KEY TO WORLD MAP

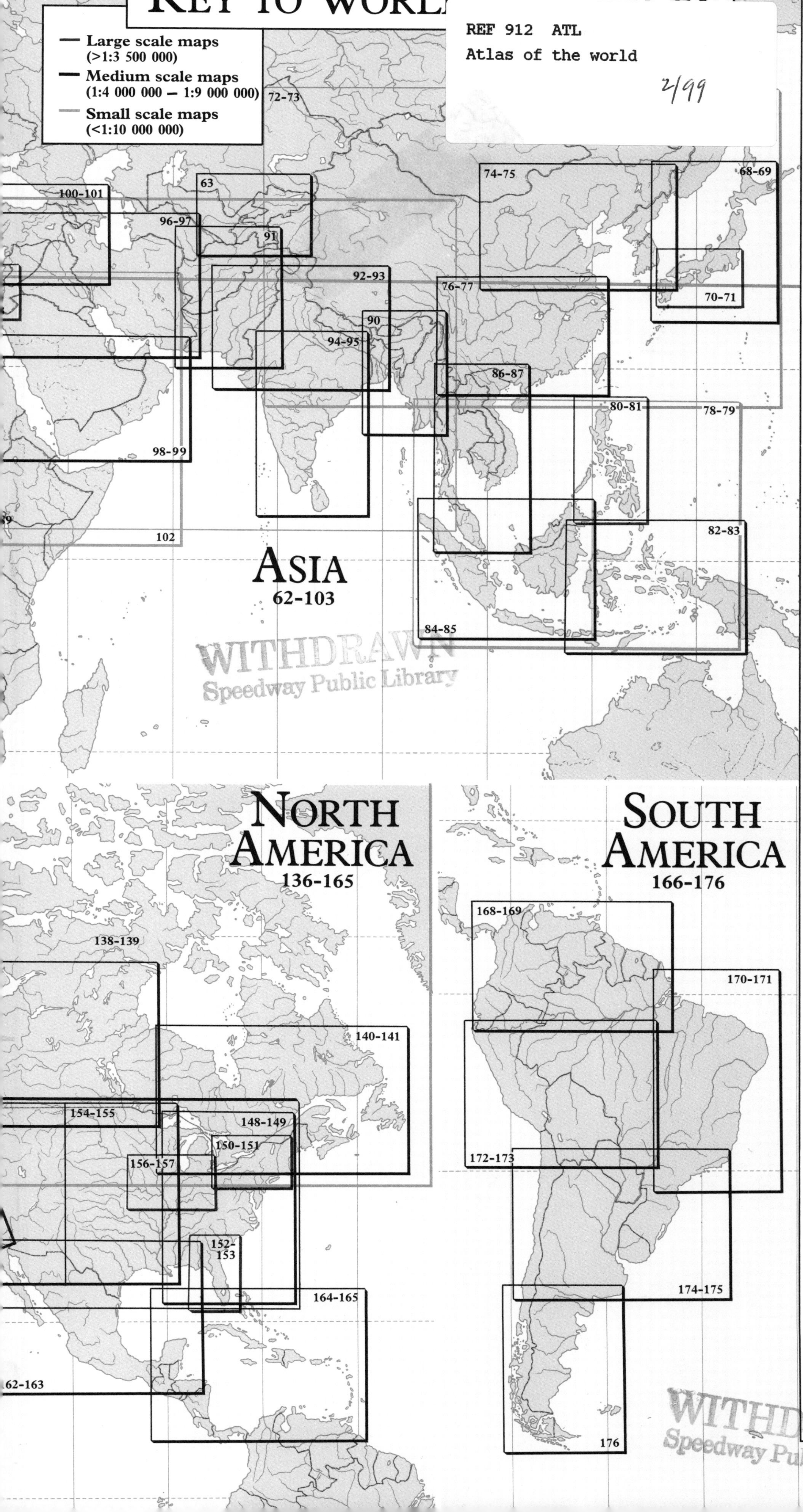

- — Large scale maps
  (>1:3 500 000)
- — Medium scale maps
  (1:4 000 000 – 1:9 000 000)
- — Small scale maps
  (<1:10 000 000)

72–73

74–75   68–69

100–101

63

96–97

91

92–93   76–77

70–71

90

94–95

86–87

80–81   78–79

98–99

102

82–83

ASIA
62–103

84–85

NORTH AMERICA
136–165

SOUTH AMERICA
166–176

138–139

140–141

168–169

170–171

154–155

148–149

150–151

156–157

152–153

172–173

164–165

174–175

62–163

176

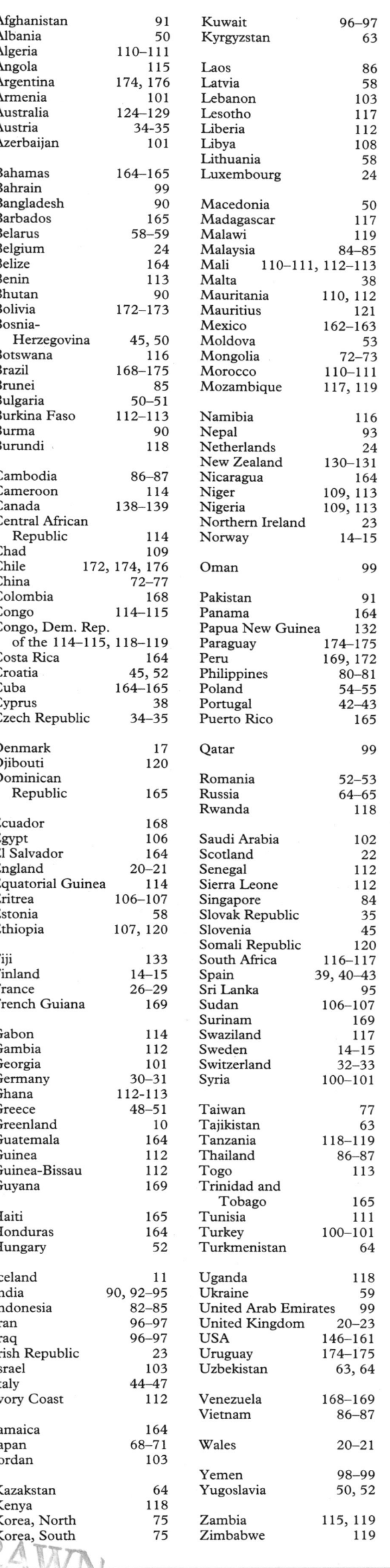

COUNTRY INDEX

ATLAS
OF THE
WORLD

OXFORD

ATLAS
OF THE
WORLD

SIXTH EDITION

Specialist Geography Consultants

The editors are grateful to the following people for acting as specialist geography consultants on the '*Introduction to World Geography*' front section:

Professor D. Brunsden, Kings College, University of London, UK

Dr C. Clarke, Oxford University, UK

Professor P. Haggett, University of Bristol, UK

Professor M-L. Hsu, University of Minnesota, Minnesota, USA

Professor K. McLachlan, Geopolitical and International Boundaries Research Centre, School of Oriental and African Studies, University of London, UK

Professor M. Monmonier, Syracuse University, New York, USA

Professor M. J. Tooley, University of St Andrews, UK

Dr T. Unwin, Royal Holloway, University of London, UK

The editors would also like to thank:

Keith Lye

Robin Scagell

Dr I. S. Evans, Durham University, UK

Introduction to World Geography

Picture Acknowledgements
Science Photo Library /BP/NRSC 9, /Earth Satellite Corporation 20, /NOAA 22 bottom left and bottom right

Illustrations
Stefan Chabluk
William Donohoe
Bernard Thornton Artists /Steve Seymour

Star charts
John Cox and Richard Monkhouse

Cartography by Philip's

Foreword

An authoritative and serious reference work, the Oxford *Atlas of the World* is one of the finest atlases available anywhere in the world. The atlas incorporates computer-derived maps which have been produced using the very latest in digital cartographic techniques.

The Oxford *Atlas of the World* has been revised and updated with the help of a panel of specialist geography consultants from the United Kingdom and the United States, whose specialties range from the history of cartography, urban and social geography, epidemiology and the European Union to biogeography and applied geomorphology. The result of their valuable input can be seen in the wealth of up-to-date maps and data contained in the "*Introduction to World Geography*" section of this atlas.

How to use the Atlas
The atlas is divided into a number of sections which are explained below.

World Statistics
Six pages of world statistics on topics such as area and population for every country in the world, city populations for the largest cities, climate statistics and physical dimensions – including the largest islands, lakes and seas, the highest mountains and the longest rivers, by continent. Also included in this section is a selection of detailed, up-to-date maps highlighting regions around the world that are currently in the news, such as the former Yugoslavia, Central East Africa, the Caucasus region of the CIS, and Israel.

Introduction to World Geography
A richly informative section comprising 48 pages of up-to-date maps, charts, graphs and clear diagrams which explain key themes about the world in which we live. The topics covered include the Solar System, oceans, climate, the environment, cities, energy and trade. Introductory text on each spread describes and explains the patterns shown by the data.

City Maps
A detailed selection of maps for 66 urban areas around the world. These are useful for planning trips abroad as well as for comparative studies of cities worldwide. Also included is a 16-page index to the city maps.

World Maps
An outstanding collection of 176 pages of distinctive Philip's cartography. The highly acclaimed physical world maps combine relief shading with layer-colored contours to give a striking visual picture of the Earth's surface. Roads, railroads, canals and airports are accurately depicted on the maps, and towns and cities are clearly marked. The maps show the recent place name changes in the countries of the former USSR. More information on the key features employed in the construction and presentation of the maps is given on the facing page.

Index
The 75,000-name index to the world maps includes geographical features as well as towns and cities, with both latitude/longitude and letter/figure grid references.

World Maps

The reference maps which form the main body of this atlas have been prepared in accordance with the highest standards of international cartography to provide an accurate and detailed representation of the Earth. The scales and projections used have been carefully chosen to give balanced coverage of the world, while emphasizing the most densely populated and economically significant regions. A hallmark of Philip's mapping is the use of hill shading and relief coloring to create a graphic impression of landforms: this makes the maps exceptionally easy to read. However, knowledge of the key features employed in the construction and presentation of the maps will enable the reader to derive the fullest benefit from the atlas.

Map Sequence

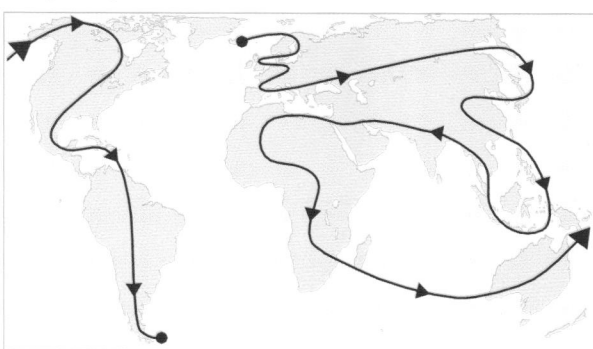

The atlas covers the Earth continent by continent: first Europe; then its land neighbor Asia (mapped north before south, in a clockwise sequence), then Africa, Australia and Oceania, North America and South America. This is the classic arrangement adopted by most cartographers since the 16th century. For each continent, there are maps at a variety of scales. First, physical relief and political maps of the whole continent; then a series of larger-scale maps of the regions within the continent, each followed, where required, by still larger-scale maps of the most important or densely populated areas. The governing principle is that by turning the pages of the atlas, the reader moves steadily from north to south through each continent, with each map overlapping its neighbors. A key map showing this sequence, and the area covered by each map, can be found on the endpapers of the atlas.

Map Presentation

With very few exceptions (e.g. for the Arctic and Antarctic), the maps are drawn with north at the top, regardless of whether they are presented upright or sideways on the page. In the borders will be found the map title; a locator diagram showing the area covered and the page numbers for maps of adjacent areas; the scale; the projection used; the degrees of latitude and longitude; and the letters and figures used in the index for locating place names and geographical features. Physical relief maps also have a height reference panel identifying the colors used for each layer of contouring.

Map Symbols

Each map contains a vast amount of detail which can only be conveyed clearly and accurately by the use of symbols. Points and circles of varying sizes locate and identify the relative importance of towns and cities; different styles of type are employed for administrative, geographical and regional place names to aid identification. A variety of pictorial symbols denote landforms such as glaciers, marshes and coral reefs, and man-made structures including roads, railroads, airports and canals. International borders are shown by red lines. Where neighboring countries are in dispute, for example in parts of the Middle East, the maps show the *de facto* boundary between nations, regardless of the legal or historical

situation. The symbols are explained on the first page of the World Maps section of the atlas.

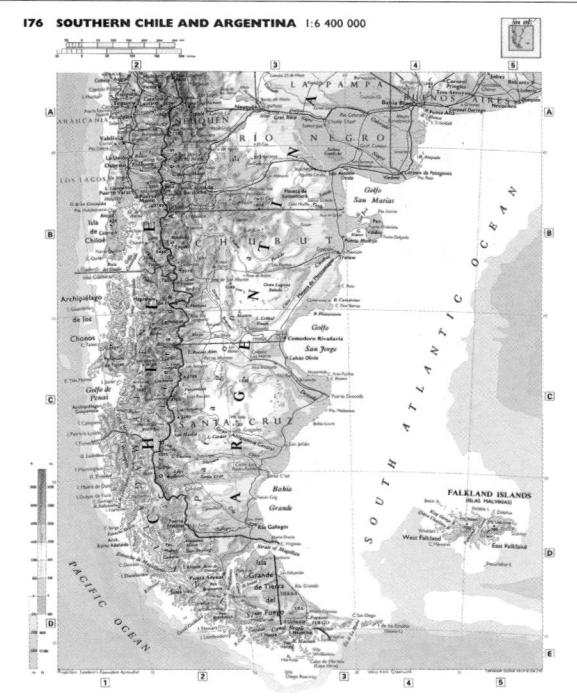

Map Scales

1:16 000 000
1 inch = 252 statute miles

The scale of each map is given in the numerical form known as the "representative fraction." The first figure is always one, signifying one unit of distance on the map; the second figure, usually in millions, is the number by which the map unit must be multiplied to give the equivalent distance on the Earth's surface. Calculations can easily be made in centimeters and kilometers, by dividing the Earth units figure by 100 000 (i.e. deleting the last five 0s). Thus 1:1 000 000 means 1 cm = 10 km. The calculation for inches and miles is more laborious, but 1 000 000 divided by 63 360 (the number of inches in a mile) shows that 1:1 000 000 means approximately 1 inch = 16 miles. The table below provides distance equivalents for scales down to 1:50 000 000.

LARGE SCALE		
1:1 000 000	1 cm = 10 km	1 inch = 16 miles
1:2 500 000	1 cm = 25 km	1 inch = 39.5 miles
1:5 000 000	1 cm = 50 km	1 inch = 79 miles
1:6 000 000	1 cm = 60 km	1 inch = 95 miles
1:8 000 000	1 cm = 80 km	1 inch = 126 miles
1:10 000 000	1 cm = 100 km	1 inch = 158 miles
1:15 000 000	1 cm = 150 km	1 inch = 237 miles
1:20 000 000	1 cm = 200 km	1 inch = 316 miles
1:50 000 000	1 cm = 500 km	1 inch = 790 miles
SMALL SCALE		

Measuring Distances

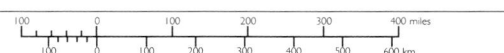

Although each map is accompanied by a scale bar, distances cannot always be measured with confidence because of the distortions involved in portraying the curved surface of the Earth on a flat page. As a general rule, the larger the map scale, the more accurate and reliable will be the distance measured. On small-scale maps such as those of the world and of entire continents, measurement may only be accurate along the "standard parallels," or central axes, and should not be attempted without considering the map projection.

Map Projections

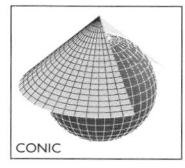  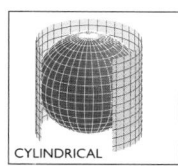

Unlike a globe, no flat map can give a true scale representation of the world in terms of area, shape and position of every region. Each of the numerous systems that have been devised for projecting the curved surface of the Earth on to a flat page involves the sacrifice of accuracy in one or more of these elements. The variations in shape and position of land masses such as Alaska, Greenland and Australia, for example, can be quite dramatic when different projections are compared.

For this atlas, the guiding principle has been to select projections that involve the least distortion of size and distance. The projection used for each map is noted in the border. Most fall into one of three categories – conic, cylindrical or azimuthal – whose basic concepts are shown above. Each involves plotting the forms of the Earth's surface on a grid of latitude and longitude lines, which may be shown as parallels, curves or radiating spokes.

Latitude and Longitude

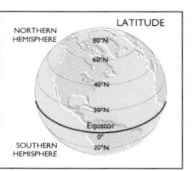 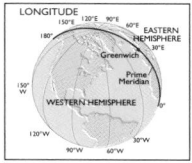 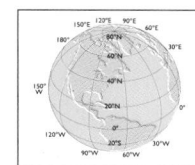

Accurate positioning of individual points on the Earth's surface is made possible by reference to the geometrical system of latitude and longitude. Latitude parallels are drawn west–east around the Earth and numbered by degrees north and south of the Equator, which is designated 0° of latitude. Longitude meridians are drawn north–south and numbered by degrees east and west of the prime meridian, 0° of longitude, which passes through Greenwich in England. By referring to these coordinates and their subdivisions of minutes (1/60th of a degree) and seconds (1/60th of a minute), any place on Earth can be located to within a few hundred yards. Latitude and longitude are indicated by blue lines on the maps; they are straight or curved according to the projection employed. Reference to these lines is the easiest way of determining the relative positions of places on different maps, and for plotting compass directions.

Name Forms

For ease of reference, both English and local name forms appear in the atlas. Oceans, seas and countries are shown in English throughout the atlas; country names may be abbreviated to their commonly accepted form (e.g. Germany, not The Federal Republic of Germany). Conventional English forms are also used for place names on the smaller-scale maps of the continents. However, local name forms are used on all large-scale and regional maps, with the English form given in brackets only for important cities – the large-scale map of Russia and Central Asia thus shows Moskva (Moscow). For countries which do not use a Roman script, place names have been transcribed according to the systems adopted by the British and US Geographic Names Authorities. For China, the Pin Yin system has been used, with some more widely known forms appearing in brackets, as with Beijing (Peking). Both English and local names appear in the index, the English form being cross-referenced to the local form.

Contents

World Statistics

Introduction to World Geography

City Maps 1–32

(Scale 1:200 000)

Index to City Maps 33–48

World Maps 1–176

The World

Europe

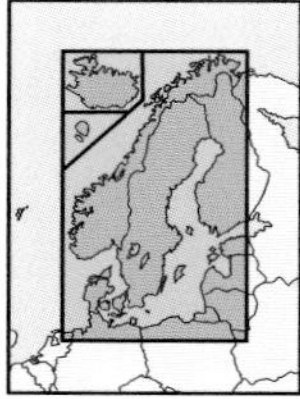

14–15

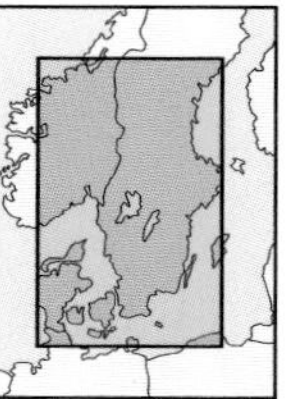

16–17

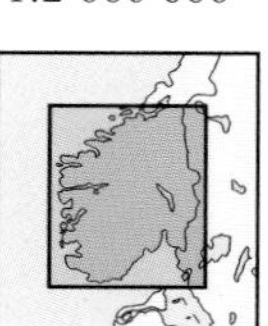

18

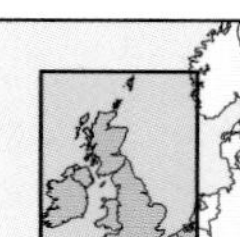

19

NOTE
The titles to the World Maps list the main countries, states and provinces covered by each map. A name given in *italics* indicates that only part of the country is shown on the map.

England and Wales
1:1 600 000
Isle of Man, Channel Islands

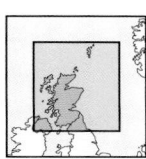

20–21

Scotland
1:1 600 000

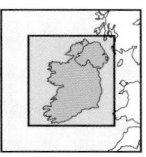

22

Ireland
1:1 600 000
Irish Republic, Northern Ireland

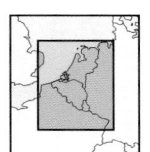

23

Netherlands, Belgium and Luxembourg
1:2 000 000

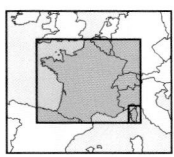

24

France
1:4 000 000

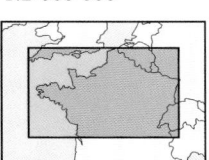

25

Northern France
1:2 000 000

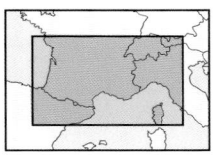

26–27

Southern France
1:2 000 000
Monaco

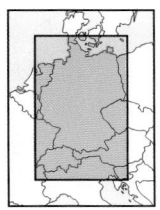

28–29

Germany
1:2 000 000

30–31

Switzerland
1:800 000
Liechtenstein

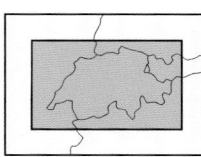

32–33

Austria, Czech Republic and Slovak Republic
1:2 000 000

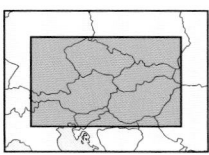

34–35

Central Europe
1:4 000 000

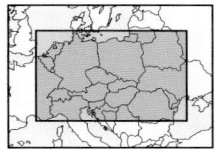

36–37

Malta, Crete, Corfu, Rhodes and Cyprus
1:800 000 / 1:1 040 000

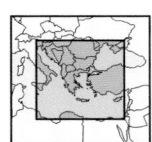

38

Balearics, Canaries and Madeira
1:800 000 / 1:1 600 000
Mallorca, Menorca, Ibiza

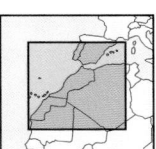

39

Eastern Spain
1:2 000 000
Andorra

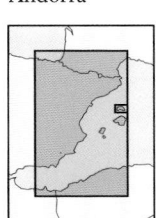

40–41

Western Spain and Portugal
1:2 000 000

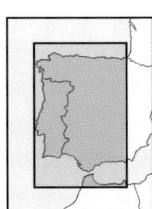

42–43

Northern Italy, Slovenia and Croatia
1:2 000 000

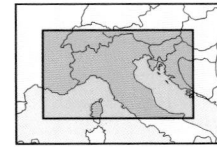

44–45

Southern Italy
1:2 000 000

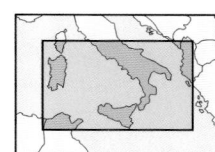

46–47

Southern Greece
1:2 000 000
Turkey

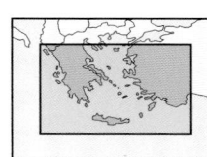

48–49

Northern Greece, Bulgaria and Yugoslavia
1:2 000 000
Macedonia

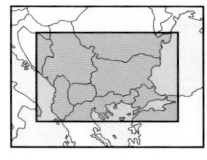

50–51

Hungary, Romania and the Lower Danube
1:2 000 000
Moldova

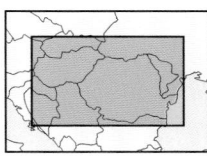

52–53

Poland and the Southern Baltic
1:2 000 000
Latvia, Lithuania

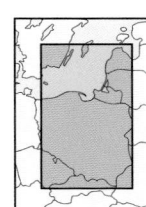

54–55

Eastern Europe and Turkey
1:8 000 000

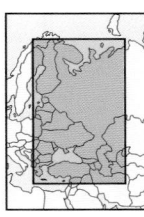

56–57

Baltic States, Belarus and Ukraine
1:4 000 000
Russia, Estonia, Latvia, Lithuania, Belarus, *Ukraine*

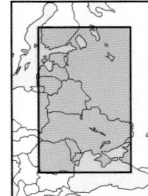

58–59

Volga Basin and the Caucasus
1:4 000 000
Russia, Georgia, Armenia, Azerbaijan

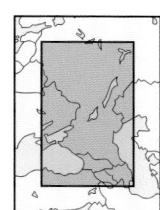

60–61

Asia

Southern Urals
1:4 000 000
Russia

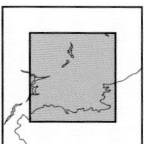

62

Central Asia
1:4 000 000
Kazakstan, *Kyrgyzstan*, Tajikistan, *Uzbekistan*

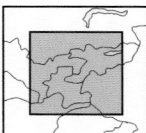

63

Russia and Central Asia
1:16 000 000
Russia, Kazakstan, Turkmenistan, Uzbekistan

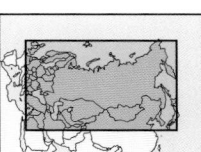

64–65

Asia: Physical   66
1:40 000 000

Asia: Political   67
1:40 000 000

Japan
1:4 000 000
Ryukyu Islands

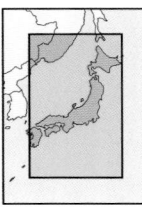

68–69

Southern Japan
1:2 000 000

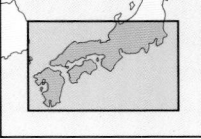

70–71

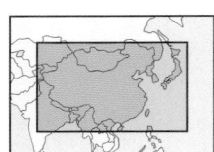
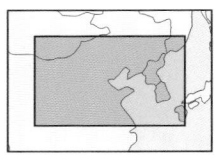

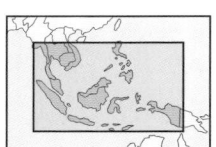
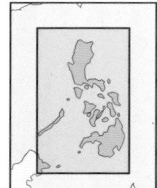
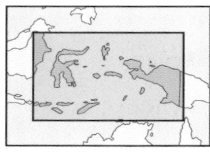
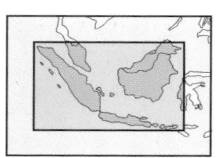
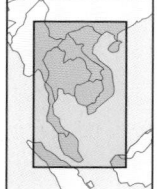
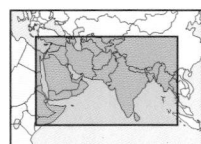

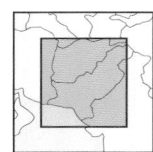
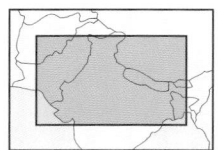
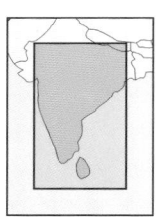
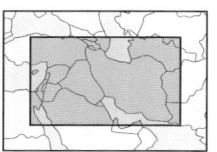
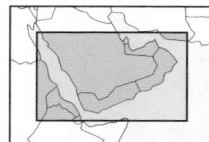
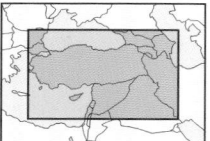
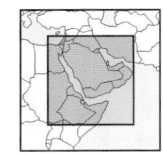
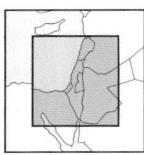
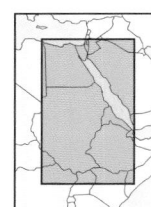
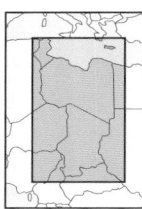
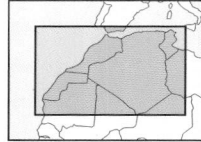
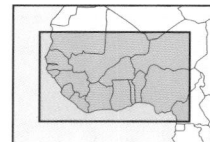
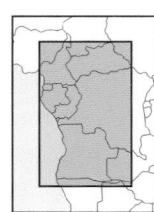
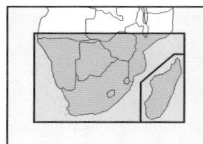
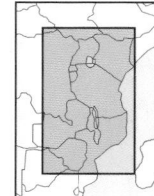
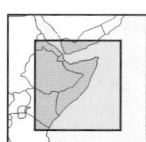
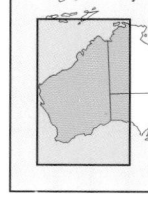

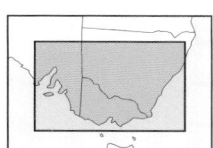
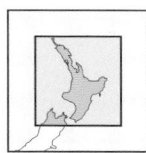

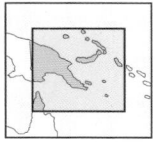
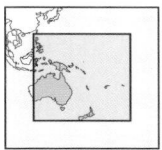
North America

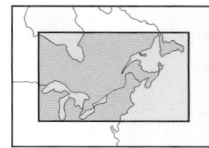
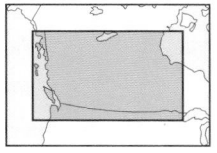
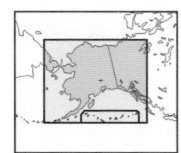
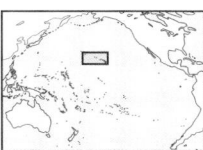
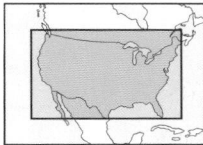

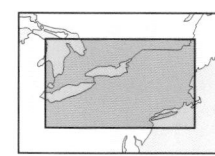
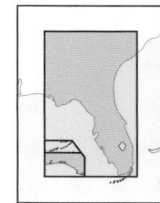
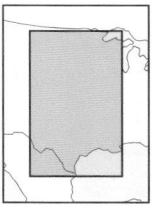
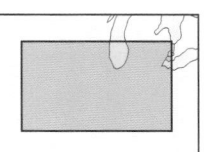
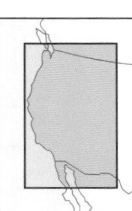
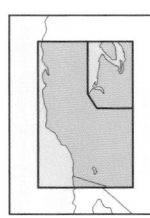
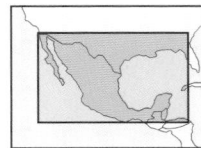

South America

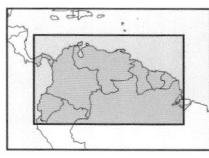

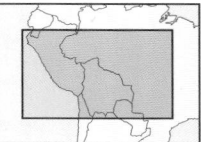

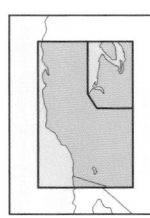
Index to World Maps

World Statistics: Countries

This alphabetical list includes all the countries and territories of the world. If a territory is not completely independent, then the country it is associated with is named. The area figures give the total area of land, inland water and ice. Units for areas and populations are thousands. The population figures are 1997 estimates. The annual income is the Gross National Product per capita in US dollars. The figures are the latest available, usually 1995.

Country/Territory	Area km^2 Thousands	Area miles^2 Thousands	Population Thousands	Capital	Annual Income US $
Adélie Land (France)	432	167	0.03	–	–
Afghanistan	652	252	23,000	Kabul	300
Albania	28.8	11.1	3,600	Tirana	670
Algeria	2,382	920	29,300	Algiers	1,600
American Samoa (US)	0.20	0.08	62	Pago Pago	2,600
Amsterdam Is. (France)	0.05	0.02	0.03	–	–
Andorra	0.45	0.17	75	Andorra-la-Vella	14,000
Angola	1,247	481	11,200	Luanda	410
Anguilla (UK)	0.1	0.04	10	The Valley	6,800
Antigua & Barbuda	0.44	0.17	66	St John's	6,390
Argentina	2,767	1,068	35,400	Buenos Aires	8,030
Armenia	29.8	11.5	3,800	Yerevan	730
Aruba (Netherlands)	0.19	0.07	70	Oranjestad	17,500
Ascension Is. (UK)	0.09	0.03	1.5	Georgetown	–
Australia	7,687	2,968	18,400	Canberra	18,720
Australian Antarctic Terr. (Aus.)	6,120	2,363	0	–	–
Austria	83.9	32.4	8,200	Vienna	26,890
Azerbaijan	86.6	33.4	7,700	Baku	480
Azores (Portugal)	2.2	0.87	238	Ponta Delgada	–
Bahamas	13.9	5.4	280	Nassau	11,940
Bahrain	0.68	0.26	605	Manama	7,840
Bangladesh	144	56	124,000	Dhaka	240
Barbados	0.43	0.17	265	Bridgetown	6,560
Belarus	207.6	80.1	10,500	Minsk	2,070
Belgium	30.5	11.8	10,200	Brussels	24,710
Belize	23	8.9	228	Belmopan	2,630
Benin	113	43	5,800	Porto-Novo	370
Bermuda (UK)	0.05	0.02	65	Hamilton	27,000
Bhutan	47	18.1	1,790	Thimphu	420
Bolivia	1,099	424	7,700	La Paz/Sucre	800
Bosnia-Herzegovina	51	20	3,600	Sarajevo	2,600
Botswana	582	225	1,500	Gaborone	3,020
Bouvet Is. (Norway)	0.05	0.02	0.02	–	–
Brazil	8,512	3,286	159,500	Brasília	3,640
British Antarctic Terr. (UK)	1,709	660	0.3	–	–
British Indian Ocean Terr. (UK)	0.08	0.03	0	–	–
Brunei	5.8	2.2	300	Bandar Seri Begawan	14,500
Bulgaria	111	43	8,600	Sofia	1,330
Burkina Faso	274	106	10,900	Ouagadougou	230
Burma (= Myanmar)	677	261	47,500	Rangoon	1,000
Burundi	27.8	10.7	6,300	Bujumbura	160
Cambodia	181	70	10,500	Phnom Penh	270
Cameroon	475	184	13,800	Yaoundé	650
Canada	9,976	3,852	30,200	Ottawa	19,380
Canary Is. (Spain)	7.3	2.8	1,494	Las Palmas/Santa Cruz	–
Cape Verde Is.	4	1.6	410	Praia	960
Cayman Is. (UK)	0.26	0.10	35	George Town	20,000
Central African Republic	623	241	3,400	Bangui	340
Chad	1,284	496	6,800	Ndjaména	180
Chatham Is. (NZ)	0.96	0.37	0.05	Waitangi	—
Chile	757	292	14,700	Santiago	4,160
China	9,597	3,705	1,210,000	Beijing	620
Christmas Is. (Australia)	0.14	0.05	2	The Settlement	–
Cocos (Keeling) Is. (Australia)	0.01	0.005	1	West Island	–
Colombia	1,139	440	35,900	Bogotá	1,910
Comoros	2.2	0.86	630	Moroni	470
Congo	342	132	2,700	Brazzaville	680
Congo (= Zaïre)	2,345	905	47,200	Kinshasa	120
Cook Is. (NZ)	0.24	0.09	20	Avarua	900
Costa Rica	51.1	19.7	3,500	San José	2,610
Croatia	56.5	21.8	4,900	Zagreb	3,250
Crozet Is. (France)	0.51	0.19	35	–	–
Cuba	111	43	11,300	Havana	1,250
Cyprus	9.3	3.6	800	Nicosia	11,500
Czech Republic	78.9	30.4	10,500	Prague	3,870
Denmark	43.1	16.6	5,400	Copenhagen	29,890
Djibouti	23.2	9	650	Djibouti	1,000
Dominica	0.75	0.29	78	Roseau	2,990
Dominican Republic	48.7	18.8	8,200	Santo Domingo	1,460
Ecuador	284	109	11,800	Quito	1,390
Egypt	1,001	387	63,000	Cairo	790
El Salvador	21	8.1	6,000	San Salvador	1,610
Equatorial Guinea	28.1	10.8	420	Malabo	380
Eritrea	94	36	3,500	Asmara	500
Estonia	44.7	17.3	1,500	Tallinn	2,860
Ethiopia	1,128	436	58,500	Addis Ababa	100
Falkland Is. (UK)	12.2	4.7	2	Stanley	–
Faroe Is. (Denmark)	1.4	0.54	45	Tórshavn	23,660
Fiji	18.3	7.1	800	Suva	2,440
Finland	338	131	5,200	Helsinki	20,580
France	552	213	58,800	Paris	24,990
French Guiana (France)	90	34.7	155	Cayenne	6,500
French Polynesia (France)	4	1.5	226	Papeete	7,500
Gabon	268	103	1,200	Libreville	3,490
Gambia, The	11.3	4.4	1,200	Banjul	320
Georgia	69.7	26.9	5,500	Tbilisi	440
Germany	357	138	82,300	Berlin/Bonn	27,510
Ghana	239	92	18,100	Accra	390
Gibraltar (UK)	0.007	0.003	28	Gibraltar Town	5,000
Greece	132	51	10,600	Athens	8,210
Greenland (Denmark)	2,176	840	57	Nuuk (Godthåb)	9,000
Grenada	0.34	0.13	99	St George's	2,980
Guadeloupe (France)	1.7	0.66	440	Basse-Terre	9,500
Guam (US)	0.55	0.21	161	Agana	6,000
Guatemala	109	42	11,300	Guatemala City	1,340
Guinea	246	95	7,500	Conakry	550
Guinea-Bissau	36.1	13.9	1,200	Bissau	250
Guyana	215	83	820	Georgetown	590
Haiti	27.8	10.7	7,400	Port-au-Prince	250
Honduras	112	43	6,300	Tegucigalpa	600
Hong Kong (China)	1.1	0.40	6,500	–	22,990
Hungary	93	35.9	10,200	Budapest	4,120
Iceland	103	40	275	Reykjavik	24,950
India	3,288	1,269	980,000	New Delhi	340
Indonesia	1,905	735	203,500	Jakarta	980
Iran	1,648	636	69,500	Tehran	4,800
Iraq	438	169	22,500	Baghdad	1,800
Ireland	70.3	27.1	3,600	Dublin	14,710
Israel	27	10.3	5,900	Jerusalem	15,920
Italy	301	116	57,800	Rome	19,020
Ivory Coast	322	125	15,100	Yamoussoukro	660
Jamaica	11	4.2	2,600	Kingston	1,510
Jan Mayen Is. (Norway)	0.38	0.15	0.06	–	–
Japan	378	146	125,900	Tokyo	39,640
Johnston Is. (US)	0.002	0.0009	1	–	–
Jordan	89.2	34.4	5,600	Amman	1,510
Kazakstan	2,717	1,049	17,000	Aqmola	1,330
Kenya	580	224	31,900	Nairobi	280
Kerguelen Is. (France)	7.2	2.8	0.7	–	–
Kermadec Is. (NZ)	0.03	0.01	0.1	–	–
Kiribati	0.72	0.28	85	Tarawa	710
Korea, North	121	47	24,500	Pyŏngyang	1,000
Korea, South	99	38.2	46,100	Seoul	9,700
Kuwait	17.8	6.9	2,050	Kuwait City	17,390
Kyrgyzstan	198.5	76.6	4,700	Bishkek	700
Laos	237	91	5,200	Vientiane	350
Latvia	65	25	2,500	Riga	2,270
Lebanon	10.4	4	3,200	Beirut	2,660
Lesotho	30.4	11.7	2,100	Maseru	770
Liberia	111	43	3,000	Monrovia	850
Libya	1,760	679	5,500	Tripoli	7,000
Liechtenstein	0.16	0.06	32	Vaduz	33,500
Lithuania	65.2	25.2	3,700	Vilnius	1,900
Luxembourg	2.6	1	400	Luxembourg	41,210
Macau (Portugal)	0.02	0.006	450	Macau	7,500
Macedonia	25.7	9.9	2,200	Skopje	860
Madagascar	587	227	15,500	Antananarivo	230
Madeira (Portugal)	0.81	0.31	253	Funchal	–
Malawi	118	46	10,300	Lilongwe	170
Malaysia	330	127	20,900	Kuala Lumpur	3,890
Maldives	0.30	0.12	275	Malé	990
Mali	1,240	479	11,000	Bamako	250
Malta	0.32	0.12	400	Valletta	11,000
Marshall Is.	0.18	0.07	60	Dalap-Uliga-Darrit	1,500
Martinique (France)	1.1	0.42	405	Fort-de-France	10,000
Mauritania	1,030	412	2,400	Nouakchott	460
Mauritius	2.0	0.72	1,200	Port Louis	3,380
Mayotte (France)	0.37	0.14	105	Mamoundzou	1,430
Mexico	1,958	756	97,400	Mexico City	3,320
Micronesia, Fed. States of	0.70	0.27	127	Palikir	1,560
Midway Is. (US)	0.005	0.002	2	–	–
Moldova	33.7	13	4,500	Chişinău	920
Monaco	0.002	0.0001	33	Monaco	16,000
Mongolia	1,567	605	2,500	Ulan Bator	310
Montserrat (UK)	0.10	0.04	12	Plymouth	4,500
Morocco	447	172	28,100	Rabat	1,110
Mozambique	802	309	19,100	Maputo	80
Namibia	825	318	1,700	Windhoek	2,000
Nauru	0.02	0.008	12	Yaren District	10,000
Nepal	141	54	22,100	Katmandu	200
Netherlands	41.5	16	15,900	Amsterdam/The Hague	24,000
Netherlands Antilles (Neths)	0.99	0.38	210	Willemstad	10,500
New Caledonia (France)	18.6	7.2	192	Nouméa	16,000
New Zealand	269	104	3,700	Wellington	14,340
Nicaragua	130	50	4,600	Managua	380
Niger	1,267	489	9,700	Niamey	220
Nigeria	924	357	118,000	Abuja	260
Niue (NZ)	0.26	0.10	2	Alofi	–
Norfolk Is. (Australia)	0.03	0.01	2	Kingston	–
Northern Mariana Is. (US)	0.48	0.18	50	Saipan	11,500
Norway	324	125	4,400	Oslo	31,250
Oman	212	82	2,400	Muscat	4,820
Pakistan	796	307	136,000	Islamabad	460
Palau	0.46	0.18	17	Koror	2,260
Panama	77.1	29.8	2,700	Panama City	2,750
Papua New Guinea	463	179	4,400	Port Moresby	1,160
Paraguay	407	157	5,200	Asunción	1,690
Peru	1,285	496	24,500	Lima	2,310
Peter 1st Is. (Norway)	0.18	0.07	0	–	–
Philippines	300	116	73,500	Manila	1,050
Pitcairn Is. (UK)	0.03	0.01	0.05	Adamstown	–
Poland	313	121	38,800	Warsaw	2,790
Portugal	92.4	35.7	10,100	Lisbon	9,740
Puerto Rico (US)	9	3.5	3,800	San Juan	7,500
Qatar	11	4.2	620	Doha	11,600
Queen Maud Land (Norway)	2,800	1,081	0	–	–
Réunion (France)	2.5	0.97	680	Saint-Denis	4,500
Romania	238	92	22,600	Bucharest	1,480
Ross Dependency (NZ)	435	168	0	–	–
Russia	17,075	6,592	147,800	Moscow	2,240
Rwanda	26.3	10.2	7,000	Kigali	180
St Helena (UK)	0.12	0.05	6	Jamestown	–
St Kitts & Nevis	0.36	0.14	42	Basseterre	4,470
St Lucia	0.62	0.24	150	Castries	3,370
St Paul Is. (France)	0.007	0.003	0	–	–
St Pierre & Miquelon (France)	0.24	0.09	7	Saint Pierre	–
St Vincent & Grenadines	0.39	0.15	114	Kingstown	2,280
San Marino	0.06	0.02	26	San Marino	20,000
São Tomé & Príncipe	0.96	0.37	135	São Tomé	350
Saudi Arabia	2,150	830	19,100	Riyadh	7,040
Senegal	197	76	8,900	Dakar	600
Seychelles	0.46	0.18	78	Victoria	6,370
Sierra Leone	71.7	27.7	4,600	Freetown	180
Singapore	0.62	0.24	3,200	Singapore	26,730
Slovak Republic	49	18.9	5,400	Bratislava	2,950
Slovenia	20.3	7.8	2,000	Ljubljana	8,200
Solomon Is.	28.9	11.2	410	Honiara	910
Somalia	638	246	9,900	Mogadishu	500
South Africa	1,220	471	42,300	C. Town/Pretoria/Bloemfontein	3,160
South Georgia (UK)	3.8	1.4	0.05	–	–
South Sandwich Is. (UK)	0.38	0.15	0	–	–
Spain	505	195	39,300	Madrid	13,580
Sri Lanka	65.6	25.3	18,700	Colombo	700
Sudan	2,506	967	31,000	Khartoum	750
Surinam	163	63	500	Paramaribo	880
Svalbard (Norway)	62.9	24.3	4	Longyearbyen	–
Swaziland	17.4	6.7	1,000	Mbabane	1,170
Sweden	450	174	8,900	Stockholm	23,750
Switzerland	41.3	15.9	7,100	Bern	40,630
Syria	185	71	15,300	Damascus	1,120
Taiwan	36	13.9	21,700	Taipei	12,000
Tajikistan	143.1	55.2	6,000	Dushanbe	340
Tanzania	945	365	31,200	Dodoma	120
Thailand	513	198	60,800	Bangkok	2,740
Togo	56.8	21.9	4,500	Lomé	310
Tokelau (NZ)	0.01	0.005	2	Nukunonu	–
Tonga	0.75	0.29	107	Nuku'alofa	1,610
Trinidad & Tobago	5.1	2	1,300	Port of Spain	3,770
Tristan da Cunha (UK)	0.11	0.04	0.33	Edinburgh	–
Tunisia	164	63	9,200	Tunis	1,820
Turkey	779	301	63,500	Ankara	2,780
Turkmenistan	488.1	188.5	4,800	Ashkhabad	920
Turks & Caicos Is. (UK)	0.43	0.17	15	Cockburn Town	5,000
Tuvalu	0.03	0.01	10	Fongafale	600
Uganda	236	91	20,800	Kampala	240
Ukraine	603.7	233.1	51,500	Kiev	1,630
United Arab Emirates	83.6	32.3	2,400	Abu Dhabi	17,400
United Kingdom	243.3	94	58,600	London	18,700
United States of America	9,373	3,619	268,000	Washington, DC	26,980
Uruguay	177	68	3,300	Montevideo	5,170
Uzbekistan	447.4	172.7	23,800	Tashkent	970
Vanuatu	12.2	4.7	175	Port-Vila	1,200
Vatican City	0.0004	0.0002	1	–	–
Venezuela	912	352	22,500	Caracas	3,020
Vietnam	332	127	77,100	Hanoi	240
Virgin Is. (UK)	0.15	0.06	13	Road Town	–
Virgin Is. (US)	0.34	0.13	105	Charlotte Amalie	12,000
Wake Is.	0.008	0.003	0.30	–	–
Wallis & Futuna Is. (France)	0.20	0.08	15	Mata-Utu	–
Western Sahara	266	103	280	El Aaiún	980
Western Samoa	2.8	1.1	175	Apia	1,120
Yemen	528	204	16,500	Sana	260
Yugoslavia	102.3	39.5	10,500	Belgrade	1,400
Zambia	753	291	9,500	Lusaka	400
Zimbabwe	391	151	12,100	Harare	540

At the time of going to press, the government of Kazakstan planned to rename the capital Aqmola to Astana.

World Statistics: Cities

This list shows the principal cities with more than 500,000 inhabitants (only cities with more than 700,000 inhabitants are included for China and India). The figures are taken from the most recent census or estimate available, and as far as possible are the population of the metropolitan area, e.g. greater New York, Mexico or Paris. All the figures are in thousands. Local name forms have been used for the smaller cities (e.g. Kraków).

Afghanistan
Kabul 1,565
Algeria
Algiers 1,722
Oran 664
Angola
Luanda 2,250
Argentina
Buenos Aires 10,990
Córdoba 1,198
Rosario 1,096
Mendoza 775
La Plata 640
San Miguel de Tucumán 622
Mar del Plata 520
Armenia
Yerevan 1,226
Australia
Sydney 3,713
Melbourne 3,189
Brisbane 1,422
Perth 1,221
Adelaide 1,071
Austria
Vienna 1,560
Azerbaijan
Baku 1,081
Bangladesh
Dhaka 7,832
Chittagong 2,041
Khulna 877
Rajshahi 517
Belarus
Minsk 1,700
Homyel 512
Belgium
Brussels 952
Benin
Cotonou 537
Bolivia
La Paz 1,126
Santa Cruz 767
Bosnia-Herzegovina
Sarajevo 526
Brazil
São Paulo 16,417
Rio de Janeiro 9,888
Salvador 2,056
Belo Horizonte 2,049
Fortaleza 1,758
Brasília 1,596
Curitiba 1,290
Recife 1,290
Nova Iguaçu 1,286
Pôrto Alegre 1,263
Belém 1,246
Manaus 1,011
Goiânia 921
Campinas 846
Guarulhos 781
São Gonçalo 748
São Luís 696
Duque de Caxias 665
Maceió 628
Santo André 614
Natal 607
Teresina 598
São Bernado de Campo 565
Osasco 563
Campo Grande 526
Bulgaria
Sofia 1,117
Burkina Faso
Ouagadougou 690
Burma (Myanmar)
Rangoon 2,513
Mandalay 533
Cambodia
Phnom Penh 920
Cameroon
Douala 884
Yaoundé 750
Canada
Toronto 4,264
Montréal 3,327
Vancouver 1,832
Ottawa–Hull 1,010
Edmonton 863
Calgary 822
Québec 672
Winnipeg 667
Hamilton 624
Central African Rep.
Bangui 706

Chad
Ndjaména 530
Chile
Santiago 5,077
China
Shanghai 15,082
Beijing 12,362
Tianjin 10,687
Hong Kong (SAR)¹ 6,205
Chongqing 3,870
Shenyang 3,762
Wuhan 3,520
Guangzhou 3,114
Harbin 2,505
Nanjing 2,211
Xi'an 2,115
Chengdu 1,933
Dalian 1,855
Changchun 1,810
Jinan 1,660
Taiyuan 1,642
Qingdao 1,584
Fuzhou, Fujian 1,380
Zibo 1,346
Zhengzhou 1,324
Lanzhou 1,296
Anshan 1,252
Fushun 1,246
Kunming 1,242
Changsha 1,198
Hangzhou 1,185
Nanchang 1,169
Shijiazhuang 1,159
Guiyang 1,131
Ürümqi 1,130
Jilin 1,118
Hefei 1,110
Tangshan 1,110
Baotou 1,033
Xuzhou, Jiangsu 937
Handan 894
Wuxi 863
Luoyang 863
Datong 845
Nanning 829
Benxi 805
Yichun, Heilongjiang 800
Huainan 769
Suzhou, Jiangsu 766
Colombia
Bogotá 5,026
Cali 1,719
Medellín 1,621
Barranquilla 1,064
Cartagena 746
Congo
Brazzaville 938
Pointe-Noire 576
Congo (Zaïre)
Kinshasa 3,804
Lubumbashi 739
Mbuji-Mayi 613
Kolwezi 544
Costa Rica
San José 1,186
Croatia
Zagreb 931
Cuba
Havana 2,143
Czech Republic
Prague 1,217
Denmark
Copenhagen 1,353
Dominican Republic
Santo Domingo 2,135
Santiago 691
Ecuador
Guayaquil 1,925
Quito 1,444
Egypt
Cairo 9,656
Alexandria 3,380
El Gîza 2,144
Shubra el Kheima 834
El Salvador
San Salvador 1,522
Ethiopia
Addis Ababa 2,316
Finland
Helsinki 525
France
Paris 9,469
Lyon 1,262
Marseille 1,087
Lille 959

Bordeaux 696
Toulouse 650
Nice 516
Georgia
Tbilisi 1,279
Germany
Berlin 3,472
Hamburg 1,706
Munich 1,245
Cologne 964
Frankfurt 652
Essen 618
Dortmund 601
Stuttgart 588
Düsseldorf 573
Bremen 549
Duisburg 536
Hanover 526
Ghana
Accra 1,781
Kumasi 540
Greece
Athens 3,097
Guatemala
Guatemala 1,814
Guinea
Conakry 1,508
Haiti
Port-au-Prince 1,402
Honduras
Tegucigalpa 739
Hungary
Budapest 1,909
India
Bombay (Mumbai) 15,093
Calcutta 11,673
Delhi 9,882
Madras (Chennai) 5,361
Hyderabad 4,280
Bangalore 4,087
Ahmadabad 3,298
Pune 2,485
Kanpur 2,111
Nagpur 1,661
Lucknow 1,642
Surat 1,517
Jaipur 1,514
Coimbatore 1,136
Vadodara 1,115
Indore 1,104
Patna 1,099
Madurai 1,094
Bhopal 1,064
Vishakhapatnam 1,052
Varanasi 1,026
Ludhiana 1,012
Agra 956
Jabalpur 887
Allahabad 858
Meerut 847
Vijayawada 845
Jamshedpur 834
Trivandrum 826
Dhanbad 818
Thane 797
Asansol 764
Nasik 722
Gwalior 720
Tiruchchirappalli 711
Amritsar 709
Indonesia
Jakarta 11,500
Surabaya 2,701
Bandung 2,368
Medan 1,910
Semarang 1,366
Palembang 1,352
Ujung Pandang 1,092
Bandar Lampung 832
Malang 763
Iran
Tehran 6,750
Mashhad 1,964
Esfahan 1,221
Tabriz 1,166
Shiraz 1,043
Ahvaz 828
Qom 780
Bakhtaran 666
Karaj 588
Iraq
Baghdad 3,841
Diyala 961
As Sulaymaniyah 952
Arbil 770

Al Mawsil 644
Kadhimain 521
Ireland
Dublin 1,024
Israel
Tel Aviv 1,880
Jerusalem 562
Italy
Rome 2,688
Milan 1,334
Naples 1,062
Turin 946
Palermo 695
Genoa 660
Ivory Coast
Abidjan 2,500
Jamaica
Kingston 644
Japan
Tokyo–Yokohama 26,836
Osaka 10,601
Nagoya 2,159
Sapporo 1,732
Kobe 1,509
Kyoto 1,452
Fukuoka 1,269
Kawasaki 1,200
Hiroshima 1,102
Kitakyushu 1,020
Sendai 951
Chiba 851
Sakai 806
Kumamoto 640
Okayama 605
Hamamatsu 561
Sagamihara 560
Funabashi 540
Kagoshima 540
Higashiosaka 515
Jordan
Amman 1,300
Az-Zarqā 609
Kazakstan
Almaty 1,151
Qaraghandy 613
Kenya
Nairobi 2,000
Mombasa 600
Korea, North
Pyŏngyang 2,639
Hamhung 775
Chŏngjin 754
Chinnampo 691
Sinŭiju 500
Korea, South
Seoul 11,641
Pusan 3,814
Taegu 2,449
Inchon 2,308
Taejŏn 1,272
Kwangju 1,258
Ulsan 967
Sŏngnam 869
Puch'on 779
Suwŏn 756
Chŏnju 563
Kyrgyzstan
Bishkek 584
Latvia
Riga 840
Lebanon
Beirut 1,500
Tripoli 500
Libya
Tripoli 960
Lithuania
Vilnius 576
Macedonia
Skopje 541
Madagascar
Antananarivo 1,053
Malaysia
Kuala Lumpur 1,145
Mali
Bamako 746
Mauritania
Nouakchott 600
Mexico
Mexico City 15,643
Guadalajara 2,847
Monterrey 2,522
Puebla 1,055
León 872
Ciudad Juárez 798
Tijuana 743

Culiacán Rosales 602
Mexicali 602
Acapulco de Juárez 592
Mérida 557
Chihuahua 530
San Luis Potosí 526
Aguascaliéntes 506
Moldova
Chişinău 700
Mongolia
Ulan Bator 619
Morocco
Casablanca 2,943
Rabat-Salé 1,220
Marrakesh 602
Fès 564
Mozambique
Maputo 2,000
Nepal
Katmandu 535
Netherlands
Amsterdam 1,100
Rotterdam 1,074
The Hague 695
Utrecht 546
New Zealand
Auckland 929
Nicaragua
Managua 974
Nigeria
Lagos 10,287
Ibadan 1,365
Ogbomosho 712
Kano 657
Norway
Oslo 714
Pakistan
Karachi 9,863
Lahore 5,085
Faisalabad 1,875
Peshawar 1,676
Gujranwala 1,663
Rawalpindi 1,290
Multan 1,257
Hyderabad 1,107
Paraguay
Asunción 945
Peru
Lima–Callao 6,601
Callao 638
Arequipa 620
Trujillo 509
Philippines
Manila 9,280
Quezon City 1,677
Davao 961
Cebu 688
Caloocan 643
Poland
Warsaw 1,638
Lódz 826
Kraków 745
Wroclaw 643
Poznań 582
Portugal
Lisbon 2,561
Oporto 1,174
Romania
Bucharest 2,061
Russia
Moscow 9,233
St Petersburg 4,883
Nizhniy Novgorod 1,425
Novosibirsk 1,418
Yekaterinburg 1,347
Samara 1,223
Omsk 1,161
Chelyabinsk 1,125
Kazan 1,092
Ufa 1,092
Perm 1,086
Rostov 1,023
Volgograd 1,000
Krasnoyarsk 914
Voronezh 905
Saratov 899
Togliatti 689
Simbirsk 670
Izhevsk 653
Krasnodar 638
Vladivostok 637
Irkutsk 632
Yaroslavl 631
Khabarovsk 609
Barnaul 596

Novokuznetsk 593
Orenburg 558
Penza 551
Tyumen 550
Tula 535
Ryazan 526
Naberezhnyye-Chelny 524
Kemerovo 513
Astrakhan 512
Saudi Arabia
Riyadh 2,000
Jedda 1,400
Mecca 618
Medina 500
Senegal
Dakar 1,729
Sierra Leone
Freetown 505
Singapore
Singapore 2,874
Somalia
Mogadishu 1,000
South Africa
Cape Town 2,350
East Rand 1,379
Johannesburg 1,196
Durban 1,137
Pretoria 1,080
West Rand 870
Port Elizabeth 853
Vanderbijlpark–Vereeniging 774
Soweto 597
Sasolburg 540
Spain
Madrid 3,041
Barcelona 1,631
Valencia 764
Sevilla 714
Zaragoza 607
Málaga 531
Sri Lanka
Colombo 1,863
Sudan
Khartoum 561
Omdurman 526
Sweden
Stockholm 1,553
Göteburg 788
Switzerland
Zürich 915
Syria
Damascus 2,230
Aleppo 1,640
Homs 644
Taiwan
Taipei 2,653
Kaohsiung 1,405
Taichung 817
Tainan 700
Panchiao 544
Tajikistan
Dushanbe 602
Tanzania
Dar-es-Salaam 1,361
Thailand
Bangkok 5,876
Togo
Lomé 590
Tunisia
Tunis 1,827
Turkey
Istanbul 7,490
Ankara 3,028
Izmir 2,333
Adana 1,472
Bursa 1,317
Konya 1,040
Gaziantep 930
Icel 908
Antalya 734
Diyarbakir 677
Kocaeli 661
Urfa 649
Kayseri 648
Manisa 641
Hatay 561
Samsun 557
Eskisehir 508
Balikesir 501
Uganda
Kampala 773
Ukraine
Kiev 2,630
Kharkiv 1,555

Dnipropetrovsk 1,147
Donetsk 1,088
Odesa 1,046
Zaporizhzhya 887
Lviv 802
Kryvyy Rih 720
Mariupol 510
Mykolayiv 508
United Kingdom
London 8,089
Birmingham 2,373
Manchester 2,353
Liverpool 852
Glasgow 832
Leeds 529
Newcastle 525
United States
New York 16,329
Los Angeles 12,410
Chicago 7,668
Philadelphia 4,949
Washington, DC 4,466
Detroit 4,307
Houston 3,653
Atlanta 3,331
Boston 3,240
Dallas 2,898
Minneapolis–St Paul 2,688
San Diego 2,632
St Louis 2,536
Phoenix 2,473
Baltimore 2,458
Pittsburgh 2,402
Cleveland 2,222
San Francisco 2,182
Seattle 2,180
Tampa 2,157
Miami 2,025
Denver 1,796
Portland (Or.) 1,676
Kansas City (Mo.) 1,647
Cincinnati 1,581
San Jose 1,557
Norfolk 1,529
Indianapolis 1,462
Milwaukee 1,456
Sacramento 1,441
San Antonio 1,437
Columbus (Oh.) 1,423
New Orleans 1,309
Charlotte 1,260
Buffalo 1,189
Salt Lake City 1,178
Hartford 1,151
Oklahoma 1,007
Jacksonville 665
Omaha 663
Memphis 614
El Paso 579
Austin 514
Nashville 505
Uruguay
Montevideo 1,326
Uzbekistan
Tashkent 2,106
Venezuela
Caracas 2,784
Maracaibo 1,364
Valencia 1,032
Maracay 800
Barquisimeto 745
Ciudad Guayana 524
Vietnam
Ho Chi Minh City 4,322
Hanoi 3,056
Haiphong 783
Yemen
Sana 972
Yugoslavia
Belgrade 1,137
Zambia
Lusaka 982
Zimbabwe
Harare 1,189
Bulawayo 622

¹ SAR = Special Administrative Region of China

World Statistics: Distances

The table shows air distances in miles and kilometers between 30 major cities. Known as "Great Circle" distances, these measure the shortest routes between the cities, which aircraft use wherever possible. The maps show the world centered on six cities, and illustrate, for example, why direct flights from Japan to northern America and Europe are across the Arctic regions. The maps have been constructed on an Azimuthal Equidistant projection, on which all distances measured through the center point are true to scale. The red lines are drawn at 5,000, 10,000 and 15,000 km from the central city.

Distances above the diagonal are in Kms; distances below the diagonal are in Miles.

	Beijing	Bombay	Buenos Aires	Cairo	Calcutta	Caracas	Chicago	Hong Kong	Honolulu	Johannesburg	Lagos	London	Los Angeles	Mexico City	Moscow	Nairobi	New York	Paris	Rio de Janeiro	Rome	Singapore	Sydney	Tokyo	Wellington
Beijing	**Beijing**	2956	11972	4688	2031	8947	6588	1220	5070	7276	7119	5057	6251	7742	3600	5727	6828	5106	10773	5049	2783	5561	1304	6700
Bombay	4757	**Bombay**	9275	2706	1034	9024	8048	2683	8024	4334	4730	4467	8700	9728	3126	2816	7793	4356	8332	3837	2432	6313	4189	7686
Buenos Aires	19268	14925	**Buenos Aires**	7341	10268	3167	5599	11481	7558	5025	4919	6917	6122	4591	8374	6463	5298	6867	1214	6929	9867	7332	11410	6202
Cairo	7544	4355	11814	**Cairo**	3541	6340	6127	5064	8838	3894	2432	2180	7580	7687	1803	2197	5605	1994	6149	1325	5137	8959	5947	10268
Calcutta	3269	1664	16524	5699	**Calcutta**	9609	7978	1653	7048	5256	5727	4946	8152	9494	3438	3839	7921	4883	9366	4486	1800	5678	3195	7055
Caracas	14399	14522	5096	10203	15464	**Caracas**	2502	10166	6009	6847	4810	4664	3612	2228	6175	7173	2131	4738	2825	5196	11407	9534	8801	8154
Chicago	10603	12953	9011	3206	12839	4027	**Chicago**	7783	4247	8689	5973	3949	1742	1694	4971	8005	711	4132	5311	4809	9369	9243	6299	8358
Hong Kong	1963	4317	18478	8150	2659	16360	12526	**Hong Kong**	5543	6669	7360	5980	7232	8775	4439	5453	8047	5984	11001	5769	1615	4582	1786	5857
Honolulu	8160	12914	12164	14223	11343	9670	6836	8921	**Honolulu**	11934	10133	7228	2558	3781	7036	10739	4958	7437	8290	8026	6721	5075	3854	4669
Johannesburg	11710	6974	8088	6267	8459	11019	13984	10732	19206	**Johannesburg**	2799	5637	10362	9063	5692	1818	7979	5426	4420	4811	5381	6860	8418	7308
Lagos	11457	7612	7916	3915	9216	7741	9612	11845	16308	4505	**Lagos**	3118	7713	6879	3886	2366	5268	2929	3750	2510	6925	9643	8376	9973
London	8138	7190	11131	3508	7961	7507	6356	9623	11632	9071	5017	**London**	5442	5552	1552	4237	3463	212	5778	889	6743	10558	5942	11691
Los Angeles	10060	14000	9852	12200	13120	5812	2804	11639	4117	16676	12414	8758	**Los Angeles**	1549	6070	9659	2446	5645	6310	6331	8776	7502	5475	6719
Mexico City	12460	15656	7389	12372	15280	3586	2726	14122	6085	14585	11071	8936	2493	**Mexico City**	6664	9207	2090	5717	4780	6365	10321	8058	7024	6897
Moscow	5794	5031	13477	2902	5534	9938	8000	7144	11323	9161	6254	2498	9769	10724	**Moscow**	3942	4666	1545	7184	1477	5237	9008	4651	10283
Nairobi	9216	4532	10402	3536	6179	11544	12883	8776	17282	2927	3807	6819	15544	14818	6344	**Nairobi**	7358	4029	5548	3350	4635	7552	6996	8490
New York	10988	12541	8526	9020	12747	3430	1145	12950	7980	12841	8477	5572	3936	3264	7510	11842	**New York**	3626	4832	4280	9531	9935	6741	8951
Paris	8217	7010	11051	3210	7858	7625	6650	9630	11968	8732	4714	342	9085	9200	2486	6485	5836	**Paris**	5708	687	6671	10539	6038	11798
Rio de Janeiro	17338	13409	1953	9896	15073	4546	8547	17704	13342	7113	6035	9299	10155	7693	11562	8928	7777	9187	**Rio de Janeiro**	5725	9763	8389	11551	7367
Rome	8126	6175	11151	2133	7219	8363	7739	9284	12916	7743	4039	1431	10188	10243	2376	5391	6888	1105	9214	**Rome**	6229	10143	6127	11523
Singapore	4478	3914	15879	8267	2897	18359	15078	2599	10816	8660	11145	10852	14123	16610	8428	7460	15339	10737	15712	10025	**Singapore**	3915	3306	5298
Sydney	8949	10160	11800	14418	9138	15343	14875	7374	8168	11040	15519	16992	12073	12969	14497	12153	15989	16962	13501	16324	6300	**Sydney**	4861	1383
Tokyo	2099	6742	18362	9571	5141	14164	10137	2874	6202	13547	13480	9562	8811	11304	7485	11260	10849	9718	18589	9861	5321	7823	**Tokyo**	5762
Wellington	10782	12370	9981	16524	11354	13122	13451	9427	7513	11761	16050	18814	10814	11100	16549	13664	14405	18987	11855	18545	8526	2226	9273	**Wellington**

Kms — above the diagonal. Miles — below the diagonal.

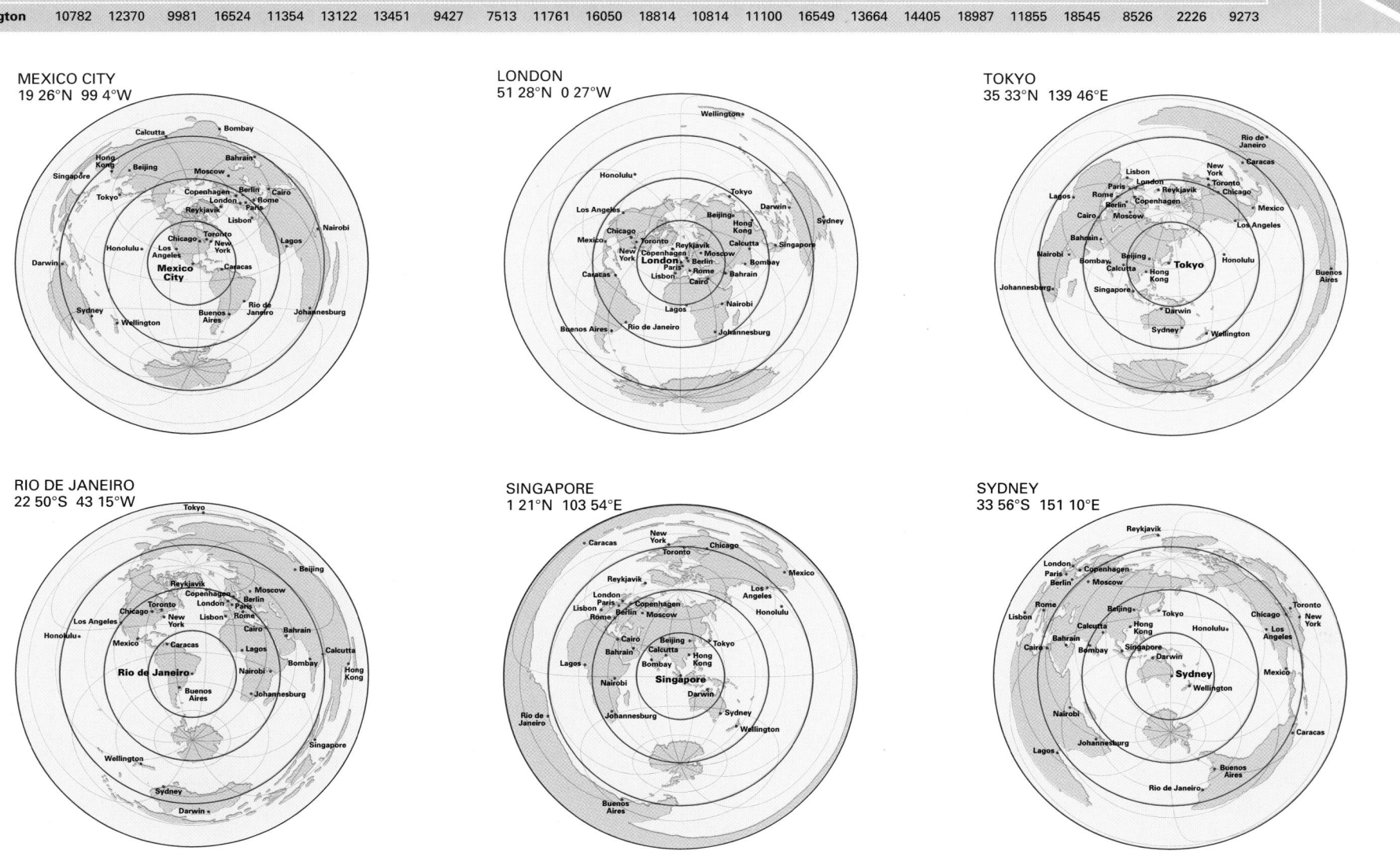

MEXICO CITY 19 26°N 99 4°W

LONDON 51 28°N 0 27°W

TOKYO 35 33°N 139 46°E

RIO DE JANEIRO 22 50°S 43 15°W

SINGAPORE 1 21°N 103 54°E

SYDNEY 33 56°S 151 10°E

World Statistics: Climate

Rainfall and temperature figures are provided for more than 70 cities around the world. As climate is affected by altitude, the height of each city is shown in feet beneath its name. For each month, the figures in blue show the total rainfall or snow in inches, and in red the average temperature in degrees Fahrenheit; the total annual rainfall and average annual temperature are at the end of the rows.

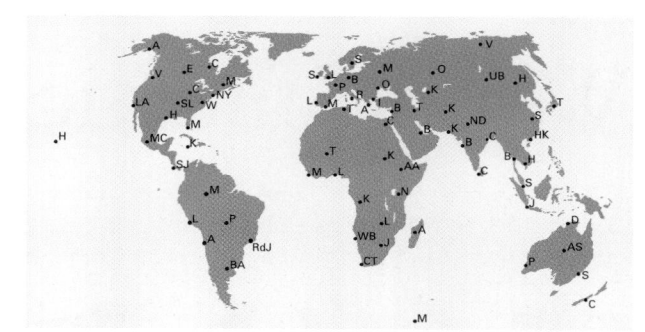

EUROPE

City	Jan.	Feb.	Mar.	Apr.	May	June	July	Aug.	Sept.	Oct.	Nov.	Dec.	Year
Athens, Greece (rain)	2.4	1.5	1.5	0.9	0.9	0.6	0.2	0.3	0.6	2	2.2	2.8	15.8
351 ft (temp)	50	50	54	61	68	77	82	82	75	68	59	52	64
Berlin, Germany (rain)	1.8	1.6	1.3	1.7	1.9	2.6	2.9	2.7	1.9	1.9	1.8	1.7	23.7
180 ft (temp)	30	32	39	48	57	63	66	64	59	48	41	34	48
Istanbul, Turkey (rain)	4.3	3.6	2.8	1.8	1.5	1.3	1.3	1.2	2.3	3.2	4.1	4.7	32.1
374 ft (temp)	41	43	45	52	61	68	73	73	68	61	54	46	57
Lisbon, Portugal (rain)	4.4	3	4.3	2.1	1.7	0.6	0.1	0.2	1.3	2.4	3.7	4.1	27.9
253 ft (temp)	52	54	57	61	63	68	72	73	70	64	57	54	63
London, UK (rain)	2.1	1.6	1.5	1.5	1.8	1.8	2.2	2.3	1.9	2.2	2.5	1.9	23.3
16 ft (temp)	39	41	45	48	54	61	64	63	59	52	46	41	52
Málaga, Spain (rain)	2.4	2	2.4	1.8	1	0.2	0	0.1	1.1	2.5	2.5	2.4	18.7
108 ft (temp)	54	54	61	63	66	84	77	79	73	68	61	55	66
Moscow, Russia (rain)	1.5	1.5	1.4	1.5	2.1	2.3	3.5	2.8	2.3	1.8	1.9	2.1	24.6
512 ft (temp)	9	14	25	43	55	61	64	63	54	43	30	19	39
Odesa, Ukraine (rain)	2.2	2.4	1.2	0.8	1.3	1.3	1.7	1.5	1.5	0.5	1.4	2.8	18.6
210 ft (temp)	27	30	36	48	59	68	72	72	64	54	48	34	51
Paris, France (rain)	2.2	1.8	1.4	1.7	2.2	2.1	2.3	2.5	2.2	2	2	2	24.4
246 ft (temp)	37	39	46	52	59	64	68	66	63	54	45	39	53
Rome, Italy (rain)	2.8	2.4	2.2	2	1.8	1.5	0.6	0.8	2.5	3.9	5.1	3.7	29.3
56 ft (temp)	46	48	52	57	64	72	77	77	72	63	55	50	61
Shannon, Irish Republic (rain)	3.7	2.6	2.2	2.1	2.4	2.2	3	3.1	3.4	3.4	3.8	4.6	36.6
7 ft (temp)	41	41	45	48	54	57	61	61	57	52	46	43	50
Stockholm, Sweden (rain)	1.7	1.2	1	1.2	1.3	1.8	2.4	3	2.4	1.9	2.1	1.9	21.8
144 ft (temp)	27	27	30	41	50	59	64	63	54	45	37	32	44

ASIA

City	Jan.	Feb.	Mar.	Apr.	May	June	July	Aug.	Sept.	Oct.	Nov.	Dec.	Year
Bahrain (rain)	0.3	0.7	0.5	0.3	<0.1	0	0	0	0	0	0.7	0.7	3.3
16 ft (temp)	63	64	70	77	84	90	91	93	88	82	75	66	79
Bangkok, Thailand (rain)	0.3	0.8	1.4	2.3	7.8	6.3	6.3	6.9	12	8.1	2.6	0.2	55
7 ft (temp)	79	82	84	86	84	84	82	82	82	82	79	77	82
Beirut, Lebanon (rain)	7.5	6.2	3.7	2.1	0.7	0.1	<0.1	<0.1	0.2	2	5.2	7.3	35
112 ft (temp)	79	82	84	86	84	84	82	82	82	82	79	77	82
Bombay, India (rain)	0.1	0.1	0.1	<0.1	0.7	19.1	24.3	13.4	10.4	2.5	0.5	0.1	71.4
36 ft (temp)	75	75	79	82	86	84	81	81	81	82	81	79	80
Calcutta, India (rain)	0.4	1.2	1.4	1.7	5.5	11.7	12.8	12.9	9.9	4.5	0.8	0.2	63
20 ft (temp)	68	72	81	86	86	86	84	84	84	82	73	66	79
Colombo, Sri Lanka (rain)	3.5	2.7	5.8	9.1	14.6	8.8	5.3	4.3	6.3	13.7	12.4	5.8	92.3
23 ft (temp)	79	79	81	82	82	81	81	81	81	81	79	79	80
Harbin, China (rain)	0.2	0.2	0.4	0.9	1.7	3.7	4.4	4.1	1.8	1.3	0.3	0.2	19.3
525 ft (temp)	0	5	23	43	55	66	72	70	57	39	21	3	38
Ho Chi Minh, Vietnam (rain)	0.6	0.1	0.5	1.7	8.7	13	12.4	10.6	13.2	10.6	4.5	2.2	78.1
30 ft (temp)	79	81	84	86	84	82	82	82	81	81	81	79	82
Hong Kong, China (rain)	1.3	1.8	2.9	5.4	11.5	15.5	15	14.2	10.1	4.5	1.7	1.2	85.2
108 ft (temp)	61	59	64	72	79	82	82	82	81	77	70	64	73
Jakarta, Indonesia (rain)	11.8	11.8	8.3	5.8	4.5	3.8	2.5	1.7	2.6	4.4	5.6	8	70.8
26 ft (temp)	79	79	81	81	81	81	81	81	81	81	81	79	80
Kabul, Afghanistan (rain)	1.2	1.4	3.7	4	0.8	0.2	0.1	0.1	<0.1	0.6	0.8	0.8	13.3
5,953 ft (temp)	27	30	43	55	64	72	77	75	68	57	45	37	54
Karachi, Pakistan (rain)	0.5	0.4	0.3	0.1	0.1	0.7	3.2	1.6	0.5	<0.1	0.1	0.2	7.8
13 ft (temp)	66	68	75	82	86	88	86	84	82	82	75	68	79
Kazalinsk, Kazakstan (rain)	0.4	0.4	0.5	0.5	0.6	0.2	0.3	0.3	0.4	0.5	0.6		4.9
207 ft (temp)	10	12	27	43	64	73	77	73	61	46	30	19	45
New Delhi, India (rain)	0.9	0.7	0.5	0.3	0.5	2.9	7.1	6.8	4.6	0.4	0.1	0.4	25.2
715 ft (temp)	57	63	73	82	91	93	88	86	84	79	68	59	77
Omsk, Russia (rain)	0.6	0.3	0.3	0.5	1.2	2	2	2	1.1	1	0.7	0.8	12.6
279 ft (temp)	-7	-1	10	30	50	61	64	61	50	34	12	0	30
Shanghai, China (rain)	1.9	2.3	3.3	3.7	3.7	7.1	5.8	5.6	5.1	2.8	2	1.4	44.7
23 ft (temp)	39	41	48	57	68	75	82	82	73	66	54	43	61
Singapore (rain)	9.9	6.8	7.6	7.4	6.8	6.8	6.7	7.7	7	8.2	10	10.1	95.1
33 ft (temp)	79	81	82	82	82	82	82	81	81	81	81	81	81
Tehran, Iran (rain)	1.8	1.5	1.8	1.4	0.5	0.1	0.1	0.1	0.1	0.3	0.8	1.2	9.8
4,002 ft (temp)	36	41	48	61	70	79	86	84	77	64	54	43	62
Tokyo, Japan (rain)	1.9	2.9	4.2	5.3	5.8	6.5	5.6	6	9.2	8.2	3.8	2.2	61.6
20 ft (temp)	37	39	45	55	63	70	77	79	73	63	52	43	58
Ulan Bator, Mongolia (rain)	<0.1	<0.1	0.1	0.2	0.4	1.1	3	2	0.9	0.2	0.2	0.1	8.2
4,346 ft (temp)	-14	-5	9	30	43	57	61	57	46	30	9	-7	26
Verkhoyansk, Russia (rain)	0.2	0.2	0.1	0.2	0.3	0.9	1.1	1	0.5	0.3	0.3	0.2	5.4
328 ft (temp)	-57	-48	-25	5	32	54	57	48	36	5	-35	-53	1

AFRICA

City	Jan.	Feb.	Mar.	Apr.	May	June	July	Aug.	Sept.	Oct.	Nov.	Dec.	Year
Addis Ababa, Ethiopia (rain)	<0.1	0.1	1	5.3	8.4	7.9	8.1	9.4	4	1.1	<0.1	0	45.4
8,036 ft (temp)	66	68	68	68	66	64	64	66	70	72	70	68	68
Antananarivo, Madagas. (rain)	11.8	11	7	2.1	0.7	0.3	0.3	0.4	0.7	2	4	5.3	53.3
4,500 ft (temp)	70	70	70	66	64	59	57	59	63	66	70	70	65
Cairo, Egypt (rain)	0.2	0.2	0.2	0.1	0.1	<0.1	0	0	<0.1	<0.1	0.2	0.2	1.1
380 ft (temp)	55	59	64	70	77	82	82	82	79	75	68	59	71
Cape Town, South Africa (rain)	0.6	0.3	0.7	1.9	3.1	3.3	3.5	2.6	1.7	1.2	0.7	0.4	20
56 ft (temp)	70	70	68	63	57	55	54	55	57	61	64	66	62
Johannesburg, S. Africa (rain)	4.5	4.3	3.5	1.5	1	0.3	0.3	0.3	0.9	2.2	4.2	4.9	28
5,461 ft (temp)	68	68	64	61	55	50	52	55	61	64	66	68	61

AFRICA (continued)

City	Jan.	Feb.	Mar.	Apr.	May	June	July	Aug.	Sept.	Oct.	Nov.	Dec.	Year
Khartoum, Sudan (rain)	<0.1	<0.1	<0.1	<0.1	0.1	0.3	2.1	2.8	0.7	0.2	<0.1	0	6.2
1,279 ft (temp)	75	77	82	88	91	93	90	88	90	90	82	77	85
Kinshasa, Congo (Zaïre) (rain)	5.3	5.7	7.7	7.7	6.2	0.3	0.1	0.1	1.2	4.7	8.7	5.6	53.4
1,066 ft (temp)	79	79	81	81	79	75	73	75	77	79	79	79	78
Lagos, Nigeria (rain)	1.1	1.8	4	5.9	10.6	18.1	11	2.5	5.5	8.1	2.7	1	72.4
10 ft (temp)	81	82	84	82	82	79	77	79	77	79	79	82	81
Lusaka, Zambia (rain)	9.1	7.5	5.6	0.7	0.1	<0.1	<0.1	0	<0.1	0.4	3.6	5.9	32.9
4,189 ft (temp)	70	72	70	70	66	61	61	64	72	75	73	72	69
Monrovia, Liberia (rain)	1.2	2.2	3.8	8.5	20.3	38.3	39.2	14.7	29.3	30.4	9.3	5.1	202.4
75 ft (temp)	79	79	81	81	79	77	75	77	77	77	79	79	78
Nairobi, Kenya (rain)	1.5	2.5	4.9	8.3	6.2	1.8	0.6	0.9	1.2	2.1	4.3	3.4	37.8
5,970 ft (temp)	66	66	66	66	64	61	61	61	64	66	64	64	64
Timbuktu, Mali (rain)	<0.1	<0.1	0.1	<0.1	0.2	0.9	3.1	3.2	1.5	0.1	<0.1	<0.1	9.1
987 ft (temp)	72	75	82	90	93	95	90	86	90	88	82	73	85
Tunis, Tunisia (rain)	2.5	2	1.6	1.4	0.7	0.3	0.1	0.3	1.3	2	1.9	2.4	16.6
216 ft (temp)	50	52	55	61	66	73	79	81	77	68	61	52	65
Walvis Bay, Namibia (rain)	<0.1	0.2	0.3	0.1	0.1	<0.1	0.1	<0.1	<0.1	<0.1	<0.1	<0.1	0.9
23 ft (temp)	66	66	66	64	63	61	59	57	57	59	63	64	62

AUSTRALIA, NEW ZEALAND AND ANTARCTICA

City	Jan.	Feb.	Mar.	Apr.	May	June	July	Aug.	Sept.	Oct.	Nov.	Dec.	Year
Alice Springs, Australia (rain)	1.7	1.3	1.1	0.4	0.6	0.5	0.3	0.3	0.3	0.7	1.2	1.5	10
1,899 ft (temp)	84	82	77	68	59	54	54	57	64	73	79	82	70
Christchurch, N. Zealand (rain)	2.2	1.7	1.9	1.9	2.6	2.6	2.7	1.9	1.8	1.7	1.9	2.2	25.1
33 ft (temp)	61	61	57	54	48	43	43	45	48	54	57	61	53
Darwin, Australia (rain)	15.2	12.3	10	3.8	0.6	0.1	<0.1	0.1	0.5	2	4.7	9.4	58.7
98 ft (temp)	84	84	84	84	82	79	77	79	82	84	86	84	83
Mawson, Antarctica (rain)	0.4	1.2	0.8	0.4	1.7	7.1	0.2	1.6	0.1	0.8	0	0	14.3
46 ft (temp)	32	23	14	7	5	3	0	0	-1	9	23	30	12
Perth, Australia (rain)	0.3	0.4	0.8	1.7	5.1	7.1	6.7	5.9	3.4	2.2	0.8	0.5	34.8
197 ft (temp)	73	73	72	66	61	57	55	55	59	61	66	72	64
Sydney, Australia (rain)	3.5	4	5.3	5	4.6	4.6	3	2.9	2.8	2.9	2.9	2.9	46.5
138 ft (temp)	72	72	70	64	59	55	54	55	59	64	66	70	63

NORTH AMERICA

City	Jan.	Feb.	Mar.	Apr.	May	June	July	Aug.	Sept.	Oct.	Nov.	Dec.	Year
Anchorage, Alaska, USA (rain)	0.8	0.7	0.6	0.4	0.5	0.7	1.6	2.6	2.6	2.2	1	0.9	14.6
131 ft (temp)	12	18	23	36	45	54	57	55	48	36	23	12	35
Chicago, Illinois, USA (rain)	2	2	2.6	2.8	3.4	3.5	3.3	3.2	3.1	2.6	2.4	2	32.9
823 ft (temp)	25	27	36	48	57	68	73	72	66	54	41	30	50
Churchill, Man., Canada (rain)	0.6	0.5	0.7	0.9	1.3	1.7	1.8	2.3	2	1.7	1.5	0.8	15.9
43 ft (temp)	-17	-14	-3	14	28	43	54	52	41	28	10	-7	19
Edmonton, Alta., Canada (rain)	1	0.7	0.7	0.9	1.7	3	3.5	3.1	1.5	0.7	0.6	1	18.5
2,217 ft (temp)	5	14	23	39	52	59	63	61	52	43	25	14	37
Honolulu, Hawaii, USA (rain)	4.1	2.6	3.1	1.9	1	0.7	0.9	1.1	1.4	1.9	2.5	4.1	25.3
39 ft (temp)	73	64	66	68	72	75	77	79	79	75	72	66	72
Houston, Texas, USA (rain)	3.5	3	3.3	3.6	4.7	4.6	3.9	3.9	4.1	3.7	3.5	4.3	46.1
39 ft (temp)	54	55	63	70	75	81	82	84	79	72	61	54	69
Kingston, Jamaica (rain)	0.9	0.6	0.9	1.2	4	3.5	1.5	3.6	3.9	7.1	2.9	1.4	31.5
112 ft (temp)	77	77	77	79	79	82	82	82	81	81	79	79	80
Los Angeles, Calif., USA (rain)	3.1	3	2.8	1	0.4	<0.1	<0.1	0.2	0.6	1.2	2.6		15
312 ft (temp)	55	57	57	61	63	66	70	72	70	64	61	57	63
Mexico City, Mexico (rain)	0.5	0.2	0.4	0.8	2.1	4.7	6.7	6	5.1	2	0.7	0.3	29.5
7,574 ft (temp)	54	55	61	64	66	66	63	64	64	61	57	55	61
Miami, Florida, USA (rain)	2.8	2.1	2.5	3.2	6.8	7	6.1	6.3	8	9.2	2.8	2	58.8
26 ft (temp)	68	68	72	73	77	81	82	82	81	77	72	70	75
Montréal, Que., Canada (rain)	2.8	2.6	2.9	2.9	2.6	3.2	3.5	3.6	3.5	3	3.2	3.4	37.3
187 ft (temp)	14	16	27	21	55	64	70	68	59	48	36	19	41
New York City, N.Y., USA (rain)	3.7	3.8	3.6	3.2	3.2	3.3	4.2	4.3	3.4	3.5	3	3.6	42.8
315 ft (temp)	30	30	37	50	61	68	73	73	70	59	45	36	53
St Louis, Mo., USA (rain)	2.3	2.5	3.5	3.8	4.5	4.5	3.5	3.4	3.2	2.9	2.8	2.5	39.4
567 ft (temp)	32	34	45	55	66	75	79	79	72	59	46	36	56
San José, Costa Rica (rain)	0.6	0.2	0.8	1.8	9	9.5	8.3	9.5	12	11.8	5.7	1.6	70.8
3,759 ft (temp)	66	66	70	70	72	70	70	70	70	68	68	66	69
Vancouver, B.C., Canada (rain)	6.1	4.5	4	2.4	2	1.8	1.3	1.6	2.6	4.5	5.9	7.2	43.8
46 ft (temp)	37	41	43	48	54	59	63	63	57	50	43	39	50
Washington, D.C., USA (rain)	3.4	3	3.6	3.3	3.7	3.9	4.4	4.3	3.7	2.9	2.6	3.1	41.9
72 ft (temp)	34	36	45	54	63	72	77	75	68	57	46	37	56

SOUTH AMERICA

City	Jan.	Feb.	Mar.	Apr.	May	June	July	Aug.	Sept.	Oct.	Nov.	Dec.	Year
Antofagasta, Chile (rain)	0	0	0	<0.1	<0.1	0.1	0.2	0.1	<0.1	0.1	<0.1	0	0.6
308 ft (temp)	70	70	68	64	61	59	57	57	59	61	64	66	63
Buenos Aires, Argentina (rain)	3.1	2.8	4.3	3.5	3	2.2	2.4	2.4	3.1	3.4	3.3	3.9	37.4
89 ft (temp)	73	73	70	63	55	48	50	52	55	59	66	72	61
Lima, Peru (rain)	0.1	<0.1	<0.1	<0.1	0.2	0.1	0.3	0.3	0.3	0.1	<0.1	0.1	1.7
394 ft (temp)	73	75	75	72	66	63	63	61	63	64	66	70	68
Manaus, Brazil (rain)	9.8	9.1	10.3	8.7	6.7	3.3	2.3	1.5	1.8	4.2	5.6	8	71.3
144 ft (temp)	82	82	82	81	82	82	82	84	84	84	82	82	83
Paraná, Brazil (rain)	11.3	9.3	9.4	4	0.5	<0.1	0.1	0.2	1.1	5	9.1	12.2	62.2
853 ft (temp)	73	73	73	73	73	70	70	72	75	75	75	73	73
Rio de Janeiro, Brazil (rain)	4.9	4.8	5.1	4.2	3.1	2.1	1.6	1.7	2.6	3.1	4.1	5.4	42.8
200 ft (temp)	79	79	77	75	72	70	70	70	72	73	77	77	74

World Statistics: Physical Dimensions

Each topic list is divided into continents and within a continent the items are listed in order of size. The order of the continents is as in the atlas, Europe through to South America. Certain lists down to this mark > are complete; below they are selective. The world top ten are shown in square brackets; in the case of mountains this has not been done because the world top 30 are all in Asia. The figures are rounded as appropriate.

World, Continents, Oceans

	km²	miles²	%
The World	509,450,000	196,672,000	
Land	149,450,000	57,688,000	29.3
Water	360,000,000	138,984,000	70.7
Asia	44,500,000	17,177,000	29.8
Africa	30,302,000	11,697,000	20.3
North America	24,241,000	9,357,000	16.2
South America	17,793,000	6,868,000	11.9
Antarctica	14,100,000	5,443,000	9.4
Europe	9,957,000	3,843,000	6.7
Australia & Oceania	8,557,000	3,303,000	5.7
Pacific Ocean	179,679,000	69,356,000	49.9
Atlantic Ocean	92,373,000	35,657,000	25.7
Indian Ocean	73,917,000	28,532,000	20.5
Arctic Ocean	14,090,000	5,439,000	3.9

Seas

Pacific

	km²	miles²
South China Sea	2,974,600	1,148,500
Bering Sea	2,268,000	875,000
Sea of Okhotsk	1,528,000	590,000
East China & Yellow	1,249,000	482,000
Sea of Japan	1,008,000	389,000
Gulf of California	162,000	62,500
Bass Strait	75,000	29,000

Atlantic

	km²	miles²
Caribbean Sea	2,766,000	1,068,000
Mediterranean Sea	2,516,000	971,000
Gulf of Mexico	1,543,000	596,000
Hudson Bay	1,232,000	476,000
North Sea	575,000	223,000
Black Sea	462,000	178,000
Baltic Sea	422,170	163,000
Gulf of St Lawrence	238,000	92,000

Indian

	km²	miles²
Red Sea	438,000	169,000
The Gulf	239,000	92,000

Mountains

Europe

		m	ft
Mont Blanc	France/Italy	4,807	15,771
Monte Rosa	Italy/Switzerland	4,634	15,203
Dom	Switzerland	4,545	14,911
Liskamm	Switzerland	4,527	14,852
Weisshorn	Switzerland	4,505	14,780
Taschorn	Switzerland	4,490	14,730
Matterhorn/Cervino	Italy/Switzerland	4,478	14,691
Mont Maudit	France/Italy	4,465	14,649
Dent Blanche	Switzerland	4,356	14,291
> Nadelhorn	Switzerland	4,327	14,196
Grandes Jorasses	France/Italy	4,208	13,806
Jungfrau	Switzerland	4,158	13,642
Barre des Ecrins	France	4,103	13,461
Gran Paradiso	Italy	4,061	13,323
Piz Bernina	Italy/Switzerland	4,049	13,284
Eiger	Switzerland	3,970	13,025
Monte Viso	Italy	3,841	12,602
Grossglockner	Austria	3,797	12,457
Wildspitze	Austria	3,772	12,382
Monte Disgrazia	Italy	3,678	12,066
Mulhacén	Spain	3,478	11,411
Pico de Aneto	Spain	3,404	11,168
Marmolada	Italy	3,342	10,964
Etna	Italy	3,340	10,958
Punta del'Argentera	Italy	3,297	10,817
Zugspitze	Germany	2,962	9,718
Musala	Bulgaria	2,925	9,596
Olympus	Greece	2,917	9,570
Triglav	Slovenia	2,863	9,393
Monte Cinto	France (Corsica)	2,710	8,891
Gerlachovka	Slovak Republic	2,655	8,711
Torre de Cerrado	Spain	2,648	8,688
Galdhöpiggen	Norway	2,468	8,100
Hvannadalshnúkur	Iceland	2,119	6,952
Kebnekaise	Sweden	2,117	6,946
Ben Nevis	UK	1,343	4,406

Asia

		m	ft
Everest	China/Nepal	8,848	29,029
K2 (Godwin Austen)	China/Kashmir	8,611	28,251
Kanchenjunga	India/Nepal	8,598	28,208
Lhotse	China/Nepal	8,516	27,939
Makalu	China/Nepal	8,481	27,824
Cho Oyu	China/Nepal	8,201	26,906
Dhaulagiri	Nepal	8,172	26,811
Manaslu	Nepal	8,156	26,758
Nanga Parbat	Kashmir	8,126	26,660
Annapurna	Nepal	8,078	26,502
Gasherbrum	China/Kashmir	8,068	26,469
Broad Peak	China/Kashmir	8,051	26,414
Xixabangma	China	8,012	26,286
Kangbachen	India/Nepal	7,902	25,925
Jannu	India/Nepal	7,902	25,925
Gayachung Kang	Nepal	7,897	25,909
Himalchuli	Nepal	7,893	25,896
Disteghil Sar	Kashmir	7,885	25,869
Nuptse	Nepal	7,879	25,849
Khunyang Chhish	Kashmir	7,852	25,761
Masherbrum	Kashmir	7,821	25,659
Nanda Devi	India	7,817	25,646
Rakaposhi	Kashmir	7,788	25,551
Batura	Kashmir	7,785	25,541
Namche Barwa	China	7,756	25,446
Kamet	India	7,756	25,446
Soltoro Kangri	Kashmir	7,742	25,400
Gurla Mandhata	China	7,728	25,354
Trivor	Pakistan	7,720	25,328
> Kongur Shan	China	7,719	25,324
Tirich Mir	Pakistan	7,690	25,229
K'ula Shan	Bhutan/China	7,543	24,747
Pik Kommunizma	Tajikistan	7,495	24,590
Elbrus	Russia	5,642	18,510
Demavend	Iran	5,604	18,386
Ararat	Turkey	5,165	16,945
Gunong Kinabalu	Malaysia (Borneo)	4,101	13,455
Yu Shan	Taiwan	3,997	13,113
Fuji-San	Japan	3,776	12,388

Africa

		m	ft
Kilimanjaro	Tanzania	5,895	19,340
Mt Kenya	Kenya	5,199	17,057
Ruwenzori (Margherita)	Uganda/Congo (Z.)	5,109	16,762
Ras Dashan	Ethiopia	4,620	15,157
Meru	Tanzania	4,565	14,977
Karisimbi	Rwanda/Congo (Z.)	4,507	14,787
Mt Elgon	Kenya/Uganda	4,321	14,176
Batu	Ethiopia	4,307	14,130
Guna	Ethiopia	4,231	13,882
Toubkal	Morocco	4,165	13,665
Irhil Mgoun	Morocco	4,071	13,356
Mt Cameroon	Cameroon	4,070	13,353
Amba Ferit	Ethiopia	3,875	13,042
Pico del Teide	Spain (Tenerife)	3,718	12,198
Thabana Ntlenyana	Lesotho	3,482	11,424
> Emi Koussi	Chad	3,415	11,204
Mt aux Sources	Lesotho/S. Africa	3,282	10,768
Mt Piton	Réunion	3,069	10,069

Oceania

		m	ft
Puncak Jaya	Indonesia	5,029	16,499
Puncak Trikora	Indonesia	4,750	15,584
Puncak Mandala	Indonesia	4,702	15,427
> Mt Wilhelm	Papua New Guinea	4,508	14,790
Mauna Kea	USA (Hawaii)	4,205	13,796
Mauna Loa	USA (Hawaii)	4,170	13,681
Mt Cook (Aoraki)	New Zealand	3,753	12,313
Mt Balbi	Solomon Is.	2,439	8,002
Orohena	Tahiti	2,241	7,352
Mt Kosciuszko	Australia	2,237	7,339

North America

		m	ft
Mt McKinley (Denali)	USA (Alaska)	6,194	20,321
Mt Logan	Canada	5,959	19,551
Citlaltepetl	Mexico	5,700	18,701
Mt St Elias	USA/Canada	5,489	18,008
Popocatepetl	Mexico	5,452	17,887
Mt Foraker	USA (Alaska)	5,304	17,401
Ixtaccihuatl	Mexico	5,286	17,342
Lucania	Canada	5,227	17,149
Mt Steele	Canada	5,073	16,644
Mt Bona	USA (Alaska)	5,005	16,420
Mt Blackburn	USA (Alaska)	4,996	16,391
Mt Sanford	USA (Alaska)	4,940	16,207
Mt Wood	Canada	4,848	15,905
Nevado de Toluca	Mexico	4,670	15,321
Mt Fairweather	USA (Alaska)	4,663	15,298
Mt Hunter	USA (Alaska)	4,442	15,573
Mt Whitney	USA	4,418	14,495
Mt Elbert	USA	4,399	14,432
Mt Harvard	USA	4,395	14,419
Mt Rainier	USA	4,392	14,409
> Blanca Peak	USA	4,372	14,344
Longs Peak	USA	4,345	14,255
Tajumulco	Guatemala	4,220	13,845
Grand Teton	USA	4,197	13,770
Mt Waddington	Canada	3,994	13,104
Mt Robson	Canada	3,954	12,972
Chirripó Grande	Costa Rica	3,837	12,589
Mt Assiniboine	Canada	3,619	11,873
Pico Duarte	Dominican Rep.	3,175	10,417

South America

		m	ft
Aconcagua	Argentina	6,960	22,834
Bonete	Argentina	6,872	22,546
Ojos del Salado	Argentina/Chile	6,863	22,516
Pissis	Argentina	6,779	22,241
Mercedario	Argentina/Chile	6,770	22,211
Huascaran	Peru	6,768	22,204
Llullaillaco	Argentina/Chile	6,723	22,057
Nudo de Cachi	Argentina	6,720	22,047
Yerupaja	Peru	6,632	21,758
N. de Tres Cruces	Argentina/Chile	6,620	21,719
Incahuasi	Argentina/Chile	6,601	21,654
Cerro Galan	Argentina	6,600	21,654
Tupungato	Argentina/Chile	6,570	21,555
> Sajama	Bolivia	6,542	21,463
Illimani	Bolivia	6,485	21,276
Coropuna	Peru	6,425	21,079
Ausangate	Peru	6,384	20,945
Cerro del Toro	Argentina	6,380	20,932
Siula Grande	Peru	6,356	20,853
Chimborazo	Ecuador	6,267	20,561
Alpamayo	Peru	5,947	19,511
Cotapaxi	Ecuador	5,896	19,344
Pico Colon	Colombia	5,800	19,029
Pico Bolivar	Venezuela	5,007	16,427

Antarctica

		m	ft
Vinson Massif		4,897	16,066
Mt Kirkpatrick		4,528	14,855
Mt Markham		4,349	14,268

Ocean Depths

Atlantic Ocean

	m	ft	
Puerto Rico (Milwaukee) Deep	9,220	30,249	[7]
Cayman Trench	7,680	25,197	
Gulf of Mexico	5,203	17,070	
Mediterranean Sea	5,121	16,801	
Black Sea	2,211	7,254	
North Sea	660	2,165	
Baltic Sea	463	1,519	

Indian Ocean

	m	ft
Java Trench	7,450	24,442
Red Sea	2,635	8,454
Persian Gulf	73	239

Pacific Ocean

	m	ft	
Mariana Trench	11,022	36,161	[1]
Tonga Trench	10,882	35,702	[2]
Japan Trench	10,554	34,626	[3]
Kuril Trench	10,542	34,587	[4]
Mindanao Trench	10,497	34,439	[5]
Kermadec Trench	10,047	32,962	[6]
New Guinea Trench	9,140	29,987	[8]
Peru–Chile Trench	8,050	26,410	[9]
Aleutian Trench	7,822	25,662	[10]
Middle American Trench	6,662	21,857	

Arctic Ocean

	m	ft
Molloy Deep	5,608	18,399

Land Lows

		m	ft
Caspian Sea	Europe	−28	−92
Dead Sea	Asia	−403	−1,322
Lake Asale	Africa	−116	−381
Lake Eyre North	Oceania	−16	−52
Death Valley	N. America	−86	−282
Valdés Peninsula	S. America	−40	−131

Rivers

Europe

		km	miles	
Volga	Caspian Sea	3,700	2,300	
Danube	Black Sea	2,850	1,770	
Ural	Caspian Sea	2,535	1,575	
Dnepr (Dnipro)	Black Sea	2,285	1,420	
Kama	Volga	2,030	1,260	
Don	Black Sea	1,990	1,240	
Petchora	Arctic Ocean	1,790	1,110	
Oka	Volga	1,480	920	
Belaya	Kama	1,420	880	
Dnister (Dniester)	Black Sea	1,400	870	
Vyatka	Kama	1,370	850	
Rhine	North Sea	1,320	820	
N. Dvina	Arctic Ocean	1,290	800	
Desna	Dnepr (Dnipro)	1,190	740	
Elbe	North Sea	1,145	710	
Wisła	Baltic Sea	1,090	675	
Loire	Atlantic Ocean	1,020	635	
W. Dvina	Baltic Sea	1,019	633	

Asia

		km	miles	
Yangtze	Pacific Ocean	6,380	3,960	[3]
Yenisey–Angara	Arctic Ocean	5,550	3,445	[5]
Huang He	Pacific Ocean	5,464	3,395	[6]
Ob–Irtysh	Arctic Ocean	5,410	3,360	[7]
Mekong	Pacific Ocean	4,500	2,795	[9]
Amur	Pacific Ocean	4,400	2,730	[10]
Lena	Arctic Ocean	4,400	2,730	
Irtysh	Ob	4,250	2,640	
Yenisey	Arctic Ocean	4,090	2,540	
Ob	Arctic Ocean	3,680	2,285	
Indus	Indian Ocean	3,100	1,925	
Brahmaputra	Indian Ocean	2,900	1,800	
Syrdarya	Aral Sea	2,860	1,775	
Salween	Indian Ocean	2,800	1,740	
Euphrates	Indian Ocean	2,700	1,675	
Vilyuy	Lena	2,650	1,645	
Kolyma	Arctic Ocean	2,600	1,615	
Amudarya	Aral Sea	2,540	1,575	
Ural	Caspian Sea	2,535	1,575	
Ganges	Indian Ocean	2,510	1,560	
Si Kiang	Pacific Ocean	2,100	1,305	
Irrawaddy	Indian Ocean	2,010	1,250	
Tarim–Yarkand	Lop Nor	2,000	1,240	
Tigris	Indian Ocean	1,900	1,180	
Angara	Yenisey	1,830	1,135	
Godavari	Indian Ocean	1,470	915	
Sutlej	Indian Ocean	1,450	900	
Yamuna	Indian Ocean	1,400	870	

Africa

		km	miles	
Nile	Mediterranean	6,670	4,140	[1]
Congo	Atlantic Ocean	4,670	2,900	[8]
Niger	Atlantic Ocean	4,180	2,595	
Zambezi	Indian Ocean	3,540	2,200	
Oubangui/Uele	Congo (Zaïre)	2,250	1,400	
Kasai	Congo (Zaïre)	1,950	1,210	
Shaballe	Indian Ocean	1,930	1,200	
Orange	Atlantic Ocean	1,860	1,155	
Cubango	Okavango Swamps	1,800	1,120	
Limpopo	Indian Ocean	1,600	995	
Senegal	Atlantic Ocean	1,600	995	
Volta	Atlantic Ocean	1,500	930	
Benue	Niger	1,350	840	

Australia

		km	miles	
Murray–Darling	Indian Ocean	3,750	2,330	
Darling	Murray	3,070	1,905	
Murray	Indian Ocean	2,575	1,600	
Murrumbidgee	Murray	1,690	1,050	

North America

		km	miles	
Mississippi–Missouri	Gulf of Mexico	6,020	3,740	[4]
Mackenzie	Arctic Ocean	4,240	2,630	
Mississippi	Gulf of Mexico	3,780	2,350	
Missouri	Mississippi	3,780	2,350	
Yukon	Pacific Ocean	3,185	1,980	
Rio Grande	Gulf of Mexico	3,030	1,880	
Arkansas	Mississippi	2,340	1,450	
Colorado	Pacific Ocean	2,330	1,445	
Red	Mississippi	2,040	1,270	
Columbia	Pacific Ocean	1,950	1,210	
Saskatchewan	Lake Winnipeg	1,940	1,205	
Snake	Columbia	1,670	1,040	
Churchill	Hudson Bay	1,600	990	
Ohio	Mississippi	1,580	980	
Brazos	Gulf of Mexico	1,400	870	
St Lawrence	Atlantic Ocean	1,170	730	

South America

		km	miles	
Amazon	Atlantic Ocean	6,450	4,010	[2]
Paraná–Plate	Atlantic Ocean	4,500	2,800	
Purus	Amazon	3,350	2,080	
Madeira	Amazon	3,200	1,990	
São Francisco	Atlantic Ocean	2,900	1,800	
Paraná	Plate	2,800	1,740	
Tocantins	Atlantic Ocean	2,750	1,710	
Paraguay	Paraná	2,550	1,580	
Orinoco	Atlantic Ocean	2,500	1,550	
Pilcomayo	Paraná	2,500	1,550	
Araguaia	Tocantins	2,250	1,400	
Juruá	Amazon	2,000	1,240	
Xingu	Amazon	1,980	1,230	
Ucayali	Amazon	1,900	1,180	
Marañón	Amazon	1,600	990	
Uruguay	Plate	1,600	990	
Magdalena	Caribbean Sea	1,540	960	

Lakes

Europe

		km²	miles²	
Lake Ladoga	Russia	17,700	6,800	
Lake Onega	Russia	9,700	3,700	
Saimaa system	Finland	8,000	3,100	
Vänern	Sweden	5,500	2,100	
Rybinskoye Res.	Russia	4,700	1,800	

Asia

		km²	miles²	
Caspian Sea	Asia	371,800	143,550	[1]
Aral Sea	Kazak./Uzbek.	33,640	13,000	[6]
Lake Baykal	Russia	30,500	11,780	[9]
Tonlé Sap	Cambodia	20,000	7,700	
Lake Balqash	Kazakstan	18,500	7,100	
Lake Dongting	China	12,000	4,600	
Lake Ysyk	Kyrgyzstan	6,200	2,400	
Lake Orumiyeh	Iran	5,900	2,300	
Lake Koko	China	5,700	2,200	
Lake Poyang	China	5,000	1,900	
Lake Khanka	China/Russia	4,400	1,700	
Lake Van	Turkey	3,500	1,400	
Lake Ubsa	China	3,400	1,300	

Africa

		km²	miles²	
Lake Victoria	E. Africa	68,000	26,000	[3]
Lake Tanganyika	C. Africa	33,000	13,000	[7]
Lake Malawi/Nyasa	E. Africa	29,600	11,430	[10]
Lake Chad	C. Africa	25,000	9,700	
Lake Turkana	Ethiopia/Kenya	8,500	3,300	
Lake Volta	Ghana	8,500	3,300	
Lake Bangweulu	Zambia	8,000	3,100	
Lake Rukwa	Tanzania	7,000	2,700	
Lake Mai-Ndombe	Congo (Zaïre)	6,500	2,500	
Lake Kariba	Zambia/Zimbabwe	5,300	2,000	
Lake Albert	Uganda/Congo (Z.)	5,300	2,000	
Lake Nasser	Egypt/Sudan	5,200	2,000	
Lake Mweru	Zambia/Congo (Z.)	4,900	1,900	
Lake Cabora Bassa	Mozambique	4,500	1,700	
Lake Kyoga	Uganda	4,400	1,700	
Lake Tana	Ethiopia	3,630	1,400	
Lake Kivu	Rwanda/Congo (Z.)	2,650	1,000	
Lake Edward	Uganda/Congo (Z.)	2,200	850	

Australia

		km²	miles²	
Lake Eyre	Australia	8,900	3,400	
Lake Torrens	Australia	5,800	2,200	
Lake Gairdner	Australia	4,800	1,900	

North America

		km²	miles²	
Lake Superior	Canada/USA	82,350	31,800	[2]
Lake Huron	Canada/USA	59,600	23,010	[4]
Lake Michigan	USA	58,000	22,400	[5]
Great Bear Lake	Canada	31,800	12,280	[8]
Great Slave Lake	Canada	28,500	11,000	
Lake Erie	Canada/USA	25,700	9,900	
Lake Winnipeg	Canada	24,400	9,400	
Lake Ontario	Canada/USA	19,500	7,500	
Lake Nicaragua	Nicaragua	8,200	3,200	
Lake Athabasca	Canada	8,100	3,100	
Smallwood Res.	Canada	6,530	2,520	
Reindeer Lake	Canada	6,400	2,500	
Lake Winnipegosis	Canada	5,400	2,100	
Nettilling Lake	Canada	5,500	2,100	
Lake Nipigon	Canada	4,850	1,900	
Lake Manitoba	Canada	4,700	1,800	

South America

		km²	miles²	
Lake Titicaca	Bolivia/Peru	8,300	3,200	
Lake Poopo	Peru	2,800	1,100	

Islands

Europe

		km²	miles²	
Great Britain	UK	229,880	88,700	[8]
Iceland	Atlantic Ocean	103,000	39,800	
Ireland	Ireland/UK	84,400	32,600	
Novaya Zemlya (N.)	Russia	48,200	18,600	
W. Spitzbergen	Norway	39,000	15,100	
Novaya Zemlya (S.)	Russia	33,200	12,800	
Sicily	Italy	25,500	9,800	
Sardinia	Italy	24,000	9,300	
N.E. Spitzbergen	Norway	15,000	5,600	
Corsica	France	8,700	3,400	
Crete	Greece	8,350	3,200	
Zealand	Denmark	6,850	2,600	

Asia

		km²	miles²	
Borneo	S. E. Asia	744,360	287,400	[3]
Sumatra	Indonesia	473,600	182,860	[6]
Honshu	Japan	230,500	88,980	[7]
Sulawesi (Celebes)	Indonesia	189,000	73,000	
Java	Indonesia	126,700	48,900	
Luzon	Philippines	104,700	40,400	
Mindanao	Philippines	101,500	39,200	
Hokkaido	Japan	78,400	30,300	
Sakhalin	Russia	74,060	28,600	
Sri Lanka	Indian Ocean	65,600	25,300	
Taiwan	Pacific Ocean	36,000	13,900	
Kyushu	Japan	35,700	13,800	
Hainan	China	34,000	13,100	
Timor	Indonesia	33,600	13,000	
Shikoku	Japan	18,800	7,300	
Halmahera	Indonesia	18,000	6,900	
Ceram	Indonesia	17,150	6,600	
Sumbawa	Indonesia	15,450	6,000	
Flores	Indonesia	15,200	5,900	
Samar	Philippines	13,100	5,100	
Negros	Philippines	12,700	4,900	
Bangka	Indonesia	12,000	4,600	
Palawan	Philippines	12,000	4,600	
Panay	Philippines	11,500	4,400	
Sumba	Indonesia	11,100	4,300	
Mindoro	Philippines	9,750	3,800	
Buru	Indonesia	9,500	3,700	
Bali	Indonesia	5,600	2,200	
Cyprus	Mediterranean	3,570	1,400	

Africa

		km²	miles²	
Madagascar	Indian Ocean	587,040	226,660	[4]
Socotra	Indian Ocean	3,600	1,400	
Réunion	Indian Ocean	2,500	965	
Tenerife	Atlantic Ocean	2,350	900	
Mauritius	Indian Ocean	1,865	720	

Oceania

		km²	miles²	
New Guinea	Indon./Pap. NG	821,030	317,000	[2]
New Zealand (S.)	Pacific Ocean	150,500	58,100	
New Zealand (N.)	Pacific Ocean	114,700	44,300	
Tasmania	Australia	67,800	26,200	
New Britain	Papua NG	37,800	14,600	
New Caledonia	Pacific Ocean	19,100	7,400	
Viti Levu	Fiji	10,500	4,100	
Hawaii	Pacific Ocean	10,450	4,000	
Bougainville	Papua NG	9,600	3,700	
Guadalcanal	Solomon Is.	6,500	2,500	
Vanua Levu	Fiji	5,550	2,100	
New Ireland	Papua NG	3,200	1,200	

North America

		km²	miles²	
Greenland	Atlantic Ocean	2,175,600	839,800	[1]
Baffin Is.	Canada	508,000	196,100	[5]
Victoria Is.	Canada	212,200	81,900	[9]
Ellesmere Is.	Canada	212,000	81,800	[10]
Cuba	Caribbean Sea	110,860	42,800	
Newfoundland	Canada	110,680	42,700	
Hispaniola	Dom. Rep./Haiti	76,200	29,400	
Banks Is.	Canada	67,000	25,900	
Devon Is.	Canada	54,500	21,000	
Melville Is.	Canada	42,400	16,400	
Vancouver Is.	Canada	32,150	12,400	
Somerset Is.	Canada	24,300	9,400	
Jamaica	Caribbean Sea	11,400	4,400	
Puerto Rico	Atlantic Ocean	8,900	3,400	
Cape Breton Is.	Canada	4,000	1,500	

South America

		km²	miles²	
Tierra del Fuego	Argentina/Chile	47,000	18,100	
Falkland Is. (E.)	Atlantic Ocean	6,800	2,600	
South Georgia	Atlantic Ocean	4,200	1,600	
Galapagos (Isabela)	Pacific Ocean	2,250	870	

World: Regions in the News

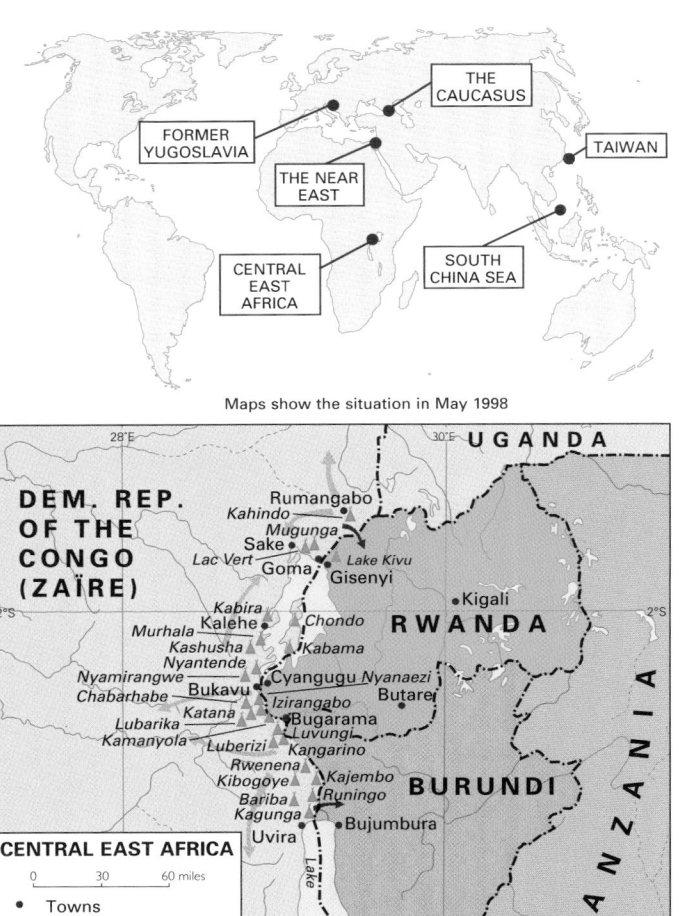

FORMER YUGOSLAVIA
THE NEAR EAST
THE CAUCASUS
TAIWAN
CENTRAL EAST AFRICA
SOUTH CHINA SEA

Maps show the situation in May 1998

CENTRAL EAST AFRICA

0   30   60 miles

- • Towns
- ▲ Camps
- ← Refugee movements
- — International boundaries
- ➤ Forced repatriation

THE NEAR EAST

0   10   20   30 miles

- —·— 1949 Armistice Line
- – – – 1974 Cease-fire Lines
- ● Efrata Main Jewish settlements in the West Bank and Gaza Strip
- ■ Halhul Main Palestinian Arab towns in the West Bank and Gaza Strip
- • 'Amman Capital cities

ISRAEL
Population: 5,696,000 (inc. East Jerusalem and Jewish settlers in the areas under Israeli administration. Jewish 82%, Arab Muslim 13.8%, Arab Christian 2.5%, Druze 1.7%)

West Bank
Population: 1,122,900 (Palestinian Arabs 97% [of whom Arab Muslim 85%, Jewish 7%, Christian 8%])

Gaza Strip
Population: 748,400 (Arab 98%)

JORDAN
Population: 5,547,000 (Arab 99% [of whom about 50% are Palestinian Arab])

LEBANON
Population: 2,971,000 (Arab 93% [of whom 83% are Lebanese Arab and 10% Palestinian Arab])

FORMER YUGOSLAVIA

0   50   100 miles

- —·— International boundaries
- – – – Republic boundaries
- - - - Province boundaries
- ■ Capital cities
- —— Dayton Peace Agreement Boundary
- Muslim–Croat Federation
- Bosnian Serb Republic

THE BREAKUP OF YUGOSLAVIA
The former country of Yugoslavia comprised six republics. In 1991 Slovenia and Croatia declared independence. Bosnia-Herzegovina followed in 1992 and Macedonia in 1993. Yugoslavia now comprises the remaining two republics, Serbia and Montenegro.

YUGOSLAVIA
Population: 10,881,000 (Serb 62.6%, Albanian 16.5%, Montenegrin 5%, Hungarian 3.3%, Muslim 3.2%)

Serbia Population: 6,060,000 (Serb 87.7%, excluding the former autonomous provinces of Kosovo and Vojvodina)
  Kosovo Population: 1,989,050
  Vojvodina Population: 2,131,900

Montenegro Population: 700,050 (Montenegrin 61.9%, Muslim 14.6%, Albanian 7%)

CROATIA
Population: 4,900,000 (Croat 78.1%, Serb 12.2%)

SLOVENIA
Population: 2,000,000 (Slovene 88%, Croat 3%, Serb 2%)

MACEDONIA (F. Y. R. O. M.)
Population: 2,173,000 (Macedonian 64%, Albanian 21.7%, Turkish 5%, Romanian 3%, Serb 2%)

BOSNIA-HERZEGOVINA
Population: 4,400,000 (Muslim 49%, Serb 31.2%, Croat 17.2%)

THE CAUCASUS

0   50   100 miles

- —·— International boundaries
- – – – Republic boundaries

Georgia, Armenia and Azerbaijan achieved independence in 1991. Abkhazia, Ajaria and South Ossetia seek independence from Georgia. Chechenia has been trying to break away from Russia since 1991, but Russia has resisted with military force. Hostility also continues between Armenia and Azerbaijan over the enclave of Nagorno-Karabakh.

COUNTRIES AND REPUBLICS OF THE CAUCASUS REGION

RUSSIA
  North Ossetia (Alania)
  Population: 695,000 (Ossetian 53%, Russian 29%, Chechen 5.2%, Armenian 1.9%)
  Chechenia Population: 1,308,000 (Chechen and Ingush 70.7%, Russian 23.1%, Armenian 1.2%)
  Ingushetia (Split from Chechenia in June 1993) Population: 250,000

GEORGIA
Population: 5,448,000 (Georgian 70.1%, Armenian 8.1%, Russian 6.3%, Azerbaijani 5.7%, Ossetian 3%, Greek 2%, Abkhazian 2%)
  Abkhazia Population: 537,500 (Georgian 45.7%, Abkhazian 17.8%, Armenian 14.6%, Russian 14.3%)
  Ajaria Population: 382,000 (Georgian 82.8%, Russian 7.7%, Armenian 4%)

ARMENIA
Population: 3,603,000 (Armenian 93%, Azerbaijani 3%)
  Nagorno-Karabakh Population: 192,400 (Armenian 76.9%, Azerbaijani 21.5%)

AZERBAIJAN
Population: 7,559,000 (Azerbaijani 83%, Russian 6%, Armenian 6%, Lezgin 2%)
  Naxçivan Population: 300,400 (Azerbaijani 95.9%)

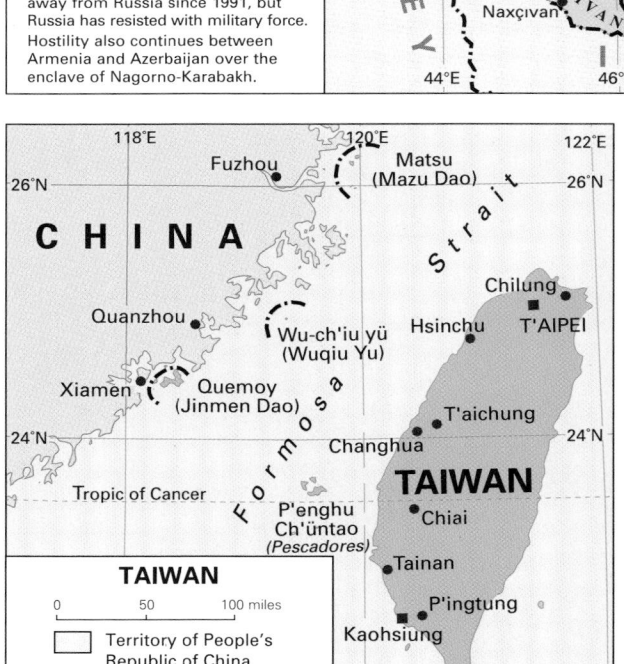

TAIWAN

0   50   100 miles

- Territory of People's Republic of China
- Territory of Republic of China (Taiwan)

SOUTH CHINA SEA

0   200   400 miles

- ▲ Philippine terr.
- ▼ Vietnamese terr.
- ▲ Chinese terr.
- ● Taiwanese terr.
- – – – Philippine claim
- - - - Vietnamese claim
- —+— Chinese claim
- —·— Malaysian claim

INTRODUCTION TO WORLD GEOGRAPHY

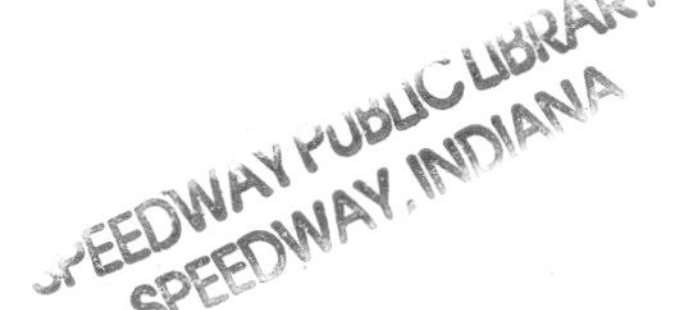

The Universe

About 15 billion years ago, time and space began with the most colossal explosion in cosmic history: the so-called "Big Bang" that is believed to have initiated the universe. According to current theory, in the first millionth of a second of its existence it expanded from a dimensionless point of infinite mass and density into a fireball about 19 billion miles across; and it has been expanding ever since.

It took almost a million years for the primal fireball to cool enough for atoms to form. They were mostly hydrogen, still the most abundant material in the universe. But the new matter was not evenly distributed around the young universe, and a few billion years later atoms in relatively dense regions began to cling together under the influence of gravity, forming distinct masses of gas separated by vast expanses of empty space. To begin with, these first proto-galaxies were dark places: the universe had cooled. But gravitational attraction continued, condensing matter into coherent lumps inside the galactic gas clouds. About three billion years later, some of these masses had contracted so much that internal pressure produced the high temperatures necessary to bring about nuclear fusion: the first stars were born.

There were several generations of stars, each feeding on the wreckage of its extinct predecessors as well as the original galactic gas swirls. With each new generation, progressively larger atoms were forged in stellar furnaces and the galaxy's range of elements, once restricted to hydrogen, grew larger. About 10 billion years after the Big Bang, a star formed on the outskirts of our galaxy with enough matter left over to create a retinue of planets. Nearly five billion years after that human beings evolved.

The Sun is one of more than 100 billion stars in the home galaxy alone. Our galaxy, in turn, forms part of a local group of approximately 30 similar structures, some much larger than our own; there are at least 100 billion other galaxies in the universe as a whole. The most distant ever observed, a highly energetic galactic core known only as quasar PC 1247 +3406, lies about 12 billion light-years away.

Life of a Star

For most of its existence, a star produces energy by the nuclear fusion of hydrogen into helium at its core. The duration of this hydrogen-burning period – known as the main sequence – depends on the star's mass; the greater the mass, the higher the core temperatures and the sooner the star's supply of hydrogen is exhausted. Dim, dwarf stars consume their hydrogen slowly, eking it out over 1,000 billion years or more. The Sun, like other stars of its mass, should spend about 10 billion years on the main sequence; since it was formed less than five billion years ago, it still has half its life left.

Once all a star's core hydrogen has been fused into helium, nuclear activity moves outward into layers of unconsumed hydrogen. For a time, energy production sharply increases: the star grows hotter and expands enormously, turning into a so-called red giant. Its energy output will increase a thousandfold, and it will swell to a hundred times its present diameter.

After a few hundred million years, helium in the core will become sufficiently compressed to initiate a new cycle of nuclear fusion: from helium to carbon. The star will contract somewhat, before beginning its last expansion, in the Sun's case engulfing the Earth and perhaps Mars. In this bloated condition, the Sun's outer layers will break off into space, leaving a tiny inner core, mainly of carbon, that shrinks progressively under the force of its own gravity: dwarf stars can attain a density more than 10,000 times that of normal matter, with crushing surface gravities to match. Gradually, the nuclear fires will die down, and the Sun will reach its terminal stage: a black dwarf, emitting insignificant amounts of energy.

However, stars more massive than the Sun may undergo another transformation. The additional mass allows gravitational collapse to continue indefinitely: eventually, all the star's remaining matter shrinks to a point, and its density approaches infinity – a state that will not permit even subatomic structures to survive.

The star has become a black hole: an anomalous "singularity" in the fabric of space and time. Although vast coruscations of radiation will be emitted by any matter falling into its grasp, the singularity itself has an escape velocity that exceeds the speed of light, and nothing can ever be released from it. Within the boundaries of the black hole, the laws of physics are suspended, but no physicist can ever observe the extraordinary events that may occur.

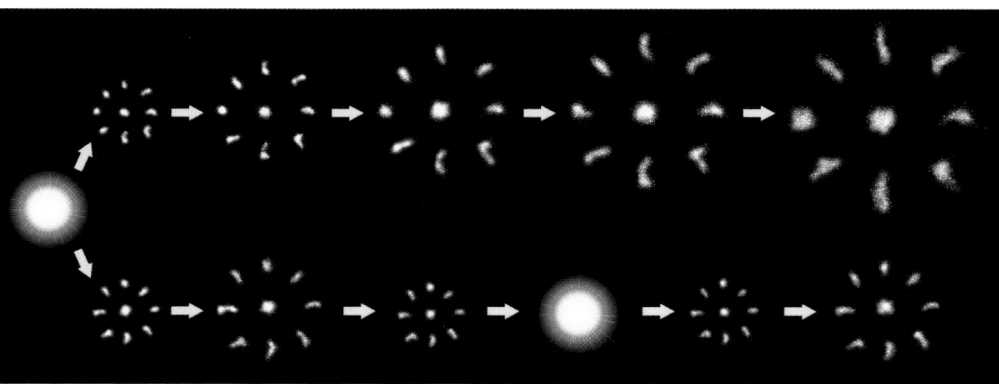

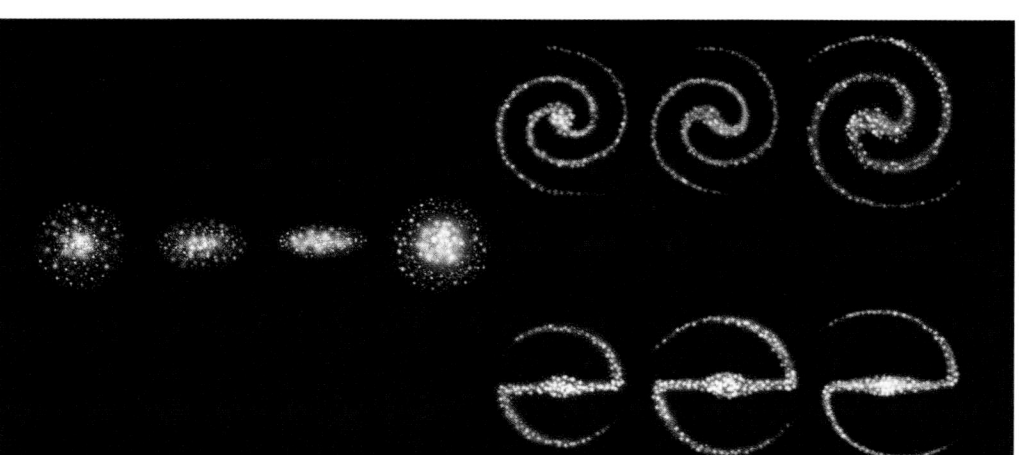

The End of the Universe

The likely fate of the universe is disputed. One theory (top left) dictates that the expansion begun at the time of the Big Bang will continue "indefinitely," with ageing galaxies moving further and further apart in an immense, dark graveyard. Alternatively, gravity may overcome the expansion (bottom left). Galaxies will fall back together until everything is again concentrated at a single point, followed by a new Big Bang and a new expansion, in an endlessly repeated cycle.

The first theory is supported by the amount of visible matter in the universe; the second assumes there is enough dark material to bring about the gravitational collapse.

Galactic Structures

Many of the universe's 100 billion galaxies show clear structural patterns, originally classified by the American astronomer Edwin Hubble in 1925. Spiral galaxies like our own (top row) have a central, almost spherical bulge and a surrounding disk composed of spiral arms. Barred spirals (bottom row) have a central bar of stars across the nucleus, with spiral arms trailing from the ends of the bar. Elliptical galaxies (far left) have a uniform appearance, ranging from a flattened disk to a near sphere. So-called SO galaxies (left row, right) have a central bulge, but no spiral arms. Most galaxies, however, have no obvious structure at all.

Galaxies also vary enormously in size, from dwarfs only 2,000 light-years across to great assemblies of stars 80 or more times larger.

The Home Galaxy

The Sun and its planets are located in one of the spiral arms, a little less than 28,000 light-years from the galactic center and orbiting around it in a period of 200 million years. The center is invisible from the Earth, masked by vast, light-absorbing clouds of interstellar dust. The galaxy is probably around 12 billion years old and, like other spiral galaxies, has three distinct regions. The central bulge is about 30,000 light-years in diameter. The disk in which the Sun is located is not much more than 1,000 light-years thick but 100,000 light-years from end to end. Around the galaxy is the halo, a spherical zone 300,000 light-years across, studded with globular star-clusters and sprinkled with individual suns.

Globular clusters

Bulge

Disk

Solar System

Star Charts

Star charts are drawn as projections of a vast, hollow sphere with the observer in the middle. Each circle below represents slightly more than one hemisphere, centered on the north and south celestial poles respectively – projections of the Earth's poles in the heavens. At the present era, the north pole is marked by the star Polaris; the south pole has no such convenient reference point.

Astronomical coordinates are normally given in terms of "Right Ascension" for longitude and "Declination" for latitude or altitude. Since the stars appear to rotate around the Earth once every 24 hours, Right Ascension is measured eastward – counterclockwise – in hours and minutes and is marked around the edge of the map. One hour is equivalent to 15 angular degrees; zero on the scale is the point at which the Sun crosses the celestial equator at the spring equinox, known to astronomers as the First Point in Aries. Unlike the Sun, stars always rise and set at the same point on the horizon. Declination measures (in degrees) a star's angular distance above or below the celestial equator and is marked on the vertical line.

To use the maps, first choose the one for your hemisphere and hold it with the month at the bottom. The stars in the lower part of the map are then due south (or north, in the southern hemisphere) at about 1 AM local time, not allowing for summer or daylight saving time. Their exact position above the horizon depends on your latitude. The closer to the Equator you live, the higher in the sky these stars will appear. Some additional stars from the map for the other hemisphere will be visible in the lower sky.

Stars near the top of the map will be below the opposite horizon at this date and time but will be visible at other times of the night and year. The sky appears to move counterclockwise around the celestial pole during the course of the day (clockwise in the southern hemisphere), so the same stars will be visible at 11 PM a month earlier.

NORTHERN HEAVENS

SOUTHERN HEAVENS

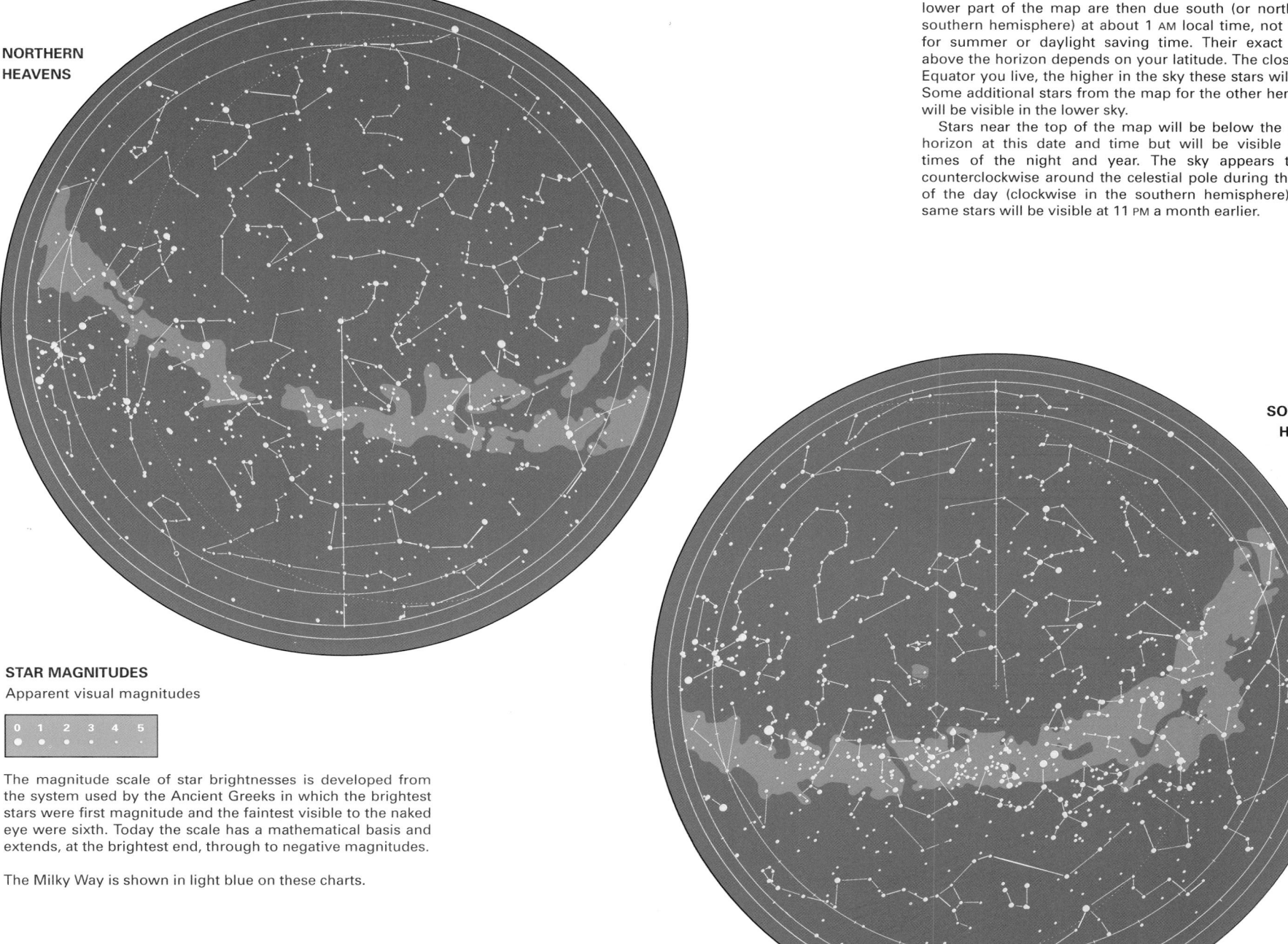

STAR MAGNITUDES

Apparent visual magnitudes

| 0 | 1 | 2 | 3 | 4 | 5 |

The magnitude scale of star brightnesses is developed from the system used by the Ancient Greeks in which the brightest stars were first magnitude and the faintest visible to the naked eye were sixth. Today the scale has a mathematical basis and extends, at the brightest end, through to negative magnitudes.

The Milky Way is shown in light blue on these charts.

THE NEAREST STARS

The 20 nearest stars, excluding the Sun, with their distance from Earth in light-years*

Star	Distance	
Proxima Centauri	4.25	Many of the nearest stars, like
Alpha Centauri A	4.3	Alpha Centauri A and B, are
Alpha Centauri B	4.3	doubles, orbiting about the
Barnard's Star	6.0	common center of gravity
Wolf 359	7.8	and to all intents and
Lalande 21185	8.3	purposes equidistant from
Sirius A	8.7	Earth. Many of them are dim
Sirius B	8.7	objects, with no name other
UV Ceti A	8.7	than the designation given
UV Ceti B	8.7	by the astronomers who
Ross 154	9.4	investigated them. However,
Ross 248	10.3	they include Sirius, the
Epsilon Eridani	10.7	brightest star in the sky,
Ross 128	10.9	and Procyon, the seventh
61 Cygni A	11.1	brightest. Both are far larger
61 Cygni B	11.1	than the Sun; of the nearest
Epsilon Indi	11.2	stars, only Epsilon Eridani is
Groombridge 34A	11.2	similar in size and luminosity.
Groombridge 34B	11.2	
L789-6	11.2	* A light-year equals approx.
Procyon A	11.4	5,900 billion miles
Procyon B	11.4	

THE CONSTELLATIONS

The constellations and their English names

Andromeda	Andromeda	Circinus	Compasses	Lacerta	Lizard	Piscis Austrinus	Southern Fish
Antlia	Air Pump	Columba	Dove	Leo	Lion	Puppis	Ship's Stern
Apus	Bird of Paradise	Coma Berenices	Berenice's Hair	Leo Minor	Little Lion	Pyxis	Mariner's Compass
Aquarius	Water Carrier	Corona Australis	Southern Crown	Lepus	Hare	Reticulum	Net
Aquila	Eagle	Corona Borealis	Northern Crown	Libra	Scales	Sagitta	Arrow
Ara	Altar	Corvus	Crow	Lupus	Wolf	Sagittarius	Archer
Aries	Ram	Crater	Cup	Lynx	Lynx	Scorpius	Scorpion
Auriga	Charioteer	Crux	Southern Cross	Lyra	Lyre	Sculptor	Sculptor
Boötes	Herdsman	Cygnus	Swan	Mensa	Table	Scutum	Shield
Caelum	Chisel	Delphinus	Dolphin	Microscopium	Microscope	Serpens	Serpent
Camelopardalis	Giraffe	Dorado	Swordfish	Monoceros	Unicorn	Sextans	Sextant
Cancer	Crab	Draco	Dragon	Musca	Fly	Taurus	Bull
Canes Venatici	Hunting Dogs	Equuleus	Little Horse	Norma	Level	Telescopium	Telescope
Canis Major	Great Dog	Eridanus	Eridanus	Octans	Octant	Triangulum	Triangle
Canis Minor	Little Dog	Fornax	Furnace	Ophiuchus	Serpent Bearer	Triangulum Australe	Southern Triangle
Capricornus	Goat	Gemini	Twins	Orion	Orion	Tucana	Toucan
Carina	Keel	Grus	Crane	Pavo	Peacock	Ursa Major	Great Bear
Cassiopeia	Cassiopeia	Hercules	Hercules	Pegasus	Winged Horse	Ursa Minor	Little Bear
Centaurus	Centaur	Horologium	Clock	Perseus	Perseus	Vela	Sails
Cepheus	Cepheus	Hydra	Water Snake	Phoenix	Phoenix	Virgo	Virgin
Cetus	Whale	Hydrus	Sea Serpent	Pictor	Easel	Volans	Flying Fish
Chamaeleon	Chameleon	Indus	Indian	Pisces	Fishes	Vulpecula	Fox

The Solar System

Lying 28,000 light-years from the center of one of billions of galaxies that comprise the observable universe, our Solar System contains nine planets and their moons, innumerable asteroids and comets, and a miscellany of dust and gas, all tethered by the immense gravitational field of the Sun, the middling-sized star whose thermonuclear furnaces provide them all with heat and light. The Solar System was formed about 4.6 billion years ago, when a spinning cloud of gas, mostly hydrogen but seeded with other, heavier elements, condensed enough to ignite a nuclear reaction and create a star. The Sun still accounts for almost 99.9% of the system's total mass; one planet, Jupiter, contains most of the remainder.

By composition as well as distance, the planetary array divides quite neatly in two: an inner system of four small, solid planets, including the Earth, and an outer system, from Jupiter to Neptune, of four much larger planets composed of lighter materials, such as gas, liquid and ice. Between the two groups lies a scattering of rocky asteroids, perhaps as many as 400,000. They may be debris left over from the inner solar system's formation. The outermost planet, Pluto, may simply be the largest of a number of bodies composed of rock and ice orbiting beyond Neptune, similarly left over from the formation of the outer solar system.

By the 1990s, however, the Solar System also included some newer anomalies: several thousand spacecraft. Most were in orbit around the Earth, but some had probed far and wide around the system. The valuable information beamed back by these robotic investigators has transformed our knowledge of our celestial environment.

Much of the early history of science is the story of people trying to make sense of the errant points of light that were all they knew of the planets. Now, men have themselves stood on the Earth's Moon; probes have landed on Mars and Venus, and orbiting radars have mapped far distant landscapes with astonishing accuracy. In the 1980s, the US *Voyagers* skimmed all four major planets of the outer system, bringing new revelations with each close approach. Only Pluto, inscrutably distant in an orbit that takes it 50 times the Earth's distance from the Sun, remains unvisited by our messengers.

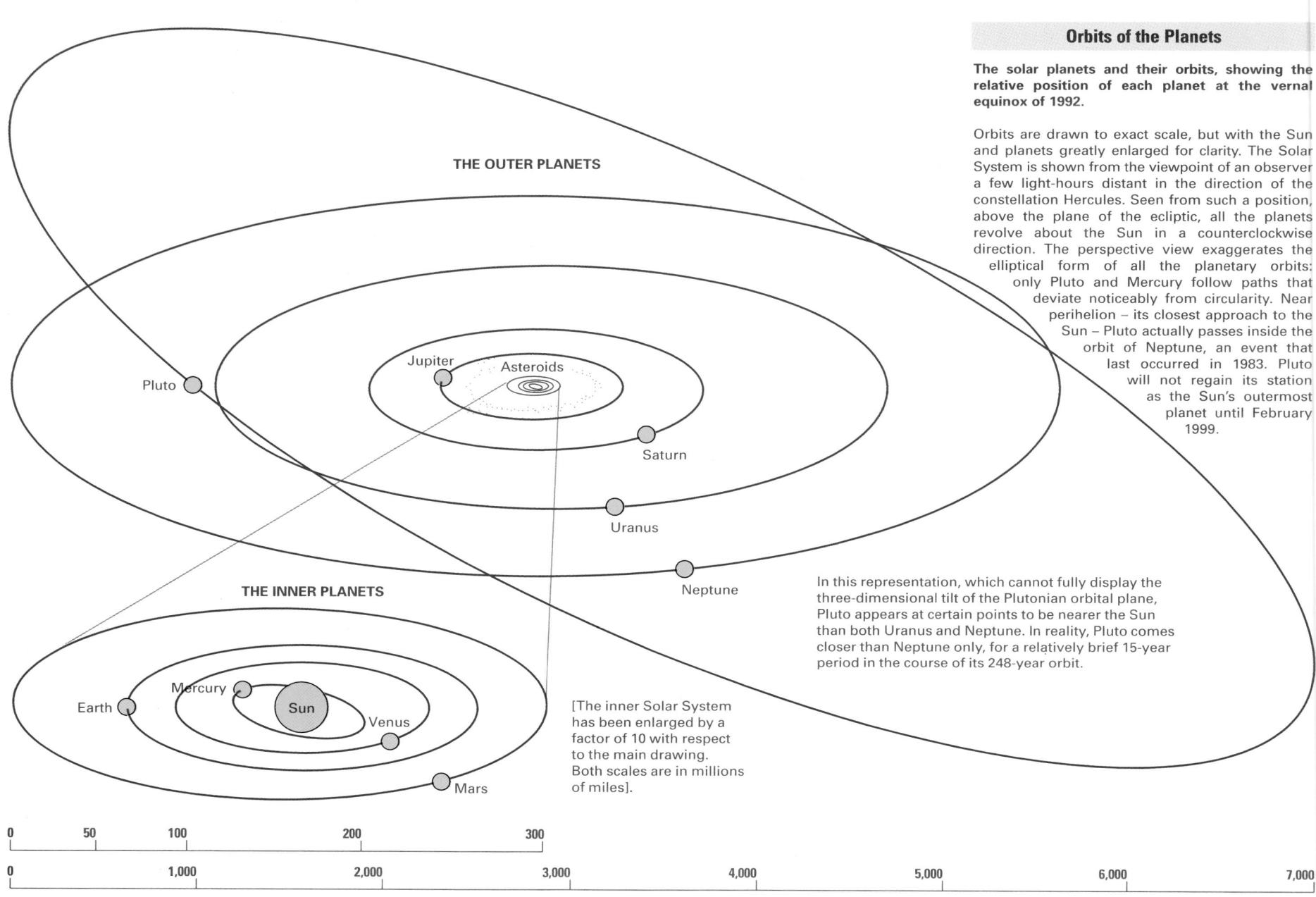

Orbits of the Planets

The solar planets and their orbits, showing the relative position of each planet at the vernal equinox of 1992.

Orbits are drawn to exact scale, but with the Sun and planets greatly enlarged for clarity. The Solar System is shown from the viewpoint of an observer a few light-hours distant in the direction of the constellation Hercules. Seen from such a position, above the plane of the ecliptic, all the planets revolve about the Sun in a counterclockwise direction. The perspective view exaggerates the elliptical form of all the planetary orbits: only Pluto and Mercury follow paths that deviate noticeably from circularity. Near perihelion – its closest approach to the Sun – Pluto actually passes inside the orbit of Neptune, an event that last occurred in 1983. Pluto will not regain its station as the Sun's outermost planet until February 1999.

THE OUTER PLANETS

Jupiter Asteroids

Pluto

Saturn

Uranus

Neptune

In this representation, which cannot fully display the three-dimensional tilt of the Plutonian orbital plane, Pluto appears at certain points to be nearer the Sun than both Uranus and Neptune. In reality, Pluto comes closer than Neptune only, for a relatively brief 15-year period in the course of its 248-year orbit.

THE INNER PLANETS

Earth Mercury Sun Venus

Mars

[The inner Solar System has been enlarged by a factor of 10 with respect to the main drawing. Both scales are in millions of miles].

0	50	100	200	300

0	1,000	2,000	3,000	4,000	5,000	6,000	7,000

Planetary Data

	Mean distance from Sun (million miles)	Mass (Earth = 1)	Period of orbit (Earth years)	Period of rotation (Earth days)	Equatorial diameter (miles)	Average density (water = 1)	Surface gravity (Earth = 1)	Escape velocity (miles/sec)	Number of known satellites
Sun	–	*332,946*	–	*25.38*	*870,000*	*1.41*	*27.9*	*383.7*	–
Mercury	36.0	0.06	0.241	58.67	3,031	5.43	0.38	2.64	0
Venus	67.2	0.8	0.615	243.00	7,519	5.24	0.90	6.44	0
Earth	93.0	1.0	1.00	1.00	7,926	5.52	1.00	6.95	1
Mars	141.6	0.1	1.88	1.02	4,222	3.93	0.38	3.13	2
Jupiter	483.6	317.8	11.86	0.41	88,732	1.33	2.69	37.03	16
Saturn	886.6	95.2	29.46	0.42	74,500	0.706	1.16	22.12	18
Uranus	1,783.0	14.5	84.01	0.45	32,600	1.25	0.93	13.11	15
Neptune	2,793.9	17.1	164.79	0.71	30,100	1.77	1.21	15.29	8
Pluto	3,666.2	0.002	247.7	6.39	1,519	1.40	0.05	0.75	1

Planetary days are given in sidereal time – that is, with respect to the stars rather than the Sun. Most of the information in the table was confirmed by spacecraft and often obtained from photographs and other data transmitted back to the Earth. In the case of Pluto, however, only earthbound observations have been made, and no spacecraft will encounter it until well into the next century. Given the planet's small size and great distance, figures for its diameter and rotation period have only recently been confirmed.

Pluto is not massive enough to account for the perturbations in the orbits of Uranus and Neptune that led to its 1930 discovery, but it is now widely believed that these perturbations can be explained away as observational errors made by the earlier observers.

The Planets

Mercury is the closest planet to the Sun and hence the fastest-moving. It is very hot with a cratered, wrinkled surface very similar to that of Earth's Moon. It is small and has no gravity, hence there is no significant atmosphere.

Venus has much the same physical dimensions as Earth. Its dense atmosphere is composed of 97% CO_2 resulting in a runaway greenhouse effect that makes the Venusian surface, at 890°F, the hottest of all the planets in the Solar System. Radar mapping shows relatively level land with volcanic regions whose sulfurous discharges explain the sulfuric acid rains reported by soft-landing space probes before they succumbed to Venus' fierce climate.

Earth seen from space is easily the most beautiful of the inner planets; it is also, and more objectively, the largest, as well as the only home of known life. Living things are the main reason why the Earth is able to retain a substantial proportion of corrosive and highly reactive oxygen in its atmosphere, a state of affairs that contradicts the laws of chemical equilibrium; the oxygen in turn supports the life that constantly regenerates it.

Mars, smaller and cooler than the Earth, is nevertheless the most likely planet other than Earth where life may have formed. Vast water channels show that it was once warmer and wetter; there may still be traces of former simple life forms, though whether life could thrive in its current cold, dry and thin atmosphere is doubtful. The ice caps are mainly frozen carbon dioxide, and whatever oxygen the planet once possessed is now locked up in the iron-bearing rock that covers its cratered surface and gives it its characteristic red hue. Mars is a dustbowl with occasional storms whirling the dust high into the air.

Jupiter masses almost three times as much as all the other planets combined; had it scooped up rather more matter during its formation, it might have evolved into a small companion star for the Sun. The planet is mostly gas, under intense pressure in the lower atmosphere above a core of fiercely compressed hydrogen and helium. The upper layers form strikingly-colored rotating belts, the outward sign of the intense storms created by Jupiter's rapid diurnal rotation. Close approaches by spacecraft have shown an orbiting ring system and discovered several previously unknown moons: Jupiter has at least 16 moons.

Saturn is structurally similar to Jupiter, rotating fast enough to produce an obvious bulge at its equator. It is composed of 89% hydrogen and 11% helium, and has wind velocities in the outer atmosphere of 1,600 feet per second. Ever since the invention of the telescope, however, Saturn's rings have been the feature that has attracted most observers. *Voyager* probes in 1980 and 1981 sent back detailed pictures that showed them to be composed of thousands of separate ringlets, each in turn made up of tiny icy particles.

Uranus was unknown to the ancients. Although it is faintly visible to the naked eye, it was not discovered until 1781. Its interior is largely water, with an atmosphere of hydrogen, helium and some methane, which gives the planet its blue-green color. Observations in 1977 suggested the presence of a faint ring system, amply confirmed when *Voyager 2* swung past the planet in 1986.

Neptune is always more than 2.5 billion miles from Earth, and despite its diameter of almost 30,000 miles, it can only be seen by telescope. Its 1846 discovery was the result of mathematical predictions by astronomers seeking to explain irregularities in the orbit of Uranus, but until *Voyager 2* closed with the planet in 1989, little was known of it. Like Uranus, it has a ring system; *Voyager*'s photographs revealed a total of eight moons.

Pluto is the most mysterious of the solar planets, if only because even the most powerful telescopes can scarcely resolve it from a point of light to a disk. It was discovered as recently as 1930, like Neptune as the result of perturbations in the orbits of the two then outermost planets. Its small size, as well as its eccentric and highly tilted orbit, has led to suggestions that it is a former satellite of Neptune, somehow liberated from its primary. In 1978 Pluto was found to have a moon of its own, Charon, apparently half the size of Pluto itself.

Mean distance from the Sun in million miles (not to scale)

Mercury — 36.0

Venus — 67.2

Earth — 93.0

Mars — 141.6

Jupiter — 483.6

Saturn — 886.6

Uranus — 1,783.0

Neptune — 2,793.9

Pluto — 3,666.2

Diagram not drawn to scale

Time and Motion

The basic unit of time measurement is the day, that is, one rotation of the Earth on its axis. Our present calendar is based on the solar year of 365.24 days, the time taken by the Earth to orbit the Sun.

Calendars based on the movements of the Sun and Moon have been used since ancient times. The average length of the year, according to the Julian Calendar introduced by Julius Caesar, was about 11 minutes too long. The cumulative error was rectified in 1582 by the Gregorian Calendar, when Pope Gregory XIII decreed that the day following 4 October was 15 October, and in that century years did not count as leap years unless they were divisible by 400. England finally adopted the reformed calendar in 1752, when it was 11 days behind the European mainland.

The rotation of the Earth on its axis causes day and night. Because the Earth rotates through 360° every 24 hours, the world is divided into 24 time zones centered on lines of longitude at 15° longitude.

The tilt of the Earth's axis, also called the obliquity of the ecliptic, accounts for the seasons which are so familiar in the middle latitudes. But geological evidence shows that, over long periods of time, climates change and the advances and retreats of the ice during the Pleistocene Ice Age may have been caused by regular variations in the Earth's tilt, its orbit around the Sun, and changes in the season when it is closest to the Sun (perihelion).

Earth Data

Aphelion (maximum distance from Sun): 94,452,780 miles

Perihelion (minimum distance from Sun): 91,342,080 miles

Angle of tilt (obliquity of the ecliptic): 23° 27′ 08″

Length of year – solar tropical (equinox to equinox): 365.24 days

Length of year: 365 days, 5 hours, 48 minutes, 46 seconds of mean solar time

Superficial area: 197,000,000 sq mi

Land surface: 57,500,000 sq mi (29.2%)

Water surface: 139,500,000 sq mi (70.8%)

Equatorial circumference: 24,903 mi

Polar circumference: 24,860 mi

Equatorial diameter: 7,926.7 mi

Polar diameter: 7,900.0 mi

Equatorial radius: 3,963.4 mi

Polar radius: 3,950.0 mi

Volume of the Earth: 260,000 x 10⁶ cu mi

Mass of the Earth: 6.5×10^{21} tons

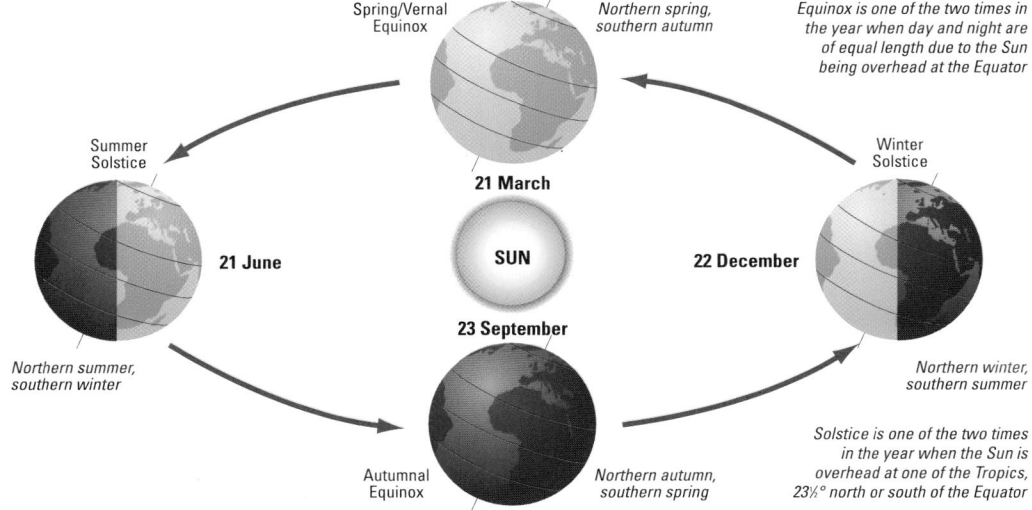

The Seasons

Seasons occur because the Earth's axis is tilted at a constant angle of 23½°. When the northern hemisphere is tilted to a maximum extent toward the Sun, on 21 June, the Sun is overhead at the Tropic of Cancer (latitude 23½° North). This is midsummer, or the summer solstice, in the northern hemisphere.

On 22 or 23 September, the Sun is overhead at the Equator, and day and night are of equal length throughout the world. This is the autumn equinox in the northern hemisphere. On 21 or 22 December, the Sun is overhead at the Tropic of Capricorn (23½° South), the winter solstice in the northern hemisphere. The overhead Sun then tracks north until, on 21 March, it is overhead at the Equator. This is the spring (vernal) equinox in the northern hemisphere.

In the southern hemisphere, the seasons are the reverse of those in the north.

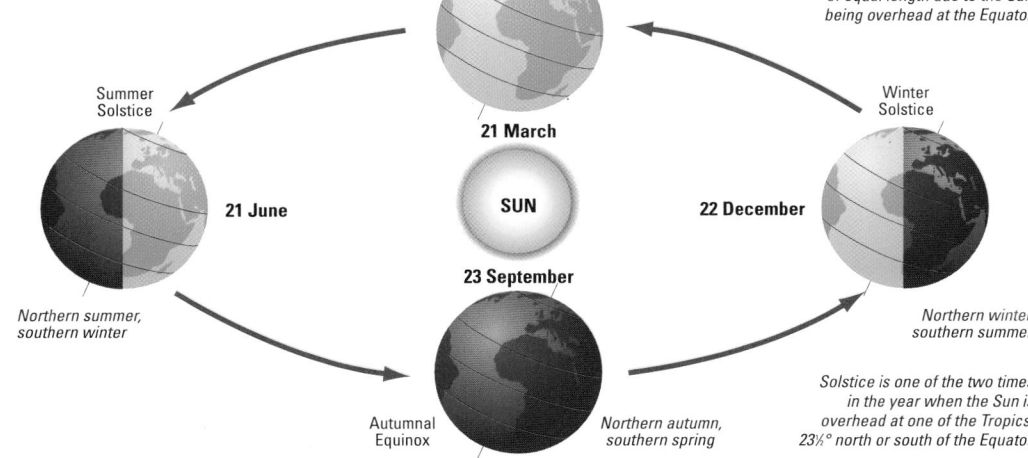

Day and Night

The Sun appears to rise in the east, reach its highest point at noon, and then set in the west, to be followed by night. In reality, it is not the Sun that is moving but the Earth rotating from west to east. The moment when the Sun's upper limb first appears above the horizon is termed sunrise; the moment when the Sun's upper limb disappears below the horizon is sunset.

At the summer solstice in the northern hemisphere (21 June), the Arctic has total daylight and the Antarctic total darkness. The opposite occurs at the winter solstice (21 or 22 December). At the Equator, the length of day and night are almost equal all year.

The Sun's Path

The diagrams on the right illustrate the apparent path of the Sun at (A) the Equator, (B) in midlatitude (45°), (C) at the Arctic Circle (66½°), and (D) at the North Pole, where there are six months of continuous daylight and six months of continuous night.

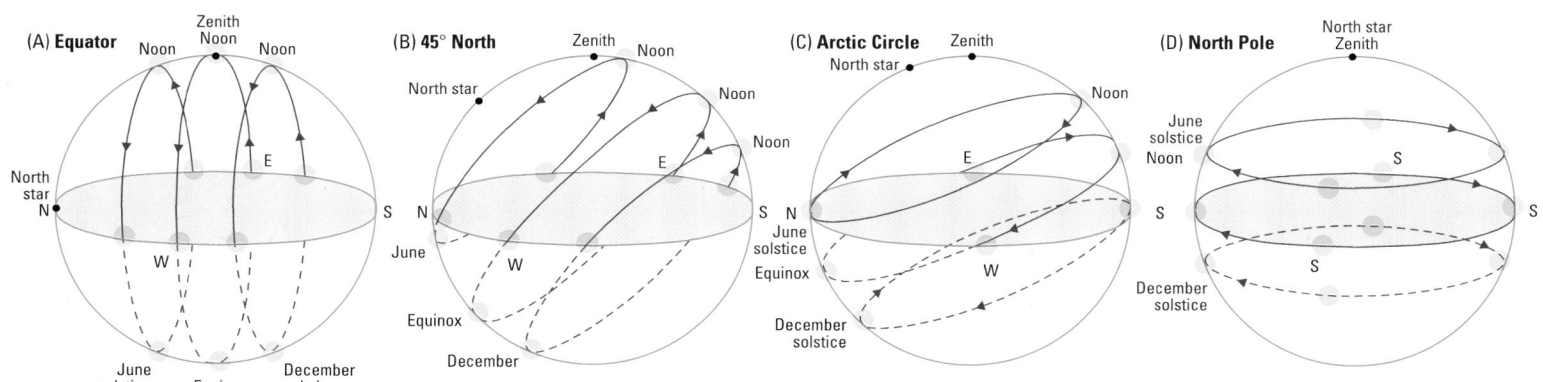

Sunrise and Sunset

The term equinox comes from two Latin words meaning "equal night." At the spring and autumn equinoxes, the Sun is vertically overhead at the Equator and all places on Earth have 12 hours of darkness and 12 of daylight. The graphs showing sunrise and sunset show that these occasions occur on 21 March and on 22 or 23 September. The graphs also show that, because the Sun remains high in the sky throughout the year, the length of the day and night at the Equator remain roughly the same throughout the year, with sunrise occurring around 6 AM and sunset at around 6 PM. The further north or south one travels, the greater the difference between the number of hours of daylight and darkness. For example, the graph, right, shows that at latitude 60°N, sunrise varies from just after 9 AM in midwinter (on 22 or 23 December) to about 2.30 AM in midsummer (around the summer solstice on 21 June). By contrast, the second graph, far right, shows that sunset at latitude 60°N occurs at about 2.45 PM in midwinter and 9.20 PM in midsummer.

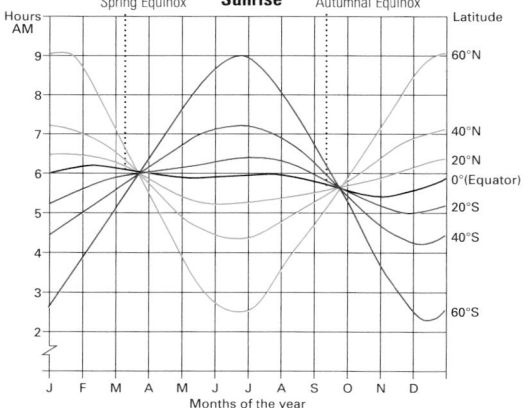

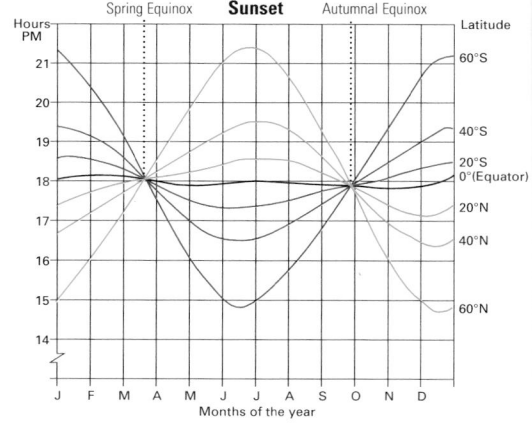

The Moon

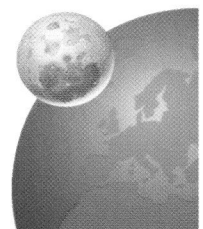

The Moon rotates more slowly than the Earth, making one complete turn on its axis in just over 27 days. Since this corresponds to its period of revolution around the Earth, the Moon always presents the same hemisphere or face to us, and we never see "the dark side." The interval between one full Moon and the next (and between new Moons) is about 29½ days – a lunar month. The apparent changes in the shape of the Moon are caused by its changing position in relation to the Earth; like the planets, it produces no light of its own and shines only by reflecting the rays of the Sun.

Phases of the Moon

Distance from Earth: 221,463 mi – 252,710 mi; Mean diameter: 2,160 mi;
Mass: approximately 1/81 that of Earth;
Surface gravity: one-sixth of Earth's; Daily range of temperature at lunar equator: 350°F;
Average orbital speed: 2,300 mph

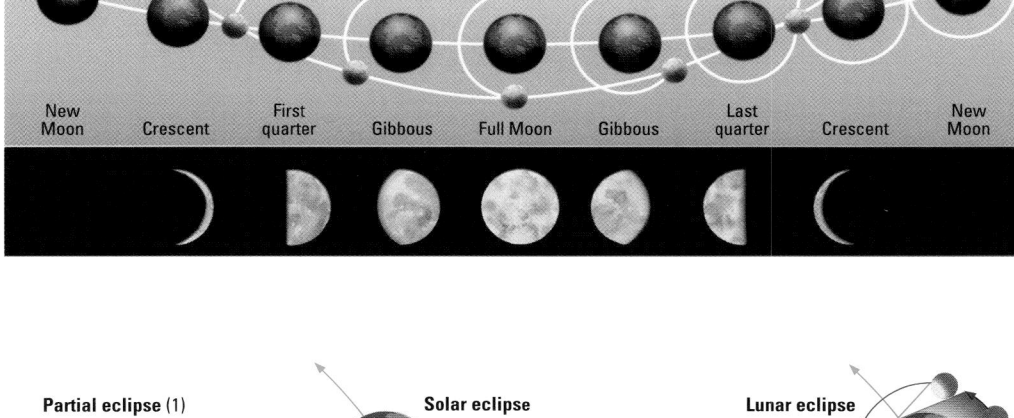

| New Moon | Crescent | First quarter | Gibbous | Full Moon | Gibbous | Last quarter | Crescent | New Moon |

Moon Data

Distance from Earth
The Moon orbits at a mean distance of 238,731 mi, at an average speed of 2,289 mph in relation to the Earth.

Size and mass
The average diameter of the Moon is 2,159.3 mi. It is 400 times smaller than the Sun but is about 400 times closer to the Earth, so we see them as the same size. The Moon has a mass of 7,975 x 10^{19} tons, with a density 3.344 times that of water.

Visibility
Only 59% of the Moon's surface is directly visible from Earth. Reflected light takes 1.25 seconds to reach Earth – compared to 8 minutes 27.3 seconds for light to reach us from the Sun.

Temperature
With the Sun overhead, the temperature on the lunar equator can reach 243°F. At night it can sink to –261°F.

Eclipses

When the Moon passes between the Sun and the Earth it causes a partial eclipse of the Sun (1) if the Earth passes through the Moon's outer shadow (P), or a total eclipse (2) if the inner cone shadow crosses the Earth's surface. In a lunar eclipse, the Earth's shadow crosses the Moon and, again, provides either a partial or total eclipse.

Eclipses of the Sun and the Moon do not occur every month because of the 5° difference between the plane of the Moon's orbit and the plane in which the Earth moves. In the 1990s only 14 lunar eclipses are possible, for example, seven partial and seven total; each is visible only from certain, and variable, parts of the world. The same period witnesses 13 solar eclipses – six partial (or annular) and seven total.

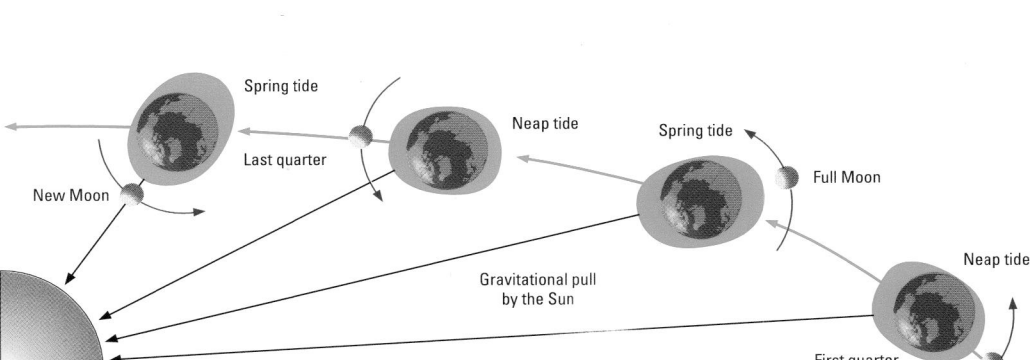

Partial eclipse (1)

Solar eclipse

Lunar eclipse

Total eclipse (2)

Tides

The daily rise and fall of the ocean's tides are the result of the gravitational pull of the Moon and that of the Sun, though the effect of the latter is only 46.6% as strong as that of the Moon. This effect is greatest on the hemisphere facing the Moon and causes a tidal "bulge." When the Sun, Earth and Moon are in line, tide-raising forces are at a maximum and Spring tides occur: high tide reaches the highest values, and low tide falls to low levels. When lunar and solar forces are least coincidental with the Sun and Moon at an angle (near the Moon's first and third quarters), Neap tides occur, which have a small tidal range.

Spring tide

Last quarter

New Moon

Neap tide

Spring tide

Full Moon

Neap tide

Gravitational pull by the Sun

First quarter

Time Zones

The Earth rotates through 360° in 24 hours, and so moves 15° every hour. The world is divided into 24 standard time zones, each centered on lines of longitude at 15° intervals. At the center of the first zone is the Prime meridian or Greenwich meridian. All places to the west of Greenwich are one hour behind for every 15° of longitude; places to the east are ahead by one hour for every 15°. When it is 12 noon at the Greenwich meridian, 180° east it is midnight of the same day – while 180° west the day is just beginning. To overcome this, the International Date Line was established, approximately following the 180° meridian. Thus, if you traveled eastward from Japan (140° East) to Samoa (170° West), you would pass from Sunday night into Sunday morning.

	Zones using GMT
	Zones slow of GMT
---	International boundaries
10	Hours slow or fast of GMT or Greenwich Mean Time
	Zones fast of GMT
	Half-hour zones
	Time zone boundaries
	International Date Line

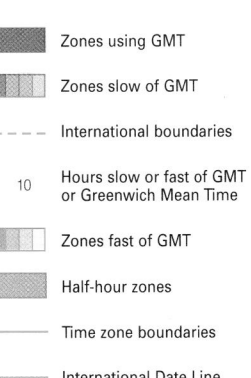

Projection: Mercator

CARTOGRAPHY BY PHILIP'S. COPYRIGHT GEORGE PHILIP LTD

Oceans

The last 40 years have been described as the "Space Age," but another exciting and perhaps even more important area of discovery, proceeding at the same time, has been the exploration of "inner space," namely the oceans which cover more than 70% of our planet. The study of the ocean floor and oceanic islands has revealed features that help to explain how continents move, and how the movements are related to earthquakes and volcanic activity.

Manned submersibles have established that life exists even in the deepest trenches, where the pressure reaches 1,000 atmospheres, the equivalent of the force of six and a half tons bearing down on every square inch. Further exploration in the pitch-black environment of the ocean ridges has revealed strange forms of marine life around scalding hot vents. The creatures include giant tubeworms, blind shrimps, and bacteria, some of which are genetically very different from any other known life forms. In 1996, an analysis of one microorganism revealed that at least half of its 1,700 or so genes were hitherto unknown. This environment, which is based on chemicals, not sunlight, may resemble the places where life on Earth first began.

Another vital area of contemporary research concerns the interactions between the oceans and the atmosphere, as exemplified in the El Niño–Southern Oscillation (ENSO), and the bearing that these have on climatic change.

Most geographers divide the world's ocean waters into four areas: the Pacific, Atlantic, Indian and Arctic oceans. The most active zone in the oceans is the sunlit upper layer, where the water is moved around by wind-blown currents. It is the home of most sea life and acts as a membrane through which the ocean breathes,

Seawater

The chemical composition of the sea, by percentage, excluding the elements of water itself

Chloride (Cl)	55.04%
Sodium (Na)	30.61%
Sulfate (SO₄)	7.69%
Magnesium (Mg)	3.69%
Calcium (Ca)	1.16%
Potassium (K)	1.10%
Bicarbonate (HCO₃)	0.41%
Bromide (Br)	0.19%
Boric Acid (H₃BO₃)	0.07%
Strontium (Sr)	0.04%
Fluoride (Fl)	0.003%
Lithium (Li)	trace
Rubidium (Rb)	trace
Phosphorus (P)	trace
Iodine (I)	trace
Barium (Ba)	trace
Arsenic (As)	trace
Cesium (Cs)	trace

Eleven constituents account for over 99% of the salt content of seawater, but seawater also contains virtually every other element. In natural conditions, its composition is broadly consistent across the world's seas and oceans; but in coastal areas especially variations are sometimes substantial. The oceans are about 35 parts water to one part salt.

Atoll Building

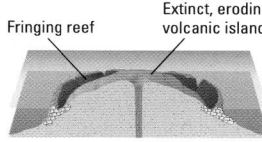

Volcano rises from ocean floor

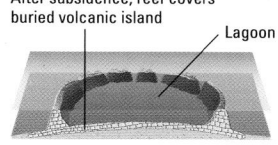

Fringing reef — Extinct, eroding volcanic island

After subsidence, reef covers buried volcanic island — Lagoon

A coral atoll usually begins existence as a bare volcanic peak, thrusting above the surface of the ocean. A colony of coral – organisms with calcium carbonate skeletons – forms itself in the shallow water around the peak. The volcano is eroded and slowly sinks, leaving the coral forming a ring of hard limestone around its remnant. In time, the barrier reef of an atoll is all that remains.

Life in the Oceans

An imaginary profile of the typical coastal and oceanic zones is shown, with a selection of the life forms that might occur in the water off the Pacific Coast of Central America. The animals illustrated are not drawn to scale as the range of sizes is too great. Most marine life is confined to the first 650 feet, the upper sunlit (photic) zone, where sunlight can still penetrate. Plant and animal plankton, the basis of life in the ocean, occur in great quantities in all zones.

In the pelagic environment (open sea), vertical gradients, including those of light, temperature and salinity, determine the distribution of organisms. From the tidal zone at the coastline, the continental shelf, geologically still part of the continental landmass, drops gently to about 650 feet – the sunlit zone. At the end of the shelf, the seabed falls away in the steeper angle of the continental slope. The subsequent descent to the deep ocean floor, known as the continental rise, is more gentle, with gradients between 1 in 100 and 1 in 700 until the abyssal plains and hills between 8,000 and 19,500 feet below the surface.

The deep sea floor contains seamounts, some of which are capped by coral reefs, ocean ridges, the longest mountain chains on Earth, and deep ocean trenches, especially in the Pacific Ocean where six trenches reach depths of more than 33,000 feet, including the Mariana Trench at 36,000 feet deep.

Each of these zones contains a distinctive community of species adapted to the different conditions of salinity, temperature and light intensity. Indeed, a few organisms have been found even in the abyssal darkness of the great ocean trenches.

absorbing great quantities of carbon dioxide and partly exchanging it for oxygen.

As the depth increases, so light fades and temperatures fall until just before 3,000 feet where there is a marked temperature change at the thermocline, the boundary between the warm surface zone and the cold deep zone. Below the thermocline, slow currents are caused by density differences between bodies of water with varying temperatures and salinity.

The El Niño Phenomenon

The importance of the ocean–atmosphere interaction is nowhere more dramatically demonstrated than the El Niño phenomenon in the southern Pacific Ocean.

Under normal conditions, shown in the diagram, top right, surface water flows eastward from South America under the influence of trade winds while, near the coast, cold, nutrient-rich water (dark blue) rises to the surface and spreads westward. In the western Pacific, sea surface temperatures reach 82°F or more and warm air rises, creating a low pressure air system and causing heavy rains. The rising warm air spreads out and some of it descends over South America and the eastern Pacific creating a high pressure air system from which winds blow westward. This rotating system is called a Walker Circulation Cell.

An El Niño event, also called an El Niño–Southern Oscillation cycle, or ENSO cycle, is characterized by a reversal of currents whereby the eastward-moving South Equatorial Current extends much further eastward and the trade winds weaken. The upwelling of cold water off South America is greatly reduced and surface water temperatures rise, causing a drastic reduction in fish life. The heaviest rainfall is over the eastern Pacific, while Southeast Asia is much drier than usual. Warm air rises in the east and spreads out, descending in the western Pacific, which then becomes a high pressure area, as shown on the second diagram, below right.

During an intense El Niño, such as in 1982–83 when sea temperatures in the eastern Pacific rose by 11°F, the effects of the current and wind reversals affect the weather around the world. In Australia and Southeast Asia, the monsoon rainfall is reduced, while, in 1983–84, a severe drought occurred in the Sahel, south of the Sahara, and also in southern Africa. The southeast coast of the United States also suffered storms and heavy rainfall, and even Europe experienced changes in weather patterns, possibly as a result of consequent changes in the course of the jet stream.

Scientists have found evidence that the frequency of the El Niño event, which normally occurs every two to seven years, may have increased in recent years with warm conditions persisting in the eastern Pacific from 1990 until mid-1995, an unprecedented length of time during the 114 years for which data exist. Another intense El Niño occurred in 1997–98, with resultant freak weather conditions across the entire Pacific region. Scientists do not know the causes of the El Niño event, though some researchers are investigating possible connections between major volcanic eruptions in the tropical Pacific region, the ENSO cycle and atmospheric circulation.

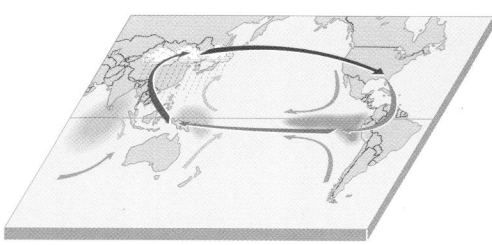

Normal year – Walker Circulation Cell

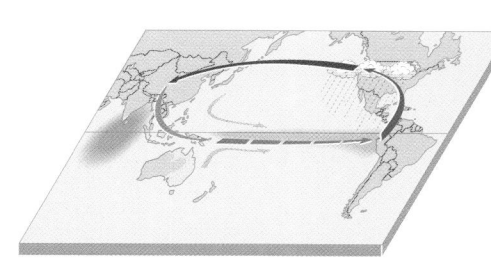

El Niño event

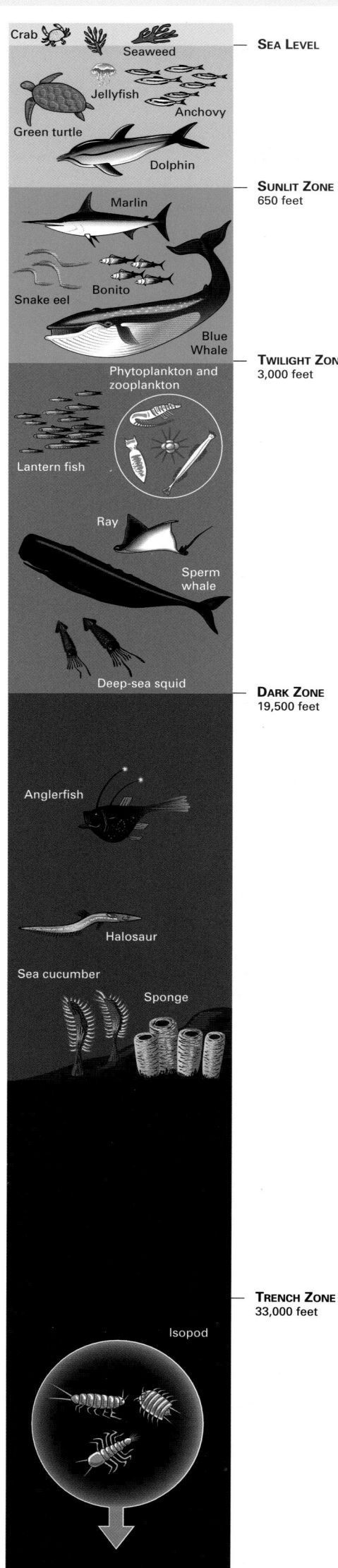

Crab — Seaweed — SEA LEVEL

Jellyfish

Green turtle — Anchovy

Dolphin

SUNLIT ZONE
650 feet

Marlin

Bonito

Snake eel — Blue Whale

TWILIGHT ZONE
3,000 feet

Phytoplankton and zooplankton

Lantern fish

Ray

Sperm whale

Deep-sea squid

DARK ZONE
19,500 feet

Anglerfish

Halosaur

Sea cucumber

Sponge

TRENCH ZONE
33,000 feet

Isopod

A R C T I C   O C E A N

Svalbard (Norw.)

Barents Sea

Novaya Zemlya

Kara Sea

Severnaya Zemlya

Laptev Sea

New Siberian Is.

East Siberian Sea

Wrangel I.

Arctic Circle

A

Murmansk

Arkhangelsk

Salekhard

Norilsk

Ob

Yenisey

Verkhoyansk

Lena

Yakutsk

Okhotsk

Magadan

Sea of Okhotsk

Bering Sea

Petropavlovsk-Kamchatskiy

International Date Line

B

NORWAY   SWEDEN   FINLAND

Oslo   Stockholm   Helsinki

ST.PETERSBURG

R U S S I A

Perm   Yekaterinburg

Tomsk   Krasnoyarsk

L. Baikal   Irkutsk   Ulan Ude

Khabarovsk

Sakhalin

Komsomolsk

Kuril Is.

DENMARK   Copenhagen   EST.

LATVIA   LITH.

Hamburg   Berlin   BELARUS

POLAND   Minsk   Kiev

MOSCOW   Kazan

Samara   Chelyabinsk   Omsk   Novosibirsk   Barnaul

Vladivostok   Sapporo

Amsterdam   Brussels   Prague   GERMANY   CZECH REP.   UKRAINE

Volga   Saratov   Irtysh   KAZAKSTAN

Ulan Bator

MONGOLIA   Harbin   Changchun   SHENYANG   NORTH KOREA   Pyongyang

JAPAN

PARIS   Vienna   AUSTRIA   SLOV.   Budapest   Volgograd   Astrakhan   Aral Sea   L. Balkhash

BEIJING   TIANJIN   SEOUL   SOUTH KOREA   TŌKYŌ

Lyons   SW.   Milan   ITALY   CROATIA   Belgrade   ROMANIA   Bucharest

Barcelona   Rome   Sofia   BULGARIA   Black Sea   GEORGIA   Tbilisi   Baku

ALB.   GREECE   ISTANBUL   Ankara   ARM.   AZER.   Yerevan

UZBEKISTAN   Samarkand   Tashkent   KYRGYZSTAN   Bishkek   Ürümqi

ÜrümqI   Alma Ata

Changchun   Dalian   Ōsaka   Kitakyūshū

PACIFIC   OCEAN

C

Algiers   TUNISIA   Tunis   MALTA   Sicily   Sardinia   Naples

Athens   Crete   CYPRUS   TURKEY   Izmir

SYRIA   Beirut   Damascus   LEB.   TEHRĀN   Mashhad   TURKMENISTAN   Ashkhabad   TAJIKISTAN   Dushanbe

CHINA   Lanzhou   Taiyuan

Xi'an   Hwang Ho   Nanjing   SHANGHAI

Benghazi   Tripoli   Mediterranean   Sea   Jerusalem   ISR.   Amman   JORDAN   IRAQ   Baghdād   IRAN   Esfahān   Shīrāz   KĀBUL   AFGHANISTAN   Islamabad   Lahore   KASHMIR   TIBET   Lhasa   Chengdu   Wuhan   CHONGQING   East China Sea   Ryukyu Is.

Bonin Is. (Japan)

LIBYA   EGYPT   CAIRO   Alexandria   KUWAIT   BAHRAIN   QATAR   Riyadh   Abu Dhabi   U.A.E.   Muscat   Karachi   New Delhi   DELHI   NEPAL   Katmandu   BHU.   Kanpur   Kunming   GUANGZHOU   Taipei   TAIWAN

Tropic of Cancer

Volcano Is. (Japan)   Marcus I. (Japan)   Wake I. (U.S.A.)

D

GERIA   Aswān   Nile   Mecca   SAUDI   ARABIA   OMAN   Arabian   Sea   Ahmadabad   BOMBAY (Mumbai)   INDIA   Nagpur   CALCUTTA   DACCA   BANGLA-DESH   BURMA   MYANMAR   Hanoi   HONG KONG   South   China   Sea   MANILA   PHILIPPINES

NIGER   CHAD   Omdurmân   Khartoum   Asmara   ERITREA   SANA   YEMEN   Aden   Socotra (Yemen)   Hyderabad   MADRAS (Chennai)   Bangalore   Andaman Is. (India)   Rangoon   THAILAND   BANGKOK   CAMBODIA   Phnom Penh   VIET-   NAM   Vientiane   Ho Chi Minh City

NORTHERN MARIANAS (U.S.A.)   GUAM (U.S.A.)   FEDERATED STATES   Yap   Truk   Pohnpei   Caroline Is.   MARSHALL IS.

Niamey   Kano   NIGERIA   Abuja   CAMEROON   Douala   Yaounde   CENTRAL AFRICAN REP.   Bangui   SUDAN   Addis Ababa   ETHIOPIA   SOMALI   REP.   Lakshadweep Is. (India)   SRI LANKA   Colombo   Nicobar Is. (India)   MALAYSIA   Medan   Kuala Lumpur   PEN. MALAYSIA   SABAH   BRUNEI   IRIAN   JAYA   PALAU   OF MICRONESIA   Gilbert Is.   NAURU   KIRIBATI

E

BENIN   Ibadan   Lagos   EQUATORIAL GUINEA   GABON   Libreville   CONGO   DEM. REP.OF THE   CONGO (Zaire)   Kisangani   Kampala   UGANDA   KENYA   L. Turkana   MADIVES   Equator   Medan   SINGAPORE   Palembang   Borneo   Banjarmasin   INDONESIA   JAKARTA   Ujung Pandang   New Ireland   New Britain   SOLOMON IS.   Santa Cruz I.   TUVALU

SÃO TOMÉ & PRÍNCIPE   Guinea   Brazzaville   Kinshasa   Kananga   RWANDA   Kigali   BURUNDI   Bujumbura   L. Victoria   Nairobi   SEYCHELLES   Amirante Is.   Diego Garcia   Chagos Arch. (U.K.)   I N D I A N   O C E A N   Sumatra   Bandung   Java   Surabaya   Timor   PAPUA NEW GUINEA   Port Moresby   C. York   Arafura Sea

CABINDA (Angola)   Luanda   Katanga   TANZANIA   Dodoma   Zanzibar   Mombasa   Dar es Salaam   Aldabra Is.   Cocos Is. (Austral.)   Christmas I. (Austral.)   Cairns   VANUATU   FIJI   Suva

Benguela   ANGOLA   Lubumbashi   COMOROS   Mayotte (Fr.)   Agalega Is. (Fr.)   Cargados Carajos   Darwin   NEW CALEDONIA (Fr.)

F

ZAMBIA   Lusaka   Lilongwe   MALAWI   L. Malawi   MOZAMBIQUE   Mozambique Channel   MADAGASCAR   Antananarivo   Rodriguez   MAURITIUS   RÉUNION (Fr.)   Port Hedland   Alice Springs   Rockhampton   Lord Howe I. (Austral.)

ZIMBABWE   Harare   Bulawayo   NAMIBIA   Windhoek   BOTSWANA   Gaborone   Pretoria   Johannesburg   SOUTH   AFRICA   Maputo   SWAZILAND   LESOTHO   Durban   Tropic of Capricorn   Geraldton   Perth   Fremantle   Kalgoorlie-Boulder   AUSTRALIA   Brisbane   Newcastle   Norfolk I. (Austral.)

Cape Town   C. of Good Hope   Port Elizabeth   Amsterdam I. (Fr.)   St.Paul (Fr.)   Great Australian Bight   Adelaide   Darling   Sydney   Canberra   Auckland   North I. (N.Z.)   NEW   ZEALAND

Melbourne   Tasman   Sea   Wellington

Tasmania   South I.   Christchurch

Prince Edward Is. (S.Africa)   Crozet Is. (Fr.)   Kerguelen (Fr.)   Hobart   Stewart I.   Bounty Is. (N.Z.)   Dunedin

Bouvet I. (Norw.)   McDonald Is. (Austral.)   Heard I. (Austral.)   Antipodes Is. (N.Z.)   Campbell I. (N.Z.)   Auckland Is. (N.Z.)   Macquarie Is. (Austral.)   **G**

OUTHERN   OCEAN

East from Greenwich

c   t   i   c   a

Antarctic Circle

Ross Sea   **H**

Hanoi ● Capital Cities

100 0   200 400 600 800 1000 1200 1400 km

100 0   200   400   600   800   1000 miles

JAPAN

PACIFIC OCEAN

Aleutian Islands (U.S.A.)

Near Is. (U.S.A.)

Komandorskiye Ostrova

Petropavlovsk Kamchatskiy

Kurilskiye Ostrova (Russia)

La Perouse Str.

Hokkaidō

Dutch Harbor

D

Gora Klyuchevskaya 4850

Sakhalin (Russia)

Unimak I.

B e r i n g   S e a

Poluostrov Kamchatka

Sakhalinskiy Zaliv

Vanino

Pribilof Is. (U.S.A.)

Ostrov Karaginskiy

Sea of Okhotsk

Amur

Khabarovsk

Kodiak I.

Bristol Bay

St. Matthew (U.S.A.)

Mys Olyutorski

Nikolayevsk

Alaska Pen.

Nunivak

Komandorskiye

Penzhinskaya G.

Gizhiginskaya Guba

Tauiskaya Guba

Udskaya Guba

Ulbanskiy Zaliv

1

G. of Alaska

Seward

Cook Inlet

St. Lawrence I. (U.S.A.)

Norton Sd.

Nome

Mys Navarin

Anadyrskiy Zaliv

Anadyr

Penzhino

Kolymskoye Nagorye

Okhotsk

Stanovoy Khrebet

14

Prince William Sd.

Cordova

Anchorage

Mt. McKinley 6194

Fairbanks

Kuskokwim

Yukon

Bering Str.

Mys Dezhneva

Chukotskoye Nagorye

Kolyma

Omolon

Aldan

Yakutsk

Prince Rupert

Mt. St. Elias 5489

ALASKA (U.S.A.)

C. Prince of Wales

C

Kotzebue Sd.

Pt. Hope

C. Lisburne

Prol iv Longa

Nizhne Kolymsk

Srednekolymsk

Indigirka

Zashiversk

Verkhoyanskiy Khrebet

Lena

Skagway

Mt. Logan 6050

Whitehorse

Rocky Mountains

Dawson

Fort Yukon

Porcupine

Koyukuk

Noatak

Chukchi Sea

Ostrov Vrangelya (Russia)

Chaunskaya G.

Russkoye Ustie

Verkhoyansk

Yana

Kazache

Zhigansk

13

Skeena

Stewart

Peel

Prudhoe Bay

C. Halkett

Fort McPherson

Pt. Barrow

46

Nizhne Kolymsk

Kolyma

S

b

i

r

Dawson Creek

Liard

Fort Good Hope

Tulita

Mackenzie

Fort Hope

Harrison B.

Herschel I.

Beaufort Sea

3767

B

Novosibirskiye Ostrova

Lyakhovskiye Ostrova

Tiksi

Olenek

Fort Vermilion

Fort Simpson

Peace

Great Bear Lake

Mackenzie Bay

C. Bathurst

C. Kellett

A R C T I C

Mendeleyev Ridge

O. Bennetta (Russia)

Bulun

O. Kotelnyy

Lena

Olenek

NORTH

Athabasca

Yellowknife

Great Slave Lake

Coppermine

Dolphin & Union Sd.

Banks I.

C. Prince Alfred

Canada Basin

O C E A N

3327

A

Laptev Sea

Anabar

Nordvik

Kotuy

Vilyuy

2

Athabasca Lake

Kugluktuk

Coronation G.

Prince Albert Pen.

M'Clure Str.

Prince Patrick I.

Melville I.

3546

Alpha Cordillera

4007

Lomonosov Ridge

4100

NORTH POLE

4484

Severnaya Zemlya

3849

Ostrova Petra

Poluostrov Taymyr

Ozero Taymyr

Khatanga

Kotuy

Gory Putorana

Nizhnyaya Tunguska

12

Hudson Bay

Chesterfield Inlet

Back

King William I.

Prince of Wales I.

Boothia Pen.

Somerset I.

Viscount Melville Sd.

Bathurst I.

North Magnetic Pole 1990

Ellef Ringnes I.

Sverdrup Is.

Axel Heiberg I.

Nansen Sd.

2104

Makarov Basin

4418

Fram Basin

3741

Nansen Basin

Nansen Cordillera

O. Oktyabrskoy Revolyutsii

O. Uedineniya

O. Ushakova

O. Vise

Golchikha

Dudinka

Igarka

Norilsk

Pyasina

Taz

Yenisey

3

Roes Welcome Sd.

Melville Pen.

Boothia

Gulf of Boothia

Prince Regent Inlet

Barrow Str.

Lancaster Sound

Devon I.

Jones Sound

Eureka

Ellesmere I. (Canada)

Alert

Lincoln Sea

C. Columbia

O. Uedineniya

Z. Vilcheka

O. Graham Bell

O. Belyy

Urengoy

Norilsk

Pyasina

11

Southampton I.

Coats I.

Mansel I.

Foxe Basin

Fury & Hecla Str.

Prince Charles I.

Bylot I.

Smith Sund

Qaanaaq

Kane Basin

Robeson Chan.

Peary Land

K. Morris Jesup

Zemlya Frantsa Iosifa

Z. Aleksandry (Russia)

Nordkapp

Nordaustlandet

Novaya Zemlya

Poluostrov Yamal

Novyy Port

Nadym

Surgut

Hudson Str.

C. Wolstenholme

Foxe Chan.

Nettilling

Baffin

C. Dyer

Upernavik

2399

K. York

Sermersuaq

Knud Rasmussen Land

Independence Fjord

McKinley Sea

A

Kong Frederik VIII.s Land

Kara Sea

Baydaratskaya Guba

Vorkuta

Salekhard

Khabarovo

Berezovo

Ob

4

Hudson Bay

Iqaluit

Frobisher B.

Cumberland Sd.

Resolution I.

Baffin Bay

Uummannaq

Qeqertarsuaq

Qeqertarsuaq

Vestspitsbergen

2571

Svalbard (Norway)

Edgeøya

Barents Sea

O. Kolguyev

Mys Kanin Nos

Pechora

Narodnaya 1894

Uralskie Gory

Tobolsk

Labrador

C. Chidley

Ungava Bay

Davis Str.

Nuuk

GREENLAND (**KALAALLIT NUNAAT**) (Denmark)

Kong Frederik IX.s Land

Kong Christian X.s Land

Kong Franz Joseph Fd.

Kong Oscar Fjord

Ittoqqortoormiit

Greenland

Nordkapp

Longyearbyen

Vardø

Varangerfjorden

Mezen

Sev. Dvina

YEKATERINBURG

PERM

60

Paamiut

Nuuk

Mt. Forel 3360

Kong Frederik VI.s Kyst

Kong Christian IX.s Land

3700

Gunnbjørn Fjeld

Kap Brewster

Sea

B

Bjørnøya

Hammerfest

Tromsø

Kolskiy Poluostrov

Murmansk

Arkhangelsk

Beloye More

Onega

UFA

5

Qaqortoq

Alluitsup Paa

Ammassalik

Kap Farvel (Nunap Isua)

Breiðafjörður

Horn

Fontur

Reykjavík

ICELAND

Öræfajökull 2119

Denmark Str.

Iceland Plateau

Jan Mayen (Norway)

Norwegian

Arctic Circle

Nordkapp

Trondheim

Lofoten

NORWAY

SWEDEN

Gulf of Bothnia

FINLAND

Ladozhskoye Ozero

Helsinki

Volga

SAMARA

Onezhskoye Ozero

RUSSIA

SARATOV

4755

3800

Sea

C

Føroyar (Den.)

Shetland Is. (U.K.)

Bergen

Oslo

STOCKHOLM

Helsinki

Chudskoye Ozero

ST. PETERBURG

VOLGOGRAD

10

Mid-Atlantic Ridge

Rockall (U.K.)

Hebrides (U.K.)

Orkney Is. (U.K.)

SCOTLAND

Edinburgh

North Sea

UNITED KINGDOM

Belfast

IRELAND

Dublin

C. Clear

ENGLAND

Skagerrak

KØBENHAVN

DENMARK

Kaliningrad

Baltic Sea

LAT.

LITH.

Riga

Vilnius

BELARUS

KYYIV

ROSTOV

WALES

D

HAMBURG

NETH.

AMSTERDAM

BERLIN

GERMANY

Elbe

POLAND

WARSZAWA

Wisła

UKRAINE

ODESA

Black Sea

ATLANTIC OCEAN

LONDON

PRAHA

6   **7**   West from Greenwich   0   East from Greenwich   **8**   **9**

CARTOGRAPHY BY PHILIP'S. COPYRIGHT REED INTERNATIONAL BOOKS LTD

Projection : Zenithal Equidistant

ft   m

80

12 000   4000

6000   2000

4500   1500

3000   1000

1200   400

600   200

0   0

500   1500

1000   3000

2000   6000

3000   9000

4000   12 000

5000   15 000

m   ft

Maximum extent of sea ice

Summer extent of sea ice

Ice caps and permanent ice shelf

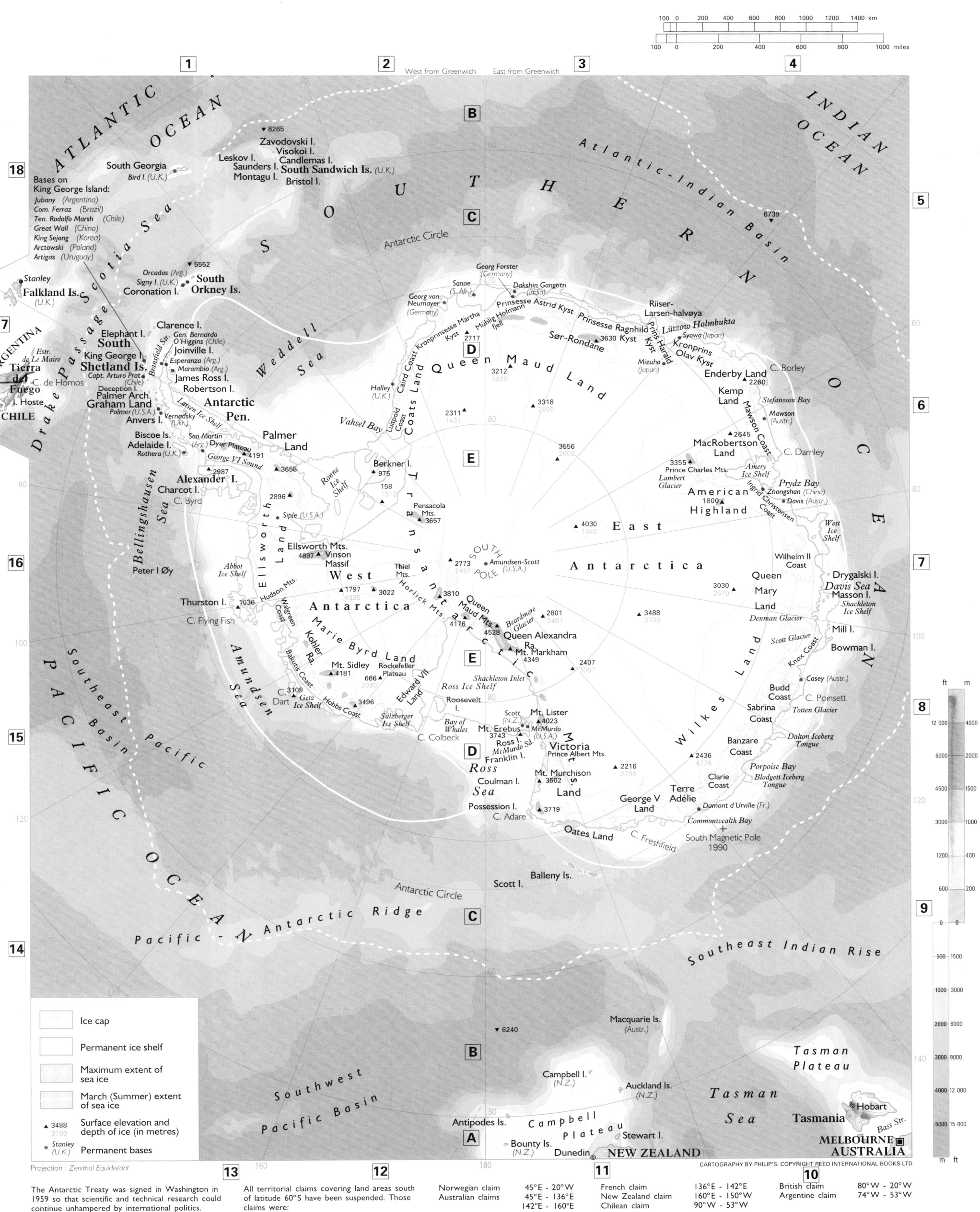

The Antarctic Treaty was signed in Washington in 1959 so that scientific and technical research could continue unhampered by international politics.

All territorial claims covering land areas south of latitude 60°S have been suspended. Those claims were:

Norwegian claim	45°E - 20°W	French claim	136°E - 142°E	British claim	80°W - 20°W
Australian claims	45°E - 136°E	New Zealand claim	160°E - 150°W	Argentine claim	74°W - 53°W
	142°E - 160°E	Chilean claim	90°W - 53°W		

Projection: Zenithal Equidistant

CARTOGRAPHY BY PHILIP'S. COPYRIGHT REED INTERNATIONAL BOOKS LTD

A
B
C
D
E
F
G
H
1
2
3
4
5
6
7
8
9
ARCTIC OCEAN
Beaufort Sea
Queen Elizabeth Islands
Baffin Bay
Davis Strait
Baffin Island
Hudson Bay
Labrador
Labrador Basin
Newfoundland
Greenland Sea
Denmark Strait
GREENLAND
ICELAND
Svalbard
Norwegian Sea
Barents Sea
Kara Sea
Novaya Zemlya
Zemlya Frantsa Iosifa
Severnaya Zemlya
RUSSIA
FINLAND
SWEDEN
North Sea
UNITED KINGDOM
IRELAND
FRANCE
SPAIN
PORTUGAL
GERMANY
POLAND
Bay of Biscay
Mediterranean Sea
MOROCCO
ALGERIA
TUNISIA
LIBYA
NIGER
MAURITANIA
WESTERN SAHARA
SENEGAL
NIGERIA
G. of Guinea
Canary Basin
Azores
Madeira
Islas Canarias
C. VERDE IS.
Cape Verde Basin
Northern Mid-Atlantic Ridge
NORTH ATLANTIC OCEAN
North American Basin
Sargasso Sea
Bermuda
BAHAMAS
West Indies
Caribbean Sea
Gulf of Mexico
CUBA
Florida
CANADA
UNITED STATES
MEXICO
VENEZUELA
COLOMBIA
Tropic of Cancer
Equator
Amazon

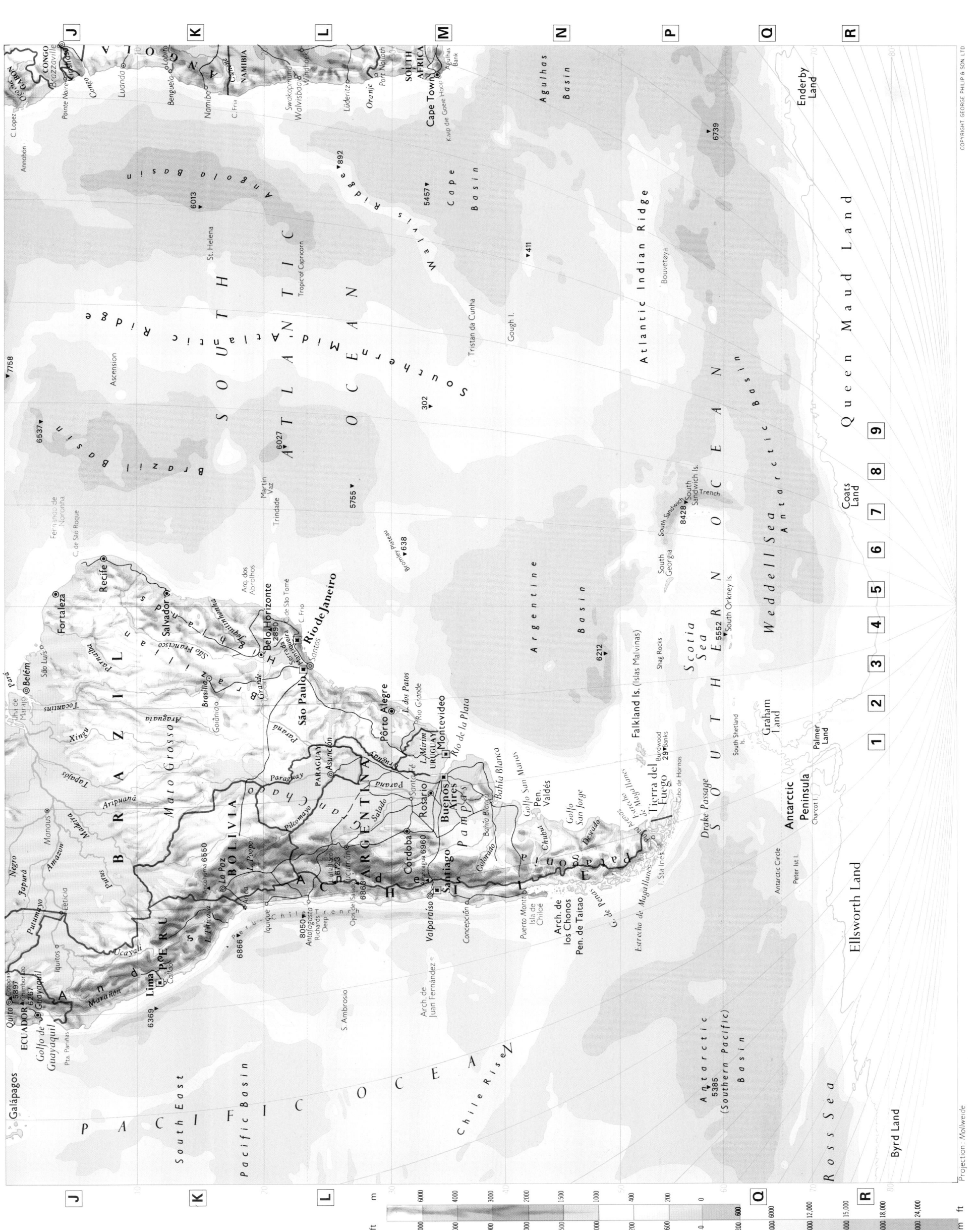

S O U T H   A T L A N T I C   O C E A N
P A C I F I C   O C E A N
S O U T H E R N   O C E A N
B R A Z I L
B O L I V I A
A R G E N T I N A
P E R U
C H I L E
A n d e s
B r a z i l i a n   H i g h l a n d s
Gran Chaco
PARAGUAY
URUGUAY
ECUADOR
ANGOLA
NAMIBIA
SOUTH AFRICA
CONGO
GABON
Brazil Basin
Angola Basin
Cape Basin
Agulhas Basin
Argentine Basin
Walvis Ridge
Southern Mid-Atlantic Ridge
Atlantic Indian Ridge
South East Pacific Basin
Chile Rise
Peru-Chile Trench
Richards Deep
Antarctic (Southern Pacific) Basin
Antarctic Basin
Bromley Plateau
Scotia Sea
Weddell Sea
Ross Sea
Drake Passage
South Sandwich Trench
Queen Maud Land
Coats Land
Enderby Land
Ellsworth Land
Byrd Land
Graham Land
Palmer Land
Antarctic Peninsula
Antarctic Circle
Tropic of Capricorn
Falkland Is. (Islas Malvinas)
South Georgia
South Sandwich Is.
South Orkney Is.
South Shetland Is.
Shag Rocks
Burdwood Banks
Tierra del Fuego
Cabo de Hornos
Estrecho de Magallanes
Punta Arenas
I. Sta Inés
Bouvetøya
Gough I.
Tristan da Cunha
St. Helena
Ascension
Trindade
Martin Vaz
Fernando de Noronha
C. de São Roque
Arq. dos Abrolhos
C. de São Tomé
C. Frio
Galápagos
Pta. Pariñas
Peter Ist I.
Charcot I.
Arch. de Juan Fernández
S. Ambrosio
Arch. de los Chonos
Pen. de Taitao
G. de Penas
Isla de Chiloé
Golfo San Jorge
Golfo San Matías
Pen. Valdés
Bahía Blanca
Río de la Plata
Pampas
Mato Grosso
Ilha de Marajó
Amazon
Madeira
Negro
Japurá
Putumayo
Ucayali
Marañón
Tapajós
Xingu
Tocantins
Pará
Araguaia
Paraná
Paraguay
Uruguay
Pilcomayo
Salado
Colorado
Chubut
Deseado
São Francisco
Parnaíba
Jequitinhonha
Aripuanã
Purus
Grande
Serra da Mantiqueira
L. Titicaca
L. Poopó
L. Mirim
L. dos Patos
Quito
Guayaquil
Golfo de Guayaquil
Lima
Callao
Iquitos
Leticia
Manaus
Belém
São Luís
Fortaleza
Recife
Salvador
Brasília
Goiânia
Belo Horizonte
Rio de Janeiro
Santos
São Paulo
Pôrto Alegre
Rio Grande
Montevideo
Buenos Aires
Rosario
Córdoba
Santa Fé
Tucumán
Asunción
La Paz
Arica
Iquique
Antofagasta
Valparaíso
Santiago
Concepción
Puerto Montt
Luanda
Lobito
Benguela
Namibe
C. Fria
Cunene
Swakopmund
Walvisbaai
Windhoek
Lüderitz
Oranje
Port Nolloth
Cape Town
Kaap die Goeie Hoop
Agulhas Bank
Kinshasa
Brazzaville
Congo
Pointe Noire
Ogooué
Annobón
C. Lopez
Aconcagua 6960
Ojos del Salado 6863
Llullaillaco 6723
Ancohuma 6550
Chimborazo 6267
Cotopaxi 5897
2890
6369
6866
8050
6537
7758
6027
5755
302
638
6013
5457
892
411
6739
6212
5552
8428
5385
29
1 2 3 4 5 6 7 8 9
J K L M N P Q R
ft m
18,000 6000
12,000 4000
9000 3000
6000 2000
4500 1500
3000 1000
1200 400
600 200
0 0
200 600
2000 6000
4000 12,000
5000 15,000
6000 18,000
8000 24,000
m ft
Projection: Mollweide
COPYRIGHT GEORGE PHILIP & SON LTD

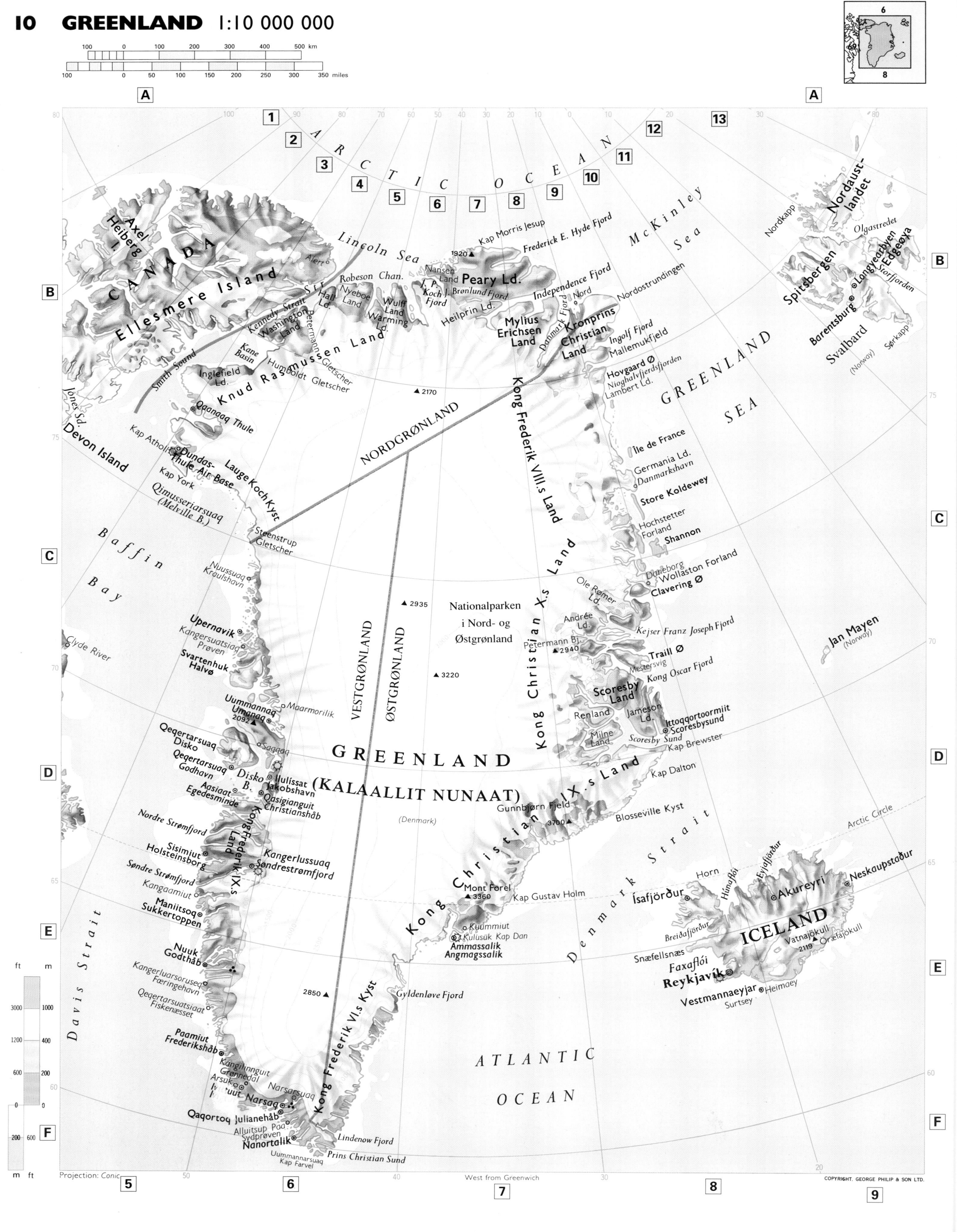

100 0 100 200 300 400 500 km
100 0 50 100 150 200 250 300 350 miles

A

6
8

A   A

1 90 80
2
3
4
5 6 7 8 9 10 11 12 13

ARCTIC OCEAN

McKinley Sea

Nordkapp   Nordaust-
landet
Olgastredet

Kap Morris Jesup   Frederick E. Hyde Fjord
Lincoln Sea   1920▲   Longyearbyen Storøya
Alert○   Nansen   Spitsbergen   ●   Edgeøya
Robeson Chan. Land Peary Ld.   Nordostrundingen
Str   Nyeboe Wulff   J.P. J. Brønlund Fjord   Nord   Barentsburg   Storfjorden
Hall Land   Land Koch   Independence Fjord   Svalbard   Sørkapp
Kennedy Strait Ld. Warming Fjord   Hellprin Ld.   (Norway)
Washington   Petermann Land   Mylius   Kronprins
Land   Erichsen   Christian   Ingolf Fjord
Kane   Land   Land   Mallemukfjeld   GREENLAND
Basin   Humboldt Gletscher   Hovgaard Ø
Inglefield Ld.   2170▲   Nioghalvfjerdsfjorden   SEA
Smith Sound   Lambert Ld.
Knud Rasmussen Land   Kong Frederik VIII.s Land
Qaanaaq Thule○   NORDGRØNLAND
Kap Atholl ○ Dundas-   NORDGRØNLAND   Île de France
Jones Sd.   Thule Air Base   Germania Ld.
Devon Island Kap York Lauge Koch Kyst   ○Danmarkshavn
Qimusseriarsuaq   Store Koldewey
(Melville B.)   Hochstetter
Steenstrup   Forland   Shannon
Gletscher   ●

Baffin   Danieborg
Bay   Nuussuaq   Wollaston Forland
Kraulshavn○   Clavering Ø
Nationalparken   Ole Rømer
i Nord- og   Ld.   Andrée   Kejser Franz Joseph Fjord   Jan Mayen
Østgrønland   Ld.   (Norway)
Upernavik●   2935▲   Petermann Bj. Mestersvig   Kong Oscar Fjord
Kangersuatsiaq   2940▲   Traill Ø
Prøven○   Scoresby   Land
Svartenhuk   Renland   Jameson
Halvø   VESTGRØNLAND ØSTGRØNLAND   3220▲   Ld. Ittoqqortoormiit
Uummannaq●   Milne   ○Scoresbysund
Umanaq   Land   Scoresby Sund
Maarmorilik○ 2092▲   ○Kap Brewster
Qeqertarsuaq   GREENLAND
Disko
Qeqertarsuaq Saqqaq○   Kap Dalton
Godhavn○ Disko○ Ilulissat   (KALAALLIT NUNAAT)
Aasiaat B.   Jakobshavn   Gunnbjørn Fjeld
Egedesminde○ Qasigiannguit   (Denmark)   3700▲   Blosseville Kyst   Arctic Circle
Christianshåb
Nordre Strømfjord   Denmark Strait
Sisimiut   Mont Forel   Horn   Húnaflói   Eyjafjörður
Holsteinsborg Kangerlussuaq   3360▲ Kap Gustav Holm   Ísafjörður   ●Akureyri   ●Neskaupstaður
Søndre Strømfjord Søndrestrømfjord   Breiðafjörður
Kangaamiut○   Kuummiut○   Vatnajökull
Maniitsoq   ○Kulusuk Kap Dan   Faxaflói   ICELAND   2119▲ Öræfajökull
Sukkertoppen   Ammassalik
Nuuk   Angmagssalik   Reykjavík Snæfellsnes
Godthåb   Vestmannaeyjar   ○Heimaey
Kangerluarsoruseq   Surtsey
Føringehavn○
Qeqertarsuatsiaat 2850▲   Gyldenløve Fjord
Fiskenæsset○
Paamiut   ATLANTIC
Frederikshåb
Kangilinnguit   OCEAN
Grønnedal
Arsuk○   Narsarsuaq
Ivittuut Narsaq
Qaqortoq Julianehåb
Alluitsup Paa   60
Sydprøven○   Lindenow Fjord
Nanortalik●
Uummannarsuaq   Prins Christian Sund
Kap Farvel

CANADA
Axel
Heiberg
I.
Ellesmere Island

Clyde River

Davis Strait

ft m
3000 1000
1200 400
600 200
60
0 0
200 600
m ft

B   B
C   C
D   D
E   E
F   F

75
70
65
60

Projection: Conic

5 6 7 8 9
50 40 West from Greenwich 30 20

COPYRIGHT. GEORGE PHILIP & SON LTD.

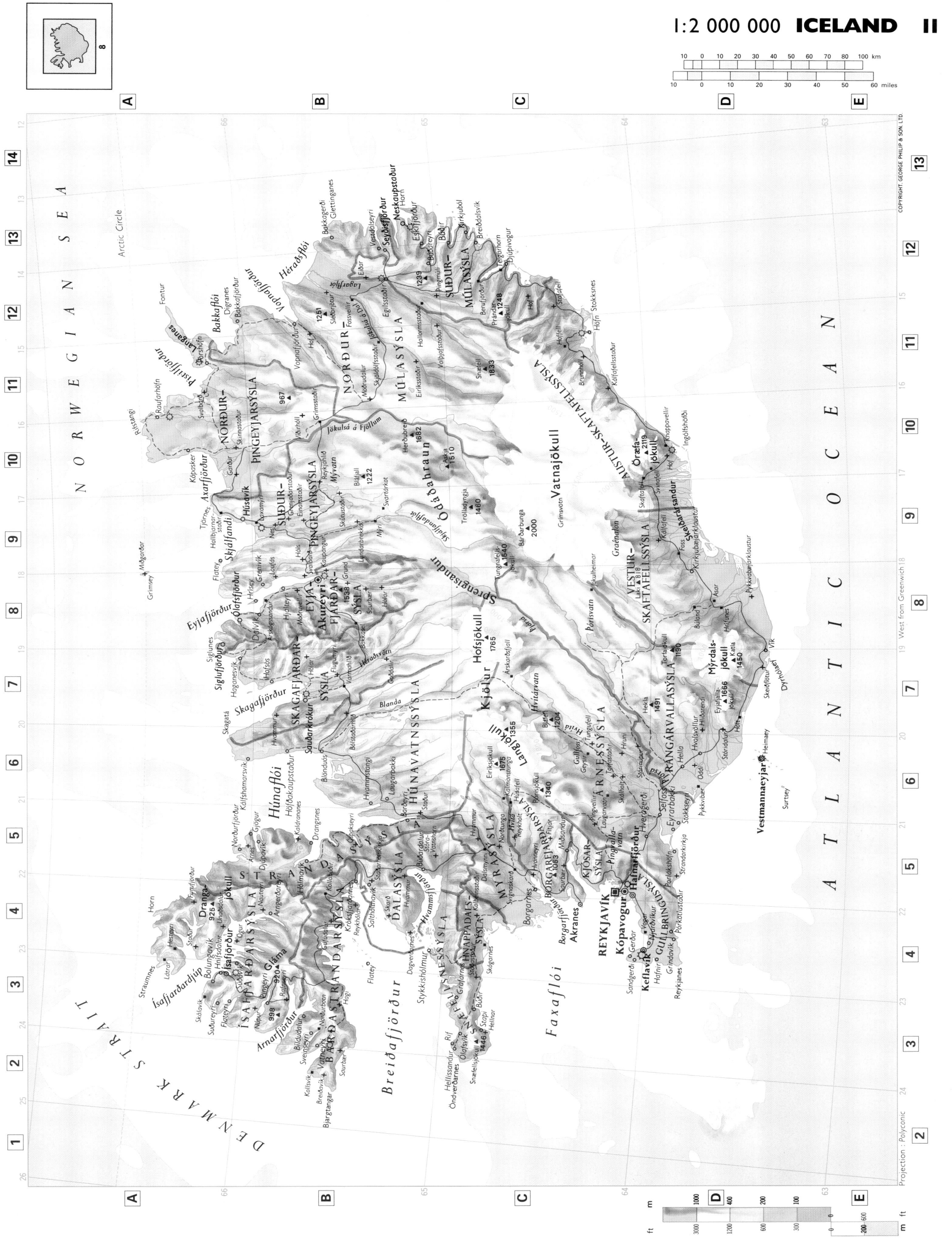

10 0 10 20 30 40 50 60 70 80 100 km
10 0 10 20 30 40 50 60 miles

Arctic Circle

N O R W E G I A N   S E A

D E N M A R K   S T R A I T

A T L A N T I C   O C E A N

Projection : Polyconic

West from Greenwich

100 0 100 200 300 400 500 600 700 800 km
100 0 100 200 300 400 500 miles
Ural Mountains
Narodnaya 1894
Telpos Iz 1617
Pechora
Ob
Kanin Pen.
Mezen
N. Dvina
Onega
Kama
Kola Pen.
White Sea
Inari
Nordkinn
North Cape
Lapland
Kebnekaise 2117
Torne
Vesterålen
Lofoten
Iceland
Arctic Circle
Hekla 1491
Öræfajökull 2119
Norwegian Sea
2851
Faroe Is.
Shetland Is.
Rockall
Orkney Is.
Scandinavia
Galdhøpiggen 2469
Ume
Indals
G. of Bothnia
Finland
L. Onega
Svir
L. Ladoga
Plain
Rybinsk Res.
Volga
Åland
G. of Finland
Neva
Mälaren
L. Chudskoye
Vänern
Vättern
Skagerrak
Kattegat
Lindesnes
Gotland
Saaremaa
G. of Riga
European
Öland
Bornholm
Baltic Sea
ATLANTIC OCEAN
British Isles
Hebrides
Ben Nevis 1347
Great Britain
North Sea
15
Ireland
Irish Sea
Jutland
Snowdon 1085
Helgoland
C. Clear
Celtic Sea
Land's End
Thames
English Channel
Channel Is.
Ushant
Brittany
North
Central Russian Uplands
W. Dvina
Oka
Niemen
Elbe
Weser
Harz
Vistula
Pripet
Don
Volga Hts.
Obshchi Syrt
Depression
Ural
Rhine
Westerwald
Taunus
Ardennes
Meuse
Seine
Hunsrück
Vosges
Black Forest
Jura
Erzgebirge
Bohemian Forest
Moravian Hts.
Oder
Sudeten
Ukraine
Donets
Tsimlyansk Res.
Caspian
-28
Caspian Sea
Loire
Bay of Biscay
4817
Gironde
Inn
Carpathians
Tatra 2655
Dniester
Dnieper
Bug
Manych
Danube
Prut
Sea of Azov
C. Finisterre
Massif Central
Puy de Sancy 1886
Mont Blanc 4807
Alps
Bakony Forest
Plain of Hungary
Kuban
Terek
Crimea
Cantabrian Mts.
Pyrenees
Cévennes
Rhône
Garonne
Drava
Tisza
Mureş
Transylvanian Alps
Str. of Kerch
Caucasus
Elbrus 5642
Douro
Old Castile
Pico de Aneto 3404
Ebro
Po
Sava
Dinaric Alps
Wallachia
Danube
Kura
C. da Roca
Tagus
Serra da Estrêla
Iberian Peninsula
G. of Lions
Ligurian Sea
Apennines
Adriatic Sea
Balkans
Black Sea
2211
New Castile
Guadiana
Sierra Morena
Guadalquivir
Corsica
Str. of Bonifacio
2914 Gran Sasso d'Italia
Tiber
Rhodope
Bosporus
Pontine Mts.
Armenia
Ararat 5165
Araks
L. Van
L. Urmia
C. de São Vicente
C. Trafalgar
Andalusia
Sierra Nevada
Mulhacén 3748
Balearic Is.
Minorca
Ibiza
Majorca
Sardinia
Vesuvius 1277
Tyrrhenian Sea
Str. of Otranto
Sea of Marmara
Dardanelles
Olympus 2917
Mt. Ida 1766
Kızıl Irmak
L. Tuz
Kurdistan
Str. of Gibraltar
Mediterranean Sea
Str. of Messina
Calabria
Ionian Sea
Ionian Is.
Pindus
Ægean Sea
Anatolia (Asia Minor)
3770 Erciyaş Dağı
Taurus Mts.
Asia
Tigris
Er Rif
Africa
C. Blanc
C. Bon
Pantelleria
Sicily
Etna 3340
Morea
Mesopotamia
Euphrates
Plateau of the Shotts
Malta
4070
C. Matapan
Crete
Rhodes
Cyprus
ft m
15 000 5000
12 000 4000
6000 2000
3000 1000
1200 400
600 200
0 0
200 600
1000 3000
2000 6000
4000 12 000
m ft
Projection: Bonne
West from Greenwich
East from Greenwich
CARTOGRAPHY BY PHILIP'S. COPYRIGHT REED INTERNATIONAL BOOKS LTD.

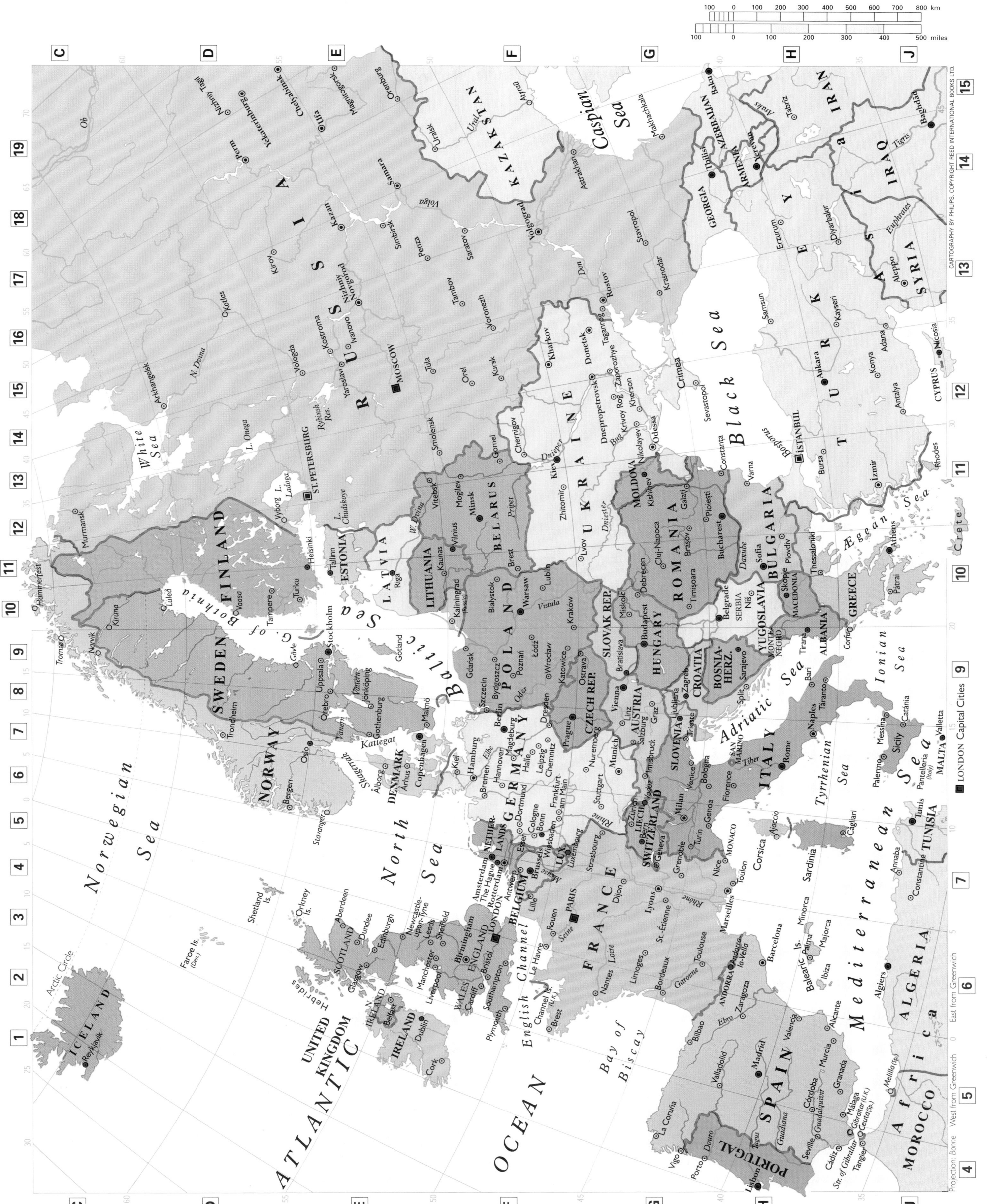
100 0 100 200 300 400 500 600 700 800 km
100 0 100 200 300 400 500 miles
Norwegian Sea
North Sea
Baltic Sea
White Sea
G. of Bothnia
Kattegat
Skagerrak
English Channel
Bay of Biscay
ATLANTIC OCEAN
Adriatic Sea
Tyrrhenian Sea
Ionian Sea
Ægean Sea
Mediterranean Sea
Black Sea
Caspian Sea
ICELAND
NORWAY
SWEDEN
FINLAND
ESTONIA
LATVIA
LITHUANIA
BELARUS
RUSSIA
UKRAINE
MOLDOVA
ROMANIA
BULGARIA
KAZAKSTAN
GEORGIA
AZERBAIJAN
ARMENIA
TURKEY
SYRIA
IRAQ
IRAN
CYPRUS
UNITED KINGDOM
IRELAND
DENMARK
NETHERLANDS
BELGIUM
LUX.
GERMANY
POLAND
CZECH REP.
SLOVAK REP.
HUNGARY
AUSTRIA
SWITZERLAND
LIECH.
SLOVENIA
CROATIA
BOSNIA-HERZ.
YUGOSLAVIA
SERBIA
MONTENEGRO
MACEDONIA
ALBANIA
GREECE
ITALY
SAN MARINO
MALTA
FRANCE
MONACO
ANDORRA
SPAIN
PORTUGAL
MOROCCO
ALGERIA
TUNISIA
Africa
LONDON Capital Cities
Projection: Bonne
West from Greenwich
East from Greenwich
CARTOGRAPHY BY PHILIPS. COPYRIGHT REED INTERNATIONAL BOOKS LTD.

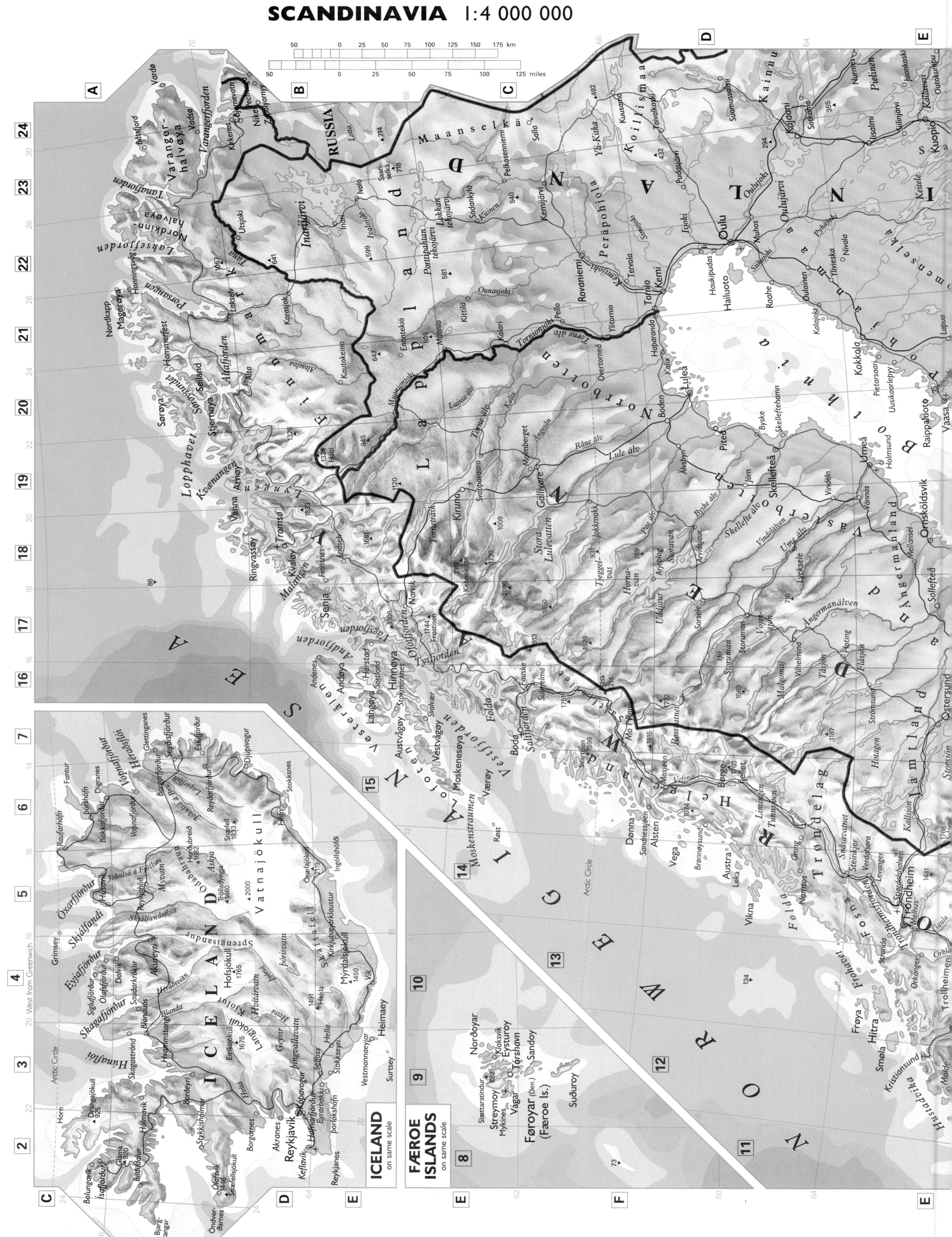

ICELAND
on same scale

FÆROE
ISLANDS
on same scale

Gulf of Finland
Gulf of Riga
BALTIC SEA
Ålands hav
Skagerrak
Kattegat
Oslofjorden
Zatoka Gdańska
Kurshskiy Zaliv
Zalew Wiślany
Deutsche Bucht
Kieler Bucht
Mecklenburger Bucht
Fehmarn Bælt
Store Bælt
Lille Bælt
ESTONIA
LATVIA
LITHUANIA
DENMARK
GERMANY
POLAND
BELARUS
Svealand
Götaland
Småland
Dalarna
Värmland
Uppland
Södermanland
Västmanland
Gästrikland
Hälsingland
Medelpad
Härjedalen
Dalsland
Bohuslän
Halland
Blekinge
Skåne
Jylland
Sjælland
Fyn
Lolland
Falster
Møn
Bornholm
Öland
Gotland
Gotska Sandön
Fårö
Saaremaa (Ösel)
Hiiumaa (Dagö)
Åland (Ahvenanmaa)
Rügen
Usedom
Fehmarn
Langeland
Læsø
Anholt
Samsø
Fanø
Rømø
Sylt
Föhr
Nordfriesische Inseln
Ostfriesische Inseln
Helgoland
Jotunheimen
Hardangervidda
Dovrefjell
Gudbrandsdalen
Østerdalen
Valdres
Telemark
Romsdalen
Rondane
Satakunta
Salpausselkä
Saimaa
Päijänne
Vänern
Vättern
Mälaren
Hjälmaren
Mjøsa
Ozero Chudskoye
Daugava
Neman
STOCKHOLM
Oslo
Helsinki (Helsingfors)
Tallinn
Riga
Vilnius
KØBENHAVN (Copenhagen)
Göteborg (Gothenburg)
Malmö
Uppsala
Västerås
Örebro
Norrköping
Linköping
Jönköping
Gävle
Sundsvall
Karlstad
Falun
Borås
Helsingborg
Kalmar
Karlskrona
Visby
Bergen
Stavanger
Kristiansand
Drammen
Turku (Åbo)
Tampere
Jyväskylä
Espoo
Vantaa
Lahti
Kotka
Pori
Tartu
Pärnu
Daugavpils
Liepāja
Ventspils
Jelgava
Kaunas
Šiauliai
Panevėžys
Klaipėda
Kaliningrad (Russia)
Gdańsk
Gdynia
Elbląg
Koszalin
Słupsk
Århus
Ålborg
Odense
Esbjerg
Kiel
Lübeck
Rostock
Flensburg
Schleswig
Holstein
Wismar
Stralsund
Greifswald
Cuxhaven
Ruhnu saar
Projection: Conical with two standard parallels
East from Greenwich
12
13
14
15
16
17
18
19
20
21
F
G
H
J
K
ft m
6000 2000
4500 1500
3000 1000
1500 500
600 200
0 0
50 150
100 300
200 600
500 1500
1000 3000
2000 6000
m ft

10 0 10 20 30 40 50 60 70 80 90 km

10 0 10 20 30 40 50 60 miles

Gulf of Bothnia

VÄSTER-NORRLANDS LÄN

NORRLANDS LÄN

JÄMTLANDS LÄN

Medelpad

Ljungan

Indalsälven

Östersund

Storsjön

Härjedalen

Hälsingland

GÄVLEBORGS LÄN

Gästrikland

Gävle

Sundsvall

Hudiksvall

Söderhamn

Ljusnan

Falun

Borlänge

KOPPARBERGS LÄN

DALARNA

Siljan

Österdalälven

Västerdalälven

UPPSALA LÄN

UPPLAND

Uppsala

STOCKHOLMS LÄN

STOCKHOLM

Södertälje

Mälaren

VÄSTMANLANDS LÄN

Västerås

Eskilstuna

SÖDERMANLANDS LÄN

Örebro

ÖREBRO LÄN

Karlstad

VÄRMLANDS LÄN

Klarälven

SØR-TRØNDELAG

Trondheim

Røros

Dovrefjell

Trollheimen

MØRE OG ROMSDAL

Kristiansund

Jotunheimen

OPPLAND

Rondane

Gudbrandsdalen

HEDMARK

Österdalen

Femunden

Elverum

Hamar

Lillehammer

Glomma

Mjøsa

BUSKERUD

Hallingdal

Numedal

AKERSHUS

Oslo

Drammen

ØSTFOLD

Fredrikstad

Halden

VESTFOLD

Tønsberg

Sandefjord

Larvik

TELEMARK

Skien

Porsgrunn

Oslofjorden

m 2000 1500 1000 500 200 0

ft 6000 4500 3000 1500 600 0

Underlined towns give their name to the
administrative area in which they stand.

East from Greenwich
COPYRIGHT GEORGE PHILIP LTD.

NORTH SEA
BALTIC SEA
ADRIATIC SEA
UNITED KINGDOM
NETHERLANDS
BELGIUM
LUXEMBOURG
DENMARK
GERMANY
FRANCE
SWITZERLAND
LIECHTENSTEIN
AUSTRIA
ITALY
SLOVENIA
CZECH R.
SAN MARINO
MONACO
AMSTERDAM
'S-Gravenhage (Den Haag)
ROTTERDAM
BRUSSEL (Bruxelles)
PARIS
BERLIN
HAMBURG
MÜNCHEN (Munich)
PRAHA (Prague)
MILANO
TORINO (Turin)
LYON
MARSEILLE
Frankfurt
Köln (Cologne)
Stuttgart
Zürich
Genève
Bern
Venézia (Venice)
Golfo di Venézia
Golfo di Génova
Ljubljana
Zagreb
Szczecin
Projection: Conical with two standard parallels

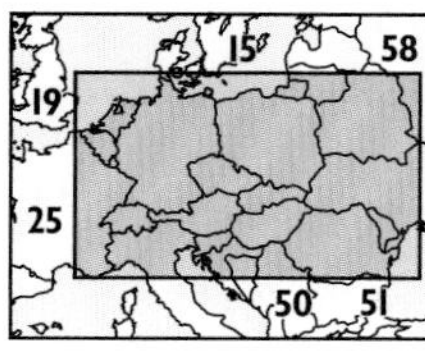

LITHUANIA

BELARUS

POLAND

UKRAINE

SLOVAK REP.

HUNGARY

ROMANIA

MOLDOVA

YUGOSLAVIA

BOSNIA-HERZEGOVINA

CROATIA

BULGARIA

Kaliningrad (Russia)

Zatoka Gdańska

Słupsk, Wejherowo, Rumia, Gdynia, Sopot, Gdańsk, Tczew, Elbląg, Lębork, Bytów, Malbork, Kwidzyn, Braniewo, Gvardeysk, Bagrationovsk, Chernyakhovsk, Gusev

Prienai, Marijampolė, Vilnius, Ashmyany, Vileyka, Smarhon, Maladzyechna, Barysaw, Krupki, Shklow, Mstsislaw, Krychaw

Varėna, Druskininkai, Lida, Nyoman, Navahrudak, Stowbtsy, Dzyarzhynsk, MINSK, Cherven, Mahilyow, Bykhaw, Cherykaw, Slawharad

Szczecinek, Chojnice, Świecie, Grudziądz, Brodnica, Rypin, Chełmno, Wałcz, Piła, Bydgoszcz, Toruń, Inowrocław, Włocławek

Olsztyn, Szczytno, Kętrzyn, Giżycko, Suwałki, Augustów, Ełk, Ostróda, Iława, Mława, Ciechanów, Działdowo, Ostrołęka

Hrodna, Sokółka, Masty, Vawkavysk, Slonim, Baranavichy, Klyetsk, Lyakhavichy, Slutsk, Salihorsk, Glusk, Babruysk, Zhlobin

Poznań, Gniezno, Września, Koło, Kutno, Płock, Legionowo, WARSZAWA (Warsaw), Mińsk Mazowiecki, Siedlce, Brest, Pinsk

Leszno, Krotoszyn, Kalisz, Konin, Turek, Łęczyca, Łowicz, Pruszków, Otwock, Łuków, Międzyrzec Podlaski, Malaryta

Wrocław, Oleśnica, Kluczbork, Piotrków Trybunalski, Radomsko, Tomaszów Mazowiecki, Radom, Puławy, Lublin, Chełm, Lyuboml, Kovel

Świdnica, Oława, Opole, Częstochowa, Myszków, Zawiercie, Kielce, Ostrowiec-Świętokrzyski, Sandomierz, Stalowa Wola, Zamość, Sokal

Kłodzko, Nysa, Tarnowskie Góry, Bytom, Sosnowiec, Katowice, Kraków, Tarnów, Rzeszów, Przemyśl, Lviv (Lvov), Ternopil

Zabrze, Gliwice, Chorzów, Tychy, Oświęcim, Bochnia, Dębica, Jarosław, Mościska, Khmelnytskyy, Vinnytsya

Ostrava, Frýdek-Místek, Karviná, Bielsko-Biała, Nowy Sącz, Jasło, Krosno, Sanok, Sambir, Drohobych

KYYIV (Kiev), Zhytomyr, Korostyshev, Fastiv, Bila Tserkva, Berdychiv, Khmelnik

Chernobyl, Korosten, Novohrad-Volynskyy, Zhmerynka, Haysyn, Uman

Olomouc, Vyškov, Přerov, Zlín, Brno, Bratislava, WIEN (Vienna)

SLOVAK REP., Žilina, Martin, Banská Bystrica, Zvolen, Poprad, Prešov, Košice, Humenné, Uzhhorod, Mukacheve, Khust

Ivano-Frankivsk, Kolomyya, Chernivtsi, CHIŞINĂU, Tiraspol, Tighina

Nitra, Levice, Lučenec, Miskolc, Nyíregyháza, Satu Mare, Baia Mare, Iaşi

BUDAPEST, Székesfehérvár, Kecskemét, Szolnok, Debrecen, Oradea, Cluj-Napoca, Târgu Mureş, Bacău, Bârlad

Győr, Tatabánya, Komárno, Esztergom, Vác, Gyöngyös, Eger, Hatvan

Szombathely, Veszprém, Nagykanizsa, Kaposvár, Pécs, Szeged, Subotica, Timişoara, Arad, Deva, Sibiu, Braşov, Ploieşti

Zalaegerszeg, Balaton, Siófok, Kiskunhalas, Baja, Mohács, Novi Sad, Reşiţa, Hunedoara, Petroşani, Târgu-Jiu, Piteşti, BUCUREŞTI (Bucharest)

ROMANIA, Craiova, Slatina, Caracal, Alexandria, Giurgiu, Ruse, Constanţa

BOSNIA-HERZEGOVINA, Banja Luka, Tuzla, Sarajevo

BEOGRAD (Belgrade), Smederevo, Požarevac, Kragujevac, Jagodina, Niš, Zaječar, Vidin

BULGARIA, Varna, Dobrich, Razgrad, Silistra

Drava, Sava, Danube (Dunărea), Vistula (Wisła), Dniester, Dnieper, Pripet Marsh

East from Greenwich

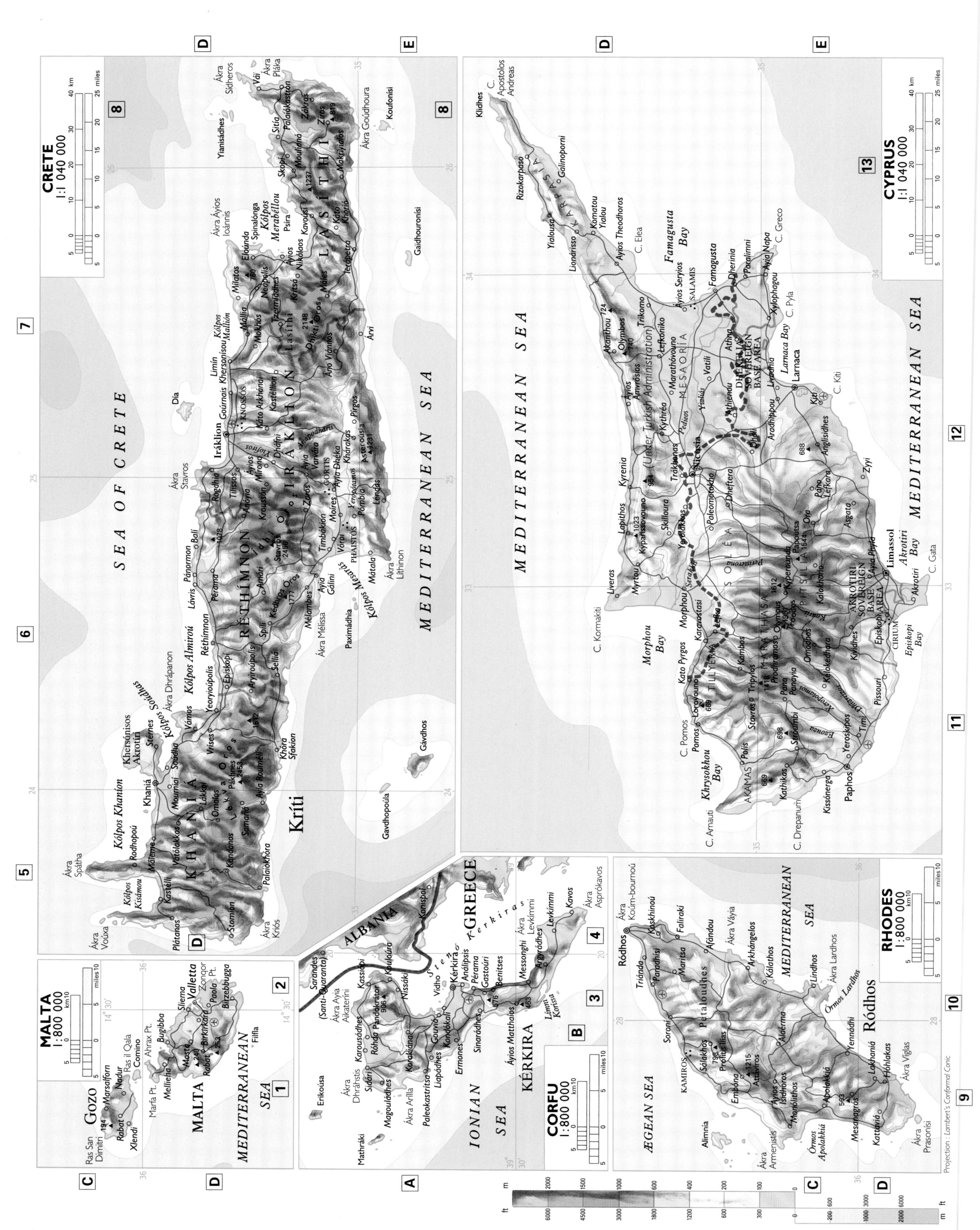
MALTA
1:800 000
CRETE
1:1 040 000
CORFU
1:800 000
RHODES
1:800 000
CYPRUS
1:1 040 000
SEA OF CRETE
MEDITERRANEAN SEA
IONIAN SEA
ÆGEAN SEA
Gozo
MALTA
Valletta
Kríti
KHANIÁ
RÉTHÍMNON
IRÁKLION
LASÍTHI
Khaniá
Réthímnon
Iráklion
KÉRKIRA
Kérkira
ALBANIA
GREECE
Ródhos
Morphou Bay
Khrysokhou Bay
Famagusta Bay
Larnaca Bay
Akrotiri Bay
Episkopi Bay
Nicosia
Famagusta
Larnaca
Limassol
Paphos
(Under Turkish Administration)
DHEKELIA SOVEREIGN BASE AREA
AKROTIRI SOVEREIGN BASE AREA
Projection : Lambert's Conformal Conic

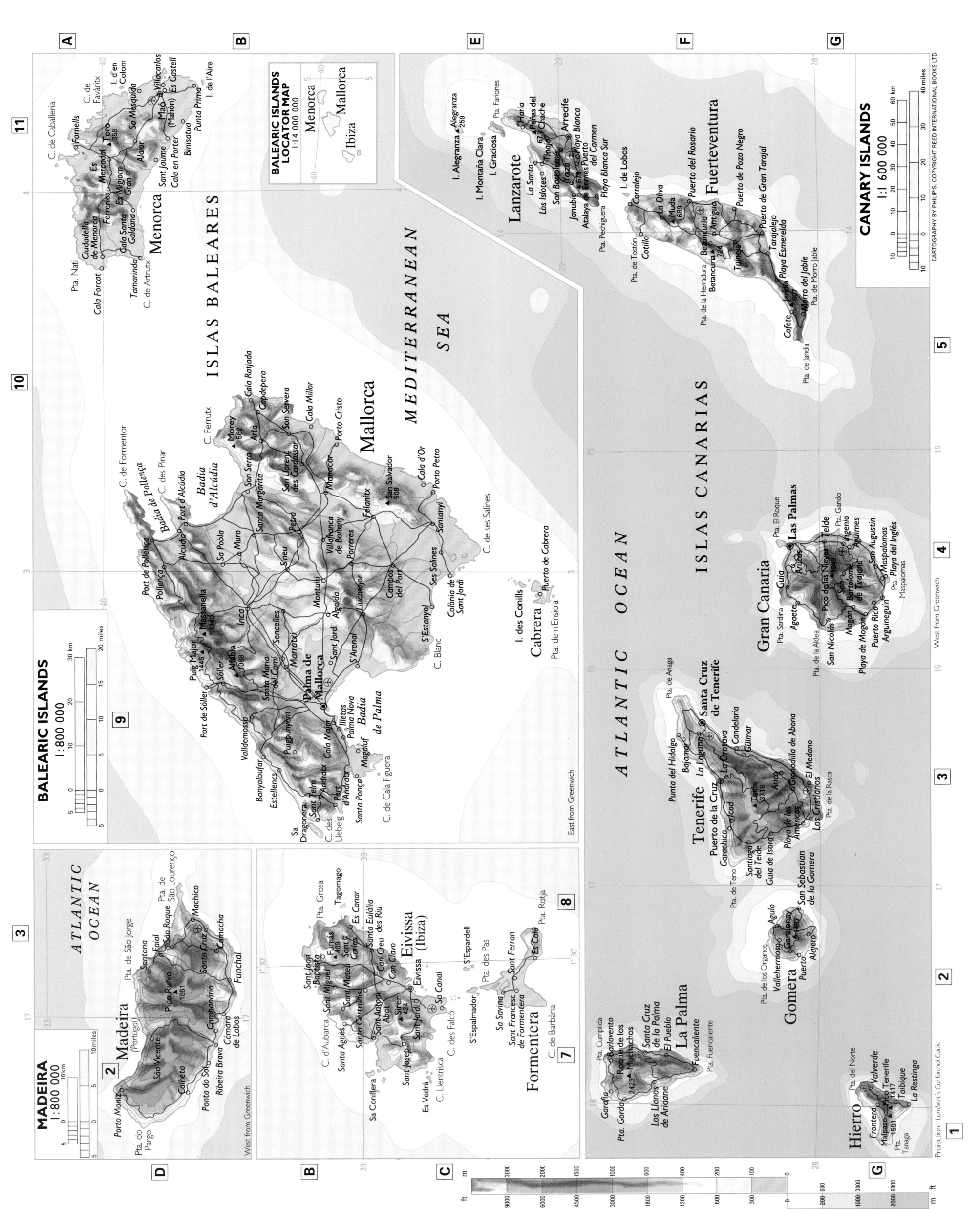

THE BALEARICS, THE CANARIES AND MADEIRA

BALEARIC ISLANDS LOCATOR MAP
1:14 000 000

Menorca
Mallorca
Ibiza

MADEIRA
1:800 000

Madeira (Portugal)

ATLANTIC OCEAN

West from Greenwich

Porto Moniz · Pta. do Pargo · São Vicente · Pico Ruivo 1861 · São Jorge · Pta. de São Lourenço · Machico · Santa Cruz · Funchal · Câmara de Lobos · Ribeira Brava · Ponta do Sol · Calheta · Caniço · Camacha

BALEARIC ISLANDS
1:800 000

ISLAS BALEARES

Menorca
Pta. Nati · C. de Caballeria · Fornells · C. de Favàritx · I. d'en Colom · Es Mercadal · Toro 358 · Villacarlos · Es Castell · Maó (Mahón) · Sa Mesquida · Binisafúa · Punta Prima · I. de l'Aire · Cala en Porter · Alaior · Sant Jaume · Cala Santa Galdana · Ferreries · Es Migjorn Gran · Cala Forcat · Ciudadella de Menorca · Tamarinda · C. de Artrux

Mallorca
C. de Formentor · Pollença · Port de Pollença · Badia de Pollença · C. des Pinar · Port d'Alcúdia · Badia d'Alcúdia · Alcúdia · Sa Pobla · Santa Margarita · Son Serra · Son Serra · Artà · Capdepera · Cala Ratjada · Cala Millor · Son Servera · Porto Cristo · Manacor · Morey 562 · C. Ferrutx · Muro · Inca · Sineu · Sencelles · Petra · Villafranca de Bonany · Porreres · Felanitx · San Salvador 509 · Cala d'Or · Porto Petro · Santanyi · Campos del Port · Llucmajor · Ses Salines · C. de ses Salines · Puig Major 1445 · Massanella 1340 · Alfabia 1068 · Sóller · Port de Sóller · Santa Maria del Camí · Marratxi · S'Arenal · Badia de Palma · Palma de Mallorca · Puigpunyent · Valldemossa · Banyalbufar · Estellencs · Andratx · Port d'Andratx · Santa Ponça · S'Estanyol · Colònia de Sant Jordi · S. Blanc · Magaluf · Palma Nova · Illetas · Cala Major · Sant Telm · Sa Dragonera · C. des Llebeig · C. de Cala Figuera

ISLAS BALEARES

Eivissa (Ibiza)
Pta. Grosa · Tagomago · Santa Eulalia del Riu · Sant Carlos · Es Canar · Es Canar · Portinatx · Sant Joan Baptista · Sant Miguel · Sant Mateu · Fornàs 409 · Santa Agnès · C. d'Aubarca · Santa Gertrudis · Sant Antoni Abat · Sant Josep · Sant Jordi · Sa Conillera · Es Vedrà · C. Llentrisca · Sirena 424 · Can Clavo · Can Creu des Riu · C. des Falcó · Eivissa

Formentera
S'Espardell · S'Espalmador · Pta. Rotja · Es Caló · Sant Ferran · Sant Francesc de Formentera · Sa Savina · Pta. des Pas · Sa Canal · C. de Barbària

EAST from Greenwich

MEDITERRANEAN SEA

Mallorca

Cabrera · I. des Conills · Puerto de Cabrera · C. Blanc · Pta. de n'Ensiola

CANARY ISLANDS
1:1 600 000

ISLAS CANARIAS

ATLANTIC OCEAN

Lanzarote
I. Alegranza · I. Montaña Clara · Alegranza 259 · I. Graciosa · Pta. Fariones · Haria · La Santa · Peñas del Chache 671 · Soo · Tinajo · Arrecife · San Bartolomé · Los Islotes · Tiza · Uga · 478 · Playa Blanca · Atalaya de Femés · Puerto del Carmen · Janubio · Playa Blanca · Pta. Pechiguera · I. de Lobos

Fuerteventura
Corralejo · La Oliva · Puerto del Rosario · Muda 689 · Betancuria · Antigua · Tuineje · Puerto de Pozo Negro · Pta. de Tostón · Cotillo · Pta. de la Herradura · Betancuria · Puerto de Gran Tarajal · Tarajalejo · Jandía Playa Esmeralda · Cofete · Pta. del Jable · Morro del Jable · 807 · Pta. de Jandia

Gran Canaria
Pta. Sardina · Guía · Gáldar · Agaete · Pico de las Nieves 1949 · San Nicolás · Arucas · Teror · Las Palmas · Telde · Ingenio · Aguimes · Virgenio · Pta. El Roque · Pta. Gando · San Agustin · Mogán · Bartolomé de Tirajana · Playa del Inglés · Maspalomas · Pta. de la Aldea · Playa de Mogán · Puerto Rico · Arguineguin · Maspalomas

West from Greenwich

Tenerife
Pta. de Anaga · Pta. del Hidalgo · Bajamar · Punta del Hidalgo · La Laguna · Santa Cruz de Tenerife · La Orotava · Puerto de la Cruz · Candelaria · Güimar · Icod · Teide 3718 · Santiago del Teide · Garachico · Guía de Isora · Arico · Pta. de Teno · Adeje · Granadilla de Abona · Playa de las Américas · El Médano · Los Cristianos · Pta. de la Rasca

Gomera
Pta. de los Organos · Vallehermoso · Agulo · Hermigua · San Sebastián de la Gomera · Garajonay 1487 · Alajeró

La Palma
Pta. Cumplida · Barlovento · Roque de los Muchachos 2423 · Santa Cruz de la Palma · El Pueblo · Garafía · Pta. Gorda · Los Llanos de Aridane · Fuencaliente · Pta. Fuencaliente

Hierro
Pta. del Norte · Frontera · Valverde · Malpaso 1501 · Pico de Tenerife 1417 · Pta. Tabaiba · La Restinga · Pta. Taniga

ISLAS CANARIAS

CARTOGRAPHY BY PHILIP'S. COPYRIGHT REED INTERNATIONAL BOOKS LTD

Projection: Lambert's Conformal Conic

10 0 10 20 30 40 50 60 70 80 90 km
10 0 10 20 30 40 50 60 miles
A
B
C
D
E
F
1
2
3
4
5
6
7
8
FRANCE
SPAIN
Golfe de Gascogne
MIDI-PYRÉNÉES
LANGUEDOC-ROUSSILLON
PYRÉNÉES-ATLANTIQUES
HAUTES-PYRÉNÉES
PYRÉNÉES-ORIENTALES
ANDORRA
Andorra La Vella
PAÍS VASCO
NAVARRA
LA RIOJA
ARAGÓN
CATALUÑA
CASTILLA Y LEÓN
Toulouse
Bayonne
Pau
Tarbes
Perpignan
Carcassonne
Narbonne
Béziers
Montauban
Albi
Santander
Bilbao
Donostia-San Sebastián
Vitoria-Gasteiz
Pamplona
Logroño
Burgos
Huesca
Zaragoza
Lleida (Lérida)
Girona
BARCELONA
L'Hospitalet de Llobregat
Tarragona
Tortosa
Teruel
Castelló de la Plana
Guadalajara
MADRID
Cuenca
Soria
Costa Brava
Costa Dorada
Costa del Azahar
BALEARES
MENORCA
MALLORCA
Palma de Mallorca
Ebro
Garonne

MEDITERRANEAN SEA
ISLAS
Golfo de Valencia
Costa Blanca
Costa del Sol
CASTILLA-LA MANCHA
CIUDAD REAL
ALBACETE
VALENCIA
ALICANTE
MURCIA
JAÉN
GRANADA
ALMERÍA
ALGERIA
ALGER (Algiers)
Valencia
Alicante
Elche
Murcia
Cartagena
Albacete
Granada
Almería
Lorca
Linares
Eivissa (Ibiza)
EIVISSA (IBIZA)
Formentera
Cabrera
B. de Palma
Oran (Ouahran)
Mostaganem
Sidi-bel-Abbès
Blida
Médéa
Tiaret
Mascara
Relizane
Ech Cheliff
Melilla (Sp.)
Nador
Alborán (Sp.)
Islas Chafarinas (Sp.)
Sierra Nevada
Sierra de Segura
Sierra de Alcaraz
Sierra de los Filabres
G. de Almería
G. de Mazarrón
Mar Menor
Golfe d'Arzew
Massif de Dahra
Massif de l'Ouarsenis
MOSTAGANEM
RELIZANE
TISSEMSILT
AÏN DEFLA
TIPASA
BLIDA
MÉDÉA
ECH CHELIFF
AÏN TÉMOUCHENT
ORAN
MASCARA
TIARET
DJELFA
C. de la Nao
C. de Palos
C. de Gata
C. des Trois Fourches
Projection : Lambert's Conformal Conic
West from Greenwich
East from Greenwich
1
2
3
4
5
6
7
F
G
H
J
K
COPYRIGHT GEORGE PHILIP LTD.

10 0 10 20 30 40 50 60 70 80 90 km
10 0 10 20 30 40 50 60 miles
1
2
3
4
5
6
7
8
A
B
C
D
E
F
Bay of Biscay
Costa Verde
Donostia-San Sebastián
Bilbao
Vitoria-Gasteiz
PAÍS VASCO
GUIPÚZCOA
VIZCAYA
Santander
CANTABRIA
Gijón
Oviedo
ASTURIAS
Picos de Europa
Cordillera Cantábrica
A Coruña (La Coruña)
Ferrol
LA CORUÑA
Santiago de Compostela
GALICIA
Lugo
LUGO
Ourense (Orense)
ORENSE
Pontevedra
PONTEVEDRA
Vigo
León
LEÓN
Ponferrada
Montes de León
Burgos
BURGOS
Logroño
LA RIOJA
Palencia
PALENCIA
Tierra de Campos
CASTILLA Y LEÓN
Valladolid
VALLADOLID
Zamora
ZAMORA
Soria
SORIA
Salamanca
SALAMANCA
Segovia
SEGOVIA
Ávila
ÁVILA
Sierra de Guadarrama
Sierra de Gredos
MADRID
Guadalajara
GUADALAJARA
La Alcarria
Toledo
TOLEDO
Talavera de la Reina
Plasencia
SPAIN
PORTUGAL
Viana do Castelo
VIANA DO CASTELO
Braga
BRAGA
VILA REAL
BRAGANÇA
Porto
PORTO
Vila Nova de Gaia
VISEU
GUARDA
Aveiro
AVEIRO
Coimbra
COIMBRA
Serra da Estrela
CASTELO BRANCO
Castelo Branco
Leiria
LEIRIA
Figueira da Foz
Douro
Duero
Ebro
Miño
Tajo
Tormes
Esla
Pisuerga
Embalse de Almendra
Embalse de Alcántara

E   **F**   **G**   **H**

1. Crimea (Ukr.)
2. Adygea (Russ.)
3. Karachey-Cherkessia (Russ.)
4. Kabardino-Balkaria (Russ.)
5. North Ossetia (Russ.)
6. Ingushetia (Russ.)
7. Chechenia (Russ.)
8. Naxçıvan (Azer.)

TURKMENISTAN

Garabogazköl Aylagy

C A S P I A N   S E A

K I R G I Z   Steppe

K A Z A K S T A N

Kaspiy Depression

KALMYKIA

Privolzhskaya Vozvyshennost

Yergeni Vozyyshennost

Volgograd (Stalingrad)

KAZAKSTAN

BAKI (Baku) Bakı Kömü

AZERBAIJAN

DAGESTAN

Caucasus Mountains

Makhachkala

Grozny

Vladikavkaz

Elbrus 5642

GEORGIA

ABKHAZIA

ARMENIA

Yerevan

Tbilisi

Tabriz

I R A N

Alborz

TEHRĀN

Qom

KURDISTAN

Kuzey Anadolu Dağları

T U R K E Y

Toros Dağları

I R A Q

Baghdād

S Y R I A

Bādiyat ash Sham

Ankara

İstanbul

B L A C K   S E A

Sea of Azov

U K R A I N E

KHARKIV (Kharkov)

KYYIV (Kiev)

Volgograd

Rostov

Donetsk

Dnipropetrovsk

MOLDOVA

R O M A N I A

BUCUREŞTI (Bucharest)

B U L G A R I A

CYPRUS

LEBANON

Dimashq (Damascus)

Bayrūt (Beirut)

M E D I T E R R A N E A N   S E A

Levant

İzmir (Smyrna)

Dhodhekánisos

Division between Greeks and Turks in Cyprus: Turks to the North.

East from Greenwich

Projection: Conical with two standard parallels

E   **F**   **H**

m / ft scale

4000 12 000
2000 6000
1000 3000
600 2000
400 1200
200 600
0
200 600
1000 3000
2000 6000
4000 12 000
6000
12 000

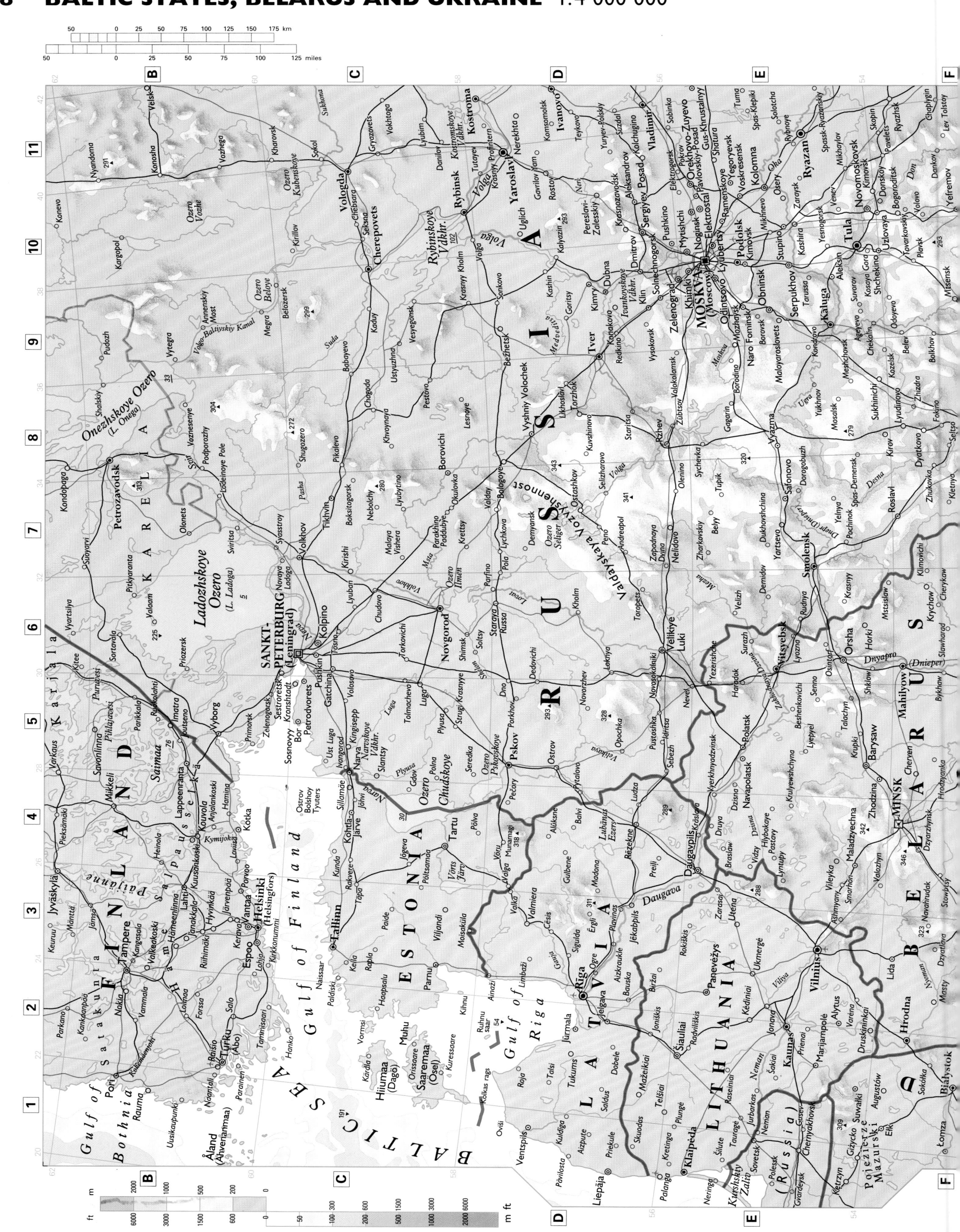

50 0 25 50 75 100 125 150 175 km
50 0 25 50 75 100 125 miles
B
C
D
E
F
11
10
9
8
7
6
5
4
3
2
1
FINLAND
ESTONIA
LATVIA
LITHUANIA
BELARUS
RUSSIA
KARELIA
Gulf of Finland
Gulf of Riga
Gulf of Bothnia
BALTIC SEA
Ladozhskoye Ozero (L. Ladoga)
Onezhskoye Ozero (L. Onega)
SANKT-PETERBURG (Leningrad)
MOSKVA (Moscow)
Helsinki (Helsingfors)
Tallinn
Riga
Vilnius
MINSK
Kaliningrad
m ft
2000 6000
1000 3000
500 1500
200 600
0 0
50
100 300
200 600
500 1500
1000 3000
2000 6000

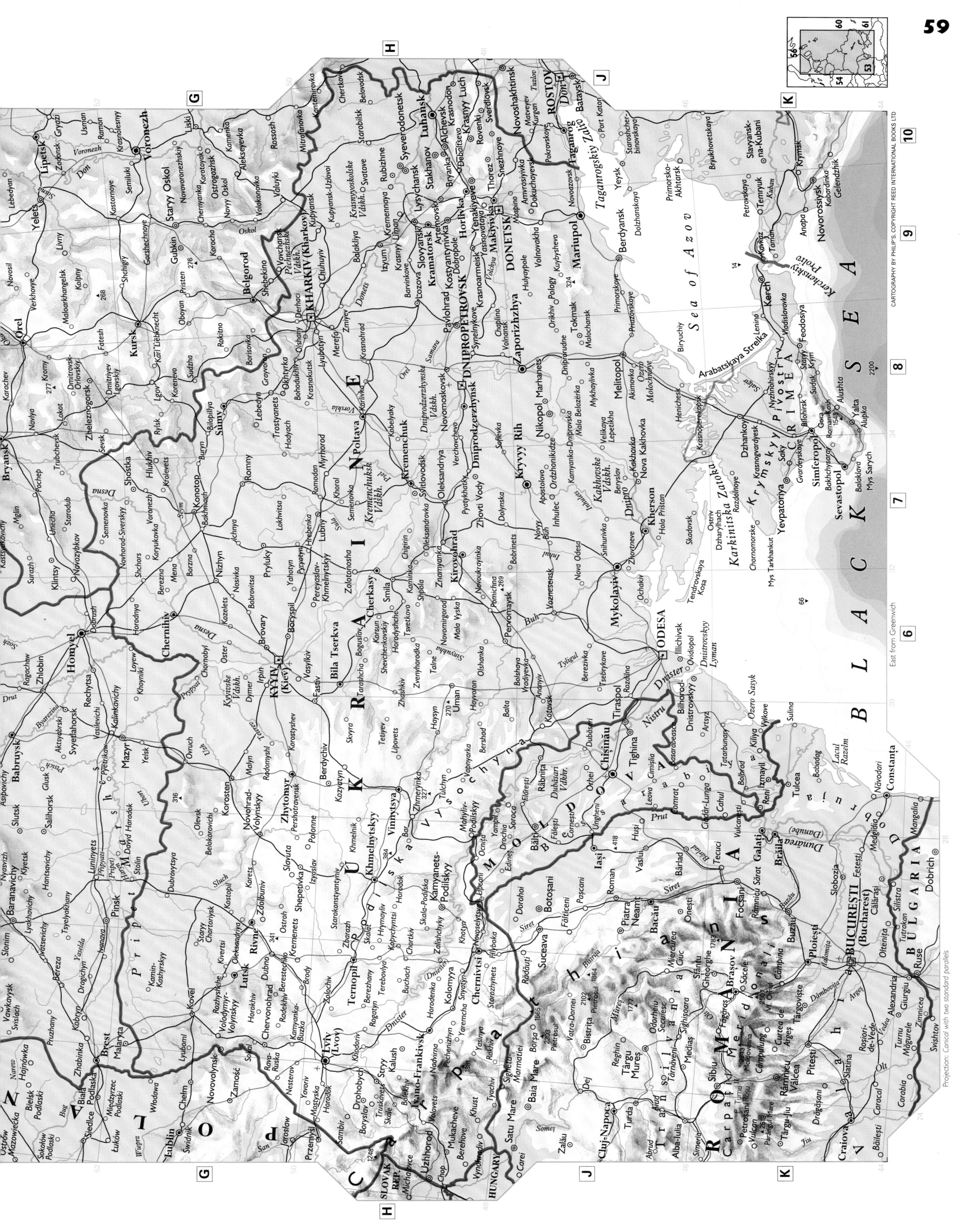
U K R A I N E
B L A C K S E A
Sea of Azov
M O L D O V A
R O M A N I A
B U L G A R I A
P O L A N D
HUNGARY
SLOVAK REP.
KYYIV (Kiev)
KHARKIV (Kharkov)
DNIPROPETROVSK
DONETSK
ODESA
Zaporizhzhya
Kryvyy Rih
Mykolayiv
Kherson
Simferopol
Sevastopol
Kerch
Feodosiya
Yalta
Yevpatoriya
Mariupol
Luhansk
Horlivka
Makiyivka
Poltava
Kremenchuk
Cherkasy
Kirovohrad
Vinnytsya
Zhytomyr
Khmelnytskyy
Ternopil
Lviv (Lvov)
Rivne
Lutsk
Ivano-Frankivsk
Chernivtsi
Uzhhorod
Chernihiv
Sumy
Bila Tserkva
Chișinău
Tiraspol
Bălți
Tighina
BUCUREȘTI (Bucharest)
Constanța
Brașov
Ploiești
Galați
Brăila
Iași
Cluj-Napoca
Homyel
Brest
Pinsk
Mazyr
Babruysk
Bryansk
Orel
Kursk
Belgorod
Voronezh
Lipetsk
Yelets
Staryy Oskol
ROSTOV
Taganrog
Novorossiysk
Krymsk
Lublin
Zamość
Przemyśl
Chełm
Dnister
Dnipro
Desna
Pripyat
Prut
Siret
Buh
Donets
Dunărea (Danube)
Don
Pripet Marshes
Carpații Meridionali
Transilvania
Krymskyy Pivostriv CRIMEA
Karkinitska Zatoka
Kerchenskiy Proliv
Taganrogskiy Zaliv
Arabatskaya Strelka
Kyyivske Vdskh.
Kremenchukske Vdskh.
Kakhovske Vdskh.
Podilska Vysochyna
Projection: Conical with two standard parallels
East from Greenwich
CARTOGRAPHY BY PHILIP'S. COPYRIGHT REED INTERNATIONAL BOOKS LTD
G
H
J
K
6
7
8
9
10

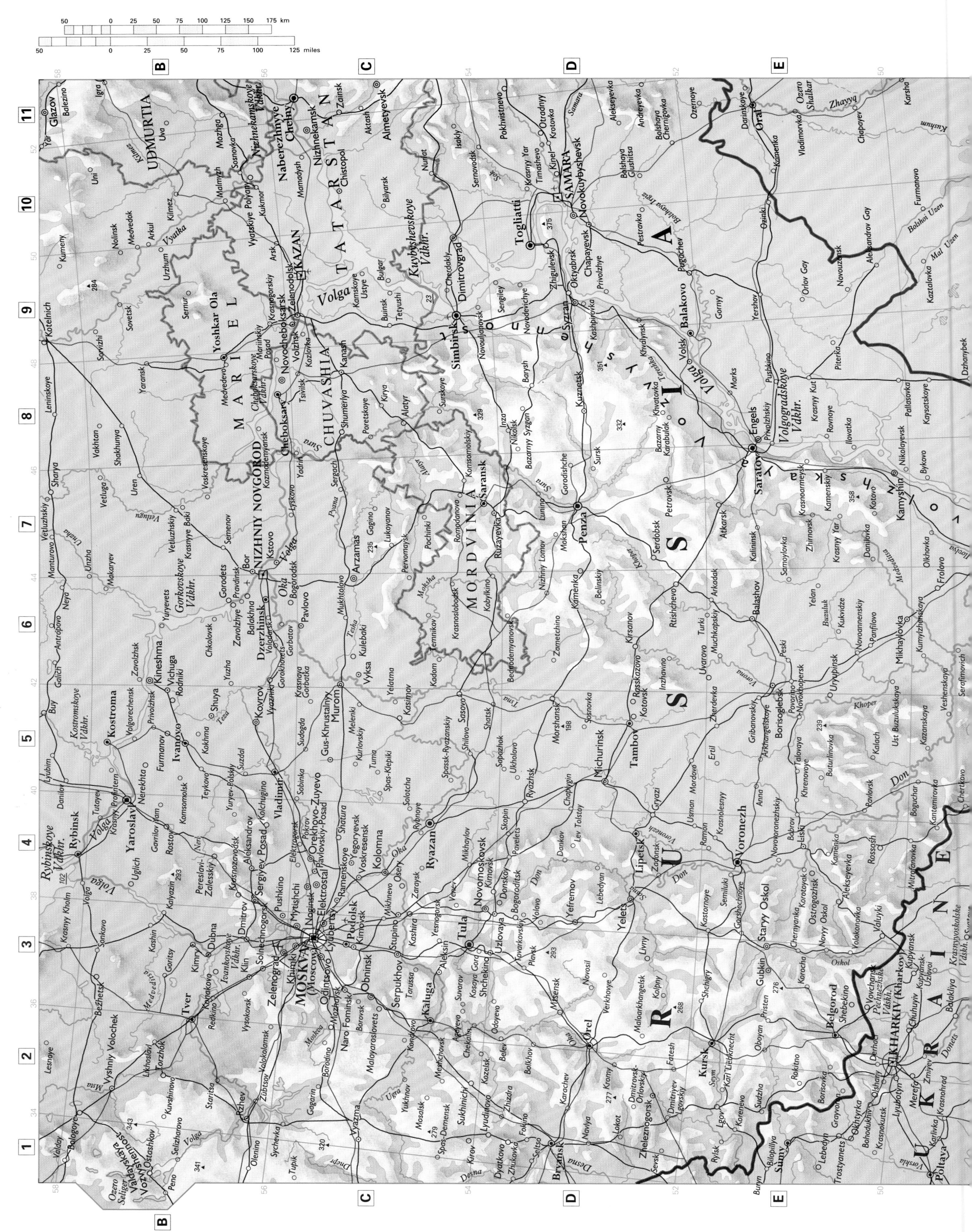

50 0 25 50 75 100 125 150 175 km
50 0 25 50 75 100 125 miles
B
C
D
E
1
2
3
4
5
6
7
8
9
10
11
R U S S I A
U K R A I N E
UDMURTIA
MARI EL
TATARSTAN
CHUVASHIA
MORDVINIA
NIZHNIY NOVGOROD
MOSKVA (Moscow)
KAZAN
SAMARA
Saratov
Engels
Voronezh
Tambov
Penza
Saransk
Ryazan
Tula
Kaluga
Orel
Kursk
Belgorod
KHARKIV (Kharkov)
Bryansk
Tver
Yaroslavl
Kostroma
Ivanovo
Vladimir
Simbirsk
Togliatti
Cheboksary
Yoshkar Ola
Lipetsk
Sumy
Poltava
Oral
Volga
Oka
Don
Rybinskoye Vdkhr.
Gorkovskoye Vdkhr.
Kuybyshevskoye Vdkhr.
Volgogradskoye Vdkhr.

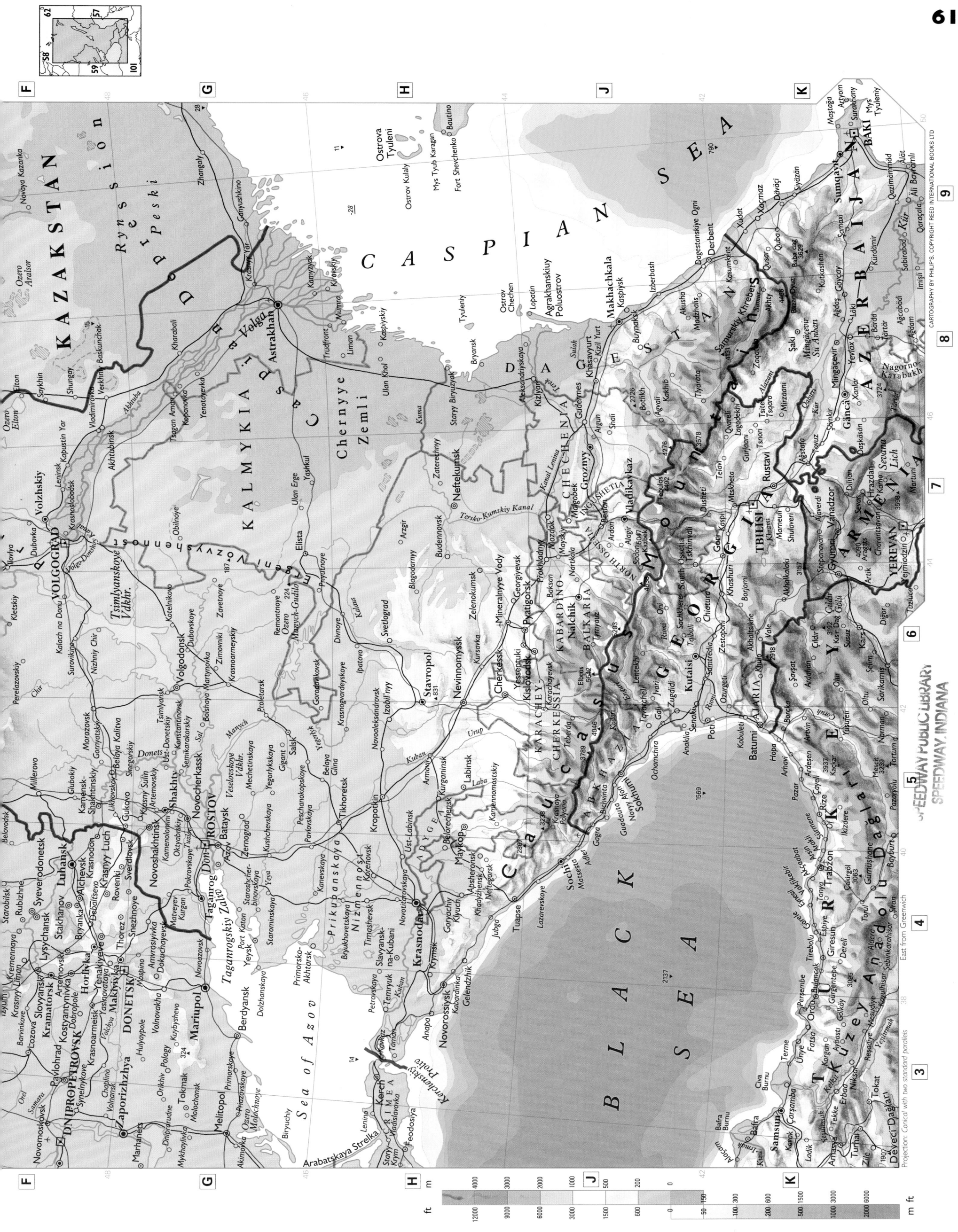
KAZAKSTAN
Caspian Depression
Ryn Peski
KALMYKIA
Chernyye Zemli
CASPIAN SEA
BLACK SEA
Sea of Azov
Taganrogskiy Zaliv
Prikubanskaya Nizmennost
Yergeni Vozvyshennost
Caucasus Mountains
GEORGIA
ARMENIA
AZERBAIJAN
TURKEY
Kuzey Anadolu Dağları
DAGESTAN
CHECHENIA
INGUSHETIA
NORTH OSSETIA
KABARDINO-BALKARIA
KARACHEY-CHERKESSIA
ADYGEA
ABKHAZIA
AJARIA
South Ossetia
CRIMEA
Kerchenskiy Proliv
Tsimlyanskoye Vdkhr.
VOLGOGRAD
Volzhskiy
Astrakhan
ROSTOV
Novocherkassk
Shakhty
Taganrog
Krasnodar
Stavropol
Elista
Grozny
Makhachkala
Vladikavkaz
Nalchik
Sochi
Novorossiysk
TBILISI
Kutaisi
Batumi
Sokhumi
YEREVAN
Gyumri
Vanadzor
Gäncä
Sumqayit
BAKI
DNIPROPETROVSK
Zaporizhzhya
DONETSK
Luhansk
Mariupol
Trabzon
Samsun
Elbrus 5642
Kazbek 5047
Sevana Lich
Agrakhanskiy Poluostrov
Ostrova Tyuleni
Nagorno-Karabakh
Mingäçevir Su Anbarı
Projection: Conical with two standard parallels
East from Greenwich
CARTOGRAPHY BY PHILIP'S. COPYRIGHT REED INTERNATIONAL BOOKS LTD

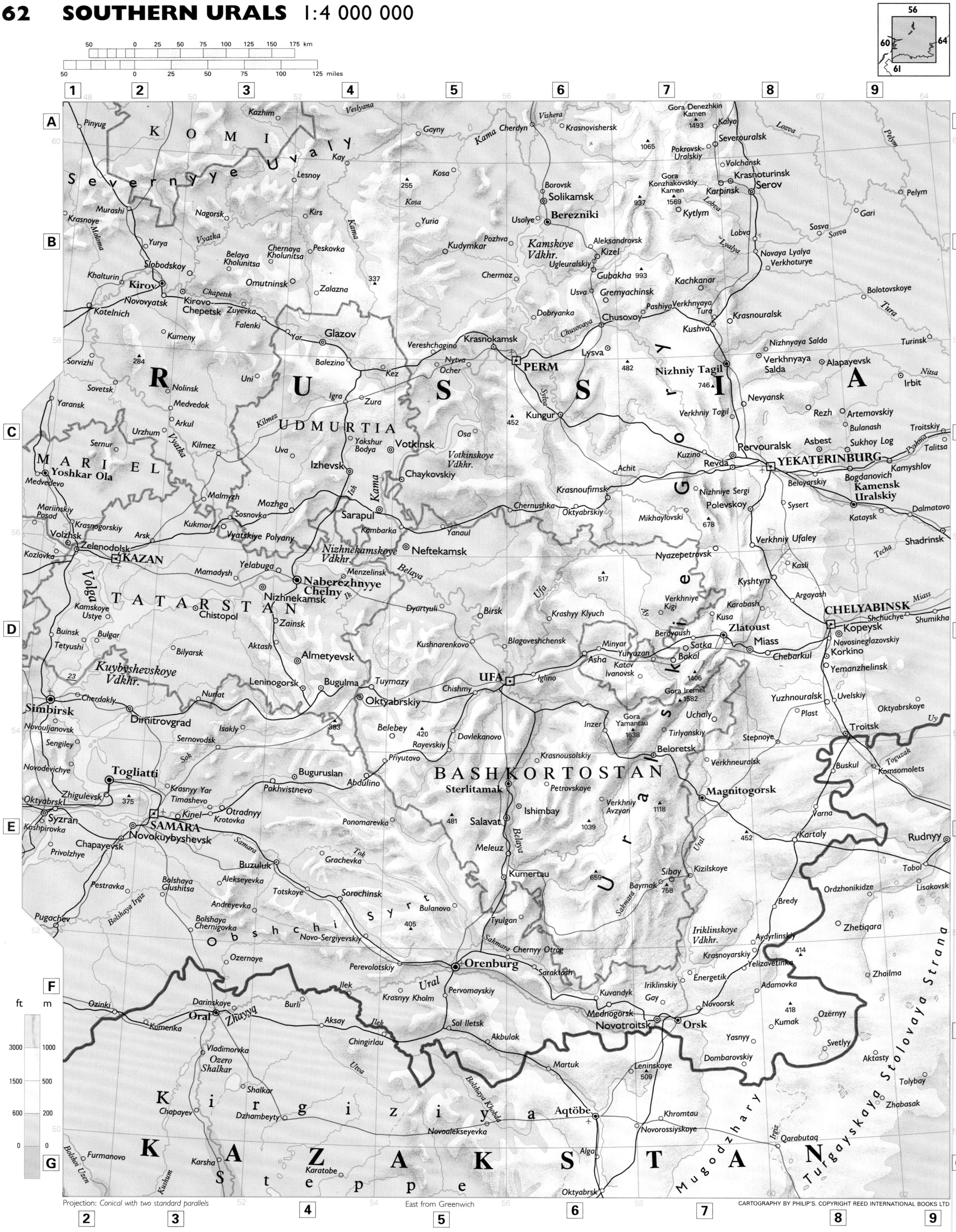

Projection: Conical with two standard parallels

East from Greenwich

CARTOGRAPHY BY PHILIP'S. COPYRIGHT REED INTERNATIONAL BOOKS LTD

50   0   25   50   75   100   125   150   175 km
50   0   25   50   75   100   125 miles

ft   m
3000   1000
1500   500
600   200
0   0

RUSSIA

KOMI

Severnyye Uvaly

UDMURTIA

MARI EL
Yoshkar Ola

TATARSTAN

KAZAN

BASHKORTOSTAN

U r a l s k i y e   G o r y

KAZAKSTAN

Kirgiziya Steppe

K a z a k   S t e p p e

Mugodzhary

Turgayskaya Stolovaya Strana

Pinyug, Kazhim, Vestyana, Gayny, Kama, Cherdyn, Vishera, Krasnovishersk, Gora Denezhkin Kamen 1493, Kalya, Lozva, Pelym, Severouralsk, Volchansk, Krasnoturinsk, Serov, Pelym, Gari

Murashi, Krasnoye, Maloma, Nagorsk, Kirs, Kosa, Yuria, Borovsk, Usolye, Solikamsk, Berezniki, Gora Konzhakovskiy Kamen 1569, Karpinsk, Kytlym, Lobva, Sosva, Sosva

Kirov, Kirovo-Chepetsk, Glazov, Balezino, Kez, Ocher, Nytva, Krasnokamsk, PERM, Kungur, Lysva, Chusovoy, Nizhniy Tagil, Verkhnyaya Salda, Alapayevsk, Irbit, Nitsa

YEKATERINBURG, Pervouralsk, Revda, Polevskoy, Sysert, Kamensk Uralskiy, Shadrinsk

Izhevsk, Votkinsk, Chaykovsky, Sarapul, Kambarka, Yanaul, Neftekamsk, Birsk, Blagoveshchensk

Naberezhnyye Chelny, Nizhnekamsk, Chistopol, Almetyevsk, Bugulma, UFA, Iglino, Zlatoust, Miass, CHELYABINSK, Kopeysk, Korkino

Simbirsk, Dimitrovgrad, Togliatti, Syzran, SAMARA, Novokuybyshevsk, Buzuluk, Sterlitamak, Salavat, Ishimbay, Meleuz, Kumertau, Beloretsk, Magnitogorsk, Troitsk

Orenburg, Sol Iletsk, Akbulak, Novotroitsk, Orsk, Mednogorsk, Kuvandyk

Oral, Zhayyq, Aqtobe, Khromtau, Rudnyy, Lisakovsk

Kuybyshevskoye Vdkhr., Nizhnekamskoye Vdkhr., Votkinskoye Vdkhr., Kamskoye Vdkhr., Iriklinskoye Vdkhr.

Gora Yamantau 1638, Gora Iremel 1582, Gora Yamantau

Volga, Kama, Belaya, Ural, Ufa, Sakmara, Samara

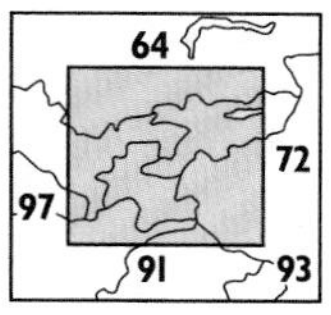

50 0 25 50 75 100 125 150 175 km
50 0 25 50 75 100 125 miles

KAZAKSTAN

KYRGYZSTAN

UZBEKISTAN

TAJIKISTAN

TURKMENISTAN

AFGHANISTAN

XINJIANG

CHINA

JAMMU AND KASHMIR

PAKISTAN

Peski Taukum

Peski Muyunkum

Kyzyl Kum

Balqash Köl

Ozero Ysyk-Köl 1609

Almaty (Alma-Ata)

Bishkek (Frunze)

Toshkent (Tashkent)

Shymkent (Chimkent)

Samarqand

Bukhoro

Dushanbe

Kashi (Kashgar)

Shache (Yarkand)

Mazār-e-Sharif

Khānābād

Feyzābād

Qyzylorda

Kunlun Shan

Hindu Kush

Pamir

Khrebet Zeravshanskiy

Khrebet Gissarskiy

Khrebet Turkestanskiy

Alayskiy Khrebet

Zaalayskiy Khrebet

Ferganskaya Dolina

Khrebet Terskey Alatau

Khrebet Kungey Alatau

Kirgizskiy Khrebet

Khrebet Talasskiy Alatau

Syrdarya

Amudarya

Pik Lenina 7134

Pik Kommunizma 7495

Pik Pobedy 7439

7719

7589

East from Greenwich

COPYRIGHT GEORGE PHILIP & SON LTD.

Projection: Conical with two standard parallels.

ft m 18 000 6000 12 000 9000 6000 4000 3000 2000 1500 1000 400 200 0

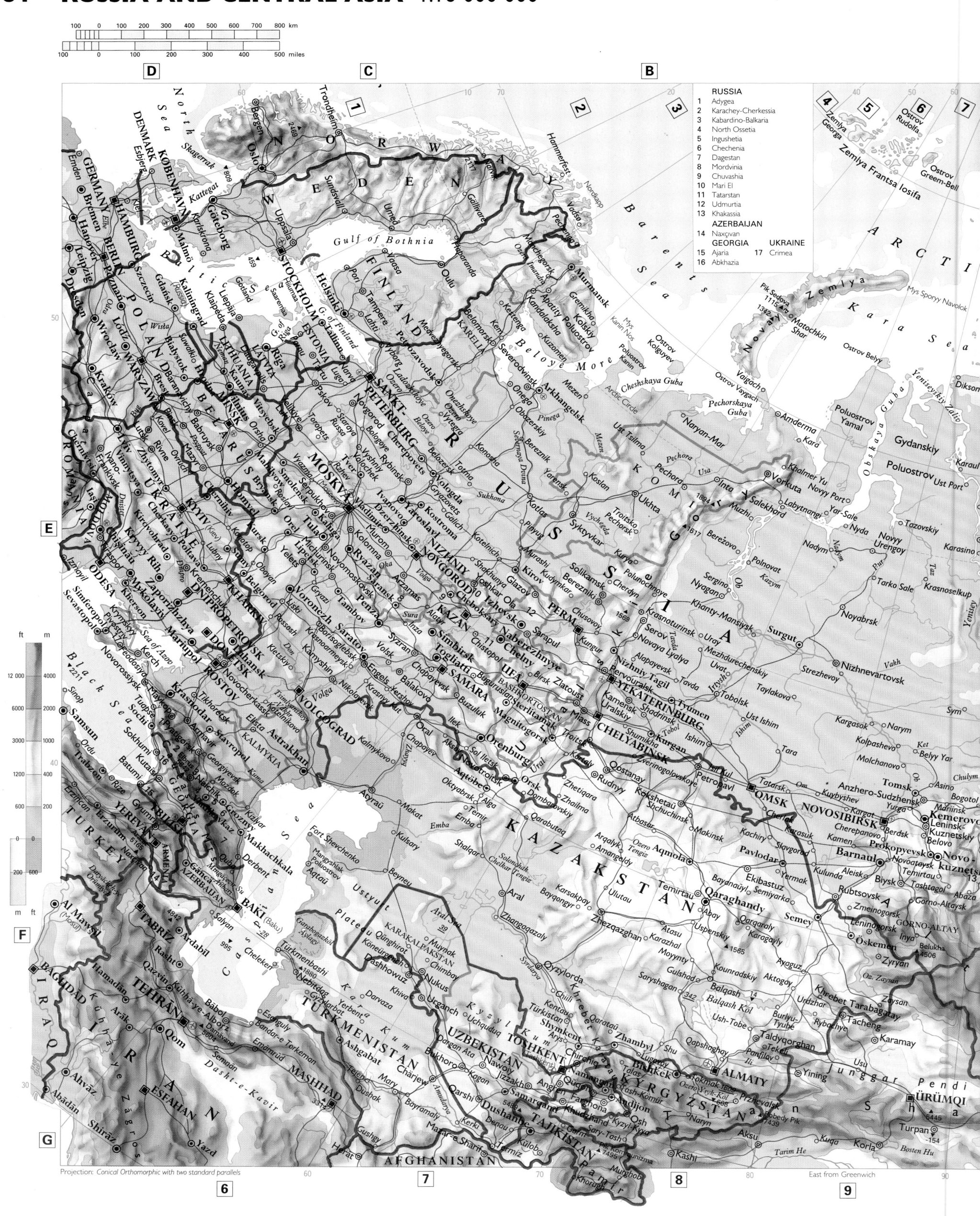
100 0 100 200 300 400 500 600 700 800 km
100 0 100 200 300 400 500 miles
D
C
B
1
2
3
4
5
6
7
E
F
G
6
7
8
9
RUSSIA
1 Adygea
2 Karachey-Cherkessia
3 Kabardino-Balkaria
4 North Ossetia
5 Ingushetia
6 Chechenia
7 Dagestan
8 Mordvinia
9 Chuvashia
10 Mari El
11 Tatarstan
12 Udmurtia
13 Khakassia
AZERBAIJAN
14 Naxçivan
GEORGIA
15 Ajaria
16 Abkhazia
UKRAINE
17 Crimea
ft m
12 000 4000
6000 2000
3000 1000
1200 400
600 200
0 0
200 600
m ft
Projection: Conical Orthomorphic with two standard parallels
East from Greenwich
NORWAY
SWEDEN
FINLAND
ESTONIA
LATVIA
LITHUANIA
BELARUS
POLAND
UKRAINE
MOLDOVA
ROMANIA
GERMANY
DENMARK
GEORGIA
ARMENIA
AZERBAIJAN
TURKEY
IRAQ
IRAN
TURKMENISTAN
UZBEKISTAN
KARAKALPAKSTAN
KYRGYZSTAN
TAJIKISTAN
AFGHANISTAN
KAZAKSTAN
RUSSIA
KOMI
KARELIA
KALMYKIA
BASHKORTOSTAN
GORNO-ALTAY
Barents Sea
Kara Sea
Beloye More
Baltic Sea
Gulf of Bothnia
North Sea
Black Sea
Sea of Azov
Caspian Sea
Aral Sea
Zemlya Frantsa Iosifa
Novaya Zemlya
Poluostrov Yamal
Gydanskiy Poluostrov
Ustyurt Plateau
Kyzyl Kum
Kara Kum
Dasht-e-Kavir
Pamir
Junggar Pendi
Arctic Circle
MOSKVA
SANKT-PETERBURG
KYYIV
MINSK
WARSZAWA
BERLIN
KØBENHAVN
STOCKHOLM
Helsinki
Tallinn
Riga
Vilnius
Oslo
NIZHNIY NOVGOROD
KAZAN
SAMARA
PERM
YEKATERINBURG
CHELYABINSK
OMSK
NOVOSIBIRSK
VOLGOGRAD
ROSTOV
DONETSK
DNIPROPETROVSK
KHARKIV
ODESA
TBILISI
YEREVAN
BAKI
TEHRAN
TOSHKENT
ALMATY
Bishkek
Dushanbe
Ashgabat
Aqmola
Qaraghandy
ÜRÜMQI
Murmansk
Arkhangelsk
Vorkuta
Surgut
Tomsk
Barnaul
Orenburg
Astrakhan
MASHHAD
ESFAHAN
TABRIZ
BAGHDAD

A
B
C
D
E
F
8
9
10
11
12
13
14
15
16
17
18
19
Chukchi Sea
Mys Dezhneva (East C.)
Uelen
St. Lawrence I. (U.S.A.)
Provideniya
Anadyrskiy Zaliv
Vankarem
Chukotskoye Nagorye
Egvekinot
Beringovskiy
Pevek
Ostrov Vrangelya
East Siberian Sea
Ostrov Genriyetty
Ostrov Zhannetty
Ostrova Delonga
Ostrov Zhokhova
Ostrov Bennetta
Ostrov Novaya Sibir
Novosibirskiye Ostrova
Ostrov Faddeyevskiy
Ostrov Belkovskiy
Ostrov Kotelnyy
Ostrov Stolbovoy
Ostrov Malyy Lyakhovskiy
Ostrov Bolshoy Lyakhovskiy
Lyakhovskiye Ostrova
Proliv Dmitriya Lapteva
Laptev Sea
OCEAN
Mys Arkticheskiy
Ostrov Shmidta
Ostrov Komsomolets
Ostrov Oktyabrskoy Revolyutsii
Ostrov Pioner
Ostrov Bolshevik
Severnaya Zemlya
Proliv Vilkitskogo
Mys Chelyuskin
Poluostrov Taymyr
Gory Byrranga
Oz. Taymyr
Ostrov Bolshoy Begichev
Nordvik
Mys Buorkhaya
Tiksi
Ust Olenek
Tit-Ary
Bulun
Kyusyur
Novorybnoye
Khatanga
Saskylakh
Zhilinda
Olenek
Volochanka
Kheta
Chernaya
Dudinka
Norilsk
Gory Putorana
Yessey
Igarka
Arctic Circle
Zhigansk
Kystatyam
Verkhoyansk
Verkhoyanskiy Khrebet
Batagai
Deputatskiy
Kazachye
Ust Kuyga
Chokurdakh
Indigirka
Druzhina
Srednekolymsk
Nizhne Kolymsk
Cherskiy
Ambarchik
Ostrova Medvezhi
Ostrov Ayon
Ust Chaun
Bilibino
Bolshoy Anyuy
Anadyr
Markovo
Yerepol
Penzhino
Kamenskoye
Omolon
Koryakskoye Nagorye
Khatyrka
Tilichiki
Korf
Kolymskoye Nagorye
Zyryanka
Khonuu
Kolyma
Ust-Nera
Pobeda
Gora Chen
Khrebet Cherskogo
Taskan
Omsukchan
Orotukan
Susuman
Yagodnoye
Oymyakon
Ust-Omchug
Palatka
Magadan
Evensk
Gizhiga
Paren
Penzhinskaya Guba
Gizhiginskaya Guba
Zaliv Shelikhova
Tigil
Palana
Sredinnyy Khrebet
Ossora
Ostrov Karaginskiy
Ust-Kamchatsk
Klyuchevskaya Gora
Poluostrov Kamchatka
Petropavlovsk-Kamchatskiy
Ust Khayryuzovo
Kirovskiy
Ust-Bolsheretsk
Komandorskiye Ostrova
Nikolskoye
Severo-Kurilsk
Ostrov Paramushir
Ostrov Onekotan
Bering Sea
Sea of Okhotsk
Sangar
Batamay
Lena
Borogontsy
Khandyga
Okhotskiy Perevoz
Ytyk Kyuyel
Yakutsk
Mayya
Amga
Namtsy
Vilyuysk
Verkhnevilyuysk
Vilyuy
Nyurba
Suntar
Pokrovsk
Mirnyy
Chernyshevskiy
Ust Maya
Ust-Mil
Aim
Maya
Nelkan
Chagda
Uchur
Ayan
Okhotsk
Ulya
Arka
Khrebet Dzhugdzur
Olekminsk
Sinsk
Tommot
Aldan
Lensk
Neryungri
Nagornyy
Stanovoy Khrebet
Tynda
Skovorodino
Zeya
Ostrov Bolshoy Shantar
Sakhalinskiy Zaliv
Nikolayevsk-na-Amur
Okha
Sakhalin
Aleksandrovsk-Sakhalinskiy
Gora Lopatina
Poronaysk
Uglegorsk
Yuzhno-Sakhalinsk
Kholmsk
Korsakov
Vanino
Tugur
Chumikan
Uda
Amgun
Amur
Khrebet Sikhote Alin
Komsomolsk
Amursk
Khabarovsk
Birobidzhan
Smidovich
Obluchye
Chegdomyn
Bureya
Norsk
Ushumun
Dzhalinda
Belogorsk
Svobodnyy
Shimanovsk
Blagoveshchensk
Zavitinsk
Raychikhinsk
Poyarkovo
Vyazemskiy
Bikin
Amgu
Terney
Dalnegorsk
Lesozavodsk
Dalnerechensk
Spassk Dalniy
Olga
Ussuriysk
Artem
Vladivostok
Nakhodka
Kraskino
Ostrov Iturup
Kurilsk
Ostrov Kunashir
Kurilskiye Ostrova
Ostrov Urup
Ostrov Simushir
Wakkanai
Rumoi
Abashiri
Kushiro
Obihiro
Hokkaidō
SAPPORO
Otaru
Hakodate
Aomori
Hachinohe
Akita
Honshū
Niigata
Kanazawa
Fuji-San
JAPAN
ŌSAKA
Sea of Japan
Tura
Noginsk
Nizhnyaya Tunguska
Turukhansk
Yukta
Yerbogachen
Kuyumba
Mutoray
Podkamennaya Tunguska
Vanavara
Yartsevo
Severo-Yeniseyskiy
RUSSIA
Vitim
Mama
Bodaybo
Chara
Kalar
Ust-Nyukzha
Olekma
Yenyuka
Kezhma
Korshunovo
Kirensk
Ust-Ilimsk
Yeniseysk
Angara
Boguchany
Strelka
Yenisey
Kondratyevo
Chuna
Zheleznogorsk-Ilimskiy
Makarovo
Ust-Kut
Magistralnyy
Kalakan
Bagdarin
Yablonovyy Khrebet
Mogocha
Achinsk
Kansk
Ilanskiy
Krasnoyarsk
Tayshet
Bratsk
Nizhneudinsk
Tulun
Artemovsk
Vostochnyy Sayan
Chernogorsk
Minusinsk
Abakan
Zima
Cheremkhovo
Usolye Sibirskoye
Angarsk
Irkutsk
Ulan Ude
Barguzin
Oz. Baykal
Munku-Sardyk
Zapadnyy Sayan
Turan
Kyzyl
Toora-Khem
Chadan
TUVA
Slyudyanka
Gusinoozersk
Kyakhta
Zakamensk
Darhan
Hövsgöl Nuur
Hatgal
Samagaltay
Tannu Ola
Uvs Nuur
Erzin
Bukachacha
Chita
Shilka
Nerchinsk
Sretensk
Aginskoye
Khilok
Olovyannaya
Borzya
Zabaykalsk
Petrovsk-Zabaykalskiy
Khapcheranga
Manzhouli
Hailar
Hulun Nur
Argun
Gulian
Da Hinggan Ling
Nenjiang
QIQIHAR
Hegang
Jiamusi
Songhua Jiang
HARBIN
Ang'angxi
Nen Jiang
Mudanjiang
Dongbei
Tao'an
JILIN
Yanji
CHANGCHUN
Siping
FUSHUN
Tonghua
SHENYANG
ANSHAN
Chŏngjin
NORTH KOREA
Dandong
Wŏnsan
Kansŏng
P'YŎNGYANG
Namp'o
SŎUL (Seoul)
INCH'ŎN
SOUTH KOREA
TAEJŎN
TAEGU
PUSAN
DALIAN
Yingkou
Chifeng
Chengde
BEIJING
Zhangjiakou
Hohhot
Baotou
Erenhot
Xilinhot
Linxi
Saynshand
Choybalsan
Tamsagbulag
Öndörhaan
Hentiyn Nuruu
Ulaanbaatar
Lun
Tsetserleg
Hangayn Nuruu
Hyargas Nuur
Har Us Nuur
Dörgön Nuur
Uliastay
Hovd
MONGOLIA
Altay
Gobi
Dalandzadgad
Hami
Gaxun Nur
CHINA
60
50
40
70
80
90
100
110
120
130
140
150
170
3800
965
374
1146
2389
1701
962
2682
3147
2959
1853
1742
1843
2562
4750
3621
3456
1780
2246
2999
1609
2640
2078
2840
1620
3491
455
1054
2800
1949
2744
3669
2290
3776
4362
3957
4266

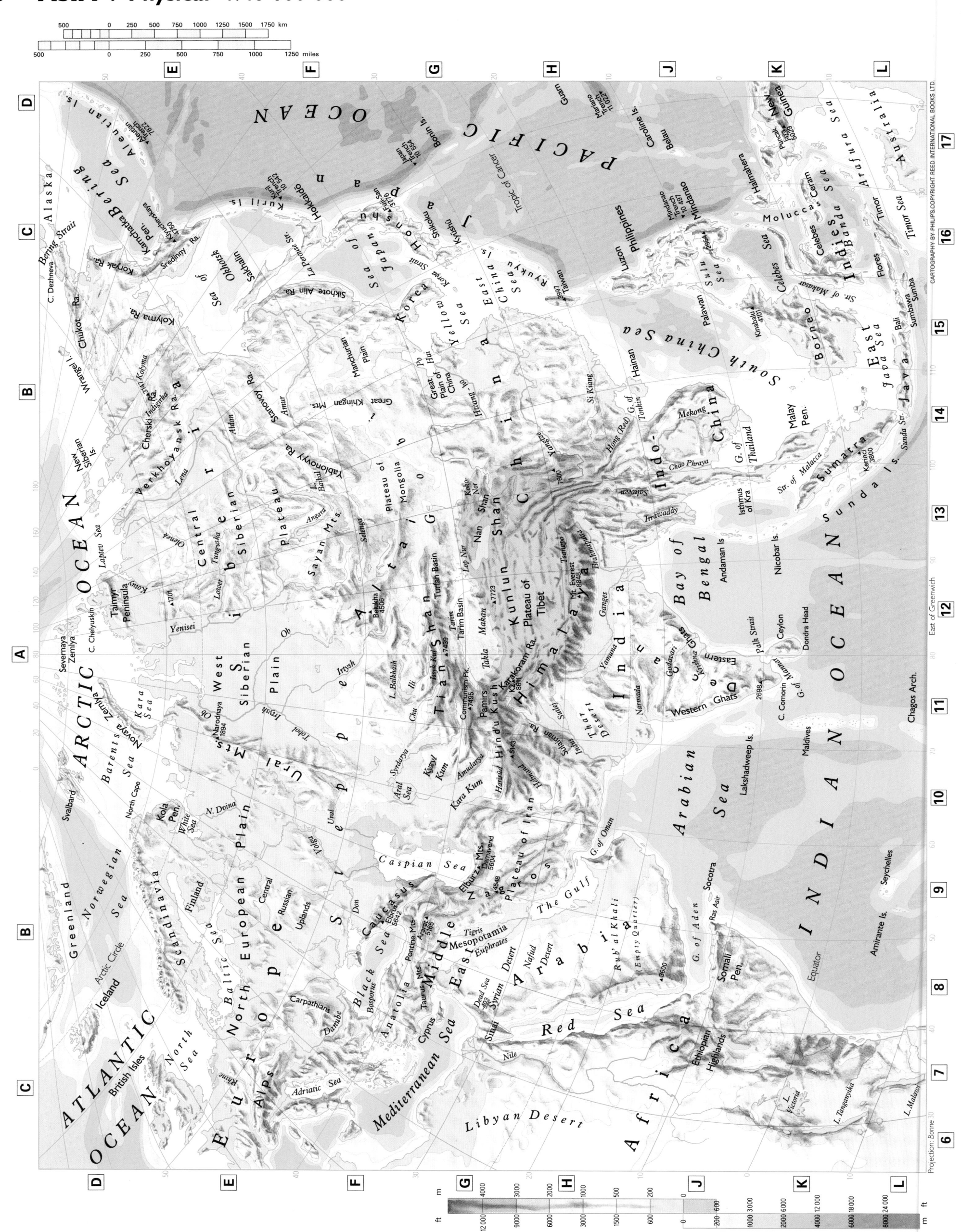

500 0 250 500 750 1000 1250 1500 1750 km

500 0 250 500 750 1000 1250 miles

CARTOGRAPHY BY PHILIPS. COPYRIGHT REED INTERNATIONAL BOOKS LTD.

PACIFIC OCEAN

ARCTIC OCEAN

ATLANTIC OCEAN

INDIAN OCEAN

Bering Strait
Bering Sea
Sea of Okhotsk
Sea of Japan

Japan
Hokkaido
Honshu
Shikoku
Kyushu

Kurll Is.
Sakhalin

Kamchatka Pen.
Koryak Ra.
Sredinny Ra.
Chukot Ra.
Kolyma Ra.
Verkhoyansk Ra.
Cherski Ra.
Indigirka
Kolyma
Lena
Aldan
Stanovoy Ra.
Amur
Yablonovy Ra.
L. Baikal
Selenga
Sayan Mts.
Angara
Yenisei

Korea
Korea Str.
Yellow Sea
East China Sea
Ryukyu Is.
Hainan
Taiwan
Luzon
Philippines
Mindanao
Mindanao Trench 10 497
Palawan
Celebes
Borneo
Sumatra
Java
Sulu Sea
Celebes Sea
Molucca Sea
Banda Sea
Halmahera
Ceram
Flores
Timor
Sumbawa
Sumba
Bali
Java Sea
Timor Sea
Arafura Sea
New Guinea
Australia

C. Dezhneva
Alaska
New Siberian Is.
Wrangel I.
Severnaya Zemlya
C. Chelyuskin
Taimyr Peninsula
Taimyr

Central Siberian Plateau

West Siberian Plain

Steppes

Ural Mts.
Ob
Irtysh
Tobol
Narodnaya 1894
L. Balkhash
Ili
Chu
Syrdarya
Aral Sea
Amudarya
Kyzyl Kum
Kara Kum

Altai
Plateau of Mongolia
Tian Shan
Tarim Basin
Turfan Basin
Lop Nor
Takla Makan
Kunlun Shan
Nan Shan
Kokor Nor
Plateau of Tibet
Himalaya
Mt. Everest 8848
Hindu Kush
Karakoram Ra.
Pamirs
Communism Pk. 7495

China
Great Plain of China
Manchurian Plain
Great Khingan Mts.
Hwang-Ho
Huang-Ho
Po Hai
Si Kiang
Yangtze
Hong (Red)
G. of Tonkin
Mekong
Indo-China
Malay Pen.
Chao Phraya
G. of Thailand
Salween
Irrawaddy
Isthmus of Kra
Str. of Malacca
Sunda Str.
Sunda Is.

Bay of Bengal
Andaman Is.
Nicobar Is.
Ceylon
Dondra Head
Palk Strait
Maldives
Lakshadweep Is.
Chagos Arch.
Amirante Is.
Seychelles
Socotra

India
Deccan
Western Ghats
Eastern Ghats
Ganges
Brahmaputra
Yamuna
Indus
Sutlej
Narmada
Godavari
Krishna
C. Comorin
Thar Desert
Sulaiman Ra.
Hamun

Arabian Sea
Red Sea
G. of Aden
G. of Oman
The Gulf
Somali Pen.
Ras Asir
Ethiopian Highlands
Africa
L. Victoria
L. Tanganyika
L. Malawi

Arabia
Rub' al Khali (Empty Quarter)
Nafud Desert
Syrian Desert
Mesopotamia
Tigris
Euphrates
Dead Sea
Sinai
Nile
Libyan Desert

Caspian Sea
Elburz Mts.
Plateau of Iran
Zagros
Caucasus
Black Sea
Anatolia
Taurus Mts.
Pontine Mts.
Araks
Cyprus
Mediterranean Sea
Bosporus

Europe
North European Plain
Russian Uplands
Central Uplands
Baltic Sea
White Sea
Kola Pen.
Scandinavia
Finland
Norwegian Sea
Greenland
Iceland
British Isles
North Sea
Carpathians
Alps
Adriatic Sea
Danube
Rhine
Volga
Don
Dnieper
N. Dvina
Novaya Zemlya
Barents Sea
Kara Sea
Laptev Sea
North Cape
Svalbard

Tropic of Cancer
Equator
Arctic Circle
East of Greenwich

Projection: Bonne

m ft
4000 12 000
3000 9000
2000 6000
1000 3000
500 1500
200 600
0
0
200 - 600
1000 3000
2000 6000
4000 12 000
6000 18 000
8000 24 000
ft m

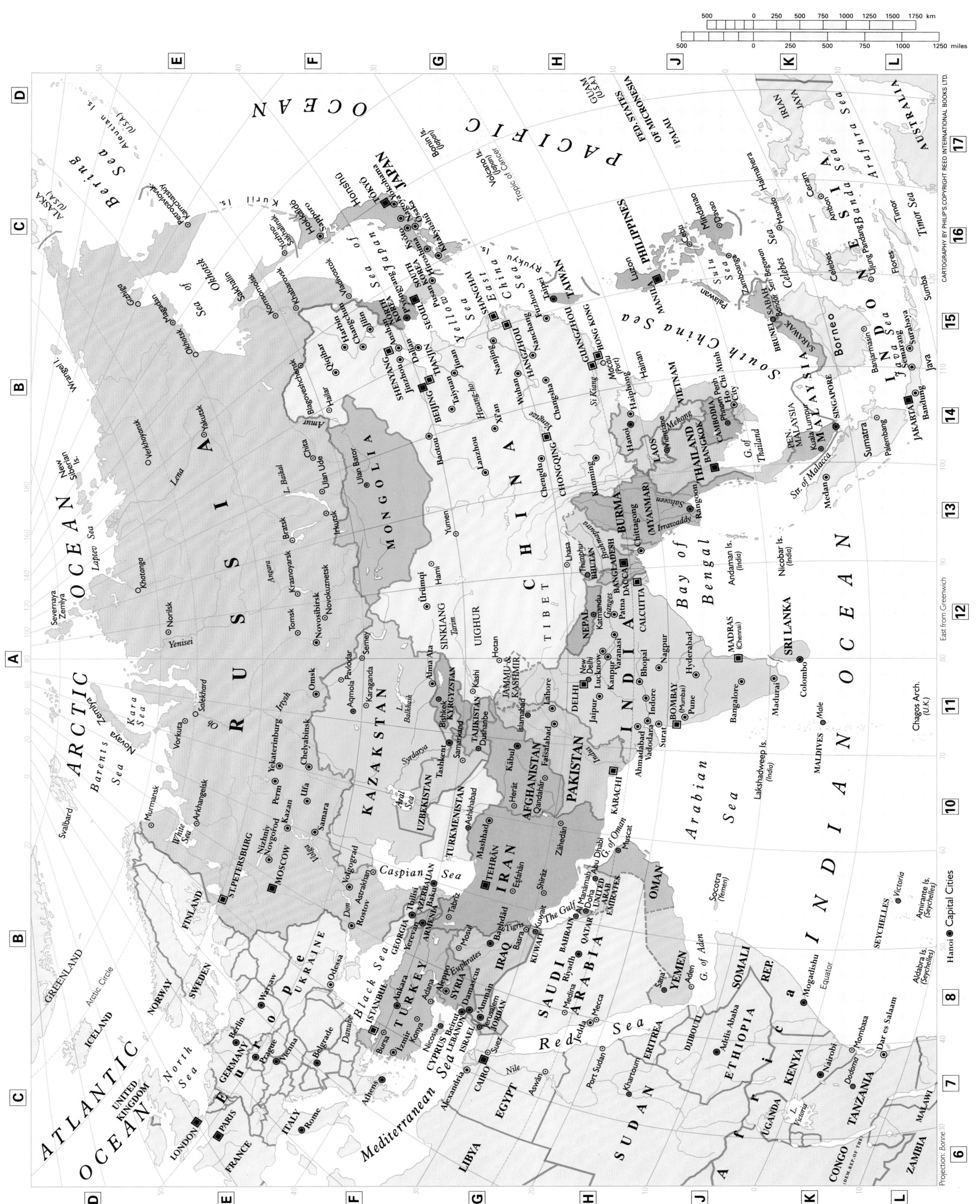
500 0 250 500 750 1000 1250 1500 1750 km
500 0 250 500 750 1000 1250 miles
ARCTIC OCEAN
PACIFIC OCEAN
INDIAN OCEAN
ATLANTIC OCEAN
R U S S I A
C H I N A
MONGOLIA
KAZAKSTAN
I N D I A
PAKISTAN
AFGHANISTAN
IRAN
IRAQ
SAUDI ARABIA
TURKEY
JAPAN
PHILIPPINES
INDONESIA
MALAYSIA
THAILAND
BURMA (MYANMAR)
VIETNAM
LAOS
CAMBODIA
TAIWAN
NORTH KOREA
SOUTH KOREA
NEPAL
BHUTAN
BANGLADESH
SRI LANKA
UZBEKISTAN
TURKMENISTAN
KYRGYZSTAN
TAJIKISTAN
OMAN
YEMEN
SYRIA
JORDAN
ISRAEL
LEBANON
CYPRUS
KUWAIT
QATAR
BAHRAIN
UNITED ARAB EMIRATES
GEORGIA
ARMENIA
AZERBAIJAN
UKRAINE
EGYPT
SUDAN
ETHIOPIA
SOMALI REP.
KENYA
TANZANIA
Bay of Bengal
Arabian Sea
South China Sea
Sea of Japan
Sea of Okhotsk
Bering Sea
Caspian Sea
Black Sea
Red Sea
Mediterranean Sea
TOKYO
BEIJING
MOSCOW
TEHRĀN
DELHI
BANGKOK
MANILA
JAKARTA
SINGAPORE
Tropic of Cancer
Equator
Arctic Circle
East from Greenwich
Projection: Bonne
Hanoi ● Capital Cities
CARTOGRAPHY BY PHILIP'S.COPYRIGHT REED INTERNATIONAL BOOKS LTD.

JAPAN 1:4 000 000

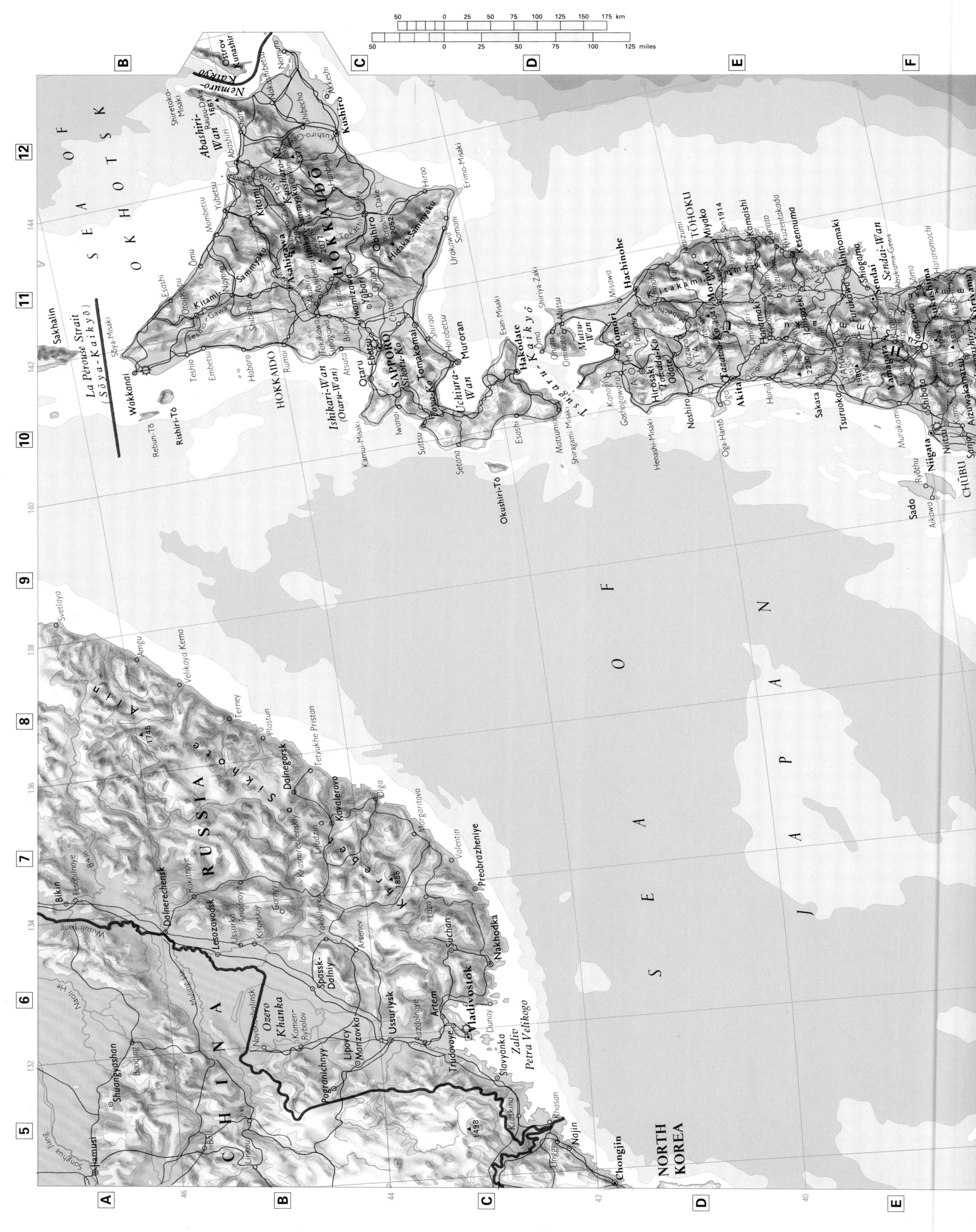

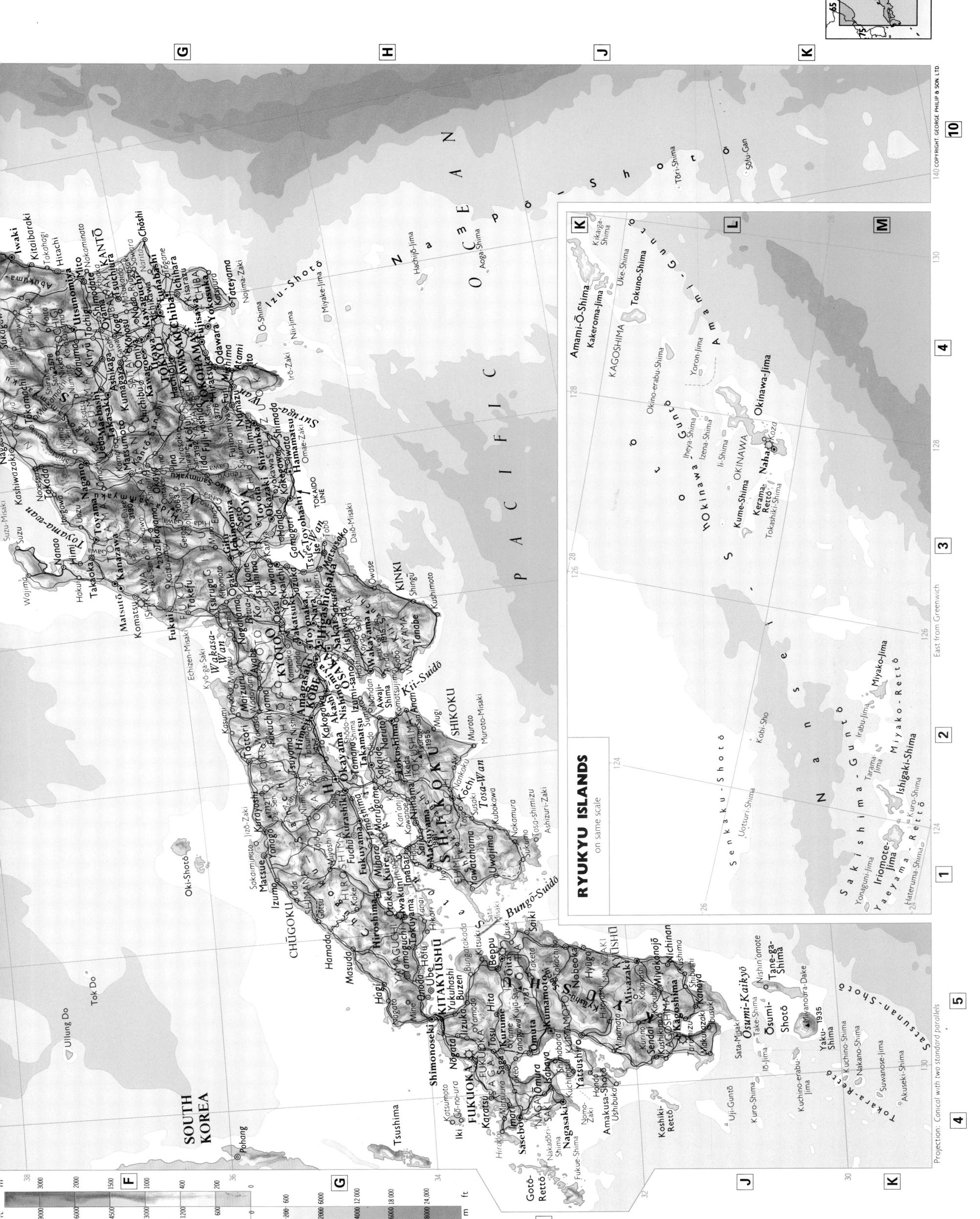

RYUKYU ISLANDS
on same scale

SOUTH KOREA

PACIFIC OCEAN

KANTŌ

TOKYO
YOKOHAMA
KAWASAKI

NAGOYA

KYOTO
KOBE
OSAKA

KINKI

SHIKOKU

KYŪSHŪ

FUKUOKA
KITAKYŪSHŪ

NAGASAKI

KAGOSHIMA

CHŪGOKU

HIROSHIMA

KAGOSHIMA

Tsushima

Tok Do

Ullung Do

Oki-Shotō

Izu-Shotō

KAGOSHIMA

OKINAWA
Naha

Amami-Ō-Shima

Miyako-Rettō

Ishigaki-Shima

Projection: Conical with two standard parallels

East from Greenwich

10 0 10 20 30 40 50 60 70 80 90 km
10 0 10 20 30 40 50 60 miles

1 **2** **3** **4** **5** **6**

130 131 132 133 134

SEA OF JAPAN

A

SOUTH KOREA

Oki-Shotō
Daimanji-San
Dōgo ▲608
Suigō

36

H O N S

B

CHŪGOKU-DISTRICT

Shimane-Hantō
Jizō-Zaki
Iwami
Matsue Sakaiminato
Hi-no-Misaki Hirata Shinji Yonago Kurayoshi Tottori
Taisha Ko Shinji Yasugi Dai-Sen Suga-no-Sen Toyooka
Izumo Daito Wakasa Hidaka

35

Ōda Kisuki TOTTORI
Yunotsu Sanbe-San Yanahara Iyo HYOGO
Gōtsu 1126 Niimi Tsuyama Yamazaki Nishiwaki
Hamada SHIMANE Shōbara OKAYAMA Kasai
Masuda Miyoshi Takahashi Bizen Tatsuno Himeji

C

Ōmi-Shima Hagi HIROSHIMA Sōja Kurashiki Shōdo-Shima Aioi Takasago Mik
Tsuno-Shima Aono-Yama ▲908 Karimui-Yama Fukuyama Kasaoka Akō Kakogawa
Mi-Shima Nagato 1339 Kake Ota-Gawa Mihara Onomichi Harima-Nada Akashi
Tsushima YAMAGUCHI HIROSHIMA Takehara In'no-shima Tonosho Awaji-Shima

34

Hibiki- Yamaguchi Mine Ōtake Kure Aki-Nada Hōjō Imabari KAGAWA Naruto Sumoto
Nada Toyoura San'yō Ogōri Hōfu Iwakuni Ondo Niihama Takamatsu
Genkai- Shimonoseki Onoda Ube Tokuyama Yanai Matsuyama Niihama Zentsuji Tokushima
Nada Ō-Shima Kudamatsu Hikari Kawanoe Iyo-mishima Saijō TOKUSHIMA Anan

Iki Shimonoseki KITAKYŪSHŪ Suō-Nada EHIME KŌCHI Kii-Suidō

D

Ō-Shima FUKUOKA Nakama Nōgata Yukuhashi Buzen Kunisaki Iyo-Nada Kōchi Tosa-yamada
Ikitsuki- Karatsu Iizuka Takawa Nakatsu Nagahama Ōzu Yawatahama Nankoku Aki
Shima FUKUOKA Usa Bungotakada Uchiko Susaki Muroto
Hirado SAGA Tosu Hita Kitsuki Uwajima Kubokawa Muroto-Misaki
Sasebo Takeo Saga Yame Beppu-Wan Tosa-Wan

33

Ōmuta Kumamoto Beppu Ōita SHIKOKU
NAGASAKI Isahaya Tamana Usuki SHIKOKU-DISTRICT
Nagasaki Unzen-Dake Shimabara KUMAMOTO Taketa Saiki Nakamura
Tachibana-Wan Uto Tsukumi Sukumo Tosa-shimizu

E

Amakusa- Hondo KUMAMOTO Takachiho Oki-no-Shima Ashizuri-Zaki
Amakusa- Shimo-Jima Yatsushiro Nobeoka
Shotō Yatsushiro-Kai MIYAZAKI Hyūga
Nada Minamata Saito Takanabe

32

Akune Izumi Ōkuchi Kobayashi KYŪSHŪ
Kami-koshiki Miyanojō Kirishima-Yama KYŪSHŪ-DISTRICT
Jima Sendai Kokubu Miyakonojō
Koshiki- Kushikino KAGOSHIMA Miyazaki Nichinan
Rettō

F

Kagoshima Kanoya
Taniyama Tarumizu Kushima
Makurazaki Kaseda Ibusuki Koyama

31

Sata-Misaki

1 **2** **3** **4** **5** **6**

Projection:
Lambert's Conformal
Conic

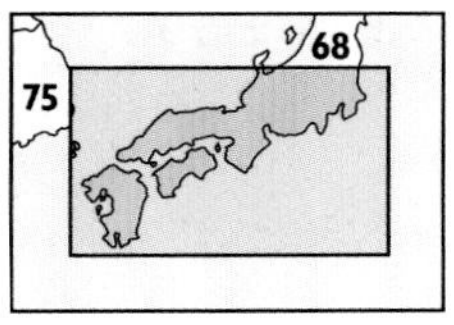

PACIFIC OCEAN

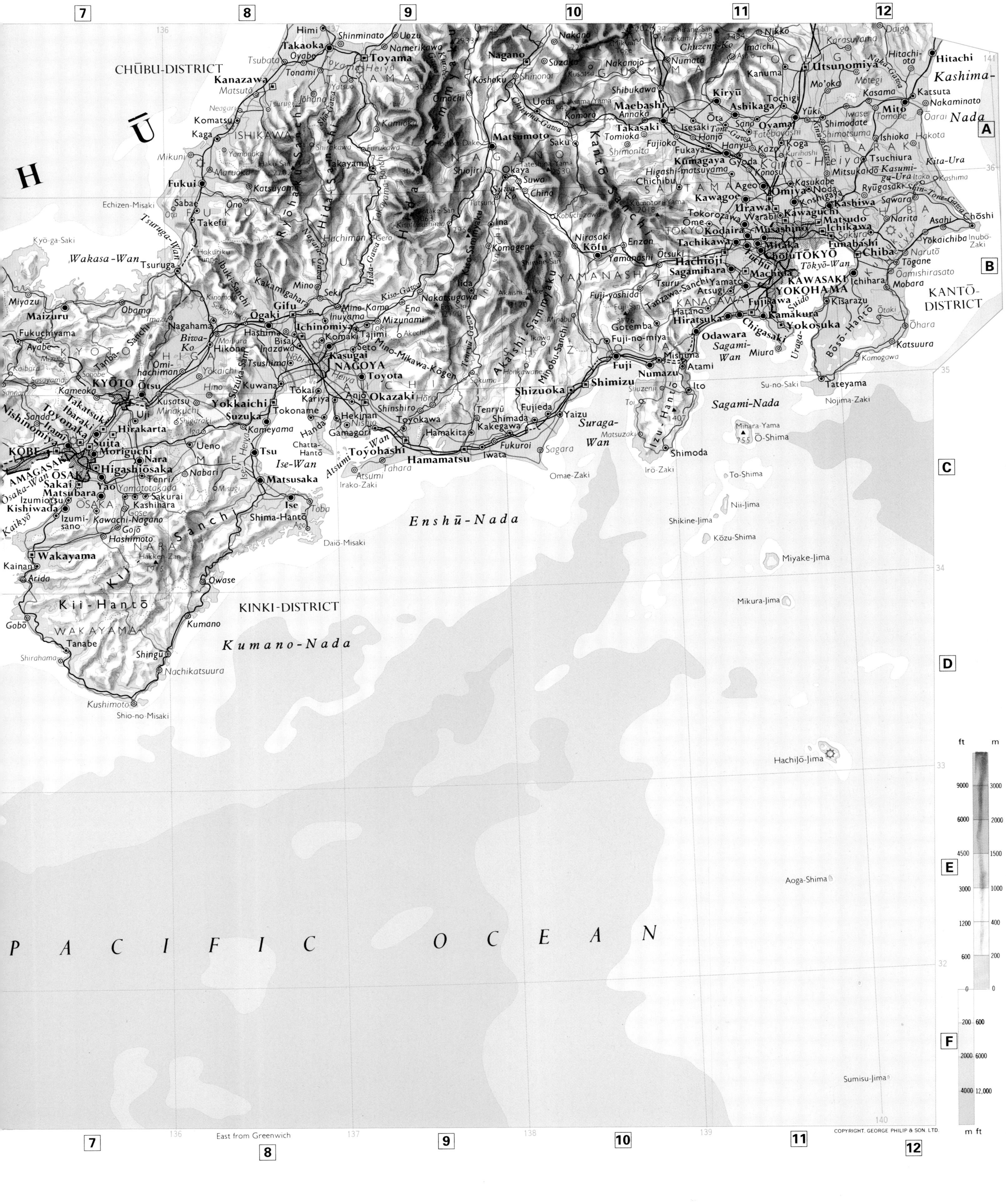

CHŪBU-DISTRICT

KANTŌ-DISTRICT

KINKI-DISTRICT

Kashima-Nada

Enshū-Nada

Kumano-Nada

Sagami-Nada

Suruga-Wan

Wakasa-Wan

Ise-Wan

Tōkyō-Wan

Biwa-Ko

TOKYO

YOKOHAMA

KAWASAKI

NAGOYA

KYŌTO

OSAKA

KOBE

AMAGASAKI

SAKAI

Wakayama

Toyama

Kanazawa

Takaoka

Fukui

Matsumoto

Nagano

Maebashi

Takasaki

Utsunomiya

Mito

Hitachi

Kiryū

Ashikaga

Kumagaya

Kawagoe

Urawa

Hachiōji

Machida

Kōfu

Hamamatsu

Shizuoka

Shimizu

Numazu

Odawara

Kamakura

Yokosuka

Hiratsuka

Fujisawa

Chiba

Funabashi

Matsudo

Ichikawa

Kashiwa

Toyohashi

Okazaki

Toyota

Ichinomiya

Ōgaki

Gifu

Kasugai

Yokkaichi

Suzuka

Tsu

Matsusaka

Ise

Higashiosaka

Suita

Moriguchi

Toyonaka

Ibaraki

Takatsuki

Matsubara

Kishiwada

Izumiotsu

Nara

Yao

Hirakata

Toyooka

Maizuru
Fukuchiyama
Ayabe
Miyazu
Komatsu
Kaga
Himi
Shinminato
Uozu
Namerikawa
Nagano
Nakano
Nikko
Imaichi
Kasama
Katsuta
Nakaminato
Ōarai
Ishioka
Tsuchiura
Kasumigaura
Narita
Chōshi
Asahi
Tōgane
Kisarazu
Tateyama
Kamogawa
Katsuura
Ōhara
Mobara
Kanuma
Mooka
Motegi
Tochigi
Sano
Ōta
Isesaki
Honjo
Fukaya
Gyōda
Kōnosu
Ōmiya
Kawaguchi
Warabi
Kodaira
Mitaka
Chōfu
Fuchū
Tachikawa
Ōme
Tokorozawa
Ageo
Kasukabe
Noda
Yūki
Shimodate
Iwase
Kita-Ura
Shin-Tone-Gawa
Itako
Ryūgasaki
Yokaichiba
Inubō-Zaki
Naruto
Ōamishirasato
Su-no-Saki
Nojima-Zaki
Izu-Hantō
Itō
Atami
Mishima
Fuji
Yaizu
Shimada
Kakegawa
Iwata
Fukuroi
Hamakita
Shinshiro
Toyokawa
Tahara
Irako-Zaki
Atsumi
Ōmae-Zaki
Shimoda
Ō-Shima
Mihara-Yama
To-Shima
Nii-Jima
Shikine-Jima
Kōzu-Shima
Miyake-Jima
Mikura-Jima
Hachijō-Jima
Aoga-Shima
Sumisu-Jima

Shio-no-Misaki
Kushimoto
Nachikatsuura
Shingū
Kumano
Owase
Tanabe
Shirahama
Gobō
Arida
Kainan
Daiō-Misaki
Toba
Shima-Hantō
Ago
Kaikyō
Izumi-sano
Kawachi-Nagano
Hashimoto
Gose
Gojō
Sakurai
Kashihara
Kusatsu
Ōtsu
Uji
Yamatotakada
Tenri
Nabari
Ueno
Kameyama
Kariya
Tōkai
Tokoname
Handa
Gamagori
Hekinan
Anjō
Nishio
Chatta-Hantō
Kuwana
Tsushima
Bisai
Inazawa
Hashima
Komaki
Seto
Tajimi
Mizunami
Ena
Nakatsugawa
Mino-Kamo
Mino
Seki
Kakamigahara
Inuyama
Kani
Tatsuno
Iida
Komagane
Ina
Okaya
Suwa
Chino
Shiojiri
Saku
Komoro
Ueda
Koshoku
Nirasaki
Enzan
Ōtsuki
Fuji-yoshida
Fujinomiya
Gotemba
Hatano
Atsugi
Yamato
Sagamihara
Chigasaki
Uraga
Miura
Kazusa
Ichihara
Kamisu
Hota
Kashima
Sawara
Omiya
Koshigaya
Higashi-matsuyama
Chichibu
Annaka
Fujioka
Shimonita
Tomioka
Tatebayashi
Koga
Kazo
Hanyū
Kanra
Tomobe
Shimotsuma
Tomobe
Tochigi
Numata
Nakanojō
Kusatsu
Suzaka
Shinonoi
Ōmachi
Takayama
Hachiman
Gujō
Gero
Kiso-Gawa
Kiso-Fukushima
Matsuto
Tsubata
Oyabe
Tonami
Johana
Inami
Mikuni
Maruoka
Katsuyama
Ōno
Takefu
Sabae
Tsuruga
Obama
Echizen-Misaki
Kyō-ga-Saki
Kinomoto
Nagahama
Hikone
Yōkaichi
Higashi-ōmi
Hino
Kōka
Shigaraki
Iga
Misugi
Yamatotakada
Tanba
Sasayama
Sonobe
Kameoka

Kashima-Nada

KANAGAWA
SAITAMA
IBARAKI
CHIBA
TOCHIGI
GUMMA
NAGANO
YAMANASHI
SHIZUOKA
AICHI
GIFU
FUKUI
ISHIKAWA
TOYAMA
KYŌTO
SHIGA
MIE
NARA
WAKAYAMA
HYŌGO

Kantō-Heiya

Akaishi-Sammyaku

Kiso-Sammyaku

Hida-Sammyaku

Kii-Hantō

Bōsō-Hantō

Tenryū-Gawa

Chikuma-Gawa

Shinano-Gawa

Kiso-Gawa

Nagara-Gawa

Kinu-Gawa

Tone-Gawa

East from Greenwich

ft m
9000 3000
6000 2000
4500 1500
3000 1000
1200 400
600 200
0 0
200 600
2000 6000
4000 12,000
m ft

KAZAKSTAN
RUSSIA
MONGOLIA
KYRGYZSTAN
XINJIANG UYGUR ZIZHIQU (SINKIANG)
XIZANG ZIZHIQU (TIBET)
QINGHAI
JAMMU & KASHMIR
NEPAL
BHUTAN
BANGLADESH
INDIA
BURMA (MYANMAR)
THAILAND (SIAM)
LAOS
VIETNAM
Qaraghandy
Semey
Rubtsovsk
Oskemen
Irkutsk
Ulaanbaatar
Almaty
Bishkek
Ürümqi
Tarim Pendi
Taklamakan Shamo
Kunlun Shan
Tanggula (Dangla) Shan
Qilian Shan
Lanzhou
Xining
Chengdu
Kunming
Lhasa
Delhi
Calcutta
Dhaka
Bay of Bengal
Hanoi
Haiphong
G. of Tonkin
Tropic of Cancer
Projection: Bonne
East from Greenwich

SEA OF JAPAN
YELLOW SEA
EAST CHINA SEA
PACIFIC OCEAN
SOUTH CHINA SEA
BEIJING
SHANGHAI
TIANJIN
NORTH KOREA
SOUTH KOREA
SŎUL (SEOUL)
P'YŎNGYANG
TŌKYŌ
ŌSAKA
NAGOYA
YOKOHAMA
KAWASAKI
KYŌTO
KŌBE
HIROSHIMA
FUKUOKA
KITAKYUSHU
SAPPORO
Hokkaidō
Kyūshū
Shikoku
Sakhalin
TAIWAN (FORMOSA)
T'AIPEI
KAOHSIUNG
HONG KONG
GUANGZHOU (CANTON)
Macau (Port.)
Hainan Dao
HARBIN
SHENYANG
CHANGCHUN
DALIAN
QINGDAO
NANJING
WUHAN
HANGZHOU
FUZHOU
Vladivostok
Khabarovsk
Manchuria
Tropic of Cancer
Korea Strait
Taiwan Strait
Ryūkyū-rettō
COPYRIGHT GEORGE PHILIP LTD.

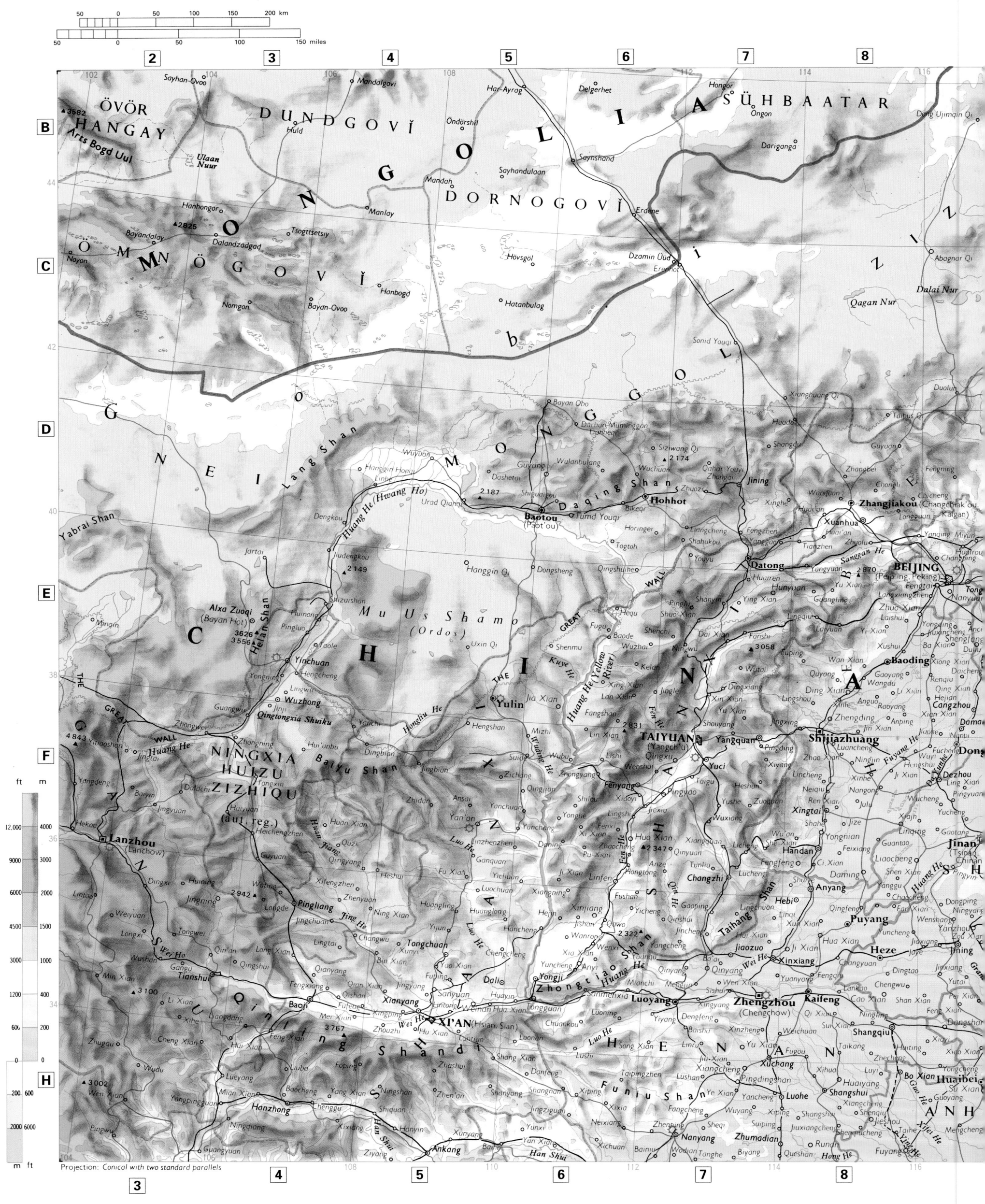

9   10   11   12   13   14   15   16

B

HIQU

HEILONGJIANG

Horqin Youyi Qianqi
Zhenlai
Baicheng
Nen Jiang
Maoxing Zhaoyuan
HARBIN
(Haerhpin)
Bin Xian
Jixi
Turiy Rog
Ozero Khanka

Hulin He
Tuquan
Tao'an
Ta'er He
Dalai
Songhua Jiang
Acheng
Shangzhi
Linkou
Mudan Jiang
RUSSIA

Tuquan
Anguang
Qian Gorlos
Fuyu
Changchunling
Yimianpo
Hailin
Maqiaohe
Pogranichnyy

Hulin He
Tongyu
Shenjingzi
Kaipan
Song hua Jiang
Yushu
Shahetun
Wuchang
Muling
Xiachengzi
Suiyang
Suifenhe
Golenki

Jarud Qi
Beizhenzhen
Nong'an
Dehui
Gangyao
Lalin He
Shulan
Shanhetun
Zhangguangcailing
Hengdaohezi
Ning'an
Dongning
C
Ussuriysk (Voroshilov)
Razdolnoye

1949
Changling
Jiutai
Wulajie
△1690
Dandanghezi
Pokrovka
Razdolnoye
44
Artem

L
Zhanyu
Changchun
Maoiin
Jilin
(Kirin)
Chilin
Jiaohe
Emu
Huadanggang
Wangqing
Mingyuegue
Yanji
Dongning
Slavyanka

2029
Linxi
Bairin Youqi
Kailu
Xiliao He
Lishu
Siping
Yitong
Panshi
Huadian
Dunhua
Daxinggou
Shixian
Tumen
Kraskino
Vladivostok
42
C

Xinkai He
Zhanyu
Shuangliao
Bamiancheng
Liaoyuan
Xifeng
Dongfeng
Huinan
Jingyu
Antu
Helong
Tumen
Hunchun
Posyet

Xar Moron He
Laoha He
Jargalang
Kangping
Zhangwu
Hailong
Sanchengzhen
Huijin
Linjiang
△1677
Musan
Puryong
Pogŏdong

Héxigten Qi
Wutonghaolai
Hure Qi
WALL
Faku
Kaiyuan
Shanchengzhen
Duhe
Baektusan
2541
Kyŏngsŏng
Chongjin
Chuuronjang

Ongniud Qi
Xiawa
Zhangwu
Liu He
Tieling
Qingyuan
Tonghua
Changbai
Huchang
Pukchong
Ondaejin
Musudan

2020
Chifeng
Heishan
Xinlitun
Xinmin
WILLOW
Fushun
Dongtua
Linjiang
△845
Kanggye
Kuup-tong
Hyesan
Kapsan
Simpungdong

1885
Beipiao
Qinghemen
SHENYANG
(Mukden)
Fushun
Hun He
Huajianzi
Ji'an
Manpojin
Pungsan
Kosŏngni
Kimchaek
(Songjin)
D

Chaoyang
Ningcheng
Benxi
Liaoyang
Qinghecheng
Yalu
Jiang
Chosan
Kuandian
Pyŏktong
Changjin-chōsuji
Changjin
Sinhung
Chanhangni
Pukchŏng
Tanchon

Chengde
Luannan
Jinzhou
Panshan
Niuzhuang
Haicheng
Lianshanguan
Supung Sk.
Chosan
Koin-dong
2522
Pujon-chōsuji
Kwangdŏch
Hapsu
Sŏhori

Longhua
Lingyuan
Jinxi
Tianzhuangtai
Anshan
Anping
Xiuyan
NORTH
Changjin
Sinhung
Hamhung
Hongwon
40

Luanping
Shangbancheng
Pingquan
Jinxi
Yingkou
Fengcheng
Taegwan
Pukchin
KOREA
Oro
Tŏkchŏn
Hamhung
Hŭngnam

Shuangshinzi
Gai Xian
Cao He
Tongjosŏn
Man

Xinglong
Zunhua
Jianchangying
Suizhong
Yingkou
Xiongyuecheng
Zhuanghe
Pikou
△1131
Gushan
Donggou
Dandong
Sinŭiju
Yalu Jiang
Sŏnchŏn
Chongju
Anju
Kŭjang
Hongwon
Tongchŏn-ni
E

Sanhe
Bodi
Fengrun
Luan Xian
Qinhuangdao
Changli
Jin Xian
Xinjin
Yongampo
Zhuanghe
Sukchŏn
Sinchang-ni
Unsan
Munchŏn
Wŏnsan
JAPAN

Xian
Fuxin
Wan
Liaodong
Fu Xian
Sŏnchŏn
Kangdong
Tangyang
Singosan
Kojŏ
Kosŏng

TIANJIN
(Tientsin, Tienching)
Tanggu
Dagu
Lulong
Leting
Sunan
Munchŏn
Anbyŏn

Wuqing
Hangu
Lüshun
DALIAN
(Luda)
Korea
Bay
P'YŎNGYANG
Chunghwa
Koksan
Pyŏngsan
1638△
Hoeyang
Kansŏng
38

Dangjiang
Oikou
Bo Hai
Nampo
Songnim
Suan
Chiha-ri
Pyŏnggang
Hwachon-chōsuji
Yangyang

zhuang
Yanshan
(Gulf of Chihli)
Cho-do
Chaeryŏng
Sinmak
Nam-chon
Chŏrwŏn
△1578
Chumunjin

guang
Qingyun
Huang He
Langkou
Penglai
Yantai
Sariwŏn
Kŭmhwa
Hongchŏn
Kangnŭng

Wudi
Zhanhua
Huang Xian
Laizhou
Wan
Weihai
Changyŏn
Haeju
Kaesŏng
Panmunjŏm
Munsan
Ŭijŏngbu
Chunchŏn
Kangnŭng

Deping
Huimin
Shanghe
Benzhou
Zhoucun
Fushan
Muping
Ongjin
Kimchŏn
Kaphwa
SŎUL
(Seoul)
Hongchŏn
Hoengsŏng
Samchŏk
Ullŭng-do

Shanghe
Qixia
923△
Wendeng
Paengnyong-do
Cease Fire Line
Yŏngdŭngpo
Ich'ŏn
Wŏnju
Yŏngwŏl
F

Qingcheng
Guangrao
Shandong Bandao
Rushan
Shidao
INCH'ŎN
Suwŏn
Osan
Ch'ungju
Chechŏn
Ulchin
36

Jiyang
Huantai
Linzi
Pingdu
Laixi
Nanhuang
Haiyang
Sŏsan
Chŏnan
Chŏngju
Yŏju
Yŏngju

Zibo
Yidu
Fangzi
Changyi
Gaomi
Jimo
Changyang
SOUTH
Hongsŏng
Chonan
Chŏngju
Yŏngdŏk

Tai Shan
1524△
Bashan
Linqu
Anqiu
Jiao Xian
Jimo
KOREA
Kongju
Nonsan
Taejŏn
Sangju
Naktong
Andong
Yŏngha

Tai'an
Laiwu
Zhucheng
Changyang
Anmyon-do
Taechŏn-ni
Chŏngŭp
Kimchŏn
Chŏngju
Pohang

Xintai
Zhuzhou
QINGDAO
(Ch'ingtao)
Kangyŏng
Iri
Kaeryŏng
Koryŏng
TAEGU
Kyongju

1108
Yishui
Wulian
Jiaozhou Wan
Puan
Kimje
Kochang
Hamyang
Miryang
Ulsan

Sishui
Mengyin
Yitie
Kunsan
Chŏnju
Namwŏn
1915△
Chinju
Masan
Tongnae

Pingyi
Ju Xian
Langcheng
△923
Chŏngŭp
Hamyang
Chinhae
PUSAN
G

Fei Xian
Linyi
Ganyu
Haizhou Wan
Sago-ri
Tamyang
Chungmu

Teng Xian
Zaozhuang
Tancheng
Lianyungang
Songjong-ni
Kwangju
Sunchŏn
Polgyo-ri
Samchonpo

Hanzhuang
Jiawang
Pei Xian
Lianyungang
(Hsinhaihen)
Mokpo
Changhŭng
Yŏsu
HUANG HAI
Korea Strait

Xuzhou
Suzhou
Shuguang
Guanyun
Chenjiagang
Haenam
Chindo
Tsushima
Iki

Suining
Suqian
Shuyang
Xiangshui
Posong
Izuhara

UI
Lingbi
Qingjiang
Binhai
(Yellow Sea)
Cheju
Cheju-do
Tsushima-kaikyō
Karatsu
Imari

Guzhen
Si Xian
Huaiyuan
Dayunhe
Hallim
1950△
Onpyong-ni
JAPAN
Sasebo
H

Bengbu
Fengyang
Huai'an
Baoying
Liuzhuang
Mosulpo
Sŏgwi-po
Nakadóri-jima
Ōmura
Isahaya

Gaoyou
Hu
Yancheng
Xinghua
Dongtai
Fukue-jima
Nagasaki
Kuchinotsu

9   10   11   12   13   14   15

East from Greenwich

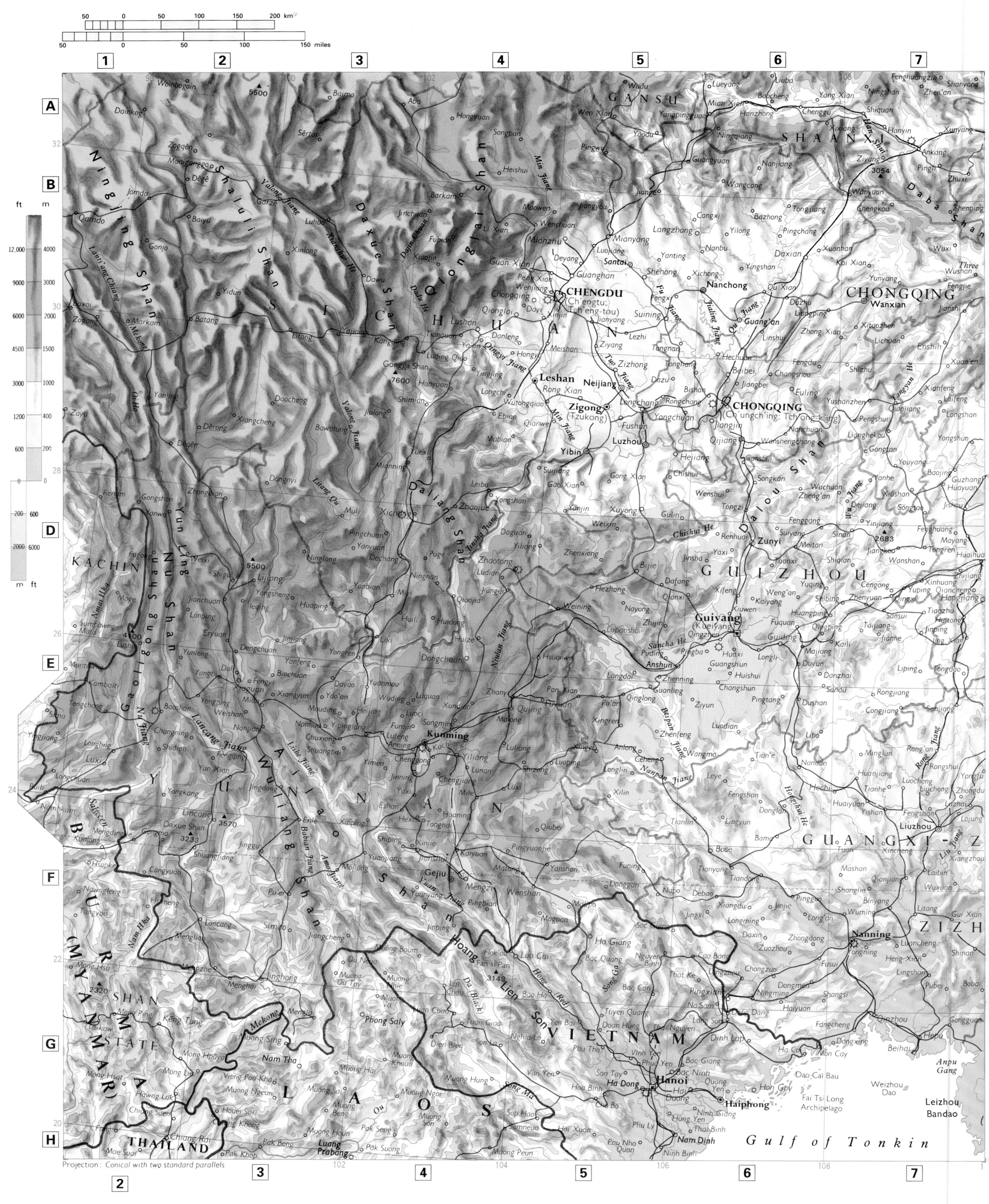
50 0 50 100 150 200 km
50 0 50 100 150 miles
1 2 3 4 5 6 7
A B C D E F G H
ft m
12,000 4000
9000 3000
6000 2000
4500 1500
3000 1000
1200 400
600 200
0 0
200 600
2000 6000
m ft
GANSU
SHAANXI
SICHUAN
CHONGQING
GUIZHOU
YUNNAN
GUANGXI-Z
KACHIN
SHAN STATE
BURMA (MYANMAR)
LAOS
VIETNAM
THAILAND
CHENGDU (Ch'engtu; Tch'eng-tou)
CHONGQING (Ch'ungch'ing; Tch'ong-k'ing)
Chongqing
Kunming
Guiyang (Kueiyang)
Nanning
Hanoi
Haiphong
Leshan
Zigong (Tzukong)
Nanchong
Neijiang
Yibin
Luzhou
Zunyi
Wanxian
Liuzhou
Ningjing Shan
Shaluli Shan
Daxue Shan
Qionglai Shan
Daliang Shan
Dalou Shan
Daba Shan
Yun Ling
Nu Shan
Gaoligong Shan
Wuliang Shan
Ailao Shan
Hoang Lien Son
Gonggga Shan 7600
Min Jiang
Yalong Jiang
Jinsha Jiang
Lancang Jiang
Mekong
Nu Jiang
Salween
Hong (Red)
Da (Black)
Gulf of Tonkin
Leizhou Bandao
Fai Tsi Long Archipelago
Weizhou Dao
Projection: Conical with two standard parallels
2 3 4 5 6 7
102 104 106 108

74
75
90
86
80
HENAN
HUBEI
ANHUI
JIANGSU
ZHEJIANG
JIANGXI
HUNAN
FUJIAN
GUANGDONG
TAIWAN
Nanyang
Bengbu
Huainan
Yangzhou
Taizhou
Nantong
NANJING (Nanking, Nanching)
Zhenjiang (Chenchiang)
Changzhou (Ch'angchou)
Wuxi (Wuhsi)
Suzhou (Suchou)
SHANGHAI (Changhai)
Chongming Dao
Hefei
Wuhu
Xiangfan
Macheng
Dabie Shan
Hankou
WUHAN (Wou-han)
Yichang (Ich'ang)
Shashi
Huangshi
Anqing
Hangzhou (Hangchou; Hangchow)
Hangzhou Wan
Zhoushan Dao
Ningbo (Ningpo)
Shaoxing
Jiujiang
Jingdezhen
Poyang Hu
Dongting Hu
Changde
Nanchang
Changsha
Xiangtan
Zhuzhou
Linchuan
Shangrao
Jinhua
Linhai
Wenzhou (Wenchow)
Hengyang
Shaoyang
Ji'an
Nanping
Sanming
Fuzhou (Foochow; Fuchou)
Matsu Tao
Ganzhou
Guilin
Shaoguan
Quanzhou (Ch'uanchou)
Zhangzhou
Xiamen (Hsiamen; Amoy)
Chao'an
Shantou (Swatow)
Wuzhou
GUANGZHOU (Kuangchou; Canton)
Foshan
Jiangmen
Shenzhen
HONG KONG
Kowloon
Macau (Macao) (Port.)
Zhanjiang
Leizhou Wan
Hsinchu
Chilung
T'AIPEI
T'aichung
Changhua
Chiai
T'ainan
Kaohsiung
Hualien
Formosa Strait
Luzon Strait
Tropic of Cancer
SOUTH CHINA SEA
Tungsha Tao
East from Greenwich
COPYRIGHT. GEORGE PHILIP & SON. LTD

100 0 100 200 300 400 500 km
100 0 50 100 150 200 250 300 350 miles
BURMA
RANGOON (YANGON)
(MYANMAR)
Insein
Thaton
Moulmein
Ma-ubin
Pyapon
G. of Martaban
Kyaikkami
Ye
Letpadan
Tharrawaddy
Thoen
Tak
Mae Sot
THAILAND
Vientiane (Viangchan)
Nong Khai
Udon Thani
Loei
Uttaradit
Sawankhalok
Phitsanulok
Phetchabun
Khon Kaen
Sakon Nakhon
Nakhon Phanom
Savannakhet
Chalyaphum
Roi Et
Nakhon Sawan
Nakhon Ratchasima
Buriram
Sisaket
Ubon Ratchathani
Khemmarat
Muang Khammouan
Natkyizin
Tavoy
Maungmagan Islands
Nam Tok
Kanchanaburi
Phra Nakhon Si Ayutthaya
Saraburi
BANGKOK
Samut Songkhram
Samut Prakan
Chon Buri
Pattaya
Sattahip
Rayong
Chanthaburi
Trat
Phet Buri
Hua Hin
Prachuap Khiri Khan
Bang Saphan
Chumphon
Mali Kyun
Kadan Kyun
Mergui
Tanintharyi
Letsök-aw Kyun
Lenya
Bokpyin
Myeik
Lambi Kyun
Kyunzu
Maliwun
Kho Khot Kra
Zadetkyi Kyun
Ranong
Phnom Dangrek
Khu Khan
Cheom Ksan
Aranyaprathet
Sisophon
Siemreab
Tonle Sap
Batdambang
Kulen
CAMBODIA
Pouthisat
Kampong Chhnang
Kompong Thom
Phnom Penh (Phnum Penh)
Ko Chang
Ko Kut
Phumi Koh Kong
Sre Ambel
Krong Kaoh Kong
Takev
Kampot
Chaak Kampong Saom
Kampong Saom
Gulf of Thailand
Hon Chong
Dao Phu Quoc
Rach Gia
Long Xuyen
My Tho
Can Tho
Sa Dec
Soc Trang
Bac Lieu
Ca Mau
Mui Ca Mau
Con Son
Ko Phangan
Ko Samui
Surat Thani
Nakhon Si Thammarat
Pak Phanang
Thung Song
Phatthalung
Thale Luang
Phangnga
Phuket
Trang
Kantang
Songkhla
Hat Yai
Satun
Tarutao
Langkawi
Pattani
Yala
Narathiwat
Tumpat
Kota Baharu
Perhentian
Redang
Kuala Terengganu
LAOS
Pakxe
Saravan
Attapu
Stoeng Treng
Kracheh
Kampong Cham
Prey Veng
Svay Rieng
Senmonorom
Ba Don
Dong Hoi
Quang Tri
Hue
Da Nang
Hoi An
VIETNAM
Chau O
Quang Ngai
Kon Tum
Plei Ku
Bong Son
Binh Dinh
Qui Nhon
Song Cau
A Yun Pa
Mui Nay
Buon Me Thuot
Nha Trang
Cam Ranh
Da Lat
Phan Rang
Mui Dinh
Bien Hoa
Phan Thiet
PHANH BHO HO CHI MINH
Vung Tau
Truong Son
Paracel Is.
SOUTH CHINA SEA
Spratly Is.
Spratly I.
Nanshan I.
Loaita I.
Itu Aba I.
Sin Cowe I.
Amboyna Cay
Mt. Mantalingajan
C. Buliluyan
Bugsuk
Balabac I.
Balabac Str.
Banggi
Kudat
Langkon
Jembongan
Gunong Kinabalu
Kota Belud
Kota Kinabalu
Papar
Beaufort
Labuan
Pulau Labuan
Ranau
Beluran
SABAH
Melalap
Tenom
Lawas
Limbang
Bandar Seri Begawan
BRUNEI
Tutong
Kuala Belait
Seria
Miri
Marudi
Niah
Tawau
ANDAMAN SEA
Strait of Malacca
Alor Setar
George Town
Pinang
Butterworth
Sungei Petani
Pasir Mas
PENINSULAR MALAYSIA
Taiping
Kuala Sepetang
Ipoh
Gunong Tahan
Dungun
Kemaman
Teluk Intan
Kampar
Kuala Lipis
Kota Kubu Baharu
Temerloh
Kuantan
Kelang
KUALA LUMPUR
Pelabuhan Kelang
Seremban
Port Dickson
Pulau Tioman
Segamat
Mersing
Melaka
Muar
Keluang
Kota Tinggi
Johor Baharu
SINGAPORE
Tenggol
MALAYSIA
We
Sabang
Banda Aceh
Meureudu
Bireuen
Lhokseumawe
Sigli
Lhokkruet
Idi
Peureulak
Calang
Takengon
Langsa
Meulaboh
ACEH
Kualasimpang
G. Leuser
Pangkalanbrandan
Ujung Raja
Belawan
MEDAN
Binjai
Tebingtinggi
Tapaktuan
Kabanjahe
Simeulue
Pematangsiantar
Tanjungbalai
Sinabang
Prapat
Danau Toba
Kepulauan Banyak
Tarutung
Rantauprapat
Sibolga
UTARA
Musala
Lahewa
Gunungsitoli
Nias
Padangsidempuan
Telukdalem
Natal
Pini
Lubuksikaping
Tanahmasa
Kepulauan Batu
Tanahbala
Bukittinggi
Payakumbuh
Padangpanjang
BARAT
Padang
Sawahlunto
Solok
Painan
Siberut
Sabulubbek
Muarabungo
Kerinci
Sungaipenuh
Sipura
Kepulauan Mentawai
Pulau Pagai Utara
Pulau Pagai Selatan
Mukomuko
Bengkulu
BENGKULU
Manna
Curup
Lubuklinggau
Tebingtinggi
Lahat
Dempo
Muaraenim
Baturaja
Martapura
Kotabumi
Menggala
Tanjungkarang
Telukbetung
LAMPUNG
Kotaagung
Kalianda
Enggano
Dumai
Bengkalis
Rupat
Batu Pahat
Bagansiapiapi
Siaksriindrapura
Pekanbaru
Bangkinang
Kampar
RIAU
Rengat
Bintan
Tanjungpinang
Kepulauan Riau
Kepulauan Lingga
Lingga
Pasirkuning
Singkep
Selat Berhala
Muaratebo
JAMBI
Jambi
Muaratembesi
Sarolangun
Bangko
PALEMBANG
Sungaigerong
Sekayu
SELATAN
Perabumulih
Muntok
Belinyu
Sungailiat
Pangkalpinang
Bangka
Selat Bangka
Toboali
Tg. Lumut
Tanjungpandan
Manggar
Dendang
Belitung
Selat Karimata
Kepulauan Karimata
Tanjung Sambar
Tanjung Puting
Laut
Kepulauan Natuna Besar
Telukbutun
Natuna Besar
Binjai
Matak
Siantan
Kepulauan Anambas
Midai
Subi
Kepulauan Natuna Selatan
Serasan
Tanjung Datu
Kepulauan Tambelan
Sambas
Singkawang
Kepulauan Badas
Mempawah
Pontianak
Lundu
Bau
Kuching
Serian
Niut
Tebakang
Ngabang
BARAT
Sanggau
Sintang
Semitau
Putussibau
Bintangor
Sarikei
Saratok
Betung
Bandar Sri Aman
Sibu
Kanowit
Kapit
Oya
Mukah
Bintulu
SARAWAK
Banjaran Dulit
Pegunungan Muller
Nangapinoh
Muarajuloi
G. Saran
Pegunungan Schwaner
Sukadana
Nangatayap
Ketapang
Sukaraja
Kendawangan
Kualajelai
Kualapembuang
Pangkalanbuun
Kumai
Semuda
Sampit
Teluk Sampit
Palangkaraya
Kualakapuas
TENGAH
KALIMANTAN
Borneo
Longnawan
Tanjungredeb
Tanjungselor
Tarakan
Tanjungbatu
Kayan
TIMUR
Sangkulirang
Bontang
Longiram
Muarakaman
Purukcahu
Tenggarong
Samarinda
Muaratewe
Kualakurun
Balikpapan
Tanahgrogot
Buntok
Tanjung
Amuntai
Barabai
Kandangan
Besar
Kepulauan Balabalangan
Banjarmasin
Kotabaru
Sebuku
Martapura
Pelaihari
Pagatan
Karambu
Pulau Laut
Jorong
Satui
Tg. Selatan
Selat Makassar
Greater Sunda Islands
JAVA SEA
Kepulauan Laut Kecil
Kepulauan Masalima
Kepulauan Masalembo
Kepulauan Karimunjawa
Bawean
Sangkapura
Tanjung Bugel
INDONESIA
Kepulauan Seribu
JAKARTA
Merak
Serang
Selat Sunda
Pulau Rakata (Krakatau)
Panaitan
Bogor
Sukabumi
Purwakarta
Jatibarang
Indramayu
Cirebon
Brebes
Tegal
Pemalang
Pekalongan
Kendal
SEMARANG
BANDUNG
Garut
Tasikmalaya
Cilacap
Teluk Pelabuhan Ratu
Slamet
TENGAH
Kebumen
Magelang
Yogyakarta
Surakarta
Lawu
Madiun
Tulungagung
Blitar
Malang
Kediri
Semeru
Bojonegoro
Tuban
Gresik
Bangkalan
Madura
Sampang
SURABAYA
Pasuruan
Probolinggo
Jember
TIMUR
Banyuwangi
Kepulauan Kangean
Kepulauan Tengah
BALI
Agung
Singaraja
Denpasar
Penida
Ampenan
Mataram
Praya
Lombok
Rinjani
Sumbawa Besar
Taliwang
Moyo
Tambora
NUSA TENGGARA BARAT
Lesser
Java (Jawa)
INDIAN OCEAN
Java Trench
Projection: Mercator
East from Greenwich

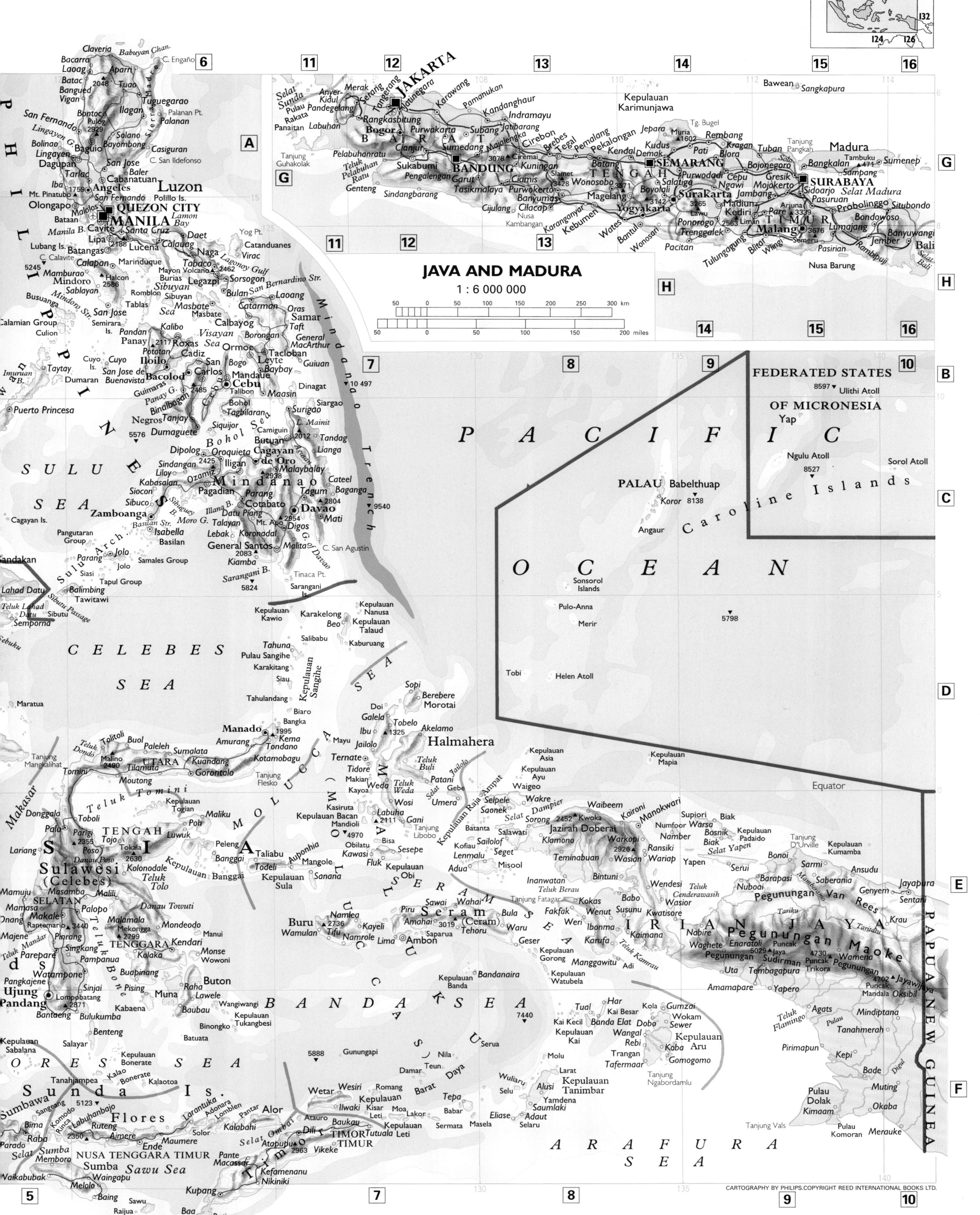

JAVA AND MADURA

1 : 6 000 000

CARTOGRAPHY BY PHILIPS.COPYRIGHT REED INTERNATIONAL BOOKS LTD.

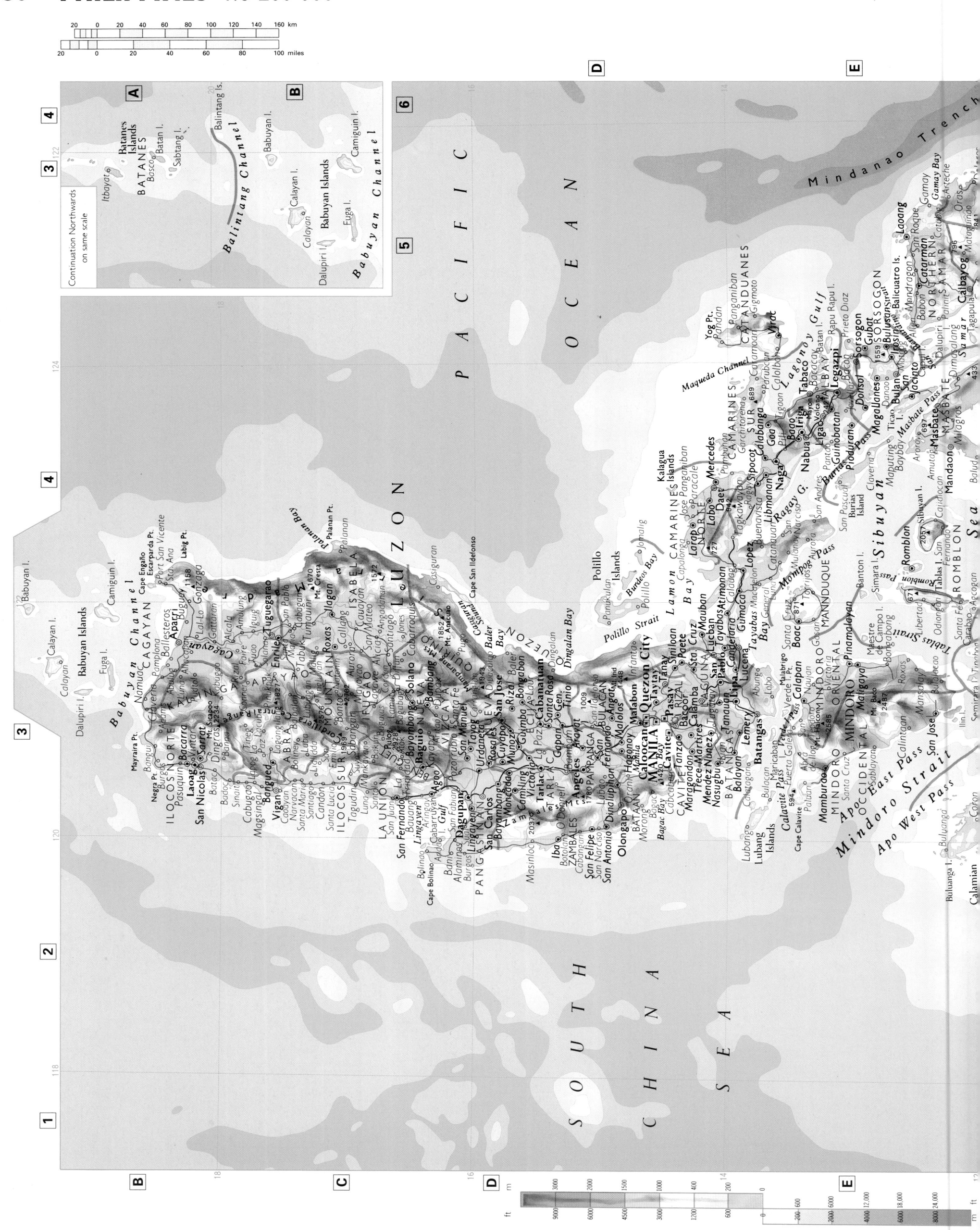

20 0 20 40 60 80 100 120 140 160 km
20 0 20 40 60 80 100 miles

A **B** **C** **D** **E**

Continuation Northwards
on same scale

Batanes Islands
Batan I.
Sabtang I.
Boscoi
Itbayat
BATANES
Balintang Is.
Balintang Channel
Camiguin I.
Babuyan I.
Babuyan Islands
Calayan I.
Fuga I.
Dalupiri
Babuyan Channel

P A C I F I C   O C E A N

Mindanao Trench

Babuyan I.
Calayan I.
Camiguin I.
Babuyan Channel
Babuyan Islands
Fuga I.
Dalupiri I.
Calayan

L U Z O N

Palanan Bay
Palanan Pt.
Palanan

Q U E Z O N

Polillo
Polillo Islands
Burdeos Bay
Polillo Strait
Johalig

CATANDUANES
Panganiban
Pandon
Yog Pt.

Gamay Bay
Laoang
NORTHERN
SAMAR
Catubig
Oras
Arteche

Maqueda Channel
CAMARINES
NORTE
Daet
Mercedes
Labo
Calabanga
Naga
Nabua
Iriga
Tabaco
Legazpi
ALBAY
Sorsogon
Donsol
Magallanes
Bulusan
Juban
MASBATE
Calbayog
SAMAR

Lamon Bay
Lagonoy Gulf

Sibuyan Sea

S O U T H

C H I N A

S E A

m
ft
9000 6000 4500 3000 1200 600 0
3000 2000 1500 1000 400 200 0
-200 -2000 -4000 -6000 -8000
600 6000 12000 18000 24000

B **C** **D** **E**

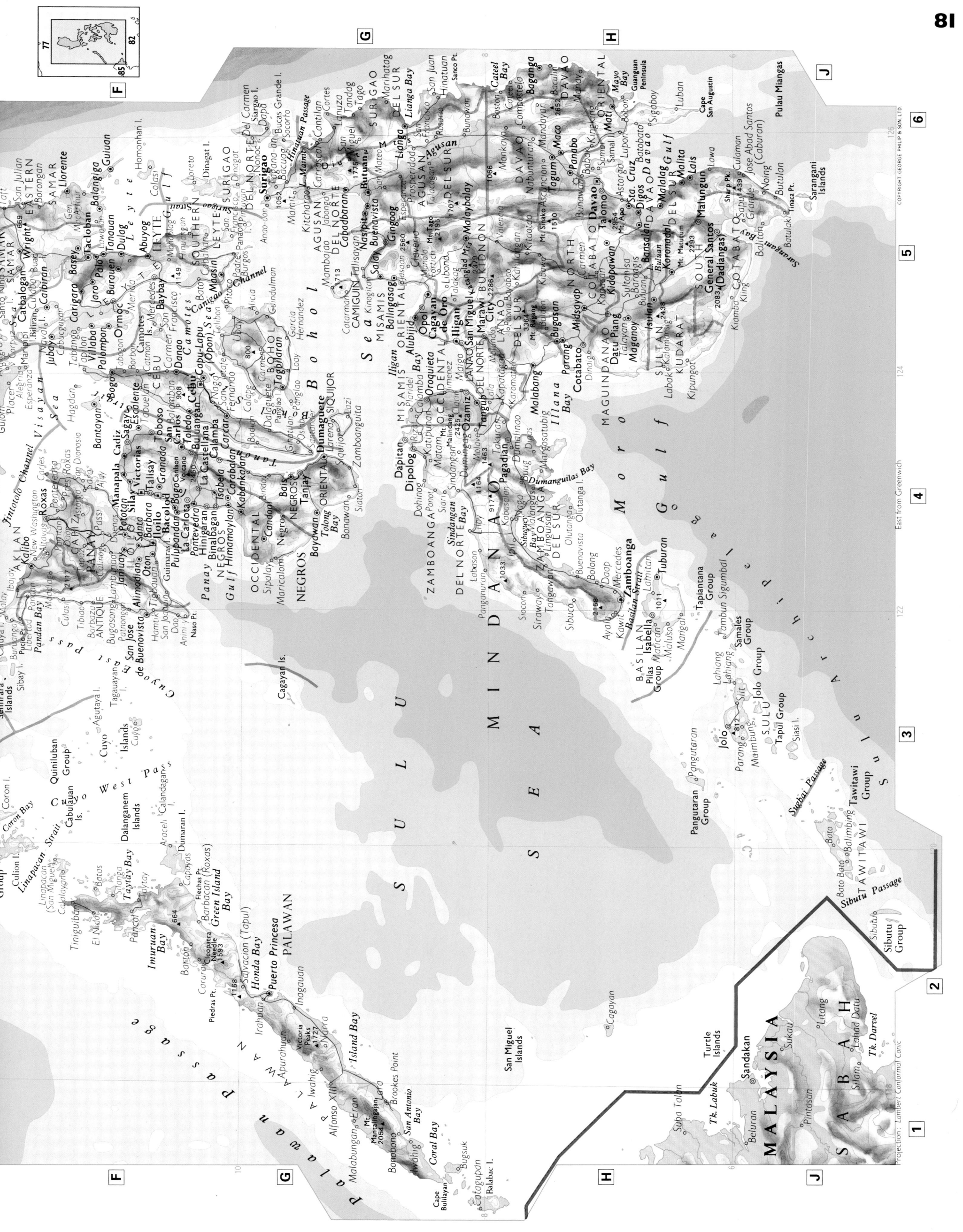

S U L U
S E A
M I N D A N A O
Moro Gulf
Sulu Archipelago
Bohol Sea
Camotes Sea
Visayan Sea
Palawan Passage
Cuyo East Pass
Cuyo West Pass
Jintotolo Channel
Canigao Channel
Hinatuan Passage
Basilan Strait
Sibutu Passage
Sugbai Passage
Leyte Gulf
Davao Gulf
Illana Bay
Panay Gulf
PALAWAN
Puerto Princesa
Honda Bay
Island Bay
Coral Bay
San Antonio Bay
Imuruan Bay
Taytay Bay
Green Island Bay
Victoria Peaks 1727
Mt. Mantalingajan 2054
Cleopatra Needle 1593
Cape Bulilayan
Balabac I.
Cuyo Islands
Cagayan Is.
Dumaran I.
Quiniluban Group
Cabulauan Is.
Dalanganem Islands
Agutaya I.
Tagauayan I.
San Miguel Islands
Cagayan
MALAYSIA
S A B A H
Sandakan
Tk. Labuk
Tk. Darvel
Turtle Islands
PANAY
ANTIQUE
AKLAN
CAPIZ
ILOILO
Iloilo
Roxas
Kalibo
San Jose de Buenovista
NEGROS OCCIDENTAL
NEGROS ORIENTAL
Bacolod
Dumaguete
SIQUIJOR
CEBU
Cebu
BOHOL
Tagbilaran
LEYTE
Tacloban
Ormoc
SOUTHERN LEYTE
Maasin
SAMAR
EASTERN SAMAR
Catbalogan
Dinagat I.
Siargao I.
SURIGAO DEL NORTE
Surigao
SURIGAO DEL SUR
AGUSAN DEL NORTE
Butuan
AGUSAN DEL SUR
CAMIGUIN
MISAMIS ORIENTAL
Cagayan de Oro
MISAMIS OCCIDENTAL
Ozamiz
ZAMBOANGA DEL NORTE
Dipolog
Dapitan
ZAMBOANGA DEL SUR
Zamboanga
Pagadian
LANAO DEL NORTE
Iligan
LANAO DEL SUR
Marawi City
BUKIDNON
Malaybalay
DAVAO
Davao
DAVAO ORIENTAL
DAVAO DEL SUR
NORTH COTABATO
MAGUINDANAO
Cotabato
SULTAN KUDARAT
SOUTH COTABATO
General Santos (Dadiangas)
Sarangani Bay
Sarangani Islands
BASILAN
Isabella
SULU
Jolo
TAWITAWI
Tawitawi Group
Sibutu Group
Pangutaran Group
Tapul Group
Pulau Miangas
Cape San Augustin
Mt. Apo 2954
East from Greenwich
Projection: Lambert Conformal Conic
COPYRIGHT GEORGE PHILIP & SON LTD

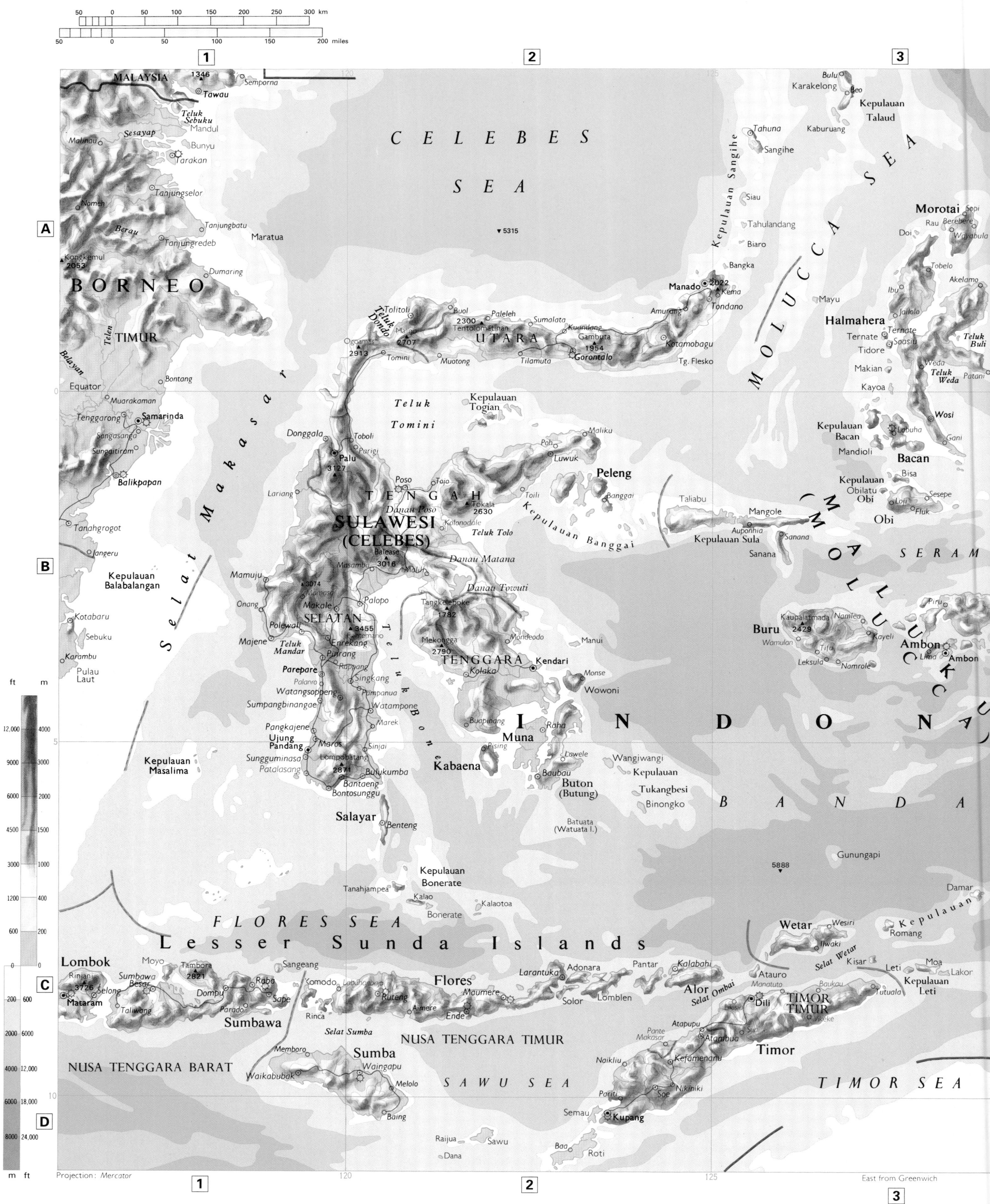

CELEBES SEA
MOLUCCA SEA
BANDA
FLORES SEA
SAWU SEA
TIMOR SEA
Selat Makasar
Lesser Sunda Islands
INDONESIA
BORNEO
TIMUR
SULAWESI (CELEBES)
UTARA
TENGAH
SELATAN
TENGGARA
MALUKU (MOLUCCAS)
NUSA TENGGARA TIMUR
NUSA TENGGARA BARAT
TIMOR TIMUR
MALAYSIA
Teluk Tomini
Teluk Bone
Halmahera
Morotai
Bacan
Obi
Buru
Ambon
Seram
Peleng
Kepulauan Banggai
Kepulauan Sula
Muna
Buton (Butung)
Kabaena
Salayar
Flores
Sumbawa
Lombok
Sumba
Timor
Wetar
Alor
Kepulauan Talaud
Kepulauan Sangihe
Manado
Ujung Pandang
Kendari
Palu
Kupang
Dili
Mataram
Samarinda
Balikpapan
Equator
Projection: Mercator
East from Greenwich

PACIFIC

OCEAN

Tobi

Helen
Atoll

Kepulauan
Asia

Kepulauan
Mapia

Kepulauan
Ayu

Gebe
Umera
Selpele
Kabarai
Wakre
Waigeo
4625
Equator

Gam
Saonek
Waibeem
Kwoka
3000
Kaironi
Warsa
Korim
Supiori

Kepulauan Raja Ampat
Batanta
Selat Dampier
Sorong
Jazirah Doberai (Vogelkop)
Manokwari
Biak
Bosnik

Salawati
Klamono
Namber
Numfoor
Biak
Kepulauan Padaido

Kofiau
Sailolof
Seget
Wersar
3100
Ransiki
Num
Selat Yapen
Tg. D'Urville
Mataboor
Kepulauan Kumamba

Adua
Ler nalu
Mogoi
Wariap
Yapen
Serui
Bonoi
Saberania
Sarmi
Ansudu

Misool
Bira
Wasian
Bintuni
Teluk
Nuboai
Barapasi

S E A
Teluk Berau
Tg. Fatagar
Saga
Babo
Wendesi
Wasior
Cenderawasih
Nabire
Tariku
Pegunungan Van Rees
Genyem
●Jayapura

Wahai
Kokas
Susunu
Kwatisore
Santani
Krau

Sawai
Binaiya
Bula
Fakfak
Weri
Wenut
Ibonma
Koimana
I R I A N   J A Y A
Taritatu

Masohi
3019
Waru
Karufa
Teluk Kamrau
Enarotali
Pegunungan
Puncak Jaya 5029
Puncak Trikora 4756
Wamena

Amahai
Haya
Geser
Manggawitu
Adi
Wanapiri
Uta
Waghete
Puncak Jaya
Pegunungan Sudirman
Pegunungan Jayawijaya
Mandala 4702

Seram (Ceram)
Kepulauan Gorong
Yapero

Kepulauan
Bandanaira
EBanda
Kepulauan Watubela
Wanapiri

Kepulauan Kai
Har
Gumzai
Kola
Teluk Flamingo
Agats
Pulau
Mindiptana

Tual
Kai Besar
Dobo
Wokam
Pirimapun
Tanahmerah

7440
Kai Kecil
Bandar Elat
Sewer
Kepulauan Aru
Kepi
Asike

Barat
Daya
Serua
Molu
Wangai
Maikoor
Kobroor
Kassue
Bade

Nila
Rebi
Koba
Muting

Teun
Trangan
Tafermaar
Gomogomo
Pulau
Yos Sudarso
Kimaän
Okaba

Tepa
Wuliaru
Larat
Tg. Ngabordamlu
Tg. Vals
Pulau Komoran
Merauke

Babar
Selu
Alusi
Yamdena

Sermata
Masela
Saumlaki
Adaut
Selaru
Eliase
Kepulauan Tanimbar

A R A F U R A   S E A

P A P U A   N E W   G U I N E A

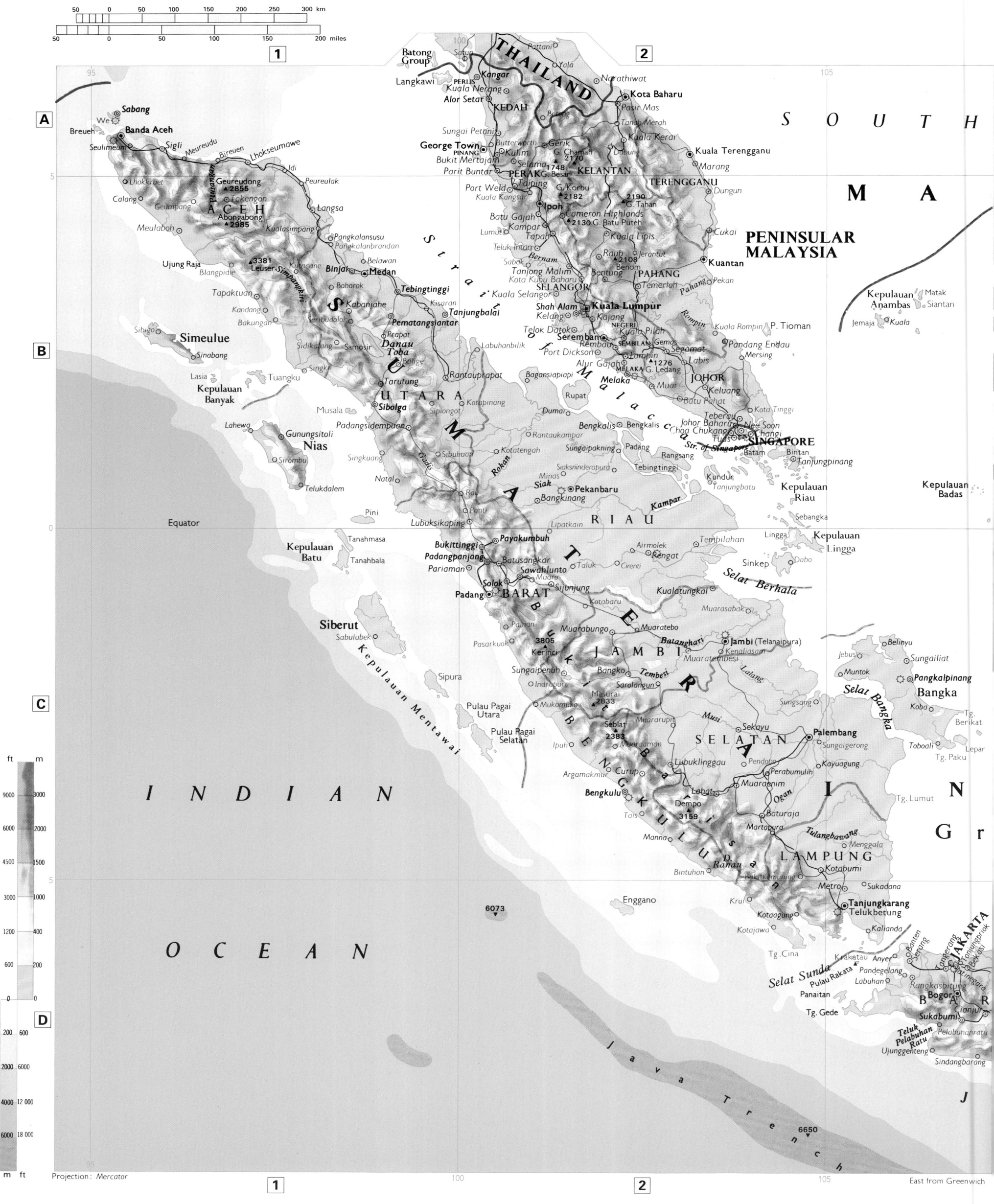

50 0 50 100 150 200 250 300 km
50 0 50 100 150 200 miles
1
2
A
B
C
D
THAILAND
Batong Group
Langkawi
Satun
Pattani
Yala
Narathiwat
PERLIS
Kangar
Kuala Nerang
Alor Setar
KEDAH
Kota Baharu
Pasir Mas
Tanah Merah
Kuala Krai
Sungai Petani
George Town
PINANG
Butterworth
Bukit Mertajam
Kulim
Gerik
G. Chamah
2170
1748
Selama
Parit Buntar
PERAK
G. Besar
KELANTAN
Kuala Terengganu
Marang
TERENGGANU
Dungun
Port Weld
Kuala Kangsar
Taiping
G. Korbu
2182
Ipoh
2190
G. Tahan
Batu Gajah
Kampar
Cameron Highlands
2130 G. Batu Puteh
Lumut
Tapah
Kuala Lipis
Teluk Intan
Bernam
Sabak
Raub
Jerantut
2108
Benom
Tanjong Malim
Kota Kubu Baharu
Bentung
PAHANG
SELANGOR
Temerloh
Pahang
Pekan
Kuantan
Cukai
PENINSULAR MALAYSIA
Kuala Selangor
Shah Alam
Kuala Lumpur
Kelang
Kajang
NEGERI SEMBILAN
Kuala Pilah
Telok Datok
Seremban
Port Dickson
Rembau
Gemas
Segamat
Rompin
Kuala Rompin
P. Tioman
Pandang Endau
Mersing
Alur Gajah
MELAKA
Melaka
1276
G. Ledang
Tampin
Labis
JOHOR
Keluang
Muar
Batu Pahat
Kota Tinggi
Johor Baharu
Teberau
Nee Soon
Changi
Choa Chukang
Tuas
SINGAPORE
Str. of Singapore
Batam
Bintan
Tanjungpinang
S O U T H
M A
Kepulauan Anambas
Matak
Siantan
Jemaja
Kuala
Kepulauan Badas
Sabang
We
Breueh
Banda Aceh
Seulimeum
Sigli
Meureudu
Bireuen
Lhokseumawe
Idi
Peureulak
Lhokkruet
Pasaman
Geureudong
2855
Takengon
Calang
Geumpang
ACEH
Abongabong
2985
Langsa
Meulaboh
Kualasimpang
Pangkalansusu
Pangkalanbrandan
Belawan
Ujung Raja
Blangpidie
3381
Leuser
Kutacane
Simpangkiri
Binjai
Medan
Tapaktuan
Boborok
Tebingtinggi
Kabanjahe
Kisaran
Tanjungbalai
Kandang
Bakungan
Seribudolok
Pematangsiantar
Sibigo
Simeulue
Sinabang
Sidikalang
Samosir
Prapat
Danau Toba
Balige
Labuhanbilik
Lasia
Tuangku
Kepulauan Banyak
Singkil
Tarutung
Rantauprapat
Bagansiapiapi
Rupat
UTARA
Musala
Sibolga
Sipiongot
Kotapinang
Dumai
Bengkalis
Lahewa
Gunungsitoli
Nias
Padangsidempuan
Rantaukampar
Sungaipakning
Padang
Rangsang
Sirombu
Singkuang
Sibuhuan
Gadis
Rokan
Kotatengah
Siaksriinderapura
Tebingtinggi
Kundur
Tanjungbatu
Kepulauan Riau
Natal
Minas
Siak
Pekanbaru
Bangkinang
Kampar
Telukdalem
Pini
Rau
Panti
Sebangka
Equator
Lubuksikaping
Lipatkain
RIAU
Lingga
Kepulauan Lingga
Tanahmasa
Kepulauan Batu
Tanahbala
Bukittinggi
Payakumbuh
Padangpanjang
Batusangkar
Airmolek
Rengat
Tembilahan
Sinkep
Dabo
Pariaman
Sawahlunto
Taluk
Cirenti
Selat Berhala
Solok
Muaro
Sijunjung
Padang
BARAT
Kualatungkal
Kotabaru
Muarasabak
Siberut
Sabulubek
Painan
Muarabungo
Muaratebo
Batanghari
Jambi (Telanaipura)
Pasarkuok
3805
Kerinci
JAMBI
Kenaliasam
Muaratembesi
Belinyu
Jebus
Sungailiat
Muntok
Pangkalpinang
Bangka
Sungaipenuh
Bangko
Tembesi
Lalang
Kepulauan Mentawai
Sipura
Indrapura
Sarolangun
Selat Bangka
Masurai
2833
Mukomuko
Koba
Tg. Berikat
Pulau Pagai Utara
Musi
Sungsang
Muararupit
Sekayu
Palembang
Pulau Pagai Selatan
Seblat
2383
Muaraaman
SELATAN
Sungaigerong
Toboali
Lepar
Tg. Paku
Ipuh
Lubuklinggau
Pendopo
Kayuagung
Argamakmur
Curup
Perabumulih
Bengkulu
Lahat
Muaraenim
Ogan
Tg. Lumut
Dempo
3159
Baturaja
Tais
Martapura
Tulangbawang
Manna
Menggala
LAMPUNG
Ranau
Kotabumi
Bintuhan
Bukitkemuning
Sukadana
Metro
Enggano
Krui
Tanjungkarang
Telukbetung
6073
Kotaagung
Kotajawa
Kalianda
Tg. Cina
INDIAN
OCEAN
Selat Sunda
Krakatau
Pulau Rakata
Anyer
Bonten
Serang
Tangerang
JAKARTA
Tanjungpriok
Bekasi
Pandeglang
Labuhan
Cikampek
Rangkasbitung
Panaitan
Tg. Gede
Bogor
Cianjur
Sukabumi
Teluk Pelabuhan Ratu
Pelabuhanratu
Ujunggenteng
Sindangbarang
Java Trench
6650
N
G
r
I
J
B
A
R
ft m
9000 3000
6000 2000
4500 1500
3000 1000
1200 400
600 200
0 0
200 600
2000 6000
4000 12 000
6000 18 000
m ft
95
100
105
5
0
Projection: Mercator
East from Greenwich

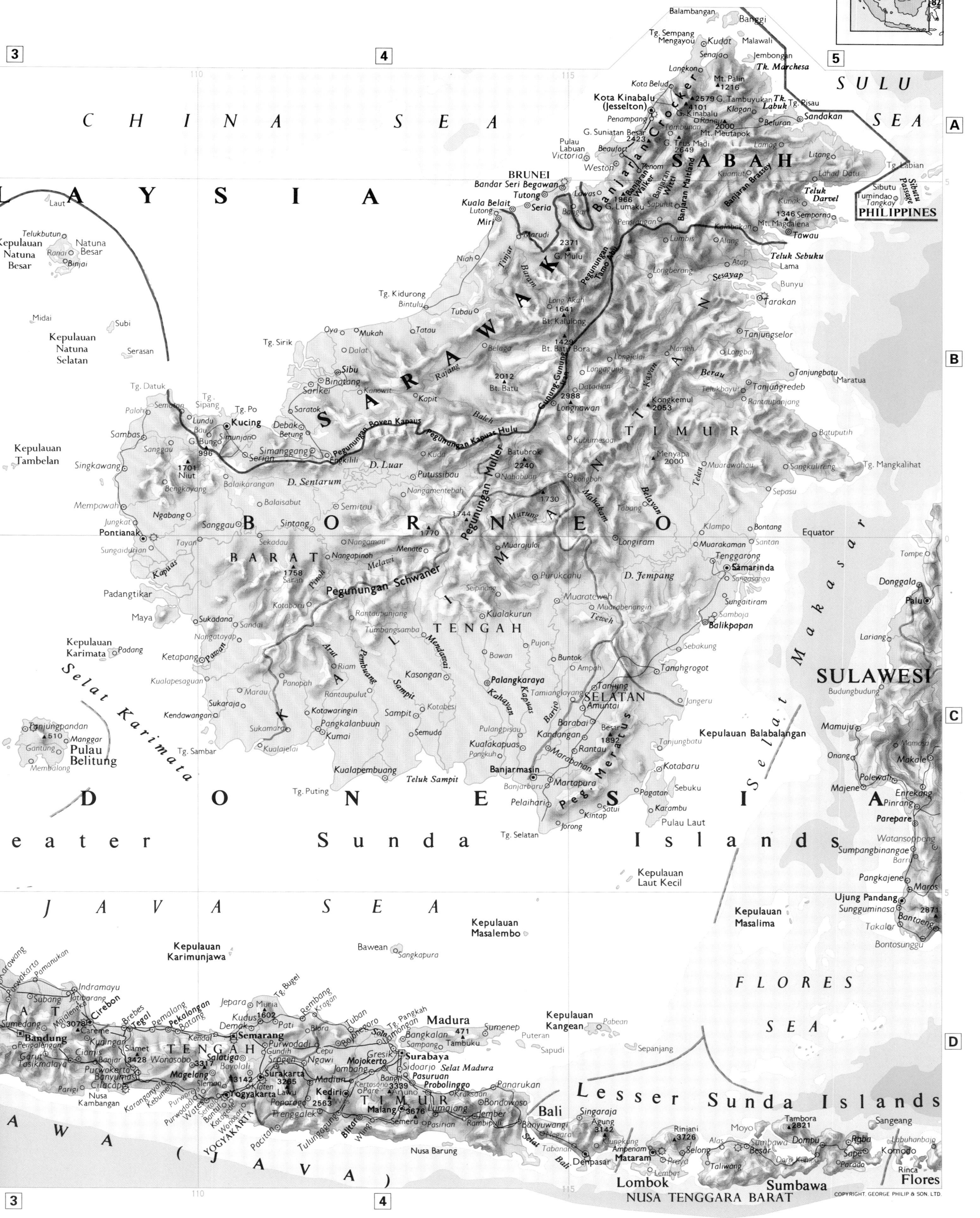

C H I N A   S E A

L A Y S I A

Telukbutun
Kepulauan
Natuna
Besar
Ranai
Binjai

Midai   Subi

Serasan

Kepulauan
Natuna
Selatan

Kepulauan
Tambelan

Tg. Datuk
Tg. Sipang
Tg. Po
Semotan
Lundu
Kucing
Bau
Simunjan
G. Bungo
996
1701
Niut
Mempawah
Ngabang
Jungkat
Pontianak
Kapuas
Sungaidurian

Kepulauan
Karimata
Padang
Padangtikar
Maya
Sukadana
Sandai

SABAH
Balambangan   Banggi
Tg. Sempang   Malawali
Mengayou   Kudat
Senajao   Jembongan
Langkon   Tk. Marchesa
Kota Belud   Mt. Palin ▲1216
Kota Kinabalu   G. Tambuyukon
(Jesselton)   2579   4101   Klagan
Penampang   G. Kinabalu   Ranau   Beluran   Sandakan
G. Suniatan Besar   2000   Tombunan
Mt. Meutapok   Lamog
2423   Trus Madi   Kuamut   Litang   Tg. Labian
Pulau   2649   Walker   Lahad Datu
Labuan   Beaufort   Tenom   Lumbis   Alang   Teluk
Victoria   Weston   1966   Sapulut   Darvel
BRUNEI   Lawas   G. Lumaku   1346 Semporna
Bandar Seri Begawan   Bangar   Mt. Magdalena
Tutong   Tawau
Kuala Belait   Teluk Sebuku
Lutong   Seria   Lama
Miri   Marudi

S A R A W A K
Niah
Tg. Kidurong
Bintulu   2371
G. Mulu
Tubau   Long Akah
1641
Oya   Mukah   Tatau   Belaga
Dalat   Bt. Kalulong
Sibu   1429
Binatang   Bt. Batu Bora
Sariker   2012   Longnawan
Saratok   Bt. Batu   2988
Kanowit   Longiram
Kapit   Batubrok
Debak   Rajang   Datadian
Betung   Baleh   Longboh   Kongkemul
Serian   Pegunungan Kapuas Hulu   2053
Simanggang   Kuda   Nahabuan   Menyapa
Engkilili   Putussibau   1730   2000
D. Luar   2240
Balaikarangan   D. Sentarum   1744
Balaisabut   Nangamentebah   1770
Semitau   Mutung   Longiram

K A L I M A N T A N   T I M U R

Berau
Tanjungselor
Tarakan
Nameh   Longbai
Tanjungbatu   Maratua
Telukbayur
Rantaupanjang
Batuputih
Tg. Mangkalihat
Muarawahau   Sangkulirang
Sepasu
Klampo   Bontang
Muarakaman   Santan
Tenggarong   Equator   Tompe
Samarinda   Donggala
Sebulu   Palu
Sungaitiram
Balikpapan   SULAWESI
Sebakung   Budungbudung
Tanahgrogot

B O R N E O

BARAT
Sintang   Nangapinoh
Sekadau   Nangamau
Tayan   Menate   1758
Sanggau   Saran
Nangatayap   Melawi
Pegunungan Schwaner   Seipinang
Kotaboru   Rantaupanjang
Tumbangsamba   Kualakurun
Rantaupulut   Mendawai   Teweh
Panopah   Bawan   Muarateweh
Riam   Buntok   Muaradenangin
Marau   Sampit   Ampah
Kendawangan   Kasongan   Pujon
Sukaraja   Kotabesi   Tamianglayang
Sukamara   Semuda   Tanjung
Kotawaringin   Pulangpisau   Amuntai
Pangkalanbuun   Kualakapuas   Barabai
Kumai   Pangkuh   Kandangan   Kotabaru
Kualajelai   Kualapembuang   Banjarmasin   Rantau
Tg. Sambar   Teluk Sampit   Banjarbaru   Marabahan
Tg. Puting   Martapura   Pagatan
Pelaihari   Sebuku
Kintap   Karambu
Jorong   Pulau Laut
Tg. Selatan

TENGAH

SELATAN
Besar
1892
Peg. Meratus

Kepulauan
Balabalangan

Kepulauan
Balalangan

Mamuju
Mamasa
Onang   Makale
Majene   Polewali
Enrekang
Pinrang
Parepare
Watansoppeng
Sumpangbinangae   Barru
Pangkajene
Maros
Ujung Pandang
Sungguminasa   2871
Takalar   Bontaeng
Bontosunggu

D O N E S I A

Tanjungpandan   510
Gantung   Manggar
Pulau   Belitung
Membalong

Selat Karimata

eater   Sunda   Islands

J A V A   S E A

Kepulauan
Laut Kecil

Kepulauan
Masalima

F L O R E S

S E A

Kepulauan
Masalembo
Bawean   Sangkapura

Kepulauan
Karimunjawa

J A V A   S E A

Karawang   Indramayu
Purwakarta   Pamanukan
Subang   Jatibarang   Cirebon
Sumedang   Majalengka   Brebes   Tegal   Pekalongan
3078   Careme   Pemalang   Batang
Bandung   Kuningan   Kendal   Kudus 1602
Pemalengan   Ciamis   Slamet   Purwodadi   Demak   Muria
Garut   Banjar   3428   Wonosobo   Gundih   Blora
Tasikmalaya   Purwokerto   3317   Salatiga   Sragen
Parigi   Banyumas   Magelang   Boyolali   Madiun
Nusa   Karangnunggal   Kebumen   3265   Klaten   Lawu
Kambangan   Purworejo   Sleman   3142   Surakarta   3339
Cilacap   Wates   Yogyakarta   Poporogo   2563
Bantul   YOGYAKARTA   Wonogiri
Pacitan   Trenggalek

TENGAH

Jepara
Rembang
Rembang   Kragan
Pati   Tuban
Blora
Cepu   Bojonegoro   Tg. Pangkah
Ngawi   Lamongan   Gresik
Jombang   Mojokerto   Surabaya
Kertosono   Bangil   Pasuruan
Kediri   Sidoarjo   Probolinggo
Pare   Malang   Lumajang
Blitar   3676   Pasirian
Tulungagung   Semeru
Wlingi   Rambipuji
Kragan

Madura
Bangkalan   471
Sampang   Tambuku
Sumenep
Pamekasan   Pabean
Selat Madura   Sapudi

T I M U R
Kepulauan
Kangean

Panarukan
Kraksaan
Bondowoso
Jember
Banyuwangi   Bali
Negara   Singaraja
Pasian   Agung
Bali   3142
Denpasar   Rinjani
Mataram   3726
Lombok   Ampenam
Praya   Selong
Kuta   Tabanan
Nusa Barung

Sumbawa
Besar   Tambora   2821
Moyo   Dompu   Raba
Sumbawa   Sape
Alas   Dara   Kore
Taliwang   Parado
Labuanbajo
Komodo
Sangeang
Flores
Rinca

L e s s e r   S u n d a   I s l a n d s

Bali

NUSA TENGGARA BARAT

A W A   (J A V A)

SULU
SEA

PHILIPPINES

Selat Makasar

50 0 50 100 150 200 km
50 0 50 100 150 miles
A
B
C
D
E
F
1
2
3
4
5
6
7
8
HAINAN
G. of Tonkin
HANOI
Haiphong
Vientiane
BANGKOK
RANGOON
Mandalay
Moulmein
Da Nang
Hue
Nanning
Zhanjiang
Haikou
CAMBODIA
THAILAND
LAOS
VIETNAM
BURMA (MYANMAR)
YUNNAN
GUANGXI ZHUANGZU ZIZHIQU
GUANGDONG
SHAN
KAYAH
KAWTHULE
TENASSERIM
Mekong
Salween
Phnom Dangrek
Cao Nguyen Khorat
Tonlé Sap
ANGKOR
G. of Martaban
Moscos Islands

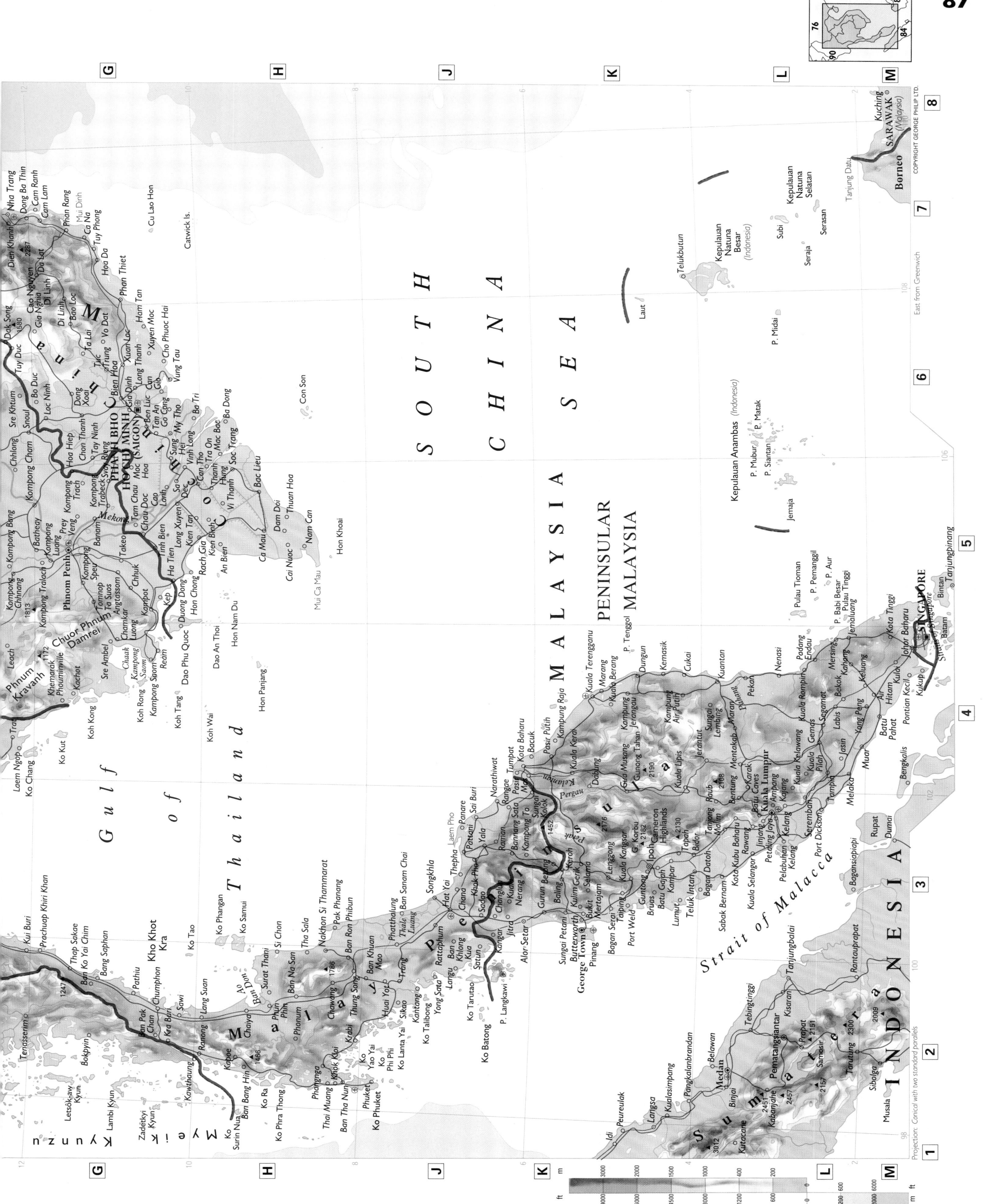

Gulf of Thailand
SOUTH CHINA SEA
MALAYSIA
PENINSULAR MALAYSIA
INDONESIA
Strait of Malacca
Sumatera
Malay Peninsula
Cochin China
Myeik Kyunzu
Kho Khot Kra
Phnom Penh
PHANH BHO HO CHI MINH (SAIGON)
Kuala Lumpur
SINGAPORE
Straits of Singapore
George Town
Medan
Ipoh
Borneo
SARAWAK (Malaysia)
Kuching
Tenasserim
Kui Buri
Prachuap Khiri Khan
Thap Sakae
Ban Ko Yai Chim
Bang Saphan
Pathiu
Chumphon
Ban Pak Chan
Kra Buri
Sawi
Lang Suan
Ranong
Kawthaung
Letsôk-aw Kyun
Bokpyin
Lambi Kyun
Zadetkyi Kyun
Ko Surin Nua
Kapoe
Ban Bang Hin
Chaiya
Ao Ban Don
Surat Thani
Ko Ra
Ko Phra Thong
Phun Phin
Ban Na San
Phanom
Si Chon
Tha Sala
Nakhon Si Thammarat
Pak Phanang
Ban Ron Phibun
Phangnga
Thai Muang
Khok Kloi
Chawang
Thung Song
Ban Tha Nun
Krabi
Phuket
Ko Phuket
Ko Yao Yai
Ko Phi Phi
Ko Lanta Yai
Huai Yot
Sikao
Trang
Ban Khuan Mao
Phatthalung
Thale Luang
Ban Sanam Chai
Kantang
Ko Talibong
Yong Sata
Rattaphum
Songkhla
Hat Yai
Langu
Ban Khlong Kua
Satun
Ko Tarutao
Ko Batong
P. Langkawi
Thepha
Laem Pho
Chana
Pattani
Panare
Sai Buri
Khok Pho
Yala
Sadao
Changlun
Kangar
Jitra
Alor Setar
Kuala Nerang
Raman
Narathiwat
Rangae
Bannang Sata
Kampong To
Pasir Mas
Tumpat
Kota Baharu
Bacuk
Sungai Kolok
Gurun
Betong
Baling
Keroh
Sungai Petani
Butterworth
Kulim
Gerik
Selama
Pinang
Bukit Mertajam
Pasir Putih
Kampung Raja
Kuala Krai
Kelantan
Pergau
Perak
Dabung
Kuala Terengganu
Marang
Kuala Berang
P. Tenggol
Bagan Serai
Taiping
Lenggong
Kuala Kangsar
Port Weld
G. Korbu
Gua Musang
Gunong Tahan
Kampung Jerangau
Dungun
Guntong
Bruas
Batu Gajah
Kampar
Cameron Highlands
Lumut
Tapah
Bidor
Teluk Intan
Kuala Lipis
Kampung Air Putih
Kemasik
Cukai
Jerantut
Bagan Datoh
Tanjong Malim
Raub
Sungai Lembing
Kuantan
Sabak Bernam
Kota Kubu Baharu
Bentung
Mentakab
Maran
Pahang
Pekan
Rawang
Kuala Selangor
Karak
Batu Caves
Jinjang
Petaling Jaya
Ampang
Kajang
Pelabuhan Kelang
Kelang
Nenasi
Kuala Kelawang
Kuala Rompin
Padang Endau
Seremban
Port Dickson
Kuala Pilah
Gemas
Segamat
Tampin
Jasin
Labis
Bekok
Mersing
Kahang
Jemaluang
Melaka
Muar
Yong Peng
Keluang
Batu Pahat
Air Hitam
Kulai
Kota Tinggi
Pontian Kecil
Johor Baharu
Kukup
Batam
Bintan
Tanjungpinang
Pulau Tioman
P. Pemanggil
P. Aur
P. Babi Besar
Pulau Tinggi
Bengkalis
Rupat
Dumai
Bagansiapiapi
Idi
Peureulak
Langsa
Kualasimpang
Pangkalanbrandan
Belawan
Binjai
Kutacane
Tebingtinggi
Kabanjahe
Pematangsiantar
Kisaran
Tanjungbalai
Prapat
Samosir
Tarutung
Rantauprapat
Sibolga
Musala
Laem Ngop
Ko Chang
Trat
Ko Kut
Koh Kong
Phnum Kravanh
Khemarak Phouminville
Kachot
Leach
Chuor Phnum Damrei
Sre Ambel
Koh Rong
Chaak Kampong Saom
Kampong Saom
Ream
Koh Tang
Dao Phu Quoc
Koh Wai
Dao An Thoi
Hon Nam Du
Hon Panjang
Chamkar Luong
Kampot
Kep
Ha Tien
Duong Dong
Hon Chong
Rach Gia
Kien Binh
An Bien
Kompong Chhnang
Kompong Bang
Kompong Tralach
Batheay
Kompong Luong
Prey Veng
Kompong Speu
Tomnop Ta Suos
Angtassom
Chhuk
Takeo
Banam
Mekong
Kompong Trabeck
Svay Rieng
Kompong Trach
Chhlong
Kompong Cham
Sre Khtum
Snoul
Tuy Duc
Dak Song
Bo Duc
Loc Ninh
Hoa Hiep
Chon Thanh
Tay Ninh
Dong Xoai
Tam Chau
Chau Doc
Moc Hoa
Tinh Bien
Long Xuyen
Cao Lanh
Kien Tan
Sa Dec
Sung Hei
My Tho
Vinh Long
Can Tho
Tra On
Mac Bac
Thanh Hung
Vi Thanh
Soc Trang
Ba Dong
Ba Tri
Bac Lieu
Ca Mau
Dam Doi
Thuan Hoa
Cai Nuoc
Nam Can
Mui Ca Mau
Hon Khoai
Con Son
Ben Luc
Tan An
Go Cong
Can Gio
Vung Tau
Gia Dinh
Bien Hoa
Long Thanh
Xuan Loc
Xuyen Moc
Cho Phuoc Hai
Ham Tan
Phan Thiet
Tuc Trung
Vo Dat
Ta Lai
Di Linh
Bao Loc
Gia Nghia
Cao Nguyen Di Linh
Da Lat
Dien Khanh
Nha Trang
Dong Ba Thin
Cam Ranh
Cam Lam
Phan Rang
Mui Dinh
Ca Na
Tuy Phong
Hoa Da
Cu Lao Hon
Catwick Is.
Laut
Telukbutun
Kepulauan Natuna Besar (Indonesia)
Kepulauan Anambas (Indonesia)
P. Mubur
P. Matak
P. Siantan
Jemaja
P. Midai
Subi
Seraja
Serasan
Kepulauan Natuna Selatan
Tanjung Datu
East from Greenwich
Projection: Conical with two standard parallels
COPYRIGHT GEORGE PHILIP LTD.

6 7 8 9

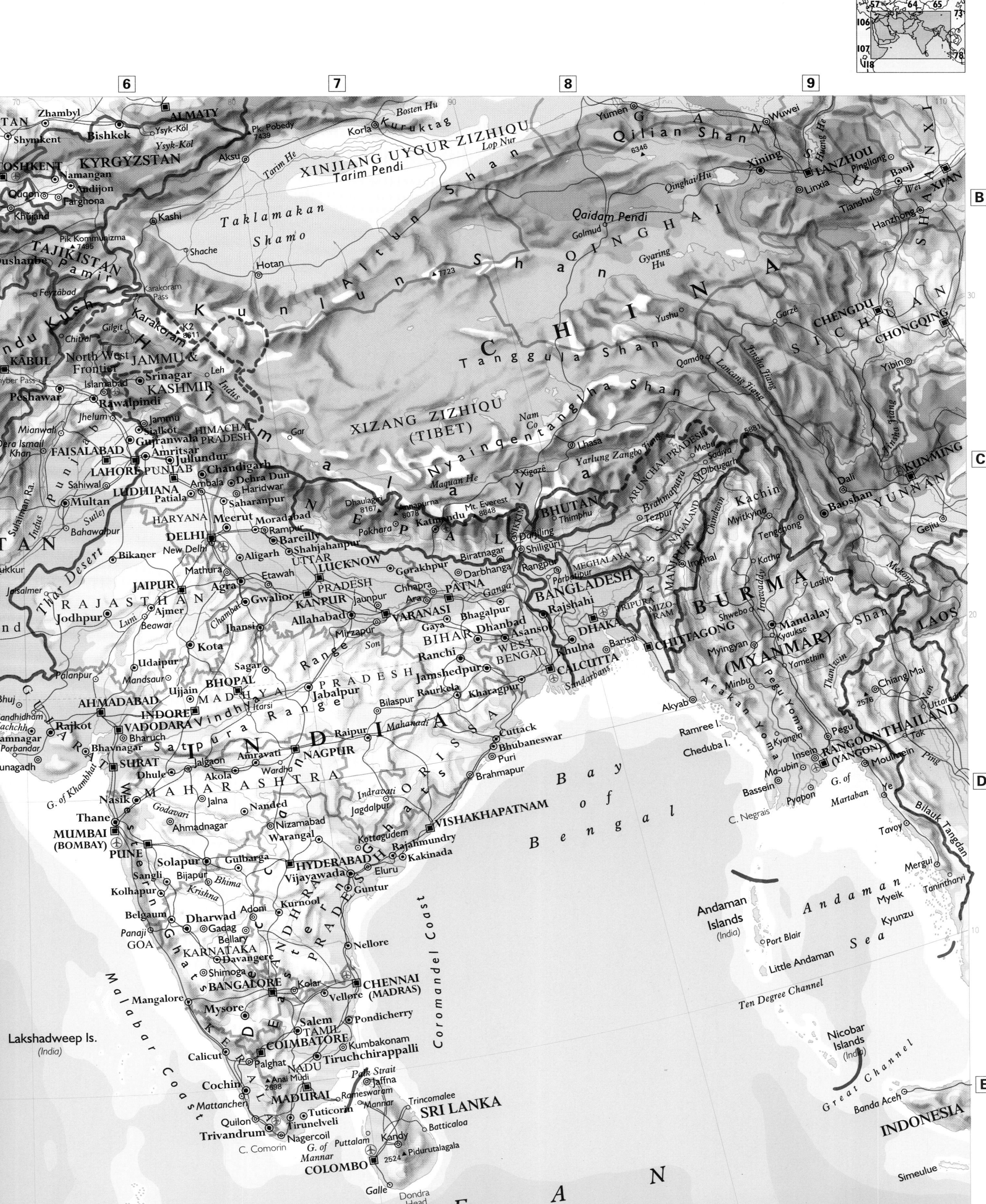

STAN
Zhambyl
ALMATY
Shymkent
Bishkek
Ysyk-Köl
KYRGYZSTAN
Pk. Pobedy 7439
TOSHKENT
Namangan
Andijon
Farghona
Quqon
Khujand
Aksu
Tarim He
Korla
Kuruktag
Bosten Hu
XINJIANG UYGUR ZIZHIQU
Tarim Pendi
Lop Nur
Yumen
Qilian Shan
Wuwei
Xining
LANZHOU
Huang He
Pingliang
Baoji
XI'AN
Linxia
Qinghai Hu
Tianshui
Wei
Hanzhong
Kashi
Taklamakan Shamo
Pik Kommunizma 7495
TAJIKISTAN
Dushanbe
Pamir
Shache
Hotan
Qaidam Pendi
Golmud
QINGHAI
Gyaring Hu
Kunlun Shan
Altun Shan
7723
6346
Feyzabad
Karakoram Pass
Karakoram
K2 8611
Hindu Kush
Gilgit
Chitral
KABUL
Khyber Pass
North West Frontier
JAMMU & KASHMIR
Srinagar
Leh
Indus
Islamabad
Peshawar
Rawalpindi
Jhelum
Jammu
Sialkot
Gujranwala
Mianwali
Dera Ismail Khan
FAISALABAD
Amritsar
LAHORE
PUNJAB
Jullundur
HIMACHAL PRADESH
Chandigarh
Dehra Dun
Haridwar
Sahiwal
LUDHIANA
Ambala
Patiala
Saharanpur
Multan
Sulaiman Ra.
Bahawalpur
HARYANA
Meerut
Moradabad
Rampur
Bareilly
Shahjahanpur
DELHI
New Delhi
Aligarh
Mathura
Agra
Etawah
UTTAR PRADESH
LUCKNOW
Gorakhpur
Bikaner
Sukkur
Thar Desert
Jaisalmer
RAJASTHAN
JAIPUR
Ajmer
Beawar
Jodhpur
Luni
Gwalior
Chambal
Jhansi
KANPUR
Jaunpur
Allahabad
VARANASI
Mirzapur
Son
Kota
Udaipur
Palanpur
Mandsaur
Sagar
Ujjain
BHOPAL
MADHYA PRADESH
Jabalpur
Indore
Itarsi
Vindhya Range
Bhuj
Kachchh
Gandhidham
Rajkot
Jamnagar
Porbandar
Junagadh
GUJARAT
AHMADABAD
VADODARA
Bharuch
Bhavnagar
SURAT
G. of Khambhat
Satpura Range
Jalgaon
Amravati
NAGPUR
Wardha
Dhule
Akola
Nasik
MAHARASHTRA
Thane
MUMBAI (BOMBAY)
PUNE
Jalna
Godavari
Ahmadnagar
Nanded
Nizamabad
Warangal
Solapur
Gulbarga
HYDERABAD
Vijayawada
Guntur
Sangli
Bijapur
Bhima
Kolhapur
Krishna
Belgaum
Dharwad
Gadag
Adoni
Kurnool
Bellary
Panaji
GOA
KARNATAKA
Davangere
Shimoga
BANGALORE
Kolar
ANDHRA PRADESH
Western Ghats
Eastern Ghats
Nellore
CHENNAI (MADRAS)
Vellore
Mangalore
Mysore
Salem
Pondicherry
TAMIL NADU
COIMBATORE
Kumbakonam
Tiruchchirappalli
Calicut
Palghat
Anai Mudi 2698
Cochin
KERALA
MADURAI
Mattancheri
Quilon
Trivandrum
Tuticorin
Tirunelveli
Nagercoil
C. Comorin
Malabar Coast
Coromandel Coast
Lakshadweep Is. (India)
Deccan
Palk Strait
Jaffna
Rameswaram
Mannar
Trincomalee
SRI LANKA
Batticaloa
G. of Mannar
Puttalam
Kandy
Pidurutalagala 2524
COLOMBO
Galle
Dondra Head
MALDIVES
INDIAN OCEAN
XIZANG ZIZHIQU (TIBET)
Tanggula Shan
Nyainqentanglha Shan
Nam Co
Gar
Lhasa
Yarlung Zangbo Jiang
Xigazê
Himalaya
Dhaulagiri 8167
Annapurna 8078
Maquan He
Mt. Everest 8848
NEPAL
Pokhara
Katmandu
SIKKIM
BHUTAN
Thimphu
Darjiling
Shiliguri
Biratnagar
Darbhanga
Chhapra
Ara
PATNA
Ganga
Bhagalpur
Gaya
BIHAR
Dhanbad
Asansol
WEST BENGAL
Ranchi
Jamshedpur
Raurkela
Kharagpur
CALCUTTA
Sandarbans
Bilaspur
Raipur
Mahanadi
ORISSA
Cuttack
Bhubaneswar
Puri
Brahmapur
Indravati
Jagdalpur
VISHAKHAPATNAM
Kottagudem
Rajahmundry
Kakinada
Eluru
Bay of Bengal
Rangpur
Parbatipur
MEGHALAYA
BANGLADESH
Rajshahi
DHAKA
Khulna
Barisal
CHITTAGONG
TRIPURA
MIZORAM
MANIPUR
Imphal
NAGALAND
ASSAM
Tezpur
Brahmaputra
ARUNACHAL PRADESH
Mebo
Sadiya
Dibrugarh
5881
Kachin
Myitkyina
Chindwin
Tengchong
Katha
Irrawaddy
BURMA (MYANMAR)
Shwebo
Mandalay
Kyaukse
Lashio
Myingyan
Yamethin
Minbu
Pegu Yoma
Arakan Yoma
Akyab
Ramree I.
Cheduba I.
Kyangin
Pegu
Insein
RANGOON (YANGON)
Bassein
Ma-ubin
Pyapon
C. Negrais
G. of Martaban
Moulmein
Ye
Tavoy
Mergui
Tanintharyi
Myeik
Kyunzu
Bilauk Taungdan
Shan
Thanlwin
THAILAND
Chiang Mai
2576
Nam
Uttaradit
Tak
Ping
LAOS
Mekong
Andaman Islands (India)
Port Blair
Little Andaman
Andaman Sea
Ten Degree Channel
Nicobar Islands (India)
Great Channel
Banda Aceh
INDONESIA
Simeulue
CHINA
Yushu
Garzê
CHENGDU
SICHUAN
CHONGQING
Qamdo
Lancang Jiang
Jinsha Jiang
Yibin
KUNMING
Dali
Baoshan
YUNNAN
Gejiu
SHAANXI

B C D E

6 7 8

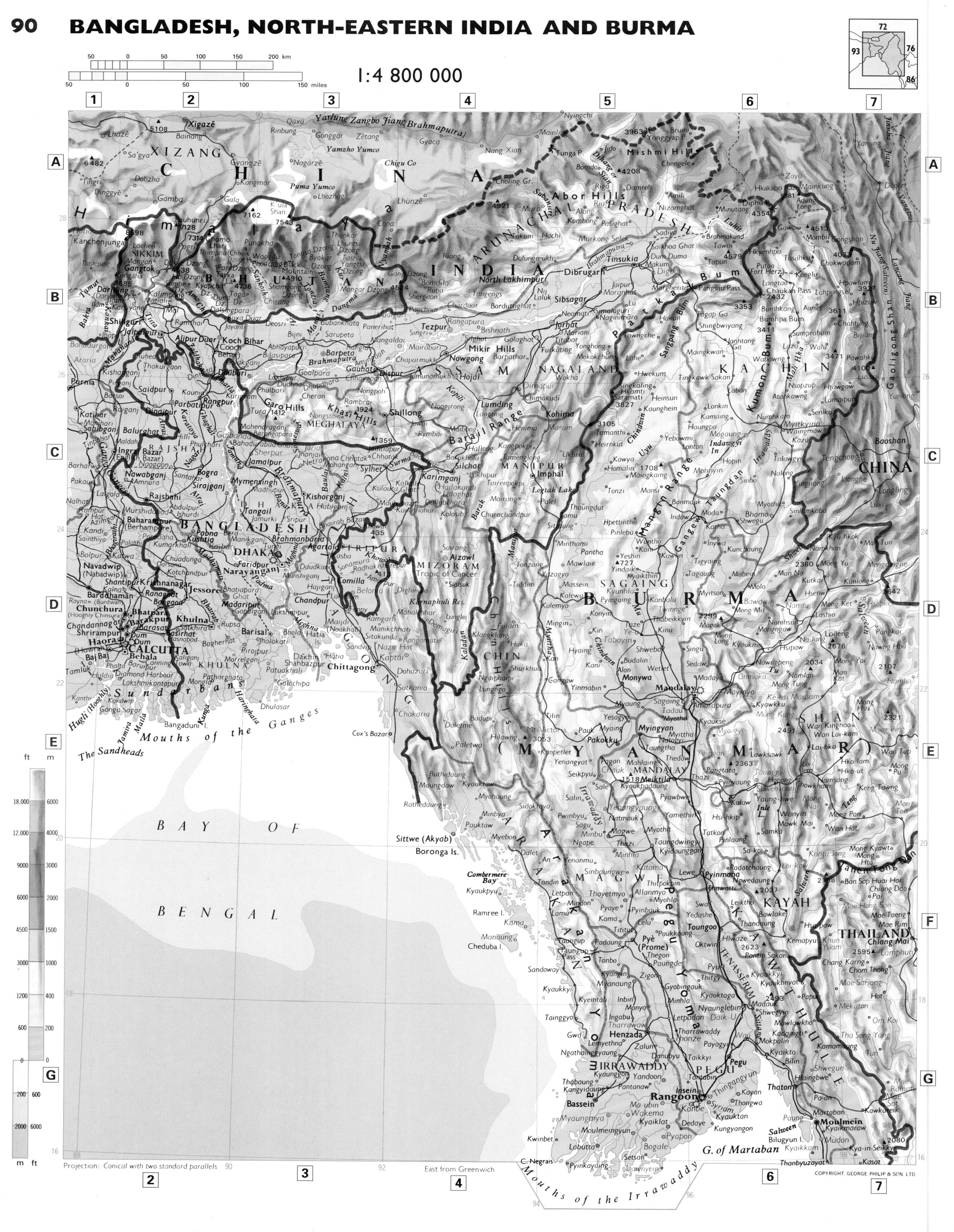
1:4 800 000
50 0 50 100 150 200 km
50 0 50 100 150 miles
XIZANG
CHINA
Himalaya
Yarlung Zangbo Jiang (Brahmaputra)
Mishmi Hills
Abor Hills
ARUNACHAL PRADESH
SIKKIM
Gangtok
Kanchenjunga
BHUTAN
INDIA
Dibrugarh
Tinsukia
North Lakhimpur
Tezpur
Mikir Hills
Nowgong
ASSAM
NAGALAND
Kohima
Shillong
Garo Hills
Khasi Hills
MEGHALAYA
Jaintia Hills
Barail Range
MANIPUR
Imphal
Silchar
Aizawl
MIZORAM
Tropic of Cancer
TRIPURA
Agartala
Comilla
Chittagong
CHITTAGONG
Cox's Bazar
BANGLADESH
DHAKA
Narayanganj
RAJSHAHI
Rajshahi
Rangpur
Dinajpur
Bogra
Mymensingh
Sylhet
KHULNA
Khulna
Jessore
Barisal
CALCUTTA
Haora
Sundarbans
Mouths of the Ganges
The Sandheads
BAY OF BENGAL
Patkai Bum
KACHIN
Kumon Bum
Myitkyina
Bhamo
Mangin Range
Gangaw Taungdan
SAGAING
BURMA
(MYANMAR)
CHIN
Chin Hills
Mandalay
Monywa
Myingyan
Meiktila
MANDALAY
MAGWE
Arakan Yoma
Sittwe (Akyab)
Boronga Is.
Combermere Bay
Ramree I.
Cheduba I.
SHAN
KAYAH
Taunggyi
Inle L.
Toungoo
Pyinmana
Pyè (Prome)
Pegu Yoma
Henzada
IRRAWADDY
Bassein
PEGU
Pegu
Rangoon
G. of Martaban
Moulmein
Thaton
TENASSERIM
DAWNA
THAILAND
Chiang Mai
Mouths of the Irrawaddy
C. Negrais
Salween
Irrawaddy
Chindwin
Brahmaputra
Projection: Conical with two standard parallels
East from Greenwich
COPYRIGHT, GEORGE PHILIP & SON, LTD.

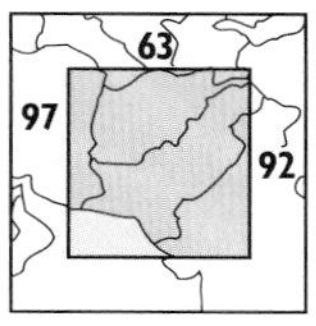

50 0 50 100 150 200 250 300 km
50 0 50 100 150 200 miles

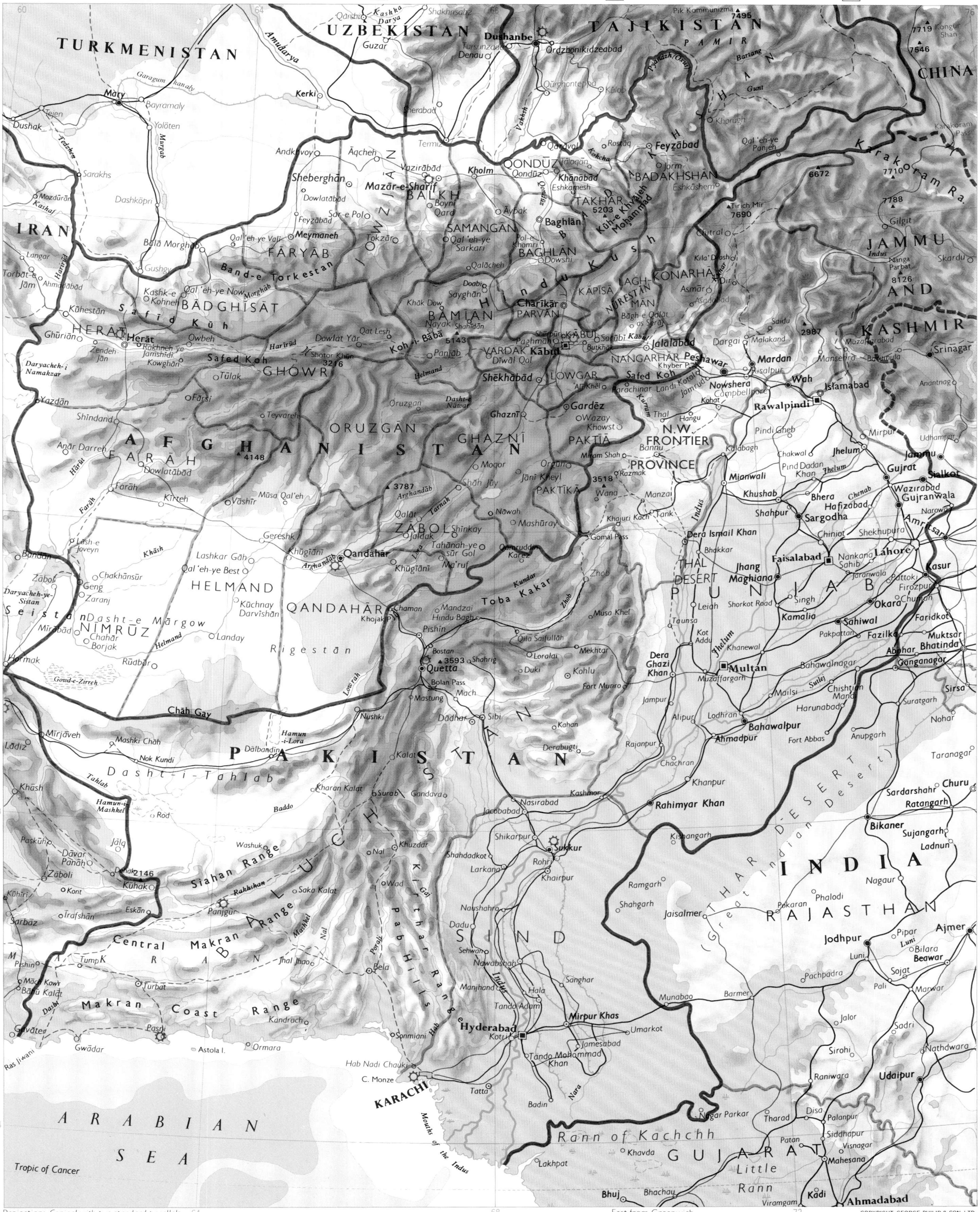

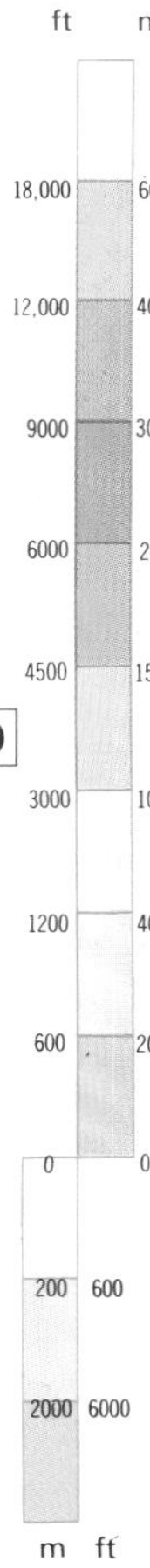

TURKMENISTAN
UZBEKISTAN
TAJIKISTAN
CHINA

Mary
Bayramaly
Dushak
Tejen
Yöloten
Kerki
Qarshi
Kashka Darya
Shakhrisabz
Guzar
Denau
Dushanbe
Ordzhonikidzeabad
Pik Kommunizma 7495
Kongur Shan 7719
7546

PAMIR
Bartang
Gunt

Andkhvoy
Sheberghan
Āqcheh
Vazirābād
Termiz
Kholm
QONDŪZ
Qondūz
Khānabād
Feyzābād
BADAKHSHAN
6672
7710
7788 Tirich Mir 7690
Gilgit
Skardu

Mazār-e-Sharīf
BALKH
SAMANGAN
Baghlān
BAGHLAN
TAKHĀR
5203
Kūh-e-Khvājeh Moḥammad

IRAN
Mashhad
Mozdūrān
Kashaf
Herāt
Ghūrīān
Zendeh Jān
Owbeh

FARYAB
Meymaneh
Qal'eh-ye Vali
Band-e Torkestan
BĀDGHĪSĀT
Safīd Kūh
GHOWR
3276

HERĀT
Safed Koh

JOWZJĀN
Sar-e Pol
Tokzār
Doabi
Sayghan
BĀMIĀN
Nāyak
Kōh-i-Bābā 5143
Charīkār
PARVĀN
KĀPISA
NŪRESTAN
LAGHMĀN
KONARHA
Jalālābād
NANGARHĀR
Peshawar
Khyber Pass
Mardan
Islamabad
Srinagar

JAMMU
8126
AND
KASHMIR

KĀBUL
VARDAK
LOWGAR
Ghaznī
GHAZNĪ
PAKTIĀ
Gardēz
Safed Koh
N.W. FRONTIER PROVINCE
Rawalpindi
Jhelum
Gujrat
Sialkot
Wazirābād
Gujranwala

AFGHANISTAN
ORUZGAN
4148
Oruzgan
3787
Arghandāb
Qalāt
ZABOL
Shinkay
Jaldak
Tarnak

FARĀH
Farāh
Dowlatābād
Mūsa Qal'eh
Vāshīr

Khāsh
Lashkar Gāh
HELMAND
Gereshk
Qandahār
Arghandāb
QANDAHĀR
Kūchnay Darvīshān
Helmand

NĪMRŪZ
Chahār Borjak
Zaranj
Dasht-e Mārgow
Rīgestān

Quetta 3593
Bostan
Shahrig
3518
Wana
Manzai
Dera Ismail Khan
Bhakkar
PUNJAB
Mianwali
Khushab
Sargodha
Faisalabad
Lahore
Kasur

BALUCHISTAN
Kalat
Mastung
Sibi
Mach
Bolan Pass
Dādhar
Kalat
Khuzdar

Siahan Range
Central Makran Range
Makran Coast Range
Panjgur
Turbat
Gwādar
Ras Jīwani
Astola I.
Ormara
Sonmiani
C. Monze
KARACHI
Tatta
Mouths of the Indus

SIND
Hyderabad
Mirpur Khas
Nawabshah
Larkana
Sukkur
Rohri
Khairpur

PAKISTAN

THAR (Great Indian) DESERT
Bahawalpur
Rahimyar Khan
Khanpur
Bikaner
RAJASTHAN
INDIA
Jaisalmer
Jodhpur
Ajmer

Multan
Dera Ghazi Khan

GUJARAT
Rann of Kachchh
Little Rann
Ahmadabad
Bhuj

ARABIAN SEA
Tropic of Cancer

East from Greenwich
Projection: Conical with two standard parallels

ft m
18,000 6000
12,000 4000
9000 3000
6000 2000
4500 1500
3000 1000
1200 400
600 200
0 0
200 600
2000 6000
m ft

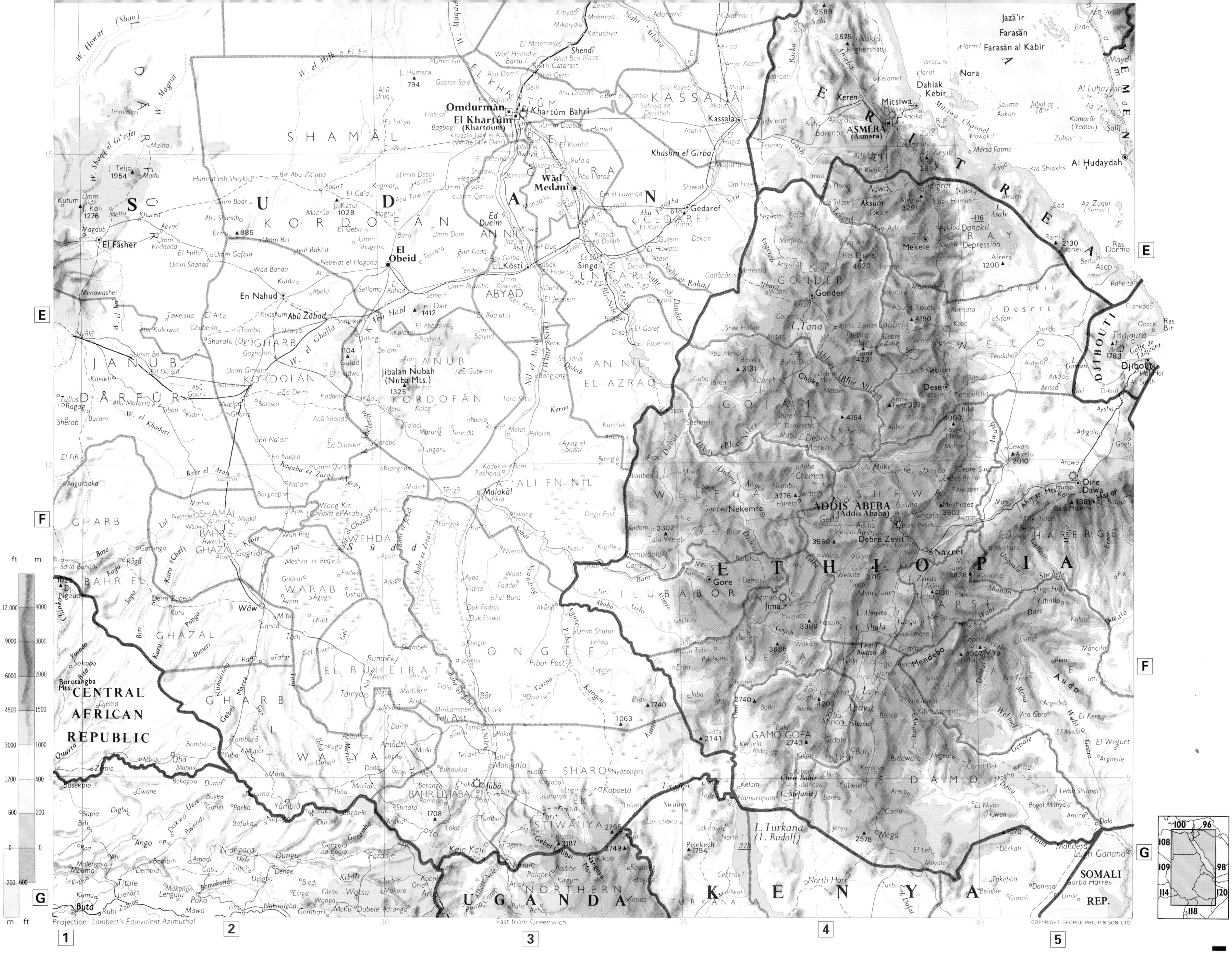

Map of Northeast Africa (Sudan, Ethiopia, Eritrea, Somalia, Uganda, Kenya)

YEMEN

Jazā'ir Farasān al Kabir
Farasān al Kabir
Al Luhayyah (Yemen)
Al Hudaydah
Kamarān (Yemen)
Az Zuqur
Jabal at Ta'ir
Dahlak Kebir

ERITREA
Mitsiwa
ASMERA (Asmara)
Keren
Nakfa
Agordat
Adi Keyih
Adi Ugri
Adwa
Aksum
Mekele
Adigrat

DJIBOUTI
Djibouti
Tadjoura
Obock
Dikhil

Danakil Desert
Danakil Depression

ETHIOPIA
Gonder
L. Tana
Debre Markos
Debre Zeyit
ADDIS ABEBA (Addis Ababa)
Nazret
Dese
Dire Dawa
HARERGE
WELO
TIGRAY
GONDER
GOJAM
SHEWA
ARSI
BALE
SIDAMO
GAMO GOFA
KEFA
ILUBABOR
WELEGA
Gore
Gimbi-Nekemte
Jima
Goba
Mega
Awasa

SUDAN
KHARTUM
El Khartûm (Khartoum)
Omdurmân
El Khartûm Bahri
Wad Medani
Kassala
Gedaref
GEDAREF
KASSALA
 AN NIL AL AZRAQ
Singa
El Kôsti
Ed Dueim
EL ABYAD
SHAMĀL KORDOFĀN
El Obeid
En Nahud
Abu Zabad
GHARB KORDOFĀN
JANŪB KORDOFĀN
Ilbalan Nubah (Nuba Mts.)
Dilling
KORDOFÁN
SHAMĀL DÁRFÚR
DÁRFÚR
El Fasher
Nyala
JANŪB DÁRFÚR
GHARB DÁRFÚR
Shendi
Atbara
AN NIL AL ABYAD
Malakâl
A'ALI EN NIL
JUNQALI
BAHR EL GHAZAL
SHAMĀL BAHR EL GHAZAL
GHARB BAHR EL GHAZAL
Wâw
Tonj
WARAB
EL BUHEIRAT
WEHDA
SHARQ EL ISTIWÁIYA
GHARB EL ISTIWÁIYA
Jûba
Torit
Bôr
Yambio
Yei

KENYA
L. Turkana (L. Rudolf)
North Horr

UGANDA
NORTHERN

SOMALI REP.

CENTRAL AFRICAN REPUBLIC

Blue Nile
White Nile

East from Greenwich

Projection: Lambert's Equivalent Azimuthal

m 4000 3000 2000 1500 1000 400 200 0 m
ft 12,000 9000 6000 4500 3000 1200 600 0 ft

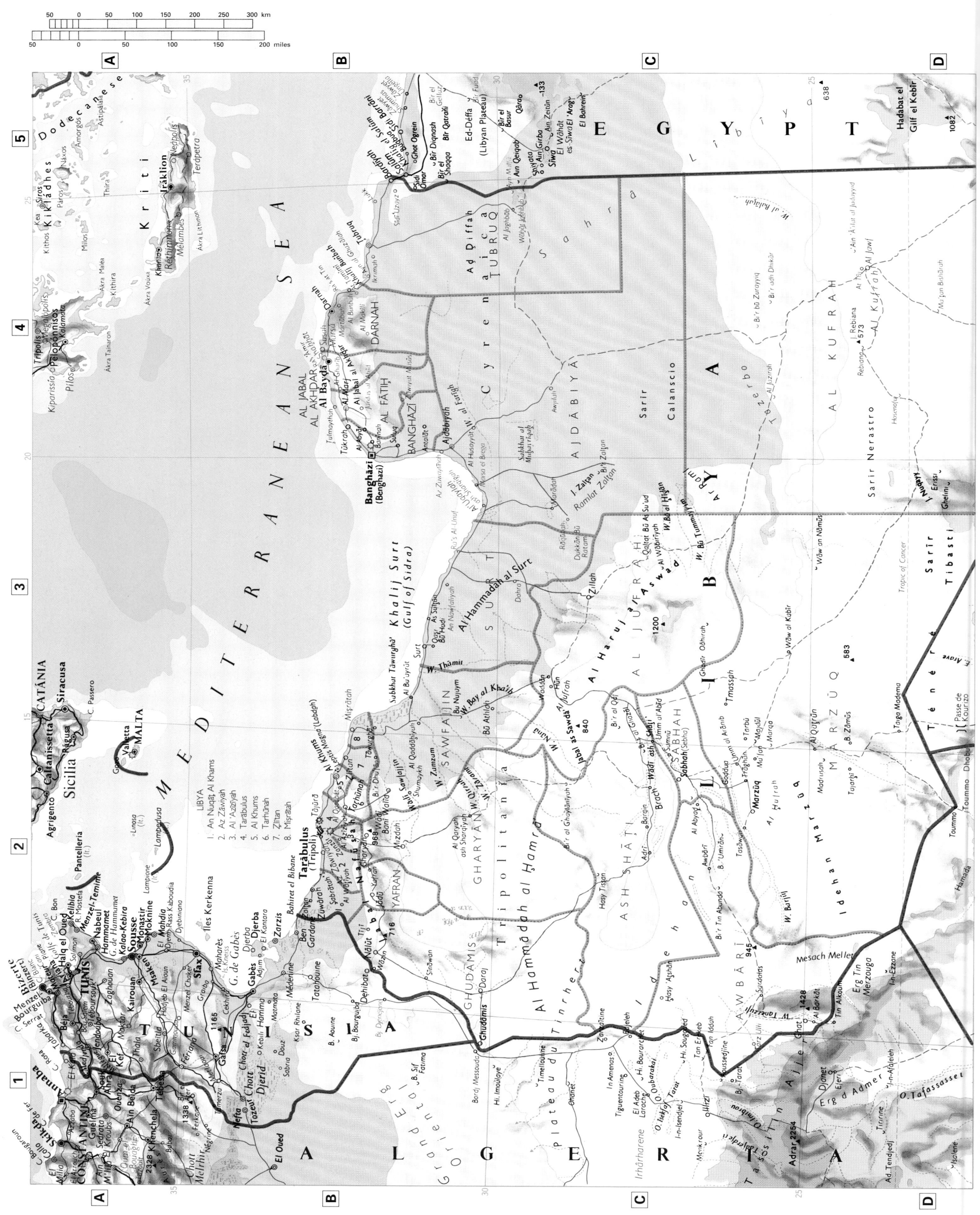
50 0 50 100 150 200 250 300 km
50 0 50 100 150 200 miles
MEDITERRANEAN SEA
Khalij Surt (Gulf of Sidra)
LIBYA
1. An Nuqāt Al Khams
2. Az Zāwiyah
3. Al 'Azīzīyah
4. Tarābulus
5. Al Khums
6. Tarhūnah
7. Zlītan
8. Misrātah
ALGERIA
TUNISIA
LIBYA
EGYPT
Sicilia
MALTA
Valletta
Siracusa
Catánia
Tarābulus (Tripoli)
Banghāzī (Benghazi)
TUNIS
Sfax
Sousse
Gabès
Djerba
Kriti
Iráklion
Pelopónnisos
Kikládhes
Dodecanese
Ed-Déffa (Libyan Plateau)
Sarir Calanscio
Sarīr Tibasti
Sarir Nerastro
Al Kufrah
Idehan Marzūq
Al Hammādah al Hamrā'
Al Harūj al Aswad
Sabhah (Sebha)
Marzūq
Ghat
Ghudāmis
Tropic of Cancer
Ténéré
Hadabat el Gilf el Kebīr
Erg d'Admer
Grand Erg Oriental
Tripolitania
Cyrenaica

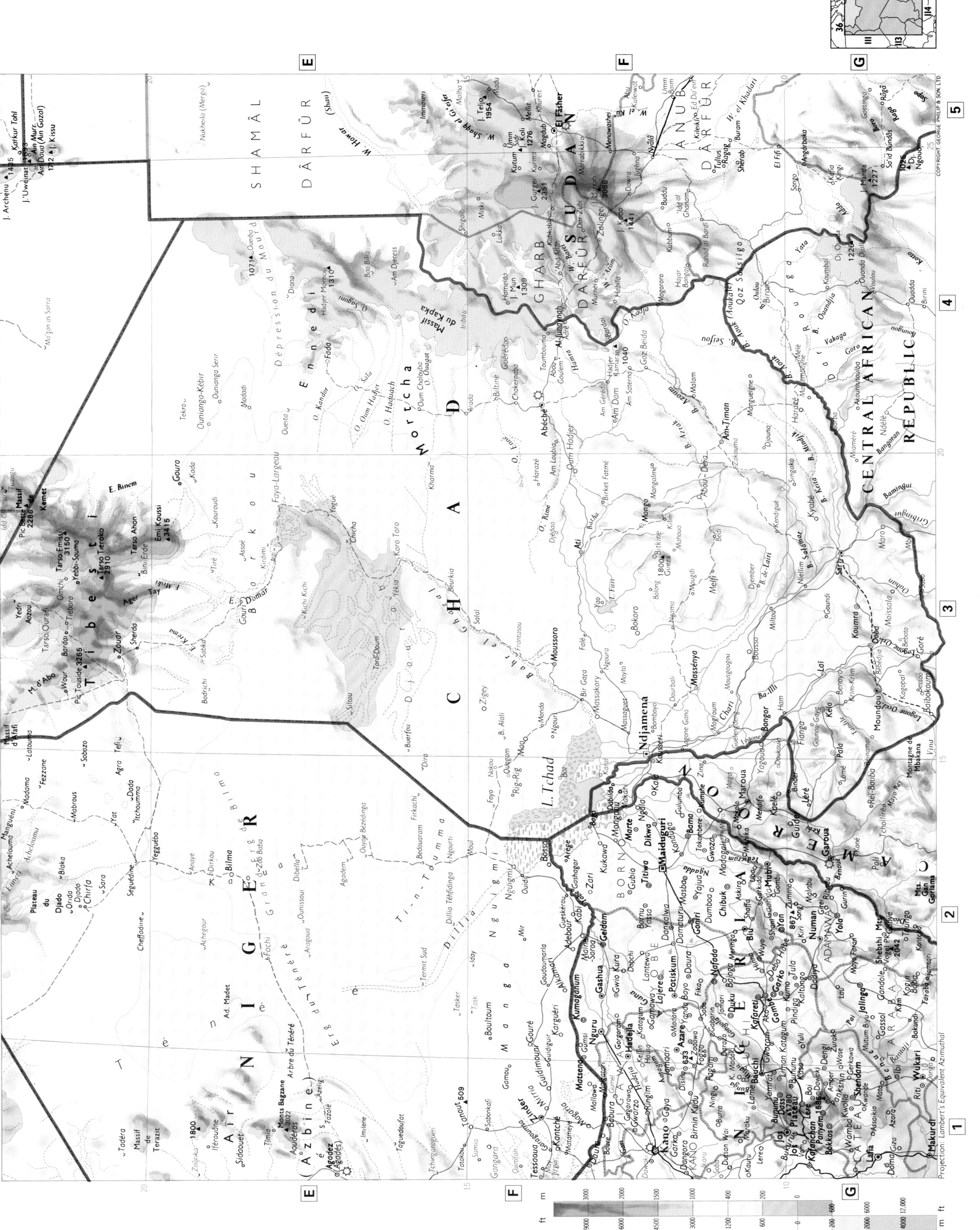
NIGER
CHAD
SUDAN
NIGERIA
CAMEROON
CENTRAL AFRICAN REPUBLIC
Tibesti
Aïr (Azbine)
Ennedi
Mortcha
Borkou
Djourab
Manga
Nguigmi
Ti-n-Toumma
Dillia
Grand Erg de Bilma
Erg du Ténéré
Dépression du Mourdi
Massif du Kapka
SHAMÂL DÂRFÛR
GHARB DÂRFÛR
JANUB DÂRFÛR
BORNO
YOBE
JIGAWA
KANO
BAUCHI
PLATEAU
ADAMAWA
TARABA
L. Tchad
Bahr el Ghazal
Ndjamena
Abéché
Moundou
Sarh
Faya-Largeau
Zinder
Agadez (Agadès)
Bilma
Kano
Maiduguri
Jos
Bauchi
Yola
Garoua
Maroua
El Fasher
Emi Koussi 3415
Pic Toussidé 3265
Tarso Emissi 3150
J. Marrah 3088
Mongo
Ati
Am-Timan
Bongor
Koumra
Doba
Moussoro
Massénya
Mao
Bol
Nyala
Zalingei
Geneina (Al Junaynah)
Dâr Rounga
Qoz Salsilgo
Projection: Lambert's Equivalent Azimuthal
COPYRIGHT. GEORGE PHILIP & SON. LTD.
ft m
9000 3000
6000 2000
4500 1500
3000 1000
1200 400
600 200
0 0
200 600
2000 6000
4000 12,000
m ft
36 37 106 107 111 113 114

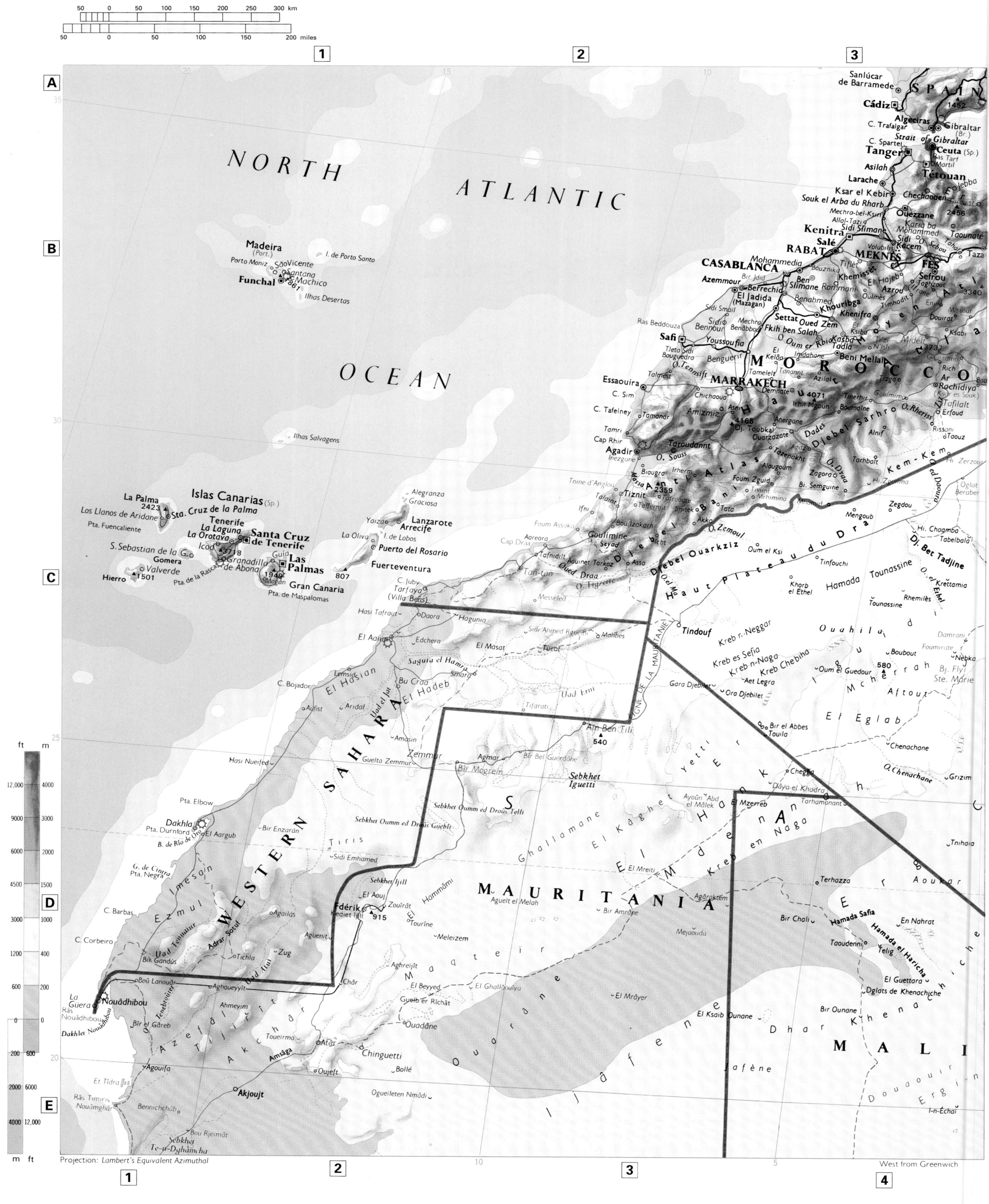

Projection: Lambert's Equivalent Azimuthal

West from Greenwich

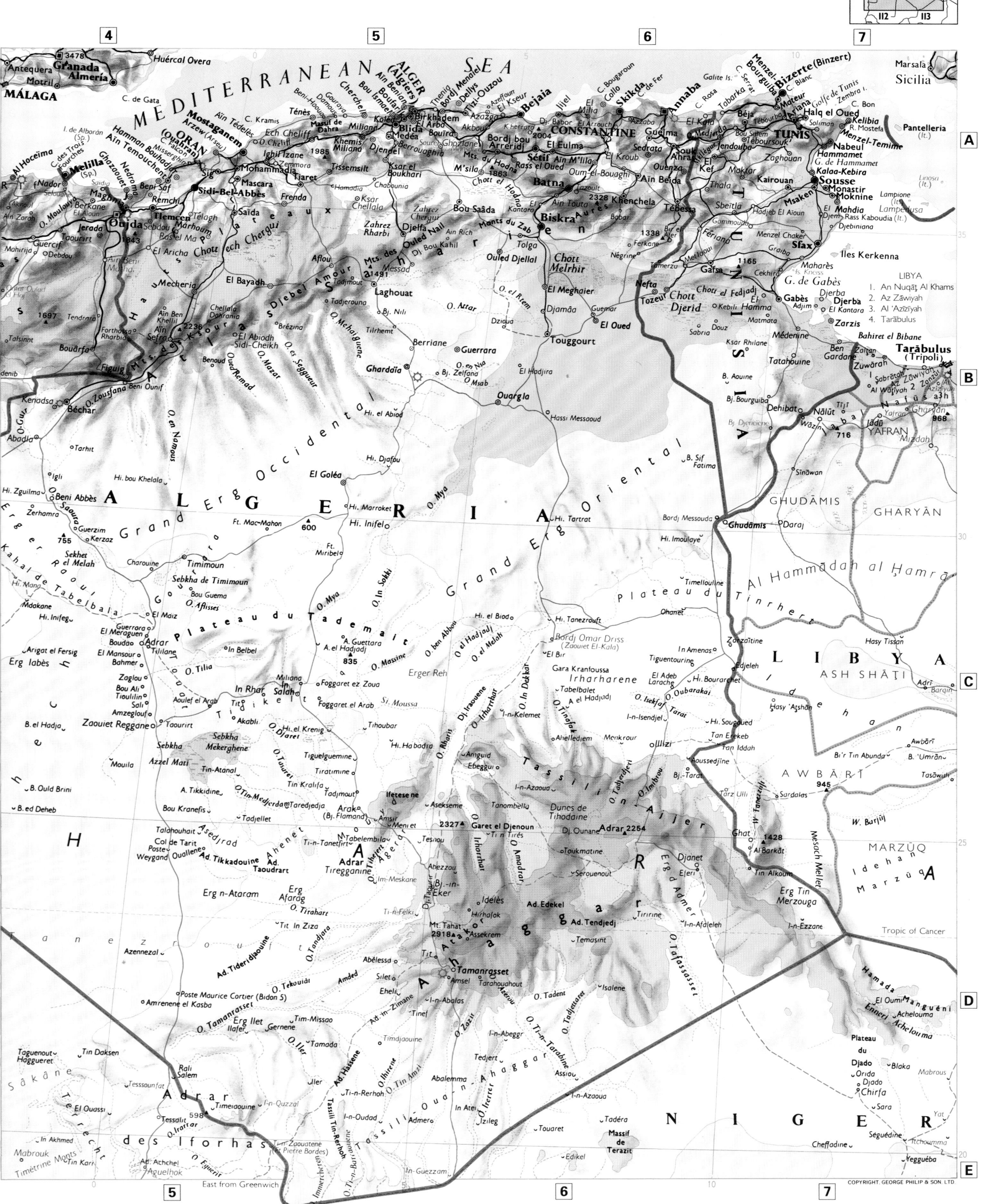
36
108
112
113
4
5
6
7
A
B
C
D
E
MEDITERRANEAN SEA
ALGERIA
TUNISIA
LIBYA
NIGER
Sicilia
Marsala
Pantelleria (It.)
Linosa (It.)
Lampione (It.)
Lampedusa
3478
Antequera
Granada
Almería
Motril
MÁLAGA
Huércal Overa
C. de Gata
I. de Alborán (Sp.)
Al Hoceïma
C. des Trois Fourches
Melilla (Sp.)
Nador
Saïdia
Selouane
Hamman Bouhadjar
Aïn Temouchent
Nedroma
Ghazaouet
Maghnia
Berkane
El Aïoun
Aïn Zorah
O. Moulouya
Oujda
Jerada
Taourirt
Guercif
Debdou
Mahirija
1843
Aïn Beni Mathar
1697
Tendrara
Talsinnt
Forthassa Rharbia
Bouârfa
Figuig
Beni Ounif
Kenadsa
Béchar
O. Guir
Abadla
O. Zousfana
Tarhit
Igli
Beni Abbès
Hi. Zguilma
Zerhamra
O. Saoura
Guerzim
Kerzaz
755
Sekhet el Melah
Kahal de Tabelbala
Erg er Raoui
Hi. Mana
Mdakane
Hi. Inifeg
Arigat el Fersig
Erg Iabès
B. el Hadjo
B. Ould Brini
B. ed Deheb
Mouila
Tlemcen
Sebdou
Telagh
Marhoum
Ras el Ma
El Aricha
Chott ech Chergui
Remchi
Beni Saf
Sidi-Bel-Abbès
Mascara
Saïda
Sig
Misserghin
ORAN (Ouahran)
C. Falcon
Arzew
Mostaganem
Aïn Tédelès
C. Kramis
Ech Cheliff
O. Cheliff
Mohammadia
Ighil Izane
Zemmora
Tiaret
Frenda
Hamadia
Ksar Chellala
Chabounia
Tissemsilt
Khemis Miliana
Miliana
1985
Ténès
Beni-Haoua
Massif de Dahra
Gouraya
Damous
Cherchell
Bou Ismaïl
Aïn Benian
Boufarik
ALGER (Algiers)
Koléa
Blida
Médéa
Berrouaghia
Birkhadem
El Arba
Bouïra
Thenia
Bordj Menaïel
Dellys
Tizi-Ouzou
Azeffoun
El Kseur
Azazga
Akbou
Bejaia
Sour el Ghozlane
Djendel
Ksar el Boukhari
M'sila
Bordj bou Arreridj
Mts. du Hodna
1863
Chott el Hodna
Kherrata
Dj. Babor
2004
Jijel
El Milia
C. Bougaroun
Collo
Skikda
C. de Fer
Annaba
El Arrouch
Azzaba
CONSTANTINE
El Eulma
Sétif
Aïn M'lila
Ras el Oued
El Kroub
Oum-el-Bouaghi
Batna
Tazoult
Aïn Touta
2328
Aurès
Khenchela
Babar
Biskra
Tolga
Kantara
Ouled Djellal
Bou Saâda
Monts du Zab
Aïn Rich
Djelfa
Zahrez Chergui
Zahrez Rharbi
Dj. Bou Kahil
Mts. des Ouled Nail
Aflou
Djebel Amour
1491
Tadjmout
Messad
Laghouat
Tadjerouna
Bj. Nili
El Bayadh
Mecheria
Hauts Plateaux
Chellala Dahrania
Aïn Ben Khellil
Aïn Sefra
2236
Mts. des Ksour
El Abiodh Sidi-Cheikh
Brézina
Benoud
Oued Remad
O. Mazar
O. es Seggueur
O. Mehaiguene
Saharan Atlas
Tilrhemt
Berriane
Ghardaïa
Guerrara
O. en Nsa
Bj. Zelfana
O. Msab
O. Attar
Ouargla
Hassi Messaoud
El Hadjira
Touggourt
Djamâa
El Meghaier
Chott Melrhir
O. el Rtem
Dzioua
Guemar
El Oued
Hi. el Abiod
Hi. Djafou
El Goléa
Grand Erg Occidental
Grand Erg Oriental
O. en Namous
Hi. bou Khelala
Hi. Marroket
Hi. Inifel
O. Mya
Ft. Mac-Mahon
600
Ft. Miribel
Charouine
Timimoun
Sebkha de Timimoun
Bou Guema
O. Aflisses
Gourara
El Maïz
Guerrara
El Meraguen
Boudao
Adrar
Tililane
El Mansour
Bahmer
Zaglou
Bou Ali
Tioulilin
Sali
Amzegloul
Zaouiet Reggane
Taourirt
Sebkha
Azzel Mati
Sebkha Mekerghene
Plateau du Tademaït
A. Guettara
A. el Hadjadj
835
In Belbel
O. Tilia
In Rhar
Aoulef el Arab
Tit
Akabli
Tidikelt
Miliana
In Salah
Foggaret ez Zoua
Foggaret el Arab
Si. Moussa
Hi. el Krenig
O. Djaret
Tihoubar
O. In Sokki
O. ben Abbou
O. el Hadjadj
O. el Melah
O. Massine
Erger Reh
Hi. Habadra
Tiguelguemine
Tiratimine
Tin Kralifa
Tadjmout
Tin-Atanal
O. Touret
A. Tikkidine
O. Tin-Medjerdam
Taredjedja
Bou Kranefis
Tadjellet
Arak (Bj. Flamand)
Amsir
Ifetesene
Asekseme
Meniet
2327
Garet el Djenoun
Talahouhait
Col de Tarit
Poste Weygand
Ouallene
Asedjrad
Ad. Tikkadouine
Ahenet
Ad. Taoudrart
Tesnou
Tin Tirés
Tin-Tanetfirt
Tabelembila
Mouydir
Adrar Tiregganine
O. Tahkjert
Agerd
Im-Meskane
Abezzou
Bj.-in-Eker
Erg n-Ataram
Erg Afarag
O. Tirahart
Tit
In Ziza
Ti-n-Felki
Idélès
Hirhafok
Mt. Tahat
2918
Assekrem
Atakor
Ahaggar
Ad. Edekel
Ad. Tendjedj
Tiririne
Temasint
Tanezrouft
Azennezal
Ad. Tideridjaouine
O. Tandjara
Abélessa
Silet
Tit
Tamanrasset
Amsel
Tarahouhaout
O. Tekouiât
Amded
Ehelil
I-n-Abalos
Ad.-n-Zimane
Tinef
Poste Maurice Cortier (Bidon 5)
Amrenene el Kasba
O. Tamanrasset
Erg Ilet
Ilafer
Gernene
Tim-Missao
O. Iler
Tamada
Timdjaouine
Ad. Hassene
O. Ihirene
O. Tin Amzi
Abalemma
O. Zazir
I-n-Abeggi
Tedjert
Assiou
O. Ti-n-Tarabine
Taguenout-Hagguerèt
Tin Daksen
Sâkâne
Rali Salem
Tesssounfat
Adrar des Iforhas
Timeïaouine
I-n-Quzzal
Jller
Ti-n-Rerhoh
Tassili Tin-Rerhoh
Tassili-Oua-n-Ahaggar
I-n-Oudad
Admero
In Atei
Izilég
I-n-Azaoua
O. Iterrer
Tessalit
598
O. Irattar
El Ouassi
Tilemsi
Terrecht
In Akhmed
Mabrouk
Timétrine Monts
Tin Karr
Ad. Achchel
Aguelhok
O. Egiaarit
Ti-n-Zaouatene (Ft. Pierre Bordes)
O. Immerchesi
O. Ti-n-Borak
In-Guezzam
Touaret
Tadéra
Massif de Terazit
Edikel
East from Greenwich
Dj. Irao uene
O. Rharis
O. Irharrhar
I-n-Kelemet
O. In Dekkar
Bordj Omar Driss (Zaouiet El-Kala)
Hi. Tanezrouft
El Bir
Hi. el Biod
Tedjert
Gara Kranfoussa
Irharharene
Tabelbalet
A. el Hadjadj
O. Tinafok
Ahelledjem
Menkrour
Amguid
Ebeggui
Tassili n Ajjer
O. Tadjedjeri
O. Imbrou
I-n-Azaoua
Tanombella
Dunes de Tihodaine
Dj. Ounane
Adrar
2254
Toukmatine
Serouenout
O. Amadror
O. Iharhar
Djanet
Ejeri
Erg d'Admer
O. Tafassasset
I-n-Afaleleh
Isalene
O. Tadent
O. Tadjettaret
Plateau du Tinrhert
Ohanet
Timelloutine
In Amenas
Tiguentourine
El Adeb Larache
Hi. Bourarchet
O. Isekfaf
O. Oubarakai
Tarat
I-n-Isendjel
Hi. Souguëd
Illizi
Tan Elekeb
Tin Iddah
Aoussedjine
Bj.-Tarat
Tarz Ulli
W. Tanezzuft
Ghat
1428
Al Barkât
Tin Alkoum
Ghudâmis
Daraj
Bordj Messouda
Hi. Imoulaye
B. Sif Fatima
Hi. Tartrat
Zarzaïtine
Edjeleh
Erg Tin Merzouga
I-n-Ezzane
Mesach Mellet
Tropic of Cancer
Hamada Manguéni
El Oumi
Achelouma
Enneri Achelouma
Plateau du Djado
Blaka
Mabrous
Orida
Djado
Chirfa
Sara
Yat
Seguédine
Tchoumma
Chefadine
Yegguéba
Galite Is.
C. Serrat
Tabarka
C. Rosa
El Kala
Guelma
Souk Ahras
Medjerda
Jendouba
Bou Salem
Teboursouk
Béja
Mateur
Menzel-Bourguiba
Bizerte (Binzert)
C. Blanc
Ariana
Golfe de Tunis
Zembra I.
C. Bon
TUNIS
Halq el Oued
Tebourba
Soliman
Kelibia
R. Mostefa
Menzel-Temime
Nabeul
Hammamet
G. de Hammamet
Zaghouan
Kalaa-Kebira
Sousse
Monastir
Moknine
Kairouan
Maktar
El Kef
Sedrata
Ouenza
Aïn Beida
Thala
Tébessa
Sbeitla
Hadjeb El Aïoun
Msaken
Mahdia
Djem
Rass Kaboudia (It.)
Djebiniana
Gammouda
Fériana
Bir el Ater
1338
Ferkane
Négrine
Tamerza
Metlaoui
Gafsa
1165
Menzel Chaker
Graiba
Sfax
Iles Kerkenna
Maharès
Cekhira
Is. Kneiss
G. de Gabès
Gabès
Djerba
Adjim
El Kantara
Zarzis
Nefta
Tozeur
Chott Djerid
Chott el Fedjadj
Kebili
El Hamma
Matmata
Douz
Sabria
Médenine
Ksar Rhilane
Tatahouine
Bahiret el Bibane
Ben Gardane
Zaltan
Zuwārah
Sabrātah
Az Zāwiyah
Al Wāţīyah
Tarābulus (Tripoli)
Al 'Azīzīyah
Jafārah
B. Aouine
Bj. Bourguiba
Bj. Djeneiene
Dehibat
Nālūt
Ţiji
Jādū
Yafran
YAFRAN
Gharyān
968
716
Wāzin
Jabal Nafūsah
Mizdah
Sinawan
LIBYA
1. An Nuqāţ Al Khams
2. Az Zāwiyah
3. Al 'Azīzīyah
4. Tarābulus
GHUDĀMIS
GHARYĀN
Al Hammādah al Ḩamrā'
ASH SHĀŢI
Idehan
Hasy Tissān
Hasy 'Aţshān
Adrī
Barqin
Awbārī
Bi'r Tin Abunda
B. 'Umrān
AWBĀRĪ
945
Sardalas
Tasāwah
W. Barjūj
MARZŪQ
Idehan Marzūq
Ahaggar
0
5
10
20
25
30
35

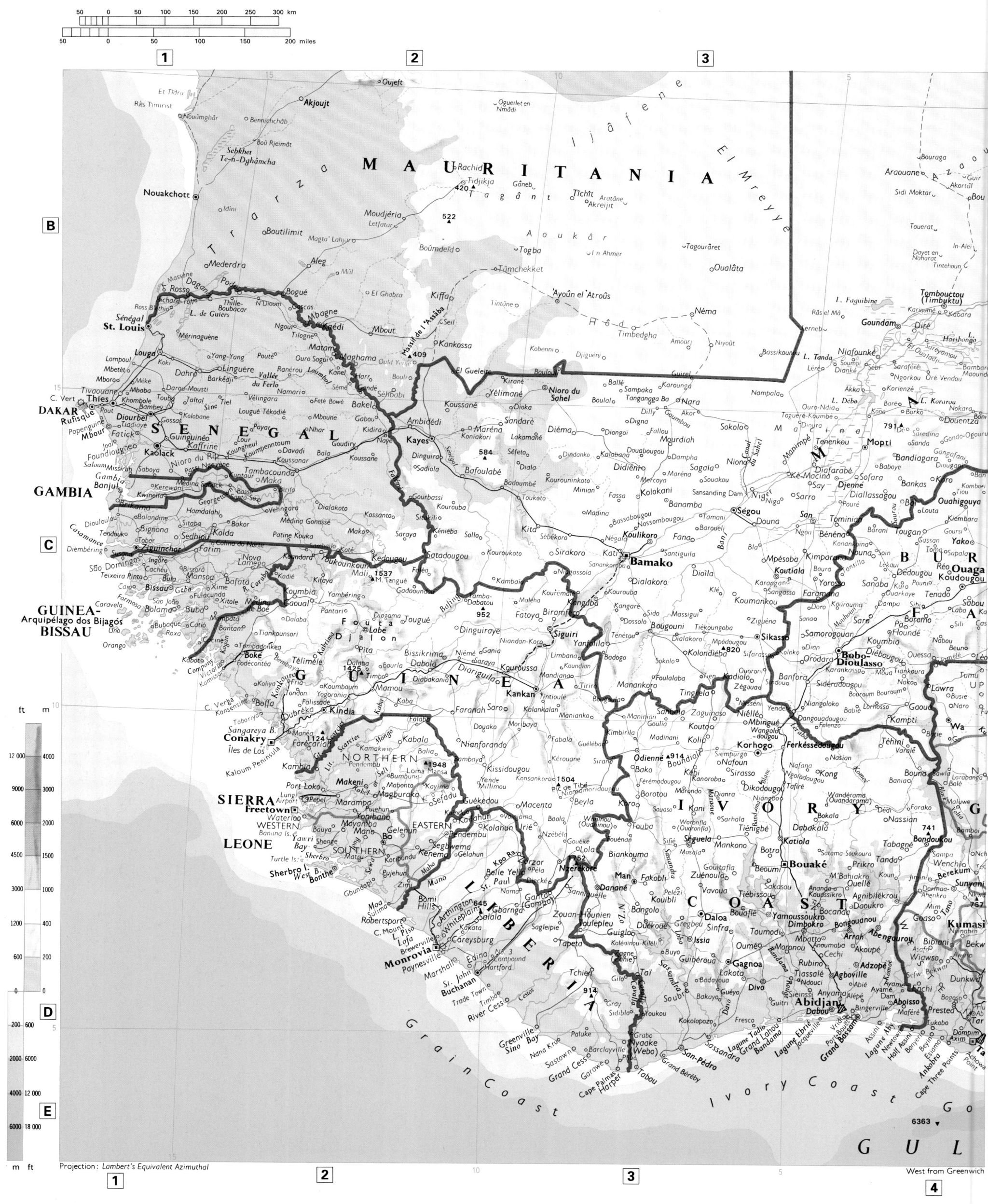

Projection: Lambert's Equivalent Azimuthal

West from Greenwich

N. E. NIGERIA
on same scale
as general map

East from Greenwich

COPYRIGHT. GEORGE PHILIP & SON LTD

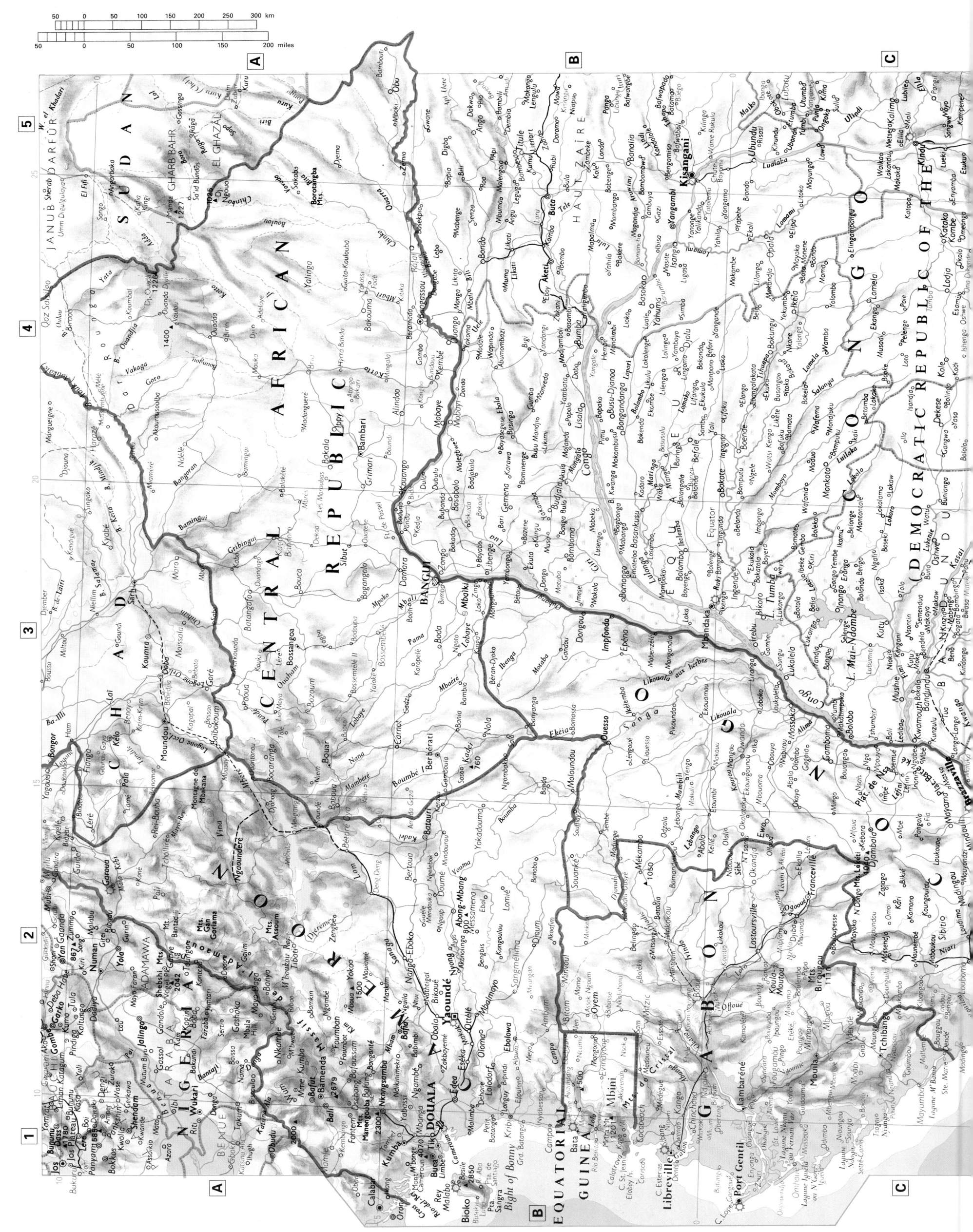
50 0 50 100 150 200 250 300 km
50 0 50 100 150 200 miles
1
2
3
4
5
A
B
C
SUDAN
CHAD
CENTRAL AFRICAN REPUBLIC
CAMEROON
NIGERIA
GABON
CONGO
CONGO (DEMOCRATIC REPUBLIC OF THE)
EQUATORIAL GUINEA
HAUT ZAÏRE
EQUATEUR
BANGUI
Yaoundé
DOUALA
Libreville
Port Gentil
Kisangani
Mbandaka
Brazzaville
Bioko
Bight of Bonny
Mbini
Congo
Equator
L. Tumba
L. Mai-Ndombe
Lambaréné
Franceville
Berbérati
Bambari
Bossangoa
Garoua
Ngaoundéré
Moundou
Sarh
Bongor
Bata
Ouesso
Impfondo
Buta
Lisala
Bumba
Yangambi
EL GHAZÂL
GHARB BAHR
JANUB DARFÛR

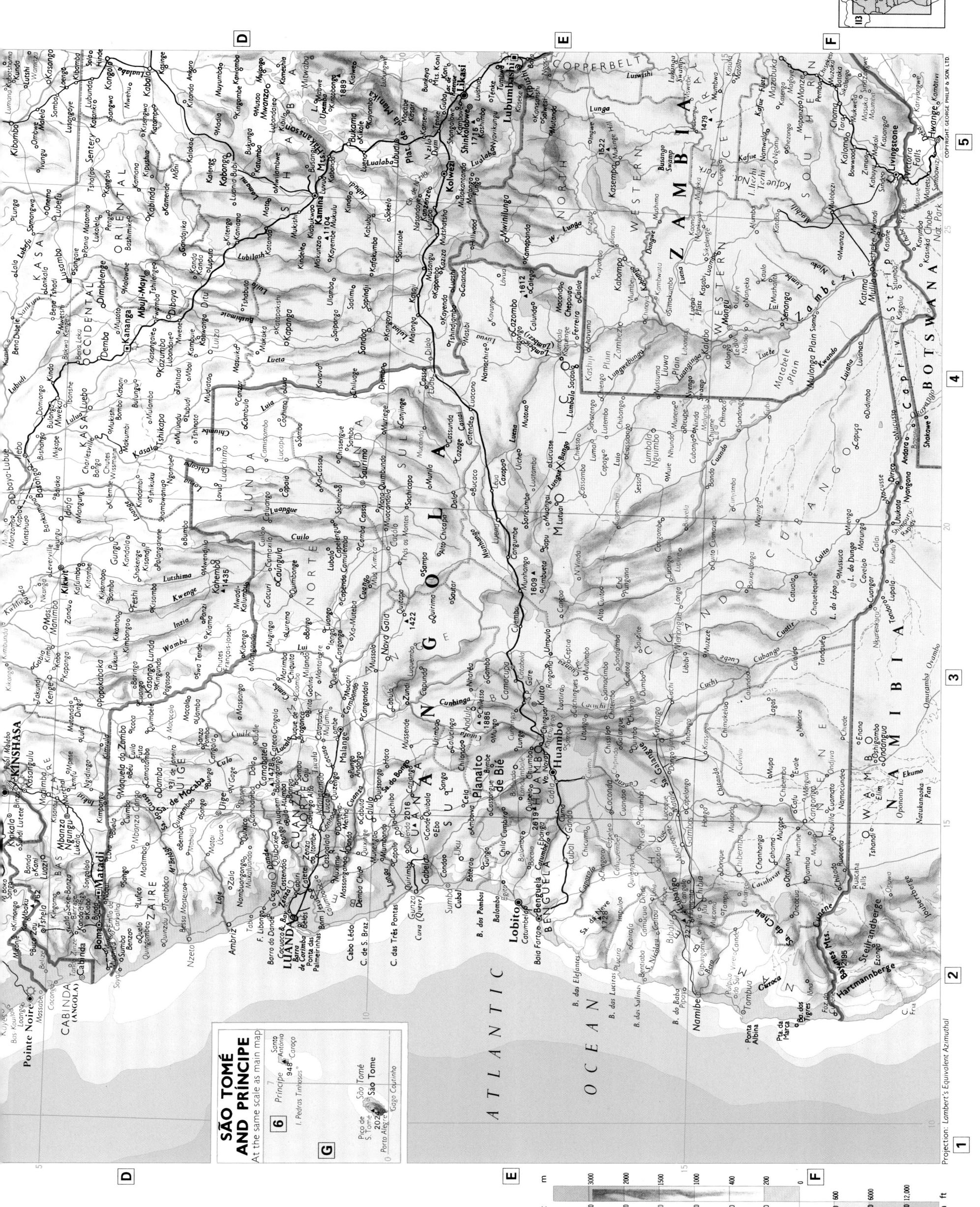

SÃO TOMÉ
AND PRÍNCIPE
At the same scale as main map

Projection: Lambert's Equivalent Azimuthal

COPYRIGHT GEORGE PHILIP & SON LTD

ATLANTIC OCEAN

ZAMBIA

BOTSWANA

NAMIBIA

ANGOLA

ZAIRE

KINSHASA

CABINDA
(ANGOLA)

Pointe Noire

50 0 50 100 150 200 250 300 km
50 0 50 100 150 200 miles
ANGOLA
CUANDO CUBANGO
NAMIBE
WESTERN
ZAMBIA
SOUTHERN
NAMIBIA
BOTSWANA
SOUTH AFRICA
NORTH WEST
FREE
NORTHERN CAPE
WESTERN CAPE
EASTERN CAPE
ATLANTIC OCEAN
Tropic of Capricorn
Caprivi Strip
Okavango Swamps
Makgadikgadi Salt Pans
Khama Country
Kalahari
Namib Desert
Kaokoveld
Ovamboland
Damaraland
Namaland
Kaukauveld
Sandveld
Ghanzi
Etosha Pan
Namaqualand
Great Karoo
Little Karoo
Kalahari Gemsbok National Park
Chobe Nat. Park
Hwange Nat. Park
Windhoek
Walvisbaai
Walvis Bay (Walvisbaai)
Swakopmund
Lüderitz
Keetmanshoop
Rehoboth
Okahandja
Otjiwarongo
Tsumeb
Gobabis
Mariental
Oranjemund
Alexander Bay
Ondangua
Oshigambo
Katima Mulilo
Livingstone
Victoria Falls
Hwange
Maun
Ghanzi
Serowe
Palapye
Mahalapye
Molepolole
Gaborone
Lobatse
Kanye
Mochudi
Tshane
Lehututu
Upington
Kimberley
Bloemfontein
Welkom
Kroonstad
Mafikeng
Vryburg
Kuruman
Springbok
Port Nolloth
Calvinia
Beaufort West
Graaff-Reinet
Cradock
Queenstown
Middelburg
De Aar
Worcester
Paarl
Stellenbosch
CAPE TOWN (Kaapstad)
Table Mt. 1086
Kaap die Goeie Hoop (Cape of Good Hope)
C. Agulhas
Oudtshoorn
George
Mosselbaai
Swellendam
Uitenhage
PORT ELIZABETH
Grahamstown
Port Alfred
Algoa Bay
Vredenburg
Saldanha
Moorreesburg
Tsodilo Hill 1300
Brandberg 2606
Erongo 2350
Hakos 2351
Groot Karasberge 2202
Hunsberge 1655
Matroosberg 2249
Atlantic
Ocean
Projection: Lambert's Equivalent Azimuthal
ft m
9000 3000
6000 2000
4500 1500
3000 1000
1200 400
600 200
0 0
200 600
2000 6000
4000 12,000
m ft
1 2 3 4
B C D E

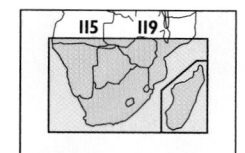

MOZAMBIQUE CHANNEL

CHANNEL

INDIAN OCEAN

ZIMBABWE

MOZAMBIQUE

MADAGASCAR

Tropic of Capricorn

East from Greenwich

MADAGASCAR

On same scale as General Map

COPYRIGHT. GEORGE PHILIP & SON. LTD.

50 0 50 100 150 200 250 300 km
50 0 50 100 150 200 miles
1
2
3
4
5
A
B
C
D
CENTRAL AFRICAN REPUBLIC
SUDAN
ETHIOPIA
SOMALI REP.
KENYA
NAIROBI
MOMBASA
Pemba I.
Zanzibar
DAR ES SALAAM
Mafia I.
UGANDA
Kampala
Entebbe
Lake Victoria
RWANDA
BURUNDI
TANZANIA
L. Tanganyika
L. Turkana (L. Rudolf)
L. Albert
L. Kyoga
L. Rukwa
Kisangani
Juba
Kigoma
Tabora
Dodoma
Arusha
Tanga
Kisumu
Mwanza
Lamu
Malindi
(DEM. REP. OF THE CONGO)

ZAMBIA
MALAWI
MOZAMBIQUE
ZIMBABWE
BOTSWANA
ANGOLA
SOUTH AFRICA
INDIAN OCEAN
Lusaka
Harare
Lilongwe
Lubumbashi
Bulawayo
Blantyre
Beira
Kitwe
Ndola
Mufulira
Luanshya
Chingola
Likasi
Kolwezi
Kabwe
Livingstone
Hwange
Gweru
Kwekwe
Kadoma
Masvingo
Mutare
Chitungwiza
Zomba
Nkhotakota
Nampula
Moçambique
Quelimane
Angoche
Lindi
Mtwara-Mikindani
Pemba
Lichinga
Tete
Chipata
Mpika
Kasama
Francistown
Serowe
Messina
Lake Kariba
L. Malawi
L. Mweru
L. Bangweulu
Zambezi
Limpopo
COPPERBELT
NORTHERN
CENTRAL
EASTERN
SOUTHERN
WESTERN
NORTH WESTERN
MASHONALAND
MATABELELAND
MIDLANDS
MASVINGO
MANICALAND
NIASSA
CABO DELGADO
NAMPULA
ZAMBÉZIA
SOFALA
MANICA
GAZA
INHAMBANE
Caprivi Strip
Projection: Lambert's Equivalent Azimuthal
East from Greenwich
COPYRIGHT, GEORGE PHILIP & SON, LTD.
ft
m
18 000
12 000
9000
6000
4500
3000
1200
600
0
200
2000
6000
4000
3000
2000
1500
1000
400
200
0
600
6000
E
F
G
1
2
3
4
5
107
114
115
116
117
120

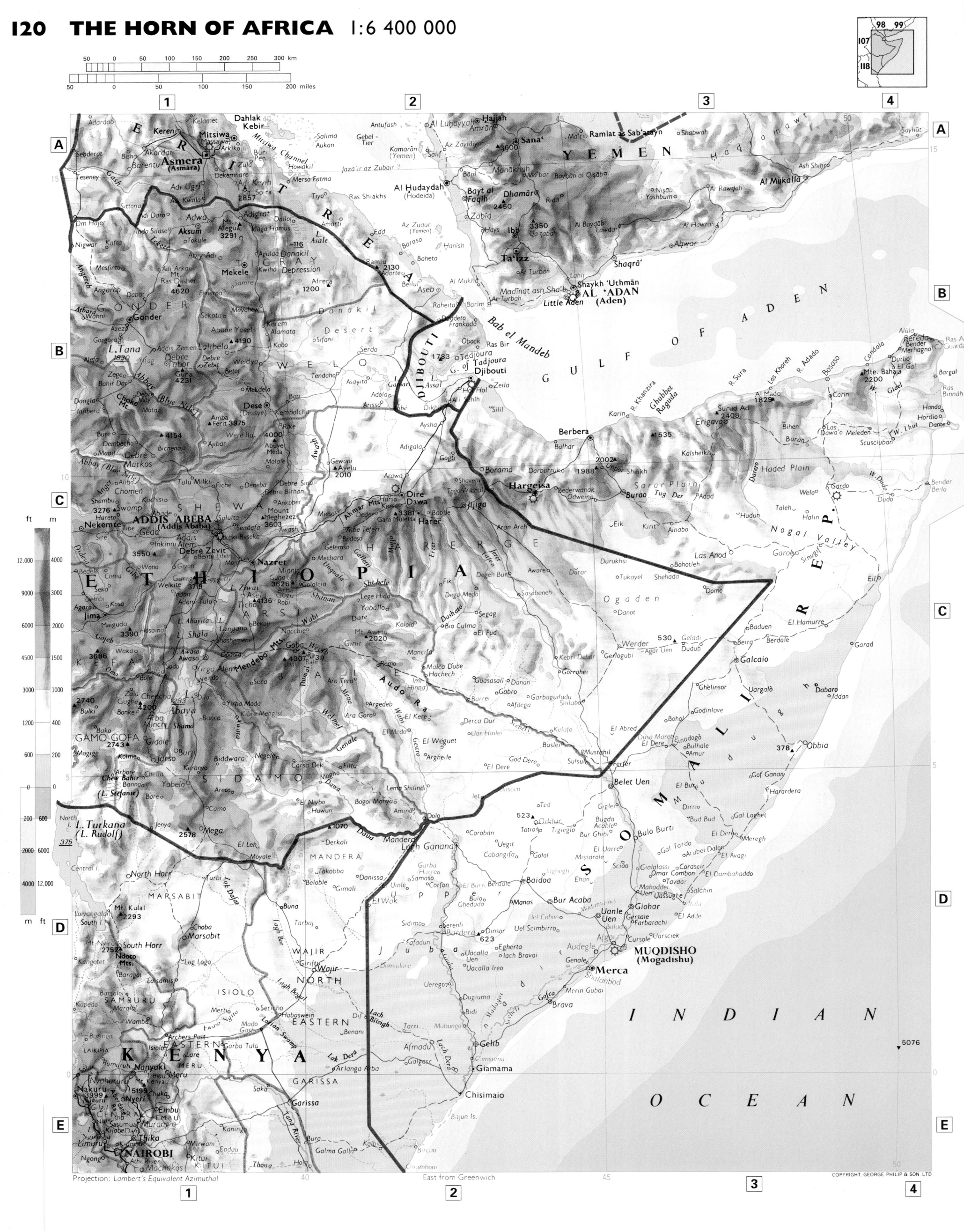

Projection: Lambert's Equivalent Azimuthal

East from Greenwich

COPYRIGHT. GEORGE PHILIP & SON. LTD

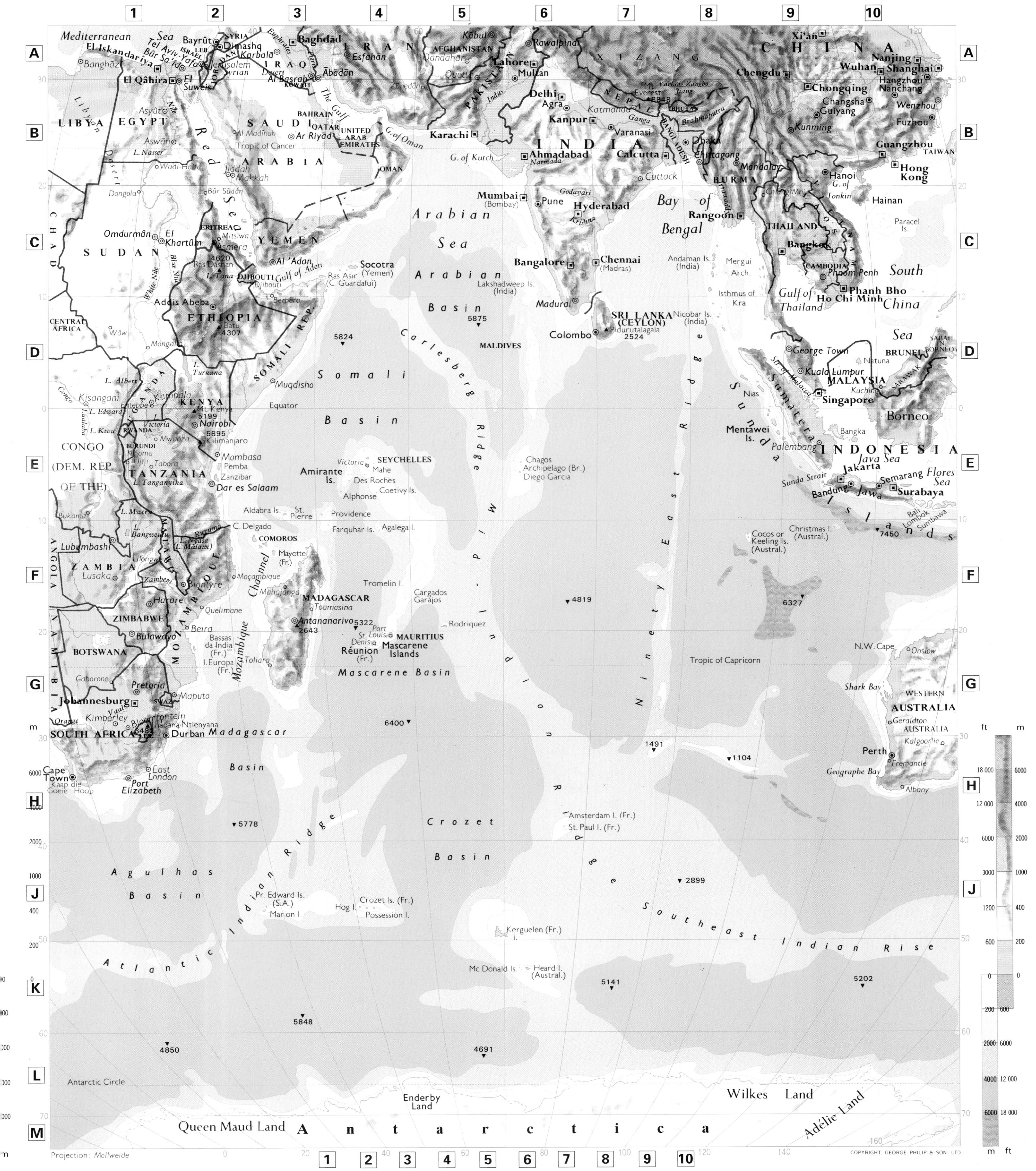

Mediterranean Sea
El Iskandarîya
Banghazî
Bayrût SYRIA Baghdad IRAN
Tel Aviv-Yafo ISRAEL LEB
Bûr Sa'îd Dimashq Karbala
El Qâhira El Suweis Jerusalem IRAQ Al Basrah Esfahân
Asyût Al Madînah KUWAIT Âbâdân Zâhedân
Kabul Rawalpindi
AFGHANISTAN
Qandahar
Quetta
Xi'an CHINA
Chengdu Wuhan Nanjing Shanghai
XIZANG
Chongqing Changsha Hangzhou Nanchang
Mt. Everest 8848 Kunming Guiyang Wenzhou Fuzhou

LIBYA EGYPT
SAUDI ARABIA
BAHRAIN QATAR
Ar Riyâd UNITED ARAB EMIRATES
OMAN
G. of Oman
Tropic of Cancer
PAKISTAN Lahore Multan
Delhi Agra
Kanpur NEPAL Katmandu BHUTAN
Yarlung Zangbo Brahmaputra
Guangzhou TAIWAN
Hong Kong

Red Sea
Dongola
Wadi Halfa
L. Nasser
Aswân
Jiddah
Mekkah
Bûr Sûdân
Karachi
INDIA
Ahmadabad
Narmada
Calcutta
Dhaka
BANGLADESH
Chittagong
Mandalay
BURMA
Hainan
G. of Tonkin
Hanoi
Paracel Is.

CHAD
SUDAN
Omdurmân El Khartûm
Blue Nile White Nile
ERITREA Mitsiwa Asmera
Ras Dashan 4620 L. Tana
DJIBOUTI Gulf of Aden
Socotra (Yemen) (C. Guardafui)
Ras Asir
YEMEN
Al 'Adan Djibouti Berbera
Mumbai (Bombay) Pune Hyderabad
Godavari Krishna
Bay of Bengal
Rangoon
THAILAND
Bangkok
Cuttack
Mergui Arch.
Andaman Is. (India)
CAMBODIA Phnom Penh
Isthmus of Kra
Gulf of Thailand
Phanh Bho Ho Chi Minh
South China Sea

CENTRAL AFRICA
Wâw
Mongalla
ETHIOPIA Addis Abeba
Batu 4307
SOMALI REP.
Muqdisho
Arabian Sea
Arabian Basin
5875 Bangalore Chennai (Madras)
Lakshadweep Is. (India)
Madurai
Colombo SRI LANKA (CEYLON)
Pidurutalagala 2524
Nicobar Is. (India)
George Town BRUNEI SABAH N. BORNEO
Kuala Lumpur Natuna
MALAYSIA Kuching SARAWAK

CONGO Kisangani L. Albert UGANDA Kampala Entebbe L. Edward RWANDA L. Kivu BURUNDI Kisoma
L. Turkana
KENYA Mt. Kenya 5199 Nairobi 5895
Victoria Mwanza Kilimanjaro
Mombasa Pemba Zanzibar
5824 Carlsberg Ridge
MALDIVES
Somali Basin
Equator
SEYCHELLES Victoria Mahé
Amirante Is. Des Roches Coetivy Is. Alphonse
Chagos Archipelago (Br.) Diego Garcia
Nias
Mentawei Is. Palembang INDONESIA
Sumatera
Strait of Malacca Singapore
Borneo Bangka
Java Sea Jakarta
Sunda Strait Bandung Jawa Semarang Flores Sea Surabaya

CONGO (DEM. REP. OF THE)
Lubumbashi
TANZANIA Tabora L. Tanganyika Dar es Salaam
Bukama L. Mweru L. Bangweulu
ANGOLA ZAMBIA Lusaka
Aldabra Is. St. Pierre Providence Farquhar Is. Agalega I.
COMOROS Mayotte (Fr.)
Tromelin I.
Cargados Garajos
Rodriquez
4819
Cocos or Keeling Is. (Austral.) Christmas I. (Austral.) 7450 6327
Sunda Islands
Lombok Bali Sumbawa

Mid
Blantyre MALAWI Lilongwe L. Nyasa (L. Malawi) Zambezi
MOZAMBIQUE Moçambique Mahajanga
MADAGASCAR Toamasina
Antananarivo 5322 2643
St. Louis Port Denis MAURITIUS Réunion (Fr.) Mascarene Islands
Mascarene Basin

NAMIBIA
ZIMBABWE Harare Bulawayo Beira
Quelimane
Mozambique Channel
Bassas da India (Fr.) I. Europa (Fr.) Toliara
Indian Ridge
Tropic of Capricorn
N.W. Cape Onslow
WESTERN AUSTRALIA

BOTSWANA Gaborone Pretoria SWAZILAND Maputo
Johannesburg Kimberley Bloemfontein Ntlenyana 3482 Durban
SOUTH AFRICA
Madagascar Basin
6400
Shark Bay
Geraldton AUSTRALIA
Kalgoorlie
Perth Fremantle
Geographe Bay
1491 1104

Cape Town Kaap die Goeie Hoop
East London Port Elizabeth
Orange Vaal
5778
Amsterdam I. (Fr.) St. Paul I. (Fr.)
Crozet Basin
Albany

Agulhas Basin
Atlantic Indian Ridge
Pr. Edward Is. (S.A.) Marion I. Hog I. Crozet Is. (Fr.) Possession I.
Kerguelen (Fr.)
2899
Southeast Indian Rise

McDonald Is. Heard I. (Austral.)
5141 5202

5848
4691 4850
Antarctic Circle
Enderby Land Wilkes Land

Queen Maud Land **Antarctica** Adélie Land

ft m
18 000 6000
12 000 4000
6000 2000
3000 1000
1200 400
600 200
0 0
200 600
2000 6000
4000 12 000
6000 18 000
m ft

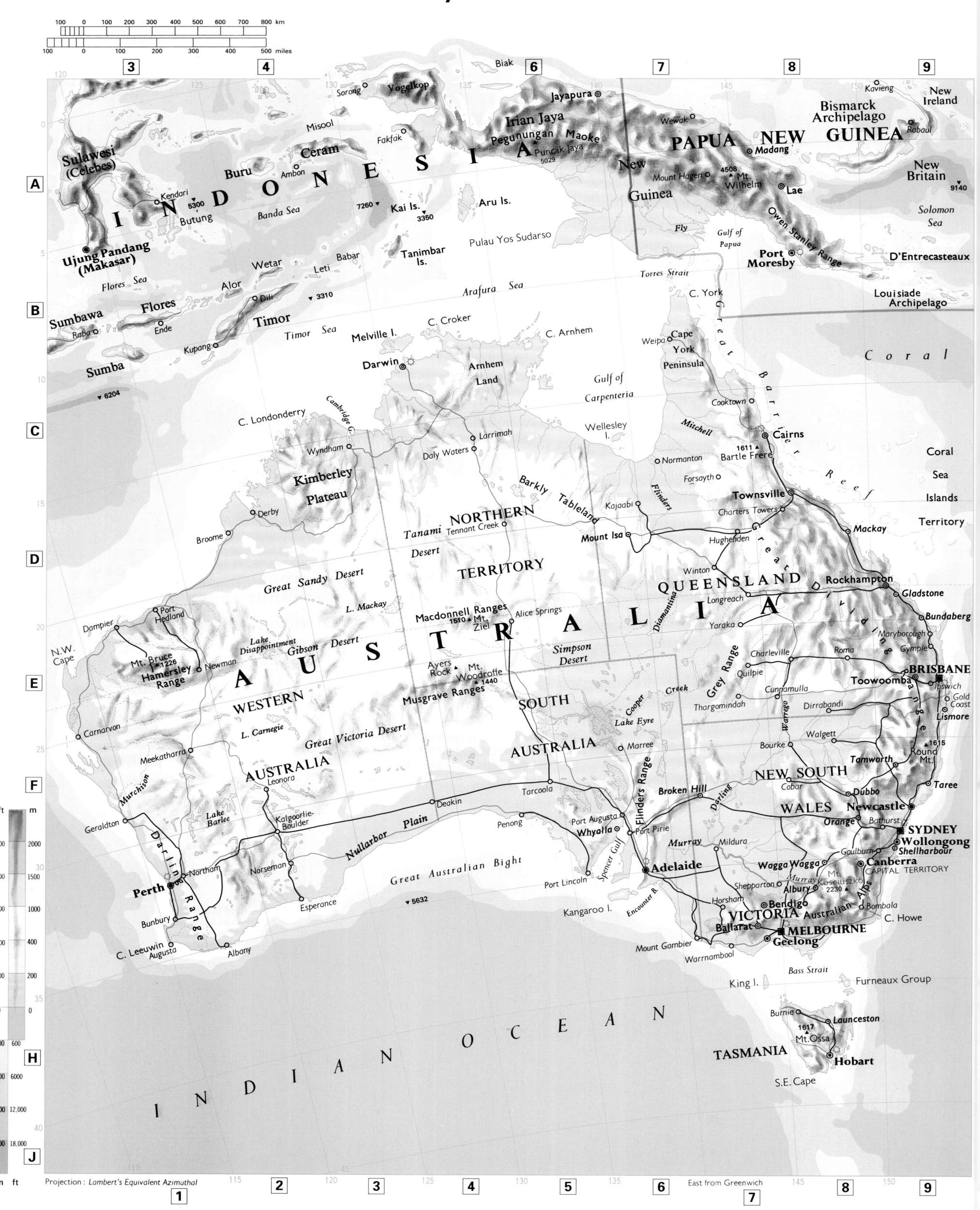

Projection: *Lambert's Equivalent Azimuthal*

East from Greenwich

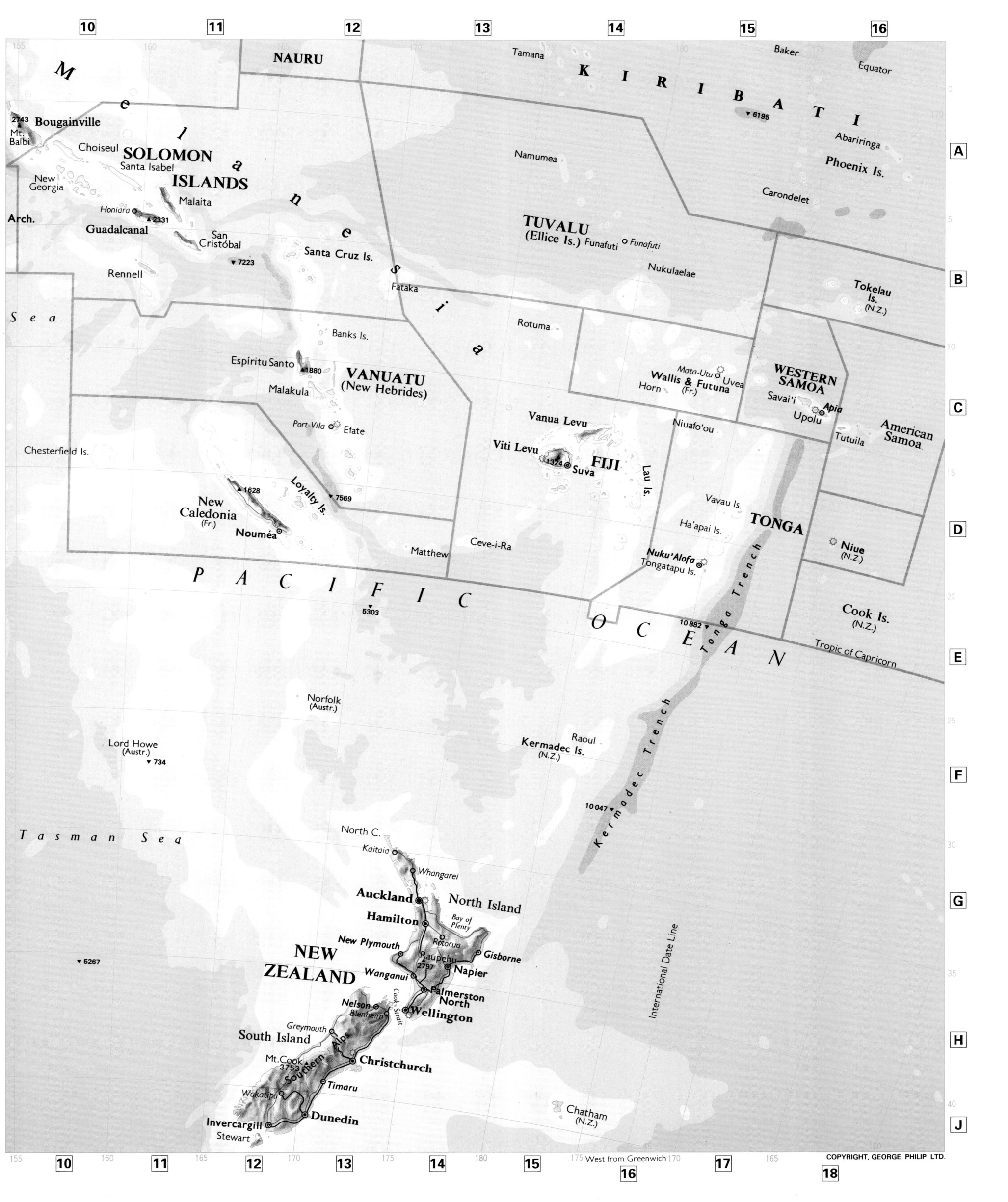

M e l a n e s i a

NAURU

K I R I B A T I

Tamana

Baker

Equator

▲2743
Mt. Balbi
Bougainville

Choiseul

Santa Isabel

SOLOMON ISLANDS

New Georgia

Arch.

Honiara ○
▲2331
Guadalcanal

San Cristóbal

Rennell

▼7223

Malaita

Santa Cruz Is.

Fataka

Namumea

Abariringa

Phoenix Is.

Carondelet

TUVALU
(Ellice Is.) Funafuti ○ Funafuti

Nukulaelae

Tokelau Is.
(N.Z.)

M

S e a

Banks Is.

Espíritu Santo ▲1880

Malakula

VANUATU
(New Hebrides)

Rotuma

Port-Vila ○ Efate

Chesterfield Is.

Matthew

Ceve-i-Ra

New Caledonia
(Fr.)

▲1628

Loyalty Is.

▼7569

Nouméa ○

Vanua Levu

Viti Levu ▲1324 ○ Suva **FIJI**

Lau Is.

Niuafo'ou

Vavau Is.

Ha'apai Is.

TONGA

Nuku'Alofa ○
Tongatapu Is.

Mata-Utu ○ Uvea
Wallis & Futuna
Horn (Fr.)

WESTERN SAMOA

Savai'i ○
Upolu ○ **Apia**

Tutuila

American Samoa

Niue
(N.Z.)

Cook Is.
(N.Z.)

Tropic of Capricorn

P A C I F I C

▼5303

O C E A N

10 882 ▼

Tonga Trench

Norfolk
(Austr.)

Raoul

Kermadec Is.
(N.Z.)

Kermadec Trench

Lord Howe
(Austr.)

▼734

10 047 ▼

T a s m a n   S e a

North C.

Kaitaia ○

▼5267

Whangarei ○

Auckland ○

Hamilton ○

New Plymouth ○

NEW ZEALAND

Wanganui ○

Nelson ○
Blenheim ○

Greymouth ○

Southern Alps

South Island

Mt.Cook
3753

Wakatipu

Invercargill ○

Stewart

Timaru ○

Dunedin ○

North Island

Bay of Plenty

Rotorua ○
Ruapehu
2797

Gisborne ○

Napier ○

Palmerston North ○

Cook Strait

Wellington ○

Christchurch ○

International Date Line

Chatham
(N.Z.)

West from Greenwich

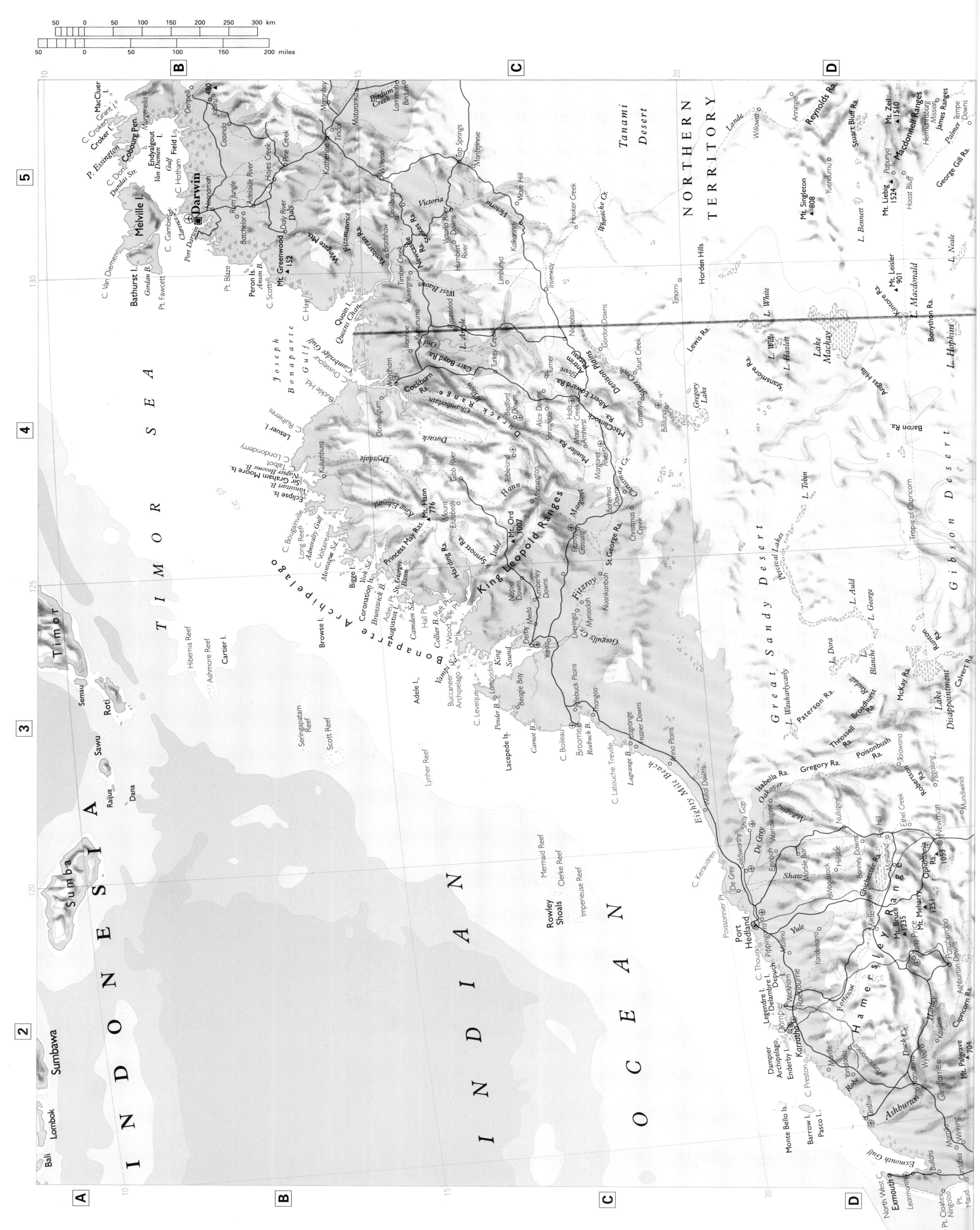
50 0 50 100 150 200 250 300 km
50 0 50 100 150 200 miles
INDONESIA
TIMOR SEA
INDIAN OCEAN
NORTHERN TERRITORY
Tanami Desert
Great Sandy Desert
Gibson Desert
Timor
Sumba
Sumbawa
Lombok
Bali
Roti
Sawu
Semau
Raijua
Dana
Darwin
Melville I.
Bathurst I.
Cobourg Pen.
Van Diemen Gulf
Joseph Bonaparte Gulf
Bonaparte Archipelago
King Leopold Ranges
Durack Range
Hamersley Range
Ophthalmia Ra.
Macdonnell Ranges
Lake Mackay
Lake Disappointment
L. Macdonald
L. Hopkins
Port Hedland
Broome
Derby
Karratha
Dampier
Roebourne
Wyndham
Kununurra
Halls Creek
Fitzroy Crossing
Newman
Tom Price
Paraburdoo
Exmouth
Onslow
Katherine
Eighty Mile Beach
Exmouth Gulf
Rowley Shoals
Scott Reef
Ashmore Reef
Cartier I.
Hibernia Reef
Browse I.
Adele I.
Lacepede Is.
Monte Bello Is.
Barrow I.
Mt. Ord 1007
Mt. Hann 776
Mt. Bruce 1235
Mt. Meharry 1251
Mt. Liebig 1524
Mt. Zeil 1510
Mt. Leisler 901
Mt. Singleton 808
Mt. Palgrave 704
Tropic of Capricorn

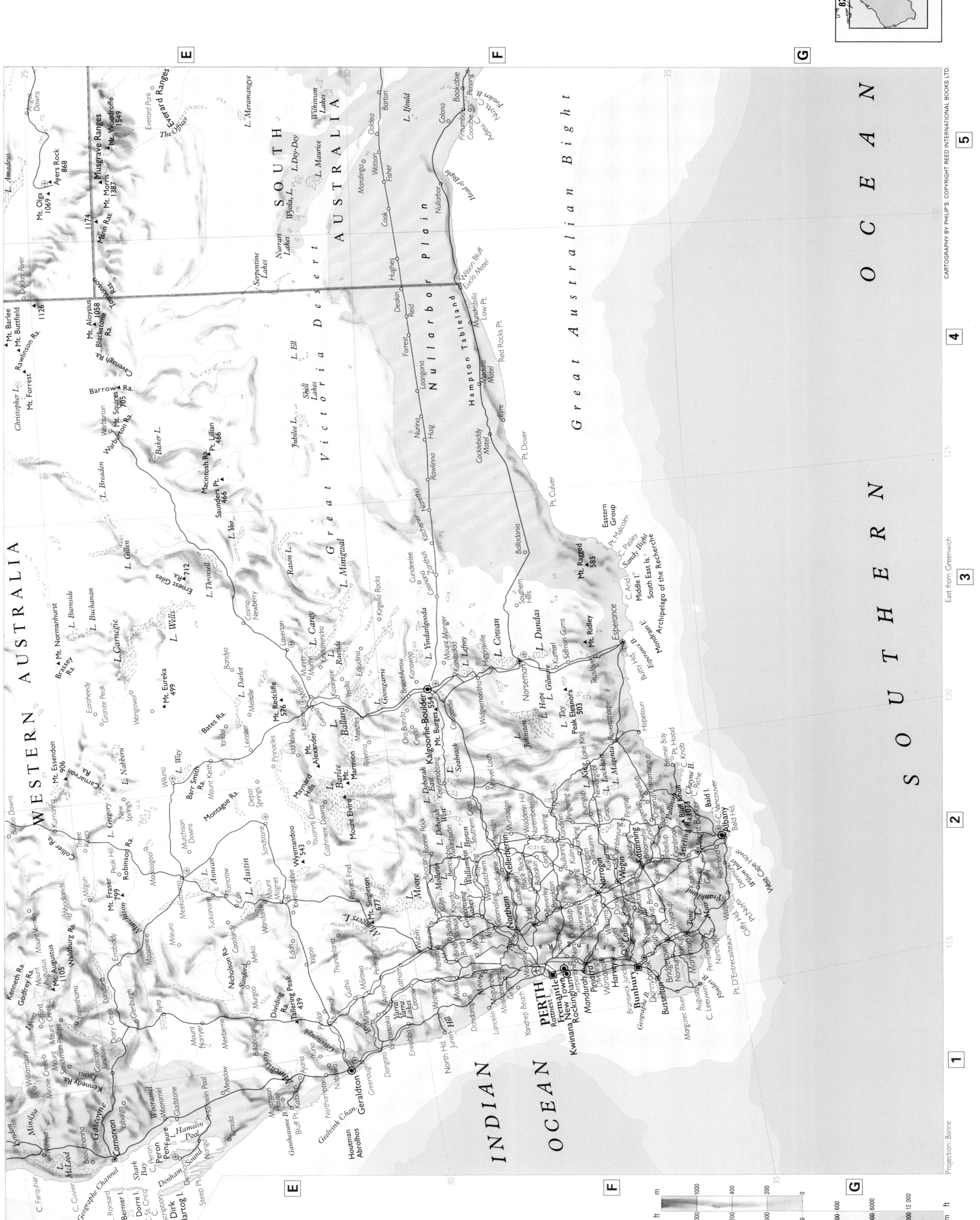

SOUTH

AUSTRALIA

WESTERN AUSTRALIA

SOUTHERN OCEAN

INDIAN

OCEAN

Great Australian Bight

Great Victoria Desert

Nullarbor Plain

Hampton Tableland

Musgrave Ranges

Mt. Woodroffe 1549
Mt. Morris 1387
Mt. Olga 1069
Ayers Rock 868
L. Amadeus

L. Meramangye
Everard Park
The Everard Ranges
The Officer
Serpentine Lakes
Wilkinson Lakes
L. Maurice
L. Dey-Dey
Wynola L.
L. Ifould
Barton
Ooldea
Colona
Coorabie Penong
Nyars Br.
Bookabie
Pintumba

Cook
Hughes
Deakin
Forrest
Reid
Wilson Bluff
Eucla Motel
Mundrabilla
Low Pt.
Red Rocks Pt.
Madura Motel
Loongana
Nurina
Haig
Cocklebiddy Motel
Rawlinna
Pt. Dover
Pt. Culver

Mt. Barlee
Mt. Buttfield 1058
Rawlinson Ra. 1126
Mt. Aloysius
Blackstone Ra.
Cavenagh Ra.
Mt. Forrest
Barrow Ra.
Mt. Squires Ra. 705
Warburton
Macintosh Ra.
Pt. Lillian 466
Saunders Pt. 466
L. Yeo
L. Throssell
Jubilee L.
Shell Lakes
L. Ell
Minigwal

Christopher L.d.
Mt. Normanhurst
L. Breaden
Baker L.
Ernest Giles
Brassey Ra.
L. Burnside
L. Buchanan
L. Gillen
Ra. 712
L. Wells
L. Carnegie
Rason L.
L. Yeo
Lake Carey
L. Minigwal
L. Rebecca

Kalgoorlie-Boulder
Esperance
Norseman
L. Cowan
L. Dundas
L. Lefroy
Mt. Ragged 585
Mt. Ridley
Eastern Group
C. Arid
Middle I.
South East Is.
Sandy Bight
Archipelago of the Recherche
C. Pasley

PERTH
Fremantle
Rockingham
Kwinana
Mandurah
Bunbury
Busselton
Albany
Geraldton
Carnarvon
Shark Bay
Hamelin Pool
Houtman Abrolhos
Geographe Channel

Projection: Bonne

CARTOGRAPHY BY PHILIP'S. COPYRIGHT REED INTERNATIONAL BOOKS LTD.

East from Greenwich

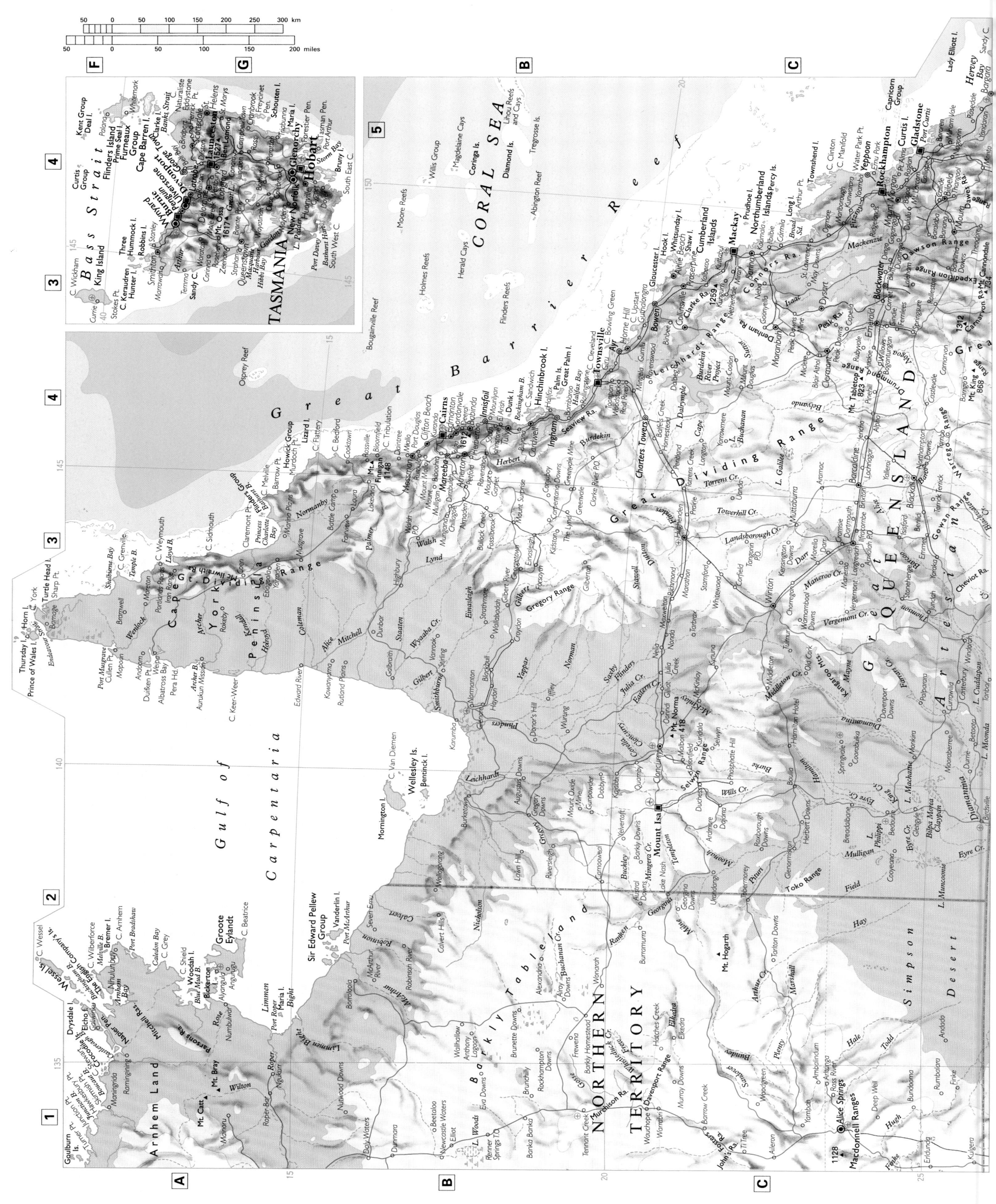

50 0 50 100 150 200 250 300 km
50 0 50 100 150 200 miles
TASMANIA
Bass Strait
King Island
Flinders Island
Furneaux Group
Cape Barren I.
Launceston
Devonport
Burnie
Hobart
Glenorchy
New Norfolk
CORAL SEA
Great Barrier Reef
Gulf of Carpentaria
Cape York Peninsula
Arnhem Land
Groote Eylandt
Wellesley Is.
Mornington I.
Sir Edward Pellew Group
NORTHERN TERRITORY
QUEENSLAND
Great Dividing Range
Great Artesian Basin
Barkly Tableland
Simpson Desert
Cairns
Townsville
Mackay
Rockhampton
Gladstone
Bowen
Charters Towers
Mount Isa
Cloncurry
Alice Springs
Macdonnell Ranges
Whitsunday I.
Cumberland Islands
Northumberland Islands
Hinchinbrook I.
Thursday I.
Prince of Wales I.
Horn I.
Turtle Head I.
Mt. Bartle Frere 1612
Mt. Norma 418
Mt. Ossa 1617

SOUTH AUSTRALIA
NEW SOUTH WALES
VICTORIA
TASMAN SEA
SOUTHERN OCEAN
Bass Strait
ADELAIDE
MELBOURNE
SYDNEY
CANBERRA
BRISBANE
Newcastle
Wollongong
Geelong
Ballarat
Bendigo
Broken Hill
Port Augusta
Whyalla
Port Pirie
Port Lincoln
Mount Gambier
Kangaroo I.
King Island
Flinders Island
Lake Eyre (North)
Lake Eyre (South)
Lake Torrens
Lake Frome
L. Gairdner
Spencer Gulf
Projection: Bonne
East from Greenwich
CARTOGRAPHY BY PHILIP'S. COPYRIGHT REED INTERNATIONAL BOOKS LTD.

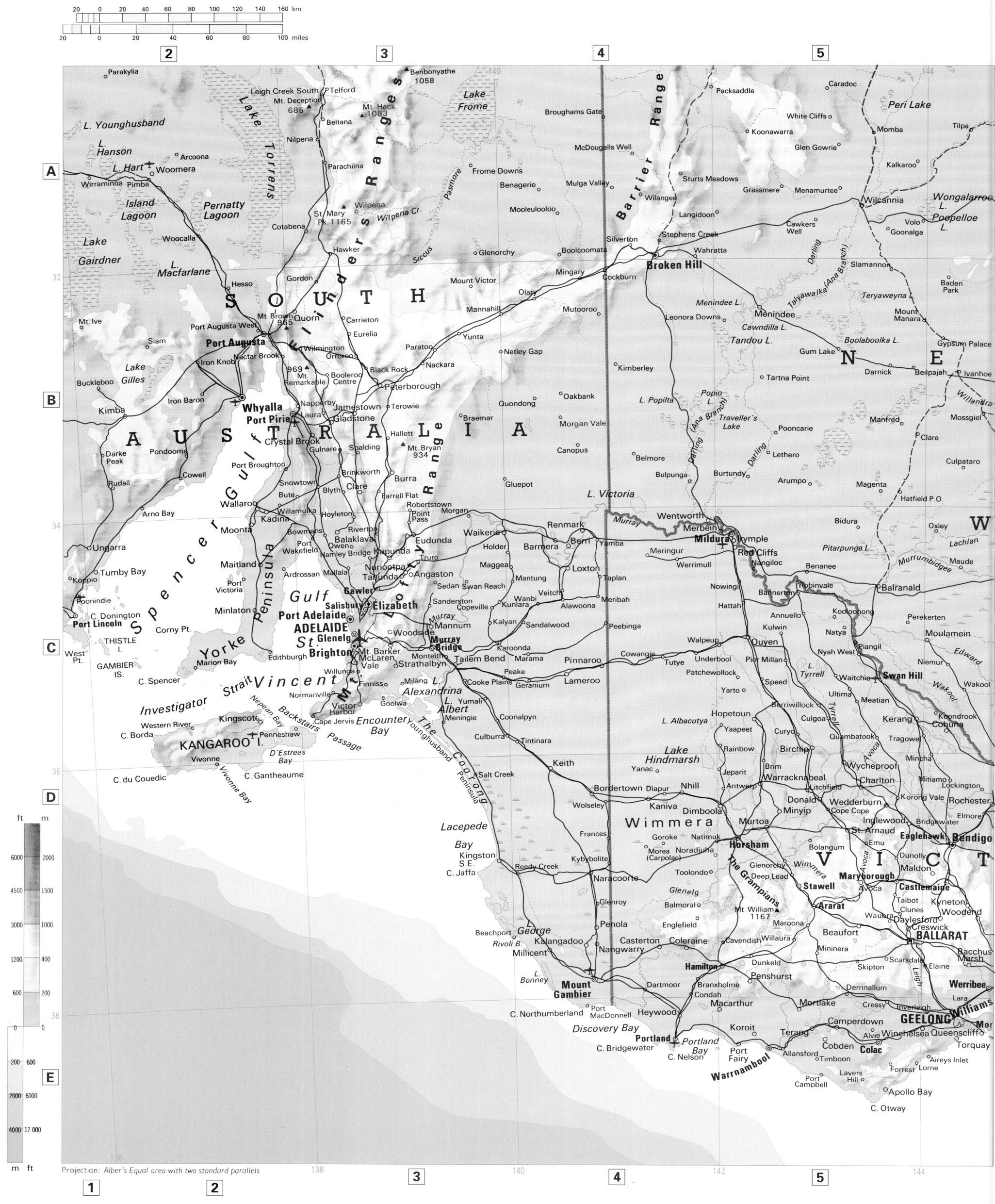

20 0 20 40 60 80 100 120 140 160 km
20 0 20 40 60 80 100 miles
SOUTH AUSTRALIA
N E W
V I C T
Spencer Gulf
Gulf St. Vincent
Investigator Strait
Backstairs Passage
Encounter Bay
Lacepede Bay
Discovery Bay
Portland Bay
Yorke Peninsula
KANGAROO I.
Flinders Ranges
Mt. Lofty Ranges
Barrier Range
Lake Torrens
Lake Frome
Lake Gairdner
L. Younghusband
L. Hanson
L. Hart
Island Lagoon
Pernatty Lagoon
L. Macfarlane
Lake Gilles
L. Victoria
Menindee L.
Cawndilla L.
Tandou L.
Peri Lake
L. Popilta
L. Albacutya
Lake Hindmarsh
L. Alexandrina
L. Tyrrell
Wimmera
The Coorong
The Grampians
Murray
Darling
Lachlan
Murrumbidgee
Woomera
Port Augusta
Whyalla
Port Pirie
Port Lincoln
Broken Hill
Wilcannia
Menindee
Wentworth
Mildura
Renmark
Berri
Barmera
Loxton
Gawler
Elizabeth
Salisbury
Port Adelaide
ADELAIDE
Glenelg
Brighton
Murray Bridge
Tailem Bend
Victor Harbor
Kingscote
Keith
Bordertown
Naracoorte
Penola
Millicent
Mount Gambier
Portland
Warrnambool
Hamilton
Horsham
Stawell
Ararat
Maryborough
Castlemaine
BALLARAT
Bendigo
GEELONG
Colac
Ouyen
Swan Hill
Kerang
Balranald
Robinvale
C. Otway
Apollo Bay
C. Nelson
C. Bridgewater
C. Northumberland
C. du Couedic
C. Borda
C. Gantheaume
C. Spencer
C. Jervis
Projection: Alber's Equal area with two standard parallels

127
127
126

6 7 8 9 10

146 148 150 152

Louth
Byrock
Carinda
Teridgerie Cr.
Gwabegar
Bohena Cr.
Turrawan
Barraba
Black Mountain
1684
Chandlers Pk.
Coffs Harbour
Dorrigo
Bellingen
Curraweena
Glenariff
Bogan
Marthaguy
Colossal
Nelgowrie
Baradine
Boggabri
Upper Manilla
Kingstown
Armidale
Uralla
Kentucky
Nambucca Heads
Macksville
Curranyalpa
Wilgaroon
Coolabah
Quambone
Coonamble
Namoi
Manilla
Attunga
Bendemeer
Walcha Road
Macleay
Smithtown
Burnamwood
Girilambone
Pine Ridge
Creek
Combara
Gunnedah
Peel
Limbri
Walcha
Booroomugga
Yearinan
Coonabarabran
Ulamambri
Liverpool
Plains
Tamworth
Range
Mt. Banda Banda
1263
Kempsey
A
Cobar
Hermidale
Gunningbar Ck.
Haddon Rig
Gulargambone
Castlereagh
Currabubula
Tia
Hastings
Rollands Plains
Barnato
Elsinore
Canbelego
Nyngan
Mullengudgery
Armatree
Binnaway
Ulinda
Werris Creek
Yarras
Wauchope
Tamarang
Quirindi
Warren
Collie
Nowendoc
Everdale
Crowl
Rest Downs
Gilgandra
Neilrex
Oakley Creek
Willow Tree
Comboyne
Port Macquarie
Kendall
Nevertire
Coolah
Murrurundi
Elandso
Buddabadah
Merrygoen
Dunedoo
Hannahs Bridge
Liverpool
Range
Ellerston
Moorland
Nymagee
Cr.
Trangie
Eumungerie
Mogriguy
Wingen
Wingham
Lansdowne
Taringo Downs
Tottenham
Bogan
Brocklehurst
Talbragar
Craboon
Barrington Tops
1555
Taree
Bobadah
Narromine
Dubbo
Minore
Tallawang
Scone
Gloucester
Albert
Geurie
Merriwa
Aberdeen
Stratford
Gilgunnia
Gungal
Muswellbrook
Wards River
Tuncurry
Forster
Toongi
Tomingley
Gulgong
Stroud Road
Yathong
Tullamore
Wellington
Denman
Ravensworth
Dungog
W
Tiarra
Tallebung
Mudgee
Baerami Creek
Bulahdelah
Mt. Hope
S
Melrose
O
Peak Hill
U
T
Yeoval
H
Lake Burrendong
Lue
Singleton
Paterson
Booral
Conoble
Wee Elwah
Matakana
Gunebang
Condobolin
Trundle
Goonumbla
Rylstone
Range
Branxton
Karuah
Trida
Roto
Ootha
Bogan Gate
Cumnock
Store Creek
Hunter
Kandos
Olinda
Coricudgy
1257
Maitland
Raymond Terrace
Thornton-Beresfield
Cessnock
Stockton
Billabong
Cr.
Gunniguldrie
Parkes
Molong
Euchareena
Putty
Kurri Kurri
Toronto
Wallsend
NEWCASTLE
Lake Cargelligo
Borenore
Ben Bullen
Blue
Belmont
552
Ural
Tullibigeal
Forbes
Orange
Portland
Morisset
Swansea
B
Marrowie Cr.
Hillston
Bucher
Spring Hill
Bathurst
Waller awang
Mts.
Lithgow
Colo
Wyong
Budgewoi
Cowl Cowl
Naradhan
Ungarie
L. Cowal
Eugowra
Canowindra
Blayney
Gosford
The Entrance
Kikoira
Tarana
Kurrajong
Rankins Springs
Pullabooka
Billimari
Carcoar
Oberon
Richmond
Windsor
Woy Woy
Merriwagga
Marsden
Caragabal
Cowra
Woodstock
Penrith
Katoomba
Hornsby
Booligal
Goolgowi
West Wyalong
Wyangala Res.
Parramatta
Fairfield
Manly
Quandialla
Grenfell
Lake Burragorang
Liverpool
SYDNEY
A
L
E
S
Bellarwi
Barmedman
Bribbaree
Koorawatha
Camden
Sutherland
Cronulla
Barellan
Mirrool
Reefton
The Oaks
Campbelltown
Griffith
Yenda
Peelwood
Picton
Helensburgh
Beabula
Hanwood
Ardlethan
Young
Frogmore
34
Hay
Willbriggie
Leeton
Temora
Boorowa
Crookwell
Mittagong
Bargo
Bulli
Woonona
Yanco
Murrumburrah
Harden
Galong
Cullarin
Range
Roslyn
Taralga
Bowral
WOLLONGONG
Port Kembla
Shellharbour
Narrandera
Cootamundra
Ganmain
Bethungra
Muttama
Binalong
Moss Vale
Robertson
Morundah
Coolamon
Goulburn
Marulan
Kiama
Kywong
Junee
Pettitts
Yass
Murrumbateman
Berry
Gerringong
Wanganella
Borea Creek
Wagga Wagga
Gundagai
Burrinjuck Res.
L. George
Nowra
Bomaderry
Yanco Cr.
Lockhart
Alfred Town
Conargo
Bundure
The Rock
Adelong
Tumut
CANBERRA
Bungendore
Wandandian
Jervis Bay (Commonwealth Territory)
C
Jerilderie
Urana
Pleasant Hills
Gilmore
A.C.T.
Queanbeyan
Braidwood
Marlow
St. Georges Hd.
Deniliquin
Oaklands
Rand
Henty
Humula
Kunama
Batlow
Royalla
Ulladulla
Finley
Berrigan
Culcairn
Rosewood
Tumbarumba
Bimberi Pk.
1910
Captains Flat
Majors Creek
Tocumwal
Holbrook
East Lynne
Mathoura
Balldale
Walla Walla
Colinton
Batemans Bay
Bateman's Bay
Nathalia
Cobram
Mulwala
Gerogery
Walwa
Adaminaby
Barnes
Numurkah
Yarrawonga
Corowa
Rutherglen
Albury
L. Eucumbene
Bredbo
Moruya
Echuca
Katamatite
Chiltern
L. Hume (Res.)
Cudgewa
Mt. Jagungal
2060
Alps
Kyabram
Yabba North
Springhurst
Wodonga
Yackandandah
Gourock
Tuross Head
ongala
Corryong
Cooma
Tatura
Wangaratta
Tallangatta
Murray
Narooma
36
Stanhope
Shepparton
Mooroopna
Everton
Beechworth
Mt. Benambra
1476
Snowy Mts.
2230
Mt. Kosciuszko
Jindabyne
Yowrie
Rock Flat
C. Dromedary
TASMAN
Rushworth
Glenrowan
Colbinabbin
Violet Town
Benalla
Ovens
Myrtleford
RANGE
Australian
Nimmitabel
Goalen Hd.
Nagambie
Euroa
Bright
Mt. Bogong
1986
Jimenbuen
Bega
Heathcote
O
R
I
A
Mount Beauty
Whitfield
GREAT DIVIDING RANGE
Tathra
Candelo
Seymour
Carisbrook
Bonnie Doon
Bowen Mts.
Bombala
Tallarook
Mansfield
Glen Valley
1836
Mt. Cobberas
Corrowidgie
Cathcart
Rowes
Broadford
Alexandra
L. Eildon
Mt. Buller
1806
Omeo
Delegate
Eden
Twofold Bay
SEA
Kilmore
Yea
Eildon
Swifts Creek
Bonang
Towamba
D
Heathcote Junc.
Snowy
Wonboyn
Green C.
Disaster B.
Whittlesea
Glenburn
GREAT
DIVIDING
Mt. Tamboritha
1646
Mt. Ellery
1297
Sunbury
Buchan
Genoa
Eltham
Healesville
Cobbannah
Club Terrace
C. Howe
Sunshine
MELBOURNE
Aberfeldy
Warburton
Gippsland
Bruthen
Nowa Nowa
Orbost
Cann River
Mallacoota
Mallacoota Inlet
Bairnsdale
town
Walhalla
Heyfield
Munro
C. Conran
Ram Head
Dandenong
Stratford
Lakes Entrance
Chelsea
Seaford
Pakenham
Hill End
Maffra
Beach
Moe
Yallourn
Traralgon
Sale
L. Wellington
ington
Drouin
Warragul
Trafalgar
Morwell
The Ninety Mile
Frankston
Hastings
FRENCH
Nyora
Mirboo North
Churchill
Boolarra
Seaspray
38
Rosebud
PHILLIP I.
Korumburra
San Remo
Leongatha
Yarram
Woodside
Koonwarra
Meeniyan
C. Woolamai
Anderson
Wonthaggi
Inverloch
Welshpool
Venus B.
Foster
Port Albert
SNAKE I.
E
C. Liptrap
Waratah B.
Wilsons Promontory

6 146 7 148 8 150 9 152 10

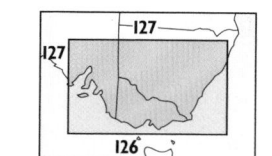

10 0 20 40 60 80 100 120 140 km
10 0 20 40 60 80 100 miles

131

1 2 3 4 5 6 7 8

C. Reinga
C. Maria van Diemen
North C.
Parengarenga Harb.
Ninety Mile Beach
Houhora
Rangaunu B.
Awanui
C. Karikari
Doubtless B.
Mangonui
Whangaroa Harb.
Kaitaia
Kaeo
Ahipara B.
Herekino
Kerikeri
Okaihau
Cavalli I.
Kohukohu
Kaikohe
Russell
Opua
Bay of Islands
C. Brett
NORTHLAND
Hokianga Harb.
Rawene
Omapere
776
Donnelly's Crossing
Aranga
Kawakawa
Kamo
Hikurangi
Poor Knights Island
Dargaville
Wairoa
Whangarei
Onerahi
Whangaruru Harb.
Te Kopuru
Ruawai
Waikieke
Waipu
Whangarei Harb.
Bream Head
Paparoa
Maungaturoto
Bream Bay
Hen & Chickens Islands
Bream Tail
Needles Point
Port Fitzroy
Wellsford
C. Rodney
Lit. Barrier I.
Great Barrier I.
Helensville
Motakana
Kawau I.
Warkworth
C. Barrier
Hauraki Gulf
Cuvier I.
AUCKLAND
Birkenhead
Takapuna
Devonport
Waiheke I.
Coromandel
Port Charles
Mercury Is.
AUCKLAND
Mt. Roskill
Howick
Coromandel
Mercury B.
Whitianga
Coromandel
Peninsula
Onehunga
835
Papatoetoe
Thames
Manukau Harb.
Manukau
Papakura
Thames
Pukekohe
Whangamata
Mayor I.
Waiuku
Mercer
Turua
Paeroa
Te Kauwhata
Waihi
Waikato
Waikare
Waihou
Ngatea
BAY OF PLENTY
White I.
C. Runaway
Hicks Bay
Huntly
Waitoa
Te Aroha
Tauranga Harb.
Matakana
Motiti I.
Te Kaha
Te Araroa
TASMAN
Glen Afton
Glen Massey
Morrinsville
Mt. Maunganui
Tauranga
Matata
Whakatane
Ohiwa Harbour
1753
Hikurangi
Waipiro
East C.
Ngaruawahia
Hamilton
Wahaoroa
Te Puke
Paengaroa
Edgecumbe
Opotiki
Ruatoria
Raglan Harb.
Frankton
Cambridge
Matamata
Kawerau
Motu
Tokomaru Bay
SEA
Raglan
Ohaupo
Leamington
Karapiro
Rotoiti
Rotoma
Te Teko
Taneatua
Aotea Harb.
Te Awamutu
Tirau
Rotorua
Mamaku
Te Karaka
Albatross Pt.
Arapuni
Kihikihi
Ngongotaha
Rotorua
Kaingaroa
East Cape
Kawhia Harb.
Otorohanga
Rotorua
L. Tarawera
Mt. Tarawera 1111
Waiotapu
Galatea
Ngatapa
Ormond
Gisborne
Tirua Pt.
Te Kuiti
Rangitoto 1165
Mangakino
Whakamaru
Kaingaroa State Forest Murupara
1403
Waikare Iti
Poverty Bay
Mokau
Aria
369
Waikite
Kaupo
Rangitaiki
Waikaremoana
Ongarue
Okahukura
Lake Taupo
Tuai
L. Waikare
iti
Wairoa
Waikokopu
North Taranaki Bight
Pukearuhe
Manunui
Taumarunui
Tokaanu
Ahimanawa Mts.
1383
Tarawera
Mohaka
Frasertown
Mahia Peninsula
Waitara
Tahora
Owhango
Rotoaira 2291
Ngauruhoe
New Plymouth
Inglewood
Whangamomona
Kaweka Ra.
Putorino
Wairoa
Nuhaka
Portland I.
TARANAKI
Okato
C. Egmont Mt. Egmont (Taranaki) 2518
Rahotu
Midhirst
National Park
Ruapehu 2796
Ohakune
Rangataua
Raetihi
Waiouru
Napier
Bay View
Taradale
Napier
Clive
C. Kidnappers
Hawke Bay
Stratford
Eltham
Normanby
Hawera
Manaia
Kaponga
Opunake
Kapuni
South Taranaki Bight
Patea
Waverley
Maxwell
Waitotara
Castlecliff
Wanganui
Turakina
Hastings
Havelock North
Opapa
Otane
Tikokino
Waipawa
HAWKES BAY
1733
Mangaweka
Apiti
Waipukurau
Marton
Halcombe
WANGANUI-MANAWATU
Bulls
Rangitikei
Hunterville
Takapau
Wanstead
Ongaonga
Herbertville
C. Turnagain
Feilding
Ashhurst
Danevirke
Woodville
Paranghau
Palmerston North
Longburn
Weber
Manawatu
Pahiatua
Foxton
Shannon
Eketahuna
Alfredton
Mauriceville
Levin
Otaki
Tinui
Castlepoint
Masterton
Mitre 1571
Carterton
Paraparaumu
Paekakariki
Greytown
Featherston
Martinborough
Flat Pt.
WELLINGTON
Up. Hutt
Petone
Lr. Hutt
Wainuiomata
L. Wairarapa
C. Palliser
Wellington
Port Nicholson
Aorangi 983 Mts.
Palliser Bay

TASMAN SEA

PACIFIC OCEAN

Bay of Plenty

Lake Taupo

2297

C. Farewell
Farewell Spit
Collingwood
Golden Bay
C. Stephens
Stephens I.
D'Urville Island
French Pass
Separation Pt.
Takaka
Kahurangi Pt.
Devil River Pk. 775
Riwaka
Motueka
Tasman Bay
Pelorus Sd.
On Charlotte Sd.
Cook Strait
Karamea
Tadmor
Mt. Richmond
Brightwater
Wakefield
Stoke
Havelock
Picton
Nelson
Tuamarina
Mt. Owen 1760
Richmond Ra.
1875
Blenheim
Renwicktown
Wairau
Mokihinui
Lyell Ra.
Murchison
L. Rotoiti
C. Campbell
Ward
Seddon
Awatere
Buller

ft m
9000 3000
6000 2000
3000 1000
1200 400
600 200
0 0
200 600
2000 6000
m ft

Projection: Conical with two standard parallels
173 174 175 East from Greenwich 176 177 178 179 COPYRIGHT GEORGE PHILIP & SON LTD.

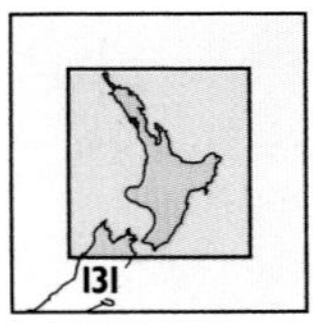

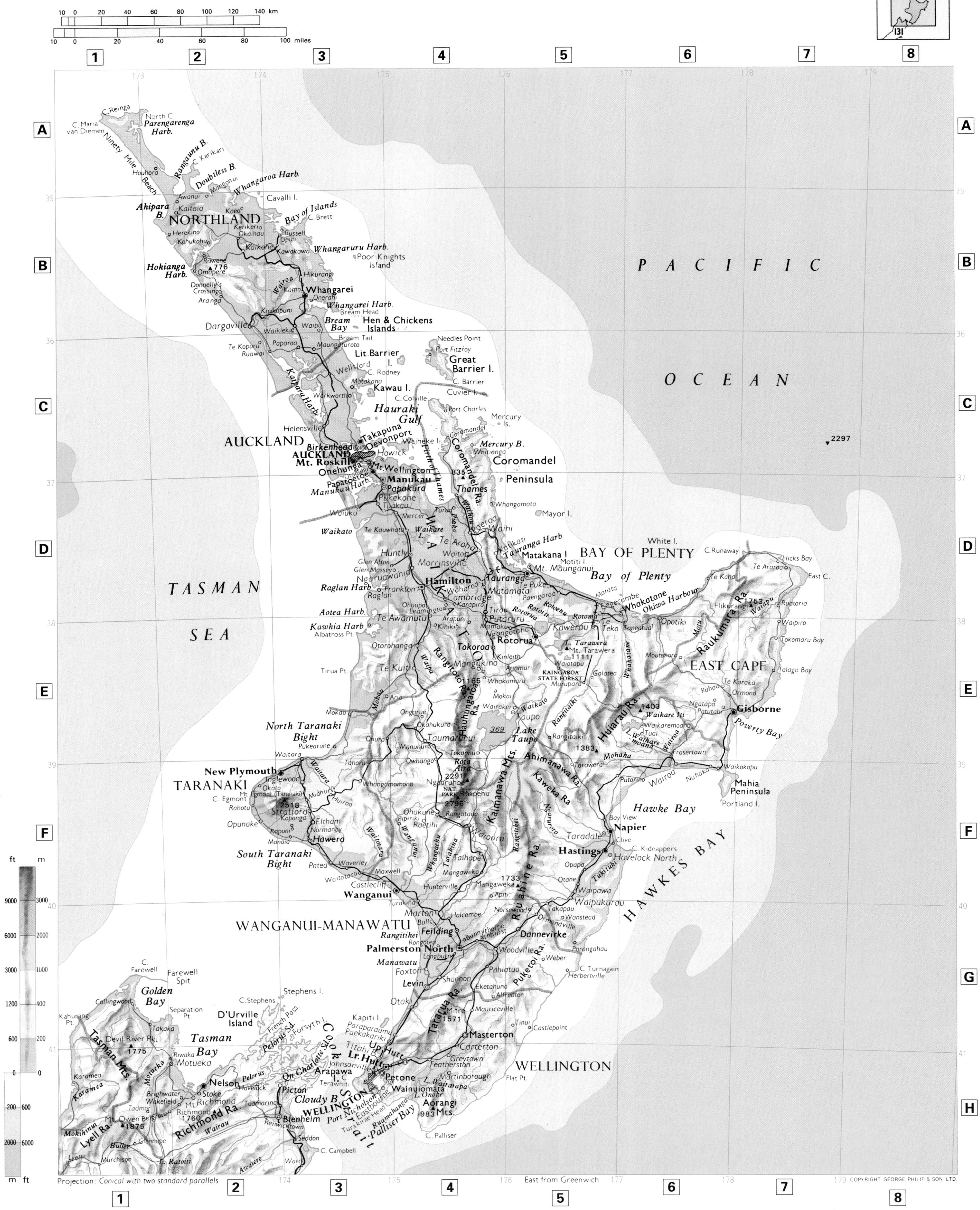

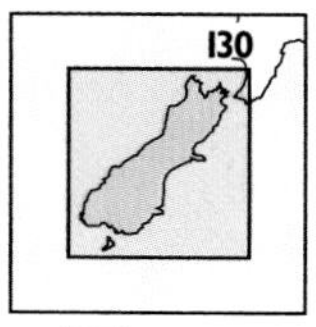

130

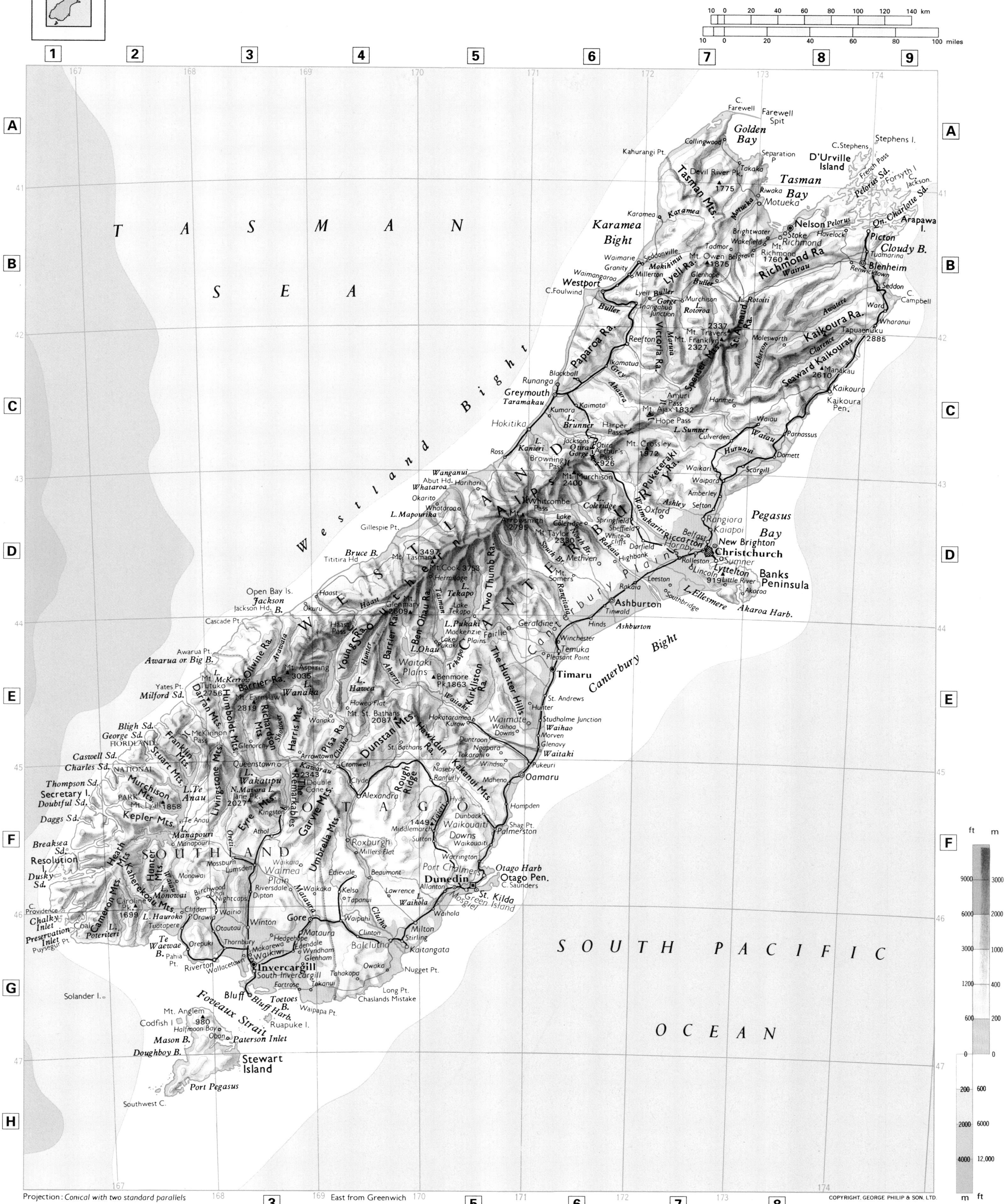

10 0 20 40 60 80 100 120 140 km
10 0 20 40 60 80 100 miles

A B C D E F G H

TASMAN SEA

Golden Bay
C. Farewell
Farewell Spit
Collingwood
Kahurangi Pt.
Takaka
Separation P
C. Stephens
Stephens I.
D'Urville Island
French Pass
Forsyth I.
Jackson
Pelorus Sd.
Qn. Charlotte Sd.
Arapawa I.

Devil River Pk.
1775
Riwaka
Motueka
Tasman Mts.
Tasman Bay
Nelson
Pelorus
Picton
Cloudy B.

Karamea
Karamea Bight
Wairau
Seddonville
Waimarie
Waimangaroa
Granity
Millerton
Mokihinui
Westport
C. Foulwind
Lyell Ra.
Brightwater
Stoke
Richmond
Wakefield
Tadmor
Mt. Owen
1875
Glenhope
Buller
Mt. Richmond
1760
Richmond Ra.
Havelock
Tuamarina
Blenheim
Renwicktown
Seddon
C. Campbell
Ward
Wharanui

Buller
Lyell
Buller Gorge
Inangahua Junction
Murchison
L. Rotoroa
L. Rotoiti
Mt. Travers
2337
Mt. Franklyn
2327
St. Arnaud Ra.
Spenser Mts.
Molesworth
Awatere
Amuri
Clarence
Kaikoura Ra.
Tapuaenuku
2885
Seaward Kaikouras
Manakau
2610
Kaikoura
Kaikoura Pen.

Paparoa Ra.
Reefton
Ikamatua
Grey
Ahaura
Maruia
Victoria Ra.
Mt. Ajax 1832
Hanmer
Amuri Pass
Hope Pass
L. Sumner
Culverden
Waiau
Waiau
Parnassus
Scargill
Domett

Blackball
Runanga
Greymouth
Taramakau
Kumara
Kaimata
Harper Pass
Mt. Crossley
1572
Puketeraki Ra.
Hurunui
Waikari
Waipara
Amberley
Sefton
Ashley

Hokitika
L. Brunner
Kanieri
Otira
Otira Gorge
Arthur's Pass
926
Browning Pass
Mt. Murchison
2400
Coleridge
Springfield
Sheffield
White Cliffs
Oxford
Rangiora
Kaiapoi
Belfast
Riccarton
Hornby
New Brighton
Christchurch
Lyttelton
Sumner
Banks Peninsula
Akaroa
Akaroa Harb.

Ross
Jackson
Wanganui
Abut Hd.
Harihari
Whataroa
Okarito
Whataroa
L. Mapourika
Gillespie Pt.
Whitcombe Pass
Mt. Arrowsmith
2795
Lake Coleridge
Mt. Taylor
2330
Darfield
Rolleston
Lincoln
919
Little River
Leeston
L. Ellesmere
Southbridge

Bruce B.
Tititira Hd.
Mt. Tasman 3497
Mt. Cook 3758
Hermitage
Mt. Somers
Methven
Rakaia
Rakaia

Open Bay Is.
Jackson
Jackson Hd.
B.
Okuru
Haast
Glenmary
2609
L. Tekapo
Tekapo
Lake Tekapo
Fairlie
Geraldine
Winchester
Temuka
Pleasant Point
Hinds
Tinwald
Ashburton

Cascade Pt.
Haast Pass
Two Thumb Ra.
L. Pukaki
Mackenzie Plains
Lake Pukaki
Burke's Pass
Cave
St. Andrews

Awarua Pt.
Awarua or Big B.
Mt. Aspiring
3035
Young & Siao
Ben Ohau Ra.
Ohau
L. Ohau
Waitaki Plains
Benmore Pk.1863
Kirkliston Ra.
The Hunter Hills
Timaru
Canterbury Bight

Yates Pt.
Milford Sd.
L. McKerrow
Tutoko
2756
Barrier Ra.
Olivine Ra.
Arawata
Dart
Mt. Earnslaw
2819
Harris Mts.
Wanaka
L. Wanaka
L. Hawea
Hawea
Haweo Flat
Mt. St. Bathans
2087
Hunter
Studholme Junction
Waimate
Waihao Downs
Waihao

Bligh Sd.
George Sd.
FIORDLAND
Caswell Sd.
Charles Sd.
Franklin Hts.
Stuart Mts.
McKinnon Pass
Glenorchy
Richardson Mts.
Pisa Ra.
Dunstan Mts.
Hawkdun Ra.
St. Bathans
Hokataramea
Kurow
Duntroon
Tokarahi
Ngapara
Waitaki
Oamaru

Thompson Sd.
Secretary I.
Doubtful Sd.
Daggs Sd.
Murchison Mts.
Mt. Lyall 1858
L. Te Anau
Te Anau
Queenstown
L. Wakatipu
2343
The Remarkables
Kawarau
Cromwell
Clyde
Rough Ridge
Nenthorn
Ranfurly
Naseby
Dunback
Windsor
Maheno
Pukeuri

Breaksea Sd.
Resolution
Dusky Sd.
Heath Mts.
Kaherekoau Mts.
Manapouri
L. Manapouri
Eyre Mts.
Garvie Mts.
2027
Jane
N. Matara L.
Double Cone
Kingston
Athol
Alexandra
Roxburgh
Millers Flat
Umbrella Mts.
Middlemarch
Hyde
Hampden
Waikouaiti Downs
Shag Pt.
Palmerston
Waikouaiti
Warrington

Providence
Chalky Inlet
Cameron Mts.
1699
Coal I.
Preservation Inlet
Puyseger Pt.
Te Waewae B.
Pahia Pt.
Riverton
Orepuki
Wallacetown
Invercargill
South Invercargill
L. Hauroka
Tuatapere
Clifden
Orawia
Otautau
Nightcaps
Mossburn
Lumsden
Birchwood
Dipton
Winton
Riversdale
Waikaia
Waimea Plain
Waikaka
Gore
Mataura
Edievale
Beaumont
Roxburgh
Tapanui
Lawrence
Waipahi
Clinton
Kelso
Waihola
Outram
Allanton
Mosgiel
Green Island
St. Kilda
Dunedin
Port Chalmers
Otago Harb.
Otago Pen.
C. Saunders

SOUTH PACIFIC OCEAN

Solander I.
Mt. Anglem 980
Codfish I.
Halfmoon Bay
Oban
Paterson Inlet
Mason B.
Doughboy B.
Stewart Island
Port Pegasus
Southwest C.

Foveaux Strait
Bluff
Bluff Hd.
Bluff Harb.
Toetoes B.
Waipapa Pt.
Ruapuke I.
Fortrose
Tokanui
Waikawa
Owaka
Tahakopa
Nugget Pt.
Long Pt.
Chaslands Mistake

Endeavour
Wyndham
Edendale
Glenham
Makarewa
Wallacetown
Balclutha
Stirling
Kaitangata
Milton

WESTLAND BIGHT
WESTLAND
CANTERBURY PLAINS
CANTERBURY
OTAGO
SOUTHLAND
Pegasus Bay

ft m
9000 3000
6000 2000
3000 1000
1200 400
600 200
0 0
200 600
2000 6000
4000 12,000
m ft

Projection: Conical with two standard parallels
East from Greenwich

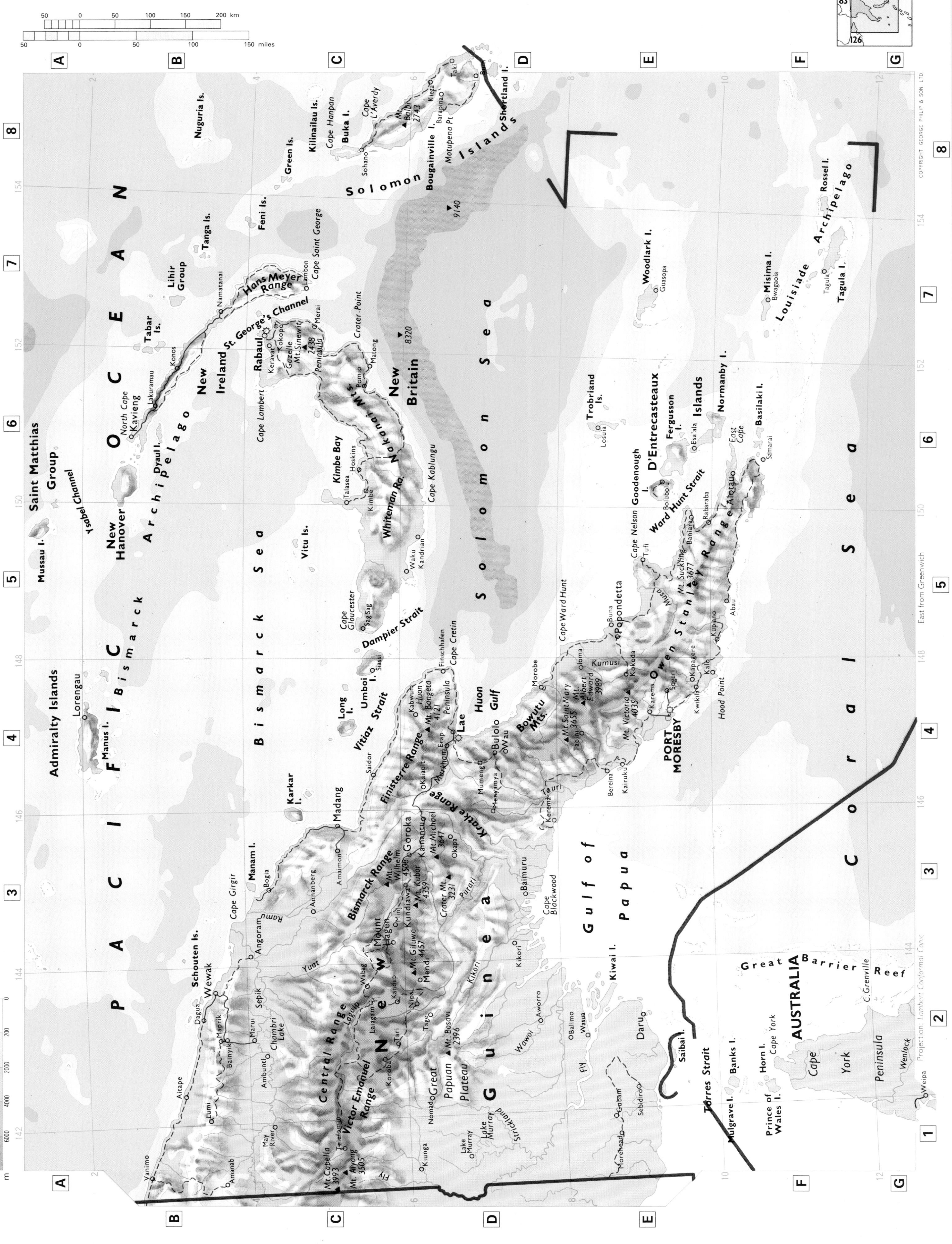

50 0 50 100 150 200 km
50 0 50 100 150 miles

83

126

COPYRIGHT GEORGE PHILIP & SON, LTD.

PACIFIC OCEAN

Admiralty Islands

Saint Matthias Group

Mussau I.

Ysabel Channel

New Hanover

New Ireland

Bismarck Archipelago

Lihir Group

Tabar Is.

Tanga Is.

Feni Is.

Nuguria Is.

Kilinailau Is.

Green Is.

Cape Hanpan

Buka I.

Cape L'Averdy

Mt. Balbi
2743

Bougainville I.

Cape Shortland I.

Motupena Pt.

Solomon Islands

9140

Kavieng
North Cape

Konos

Lakuramau

Namatanai

Hans Meyer Range

St. George's Channel

Cape Saint George

Rabaul

Gazelle Peninsula

Keravat

Kokopo

Mt. Sinewit
2438

Merai

Crater Point

Cape Lambert

Cape Gloucester

Sag Sag

Dampier Strait

Kimbe Bay

Hoskins

Talasea

Kimbe

Bomio

Matong

Nakanai Mts.

Whiteman Ra.

8320

New Britain

Cape Kabiungu

Solomon Sea

Trobriand Is.

Losuia

Woodlark I.

Gusiopa

Misima I.

Bwagaoia

Rossel I.

Tagula

Louisiade Archipelago

Tagula I.

Coral Sea

Bismarck Sea

Vitu Is.

Long I.

Umboi I.

Vitiaz Strait

Siassi

Siador

Madang

Karkar I.

Manam I.

Bogia

Cape Girgir

Schouten Is.

Wewak

Angoram

Sepik

Ramu

Yuat

Maprik

Ambunti

Chambri Lake

Baihyik

Dreikikir

Aitape

Vanimo

Ulumi

Amanab

Mt. Capella
3953

Mt. Aiyung
3605

Fly

May River

Telefomin

Victor Emanuel Range

Central Range

Koroba

Tari

Lagaip

Laiagam

Mendi
2396

Mt. Bosavi

Mt. Giluwe
4457

Mt. Hagen

Mount Hagen

Wabag

Kandep

Nipa

Tago

Kikori

Great Papuan Plateau

Lake Murray

Strickland

Okunga

Morehead

Sebidiro

Gogal

Nomad

Wawoi

Balimo

Wasua

Aworro

Daru

Kiwai I.

Kikori

Cape Blackwood

Baimuru

Kerema

Kerema

Keremu

Purari

Okapa

Mt. Michael
3647

Crater Mt.
3231

Kainantu

Goroka

Mt. Wilhelm
4508

Mt. Kubor
4359

Kundiawa

Minj

Bismarck Range

NEW GUINEA

Kratke Range

Finisterre Range

Saidor

Amaimon

Annanberg

Bulolo Range

Kabwum

Huon

Mt. Bangeta
4121

Huon Peninsula

Finschhafen

Markham

Lae

Wau

Bulolo

Mumeng

Menyamya

Wau

Morobe

Cape Cretin

Huon Gulf

Bowutu Mts.

Tapini

Mt. Saint Mary
3653

Mt. Albert Edward
3989

Otona

Kokoda

Kumusi

Mt. Victoria
4035

Mt. Suckling
3677

Owen Stanley Range

Tauri

Chemyamya

Bereina

Kairuku

Abau

Karema

Sogeri

Kwikila

Okapanere

Kalo

Hood Point

PORT MORESBY

Abuari

Cape Ward Hunt

Buna

Kumusi

Popondetta

Tufi

Cape Nelson

Goodenough I.

Bolubolu

Esa'ala

Rabaraba

Baniara

Mt. Suckling

Ward Hunt Strait

D'Entrecasteaux Islands

Fergusson I.

Normanby I.

Basilaki I.

East Cape

Samarai

Misö

Gulf of Papua

Cape York

Prince of Wales I.

Banks I.

Mulgrave I.

Saibai I.

Torres Strait

Horn I.

Weipa

C. Grenville

AUSTRALIA

Great Barrier Reef

Cape York Peninsula

Wenlock

East from Greenwich

Projection: Lambert Conformal Conic

m   ft
4000 12,000
2000 6000
1000 3000
400 1200
200 600
0   0

600 2000
2000 6000
4000 12,000
6000 18,000
ft   m

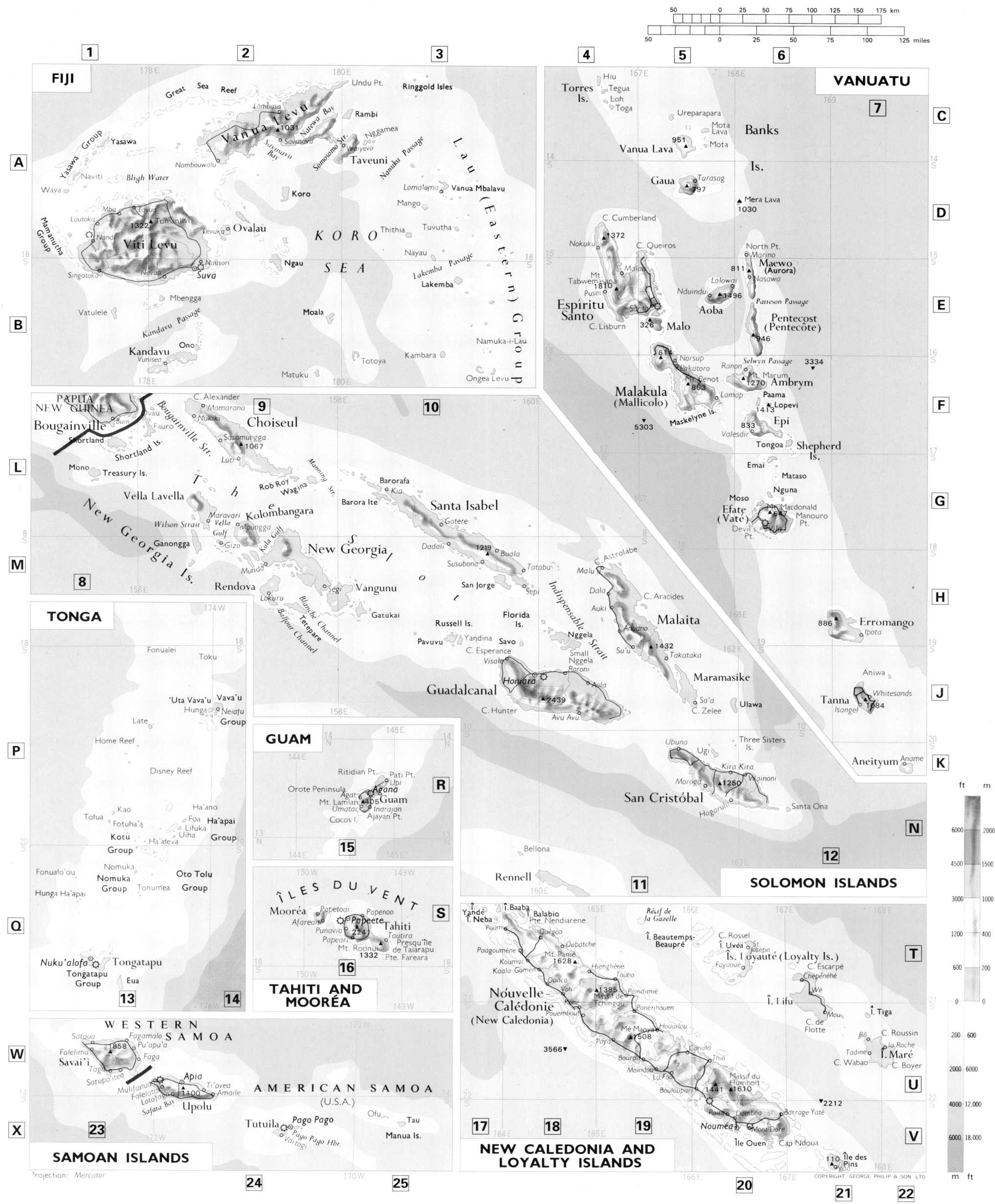

50 0 25 50 75 100 125 150 175 km
50 0 25 50 75 100 125 miles

FIJI

1 **2** **3** **4** **5** **6**

A

Great Sea Reef
Undu Pt.
Ringgold Isles
Yasawa Group
Yasawa
Lombasa
Vanua Levu
Nutewa Bay
Rambi
1031
Savusavu
Nggamea
Naviti
Nambouwalu
Savusavu Bay
Somosomo Str.
Waiyevo
Taveuni
Namuka Passage
Waya
Bligh Water
Koro
Lomaloma
Vanua Mbalavu
Mamanutha Group
Lautoka
Tavua
Tomaniivi
1322
Levuka
Ovalau
Mango
Lau (Eastern) Group
Mba
Viti Levu
Nausori
Ngau
Thithia
Tuvutha
Nayau
B
Singatoka
Navua
Suva
KORO SEA
Lakemba Passage
Lakemba
Vatulele
Mbengga
Moala
Kandavu Passage
Namuka-i-Lau
Kandavu
Ono
Totoya
Kambara
Vunisea
Matuku
Ongea Levu

VANUATU

7

C
Torres Is.
Hiu
Tegua
Loh
Toga
Ureparapara
Banks
Mota Lava
Vanua Lava
951
Mota
Is.
Gaua
Tarasag
797
Mera Lava
1030
D
C. Cumberland
North Pt.
Nokuku
1372
Marino
Maewo (Aurora)
C. Queiros
811
Mt. Tabwemasana
Malao
1810
Nasawa
Espiritu Santo
Pusei
Santo
Nduindui
Lolowai
1496
Aoba
Patteson Passage
E
326
Malo
C. Lisburn
Norsup
1617
Ranon
Mt. Marum
3334
Vunkatoro
Mt. Benot
1270
Ambrym
863
Lamap
Paama
Malakula (Mallicolo)
1413
Lopevi
Maskelyne Is.
833
Epi
5303
Valesdir
Tongoa
Shepherd Is.
F
Emai
Mataso
Nguna
Moso
Efate (Vaté)
Macdonald
647
Manouro Pt.
Devil's Pt.

PAPUA NEW GUINEA

9 **10**

L
Bougainville
C. Alexander
Mamarana
Buin
Vau
Nukiki
Choiseul
Shortland
Shortland Is.
Sasamungga
1067
Mono
Treasury Is.
Luti
M
Vella Lavella
Rob Roy
Wagina
Barorafa
Kia
Santa Isabel
Maravari
Vella
Kolombangara
Barora Ite
Gatere
Wilson Strait
Vounga
Gizo
Kolo Gull
Ganongga
New Georgia
Dadali
1219
Buala
Tataba
8
Rendova
Mundo
Segi
Vangunu
San Jorge
Sepi
Lokuru
Gatukai
Susubona
H
Blanche Channel
Tetepare
Florida Is.
Malu
Dala
C. Aracides
886
Erromango
Tepare
Balfour Channel
Russell Is.
Yandina
Savo
Auki
Malaita
Ipota
Pavuvu
Nggela
Su'u
1432
J
Guadalcanal
C. Esperance
Visale
Small Nggela
Baroni
Takataka
Tanna
1684
Honiara
2439
Aola
Maramasike
Isangel
C. Hunter
Avu Avu
Sa'a
Ulawa
C. Zelee
Aniwa
Whitesands
P
Ubuna
Ugi
Three Sisters Is.
Moroga
Kira Kira
Wainoni
Aneityum
Aname
San Cristóbal
1250
Hogarhu
Santa Ona

GUAM

15

Ritidian Pt.
Pati Pt.
Upi
Orote Peninsula
Agat
Agana
R
Mt. Lamlam
405
Guam
Umatac
Inarajan
Cocos I.
Ajayan Pt.

TONGA

Fonualei
Toku
'Uta Vava'u
Vava'u
Hunga
Neiafu
Group
Late
Home Reef
Disney Reef
P
Tofua
Kao
Ha'ano
Fotuha'a
Foa
Lifuka
Ha'apai
Kotu
Uiha
Group
Group
Fonuafo'ou
Ha'afeva
Nomuka
Oto Tolu
Q
Hunga Ha'apai
Nomuka Group
Tonumea
Group
Nuku'alofa
Tongatapu
Tongatapu Group
Eua
13

ÎLES DU VENT

Moorea
Papetoai
Papenoo
Afareaitu
Papeete
S
Punavia
274
Tahiti
Papeari
Tautira
Presqu'île de Taiarapu
Mt. Roonui
1332
Pte. Fareara
16

TAHITI AND MOORÉA

SOLOMON ISLANDS

12

Rennell
Bellona
11

NEW CALEDONIA AND LOYALTY ISLANDS

Î. Yandé
Î. Baaba
Î. Neba
Balabio
Pte. Nendiarene
Poum
Récif de la Gazelle
C. Rossel
Î. Beautemps-Beaupré
T
Paagoumène
Ouégoa
Koumac
Oubatche
Î. Uvéa (Loyalty Is.)
Mt. Panié
1628
St. Joseph
Kaala-Gomen
Hienghène
Touho
Fayaoué
Voh
1385
Poindimié
Î. Lifou
Ouaco
Massif de Tchingou
Ponérihouen
Wé
U
Nouvelle-Calédonie (New Caledonia)
Pouembout
Me Maoya
1508
Houailou
Chépénéhé
Poya
Canala
Î. Tiga
Bourail
Thio
C. de Flotte
Mou
3566
Moindou
La Foa
Massif du Humbolt
Ro
Tadine
C. Roussin
1441
1610
Î. Maré
V
Boulouparis
Païta
2212
Dumbéa
Barrage Yaté
C. Wabao
C. Boyer
Nouméa
Mont Dore
Île Ouen
Cap Ndoua
17 **18** **19**
Île des Pins
110
Vao
20 **21** **22**

WESTERN SAMOA

W
Sataua
Fagamalo
Pu'apu'a
Falelima
1858
Savai'i
Taga
Faga
Satupa'itea
Apia
Ti'avea
X
Mulifanua
Falefa
Amaile
AMERICAN SAMOA (U.S.A.)
1100
Lotofaga
Upolu
Safata Bay
Tutuila
Pago Pago
Ofu
Tau
Pago Pago Hbr.
Manua Is.
Vaitogi

SAMOAN ISLANDS

Projection: Mercator

23 **24** **25**

COPYRIGHT. GEORGE PHILIP & SON, LTD.

ft m
6000 2000
4500 1500
3000 1000
1200 400
600 200
0 0
200 600
2000 6000
4000 12,000
6000 18,000
m ft

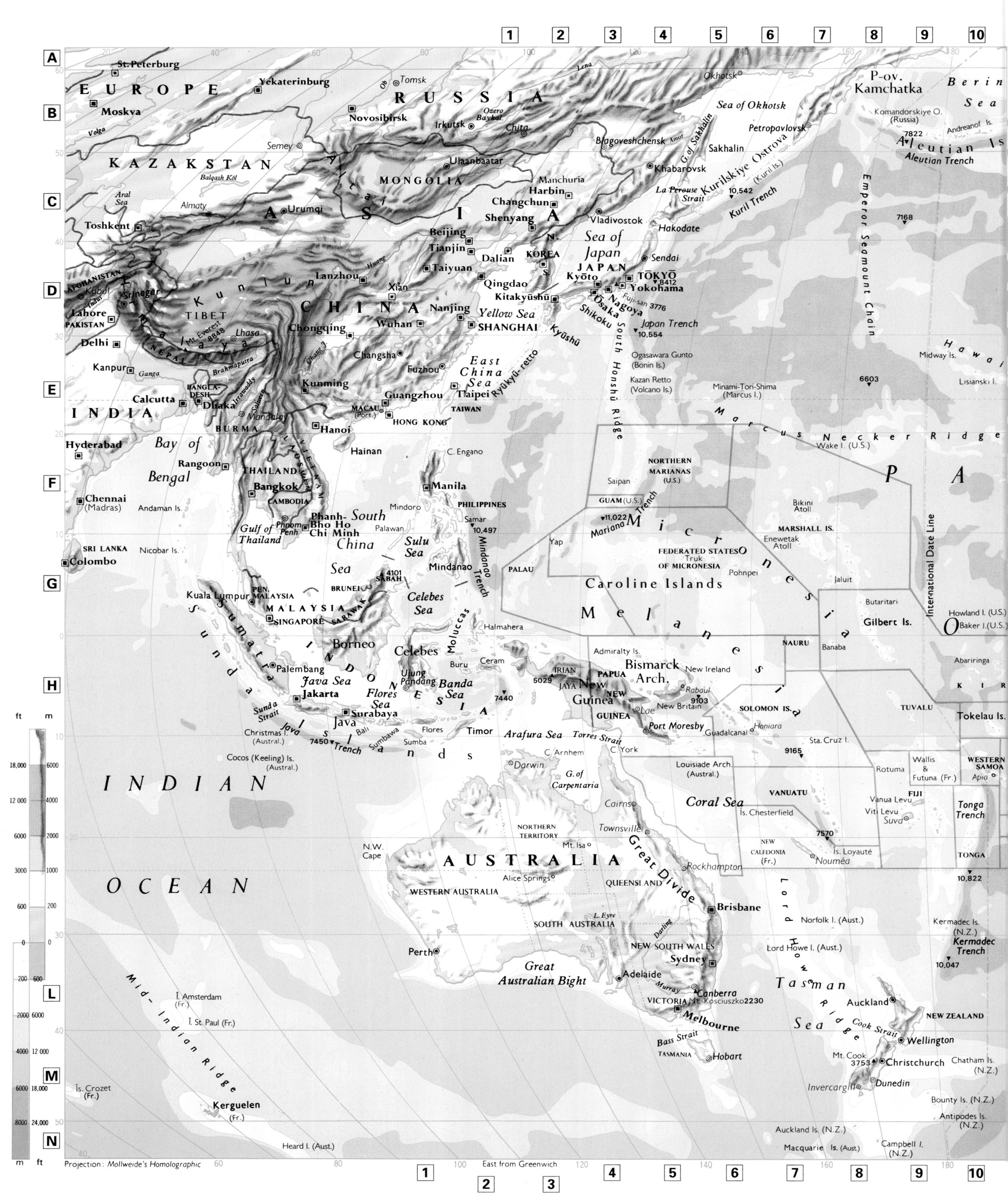

EUROPE
RUSSIA
KAZAKSTAN
MONGOLIA
ASIA
CHINA
TIBET
INDIA
JAPAN
St.Peterburg
Moskva
Volga
Yekaterinburg
Novosibirsk
Tomsk
Irkutsk
Ozero Baykal
Chita
Okhotsk
Sea of Okhotsk
P-ov. Kamchatka
Bering Sea
Petropavlovsk
Komandorskiye O. (Russia)
Andreanof Is.
Aleutian Is
Aleutian Trench
7822
Semey
Balqash Köl
Aral Sea
Almaty
Urumqi
Ulaanbaatar
Altai
Blagoveshchensk
Amur
G. of Sakhalin
Sakhalin
Khabarovsk
Kurilskiye Ostrova (Kuril Is.)
Manchuria
Harbin
Changchun
Shenyang
Vladivostok
La Perouse Strait
10,542
Kuril Trench
Hakodate
Emperor Seamount Chain
7168
Toshkent
Beijing
Tianjin
Dalian
N. KOREA
S. KOREA
Sea of Japan
Sendai
Taiyuan
Lanzhou
Huang
AFGHANISTAN
Kabul
Srinagar
Kunlun
Qingdao
Kyoto
TŌKYŌ
8412
Yokohama
Fuji-san 3776
Lahore
Indus
PAKISTAN
Xi'an
Nanjing
Kitakyūshū
Osaka
Nagoya
Shikoku
Yellow Sea
SHANGHAI
Wuhan
Chongqing
Japan Trench
10,554
Delhi
Himalaya
NEPAL
Mt. Everest 8848
Lhasa
Kyūshū
South Honshū Ridge
Changsha
Chang J.
Brahmaputra
Ganga
Kanpur
East China Sea
Ryūkyū-retto
Fuzhou
Ogasawara Gunto (Bonin Is.)
Kazan Retto (Volcano Is.)
Minami-Tori-Shima (Marcus I.)
6603
Midway Is.
Hawai
Lisianski I.
BANGLADESH
Dhaka
Calcutta
Kunming
Irrawaddy
Salween
Mandalay
BURMA
Guangzhou
Taipei
TAIWAN
MACAU (Port.)
HONG KONG
Hanoi
Marcus Necker Ridge
Wake I. (U.S.)
Hyderabad
Bay of Bengal
Rangoon
Hainan
C. Engano
LAOS
VIETNAM
Mekong
THAILAND
Bangkok
Manila
NORTHERN MARIANAS (U.S.)
Saipan
PACIFIC OCEAN
Chennai (Madras)
Andaman Is.
CAMBODIA
Mindoro
PHILIPPINES
GUAM (U.S.)
Mariana Trench
11,022
Bikini Atoll
Gulf of Thailand
Phnom Penh
Phanh-Bho Ho Chi Minh
South China Sea
Palawan
Samar
10,497
Mindanao Trench
MARSHALL IS.
Enewetak Atoll
Micronesia
SRI LANKA
Colombo
Nicobar Is.
Sulu Sea
Mindanao
Yap
FEDERATED STATES OF MICRONESIA
Truk
Pohnpei
PALAU
Jaluit
Kuala Lumpur
PEN. MALAYSIA
BRUNEI
SABAH
4101
Celebes Sea
Caroline Islands
Butaritari
MALAYSIA
SARAWAK
SINGAPORE
Halmahera
Melanesia
Gilbert Is.
International Date Line
Howland I. (U.S.)
Baker I.(U.S.)
Sumatra
Borneo
Moluccas
Celebes
Buru
Ceram
Admiralty Is.
NAURU
Banaba
Abariringa
INDONESIA
Palembang
Java Sea
Jakarta
Ujung Pandang
Banda Sea
Flores Sea
IRIAN JAYA
PAPUA NEW GUINEA
New Guinea
Bismarck Arch.
New Ireland
Rabaul
9103
New Britain
Lae
SOLOMON IS.
TUVALU
KIRIBATI
Tokelau Is.
Sunda Strait
Java
Surabaya
Bali
7440
5029
Port Moresby
Guadalcanal
Honiara
Christmas I. (Austral.)
Java Trench
7450
Sumbawa
Sumba
Flores
Timor
Arafura Sea
Torres Strait
Sunda Islands
Sta. Cruz I.
9165
C. Arnhem
C. York
Louisiade Arch. (Austral.)
Rotuma
Wallis & Futuna (Fr.)
WESTERN SAMOA
Apia
Cocos (Keeling) Is. (Austral.)
Darwin
G. of Carpentaria
INDIAN OCEAN
Coral Sea
VANUATU
FIJI
Vanua Levu
Viti Levu
Suva
Tonga Trench
Cairns
Is. Chesterfield
Townsville
NORTHERN TERRITORY
Mt. Isa
NEW CALEDONIA (Fr.)
7570
Is. Loyauté
Nouméa
N.W. Cape
AUSTRALIA
Great Divide
Rockhampton
TONGA
10,822
Alice Springs
WESTERN AUSTRALIA
QUEENSLAND
Lord Howe Rise
Brisbane
L. Eyre
SOUTH AUSTRALIA
Darling
Norfolk I. (Aust.)
Kermadec Is. (N.Z.)
Kermadec Trench
NEW SOUTH WALES
Lord Howe I. (Aust.)
Perth
Sydney
10,047
Great Australian Bight
Adelaide
Murray
Canberra
Mt. Kosciuszko 2230
VICTORIA
Melbourne
Tasman Sea
Auckland
Î. Amsterdam (Fr.)
Î. St. Paul (Fr.)
Mid-Indian Ridge
Cook Strait
NEW ZEALAND
Wellington
Bass Strait
TASMANIA
Hobart
Mt. Cook 3753
Christchurch
Chatham Is. (N.Z.)
Îs. Crozet (Fr.)
Kerguelen (Fr.)
Invercargill
Dunedin
Bounty Is. (N.Z.)
Antipodes Is. (N.Z.)
Heard I. (Aust.)
Auckland Is. (N.Z.)
Macquarie Is. (Aust.)
Campbell I. (N.Z.)
East from Greenwich
Projection: Mollweide's Homolographic
ft m
18,000 6000
12 000 4000
6000 2000
3000 1000
600 200
0 0
200 600
2000 6000
4000 12 000
6000 18,000
8000 24,000
m ft

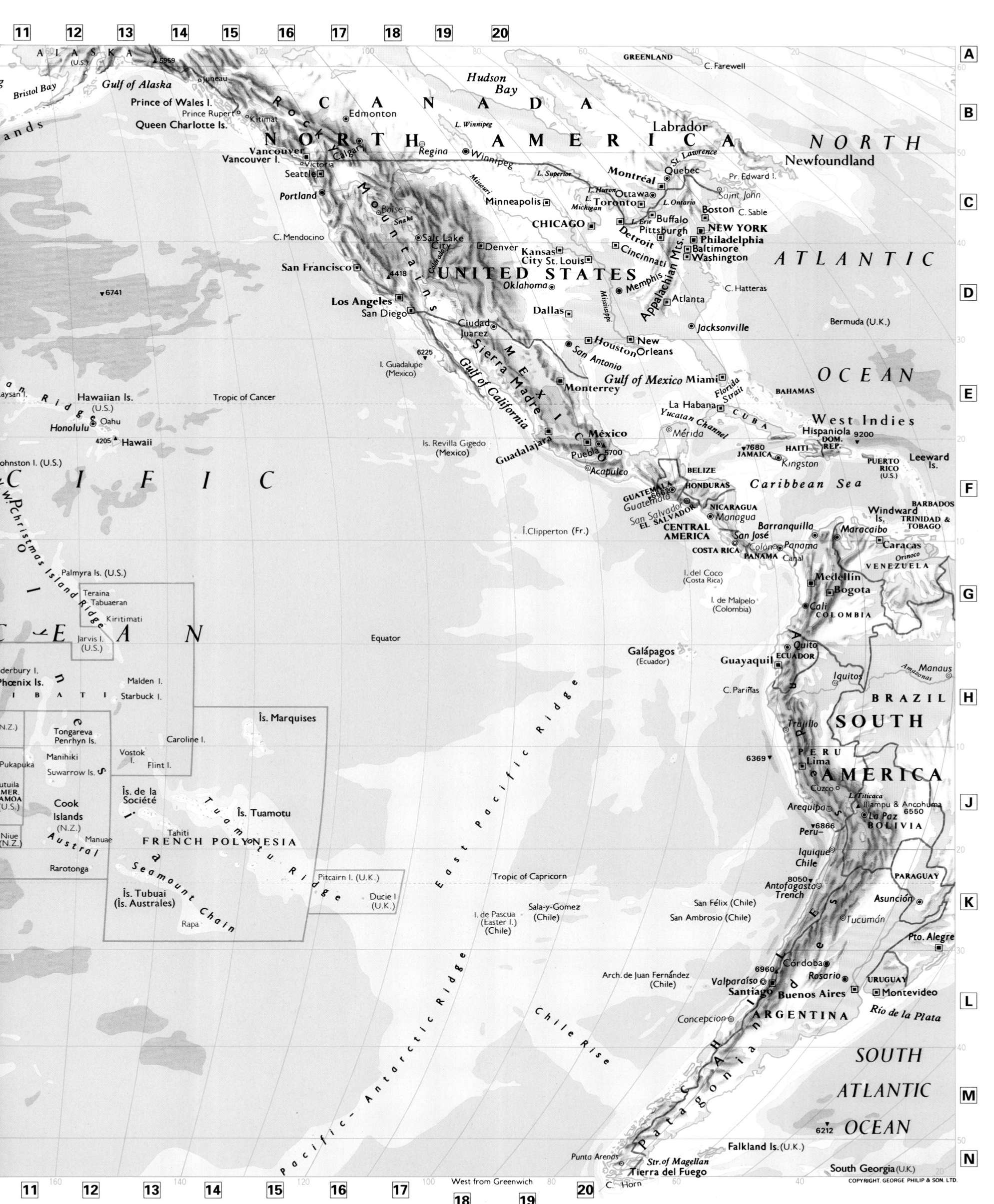

GREENLAND
C. Farewell

Bristol Bay
Gulf of Alaska
Juneau
Hudson Bay
NORTH

A L A S K A (U.S.)
5959

Prince of Wales I.
Prince Rupert
Queen Charlotte Is.
Kitimat
Edmonton
C A N A D A
Labrador

Vancouver
Vancouver I.
Victoria
Seattle
Portland
Calgary
Regina
Winnipeg
L. Winnipeg
NORTH AMERICA
Newfoundland
Montréal
Quebec
Pr. Edward I.
Saint John
St. Lawrence

Boise
Minneapolis
L. Superior
L. Huron
Ottawa
Toronto
L. Ontario
Boston
C. Sable
Buffalo
Michigan
CHICAGO
Detroit
Pittsburgh
NEW YORK
Philadelphia

C. Mendocino
Salt Lake City
Denver
Kansas City
St. Louis
Cincinnati
Baltimore
Washington
ATLANTIC

San Francisco
4418
UNITED STATES
Oklahoma
Memphis
Atlanta
C. Hatteras

6741
Los Angeles
San Diego
Ciudad Juarez
Dallas
Jacksonville
Bermuda (U.K.)

Hawaiian Is. (U.S.)
6225
I. Guadalupe (Mexico)
Sierra Madre
San Antonio
Houston
New Orleans
Gulf of Mexico
Miami
Florida Strait
BAHAMAS
OCEAN

Tropic of Cancer
Honolulu
Oahu
Monterrey
La Habana
CUBA
West Indies

4205
Hawaii
Is. Revilla Gigedo (Mexico)
México
Mérida
Yucatan Channel
Hispaniola
DOM. REP.
9200

Johnston I. (U.S.)
Guadalajara
Puebla
5700
Acapulco
BELIZE
7680
JAMAICA
HAITI
Kingston
PUERTO RICO (U.S.)
Leeward Is.

P A C I F I C
GUATEMALA
8862
HONDURAS
Caribbean Sea
BARBADOS

N.W. Christmas I.
Palmyra Is. (U.S.)
Teraina
Tabuaeran
Kiritimati
GUATEMALA
San Salvador
EL SALVADOR
NICARAGUA
Managua
CENTRAL AMERICA
San José
COSTA RICA
Barranquilla
San Andrés
PANAMA
Panama Canal
Windward Is.
TRINIDAD & TOBAGO
Maracaibo
Caracas
VENEZUELA

Jarvis I. (U.S.)
O C E A N
I. Clipperton (Fr.)
I. del Coco (Costa Rica)
Colón
Panama
Medellín
Bogota
Orinoco

Phoenix Is.
KIRIBATI
Equator
I. de Malpelo (Colombia)
Cali
COLOMBIA

Enderbury I.
Malden I.
Starbuck I.
Galápagos (Ecuador)
Guayaquil
Quito
ECUADOR
Manaus

Tongareva
Penrhyn Is.
Vostok I.
Flint I.
Caroline I.
C. Pariñas
Iquitos
Amazonas
BRAZIL

Pukapuka
Manihiki
Suwarrow Is.
Îs. Marquises
Trujillo
SOUTH

AMER. SAMOA (U.S.)
Îs. de la Société
Îs. Tuamotu
6369
PERU
Lima
AMERICA

Niue (N.Z.)
Cook Islands (N.Z.)
Tahiti
FRENCH POLYNESIA
Cuzco
L. Titicaca
Illampu & Ancohuma
6550

Manuae
Austral
Arequipa
La Paz
BOLIVIA

Rarotonga
Seamount Chain
6866
Peru
Iquique
Chile

Îs. Tubuai (Îs. Australes)
Pitcairn I. (U.K.)
Tropic of Capricorn
PARAGUAY

Rapa
Ducie I. (U.K.)
8050
Antofagasta
Trench

I. de Pascua (Easter I.) (Chile)
Sala-y-Gomez (Chile)
San Félix (Chile)
San Ambrosio (Chile)
Tucumán
Asunción

Arch. de Juan Fernández (Chile)
6960
Córdoba
URUGUAY
Pto. Alegre

East Pacific Ridge
Valparaíso
Rosario
Montevideo

Santiago
Buenos Aires
Río de la Plata

Concepción
ARGENTINA
SOUTH

Chile Rise
ATLANTIC

Pacific–Antarctic Ridge
Patagonia
6212
OCEAN

Punta Arenas
Str. of Magellan
Tierra del Fuego
Falkland Is. (U.K.)
South Georgia (U.K.)
C. Horn

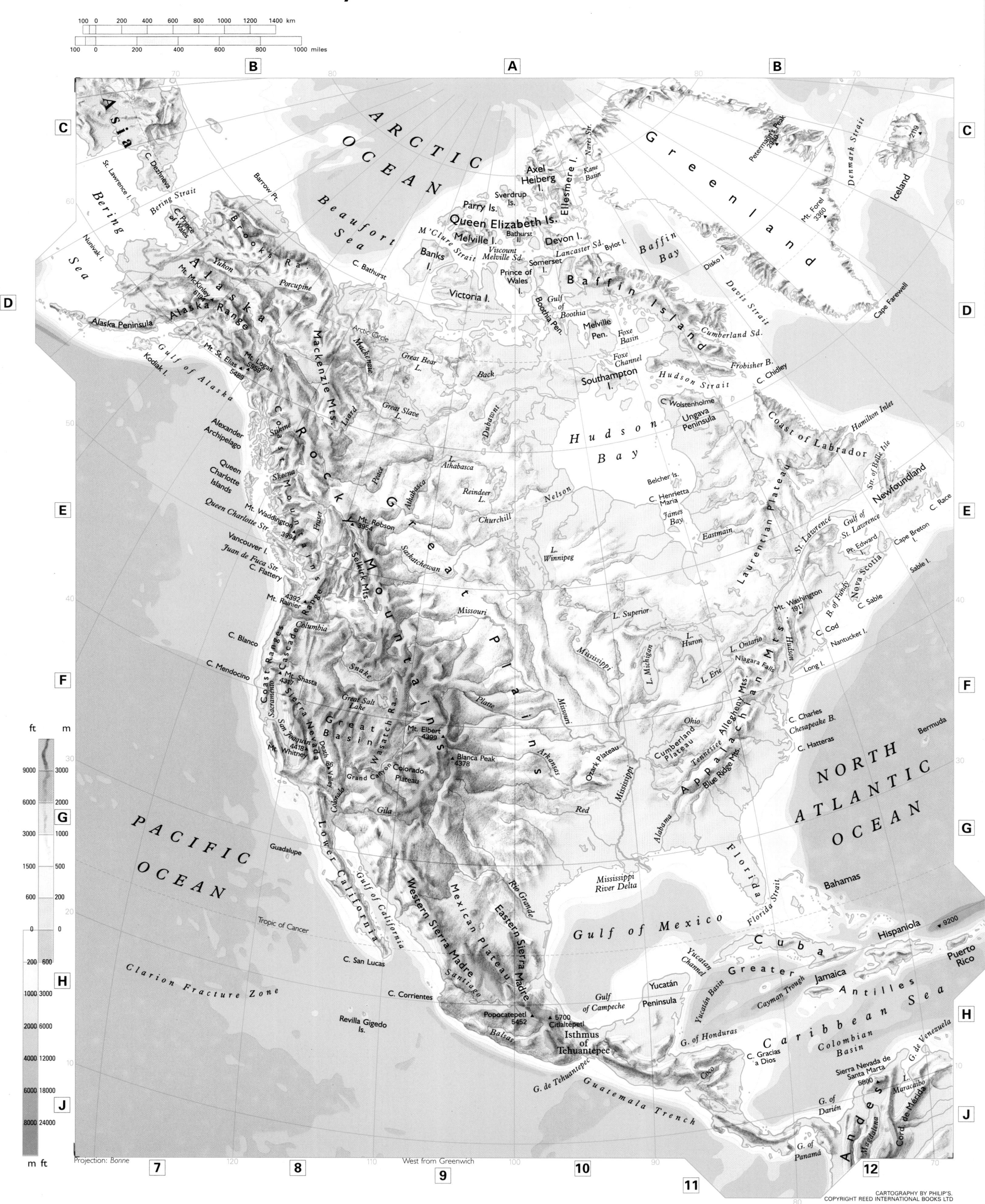
ARCTIC OCEAN
Beaufort Sea
Greenland
Baffin Bay
Baffin Island
Queen Elizabeth Is.
Hudson Bay
Hudson Strait
Alaska
Alaska Range
Alaska Peninsula
Bering Sea
Bering Strait
Asia
Gulf of Alaska
Rocky Mountains
Coast Mountains
Mackenzie Mts.
Great Plains
Great Basin
Sierra Nevada
Cascade Range
Coast Ranges
Colorado Plateau
Mexican Plateau
Western Sierra Madre
Eastern Sierra Madre
Appalachian Mts.
Laurentian Plateau
Coast of Labrador
Newfoundland
Nova Scotia
PACIFIC OCEAN
NORTH ATLANTIC OCEAN
Gulf of Mexico
Caribbean Sea
Greater Antilles
Cuba
Hispaniola
Jamaica
Puerto Rico
Bahamas
Bermuda
Yucatán Peninsula
Isthmus of Tehuantepec
Lower California
Gulf of California
Guatemala Trench
Clarion Fracture Zone
Tropic of Cancer
Arctic Circle
Andes
Iceland
Mt. McKinley 6194
Mt. Logan 5959
Mt. St. Elias 5489
Mt. Robson 3954
Mt. Waddington 3994
Mt. Rainier 4392
Mt. Shasta 4317
Mt. Whitney 4418
Mt. Elbert 4399
Blanca Peak 4378
Popocatepetl 5452
Citlaltépetl 5700
Mt. Washington 1917
Sierra Nevada de Santa Marta 5800
Petermann's Peak 2940
Mt. Forel 3360
Projection: Bonne
West from Greenwich
CARTOGRAPHY BY PHILIP'S.
COPYRIGHT REED INTERNATIONAL BOOKS LTD

100 0   200   400   600   800   1000 1200 1400 km

100 0   200   400   600   800   1000 miles

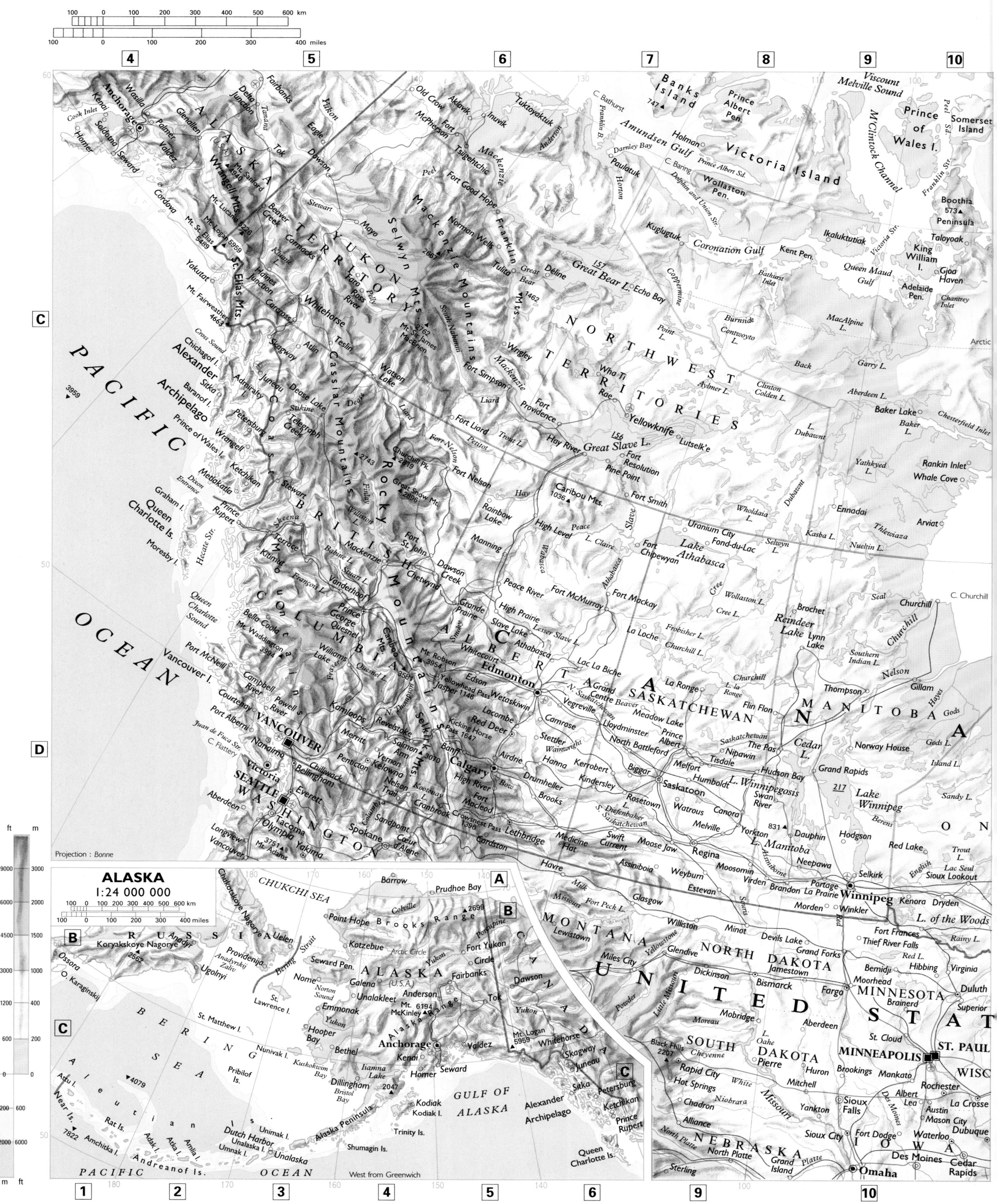

100 0 100 200 300 400 500 600 km
100 0 100 200 300 400 miles
PACIFIC OCEAN
ALASKA
YUKON TERRITORY
NORTHWEST TERRITORIES
BRITISH COLUMBIA
ALBERTA
SASKATCHEWAN
MANITOBA
WASHINGTON
MONTANA
NORTH DAKOTA
SOUTH DAKOTA
NEBRASKA
MINNESOTA
IOWA
UNITED STATES
Banks Island
Victoria Island
Prince of Wales I.
Somerset Island
Prince Albert Pen.
Wollaston Pen.
Boothia Peninsula
King William I.
Viscount Melville Sound
M'Clintock Channel
Amundsen Gulf
Coronation Gulf
Queen Maud Gulf
Great Bear L.
Great Slave L.
Lake Athabasca
Reindeer Lake
L. Winnipegosis
L. Manitoba
Lake Winnipeg
L. of the Woods
Mackenzie Mountains
Selwyn Mts.
St. Elias Mts.
Wrangell Mts.
Cassiar Mountains
Rocky Mountains
Coast Mountains
Columbia Mountains
Selkirk Mts.
Caribou Mts.
Mt. Logan 5959
Mt. St. Elias 5489
Mt. Fairweather 4663
Mt. Robson 3954
Mt. Waddington 3994
Alexander Archipelago
Queen Charlotte Is.
Vancouver I.
Anchorage
Fairbanks
Inuvik
Tuktoyaktuk
Dawson
Whitehorse
Yellowknife
Prince Rupert
Prince George
Kamloops
VANCOUVER
Victoria
SEATTLE
Edmonton
Calgary
Red Deer
Lethbridge
Medicine Hat
Saskatoon
Regina
Moose Jaw
Prince Albert
The Pas
Flin Flon
Thompson
Churchill
Winnipeg
Brandon
Kenora
Spokane
Yakima
Tacoma
Olympia
Fargo
Bismarck
Pierre
Rapid City
Sioux Falls
Omaha
Des Moines
Duluth
MINNEAPOLIS
ST. PAUL
Projection : Bonne
ALASKA
1:24 000 000
100 0 100 200 300 400 500 600 km
100 0 100 200 300 400 miles
CHUKCHI SEA
BERING SEA
GULF OF ALASKA
PACIFIC OCEAN
RUSSIA
CANADA
Brooks Range
Alaska Range
Mt. McKinley 6194
Barrow
Prudhoe Bay
Nome
Kodiak I.
Aleutian Is.
Alaska Peninsula
West from Greenwich

146
147

11
12
13
14
15
16
B
C
D
E
Devon I.
Lancaster Sound
Baffin Bay
Nunavik
Uummannaq
Ilulissat
Qasigiannguit
Ammassalik
GREENLAND
(KALAALLIT NUNAAT)
(Denmark)
Bylot I.
Arctic Bay
Borden Pen.
Pond Inlet
Brodeur Peninsula
Clyde River
C. Adair
C. Raper
Qeqertarsuaq
Davis Strait
Kangerlussuaq
Kong Frederik VI's Kyst
Gulf of Boothia
Fury and Hecla Str.
Igloolik
Baffin Island
Home B.
Broughton Island
Sisimiut
Maniitsoq
Nuuk
Simpson Pen.
Pelly Bay
Melville Peninsula
Sanirajak
Foxe
Prince Charles I.
Air Force I.
Cumberland Peninsula
C. Dyer
Pangnirtung
Hoare B.
Mercy C.
Cumberland Sd.
Qeqertarsuatsiaat
Paamiut
Arsuk
Qaqortoq
Alluitsup Paa
Nanortalik
ATLANTIC
OCEAN
Rae Isthmus
Repulse Bay
Circle
Basin
Nettilling L.
NUNAVUT
C. Dorchester
Amadjuak L.
Wager B.
Roes Welcome Sd.
Foxe Channel
Foxe Pen.
Cape Dorset
Iqaluit
Meta Incognita Peninsula
Kimmirut
Hall Peninsula
Frobisher Bay
Resolution I.
Southampton I.
Salliq
Bell Pen.
Salisbury I.
Nottingham I.
Hudson Strait
Labrador Sea
Igluligaarjuk
Coats I.
Mansel I.
Ivujivik
Salluit
Kangiqsujuaq
Quaqtaq
C. Chidley
Akpatok I.
Péninsule d'Ungava
Kangirsuk
Ungava Bay
Kangiqsualujjuaq
Hebron
Puvirnituq
L. Payne
Arnaud
Hudson Bay
Ottawa Is.
Kuujjuaq
Nain
NEWFOUNDLAND
Inukjuak
Feuilles
Koksoak
Baleine
George
Hopedale
C. Harrison
Rigolet
Cartwright
L. Minto
Mélèzes
Caniapiscau
Sleeper Is.
King George Is.
Baker's Dozen Is.
Port Hope Simpson
Belle Isle
C. Tatnam
Smallwood Res.
Labrador
North West River
Happy Valley
Goose Bay
C. Bauld
St. Anthony
Schefferville
L. Petitsikapau
L. à l'Eau Claire
L. Bienville
Belcher Is.
Grande Baleine
Kuujjuarapik
Esker
Churchill Falls
Churchill
Str. of Belle Isle
Notre Dame B.
C. Henrietta Maria
Peawanuck
Winisk
Severn
Rés. de Caniapiscau
Kanaaupscow
Labrador City
Fermont
L. Ashuanipi
Natashquan
St-Augustin
Baie Verte
Lewisporte
Gander
Bonavista
Pte. Louis XIV
Chisasibi
La Grande
QUÉBEC
Gagnon
Moisie
Romaine
Deer Lake
Grand Falls
Trinity B.
Carbonear
St. John's
Big Trout L.
James Bay
Akimiski I.
Wemindji
Rés. Manicouagan
Havre-St-Pierre
Î. d'Anticosti
Corner Brook
Newfoundland
Placentia
C. Race
Attawapiskat
Stephenville
Marystown
Placentia B.
Fort Albany
Charlton I.
Eastmain
Waskaganish
Rupert
L. Mistassini
L. Albanel
Sept-Îles
Port-Cartier
Gulf of St. Lawrence
C. Ray
Channel-Port aux Basques
Cabot Str.
ST.-PIERRE et MIQUELON (Fr)
ONTARIO
Albany
Moosonee
Nottaway
Harricana
Missinaibi
Péribonca
Baie Comeau
St. Lawrence
Matane
Gaspé
Pén. de Gaspé
Îs. de la Madeleine
North C.
Cape Breton I.
Glace Bay
L. St. Joseph
Nakina
Kenogami
Matagami
Chibougamau
Rimouski
Campbellton
Bathurst
PR. EDWARD I.
Sydney
L. Nipigon
Hearst
L. Matagami
Dolbeau
L. St-Jean
Chicoutimi
Rivière-du-Loup
Edmundston
Chatham
Summerside
Charlottetown
Port Hawkesbury
Geraldton
Kapuskasing
Cochrane
Abitibi L.
Amos
Rés. Gouin
Roberval
Jonquière
Northumberland Str.
Antigonish
Nipigon
Marathon
Oba
Mattagami
Timmins
Grand Falls
NEW BRUNSWICK
NOVA SCOTIA
New Glasgow
Truro
Thunder Bay
Kirkland Lake
Rouyn-Noranda
Val-d'Or
La Tuque
Québec
Lévis
Woodstock
Moncton
Amherst
Kentville
Dartmouth
Sable I. (Nova Scotia)
Chapleau
New Liskeard
Rés. Cabonga
Mont-Laurier
Shawinigan
Trois-Rivières
Thetford Mines
Fredericton
Saint John
Halifax
Lake Superior
Wawa
Joliette
St-Hyacinthe
MAINE
B. of Fundy
Digby
Bridgewater
Liverpool
Houghton
Sault Ste. Marie
Elliot Lake
Sudbury
Outaouais
MONTRÉAL
Sherbrooke
Bangor
Yarmouth
C. Sable
Ironwood
Marquette
MICHIGAN
L. Nipissing
North Bay
Hull
Granby
L. Champlain
Augusta
Lewiston
Manistique
Manitoulin I.
Parry Sound
Georgian Bay
Pembroke
Ottawa
Cornwall
Montpelier
Portland
Rhinelander
Escanaba
Menominee
Petoskey
Lake Huron
Huntsville
Burlington
VERMONT
NEW HAMPSHIRE
Traverse City
Barrie
Peterborough
Kingston
Concord
Wausau
WISCONSIN
Green Bay
L. Michigan
Cadillac
Owen Sound
Oshawa
L. Ontario
Belleville
Manchester
Appleton
TORONTO
Syracuse
MASS.
BOSTON
C. Cod
Sheboygan
Saginaw
Kitchener
Hamilton
Albany
Springfield
Providence
MILWAUKEE
Grand Rapids
Flint
London
Niagara Falls
Rochester
NEW YORK
HARTFORD
R.I.
Madison
Sarnia
BUFFALO
Elmira
CONN.
New Haven
Racine
Kenosha
Lansing
L. Erie
Binghamton
Rockford
DETROIT
Windsor
Erie
Jamestown
Scranton
Bridgeport
CHICAGO
Toledo
CLEVELAND
Newark
NEW YORK
Gary
South Bend
ILLINOIS
INDIANA
OHIO
PENNSYLVANIA
N.J.
Allentown
Trenton
1899
2136
2850
2591
3809
1652
257
1135
814
1190
6309
183
177
174
75
90
80
70
60
50
40
30
West from Greenwich
COPYRIGHT GEORGE PHILIP LTD.

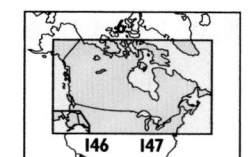

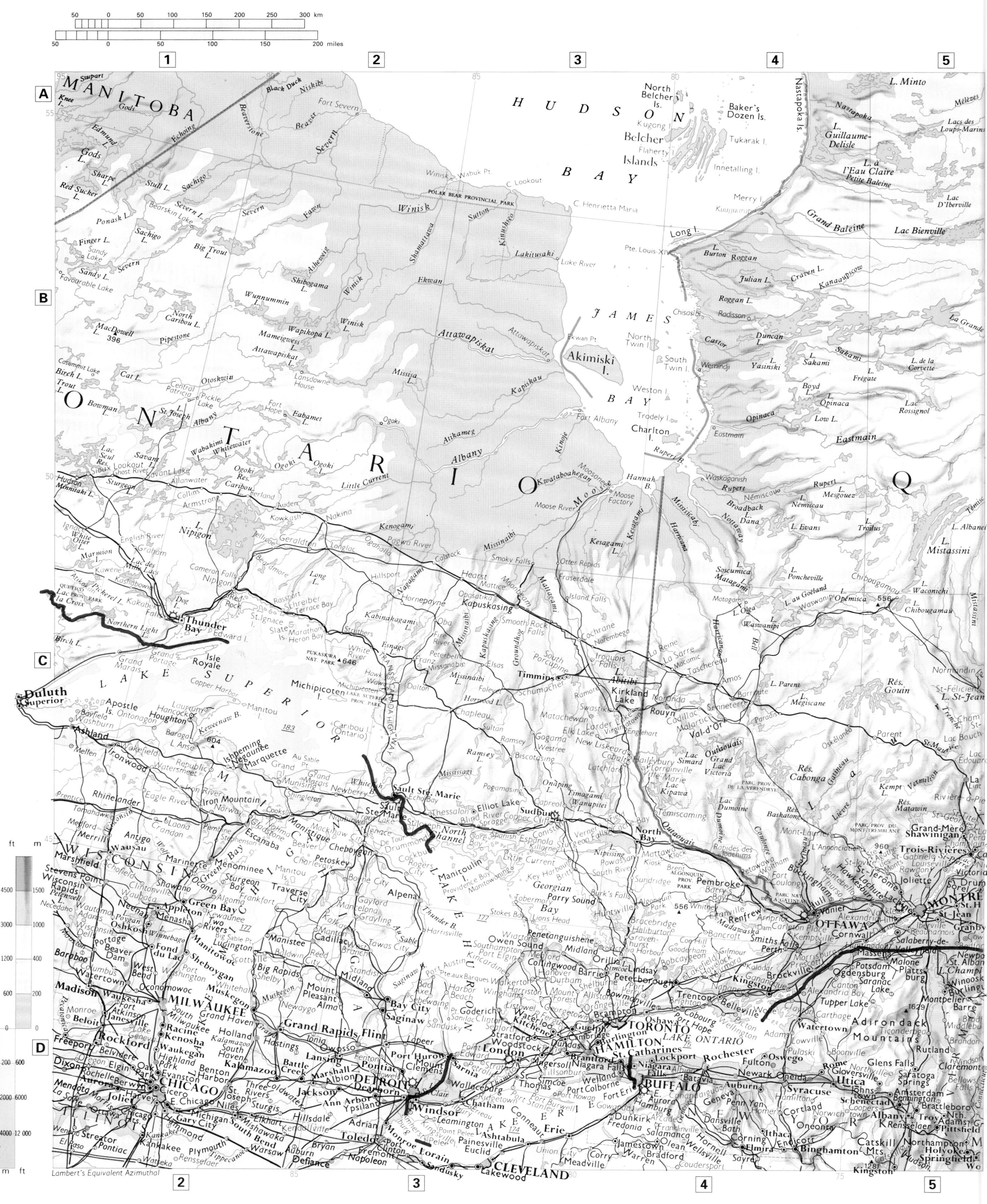

50 0 50 100 150 200 250 300 km
50 0 50 100 150 200 miles
1
2
3
4
5
A
B
C
D
MANITOBA
HUDSON BAY
JAMES BAY
ONTARIO
Q
North Belcher Is.
Baker's Dozen Is.
Belcher Islands
Nastapoka Is.
L. Minto
L. Guillaume-Delisle
L. à l'Eau Claire
Petite Baleine
Lac D'Iberville
Grand Baleine
Lac Bienville
Long I.
Pte. Louis-XIV
L. Burton
Roggan
Julian L.
Craven L.
Chisasibi
Radisson
La Grande
Akimiski I.
North Twin I.
South Twin I.
Weston I.
Trodely I.
Charlton I.
Eastmain
Opinaca
Rupert B.
Fort Severn
Winisk
C. Lookout
POLAR BEAR PROVINCIAL PARK
C. Henrietta Maria
Lake River
Attawapiskat
Albany
Fort Albany
Moosonee
Moose Factory
Moose River
Hannah B.
Kesagami L.
Missinaibi
Kenogami
Little Current
Ogoki
L. Nipigon
Nipigon
Thunder Bay
Isle Royale
LAKE SUPERIOR
Michipicoten I.
Duluth
Superior
Apostle Is.
Houghton
Hancock
Ashland
Marquette
Ishpeming
Hearst
Kapuskasing
Cochrane
Timmins
Kirkland Lake
Rouyn
Val-d'Or
Amos
Matagami
Chibougamau
L. Mistassini
L. Evans
Waskaganish
Nemiscau
Rés. Gouin
Rés. Cabonga
L. St-Jean
Sault Ste. Marie
Elliot Lake
Sudbury
North Bay
Manitoulin I.
LAKE HURON
Georgian Bay
Parry Sound
Owen Sound
Pembroke
OTTAWA
Hull
MONTRÉAL
Trois-Rivières
Shawinigan
Kingston
Belleville
Peterborough
Barrie
TORONTO
HAMILTON
LAKE ONTARIO
St. Catharines
Niagara Falls
BUFFALO
Rochester
Syracuse
Utica
Albany
Binghamton
Adirondack Mountains
Watertown
London
Kitchener
Guelph
Brantford
Sarnia
Port Huron
DETROIT
Windsor
Toledo
CLEVELAND
Lorain
Sandusky
LAKE ERIE
Erie
Flint
Saginaw
Bay City
Lansing
Grand Rapids
Kalamazoo
MILWAUKEE
CHICAGO
Madison
Rockford
Green Bay
Appleton
Oshkosh
Sheboygan
Manitowoc
WISCONSIN
MICHIGAN
ILLINOIS
NEW YORK
Lambert's Equivalent Azimuthal
ft m
4500 1500
3000 1000
1200 400
600 200
0 0
200 600
2000 6000
4000 12 000
m ft

139
143
154
148
6
7
8
9
A
B
C
D
NEWFOUNDLAND
Labrador
QUEBEC
NEW BRUNSWICK
NOVA SCOTIA
PRINCE EDWARD ISLAND
MAINE
NEWFOUNDLAND
ATLANTIC OCEAN
GULF OF ST. LAWRENCE
St. Lawrence
Cabot Strait
Str. of Belle Isle
Bay of Fundy
Northumberland Str.
Î. d'Anticosti
Dét. de Jacques-Cartier
Dét. d'Honguedo
Pén. de Gaspé
Mts. Chic-Chocs
Îs. de la Madeleine (Quebec)
Cape Breton Island
Sable I. (Nova Scotia)
SAINT-PIERRE ET MIQUELON (Fr.)
Miquelon
Langlade
St. Pierre
Avalon Peninsula
Long Range Mts.
Mealy Mts.
Smallwood Res.
Churchill Falls
Happy Valley-Goose Bay
L. Melville
Labrador City
Kaniapiskau Lake
Rés. Manicouagan
Sept-Îles
Port-Cartier
Rimouski
Rivière-du-Loup
QUEBEC
Lévis
Chicoutimi
Jonquière
Sherbrooke
Thetford Mines
Edmundston
Campbellton
Dalhousie
Bathurst
Chatham
Moncton
Fredericton
Saint John
Summerside
Charlottetown
Amherst
Truro
New Glasgow
Halifax
Dartmouth
Yarmouth
Sydney
Glace Bay
Sydney Mines
New Waterford
Corner Brook
Gander
St. John's
Grand Falls
Bangor
Augusta
Portland
Lewiston
Auburn
Manchester
BOSTON
Brockton
Lowell
Portsmouth
Gaspé
Belle I.
Fogo I.
Bonavista
Placentia
Stephenville
Channel-Port aux Basques
Presque Isle
Houlton
Old Town
Skowhegan
Waterville
Rockland
Bath
Brunswick
Biddeford
Saco
Sanford
Rochester
Concord
Nashua
Lawrence
Haverhill
Gloucester
Lynn
Waltham
Worcester
Fitchburg
Keene
Plessisville
Magog
Coaticook
Lac-Mégantic
Asbestos
Alma
Dolbeau
Montmagny
Matane
Amqui
Mont-Joli
Trois-Pistoles
Chandler
Percé
Kentville
Windsor
Digby
Lunenburg
Bridgewater
Liverpool
Antigonish
Inverness
Louisbourg
Canso
Nain
Hopedale
Cartwright
Battle Harbour
Schefferville
Gagnon

West from Greenwich

Projection: Lambert's Equivalent Azimuthal

West from Greenwich

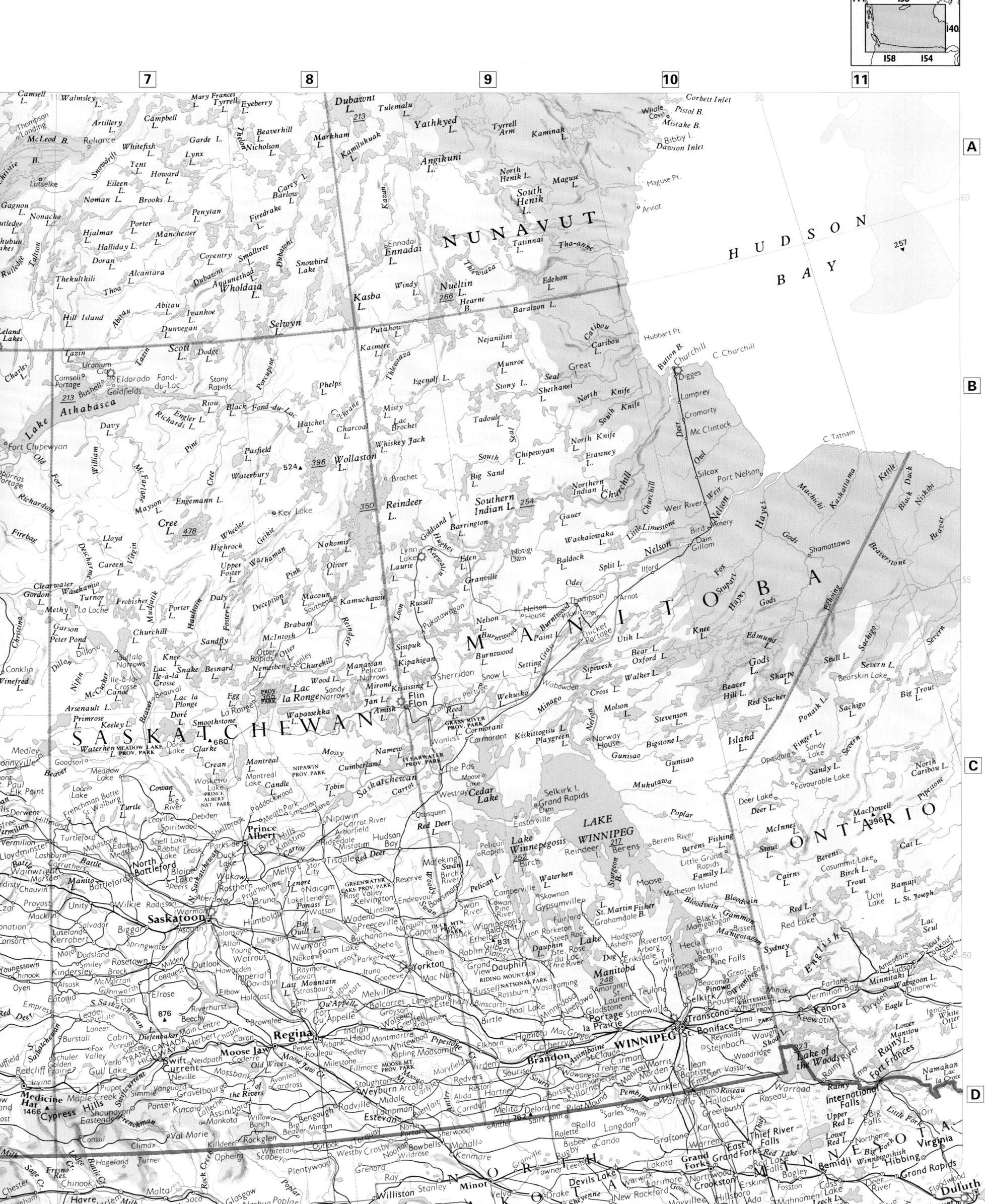

144
138
140
158
154
7
8
9
10
11
A
B
C
D
NUNAVUT
HUDSON
BAY
MANITOBA
SASKATCHEWAN
ONTARIO
NORTH DAKOTA
MINNESOTA
Dubawnt L.
Yathkyed L.
Tulemalu L.
Kaminak L.
Tyrrell Arm
Angikuni L.
Kamilukuak L.
Markham L.
South Henik L.
North Henik L.
Maguse L.
Maguse Pt.
Arviat
Whale Cove
Corbett Inlet
Pistol B.
Mistake B.
Bibby I.
Dawson Inlet
Ennadai L.
Tatinnai L.
Tha-anne
Thlewiaza
Nueltin L.
Hearne B.
Edehon L.
Baralzon L.
Kasba L.
Windy L.
Snowbird Lake
Wholdaia L.
Selwyn L.
Scott L.
Dodge L.
Tazin L.
Uranium City
Eldorado
Goldfields
Fond-du-Lac
Stony Rapids
Camsell Portage
Bushell
Lake Athabasca
Fort Chipewyan
Wollaston L.
Reindeer L.
Southern Indian L.
Brochet
Key Lake
Cree L.
Lynn Lake
Churchill
C. Churchill
Hubbart Pt.
Button B.
Digges
Lamprey
Cromarty
McClintock
C. Tatnam
Port Nelson
Weir River
Amery
Gillam
Split L.
Nelson
Hayes
Gods L.
Thompson
Flin Flon
The Pas
Cedar Lake
Grand Rapids
Lake Winnipegosis
LAKE WINNIPEG
Lake Manitoba
Dauphin
Brandon
Portage la Prairie
WINNIPEG
St. Boniface
Selkirk
Kenora
Lake of the Woods
Fort Frances
International Falls
Regina
Saskatoon
Prince Albert
North Battleford
Lloydminster
Moose Jaw
Swift Current
Medicine Hat
Cypress Hills
Yorkton
Weyburn
Estevan
Minot
Williston
Grand Forks
Devils Lake
Thief River Falls
Bemidji
Hibbing
Virginia
Duluth
Grand Rapids
Lac la Ronge
Havre
Glasgow
Fort Peck Res.
Sakakawea
Lake Athabasca
Red Lake
Sioux Lookout
Dryden
Big Trout L.
Sandy Lake
Island L.

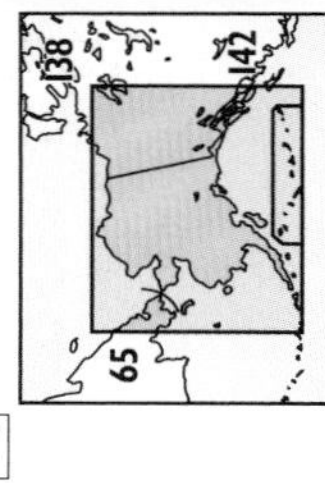

50 0 100 200 300 400 km
50 0 50 100 150 200 250 miles

Continuation Westwards on same scale

CARTOGRAPHY BY PHILIP'S. COPYRIGHT REED INTERNATIONAL BOOKS LTD

Projection: Bipolar oblique conic conformal

ARCTIC OCEAN

BEAUFORT SEA

CHUKCHI SEA

BERING SEA

GULF OF ALASKA

PACIFIC OCEAN

CANADA — UNITED STATES
YUKON — ALASKA
BRITISH COLUMBIA — ALASKA

NORTH WEST TERRITORIES

RUSSIA — UNITED STATES

Franklin Mts.
Richardson Mts.
British Mts.
Mackenzie Mts.
Selwyn Mts.
Ogilvie Mts.
Yukon Mountains
Pelly Mts.
Big Salmon Mts.
Dawson Range
St. Elias Mts.
Wrangell Mts.
Chugach Mts.
Brooks Range
Endicott Mts.
Philip Smith Mts.
Schwatka Mts.
Baird Mts.
De Long Mts.
Ray Mts.
Kaiyuh Mts.
White Mts.
Kuskokwim Mountains
Alaska Range
Aleutian Range
Kenai Mts.

ALEUTIAN ISLANDS
ANDREANOF ISLANDS
RAT ISLANDS
NEAR ISLANDS
FOX ISLANDS
ISLANDS OF THE FOUR MTS.

ALEXANDER ARCHIPELAGO

Barrow
Prudhoe Bay
Kotzebue
Nome
Fairbanks
Anchorage
Juneau
Ketchikan
Seward
Valdez
Cordova
Bethel
Dillingham
Kodiak
Unalaska
Dutch Harbor

Mt. McKinley
Mt. Foraker
Mt. St. Elias 5489
Mt. Fairweather 4663
Mt. Marcus Baker
Denali Nat. Park and Preserve
Wrangell-St. Elias Nat. Park and Preserve
Gates of the Arctic Nat. Park and Preserve
Katmai Nat. Park and Preserve
Lake Clark Nat. Park and Preserve
Kenai Fjords Nat. Park
Glacier Bay Nat. Park
Tongass National Forest
Chugach Nat. Forest
Kobuk Valley Nat. Park
Noatak National Preserve
Yukon-Charley Rivers Nat. Preserve
Arctic National Wildlife Refuge
Yukon Flats Nat. Wildlife Refuge

Yukon
Kuskokwim
Tanana
Koyukuk
Porcupine
Colville
Noatak
Kobuk
Copper

Norton Sound
Bristol Bay
Cook Inlet
Kotzebue Sound
Bering Strait

St. Lawrence I.
Nunivak I.
St. Matthew I.
Pribilof Is.
Nelson I.
Kodiak I.
Afognak I.
Unimak Island
Unalaska I.
Umnak I.
Montague I.
Chichagof I.
Baranof I.
Admiralty I.
Prince of Wales I.

Seward Peninsula
Alaska Peninsula
Kenai Peninsula

Shishaldin Volcano 2857
Makushin Volcano 2036
Pavlof Volcano
Redoubt Volcano 3108
Mt. Spurr

Chukotskiy Poluostrov
Anadyrskiy Zaliv
Providentiya

West from Greenwich
East from Greenwich

Permanent Ice
m 3000 2000 1500 1000 400 200 0
ft 9000 6000 4500 3000 1200 600 0

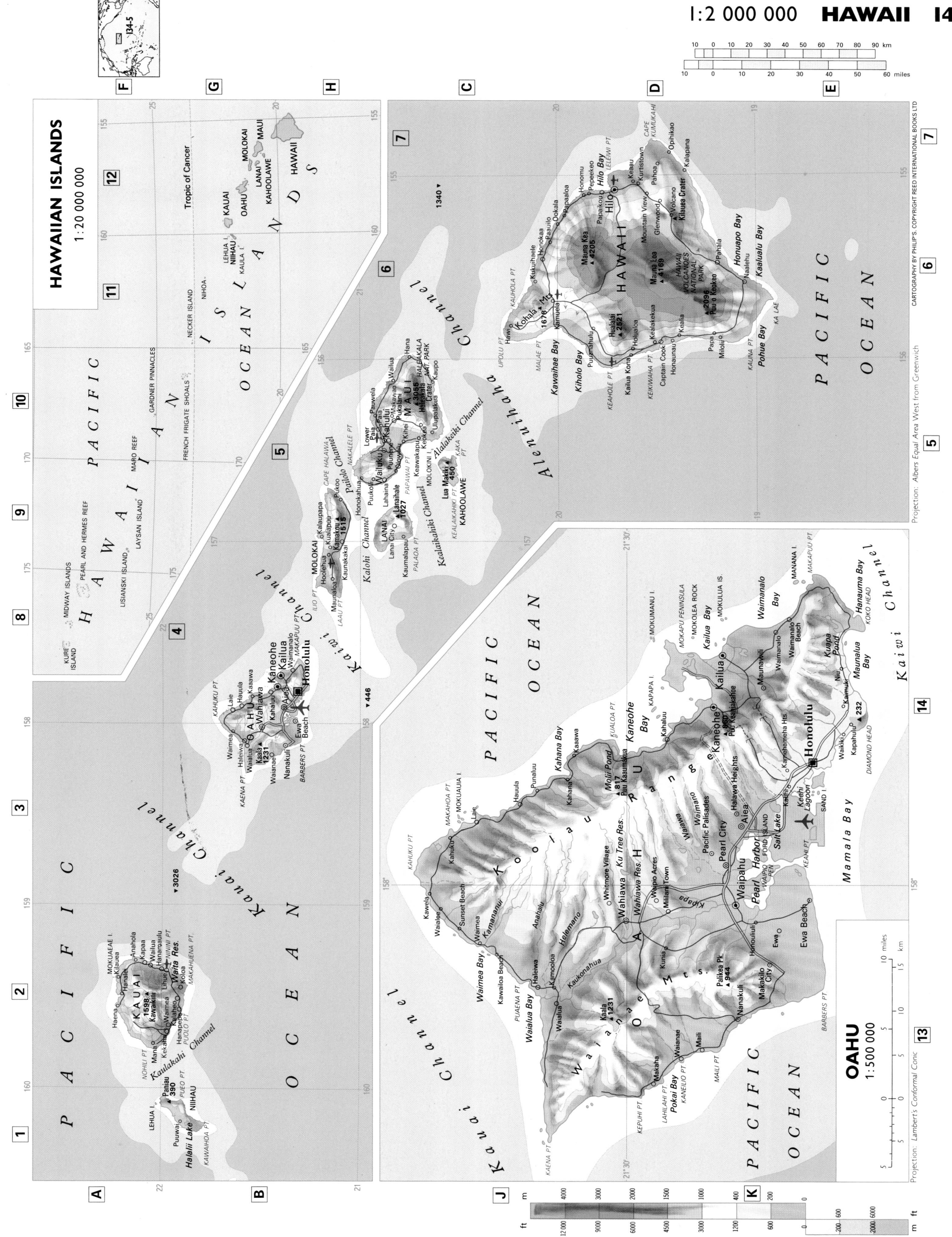

HAWAIIAN ISLANDS
1:20 000 000

OCEAN

Tropic of Cancer

KAUAI
NIIHAU OAHU MOLOKAI
LANAI MAUI
KAHOOLAWE
HAWAII

I S L A N D S

LEHUA I.
NIIHAU
KAULA I.
NIHOA

PACIFIC

PACIFIC

OCEAN

NECKER ISLAND

GARDNER PINNACLES

FRENCH FRIGATE SHOALS

H A W A I I A N

PEARL AND HERMES REEF

MARO REEF

LISIANSKI ISLAND

LAYSAN ISLAND

MIDWAY ISLANDS

KURE ISLAND

Projection: Albers Equal Area Area West from Greenwich

CARTOGRAPHY BY PHILIP'S, COPYRIGHT REED INTERNATIONAL BOOKS LTD

HAWAII

HAWAII VOLCANOES NATIONAL PARK

Mauna Kea ▲4205
Mauna Loa ▲4169
Hualalai ▲2521
Puʻu o Keokeo ▲2096
Kohala Mts ▲1676

Hilo Bay
Hilo

Honuapo Bay
Kaalualu Bay
Pohue Bay

PACIFIC

OCEAN

KAUAI

KAUAI ▲1598
Waimea
Lihue

NIIHAU
Puuwai ▲390
Halalii Lake

LEHUA I.

Kaulakahi Channel

Kauai Channel

PACIFIC

OCEAN

OAHU
Honolulu
Kaneohe
Kailua
Waialua
Kaala ▲1231

MOLOKAI
Kaunakakai ▲1515

MAUI
HALEAKALA
▲3055
Kahului
Wailuku

LANAI
Lana City ▲1027

KAHOOLAWE
Lua Makika ▲450

Kaiwi Channel

Kalohi Channel

Pailolo Channel

Alalakeiki Channel

Alenuihaha Channel

Kaiwi Channel

OAHU
1:500 000

Projection: Lambert's Conformal Conic

Honolulu

Pearl Harbor

Koolau Range

Waianae Mts

Diamond Head ▲232
Koko Head

Hanauma Bay
Waimanalo Bay
Kailua Bay
Kaneohe Bay
Kahana Bay

Kaala ▲1231

Palikea Pk ▲944

PACIFIC OCEAN

Mamala Bay

Kaiwi Channel

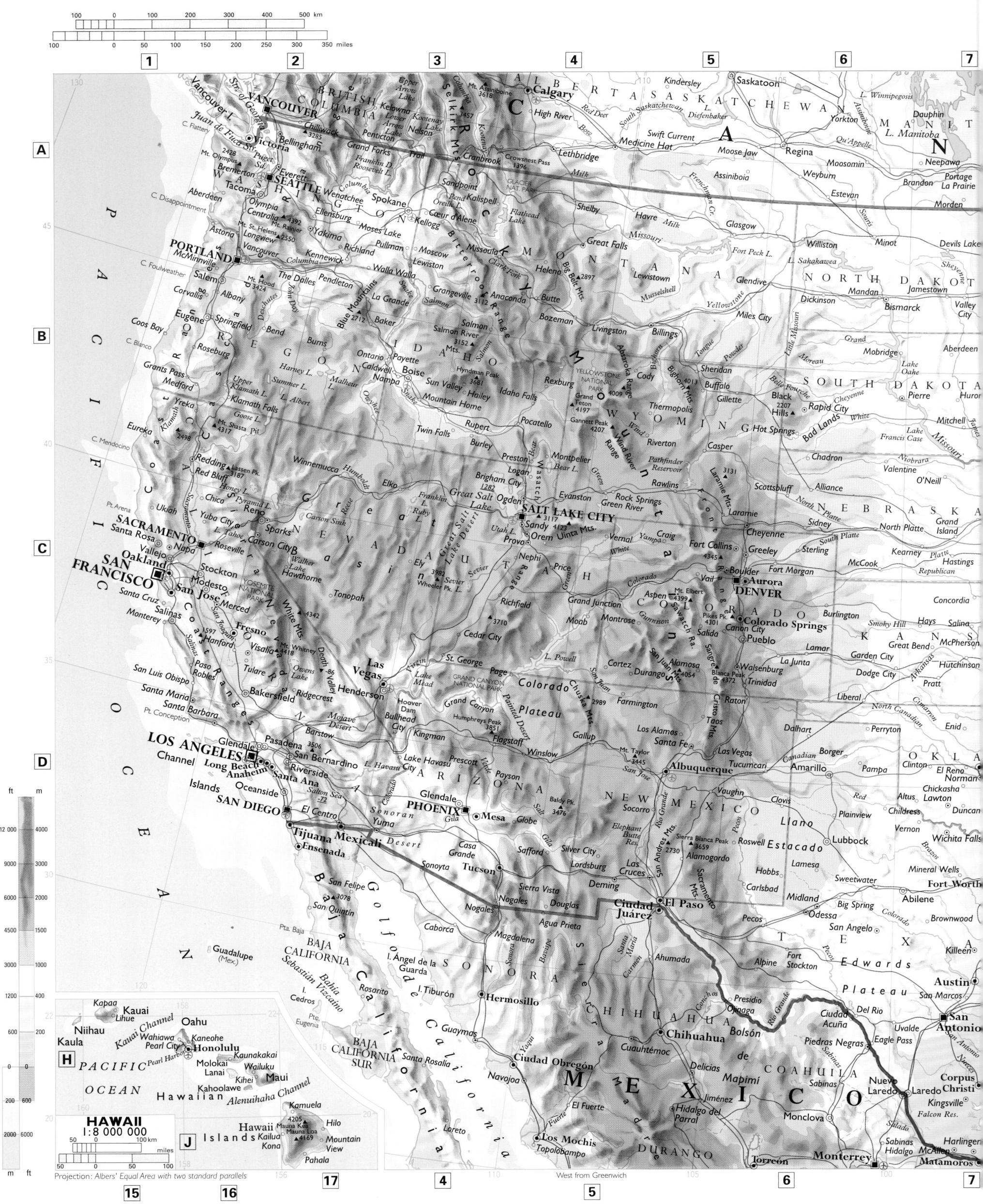
100 0 100 200 300 400 500 km
100 0 50 100 150 200 250 300 350 miles
1 2 3 4 5 6 7
A B C D
BRITISH COLUMBIA
ALBERTA
SASKATCHEWAN
MANITOBA
CANADA
Vancouver I.
VANCOUVER
Victoria
Str. of Georgia
Juan de Fuca Str.
C. Flattery
Calgary
High River
Kindersley
Saskatoon
L. Winnipegosis
Dauphin
L. Manitoba
Yorkton
Regina
Moose Jaw
Swift Current
Medicine Hat
Lethbridge
Assiniboia
Moosomin
Weyburn
Estevan
Brandon
Portage La Prairie
Neepawa
Morden
Kelowna
Penticton
Nelson
Kootenay Lake
Chilliwack
Bellingham
Grand Forks
Trail
Cranbrook
Crowsnest Pass 1396
Mt. Assiniboine 3618
Selkirk Mts.
Mt. Olympus 2428
Bremerton
Everett
SEATTLE
Tacoma
Olympia
Aberdeen
C. Disappointment
Centralia
Mt. Rainier 4392
Mt. St. Helens 2550
Longview
Astoria
WASHINGTON
Wenatchee
Spokane
Ellensburg
Yakima
Moses Lake
Kellogg
Cœur d'Alene
Sandpoint
Kalispell
Flathead Lake
GLACIER NAT. PARK
Shelby
Havre
Milk
Glasgow
Missouri
Fort Peck L.
Williston
Minot
Devils Lake
L. Sakakawea
NORTH DAKOTA
Jamestown
Valley City
Bismarck
Mandan
Dickinson
Richland
Kennewick
Pullman
Moscow
Lewiston
Walla Walla
Missoula
Great Falls
Helena
MONTANA
Lewistown
Glendive
Miles City
Big Belt Mts. 2897
PORTLAND
McMinnville
Vancouver
C. Foulweather
Salem
Corvallis
Albany
Eugene
Springfield
The Dalles
Pendleton
Mt. Hood 3424
Coos Bay
C. Blanco
Roseburg
Grants Pass
Medford
Bend
Blue Mountains 2712
La Grande
Baker
Burns
OREGON
Ontario
Caldwell
Nampa
Payette
Boise
Grangeville
Salmon River Mts. 3152
Salmon
Anaconda
Butte
Bozeman
Livingston
Billings
Bitterroot Range
IDAHO
Sun Valley
Hailey
Hyndman Peak 3681
Mountain Home
Idaho Falls
Rexburg
YELLOWSTONE NATIONAL PARK
Grand Teton 4197
Cody
Absaroka Range
Bighorn Mts. 4013
Sheridan
Buffalo
Gillette
Thermopolis
WYOMING
Gannett Peak 4207
Wind River Range
Riverton
Casper
Pathfinder Reservoir
Rawlins
Rock Springs
Green River
Evanston
Laramie
Cheyenne
Laramie Mts. 3131
Mobridge
Aberdeen
SOUTH DAKOTA
Pierre
Huron
Black Hills 2207
Rapid City
Hot Springs
Bad Lands
Mitchell
Lake Francis Case
Chadron
Niobrara
Valentine
O'Neill
Alliance
Scottsbluff
NEBRASKA
North Platte
Sidney
Grand Island
Kearney
Hastings
McCook
Upper Klamath L.
Klamath Falls
Goose L.
Yreka
Mt. Shasta 4317
Eureka
C. Mendocino
Redding
Lassen Pk. 3187
Red Bluff
Chico
Ukiah
Pt. Arena
Yuba City
Reno
Sparks
Carson City
L. Tahoe
Roseville
SACRAMENTO
Santa Rosa
Napa
Vallejo
Oakland
SAN FRANCISCO
Stockton
Modesto
San Jose
Merced
Santa Cruz
Salinas
Monterey
Fresno
Visalia
Hanford
Tulare
Mt. Whitney 4418
Paso Robles
San Luis Obispo
Santa Maria
Santa Barbara
Pt. Conception
Bakersfield
Ridgecrest
Coast Ranges
Sierra Nevada
YOSEMITE NATIONAL PARK
Winnemucca
Humboldt
Elko
NEVADA
Great Basin
Walker Lake
Hawthorne
Tonopah
White Mts. 4342
Owens Lake
Death Valley -86
Ely
Wheeler Pk. 3982
Sevier L.
Great Salt Lake
Great Salt Lake Desert
Brigham City
Ogden
Logan
Preston
Montpelier
Bear L.
Pocatello
Twin Falls
Rupert
Burley
SALT LAKE CITY
Sandy 3117
Orem
Uinta Mts. 4123
Provo
Utah L.
Vernal
Nephi
Price
Richfield
UTAH
Cedar City
St. George
Las Vegas
Henderson
Hoover Dam
Bullhead City
Lake Mead
Kingman
GRAND CANYON NATIONAL PARK
Grand Canyon
Humphreys Peak 3851
Flagstaff
Colorado Plateau
Painted Desert
Chuska Mts. 2989
Farmington
Gallup
Winslow
Page
Craig
Fort Collins
Greeley
Boulder
Aurora
DENVER
Vail
Mt. Elbert 4399
Aspen
Grand Junction
Montrose
Moab
COLORADO
Pikes Pk. 4301
Colorado Springs
Cañon City
Pueblo
Salida
Walsenburg
Trinidad
Alamosa
San Juan Mts. 4054
Blanca Peak 4372
Cortez
Durango
Raton
Sterling
Fort Morgan
Burlington
Lamar
La Junta
KANSAS
Hays
Salina
Great Bend
McPherson
Garden City
Dodge City
Hutchinson
Pratt
Liberal
Concordia
LOS ANGELES
Glendale
Pasadena
San Bernardino
Riverside
Long Beach
Anaheim
Santa Ana
Oceanside
SAN DIEGO
Channel Islands
Barstow
Mojave Desert
Salton Sea -72
El Centro
Tijuana
Mexicali
Ensenada
Yuma
Lake Havasu City
Prescott
Payson
ARIZONA
Glendale
PHOENIX
Mesa
Globe
Baldy Pk. 3476
Sonoran Desert
Casa Grande
Tucson
Safford
Sierra Vista
Nogales
Douglas
Agua Prieta
Los Alamos
Santa Fe
Taos
Mt. Taylor 3445
Albuquerque
Las Vegas
Tucumcari
NEW MEXICO
Socorro
Elephant Butte Res.
Silver City
Lordsburg
Deming
Las Cruces
Sierra Blanca Peak 3659
Alamogordo
Roswell
Carlsbad
Hobbs
Clovis
Ciudad Juárez
El Paso
Dalhart
Amarillo
Borger
Pampa
Perryton
Enid
Clinton
El Reno
Norman
Chickasha
Lawton
Altus
Childress
Duncan
Vernon
Wichita Falls
Plainview
Llano Estacado
Lubbock
Lamesa
Sweetwater
Mineral Wells
Fort Worth
Abilene
Big Spring
Midland
Odessa
Pecos
San Angelo
Brownwood
Killeen
TEXAS
Edwards Plateau
Fort Stockton
Alpine
Austin
San Marcos
San Antonio
Del Rio
Uvalde
Eagle Pass
Ciudad Acuña
Piedras Negras
Nuevo Laredo
Laredo
Corpus Christi
Kingsville
Falcon Res.
Harlingen
McAllen
Matamoros
Guadalupe (Mex.)
Pta. Baja
San Quintín
San Felipe
BAJA CALIFORNIA
Sebastián Vizcaíno Bahía
I. Cedros
Pte. Eugenia
BAJA CALIFORNIA SUR
Santa Rosalía
Loreto
Golfo de California
I. Ángel de la Guarda
I. Tiburón
Caborca
Magdalena
Hermosillo
SONORA
Guaymas
Ciudad Obregón
Navojoa
El Fuerte
Los Mochis
Topolobampo
Sierra Madre
CHIHUAHUA
Ahumada
Presidio
Ojinaga
Chihuahua
Cuauhtémoc
Delicias
Bolsón de Mapimí
Jiménez
Hidalgo del Parral
COAHUILA
Sabinas
Monclova
Sabinas Hidalgo
Monterrey
Torreón
DURANGO
MEXICO
PACIFIC OCEAN
HAWAII 1:8 000 000
50 0 100 km
50 0 50 100 miles
H J
Kauai
Kapaa
Lihue
Niihau
Kaula
Kauai Channel
Oahu
Wahiawa
Pearl City
Kaneohe
Honolulu
Pearl Harbor
Molokai
Lanai
Kaunakakai
Wailuku
Kihei
Maui
Kahoolawe
Alenuihaha Channel
Hawaiian Islands
Kamuela
Hawaii
Mauna Kea 4205
Mauna Loa 4169
Hilo
Kailua Kona
Mountain View
Pahala
ft m
12 000 4000
9000 3000
6000 2000
4500 1500
3000 1000
1200 400
600 200
0 0
200 600
2000 6000
m ft
Projection: Albers' Equal Area with two standard parallels
West from Greenwich
15 16 17 4 5 6 7

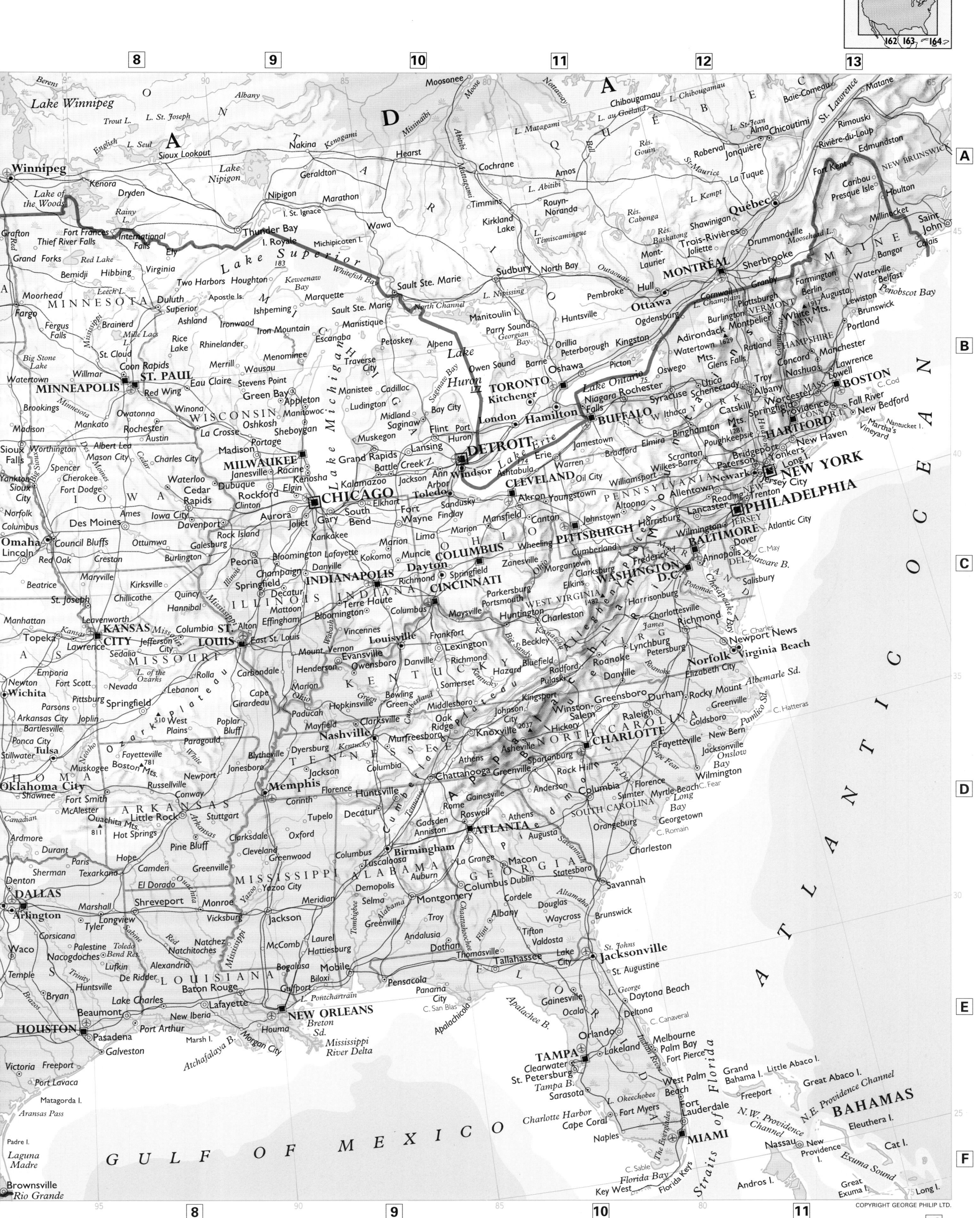
138
139
162
163
164
8
9
10
11
12
13
A
B
C
D
E
F
Berens
Lake Winnipeg
Albany
Trout L.
L. St. Joseph
English
L. Seul
Sioux Lookout
Moosonee
Moose
Abitibi
Nottaway
L. Chibougamau
Chibougamau
L. au Goéland
Matane
Baie-Comeau
St. Lawrence
L. Matagami
Nakina
Kenogami
Missinaibi
Hearst
Rés. Gouin
L. St-Jean
Alma
Chicoutimi
Rimouski
Rivière-du-Loup
Edmundston
NEW BRUNSWICK
Winnipeg
Kenora
Dryden
Lake of the Woods
Rainy L.
Lake Nipigon
Geraldton
Cochrane
Mattagami
Amos
L. Abitibi
Roberval
Jonquière
St.-Maurice
La Tuque
Fort Kent
Caribou
Presque Isle
Houlton
Nipigon
Marathon
I. St. Ignace
Timmins
Rouyn-Noranda
Kirkland Lake
Rés. Cabonga
L. Kempt
Québec
Shawinigan
Millinocket
Saint John
Calais
Grafton
Red
Thief River Falls
Fort Frances
International Falls
Ely
Thunder Bay
I. Royale
Wawa
Michipicoten I.
L. Témiscamingue
Rés. Baskatong
Trois-Rivières
Joliette
Drummondville
Moosehead L.
MAINE
Grand Forks
Red Lake
Virginia
Lake Superior
183
Mont-Laurier
Bangor
Belfast
Penobscot Bay
Bemidji
Hibbing
Two Harbors
Houghton
Keweenaw Bay
Whitefish Bay
Sault Ste. Marie
Sudbury
North Bay
Outaouais
MONTREAL
Sherbrooke
Farmington
Waterville
Leech L.
Moorhead
MINNESOTA
Duluth
Superior
Apostle Is.
Marquette
Sault Ste. Marie
North Channel
L. Nipissing
Pembroke
Hull
Ottawa
Cornwall
L. Champlain
Granby
Plattsburgh
Berlin
Augusta
Lewiston
Brunswick
Fargo
Ishpeming
Ashland
Ironwood
Iron Mountain
Manistique
Manitoulin I.
Huntsville
Ogdensburg
Burlington
VERMONT
White Mts.
1917
Fergus Falls
Brainerd
Mille Lacs L.
Mississippi
Rice Lake
Rhinelander
Escanaba
Petoskey
Alpena
Parry Sound
Georgian Bay
Orillia
Peterborough
Kingston
Adirondack Mts.
Montpelier
1629
Rutland
NEW HAMPSHIRE
Portland
Big Stone Lake
St. Cloud
Coon Rapids
Merrill
Wausau
Menominee
Traverse City
Lake Huron
Owen Sound
Barrie
Oshawa
Picton
Watertown
Glens Falls
Concord
Manchester
Lawrence
Lowell
Nashua
MASS.
BOSTON
C. Cod
Watertown
Willmar
MINNEAPOLIS
ST. PAUL
Eau Claire
Stevens Point
Green Bay
Appleton
Manistee
Cadillac
Saginaw Bay
177
TORONTO
Kitchener
Lake Ontario
75
Rochester
Oswego
Syracuse
Utica
Schenectady
Troy
Albany
Worcester
Fall River
New Bedford
Nantucket I.
Martha's Vineyard
Red Wing
Winona
Brookings
Minnesota
Mankato
Owatonna
Rochester
Austin
WISCONSIN
Manitowoc
Oshkosh
Sheboygan
La Crosse
Ludington
Midland
Saginaw
Bay City
Muskegon
Flint
Port Huron
London
Hamilton
Niagara Falls
BUFFALO
NEW YORK
Ithaca
Catskill Mts.
1281
Springfield
Providence
CONN.
R.I.
HARTFORD
Madison
Worthington
Albert Lea
Mason City
Charles City
Madison
MILWAUKEE
Portage
Lansing
Grand Rapids
Lake Michigan
DETROIT
Lake Erie
Jamestown
Bradford
Elmira
Binghamton
Poughkeepsie
New Haven
Sioux Falls
Des Moines
Spencer
Cherokee
Fort Dodge
Cedar
Janesville
Racine
Kenosha
Battle Creek
Kalamazoo
Ann Arbor
Windsor
Ashtabula
174
Erie
Warren
Scranton
Wilkes-Barre
Bridgeport
Yonkers
Long I.
Paterson
Newark
NEW YORK
Jersey City
Yankton
Sioux City
Big Sioux
Waterloo
Cedar Rapids
Dubuque
Elgin
Rockford
Clinton
IOWA
CHICAGO
Elkhart
Jackson
Toledo
Sandusky
CLEVELAND
Akron
Youngstown
Oil City
Williamsport
PENNSYLVANIA
Allentown
Reading
Trenton
Lancaster
PHILADELPHIA
Norfolk
Columbus
Ames
Iowa City
Davenport
Aurora
Joliet
Gary
South Bend
Fort Wayne
Findlay
Mansfield
Canton
Johnstown
Altoona
Harrisburg
NEW JERSEY
Atlantic City
Omaha
Lincoln
Council Bluffs
Red Oak
Creston
Ottumwa
Burlington
Rock Island
Galesburg
Kankakee
Marion
Lima
Marion
OHIO
Wheeling
PITTSBURGH
Wilmington
BALTIMORE
Dover
C. May
Delaware B.
Beatrice
Maryville
Kirksville
Peoria
Bloomington
Champaign
Lafayette
Kokomo
Muncie
Danville
COLUMBUS
Zanesville
Cumberland
Frederick
Annapolis
MARYLAND
DEL.
Springfield
Decatur
INDIANAPOLIS
INDIANA
Dayton
Springfield
Richmond
CINCINNATI
Morgantown
Clarksburg
WASHINGTON D.C.
Potomac
Salisbury
Chesapeake Bay
St. Joseph
Chillicothe
Quincy
Hannibal
ILLINOIS
Mattoon
Effingham
Terre Haute
Bloomington
Columbus
Parkersburg
Portsmouth
Elkins
WEST VIRGINIA
1482
Harrisonburg
Manhattan
Leavenworth
Topeka
Kansas
KANSAS CITY
Missouri
Jefferson City
Columbia
Alton
ST. LOUIS
East St. Louis
Vincennes
Maysville
Huntington
Charleston
Charlottesville
Richmond
James
C. Charles
Newport News
Lawrence
Sedalia
MISSOURI
Wabash
Louisville
Frankfort
Lexington
Beckley
Lynchburg
Petersburg
Norfolk
Virginia Beach
Emporia
L. of the Ozarks
Rolla
Carbondale
Mount Vernon
Evansville
Owensboro
Henderson
Danville
Richmond
KENTUCKY
Big Sandy
Bluefield
Hazard
Roanoke
Roanoke
Elizabeth City
Albemarle Sd.
Newton
Wichita
Fort Scott
Nevada
Lebanon
Cape Girardeau
Marion
Ohio
Somerset
Pulaski
Radford
Danville
Pittsburg
Springfield
Parsons
Joplin
Ozark Plateau
510
West Plains
Poplar Bluff
Paducah
Hopkinsville
Bowling Green
Green
Cumberland
Middlesboro
Kingsport
Johnson City
Winston-Salem
Greensboro
Durham
Raleigh
Rocky Mount
Greenville
Pamlico Sd.
C. Hatteras
Arkansas City
Bartlesville
Ponca City
Stillwater
Tulsa
Neosho
Fayetteville
Boston Mts.
781
Paragould
White
Blytheville
Dyersburg
Mayfield
Clarksville
Nashville
Kentucky
Oak Ridge
Knoxville
2037
Hickory
Asheville
NORTH CAROLINA
CHARLOTTE
Goldsboro
New Bern
Jacksonville
Onslow Bay
Muskogee
Fort Smith
Russellville
Newport
Jonesboro
TENNESSEE
Murfreesboro
Columbia
Jackson
Athens
Chattanooga
Spartanburg
Greenville
Rock Hill
Fayetteville
Cape Fear
Wilmington
Oklahoma City
Shawnee
McAlester
Ouachita Mts.
811
Conway
Little Rock
Arkansas
Memphis
Corinth
Florence
Huntsville
Gainesville
Anderson
Columbia
Florence
Sumter
Myrtle Beach
C. Fear
Long Bay
ARKANSAS
Hot Springs
Stuttgart
Clarksdale
Tupelo
Decatur
Rome
Roswell
Athens
SOUTH CAROLINA
Georgetown
Canadian
Ardmore
Durant
Paris
Pine Bluff
Oxford
Cleveland
Greenwood
Columbus
Birmingham
Gadsden
Anniston
ATLANTA
Augusta
Savannah
Orangeburg
C. Romain
Charleston
Hope
Camden
El Dorado
Greenville
Tuscaloosa
ALABAMA
La Grange
Macon
GEORGIA
Statesboro
Sherman
Texarkana
Ouachita
MISSISSIPPI
Yazoo
Yazoo City
Demopolis
Auburn
Columbus
Dublin
Denton
DALLAS
Arlington
Marshall
Shreveport
Monroe
Vicksburg
Jackson
Meridian
Selma
Montgomery
Cordele
Altamaha
Douglas
Savannah
Brunswick
Longview
Tyler
Sabine
Red
Natchez
Tombigbee
Alabama
Greenville
Troy
Albany
Waycross
Corsicana
Palestine
Toledo Bend Res.
Natchitoches
McComb
Laurel
Hattiesburg
Andalusia
Chattahoochee
Flint
Tifton
Valdosta
Waco
Nacogdoches
Lufkin
Alexandria
Dothan
Thomasville
Tallahassee
Lake City
St. Johns
Jacksonville
St. Augustine
Temple
Trinity
Huntsville
De Ridder
LOUISIANA
Bogalusa
Mobile
Biloxi
Gulfport
Pensacola
FLORIDA
Brazos
Bryan
Lake Charles
Baton Rouge
Lafayette
L. Pontchartrain
Panama City
C. San Blas
Apalachicola
Apalachee B.
Gainesville
L. George
Daytona Beach
Beaumont
New Iberia
NEW ORLEANS
Breton Sd.
Ocala
Deltona
C. Canaveral
HOUSTON
Pasadena
Port Arthur
Marsh I.
Houma
Mississippi River Delta
Orlando
Melbourne
Palm Bay
Indian River
Galveston
Atchafalaya B.
Morgan City
TAMPA
Clearwater
St. Petersburg
Tampa B.
Lakeland
Fort Pierce
Victoria
Freeport
Sarasota
West Palm Beach
Grand Bahama I.
Little Abaco I.
Great Abaco I.
Port Lavaca
L. Okeechobee
Freeport
N.E. Providence Channel
Matagorda I.
Aransas Pass
Charlotte Harbor
Cape Coral
Fort Myers
The Everglades
Fort Lauderdale
Straits of Florida
N.W. Providence Channel
BAHAMAS
Eleuthera I.
GULF OF MEXICO
Naples
MIAMI
Nassau
New Providence I.
Padre I.
Laguna Madre
C. Sable
Florida Bay
Cat I.
Exuma Sound
Key West
Florida Keys
Andros I.
Great Exuma I.
Long I.
Brownsville
Rio Grande
ATLANTIC OCEAN
CANADA
ONTARIO
QUÉBEC
MICHIGAN
KANSAS
OKLAHOMA
TEXAS
VIRGINIA
Appalachian Mts.
Cumberland Plateau
Allegheny Mts.
Blue Ridge
Piedmont
95
90
85
80
75
70
65
45
40
35
30
25

50 0 50 100 150 200 km

50 0 50 100 150 miles

LAKE SUPERIOR
LAKE HURON
Georgian Bay
LAKE ONTARIO
LAKE ERIE
ONTARIO
QUÉBEC
CANADA
MICHIGAN
NEW YORK
PENNSYLVANIA
OHIO
INDIANA
KENTUCKY
WEST VIRGINIA
VIRGINIA
MARYLAND
NEW JERSEY
Chesapeake Bay
TORONTO
MONTRÉAL
Ottawa
Québec
BOSTON
NEW YORK
PHILADELPHIA
BALTIMORE
WASHINGTON D.C.
PITTSBURGH
CLEVELAND
DETROIT
CHICAGO
MILWAUKEE
COLUMBUS
INDIANAPOLIS
Buffalo
Cincinnati
Louisville

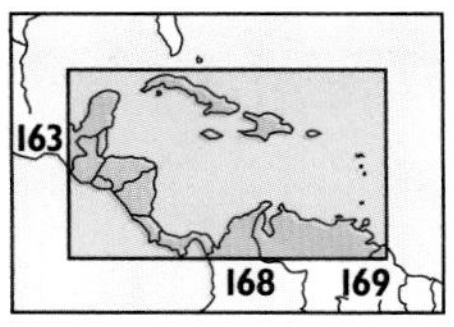

AMAS

50 0 50 100 150 200 250 300 km
50 0 50 100 150 200 miles

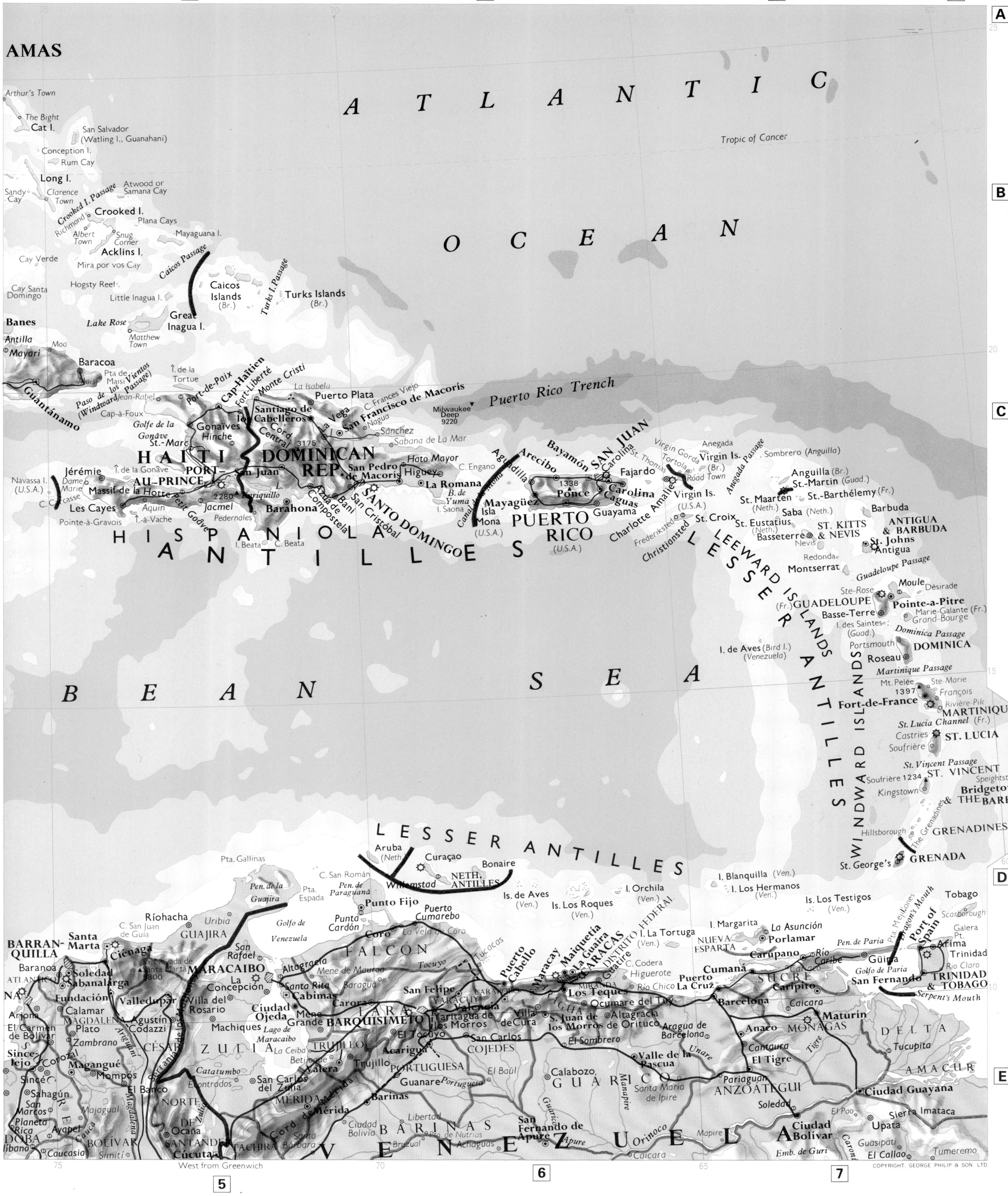

5   6   7   8

A T L A N T I C

O C E A N

Tropic of Cancer

Arthur's Town
The Bight
Cat I.
San Salvador
(Watling I., Guanahani)
Conception I.
Rum Cay
Sandy Cay
Long I.
Clarence Town
Cay Verde
Albert Town
Snug Corner
Crooked I.
Plana Cays
Acklins I.
Mira por vos Cay
Mayaguana I.
Hogsty Reef
Little Inagua I.
Caicos Islands (Br.)
Turks Islands (Br.)
Cay Santa Domingo
Lake Rose
Great Inagua I.
Matthew Town
Caicos Passage
Turks I. Passage

Banes
Antilla
Mayari
Moa
Baracoa
Pta de los Vientos
Maisi
Pta. de Maisi
Î. de la Tortue
Port-de-Paix
Cap-Haïtien
Fort-Liberté
Monte Cristi
La Isabela
Puerto Plata
S. Frances Viejo
San Francisco de Macoris
Nagua
Sánchez
Sabana de la Mar
Puerto Rico Trench
Milwaukee Deep 9220

Guantánamo
Paso de los (Windward) Passage
Jean-Rabel
Cap-à-Foux
Golfe de la Gonâve
Gonaïves
Hinche
St.-Marc
Cord. Central
3175
Santiago de los Cabelleros
La Vega

Puerto Rico Trench

HAITI
DOMINICAN REP
Jérémie
Î. de la Gonâve
PORT-AU-PRINCE
San Juan
2280
Hato Mayor
Higüey
C. Engano
Dame Marie
Massif de la Hotte
Aquin
Jacmel
Pedernales
Barahona
Compostela
San Cristóbal
SANTO DOMINGO
La Romana
San Pedro de Macoris
B. de Yuma
I. Saona
Bayamón
SAN JUAN
Arecibo
Aguadilla
Carolina
Fajardo
Caguas
Ponce
Mayagüez
1338
Isla Mona (U.S.A.)
Guayama
PUERTO RICO (U.S.A.)

Les Cayes
Pointe-à-Gravois
Î.-à-Vache
C. Tiburon
HISPANIOLA
C. Beata
I. Beata

Virgin Is.
Anegada
Virgin Gorda
Tortola (Br.)
Road Town
St. Thomas
Virgin Is. (U.S.A.)
Charlotte Amalie
St. Croix
Frederiksted
Christiansted
Anegada Passage
Sombrero (Anguilla)
Anguilla (Br.)
St.-Martin (Guad.)
St. Maarten (Neth.)
St.-Barthélemy (Fr.)
Saba (Neth.)
St. Eustatius (Neth.)
Basseterre
Nevis
ST. KITTS & NEVIS
Redonda
Barbuda
ANTIGUA & BARBUDA
St. Johns
Antigua
Montserrat
Guadeloupe Passage

A N T I L L E S

LEEWARD ISLANDS

Moule
Désirade
Ste-Rose
GUADELOUPE (Fr.)
Basse-Terre
Pointe-à-Pitre
Marie-Galante (Fr.)
Grand-Bourge
I. des Saintes (Guad.)
Dominica Passage
Portsmouth
DOMINICA
Roseau
Martinique Passage
I. de Aves (Bird I.) (Venezuela)

B E A N   S E A

Mt. Pelée 1397
Ste-Marie
François
Rivière-Pilc
Fort-de-France
MARTINIQUE (Fr.)
St. Lucia Channel (Fr.)
Castries
ST. LUCIA
Soufrière
St. Vincent Passage
Soufrière 1234
ST. VINCENT
Speightstown
Kingstown
Bridgetown & THE BARBADOS

WINDWARD ISLANDS
LESSER ANTILLES

LESSER ANTILLES
Aruba (Neth.)
Curaçao
Bonaire
Willemstad
NETH. ANTILLES
Is. de Aves (Ven.)
I. Orchila (Ven.)
I. Blanquilla (Ven.)
I. Los Hermanos (Ven.)
Hillsborough
The Grenadines
GRENADINES
St. George's
GRENADA

Pta. Gallinas
Pen. de la Guajira
Pta. Espada
C. San Román
Pta. de Paraguaná
Punto Fijo
Punta Cardón
Coro
La Vela de Coro
Is. Los Roques (Ven.)
I. La Tortuga (Ven.)
I. Margarita
La Asunción
NUEVA ESPARTA
Porlamar
I. Los Testigos (Ven.)
Tobago
Scarborough
Galera Pt.
Port of Spain

BARRANQUILLA
Santa Marta
Ciénaga
Ríohacha
Uribia
San Juan de Guia
GUAJIRA
Golfo de Venezuela
FALCON
Cumarebo
Tucacas
Puerto Cabello
Maracay
Maiquetía
La Guaira
CARACAS
DISTRITO FEDERAL
C. Codera
Río Chico
La Asunción
Carúpano
Río Caribe
Güiria
Pen. de Paria
Pta Mejillones
Dragon's Mouth
Trinidad
TRINIDAD & TOBAGO
San Fernando
Serpent's Mouth

Baranoa
Soledad
Sabanalarga
Fundación
Calamar
Plato
MAGDALENA
Zambrano
Valledupar
Villa del Rosario
La Concepción
MARACAIBO
Santa Rita
Cabimas
Maracaibo
Mene Grande
Ojeda
TRUJILLO
Barquisimeto
San Felipe
YARACUY
CARABOBO
ARAGUA
Villa de Cura
Cagua
Ocumare del Tuy
MIRANDA
Los Teques
San Juan de los Morros
Altagracia de Orituco
Aragua de Barcelona
Barcelona
Anaco
MONAGAS
Maturín
DELTA AMACUR
Tucupita

Sincelejo
Ocaña
César
ZULIA
La Ceiba
Trujillo
Betijoque
Machiques
MÉRIDA
Mérida
Barinas
Guanare
PORTUGUESA
Acarigua
El Tocuyo
LARA
San Carlos
COJEDES
El Baúl
Calabozo
GUÁRICO
Santa María de Ipire
Valle de la Pascua
Zaraza
El Sombrero
Pariaguán
ANZOÁTEGUI
El Tigre
Soledad
El Pao
Ciudad Guayana
Sierra Imataca
Upata

Magangué
Mompós
El Banco
Agustín Codazzi
NORTE
Cúcuta
San Cristóbal
TÁCHIRA
Cord. de Mérida
San Antonio
Barinas
BARINAS
Ciudad Bolivia
Libertad
San Fernando de Apure
APURE
Achaguas
Río Apure
Caicara
Cabruta
VENEZUELA
Orinoco
Ciudad Bolívar
BOLÍVAR
Caroní
Emb. de Guri
Guasipati
Tumeremo
El Callao

BOLÍVAR
Simití
Ayapel
Planeta Rica

ATLANTICO

West from Greenwich

5   6   7

COPYRIGHT GEORGE PHILIP & SON LTD

100 0 200 400 600 800 1000 1200 1400 km
100 0 200 400 600 800 1000 miles
Tropic of Cancer
Turks & Caicos Is.
Yucatan Channel
Cuba
Gulf of Campeche
Yucatán Peninsula
Greater Antilles
Hispaniola
9200
Puerto Rico
NORTH ATLANTIC OCEAN
Jamaica
Isthmus of Tehuantepec
G. de Honduras
Lesser Antilles
Guadeloupe
Dominica
Martinique
C. Gracias a Dios
Caribbean Sea
St. Lucia
St. Vincent
Barbados
Guatemala Trench
Coco
L. Nicaragua
C. de la Aguja
5800
Sierra Nevada de Santa Marta
L. Maracaibo
I. Margarita
Grenada
Tobago
Trinidad
Panama Canal
G. of Darién
Orinoco
Cord. de Mérida
Llanos
Magdalena
Meta
Gulf of Panamá
Cordillera Occidental
Cordillera Central
Cordillera Oriental
Guiana Highlands
2810
Mt. Roraima
Sierra Pacaraima
Essequibo
Courantyne
C. Orange
Serra Tumucumaque
Guaviare
Branco
C. de San Francisco
Caquetá
Negro
Equator
Cotopaxi 5897
Chimborazo 6267
Japurá
Putumayo
Marajó I.
Amazon
Galapagos Is.
Napo
Amazon
G. of Guayaquil
Marañón
Tocantins
Pta. Pariñas
Pta. Negra
Xingu
Parnaíba
C. de São Roque
Juruá
Selvas
Purus
Madeira
Tapajós
Aripuanã
Roosevelt
Ucayali
Huascarán 6768
Teles Pires
Plat. of Borborema
São Francisco
Araguaia
Chile Peru Trench
Madre de Dios
Arinos
Guaporé
Brazilian Highlands
Plateau of Mato Grosso
Mamoré
Chincha Alta
L. Titicaca
Nevada Ancohuma 6550
Bolivian Plateau
PACIFIC OCEAN
L. de Poopó
Abrolhos Bank
Gran Chaco
Paraguay
Paraná
2890
Pico da Bandeira
Serra da Mantiqueira
8050
Pilcomayo
C. Frio
Tropic of Capricorn
Atacama Desert
Cerro Ojos del Salado 6863
Iguaçu Falls
San Félix
San Ambrosio
Uruguay
Serra do Mar
Salado
Paraná
Salinas Grandes
Entre Ríos
L. Mar Chiquita
Sierra de Córdoba
L. dos Patos
Mt. Aconcagua 6960
Andes
Pampas
Arch. de Juan Fernández
Río de la Plata
SOUTH ATLANTIC OCEAN
Colorado
Bahía Blanca
Negro
G. San Matías
Chile Rise
Chiloé I.
40
Valdés Peninsula
Chubut
Patagonia
Argentine Basin
Chonos Archipelago
Mte. San Valentín 4058
Gulf of San Jorge
Taitao Peninsula
Gulf of Penas
6212
Wellington I.
Falkland Is.
Madre de Dios I.
West Falkland
East Falkland
Magellan's Str.
Tierra del Fuego
Santa Inés I.
Canal Cockburn
Staten I.
South Georgia
Canal Beagle
C. Horn
ft m
12000 4000
9000 3000
6000 2000
3000 1000
1500 500
600 200
0 0
200 600
1000 3000
2000 6000
4000 12000
6000 18000
8000 24000
m ft
Projection: Lambert's Azimuthal Equal Area
West from Greenwich
CARTOGRAPHY BY PHILIP'S. COPYRIGHT REED INTERNATIONAL BOOKS LTD

100 0 200 400 600 800 1000 1200 1400 km

100 0 200 400 600 800 1000 miles

1   **2**   **3**   **4**   **5**   **6**   **7**

Tropic of Cancer

A

Havana *C U B A*   BAHAMAS   Turks & Caicos Is. (U.K.)

N O R T H

Virgin Is. (U.K.)

HAITI   DOMINICAN REP.   San Juan

Port-au-Prince

JAMAICA   Kingston   PUERTO RICO (U.S.A.)   ST. KITTS & NEVIS   ANTIGUA & BARBUDA

A T L A N T I C

Basse-Terre   GUADELOUPE (Fr.)

DOMINICA

Fort-de-France   MARTINIQUE (Fr.)

C a r i b b e a n   S e a   Castries   ST. LUCIA

ST. VINCENT   BARBADOS

B   Kingstown   Bridgetown   **B**

O C E A N

MEXICO

BELIZE

GUATEMALA   HONDURAS   Tegucigalpa

Guatemala   NICARAGUA

San Salvador   Managua

EL SALVADOR

COSTA RICA   San José

Aruba   Curaçao   GRENADA   St. George's   TRINIDAD & TOBAGO

Port of Spain

C. de la Aguja

Barranquilla   Maracaibo   Caracas

PANAMÁ   Panamá   Cartagena   Barquisimeto   Valencia

G. of Darién   Cúcuta   San Cristóbal   Ciudad Guayana

C   Gulf of Panamá   Medellín   Bucaramanga   **VENEZUELA**   Georgetown   **GUYANA**   Paramaribo   **C**

Cali   Bogotá   **SURINAM**   Cayenne   C. Orange

FRENCH GUIANA

COLOMBIA   RORAIMA   Essequibo   AMAPÁ

Galapagos Is. (Ecuador)   Quito   Napo   Putumayo   Japurá   Branco   Equator

ECUADOR   Amazon   Marajó I.   Belém

Guayaquil   Iquitos   Manaus   Santarém   São Luís

G. of Guayaquil   Marañón   **AMAZONAS**   *P A R Á*   Fortaleza

Chiclayo   Juruá   Purus   Madeira   Tapajós   Xingu   Tocantins   **MARANHÃO**   Teresina   C. de São Roque

D   Trujillo   *A C R E*   **CEARÁ**   RIO G. DO NORTE   Natal   **D**

Chimbote   Pôrto Velho   Parnaíba   PARAÍBA   Campina Grande

PERÚ   RONDÔNIA   **PIAUÍ**   PERNAMBUCO   Recife

Callao   Madre de Dios   **B R A Z I L**   ALAGOAS   Maceió

LIMA   Cuzco   Mamoré   **MATO GROSSO**   TOCANTINS   São Francisco   SERGIPE

L. Titicaca   **BOLIVIA**   GOIÁS   **B A H Í A**   Aracaju

E   Arequipa   La Paz   Cochabamba   DIS. FED.   Brasília   Salvador   **E**

Santa Cruz   Cuiabá   Goiânia

Sucre   Paraguay   **MINAS GERAIS**

Iquique   **MATO GROSSO DO SUL**   Ribeirão Prêto   Belo Horizonte   ESPÍRITO SANTO

Juiz de Fora   Vitória

Antofagasta   Paraná   **SÃO PAULO**   Campinas   Campos

Tropic of Capricorn   **PARAGUAY**   *P A R A N Á*   SÃO PAULO   R. DE J.   Niterói

Pilcomayo   Asunción   Curitiba   RIO DE JANEIRO

San Félix (Chile)   San Ambrosio (Chile)   Salta   SANTA CATARINA

F   San Miguel de Tucumán   Resistencia   Uruguay   **F**

Corrientes   RIO GRANDE DO SUL

Salado   Pôrto Alegre

Arch. de Juan Fernández (Chile)   Córdoba   Santa Fe   Pelotas

O C E A N   San Juan   Paraná   **URUGUAY**

Viña del Mar   Mendoza   Rosario

Valparaíso   Montevideo

SANTIAGO   **BUENOS AIRES**

Talca   *A R G E N T I N A*   La Plata   Río de la Plata

Concepción   Bahía Blanca   Mar del Plata

G   Colorado   *S O U T H*   **G**

Valdivia   *C H I L E*   Negro   Viedma

Puerto Montt   Chubut   *A T L A N T I C*

Comodoro Rivadavia   Gulf of San Jorge

P A C I F I C   Gulf of Penas   *O C E A N*

H   West Falkland   FALKLAND IS. (U.K.)   **H**

Magellan's Str.   Stanley   East Falkland

Punta Arenas   Tierra del Fuego

South Georgia (U.K.)

C. Horn

1   **2**   **3**   **4**   **5**   **6**   **7**

Projection: Lambert's Azimuthal Equal Area

■ LIMA   Capital Cities

West from Greenwich

CARTOGRAPHY BY PHILIP'S.
COPYRIGHT REED INTERNATIONAL BOOKS LTD

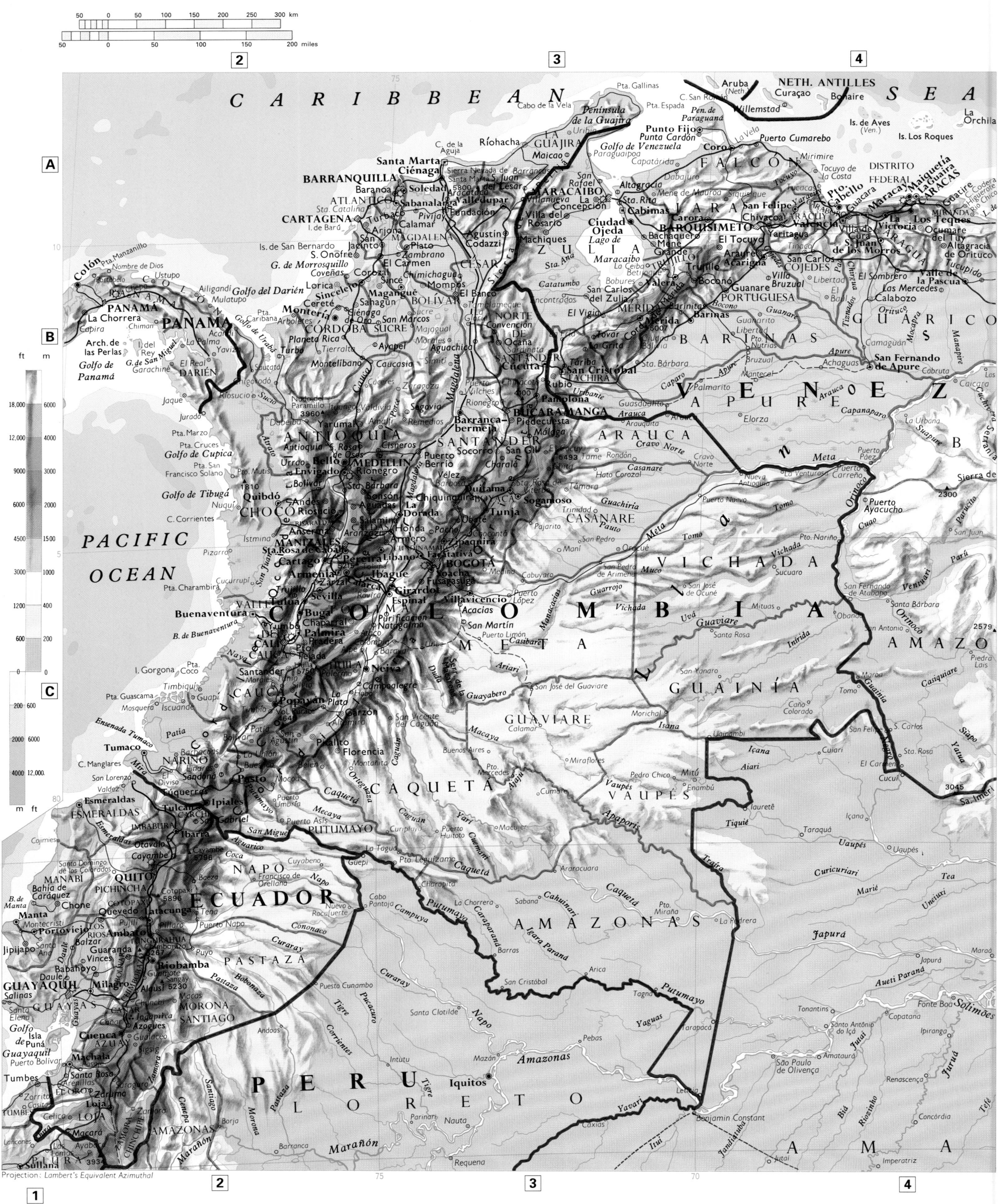

50 0 50 100 150 200 250 300 km
50 0 50 100 150 200 miles
2
3
4
A
B
C
1
CARIBBEAN SEA
PACIFIC OCEAN
NETH. ANTILLES
Aruba (Neth.)
Curaçao
Bonaire
Willemstad
Pta. Gallinas
Cabo de la Vela
Península de la Guajira
Pta. Espada
Pen. de Paraguaná
Punto Fijo
Punta Cardón
Golfo de Venezuela
Coro
Puerto Cumarebo
Is. de Aves (Ven.)
Is. Los Roques
La Orchila
DISTRITO FEDERAL
Maiquetía
La Guaira
CARACAS
Los Teques
Maracay
Valencia
Pto. Cabello
San Felipe
Barquisimeto
FALCÓN
LARA
ZULIA
MARACAIBO
Lago de Maracaibo
Ciudad Ojeda
Cabimas
Machiques
Mérida
Barinas
Trujillo
Valera
Guanare
Acarigua
Araure
San Carlos
COJEDES
PORTUGUESA
BARINAS
GUÁRICO
San Fernando de Apure
APURE
Calabozo
Valle de la Pascua
San Cristóbal
TÁCHIRA
Cúcuta
Pamplona
Ríohacha
LA GUAJIRA
Maicao
Santa Marta
Ciénaga
BARRANQUILLA
ATLÁNTICO
Soledad
Sabanalarga
CARTAGENA
Valledupar
Sierra Nevada de Santa Marta
5800
CESAR
MAGDALENA
BOLÍVAR
Sincelejo
SUCRE
Montería
CÓRDOBA
Mompós
El Banco
Ocaña
NORTE SANTANDER
Bucaramanga
SANTANDER
Barrancabermeja
Colón
PANAMÁ
PANAMA
La Chorrera
Golfo de Panamá
Arch. de las Perlas
DARIÉN
Golfo del Darién
Golfo de Urabá
Turbo
ANTIOQUIA
MEDELLÍN
Bello
Envigado
Rionegro
CHOCÓ
Quibdó
Golfo de Tibugá
Golfo de Cupica
C. Corrientes
Manizales
Pereira
Armenia
Ibagué
BOGOTÁ
Tunja
BOYACÁ
Sogamoso
Duitama
Zipaquirá
Girardot
Villavicencio
CASANARE
ARAUCA
Arauca
VICHADA
Puerto Carreño
Puerto Ayacucho
COLOMBIA
VENEZUELA
Buenaventura
B. de Buenaventura
VALLE
Tuluá
Buga
Palmira
Cali
CAUCA
Popayán
HUILA
Neiva
Garzón
Florencia
CAQUETÁ
META
GUAVIARE
GUAINÍA
VAUPÉS
Mitú
AMAZONAS
AMAZONAS
Tumaco
NARIÑO
Pasto
Ipiales
Tulcán
Esmeraldas
ESMERALDAS
Ibarra
Cayambe 5790
QUITO
PICHINCHA
ECUADOR
NAPO
Cotopaxi 5896
Latacunga
Ambato
Chimborazo 6267
Riobamba
Sangay 5230
PASTAZA
MORONA-SANTIAGO
Manta
Portoviejo
MANABÍ
GUAYAQUIL
GUAYAS
Golfo de Guayaquil
Isla de Puná
Cuenca
AZUAY
Machala
Loja
LOJA
Tumbes
Sullana
PIURA
PERU
LORETO
Iquitos
Amazonas
Marañón
Putumayo
Caquetá
Japurá
Solimões
Leticia
Benjamin Constant
Tabatinga
Yavari
Sa. Imeri
3045
AMAZO
Projection: Lambert's Equivalent Azimuthal
ft m
18,000 6000
12,000 4000
9000 3000
6000 2000
4500 1500
3000 1000
1200 400
600 200
0 0
200 600
2000 6000
4000 12,000
m ft
75
70
80
10
5
0

165
164
172
173
170

5
6
7

A
B
C
D

La Blanquilla (Ven.)
Los Hermanos (Ven.)
St. George's
GRENADA
Is. Los Testigos (Ven.)
Tobago
Scarborough
I. La Tortuga (Ven.)
NUEVA ESPARTA
Margarita
La Asunción
Porlamar
I. Coche
Pen. de Araya
Cumaná
Carúpano
Río Caribe
Pen. de Paria
Güiria
Dragon's Mouth
Boca del Dragon
Port of Spain
Arima
TRINIDAD AND TOBAGO
Trinidad
Rio Claro
Galeota Point
San Fernando
Golfo de Paria
Serpent's Mouth
Boca de la Sierpe
Puerto La Cruz
Barcelona
SUCRE
2596
Caripito
Maturín
Aragua de Barcelona
Anaco
Cantaura
MONAGAS
Zaraza
ANZOÁTEGUI
El Tigre
Tigre
Guanipa
Temblador
DELTA
Tucupita
Barrancas
Pariaguán
Orinoco
Ciudad Guayana
Pto. Ordaz
Soledad
Ciudad Bolívar
Upata
El Palmar
AMACURO
Boca Grande
I. Corocoro
Morawhanna
Mabaruma
BARIMA-WAINI
Matthew's Ridge
Tumeremo
El Callao
Guasipati
Embalse de Guri
El Dorado
Kokerite
POMEROON-SUPERNAAM
Anna Regina
ESSEQUIBO ISLANDS-WEST DEMERARA
Georgetown
DEMERARA-MAHAICA
MAHAICA-BERBICE
New Amsterdam
Port Mourant
Cuyuni
CUYUNI-MAZARUNI
Mazaruni
Bartica
Linden
Skeldon
GUYANA
Serranía Turagua
B O L Í V A R
Angel Falls
2560
La Gran Sabana
Mt. Roraima
2772
Pakaraima Mts.
Kaieteur Falls
POTARO-SIPARUNI
Essequibo
UPPER DEMERARA-BERBICE
Nieuw Nickerie
Wageningen
Paramaribo
Totness
CORONIE
Groningen
Nieuw Amsterdam
Moengo
Albina
COMMEWIJNE
BROKOPONDO
SARAMACCA
Prof. Dr. Ir. W. J. Van Blommestein Meer
Wilhelmina Geb.
Julianatop
1280
SURINAM
NICKERIE
MAROWIJNE
Tapanahoni
Lawa
Litani
FRENCH GUIANA
Cayenne
Kourou
Iles du Salut
I. du Diable
St. Laurent
Mana
St. Georges
Oiapoque
Cabo Orange
Oyapock
Camopi
Serra Pacaraima
Sa. Tepequem
Uraricoera
Boa Vista
Lethem
Tacutu
Rupununi
UPPER TAKUTU-UPPER ESSEQUIBO
Kamoa Mts.
EAST BERBICE-CORENTYNE
Serra Acaraí
734
Serra Tumucumaque
690
Serra Parima
Serra do Apiaú
Serra do Mucajaí
Caracaraí
R O R A I M A
Serra Tabatinga
Branco
Jari
A M A P Á
Amapá
Araguari
Serra do Navio
Macapá
Canal do Norte
I. Caviana
Ilha de Marajó
Almeirim
Pôrto de Moz
Negro
B R A Z I L
Barcelos
Moura
Represa de Balbina
Uatumã
Nhamundá
Trombetas
Óbidos
Alenquer
Monte Alegre
Amazonas
Santarém
Faro
Juruti
Parintins
Urucará
MANAUS
Manacapuru
Codajás
Coari
Tefé
(Amazonas)
Autazes
Maués
Itaituba
Tapajós
Altamira
Xingu
Iriri
P A R Á
Madeira
Purus
Borba
Novo Aripuanã
A M A Z O N A S

65
60
55
50
10
5
0

West from Greenwich

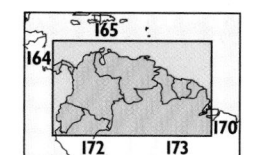

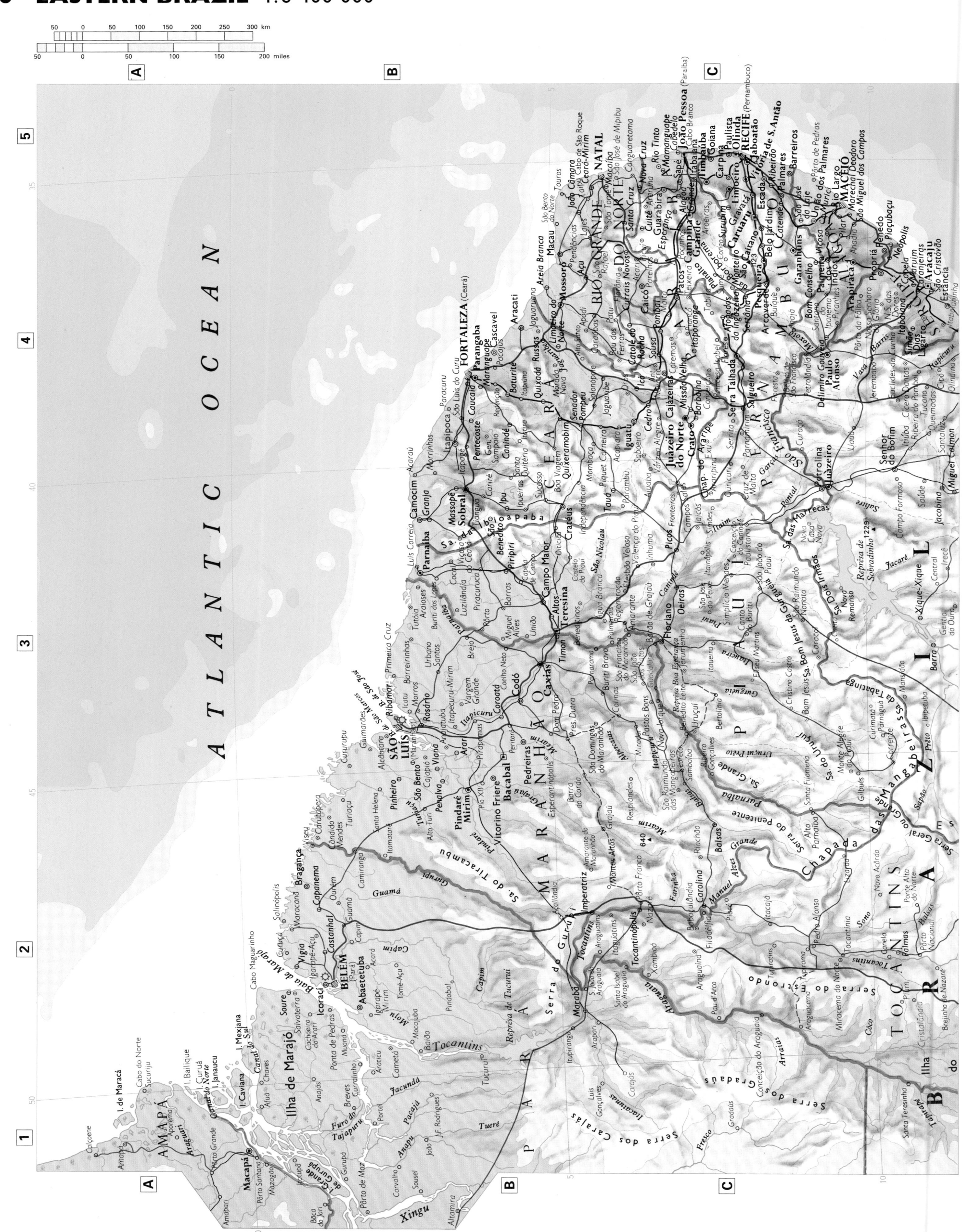

50 0 50 100 150 200 250 300 km
50 0 50 100 150 200 miles
1
2
3
4
5
A
B
C
ATLANTIC OCEAN
AMAPÁ
Macapá
I. de Maracá
Cabo do Norte
I. Caviana
I. Mexiana
Ilha de Marajó
Soure
Baía de Marajó
BELÉM
(Pará)
Abaetetuba
Vigia
Castanhal
Capanema
Bragança
Tocantins
Xingu
Altamira
Marabá
Imperatriz
Tocantinópolis
Carolina
Balsas
Serra do Gurupi
Sa. do Tiracambu
Serra dos Carajás
Serra dos Gradaús
Serra do Estrondo
Chapada das Mangabeiras
Serra Geral
P A R Á
M A R A N H Ã O
P I A U Í
C E A R Á
RIO GRANDE DO NORTE
P A R A Í B A
P E R N A M B U C O
A L A G O A S
S E R G I P E
B R A Z I L
T O C A N T I N S
SÃO LUÍS
Bacabal
Pedreiras
Caxias
Codó
Teresina
Timon
Floriano
Oeiras
Picos
Parnaíba
Sobral
Camocim
FORTALEZA (Ceará)
Maranguape
Quixadá
Crato
Juàzeiro do Norte
Iguatu
Mossoró
NATAL
João Pessoa (Paraíba)
Campina Grande
RECIFE (Pernambuco)
Olinda
Caruaru
Garanhuns
Arcoverde
Serra Talhada
Petrolina
Juàzeiro
Paulo Afonso
MACEIÓ
Aracaju
Propriá
Penedo
Senhor do Bonfim
Xique-Xique
Barra
Ilha do Bananal
Palmas
Pôrto Nacional
Represa de Sobradinho
Represa de Tucuruí
Represa Boa Esperança

Projection: Lambert's Equivalent Azimuthal

West from Greenwich

m	ft
2000	6000
1500	4500
1000	3000
400	1200
200	600
0	0
200	600
2000	6000
4000	12,000

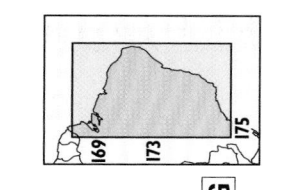

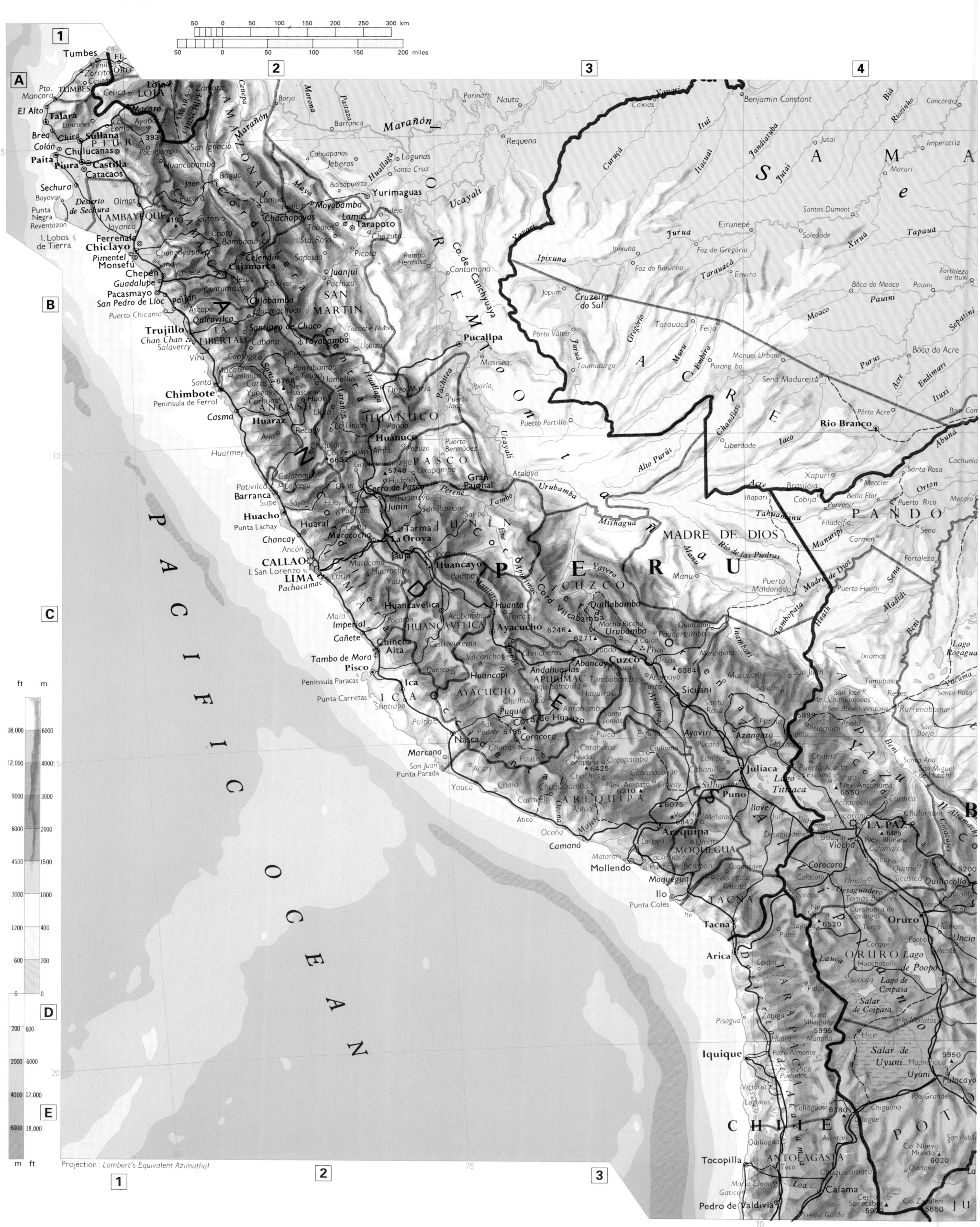

50 0 50 100 150 200 250 300 km
50 0 50 100 150 200 miles
PACIFIC OCEAN
PERU
CHILE
ACRE
AMAZONAS
SAN MARTIN
LA LIBERTAD
ANCASH
HUANUCO
PASCO
JUNIN
LIMA
HUANCAVELICA
AYACUCHO
APURIMAC
CUZCO
MADRE DE DIOS
PANDO
LA PAZ
ORURO
POTOSI
ANTOFAGASTA
TARAPACA
TACNA
MOQUEGUA
AREQUIPA
PUNO
ICA
PIURA
TUMBES
LAMBAYEQUE
CAJAMARCA
LORETO
Tumbes
Talara
Sullana
Piura
Paita
Chiclayo
Cajamarca
Trujillo
Chimbote
Huaraz
Huánuco
Pucallpa
Cerro de Pasco
La Oroya
Huancayo
CALLAO
LIMA
Huancavelica
Ayacucho
Pisco
Ica
Nasca
Cuzco
Abancay
Juliaca
Puno
Arequipa
Mollendo
Moquegua
Ilo
Tacna
Arica
Iquique
Tocopilla
Calama
LA PAZ
Oruro
Uyuni
Rio Branco
Cobija
Puerto Maldonado
Yurimaguas
Tarapoto
Moyobamba
Chachapoyas
Iquitos
Cruzeiro do Sul
Lago Titicaca
Lago de Poopó
Salar de Uyuni
Salar de Coipasa
Marañón
Ucayali
Huallaga
Urubamba
Madre de Dios
Juruá
Purus
6768
6632
6271
6384
6425
6310
6075
6485
6520
5995
6180
6020
Projection: Lambert's Equivalent Azimuthal

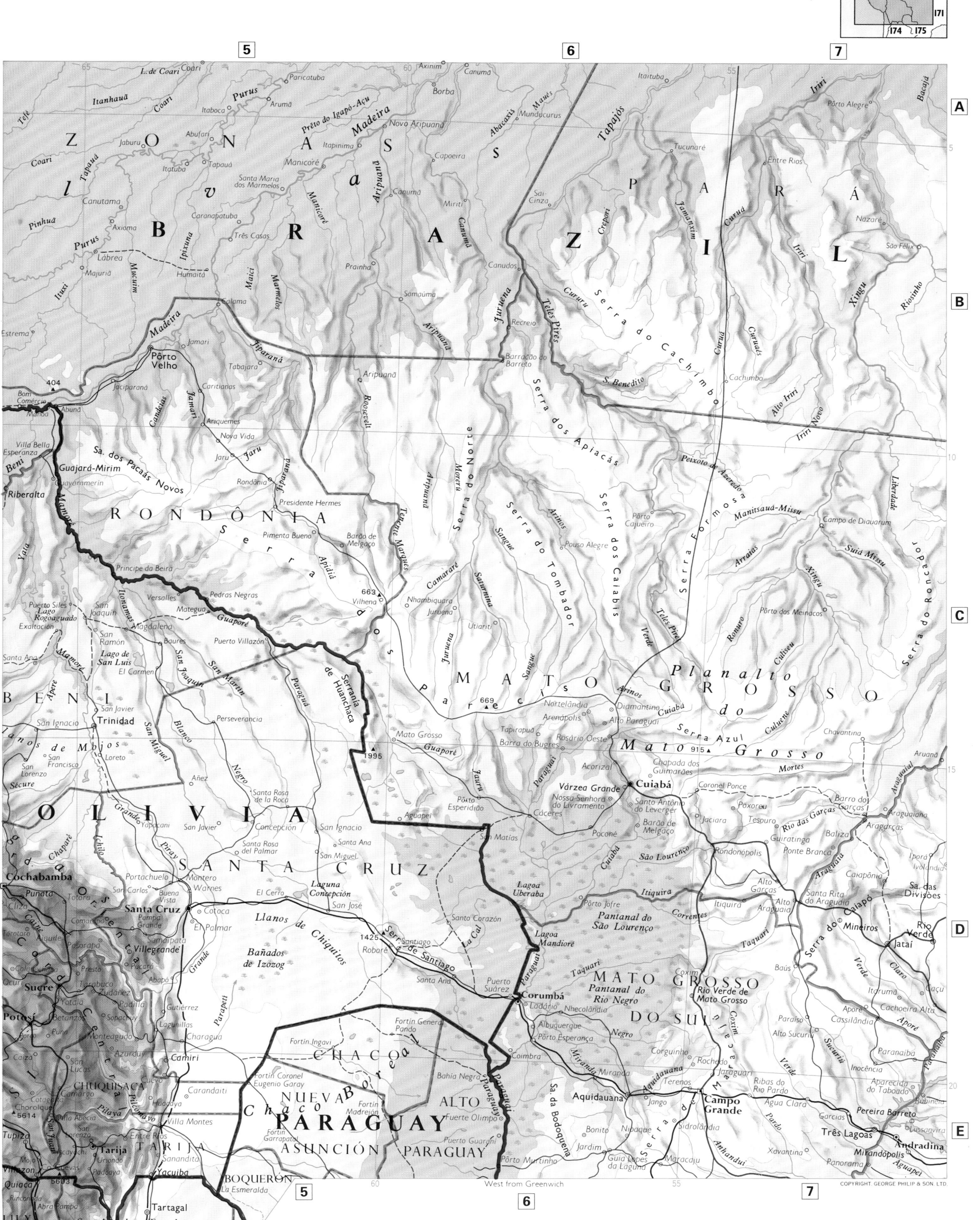
168
169
170
171
174
175
5
6
7
A
B
C
D
E
West from Greenwich
BRAZIL
BOLIVIA
PARAGUAY
Amazonas
Pará
Rondônia
Mato Grosso
Mato Grosso do Sul
Planalto do Mato Grosso
Beni
Santa Cruz
Chuquisaca
Tarija
Salta
Jujuy
Chaco
Nueva Asunción
Alto Paraguay
Boquerón
Chaco Boreal
Cochabamba
Sucre
Potosí
Tarija
Santa Cruz
Trinidad
Riberalta
Guajará-Mirim
Pôrto Velho
Cuiabá
Várzea Grande
Corumbá
Campo Grande
Aquidauana
Três Lagoas
Andradina
Jataí
Rio Verde
Humaitá
Manicoré
Borba
Novo Aripuanã
Lábrea
Coari
Purus
Madeira
Tapajós
Juruena
Teles Pires
Xingu
Iriri
Guaporé
Mamoré
Paraguai
Aripuanã
Serra do Cachimbo
Serra dos Apiacás
Serra dos Caiabis
Serra do Tombador
Serra Formosa
Serra dos Pacaás Novos
Serra dos Parecis
Serra do Norte
Serra do Roncador
Serra Azul
Serra de Maracaju
Serranía de Huanchaca
Serra de Santiago
Cordillera Oriental
Cordillera Central
Llanos de Mojos
Llanos de Chiquitos
Bañados de Izozog
Pantanal do São Lourenço
Pantanal do Rio Negro
Lagoa Uberaba
Lagoa Mandioré
Lago Rogoaguado
Lago de San Luis
Vilhena
Cáceres
Poconé
Rondonópolis
Coxim
Camiri
Yacuiba
Tartagal
Villazón
Tupiza
Villa Montes
Puerto Suárez
Roboré
San José
Concepción
Montero
Warnes
Vallegrande
San Ignacio
San Matías
Magdalena
Mato Grosso
Diamantino
Rosário Oeste
Barra do Bugres
Alto Araguaia
Aragarças
Barra do Garças
Mineiros
Paranaíba
Ponta Porã
Miranda
Maracaju
Nioaque
Bela Vista
Pôrto Murtinho
Fuerte Olimpo
Bahía Negra
Fortín Ravelo
Fortín General Pando
404
663
669
915
1425
1995
5603
5614
5
10
15
20
55
60
65

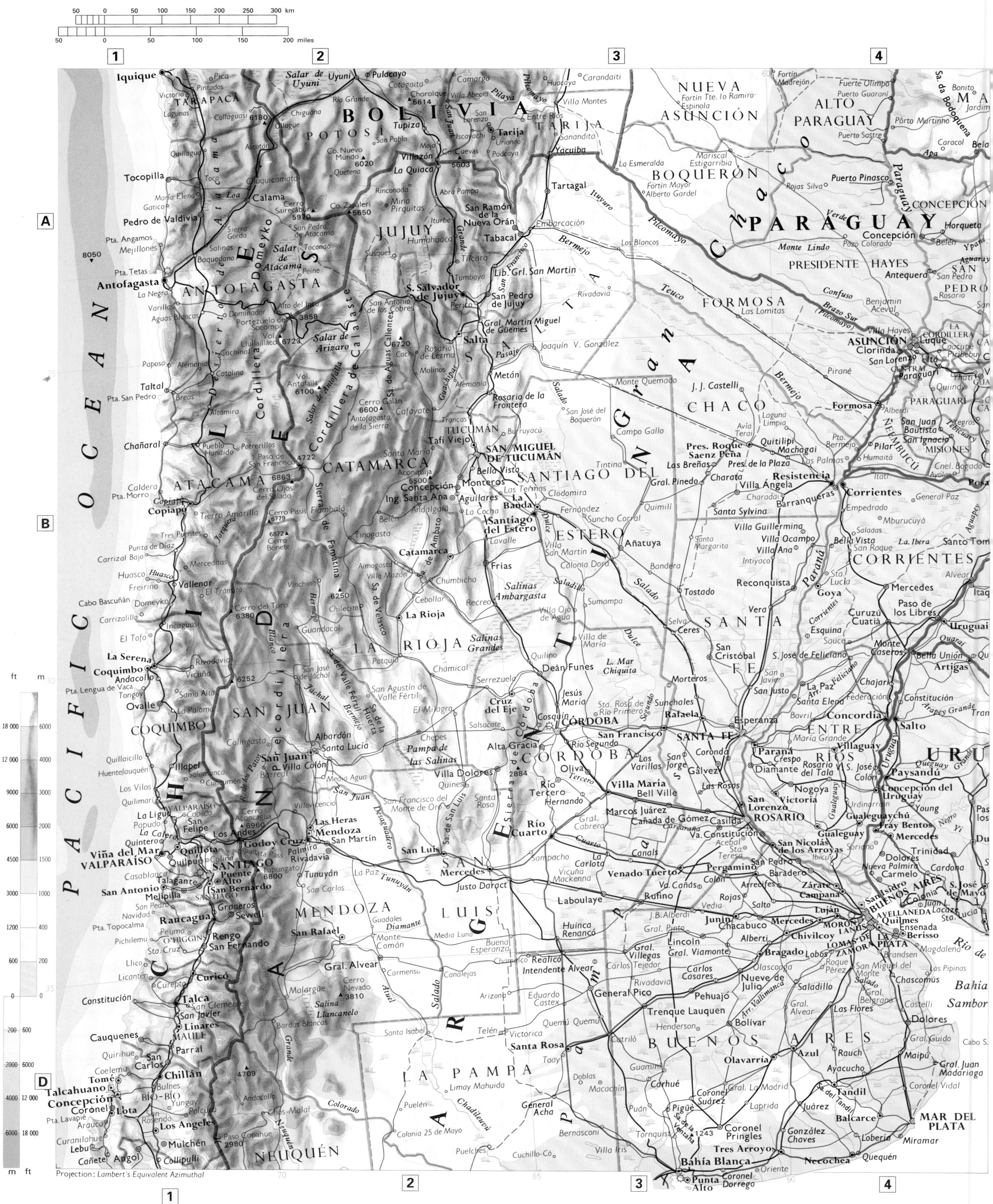

50 0 50 100 150 200 250 300 km
50 0 50 100 150 200 miles
1 2 3 4
A B D
PACIFIC OCEAN
BOLIVIA
PARAGUAY
CHILE
ARGENTINA
Gran Chaco
ANDES
Cordillera Domeyko
Desierto de Atacama
Cordillera de Calalaste
Cordillera de San Buenaventura
Salar de Uyuni
Salar de Atacama
Salar de Arizaro
Salinas Grandes
Salinas Ambargasta
Salina Llancanelo
Pampa de las Salinas
Iquique
TARAPACÁ
POTOSÍ
TARIJA
NUEVA ASUNCIÓN
ALTO PARAGUAY
BOQUERÓN
PRESIDENTE HAYES
SAN PEDRO
CONCEPCIÓN
ASUNCIÓN
PARAGUARÍ
MISIONES
ÑEEMBUCÚ
Tocopilla
Calama
Pedro de Valdivia
ANTOFAGASTA
Antofagasta
Mejillones
Taltal
Chañaral
ATACAMA
Copiapó
Caldera
Vallenar
Huasco
La Serena
Coquimbo
Ovalle
COQUIMBO
Illapel
VALPARAÍSO
Viña del Mar
Valparaíso
Quillota
Los Andes
SANTIAGO
San Antonio
Rancagua
O'HIGGINS
San Fernando
Curicó
Talca
MAULE
Linares
Cauquenes
Chillán
Talcahuano
Concepción
BÍO-BÍO
Coronel
Lota
Los Angeles
Mulchén
Angol
Collipulli
JUJUY
S. Salvador de Jujuy
SALTA
Salta
Tartagal
San Ramón de la Nueva Orán
Formosa
FORMOSA
CHACO
Resistencia
Corrientes
CORRIENTES
TUCUMÁN
SAN MIGUEL DE TUCUMÁN
CATAMARCA
Catamarca
SANTIAGO DEL ESTERO
Santiago del Estero
LA RIOJA
La Rioja
SAN JUAN
San Juan
MENDOZA
Mendoza
San Rafael
SAN LUIS
San Luis
CÓRDOBA
Córdoba
Río Cuarto
Villa María
SANTA FE
Santa Fe
Rosario
ENTRE RÍOS
Paraná
Concordia
BUENOS AIRES
Buenos Aires
La Plata
Mar del Plata
Bahía Blanca
Tandil
Olavarría
LA PAMPA
Santa Rosa
NEUQUÉN
Aconcagua 6960
Río de la Plata
URUGUAY
Salto
Paysandú
Projection: Lambert's Equivalent Azimuthal

172
173
171
176
5
6
7
A
B
C
D
BELO HORIZONTE
N. Lima
Itabirito
Congonhas
Cons. Lafaiete
Ouro Prêto
Ponte Nova
Vitória
Itaquari
Vila Velha
Guarapari
Pico da Bandeira 2890
Castelo
Cachoeiro de Itapemirim
Carangola
Alegre
Muriaé
Ubá
Itaperuna
Leopoldina
Cambuci
Guarus
CAMPOS
Cabo de São Tomé
Oliveira
Campo Belo
São João del Rei
Barbacena
Cataguases
Santos Dumont
Juiz de Fora
Três Rios
Além Paraíba
Paraíba do Sul
RIO DE JANEIRO
Nova Friburgo
Petrópolis
Macaé
DUQUE DE CAXIAS
SÃO GONÇALO
NITERÓI
RIO DE JANEIRO
Cabo Frio
La. de Araruama
Tropic of Capricorn
Passos
Batatais
São Seb. do Paraíso
Represa de Furnas
Três Pontas
Lavras
Guaxupé
Alfenas
Varginha
Três Corações
São Lourenço
Poços de Caldas
Pouso Alegre
Itajubá
Mantiqueira
2787
Volta Redonda
Barra do Piraí
Barra Mansa
Nova Iguaçu
Ouro Fino
Mogi-Mirim
Cruzeiro
Guaratinguetá
Bragança Paulista
Taubaté
S. J. dos Campos
Jacareí
Mogi das Cruzes
Angra dos Reis
Ilha Grande
Baía da Ilha Grande
Pta. de Juatinga
SÃO PAULO
SANTO ANDRÉ
São Bernardo do Campo
São Vicente
SANTOS
Guarujá
Ilha de São Sebastião
Pta. do Boi
Itanhaém
Jundiaí
Itu
Sorocaba
CAMPINAS
Americana
Limeira
Piracicaba
Rio Claro
Araras
Pinhal
São João da Boa Vista
Casa Branca
Mococa
Ribeirão Prêto
Jaboticabal
Bebedouro
Olímpia
S. José do Rio Prêto
Mirassol
Catanduva
Taquaritinga
Novo Horizonte
Araraquara
São Carlos
Jaú
Barirí
Pirajuí
Lins
Penápolis
Birigui
Araçatuba
Andradina
Três Lagoas
Mirandópolis
Adamantina
Pres. Epitácio
Panorama
Tupã
Marília
Martinópolis
Presidente Prudente
Santo Anastácio
Rancharia
Paraguaçu Paulista
Garça
Bauru
Assis
Porecatu
Cambará
Ourinhos
Santa Cruz do Rio Pardo
Botucatu
Avaré
Tatuí
Tietê
Itapetininga
Itapeva
Itararé
Jaguariaíva
Paranapiacaba
Juquiá
Registro
Apiaí
Iguape
Ilha Comprida
Ilha do Cardoso
Cornélio Procópio
Jacarèzinho
Londrina
Rolândia
Arapongas
Apucarana
Maringá
Mandaguari
Nova Esperança
Paranavaí
Sertanópolis
Centenário do Sul
Cianorte
Umuarama
Cruzeiro do Oeste
Goio Erê
Campo Mourão
Cândido de Abreu
PARANÁ
BRAZIL
Castro
Ponta Grossa
Serra do Mar
1889
CURITIBA
Antonina
Paranaguá
Guaratuba
Palmeira
Lapa
Irati
Guarapuava
Prudentópolis
Laranjeiras do Sul
Pitanga
Cascavel
Represa de Itaipú
Foz do Iguaçu
Ciudad del Este
Cat. del Iguaçu
Guaíra
Pto. Mendes
União da Vitória
Rio Negro
Mafra
Pto. União
São Francisco do Sul
Joinville
Caçador
Blumenau
Itajaí
Brusque
Santa Cecília
SANTA CATARINA
Campos Novos
Rio do Sul
Joaçaba
Chapecó
Palmas
Clevelândia
1340
Ilha de Santa Catarina
Florianópolis
Lajes
1808
Tubarão
Laguna
Cabo Santa Marta Grande
Criciúma
Araranguá
Vacaria
Erechim
Passo Fundo
Caràzinho
Cruz Alta
Guaporé
Bento Gonçalves
Caxias do Sul
RIO GRANDE
Santo Ângelo
São Luís Gonzaga
São Borja
Santiago
Santa Maria
Santa Cruz do Sul
Montenegro
Nôvo Hamburgo
Taquara
São Leopoldo
Osório
Canoas
PÔRTO ALEGRE
Viamão
Rio Pardo
Cachoeira do Sul
DO SUL
Alegrete
Rosário do Sul
São Gabriel
Caçapava do Sul
Camaquã
Encantadas
Santana do Livramento
Rivera
Dom Pedrito
Bagé
Canguçu
Lagoa dos Patos
Mostardas
Pelotas
Rio Grande
Jaguarão
Lagoa Mirim
Lagoa Mangueira
Santa Vitória do Palmar
Melo
Río Branco
Tacuarembó
URUGUAY
San Gregorio
Sta. Clara de Olimar
Treinta y Tres
José Batlle y Ordóñez
Lascano
Sarandí del Yi
Florida
Aigua
Castillos
Minas
Rocha
Pando
San Carlos
Maldonado
MONTEVIDEO
Río de la Plata
Tala
Canelones
Las Piedras
Antonio
MATO GROSSO DO SUL
Xavantina
Nioaque
Guia Lopes da Laguna
Maracaju
Rio Brilhante
Dourados
Ponta Porã
Pedro Juan Caballero
Amambaí
AMAMBAY
CANINDEYÚ
Igatimi
ALTO PARANÁ
Hernandarias
Irala
Abaí
Bernardo de Irigoyen
Eldorado
San Pedro
ITAPUÁ
MISIONES
Encarnación
Posadas
Obera
Candelaria
Leandro N. Alem
Santa Rosa
Apóstoles
San Javier
Corpus
Gral. Artigas
Carmen
Atlantic Ocean
ATLANTIC OCEAN
5304
West from Greenwich
55
50
45
40
25
30
35
COPYRIGHT, GEORGE PHILIP & SON, LTD

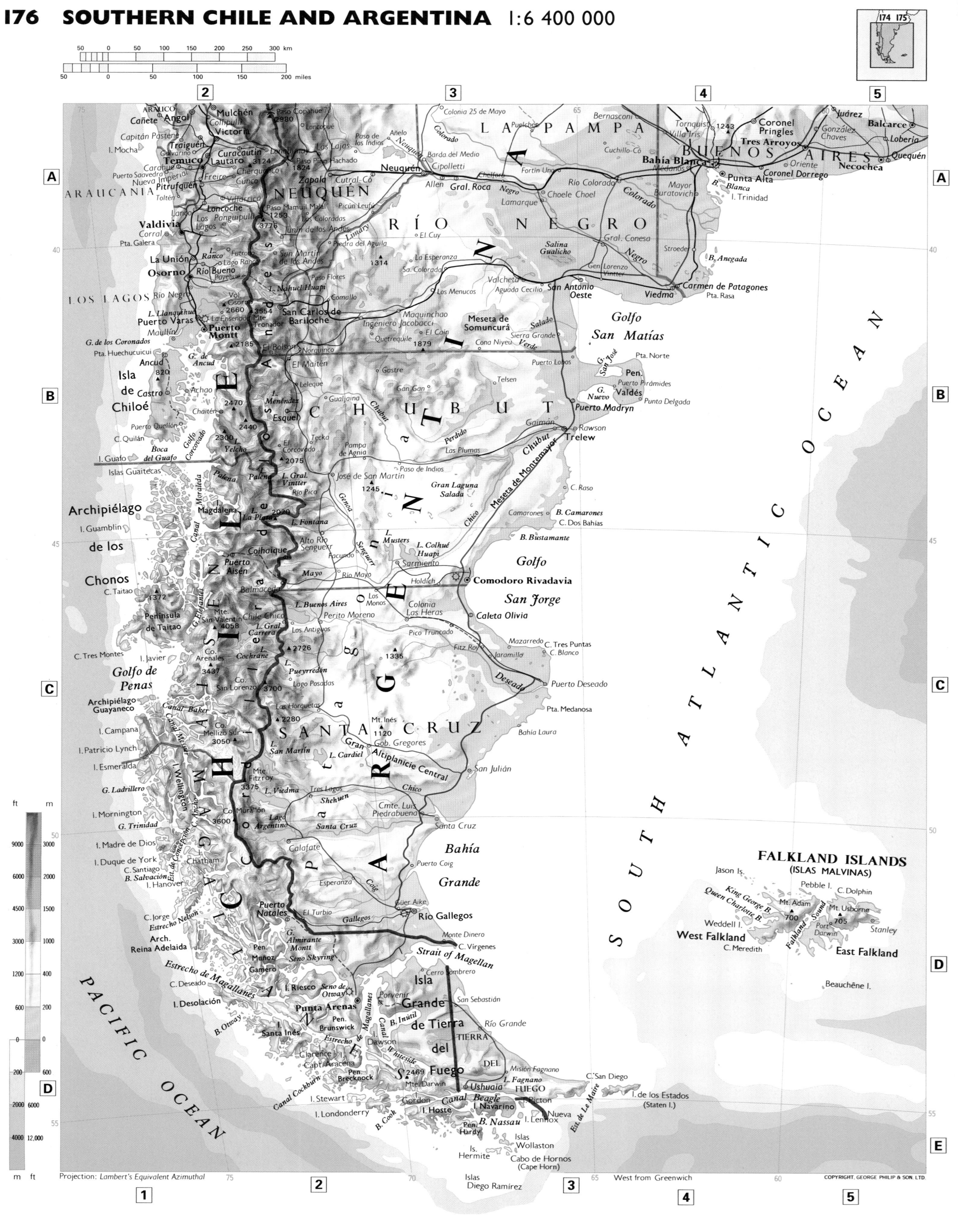

Projection: Lambert's Equivalent Azimuthal

West from Greenwich

COPYRIGHT. GEORGE PHILIP & SON. LTD.

INDEX

The index contains the names of all the principal places and features shown on the World Maps. Each name is followed by an additional entry in italics giving the country or region within which it is located. The alphabetical order of names composed of two or more words is governed primarily by the first word and then by the second. This is an example of the rule:

Mīr Kūh, *Iran*	**97 E8**	26 22N	58 55E
Mīr Shahdād, *Iran*	**97 E8**	26 15N	58 29E
Mira, *Italy*	**45 C9**	45 26N	12 8E
Mira por vos Cay, *Bahamas*	**165 B5**	22 9N	74 30W
Mīrābād, *Afghan*	**91 C1**	30 25N	61 50E

Physical features composed of a proper name (Erie) and a description (Lake) are positioned alphabetically by the proper name. The description is positioned after the proper name and is usually abbreviated:

Erie, L., *N. Amer.*	**150 D4**	42 15N	81 0W

Where a description forms part of a settlement or administrative name however, it is always written in full and put in its true alphabetic position:

Mount Morris, *U.S.A.*	**150 D7**	42 44N	77 52W

Names beginning with M' and Mc are indexed as if they were spelled Mac. Names beginning St. are alphabetised under Saint, but Sankt, Sint, Sant', Santa and San are all spelt in full and are alphabetised accordingly. If the same place name occurs two or more times in the index and all are in the same country, each is followed by the name of the administrative subdivision in which it is located. The names are placed in the alphabetical order of the subdivisions. For example:

Jackson, *Ky., U.S.A.*	**148 G4**	37 33N	83 23W
Jackson, *Mich., U.S.A.*	**157 D12**	42 15N	84 24W
Jackson, *Minn., U.S.A.*	**154 D7**	43 37N	95 1W

The number in bold type which follows each name in the index refers to the number of the map page where that feature or place will be found. This is usually the largest scale at which the place or feature appears.

The letter and figure which are in bold type immediately after the page number give the grid square on the map page, within which the feature is situated. The letter represents the latitude and the figure the longitude.

In some cases the feature itself may fall within the specified square, while the name is outside. This is usually the case only with features which are larger than a grid square.

For a more precise location the geographical coordinates which follow the letter/figure references give the latitude and the longitude of each place. The first set of figures represent the latitude which is the distance north or south of the Equator measured as an angle at the centre of the earth. The Equator is latitude 0°, the North Pole is 90°N, and the South Pole 90°S.

The second set of figures represent the longitude, which is the distance East or West of the prime meridian, which runs through Greenwich, England. Longitude is also measured as an angle at the centre of the earth and is given East or West of the prime meridian, from 0° to 180° in either direction.

The unit of measurement for latitude and longitude is the degree, which is subdivided into 60 minutes. Each index entry states the position of a place in degrees and minutes, a space being left between the degrees and the minutes.

The latitude is followed by N(orth) or S(outh) and the longitude by E(ast) or W(est).

Rivers are indexed to their mouths or confluences, and carry the symbol ⇢ after their names. A solid square ■ follows the name of a country, while an open square □ refers to a first order administrative area.

ABBREVIATIONS USED IN THE INDEX

A.C.T. – Australian Capital Territory
Afghan. – Afghanistan
Ala. – Alabama
Alta. – Alberta
Amer. – America(n)
Arch. – Archipelago
Ariz. – Arizona
Ark. – Arkansas
Atl. Oc. – Atlantic Ocean
B. – Baie, Bahía, Bay, Bucht, Bugt
B.C. – British Columbia
Bangla. – Bangladesh
Barr. – Barrage
Bos.-H. – Bosnia-Herzegovina
C. – Cabo, Cap, Cape, Coast
C.A.R. – Central African Republic
C. Prov. – Cape Province
Calif. – California
Cent. – Central
Chan. – Channel
Colo. – Colorado
Conn. – Connecticut
Cord. – Cordillera
Cr. – Creek
Czech. – Czech Republic
D.C. – District of Columbia
Del. – Delaware
Dep. – Dependency
Des. – Desert
Dist. – District
Dj. – Djebel
Domin. – Dominica
Dom. Rep. – Dominican Republic
E. – East

E. Salv. – El Salvador
Eq. Guin. – Equatorial Guinea
Fla. – Florida
Falk. Is. – Falkland Is.
G. – Golfe, Golfo, Gulf, Guba, Gebel
Ga. – Georgia
Gt. – Great, Greater
Guinea-Biss. – Guinea-Bissau
H.K. – Hong Kong
H.P. – Himachal Pradesh
Hants. – Hampshire
Harb. – Harbor, Harbour
Hd. – Head
Hts. – Heights
I. (s). – Île, Ilha, Insel, Isla, Island, Isle
Ill. – Illinois
Ind. – Indiana
Ind. Oc. – Indian Ocean
Ivory C. – Ivory Coast
J. – Jabal, Jebel, Jazira
Junc. – Junction
K. – Kap, Kapp
Kans. – Kansas
Kep. – Kepulauan
Ky. – Kentucky
L. – Lac, Lacul, Lago, Lagoa, Lake, Limni, Loch, Lough
La. – Louisiana
Liech. – Liechtenstein
Lux. – Luxembourg
Mad. P. – Madhya Pradesh
Madag. – Madagascar
Man. – Manitoba
Mass. – Massachusetts

Md. – Maryland
Me. – Maine
Medit. S. – Mediterranean Sea
Mich. – Michigan
Minn. – Minnesota
Miss. – Mississippi
Mo. – Missouri
Mont. – Montana
Mozam. – Mozambique
Mt. (e) – Mont, Monte, Monti, Montaña, Mountain
N. – Nord, Norte, North, Northern, Nouveau
N.B. – New Brunswick
N.C. – North Carolina
N. Cal. – New Caledonia
N. Dak. – North Dakota
N.H. – New Hampshire
N.I. – North Island
N.J. – New Jersey
N. Mex. – New Mexico
N.S. – Nova Scotia
N.S.W. – New South Wales
N.W.T. – North West Territory
N.Y. – New York
N.Z. – New Zealand
Nebr. – Nebraska
Neths. – Netherlands
Nev. – Nevada
Nfld. – Newfoundland
Nic. – Nicaragua
O. – Oued, Ouadi
Occ. – Occidentale
Okla. – Oklahoma
Ont. – Ontario
Or. – Orientale

Oreg. – Oregon
Os. – Ostrov
Oz. – Ozero
P. – Pass, Passo, Pasul, Pulau
P.E.I. – Prince Edward Island
Pa. – Pennsylvania
Pac. Oc. – Pacific Ocean
Papua N.G. – Papua New Guinea
Pass. – Passage
Pen. – Peninsula, Péninsule
Phil. – Philippines
Pk. – Park, Peak
Plat. – Plateau
Prov. – Province, Provincial
Pt. – Point
Pta. – Ponta, Punta
Pte. – Pointe
Qué. – Québec
Queens. – Queensland
R. – Rio, River
R.I. – Rhode Island
Ra. (s). – Range(s)
Raj. – Rajasthan
Reg. – Region
Rep. – Republic
Res. – Reserve, Reservoir
S. – San, South, Sea
Si. Arabia – Saudi Arabia
S.C. – South Carolina
S. Dak. – South Dakota
S.I. – South Island
S. Leone – Sierra Leone
Sa. – Serra, Sierra
Sask. – Saskatchewan
Scot. – Scotland
Sd. – Sound

Sev. – Severnaya
Sib. – Siberia
Sprs. – Springs
St. – Saint
Sta. – Santa, Station
Ste. – Sainte
Sto. – Santo
Str. – Strait, Stretto
Switz. – Switzerland
Tas. – Tasmania
Tenn. – Tennessee
Tex. – Texas
Tg. – Tanjung
Trin. & Tob. – Trinidad & Tobago
U.A.E. – United Arab Emirates
U.K. – United Kingdom
U.S.A. – United States of America
Ut. P. – Uttar Pradesh
Va. – Virginia
Vdkhr. – Vodokhranilishche
Vf. – Vírful
Vic. – Victoria
Vol. – Volcano
Vt. – Vermont
W. – Wadi, West
W. Va. – West Virginia
Wash. – Washington
Wis. – Wisconsin
Wlkp. – Wielkopolski
Wyo. – Wyoming
Yorks. – Yorkshire
Yug. – Yugoslavia

178 A Baña

A

A Baña, *Spain* **42 C2** 42 58N 8 46W
A Cañiza, *Spain* **42 C2** 42 13N 8 16W
A Coruña, *Spain* **42 B2** 43 20N 8 25W
A Estrada, *Spain* **42 C2** 42 43N 8 27W
A Fonsagrada, *Spain* **42 B3** 43 8N 7 4W
A Guarda, *Spain* **42 D2** 41 56N 8 52W
A Gudiña, *Spain* **42 C3** 42 4N 7 8W
A Rúa, *Spain* **42 C3** 42 24N 7 6W
Aachen, *Germany* **30 E2** 50 45N 6 6 E
Aadorf, *Switz.* **33 B7** 47 30N 8 55 E
Aalborg = Ålborg, *Denmark* **17 G3** 57 2N 9 54 E
Aalen, *Germany* **31 G6** 48 51N 10 6 E
A'âli en Nîl □, *Sudan* **107 F3** 9 30N 33 0 E
Aalst, *Belgium* **24 D4** 50 56N 4 2 E
Aalten, *Neths.* **24 C6** 51 56N 6 35 E
Aalter, *Belgium* **24 C3** 51 5N 3 28 E
Äänekoski, *Finland* **15 E21** 62 36N 25 44 E
Aarau, *Switz.* **32 B6** 47 23N 8 4 E
Aarberg, *Switz.* **32 B4** 47 2N 7 16 E
Aare →, *Switz.* **32 A6** 47 33N 8 14 E
Aargau □, *Switz.* **32 B6** 47 26N 8 10 E
Aarhus = Århus, *Denmark* **17 H4** 56 8N 10 11 E
Aarschot, *Belgium* **24 D4** 50 59N 4 49 E
Aarwangen, *Switz.* **32 B5** 47 15N 7 46 E
Aasiaat = Egedesminde, *Greenland* **10 D5** 68 43N 52 56W
Aba, *China* **76 A3** 32 59N 101 42 E
Aba, *Dem. Rep. of the Congo* **118 B3** 3 58N 30 17 E
Aba, *Nigeria* **113 D6** 5 10N 7 19 E
Âbâ, Jazîrat, *Sudan* **107 E3** 13 30N 32 31 E
Abacaxis →, *Brazil* **169 D6** 3 54 S 58 47W
Ābādān, *Iran* **97 D6** 30 22N 48 20 E
Abade, *Ethiopia* **107 F4** 9 22N 38 3 E
Ābādeh, *Iran* **97 D7** 31 8N 52 40 E
Abadin, *Spain* **42 B3** 43 21N 7 29W
Abadla, *Algeria* **111 B4** 31 2N 2 45W
Abaeté, *Brazil* **171 E2** 19 9 S 45 27W
Abaeté →, *Brazil* **171 E2** 18 2 S 45 12W
Abaetetuba, *Brazil* **170 B2** 1 40 S 48 50W
Abagnar Qi, *China* **74 C9** 43 52N 116 2 E
Abai, *Paraguay* **175 B4** 25 58 S 55 54W
Abak, *Nigeria* **113 E6** 4 58N 7 50 E
Abakaliki, *Nigeria* **113 D6** 6 22N 8 2 E
Abakan, *Russia* **65 D10** 53 40N 91 10 E
Abalemma, *Niger* **113 B6** 16 12N 7 50 E
Abana, *Turkey* **100 B6** 41 59N 34 1 E
Abancay, *Peru* **172 C3** 13 35 S 72 55W
Abano Terme, *Italy* **45 C8** 45 22N 11 46 E
Abapó, *Bolivia* **173 D5** 18 48 S 63 25W
Abarán, *Spain* **41 G3** 38 12N 1 23W
Abariringa, *Kiribati* **134 H10** 2 50 S 171 40W
Abarqū, *Iran* **97 D7** 31 10N 53 20 E
Abashiri, *Japan* **68 C12** 44 0N 144 15 E
Abashiri-Wan, *Japan* **68 C12** 44 0N 144 30 E
Abau, *Papua N. G.* **132 F5** 10 11 S 148 46 E
Abaújszántó, *Hungary* **52 B6** 48 16N 21 12 E
Abava →, *Latvia* **54 A8** 57 6N 21 54 E
Abay, *Kazakstan* **64 E8** 49 38N 72 53 E
Abaya, L., *Ethiopia* **107 F4** 6 30N 37 50 E
Abaza, *Russia* **64 D10** 52 39N 90 6 E
Abbadia San Salvatore, *Italy* **45 F8** 42 53N 11 41 E
'Abbāsābād, *Iran* **97 C8** 33 34N 58 23 E
Abbay = Nîl el Azraq →, *Sudan* **107 D3** 15 38N 32 31 E
Abbaye, Pt., *U.S.A.* **148 B1** 46 58N 88 8W
Abbé, L., *Ethiopia* **107 E5** 11 8N 41 47 E
Abbeville, *France* **27 B8** 50 6N 1 49 E
Abbeville, *Ala., U.S.A.* **152 D4** 31 34N 85 15W
Abbeville, *Ga., U.S.A.* **152 D6** 31 59N 83 18W
Abbeville, *La., U.S.A.* **155 L8** 29 58N 92 8W
Abbeville, *S.C., U.S.A.* **152 A7** 34 11N 82 23W
Abbiategrasso, *Italy* **44 C5** 45 24N 8 54 E
Abbieglassie, *Australia* **127 D4** 27 15 S 147 28 E
Abbot Ice Shelf, *Antarctica* **7 D16** 73 0 S 92 0W
Abbotsford, *Canada* **142 D4** 49 5N 122 20W
Abbotsford, *U.S.A.* **154 C9** 44 57N 90 19W
Abbottabad, *Pakistan* **92 B5** 34 10N 73 15 E
Abd al Kūrī, *Ind. Oc.* **99 D6** 12 5N 52 20 E
Ābdar, *Iran* **97 D7** 30 16N 55 19 E
'Abdolābād, *Iran* **97 C8** 34 12N 56 30 E
Abdulino, *Russia* **62 E4** 53 42N 53 40 E
Abéché, *Chad* **109 F4** 13 50N 20 35 E
Abejar, *Spain* **40 D2** 41 48N 2 47W
Abekr, *Sudan* **107 E2** 12 45N 28 50 E
Abêlessa, *Algeria* **111 D5** 22 58N 4 47 E
Abengourou, *Ivory C.* **112 D4** 6 42N 3 27W
Abenójar, *Spain* **43 G6** 38 53N 4 21W
Åbenrå, *Denmark* **17 J3** 55 3N 9 25 E
Abensberg, *Germany* **31 G7** 48 48N 11 51 E
Abeokuta, *Nigeria* **113 D5** 7 3N 3 19 E
Aber, *Uganda* **118 B3** 2 12N 32 25 E
Aberaeron, *U.K.* **21 E3** 52 15N 4 15W
Aberayron = Aberaeron, *U.K.* **21 E3** 52 15N 4 15W
Aberchirder, *U.K.* **22 D6** 57 34N 2 37W
Abercorn = Mbala, *Zambia* **119 D3** 8 46 S 31 24 E
Abercorn, *Australia* **127 D5** 25 12 S 151 5 E
Aberdare, *U.K.* **21 F4** 51 43N 3 27W
Aberdare Ra., *Kenya* **118 C4** 0 15 S 36 50 E
Aberdeen, *Australia* **129 B9** 32 9 S 150 56 E
Aberdeen, *Canada* **143 C7** 52 20N 106 8W
Aberdeen, *S. Africa* **116 E3** 32 28 S 24 2 E
Aberdeen, *U.K.* **22 D6** 57 9N 2 5W
Aberdeen, *Ala., U.S.A.* **149 J1** 33 49N 88 33W
Aberdeen, *Idaho, U.S.A.* **158 E7** 42 57N 112 50W
Aberdeen, *Ohio, U.S.A.* **157 F13** 38 39N 83 46W
Aberdeen, *S. Dak., U.S.A.* **154 C5** 45 28N 98 29W
Aberdeen, *Wash., U.S.A.* **160 D3** 46 59N 123 50W
Aberdeen, City of □, *U.K.* **22 D6** 57 10N 2 10W
Aberdeenshire □, *U.K.* **22 D6** 57 17N 2 36W
Aberdovey = Aberdyfi, *U.K.* **21 E3** 52 33N 4 3W
Aberdyfi, *U.K.* **21 E3** 52 33N 4 3W
Aberfeldy, *Australia* **129 D7** 37 42 S 146 22 E
Aberfeldy, *U.K.* **22 E5** 56 37N 3 51W
Abergavenny, *U.K.* **21 F4** 51 49N 3 1W
Abergele, *U.K.* **20 D4** 53 17N 3 35W
Abernathy, *U.S.A.* **155 J4** 33 50N 101 51W
Abert, L., *U.S.A.* **158 E3** 42 38N 120 14W
Aberystwyth, *U.K.* **21 E3** 52 25N 4 5W
Abha, *Si. Arabia* **106 D5** 18 0N 42 34 E
Abhar, *Iran* **97 B6** 36 9N 49 13 E
Abhayapuri, *India* **90 B3** 26 24N 90 38 E
Abia □, *Nigeria* **113 D6** 5 30N 7 35 E
Abide, *Turkey* **49 C11** 38 55N 29 20 E
Abidiya, *Sudan* **106 D3** 18 18N 34 3 E
Abidjan, *Ivory C.* **112 D4** 5 26N 3 58W
Abilene, *Kans., U.S.A.* **154 F6** 38 55N 97 13W
Abilene, *Tex., U.S.A.* **155 J5** 32 28N 99 43W
Abingdon, *U.K.* **21 F6** 51 40N 1 17W
Abingdon, *Ill., U.S.A.* **156 D6** 40 48N 90 24W
Abingdon, *Va., U.S.A.* **149 G5** 36 43N 81 59W
Abington Reef, *Australia* **126 B4** 18 0 S 149 35 E
Abitau →, *Canada* **143 B7** 59 53N 109 3W
Abitau L., *Canada* **143 A7** 60 27N 107 15W
Abitibi L., *Canada* **140 C4** 48 40N 79 40W
Abiy Adi, *Ethiopia* **107 E4** 13 39N 39 3 E
Abkhaz Republic □ = Abkhazia □, *Georgia* **61 J5** 43 12N 41 5 E
Abkhazia □, *Georgia* **61 J5** 43 12N 41 5 E
Abminga, *Australia* **127 D1** 26 8 S 134 51 E
Abnûb, *Egypt* **106 B3** 27 18N 31 4 E
Åbo = Turku, *Finland* **15 F20** 60 30N 22 19 E
Abo, Massif d', *Chad* **109 D3** 21 41N 16 8 E
Abocho, *Nigeria* **113 D6** 7 35N 6 56 E
Abohar, *India* **92 D6** 30 10N 74 10 E
Aboisso, *Ivory C.* **112 D4** 5 30N 3 5W
Abolo, *Congo* **114 B2** 0 8N 14 16 E
Abomey, *Benin* **113 D5** 7 10N 2 5 E
Abong-Mbang, *Cameroon* **114 B2** 4 0N 13 8 E
Abongabong, *Indonesia* **84 B1** 4 15N 96 48 E
Abonnema, *Nigeria* **113 E6** 4 41N 6 49 E
Abony, *Hungary* **52 C5** 47 12N 20 3 E
Aboso, *Ghana* **112 D4** 5 23N 1 57W
Abou-Deïa, *Chad* **109 F3** 11 20N 19 20 E
Abou Goulem, *Chad* **109 F4** 13 37N 21 38 E
Aboyne, *U.K.* **22 D6** 57 4N 2 47W
Abra □, *Phil.* **80 C3** 17 35N 120 45 E
Abra de Ilog, *Phil.* **80 E3** 13 27N 120 44 E
Abra Pampa, *Argentina* **174 A2** 22 43 S 65 42W
Abrantes, *Portugal* **43 F2** 39 24N 8 7W
Abreojos, Pta., *Mexico* **162 B2** 26 50N 113 40W
Abri, *Esh Shamâliya, Sudan* **106 C3** 20 50N 30 27 E
Abri, *Janub Kordofân, Sudan* **107 E3** 11 40N 30 21 E
Abrolhos, Banka, *Brazil* **171 E4** 18 0 S 38 0W
Abrud, *Romania* **52 D8** 46 19N 23 5 E
Abruzzo □, *Italy* **45 F10** 42 15N 14 0 E
Absaroka Range, *U.S.A.* **158 D9** 44 45N 109 50W
Abtenau, *Austria* **34 D6** 47 33N 13 21 E
Abū al Khaşīb, *Iraq* **97 D6** 30 25N 48 0 E
Abū 'Alī, *Si. Arabia* **97 E6** 27 20N 49 27 E
Abū 'Alī →, *Lebanon* **103 A4** 34 25N 35 50 E
Abu 'Arīsh, *Si. Arabia* **98 C3** 16 53N 42 48 E
Abû Ballas, *Egypt* **106 C2** 24 26N 27 36 E
Abu Deleiq, *Sudan* **107 D3** 15 57N 33 48 E
Abu Dhabi = Abū Z̧āby, *U.A.E.* **97 E7** 24 28N 54 22 E
Abû Dis, *Sudan* **106 D3** 19 12N 33 38 E
Abû Dom, *Sudan* **107 D3** 16 18N 32 25 E
Abū Du'ān, *Syria* **101 D8** 36 25N 38 15 E
Abu el Gairi, W. →, *Egypt* **103 F2** 29 35N 33 30 E
Abû Gabra, *Sudan* **107 E2** 11 2N 26 50 E
Abu Ga'da, W. →, *Egypt* **103 F1** 29 15N 32 53 E
Abû Gubeiha, *Sudan* **107 E3** 11 30N 31 15 E
Abu Habl, Khawr →, *Sudan* **107 E3** 12 37N 31 0 E
Abū Ḩadrīyah, *Si. Arabia* **97 E6** 27 20N 48 58 E
Abu Hamed, *Sudan* **106 D3** 19 32N 33 13 E
Abu Haraz, *An Nîl el Azraq, Sudan* **107 E3** 14 35N 33 30 E
Abû Haraz, *Esh Shamâliya, Sudan* **106 D3** 19 8N 32 18 E
Abû Higar, *Sudan* **107 E3** 12 50N 33 59 E
Abū Kamāl, *Syria* **101 E9** 34 30N 41 0 E
Abū Madd, Ra's, *Si. Arabia* **96 E3** 24 50N 37 7 E
Abu Matariq, *Sudan* **107 E2** 10 59N 26 9 E
Abu Qir, *Egypt* **106 H7** 31 18N 30 0 E
Abu Qireiya, *Egypt* **106 C4** 24 5N 35 28 E
Abu Qurqâs, *Egypt* **106 J7** 28 1N 30 44 E
Abū Rașāș, Ra's, *Oman* **99 B7** 20 10N 58 38 E
Abû Rubayq, *Si. Arabia* **98 B2** 23 44N 39 42 E
Abu Şafāt, W. →, *Jordan* **103 E5** 30 24N 36 7 E
Abû Simbel, *Egypt* **106 C3** 22 18N 31 40 E
Abū Şukhayr, *Iraq* **101 G11** 31 54N 44 30 E
Abu Tig, *Egypt* **106 B3** 27 4N 31 15 E
Abu Tiga, *Sudan* **107 E3** 12 47N 34 12 E
Abû Zabad, *Sudan* **107 E2** 12 25N 29 10 E
Abū Z̧āby, *U.A.E.* **97 E7** 24 28N 54 22 E
Abū Zeydābād, *Iran* **97 C6** 33 54N 51 45 E
Abufari, *Brazil* **173 B5** 5 25 S 62 59W
Abuja, *Nigeria* **113 D6** 9 16N 7 2 E
Abukuma-Gawa →, *Japan* **68 E10** 38 6N 140 52 E
Abukuma-Sammyaku, *Japan* **68 F10** 37 30N 140 45 E
Abulug, *Phil.* **80 B3** 18 27N 121 27 E
Abumombazi, *Dem. Rep. of the Congo* **114 B4** 3 42N 22 10 E
Abunã, *Brazil* **173 B4** 9 40 S 65 20W
Abunã →, *Brazil* **173 B4** 9 41 S 65 20W
Abung, *Phil.* **80 E3** 13 46N 121 26 E
Aburatsu, *Japan* **70 F3** 31 34N 131 24 E
Aburo, *Dem. Rep. of the Congo* **118 B3** 2 4N 30 53 E
Abut Hd., *N.Z.* **131 D5** 43 7 S 170 15 E
Abwong, *Sudan* **107 F3** 9 2N 32 14 E
Åby, *Sweden* **17 F10** 58 40N 16 10 E
Aby, Lagune, *Ivory C.* **112 D4** 5 15N 3 14W
Åbybro, *Denmark* **17 G3** 57 10N 9 44 E
Acacías, *Colombia* **168 C3** 3 59N 73 46W
Acajutla, *El Salv.* **164 D2** 13 36N 89 50W
Açailândia, *Brazil* **170 C2** 5 0 S 47 30W
Acámbaro, *Mexico* **162 D4** 20 0N 100 40W
Acanthus, *Greece* **50 F7** 40 27N 23 47 E
Acaponeta, *Mexico* **162 C3** 22 30N 105 20W
Acapulco, *Mexico* **163 D5** 16 51N 99 56W
Acarai, Serra, *Brazil* **169 C6** 1 50N 57 50W
Acaraú, *Brazil* **170 B3** 2 53 S 40 7W
Acari, *Brazil* **170 C4** 6 31 S 36 38W
Acarí, *Peru* **172 D3** 15 25 S 74 36W
Acarigua, *Venezuela* **168 B4** 9 33N 69 12W
Acatlán, *Mexico* **163 D5** 18 10N 98 3W
Acayucan, *Mexico* **163 D6** 17 59N 94 58W
Accéglio, *Italy* **44 D4** 44 28N 7 0 E
Accomac, *U.S.A.* **148 G8** 37 43N 75 40W
Accous, *France* **28 E3** 43 0N 0 36W
Accra, *Ghana* **113 D4** 5 35N 0 6W
Accrington, *U.K.* **20 D5** 53 45N 2 22W
Acebal, *Argentina* **174 C3** 33 20 S 60 50W
Aceh □, *Indonesia* **84 B1** 4 15N 97 30 E
Acerra, *Italy* **47 B7** 40 57N 14 22 E
Aceuchal, *Spain* **43 G4** 38 39N 6 30W
Achacachi, *Bolivia* **172 D4** 16 3 S 68 43W
Achaguas, *Venezuela* **168 B4** 7 46N 68 14W
Achalpur, *India* **94 D3** 21 22N 77 32 E
Achao, *Chile* **176 B2** 42 28 S 73 30W
Acheng, *China* **75 B14** 45 30N 126 58 E
Achenkirch, *Austria* **34 D4** 47 32N 11 45 E
Achensee, *Austria* **34 D4** 47 26N 11 45 E
Acher, *India* **92 H5** 23 10N 72 32 E
Achern, *Germany* **31 G4** 48 37N 8 4 E
Acheron →, *N.Z.* **131 C8** 42 16 S 173 4 E
Achill Hd., *Ireland* **23 C1** 53 58N 10 15W
Achill I., *Ireland* **23 C1** 53 58N 10 1W
Achim, *Germany* **30 B5** 53 1N 9 3 E
Achinsk, *Russia* **65 D10** 56 20N 90 20 E
Achisay = Ashchysay, *Kazakstan* **63 B4** 43 35N 68 53 E
Achit, *Russia* **62 C6** 56 48N 57 54 E
Achol, *Sudan* **107 F3** 6 35N 31 32 E
Acıgöl, *Turkey* **49 D11** 37 50N 29 50 E
Acıpayam, *Turkey* **49 D11** 37 26N 29 22 E
Acireale, *Italy* **47 E8** 37 37N 15 10 E
Ackerman, *U.S.A.* **155 J10** 33 19N 89 11W
Ackley, *U.S.A.* **156 B3** 42 33N 93 3W
Acklins I., *Bahamas* **165 B5** 22 30N 74 0W
Acme, *Canada* **142 C6** 51 33N 113 30W
Acobamba, *Peru* **172 C3** 12 52 S 74 35W
Acomayo, *Peru* **172 C3** 13 55 S 71 38W
Aconcagua, Cerro, *Argentina* **174 C2** 32 39 S 70 0W
Aconquija, Mt., *Argentina* **174 B2** 27 0 S 66 0W
Acopiara, *Brazil* **170 C4** 6 6 S 39 27W
Açores, Is. dos = Azores, *Atl. Oc.* **8 E6** 38 44N 29 0W
Acorizal, *Brazil* **173 D6** 15 12 S 56 22W
Acquapendente, *Italy* **45 F8** 42 44N 11 52 E
Acquasanta Terme, *Italy* **45 F10** 42 46N 13 24 E
Acquasparta, *Italy* **45 F9** 42 41N 12 33 E
Acquaviva delle Fonti, *Italy* **47 B9** 40 54N 16 50 E
Ácqui Terme, *Italy* **44 D5** 44 41N 8 28 E
Acraman, L., *Australia* **127 E2** 32 2 S 135 23 E
Acre = 'Akko, *Israel* **103 C4** 32 55N 35 4 E
Acre □, *Brazil* **172 B3** 9 1 S 71 0W
Acre →, *Brazil* **172 B4** 8 45 S 67 22W
Acri, *Italy* **47 C9** 39 29N 16 23 E
Acs, *Hungary* **52 C3** 47 42N 18 2 E
Actium, *Greece* **48 C2** 38 57N 20 45 E
Acton, *Canada* **150 C4** 43 38N 80 3W
Açu, *Brazil* **170 C4** 5 34 S 36 54W
Acworth, *U.S.A.* **152 A5** 34 4N 84 41W
Ad Dahnā, *Si. Arabia* **99 A5** 24 30N 48 10 E
Ad Ḑāli', *Yemen* **98 D4** 13 42N 44 44 E
Ad Dammām, *Si. Arabia* **97 E6** 26 20N 50 5 E
Ad Darb, *Si. Arabia* **98 C3** 18 2N 43 7 E
Ad Dawhah, *Qatar* **97 E6** 25 15N 51 35 E
Ad Dawr, *Iraq* **101 E10** 34 27N 43 47 E
Ad Ḑiffah, *Libya* **108 B4** 30 30N 24 30 E
Ad Dilam, *Si. Arabia* **98 B4** 23 55N 47 10 E
Ad Dir'īyah, *Si. Arabia* **96 E5** 24 44N 46 35 E
Ad Dīwānīyah, *Iraq* **101 F11** 32 0N 45 0 E
Ad Dujayl, *Iraq* **101 F11** 33 51N 44 14 E
Ada, *Ghana* **113 D5** 5 44N 0 40 E
Ada, *Serbia, Yug.* **52 E5** 45 49N 20 9 E
Ada, *Minn., U.S.A.* **154 B6** 47 18N 96 31W
Ada, *Ohio, U.S.A.* **157 D13** 40 46N 83 49W
Ada, *Okla., U.S.A.* **155 H6** 34 46N 96 41W
Adad, *Somali Rep.* **120 C3** 9 27N 46 49 E
Adaja →, *Spain* **42 D6** 41 32N 4 52W
Adak, *U.S.A.* **144 L3** 51 45N 176 45W
Adak I., *U.S.A.* **144 L3** 51 45N 176 45W
Ådalsbruk, *Norway* **18 D8** 60 43N 11 19 E
Adam, *Oman* **99 B7** 22 15N 57 28 E
Adam, Mt., *Falk. Is.* **176 D4** 51 34 S 60 4W
Adamantina, *Brazil* **171 F1** 21 42 S 51 4W
Adamaoua, Massif de l', *Cameroon* **113 D7** 7 20N 12 20 E
Adamawa □, *Nigeria* **113 D7** 9 20N 12 30 E
Adamawa Highlands = Adamaoua, Massif de l', *Cameroon* **113 D7** 7 20N 12 20 E
Adamello, Mte., *Italy* **44 B7** 46 9N 10 30 E
Adami Tulu, *Ethiopia* **107 F4** 7 53N 38 41 E
Adaminaby, *Australia* **129 D8** 36 0 S 148 45 E
Adamovka, *Russia* **62 F7** 51 32N 59 56 E
Adams, *Phil.* **80 B3** 18 28N 120 54 E
Adams, *Mass., U.S.A.* **151 D11** 42 38N 73 7W
Adams, *N.Y., U.S.A.* **151 C8** 43 49N 76 1W
Adams, *Wis., U.S.A.* **154 D10** 43 57N 89 49W
Adam's Bridge, *Sri Lanka* **95 K4** 9 15N 79 40 E
Adams L., *Canada* **142 C5** 51 10N 119 40W
Adams Mt., *U.S.A.* **160 D5** 46 12N 121 30W
Adam's Peak, *Sri Lanka* **95 L5** 6 48N 80 30 E
Adamuz, *Spain* **43 G6** 38 2N 4 32W
Adana, *Turkey* **100 D6** 37 0N 35 16 E
Adanero, *Spain* **42 E6** 40 56N 4 36W
Adapazarı, *Turkey* **100 B4** 40 48N 30 25 E
Adarama, *Sudan* **107 D3** 17 10N 34 52 E
Adare, C., *Antarctica* **7 D11** 71 0 S 171 0 E
Adaut, *Indonesia* **83 C4** 8 8 S 131 7 E
Adavale, *Australia* **127 D3** 25 52 S 144 32 E
Adda →, *Italy* **44 C6** 45 8N 9 53 E
Addis Ababa = Addis Abeba, *Ethiopia* **107 F4** 9 2N 38 42 E
Addis Abeba, *Ethiopia* **107 F4** 9 2N 38 42 E
Addis Alem, *Ethiopia* **107 F4** 9 0N 38 17 E
Addison, *Ill., U.S.A.* **157 C8** 41 55N 88 0W
Addison, *N.Y., U.S.A.* **150 D7** 42 1N 77 14W
Addo, *S. Africa* **116 E4** 33 32 S 25 45 E
Addyston, *U.S.A.* **157 E12** 39 8N 84 43W
Adebour, *Niger* **109 F2** 13 17N 11 50 E
Ādeh, *Iran* **96 B5** 37 42N 45 11 E
Adel, *Ga., U.S.A.* **152 D6** 31 8N 83 25W
Adel, *Iowa, U.S.A.* **156 C2** 41 37N 94 1W
Adelaide, *Australia* **128 C3** 34 52 S 138 30 E
Adelaide, *Bahamas* **164 A4** 25 4N 77 31W
Adelaide, *S. Africa* **116 E4** 32 42 S 26 20 E
Adelaide I., *Antarctica* **7 C17** 67 15 S 68 30W
Adelaide Pen., *Canada* **138 B10** 68 15N 97 30W
Adelaide River, *Australia* **126 B5** 13 15 S 131 7 E
Adelanto, *U.S.A.* **161 L9** 34 35N 117 22W
Adelboden, *Switz.* **32 D5** 46 29N 7 33 E
Adele I., *Australia* **124 C3** 15 32 S 123 9 E
Adélie, Terre, *Antarctica* **7 C10** 68 0 S 140 0 E
Adélie Land = Adélie, Terre, *Antarctica* **7 C10** 68 0 S 140 0 E
Ademuz, *Spain* **40 E3** 40 5N 1 13W
Aden = Al 'Adan, *Yemen* **98 D4** 12 45N 45 0 E
Aden, G. of, *Asia* **102 E4** 12 30N 47 30 E
Adendorp, *S. Africa* **116 E3** 32 15 S 24 30 E
Adh Dhayd, *U.A.E.* **97 E7** 25 17N 55 53 E
Adhoi, *India* **92 H4** 23 26N 70 32 E
Adi, *Indonesia* **83 B4** 4 15 S 133 30 E
Adi Daro, *Ethiopia* **107 E4** 14 20N 38 14 E
Adi Keyih, *Eritrea* **107 E4** 14 51N 39 22 E
Adi Kwala, *Eritrea* **107 E4** 14 38N 38 48 E
Adi Ugri, *Eritrea* **107 E4** 14 58N 38 48 E
Adieu, C., *Australia* **125 F5** 32 0 S 132 10 E
Adieu Pt., *Australia* **124 C3** 15 14 S 124 35 E
Adigala, *Ethiopia* **107 E5** 10 24N 42 15 E
Adige →, *Italy* **45 C9** 45 9N 12 20 E
Adigrat, *Ethiopia* **107 E4** 14 20N 39 26 E
Adıgüzel Barajı, *Turkey* **49 C11** 38 13N 29 14 E
Adilabad, *India* **94 E4** 19 33N 78 20 E
Adilcevaz, *Turkey* **101 C10** 38 47N 42 43 E
Adin, *U.S.A.* **158 F3** 41 12N 120 57W
Adirondack Mts., *U.S.A.* **151 C10** 44 0N 74 0W
Adıyaman, *Turkey* **101 D8** 37 45N 38 16 E
Adjim, *Tunisia* **108 B2** 33 47N 10 50 E
Adjohon, *Benin* **113 D5** 6 41N 2 32 E
Adjud, *Romania* **53 D12** 46 7N 27 10 E
Adjumani, *Uganda* **118 B3** 3 20N 31 50 E
Adlavik Is., *Canada* **141 A8** 55 2N 57 45W
Adler, *Russia* **61 J4** 43 28N 39 52 E
Adliswil, *Switz.* **33 B7** 47 19N 8 32 E
Admer, *Algeria* **111 D6** 20 21N 5 27 E
Admer, Erg d', *Algeria* **111 D6** 24 0N 9 5 E
Admiralty G., *Australia* **124 B4** 14 20 S 125 55 E
Admiralty I., *U.S.A.* **138 C6** 57 30N 134 30W
Admiralty Inlet, *U.S.A.* **158 C2** 48 8N 122 58W
Admiralty Is., *Papua N. G.* **132 B4** 2 0 S 147 0 E
Ado, *Nigeria* **113 D5** 6 36N 2 56 E
Ado-Ekiti, *Nigeria* **113 D6** 7 38N 5 12 E
Adok, *Sudan* **107 F3** 8 10N 30 20 E
Adola, *Ethiopia* **107 E5** 11 14N 41 44 E
Adonara, *Indonesia* **82 C2** 8 15 S 123 5 E
Adoni, *India* **95 G3** 15 33N 77 18 E
Adony, *Hungary* **52 C3** 47 6N 18 52 E
Adour →, *France* **28 E2** 43 32N 1 32W
Adra, *India* **93 H12** 23 30N 86 42 E
Adra, *Spain* **43 J7** 36 43N 3 3W
Adrano, *Italy* **47 E7** 37 40N 14 50 E
Adrar, *Algeria* **111 C4** 27 51N 0 11W
Adrasman, *Tajikistan* **63 C4** 40 38N 69 58 E
Adré, *Chad* **109 F4** 13 40N 22 20 E
Adrī, *Libya* **108 C2** 27 32N 13 2 E
Ádria, *Italy* **45 C9** 45 3N 12 3 E
Adrian, *Ga., U.S.A.* **152 C7** 32 33N 82 35W
Adrian, *Mich., U.S.A.* **157 C12** 41 54N 84 2W
Adrian, *Mo., U.S.A.* **156 F2** 38 24N 94 21W
Adrian, *Tex., U.S.A.* **155 H3** 35 16N 102 40W
Adriatic Sea, *Medit. S.* **12 G9** 43 0N 16 0 E
Adua, *Indonesia* **83 B3** 1 45 S 129 50 E
Adula, *Switz.* **33 D8** 46 30N 9 3 E
Adung Long, *Burma* **90 A6** 28 7N 97 42 E
Adur, *India* **95 K3** 9 8N 76 40 E
Adwa, *Ethiopia* **107 E4** 14 15N 38 52 E
Adygea □, *Russia* **61 H5** 45 0N 40 0 E
Adzhar Republic □ = Ajaria □, *Georgia* **61 K6** 41 30N 42 0 E
Adzopé, *Ivory C.* **112 D4** 6 7N 3 49W
Ægean Sea, *Medit. S.* **49 C7** 38 30N 25 0 E
Aerhtai Shan, *Mongolia* **72 B4** 46 40N 92 45 E
Ærø, *Denmark* **17 K4** 54 52N 10 25 E
Ærøskøbing, *Denmark* **17 K4** 54 53N 10 24 E
Aesch, *Switz.* **32 B5** 47 28N 7 36 E
Aëtós, *Greece* **48 D3** 37 15N 21 50 E
Afafi, Massif d', *Niger* **109 D3** 22 11N 15 10 E
'Afak, *Iraq* **101 F11** 32 4N 45 15 E
Afándou, *Greece* **38 C10** 36 18N 28 12 E
Afarag, Erg, *Algeria* **111 D5** 23 50N 2 47 E
Åfarnes, *Norway* **18 B4** 62 40N 7 32 E
Afdega, *Ethiopia* **120 C2** 6 4N 43 30 E
Affreville = Khemis Miliana, *Algeria* **111 A5** 36 11N 2 14 E
Affton, *U.S.A.* **156 F6** 38 33N 90 20W
Afghanistan ■, *Asia* **91 B2** 33 0N 65 0 E
Afgoi, *Somali Rep.* **120 D2** 2 7N 44 59 E
'Afif, *Si. Arabia* **98 B3** 23 53N 42 56 E
Afikpo, *Nigeria* **113 D6** 5 53N 7 54 E
Aflou, *Algeria* **111 B5** 34 7N 2 3 E
Afmadu, *Somali Rep.* **120 D2** 0 31N 42 4 E
Afogados da Ingàzeira, *Brazil* **170 C4** 7 45 S 37 39W
Afragóla, *Italy* **47 B7** 40 55N 14 18 E
Afrera, *Ethiopia* **107 E5** 13 16N 41 5 E
'Afrīn, *Syria* **100 D7** 36 32N 36 50 E
Afşin, *Turkey* **100 C7** 38 14N 36 55 E
Afton, *U.S.A.* **151 D9** 42 14N 75 32W
Aftout, *Algeria* **110 C4** 26 50N 3 45W
Afuá, *Brazil* **169 D7** 0 15 S 50 20W
'Afula, *Israel* **103 C4** 32 37N 35 17 E
Afyon, *Turkey* **57 G5** 38 45N 30 33 E
Afyon □, *Turkey* **49 C12** 38 25N 30 30 E
Afyonkarahisar = Afyon, *Turkey* **57 G5** 38 45N 30 33 E
Aga, *Egypt* **106 H7** 30 55N 31 10 E
Agadès = Agadez, *Niger* **109 E1** 16 58N 7 59 E
Agadez, *Niger* **109 E1** 16 58N 7 59 E
Agadir, *Morocco* **110 B3** 30 28N 9 55W
Agaete, *Canary Is.* **39 F4** 28 6N 15 43W
Agailás, *Mauritania* **110 D2** 22 37N 14 22W
Agana, *Guam* **133 R15** 13 28N 144 45 E
Ağapınar, *Turkey* **49 B12** 39 48N 30 47 E
Agar, *India* **92 H7** 23 40N 76 2 E
Agaro, *Ethiopia* **107 F4** 7 50N 36 38 E
Agartala, *India* **90 D3** 23 50N 91 23 E
Agăş, *Romania* **53 D11** 46 28N 26 15 E
Agassiz, *Canada* **142 D4** 49 14N 121 46W
Agats, *Indonesia* **83 C5** 5 33 S 138 0 E
Agattu I., *U.S.A.* **144 K2** 52 25N 173 35 E
Agbélouvé, *Togo* **113 D5** 6 35N 1 14 E
Agboville, *Ivory C.* **112 D4** 5 55N 4 15W
Ağcabädi, *Azerbaijan* **61 K8** 40 5N 47 27 E
Agcogan, *Phil.* **80 E3** 12 4N 121 57 E
Ağdam, *Azerbaijan* **61 L8** 40 0N 46 58 E
Ağdaş, *Azerbaijan* **61 K8** 40 44N 47 22 E
Agde, *France* **28 E7** 43 19N 3 28 E
Agde, C. d', *France* **28 E7** 43 16N 3 28 E
Agdz, *Morocco* **110 B3** 30 47N 6 30W
Agdzhabedi = Ağcabädi, *Azerbaijan* **61 K8** 40 5N 47 27 E
Agen, *France* **28 D4** 44 12N 0 38 E
Ageo, *Japan* **71 B11** 35 58N 139 36 E
Ager Tay, *Chad* **109 E3** 20 0N 17 41 E
Agerbæk, *Denmark* **17 J2** 55 36N 8 48 E
Agersø, *Denmark* **17 J5** 55 13N 11 12 E
Ageyevo, *Russia* **58 E9** 54 10N 36 27 E
Āgh Kand, *Iran* **97 B6** 37 15N 48 4 E
Aghireşu, *Romania* **53 D8** 46 53N 23 15 E
Aghoueyyît, *Mauritania* **110 D1** 21 10N 15 6W
Aginskoye, *Russia* **65 D12** 51 6N 114 32 E
Ağlasun, *Turkey* **49 D12** 37 39N 30 31 E
Agly →, *France* **28 F7** 42 46N 3 3 E
Agnibilékrou, *Ivory C.* **112 D4** 7 10N 3 11W
Agnita, *Romania* **53 E9** 45 59N 24 40 E
Agnone, *Italy* **45 G11** 41 48N 14 22 E
Ago, *Japan* **71 C8** 34 20N 136 51 E
Agofie, *Ghana* **113 D5** 8 27N 0 15 E

Al Kiswah, *Syria* **103 B5** 33 23N 36 14 E
Al Kūfah, *Iraq* **101 F11** 32 2N 44 24 E
Al Kufrah, *Libya* **108 D4** 24 17N 23 15 E
Al Kuhayfīyah, *Si. Arabia* **96 E4** 27 12N 43 3 E
Al Kūt, *Iraq* **101 F11** 32 30N 46 0 E
Al Kuwayt, *Kuwait* **96 D5** 29 30N 48 0 E
Al Labwah, *Lebanon* **103 A5** 34 11N 36 20 E
Al Lādhiqīyah, *Syria* **100 E6** 35 30N 35 45 E
Al Līth, *Si. Arabia* **98 B3** 20 9N 40 15 E
Al Liwā', *Oman* **97 E8** 24 31N 56 36 E
Al Luḩayyah, *Yemen* **98 D3** 15 45N 42 40 E
Al Madīnah, *Iraq* **96 D5** 30 57N 47 16 E
Al Madīnah, *Si. Arabia* **96 E3** 24 35N 39 52 E
Al Mafraq, *Jordan* **103 C5** 32 17N 36 14 E
Al Maghārīm, *Yemen* **98 D4** 15 1N 47 49 E
Al Maḩmūdīyah, *Iraq* **101 F11** 33 3N 44 21 E
Al Majma'ah, *Si. Arabia* **96 E5** 25 57N 45 22 E
Al Makhruq, W. →, *Jordan* **103 D6** 31 28N 37 0 E
Al Makḩūl, *Si. Arabia* **96 E4** 26 37N 42 39 E
Al Makīlī, *Libya* **108 B4** 32 10N 22 17 E
Al Manā'if, *Si. Arabia* **99 B5** 23 49N 51 20 E
Al Manāmah, *Bahrain* **97 E6** 26 10N 50 30 E
Al Manşūrī, *Yemen* **98 D4** 14 17N 45 16 E
Al Maqwa', *Kuwait* **96 D5** 29 10N 47 59 E
Al Marj, *Libya* **108 B4** 32 25N 20 30 E
Al Maţlā, *Kuwait* **96 D5** 29 24N 47 40 E
Al Mawjib, W. →, *Jordan* **103 D4** 31 28N 35 36 E
Al Mawşil, *Iraq* **101 D10** 36 15N 43 5 E
Al Mayādin, *Syria* **101 E9** 35 1N 40 27 E
Al Mazār, *Jordan* **103 D4** 31 4N 35 41 E
Al Midhnab, *Si. Arabia* **96 E5** 25 50N 44 18 E
Al Mīfá, *Si. Arabia* **98 C3** 18 54N 41 57 E
Al Minā', *Lebanon* **103 A4** 34 24N 35 49 E
Al Miqdādīyah, *Iraq* **101 E11** 34 0N 45 0 E
Al Mubarraz, *Si. Arabia* **97 E6** 25 30N 49 40 E
Al Muḑaybī, *Oman* **99 B7** 22 34N 58 7 E
Al Mughayrā', *U.A.E.* **97 E7** 24 5N 53 32 E
Al Muḩarraq, *Bahrain* **97 E6** 26 15N 50 40 E
Al Mukallā, *Yemen* **99 D5** 14 33N 49 2 E
Al Mukhā, *Yemen* **98 D3** 13 18N 43 15 E
Al Muladdah, *Oman* **99 B7** 23 45N 57 34 E
Al Musayjīd, *Si. Arabia* **96 E3** 24 5N 39 5 E
Al Musayyib, *Iraq* **101 F11** 32 49N 44 20 E
Al Muwayliḩ, *Si. Arabia* **96 E2** 27 40N 35 30 E
Al Owuho = Otukpa, *Nigeria* **113 D6** 7 9N 7 41 E
Al Qaddāhīyah, *Libya* **108 B3** 31 15N 15 9 E
Al Qadīmah, *Si. Arabia* **98 B2** 22 20N 39 13 E
Al Qaḩmah, *Si. Arabia* **98 C3** 18 0N 41 41 E
Al Qā'im, *Iraq* **101 E9** 34 21N 41 7 E
Al Qalībah, *Si. Arabia* **96 D3** 28 24N 37 42 E
Al Qāmishlī, *Syria* **101 D9** 37 2N 41 14 E
Al Qaryah ash Sharqīyah, *Libya* **108 B2** 30 28N 13 40 E
Al Qaryatayn, *Syria* **103 A6** 34 12N 37 13 E
Al Qaşabāt, *Libya* **108 B2** 32 39N 14 1 E
Al Qaţ'ā, *Syria* **101 E9** 34 40N 40 48 E
Al Qaţīf, *Si. Arabia* **97 E6** 26 35N 50 0 E
Al Qaţn, *Yemen* **99 D5** 15 51N 48 26 E
Al Qatrānah, *Jordan* **103 D5** 31 12N 36 6 E
Al Qaţrūn, *Libya* **108 D3** 24 56N 15 3 E
Al Qayşūmah, *Si. Arabia* **96 D5** 28 20N 46 7 E
Al Qiblīyah, *Oman* **99 C7** 17 30N 56 20 E
Al Quds = Jerusalem, *Israel* **103 D4** 31 47N 35 10 E
Al Qunayţirah, *Syria* **103 C4** 32 55N 35 45 E
Al Qunfudhah, *Si. Arabia* **98 C3** 19 3N 41 4 E
Al Qurḩ, *Yemen* **99 C5** 16 44N 51 29 E
Al Qurnah, *Iraq* **96 D5** 31 1N 47 25 E
Al Quşayr, *Iraq* **96 D5** 30 39N 45 50 E
Al Quşayr, *Syria* **103 A5** 34 31N 36 34 E
Al Qutayfah, *Syria* **103 B5** 33 44N 36 36 E
Al Quway'īyah, *Si. Arabia* **98 A4** 24 3N 45 15 E
Al 'Ubaylah, *Si. Arabia* **99 B5** 21 59N 50 57 E
Al 'Uḑaylīyah, *Si. Arabia* **97 E6** 25 8N 49 18 E
Al 'Ulā, *Si. Arabia* **96 E3** 26 35N 38 0 E
Al 'Ulayyah, *Si. Arabia* **98 C3** 19 39N 41 54 E
Al Uqaylah ash Sharqīgah, *Libya* **108 B3** 30 12N 19 10 E
Al Uqayr, *Si. Arabia* **97 E6** 25 40N 50 15 E
Al 'Uwaynid, *Si. Arabia* **96 E5** 24 50N 46 0 E
Al 'Uwayqīlah, *Si. Arabia* **96 D4** 30 30N 42 10 E
Al 'Uyūn, *Ḩijāz, Si. Arabia* **96 E3** 24 33N 39 35 E
Al 'Uyūn, *Najd, Si. Arabia* **96 E4** 26 30N 43 50 E
Al Wajh, *Si. Arabia* **96 E3** 26 10N 36 30 E
Al Wakrah, *Qatar* **97 E6** 25 10N 51 40 E
Al Wannān, *Si. Arabia* **97 E6** 26 55N 48 24 E
Al Waqbah, *Si. Arabia* **96 D5** 28 48N 45 33 E
Al Wari'āh, *Si. Arabia* **96 E5** 27 51N 47 25 E
Al Wātīyah, *Libya* **108 B2** 32 28N 11 57 E
Al Wusayl, *Qatar* **97 E6** 25 29N 51 29 E
Ala, *Italy* **44 C8** 45 45N 11 0 E
Ala Dağları, *Turkey* **101 C10** 39 15N 43 33 E
Ala Tau Shankou = Dzungarian Gates, *Kazakstan* **72 B3** 45 0N 82 0 E
Alabama □, *U.S.A.* **149 J2** 33 0N 87 0W
Alabama →, *U.S.A.* **149 K2** 31 8N 87 57W
Alaca, *Turkey* **100 B6** 40 10N 34 51 E
Alacaatlı, *Turkey* **49 B10** 39 15N 28 3 E
Alaçam, *Turkey* **100 B6** 41 36N 35 36 E
Alaçam Dağları, *Turkey* **49 B10** 39 18N 28 49 E
Alaçatı, *Turkey* **49 C8** 38 15N 26 22 E
Alachua, *U.S.A.* **152 F7** 29 47N 82 30W
Alaejos, *Spain* **42 D5** 41 18N 5 13W
Alaérma, *Greece* **38 C9** 36 9N 27 57 E
Alagir, *Russia* **61 J7** 43 3N 44 14 E
Alagna Valsésia, *Italy* **44 C4** 45 51N 7 56 E
Alagoa Grande, *Brazil* **170 C4** 7 3 S 35 35W
Alagoas □, *Brazil* **170 C4** 9 0 S 36 0W
Alagoinhas, *Brazil* **171 D4** 12 7 S 38 20W
Alagón, *Spain* **40 D3** 41 46N 1 12W
Alagón →, *Spain* **42 F4** 39 44N 6 53W
Alaior, *Spain* **39 B11** 39 57N 4 8 E
Alajero, *Canary Is.* **39 F2** 28 3N 17 13W
Alajuela, *Costa Rica* **164 D3** 10 2N 84 8W
Alakamisy, *Madag.* **117 C8** 21 19 S 47 14 E
Alakanuk, *U.S.A.* **144 E6** 62 41N 164 37W
Alakurtti, *Russia* **56 A5** 67 0N 30 30 E
Alalakeiki Channel, *U.S.A.* **145 C5** 20 30N 156 30W
Alalapura, *Surinam* **169 C6** 2 20N 56 25W
Alalaú →, *Brazil* **169 D5** 0 30 S 61 9W
Alameda, *Calif., U.S.A.* **160 H4** 37 46N 122 15W
Alameda, *N. Mex., U.S.A.* **159 J10** 35 11N 106 37W
Alaminos, *Phil.* **80 C2** 16 10N 119 59 E
Alamo, *Ga., U.S.A.* **152 C7** 32 9N 82 47W
Alamo, *Nev., U.S.A.* **161 J11** 36 21N 115 10W
Alamo Crossing, *U.S.A.* **161 L13** 34 16N 113 33W
Alamogordo, *U.S.A.* **159 K11** 32 54N 105 57W
Alamos, *Mexico* **162 B3** 27 0N 109 0W
Alamosa, *U.S.A.* **159 H11** 37 28N 105 52W
Åland, *Finland* **15 F19** 60 15N 20 0 E
Aland, *India* **94 F3** 17 36N 76 35 E
Alandroal, *Portugal* **43 G3** 38 41N 7 24W
Ålands hav, *Sweden* **15 F18** 60 0N 19 30 E
Alandur, *India* **95 H5** 13 0N 80 15 E
Alange, Presa de, *Spain* **43 G4** 38 45N 6 18W
Alania = North Ossetia □, *Russia* **61 J7** 43 30N 44 30 E
Alanís, *Spain* **43 G5** 38 3N 5 43W
Alanya, *Turkey* **100 D5** 36 38N 32 0 E
Alaotra, Farihin', *Madag.* **117 B8** 17 30 S 48 30 E
Alapaha, *U.S.A.* **152 D6** 31 23N 83 13W
Alapayevsk, *Russia* **62 C8** 57 52N 61 42 E
Alar del Rey, *Spain* **42 C6** 42 38N 4 20W
Alaraz, *Spain* **42 E5** 40 45N 5 17W
Alarcón, Embalse de, *Spain* **40 F2** 39 36N 2 10W
Alaşehir, *Turkey* **57 G4** 38 23N 28 30 E
Alaska □, *U.S.A.* **138 B5** 64 0N 154 0W
Alaska, G. of, *Pac. Oc.* **144 G11** 58 0N 145 0W
Alaska Peninsula, *U.S.A.* **138 C4** 56 0N 159 0W
Alaska Range, *U.S.A.* **144 F10** 62 50N 151 0W
Alássio, *Italy* **44 E5** 44 0N 8 10 E
Ālāt, *Azerbaijan* **61 L9** 39 58N 49 25 E
Alatri, *Italy* **45 G10** 41 43N 13 21 E
Alatyr, *Russia* **60 C8** 54 55N 46 35 E
Alatyr →, *Russia* **60 C8** 54 52N 46 36 E
Alausi, *Ecuador* **168 D2** 2 0 S 78 50W
Álava □, *Spain* **40 C2** 42 48N 2 28W
Alava, C., *U.S.A.* **158 B1** 48 10N 124 44W
Alaverdi, *Armenia* **61 K7** 41 15N 44 37 E
Alavus, *Finland* **15 E20** 62 35N 23 36 E
Alawoona, *Australia* **128 C4** 34 45 S 140 30 E
'Alayh, *Lebanon* **103 B4** 33 46N 35 33 E
Alaykel, *Kyrgyzstan* **63 C7** 40 15N 74 25 E
Alayskiy Khrebet, *Kyrgyzstan* **63 D6** 39 45N 72 0 E
Alazani →, *Azerbaijan* **61 K8** 41 5N 46 40 E
Alba, *Italy* **44 D5** 44 42N 8 2 E
Alba □, *Romania* **53 D8** 46 10N 23 30 E
Alba Adriática, *Italy* **45 F10** 42 50N 13 56 E
Alba de Tormes, *Spain* **42 E5** 40 50N 5 30W
Alba-Iulia, *Romania* **53 D8** 46 8N 23 39 E
Albac, *Romania* **52 D7** 46 28N 22 58 E
Albacete, *Spain* **41 F3** 39 0N 1 50W
Albacete □, *Spain* **41 G3** 38 50N 2 0W
Albacutya, L., *Australia* **128 C4** 35 45 S 141 58 E
Ålbæk, *Denmark* **17 G4** 57 36N 10 25 E
Ålbæk Bugt, *Denmark* **17 G4** 57 35N 10 40 E
Albaida, *Spain* **41 G4** 38 51N 0 31W
Albalate de las Nogueras, *Spain* **40 E2** 40 22N 2 18W
Albalate del Arzobispo, *Spain* **40 D4** 41 6N 0 31W
Alban, *France* **28 E6** 43 53N 2 28 E
Albania ■, *Europe* **50 E4** 41 0N 20 0 E
Albano Laziale, *Italy* **45 G9** 41 44N 12 39 E
Albany, *Australia* **125 G2** 35 1 S 117 58 E
Albany, *Ga., U.S.A.* **152 D5** 31 35N 84 10W
Albany, *Ind., U.S.A.* **157 D11** 40 18N 85 14W
Albany, *Minn., U.S.A.* **154 C7** 45 38N 94 34W
Albany, *Mo., U.S.A.* **156 D2** 40 15N 94 20W
Albany, *N.Y., U.S.A.* **151 D11** 42 39N 73 45W
Albany, *Oreg., U.S.A.* **158 D2** 44 38N 123 6W
Albany, *Tex., U.S.A.* **155 J5** 32 44N 99 18W
Albany, *Wis., U.S.A.* **156 B7** 42 43N 89 26W
Albany →, *Canada* **140 B3** 52 17N 81 31W
Albardón, *Argentina* **174 C2** 31 20 S 68 30W
Albarracín, *Spain* **40 E3** 40 25N 1 26W
Albarracín, Sierra de, *Spain* **40 E3** 40 30N 1 30W
Albatera, *Spain* **41 G4** 38 11N 0 52W
Albatross B., *Australia* **126 A3** 12 45 S 141 30 E
Albatross Pt., *N.Z.* **130 E3** 38 7 S 174 44 E
Albay □, *Phil.* **80 E4** 13 13N 123 33 E
Albegna →, *Italy* **45 F8** 42 30N 11 11 E
Albemarle, *U.S.A.* **149 H5** 35 21N 80 11W
Albemarle Sd., *U.S.A.* **149 H7** 36 5N 76 0W
Albenga, *Italy* **44 D5** 44 3N 8 13 E
Alberche →, *Spain* **42 F6** 39 58N 4 46W
Alberdi, *Paraguay* **174 B4** 26 14 S 58 20W
Alberes, Mts., *France* **28 F6** 42 28N 2 56 E
Ålberga, *Sweden* **17 F10** 58 44N 16 35 E
Albersdorf, *Germany* **30 A5** 54 8N 9 17 E
Albert, *Australia* **129 B7** 32 22 S 147 30 E
Albert, *France* **27 C9** 50 0N 2 38 E
Albert, L., *Australia* **128 C3** 35 30 S 139 10 E
Albert Canyon, *Canada* **142 C5** 51 8N 117 41W
Albert Edward, Mt., *Papua N. G.* **132 E4** 8 20 S 147 24 E
Albert Edward Ra., *Australia* **124 C4** 18 17 S 127 57 E
Albert L., *Africa* **118 B3** 1 30N 31 0 E
Albert Lea, *U.S.A.* **154 D8** 43 39N 93 22W
Albert Nile →, *Uganda* **118 B3** 3 36N 32 2 E
Albert Town, *Bahamas* **165 B5** 22 37N 74 33W
Alberta □, *Canada* **142 C6** 54 40N 115 0W
Alberti, *Argentina* **174 D3** 35 1 S 60 16W
Albertinia, *S. Africa* **116 E3** 34 11 S 21 34 E
Albertirsa, *Hungary* **52 C4** 47 14N 19 37 E
Alberton, *Canada* **141 C7** 46 50N 64 0W
Albertville = Kalemie, *Dem. Rep. of the Congo* **118 D2** 5 55 S 29 9 E
Albertville, *France* **29 C10** 45 40N 6 22 E
Albi, *France* **28 E6** 43 56N 2 9 E
Albia, *U.S.A.* **156 C4** 41 2N 92 48W
Albina, *Surinam* **169 B7** 5 37N 54 15W
Albina, Ponta, *Angola* **115 F2** 15 52 S 11 44 E
Albino, *Italy* **44 C6** 45 46N 9 47 E
Albion, *Idaho, U.S.A.* **158 E7** 42 25N 113 35W
Albion, *Ill., U.S.A.* **157 F8** 38 23N 88 4W
Albion, *Ind., U.S.A.* **157 C11** 41 24N 85 25W
Albion, *Mich., U.S.A.* **157 B12** 42 15N 84 45W
Albion, *Nebr., U.S.A.* **154 E6** 41 42N 98 0W
Albion, *Pa., U.S.A.* **150 E4** 41 53N 80 22W
Albocácer, *Spain* **40 E5** 40 21N 0 1 E
Albolote, *Spain* **43 H7** 37 14N 3 39W
Alborán, *Medit. S.* **43 K7** 35 57N 3 0W
Alborea, *Spain* **41 F3** 39 17N 1 24W
Ålborg, *Denmark* **17 G3** 57 2N 9 54 E
Ålborg Bugt, *Denmark* **17 H4** 56 50N 10 35 E
Alborz, Reshteh-ye Kūhhā-ye, *Iran* **97 C7** 36 0N 52 0 E
Albox, *Spain* **41 H2** 37 23N 2 8W
Albreda, *Canada* **142 C5** 52 35N 119 10W
Albufeira, *Portugal* **43 H2** 37 5N 8 15W
Albula →, *Switz.* **33 C8** 46 38N 9 28 E
Albuñol, *Spain* **43 J7** 36 48N 3 11W
Albuquerque, *Brazil* **173 D6** 19 23 S 57 26W
Albuquerque, *U.S.A.* **159 J10** 35 5N 106 39W
Albuquerque, Cayos de, *Caribbean* **164 D3** 12 10N 81 50W
Alburg, *U.S.A.* **151 B11** 44 59N 73 18W
Alburno, Mte., *Italy* **47 B8** 40 33N 15 17 E
Alburquerque, *Spain* **43 F4** 39 15N 6 59W
Albury, *Australia* **129 D7** 36 3 S 146 56 E
Alcácer do Sal, *Portugal* **43 G2** 38 22N 8 33W
Alcáçovas, *Portugal* **43 G2** 38 23N 8 9W
Alcala, *Phil.* **80 C3** 17 54N 121 39 E
Alcalá de Chivert, *Spain* **40 E5** 40 19N 0 13 E
Alcalá de Guadaira, *Spain* **43 H5** 37 20N 5 50W
Alcalá de Henares, *Spain* **42 E7** 40 28N 3 22W
Alcalá de los Gazules, *Spain* **43 J5** 36 29N 5 43W
Alcalá del Júcar, *Spain* **41 F3** 39 12N 1 26W
Alcalá del Río, *Spain* **43 H5** 37 31N 5 59W
Alcalá del Valle, *Spain* **43 J5** 36 54N 5 10W
Alcalá la Real, *Spain* **43 H7** 37 27N 3 57W
Álcamo, *Italy* **46 E5** 37 59N 12 55 E
Alcanadre, *Spain* **40 C2** 42 24N 2 7W
Alcanadre →, *Spain* **40 D4** 41 43N 0 12W
Alcanar, *Spain* **40 E5** 40 33N 0 28 E
Alcanede, *Portugal* **43 F2** 39 25N 8 49W
Alcanena, *Portugal* **43 F2** 39 27N 8 40W
Alcañices, *Spain* **42 D4** 41 41N 6 21W
Alcañiz, *Spain* **40 D4** 41 2N 0 8W
Alcântara, *Brazil* **170 B3** 2 20 S 44 30W
Alcántara, *Spain* **42 F4** 39 41N 6 57W
Alcántara, Embalse de, *Spain* **42 F4** 39 44N 6 50W
Alcantara L., *Canada* **143 A7** 60 57N 108 9W
Alcantarilla, *Spain* **41 H3** 37 59N 1 12W
Alcaracejos, *Spain* **43 G6** 38 24N 4 58W
Alcaraz, *Spain* **41 G2** 38 40N 2 29W
Alcaraz, Sierra de, *Spain* **41 G2** 38 40N 2 20W
Alcaudete, *Spain* **43 H6** 37 35N 4 5W
Alcázar de San Juan, *Spain* **43 F7** 39 24N 3 12W
Alchevsk, *Ukraine* **59 H10** 48 30N 38 45 E
Alcira = Alzira, *Spain* **41 F4** 39 9N 0 30W
Alcoa, *U.S.A.* **149 H4** 35 48N 83 59W
Alcobaça, *Portugal* **43 F2** 39 32N 8 58W
Alcobendas, *Spain* **42 E7** 40 32N 3 38W
Alcolea del Pinar, *Spain* **40 D2** 41 2N 2 28W
Alcoma, *U.S.A.* **153 H8** 27 54N 81 29W
Alcora, *Spain* **40 E4** 40 5N 0 14W
Alcorcón, *Spain* **42 E7** 40 20N 3 50W
Alcoutim, *Portugal* **43 H3** 37 25N 7 28W
Alcova, *U.S.A.* **158 E10** 42 34N 106 43W
Alcoy, *Spain* **41 G4** 38 43N 0 30W
Alcubierre, Sierra de, *Spain* **40 D4** 41 45N 0 22W
Alcublas, *Spain* **40 F4** 39 48N 0 43W
Alcúdia, *Spain* **39 B10** 39 51N 3 7 E
Alcúdia, B. d', *Spain* **39 B10** 39 47N 3 15 E
Alcudia, Sierra de la, *Spain* **43 G6** 38 34N 4 30W
Aldabra Is., *Seychelles* **121 E3** 9 22 S 46 28 E
Aldama, *Mexico* **163 C5** 23 0N 98 4W
Aldan, *Russia* **65 D13** 58 40N 125 30 E
Aldan →, *Russia* **65 C13** 63 28N 129 35 E
Aldea, Pta. de la, *Canary Is.* **39 G4** 28 0N 15 50W
Aldeburgh, *U.K.* **21 E9** 52 10N 1 37 E
Alden, *Norway* **18 C1** 61 19N 4 45 E
Alder, *U.S.A.* **158 D7** 45 19N 112 6W
Alder Pk., *U.S.A.* **160 K5** 35 53N 121 22W
Alderney, *U.K.* **21 H5** 49 42N 2 11W
Aldershot, *U.K.* **21 F7** 51 15N 0 44W
Åled, *Sweden* **17 H6** 56 44N 12 57 E
Aledo, *U.S.A.* **156 C6** 41 12N 90 45W
Alefa, *Ethiopia* **107 E4** 11 55N 36 55 E
Aleg, *Mauritania* **112 B2** 17 3N 13 55W
Alegranza, *Canary Is.* **39 E6** 29 23N 13 32W
Alegranza, I., *Canary Is.* **39 E6** 29 23N 13 32W
Alegre, *Brazil* **171 F3** 20 50 S 41 30W
Alegrete, *Brazil* **175 B4** 29 40 S 56 0W
Alegria, *Phil.* **81 F5** 11 47N 124 3 E
Aleisk, *Russia* **64 D9** 52 40N 83 0 E
Aleknagik, *U.S.A.* **144 G8** 59 17N 158 36W
Aleksandriya = Oleksandriya, *Kirovohrad, Ukraine* **59 H7** 48 42N 33 3 E
Aleksandriya = Oleksandriya, *Rivne, Ukraine* **59 G4** 50 37N 26 19 E
Aleksandriyskaya, *Russia* **61 J8** 43 58N 47 14 E
Aleksandrov, *Russia* **58 D10** 56 23N 38 44 E
Aleksandrov Gay, *Russia* **60 E9** 50 9N 48 34 E
Aleksandrovac, *Serbia, Yug.* **50 C5** 43 28N 21 3 E
Aleksandrovac, *Serbia, Yug.* **50 B5** 44 28N 21 13 E
Aleksandrovka = Oleksandrovka, *Ukraine* **59 H7** 48 55N 32 20 E
Aleksandrovo, *Bulgaria* **51 C8** 43 14N 24 51 E
Aleksandrovsk, *Russia* **62 B6** 59 9N 57 33 E
Aleksandrovsk-Sakhalinskiy, *Russia* **65 D15** 50 50N 142 20 E
Aleksandrów Kujawski, *Poland* **55 F5** 52 53N 18 43 E
Aleksandrów Łódźki, *Poland* **55 G6** 51 49N 19 17 E
Alekseyevka, *Samara, Russia* **60 D10** 52 35N 51 17 E
Alekseyevka, *Voronezh, Russia* **59 G10** 50 43N 38 40 E
Aleksin, *Russia* **58 E9** 54 31N 37 9 E
Aleksinac, *Serbia, Yug.* **50 C5** 43 31N 21 42 E
Além Paraíba, *Brazil* **171 F3** 21 52 S 42 41W
Alemania, *Argentina* **174 B2** 25 40 S 65 30W
Alemania, *Chile* **174 B2** 25 10 S 69 55W
Alençon, *France* **26 D7** 48 27N 0 4 E
Alenuihaha Channel, *U.S.A.* **146 H17** 20 30N 156 0W
Aleppo = Ḩalab, *Syria* **100 D7** 36 10N 37 15 E
Aléria, *France* **29 F13** 42 5N 9 26 E
Alert Bay, *Canada* **142 C3** 50 30N 126 55W
Alès, *France* **29 D8** 44 9N 4 5 E
Aleşd, *Romania* **52 C7** 47 3N 22 22 E
Alessándria, *Italy* **44 D5** 44 54N 8 37 E
Ålestrup, *Denmark* **17 H3** 56 42N 9 29 E
Ålesund, *Norway* **15 E12** 62 28N 6 12 E
Alet-les-Bains, *France* **28 F6** 42 59N 2 14 E
Aletschhorn, *Switz.* **32 D6** 46 28N 8 0 E
Aleutian Is., *Pac. Oc.* **144 L3** 52 0N 175 0W
Aleutian Range, *U.S.A.* **144 G9** 60 0N 154 0W
Aleutian Trench, *Pac. Oc.* **134 C10** 48 0N 180 0 E
Alexander, *Ga., U.S.A.* **152 B8** 33 1N 81 53W
Alexander, *N. Dak., U.S.A.* **154 B3** 47 51N 103 39W
Alexander, Mt., *Australia* **125 E3** 28 58 S 120 16 E
Alexander Arch., *U.S.A.* **142 B2** 56 0N 136 0W
Alexander Bay, *S. Africa* **116 D2** 28 40 S 16 30 E
Alexander City, *U.S.A.* **152 C4** 32 56N 85 58W
Alexander I., *Antarctica* **7 C17** 69 0 S 70 0W
Alexandra, *Australia* **129 D6** 37 8 S 145 40 E
Alexandra, *N.Z.* **131 F4** 45 14 S 169 25 E
Alexandra Falls, *Canada* **142 A5** 60 29N 116 18W
Alexandria = El Iskandarîya, *Egypt* **106 H7** 31 13N 29 58 E
Alexandria, *Australia* **126 B2** 19 5 S 136 40 E
Alexandria, *B.C., Canada* **142 C4** 52 35N 122 27W
Alexandria, *Ont., Canada* **140 C5** 45 19N 74 38W
Alexandria, *Romania* **53 G10** 43 57N 25 24 E
Alexandria, *S. Africa* **116 E4** 33 38 S 26 28 E
Alexandria, *U.K.* **22 F4** 55 59N 4 35W
Alexandria, *Ind., U.S.A.* **157 D11** 40 16N 85 41W
Alexandria, *Ky., U.S.A.* **157 F12** 38 58N 84 23W
Alexandria, *La., U.S.A.* **155 K8** 31 18N 92 27W
Alexandria, *Minn., U.S.A.* **154 C7** 45 53N 95 22W
Alexandria, *Mo., U.S.A.* **156 D5** 40 27N 91 28W
Alexandria, *S. Dak., U.S.A.* **154 D6** 43 39N 97 47W
Alexandria, *Va., U.S.A.* **148 F7** 38 48N 77 3W
Alexandria Bay, *U.S.A.* **151 B9** 44 20N 75 55W
Alexandrina, L., *Australia* **128 C3** 35 25 S 139 10 E
Alexandroúpolis, *Greece* **51 F9** 40 50N 25 54 E
Alexis, *U.S.A.* **156 C6** 41 4N 90 33W
Alexis →, *Canada* **141 B8** 52 33N 56 8W
Alexis Creek, *Canada* **142 C4** 52 10N 123 20W
Alfabia, *Spain* **39 B9** 39 44N 2 44 E
Alfambra, *Spain* **40 E3** 40 33N 1 5W
Alfândega da Fé, *Portugal* **42 D4** 41 20N 6 59W
Alfaro, *Spain* **40 C3** 42 10N 1 50W
Alfatar, *Bulgaria* **51 C11** 43 59N 27 13 E
Alfaz del Pi, *Spain* **41 G4** 38 35N 0 5W
Alfeld, *Germany* **30 D5** 51 59N 9 50 E
Alfenas, *Brazil* **175 A6** 21 20 S 46 10W
Alfiós →, *Greece* **48 D3** 37 40N 21 33 E
Alföld, *Hungary* **52 D5** 46 30N 20 0 E
Alfonsine, *Italy* **45 D9** 44 30N 12 3 E
Alfonso XIII, *Phil.* **81 G1** 9 15N 117 59 E
Alford, *Aberds., U.K.* **22 D6** 57 14N 2 41W
Alford, *Lincs., U.K.* **20 D8** 53 15N 0 10 E
Alford, *U.S.A.* **152 E4** 30 42N 85 24W
Ålfotbreen, *Norway* **18 C2** 61 45N 5 39 E
Ålfoten, *Norway* **18 C1** 61 51N 5 41 E
Alfred, *Maine, U.S.A.* **151 C14** 43 29N 70 43W
Alfred, *N.Y., U.S.A.* **150 D7** 42 16N 77 48W
Alfred Town, *Australia* **129 C7** 35 8 S 147 30 E
Alfredton, *N.Z.* **130 G4** 40 41 S 175 54 E
Alfreton, *U.K.* **20 D6** 53 6N 1 24W
Alfta, *Sweden* **16 C10** 61 21N 16 4 E
Alga, *Kazakstan* **57 E10** 49 53N 57 20 E
Algaida, *Spain* **39 B9** 39 33N 2 53 E
Algar, *Spain* **43 J5** 36 40N 5 39W
Ålgård, *Norway* **15 G11** 58 46N 5 53 E
Algarinejo, *Spain* **43 H6** 37 19N 4 9W
Algarve, *Portugal* **43 J2** 36 58N 8 20W
Algeciras, *Spain* **43 J5** 36 9N 5 28W
Algemesí, *Spain* **41 F4** 39 11N 0 27W
Alger, *Algeria* **111 A5** 36 42N 3 8 E
Algeria ■, *Africa* **111 C5** 28 30N 2 0 E
Alghero, *Italy* **46 B1** 40 33N 8 19 E
Älghult, *Sweden* **17 G9** 57 0N 15 35 E
Algiers = Alger, *Algeria* **111 A5** 36 42N 3 8 E
Algoa B., *S. Africa* **116 E4** 33 50 S 25 45 E
Algodonales, *Spain* **43 J5** 36 54N 5 24W
Algodor →, *Spain* **42 F7** 39 55N 3 53W
Algoma, *U.S.A.* **148 C2** 44 36N 87 26W
Algona, *U.S.A.* **156 A2** 43 4N 94 14W
Algonac, *U.S.A.* **150 D2** 42 37N 82 32W
Alhama de Almería, *Spain* **43 J8** 36 57N 2 34W
Alhama de Aragón, *Spain* **40 D3** 41 18N 1 54W
Alhama de Granada, *Spain* **43 H7** 37 0N 3 59W
Alhama de Murcia, *Spain* **41 H3** 37 51N 1 25W
Alhambra, *Calif., U.S.A.* **161 L8** 34 8N 118 6W
Alhambra, *Ill., U.S.A.* **156 F7** 38 52N 89 45W
Alhaurín el Grande, *Spain* **43 J6** 36 39N 4 41W
Alhucemas = Al Hoceïma, *Morocco* **110 A4** 35 8N 3 58W
'Alī al Gharbī, *Iraq* **101 F12** 32 30N 46 45 E
Alī ash Sharqī, *Iraq* **101 F12** 32 7N 46 44 E
Äli Bayramlı, *Azerbaijan* **61 L9** 39 59N 48 52 E
'Alī Khēl, *Afghan.* **91 B3** 33 57N 69 43 E
Ali Sahîh, *Djibouti* **107 E5** 11 10N 42 44 E
Alī Shāh, *Iran* **96 B5** 38 9N 45 50 E
Ália, *Italy* **46 E6** 37 47N 13 43 E
'Alīābād, *Khorāsān, Iran* **97 C8** 32 30N 57 30 E
'Alīābād, *Kordestān, Iran* **96 C5** 35 4N 46 58 E
'Alīābād, *Yazd, Iran* **97 D7** 31 41N 53 49 E
Aliaga, *Spain* **40 E4** 40 40N 0 42W
Aliağa, *Turkey* **49 C8** 38 47N 26 59 E
Aliákmon →, *Greece* **50 F6** 40 30N 22 36 E
Alibag, *India* **94 E1** 18 38N 72 56 E
Alibo, *Ethiopia* **107 F4** 9 52N 37 5 E
Alibunar, *Serbia, Yug.* **52 E5** 45 5N 20 57 E
Alicante, *Spain* **41 G4** 38 23N 0 30W
Alicante □, *Spain* **41 G4** 38 30N 0 37W
Alice, *S. Africa* **116 E4** 32 48 S 26 55 E
Alice, *U.S.A.* **155 M5** 27 45N 98 5W
Alice →, *Queens., Australia* **126 C3** 24 2 S 144 50 E
Alice →, *Queens., Australia* **126 B3** 15 35 S 142 20 E
Alice, Punta, *Italy* **47 C10** 39 24N 17 9 E
Alice Arm, *Canada* **142 B3** 55 29N 129 31W
Alice Downs, *Australia* **124 C4** 17 45 S 127 56 E
Alice Springs, *Australia* **126 C1** 23 40 S 133 50 E
Alicedale, *S. Africa* **116 E4** 33 15 S 26 4 E
Aliceville, *U.S.A.* **149 J1** 33 8N 88 9W
Alicia, *Phil.* **81 G5** 9 54N 124 26 E
Alick Cr. →, *Australia* **126 C3** 20 55 S 142 20 E
Alicudi, *Italy* **47 D7** 38 33N 14 20 E
Alida, *Canada* **143 D8** 49 25N 101 55W
Aligarh, *Raj., India* **92 G7** 25 55N 76 15 E
Aligarh, *Ut. P., India* **92 F8** 27 55N 78 10 E
Alīgūdarz, *Iran* **97 C6** 33 25N 49 45 E
Alijó, *Portugal* **42 D3** 41 16N 7 27W
Alimnía, *Greece* **38 C9** 36 16N 27 43 E
Alimodian, *Phil.* **81 F4** 10 49N 122 26 E
Alindao, *C.A.R.* **114 A4** 5 2N 21 13 E
Alingsås, *Sweden* **17 G6** 57 56N 12 31 E
Alipur, *Pakistan* **92 E4** 29 25N 70 55 E
Alipur Duar, *India* **90 B2** 26 30N 89 35 E
Aliquippa, *U.S.A.* **150 F4** 40 37N 80 15W
Aliste →, *Spain* **42 D5** 41 34N 5 58W
Alitus = Alytus, *Lithuania* **15 J21** 54 24N 24 3 E
Alivérion, *Greece* **48 C6** 38 24N 24 2 E
Aliwal North, *S. Africa* **116 E4** 30 45 S 26 45 E
Alix, *Canada* **142 C6** 52 24N 113 11W
Aljezur, *Portugal* **43 H2** 37 18N 8 49W
Aljustrel, *Portugal* **43 H2** 37 55N 8 10W
Alkamari, *Niger* **109 F2** 13 27N 11 10 E
Alkmaar, *Neths.* **24 B4** 52 37N 4 45 E
All American Canal, *U.S.A.* **159 K6** 32 45N 115 15W
Allacapan, *Phil.* **80 B3** 18 15N 121 35 E
Allada, *Benin* **113 D5** 6 41N 2 9 E
Allah Dad, *Pakistan* **92 G2** 25 38N 67 34 E
Allahabad, *India* **93 G9** 25 25N 81 58 E
Allal Tazi, *Morocco* **110 B3** 34 30N 6 20W
Allan, *Canada* **143 C7** 51 53N 106 4W
Allanche, *France* **28 C6** 45 14N 2 57 E
Allanmyo, *Burma* **90 F5** 19 30N 95 17 E
Allanridge, *S. Africa* **116 D4** 27 45 S 26 40 E
Allansford, *Australia* **128 E5** 38 26 S 142 39 E

Allanton, *N.Z.* **131 F5** 45 55 S 170 15 E
Allanwater, *Canada* **140 B1** 50 14N 90 10W
Allaqi, Wadi →, *Egypt* **106 C3** 23 7N 32 47 E
Allariz, *Spain* **42 C3** 42 11N 7 50W
Allassac, *France* **28 C5** 45 15N 1 29 E
Allatoona L., *U.S.A.* **152 A5** 34 10N 84 44W
Ålleberg, *Sweden* **17 F7** 58 8N 13 36 E
Allegan, *U.S.A.* **157 B11** 42 32N 85 51W
Allegany, *U.S.A.* **150 D6** 42 6N 78 30W
Allegheny →, *U.S.A.* **150 F5** 40 27N 80 1W
Allegheny Mts., *U.S.A.* **136 F11** 38 15N 80 10W
Allegheny Plateau, *U.S.A.* **148 G6** 38 0N 80 0W
Allegheny Reservoir, *U.S.A.* **150 E6** 41 50N 79 0W
Allègre, *France* **28 C7** 45 12N 3 41 E
Allen, *Argentina* **176 A3** 38 58 S 67 50W
Allen, *Phil.* **80 E5** 12 30N 124 17 E
Allen, Bog of, *Ireland* **23 C5** 53 15N 7 0W
Allen, L., *Ireland* **23 B3** 54 8N 8 4W
Allendale, *U.S.A.* **152 B9** 33 1N 81 18W
Allende, *Mexico* **162 B4** 28 20N 100 50W
Allentown, *U.S.A.* **151 F9** 40 37N 75 29W
Allentsteig, *Austria* **34 C8** 48 41N 15 20 E
Alleppey, *India* **95 K3** 9 30N 76 28 E
Allepuz, *Spain* **40 E4** 40 29N 0 44W
Aller →, *Germany* **30 C5** 52 56N 9 12 E
Alliance, *Surinam* **169 B7** 5 50N 54 50W
Alliance, *Nebr., U.S.A.* **154 D3** 42 6N 102 52W
Alliance, *Ohio, U.S.A.* **150 F3** 40 55N 81 6W
Allier □, *France* **27 F9** 46 25N 2 40 E
Allier →, *France* **27 F10** 46 57N 3 4 E
Allinge, *Denmark* **17 J8** 55 17N 14 50 E
Allison, *U.S.A.* **156 B4** 42 45N 92 48W
Alliston, *Canada* **140 D4** 44 9N 79 52W
Alloa, *U.K.* **22 E5** 56 7N 3 47W
Allones, *France* **26 D8** 48 20N 1 40 E
Allora, *Australia* **127 D5** 28 2 S 152 0 E
Allos, *France* **29 D10** 44 15N 6 38 E
Alluitsup Paa = Sydprøven, *Greenland* **10 E6** 60 30N 45 35W
Alma, *Canada* **141 C5** 48 35N 71 40W
Alma, *Ga., U.S.A.* **152 D7** 31 33N 82 28W
Alma, *Kans., U.S.A.* **154 F6** 39 1N 96 17W
Alma, *Mich., U.S.A.* **148 D3** 43 23N 84 39W
Alma, *Nebr., U.S.A.* **154 E5** 40 6N 99 22W
Alma, *Wis., U.S.A.* **154 C9** 44 20N 91 55W
Alma Ata = Almaty, *Kazakstan* **63 B8** 43 15N 76 57 E
Almacelles, *Spain* **40 D5** 41 43N 0 27 E
Almada, *Portugal* **43 G1** 38 40N 9 9W
Almaden, *Australia* **126 B3** 17 22 S 144 40 E
Almadén, *Spain* **43 G6** 38 49N 4 52W
Almagro I., *Phil.* **81 F5** 11 56N 124 18 E
Almalyk = Olmaliq, *Uzbekistan* **63 C4** 40 50N 69 35 E
Almanor, L., *U.S.A.* **158 F3** 40 14N 121 9W
Almansa, *Spain* **41 G3** 38 51N 1 5W
Almanza, *Spain* **42 C5** 42 39N 5 3W
Almanzor, Pico, *Spain* **42 E5** 40 15N 5 18W
Almanzora →, *Spain* **41 H3** 37 14N 1 46W
Almas, *Brazil* **171 D2** 11 33 S 47 9W
Almaş, Munţii, *Romania* **52 F7** 44 49N 22 12 E
Almassora, *Spain* **40 F4** 39 57N 0 3W
Almaty, *Kazakstan* **63 B8** 43 15N 76 57 E
Almazán, *Spain* **40 D2** 41 30N 2 30W
Almeirim, *Brazil* **169 D7** 1 30 S 52 34W
Almeirim, *Portugal* **43 F2** 39 12N 8 37W
Almelo, *Neths.* **24 B6** 52 22N 6 42 E
Almenar de Soria, *Spain* **40 D2** 41 43N 2 12W
Almenara, *Brazil* **171 E3** 16 11 S 40 42W
Almenara, *Spain* **40 F4** 39 46N 0 14W
Almenara, Sierra de la, *Spain* **41 H3** 37 34N 1 32W
Almendra, Embalse de, *Spain* **42 D4** 41 10N 6 5W
Almendralejo, *Spain* **43 G4** 38 41N 6 26W
Almere-Stad, *Neths.* **24 B5** 52 20N 5 15 E
Almería, *Spain* **43 J8** 36 52N 2 27W
Almería □, *Spain* **41 H2** 37 20N 2 20W
Almería, G. de, *Spain* **41 J2** 36 41N 2 28W
Almetyevsk, *Russia* **60 C11** 54 53N 52 20 E
Älmhult, *Sweden* **17 H8** 56 33N 14 8 E
Almirante, *Panama* **164 E3** 9 10N 82 30W
Almirante Montt, G., *Chile* **176 D2** 51 52 S 72 50W
Almirópótamos, *Greece* **48 C6** 38 16N 24 11 E
Almirós, *Greece* **48 B4** 39 11N 22 45 E
Almiroú, Kólpos, *Greece* **38 D6** 35 23N 24 20 E
Almodôvar, *Portugal* **43 H2** 37 31N 8 2W
Almodóvar del Campo, *Spain* **43 G6** 38 43N 4 10W
Almodóvar del Río, *Spain* **43 H5** 37 48N 5 1W
Almon, *U.S.A.* **152 B6** 33 37N 83 56W
Almont, *U.S.A.* **150 D1** 42 55N 83 3W
Almonte, *Canada* **151 A8** 45 14N 76 12W
Almonte, *Spain* **43 H4** 37 13N 6 38W
Almora, *India* **93 E8** 29 38N 79 40 E
Almoradí, *Spain* **41 G4** 38 7N 0 46W
Almorox, *Spain* **42 E6** 40 14N 4 24W
Almoustarat, *Mali* **113 B5** 17 35N 0 8 E
Älmsta, *Sweden* **16 E12** 59 58N 18 50 E
Almudévar, *Spain* **40 C4** 42 3N 0 35W
Almuñécar, *Spain* **43 J7** 36 43N 3 41W
Almunge, *Sweden* **16 E12** 59 53N 18 3 E
Almuradiel, *Spain* **43 G7** 38 32N 3 28W
Alness, *U.K.* **22 D4** 57 41N 4 16W
Alnif, *Morocco* **110 B3** 31 10N 5 8W
Alnmouth, *U.K.* **20 B6** 55 24N 1 37W
Alnwick, *U.K.* **20 B6** 55 24N 1 42W
Aloi, *Uganda* **118 B3** 2 16N 33 10 E
Alon, *Burma* **90 D5** 22 12N 95 5 E
Alor, *Indonesia* **82 C2** 8 15 S 124 30 E
Alor Setar, *Malaysia* **87 J3** 6 7N 100 22 E
Álora, *Spain* **43 J6** 36 49N 4 46W
Alosno, *Spain* **43 H3** 37 33N 7 7W
Alotau, *Papua N. G.* **132 F6** 10 16 S 150 30 E
Alougoum, *Morocco* **110 B3** 30 17N 6 56W
Aloysius, Mt., *Australia* **125 E4** 26 0 S 128 38 E
Alpaugh, *U.S.A.* **160 K7** 35 53N 119 29W
Alpedrinha, *Portugal* **42 E3** 40 6N 7 27W
Alpena, *U.S.A.* **148 C4** 45 4N 83 27W
Alpercatas →, *Brazil* **170 C3** 6 2 S 44 19W
Alpes-de-Haute-Provence □, *France* **29 D10** 44 8N 6 10 E
Alpes-Maritimes □, *France* **29 E11** 43 55N 7 10 E
Alpha, *Australia* **126 C4** 23 39 S 146 37 E
Alpha, *U.S.A.* **156 C6** 41 12N 90 23W
Alphen aan den Rijn, *Neths.* **24 B4** 52 7N 4 40 E
Alphonse, *Seychelles* **121 E4** 7 0 S 52 45 E
Alpiarça, *Portugal* **43 F2** 39 15N 8 35W
Alpine, *Ariz., U.S.A.* **159 K9** 33 51N 109 9W
Alpine, *Calif., U.S.A.* **161 N10** 32 50N 116 46W
Alpine, *Tex., U.S.A.* **155 K3** 30 22N 103 40W
Alpnach, *Switz.* **33 C6** 46 57N 8 17 E
Alps, *Europe* **12 F7** 46 30N 9 30 E
Alpu, *Turkey* **100 C4** 39 46N 30 58 E
Alqueta, Barragem do, *Portugal* **43 G3** 38 20N 7 25W
Alrø, *Denmark* **17 J4** 55 52N 10 5 E
Alroy Downs, *Australia* **126 B2** 19 20 S 136 5 E
Als, *Denmark* **17 K3** 54 59N 9 55 E
Alsace, *France* **27 D14** 48 15N 7 25 E
Alsask, *Canada* **143 C7** 51 21N 109 59W
Alsasua, *Spain* **40 C2** 42 54N 2 10W
Alsfeld, *Germany* **30 E5** 50 44N 9 16 E
Alsten, *Norway* **14 D15** 65 58N 12 40 E
Alstermo, *Sweden* **17 H9** 56 58N 15 38 E
Alston, *U.K.* **20 C5** 54 49N 2 25W
Alta, *Norway* **14 B20** 69 57N 23 10 E
Alta, Sierra, *Spain* **40 E3** 40 31N 1 30W
Alta Gracia, *Argentina* **174 C3** 31 40 S 64 30W
Alta Lake, *Canada* **142 C4** 50 10N 123 0W
Alta Sierra, *U.S.A.* **161 K8** 35 42N 118 33W
Altaelva →, *Norway* **14 B20** 69 54N 23 17 E
Altafjorden, *Norway* **14 A20** 70 5N 23 5 E
Altagracia, *Venezuela* **168 A3** 10 45N 71 30W
Altagracia de Orituco, *Venezuela* **168 B4** 9 52N 66 23W
Altai = Aerhtai Shan, *Mongolia* **72 B4** 46 40N 92 45 E
Altamachi →, *Bolivia* **172 D4** 16 8 S 66 50W
Altamaha →, *U.S.A.* **152 D8** 31 20N 81 20W
Altamira, *Brazil* **169 D7** 3 12 S 52 10W
Altamira, *Chile* **174 B2** 25 47 S 69 51W
Altamira, *Colombia* **168 C2** 2 3N 75 47W
Altamira, *Mexico* **163 C5** 22 24N 97 55W
Altamira, Cuevas de, *Spain* **42 B6** 43 20N 4 5W
Altamont, *Ill., U.S.A.* **157 E8** 39 4N 88 45W
Altamont, *N.Y., U.S.A.* **151 D10** 42 43N 74 3W
Altamura, *Italy* **47 B9** 40 49N 16 33 E
Altanbulag, *Mongolia* **72 A5** 50 16N 106 30 E
Altar, *Mexico* **162 A2** 30 40N 111 50W
Altata, *Mexico* **162 C3** 24 30N 108 0W
Altavas, *Phil.* **81 F4** 11 32N 122 29 E
Altavista, *U.S.A.* **148 G6** 37 6N 79 17W
Altay, *China* **72 B3** 47 48N 88 10 E
Altdorf, *Switz.* **33 C7** 46 52N 8 36 E
Alte Mellum, *Germany* **30 B4** 53 43N 8 10 E
Altea, *Spain* **41 G4** 38 38N 0 2W
Altenberg, *Germany* **30 E9** 50 45N 13 45 E
Altenbruch, *Germany* **30 B4** 53 49N 8 46 E
Altenburg, *Germany* **30 E8** 50 59N 12 25 E
Altenkirchen, *Mecklenburg-Vorpommern, Germany* **30 A9** 54 38N 13 22 E
Altenkirchen, *Rhld-Pfz., Germany* **30 E3** 50 41N 7 39 E
Altenmarkt, *Austria* **34 D7** 47 43N 14 39 E
Alter do Chão, *Portugal* **43 F3** 39 12N 7 40W
Altha, *U.S.A.* **152 E4** 30 34N 85 8W
Altınoluk, *Turkey* **49 B8** 39 34N 26 45 E
Altınova, *Turkey* **49 B8** 39 12N 26 47 E
Altıntaş, *Turkey* **49 B12** 39 4N 30 7 E
Altınyaka, *Turkey* **49 E12** 36 33N 30 20 E
Altınyayla, *Turkey* **49 D11** 37 0N 29 33 E
Altiplano, *Bolivia* **172 D4** 17 0 S 68 0W
Altkirch, *France* **27 E14** 47 37N 7 15 E
Altmark, *Germany* **30 C7** 52 45N 11 30 E
Altmühl →, *Germany* **31 G7** 48 54N 11 52 E
Altmunster, *Austria* **34 D6** 47 54N 13 45 E
Alto Adige = Trentino-Alto Adige □, *Italy* **45 B8** 46 30N 11 20 E
Alto Araguaia, *Brazil* **173 D7** 17 15 S 53 20W
Alto Cuchumatanes = Cuchumatanes, Sierra de los, *Guatemala* **164 C1** 15 35N 91 25W
Alto Cuito, *Angola* **115 E3** 13 27 S 18 49 E
Alto del Inca, *Chile* **174 A2** 24 10 S 68 10W
Alto Garças, *Brazil* **173 D7** 16 56 S 53 32W
Alto Iriri →, *Brazil* **173 B7** 8 50 S 53 25W
Alto Ligonha, *Mozam.* **119 F4** 15 30 S 38 11 E
Alto Molocue, *Mozam.* **119 F4** 15 50 S 37 35 E
Alto Paraguai, *Brazil* **173 C6** 14 30 S 56 31W
Alto Paraguay □, *Paraguay* **174 A4** 21 0 S 58 30W
Alto Paraná □, *Paraguay* **175 B5** 25 30 S 54 50W
Alto Parnaíba, *Brazil* **170 C2** 9 6 S 45 57W
Alto Purús →, *Peru* **172 B3** 9 12 S 70 28W
Alto Río Senguerr, *Argentina* **176 C2** 45 2 S 70 50W
Alto Santo, *Brazil* **170 C4** 5 31 S 38 15W
Alto Sucuriú, *Brazil* **173 D7** 19 19 S 52 47W
Alto Turi, *Brazil* **170 B2** 2 54 S 45 38W
Alton, *Canada* **150 C4** 43 54N 80 5W
Alton, *U.K.* **21 F7** 51 9N 0 59W
Alton, *U.S.A.* **156 F6** 38 53N 90 11W
Alton Downs, *Australia* **127 D2** 26 7 S 138 57 E
Altoona, *Ala., U.S.A.* **152 A3** 34 2N 86 20W
Altoona, *Iowa, U.S.A.* **156 C3** 41 39N 93 28W
Altoona, *Pa., U.S.A.* **150 F6** 40 31N 78 24W
Altos, *Brazil* **170 C3** 5 3 S 42 28W
Altötting, *Germany* **31 G8** 48 12N 12 39 E
Altstätten, *Switz.* **33 B9** 47 22N 9 33 E
Altūn Kūprī, *Iraq* **101 E11** 35 45N 44 9 E
Altun Shan, *China* **72 C3** 38 30N 88 0 E
Alturas, *U.S.A.* **158 F3** 41 29N 120 32W
Altus, *U.S.A.* **155 H5** 34 38N 99 20W
Alubijid, *Phil.* **81 G5** 8 35N 124 29 E
Alucra, *Turkey* **101 B8** 40 22N 38 47 E
Alūksne, *Latvia* **15 H22** 57 24N 27 3 E
Alùla, *Somali Rep.* **120 B4** 11 50N 50 45 E
Alunda, *Sweden* **16 D12** 60 4N 18 5 E
Alunite, *U.S.A.* **161 K12** 35 59N 114 55W
Alupka, *Ukraine* **59 K8** 44 23N 34 2 E
Alur Gajah, *Malaysia* **84 B2** 2 23N 102 13 E
Alushta, *Ukraine* **59 K8** 44 40N 34 25 E
Alusi, *Indonesia* **83 C4** 7 35 S 131 40 E
Alustante, *Spain* **40 E3** 40 36N 1 40W
Al'Uzayr, *Iraq* **96 D5** 31 19N 47 25 E
Alva, *U.S.A.* **155 G5** 36 48N 98 40W
Alvaiázere, *Portugal* **42 F2** 39 49N 8 23W
Älvängen, *Sweden* **17 G6** 57 58N 12 8 E
Alvarado, *Mexico* **163 D5** 18 40N 95 50W
Alvarado, *U.S.A.* **155 J6** 32 24N 97 13W
Alvarães, *Brazil* **169 D5** 3 12 S 64 50W
Alvaro Obregón, Presa, *Mexico* **162 B3** 27 55N 109 52W
Alvdal, *Norway* **18 B7** 62 6N 10 37 E
Älvdalen, *Sweden* **16 C8** 61 13N 14 4 E
Alvear, *Argentina* **174 B4** 29 5 S 56 30W
Alverca, *Portugal* **43 G1** 38 56N 9 1W
Alvesta, *Sweden* **17 H8** 56 54N 14 35 E
Alvie, *Australia* **128 E5** 38 14 S 143 30 E
Ålvik, *Norway* **18 D3** 60 26N 6 26 E
Alvin, *S.C., U.S.A.* **152 B10** 33 22N 79 48W
Alvin, *Tex., U.S.A.* **155 L7** 29 26N 95 15W
Alvinston, *Canada* **150 D3** 42 49N 81 52W
Alvito, *Portugal* **43 G3** 38 15N 7 58W
Älvkarleby, *Sweden* **16 D11** 60 34N 17 26 E
Älvros, *Sweden* **16 B8** 62 3N 14 38 E
Älvsborgs län □, *Sweden* **17 F6** 58 30N 12 30 E
Älvsbyn, *Sweden* **14 D19** 65 40N 21 0 E
Ålvundeid, *Norway* **18 B5** 62 45N 8 33 E
Alwar, *India* **92 F7** 27 38N 76 34 E
Alwaye, *India* **95 J3** 10 8N 76 24 E
Alxa Zuoqi, *China* **74 E3** 38 50N 105 40 E
Alyata = Älät, *Azerbaijan* **61 L9** 39 58N 49 25 E
Alyth, *U.K.* **22 E5** 56 38N 3 13W
Alytus, *Lithuania* **15 J21** 54 24N 24 3 E
Alzada, *U.S.A.* **154 C2** 45 2N 104 25W
Alzey, *Germany* **31 F4** 49 45N 8 7 E
Alzira, *Spain* **41 F4** 39 9N 0 30W
Am Dam, *Chad* **109 F4** 12 40N 20 35 E
Am Géréda, *Chad* **109 F4** 12 53N 21 14 E
Am-Timan, *Chad* **109 F4** 11 0N 20 10 E
Amacuro □, *Venezuela* **169 B5** 8 50N 61 5W
Amadeus, L., *Australia* **125 D5** 24 54 S 131 0 E
Amadi, *Dem. Rep. of the Congo* **118 B2** 3 40N 26 40 E
Amâdi, *Sudan* **107 F3** 5 29N 30 25 E
Amadjuak L., *Canada* **139 B12** 65 0N 71 8W
Amadora, *Portugal* **43 G1** 38 45N 9 13W
Amagasaki, *Japan* **71 C7** 34 42N 135 20 E
Amager, *Denmark* **17 J6** 55 37N 12 35 E
Amagi, *Japan* **70 D2** 33 25N 130 39 E
Amahai, *Indonesia* **83 B3** 3 20 S 128 55 E
Amaimon, *Papua N. G.* **132 C3** 5 12 S 145 30 E
Amakusa-Nada, *Japan* **70 E2** 32 35N 130 5 E
Amakusa-Shotō, *Japan* **70 E2** 32 15N 130 10 E
Åmål, *Sweden* **16 E6** 59 3N 12 42 E
Amalapuram, *India* **95 F5** 16 35N 81 55 E
Amalfi, *Colombia* **168 B2** 6 55N 75 4W
Amalfi, *Italy* **47 B7** 40 38N 14 36 E
Amaliás, *Greece* **48 D3** 37 47N 21 22 E
Amalner, *India* **94 D2** 21 5N 75 5 E
Amambaí, *Brazil* **175 A4** 23 5 S 55 13W
Amambaí →, *Brazil* **175 A5** 23 22 S 53 56W
Amambay □, *Paraguay* **175 A4** 23 0 S 56 0W
Amambay, Cordillera de, *S. Amer.* **175 A4** 23 0 S 55 45W
Amami-Guntō, *Japan* **69 L4** 27 16N 129 21 E
Amami-Ō-Shima, *Japan* **69 L4** 28 0N 129 0 E
Amana →, *Venezuela* **169 B5** 9 45N 62 39W
Amaná, L., *Brazil* **169 D5** 2 35 S 64 40W
Amanab, *Papua N. G.* **132 B1** 3 40 S 141 14 E
Amanda Park, *U.S.A.* **160 C3** 47 28N 123 55W
Amangeldy, *Kazakstan* **64 D7** 50 10N 65 10 E
Amantea, *Italy* **47 C9** 39 8N 16 4 E
Amapá, *Brazil* **169 C7** 2 5N 50 50W
Amapá □, *Brazil* **169 C7** 1 40N 52 0W
Amapari, *Brazil* **169 C7** 0 37N 51 39W
Amara, *Sudan* **107 E3** 10 25N 34 10 E
Amarante, *Brazil* **170 C3** 6 14 S 42 50W
Amarante, *Portugal* **42 D2** 41 16N 8 5W
Amarante do Maranhão, *Brazil* **170 C2** 5 36 S 46 45W
Amaranth, *Canada* **143 C9** 50 36N 98 43W
Amarapura, *Burma* **90 E6** 21 54N 96 3 E
Amaravati →, *India* **95 J4** 11 0N 78 15 E
Amareleja, *Portugal* **43 G3** 38 12N 7 13W
Amargosa, *Brazil* **171 D4** 13 2 S 39 36W
Amargosa →, *U.S.A.* **161 J10** 36 14N 116 51W
Amargosa Range, *U.S.A.* **161 J10** 36 20N 116 45W
Amári, *Greece* **38 D6** 35 13N 24 40 E
Amarillo, *U.S.A.* **155 H4** 35 13N 101 50W
Amarnath, *India* **94 E1** 19 12N 73 22 E
Amaro, Mte., *Italy* **45 F11** 42 5N 14 5 E
Amaro Leite, *Brazil* **171 D2** 13 58 S 49 9W
Amarpur, *India* **93 G12** 25 5N 87 0 E
Amasra, *Turkey* **100 B5** 41 45N 32 23 E
Amassama, *Nigeria* **113 D6** 5 1N 6 2 E
Amasya, *Turkey* **100 B6** 40 40N 35 50 E
Amataurá, *Brazil* **168 D4** 3 29 S 68 6W
Amatikulu, *S. Africa* **117 D5** 29 3 S 31 33 E
Amatitlán, *Guatemala* **164 D1** 14 29N 90 38W
Amatrice, *Italy* **45 F10** 42 38N 13 17 E
Amay, *Belgium* **24 D5** 50 33N 5 19 E
Amazon = Amazonas →, *S. Amer.* **169 D8** 0 5 S 50 0W
Amazonas □, *Brazil* **173 B5** 5 0 S 65 0W
Amazonas □, *Peru* **172 B2** 5 0 S 78 0W
Amazonas □, *Venezuela* **168 C4** 3 30N 66 0W
Amazonas →, *S. Amer.* **169 D8** 0 5 S 50 0W
Ambad, *India* **94 E2** 19 38N 75 50 E
Ambahakily, *Madag.* **117 C7** 21 36 S 43 41 E
Ambala, *India* **92 D7** 30 23N 76 56 E
Ambalangoda, *Sri Lanka* **95 L5** 6 15N 80 5 E
Ambalapulai, *India* **95 K3** 9 25N 76 25 E
Ambalavao, *Madag.* **117 C8** 21 50 S 46 56 E
Ambalindum, *Australia* **126 C2** 23 23 S 135 0 E
Ambam, *Cameroon* **114 B2** 2 20N 11 15 E
Ambanja, *Madag.* **117 A8** 13 40 S 48 27 E
Ambarchik, *Russia* **65 C17** 69 40N 162 20 E
Ambarijeby, *Madag.* **117 A8** 14 56 S 47 41 E
Ambaro, Helodranon', *Madag.* **117 A8** 13 23 S 48 38 E
Ambasamudram, *India* **95 K3** 8 43N 77 25 E
Ambato, *Ecuador* **168 D2** 1 5 S 78 42W
Ambato, Sierra de, *Argentina* **174 B2** 28 25 S 66 10W
Ambato Boeny, *Madag.* **117 B8** 16 28 S 46 43 E
Ambatofinandrahana, *Madag.* **117 C8** 20 33 S 46 48 E
Ambatolampy, *Madag.* **117 B8** 19 20 S 47 35 E
Ambatondrazaka, *Madag.* **117 B8** 17 55 S 48 28 E
Ambatosoratra, *Madag.* **117 B8** 17 37 S 48 31 E
Ambelón, *Greece* **48 B4** 39 45N 22 22 E
Ambenja, *Madag.* **117 B8** 15 17 S 46 58 E
Amberg, *Germany* **31 F7** 49 26N 11 52 E
Ambergris Cay, *Belize* **163 D7** 18 0N 88 0W
Ambérieu-en-Bugey, *France* **29 C9** 45 57N 5 20 E
Amberley, *N.Z.* **131 D7** 43 9 S 172 44 E
Ambert, *France* **28 C7** 45 33N 3 44 E
Ambidédi, *Mali* **112 C2** 14 35N 11 47W
Ambikapur, *India* **93 H10** 23 15N 83 15 E
Ambikol, *Sudan* **106 C3** 21 20N 30 50 E
Ambilobé, *Madag.* **117 A8** 13 10 S 49 3 E
Ambinanindrano, *Madag.* **117 C8** 20 5 S 48 23 E
Amble, *U.K.* **20 B6** 55 20N 1 36W
Ambler, *U.S.A.* **144 C8** 67 5N 157 52W
Ambleside, *U.K.* **20 C5** 54 26N 2 58W
Ambo, *Ethiopia* **107 E4** 12 20N 37 30 E
Ambo, *Peru* **172 C2** 10 5 S 76 10W
Ambodifototra, *Madag.* **117 B8** 16 59 S 49 52 E
Ambodilazana, *Madag.* **117 B8** 18 6 S 49 10 E
Ambohimahasoa, *Madag.* **117 C8** 21 7 S 47 13 E
Ambohimanga, *Madag.* **117 C8** 20 52 S 47 36 E
Ambohitra, *Madag.* **117 A8** 12 30 S 49 10 E
Amboise, *France* **26 E8** 47 24N 1 2 E
Ambon, *Indonesia* **82 B3** 3 43 S 128 12 E
Amboseli, L., *Kenya* **118 C4** 2 40 S 37 10 E
Ambositra, *Madag.* **117 C8** 20 31 S 47 25 E
Ambovombé, *Madag.* **117 D8** 25 11 S 46 5 E
Amboy, *Calif., U.S.A.* **161 L11** 34 33N 115 45W
Amboy, *Ill., U.S.A.* **156 C7** 41 44N 89 20W
Ambridge, *U.S.A.* **150 F4** 40 36N 80 14W
Ambriz, *Angola* **115 D2** 7 48 S 13 8 E
Ambrose, *U.S.A.* **152 D6** 31 36N 83 1W
Ambrym, *Vanuatu* **133 F6** 16 15 S 168 10 E
Ambunti, *Papua N. G.* **132 C2** 4 13 S 142 52 E
Ambur, *India* **95 H4** 12 48N 78 43 E
Amby, *Australia* **127 D4** 26 30 S 148 11 E
Amchitka I., *U.S.A.* **138 C1** 51 32N 179 0 E
Amderma, *Russia* **64 C7** 69 45N 61 30 E
Ameca, *Mexico* **162 C4** 20 30N 104 0W
Ameca →, *Mexico* **162 C3** 20 40N 105 15W
Amecameca, *Mexico* **163 D5** 19 7N 98 46W
Ameland, *Neths.* **24 A5** 53 27N 5 45 E
Amélia, *Italy* **45 F9** 42 33N 12 25 E
Amelia City, *U.S.A.* **152 E8** 30 35N 81 28W
Amelia I., *U.S.A.* **152 E8** 30 40N 81 25W
Amendolara, *Italy* **47 C9** 39 57N 16 35 E
American Falls, *U.S.A.* **158 E7** 42 47N 112 51W
American Falls Reservoir, *U.S.A.* **158 E7** 42 47N 112 52W
American Highland, *Antarctica* **7 D6** 73 0 S 75 0 E
American Samoa ■, *Pac. Oc.* **133 X24** 14 20 S 170 40W
Americana, *Brazil* **175 A6** 22 45 S 47 20W
Americus, *U.S.A.* **152 C5** 32 4N 84 14W
Amersfoort, *Neths.* **24 B5** 52 9N 5 23 E
Amersfoort, *S. Africa* **117 D4** 26 59 S 29 53 E
Amery, *Australia* **125 F2** 31 9 S 117 5 E
Amery, *Canada* **143 B10** 56 34N 94 3W
Amery Ice Shelf, *Antarctica* **7 C6** 69 30 S 72 0 E
Ames, *Spain* **42 C2** 42 54N 8 38W
Ames, *U.S.A.* **156 C3** 42 2N 93 37W
Amesbury, *U.S.A.* **151 D14** 42 51N 70 56W
Amfíklia, *Greece* **48 C4** 38 38N 22 35 E
Amfilokhía, *Greece* **48 C3** 38 52N 21 9 E
Amfípolis, *Greece* **50 F7** 40 48N 23 52 E
Ámfissa, *Greece* **48 C4** 38 32N 22 22 E
Amga, *Russia* **65 C14** 60 50N 132 0 E
Amga →, *Russia* **65 C14** 62 38N 134 32 E
Amgu, *Russia* **68 B8** 45 45N 137 15 E
Amgun →, *Russia* **65 D14** 52 56N 139 38 E
Amherst, *Canada* **141 C7** 45 48N 64 8W
Amherst, *Mass., U.S.A.* **151 D12** 42 23N 72 31W
Amherst, *N.Y., U.S.A.* **150 D6** 42 59N 78 48W
Amherst, *Ohio, U.S.A.* **150 E2** 41 24N 82 14W
Amherst, *Tex., U.S.A.* **155 J3** 34 1N 102 25W
Amherst I., *Canada* **151 B8** 44 8N 76 43W
Amherstburg, *Canada* **140 D3** 42 6N 83 6W
Amiata, Mte., *Italy* **45 F8** 42 53N 11 37 E
Amiens, *France* **27 C9** 49 54N 2 16 E
Amili, *India* **90 A5** 28 25N 95 52 E
Amindaion, *Greece* **50 F5** 40 42N 21 42 E
Åminne, *Sweden* **17 G7** 57 7N 14 0 E
Amīrābād, *Iran* **96 C5** 33 20N 46 16 E
Amirante Is., *Seychelles* **121 E4** 6 0 S 53 0 E
Amisk L., *Canada* **143 C8** 54 35N 102 15W
Amistad, Presa de la, *Mexico* **162 B4** 29 24N 101 0W
Amite, *U.S.A.* **155 K9** 30 44N 90 30W
Amizmiz, *Morocco* **110 B3** 31 12N 8 15W
Åmli, *Norway* **18 F5** 58 45N 8 32 E
Amlwch, *U.K.* **20 D3** 53 24N 4 20W
Amm Adam, *Sudan* **107 D4** 16 20N 36 1 E
'Ammān, *Jordan* **103 D4** 31 57N 35 52 E
'Ammān □, *Jordan* **103 D5** 31 40N 36 30 E
Ammanford, *U.K.* **21 F4** 51 48N 3 59W
Ammassalik = Angmagssalik, *Greenland* **10 D7** 65 40N 37 20W
Ammerån →, *Sweden* **16 A10** 63 9N 16 13 E
Ammersee, *Germany* **31 G7** 48 0N 11 7 E
Amnat Charoen, *Thailand* **86 E5** 15 51N 104 38 E
Amo Jiang →, *China* **76 F3** 23 0N 101 50 E
Āmol, *Iran* **97 B7** 36 23N 52 20 E
Amoret, *U.S.A.* **156 F2** 38 15N 94 35W
Amorgós, *Greece* **49 E7** 36 50N 25 58 E
Amory, *U.S.A.* **149 J1** 33 59N 88 29W
Amos, *Canada* **140 C4** 48 35N 78 5W
Åmot, *Buskerud, Norway* **15 G13** 59 57N 9 54 E
Åmot, *Oppland, Norway* **18 D7** 61 0N 10 2 E
Åmot, *Telemark, Norway* **18 E5** 59 34N 8 0 E
Åmotfors, *Sweden* **16 E6** 59 47N 12 22 E
Åmotsdal, *Norway* **18 E5** 59 37N 8 26 E
Amour, Djebel, *Algeria* **111 B5** 33 42N 1 37 E
Amoy = Xiamen, *China* **77 E12** 24 25N 118 4 E
Ampang, *Malaysia* **87 L3** 3 8N 101 45 E
Ampanihy, *Madag.* **117 C7** 24 40 S 44 45 E
Ampasindava, Helodranon', *Madag.* **117 A8** 13 40 S 48 15 E
Ampasindava, Saikanosy, *Madag.* **117 A8** 13 42 S 47 55 E
Ampato, Nevado, *Peru* **172 D3** 15 40 S 71 56W
Ampenan, *Indonesia* **85 D5** 8 35 S 116 13 E
Amper, *Nigeria* **113 D6** 9 25N 9 40 E
Amper →, *Germany* **31 G7** 48 29N 11 55 E
Ampezzo, *Italy* **45 B9** 46 25N 12 48 E
Amposta, *Spain* **40 E5** 40 43N 0 34 E
Ampotaka, *Madag.* **117 D7** 25 3 S 44 41 E
Ampoza, *Madag.* **117 C7** 22 20 S 44 44 E
Amqui, *Canada* **141 C6** 48 28N 67 27W
'Amrān, *Yemen* **98 D3** 15 41N 43 55 E
Amravati, *India* **94 D3** 20 55N 77 45 E
Amreli, *India* **92 J4** 21 35N 71 17 E
Amrenene el Kasba, *Algeria* **111 D5** 22 10N 0 30 E
Amriswil, *Switz.* **33 A8** 47 33N 9 18 E
Amritsar, *India* **92 D6** 31 35N 74 57 E
Amroha, *India* **93 E8** 28 53N 78 30 E
Amrum, *Germany* **30 A4** 54 38N 8 22 E
Amsel, *Algeria* **111 D6** 22 47N 5 29 E
Amsterdam, *Neths.* **24 B4** 52 23N 4 54 E
Amsterdam, *U.S.A.* **151 D10** 42 56N 74 11W
Amsterdam, I., *Ind. Oc.* **121 H6** 38 30 S 77 30 E
Amstetten, *Austria* **34 C7** 48 7N 14 51 E
Amudarya →, *Uzbekistan* **64 E6** 43 58N 59 34 E
Amukta Pass, *U.S.A.* **144 L5** 52 0N 171 0W
Amulung, *Phil.* **80 C3** 17 50N 121 43 E
Amundsen Gulf, *Canada* **138 A7** 71 0N 124 0W
Amundsen Sea, *Antarctica* **7 D15** 72 0 S 115 0W
Amungen, *Sweden* **16 C9** 61 10N 15 40 E
Amuntai, *Indonesia* **85 C5** 2 28 S 115 25 E
Amur, *Somali Rep.* **120 C3** 5 16N 46 30 E

Amur →, Russia **65 D15** 52 56N 141 10 E
Amurang, Indonesia **82 A2** 1 5N 124 40 E
Amuri Pass, N.Z. **131 C7** 42 31 S 172 11 E
Amurrio, Spain **40 B1** 43 3N 3 0W
Amursk, Russia **65 D14** 50 14N 136 54 E
Amusco, Spain **42 C6** 42 10N 4 28W
Amutag, Phil. **80 E4** 12 23N 123 16 E
Amvrakikós Kólpos, Greece **48 C2** 39 0N 20 55 E
Amvrosiyivka, Ukraine **59 J10** 47 43N 38 30 E
Amyderya = Amudarya →, Uzbekistan **64 E6** 43 58N 59 34 E
Amzeglouf, Algeria **111 C5** 26 50N 0 1 E
An, Burma **90 F5** 19 48N 94 0 E
An Bien, Vietnam **87 H5** 9 45N 105 0 E
An Hoa, Vietnam **86 E7** 15 40N 108 5 E
An Nabatīyah at Tahta, Lebanon **103 B4** 33 23N 35 27 E
An Nabk, Si. Arabia **96 D3** 31 20N 37 20 E
An Nabk, Syria **103 A5** 34 2N 36 44 E
An Nabk Abū Qaşr, Si. Arabia **96 D3** 30 21N 38 34 E
An Nafūd, Si. Arabia **96 D4** 28 15N 41 0 E
An Najaf, Iraq **101 G11** 32 3N 44 15 E
An Nāşirīyah, Iraq **96 D5** 31 0N 46 15 E
An Nawfaliyah, Libya **108 B3** 30 54N 17 58 E
An Nhon, Vietnam **86 F7** 13 55N 109 7 E
An Nîl □, Sudan **106 D3** 19 30N 33 0 E
An Nîl el Abyad □, Sudan **107 E3** 14 0N 32 15 E
An Nîl el Azraq □, Sudan **107 E3** 11 30N 34 30 E
An Nimāş, Si. Arabia **98 C3** 19 7N 42 8 E
An Nu'ayrīyah, Si. Arabia **97 E6** 27 30N 48 30 E
An Nu'mānīyah, Iraq **101 F11** 32 32N 45 25 E
An Nuwayb'ī, W. →, Si. Arabia **103 F3** 29 18N 34 57 E
An Thoi, Dao, Vietnam **87 H5** 9 58N 104 0 E
An Uaimh, Ireland **23 C5** 53 39N 6 41W
Åna-Sira, Norway **18 F3** 58 17N 6 25 E
Anabar →, Russia **65 B12** 73 8N 113 36 E
'Anabtā, West Bank **103 C4** 32 19N 35 7 E
Anabuki, Japan **70 C6** 34 2N 134 11 E
Anaco, Venezuela **169 B5** 9 27N 64 28W
Anaconda, U.S.A. **158 C7** 46 8N 112 57W
Anacortes, U.S.A. **160 B4** 48 30N 122 37W
Anacuao, Mt., Phil. **80 C3** 16 16N 121 53 E
Anadarko, U.S.A. **155 H5** 35 4N 98 15W
Anadia, Brazil **170 C4** 9 42 S 36 18W
Anadia, Portugal **42 E2** 40 26N 8 27W
Anadolu, Turkey **100 C5** 39 0N 30 0 E
Anadyr, Russia **65 C18** 64 35N 177 20 E
Anadyr →, Russia **65 C18** 64 55N 176 5 E
Anadyrskiy Zaliv, Russia **65 C19** 64 0N 180 0 E
Anáfi, Greece **49 E7** 36 22N 25 48 E
Anafópoulo, Greece **49 E7** 36 17N 25 50 E
Anaga, Pta. de, Canary Is. **39 F3** 28 34N 16 9W
Anagni, Italy **45 G10** 41 44N 13 9 E
'Ānah, Iraq **101 E10** 34 25N 42 0 E
Anahalu →, U.S.A. **145 J13** 21 37N 158 6W
Anaheim, U.S.A. **161 M9** 33 50N 117 55W
Anahim Lake, Canada **142 C3** 52 28N 125 18W
Anahola, U.S.A. **145 A2** 22 9N 159 19W
Anáhuac, Mexico **162 B4** 27 14N 100 9W
Anai Mudi, India **95 J3** 10 12N 77 4 E
Anaimalai Hills, India **95 J3** 10 20N 76 40 E
Anajás, Brazil **170 B2** 0 59 S 49 57W
Anajatuba, Brazil **170 B3** 3 16 S 44 37W
Anakapalle, India **94 F6** 17 42N 83 6 E
Anakie, Australia **126 C4** 23 32 S 147 45 E
Anaklia, Georgia **61 J5** 42 22N 41 35 E
Anaktuvuk Pass, U.S.A. **144 B10** 68 8N 151 45W
Analalava, Madag. **117 A8** 14 35 S 48 0 E
Análipsis, Greece **38 A3** 39 36N 19 55 E
Anamã, Brazil **169 D5** 3 35 S 61 22W
Anambar →, Pakistan **92 D3** 30 15N 68 50 E
Anambas, Kepulauan, Indonesia **87 L6** 3 20N 106 30 E
Anambas Is. = Anambas, Kepulauan, Indonesia **87 L6** 3 20N 106 30 E
Anambra □, Nigeria **113 D6** 6 20N 7 0 E
Aname, Vanuatu **133 K7** 20 8 S 169 47 E
Anamoose, U.S.A. **154 B4** 47 53N 100 15W
Anamosa, U.S.A. **156 B5** 42 7N 91 17W
Anamur, Turkey **100 D5** 36 8N 32 58 E
Anamur Burnu, Turkey **100 D5** 36 2N 32 47 E
Anan, Japan **70 D6** 33 54N 134 40 E
Anand, India **92 H5** 22 32N 72 59 E
Anandpur, India **94 D8** 21 16N 86 13 E
Anánes, Greece **48 E6** 36 33N 24 9 E
Anantapur, India **95 G3** 14 39N 77 42 E
Anantnag, India **93 C6** 33 45N 75 10 E
Ananyiv, Ukraine **59 J5** 47 44N 29 58 E
Anao-aon, Phil. **81 G5** 9 47N 125 25 E
Anapa, Russia **59 K9** 44 55N 37 25 E
Anapodháris →, Greece **38 E7** 34 59N 25 20 E
Anápolis, Brazil **171 E2** 16 15 S 48 50W
Anapu →, Brazil **169 D7** 1 53 S 50 53W
Anār, Iran **97 D7** 30 55N 55 13 E
Anār Darreh, Afghan. **91 B1** 32 46N 61 39 E
Anārak, Iran **97 C7** 33 25N 53 40 E
Anarisfjällen, Sweden **16 A7** 63 6N 13 10 E
Anatolia = Anadolu, Turkey **100 C5** 39 0N 30 0 E
Anatone, U.S.A. **158 C5** 46 8N 117 8W
Anatsogno, Madag. **117 C7** 23 33 S 43 46 E
Añatuya, Argentina **174 B3** 28 20 S 62 50W
Anauá →, Brazil **169 C5** 0 58N 61 21W
Anaunethad L., Canada **143 A8** 60 55N 104 25W
Anavilhanas, Arquipélago das, Brazil **169 D5** 2 42 S 60 45W
Anaye, Niger **109 E2** 19 15N 12 50 E
Anbyŏn, N. Korea **75 E14** 39 1N 127 35 E
Ancares, Sierra de, Spain **42 C4** 42 51N 6 52W
Ancash □, Peru **172 B2** 9 30 S 77 45W
Ancenis, France **26 E5** 47 21N 1 10W
Ancho, Canal, Chile **176 D2** 50 0 S 74 20W
Anchor Bay, U.S.A. **160 G3** 38 48N 123 34W
Anchorage, U.S.A. **138 B5** 61 13N 149 54W
Anci, China **74 E9** 39 20N 116 40 E
Ancohuma, Nevada, Bolivia **172 D4** 16 0 S 68 50W
Ancón, Peru **172 C2** 11 50 S 77 10W
Ancona, Italy **45 E10** 43 38N 13 30 E
Ancud, Chile **176 B2** 42 0 S 73 50W
Ancud, G. de, Chile **176 B2** 42 0 S 73 0W
Ancy-le-Franc, France **27 E11** 47 46N 4 10 E
Anda, China **73 B7** 46 24N 125 19 E
Anda, Phil. **80 C2** 16 17N 119 57 E
Andacollo, Argentina **174 D1** 37 10 S 70 42W
Andacollo, Chile **174 C1** 30 5 S 71 10W
Andado, Australia **126 D2** 25 25 S 135 15 E
Andahuaylas, Peru **172 C3** 13 40 S 73 25W
Andalgalá, Argentina **174 B2** 27 40 S 66 30W
Åndalsnes, Norway **15 E12** 62 35N 7 43 E
Andalucía □, Spain **43 H6** 37 35N 5 0W
Andalusia, U.S.A. **149 K2** 31 18N 86 29W
Andalusia □ = Andalucía □, Spain **43 H6** 37 35N 5 0W
Andaman Is., Ind. Oc. **121 C8** 12 30N 92 30 E
Andara, Namibia **116 B3** 18 2 S 21 9 E
Andaraí, Brazil **171 D3** 12 48 S 41 20W
Andeer, Switz. **33 C8** 46 36N 9 26 E
Andelfingen, Switz. **33 A7** 47 36N 8 41 E
Andelot-Blancheville, France **27 D12** 48 15N 5 18 E
Andenes, Norway **14 B17** 69 19N 16 18 E
Andenne, Belgium **24 D5** 50 28N 5 5 E
Andéranboukane, Mali **113 B5** 15 26N 3 2 E
Andermatt, Switz. **33 C7** 46 38N 8 35 E
Andernach, Germany **30 E3** 50 26N 7 24 E
Andernos-les-Bains, France **28 D2** 44 44N 1 6W
Anderslöv, Sweden **17 J7** 55 26N 13 19 E
Anderson, Alaska, U.S.A. **144 D10** 64 25N 149 15W
Anderson, Calif., U.S.A. **158 F2** 40 27N 122 18W
Anderson, Ind., U.S.A. **157 D11** 40 10N 85 41W
Anderson, Mo., U.S.A. **155 G7** 36 39N 94 27W
Anderson, S.C., U.S.A. **149 H4** 34 31N 82 39W
Anderson →, Canada **138 B7** 69 42N 129 0W
Andersonville, U.S.A. **152 C5** 32 12N 84 9W
Anderstorp, Sweden **17 G7** 57 19N 13 39 E
Andes, Cord. de los, S. Amer. **172 E4** 20 0 S 68 0W
Andfjorden, Norway **14 B17** 69 10N 16 20 E
Andhra, L., India **94 E1** 18 54N 73 32 E
Andhra Pradesh □, India **95 F4** 18 0N 79 0 E
Andijon, Uzbekistan **63 C6** 41 10N 72 15 E
Andikíthira, Greece **48 F5** 35 52N 23 15 E
Andīmeshk, Iran **97 C6** 32 27N 48 21 E
Andímilos, Greece **48 E6** 36 47N 24 12 E
Andíparos, Greece **49 D7** 37 0N 25 3 E
Andípaxoi, Greece **48 B2** 39 9N 20 13 E
Andípsara, Greece **49 C7** 38 30N 25 29 E
Andírrion, Greece **48 C3** 38 20N 21 46 E
Andizhan = Andijon, Uzbekistan **63 C6** 41 10N 72 15 E
Andkhvoy, Afghan. **91 A2** 36 52N 65 8 E
Andoain, Spain **40 B2** 43 13N 2 1W
Andoany, Madag. **117 A8** 13 25 S 48 16 E
Andoas, Peru **168 D2** 2 55 S 76 25W
Andol, India **94 F4** 17 51N 78 4 E
Andong, S. Korea **75 F15** 36 40N 128 43 E
Andongwei, China **75 G10** 35 6N 119 20 E
Andorra, Spain **40 E4** 40 59N 0 28W
Andorra ■, Europe **28 F5** 42 30N 1 30 E
Andorra La Vella, Andorra **28 F5** 42 31N 1 32 E
Andover, U.K. **21 F6** 51 12N 1 29W
Andover, Mass., U.S.A. **151 D13** 42 40N 71 8W
Andover, N.Y., U.S.A. **150 D7** 42 10N 77 48W
Andover, Ohio, U.S.A. **150 E4** 41 36N 80 34W
Andøya, Norway **14 B16** 69 10N 15 50 E
Andradina, Brazil **171 F1** 20 54 S 51 23W
Andrahary, Mt., Madag. **117 A8** 13 37 S 49 17 E
Andramasina, Madag. **117 B8** 19 11 S 47 35 E
Andranopasy, Madag. **117 C7** 21 17 S 43 44 E
Andratx, Spain **39 B9** 39 39N 2 25 E
Andreanof Is., U.S.A. **144 L4** 51 30N 176 0W
Andreapol, Russia **58 D7** 56 40N 32 17 E
Andrée Land, Greenland **10 C8** 73 40N 26 0W
Andrewilla, Australia **127 D2** 26 31 S 139 17 E
Andrews, S.C., U.S.A. **149 J6** 33 27N 79 34W
Andrews, Tex., U.S.A. **155 J3** 32 19N 102 33W
Andreyevka, Russia **60 D10** 52 19N 51 55 E
Ándria, Italy **47 A9** 41 13N 16 17 E
Andriba, Madag. **117 B8** 17 30 S 46 58 E
Andrijevica, Montenegro, Yug. **50 D3** 42 45N 19 48 E
Andrítsaina, Greece **48 D3** 37 29N 21 52 E
Androka, Madag. **117 C7** 24 58 S 44 2 E
Andropov = Rybinsk, Russia **58 C10** 58 5N 38 50 E
Ándros, Greece **48 D6** 37 50N 24 57 E
Andros I., Bahamas **164 B4** 24 30N 78 0W
Andros Town, Bahamas **164 B4** 24 43N 77 47W
Andrychów, Poland **55 J6** 49 51N 19 18 E
Andselv, Norway **14 B18** 69 4N 18 34 E
Andújar, Spain **43 G6** 38 3N 4 5W
Andulo, Angola **115 E3** 11 25 S 16 45 E
Åneby, Norway **18 D7** 60 5N 10 51 E
Aneby, Sweden **17 G8** 57 48N 14 49 E
Anegada, B., Argentina **176 B4** 40 20 S 62 20W
Anegada I., Virgin Is. **165 C7** 18 45N 64 20W
Anegada Passage, W. Indies **165 C7** 18 15N 63 45W
Aného, Togo **113 D5** 6 12N 1 34 E
Aneityum, Vanuatu **133 K7** 20 12 S 169 45 E
Añelo, Argentina **176 A3** 38 20 S 68 45W
Anenni-Noi, Moldova **53 D14** 46 53N 29 15 E
Anergane, Morocco **110 B3** 31 4N 7 14W
Aneto, Pico de, Spain **40 C5** 42 37N 0 40 E
Añez, Bolivia **173 D5** 15 40 S 63 10W
Anfu, China **77 D10** 27 21N 114 40 E
Ang Thong, Thailand **86 E3** 14 35N 100 31 E
Angadanan, Phil. **80 C3** 16 45N 121 45 E
Angamos, Punta, Chile **174 A1** 23 1 S 70 32W
Angara →, Russia **65 D10** 58 5N 94 20 E
Angarab, Ethiopia **107 E4** 13 11N 37 7 E
Angarsk, Russia **65 D11** 52 30N 104 0 E
Angas Downs, Australia **125 E5** 25 2 S 132 14 E
Angas Hills, Australia **124 D4** 23 0 S 127 50 E
Angaston, Australia **128 C3** 34 30 S 139 8 E
Angat, Phil. **80 D3** 14 56N 121 2 E
Ånge, Sweden **16 B9** 62 31N 15 35 E
Ángel, Salto = Angel Falls, Venezuela **169 B5** 5 57N 62 30W
Ángel de la Guarda, I., Mexico **162 B2** 29 30N 113 30W
Angel Falls, Venezuela **169 B5** 5 57N 62 30W
Angeles, Phil. **80 D3** 15 9N 120 33 E
Ängelholm, Sweden **17 H6** 56 15N 12 58 E
Angellala, Australia **127 D4** 26 24 S 146 54 E
Angels Camp, U.S.A. **160 G6** 38 4N 120 32W
Ängelsberg, Sweden **16 E10** 59 58N 16 0 E
Anger →, Ethiopia **107 F4** 9 37N 36 6 E
Angereb →, Ethiopia **107 E4** 13 45N 36 40 E
Ångermanälven →, Sweden **16 B11** 62 40N 18 0 E
Ångermanland, Sweden **14 E18** 63 36N 17 45 E
Angermünde, Germany **30 B9** 53 0N 14 0 E
Angers, Canada **151 A9** 45 31N 75 29W
Angers, France **26 E6** 47 30N 0 35W
Angerville, France **27 D9** 48 19N 2 0 E
Ängesån →, Sweden **14 C20** 66 16N 22 47 E
Angical, Brazil **171 D3** 12 0 S 44 42W
Angikuni L., Canada **143 A9** 62 0N 100 0W
Angkor, Cambodia **86 F4** 13 22N 103 50 E
Anglem Mt., N.Z. **131 G2** 46 45 S 167 53 E
Anglès, Spain **40 D7** 41 57N 2 38 E
Anglesey, U.K. **20 D3** 53 17N 4 20W
Anglesey, Isle of □, U.K. **20 D3** 53 16N 4 18W
Anglet, France **28 E2** 43 29N 1 31W
Angleton, U.S.A. **155 L7** 29 10N 95 26W
Anglin →, France **28 B4** 46 42N 0 52 E
Anglisidhes, Cyprus **38 E12** 34 51N 33 27 E
Anglure, France **27 D10** 48 35N 3 50 E
Angmagssalik, Greenland **10 D7** 65 40N 37 20W
Ango, Dem. Rep. of the Congo **118 B2** 4 10N 26 5 E
Angoche, Mozam. **119 F4** 16 8 S 39 55 E
Angoche, I., Mozam. **119 F4** 16 20 S 39 50 E
Angol, Chile **174 D1** 37 56 S 72 45W
Angola, Ind., U.S.A. **157 C12** 41 38N 85 0W
Angola, N.Y., U.S.A. **150 D5** 42 38N 79 2W
Angola ■, Africa **115 E3** 12 0 S 18 0 E
Angoon, U.S.A. **142 B2** 57 30N 134 35W
Angoram, Papua N. G. **132 C3** 4 4 S 144 4 E
Angoulême, France **28 C4** 45 39N 0 10 E
Angoumois, France **28 C4** 45 50N 0 25 E
Angra dos Reis, Brazil **175 A7** 23 0 S 44 10W
Angren, Uzbekistan **63 C5** 41 1N 70 12 E
Angtassom, Cambodia **87 G5** 11 1N 104 41 E
Angu, Dem. Rep. of the Congo **118 B1** 3 25N 24 28 E
Anguang, China **75 B12** 45 15N 123 45 E
Anguilla ■, W. Indies **165 C7** 18 14N 63 5W
Anguo, China **74 E8** 38 28N 115 15 E
Angurugu, Australia **126 A2** 14 0 S 136 25 E
Angus □, U.K. **22 E6** 56 46N 2 56W
Anhanduí →, Brazil **175 A5** 21 46 S 52 9W
Anholt, Denmark **17 H5** 56 42N 11 33 E
Anhua, China **77 C8** 28 23N 111 12 E
Anhui □, China **77 B11** 32 0N 117 0 E
Anhwei □ = Anhui □, China **77 B11** 32 0N 117 0 E
Anichab, Namibia **116 C1** 21 0 S 14 46 E
Anicuns, Brazil **171 E2** 16 28 S 49 58W
Ánidhros, Greece **49 E7** 36 38N 25 43 E
Anié, Togo **113 D5** 7 42N 1 8 E
Animas, U.S.A. **159 L9** 31 57N 108 48W
Anina, Romania **52 E6** 45 6N 21 51 E
Anini-y, Phil. **81 F3** 10 25N 121 55 E
Aninoasa, Romania **53 F9** 44 47N 24 10 E
Anita, U.S.A. **156 C2** 41 27N 94 46W
Anivorano, Madag. **117 B8** 18 44 S 48 58 E
Aniwa, Vanuatu **133 J7** 19 17 S 169 35 E
Anjalankoski, Finland **15 F22** 60 45N 26 51 E
Anjangaon, India **94 D3** 21 10N 77 20 E
Anjar, India **92 H4** 23 6N 70 10 E
Anjidiv I., India **95 G2** 14 40N 74 10 E
Anjō, Japan **71 C9** 34 57N 137 5 E
Anjou, France **26 E6** 47 20N 0 15W
Anjozorobe, Madag. **117 B8** 18 22 S 47 52 E
Anju, N. Korea **75 E13** 39 36N 125 40 E
Anka, Nigeria **113 C6** 12 13N 5 58 E
Ankaboa, Tanjona, Madag. **117 C7** 21 58 S 43 20 E
Ankang, China **74 H5** 32 40N 109 1 E
Ankara, Turkey **100 C5** 39 57N 32 54 E
Ankaramena, Madag. **117 C8** 21 57 S 46 39 E
Ankarsrum, Sweden **17 G10** 57 41N 16 20 E
Ankazoabo, Madag. **117 C7** 22 18 S 44 31 E
Ankazobe, Madag. **117 B8** 18 20 S 47 10 E
Ankeny, U.S.A. **156 C3** 41 44N 93 36W
Ankisabe, Madag. **117 B8** 19 17 S 46 29 E
Anklam, Germany **30 B9** 53 51N 13 41 E
Ankleshwar, India **94 D1** 21 38N 73 3 E
Ankober, Ethiopia **107 F4** 9 35N 39 40 E
Ankona, U.S.A. **153 H9** 27 21N 80 17W
Ankoro, Dem. Rep. of the Congo **118 D2** 6 45 S 26 55 E
Anlong, China **76 E5** 25 2N 105 27 E
Anlu, China **77 B9** 31 15N 113 45 E
Anmyŏn-do, S. Korea **75 F14** 36 25N 126 25 E
Ånn, Sweden **16 A6** 63 16N 12 34 E
Ann, C., U.S.A. **151 D14** 42 38N 70 35W
Ann Arbor, U.S.A. **157 B13** 42 17N 83 45W
Anna, Russia **60 E5** 51 28N 40 23 E
Anna, Ill., U.S.A. **155 G10** 37 28N 89 15W
Anna, Ohio, U.S.A. **157 D12** 40 24N 84 11W
Anna Plains, Australia **124 C3** 19 17 S 121 37 E
Anna Regina, Guyana **169 B6** 7 10N 58 30W
Annaba, Algeria **111 A6** 36 50N 7 46 E
Annaberg-Buchholz, Germany **30 E9** 50 34N 13 0 E
Annaka, Japan **71 A10** 36 19N 138 54 E
Annalee →, Ireland **23 B4** 54 2N 7 24W
Annam, Vietnam **86 E7** 16 0N 108 0 E
Annamitique, Chaîne, Asia **86 D6** 17 0N 106 0 E
Annan, U.K. **22 G5** 54 59N 3 16W
Annan →, U.K. **22 G5** 54 58N 3 16W
Annanberg, Papua N. G. **132 C3** 4 52 S 144 42 E
Annapolis, U.S.A. **148 F7** 38 59N 76 30W
Annapolis Royal, Canada **141 D6** 44 44N 65 32W
Annapurna, Nepal **93 E10** 28 34N 83 50 E
Annean, L., Australia **125 E2** 26 54 S 118 14 E
Anneberg, Sweden **17 G8** 57 44N 14 49 E
Annecy, France **29 C10** 45 55N 6 8 E
Annecy, L. d', France **29 C10** 45 52N 6 10 E
Annemasse, France **27 F13** 46 12N 6 16 E
Annenskiy Most, Russia **58 B9** 60 45N 37 10 E
Annette I., U.S.A. **144 J15** 55 9N 131 28W
Annette Island Indian Reservation, U.S.A. **144 J15** 55 5N 131 30W
Anning, China **76 E4** 24 55N 102 26 E
Anningie, Australia **124 D5** 21 50 S 133 7 E
Anniston, U.S.A. **152 B4** 33 39N 85 50W
Annobón, Atl. Oc. **105 G4** 1 25 S 5 36 E
Annonay, France **29 C8** 45 15N 4 40 E
Annot, France **29 E10** 43 58N 6 38 E
Annotto Bay, Jamaica **164 C4** 18 17N 76 45W
Ånnsjön, Sweden **16 A6** 63 19N 12 34 E
Annuello, Australia **128 C5** 34 53 S 142 55 E
Annville, U.S.A. **151 F8** 40 20N 76 31W
Annweiler, Germany **31 F3** 49 12N 7 57 E
Áno Arkhánai, Greece **49 F7** 35 16N 25 11 E
Áno Porróia, Greece **50 E7** 41 17N 23 2 E
Áno Síros, Greece **48 D6** 37 29N 24 56 E
Áno Viánnos, Greece **38 D7** 35 2N 25 21 E
Anoano, Solomon Is. **133 M11** 8 59 S 160 46 E
Anoka, U.S.A. **147 A8** 45 12N 93 23W
Anorotsangana, Madag. **117 A8** 13 56 S 47 55 E
Anóyia, Greece **38 D6** 35 16N 24 52 E
Anping, Hebei, China **74 E8** 38 15N 115 30 E
Anping, Liaoning, China **75 D12** 41 5N 123 30 E
Anpu Gang, China **76 G7** 21 25N 109 50 E
Anqing, China **77 B11** 30 30N 117 3 E
Anqiu, China **75 F10** 36 25N 119 10 E
Anren, China **77 D9** 26 43N 113 18 E
Ansager, Denmark **17 J2** 55 43N 8 45 E
Ansai, China **74 F5** 36 50N 109 20 E
Ansbach, Germany **31 F6** 49 28N 10 34 E
Anseba →, Eritrea **107 D4** 16 0N 38 30 E
Anserma, Colombia **168 B2** 5 13N 75 48W
Ansfelden, Austria **34 C7** 48 12N 14 17 E
Anshan, China **75 D12** 41 5N 122 58 E
Anshun, China **76 D5** 26 18N 105 57 E
Ansião, Portugal **42 F2** 39 56N 8 27W
Ansley, U.S.A. **154 E5** 41 18N 99 23W
Ansó, Spain **40 C4** 42 51N 0 48W
Ansoain, Spain **40 C3** 42 50N 1 38W
Anson, U.S.A. **155 J5** 32 45N 99 54W
Anson B., Australia **124 B5** 13 20 S 130 6 E
Ansongo, Mali **113 B5** 15 25N 0 35 E
Ansonia, Conn., U.S.A. **151 E11** 41 21N 73 5W
Ansonia, Ohio, U.S.A. **157 D12** 40 13N 84 38W
Anstruther, U.K. **22 E6** 56 14N 2 41W
Ansudu, Indonesia **83 B5** 2 11 S 139 22 E
Antabamba, Peru **172 C3** 14 40 S 73 0W
Antakya, Turkey **100 D7** 36 14N 36 10 E
Antalaha, Madag. **117 A9** 14 57 S 50 20 E
Antalya, Turkey **100 D4** 36 52N 30 45 E
Antalya □, Turkey **49 E12** 36 30N 30 0 E
Antalya Körfezi, Turkey **100 D4** 36 15N 31 30 E
Antananarivo, Madag. **117 B8** 18 55 S 47 31 E
Antananarivo □, Madag. **117 B8** 19 0 S 47 0 E
Antanimbaribe, Madag. **117 C7** 21 30 S 44 48 E
Antarctic Pen., Antarctica **7 C18** 67 0 S 60 0W
Antarctica **7 E3** 90 0 S 0 0 E
Antelope, Zimbabwe **119 G2** 21 2 S 28 31 E
Antenor Navarro, Brazil **170 C4** 6 44 S 38 27W
Antequera, Paraguay **174 A4** 24 8 S 57 7W
Antequera, Spain **43 H6** 37 5N 4 33W
Antero, Mt., U.S.A. **159 G10** 38 41N 106 15W
Anthemoús, Greece **50 F7** 40 31N 23 15 E
Anthony, Fla., U.S.A. **153 F7** 29 18N 82 7W
Anthony, Kans., U.S.A. **155 G5** 37 9N 98 2W
Anthony, N. Mex., U.S.A. **159 K10** 32 0N 106 36W
Anthony Lagoon, Australia **126 B2** 18 0 S 135 30 E
Anti Atlas, Morocco **110 C3** 30 0N 8 30W
Anti-Lebanon = Ash Sharqi, Al Jabal, Lebanon **103 B5** 33 40N 36 10 E
Antibes, France **29 E11** 43 34N 7 6 E
Antibes, C. d', France **29 E11** 43 31N 7 7 E
Anticosti, Î. d', Canada **141 C7** 49 30N 63 0W
Antifer, C. d', France **26 C7** 49 41N 0 10 E
Antigo, U.S.A. **154 C10** 45 9N 89 9W
Antigonish, Canada **141 C7** 45 38N 61 58W
Antigua, Canary Is. **39 F5** 28 24N 14 1W
Antigua, W. Indies **165 C7** 17 0N 61 50W
Antigua & Barbuda ■, W. Indies **165 C7** 17 20N 61 48W
Antigua Guatemala, Guatemala **164 D1** 14 34N 90 41W
Antilla, Cuba **164 B4** 20 40N 75 50W
Antimony, U.S.A. **159 G8** 38 7N 112 0W
Antioch, U.S.A. **160 G5** 38 1N 121 48W
Antioche, Pertuis d', France **28 B2** 46 6N 1 20W
Antioquia, Colombia **168 B2** 6 40N 75 55W
Antioquia □, Colombia **168 B2** 7 0N 75 30W
Antipodes Is., Pac. Oc. **134 M9** 49 45 S 178 40 E
Antique □, Phil. **81 F4** 11 10N 122 5 E
Antler, U.S.A. **154 A4** 48 59N 101 17W
Antler →, Canada **143 D8** 49 8N 101 0W
Antlers, U.S.A. **155 H7** 34 14N 95 37W
Antofagasta, Chile **174 A1** 23 50 S 70 30W
Antofagasta □, Chile **174 A2** 24 0 S 69 0W
Antofagasta de la Sierra, Argentina **174 B2** 26 5 S 67 20W
Antofalla, Argentina **174 B2** 25 30 S 68 5W
Antofalla, Salar de, Argentina **174 B2** 25 40 S 67 45W
Anton, U.S.A. **155 J3** 33 49N 102 10W
Anton Chico, U.S.A. **159 J11** 35 12N 105 9W
Antongila, Helodrano, Madag. **117 B8** 15 30 S 49 50 E
Antonibé, Madag. **117 B8** 15 7 S 47 24 E
Antonibé, Presqu'île d', Madag. **117 A8** 14 55 S 47 20 E
Antonina, Brazil **175 B6** 25 26 S 48 42W
Antonito, U.S.A. **159 H10** 37 5N 106 0W
Antrain, France **26 D5** 48 28N 1 30W
Antrim, U.K. **23 B5** 54 43N 6 14W
Antrim □, U.K. **23 B5** 54 56N 6 25W
Antrim, Mts. of, U.K. **23 A5** 55 3N 6 14W
Antrim Plateau, Australia **124 C4** 18 8 S 128 20 E
Antrodoco, Italy **45 F10** 42 25N 13 5 E
Antropovo, Russia **60 A6** 58 24N 43 6 E
Antsalova, Madag. **117 B7** 18 40 S 44 37 E
Antsirabe, Madag. **117 B8** 19 55 S 47 2 E
Antsiranana, Madag. **117 A8** 12 25 S 49 20 E
Antsohihy, Madag. **117 A8** 14 50 S 47 59 E
Antsohimbondrona Seranana, Madag. **117 A8** 13 7 S 48 48 E
Antu, China **75 C15** 42 30N 128 20 E
Antufash, Yemen **98 D3** 15 42N 42 25 E
Antwerp = Antwerpen, Belgium **24 C4** 51 13N 4 25 E
Antwerp, Australia **128 D5** 36 17 S 142 4 E
Antwerp, N.Y., U.S.A. **151 B9** 44 12N 75 37W
Antwerp, Ohio, U.S.A. **157 C12** 41 11N 84 45W
Antwerpen, Belgium **24 C4** 51 13N 4 25 E
Antwerpen □, Belgium **24 C4** 51 15N 4 40 E
Anupgarh, India **92 E5** 29 10N 73 10 E
Anuradhapura, Sri Lanka **95 K5** 8 22N 80 28 E
Anveh, Iran **97 E7** 27 23N 54 11 E
Anvers = Antwerpen, Belgium **24 C4** 51 13N 4 25 E
Anvers I., Antarctica **7 C17** 64 30 S 63 40W
Anxi, Fujian, China **77 E12** 25 2N 118 12 E
Anxi, Gansu, China **72 B4** 40 30N 95 43 E
Anxiang, China **77 C9** 29 27N 112 11 E
Anxious B., Australia **127 E1** 33 24 S 134 45 E
Anyama, Ivory C. **112 D4** 5 30N 4 3W
Anyang, China **74 F8** 36 5N 114 21 E
Anyer, Indonesia **84 D3** 6 4 S 105 53 E
Anyi, Jiangxi, China **77 C10** 28 49N 115 25 E
Anyi, Shanxi, China **74 G6** 35 2N 111 2 E
Anyuan, China **77 E10** 25 9N 115 21 E
Anza, U.S.A. **161 M10** 33 35N 116 39W
Anzawr, Oman **99 C6** 17 28N 52 50 E
Anze, China **74 F7** 36 10N 112 12 E
Anzhero-Sudzhensk, Russia **64 D9** 56 10N 86 0 E
Ánzio, Italy **46 A5** 41 27N 12 37 E
Anzoátegui □, Venezuela **169 B5** 9 0N 64 30W
Aoba, Vanuatu **133 E5** 15 25 S 167 50 E
Aoga-Shima, Japan **71 E11** 32 28N 139 46 E
Aoiz, Spain **40 C3** 42 46N 1 22W
Aomori, Japan **68 D10** 40 45N 140 45 E
Aomori □, Japan **68 D10** 40 45N 140 40 E
Aonla, India **93 E8** 28 16N 79 11 E

186 Ba Don

B

Ba Don, *Vietnam* **86 D6** 17 45N 106 26 E
Ba Dong, *Vietnam* **87 H6** 9 40N 106 33 E
Ba Ngoi = Cam Lam, *Vietnam* **87 G7** 11 54N 109 10 E
Ba Tri, *Vietnam* **87 G6** 10 2N 106 36 E
Ba Xian, *China* **74 E9** 39 8N 116 22 E
Baa, *Indonesia* **82 D2** 10 50 S 123 0 E
Baaba, I., *N. Cal.* **133 T18** 20 3 S 164 59 E
Baamonde, *Spain* **42 B3** 43 7N 7 44W
Baao, *Phil.* **80 E4** 13 27N 123 22 E
Baar, *Switz.* **33 B7** 47 12N 8 32 E
Baarle-Nassau, *Belgium* **24 C4** 51 27N 4 56 E
Bab el Mandeb, *Red Sea* **98 D3** 12 35N 43 25 E
Baba, *Bulgaria* **50 D7** 42 44N 23 59 E
Baba Burnu, *Turkey* **49 B8** 39 29N 26 2 E
Baba dag, *Azerbaijan* **61 K9** 41 0N 48 19 E
Bābā Kalū, *Iran* **97 D6** 30 7N 50 49 E
Babaçulândia, *Brazil* **170 C2** 7 13 S 47 46W
Babadag, *Romania* **53 F13** 44 53N 28 44 E
Babadağ, *Turkey* **49 D10** 37 49N 28 52 E
Babadayhan, *Turkmenistan* **64 F7** 37 42N 60 23 E
Babaeski, *Turkey* **51 E11** 41 26N 27 6 E
Babahoyo, *Ecuador* **168 D2** 1 40 S 79 30W
Babak, *Phil.* **81 H5** 7 8N 125 41 E
Babakin, *Australia* **125 F2** 32 7 S 118 1 E
Babana, *Nigeria* **113 C5** 10 31N 3 46 E
Babar, *Algeria* **111 A6** 35 10N 7 6 E
Babar, *Indonesia* **83 C3** 8 0 S 129 30 E
Babar, *Pakistan* **92 D3** 31 7N 69 32 E
Babarkach, *Pakistan* **92 E3** 29 45N 68 0 E
Babayevo, *Russia* **58 C8** 59 24N 35 55 E
Babb, *U.S.A.* **158 B7** 48 51N 113 27W
Babenhausen, *Germany* **31 F4** 49 57N 8 57 E
Băbeni, *Romania* **53 F9** 44 59N 24 11 E
Babi Besar, Pulau, *Malaysia* **87 L4** 2 25N 103 59 E
Babia Gora, *Europe* **55 J6** 49 38N 19 38 E
Babian Jiang →, *China* **76 F3** 22 55N 101 47 E
Babile, *Ethiopia* **107 F5** 9 16N 42 11 E
Babimost, *Poland* **55 F2** 52 10N 15 49 E
Babinda, *Australia* **126 B4** 17 20 S 145 56 E
Babine, *Canada* **142 B3** 55 22N 126 37W
Babine →, *Canada* **142 B3** 55 45N 127 44W
Babine L., *Canada* **142 C3** 54 48N 126 0W
Babo, *Indonesia* **83 B4** 2 30 S 133 30 E
Babócsa, *Hungary* **52 D2** 46 2N 17 21 E
Bābol, *Iran* **97 B7** 36 40N 52 50 E
Bābol Sar, *Iran* **97 B7** 36 45N 52 45 E
Baborów, *Poland* **55 H5** 50 7N 18 1 E
Baboua, *C.A.R.* **114 A2** 5 49N 14 58 E
Babruysk, *Belarus* **59 F5** 53 10N 29 15 E
Babson Park, *U.S.A.* **153 H8** 27 49N 81 32W
Babuna, *Macedonia* **50 E5** 41 30N 21 40 E
Babura, *Nigeria* **113 C6** 12 51N 8 59 E
Babusar Pass, *Pakistan* **93 B5** 35 12N 73 59 E
Babušnica, *Serbia, Yug.* **50 C6** 43 7N 22 27 E
Babuyan Chan., *Phil.* **80 B3** 18 40N 121 30 E
Babuyan I., *Phil.* **80 B3** 19 32N 121 57 E
Babuyan Is., *Phil.* **80 B3** 19 15N 121 40 E
Babylon, *Iraq* **101 F11** 32 34N 44 22 E
Bač, *Serbia, Yug.* **52 E4** 45 29N 19 17 E
Bâc →, *Moldova* **53 D14** 46 55N 29 26 E
Bac Can, *Vietnam* **86 A5** 22 8N 105 49 E
Bac Giang, *Vietnam* **86 B6** 21 16N 106 11 E
Bac Ninh, *Vietnam* **86 B6** 21 13N 106 4 E
Bac Phan, *Vietnam* **86 B5** 22 0N 105 0 E
Bac Quang, *Vietnam* **86 A5** 22 30N 104 48 E
Bacabal, *Brazil* **170 B3** 4 15 S 44 45W
Bacacay, *Phil.* **80 E4** 13 18N 123 47 E
Bacajá →, *Brazil* **169 D7** 3 25 S 51 50W
Bacalar, *Mexico* **163 D7** 18 50N 87 27W
Bacan, Kepulauan, *Indonesia* **82 B3** 0 35 S 127 30 E
Bacan, Pulau, *Indonesia* **82 B3** 0 50 S 127 30 E
Bacarra, *Phil.* **80 B3** 18 15N 120 37 E
Bacău, *Romania* **53 D11** 46 35N 26 55 E
Bacău □, *Romania* **53 D11** 46 30N 26 45 E
Baccarat, *France* **27 D13** 48 28N 6 42 E
Bacchus Marsh, *Australia* **128 D6** 37 43 S 144 27 E
Bacerac, *Mexico* **162 A3** 30 18N 108 50W
Băceşti, *Romania* **53 D12** 46 50N 27 11 E
Bach Long Vi, Dao, *Vietnam* **86 B6** 20 10N 107 40 E
Bachaquero, *Venezuela* **168 B3** 9 56N 71 8W
Bacharach, *Germany* **31 E3** 50 3N 7 44 E
Bachelina, *Russia* **64 D7** 57 45N 67 20 E
Bachuma, *Ethiopia* **107 F4** 6 48N 35 53 E
Bačina, *Serbia, Yug.* **50 C5** 43 42N 21 23 E
Back →, *Canada* **138 B9** 65 10N 104 0W
Bačka Palanka, *Serbia, Yug.* **52 E4** 45 17N 19 27 E
Bačka Topola, *Serbia, Yug.* **52 E4** 45 49N 19 39 E
Bäckebo, *Sweden* **17 H10** 56 53N 16 4 E
Bäckefors, *Sweden* **17 F6** 58 48N 12 9 E
Bäckhammar, *Sweden* **16 E8** 59 10N 14 11 E
Bački Petrovac, *Serbia, Yug.* **52 E4** 45 29N 19 32 E
Backnang, *Germany* **31 G5** 48 56N 9 26 E
Backstairs Passage, *Australia* **128 C3** 35 40 S 138 5 E
Baco, Mt., *Phil.* **80 E3** 12 49N 121 10 E
Bacolod, *Phil.* **81 F4** 10 40N 122 57 E
Bacon, *Phil.* **80 E5** 13 3N 124 3 E
Baconton, *U.S.A.* **152 D5** 31 23N 84 10W
Bacoor, *Phil.* **80 D3** 14 28N 120 56 E
Bacqueville-en-Caux, *France* **26 C8** 49 47N 1 0 E
Bács-Kiskun □, *Hungary* **52 D4** 46 43N 19 30 E
Bácsalmás, *Hungary* **52 D4** 46 8N 19 17 E
Bacuag, *Phil.* **81 G5** 9 36N 125 38 E
Bacuk, *Malaysia* **87 J4** 6 4N 102 25 E
Baculin, *Phil.* **81 H6** 7 27N 126 35 E
Bād, *Iran* **97 C7** 33 41N 52 1 E
Bad →, *U.S.A.* **154 C4** 44 21N 100 22W
Bad Aussee, *Austria* **34 D6** 47 43N 13 45 E
Bad Axe, *U.S.A.* **150 C2** 43 48N 83 0W
Bad Bergzabern, *Germany* **31 F3** 49 6N 7 59 E
Bad Berleburg, *Germany* **30 D4** 51 2N 8 26 E
Bad Bevensen, *Germany* **30 B6** 53 5N 10 35 E
Bad Bramstedt, *Germany* **30 B5** 53 55N 9 53 E
Bad Brückenau, *Germany* **31 E5** 50 18N 9 47 E
Bad Doberan, *Germany* **30 A7** 54 6N 11 53 E
Bad Driburg, *Germany* **30 D5** 51 43N 9 1 E
Bad Ems, *Germany* **31 E3** 50 20N 7 43 E
Bad Frankenhausen, *Germany* **30 D7** 51 21N 11 5 E
Bad Freienwalde, *Germany* **30 C10** 52 46N 14 1 E
Bad Goisern, *Austria* **34 D6** 47 38N 13 38 E
Bad Harzburg, *Germany* **30 D6** 51 52N 10 34 E
Bad Hersfeld, *Germany* **30 E5** 50 52N 9 42 E
Bad Hofgastein, *Austria* **34 D6** 47 17N 13 6 E
Bad Homburg, *Germany* **31 E4** 50 13N 8 38 E
Bad Honnef, *Germany* **30 E3** 50 38N 7 13 E
Bad Iburg, *Germany* **30 C4** 52 10N 8 3 E
Bad Ischl, *Austria* **34 D6** 47 44N 13 38 E
Bad Kissingen, *Germany* **31 E6** 50 11N 10 4 E
Bad Königshofen, *Germany* **31 E6** 50 17N 10 28 E
Bad Kreuznach, *Germany* **31 F3** 49 50N 7 51 E
Bad Krozingen, *Germany* **31 H3** 47 54N 7 42 E
Bad Laasphe, *Germany* **30 E4** 50 56N 8 25 E
Bad Lands, *U.S.A.* **154 D3** 43 40N 102 10W
Bad Langensalza, *Germany* **30 D6** 51 5N 10 38 E
Bad Lauterberg, *Germany* **30 D6** 51 38N 10 28 E
Bad Leonfelden, *Austria* **34 C7** 48 31N 14 18 E
Bad Liebenwerda, *Germany* **30 D9** 51 31N 13 24 E
Bad Mergentheim, *Germany* **31 F5** 49 28N 9 42 E
Bad Münstereifel, *Germany* **30 E2** 50 33N 6 46 E
Bad Nauheim, *Germany* **31 E4** 50 21N 8 43 E
Bad Neuenahr-Ahrweiler, *Germany* **30 E3** 50 32N 7 5 E
Bad Neustadt, *Germany* **31 E6** 50 18N 10 13 E
Bad Oeynhausen, *Germany* **30 C4** 52 12N 8 46 E
Bad Oldesloe, *Germany* **30 B6** 53 48N 10 22 E
Bad Orb, *Germany* **31 E5** 50 12N 9 22 E
Bad Pyrmont, *Germany* **30 D5** 51 59N 9 16 E
Bad Ragaz, *Switz.* **33 C9** 47 0N 9 30 E
Bad Reichenhall, *Germany* **31 H8** 47 43N 12 54 E
Bad Säckingen, *Germany* **31 H3** 47 33N 7 56 E
Bad Salzuflen, *Germany* **30 C4** 52 5N 8 45 E
Bad Salzungen, *Germany* **30 E6** 50 48N 10 14 E
Bad Schwartau, *Germany* **30 B6** 53 55N 10 41 E
Bad Segeberg, *Germany* **30 B6** 53 56N 10 17 E
Bad St. Leonhard, *Austria* **34 E7** 46 58N 14 47 E
Bad Tölz, *Germany* **31 H7** 47 45N 11 34 E
Bad Urach, *Germany* **31 G5** 48 29N 9 23 E
Bad Vöslau, *Austria* **35 D9** 47 58N 16 12 E
Bad Waldsee, *Germany* **31 H5** 47 55N 9 45 E
Bad Wildungen, *Germany* **30 D5** 51 6N 9 7 E
Bad Wimpfen, *Germany* **31 F5** 49 13N 9 11 E
Bad Windsheim, *Germany* **31 F6** 49 36N 10 25 E
Bad Zwischenahn, *Germany* **30 B4** 53 12N 8 1 E
Badagara, *India* **95 J2** 11 35N 75 40 E
Badagri, *Nigeria* **113 D5** 6 25N 2 55 E
Badajós, L., *Brazil* **169 D5** 3 15 S 62 50W
Badajoz, *Spain* **43 G4** 38 50N 6 59W
Badajoz □, *Spain* **43 G4** 38 40N 6 30W
Badakhshān □, *Afghan.* **91 A3** 36 30N 71 0 E
Badalona, *Spain* **40 D7** 41 26N 2 15 E
Badalzai, *Afghan.* **92 E1** 29 50N 65 35 E
Badampahar, *India* **94 C8** 22 10N 86 10 E
Badanah, *Si. Arabia* **96 D4** 30 58N 41 30 E
Badarinath, *India* **93 D8** 30 45N 79 30 E
Badas, Kepulauan, *Indonesia* **84 B3** 0 45N 107 5 E
Baddo →, *Pakistan* **91 D2** 28 0N 64 20 E
Bade, *Indonesia* **83 C5** 7 10 S 139 35 E
Baden, *Austria* **35 C9** 48 1N 16 13 E
Baden, *Switz.* **33 B6** 47 28N 8 18 E
Baden-Baden, *Germany* **31 G4** 48 44N 8 13 E
Baden Park, *Australia* **128 B6** 32 8 S 144 12 E
Baden-Württemberg □, *Germany* **31 G4** 48 20N 8 40 E
Badgastein, *Austria* **34 D6** 47 7N 13 9 E
Badger, *Canada* **141 C8** 49 0N 56 4W
Badger, *U.S.A.* **160 J7** 36 38N 119 1W
Bādghīsāt □, *Afghan.* **91 B1** 35 0N 63 0 E
Badgom, *India* **93 B6** 34 1N 74 45 E
Badia Polésine, *Italy* **45 C8** 45 5N 11 29 E
Badian, *Phil.* **81 G4** 9 55N 123 24 E
Badin, *Pakistan* **91 D3** 24 38N 68 54 E
Badnera, *India* **94 D3** 20 48N 77 44 E
Badoc, *Phil.* **80 C3** 17 56N 120 28 E
Badogo, *Mali* **112 C3** 11 2N 8 13W
Badong, *China* **77 B8** 31 1N 110 23 E
Badr Ḩunayn, *Si. Arabia* **98 B2** 23 44N 38 46 E
Badrah, *Iraq* **101 F11** 33 6N 45 58 E
Baduen, *Somali Rep.* **120 C3** 7 15N 47 40 E
Badulla, *Sri Lanka* **95 L5** 7 1N 81 7 E
Badupi, *Burma* **90 E4** 21 36N 93 27 E
Baena, *Spain* **43 H6** 37 37N 4 20W
Baerami Creek, *Australia* **129 B9** 32 27 S 150 27 E
Baeza, *Ecuador* **168 D2** 0 25 S 77 53W
Baeza, *Spain* **43 H7** 37 57N 3 25W
Bafang, *Cameroon* **113 D7** 5 9N 10 11 E
Bafatá, *Guinea-Biss.* **112 C2** 12 8N 14 40W
Baffin B., *Canada* **6 B4** 72 0N 64 0W
Baffin I., *Canada* **139 B12** 68 0N 75 0W
Bafia, *Cameroon* **113 E7** 4 40N 11 10 E
Bafilo, *Togo* **113 D5** 9 22N 1 22 E
Bafing →, *Mali* **112 C2** 13 49N 10 50W
Bafliyūn, *Syria* **96 B3** 36 37N 36 59 E
Bafoulabé, *Mali* **112 C2** 13 50N 10 55W
Bafoussam, *Cameroon* **113 D7** 5 28N 10 25 E
Bāfq, *Iran* **97 D7** 31 40N 55 25 E
Bafra, *Turkey* **100 B6** 41 34N 35 54 E
Bafra Burnu, *Turkey* **100 B7** 41 45N 36 2 E
Bāft, *Iran* **97 D8** 29 15N 56 38 E
Bafut, *Cameroon* **113 D7** 6 6N 10 2 E
Bafwasende, *Dem. Rep. of the Congo* **118 B2** 1 3N 27 5 E
Bagabag, *Phil.* **80 C3** 16 30N 121 15 E
Bagac, *Phil.* **80 D3** 14 36N 120 23 E
Bagac Bay, *Phil.* **80 D3** 14 36N 120 20 E
Bagalkot, *India* **95 F2** 16 10N 75 40 E
Bagamoyo, *Tanzania* **118 D4** 6 28 S 38 55 E
Bagamoyo □, *Tanzania* **118 D4** 6 20 S 38 30 E
Bagan Datoh, *Malaysia* **87 L3** 3 59N 100 47 E
Bagan Serai, *Malaysia* **87 K3** 5 1N 100 32 E
Baganga, *Phil.* **81 H6** 7 34N 126 33 E
Bagani, *Namibia* **116 B3** 18 7 S 21 41 E
Bagansiapiapi, *Indonesia* **84 B2** 2 12N 100 50 E
Bagasra, *India* **92 J4** 21 30N 71 0 E
Bagata, *Dem. Rep. of the Congo* **114 C3** 3 44 S 17 57 E
Bagawi, *Sudan* **107 E3** 12 20N 34 18 E
Bagdad, *Calif., U.S.A.* **161 L11** 34 35N 115 53W
Bagdad, *Fla., U.S.A.* **153 E2** 30 36N 87 2W
Bagdarin, *Russia* **65 D12** 54 26N 113 36 E
Bagé, *Brazil* **175 C5** 31 20 S 54 15W
Bagenalstown = Muine Bheag, *Ireland* **23 D5** 52 42N 6 58W
Baggs, *U.S.A.* **158 F10** 41 2N 107 39W
Bagh, *Pakistan* **93 C5** 33 59N 73 45 E
Baghdād, *Iraq* **101 F11** 33 20N 44 30 E
Bagherhat, *Bangla.* **90 D2** 22 40N 89 47 E
Bagheria, *Italy* **46 D6** 38 5N 13 30 E
Baghlān, *Afghan.* **91 A3** 36 12N 69 0 E
Baghlān □, *Afghan.* **91 B3** 36 0N 68 30 E
Bagley, *U.S.A.* **154 B7** 47 32N 95 24W
Bagn, *Norway* **18 D6** 60 49N 9 34 E
Bagnara Cálabra, *Italy* **47 D8** 38 17N 15 48 E
Bagnasco, *Italy* **44 D5** 44 18N 8 2 E
Bagnell Dam, *U.S.A.* **156 F4** 38 14N 92 36W
Bagnères-de-Bigorre, *France* **28 E4** 43 5N 0 9 E
Bagnères-de-Luchon, *France* **28 F4** 42 47N 0 38 E
Bagni di Lucca, *Italy* **44 D7** 44 1N 10 35 E
Bagno di Romagna, *Italy* **45 E8** 43 50N 11 57 E
Bagnoles-de-l'Orne, *France* **26 D6** 48 32N 0 25W
Bagnols-sur-Cèze, *France* **29 D8** 44 10N 4 36 E
Bagnorégio, *Italy* **45 F9** 42 37N 12 5 E
Bago, *Phil.* **81 F4** 10 32N 122 50 E
Bagotville, *Canada* **141 C5** 48 22N 70 54W
Bagrationovsk, *Russia* **15 J19** 54 23N 20 39 E
Bagrdan, *Serbia, Yug.* **50 B5** 44 5N 21 11 E
Bagua, *Peru* **172 B2** 5 35 S 78 22W
Baguio, *Phil.* **80 C3** 16 26N 120 34 E
Bağyurdu, *Turkey* **49 C9** 38 25N 27 41 E
Bahabón de Esgueva, *Spain* **42 D7** 41 52N 3 43W
Bahadurabad Ghat, *Bangla.* **90 C2** 25 11N 89 44 E
Bahadurgarh, *India* **92 E7** 28 40N 76 57 E
Bahama, Canal Viejo de, *W. Indies* **164 B4** 22 10N 77 30W
Bahamas ■, *N. Amer.* **165 B5** 24 0N 75 0W
Bahār, *Iran* **101 E13** 34 54N 48 26 E
Baharampur, *India* **93 G13** 24 2N 88 27 E
Baharîya, El Wâhât al, *Egypt* **106 J6** 28 0N 28 50 E
Bahawalnagar, *Pakistan* **91 C4** 30 0N 73 15 E
Bahawalpur, *Pakistan* **91 C3** 29 24N 71 40 E
Bahçe, *Turkey* **100 D7** 37 13N 36 34 E
Bahçecik, *Turkey* **51 F13** 40 41N 29 44 E
Baheri, *India* **93 E8** 28 45N 79 34 E
Bahi, *Tanzania* **118 D4** 5 58 S 35 21 E
Bahi Swamp, *Tanzania* **118 D4** 6 10 S 35 0 E
Bahía = Salvador, *Brazil* **171 D4** 13 0 S 38 30W
Bahía □, *Brazil* **171 D3** 12 0 S 42 0W
Bahía, Is. de la, *Honduras* **164 C2** 16 45N 86 15W
Bahía Blanca, *Argentina* **174 D3** 38 35 S 62 13W
Bahía de Caráquez, *Ecuador* **168 D1** 0 40 S 80 27W
Bahía Honda, *Cuba* **164 B3** 22 54N 83 10W
Bahía Laura, *Argentina* **176 C3** 48 10 S 66 30W
Bahía Negra, *Paraguay* **173 E6** 20 5 S 58 5W
Bahir Dar, *Ethiopia* **107 E4** 11 37N 37 10 E
Bahlah, *Oman* **99 B7** 22 58N 57 18 E
Bahmanzād, *Iran* **97 D6** 31 15N 51 47 E
Bahmer, *Algeria* **111 C4** 27 32N 0 10W
Bahr Aouk →, *C.A.R.* **114 A3** 8 40N 19 0 E
Bahr el Ahmar □, *Sudan* **106 D4** 20 0N 35 0 E
Bahr el Ghazâl □, *Sudan* **107 F2** 7 0N 28 0 E
Bahr Salamat →, *Chad* **109 G3** 9 20N 18 0 E
Bahr Yûsef →, *Egypt* **106 J7** 28 25N 30 35 E
Bahra el Burullus, *Egypt* **106 H7** 31 28N 30 48 E
Bahraich, *India* **93 F9** 27 38N 81 37 E
Bahrain ■, *Asia* **97 E6** 26 0N 50 35 E
Bahror, *India* **92 F7** 27 51N 76 20 E
Bāhū Kalāt, *Iran* **97 E9** 25 43N 61 25 E
Bai, *Mali* **112 C4** 13 35N 3 28W
Bai Bung, Mui = Ca Mau, Mui, *Vietnam* **87 H5** 8 38N 104 44 E
Bai Duc, *Vietnam* **86 C5** 18 3N 105 49 E
Bai Thuong, *Vietnam* **86 C5** 19 54N 105 23 E
Baia de Aramă, *Romania* **52 E7** 45 0N 22 50 E
Baía Farta, *Angola* **115 E2** 12 40 S 13 11 E
Baia Mare, *Romania* **53 C8** 47 40N 23 35 E
Baia-Sprie, *Romania* **53 C8** 47 41N 23 43 E
Baião, *Brazil* **170 B2** 2 40 S 49 40W
Baïbokoum, *Chad* **109 G3** 7 46N 15 43 E
Baicheng, *China* **75 B12** 45 38N 122 42 E
Băicoi, *Romania* **53 E10** 45 3N 25 52 E
Baidoa, *Somali Rep.* **120 D2** 3 8N 43 30 E
Baie Comeau, *Canada* **141 C6** 49 12N 68 10W
Baie-St-Paul, *Canada* **141 C5** 47 28N 70 32W
Baie Trinité, *Canada* **141 C6** 49 25N 67 20W
Baie Verte, *Canada* **141 C8** 49 55N 56 12W
Baignes-Ste-Radegonde, *France* **28 C3** 45 23N 0 25W
Baigneux-les-Juifs, *France* **27 E11** 47 31N 4 39 E
Baihe, *China* **74 H6** 32 50N 110 5 E
Ba'ījī, *Iraq* **101 E10** 35 0N 43 30 E
Baikal, L. = Baykal, Oz., *Russia* **65 D11** 53 0N 108 0 E
Bailadila, Mt., *India* **94 E5** 18 43N 81 15 E
Baile Atha Cliath = Dublin, *Ireland* **23 C5** 53 21N 6 15W
Băile Govora, *Romania* **53 E9** 45 5N 24 11 E
Băile Herculane, *Romania* **52 F7** 44 53N 22 26 E
Băile Olăneşti, *Romania* **53 E9** 45 12N 24 14 E
Băile Tuşnad, *Romania* **53 D10** 46 9N 25 51 E
Bailei, *Ethiopia* **107 F5** 6 44N 40 18 E
Bailén, *Spain* **43 G7** 38 8N 3 48W
Băileşti, *Romania* **53 F8** 44 1N 23 20 E
Bailhongal, *India* **95 G2** 15 55N 74 53 E
Bailique, Ilha, *Brazil* **170 A2** 1 2N 49 58W
Bailundo, *Angola* **115 E3** 12 10 S 15 50 E
Baima, *China* **76 A3** 33 0N 100 26 E
Baimuru, *Papua N. G.* **132 D3** 7 35 S 144 51 E
Bain-de-Bretagne, *France* **26 E5** 47 50N 1 40W
Bainbridge, *Ga., U.S.A.* **152 E5** 30 55N 84 35W
Bainbridge, *Ind., U.S.A.* **157 E10** 39 46N 86 49W
Bainbridge, *N.Y., U.S.A.* **151 D9** 42 18N 75 29W
Bainbridge, *Ohio, U.S.A.* **157 E13** 39 14N 83 16W
Baing, *Indonesia* **82 D2** 10 14 S 120 34 E
Bainiu, *China* **74 H7** 32 50N 112 15 E
Bainville, *U.S.A.* **154 A2** 48 8N 104 13W
Bainyik, *Papua N. G.* **132 B2** 3 40 S 143 4 E
Baiona, *Spain* **42 C2** 42 6N 8 52W
Bā'ir, *Jordan* **103 E5** 30 45N 36 55 E
Baird, *U.S.A.* **155 J5** 32 24N 99 24W
Bairin Youqi, *China* **75 C10** 43 30N 118 35 E
Bairin Zuoqi, *China* **75 C10** 43 58N 119 15 E
Bairnsdale, *Australia* **129 D7** 37 48 S 147 36 E
Bais, *Phil.* **81 G4** 9 35N 123 7 E
Baisha, *China* **74 G7** 34 20N 112 32 E
Baissa, *Nigeria* **113 D7** 7 14N 10 38 E
Baitadi, *Nepal* **93 E9** 29 35N 80 25 E
Baixa Grande, *Brazil* **171 D3** 11 57 S 40 11W
Baiyin, *China* **74 F3** 36 45N 104 14 E
Baiyü, *China* **76 B2** 31 16N 98 50 E
Baiyu Shan, *China* **74 F4** 37 15N 107 30 E
Baiyuda, *Sudan* **106 D3** 17 35N 32 7 E
Baj Baj, *India* **93 H13** 22 30N 88 5 E
Baja, *Hungary* **52 D3** 46 12N 18 59 E
Baja, Pta., *Mexico* **162 B1** 29 50N 116 0W
Baja California, *Mexico* **162 A1** 31 10N 115 12W
Baja California □, *Mexico* **162 B2** 30 0N 115 0W
Baja California Sur □, *Mexico* **162 B2** 25 50N 111 50W
Bajamar, *Canary Is.* **39 F3** 28 33N 16 20W
Bajana, *India* **92 H4** 23 7N 71 49 E
Bājgīrān, *Iran* **97 B8** 37 36N 58 24 E
Bājil, *Yemen* **98 D3** 15 4N 43 17 E
Bajimba, Mt., *Australia* **127 D5** 29 17 S 152 6 E
Bajina Bašta, *Serbia, Yug.* **50 C3** 43 58N 19 35 E
Bajmok, *Serbia, Yug.* **52 E4** 45 57N 19 24 E
Bajo Nuevo, *Caribbean* **164 C4** 15 40N 78 50W
Bajoga, *Nigeria* **113 C7** 10 57N 11 20 E
Bajool, *Australia* **126 C5** 23 40 S 150 35 E
Bak, *Hungary* **52 D1** 46 43N 16 51 E
Bakal, *Russia* **62 D7** 54 56N 58 48 E
Bakala, *C.A.R.* **114 A4** 6 15N 20 20 E
Bakanas, *Kazakstan* **63 A8** 44 50N 76 15 E
Bakar, *Croatia* **45 C11** 45 18N 14 32 E
Bakbakty, *Kazakstan* **63 A8** 44 35N 76 40 E
Bakel, *Senegal* **112 C2** 14 56N 12 20W
Baker, *Calif., U.S.A.* **161 K10** 35 16N 116 4W
Baker, *Fla., U.S.A.* **153 E3** 30 48N 86 41W
Baker, *Mont., U.S.A.* **154 B2** 46 22N 104 17W
Baker, *Oreg., U.S.A.* **158 D5** 44 47N 117 50W
Baker, Canal, *Chile* **176 C2** 47 45 S 74 45W
Baker, L., *Canada* **138 B10** 64 0N 96 0W
Baker Hill, *U.S.A.* **152 D4** 31 47N 85 18W
Baker I., *Pac. Oc.* **134 G10** 0 10N 176 35W
Baker L., *Australia* **125 E4** 26 54 S 126 5 E
Baker Lake, *Canada* **138 B10** 64 20N 96 3W
Baker Mt., *U.S.A.* **158 B3** 48 50N 121 49W
Bakers Creek, *Australia* **126 C4** 21 13 S 149 7 E
Baker's Dozen Is., *Canada* **140 A4** 56 45N 78 45W
Bakersfield, *Calif., U.S.A.* **161 K8** 35 23N 119 1W
Bakersfield, *Vt., U.S.A.* **151 B12** 44 45N 72 48W
Bakhchysaray, *Ukraine* **59 K7** 44 40N 33 45 E
Bakhmach, *Ukraine* **59 G7** 51 10N 32 45 E
Bākhtarān, *Iran* **101 E12** 34 23N 47 0 E
Bākhtarān □, *Iran* **96 C5** 34 0N 46 30 E
Bakı, *Azerbaijan* **61 K9** 40 29N 49 56 E
Bakır →, *Turkey* **49 C9** 38 55N 27 0 E
Bakırdaği, *Turkey* **100 C6** 38 13N 35 46 E
Bakkafjörður, *Iceland* **11 A12** 66 2N 14 48W
Bakkaflói, *Iceland* **11 A12** 66 10N 14 45W
Bakkasel, *Iceland* **11 B8** 65 28N 18 40W
Baklan, *Turkey* **49 C11** 38 0N 29 36 E
Bakony, *Hungary* **52 C2** 47 10N 17 30 E
Bakony Forest = Bakony, *Hungary* **52 C2** 47 10N 17 30 E
Bakori, *Nigeria* **113 C6** 11 34N 7 25 E
Bakouma, *C.A.R.* **114 A4** 5 40N 22 56 E
Bakpakty = Bakbakty, *Kazakstan* **63 A8** 44 35N 76 40 E
Baksan, *Russia* **61 J6** 43 42N 43 32 E
Baku = Bakı, *Azerbaijan* **61 K9** 40 29N 49 56 E
Bakun, *Phil.* **80 C3** 16 48N 120 40 E
Bakutis Coast, *Antarctica* **7 D15** 74 0 S 120 0W
Bakwa-Kenge, *Dem. Rep. of the Congo* **115 C4** 4 51 S 22 4 E
Baky = Bakı, *Azerbaijan* **61 K9** 40 29N 49 56 E
Bala, *Canada* **150 A5** 45 1N 79 37W
Bâlâ, *Turkey* **100 C5** 39 32N 33 6 E
Bala, *U.K.* **20 E4** 52 54N 3 36W
Bala, L., *U.K.* **20 E4** 52 53N 3 37W
Bālā Morghāb, *Afghan.* **91 B1** 35 35N 63 20 E
Balabac I., *Phil.* **81 H1** 8 0N 117 0 E
Balabagh, *Afghan.* **92 B4** 34 25N 70 12 E
Ba'labakk, *Lebanon* **103 B5** 34 0N 36 10 E
Balabalangan, Kepulauan, *Indonesia* **85 C5** 2 20 S 117 30 E
Balabio, I., *N. Cal.* **133 T18** 20 7 S 164 11 E
Bălăciţa, *Romania* **53 F8** 44 23N 23 8 E
Balad, *Iraq* **101 F11** 34 1N 44 9 E
Balad Rūz, *Iraq* **101 F11** 33 42N 45 5 E
Bālādeh, *Fārs, Iran* **97 D6** 29 17N 51 56 E
Bālādeh, *Māzandaran, Iran* **97 B6** 36 12N 51 48 E
Balaghat, *India* **94 D5** 21 49N 80 12 E
Balaghat Ra., *India* **94 E3** 18 50N 76 30 E
Balaguer, *Spain* **40 D5** 41 50N 0 50 E
Balakété, *C.A.R.* **114 A3** 6 56N 19 54 E
Balakhna, *Russia* **60 B6** 56 25N 43 32 E
Balaklava, *Australia* **128 C3** 34 7 S 138 22 E
Balaklava, *Ukraine* **59 K7** 44 30N 33 30 E
Balakliya, *Ukraine* **59 H9** 49 28N 36 55 E
Balakovo, *Russia* **60 D8** 52 4N 47 55 E
Balamban, *Phil.* **81 F4** 10 30N 123 43 E
Balambangan, *Malaysia* **85 A5** 7 17N 116 55 E
Bălan, *Romania* **53 D10** 46 39N 25 49 E
Balancán, *Mexico* **163 D6** 17 48N 91 32W
Balangiga, *Phil.* **81 F5** 11 7N 125 23 E
Balangir, *India* **94 D6** 20 43N 83 35 E
Balapur, *India* **94 D3** 20 40N 76 45 E
Balashov, *Russia* **60 E6** 51 30N 43 10 E
Balasinor, *India* **92 H5** 22 57N 73 23 E
Balasore = Baleshwar, *India* **94 D8** 21 35N 87 3 E
Balassagyarmat, *Hungary* **52 B4** 48 4N 19 15 E
Balât, *Egypt* **106 B2** 25 36N 29 19 E
Balaton, *Hungary* **52 D2** 46 50N 17 40 E
Balatonboglár, *Hungary* **52 D2** 46 46N 17 40 E
Balatonfüred, *Hungary* **52 D2** 46 58N 17 54 E
Balatonszentgyörgy, *Hungary* **52 D2** 46 41N 17 19 E
Balayan, *Phil.* **80 E3** 13 57N 120 44 E
Balazote, *Spain* **41 G2** 38 54N 2 9W
Balbalan, *Phil.* **80 C3** 17 27N 121 12 E
Balbi, Mt., *Papua N. G.* **132 C8** 5 55 S 154 58 E
Balbieriškis, *Lithuania* **54 D10** 54 32N 23 53 E
Balbigny, *France* **29 C8** 45 49N 4 11 E
Balbina, Reprêsa de, *Brazil* **169 D6** 2 0 S 59 30W
Balboa, *Panama* **164 E4** 8 57N 79 34W
Balbriggan, *Ireland* **23 C5** 53 37N 6 11W
Balcarce, *Argentina* **174 D4** 38 0 S 58 10W
Balcarres, *Canada* **143 C8** 50 50N 103 35W
Bălceşti, *Romania* **53 F8** 44 37N 23 57 E
Balchik, *Bulgaria* **51 C12** 43 28N 28 11 E
Balclutha, *N.Z.* **131 G4** 46 15 S 169 45 E
Balçova, *Turkey* **49 C9** 38 22N 27 4 E
Bald Hd., *Australia* **125 G2** 35 6 S 118 1 E
Bald I., *Australia* **125 F2** 34 57 S 118 27 E
Bald Knob, *U.S.A.* **155 H9** 35 19N 91 34W
Baldock L., *Canada* **143 B9** 56 33N 97 57W
Baldwin, *Fla., U.S.A.* **152 E8** 30 18N 81 59W
Baldwin, *Mich., U.S.A.* **148 D3** 43 54N 85 51W
Baldwinsville, *U.S.A.* **151 C8** 43 10N 76 20W
Baldy Peak, *U.S.A.* **159 K9** 33 54N 109 34W
Bale, *Croatia* **45 C10** 45 4N 13 46 E
Bale □, *Ethiopia* **107 F5** 6 20N 41 30 E
Baleares, Is., *Spain* **39 B10** 39 30N 3 0 E
Balearic Is. = Baleares, Is., *Spain* **39 B10** 39 30N 3 0 E
Baleia, Pta. da, *Brazil* **171 E4** 17 40 S 39 7W
Baleine = Whale →, *Canada* **141 A6** 58 15N 67 40W
Băleni, *Romania* **53 E12** 45 48N 27 51 E
Baler, *Phil.* **80 D3** 15 46N 121 34 E
Baler Bay, *Phil.* **80 D3** 15 50N 121 35 E
Balerna, *Switz.* **33 E8** 45 52N 9 0 E
Baleshare, *U.K.* **22 D1** 57 31N 7 22W
Baleshwar, *India* **94 D8** 21 35N 87 3 E
Balestrand, *Norway* **18 C3** 61 11N 6 31 E
Balezino, *Russia* **60 B11** 58 2N 53 6 E
Balfate, *Honduras* **164 C2** 15 48N 86 25W
Balfe's Creek, *Australia* **126 C4** 20 12 S 145 55 E

Bardejov 187

Balfour Channel, Solomon Is. 133 M9 8 43 S 157 27 E
Balharshah, India 94 E4 19 50N 79 23 E
Bali, Cameroon 113 D7 5 54N 10 0 E
Balí, Greece 38 D6 35 25N 24 47 E
Bali, Indonesia 85 D5 8 20 S 115 0 E
Bali □, Indonesia 85 D4 8 20 S 115 0 E
Bali, Selat, Indonesia 85 D4 8 18 S 114 25 E
Balicuatro Is., Phil. 80 E5 12 39N 124 24 E
Baligród, Poland 55 J9 49 20N 22 17 E
Balikeşir, Turkey 57 G4 39 39N 27 53 E
Balikeşir □, Turkey 49 B9 39 45N 28 0 E
Balıklıçeşme, Turkey 51 F11 40 18N 27 5 E
Balikpapan, Indonesia 85 C5 1 10 S 116 55 E
Balimbing, Phil. 81 J2 5 5N 119 58 E
Balimo, Papua N. G. 132 E2 8 6 S 142 57 E
Baling, Malaysia 87 K3 5 41N 100 55 E
Balingen, Germany 31 G4 48 16N 8 51 E
Balinţ, Romania 52 E6 45 48N 21 54 E
Balintang Channel, Phil. 80 B3 19 49N 121 40 E
Balintang Is., Phil. 80 B4 19 58N 122 9 E
Baliton, Phil. 81 J5 5 44N 125 14 E
Baliza, Brazil 173 D7 16 0 S 52 20W
Baljurshi, Si. Arabia 98 C3 19 51N 41 33 E
Balkan Mts. = Stara Planina, Bulgaria 50 C7 43 15N 23 0 E
Balkh □, Afghan. 91 A2 36 50N 67 0 E
Balkhash = Balqash, Kazakstan 64 E8 46 50N 74 50 E
Balkhash, Ozero = Balqash Köl, Kazakstan 64 E8 46 0N 74 50 E
Ballachulish, U.K. 22 E3 56 41N 5 8W
Balladonia, Australia 125 F3 32 27 S 123 51 E
Ballaghaderreen, Ireland 23 C3 53 55N 8 34W
Ballara, Australia 128 B4 32 19 S 140 45 E
Ballarat, Australia 127 F3 37 33 S 143 50 E
Ballard, L., Australia 125 E3 29 20 S 120 40 E
Ballater, U.K. 22 D5 57 3N 3 3W
Balldale, Australia 129 C7 35 50 S 146 33 E
Ballenas, Canal de, Mexico 162 B2 29 10N 113 45W
Balleny Is., Antarctica 7 C11 66 30 S 163 0 E
Balleroy, France 26 C6 49 11N 0 50W
Ballerup, Denmark 17 J6 55 44N 12 21 E
Ballesteros, Phil. 80 B3 18 25N 121 31 E
Ballı, Turkey 51 F11 40 50N 27 3 E
Ballia, India 93 G11 25 46N 84 12 E
Ballidu, Australia 125 F2 30 35 S 116 45 E
Ballina, Australia 127 D5 28 50 S 153 31 E
Ballina, Ireland 23 B2 54 7N 9 9W
Ballinasloe, Ireland 23 C3 53 20N 8 13W
Ballinger, U.S.A. 155 K5 31 45N 99 57W
Ballinrobe, Ireland 23 C2 53 38N 9 13W
Ballinskelligs B., Ireland 23 E1 51 48N 10 13W
Ballon, France 26 D7 48 10N 0 14 E
Ballsh, Albania 50 F3 40 36N 19 44 E
Ballycastle, U.K. 23 A5 55 12N 6 15W
Ballyclare, U.K. 23 B5 54 46N 6 0W
Ballyhaunis, Ireland 23 C3 53 46N 8 46W
Ballymena, U.K. 23 B5 54 52N 6 17W
Ballymoney, U.K. 23 A5 55 5N 6 31W
Ballymote, Ireland 23 B3 54 5N 8 31W
Ballynahinch, U.K. 23 B6 54 24N 5 54W
Ballyquintin Pt., U.K. 23 B6 54 20N 5 30W
Ballyshannon, Ireland 23 B3 54 30N 8 11W
Balmaceda, Chile 176 C2 46 0 S 71 50W
Balmaseda, Spain 40 B1 43 11N 3 12W
Balmazújváros, Hungary 52 C6 47 37N 21 21 E
Balmhorn, Switz. 32 D5 46 26N 7 42 E
Balmoral, Australia 128 D4 37 15 S 141 48 E
Balmorhea, U.S.A. 155 K3 30 59N 103 45W
Balombo, Angola 115 E2 12 21 S 14 46 E
Balonne →, Australia 127 D4 28 47 S 147 56 E
Balqash, Kazakstan 64 E8 46 50N 74 50 E
Balqash Köl, Kazakstan 64 E8 46 0N 74 50 E
Balrampur, India 93 F10 27 30N 82 20 E
Balranald, Australia 128 C5 34 38 S 143 33 E
Balş, Romania 53 F9 44 22N 24 5 E
Balsapuerto, Peru 172 B2 5 48 S 76 33W
Balsas, Mexico 163 D5 18 0N 99 40W
Balsas →, Maranhão, Brazil 170 C3 7 15 S 44 35W
Balsas →, Tocantins, Brazil 170 C2 9 58 S 47 52W
Balsas →, Mexico 162 D4 17 55N 102 10W
Bålsta, Sweden 16 E11 59 35N 17 30 E
Balsthal, Switz. 32 B5 47 19N 7 41 E
Balston Spa, U.S.A. 151 D11 43 0N 73 52W
Balta, Romania 52 F7 44 54N 22 38 E
Balta, Ukraine 59 H5 48 2N 29 45 E
Balta, U.S.A. 154 A4 48 10N 100 2W
Baltanás, Spain 42 D6 41 56N 4 15W
Bălţi, Moldova 53 C12 47 48N 27 58 E
Baltic Sea, Europe 15 H18 57 0N 19 0 E
Baltîm, Egypt 106 H7 31 35N 31 10 E
Baltimore, Ireland 23 E2 51 29N 9 22W
Baltimore, U.S.A. 148 F7 39 17N 76 37W
Baltit, Pakistan 93 A6 36 15N 74 40 E
Baltiysk, Russia 15 J18 54 41N 19 58 E
Baltrum, Germany 30 B3 53 43N 7 24 E
Baluchistan □, Pakistan 91 D2 27 30N 65 0 E
Balud, Phil. 80 E4 12 2N 123 12 E
Balurghat, India 93 G13 25 15N 88 44 E
Balvi, Latvia 15 H22 57 8N 27 15 E
Balya, Turkey 49 B9 39 44N 27 35 E
Balzar, Ecuador 168 D2 2 2 S 79 54W
Bam, Iran 97 D8 29 7N 58 14 E
Bama, China 76 E6 24 8N 107 12 E
Bama, Nigeria 113 C7 11 33N 13 41 E
Bamako, Mali 112 C3 12 34N 7 55W
Bamba, Dem. Rep. of the Congo 115 D3 5 45 S 18 23 E
Bamba, Mali 113 B4 17 5N 1 24W
Bambam, Phil. 80 D3 15 40N 120 20 E
Bambamarca, Peru 172 B2 6 36 S 78 32W
Bambang, Phil. 80 C3 16 23N 121 6 E
Bambari, C.A.R. 114 A4 5 40N 20 35 E
Bambaroo, Australia 126 B4 18 50 S 146 10 E
Bamberg, Germany 31 F6 49 54N 10 54 E
Bamberg, U.S.A. 152 B9 33 18N 81 2W
Bambesi, Ethiopia 107 F3 9 45N 34 40 E
Bambey, Senegal 112 C1 14 42N 16 28W
Bambili, Dem. Rep. of the Congo 118 B2 3 40N 26 0 E
Bambuí, Brazil 171 F2 20 1 S 45 58W
Bamenda, Cameroon 113 D7 5 57N 10 11 E
Bamfield, Canada 142 D3 48 45N 125 10W
Bāmīān □, Afghan. 91 B2 35 0N 67 0 E
Bamiancheng, China 75 C13 43 15N 124 2 E
Bamingui, C.A.R. 114 A4 7 34N 20 11 E
Bamkin, Cameroon 113 D7 6 3N 11 27 E
Bampūr, Iran 97 E9 27 15N 60 21 E
Ban Ban, Laos 86 C4 19 31N 103 30 E
Ban Bang Hin, Thailand 87 H2 9 32N 98 35 E
Ban Chiang Klang, Thailand 86 C3 19 25N 100 55 E
Ban Chik, Laos 86 D4 17 15N 102 22 E
Ban Choho, Thailand 86 E4 15 2N 102 9 E
Ban Dan Lan Hoi, Thailand 86 D2 17 0N 99 35 E
Ban Don = Surat Thani, Thailand 87 H2 9 6N 99 20 E
Ban Don, Vietnam 86 F6 12 53N 107 48 E
Ban Don, Ao →, Thailand 87 H2 9 20N 99 25 E
Ban Dong, Thailand 86 C3 19 30N 100 59 E
Ban Hong, Thailand 86 C2 18 18N 98 50 E
Ban Kaeng, Thailand 86 D3 17 29N 100 7 E
Ban Kantang, Thailand 87 J2 7 25N 99 31 E
Ban Keun, Laos 86 C4 18 22N 102 35 E
Ban Khai, Thailand 86 F3 12 46N 101 18 E
Ban Kheun, Laos 86 B3 20 13N 101 7 E
Ban Khlong Kua, Thailand 87 J3 6 57N 100 8 E
Ban Khuan Mao, Thailand 87 J2 7 50N 99 37 E
Ban Ko Yai Chim, Thailand 87 G2 11 17N 99 26 E
Ban Kok, Thailand 86 D4 16 40N 103 40 E
Ban Laem, Thailand 86 F2 13 13N 99 59 E
Ban Lao Ngam, Laos 86 E6 15 28N 106 10 E
Ban Le Kathe, Thailand 86 E2 15 49N 98 53 E
Ban Mae Chedi, Thailand 86 C2 19 11N 99 31 E
Ban Mae Laeng, Thailand 86 B2 20 1N 99 17 E
Ban Mae Sariang, Thailand 86 C1 18 10N 97 56 E
Ban Mê Thuôt = Buon Ma Thuot, Vietnam 86 F7 12 40N 108 3 E
Ban Mi, Thailand 86 E3 15 3N 100 32 E
Ban Muong Mo, Laos 86 C4 19 4N 103 58 E
Ban Na Mo, Laos 86 D5 17 7N 105 40 E
Ban Na San, Thailand 87 H2 8 53N 99 52 E
Ban Na Tong, Laos 86 B3 20 56N 101 47 E
Ban Nam Bac, Laos 86 B4 20 38N 102 20 E
Ban Nam Ma, Laos 86 A3 22 2N 101 37 E
Ban Ngang, Laos 86 E6 15 59N 106 11 E
Ban Nong Bok, Laos 86 D5 17 5N 104 48 E
Ban Nong Boua, Laos 86 E6 15 40N 106 33 E
Ban Nong Pling, Thailand 86 E3 15 40N 100 10 E
Ban Pak Chan, Thailand 87 G2 10 32N 98 51 E
Ban Phai, Thailand 86 D4 16 4N 102 44 E
Ban Pong, Thailand 86 F2 13 50N 99 55 E
Ban Ron Phibun, Thailand 87 H2 8 9N 99 51 E
Ban Sanam Chai, Thailand 87 J3 7 33N 100 25 E
Ban Sangkha, Thailand 86 E4 14 37N 103 52 E
Ban Tak, Thailand 86 D2 17 2N 99 4 E
Ban Tako, Thailand 86 E4 14 5N 102 40 E
Ban Tha Dua, Thailand 86 D2 17 59N 98 39 E
Ban Tha Li, Thailand 86 D3 17 37N 101 25 E
Ban Tha Nun, Thailand 87 H2 8 12N 98 18 E
Ban Thahine, Laos 86 E5 14 12N 105 33 E
Ban Xien Kok, Laos 86 B3 20 54N 100 39 E
Ban Yen Nhan, Vietnam 86 B6 20 57N 106 2 E
Banā, W. →, Yemen 98 D4 13 3N 45 24 E
Banaba, Kiribati 134 H8 0 45 S 169 50 E
Bañalbufar, Spain 39 B9 39 42N 2 31 E
Banalia, Dem. Rep. of the Congo 118 B2 1 32N 25 5 E
Banam, Cambodia 87 G5 11 20N 105 17 E
Banamba, Mali 112 C3 13 29N 7 22W
Banana, Australia 126 C5 24 28 S 150 8 E
Bananal, I. do, Brazil 171 D1 11 30 S 50 30W
Banaras = Varanasi, India 93 G10 25 22N 83 0 E
Banas →, Gujarat, India 92 H4 23 45N 71 25 E
Banas →, Mad. P., India 93 G9 24 15N 81 30 E
Bânâs, Ras, Egypt 106 C4 23 57N 35 59 E
Banaz, Turkey 49 C11 38 46N 29 46 E
Banaz →, Turkey 49 C11 38 12N 29 14 E
Banbān, Si. Arabia 96 E5 25 1N 46 35 E
Banbridge, U.K. 23 B5 54 22N 6 16W
Banbury, U.K. 21 E6 52 4N 1 20W
Banchory, U.K. 22 D6 57 3N 2 29W
Bancroft, Canada 140 C4 45 3N 77 51W
Band, Romania 53 D9 46 30N 24 25 E
Band Bonī, Iran 97 E8 25 30N 59 33 E
Band-e Torkestān, Afghan. 91 B2 35 30N 64 0 E
Band Qīr, Iran 97 D6 31 39N 48 53 E
Banda, Cameroon 114 B2 3 58N 14 32 E
Banda, India 93 G9 25 30N 80 26 E
Banda, Kepulauan, Indonesia 83 B3 4 37 S 129 50 E
Banda Aceh, Indonesia 84 A1 5 35N 95 20 E
Banda Banda, Mt., Australia 129 A10 31 10 S 152 28 E
Banda Elat, Indonesia 83 C4 5 40 S 133 5 E
Banda Is. = Banda, Kepulauan, Indonesia 83 B3 4 37 S 129 50 E
Banda Sea, Indonesia 82 C3 6 0 S 130 0 E
Bandai-San, Japan 68 F10 37 36N 140 4 E
Bandak, Norway 18 E5 59 23N 8 29 E
Bandama →, Ivory C. 112 D3 6 32N 5 30W
Bandān, Iran 97 D9 31 23N 60 44 E
Bandanaira, Indonesia 83 B3 4 32 S 129 54 E
Bandanwara, India 92 F6 26 9N 74 38 E
Bandar = Machilipatnam, India 95 F5 16 12N 81 8 E
Bandār 'Abbās, Iran 97 E8 27 15N 56 15 E
Bandar-e Anzalī, Iran 97 B6 37 30N 49 30 E
Bandar-e Bushehr = Būshehr, Iran 97 D6 28 55N 50 55 E
Bandar-e Chārak, Iran 97 E7 26 45N 54 20 E
Bandar-e Deylam, Iran 97 D6 30 5N 50 10 E
Bandar-e Khomeynı, Iran 97 D6 30 30N 49 5 E
Bandar-e Lengeh, Iran 97 E7 26 35N 54 58 E
Bandar-e Maqām, Iran 97 E7 26 56N 53 29 E
Bandar-e Ma'shur, Iran 97 D6 30 35N 49 10 E
Bandar-e Nakhīlū, Iran 97 E7 26 58N 53 30 E
Bandar-e Rīg, Iran 97 D6 29 29N 50 38 E
Bandar-e Torkeman, Iran 97 B7 37 0N 54 10 E
Bandar Maharani = Muar, Malaysia 87 L4 2 3N 102 34 E
Bandar Penggaram = Batu Pahat, Malaysia 87 M4 1 50N 102 56 E
Bandar Seri Begawan, Brunei 85 B4 4 52N 115 0 E
Bandar Sri Aman, Malaysia 85 B4 1 15N 111 32 E
Bandawe, Malawi 119 E3 11 58 S 34 5 E
Bande, Spain 42 C3 42 3N 7 58W
Bandeira, Pico da, Brazil 171 F3 20 26 S 41 47W
Bandeirante, Brazil 171 D1 13 41 S 50 48W
Bandera, Argentina 174 B3 28 55 S 62 20W
Bandera, U.S.A. 155 L5 29 44N 99 5W
Banderas, B. de, Mexico 162 C3 20 40N 105 30W
Bandia →, India 94 E5 19 2N 80 28 E
Bandiagara, Mali 112 C4 14 12N 3 29W
Bandırma, Turkey 51 F11 40 20N 28 0 E
Bandol, France 29 E9 43 8N 5 46 E
Bandon, Ireland 23 E3 51 44N 8 44W
Bandon →, Ireland 23 E3 51 43N 8 37W
Bandoua, C.A.R. 114 B4 4 39N 21 42 E
Bandula, Mozam. 119 F3 19 0 S 33 7 E
Bandundu, Dem. Rep. of the Congo 114 C3 3 15 S 17 22 E
Bandung, Indonesia 85 D3 6 54 S 107 36 E
Bandya, Australia 125 E3 27 40 S 122 5 E
Băneasa, Romania 53 E12 45 56N 27 55 E
Bāneh, Iran 101 E11 35 59N 45 53 E
Bañeres, Spain 41 G4 38 44N 0 38W
Banes, Cuba 165 B4 21 0N 75 42W
Banff, Canada 142 C5 51 10N 115 34W
Banff, U.K. 22 D6 57 40N 2 33W
Banff Nat. Park, Canada 142 C5 51 30N 116 15W
Banfora, Burkina Faso 112 C4 10 40N 4 40W
Bang Fai →, Laos 86 D5 16 57N 104 45 E
Bang Hieng →, Laos 86 D5 16 10N 105 10 E
Bang Krathum, Thailand 86 D3 16 34N 100 18 E
Bang Lamung, Thailand 86 F3 13 3N 100 56 E
Bang Mun Nak, Thailand 86 D3 16 2N 100 23 E
Bang Pa In, Thailand 86 E3 14 14N 100 35 E
Bang Rakam, Thailand 86 D3 16 45N 100 7 E
Bang Saphan, Thailand 87 G2 11 14N 99 28 E
Bangala Dam, Zimbabwe 119 G3 21 7 S 31 25 E
Bangalore, India 95 H3 12 59N 77 40 E
Bangangté, Cameroon 113 D7 5 8N 10 32 E
Bangaon, India 93 H13 23 0N 88 47 E
Bangassou, C.A.R. 114 B4 4 55N 23 7 E
Bangeta, Mt., Papua N. G. 132 D4 6 21 S 147 3 E
Banggai, Kepulauan, Indonesia 82 B2 1 40 S 123 30 E
Banggai Arch. = Banggai, Kepulauan, Indonesia 82 B2 1 40 S 123 30 E
Banggi, Malaysia 85 A5 7 17N 117 12 E
Banghāzī, Libya 108 B4 32 11N 20 3 E
Banghāzī □, Libya 108 B4 32 7N 20 4 E
Bangil, Indonesia 85 D4 7 36 S 112 50 E
Bangjang, Sudan 107 E3 11 23N 32 41 E
Bangka, Sulawesi, Indonesia 82 A3 1 50N 125 5 E
Bangka, Sumatera, Indonesia 84 C3 2 0 S 105 50 E
Bangka, Selat, Indonesia 84 C3 2 30 S 105 30 E
Bangkalan, Indonesia 85 D4 7 2 S 112 46 E
Bangkinang, Indonesia 84 B2 0 18N 101 5 E
Bangko, Indonesia 84 C2 2 5 S 102 9 E
Bangkok, Thailand 86 F3 13 45N 100 35 E
Bangladesh ■, Asia 90 D3 24 0N 90 0 E
Bangolo, Ivory C. 112 D3 7 1N 7 29W
Bangong Co, India 93 B8 35 50N 79 20 E
Bangor, Down, U.K. 23 B6 54 40N 5 40W
Bangor, Gwynedd, U.K. 20 D3 53 14N 4 8W
Bangor, Maine, U.S.A. 141 D6 44 48N 68 46W
Bangor, Mich., U.S.A. 157 B10 42 18N 86 7W
Bangor, Pa., U.S.A. 151 F9 40 52N 75 13W
Bangu, Dem. Rep. of the Congo 114 C3 0 3 S 19 12 E
Bangued, Phil. 80 C3 17 40N 120 37 E
Bangui, C.A.R. 114 B3 4 23N 18 35 E
Bangui, Phil. 80 B3 18 32N 120 46 E
Banguru, Dem. Rep. of the Congo 118 B2 0 30N 27 10 E
Bangweulu, L., Zambia 119 E3 11 0 S 30 0 E
Bangweulu Swamp, Zambia 119 E3 11 20 S 30 15 E
Bani, Dom. Rep. 165 C5 18 16N 70 22W
Bani, Phil. 80 C2 16 11N 119 52 E
Bani →, Mali 112 C4 14 30N 4 12W
Bani, Djebel, Morocco 110 C3 29 16N 8 0W
Bani Bangou, Niger 113 B5 15 3N 2 42 E
Banī Sa'd, Iraq 101 F11 33 34N 44 32 E
Banī Sār, Si. Arabia 98 B3 20 6N 41 27 E
Banī Walīd, Libya 108 B2 31 36N 13 53 E
Bania, Ivory C. 112 D4 9 4N 3 6W
Baniara, Papua N. G. 132 E5 9 44 S 149 54 E
Banihal Pass, India 93 C6 33 30N 75 12 E
Banīnah, Libya 108 B4 32 0N 20 12 E
Bāniyās, Syria 100 E6 35 10N 36 0 E
Banja Luka, Bos.-H. 52 F2 44 49N 17 11 E
Banjar, Indonesia 85 D3 7 24 S 108 30 E
Banjarmasin, Indonesia 85 C4 3 20 S 114 35 E
Banjarnegara, Indonesia 85 D3 7 24 S 109 42 E
Banjul, Gambia 112 C1 13 28N 16 40W
Banka Banka, Australia 126 B1 18 50 S 134 0 E
Bankeryd, Sweden 17 G8 57 53N 14 6 E
Banket, Zimbabwe 119 F3 17 27 S 30 19 E
Bankilaré, Niger 113 C5 14 35N 0 44 E
Bankipore, India 93 G11 25 35N 85 10 E
Banks, U.S.A. 152 D4 31 49N 85 51W
Banks, Is., Vanuatu 133 C5 13 50 S 167 30 E
Banks I., B.C., Canada 142 C3 53 20N 130 0W
Banks I., N.W.T., Canada 138 A7 73 15N 121 30W
Banks I., Papua N. G. 132 F2 10 10 S 142 15 E
Banks L., U.S.A. 152 D6 31 2N 83 6W
Banks Pen., N.Z. 131 D8 43 45 S 173 15 E
Banks Str., Australia 126 G4 40 40 S 148 10 E
Bankura, India 93 H12 23 11N 87 18 E
Bankya, Bulgaria 50 D7 42 43N 23 8 E
Bann →, Arm., U.K. 23 B5 54 30N 6 31W
Bann →, L'derry., U.K. 23 A5 55 8N 6 41W
Banna, Phil. 80 C3 17 59N 120 39 E
Bannalec, France 26 E3 47 57N 3 42W
Bannang Sata, Thailand 87 J3 6 16N 101 16 E
Bannerton, Australia 128 C5 34 42 S 142 47 E
Banning, U.S.A. 161 M10 33 56N 116 53W
Banningville = Bandundu, Dem. Rep. of the Congo 114 C3 3 15 S 17 22 E
Bannockburn, Canada 150 B7 44 39N 77 33W
Bannockburn, U.K. 22 E5 56 5N 3 55W
Bannockburn, Zimbabwe 119 G2 20 17 S 29 48 E
Bannu, Pakistan 91 B3 33 0N 70 18 E
Bañolas = Banyoles, Spain 40 C7 42 16N 2 44 E
Banon, France 29 D9 44 2N 5 38 E
Baños de la Encina, Spain 43 G7 38 10N 3 46W
Baños de Molgas, Spain 42 C3 42 15N 7 40W
Bánovce nad Bebravou, Slovak Rep. 35 C11 48 44N 18 16 E
Banovići, Bos.-H. 52 F3 44 25N 18 32 E
Bansilan □, Phil. 81 H3 6 40N 121 40 E
Banská Bystrica, Slovak Rep. 35 C12 48 46N 19 14 E
Banská Štiavnica, Slovak Rep. 35 C11 48 25N 18 55 E
Bansko, Bulgaria 50 E7 41 52N 23 28 E
Banskobystrický □, Slovak Rep. 35 C12 48 20N 19 0 E
Banswara, India 92 H6 23 32N 74 24 E
Bantayan, Phil. 81 F4 11 10N 123 43 E
Bantayan I., Phil. 81 F4 11 13N 123 44 E
Banten, Indonesia 84 D3 6 5 S 106 8 E
Banton I., Phil. 80 E4 12 56N 122 4 E
Bantry, Ireland 23 E2 51 41N 9 27W
Bantry B., Ireland 23 E2 51 37N 9 44W
Bantul, Indonesia 85 D4 7 55 S 110 19 E
Bantva, India 92 J4 21 29N 70 12 E
Bantval, India 95 H2 12 55N 75 0 E
Banya, Bulgaria 51 D8 42 33N 24 50 E
Banyak, Kepulauan, Indonesia 84 B1 2 10N 97 10 E
Banyo, Cameroon 113 D7 6 52N 11 45 E
Banyoles, Spain 40 C7 42 16N 2 44 E
Banyuls-sur-Mer, France 28 F7 42 28N 3 8 E
Banyumas, Indonesia 85 D3 7 32 S 109 18 E
Banyuwangi, Indonesia 85 D4 8 13 S 114 21 E
Banzare Coast, Antarctica 7 C9 68 0 S 125 0 E
Banzyville = Mobayi, Dem. Rep. of the Congo 114 B4 4 15N 21 8 E
Bao Ha, Vietnam 86 A5 22 11N 104 21 E
Bao Lac, Vietnam 86 A5 22 57N 105 40 E
Bao Loc, Vietnam 87 G6 11 32N 107 48 E
Bao'an = Shenzhen, China 77 F10 22 27N 114 10 E
Baocheng, China 74 H4 33 12N 106 56 E
Baode, China 74 E6 39 1N 111 5 E
Baodi, China 75 E9 39 38N 117 20 E
Baoding, China 74 E8 38 50N 115 28 E
Baoji, China 74 G4 34 20N 107 5 E
Baojing, China 76 C7 28 45N 109 41 E
Baokang, China 77 B8 31 54N 111 12 E
Baoro, C.A.R. 114 A3 5 40N 15 58 E
Baoshan, Shanghai, China 77 B13 31 27N 121 26 E
Baoshan, Yunnan, China 76 E2 25 10N 99 5 E
Baotou, China 74 D6 40 32N 110 2 E
Baoying, China 75 H10 33 17N 119 20 E
Bap, India 92 F5 27 23N 72 18 E
Bapatla, India 95 G5 15 55N 80 30 E
Bapaume, France 27 B9 50 7N 2 50 E
Bāqerābād, Iran 97 C6 33 2N 51 58 E
Ba'qūbah, Iraq 101 F11 33 45N 44 50 E
Baquedano, Chile 174 A2 23 20 S 69 52W
Bar, Montenegro, Yug. 50 D3 42 8N 19 6 E
Bar, Ukraine 59 H4 49 4N 27 40 E
Bar Bigha, India 93 G11 25 21N 85 47 E
Bar Harbor, U.S.A. 141 D6 44 23N 68 13W
Bar-le-Duc, France 27 D12 48 47N 5 10 E
Bar-sur-Aube, France 27 D11 48 14N 4 40 E
Bar-sur-Seine, France 27 D11 48 7N 4 20 E
Băra, Romania 53 C12 47 2N 27 3 E
Barabai, Indonesia 85 C5 2 32 S 115 34 E
Baraboo, U.S.A. 154 D10 43 28N 89 45W
Baracoa, Cuba 165 B5 20 20N 74 30W
Baradā →, Syria 103 B5 33 33N 36 34 E
Baradero, Argentina 174 C4 33 52 S 59 29W
Baradine, Australia 129 A8 30 56 S 149 4 E
Baraga, U.S.A. 154 B10 46 47N 88 30W
Bărăganul, Romania 53 F12 44 49N 27 31 E
Barahona, Dom. Rep. 165 C5 18 13N 71 7W
Barahona, Spain 40 D2 41 17N 2 39W
Baraka →, Sudan 106 D4 18 13N 37 35 E
Barakaldo, Spain 40 B2 43 18N 2 59W
Barakot, India 93 J11 21 33N 84 59 E
Barakpur, India 93 H13 22 44N 88 30 E
Barakula, Australia 127 D5 26 30 S 150 33 E
Baralaba, Australia 126 C4 24 13 S 149 50 E
Baralla, Spain 42 C3 42 53N 7 15W
Baralzon L., Canada 143 B9 60 0N 98 3W
Baram →, Malaysia 85 B4 4 35N 113 58 E
Baramati, India 94 E2 18 11N 74 33 E
Baramba, India 94 D7 20 25N 85 23 E
Barameiya, Sudan 106 D4 18 32N 36 38 E
Baramula, India 93 B6 34 15N 74 20 E
Baran, India 92 G7 25 9N 76 40 E
Barañain, Spain 40 C3 42 48N 1 40W
Baranavichy, Belarus 59 F4 53 10N 26 0 E
Baranoa, Colombia 168 A3 10 48N 74 55W
Baranof, U.S.A. 144 H14 57 5N 134 50W
Baranof I., U.S.A. 142 B1 57 0N 135 0W
Baranów Sandomierski, Poland 55 H8 50 29N 21 30 E
Baranya □, Hungary 52 E3 46 0N 18 15 E
Barão de Cocais, Brazil 171 E3 19 56 S 43 28W
Barão de Grajaú, Brazil 170 C3 6 45 S 43 1W
Barão de Melgaço, Mato Grosso, Brazil 173 D6 16 14 S 55 52W
Barão de Melgaço, Rondônia, Brazil 173 C5 11 50 S 60 45W
Baraolt, Romania 53 D10 46 5N 25 34 E
Barapasi, Indonesia 83 B5 2 15 S 137 5 E
Barapina, Papua N. G. 132 D8 6 21 S 155 25 E
Barasat, India 93 H13 22 46N 88 31 E
Barat Daya, Kepulauan, Indonesia 82 C3 7 30 S 128 0 E
Barataria B., U.S.A. 155 L10 29 20N 89 55W
Baraut, India 92 E7 29 13N 77 7 E
Baraya, Colombia 168 C2 3 10N 75 4W
Barbacan, Phil. 81 F2 10 20N 119 21 E
Barbacena, Brazil 171 F3 21 15 S 43 56W
Barbacoas, Colombia 168 C2 1 45N 78 0W
Barbacoas, Venezuela 168 B4 9 29N 66 58W
Barbados ■, W. Indies 165 D8 13 10N 59 30W
Barbalha, Brazil 170 C4 7 19 S 39 17W
Barban, Croatia 45 C11 45 5N 14 2 E
Barbària, C. de, Spain 39 C7 38 39N 1 24 E
Barbaros, Turkey 51 F11 40 54N 27 27 E
Barbastro, Spain 40 C5 42 2N 0 5 E
Barbate = Barbate de Franco, Spain 43 J5 36 13N 5 56W
Barbate de Franco, Spain 43 J5 36 13N 5 56W
Barbaza, Phil. 81 F4 11 12N 122 2 E
Barberino di Mugello, Italy 45 E8 44 0N 11 15 E
Barbers Pt., U.S.A. 145 K13 21 18N 158 7W
Barberton, S. Africa 117 D5 25 42 S 31 2 E
Barberton, U.S.A. 150 E3 41 0N 81 39W
Barberville, U.S.A. 153 F8 29 11N 81 26W
Barbezieux-St-Hilaire, France 28 C3 45 28N 0 9W
Barbosa, Colombia 168 B3 5 57N 73 37W
Barbourville, U.S.A. 149 G4 36 52N 83 53W
Barbuda, W. Indies 165 C7 17 30N 61 40W
Bârca, Romania 53 G8 43 59N 23 36 E
Barcaldine, Australia 126 C4 23 43 S 145 6 E
Barcarrota, Spain 43 G4 38 31N 6 51W
Barcellona Pozzo di Gotto, Italy 47 D8 38 9N 15 13 E
Barcelona, Spain 40 D7 41 21N 2 10 E
Barcelona, Venezuela 169 A5 10 10N 64 40W
Barcelona □, Spain 40 D7 41 30N 2 0 E
Barcelonette, France 29 D10 44 23N 6 40 E
Barcelos, Brazil 169 D5 1 0 S 63 0W
Barcin, Poland 55 F4 52 52N 17 55 E
Barcoo →, Australia 126 D3 25 30 S 142 50 E
Barcs, Hungary 52 E2 45 58N 17 28 E
Barczewo, Poland 54 E7 53 50N 20 42 E
Bärdä, Azerbaijan 61 K8 40 25N 47 10 E
Barda del Medio, Argentina 176 A3 38 45 S 68 11W
Bardaï, Chad 109 D3 21 25N 17 0 E
Bardas Blancas, Argentina 174 D2 35 49 S 69 45W
Barddhaman, India 93 H12 23 14N 87 39 E
Bardejov, Slovak Rep. 35 B14 49 18N 21 15 E

188 Bardera

Bardera, *Somali Rep.* **120 D2** 2 20N 42 27 E
Bárðarbunga, *Iceland* **11 C9** 64 38N 17 32W
Barðastrandarsýsla □, *Iceland* **11 B3** 65 40N 23 0W
Bardi, *Italy* **44 D6** 44 38N 9 44 E
Bardīyah, *Libya* **108 B5** 31 45N 25 5 E
Bardoli, *India* **94 D1** 21 12N 73 5 E
Bardolino, *Italy* **44 C7** 45 33N 10 43 E
Bardonécchia, *Italy* **44 C3** 45 5N 6 42 E
Bardsey I., *U.K.* **20 E3** 52 45N 4 47W
Bardstown, *U.S.A.* **157 G11** 37 49N 85 28W
Bareilly, *India* **93 E8** 28 22N 79 27 E
Barellan, *Australia* **129 C7** 34 16 S 146 24 E
Barentin, *France* **26 C7** 49 33N 0 58 E
Barenton, *France* **26 D6** 48 38N 0 50W
Barents Sea, *Arctic* **6 B9** 73 0N 39 0 E
Barentu, *Eritrea* **107 D4** 15 2N 37 35 E
Barfleur, *France* **26 C5** 49 40N 1 17W
Barfleur, Pte. de, *France* **26 C5** 49 42N 1 16W
Barga, *Italy* **44 D7** 44 4N 10 29 E
Bargal, *Somali Rep.* **120 B4** 11 25N 51 0 E
Bargara, *Australia* **126 C5** 24 50 S 152 25 E
Bargas, *Spain* **42 F6** 39 56N 4 3W
Bârgăului Bistriţa, *Romania* **53 C9** 47 13N 24 46 E
Barge, *Italy* **44 D4** 44 43N 7 20 E
Bargnop, *Sudan* **107 F2** 9 32N 28 25 E
Bargo, *Australia* **129 C9** 34 18 S 150 35 E
Bargteheide, *Germany* **30 B6** 53 44N 10 14 E
Barguzin, *Russia* **65 D11** 53 37N 109 37 E
Barh, *India* **93 G11** 25 29N 85 46 E
Barhaj, *India* **93 F10** 26 18N 83 44 E
Barhi, *India* **93 G11** 24 15N 85 25 E
Bari, *India* **92 F7** 26 39N 77 39 E
Bari, *Italy* **47 A9** 41 8N 16 51 E
Bari Doab, *Pakistan* **92 D5** 30 20N 73 0 E
Bari Sardo, *Italy* **46 C2** 39 50N 9 38 E
Bariadi □, *Tanzania* **118 C3** 2 45 S 34 40 E
Barīm, *Yemen* **98 D3** 12 39N 43 25 E
Barima →, *Guyana* **169 B5** 8 33N 60 25W
Barinas, *Venezuela* **168 B3** 8 36N 70 15W
Barinas □, *Venezuela* **168 B4** 8 10N 69 50W
Baring, *U.S.A.* **156 D4** 40 15N 92 12W
Baring, C., *Canada* **138 B8** 70 0N 117 30W
Baringa, *Dem. Rep. of the Congo* **114 B4** 0 45N 20 52 E
Baringo, *Kenya* **118 B4** 0 47N 36 16 E
Baringo □, *Kenya* **118 B4** 0 55N 36 0 E
Baringo, L., *Kenya* **118 B4** 0 47N 36 16 E
Barinitas, *Venezuela* **168 B3** 8 45N 70 25W
Baripada, *India* **94 D8** 21 57N 86 45 E
Bariri, *Brazil* **171 F2** 22 4 S 48 44W
Bâris, *Egypt* **106 C3** 24 42N 30 31 E
Barisal, *Bangla.* **90 D3** 22 45N 90 20 E
Barito →, *Indonesia* **85 C4** 4 0 S 114 50 E
Barjac, *France* **29 D8** 44 20N 4 22 E
Barjols, *France* **29 E10** 43 34N 6 2 E
Barjūj, Wadi →, *Libya* **108 C2** 25 26N 12 12 E
Bark L., *Canada* **150 A7** 45 27N 77 51W
Barka = Baraka →, *Sudan* **106 D4** 18 13N 37 35 E
Barkald, *Norway* **18 C7** 61 59N 10 53 E
Barkam, *China* **76 B4** 31 51N 102 28 E
Barker, *U.S.A.* **150 C6** 43 20N 78 33W
Barkley Sound, *Canada* **142 D3** 48 50N 125 10W
Barkly Downs, *Australia* **126 C2** 20 30 S 138 30 E
Barkly East, *S. Africa* **116 E4** 30 58 S 27 33 E
Barkly Tableland, *Australia* **126 B2** 17 50 S 136 40 E
Barkly West, *S. Africa* **116 D3** 28 5 S 24 31 E
Barkol, Wadi →, *Sudan* **106 D3** 17 40N 32 0 E
Barkol Kazak Zizhixian, *China* **72 B4** 43 37N 93 2 E
Barksdale, *U.S.A.* **155 L4** 29 44N 100 2W
Barla Dağı, *Turkey* **49 C12** 38 5N 30 40 E
Bârlad, *Romania* **53 D12** 46 15N 27 38 E
Bârlad →, *Romania* **53 E12** 45 38N 27 32 E
Barlee, L., *Australia* **125 E2** 29 15 S 119 30 E
Barlee, Mt., *Australia* **125 D4** 24 38 S 128 13 E
Barletta, *Italy* **47 A9** 41 19N 16 17 E
Barlinek, *Poland* **55 F2** 53 0N 15 15 E
Barlovento, *Canary Is.* **39 F2** 28 48N 17 48W
Barlow L., *Canada* **143 A8** 62 0N 103 0W
Barmedman, *Australia* **129 C7** 34 9 S 147 21 E
Barmer, *India* **92 G4** 25 45N 71 20 E
Barmera, *Australia* **128 C4** 34 15 S 140 28 E
Barmouth, *U.K.* **20 E3** 52 44N 4 4W
Barmstedt, *Germany* **30 B5** 53 47N 9 46 E
Barnagar, *India* **92 H6** 23 7N 75 19 E
Barnard Castle, *U.K.* **20 C6** 54 33N 1 55W
Barnato, *Australia* **129 A6** 31 38 S 145 0 E
Barnaul, *Russia* **64 D9** 53 20N 83 40 E
Barnesville, *U.S.A.* **152 B5** 33 3N 84 9W
Barnet, *U.K.* **21 F7** 51 38N 0 9W
Barneveld, *Neths.* **24 B5** 52 7N 5 36 E
Barneveld, *U.S.A.* **151 C9** 43 16N 75 14W
Barneville-Cartevert, *France* **26 C5** 49 23N 1 46W
Barney, *U.S.A.* **152 D6** 31 1N 83 31W
Barngo, *Australia* **126 D4** 25 3 S 147 20 E
Barnhart, *U.S.A.* **155 K4** 31 8N 101 10W
Barnsley, *U.K.* **20 D6** 53 34N 1 27W
Barnstaple, *U.K.* **21 F3** 51 5N 4 4W
Barnstaple Bay = Bideford Bay, *U.K.* **21 F3** 51 5N 4 20W
Barnsville, *U.S.A.* **154 B6** 46 43N 96 28W
Barnwell, *U.S.A.* **152 B9** 33 15N 81 23W
Baro, *Nigeria* **113 D6** 8 35N 6 18 E
Baro →, *Ethiopia* **107 F3** 8 26N 33 13 E
Baroda = Vadodara, *India* **92 H5** 22 20N 73 10 E
Baroda, *India* **92 G7** 25 29N 76 35 E
Baroe, *S. Africa* **116 E3** 33 13 S 24 33 E
Baron Ra., *Australia* **124 D4** 23 30 S 127 45 E
Barora Ite, *Solomon Is.* **133 L10** 7 36 S 158 24 E
Barorafa, *Solomon Is.* **133 L10** 7 30 S 158 20 E
Barpali, *India* **94 D6** 21 11N 83 35 E
Barpathar, *India* **90 B4** 26 17N 93 53 E
Barpeta, *India* **90 B3** 26 20N 91 10 E
Barqin, *Libya* **108 C2** 27 33N 13 34 E
Barques, Pt. Aux, *U.S.A.* **148 C4** 44 4N 82 58W
Barquísimeto, *Venezuela* **168 A4** 10 4N 69 19W
Barra, *Brazil* **170 D3** 11 5 S 43 10W
Barra, *U.K.* **22 E1** 57 0N 7 29W
Barra, Sd. of, *U.K.* **22 D1** 57 4N 7 25W
Barra da Estiva, *Brazil* **171 D3** 13 38 S 41 19W
Barra de Navidad, *Mexico* **162 D4** 19 12N 104 41W
Barra do Corda, *Brazil* **170 C2** 5 30 S 45 10W
Barra do Dande, *Angola* **115 D2** 8 28 S 13 22 E
Barra do Mendes, *Brazil* **171 D3** 11 43 S 42 4W
Barra do Piraí, *Brazil* **171 F3** 22 30 S 43 50W
Barra Falsa, Pta. da, *Mozam.* **117 C6** 22 58 S 35 37 E
Barra Hd., *U.K.* **22 E1** 56 47N 7 40W
Barra Mansa, *Brazil* **171 F3** 22 35 S 44 12W
Barraba, *Australia* **129 A9** 30 21 S 150 35 E
Barracão do Barreto, *Brazil* **173 B6** 8 48 S 58 24W
Barrackpur = Barakpur, *India* **93 H13** 22 44N 88 30 E
Barrafranca, *Italy* **47 E7** 37 22N 14 12 E
Barraigh = Barra, *U.K.* **22 E1** 57 0N 7 29W
Barranca, *Lima, Peru* **172 C2** 10 45 S 77 50W
Barranca, *Loreto, Peru* **168 D2** 4 50 S 76 50W
Barrancabermeja, *Colombia* **168 B3** 7 0N 73 50W
Barrancas, *Colombia* **168 A3** 10 57N 72 50W
Barrancas, *Venezuela* **169 B5** 8 55N 62 5W
Barrancos, *Portugal* **43 G4** 38 10N 6 58W
Barranqueras, *Argentina* **174 B4** 27 30 S 59 0W
Barranquilla, *Colombia* **168 A3** 11 0N 74 50W
Barras, *Brazil* **170 B3** 4 15 S 42 18W
Barras, *Colombia* **168 D3** 1 45 S 73 13W
Barraute, *Canada* **140 C4** 48 26N 77 38W
Barre, *Mass., U.S.A.* **151 D12** 42 25N 72 6W
Barre, *Vt., U.S.A.* **151 B12** 44 12N 72 30W
Barre do Bugres, *Brazil* **173 D6** 15 0 S 57 11W
Barreal, *Argentina* **174 C2** 31 33 S 69 28W
Barrei, *Ethiopia* **120 C2** 6 10N 42 49 E
Barreiras, *Brazil* **171 D3** 12 8 S 45 0W
Barreirinha, *Brazil* **169 D6** 2 47 S 57 3W
Barreirinhas, *Brazil* **170 B3** 2 30 S 42 50W
Barreiro, *Portugal* **43 G1** 38 40N 9 6W
Barreiros, *Brazil* **170 C4** 8 49 S 35 12W
Barrême, *France* **29 E10** 43 57N 6 23 E
Barren, Nosy, *Madag.* **117 B7** 18 25 S 43 40 E
Barren Is., *U.S.A.* **144 G9** 58 55N 152 15W
Barretos, *Brazil* **171 F2** 20 30 S 48 35W
Barrhead, *Canada* **142 C6** 54 10N 114 24W
Barrie, *Canada* **140 D4** 44 24N 79 40W
Barrier, C., *N.Z.* **130 C4** 36 25 S 175 32 E
Barrier Ra., *Australia* **128 A4** 31 0 S 141 30 E
Barrier Ra., *Otago, N.Z.* **131 E4** 44 15 S 169 32 E
Barrier Ra., *W. Coast, N.Z.* **131 E3** 44 30 S 168 30 E
Barrière, *Canada* **142 C4** 51 12N 120 7W
Barrineau Park, *U.S.A.* **153 E2** 30 42N 87 26W
Barrington, *U.S.A.* **151 E13** 41 44N 71 18W
Barrington L., *Canada* **143 B8** 56 55N 100 15W
Barrington Tops, *Australia* **129 B9** 32 6 S 151 28 E
Barringun, *Australia* **127 D4** 29 1 S 145 41 E
Barro do Garças, *Brazil* **173 D7** 15 54 S 52 16W
Barrow, *U.S.A.* **138 A4** 71 18N 156 47W
Barrow →, *Ireland* **23 D5** 52 25N 6 58W
Barrow Creek, *Australia* **126 C1** 21 30 S 133 55 E
Barrow I., *Australia* **124 D2** 20 45 S 115 20 E
Barrow-in-Furness, *U.K.* **20 C4** 54 7N 3 14W
Barrow Pt., *Australia* **126 A3** 14 20 S 144 40 E
Barrow Pt., *U.S.A.* **136 B4** 71 24N 156 29W
Barrow Ra., *Australia* **125 E4** 26 0 S 127 40 E
Barrow Str., *Canada* **6 B3** 74 20N 95 0W
Barruecopardo, *Spain* **42 D4** 41 4N 6 40W
Barruelo de Santullán, *Spain* **42 C6** 42 54N 4 17W
Barry, *U.K.* **21 F4** 51 24N 3 16W
Barry, *U.S.A.* **156 E5** 39 42N 91 2W
Barry's Bay, *Canada* **140 C4** 45 29N 77 41W
Barsalogho, *Burkina Faso* **113 C4** 13 25N 1 3W
Barsat, *Pakistan* **93 A5** 36 10N 72 45 E
Barsham, *Syria* **101 E9** 35 21N 40 33 E
Barsi, *India* **94 E2** 18 10N 75 50 E
Barsinghausen, *Germany* **30 C5** 52 18N 9 28 E
Barstow, *Calif., U.S.A.* **161 L9** 34 54N 117 1W
Barstow, *Tex., U.S.A.* **155 K3** 31 28N 103 24W
Barth, *Germany* **30 A8** 54 22N 12 42 E
Barthélemy, Col, *Vietnam* **86 C5** 19 26N 104 6 E
Bartica, *Guyana* **169 B6** 6 25N 58 40W
Bartin, *Turkey* **100 B5** 41 38N 32 21 E
Bartle Frere, *Australia* **122 D8** 17 27 S 145 50 E
Bartlesville, *U.S.A.* **155 G7** 36 45N 95 59W
Bartlett, *Calif., U.S.A.* **160 J8** 36 29N 118 2W
Bartlett, *Tex., U.S.A.* **155 K6** 30 48N 97 26W
Bartlett, L., *Canada* **142 A5** 63 5N 118 20W
Bartolomeu Dias, *Mozam.* **119 G4** 21 10 S 35 8 E
Barton, *Australia* **125 F5** 30 31 S 132 39 E
Barton, *Phil.* **81 F2** 10 24N 119 8 E
Barton upon Humber, *U.K.* **20 D7** 53 41N 0 25W
Bartonville, *U.S.A.* **156 D7** 40 39N 89 39W
Bartoszyce, *Poland* **54 D7** 54 15N 20 55 E
Bartow, *Fla., U.S.A.* **149 M5** 27 54N 81 50W
Bartow, *Ga., U.S.A.* **152 C7** 32 53N 82 29W
Barú, I. de, *Colombia* **168 A2** 10 15N 75 35W
Barú, Volcan, *Panama* **164 E3** 8 55N 82 35W
Barumba, *Dem. Rep. of the Congo* **118 B1** 1 3N 23 37 E
Baruth, *Germany* **30 C9** 52 4N 13 30 E
Barvinkove, *Ukraine* **59 H9** 48 57N 37 0 E
Barwani, *India* **92 H6** 22 2N 74 57 E
Barwice, *Poland* **54 E3** 53 44N 16 21 E
Barwick, *U.S.A.* **152 E6** 30 54N 83 44W
Barycz →, *Poland* **55 G3** 51 42N 16 15 E
Barysaw, *Belarus* **58 E5** 54 17N 28 28 E
Barysh, *Russia* **60 D8** 53 39N 47 8 E
Barzān, *Iraq* **96 B5** 36 55N 44 3 E
Bârzava, *Romania* **52 D6** 46 7N 21 59 E
Bas-Rhin □, *France* **27 D14** 48 40N 7 30 E
Bašaid, *Serbia, Yug.* **52 E5** 45 38N 20 25 E
Bāsa'idū, *Iran* **97 E7** 26 35N 55 20 E
Basal, *Pakistan* **92 C5** 33 33N 72 13 E
Basankusa, *Dem. Rep. of the Congo* **114 B3** 1 5N 19 50 E
Basarabeasca, *Moldova* **53 D13** 46 21N 28 58 E
Basarabi, *Romania* **53 F13** 44 10N 28 26 E
Basauri, *Spain* **40 B2** 43 13N 2 53W
Basawa, *Afghan.* **92 B4** 34 15N 70 50 E
Basco, *Phil.* **80 A3** 20 27N 121 58 E
Bascuñán, C., *Chile* **174 B1** 28 52 S 71 35W
Basel, *Switz.* **32 A5** 47 35N 7 35 E
Basel-Stadt □, *Switz.* **32 A5** 47 35N 7 35 E
Baselland □, *Switz.* **32 B5** 47 26N 7 45 E
Basento →, *Italy* **47 B9** 40 20N 16 49 E
Basey, *Phil.* **81 F5** 11 17N 125 4 E
Bāshī, *Iran* **97 D6** 28 41N 51 4 E
Bashkir Republic = Bashkortostan □, *Russia* **62 E6** 54 0N 57 0 E
Bashkortostan □, *Russia* **62 E6** 54 0N 57 0 E
Basilaki I., *Papua N. G.* **132 F6** 10 35 S 151 0 E
Basilan, *Phil.* **81 H4** 6 35N 122 0 E
Basilan □, *Phil.* **81 H4** 6 33N 122 4 E
Basilan Str., *Phil.* **81 H4** 6 50N 122 0 E
Basildon, *U.K.* **21 F8** 51 34N 0 28 E
Basilicata □, *Italy* **47 B9** 40 30N 16 30 E
Basim = Washim, *India* **94 D3** 20 3N 77 0 E
Basin, *U.S.A.* **158 D9** 44 23N 108 2W
Basinger, *U.S.A.* **153 H8** 27 23N 81 2W
Basingstoke, *U.K.* **21 F6** 51 15N 1 5W
Basirhat, *Bangla.* **90 D2** 22 40N 88 54 E
Baška, *Croatia* **45 D11** 44 58N 14 45 E
Başkale, *Turkey* **101 C10** 38 2N 43 59 E
Baskatong, Rés., *Canada* **140 C4** 46 46N 75 50W
Basle = Basel, *Switz.* **32 A5** 47 35N 7 35 E
Başmakçı, *Turkey* **49 D12** 37 54N 30 1 E
Basmat, *India* **94 E3** 19 15N 77 12 E
Basoda, *India* **92 H7** 23 52N 77 54 E
Basodino, *Switz.* **33 D6** 46 25N 8 28 E
Basoka, *Dem. Rep. of the Congo* **118 B1** 1 16N 23 40 E
Basongo, *Dem. Rep. of the Congo* **115 C4** 4 15 S 20 20 E
Basque, Pays, *France* **28 E2** 43 15N 1 20W
Basque Provinces = País Vasco □, *Spain* **40 C2** 42 50N 2 45W
Basra = Al Başrah, *Iraq* **96 D5** 30 30N 47 50 E
Bass Str., *Australia* **126 F4** 39 15 S 146 30 E
Bassano, *Canada* **142 C6** 50 48N 112 20W
Bassano del Grappa, *Italy* **45 C8** 45 46N 11 44 E
Bassar, *Togo* **113 D5** 9 19N 0 57 E
Basse-Normandie □, *France* **26 D6** 48 45N 0 30W
Basse Santa-Su, *Gambia* **112 C2** 13 13N 14 15W
Basse-Terre, *Guadeloupe* **165 C7** 16 0N 61 44W
Bassecourt, *Switz.* **32 B4** 47 20N 7 15 E
Bassein, *Burma* **90 G5** 16 45N 94 30 E
Bassein, *India* **94 E1** 19 26N 72 48 E
Basseterre, *St. Kitts & Nevis* **165 C7** 17 17N 62 43W
Bassett, *Nebr., U.S.A.* **154 D5** 42 35N 99 32W
Bassett, *Va., U.S.A.* **149 G6** 36 46N 79 59W
Bassi, *India* **92 D7** 30 44N 76 21 E
Bassigny, *France* **27 E12** 48 0N 5 30 E
Bassikounou, *Mauritania* **112 B3** 15 55N 6 1W
Bassum, *Germany* **30 C4** 52 50N 8 40 E
Båstad, *Sweden* **17 H6** 56 25N 12 51 E
Bastak, *Iran* **97 E7** 27 15N 54 25 E
Baştām, *Iran* **97 B7** 36 29N 55 4 E
Bastar, *India* **94 E5** 19 15N 81 40 E
Bastelica, *France* **29 F13** 42 1N 9 3 E
Basti, *India* **93 F10** 26 52N 82 55 E
Bastia, *France* **29 F13** 42 40N 9 30 E
Bastogne, *Belgium* **25 D5** 50 1N 5 43 E
Bastrop, *U.S.A.* **155 K6** 30 7N 97 19W
Bat Yam, *Israel* **103 C3** 32 2N 34 44 E
Bata, *Eq. Guin.* **114 B1** 1 57N 9 50 E
Bata, *Romania* **52 D7** 46 1N 22 4 E
Bataan, *Phil.* **80 D3** 14 40N 120 25 E
Batabanó, *Cuba* **164 B3** 22 40N 82 20W
Batabanó, G. de, *Cuba* **164 B3** 22 30N 82 30W
Batac, *Phil.* **80 B3** 18 3N 120 34 E
Batagai, *Russia* **65 C14** 67 38N 134 38 E
Batajnica, *Serbia, Yug.* **50 B4** 44 54N 20 17 E
Batak, *Bulgaria* **51 E8** 41 57N 24 12 E
Batalha, *Portugal* **42 F2** 39 40N 8 50W
Batam, *Indonesia* **84 B2** 1 5N 104 3 E
Batama, *Dem. Rep. of the Congo* **118 B2** 0 58N 26 33 E
Batamay, *Russia* **65 C13** 63 30N 129 15 E
Batan I., *Phil.* **80 A3** 20 30N 121 50 E
Batanes □, *Phil.* **80 A3** 20 40N 121 55 E
Batanes Is., *Phil.* **80 A3** 20 25N 121 50 E
Batang, *China* **76 B2** 30 1N 99 0 E
Batang, *Indonesia* **85 D3** 6 55 S 109 45 E
Batangafo, *C.A.R.* **114 A3** 7 25N 18 20 E
Batangas, *Phil.* **80 E3** 13 35N 121 10 E
Batangas □, *Phil.* **80 E3** 13 15N 121 5 E
Batanghari, *Indonesia* **84 C2** 1 36 S 103 37 E
Batanta, *Indonesia* **83 B4** 0 55 S 130 40 E
Batas, *Phil.* **81 F2** 11 10N 119 37 E
Batas I., *Phil.* **81 F2** 11 10N 119 36 E
Batatais, *Brazil* **175 A6** 20 54 S 47 37W
Batavia, *Ill., U.S.A.* **157 C8** 41 51N 88 19W
Batavia, *N.Y., U.S.A.* **150 D6** 43 0N 78 11W
Batavia, *Ohio, U.S.A.* **157 E12** 39 5N 84 11W
Bataysk, *Russia* **59 J10** 47 3N 39 45 E
Batchelor, *Australia* **124 B5** 13 4 S 131 1 E
Batdambang, *Cambodia* **86 F4** 13 7N 103 12 E
Batéké, Plateau, *Congo* **114 C3** 3 30 S 15 45 E
Bateman's B., *Australia* **129 C9** 35 40 S 150 12 E
Batemans Bay, *Australia* **129 C9** 35 44 S 150 11 E
Bates Ra., *Australia* **125 E3** 27 27 S 121 5 E
Batesburg, *U.S.A.* **152 B9** 33 54N 81 33W
Batesville, *Ark., U.S.A.* **155 H9** 35 46N 91 39W
Batesville, *Ind., U.S.A.* **157 E11** 39 18N 85 13W
Batesville, *Miss., U.S.A.* **155 H10** 34 19N 89 57W
Batesville, *Tex., U.S.A.* **155 L5** 28 58N 99 37W
Bath, *U.K.* **21 F5** 51 23N 2 22W
Bath, *Maine, U.S.A.* **141 D6** 43 55N 69 49W
Bath, *N.Y., U.S.A.* **150 D7** 42 20N 77 19W
Bath, *S.C., U.S.A.* **152 B9** 33 31N 81 51W
Bath & North East Somerset □, *U.K.* **21 F5** 51 21N 2 27W
Batheay, *Cambodia* **87 G5** 11 59N 104 57 E
Bathurst = Banjul, *Gambia* **112 C1** 13 28N 16 40W
Bathurst, *Australia* **129 B8** 33 25 S 149 31 E
Bathurst, *Canada* **141 C6** 47 37N 65 43W
Bathurst, *S. Africa* **116 E4** 33 30 S 26 50 E
Bathurst, C., *Canada* **138 A7** 70 34N 128 0W
Bathurst B., *Australia* **126 A3** 14 16 S 144 25 E
Bathurst Harb., *Australia* **126 G4** 43 15 S 146 10 E
Bathurst I., *Australia* **124 B5** 11 30 S 130 10 E
Bathurst I., *Canada* **6 B2** 76 0N 100 30W
Bathurst Inlet, *Canada* **138 B9** 66 50N 108 1W
Batie, *Burkina Faso* **112 D4** 9 53N 2 53W
Batlow, *Australia* **129 C8** 35 31 S 148 9 E
Batman, *Turkey* **101 D9** 37 55N 41 5 E
Batna, *Algeria* **111 A6** 35 34N 6 15 E
Batnfjordsøra, *Norway* **18 B4** 62 53N 7 42 E
Bato, *Leyte, Phil.* **81 F5** 10 13N 124 48 E
Bato, *Sulu, Phil.* **81 J3** 5 15N 120 3 E
Bato Bato, *Phil.* **81 J2** 5 6N 119 49 E
Batoala, *Gabon* **114 B2** 0 48N 13 27 E
Batobato, *Phil.* **81 H6** 6 50N 126 5 E
Batočina, *Serbia, Yug.* **50 B5** 44 7N 21 5 E
Batoka, *Zambia* **119 F2** 16 45 S 27 15 E
Baton Rouge, *U.S.A.* **155 K9** 30 27N 91 11W
Batong, Ko, *Thailand* **87 J2** 6 32N 99 12 E
Bátonyterenye, *Hungary* **52 C4** 47 59N 19 50 E
Batopilas, *Mexico* **162 B3** 27 0N 107 45W
Batouri, *Cameroon* **114 B2** 4 30N 14 25 E
Båtsfjord, *Norway* **14 A23** 70 38N 29 39 E
Battambang = Batdambang, *Cambodia* **86 F4** 13 7N 103 12 E
Batticaloa, *Sri Lanka* **95 L5** 7 43N 81 45 E
Battipáglia, *Italy* **47 B7** 40 37N 14 58 E
Battle, *U.K.* **21 G8** 50 55N 0 30 E
Battle →, *Canada* **143 C7** 52 43N 108 15W
Battle Camp, *Australia* **126 B3** 15 20 S 144 40 E
Battle Creek, *U.S.A.* **157 B11** 42 19N 85 11W
Battle Ground, *U.S.A.* **160 E4** 45 47N 122 32W
Battle Harbour, *Canada* **141 B8** 52 16N 55 35W
Battle Lake, *U.S.A.* **154 B7** 46 17N 95 43W
Battle Mountain, *U.S.A.* **158 F5** 40 38N 116 56W
Battlefields, *Zimbabwe* **119 F2** 18 37 S 29 47 E
Battleford, *Canada* **143 C7** 52 45N 108 15W
Battonya, *Hungary* **52 D6** 46 16N 21 3 E
Batu, Bukit, *Malaysia* **85 B4** 2 16N 113 43 E
Batu, Kepulauan, *Indonesia* **84 C1** 0 30 S 98 25 E
Batu, Mt., *Ethiopia* **107 F4** 6 55N 39 45 E
Batu Bora, Bukit, *Malaysia* **85 B4** 2 43N 114 43 E
Batu Caves, *Malaysia* **87 L3** 3 15N 101 40 E
Batu Gajah, *Malaysia* **87 K3** 4 28N 101 3 E
Batu Is. = Batu, Kepulauan, *Indonesia* **84 C1** 0 30 S 98 25 E
Batu Pahat, *Malaysia* **87 M4** 1 50N 102 56 E
Batu Puteh, Gunong, *Malaysia* **84 B2** 4 15N 101 31 E
Batuata, *Indonesia* **82 C2** 6 12 S 122 42 E
Batulaki, *Phil.* **81 J5** 5 34N 125 19 E
Batumi, *Georgia* **61 K5** 41 39N 41 44 E
Baturaja, *Indonesia* **84 C2** 4 11 S 104 15 E
Baturité, *Brazil* **170 B4** 4 28 S 38 45W
Batusangkar, *Indonesia* **84 C2** 0 27 S 100 35 E
Bau, *Malaysia* **85 B4** 1 25N 110 9 E
Bauang, *Phil.* **80 C3** 16 31N 120 20 E
Baubau, *Indonesia* **82 C2** 5 25 S 122 38 E
Bauchi, *Nigeria* **113 C6** 10 22N 9 48 E
Bauchi □, *Nigeria* **113 C7** 10 30N 10 0 E
Baud, *France* **26 E3** 47 52N 3 1W
Baudette, *U.S.A.* **154 A7** 48 43N 94 36W
Bauer, C., *Australia* **127 E1** 32 44 S 134 4 E
Bauhinia Downs, *Australia* **126 C4** 24 35 S 149 18 E
Baukau, *Indonesia* **82 C3** 8 27 S 126 27 E
Bauma, *Switz.* **33 B7** 47 23N 8 53 E
Baume-les-Dames, *France* **27 E13** 47 22N 6 22 E
Baunatal, *Germany* **30 D5** 51 14N 9 24 E
Baunei, *Italy* **46 B2** 40 2N 9 40 E
Baures, *Bolivia* **173 C5** 13 35 S 63 35W
Bauru, *Brazil* **175 A6** 22 10 S 49 0W
Baús, *Brazil* **173 D7** 18 22 S 52 47W
Bauska, *Latvia* **15 H21** 56 24N 24 15 E
Bautino, *Kazakstan* **61 H10** 44 35N 50 14 E
Bautzen, *Germany* **30 D10** 51 10N 14 26 E
Bavānāt, *Iran* **97 D7** 30 28N 53 27 E
Bavanište, *Serbia, Yug.* **52 F5** 44 49N 20 55 E
Bavaria = Bayern □, *Germany* **31 G7** 48 50N 12 0 E
Båven, *Sweden* **16 E10** 59 0N 16 56 E
Bavi Sadri, *India* **92 G6** 24 28N 74 30 E
Bavispe →, *Mexico* **162 B3** 29 30N 109 11W
Bawdwin, *Burma* **90 D6** 23 5N 97 20 E
Bawean, *Indonesia* **85 D4** 5 46 S 112 35 E
Bawku, *Ghana* **113 C4** 11 3N 0 19W
Bawlake, *Burma* **90 F6** 19 11N 97 21 E
Bawolung, *China* **76 C3** 28 50N 101 16 E
Baxley, *U.S.A.* **152 D7** 31 47N 82 21W
Baxoi, *China* **76 B1** 30 1N 96 50 E
Baxter, *U.S.A.* **156 C3** 41 49N 93 9W
Baxter Springs, *U.S.A.* **155 G7** 37 2N 94 44W
Bay Bulls, *Canada* **141 C9** 47 19N 52 50W
Bay City, *Mich., U.S.A.* **148 D4** 43 36N 83 54W
Bay City, *Oreg., U.S.A.* **158 D2** 45 31N 123 53W
Bay City, *Tex., U.S.A.* **155 L7** 28 59N 95 58W
Bay de Verde, *Canada* **141 C9** 48 5N 52 54W
Bay Minette, *U.S.A.* **149 K2** 30 53N 87 46W
Bay St. Louis, *U.S.A.* **155 K10** 30 19N 89 20W
Bay Springs, *U.S.A.* **155 K10** 31 59N 89 17W
Bay View, *N.Z.* **130 F5** 39 25 S 176 50 E
Baya, *Dem. Rep. of the Congo* **119 E2** 11 53 S 27 25 E
Bayambang, *Phil.* **80 D3** 15 49N 120 27 E
Bayamo, *Cuba* **164 B4** 20 20N 76 40W
Bayamón, *Puerto Rico* **165 C6** 18 24N 66 10W
Bayan Har Shan, *China* **72 C4** 34 0N 98 0 E
Bayan Hot = Alxa Zuoqi, *China* **74 E3** 38 50N 105 40 E
Bayan Obo, *China* **74 D5** 41 52N 109 59 E
Bayan-Ovoo, *Mongolia* **74 C4** 42 55N 106 5 E
Bayana, *India* **92 F7** 26 55N 77 18 E
Bayanaūyl, *Kazakstan* **64 D8** 50 45N 75 45 E
Bayandalay, *Mongolia* **74 C2** 43 30N 103 29 E
Bayanhongor, *Mongolia* **72 B5** 46 8N 102 43 E
Bayard, *U.S.A.* **154 E3** 41 45N 103 20W
Bayawan, *Phil.* **81 G4** 9 46N 122 45 E
Baybay, *Phil.* **81 F5** 10 40N 124 55 E
Bayburt, *Turkey* **101 B9** 40 15N 40 20 E
Bayerische Alpen, *Germany* **31 H7** 47 35N 11 30 E
Bayerischer Wald, *Germany* **31 G8** 48 56N 12 50 E
Bayern □, *Germany* **31 G7** 48 50N 12 0 E
Bayeux, *France* **26 C6** 49 17N 0 42W
Bayfield, *Canada* **150 C3** 43 34N 81 42W
Bayfield, *U.S.A.* **154 B9** 46 49N 90 49W
Bayḥān al Qisāb, *Yemen* **98 D4** 15 48N 45 44 E
Bayındır, *Turkey* **49 C9** 38 13N 27 39 E
Baykadam, *Kazakstan* **63 B4** 43 48N 69 58 E
Baykal, Oz., *Russia* **65 D11** 53 0N 108 0 E
Baykonur = Bayqongyr, *Kazakstan* **64 E7** 47 48N 65 50 E
Baymak, *Russia* **62 E7** 52 36N 58 19 E
Baynes Mts., *Namibia* **116 B1** 17 15 S 13 0 E
Bayombong, *Phil.* **80 C3** 16 30N 121 10 E
Bayon, *France* **27 D13** 48 30N 6 20 E
Bayona = Baiona, *Spain* **42 C2** 42 6N 8 52W
Bayonne, *France* **28 E2** 43 30N 1 28W
Bayonne, *U.S.A.* **151 F10** 40 40N 74 7W
Bayou George, *U.S.A.* **152 E4** 30 16N 85 33W
Bayovar, *Peru* **172 B1** 5 50 S 81 0W
Bayport, *U.S.A.* **153 G7** 28 32N 82 39W
Bayqongyr, *Kazakstan* **64 E7** 47 48N 65 50 E
Bayram-Ali = Bayramaly, *Turkmenistan* **64 F7** 37 37N 62 10 E
Bayramaly, *Turkmenistan* **64 F7** 37 37N 62 10 E
Bayramiç, *Turkey* **49 B8** 39 48N 26 36 E
Bayreuth, *Germany* **31 F7** 49 56N 11 35 E
Bayrischzell, *Germany* **31 H8** 47 41N 12 0 E
Bayrūt, *Lebanon* **103 B4** 33 53N 35 31 E
Bayshore, *U.S.A.* **153 J8** 26 43N 81 50W
Baysun, *Uzbekistan* **63 D3** 38 12N 67 12 E
Bayt al Faqīh, *Yemen* **98 D3** 14 31N 43 19 E
Bayt Lahm, *West Bank* **103 D4** 31 43N 35 12 E
Baytown, *U.S.A.* **155 L7** 29 43N 94 59W
Bayzhansay, *Kazakstan* **63 B4** 43 14N 69 54 E
Bayzo, *Niger* **113 C5** 13 52N 4 35 E
Baza, *Spain* **43 H8** 37 30N 2 47W
Bazar Dyuzi, *Russia* **61 K8** 41 12N 47 50 E
Bazardüzü = Bazar Dyuzi, *Russia* **61 K8** 41 12N 47 50 E
Bazarny Karabulak, *Russia* **60 D8** 52 20N 46 29 E
Bazarnyy Syzgan, *Russia* **60 D8** 53 45N 46 40 E
Bazaruto, I. do, *Mozam.* **117 C6** 21 40 S 35 28 E
Bazas, *France* **28 D3** 44 27N 0 13W
Bazhong, *China* **76 B6** 31 52N 106 46 E
Bazmān, Kūh-e, *Iran* **97 D9** 28 4N 60 1 E
Beabula, *Australia* **129 C6** 34 26 S 145 9 E
Beach, *U.S.A.* **154 B3** 46 58N 104 0W

Bengal, Bay of 189

Beach City, *U.S.A.* **150 F3** 40 39N 81 35W
Beachport, *Australia* **128 D4** 37 29S 140 0 E
Beachy Hd., *U.K.* **21 G8** 50 44N 0 15 E
Beacon, *Australia* **125 F2** 30 26S 117 52 E
Beacon, *U.S.A.* **151 E11** 41 30N 73 58W
Beacon Hill, *U.S.A.* **152 F4** 29 55N 85 23W
Beaconia, *Canada* **143 C9** 50 25N 96 31W
Beagle, Canal, *S. Amer.* **176 E3** 55 0S 68 30W
Beagle Bay, *Australia* **124 C3** 16 58S 122 40 E
Bealanana, *Madag.* **117 A8** 14 33S 48 44 E
Beamsville, *Canada* **150 C5** 43 12N 79 28W
Bear →, *U.S.A.* **160 G5** 38 56N 121 36W
Béar, C., *France* **28 F7** 42 31N 3 8 E
Bear I., *Ireland* **23 E2** 51 38N 9 50W
Bear L., *B.C., Canada* **142 B3** 56 10N 126 52W
Bear L., *Man., Canada* **143 B9** 55 8N 96 0W
Bear L., *U.S.A.* **158 F8** 41 59N 111 21W
Bearcreek, *U.S.A.* **158 D9** 45 11N 109 6W
Beardmore, *Canada* **140 C2** 49 36N 87 57W
Beardmore Glacier, *Antarctica* **7 E11** 84 30S 170 0 E
Beardstown, *U.S.A.* **156 E6** 40 1N 90 26W
Béarn, *France* **28 E3** 43 20N 0 30W
Bearpaw Mts., *U.S.A.* **158 B9** 48 12N 109 30W
Bearskin Lake, *Canada* **140 B1** 53 58N 91 2W
Beas de Segura, *Spain* **43 G8** 38 15N 2 53W
Beasain, *Spain* **40 B2** 43 3N 2 11W
Beata, C., *Dom. Rep.* **165 C5** 17 40N 71 30W
Beata, I., *Dom. Rep.* **165 C5** 17 34N 71 31W
Beatrice, *U.S.A.* **154 E6** 40 16N 96 45W
Beatrice, *Zimbabwe* **119 F3** 18 15S 30 55 E
Beatrice, C., *Australia* **126 A2** 14 20S 136 55 E
Beatton →, *Canada* **142 B4** 56 15N 120 45W
Beatton River, *Canada* **142 B4** 57 26N 121 20W
Beatty, *U.S.A.* **160 J10** 36 54N 116 46W
Beaucaire, *France* **29 E8** 43 48N 4 39 E
Beauce, Plaine de la, *France* **27 D8** 48 10N 1 45 E
Beauceville, *Canada* **141 C5** 46 13N 70 46W
Beauchêne, I., *Falk. Is.* **176 D5** 52 55S 59 15W
Beaudesert, *Australia* **127 D5** 27 59S 153 0 E
Beaufort, *Australia* **128 D5** 37 25S 143 25 E
Beaufort, *France* **29 C10** 45 44N 6 34 E
Beaufort, *Malaysia* **85 A5** 5 30N 115 40 E
Beaufort, *N.C., U.S.A.* **149 H7** 34 43N 76 40W
Beaufort, *S.C., U.S.A.* **152 C9** 32 26N 80 40W
Beaufort Sea, *Arctic* **6 B1** 72 0N 140 0W
Beaufort West, *S. Africa* **116 E3** 32 18S 22 36 E
Beaugency, *France* **27 E8** 47 47N 1 38 E
Beauharnois, *Canada* **140 C5** 45 20N 73 52W
Beaujeu, *France* **27 F11** 46 10N 4 35 E
Beaujolais, *France* **27 F11** 46 0N 4 22 E
Beaulieu →, *Canada* **142 A6** 62 3N 113 11W
Beaulieu-sur-Dordogne, *France* **28 D5** 44 58N 1 50 E
Beaulieu-sur-Mer, *France* **29 E11** 43 42N 7 20 E
Beauly, *U.K.* **22 D4** 57 30N 4 28W
Beauly →, *U.K.* **22 D4** 57 29N 4 27W
Beaumaris, *U.K.* **20 D3** 53 16N 4 6W
Beaumont, *Belgium* **24 D4** 50 15N 4 14 E
Beaumont, *France* **28 D4** 44 45N 0 46 E
Beaumont, *N.Z.* **131 F4** 45 50S 169 33 E
Beaumont, *U.S.A.* **155 K7** 30 5N 94 6W
Beaumont-de-Lomagne, *France* **28 E5** 43 53N 1 0 E
Beaumont-le-Roger, *France* **26 C7** 49 4N 0 47 E
Beaumont-sur-Sarthe, *France* **26 D7** 48 13N 0 8 E
Beaune, *France* **27 E11** 47 2N 4 50 E
Beaune-la-Rolande, *France* **27 D9** 48 4N 2 25 E
Beaupréau, *France* **26 E6** 47 12N 1 0W
Beauraing, *Belgium* **24 D4** 50 7N 4 57 E
Beaurepaire, *France* **29 C9** 45 22N 5 1 E
Beauséjour, *Canada* **143 C9** 50 5N 96 35W
Beautemps-Beaupré, I., *N. Cal.* **133 K4** 20 24S 166 9 E
Beauvais, *France* **27 C9** 49 25N 2 8 E
Beauval, *Canada* **143 B7** 55 9N 107 37W
Beauvoir-sur-Mer, *France* **26 F4** 46 55N 2 2W
Beauvoir-sur-Niort, *France* **28 B3** 46 12N 0 30W
Beaver, *Okla., U.S.A.* **155 G4** 36 49N 100 31W
Beaver, *Pa., U.S.A.* **150 F4** 40 42N 80 19W
Beaver, *Utah, U.S.A.* **159 G7** 38 17N 112 38W
Beaver →, *B.C., Canada* **142 B4** 59 52N 124 20W
Beaver →, *Ont., Canada* **140 A2** 55 55N 87 48W
Beaver →, *Sask., Canada* **143 B7** 55 26N 107 45W
Beaver City, *U.S.A.* **154 E5** 40 8N 99 50W
Beaver Dam, *U.S.A.* **154 D10** 43 28N 88 50W
Beaver Falls, *U.S.A.* **150 F4** 40 46N 80 20W
Beaver Hill L., *Canada* **143 C10** 54 5N 94 50W
Beaver I., *U.S.A.* **148 C3** 45 40N 85 33W
Beavercreek, *U.S.A.* **157 E12** 39 43N 84 11W
Beaverhill L., *Alta., Canada* **142 C6** 53 27N 112 32W
Beaverhill L., *N.W.T., Canada* **143 A8** 63 2N 104 22W
Beaverlodge, *Canada* **142 B5** 55 11N 119 29W
Beavermouth, *Canada* **142 C5** 51 32N 117 23W
Beaverstone →, *Canada* **140 B2** 54 59N 89 25W
Beaverton, *Canada* **150 B5** 44 26N 79 9W
Beaverton, *U.S.A.* **160 E4** 45 29N 122 48W
Beaverville, *U.S.A.* **157 D9** 40 57N 87 39W
Beawar, *India* **92 F6** 26 3N 74 18 E
Bebedouro, *Brazil* **175 A6** 21 0S 48 25W
Beboa, *Madag.* **117 B7** 17 22S 44 33 E
Bebra, *Germany* **30 E5** 50 58N 9 48 E
Beccles, *U.K.* **21 E9** 52 27N 1 35 E
Bečej, *Serbia, Yug.* **52 E5** 45 36N 20 3 E
Beceni, *Romania* **53 E11** 45 23N 26 48 E
Becerreá, *Spain* **42 C3** 42 51N 7 10W
Béchar, *Algeria* **111 B4** 31 38N 2 18W
Bechyně, *Czech Rep.* **34 B7** 49 17N 14 29 E
Beckley, *U.S.A.* **148 G5** 37 47N 81 11W
Beckum, *Germany* **30 D4** 51 45N 8 3 E
Beclean, *Romania* **53 C9** 47 11N 24 11 E
Bečov nad Teplou, *Czech Rep.* **34 A5** 50 5N 12 49 E
Bečva →, *Czech Rep.* **35 B10** 49 31N 17 20 E
Bédar, *Spain* **41 H3** 37 11N 1 59W
Bédarieux, *France* **28 E7** 43 37N 3 10 E
Beddouza, Ras, *Morocco* **110 B3** 32 33N 9 9W
Bedel, Pereval, *Kyrgyzstan* **63 C9** 41 26N 78 26 E
Bedele, *Ethiopia* **107 F4** 8 31N 36 23 E
Bederkesa, *Germany* **30 B4** 53 37N 8 50 E
Bederwanak, *Somali Rep.* **120 C2** 9 34N 44 23 E
Bedeso, *Ethiopia* **107 F5** 9 58N 40 52 E
Bedford, *Canada* **140 C5** 45 7N 72 59W
Bedford, *S. Africa* **116 E4** 32 40S 26 10 E
Bedford, *U.K.* **21 E7** 52 8N 0 28W
Bedford, *Ind., U.S.A.* **157 F10** 38 52N 86 29W
Bedford, *Iowa, U.S.A.* **156 D2** 40 40N 94 44W
Bedford, *Ky., U.S.A.* **157 F11** 38 36N 85 19W
Bedford, *Ohio, U.S.A.* **150 E3** 41 23N 81 32W
Bedford, *Pa., U.S.A.* **150 F6** 40 1N 78 30W
Bedford, *Va., U.S.A.* **148 G6** 37 20N 79 31W
Bedford, C., *Australia* **126 B4** 15 14S 145 21 E
Bedford Downs, *Australia* **124 C4** 17 19S 127 20 E
Bedfordshire □, *U.K.* **21 E7** 52 4N 0 28W
Bedi, *Chad* **109 F3** 11 6N 18 33 E
Będków, *Poland* **55 G6** 51 36N 19 44 E
Bednja →, *Croatia* **45 B13** 46 20N 16 52 E
Bednodemyanovsk, *Russia* **60 D6** 53 55N 43 15 E
Bedónia, *Italy* **44 D6** 44 30N 9 38 E
Bedourie, *Australia* **126 C2** 24 30S 139 30 E
Bedretto, *Switz.* **33 C7** 46 31N 8 31 E
Bedum, *Neths.* **24 A6** 53 18N 6 36 E
Będzin, *Poland* **55 H6** 50 19N 19 7 E
Bee Ridge, *U.S.A.* **153 H7** 27 17N 82 29W
Beech Fork →, *U.S.A.* **157 G11** 37 46N 85 41W
Beech Grove, *U.S.A.* **157 E10** 39 44N 86 3W
Beecher, *U.S.A.* **157 C9** 41 21N 87 38W
Beechworth, *Australia* **129 D7** 36 22S 146 43 E
Beechy, *Canada* **143 C7** 50 53N 107 24W
Beelitz, *Germany* **30 C8** 52 14N 12 58 E
Beenleigh, *Australia* **127 D5** 27 43S 153 10 E
Be'er Menuḥa, *Israel* **96 D2** 30 19N 35 8 E
Be'er Sheva, *Israel* **103 D3** 31 15N 34 48 E
Beersheba = Be'er Sheva, *Israel* **103 D3** 31 15N 34 48 E
Beeskow, *Germany* **30 C10** 52 10N 14 15 E
Beeston, *U.K.* **20 E6** 52 56N 1 14W
Beetaloo, *Australia* **126 B1** 17 15S 133 50 E
Beetzendorf, *Germany* **30 C7** 52 42N 11 6 E
Beeville, *U.S.A.* **155 L6** 28 24N 97 45W
Befale, *Dem. Rep. of the Congo* **114 B4** 0 25N 20 45 E
Befandriana, *Madag.* **117 C7** 21 55S 44 0 E
Befotaka, *Madag.* **117 C8** 23 49S 47 0 E
Bega, *Australia* **129 D8** 36 41S 149 51 E
Bega, Canalul, *Romania* **52 E5** 45 37N 20 46 E
Bégard, *France* **26 D3** 48 38N 3 18W
Beğendik, *Turkey* **51 F10** 40 55N 26 34 E
Begndal, *Norway* **18 D6** 60 49N 9 46 E
Begusarai, *India* **93 G12** 25 24N 86 9 E
Behābād, *Iran* **97 C8** 32 24N 59 47 E
Behara, *Madag.* **117 C8** 24 55S 46 20 E
Behbehān, *Iran* **97 D6** 30 30N 50 15 E
Behshahr, *Iran* **97 B7** 36 45N 53 35 E
Bei Jiang →, *China* **77 F9** 23 2N 112 58 E
Bei'an, *China* **73 B7** 48 10N 126 20 E
Beihai, *China* **76 G7** 21 28N 109 6 E
Beijing, *China* **74 E9** 39 55N 116 20 E
Beijing □, *China* **74 E9** 39 55N 116 20 E
Beilen, *Neths.* **24 B6** 52 52N 6 27 E
Beiliu, *China* **77 F8** 22 41N 110 21 E
Beilngries, *Germany* **31 F7** 49 2N 11 28 E
Beilpajah, *Australia* **128 B5** 32 54S 143 52 E
Beilul, *Eritrea* **107 E5** 13 2N 42 20 E
Beinn na Faoghla = Benbecula, *U.K.* **22 D1** 57 26N 7 21W
Beipiao, *China* **75 D11** 41 52N 120 32 E
Beira, *Mozam.* **119 F3** 19 50S 34 52 E
Beira, *Somali Rep.* **120 C3** 6 57N 47 19 E
Beirut = Bayrūt, *Lebanon* **103 B4** 33 53N 35 31 E
Beitaolaizhao, *China* **75 B13** 44 58N 125 58 E
Beitbridge, *Zimbabwe* **119 G3** 22 12S 30 0 E
Beiuş, *Romania* **52 D7** 46 40N 22 21 E
Beizhen, *Liaoning, China* **75 D11** 41 38N 121 54 E
Beizhen, *Shandong, China* **75 F10** 37 20N 118 2 E
Beizhengzhen, *China* **75 B12** 44 31N 123 30 E
Beja, *Portugal* **43 G3** 38 2N 7 53W
Béja, *Tunisia* **108 A1** 36 43N 9 12 E
Beja □, *Portugal* **43 H3** 37 55N 7 55W
Bejaia, *Algeria* **111 A6** 36 42N 5 2 E
Béjar, *Spain* **42 E5** 40 23N 5 46W
Bejestān, *Iran* **97 C8** 34 30N 58 5 E
Bekabad, *Uzbekistan* **63 C4** 40 13N 69 14 E
Bekasi, *Indonesia* **84 D3** 6 14S 106 59 E
Bekçiler, *Turkey* **49 E11** 36 56N 29 44 E
Békés, *Hungary* **52 D6** 46 47N 21 9 E
Békés □, *Hungary* **52 D6** 46 45N 21 0 E
Békéscsaba, *Hungary* **52 D6** 46 40N 21 5 E
Bekilli, *Turkey* **49 C11** 38 17N 29 27 E
Bekily, *Madag.* **117 C8** 24 13S 45 19 E
Bekoji, *Ethiopia* **107 F4** 7 40N 39 17 E
Bekok, *Malaysia* **87 L4** 2 20N 103 7 E
Bekwai, *Ghana* **113 D4** 6 30N 1 34W
Bela, *India* **93 G10** 25 50N 82 0 E
Bela, *Pakistan* **91 D2** 26 12N 66 20 E
Bela Crkva, *Serbia, Yug.* **52 F6** 44 55N 21 27 E
Bela Palanka, *Serbia, Yug.* **50 C6** 43 13N 22 17 E
Bela Vista, *Brazil* **174 A4** 22 12S 56 20W
Bela Vista, *Mozam.* **117 D5** 26 10S 32 44 E
Bélâbre, *France* **28 B5** 46 34N 1 8 E
Belaga, *Malaysia* **85 B4** 2 42N 113 47 E
Belalcázar, *Spain* **43 G5** 38 35N 5 10W
Belanovica, *Serbia, Yug.* **50 B4** 44 15N 20 23 E
Belarus ■, *Europe* **58 F4** 53 30N 27 0 E
Belas, *Angola* **115 D2** 8 55S 13 9 E
Belau = Palau ■, *Pac. Oc.* **134 G5** 7 30N 134 30 E
Belavenona, *Madag.* **117 C8** 24 50S 47 4 E
Belawan, *Indonesia* **84 B1** 3 33N 98 32 E
Belaya, *Ethiopia* **107 E4** 11 25N 36 8 E
Belaya →, *Russia* **62 D6** 54 40N 56 0 E
Belaya Glina, *Russia* **61 G5** 46 5N 40 48 E
Belaya Kalitva, *Russia* **61 F5** 48 13N 40 50 E
Belaya Kholunitsa, *Russia* **62 B3** 58 51N 50 53 E
Belaya Tserkov = Bila Tserkva, *Ukraine* **59 H6** 49 45N 30 10 E
Belayan →, *Indonesia* **85 C5** 0 14S 116 36 E
Belceşti, *Romania* **53 C12** 47 19N 27 7 E
Bełchatów, *Poland* **55 G6** 51 21N 19 22 E
Belcher Is., *Canada* **139 C12** 56 15N 78 45W
Belchite, *Spain* **40 D4** 41 18N 0 43W
Belden, *U.S.A.* **160 E5** 40 2N 121 17W
Belebey, *Russia* **62 D5** 54 7N 54 7 E
Belém, *Brazil* **170 B2** 1 20S 48 30W
Belém de São Francisco, *Brazil* **170 C4** 8 46S 38 58W
Belén, *Argentina* **174 B2** 27 40S 67 5W
Belén, *Colombia* **168 C2** 1 26N 75 56W
Belén, *Paraguay* **174 A4** 23 30S 57 6W
Belen, *U.S.A.* **159 J10** 34 40N 106 46W
Belene, *Bulgaria* **51 C9** 43 39N 25 10 E
Beleni, *Turkey* **100 D7** 36 31N 36 10 E
Bélesta, *France* **28 F5** 42 55N 1 56 E
Belet Uen, *Somali Rep.* **120 D3** 4 30N 45 5 E
Belev, *Russia* **58 F9** 53 50N 36 5 E
Belevi, *Turkey* **49 C9** 38 0N 27 28 E
Belfair, *U.S.A.* **160 C4** 47 27N 122 50W
Belfast, *N.Z.* **131 D7** 43 27S 172 39 E
Belfast, *S. Africa* **117 D5** 25 42S 30 2 E
Belfast, *U.K.* **23 B6** 54 37N 5 56W
Belfast, *Maine, U.S.A.* **141 D6** 44 26N 69 1W
Belfast, *N.Y., U.S.A.* **150 D6** 42 21N 78 7W
Belfast L., *U.K.* **23 B6** 54 40N 5 50W
Belfield, *U.S.A.* **154 B3** 46 53N 103 12W
Belfort, *France* **27 E13** 47 38N 6 50 E
Belfort, Territoire de □, *France* **27 E13** 47 40N 6 55 E
Belfry, *U.S.A.* **158 D9** 45 9N 109 1W
Belgaum, *India* **95 G2** 15 55N 74 35 E
Belgioioso, *Italy* **44 C6** 45 10N 9 19 E
Belgium ■, *Europe* **24 D4** 50 30N 5 0 E
Belgodère, *France* **29 F13** 42 35N 9 1 E
Belgorod, *Russia* **59 G9** 50 35N 36 35 E
Belgorod-Dnestrovskiy = Bilhorod-Dnistrovskyy, *Ukraine* **59 J6** 46 11N 30 23 E
Belgrade = Beograd, *Serbia, Yug.* **50 B4** 44 50N 20 37 E
Belgrade, *U.S.A.* **158 D8** 45 47N 111 11W
Belgrove, *N.Z.* **131 B7** 41 27S 172 59 E
Belhaven, *U.S.A.* **149 H7** 35 33N 76 37W
Beli Drim →, *Europe* **50 D4** 42 6N 20 25 E
Beli Manastir, *Croatia* **52 E3** 45 45N 18 36 E
Beli Timok →, *Serbia, Yug.* **50 C6** 43 53N 22 14 E
Bélice →, *Italy* **46 E5** 37 35N 12 55 E
Belinga, *Gabon* **114 B2** 1 10N 13 2 E
Belinskiy, *Russia* **60 D6** 53 0N 43 25 E
Belinyu, *Indonesia* **84 C3** 1 35S 105 50 E
Beliton Is. = Belitung, *Indonesia* **85 C3** 3 10S 107 50 E
Belitung, *Indonesia* **85 C3** 3 10S 107 50 E
Beliu, *Romania* **52 D6** 46 30N 22 0 E
Belize ■, *Cent. Amer.* **163 D7** 17 0N 88 30W
Belize City, *Belize* **163 D7** 17 25N 88 0W
Beljakovci, *Macedonia* **50 D5** 42 6N 21 59 E
Beljanica, *Serbia, Yug.* **50 B5** 44 8N 21 43 E
Belkovskiy, Ostrov, *Russia* **65 B14** 75 32N 135 44 E
Bell, *U.S.A.* **152 F7** 29 45N 82 52W
Bell →, *Canada* **140 C4** 49 48N 77 38W
Bell Bay, *Australia* **126 G4** 41 6S 146 53 E
Bell I., *Canada* **141 B8** 50 46N 55 35W
Bell-Irving →, *Canada* **142 B3** 56 12N 129 5W
Bell Peninsula, *Canada* **139 B11** 63 50N 82 0W
Bell Ville, *Argentina* **174 C3** 32 40S 62 40W
Bella, *Italy* **47 B8** 40 45N 15 32 E
Bella Bella, *Canada* **142 C3** 52 10N 128 10W
Bella Coola, *Canada* **142 C3** 52 25N 126 40W
Bella Flor, *Bolivia* **172 C4** 11 9S 67 49W
Bella Unión, *Uruguay* **174 C4** 30 15S 57 40W
Bella Vista, *Corrientes, Argentina* **174 B4** 28 33S 59 0W
Bella Vista, *Tucuman, Argentina* **174 B2** 27 10S 65 25W
Bellac, *France* **28 B5** 46 7N 1 3 E
Bellágio, *Italy* **44 C6** 45 59N 9 15 E
Bellaire, *U.S.A.* **150 F4** 40 1N 80 45W
Bellária, *Italy* **45 D9** 44 9N 12 28 E
Bellary, *India* **95 G3** 15 10N 76 56 E
Bellata, *Australia* **127 D4** 29 53S 149 46 E
Belle, *U.S.A.* **156 F5** 38 17N 91 43W
Belle Fourche, *U.S.A.* **154 C3** 44 40N 103 51W
Belle Fourche →, *U.S.A.* **154 C3** 44 26N 102 18W
Belle Glade, *U.S.A.* **149 M5** 26 41N 80 40W
Belle-Île, *France* **26 E3** 47 20N 3 10W
Belle Isle, *Canada* **141 B8** 51 57N 55 25W
Belle Isle, *U.S.A.* **153 G8** 28 27N 81 21W
Belle Isle, Str. of, *Canada* **141 B8** 51 30N 56 30W
Belle Plaine, *Iowa, U.S.A.* **156 C4** 41 54N 92 17W
Belle Plaine, *Minn., U.S.A.* **154 C8** 44 37N 93 46W
Belle Rive, *U.S.A.* **157 F8** 38 14N 88 45W
Belle Yella, *Liberia* **112 D3** 7 24N 10 0W
Belledonne, *France* **29 C10** 45 20N 6 10 E
Belledune, *Canada* **141 C6** 47 55N 65 50W
Bellefontaine, *U.S.A.* **157 D13** 40 22N 83 46W
Bellefonte, *U.S.A.* **150 F7** 40 55N 77 47W
Bellegarde, *France* **27 E9** 47 59N 2 26 E
Bellegarde-en-Marche, *France* **28 C6** 45 59N 2 18 E
Bellegarde-sur-Valserine, *France* **27 F12** 46 4N 5 50 E
Bellême, *France* **26 D7** 48 22N 0 34 E
Belleoram, *Canada* **141 C8** 47 31N 55 25W
Belleview, *U.S.A.* **153 F7** 29 4N 82 3W
Belleville, *Canada* **140 D4** 44 10N 77 23W
Belleville, *France* **27 F11** 46 7N 4 45 E
Belleville, *Ill., U.S.A.* **156 F7** 38 31N 89 59W
Belleville, *Kans., U.S.A.* **154 F6** 39 50N 97 38W
Belleville, *N.Y., U.S.A.* **151 C8** 43 46N 76 10W
Belleville-sur-Vie, *France* **26 F5** 46 46N 1 25W
Bellevue, *Canada* **142 D6** 49 35N 114 22W
Bellevue, *Idaho, U.S.A.* **158 E6** 43 28N 114 16W
Bellevue, *Iowa, U.S.A.* **156 B6** 42 16N 90 26W
Bellevue, *Mich., U.S.A.* **157 B11** 42 27N 85 1W
Bellevue, *Ohio, U.S.A.* **150 E2** 41 17N 82 51W
Bellevue, *Wash., U.S.A.* **160 C4** 47 37N 122 12W
Belley, *France* **29 C9** 45 46N 5 41 E
Bellflower, *U.S.A.* **156 F5** 39 0N 91 21W
Bellin = Kangirsuk, *Canada* **139 C13** 60 0N 70 0W
Bellinge, *Denmark* **17 J4** 55 20N 10 20 E
Bellingen, *Australia* **129 A10** 30 25S 152 50 E
Bellingham, *U.S.A.* **160 B4** 48 46N 122 29W
Bellingshausen Sea, *Antarctica* **7 C17** 66 0S 80 0W
Bellinzona, *Switz.* **33 D8** 46 11N 9 1 E
Bello, *Colombia* **168 B2** 6 20N 75 33W
Bellona, *Solomon Is.* **133 N10** 11 17S 159 47 E
Bellows Falls, *U.S.A.* **151 C12** 43 8N 72 27W
Bellpat, *Pakistan* **92 E3** 29 0N 68 5 E
Bellpuig d'Urgell, *Spain* **40 D6** 41 37N 1 1 E
Belluno, *Italy* **45 B9** 46 9N 12 13 E
Bellville, *Ga., U.S.A.* **152 C8** 32 9N 81 59W
Bellville, *Tex., U.S.A.* **155 L6** 29 57N 96 15W
Bellwood, *U.S.A.* **150 F6** 40 36N 78 20W
Bélmez, *Spain* **43 G5** 38 17N 5 17W
Belmond, *U.S.A.* **156 B3** 42 51N 93 37W
Belmont, *Australia* **129 B9** 33 4S 151 42 E
Belmont, *Canada* **150 D3** 42 53N 81 5W
Belmont, *S. Africa* **116 D3** 29 28S 24 22 E
Belmont, *U.S.A.* **150 D6** 42 14N 78 2W
Belmonte, *Brazil* **171 E4** 16 0S 39 0W
Belmonte, *Portugal* **42 E3** 40 21N 7 20W
Belmonte, *Spain* **41 F2** 39 34N 2 43W
Belmopan, *Belize* **163 D7** 17 18N 88 30W
Belmore, *Australia* **128 B4** 33 34S 141 13 E
Belmullet, *Ireland* **23 B2** 54 14N 9 58W
Belo Horizonte, *Brazil* **171 E3** 19 55S 43 56W
Belo Jardim, *Brazil* **170 C4** 8 20S 36 26W
Belo-sur-Mer, *Madag.* **117 C7** 20 42S 44 0 E
Belo-Tsiribihina, *Madag.* **117 B7** 19 40S 44 30 E
Belogorsk = Bilohirsk, *Ukraine* **59 K8** 45 3N 34 35 E
Belogorsk, *Russia* **65 D13** 51 0N 128 20 E
Belogradchik, *Bulgaria* **50 C6** 43 53N 22 42 E
Belogradets, *Bulgaria* **51 C11** 43 22N 27 18 E
Beloha, *Madag.* **117 D8** 25 10S 45 3 E
Beloit, *Kans., U.S.A.* **154 F5** 39 28N 98 6W
Beloit, *Wis., U.S.A.* **156 B7** 42 31N 89 2W
Belokorovichi, *Ukraine* **59 G5** 51 7N 28 2 E
Belomorsk, *Russia* **56 B5** 64 35N 34 54 E
Belonia, *India* **90 D3** 23 15N 91 30 E
Belopolye = Bilopillya, *Ukraine* **59 G8** 51 14N 34 20 E
Belorechensk, *Russia* **61 H4** 44 46N 39 52 E
Beloretsk, *Russia* **62 E7** 53 58N 58 24 E
Belorussia = Belarus ■, *Europe* **58 F4** 53 30N 27 0 E
Beloslav, *Bulgaria* **51 C11** 43 11N 27 42 E
Belovo, *Bulgaria* **51 D8** 42 13N 24 1 E
Belovo, *Russia* **64 D9** 54 30N 86 0 E
Belovodsk, *Ukraine* **59 H10** 49 13N 39 36 E
Beloyarskiy, *Russia* **62 C8** 56 45N 61 24 E
Beloye, Ozero, *Russia* **58 B9** 60 10N 37 35 E
Beloye More, *Russia* **56 A6** 66 30N 38 0 E
Belozem, *Bulgaria* **51 D9** 42 12N 25 2 E
Belozersk, *Russia* **58 B9** 60 1N 37 45 E
Belpasso, *Italy* **47 E7** 37 35N 14 58 E
Beltana, *Australia* **128 A3** 30 48S 138 25 E
Belterra, *Brazil* **169 D7** 2 45S 55 0W
Beltinci, *Slovenia* **45 B13** 46 37N 16 20 E
Belton, *S.C., U.S.A.* **149 H4** 34 31N 82 30W
Belton, *Tex., U.S.A.* **155 K6** 31 3N 97 28W
Belton Res., *U.S.A.* **155 K6** 31 8N 97 32W
Beltsy = Bălţi, *Moldova* **53 C12** 47 48N 27 58 E
Belturbet, *Ireland* **23 B4** 54 6N 7 26W
Belukha, *Russia* **64 E9** 49 50N 86 50 E
Beluša, *Slovak Rep.* **35 B11** 49 5N 18 17 E
Belušić, *Serbia, Yug.* **50 C5** 43 50N 21 10 E
Belvedere Marittimo, *Italy* **47 C8** 39 37N 15 52 E
Belvès, *France* **28 D5** 44 46N 1 0 E
Belvidere, *Ill., U.S.A.* **154 D10** 42 15N 88 50W
Belvidere, *N.J., U.S.A.* **151 F9** 40 50N 75 5W
Belvis de la Jara, *Spain* **42 F6** 39 45N 4 57W
Belyando →, *Australia* **126 C4** 21 38S 146 50 E
Belyy, *Russia* **58 E7** 55 49N 33 3 E
Belyy, Ostrov, *Russia* **64 B8** 73 30N 71 0 E
Belyy Yar, *Russia* **64 D9** 58 26N 84 39 E
Belyye Vody, *Kazakstan* **63 B4** 42 25N 69 50 E
Bełżec, *Poland* **55 H10** 50 23N 23 26 E
Belzig, *Germany* **30 C8** 52 8N 12 35 E
Belzoni, *U.S.A.* **155 J9** 33 11N 90 29W
Bełżyce, *Poland* **55 G9** 51 11N 22 17 E
Bemaraha, Lembalemban' i, *Madag.* **117 B7** 18 40S 44 45 E
Bemarivo, *Madag.* **117 C7** 21 45S 44 45 E
Bemarivo →, *Madag.* **117 B8** 15 27S 47 40 E
Bemavo, *Madag.* **117 C8** 21 33S 45 25 E
Bembéréke, *Benin* **113 C5** 10 11N 2 43 E
Bembesi, *Zimbabwe* **119 G2** 20 0S 28 58 E
Bembesi →, *Zimbabwe* **119 F2** 18 57S 27 47 E
Bembézar →, *Spain* **43 H5** 37 45N 5 13W
Bembibre, *Spain* **42 C4** 42 37N 6 25W
Bement, *U.S.A.* **157 E8** 39 55N 88 34W
Bemidji, *U.S.A.* **154 B7** 47 28N 94 53W
Ben, *Iran* **97 C6** 32 32N 50 45 E
Ben Bullen, *Australia* **129 B9** 33 12S 150 2 E
Ben Cruachan, *U.K.* **22 E3** 56 26N 5 8W
Ben Dearg, *U.K.* **22 D4** 57 47N 4 56W
Ben Gardane, *Tunisia* **108 B2** 33 11N 11 11 E
Ben Hope, *U.K.* **22 C4** 58 25N 4 36W
Ben Lawers, *U.K.* **22 E4** 56 32N 4 14W
Ben Lomond, *N.S.W., Australia* **127 E5** 30 1S 151 43 E
Ben Lomond, *Tas., Australia* **126 G4** 41 38S 147 42 E
Ben Lomond, *U.K.* **22 E4** 56 11N 4 38W
Ben Luc, *Vietnam* **87 G6** 10 39N 106 29 E
Ben Macdhui, *U.K.* **22 D5** 57 4N 3 40W
Ben Mhor, *U.K.* **22 D1** 57 15N 7 18W
Ben More, *Arg. & Bute, U.K.* **22 E2** 56 26N 6 1W
Ben More, *Stirl., U.K.* **22 E4** 56 23N 4 32W
Ben More Assynt, *U.K.* **22 C4** 58 8N 4 52W
Ben Nevis, *U.K.* **22 E3** 56 48N 5 1W
Ben Ohau Ra., *N.Z.* **131 E5** 44 1S 170 4 E
Ben Quang, *Vietnam* **86 D6** 17 3N 106 55 E
Ben Slimane, *Morocco* **110 B3** 33 38N 7 7W
Ben Vorlich, *U.K.* **22 E4** 56 21N 4 14W
Ben Wyvis, *U.K.* **22 D4** 57 40N 4 35W
Bena, *Nigeria* **113 C6** 11 20N 5 50 E
Bena Dibele, *Dem. Rep. of the Congo* **115 C4** 4 4S 22 50 E
Bena-Leka, *Dem. Rep. of the Congo* **115 D4** 5 8S 22 10 E
Bena-Tshadi, *Dem. Rep. of the Congo* **115 C4** 4 40S 22 49 E
Benāb, *Iran* **101 D12** 37 20N 46 4 E
Benadir, *Somali Rep.* **120 D2** 1 30N 44 30 E
Benagerie, *Australia* **128 A4** 31 25S 140 22 E
Benahmed, *Morocco* **110 B3** 33 4N 7 9W
Benalla, *Australia* **129 D7** 36 30S 146 0 E
Benalmádena, *Spain* **43 J6** 36 36N 4 34W
Benambra, Mt., *Australia* **129 D7** 36 31S 147 34 E
Benares = Varanasi, *India* **93 G10** 25 22N 83 0 E
Bénat, C., *France* **29 E10** 43 5N 6 22 E
Benavente, *Portugal* **43 G2** 38 59N 8 49W
Benavente, *Spain* **42 C5** 42 2N 5 43W
Benavides, *U.S.A.* **155 M5** 27 36N 98 25W
Benavides de Órbigo, *Spain* **42 C5** 42 30N 5 54W
Benbecula, *U.K.* **22 D1** 57 26N 7 21W
Benbonyathe, *Australia* **128 A3** 30 25S 139 11 E
Bencubbin, *Australia* **125 F2** 30 48S 117 52 E
Bend, *U.S.A.* **158 D3** 44 4N 121 19W
Bendela, *Dem. Rep. of the Congo* **114 C3** 3 18S 17 36 E
Bender Beila, *Somali Rep.* **120 C4** 9 30N 50 48 E
Bender Merchagno, *Somali Rep.* **120 B4** 11 41N 50 34 E
Bendering, *Australia* **125 F2** 32 23S 118 18 E
Bendery = Tighina, *Moldova* **53 D14** 46 50N 29 30 E
Bendigo, *Australia* **128 D6** 36 40S 144 15 E
Bendorf, *Germany* **30 E3** 50 25N 7 35 E
Benē Beraq, *Israel* **103 C3** 32 6N 34 51 E
Beneditinos, *Brazil* **170 C3** 5 27S 42 22W
Benedito Leite, *Brazil* **170 C3** 7 13S 44 34W
Bénéna, *Mali* **112 C4** 13 9N 4 17W
Benenitra, *Madag.* **117 C8** 23 27S 45 5 E
Benešov, *Czech Rep.* **34 B7** 49 46N 14 41 E
Benevento, *Italy* **47 A7** 41 8N 14 45 E
Benevolence, *U.S.A.* **152 D5** 31 53N 84 44W
Benfeld, *France* **27 D14** 48 22N 7 34 E
Benga, *Mozam.* **119 F3** 16 11S 33 40 E
Bengal, Bay of, *Ind. Oc.* **66 H12** 15 0N 90 0 E

192 Blairsville

Blairsville, *U.S.A.* **150 F5** 40 26N 79 16W
Blaj, *Romania* **53 D8** 46 10N 23 57 E
Blake Pt., *U.S.A.* **154 A10** 48 11N 88 25W
Blakely, *U.S.A.* **152 D5** 31 23N 84 56W
Blakesburg, *U.S.A.* **156 D4** 40 58N 92 38W
Blakstad, *Norway* **18 F5** 58 30N 8 39 E
Blâmont, *France* **27 D13** 48 35N 6 50 E
Blanc, C., *Spain* **39 B9** 39 21N 2 51 E
Blanc, C., *Tunisia* **108 A1** 37 15N 9 56 E
Blanc, Mont, *Alps* **29 C10** 45 48N 6 50 E
Blanca, B., *Argentina* **176 A4** 39 10 S 61 30W
Blanca Peak, *U.S.A.* **159 H11** 37 35N 105 29W
Blanchard, *U.S.A.* **155 H6** 35 8N 97 39W
Blanchardville, *U.S.A.* **156 B7** 42 49N 89 52W
Blanche, C., *Australia* **127 E1** 33 1 S 134 9 E
Blanche, L., *S. Austral., Australia* **127 D2** 29 15 S 139 40 E
Blanche, L., *W. Austral., Australia* **124 D3** 22 25 S 123 17 E
Blanche Channel, *Solomon Is.* **133 M9** 8 30 S 157 30 E
Blanchester, *U.S.A.* **157 E13** 39 17N 83 59W
Blanco, *S. Africa* **116 E3** 33 55 S 22 23 E
Blanco, *U.S.A.* **155 K5** 30 6N 98 25W
Blanco →, *Argentina* **174 C2** 30 20 S 68 42W
Blanco, C., *Costa Rica* **164 E2** 9 34N 85 8W
Blanco, C., *U.S.A.* **158 E1** 42 51N 124 34W
Blanda →, *Iceland* **11 B6** 65 37N 20 9W
Blandford Forum, *U.K.* **21 G5** 50 51N 2 9W
Blanding, *U.S.A.* **159 H9** 37 37N 109 29W
Blandinsville, *U.S.A.* **156 D6** 40 33N 90 52W
Blanes, *Spain* **40 D7** 41 40N 2 48 E
Blangy-sur-Bresle, *France* **27 C8** 49 55N 1 37 E
Blanice →, *Czech Rep.* **34 B7** 49 10N 14 5 E
Blankaholm, *Sweden* **17 G10** 57 36N 16 31 E
Blankenberge, *Belgium* **24 C3** 51 20N 3 9 E
Blankenburg, *Germany* **30 D6** 51 47N 10 57 E
Blanquefort, *France* **28 D3** 44 55N 0 38W
Blanquillo, *Uruguay* **175 C4** 32 53 S 55 37W
Blansko, *Czech Rep.* **35 B9** 49 22N 16 40 E
Blantyre, *Malawi* **119 F4** 15 45 S 35 0 E
Blarney, *Ireland* **23 E3** 51 56N 8 33W
Blåsjø, *Norway* **18 E3** 59 20N 6 50 E
Błaszki, *Poland* **55 G5** 51 38N 18 30 E
Blatná, *Czech Rep.* **34 B6** 49 25N 13 52 E
Blato, *Croatia* **45 F13** 42 56N 16 48 E
Blatten, *Switz.* **32 D5** 46 20N 7 50 E
Blaubeuren, *Germany* **31 G5** 48 24N 9 46 E
Blaustein, *Germany* **31 G5** 48 25N 9 53 E
Blåvands Huk, *Denmark* **17 J2** 55 33N 8 4 E
Blaydon, *U.K.* **20 C6** 54 58N 1 42W
Blaye, *France* **28 C3** 45 8N 0 40W
Blaye-les-Mines, *France* **28 D6** 44 1N 2 8 E
Blayney, *Australia* **129 B8** 33 32 S 149 14 E
Blaze, Pt., *Australia* **124 B5** 12 56 S 130 11 E
Błażowa, *Poland* **55 J9** 49 53N 22 7 E
Bleckede, *Germany* **30 B6** 53 17N 10 43 E
Bled, *Slovenia* **45 B11** 46 27N 14 7 E
Blefjell, *Norway* **18 E6** 59 48N 9 10 E
Bleiburg, *Austria* **34 E7** 46 35N 14 49 E
Blejeşti, *Romania* **53 F10** 44 19N 25 27 E
Blekinge, *Sweden* **15 H16** 56 25N 15 20 E
Blekinge län □, *Sweden* **17 H9** 56 20N 15 20 E
Blenheim, *Canada* **150 D3** 42 20N 82 0W
Blenheim, *N.Z.* **131 B8** 41 38 S 173 57 E
Bléone →, *France* **29 D10** 44 5N 6 0 E
Blérancourt, *France* **27 C10** 49 31N 3 9 E
Bletchley, *U.K.* **21 F7** 51 59N 0 44W
Blida, *Algeria* **111 A5** 36 30N 2 49 E
Blidet Amor, *Algeria* **111 B6** 32 59N 5 58 E
Blidö, *Sweden* **16 E12** 59 37N 18 53 E
Blidsberg, *Sweden* **17 G7** 57 56N 13 30 E
Blieskastel, *Germany* **31 F3** 49 14N 7 12 E
Bligh Sound, *N.Z.* **131 E2** 44 47 S 167 32 E
Bligh Water, *Fiji* **133 A2** 17 0 S 178 0 E
Blind River, *Canada* **140 C3** 46 10N 82 58W
Blinishti, *Albania* **50 E3** 41 52N 19 59 E
Blinnenhorn, *Switz.* **33 D6** 46 26N 8 19 E
Blissfield, *U.S.A.* **157 C13** 41 50N 83 52W
Blitar, *Indonesia* **85 D4** 8 5 S 112 11 E
Blitchton, *U.S.A.* **152 C8** 32 12N 81 26W
Blitta, *Togo* **113 D5** 8 23N 1 6 E
Block I., *U.S.A.* **151 E13** 41 11N 71 35W
Block Island Sd., *U.S.A.* **151 E13** 41 15N 71 40W
Blockton, *U.S.A.* **156 D2** 40 37N 94 29W
Blodgett Iceberg Tongue, *Antarctica* **7 C9** 66 8 S 130 35 E
Bloemfontein, *S. Africa* **116 D4** 29 6 S 26 7 E
Bloemhof, *S. Africa* **116 D4** 27 38 S 25 32 E
Blois, *France* **26 E8** 47 35N 1 20 E
Blomskog, *Sweden* **16 E6** 59 16N 12 2 E
Blomstermåla, *Sweden* **17 H10** 56 59N 16 21 E
Blomvåg, *Norway* **18 D1** 60 32N 4 50 E
Blönduós, *Iceland* **11 B6** 65 40N 20 12W
Błonie, *Poland* **55 F7** 52 12N 20 37 E
Bloodvein →, *Canada* **143 C9** 51 47N 96 43W
Bloody Foreland, *Ireland* **23 A3** 55 10N 8 17W
Bloomer, *U.S.A.* **154 C9** 45 6N 91 29W
Bloomfield, *Australia* **126 B4** 15 56 S 145 22 E
Bloomfield, *Canada* **150 C7** 43 59N 77 14W
Bloomfield, *Ind., U.S.A.* **157 E10** 39 1N 86 57W
Bloomfield, *Iowa, U.S.A.* **156 D4** 40 45N 92 25W
Bloomfield, *Ky., U.S.A.* **157 G11** 37 55N 85 19W
Bloomfield, *N. Mex., U.S.A.* **159 H10** 36 43N 107 59W
Bloomfield, *Nebr., U.S.A.* **154 D6** 42 36N 97 39W
Bloomingburg, *U.S.A.* **157 E13** 39 36N 83 24W
Bloomington, *Ill., U.S.A.* **156 D8** 40 28N 89 0W
Bloomington, *Ind., U.S.A.* **157 E10** 39 10N 86 32W
Bloomington, *Minn., U.S.A.* **154 C8** 44 50N 93 17W
Bloomington, *Wis., U.S.A.* **156 B6** 42 53N 90 55W
Bloomsburg, *U.S.A.* **151 F8** 41 0N 76 27W
Blora, *Indonesia* **85 D4** 6 57 S 111 25 E
Blossburg, *U.S.A.* **150 E7** 41 41N 77 4W
Blosseville Kyst, *Greenland* **10 D8** 68 50N 26 30W
Blouberg, *S. Africa* **117 C4** 23 8 S 28 59 E
Blountstown, *U.S.A.* **152 E4** 30 27N 85 3W
Bludenz, *Austria* **34 D2** 47 10N 9 50 E
Blue →, *U.S.A.* **157 F10** 38 11N 86 19W
Blue Cypress L., *U.S.A.* **153 H9** 27 44N 80 45W
Blue Island, *U.S.A.* **148 E2** 41 40N 87 40W
Blue Lake, *U.S.A.* **158 F2** 40 53N 123 59W
Blue Mesa Reservoir, *U.S.A.* **159 G10** 38 28N 107 20W
Blue Mound, *U.S.A.* **156 E7** 39 42N 89 7W
Blue Mts., *Australia* **129 B9** 33 40 S 150 0 E
Blue Mts., *Oreg., U.S.A.* **158 D4** 45 15N 119 0W
Blue Mts., *Pa., U.S.A.* **151 F8** 40 30N 76 30W
Blue Mud B., *Australia* **126 A2** 13 30 S 136 0 E
Blue Nile = Nîl el Azraq →, *Sudan* **107 D3** 15 38N 32 31 E
Blue Rapids, *U.S.A.* **154 F6** 39 41N 96 39W
Blue Ridge Mts., *U.S.A.* **149 G5** 36 30N 80 15W
Blue Springs, *U.S.A.* **156 E2** 39 1N 94 17W
Blueberry →, *Canada* **142 B4** 56 45N 120 49W
Bluefield, *U.S.A.* **148 G5** 37 15N 81 17W
Bluefields, *Nic.* **164 D3** 12 20N 83 50W
Blueskin B., *N.Z.* **131 F5** 45 44 S 170 38 E
Bluff, *Australia* **126 C4** 23 35 S 149 4 E
Bluff, *N.Z.* **131 G3** 46 37 S 168 20 E
Bluff, *U.S.A.* **159 H9** 37 17N 109 33W
Bluff Harbour, *N.Z.* **131 G3** 46 36 S 168 21 E
Bluff Knoll, *Australia* **125 F2** 34 24 S 118 15 E
Bluff Pt., *Australia* **125 E1** 27 50 S 114 5 E
Bluffs, *U.S.A.* **156 E6** 39 45N 90 32W
Bluffton, *Ga., U.S.A.* **152 D5** 31 31N 84 52W
Bluffton, *Ind., U.S.A.* **157 D11** 40 44N 85 11W
Bluffton, *Ohio, U.S.A.* **157 D13** 40 54N 83 54W
Bluffton, *S.C., U.S.A.* **152 C9** 32 14N 80 52W
Bluford, *U.S.A.* **157 F8** 38 20N 88 45W
Blumenau, *Brazil* **175 B6** 27 0 S 49 0W
Blümisalphorn, *Switz.* **32 D5** 46 28N 7 47 E
Blunt, *U.S.A.* **154 C5** 44 31N 99 59W
Bly, *U.S.A.* **158 E3** 42 24N 121 3W
Blyth, *Canada* **150 C3** 43 44N 81 26W
Blyth, *U.K.* **20 B6** 55 8N 1 31W
Blythe, *Calif., U.S.A.* **161 M12** 33 37N 114 36W
Blythe, *Ga., U.S.A.* **152 B7** 33 17N 82 12W
Bø, *Norway* **18 E6** 59 25N 9 3 E
Bo, *S. Leone* **112 D2** 7 55N 11 50W
Bo Duc, *Vietnam* **87 G6** 11 58N 106 50 E
Bo Hai, *China* **75 E10** 39 0N 119 0 E
Bō-no-Misaki, *Japan* **70 F2** 31 15N 130 13 E
Bo Xian, *China* **74 H8** 33 55N 115 41 E
Boa Esperança, *Brazil* **169 C5** 3 21N 61 23W
Boa Esperança, Represa, *Brazil* **170 C3** 6 50 S 43 50W
Boa Nova, *Brazil* **171 D3** 14 22 S 40 10W
Boa Viagem, *Brazil* **170 C4** 5 7 S 39 44W
Boa Vista, *Brazil* **169 C5** 2 48N 60 30W
Boa Vista, *C. Verde Is.* **8 G6** 16 0N 22 50W
Boac, *Phil.* **80 E3** 13 27N 121 50 E
Boaco, *Nic.* **164 D2** 12 29N 85 35W
Bo'ai, *China* **74 G7** 35 10N 113 3 E
Boal, *Spain* **42 B4** 43 25N 6 49W
Boali, *C.A.R.* **114 B3** 4 48N 18 7 E
Boardman, *U.S.A.* **150 E4** 41 2N 80 40W
Boatman, *Australia* **127 D4** 27 16 S 146 55 E
Bobadah, *Australia* **129 B7** 32 19 S 146 41 E
Bobai, *China* **76 F7** 22 17N 109 59 E
Bobbili, *India* **94 E6** 18 35N 83 30 E
Bòbbio, *Italy* **44 D6** 44 46N 9 23 E
Bobcaygeon, *Canada* **140 D4** 44 33N 78 33W
Böblingen, *Germany* **31 G5** 48 40N 9 1 E
Bobo-Dioulasso, *Burkina Faso* **112 C4** 11 8N 4 13W
Bobolice, *Poland* **54 E3** 53 58N 16 37 E
Bobon, *Davao, Phil.* **81 H6** 6 53N 126 19 E
Bobon, *Samar, Phil.* **80 E5** 12 32N 124 34 E
Bobonaza →, *Ecuador* **168 D2** 2 36 S 76 38W
Boboshevo, *Bulgaria* **50 D7** 42 9N 23 0 E
Bobov Dol, *Bulgaria* **50 D6** 42 20N 23 0 E
Bóbr →, *Poland* **55 F2** 52 4N 15 4 E
Bobraomby, Tanjon' i, *Madag.* **117 A8** 12 40 S 49 10 E
Bobrinets, *Ukraine* **59 H7** 48 4N 32 5 E
Bobrov, *Russia* **60 E5** 51 5N 40 2 E
Bobrovitsa, *Ukraine* **59 G6** 50 45N 31 23 E
Bobruysk = Babruysk, *Belarus* **59 F5** 53 10N 29 15 E
Bobures, *Venezuela* **168 B3** 9 15N 71 11W
Boca de Drago, *Venezuela* **169 A5** 11 0N 61 50W
Bôca do Acre, *Brazil* **172 B4** 8 50 S 67 27W
Bôca do Jari, *Brazil* **169 D7** 1 7 S 51 58W
Bôca do Moaco, *Brazil* **172 B4** 7 41 S 68 17W
Boca Grande, *U.S.A.* **153 J7** 26 45N 82 16W
Boca Grande, *Venezuela* **169 B5** 8 40N 60 40W
Boca Raton, *U.S.A.* **149 M5** 26 21N 80 5W
Bocaiúva, *Brazil* **171 E3** 17 7 S 43 49W
Bocanda, *Ivory C.* **112 D4** 7 5N 4 31W
Bocaranga, *C.A.R.* **114 A3** 7 0N 15 35 E
Bocas del Toro, *Panama* **164 E3** 9 15N 82 20W
Boceguillas, *Spain* **42 D7** 41 20N 3 39W
Bochnia, *Poland* **55 J7** 49 58N 20 27 E
Bocholt, *Germany* **30 D2** 51 50N 6 36 E
Bochum, *Germany* **30 D3** 51 28N 7 13 E
Bockenem, *Germany* **30 C6** 52 1N 10 8 E
Boćki, *Poland* **55 F10** 52 39N 23 3 E
Bocognano, *France* **29 F13** 42 5N 9 4 E
Boconó, *Venezuela* **168 B3** 9 15N 70 16W
Boconó →, *Venezuela* **168 B4** 8 43N 69 34W
Bocoyna, *Mexico* **162 B3** 27 52N 107 35W
Bocşa, *Romania* **52 E6** 45 21N 21 47 E
Boda, *C.A.R.* **114 B3** 4 19N 17 26 E
Böda, *Sweden* **17 G11** 57 15N 17 3 E
Boda, *Kopparberg, Sweden* **16 C9** 61 1N 15 13 E
Boda, *Västernorrland, Sweden* **16 B10** 62 52N 16 39 E
Bodafors, *Sweden* **17 G8** 57 48N 14 23 E
Bodaybo, *Russia* **65 D12** 57 50N 114 0 E
Boddam, *U.K.* **22 B7** 59 56N 1 17W
Boddington, *Australia* **125 F2** 32 50 S 116 30 E
Bodega Bay, *U.S.A.* **160 G3** 38 20N 123 3W
Boden, *Sweden* **14 D19** 65 50N 21 42 E
Bodensee, *Europe* **33 A8** 47 35N 9 25 E
Bodenteich, *Germany* **30 C6** 52 50N 10 42 E
Bodhan, *India* **94 E3** 18 40N 77 44 E
Bodinayakkanur, *India* **95 J3** 10 2N 77 10 E
Bodinga, *Nigeria* **113 C6** 12 58N 5 10 E
Bodio, *Switz.* **33 D7** 46 23N 8 55 E
Bodmin, *U.K.* **21 G3** 50 28N 4 43W
Bodmin Moor, *U.K.* **21 G3** 50 33N 4 36W
Bodø, *Norway* **14 C16** 67 17N 14 24 E
Bodoquena, Serra da, *Brazil* **173 E6** 21 0 S 56 50W
Bodoupa, *C.A.R.* **114 A3** 5 43N 17 36 E
Bodrog →, *Hungary* **52 B6** 48 11N 21 22 E
Bodrum, *Turkey* **49 D9** 37 3N 27 30 E
Bódva →, *Hungary* **52 B5** 48 19N 20 45 E
Boembé, *Congo* **114 C3** 2 54 S 15 39 E
Boën, *France* **29 C8** 45 44N 4 1 E
Boende, *Dem. Rep. of the Congo* **114 C4** 0 24 S 21 12 E
Boerne, *U.S.A.* **155 L5** 29 47N 98 44W
Boffa, *Guinea* **112 C2** 10 16N 14 3W
Bogale, *Burma* **90 G5** 16 17N 95 24 E
Bogalusa, *U.S.A.* **155 K10** 30 47N 89 52W
Bogan →, *Australia* **129 A7** 29 59 S 146 17 E
Bogan Gate, *Australia* **129 B7** 33 7 S 147 49 E
Bogangolo, *C.A.R.* **114 A3** 5 34N 18 15 E
Bogantungan, *Australia* **126 C4** 23 41 S 147 17 E
Bogata, *U.S.A.* **155 J7** 33 28N 95 13W
Bogatić, *Serbia, Yug.* **50 B3** 44 51N 19 30 E
Boğazkale, *Turkey* **100 B6** 40 2N 34 37 E
Boğazlıyan, *Turkey* **100 C6** 39 11N 35 14 E
Bogdanovich, *Russia* **62 C9** 56 47N 62 1 E
Bogen, *Sweden* **16 D6** 60 4N 12 33 E
Bogense, *Denmark* **17 J4** 55 34N 10 5 E
Bogetići, *Montenegro, Yug.* **50 D2** 42 41N 18 58 E
Boggabilla, *Australia* **127 D5** 28 36 S 150 24 E
Boggabri, *Australia* **129 A9** 30 45 S 150 5 E
Boggeragh Mts., *Ireland* **23 D3** 52 2N 8 55W
Bogia, *Papua N. G.* **132 C3** 4 9 S 145 0 E
Bognor Regis, *U.K.* **21 G7** 50 47N 0 40W
Bogo, *Phil.* **81 F4** 11 3N 124 0 E
Bogodukhov = Bohodukhiv, *Ukraine* **59 G8** 50 9N 35 33 E
Bogong, Mt., *Australia* **129 D7** 36 47 S 147 17 E
Bogor, *Indonesia* **84 D3** 6 36 S 106 48 E
Bogoroditsk, *Russia* **58 F10** 53 47N 38 8 E
Bogorodsk, *Russia* **60 B6** 56 4N 43 30 E
Bogoso, *Ghana* **112 D4** 5 38N 2 3W
Bogotá, *Colombia* **168 C3** 4 34N 74 0W
Bogotol, *Russia* **64 D9** 56 15N 89 50 E
Bogra, *Bangla.* **90 C2** 24 51N 89 22 E
Boguchany, *Russia* **65 D10** 58 40N 97 30 E
Boguchar, *Russia* **60 F5** 49 55N 40 32 E
Bogué, *Mauritania* **112 B2** 16 45N 14 10W
Boguslav, *Ukraine* **59 H6** 49 33N 30 56 E
Boguszów-Gorce, *Poland* **55 H3** 50 45N 16 12 E
Bohain-en-Vermandois, *France* **27 C10** 49 59N 3 28 E
Bohemia Downs, *Australia* **124 C4** 18 53 S 126 14 E
Bohemian Forest = Böhmerwald, *Germany* **31 F9** 49 8N 13 14 E
Bohena Cr. →, *Australia* **127 E4** 30 17 S 149 42 E
Bohinjska Bistrica, *Slovenia* **45 B11** 46 17N 14 1 E
Böhmerwald, *Germany* **31 F9** 49 8N 13 14 E
Bohmte, *Germany* **30 C4** 52 22N 8 19 E
Bohodukhiv, *Ukraine* **59 G8** 50 9N 35 33 E
Bohol, *Phil.* **81 G5** 9 50N 124 10 E
Bohol, *Somali Rep.* **120 C3** 5 45N 46 9 E
Bohol Sea, *Phil.* **81 G5** 9 0N 124 0 E
Bohol Str., *Phil.* **81 G4** 9 45N 123 40 E
Böhönye, *Hungary* **52 D2** 46 25N 17 28 E
Bohotleh, *Somali Rep.* **120 C3** 8 20N 46 25 E
Bohuslän, *Sweden* **17 F5** 58 25N 12 0 E
Boi, *Nigeria* **113 D6** 9 35N 9 27 E
Boi, Pta. de, *Brazil* **175 A6** 23 55 S 45 15W
Boiaçu, *Brazil* **169 D5** 0 27 S 61 46W
Boileau, C., *Australia* **124 C3** 17 40 S 122 7 E
Boipeba, I. de, *Brazil* **171 D4** 13 39 S 38 55W
Boiro, *Spain* **42 C2** 42 39N 8 54W
Bois →, *Brazil* **171 E1** 18 35 S 50 2W
Boise, *U.S.A.* **158 E5** 43 37N 116 13W
Boise City, *U.S.A.* **155 G3** 36 44N 102 31W
Boissevain, *Canada* **143 D8** 49 15N 100 5W
Bóite →, *Italy* **45 B9** 46 5N 12 5 E
Boitzenburg, *Germany* **30 B9** 53 16N 13 35 E
Boizenburg, *Germany* **30 B6** 53 23N 10 43 E
Bojador C., *W. Sahara* **110 C2** 26 0N 14 30W
Bojana →, *Albania* **50 E3** 41 52N 19 22 E
Bojano, *Italy* **47 A7** 41 29N 14 29 E
Bojanowo, *Poland* **55 G3** 51 43N 16 42 E
Bøjden, *Denmark* **17 J4** 55 6N 10 7 E
Bojnūrd, *Iran* **97 B8** 37 30N 57 20 E
Bojonegoro, *Indonesia* **85 D4** 7 11 S 111 54 E
Boju, *Nigeria* **113 D6** 7 22N 7 55 E
Boka, *Serbia, Yug.* **52 E5** 45 22N 20 52 E
Boka Kotorska, *Montenegro, Yug.* **50 D2** 42 23N 18 32 E
Bokada, *Dem. Rep. of the Congo* **114 B3** 4 8N 19 23 E
Bokala, *Ivory C.* **112 D4** 8 31N 4 33W
Bokatola, *Dem. Rep. of the Congo* **114 C3** 0 38 S 18 46 E
Boké, *Guinea* **112 C2** 10 56N 14 17W
Bokhara →, *Australia* **127 D4** 29 55 S 146 42 E
Bokkos, *Nigeria* **113 D6** 9 17N 9 1 E
Boknafjorden, *Norway* **15 G11** 59 14N 5 40 E
Bokombayevskoye, *Kyrgyzstan* **63 B8** 42 10N 76 55 E
Bokoro, *Chad* **109 F3** 12 25N 17 14 E
Bokote, *Dem. Rep. of the Congo* **114 C4** 0 12 S 21 8 E
Bokpyin, *Burma* **87 G2** 11 18N 98 42 E
Boksitogorsk, *Russia* **58 C7** 59 22N 33 50 E
Bokungu, *Dem. Rep. of the Congo* **114 C4** 0 35 S 22 50 E
Bol, *Chad* **109 F2** 13 30N 14 40 E
Bol, *Croatia* **45 E13** 43 18N 16 38 E
Bolama, *Guinea-Biss.* **112 C1** 11 30N 15 30W
Bolan Pass, *Pakistan* **91 C2** 29 50N 67 20 E
Bolangum, *Australia* **128 D5** 36 42 S 142 54 E
Bolaños →, *Mexico* **162 C4** 21 14N 104 8W
Bolaños de Calatrava, *Spain* **43 G7** 38 54N 3 40W
Bolayır, *Turkey* **51 F10** 40 31N 26 45 E
Bolbec, *France* **26 C7** 49 30N 0 30 E
Boldājī, *Iran* **97 D6** 31 56N 51 3 E
Boldeşti-Scăeni, *Romania* **53 E11** 45 3N 26 2 E
Bole, *China* **72 B3** 45 11N 81 37 E
Bole, *Ethiopia* **107 F4** 6 36N 37 20 E
Bolekhiv, *Ukraine* **59 H2** 49 0N 23 57 E
Bolesławiec, *Poland* **55 G2** 51 17N 15 37 E
Bolgatanga, *Ghana* **113 C4** 10 44N 0 53W
Bolgrad = Bolhrad, *Ukraine* **59 K5** 45 40N 28 32 E
Bolhrad, *Ukraine* **59 K5** 45 40N 28 32 E
Boli, *Sudan* **107 F2** 6 2N 28 48 E
Bolinao, *Phil.* **80 C2** 16 23N 119 54 E
Bolinao C., *Phil.* **80 C2** 16 23N 119 55 E
Boliney, *Phil.* **80 C3** 17 24N 120 48 E
Bolingbroke, *U.S.A.* **152 C6** 32 57N 83 48W
Bolintin-Vale, *Romania* **53 F10** 44 27N 25 46 E
Bolívar, *Argentina* **174 D3** 36 15 S 60 53W
Bolívar, *Antioquía, Colombia* **168 B2** 5 50N 76 1W
Bolívar, *Cauca, Colombia* **168 C2** 2 0N 77 0W
Bolívar, *Peru* **172 B2** 7 18 S 77 48W
Bolivar, *Mo., U.S.A.* **155 G8** 37 37N 93 25W
Bolivar, *Tenn., U.S.A.* **155 H10** 35 12N 89 0W
Bolívar □, *Colombia* **168 B3** 9 0N 74 40W
Bolívar □, *Ecuador* **168 D2** 1 15 S 79 5W
Bolívar □, *Venezuela* **169 B5** 6 20N 63 30W
Bolivia ■, *S. Amer.* **173 D5** 17 6 S 64 0W
Bolivian Plateau, *S. Amer.* **166 E4** 20 0 S 67 30W
Boljevac, *Serbia, Yug.* **50 C5** 43 51N 21 58 E
Bolkhov, *Russia* **58 F9** 53 25N 36 0 E
Bolków, *Poland* **55 H3** 50 55N 16 6 E
Bollebygd, *Sweden* **17 G6** 57 40N 12 35 E
Bollène, *France* **29 D8** 44 18N 4 45 E
Bollnäs, *Sweden* **16 C10** 61 21N 16 24 E
Bollon, *Australia* **127 D4** 28 2 S 147 29 E
Bollstabruk, *Sweden* **16 B11** 62 59N 17 40 E
Bolmen, *Sweden* **17 H7** 56 55N 13 40 E
Bolobo, *Dem. Rep. of the Congo* **114 C3** 2 6 S 16 20 E
Bologna, *Italy* **45 D8** 44 29N 11 20 E
Bologoye, *Russia* **58 D8** 57 55N 34 5 E
Bolomba, *Dem. Rep. of the Congo* **114 B3** 0 35N 19 0 E
Bolonchenticul, *Mexico* **163 D7** 20 0N 89 49W
Bolong, *Phil.* **81 H4** 7 6N 122 14 E
Bolótana, *Italy* **46 B1** 40 20N 8 52 E
Bolotovskoye, *Russia* **62 B9** 58 31N 62 28 E
Boloven, Cao Nguyen, *Laos* **86 E6** 15 10N 106 30 E
Bolpur, *India* **93 H12** 23 40N 87 45 E
Bolsena, *Italy* **45 F8** 42 39N 11 59 E
Bolsena, L. di, *Italy* **45 F8** 42 36N 11 56 E
Bolshaya Chernigovka, *Russia* **60 D10** 52 6N 50 52 E
Bolshaya Glushitsa, *Russia* **60 D10** 52 28N 50 30 E
Bolshaya Khobda →, *Kazakstan* **62 F5** 50 56N 54 34 E
Bolshaya Martynovka, *Russia* **61 G5** 47 19N 41 37 E
Bolshaya Vradiyevka, *Ukraine* **59 J6** 47 50N 30 40 E
Bolshevik, Ostrov, *Russia* **65 B11** 78 30N 102 0 E
Bolshezemelskaya Tundra, *Russia* **56 A10** 67 0N 56 0 E
Bolshoi Kavkas = Caucasus Mountains, *Eurasia* **61 J7** 42 50N 44 0 E
Bolshoy Anyuy →, *Russia* **65 C17** 68 30N 160 49 E
Bolshoy Begichev, Ostrov, *Russia* **65 B12** 74 20N 112 30 E
Bolshoy Lyakhovskiy, Ostrov, *Russia* **65 B15** 73 35N 142 0 E
Bolshoy Tokmak = Tokmak, *Ukraine* **59 J8** 47 16N 35 42 E
Bolshoy Tyuters, Ostrov, *Russia* **15 G22** 59 51N 27 13 E
Bólstaðarhlíð, *Iceland* **11 B7** 65 31N 19 49W
Bolsward, *Neths.* **24 A5** 53 3N 5 32 E
Bolt Head, *U.K.* **21 G4** 50 12N 3 48W
Boltaña, *Spain* **40 C5** 42 28N 0 4 E
Boltigen, *Switz.* **32 C4** 46 38N 7 24 E
Bolton, *Canada* **150 C5** 43 54N 79 45W
Bolton, *U.K.* **20 D5** 53 35N 2 26W
Bolu, *Turkey* **100 B4** 40 45N 31 35 E
Bolubolu, *Papua N. G.* **132 E6** 9 21 S 150 20 E
Bolungavík, *Iceland* **11 A3** 66 9N 23 15W
Boluo, *China* **77 F10** 23 3N 114 21 E
Bolvadin, *Turkey* **100 C4** 38 45N 31 4 E
Bolzano, *Italy* **45 B8** 46 31N 11 22 E
Bom Comércio, *Brazil* **173 B4** 9 45 S 65 54W
Bom Conselho, *Brazil* **170 C4** 9 10 S 36 41W
Bom Despacho, *Brazil* **171 E2** 19 43 S 45 15W
Bom Jesus, *Brazil* **170 C3** 9 4 S 44 22W
Bom Jesus da Gurguéia, Serra, *Brazil* **170 C3** 9 0 S 43 0W
Bom Jesus da Lapa, *Brazil* **171 D3** 13 15 S 43 25W
Boma, *Dem. Rep. of the Congo* **115 D2** 5 50 S 13 4 E
Bomaderry, *Australia* **129 C9** 34 52 S 150 37 E
Bomandjokou, *Congo* **114 B2** 0 34N 14 23 E
Bomassa, *Congo* **114 B3** 2 12N 16 12 E
Bombala, *Australia* **129 D8** 36 56 S 149 15 E
Bombarral, *Portugal* **43 F1** 39 15N 9 9W
Bombay = Mumbai, *India* **94 E1** 18 55N 72 50 E
Bomboma, *Dem. Rep. of the Congo* **114 B3** 2 25N 18 55 E
Bombombwa, *Dem. Rep. of the Congo* **118 B2** 1 40N 25 40 E
Bomi Hills, *Liberia* **112 D2** 7 1N 10 38W
Bomili, *Dem. Rep. of the Congo* **118 B2** 1 45N 27 5 E
Bømlo, *Norway* **15 G11** 59 37N 5 13 E
Bomokandi →, *Dem. Rep. of the Congo* **118 B2** 3 39N 26 8 E
Bomongo, *Dem. Rep. of the Congo* **114 B3** 1 27N 18 21 E
Bomu →, *C.A.R.* **114 B4** 4 40N 22 30 E
Bon, C., *Tunisia* **108 A2** 37 1N 11 2 E
Bon Sar Pa, *Vietnam* **86 F6** 12 24N 107 35 E
Bonaduz, *Switz.* **33 C8** 46 49N 9 25 E
Bonaire, *Neth. Ant.* **165 D6** 12 10N 68 15W
Bonaire, *U.S.A.* **152 C6** 32 33N 83 36W
Bonang, *Australia* **129 D8** 37 11 S 148 41 E
Bonanza, *Nic.* **164 D3** 13 54N 84 35W
Bonaparte Arch., *Australia* **124 B3** 14 0 S 124 30 E
Boñar, *Spain* **42 C5** 42 52N 5 19W
Bonaventure, *Canada* **141 C6** 48 5N 65 32W
Bonavista, *Canada* **141 C9** 48 40N 53 5W
Bonavista, C., *Canada* **141 C9** 48 42N 53 5W
Bonawan, *Phil.* **81 G4** 9 8N 122 55 E
Bondeno, *Italy* **45 D8** 44 53N 11 25 E
Bondo, *Dem. Rep. of the Congo* **114 B4** 3 55N 23 53 E
Bondoukou, *Ivory C.* **112 D4** 8 2N 2 47W
Bondowoso, *Indonesia* **85 D4** 7 55 S 113 49 E
Bone, Teluk, *Indonesia* **82 B2** 4 10 S 120 50 E
Bonerate, *Indonesia* **82 C2** 7 25 S 121 5 E
Bonerate, Kepulauan, *Indonesia* **82 C2** 6 30 S 121 10 E
Bo'ness, *U.K.* **22 E5** 56 1N 3 37W
Bonete, Cerro, *Argentina* **174 B2** 27 55 S 68 40W
Bong Son = Hoai Nhon, *Vietnam* **86 E7** 14 28N 109 1 E
Bongabon, *Phil.* **80 D3** 15 38N 121 8 E
Bongabong, *Phil.* **80 E3** 12 45N 121 29 E
Bongandanga, *Dem. Rep. of the Congo* **114 B4** 1 24N 21 3 E
Bongo, *Dem. Rep. of the Congo* **114 C3** 1 47 S 17 41 E
Bongor, *Chad* **109 F3** 10 35N 15 20 E
Bongouanou, *Ivory C.* **112 D4** 6 42N 4 15W
Bonham, *U.S.A.* **155 J6** 33 35N 96 11W
Bonifacio, *France* **29 G13** 41 24N 9 10 E
Bonifacio, Bouches de, *Medit. S.* **46 A2** 41 12N 9 15 E
Bonifay, *U.S.A.* **152 E4** 30 47N 85 41W
Bonin Is. = Ogasawara Gunto, *Pac. Oc.* **134 E6** 27 0N 142 0 E
Bonita Springs, *U.S.A.* **153 J8** 26 21N 81 47W
Bonke, *Ethiopia* **107 F4** 6 5N 37 16 E
Bonn, *Germany* **30 E3** 50 46N 7 6 E
Bonnat, *France* **27 F8** 46 20N 1 54 E
Bonne Terre, *U.S.A.* **155 G9** 37 55N 90 33W
Bonneau, *U.S.A.* **152 B10** 33 16N 79 58W
Bonners Ferry, *U.S.A.* **158 B5** 48 42N 116 19W
Bonnétable, *France* **26 D7** 48 11N 0 25 E
Bonneval, *Eure-et-Loir, France* **26 D8** 48 11N 1 24 E
Bonneval, *Savoie, France* **29 C11** 45 22N 7 3 E
Bonneville, *France* **27 F13** 46 4N 6 24 E

Branchville 193

Bonney, L., *Australia* **128 D4** 37 50 S 140 20 E
Bonnie Doon, *Australia* **129 D6** 37 2 S 145 53 E
Bonnie Downs, *Australia* **126 C3** 22 7 S 143 50 E
Bonnie Rock, *Australia* **125 F2** 30 29 S 118 22 E
Bonny, *Nigeria* **113 E6** 4 25N 7 13 E
Bonny →, *Nigeria* **113 E6** 4 20N 7 10 E
Bonny, Bight of, *Africa* **113 E6** 3 30N 9 20 E
Bonny-sur-Loire, *France* **27 E9** 47 33N 2 50 E
Bonnyrigg, *U.K.* **22 F5** 55 53N 3 6W
Bonnyville, *Canada* **143 C6** 54 20N 110 45W
Bono, *Italy* **46 B2** 40 25N 9 2 E
Bonobono, *Phil.* **81 G1** 8 40N 117 36 E
Bonoi, *Indonesia* **83 B5** 1 45 S 137 41 E
Bonorva, *Italy* **46 B1** 40 25N 8 46 E
Bonsall, *U.S.A.* **161 M9** 33 16N 117 14W
Bontang, *Indonesia* **85 B5** 0 10N 117 30 E
Bonthain, *Indonesia* **82 C1** 5 34 S 119 56 E
Bonthe, *S. Leone* **112 D2** 7 30N 12 33W
Bontoc, *Phil.* **80 C3** 17 7N 120 58 E
Bontosunggu, *Indonesia* **82 C1** 5 41 S 119 42 E
Bonyeri, *Ghana* **112 D4** 5 1N 2 46W
Bonyhád, *Hungary* **52 D3** 46 18N 18 32 E
Bonython Ra., *Australia* **124 D4** 23 40 S 128 45 E
Bookabie, *Australia* **125 F5** 31 50 S 132 41 E
Booker, *U.S.A.* **155 G4** 36 27N 100 32W
Boolaboolka L., *Australia* **128 B5** 32 38 S 143 10 E
Boolarra, *Australia* **129 E7** 38 20 S 146 20 E
Boolcoomata, *Australia* **128 A4** 31 57 S 140 33 E
Booleroo Centre, *Australia* **128 B3** 32 53 S 138 21 E
Booligal, *Australia* **129 B6** 33 58 S 144 53 E
Boonah, *Australia* **127 D5** 27 58 S 152 41 E
Boone, *Iowa, U.S.A.* **156 B3** 42 4N 93 53W
Boone, *N.C., U.S.A.* **149 G5** 36 13N 81 41W
Booneville, *Ark., U.S.A.* **155 H8** 35 8N 93 55W
Booneville, *Miss., U.S.A.* **149 H1** 34 39N 88 34W
Boonville, *Calif., U.S.A.* **160 F3** 39 1N 123 22W
Boonville, *Ind., U.S.A.* **157 F9** 38 3N 87 16W
Boonville, *Mo., U.S.A.* **156 F4** 38 58N 92 44W
Boonville, *N.Y., U.S.A.* **151 C9** 43 29N 75 20W
Booral, *Australia* **129 B9** 32 30 S 151 56 E
Boorindal, *Australia* **127 E4** 30 22 S 146 11 E
Booroomugga, *Australia* **129 A7** 31 17 S 146 27 E
Boorowa, *Australia* **129 C8** 34 28 S 148 44 E
Boothia, Gulf of, *Canada* **139 A11** 71 0N 90 0W
Boothia Pen., *Canada* **138 A10** 71 0N 94 0W
Bootle, *U.K.* **20 D4** 53 28N 3 1W
Booué, *Gabon* **114 C2** 0 5 S 11 55 E
Boppard, *Germany* **31 E3** 50 13N 7 35 E
Boquerón □, *Paraguay* **173 E6** 23 0 S 60 0W
Boquete, *Panama* **164 E3** 8 46N 82 27W
Boquilla, Presa de la, *Mexico* **162 B3** 27 40N 105 30W
Boquillas del Carmen, *Mexico* **162 B4** 29 17N 102 53W
Bor, *Czech Rep.* **34 B5** 49 41N 12 45 E
Bor, *Russia* **60 B7** 56 28N 43 59 E
Bor, *Serbia, Yug.* **50 B6** 44 5N 22 7 E
Bôr, *Sudan* **107 F3** 6 10N 31 40 E
Bor, *Sweden* **17 G8** 57 9N 14 10 E
Bor, *Turkey* **100 D6** 37 54N 34 32 E
Bor Mashash, *Israel* **103 D3** 31 7N 34 50 E
Borah Peak, *U.S.A.* **158 D7** 44 8N 113 47W
Borama, *Somali Rep.* **120 C2** 9 55N 43 7 E
Borang, *Sudan* **107 G3** 4 50N 30 59 E
Borangapara, *India* **90 C3** 25 14N 90 14 E
Borås, *Sweden* **17 G6** 57 43N 12 56 E
Borāzjān, *Iran* **97 D6** 29 22N 51 10 E
Borba, *Brazil* **169 D6** 4 12 S 59 34W
Borba, *Portugal* **43 G3** 38 50N 7 26W
Borbon, *Phil.* **81 F5** 10 50N 124 2 E
Borborema, Planalto da, *Brazil* **170 C4** 7 0 S 37 0W
Borcea, *Romania* **53 F12** 44 20N 27 45 E
Borçka, *Turkey* **101 B9** 41 25N 41 41 E
Bord Khūn-e Now, *Iran* **97 D6** 28 3N 51 28 E
Borda, C., *Australia* **128 C2** 35 45 S 136 34 E
Bordeaux, *France* **28 D3** 44 50N 0 36W
Borden, *Australia* **125 F2** 34 3 S 118 12 E
Borden, *Canada* **141 C7** 46 18N 63 47W
Borden I., *Canada* **6 B2** 78 30N 111 30W
Borden Springs, *U.S.A.* **152 B4** 33 56N 85 28W
Bordertown, *Australia* **128 D4** 36 19 S 140 45 E
Borðeyri, *Iceland* **11 B5** 65 12N 21 6W
Bordighera, *Italy* **44 E4** 43 46N 7 39 E
Bordj bou Arreridj, *Algeria* **111 A5** 36 4N 4 45 E
Bordj Bourguiba, *Tunisia* **108 B2** 32 12N 10 2 E
Bordj Fly Ste. Marie, *Algeria* **110 C4** 27 19N 2 32W
Bordj-in-Eker, *Algeria* **111 D6** 24 9N 5 3 E
Bordj Menaiel, *Algeria* **111 A5** 36 46N 3 43 E
Bordj Messouda, *Algeria* **111 B6** 30 12N 9 25 E
Bordj Nili, *Algeria* **111 B5** 33 28N 3 2 E
Bordj Omar Driss, *Algeria* **111 C6** 28 10N 6 40 E
Bordj Sif Fatima, *Algeria* **111 B6** 31 6N 8 41 E
Bordj-Tarat, *Algeria* **111 C6** 25 55N 9 3 E
Bordj Zelfana, *Algeria* **111 B5** 32 27N 4 15 E
Bordoba, *Kyrgyzstan* **63 D6** 39 31N 73 16 E
Borea Creek, *Australia* **129 C7** 35 5 S 146 35 E
Borehamwood, *U.K.* **21 F7** 51 40N 0 15W
Borek Wielkopolski, *Poland* **55 G4** 51 54N 17 11 E
Boremore, *Australia* **129 B8** 33 15 S 149 0 E
Boren Kapuas, Pegunungan, *Malaysia* **85 B4** 1 25N 113 15 E
Borensberg, *Sweden* **17 F9** 58 34N 15 17 E
Borgå = Porvoo, *Finland* **15 F21** 60 24N 25 40 E
Borgarfjarðarsýsla □, *Iceland* **11 C5** 64 30N 21 30W
Borgarfjörður, *Iceland* **11 C4** 64 30N 22 0W
Borgarfjörður, *Iceland* **11 B13** 65 31N 13 49W
Borgarnes, *Iceland* **11 C5** 64 32N 21 55W
Børgefjellet, *Norway* **14 D15** 65 20N 13 45 E
Borger, *Neths.* **24 B6** 52 54N 6 44 E
Borger, *U.S.A.* **155 H4** 35 39N 101 24W
Borgholm, *Sweden* **17 H10** 56 52N 16 39 E
Bórgia, *Italy* **47 D9** 38 49N 16 30 E
Borgo San Dalmazzo, *Italy* **44 D4** 44 20N 7 30 E
Borgo San Lorenzo, *Italy* **45 E8** 43 57N 11 23 E
Borgo Val di Taro, *Italy* **44 D6** 44 29N 9 46 E
Borgo Valsugana, *Italy* **45 B8** 46 3N 11 27 E
Borgomanero, *Italy* **44 C5** 45 42N 8 28 E
Borgorose, *Italy* **45 F10** 42 11N 13 13 E
Borgosésia, *Italy* **44 C5** 45 43N 8 16 E
Borgund, *Norway* **18 C4** 61 3N 7 48 E
Borikhane, *Laos* **86 C4** 18 33N 103 43 E
Borisoglebsk, *Russia* **60 E6** 51 27N 42 5 E
Borisov = Barysaw, *Belarus* **58 E5** 54 17N 28 28 E
Borisovka, *Kazakstan* **63 B4** 43 15N 68 10 E
Borisovka, *Russia* **59 G9** 50 36N 36 1 E
Borja, *Peru* **168 D2** 4 20 S 77 40W
Borja, *Spain* **40 D3** 41 48N 1 34W
Borjas Blancas = Les Borges Blanques, *Spain* **40 D5** 41 31N 0 52 E
Borjomi, *Georgia* **61 K6** 41 48N 43 28 E
Borken, *Germany* **30 D9** 51 40N 13 10 E
Børkop, *Denmark* **17 J3** 55 39N 9 39 E
Borkou, *Chad* **109 E3** 18 15N 18 50 E
Borkum, *Germany* **30 B2** 53 34N 6 40 E
Borlänge, *Sweden* **16 D9** 60 29N 15 26 E
Borley, C., *Antarctica* **7 C5** 66 15 S 52 30 E
Borlu, *Turkey* **49 C10** 38 44N 28 27 E
Bormida →, *Italy* **44 D5** 44 23N 8 13 E
Bórmio, *Italy* **44 B7** 46 28N 10 22 E
Borna, *Germany* **30 D8** 51 7N 12 29 E
Borne Sulinowo, *Poland* **54 E3** 53 32N 16 36 E
Borneo, *E. Indies* **85 B4** 1 0N 115 0 E
Bornholm, *Denmark* **17 J8** 55 10N 15 0 E
Bornholms Amtskommune □, *Denmark* **17 J8** 55 5N 15 0 E
Bornholmsgattet, *Europe* **17 J8** 55 15N 14 20 E
Borno □, *Nigeria* **113 C7** 11 30N 13 0 E
Bornos, *Spain* **43 J5** 36 48N 5 42W
Bornova, *Turkey* **49 C9** 38 27N 27 14 E
Bornu Yassa, *Nigeria* **113 C7** 12 14N 12 25 E
Borobudur, *Indonesia* **85 D4** 7 36 S 110 13 E
Borodino, *Russia* **58 E8** 55 31N 35 40 E
Borogontsy, *Russia* **65 C14** 62 42N 131 8 E
Boromo, *Burkina Faso* **112 C4** 11 45N 2 58W
Boron, *U.S.A.* **161 L9** 35 0N 117 39W
Borongа Is., *Burma* **90 F4** 19 58N 93 6 E
Borongan, *Phil.* **81 F5** 11 37N 125 26 E
Bororen, *Australia* **126 C5** 24 13 S 151 33 E
Borotangba Mts., *C.A.R.* **107 F2** 6 30N 25 0 E
Borovan, *Bulgaria* **50 C7** 43 27N 23 45 E
Borovichi, *Russia* **58 C7** 58 25N 33 55 E
Borovsk, *Berezniki, Russia* **62 B6** 59 43N 56 40 E
Borovsk, *Moskva, Russia* **58 E9** 55 12N 36 24 E
Borrby, *Sweden* **17 J8** 55 27N 14 10 E
Borrego Springs, *U.S.A.* **161 M10** 33 15N 116 23W
Borriol, *Spain* **40 E4** 40 4N 0 4W
Borroloola, *Australia* **126 B2** 16 4 S 136 17 E
Borşa, *Cluj, Romania* **53 D8** 46 56N 23 40 E
Borşa, *Maramureş, Romania* **53 C9** 47 41N 24 50 E
Borsec, *Romania* **53 D10** 46 57N 25 34 E
Borsod-Abaúj-Zemplén □, *Hungary* **52 B6** 48 20N 21 0 E
Bort-les-Orgues, *France* **28 C6** 45 24N 2 29 E
Borth, *U.K.* **21 E3** 52 29N 4 2W
Börtnan, *Sweden* **16 B7** 62 45N 13 50 E
Borūjerd, *Iran* **97 C6** 33 55N 48 50 E
Boryslav, *Ukraine* **59 H2** 49 18N 23 28 E
Boryspil, *Ukraine* **59 G6** 50 21N 30 59 E
Borzhomi = Borjomi, *Georgia* **61 K6** 41 48N 43 28 E
Borzna, *Ukraine* **59 G7** 51 18N 32 26 E
Borzya, *Russia* **65 D12** 50 24N 116 31 E
Bosa, *Italy* **46 B1** 40 18N 8 30 E
Bosaga, *Turkmenistan* **63 E2** 37 33N 65 41 E
Bosanska Dubica, *Bos.-H.* **45 C13** 45 10N 16 50 E
Bosanska Gradiška, *Bos.-H.* **52 E2** 45 10N 17 15 E
Bosanska Kostajnica, *Bos.-H.* **45 C13** 45 11N 16 33 E
Bosanska Krupa, *Bos.-H.* **45 D13** 44 53N 16 10 E
Bosanski Brod, *Bos.-H.* **52 E2** 45 10N 18 0 E
Bosanski Novi, *Bos.-H.* **45 C13** 45 2N 16 22 E
Bosanski Petrovac, *Bos.-H.* **45 D13** 44 35N 16 21 E
Bosanski Šamac, *Bos.-H.* **52 E3** 45 3N 18 29 E
Bosansko Grahovo, *Bos.-H.* **45 D13** 44 12N 16 26 E
Bosaso, *Somali Rep.* **120 B3** 11 12N 49 18 E
Bosavi, Mt., *Papua N. G.* **132 D2** 6 30 S 142 49 E
Boscastle, *U.K.* **21 G3** 50 41N 4 42W
Boscobel, *U.S.A.* **156 A6** 43 8N 90 42W
Bose, *China* **76 F6** 23 53N 106 35 E
Boshan, *China* **75 F9** 36 28N 117 49 E
Boshof, *S. Africa* **116 D4** 28 31 S 25 13 E
Boshrūyeh, *Iran* **97 C8** 33 50N 57 30 E
Bosilegrad, *Serbia, Yug.* **50 D6** 42 30N 22 27 E
Boskovice, *Czech Rep.* **35 B9** 49 29N 16 40 E
Bosna →, *Bos.-H.* **52 E3** 45 4N 18 29 E
Bosna i Hercegovina = Bosnia-Herzegovina ■, *Europe* **52 G2** 44 0N 18 0 E
Bosnia-Herzegovina ■, *Europe* **52 G2** 44 0N 18 0 E
Bosnik, *Indonesia* **83 B5** 1 5 S 136 10 E
Bōsō-Hantō, *Japan* **71 B12** 35 20N 140 20 E
Bosobolo, *Dem. Rep. of the Congo* **114 B3** 4 15N 19 50 E
Bosporus = İstanbul Boğazı, *Turkey* **51 E13** 41 10N 29 10 E
Bossangoa, *C.A.R.* **114 A3** 6 35N 17 30 E
Bossembélé, *C.A.R.* **114 A3** 5 25N 17 40 E
Bossembélé II, *C.A.R.* **114 A3** 5 41N 16 38 E
Bossier City, *U.S.A.* **155 J8** 32 31N 93 44W
Bosso, *Niger* **109 F2** 13 43N 13 19 E
Bostānābād, *Iran* **101 D12** 37 50N 46 50 E
Bosten Hu, *China* **72 B3** 41 55N 87 40 E
Boston, *Phil.* **81 H6** 7 52N 126 22 E
Boston, *U.K.* **20 E7** 52 59N 0 2W
Boston, *Ga., U.S.A.* **152 E6** 30 47N 83 47W
Boston, *Mass., U.S.A.* **151 D13** 42 22N 71 4W
Boston Bar, *Canada* **142 D4** 49 52N 121 30W
Bostwick, *Fla., U.S.A.* **152 F8** 29 46N 81 38W
Bostwick, *Ga., U.S.A.* **152 B6** 33 44N 83 31W
Bosusulu, *Dem. Rep. of the Congo* **114 B4** 0 50N 20 45 E
Bosut →, *Croatia* **52 E3** 45 20N 18 45 E
Boswell, *Canada* **142 D5** 49 28N 116 45W
Boswell, *Ind., U.S.A.* **157 D9** 40 31N 87 23W
Boswell, *Okla., U.S.A.* **155 H7** 34 2N 95 52W
Boswell, *Pa., U.S.A.* **150 F5** 40 10N 79 2W
Bosworth, *U.S.A.* **156 E3** 39 28N 93 20W
Botad, *India* **92 H4** 22 15N 71 40 E
Botan →, *Turkey* **101 D10** 37 57N 42 2 E
Botany B., *Australia* **127 E5** 34 0 S 151 14 E
Botene, *Laos* **86 D3** 17 35N 101 12 E
Botev, *Bulgaria* **51 D8** 42 44N 24 52 E
Botevgrad, *Bulgaria* **50 D7** 42 55N 23 47 E
Bothaville, *S. Africa* **116 D4** 27 23 S 26 34 E
Bothnia, G. of, *Europe* **14 E19** 63 0N 20 15 E
Bothwell, *Australia* **126 G4** 42 20 S 147 1 E
Bothwell, *Canada* **150 D3** 42 38N 81 52W
Boticas, *Portugal* **42 D3** 41 41N 7 40W
Botletle →, *Botswana* **116 C3** 20 10 S 23 15 E
Botlikh, *Russia* **61 J8** 42 39N 46 11 E
Botna →, *Moldova* **53 D14** 46 45N 29 34 E
Botolan, *Phil.* **80 D3** 15 17N 120 1 E
Botoroaga, *Romania* **53 F10** 44 8N 25 32 E
Botoşani, *Romania* **53 C11** 47 42N 26 41 E
Botoşani □, *Romania* **53 C11** 47 50N 26 50 E
Botricello, *Italy* **47 D9** 38 56N 16 51 E
Botro, *Ivory C.* **112 D3** 7 51N 5 19W
Botswana ■, *Africa* **116 C3** 22 0 S 24 0 E
Bottineau, *U.S.A.* **154 A4** 48 50N 100 27W
Bottnaryd, *Sweden* **17 G7** 57 47N 13 50 E
Bottrop, *Germany* **30 D2** 51 31N 6 58 E
Botucatu, *Brazil* **175 A6** 22 55 S 48 30W
Botwood, *Canada* **141 C8** 49 6N 55 23W
Bou Alam, *Algeria* **111 B5** 33 50N 1 26 E
Bou Ali, *Algeria* **111 C4** 27 11N 0 4W
Bou Djébéha, *Mali* **112 B4** 18 25N 2 45W
Bou Guema, *Algeria* **111 C5** 28 49N 0 19 E
Bou Ismael, *Algeria* **111 A5** 36 38N 2 42 E
Bou Izakarn, *Morocco* **110 C3** 29 12N 9 46W
Boû Lanouâr, *Mauritania* **110 D1** 21 12N 16 34W
Bou Saâda, *Algeria* **111 A5** 35 11N 4 9 E
Bou Salem, *Tunisia* **108 A1** 36 45N 9 2 E
Bouaflé, *Ivory C.* **112 D3** 7 1N 5 47W
Bouaké, *Ivory C.* **112 D3** 7 40N 5 2W
Bouanga, *Congo* **114 C3** 2 7 S 16 8 E
Bouar, *C.A.R.* **114 A3** 6 0N 15 40 E
Bouârfa, *Morocco* **110 B4** 32 32N 1 58W
Bouca, *C.A.R.* **114 A3** 6 45N 18 25 E
Boucaut B., *Australia* **126 A1** 12 0 S 134 25 E
Bouches-du-Rhône □, *France* **29 E9** 43 37N 5 2 E
Bouda, *Algeria* **111 C4** 27 50N 0 27W
Boudenib, *Morocco* **110 B4** 31 59N 3 31W
Boudry, *Switz.* **32 C3** 46 57N 6 50 E
Boufarik, *Algeria* **111 A5** 36 34N 2 58 E
Bougainville, C., *Australia* **124 B4** 13 57 S 126 4 E
Bougainville I., *Solomon Is.* **133 L8** 6 0 S 155 0 E
Bougainville Reef, *Australia* **126 B4** 15 30 S 147 5 E
Bougainville Str., *Solomon Is.* **133 L9** 6 40 S 156 10 E
Bougaroun, C., *Algeria* **111 A6** 37 6N 6 30 E
Bougie = Bejaia, *Algeria* **111 A6** 36 42N 5 2 E
Bougouni, *Mali* **112 C3** 11 30N 7 20W
Bouillon, *Belgium* **25 J6** 49 44N 5 3 E
Bouïra, *Algeria* **111 A5** 36 20N 3 59 E
Boulazac, *France* **28 C4** 45 10N 0 47 E
Boulder, *Colo., U.S.A.* **154 E2** 40 1N 105 17W
Boulder, *Mont., U.S.A.* **158 C7** 46 14N 112 7W
Boulder City, *U.S.A.* **161 K12** 35 59N 114 50W
Boulder Creek, *U.S.A.* **160 H4** 37 7N 122 7W
Boulder Dam = Hoover Dam, *U.S.A.* **161 K12** 36 1N 114 44W
Boulembo, *Gabon* **114 C2** 1 26 S 12 0 E
Bouli, *Mauritania* **112 B2** 15 17N 12 18W
Boulia, *Australia* **126 C2** 22 52 S 139 51 E
Bouligny, *France* **27 C12** 49 17N 5 45 E
Boulogne →, *France* **26 E5** 47 12N 1 47W
Boulogne-sur-Gesse, *France* **28 E4** 43 18N 0 38 E
Boulogne-sur-Mer, *France* **27 B8** 50 42N 1 36 E
Bouloire, *France* **26 E7** 47 59N 0 45 E
Bouloupari, *N. Cal.* **133 U20** 21 52 S 166 4 E
Boulsa, *Burkina Faso* **113 C4** 12 39N 0 34W
Boultoum, *Niger* **109 F2** 14 45N 10 25 E
Boumalne, *Morocco* **110 B3** 31 25N 6 0W
Boun Neua, *Laos* **86 B3** 21 38N 101 54 E
Boun Tai, *Laos* **86 B3** 21 23N 101 58 E
Bouna, *Ivory C.* **112 D4** 9 10N 3 0W
Boundary Peak, *U.S.A.* **160 H8** 37 51N 118 21W
Boundiali, *Ivory C.* **112 D3** 9 30N 6 20W
Bountiful, *U.S.A.* **158 F8** 40 53N 111 53W
Bounty Is., *Pac. Oc.* **134 M9** 48 0 S 178 30 E
Bourail, *N. Cal.* **133 U19** 21 34 S 165 30 E
Bourbeuse →, *U.S.A.* **156 F6** 38 24N 90 53W
Bourbon, *U.S.A.* **157 C10** 41 18N 86 7W
Bourbon-Lancy, *France* **27 F10** 46 37N 3 45 E
Bourbon-l'Archambault, *France* **27 F10** 46 36N 3 4 E
Bourbonnais, *France* **27 F10** 46 28N 3 0 E
Bourbonne-les-Bains, *France* **27 E12** 47 54N 5 45 E
Bourbourg, *France* **27 B9** 50 56N 2 12 E
Bourem, *Mali* **113 B4** 17 0N 0 24W
Bourg, *France* **28 C3** 45 3N 0 34W
Bourg-Argental, *France* **29 C8** 45 18N 4 32 E
Bourg-de-Péage, *France* **29 C9** 45 2N 5 3 E
Bourg-en-Bresse, *France* **27 F12** 46 13N 5 12 E
Bourg-Lastic, *France* **28 C6** 45 39N 2 35 E
Bourg-Madame, *France* **28 F5** 42 26N 1 55 E
Bourg-St-Andéol, *France* **29 D8** 44 23N 4 39 E
Bourg-St-Maurice, *France* **29 C10** 45 35N 6 46 E
Bourg-St.-Pierre, *Switz.* **32 E4** 45 57N 7 12 E
Bourganeuf, *France* **28 C5** 45 57N 1 45 E
Bourges, *France* **27 E9** 47 9N 2 25 E
Bourget, *Canada* **151 A9** 45 26N 75 9W
Bourget, L. du, *France* **29 C9** 45 44N 5 52 E
Bourgneuf, B. de, *France* **26 E4** 47 3N 2 10W
Bourgneuf-en-Retz, *France* **26 E5** 47 2N 1 58W
Bourgogne, *France* **27 F11** 47 0N 4 50 E
Bourgoin-Jallieu, *France* **29 C9** 45 36N 5 17 E
Bourgueil, *France* **26 E7** 47 17N 0 10 E
Bourke, *Australia* **127 E4** 30 8 S 145 55 E
Bourne, *U.K.* **20 E7** 52 47N 0 22W
Bournemouth, *U.K.* **21 G6** 50 43N 1 52W
Bournemouth □, *U.K.* **21 G6** 50 43N 1 52W
Bouse, *U.S.A.* **161 M13** 33 56N 114 0W
Boussac, *France* **27 F9** 46 22N 2 13 E
Bousso, *Chad* **109 F3** 10 34N 16 52 E
Boutilimit, *Mauritania* **112 B2** 17 45N 14 40W
Boutonne →, *France* **28 C3** 45 54N 0 50W
Bouvet I. = Bouvetøya, *Antarctica* **9 P9** 54 26 S 3 24 E
Bouvetøya, *Antarctica* **9 P9** 54 26 S 3 24 E
Bouxwiller, *France* **27 D14** 48 49N 7 27 E
Bouznika, *Morocco* **110 B3** 33 46N 7 6W
Bouzonville, *France* **27 C13** 49 17N 6 32 E
Bova Marina, *Italy* **47 E8** 37 56N 15 55 E
Bovalino Marina, *Italy* **47 D9** 38 10N 16 10 E
Bovec, *Slovenia* **45 B10** 46 20N 13 33 E
Bøverdal, *Norway* **18 C5** 61 44N 8 20 E
Bøverfjorden, *Norway* **18 A5** 63 1N 8 32 E
Bovill, *U.S.A.* **158 C5** 46 51N 116 24W
Bovino, *Italy* **47 A8** 41 15N 15 20 E
Bow Island, *Canada* **142 D6** 49 50N 111 23W
Bowbells, *U.S.A.* **154 A3** 48 48N 102 15W
Bowdle, *U.S.A.* **154 C5** 45 27N 99 39W
Bowdon, *U.S.A.* **152 B4** 33 32N 85 15W
Bowdon Junction, *U.S.A.* **152 B4** 33 40N 85 9W
Bowelling, *Australia* **125 F2** 33 25 S 116 30 E
Bowen, *Australia* **126 C4** 20 0 S 148 16 E
Bowen Mts., *Australia* **129 D7** 37 0 S 147 50 E
Bowie, *Ariz., U.S.A.* **159 K9** 32 19N 109 29W
Bowie, *Tex., U.S.A.* **155 J6** 33 34N 97 51W
Bowkān, *Iran* **101 D12** 36 31N 46 12 E
Bowland, Forest of, *U.K.* **20 D5** 54 0N 2 30W
Bowling Green, *Fla., U.S.A.* **153 H8** 27 38N 81 50W
Bowling Green, *Ky., U.S.A.* **148 G2** 36 59N 86 27W
Bowling Green, *Mo., U.S.A.* **156 E5** 39 21N 91 12W
Bowling Green, *Ohio, U.S.A.* **157 C13** 41 23N 83 39W
Bowling Green, C., *Australia* **126 B4** 19 19 S 147 25 E
Bowman, *N. Dak., U.S.A.* **154 B3** 46 11N 103 24W
Bowman, *S.C., U.S.A.* **152 B9** 33 21N 80 41W
Bowman I., *Antarctica* **7 C8** 65 0 S 104 0 E
Bowmans, *Australia* **128 C3** 34 10 S 138 17 E
Bowmanville, *Canada* **140 D4** 43 55N 78 41W
Bowmore, *U.K.* **22 F2** 55 45N 6 17W
Bowral, *Australia* **129 C9** 34 26 S 150 27 E
Bowraville, *Australia* **127 E5** 30 37 S 152 52 E
Bowron →, *Canada* **142 C4** 54 3N 121 50W
Bowser L., *Canada* **142 B3** 56 30N 129 30W
Bowsman, *Canada* **143 C8** 52 14N 101 12W
Bowutu Mts., *Papua N. G.* **132 D4** 7 45 S 147 10 E
Bowwood, *Zambia* **119 F2** 17 5 S 26 20 E
Boxholm, *Sweden* **17 F9** 58 12N 15 3 E
Boxmeer, *Neths.* **24 C5** 51 38N 5 56 E
Boxtel, *Neths.* **24 C5** 51 36N 5 20 E
Boyabat, *Turkey* **100 B6** 41 28N 34 47 E
Boyabo, *Dem. Rep. of the Congo* **114 B3** 3 43N 18 46 E
Boyaca = Casanare □, *Colombia* **168 B3** 5 30N 72 0W
Boyalıca, *Turkey* **51 F13** 40 29N 29 33 E
Boyce, *U.S.A.* **155 K8** 31 23N 92 40W
Boyd, *U.S.A.* **152 E6** 30 11N 83 37W
Boyer →, *Canada* **142 B5** 58 27N 115 57W
Boyer, C., *N. Cal.* **133 U22** 21 37 S 168 6 E
Boykin, *U.S.A.* **152 D5** 31 6N 84 41W
Boyle, *Ireland* **23 C3** 53 59N 8 18W
Boyne →, *Ireland* **23 C5** 53 43N 6 15W
Boyne City, *U.S.A.* **148 C3** 45 13N 85 1W
Boyni Qara, *Afghan.* **91 A2** 36 20N 67 0 E
Boynitsa, *Bulgaria* **50 C6** 43 58N 22 32 E
Boynton Beach, *U.S.A.* **149 M5** 26 32N 80 4W
Boyolali, *Indonesia* **85 D4** 7 32 S 110 35 E
Boyoma, Chutes, *Dem. Rep. of the Congo* **114 B5** 0 35N 25 23 E
Boyup Brook, *Australia* **125 F2** 33 50 S 116 23 E
Boz Burun, *Turkey* **51 F12** 40 32N 28 46 E
Boz Dağ, *Turkey* **49 D11** 37 18N 29 11 E
Boz Dağları, *Turkey* **49 C10** 38 20N 28 0 E
Bozburun, *Turkey* **49 E10** 36 43N 28 4 E
Bozcaada, *Turkey* **49 B8** 39 50N 26 4 E
Bozdoğan, *Turkey* **49 D10** 37 40N 28 17 E
Bozeman, *U.S.A.* **158 D8** 45 41N 111 2W
Bozen = Bolzano, *Italy* **45 B8** 46 31N 11 22 E
Bozene, *Dem. Rep. of the Congo* **114 B3** 2 56N 19 12 E
Boževac, *Serbia, Yug.* **50 B5** 44 32N 21 24 E
Bozkır, *Turkey* **100 D5** 37 11N 32 14 E
Bozkurt, *Turkey* **49 D11** 37 50N 29 37 E
Bozouls, *France* **28 D6** 44 28N 2 43 E
Bozoum, *C.A.R.* **114 A3** 6 25N 16 35 E
Bozova, *Antalya, Turkey* **49 D12** 37 13N 30 18 E
Bozova, *Sanliurfa, Turkey* **101 D8** 37 21N 38 32 E
Bozovici, *Romania* **52 F7** 44 56N 22 0 E
Bozüyük, *Turkey* **49 B12** 39 54N 30 3 E
Bra, *Italy* **44 D4** 44 42N 7 51 E
Braås, *Sweden* **17 G9** 57 4N 15 3 E
Brabant □, *Belgium* **24 D4** 50 46N 4 30 E
Brabant L., *Canada* **143 B8** 55 58N 103 43W
Brabrand, *Denmark* **17 H4** 56 9N 10 7 E
Brač, *Croatia* **45 E13** 43 20N 16 40 E
Bracadale, L., *U.K.* **22 D2** 57 20N 6 30W
Bracciano, *Italy* **45 F9** 42 6N 12 10 E
Bracciano, L. di, *Italy* **45 F9** 42 7N 12 14 E
Bracebridge, *Canada* **140 C4** 45 2N 79 19W
Brach, *Libya* **108 C2** 27 31N 14 20 E
Bracieux, *France* **26 E8** 47 30N 1 30 E
Bräcke, *Sweden* **16 B9** 62 45N 15 26 E
Brackettville, *U.S.A.* **155 L4** 29 19N 100 25W
Brački Kanal, *Croatia* **45 E13** 43 24N 16 40 E
Bracknell, *U.K.* **21 F7** 51 25N 0 43W
Bracknell Forest □, *U.K.* **21 F7** 51 25N 0 44W
Brad, *Romania* **52 D7** 46 10N 22 50 E
Brădano →, *Italy* **47 B9** 40 23N 16 51 E
Bradenton, *U.S.A.* **149 M4** 27 30N 82 34W
Bradford, *Canada* **150 B5** 44 7N 79 34W
Bradford, *U.K.* **20 D6** 53 47N 1 45W
Bradford, *Ill., U.S.A.* **156 C7** 41 11N 89 39W
Bradford, *Ohio, U.S.A.* **157 D12** 40 8N 84 27W
Bradford, *Pa., U.S.A.* **150 E6** 41 58N 78 38W
Bradford, *Vt., U.S.A.* **151 C12** 43 59N 72 9W
Bradley, *Ark., U.S.A.* **155 J8** 33 6N 93 39W
Bradley, *Calif., U.S.A.* **160 K6** 35 52N 120 48W
Bradley, *Fla., U.S.A.* **153 H8** 27 48N 81 59W
Bradley, *Ill., U.S.A.* **157 C9** 41 9N 87 52W
Bradley, *S. Dak., U.S.A.* **154 C6** 45 5N 97 39W
Bradley Institute, *Zimbabwe* **119 F3** 17 7 S 31 25 E
Bradore Bay, *Canada* **141 B8** 51 27N 57 18W
Bradshaw, *Australia* **124 C5** 15 21 S 130 16 E
Brady, *U.S.A.* **155 K5** 31 9N 99 20W
Brædstrup, *Denmark* **17 J3** 55 58N 9 37 E
Braemar, *Australia* **128 B3** 33 12 S 139 35 E
Braemar, *U.K.* **22 D5** 57 0N 3 23W
Braeside, *Canada* **151 A8** 45 28N 76 24W
Braga, *Portugal* **42 D2** 41 35N 8 25W
Braga □, *Portugal* **42 D2** 41 30N 8 30W
Bragadiru, *Romania* **53 G10** 43 46N 25 31 E
Bragado, *Argentina* **174 D3** 35 2 S 60 27W
Bragança, *Brazil* **170 B2** 1 0 S 47 2W
Bragança, *Portugal* **42 D4** 41 48N 6 50W
Bragança □, *Portugal* **42 D4** 41 30N 6 45W
Bragança Paulista, *Brazil* **175 A6** 22 55 S 46 32W
Brahmanbaria, *Bangla.* **90 D3** 23 58N 91 15 E
Brahmani →, *India* **94 D8** 20 39N 86 46 E
Brahmapur, *India* **94 E7** 19 15N 84 54 E
Brahmaputra →, *India* **90 D2** 23 58N 89 50 E
Braich-y-pwll, *U.K.* **20 E3** 52 47N 4 46W
Braidwood, *Australia* **129 C8** 35 27 S 149 49 E
Brăila, *Romania* **53 E12** 45 19N 27 59 E
Brăila □, *Romania* **53 E12** 45 5N 27 30 E
Brainerd, *U.S.A.* **154 B7** 46 22N 94 12W
Braintree, *U.K.* **21 F8** 51 53N 0 34 E
Braintree, *U.S.A.* **151 D14** 42 13N 71 0W
Brak →, *S. Africa* **116 D3** 29 35 S 22 55 E
Brake, *Germany* **30 B4** 53 20N 8 28 E
Brakel, *Germany* **30 D5** 51 42N 9 11 E
Bräkne-Hoby, *Sweden* **17 H9** 56 14N 15 6 E
Brakwater, *Namibia* **116 C2** 22 28 S 17 3 E
Brålanda, *Sweden* **17 F6** 58 34N 12 21 E
Bralorne, *Canada* **142 C4** 50 50N 122 50W
Bramberg, *Germany* **31 E6** 50 6N 10 40 E
Bramdrupdam, *Denmark* **17 J3** 55 31N 9 28 E
Bramming, *Denmark* **17 J2** 55 28N 8 42 E
Brämön, *Sweden* **16 B11** 62 14N 17 40 E
Brampton, *Canada* **140 D4** 43 45N 79 45W
Brampton, *U.K.* **20 C5** 54 57N 2 44W
Bramsche, *Germany* **30 C3** 52 24N 7 58 E
Bramwell, *Australia* **126 A3** 12 8 S 142 37 E
Branchville, *U.S.A.* **152 B9** 33 15N 80 49W

194 Branco

Branco →, Brazil 169 D5 1 20 S 61 50W
Branco, C., Brazil 170 C5 7 9 S 34 47W
Brandbu, Norway 18 D7 60 26N 10 28 E
Brande, Denmark 17 J3 55 57N 9 8 E
Brandenburg =
Neubrandenburg, Germany 30 B9 53 33N 13 15 E
Brandenburg, Germany 30 C8 52 25N 12 33 E
Brandenburg, U.S.A. 157 G10 38 0N 86 10W
Brandenburg □, Germany 30 C9 52 50N 13 0 E
Brandfort, S. Africa 116 D4 28 40 S 26 30 E
Brando, France 29 F13 42 47N 9 27 E
Brandon, Canada 143 D9 49 50N 99 57W
Brandon, Fla., U.S.A. 153 H7 27 56N 82 17W
Brandon, Vt., U.S.A. 151 C11 43 48N 73 4W
Brandon B., Ireland 23 D1 52 17N 10 8W
Brandon Mt., Ireland 23 D1 52 15N 10 15W
Brandsen, Argentina 174 D4 35 10 S 58 15W
Brandvlei, S. Africa 116 E3 30 25 S 20 30 E
Brandýs nad Labem,
Czech Rep. 34 A7 50 10N 14 40 E
Brăneşti, Romania 53 F11 44 27N 26 20 E
Branford, Conn., U.S.A. 151 E12 41 17N 72 49W
Branford, Fla., U.S.A. 152 F7 29 58N 82 56W
Braniewo, Poland 54 D6 54 25N 19 50 E
Bransfield Str., Antarctica 7 C18 63 0 S 59 0W
Brańsk, Poland 55 F9 52 45N 22 50 E
Branson, Colo., U.S.A. 155 G3 37 1N 103 53W
Branson, Mo., U.S.A. 155 G8 36 39N 93 13W
Brantford, Canada 140 D3 43 10N 80 15W
Brantley, U.S.A. 152 D3 31 35N 86 16W
Brantôme, France 28 C4 45 22N 0 39 E
Branxholme, Australia 128 D4 37 52 S 141 49 E
Branxton, Australia 129 B9 32 38 S 151 21 E
Branzi, Italy 44 B6 46 1N 9 46 E
Bras d'Or L., Canada 141 C7 45 50N 60 50W
Brasil, Planalto, Brazil 166 E6 18 0 S 46 30W
Brasiléia, Brazil 172 C4 11 0 S 68 45W
Brasília, Brazil 171 E2 15 47 S 47 55W
Brasília Legal, Brazil 169 D6 3 49 S 55 36W
Braskereidfoss, Norway 18 D8 60 44N 11 46 E
Braslaw, Belarus 15 J22 55 38N 27 0 E
Braslovče, Slovenia 45 B12 46 21N 15 3 E
Braşov, Romania 53 E10 45 38N 25 35 E
Braşov □, Romania 53 E10 45 45N 25 15 E
Brass, Nigeria 113 E6 4 35N 6 14 E
Brass →, Nigeria 113 E6 4 15N 6 13 E
Brassac-les-Mines, France 28 C7 45 24N 3 20 E
Brasschaat, Belgium 24 C4 51 19N 4 27 E
Brassey, Banjaran, Malaysia 85 B5 5 0N 117 15 E
Brassey Ra., Australia 125 E3 25 8 S 122 15 E
Brasstown Bald, U.S.A. 149 H4 34 53N 83 49W
Brastad, Sweden 17 F5 58 23N 11 30 E
Brastavăţu, Romania 53 G9 43 55N 24 24 E
Bratan = Morozov, Bulgaria 51 D9 42 30N 25 10 E
Brateş, Romania 53 E11 45 50N 26 4 E
Bratislava, Slovak Rep. 35 C10 48 10N 17 7 E
Bratislavský □, Slovak Rep. 35 C10 48 15N 17 20 E
Bratsigovo, Bulgaria 51 D8 42 1N 24 22 E
Bratsk, Russia 65 D11 56 10N 101 30 E
Bratt, U.S.A. 153 E2 30 58N 87 26W
Brattleboro, U.S.A. 151 D12 42 51N 72 34W
Brattvåg, Norway 18 B3 62 37N 6 25 E
Bratunac, Bos.-H. 52 F4 44 13N 19 21 E
Braunau, Austria 34 C6 48 15N 13 3 E
Braunschweig, Germany 30 C6 52 15N 10 31 E
Braunton, U.K. 21 F3 51 7N 4 10W
Brava, Somali Rep. 120 D2 1 20N 44 8 E
Bravicea, Moldova 53 C13 47 22N 28 27 E
Bråviken, Sweden 17 F10 58 38N 16 32 E
Bravo del Norte →, Mexico 162 B5 25 57N 97 9W
Bravo del Norte, Rio → =
Grande, Rio →, U.S.A. 155 N6 25 58N 97 9W
Brawley, U.S.A. 161 N11 32 59N 115 31W
Bray, Ireland 23 C5 53 13N 6 7W
Bray, Mt., Australia 126 A1 14 0 S 134 30 E
Bray-sur-Seine, France 27 D10 48 25N 3 14 E
Braymer, U.S.A. 156 E3 39 35N 93 48W
Brazeau →, Canada 142 C5 52 55N 115 14W
Brazil, U.S.A. 157 E9 39 32N 87 8W
Brazil ■, S. Amer. 171 D2 12 0 S 50 0W
Brazilian Highlands = Brasil,
Planalto, Brazil 166 E6 18 0 S 46 30W
Brazo Sur →, S. Amer. 174 B4 25 21 S 57 42W
Brazos →, U.S.A. 155 L7 28 53N 95 23W
Brazzaville, Congo 115 C3 4 9 S 15 12 E
Brčko, Bos.-H. 52 F3 44 54N 18 46 E
Brda →, Poland 55 E5 53 8N 18 8 E
Brdy, Czech Rep. 34 B6 49 43N 13 55 E
Brea, Peru 172 A1 4 40 S 81 7W
Breadalbane, Australia 126 C2 23 50 S 139 35 E
Breaden, L., Australia 125 E4 25 51 S 125 28 E
Breaksea Sd., N.Z. 131 F1 45 35 S 166 35 E
Bream B., N.Z. 130 B3 35 56 S 174 28 E
Bream Hd., N.Z. 130 B3 35 51 S 174 36 E
Bream Tail, N.Z. 130 C3 36 3 S 174 36 E
Breas, Chile 174 B1 25 29 S 70 24W
Breaza, Romania 53 E10 45 11N 25 40 E
Brebes, Indonesia 85 D3 6 52 S 109 3 E
Brechin, Canada 150 B5 44 32N 79 10W
Brechin, U.K. 22 E6 56 44N 2 39W
Brecht, Belgium 24 C4 51 21N 4 38 E
Breckenridge, Colo., U.S.A. 158 G10 39 29N 106 3W
Breckenridge, Minn., U.S.A. 154 B6 46 16N 96 35W
Breckenridge, Mo., U.S.A. 156 E3 39 46N 93 48W
Breckenridge, Tex., U.S.A. 155 J5 32 45N 98 54W
Breckland, U.K. 21 E8 52 30N 0 40 E
Brecknock, Pen., Chile 176 D2 54 35 S 71 30W
Břeclav, Czech Rep. 35 C9 48 46N 16 53 E
Brecon, U.K. 21 F4 51 57N 3 23W
Brecon Beacons, U.K. 21 F4 51 53N 3 26W
Breda, Neths. 24 C4 51 35N 4 45 E
Bredaryd, Sweden 17 G7 57 10N 13 45 E
Bredasdorp, S. Africa 116 E3 34 33 S 20 2 E
Bredbo, Australia 129 C8 35 58 S 149 10 E
Bredebro, Denmark 17 J2 55 4N 8 50 E
Bredstedt, Germany 30 A4 54 37N 8 55 E
Bredy, Russia 62 E8 52 26N 60 21 E
Bree, Belgium 24 C5 51 8N 5 35 E
Bregalnica →, Macedonia 50 E6 41 43N 22 9 E
Bregenz, Austria 34 D2 47 30N 9 45 E
Bregovo, Bulgaria 50 B6 44 9N 22 39 E
Bréhal, France 26 D5 48 53N 1 30W
Bréhat, Î. de, France 26 D4 48 51N 3 0W
Breiðafjörður, Iceland 11 B3 65 15N 23 15W
Breiðavík, Iceland 11 B2 65 33N 24 21W
Breiðdalsvík, Iceland 11 C13 64 44N 14 0W
Breil-sur-Roya, France 29 E11 43 56N 7 31 E
Breim, Norway 18 C3 61 44N 6 25 E
Breisach, Germany 31 G3 48 1N 7 36 E
Brejinho de Nazaré, Brazil 170 D2 11 1 S 48 34W
Brejo, Brazil 170 B3 3 41 S 42 47W
Brekke, Norway 18 C2 61 1N 5 26 E
Brekken, Norway 18 B8 62 40N 11 51 E
Brekkestø, Norway 18 F5 58 11N 8 22 E
Bremanger, Norway 18 C1 61 51N 4 58 E
Bremangerlandet, Norway 18 C1 61 51N 5 0 E
Bremen, Germany 30 B4 53 4N 8 47 E
Bremen, U.S.A. 152 B4 33 43N 85 9W
Bremen □, Germany 30 B4 53 4N 8 50 E
Bremer I., Australia 126 A2 12 5 S 136 45 E
Bremerhaven, Germany 30 B4 53 33N 8 36 E
Bremerton, U.S.A. 160 C4 47 34N 122 38W
Bremervörde, Germany 30 B5 53 29N 9 8 E
Bremnes, Norway 18 E2 59 47N 5 8 E
Bremsnes, Norway 18 A4 63 6N 7 40 E
Brenes, Spain 43 H5 37 32N 5 54W
Brenham, U.S.A. 155 K6 30 10N 96 24W
Brenne, France 28 B5 46 44N 1 14 E
Brennerpass, Austria 34 D4 47 2N 11 30 E
Brennhaug, Norway 18 C6 61 54N 9 21 E
Breno, Italy 44 C7 45 57N 10 18 E
Brent, Canada 140 C4 46 2N 78 29W
Brenta →, Italy 45 C9 45 11N 12 18 E
Brentwood, U.K. 21 F8 51 37N 0 19 E
Brentwood, U.S.A. 151 F11 40 47N 73 15W
Bréscia, Italy 44 C7 45 33N 10 15 E
Breskens, Neths. 24 C3 51 23N 3 33 E
Breslau = Wrocław, Poland 55 G4 51 5N 17 5 E
Bresle →, France 26 B8 50 4N 1 22 E
Bressanone, Italy 45 B8 46 43N 11 39 E
Bressay, U.K. 22 A7 60 9N 1 6W
Bresse, France 27 F12 46 50N 5 10 E
Bressuire, France 26 F6 46 51N 0 30W
Brest, Belarus 59 F2 52 10N 23 40 E
Brest, France 26 D2 48 24N 4 31W
Brest-Litovsk = Brest,
Belarus 59 F2 52 10N 23 40 E
Bretagne, France 26 D3 48 10N 3 0W
Bretçu, Romania 53 D11 46 7N 26 18 E
Bretenoux, France 28 D5 44 54N 1 51 E
Breteuil, Eure, France 26 D7 48 50N 0 57 E
Breteuil, Oise, France 27 C9 49 38N 2 18 E
Breton, Canada 142 C6 53 7N 114 28W
Breton, Pertuis, France 28 B2 46 17N 1 25W
Breton Sd., U.S.A. 155 L10 29 35N 89 15W
Brett, C., N.Z. 130 B3 35 10 S 174 20 E
Bretten, Germany 31 F4 49 2N 8 42 E
Breuil-Cervínia, Italy 44 C4 45 56N 7 38 E
Brevard, U.S.A. 149 H4 35 14N 82 44W
Breves, Brazil 170 B1 1 40 S 50 29W
Brevig Mission, U.S.A. 144 D6 65 20N 166 29W
Brevik, Norway 18 E6 59 4N 9 42 E
Brewarrina, Australia 127 E4 30 0 S 146 51 E
Brewer, U.S.A. 141 D6 44 48N 68 46W
Brewer, Mt., U.S.A. 160 J8 36 44N 118 28W
Brewster, N.Y., U.S.A. 151 E11 41 23N 73 37W
Brewster, Wash., U.S.A. 158 B4 48 6N 119 47W
Brewster, Kap, Greenland 10 C8 70 7N 22 0W
Brewton, Ala., U.S.A. 149 K2 31 7N 87 4W
Brewton, Ga., U.S.A. 152 C7 32 36N 82 48W
Breyten, S. Africa 117 D5 26 16 S 30 0 E
Breza, Bos.-H. 52 F3 44 2N 18 16 E
Brezhnev = Naberezhnyye
Chelny, Russia 60 C11 55 42N 52 19 E
Brežice, Slovenia 45 C12 45 54N 15 35 E
Brézina, Algeria 111 B5 33 4N 1 14 E
Březnice, Czech Rep. 34 B6 49 32N 13 57 E
Breznik, Bulgaria 50 D6 42 44N 22 55 E
Brezno, Slovak Rep. 35 C12 48 50N 19 40 E
Brezoi, Romania 53 E9 45 21N 24 15 E
Brezovo, Bulgaria 51 D9 42 21N 25 5 E
Bria, C.A.R. 114 A4 6 30N 21 58 E
Briançon, France 29 D10 44 54N 6 39 E
Briare, France 27 E9 47 38N 2 45 E
Briático, Italy 47 D9 38 43N 16 2 E
Bribbaree, Australia 129 C7 34 10 S 147 51 E
Bribie I., Australia 127 D5 27 0 S 153 10 E
Briceni, Moldova 53 B12 48 22N 27 6 E
Bricquebec, France 26 C5 49 28N 1 38W
Bridgeboro, U.S.A. 152 D6 31 24N 83 59W
Bridgehampton, U.S.A. 151 F12 40 56N 72 19W
Bridgend, U.K. 21 F4 51 30N 3 34W
Bridgend □, U.K. 21 F4 51 36N 3 36W
Bridgeport, Calif., U.S.A. 160 G7 38 15N 119 14W
Bridgeport, Conn., U.S.A. 151 E11 41 11N 73 12W
Bridgeport, Nebr., U.S.A. 154 E3 41 40N 103 6W
Bridgeport, Tex., U.S.A. 155 J6 33 13N 97 45W
Bridger, U.S.A. 158 D9 45 18N 108 55W
Bridgeton, U.S.A. 148 F8 39 26N 75 14W
Bridgetown, Australia 125 F2 33 58 S 116 7 E
Bridgetown, Barbados 165 D8 13 5N 59 30W
Bridgetown, Canada 141 D6 44 55N 65 18W
Bridgewater, Australia 128 D5 36 36 S 143 59 E
Bridgewater, Canada 141 D7 44 25N 64 31W
Bridgewater, Mass., U.S.A. 151 E14 41 59N 70 58W
Bridgewater, S. Dak.,
U.S.A. 154 D6 43 33N 97 30W
Bridgewater, C., Australia 128 E4 38 23 S 141 23 E
Bridgman, U.S.A. 157 C10 41 57N 86 33W
Bridgnorth, U.K. 21 E5 52 32N 2 25W
Bridgton, U.S.A. 151 B14 44 3N 70 42W
Bridgwater, U.K. 21 F5 51 8N 2 59W
Bridgwater B., U.K. 21 F4 51 15N 3 15W
Bridlington, U.K. 20 C7 54 5N 0 12W
Bridlington B., U.K. 20 C7 54 4N 0 10W
Bridport, Australia 126 G4 40 59 S 147 23 E
Bridport, U.K. 21 G5 50 44N 2 45W
Briec, France 26 D2 48 6N 4 0W
Brienne-le-Château, France 27 D11 48 24N 4 30 E
Brienon-sur-Armançon,
France 27 E10 47 59N 3 38 E
Brienz, Switz. 32 C6 46 46N 8 2 E
Brienzersee, Switz. 32 C5 46 44N 7 53 E
Brier Cr. →, U.S.A. 152 C8 32 44N 81 26W
Brig, Switz. 32 D5 46 18N 7 59 E
Brigg, U.K. 20 D7 53 34N 0 28W
Briggsdale, U.S.A. 154 E2 40 38N 104 20W
Brigham City, U.S.A. 158 F7 41 31N 112 1W
Bright, Australia 129 D7 36 42 S 146 56 E
Brighton, Australia 128 C3 35 5 S 138 30 E
Brighton, Canada 140 D4 44 2N 77 44W
Brighton, U.K. 21 G7 50 49N 0 7W
Brighton, Colo., U.S.A. 154 F2 39 59N 104 49W
Brighton, Fla., U.S.A. 153 H8 27 14N 81 6W
Brighton, Ill., U.S.A. 156 E6 39 2N 90 8W
Brighton, Iowa, U.S.A. 156 C5 41 10N 91 49W
Brighton Seminole Indian
Reservation, U.S.A. 153 H8 27 0N 81 15W
Brightwater, N.Z. 131 B8 41 22 S 173 9 E
Brignogan-Plage, France 26 D2 48 40N 4 20W
Brignoles, France 29 E10 43 25N 6 5 E
Brihuega, Spain 40 E2 40 45N 2 52W
Brikama, Gambia 112 C1 13 15N 16 45W
Brilliant, Canada 142 D5 49 19N 117 38W
Brilliant, U.S.A. 150 F4 40 15N 80 39W
Brilon, Germany 30 D4 51 23N 8 35 E
Brim, Australia 128 D5 36 3 S 142 27 E
Brimfield, U.S.A. 156 D7 40 50N 89 53W
Bríndisi, Italy 47 B10 40 39N 17 55 E
Brinje, Croatia 45 D12 44 59N 15 9 E
Brinkley, U.S.A. 155 H9 34 53N 91 12W
Brinkworth, Australia 128 B3 33 42 S 138 26 E
Brinnon, U.S.A. 160 C4 47 41N 122 54W
Brinson, U.S.A. 152 E5 30 59N 84 44W
Brion, I., Canada 141 C7 47 46N 61 26W
Brionne, France 26 C7 49 11N 0 43 E
Brionski, Croatia 45 D10 44 55N 13 45 E
Brioude, France 28 C7 45 18N 3 24 E
Briouze, France 26 D6 48 42N 0 23W
Brisbane, Australia 127 D5 27 25 S 153 2 E
Brisbane →, Australia 127 D5 27 24 S 153 9 E
Brisighella, Italy 45 D8 44 13N 11 46 E
Bristol, U.K. 21 F5 51 26N 2 35W
Bristol, Conn., U.S.A. 151 E12 41 40N 72 57W
Bristol, Fla., U.S.A. 152 E5 30 26N 84 59W
Bristol, Pa., U.S.A. 151 F10 40 6N 74 51W
Bristol, R.I., U.S.A. 151 E13 41 40N 71 16W
Bristol, S. Dak., U.S.A. 154 C6 45 21N 97 45W
Bristol, Tenn., U.S.A. 149 G4 36 36N 82 11W
Bristol, City of □, U.K. 21 F5 51 27N 2 36W
Bristol B., U.S.A. 138 C4 58 0N 160 0W
Bristol Channel, U.K. 21 F3 51 18N 4 30W
Bristol I., Antarctica 7 B1 58 45 S 28 0W
Bristol L., U.S.A. 159 J5 34 23N 116 50W
Bristow, U.S.A. 155 H6 35 50N 96 23W
British Columbia □, Canada 142 C3 55 0N 125 15W
British Isles, Europe 12 E5 54 0N 4 0W
Brits, S. Africa 117 D4 25 37 S 27 48 E
Britstown, S. Africa 116 E3 30 37 S 23 30 E
Britt, Canada 140 C3 45 46N 80 34W
Britt, U.S.A. 156 A3 43 6N 93 48W
Brittany = Bretagne, France 26 D3 48 10N 3 0W
Britton, U.S.A. 154 C6 45 48N 97 45W
Brive-la-Gaillarde, France 28 C5 45 10N 1 32 E
Briviesca, Spain 42 C7 42 32N 3 19W
Brixen = Bressanone, Italy 45 B8 46 43N 11 39 E
Brixham, U.K. 21 G4 50 23N 3 31W
Brixton, Australia 126 C3 23 32 S 144 57 E
Brlik = Birlik, Kazakstan 63 A6 44 5N 73 31 E
Brlik, Kazakstan 63 B6 43 40N 73 49 E
Brnaze, Croatia 45 E13 43 41N 16 40 E
Brno, Czech Rep. 35 B9 49 10N 16 35 E
Broach = Bharuch, India 94 D1 21 47N 73 0 E
Broad →, Ga., U.S.A. 152 B7 33 59N 82 39W
Broad →, S.C., U.S.A. 149 J5 34 1N 81 4W
Broad Arrow, Australia 125 F3 30 23 S 121 15 E
Broad B., U.K. 22 C2 58 14N 6 18W
Broad Haven, Ireland 23 B2 54 20N 9 55W
Broad Law, U.K. 22 F5 55 30N 3 21W
Broad Sd., Australia 126 C4 22 0 S 149 45 E
Broadford, Australia 129 D6 37 14 S 145 4 E
Broadhurst, U.S.A. 152 D8 31 28N 81 55W
Broadhurst Ra., Australia 124 D3 22 30 S 122 30 E
Broads, The, U.K. 20 E9 52 45N 1 30 E
Broadus, U.S.A. 154 C2 45 27N 105 25W
Broadview, Canada 143 C8 50 22N 102 35W
Broager, Denmark 17 K3 54 53N 9 40 E
Broby, Sweden 17 H8 56 15N 14 4 E
Broceni, Latvia 54 B9 56 42N 22 32 E
Brochet, Canada 143 B8 57 53N 101 40W
Brochet, L., Canada 143 B8 58 36N 101 35W
Brock, Canada 143 C7 51 26N 108 43W
Brocken, Germany 30 D6 51 47N 10 37 E
Brocklehurst, Australia 129 B8 32 9 S 148 38 E
Brockport, U.S.A. 150 C7 43 13N 77 56W
Brockton, U.S.A. 151 D13 42 5N 71 1W
Brockville, Canada 140 D4 44 35N 75 41W
Brockway, Mont., U.S.A. 154 B2 47 18N 105 45W
Brockway, Pa., U.S.A. 150 E6 41 15N 78 47W
Brocton, U.S.A. 150 D5 42 23N 79 26W
Brod, Macedonia 50 E5 41 32N 21 17 E
Brodarevo, Serbia, Yug. 50 C3 43 14N 19 44 E
Brodeur Pen., Canada 139 A11 72 30N 88 10W
Brodhead, U.S.A. 156 B7 42 37N 89 22W
Brodick, U.K. 22 F3 55 35N 5 9W
Brodnica, Poland 55 E6 53 15N 19 25 E
Brody, Ukraine 59 G3 50 5N 25 10 E
Brogan, U.S.A. 158 D5 44 15N 117 31W
Broglie, France 26 C7 49 2N 0 30 E
Brok, Poland 55 F8 52 43N 21 52 E
Broken Arrow, U.S.A. 155 G7 36 3N 95 48W
Broken Bow, Nebr., U.S.A. 154 E5 41 24N 99 38W
Broken Bow, Okla., U.S.A. 155 H7 34 2N 94 44W
Broken Hill = Kabwe,
Zambia 119 E2 14 30 S 28 29 E
Broken Hill, Australia 128 A4 31 58 S 141 29 E
Brokind, Sweden 17 F9 58 13N 15 42 E
Brokopondo, Surinam 169 B7 5 3N 54 59W
Brokopondo □, Surinam 169 C6 4 30N 55 30W
Bromley, U.K. 21 F8 51 24N 0 2 E
Bromölla, Sweden 17 H8 56 5N 14 28 E
Bromsgrove, U.K. 21 E5 52 21N 2 2W
Bronaugh, U.S.A. 156 G2 37 41N 94 28W
Brønderslev, Denmark 17 G3 57 16N 9 57 E
Brong-Ahafo □, Ghana 112 D4 7 50N 2 0W
Broni, Italy 44 C6 45 4N 9 16 E
Bronkhorstspruit, S. Africa 117 D4 25 46 S 28 45 E
Brønnøysund, Norway 14 D15 65 28N 12 14 E
Bronson, Fla., U.S.A. 153 F7 29 27N 82 39W
Bronson, Mich., U.S.A. 157 C11 41 52N 85 12W
Bronte, Italy 47 E7 37 47N 14 50 E
Bronte, U.S.A. 155 K4 31 53N 100 18W
Bronte Park, Australia 126 G4 42 8 S 146 30 E
Bronwood, U.S.A. 152 D5 31 50N 84 22W
Brook Park, U.S.A. 150 E4 41 24N 80 51W
Brookes Point, Phil. 81 G1 8 47N 117 50 E
Brookfield, U.S.A. 156 E3 39 47N 93 4W
Brookhaven, U.S.A. 155 K9 31 35N 90 26W
Brookings, Oreg., U.S.A. 158 E1 42 3N 124 17W
Brookings, S. Dak., U.S.A. 154 C6 44 19N 96 48W
Brooklet, U.S.A. 152 C8 32 23N 81 40W
Brooklin, Canada 150 C6 43 55N 78 55W
Brooklyn, U.S.A. 156 C4 41 44N 92 27W
Brooklyn Park, U.S.A. 154 C8 45 6N 93 23W
Brookmere, Canada 142 D4 49 52N 120 53W
Brooks, Canada 142 C6 50 35N 111 55W
Brooks B., Canada 142 C3 50 15N 127 55W
Brooks L., Canada 143 A7 61 55N 106 35W
Brooks Range, U.S.A. 144 C10 68 0N 152 0W
Brookston, U.S.A. 157 D10 40 36N 86 52W
Brooksville, Fla., U.S.A. 149 L4 28 33N 82 23W
Brooksville, Ky., U.S.A. 157 F12 38 41N 84 4W
Brookville, U.S.A. 157 E12 39 25N 85 1W
Brooloo, Australia 127 D5 26 30 S 152 43 E
Broom, L., U.K. 22 D3 57 55N 5 15W
Broome, Australia 124 C3 18 0 S 122 15 E
Broomehill, Australia 125 F2 33 51 S 117 39 E
Broons, France 26 D4 48 20N 2 16W
Brora, U.K. 22 C5 58 0N 3 52W
Brora →, U.K. 22 C5 58 0N 3 51W
Brørup, Denmark 17 J2 55 29N 9 1 E
Brösarp, Sweden 17 J8 55 43N 14 6 E
Brosna →, Ireland 23 C4 53 14N 7 58W
Broşteni, Mehedinţi,
Romania 52 F7 44 45N 22 59 E
Broşteni, Suceava, Romania 53 C10 47 14N 25 43 E
Brostrud, Norway 18 D5 60 18N 8 34 E
Brotas de Macaúbas, Brazil 171 D3 12 0 S 42 38W
Brothers, U.S.A. 158 E3 43 49N 120 36W
Brøttum, Norway 18 C7 61 2N 10 34 E
Brou, France 26 D8 48 13N 1 11 E
Brouage, France 28 C2 45 52N 1 4W
Brough, U.K. 20 C5 54 32N 2 18W
Brough Hd., U.K. 22 B5 59 8N 3 20W
Broughams Gate, Australia 128 A4 30 51 S 140 59 E
Broughton, U.S.A. 157 G8 37 56N 88 27W
Broughton Island, Canada 139 B13 67 33N 63 0W
Broumov, Czech Rep. 35 A9 50 35N 16 20 E
Brovary, Ukraine 59 G6 50 34N 30 48 E
Brovst, Denmark 17 G3 57 6N 9 31 E
Browerville, U.S.A. 154 B7 46 5N 94 52W
Brown, Mt., Australia 128 B3 32 30 S 138 0 E
Brown, Pt., Australia 127 E1 32 32 S 133 50 E
Brown Willy, U.K. 21 G3 50 35N 4 37W
Brownfield, U.S.A. 155 J3 33 11N 102 17W
Browning, Ill., U.S.A. 156 D6 40 8N 90 22W
Browning, Mo., U.S.A. 156 D3 40 3N 93 12W
Browning, Mont., U.S.A. 158 B7 48 34N 113 1W
Browning Pass, N.Z. 131 C6 42 55 S 171 22 E
Brownlee, Canada 143 C7 50 43N 106 1W
Brownsburg, U.S.A. 157 E10 39 51N 86 24W
Brownstown, U.S.A. 157 F10 38 53N 86 3W
Brownsville, Oreg., U.S.A. 158 D2 44 24N 122 59W
Brownsville, Tenn., U.S.A. 155 H10 35 36N 89 16W
Brownsville, Tex., U.S.A. 155 N6 25 54N 97 30W
Brownsweg, Surinam 169 B6 5 5N 55 15W
Brownwood, U.S.A. 155 K5 31 43N 98 59W
Brownwood, L., U.S.A. 155 K5 31 51N 98 35W
Browse I., Australia 124 B3 14 7 S 123 33 E
Broxton, U.S.A. 152 D7 31 38N 82 53W
Broye →, Switz. 32 C3 46 52N 6 58 E
Bru, Norway 18 C2 61 32N 5 11 E
Bruas, Malaysia 87 K3 4 30N 100 47 E
Bruay-la-Buissière, France 27 B9 50 29N 2 33 E
Bruce, U.S.A. 152 E4 30 28N 85 58W
Bruce, Mt., Australia 124 D2 22 37 S 118 8 E
Bruce B., N.Z. 131 D4 43 35 S 169 42 E
Bruce Pen., Canada 150 B3 45 0N 81 30W
Bruce Rock, Australia 125 F2 31 52 S 118 8 E
Bruche →, France 27 D14 48 34N 7 43 E
Bruchsal, Germany 31 F4 49 7N 8 35 E
Bruck an der Leitha, Austria 35 C9 48 1N 16 47 E
Bruck an der Mur, Austria 34 D8 47 24N 15 16 E
Brue →, U.K. 21 F5 51 13N 2 59W
Bruflat, Norway 18 D6 60 53N 9 37 E
Bruges = Brugge, Belgium 24 C3 51 13N 3 13 E
Brugg, Switz. 32 B6 47 29N 8 11 E
Brugge, Belgium 24 C3 51 13N 3 13 E
Brûlé, Canada 142 C5 53 15N 117 58W
Brûlon, France 26 E6 47 58N 0 15W
Brumado, Brazil 171 D3 14 14 S 41 40W
Brumado →, Brazil 171 D3 14 13 S 41 40W
Brumath, France 27 D14 48 43N 7 40 E
Brumunddal, Norway 15 F14 60 53N 10 56 E
Brunchilly, Australia 126 B1 18 50 S 134 30 E
Brundidge, U.S.A. 152 D4 31 43N 85 49W
Bruneau, U.S.A. 158 E6 42 53N 115 48W
Bruneau →, U.S.A. 158 E6 42 56N 115 57W
Bruneck = Brunico, Italy 45 B8 46 48N 11 56 E
Brunei = Bandar Seri
Begawan, Brunei 85 B4 4 52N 115 0 E
Brunei ■, Asia 85 B4 4 50N 115 0 E
Brunette Downs, Australia 126 B2 18 40 S 135 55 E
Brunflo, Sweden 16 A8 63 5N 14 50 E
Brunico, Italy 45 B8 46 48N 11 56 E
Brünig, P., Switz. 32 C6 46 46N 8 8 E
Brunna, Sweden 16 E11 59 52N 17 25 E
Brunnen, Switz. 33 C7 46 59N 8 37 E
Brunner, L., N.Z. 131 C6 42 37 S 171 27 E
Brunnhóll, Iceland 11 C11 64 17N 15 26W
Bruno, Canada 143 C7 52 20N 105 30W
Brunsbüttel, Germany 30 B5 53 53N 9 6 E
Brunssum, Neths. 24 D5 50 57N 5 59 E
Brunswick = Braunschweig,
Germany 30 C6 52 15N 10 31 E
Brunswick, Ga., U.S.A. 152 D8 31 10N 81 30W
Brunswick, Maine, U.S.A. 141 D6 43 55N 69 58W
Brunswick, Md., U.S.A. 148 F7 39 19N 77 38W
Brunswick, Mo., U.S.A. 156 E3 39 26N 93 8W
Brunswick, Ohio, U.S.A. 150 E3 41 14N 81 51W
Brunswick, Pen. de, Chile 176 D2 53 30 S 71 30W
Brunswick B., Australia 124 C3 15 15 S 124 50 E
Brunswick Junction,
Australia 125 F2 33 15 S 115 50 E
Bruntál, Czech Rep. 35 B10 49 59N 17 27 E
Bruny I., Australia 126 G4 43 20 S 147 15 E
Brus Laguna, Honduras 164 C3 15 47N 84 35W
Brusartsi, Bulgaria 50 C7 43 40N 23 5 E
Brush, U.S.A. 154 E3 40 15N 103 37W
Brushton, U.S.A. 151 B10 44 50N 74 31W
Brusio, Switz. 33 D10 46 14N 10 8 E
Brusque, Brazil 175 B6 27 5 S 49 0W
Brussel, Belgium 24 D4 50 51N 4 21 E
Brussels = Brussel, Belgium 24 D4 50 51N 4 21 E
Brussels, Canada 150 C3 43 44N 81 15W
Brusy, Poland 54 E4 53 53N 17 43 E
Bruthen, Australia 129 D7 37 42 S 147 50 E
Bruvoll, Norway 18 D8 60 27N 11 29 E
Bruxelles = Brussel, Belgium 24 D4 50 51N 4 21 E
Bruyères, France 27 D13 48 10N 6 40 E
Bruz, France 26 D5 48 1N 1 46W
Brwinów, Poland 55 F7 52 9N 20 40 E
Bryagovo, Bulgaria 51 E9 41 58N 25 8 E
Bryan, Ohio, U.S.A. 157 C12 41 28N 84 33W
Bryan, Tex., U.S.A. 155 K6 30 40N 96 22W
Bryan, Mt., Australia 128 B3 33 30 S 139 0 E
Bryanka, Ukraine 59 H10 48 32N 38 45 E
Bryansk, Bryansk, Russia 59 F8 53 13N 34 25 E
Bryansk, Dagestan, Russia 61 H8 44 20N 47 10 E
Bryanskoye = Bryansk,
Russia 61 H8 44 20N 47 10 E

Bussolengo 195

Bryant, *U.S.A.* **154 C6** 44 35N 97 28W
Bryne, *Norway* **15 G11** 58 44N 5 38 E
Bryson City, *U.S.A.* **149 H4** 35 26N 83 27W
Bryukhovetskaya, *Russia* **59 K10** 45 48N 39 0 E
Brza Palanka, *Serbia, Yug.* **50 B6** 44 28N 22 27 E
Brzeg, *Poland* **55 H4** 50 52N 17 30 E
Brzeg Dolny, *Poland* **55 G3** 51 16N 16 41 E
Brześć Kujawski, *Poland* **55 F5** 52 36N 18 55 E
Brzesko, *Poland* **55 J7** 49 59N 20 34 E
Brzeziny, *Poland* **55 G6** 51 49N 19 42 E
Brzozów, *Poland* **55 J9** 49 41N 22 3 E
Bsharri, *Lebanon* **103 A5** 34 15N 36 0 E
Bū Athlah, *Libya* **108 B3** 30 9N 15 39 E
Bū Baqarah, *U.A.E.* **97 E8** 25 35N 56 25 E
Bu Craa, *W. Sahara* **110 C2** 26 45N 12 50W
Bū Ḩasā, *U.A.E.* **97 F7** 23 30N 53 20 E
Bua, *Sweden* **17 G6** 57 14N 12 7 E
Bua Yai, *Thailand* **86 E4** 15 33N 102 26 E
Buad I., *Phil.* **81 F5** 11 40N 124 15 E
Buala, *Solomon Is.* **133 M10** 8 10 S 159 35 E
Buapinang, *Indonesia* **82 B2** 4 40 S 121 30 E
Buba, *Guinea-Biss.* **112 C2** 11 40N 14 59W
Bubanda, *Dem. Rep. of the Congo* **114 B3** 4 14N 19 38 E
Bubanza, *Burundi* **118 C2** 3 6 S 29 23 E
Būbiyān, *Kuwait* **97 D6** 29 45N 48 15 E
Buca, *Turkey* **49 C9** 38 22N 27 11 E
Bucak, *Turkey* **49 D12** 37 28N 30 36 E
Bucaramanga, *Colombia* **168 B3** 7 0N 73 0W
Bucas Grande I., *Phil.* **81 G5** 9 40N 125 57 E
Buccaneer Arch., *Australia* **124 C3** 16 7 S 123 20 E
Buccino, *Italy* **47 B8** 40 38N 15 22 E
Bucecea, *Romania* **53 C11** 47 47N 26 28 E
Buchach, *Ukraine* **59 H3** 49 5N 25 25 E
Buchan, *Australia* **129 D8** 37 30 S 148 12 E
Buchan, *U.K.* **22 D6** 57 32N 2 21W
Buchan Ness, *U.K.* **22 D7** 57 29N 1 46W
Buchanan, *Canada* **143 C8** 51 40N 102 45W
Buchanan, *Liberia* **112 D2** 5 57N 10 2W
Buchanan, *Ga., U.S.A.* **152 B4** 33 48N 85 11W
Buchanan, *Mich., U.S.A.* **157 C10** 41 50N 86 22W
Buchanan, L., *Queens., Australia* **126 C4** 21 35 S 145 52 E
Buchanan, L., *W. Austral., Australia* **125 E3** 25 33 S 123 2 E
Buchanan, L., *U.S.A.* **155 K5** 30 45N 98 25W
Buchanan Cr. →, *Australia* **126 B2** 19 13 S 136 33 E
Buchans, *Canada* **141 C8** 48 50N 56 52W
Bucharest = București, *Romania* **53 F11** 44 27N 26 10 E
Buchen, *Germany* **31 F5** 49 32N 9 20 E
Buchholz, *Germany* **30 B5** 53 19N 9 52 E
Buchloe, *Germany* **31 G6** 48 1N 10 44 E
Buchon, Pt., *U.S.A.* **160 K6** 35 15N 120 54W
Buchs, *Switz.* **33 B8** 47 10N 9 28 E
Buciumi, *Romania* **52 C8** 47 3N 23 1 E
Bückeburg, *Germany* **30 C5** 52 16N 9 7 E
Buckeye, *U.S.A.* **159 K7** 33 22N 112 35W
Buckhannon, *U.S.A.* **148 F5** 39 0N 80 8W
Buckhaven, *U.K.* **22 E5** 56 11N 3 3W
Buckie, *U.K.* **22 D6** 57 41N 2 58W
Buckingham, *Canada* **140 C4** 45 37N 75 24W
Buckingham, *U.K.* **21 F7** 51 59N 0 57W
Buckingham B., *Australia* **126 A2** 12 10 S 135 40 E
Buckingham Canal, *India* **95 H5** 14 0N 80 5 E
Buckinghamshire □, *U.K.* **21 F7** 51 53N 0 55W
Buckland, *U.S.A.* **157 D12** 40 37N 84 16W
Buckle Hd., *Australia* **124 B4** 14 26 S 127 52 E
Buckleboo, *Australia* **128 B2** 32 54 S 136 12 E
Buckley, *U.K.* **20 D4** 53 10N 3 5W
Buckley, *Ill., U.S.A.* **157 D8** 40 36N 88 2W
Buckley, *Wash., U.S.A.* **158 C2** 47 10N 122 2W
Buckley →, *Australia* **126 C2** 20 10 S 138 49 E
Bucklin, *Kans., U.S.A.* **155 G5** 37 33N 99 38W
Bucklin, *Mo., U.S.A.* **156 E4** 39 47N 92 53W
Bucks L., *U.S.A.* **160 F5** 39 54N 121 12W
Buco Zau, *Angola* **115 C2** 4 46 S 12 33 E
Bucquoy, *France* **27 B9** 50 9N 2 43 E
Buctouche, *Canada* **141 C7** 46 30N 64 45W
București, *Romania* **53 F11** 44 27N 26 10 E
Bucyrus, *U.S.A.* **157 D14** 40 48N 82 59W
Bud, *Norway* **18 B3** 62 55N 6 55 E
Budacu, Vf., *Romania* **53 C10** 47 7N 25 41 E
Budalin, *Burma* **90 D5** 22 20N 95 10 E
Budaörs, *Hungary* **52 C3** 47 27N 18 58 E
Budapest, *Hungary* **52 C4** 47 29N 19 5 E
Budapest □, *Hungary* **52 C4** 47 29N 19 5 E
Budaun, *India* **93 E8** 28 5N 79 10 E
Budd Coast, *Antarctica* **7 C8** 68 0 S 112 0 E
Buddabadah, *Australia* **129 A7** 31 56 S 147 14 E
Buddusò, *Italy* **46 B2** 40 35N 9 15 E
Bude, *U.K.* **21 G3** 50 49N 4 34W
Budennovsk, *Russia* **61 H7** 44 50N 44 10 E
Budești, *Romania* **53 F11** 44 13N 26 30 E
Budge Budge = Baj Baj, *India* **93 H13** 22 30N 88 5 E
Budgewoi, *Australia* **129 B9** 33 13 S 151 34 E
Búðardalur, *Iceland* **11 B5** 65 7N 21 46W
Búðir, *Snæfellsnessýsla, Iceland* **11 C3** 64 49N 23 23W
Búðir, *Suður-Múlasýsla, Iceland* **11 C12** 64 56N 14 1W
Budia, *Spain* **40 E2** 40 38N 2 46W
Büdingen, *Germany* **31 E5** 50 16N 9 7 E
Budjala, *Dem. Rep. of the Congo* **114 B3** 2 50N 19 40 E
Budoni, *Italy* **46 B2** 40 40N 9 45 E
Búdrio, *Italy* **45 D8** 44 32N 11 32 E
Budva, *Montenegro, Yug.* **50 D2** 42 17N 18 50 E
Budzyń, *Poland* **55 F3** 52 54N 16 59 E
Bue, *Norway* **18 F2** 58 40N 5 58 E
Buea, *Cameroon* **113 E6** 4 10N 9 9 E
Buellton, *U.S.A.* **161 L6** 34 37N 120 12W
Buena Park, *U.S.A.* **161 M9** 33 52N 117 59W
Buena Vista, *Bolivia* **173 D5** 17 27 S 63 40W
Buena Vista, *Colo., U.S.A.* **159 G10** 38 51N 106 8W
Buena Vista, *Ga., U.S.A.* **152 C3** 32 19N 84 31W
Buena Vista, *Va., U.S.A.* **148 G6** 37 44N 79 21W
Buena Vista L., *U.S.A.* **161 K7** 35 12N 119 18W
Buenaventura, *Colombia* **168 C2** 3 53N 77 4W
Buenaventura, *Mexico* **162 B3** 29 50N 107 30W
Buenaventura, B. de, *Colombia* **168 C2** 3 48N 77 17W
Buenavista, *Luzon, Phil.* **80 E4** 13 35N 122 34 E
Buenavista, *Mindanao, Phil.* **81 G5** 8 59N 125 24 E
Buenavista, *Zamboanga del S., Phil.* **81 H4** 7 15N 122 16 E
Buendía, Embalse de, *Spain* **40 E2** 40 25N 2 43W
Buenópolis, *Brazil* **171 E3** 17 54 S 44 11W
Buenos Aires, *Argentina* **174 C4** 34 30 S 58 20W
Buenos Aires, *Colombia* **168 C3** 1 36N 73 18W
Buenos Aires, *Costa Rica* **164 E3** 9 10N 83 20W
Buenos Aires □, *Argentina* **174 D4** 36 30 S 60 0W
Buenos Aires, L., *Chile* **176 C2** 46 35 S 72 30W
Buesaco, *Colombia* **168 C2** 1 23N 77 9W
Buffalo, *Mo., U.S.A.* **155 G8** 37 39N 93 6W
Buffalo, *N.Y., U.S.A.* **150 D6** 42 53N 78 53W
Buffalo, *Okla., U.S.A.* **155 G5** 36 50N 99 38W
Buffalo, *S. Dak., U.S.A.* **154 C3** 45 35N 103 33W
Buffalo, *Wyo., U.S.A.* **158 D10** 44 21N 106 42W
Buffalo →, *Canada* **142 A5** 60 5N 115 5W
Buffalo Head Hills, *Canada* **142 B5** 57 25N 115 55W
Buffalo L., *Canada* **142 C6** 52 27N 112 54W
Buffalo Narrows, *Canada* **143 B7** 55 51N 108 29W
Buffels →, *S. Africa* **116 D2** 29 36 S 17 3 E
Buford, *U.S.A.* **149 H4** 34 10N 84 0W
Bug → = Buh →, *Ukraine* **59 J6** 46 59N 31 58 E
Bug →, *Poland* **55 F8** 52 31N 21 5 E
Buga, *Colombia* **168 C2** 4 0N 76 15W
Buganda, *Uganda* **118 C3** 0 0 31 30 E
Buganga, *Uganda* **118 C3** 0 3 S 32 0 E
Bugasan, *Phil.* **81 H5** 7 27N 124 14 E
Bugasong, *Phil.* **81 F4** 11 3N 122 4 E
Bugeat, *France* **28 C5** 45 36N 1 55 E
Bugel, Tanjung, *Indonesia* **85 D4** 6 26 S 111 3 E
Bugibba, *Malta* **38 D1** 35 57N 14 25 E
Bugojno, *Bos.-H.* **52 F2** 44 2N 17 25 E
Bugsuk, *Phil.* **81 G1** 8 15N 117 15 E
Buguey, *Phil.* **80 B3** 18 17N 121 50 E
Bugulma, *Russia* **62 D4** 54 33N 52 48 E
Buguma, *Nigeria* **113 E6** 4 42N 6 55 E
Bugun Shara, *Mongolia* **72 B5** 49 0N 104 0 E
Buguruslan, *Russia* **62 E4** 53 39N 52 26 E
Buh →, *Ukraine* **59 J6** 46 59N 31 58 E
Buharkent, *Turkey* **49 D10** 37 58N 28 44 E
Buheirat-Murrat-el-Kubra, *Egypt* **106 H8** 30 18N 32 26 E
Bühl, *Germany* **31 G4** 48 40N 8 8 E
Buhl, *Idaho, U.S.A.* **158 E6** 42 36N 114 46W
Buhl, *Minn., U.S.A.* **154 B8** 47 30N 92 46W
Buhuşi, *Romania* **53 D11** 46 41N 26 45 E
Buick, *U.S.A.* **155 G9** 37 38N 91 2W
Builth Wells, *U.K.* **21 E4** 52 9N 3 25W
Buin, *Papua N. G.* **133 L8** 6 48 S 155 42 E
Buinsk, *Russia* **60 C9** 55 0N 48 18 E
Buíque, *Brazil* **170 C4** 8 37 S 37 9W
Buir Nur, *Mongolia* **73 B6** 47 50N 117 42 E
Buis-les-Baronnies, *France* **29 D9** 44 17N 5 16 E
Buitrago = Buitrago del Lozoya, *Spain* **42 E7** 40 58N 3 38W
Buitrago del Lozoya, *Spain* **42 E7** 40 58N 3 38W
Bujalance, *Spain* **43 H6** 37 54N 4 23W
Bujanovac, *Serbia, Yug.* **50 D5** 42 28N 21 44 E
Bujaraloz, *Spain* **40 D4** 41 29N 0 10W
Buje, *Croatia* **45 C10** 45 24N 13 39 E
Bujumbura, *Burundi* **118 C2** 3 16 S 29 18 E
Bük, *Hungary* **52 C1** 47 22N 16 45 E
Buk, *Poland* **55 F3** 52 21N 16 30 E
Buka I., *Papua N. G.* **132 C8** 5 10 S 154 35 E
Bukachacha, *Russia* **65 D12** 52 55N 116 50 E
Bukama, *Dem. Rep. of the Congo* **119 D2** 9 10 S 25 50 E
Bukavu, *Dem. Rep. of the Congo* **118 C2** 2 20 S 28 52 E
Bukene, *Tanzania* **118 C3** 4 15 S 32 48 E
Bukhara = Bukhoro, *Uzbekistan* **63 D2** 39 48N 64 25 E
Bukhoro, *Uzbekistan* **63 D2** 39 48N 64 25 E
Bukidnon □, *Phil.* **81 H5** 8 0N 125 0 E
Bukima, *Tanzania* **118 C3** 1 50 S 33 25 E
Bukit Mertajam, *Malaysia* **87 K3** 5 22N 100 28 E
Bukittinggi, *Indonesia* **84 C2** 0 20 S 100 20 E
Bükk, *Hungary* **52 B5** 48 0N 20 30 E
Bukkapatnam, *India* **95 G3** 14 14N 77 46 E
Bukoba, *Tanzania* **118 C3** 1 20 S 31 49 E
Bukoba □, *Tanzania* **118 C3** 1 30 S 32 0 E
Bukuru, *Nigeria* **113 D6** 9 42N 8 48 E
Bukuya, *Uganda* **118 B3** 0 40N 31 52 E
Bula, *Guinea-Biss.* **112 C1** 12 7N 15 43W
Bula, *Indonesia* **83 B4** 3 6 S 130 30 E
Bulacan, *Phil.* **80 E3** 13 40N 120 21 E
Bulacan □, *Phil.* **80 D3** 15 0N 121 5 E
Bülach, *Switz.* **33 A7** 47 31N 8 32 E
Bulahdelah, *Australia* **129 B10** 32 23 S 152 13 E
Bulalacao, *Phil.* **80 E3** 12 31N 121 26 E
Bulan, *Phil.* **80 E4** 12 40N 123 52 E
Bulanash, *Russia* **62 C9** 57 16N 62 0 E
Bulancak, *Turkey* **101 B8** 40 56N 38 14 E
Búland, *Iceland* **11 D8** 63 46N 18 30W
Bulandshahr, *India* **92 E7** 28 28N 77 51 E
Bulanık, *Turkey* **101 C10** 39 4N 42 14 E
Bulanovo, *Russia* **62 E5** 52 27N 55 10 E
Bûlâq, *Egypt* **106 B3** 25 10N 30 38 E
Bulawayo, *Zimbabwe* **119 G2** 20 7 S 28 32 E
Buldan, *Turkey* **49 C10** 38 2N 28 50 E
Buldana, *India* **94 D3** 20 30N 76 18 E
Buldon, *Phil.* **81 H5** 7 33N 124 25 E
Bulgar, *Russia* **60 C9** 54 57N 49 4 E
Bulgaria ■, *Europe* **51 D9** 42 35N 25 30 E
Bulgheria, Monte, *Italy* **47 B8** 40 4N 15 26 E
Bulgroo, *Australia* **127 D3** 25 47 S 143 58 E
Bulgunnia, *Australia* **127 E1** 30 10 S 134 53 E
Bulgurca, *Turkey* **49 C9** 38 9N 27 9 E
Bulhale, *Somali Rep.* **120 C3** 5 20N 46 29 E
Bulhar, *Somali Rep.* **120 B2** 10 25N 44 30 E
Buli, Teluk, *Indonesia* **82 A3** 1 5N 128 25 E
Buliluyan, C., *Phil.* **81 G1** 8 20N 117 15 E
Bulki, *Ethiopia* **107 F4** 6 11N 36 31 E
Bulkley →, *Canada* **142 B3** 55 15N 127 40W
Bull Shoals L., *U.S.A.* **155 G8** 36 22N 92 35W
Bullaque →, *Spain* **43 G6** 38 59N 4 17W
Bullara, *Australia* **124 D1** 22 40 S 114 3 E
Bullard, *U.S.A.* **152 C6** 32 38N 83 30W
Bullaring, *Australia* **125 F2** 32 30 S 117 45 E
Bullas, *Spain* **41 G3** 38 2N 1 40W
Bulle, *Switz.* **32 C4** 46 37N 7 3 E
Buller →, *N.Z.* **131 B6** 41 44 S 171 36 E
Buller, Mt., *Australia* **129 D7** 37 10 S 146 28 E
Buller Gorge, *N.Z.* **131 B7** 41 40 S 172 10 E
Bulli, *Australia* **129 C9** 34 15 S 150 57 E
Büllingen, *Belgium* **24 D6** 50 25N 6 16 E
Bullock Creek, *Australia* **126 B3** 17 43 S 144 31 E
Bulloo →, *Australia* **127 D3** 28 43 S 142 30 E
Bulloo Downs, *Queens., Australia* **127 D3** 28 31 S 142 57 E
Bulloo Downs, *W. Austral., Australia* **124 D2** 24 0 S 119 32 E
Bulloo L., *Australia* **127 D3** 28 43 S 142 25 E
Bulls, *N.Z.* **130 G4** 40 10 S 175 24 E
Bully-les-Mines, *France* **27 B9** 50 27N 2 44 E
Bulnes, *Chile* **174 D1** 36 42 S 72 19W
Bulo Burti, *Somali Rep.* **120 D3** 3 50N 45 33 E
Bulo Ghedudo, *Somali Rep.* **120 D2** 2 52N 43 1 E
Bulolo, *Papua N. G.* **132 D4** 7 10 S 146 40 E
Bulongo, *Dem. Rep. of the Congo* **115 C4** 4 45 S 21 30 E
Bulpunga, *Australia* **128 B4** 33 47 S 141 45 E
Bulqiza, *Albania* **50 E4** 41 30N 20 21 E
Bulsar = Valsad, *India* **94 D1** 20 40N 72 58 E
Bultfontein, *S. Africa* **116 D4** 28 18 S 26 10 E
Buluan, L., *Phil.* **81 H5** 6 40N 124 49 E
Buluangan, *Phil.* **81 F4** 10 24N 123 20 E
Bulukumba, *Indonesia* **82 C2** 5 33 S 120 11 E
Bulun, *Russia* **65 B13** 70 37N 127 30 E
Bulunghur, *Uzbekistan* **63 D3** 39 46N 67 16 E
Bulungu, *Dem. Rep. of the Congo* **115 D4** 6 4 S 21 54 E
Bulusan, *Phil.* **80 E5** 12 45N 124 8 E
Bumba, *Dem. Rep. of the Congo* **114 B4** 2 13N 22 30 E
Bumbeşti-Jiu, *Romania* **53 E8** 45 10N 23 24 E
Bumbiri I., *Tanzania* **118 C3** 1 40 S 31 55 E
Bumhkang, *Burma* **90 B6** 26 51N 97 40 E
Bumhpa Bum, *Burma* **90 B6** 26 51N 97 14 E
Bumi →, *Zimbabwe* **119 F2** 17 0 S 28 20 E
Bumtang →, *Bhutan* **90 B3** 26 56N 90 53 E
Buna, *Kenya* **118 B4** 2 58N 39 30 E
Buna, *Papua N. G.* **132 E5** 8 42 S 148 27 E
Bunawan, *Agusan del S., Phil.* **81 G5** 8 12N 125 57 E
Bunawan, *Davao del S., Phil.* **81 H5** 7 14N 125 38 E
Bunazi, *Tanzania* **118 C3** 1 3 S 31 23 E
Bunbah, Khalīj, *Libya* **108 B4** 32 20N 23 15 E
Bunbury, *Australia* **125 F2** 33 20 S 115 35 E
Bunclody, *Ireland* **23 D5** 52 39N 6 40W
Buncrana, *Ireland* **23 A4** 55 8N 7 27W
Bundaberg, *Australia* **127 C5** 24 54 S 152 22 E
Bünde, *Germany* **30 C4** 52 11N 8 35 E
Bundey →, *Australia* **126 C2** 21 46 S 135 37 E
Bundi, *India* **92 G6** 25 30N 75 35 E
Bundooma, *Australia* **126 C1** 24 54 S 134 16 E
Bundoran, *Ireland* **23 B3** 54 28N 8 16W
Bundukia, *Sudan* **107 F3** 5 14N 30 55 E
Bundure, *Australia* **129 C7** 35 10 S 146 1 E
Bung Kan, *Thailand* **86 C4** 18 23N 103 37 E
Bungatakada, *Japan* **70 D3** 33 35N 131 25 E
Bungay, *U.K.* **21 E9** 52 27N 1 28 E
Bungendore, *Australia* **129 C8** 35 14 S 149 30 E
Bungil Cr. →, *Australia* **126 D4** 27 5 S 149 5 E
Bungo, Gunong, *Malaysia* **85 B4** 1 16N 110 9 E
Bungo-Suidō, *Japan* **70 E4** 33 0N 132 15 E
Bungoma, *Kenya* **118 B3** 0 34N 34 34 E
Bungu, *Tanzania* **118 D4** 7 35 S 39 0 E
Bunia, *Dem. Rep. of the Congo* **118 B3** 1 35N 30 20 E
Bunji, *Pakistan* **93 B6** 35 45N 74 40 E
Bunker Hill, *Ill., U.S.A.* **156 E7** 39 3N 89 57W
Bunker Hill, *Ind., U.S.A.* **157 D10** 40 40N 86 6W
Bunkie, *U.S.A.* **155 K8** 30 57N 92 11W
Bunnell, *U.S.A.* **149 L5** 29 28N 81 16W
Bunnythorpe, *N.Z.* **130 G4** 40 16 S 175 39 E
Buñol, *Spain* **41 F4** 39 25N 0 47W
Buntok, *Indonesia* **85 C4** 1 40 S 114 58 E
Bununu, *Nigeria* **113 D6** 9 51N 9 32 E
Bununu Dass, *Nigeria* **113 C6** 10 5N 9 31 E
Bünyan, *Turkey* **100 C6** 38 51N 35 51 E
Bunyu, *Indonesia* **85 B5** 3 35N 117 50 E
Bunza, *Nigeria* **113 C5** 12 8N 4 0 E
Buol, *Indonesia* **82 A2** 1 15N 121 32 E
Buon Brieng, *Vietnam* **86 F7** 13 9N 108 12 E
Buon Ma Thuot, *Vietnam* **86 F7** 12 40N 108 3 E
Buong Long, *Cambodia* **86 F6** 13 44N 106 59 E
Buorkhaya, Mys, *Russia* **65 B14** 71 50N 132 40 E
Buqayq, *Si. Arabia* **97 E6** 26 0N 49 45 E
Buqbuq, *Egypt* **106 A2** 31 29N 25 29 E
Bur Acaba, *Somali Rep.* **120 D2** 3 12N 44 20 E
Bûr Fuad, *Egypt* **106 H8** 31 15N 32 20 E
Bur Ghibi, *Somali Rep.* **120 D3** 3 56N 45 7 E
Bûr Safâga, *Egypt* **106 B3** 26 43N 33 57 E
Bûr Sa'îd, *Egypt* **106 H8** 31 16N 32 18 E
Bûr Sûdân, *Sudan* **106 D4** 19 32N 37 9 E
Bûr Taufiq, *Egypt* **106 J8** 29 54N 32 32 E
Bura, *Kenya* **118 C4** 1 4 S 39 58 E
Buran, *Somali Rep.* **120 B3** 10 14N 48 44 E
Burao, *Somali Rep.* **120 C3** 9 32N 45 32 E
Burāq, *Syria* **103 B5** 33 11N 36 29 E
Buras, *U.S.A.* **155 L10** 29 22N 89 32W
Burauen, *Phil.* **81 F5** 10 58N 124 53 E
Buraydah, *Si. Arabia* **96 E5** 26 20N 44 8 E
Burbank, *U.S.A.* **161 L8** 34 11N 118 19W
Burcher, *Australia* **129 B7** 33 30 S 147 16 E
Burdekin →, *Australia* **126 B4** 19 38 S 147 25 E
Burdeos Bay, *Phil.* **80 D4** 14 44N 122 6 E
Burdett, *Canada* **142 D6** 49 50N 111 32W
Burdur, *Turkey* **57 G5** 37 45N 30 17 E
Burdur □, *Turkey* **49 D12** 37 45N 30 0 E
Burdur Gölü, *Turkey* **49 D12** 37 44N 30 10 E
Burdwan = Barddhaman, *India* **93 H12** 23 14N 87 39 E
Bure, *Ethiopia* **107 E4** 10 40N 37 4 E
Bure →, *U.K.* **20 E9** 52 38N 1 43 E
Büren, *Germany* **30 D4** 51 33N 8 35 E
Bureya →, *Russia* **65 E13** 49 27N 129 30 E
Burford, *Canada* **150 C4** 43 7N 80 27W
Burg, *Germany* **30 C7** 52 16N 11 51 E
Burg auf Fehmarn, *Germany* **30 A7** 54 28N 11 9 E
Burg el Arab, *Egypt* **106 H6** 30 54N 29 32 E
Burg et Tuyur, *Sudan* **106 C2** 20 55N 27 56 E
Burg Stargard, *Germany* **30 B9** 53 29N 13 18 E
Burgas, *Bulgaria* **51 D11** 42 33N 27 29 E
Burgas □, *Bulgaria* **51 D10** 42 30N 26 50 E
Burgaski Zaliv, *Bulgaria* **51 D11** 42 30N 27 39 E
Burgdorf, *Germany* **30 C6** 52 27N 10 1 E
Burgdorf, *Switz.* **32 B5** 47 3N 7 37 E
Burgenland □, *Austria* **35 D9** 47 20N 16 20 E
Burgeo, *Canada* **141 C8** 47 37N 57 38W
Burgersdorp, *S. Africa* **116 E4** 31 0 S 26 20 E
Burges, Mt., *Australia* **125 F3** 30 50 S 121 5 E
Burghausen, *Germany* **31 G8** 48 9N 12 49 E
Búrgio, *Italy* **46 E6** 37 36N 13 17 E
Bürglen, *Switz.* **33 C7** 46 53N 8 40 E
Burglengenfeld, *Germany* **31 F8** 49 12N 12 2 E
Burgohondo, *Spain* **42 E6** 40 26N 4 47W
Burgos, *Ilocos N., Phil.* **80 B3** 18 31N 120 39 E
Burgos, *Pangasinan, Phil.* **80 C2** 16 4N 119 52 E
Burgos, *Spain* **42 C7** 42 21N 3 41W
Burgos □, *Spain* **42 C7** 42 21N 3 42W
Burgstädt, *Germany* **30 E8** 50 54N 12 49 E
Burgsvik, *Sweden* **17 G12** 57 3N 18 19 E
Burguillos del Cerro, *Spain* **43 G4** 38 23N 6 35W
Burgundy = Bourgogne, *France* **27 F11** 47 0N 4 50 E
Burhaniye, *Turkey* **49 B8** 39 30N 26 58 E
Burhanpur, *India* **94 D3** 21 18N 76 14 E
Buri Pen., *Eritrea* **107 D4** 15 25N 39 55 E
Burias, *Phil.* **80 E4** 12 55N 123 5 E
Burias Pass, *Phil.* **80 E4** 13 0N 123 15 E
Burica, Pta., *Costa Rica* **164 E3** 8 3N 82 51W
Burigi, L., *Tanzania* **118 C3** 2 2 S 31 22 E
Burin, *Canada* **141 C8** 47 1N 55 14W
Buriram, *Thailand* **86 E4** 15 0N 103 0 E
Buriti Alegre, *Brazil* **171 E2** 18 9 S 49 3W
Buriti Bravo, *Brazil* **170 C3** 5 50 S 43 50W
Buriti dos Lopes, *Brazil* **170 B3** 3 10 S 41 52W
Burj Sāfitā, *Syria* **100 E7** 34 48N 36 7 E
Burji, *Ethiopia* **107 F4** 5 29N 37 51 E
Burkburnett, *U.S.A.* **155 H5** 34 6N 98 34W
Burke, *U.S.A.* **158 C6** 47 31N 115 49W
Burke →, *Australia* **126 C2** 23 12 S 139 33 E
Burketown, *Australia* **126 B2** 17 45 S 139 33 E
Burkettsville, *U.S.A.* **157 D12** 40 21N 84 39W
Burkina Faso ■, *Africa* **112 C4** 12 0N 1 0W
Burk's Falls, *Canada* **140 C4** 45 37N 79 24W
Burlada, *Spain* **40 C3** 42 49N 1 36W
Burley, *U.S.A.* **158 E7** 42 32N 113 48W
Burli, *Kazakstan* **62 F4** 51 25N 52 40 E
Burlingame, *U.S.A.* **160 H4** 37 35N 122 21W
Burlington, *Canada* **150 C5** 43 18N 79 45W
Burlington, *Colo., U.S.A.* **154 F3** 39 18N 102 16W
Burlington, *Ill., U.S.A.* **157 B8** 42 3N 88 33W
Burlington, *Iowa, U.S.A.* **156 D5** 40 49N 91 14W
Burlington, *Kans., U.S.A.* **154 F7** 38 12N 95 45W
Burlington, *Ky., U.S.A.* **157 E12** 39 2N 84 43W
Burlington, *N.C., U.S.A.* **149 G6** 36 6N 79 26W
Burlington, *N.J., U.S.A.* **151 F10** 40 4N 74 51W
Burlington, *Vt., U.S.A.* **151 B11** 44 29N 73 12W
Burlington, *Wash., U.S.A.* **160 B4** 48 28N 122 20W
Burlington, *Wis., U.S.A.* **148 D1** 42 41N 88 17W
Burlyu-Tyube, *Kazakstan* **64 E8** 46 30N 79 10 E
Burma ■, *Asia* **90 E6** 21 0N 96 30 E
Burnaby I., *Canada* **142 C2** 52 25N 131 19W
Burnamwood, *Australia* **129 A6** 31 7 S 144 53 E
Burnet, *U.S.A.* **155 K5** 30 45N 98 14W
Burney, *U.S.A.* **158 F3** 40 53N 121 40W
Burngup, *Australia* **125 F2** 33 2 S 118 42 E
Burnham, *U.S.A.* **150 F7** 40 38N 77 34W
Burnham-on-Sea, *U.K.* **21 F5** 51 14N 3 0W
Burnie, *Australia* **126 G4** 41 4 S 145 56 E
Burnley, *U.K.* **20 D5** 53 47N 2 14W
Burnoye, *Kazakstan* **63 B5** 42 36N 70 47 E
Burns, *Oreg., U.S.A.* **158 E4** 43 35N 119 3W
Burns, *Wyo., U.S.A.* **154 E2** 41 12N 104 21W
Burns Lake, *Canada* **142 C3** 54 20N 125 45W
Burnside →, *Canada* **138 B9** 66 51N 108 4W
Burnside, L., *Australia* **125 E3** 25 22 S 123 0 E
Burnsville, *U.S.A.* **154 C8** 44 47N 93 17W
Burnt River, *Canada* **150 B6** 44 41N 78 42W
Burntwood →, *Canada* **143 B9** 56 8N 96 34W
Burntwood L., *Canada* **143 B8** 55 22N 100 26W
Burqān, *Kuwait* **96 D5** 29 0N 47 57 E
Burra, *Australia* **128 B3** 33 40 S 138 55 E
Burragorang, L., *Australia* **129 B9** 33 52 S 150 37 E
Burramurra, *Australia* **126 C2** 20 25 S 137 15 E
Burray, *U.K.* **22 C6** 58 51N 2 54W
Burreli, *Albania* **50 E4** 41 36N 20 1 E
Burren Junction, *Australia* **127 E4** 30 7 S 148 59 E
Burrendong, L., *Australia* **129 B8** 32 45 S 149 10 E
Burrendong Dam, *Australia* **127 E4** 32 39 S 149 6 E
Burriana, *Spain* **40 F4** 39 50N 0 4W
Burrinjuck Res., *Australia* **129 C8** 35 0 S 148 36 E
Burro, Serranías del, *Mexico* **162 B4** 29 0N 102 0W
Burrow Hd., *U.K.* **22 G4** 54 41N 4 24W
Burruyacú, *Argentina* **174 B3** 26 30 S 64 40W
Burry Port, *U.K.* **21 F3** 51 41N 4 15W
Bursa, *Turkey* **51 F13** 40 15N 29 5 E
Burseryd, *Sweden* **17 G7** 57 12N 13 17 E
Burstall, *Canada* **143 C7** 50 39N 109 54W
Burton, *U.S.A.* **157 B13** 43 0N 83 40W
Burton L., *Canada* **140 B4** 54 45N 78 20W
Burton upon Trent, *U.K.* **20 E6** 52 48N 1 38W
Burtundy, *Australia* **128 B5** 33 45 S 142 15 E
Buru, *Indonesia* **82 B3** 3 30 S 126 30 E
Buruanga, *Phil.* **81 F3** 11 51N 121 53 E
Burullus, Bahra el, *Egypt* **106 H7** 31 25N 31 0 E
Burûm, *Yemen* **99 D5** 14 22N 48 59 E
Burûn, Râs, *Egypt* **103 D2** 31 14N 33 7 E
Burunday, *Kazakstan* **63 B8** 43 20N 76 51 E
Burundi ■, *Africa* **118 C3** 3 15 S 30 0 E
Bururi, *Burundi* **118 C2** 3 57 S 29 37 E
Burutu, *Nigeria* **113 D6** 5 20N 5 29 E
Burwell, *U.S.A.* **154 E5** 41 47N 99 8W
Burwick, *U.K.* **22 C5** 58 45N 2 58W
Bury, *U.K.* **20 D5** 53 35N 2 17W
Bury St. Edmunds, *U.K.* **21 E8** 52 15N 0 43 E
Buryatia □, *Russia* **65 D12** 53 0N 110 0 E
Buryn, *Ukraine* **59 G7** 51 13N 33 50 E
Burzenin, *Poland* **55 G5** 51 28N 18 47 E
Busalla, *Italy* **44 D5** 44 34N 8 57 E
Busango Swamp, *Zambia* **119 E2** 14 15 S 25 45 E
Buşayrah, *Syria* **101 E9** 35 9N 40 26 E
Buşayyah, *Iraq* **96 D5** 30 0N 46 10 E
Busca, *Italy* **44 D4** 44 31N 7 29 E
Bushati, *Albania* **50 E3** 41 58N 19 34 E
Būshehr, *Iran* **97 D6** 28 55N 50 55 E
Būshehr □, *Iran* **97 D6** 28 20N 51 45 E
Bushell, *Canada* **143 B7** 59 31N 108 45W
Bushenyi, *Uganda* **118 C3** 0 35 S 30 10 E
Bushire = Būshehr, *Iran* **97 D6** 28 55N 50 55 E
Bushnell, *Fla., U.S.A.* **153 G7** 28 40N 82 7W
Bushnell, *Ill., U.S.A.* **154 E9** 40 33N 90 31W
Bushnell, *Nebr., U.S.A.* **154 E3** 41 14N 103 54W
Busia □, *Kenya* **118 B3** 0 25N 34 6 E
Busie, *Ghana* **112 C4** 10 29N 2 22W
Businga, *Dem. Rep. of the Congo* **114 B4** 3 16N 20 59 E
Buskerud □, *Norway* **18 D5** 60 20N 9 0 E
Busko-Zdrój, *Poland* **55 H7** 50 28N 20 42 E
Buskul, *Kazakstan* **62 E8** 53 45N 61 12 E
Buslei, *Ethiopia* **120 C2** 5 28N 44 25 E
Busovača, *Bos.-H.* **52 F2** 44 6N 17 53 E
Buşra ash Shām, *Syria* **103 C5** 32 30N 36 25 E
Busselton, *Australia* **125 F2** 33 42 S 115 15 E
Busseto, *Italy* **44 D7** 44 59N 10 2 E
Bussière-Badil, *France* **28 C4** 45 39N 0 36 E
Bussigny, *Switz.* **32 C3** 46 33N 6 33 E
Bussolengo, *Italy* **44 C7** 45 28N 10 51 E

C

Carbonara, C. 197

Camagüey, *Cuba* **164 B4** 21 20N 78 0W
Camaiore, *Italy* **44 E7** 43 56N 10 18 E
Camak, *U.S.A.* **152 B7** 33 27N 82 39W
Camamu, *Brazil* **171 D4** 13 57 S 39 7W
Camaná, *Peru* **172 D3** 16 30 S 72 50W
Camanche, *U.S.A.* **156 C6** 41 47N 90 15W
Camanche Reservoir, *U.S.A.* **160 G6** 38 14N 121 1W
Camanongue, *Angola* **115 E4** 11 24 S 20 17 E
Camaquã →, *Brazil* **175 C5** 31 17 S 51 47W
Câmara de Lobos, *Madeira* . **39 D3** 32 39N 16 59W
Camararé →, *Brazil* **173 C6** 12 15 S 58 55W
Camarat, C., *France* **29 E10** 43 12N 6 41 E
Camarès, *France* **28 E6** 43 49N 2 53 E
Camaret-sur-Mer, *France* **26 D2** 48 16N 4 37W
Camargo, *Bolivia* **173 E4** 20 38 S 65 15W
Camargue, *France* **29 E8** 43 34N 4 34 E
Camarillo, *U.S.A.* **161 L7** 34 13N 119 2W
Camariñas, *Spain* **42 B1** 43 8N 9 12W
Camarines Norte □, *Phil.* **80 D4** 14 10N 122 45 E
Camarines Sur □, *Phil.* **80 E4** 13 40N 123 20 E
Camarón, C., *Honduras* **164 C2** 16 0N 85 5W
Camarones, *Argentina* **176 B3** 44 50 S 65 40W
Camarones, B., *Argentina* **176 B3** 44 45 S 65 35W
Camas, *Spain* **43 H4** 37 24N 6 2W
Camas, *U.S.A.* **160 E4** 45 35N 122 24W
Camas Valley, *U.S.A.* **158 E2** 43 2N 123 40W
Camaxilo, *Angola* **115 D3** 8 21 S 18 56 E
Cambamba, *Angola* **115 D2** 8 53 S 14 44 E
Cambará, *Brazil* **175 A5** 23 2 S 50 5W
Cambay = Khambhat, *India* **92 H5** 22 23N 72 33 E
Cambay, G. of = Khambhat, G. of, *India* **92 J5** 20 45N 72 30 E
Cambil, *Spain* **43 H7** 37 40N 3 33W
Cambo-les-Bains, *France* **28 E2** 43 22N 1 23W
Cambodia ■, *Asia* **86 F5** 12 15N 105 0 E
Camborne, *U.K.* **21 G2** 50 12N 5 19W
Cambrai, *France* **27 B10** 50 11N 3 14 E
Cambre, *Spain* **42 B2** 43 17N 8 20W
Cambria, *U.S.A.* **160 K5** 35 34N 121 5W
Cambrian Mts., *U.K.* **21 E4** 52 3N 3 57W
Cambridge, *Canada* **140 D3** 43 23N 80 15W
Cambridge, *Jamaica* **164 C4** 18 18N 77 54W
Cambridge, *N.Z.* **130 D4** 37 54 S 175 29 E
Cambridge, *U.K.* **21 E8** 52 12N 0 8 E
Cambridge, *Idaho, U.S.A.* **158 D5** 44 34N 116 41W
Cambridge, *Ill., U.S.A.* **156 C6** 41 18N 90 12W
Cambridge, *Iowa, U.S.A.* **156 C3** 41 54N 93 32W
Cambridge, *Mass., U.S.A.* **151 D13** 42 22N 71 6W
Cambridge, *Md., U.S.A.* **148 F7** 38 34N 76 5W
Cambridge, *Minn., U.S.A.* **154 C8** 45 34N 93 13W
Cambridge, *N.Y., U.S.A.* **151 C11** 43 2N 73 22W
Cambridge, *Nebr., U.S.A.* **154 E4** 40 17N 100 10W
Cambridge, *Ohio, U.S.A.* **150 F3** 40 2N 81 35W
Cambridge City, *U.S.A.* **157 E11** 39 49N 85 10W
Cambridge G., *Australia* **124 B4** 14 55 S 128 15 E
Cambridge Springs, *U.S.A.* **150 E4** 41 48N 80 4W
Cambridgeshire □, *U.K.* **21 E7** 52 25N 0 7W
Cambrils, *Spain* **40 D6** 41 8N 1 3 E
Cambuci, *Brazil* **171 F3** 21 35 S 41 55W
Cambundi-Catembo, *Angola* **115 E3** 10 10 S 17 35 E
Camden, *Australia* **129 C9** 34 1 S 150 43 E
Camden, *Ala., U.S.A.* **149 K2** 31 59N 87 17W
Camden, *Ark., U.S.A.* **155 J8** 33 35N 92 50W
Camden, *Maine, U.S.A.* **141 D6** 44 13N 69 4W
Camden, *N.J., U.S.A.* **151 G9** 39 56N 75 7W
Camden, *Ohio, U.S.A.* **157 E12** 39 38N 84 39W
Camden, *S.C., U.S.A.* **149 H5** 34 16N 80 36W
Camden Bay, *U.S.A.* **144 A11** 70 10N 145 15W
Camden Sd., *Australia* **124 C3** 15 27 S 124 25 E
Camdenton, *U.S.A.* **155 F8** 38 1N 92 45W
Çameli, *Turkey* **49 D11** 37 5N 29 24 E
Camenca, *Moldova* **53 B13** 48 3N 28 42 E
Camerino, *Italy* **45 E10** 43 8N 13 4 E
Cameron, *Ariz., U.S.A.* **159 J8** 35 53N 111 25W
Cameron, *La., U.S.A.* **155 L8** 29 48N 93 20W
Cameron, *Mo., U.S.A.* **156 E2** 39 44N 94 14W
Cameron, *S.C., U.S.A.* **153 B9** 33 34N 80 43W
Cameron, *Tex., U.S.A.* **155 K6** 30 51N 96 59W
Cameron Falls, *Canada* **140 C2** 49 8N 88 19W
Cameron Highlands, *Malaysia* **87 K3** 4 27N 101 22 E
Cameron Hills, *Canada* **142 B5** 59 48N 118 0W
Cameron Mts., *N.Z.* **131 G2** 46 1 S 167 0 E
Cameroon ■, *Africa* **114 A2** 6 0N 12 30 E
Camerota, *Italy* **47 B8** 40 2N 15 22 E
Cameroun →, *Cameroon* **113 E6** 4 0N 9 35 E
Cameroun, Mt., *Cameroon* **113 E6** 4 13N 9 10 E
Cametá, *Brazil* **170 B2** 2 12 S 49 30W
Çamiçi Gölü, *Turkey* **49 D9** 37 29N 27 28 E
Camiguin □, *Phil.* **81 G5** 9 11N 124 42 E
Camiguin I., *Phil.* **80 B3** 18 56N 121 55 E
Camiling, *Phil.* **80 D3** 15 42N 120 24 E
Camilla, *U.S.A.* **152 D5** 31 14N 84 12W
Caminha, *Portugal* **42 D2** 41 50N 8 50W
Camino, *U.S.A.* **160 G6** 38 44N 120 41W
Camira Creek, *Australia* **127 D5** 29 15 S 152 58 E
Camiranga, *Brazil* **170 B2** 1 48 S 46 17W
Camiri, *Bolivia* **173 E5** 20 3 S 63 31W
Camissombo, *Angola* **115 D4** 8 7 S 20 38 E
Cammal, *U.S.A.* **150 E7** 41 24N 77 28W
Cammarata, *Italy* **46 E6** 37 38N 13 38 E
Camocim, *Brazil* **170 B3** 2 55 S 40 50W
Camooweal, *Australia* **126 B2** 19 56 S 138 7 E
Camopi, *Fr. Guiana* **169 C7** 3 12N 52 17W
Camopi →, *Fr. Guiana* **169 C7** 3 10N 52 20W
Camotes Is., *Phil.* **81 F5** 10 40N 124 24 E
Camotes Sea, *Phil.* **81 F5** 10 30N 124 15 E
Camp Crook, *U.S.A.* **154 C3** 45 33N 103 59W
Camp Hill, *U.S.A.* **152 C4** 32 48N 85 39W
Camp Nelson, *U.S.A.* **161 J8** 36 8N 118 39W
Camp Point, *U.S.A.* **156 D5** 40 3N 91 4W
Camp Wood, *U.S.A.* **155 L5** 29 40N 100 1W
Campagna, *Italy* **47 B8** 40 40N 15 6 E
Campana, *Argentina* **174 C4** 34 10 S 58 55W
Campana, I., *Chile* **176 C1** 48 20 S 75 20W
Campanário, *Madeira* **39 D2** 32 39N 17 2W
Campanario, *Spain* **43 G5** 38 52N 5 36W
Campânia □, *Italy* **47 B7** 41 0N 14 30 E
Campbell, *S. Africa* **116 D3** 28 48 S 23 44 E
Campbell, *Calif., U.S.A.* **160 H5** 37 17N 121 57W
Campbell, *Ohio, U.S.A.* **150 E4** 41 5N 80 37W
Campbell, C., *N.Z.* **131 B9** 41 47 S 174 18 E
Campbell I., *Pac. Oc.* **134 N8** 52 30 S 169 0 E
Campbell L., *Canada* **143 A7** 63 14N 106 55W
Campbell River, *Canada* **142 C3** 50 5N 125 20W
Campbell Town, *Australia* **126 G4** 41 52 S 147 30 E
Campbellford, *Canada* **150 B7** 44 18N 77 48W
Campbellpur, *Pakistan* **92 C5** 33 46N 72 26 E
Campbellsburg, *U.S.A.* **157 F10** 38 39N 86 16W
Campbellsville, *U.S.A.* **148 G3** 37 21N 85 20W
Campbellton, *Canada* **141 C6** 47 57N 66 43W
Campbellton, *U.S.A.* **152 E4** 30 57N 85 24W
Campbelltown, *Australia* **129 C9** 34 4 S 150 49 E
Campbeltown, *U.K.* **22 F3** 55 26N 5 36W
Campeche, *Mexico* **163 D6** 19 50N 90 32W
Campeche □, *Mexico* **163 D6** 19 50N 90 32W
Campeche, B. de, *Mexico* **163 D6** 19 30N 93 0W
Campello , *Spain* **41 G4** 38 26N 0 24W
Câmpeni, *Romania* **52 D8** 46 22N 23 3 E
Camperdown, *Australia* **128 E5** 38 14 S 143 9 E
Camperville, *Canada* **143 C8** 51 59N 100 9W
Campi Salentina, *Italy* **47 B11** 40 24N 18 1 E
Câmpia Turzii, *Romania* **53 D8** 46 34N 23 53 E
Campidano, *Italy* **46 C1** 39 30N 8 47 E
Campíglia Maríttima, *Italy* **44 E7** 43 4N 10 37 E
Campillo de Altobuey, *Spain* **41 F3** 39 36N 1 49W
Campillos, *Spain* **43 H6** 37 4N 4 51W
Câmpina, *Romania* **53 E10** 45 10N 25 45 E
Campina Grande, *Brazil* **170 C4** 7 20 S 35 47W
Campina Verde, *Brazil* **171 E2** 19 31 S 49 28W
Campinas, *Brazil* **175 A6** 22 50 S 47 0W
Campli, *Italy* **45 F10** 42 43N 13 41 E
Campo, *Cameroon* **114 B1** 2 22N 9 50 E
Campo, *Spain* **40 C5** 42 25N 0 24 E
Campo Belo, *Brazil* **171 F2** 20 52 S 45 16W
Campo de Criptana, *Spain* **43 F7** 39 24N 3 7W
Campo de Diauarum, *Brazil* **173 C7** 11 12 S 53 14W
Campo de Gibraltar, *Spain* **43 J5** 36 15N 5 25W
Campo Flórido, *Brazil* **171 E2** 19 47 S 48 35W
Campo Formoso, *Brazil* **170 D3** 10 30 S 40 20W
Campo Grande, *Brazil* **173 E7** 20 25 S 54 40W
Campo Maíor, *Brazil* **170 B3** 4 50 S 42 12W
Campo Maior, *Portugal* **43 F3** 39 2N 7 7W
Campo Mourão, *Brazil* **175 A5** 24 3 S 52 22W
Campo Tencia, *Switz.* **33 D7** 46 26N 8 43 E
Campo Túres, *Italy* **45 B8** 46 53N 11 55 E
Campoalegre, *Colombia* **168 C2** 2 41N 75 20W
Campobasso, *Italy* **47 A7** 41 34N 14 39 E
Campobello di Licata, *Italy* **46 E6** 37 15N 13 55 E
Campobello di Mazara, *Italy* **46 E5** 37 38N 12 45 E
Campofelice di Roccella, *Italy* **46 E6** 37 59N 13 53 E
Campomarino, *Italy* **45 G12** 41 57N 15 2 E
Camporeale, *Italy* **46 E6** 37 54N 13 6 E
Camporrobles, *Spain* **40 F3** 39 39N 1 24W
Campos, *Brazil* **171 F3** 21 50 S 41 20W
Campos Altos, *Brazil* **171 E2** 19 47 S 46 10W
Campos Belos, *Brazil* **171 D2** 13 10 S 47 3W
Campos del Puerto, *Spain* **39 B10** 39 26N 3 1 E
Campos Novos, *Brazil* **175 B5** 27 21 S 51 50W
Campos Sales, *Brazil* **170 C3** 7 4 S 40 23W
Camprodón, *Spain* **40 C7** 42 19N 2 23 E
Campton, *Fla., U.S.A.* **153 E3** 30 53N 86 31W
Campton, *Ga., U.S.A.* **152 B6** 33 52N 83 43W
Campton, *Ky., U.S.A.* **157 G13** 37 44N 83 33W
Camptonville, *U.S.A.* **160 F5** 39 27N 121 3W
Câmpulung, *Argeş, Romania* **53 E10** 45 17N 25 3 E
Câmpulung, *Suceava, Romania* **53 C10** 47 32N 25 30 E
Câmpuri, *Romania* **53 D11** 46 0N 26 50 E
Campuya →, *Peru* **168 D3** 1 40 S 73 30W
Campville, *U.S.A.* **152 F7** 29 40N 82 7W
Camrose, *Canada* **142 C6** 53 0N 112 50W
Camsell Portage, *Canada* **143 B7** 59 37N 109 15W
Çamyuva, *Turkey* **49 E12** 36 30N 30 30 E
Çan, *Turkey* **51 F11** 40 2N 27 3 E
Can Clavo, *Spain* **39 C7** 38 57N 1 27 E
Can Creu, *Spain* **39 C7** 38 58N 1 28 E
Can Gio, *Vietnam* **87 G6** 10 25N 106 58 E
Can Tho, *Vietnam* **87 G5** 10 2N 105 46 E
Canaan, *U.S.A.* **151 D11** 42 2N 73 20W
Canada ■, *N. Amer.* **138 C10** 60 0N 100 0W
Cañada de Gómez, *Argentina* **174 C3** 32 40 S 61 30W
Canadian, *U.S.A.* **155 H4** 35 55N 100 23W
Canadian →, *U.S.A.* **155 H7** 35 28N 95 3W
Canadys, *U.S.A.* **152 B9** 33 3N 80 37W
Çanakkale, *Turkey* **51 F10** 40 8N 26 24 E
Çanakkale □, *Turkey* **51 F10** 40 10N 26 25 E
Çanakkale Boğazı, *Turkey* **51 F10** 40 17N 26 32 E
Canal Flats, *Canada* **142 C5** 50 10N 115 48W
Canal Point, *U.S.A.* **153 J9** 26 52N 80 38W
Canala, *N. Cal.* **133 U19** 21 32 S 165 57 E
Canalejas, *Argentina* **174 D2** 35 15 S 66 34W
Canals, *Argentina* **174 C3** 33 35 S 62 53W
Canals, *Spain* **41 G4** 38 58N 0 35W
Canandaigua, *U.S.A.* **150 D7** 42 54N 77 17W
Cananea, *Mexico* **162 A2** 31 0N 110 20W
Cañar, *Ecuador* **168 D2** 2 33 S 78 56W
Cañar □, *Ecuador* **168 D2** 2 30 S 79 0W
Canarias, Is., *Atl. Oc.* **39 F4** 28 30N 16 0W
Canarreos, Arch. de los, *Cuba* **164 B3** 21 35N 81 40W
Canary Is. = Canarias, Is., *Atl. Oc.* **39 F4** 28 30N 16 0W
Canastra, Serra da, *Brazil* **171 F2** 20 0 S 46 20W
Canatlán, *Mexico* **162 C4** 24 31N 104 47W
Canaveral, C., *U.S.A.* **149 L5** 28 27N 80 32W
Canaveral National Seashore, *U.S.A.* **153 G9** 28 28N 80 34W
Cañaveruelas, *Spain* **40 E2** 40 24N 2 38W
Canavieiras, *Brazil* **171 E4** 15 39 S 39 0W
Canbelego, *Australia* **129 A7** 31 32 S 146 18 E
Canberra, *Australia* **129 C8** 35 15 S 149 8 E
Canby, *Calif., U.S.A.* **158 F3** 41 27N 120 52W
Canby, *Minn., U.S.A.* **154 C6** 44 43N 96 16W
Canby, *Oreg., U.S.A.* **160 E4** 45 16N 122 42W
Cancale, *France* **26 D5** 48 40N 1 50W
Canche →, *France* **27 B8** 50 31N 1 39 E
Canchyuaya, Cordillera de, *Peru* **172 B3** 7 30 S 74 0W
Cancún, *Mexico* **163 C7** 21 8N 86 44W
Candala, *Somali Rep.* **120 B3** 11 30N 49 58 E
Candanchu, *Spain* **40 C4** 42 47N 0 32W
Candarave, *Peru* **172 D3** 17 15 S 70 13W
Çandarli, *Turkey* **49 C8** 38 56N 26 56 E
Çandarli Körfezi, *Turkey* **49 C8** 38 52N 26 55 E
Candas, *Spain* **42 B5** 43 35N 5 45W
Candé, *France* **26 E5** 47 34N 1 2W
Candeias →, *Brazil* **173 B5** 8 39 S 63 31W
Candela, *Italy* **47 A8** 41 8N 15 31 E
Candelaria, *Argentina* **175 B4** 27 29 S 55 44W
Candelaria, *Canary Is.* **39 F3** 28 22N 16 22W
Candelaria, *Phil.* **80 E3** 15 36N 121 25 E
Candeleda, *Spain* **42 E5** 40 10N 5 14W
Candelo, *Australia* **129 D8** 36 47 S 149 43 E
Candia = Iráklion, *Greece* **38 D7** 35 20N 25 12 E
Candia, Sea of = Crete, Sea of, *Greece* **49 E7** 36 0N 25 0 E
Cândido de Abreu, *Brazil* **171 F1** 24 35 S 51 20W
Cândido Mendes, *Brazil* **170 B2** 1 27 S 45 43W
Candle, *U.S.A.* **144 D7** 65 55N 161 56W
Candle L., *Canada* **143 C7** 53 50N 105 18W
Candlemas I., *Antarctica* **7 B1** 57 3 S 26 40W
Cando, *U.S.A.* **154 A5** 48 32N 99 12W
Candon, *Phil.* **80 C3** 17 12N 120 27 E
Candoni, *Phil.* **81 G4** 9 48N 122 30 E
Canea = Khaniá, *Greece* **38 D6** 35 30N 24 4 E
Canela, *Brazil* **170 D2** 10 15 S 48 25W
Canelli, *Italy* **44 D5** 44 43N 8 17 E
Canelones, *Uruguay* **175 C4** 34 32 S 56 17W
Canet-Plage, *France* **28 F7** 42 41N 3 2 E
Cañete, *Chile* **174 D1** 37 50 S 73 30W
Cañete, *Peru* **172 C2** 13 8 S 76 30W
Cañete, *Spain* **40 E3** 40 3N 1 54W
Cañete de las Torres, *Spain* **43 H6** 37 53N 4 19W
Cangamba, *Angola* **115 E3** 13 40 S 19 54 E
Cangandala, *Angola* **115 D3** 9 45 S 16 33 E
Cangas, *Spain* **42 C2** 42 16N 8 47W
Cangas de Narcea, *Spain* **42 B4** 43 10N 6 32W
Cangas de Onís, *Spain* **42 B5** 43 21N 5 8W
Cangoa, *Angola* **115 E3** 13 8 S 18 30 E
Cangombo, *Angola* **115 E3** 14 24 S 19 59 E
Cangongo, *Angola* **115 D3** 9 24 S 17 30 E
Canguaretama, *Brazil* **170 C4** 6 20 S 35 5W
Canguçu, *Brazil* **175 C5** 31 22 S 52 43W
Cangxi, *China* **76 B5** 31 47N 105 59 E
Cangyuan, *China* **76 F2** 23 12N 99 14 E
Cangzhou, *China* **74 E9** 38 19N 116 52 E
Canhoca, *Angola* **115 D2** 9 15 S 14 41 E
Caniapiscau = Kaniapiskau →, *Canada* **141 A6** 56 40N 69 30W
Caniapiscau, L. = Kaniapiskau L., *Canada* **141 B6** 54 10N 69 55W
Canicattì, *Italy* **46 E6** 37 21N 13 51 E
Canicattini Bagni, *Italy* **47 E8** 37 2N 15 4 E
Canigao Channel, *Phil.* **81 F5** 10 15N 124 42 E
Caniles, *Spain* **43 H8** 37 26N 2 43W
Canim Lake, *Canada* **142 C4** 51 47N 120 54W
Canindé, *Brazil* **170 B4** 4 22 S 39 19W
Canindé →, *Brazil* **170 C3** 6 15 S 42 52W
Canindeyu □, *Paraguay* **175 A5** 24 10 S 55 0W
Canino, *Italy* **45 F8** 42 28N 11 45 E
Canisteo, *U.S.A.* **150 D7** 42 16N 77 36W
Canisteo →, *U.S.A.* **150 D7** 42 7N 77 8W
Cañitas, *Mexico* **162 C4** 23 36N 102 43W
Cañizal, *Spain* **42 D5** 41 12N 5 22W
Canjáyar, *Spain* **43 H8** 37 1N 2 44W
Canjinge, *Angola* **115 E4** 10 12 S 21 17 E
Çankırı, *Turkey* **100 B5** 40 40N 33 37 E
Cankuzo, *Burundi* **118 C3** 3 10 S 30 31 E
Canmore, *Canada* **142 C5** 51 7N 115 18W
Cann River, *Australia* **129 D8** 37 35 S 149 7 E
Canna, *U.K.* **22 D2** 57 3N 6 33W
Cannanore, *India* **95 J2** 11 53N 75 27 E
Cannelton, *U.S.A.* **157 G10** 37 55N 86 45W
Cannes, *France* **29 E11** 43 32N 7 1 E
Canning Town = Port Canning, *India* **93 H13** 22 23N 88 40 E
Cannington, *Canada* **150 B5** 44 20N 79 2W
Cannóbio, *Italy* **44 B5** 46 4N 8 42 E
Cannock, *U.K.* **21 E5** 52 41N 2 1W
Cannon Ball →, *U.S.A.* **154 B4** 46 20N 100 38W
Cannondale Mt., *Australia* **126 D4** 25 13 S 148 57 E
Caño Colorado, *Colombia* **168 C4** 2 18N 68 22W
Canoas, *Brazil* **175 B5** 29 56 S 51 11W
Canoe L., *Canada* **143 B7** 55 10N 108 15W
Canon City, *U.S.A.* **154 F2** 38 27N 105 14W
Canoochee →, *U.S.A.* **152 D8** 31 59N 81 19W
Canopus, *Australia* **128 B4** 33 29 S 140 42 E
Canora, *Canada* **143 C8** 51 40N 102 30W
Canosa di Púglia, *Italy* **47 A9** 41 13N 16 4 E
Canowindra, *Australia* **129 B8** 33 35 S 148 38 E
Canso, *Canada* **141 C7** 45 20N 61 0W
Canta, *Peru* **172 C2** 11 29 S 76 37W
Cantabria □, *Spain* **42 B7** 43 10N 4 0W
Cantabria, Sierra de, *Spain* **40 C2** 42 40N 2 30W
Cantabrian Mts. = Cantábrica, Cordillera, *Spain* **42 C5** 43 0N 5 10W
Cantábrica, Cordillera, *Spain* **42 C5** 43 0N 5 10W
Cantal □, *France* **28 C6** 45 5N 2 45 E
Cantal, Plomb du, *France* **28 C6** 45 3N 2 45 E
Cantanhede, *Portugal* **42 E2** 40 20N 8 36W
Cantaura, *Venezuela* **169 B5** 9 19N 64 21W
Cantavieja, *Spain* **40 E4** 40 31N 0 25W
Čantavir, *Serbia, Yug.* **52 E4** 45 55N 19 46 E
Cantemir, *Moldova* **53 D13** 46 17N 28 14 E
Canterbury, *Australia* **126 D3** 25 23 S 141 53 E
Canterbury, *U.K.* **21 F9** 51 16N 1 6 E
Canterbury □, *N.Z.* **131 D6** 43 45 S 171 19 E
Canterbury Bight, *N.Z.* **131 E6** 44 16 S 171 55 E
Canterbury Plains, *N.Z.* **131 D6** 43 55 S 171 22 E
Cantil, *U.S.A.* **161 K9** 35 18N 117 58W
Cantilan, *Phil.* **81 G5** 9 20N 125 58 E
Cantillana, *Spain* **43 H5** 37 36N 5 50W
Canto do Buriti, *Brazil* **170 C3** 8 7 S 42 58W
Canton = Guangzhou, *China* **77 F9** 23 5N 113 10 E
Canton, *Ga., U.S.A.* **149 H3** 34 14N 84 29W
Canton, *Ill., U.S.A.* **156 D6** 40 33N 90 2W
Canton, *Miss., U.S.A.* **155 J9** 32 37N 90 2W
Canton, *Mo., U.S.A.* **156 D5** 40 8N 91 32W
Canton, *N.Y., U.S.A.* **151 B9** 44 36N 75 10W
Canton, *Ohio, U.S.A.* **150 F3** 40 48N 81 23W
Canton, *Okla., U.S.A.* **155 G5** 36 3N 98 35W
Canton, *S. Dak., U.S.A.* **154 D6** 43 18N 96 35W
Canton L., *U.S.A.* **155 G5** 36 6N 98 35W
Cantonment, *U.S.A.* **153 E2** 30 37N 87 20W
Cantù, *Italy* **44 C6** 45 44N 9 8 E
Cantwell, *U.S.A.* **144 E10** 63 24N 148 57W
Canudos, *Brazil* **173 B6** 7 13 S 58 5W
Canumã, *Amazonas, Brazil* **169 D6** 4 2 S 59 4W
Canumã, *Amazonas, Brazil* **173 B5** 6 8 S 60 10W
Canumã →, *Brazil* **173 A6** 3 55 S 59 10W
Canutama, *Brazil* **173 B5** 6 30 S 64 20W
Canutillo, *U.S.A.* **159 L10** 31 55N 106 36W
Canvey, *U.K.* **21 F8** 51 31N 0 37 E
Canyon, *Tex., U.S.A.* **155 H4** 34 59N 101 55W
Canyon, *Wyo., U.S.A.* **158 D8** 44 43N 110 36W
Canyonlands National Park, *U.S.A.* **159 G9** 38 15N 110 0W
Canyonville, *U.S.A.* **158 E2** 42 56N 123 17W
Cao Bang, *Vietnam* **86 A6** 22 40N 106 15 E
Cao He →, *China* **75 D13** 40 10N 124 32 E
Cao Lanh, *Vietnam* **87 G5** 10 27N 105 38 E
Cao Xian, *China* **74 G8** 34 50N 115 35 E
Caoayan, *Phil.* **80 C3** 17 37N 120 23 E
Cáorle, *Italy* **45 C9** 45 36N 12 53 E
Cap-aux-Meules, *Canada* **141 C7** 47 23N 61 52W
Cap-Chat, *Canada* **141 C6** 49 6N 66 40W
Cap-de-la-Madeleine, *Canada* **140 C5** 46 22N 72 31W
Cap-Haïtien, *Haiti* **165 C5** 19 40N 72 20W
Capáccio, *Italy* **47 B8** 40 25N 15 5 E
Capaci, *Italy* **46 D6** 38 10N 13 14 E
Capaia, *Angola* **115 D4** 8 27 S 20 13 E
Capalonga, *Phil.* **80 D4** 14 20N 122 30 E
Capanaparo →, *Venezuela* **168 B4** 7 1N 67 7W
Capanema, *Brazil* **170 B2** 1 12 S 47 11W
Capánnori, *Italy* **44 E7** 43 50N 10 34 E
Caparo →, *Barinas, Venezuela* **168 B3** 7 46N 70 23W
Caparo →, *Bolívar, Venezuela* **169 B5** 7 30N 64 0W
Capatárida, *Venezuela* **168 A3** 11 11N 70 37W
Capayas, *Phil.* **81 F2** 10 28N 119 39 E
Capbreton, *France* **28 E2** 43 39N 1 26W
Capdenac, *France* **28 D6** 44 34N 2 5 E
Capdepera, *Spain* **40 F8** 39 42N 3 26 E
Cape →, *Australia* **126 C4** 20 59 S 146 51 E
Cape Barren I., *Australia* **126 G4** 40 25 S 148 15 E
Cape Breton Highlands Nat. Park, *Canada* **141 C7** 46 50N 60 40W
Cape Breton I., *Canada* **141 C7** 46 0N 60 30W
Cape Canaveral, *U.S.A.* **153 G9** 28 24N 80 36W
Cape Charles, *U.S.A.* **148 G8** 37 16N 76 1W
Cape Coast, *Ghana* **113 D4** 5 5N 1 15W
Cape Coral, *U.S.A.* **149 M5** 26 33N 81 57W
Cape Dorset, *Canada* **139 B12** 64 14N 76 32W
Cape Fear →, *U.S.A.* **149 H6** 33 53N 78 1W
Cape Girardeau, *U.S.A.* **155 G10** 37 19N 89 32W
Cape Jervis, *Australia* **128 C3** 35 40 S 138 5 E
Cape May, *U.S.A.* **148 F8** 38 56N 74 56W
Cape May Point, *U.S.A.* **147 C12** 38 56N 74 58W
Cape Palmas, *Liberia* **112 E3** 4 25N 7 49W
Cape Pole, *U.S.A.* **144 J14** 55 58N 133 48W
Cape Tormentine, *Canada* **141 C7** 46 8N 63 47W
Cape Town, *S. Africa* **116 E2** 33 55 S 18 22 E
Cape Verde Is. ■, *Atl. Oc.* **8 G6** 17 10N 25 20W
Cape Vincent, *U.S.A.* **151 B8** 44 8N 76 20W
Cape Yakataga, *U.S.A.* **144 F12** 60 4N 142 26W
Cape York Peninsula, *Australia* **126 A3** 12 0 S 142 30 E
Capela, *Brazil* **170 D4** 10 30 S 37 0W
Capela de Campo, *Brazil* **170 B3** 4 40 S 41 55W
Capele, *Angola* **115 E2** 13 39 S 14 53 E
Capelinha, *Brazil* **171 E3** 17 42 S 42 31W
Capella, *Australia* **126 C4** 23 2 S 148 1 E
Capella, Mt., *Papua N. G.* **132 C1** 5 4 S 141 8 E
Capenda Camulemba, *Angola* **115 D3** 9 24 S 18 27 E
Capendu, *France* **28 E6** 43 11N 2 31 E
Capestang, *France* **28 E7** 43 20N 3 2 E
Capim, *Brazil* **170 B2** 1 41 S 47 47W
Capim →, *Brazil* **170 B2** 1 40 S 47 47W
Capinópolis, *Brazil* **171 E2** 18 41 S 49 35W
Capinota, *Bolivia* **172 D4** 17 43 S 66 14W
Capistrello, *Italy* **45 G10** 41 57N 13 23 E
Capitan, *U.S.A.* **159 K11** 33 35N 105 35W
Capitán Aracena, I., *Chile* **176 D2** 54 10 S 71 20W
Capitán Pastene, *Chile* **176 A2** 38 13 S 73 1W
Capitola, *Calif., U.S.A.* **160 J5** 36 59N 121 57W
Capitola, *Fla., U.S.A.* **152 E5** 30 27N 84 5W
Capivara, Serra da, *Brazil* **171 D3** 14 35 S 45 0W
Capiz □, *Phil.* **81 F4** 11 35N 122 30 E
Capizzi, *Italy* **47 E7** 37 51N 14 29 E
Capoche →, *Mozam.* **119 F3** 15 35 S 33 0 E
Capoeira, *Brazil* **173 B6** 5 37 S 59 33W
Capolo, *Angola* **115 E2** 10 22 S 14 7 E
Capoterra, *Italy* **46 C1** 39 11N 8 58 E
Cappadocia, *Turkey* **100 C6** 39 0N 35 0 E
Capraia, *Italy* **44 E6** 43 2N 9 50 E
Caprara, Pta., *Italy* **46 A1** 41 7N 8 19 E
Caprarola, *Italy* **45 F9** 42 19N 12 14 E
Capreol, *Canada* **140 C3** 46 43N 80 56W
Caprera, *Italy* **46 A2** 41 12N 9 28 E
Capri, *Italy* **47 B7** 40 33N 14 14 E
Capricorn Group, *Australia* **126 C5** 23 30 S 151 55 E
Capricorn Ra., *Australia* **124 D2** 23 20 S 116 50 E
Caprino Veronese, *Italy* **44 C7** 45 36N 10 47 E
Caprivi Strip, *Namibia* **116 B3** 18 0 S 23 0 E
Captain Cook, *U.S.A.* **145 D6** 19 30N 155 55W
Captainganj, *India* **93 F10** 26 55N 83 45 E
Captain's Flat, *Australia* **129 C8** 35 35 S 149 27 E
Captieux, *France* **28 D3** 44 18N 0 16W
Captiva, *U.S.A.* **153 J7** 26 31N 82 11W
Capul I., *Phil.* **80 E5** 12 26N 124 10 E
Caquetá □, *Colombia* **168 C3** 1 0N 74 0W
Caquetá →, *Colombia* **168 D4** 1 15 S 69 15W
Carabalan, *Phil.* **81 F4** 10 6N 122 57 E
Carabao I., *Phil.* **80 E3** 12 4N 121 56 E
Carabobo, *Venezuela* **168 A4** 10 2N 68 5W
Carabobo □, *Venezuela* **168 A4** 10 10N 68 5W
Caracal, *Romania* **53 F9** 44 8N 24 22 E
Caracaraí, *Brazil* **169 C5** 1 50N 61 8W
Caracas, *Venezuela* **168 A4** 10 30N 66 55W
Caracol, *Brazil* **170 C3** 9 15 S 43 22W
Caracollo, *Bolivia* **172 D4** 17 39 S 67 10W
Caradoc, *Australia* **128 A5** 30 35 S 143 5 E
Caragabal, *Australia* **129 B7** 33 49 S 147 45 E
Caráglio, *Italy* **44 D4** 44 25N 7 26 E
Carahue, *Chile* **176 A2** 38 43 S 73 12W
Caraí, *Brazil* **171 E3** 17 12 S 41 42W
Carajás, Serra dos, *Brazil* **170 C1** 6 0 S 51 30W
Caramoan, *Phil.* **80 E4** 13 46N 123 52 E
Caranapatuba, *Brazil* **173 B5** 6 38 S 62 34W
Carandaiti, *Bolivia* **173 E5** 20 45 S 63 4W
Carangola, *Brazil* **171 F3** 20 44 S 42 5W
Carani, *Australia* **125 F2** 30 57 S 116 28 E
Caransebeş, *Romania* **52 E7** 45 28N 22 18 E
Carantec, *France* **26 D3** 48 40N 3 55W
Caraparaná →, *Colombia* **168 D3** 1 45 S 73 13W
Caras, *Peru* **172 B2** 9 3 S 77 47W
Caraş Severin □, *Romania* **52 E7** 45 10N 22 10 E
Caraşova, *Romania* **52 E6** 45 11N 21 51 E
Caratasca, L., *Honduras* **164 C3** 15 20N 83 40W
Caratinga, *Brazil* **171 E3** 19 50 S 42 10W
Caraúbas, *Brazil* **170 C4** 5 43 S 37 33W
Caravaca = Caravaca de la Cruz, *Spain* **41 G3** 38 8N 1 52W
Caravaca de la Cruz, *Spain* **41 G3** 38 8N 1 52W
Caravággio, *Italy* **44 C6** 45 30N 9 38 E
Caravelas, *Brazil* **171 E4** 17 45 S 39 15W
Caraveli, *Peru* **172 D3** 15 45 S 73 25W
Caràzinho, *Brazil* **175 B5** 28 16 S 52 46W
Carballino = O Carballiño, *Spain* **42 C2** 42 26N 8 5W
Carballo, *Spain* **42 B2** 43 13N 8 41W
Carberry, *Canada* **143 D9** 49 50N 99 25W
Carbó, *Mexico* **162 B2** 29 42N 110 58W
Carbon, *Canada* **142 C6** 51 30N 113 9W
Carbonara, C., *Italy* **46 C2** 39 6N 9 31 E

Chantilly 199

Caucaia, *Brazil* **170 B4** 3 40 S 38 35W
Caucasia, *Colombia* **168 B2** 8 0N 75 12W
Caucasus Mountains, *Eurasia* **61 J7** 42 50N 44 0 E
Caudete, *Spain* **41 G3** 38 42N 1 2W
Caudry, *France* **27 B10** 50 7N 3 22 E
Caulnes, *France* **26 D4** 48 18N 2 10W
Caulónia, *Italy* **47 D9** 38 23N 16 24 E
Caungula, *Angola* **115 D3** 8 26 S 18 38 E
Cauquenes, *Chile* **174 D1** 36 0 S 72 22W
Caura →, *Venezuela* **169 B5** 7 38N 64 53W
Caurés →, *Brazil* **169 D5** 1 21 S 62 20W
Cauresi →, *Mozam.* **119 F3** 17 8 S 33 0 E
Căuşani, *Moldova* **53 D14** 46 38N 29 25 E
Causapscal, *Canada* **141 C6** 48 19N 67 12W
Caussade, *France* **28 D5** 44 10N 1 33 E
Causse-Méjean, *France* **28 D7** 44 18N 3 42 E
Cauterets, *France* **28 F3** 42 52N 0 8W
Cautín □, *Chile* **176 A2** 39 0 S 72 30W
Caux, Pays de, *France* **26 C7** 49 38N 0 35 E
Cava de' Tirreni, *Italy* **47 B7** 40 42N 14 42 E
Cávado →, *Portugal* **42 D2** 41 32N 8 48W
Cavaillon, *France* **29 E9** 43 50N 5 2 E
Cavalaire-sur-Mer, *France* **29 E10** 43 10N 6 33 E
Cavalcante, *Brazil* **171 D2** 13 48 S 47 30W
Cavalese, *Italy* **45 B8** 46 17N 11 27 E
Cavalier, *U.S.A.* **154 A6** 48 48N 97 37W
Cavalla = Cavally →, *Africa* **112 E3** 4 22N 7 32W
Cavalleria, C. de, *Spain* **39 A11** 40 5N 4 5 E
Cavalli Is., *N.Z.* **130 B2** 35 0 S 173 58 E
Cavallo, I. de, *France* **29 G13** 41 22N 9 16 E
Cavally →, *Africa* **112 E3** 4 22N 7 32W
Cavan, *Ireland* **23 B4** 54 0N 7 22W
Cavan □, *Ireland* **23 C4** 54 1N 7 16W
Cavárzere, *Italy* **45 C9** 45 8N 12 5 E
Çavdarhisar, *Turkey* **49 B11** 39 12N 29 37 E
Çavdır, *Turkey* **49 D11** 37 10N 29 42 E
Cave City, *U.S.A.* **148 G3** 37 8N 85 58W
Cave Spring, *U.S.A.* **152 A4** 34 6N 85 20W
Cavenagh Ra., *Australia* **125 E4** 26 12 S 127 55 E
Cavendish, *Australia* **128 D5** 37 31 S 142 2 E
Caviana, I., *Brazil* **169 C7** 0 10N 50 10W
Cavite, *Phil.* **80 D3** 14 29N 120 55 E
Cavite □, *Phil.* **80 D3** 14 15N 120 50 E
Cavnic, *Romania* **53 C8** 47 40N 23 52 E
Cavour, *Italy* **44 D4** 44 47N 7 22 E
Cavtat, *Croatia* **50 D2** 42 35N 18 13 E
Cawkers Well, *Australia* **128 A5** 31 41 S 142 57 E
Cawndilla L., *Australia* **128 B5** 32 30 S 142 15 E
Cawnpore = Kanpur, *India* **93 F9** 26 28N 80 20 E
Caxias, *Brazil* **170 B3** 4 55 S 43 20W
Caxias do Sul, *Brazil* **175 B5** 29 10 S 51 10W
Caxito, *Angola* **115 D2** 8 30 S 13 30 E
Caxopa, *Angola* **115 E4** 11 52 S 20 52 E
Çay, *Turkey* **100 C4** 38 35N 31 1 E
Cay Sal Bank, *Bahamas* **164 B4** 23 45N 80 0W
Cayambe, *Napo, Ecuador* **168 C2** 0 2N 77 59W
Cayambe, *Quito, Ecuador* **168 C2** 0 3N 78 8W
Cayce, *U.S.A.* **152 B9** 33 59N 81 4W
Çaycuma, *Turkey* **100 B5** 41 25N 32 4 E
Çayeli, *Turkey* **101 B9** 41 5N 40 45 E
Cayenne, *Fr. Guiana* **169 B7** 5 5N 52 18W
Cayenne □, *Fr. Guiana* **169 C7** 5 0N 53 0W
Caygören Barajı, *Turkey* **49 B10** 39 15N 28 12 E
Cayiralan, *Turkey* **100 C6** 39 17N 35 38 E
Caylus, *France* **28 D5** 44 15N 1 47 E
Cayman Brac, *Cayman Is.* **164 C4** 19 43N 79 49W
Cayman Is. ■, *W. Indies* **164 C3** 19 40N 80 30W
Cayo Romano, *Cuba* **165 B4** 22 0N 78 0W
Cayres, *France* **28 D7** 44 55N 3 48 E
Cayuga, *Canada* **150 D5** 42 59N 79 50W
Cayuga, *Ind., U.S.A.* **157 E9** 39 57N 87 28W
Cayuga, *N.Y., U.S.A.* **151 D8** 42 54N 76 44W
Cayuga L., *U.S.A.* **151 D8** 42 41N 76 41W
Cazaje, *Angola* **115 E4** 11 2 S 20 45 E
Cazalla de la Sierra, *Spain* **43 H5** 37 56N 5 45W
Căzăneşti, *Romania* **53 F12** 44 36N 27 3 E
Cazaubon, *France* **28 E3** 43 56N 0 3W
Cazaux et de Sanguinet, Étang de, *France* **28 D2** 44 29N 1 10W
Cazères, *France* **28 E5** 43 13N 1 5 E
Cazin, *Bos.-H.* **45 D12** 44 57N 15 57 E
Čazma, *Croatia* **45 C13** 45 45N 16 39 E
Cazombo, *Angola* **115 E4** 11 54 S 22 56 E
Cazorla, *Spain* **43 H7** 37 55N 3 2W
Cazorla, *Venezuela* **168 B4** 8 1N 67 0W
Cazorla, Sierra de, *Spain* **43 G8** 38 5N 2 55W
Cea →, *Spain* **42 C5** 42 0N 5 36W
Ceamurlia de Jos, *Romania* **53 F13** 44 43N 28 47 E
Ceanannus Mor, *Ireland* **23 C5** 53 44N 6 53W
Ceará = Fortaleza, *Brazil* **170 B4** 3 45 S 38 35W
Ceará □, *Brazil* **170 C4** 5 0 S 40 0W
Ceará Mirim, *Brazil* **170 C4** 5 38 S 35 25W
Ceauru, L., *Romania* **53 F8** 44 58N 23 11 E
Cebaco, I. de, *Panama* **164 E3** 7 33N 81 9W
Cebollar, *Argentina* **174 B2** 29 10 S 66 35W
Cebollera, Sierra de, *Spain* **40 D2** 42 0N 2 30W
Cebreros, *Spain* **42 E6** 40 27N 4 28W
Cebu, *Phil.* **81 F4** 10 18N 123 54 E
Čečava, *Bos.-H.* **52 F2** 44 42N 17 44 E
Ceccano, *Italy* **46 A6** 41 34N 13 20 E
Cece, *Hungary* **52 D3** 46 46N 18 39 E
Cechi, *Ivory C.* **112 D4** 6 15N 4 25W
Cecil, *U.S.A.* **152 D6** 31 3N 83 24W
Cecil Plains, *Australia* **127 D5** 27 30 S 151 11 E
Cécina, *Italy* **44 E7** 43 19N 10 31 E
Cécina →, *Italy* **44 E7** 43 18N 10 29 E
Ceclavín, *Spain* **42 F4** 39 50N 6 45W
Cedar →, *U.S.A.* **156 C5** 41 17N 91 21W
Cedar City, *U.S.A.* **159 H7** 37 41N 113 4W
Cedar Creek Reservoir, *U.S.A.* **155 J6** 32 11N 96 4W
Cedar Falls, *Iowa, U.S.A.* **156 B4** 42 32N 92 27W
Cedar Falls, *Wash., U.S.A.* **160 C5** 47 25N 121 45W
Cedar Grove, *U.S.A.* **157 E12** 39 22N 84 56W
Cedar Key, *U.S.A.* **149 L4** 29 8N 83 2W
Cedar L., *Canada* **143 C9** 53 10N 100 0W
Cedar Lake, *U.S.A.* **157 C9** 41 22N 87 26W
Cedar Point, *U.S.A.* **157 C13** 41 44N 83 21W
Cedar Rapids, *U.S.A.* **156 C5** 41 59N 91 40W
Cedartown, *U.S.A.* **152 A4** 34 1N 85 15W
Cedarvale, *Canada* **142 B3** 55 1N 128 22W
Cedarville, *S. Africa* **117 E4** 30 23 S 29 3 E
Cedarville, *Calif., U.S.A.* **158 F3** 41 32N 120 10W
Cedarville, *Ill., U.S.A.* **156 B7** 42 23N 89 38W
Cedarville, *Ohio, U.S.A.* **157 E13** 39 44N 83 49W
Cedeira, *Spain* **42 B2** 43 39N 8 2W
Cedral, *Mexico* **162 C4** 23 50N 100 42W
Cedrino →, *Italy* **46 B2** 40 11N 9 24 E
Cedro, *Brazil* **170 C4** 6 34 S 39 3W
Cedros, I. de, *Mexico* **162 B1** 28 10N 115 20W
Ceduna, *Australia* **127 E1** 32 7 S 133 46 E
Cedynia, *Poland* **55 F1** 52 53N 14 12 E
Cée, *Spain* **42 C1** 42 57N 9 10W
Cefalù, *Italy* **47 D7** 38 2N 14 1 E
Cega →, *Spain* **42 D6** 41 33N 4 46W
Cegléd, *Hungary* **52 C4** 47 11N 19 47 E
Céglie Messápico, *Italy* **47 B10** 40 39N 17 31 E
Cehegín, *Spain* **41 G3** 38 6N 1 48W
Ceheng, *China* **76 E5** 24 58N 105 48 E
Cehu-Silvaniei, *Romania* **53 C8** 47 24N 23 9 E
Ceica, *Romania* **52 D7** 46 53N 22 10 E
Ceira →, *Portugal* **42 E2** 40 13N 8 16W
Cekhira, *Tunisia* **108 B2** 34 20N 10 5 E
Cela, *Angola* **115 E3** 11 25 S 15 7 E
Čelákovice, *Czech Rep.* **34 A7** 50 10N 14 46 E
Celano, *Italy* **45 F10** 42 5N 13 33 E
Celanova, *Spain* **42 C3** 42 9N 7 58W
Celaya, *Mexico* **162 C4** 20 31N 100 37W
Celebes = Sulawesi □, *Indonesia* **82 B2** 2 0 S 120 0 E
Celebes Sea, *Indonesia* **82 A2** 3 0N 123 0 E
Celendín, *Peru* **172 B2** 6 52 S 78 10W
Čelić, *Bos.-H.* **52 F3** 44 43N 18 49 E
Celica, *Ecuador* **168 D2** 4 7 S 79 59W
Celina, *U.S.A.* **157 D12** 40 33N 84 35W
Celinac, *Bos.-H.* **52 F2** 44 44N 17 22 E
Celje, *Slovenia* **45 B12** 46 16N 15 18 E
Celldömölk, *Hungary* **52 C2** 47 16N 17 10 E
Celle, *Germany* **30 C6** 52 37N 10 4 E
Celorico da Beira, *Portugal* **42 E3** 40 38N 7 24W
Çeltikçi, *Turkey* **49 D12** 37 32N 30 29 E
Cement, *U.S.A.* **155 H5** 34 56N 98 8W
Çemişgezek, *Turkey* **101 C8** 39 3N 38 56 E
Cenepa →, *Peru* **168 D2** 4 40 S 78 10W
Cengong, *China* **76 D7** 27 13N 108 44 E
Ceno →, *Italy* **44 D7** 44 43N 10 5 E
Centallo, *Italy* **44 D4** 44 30N 7 35 E
Centelles, *Spain* **40 D7** 41 50N 2 14 E
Centenário do Sul, *Brazil* **171 F1** 22 48 S 51 36W
Center, *Ga., U.S.A.* **152 A6** 34 3N 83 25W
Center, *N. Dak., U.S.A.* **154 B4** 47 7N 101 18W
Center, *Tex., U.S.A.* **155 K7** 31 48N 94 11W
Center Hill, *U.S.A.* **153 G7** 28 38N 82 3W
Center Point, *U.S.A.* **156 B5** 42 12N 91 46W
Centerfield, *U.S.A.* **159 G8** 39 8N 111 49W
Centerville, *Calif., U.S.A.* **160 J7** 36 44N 119 30W
Centerville, *Iowa, U.S.A.* **156 D4** 40 44N 92 52W
Centerville, *Mich., U.S.A.* **157 C11** 41 55N 85 32W
Centerville, *Pa., U.S.A.* **150 F5** 40 3N 79 59W
Centerville, *S. Dak., U.S.A.* **154 D6** 43 7N 96 58W
Centerville, *Tenn., U.S.A.* **149 H2** 35 47N 87 28W
Centerville, *Tex., U.S.A.* **155 K7** 31 16N 95 59W
Cento, *Italy* **45 D8** 44 43N 11 17 E
Central, *Brazil* **170 D3** 11 8 S 42 8W
Central, *Alaska, U.S.A.* **144 D11** 65 35N 144 48W
Central, *N. Mex., U.S.A.* **159 K9** 32 47N 108 9W
Central □, *Kenya* **118 C4** 0 30 S 37 30 E
Central □, *Malawi* **119 E3** 13 30 S 33 30 E
Central □, *Zambia* **119 E2** 14 25 S 28 50 E
Central, Cordillera, *Bolivia* **173 D5** 18 30 S 64 55W
Central, Cordillera, *Colombia* **168 C3** 5 0N 75 0W
Central, Cordillera, *Costa Rica* **164 D3** 10 10N 84 5W
Central, Cordillera, *Dom. Rep.* **165 C5** 19 15N 71 0W
Central, Cordillera, *Peru* **172 B2** 7 0 S 77 30W
Central, Cordillera, *Phil.* **80 C3** 17 20N 120 57 E
Central African Rep. ■, *Africa* **114 A4** 7 0N 20 0 E
Central City, *Ky., U.S.A.* **148 G2** 37 18N 87 7W
Central City, *Nebr., U.S.A.* **154 E6** 41 7N 98 0W
Central I., *Kenya* **118 B4** 3 30N 36 0 E
Central Makran Range, *Pakistan* **91 D2** 26 30N 64 15 E
Central Patricia, *Canada* **140 B1** 51 30N 90 9W
Central Ra., *Papua N. G.* **132 C2** 5 0 S 143 0 E
Central Russian Uplands, *Europe* **12 E13** 54 0N 36 0 E
Central Siberian Plateau, *Russia* **66 C14** 65 0N 105 0 E
Centralhatchee, *U.S.A.* **152 B4** 33 22N 85 6W
Centralia, *Ill., U.S.A.* **156 F7** 38 32N 89 8W
Centralia, *Mo., U.S.A.* **156 E4** 39 13N 92 8W
Centralia, *Wash., U.S.A.* **160 D4** 46 43N 122 58W
Centre, *U.S.A.* **152 A4** 34 9N 85 41W
Centreville, *Ala., U.S.A.* **149 J2** 32 57N 87 8W
Centreville, *Miss., U.S.A.* **155 K9** 31 5N 91 4W
Century, *U.S.A.* **153 E2** 30 58N 87 16W
Čeotina →, *Bos.-H.* **50 C2** 43 36N 18 50 E
Cephalonia = Kefallinía, *Greece* **48 C2** 38 20N 20 30 E
Čepin, *Croatia* **52 E3** 45 32N 18 34 E
Ceprano, *Italy* **46 A6** 41 33N 13 31 E
Ceptura, *Romania* **53 E11** 45 1N 26 21 E
Cepu, *Indonesia* **85 D4** 7 9 S 111 35 E
Ceram = Seram, *Indonesia* **83 B3** 3 10 S 129 0 E
Ceram Sea = Seram Sea, *Indonesia* **82 B3** 2 30 S 128 30 E
Cerbère, *France* **28 F7** 42 26N 3 10 E
Cerbicales, Is., *France* **29 G13** 41 33N 9 22 E
Cercal, *Portugal* **43 H2** 37 48N 8 40W
Cerdaña, *Spain* **40 C6** 42 22N 1 35 E
Cère →, *France* **28 D5** 44 55N 1 49 E
Cerea, *Italy* **45 C8** 45 12N 11 13 E
Ceredigion □, *U.K.* **21 E3** 52 16N 4 15W
Ceres, *Argentina* **174 B3** 29 55 S 61 55W
Ceres, *Brazil* **171 E2** 15 17 S 49 35W
Ceres, *S. Africa* **116 E2** 33 21 S 19 18 E
Ceres, *U.S.A.* **160 H6** 37 35N 120 57W
Céret, *France* **28 F6** 42 30N 2 42 E
Cereté, *Colombia* **168 B2** 8 53N 75 48W
Cergy, *France* **27 C9** 49 2N 2 4 E
Cerignola, *Italy* **47 A8** 41 17N 15 53 E
Cerigo = Kíthira, *Greece* **48 E5** 36 8N 23 0 E
Cérilly, *France* **27 F9** 46 37N 2 50 E
Cerisiers, *France* **27 D10** 48 8N 3 30 E
Cerizay, *France* **26 F6** 46 50N 0 40W
Çerkeş, *Turkey* **100 B5** 40 49N 32 52 E
Çerkezköy, *Turkey* **51 E12** 41 17N 28 0 E
Cerknica, *Slovenia* **45 C11** 45 48N 14 21 E
Cerkovica, *Bulgaria* **51 C8** 43 41N 24 50 E
Cermerno, *Serbia, Yug.* **50 C4** 43 35N 20 25 E
Çermik, *Turkey* **101 C8** 38 8N 39 26 E
Cerna, *Romania* **53 E13** 45 4N 28 17 E
Cerna →, *Romania* **53 F8** 44 38N 23 58 E
Cernavodă, *Romania* **53 F13** 44 22N 28 3 E
Cernay, *France* **27 E14** 47 44N 7 10 E
Cernik, *Croatia* **52 E2** 45 17N 17 22 E
Cerralvo, I., *Mexico* **162 C3** 24 20N 109 45W
Cërrik, *Albania* **50 E3** 41 2N 19 58 E
Cerritos, *Mexico* **162 C4** 22 27N 100 20W
Cerro Gordo, *U.S.A.* **157 E8** 39 53N 88 44W
Cerro Sombrero, *Chile* **176 D3** 52 45 S 69 15W
Certaldo, *Italy* **44 E8** 43 33N 11 2 E
Cervaro →, *Italy* **47 A8** 41 30N 15 52 E
Cervati, Monte, *Italy* **47 B8** 40 17N 15 29 E
Cervera, *Spain* **40 D6** 41 40N 1 16 E
Cervera de Pisuerga, *Spain* **42 C6** 42 51N 4 30W
Cervera del Río Alhama, *Spain* **40 C3** 42 2N 1 58W
Cervéteri, *Italy* **45 F9** 42 0N 12 6 E
Cérvia, *Italy* **45 D9** 44 15N 12 22 E
Cervignano del Friuli, *Italy* **45 C10** 45 49N 13 20 E
Cervinara, *Italy* **47 A7** 41 1N 14 37 E
Cervione, *France* **29 F13** 42 20N 9 29 E
Cervo, *Spain* **42 B3** 43 40N 7 24W
César □, *Colombia* **168 B3** 9 0N 73 30W
Cesarò, *Italy* **47 E7** 37 50N 14 38 E
Cesena, *Italy* **45 D9** 44 8N 12 15 E
Cesenático, *Italy* **45 D9** 44 12N 12 24 E
Cēsis, *Latvia* **15 H21** 57 18N 25 15 E
Česká Lípa, *Czech Rep.* **34 A7** 50 45N 14 30 E
Česká Třebová, *Czech Rep.* **35 B9** 49 54N 16 27 E
České Budějovice, *Czech Rep.* **34 C7** 48 55N 14 25 E
České Velenice, *Czech Rep.* **34 C7** 48 45N 14 57 E
Českomoravská Vrchovina, *Czech Rep.* **34 B8** 49 30N 15 40 E
Český Brod, *Czech Rep.* **34 A7** 50 4N 14 52 E
Český Krumlov, *Czech Rep.* **34 C7** 48 43N 14 21 E
Český Těšín, *Czech Rep.* **35 B11** 49 45N 18 39 E
Česma →, *Croatia* **45 C13** 45 35N 16 29 E
Çeşme, *Turkey* **49 C8** 38 20N 26 23 E
Cessnock, *Australia* **129 B9** 32 50 S 151 21 E
Cesson-Sévigné, *France* **26 D5** 48 7N 1 36W
Cestas, *France* **28 D3** 44 44N 0 41W
Cestos →, *Liberia* **112 D3** 5 40N 9 10W
Cetate, *Romania* **52 F8** 44 7N 23 2 E
Cetin Grad, *Croatia* **45 C12** 45 9N 15 45 E
Cetina →, *Croatia* **45 E13** 43 26N 16 42 E
Cetinje, *Montenegro, Yug.* **50 D2** 42 23N 18 59 E
Cetraro, *Italy* **47 C8** 39 31N 15 55 E
Ceuta, *N. Afr.* **110 A3** 35 52N 5 18W
Ceva, *Italy* **44 D5** 44 23N 8 2 E
Cévennes, *France* **28 D7** 44 10N 3 50 E
Ceyhan, *Turkey* **100 D6** 37 4N 35 47 E
Ceyhan →, *Turkey* **100 D6** 36 38N 35 40 E
Ceylânpınar, *Turkey* **101 D9** 36 50N 40 2 E
Ceylon = Sri Lanka ■, *Asia* **95 L5** 7 30N 80 50 E
Cèze →, *France* **29 D8** 44 6N 4 43 E
Cha-am, *Thailand* **86 F2** 12 48N 99 58 E
Cha Pa, *Vietnam* **86 A4** 22 20N 103 47 E
Chá Pungana, *Angola* **115 E3** 13 44 S 18 39 E
Chabanais, *France* **28 C4** 45 52N 0 48 E
Chabeuil, *France* **29 D9** 44 54N 5 3 E
Chablais, *France* **27 F13** 46 20N 6 36 E
Chablis, *France* **27 E10** 47 47N 3 48 E
Chabounia, *Algeria* **111 A5** 35 30N 2 38 E
Chacabuco, *Argentina* **174 C3** 34 40 S 60 27W
Chachapoyas, *Peru* **172 B2** 6 15 S 77 50W
Chachasp, *Peru* **172 D3** 15 30 S 72 15W
Chachoengsao, *Thailand* **86 F3** 13 42N 101 5 E
Chachro, *Pakistan* **92 G4** 25 5N 70 15 E
Chaco □, *Argentina* **174 B3** 26 30 S 61 0W
Chaco □, *Paraguay* **174 B4** 26 0 S 60 0W
Chad ■, *Africa* **109 F3** 15 0N 17 15 E
Chad, L. = Tchad, L., *Chad* **109 F2** 13 30N 14 30 E
Chadan, *Russia* **65 D10** 51 17N 91 35 E
Chadileuvú →, *Argentina* **174 D2** 37 46 S 66 0W
Chadiza, *Zambia* **119 E3** 14 45 S 32 27 E
Chadron, *U.S.A.* **154 D3** 42 50N 103 0W
Chadyr-Lunga = Ciadâr-Lunga, *Moldova* **53 D13** 46 3N 28 51 E
Chae Hom, *Thailand* **86 C2** 18 43N 99 35 E
Chaem →, *Thailand* **86 C2** 18 11N 98 38 E
Chaeryŏng, *N. Korea* **75 E13** 38 24N 125 36 E
Chagda, *Russia* **65 D14** 58 45N 130 38 E
Chagny, *France* **27 F11** 46 57N 4 45 E
Chagoda, *Russia* **58 C8** 59 10N 35 15 E
Chagos Arch., *Ind. Oc.* **121 E6** 6 0 S 72 0 E
Chāh Ākhvor, *Iran* **97 C8** 32 41N 59 40 E
Chāh Bahār, *Iran* **97 E9** 25 20N 60 40 E
Chāh-e-Malek, *Iran* **97 D8** 28 35N 59 7 E
Chāh Gay Hills, *Afghan.* **91 C2** 29 30N 64 0 E
Chāh Kavīr, *Iran* **97 D7** 31 45N 54 52 E
Chahār Borjak, *Afghan.* **91 C1** 30 17N 62 3 E
Chahtung, *Burma* **90 B7** 26 41N 98 10 E
Chaillé-les-Marais, *France* **28 B2** 46 25N 1 2W
Chainat, *Thailand* **86 E3** 15 11N 100 8 E
Chaires, *U.S.A.* **152 E5** 30 26N 84 7W
Chaitén, *Chile* **176 B2** 42 55 S 72 43W
Chaiya, *Thailand* **87 H2** 9 23N 99 14 E
Chaj Doab, *Pakistan* **92 C5** 32 15N 73 0 E
Chajari, *Argentina* **174 C4** 30 42 S 58 0W
Chakaria, *Bangla.* **90 E4** 21 45N 92 5 E
Chake Chake, *Tanzania* **118 D4** 5 15 S 39 45 E
Chakhānsūr, *Afghan.* **91 C1** 31 10N 62 0 E
Chakonipau, L., *Canada* **141 A6** 56 18N 68 30W
Chakradharpur, *India* **93 H11** 22 45N 85 40 E
Chakwadam, *Burma* **90 B7** 27 29N 98 31 E
Chakwal, *Pakistan* **91 B4** 32 56N 72 53 E
Chala, *Peru* **172 D3** 15 48 S 74 20W
Chalais, *France* **28 C4** 45 16N 0 3 E
Chalakudi, *India* **95 J3** 10 18N 76 20 E
Chalchihuites, *Mexico* **162 C4** 23 29N 103 53W
Chalcis = Khalkís, *Greece* **48 C5** 38 27N 23 42 E
Châlette-sur-Loing, *France* **27 D9** 48 1N 2 44 E
Chaleur B., *Canada* **141 C6** 47 55N 65 30W
Chalfant, *U.S.A.* **160 H8** 37 32N 118 21W
Chalhuanca, *Peru* **172 C3** 14 15 S 73 15W
Chalindrey, *France* **27 E12** 47 43N 5 26 E
Chaling, *China* **77 D9** 26 58N 113 30 E
Chalisgaon, *India* **94 D2** 20 30N 75 10 E
Chalkar = Shalkar, *Kazakstan* **62 F3** 50 40N 51 53 E
Chalkar, Ozero = Shalkar, Ozero, *Kazakstan* **62 F3** 50 35N 51 47 E
Chalky Inlet, *N.Z.* **131 G1** 46 3 S 166 31 E
Chalkyitsik, *U.S.A.* **144 C12** 66 39N 143 43W
Challans, *France* **26 F5** 46 50N 1 52W
Challapata, *Bolivia* **172 D4** 18 53 S 66 50W
Challis, *U.S.A.* **158 D6** 44 30N 114 14W
Chalna, *India* **93 H13** 22 36N 89 35 E
Chalon-sur-Saône, *France* **27 F11** 46 48N 4 50 E
Chalonnes-sur-Loire, *France* **26 E6** 47 20N 0 45W
Châlons-en-Champagne, *France* **27 D11** 48 58N 4 20 E
Châlus, *France* **28 C4** 45 39N 0 58 E
Chalyaphum, *Thailand* **86 E4** 15 48N 102 2 E
Cham, *Germany* **31 F8** 49 13N 12 39 E
Cham, *Switz.* **33 B6** 47 11N 8 28 E
Cham, Cu Lao, *Vietnam* **86 E7** 15 57N 108 30 E
Chama, *U.S.A.* **159 H10** 36 54N 106 35W
Chamah, Gunong, *Malaysia* **84 A2** 5 13N 101 35 E
Chaman, *Pakistan* **91 C2** 30 58N 66 25 E
Chamba, *India* **92 C7** 32 35N 76 10 E
Chamba, *Tanzania* **119 E4** 11 37 S 37 0 E
Chambal →, *India* **93 F8** 26 29N 79 15 E
Chamberlain, *U.S.A.* **154 D5** 43 49N 99 20W
Chamberlain →, *Australia* **124 C4** 15 30 S 127 54 E
Chambers, *U.S.A.* **159 J9** 35 11N 109 26W
Chambersburg, *U.S.A.* **148 F7** 39 56N 77 40W
Chambéry, *France* **29 C9** 45 34N 5 55 E
Chamblee, *U.S.A.* **152 B5** 33 53N 84 18W
Chambly, *Canada* **151 A11** 45 27N 73 17W
Chambord, *Canada* **141 C5** 48 25N 72 6W
Chamboulive, *France* **28 C5** 45 26N 1 42 E
Chambri L., *Papua N. G.* **132 C2** 4 15 S 143 10 E
Chamchamal, *Iraq* **101 E11** 35 32N 44 50 E
Chamela, *Mexico* **162 D3** 19 32N 105 5W
Chamical, *Argentina* **174 C2** 30 22 S 66 27W
Chamkar Luong, *Cambodia* **87 G4** 11 0N 103 45 E
Chamois, *U.S.A.* **156 F5** 38 41N 91 46W
Chamonix-Mont Blanc, *France* **29 C10** 45 55N 6 51 E
Champa, *India* **93 H10** 22 2N 82 43 E
Champagne, *Canada* **142 A1** 60 49N 136 30W
Champagne, *France* **27 D11** 48 40N 4 20 E
Champagnole, *France* **27 F12** 46 45N 5 55 E
Champaign, *U.S.A.* **157 D8** 40 7N 88 15W
Champassak, *Laos* **86 E5** 14 53N 105 52 E
Champaubert, *France* **27 D10** 48 50N 3 45 E
Champdeniers-St-Denis, *France* **28 B3** 46 29N 0 25W
Champeix, *France* **28 C7** 45 37N 3 8 E
Champlain, *Canada* **148 B9** 46 27N 72 24W
Champlain, *U.S.A.* **151 B11** 44 59N 73 27W
Champlain, L., *U.S.A.* **151 B11** 44 40N 73 20W
Champlitte, *France* **27 E12** 47 32N 5 31 E
Champotón, *Mexico* **163 D6** 19 20N 90 50W
Chamrajnagar, *India* **95 J3** 11 52N 76 52 E
Chamusca, *Portugal* **43 F2** 39 21N 8 29W
Chan Chan, *Peru* **172 B2** 8 7 S 79 0W
Chana, *Thailand* **87 J3** 6 55N 100 44 E
Chañaral, *Chile* **174 B1** 26 23 S 70 40W
Chanārān, *Iran* **97 B8** 36 39N 59 6 E
Chanasma, *India* **92 H5** 23 44N 72 5 E
Chancay, *Peru* **172 C2** 11 32 S 77 25W
Chancy, *Switz.* **32 D1** 46 8N 5 58 E
Chandalar →, *U.S.A.* **144 C11** 66 37N 146 0W
Chandannagar, *India* **93 H13** 22 52N 88 24 E
Chandausi, *India* **93 E8** 28 27N 78 49 E
Chandeleur Is., *U.S.A.* **155 L10** 29 55N 88 57W
Chandeleur Sd., *U.S.A.* **155 L10** 29 55N 89 0W
Chandigarh, *India* **92 D7** 30 43N 76 47 E
Chandler, *Australia* **127 D1** 27 0 S 133 19 E
Chandler, *Canada* **141 C7** 48 18N 64 46W
Chandler, *Ariz., U.S.A.* **159 K8** 33 18N 111 50W
Chandler, *Okla., U.S.A.* **155 H6** 35 42N 96 53W
Chandlers Pk., *Australia* **129 A9** 30 15 S 151 48 E
Chandless →, *Brazil* **172 B4** 9 8 S 69 51W
Chandpur, *Bangla.* **90 D3** 23 8N 90 45 E
Chandpur, *India* **92 E8** 29 8N 78 19 E
Chandrapur, *India* **94 E4** 19 57N 79 25 E
Chānf, *Iran* **97 E9** 26 38N 60 29 E
Chang, *Pakistan* **92 F3** 26 59N 68 30 E
Chang, Ko, *Thailand* **87 G4** 12 0N 102 23 E
Ch'ang Chiang = Chang Jiang →, *China* **77 B13** 31 48N 121 10 E
Chang Jiang →, *China* **77 B13** 31 48N 121 10 E
Changa, *India* **93 C7** 33 53N 77 35 E
Changanacheri, *India* **95 K3** 9 25N 76 31 E
Changane →, *Mozam.* **117 C5** 24 30 S 33 30 E
Changbai, *China* **75 D15** 41 25N 128 5 E
Changbai Shan, *China* **75 C15** 42 20N 129 0 E
Changchiak'ou = Zhangjiakou, *China* **74 D8** 40 48N 114 55 E
Ch'angchou = Changzhou, *China* **77 B12** 31 47N 119 58 E
Changchun, *China* **75 C13** 43 57N 125 17 E
Changchunling, *China* **75 B13** 45 18N 125 27 E
Changde, *China* **77 C8** 29 4N 111 35 E
Changdo-ri, *N. Korea* **75 E14** 38 30N 127 40 E
Changfeng, *China* **77 A11** 32 28N 117 10 E
Changhai = Shanghai, *China* **77 B13** 31 15N 121 26 E
Changhua, *China* **77 B12** 30 12N 119 12 E
Changhua, *Taiwan* **77 E13** 24 2N 120 30 E
Changhŭng, *S. Korea* **75 G14** 34 41N 126 52 E
Changhŭngni, *N. Korea* **75 D15** 40 24N 128 19 E
Changi, *Malaysia* **84 B2** 1 23N 103 59 E
Changjiang, *China* **86 C7** 19 20N 108 55 E
Changjin, *N. Korea* **75 D14** 40 23N 127 15 E
Changjin-chŏsuji, *N. Korea* **75 D14** 40 30N 127 15 E
Changle, *China* **77 E12** 25 59N 119 27 E
Changli, *China* **75 E10** 39 40N 119 13 E
Changling, *China* **75 B12** 44 20N 123 58 E
Changlun, *Malaysia* **87 J3** 6 25N 100 26 E
Changning, *Hunan, China* **77 D9** 26 28N 112 22 E
Changning, *Yunnan, China* **76 E2** 24 45N 99 30 E
Changping, *China* **74 D9** 40 14N 116 12 E
Changsha, *China* **77 C9** 28 12N 113 0 E
Changshan, *China* **77 C12** 28 55N 118 27 E
Changshou, *China* **76 C6** 29 51N 107 8 E
Changshu, *China* **77 B13** 31 38N 120 43 E
Changshun, *China* **76 D6** 26 3N 106 25 E
Changtai, *China* **77 E11** 24 35N 117 42 E
Changting, *China* **77 E11** 25 50N 116 22 E
Changwu, *China* **74 G4** 35 10N 107 45 E
Changxing, *China* **77 B12** 31 0N 119 55 E
Changyang, *China* **77 B8** 30 30N 111 10 E
Changyi, *China* **75 F10** 36 40N 119 30 E
Changyŏn, *N. Korea* **75 E13** 38 15N 125 6 E
Changyuan, *China* **74 G8** 35 15N 114 42 E
Changzhi, *China* **74 F7** 36 10N 113 6 E
Changzhou, *China* **77 B12** 31 47N 119 58 E
Chanhanga, *Angola* **115 F2** 16 0 S 14 8 E
Chanlar = Xanlar, *Azerbaijan* **61 K8** 40 37N 46 12 E
Channapatna, *India* **95 H3** 12 40N 77 15 E
Channel Is., *U.K.* **21 H5** 49 19N 2 24W
Channel Is., *U.S.A.* **161 M7** 33 40N 119 15W
Channel-Port aux Basques, *Canada* **141 C8** 47 30N 59 9W
Channing, *Mich., U.S.A.* **148 B1** 46 9N 88 5W
Channing, *Tex., U.S.A.* **155 H3** 35 41N 102 20W
Chantada, *Spain* **42 C3** 42 36N 7 46W
Chanthaburi, *Thailand* **86 F4** 12 38N 102 12 E
Chantilly, *France* **27 C9** 49 12N 2 29 E

Cittadella 201

Chigorodó, *Colombia* **168 B2** 7 41N 76 42W
Chiguana, *Bolivia* **174 A2** 21 0S 67 58W
Chigwell, *U.K.* **21 F8** 51 37N 0 5 E
Chiha-ri, *N. Korea* **75 E14** 38 40N 126 30 E
Chihli, G. of = Bo Hai, *China* **75 E10** 39 0N 119 0 E
Chihuahua, *Mexico* **162 B3** 28 40N 106 3W
Chihuahua □, *Mexico* **162 B3** 28 40N 106 3W
Chiili, *Kazakstan* **63 A3** 44 20N 66 15 E
Chik Bollapur, *India* **95 H3** 13 25N 77 45 E
Chikhli, *India* **94 D3** 20 20N 76 18 E
Chikmagalur, *India* **95 H2** 13 15N 75 45 E
Chikodi, *India* **95 F2** 16 26N 74 38 E
Chikugo, *Japan* **70 D2** 33 14N 130 28 E
Chikuma-Gawa →, *Japan* **71 A10** 36 59N 138 35 E
Chikwawa, *Malawi* **119 F3** 16 2S 34 50 E
Chilac, *Mexico* **163 D5** 18 20N 97 24W
Chilako →, *Canada* **142 C4** 53 53N 122 57W
Chilam Chavki, *Pakistan* **93 B6** 35 5N 75 5 E
Chilanga, *Zambia* **119 F2** 15 33S 28 16 E
Chilapa, *Mexico* **163 D5** 17 40N 99 11W
Chilas, *Pakistan* **93 B6** 35 25N 74 5 E
Chilcotin →, *Canada* **142 C4** 51 44N 122 23W
Childers, *Australia* **127 D5** 25 15S 152 17 E
Childersburg, *U.S.A.* **152 B3** 33 16N 86 21W
Childress, *U.S.A.* **155 H4** 34 25N 100 13W
Chile ■, *S. Amer.* **176 A2** 35 0S 72 0W
Chile Chico, *Chile* **176 C2** 46 33S 71 44W
Chile Rise, *Pac. Oc.* **135 L18** 38 0S 92 0W
Chilecito, *Argentina* **174 B2** 29 10S 67 30W
Chilete, *Peru* **172 B2** 7 10S 78 50W
Chilhowee, *U.S.A.* **156 F3** 38 36N 93 51W
Chilia, Braţul →, *Romania* **53 E14** 45 14N 29 42 E
Chilik = Shelek, *Kazakstan* **63 B9** 43 33N 78 17 E
Chililabombwe, *Zambia* **119 E2** 12 18S 27 43 E
Chilin = Jilin, *China* **75 C14** 43 44N 126 30 E
Chilka L., *India* **94 E7** 19 40N 85 25 E
Chilko →, *Canada* **142 C4** 52 0N 123 40W
Chilko, L., *Canada* **142 C4** 51 20N 124 10W
Chillagoe, *Australia* **126 B3** 17 7S 144 33 E
Chillán, *Chile* **174 D1** 36 40S 72 10W
Chillicothe, *Ill., U.S.A.* **156 D7** 40 55N 89 29W
Chillicothe, *Mo., U.S.A.* **156 E3** 39 48N 93 33W
Chillicothe, *Ohio, U.S.A.* **148 F4** 39 20N 82 59W
Chilliwack, *Canada* **142 D4** 49 10N 121 54W
Chilo, *India* **92 F5** 27 25N 73 32 E
Chiloane, I., *Mozam.* **117 C5** 20 40S 34 55 E
Chiloé □, *Chile* **176 B2** 43 0S 73 0W
Chiloé, I. de, *Chile* **176 B2** 42 30S 73 50W
Chilonda, *Angola* **115 E3** 11 19S 16 12 E
Chilpancingo, *Mexico* **163 D5** 17 30N 99 30W
Chiltern, *Australia* **129 D7** 36 10S 146 36 E
Chiltern Hills, *U.K.* **21 F7** 51 40N 0 53W
Chilton, *U.S.A.* **148 C1** 44 2N 88 10W
Chiluage, *Angola* **115 D4** 9 30S 21 50 E
Chilubi, *Zambia* **119 E2** 11 5S 29 58 E
Chilubula, *Zambia* **119 E3** 10 14S 30 51 E
Chilumba, *Malawi* **119 E3** 10 28S 34 12 E
Chilung, *Taiwan* **77 E13** 25 3N 121 45 E
Chilwa, L., *Malawi* **119 F4** 15 15S 35 40 E
Chimaltitán, *Mexico* **162 C4** 21 46N 103 50W
Chimán, *Panama* **164 E4** 8 45N 78 40W
Chimay, *Belgium* **24 D4** 50 3N 4 20 E
Chimbay, *Uzbekistan* **64 E6** 42 57N 59 47 E
Chimborazo, *Ecuador* **168 D2** 1 29S 78 55W
Chimborazo □, *Ecuador* **168 D2** 1 0S 78 40W
Chimbote, *Peru* **172 B2** 9 0S 78 35W
Chimion, *Uzbekistan* **63 C5** 40 15N 71 32 E
Chimkent = Shymkent, *Kazakstan* **63 B4** 42 18N 69 36 E
Chimoio, *Mozam.* **119 F3** 19 4S 33 30 E
Chimpembe, *Zambia* **119 D2** 9 31S 29 33 E
Chin □, *Burma* **90 E4** 22 0N 93 0 E
Chin Hills, *Burma* **90 D4** 22 30N 93 30 E
Chin Ling Shan = Qinling Shandi, *China* **74 H5** 33 50N 108 10 E
China, *Mexico* **163 B5** 25 40N 99 20W
China ■, *Asia* **74 E3** 30 0N 110 0 E
China Lake, *U.S.A.* **161 K9** 35 44N 117 37W
Chinacota, *Colombia* **168 B3** 7 37N 72 36W
Chinan = Jinan, *China* **74 F9** 36 38N 117 1 E
Chinandega, *Nic.* **164 D2** 12 35N 87 12W
Chinati Peak, *U.S.A.* **155 L2** 29 57N 104 29W
Chincha Alta, *Peru* **172 C2** 13 25S 76 7W
Chinchilla, *Australia* **127 D5** 26 45S 150 38 E
Chinchilla de Monte Aragón, *Spain* **41 G3** 38 53N 1 40W
Chinchorro, Banco, *Mexico* **163 D7** 18 35N 87 20W
Chinchou = Jinzhou, *China* **75 D11** 41 5N 121 3 E
Chinchoua, *Gabon* **114 B1** 0 1N 9 48 E
Chincoteague, *U.S.A.* **148 G8** 37 56N 75 23W
Chinde, *Mozam.* **119 F4** 18 35S 36 30 E
Chindo, *S. Korea* **75 G14** 34 28N 126 15 E
Chindwin →, *Burma* **90 E5** 21 26N 95 15 E
Chineni, *India* **93 C6** 33 2N 75 15 E
Chinga, *Mozam.* **119 F4** 15 13S 38 35 E
Chingirlau, *Kazakstan* **62 F5** 51 7N 54 7 E
Chingola, *Zambia* **119 E2** 12 31S 27 53 E
Chingole, *Malawi* **119 E3** 13 4S 34 17 E
Chingoroi, *Angola* **115 E2** 13 37S 14 1 E
Ch'ingtao = Qingdao, *China* **75 F11** 36 5N 120 20 E
Chinguar, *Angola* **115 E3** 12 25S 16 45 E
Chinguetti, *Mauritania* **110 D2** 20 25N 12 24W
Chingune, *Mozam.* **117 C5** 20 33S 34 58 E
Chinhae, *S. Korea* **75 G15** 35 9N 128 47 E
Chinhanguanine, *Mozam.* **117 D5** 25 21S 32 30 E
Chinhoyi, *Zimbabwe* **119 F3** 17 20S 30 8 E
Chiniot, *Pakistan* **91 C4** 31 45N 73 0 E
Chínipas, *Mexico* **162 B3** 27 22N 108 32W
Chinju, *S. Korea* **75 G15** 35 12N 128 2 E
Chinle, *U.S.A.* **159 H9** 36 9N 109 33W
Chinmen, *Taiwan* **77 E13** 24 26N 118 19 E
Chinmen Tao, *Taiwan* **77 E13** 24 27N 118 23 E
Chinnamanur, *India* **95 K3** 9 50N 77 24 E
Chinnampo = Nampo, *N. Korea* **75 E13** 38 52N 125 10 E
Chinnur, *India* **94 E4** 18 57N 79 49 E
Chino, *Japan* **71 B10** 35 59N 138 9 E
Chino, *U.S.A.* **161 L9** 34 1N 117 41W
Chino Valley, *U.S.A.* **159 J7** 34 45N 112 27W
Chinon, *France* **26 E7** 47 10N 0 15 E
Chinook, *Canada* **143 C6** 51 28N 110 59W
Chinook, *U.S.A.* **158 B9** 48 35N 109 14W
Chinsali, *Zambia* **119 E3** 10 30S 32 2 E
Chintamani, *India* **95 H4** 13 26N 78 3 E
Chióggia, *Italy* **45 C9** 45 13N 12 17 E
Chipata, *Zambia* **119 E3** 13 38S 32 28 E
Chiperceni, *Moldova* **53 C13** 47 31N 28 50 E
Chipewyan L., *Canada* **143 B9** 58 0N 98 27W
Chipinge, *Zimbabwe* **119 G3** 20 13S 32 28 E
Chipiona, *Spain* **43 J4** 36 44N 6 26W
Chipley, *U.S.A.* **152 E4** 30 47N 85 32W
Chiplun, *India* **94 F1** 17 31N 73 34 E
Chipman, *Canada* **141 C6** 46 6N 65 53W
Chipoka, *Malawi* **119 E3** 13 57S 34 28 E
Chipola →, *U.S.A.* **152 E4** 30 1N 85 5W
Chippenham, *U.K.* **21 F5** 51 27N 2 6W
Chippewa →, *U.S.A.* **154 C8** 44 25N 92 5W
Chippewa Falls, *U.S.A.* **154 C9** 44 56N 91 24W
Chipping Norton, *U.K.* **21 F6** 51 56N 1 32W
Chiprovtsi, *Bulgaria* **50 C6** 43 24N 22 52 E
Chiquián, *Peru* **172 C2** 10 10S 77 0W
Chiquimula, *Guatemala* **164 D2** 14 51N 89 37W
Chiquinquira, *Colombia* **168 B3** 5 37N 73 50W
Chiquitos, Llanos de, *Bolivia* **173 D5** 18 0S 61 30W
Chir →, *Russia* **61 F6** 48 30N 43 0 E
Chirala, *India* **95 G5** 15 50N 80 26 E
Chiramba, *Mozam.* **119 F3** 16 55S 34 39 E
Chiran, *Japan* **70 F2** 31 22N 130 27 E
Chirawa, *India* **92 E6** 28 14N 75 42 E
Chirayinkil, *India* **95 K3** 8 41N 76 49 E
Chirchiq, *Uzbekistan* **63 C4** 41 29N 69 35 E
Chiredzi, *Zimbabwe* **117 C5** 21 0S 31 38 E
Chirfa, *Niger* **109 D2** 20 55N 12 22 E
Chirgua →, *Venezuela* **168 B4** 8 54N 67 58W
Chiricahua Peak, *U.S.A.* **159 L9** 31 51N 109 18W
Chiriquí, G. de, *Panama* **164 E3** 8 0N 82 10W
Chiriquí, L. de, *Panama* **164 E3** 9 10N 82 0W
Chirivira Falls, *Zimbabwe* **119 G3** 21 10S 32 12 E
Chirnogi, *Romania* **53 F11** 44 7N 26 32 E
Chirpan, *Bulgaria* **51 D9** 42 10N 25 19 E
Chirripó Grande, Cerro, *Costa Rica* **164 E3** 9 29N 83 29W
Chisamba, *Zambia* **119 E2** 14 55S 28 20 E
Chisasibi, *Canada* **140 B4** 53 50N 79 0W
Ch'ishan, *Taiwan* **77 F13** 22 44N 120 31 E
Chishmy, *Russia* **62 D5** 54 35N 55 23 E
Chisholm, *Canada* **142 C6** 54 55N 114 10W
Chishtian Mandi, *Pakistan* **92 E5** 29 50N 72 55 E
Chishui, *China* **76 C5** 28 30N 105 42 E
Chishui He →, *China* **76 C5** 28 49N 105 50 E
Chisimaio, *Somali Rep.* **120 E2** 0 22S 42 32 E
Chisimba Falls, *Zambia* **119 E3** 10 12S 30 56 E
Chişinău, *Moldova* **53 C13** 47 2N 28 50 E
Chişineu Criş, *Romania* **52 D6** 46 32N 21 37 E
Chisone →, *Italy* **44 D4** 44 49N 7 25 E
Chisos Mts., *U.S.A.* **155 L3** 29 5N 103 15W
Chistochina, *U.S.A.* **144 E11** 62 34N 144 40W
Chistopol, *Russia* **60 C10** 55 25N 50 38 E
Chita, *Colombia* **168 B3** 6 11N 72 28W
Chita, *Russia* **65 D12** 52 0N 113 35 E
Chitado, *Angola* **115 F2** 17 10S 14 8 E
Chitapur, *India* **94 F3** 17 10N 77 5 E
Chitembo, *Angola* **115 E3** 13 30S 16 50 E
Chitipa, *Malawi* **119 D3** 9 41S 33 19 E
Chitose, *Japan* **68 C10** 42 49N 141 39 E
Chitrakot, *India* **94 E5** 19 10N 81 40 E
Chitral, *Pakistan* **91 B3** 35 50N 71 56 E
Chitravati →, *India* **95 G4** 14 45N 78 15 E
Chitré, *Panama* **164 E3** 7 59N 80 27W
Chittagong, *Bangla.* **90 D3** 22 19N 91 48 E
Chittagong □, *Bangla.* **90 C3** 24 5N 91 0 E
Chittaurgarh, *India* **92 G6** 24 52N 74 38 E
Chittoor, *India* **95 H4** 13 15N 79 5 E
Chittur, *India* **95 J3** 10 40N 76 45 E
Chitungwiza, *Zimbabwe* **119 F3** 18 0S 31 6 E
Chiumba, *Angola* **115 E3** 12 29S 16 8 E
Chiume, *Angola* **115 F4** 15 3S 21 14 E
Chiusi, *Italy* **45 E8** 43 1N 11 57 E
Chiva, *Spain* **41 F4** 39 27N 0 41W
Chivacoa, *Venezuela* **168 A4** 10 10N 68 54W
Chivasso, *Italy* **44 C4** 45 11N 7 53 E
Chivay, *Peru* **172 D3** 15 40S 71 35W
Chivhu, *Zimbabwe* **119 F3** 19 2S 30 52 E
Chivilcoy, *Argentina* **174 C4** 34 55S 60 0W
Chiwanda, *Tanzania* **119 E3** 11 23S 34 55 E
Chixi, *China* **77 G9** 22 0N 112 58 E
Chizera, *Zambia* **119 E2** 13 10S 25 0 E
Chkalov = Orenburg, *Russia* **62 F5** 51 45N 55 6 E
Chkolovsk, *Russia* **60 B6** 56 50N 43 10 E
Chloride, *U.S.A.* **161 K12** 35 25N 114 12W
Chlumec nad Cidlinou, *Czech Rep.* **34 A8** 50 9N 15 29 E
Chmielnik, *Poland* **55 H7** 50 37N 20 43 E
Cho Bo, *Vietnam* **86 B5** 20 46N 105 10 E
Cho-do, *N. Korea* **75 E13** 38 30N 124 40 E
Cho Phuoc Hai, *Vietnam* **87 G6** 10 26N 107 18 E
Choa Chukang, *Malaysia* **84 B2** 1 22N 103 41 E
Choba, *Kenya* **118 B4** 2 30N 38 5 E
Chobe National Park, *Botswana* **116 B4** 18 0S 25 0 E
Chochiwŏn, *S. Korea* **75 F14** 36 37N 127 18 E
Chocianów, *Poland* **55 G2** 51 27N 15 55 E
Chociwel, *Poland* **54 E2** 53 29N 15 21 E
Chocó □, *Colombia* **168 B2** 6 0N 77 0W
Chocontá, *Colombia* **168 B3** 5 9N 73 41W
Choctawatchee →, *U.S.A.* **152 E3** 30 25N 86 8W
Choctawhatchee B., *U.S.A.* **147 D9** 30 20N 86 20W
Chodavaram, *India* **94 F6** 17 50N 82 57 E
Chodecz, *Poland* **55 F6** 52 24N 19 2 E
Chodov, *Czech Rep.* **34 A5** 50 15N 12 45 E
Chodziez, *Poland* **55 F3** 52 58N 16 58 E
Choele Choel, *Argentina* **176 A3** 39 11S 65 40W
Chōfu, *Japan* **71 B11** 35 39N 139 33 E
Choiseul, *Solomon Is.* **133 L9** 7 0S 156 40 E
Choix, *Mexico* **162 B3** 26 40N 108 23W
Chojna, *Poland* **55 F1** 52 58N 14 25 E
Chojnice, *Poland* **54 E4** 53 42N 17 32 E
Chojnów, *Poland* **55 G2** 51 18N 15 58 E
Chōkai-San, *Japan* **68 E10** 39 6N 140 3 E
Choke, *Ethiopia* **107 E4** 11 18N 37 15 E
Chokoloskee, *U.S.A.* **153 K8** 25 49N 81 22W
Chokurdakh, *Russia* **65 B15** 70 38N 147 55 E
Cholame, *U.S.A.* **160 K6** 35 44N 120 18W
Cholet, *France* **26 E6** 47 4N 0 52W
Cholpon-Ata, *Kyrgyzstan* **63 B8** 42 40N 77 6 E
Choluteca, *Honduras* **164 D2** 13 20N 87 14W
Choluteca →, *Honduras* **164 D2** 13 0N 87 20W
Chom Bung, *Thailand* **86 F2** 13 37N 99 36 E
Chom Thong, *Thailand* **86 C2** 18 25N 98 41 E
Choma, *Zambia* **119 F2** 16 48S 26 59 E
Chomen Swamp, *Ethiopia* **107 F4** 9 20N 37 10 E
Chomun, *India* **92 F6** 27 15N 75 40 E
Chomutov, *Czech Rep.* **34 A6** 50 28N 13 23 E
Chon Buri, *Thailand* **86 F3** 13 21N 101 1 E
Chon Thanh, *Vietnam* **87 G6** 11 24N 106 36 E
Chonan, *S. Korea* **75 F14** 36 48N 127 9 E
Chone, *Ecuador* **168 D2** 0 40S 80 0W
Chong Kai, *Cambodia* **86 F4** 13 57N 103 35 E
Chong Mek, *Thailand* **86 E5** 15 10N 105 27 E
Chong'an, *China* **77 D12** 27 45N 118 0 E
Chongde, *China* **77 B13** 30 32N 120 26 E
Chŏngdo, *S. Korea* **75 G15** 35 38N 128 42 E
Chŏngha, *S. Korea* **75 F15** 36 12N 129 21 E
Chŏngjin, *N. Korea* **75 D15** 41 47N 129 50 E
Chŏngju, *N. Korea* **75 E13** 39 40N 125 5 E
Chŏngju, *S. Korea* **75 F14** 36 39N 127 27 E
Chongli, *China* **74 D8** 40 58N 115 15 E
Chongming, *China* **77 B13** 31 38N 121 23 E
Chongming Dao, *China* **77 B13** 31 40N 121 30 E
Chongoyape, *Peru* **172 B2** 6 35S 79 25W
Chongqing, Chongqing, *China* **76 C6** 29 35N 106 25 E
Chongqing, *Sichuan, China* **76 B4** 30 38N 103 40 E
Chongqing □, *China* **76 B7** 30 0N 108 0 E
Chongren, *China* **77 D11** 27 46N 116 3 E
Chŏngŭp, *S. Korea* **75 G14** 35 35N 126 50 E
Chongzuo, *China* **76 F6** 22 23N 107 20 E
Chŏnju, *S. Korea* **75 G14** 35 50N 127 4 E
Chonos, Arch. de los, *Chile* **176 C2** 45 0S 75 0W
Chop, *Ukraine* **59 H2** 48 26N 22 12 E
Chopda, *India* **94 D2** 21 20N 75 15 E
Chopim →, *Brazil* **175 B5** 25 35S 53 5W
Chorbat La, *India* **93 B7** 34 42N 76 37 E
Chorley, *U.K.* **20 D5** 53 39N 2 38W
Chornobyl, *Ukraine* **59 G6** 51 20N 30 15 E
Chornomorske, *Ukraine* **59 K7** 45 31N 32 40 E
Chorolque, Cerro, *Bolivia* **174 A2** 20 59S 66 5W
Choroszcz, *Poland* **55 E9** 53 10N 22 59 E
Chorregon, *Australia* **126 C3** 22 40S 143 32 E
Chortkiv, *Ukraine* **59 H3** 49 2N 25 46 E
Chŏrwŏn, *S. Korea* **75 E14** 38 15N 127 10 E
Chorzele, *Poland* **55 E7** 53 15N 20 55 E
Chorzów, *Poland* **55 H5** 50 18N 18 57 E
Chos-Malal, *Argentina* **174 D1** 37 20S 70 15W
Chosan, *N. Korea* **75 D13** 40 50N 125 47 E
Chōshi, *Japan* **71 B12** 35 45N 140 51 E
Choszczno, *Poland* **55 E2** 53 7N 15 25 E
Chota, *Peru* **172 B2** 6 33S 78 39W
Choteau, *U.S.A.* **158 C7** 47 49N 112 11W
Chotěboř, *Czech Rep.* **34 B8** 49 43N 15 40 E
Chotila, *India* **92 H4** 22 23N 71 15 E
Chowchilla, *U.S.A.* **160 H6** 37 7N 120 16W
Chowkham, *Burma* **90 E6** 20 52N 97 28 E
Choybalsan, *Mongolia* **73 B6** 48 4N 114 30 E
Chrisman, *U.S.A.* **157 E9** 39 48N 87 41W
Christchurch, *N.Z.* **131 D7** 43 33S 172 47 E
Christchurch, *U.K.* **21 G6** 50 44N 1 47W
Christian I., *Canada* **150 B4** 44 50N 80 12W
Christian Sd., *U.S.A.* **144 J14** 55 56N 134 40W
Christiana, *S. Africa* **116 D4** 27 52S 25 8 E
Christiansfeld, *Denmark* **17 J3** 55 21N 9 29 E
Christianshåb, *Greenland* **10 D5** 68 50N 51 18W
Christiansted, *Virgin Is.* **165 C7** 17 45N 64 42W
Christie B., *Canada* **143 A6** 62 32N 111 10W
Christina →, *Canada* **143 B6** 56 40N 111 3W
Christmas Cr. →, *Australia* **124 C4** 18 29S 125 23 E
Christmas Creek, *Australia* **124 C4** 18 29S 125 23 E
Christmas I. = Kiritimati, *Kiribati* **135 G12** 1 58N 157 27W
Christmas I., *Ind. Oc.* **121 F9** 10 30S 105 40 E
Christopher L., *Australia* **125 D4** 24 49S 127 42 E
Chrudim, *Czech Rep.* **34 B8** 49 58N 15 43 E
Chrzanów, *Poland* **55 H6** 50 10N 19 21 E
Chtimba, *Malawi* **119 E3** 10 35S 34 13 E
Chu = Shu, *Kazakstan* **63 B6** 43 36N 73 42 E
Chu → = Shu →, *Kazakstan* **63 A3** 45 0N 67 44 E
Chu →, *Vietnam* **86 C5** 19 53N 105 45 E
Chu Chua, *Canada* **142 C4** 51 22N 120 10W
Chu Lai, *Vietnam* **86 E7** 15 28N 108 45 E
Chu Xian, *China* **77 A12** 32 19N 118 20 E
Chuadanga, *Bangla.* **90 D2** 23 38N 88 51 E
Ch'uanchou = Quanzhou, *China* **77 E12** 24 55N 118 34 E
Chuankou, *China* **74 G6** 34 20N 110 59 E
Chuathbaluk, *U.S.A.* **144 F8** 61 40N 159 15W
Chūbu □, *Japan* **71 A9** 36 45N 137 30 E
Chubut □, *Argentina* **176 B3** 43 30S 69 0W
Chubut →, *Argentina* **176 B3** 43 20S 65 5W
Chuchi L., *Canada* **142 B4** 55 12N 124 30W
Chudovo, *Russia* **58 C6** 59 10N 31 41 E
Chudskoye, Oz., *Russia* **15 G22** 58 13N 27 30 E
Chugach National Forest, *U.S.A.* **144 G9** 58 15N 152 45W
Chūgoku □, *Japan* **70 C5** 35 0N 133 0 E
Chūgoku-Sanchi, *Japan* **70 C5** 35 0N 133 0 E
Chuguyev = Chuhuyiv, *Ukraine* **59 H9** 49 55N 36 45 E
Chugwater, *U.S.A.* **154 E2** 41 46N 104 50W
Chuhuyiv, *Ukraine* **59 H9** 49 55N 36 45 E
Chukchi Sea, *Russia* **65 C19** 68 0N 175 0W
Chukotskoye Nagorye, *Russia* **65 C18** 68 0N 175 0 E
Chula, *Ga., U.S.A.* **152 D6** 31 33N 83 32W
Chula, *Mo., U.S.A.* **156 E3** 39 55N 93 29W
Chula Vista, *U.S.A.* **161 N9** 32 39N 117 5W
Chulakkurgan, *Kazakstan* **63 B4** 43 46N 69 9 E
Chulucanas, *Peru* **172 B1** 5 8S 80 10W
Chulumani, *Bolivia* **172 D4** 16 24S 67 31W
Chulym →, *Russia* **64 D9** 57 43N 83 51 E
Chum Phae, *Thailand* **86 D4** 16 40N 102 6 E
Chum Saeng, *Thailand* **86 E3** 15 55N 100 15 E
Chuma, *Bolivia* **172 D4** 15 24S 68 56W
Chumar, *India* **93 C8** 32 40N 78 35 E
Chumbicha, *Argentina* **174 B2** 29 0S 66 10W
Chumerna, *Bulgaria* **51 D9** 42 45N 25 55 E
Chumikan, *Russia* **65 D14** 54 40N 135 10 E
Chumphon, *Thailand* **87 G2** 10 35N 99 14 E
Chumpi, *Peru* **172 D3** 15 4S 73 46W
Chumuare, *Mozam.* **119 E3** 14 31S 31 50 E
Chumunjin, *S. Korea* **75 F15** 37 55N 128 54 E
Chuna →, *Russia* **65 D10** 57 47N 94 37 E
Chun'an, *China* **77 C12** 29 35N 119 3 E
Chunchŏn, *S. Korea* **75 F14** 37 58N 127 44 E
Chunchura, *India* **93 H13** 22 53N 88 27 E
Chunga, *Zambia* **119 F2** 15 0S 26 2 E
Chunggang-ŭp, *N. Korea* **75 D14** 41 48N 126 48 E
Chunghwa, *N. Korea* **75 E13** 38 52N 125 47 E
Chungju, *S. Korea* **75 F14** 36 58N 127 58 E
Chungking = Chongqing, *China* **76 C6** 29 35N 106 25 E
Chungli, *Taiwan* **77 E13** 24 57N 121 13 E
Chungmu, *S. Korea* **75 G15** 34 50N 128 20 E
Chungt'iaoshan = Zhongtiao Shan, *China* **74 G6** 35 0N 111 10 E
Chungyang Shanmo, *Taiwan* **77 F13** 23 10N 121 0 E
Chunian, *Pakistan* **92 D6** 30 57N 74 0 E
Chunya, *Tanzania* **119 D3** 8 30S 33 27 E
Chunya □, *Tanzania* **118 D3** 7 48S 33 0 E
Chunyang, *China* **75 C15** 43 38N 129 23 E
Chuquibamba, *Peru* **172 D3** 15 47S 72 44W
Chuquibambilla, *Peru* **172 C3** 14 7S 72 41W
Chuquicamata, *Chile* **174 A2** 22 15S 69 0W
Chuquisaca □, *Bolivia* **173 E5** 20 30S 63 30W
Chur, *Switz.* **33 C9** 46 52N 9 32 E
Churachandpur, *India* **90 C4** 24 20N 93 40 E
Churchill, *Canada* **143 B10** 58 47N 94 11W
Churchill →, *Man., Canada* **143 B10** 58 47N 94 12W
Churchill →, *Nfld., Canada* **141 B7** 53 19N 60 10W
Churchill, C., *Canada* **143 B10** 58 46N 93 12W
Churchill Falls, *Canada* **141 B7** 53 36N 64 19W
Churchill L., *Canada* **143 B7** 55 55N 108 20W
Churchill Pk., *Canada* **142 B3** 58 10N 125 10W
Churdan, *U.S.A.* **156 B2** 42 9N 94 29W
Churfisten, *Switz.* **33 B8** 47 8N 9 17 E
Churu, *India* **92 E6** 28 20N 74 50 E
Churubusco, *U.S.A.* **157 C11** 41 14N 85 19W
Churún Merú = Angel Falls, *Venezuela* **169 B5** 5 57N 62 30W
Churwalden, *Switz.* **33 C9** 46 47N 9 33 E
Chushal, *India* **93 C8** 33 40N 78 40 E
Chusovaya →, *Russia* **62 B6** 58 12N 56 54 E
Chusovoy, *Russia* **62 B6** 58 22N 57 50 E
Chust, *Uzbekistan* **63 C5** 41 0N 71 13 E
Chuuronjang, *N. Korea* **75 D15** 41 35N 129 40 E
Chuvash Republic □ = Chuvashia □, *Russia* **60 C8** 55 30N 47 0 E
Chuvashia □, *Russia* **60 C8** 55 30N 47 0 E
Chuwārtah, *Iraq* **96 C5** 35 43N 45 34 E
Chuxiong, *China* **76 E3** 25 2N 101 28 E
Ci Xian, *China* **74 F8** 36 20N 114 25 E
Ciacova, *Romania* **52 E6** 45 35N 21 10 E
Ciadâr-Lunga, *Moldova* **53 D13** 46 3N 28 51 E
Ciamis, *Indonesia* **85 D3** 7 20S 108 21 E
Cianjur, *Indonesia* **84 D3** 6 49S 107 8 E
Cibola, *U.S.A.* **161 M12** 33 17N 114 42W
Cicero, *U.S.A.* **148 E2** 41 48N 87 48W
Cicero, *Ill., U.S.A.* **157 C9** 41 51N 87 45W
Cícero Dantas, *Brazil* **170 D4** 10 36S 38 23W
Cidacos →, *Spain* **40 C3** 42 21N 1 38W
Cide, *Turkey* **100 B5** 41 53N 33 1 E
Ciechanów, *Poland* **55 F7** 52 52N 20 38 E
Ciechanów □, *Poland* **55 E7** 53 0N 20 30 E
Ciechanowiec, *Poland* **55 F9** 52 40N 22 31 E
Ciechocinek, *Poland* **55 F5** 52 53N 18 45 E
Ciego de Avila, *Cuba* **164 B4** 21 50N 78 50W
Ciénaga, *Colombia* **168 A3** 11 1N 74 15W
Ciénaga de Oro, *Colombia* **168 B2** 8 53N 75 37W
Cienfuegos, *Cuba* **164 B3** 22 10N 80 30W
Cierp, *France* **28 F4** 42 55N 0 40 E
Cíes, Is., *Spain* **42 C2** 42 12N 8 55W
Cieszanów, *Poland* **55 H10** 50 14N 23 8 E
Cieszyn, *Poland* **55 J5** 49 45N 18 35 E
Cieza, *Spain* **41 G3** 38 17N 1 23W
Çifteler, *Turkey* **100 C4** 39 22N 31 2 E
Cifuentes, *Spain* **40 E2** 40 47N 2 37W
Cihanbeyli, *Turkey* **100 C5** 38 40N 32 55 E
Cihuatlán, *Mexico* **162 D4** 19 14N 104 35W
Cijara, Embalse de, *Spain* **43 F6** 39 18N 4 52W
Cilacap, *Indonesia* **85 D3** 7 43S 109 0 E
Çıldır, *Turkey* **101 B10** 41 7N 43 8 E
Çıldır Gölü, *Turkey* **101 B10** 41 5N 43 15 E
Cili, *China* **77 C8** 29 30N 111 8 E
Cilibia, *Romania* **53 E12** 45 4N 27 4 E
Cilicia, *Turkey* **100 D5** 36 30N 33 40 E
Cill Chainnigh = Kilkenny, *Ireland* **23 D4** 52 39N 7 15W
Cilo Dağı, *Turkey* **101 D10** 37 28N 43 55 E
Cima, *U.S.A.* **161 K11** 35 14N 115 30W
Cimarron, *Kans., U.S.A.* **155 G4** 37 48N 100 21W
Cimarron, *N. Mex., U.S.A.* **155 G2** 36 31N 104 55W
Cimarron →, *U.S.A.* **155 G6** 36 10N 96 17W
Cimişlia, *Moldova* **53 D13** 46 34N 28 44 E
Cimone, Mte., *Italy* **44 D7** 44 12N 10 42 E
Çinar, *Turkey* **101 D9** 37 46N 40 19 E
Çınarcık, *Turkey* **51 F13** 40 38N 29 5 E
Cinca →, *Spain* **40 D5** 41 26N 0 21 E
Cincar, *Bos.-H.* **52 G2** 43 55N 17 5 E
Cincinnati, *Iowa, U.S.A.* **156 D4** 40 38N 92 56W
Cincinnati, *Ohio, U.S.A.* **157 E12** 39 6N 84 31W
Çine, *Turkey* **49 D10** 37 37N 28 2 E
Ciney, *Belgium* **24 D5** 50 18N 5 5 E
Cíngoli, *Italy* **45 E10** 43 23N 13 10 E
Cinigiano, *Italy* **45 F8** 42 53N 11 24 E
Cinto, Mte., *France* **29 F12** 42 24N 8 54 E
Cintruénigo, *Spain* **40 C3** 42 5N 1 49W
Ciocile, *Romania* **53 F12** 44 49N 27 14 E
Ciolăneşti din Deal, *Romania* **53 F10** 44 19N 25 5 E
Ciorani, *Romania* **53 F11** 44 45N 26 25 E
Čiovo, *Croatia* **45 E13** 43 30N 16 17 E
Cipó, *Brazil* **170 D4** 11 6S 38 31W
Circeo, Mte., *Italy* **46 A6** 41 14N 13 3 E
Çirçir, *Turkey* **100 C7** 40 5N 36 47 E
Circle, *Alaska, U.S.A.* **138 B5** 65 50N 144 4W
Circle, *Mont., U.S.A.* **154 B2** 47 25N 105 35W
Circleville, *Ohio, U.S.A.* **148 F4** 39 36N 82 57W
Circleville, *Utah, U.S.A.* **159 G7** 38 10N 112 16W
Cirebon, *Indonesia* **85 D3** 6 45S 108 32 E
Ciremai, *Indonesia* **85 D3** 6 55S 108 27 E
Cirencester, *U.K.* **21 F6** 51 43N 1 57W
Cireşu, *Romania* **52 F7** 44 47N 22 31 E
Cirey-sur-Vezouze, *France* **27 D13** 48 35N 6 57 E
Ciriè, *Italy* **44 C4** 45 14N 7 36 E
Cirium, *Cyprus* **38 E11** 34 40N 32 53 E
Cirò, *Italy* **47 C10** 39 23N 17 4 E
Cirò Marina, *Italy* **47 C10** 39 22N 17 8 E
Ciron →, *France* **28 D3** 44 36N 0 18W
Cisco, *U.S.A.* **155 J5** 32 23N 98 59W
Cislău, *Romania* **53 E11** 45 14N 26 20 E
Cisna, *Poland* **55 J9** 49 12N 22 20 E
Cisnădie, *Romania* **53 E9** 45 42N 24 9 E
Cisne, *U.S.A.* **157 F8** 38 31N 88 26W
Cisneros, *Colombia* **168 B2** 6 33N 75 4W
Cissna Park, *U.S.A.* **157 D9** 40 34N 87 54W
Cisterna di Latina, *Italy* **46 A5** 41 35N 12 49 E
Cisternino, *Italy* **47 B10** 40 44N 17 25 E
Cistierna, *Spain* **42 C5** 42 48N 5 7W
Citaré →, *Brazil* **169 C7** 1 11N 54 41W
Citeli-Ckaro = Tsiteli-Tsqaro, *Georgia* **61 K8** 41 33N 46 0 E
Citlaltépetl, *Mexico* **163 D5** 19 0N 97 20W
Citra, *U.S.A.* **153 F7** 29 25N 82 7W
Citrus Heights, *U.S.A.* **160 G5** 38 42N 121 17W
Citrus Springs, *U.S.A.* **153 F7** 29 2N 82 27W
Citrusdal, *S. Africa* **116 E2** 32 35S 19 0 E
Città della Pieve, *Italy* **45 F9** 42 57N 12 1 E
Città di Castello, *Italy* **45 E9** 43 27N 12 14 E
Città Sant' Angelo, *Italy* **45 F11** 42 32N 14 5 E
Cittadella, *Italy* **45 C8** 45 39N 11 47 E

Coronel Ponce 203

Colombier, *Switz.* **32 C3** 46 58N 6 53 E
Colombo, *Sri Lanka* **95 L4** 6 56N 79 58 E
Colome, *U.S.A.* **154 D5** 43 16N 99 43W
Colomiers, *France* **28 E5** 43 36N 1 21 E
Colón, *Argentina* **174 C4** 32 12 S 58 10W
Colón, *Cuba* **164 B3** 22 42N 80 54W
Colón, *Panama* **164 E4** 9 20N 79 54W
Colón, *Peru* **172 B1** 5 0 S 81 0W
Colona, *Australia* **125 F5** 31 38 S 132 4 E
Colonia de San Jordi, *Spain* **39 B9** 39 19N 2 59 E
Colonia del Sacramento, *Uruguay* **174 C4** 34 25 S 57 50W
Colonia Dora, *Argentina* **174 B3** 28 34 S 62 59W
Colonial Heights, *U.S.A.* **148 G7** 37 15N 77 25W
Colonna, C., *Italy* **47 C10** 39 2N 17 12 E
Colonsay, *Canada* **143 C7** 51 59N 105 52W
Colonsay, *U.K.* **22 E2** 56 5N 6 12W
Colorado □, *U.S.A.* **159 G10** 39 30N 105 30W
Colorado →, *Argentina* **176 A4** 39 50 S 62 8W
Colorado →, *N. Amer.* **159 L6** 31 45N 114 40W
Colorado →, *U.S.A.* **155 L7** 28 36N 95 59W
Colorado City, *U.S.A.* **155 J4** 32 24N 100 52W
Colorado Desert, *U.S.A.* **146 D3** 34 20N 116 0W
Colorado Plateau, *U.S.A.* **159 H8** 37 0N 111 0W
Colorado River Aqueduct, *U.S.A.* **161 L12** 34 17N 114 10W
Colorado Springs, *U.S.A.* **154 F2** 38 50N 104 49W
Colorno, *Italy* **44 D7** 44 56N 10 23 E
Colotlán, *Mexico* **162 C4** 22 6N 103 16W
Colquechaca, *Bolivia* **173 D4** 18 40 S 66 1W
Colquitt, *U.S.A.* **152 D5** 31 10N 84 44W
Colton, *N.Y., U.S.A.* **151 B10** 44 33N 74 56W
Colton, *Wash., U.S.A.* **158 C5** 46 34N 117 8W
Columbia, *Ala., U.S.A.* **152 D4** 31 18N 85 7W
Columbia, *Ill., U.S.A.* **156 F6** 38 27N 90 12W
Columbia, *La., U.S.A.* **155 J8** 32 6N 92 5W
Columbia, *Miss., U.S.A.* **155 K10** 31 15N 89 50W
Columbia, *Mo., U.S.A.* **156 F4** 38 57N 92 20W
Columbia, *Pa., U.S.A.* **151 F8** 40 2N 76 30W
Columbia, *S.C., U.S.A.* **152 A8** 34 0N 81 2W
Columbia, *Tenn., U.S.A.* **149 H2** 35 37N 87 2W
Columbia →, *U.S.A.* **158 C1** 46 15N 124 5W
Columbia, C., *Canada* **6 A4** 83 0N 70 0W
Columbia, District of □, *U.S.A.* **148 F7** 38 55N 77 0W
Columbia, Mt., *Canada* **142 C5** 52 8N 117 20W
Columbia Basin, *U.S.A.* **158 C4** 46 45N 119 5W
Columbia Falls, *U.S.A.* **158 B6** 48 23N 114 11W
Columbia Heights, *U.S.A.* **154 C8** 45 3N 93 15W
Columbiana, *U.S.A.* **150 F4** 40 53N 80 42W
Columbretes, Is., *Spain* **40 F5** 39 50N 0 50 E
Columbus, *Ga., U.S.A.* **152 C5** 32 28N 84 59W
Columbus, *Ind., U.S.A.* **157 E11** 39 13N 85 55W
Columbus, *Kans., U.S.A.* **155 G7** 37 10N 94 50W
Columbus, *Miss., U.S.A.* **149 J1** 33 30N 88 25W
Columbus, *Mont., U.S.A.* **158 D9** 45 38N 109 15W
Columbus, *N. Dak., U.S.A.* **154 A3** 48 54N 102 47W
Columbus, *N. Mex., U.S.A.* **159 L10** 31 50N 107 38W
Columbus, *Nebr., U.S.A.* **154 E6** 41 26N 97 22W
Columbus, *Ohio, U.S.A.* **157 E13** 39 58N 83 0W
Columbus, *Tex., U.S.A.* **155 L6** 29 42N 96 33W
Columbus, *Wis., U.S.A.* **154 D10** 43 21N 89 1W
Columbus Grove, *U.S.A.* **157 D12** 40 55N 84 4W
Columbus Junction, *U.S.A.* **156 C5** 41 17N 91 22W
Colunga, *Spain* **42 B5** 43 29N 5 16W
Colusa, *U.S.A.* **160 F4** 39 13N 122 1W
Colville, *U.S.A.* **158 B5** 48 33N 117 54W
Colville →, *U.S.A.* **138 A4** 70 25N 150 30W
Colville, C., *N.Z.* **130 C4** 36 29 S 175 21 E
Colwyn Bay, *U.K.* **20 D4** 53 18N 3 44W
Coma, *Ethiopia* **107 F4** 8 29N 36 53 E
Comácchio, *Italy* **45 D9** 44 42N 12 11 E
Comalcalco, *Mexico* **163 D6** 18 16N 93 13W
Comallo, *Argentina* **176 B2** 41 0S 70 5W
Comana, *Romania* **53 F11** 44 10N 26 10 E
Comanche, *Okla., U.S.A.* **155 H6** 34 22N 97 58W
Comanche, *Tex., U.S.A.* **155 K5** 31 54N 98 36W
Comandante Luis Piedrabuena, *Argentina* **176 C3** 49 59 S 68 54W
Comănești, *Romania* **53 D11** 46 25N 26 26 E
Comarapa, *Bolivia* **173 D5** 17 54 S 64 29W
Comarnic, *Romania* **53 E10** 45 15N 25 38 E
Comayagua, *Honduras* **164 D2** 14 25N 87 37W
Combahee →, *U.S.A.* **149 J5** 32 30N 80 31W
Combara, *Australia* **129 A8** 31 10 S 148 22 E
Combeaufontaine, *France* **27 E12** 47 38N 5 54 E
Comber, *Canada* **150 D2** 42 14N 82 33W
Comber, *U.K.* **23 B6** 54 33N 5 45W
Combermere Bay, *Burma* **90 F4** 19 37N 93 34 E
Comblain-au-Pont, *Belgium* **24 D5** 50 29N 5 35 E
Combourg, *France* **26 D5** 48 25N 1 46W
Comboyne, *Australia* **129 A10** 31 34 S 152 27 E
Combrailles, *France* **27 F9** 46 8N 2 8 E
Combronde, *France* **28 C7** 45 58N 3 5 E
Comer, *Ala., U.S.A.* **152 C4** 32 2N 85 23W
Comer, *Ga., U.S.A.* **152 A6** 34 4N 83 8W
Comeragh Mts., *Ireland* **23 D4** 52 18N 7 34W
Comet, *Australia* **126 C4** 23 36 S 148 38 E
Comilla, *Bangla.* **90 D3** 23 28N 91 10 E
Comino, *Malta* **38 C1** 36 2N 14 20 E
Comino, C., *Italy* **46 B2** 40 32N 9 49 E
Cómiso, *Italy* **47 F7** 36 56N 14 36 E
Comitán, *Mexico* **163 D6** 16 18N 92 9W
Commentry, *France* **27 F9** 46 20N 2 46 E
Commerce, *Ga., U.S.A.* **149 H4** 34 12N 83 28W
Commerce, *Tex., U.S.A.* **155 J7** 33 15N 95 54W
Commercy, *France* **27 D12** 48 43N 5 34 E
Commewijne □, *Surinam* **169 B7** 5 25N 54 45W
Committee B., *Canada* **139 B11** 68 30N 86 30W
Commonwealth B., *Antarctica* **7 C10** 67 0 S 144 0 E
Commoron Cr. →, *Australia* **127 D5** 28 22 S 150 8 E
Communism Pk. = Kommunizma, Pik, *Tajikistan* **63 D6** 39 0N 72 2 E
Como, *Italy* **44 C6** 45 47N 9 5 E
Como, L. di, *Italy* **44 B6** 46 0N 9 11 E
Comodoro Rivadavia, *Argentina* **176 C3** 45 50 S 67 40W
Comorâște, *Romania* **52 E6** 45 10N 21 35 E
Comorin, C., *India* **95 K3** 8 3N 77 40 E
Comoro Is. = Comoros ■, *Ind. Oc.* **121 F3** 12 10 S 44 15 E
Comoros ■, *Ind. Oc.* **121 F3** 12 10 S 44 15 E
Comox, *Canada* **142 D4** 49 42N 124 55W
Compass Lake, *U.S.A.* **152 E4** 30 36N 85 24W
Compiègne, *France* **27 C9** 49 24N 2 50 E
Comporta, *Portugal* **43 G2** 38 22N 8 46W
Compostela, *Mexico* **162 C4** 21 15N 104 53W
Compostela, *Phil.* **81 H6** 7 40N 126 2 E
Comprida, I., *Brazil* **175 A6** 24 50 S 47 42W
Compton, *U.S.A.* **161 M8** 33 54N 118 13W
Compton Downs, *Australia* **127 E4** 30 28 S 146 30 E
Comrat, *Moldova* **53 D13** 46 18N 28 40 E
Con Cuong, *Vietnam* **86 C5** 19 2N 104 54 E
Con Son, *Vietnam* **87 H6** 8 41N 106 37 E
Cona Niyeu, *Argentina* **176 B3** 41 58 S 67 0W
Conakry, *Guinea* **112 D2** 9 29N 13 49W
Conara Junction, *Australia* **126 G4** 41 50 S 147 26 E
Conargo, *Australia* **129 C6** 35 16 S 145 10 E
Concarneau, *France* **26 E3** 47 52N 3 56W
Conceição, *Brazil* **170 C4** 7 33 S 38 31W
Conceição, *Mozam.* **119 F4** 18 47 S 36 7 E
Conceição da Barra, *Brazil* **171 E4** 18 35 S 39 45W
Conceição do Araguaia, *Brazil* **170 C2** 8 0 S 49 2W
Conceição do Canindé, *Brazil* **170 C3** 7 54 S 41 34W
Concepción, *Argentina* **174 B2** 27 20 S 65 35W
Concepción, *Bolivia* **173 D5** 16 15 S 62 8W
Concepción, *Chile* **174 D1** 36 50 S 73 0W
Concepción, *Mexico* **163 D6** 18 15N 90 5W
Concepción, *Paraguay* **174 A4** 23 22 S 57 26W
Concepción, *Peru* **172 C2** 11 54 S 75 19W
Concepción □, *Chile* **174 D1** 37 0 S 72 30W
Concepción →, *Mexico* **162 A2** 30 32N 113 2W
Concepción, Est. de, *Chile* **176 D2** 50 30 S 74 55W
Concepción, L., *Bolivia* **173 D5** 17 20 S 61 20W
Concepción, Punta, *Mexico* **162 B2** 26 55N 111 59W
Concepción del Oro, *Mexico* **162 C4** 24 40N 101 30W
Concepción del Uruguay, *Argentina* **174 C4** 32 35 S 58 20W
Conception, Pt., *U.S.A.* **161 L6** 34 27N 120 28W
Conception B., *Namibia* **116 C1** 23 55 S 14 22 E
Conception I., *Bahamas* **165 B4** 23 52N 75 9W
Concession, *Zimbabwe* **119 F3** 17 27 S 30 56 E
Conchas Dam, *U.S.A.* **155 H2** 35 22N 104 11W
Conche, *Canada* **141 B8** 50 55N 55 58W
Conches-en-Ouche, *France* **26 D7** 48 58N 0 56 E
Concho, *U.S.A.* **159 J9** 34 28N 109 36W
Concho →, *U.S.A.* **155 K5** 31 34N 99 43W
Conchos →, *Chihuahua, Mexico* **162 B4** 29 32N 105 0W
Conchos →, *Tamaulipas, Mexico* **163 B5** 25 9N 98 35W
Concord, *Calif., U.S.A.* **160 H4** 37 59N 122 2W
Concord, *Ga., U.S.A.* **152 B5** 33 5N 84 27W
Concord, *Mich., U.S.A.* **157 B12** 42 11N 84 38W
Concord, *N.C., U.S.A.* **149 H5** 35 25N 80 35W
Concord, *N.H., U.S.A.* **151 C13** 43 12N 71 32W
Concordia, *Argentina* **174 C4** 31 20 S 58 2W
Concórdia, *Brazil* **168 D4** 4 36 S 66 36W
Concordia, *Mexico* **162 C3** 23 18N 106 2W
Concordia, *Kans., U.S.A.* **154 F6** 39 34N 97 40W
Concordia, *Mo., U.S.A.* **156 F3** 38 59N 93 34W
Concrete, *U.S.A.* **158 B3** 48 32N 121 45W
Condah, *Australia* **128 D4** 37 57 S 141 44 E
Condamine, *Australia* **127 D5** 26 56 S 150 9 E
Condat, *France* **28 C6** 45 21N 2 46 E
Condé, *Angola* **115 E2** 10 50 S 14 37 E
Conde, *Brazil* **171 D4** 11 49 S 37 37W
Conde, *U.S.A.* **154 C5** 45 9N 98 6W
Condé-sur-Noireau, *France* **26 D6** 48 51N 0 33W
Condeúba, *Brazil* **171 D3** 14 52 S 42 0W
Condobolin, *Australia* **127 E4** 33 4 S 147 6 E
Condom, *France* **28 E4** 43 57N 0 22 E
Condon, *U.S.A.* **158 D3** 45 14N 120 11W
Conegliano, *Italy* **45 C9** 45 53N 12 18 E
Conejera, I. = Conills, I. des, *Spain* **39 B9** 39 11N 2 58 E
Conejos, *Mexico* **162 B4** 26 14N 103 53W
Confolens, *France* **28 B4** 46 2N 0 40 E
Confuso →, *Paraguay* **174 B4** 25 9 S 57 34W
Congaree →, *U.S.A.* **152 B9** 33 44N 80 38W
Congaz, *Moldova* **53 D13** 46 7N 28 36 E
Congjiang, *China* **76 E7** 25 43N 108 52 E
Congleton, *U.K.* **20 D5** 53 10N 2 13W
Congo, *Brazil* **170 C4** 7 48 S 36 40W
Congo (Kinshasa) = Congo, Dem. Rep. of the ■, *Africa* **115 C4** 3 0 S 23 0 E
Congo ■, *Africa* **114 C3** 1 0 S 16 0 E
Congo →, *Africa* **115 D2** 6 4 S 12 24 E
Congo, Dem. Rep. of the ■, *Africa* **115 C4** 3 0 S 23 0 E
Congo Basin, *Africa* **104 G6** 0 10 S 24 30 E
Congonhas, *Brazil* **171 F3** 20 30 S 43 52W
Congress, *U.S.A.* **159 J7** 34 9N 112 51W
Conil = Conil de la Frontera, *Spain* **43 J4** 36 17N 6 10W
Conil de la Frontera, *Spain* **43 J4** 36 17N 6 10W
Conills, I. des, *Spain* **39 B9** 39 11N 2 58 E
Coniston, *Canada* **140 C3** 46 29N 80 51W
Conjeeveram = Kanchipuram, *India* **95 H4** 12 52N 79 45 E
Conjuboy, *Australia* **126 B3** 18 35 S 144 35 E
Conklin, *Canada* **143 B6** 55 38N 111 5W
Conlea, *Australia* **127 E3** 30 7 S 144 35 E
Conn, L., *Ireland* **23 B2** 54 3N 9 15W
Connacht □, *Ireland* **23 C2** 53 43N 9 12W
Conneaut, *U.S.A.* **150 E4** 41 57N 80 34W
Connecticut □, *U.S.A.* **151 E12** 41 30N 72 45W
Connecticut →, *U.S.A.* **151 E12** 41 16N 72 20W
Connell, *U.S.A.* **158 C4** 46 40N 118 52W
Connellsville, *U.S.A.* **150 F5** 40 1N 79 35W
Connemara, *Ireland* **23 C2** 53 29N 9 45W
Connemaugh →, *U.S.A.* **150 F5** 40 28N 79 19W
Conner, *Phil.* **80 C3** 17 48N 121 19 E
Connerré, *France* **26 D7** 48 3N 0 30 E
Connersville, *U.S.A.* **157 E11** 39 39N 85 8W
Connors Ra., *Australia* **126 C4** 21 40 S 149 10 E
Conoble, *Australia* **129 B6** 32 55 S 144 33 E
Cononaco →, *Ecuador* **168 D2** 1 32 S 75 35W
Conques, *France* **28 D6** 44 36N 2 23 E
Conquest, *Canada* **143 C7** 51 32N 107 14W
Conrad, *Iowa, U.S.A.* **156 B4** 42 14N 92 52W
Conrad, *Mont., U.S.A.* **158 B8** 48 10N 111 57W
Conran, C., *Australia* **129 D8** 37 49 S 148 44 E
Conroe, *U.S.A.* **155 K7** 30 19N 95 27W
Conselheiro Lafaiete, *Brazil* **171 F3** 20 40 S 43 48W
Conselheiro Pena, *Brazil* **171 E3** 19 10 S 41 30W
Conselve, *Italy* **45 C8** 45 14N 11 52 E
Consett, *U.K.* **20 C6** 54 51N 1 50W
Consort, *Canada* **143 C6** 52 1N 110 46W
Constance = Konstanz, *Germany* **31 H5** 47 40N 9 10 E
Constance, L. = Bodensee, *Europe* **33 A8** 47 35N 9 25 E
Constanța, *Romania* **53 F13** 44 14N 28 38 E
Constanța □, *Romania* **53 F13** 44 15N 28 15 E
Constantina, *Spain* **43 H5** 37 51N 5 40W
Constantine, *Algeria* **111 A6** 36 25N 6 42 E
Constantine, *U.S.A.* **157 C11** 41 50N 85 40W
Constantine, C., *U.S.A.* **144 G8** 58 24N 158 54W
Constitución, *Chile* **174 D1** 35 20 S 72 30W
Constitución, *Uruguay* **174 C4** 31 0 S 57 50W
Consuegra, *Spain* **43 F7** 39 28N 3 36W
Consul, *Canada* **143 D7** 49 20N 109 30W
Contact, *U.S.A.* **158 F6** 41 46N 114 45W
Contai, *India* **93 J12** 21 54N 87 46 E
Contamana, *Peru* **172 B3** 7 19 S 74 55W
Contarina, *Italy* **45 C9** 45 1N 12 13 E
Contas →, *Brazil* **171 D4** 14 17 S 39 1W
Contes, *France* **29 E11** 43 49N 7 19 E
Conthey, *Switz.* **32 D4** 46 14N 7 18 E
Continental, *U.S.A.* **157 C12** 41 6N 84 16W
Contoocook, *U.S.A.* **151 C13** 43 13N 71 45W
Contra Costa, *Mozam.* **117 D5** 25 9 S 33 30 E
Contres, *France* **26 E8** 47 24N 1 26 E
Contrexéville, *France* **27 D12** 48 10N 5 53 E
Controller B., *U.S.A.* **144 F11** 60 7N 144 15W
Contumaza, *Peru* **172 B2** 7 23 S 78 57W
Convención, *Colombia* **168 B3** 8 28N 73 21W
Conversano, *Italy* **47 B10** 40 58N 17 7 E
Converse, *U.S.A.* **157 D11** 40 35N 85 52W
Convoy, *U.S.A.* **157 D12** 40 55N 84 43W
Conway = Conwy, *U.K.* **20 D4** 53 17N 3 50W
Conway = Conwy →, *U.K.* **20 D4** 53 17N 3 50W
Conway, *Ark., U.S.A.* **155 H8** 35 5N 92 26W
Conway, *Fla., U.S.A.* **153 G8** 28 28N 81 22W
Conway, *N.H., U.S.A.* **151 C13** 43 59N 71 7W
Conway, *S.C., U.S.A.* **149 J6** 33 51N 79 3W
Conway, L., *Australia* **127 D2** 28 17 S 135 35 E
Conwy, *U.K.* **20 D4** 53 17N 3 50W
Conwy □, *U.K.* **20 D4** 53 10N 3 44W
Conwy →, *U.K.* **20 D4** 53 17N 3 50W
Conyers, *U.S.A.* **152 B5** 33 40N 84 1W
Coober Pedy, *Australia* **127 D1** 29 1 S 134 43 E
Cooch Behar = Koch Bihar, *India* **90 B2** 26 22N 89 29 E
Coodardy, *Australia* **125 E2** 27 15 S 117 39 E
Cook, *Australia* **125 F5** 30 37 S 130 25 E
Cook, *U.S.A.* **154 B8** 47 49N 92 39W
Cook, B., *Chile* **176 E3** 55 10 S 70 0W
Cook, Mt., *N.Z.* **131 D5** 43 36 S 170 9 E
Cook Inlet, *U.S.A.* **138 C4** 60 0N 152 0W
Cook Is., *Pac. Oc.* **135 J12** 17 0 S 160 0W
Cook Strait, *N.Z.* **130 H3** 41 15 S 174 29 E
Cooke Plains, *Australia* **128 C3** 35 23 S 139 34 E
Cookeville, *U.S.A.* **149 G3** 36 10N 85 30W
Cookhouse, *S. Africa* **116 E4** 32 44 S 25 47 E
Cooks Hammock, *U.S.A.* **152 F6** 29 56N 83 17W
Cookshire, *Canada* **151 A13** 45 25N 71 38W
Cookstown, *U.K.* **23 B5** 54 39N 6 45W
Cooksville, *Canada* **150 C5** 43 36N 79 35W
Cooktown, *Australia* **126 B4** 15 30 S 145 16 E
Coolabah, *Australia* **129 A7** 31 1 S 146 43 E
Cooladdi, *Australia* **127 D4** 26 37 S 145 23 E
Coolah, *Australia* **129 A8** 31 48 S 149 41 E
Coolamon, *Australia* **127 E4** 34 46 S 147 8 E
Coolangatta, *Australia* **127 D5** 28 11 S 153 29 E
Coolgardie, *Australia* **125 F3** 30 55 S 121 8 E
Coolibah, *Australia* **124 C5** 15 33 S 130 56 E
Coolidge, *Ariz., U.S.A.* **159 K8** 32 59N 111 31W
Coolidge, *Ga., U.S.A.* **152 D6** 31 1N 83 52W
Coolidge Dam, *U.S.A.* **159 K8** 33 0N 110 20W
Cooma, *Australia* **129 D8** 36 12 S 149 8 E
Coon Rapids, *Iowa, U.S.A.* **156 C2** 41 53N 94 41W
Coon Rapids, *Minn., U.S.A.* **154 C8** 45 9N 93 19W
Coonabarabran, *Australia* **129 A8** 31 14 S 149 18 E
Coonalpyn, *Australia* **128 C3** 35 43 S 139 52 E
Coonamble, *Australia* **129 A8** 30 56 S 148 27 E
Coonana, *Australia* **125 F3** 31 0 S 123 0 E
Coondapoor, *India* **95 H2** 13 42N 74 40 E
Coongie, *Australia* **127 D3** 27 9 S 140 8 E
Coongoola, *Australia* **127 D4** 27 43 S 145 51 E
Cooninie, L., *Australia* **127 D2** 26 4 S 139 59 E
Coonoor, *India* **95 J3** 11 21N 76 45 E
Cooper, *U.S.A.* **155 J7** 33 23N 95 42W
Cooper →, *U.S.A.* **152 C10** 32 50N 79 56W
Cooper Cr. →, *N. Terr., Australia* **122 C5** 12 7 S 132 41 E
Cooper Cr. →, *S. Austral., Australia* **127 D2** 28 29 S 137 46 E
Cooperstown, *N. Dak., U.S.A.* **154 B5** 47 27N 98 8W
Cooperstown, *N.Y., U.S.A.* **151 D10** 42 42N 74 56W
Coopersville, *U.S.A.* **157 A11** 43 4N 85 57W
Coorabie, *Australia* **125 F5** 31 54 S 132 18 E
Coorabulka, *Australia* **126 C3** 23 41 S 140 20 E
Coorow, *Australia* **125 E2** 29 53 S 116 2 E
Cooroy, *Australia* **127 D5** 26 22 S 152 54 E
Coos Bay, *U.S.A.* **158 E1** 43 22N 124 13W
Cootamundra, *Australia* **129 C8** 34 36 S 148 1 E
Cootehill, *Ireland* **23 B4** 54 4N 7 5W
Cooyar, *Australia* **127 D5** 26 59 S 151 51 E
Cooyeana, *Australia* **126 C2** 24 29 S 138 45 E
Copahue Paso, *Argentina* **174 D1** 37 49 S 71 8W
Copainalá, *Mexico* **163 D6** 17 8N 93 11W
Copalnic Mănăștur, *Romania* **53 C8** 47 30N 23 41 E
Copán, *Honduras* **164 D2** 14 50N 89 9W
Copatana, *Brazil* **168 D4** 2 48 S 67 4W
Cope, *Colo., U.S.A.* **154 F3** 39 40N 102 51W
Cope, *S.C., U.S.A.* **152 B9** 33 23N 81 0W
Cope, C., *Spain* **41 H3** 37 26N 1 28W
Cope Cope, *Australia* **128 D5** 36 27 S 143 5 E
Copeland, *U.S.A.* **153 K8** 25 57N 81 22W
Copenhagen = København, *Denmark* **17 J6** 55 41N 12 34 E
Copertino, *Italy* **47 B11** 40 16N 18 3 E
Copeville, *Australia* **128 C3** 34 47 S 139 51 E
Copiapó, *Chile* **174 B1** 27 30 S 70 20W
Copiapó →, *Chile* **174 B1** 27 19 S 70 56W
Copley, *Australia* **128 A3** 30 36 S 138 26 E
Copp L., *Canada* **142 A6** 60 14N 114 40W
Copparo, *Italy* **45 D8** 44 54N 11 49 E
Copperame →, *Surinam* **169 B6** 5 48N 55 55W
Copper Cliff, *Canada* **140 C3** 46 28N 81 4W
Copper Harbor, *U.S.A.* **148 B2** 47 28N 87 53W
Copper Queen, *Zimbabwe* **119 F2** 17 29 S 29 18 E
Copperbelt □, *Zambia* **119 E2** 13 15 S 27 30 E
Coppermine = Kugluktuk, *Canada* **138 B8** 67 50N 115 5W
Coppermine →, *Canada* **138 B8** 67 49N 116 4W
Copperopolis, *U.S.A.* **160 H6** 37 58N 120 38W
Copșa Mică, *Romania* **53 D9** 46 7N 24 15 E
Coquet →, *U.K.* **20 B6** 55 20N 1 32W
Coquilhatville = Mbandaka, *Dem. Rep. of the Congo* **114 B3** 0 1N 18 18 E
Coquille, *U.S.A.* **158 E1** 43 11N 124 11W
Coquimbo, *Chile* **174 C1** 30 0 S 71 20W
Coquimbo □, *Chile* **174 C1** 31 0 S 71 0W
Corabia, *Romania* **53 G9** 43 48N 24 30 E
Coração de Jesus, *Brazil* **171 E3** 16 43 S 44 22W
Coracora, *Peru* **172 D3** 15 5 S 73 45W
Coral Bay, *Phil.* **81 G1** 8 25N 117 20 E
Coral Gables, *U.S.A.* **149 N5** 25 45N 80 16W
Coral Sea, *Pac. Oc.* **134 J7** 15 0 S 150 0 E
Coral Springs, *U.S.A.* **149 M5** 26 16N 80 13W
Coralville, *U.S.A.* **156 C5** 41 40N 91 35W
Coralville Res., *U.S.A.* **156 C5** 41 50N 91 40W
Corantijn →, *Surinam* **169 B6** 5 50N 57 8W
Coraopolis, *U.S.A.* **150 F4** 40 31N 80 10W
Corato, *Italy* **47 A9** 41 9N 16 25 E
Corbeil-Essonnes, *France* **27 D9** 48 36N 2 26 E
Corbie, *France* **27 C9** 49 54N 2 30 E
Corbières, *France* **28 F6** 42 55N 2 35 E
Corbigny, *France* **27 E10** 47 16N 3 40 E
Corbin, *U.S.A.* **148 G3** 36 57N 84 6W
Corbones →, *Spain* **43 H5** 37 36N 5 39W
Corbu, *Romania* **53 F13** 44 25N 28 39 E
Corby, *U.K.* **21 E7** 52 30N 0 41W
Corcaigh = Cork, *Ireland* **23 E3** 51 54N 8 29W
Corcoran, *U.S.A.* **160 J7** 36 6N 119 33W
Cordele, *U.S.A.* **152 D6** 31 58N 83 47W
Cordell, *U.S.A.* **155 H5** 35 17N 98 59W
Cordenòns, *Italy* **45 C9** 45 59N 12 42 E
Cordes, *France* **28 D5** 44 5N 1 57 E
Cordisburgo, *Brazil* **171 E3** 19 7 S 44 21W
Córdoba, *Argentina* **174 C3** 31 20 S 64 10W
Córdoba, *Mexico* **163 D5** 18 50N 97 0W
Córdoba, *Spain* **43 H6** 37 50N 4 50W
Córdoba □, *Argentina* **174 C3** 31 22 S 64 15W
Córdoba □, *Colombia* **168 B2** 8 20N 75 40W
Córdoba □, *Spain* **43 G6** 38 5N 5 0W
Córdoba, Sierra de, *Argentina* **174 C3** 31 10 S 64 25W
Cordon, *Phil.* **80 C3** 16 42N 121 32 E
Cordova, *Ala., U.S.A.* **149 J2** 33 46N 87 11W
Cordova, *Alaska, U.S.A.* **138 B5** 60 33N 145 45W
Cordova, *Ill., U.S.A.* **156 C6** 41 41N 90 19W
Corella, *Spain* **40 C3** 42 7N 1 48W
Corella →, *Australia* **126 B3** 19 34 S 140 47 E
Coremas, *Brazil* **170 C4** 7 1 S 37 58W
Corentyne →, *Guyana* **169 B6** 5 50N 57 8W
Corfield, *Australia* **126 C3** 21 40 S 143 21 E
Corfu, Str of, *Greece* **38 A4** 39 34N 20 0 E
Corgo = O Corgo, *Spain* **42 C3** 42 56N 7 25W
Corguinho, *Brazil* **173 D7** 19 53 S 54 52W
Cori, *Italy* **46 A5** 41 39N 12 55 E
Coria, *Spain* **42 F4** 39 58N 6 33W
Coria del Río, *Spain* **43 H4** 37 16N 6 3W
Coricudgy, *Australia* **129 B9** 32 51 S 150 24 E
Corigliano Cálabro, *Italy* **47 C9** 39 36N 16 31 E
Coringa Is., *Australia* **126 B4** 16 58 S 149 58 E
Corinna, *Australia* **126 G4** 41 35 S 145 10 E
Corinth = Kórinthos, *Greece* **48 D4** 37 56N 22 55 E
Corinth, *Ga., U.S.A.* **152 B5** 33 14N 84 57W
Corinth, *Ky., U.S.A.* **157 F12** 38 30N 84 34W
Corinth, *Miss., U.S.A.* **149 H1** 34 56N 88 31W
Corinth, *N.Y., U.S.A.* **151 C11** 43 15N 73 49W
Corinth, G. of = Korinthiakós Kólpos, *Greece* **48 C4** 38 16N 22 30 E
Corinth Canal, *Greece* **48 D4** 37 58N 23 0 E
Corinto, *Brazil* **171 E3** 18 20 S 44 30W
Corinto, *Nic.* **164 D2** 12 30N 87 10W
Cork, *Ireland* **23 E3** 51 54N 8 29W
Cork □, *Ireland* **23 E3** 51 57N 8 40W
Cork Harbour, *Ireland* **23 E3** 51 47N 8 16W
Corlay, *France* **26 D3** 48 20N 3 5W
Corleone, *Italy* **46 E6** 37 49N 13 18 E
Corleto Perticara, *Italy* **47 B9** 40 23N 16 2 E
Çorlu, *Turkey* **51 E11** 41 11N 27 49 E
Cormack L., *Canada* **142 A4** 60 56N 121 37W
Cormòns, *Italy* **45 C10** 45 58N 13 28 E
Cormorant, *Canada* **143 C8** 54 14N 100 35W
Cormorant L., *Canada* **143 C8** 54 15N 100 50W
Corn Is. = Maiz, Is. del, *Nic.* **164 D3** 12 15N 83 4W
Cornélio Procópio, *Brazil* **175 A5** 23 7 S 50 40W
Cornell, *Ill., U.S.A.* **157 D8** 41 0N 88 44W
Cornell, *Wis., U.S.A.* **154 C9** 45 10N 91 9W
Corner Brook, *Canada* **141 C8** 48 57N 57 58W
Cornești, *Moldova* **53 C13** 47 21N 28 1 E
Corníglio, *Italy* **44 D7** 44 29N 10 5 E
Corning, *Ark., U.S.A.* **155 G9** 36 25N 90 35W
Corning, *Calif., U.S.A.* **158 G2** 39 56N 122 11W
Corning, *Iowa, U.S.A.* **156 D2** 40 59N 94 44W
Corning, *N.Y., U.S.A.* **150 D7** 42 9N 77 3W
Corno Grande, *Italy* **45 F10** 42 28N 13 34 E
Cornwall, *Canada* **140 C5** 45 2N 74 44W
Cornwall □, *U.K.* **21 G3** 50 26N 4 40W
Cornwell, *U.S.A.* **153 H8** 27 23N 81 6W
Corny Pt., *Australia* **128 C2** 34 55 S 137 0 E
Coro, *Venezuela* **168 A4** 11 25N 69 41W
Coroaci, *Brazil* **171 E3** 18 35 S 42 17W
Coroatá, *Brazil* **170 B3** 4 8 S 44 0W
Coroban, *Somali Rep.* **120 D2** 3 58N 42 44 E
Corocoro, *Bolivia* **172 D4** 17 15 S 68 28W
Corocoro, I., *Venezuela* **169 B5** 8 30N 60 10W
Coroico, *Bolivia* **172 D4** 16 0 S 67 50W
Coromandel, *Brazil* **171 E2** 18 28 S 47 13W
Coromandel, *N.Z.* **130 C4** 36 45 S 175 31 E
Coromandel Coast, *India* **95 H5** 12 30N 81 0 E
Coromandel Pen., *N.Z.* **130 D4** 37 0 S 175 45 E
Coromandel Ra., *N.Z.* **130 D4** 37 0 S 175 40 E
Coron, *Phil.* **80 E3** 12 1N 120 12 E
Coron Bay, *Phil.* **81 F3** 11 54N 120 8 E
Coron I., *Phil.* **81 F3** 11 55N 120 14 E
Corona, *Australia* **127 E3** 31 16 S 141 24 E
Corona, *Calif., U.S.A.* **161 M9** 33 53N 117 34W
Corona, *N. Mex., U.S.A.* **159 J11** 34 15N 105 36W
Coronado, *U.S.A.* **161 N9** 32 41N 117 11W
Coronado, B. de, *Costa Rica* **164 E3** 9 0N 83 40W
Coronados, G. de los, *Chile* **176 B2** 41 40 S 74 0W
Coronados, Is. los, *U.S.A.* **161 N9** 32 25N 117 15W
Coronation, *Canada* **142 C6** 52 5N 111 27W
Coronation Gulf, *Canada* **138 B9** 68 25N 110 0W
Coronation I., *Antarctica* **7 C18** 60 45 S 46 0W
Coronation I., *U.S.A.* **142 B2** 55 52N 134 20W
Coronation Is., *Australia* **124 B3** 14 57 S 124 55 E
Coronda, *Argentina* **174 C3** 31 58 S 60 56W
Coronel, *Chile* **174 D1** 37 0 S 73 10W
Coronel Bogado, *Paraguay* **174 B4** 27 11 S 56 18W
Coronel Dorrego, *Argentina* **174 D3** 38 40 S 61 10W
Coronel Fabriciano, *Brazil* **171 E3** 19 31 S 42 38W
Coronel Murta, *Brazil* **171 E3** 16 37 S 42 11W
Coronel Oviedo, *Paraguay* **174 B4** 25 24 S 56 30W
Coronel Ponce, *Brazil* **173 D6** 15 34 S 55 1W

204 Coronel Pringles

Coronel Pringles, *Argentina* **174 D3** 38 0 S 61 30W
Coronel Suárez, *Argentina* . **174 D3** 37 30 S 61 52W
Coronel Vidal, *Argentina* ... **174 D4** 37 28 S 57 45W
Corongo, *Peru* .... **172 B2** 8 30 S 77 53W
Coronie □, *Surinam* **169 B6** 5 55N 56 20W
Coropuna, Nevado, *Peru* ... **172 D3** 15 30 S 72 41W
Çorovoda, *Albania* .... **50 F4** 40 31N 20 14 E
Corowa, *Australia* .... **129 C7** 35 58 S 146 21 E
Corozal, *Belize* **163 D7** 18 23N 88 23W
Corozal, *Colombia* **168 B2** 9 19N 75 18W
Corps, *France* .... **29 D9** 44 50N 5 56 E
Corpus, *Argentina* .... **175 B4** 27 10 S 55 30W
Corpus Christi, *U.S.A.* **155 M6** 27 47N 97 24W
Corpus Christi, L., *U.S.A.* . **155 L6** 28 2N 97 52W
Corque, *Bolivia* .... **172 D4** 18 20 S 67 41W
Corral, *Chile* .... **176 A2** 39 52 S 73 26W
Corral de Almaguer, *Spain* . **42 F7** 39 45N 3 10W
Corralejo, *Canary Is.* **39 F6** 28 43N 13 53W
Corraun Pen., *Ireland* **23 C2** 53 54N 9 54W
Corréggio, *Italy* .... **44 D7** 44 46N 10 47 E
Corrente, *Brazil* .... **170 D2** 10 27 S 45 10W
Corrente →, *Brazil* .... **171 D3** 13 8 S 43 28W
Correntes →, *Brazil* ...... **173 D6** 17 38 S 55 8W
Correntes, C. das, *Mozam.* . **117 C6** 24 6 S 35 34 E
Correntina, *Brazil* .... **171 D3** 13 20 S 44 39W
Corrèze □, *France* .... **28 C5** 45 20N 1 45 E
Corrèze →, *France* .... **28 C5** 45 10N 1 28 E
Corrib, L., *Ireland* **23 C2** 53 27N 9 16W
Corridónia, *Italy* .... **45 E10** 43 15N 13 30 E
Corrientes, *Argentina* ..... **174 B4** 27 30 S 58 45W
Corrientes □, *Argentina* ... **174 B4** 28 0 S 57 0W
Corrientes →, *Argentina* .. **174 C4** 30 42 S 59 38W
Corrientes →, *Peru* .... **168 D3** 3 43 S 74 35W
Corrientes, C., *Colombia* .. **168 B2** 5 30N 77 34W
Corrientes, C., *Cuba* **164 B3** 21 43N 84 30W
Corrientes, C., *Mexico* **162 C3** 20 25N 105 42W
Corrigan, *U.S.A.* **155 K7** 31 0N 94 52W
Corrigin, *Australia* .... **125 F2** 32 20 S 117 53 E
Corrowidgie, *Australia* .... **129 D8** 36 56 S 148 50 E
Corry, *U.S.A.* **150 E5** 41 55N 79 39W
Corryong, *Australia* .... **129 D7** 36 12 S 147 53 E
Corse, *France* .... **29 G13** 42 0N 9 0 E
Corse, C., *France* .... **29 E13** 43 1N 9 25 E
Corse-du-Sud □, *France* ... **29 G13** 41 45N 9 0 E
Corsica = Corse, *France* ... **29 G13** 42 0N 9 0 E
Corsicana, *U.S.A.* **155 J6** 32 6N 96 28W
Corte, *France* .... **29 F13** 42 19N 9 11 E
Corte Pinto, *Portugal* **43 H3** 37 42N 7 29W
Cortegana, *Spain* **43 H4** 37 52N 6 49W
Cortes, *Phil.* .... **81 G6** 9 17N 126 11 E
Cortez, *U.S.A.* **159 H9** 37 21N 108 35W
Cortina d'Ampezzo, *Italy* .. **45 B9** 46 32N 12 8 E
Cortland, *U.S.A.* **151 D8** 42 36N 76 11W
Cortona, *Italy* .... **45 E8** 43 16N 11 59 E
Coruche, *Portugal* **43 G2** 38 57N 8 30W
Çoruh →, *Turkey* .... **61 K5** 41 38N 41 38 E
Çorum, *Turkey* .... **100 B6** 40 30N 34 57 E
Corumbá, *Brazil* .... **173 D6** 19 0 S 57 30W
Corumbá →, *Brazil* ...... **171 E2** 18 19 S 48 55W
Corumbá de Goiás, *Brazil* .. **171 E2** 16 0 S 48 50W
Corumbaíba, *Brazil* .... **171 E2** 18 9 S 48 34W
Corund, *Romania* .... **53 D10** 46 30N 25 13 E
Corunna = A Coruña, *Spain* **42 B2** 43 20N 8 25W
Corunna, *U.S.A.* **157 B12** 42 59N 84 7W
Corvallis, *U.S.A.* **158 D2** 44 34N 123 16W
Corvette, L. de la, *Canada* . **140 B5** 53 25N 74 3W
Corydon, *Ind., U.S.A.* **157 F10** 38 13N 86 7W
Corydon, *Iowa, U.S.A.* **156 D3** 40 46N 93 19W
Corydon, *Ky., U.S.A.* **157 G9** 37 44N 87 43W
Cosalá, *Mexico* **162 C3** 24 28N 106 40W
Cosamaloapan, *Mexico* **163 D5** 18 23N 95 50W
Cosenza, *Italy* .... **47 C9** 39 18N 16 15 E
Coşereni, *Romania* .... **53 F11** 44 38N 26 35 E
Coshocton, *U.S.A.* **150 F3** 40 16N 81 51W
Cosmo Newberry, *Australia* **125 E3** 28 0 S 122 54 E
Cosne-Cours-sur-Loire, *France* .... **27 E9** 47 24N 2 54 E
Coso Junction, *U.S.A.* **161 J9** 36 3N 117 57W
Coso Pk., *U.S.A.* **161 J9** 36 13N 117 44W
Cospeito, *Spain* **42 B3** 43 12N 7 34W
Cosquín, *Argentina* .... **174 C3** 31 15 S 64 30W
Cossato, *Italy* .... **44 C5** 45 34N 8 10 E
Cossé-le-Vivien, *France* **26 E6** 47 57N 0 54W
Cosson →, *France* .... **26 E8** 47 30N 1 15 E
Costa Blanca, *Spain* **41 G4** 38 25N 0 10W
Costa Brava, *Spain* .... **40 D8** 41 30N 3 0 E
Costa del Sol, *Spain* **43 J6** 36 30N 4 30W
Costa Dorada, *Spain* ...... **40 D6** 41 12N 1 15 E
Costa Mesa, *U.S.A.* **161 M9** 33 38N 117 55W
Costa Rica ■, *Cent. Amer.* . **164 E3** 10 0N 84 0W
Costa Smeralda, *Italy* ..... **46 A2** 41 5N 9 35 E
Costeşti, *Romania* .... **53 F9** 44 40N 24 53 E
Costigliole d'Asti, *Italy* .... **44 D5** 44 47N 8 11 E
Costilla, *U.S.A.* **159 H11** 36 59N 105 32W
Cosumnes →, *U.S.A.* **160 G5** 38 16N 121 26W
Coswig, *Sachsen, Germany* . **30 D9** 51 7N 13 34 E
Coswig, *Sachsen-Anhalt, Germany* .... **30 D8** 51 53N 12 27 E
Cotabato, *Phil.* .... **81 H5** 7 14N 124 15 E
Cotabena, *Australia* .... **128 A3** 31 42 S 138 11 E
Cotacajes →, *Bolivia* ..... **172 D4** 16 0 S 67 1W
Cotagaita, *Bolivia* .... **174 A2** 20 45 S 65 40W
Cotahuasi, *Peru* .... **172 D3** 15 12 S 72 50W
Côte d'Azur, *France* .... **29 E11** 43 25N 7 10 E
Côte-d'Ivoire = Ivory Coast ■, *Africa* **112 D4** 7 30N 5 0W
Côte d'Or, *France* .... **27 E11** 47 10N 4 50 E
Côte-d'Or □, *France* ...... **27 E11** 47 30N 4 50 E
Coteau des Prairies, *U.S.A.* **154 C6** 45 20N 97 50W
Coteau du Missouri, *U.S.A.* **154 B4** 47 0N 100 0W
Coteau Landing, *Canada* ... **151 A10** 45 15N 74 13W
Cotegipe, *Brazil* **171 D3** 12 2S 44 15W
Cotentin, *France* **26 C5** 49 15N 1 30W
Côtes-d'Armor □, *France* .. **26 D4** 48 25N 2 40W
Côtes de Meuse, *France* .... **27 C12** 49 15N 5 22 E
Côtes-du-Nord = Côtes-d'Armor □, *France* **26 D4** 48 25N 2 40W
Cotiella, *Spain* .... **40 C5** 42 31N 0 19 E
Cotillo, *Canary Is.* **39 F5** 28 41N 14 1W
Cotiujeni, *Moldova* .... **53 C13** 47 51N 28 33 E
Cotoca, *Bolivia* .... **173 D5** 17 49 S 63 3W
Cotonou, *Benin* .... **113 D5** 6 20N 2 25 E
Cotopaxi, *Ecuador* .... **168 D2** 0 40 S 78 30W
Cotopaxi □, *Ecuador* ...... **168 D2** 0 5 S 78 55W
Cotronei, *Italy* .... **47 C9** 39 9N 16 47 E
Cotswold Hills, *U.K.* **21 F5** 51 42N 2 10W
Cottage Grove, *U.S.A.* **158 E2** 43 48N 123 3W
Cottageville, *U.S.A.* **152 C9** 32 56N 80 29W
Cottbus, *Germany* .... **30 D10** 51 45N 14 20 E
Cotton, *U.S.A.* **152 D5** 31 10N 84 4W
Cottondale, *U.S.A.* **152 E4** 30 48N 85 23W
Cottonwood, *Ala., U.S.A.* . **152 D4** 31 3N 85 18W
Cottonwood, *Ariz., U.S.A.* . **159 J7** 34 45N 112 1W
Cotulla, *U.S.A.* **155 L5** 28 26N 99 14W
Coubre, Pte. de la, *France* . **28 C2** 45 42N 1 15W
Couches, *France* .... **27 F11** 46 53N 4 30 E
Couço, *Portugal* **43 G2** 38 59N 8 17W
Coudersport, *U.S.A.* **150 E6** 41 46N 78 1W
Couedic, C. du, *Australia* .. **128 D2** 36 5 S 136 40 E
Couëron, *France* **26 E5** 47 13N 1 44W
Couesnon →, *France* **26 D5** 48 38N 1 32W
Couhé, *France* .... **28 B4** 46 17N 0 11 E
Coulanges-sur-Yonne, *France* **27 E10** 47 31N 3 33 E
Coulee City, *U.S.A.* **158 C4** 47 37N 119 17W
Coulman I., *Antarctica* ..... **7 D11** 73 35 S 170 0 E
Coulommiers, *France* ...... **27 D10** 48 50N 3 3 E
Coulon →, *France* .... **29 E9** 43 51N 5 6 E
Coulonge →, *Canada* **140 C4** 45 52N 76 46W
Coulonges-sur-l'Autize, *France* **28 B3** 46 29N 0 36W
Coulounieix-Chamiers, *France* .... **28 C4** 45 11N 0 42 E
Coulterville, *Calif., U.S.A.* . **160 H6** 37 43N 120 12W
Coulterville, *Ill., U.S.A.* ... **156 F7** 38 11N 89 36W
Council, *Ga., U.S.A.* **152 E7** 30 37N 82 31W
Council, *Idaho, U.S.A.* **158 D5** 44 44N 116 26W
Council Bluffs, *U.S.A.* **154 E7** 41 16N 95 52W
Council Grove, *U.S.A.* **154 F6** 38 40N 96 29W
Coupeville, *U.S.A.* **160 B4** 48 13N 122 41W
Courantyne →, *S. Amer.* .. **169 B6** 5 55N 57 5W
Courcelles, *Belgium* .... **24 D4** 50 28N 4 22 E
Courçon, *France* **28 B3** 46 15N 0 50W
Courmayeur, *Italy* .... **44 C3** 45 47N 6 58 E
Couronne, C., *France* ..... **29 E9** 43 19N 5 3 E
Cours-la-Ville, *France* ..... **27 F11** 46 7N 4 19 E
Coursan, *France* .... **28 E7** 43 14N 3 4 E
Courseulles-sur-Mer, *France* **26 C6** 49 20N 0 29W
Courtenay, *Canada* **142 D4** 49 45N 125 0W
Courtenay, *France* .... **27 D10** 48 2N 3 3 E
Courtland, *U.S.A.* **160 G5** 38 20N 121 34W
Courtrai = Kortrijk, *Belgium* **24 D3** 50 50N 3 17 E
Courtright, *Canada* **150 D2** 42 49N 82 28W
Coushatta, *U.S.A.* **155 J8** 32 1N 93 21W
Coutances, *France* **26 C5** 49 3N 1 28W
Coutras, *France* **28 C3** 45 3N 0 8W
Coutts, *Canada* **142 D6** 49 0N 111 57W
Couvet, *Switz.* .... **32 C3** 46 57N 6 38 E
Couvin, *Belgium* .... **24 D4** 50 3N 4 29 E
Covarrubias, *Spain* **42 C7** 42 4N 3 31W
Covasna, *Romania* .... **53 E11** 45 50N 26 10 E
Covasna □, *Romania* ...... **53 E10** 45 50N 26 0 E
Coveñas, *Colombia* **168 B2** 9 24N 75 44W
Coventry, *U.K.* **21 E6** 52 25N 1 28W
Coventry L., *Canada* **143 A7** 61 15N 106 15W
Coverdale, *U.S.A.* **152 D6** 31 38N 83 58W
Covilhã, *Portugal* **42 E3** 40 17N 7 31W
Covington, *Ga., U.S.A.* **152 B6** 33 36N 83 51W
Covington, *Ind., U.S.A.* **157 D9** 40 9N 87 24W
Covington, *Ky., U.S.A.* **157 E12** 39 5N 84 31W
Covington, *Ohio, U.S.A.* ... **157 D12** 40 7N 84 21W
Covington, *Okla., U.S.A.* .. **155 G6** 36 18N 97 35W
Covington, *Tenn., U.S.A.* .. **155 H10** 35 34N 89 39W
Cowal, L., *Australia* ...... **129 B7** 33 40 S 147 25 E
Cowan, *Canada* **143 C8** 52 5N 100 45W
Cowan, L., *Australia* ...... **125 F3** 31 45 S 121 45 E
Cowan L., *Canada* **143 C7** 54 0N 107 15W
Cowangie, *Australia* ...... **128 C4** 35 12 S 141 26 E
Cowansville, *Canada* **151 A12** 45 14N 72 46W
Coward, *U.S.A.* **152 B10** 33 58N 79 45W
Cowarie, *Australia* .... **127 D2** 27 45 S 138 15 E
Cowcowing Lakes, *Australia* **125 F2** 30 55 S 117 20 E
Cowden, *U.S.A.* **157 E8** 39 15N 88 52W
Cowdenbeath, *U.K.* **22 E5** 56 7N 3 21W
Cowell, *Australia* .... **128 B2** 33 39 S 136 56 E
Cowes, *U.K.* **21 G6** 50 45N 1 18W
Cowl Cowl, *Australia* ..... **129 B6** 33 36 S 145 18 E
Cowlitz →, *U.S.A.* **160 D4** 46 6N 122 55W
Cowra, *Australia* .... **129 B8** 33 49 S 148 42 E
Cox, *Spain* **41 G4** 38 8N 0 53W
Cox, *U.S.A.* **152 D8** 31 27N 81 34W
Coxilha Grande, *Brazil* .... **175 B5** 28 18 S 51 30W
Coxim, *Brazil* .... **173 D7** 18 30 S 54 55W
Coxim →, *Brazil* .... **173 D7** 18 34 S 54 46W
Cox's Bazar, *Bangla.* ...... **90 E3** 21 26N 91 59 E
Cox's Cove, *Canada* **141 C8** 49 7N 58 5W
Coyame, *Mexico* **162 B3** 29 28N 105 6W
Coyote Wells, *U.S.A.* **161 N11** 32 44N 115 58W
Coyuca de Benítez, *Mexico* . **163 D4** 17 1N 100 8W
Coyuca de Catalan, *Mexico* . **162 D4** 18 18N 100 41W
Cozad, *U.S.A.* **154 E5** 40 52N 99 59W
Cozes, *France* **28 C3** 45 34N 0 49W
Cozumel, *Mexico* **163 C7** 20 31N 86 55W
Cozumel, I. de, *Mexico* **163 C7** 20 30N 86 40W
Craboon, *Australia* .... **129 B8** 32 3 S 149 30 E
Cracow = Kraków, *Poland* . **55 H6** 50 4N 19 57 E
Cracow, *Australia* .... **127 D5** 25 17 S 150 17 E
Cradock, *S. Africa* .... **116 E4** 32 8 S 25 36 E
Craig, *Alaska, U.S.A.* **142 B2** 55 29N 133 9W
Craig, *Colo., U.S.A.* **158 F10** 40 31N 107 33W
Craigavon, *U.K.* **23 B5** 54 27N 6 23W
Craigmore, *Zimbabwe* ..... **119 G3** 20 28 S 32 50 E
Crailsheim, *Germany* ...... **31 F6** 49 8N 10 5 E
Craiova, *Romania* .... **53 F8** 44 21N 23 48 E
Cramsie, *Australia* .... **126 C3** 23 20 S 144 15 E
Cranberry Portage, *Canada* . **143 C8** 54 35N 101 23W
Cranbrook, *Tas., Australia* . **126 G4** 42 0 S 148 5 E
Cranbrook, *W. Austral., Australia* .... **125 F2** 34 18 S 117 33 E
Cranbrook, *Canada* **142 D5** 49 30N 115 46W
Crandon, *U.S.A.* **154 C10** 45 34N 88 54W
Crane, *Oreg., U.S.A.* **158 E4** 43 25N 118 35W
Crane, *Tex., U.S.A.* **155 K3** 31 24N 102 21W
Cranston, *U.S.A.* **151 E13** 41 47N 71 26W
Craon, *France* **26 E6** 47 50N 0 58W
Craonne, *France* .... **27 C10** 49 27N 3 46 E
Craponne-sur-Arzon, *France* **28 C7** 45 19N 3 51 E
Crasna, *Romania* .... **53 D12** 46 32N 27 51 E
Crasna →, *Romania* ...... **52 C7** 47 44N 22 35 E
Crasnei, Munţii, *Romania* .. **53 C8** 47 0N 23 20 E
Crater L., *U.S.A.* **158 E2** 42 56N 122 6W
Crater Mt., *Papua N. G.* ... **132 D3** 6 37 S 145 7 E
Crater Pt., *Papua N. G.* ... **132 C7** 5 25 S 152 9 E
Crateús, *Brazil* .... **170 C3** 5 10 S 40 39W
Crati →, *Italy* .... **47 C9** 39 43N 16 31 E
Crato, *Brazil* .... **170 C4** 7 10 S 39 25W
Crato, *Portugal* **43 F3** 39 16N 7 39W
Cravo Norte, *Colombia* **168 B3** 6 18N 70 12W
Cravo Norte →, *Colombia* . **168 B3** 6 18N 70 12W
Crawford, *Ala., U.S.A.* **152 C4** 32 27N 85 11W
Crawford, *Nebr., U.S.A.* ... **154 D3** 42 41N 103 25W
Crawfordsville, *U.S.A.* **157 D10** 40 2N 86 54W
Crawfordville, *Fla., U.S.A.* . **152 E5** 30 11N 84 23W
Crawfordville, *Ga., U.S.A.* . **152 B7** 33 33N 82 54W
Crawley, *U.K.* **21 F7** 51 7N 0 11W
Crazy Mts., *U.S.A.* **158 C8** 46 12N 110 20W
Crean L., *Canada* **143 C7** 54 5N 106 9W
Crécy-en-Ponthieu, *France* . **27 B8** 50 15N 1 53 E
Crediton, *Canada* **150 C3** 43 17N 81 33W
Credo, *Australia* .... **125 F3** 30 28 S 120 45 E
Cree →, *Canada* **143 B7** 58 57N 105 47W
Cree →, *U.K.* **22 G4** 54 55N 4 25W
Cree L., *Canada* **143 B7** 57 30N 106 30W
Creede, *U.S.A.* **159 H10** 37 51N 106 56W
Creel, *Mexico* **162 B3** 27 45N 107 38W
Creighton, *U.S.A.* **154 D6** 42 28N 97 54W
Creil, *France* .... **27 C9** 49 15N 2 29 E
Crema, *Italy* .... **44 C6** 45 22N 9 41 E
Cremona, *Italy* .... **44 C7** 45 7N 10 2 E
Crepaja, *Serbia, Yug.* ..... **52 E5** 45 1N 20 38 E
Crepori →, *Brazil* .... **173 B6** 5 42 S 57 8W
Crépy, *France* .... **27 C10** 49 35N 3 32 E
Crépy-en-Valois, *France* ... **27 C9** 49 14N 2 54 E
Cres, *Croatia* .... **45 D11** 44 58N 14 25 E
Cresbard, *U.S.A.* **154 C5** 45 10N 98 57W
Crescent, *Okla., U.S.A.* ... **155 H6** 35 57N 97 36W
Crescent, *Oreg., U.S.A.* ... **158 E3** 43 28N 121 42W
Crescent Beach, *U.S.A.* **152 E8** 29 46N 81 15W
Crescent City, *Calif., U.S.A.* **158 F1** 41 45N 124 12W
Crescent City, *Fla., U.S.A.* . **153 F8** 29 26N 81 31W
Crescent L., *U.S.A.* **153 F8** 29 28N 81 30W
Crescentino, *Italy* .... **44 C5** 45 11N 8 6 E
Crespo, *Argentina* .... **174 C3** 32 2 S 60 19W
Cressy, *Australia* .... **128 E5** 38 2 S 143 40 E
Crest, *France* .... **29 D9** 44 44N 5 2 E
Cresta, Mt., *Phil.* .... **80 C4** 17 17N 122 6 E
Crested Butte, *U.S.A.* **159 G10** 38 52N 106 59W
Crestline, *Calif., U.S.A.* ... **161 L9** 34 14N 117 18W
Crestline, *Ohio, U.S.A.* ... **150 F2** 40 47N 82 44W
Creston, *Canada* **142 D5** 49 10N 116 31W
Creston, *Calif., U.S.A.* **160 K6** 35 32N 120 33W
Creston, *Iowa, U.S.A.* **156 C2** 41 4N 94 22W
Creston, *Wash., U.S.A.* ... **158 C4** 47 46N 118 31W
Crestview, *Calif., U.S.A.* .. **160 H8** 37 46N 118 58W
Crestview, *Fla., U.S.A.* **149 K2** 30 46N 86 34W
Creswick, *Australia* ...... **128 D5** 37 25 S 143 58 E
Crêt de la Neige, *France* ... **27 F12** 46 16N 5 58 E
Crete = Kríti, *Greece* ..... **38 D7** 35 15N 25 0 E
Crete, *U.S.A.* **154 E6** 40 38N 96 58W
Crete, Sea of, *Greece* ..... **49 E7** 36 0N 25 0 E
Créteil, *France* .... **27 D9** 48 47N 2 28 E
Cretin, C., *Papua N. G.* ... **132 D4** 6 40 S 147 53 E
Creus, C. de, *Spain* ...... **40 C8** 42 20N 3 19 E
Creuse □, *France* .... **27 F9** 46 10N 2 0 E
Creuse →, *France* .... **28 B4** 47 0N 0 34 E
Creutzwald, *France* .... **27 C13** 49 12N 6 42 E
Creuzburg, *Germany* ...... **30 D6** 51 3N 10 14 E
Crèvecœur-le-Grand, *France* **27 C9** 49 37N 2 5 E
Crevillente, *Spain* **41 G4** 38 12N 0 48W
Crewe, *U.K.* **20 D5** 53 6N 2 26W
Crewkerne, *U.K.* **21 G5** 50 53N 2 48W
Criciúma, *Brazil* .... **175 B6** 28 40 S 49 23W
Cricova, *Moldova* .... **53 C13** 47 8N 28 52 E
Cridersville, *U.S.A.* **157 D12** 40 39N 84 9W
Crieff, *U.K.* **22 E5** 56 22N 3 50W
Crikvenica, *Croatia* ...... **45 C11** 45 11N 14 40 E
Crimea □, *Ukraine* .... **59 K8** 45 30N 33 10 E
Crimean Pen. = Krymskyy Pivostriv, *Ukraine* ...... **59 K8** 45 0N 34 0 E
Crimmitschau, *Germany* ... **30 E8** 50 48N 12 24 E
Cristal, Mts. de, *Gabon* .... **114 B2** 0 30N 10 30 E
Cristalândia, *Brazil* ...... **170 D2** 10 36 S 49 11W
Cristino Castro, *Brazil* .... **170 C3** 8 49 S 44 13W
Cristuru Secuiesc, *Romania* . **53 D10** 46 17N 25 2 E
Crişul Alb →, *Romania* ... **52 D6** 46 42N 21 17 E
Crişul Negru →, *Romania* . **52 D6** 46 42N 21 16 E
Crişul Repede →, *Romania* **52 D5** 46 55N 20 59 E
Crittenden, *U.S.A.* **157 F12** 38 47N 84 36W
Criuleni, *Moldova* .... **53 C14** 47 13N 29 10 E
Crivitz, *Germany* .... **30 B7** 53 34N 11 39 E
Crixás, *Brazil* .... **171 D2** 14 27 S 49 58W
Crna →, *Macedonia* ...... **50 E5** 41 33N 21 59 E
Crna Gora = Montenegro □, *Yugoslavia* .... **50 D3** 42 40N 19 20 E
Crna Gora, *Macedonia* .... **50 D5** 42 10N 21 30 E
Crna Reka = Crna →, *Macedonia* .... **50 E5** 41 33N 21 59 E
Crna Trava, *Serbia, Yug.* ... **50 D6** 42 49N 22 19 E
Crni Drim →, *Macedonia* . **50 E4** 41 17N 20 40 E
Crni Timok →, *Serbia, Yug.* **50 C6** 43 53N 22 15 E
Crnoljeva Planina, *Serbia, Yug.* .... **50 D5** 42 20N 21 0 E
Črnomelj, *Slovenia* .... **45 C12** 45 33N 15 10 E
Croagh Patrick, *Ireland* **23 C2** 53 46N 9 40W
Croatia ■, *Europe* .... **45 C13** 45 20N 16 0 E
Crocker, *U.S.A.* **156 G4** 37 57N 92 16W
Crocker, Banjaran, *Malaysia* **85 A5** 5 40N 116 30 E
Crockett, *U.S.A.* **155 K7** 31 19N 95 27W
Crocodile = Krokodil →, *Mozam.* .... **117 D5** 25 14 S 32 18 E
Crocodile Is., *Australia* .... **126 A1** 12 3 S 134 58 E
Crocq, *France* .... **28 C6** 45 52N 2 21 E
Crodo, *Italy* .... **44 B5** 46 13N 8 19 E
Crohy Hd., *Ireland* **23 B3** 54 55N 8 26W
Croisette, C., *France* ...... **29 E9** 43 14N 5 22 E
Croisic, Pte. du, *France* ... **26 E4** 47 19N 2 31W
Croix, L. La, *Canada* **140 C1** 48 20N 92 15W
Croker, C., *Australia* ..... **124 B5** 10 58 S 132 35 E
Croker I., *Australia* ...... **124 B5** 11 12 S 132 32 E
Cromarty, *Canada* **143 B10** 58 3N 94 9W
Cromarty, *U.K.* **22 D4** 57 40N 4 2W
Cromer, *U.K.* .... **20 E9** 52 56N 1 17 E
Cromwell, *N.Z.* .... **131 F4** 45 3 S 169 14 E
Cronat, *France* .... **27 F10** 46 43N 3 40 E
Cronulla, *Australia* .... **129 C9** 34 3 S 151 8 E
Crook, *U.K.* **20 C6** 54 43N 1 45W
Crooked →, *Canada* **142 C4** 54 50N 122 54W
Crooked →, *U.S.A.* **158 D3** 44 32N 121 16W
Crooked Creek, *U.S.A.* **144 F8** 61 52N 158 7W
Crooked I., *Bahamas* **165 B5** 22 50N 74 10W
Crooked Island Passage, *Bahamas* **165 B5** 23 0N 74 30W
Crooked L., *U.S.A.* **153 H8** 27 48N 81 35W
Crookston, *Minn., U.S.A.* .. **154 B6** 47 47N 96 37W
Crookston, *Nebr., U.S.A.* .. **154 D4** 42 56N 100 45W
Crooksville, *U.S.A.* **148 F4** 39 46N 82 6W
Crookwell, *Australia* ...... **129 C8** 34 28 S 149 24 E
Crosby, *U.K.* **20 D4** 53 30N 3 3W
Crosby, *Minn., U.S.A.* **154 B8** 46 29N 93 58W
Crosby, *N. Dak., U.S.A.* ... **143 D8** 48 55N 103 18W
Crosby, *Pa., U.S.A.* **150 E6** 41 45N 78 23W
Crosbyton, *U.S.A.* **155 J4** 33 40N 101 14W
Crosía, *Italy* .... **47 C9** 39 35N 16 45 E
Cross →, *Nigeria* .... **113 E6** 4 42N 8 21 E
Cross City, *U.S.A.* **153 F6** 29 38N 83 7W
Cross Fell, *U.K.* **20 C5** 54 43N 2 28W
Cross L., *Canada* **143 C9** 54 45N 97 30W
Cross Plains, *U.S.A.* **155 J5** 32 8N 99 11W
Cross River □, *Nigeria* .... **113 D6** 6 0N 8 0 E
Cross Sound, *U.S.A.* **138 C6** 58 0N 135 0W
Cross Timbers, *U.S.A.* **156 F3** 38 1N 93 14W
Crossett, *U.S.A.* **155 J9** 33 8N 91 58W
Crossfield, *Canada* **142 C6** 51 25N 114 0W
Crosshaven, *Ireland* **23 E3** 51 47N 8 17W
Crossley, Mt., *N.Z.* .... **131 C7** 42 50 S 172 5 E
Crossville, *U.S.A.* **157 F8** 38 10N 88 4W
Croton-on-Hudson, *U.S.A.* . **151 E11** 41 12N 73 55W
Crotone, *Italy* .... **47 C10** 39 5N 17 8 E
Crow →, *Canada* **142 B4** 59 41N 124 20W
Crow Agency, *U.S.A.* **158 D10** 45 36N 107 28W
Crow Hd., *Ireland* **23 E1** 51 35N 10 9W
Crowell, *U.S.A.* **155 J5** 33 59N 99 43W
Crowl Creek, *Australia* .... **129 B6** 32 0 S 145 30 E
Crowley, *U.S.A.* **155 K8** 30 13N 92 22W
Crowley, L., *U.S.A.* **160 H8** 37 35N 118 42W
Crown Point, *U.S.A.* **157 C9** 41 25N 87 22W
Crows Landing, *U.S.A.* **160 H5** 37 23N 121 6W
Crows Nest, *Australia* ..... **127 D5** 27 16 S 152 4 E
Crowsnest Pass, *Canada* ... **142 D6** 49 40N 114 40W
Croydon, *Australia* .... **126 B3** 18 13 S 142 14 E
Croydon, *U.K.* **21 F7** 51 22N 0 5W
Crozet Is., *Ind. Oc.* .... **121 J4** 46 27 S 52 0 E
Crozon, *France* **26 D2** 48 15N 4 30W
Cruz, C., *Cuba* **164 C4** 19 50N 77 50W
Cruz Alta, *Brazil* .... **175 B5** 28 45 S 53 40W
Cruz das Almas, *Brazil* .... **171 D4** 12 0 S 39 6W
Cruz de Incio, *Spain* **42 C3** 42 39N 7 21W
Cruz de Malta, *Brazil* ..... **170 C3** 8 15 S 40 20W
Cruz del Eje, *Argentina* .... **174 C3** 30 45 S 64 50W
Cruzeiro, *Brazil* .... **171 F3** 22 33 S 45 0W
Cruzeiro do Oeste, *Brazil* .. **175 A5** 23 46 S 53 4W
Cruzeiro do Sul, *Brazil* .... **172 B3** 7 35 S 72 35W
Cry L., *Canada* **142 B3** 58 45N 129 0W
Crystal B., *U.S.A.* **153 G7** 28 50N 82 45W
Crystal Bay, *U.S.A.* **160 F7** 39 15N 120 0W
Crystal Brook, *Australia* ... **128 B3** 33 21 S 138 12 E
Crystal City, *Mo., U.S.A.* .. **156 F6** 38 13N 90 23W
Crystal City, *Tex., U.S.A.* . **155 L5** 28 41N 99 50W
Crystal Falls, *U.S.A.* **148 B1** 46 5N 88 20W
Crystal Lake, *Fla., U.S.A.* . **152 E4** 30 26N 85 42W
Crystal Lake, *Ill., U.S.A.* .. **157 B8** 42 14N 88 19W
Crystal River, *U.S.A.* **149 L4** 28 54N 82 35W
Crystal Springs, *U.S.A.* **155 K9** 31 59N 90 21W
Csenger, *Hungary* .... **52 C7** 47 50N 22 41 E
Csongrád, *Hungary* .... **52 D5** 46 43N 20 12 E
Csongrád □, *Hungary* ..... **52 D5** 46 32N 20 15 E
Csorna, *Hungary* .... **52 C2** 47 38N 17 18 E
Csurgo, *Hungary* .... **52 D2** 46 16N 17 9 E
Cu Lao Hon, *Vietnam* ..... **87 G7** 10 54N 108 18 E
Cua Rao, *Vietnam* .... **86 C5** 19 16N 104 27 E
Cuácua →, *Mozam.* ...... **119 F4** 17 54 S 37 0 E
Cuamato, *Angola* .... **115 F3** 17 2 S 15 7 E
Cuamba, *Mozam.* .... **119 E4** 14 45 S 36 22 E
Cuando →, *Angola* ...... **115 F4** 17 30 S 23 15 E
Cuando Cubango □, *Angola* **115 F4** 16 25 S 20 0 E
Cuangar, *Angola* .... **115 F3** 17 36 S 18 39 E
Cuango, *Angola* .... **115 D3** 6 15 S 16 42 E
Cuanza →, *Angola* .... **115 D2** 9 2 S 13 30 E
Cuanza Norte □, *Angola* ... **115 D2** 8 50 S 14 30 E
Cuanza Sul □, *Angola* .... **115 E2** 10 50 S 14 50 E
Cuarto →, *Argentina* ..... **174 C3** 33 25 S 63 2W
Cuatrociénegas, *Mexico* **162 B4** 26 59N 102 5W
Cuauhtémoc, *Mexico* **162 B3** 28 25N 106 52W
Cuba, *Portugal* **43 G3** 38 10N 7 54W
Cuba, *Mo., U.S.A.* **156 F5** 38 4N 91 24W
Cuba, *N. Mex., U.S.A.* **159 J10** 36 1N 107 4W
Cuba, *N.Y., U.S.A.* **150 D6** 42 13N 78 17W
Cuba ■, *W. Indies* **164 B4** 22 0N 79 0W
Cuba City, *U.S.A.* **156 B6** 42 36N 90 26W
Cubal, *Angola* .... **115 E2** 12 26 S 14 3 E
Cuballing, *Australia* ...... **125 F2** 32 50 S 117 10 E
Cubango →, *Africa* ...... **115 F4** 18 50 S 22 25 E
Cubanja, *Angola* .... **115 E4** 14 49 S 21 20 E
Cubia, *Angola* .... **115 F4** 15 58 S 21 42 E
Çubuk, *Turkey* .... **100 B5** 40 14N 33 3 E
Cuchi, *Angola* .... **115 E3** 14 37 S 16 58 E
Cuchillo-Có, *Argentina* .... **176 A4** 38 20 S 64 37W
Cuchivero →, *Venezuela* .. **168 B4** 7 40N 65 57W
Cuchumatanes, Sierra de los, *Guatemala* **164 C1** 15 35N 91 25W
Cuckfield, *U.K.* **21 F7** 51 1N 0 8W
Cucuí, *Brazil* **168 C4** 1 12N 66 50W
Cucurpe, *Mexico* **162 A2** 30 20N 110 43W
Cucurrupí, *Colombia* **168 C2** 4 23N 76 56W
Cúcuta, *Colombia* **168 B3** 7 54N 72 31W
Cudahy, *U.S.A.* **157 B9** 42 58N 87 52W
Cudalbi, *Romania* .... **53 E12** 45 46N 27 41 E
Cuddalore, *India* .... **95 J4** 11 46N 79 45 E
Cuddapah, *India* .... **95 G4** 14 30N 78 47 E
Cuddapan, L., *Australia* ... **126 D3** 25 45 S 141 26 E
Cudgewa, *Australia* ...... **129 D7** 36 10 S 147 42 E
Cudillero, *Spain* **42 B4** 43 33N 6 9W
Cue, *Australia* .... **125 E2** 27 25 S 117 54 E
Cuéllar, *Spain* **42 D6** 41 23N 4 21W
Cuemba, *Angola* .... **115 E3** 11 50 S 17 42 E
Cuenca, *Ecuador* .... **168 D2** 2 50 S 79 9W
Cuenca, *Spain* **40 E2** 40 5N 2 10W
Cuenca □, *Spain* **40 F3** 40 0N 2 0W
Cuenca, Serranía de, *Spain* . **40 F3** 39 55N 1 50W
Cuerdo del Pozo, Embalse de la, *Spain* **40 D2** 41 51N 2 44W
Cuernavaca, *Mexico* **163 D5** 18 55N 99 15W
Cuero, *U.S.A.* **155 L6** 29 6N 97 17W
Cuers, *France* .... **29 E10** 43 14N 6 5 E
Cuervo, *U.S.A.* **155 H2** 35 2N 104 25W
Cuevas, Cerro, *Bolivia* .... **173 E4** 22 0 S 65 12W
Cuevas del Almanzora, *Spain* **41 H3** 37 18N 1 58W
Cuevo, *Bolivia* .... **173 E5** 20 15 S 63 30W
Cugir, *Romania* .... **53 E8** 45 48N 23 25 E
Cugnaux, *France* .... **28 E5** 43 32N 1 20 E
Cuhai-Bakony →, *Hungary* **52 C2** 47 35N 17 54 E
Cuiabá, *Brazil* .... **173 D6** 15 30 S 56 0W
Cuiabá →, *Brazil* .... **173 D6** 17 5 S 56 36W
Cuijk, *Neths.* .... **24 C5** 51 44N 5 50 E
Cuilco, *Guatemala* **164 C1** 15 24N 91 58W
Cuillin Hills, *U.K.* **22 D2** 57 13N 6 15W
Cuillin Sd., *U.K.* **22 D2** 57 4N 6 20W
Cuima, *Angola* .... **115 E3** 13 25 S 15 45 E
Cuiseaux, *France* .... **27 F12** 46 30N 5 22 E
Cuité, *Brazil* .... **170 C4** 6 29 S 36 9W
Cuito →, *Angola* .... **115 F4** 18 1 S 20 48 E

D

206 Dante

Dante, *Somali Rep.* .... **120 B4** 10 25N 51 16 E
Danube = Dunărea ⟶,
Europe .... **53 E14** 45 20N 29 40 E
Danubyu, *Burma* .... **90 G5** 17 15N 95 35 E
Danukandi, *Bangla.* .... **90 D3** 23 32N 90 43 E
Danvers, *U.S.A.* **151 D14** 42 34N 70 56W
Danville, *Ga., U.S.A.* **152 C6** 32 37N 83 15W
Danville, *Ill., U.S.A.* **157 D9** 40 8N 87 37W
Danville, *Ind., U.S.A.* **157 E10** 39 46N 86 32W
Danville, *Ky., U.S.A.* **157 G12** 37 39N 84 46W
Danville, *Va., U.S.A.* **149 G6** 36 36N 79 23W
Danyang, *China* .... **77 B12** 32 0N 119 31 E
Danzhai, *China* .... **76 D6** 26 11N 107 48 E
Danzig = Gdańsk, *Poland* . **54 D5** 54 22N 18 40 E
Dao, *Antique, Phil.* .... **81 F3** 10 30N 121 57 E
Dao, *Capiz, Phil.* .... **81 F4** 11 24N 122 41 E
Dão ⟶, *Portugal* **42 E2** 40 20N 8 11W
Dao Xian, *China* .... **77 E8** 25 36N 111 31 E
Daocheng, *China* .... **76 C3** 29 0N 100 10 E
Daora, *W. Sahara* **110 C2** 27 5N 12 59W
Daoud = Aïn Beïda, *Algeria* **111 A6** 35 50N 7 29 E
Dapa, *Phil.* .... **81 G6** 9 46N 126 3 E
Dapaong, *Togo* .... **113 C5** 10 55N 0 16 E
Dapitan, *Phil.* .... **81 G4** 8 39N 123 25 E
Daqing Shan, *China* .... **74 D6** 40 40N 111 0 E
Daqu Shan, *China* .... **77 B14** 30 25N 122 20 E
Dar Banda, *Africa* .... **104 F6** 8 0N 23 0 E
Dar el Beida = Casablanca,
Morocco **110 B3** 33 36N 7 36W
Dar es Salaam, *Tanzania* ... **118 D4** 6 50 S 39 12 E
Dar Mazār, *Iran* .... **97 D8** 29 14N 57 20 E
Dar'ā, *Syria* .... **103 C5** 32 36N 36 7 E
Dar'ā □, *Syria* .... **103 C5** 32 55N 36 10 E
Dārāb, *Iran* .... **97 D7** 28 50N 54 30 E
Darabani, *Romania* .... **53 B11** 48 10N 26 39 E
Daraj, *Libya* .... **108 B2** 30 10N 10 28 E
Dārān, *Iran* .... **97 C6** 32 59N 50 24 E
Daraut Kurgan, *Kyrgyzstan* **63 D6** 39 33N 72 11 E
Daravica, *Serbia, Yug.* ..... **50 D4** 42 32N 20 8 E
Daraw, *Egypt* .... **106 C3** 24 22N 32 51 E
Dārayyā, *Syria* .... **103 B5** 33 28N 36 15 E
Darazo, *Nigeria* .... **113 C7** 11 1N 10 24 E
Darband, *Pakistan* .... **92 B5** 34 20N 72 50 E
Darband, Kūh-e, *Iran* ..... **97 D8** 31 34N 57 8 E
Darbhanga, *India* .... **93 F11** 26 15N 85 55 E
Darburruk, *Somali Rep.* ... **120 C2** 9 44N 44 31 E
Darby, *U.S.A.* **158 C6** 46 1N 114 11W
Darby, C., *U.S.A.* **144 D7** 64 19N 162 47W
Darda, *Croatia* .... **52 E3** 45 40N 18 41 E
Dardanelle, *Ark., U.S.A.* .. **155 H8** 35 13N 93 9W
Dardanelle, *Calif., U.S.A.* . **160 G7** 38 20N 119 50W
Dardanelles = Çanakkale
Boğazı, *Turkey* .... **51 F10** 40 17N 26 32 E
Darende, *Turkey* .... **100 C7** 38 31N 37 30 E
Dārestān, *Iran* .... **97 D8** 29 9N 58 42 E
Darfield, *N.Z.* .... **131 D7** 43 29 S 172 7 E
Darfo, *Italy* .... **44 C7** 45 53N 10 11 E
Dārfūr, *Sudan* .... **104 E6** 13 40N 24 0 E
Dargai, *Pakistan* .... **91 B3** 34 25N 71 55 E
Dargan Ata, *Uzbekistan* ... **64 E7** 40 29N 62 10 E
Dargaville, *N.Z.* .... **130 B2** 35 57 S 173 52 E
Darhan, *Mongolia* .... **72 B5** 49 37N 106 21 E
Darhan Muminggan
Lianheqi, *China* .... **74 D6** 41 40N 110 28 E
Dari, *Sudan* .... **107 F3** 5 48N 30 26 E
Darıca, *Turkey* .... **100 B3** 40 45N 29 23 E
Darien, *U.S.A.* **152 D8** 31 23N 81 26W
Darién, G. del, *Colombia* .. **168 B2** 9 0N 77 0W
Darién, Serranía del,
Colombia **168 B2** 8 30N 77 30W
Dariganga, *Mongolia* ..... **74 B7** 45 21N 113 45 E
Darinskoye, *Kazakstan* ... **60 E10** 51 20N 51 44 E
Darjeeling = Darjiling, *India* **93 F13** 27 3N 88 18 E
Darjiling, *India* .... **93 F13** 27 3N 88 18 E
Dark Cove, *Canada* **141 C9** 48 47N 54 13W
Darkan, *Australia* .... **125 F2** 33 20 S 116 43 E
Darke Peak, *Australia* ..... **128 B2** 33 27 S 136 12 E
Darkhazīneh, *Iran* .... **97 D6** 31 54N 48 39 E
Darkot Pass, *Pakistan* ..... **93 A5** 36 45N 73 26 E
Darling ⟶, *Australia* ..... **128 C4** 34 4 S 141 54 E
Darling Downs, *Australia* .. **127 D5** 27 30 S 150 30 E
Darling Ra., *Australia* ..... **125 F2** 32 30 S 116 0 E
Darlington, *U.K.* **20 C6** 54 32N 1 33W
Darlington, *Fla., U.S.A.* ... **152 E3** 30 57N 86 3W
Darlington, *S.C., U.S.A.* ... **149 H6** 34 18N 79 52W
Darlington, *Wis., U.S.A.* ... **156 B6** 42 41N 90 7W
Darlington □, *U.K.* **20 C6** 54 32N 1 33W
Darlington, L., *S. Africa* .. **116 E4** 33 10 S 25 9 E
Darlot, L., *Australia* ..... **125 E3** 27 48 S 121 35 E
Darłowo, *Poland* .... **54 D3** 54 25N 16 25 E
Dărmănești, *Bacău, Romania* **53 D11** 46 21N 26 33 E
Dărmănești, *Suceava,*
Romania .... **53 C11** 47 44N 26 9 E
Darmstadt, *Germany* ..... **31 F4** 49 51N 8 39 E
Darnah, *Libya* .... **108 B4** 32 45N 22 45 E
Darnah □, *Libya* .... **108 B4** 31 0N 23 40 E
Darnall, *S. Africa* .... **117 D5** 29 23 S 31 18 E
Darney, *France* .... **27 D13** 48 5N 6 2 E
Darnick, *Australia* .... **128 B5** 32 48 S 143 38 E
Darnley, C., *Antarctica* .... **7 C6** 68 0 S 69 0 E
Darnley B., *Canada* **138 B7** 69 30N 123 30W
Daroca, *Spain* **40 D3** 41 9N 1 25W
Darr, *Australia* .... **126 C3** 23 13 S 144 7 E
Darr ⟶, *Australia* .... **126 C3** 23 39 S 143 50 E
Darran Mts., *N.Z.* .... **131 E2** 44 37 S 167 59 E
Darrington, *U.S.A.* **158 B3** 48 15N 121 36W
Darsana, *Bangla.* .... **90 D2** 23 35N 88 48 E
Darsi, *India* .... **95 G4** 15 46N 79 44 E
Darsser Ort, *Germany* ..... **30 A8** 54 28N 12 32 E
Dart ⟶, *U.K.* **21 G4** 50 24N 3 39W
Dart, C., *Antarctica* .... **7 D14** 73 6 S 126 20W
Dartford, *U.K.* .... **21 F8** 51 26N 0 13 E
Dartmoor, *Australia* .... **128 D4** 37 56 S 141 19 E
Dartmoor, *U.K.* **21 G4** 50 38N 3 57W
Dartmouth, *Australia* ..... **126 C3** 23 31 S 144 44 E
Dartmouth, *Canada* **141 D7** 44 40N 63 30W
Dartmouth, *U.K.* **21 G4** 50 21N 3 36W
Dartmouth, L., *Australia* .. **127 D4** 26 4 S 145 18 E
Dartuch, C., *Spain* .... **39 B10** 39 55N 3 49 E
Daru, *Papua N. G.* .... **132 E2** 9 3 S 143 13 E
Daruvar, *Croatia* .... **52 E2** 45 35N 17 14 E
Darvaza, *Turkmenistan* ... **64 E6** 40 11N 58 24 E
Darvel, Teluk = Lahad
Datu, Teluk, *Malaysia* ... **85 B5** 4 50N 118 20 E
Darwen, *U.K.* **20 D5** 53 42N 2 29W
Darwha, *India* .... **94 D3** 20 15N 77 45 E
Darwin, *Australia* .... **124 B5** 12 25 S 130 51 E
Darwin, *U.S.A.* **161 J9** 36 15N 117 35W
Darwin, Mt., *Chile* .... **176 D3** 54 47 S 69 55W
Darwin River, *Australia* .... **124 B5** 12 50 S 130 58 E
Daryapur, *India* .... **94 D3** 20 55N 77 20 E
Daryoi Amu =
Amudarya ⟶, *Uzbekistan* **64 E6** 43 58N 59 34 E
Dās, *U.A.E.* .... **97 E7** 25 20N 53 30 E
Dasher, *U.S.A.* **152 E6** 30 45N 83 13W
Dashetai, *China* .... **74 D5** 41 0N 109 5 E
Dashhowuz, *Turkmenistan* . **64 E6** 41 49N 59 58 E
Dashkesan = Daşkäsän,
Azerbaijan .... **61 K7** 40 25N 46 0 E
Dasht, *Iran* .... **97 B8** 37 17N 56 7 E
Dasht ⟶, *Pakistan* .... **91 D1** 25 10N 61 40 E
Dasht-i-Nawar, *Afghan.* .... **92 C3** 33 52N 68 0 E
Daska, *Pakistan* .... **92 C6** 32 20N 74 20 E
Daşkäsän, *Azerbaijan* ..... **61 K7** 40 25N 46 0 E
Dassa, *Benin* .... **113 D5** 7 46N 2 14 E
Datça, *Turkey* .... **49 E9** 36 46N 27 40 E
Datia, *India* .... **93 G8** 25 39N 78 27 E
Datian, *China* .... **77 E11** 25 40N 117 50 E
Datong, *Anhui, China* ..... **77 B11** 30 48N 117 44 E
Datong, *Shanxi, China* ..... **74 D7** 40 6N 113 18 E
Dattapur = Dhamangaon,
India .... **94 D4** 20 45N 78 15 E
Datteln, *Germany* .... **30 D3** 51 39N 7 21 E
Datu, Tanjung, *Indonesia* .. **85 B3** 2 5N 109 39 E
Datu Piang, *Phil.* .... **81 H5** 7 2N 124 30 E
Datuk, Tanjong, *Malaysia* .. **85 B3** 2 5N 109 39 E
Daugava ⟶, *Latvia* .... **15 H21** 57 4N 24 3 E
Daugavpils, *Latvia* .... **15 J22** 55 53N 26 32 E
Daulatabad, *India* .... **94 E2** 19 57N 75 15 E
Daule, *Ecuador* .... **168 D2** 1 56 S 79 56W
Daule ⟶, *Ecuador* .... **168 D2** 2 10 S 79 52W
Daulpur, *India* .... **92 F7** 26 45N 77 59 E
Daun, *Germany* .... **31 E2** 50 11N 6 49 E
Daund, *India* .... **94 E2** 18 26N 74 40 E
Dauphin, *Canada* **143 C8** 51 9N 100 5W
Dauphin I., *U.S.A.* **149 K1** 30 15N 88 11W
Dauphin L., *Canada* **143 C9** 51 20N 99 45W
Dauphiné, *France* .... **29 C9** 45 15N 5 25 E
Daura, *Borno, Nigeria* ..... **113 C7** 11 31N 11 24 E
Daura, *Kaduna, Nigeria* .... **113 C6** 13 2N 8 21 E
Dausa, *India* .... **92 F7** 26 52N 76 20 E
Däväçi, *Azerbaijan* .... **61 K9** 41 15N 48 57 E
Davangere, *India* .... **95 G2** 14 25N 75 55 E
Davao, *Phil.* .... **81 H5** 7 0N 125 40 E
Davao □, *Phil.* .... **81 H5** 7 0N 125 55 E
Davao, G. of, *Phil.* .... **81 H5** 6 30N 125 48 E
Davao del Sur □, *Phil.* .... **81 H5** 6 30N 125 25 E
Davao Oriental □, *Phil.* ... **81 H6** 7 10N 126 30 E
Dāvar Panāh, *Iran* .... **97 E9** 27 25N 62 15 E
Davenport, *Calif., U.S.A.* .. **160 H4** 37 1N 122 12W
Davenport, *Fla., U.S.A.* ... **153 G8** 28 10N 81 36W
Davenport, *Iowa, U.S.A.* .. **156 C6** 41 32N 90 35W
Davenport, *Wash., U.S.A.* . **158 C4** 47 39N 118 9W
Davenport Downs, *Australia* **126 C3** 24 8 S 141 7 E
Davenport Ra., *Australia* .. **126 C1** 20 28 S 134 0 E
Daventry, *U.K.* **21 E6** 52 16N 1 10W
David, *Panama* **164 E3** 8 30N 82 30W
David City, *U.S.A.* **154 E6** 41 15N 97 8W
David Gorodok = Davyd
Haradok, *Belarus* .... **59 F4** 52 4N 27 8 E
Davidson, *Canada* **143 C7** 51 16N 105 59W
Davis, *U.S.A.* **160 G5** 38 33N 121 44W
Davis Dam, *U.S.A.* **161 K12** 35 11N 114 34W
Davis Inlet, *Canada* **141 A7** 55 50N 60 59W
Davis Mts., *U.S.A.* **155 K2** 30 50N 103 55W
Davis Sea, *Antarctica* ..... **7 C7** 66 0 S 92 0 E
Davis Str., *N. Amer.* **139 B14** 65 0N 58 0W
Davisboro, *U.S.A.* **152 C7** 32 59N 82 36W
Davlekanovo, *Russia* .... **62 D5** 54 13N 55 3 E
Davos, *Switz.* .... **33 C9** 46 48N 9 49 E
Davutlar, *Turkey* .... **49 D9** 37 43N 27 17 E
Davy L., *Canada* **143 B7** 58 53N 108 18W
Davyd Haradok, *Belarus* ... **59 F4** 52 4N 27 8 E
Dawa ⟶, *Ethiopia* .... **107 G5** 4 11N 42 6 E
Dawaki, *Bauchi, Nigeria* ... **113 D6** 9 25N 9 33 E
Dawaki, *Kano, Nigeria* .... **113 C6** 12 5N 8 23 E
Dawei, *Burma* .... **86 E2** 14 2N 98 12 E
Dawes Ra., *Australia* ..... **126 C5** 24 40 S 150 40 E
Dawlish, *U.K.* **21 G4** 50 35N 3 28W
Dawna Range, *Burma* ..... **90 G7** 16 30N 98 30 E
Dawnyein, *Burma* .... **90 G5** 15 54N 95 36 E
Dawqah, *Si. Arabia* .... **98 C3** 19 36N 40 54 E
Dawros Hd., *Ireland* **23 B3** 54 50N 8 33W
Dawson, *Canada* **138 B6** 64 10N 139 30W
Dawson, *Ga., U.S.A.* **152 D5** 31 46N 84 27W
Dawson, *N. Dak., U.S.A.* .. **154 B5** 46 52N 99 45W
Dawson, I., *Chile* .... **176 D2** 53 50 S 70 50W
Dawson Creek, *Canada* **142 B4** 55 45N 120 15W
Dawson Inlet, *Canada* **143 A10** 61 50N 93 25W
Dawson Ra., *Australia* ..... **126 C4** 24 30 S 149 48 E
Dawu, *China* .... **76 B3** 30 55N 101 10 E
Dawwah, *Oman* .... **99 B7** 20 33N 58 48 E
Dax, *France* **28 E2** 43 44N 1 3W
Daxian, *China* .... **76 B6** 31 15N 107 23 E
Daxin, *China* .... **76 F6** 22 50N 107 11 E
Daxindian, *China* .... **75 F11** 37 30N 120 50 E
Daxinggou, *China* .... **75 C15** 43 25N 129 40 E
Daxue Shan, *Sichuan, China* **76 B3** 30 30N 101 30 E
Daxue Shan, *Yunnan, China* **76 F2** 23 42N 99 48 E
Day, *U.S.A.* **152 E6** 30 12N 83 17W
Dayao, *China* .... **76 E3** 25 43N 101 20 E
Daye, *China* .... **77 B10** 30 6N 114 58 E
Dayi, *China* .... **76 B4** 30 41N 103 29 E
Daylesford, *Australia* ..... **128 D6** 37 21 S 144 9 E
Dayong, *China* .... **77 C8** 29 11N 110 30 E
Dayr az Zawr, *Syria* .... **101 E9** 35 20N 40 5 E
Daysland, *Canada* **142 C6** 52 50N 112 20W
Dayton, *Iowa, U.S.A.* **156 B2** 42 14N 94 6W
Dayton, *Ky., U.S.A.* **157 E12** 39 47N 84 28W
Dayton, *Nev., U.S.A.* **160 F7** 39 14N 119 36W
Dayton, *Ohio, U.S.A.* **148 F3** 39 45N 84 12W
Dayton, *Pa., U.S.A.* **150 F5** 40 53N 79 15W
Dayton, *Tenn., U.S.A.* **149 H3** 35 30N 85 1W
Dayton, *Wash., U.S.A.* **158 C4** 46 19N 117 59W
Daytona Beach, *U.S.A.* **149 L5** 29 13N 81 1W
Dayu, *China* .... **77 E10** 25 24N 114 22 E
Dayville, *U.S.A.* **158 D4** 44 28N 119 32W
Dazhu, *China* .... **76 B6** 30 41N 107 15 E
Dazkırı, *Turkey* .... **49 D11** 37 57N 29 52 E
Dazu, *China* .... **76 C5** 29 40N 105 42 E
De Aar, *S. Africa* .... **116 E3** 30 39 S 24 0 E
De Armanville, *U.S.A.* **152 B4** 33 38N 85 45W
De Bary, *U.S.A.* **153 G8** 28 54N 81 18W
De Forest, *U.S.A.* **156 A7** 43 15N 89 20W
De Funiak Springs, *U.S.A.* . **152 E3** 30 43N 86 7W
De Grey, *Australia* .... **124 D2** 20 12 S 119 12 E
De Grey ⟶, *Australia* .... **124 D2** 20 12 S 119 13 E
De Haan, *Belgium* .... **24 C3** 51 16N 3 2 E
De Kalb, *U.S.A.* **154 E10** 41 56N 88 46W
De Land, *U.S.A.* **149 L5** 29 2N 81 18W
De Leon, *U.S.A.* **155 J5** 32 7N 98 32W
De Leon Springs, *U.S.A.* .. **153 F8** 29 7N 81 21W
De Panne, *Belgium* .... **24 C2** 51 6N 2 34 E
De Pere, *U.S.A.* **148 C1** 44 27N 88 4W
De Queen, *U.S.A.* **155 H7** 34 2N 94 21W
De Quincy, *U.S.A.* **155 K8** 30 27N 93 26W
De Ridder, *U.S.A.* **155 K8** 30 51N 93 17W
De Smet, *U.S.A.* **154 C6** 44 23N 97 33W
De Soto, *U.S.A.* **156 F6** 38 8N 90 34W
De Soto City, *U.S.A.* **153 H8** 27 27N 81 24W
De Tour Village, *U.S.A.* ... **148 C4** 46 0N 83 56W
De Witt, *Ark., U.S.A.* **155 H9** 34 18N 91 20W
De Witt, *Iowa, U.S.A.* **156 C6** 41 49N 90 33W
De Witt, *Mich., U.S.A.* **157 B12** 42 51N 84 34W
Dead L., *U.S.A.* **152 E4** 30 10N 85 10W
Dead Sea, *Asia* .... **103 D4** 31 30N 35 30 E
Deadhorse, *U.S.A.* **144 A10** 70 11N 148 27W
Deadman B., *U.S.A.* **153 F6** 29 30N 83 30W
Deadwood, *U.S.A.* **154 C3** 44 23N 103 44W
Deadwood L., *Canada* **142 B3** 59 10N 128 30W
Deakin, *Australia* .... **125 F4** 30 46 S 128 58 E
Deal, *U.K.* .... **21 F9** 51 13N 1 25 E
Deal I., *Australia* .... **126 F4** 39 30 S 147 20 E
Dealesville, *S. Africa* ..... **116 D4** 28 41 S 25 44 E
De'an, *China* .... **77 C10** 29 21N 115 46 E
Dean, Forest of, *U.K.* **21 F5** 51 45N 2 33W
Deán Funes, *Argentina* .... **174 C3** 30 20 S 64 20W
Dearborn, *Mich., U.S.A.* ... **140 D3** 42 19N 83 11W
Dearborn, *Mo., U.S.A.* **156 E2** 39 32N 94 46W
Dease ⟶, *Canada* **142 B3** 59 56N 128 32W
Dease Inlet, *U.S.A.* **144 A9** 70 30N 155 0W
Dease L., *Canada* **142 B2** 58 40N 130 5W
Dease Lake, *Canada* **142 B2** 58 25N 130 6W
Death Valley, *U.S.A.* **161 J10** 36 15N 116 50W
Death Valley Junction,
U.S.A. **161 J10** 36 20N 116 25W
Death Valley National
Monument, *U.S.A.* **161 J10** 36 45N 117 15W
Deauville, *France* .... **26 C7** 49 23N 0 2 E
Deba, *Spain* **40 B2** 43 18N 2 21W
Deba Habe, *Nigeria* .... **113 C7** 10 14N 11 20 E
Debak, *Malaysia* .... **85 B4** 1 34N 111 25 E
Debaltsevo, *Ukraine* .... **59 H10** 48 22N 38 26 E
Debao, *China* .... **76 F6** 23 21N 106 46 E
Debar, *Macedonia* .... **50 E4** 41 31N 20 30 E
Debden, *Canada* **143 C7** 53 30N 106 50W
Debdou, *Morocco* **111 B4** 33 59N 3 0W
Dębica, *Poland* .... **55 H8** 50 2N 21 25 E
Dęblin, *Poland* .... **55 G8** 51 34N 21 50 E
Dębno, *Poland* .... **55 F1** 52 44N 14 41 E
Débo, L., *Mali* **112 B4** 15 14N 4 15W
Debolt, *Canada* **142 B5** 55 12N 118 1W
Deborah East, L., *Australia* **125 F2** 30 45 S 119 0 E
Deborah West, L., *Australia* **125 F2** 30 45 S 118 50 E
Debrc, *Serbia, Yug.* .... **50 B3** 44 38N 19 53 E
Debre Birhan, *Ethiopia* .... **107 F4** 9 41N 39 31 E
Debre Markos, *Ethiopia* ... **107 E4** 10 20N 37 40 E
Debre May, *Ethiopia* .... **107 E4** 11 20N 37 25 E
Debre Sina, *Ethiopia* .... **107 F4** 9 51N 39 50 E
Debre Tabor, *Ethiopia* ..... **107 E4** 11 50N 38 26 E
Debre Zebit, *Ethiopia* ..... **107 E4** 11 48N 38 30 E
Debrecen, *Hungary* .... **52 C6** 47 33N 21 42 E
Debrzno, *Poland* .... **54 E4** 53 31N 17 14 E
Dečani, *Serbia, Yug.* .... **50 D4** 42 30N 20 18 E
Decatur, *Ala., U.S.A.* **149 H2** 34 36N 86 59W
Decatur, *Ga., U.S.A.* **152 B5** 33 47N 84 18W
Decatur, *Ill., U.S.A.* **156 E8** 39 51N 88 57W
Decatur, *Ind., U.S.A.* **157 D12** 40 50N 84 56W
Decatur, *Mich., U.S.A.* **157 B11** 42 7N 85 58W
Decatur, *Tex., U.S.A.* **155 J6** 33 14N 97 35W
Decazeville, *France* .... **28 D6** 44 34N 2 15 E
Deccan, *India* .... **94 F4** 18 0N 79 0 E
Deception, Mt., *Australia* .. **128 A3** 30 42 S 138 16 E
Deception L., *Canada* **143 B8** 56 33N 104 13W
Dechang, *China* .... **76 D4** 27 25N 102 11 E
Děčín, *Czech Rep.* .... **34 A7** 50 47N 14 12 E
Decize, *France* .... **27 F10** 46 50N 3 28 E
Deckerville, *U.S.A.* **150 C2** 43 32N 82 44W
Decollatura, *Italy* .... **47 C9** 39 3N 16 21 E
Decorah, *U.S.A.* **154 D9** 43 18N 91 48W
Deda, *Romania* .... **53 D9** 46 56N 24 50 E
Dedaye, *Burma* .... **90 G5** 16 24N 95 53 E
Dedéagach =
Alexandroúpolis, *Greece* . **51 F9** 40 50N 25 54 E
Dedham, *U.S.A.* **151 D13** 42 15N 71 10W
Dédougou, *Burkina Faso* ... **112 C4** 12 30N 3 25W
Dedovichi, *Russia* .... **58 D5** 57 32N 29 56 E
Deduru Oya, *Sri Lanka* .... **95 L4** 7 32N 79 50 E
Dedza, *Malawi* .... **119 E3** 14 20 S 34 20 E
Dee ⟶, *Aberds., U.K.* **22 D6** 57 9N 2 5W
Dee ⟶, *Dumf. & Gall.,*
U.K. **22 G4** 54 51N 4 3W
Dee ⟶, *Wales, U.K.* **20 D4** 53 22N 3 17W
Deep B., *Canada* **142 A5** 61 15N 116 35W
Deep Lead, *Australia* ..... **128 D5** 37 0 S 142 43 E
Deep River, *U.S.A.* **156 C4** 41 35N 92 22W
Deep Well, *Australia* ..... **126 C1** 24 20 S 134 0 E
Deepstep, *U.S.A.* **152 B7** 33 1N 82 58W
Deepwater, *Australia* ..... **127 D5** 29 25 S 151 51 E
Deepwater, *U.S.A.* **156 F3** 38 16N 93 47W
Deer ⟶, *Canada* **143 B10** 58 23N 94 13W
Deer Lake, *Nfld., Canada* .. **141 C8** 49 11N 57 27W
Deer Lake, *Ont., Canada* .. **143 C10** 52 36N 94 20W
Deer Lodge, *U.S.A.* **158 C7** 46 24N 112 44W
Deer Park, *Fla., U.S.A.* ... **153 G9** 28 6N 80 54W
Deer Park, *Ohio, U.S.A.* .. **157 E12** 39 13N 84 23W
Deer Park, *Wash., U.S.A.* . **158 C5** 47 57N 117 28W
Deer River, *U.S.A.* **154 B8** 47 20N 93 48W
Deeral, *Australia* .... **126 B4** 17 14 S 145 55 E
Deerdepoort, *S. Africa* .... **116 C4** 24 37 S 26 27 E
Deerfield Beach, *U.S.A.* ... **153 J9** 26 19N 80 6W
Deferiet, *U.S.A.* **151 B9** 44 2N 75 41W
Defiance, *U.S.A.* **157 C12** 41 17N 84 22W
Dêgê, *China* .... **76 B2** 31 44N 98 39 E
Degebe ⟶, *Portugal* **43 G3** 38 13N 7 29W
Degeberga, *Sweden* .... **17 J8** 55 51N 14 5 E
Degeh Bur, *Ethiopia* .... **120 C2** 8 11N 43 31 E
Degema, *Nigeria* .... **113 E6** 4 50N 6 48 E
Degerfors, *Sweden* .... **16 E8** 59 15N 14 27 E
Degerhamn, *Sweden* .... **17 H10** 56 20N 16 24 E
Degersheim, *Switz.* .... **33 B8** 47 23N 9 12 E
Deggendorf, *Germany* ..... **31 G8** 48 50N 12 57 E
Değirmendere, *Turkey* ..... **51 F13** 40 42N 29 47 E
Deh Bīd, *Iran* .... **97 D7** 30 39N 53 11 E
Deh-e Shīr, *Iran* .... **97 D7** 31 29N 53 45 E
Dehaj, *Iran* .... **97 D7** 30 42N 54 53 E
Dehak, *Iran* .... **91 D1** 27 11N 62 37 E
Dehdez, *Iran* .... **97 D6** 31 43N 50 17 E
Dehestān, *Iran* .... **97 D7** 28 30N 55 35 E
Dehgolān, *Iran* .... **101 E12** 35 17N 47 25 E
Dehibat, *Tunisia* .... **108 B2** 32 0N 10 47 E
Dehiwala, *Sri Lanka* .... **95 L4** 6 50N 79 51 E
Dehlorān, *Iran* .... **101 F12** 32 41N 47 16 E
Dehnow-e Kūhestān, *Iran* .. **97 E8** 27 58N 58 32 E
Dehra Dun, *India* .... **92 D8** 30 20N 78 4 E
Dehri, *India* .... **93 G11** 24 50N 84 15 E
Dehua, *China* .... **77 E12** 25 26N 118 14 E
Dehui, *China* .... **75 B13** 44 30N 125 40 E
Deinze, *Belgium* .... **24 D3** 50 59N 3 32 E
Dej, *Romania* .... **53 C8** 47 10N 23 52 E
Deje, *Sweden* .... **16 E7** 59 35N 13 29 E
Dejiang, *China* .... **76 C7** 28 18N 108 7 E
Dekemhare, *Eritrea* .... **107 D4** 15 6N 39 0 E
Dekese,
Dem. Rep. of the Congo . **114 C4** 3 24 S 21 24 E
Dekhkanabad, *Uzbekistan* .. **63 D3** 38 21N 66 30 E
Dekoa, *C.A.R.* .... **114 A3** 6 19N 19 4 E
Del Carmen, *Phil.* .... **81 G6** 9 50N 126 0 E
Del Mar, *U.S.A.* **161 N9** 32 58N 117 16W
Del Norte, *U.S.A.* **159 H10** 37 41N 106 21W
Del Rio, *U.S.A.* **155 L4** 29 22N 100 54W
Delai, *Sudan* .... **106 D4** 17 21N 36 6 E
Delano, *U.S.A.* **161 K7** 35 46N 119 15W
Delareyville, *S. Africa* ..... **116 D4** 26 41 S 25 26 E
Delavan, *Ill., U.S.A.* **156 D7** 40 22N 89 33W
Delavan, *Wis., U.S.A.* **154 D10** 42 38N 88 39W
Delaware, *U.S.A.* **157 D13** 40 18N 83 4W
Delaware □, *U.S.A.* **148 F8** 39 0N 75 20W
Delaware ⟶, *U.S.A.* **148 F8** 39 15N 75 20W
Delaware B., *U.S.A.* **147 C12** 39 0N 75 10W
Delbrück, *Germany* .... **30 D4** 51 46N 8 34 E
Delčevo, *Macedonia* .... **50 E6** 41 58N 22 46 E
Delegate, *Australia* .... **129 D8** 37 4 S 148 56 E
Delémont, *Switz.* .... **32 B4** 47 22N 7 20 E
Delft, *Neths.* .... **24 B4** 52 1N 4 22 E
Delft I., *Sri Lanka* .... **95 K4** 9 30N 79 40 E
Delfzijl, *Neths.* .... **24 A6** 53 20N 6 55 E
Delgado, C., *Mozam.* .... **119 E5** 10 45 S 40 40 E
Delgerhet, *Mongolia* .... **74 B6** 45 50N 110 30 E
Delgo, *Sudan* .... **106 C3** 20 6N 30 40 E
Delhi, *Canada* **150 D4** 42 51N 80 30W
Delhi, *India* .... **92 E7** 28 38N 77 17 E
Delhi, *U.S.A.* **151 D10** 42 17N 74 55W
Deli Jovan, *Serbia, Yug.* ... **50 B6** 44 13N 22 9 E
Delia, *Canada* **142 C6** 51 38N 112 23W
Delice, *Turkey* .... **100 C6** 39 54N 34 2 E
Delice ⟶, *Turkey* .... **57 G5** 39 45N 34 15 E
Delicias, *Mexico* **162 B3** 28 10N 105 30W
Delījān, *Iran* .... **97 C6** 33 59N 50 40 E
Déline, *Canada* **138 B7** 65 10N 123 30W
Delitzsch, *Germany* .... **30 D8** 51 31N 12 20 E
Dell City, *U.S.A.* **159 L11** 31 56N 105 12W
Dell Rapids, *U.S.A.* **154 D6** 43 50N 96 43W
Delle, *France* .... **27 E14** 47 30N 7 2 E
Dellys, *Algeria* .... **111 A5** 36 57N 3 57 E
Delmar, *Iowa, U.S.A.* **156 C6** 42 0N 90 37W
Delmar, *N.Y., U.S.A.* **151 D11** 42 37N 73 47W
Delmenhorst, *Germany* ..... **30 B4** 53 3N 8 37 E
Delmiro Gouveia, *Brazil* ... **170 C4** 9 24 S 38 6W
Delnice, *Croatia* .... **45 C11** 45 23N 14 50 E
Delong, Ostrova, *Russia* ... **65 B15** 76 40N 149 20 E
Deloraine, *Australia* .... **126 G4** 41 30 S 146 40 E
Deloraine, *Canada* **143 D8** 49 15N 100 29W
Delphi, *Greece* .... **48 C4** 38 28N 22 30 E
Delphi, *U.S.A.* **157 D10** 40 36N 86 41W
Delphos, *U.S.A.* **157 D12** 40 51N 84 21W
Delportshoop, *S. Africa* .... **116 D3** 28 22 S 24 20 E
Delray Beach, *U.S.A.* **149 M5** 26 28N 80 4W
Delsbo, *Sweden* .... **16 C10** 61 48N 16 32 E
Delta, *Ala., U.S.A.* **152 B4** 33 26N 85 42W
Delta, *Colo., U.S.A.* **159 G9** 38 44N 108 4W
Delta, *Utah, U.S.A.* **158 G7** 39 21N 112 35W
Delta □, *Nigeria* .... **113 D6** 5 30N 6 0 E
Delta Amacuro □, *Venezuela* **169 B5** 8 30N 61 30W
Delta Junction, *U.S.A.* **144 D11** 64 2N 145 44W
Deltona, *U.S.A.* **153 G8** 28 54N 81 16W
Delungra, *Australia* .... **127 D5** 29 39 S 150 51 E
Delvinákion, *Greece* .... **48 B2** 39 57N 20 32 E
Delvinë, *Albania* .... **50 G4** 39 59N 20 6 E
Demak, *Indonesia* .... **85 D4** 6 53 S 110 38 E
Demanda, Sierra de la, *Spain* **40 C2** 42 15N 3 0W
Demavand = Damāvand,
Iran .... **97 C7** 35 47N 52 0 E
Demba,
Dem. Rep. of the Congo . **115 D4** 5 28 S 22 15 E
Demba Chio, *Angola* .... **115 D2** 9 41 S 13 41 E
Dembecha, *Ethiopia* .... **107 E4** 10 32N 37 30 E
Dembi, *Ethiopia* .... **107 F4** 8 5N 36 25 E
Dembia,
Dem. Rep. of the Congo . **118 B2** 3 33N 25 48 E
Dembidolo, *Ethiopia* .... **107 F3** 8 34N 34 50 E
Demer ⟶, *Belgium* .... **24 D4** 50 57N 4 42 E
Demetrias, *Greece* .... **48 B5** 39 22N 23 1 E
Demidov, *Russia* .... **58 E6** 55 16N 31 30 E
Deming, *N. Mex., U.S.A.* .. **159 K10** 32 16N 107 46W
Deming, *Wash., U.S.A.* **160 B4** 48 50N 122 13W
Demini ⟶, *Brazil* .... **169 D5** 0 46 S 62 56W
Demirci, *Turkey* .... **49 B10** 39 2N 28 38 E
Demirköprü Barajı, *Turkey* . **49 C10** 38 42N 28 25 E
Demirköy, *Turkey* .... **51 E11** 41 49N 27 45 E
Demmin, *Germany* .... **30 B9** 53 54N 13 2 E
Demnate, *Morocco* **110 B3** 31 44N 6 59W
Demonte, *Italy* .... **44 D4** 44 19N 7 17 E
Demopolis, *U.S.A.* **149 J2** 32 31N 87 50W
Dempo, *Indonesia* .... **84 C2** 4 2 S 103 15 E
Demyansk, *Russia* .... **58 D7** 57 40N 32 27 E
Den Burg, *Neths.* .... **24 A4** 53 3N 4 47 E
Den Chai, *Thailand* .... **86 D3** 17 59N 100 4 E
Den Haag = 's-Gravenhage,
Neths. .... **24 B4** 52 7N 4 17 E
Den Helder, *Neths.* .... **24 B4** 52 57N 4 45 E
Den Oever, *Neths.* .... **24 B5** 52 56N 5 2 E
Denain, *France* .... **27 B10** 50 20N 3 22 E
Denair, *U.S.A.* **160 H6** 37 32N 120 48W
Denau, *Uzbekistan* .... **63 D3** 38 16N 67 54 E
Denbigh, *U.K.* **20 D4** 53 12N 3 25W
Denbigh, C., *U.S.A.* **144 D7** 64 23N 161 32W
Denbighshire □, *U.K.* **20 D4** 53 8N 3 22W
Dendang, *Indonesia* .... **85 C3** 3 7 S 107 56 E
Dendermonde, *Belgium* ..... **24 C4** 51 2N 4 5 E
Deneba, *Ethiopia* .... **107 F4** 9 47N 39 10 E
Denezhkin Kamen, Gora,
Russia .... **62 A7** 60 25N 59 32 E
Deng Deng, *Cameroon* ..... **114 A2** 5 12N 13 31 E
Deng Xian, *China* .... **77 A9** 32 34N 112 4 E
Dengchuan, *China* .... **76 E3** 25 59N 100 3 E
Denge, *Nigeria* .... **113 C6** 12 52N 5 21 E
Dengfeng, *China* .... **74 G7** 34 25N 113 2 E
Dengi, *Nigeria* .... **113 D6** 9 25N 9 55 E
Dengkou, *China* .... **74 D4** 40 18N 106 55 E

Djebiniana 207

Denham, *Australia* 125 E1 25 56 S 113 31 E
Denham Ra., *Australia* 126 C4 21 55 S 147 46 E
Denham Sd., *Australia* 125 E1 25 45 S 113 15 E
Denia, *Spain* 41 G5 38 49N 0 8 E
Denial B., *Australia* 127 E1 32 14 S 133 32 E
Deniliquin, *Australia* 129 C6 35 30 S 144 58 E
Denison, *Iowa, U.S.A.* 154 E7 42 1N 95 21W
Denison, *Tex., U.S.A.* 155 J6 33 45N 96 33W
Denison Plains, *Australia* 124 C4 18 35 S 128 0 E
Denizli, *Turkey* 57 G4 37 42N 29 2 E
Denizli □, *Turkey* 49 D11 37 45N 29 5 E
Denman, *Australia* 129 B9 32 24 S 150 42 E
Denman Glacier, *Antarctica* 7 C7 66 45 S 99 25 E
Denmark, *Australia* 125 F2 34 59 S 117 25 E
Denmark, *U.S.A.* 152 B9 33 19N 81 9W
Denmark ■, *Europe* 17 J3 55 45N 10 0 E
Denmark Str., *Atl. Oc.* 8 B6 66 0N 30 0W
Dennison, *U.S.A.* 150 F3 40 24N 81 19W
Denny, *U.K.* 22 E5 56 1N 3 55W
Denpasar, *Indonesia* 85 D5 8 45 S 115 14 E
Denton, *Ga., U.S.A.* 152 D7 31 44N 82 42W
Denton, *Mont., U.S.A.* 158 C9 47 19N 109 57W
Denton, *Tex., U.S.A.* 155 J6 33 13N 97 8W
D'Entrecasteaux, Pt., *Australia* 125 F2 34 50 S 115 57 E
D'Entrecasteaux Is., *Papua N. G.* 132 E6 9 0 S 151 0 E
Dents du Midi, *Switz.* 32 D3 46 10N 6 56 E
Dentsville, *U.S.A.* 152 A9 34 4N 80 58W
Denu, *Ghana* 113 D5 6 4N 1 8 E
Denver, *Colo., U.S.A.* 154 F2 39 44N 104 59W
Denver, *Ind., U.S.A.* 157 D10 40 52N 86 5W
Denver, *Iowa, U.S.A.* 156 B4 42 40N 92 20W
Denver City, *U.S.A.* 155 J3 32 58N 102 50W
Deoband, *India* 92 E7 29 42N 77 43 E
Deobhog, *India* 94 E6 19 53N 82 44 E
Deogarh, *India* 94 D7 21 32N 84 45 E
Deoghar, *India* 93 G12 24 30N 86 42 E
Deolali, *India* 94 E1 19 58N 73 50 E
Deoli = Devli, *India* 92 G6 25 50N 75 20 E
Déols, *France* 27 F8 46 50N 1 43 E
Deoria, *India* 93 F10 26 31N 83 48 E
Deosai Mts., *Pakistan* 93 B6 35 40N 75 0 E
Deping, *China* 75 F9 37 25N 116 58 E
Deposit, *U.S.A.* 151 D9 42 4N 75 25W
Depot Springs, *Australia* 125 E3 27 55 S 120 3 E
Deputatskiy, *Russia* 65 C14 69 18N 139 54 E
Dêqên, *China* 76 C2 28 34N 98 51 E
Deqing, *China* 77 F8 23 8N 111 42 E
Dera Ghazi Khan, *Pakistan* 91 C3 30 5N 70 43 E
Dera Ismail Khan, *Pakistan* 91 C3 31 50N 70 50 E
Derbent, *Russia* 61 J9 42 5N 48 15 E
Derbent, *Turkey* 49 C10 38 11N 28 33 E
Derby, *Australia* 124 C3 17 18 S 123 38 E
Derby, *U.K.* 20 E6 52 56N 1 28W
Derby, *Conn., U.S.A.* 151 E11 41 19N 73 5W
Derby, *N.Y., U.S.A.* 150 D6 42 41N 78 58W
Derby City □, *U.K.* 20 E6 52 56N 1 28W
Derbyshire □, *U.K.* 20 D6 53 11N 1 38W
Derecske, *Hungary* 52 C6 47 20N 21 33 E
Dereköy, *Turkey* 51 E11 41 55N 27 21 E
Dereli, *Turkey* 101 B8 40 44N 38 26 E
Derg →, *U.K.* 23 B4 54 44N 7 26W
Derg, L., *Ireland* 23 D3 53 0N 8 20W
Dergachi = Derhaci, *Ukraine* 59 G9 50 9N 36 11 E
Derhaci, *Ukraine* 59 G9 50 9N 36 11 E
Derik, *Turkey* 101 D9 37 21N 40 18 E
Derinkuyu, *Turkey* 100 C6 38 22N 34 45 E
Dermantsi, *Bulgaria* 51 C8 43 8N 24 17 E
Dernieres, Isles, *U.S.A.* 155 L9 29 2N 90 50W
Dêrong, *China* 76 C2 28 44N 99 9 E
Derrinallum, *Australia* 128 D5 37 57 S 143 15 E
Derry = Londonderry, *U.K.* 23 B4 55 0N 7 20W
Derry = Londonderry □, *U.K.* 23 B4 55 0N 7 20W
Derryveagh Mts., *Ireland* 23 B3 54 56N 8 11W
Derudub, *Sudan* 106 D4 17 31N 36 7 E
Derval, *France* 26 E5 47 40N 1 41W
Dervéni, *Greece* 48 C4 38 8N 22 25 E
Derventa, *Bos.-H.* 52 F2 44 59N 17 55 E
Derwent, *Canada* 143 C6 53 41N 110 58W
Derwent →, *Cumb., U.K.* 20 C4 54 39N 3 33W
Derwent →, *Derby, U.K.* 20 E6 52 57N 1 28W
Derwent →, *N. Yorks., U.K.* 20 D7 53 45N 0 58W
Derwent Water, *U.K.* 20 C4 54 35N 3 9W
Des Moines, *Iowa, U.S.A.* 156 C3 41 35N 93 37W
Des Moines, *N. Mex., U.S.A.* 155 G3 36 46N 103 50W
Des Moines →, *U.S.A.* 154 E9 40 23N 91 25W
Des Plaines, *U.S.A.* 157 B9 42 3N 87 52W
Des Plaines →, *U.S.A.* 157 C8 41 23N 88 15W
Desa, *Romania* 52 G8 43 52N 23 2 E
Desaguadero →, *Argentina* 174 C2 34 30 S 66 46W
Desaguadero →, *Bolivia* 172 D4 16 35 S 69 5W
Descanso, Pta., *Mexico* 161 N9 32 21N 117 3W
Descartes, *France* 28 B4 46 59N 0 42 E
Deschaillons, *Canada* 141 C5 46 32N 72 7W
Descharme →, *Canada* 143 B7 56 51N 109 13W
Deschutes →, *U.S.A.* 158 D3 45 38N 120 55W
Dese, *Ethiopia* 107 E4 11 5N 39 40 E
Deseado, C., *Chile* 176 D2 52 45 S 74 42W
Desenzano del Garda, *Italy* 44 C7 45 28N 10 32 E
Desert Center, *U.S.A.* 161 M11 33 43N 115 24W
Desert Hot Springs, *U.S.A.* 161 M10 33 58N 116 30W
Deset, *Norway* 18 C8 61 20N 11 26 E
Désirade, I., *Guadeloupe* 165 C7 16 18N 61 3W
Deskenatlata L., *Canada* 142 A6 60 55N 112 3W
Desna →, *Ukraine* 59 G6 50 33N 30 32 E
Desnătui →, *Romania* 53 G8 43 53N 23 35 E
Desolación, I., *Chile* 176 D2 53 0 S 74 0W
Despeñaperros, Paso, *Spain* 43 G7 38 24N 3 30W
Despotovac, *Serbia, Yug.* 50 B5 44 6N 21 30 E
Dessau, *Germany* 30 D8 51 51N 12 14 E
Dessye = Dese, *Ethiopia* 107 E4 11 5N 39 40 E
Destin, *U.S.A.* 153 E3 30 24N 86 30W
D'Estrees B., *Australia* 128 C2 35 55 S 137 45 E
Desuri, *India* 92 G5 25 18N 73 35 E
Desvres, *France* 27 B8 50 40N 1 48 E
Det Udom, *Thailand* 86 E5 14 54N 105 5 E
Deta, *Romania* 52 E6 45 24N 21 13 E
Dete, *Zimbabwe* 119 F2 18 38 S 26 50 E
Đetinja →, *Serbia, Yug.* 50 C4 43 51N 20 5 E
Detmold, *Germany* 30 D4 51 56N 8 52 E
Detour, Pt., *U.S.A.* 148 C2 45 40N 86 40W
Detroit, *Mich., U.S.A.* 140 D3 42 20N 83 3W
Detroit, *Tex., U.S.A.* 155 J7 33 40N 95 16W
Detroit Lakes, *U.S.A.* 154 B7 46 49N 95 51W
Detva, *Slovak Rep.* 35 C12 48 34N 19 25 E
Deurne, *Neths.* 24 C5 51 27N 5 49 E
Deutsche Bucht, *Germany* 30 A4 54 15N 8 0 E
Deutschlandsberg, *Austria* 34 E8 46 49N 15 14 E
Deux-Sèvres □, *France* 26 F6 46 35N 0 20W
Deva, *Romania* 52 E7 45 53N 22 55 E
Devakottai, *India* 95 K4 9 55N 78 45 E
Devaprayag, *India* 93 D8 30 13N 78 35 E
Dévaványa, *Hungary* 52 C5 47 2N 20 59 E
Deveci Dağları, *Turkey* 100 B7 40 6N 36 15 E
Devecikonağı, *Turkey* 51 G12 39 55N 28 34 E
Devecser, *Hungary* 52 C2 47 6N 17 26 E
Develi, *Turkey* 100 C6 38 23N 35 29 E
Deventer, *Neths.* 24 B6 52 15N 6 10 E
Devereux, *U.S.A.* 152 B6 33 13N 83 5W
Deveron →, *U.K.* 22 D6 57 41N 2 32W
Devesel, *Romania* 52 F7 44 28N 22 41 E
Devgad I., *India* 95 G2 14 48N 74 5 E
Devgadh Bariya, *India* 92 H5 22 40N 73 55 E
Devil River Pk., *N.Z.* 131 A7 40 56 S 172 37 E
Devils Den, *U.S.A.* 160 K7 35 46N 119 58W
Devils Lake, *U.S.A.* 154 A5 48 7N 98 52W
Devils Paw, *Canada* 142 B2 58 47N 134 0W
Devil's Pt., *Sri Lanka* 95 K5 9 26N 80 6 E
Devil's Pt., *Vanuatu* 133 G6 17 44 S 168 11 E
Devin, *Bulgaria* 51 E8 41 44N 24 24 E
Devizes, *U.K.* 21 F6 51 22N 1 58W
Devli, *India* 92 G6 25 50N 75 20 E
Devnya, *Bulgaria* 51 C11 43 13N 27 33 E
Devolii →, *Albania* 50 F4 40 57N 20 15 E
Devon, *Canada* 142 C6 53 24N 113 44W
Devon □, *U.K.* 21 G4 50 50N 3 40W
Devon I., *Canada* 6 B3 75 10N 85 0W
Devonport, *Australia* 126 G4 41 10 S 146 22 E
Devonport, *N.Z.* 130 C3 36 49 S 174 49 E
Devrek, *Turkey* 100 B4 41 13N 31 57 E
Devrekâni, *Turkey* 100 B5 41 36N 33 50 E
Devrez →, *Turkey* 100 B6 41 6N 34 25 E
Dewas, *India* 92 H7 22 59N 76 3 E
Dewetsdorp, *S. Africa* 116 D4 29 33 S 26 39 E
Dewey Rose, *U.S.A.* 152 A7 34 10N 82 57W
Dexing, *China* 77 C11 28 46N 117 30 E
Dexter, *Ga., U.S.A.* 152 C6 32 27N 83 4W
Dexter, *Mich., U.S.A.* 157 B13 42 20N 83 53W
Dexter, *Mo., U.S.A.* 155 G10 36 48N 89 57W
Dexter, *N. Mex., U.S.A.* 155 J2 33 12N 104 22W
Dey-Dey, L., *Australia* 125 E5 29 12 S 131 4 E
Deyang, *China* 76 B5 31 3N 104 27 E
Deyhūk, *Iran* 97 C8 33 15N 57 30 E
Deyyer, *Iran* 97 E6 27 55N 51 55 E
Dezadeash L., *Canada* 142 A1 60 28N 136 58W
Dezfūl, *Iran* 97 C6 32 20N 48 30 E
Dezhneva, Mys, *Russia* 65 C19 66 5N 169 40W
Dezhou, *China* 74 F9 37 26N 116 18 E
Dháfni, *Kríti, Greece* 38 D7 35 13N 25 3 E
Dháfni, *Pelóponnisos, Greece* 48 D4 37 48N 22 1 E
Dhahaban, *Si. Arabia* 98 B2 21 58N 39 3 E
Dhahiriya = Aẓ Ẓāhirīyah, *West Bank* 103 D3 31 25N 34 58 E
Dhahran = Aẓ Ẓahrān, *Si. Arabia* 97 E6 26 10N 50 7 E
Dhaka, *Bangla.* 90 D3 23 43N 90 26 E
Dhaka □, *Bangla.* 90 C3 24 25N 90 25 E
Dhali, *Cyprus* 38 D12 35 1N 33 25 E
Dhamangaon, *India* 94 D4 20 45N 78 15 E
Dhamar, *Yemen* 98 D4 14 30N 44 20 E
Dhamási, *Greece* 48 B4 39 43N 22 11 E
Dhampur, *India* 93 E8 29 19N 78 33 E
Dhamtari, *India* 94 D5 20 42N 81 35 E
Dhanbad, *India* 93 H12 23 50N 86 30 E
Dhankuta, *Nepal* 93 F12 26 55N 87 40 E
Dhanora, *India* 94 D5 20 20N 80 22 E
Dhar, *India* 92 H6 22 35N 75 26 E
Dharampur, *Gujarat, India* 94 D1 20 32N 73 17 E
Dharampur, *Mad. P., India* 92 H6 22 13N 75 18 E
Dharamsala = Dharmsala, *India* 92 C7 32 16N 76 23 E
Dharapuram, *India* 95 J3 10 45N 77 34 E
Dharmapuri, *India* 95 H4 12 10N 78 10 E
Dharmavaram, *India* 95 G3 14 29N 77 44 E
Dharmsala, *India* 92 C7 32 16N 76 23 E
Dharwad, *India* 95 G2 15 22N 75 15 E
Dhaulagiri, *Nepal* 93 E10 28 39N 83 28 E
Dhebar, L., *India* 92 G6 24 10N 74 0 E
Dheftera, *Cyprus* 38 D12 35 5N 33 16 E
Dhenkanal, *India* 94 D7 20 45N 85 35 E
Dhenoúsa, *Greece* 49 D7 37 8N 25 48 E
Dherinia, *Cyprus* 38 D12 35 3N 33 57 E
Dheskáti, *Greece* 50 G5 39 55N 21 49 E
Dhespotikó, *Greece* 48 E6 36 57N 24 58 E
Dhestina, *Greece* 48 C4 38 25N 22 31 E
Dhiarrizos →, *Cyprus* 38 E11 34 41N 32 34 E
Dhībān, *Jordan* 103 D4 31 30N 35 46 E
Dhidhimótikhon, *Greece* 51 E10 41 22N 26 29 E
Dhíkti Óros, *Greece* 38 D7 35 8N 25 30 E
Dhilianáta, *Greece* 48 C2 38 15N 20 34 E
Dhílos, *Greece* 49 D7 37 23N 25 15 E
Dhimitsána, *Greece* 48 D4 37 36N 22 3 E
Dhírfis = Dhírfis Óros, *Greece* 48 C5 38 40N 23 54 E
Dhírfis Óros, *Greece* 48 C5 38 40N 23 54 E
Dhodhekánisos, *Greece* 49 E8 36 35N 27 0 E
Dhodhekánisos □, *Greece* 49 E8 36 35N 27 0 E
Dhokós, *Greece* 48 D5 37 20N 23 20 E
Dholiana, *Greece* 48 B2 39 54N 20 32 E
Dholka, *India* 92 H5 22 44N 72 29 E
Dhomokós, *Greece* 48 B4 39 10N 22 18 E
Dhoraji, *India* 92 J4 21 45N 70 37 E
Dhoxáton, *Greece* 51 E8 41 9N 24 16 E
Dhragonísi, *Greece* 49 D7 37 27N 25 29 E
Dhráhstis, Ákra, *Greece* 38 A3 39 48N 19 40 E
Dhrangadhra, *India* 92 H4 22 59N 71 31 E
Dhrápanon, Ákra, *Greece* 38 D6 35 28N 24 14 E
Dhriopís, *Greece* 48 D6 37 25N 24 26 E
Dhrol, *India* 92 H4 22 33N 70 25 E
Dhubāb, *Yemen* 98 D3 12 56N 43 25 E
Dhuburi, *India* 90 B2 26 2N 89 59 E
Dhulasar, *Bangla.* 90 E3 21 52N 90 14 E
Dhule, *India* 94 D2 20 58N 74 50 E
Dhupdhara, *India* 90 B3 26 10N 91 4 E
Di Linh, *Vietnam* 87 G7 11 35N 108 4 E
Di Linh, Cao Nguyen, *Vietnam* 87 G7 11 30N 108 0 E
Día, *Greece* 38 D7 35 28N 25 14 E
Diablo, Mt., *U.S.A.* 160 H5 37 53N 121 56W
Diablo Range, *U.S.A.* 160 J5 37 20N 121 25W
Diafarabé, *Mali* 112 C4 14 9N 4 57W
Diagonal, *U.S.A.* 156 D2 40 49N 94 20W
Diala, *Mali* 112 C3 14 10N 9 58W
Dialakoro, *Mali* 112 C3 12 18N 7 54W
Diallassagou, *Mali* 112 C4 13 47N 3 41W
Diamante, *Argentina* 174 C3 32 5 S 60 40W
Diamante, *Italy* 47 C8 39 41N 15 49 E
Diamante →, *Argentina* 174 C2 34 30 S 66 46W
Diamantina, *Brazil* 171 E3 18 17 S 43 40W
Diamantina →, *Australia* 127 D2 26 45 S 139 10 E
Diamantino, *Brazil* 173 C6 14 30 S 56 30W
Diamond Bar, *U.S.A.* 161 L9 34 1N 117 48W
Diamond Harbour, *India* 93 H13 22 11N 88 14 E
Diamond Head, *U.S.A.* 145 K14 21 16N 157 49W
Diamond Is., *Australia* 126 B5 17 25 S 151 5 E
Diamond Mts., *U.S.A.* 158 G6 39 50N 115 30W
Diamond Springs, *U.S.A.* 160 G6 38 42N 120 49W
Diamondville, *U.S.A.* 158 F8 41 47N 110 32W
Dianalund, *Denmark* 17 J5 55 32N 11 30 E
Dianbai, *China* 77 G8 21 33N 111 0 E
Diancheng, *China* 77 G8 21 30N 111 4 E
Diano Marina, *Italy* 44 E5 43 54N 8 5 E
Dianópolis, *Brazil* 171 D2 11 38 S 46 50W
Dianra, *Ivory C.* 112 D3 8 45N 6 14W
Diapaga, *Burkina Faso* 113 C5 12 5N 1 46 E
Diapangou, *Burkina Faso* 113 C5 12 5N 0 10 E
Diapur, *Australia* 128 D4 36 19 S 141 29 E
Diariguila, *Guinea* 112 C2 10 35N 10 2W
Dībā, *Oman* 97 E8 25 45N 56 16 E
Dibaya, *Dem. Rep. of the Congo* 115 D4 6 30 S 22 57 E
Dibaya-Lubue, *Dem. Rep. of the Congo* 115 C3 4 12 S 19 54 E
Dibbi, *Ethiopia* 107 G5 4 10N 41 52 E
Dibete, *Botswana* 116 C4 23 45 S 26 32 E
Dibrugarh, *India* 90 B5 27 29N 94 55 E
Dickeyville, *U.S.A.* 156 B6 42 38N 90 36W
Dickinson, *U.S.A.* 154 B3 46 53N 102 47W
Dickson = Dikson, *Russia* 64 B9 73 40N 80 5 E
Dickson, *U.S.A.* 149 G2 36 5N 87 23W
Dickson City, *U.S.A.* 151 E9 41 29N 75 40W
Dicle Nehri →, *Turkey* 101 D9 37 44N 41 10 E
Dicomano, *Italy* 45 E8 43 53N 11 31 E
Didesa, W. →, *Ethiopia* 107 E4 10 2N 35 32 E
Didiéni, *Mali* 112 C3 13 53N 8 6W
Didsbury, *Canada* 142 C6 51 35N 114 10W
Didwana, *India* 92 F6 27 23N 74 36 E
Die, *France* 29 D9 44 47N 5 22 E
Diébougou, *Burkina Faso* 112 C4 11 0N 3 15W
Diefenbaker L., *Canada* 143 C7 51 0N 106 55W
Diego Garcia, *Ind. Oc.* 121 E6 7 50 S 72 50 E
Diekirch, *Lux.* 24 E6 49 52N 6 10 E
Diéma, *Mali* 112 C3 14 32N 9 12W
Diembéring, *Senegal* 112 C1 12 29N 16 47W
Dien Ban, *Vietnam* 86 E7 15 53N 108 16 E
Dien Bien, *Vietnam* 86 B4 21 20N 103 0 E
Dien Khanh, *Vietnam* 87 F7 12 15N 109 6 E
Diepholz, *Germany* 30 C4 52 37N 8 22 E
Diepoldsau, *Switz.* 33 B9 47 23N 9 40 E
Dieppe, *France* 26 C8 49 54N 1 4 E
Dierks, *U.S.A.* 155 H8 34 7N 94 1W
Diessenhofen, *Switz.* 33 A7 47 42N 8 46 E
Diest, *Belgium* 24 D5 50 58N 5 4 E
Dieterich, *U.S.A.* 157 E8 39 4N 88 23W
Dietikon, *Switz.* 33 B6 47 24N 8 24 E
Dieulefit, *France* 29 D9 44 32N 5 4 E
Dieuze, *France* 27 D13 48 49N 6 43 E
Differdange, *Lux.* 24 E5 49 31N 5 54 E
Diffun, *Phil.* 80 C3 16 36N 121 33 E
Dig, *India* 92 F7 27 28N 77 20 E
Digba, *Dem. Rep. of the Congo* 118 B2 4 25N 25 48 E
Digboi, *India* 90 B5 27 23N 95 38 E
Digby, *Canada* 141 D6 44 38N 65 50W
Digges, *Canada* 143 B10 58 40N 94 0W
Dighinala, *Bangla.* 90 D4 23 15N 92 5 E
Dighton, *U.S.A.* 154 F4 38 29N 100 28W
Diglur, *India* 94 E3 18 34N 77 33 E
Digne-les-Bains, *France* 29 D10 44 5N 6 12 E
Digoin, *France* 27 F11 46 29N 4 1 E
Digor, *Turkey* 101 B10 40 22N 43 25 E
Digos, *Phil.* 81 H5 6 45N 125 20 E
Digranes, *Iceland* 11 A12 66 4N 14 44W
Digras, *India* 94 D3 20 6N 77 45 E
Digul →, *Indonesia* 83 C5 7 7 S 138 42 E
Dīhōk, *Iraq* 101 D10 36 50N 43 1 E
Dijlah, Nahr →, *Asia* 96 D5 31 0N 47 25 E
Dijon, *France* 27 E12 47 20N 5 3 E
Dikala, *Sudan* 107 G3 4 45N 31 28 E
Dikili, *Turkey* 49 B8 39 4N 26 53 E
Dikkil, *Djibouti* 107 E5 11 8N 42 20 E
Dikomu di Kai, *Botswana* 116 C3 24 58 S 24 36 E
Diksmuide, *Belgium* 24 C2 51 2N 2 52 E
Dikson, *Russia* 64 B9 73 40N 80 5 E
Dikwa, *Nigeria* 113 C7 12 4N 13 30 E
Dila, *Ethiopia* 107 F4 6 21N 38 22 E
Dili, *Indonesia* 82 C3 8 39 S 125 34 E
Dilijan, *Armenia* 61 K7 40 46N 44 57 E
Dilizhan = Dilijan, *Armenia* 61 K7 40 46N 44 57 E
Dilj, *Croatia* 52 E3 45 29N 18 1 E
Dillard, *U.S.A.* 156 G5 37 44N 91 13W
Dillenburg, *Germany* 30 E4 50 43N 8 17 E
Dilley, *U.S.A.* 155 L5 28 40N 99 10W
Dilling, *Sudan* 107 E2 12 3N 29 35 E
Dillingen, *Bayern, Germany* 31 G6 48 36N 10 30 E
Dillingen, *Saarland, Germany* 31 F2 49 22N 6 43 E
Dillingham, *U.S.A.* 138 C4 59 3N 158 28W
Dillon, *Canada* 143 B7 55 56N 108 35W
Dillon, *Mont., U.S.A.* 158 D7 45 13N 112 38W
Dillon, *S.C., U.S.A.* 149 H6 34 25N 79 22W
Dillon →, *Canada* 143 B7 55 56N 108 56W
Dillsboro, *U.S.A.* 157 E11 39 1N 85 4W
Dilolo, *Dem. Rep. of the Congo* 115 E4 10 28 S 22 18 E
Dilston, *Australia* 126 G4 41 22 S 147 10 E
Dimapur, *India* 90 C4 25 54N 93 45 E
Dimas, *Mexico* 162 C3 23 43N 106 47W
Dimasalang, *Phil.* 80 E4 12 12N 123 51 E
Dimashq, *Syria* 103 B5 33 30N 36 18 E
Dimashq □, *Syria* 103 B5 33 30N 36 30 E
Dimbaza, *S. Africa* 117 E4 32 50 S 27 14 E
Dimbelenge, *Dem. Rep. of the Congo* 115 D4 5 33 S 23 7 E
Dimbokro, *Ivory C.* 112 D4 6 45N 4 46W
Dimboola, *Australia* 128 D5 36 28 S 142 7 E
Dîmboviţa = Dâmboviţa →, *Romania* 53 F11 44 12N 26 26 E
Dimbulah, *Australia* 126 B4 17 8 S 145 4 E
Dimitrovgrad, *Bulgaria* 51 D9 42 5N 25 35 E
Dimitrovgrad, *Russia* 60 C9 54 14N 49 39 E
Dimitrovgrad, *Serbia, Yug.* 50 C6 43 2N 22 48 E
Dimitrovo = Pernik, *Bulgaria* 50 D7 42 35N 23 2 E
Dimmitt, *U.S.A.* 155 H3 34 33N 102 19W
Dimo, *Sudan* 107 F2 5 19N 29 10 E
Dimona, *Israel* 103 D4 31 2N 35 1 E
Dimovo, *Bulgaria* 50 C6 43 43N 22 50 E
Dinagat, *Phil.* 81 F5 10 10N 125 40 E
Dinaig, *Phil.* 81 H5 7 11N 124 10 E
Dinajpur, *Bangla.* 90 C2 25 33N 88 43 E
Dinalupihan, *Phil.* 80 D3 14 52N 120 28 E
Dinan, *France* 26 D4 48 28N 2 2W
Dīnān Āb, *Iran* 97 C8 32 4N 56 49 E
Dinant, *Belgium* 24 D4 50 16N 4 55 E
Dinapur, *India* 93 G11 25 38N 85 5 E
Dinar, *Turkey* 49 C12 38 5N 30 10 E
Dīnār, Kūh-e, *Iran* 97 D6 30 42N 51 46 E
Dinara Planina, *Croatia* 45 D13 44 0N 16 30 E
Dinard, *France* 26 D4 48 38N 2 6W
Dinaric Alps = Dinara Planina, *Croatia* 45 D13 44 0N 16 30 E
Dinas, *Phil.* 81 H4 7 38N 123 20 E
Dinder, Nahr ed →, *Sudan* 107 E3 14 6N 33 40 E
Dindi →, *India* 95 F4 16 24N 78 15 E
Dindigul, *India* 95 J4 10 25N 78 0 E
Ding Xian, *China* 74 E8 38 30N 114 59 E
Dingalan, *Phil.* 80 D3 15 18N 121 25 E
Dingalan Bay, *Phil.* 80 D3 15 18N 121 25 E
Dingbian, *China* 74 F4 37 35N 107 32 E
Dingelstädt, *Germany* 30 D6 51 18N 10 19 E
Dinghai, *China* 77 B14 30 1N 122 6 E
Dingle, *Ireland* 23 D1 52 9N 10 17W
Dingle, *Sweden* 17 F5 58 32N 11 35 E
Dingle B., *Ireland* 23 D1 52 3N 10 20W
Dingmans Ferry, *U.S.A.* 151 E10 41 13N 74 55W
Dingnan, *China* 77 E10 24 45N 115 0 E
Dingo, *Australia* 126 C4 23 38 S 149 19 E
Dingolfing, *Germany* 31 G8 48 37N 12 30 E
Dingras, *Phil.* 80 B3 18 6N 120 42 E
Dingtao, *China* 74 G8 35 5N 115 35 E
Dinguiraye, *Guinea* 112 C2 11 18N 10 49W
Dingwall, *U.K.* 22 D4 57 36N 4 26W
Dingxi, *China* 74 G3 35 30N 104 33 E
Dingxiang, *China* 74 E7 38 30N 112 58 E
Dingyuan, *China* 77 A11 32 32N 117 41 E
Dinh, Mui, *Vietnam* 87 G7 11 22N 109 1 E
Dinh Lap, *Vietnam* 86 B6 21 33N 107 6 E
Dinhata, *India* 90 B2 26 8N 89 27 E
Dinokwe, *Botswana* 116 C4 23 29 S 26 37 E
Dinosaur National Monument, *U.S.A.* 158 F9 40 30N 108 45W
Dinsor, *Somali Rep.* 120 D2 2 24N 42 59 E
Dinuba, *U.S.A.* 160 J7 36 32N 119 23W
Diö, *Sweden* 17 H8 56 37N 14 15 E
Diomede, *U.S.A.* 144 D5 65 47N 169 0W
Diósgyör, *Hungary* 52 B5 48 7N 20 43 E
Diosig, *Romania* 52 C7 47 18N 22 2 E
Diourbel, *Senegal* 112 C1 14 39N 16 12W
Dipaculao, *Phil.* 80 D3 15 51N 121 32 E
Diphu Pass, *India* 90 A6 28 9N 97 20 E
Diplo, *Pakistan* 92 G3 24 35N 69 35 E
Dipolog, *Phil.* 81 G4 8 36N 123 20 E
Dipton, *N.Z.* 131 F3 45 54 S 168 22 E
Dir, *Pakistan* 91 B3 35 8N 71 59 E
Diré, *Mali* 112 B4 16 20N 3 25W
Dire Dawa, *Ethiopia* 107 F5 9 35N 41 45 E
Diriamba, *Nic.* 164 D2 11 51N 86 19W
Dirico, *Angola* 115 F4 17 50 S 20 42 E
Dirk Hartog I., *Australia* 125 E1 25 50 S 113 5 E
Dirkou, *Niger* 109 E2 19 1N 12 53 E
Dirranbandi, *Australia* 127 D4 28 33 S 148 17 E
Dirs, *Si. Arabia* 98 C3 18 32N 42 5 E
Disa, *India* 92 G5 24 18N 72 10 E
Disa, *Sudan* 107 E3 12 5N 34 15 E
Disappointment, C., *U.S.A.* 158 C2 46 18N 124 5W
Disappointment, L., *Australia* 124 D3 23 20 S 122 40 E
Disaster B., *Australia* 129 D8 37 15 S 149 58 E
Discovery B., *Australia* 128 E4 38 10 S 140 40 E
Disentis, *Switz.* 33 C7 46 42N 8 50 E
Dishna, *Egypt* 106 B3 26 9N 32 32 E
Disina, *Nigeria* 113 C6 11 35N 9 50 E
Disko, *Greenland* 10 D5 69 45N 53 30W
Disko Bugt, *Greenland* 10 D5 69 10N 52 0W
Disna = Dzisna →, *Belarus* 58 E5 55 34N 28 12 E
Disney Reef, *Tonga* 133 P13 19 17 S 174 7W
Diss, *U.K.* 21 E9 52 23N 1 7 E
Disteghil Sar, *Pakistan* 93 A6 36 20N 75 12 E
Distrito Federal □, *Brazil* 171 E2 15 45 S 47 45W
Distrito Federal □, *Venezuela* 168 A4 10 30N 66 55W
Disūq, *Egypt* 106 H7 31 8N 30 35 E
Ditu, *Dem. Rep. of the Congo* 115 D4 5 23 S 21 27 E
Diu, *India* 92 J4 20 45N 70 58 E
Diuata Mts., *Phil.* 81 G5 9 0N 125 50 E
Dīvāndarreh, *Iran* 101 E12 35 55N 47 2 E
Dives →, *France* 26 C6 49 18N 0 7W
Dives-sur-Mer, *France* 26 C6 49 18N 0 8W
Divi Pt., *India* 95 G5 15 59N 81 9 E
Divichi = Däväçi, *Azerbaijan* 61 K9 41 15N 48 57 E
Divide, *U.S.A.* 158 D7 45 45N 112 45W
Dividing Ra., *Australia* 125 E2 27 45 S 116 0 E
Divinópolis, *Brazil* 171 F3 20 10 S 44 54W
Divisões, Serra dos, *Brazil* 171 E1 17 0 S 51 0W
Divjake, *Albania* 50 F3 41 0N 19 32 E
Divnoye, *Russia* 61 H6 45 55N 43 21 E
Divo, *Ivory C.* 112 D3 5 48N 5 15W
Divriği, *Turkey* 101 C8 39 22N 38 7 E
Dīwāl Kol, *Afghan.* 91 B2 34 23N 67 52 E
Dix →, *U.S.A.* 157 G12 37 49N 84 43W
Dixie Mt., *U.S.A.* 160 F6 39 55N 120 16W
Dixie Union, *U.S.A.* 152 D7 31 20N 82 28W
Dixon, *Calif., U.S.A.* 160 G5 38 27N 121 49W
Dixon, *Ill., U.S.A.* 156 C7 41 50N 89 29W
Dixon, *Iowa, U.S.A.* 156 C6 41 45N 90 47W
Dixon, *Mo., U.S.A.* 156 G4 37 59N 92 6W
Dixon, *Mont., U.S.A.* 158 C6 47 19N 114 19W
Dixon, *N. Mex., U.S.A.* 159 H11 36 12N 105 53W
Dixon Entrance, *U.S.A.* 142 C2 54 30N 132 0W
Dixonville, *Canada* 142 B5 56 32N 117 40W
Diyadin, *Turkey* 101 C10 39 33N 43 40 E
Diyālā →, *Iraq* 101 F11 33 14N 44 31 E
Diyarbakır, *Turkey* 101 D9 37 55N 40 18 E
Djado, *Niger* 109 D2 21 4N 12 14 E
Djado, Plateau du, *Niger* 109 D2 21 29N 12 21 E
Djakarta = Jakarta, *Indonesia* 84 D3 6 9 S 106 49 E
Djamâa, *Algeria* 111 B6 33 32N 5 59 E
Djamba, *Angola* 115 F2 16 45 S 13 58 E
Djambala, *Congo* 114 C2 2 32 S 14 30 E
Djanet, *Algeria* 111 D6 24 35N 9 32 E
Djaul I., *Papua N. G.* 132 B6 2 58 S 150 57 E
Djawa = Jawa, *Indonesia* 85 D4 7 0 S 110 0 E
Djebiniana, *Tunisia* 108 A2 35 1N 11 0 E

208 Djédaa

Djédaa, *Chad* **109 F3** 13 31N 18 34 E
Djelfa, *Algeria* **111 B5** 34 40N 3 15 E
Djema, *C.A.R.* **114 A5** 6 3N 25 15 E
Djember, *Chad* **109 F3** 10 25N 17 50 E
Djendel, *Algeria* **111 A5** 36 15N 2 25 E
Djeneïene, *Tunisia* **108 B2** 31 45N 10 9 E
Djenné, *Mali* **112 C4** 14 0N 4 30W
Djenoun, Garet el, *Algeria* **111 C6** 25 4N 5 31 E
Djerba, *Tunisia* **108 B2** 33 52N 10 51 E
Djerba, I. de, *Tunisia* **108 B2** 33 50N 10 48 E
Djerid, Chott, *Tunisia* **108 B1** 33 42N 8 30 E
Djiba, *Gabon* **114 C2** 1 20 S 13 9 E
Djibo, *Burkina Faso* **113 C4** 14 9N 1 35W
Djibouti, *Djibouti* **107 E5** 11 30N 43 5 E
Djibouti ■, *Africa* **107 E5** 12 0N 43 0 E
Djolu, *Dem. Rep. of the Congo* **114 B4** 0 35N 22 5 E
Djougou, *Benin* **113 D5** 9 40N 1 45 E
Djoum, *Cameroon* **114 B2** 2 41N 12 35 E
Djourab, Erg du, *Chad* **109 E3** 16 40N 18 50 E
Djugu, *Dem. Rep. of the Congo* **118 B3** 1 55N 30 35 E
Djúpavík, *Iceland* **11 B5** 65 57N 21 34W
Djúpivogur, *Iceland* **11 C12** 64 39N 14 17W
Djupvasshytta, *Norway* **18 B4** 62 2N 7 16 E
Djurås, *Sweden* **16 D9** 60 34N 15 8 E
Djursland, *Denmark* **17 H4** 56 27N 10 45 E
Dmitriya Lapteva, Proliv, *Russia* **65 B15** 73 0N 140 0 E
Dmitriyev Lgovskiy, *Russia* **59 F8** 52 10N 35 0 E
Dmitrov, *Russia* **58 D9** 56 25N 37 32 E
Dmitrovsk-Orlovskiy, *Russia* **59 F8** 52 29N 35 10 E
Dnepr → = Dnipro →, *Ukraine* **59 J7** 46 30N 32 18 E
Dneprodzerzhinsk = Dniprodzerzhynsk, *Ukraine* **59 H8** 48 32N 34 37 E
Dneprodzerzhinskoye Vdkhr. = Dniprodzerzhynske Vdskh., *Ukraine* **59 H8** 48 49N 34 8 E
Dnepropetrovsk = Dnipropetrovsk, *Ukraine* **59 H8** 48 30N 35 0 E
Dneprorudnoye = Dniprorudne, *Ukraine* **59 J8** 47 21N 34 58 E
Dnestr → = Dnister →, *Europe* **59 J6** 46 18N 30 17 E
Dnestrovski = Belgorod, *Russia* **59 G9** 50 35N 36 35 E
Dnieper = Dnipro →, *Ukraine* **59 J7** 46 30N 32 18 E
Dniester = Dnister →, *Europe* **59 J6** 46 18N 30 17 E
Dnipro →, *Ukraine* **59 J7** 46 30N 32 18 E
Dniprodzerzhynsk, *Ukraine* **59 H8** 48 32N 34 37 E
Dniprodzerzhynske Vdskh., *Ukraine* **59 H8** 48 49N 34 8 E
Dnipropetrovsk, *Ukraine* **59 H8** 48 30N 35 0 E
Dniprorudne, *Ukraine* **59 J8** 47 21N 34 58 E
Dnister →, *Europe* **59 J6** 46 18N 30 17 E
Dnistrovskyy Lyman, *Ukraine* **59 J6** 46 15N 30 17 E
Dno, *Russia* **58 D5** 57 50N 29 58 E
Dnyapro = Dnipro →, *Ukraine* **59 J7** 46 30N 32 18 E
Doabi, *Afghan.* **91 A3** 36 1N 69 32 E
Doan Hung, *Vietnam* **86 B5** 21 30N 105 10 E
Doba, *Chad* **109 G3** 8 40N 16 50 E
Dobbiaco, *Italy* **45 B9** 46 44N 12 14 E
Dobbyn, *Australia* **126 B3** 19 44 S 140 2 E
Dobczyce, *Poland* **55 J7** 49 52N 20 5 E
Dobele, *Latvia* **15 H20** 56 37N 23 16 E
Dobele □, *Latvia* **54 B10** 56 35N 23 5 E
Döbeln, *Germany* **30 D9** 51 6N 13 7 E
Doberai, Jazirah, *Indonesia* **83 B4** 1 25 S 133 0 E
Dobiegniew, *Poland* **55 F2** 52 59N 15 45 E
Doblas, *Argentina* **174 D3** 37 5 S 64 0W
Dobo, *Indonesia* **83 C4** 5 45 S 134 15 E
Doboj, *Bos.-H.* **52 F3** 44 46N 18 4 E
Dobra, *Konin, Poland* **55 G5** 51 55N 18 37 E
Dobra, *Szczecin, Poland* **54 E2** 53 34N 15 20 E
Dobra, *Dîmbovita, Romania* **53 F10** 44 52N 25 40 E
Dobra, *Hunedoara, Romania* **52 E7** 45 54N 22 36 E
Dobre Miasto, *Poland* **54 E7** 53 58N 20 26 E
Dobreşti, *Romania* **52 D7** 46 51N 22 18 E
Dobreta-Turnu Severin, *Romania* **52 F7** 44 39N 22 41 E
Dobrich, *Bulgaria* **51 C11** 43 37N 27 49 E
Dobrinishta, *Bulgaria* **50 E7** 41 49N 23 34 E
Dobříš, *Czech Rep.* **34 B7** 49 46N 14 10 E
Dobrodzień, *Poland* **55 H5** 50 45N 18 25 E
Dobropole, *Ukraine* **59 H9** 48 25N 37 2 E
Dobruja, *Europe* **53 F13** 44 30N 28 15 E
Dobrush, *Belarus* **59 F6** 52 25N 31 22 E
Dobryanka, *Russia* **62 B6** 58 27N 56 25 E
Dobrzany, *Poland* **54 E2** 53 22N 15 25 E
Dobrzyń nad Wisłą, *Poland* **55 F6** 52 39N 19 22 E
Dobtong, *Sudan* **107 F3** 6 25N 31 40 E
Doc, Mui, *Vietnam* **86 D6** 17 58N 106 30 E
Doce →, *Brazil* **171 E4** 19 37 S 39 49W
Docksta, *Sweden* **16 A12** 63 3N 18 18 E
Doctors Inlet, *U.S.A.* **152 E8** 30 6N 81 47W
Doda, *India* **93 C6** 33 10N 75 34 E
Dodecanese = Dhodhekánisos, *Greece* **49 E8** 36 35N 27 0 E
Dodge Center, *U.S.A.* **154 C8** 44 2N 92 52W
Dodge City, *U.S.A.* **155 G5** 37 45N 100 1W
Dodge L., *Canada* **143 B7** 59 50N 105 36W
Dodgeville, *U.S.A.* **156 B6** 42 58N 90 8W
Dodo, *Sudan* **107 F2** 5 10N 29 57 E
Dodola, *Ethiopia* **107 F4** 6 59N 39 11 E
Dodoma, *Tanzania* **118 D4** 6 8 S 35 45 E
Dodoma □, *Tanzania* **118 D4** 6 0 S 36 0 E
Dodona, *Greece* **48 B2** 39 40N 20 46 E
Dodsland, *Canada* **143 C7** 51 50N 108 45W
Dodson, *U.S.A.* **158 B9** 48 24N 108 15W
Dodurga, *Turkey* **49 B11** 39 49N 29 57 E
Doerun, *U.S.A.* **152 D6** 31 19N 83 55W
Doesburg, *Neths.* **24 B6** 52 1N 6 9 E
Doetinchem, *Neths.* **24 C6** 51 59N 6 18 E
Dog Creek, *Canada* **142 C4** 51 35N 122 14W
Dog I., *U.S.A.* **152 F5** 29 48N 84 36W
Dog L., *Man., Canada* **143 C9** 51 2N 98 31W
Dog L., *Ont., Canada* **140 C2** 48 48N 89 30W
Doğanşehir, *Turkey* **100 C7** 38 5N 37 53 E
Dogliani, *Italy* **44 D4** 44 32N 7 56 E
Dōgo, *Japan* **70 A5** 36 15N 133 16 E
Dōgo-San, *Japan* **70 B5** 35 2N 133 13 E
Dogondoutchi, *Niger* **113 C5** 13 38N 4 2 E
Dogran, *Pakistan* **92 D5** 31 48N 73 35 E
Doğubayazıt, *Turkey* **101 C11** 39 31N 44 5 E
Doguéraoua, *Niger* **113 C6** 14 0N 5 31 E
Doha = Ad Dawhah, *Qatar* **97 E6** 25 15N 51 35 E
Dohinog, *Phil.* **81 G4** 8 32N 123 12 E
Doi, *Indonesia* **82 A3** 2 14N 127 49 E
Doi Luang, *Thailand* **86 C3** 18 30N 101 0 E
Doi Saket, *Thailand* **86 C2** 18 52N 99 9 E
Doig →, *Canada* **142 B4** 56 25N 120 40W
Dois Irmãos, Sa., *Brazil* **170 C3** 9 0 S 42 30W
Dojransko Jezero, *Macedonia* **50 E6** 41 13N 22 44 E
Dokka, *Norway* **18 D7** 60 49N 10 7 E
Dokka →, *Norway* **18 D7** 60 49N 10 7 E
Dokkum, *Neths.* **24 A5** 53 20N 5 59 E
Dokri, *Pakistan* **92 F3** 27 25N 68 7 E
Dokuchayevsk, *Ukraine* **59 J9** 47 44N 37 40 E
Dol-de-Bretagne, *France* **26 D5** 48 34N 1 47W
Dolak, Pulau, *Indonesia* **83 C5** 8 0 S 138 30 E
Doland, *U.S.A.* **154 C5** 44 54N 98 6W
Dolbeau, *Canada* **141 C5** 48 53N 72 18W
Dole, *France* **27 E12** 47 7N 5 31 E
Doleib, Wadi →, *Sudan* **107 E3** 12 10N 33 15 E
Dolenji Logatec, *Slovenia* **45 C11** 45 56N 14 15 E
Doles, *U.S.A.* **152 D6** 31 42N 83 53W
Dolgellau, *U.K.* **20 E4** 52 45N 3 53W
Dolgelley = Dolgellau, *U.K.* **20 E4** 52 45N 3 53W
Dolhasca, *Romania* **53 C11** 47 26N 26 36 E
Dolianova, *Italy* **46 C2** 39 22N 9 10 E
Dolinskaya = Dolynska, *Ukraine* **59 H7** 48 6N 32 46 E
Dolj □, *Romania* **53 F8** 44 10N 23 30 E
Dollard, *Neths.* **24 A7** 53 20N 7 10 E
Dolna Banya, *Bulgaria* **50 D7** 42 18N 23 44 E
Dolni Chiflik, *Bulgaria* **51 D11** 42 59N 27 43 E
Dolni Dŭbnik, *Bulgaria* **51 C8** 43 24N 24 26 E
Dolný Kubín, *Slovak Rep.* **35 B12** 49 12N 19 18 E
Dolo, *Ethiopia* **107 G5** 4 11N 42 3 E
Dolo, *Italy* **45 C9** 45 25N 12 5 E
Dolomites = Dolomiti, *Italy* **45 B8** 46 23N 11 51 E
Dolomiti, *Italy* **45 B8** 46 23N 11 51 E
Dolores, *Argentina* **174 D4** 36 20 S 57 40W
Dolores, *Phil.* **80 E5** 12 2N 125 29 E
Dolores, *Uruguay* **174 C4** 33 34 S 58 15W
Dolores, *U.S.A.* **159 H9** 37 28N 108 30W
Dolores →, *U.S.A.* **159 G9** 38 49N 109 17W
Dolovo, *Serbia, Yug.* **52 F5** 44 55N 20 52 E
Dolphin, C., *Falk. Is.* **176 D5** 51 10 S 59 0W
Dolphin and Union Str., *Canada* **138 B8** 69 5N 114 45W
Dolsk, *Poland* **55 G4** 51 59N 17 3 E
Dolton, *U.S.A.* **157 C9** 41 38N 87 36W
Dolynska, *Ukraine* **59 H7** 48 6N 32 46 E
Dolzhanskaya, *Russia* **59 J9** 46 37N 37 48 E
Dom, *Switz.* **32 D5** 46 6N 7 50 E
Dom Joaquim, *Brazil* **171 E3** 18 57 S 43 16W
Dom Pedrito, *Brazil* **175 C5** 31 0 S 54 40W
Dom Pedro, *Brazil* **170 B3** 4 59 S 44 27W
Doma, *Nigeria* **113 D6** 8 25N 8 18 E
Domaniç, *Turkey* **49 B11** 39 48N 29 36 E
Domasi, *Malawi* **119 F4** 15 15 S 35 22 E
Domat Ems, *Switz.* **33 C8** 46 50N 9 27 E
Domažlice, *Czech Rep.* **34 B5** 49 28N 12 58 E
Dombarovskiy, *Russia* **62 F7** 50 46N 59 32 E
Dombås, *Norway* **15 E13** 62 4N 9 8 E
Dombasle-sur-Meurthe, *France* **27 D13** 48 38N 6 21 E
Dombes, *France* **29 C9** 45 58N 5 0 E
Dombóvár, *Hungary* **52 D3** 46 21N 18 9 E
Dombrád, *Hungary* **52 B6** 48 13N 21 54 E
Domel I. = Letsôk-aw Kyun, *Burma* **87 G2** 11 30N 98 25 E
Domérat, *France* **27 F9** 46 21N 2 32 E
Domett, *N.Z.* **131 C8** 42 53 S 173 12 E
Domeyko, *Chile* **174 B1** 29 0 S 71 0W
Domeyko, Cordillera, *Chile* **174 A2** 24 30 S 69 0W
Domfront, *France* **26 D6** 48 37N 0 40W
Dominador, *Chile* **174 A2** 24 21 S 69 20W
Dominica ■, *W. Indies* **165 C7** 15 20N 61 20W
Dominica Passage, *W. Indies* **165 C7** 15 10N 61 20W
Dominican Rep. ■, *W. Indies* **165 C5** 19 0N 70 30W
Domiongo, *Dem. Rep. of the Congo* **115 C4** 4 37 S 21 15 E
Dömitz, *Germany* **30 B7** 53 8N 11 15 E
Domme, *France* **28 D5** 44 48N 1 12 E
Domneşti, *Romania* **53 E9** 45 12N 24 50 E
Domo, *Ethiopia* **120 C3** 7 50N 47 10 E
Domodóssola, *Italy* **44 B5** 46 7N 8 17 E
Dompaire, *France* **27 D13** 48 14N 6 14 E
Dompierre-sur-Besbre, *France* **27 F10** 46 31N 3 41 E
Dompim, *Ghana* **112 D4** 5 10N 2 5W
Dompu, *Indonesia* **82 C1** 8 32 S 118 28 E
Domrémy-la-Pucelle, *France* **27 D12** 48 26N 5 40 E
Domville, Mt., *Australia* **127 D5** 28 1 S 151 15 E
Domvraína, *Greece* **48 C4** 38 15N 22 59 E
Domžale, *Slovenia* **45 B11** 46 9N 14 35 E
Don →, *India* **95 F3** 16 20N 76 15 E
Don →, *Russia* **59 J10** 47 4N 39 18 E
Don →, *Aberds., U.K.* **22 D6** 57 11N 2 5W
Don →, *S. Yorks., U.K.* **20 D7** 53 41N 0 52W
Don, C., *Australia* **124 B5** 11 18 S 131 46 E
Don Benito, *Spain* **43 G5** 38 53N 5 51W
Don Martín, Presa de, *Mexico* **162 B4** 27 30N 100 50W
Dona Ana = Nhamaabué, *Mozam.* **119 F4** 17 25 S 35 5 E
Doña Mencía, *Spain* **43 H6** 37 33N 4 21W
Donaghadee, *U.K.* **23 B6** 54 39N 5 33W
Donald, *Australia* **128 D5** 36 23 S 143 0 E
Donalda, *Canada* **142 C6** 52 35N 112 34W
Donaldsonville, *U.S.A.* **155 K9** 30 6N 90 59W
Donalsonville, *U.S.A.* **152 D5** 31 3N 84 53W
Donau = Dunărea →, *Europe* **53 E14** 45 20N 29 40 E
Donau →, *Austria* **35 C10** 48 10N 17 0 E
Donaueschingen, *Germany* **31 H4** 47 56N 8 29 E
Donauwörth, *Germany* **31 G6** 48 43N 10 47 E
Doncaster, *U.K.* **20 D6** 53 32N 1 6W
Dondo, *Angola* **115 D2** 9 45 S 14 25 E
Dondo, *Dem. Rep. of the Congo* **114 B4** 4 11N 21 39 E
Dondo, *Mozam.* **119 F3** 19 33 S 34 46 E
Dondo, Teluk, *Indonesia* **82 A2** 0 50N 120 30 E
Dondra Head, *Sri Lanka* **95 M5** 5 55N 80 40 E
Donduşeni, *Moldova* **53 B12** 48 14N 27 36 E
Donegal, *Ireland* **23 B3** 54 39N 8 5W
Donegal □, *Ireland* **23 B4** 54 53N 8 0W
Donegal B., *Ireland* **23 B3** 54 31N 8 49W
Donets →, *Russia* **61 G5** 47 33N 40 55 E
Donetsk, *Ukraine* **59 J9** 48 0N 37 45 E
Dong Ba Thin, *Vietnam* **87 F7** 12 8N 109 13 E
Dong Dang, *Vietnam* **86 B6** 21 54N 106 42 E
Dong Giam, *Vietnam* **86 C5** 19 25N 105 31 E
Dong Ha, *Vietnam* **86 D6** 16 55N 107 8 E
Dong Hene, *Laos* **86 D5** 16 40N 105 18 E
Dong Hoi, *Vietnam* **86 D6** 17 29N 106 36 E
Dong Jiang →, *China* **77 F10** 23 6N 114 0 E
Dong Khe, *Vietnam* **86 A6** 22 26N 106 27 E
Dong Ujimqin Qi, *China* **74 B9** 45 32N 116 55 E
Dong Van, *Vietnam* **86 A5** 23 16N 105 22 E
Dong Xoai, *Vietnam* **87 G6** 11 32N 106 55 E
Donga, *Nigeria* **113 D7** 7 45N 10 2 E
Dong'an, *China* **77 D8** 26 23N 111 12 E
Dongara, *Australia* **125 E1** 29 14 S 114 57 E
Dongargarh, *India* **94 D5** 21 10N 80 40 E
Dongbei, *China* **75 D13** 45 0N 125 0 E
Dongchuan, *China* **76 D4** 26 8N 103 1 E
Donges, *France* **26 E4** 47 18N 2 4W
Dongfang, *China* **86 C7** 18 50N 108 33 E
Dongfeng, *China* **75 C13** 42 40N 125 34 E
Donggala, *Indonesia* **82 B1** 0 30 S 119 40 E
Donggan, *China* **76 F5** 23 22N 105 9 E
Donggou, *China* **75 E13** 39 52N 124 10 E
Dongguan, *China* **77 F9** 22 58N 113 44 E
Dongguang, *China* **74 F9** 37 50N 116 30 E
Donghai Dao, *China* **77 G8** 21 0N 110 15 E
Dongjingcheng, *China* **75 B15** 44 5N 129 10 E
Donglan, *China* **76 E6** 24 30N 107 21 E
Dongliu, *China* **77 B11** 30 13N 116 55 E
Dongmen, *China* **76 F6** 22 20N 107 48 E
Dongning, *China* **75 B16** 44 2N 131 5 E
Dongnyi, *China* **76 C3** 28 3N 100 15 E
Dongo, *Angola* **115 E3** 14 36 S 15 48 E
Dongola, *Sudan* **106 D3** 19 9N 30 22 E
Dongou, *Congo* **114 B3** 2 0N 18 5 E
Dongping, *China* **74 G9** 35 55N 116 20 E
Dongshan, *China* **77 F11** 23 43N 117 30 E
Dongsheng, *China* **74 E6** 39 50N 110 0 E
Dongtai, *China* **75 H11** 32 51N 120 21 E
Dongting Hu, *China* **77 C9** 29 18N 112 45 E
Dongxiang, *China* **77 C11** 28 11N 116 34 E
Dongxing, *China* **76 G7** 21 34N 108 0 E
Dongyang, *China* **77 C13** 29 13N 120 15 E
Dongzhi, *China* **77 B11** 30 9N 117 0 E
Donington, C., *Australia* **128 C2** 34 45 S 136 0 E
Doniphan, *U.S.A.* **155 G9** 36 37N 90 50W
Donja Stubica, *Croatia* **45 C12** 45 59N 15 59 E
Donji Dušnik, *Serbia, Yug.* **50 C6** 43 12N 22 5 E
Donji Miholjac, *Croatia* **52 E3** 45 45N 18 10 E
Donji Milanovac, *Serbia, Yug.* **50 B6** 44 28N 22 6 E
Donji Vakuf, *Bos.-H.* **52 F2** 44 8N 17 24 E
Dønna, *Norway* **14 C15** 66 6N 12 30 E
Donna, *U.S.A.* **155 M5** 26 9N 98 4W
Donnaconna, *Canada* **141 C5** 46 41N 71 41W
Donnelly's Crossing, *N.Z.* **130 B2** 35 42 S 173 38 E
Donnybrook, *Australia* **125 F2** 33 34 S 115 48 E
Donnybrook, *S. Africa* **117 D4** 29 59 S 29 48 E
Donora, *U.S.A.* **150 F5** 40 11N 79 52W
Donor's Hill, *Australia* **126 B3** 18 42 S 140 33 E
Donostia = Donostia-San Sebastián, *Spain* **40 B3** 43 17N 1 58W
Donostia-San Sebastián, *Spain* **40 B3** 43 17N 1 58W
Donque, *Angola* **115 F2** 15 28 S 14 6 E
Donskoy, *Russia* **58 F10** 53 55N 38 15 E
Donsol, *Phil.* **80 E4** 12 54N 123 36 E
Donzère, *France* **29 D8** 44 28N 4 43 E
Donzy, *France* **27 E10** 47 20N 3 6 E
Doon →, *U.K.* **22 F4** 55 27N 4 39W
Dora, L., *Australia* **124 D3** 22 0 S 123 0 E
Dora Báltea →, *Italy* **44 C5** 45 11N 8 3 E
Dora Ripária →, *Italy* **44 C4** 45 5N 7 44 E
Doran L., *Canada* **143 A7** 61 13N 108 6W
Doraville, *U.S.A.* **152 B5** 33 54N 84 17W
Dorchester, *U.K.* **21 G5** 50 42N 2 27W
Dorchester, C., *Canada* **139 B12** 65 27N 77 27W
Dordogne □, *France* **28 C4** 45 5N 0 40 E
Dordogne →, *France* **28 C3** 45 2N 0 36W
Dordrecht, *Neths.* **24 C4** 51 48N 4 39 E
Dordrecht, *S. Africa* **116 E4** 31 20 S 27 3 E
Dore →, *France* **28 C7** 45 50N 3 35 E
Dore, Mts., *France* **28 C6** 45 32N 2 50 E
Doré L., *Canada* **143 C7** 54 46N 107 17W
Doré Lake, *Canada* **143 C7** 54 38N 107 36W
Dores do Indaiá, *Brazil* **171 E2** 19 27 S 45 36W
Dorfen, *Germany* **31 G8** 48 15N 12 8 E
Dorgali, *Italy* **46 B2** 40 17N 9 35 E
Dori, *Burkina Faso* **113 C4** 14 3N 0 2W
Doring →, *S. Africa* **116 E2** 31 54 S 18 39 E
Doringbos, *S. Africa* **116 E2** 31 59 S 19 16 E
Dorion, *Canada* **140 C5** 45 23N 74 3W
Dormaa-Ahenkro, *Ghana* **112 D4** 7 15N 2 52W
Dormans, *France* **27 C10** 49 4N 3 38 E
Dormo, Ras, *Eritrea* **107 E5** 13 14N 42 35 E
Dornach, *Switz.* **32 B5** 47 29N 7 37 E
Dornbirn, *Austria* **34 D2** 47 25N 9 45 E
Dornes, *France* **27 F10** 46 48N 3 18 E
Dorneşti, *Romania* **53 C11** 47 52N 26 1 E
Dornie, *U.K.* **22 D3** 57 17N 5 31W
Dornoch, *U.K.* **22 D4** 57 53N 4 2W
Dornoch Firth, *U.K.* **22 D4** 57 51N 4 4W
Dornogovĭ □, *Mongolia* **74 C6** 44 0N 110 0 E
Doro, *Mali* **113 B4** 16 9N 0 51W
Dorog, *Hungary* **52 C3** 47 42N 18 45 E
Dorogobuzh, *Russia* **58 E7** 54 50N 33 18 E
Dorohoi, *Romania* **53 C11** 47 56N 26 23 E
Döröö Nuur, *Mongolia* **72 B4** 48 0N 93 0 E
Dorr, *Iran* **97 C6** 33 17N 50 38 E
Dorre I., *Australia* **125 E1** 25 13 S 113 12 E
Dorrigo, *Australia* **129 A10** 30 20 S 152 44 E
Dorris, *U.S.A.* **158 F3** 41 58N 121 55W
Dorset, *Canada* **150 A6** 45 14N 78 54W
Dorset, *U.S.A.* **150 E4** 41 4N 80 40W
Dorset □, *U.K.* **21 G5** 50 45N 2 26W
Dorsten, *Germany* **30 D2** 51 40N 6 58 E
Dortmund, *Germany* **30 D3** 51 30N 7 28 E
Dortmund-Ems-Kanal →, *Germany* **30 D3** 51 50N 7 26 E
Dörtyol, *Turkey* **100 D7** 36 50N 36 13 E
Dorum, *Germany* **30 B4** 53 41N 8 34 E
Doruma, *Dem. Rep. of the Congo* **118 B2** 4 42N 27 33 E
Dorūneh, *Iran* **97 C8** 35 10N 57 18 E
Dos Bahías, C., *Argentina* **176 B3** 44 58 S 65 32W
Dos Hermanas, *Spain* **43 H5** 37 16N 5 55W
Dos Palos, *U.S.A.* **160 J6** 36 59N 120 37W
Döşemealtı, *Turkey* **49 D12** 37 4N 30 36 E
Dosso, *Niger* **113 C5** 13 0N 3 13 E
Dot Lake, *U.S.A.* **144 E11** 63 40N 144 4W
Dothan, *U.S.A.* **152 D4** 31 13N 85 24W
Doty, *U.S.A.* **160 D3** 46 38N 123 17W
Douai, *France* **27 B10** 50 21N 3 4 E
Douala, *Cameroon* **113 E6** 4 0N 9 45 E
Douarnenez, *France* **26 D2** 48 6N 4 21W
Double Island Pt., *Australia* **127 D5** 25 56 S 153 11 E
Doubrava →, *Czech Rep.* **34 A8** 50 2N 15 20 E
Doubs □, *France* **27 E13** 47 10N 6 20 E
Doubs →, *France* **27 F12** 46 53N 5 1 E
Doubtful Sd., *N.Z.* **131 F1** 45 20 S 166 49 E
Doubtless B., *N.Z.* **130 A2** 34 55 S 173 26 E
Doudeville, *France* **26 C7** 49 43N 0 47 E
Doué-la-Fontaine, *France* **26 E6** 47 11N 0 16W
Douentza, *Mali* **112 C4** 14 58N 2 48W
Doughboy B., *N.Z.* **131 H2** 47 2 S 167 40 E
Douglas, *S. Africa* **116 D3** 29 4 S 23 46 E
Douglas, *U.K.* **20 C3** 54 10N 4 28W
Douglas, *Alaska, U.S.A.* **142 B2** 58 17N 134 24W
Douglas, *Ariz., U.S.A.* **159 L9** 31 21N 109 33W
Douglas, *Ga., U.S.A.* **152 D7** 31 31N 82 51W
Douglas, *Wyo., U.S.A.* **154 D2** 42 45N 105 24W
Douglas C., *U.S.A.* **144 G9** 58 51N 153 15W
Douglastown, *Canada* **141 C7** 48 46N 64 24W
Douglasville, *U.S.A.* **152 B5** 33 45N 84 45W
Douirat, *Morocco* **110 B4** 33 2N 4 11W
Doukáton, Ákra, *Greece* **48 C2** 38 34N 20 30 E
Doulevant-le-Château, *France* **27 D11** 48 23N 4 55 E
Doullens, *France* **27 B9** 50 10N 2 20 E
Doumé, *Cameroon* **114 B2** 4 15N 13 25 E
Douna, *Mali* **112 C3** 13 13N 6 0W
Dounguila, *Congo* **114 C2** 2 53 S 11 58 E
Dounreay, *U.K.* **22 C5** 58 35N 3 44W
Dourada, Serra, *Brazil* **171 D2** 13 10 S 48 45W
Dourados, *Brazil* **175 A5** 22 9 S 54 50W
Dourados →, *Brazil* **175 A5** 21 58 S 54 18W
Dourdan, *France* **27 D9** 48 30N 2 1 E
Douro →, *Europe* **42 D2** 41 8N 8 40W
Douvaine, *France* **27 F13** 46 19N 6 16 E
Douvres-la-Délivrande, *France* **26 C6** 49 17N 0 23W
Douz, *Tunisia* **108 B1** 33 25N 9 0 E
Douze →, *France* **28 E3** 43 54N 0 30W
Dove →, *U.K.* **20 E6** 52 51N 1 36W
Dove Creek, *U.S.A.* **159 H9** 37 46N 108 54W
Dover, *Australia* **126 G4** 43 18 S 147 2 E
Dover, *U.K.* **21 F9** 51 7N 1 19 E
Dover, *Del., U.S.A.* **148 F8** 39 10N 75 32W
Dover, *Ky., U.S.A.* **157 F13** 38 43N 83 52W
Dover, *N.H., U.S.A.* **151 C14** 43 12N 70 56W
Dover, *N.J., U.S.A.* **151 F10** 40 53N 74 34W
Dover, *Ohio, U.S.A.* **150 F3** 40 32N 81 29W
Dover, Pt., *Australia* **125 F4** 32 32 S 125 32 E
Dover, Str. of, *Europe* **21 G9** 51 0N 1 30 E
Dover-Foxcroft, *U.S.A.* **141 C6** 45 11N 69 13W
Dover Plains, *U.S.A.* **151 E11** 41 43N 73 35W
Dovey = Dyfi →, *U.K.* **21 E3** 52 32N 4 3W
Dovre, *Norway* **18 C6** 61 58N 9 15 E
Dovrefjell, *Norway* **15 E13** 62 15N 9 33 E
Dow Rūd, *Iran* **97 C6** 33 28N 49 4 E
Dowa, *Malawi* **119 E3** 13 38 S 33 58 E
Dowagiac, *U.S.A.* **157 C10** 41 59N 86 6W
Dowgha'i, *Iran* **97 B8** 36 54N 58 32 E
Dowlat Yār, *Afghan.* **91 B2** 34 30N 65 45 E
Dowlatābād, *Farāh, Afghan.* **91 B1** 32 47N 62 40 E
Dowlatābād, *Fāryāb, Afghan.* **91 A2** 36 26N 64 55 E
Dowlatābād, *Iran* **97 D8** 28 20N 56 40 E
Dowling Park, *U.S.A.* **152 E6** 30 15N 83 15W
Down □, *U.K.* **23 B5** 54 23N 6 2W
Downers Grove, *U.S.A.* **157 C8** 41 48N 88 1W
Downey, *Calif., U.S.A.* **161 M8** 33 56N 118 7W
Downey, *Idaho, U.S.A.* **158 E7** 42 26N 112 7W
Downham Market, *U.K.* **21 E8** 52 37N 0 23 E
Downieville, *U.S.A.* **160 F6** 39 34N 120 50W
Downing, *U.S.A.* **156 D4** 40 29N 92 22W
Downpatrick, *U.K.* **23 B6** 54 20N 5 43W
Downpatrick Hd., *Ireland* **23 B2** 54 20N 9 21W
Dowsārī, *Iran* **97 D8** 28 25N 57 59 E
Dowshī, *Afghan.* **91 B3** 35 35N 68 43 E
Doyle, *U.S.A.* **160 E6** 40 2N 120 6W
Doylestown, *U.S.A.* **151 F9** 40 21N 75 10W
Draa, C., *Morocco* **110 C2** 28 47N 11 0W
Draa, Oued →, *Morocco* **110 C2** 28 40N 11 10W
Drac →, *France* **29 C9** 45 12N 5 42 E
Dračevo, *Macedonia* **50 E5** 41 56N 21 31 E
Drachten, *Neths.* **24 A6** 53 7N 6 5 E
Drăgăneşti, *Moldova* **53 C13** 47 43N 28 15 E
Drăgăneşti-Olt, *Romania* **53 F9** 44 9N 24 32 E
Drăgăneşti-Vlaşca, *Romania* **53 F10** 44 5N 25 33 E
Dragaš, *Serbia, Yug.* **50 D4** 42 5N 20 41 E
Drăgăşani, *Romania* **53 F9** 44 39N 24 17 E
Dragichyn, *Belarus* **59 F3** 52 15N 25 8 E
Dragocvet, *Serbia, Yug.* **50 C5** 43 58N 21 15 E
Dragovishtitsa, *Bulgaria* **50 D6** 42 22N 22 39 E
Draguignan, *France* **29 E10** 43 32N 6 27 E
Drain, *U.S.A.* **158 E2** 43 40N 123 19W
Drake, *Australia* **127 D5** 28 55 S 152 25 E
Drake, *U.S.A.* **154 B4** 47 55N 100 23W
Drake Passage, *S. Ocean* **7 B17** 58 0 S 68 0W
Drakensberg, *S. Africa* **117 E4** 31 0 S 28 0 E
Dráma, *Greece* **51 E8** 41 9N 24 10 E
Dráma □, *Greece* **51 E8** 41 20N 24 0 E
Drammen, *Norway* **15 G14** 59 42N 10 12 E
Drangajökull, *Iceland* **11 A4** 66 9N 22 15W
Drangedal, *Norway* **18 E6** 59 6N 9 3 E
Drangsnes, *Iceland* **11 B5** 65 41N 21 27W
Dranov, Ostrov, *Romania* **53 F14** 44 55N 29 30 E
Dras, *India* **93 B6** 34 25N 75 48 E
Drau = Drava →, *Croatia* **52 E3** 45 33N 18 55 E
Drava →, *Croatia* **52 E3** 45 33N 18 55 E
Dravograd, *Slovenia* **45 B12** 46 36N 15 5 E
Drawa →, *Poland* **55 F2** 52 52N 15 59 E
Drawno, *Poland* **55 E2** 53 13N 15 46 E
Drawsko Pomorskie, *Poland* **54 E2** 53 35N 15 50 E
Drayton Plains, *U.S.A.* **157 B13** 42 42N 83 23W
Drayton Valley, *Canada* **142 C6** 53 12N 114 58W
Dreieich, *Germany* **31 E4** 50 1N 8 41 E
Dren, *Serbia, Yug.* **50 C4** 43 8N 20 46 E
Drenthe □, *Neths.* **24 B6** 52 52N 6 40 E
Drepanum, C., *Cyprus* **38 E11** 34 54N 32 19 E
Dresden, *Canada* **150 D2** 42 35N 82 11W
Dresden, *Germany* **30 D9** 51 3N 13 44 E
Dreux, *France* **26 D8** 48 44N 1 23 E
Drevsjø, *Norway* **18 C9** 61 53N 12 1 E
Drexel, *U.S.A.* **157 E12** 39 45N 84 18W
Drezdenko, *Poland* **55 F2** 52 50N 15 49 E
Driffield, *U.K.* **20 C7** 54 0N 0 26W
Driftwood, *U.S.A.* **150 E6** 41 20N 78 8W
Driggs, *U.S.A.* **158 E8** 43 44N 111 6W
Drin i Zi →, *Albania* **50 E4** 41 37N 20 28 E

East Grand Forks 209

Drina →, *Bos.-H.* **50 B3** 44 53N 19 21 E
Drincea →, *Romania* **52 F7** 44 20N 22 55 E
Drini →, *Albania* **50 D3** 42 1N 19 38 E
Drinjača →, *Bos.-H.* **52 F4** 44 15N 19 8 E
Drissa = Vyerkhnyadzvinsk,
Belarus **58 E4** 55 45N 27 58 E
Driva →, *Norway* **18 B6** 62 41N 9 31 E
Drivstua, *Norway* **18 B6** 62 26N 9 47 E
Drniš, *Croatia* **45 E13** 43 51N 16 10 E
Drøbak, *Norway* **15 G14** 59 39N 10 39 E
Drobin, *Poland* **55 F6** 52 42N 19 58 E
Drochia, *Moldova* **53 B12** 48 2N 27 48 E
Drogheda, *Ireland* **23 C5** 53 43N 6 22W
Drogichin = Dragichyn,
Belarus **59 F3** 52 15N 25 8 E
Drogobych = Drohobych,
Ukraine **59 H2** 49 20N 23 30 E
Drohiczyn, *Poland* **55 F9** 52 24N 22 39 E
Drohobych, *Ukraine* **59 H2** 49 20N 23 30 E
Droichead Atha =
Drogheda, *Ireland* **23 C5** 53 43N 6 22W
Droichead Nua, *Ireland* **23 C5** 53 11N 6 48W
Droitwich, *U.K.* **21 E5** 52 16N 2 8W
Drôme □, *France* **29 D9** 44 38N 5 15 E
Drôme →, *France* **29 D8** 44 46N 4 46 E
Dromedary, C., *Australia* **129 D9** 36 17 S 150 10 E
Dromore, *U.K.* **23 B4** 54 31N 7 28W
Dromore West, *Ireland* **23 B3** 54 15N 8 52W
Dronero, *Italy* **44 D4** 44 28N 7 22 E
Dronfield, *Australia* **126 C3** 21 12 S 140 3 E
Dronfield, *U.K.* **20 D6** 53 19N 1 27W
Dronne →, *France* **28 C3** 45 2N 0 9W
Dronninglund, *Denmark* **17 G4** 57 10N 10 19 E
Dronten, *Neths.* **24 B5** 52 32N 5 43 E
Dropt →, *France* **28 D3** 44 35N 0 6W
Drosendorf, *Austria* **34 C8** 48 52N 15 37 E
Droué, *France* **26 D8** 48 3N 1 6 E
Drouin, *Australia* **129 E6** 38 10 S 145 53 E
Drumbo, *Canada* **150 C4** 43 16N 80 35W
Drumheller, *Canada* **142 C6** 51 25N 112 40W
Drummond, *U.S.A.* **158 C7** 46 40N 113 9W
Drummond I., *U.S.A.* **140 C3** 46 1N 83 39W
Drummond Pt., *Australia* **127 E2** 34 9 S 135 16 E
Drummond Ra., *Australia* **126 C4** 23 45 S 147 10 E
Drummondville, *Canada* **140 C5** 45 55N 72 25W
Drumright, *U.S.A.* **155 H6** 35 59N 96 36W
Druskininkai, *Lithuania* **15 J20** 54 3N 23 58 E
Drut →, *Belarus* **59 F6** 53 8N 30 5 E
Druya, *Belarus* **58 E4** 55 45N 27 28 E
Druzhba, *Bulgaria* **51 C12** 43 15N 28 1 E
Druzhina, *Russia* **65 C15** 68 14N 145 18 E
Drvar, *Bos.-H.* **45 D13** 44 21N 16 23 E
Drvenik, *Croatia* **45 E13** 43 27N 16 3 E
Drwęca →, *Poland* **55 E5** 53 0N 18 42 E
Dry Tortugas, *U.S.A.* **164 B3** 24 38N 82 55W
Dryanovo, *Bulgaria* **51 D9** 42 59N 25 28 E
Dryden, *Canada* **143 D10** 49 47N 92 50W
Dryden, *U.S.A.* **155 K3** 30 3N 102 7W
Drygalski I., *Antarctica* **7 C7** 66 0 S 92 0 E
Drysdale →, *Australia* **124 B4** 13 59 S 126 51 E
Drysdale I., *Australia* **126 A2** 11 41 S 136 0 E
Drzewica, *Poland* **55 G7** 51 27N 20 29 E
Drzewiczka →, *Poland* **55 G7** 51 36N 20 36 E
Dschang, *Cameroon* **113 D7** 5 32N 10 3 E
Du Bois, *U.S.A.* **150 E6** 41 8N 78 46W
Du Quoin, *U.S.A.* **156 G7** 38 1N 89 14W
Duanesburg, *U.S.A.* **151 D10** 42 45N 74 11W
Duaringa, *Australia* **126 C4** 23 42 S 149 42 E
Ḍubā, *Si. Arabia* **96 E2** 27 10N 35 40 E
Dubai = Dubayy, *U.A.E.* **97 E7** 25 18N 55 20 E
Dubăsari, *Moldova* **53 C14** 47 15N 29 10 E
Dubăsari Vdkhr., *Moldova* **53 C13** 47 30N 29 0 E
Dubawnt →, *Canada* **143 A8** 64 33N 100 6W
Dubawnt, L., *Canada* **143 A8** 63 4N 101 42W
Dubayy, *U.A.E.* **97 E7** 25 18N 55 20 E
Dubbo, *Australia* **129 B8** 32 11 S 148 35 E
Dubele,
Dem. Rep. of the Congo **118 B2** 2 56N 29 35 E
Dübendorf, *Switz.* **33 B7** 47 24N 8 37 E
Dubica, *Croatia* **45 C13** 45 11N 16 48 E
Dublin, *Ireland* **23 C5** 53 21N 6 15W
Dublin, *Ga., U.S.A.* **152 C7** 32 32N 82 54W
Dublin, *Tex., U.S.A.* **155 J5** 32 5N 98 21W
Dublin □, *Ireland* **23 C5** 53 24N 6 20W
Dubna, *Russia* **58 D9** 56 44N 37 10 E
Dubnica nad Váhom,
Slovak Rep. **35 C11** 48 58N 18 11 E
Dubno, *Ukraine* **59 G3** 50 25N 25 45 E
Dubois, *Idaho, U.S.A.* **158 D7** 44 10N 112 14W
Dubois, *Ind., U.S.A.* **157 F10** 38 27N 86 48W
Dubossary = Dubăsari,
Moldova **53 C14** 47 15N 29 10 E
Dubossary Vdkhr. =
Dubăsari Vdkhr., *Moldova* **53 C13** 47 30N 29 0 E
Dubovka, *Russia* **61 F7** 49 5N 44 50 E
Dubovskoye, *Russia* **61 G6** 47 28N 42 46 E
Dubrajpur, *India* **93 H12** 23 48N 87 25 E
Dubréka, *Guinea* **112 D2** 9 46N 13 31W
Dubrovitsa = Dubrovytsya,
Ukraine **59 G4** 51 31N 26 35 E
Dubrovnik, *Croatia* **50 D2** 42 39N 18 6 E
Dubrovytsya, *Ukraine* **59 G4** 51 31N 26 35 E
Dubulu,
Dem. Rep. of the Congo **114 B4** 4 18N 20 16 E
Dubuque, *U.S.A.* **156 B6** 42 30N 90 41W
Dubysa →, *Lithuania* **54 C10** 55 5N 23 26 E
Duchang, *China* **77 C11** 29 18N 116 12 E
Duchesne, *U.S.A.* **158 F8** 40 10N 110 24W
Duchess, *Australia* **126 C2** 21 20 S 139 50 E
Ducie I., *Pac. Oc.* **135 K15** 24 40 S 124 48W
Duck Cr. →, *Australia* **124 D2** 22 37 S 116 53 E
Duck Lake, *Canada* **143 C7** 52 50N 106 16W
Duck Mountain Prov. Park,
Canada **143 C8** 51 45N 101 0W
Duckwall, Mt., *U.S.A.* **160 H6** 37 58N 120 7W
Duderstadt, *Germany* **30 D6** 51 31N 10 15 E
Dudhnai, *India* **90 C3** 25 59N 90 47 E
Düdingen, *Switz.* **32 C4** 46 52N 7 12 E
Dudinka, *Russia* **65 C9** 69 30N 86 13 E
Dudley, *U.K.* **21 E5** 52 31N 2 5W
Dudley, *U.S.A.* **152 C6** 32 32N 83 5W
Dudna →, *India* **94 E3** 19 17N 76 54 E
Dudo, *Somali Rep.* **120 C4** 9 20N 50 12 E
Dudub, *Ethiopia* **120 C3** 6 55N 46 43 E
Duenas, *Phil.* **81 F4** 11 4N 122 37 E
Dueñas, *Spain* **42 D6** 41 52N 4 33W
Dueré, *Brazil* **171 D2** 11 20 S 49 17W
Duero = Douro →, *Europe* **42 D2** 41 8N 8 40W
Dūfah, W. →, *Si. Arabia* **98 C3** 18 45N 41 49 E
Dufftown, *U.K.* **22 D5** 57 27N 3 8W
Dufourspitz, *Switz.* **32 E5** 45 56N 7 52 E
Dugger, *U.S.A.* **157 E9** 39 4N 87 18W
Dugi Otok, *Croatia* **45 D11** 44 0N 15 3 E
Dugiuma, *Somali Rep.* **120 D2** 1 15N 42 34 E
Dugo Selo, *Croatia* **45 C13** 45 51N 16 18 E
Duifken Pt., *Australia* **126 A3** 12 33 S 141 38 E
Duisburg, *Germany* **30 D2** 51 26N 6 45 E
Duitama, *Colombia* **168 B3** 5 50N 73 2W
Duiwelskloof, *S. Africa* **117 C5** 23 42 S 30 10 E
Dukati, *Albania* **50 F3** 40 16N 19 32 E
Dūkdamīn, *Iran* **97 C8** 35 59N 57 43 E
Duke I., *U.S.A.* **142 C2** 54 50N 131 20W
Dukelský Průsmyk,
Slovak Rep. **35 B14** 49 25N 21 42 E
Dukhān, *Qatar* **97 E6** 25 25N 50 50 E
Dukhovshchina, *Russia* **58 E7** 55 15N 32 27 E
Duki, *Pakistan* **91 C3** 30 14N 68 25 E
Dukla, *Poland* **55 J8** 49 30N 21 35 E
Duku, *Bauchi, Nigeria* **113 C7** 10 43N 10 43 E
Duku, *Sokoto, Nigeria* **113 C5** 11 11N 4 55 E
Dulag, *Phil.* **81 F5** 10 57N 125 2 E
Dulce →, *Argentina* **174 C3** 30 32 S 62 33W
Dulce, G., *Costa Rica* **164 E3** 8 40N 83 20W
Dulf, *Iraq* **96 C5** 35 7N 45 51 E
Dŭlgopol, *Bulgaria* **51 C11** 43 3N 27 22 E
Duliu, *China* **74 E9** 39 2N 116 55 E
Dullewala, *Pakistan* **92 D4** 31 50N 71 25 E
Dülmen, *Germany* **30 D3** 51 49N 7 17 E
Dulovo, *Bulgaria* **51 C11** 43 48N 27 9 E
Dulpetorpet, *Norway* **18 D9** 60 34N 12 19 E
Dulq Maghār, *Syria* **101 D8** 36 22N 38 39 E
Dululu, *Australia* **126 C5** 23 48 S 150 15 E
Duluth, *Ga., U.S.A.* **152 A5** 34 0N 84 9W
Duluth, *Minn., U.S.A.* **154 B8** 46 47N 92 6W
Dum Dum, *India* **93 H13** 22 39N 88 33 E
Dūmā, *Syria* **103 B5** 33 34N 36 24 E
Dumaguete, *Phil.* **81 G4** 9 17N 123 15 E
Dumai, *Indonesia* **84 B2** 1 35N 101 28 E
Dumalinao, *Phil.* **81 H4** 7 49N 123 23 E
Dumanguilas Bay, *Phil.* **81 H4** 7 34N 123 4 E
Dumaran, *Phil.* **81 F2** 10 33N 119 50 E
Dumas, *Ark., U.S.A.* **155 J9** 33 53N 91 29W
Dumas, *Tex., U.S.A.* **155 H4** 35 52N 101 58W
Dumbarton, *U.K.* **22 F4** 55 57N 4 33W
Dumbea, *N. Cal.* **133 V20** 22 10 S 166 27 E
Ďumbier, *Slovak Rep.* **35 C12** 48 56N 19 38 E
Dumbleyung, *Australia* **125 F2** 33 17 S 117 42 E
Dumbo, *Angola* **115 E3** 14 6 S 17 24 E
Dumbrăveni, *Romania* **53 D9** 46 14N 24 34 E
Dumfries, *U.K.* **22 F5** 55 4N 3 37W
Dumfries & Galloway □,
U.K. **22 F5** 55 9N 3 58W
Dumingag, *Phil.* **81 G4** 8 20N 123 20 E
Dumitreşti, *Romania* **53 E11** 45 33N 26 55 E
Dumka, *India* **93 G12** 24 12N 87 15 E
Dumlupınar, *Turkey* **49 C12** 38 53N 30 0 E
Dümmer, *Germany* **30 C4** 52 31N 8 20 E
Dumoine →, *Canada* **140 C4** 46 13N 77 51W
Dumoine L., *Canada* **140 C4** 46 55N 77 55W
Dumraon, *India* **93 G11** 25 33N 84 8 E
Dumyât, *Egypt* **106 H7** 31 24N 31 48 E
Dumyât, Masabb, *Egypt* **106 H7** 31 28N 31 51 E
Dún Dealgan = Dundalk,
Ireland **23 B5** 54 1N 6 24W
Dun Laoghaire, *Ireland* **23 C5** 53 17N 6 8W
Dun-le-Palestel, *France* **27 F8** 46 18N 1 39 E
Dun-sur-Auron, *France* **27 F9** 46 53N 2 33 E
Dun-sur-Meuse, *France* **27 C12** 49 23N 5 11 E
Duna = Dunărea →,
Europe **53 E14** 45 20N 29 40 E
Duna →, *Hungary* **52 E3** 45 51N 18 48 E
Duna-völgyi-főcsatorna,
Hungary **52 D4** 46 40N 19 14 E
Dunaföldvár, *Hungary* **52 D3** 46 50N 18 57 E
Dunaj = Dunărea →,
Europe **53 E14** 45 20N 29 40 E
Dunaj →, *Slovak Rep.* **35 D11** 47 50N 18 50 E
Dunajec →, *Poland* **55 H7** 50 15N 20 44 E
Dunajská Streda,
Slovak Rep. **35 C10** 48 0N 17 37 E
Dunakeszi, *Hungary* **52 C4** 47 37N 19 8 E
Dunapataj, *Hungary* **52 D4** 46 39N 19 4 E
Dunărea →, *Europe* **53 E14** 45 20N 29 40 E
Dunaszekcső, *Hungary* **52 D3** 46 6N 18 45 E
Dunaújváros, *Hungary* **52 D3** 46 58N 18 57 E
Dunav = Dunărea →,
Europe **53 E14** 45 20N 29 40 E
Dunavaţu de Jos, *Romania* **53 F14** 44 59N 29 13 E
Dunavtsi, *Bulgaria* **50 C6** 43 57N 22 53 E
Dunay, *Russia* **68 C6** 42 52N 132 22 E
Dunback, *N.Z.* **131 F5** 45 23 S 170 36 E
Dunbar, *Australia* **126 B3** 16 0 S 142 22 E
Dunbar, *U.K.* **22 E6** 56 0N 2 31W
Dunblane, *U.K.* **22 E5** 56 11N 3 58W
Duncan, *Canada* **142 D4** 48 45N 123 40W
Duncan, *Ariz., U.S.A.* **159 K9** 32 43N 109 6W
Duncan, *Okla., U.S.A.* **155 H6** 34 30N 97 57W
Duncan, L., *Canada* **140 B4** 53 29N 77 58W
Duncan L., *Canada* **142 A6** 62 51N 113 58W
Duncan Town, *Bahamas* **164 B4** 22 15N 75 45W
Duncannon, *U.S.A.* **150 F7** 40 23N 77 2W
Duncansby Head, *U.K.* **22 C5** 58 38N 3 1W
Dundaga, *Latvia* **54 A9** 57 31N 22 21 E
Dundalk, *Canada* **150 B4** 44 10N 80 24W
Dundalk, *Ireland* **23 B5** 54 1N 6 24W
Dundalk Bay, *Ireland* **23 C5** 53 55N 6 15W
Dundas, *Canada* **140 D4** 43 17N 79 59W
Dundas, *Greenland* **10 B4** 77 0N 69 0W
Dundas, L., *Australia* **125 F3** 32 35 S 121 50 E
Dundas I., *Canada* **142 C2** 54 30N 130 50W
Dundas Str., *Australia* **124 B5** 11 15 S 131 35 E
Dundee, *S. Africa* **117 D5** 28 11 S 30 15 E
Dundee, *U.K.* **22 E6** 56 28N 2 59W
Dundee, *U.S.A.* **157 C13** 41 57N 83 40W
Dundee City □, *U.K.* **22 E6** 56 30N 2 58W
Dundgovĭ □, *Mongolia* **74 B4** 45 10N 106 0 E
Dundoo, *Australia* **127 D3** 27 40 S 144 37 E
Dundrum, *U.K.* **23 B6** 54 16N 5 52W
Dundrum B., *U.K.* **23 B6** 54 13N 5 47W
Dundwara, *India* **93 F8** 27 48N 79 9 E
Dunedin, *N.Z.* **131 F5** 45 50 S 170 33 E
Dunedin, *U.S.A.* **149 L4** 28 1N 82 47W
Dunedin →, *Canada* **142 B4** 59 30N 124 5W
Dunfermline, *U.K.* **22 E5** 56 5N 3 27W
Dungannon, *Canada* **150 C3** 43 51N 81 36W
Dungannon, *U.K.* **23 B5** 54 31N 6 46W
Dungarpur, *India* **92 H5** 23 52N 73 45 E
Dungarvan, *Ireland* **23 D4** 52 5N 7 37W
Dungarvan Harbour, *Ireland* **23 D4** 52 4N 7 35W
Dungeness, *U.K.* **21 G8** 50 54N 0 59 E
Dungo, L. do, *Angola* **115 F3** 17 15 S 19 0 E
Dungog, *Australia* **129 B9** 32 22 S 151 46 E
Dungu,
Dem. Rep. of the Congo **118 B2** 3 40N 28 32 E
Dungun, *Malaysia* **87 K4** 4 45N 103 25 E
Dungunāb, *Sudan* **106 C4** 21 10N 37 9 E
Dungunāb, Khalij, *Sudan* **106 C4** 21 5N 37 12 E
Dunhinda Falls, *Sri Lanka* **95 L5** 7 5N 81 6 E
Dunhua, *China* **75 C15** 43 20N 128 14 E
Dunhuang, *China* **72 B4** 40 8N 94 36 E
Dunk I., *Australia* **126 B4** 17 59 S 146 29 E
Dunkeld, *Australia* **128 D5** 37 40 S 142 22 E
Dunkeld, *U.K.* **22 E5** 56 34N 3 35W
Dunkerque, *France* **27 A9** 51 2N 2 20 E
Dunkery Beacon, *U.K.* **21 F4** 51 9N 3 36W
Dunkirk = Dunkerque,
France **27 A9** 51 2N 2 20 E
Dunkirk, *U.S.A.* **150 D5** 42 29N 79 20W
Dunkuj, *Sudan* **107 E3** 12 50N 32 49 E
Dunkwa, *Central, Ghana* **112 D4** 6 0N 1 47W
Dunkwa, *Central, Ghana* **113 D4** 5 30N 1 0W
Dunlap, *U.S.A.* **154 E7** 41 51N 95 36W
Dúnleary = Dun Laoghaire,
Ireland **23 C5** 53 17N 6 8W
Dunleer, *Ireland* **23 C5** 53 50N 6 24W
Dunmanus B., *Ireland* **23 E2** 51 31N 9 50W
Dunmanway, *Ireland* **23 E2** 51 43N 9 6W
Dunmara, *Australia* **126 B1** 16 42 S 133 25 E
Dunmore, *U.S.A.* **151 E9** 41 25N 75 38W
Dunmore Hd., *Ireland* **23 D1** 52 10N 10 35W
Dunmore Town, *Bahamas* **164 A4** 25 30N 76 39W
Dunn, *U.S.A.* **149 H6** 35 19N 78 37W
Dunnellon, *U.S.A.* **149 L4** 29 3N 82 28W
Dunnet Hd., *U.K.* **22 C5** 58 40N 3 21W
Dunning, *U.S.A.* **154 E4** 41 50N 100 6W
Dunnville, *Canada* **150 D5** 42 54N 79 36W
Dunolly, *Australia* **128 D5** 36 51 S 143 44 E
Dunoon, *U.K.* **22 F4** 55 57N 4 56W
Dunqul, *Egypt* **106 C3** 23 26N 31 37 E
Duns, *U.K.* **22 F6** 55 47N 2 20W
Dunseith, *U.S.A.* **154 A4** 48 50N 100 3W
Dunsmuir, *U.S.A.* **158 F2** 41 13N 122 16W
Dunstable, *U.K.* **21 F7** 51 53N 0 32W
Dunstan Mts., *N.Z.* **131 E4** 44 53 S 169 35 E
Dunster, *Canada* **142 C5** 53 8N 119 50W
Duntroon, *N.Z.* **131 E5** 44 51 S 170 40 E
Dunvegan L., *Canada* **143 A7** 60 8N 107 10W
Duolun, *China* **74 C9** 42 12N 116 28 E
Duong Dong, *Vietnam* **87 G4** 10 13N 103 58 E
Dupax, *Phil.* **80 C3** 16 17N 121 5 E
Dupree, *U.S.A.* **154 C4** 45 4N 101 35W
Dupuyer, *U.S.A.* **158 B7** 48 13N 112 30W
Duqm, *Oman* **99 C7** 19 39N 57 42 E
Duque de Caxias, *Brazil* **171 F3** 22 45 S 43 19W
Duque de York, I., *Chile* **176 D1** 50 37 S 75 25W
Durack →, *Australia* **124 C4** 15 33 S 127 52 E
Durack Ra., *Australia* **124 C4** 16 50 S 127 40 E
Durağan, *Turkey* **100 B6** 41 25N 35 3 E
Durak, *Turkey* **49 B10** 39 42N 28 17 E
Durance →, *France* **29 E8** 43 55N 4 45 E
Durand, *Ga., U.S.A.* **152 C5** 32 54N 84 51W
Durand, *Ill., U.S.A.* **156 B7** 42 26N 89 20W
Durand, *Mich., U.S.A.* **157 B13** 42 55N 83 59W
Durango, *Mexico* **162 C4** 24 3N 104 39W
Durango, *U.S.A.* **159 H10** 37 16N 107 53W
Durango □, *Mexico* **162 C4** 25 0N 105 0W
Duranillin, *Australia* **125 F2** 33 30 S 116 45 E
Durankulak, *Bulgaria* **51 C12** 43 41N 28 32 E
Durant, *Iowa, U.S.A.* **156 C6** 41 36N 90 54W
Durant, *Okla., U.S.A.* **155 J6** 33 59N 96 25W
Duratón →, *Spain* **42 D6** 41 37N 4 7W
Durazno, *Uruguay* **174 C4** 33 25 S 56 31W
Durazzo = Durrësi, *Albania* **50 E3** 41 19N 19 28 E
Durban, *France* **28 F6** 42 59N 2 49 E
Durban, *S. Africa* **117 D5** 29 49 S 31 1 E
Durbo, *Somali Rep.* **120 B4** 11 37N 50 20 E
Durbuy, *Belgium* **24 D5** 50 21N 5 28 E
Dúrcal, *Spain* **43 J7** 36 59N 3 34W
Đurđevac, *Croatia* **52 D2** 46 2N 17 3 E
Düren, *Germany* **30 E2** 50 48N 6 29 E
Durg, *India* **94 D5** 21 15N 81 22 E
Durgapur, *India* **93 H12** 23 30N 87 20 E
Durham, *Canada* **140 D3** 44 10N 80 49W
Durham, *U.K.* **20 C6** 54 47N 1 34W
Durham, *Calif., U.S.A.* **160 F5** 39 39N 121 48W
Durham, *N.C., U.S.A.* **149 H6** 35 59N 78 54W
Durham □, *U.K.* **20 C6** 54 42N 1 45W
Durham Downs, *Australia* **127 D4** 26 6 S 149 5 E
Durleşti, *Moldova* **53 C13** 47 1N 28 46 E
Ḍurmā, *Si. Arabia* **98 A4** 24 37N 46 8 E
Durmitor, *Montenegro, Yug.* **50 C2** 43 10N 19 0 E
Durness, *U.K.* **22 C4** 58 34N 4 45W
Durrësi, *Albania* **50 E3** 41 19N 19 28 E
Durrie, *Australia* **126 D3** 25 40 S 140 15 E
Durrow, *Ireland* **23 D4** 52 51N 7 24W
Dursey I., *Ireland* **23 E1** 51 36N 10 12W
Dursunbey, *Turkey* **49 B10** 39 35N 28 37 E
Durtal, *France* **26 E6** 47 40N 0 18W
Duru,
Dem. Rep. of the Congo **118 B2** 4 14N 28 50 E
Duru Gölü, *Turkey* **51 E12** 41 20N 28 35 E
Durusu, *Turkey* **51 E12** 41 17N 28 41 E
Durūz, Jabal ad, *Jordan* **103 C5** 32 35N 36 40 E
D'Urville, Tanjung,
Indonesia **83 B5** 1 28 S 137 54 E
D'Urville I., *N.Z.* **131 A8** 40 50 S 173 55 E
Duryea, *U.S.A.* **151 E9** 41 20N 75 45W
Dusa Mareb, *Somali Rep.* **120 C3** 5 30N 46 15 E
Dûsh, *Egypt* **106 C3** 24 35N 30 41 E
Dushak, *Turkmenistan* **64 F7** 37 13N 60 1 E
Dushan, *China* **76 E6** 25 48N 107 30 E
Dushanbe, *Tajikistan* **63 D4** 38 33N 68 48 E
Dusheti, *Georgia* **61 J7** 42 10N 44 42 E
Dusky Sd., *N.Z.* **131 F1** 45 47 S 166 30 E
Dussejour, C., *Australia* **124 B4** 14 45 S 128 13 E
Düsseldorf, *Germany* **30 D2** 51 14N 6 47 E
Duszniki-Zdrój, *Poland* **55 H3** 50 24N 16 24 E
Dutch Harbor, *U.S.A.* **138 C3** 53 53N 166 32W
Dutlwe, *Botswana* **116 C3** 23 58 S 23 46 E
Dutsan Wai, *Nigeria* **113 C6** 10 50N 8 10 E
Dutton, *Canada* **150 D3** 42 39N 81 30W
Dutton →, *Australia* **126 C3** 20 44 S 143 10 E
Duved, *Sweden* **16 A6** 63 24N 12 55 E
Düvertepe, *Turkey* **49 B10** 39 14N 28 27 E
Duyun, *China* **76 D6** 26 18N 107 29 E
Düzağac, *Turkey* **49 C12** 38 48N 30 10 E
Düzce, *Turkey* **100 B4** 40 50N 31 10 E
Duzdab = Zāhedān, *Iran* **97 D9** 29 30N 60 50 E
Dve Mogili, *Bulgaria* **51 C9** 43 35N 25 55 E
Dvina, Severnaya →, *Russia* **56 B7** 64 32N 40 30 E
Dvinsk = Daugavpils, *Latvia* **15 J22** 55 53N 26 32 E
Dvinskaya Guba, *Russia* **65 B6** 65 0N 39 0 E
Dvor, *Croatia* **45 C13** 45 4N 16 22 E
Dvůr Králové nad Labem,
Czech Rep. **34 A8** 50 27N 15 50 E
Dwarka, *India* **92 H3** 22 18N 69 8 E
Dwellingup, *Australia* **125 F2** 32 43 S 116 4 E
Dwight, *Canada* **150 A5** 45 20N 79 1W
Dwight, *U.S.A.* **157 C8** 41 5N 88 26W
Dyatkovo, *Russia* **58 F8** 53 40N 34 27 E
Dyatlovo = Dzyatlava,
Belarus **58 F3** 53 28N 25 28 E
Dyce, *U.K.* **22 D6** 57 13N 2 12W
Dyer, *U.S.A.* **157 G10** 37 24N 86 13W
Dyer, C., *Canada* **139 B13** 66 40N 61 0W
Dyer Plateau, *Antarctica* **7 D17** 70 45 S 65 30W
Dyerbeldzhin, *Kyrgyzstan* **63 C7** 41 13N 74 54 E
Dyersburg, *U.S.A.* **155 G10** 36 3N 89 23W
Dyersville, *U.S.A.* **156 B5** 42 29N 91 8W
Dyfi →, *U.K.* **21 E3** 52 32N 4 3W
Dyje →, *Czech Rep.* **35 C9** 48 37N 16 56 E
Dymer, *Ukraine* **59 G6** 50 47N 30 18 E
Dynevor Downs, *Australia* **127 D3** 28 10 S 144 20 E
Dynów, *Poland* **55 J9** 49 50N 22 11 E
Dyranut, *Norway* **18 D4** 60 22N 7 31 E
Dyrhólaey, *Iceland* **11 D7** 63 24N 19 8W
Dyrnes, *Norway* **18 A4** 63 25N 7 52 E
Dysart, *Canada* **143 C8** 50 57N 104 2W
Dyurtyuli, *Russia* **62 D5** 55 9N 54 40 E
Dzamin Üüd, *Mongolia* **74 C6** 43 50N 111 58 E
Dzerzhinsk, *Russia* **60 B6** 56 14N 43 30 E
Dzhalal-Abad = Jalal-Abad,
Kyrgyzstan **63 C6** 40 56N 73 0 E
Dzhalinda, *Russia* **65 D13** 53 26N 124 0 E
Dzhambeyty, *Kazakstan* **62 F4** 50 16N 52 35 E
Dzhambul = Zhambyl,
Kazakstan **63 B5** 42 54N 71 22 E
Dzhambul, Gora =
Zhambyl, Gora, *Kazakstan* **63 A6** 44 54N 73 0 E
Dzhankoy, *Ukraine* **59 K8** 45 40N 34 20 E
Dzhanybek, *Kazakstan* **60 F8** 49 25N 46 50 E
Dzharkurgan = Jarqūrghon,
Uzbekistan **63 E3** 37 31N 67 25 E
Dzharylhach, Ostriv, *Ukraine* **59 J7** 46 2N 32 55 E
Dzhetym, Khrebet,
Kyrgyzstan **63 C8** 41 30N 77 0 E
Dzhezkazgan =
Zhezqazghan, *Kazakstan* **64 E7** 47 44N 67 40 E
Dzhizak = Jizzakh,
Uzbekistan **63 C3** 40 6N 67 50 E
Dzhugdzur, Khrebet, *Russia* **65 D14** 57 30N 138 0 E
Dzhuma, *Uzbekistan* **63 D3** 39 42N 66 40 E
Dzhumgoltau, Khrebet,
Kyrgyzstan **63 B7** 42 15N 74 30 E
Dzhungarskiye Vorota =
Dzungarian Gates,
Kazakstan **72 B3** 45 0N 82 0 E
Dzhvari = Jvari, *Georgia* **61 J6** 42 42N 42 4 E
Działdowo, *Poland* **55 E7** 53 15N 20 15 E
Działoszyce, *Poland* **55 H7** 50 22N 20 20 E
Działoszyn, *Poland* **55 G5** 51 6N 18 50 E
Dzierzgoń, *Poland* **54 E6** 53 58N 19 20 E
Dzierżoniów, *Poland* **55 H3** 50 45N 16 39 E
Dzilam de Bravo, *Mexico* **163 C7** 21 24N 88 53W
Dzioua, *Algeria* **111 B6** 33 14N 5 14 E
Dzisna, *Belarus* **58 E5** 55 34N 28 12 E
Dzisna →, *Belarus* **58 E5** 55 34N 28 12 E
Dziwnów, *Poland* **54 D1** 54 2N 14 45 E
Dzungaria = Junggar Pendi,
China **72 B3** 44 30N 86 0 E
Dzungarian Gates, *Kazakstan* **72 B3** 45 0N 82 0 E
Dzuumod, *Mongolia* **72 B5** 47 45N 106 58 E
Dzyarzhynsk, *Belarus* **58 F4** 53 40N 27 1 E
Dzyatlava, *Belarus* **58 F3** 53 28N 25 28 E

E

Eabamet, L., *Canada* **140 B2** 51 30N 87 46W
Eads, *U.S.A.* **154 F3** 38 29N 102 47W
Eagle, *U.S.A.* **158 G10** 39 39N 106 50W
Eagle →, *Canada* **141 B8** 53 36N 57 26W
Eagle Butte, *U.S.A.* **154 C4** 45 0N 101 10W
Eagle Cr. →, *U.S.A.* **157 F11** 38 36N 85 4W
Eagle Grove, *U.S.A.* **156 B3** 42 40N 93 54W
Eagle L., *Calif., U.S.A.* **158 F3** 40 39N 120 45W
Eagle L., *Maine, U.S.A.* **141 C6** 46 20N 69 22W
Eagle Lake, *U.S.A.* **155 L6** 29 35N 96 20W
Eagle Mountain, *U.S.A.* **161 M11** 33 49N 115 27W
Eagle Nest, *U.S.A.* **159 H11** 36 33N 105 16W
Eagle Pass, *U.S.A.* **155 L4** 28 43N 100 30W
Eagle Pk., *U.S.A.* **160 G7** 38 10N 119 25W
Eagle Pt., *Australia* **124 C3** 16 11 S 124 23 E
Eagle River, *U.S.A.* **154 C10** 45 55N 89 15W
Eagleville, *U.S.A.* **156 D3** 40 28N 93 59W
Ealing, *U.K.* **21 F7** 51 31N 0 20W
Earaheedy, *Australia* **125 E3** 25 34 S 121 29 E
Earl Grey, *Canada* **143 C8** 50 57N 104 43W
Earle, *U.S.A.* **155 H9** 35 16N 90 28W
Earlimart, *U.S.A.* **161 K7** 35 53N 119 16W
Earlville, *U.S.A.* **157 C8** 41 35N 88 55W
Early Branch, *U.S.A.* **152 C9** 32 45N 80 55W
Earn →, *U.K.* **22 E5** 56 21N 3 18W
Earn, L., *U.K.* **22 E4** 56 23N 4 13W
Earnslaw, Mt., *N.Z.* **131 E3** 44 32 S 168 27 E
Earth, *U.S.A.* **155 H3** 34 14N 102 24W
Easley, *U.S.A.* **149 H4** 34 50N 82 36W
East Angus, *Canada* **141 C5** 45 30N 71 40W
East Aurora, *U.S.A.* **150 D6** 42 46N 78 37W
East Ayrshire □, *U.K.* **22 F4** 55 26N 4 11W
East B., *Fla., U.S.A.* **152 E4** 30 5N 85 32W
East B., *La., U.S.A.* **155 L10** 29 0N 89 15W
East Beskids = Vychodné
Beskydy, *Europe* **35 B15** 49 20N 22 0 E
East Brady, *U.S.A.* **150 F5** 40 59N 79 36W
East C., *N.Z.* **130 D7** 37 42 S 178 35 E
East C., *Papua N. G.* **132 F6** 10 13 S 150 53 E
East Chicago, *U.S.A.* **157 C9** 41 38N 87 27W
East China Sea, *Asia* **73 D7** 30 0N 126 0 E
East Coast Bays, *N.Z.* **130 C3** 36 46 S 174 46 E
East Coulee, *Canada* **142 C6** 51 23N 112 27W
East Dereham, *U.K.* **21 E8** 52 41N 0 57 E
East Dublin, *U.S.A.* **152 C7** 32 32N 82 52W
East Dubuque, *U.S.A.* **156 B6** 42 30N 90 39W
East Dunbartonshire □,
U.K. **22 F4** 55 57N 4 13W
East Falkland, *Falk. Is.* **176 D5** 51 30 S 58 30W
East Grand Forks, *U.S.A.* **154 B6** 47 56N 97 1W

210 East Greenwich

East Greenwich, *U.S.A.* **151 E13** 41 40N 71 27W
East Grinstead, *U.K.* **21 F8** 51 7N 0 0 E
East Hartford, *U.S.A.* **151 E12** 41 46N 72 39W
East Helena, *U.S.A.* **158 C8** 46 35N 111 56W
East Indies, *Asia* **66 K15** 0 0 120 0 E
East Jordan, *U.S.A.* **148 C3** 45 10N 85 7W
East Kilbride, *U.K.* **22 F4** 55 47N 4 11W
East Lansing, *U.S.A.* **157 B12** 42 44N 84 29W
East Liverpool, *U.S.A.* **150 F4** 40 37N 80 35W
East London, *S. Africa* **117 E4** 33 0 S 27 55 E
East Lothian □, *U.K.* **22 F6** 55 58N 2 44W
East Lynne, *Australia* **129 C9** 35 35 S 150 16 E
East Main = Eastmain, *Canada* **140 B4** 52 10N 78 30W
East Moline, *U.S.A.* **156 C6** 41 32N 90 26W
East Naples, *U.S.A.* **153 J8** 26 8N 81 46W
East Orange, *U.S.A.* **151 F10** 40 46N 74 13W
East Pacific Ridge, *Pac. Oc.* **135 J17** 15 0 S 110 0W
East Palatka, *U.S.A.* **152 F5** 29 39N 81 36W
East Palestine, *U.S.A.* **150 F4** 40 50N 80 33W
East Peoria, *U.S.A.* **156 D7** 40 40N 89 34W
East Pine, *Canada* **142 B4** 55 48N 120 12W
East Point, *U.S.A.* **152 B5** 33 41N 84 27W
East Providence, *U.S.A.* **151 E13** 41 49N 71 23W
East Pt., *Canada* **141 C7** 46 27N 61 58W
East Renfrewshire □, *U.K.* **22 F4** 55 46N 4 21W
East Retford = Retford, *U.K.* **20 D7** 53 19N 0 56W
East Riding of Yorkshire □, *U.K.* **20 D7** 53 55N 0 30W
East St. Louis, *U.S.A.* **156 F6** 38 37N 90 9W
East Schelde → = Oosterschelde, *Neths.* **24 C4** 51 33N 4 0 E
East Siberian Sea, *Russia* **65 B17** 73 0N 160 0 E
East Stroudsburg, *U.S.A.* **151 E9** 41 1N 75 11W
East Sussex □, *U.K.* **21 G8** 50 56N 0 19 E
East Tawas, *U.S.A.* **148 C4** 44 17N 83 29W
East Tohopekaliga Lake, *U.S.A.* **153 G8** 28 18N 81 15W
East Toorale, *Australia* **127 E4** 30 27 S 145 28 E
East Troy, *U.S.A.* **157 B8** 42 47N 88 24W
East Walker →, *U.S.A.* **160 G7** 38 52N 119 10W
Eastbourne, *N.Z.* **130 H3** 41 19 S 174 55 E
Eastbourne, *U.K.* **21 G8** 50 46N 0 18 E
Eastend, *Canada* **143 D7** 49 32N 108 50W
Easter Islands = Pascua, I. de, *Pac. Oc.* **135 K17** 27 0 S 109 0W
Eastern □, *Kenya* **118 C4** 0 0 38 30 E
Eastern □, *Uganda* **118 B3** 1 50N 33 45 E
Eastern Cape □, *S. Africa* **116 E4** 32 0 S 26 0 E
Eastern Cr. →, *Australia* **126 C3** 20 40 S 141 35 E
Eastern Ghats, *India* **95 H4** 14 0N 78 50 E
Eastern Group = Lau Group, *Fiji* **133 A3** 17 0 S 178 30W
Eastern Group, *Australia* **125 F3** 33 30 S 124 30 E
Eastern Province □, *S. Leone* **112 D2** 8 15N 11 0W
Eastern Samar □, *Phil.* **81 F5** 11 40N 125 40 E
Eastern Transvaal = Mpumalanga □, *S. Africa* **117 B5** 26 0 S 30 0 E
Easterville, *Canada* **143 C9** 53 8N 99 49W
Easthampton, *U.S.A.* **151 D12** 42 16N 72 40W
Eastland, *U.S.A.* **155 J5** 32 24N 98 49W
Eastleigh, *U.K.* **21 G6** 50 58N 1 21W
Eastmain, *Canada* **140 B4** 52 10N 78 30W
Eastmain →, *Canada* **140 B4** 52 27N 78 26W
Eastman, *Canada* **151 A12** 45 18N 72 19W
Eastman, *Ga., U.S.A.* **152 C6** 32 12N 83 11W
Eastman, *Wis., U.S.A.* **156 A5** 43 10N 91 1W
Easton, *Md., U.S.A.* **148 F7** 38 47N 76 5W
Easton, *Pa., U.S.A.* **151 F9** 40 41N 75 13W
Easton, *Wash., U.S.A.* **160 C5** 47 14N 121 11W
Eastover, *U.S.A.* **152 B9** 33 52N 80 41W
Eastpoint, *U.S.A.* **152 F5** 29 44N 84 53W
Eastport, *U.S.A.* **141 D6** 44 56N 67 0W
Eastsound, *U.S.A.* **160 B4** 48 42N 122 55W
Eaton, *Colo., U.S.A.* **154 E2** 40 32N 104 42W
Eaton, *Ohio, U.S.A.* **157 E12** 39 45N 84 38W
Eaton Rapids, *U.S.A.* **157 B12** 42 31N 84 39W
Eatonia, *Canada* **143 C7** 51 13N 109 25W
Eatonton, *U.S.A.* **152 B6** 33 20N 83 23W
Eatontown, *U.S.A.* **151 F10** 40 19N 74 4W
Eatonville, *U.S.A.* **160 D4** 46 52N 122 16W
Eau Claire, *Fr. Guiana* **169 C7** 3 30N 53 40W
Eau Claire, *U.S.A.* **154 C9** 44 49N 91 30W
Eauze, *France* **28 E4** 43 53N 0 7 E
Ebagoola, *Australia* **126 A3** 14 15 S 143 12 E
Eban, *Nigeria* **113 D5** 9 40N 4 50 E
Ebangalakata, *Dem. Rep. of the Congo* **114 C4** 0 29 S 21 29 E
Ebbw Vale, *U.K.* **21 F4** 51 46N 3 12W
Ebebiyín, *Eq. Guin.* **114 B2** 2 9N 11 20 E
Ebeggui, *Algeria* **111 C6** 26 2N 6 0 E
Ebel, *Gabon* **114 B2** 0 7N 11 5 E
Ebeltoft, *Denmark* **17 H4** 56 12N 10 41 E
Ebeltoft Vig, *Denmark* **17 H4** 56 10N 10 35 E
Ebensburg, *U.S.A.* **150 F6** 40 29N 78 44W
Ebensee, *Austria* **34 D6** 47 48N 13 46 E
Eber Gölü, *Turkey* **100 C4** 38 38N 31 11 E
Eberbach, *Germany* **31 F4** 49 28N 8 59 E
Eberswalde-Finow, *Germany* **30 C9** 52 50N 13 49 E
Ebetsu, *Japan* **68 C10** 43 7N 141 34 E
Ebian, *China* **76 C4** 29 11N 103 13 E
Ebikon, *Switz.* **33 B6** 47 5N 8 21 E
Ebingen, *Germany* **31 G5** 48 13N 9 1 E
Ebino, *Japan* **70 E2** 32 2N 130 48 E
Ebnat-Kappel, *Switz.* **33 B8** 47 16N 9 7 E
Éboli, *Italy* **47 B8** 40 39N 15 2 E
Ebolowa, *Cameroon* **113 E7** 2 55N 11 10 E
Ebrach, *Germany* **31 F6** 49 51N 10 29 E
Ébrié, Lagune, *Ivory C.* **112 D4** 5 12N 4 26W
Ebro →, *Spain* **40 E5** 40 43N 0 54 E
Ebro, Embalse del, *Spain* **42 C7** 43 0N 3 58W
Ebstorf, *Germany* **30 B6** 53 2N 10 24 E
Eceabat, *Turkey* **51 F10** 40 11N 26 21 E
Ech Cheliff, *Algeria* **111 A5** 36 10N 1 20 E
Echallens, *Switz.* **32 C3** 46 38N 6 38 E
Echechonnee →, *U.S.A.* **152 C6** 32 39N 83 36W
Echeng, *China* **77 B10** 30 23N 114 50 E
Echigo-Sammyaku, *Japan* **69 F9** 36 50N 139 50 E
Échirolles, *France* **29 C9** 45 8N 5 44 E
Echizen-Misaki, *Japan* **71 B7** 35 59N 135 57 E
Echmiadzin = Yejmiadzin, *Armenia* **61 K7** 40 12N 44 19 E
Echo, *U.S.A.* **152 D4** 31 29N 85 28W
Echo Bay, *N.W.T., Canada* **138 B8** 66 5N 117 55W
Echo Bay, *Ont., Canada* **140 C3** 46 29N 84 4W
Echoing →, *Canada* **143 B10** 55 51N 92 5W
Echternach, *Lux.* **24 E6** 49 49N 6 25 E
Echuca, *Australia* **129 D6** 36 10 S 144 20 E
Ecija, *Spain* **43 H5** 37 30N 5 10W
Eckental, *Germany* **31 F7** 49 35N 11 12 E
Eckernförde, *Germany* **30 A5** 54 28N 9 50 E
Eclectic, *U.S.A.* **152 C3** 32 38N 86 2W
Eclipse Is., *Australia* **124 B4** 13 54 S 126 19 E
Écommoy, *France* **26 E7** 47 50N 0 17 E
Econfina, *U.S.A.* **152 E4** 30 22N 85 35W
Ecoporanga, *Brazil* **171 E3** 18 23 S 40 50W
Écouché, *France* **26 D6** 48 42N 0 10W
Ecuador ■, *S. Amer.* **168 D2** 2 0 S 78 0W
Écueillé, *France* **26 E8** 47 5N 1 21 E
Ed, *Sweden* **17 F5** 58 55N 11 55 E
Ed Dabbura, *Sudan* **106 D3** 17 40N 34 15 E
Ed Dâmer, *Sudan* **106 D3** 17 27N 34 0 E
Ed Debba, *Sudan* **106 D3** 18 0N 30 51 E
Ed-Déffa, *Egypt* **106 A2** 30 40N 26 30 E
Ed Deim, *Sudan* **107 E2** 10 10N 28 20 E
Ed Dueim, *Sudan* **107 E3** 14 0N 32 10 E
Edah, *Australia* **125 E2** 28 16 S 117 10 E
Edam, *Canada* **143 C7** 53 11N 108 46W
Edam, *Neths.* **24 B5** 52 31N 5 3 E
Edane, *Sweden* **16 E6** 59 38N 12 49 E
Edapally, *India* **95 J4** 11 19N 78 3 E
Eday, *U.K.* **22 B6** 59 11N 2 47W
Edd, *Eritrea* **107 E5** 14 0N 41 38 E
Eddrachillis B., *U.K.* **22 C3** 58 17N 5 14W
Eddystone Pt., *Australia* **126 G4** 40 59 S 148 20 E
Eddyville, *U.S.A.* **156 C4** 41 9N 92 38W
Ede, *Neths.* **24 B5** 52 4N 5 40 E
Ede, *Nigeria* **113 D5** 7 45N 4 29 E
Edéa, *Cameroon* **113 E7** 3 51N 10 9 E
Edebäck, *Sweden* **16 D7** 60 4N 13 32 E
Edehon L., *Canada* **143 A9** 60 25N 97 15W
Edekel, Adrar, *Algeria* **111 D6** 23 56N 6 47 E
Edelény, *Hungary* **52 B5** 48 18N 20 44 E
Eden, *Australia* **129 D8** 37 3 S 149 55 E
Eden, *N.C., U.S.A.* **149 G6** 36 29N 79 53W
Eden, *N.Y., U.S.A.* **150 D6** 42 39N 78 55W
Eden, *Tex., U.S.A.* **155 K5** 31 13N 99 51W
Eden, *Wyo., U.S.A.* **158 E9** 42 3N 109 26W
Eden →, *U.K.* **20 C4** 54 57N 3 1W
Eden L., *Canada* **143 B8** 56 38N 100 15W
Edenburg, *S. Africa* **116 D4** 29 43 S 25 58 E
Edendale, *N.Z.* **131 G3** 46 19 S 168 48 E
Edendale, *S. Africa* **117 D5** 29 39 S 30 18 E
Edenderry, *Ireland* **23 C4** 53 21N 7 4W
Edenton, *U.S.A.* **149 G7** 36 4N 76 39W
Edenville, *S. Africa* **117 D4** 27 37 S 27 34 E
Eder →, *Germany* **30 D5** 51 12N 9 28 E
Eder-Stausee, *Germany* **30 D4** 51 10N 8 57 E
Edewecht, *Germany* **30 B3** 53 8N 7 58 E
Edgar, *U.S.A.* **154 E6** 40 22N 97 58W
Edgartown, *U.S.A.* **151 E14** 41 23N 70 31W
Edge Hill, *U.K.* **21 E6** 52 8N 1 26W
Edgecumbe, *N.Z.* **130 D5** 37 59 S 176 47 E
Edgefield, *U.S.A.* **152 B9** 33 47N 81 56W
Edgeley, *U.S.A.* **154 B5** 46 22N 98 43W
Edgemont, *U.S.A.* **154 D3** 43 18N 103 50W
Edgeøya, *Svalbard* **6 B9** 77 45N 22 30 E
Edgerton, *Ohio, U.S.A.* **157 C12** 41 27N 84 45W
Edgerton, *Wis., U.S.A.* **156 B7** 42 50N 89 4W
Edgewater, *U.S.A.* **153 G9** 28 59N 80 54W
Edgewood, *U.S.A.* **157 F8** 38 55N 88 40W
Édhessa, *Greece* **50 F6** 40 48N 22 5 E
Edievale, *N.Z.* **131 F4** 45 49 S 169 22 E
Edina, *Liberia* **112 D2** 6 0N 10 10W
Edina, *U.S.A.* **156 D4** 40 10N 92 11W
Edinburg, *Ill., U.S.A.* **156 E7** 39 39N 89 23W
Edinburg, *Ind., U.S.A.* **157 E11** 39 21N 85 58W
Edinburg, *Tex., U.S.A.* **155 M5** 26 18N 98 10W
Edinburgh, *U.K.* **22 F5** 55 57N 3 13W
Edineţ, *Moldova* **53 B12** 48 9N 27 18 E
Edirne, *Turkey* **51 E10** 41 40N 26 34 E
Edirne □, *Turkey* **51 E10** 41 12N 26 30 E
Edison, *Ga., U.S.A.* **152 D5** 31 34N 84 44W
Edison, *Wash., U.S.A.* **160 B4** 48 33N 122 27W
Edisto →, *U.S.A.* **152 C9** 32 29N 80 21W
Edisto Beach, *U.S.A.* **152 C9** 32 29N 80 20W
Edisto I., *U.S.A.* **152 C9** 32 35N 80 20W
Edithburgh, *Australia* **128 C2** 35 5 S 137 43 E
Edjeleh, *Algeria* **111 C6** 28 38N 9 50 E
Edjudina, *Australia* **125 E3** 29 48 S 122 23 E
Edmeston, *U.S.A.* **151 D9** 42 42N 75 15W
Edmond, *U.S.A.* **155 H6** 35 39N 97 29W
Edmonds, *U.S.A.* **160 C4** 47 49N 122 23W
Edmonton, *Australia* **126 B4** 17 2 S 145 46 E
Edmonton, *Canada* **142 C6** 53 30N 113 30W
Edmund L., *Canada* **143 C10** 54 45N 93 17W
Edmundston, *Canada* **141 C6** 47 23N 68 20W
Edna, *U.S.A.* **155 L6** 28 59N 96 39W
Edna Bay, *U.S.A.* **142 B2** 55 55N 133 40W
Edo □, *Nigeria* **113 D6** 6 30N 6 0 E
Edolo, *Italy* **44 B7** 46 10N 10 21 E
Edøy, *Norway* **18 A5** 63 18N 8 10 E
Edremit, *Turkey* **49 B9** 39 34N 27 0 E
Edremit Körfezi, *Turkey* **49 B8** 39 30N 26 45 E
Edsbro, *Sweden* **16 E12** 59 54N 18 29 E
Edsbyn, *Sweden* **16 C9** 61 23N 15 49 E
Edson, *Canada* **142 C5** 53 35N 116 28W
Eduardo Castex, *Argentina* **174 D3** 35 50 S 64 18W
Edward →, *Australia* **128 C5** 35 5 S 143 30 E
Edward, L., *Africa* **118 C2** 0 25 S 29 40 E
Edward I., *Canada* **140 C2** 48 22N 88 37W
Edward River, *Australia* **126 A3** 14 59 S 141 26 E
Edward VII Land, *Antarctica* **7 E13** 80 0 S 150 0W
Edwards, *U.S.A.* **161 L9** 34 55N 117 51W
Edwards →, *U.S.A.* **156 C6** 41 9N 90 59W
Edwards Plateau, *U.S.A.* **155 K4** 30 45N 101 20W
Edwardsburg, *U.S.A.* **157 C10** 41 48N 86 6W
Edwardsport, *U.S.A.* **157 F9** 38 49N 87 15W
Edwardsville, *Ill., U.S.A.* **156 F7** 38 49N 89 58W
Edwardsville, *Pa., U.S.A.* **151 E9** 41 15N 75 56W
Edzo, *Canada* **142 A5** 62 49N 116 4W
Eeklo, *Belgium* **24 C3** 51 11N 3 33 E
Eel →, *Ind., U.S.A.* **157 E10** 39 7N 86 57W
Eel →, *Ind., U.S.A.* **157 D10** 40 45N 86 22W
Efate, I., *Vanuatu* **133 G6** 17 40 S 168 25 E
Eferding, *Austria* **34 C7** 48 18N 14 1 E
Eferi, *Algeria* **111 D6** 24 30N 9 28 E
Effingham, *Ill., U.S.A.* **157 E8** 39 7N 88 33W
Effingham, *S.C., U.S.A.* **152 A10** 34 5N 79 46W
Effretikon, *Switz.* **33 B7** 47 25N 8 42 E
Eforie, *Romania* **53 F13** 44 1N 28 37 E
Efteløt, *Norway* **18 E6** 59 33N 9 49 E
Ega →, *Spain* **40 C3** 42 19N 1 55W
Égadi, Ísole, *Italy* **46 E5** 37 55N 12 16 E
Eganville, *Canada* **140 C4** 45 32N 77 5W
Egedesminde, *Greenland* **10 D5** 68 43N 52 56W
Egegik, *U.S.A.* **144 G8** 58 13N 157 22W
Egeland, *U.S.A.* **154 A5** 48 38N 99 6W
Egenolf L., *Canada* **143 B9** 59 3N 100 0W
Eger = Cheb, *Czech Rep.* **34 A5** 50 9N 12 28 E
Eger, *Hungary* **52 C5** 47 53N 20 27 E
Eger →, *Hungary* **52 C5** 47 38N 20 50 E
Egersund, *Norway* **15 G12** 58 26N 6 1 E
Egg L., *Canada* **143 B7** 55 5N 105 30W
Eggedal, *Norway* **18 D6** 60 14N 9 22 E
Eggenburg, *Austria* **34 C8** 48 38N 15 50 E
Eggenfelden, *Germany* **31 G8** 48 23N 12 46 E
Eggiwil, *Switz.* **32 C5** 46 52N 7 47 E
Egherta, *Somali Rep.* **120 D2** 2 4N 43 11 E
Éghezée, *Belgium* **24 D4** 50 35N 4 55 E
Egilsstaðir, *Iceland* **11 B12** 65 16N 14 25W
Eginbah, *Australia* **124 D2** 20 53 S 119 47 E
Egito, *Angola* **115 E2** 12 4 S 13 58 E
Égletons, *France* **28 C6** 45 24N 2 3 E
Eglisau, *Switz.* **33 A7** 47 35N 8 31 E
Egmont, C., *N.Z.* **130 F2** 39 16 S 173 45 E
Egmont, Mt., *N.Z.* **130 F3** 39 17 S 174 5 E
Eğridir, *Turkey* **100 D4** 37 52N 30 51 E
Eğridir Gölü, *Turkey* **100 D4** 37 53N 30 50 E
Egtved, *Denmark* **17 J3** 55 38N 9 18 E
Éguas →, *Brazil* **171 D3** 13 26 S 44 14W
Egume, *Nigeria* **113 D6** 7 30N 7 14 E
Éguzon-Chantôme, *France* **27 F8** 46 27N 1 33 E
Egvekinot, *Russia* **65 C19** 66 19N 179 50W
Egyek, *Hungary* **52 C5** 47 39N 20 52 E
Egypt ■, *Africa* **106 J7** 28 0N 31 0 E
Eha Amufu, *Nigeria* **113 D6** 6 30N 7 46 E
Ehime □, *Japan* **70 D4** 33 30N 132 40 E
Ehingen, *Germany* **31 G5** 48 16N 9 43 E
Ehrenberg, *U.S.A.* **161 M12** 33 36N 114 31W
Ehrhardt, *U.S.A.* **152 B9** 33 6N 81 1W
Ehrwald, *Austria* **34 D3** 47 24N 10 56 E
Eibar, *Spain* **40 B2** 43 11N 2 28W
Eichstätt, *Germany* **31 G7** 48 54N 11 11 E
Eide, *Hordaland, Norway* **18 D3** 60 31N 6 44 E
Eide, *Møre og Romsdal, Norway* **18 B4** 62 55N 7 27 E
Eider →, *Germany* **30 A4** 54 19N 8 57 E
Eiðar, *Iceland* **11 B12** 65 22N 14 21W
Eidsbugarden, *Norway* **18 C5** 61 23N 8 16 E
Eidsbygda, *Norway* **18 B4** 62 36N 7 30 E
Eidsdal, *Norway* **18 B4** 62 16N 7 10 E
Eidsvåg, *Norway* **18 B5** 62 46N 8 2 E
Eidsvold, *Australia* **127 D5** 25 25 S 151 12 E
Eidsvoll, *Norway* **15 F14** 60 19N 11 14 E
Eifel, *Germany* **31 E2** 50 15N 6 50 E
Eiffel Flats, *Zimbabwe* **119 F3** 18 20 S 30 0 E
Eiger, *Switz.* **44 B5** 46 34N 8 1 E
Eigg, *U.K.* **22 E2** 56 54N 6 10W
Eighty Mile Beach, *Australia* **124 C3** 19 30 S 120 40 E
Eikefjord, *Norway* **18 C2** 61 35N 5 27 E
Eikelandsosen, *Norway* **18 D2** 60 15N 5 43 E
Eiken, *Norway* **18 F4** 58 29N 7 14 E
Eikeren, *Norway* **18 E6** 59 38N 9 58 E
Eikesdal, *Norway* **18 B5** 62 28N 8 12 E
Eil, *Somali Rep.* **120 C3** 8 0N 49 50 E
Eil, L., *U.K.* **22 E3** 56 51N 5 16W
Eildon, *Australia* **129 D6** 37 14 S 145 55 E
Eildon, L., *Australia* **127 F4** 37 10 S 146 0 E
Eileen L., *Canada* **143 A7** 62 16N 107 37W
Eilenburg, *Germany* **30 D8** 51 27N 12 36 E
Ein el Luweiqa, *Sudan* **107 E3** 14 5N 33 50 E
Eina, *Norway* **18 D7** 60 38N 10 35 E
Einarsstaðir, *Iceland* **11 B9** 65 44N 17 24W
Einasleigh, *Australia* **126 B3** 18 32 S 144 5 E
Einasleigh →, *Australia* **126 B3** 17 30 S 142 17 E
Einbeck, *Germany* **30 D5** 51 49N 9 53 E
Eindhoven, *Neths.* **24 C5** 51 26N 5 28 E
Einsiedeln, *Switz.* **33 B7** 47 7N 8 46 E
Eire = Ireland ■, *Europe* **23 C4** 53 50N 7 52W
Eiríksjökull, *Iceland* **11 C6** 64 46N 20 24W
Eiríksstaðir, *Iceland* **11 B11** 65 7N 15 25W
Eirunepé, *Brazil* **172 B4** 6 35 S 69 53W
Eisenach, *Germany* **30 E6** 50 58N 10 19 E
Eisenberg, *Germany* **30 E7** 50 58N 11 54 E
Eisenerz, *Austria* **34 D7** 47 32N 14 54 E
Eisenhüttenstadt, *Germany* **30 C10** 52 9N 14 38 E
Eisenkappel, *Austria* **34 E7** 46 29N 14 36 E
Eisenstadt, *Austria* **35 D9** 47 51N 16 31 E
Eisfeld, *Germany* **31 E6** 50 25N 10 54 E
Eisleben, *Germany* **30 D7** 51 32N 11 32 E
Eislingen, *Germany* **31 G5** 48 41N 9 42 E
Eivindvik, *Norway* **18 D2** 60 59N 5 5 E
Eivissa, *Spain* **39 C7** 38 54N 1 26 E
Eixe, Serra do, *Spain* **42 C4** 42 24N 6 54W
Ejea de los Caballeros, *Spain* **40 C3** 42 7N 1 9W
Ejutla, *Mexico* **163 D5** 16 34N 96 44W
Ekalaka, *U.S.A.* **154 C2** 45 53N 104 33W
Ekalla, *Gabon* **114 C2** 1 27 S 14 0 E
Ekanga, *Dem. Rep. of the Congo* **114 C4** 2 23 S 23 14 E
Ekawasaki, *Japan* **70 D4** 33 13N 132 46 E
Ekenässjön, *Sweden* **17 G9** 57 28N 15 1 E
Ekerö, *Sweden* **16 E11** 59 16N 17 45 E
Eket, *Nigeria* **113 E6** 4 38N 7 56 E
Eketahuna, *N.Z.* **130 G4** 40 38 S 175 43 E
Ekhínos, *Greece* **51 E9** 41 16N 25 1 E
Ekibastuz, *Kazakstan* **64 D8** 51 50N 75 10 E
Ekoli, *Dem. Rep. of the Congo* **114 C4** 0 23 S 24 13 E
Ekoln, *Sweden* **16 E11** 59 45N 17 37 E
Ekshärad, *Sweden* **16 D7** 60 10N 13 30 E
Eksjö, *Sweden* **17 G8** 57 40N 14 58 E
Ekwan →, *Canada* **140 B3** 53 12N 82 15W
Ekwan Pt., *Canada* **140 B3** 53 16N 82 7W
Ekwok, *U.S.A.* **144 G8** 59 22N 157 30W
El Aaiún, *W. Sahara* **110 C2** 27 9N 13 12W
El Aargub, *Mauritania* **110 D1** 23 37N 15 52W
El Abiodh-Sidi-Cheikh, *Algeria* **111 B5** 32 53N 0 31 E
El Adde, *Somali Rep.* **120 D3** 2 35N 46 9 E
El 'Agrûd, *Egypt* **103 E3** 30 14N 34 24 E
El Aïoun, *Morocco* **111 B4** 34 33N 2 30W
El 'Aiyat, *Egypt* **106 J7** 29 36N 31 15 E
El Alamein, *Egypt* **106 H6** 30 48N 28 58 E
El Alto, *Peru* **172 A1** 4 15 S 81 14W
El 'Aqaba, W. →, *Egypt* **103 E2** 30 7N 33 54 E
El 'Arag, *Egypt* **106 B2** 28 40N 26 20 E
El Arahal, *Spain* **43 H5** 37 15N 5 33W
El Aricha, *Algeria* **111 B4** 34 13N 1 10W
El Arîḥā, *West Bank* **103 D4** 31 52N 35 27 E
El Arish, *Australia* **126 B4** 17 35 S 146 1 E
El 'Arîsh, *Egypt* **103 D2** 31 8N 33 50 E
El 'Arîsh, W. →, *Egypt* **103 D2** 31 8N 33 47 E
El Arrouch, *Algeria* **111 A6** 36 37N 6 53 E
El Asnam = Ech Cheliff, *Algeria* **111 A5** 36 10N 1 20 E
El Astillero, *Spain* **42 B7** 43 24N 3 49W
El Badâri, *Egypt* **106 B3** 27 4N 31 25 E
El Bahrein, *Egypt* **106 B2** 28 30N 26 25 E
El Ballâs, *Egypt* **106 B3** 26 2N 32 43 E
El Balyana, *Egypt* **106 B3** 26 10N 32 3 E
El Banco, *Colombia* **168 B3** 9 0N 73 58W
El Baqeir, *Sudan* **106 D3** 18 40N 33 40 E
El Barco de Ávila, *Spain* **42 E5** 40 21N 5 31W
El Barco de Valdeorras = O Barco, *Spain* **42 C4** 42 23N 6 58W
El Bauga, *Sudan* **106 D3** 18 18N 33 52 E
El Baúl, *Venezuela* **168 B4** 8 57N 68 17W
El Bawiti, *Egypt* **106 J6** 28 25N 28 45 E
El Bayadh, *Algeria* **111 B5** 33 40N 1 1 E
El Bierzo, *Spain* **42 C4** 42 45N 6 30W
El Bluff, *Nic.* **164 D3** 11 59N 83 40W
El Bolsón, *Argentina* **176 B2** 41 55 S 71 30W
El Bonillo, *Spain* **41 G2** 38 57N 2 35W
El Brûk, W. →, *Egypt* **103 E2** 30 15N 33 50 E
El Buheirat □, *Sudan* **107 F3** 7 0N 30 0 E
El Bur, *Somali Rep.* **120 D3** 4 40N 46 37 E
El Burgo de Osma, *Spain* **40 D1** 41 35N 3 4W
El Caín, *Argentina* **176 B3** 41 38 S 68 19W
El Cajon, *U.S.A.* **161 N10** 32 48N 116 58W
El Callao, *Venezuela* **169 B5** 7 18N 61 50W
El Campo, *U.S.A.* **155 L6** 29 12N 96 16W
El Carmen, *Bolivia* **173 C5** 13 40 S 63 55W
El Carmen, *Venezuela* **168 C4** 1 16N 66 52W
El Centro, *U.S.A.* **161 N11** 32 48N 115 34W
El Cerro, *Bolivia* **173 D5** 17 30 S 61 40W
El Cerro de Andévalo, *Spain* **43 H4** 37 45N 6 57W
El Cocuy, *Colombia* **168 B3** 6 25N 72 27W
El Compadre, *Mexico* **161 N10** 32 20N 116 14W
El Corcovado, *Argentina* **176 B2** 43 25 S 71 35W
El Coronil, *Spain* **43 H5** 37 5N 5 38W
El Cuy, *Argentina* **176 A3** 39 55 S 68 25W
El Cuyo, *Mexico* **163 C7** 21 30N 87 40W
El Dab'a, *Egypt* **106 H6** 31 0N 28 27 E
El Daheir, *Egypt* **103 D3** 31 13N 34 10 E
El Dambahaddo, *Somali Rep.* **120 D3** 3 17N 46 40 E
El Deir, *Egypt* **106 B3** 25 25N 32 20 E
El Dere, *Ethiopia* **120 C2** 5 6N 43 5 E
El Dere, *Somali Rep.* **120 D3** 3 50N 47 8 E
El Dere, *Somali Rep.* **120 C3** 5 22N 46 11 E
El Descanso, *Mexico* **161 N10** 32 12N 116 58W
El Desemboque, *Mexico* **162 A2** 30 30N 112 57W
El Dilingat, *Egypt* **106 H7** 30 50N 30 31 E
El Diviso, *Colombia* **168 C2** 1 22N 78 14W
El Djem, *Tunisia* **108 A2** 35 18N 10 42 E
El Djouf, *Mauritania* **104 D3** 20 0N 9 0W
El Dorado, *Ark., U.S.A.* **155 J8** 33 12N 92 40W
El Dorado, *Kans., U.S.A.* **155 G6** 37 49N 96 52W
El Dorado, *Venezuela* **169 B5** 6 55N 61 37W
El Eglab, *Algeria* **110 C4** 26 20N 4 30W
El Ejido, *Spain* **43 J8** 36 47N 2 49W
El Escorial, *Spain* **42 E6** 40 35N 4 7W
El Espinar, *Spain* **42 D6** 41 43N 4 15W
El Eulma, *Algeria* **111 A6** 36 9N 5 42 E
El Faiyûm, *Egypt* **106 J7** 29 19N 30 50 E
El Fâsher, *Sudan* **107 E2** 13 33N 25 26 E
El Fashn, *Egypt* **106 J7** 28 50N 30 54 E
El Ferrol = Ferrol, *Spain* **42 B2** 43 29N 8 15W
El Fifi, *Sudan* **107 E2** 10 4N 25 0 E
El Fud, *Ethiopia* **120 C2** 7 15N 42 52 E
El Fuerte, *Mexico* **162 B3** 26 30N 108 40W
El Gal, *Somali Rep.* **120 B4** 10 58N 50 20 E
El Gebir, *Sudan* **107 E2** 13 40N 29 40 E
El Gedida, *Egypt* **106 B2** 25 40N 28 30 E
El Geneina = Al Junaynah, *Sudan* **109 F4** 13 27N 22 45 E
El Geteina, *Sudan* **107 E3** 14 50N 32 27 E
El Gezira □, *Sudan* **107 E3** 15 0N 33 0 E
El Gîza, *Egypt* **106 J7** 30 0N 31 10 E
El Goléa, *Algeria* **111 B5** 30 30N 2 50 E
El Grau, *Spain* **41 G4** 39 0N 0 7W
El Hadeb, *W. Sahara* **110 C2** 25 51N 13 0W
El Hadjira, *Algeria* **111 B6** 32 36N 5 30 E
El Hagiz, *Sudan* **107 D4** 15 15N 35 50 E
El Hajeb, *Morocco* **110 B3** 33 43N 5 13W
El Hammam, *Egypt* **106 H6** 30 52N 29 25 E
El Hammâmi, *Mauritania* **110 D2** 22 3N 11 30W
El Hamurre, *Somali Rep.* **120 C3** 7 13N 48 54 E
El Hank, *Mauritania* **110 D3** 24 30N 7 0W
El Hasian, *W. Sahara* **110 C2** 26 20N 14 0W
El Hawata, *Sudan* **107 E3** 13 25N 34 42 E
El Heiz, *Egypt* **106 B2** 27 50N 28 40 E
El 'Idisât, *Egypt* **106 B3** 25 30N 32 35 E
El Iskandarîya, *Egypt* **106 H7** 31 13N 29 58 E
El Jadida, *Morocco* **110 B3** 33 11N 8 17W
El Jebelein, *Sudan* **107 E3** 12 40N 32 55 E
El Kab, *Sudan* **106 D3** 19 27N 32 46 E
El Kabrît, G., *Egypt* **103 F2** 29 42N 33 16 E
El Kala, *Algeria* **111 A6** 36 50N 8 30 E
El Kalâa, *Morocco* **110 B3** 32 4N 7 27W
El Kamlin, *Sudan* **107 D3** 15 3N 33 11 E
El Kantara, *Algeria* **111 A6** 35 14N 5 45 E
El Kantara, *Tunisia* **108 B2** 33 45N 10 58 E
El Karaba, *Sudan* **106 D3** 18 32N 33 41 E
El Kef, *Tunisia* **108 A1** 36 12N 8 47 E
El Khandaq, *Sudan* **106 D3** 18 30N 30 30 E
El Khârga, *Egypt* **106 B3** 25 30N 30 33 E
El Khartûm, *Sudan* **107 D3** 15 31N 32 35 E
El Khartûm □, *Sudan* **107 D3** 16 0N 33 0 E
El Khartûm Bahrî, *Sudan* **107 D3** 15 40N 32 31 E
El Khroub, *Algeria* **111 A6** 36 10N 6 55 E
El Kseur, *Algeria* **111 A5** 36 46N 4 49 E
El Ksiba, *Morocco* **110 B3** 32 45N 6 1W
El Kuntilla, *Egypt* **103 E3** 30 1N 34 45 E
El Laqâwa, *Sudan* **107 E2** 11 25N 29 1 E
El Laqeita, *Egypt* **106 B3** 25 50N 33 15 E
El Leiya, *Sudan* **107 D4** 16 15N 35 28 E
El Mafâza, *Sudan* **107 E3** 13 38N 34 30 E
El Mahalla el Kubra, *Egypt* **106 H7** 31 0N 31 0 E
El Mahârîq, *Egypt* **106 B3** 25 35N 30 35 E
El Mahmûdîya, *Egypt* **106 H7** 31 10N 30 32 E
El Maitén, *Argentina* **176 B2** 42 3 S 71 10W
El Maiz, *Algeria* **111 C4** 28 19N 0 9W
El-Maks el-Bahari, *Egypt* **106 C3** 24 30N 30 40 E
El Manshâh, *Egypt* **106 B3** 26 26N 31 50 E
El Mansour, *Algeria* **111 C4** 27 47N 0 14W
El Mansûra, *Egypt* **106 H7** 31 0N 31 19 E
El Mantico, *Venezuela* **169 B5** 7 38N 62 45W
El Manzala, *Egypt* **106 H7** 31 10N 31 50 E
El Marâgha, *Egypt* **106 B3** 26 35N 31 10 E
El Masid, *Sudan* **107 D3** 15 15N 33 0 E
El Masnou, *Spain* **40 D7** 41 28N 2 20 E
El Matariya, *Egypt* **106 H8** 31 15N 32 0 E
El Medano, *Canary Is.* **39 F3** 28 3N 16 32W
El Meghaier, *Algeria* **111 B6** 33 55N 5 58 E
El Meraguen, *Algeria* **111 C4** 28 0N 0 7W

Eppingen 211

El Metemma, *Sudan* **107 D3** 16 50N 33 10 E
El Miamo, *Venezuela* **169 B5** 7 39N 61 46W
El Milagro, *Argentina* **174 C2** 30 59 S 65 59W
El Milia, *Algeria* **111 A6** 36 51N 6 13 E
El Minyâ, *Egypt* **106 J7** 28 7N 30 33 E
El Monte, *U.S.A.* **161 L8** 34 4N 118 1W
El Montseny, *Spain* **40 D7** 41 55N 2 25 E
El Mreyye, *Mauritania* **112 B3** 18 0N 6 0W
El Nido, *Phil.* **81 F2** 11 10N 119 25 E
El Obeid, *Sudan* **107 E3** 13 8N 30 10 E
El Odaiya, *Sudan* **107 E2** 12 8N 28 12 E
El Oro, *Mexico* **163 D4** 19 48N 100 8W
El Oro □, *Ecuador* **168 D2** 3 30 S 79 50W
El Oued, *Algeria* **111 B6** 33 20N 6 58 E
El Palmar, *Bolivia* **173 D5** 17 50 S 63 9W
El Palmar, *Venezuela* **169 B5** 7 58N 61 53W
El Palmito, Presa, *Mexico* **162 B3** 25 40N 105 30W
El Paso, *Ill., U.S.A.* **156 D7** 40 44N 89 1W
El Paso, *Tex., U.S.A.* **159 L10** 31 45N 106 29W
El Paso Robles, *U.S.A.* **160 K6** 35 38N 120 41W
El Pedernoso, *Spain* **41 F2** 39 29N 2 45W
El Pedroso, *Spain* **43 H5** 37 51N 5 45W
El Pobo de Dueñas, *Spain* **40 E3** 40 46N 1 39W
El Portal, *U.S.A.* **160 H7** 37 41N 119 47W
El Porvenir, *Mexico* **162 A3** 31 15N 105 51W
El Prat de Llobregat, *Spain* **40 D7** 41 18N 2 3 E
El Progreso, *Honduras* **164 C2** 15 26N 87 51W
El Pueblito, *Mexico* **162 B3** 29 3N 105 4W
El Pueblo, *Canary Is.* **39 F2** 28 36N 17 47W
El Puente del Arzobispo, *Spain* **42 F5** 39 48N 5 10W
El Puerto de Santa María, *Spain* **43 J4** 36 36N 6 13W
El Qâhira, *Egypt* **106 H7** 30 1N 31 14 E
El Qantara, *Egypt* **103 E1** 30 51N 32 20 E
El Qasr, *Egypt* **106 B2** 25 44N 28 42 E
El Quseima, *Egypt* **103 E3** 30 40N 34 15 E
El Qusîya, *Egypt* **106 B3** 27 29N 30 44 E
El Râshda, *Egypt* **106 B2** 25 36N 28 57 E
El Reno, *U.S.A.* **155 H6** 35 32N 97 57W
El Rîdisiya, *Egypt* **106 C3** 24 56N 32 51 E
El Rio, *U.S.A.* **161 L7** 34 14N 119 10W
El Ronquillo, *Spain* **43 H4** 37 44N 6 10W
El Roque, Pta., *Canary Is.* **39 F4** 28 10N 15 25W
El Rosarito, *Mexico* **162 B2** 28 38N 114 4W
El Rubio, *Spain* **43 H5** 37 22N 5 0W
El Saff, *Egypt* **106 J7** 29 34N 31 16 E
El Saheira, W. →, *Egypt* **103 E2** 30 5N 33 25 E
El Salto, *Mexico* **162 C3** 23 47N 105 22W
El Salvador ■, *Cent. Amer.* **164 D2** 13 50N 89 0W
El Sauce, *Nic.* **164 D2** 13 0N 86 40W
El Saucejo, *Spain* **43 H5** 37 4N 5 6W
El Shallal, *Egypt* **106 C3** 24 0N 32 53 E
El Simbillawein, *Egypt* **106 H7** 30 48N 31 13 E
El Sombrero, *Venezuela* **168 B4** 9 23N 67 3W
El Suweis, *Egypt* **106 J8** 29 58N 32 31 E
El Tamarâni, W. →, *Egypt* **103 E3** 30 7N 34 43 E
El Thamad, *Egypt* **103 F3** 29 40N 34 28 E
El Tigre, *Venezuela* **169 B5** 8 44N 64 15W
El Tîh, G., *Egypt* **103 F2** 29 40N 33 50 E
El Tîna, Khalîg, *Egypt* **103 D1** 31 10N 32 40 E
El Tocuyo, *Venezuela* **168 B4** 9 47N 69 48W
El Tofo, *Chile* **174 B1** 29 22 S 71 18W
El Tránsito, *Chile* **174 B1** 28 52 S 70 17W
El Tûr, *Egypt* **106 J8** 28 14N 33 36 E
El Turbio, *Argentina* **176 D2** 51 45 S 72 5W
El Uinle, *Somali Rep.* **120 D2** 3 4N 41 42 E
El Uqsur, *Egypt* **106 B3** 25 41N 32 38 E
El Venado, *Mexico* **162 C4** 22 56N 101 10W
El Vendrell, *Spain* **40 D6** 41 10N 1 30 E
El Vigía, *Venezuela* **168 B3** 8 38N 71 39W
El Viso del Alcor, *Spain* **43 H5** 37 23N 5 43W
El Wabeira, *Egypt* **103 F2** 29 34N 33 6 E
El Wak, *Kenya* **118 B5** 2 49N 40 56 E
El Wak, *Somali Rep.* **120 D2** 2 44N 41 1 E
El Waqf, *Egypt* **106 B3** 25 45N 32 15 E
El Wâsta, *Egypt* **106 J7** 29 19N 31 12 E
El Weguet, *Ethiopia* **107 F5** 5 28N 42 17 E
El Wuz, *Sudan* **107 D3** 15 5N 30 7 E
Elafónisos, *Greece* **48 E4** 36 29N 22 56 E
Elaine, *Australia* **128 D6** 37 44 S 144 2 E
Elamanchili, *India* **94 F6** 17 33N 82 50 E
Élancourt, *France* **27 D8** 48 47N 1 58 E
Elands, *Australia* **129 A10** 31 37 S 152 20 E
Élassa, *Greece* **49 F8** 35 18N 26 21 E
Elassón, *Greece* **48 B4** 39 53N 22 12 E
Elat, *Israel* **103 F3** 29 30N 34 56 E
Eláthia, *Greece* **48 C4** 38 37N 22 46 E
Elâzığ, *Turkey* **101 C8** 38 37N 39 14 E
Elba, *Italy* **44 F7** 42 46N 10 17 E
Elba, *U.S.A.* **152 D3** 31 25N 86 4W
Elbasani, *Albania* **50 E4** 41 9N 20 9 E
Elbe, *U.S.A.* **160 D4** 46 45N 122 10W
Elbe →, *Europe* **30 B4** 53 50N 9 0 E
Elbe-Seitenkanal, *Germany* **30 C6** 52 45N 10 32 E
Elberfeld, *U.S.A.* **157 F9** 38 10N 87 27W
Elbert, Mt., *U.S.A.* **159 G10** 39 7N 106 27W
Elberta, *U.S.A.* **148 C2** 44 37N 86 14W
Elberton, *U.S.A.* **152 A7** 34 7N 82 52W
Elbeuf, *France* **26 C8** 49 17N 1 2 E
Elbing = Elbląg, *Poland* **54 D6** 54 10N 19 25 E
Elbistan, *Turkey* **100 C7** 38 13N 37 15 E
Elbląg, *Poland* **54 D6** 54 10N 19 25 E
Elbląg □, *Poland* **54 D6** 54 15N 19 30 E
Elbow, *Canada* **143 C7** 51 7N 106 35W
Elbrus, *Asia* **61 J6** 43 21N 42 30 E
Elburn, *U.S.A.* **157 C8** 41 54N 88 28W
Elburz Mts. = Alborz, Reshteh-ye Kūhhā-ye, *Iran* **97 C7** 36 0N 52 0 E
Elche, *Spain* **41 G4** 38 15N 0 42W
Elche de la Sierra, *Spain* **41 G2** 38 27N 2 3W
Elcho I., *Australia* **126 A2** 11 55 S 135 45 E
Elda, *Spain* **41 G4** 38 29N 0 47W
Elde →, *Germany* **30 B7** 53 7N 11 15 E
Eldon, *Mo., U.S.A.* **156 F4** 38 21N 92 35W
Eldon, *Wash., U.S.A.* **160 C3** 47 33N 123 3W
Eldora, *U.S.A.* **156 B3** 42 22N 93 5W
Eldorado, *Argentina* **175 B5** 26 28 S 54 43W
Eldorado, *Canada* **143 B7** 59 35N 108 30W
Eldorado, *Mexico* **162 C3** 24 20N 107 22W
Eldorado, *Ill., U.S.A.* **157 G8** 37 49N 88 26W
Eldorado, *Tex., U.S.A.* **155 K4** 30 52N 100 36W
Eldorado Springs, *U.S.A.* **155 G8** 37 52N 94 1W
Eldorendo, *U.S.A.* **152 D5** 31 3N 84 39W
Eldoret, *Kenya* **118 B4** 0 30N 35 17 E
Eldred, *U.S.A.* **150 E6** 41 58N 78 23W
Eldridge, *U.S.A.* **156 C6** 41 39N 90 35W
Elea, C., *Cyprus* **38 D13** 35 19N 34 4 E
Electra, *U.S.A.* **155 H5** 34 2N 98 55W
Elefantes →, *Mozam.* **117 C5** 24 10 S 32 40 E
Elefantes, G., *Chile* **176 C2** 46 28 S 73 49W
Elektrogorsk, *Russia* **58 E10** 55 56N 38 50 E
Elektrostal, *Russia* **58 E10** 55 41N 38 32 E
Elele, *Nigeria* **113 D6** 5 5N 6 50 E
Elena, *Bulgaria* **51 D9** 42 55N 25 53 E
Elephant Butte Reservoir, *U.S.A.* **159 K10** 33 9N 107 11W
Elephant I., *Antarctica* **7 C18** 61 0 S 55 0W
Elephant Pass, *Sri Lanka* **95 K5** 9 35N 80 25 E
Elesbão Veloso, *Brazil* **170 C3** 6 13 S 42 8W
Eleshnitsa, *Bulgaria* **50 E7** 41 52N 23 36 E
Eleşkirt, *Turkey* **101 C10** 39 50N 42 50 E
Eleuthera, *Bahamas* **164 B4** 25 0N 76 20W
Elevsís, *Greece* **48 C5** 38 4N 23 26 E
Elevtheroúpolis, *Greece* **51 F8** 40 52N 24 20 E
Elgå, *Norway* **18 B8** 62 10N 11 56 E
Elgepiggen, *Norway* **18 B8** 62 10N 11 21 E
Elgeyo-Marakwet □, *Kenya* **118 B4** 0 45N 35 30 E
Elgg, *Switz.* **33 B7** 47 29N 8 52 E
Elgin, *N.B., Canada* **141 C6** 45 48N 65 10W
Elgin, *Ont., Canada* **151 B8** 44 36N 76 13W
Elgin, *U.K.* **22 D5** 57 39N 3 19W
Elgin, *Ill., U.S.A.* **157 B8** 42 2N 88 17W
Elgin, *N. Dak., U.S.A.* **154 B4** 46 24N 101 51W
Elgin, *Nebr., U.S.A.* **154 E5** 41 59N 98 5W
Elgin, *Nev., U.S.A.* **159 H6** 37 21N 114 32W
Elgin, *Oreg., U.S.A.* **158 D5** 45 34N 117 55W
Elgin, *S.C., U.S.A.* **152 A9** 34 10N 80 48W
Elgin, *Tex., U.S.A.* **155 K6** 30 21N 97 22W
Elgoibar, *Spain* **40 B2** 43 13N 2 24W
Elgon, Mt., *Africa* **118 B3** 1 10N 34 30 E
Eliase, *Indonesia* **83 C4** 8 21 S 130 48 E
Elida, *U.S.A.* **155 J3** 33 57N 103 39W
Elikón, *Greece* **48 C4** 38 18N 22 45 E
Elim, *S. Africa* **116 E2** 34 35 S 19 45 E
Elim Indian Reservation, *U.S.A.* **144 D7** 64 40N 162 0W
Elin Pelin, *Bulgaria* **50 D7** 42 40N 23 36 E
Elisabethville = Lubumbashi, *Dem. Rep. of the Congo* **119 E2** 11 40 S 27 28 E
Eliseu Martins, *Brazil* **170 C3** 8 13 S 43 42W
Elista, *Russia* **61 G7** 46 16N 44 14 E
Elizabeth, *Australia* **128 C3** 34 42 S 138 41 E
Elizabeth, *Ill., U.S.A.* **156 B6** 42 19N 90 13W
Elizabeth, *N.J., U.S.A.* **151 F10** 40 40N 74 13W
Elizabeth City, *U.S.A.* **149 G7** 36 18N 76 14W
Elizabethton, *U.S.A.* **149 G4** 36 21N 82 13W
Elizabethtown, *Ky., U.S.A.* **148 G3** 37 42N 85 52W
Elizabethtown, *N.Y., U.S.A.* **151 B11** 44 13N 73 36W
Elizabethtown, *Pa., U.S.A.* **151 F8** 40 9N 76 36W
Elizondo, *Spain* **40 B3** 43 12N 1 30W
Ełk, *Poland* **54 E9** 53 50N 22 21 E
Ełk →, *Poland* **54 E9** 53 41N 22 28 E
Elk City, *U.S.A.* **155 H5** 35 25N 99 25W
Elk Creek, *U.S.A.* **160 F4** 39 36N 122 32W
Elk Grove, *U.S.A.* **160 G5** 38 25N 121 22W
Elk Island Nat. Park, *Canada* **142 C6** 53 35N 112 59W
Elk Lake, *Canada* **140 C3** 47 40N 80 25W
Elk Point, *Canada* **143 C6** 53 54N 110 55W
Elk River, *Idaho, U.S.A.* **158 C5** 46 47N 116 11W
Elk River, *Minn., U.S.A.* **154 C8** 45 18N 93 35W
Elkader, *U.S.A.* **156 B5** 42 51N 91 24W
Elkedra, *Australia* **126 C2** 21 9 S 135 33 E
Elkedra →, *Australia* **126 C2** 21 8 S 136 22 E
Elkhart, *Ind., U.S.A.* **157 C11** 41 41N 85 58W
Elkhart, *Kans., U.S.A.* **155 G4** 37 0N 101 54W
Elkhart →, *U.S.A.* **157 C11** 41 41N 85 58W
Elkhorn, *Canada* **143 D8** 49 59N 101 14W
Elkhorn, *U.S.A.* **157 B8** 42 40N 88 33W
Elkhorn →, *U.S.A.* **154 E6** 41 8N 96 19W
Elkhovo, *Bulgaria* **51 D10** 42 10N 26 35 E
Elkin, *U.S.A.* **149 G5** 36 15N 80 51W
Elkins, *U.S.A.* **148 F6** 38 55N 79 51W
Elko, *Canada* **142 D5** 49 20N 115 10W
Elko, *Ga., U.S.A.* **152 C6** 32 20N 83 42W
Elko, *Nev., U.S.A.* **158 F6** 40 50N 115 46W
Ell, L., *Australia* **125 E4** 29 13 S 127 46 E
Ellaville, *U.S.A.* **152 C5** 32 14N 84 19W
Ellef Ringnes I., *Canada* **6 B2** 78 30N 102 2W
Ellendale, *Australia* **124 C3** 17 56 S 124 48 E
Ellendale, *U.S.A.* **154 B5** 46 0N 98 32W
Ellensburg, *U.S.A.* **158 C3** 46 59N 120 34W
Ellenton, *U.S.A.* **152 D6** 31 11N 83 35W
Ellenville, *U.S.A.* **151 E10** 41 43N 74 24W
Ellerston, *Australia* **129 A9** 31 49 S 151 20 E
Ellery, Mt., *Australia* **129 D8** 37 28 S 148 47 E
Ellesmere, L., *N.Z.* **131 H7** 47 47 S 172 28 E
Ellesmere I., *Canada* **6 B4** 79 30N 80 0W
Ellesmere Port, *U.K.* **20 D5** 53 17N 2 54W
Ellettsville, *U.S.A.* **157 E10** 39 14N 86 38W
Ellice Is. = Tuvalu ■, *Pac. Oc.* **134 H9** 8 0 S 178 0 E
Ellinwood, *U.S.A.* **154 F5** 38 21N 98 35W
Elliot, *Australia* **126 B1** 17 33 S 133 32 E
Elliot, *S. Africa* **117 E4** 31 22 S 27 48 E
Elliot Lake, *Canada* **140 C3** 46 25N 82 35W
Elliotdale = Xhora, *S. Africa* **117 E4** 31 55 S 28 38 E
Elliott, *U.S.A.* **152 A9** 34 6N 80 10W
Elliott Key, *U.S.A.* **153 K9** 25 27N 80 12W
Ellis, *U.S.A.* **154 F5** 38 56N 99 34W
Elliston, *Australia* **127 E1** 33 39 S 134 53 E
Ellisville, *U.S.A.* **155 K10** 31 36N 89 12W
Ellon, *U.K.* **22 D6** 57 22N 2 4W
Ellore = Eluru, *India* **94 F5** 16 48N 81 8 E
Elloree, *U.S.A.* **152 B9** 33 32N 80 34W
Ells →, *Canada* **142 B6** 57 18N 111 40W
Ellsworth, *U.S.A.* **154 F5** 38 44N 98 14W
Ellsworth Land, *Antarctica* **7 D16** 76 0 S 89 0W
Ellsworth Mts., *Antarctica* **7 D16** 78 30 S 85 0W
Ellwangen, *Germany* **31 G6** 48 57N 10 8 E
Ellwood City, *U.S.A.* **150 F4** 40 52N 80 17W
Ellzey, *U.S.A.* **153 F7** 29 19N 82 48W
Elm, *Switz.* **33 C8** 46 54N 9 10 E
Elma, *Canada* **143 D9** 49 52N 95 55W
Elma, *U.S.A.* **160 D3** 47 0N 123 25W
Elmadağ, *Turkey* **100 C5** 39 55N 33 14 E
Elmalı, *Turkey* **57 G4** 36 44N 29 56 E
Elmer, *U.S.A.* **156 E4** 39 57N 92 39W
Elmhurst, *U.S.A.* **157 C9** 41 53N 87 56W
Elmina, *Ghana* **113 D4** 5 5N 1 21W
Elmira, *Canada* **150 C4** 43 36N 80 33W
Elmira, *U.S.A.* **150 D8** 42 6N 76 48W
Elmodel, *U.S.A.* **152 D5** 31 21N 84 29W
Elmore, *Australia* **128 D6** 36 30 S 144 37 E
Elmore, *Ala., U.S.A.* **152 C3** 32 32N 86 19W
Elmore, *Calif., U.S.A.* **161 M11** 33 7N 115 49W
Elmore, *Minn., U.S.A.* **157 C13** 41 29N 83 18W
Elmshorn, *Germany* **30 B5** 53 43N 9 40 E
Elmvale, *Canada* **150 B5** 44 35N 79 52W
Elmwood, *U.S.A.* **156 D7** 40 47N 89 58W
Elne, *France* **28 F6** 42 36N 2 58 E
Elnesvågen, *Norway* **18 B4** 62 52N 7 10 E
Elnora, *U.S.A.* **157 F9** 38 53N 87 5W
Elora, *Canada* **150 C4** 43 41N 80 26W
Elorza, *Venezuela* **168 B4** 7 3N 69 31W
Elos, *Greece* **48 E4** 36 46N 22 43 E
Eloúnda, *Greece* **38 D7** 35 16N 25 42 E
Eloy, *U.S.A.* **159 K8** 32 45N 111 33W
Éloyes, *France* **27 D13** 48 6N 6 36 E
Elrose, *Canada* **143 C7** 51 12N 108 0W
Elsas, *Canada* **140 C3** 48 32N 82 55W
Elsdorf, *Germany* **30 E2** 50 55N 6 34 E
Elsie, *U.S.A.* **160 E3** 45 52N 123 36W
Elsinore = Helsingør, *Denmark* **17 H6** 56 2N 12 35 E
Elsinore, *Australia* **129 A6** 31 35 S 145 11 E
Elsinore, *U.S.A.* **159 G7** 38 41N 112 9W
Elster →, *Germany* **30 D7** 51 25N 11 57 E
Elsterwerda, *Germany* **30 D9** 51 27N 13 31 E
Eltham, *Australia* **129 D6** 37 43 S 145 12 E
Eltham, *N.Z.* **130 F3** 39 26 S 174 19 E
Elton, *Russia* **61 F8** 49 5N 46 52 E
Elton, Ozero, *Russia* **61 F8** 49 5N 46 42 E
Eltville, *Germany* **31 E4** 50 2N 8 7 E
Eluru, *India* **94 F5** 16 48N 81 8 E
Elvas, *Portugal* **43 G3** 38 50N 7 10W
Elven, *France* **26 E4** 47 44N 2 36W
Elverum, *Norway* **15 F14** 60 53N 11 34 E
Elvire →, *Australia* **124 C4** 17 51 S 128 11 E
Elvo →, *Italy* **44 C5** 45 23N 8 21 E
Elwood, *Ill., U.S.A.* **157 C8** 41 24N 88 7W
Elwood, *Ind., U.S.A.* **157 D11** 40 17N 85 50W
Elwood, *Nebr., U.S.A.* **154 E5** 40 36N 99 52W
Elx = Elche, *Spain* **41 G4** 38 15N 0 42W
Ely, *U.K.* **21 E8** 52 24N 0 16 E
Ely, *Minn., U.S.A.* **154 B9** 47 55N 91 51W
Ely, *Nev., U.S.A.* **158 G6** 39 15N 114 54W
Elyria, *U.S.A.* **150 E2** 41 22N 82 7W
Elyrus, *Greece* **48 F5** 35 15N 23 45 E
Elz →, *Germany* **31 G3** 48 18N 7 44 E
Emådalen, *Sweden* **16 C8** 61 20N 14 44 E
Emai, *Vanuatu* **133 G6** 17 4 S 168 24 E
Emāmrūd, *Iran* **97 B7** 36 30N 55 0 E
Emån →, *Sweden* **17 G10** 57 8N 16 30 E
Emba, *Kazakstan* **64 E6** 48 50N 58 8 E
Emba →, *Kazakstan* **57 E9** 46 55N 53 28 E
Embarcación, *Argentina* **174 A3** 23 10 S 64 0W
Embarras →, *U.S.A.* **157 F9** 38 39N 87 37W
Embarras Portage, *Canada* **143 B6** 58 27N 111 28W
Embetsu, *Japan* **68 B10** 44 44N 141 47 E
Embi = Emba, *Kazakstan* **64 E6** 48 50N 58 8 E
Embi → = Emba →, *Kazakstan* **57 E9** 46 55N 53 28 E
Embira →, *Brazil* **172 B3** 7 19 S 70 15W
Embóna, *Greece* **38 C9** 36 13N 27 51 E
Embrach, *Switz.* **33 B7** 47 30N 8 36 E
Embrun, *France* **29 D10** 44 34N 6 30 E
Embu, *Kenya* **118 C4** 0 32 S 37 38 E
Embu □, *Kenya* **118 C4** 0 30 S 37 35 E
Emden, *Germany* **30 B3** 53 21N 7 12 E
Emecik, *Turkey* **49 E9** 36 46N 27 49 E
Emerald, *Australia* **126 C4** 23 32 S 148 10 E
Emerson, *Canada* **143 D9** 49 0N 97 10W
Emerson, *U.S.A.* **152 A5** 34 8N 84 45W
Emery, *U.S.A.* **159 G8** 38 55N 111 15W
Emet, *Turkey* **49 B11** 39 20N 29 15 E
Emi Koussi, *Chad* **109 E3** 19 45N 18 55 E
Emília-Romagna □, *Italy* **44 D8** 44 45N 11 0 E
Emilius, Mte., *Italy* **44 C4** 45 45N 7 20 E
Eminabad, *Pakistan* **92 C6** 32 2N 74 8 E
Emine, Nos, *Bulgaria* **51 D11** 42 40N 27 56 E
Eminence, *U.S.A.* **157 F11** 38 22N 85 11W
Emirdağ, *Turkey* **100 C4** 39 2N 31 8 E
Emlenton, *U.S.A.* **150 E5** 41 11N 79 43W
Emlichheim, *Germany* **30 C2** 52 37N 6 51 E
Emmaboda, *Sweden* **17 H9** 56 37N 15 32 E
Emmalane, *U.S.A.* **152 C7** 32 46N 82 0W
Emme →, *Switz.* **32 B5** 47 14N 7 32 E
Emmeloord, *Neths.* **24 B5** 52 44N 5 46 E
Emmen, *Neths.* **24 B6** 52 48N 6 57 E
Emmen, *Switz.* **31 H4** 47 5N 8 18 E
Emmendingen, *Germany* **31 G3** 48 6N 7 51 E
Emmental, *Switz.* **32 C4** 46 55N 7 40 E
Emmerich, *Germany* **30 D2** 51 50N 6 14 E
Emmet, *Australia* **126 C3** 24 45 S 144 30 E
Emmetsburg, *U.S.A.* **156 A2** 43 7N 94 41W
Emmett, *U.S.A.* **158 E5** 43 52N 116 30W
Emmonak, *U.S.A.* **144 E6** 62 46N 164 30W
Emőd, *Hungary* **52 C5** 47 57N 20 47 E
Emona, *Bulgaria* **51 D11** 42 43N 27 53 E
Empalme, *Mexico* **162 B2** 28 1N 110 49W
Empangeni, *S. Africa* **117 D5** 28 50 S 31 52 E
Empedrado, *Argentina* **174 B4** 28 0 S 58 46W
Emperor Seamount Chain, *Pac. Oc.* **134 D9** 40 0N 170 0 E
Empire, *U.S.A.* **152 C6** 32 21N 83 18W
Empoli, *Italy* **44 E7** 43 43N 10 57 E
Emporia, *Kans., U.S.A.* **154 F6** 38 25N 96 11W
Emporia, *Va., U.S.A.* **149 G7** 36 42N 77 32W
Emporium, *U.S.A.* **150 E6** 41 31N 78 14W
Empress, *Canada* **143 C7** 50 57N 110 0W
Empty Quarter = Rub' al Khālī, *Si. Arabia* **99 C5** 18 0N 48 0 E
Ems →, *Germany* **30 B3** 53 20N 7 12 E
Emsdale, *Canada* **150 A5** 45 32N 79 19W
Emsdetten, *Germany* **30 C3** 52 10N 7 32 E
Emu, *Australia* **128 D5** 36 44 S 143 26 E
Emu, *China* **75 C15** 43 40N 128 6 E
Emu Park, *Australia* **126 C5** 23 13 S 150 50 E
'En 'Avrona, *Israel* **103 F4** 29 43N 35 0 E
En Nahud, *Sudan* **107 E2** 12 45N 28 25 E
Ena, *Japan* **71 B9** 35 25N 137 25 E
Ena-San, *Japan* **71 B9** 35 26N 137 36 E
Enambú, *Colombia* **168 C3** 1 1N 70 17W
Enana, *Namibia* **116 B2** 17 30 S 16 23 E
Enånger, *Sweden* **16 C11** 61 30N 17 9 E
Enaratoli, *Indonesia* **83 B5** 3 55 S 136 21 E
Enard B., *U.K.* **22 C3** 58 5N 5 20W
Enare = Inarijärvi, *Finland* **14 B22** 69 0N 28 0 E
Encantadas, Serra, *Brazil* **175 C5** 30 40 S 53 0W
Encarnación, *Paraguay* **175 B4** 27 15 S 55 50W
Encarnación de Diaz, *Mexico* **162 C4** 21 30N 102 13W
Enchi, *Ghana* **112 D4** 5 53N 2 48W
Encinal, *U.S.A.* **155 L5** 28 2N 99 21W
Encinitas, *U.S.A.* **161 M9** 33 3N 117 17W
Encino, *U.S.A.* **159 J11** 34 39N 105 28W
Encontrados, *Venezuela* **168 B3** 9 3N 72 14W
Encounter B., *Australia* **128 C3** 35 45 S 138 45 E
Encruzilhada, *Brazil* **171 E3** 15 31 S 40 54W
Encs, *Hungary* **52 B6** 48 20N 21 8 E
Ende, *Indonesia* **82 C2** 8 45 S 121 40 E
Endeavour, *Canada* **143 C8** 52 10N 102 39W
Endeavour Str., *Australia* **126 A3** 10 45 S 142 0 E
Endelave, *Denmark* **17 J4** 55 46N 10 18 E
Enden, *Norway* **18 C7** 61 47N 10 15 E
Enderbury I., *Kiribati* **134 H10** 3 8 S 171 5W
Enderby, *Canada* **142 C5** 50 35N 119 10W
Enderby I., *Australia* **124 D2** 20 35 S 116 30 E
Enderby Land, *Antarctica* **7 C5** 66 0 S 53 0 E
Enderlin, *U.S.A.* **154 B6** 46 38N 97 36W
Endicott, *N.Y., U.S.A.* **151 D8** 42 6N 76 4W
Endicott, *Wash., U.S.A.* **158 C5** 46 56N 117 41W
Endimari →, *Brazil* **172 B4** 8 46 S 66 7W
Endyalgout I., *Australia* **124 B5** 11 40 S 132 35 E
Ene →, *Peru* **172 C3** 11 10 S 74 18W
Energetik, *Russia* **62 F7** 51 45N 58 45 E
Enewetak Atoll, *Pac. Oc.* **134 F8** 11 30N 162 15 E
Enez, *Turkey* **51 F10** 40 45N 26 5 E
Enfield, *U.S.A.* **157 F8** 38 6N 88 20W
Engadin, *Switz.* **31 J6** 46 45N 10 10 E
Engan, *Norway* **18 A5** 63 8N 8 31 E
Engaño, C., *Dom. Rep.* **165 C6** 18 30N 68 20W
Engaño, C., *Phil.* **80 B4** 18 35N 122 23 E
Engcobo, *S. Africa* **117 E4** 31 37 S 28 0 E
Engelberg, *Switz.* **33 C6** 46 48N 8 26 E
Engels, *Russia* **60 E8** 51 28N 46 6 E
Engemann L., *Canada* **143 B7** 58 0N 106 55W
Engerdal, *Norway* **18 C8** 61 45N 11 58 E
Enggano, *Indonesia* **84 D2** 5 20 S 102 40 E
Engil, *Morocco* **110 B4** 33 12N 4 32W
Engkilili, *Malaysia* **85 B4** 1 3N 111 42 E
England, *U.S.A.* **155 H9** 34 33N 91 58W
England □, *U.K.* **20 D7** 53 0N 2 0W
Englee, *Canada* **141 B8** 50 45N 56 5W
Englefield, *Australia* **128 D4** 37 21 S 141 48 E
Englehart, *Canada* **140 C4** 47 49N 79 52W
Engler L., *Canada* **143 B7** 59 8N 106 52W
Englewood, *Colo., U.S.A.* **154 F2** 39 39N 104 59W
Englewood, *Fla., U.S.A.* **153 J7** 26 58N 82 21W
Englewood, *Kans., U.S.A.* **155 G5** 37 2N 99 59W
Englewood, *Ohio, U.S.A.* **157 E12** 39 53N 84 18W
English, *U.S.A.* **157 F10** 38 20N 86 28W
English →, *Canada* **143 C10** 50 35N 93 30W
English →, *U.S.A.* **156 C5** 41 29N 91 32W
English Bazar = Ingraj Bazar, *India* **93 G13** 24 58N 88 10 E
English Channel, *Europe* **21 G6** 50 0N 2 0W
English River, *Canada* **140 C1** 49 14N 91 0W
Engures ezers, *Latvia* **54 A10** 57 16N 23 6 E
Enguri →, *Georgia* **61 J5** 42 27N 41 38 E
Enid, *U.S.A.* **155 G6** 36 24N 97 53W
Enipévs →, *Greece* **48 B4** 39 22N 22 17 E
Enkhuizen, *Neths.* **24 B5** 52 42N 5 17 E
Enköping, *Sweden* **16 E11** 59 37N 17 4 E
Enle, *China* **76 F3** 24 0N 101 9 E
Enna, *Italy* **47 E7** 37 34N 14 16 E
Ennadai, *Canada* **143 A8** 61 8N 100 53W
Ennadai L., *Canada* **143 A8** 61 0N 101 0W
Ennedi, *Chad* **109 E4** 17 15N 22 0 E
Enngonia, *Australia* **127 D4** 29 21 S 145 50 E
Ennigerloh, *Germany* **30 D4** 51 50N 8 2 E
Ennis, *Ireland* **23 D3** 52 51N 8 59W
Ennis, *Mont., U.S.A.* **158 D8** 45 21N 111 44W
Ennis, *Tex., U.S.A.* **155 J6** 32 20N 96 38W
Enniscorthy, *Ireland* **23 D5** 52 30N 6 34W
Enniskillen, *U.K.* **23 B4** 54 21N 7 39W
Ennistimon, *Ireland* **23 D2** 52 57N 9 17W
Enns, *Austria* **34 C7** 48 12N 14 28 E
Enns →, *Austria* **34 C7** 48 14N 14 32 E
Enontekiö, *Finland* **14 B20** 68 23N 23 37 E
Enping, *China* **77 F9** 22 16N 112 21 E
Enrekang, *Indonesia* **82 B1** 3 34 S 119 47 E
Enrile, *Phil.* **80 C3** 17 34N 121 42 E
Enriquillo, L., *Dom. Rep.* **165 C5** 18 20N 72 5W
Enschede, *Neths.* **24 B6** 52 13N 6 53 E
Ensenada, *Argentina* **174 C4** 34 55 S 57 55W
Ensenada, *Mexico* **162 A1** 31 50N 116 50W
Enshi, *China* **76 B7** 30 18N 109 29 E
Enshū-Nada, *Japan* **71 C9** 34 27N 137 38 E
Ensiola, Pta., *Spain* **39 B9** 39 7N 2 55 E
Ensisheim, *France* **27 E14** 47 50N 7 20 E
Ensley, *U.S.A.* **153 E2** 30 31N 87 16W
Entebbe, *Uganda* **118 B3** 0 4N 32 28 E
Enterprise, *Canada* **142 A5** 60 47N 115 45W
Enterprise, *Ala., U.S.A.* **152 D4** 31 19N 85 51W
Enterprise, *Oreg., U.S.A.* **158 D5** 45 25N 117 17W
Enterprise, *Utah, U.S.A.* **159 H7** 37 34N 113 43W
Entlebuch, *Switz.* **32 C6** 46 59N 8 4 E
Entraygues-sur-Truyère, *France* **28 D6** 44 38N 2 35 E
Entre Ríos, *Bolivia* **174 A3** 21 30 S 64 25W
Entre Rios, *Bahia, Brazil* **171 D4** 11 56 S 38 5W
Entre Rios, *Pará, Brazil* **173 B7** 5 24 S 54 21W
Entre Ríos □, *Argentina* **174 C4** 30 30 S 58 30W
Entrepeñas, Embalse de, *Spain* **40 E2** 40 34N 2 42W
Entroncamento, *Portugal* **43 F2** 39 28N 8 28W
Enugu, *Nigeria* **113 D6** 6 20N 7 30 E
Enugu □, *Nigeria* **113 D6** 6 30N 7 45 E
Enugu Ezike, *Nigeria* **113 D6** 7 0N 7 29 E
Enumclaw, *U.S.A.* **160 C5** 47 12N 121 59W
Envermeu, *France* **26 C8** 49 53N 1 15 E
Envigado, *Colombia* **168 B2** 6 10N 75 35W
Enviken, *Sweden* **16 D9** 60 49N 15 46 E
Envira, *Brazil* **172 B3** 7 18 S 70 13W
Enying, *Hungary* **52 D3** 46 56N 18 15 E
Enza →, *Italy* **44 D7** 44 54N 10 31 E
Enzan, *Japan* **71 B10** 35 42N 138 44 E
Eólie, Ís., *Italy* **47 D7** 38 30N 14 57 E
Epanomí, *Greece* **50 F6** 40 25N 22 59 E
Epe, *Neths.* **24 B5** 52 21N 5 59 E
Epe, *Nigeria* **113 D5** 6 36N 3 59 E
Epéna, *Congo* **114 B3** 1 22N 17 29 E
Épernay, *France* **27 C10** 49 3N 3 56 E
Épernon, *France* **27 D8** 48 35N 1 40 E
Ephesus, *Turkey* **49 D9** 37 55N 27 22 E
Ephraim, *U.S.A.* **158 G8** 39 22N 111 35W
Ephrata, *U.S.A.* **158 C4** 47 19N 119 33W
Epi, *Vanuatu* **133 F6** 16 43 S 168 15 E
Epidaurus Limera, *Greece* **48 E5** 36 46N 23 0 E
Épila, *Spain* **40 D3** 41 36N 1 17W
Épinac, *France* **27 F11** 46 59N 4 31 E
Épinal, *France* **27 D13** 48 10N 6 27 E
Epira, *Guyana* **169 B6** 5 5N 57 20W
Episkopi, *Cyprus* **38 E11** 34 40N 32 54 E
Episkopí, *Greece* **38 D6** 35 20N 24 20 E
Episkopi Bay, *Cyprus* **38 E11** 34 35N 32 50 E
Epitálion, *Greece* **48 D3** 37 37N 21 30 E
Eppan = Appiano, *Italy* **45 B8** 46 28N 11 15 E
Eppingen, *Germany* **31 F4** 49 8N 8 53 E

Figueiró dos Vinhos 213

Faga, W. Samoa 133 W23 13 39 S 172 8W
Fagam, Nigeria 113 C7 11 1N 10 1 E
Fagamalo, W. Samoa 133 W23 13 25 S 172 21W
Făgăras, Romania 53 E9 45 48N 24 58 E
Făgăras, Munţii, Romania 53 E9 45 40N 24 40 E
Fågelmara, Sweden 17 H9 56 16N 15 58 E
Fagerheim, Norway 18 D4 60 26N 7 46 E
Fagerhult, Sweden 17 G9 57 8N 15 40 E
Fagernes, Norway 18 D6 60 59N 9 14 E
Fagersta, Sweden 16 D9 60 1N 15 46 E
Făget, Romania 52 E7 45 52N 22 10 E
Făget, Munţii, Romania 53 C8 47 40N 23 10 E
Fagnano, L., Argentina 176 D3 54 30 S 68 0W
Fagnières, France 27 D11 48 58N 4 20 E
Fahlīān, Iran 97 D6 30 11N 51 28 E
Fahr, Yemen 99 D6 12 26N 54 8 E
Fahraj, Kermān, Iran 97 D8 29 0N 59 0 E
Fahraj, Yazd, Iran 97 D7 31 46N 54 36 E
Faial, Madeira 39 D3 32 47N 16 53W
Faido, Switz. 33 D7 46 29N 8 48 E
Fair Hd., U.K. 23 A5 55 14N 6 9W
Fair Oaks, U.S.A. 160 G5 38 39N 121 16W
Fairbank, U.S.A. 159 L8 31 43N 110 11W
Fairbanks, Alaska, U.S.A. 138 B5 64 51N 147 43W
Fairbanks, Fla., U.S.A. 152 F7 29 44N 82 16W
Fairborn, U.S.A. 157 E12 39 49N 84 2W
Fairburn, U.S.A. 152 B5 33 34N 84 35W
Fairbury, Ill., U.S.A. 157 D8 40 45N 88 31W
Fairbury, Nebr., U.S.A. 154 E6 40 8N 97 11W
Faire, Phil. 80 C3 17 53N 121 34 E
Fairfax, Ala., U.S.A. 152 C4 32 48N 85 11W
Fairfax, Ohio, U.S.A. 157 E13 39 5N 83 37W
Fairfax, Okla., U.S.A. 155 G6 36 34N 96 42W
Fairfax, S.C., U.S.A. 152 C7 32 59N 81 15W
Fairfield, Australia 129 B9 33 53 S 150 57 E
Fairfield, Ala., U.S.A. 149 J2 33 29N 86 55W
Fairfield, Calif., U.S.A. 160 G4 38 15N 122 3W
Fairfield, Conn., U.S.A. 151 E11 41 9N 73 16W
Fairfield, Idaho, U.S.A. 158 E6 43 21N 114 44W
Fairfield, Ill., U.S.A. 157 F8 38 23N 88 22W
Fairfield, Iowa, U.S.A. 156 D5 40 56N 91 57W
Fairfield, Mont., U.S.A. 158 C8 47 37N 111 59W
Fairfield, Ohio, U.S.A. 157 E12 39 21N 84 34W
Fairfield, Tex., U.S.A. 155 K7 31 44N 96 10W
Fairford, Canada 143 C9 51 37N 98 38W
Fairhope, U.S.A. 149 K2 30 31N 87 54W
Fairlie, N.Z. 131 E5 44 5 S 170 49 E
Fairmead, U.S.A. 160 H6 37 5N 120 10W
Fairmont, Minn., U.S.A. 154 D7 43 39N 94 28W
Fairmont, W. Va., U.S.A. 148 F5 39 29N 80 9W
Fairmount, U.S.A. 161 L8 34 45N 118 26W
Fairplay, U.S.A. 159 G11 39 15N 106 2W
Fairport, U.S.A. 150 C7 43 6N 77 27W
Fairport Harbor, U.S.A. 150 E3 41 45N 81 17W
Fairview, Australia 126 B3 15 31 S 144 17 E
Fairview, Canada 142 B5 56 5N 118 25W
Fairview, Mont., U.S.A. 154 B2 47 51N 104 3W
Fairview, Okla., U.S.A. 155 G5 36 16N 98 29W
Fairview, Utah, U.S.A. 158 G8 39 50N 111 0W
Fairweather, Mt., U.S.A. 138 C6 58 55N 137 32W
Faisalabad, Pakistan 91 C4 31 30N 73 5 E
Faith, U.S.A. 154 C3 45 2N 102 2W
Faizabad, India 93 F10 26 45N 82 10 E
Faizpur, India 94 D2 21 14N 75 49 E
Fajardo, Puerto Rico 165 C6 18 20N 65 39W
Fakam, Yemen 98 C3 16 38N 43 49 E
Fakenham, U.K. 20 E8 52 51N 0 51 E
Fåker, Sweden 16 A8 63 0N 14 34 E
Fakfak, Indonesia 83 B4 3 0 S 132 15 E
Fakiya, Bulgaria 51 D11 42 10N 27 6 E
Fakobli, Ivory C. 112 D3 7 23N 7 23W
Fakse, Denmark 17 J6 55 15N 12 8 E
Fakse Bugt, Denmark 17 J6 55 11N 12 15 E
Fakse Ladeplads, Denmark 17 J6 55 11N 12 9 E
Faku, China 75 C12 42 32N 123 21 E
Falaise, France 26 D6 48 54N 0 12W
Falaise, Mui, Vietnam 86 C5 19 6N 105 45 E
Falakrón Óros, Greece 50 E7 41 15N 23 58 E
Falam, Burma 90 D4 23 0N 93 45 E
Falces, Spain 40 C3 42 24N 1 48W
Fălciu, Romania 53 D13 46 17N 28 7 E
Falcón □, Venezuela 168 A4 11 0N 69 50W
Falcon, C., Algeria 111 A4 35 50N 0 50W
Falcón, C., Spain 39 C7 38 50N 1 23 E
Falcon Dam, U.S.A. 155 M5 26 50N 99 20W
Falconara Maríttima, Italy 45 E10 43 37N 13 24 E
Falcone, C. del, Italy 46 B1 40 58N 8 12 E
Falconer, U.S.A. 150 D5 42 7N 79 13W
Faléa, Mali 112 C2 12 16N 11 17W
Falelatai, W. Samoa 133 W24 13 55 S 171 59W
Falelima, W. Samoa 133 W23 13 32 S 172 41W
Falenki, Russia 62 B3 58 22N 51 35 E
Falerum, Sweden 17 F10 58 8N 16 13 E
Faleshty = Făleşti, Moldova 53 C12 47 32N 27 44 E
Făleşti, Moldova 53 C12 47 32N 27 44 E
Falfurrias, U.S.A. 155 M5 27 14N 98 9W
Falher, Canada 142 B5 55 44N 117 15W
Falirakí, Greece 38 C10 36 22N 28 12 E
Falkenberg, Germany 30 D9 51 35N 13 14 E
Falkenberg, Sweden 17 H6 56 54N 12 30 E
Falkensee, Germany 30 C9 52 34N 13 4 E
Falkirk, U.K. 22 F5 56 0N 3 47W
Falkirk □, U.K. 22 F5 55 58N 3 49W
Falkland, U.K. 22 E5 56 16N 3 12W
Falkland, East, I., Falk. Is. 176 D5 51 40 S 58 30W
Falkland, West, I., Falk. Is. 176 D4 51 40 S 60 0W
Falkland Is. □, Atl. Oc. 176 D5 51 30 S 59 0W
Falkland Sd., Falk. Is. 176 D5 52 0 S 60 0W
Falkonéra, Greece 48 E5 36 50N 23 52 E
Falköping, Sweden 17 F7 58 12N 13 33 E
Fall River, U.S.A. 151 E13 41 43N 71 10W
Fall River Mills, U.S.A. 158 F3 41 3N 121 26W
Fallbrook, U.S.A. 159 K5 33 25N 117 12W
Fallbrook, Calif., U.S.A. 161 M9 33 23N 117 15W
Fallon, Mont., U.S.A. 154 B2 46 50N 105 8W
Fallon, Nev., U.S.A. 158 G4 39 28N 118 47W
Falls City, Nebr., U.S.A. 154 E7 40 3N 95 36W
Falls City, Oreg., U.S.A. 158 D2 44 52N 123 26W
Falls Creek, U.S.A. 150 E6 41 9N 78 48W
Falmouth, Jamaica 164 C4 18 30N 77 40W
Falmouth, U.K. 21 G2 50 9N 5 5W
Falmouth, Fla., U.S.A. 152 E6 30 21N 83 8W
Falmouth, Ky., U.S.A. 157 F12 38 41N 84 20W
False B., S. Africa 116 E2 34 15 S 18 40 E
False Divi Pt., India 95 G5 15 43N 80 50 E
False Pass, U.S.A. 144 J7 54 51N 163 25W
Falso, C., Honduras 164 C3 15 12N 83 21W
Falster, Denmark 17 K5 54 45N 11 55 E
Falsterbo, Sweden 15 J15 55 23N 12 50 E
Fălticeni, Romania 53 C11 47 21N 26 20 E
Falun, Sweden 16 D9 60 37N 15 37 E
Famagusta, Cyprus 38 D12 35 8N 33 55 E
Famagusta Bay, Cyprus 38 D13 35 15N 34 0 E
Famatina, Sierra de, Argentina 174 B2 27 30 S 68 0W
Family L., Canada 143 C9 51 54N 95 27W
Famoso, U.S.A. 161 K7 35 37N 119 12W
Fan Xian, China 74 G8 35 55N 115 38 E
Fana, Mali 112 C3 13 0N 6 56W
Fanad Hd., Ireland 23 A4 55 17N 7 38W
Fanahammaren, Norway 18 D2 60 16N 5 20 E
Fanárion, Greece 48 B3 39 24N 21 47 E
Fandriana, Madag. 117 C8 20 14 S 47 21 E
Fang, Thailand 86 C2 19 55N 99 13 E
Fang Xian, China 77 A8 32 3N 110 40 E
Fangchang, China 77 B12 31 5N 118 4 E
Fangcheng, Guangxi Zhuangzu, China 76 G7 21 42N 108 21 E
Fangcheng, Henan, China 74 H7 33 18N 112 59 E
Fangliao, Taiwan 77 F13 22 22N 120 38 E
Fangshan, China 74 E6 38 3N 111 25 E
Fangzi, China 75 F10 36 33N 119 10 E
Fani i Madh →, Albania 50 E4 41 56N 20 16 E
Fanjiatun, China 75 C13 43 40N 125 15 E
Fannich, L., U.K. 22 D4 57 38N 4 59W
Fannrem, Norway 18 A6 63 16N 9 50 E
Fannūj, Iran 97 E8 26 35N 59 38 E
Fanny Bay, Canada 142 D4 49 37N 124 48W
Fanø, Denmark 17 J2 55 25N 8 25 E
Fano, Italy 45 E10 43 50N 13 1 E
Fanshaw, U.S.A. 142 B2 57 11N 133 30W
Fanshi, China 74 E7 39 12N 113 20 E
Fao = Al Fāw, Iraq 97 D6 30 0N 48 30 E
Faqirwali, Pakistan 92 E5 29 27N 73 0 E
Fara in Sabina, Italy 45 F9 42 12N 12 43 E
Faradje, Dem. Rep. of the Congo 118 B2 3 50N 29 45 E
Farafangana, Madag. 117 C8 22 49 S 47 50 E
Farâfra, El Wâhât el-, Egypt 106 B2 27 15N 28 20 E
Farāh, Afghan. 91 B1 32 20N 62 7 E
Farāh □, Afghan. 91 B1 32 25N 62 10 E
Farahalana, Madag. 117 A9 14 26 S 50 10 E
Faraid, Gebel, Egypt 106 C4 23 33N 35 19 E
Faramana, Burkina Faso 112 C4 11 56N 4 45W
Faranah, Guinea 112 C2 10 3N 10 45W
Farap, Turkmenistan 63 D1 39 9N 63 36 E
Farasān, Jazā'ir, Si. Arabia 98 C3 16 45N 41 55 E
Farasan Is. = Farasān, Jazā'ir, Si. Arabia 98 C3 16 45N 41 55 E
Faratsiho, Madag. 117 B8 19 24 S 46 57 E
Farbarachi, Somali Rep. 120 D3 2 30N 45 30 E
Fardes →, Spain 43 H7 37 35N 3 0W
Fareham, U.K. 21 G6 50 51N 1 11W
Farewell, C., N.Z. 131 A7 40 29 S 172 43 E
Farewell C. = Farvel, Kap, Greenland 10 F6 59 48N 43 55W
Farewell Spit, N.Z. 131 A8 40 35 S 173 0 E
Färgelanda, Sweden 17 F5 58 34N 12 0 E
Farghona, Uzbekistan 63 C5 40 23N 71 19 E
Farghonskaya Dolina, Uzbekistan 63 C5 40 50N 71 30 E
Fargo, Ga., U.S.A. 152 E7 30 41N 82 34W
Fargo, N. Dak., U.S.A. 154 B6 46 53N 96 48W
Fār'iah, W. al →, West Bank 103 C4 32 12N 35 27 E
Faribault, U.S.A. 154 C8 44 18N 93 16W
Faridkot, India 92 D6 30 44N 74 45 E
Faridpur, Bangla. 90 D2 23 15N 89 55 E
Färila, Sweden 16 C9 61 48N 15 50 E
Farim, Guinea-Biss. 112 C1 12 27N 15 9W
Farīmān, Iran 97 C8 35 40N 59 49 E
Farina, Australia 127 E2 30 3 S 138 15 E
Farinha →, Brazil 170 C2 6 51 S 47 30W
Fariones, Pta., Canary Is. 39 E6 29 13N 13 28W
Fâriskûr, Egypt 106 H7 31 20N 31 43 E
Färjestaden, Sweden 17 H10 56 39N 16 27 E
Farkadhón, Greece 48 B4 39 36N 22 4 E
Farmakonisi, Greece 49 D9 37 17N 27 5 E
Farmer City, U.S.A. 157 D8 40 15N 88 39W
Farmersburg, U.S.A. 157 E9 39 15N 87 23W
Farmerville, U.S.A. 155 J8 32 47N 92 24W
Farmington, Calif., U.S.A. 160 H6 37 55N 120 59W
Farmington, Ga., U.S.A. 152 B6 33 47N 83 26W
Farmington, Ill., U.S.A. 156 D7 40 42N 90 0W
Farmington, Iowa, U.S.A. 156 D5 40 38N 91 44W
Farmington, Mo., U.S.A. 156 G6 37 47N 90 25W
Farmington, N.H., U.S.A. 151 C13 43 24N 71 4W
Farmington, N. Mex., U.S.A. 159 H9 36 44N 108 12W
Farmington, Utah, U.S.A. 158 F8 41 0N 111 12W
Farmington →, U.S.A. 151 E12 41 51N 72 38W
Farmland, U.S.A. 157 D11 40 15N 85 5W
Farmville, U.S.A. 148 G6 37 18N 78 24W
Färnäs, Sweden 16 D8 61 0N 14 39 E
Farne Is., U.K. 20 B6 55 38N 1 37W
Farnham, Canada 151 A12 45 17N 72 59W
Faro, Brazil 169 D6 2 10 S 56 39W
Faro, Portugal 43 H3 37 2N 7 55W
Fårö, Sweden 15 H18 57 55N 19 5 E
Faro □, Portugal 43 H2 37 12N 8 10W
Fårösund, Sweden 17 G13 57 52N 19 2 E
Farquhar, C., Australia 125 D1 23 50 S 113 36 E
Farquhar Is., Seychelles 121 F4 11 0 S 52 0 E
Farrars Cr. →, Australia 126 D3 25 35 S 140 43 E
Farrāshband, Iran 97 D7 28 57N 52 5 E
Farrell, U.S.A. 150 E4 41 13N 80 30W
Farrell Flat, Australia 128 B3 33 48 S 138 48 E
Farrokhī, Iran 97 C8 33 50N 59 31 E
Farruch, C., Spain 39 B10 39 47N 3 21 E
Farrukhabad-cum-Fatehgarh, India 93 F8 27 30N 79 32 E
Fārs □, Iran 97 D7 29 30N 55 0 E
Fársala, Greece 48 B4 39 17N 22 23 E
Fārsī, Afghan. 91 B1 33 47N 63 15 E
Farsø, Denmark 17 H3 56 46N 9 19 E
Farsund, Norway 15 G12 58 5N 6 55 E
Fartak, Râs, Si. Arabia 96 D2 28 5N 34 34 E
Fartak, Ra's, Yemen 99 D6 15 38N 52 15 E
Fârţăneşti, Romania 53 E12 45 49N 27 59 E
Fartura, Serra da, Brazil 175 B5 26 21 S 52 52W
Faru, Nigeria 113 C6 12 48N 6 12 E
Farūj, Iran 97 B8 37 14N 58 14 E
Fårup, Denmark 17 H3 56 33N 9 51 E
Farvel, Kap, Greenland 10 F6 59 48N 43 55W
Farwell, U.S.A. 155 H3 34 23N 103 2W
Fāryāb □, Afghan. 91 B2 36 0N 65 0 E
Fasā, Iran 97 D7 29 0N 53 39 E
Fasano, Italy 47 B10 40 50N 17 22 E
Fashoda, Sudan 107 F3 9 50N 32 2 E
Fastiv, Ukraine 59 G5 50 7N 29 57 E
Fastov = Fastiv, Ukraine 59 G5 50 7N 29 57 E
Fatagar, Tanjung, Indonesia 83 B4 2 46 S 131 57 E
Fatehgarh, India 93 F8 27 25N 79 35 E
Fatehpur, Raj., India 92 F6 28 0N 74 40 E
Fatehpur, Ut. P., India 93 G9 25 56N 81 13 E
Fatesh, Russia 59 F8 52 8N 35 57 E
Fatick, Senegal 112 C1 14 19N 16 27W
Fatima, Canada 141 C7 47 24N 61 53W
Fátima, Portugal 43 F2 39 37N 8 39W
Fatoya, Guinea 112 C3 11 37N 9 10W
Fatsa, Turkey 100 B7 41 2N 37 31 E
Faucille, Col de la, France 27 F13 46 22N 6 2 E
Faulkton, U.S.A. 154 C5 45 2N 99 8W
Faulquemont, France 27 C13 49 3N 6 36 E
Faure I., Australia 125 E1 25 52 S 113 50 E
Făurei, Romania 53 E12 45 6N 27 19 E
Fauresmith, S. Africa 116 D4 29 44 S 25 17 E
Fauro, Solomon Is. 133 L9 6 55 S 156 7 E
Fauske, Norway 14 C16 67 17N 15 25 E
Fåvang, Norway 18 C7 61 27N 10 11 E
Favara, Italy 46 E6 37 19N 13 39 E
Favaritx, C., Spain 39 B11 40 0N 4 15 E
Faverges, France 29 C10 45 45N 6 17 E
Favignana, Italy 46 E5 37 56N 12 20 E
Favignana, I., Italy 46 E5 37 56N 12 19 E
Favourable Lake, Canada 140 B1 52 50N 93 39W
Fawn →, Canada 140 A2 55 20N 87 35W
Fawnskin, U.S.A. 161 L10 34 16N 116 56W
Faxaflói, Iceland 11 C3 64 29N 23 0W
Faxälven →, Sweden 16 A10 63 13N 17 13 E
Faya-Largeau, Chad 109 E3 17 58N 19 6 E
Fayaoué, Vanuatu 133 K4 20 38 S 166 33 E
Fayd, Si. Arabia 96 E4 27 1N 42 52 E
Fayence, France 29 E10 43 38N 6 42 E
Fayette, Ala., U.S.A. 149 J2 33 41N 87 50W
Fayette, Iowa, U.S.A. 156 B5 42 51N 91 48W
Fayette, Mo., U.S.A. 156 E4 39 9N 92 41W
Fayette, Ohio, U.S.A. 157 C12 41 40N 84 20W
Fayetteville, Ark., U.S.A. 155 G7 36 4N 94 10W
Fayetteville, Ga., U.S.A. 152 B5 33 27N 84 27W
Fayetteville, N.C., U.S.A. 149 H6 35 3N 78 53W
Fayetteville, Tenn., U.S.A. 149 H2 35 9N 86 34W
Fayón, Spain 40 D5 41 15N 0 20 E
Fazenda Libongo, Angola 115 D2 8 24 S 13 24 E
Fazenda Nova, Brazil 171 E1 16 11 S 50 48W
Fazilka, India 92 D6 30 27N 74 2 E
Fazilpur, Pakistan 92 E4 29 18N 70 29 E
Fdérik, Mauritania 110 D2 22 40N 12 45W
Feale →, Ireland 23 D2 52 27N 9 37W
Fear, C., U.S.A. 149 J7 33 50N 77 58W
Feather →, U.S.A. 158 G3 38 47N 121 36W
Feather Falls, U.S.A. 160 F5 39 36N 121 16W
Featherston, N.Z. 130 H4 41 6 S 175 20 E
Featherstone, Zimbabwe 119 F3 18 42 S 30 55 E
Fécamp, France 26 C7 49 45N 0 22 E
Feda, Norway 18 F3 58 17N 6 50 E
Fedala = Mohammedia, Morocco 110 B3 33 44N 7 21W
Federación, Argentina 174 C4 31 0 S 57 55W
Fedeshkūh, Iran 97 D7 28 49N 53 50 E
Fedjadj, Chott el, Tunisia 108 B1 33 52N 9 14 E
Fedje, Norway 18 D1 60 47N 4 43 E
Fehérgyarmat, Hungary 52 C7 47 58N 22 30 E
Fehmarn, Germany 30 A7 54 27N 11 7 E
Fehmarn Bælt, Europe 17 K5 54 35N 11 20 E
Fehmarn Belt = Fehmarn Bælt, Europe 17 K5 54 35N 11 20 E
Fei Xian, China 75 G9 35 18N 117 59 E
Feijó, Brazil 172 B3 8 9 S 70 21W
Feilding, N.Z. 130 G4 40 13 S 175 35 E
Feira de Santana, Brazil 171 D4 12 15 S 38 57W
Feiring, Norway 18 D8 60 30N 11 10 E
Feixiang, China 74 F8 36 30N 114 45 E
Fejér □, Hungary 52 C3 47 9N 18 30 E
Fejø, Denmark 17 K5 54 55N 11 30 E
Feke, Turkey 100 D6 37 48N 35 56 E
Fekete →, Hungary 52 E3 45 47N 18 15 E
Felanitx, Spain 39 B10 39 28N 3 9 E
Felda, U.S.A. 153 J8 26 34N 81 26W
Feldbach, Austria 34 E8 46 57N 15 52 E
Feldberg, Baden-W., Germany 31 H3 47 52N 8 0 E
Feldberg, Mecklenburg-Vorpommern, Germany 30 B9 53 20N 13 25 E
Feldkirch, Austria 34 D2 47 15N 9 37 E
Feldkirchen, Austria 34 E7 46 44N 14 6 E
Felicity, U.S.A. 157 F12 38 51N 84 6W
Felipe Carrillo Puerto, Mexico 163 D7 19 38N 88 3W
Felixlândia, Brazil 171 E3 18 47 S 44 55W
Felixstowe, U.K. 21 F9 51 58N 1 23 E
Felletin, France 28 C6 45 53N 2 11 E
Fellingsbro, Sweden 16 E9 59 26N 15 37 E
Fellsmere, U.S.A. 153 H9 27 46N 80 36W
Felton, U.S.A. 160 H4 37 3N 122 4W
Feltre, Italy 45 B8 46 1N 11 54 E
Femer Bælt = Fehmarn Bælt, Europe 17 K5 54 35N 11 20 E
Femø, Denmark 17 K5 54 58N 11 35 E
Femunden, Norway 15 E14 62 10N 11 53 E
Fen He →, China 74 G6 35 36N 110 42 E
Fene, Spain 42 B2 43 27N 8 9W
Fenelon Falls, Canada 150 B6 44 32N 78 45W
Fener Burnu, Turkey 49 E9 36 58N 27 18 E
Feneroa, Ethiopia 107 E4 13 5N 39 3 E
Feng Xian, Jiangsu, China 74 G9 34 43N 116 35 E
Feng Xian, Shaanxi, China 74 H4 33 54N 106 40 E
Fengári, Greece 51 F9 40 25N 25 32 E
Fengcheng, Jiangxi, China 77 C10 28 12N 115 48 E
Fengcheng, Liaoning, China 75 D13 40 28N 124 5 E
Fengdu, China 76 C6 29 55N 107 41 E
Fengfeng, China 74 F8 36 28N 114 8 E
Fenggang, China 76 D6 27 57N 107 47 E
Fenghua, China 77 C13 29 40N 121 25 E
Fenghuang, China 76 D7 27 57N 109 29 E
Fenghuangzui, China 76 A7 33 30N 109 23 E
Fengjie, China 76 B7 31 5N 109 36 E
Fengkai, China 77 F8 23 24N 111 30 E
Fengle, China 77 B9 31 29N 112 29 E
Fengning, China 74 D9 41 10N 116 33 E
Fengqing, China 76 E2 24 38N 99 55 E
Fengqiu, China 74 G8 35 2N 114 25 E
Fengrun, China 75 E10 39 48N 118 8 E
Fengshan, Guangxi Zhuangzu, China 76 E7 24 39N 109 15 E
Fengshan, Guangxi Zhuangzu, China 76 E6 24 31N 107 3 E
Fengtai, Anhui, China 77 A11 32 50N 116 40 E
Fengtai, Beijing, China 74 E9 39 50N 116 18 E
Fengxian, China 77 B13 30 55N 121 26 E
Fengxiang, China 74 G4 34 29N 107 25 E
Fengxin, China 77 C10 28 41N 115 18 E
Fengyang, China 75 H9 32 51N 117 29 E
Fengyi, China 76 E3 25 37N 100 20 E
Fengzhen, China 74 D7 40 25N 113 2 E
Feni Is., Papua N. G. 132 C7 4 0 S 153 40 E
Fennimore, U.S.A. 156 B6 42 59N 90 39W
Fenny, Bangla. 90 D3 22 55N 91 32 E
Feno, C. de, France 29 G12 41 58N 8 33 E
Fenoarivo Afovoany, Madag. 117 B8 18 26 S 46 34 E
Fenoarivo Atsinanana, Madag. 117 B8 17 22 S 49 25 E
Fens, The, U.K. 20 E7 52 38N 0 2W
Fensmark, Denmark 17 J5 55 17N 11 48 E
Fenton, U.S.A. 157 B13 42 48N 83 42W
Fenxi, China 74 F6 36 40N 111 31 E
Fenyang, China 74 F6 37 18N 111 48 E
Fenyi, China 77 D10 27 45N 114 47 E
Feodosiya, Ukraine 59 K8 45 2N 35 16 E
Fer, C. de, Algeria 111 A6 37 3N 7 10 E
Ferdows, Iran 97 C8 33 58N 58 2 E
Fère-Champenoise, France 27 D10 48 45N 3 59 E
Fère-en-Tardenois, France 27 C10 49 10N 3 30 E
Ferentino, Italy 45 G10 41 42N 13 15 E
Ferfer, Somali Rep. 120 C3 5 4N 45 9 E
Fergana = Farghona, Uzbekistan 63 C5 40 23N 71 19 E
Ferganskaya Dolina = Farghonskaya Dolina, Uzbekistan 63 C5 40 50N 71 30 E
Ferganskiy Khrebet, Kyrgyzstan 63 C6 41 0N 73 50 E
Fergus, Canada 140 D3 43 43N 80 24W
Fergus Falls, U.S.A. 154 B6 46 17N 96 4W
Ferguson, U.S.A. 156 F6 38 45N 90 18W
Fergusson I., Papua N. G. 132 E6 9 30 S 150 45 E
Fériana, Tunisia 108 B1 34 59N 8 33 E
Feričanci, Croatia 52 E2 45 32N 18 0 E
Ferkane, Algeria 111 B6 34 37N 7 26 E
Ferkéssédougou, Ivory C. 112 D3 9 35N 5 6W
Ferlach, Austria 34 E7 46 32N 14 18 E
Ferland, Canada 140 B2 50 19N 88 27W
Ferlo, Vallée du, Senegal 112 B2 15 15N 14 15W
Fermanagh □, U.K. 23 B4 54 21N 7 40W
Fermo, Italy 45 E10 43 9N 13 43 E
Fermoselle, Spain 42 D4 41 19N 6 27W
Fermoy, Ireland 23 D3 52 9N 8 16W
Fernán Nuñéz, Spain 43 H6 37 40N 4 44W
Fernández, Argentina 174 B3 27 55 S 63 50W
Fernandina Beach, U.S.A. 152 E8 30 40N 81 27W
Fernando de Noronha, Brazil 170 B5 4 0 S 33 10W
Fernando Póo = Bioko, Eq. Guin. 113 E6 3 30N 8 40 E
Fernandópolis, Brazil 171 F1 20 16 S 50 14W
Ferndale, Calif., U.S.A. 158 F1 40 35N 124 16W
Ferndale, Wash., U.S.A. 160 B4 48 51N 122 36W
Fernie, Canada 142 D5 49 30N 115 5W
Fernlees, Australia 126 C4 23 51 S 148 7 E
Fernley, U.S.A. 158 G4 39 36N 119 15W
Feroke, India 95 J2 11 9N 75 46 E
Ferozepore = Firozpur, India 92 D6 30 55N 74 40 E
Férrai, Greece 51 F10 40 53N 26 10 E
Ferrandina, Italy 47 B9 40 29N 16 27 E
Ferrara, Italy 45 D8 44 50N 11 35 E
Ferrato, C., Italy 46 C2 39 18N 9 38 E
Ferreira do Alentejo, Portugal 43 G2 38 4N 8 6W
Ferreñafe, Peru 172 B2 6 42 S 79 50W
Ferrerías, Spain 39 B11 39 59N 4 1 E
Ferret, C., France 28 D2 44 38N 1 15W
Ferrette, France 27 E14 47 30N 7 20 E
Ferriday, U.S.A. 155 K9 31 38N 91 33W
Ferriere, Italy 44 D6 44 40N 9 30 E
Ferrières, France 27 D9 48 5N 2 48 E
Ferro, Capo, Italy 46 A2 41 9N 9 31 E
Ferrol, Spain 42 B2 43 29N 8 15W
Ferrol, Pen. de, Peru 172 B2 9 10 S 78 35W
Ferron, U.S.A. 159 G8 39 5N 111 8W
Ferros, Brazil 171 E3 19 14 S 43 2W
Ferryland, Canada 141 C9 47 2N 52 53W
Ferrysburg, U.S.A. 157 A10 43 5N 86 13W
Fertile, U.S.A. 154 B6 47 32N 96 17W
Fertőszentmiklós, Hungary 52 C1 47 35N 16 53 E
Fès, Morocco 110 B4 34 0N 5 0W
Feshi, Dem. Rep. of the Congo 115 D3 6 8 S 18 10 E
Fessenden, U.S.A. 154 B5 47 39N 99 38W
Festøy, Norway 18 B3 62 22N 6 19 E
Festus, U.S.A. 156 F6 38 13N 90 24W
Feteşti, Romania 53 F12 44 22N 27 51 E
Fethiye, Turkey 49 E11 36 36N 29 6 E
Fethiye Körfezi, Turkey 49 E10 36 40N 28 50 E
Fetlar, U.K. 22 A8 60 36N 0 52W
Fetsund, Norway 18 E8 59 56N 11 10 E
Feuerthalen, Switz. 33 A7 47 37N 8 38 E
Feuilles →, Canada 139 C12 58 47N 70 4W
Feurs, France 29 C8 45 45N 4 13 E
Fevik, Norway 18 F5 58 22N 8 39 E
Feyẕābād, Badākhshān, Afghan. 91 A3 37 7N 70 33 E
Feyẕābād, Fāryāb, Afghan. 91 A2 36 17N 64 52 E
Fiambalá, Argentina 174 B2 27 45 S 67 37W
Fianarantsoa, Madag. 117 C8 21 26 S 47 5 E
Fianarantsoa □, Madag. 117 B8 19 30 S 47 0 E
Fianga, Cameroon 109 G3 9 55N 15 9 E
Fichtelgebirge, Germany 31 E7 50 2N 11 55 E
Ficksburg, S. Africa 117 D4 28 51 S 27 53 E
Fidenza, Italy 44 D7 44 52N 10 3 E
Fiditi, Nigeria 113 D5 7 45N 3 53 E
Fidjeland, Norway 18 F3 58 57N 6 56 E
Field, Canada 140 C3 46 31N 80 1W
Field →, Australia 126 C2 23 48 S 138 0 E
Field I., Australia 124 B5 12 5 S 132 23 E
Fieni, Romania 53 E10 45 8N 25 25 E
Fieri, Albania 50 F3 40 43N 19 33 E
Fierzë, Albania 50 D4 42 15N 20 1 E
Fiesch, Switz. 32 D6 46 25N 8 12 E
Fife □, U.K. 22 E5 56 16N 3 1W
Fife Ness, U.K. 22 E6 56 17N 2 35W
Fifth Cataract, Sudan 106 D3 18 22N 33 50 E
Figari, France 29 G13 41 29N 9 7 E
Figeac, France 28 D6 44 37N 2 2 E
Figeholm, Sweden 17 G10 57 22N 16 33 E
Figline Valdarno, Italy 45 E8 43 37N 11 28 E
Figtree, Zimbabwe 119 G2 20 22 S 28 20 E
Figueira Castelo Rodrigo, Portugal 42 E4 40 57N 6 58W
Figueira da Foz, Portugal 42 E2 40 7N 8 54W
Figueiró dos Vinhos, Portugal 42 F2 39 55N 8 16W

216 Futago-Yama

Futago-Yama, *Japan* **70 D3** 33 35N 131 36 E
Futog, *Yugoslavia* **52 E4** 45 15N 19 42 E
Futrono, *Chile* **176 B2** 40 8 S 72 24W
Futuna, *Wall. & F. Is.* **134 J9** 14 25 S 178 20 E
Fuwa, *Egypt* **106 H7** 31 12N 30 33 E
Fuxin, *China* **75 C11** 42 5N 121 48 E
Fuyang, *Anhui, China* **74 H8** 33 0N 115 48 E
Fuyang, *Zhejiang, China* **77 B12** 30 5N 119 57 E
Fuyang He ⟶, *China* **74 E9** 38 12N 117 0 E
Fuying Dao, *China* **77 D13** 26 34N 120 9 E
Fuyu, *China* **75 B13** 45 12N 124 43 E
Fuyuan, *China* **76 E5** 25 40N 104 16 E
Füzesgyarmat, *Hungary* **52 C6** 47 6N 21 14 E
Fuzhou, *China* **77 D12** 26 5N 119 16 E
Fylde, *U.K.* **20 D5** 53 50N 2 58W
Fyn, *Denmark* **17 J4** 55 20N 10 30 E
Fyne, L., *U.K.* **22 F3** 55 59N 5 23W
Fyns Amtskommune □, *Denmark* **17 J4** 55 15N 10 30 E
Fynshav, *Denmark* **17 K3** 54 59N 9 59 E
Fyresdal, *Norway* **18 E5** 59 11N 8 5 E
Fyresvatn, *Norway* **18 E5** 59 6N 8 10 E

G

Gaanda, *Nigeria* **113 C7** 10 10N 12 27 E
Gabarin, *Nigeria* **113 C7** 11 8N 10 27 E
Gabas ⟶, *France* **28 E3** 43 46N 0 42W
Gabbettville, *U.S.A.* **152 C4** 32 57N 85 8W
Gabela, *Angola* **115 E2** 11 0S 14 24 E
Gabès, *Tunisia* **108 B2** 33 53N 10 2 E
Gabès, G. de, *Tunisia* **108 B2** 34 0N 10 30 E
Gabgaba, W. ⟶, *Egypt* **106 C3** 22 10N 33 5 E
Gąbin, *Poland* **55 F6** 52 23N 19 41 E
Gabon ■, *Africa* **114 C2** 0 10 S 10 0 E
Gaborone, *Botswana* **116 C4** 24 45 S 25 57 E
Gabriels, *U.S.A.* **151 B10** 44 26N 74 12W
Gābrīk, *Iran* **97 E8** 25 44N 58 28 E
Gabro, *Ethiopia* **120 C2** 6 18N 43 16 E
Gabrovo, *Bulgaria* **51 D9** 42 52N 25 19 E
Gacé, *France* **26 D7** 48 49N 0 20 E
Gāch Sār, *Iran* **97 B6** 36 7N 51 19 E
Gachsārān, *Iran* **97 D6** 30 15N 50 45 E
Gacko, *Bos.-H.* **50 C2** 43 10N 18 33 E
Gadag, *India* **95 G2** 15 30N 75 45 E
Gadamai, *Sudan* **107 D4** 17 11N 36 10 E
Gadap, *Pakistan* **92 G2** 25 5N 67 28 E
Gadarwara, *India* **93 H8** 22 50N 78 50 E
Gadebusch, *Germany* **30 B7** 53 42N 11 7 E
Gadein, *Sudan* **107 F2** 8 10N 28 45 E
Gadhada, *India* **92 J4** 22 0N 71 35 E
Gadmen, *Switz.* **33 C6** 46 45N 8 16 E
Gádor, Sierra de, *Spain* **43 J8** 36 57N 2 45W
Gadsden, *Ala., U.S.A.* **152 A3** 34 1N 86 1W
Gadsden, *Ariz., U.S.A.* **159 K6** 32 33N 114 47W
Gadsden, *S.C., U.S.A.* **152 B9** 33 51N 80 46W
Gadwal, *India* **95 F3** 16 10N 77 50 E
Gadyach = Hadyach, *Ukraine* **59 G8** 50 21N 34 0 E
Gadzi, *C.A.R.* **114 B3** 4 47N 16 42 E
Găeşti, *Romania* **53 F10** 44 48N 25 19 E
Gaeta, *Italy* **46 A6** 41 12N 13 35 E
Gaeta, G. di, *Italy* **46 A6** 41 6N 13 30 E
Gaffney, *U.S.A.* **149 H5** 35 5N 81 39W
Gafsa, *Tunisia* **108 B1** 34 24N 8 43 E
Gagarin, *Russia* **58 E8** 55 38N 35 0 E
Gagetown, *Canada* **141 C6** 45 46N 66 10W
Gaggenau, *Germany* **31 G4** 48 48N 8 18 E
Gagino, *Russia* **60 C7** 55 15N 45 1 E
Gagliano del Capo, *Italy* **47 C11** 39 50N 18 22 E
Gagnef, *Sweden* **16 D9** 60 36N 15 5 E
Gagnoa, *Ivory C.* **112 D3** 6 56N 5 16W
Gagnon, *Canada* **141 B6** 51 50N 68 5W
Gagnon, L., *Canada* **143 A6** 62 3N 110 27W
Gagra, *Georgia* **61 J5** 43 20N 40 10 E
Gahini, *Rwanda* **118 C3** 1 50 S 30 30 E
Gahmar, *India* **93 G10** 25 27N 83 49 E
Gai Xian, *China* **75 D12** 40 22N 122 20 E
Gaibanda, *Bangla.* **90 C2** 25 20N 89 36 E
Gaïdhouronísi, *Greece* **38 E7** 34 53N 25 41 E
Gail, *U.S.A.* **155 J4** 32 46N 101 27W
Gail ⟶, *Austria* **34 E6** 46 36N 13 53 E
Gaillac, *France* **28 E5** 43 54N 1 54 E
Gaillimh = Galway, *Ireland* **23 C2** 53 17N 9 3W
Gaillon, *France* **26 C8** 49 10N 1 20 E
Gaimán, *Argentina* **176 B3** 43 10 S 65 25W
Gaines, *U.S.A.* **150 E7** 41 46N 77 35W
Gainesville, *Fla., U.S.A.* **153 F7** 29 40N 82 20W
Gainesville, *Ga., U.S.A.* **149 H4** 34 18N 83 50W
Gainesville, *Mo., U.S.A.* **155 G8** 36 36N 92 26W
Gainesville, *Tex., U.S.A.* **155 J6** 33 38N 97 8W
Gainsborough, *U.K.* **20 D7** 53 24N 0 46W
Gairdner, L., *Australia* **128 A2** 31 30 S 136 0 E
Gairloch, L., *U.K.* **22 D3** 57 43N 5 45W
Gais, *Switz.* **33 B8** 47 22N 9 27 E
Gaj, *Croatia* **52 E2** 45 28N 17 3 E
Gakona, *U.S.A.* **144 E11** 62 18N 145 18W
Gakuch, *Pakistan* **93 A5** 36 7N 73 45 E
Gal Laghet, *Somali Rep.* **120 D3** 4 9N 47 10 E
Gal Oya Res., *Sri Lanka* **95 L5** 7 5N 81 30 E
Gal Tardo, *Somali Rep.* **120 D3** 3 34N 45 58 E
Galachipa, *Bangla.* **90 D3** 22 8N 90 26 E
Galán, Cerro, *Argentina* **174 B2** 25 55 S 66 52W
Galana ⟶, *Kenya* **118 C5** 3 9S 40 8 E
Galangue, *Angola* **115 E3** 13 42 S 16 9 E
Galangue, Serra, *Angola* **115 E3** 14 18 S 15 52 E
Galanta, *Slovak Rep.* **35 C10** 48 11N 17 45 E
Galapagar, *Spain* **42 E7** 40 36N 3 58W
Galápagos, *Pac. Oc.* **135 H18** 0 0 91 0W
Galashiels, *U.K.* **22 F6** 55 37N 2 49W
Galatás, *Greece* **48 D5** 37 30N 23 26 E
Galatea, *N.Z.* **130 E5** 38 24 S 176 45 E
Galați, *Romania* **53 E13** 45 27N 28 2 E
Galați □, *Romania* **53 E12** 45 45N 27 30 E
Galatia, *Turkey* **100 C5** 39 30N 33 0 E
Galatina, *Italy* **47 B11** 40 10N 18 10 E
Galátone, *Italy* **47 B11** 40 9N 18 4 E
Galax, *U.S.A.* **149 G5** 36 40N 80 56W
Galaxídhion, *Greece* **48 C4** 38 22N 22 23 E
Galbraith, *Australia* **126 B3** 16 25 S 141 30 E
Galcaio, *Somali Rep.* **102 F4** 6 30N 47 30 E
Galdhøpiggen, *Norway* **15 F12** 61 38N 8 18 E
Galeana, *Mexico* **162 C4** 24 50N 100 4W
Galela, *Indonesia* **82 A3** 1 50N 127 49 E
Galena, *U.S.A.* **156 B6** 42 25N 90 26W
Galera, *Spain* **41 H2** 37 45N 2 33W
Galera, Pta., *Chile* **176 A2** 39 59 S 73 43W
Galera Point, *Trin. & Tob.* **165 D7** 10 8N 61 0W
Galesburg, *Ill., U.S.A.* **156 D6** 40 57N 90 22W
Galesburg, *Mich., U.S.A.* **157 B11** 42 17N 85 26W
Galeton, *U.S.A.* **150 E7** 41 44N 77 39W
Galgasc, *Somali Rep.* **120 D2** 0 11N 41 38 E
Galheirão ⟶, *Brazil* **171 D2** 12 23 S 45 5W
Galheiros, *Brazil* **171 D2** 13 18 S 46 25W
Gali, *Georgia* **61 J5** 42 37N 41 46 E
Galicea Mare, *Romania* **53 F8** 44 4N 23 19 E
Galich, *Russia* **60 A6** 58 22N 42 24 E
Galiche, *Bulgaria* **50 C7** 43 34N 23 53 E
Galicia □, *Spain* **42 C3** 42 43N 7 45W
Galien, *U.S.A.* **157 C10** 41 48N 86 30W
Galilee = Hagalil, *Israel* **103 C4** 32 53N 35 18 E
Galilee, L., *Australia* **126 C4** 22 20 S 145 50 E
Galilee, Sea of = Yam Kinneret, *Israel* **103 C4** 32 45N 35 35 E
Galinoporni, *Cyprus* **38 D13** 35 31N 34 18 E
Galion, *U.S.A.* **150 F2** 40 44N 82 47W
Galite, Îs. de la, *Tunisia* **111 A6** 37 30N 8 59 E
Galiuro Mts., *U.S.A.* **159 K8** 32 30N 110 20W
Gallabat, *Sudan* **107 E4** 12 58N 36 11 E
Gallan Hd., *U.K.* **22 C1** 58 15N 7 2W
Gallarate, *Italy* **44 C5** 45 40N 8 48 E
Gallatin, *Mo., U.S.A.* **156 E3** 39 55N 93 58W
Gallatin, *Tenn., U.S.A.* **149 G2** 36 24N 86 27W
Galle, *Sri Lanka* **95 L5** 6 5N 80 10 E
Gállego ⟶, *Spain* **40 D4** 41 39N 0 51W
Gallegos ⟶, *Argentina* **176 D3** 51 35 S 69 0W
Galley Hd., *Ireland* **23 E3** 51 32N 8 55W
Galliate, *Italy* **44 C5** 45 29N 8 42 E
Gallinas, Pta., *Colombia* **168 A3** 12 28N 71 40W
Gallipoli = Gelibolu, *Turkey* **51 F10** 40 28N 26 43 E
Gallípoli, *Italy* **47 B10** 40 3N 17 58 E
Gallipolis, *U.S.A.* **148 F4** 38 49N 82 12W
Gällivare, *Sweden* **14 C19** 67 9N 20 40 E
Gallneukirchen, *Austria* **34 C7** 48 21N 14 25 E
Gällö, *Sweden* **16 B9** 62 55N 15 13 E
Gallo, C., *Italy* **46 D6** 38 13N 13 19 E
Gallocanta, L. de, *Spain* **40 E3** 40 58N 1 30W
Galloway, *U.K.* **22 F4** 55 1N 4 29W
Galloway, Mull of, *U.K.* **22 G4** 54 39N 4 52W
Gallup, *U.S.A.* **159 J9** 35 32N 108 45W
Gallur, *Spain* **40 D3** 41 52N 1 19W
Gallyaaral, *Uzbekistan* **63 C3** 40 2N 67 35 E
Galong, *Australia* **129 C8** 34 37 S 148 34 E
Galt, *Calif., U.S.A.* **160 G5** 38 15N 121 18W
Galt, *Mo., U.S.A.* **156 D3** 40 8N 93 23W
Galten, *Denmark* **17 H3** 56 9N 9 54 E
Galtür, *Austria* **34 E3** 46 58N 10 11 E
Galty Mts., *Ireland* **23 D3** 52 22N 8 10W
Galtymore, *Ireland* **23 D3** 52 21N 8 11W
Galva, *U.S.A.* **156 C6** 41 10N 90 3W
Galvarino, *Chile* **176 A2** 38 24 S 72 47W
Galve de Sorbe, *Spain* **40 D1** 41 13N 3 10W
Galveston, *Ind., U.S.A.* **157 D10** 40 35N 86 11W
Galveston, *Tex., U.S.A.* **155 L7** 29 18N 94 48W
Galveston B., *U.S.A.* **155 L7** 29 36N 94 50W
Gálvez, *Argentina* **174 C3** 32 0 S 61 14W
Galway, *Ireland* **23 C2** 53 17N 9 3W
Galway □, *Ireland* **23 C2** 53 22N 9 1W
Galway B., *Ireland* **23 C2** 53 13N 9 10W
Gam, *Indonesia* **83 B4** 0 27 S 130 36 E
Gam ⟶, *Vietnam* **86 B5** 21 55N 105 12 E
Gamagori, *Japan* **71 C9** 34 50N 137 14 E
Gamari, L., *Ethiopia* **107 E5** 11 32N 41 40 E
Gamawa, *Nigeria* **113 C7** 12 10N 10 31 E
Gamay, *Phil.* **80 E5** 12 23N 125 18 E
Gamay Bay, *Phil.* **80 E5** 12 21N 125 21 E
Gamba, *Angola* **115 E3** 11 42 S 17 14 E
Gambaga, *Ghana* **113 C4** 10 30N 0 28W
Gambat, *Pakistan* **92 F3** 27 17N 68 26 E
Gambela, *Ethiopia* **107 F3** 8 14N 34 38 E
Gambia ■, *W. Afr.* **112 C1** 13 25N 16 0W
Gambia ⟶, *W. Afr.* **112 C1** 13 28N 16 34W
Gambier, C., *Australia* **124 B5** 11 56 S 130 57 E
Gambier Is., *Australia* **128 C2** 35 3 S 136 30 E
Gambo, *C.A.R.* **114 B4** 4 39N 22 16 E
Gamboli, *Pakistan* **92 E3** 29 53N 68 24 E
Gamboma, *Congo* **114 C3** 1 55 S 15 52 E
Gamboula, *C.A.R.* **114 B3** 4 8N 15 9 E
Gambuta, *Indonesia* **82 A2** 0 30N 123 20 E
Gamerco, *U.S.A.* **159 J9** 35 34N 108 46W
Gamla Uppsala, *Sweden* **16 E11** 59 54N 17 40 E
Gamlakarleby = Kokkola, *Finland* **14 E20** 63 50N 23 8 E
Gamleby, *Sweden* **17 G10** 57 54N 16 24 E
Gammon ⟶, *Canada* **143 C9** 51 24N 95 44W
Gammouda, *Tunisia* **108 A1** 35 3N 9 39 E
Gamoda-Saki, *Japan* **70 D6** 33 50N 134 45 E
Gamu-Gofa □, *Ethiopia* **107 F4** 5 40N 36 40 E
Gan, *France* **28 E3** 43 12N 0 27W
Gan Gan, *Argentina* **176 B3** 42 30 S 68 10W
Gan Goriama, Mts., *Cameroon* **113 D7** 7 44N 12 45 E
Gan Jiang ⟶, *China* **77 C11** 29 15N 116 0 E
Ganado, *Ariz., U.S.A.* **159 J9** 35 43N 109 33W
Ganado, *Tex., U.S.A.* **155 L6** 29 2N 96 31W
Gananoque, *Canada* **140 D4** 44 20N 76 10W
Ganāveh, *Iran* **97 D6** 29 35N 50 35 E
Gäncä, *Azerbaijan* **61 K8** 40 45N 46 20 E
Gancheng, *China* **86 C7** 18 51N 108 37 E
Gand = Gent, *Belgium* **24 C3** 51 2N 3 42 E
Ganda, *Angola* **115 E2** 13 3 S 14 35 E
Gandak ⟶, *India* **93 G11** 25 39N 85 13 E
Gandara, *Phil.* **80 E5** 12 1N 124 49 E
Gandava, *Pakistan* **91 C2** 28 32N 67 32 E
Gander, *Canada* **141 C9** 48 58N 54 35W
Gander L., *Canada* **141 C9** 48 58N 54 35W
Ganderkesee, *Germany* **30 B4** 53 2N 8 32 E
Ganderowe Falls, *Zimbabwe* **119 F2** 17 20 S 29 10 E
Gandesa, *Spain* **40 D5** 41 3N 0 26 E
Gandhi Sagar, *India* **92 G6** 24 40N 75 40 E
Gandi, *Nigeria* **113 C6** 12 55N 5 49 E
Gandía, *Spain* **41 G4** 38 58N 0 9W
Gandino, *Italy* **44 C6** 45 49N 9 54 E
Gando, Pta., *Canary Is.* **39 G4** 27 55N 15 22W
Gandole, *Nigeria* **113 D7** 8 28N 11 35 E
Gandu, *Brazil* **171 D4** 13 45 S 39 30W
Ganedidalem = Gani, *Indonesia* **82 B3** 0 48 S 128 14 E
Ganetti, *Sudan* **106 D3** 18 0N 31 10 E
Ganga ⟶, *India* **93 H14** 23 20N 90 30 E
Ganganagar, *India* **92 E5** 29 56N 73 56 E
Gangapur, *India* **92 F7** 26 32N 76 49 E
Gangara, *Niger* **109 F1** 14 35N 8 29 E
Gangaw, *Burma* **90 D5** 22 5N 94 5 E
Gangawati, *India* **95 G3** 15 30N 76 36 E
Ganges = Ganga ⟶, *India* **93 H14** 23 20N 90 30 E
Ganges, *France* **28 E7** 43 56N 3 42 E
Ganges, Mouths of the, *India* **93 J14** 21 30N 90 0 E
Gånghester, *Sweden* **17 G7** 57 42N 13 1 E
Gangi, *Italy* **47 E7** 37 48N 14 12 E
Gângiova, *Romania* **53 G8** 43 54N 23 50 E
Gangoh, *India* **92 E7** 29 46N 77 18 E
Gangtok, *India* **90 B2** 27 20N 88 37 E
Gangu, *China* **74 G3** 34 40N 105 15 E
Gangyao, *China* **75 B14** 44 12N 126 37 E
Gani, *Indonesia* **82 B3** 0 48 S 128 14 E
Ganj, *India* **93 F8** 27 45N 78 57 E
Gannat, *France* **27 F10** 46 7N 3 11 E
Gannett Peak, *U.S.A.* **158 E9** 43 11N 109 39W
Gannvalley, *U.S.A.* **154 C5** 44 2N 98 59W
Ganongga, *Solomon Is.* **133 M9** 8 5 S 156 35 E
Ganquan, *China* **74 F5** 36 20N 109 20 E
Gänserdorf, *Austria* **35 C9** 48 20N 16 43 E
Ganshui, *China* **76 C6** 28 40N 106 40 E
Gansu □, *China* **74 G3** 36 0N 104 0 E
Ganta, *Liberia* **112 D3** 7 15N 8 59W
Gantheaume, C., *Australia* **128 D2** 36 4 S 137 32 E
Gantheaume B., *Australia* **125 E1** 27 40 S 114 10 E
Gantsevichi = Hantsavichy, *Belarus* **59 F4** 52 49N 26 30 E
Ganyem = Genyem, *Indonesia* **83 B6** 2 46 S 140 12 E
Ganyu, *China* **75 G10** 34 50N 119 8 E
Ganyushkino, *Kazakstan* **61 G9** 46 35N 49 20 E
Ganzhou, *China* **77 E10** 25 51N 114 56 E
Gao, *Mali* **113 B5** 18 0N 1 0 E
Gao Xian, *China* **76 C5** 28 21N 104 32 E
Gao'an, *China* **77 C10** 28 26N 115 17 E
Gaohe, *China* **77 F9** 22 46N 112 57 E
Gaohebu, *China* **77 B11** 30 43N 116 49 E
Gaokeng, *China* **77 D9** 27 40N 113 58 E
Gaolan Dao, *China* **77 G9** 21 55N 113 10 E
Gaoligong Shan, *China* **76 E2** 24 45N 98 45 E
Gaomi, *China* **75 F10** 36 20N 119 42 E
Gaoping, *China* **74 G7** 35 45N 112 55 E
Gaotang, *China* **74 F9** 36 50N 116 15 E
Gaoua, *Burkina Faso* **112 C4** 10 20N 3 8W
Gaoual, *Guinea* **112 C2** 11 45N 13 25W
Gaoxiong = Kaohsiung, *Taiwan* **77 F13** 22 35N 120 16 E
Gaoyang, *China* **74 E8** 38 40N 115 45 E
Gaoyou, *China* **77 A12** 32 47N 119 26 E
Gaoyou Hu, *China* **75 H10** 32 45N 119 20 E
Gaoyuan, *China* **75 F9** 37 8N 117 58 E
Gaozhou, *China* **77 G8** 21 58N 110 50 E
Gap, *France* **29 D10** 44 33N 6 5 E
Gapan, *Phil.* **80 D3** 15 19N 120 57 E
Gar, *China* **72 C2** 32 10N 79 58 E
Garabekewül, *Turkmenistan* **63 D2** 38 30N 64 8 E
Garabogazköl Aylagy, *Turkmenistan* **57 F9** 41 0N 53 30 E
Garachico, *Canary Is.* **39 F3** 28 22N 16 46W
Garachiné, *Panama* **164 E4** 8 0N 78 12W
Garad, *Somali Rep.* **120 C3** 6 57N 49 24 E
Garafia, *Canary Is.* **39 F2** 28 48N 17 57W
Garajonay, *Canary Is.* **39 F2** 28 7N 17 14W
Garamätnyyaz, *Turkmenistan* **63 E2** 37 45N 64 34 E
Garanhuns, *Brazil* **170 C4** 8 50 S 36 30W
Garawe, *Liberia* **112 E3** 4 35N 8 0W
Garba Harre, *Somali Rep.* **120 D2** 3 19N 42 13 E
Garba Tula, *Kenya* **118 B4** 0 30N 38 32 E
Garbagududu, *Ethiopia* **120 C2** 6 12N 43 50 E
Garber, *U.S.A.* **155 G6** 36 26N 97 35W
Garberville, *U.S.A.* **158 F2** 40 6N 123 48W
Garbsen, *Germany* **30 C5** 52 26N 9 31 E
Garça, *Brazil* **171 F2** 22 14 S 49 37W
Garças ⟶, *Mato Grosso, Brazil* **173 D7** 15 54 S 52 16W
Garças ⟶, *Pernambuco, Brazil* **170 C4** 8 43 S 39 41W
Garchitorena, *Phil.* **80 E4** 13 52N 123 40 E
Garcia Hernandez, *Phil.* **81 G5** 9 37N 124 18 E
Garcias, *Brazil* **173 E7** 20 34 S 52 13W
Gard □, *France* **29 D8** 44 2N 4 10 E
Gard ⟶, *France* **29 E8** 43 51N 4 37 E
Garda, L. di, *Italy* **44 C7** 45 40N 10 41 E
Gardanne, *France* **29 E9** 43 27N 5 27 E
Gårdby, *Sweden* **17 H10** 56 36N 16 38 E
Garde L., *Canada* **143 A7** 62 50N 106 13W
Gardelegen, *Germany* **30 C7** 52 32N 11 24 E
Garden City, *Ga., U.S.A.* **152 C8** 32 6N 81 9W
Garden City, *Kans., U.S.A.* **155 G4** 37 58N 100 53W
Garden City, *Mo., U.S.A.* **156 F2** 38 34N 94 12W
Garden City, *Tex., U.S.A.* **155 K4** 31 52N 101 29W
Garden Grove, *U.S.A.* **161 M9** 33 47N 117 55W
Gardēz, *Afghan.* **91 B3** 33 37N 69 9 E
Gardhíki, *Greece* **48 C3** 38 50N 21 55 E
Garðour, *Iceland* **11 A10** 66 4N 16 46W
Gardi, *U.S.A.* **152 D8** 31 32N 81 48W
Gardiner, *U.S.A.* **158 D8** 45 2N 110 22W
Gardiners I., *U.S.A.* **151 E12** 41 6N 72 6W
Gardner, *Fla., U.S.A.* **153 H8** 27 21N 81 48W
Gardner, *Ill., U.S.A.* **157 C8** 41 12N 88 17W
Gardner, *Mass., U.S.A.* **151 D13** 42 34N 71 59W
Gardner Canal, *Canada* **142 C3** 53 27N 128 8W
Gardner Pinnacles, *U.S.A.* **145 G10** 25 0N 167 55W
Gardnerville, *U.S.A.* **160 G7** 38 56N 119 45W
Gardno, Jezioro, *Poland* **54 D4** 54 40N 17 7 E
Gardo, *Somali Rep.* **120 C3** 9 30N 49 6 E
Gardone Val Trómpia, *Italy* **44 C7** 45 41N 10 11 E
Gárdony, *Hungary* **52 C3** 47 12N 18 39 E
Gare Tigre, *Fr. Guiana* **169 C7** 4 58N 53 9W
Garešnica, *Croatia* **45 C13** 45 36N 16 56 E
Garéssio, *Italy* **44 D5** 44 12N 8 2 E
Garey, *U.S.A.* **161 L6** 34 53N 120 19W
Garfield, *U.S.A.* **158 C5** 47 1N 117 9W
Garforth, *U.K.* **20 D6** 53 47N 1 24W
Gargaliánoi, *Greece* **48 D3** 37 4N 21 38 E
Gargan, Mt., *France* **28 C5** 45 37N 1 39 E
Gargouna, *Mali* **113 B5** 15 56N 0 13 E
Gargždai, *Lithuania* **54 C8** 55 43N 21 24 E
Garhshankar, *India* **92 D7** 31 13N 76 11 E
Gari, *Russia* **62 B9** 59 26N 62 21 E
Garibaldi Prov. Park, *Canada* **142 D4** 49 50N 122 40W
Garies, *S. Africa* **116 E2** 30 32 S 17 59 E
Garigliano ⟶, *Italy* **46 A6** 41 13N 13 45 E
Garissa, *Kenya* **118 C4** 0 25 S 39 40 E
Garissa □, *Kenya* **118 C5** 0 20 S 40 0 E
Garkida, *Nigeria* **113 C7** 10 27N 12 36 E
Garko, *Nigeria* **113 C6** 11 45N 8 53 E
Garland, *Tex., U.S.A.* **155 J6** 32 55N 96 38W
Garland, *Utah, U.S.A.* **158 F7** 41 47N 112 10W
Garlasco, *Italy* **44 C5** 45 12N 8 55 E
Garliava, *Lithuania* **54 D10** 54 49N 23 52 E
Garlin, *France* **28 E3** 43 33N 0 16W
Garm, *Tajikistan* **63 D5** 39 0N 70 20 E
Garmāb, *Iran* **97 C8** 35 25N 56 45 E
Garmisch-Partenkirchen, *Germany* **31 H7** 47 30N 11 6 E
Garmsār, *Iran* **97 C7** 35 20N 52 25 E
Garner, *U.S.A.* **156 A3** 43 6N 93 36W
Garnett, *U.S.A.* **154 F7** 38 17N 95 14W
Garo Hills, *India* **93 G14** 25 30N 90 30 E
Garoe, *Somali Rep.* **120 C3** 8 25N 48 33 E
Garonne ⟶, *France* **28 C3** 45 2N 0 36W
Garonne, Canal Latéral à la ⟶, *France* **28 D4** 44 15N 0 18 E
Garoua, *Cameroon* **113 D7** 9 19N 13 21 E
Garpenberg, *Sweden* **16 D10** 60 19N 16 12 E
Garphyttan, *Sweden* **16 E8** 59 18N 14 56 E
Garrel, *Germany* **30 C4** 52 57N 8 1 E
Garrett, *U.S.A.* **157 C11** 41 21N 85 8W
Garrigue = Garrigues, *France* **28 E7** 43 40N 3 55 E
Garrigues, *France* **28 E7** 43 40N 3 55 E
Garrison, *Ky., U.S.A.* **157 F13** 38 36N 83 10W
Garrison, *Mont., U.S.A.* **158 C7** 46 31N 112 49W
Garrison, *N. Dak., U.S.A.* **154 B4** 47 40N 101 25W
Garrison, *Tex., U.S.A.* **155 K7** 31 49N 94 30W
Garrison Res. = Sakakawea, L., *U.S.A.* **154 B4** 47 30N 101 25W
Garron Pt., *U.K.* **23 A6** 55 3N 5 59W
Garrovillas, *Spain* **43 F4** 39 40N 6 33W
Garrucha, *Spain* **41 H3** 37 11N 1 49W
Garry ⟶, *U.K.* **22 E5** 56 44N 3 47W
Garry, L., *Canada* **138 B9** 65 58N 100 18W
Garsen, *Kenya* **118 C5** 2 20 S 40 5 E
Gärsnäs, *Sweden* **17 J8** 55 32N 14 10 E
Garson L., *Canada* **143 B6** 56 19N 110 2W
Gartempe ⟶, *France* **28 B4** 46 47N 0 49 E
Gartz, *Germany* **30 B10** 53 13N 14 22 E
Garu, *Ghana* **113 C4** 10 55N 0 11W
Garub, *Namibia* **116 D2** 26 37 S 16 0 E
Garut, *Indonesia* **85 D3** 7 14 S 107 53 E
Garvão, *Portugal* **43 H2** 37 42N 8 21W
Garvie Mts., *N.Z.* **131 F3** 45 30 S 168 50 E
Garwa = Garoua, *Cameroon* **113 D7** 9 19N 13 21 E
Garwa, *India* **93 G10** 24 11N 83 47 E
Garwolin, *Poland* **55 G8** 51 55N 21 38 E
Gary, *U.S.A.* **157 C9** 41 36N 87 20W
Garz, *Germany* **30 A9** 54 19N 13 20 E
Garzê, *China* **76 B3** 31 38N 100 1 E
Garzón, *Colombia* **168 C2** 2 10N 75 40W
Gas City, *U.S.A.* **157 D11** 40 29N 85 37W
Gas-San, *Japan* **68 E10** 38 32N 140 1 E
Gasan, *Phil.* **80 E3** 13 19N 121 51 E
Gasan Kuli = Esenguly, *Turkmenistan* **64 F6** 37 37N 53 59 E
Gascogne, *France* **28 E4** 43 45N 0 20 E
Gascogne, G. de, *Europe* **28 E2** 44 0N 2 0W
Gasconade, *U.S.A.* **156 F5** 38 40N 91 34W
Gasconade ⟶, *U.S.A.* **156 F5** 38 41N 91 33W
Gascony = Gascogne, *France* **28 E4** 43 45N 0 20 E
Gascoyne ⟶, *Australia* **125 D1** 24 52 S 113 37 E
Gascoyne Junc. T.O., *Australia* **125 E2** 25 2 S 115 17 E
Gascueña, *Spain* **40 E2** 40 18N 2 31W
Gash, Wadi ⟶, *Ethiopia* **107 D4** 16 48N 35 51 E
Gashaka, *Nigeria* **113 D7** 7 20N 11 29 E
Gasherbrum, *Pakistan* **93 B7** 35 40N 76 40 E
Gashua, *Nigeria* **113 C7** 12 54N 11 0 E
Gasparilla I., *U.S.A.* **153 J7** 26 46N 82 16W
Gaspé, *Canada* **141 C7** 48 52N 64 30W
Gaspé, C. de, *Canada* **141 C7** 48 48N 64 7W
Gaspé, Pén. de, *Canada* **141 C6** 48 45N 65 40W
Gaspésie, Parc Prov. de la, *Canada* **141 C6** 48 55N 65 50W
Gassaway, *U.S.A.* **148 F5** 38 41N 80 47W
Gassol, *Nigeria* **113 D7** 8 34N 10 25 E
Gasteiz = Vitoria-Gasteiz, *Spain* **40 C2** 42 50N 2 41W
Gaston, *U.S.A.* **152 B9** 33 49N 81 5W
Gastonia, *U.S.A.* **149 H5** 35 16N 81 11W
Gastoúni, *Greece* **48 D3** 37 51N 21 15 E
Gastoúri, *Greece* **48 B1** 39 34N 19 54 E
Gastre, *Argentina* **176 B3** 42 20 S 69 15W
Gästrikland, *Sweden* **16 D10** 60 45N 16 40 E
Gata, C., *Cyprus* **38 E12** 34 34N 33 2 E
Gata, C. de, *Spain* **41 J2** 36 41N 2 13W
Gata, Sierra de, *Spain* **42 E4** 40 20N 6 45W
Gataga ⟶, *Canada* **142 B3** 58 35N 126 59W
Gătaia, *Romania* **52 E6** 45 26N 21 30 E
Gatchina, *Russia* **58 C6** 59 35N 30 9 E
Gatehouse of Fleet, *U.K.* **22 G4** 54 53N 4 12W
Gates, *U.S.A.* **150 C7** 43 9N 77 42W
Gateshead, *U.K.* **20 C6** 54 57N 1 35W
Gatesville, *U.S.A.* **155 K6** 31 26N 97 45W
Gaths, *Zimbabwe* **119 G3** 20 2 S 30 32 E
Gatico, *Chile* **174 A1** 22 29 S 70 20W
Gâtinais, *France* **27 D9** 48 5N 2 40 E
Gâtine, Hauteurs de, *France* **28 B3** 46 35N 0 45W
Gatineau ⟶, *Canada* **140 C4** 45 27N 75 42W
Gatineau, Parc de la, *Canada* **140 C4** 45 40N 76 0W
Gattaran, *Phil.* **80 B3** 18 4N 121 38 E
Gattinara, *Italy* **44 C5** 45 37N 8 22 E
Gatukai, *Solomon Is.* **133 M10** 8 45 S 158 15 E
Gatun, L., *Panama* **164 E4** 9 7N 79 56W
Gatyana, *S. Africa* **117 E4** 32 16 S 28 31 E
Gau, *Fiji* **133 B2** 18 2 S 179 18 E
Gaua, *Vanuatu* **133 D5** 14 15 S 167 30 E
Gaucín, *Spain* **43 J5** 36 31N 5 19W
Gauer L., *Canada* **143 B9** 57 0N 97 50W
Gauhati, *India* **93 F14** 26 10N 91 45 E
Gauja ⟶, *Latvia* **15 H21** 57 10N 24 16 E
Gaula ⟶, *Norway* **14 E14** 63 21N 10 14 E
Gaupne, *Norway* **18 C4** 61 25N 7 18 E
Gaurdak = Gowurdak, *Turkmenistan* **63 E3** 37 50N 66 4 E
Gausta, *Norway* **15 G13** 59 48N 8 40 E
Gauteng □, *S. Africa* **117 D4** 26 0S 28 0 E
Gāv Koshī, *Iran* **97 D8** 28 38N 57 12 E
Gāvakān, *Iran* **97 D7** 29 37N 53 10 E
Gavarnie, *France* **28 F3** 42 44N 0 1W
Gavāter, *Iran* **97 E9** 25 10N 61 31 E
Gāvbandī, *Iran* **97 E7** 27 12N 53 4 E
Gavdhopoúla, *Greece* **38 E6** 34 56N 24 0 E
Gávdhos, *Greece* **38 E6** 34 50N 24 5 E
Gavi, *Italy* **44 D5** 44 41N 8 49 E
Gavião, *Portugal* **43 F3** 39 28N 7 56W
Gaviota, *U.S.A.* **161 L6** 34 29N 120 13W
Gävle, *Sweden* **16 D11** 60 40N 17 9 E
Gävleborgs län □, *Sweden* **16 C10** 61 30N 16 15 E
Gävlebukten, *Sweden* **16 D11** 60 40N 17 20 E
Gavorrano, *Italy* **44 F7** 42 55N 10 54 E
Gavray, *France* **26 D5** 48 55N 1 20W
Gavrilov Yam, *Russia* **58 D10** 57 18N 39 49 E

Girona 217

Gávrion, *Greece* **48 D6** 37 54N 24 44 E
Gawachab, *Namibia* **116 D2** 27 4 S 17 55 E
Gawai, *Burma* **90 B6** 27 56N 97 30 E
Gawilgarh Hills, *India* **94 D3** 21 15N 76 45 E
Gawler, *Australia* **128 C3** 34 30 S 138 42 E
Gaxun Nur, *China* **72 B5** 42 22N 100 30 E
Gay, *Russia* **62 F7** 51 27N 58 27 E
Gay, *U.S.A.* **152 B5** 33 6N 84 35W
Gaya, *India* **93 G11** 24 47N 85 4 E
Gaya, *Niger* **113 C5** 11 52N 3 28 E
Gaya, *Nigeria* **113 C6** 11 57N 9 0 E
Gaylord, *U.S.A.* **148 C3** 45 2N 84 41W
Gayndah, *Australia* **127 D5** 25 35 S 151 32 E
Gayny, *Russia* **62 A5** 60 18N 54 19 E
Gaysin = Haysyn, *Ukraine* **59 H5** 48 57N 29 25 E
Gayvoron = Hayvoron, *Ukraine* **59 H5** 48 22N 29 52 E
Gaza, *Gaza Strip* **103 D3** 31 30N 34 28 E
Gaza □, *Mozam.* **117 C5** 23 10 S 32 45 E
Gaza Strip □, *Asia* **103 D3** 31 29N 34 25 E
Gazaoua, *Niger* **109 F1** 13 32N 7 55 E
Gāzbor, *Iran* **97 D8** 28 5N 58 51 E
Gazelle Pen., *Papua N. G.* **132 C7** 4 40 S 152 0 E
Gazi, *Dem. Rep. of the Congo* **118 B1** 1 3N 24 30 E
Gaziantep, *Turkey* **100 D7** 37 6N 37 23 E
Gazipaşa, *Turkey* **100 D5** 36 16N 32 18 E
Gbarnga, *Liberia* **112 D3** 7 19N 9 13W
Gbekebo, *Nigeria* **113 D5** 6 20N 4 56 E
Gboko, *Nigeria* **113 D6** 7 17N 9 4 E
Gbongan, *Nigeria* **113 D5** 7 28N 4 20 E
Gcuwa, *S. Africa* **117 E4** 32 20 S 28 11 E
Gdańsk, *Poland* **54 D5** 54 22N 18 40 E
Gdańsk □, *Poland* **54 D5** 54 10N 18 30 E
Gdańska, Zatoka, *Poland* **54 D6** 54 30N 19 20 E
Gdov, *Russia* **15 G22** 58 48N 27 55 E
Gdynia, *Poland* **54 D5** 54 35N 18 33 E
Gebe, *Indonesia* **83 A3** 0 5N 129 25 E
Gebeciler, *Turkey* **49 C12** 38 46N 30 46 E
Gebeit Mine, *Sudan* **106 C4** 21 3N 36 29 E
Gebze, *Turkey* **51 F13** 40 47N 29 25 E
Gecha, *Ethiopia* **107 F4** 7 30N 35 18 E
Gedaref, *Sudan* **107 E4** 14 2N 35 28 E
Gede, Tanjung, *Indonesia* **84 D3** 6 46 S 105 12 E
Gediz, *Turkey* **49 B11** 39 1N 29 24 E
Gediz →, *Turkey* **49 C8** 38 35N 26 48 E
Gedo, *Ethiopia* **107 F4** 9 2N 37 25 E
Gèdre, *France* **28 F4** 42 47N 0 2 E
Gedser, *Denmark* **17 K5** 54 35N 11 55 E
Geegully Cr. →, *Australia* **124 C3** 18 32 S 123 41 E
Geel, *Belgium* **24 C4** 51 10N 4 59 E
Geelong, *Australia* **128 E6** 38 10 S 144 22 E
Geelvink Chan., *Australia* **125 E1** 28 30 S 114 0 E
Geesthacht, *Germany* **30 B6** 53 26N 10 22 E
Geidam, *Nigeria* **113 C7** 12 57N 11 57 E
Geikie →, *Canada* **143 B8** 57 45N 103 52W
Geilenkirchen, *Germany* **30 E2** 50 57N 6 8 E
Geili, *Sudan* **107 D3** 16 1N 32 37 E
Geilo, *Norway* **18 D5** 60 32N 8 14 E
Geisingen, *Germany* **31 H4** 47 54N 8 38 E
Geislingen, *Germany* **31 G5** 48 37N 9 50 E
Geita, *Tanzania* **118 C3** 2 48 S 32 12 E
Geita □, *Tanzania* **118 C3** 2 50 S 32 10 E
Geitastrand, *Norway* **18 A6** 63 22N 9 56 E
Geithus, *Norway* **18 E6** 59 57N 9 59 E
Gejiu, *China* **76 F4** 23 20N 103 10 E
Gel →, *Sudan* **107 F2** 7 5N 29 10 E
Gel River, *Sudan* **107 F2** 7 5N 29 10 E
Gela, *Italy* **47 E7** 37 4N 14 15 E
Gela, G. di, *Italy* **47 F7** 37 0N 14 20 E
Geladi, *Ethiopia* **120 C3** 6 59N 46 30 E
Gelderland □, *Neths.* **24 B6** 52 5N 6 10 E
Geldern, *Germany* **30 D2** 51 31N 6 20 E
Geldrop, *Neths.* **24 C5** 51 25N 5 32 E
Geleen, *Neths.* **24 D5** 50 57N 5 49 E
Gelehun, *S. Leone* **112 D2** 8 20N 11 40W
Gelembe, *Turkey* **49 B9** 39 10N 27 50 E
Gelendost, *Turkey* **100 C4** 38 7N 31 1 E
Gelendzhik, *Russia* **59 K10** 44 33N 38 10 E
Gelib, *Somali Rep.* **120 D2** 0 29N 42 46 E
Gelibolu, *Turkey* **51 F10** 40 28N 26 43 E
Gelibolu Yarımadası, *Turkey* **51 F10** 40 20N 26 30 E
Gelidonya Burnu, *Turkey* **100 D4** 36 12N 30 24 E
Gelnhausen, *Germany* **31 E5** 50 11N 9 11 E
Gelnica, *Slovak Rep.* **35 C13** 48 51N 20 55 E
Gelsenkirchen, *Germany* **30 D3** 51 32N 7 6 E
Gelting, *Germany* **30 A5** 54 45N 9 53 E
Gemas, *Malaysia* **87 L4** 2 37N 102 36 E
Gembloux, *Belgium* **24 D4** 50 34N 4 43 E
Gemena, *Dem. Rep. of the Congo* **114 B3** 3 13N 19 48 E
Gemerek, *Turkey* **100 C7** 39 15N 36 10 E
Gemla, *Sweden* **17 H8** 56 52N 14 39 E
Gemlik, *Turkey* **51 F13** 40 26N 29 9 E
Gemlik Körfezi, *Turkey* **51 F12** 40 25N 28 55 E
Gemona del Friuli, *Italy* **45 B10** 46 16N 13 9 E
Gemsa, *Egypt* **106 B3** 27 39N 33 35 E
Gemünden, *Germany* **31 E5** 50 3N 9 42 E
Genale, *Ethiopia* **107 F4** 6 0N 39 30 E
Genale, *Somali Rep.* **120 D2** 1 48N 44 42 E
Genç, *Turkey* **101 C9** 38 44N 40 34 E
Gençay, *France* **28 B4** 46 23N 0 23 E
Geneina, Gebel, *Egypt* **106 J8** 29 2N 33 55 E
General Acha, *Argentina* **174 D3** 37 20 S 64 38W
General Alvear, *Buenos Aires, Argentina* **174 D4** 36 0 S 60 0W
General Alvear, *Mendoza, Argentina* **174 D2** 35 0 S 67 40W
General Artigas, *Paraguay* **174 B4** 26 52 S 56 16W
General Belgrano, *Argentina* **174 D4** 36 35 S 58 47W
General Cabrera, *Argentina* **174 C3** 32 53 S 63 52W
General Carrera, L., *Chile* **176 C2** 46 35 S 72 0W
General Cepeda, *Mexico* **162 B4** 25 23N 101 27W
General Conesa, *Argentina* **176 B4** 40 6 S 64 25W
General Eugenio A. Garay, *Paraguay* **173 E5** 20 31 S 62 8W
General Guido, *Argentina* **174 D4** 36 40 S 57 50W
General Juan Madariaga, *Argentina* **174 D4** 37 0 S 57 0W
General La Madrid, *Argentina* **174 D3** 37 17 S 61 20W
General Lorenzo Vintter, *Argentina* **176 B4** 40 45 S 64 26W
General Luna, *Phil.* **80 E4** 13 41N 122 10 E
General MacArthur, *Phil.* **81 F5** 11 18N 125 28 E
General Martin Miguel de Güemes, *Argentina* **174 A3** 24 50 S 65 0W
General Paz, *Argentina* **174 B4** 27 45 S 57 36W
General Pico, *Argentina* **174 D3** 35 45 S 63 50W
General Pinedo, *Argentina* **174 B3** 27 15 S 61 20W
General Pinto, *Argentina* **174 C3** 34 45 S 61 50W
General Sampaio, *Brazil* **170 B4** 4 2 S 39 29W
General Santos, *Phil.* **81 H5** 6 5N 125 14 E
General Tinio, *Phil.* **80 D3** 15 39N 121 10 E
General Toshevo, *Bulgaria* **51 C12** 43 42N 28 6 E
General Trevino, *Mexico* **163 B5** 26 14N 99 29W
General Trías, *Mexico* **162 B3** 28 21N 106 22W
General Viamonte, *Argentina* **174 D3** 35 1 S 61 3W
General Villegas, *Argentina* **174 D3** 35 5 S 63 0W
General Vintter, L., *Argentina* **176 B2** 43 55 S 71 40W
Generoso, Mte., *Switz.* **33 E8** 45 56N 9 2 E
Genesee, *Idaho, U.S.A.* **158 C5** 46 33N 116 56W
Genesee, *Pa., U.S.A.* **150 E7** 41 59N 77 54W
Genesee →, *U.S.A.* **150 C7** 43 16N 77 36W
Geneseo, *Ill., U.S.A.* **156 C6** 41 27N 90 9W
Geneseo, *Kans., U.S.A.* **154 F5** 38 31N 98 10W
Geneseo, *N.Y., U.S.A.* **150 D7** 42 48N 77 49W
Geneva = Genève, *Switz.* **32 D2** 46 12N 6 9 E
Geneva, *Ala., U.S.A.* **152 D4** 31 2N 85 52W
Geneva, *Ga., U.S.A.* **152 C5** 32 35N 84 33W
Geneva, *Ill., U.S.A.* **157 C8** 41 53N 88 18W
Geneva, *Ind., U.S.A.* **157 D12** 40 36N 84 58W
Geneva, *N.Y., U.S.A.* **150 D8** 42 52N 76 59W
Geneva, *Nebr., U.S.A.* **154 E6** 40 32N 97 36W
Geneva, *Ohio, U.S.A.* **150 E4** 41 48N 80 57W
Geneva, L. = Léman, L., *Europe* **32 D3** 46 26N 6 30 E
Geneva, L., *U.S.A.* **157 B8** 42 38N 88 30W
Genève, *Switz.* **32 D2** 46 12N 6 9 E
Genève □, *Switz.* **32 D2** 46 10N 6 10 E
Geng, *Afghan.* **91 C1** 31 22N 61 28 E
Gengenbach, *Germany* **31 G4** 48 24N 8 1 E
Gengma, *China* **76 F2** 23 32N 99 20 E
Genichesk = Henichesk, *Ukraine* **59 J8** 46 12N 34 50 E
Genil →, *Spain* **43 H5** 37 42N 5 19W
Genk, *Belgium* **24 D5** 50 58N 5 32 E
Genkai-Nada, *Japan* **70 D2** 34 0N 130 0 E
Genlis, *France* **27 E12** 47 11N 5 12 E
Gennargentu, Mti. del, *Italy* **46 B2** 40 1N 9 19 E
Gennes, *France* **26 E6** 47 20N 0 17W
Genoa = Génova, *Italy* **44 D5** 44 25N 8 57 E
Genoa, *Australia* **129 D8** 37 29 S 149 35 E
Genoa, *Ill., U.S.A.* **157 B8** 42 6N 88 42W
Genoa, *N.Y., U.S.A.* **151 D8** 42 40N 76 32W
Genoa, *Nebr., U.S.A.* **154 E6** 41 27N 97 44W
Genoa, *Nev., U.S.A.* **160 F7** 39 2N 119 50W
Genoa →, *Argentina* **176 B2** 44 55 S 70 5W
Genoa City, *U.S.A.* **157 B8** 42 30N 88 20W
Génova, *Italy* **44 D5** 44 25N 8 57 E
Génova, G. di, *Italy* **44 E6** 44 0N 9 0 E
Genriyetty, Ostrov, *Russia* **65 B16** 77 6N 156 30 E
Gent, *Belgium* **24 C3** 51 2N 3 42 E
Genthin, *Germany* **30 C8** 52 25N 12 9 E
Gentio do Ouro, *Brazil* **170 D3** 11 25 S 42 30W
Genyem, *Indonesia* **83 B6** 2 46 S 140 12 E
Genzano di Lucánia, *Italy* **47 B9** 40 51N 16 2 E
Genzano di Roma, *Italy* **45 G9** 41 42N 12 41 E
Geoagiu, *Romania* **53 E8** 45 55N 23 12 E
Geographe B., *Australia* **125 F2** 33 30 S 115 15 E
Geographe Chan., *Australia* **125 D1** 24 30 S 113 0 E
Geokchay = Göyçay, *Azerbaijan* **61 K8** 40 42N 47 43 E
George, Zemlya, *Russia* **64 A5** 80 30N 49 0 E
George, *S. Africa* **116 E3** 33 58 S 22 29 E
George →, *Canada* **141 A6** 58 49N 66 10W
George, L., *N.S.W., Australia* **129 C8** 35 10 S 149 25 E
George, L., *S. Austral., Australia* **128 D4** 37 25 S 140 0 E
George, L., *W. Austral., Australia* **124 D3** 22 45 S 123 40 E
George, L., *Uganda* **118 B3** 0 5N 30 10 E
George, L., *Fla., U.S.A.* **149 L5** 29 17N 81 36W
George, L., *N.Y., U.S.A.* **151 C11** 43 37N 73 33W
George Gill Ra., *Australia* **124 D5** 24 22 S 131 45 E
George River = Kangiqsualujjuaq, *Canada* **139 C13** 58 30N 65 59W
George Sound, *N.Z.* **131 E2** 44 52 S 167 25 E
George Town, *Bahamas* **164 B4** 23 33N 75 47W
George Town, *Malaysia* **87 K3** 5 25N 100 15 E
George V Land, *Antarctica* **7 C10** 69 0 S 148 0 E
George VI Sound, *Antarctica* **7 D17** 71 0 S 68 0W
George West, *U.S.A.* **155 L5** 28 20N 98 7W
Georgetown, *Australia* **126 B3** 18 17 S 143 33 E
Georgetown, *Ont., Canada* **140 D4** 43 40N 79 56W
Georgetown, *P.E.I., Canada* **141 C7** 46 13N 62 24W
Georgetown, *Cayman Is.* **164 C3** 19 20N 81 24W
Georgetown, *Gambia* **112 C2** 13 30N 14 47W
Georgetown, *Guyana* **169 B6** 6 50N 58 12W
Georgetown, *Calif., U.S.A.* **160 G6** 38 54N 120 50W
Georgetown, *Colo., U.S.A.* **158 G11** 39 42N 105 42W
Georgetown, *Fla., U.S.A.* **153 F8** 29 23N 81 38W
Georgetown, *Ga., U.S.A.* **152 D4** 31 53N 85 6W
Georgetown, *Ill., U.S.A.* **157 E9** 39 59N 87 38W
Georgetown, *Ky., U.S.A.* **148 F3** 38 13N 84 33W
Georgetown, *Ohio, U.S.A.* **157 F13** 38 52N 83 54W
Georgetown, *S.C., U.S.A.* **149 J6** 33 23N 79 17W
Georgetown, *Tex., U.S.A.* **155 K6** 30 38N 97 41W
Georgia □, *U.S.A.* **152 C6** 32 50N 83 15W
Georgia ■, *Asia* **61 J6** 42 0N 43 0 E
Georgia, Str. of, *Canada* **142 D4** 49 25N 124 0W
Georgian B., *Canada* **140 C3** 45 15N 81 0W
Georgina →, *Australia* **126 C2** 23 30 S 139 47 E
Georgina Downs, *Australia* **126 C2** 21 10 S 137 40 E
Georgiu-Dezh = Liski, *Russia* **59 G10** 51 3N 39 30 E
Georgiyevka, *Kazakstan* **63 B7** 43 3N 74 43 E
Georgiyevsk, *Russia* **61 H6** 44 12N 43 28 E
Georgsmarienhütte, *Germany* **30 C4** 52 13N 8 3 E
Gera, *Germany* **30 E8** 50 53N 12 4 E
Geraardsbergen, *Belgium* **24 D3** 50 45N 3 53 E
Geral, Serra, *Bahia, Brazil* **171 D3** 14 0 S 41 0W
Geral, Serra, *Goiás, Brazil* **170 D2** 11 15 S 46 30W
Geral, Serra, *Sta. Catarina, Brazil* **175 B6** 26 25 S 50 0W
Geral de Goiás, Serra, *Brazil* **171 D2** 12 0 S 46 0W
Geral do Paraná Serra, *Brazil* **171 E2** 15 0 S 47 30W
Gerald, *U.S.A.* **156 F5** 38 24N 91 20W
Geraldine, *N.Z.* **131 E6** 44 5 S 171 15 E
Geraldine, *U.S.A.* **158 C8** 47 36N 110 16W
Geraldton, *Australia* **125 E1** 28 48 S 114 32 E
Geraldton, *Canada* **140 C2** 49 44N 86 59W
Geranium, *Australia* **128 C4** 35 23 S 140 11 E
Gérardmer, *France* **27 D13** 48 3N 6 50 E
Gercüş, *Turkey* **101 D9** 37 34N 41 23 E
Gerðar, *Iceland* **11 C4** 64 4N 22 39W
Gerdine, Mt., *U.S.A.* **144 F9** 61 35N 152 27W
Gerede, *Turkey* **100 B5** 40 45N 32 10 E
Gerês, Sierra do, *Portugal* **42 D3** 41 48N 8 0W
Gereshk, *Afghan.* **91 C2** 31 47N 64 35 E
Geretsried, *Germany* **31 H7** 47 51N 11 28 E
Gérgal, *Spain* **41 H2** 37 7N 2 31W
Gerik, *Malaysia* **87 K3** 5 50N 101 15 E
Gering, *U.S.A.* **154 E3** 41 50N 103 40W
Gerlach, *U.S.A.* **158 F4** 40 39N 119 21W
Gerlachovský štit, *Slovak Rep.* **35 B13** 49 11N 20 7 E
Gerlogubi, *Ethiopia* **120 C3** 6 53N 45 3 E
German Planina, *Macedonia* **50 D6** 42 20N 22 0 E
Germania Land, *Greenland* **10 B9** 77 5N 19 30W
Germansen Landing, *Canada* **142 B4** 55 43N 124 40W
Germantown, *U.S.A.* **157 E12** 39 38N 84 22W
Germany ■, *Europe* **30 E6** 51 0N 10 0 E
Germencik, *Turkey* **49 D9** 37 52N 27 37 E
Germering, *Germany* **31 G7** 48 8N 11 22 E
Germersheim, *Germany* **31 F4** 49 12N 8 22 E
Germī, *Iran* **101 C13** 39 1N 48 3 E
Germiston, *S. Africa* **117 D4** 26 15 S 28 10 E
Gernika-Lumo, *Spain* **40 B2** 43 19N 2 40W
Gernsheim, *Germany* **31 F4** 49 45N 8 30 E
Gero, *Japan* **71 B9** 35 48N 137 14 E
Gerolzhofen, *Germany* **31 F6** 49 54N 10 21 E
Gerona = Girona, *Spain* **40 D7** 41 58N 2 46 E
Gerrard, *Canada* **142 C5** 50 30N 117 17W
Gerringong, *Australia* **129 C9** 34 46 S 150 47 E
Gers □, *France* **28 E4** 43 35N 0 30 E
Gers →, *France* **28 D4** 44 9N 0 39 E
Gersfeld, *Germany* **30 E5** 50 27N 9 56 E
Gersoppa Falls, *India* **95 G2** 14 12N 74 46 E
Gersthofen, *Germany* **31 G6** 48 25N 10 53 E
Gerzat, *France* **28 C7** 45 48N 3 8 E
Gerze, *Turkey* **100 B6** 41 48N 35 12 E
Geseke, *Germany* **30 D4** 51 38N 8 31 E
Geser, *Indonesia* **83 B4** 3 50 S 130 54 E
Gesso →, *Italy* **44 D4** 44 24N 7 33 E
Gestro, Wabi →, *Ethiopia* **107 G5** 4 12N 42 2 E
Getafe, *Spain* **42 E7** 40 18N 3 44W
Gethsémani, *Canada* **141 B7** 50 13N 60 40W
Getinge, *Sweden* **17 H6** 56 49N 12 44 E
Gettysburg, *Pa., U.S.A.* **148 F7** 39 50N 77 14W
Gettysburg, *S. Dak., U.S.A.* **154 C5** 45 1N 99 57W
Getxo, *Spain* **40 B2** 43 21N 2 59W
Getz Ice Shelf, *Antarctica* **7 D14** 75 0 S 130 0W
Geureudong, Mt., *Indonesia* **84 B1** 4 13N 96 42 E
Geurie, *Australia* **129 B8** 32 22 S 148 50 E
Gevaş, *Turkey* **101 C10** 38 15N 43 6 E
Gévaudan, *France* **28 D7** 44 40N 3 40 E
Gevgelija, *Macedonia* **50 E6** 41 9N 22 30 E
Gévora →, *Spain* **43 G4** 38 53N 6 57W
Gex, *France* **27 F13** 46 21N 6 3 E
Geyikli, *Turkey* **49 B8** 39 48N 26 12 E
Geyser, *U.S.A.* **158 C8** 47 16N 110 30W
Geyserville, *U.S.A.* **160 G4** 38 42N 122 54W
Geysir, *Iceland* **11 C6** 64 19N 20 18W
Geyve, *Turkey* **100 B4** 40 30N 30 18 E
Ghâbat el Arab = Wang Kai, *Sudan* **107 F2** 9 3N 29 23 E
Ghaghara →, *India* **93 G11** 25 45N 84 40 E
Ghalat, *Oman* **99 B7** 21 6N 58 53 E
Ghalla, Wadi el →, *Sudan* **107 E2** 10 25N 27 32 E
Ghallamane, *Mauritania* **110 D3** 23 15N 10 0W
Ghana ■, *W. Afr.* **113 D4** 8 0N 1 0W
Ghansor, *India* **93 H9** 22 39N 80 1 E
Ghanzi, *Botswana* **116 C3** 21 50 S 21 34 E
Ghanzi □, *Botswana* **116 C3** 21 50 S 21 45 E
Gharb el Istiwa'iya □, *Sudan* **107 G3** 5 0N 30 0 E
Gharbîya, Es Sahrâ el, *Egypt* **106 B2** 27 40N 26 30 E
Ghard Abû Muharik, *Egypt* **106 B3** 26 50N 30 0 E
Ghardaïa, *Algeria* **111 B5** 32 20N 3 37 E
Ghârib, G., *Egypt* **106 J8** 28 6N 32 54 E
Gharm, W. →, *Oman* **99 C7** 19 57N 57 38 E
Gharyān, *Libya* **108 B2** 32 10N 13 0 E
Gharyān □, *Libya* **108 B2** 30 35N 13 0 E
Ghat, *Libya* **108 D2** 24 59N 10 11 E
Ghatal, *India* **93 H12** 22 40N 87 46 E
Ghatampur, *India* **93 F9** 26 8N 80 13 E
Ghatere, *Solomon Is.* **133 L10** 7 55 S 159 0 E
Ghatprabha →, *India* **95 F2** 16 15N 75 20 E
Ghaţţī, *Si. Arabia* **96 D3** 31 16N 37 31 E
Ghawdex = Gozo, *Malta* **38 C1** 36 3N 14 13 E
Ghayl, *Si. Arabia* **98 B4** 21 40N 46 20 E
Ghayl Bā Wazīr, *Yemen* **99 D5** 14 47N 49 22 E
Ghazal, Bahr el →, *Chad* **109 F3** 13 0N 15 47 E
Ghazâl, Bahr el →, *Sudan* **107 F3** 9 31N 30 25 E
Ghazaouet, *Algeria* **111 A4** 35 8N 1 50W
Ghaziabad, *India* **92 E7** 28 42N 77 26 E
Ghazipur, *India* **93 G10** 25 38N 83 35 E
Ghaznī, *Afghan.* **91 B3** 33 30N 68 28 E
Ghaznī □, *Afghan.* **91 B3** 32 10N 68 20 E
Ghedi, *Italy* **44 C7** 45 24N 10 16 E
Ghelari, *Romania* **52 E7** 45 38N 22 45 E
Ghèlinsor, *Somali Rep.* **120 C3** 6 28N 46 39 E
Ghent = Gent, *Belgium* **24 C3** 51 2N 3 42 E
Gheorghe Gheorghiu-Dej = Oneşti, *Romania* **53 D11** 46 17N 26 47 E
Gheorgheni, *Romania* **53 D10** 46 43N 25 41 E
Gherla, *Romania* **53 C8** 47 2N 23 57 E
Ghidigeni, *Romania* **53 D12** 46 3N 27 30 E
Ghilarza, *Italy* **46 B1** 40 7N 8 50 E
Ghimeş-Făget, *Romania* **53 D11** 46 35N 26 2 E
Ghisonaccia, *France* **29 F13** 42 1N 9 26 E
Ghisoni, *France* **29 F13** 42 7N 9 12 E
Ghizao, *Afghan.* **92 C1** 33 20N 65 44 E
Ghizar →, *Pakistan* **93 A5** 36 15N 73 43 E
Ghod →, *India* **94 E2** 18 30N 74 35 E
Ghogha, *India* **92 J5** 21 40N 72 20 E
Ghot Ogrein, *Egypt* **106 A2** 31 10N 25 20 E
Ghotaru, *India* **92 F4** 27 20N 70 1 E
Ghotki, *Pakistan* **92 E3** 28 5N 69 21 E
Ghowr □, *Afghan.* **91 B2** 34 0N 64 20 E
Ghudaf, W. al →, *Iraq* **101 F10** 32 56N 43 30 E
Ghudāmis, *Libya* **108 B1** 30 11N 9 29 E
Ghughri, *India* **93 H9** 22 39N 80 41 E
Ghugus, *India* **94 E4** 19 58N 79 12 E
Ghulam Mohammad Barrage, *Pakistan* **92 G3** 25 30N 68 20 E
Ghurayrah, *Si. Arabia* **98 C3** 18 37N 42 41 E
Ghūrīān, *Afghan.* **91 B1** 34 17N 61 25 E
Gia Dinh, *Vietnam* **87 G6** 10 49N 106 42 E
Gia Lai = Plei Ku, *Vietnam* **86 F7** 13 57N 108 0 E
Gia Nghia, *Vietnam* **87 G6** 11 58N 107 42 E
Gia Ngoc, *Vietnam* **86 E7** 14 50N 108 58 E
Gia Vuc, *Vietnam* **86 E7** 14 42N 108 34 E
Giamama, *Somali Rep.* **120 D2** 0 4N 42 44 E
Giannutri, *Italy* **44 F8** 42 15N 11 6 E
Giant Forest, *U.S.A.* **160 J8** 36 36N 118 43W
Giant Mts. = Krkonoše, *Czech Rep.* **34 A8** 50 50N 15 35 E
Giants Causeway, *U.K.* **23 A5** 55 16N 6 29W
Giarabub = Al Jaghbūb, *Libya* **108 C4** 29 42N 24 38 E
Giarre, *Italy* **47 E8** 37 43N 15 11 E
Giaveno, *Italy* **44 C4** 45 2N 7 21 E
Gibara, *Cuba* **164 B4** 21 9N 76 11W
Gibb River, *Australia* **124 C4** 16 26 S 126 26 E
Gibbon, *U.S.A.* **154 E5** 40 45N 98 51W
Gibe →, *Ethiopia* **107 F4** 7 20N 37 36 E
Gibellina Nuova, *Italy* **46 E5** 37 47N 12 58 E
Gibraleón, *Spain* **43 H4** 37 23N 6 58W
Gibraltar ■, *Europe* **43 J5** 36 7N 5 22W
Gibraltar, Str. of, *Medit. S.* **43 K5** 35 55N 5 40W
Gibson, *U.S.A.* **152 B7** 33 14N 82 36W
Gibson City, *U.S.A.* **157 D8** 40 28N 88 22W
Gibson Desert, *Australia* **124 D4** 24 0 S 126 0 E
Gibsonburg, *U.S.A.* **157 C13** 41 23N 83 19W
Gibsons, *Canada* **142 D4** 49 24N 123 32W
Gibsonton, *U.S.A.* **153 H7** 27 51N 82 23W
Gibsonville, *U.S.A.* **160 F6** 39 46N 120 54W
Giddalur, *India* **95 G4** 15 20N 78 57 E
Giddings, *U.S.A.* **155 K6** 30 11N 96 56W
Gidole, *Ethiopia* **107 F4** 5 40N 37 25 E
Gien, *France* **27 E9** 47 40N 2 36 E
Giengen, *Germany* **31 G6** 48 37N 10 14 E
Giessen, *Germany* **30 E4** 50 34N 8 41 E
Gīfān, *Iran* **97 B8** 37 54N 57 28 E
Gifatin, Geziret, *Egypt* **106 B3** 27 10N 33 50 E
Gifford, *U.S.A.* **153 H9** 27 40N 80 25W
Gifford Creek, *Australia* **124 D2** 24 3 S 116 16 E
Gifhorn, *Germany* **30 C6** 52 30N 10 33 E
Gifu, *Japan* **71 B8** 35 30N 136 45 E
Gifu □, *Japan* **71 B9** 35 40N 137 0 E
Gigant, *Russia* **61 G5** 46 28N 41 20 E
Giganta, Sa. de la, *Mexico* **162 B2** 25 30N 111 30W
Gigen, *Bulgaria* **51 C8** 43 40N 24 28 E
Gigha, *U.K.* **22 F3** 55 42N 5 44W
Giglei, *Somali Rep.* **120 C3** 5 25N 45 20 E
Gíglio, *Italy* **44 F7** 42 20N 10 52 E
Gigmoto, *Phil.* **80 E5** 13 47N 124 23 E
Gignac, *France* **28 E7** 43 39N 3 32 E
Gigüela →, *Spain* **43 F7** 39 8N 3 44W
Gijón, *Spain* **42 B5** 43 32N 5 42W
Gil I., *Canada* **142 C3** 53 12N 129 15W
Gila →, *U.S.A.* **159 K6** 32 43N 114 33W
Gila Bend, *U.S.A.* **159 K7** 32 57N 112 43W
Gila Bend Mts., *U.S.A.* **159 K7** 33 10N 113 0W
Gīlān □, *Iran* **97 B6** 37 0N 50 0 E
Gilău, *Romania* **53 D8** 46 45N 23 23 E
Gilbert, *U.S.A.* **153 B9** 33 56N 81 24W
Gilbert →, *Australia* **126 B3** 16 35 S 141 15 E
Gilbert Is., *Kiribati* **134 G9** 1 0N 172 0 E
Gilbert Plains, *Canada* **143 C8** 51 9N 100 28W
Gilbert River, *Australia* **126 B3** 18 9 S 142 52 E
Gilberton, *Australia* **126 B3** 19 16 S 143 35 E
Gilbués, *Brazil* **170 C2** 9 50 S 45 21W
Gilf el Kebîr, Hadabat el, *Egypt* **106 C2** 23 50N 25 50 E
Gilford I., *Canada* **142 C3** 50 40N 126 30W
Gilgandra, *Australia* **129 A8** 31 43 S 148 39 E
Gilgil, *Kenya* **118 C4** 0 30 S 36 20 E
Gilgit, *India* **93 B6** 35 50N 74 15 E
Gilgit →, *Pakistan* **93 B6** 35 44N 74 37 E
Gilgunnia, *Australia* **129 B7** 32 26 S 146 2 E
Giljeva Planina, *Serbia, Yug.* **50 C3** 43 9N 20 0 E
Gillam, *Canada* **143 B10** 56 20N 94 40W
Gilleleje, *Denmark* **17 H6** 56 8N 12 19 E
Gillen, L., *Australia* **125 E3** 26 11 S 124 38 E
Gilles, L., *Australia* **128 B2** 32 50 S 136 45 E
Gillespie, *U.S.A.* **156 E7** 39 8N 89 49W
Gillespies Pt., *N.Z.* **131 D4** 43 24 S 169 49 E
Gillette, *U.S.A.* **154 C2** 44 18N 105 30W
Gilliat, *Australia* **126 C3** 20 40 S 141 28 E
Gillingham, *U.K.* **21 F8** 51 23N 0 33 E
Gilman, *U.S.A.* **157 D9** 40 46N 88 0W
Gilman City, *U.S.A.* **156 D3** 40 8N 93 53W
Gilmer, *U.S.A.* **155 J7** 32 44N 94 57W
Gilmore, *Australia* **129 C8** 35 20 S 148 12 E
Gilmore, L., *Australia* **125 F3** 32 29 S 121 37 E
Gilmour, *Canada* **140 D4** 44 48N 77 37W
Gilo →, *Ethiopia* **107 F3** 8 10N 33 15 E
Gilort →, *Romania* **53 F8** 44 38N 23 32 E
Gilroy, *U.S.A.* **160 H5** 37 1N 121 34W
Giluwe, Mt., *Papua N. G.* **132 D2** 6 8 S 143 52 E
Gimbi, *Ethiopia* **107 F4** 9 3N 35 42 E
Gimli, *Canada* **143 C9** 50 40N 97 0W
Gimo, *Sweden* **16 D12** 60 11N 18 12 E
Gimone →, *France* **28 E5** 44 0N 1 6 E
Gimont, *France* **28 E4** 43 38N 0 52 E
Gin →, *Sri Lanka* **95 L5** 6 5N 80 7 E
Gin Gin, *Australia* **127 D5** 25 0 S 151 58 E
Ginâh, *Egypt* **106 B3** 25 21N 30 30 E
Ginatilan, *Phil.* **81 G4** 9 34N 123 19 E
Gindie, *Australia* **126 C4** 23 44 S 148 8 E
Gingin, *Australia* **125 F2** 31 22 S 115 54 E
Gingoog, *Phil.* **81 G5** 8 50N 125 7 E
Ginir, *Ethiopia* **107 F5** 7 6N 40 40 E
Ginosa, *Italy* **47 B9** 40 35N 16 45 E
Ginzo de Limia = Xinzo de Limia, *Spain* **42 C3** 42 3N 7 47W
Giohar, *Somali Rep.* **120 D3** 2 48N 45 30 E
Gióia, G. di, *Italy* **47 D8** 38 30N 15 45 E
Gióia del Colle, *Italy* **47 B9** 40 48N 16 55 E
Gióia Táuro, *Italy* **47 D8** 38 25N 15 54 E
Gioiosa Iónica, *Italy* **47 D9** 38 20N 16 18 E
Gioiosa Marea, *Italy* **47 D7** 38 10N 14 54 E
Gióna, Óros, *Greece* **48 C4** 38 38N 22 14 E
Giovi, Passo dei, *Italy* **44 D5** 44 33N 8 57 E
Giovinazzo, *Italy* **47 A9** 41 11N 16 40 E
Gir Hills, *India* **92 J4** 21 0N 71 0 E
Girab, *India* **92 F4** 26 2N 70 38 E
Girâfi, W. →, *Egypt* **103 F3** 29 58N 34 39 E
Giraltovce, *Slovak Rep.* **35 B14** 49 7N 21 32 E
Girard, *Ga., U.S.A.* **152 B8** 33 3N 81 43W
Girard, *Ill., U.S.A.* **156 E7** 39 27N 89 47W
Girard, *Kans., U.S.A.* **155 G7** 37 31N 94 51W
Girard, *Ohio, U.S.A.* **150 E4** 41 9N 80 42W
Girard, *Pa., U.S.A.* **150 D4** 42 0N 80 19W
Girardot, *Colombia* **168 C3** 4 18N 74 48W
Girdle Ness, *U.K.* **22 D6** 57 9N 2 3W
Giresun, *Turkey* **101 B8** 40 55N 38 30 E
Girga, *Egypt* **106 B3** 26 17N 31 55 E
Girgir, C., *Papua N. G.* **132 B3** 3 50 S 144 35 E
Giridih, *India* **93 G12** 24 10N 86 21 E
Girifalco, *Italy* **47 D9** 38 49N 16 25 E
Girilambone, *Australia* **129 A7** 31 16 S 146 57 E
Girne = Kyrenia, *Cyprus* **38 D12** 35 20N 33 20 E
Giro, *Nigeria* **113 C5** 11 7N 4 42 E
Giromagny, *France* **27 E13** 47 45N 6 50 E
Girona, *Spain* **40 D7** 41 58N 2 46 E

218 Girona

Girona □, *Spain* **40 C7** 42 11N 2 30 E
Gironde □, *France* **28 D3** 44 45N 0 30W
Gironde →, *France* **28 C2** 45 32N 1 7W
Gironella, *Spain* **40 C6** 42 2N 1 53 E
Giru, *Australia* **126 B4** 19 30 S 147 5 E
Girvan, *U.K.* **22 F4** 55 14N 4 51W
Gisborne, *N.Z.* **130 E7** 38 39 S 178 5 E
Gisenyi, *Rwanda* **118 C2** 1 41 S 29 15 E
Gislaved, *Sweden* **17 G7** 57 19N 13 32 E
Gisors, *France* **27 C8** 49 15N 1 47 E
Gissarskiy, Khrebet, *Tajikistan* **63 D4** 39 0N 69 0 E
Giswil, *Switz.* **32 C6** 46 50N 8 11 E
Gitega, *Burundi* **118 C2** 3 26 S 29 56 E
Giuba →, *Somali Rep.* **120 D2** 1 30N 42 35 E
Giubiasco, *Switz.* **33 D8** 46 11N 9 1 E
Giugliano in Campania, *Italy* **47 B7** 40 56N 14 12 E
Giulianova, *Italy* **45 F10** 42 45N 13 57 E
Giurgeni, *Romania* **53 F12** 44 45N 27 48 E
Giurgiu, *Romania* **53 G10** 43 52N 25 57 E
Giurgiu □, *Romania* **53 F10** 44 20N 26 0 E
Giurgiuleşti, *Moldova* **53 E13** 45 29N 28 12 E
Give, *Denmark* **17 J3** 55 51N 9 13 E
Givet, *France* **27 B11** 50 8N 4 49 E
Givors, *France* **29 C8** 45 35N 4 45 E
Givry, *France* **27 F11** 46 41N 4 46 E
Giyon, *Ethiopia* **107 F4** 8 33N 38 1 E
Giza = El Gîza, *Egypt* **106 J7** 30 0N 31 10 E
Gizhduvan, *Uzbekistan* **63 C2** 40 6N 64 41 E
Gizhiga, *Russia* **65 C17** 62 3N 160 30 E
Gizhiginskaya Guba, *Russia* **65 C16** 61 0N 158 0 E
Gizo, *Solomon Is.* **133 M9** 8 7 S 156 50 E
Giżycko, *Poland* **54 D8** 54 2N 21 48 E
Gizzeria, *Italy* **47 D9** 38 59N 16 12 E
Gjegjani, *Albania* **50 E4** 41 58N 20 3 E
Gjendesheim, *Norway* **18 C5** 61 30N 8 48 E
Gjerstad, *Norway* **18 F6** 58 54N 9 1 E
Gjevilvatnet, *Norway* **18 B6** 62 43N 9 23 E
Gjirokastra, *Albania* **50 F4** 40 7N 20 10 E
Gjoa Haven, *Canada* **138 B10** 68 20N 96 8W
Gjögur, *Iceland* **11 B5** 65 59N 21 21W
Gjøra, *Norway* **18 B6** 62 33N 9 4 E
Gjøvik, *Norway* **15 F14** 60 47N 10 43 E
Gjuhës, Kep i, *Albania* **50 F3** 40 28N 19 15 E
Glace Bay, *Canada* **141 C8** 46 11N 59 58W
Glacier Bay, *U.S.A.* **142 B1** 58 40N 136 0W
Glacier Bay National Park, *U.S.A.* **144 G13** 58 45N 136 30W
Glacier Nat. Park, *Canada* **142 C5** 51 15N 117 30W
Glacier Park, *U.S.A.* **158 B7** 48 30N 113 18W
Glacier Peak, *U.S.A.* **158 B3** 48 7N 121 7W
Gladewater, *U.S.A.* **155 J7** 32 33N 94 56W
Gladstone, *Queens., Australia* **126 C5** 23 52 S 151 16 E
Gladstone, *S. Austral., Australia* **128 B3** 33 15 S 138 22 E
Gladstone, *W. Austral., Australia* **125 E1** 25 57 S 114 17 E
Gladstone, *Canada* **143 C9** 50 13N 98 57W
Gladstone, *Mich., U.S.A.* **148 C2** 45 51N 87 1W
Gladstone, *Mo., U.S.A.* **156 E2** 39 13N 94 35W
Gladwin, *U.S.A.* **148 D3** 43 59N 84 29W
Gladys L., *Canada* **142 B2** 59 50N 133 0W
Glafsfjorden, *Sweden* **16 E6** 59 30N 12 37 E
Głagów Małapolski, *Poland* **55 H8** 50 10N 21 56 E
Glåma = Glomma →, *Norway* **15 G14** 59 12N 10 57 E
Gláma, *Iceland* **11 B4** 65 48N 23 0W
Glamis, *U.S.A.* **161 N11** 32 55N 115 5W
Glamoč, *Bos.-H.* **45 D13** 44 3N 16 51 E
Glåmos, *Norway* **18 B8** 62 41N 11 25 E
Glamsbjerg, *Denmark* **17 J4** 55 17N 10 6 E
Glarner Alpen, *Switz.* **33 C8** 46 50N 9 0 E
Glärnisch, *Switz.* **33 C8** 47 0N 9 0 E
Glarus, *Switz.* **33 B8** 47 3N 9 4 E
Glarus □, *Switz.* **33 C8** 47 0N 9 5 E
Glasco, *Kans., U.S.A.* **154 F6** 39 22N 97 50W
Glasco, *N.Y., U.S.A.* **151 D11** 42 3N 73 57W
Glasgow, *U.K.* **22 F4** 55 51N 4 15W
Glasgow, *Ky., U.S.A.* **148 G3** 37 0N 85 55W
Glasgow, *Mo., U.S.A.* **156 E4** 39 14N 92 51W
Glasgow, *Mont., U.S.A.* **158 B10** 48 12N 106 38W
Glastonbury, *U.K.* **21 F5** 51 9N 2 43W
Glastonbury, *U.S.A.* **151 E12** 41 43N 72 37W
Glatt →, *Switz.* **33 B7** 47 28N 8 32 E
Glattfelden, *Switz.* **33 A7** 47 33N 8 30 E
Glauchau, *Germany* **30 E8** 50 49N 12 31 E
Glava, *Sweden* **16 E6** 59 33N 12 35 E
Glavice, *Croatia* **45 E13** 43 43N 16 41 E
Glazov, *Russia* **60 A11** 58 9N 52 40 E
Gleisdorf, *Austria* **34 D8** 47 6N 15 44 E
Gleiwitz = Gliwice, *Poland* **55 H5** 50 22N 18 41 E
Glen, *U.S.A.* **151 B13** 44 7N 71 11W
Glen Affric, *U.K.* **22 D3** 57 17N 5 1W
Glen Afton, *N.Z.* **130 D4** 37 37 S 175 4 E
Glen Canyon Dam, *U.S.A.* **159 H8** 36 57N 111 29W
Glen Canyon National Recreation Area, *U.S.A.* **159 H8** 37 15N 111 0W
Glen Coe, *U.K.* **22 E3** 56 40N 5 0W
Glen Cove, *U.S.A.* **151 F11** 40 52N 73 38W
Glen Garry, *U.K.* **22 D3** 57 3N 5 7W
Glen Gowrie, *Australia* **128 A5** 31 4 S 143 10 E
Glen Innes, *Australia* **127 D5** 29 44 S 151 44 E
Glen Lyon, *U.S.A.* **151 E8** 41 10N 76 5W
Glen Massey, *N.Z.* **130 D4** 37 38 S 175 2 E
Glen Mor, *U.K.* **22 D4** 57 9N 4 37W
Glen Moriston, *U.K.* **22 D4** 57 11N 4 52W
Glen Spean, *U.K.* **22 E4** 56 53N 4 40W
Glen Ullin, *U.S.A.* **154 B4** 46 49N 101 50W
Glen Valley, *Australia* **129 D7** 36 54 S 147 28 E
Glénan, Îs. de, *France* **26 E3** 47 42N 4 0W
Glenariff, *Australia* **129 A7** 30 50 S 146 33 E
Glenavy, *N.Z.* **131 E6** 44 54 S 171 7 E
Glenburgh, *Australia* **125 E2** 25 26 S 116 6 E
Glenburn, *Australia* **129 D6** 37 27 S 145 26 E
Glencoe, *Canada* **150 D3** 42 45N 81 43W
Glencoe, *S. Africa* **117 D5** 28 11 S 30 11 E
Glencoe, *Ala., U.S.A.* **152 B4** 33 57N 85 56W
Glencoe, *Minn., U.S.A.* **154 C7** 44 46N 94 9W
Glendale, *Ariz., U.S.A.* **159 K7** 33 32N 112 11W
Glendale, *Calif., U.S.A.* **161 L8** 34 9N 118 15W
Glendale, *Fla., U.S.A.* **152 E3** 30 52N 86 7W
Glendale, *Oreg., U.S.A.* **158 E2** 42 44N 123 26W
Glendale, *Zimbabwe* **119 F3** 17 22 S 31 5 E
Glendive, *U.S.A.* **154 B2** 47 7N 104 43W
Glendo, *U.S.A.* **154 D2** 42 30N 105 2W
Glenelg, *Australia* **128 C3** 34 58 S 138 31 E
Glenelg →, *Australia* **128 E4** 38 4 S 140 59 E
Glenflorrie, *Australia* **124 D2** 22 55 S 115 59 E
Glengarriff, *Ireland* **23 E2** 51 45N 9 34W
Glengyle, *Australia* **126 C2** 24 48 S 139 37 E
Glenham, *N.Z.* **131 G3** 46 26 S 168 52 E
Glenhope, *N.Z.* **131 B7** 41 40 S 172 39 E
Glenmary, Mt., *N.Z.* **131 D4** 43 55 S 169 55 E
Glenmora, *U.S.A.* **155 K8** 30 59N 92 35W
Glenmorgan, *Australia* **127 D4** 27 14 S 149 42 E
Glenn, *U.S.A.* **160 F4** 39 31N 122 1W
Glennallen, *U.S.A.* **144 E11** 62 7N 145 33W
Glennamaddy, *Ireland* **23 C3** 53 37N 8 33W
Glenns Ferry, *U.S.A.* **158 E6** 42 57N 115 18W
Glennville, *U.S.A.* **152 D8** 31 56N 81 56W
Glenorchy, *S. Austral., Australia* **128 A3** 31 55 S 139 46 E
Glenorchy, *Tas., Australia* **126 G4** 42 49 S 147 18 E
Glenorchy, *Vic., Australia* **128 D5** 36 55 S 142 41 E
Glenorchy, *N.Z.* **131 E3** 44 51 S 168 24 E
Glenore, *Australia* **126 B3** 17 50 S 141 12 E
Glenormiston, *Australia* **126 C2** 22 55 S 138 50 E
Glenreagh, *Australia* **127 E5** 30 2 S 153 1 E
Glenrock, *U.S.A.* **158 E11** 42 52N 105 52W
Glenrothes, *U.K.* **22 E5** 56 12N 3 10W
Glenrowan, *Australia* **129 D7** 36 29 S 146 13 E
Glenroy, *Australia* **128 D4** 37 13 S 140 48 E
Glens Falls, *U.S.A.* **151 C11** 43 19N 73 39W
Glenties, *Ireland* **23 B3** 54 49N 8 16W
Glenville, *U.S.A.* **148 F5** 38 56N 80 50W
Glenwood, *Alta., Canada* **142 D6** 49 21N 113 31W
Glenwood, *Nfld., Canada* **141 C9** 49 0N 54 58W
Glenwood, *Ala., U.S.A.* **152 D3** 31 40N 86 10W
Glenwood, *Ark., U.S.A.* **155 H8** 34 20N 93 33W
Glenwood, *Ga., U.S.A.* **152 C7** 32 11N 82 40W
Glenwood, *Hawaii, U.S.A.* **146 J17** 19 29N 155 9W
Glenwood, *Iowa, U.S.A.* **154 E7** 41 3N 95 45W
Glenwood, *Minn., U.S.A.* **154 C7** 45 39N 95 23W
Glenwood, *Wash., U.S.A.* **160 D5** 46 1N 121 17W
Glenwood Springs, *U.S.A.* **158 G10** 39 33N 107 19W
Gletsch, *Switz.* **33 C6** 46 34N 8 22 E
Glettinganes, *Iceland* **14 D7** 65 30N 13 37W
Glifádha, *Greece* **48 D5** 37 52N 23 45 E
Glimåkra, *Sweden* **17 H8** 56 19N 14 7 E
Glina, *Croatia* **45 C13** 45 20N 16 6 E
Glinojeck, *Poland* **55 F7** 52 49N 20 21 E
Glittertind, *Norway* **18 C5** 61 40N 8 32 E
Gliwice, *Poland* **55 H5** 50 22N 18 41 E
Globe, *U.S.A.* **159 K8** 33 24N 110 47W
Glodeanu Siliştea, *Romania* **53 F11** 44 50N 26 48 E
Glodeni, *Moldova* **53 C12** 47 45N 27 31 E
Glödnitz, *Austria* **34 E7** 46 53N 14 7 E
Gloggnitz, *Austria* **34 D8** 47 41N 15 56 E
Głogów, *Poland* **55 G3** 51 37N 16 5 E
Głogówek, *Poland* **55 H4** 50 21N 17 53 E
Glomma →, *Norway* **15 G14** 59 12N 10 57 E
Glorieuses, Is., *Ind. Oc.* **117 A8** 11 30 S 47 20 E
Glóssa, *Greece* **48 B5** 39 10N 23 45 E
Glossop, *U.K.* **20 D6** 53 27N 1 56W
Gloucester, *Australia* **129 B9** 32 0 S 151 59 E
Gloucester, *U.K.* **21 F5** 51 53N 2 15W
Gloucester, *U.S.A.* **151 D14** 42 37N 70 40W
Gloucester, C., *Papua N. G.* **132 C5** 5 26 S 148 21 E
Gloucester I., *Australia* **126 C4** 20 0 S 148 30 E
Gloucestershire □, *U.K.* **21 F5** 51 46N 2 15W
Gloversville, *U.S.A.* **151 C10** 43 3N 74 21W
Glovertown, *Canada* **141 C9** 48 40N 54 3W
Głowno, *Poland* **55 G6** 51 59N 19 42 E
Głubczyce, *Poland* **55 H4** 50 13N 17 52 E
Glubokiy, *Russia* **61 F5** 48 35N 40 25 E
Glubokoye = Hlybokaye, *Belarus* **58 E4** 55 10N 27 45 E
Głuchołazy, *Poland* **55 H4** 50 19N 17 24 E
Glücksburg, *Germany* **30 A5** 54 50N 9 33 E
Glückstadt, *Germany* **30 B5** 53 45N 9 25 E
Gluepot, *Australia* **128 B4** 33 45 S 140 0 E
Glukhov = Hlukhiv, *Ukraine* **59 G7** 51 40N 33 58 E
Glusk, *Belarus* **59 F5** 52 53N 28 41 E
Głuszyca, *Poland* **55 H3** 50 41N 16 23 E
Glyngøre, *Denmark* **17 H2** 56 46N 8 52 E
Gmünd, *Kärnten, Austria* **34 E6** 46 54N 13 31 E
Gmünd, *Niederösterreich, Austria* **34 C8** 48 45N 15 0 E
Gmunden, *Austria* **34 D6** 47 55N 13 48 E
Gnali, *Gabon* **114 C2** 2 34 S 11 18 E
Gnarp, *Sweden* **16 B11** 62 3N 17 16 E
Gnesta, *Sweden* **16 E11** 59 3N 17 17 E
Gniew, *Poland* **54 E5** 53 50N 18 50 E
Gniewkowo, *Poland* **55 F5** 52 54N 18 25 E
Gniezno, *Poland* **55 F4** 52 30N 17 35 E
Gnjilane, *Serbia, Yug.* **50 D5** 42 28N 21 29 E
Gnoien, *Germany* **30 B8** 53 58N 12 41 E
Gnosjö, *Sweden* **17 G7** 57 22N 13 43 E
Gnowangerup, *Australia* **125 F2** 33 58 S 117 59 E
Go Cong, *Vietnam* **87 G6** 10 22N 106 40 E
Gō-no-ura, *Japan* **70 D1** 33 44N 129 40 E
Goa, *India* **95 G1** 15 33N 73 59 E
Goa, *Phil.* **80 E4** 13 42N 123 29 E
Goa □, *India* **95 G1** 15 33N 73 59 E
Goalen Hd., *Australia* **129 D9** 36 33 S 150 4 E
Goalpara, *India* **90 B3** 26 10N 90 40 E
Goalundo Ghat, *Bangla.* **93 H13** 23 50N 89 47 E
Goaso, *Ghana* **112 D4** 6 48N 2 30W
Goat Fell, *U.K.* **22 F3** 55 38N 5 11W
Goba, *Ethiopia* **107 F4** 7 1N 39 59 E
Goba, *Mozam.* **117 D5** 26 15 S 32 13 E
Gobabis, *Namibia* **116 C2** 22 30 S 19 0 E
Göbel, *Turkey* **51 F12** 40 0N 28 9 E
Gobernador Gregores, *Argentina* **176 C2** 48 46 S 70 15W
Gobi, *Asia* **74 C6** 44 0N 110 0 E
Gobichettipalayam, *India* **95 J3** 11 31N 77 21 E
Gobles, *U.S.A.* **157 B11** 42 22N 85 53W
Gobō, *Japan* **71 D7** 33 53N 135 10 E
Gobo, *Sudan* **107 F3** 5 40N 31 10 E
Göçbeyli, *Turkey* **49 B9** 39 13N 27 25 E
Goch, *Germany* **30 D2** 51 41N 6 9 E
Gochas, *Namibia* **116 C2** 24 59 S 18 55 E
Godavari →, *India* **94 F6** 16 25N 82 18 E
Godavari Point, *India* **94 F6** 17 0N 82 20 E
Godbout, *Canada* **141 C6** 49 20N 67 38W
Godda, *India* **93 G12** 24 50N 87 13 E
Goddua, *Libya* **108 C2** 26 26N 14 19 E
Godech, *Bulgaria* **50 C7** 43 1N 23 4 E
Goderich, *Canada* **140 D3** 43 45N 81 41W
Goderville, *France* **26 C7** 49 38N 0 22 E
Godfrey, *Ga., U.S.A.* **152 B6** 33 27N 83 30W
Godfrey, *Ill., U.S.A.* **156 F6** 38 58N 90 11W
Godhavn, *Greenland* **10 D5** 69 15N 53 38W
Goðdalir, *Iceland* **11 B7** 65 20N 19 6W
Godhra, *India* **92 H5** 22 49N 73 40 E
Godinlave, *Somali Rep.* **120 C3** 5 54N 46 38 E
Gödöllő, *Hungary* **52 C4** 47 38N 19 25 E
Godoy Cruz, *Argentina* **174 C2** 32 56 S 68 52W
Gods →, *Canada* **143 B10** 56 22N 92 51W
Gods L., *Canada* **143 C10** 54 40N 94 15W
Godthåb = Nuuk, *Greenland* **10 E5** 64 10N 51 35W
Godwin Austen = K2, *Pakistan* **93 B7** 35 58N 76 32 E
Goeie Hoop, Kaap die = Good Hope, C. of, *S. Africa* **116 E2** 34 24 S 18 30 E
Goéland, L. au, *Canada* **140 C4** 49 50N 76 48W
Goeree, *Neths.* **24 C3** 51 50N 4 0 E
Goes, *Neths.* **24 C3** 51 30N 3 55 E
Gogama, *Canada* **140 C3** 47 35N 81 43W
Gogango, *Australia* **126 C5** 23 40 S 150 2 E
Gogebic, L., *U.S.A.* **154 B10** 46 30N 89 35W
Gogolin, *Poland* **55 H5** 50 30N 18 0 E
Gogra = Ghaghara →, *India* **93 G11** 25 45N 84 40 E
Gogriâl, *Sudan* **107 F2** 8 30N 28 8 E
Goiana, *Brazil* **170 C5** 7 33 S 34 59W
Goianésia, *Brazil* **171 E2** 15 18 S 49 7W
Goiânia, *Brazil* **171 E2** 16 43 S 49 20W
Goiás, *Brazil* **171 E1** 15 55 S 50 10W
Goiás □, *Brazil* **170 D2** 12 10 S 48 0W
Goiatuba, *Brazil* **171 E2** 18 1 S 49 23W
Goio-Ere, *Brazil* **175 A5** 24 12 S 53 1W
Góis, *Portugal* **42 E2** 40 10N 8 6W
Gojam □, *Ethiopia* **107 E4** 10 55N 36 30 E
Gojeb, Wabi →, *Ethiopia* **107 F4** 7 12N 36 40 E
Gojō, *Japan* **71 C7** 34 21N 135 42 E
Gojra, *Pakistan* **92 D5** 31 10N 72 40 E
Gokak, *India* **95 F2** 16 11N 74 52 E
Gokarannath, *India* **93 F9** 27 57N 80 39 E
Gokarn, *India* **95 G2** 14 33N 74 17 E
Gökçeada, *Turkey* **51 F9** 40 10N 25 50 E
Gökçedağ, *Turkey* **49 B10** 39 33N 28 56 E
Gökçen, *Turkey* **49 C9** 38 7N 27 53 E
Gökçeören, *Turkey* **49 C10** 38 37N 28 35 E
Gökçeyazı, *Turkey* **49 B9** 39 40N 27 40 E
Gökırmak →, *Turkey* **100 B6** 41 25N 35 8 E
Gökova, *Turkey* **49 D10** 37 1N 28 17 E
Gökova Körfezi, *Turkey* **49 E9** 36 55N 27 50 E
Göksu →, *Turkey* **100 D6** 36 19N 34 5 E
Göksun, *Turkey* **100 C7** 38 2N 36 30 E
Göktepe, *Turkey* **49 D10** 37 25N 28 34 E
Gokurt, *Pakistan* **92 E2** 29 40N 67 26 E
Gol, *Norway* **18 D5** 60 42N 8 55 E
Gola, *India* **93 E9** 28 3N 80 32 E
Golaghat, *India* **90 B5** 26 30N 94 0 E
Golakganj, *India* **93 F13** 26 8N 89 52 E
Golan Heights = Hagolan, *Syria* **103 C4** 33 0N 35 45 E
Gołańcz, *Poland* **55 F4** 52 57N 17 18 E
Golāshkerd, *Iran* **97 E8** 27 59N 57 16 E
Golaya Pristen = Hola Pristan, *Ukraine* **59 J7** 46 29N 32 32 E
Gölbaşı, *Adiyaman, Turkey* **100 D7** 37 43N 37 25 E
Gölbaşı, *Ankara, Turkey* **100 C5** 39 47N 32 49 E
Golchikha, *Russia* **6 B12** 71 45N 83 30 E
Golconda, *U.S.A.* **158 F5** 40 58N 117 30W
Gölcük, *Kocaeli, Turkey* **51 F13** 40 42N 29 48 E
Gölcük, *Niğde, Turkey* **100 C6** 38 14N 34 47 E
Gold Beach, *U.S.A.* **158 E1** 42 25N 124 25W
Gold Coast, *Australia* **127 D5** 28 0 S 153 25 E
Gold Coast, *W. Afr.* **113 E4** 4 0N 1 40W
Gold Hill, *U.S.A.* **158 E2** 42 26N 123 3W
Goldach, *Switz.* **33 B8** 47 28N 9 28 E
Gołdap, *Poland* **54 D9** 54 19N 22 18 E
Goldau, *Switz.* **33 B7** 47 3N 8 33 E
Goldberg, *Germany* **30 B8** 53 35N 12 4 E
Golden, *Canada* **142 C5** 51 20N 116 59W
Golden, *Colo., U.S.A.* **154 F2** 39 42N 105 15W
Golden, *Ill., U.S.A.* **156 D5** 40 7N 91 1W
Golden B., *N.Z.* **131 A7** 40 40 S 172 50 E
Golden Gate, *U.S.A.* **158 H2** 37 54N 122 30W
Golden Hinde, *Canada* **142 D3** 49 40N 125 44W
Golden Lake, *Canada* **150 A7** 45 34N 77 21W
Golden Prairie, *Canada* **143 C7** 50 13N 109 37W
Golden Rock, *India* **95 J4** 10 45N 78 48 E
Golden Vale, *Ireland* **23 D3** 52 33N 8 17W
Goldendale, *U.S.A.* **158 D3** 45 49N 120 50W
Goldfield, *U.S.A.* **159 H5** 37 42N 117 14W
Goldfields, *Canada* **143 B7** 59 28N 108 29W
Goldsand L., *Canada* **143 B8** 57 2N 101 8W
Goldsboro, *U.S.A.* **149 H7** 35 23N 77 59W
Goldsmith, *U.S.A.* **155 K3** 31 59N 102 37W
Goldsworthy, *Australia* **124 D2** 20 21 S 119 30 E
Goldthwaite, *U.S.A.* **155 K5** 31 27N 98 34W
Golegã, *Portugal* **43 F2** 39 24N 8 29W
Goleniów, *Poland* **54 E1** 53 35N 14 50 E
Golestānak, *Iran* **97 D7** 30 36N 54 14 E
Goleta, *U.S.A.* **161 L7** 34 27N 119 50W
Golfito, *Costa Rica* **164 E3** 8 41N 83 5W
Golfo Aranci, *Italy* **46 B2** 40 59N 9 38 E
Gölgeli Dağları, *Turkey* **49 D10** 37 10N 28 55 E
Gölhisar, *Turkey* **49 D11** 37 8N 29 31 E
Goliad, *U.S.A.* **155 L6** 28 40N 97 23W
Golija, *Montenegro, Yug.* **50 C2** 43 5N 18 45 E
Golija, *Serbia, Yug.* **50 C4** 43 22N 20 15 E
Golina, *Poland* **55 F5** 52 15N 18 4 E
Gölköy, *Turkey* **100 B7** 40 41N 37 37 E
Göllersdorf, *Austria* **34 C9** 48 29N 16 7 E
Gölmarmara, *Turkey* **49 C9** 38 42N 27 55 E
Golo →, *France* **29 F13** 42 31N 9 32 E
Golol, *Somali Rep.* **120 D2** 3 38N 43 49 E
Gölova, *Turkey* **49 E12** 36 48N 30 5 E
Golovin, *U.S.A.* **144 D7** 64 33N 163 2W
Golpāyegān, *Iran* **97 C6** 33 27N 50 18 E
Gölpazarı, *Turkey* **100 B4** 40 16N 30 18 E
Golra, *Pakistan* **92 C5** 33 37N 72 56 E
Golspie, *U.K.* **22 D5** 57 58N 3 59W
Golub-Dobrzyń, *Poland* **55 E6** 53 7N 19 2 E
Golubac, *Serbia, Yug.* **50 B5** 44 38N 21 38 E
Golungo Alto, *Angola* **115 D2** 9 8S 14 46 E
Golyam Perelik, *Bulgaria* **51 E8** 41 36N 24 33 E
Golyama Kamchiya →, *Bulgaria* **51 C11** 43 10N 27 55 E
Goma, *Dem. Rep. of the Congo* **118 C2** 1 37 S 29 10 E
Gomati →, *India* **93 G10** 25 32N 83 11 E
Gombari, *Dem. Rep. of the Congo* **118 B2** 2 45N 29 3 E
Gombe, *Nigeria* **113 C7** 10 19N 11 2 E
Gömbe, *Turkey* **49 E11** 36 33N 29 38 E
Gombe →, *Tanzania* **118 C3** 4 38 S 31 40 E
Gombi, *Nigeria* **113 C7** 10 12N 12 30 E
Gomel = Homyel, *Belarus* **59 F6** 52 28N 31 0 E
Gomera, *Canary Is.* **39 F2** 28 7N 17 14W
Gómez Palacio, *Mexico* **162 B4** 25 40N 104 0W
Gomīshān, *Iran* **97 B7** 37 4N 54 6 E
Gommern, *Germany* **30 C7** 52 4N 11 50 E
Gomogomo, *Indonesia* **83 C4** 6 39 S 134 43 E
Gomotartsi, *Bulgaria* **50 B6** 44 6N 22 57 E
Gompa = Ganta, *Liberia* **112 D3** 7 15N 8 59W
Gomphi, *Greece* **48 B3** 39 26N 21 36 E
Goms, *Switz.* **32 D6** 46 27N 8 15 E
Gonābād, *Iran* **97 C8** 34 15N 58 45 E
Gonaïves, *Haiti* **165 C5** 19 20N 72 42W
Gonâve, G. de la, *Haiti* **165 C5** 19 29N 72 42W
Gonâve, I. de la, *Haiti* **165 C5** 18 45N 73 0W
Gonbad-e-Kāvūs, *Iran* **97 B7** 37 20N 55 25 E
Gönc, *Hungary* **52 B6** 48 28N 21 14 E
Gonda, *India* **93 F9** 27 9N 81 58 E
Gondal, *India* **92 J4** 21 58N 70 52 E
Gonder, *Ethiopia* **107 E4** 12 39N 37 30 E
Gonder □, *Ethiopia* **107 E4** 12 55N 37 30 E
Gondia, *India* **94 D5** 21 23N 80 10 E
Gondola, *Mozam.* **119 F3** 19 10 S 33 37 E
Gondomar, *Portugal* **42 D2** 41 10N 8 35W
Gondrecourt-le-Château, *France* **27 D12** 48 31N 5 30 E
Gönen, *Balıkesir, Turkey* **51 F11** 40 6N 27 39 E
Gönen, *Isparta, Turkey* **49 D12** 37 57N 30 31 E
Gönen →, *Turkey* **51 F11** 40 6N 27 39 E
Gong Xian, *China* **76 C5** 28 23N 104 47 E
Gong'an, *China* **77 B9** 30 7N 112 12 E
Gongcheng, *China* **77 E8** 24 50N 110 49 E
Gongga Shan, *China* **76 C3** 29 40N 101 55 E
Gongguan, *China* **76 G7** 21 48N 109 36 E
Gonghe, *China* **72 C5** 36 18N 100 32 E
Gongola →, *Nigeria* **113 D7** 9 30N 12 4 E
Gongolgon, *Australia* **127 E4** 30 21 S 146 54 E
Gongshan, *China* **76 D2** 27 43N 98 29 E
Gongtan, *China* **76 C7** 28 55N 108 20 E
Goniadz, *Poland* **54 E9** 53 30N 22 44 E
Goniri, *Nigeria* **113 C7** 11 30N 12 15 E
Gonjo, *China* **76 B2** 30 52N 98 17 E
Gonnesa, *Italy* **46 C1** 39 16N 8 28 E
Gónnos, *Greece* **48 B4** 39 52N 22 29 E
Gonnosfanádiga, *Italy* **46 C1** 39 29N 8 39 E
Gonzaga, *Phil.* **80 B4** 18 16N 122 0 E
Gonzales, *Calif., U.S.A.* **160 J5** 36 30N 121 26W
Gonzales, *Tex., U.S.A.* **155 L6** 29 30N 97 27W
González Chaves, *Argentina* **174 D3** 38 2 S 60 5W
Good Hope, C. of, *S. Africa* **116 E2** 34 24 S 18 30 E
Goodenough I., *Papua N. G.* **132 E6** 9 20 S 150 15 E
Gooderham, *Canada* **140 D4** 44 54N 78 21W
Goodeve, *Canada* **143 C8** 51 4N 103 10W
Gooding, *U.S.A.* **158 E6** 42 56N 114 43W
Goodland, *U.S.A.* **154 F4** 39 21N 101 43W
Goodnews Bay, *U.S.A.* **144 G7** 59 7N 161 35W
Goodnight, *U.S.A.* **155 H4** 35 2N 101 11W
Goodooga, *Australia* **127 D4** 29 3 S 147 28 E
Goodsoil, *Canada* **143 C7** 54 24N 109 13W
Goodsprings, *U.S.A.* **159 J6** 35 50N 115 26W
Goodwater, *U.S.A.* **152 B3** 33 4N 86 3W
Goole, *U.K.* **20 D7** 53 42N 0 53W
Goolgowi, *Australia* **129 B6** 33 58 S 145 41 E
Goolwa, *Australia* **128 C3** 35 30 S 138 47 E
Goomalling, *Australia* **125 F2** 31 15 S 116 49 E
Goombalie, *Australia* **127 D4** 29 59 S 145 26 E
Goonalga, *Australia* **128 A5** 31 45 S 143 37 E
Goonda, *Mozam.* **119 F3** 19 48 S 33 57 E
Goondiwindi, *Australia* **127 D5** 28 30 S 150 21 E
Goongarrie, L., *Australia* **125 F3** 30 3 S 121 9 E
Goonumbla, *Australia* **129 B8** 32 59 S 148 11 E
Goonyella, *Australia* **126 C4** 21 47 S 147 58 E
Gooray, *Australia* **127 D5** 28 25 S 150 2 E
Goose →, *Canada* **141 B7** 53 20N 60 35W
Goose Creek, *U.S.A.* **152 C9** 32 59N 80 2W
Goose L., *U.S.A.* **158 F3** 41 56N 120 26W
Gooty, *India* **95 G3** 15 7N 77 41 E
Gopalganj, *Bangla.* **90 D2** 23 1N 89 50 E
Gopalganj, *India* **93 F11** 26 28N 84 30 E
Goppenstein, *Switz.* **32 D5** 46 23N 7 46 E
Göppingen, *Germany* **31 G5** 48 42N 9 39 E
Gor, *Spain* **43 H8** 37 23N 2 58W
Góra, *Leszno, Poland* **55 G3** 51 40N 16 31 E
Góra, *Płock, Poland* **55 F7** 52 39N 20 6 E
Góra Kalwaria, *Poland* **55 G8** 51 59N 21 14 E
Gorakhpur, *India* **93 F10** 26 47N 83 23 E
Goražde, *Bos.-H.* **52 G3** 43 38N 18 58 E
Gorbatov, *Russia* **60 B6** 56 12N 43 2 E
Gorbea, Peña, *Spain* **40 B2** 43 1N 2 50W
Gorda, *U.S.A.* **160 K5** 35 53N 121 26W
Gorda, Pta., *Canary Is.* **39 F2** 28 45N 18 0W
Gorda, Pta., *Nic.* **164 D3** 14 20N 83 10W
Gordan B., *Australia* **124 B5** 11 35 S 130 10 E
Gördes, *Turkey* **49 C10** 38 54N 28 17 E
Gordon, *Ga., U.S.A.* **152 C6** 32 54N 83 20W
Gordon, *Nebr., U.S.A.* **154 D3** 42 48N 102 12W
Gordon →, *Australia* **126 G4** 42 27 S 145 30 E
Gordon, I., *Chile* **176 D3** 54 55 S 69 30W
Gordon Downs, *Australia* **124 C4** 18 48 S 128 33 E
Gordon L., *Alta., Canada* **143 B6** 56 30N 110 25W
Gordon L., *N.W.T., Canada* **142 A6** 63 5N 113 11W
Gordonvale, *Australia* **126 B4** 17 5 S 145 50 E
Gore, *Australia* **127 D5** 28 17 S 151 30 E
Goré, *Chad* **109 G3** 7 59N 16 31 E
Gore, *Ethiopia* **107 F4** 8 12N 35 32 E
Gore, *N.Z.* **131 G3** 46 5 S 168 58 E
Gore Bay, *Canada* **140 C3** 45 57N 82 28W
Görele, *Turkey* **101 B8** 41 2N 39 0 E
Goreme, *Turkey* **100 C6** 38 35N 34 52 E
Gorey, *Ireland* **23 D5** 52 41N 6 18W
Gorg, *Iran* **97 D8** 29 29N 59 43 E
Gorgān, *Iran* **97 B7** 36 50N 54 29 E
Gorgona, *Italy* **44 E6** 43 26N 9 54 E
Gorgora, *Ethiopia* **107 E4** 12 15N 37 17 E
Gorham, *U.S.A.* **151 B13** 44 23N 71 10W
Gori, *Georgia* **61 J7** 42 0N 44 7 E
Gorin, *U.S.A.* **156 D4** 40 22N 92 1W
Gorinchem, *Neths.* **24 C4** 51 50N 4 59 E
Gorinhatã, *Brazil* **171 E2** 19 15 S 49 45W
Goris, *Armenia* **101 C12** 39 31N 46 22 E
Goritsy, *Russia* **58 D9** 57 4N 36 43 E
Gorízia, *Italy* **45 C10** 45 56N 13 37 E
Gorj □, *Romania* **53 E8** 45 5N 23 25 E
Gorki = Horki, *Belarus* **58 E6** 54 17N 30 59 E
Gorki = Nizhniy Novgorod, *Russia* **60 B7** 56 20N 44 0 E
Gorkiy = Nizhniy Novgorod, *Russia* **60 B7** 56 20N 44 0 E
Gorkovskoye Vdkhr., *Russia* **60 B6** 57 2N 43 4 E
Gorlice, *Poland* **55 J8** 49 35N 21 11 E
Görlitz, *Germany* **30 D10** 51 9N 14 58 E
Gorlovka = Horlivka, *Ukraine* **59 H10** 48 19N 38 5 E
Gorman, *Calif., U.S.A.* **161 L8** 34 47N 118 51W
Gorman, *Tex., U.S.A.* **155 J5** 32 12N 98 41W
Gorna Dzhumayo = Blagoevgrad, *Bulgaria* **50 D7** 42 2N 23 5 E

220 Greenacres City

Greenacres City, *U.S.A.* **153 J9** 26 38N 80 7W
Greenbank, *U.S.A.* **160 B4** 48 6N 122 34W
Greenbush, *Mich., U.S.A.* **150 B1** 44 35N 83 19W
Greenbush, *Minn., U.S.A.* **154 A6** 48 42N 96 11W
Greencastle, *U.S.A.* **157 E10** 39 38N 86 52W
Greene, *Iowa, U.S.A.* **156 B4** 42 54N 92 48W
Greene, *N.Y., U.S.A.* **151 D9** 42 20N 75 46W
Greenfield, *Calif., U.S.A.* **160 J5** 36 19N 121 15W
Greenfield, *Calif., U.S.A.* **161 K8** 35 15N 119 0W
Greenfield, *Ill., U.S.A.* **156 E6** 39 21N 90 12W
Greenfield, *Ind., U.S.A.* **157 E11** 39 47N 85 46W
Greenfield, *Iowa, U.S.A.* **156 C2** 41 18N 94 28W
Greenfield, *Mass., U.S.A.* **151 D12** 42 35N 72 36W
Greenfield, *Mo., U.S.A.* **155 G8** 37 25N 93 51W
Greenfield, *Ohio, U.S.A.* **157 E13** 39 21N 83 23W
Greenfield Park, *Canada* **151 A11** 45 29N 73 29W
Greenland ■, *N. Amer.* **10 D6** 66 0N 45 0W
Greenland Sea, *Arctic* **10 B10** 73 0N 10 0W
Greenock, *U.K.* **22 F4** 55 57N 4 46W
Greenore, *Ireland* **23 B5** 54 2N 6 8W
Greenore Pt., *Ireland* **23 D5** 52 14N 6 19W
Greenough →, *Australia* **125 E1** 28 51 S 114 38 E
Greenport, *U.S.A.* **151 E12** 41 6N 72 22W
Greensboro, *Fla., U.S.A.* **152 E5** 30 34N 84 45W
Greensboro, *Ga., U.S.A.* **152 B6** 33 35N 83 11W
Greensboro, *N.C., U.S.A.* **149 G6** 36 4N 79 48W
Greensburg, *Ind., U.S.A.* **157 E11** 39 20N 85 29W
Greensburg, *Kans., U.S.A.* **155 G5** 37 36N 99 18W
Greensburg, *Pa., U.S.A.* **150 F5** 40 18N 79 33W
Greenstone Pt., *U.K.* **22 D3** 57 55N 5 37W
Greentown, *U.S.A.* **157 D11** 40 29N 85 58W
Greenup, *U.S.A.* **157 E8** 39 15N 88 10W
Greenville, *Liberia* **112 D3** 5 1N 9 6W
Greenville, *Ala., U.S.A.* **149 K2** 31 50N 86 38W
Greenville, *Calif., U.S.A.* **160 E6** 40 8N 120 57W
Greenville, *Fla., U.S.A.* **152 E6** 30 28N 83 38W
Greenville, *Ga., U.S.A.* **152 B5** 33 2N 84 43W
Greenville, *Ill., U.S.A.* **156 F7** 38 53N 89 25W
Greenville, *Ind., U.S.A.* **157 F11** 38 22N 85 59W
Greenville, *Maine, U.S.A.* **141 C6** 45 28N 69 35W
Greenville, *Mich., U.S.A.* **157 A11** 43 11N 85 15W
Greenville, *Miss., U.S.A.* **155 J9** 33 24N 91 4W
Greenville, *N.C., U.S.A.* **149 H7** 35 37N 77 23W
Greenville, *Ohio, U.S.A.* **157 D12** 40 6N 84 38W
Greenville, *Pa., U.S.A.* **150 E4** 41 24N 80 23W
Greenville, *S.C., U.S.A.* **149 H4** 34 51N 82 24W
Greenville, *Tenn., U.S.A.* **149 G4** 36 13N 82 51W
Greenville, *Tex., U.S.A.* **155 J6** 33 8N 96 7W
Greenwater Lake Prov. Park, *Canada* **143 C8** 52 32N 103 30W
Greenwich, *U.K.* **21 F8** 51 29N 0 1 E
Greenwich, *Conn., U.S.A.* **151 E11** 41 2N 73 38W
Greenwich, *N.Y., U.S.A.* **151 C11** 43 5N 73 30W
Greenwich, *Ohio, U.S.A.* **150 E2** 41 2N 82 31W
Greenwood, *Canada* **142 D5** 49 10N 118 40W
Greenwood, *Fla., U.S.A.* **152 E4** 30 52N 85 10W
Greenwood, *Ind., U.S.A.* **157 E10** 39 37N 86 7W
Greenwood, *Miss., U.S.A.* **155 J9** 33 31N 90 11W
Greenwood, *S.C., U.S.A.* **149 H4** 34 12N 82 10W
Greenwood, Mt., *Australia* **124 B5** 13 48 S 130 4 E
Greenwood L., *U.S.A.* **152 A8** 34 11N 81 54W
Gregório →, *Brazil* **172 B3** 6 50 S 70 46W
Gregory, *U.S.A.* **154 D5** 43 14N 99 20W
Gregory →, *Australia* **126 B2** 17 53 S 139 17 E
Gregory, L., *S. Austral., Australia* **127 D2** 28 55 S 139 0 E
Gregory, L., *W. Austral., Australia* **125 E2** 25 38 S 119 58 E
Gregory Downs, *Australia* **126 B2** 18 35 S 138 45 E
Gregory L., *Australia* **124 D4** 20 0 S 127 40 E
Gregory Ra., *Queens., Australia* **126 B3** 19 30 S 143 40 E
Gregory Ra., *W. Austral., Australia* **124 D3** 21 20 S 121 12 E
Greiffenberg, *Germany* **30 B9** 53 5N 13 57 E
Greifswald, *Germany* **30 A9** 54 5N 13 23 E
Greifswalder Bodden, *Germany* **30 A9** 54 12N 13 35 E
Grein, *Austria* **34 C7** 48 14N 14 51 E
Greiz, *Germany* **30 E8** 50 39N 12 10 E
Gremikha, *Russia* **56 A6** 67 59N 39 47 E
Gremyachinsk, *Russia* **62 B6** 58 34N 57 51 E
Grenå, *Denmark* **17 H4** 56 25N 10 53 E
Grenada, *U.S.A.* **155 J10** 33 47N 89 49W
Grenada ■, *W. Indies* **165 D7** 12 10N 61 40W
Grenade, *France* **28 E5** 43 47N 1 17 E
Grenadines, *W. Indies* **165 D7** 12 40N 61 20W
Grenchen, *Switz.* **32 B4** 47 12N 7 24 E
Grenen, *Denmark* **17 G4** 57 44N 10 40 E
Grenfell, *Australia* **129 B8** 33 52 S 148 8 E
Grenfell, *Canada* **143 C8** 50 30N 102 56W
Grenivík, *Iceland* **11 B8** 65 57N 18 11W
Grenjaðarstaður, *Iceland* **11 B9** 65 49N 17 21W
Grenoble, *France* **29 C9** 45 12N 5 42 E
Grenora, *U.S.A.* **154 A3** 48 37N 103 56W
Grenville, C., *Australia* **126 A3** 12 0 S 143 13 E
Grenville Chan., *Canada* **142 C3** 53 40N 129 46W
Gréoux-les-Bains, *France* **29 E9** 43 45N 5 52 E
Gresham, *U.S.A.* **160 E4** 45 30N 122 26W
Gresik, *Indonesia* **85 D4** 7 13 S 112 38 E
Gresston, *U.S.A.* **152 C6** 32 17N 83 15W
Gretna, *U.K.* **22 F5** 55 0N 3 3W
Gretna, *U.S.A.* **152 E5** 30 37N 84 40W
Greven, *Germany* **30 C3** 52 6N 7 37 E
Grevená, *Greece* **50 F5** 40 4N 21 25 E
Grevená □, *Greece* **50 F5** 40 2N 21 25 E
Grevenbroich, *Germany* **30 D2** 51 5N 6 35 E
Grevenmacher, *Lux.* **24 E6** 49 41N 6 26 E
Grevesmühlen, *Germany* **30 B7** 53 52N 11 12 E
Grevestrand, *Denmark* **17 J6** 55 36N 12 19 E
Grey →, *N.Z.* **131 C6** 42 27 S 171 12 E
Grey, C., *Australia* **126 A2** 13 0 S 136 35 E
Grey Ra., *Australia* **127 D3** 27 0 S 143 30 E
Grey Res., *Canada* **141 C8** 48 20N 56 30W
Greybull, *U.S.A.* **158 D9** 44 30N 108 3W
Greymouth, *N.Z.* **131 C6** 42 29 S 171 13 E
Greystones, *Ireland* **23 C5** 53 9N 6 5W
Greytown, *N.Z.* **130 H4** 41 5 S 175 29 E
Greytown, *S. Africa* **117 D5** 29 1 S 30 36 E
Gribanovskiy, *Russia* **60 E5** 51 28N 41 50 E
Gribbell I., *Canada* **142 C3** 53 23N 129 0W
Gridley, *U.S.A.* **160 F5** 39 22N 121 42W
Griekwastad, *S. Africa* **116 D3** 28 49 S 23 15 E
Griesheim, *Germany* **31 F4** 49 51N 8 33 E
Grieskirchen, *Austria* **34 C6** 48 16N 13 48 E
Griffin, *U.S.A.* **152 B5** 33 15N 84 16W
Griffin, L., *U.S.A.* **153 G8** 28 52N 81 51W
Griffith, *Australia* **129 C7** 34 18 S 146 2 E
Grignols, *France* **28 D3** 44 23N 0 2W
Grigoriopol, *Moldova* **53 C14** 47 9N 29 18 E
Grimari, *C.A.R.* **114 A4** 5 43N 20 6 E
Grimaylov = Hrymayliv, *Ukraine* **59 H4** 49 20N 26 5 E
Grimes, *U.S.A.* **160 F5** 39 4N 121 54W
Grimma, *Germany* **30 D8** 51 14N 12 43 E
Grimmen, *Germany* **30 A9** 54 7N 13 3 E
Grimsay, *U.K.* **22 D1** 57 29N 7 14W
Grimsby, *Canada* **150 C5** 43 12N 79 34W
Grimsby, *U.K.* **20 D7** 53 34N 0 5W
Grimselpass, *Switz.* **33 C6** 46 34N 8 23 E
Grímsey, *Iceland* **11 A9** 66 33N 17 58W
Grimshaw, *Canada* **142 B5** 56 10N 117 40W
Grimslöv, *Sweden* **17 H8** 56 44N 14 34 E
Grímsstaðir, *Iceland* **11 B10** 65 39N 16 7W
Grimstad, *Norway* **15 G13** 58 20N 8 35 E
Grímsvötn, *Iceland* **11 C9** 64 26N 17 22W
Grindavík, *Iceland* **11 D4** 63 50N 22 26W
Grindelwald, *Switz.* **32 C6** 46 38N 8 2 E
Grindsted, *Denmark* **17 J2** 55 46N 8 55 E
Grindu, *Romania* **53 F11** 44 44N 26 50 E
Grinnell, *U.S.A.* **156 C4** 41 45N 92 43W
Grintavec, *Slovenia* **45 B11** 46 22N 14 32 E
Gris-Nez, C., *France* **27 B8** 50 52N 1 35 E
Grisolles, *France* **28 E5** 43 49N 1 19 E
Grisons = Graubünden □, *Switz.* **33 C9** 46 45N 9 30 E
Grisslehamn, *Sweden* **16 D12** 60 5N 18 49 E
Grmeč Planina, *Bos.-H.* **45 D13** 44 43N 16 16 E
Groais I., *Canada* **141 B8** 50 55N 55 35W
Grobiņa, *Latvia* **54 B8** 56 33N 21 10 E
Groblersdal, *S. Africa* **117 D4** 25 15 S 29 25 E
Grobming, *Austria* **34 D6** 47 27N 13 54 E
Grocka, *Serbia, Yug.* **50 B4** 44 40N 20 42 E
Gródek, *Poland* **55 E10** 53 6N 23 40 E
Grodków, *Poland* **55 H4** 50 43N 17 21 E
Grodno = Hrodna, *Belarus* **58 F2** 53 42N 23 52 E
Grodzisk Mazowiecki, *Poland* **55 F7** 52 7N 20 37 E
Grodzisk Wielkopolski, *Poland* **55 F3** 52 15N 16 22 E
Grodzyanka = Hrodzyanka, *Belarus* **58 F5** 53 31N 28 42 E
Groesbeck, *U.S.A.* **155 K6** 30 48N 96 31W
Groix, *France* **26 E3** 47 38N 3 29W
Groix, Î. de, *France* **26 E3** 47 38N 3 28W
Grójec, *Poland* **55 G7** 51 50N 20 58 E
Gronau, *Niedersachsen, Germany* **30 C5** 52 5N 9 47 E
Gronau, *Nordrhein-Westfalen, Germany* **30 C3** 52 12N 7 2 E
Grong, *Norway* **14 D15** 64 25N 12 8 E
Grönhögen, *Sweden* **17 H10** 56 16N 16 24 E
Groningen, *Neths.* **24 A6** 53 15N 6 35 E
Groningen, *Surinam* **169 B6** 5 48N 55 28W
Groningen □, *Neths.* **24 A6** 53 16N 6 40 E
Grønnedal, *Greenland* **10 E6** 61 20N 47 57W
Groom, *U.S.A.* **155 H4** 35 12N 101 6W
Groot →, *S. Africa* **116 E3** 33 45 S 24 36 E
Groot Berg →, *S. Africa* **116 E2** 32 47 S 18 8 E
Groot-Brakrivier, *S. Africa* **116 E3** 34 2 S 22 18 E
Groot-Kei →, *S. Africa* **117 E4** 32 41 S 28 22 E
Groot Vis →, *S. Africa* **116 E4** 33 28 S 27 5 E
Groote Eylandt, *Australia* **126 A2** 14 0 S 136 40 E
Grootfontein, *Namibia* **116 B2** 19 31 S 18 6 E
Grootlaagte →, *Africa* **116 C3** 20 55 S 21 27 E
Grootvloer, *S. Africa* **116 E3** 30 0 S 20 40 E
Gros C., *Canada* **142 A6** 61 59N 113 32W
Grósio, *Italy* **44 B7** 46 18N 10 16 E
Grosne →, *France* **27 F11** 46 42N 4 56 E
Grossa, Pta., *Spain* **39 B8** 39 6N 1 36 E
Grossenbrode, *Germany* **30 A7** 54 21N 11 4 E
Grossenhain, *Germany* **30 D9** 51 17N 13 32 E
Grosser Arber, *Germany* **31 F9** 49 6N 13 8 E
Grosser Plöner See, *Germany* **30 A6** 54 10N 10 22 E
Grosseto, *Italy* **45 F8** 42 46N 11 8 E
Grossgerungs, *Austria* **34 C7** 48 34N 14 57 E
Grossglockner, *Austria* **34 D5** 47 5N 12 40 E
Groswater B., *Canada* **141 B8** 54 20N 57 40W
Grotli, *Norway* **18 B4** 62 2N 7 42 E
Groton, *Conn., U.S.A.* **151 E12** 41 21N 72 5W
Groton, *S. Dak., U.S.A.* **154 C5** 45 27N 98 6W
Grottáglie, *Italy* **47 B10** 40 32N 17 26 E
Grottaminarda, *Italy* **47 A8** 41 4N 15 2 E
Grottammare, *Italy* **45 F10** 42 59N 13 52 E
Grouard Mission, *Canada* **142 B5** 55 33N 116 9W
Grouin, Pte. du, *France* **26 D5** 48 43N 1 51W
Groundhog →, *Canada* **140 C3** 48 45N 82 58W
Grouse Creek, *U.S.A.* **158 F7** 41 42N 113 53W
Grouw, *Neths.* **24 A5** 53 5N 5 51 E
Grovania, *U.S.A.* **152 C6** 32 22N 83 40W
Grove City, *Ohio, U.S.A.* **157 E13** 39 53N 83 6W
Grove City, *Pa., U.S.A.* **150 E4** 41 10N 80 5W
Groveland, *Calif., U.S.A.* **160 H6** 37 50N 120 14W
Groveland, *Fla., U.S.A.* **153 G8** 28 34N 81 51W
Grover City, *U.S.A.* **161 K6** 35 7N 120 37W
Grover Hill, *U.S.A.* **157 C12** 41 1N 84 29W
Groveton, *N.H., U.S.A.* **151 B13** 44 36N 71 31W
Groveton, *Tex., U.S.A.* **155 K7** 31 4N 95 8W
Grovetown, *U.S.A.* **152 B7** 33 27N 82 12W
Grožnjan, *Croatia* **45 C10** 45 22N 13 43 E
Groznyy, *Russia* **61 J7** 43 20N 45 45 E
Grua, *Norway* **18 D7** 60 16N 10 40 E
Grubišno Polje, *Croatia* **52 E2** 45 44N 17 12 E
Grudovo, *Bulgaria* **51 D11** 42 21N 27 10 E
Grudusk, *Poland* **55 E7** 53 3N 20 38 E
Grudziądz, *Poland* **54 E5** 53 30N 18 47 E
Gruinard B., *U.K.* **22 D3** 57 56N 5 35W
Gruissan, *France* **28 E7** 43 8N 3 7 E
Grumo Áppula, *Italy* **47 A9** 41 1N 16 42 E
Grums, *Sweden* **16 E7** 59 22N 13 5 E
Grünberg, *Germany* **30 E4** 50 35N 8 58 E
Grund, *Iceland* **11 B8** 65 31N 18 9W
Gründau, *Germany* **31 E5** 50 10N 9 9 E
Grundy Center, *U.S.A.* **156 B4** 42 22N 92 47W
Grungedal, *Norway* **18 E4** 59 44N 7 43 E
Grünstadt, *Germany* **31 F4** 49 34N 8 9 E
Gruvberget, *Sweden* **16 C10** 61 6N 16 10 E
Gruver, *U.S.A.* **155 G4** 36 16N 101 24W
Gruyères, *Switz.* **32 C4** 46 35N 7 4 E
Gruža, *Serbia, Yug.* **50 C4** 43 54N 20 46 E
Gryazi, *Russia* **59 F10** 52 30N 39 58 E
Gryazovets, *Russia* **58 C11** 58 50N 40 10 E
Grybów, *Poland* **55 J7** 49 36N 20 55 E
Grycksbo, *Sweden* **16 D9** 60 40N 15 29 E
Gryfice, *Poland* **54 E2** 53 55N 15 13 E
Gryfino, *Poland* **55 E1** 53 16N 14 29 E
Gryfów Śląski, *Poland* **55 G2** 51 2N 15 24 E
Grythyttan, *Sweden* **16 E8** 59 41N 14 32 E
Gstaad, *Switz.* **32 D4** 46 28N 7 18 E
Gua Musang, *Malaysia* **87 K3** 4 53N 101 58 E
Guacanayabo, G. de, *Cuba* **164 B4** 20 40N 77 20W
Guacara, *Venezuela* **168 A4** 10 14N 67 53W
Guachípas →, *Argentina* **174 B2** 25 40 S 65 30W
Guachiría →, *Colombia* **168 B3** 5 27N 70 36W
Guadajoz →, *Spain* **43 H6** 37 50N 4 51W
Guadalajara, *Mexico* **162 C4** 20 40N 103 20W
Guadalajara, *Spain* **40 E1** 40 37N 3 12W
Guadalajara □, *Spain* **40 E2** 40 47N 2 30W
Guadalcanal, *Solomon Is.* **133 M11** 9 32 S 160 12 E
Guadalcanal, *Spain* **43 G5** 38 5N 5 52W
Guadalén →, *Spain* **43 G7** 38 5N 3 32W
Guadales, *Argentina* **174 C2** 34 30 S 67 55W
Guadalete →, *Spain* **43 J4** 36 35N 6 13W
Guadalimar →, *Spain* **43 G7** 38 5N 3 28W
Guadalmena →, *Spain* **43 G8** 38 19N 2 56W
Guadalmez →, *Spain* **43 G5** 38 46N 5 4W
Guadalope →, *Spain* **40 D4** 41 15N 0 3W
Guadalquivir →, *Spain* **43 J4** 36 47N 6 22W
Guadalupe = Guadeloupe ■, *W. Indies* **165 C7** 16 20N 61 40W
Guadalupe, *Brazil* **170 C3** 6 44 S 43 47W
Guadalupe, *Mexico* **161 N10** 32 4N 116 32W
Guadalupe, *Spain* **43 F5** 39 27N 5 17W
Guadalupe, *U.S.A.* **161 L6** 34 59N 120 33W
Guadalupe →, *Mexico* **161 N10** 32 6N 116 51W
Guadalupe →, *U.S.A.* **155 L6** 28 27N 96 47W
Guadalupe, Sierra de, *Spain* **43 F5** 39 28N 5 30W
Guadalupe Bravos, *Mexico* **162 A3** 31 20N 106 10W
Guadalupe I., *Pac. Oc.* **136 G8** 29 0N 118 50W
Guadalupe Peak, *U.S.A.* **159 L11** 31 50N 104 52W
Guadalupe y Calvo, *Mexico* **162 B3** 26 6N 106 58W
Guadarrama, Sierra de, *Spain* **42 E7** 41 0N 4 0W
Guadauta, *Georgia* **61 J5** 43 7N 40 32 E
Guadeloupe ■, *W. Indies* **165 C7** 16 20N 61 40W
Guadeloupe Passage, *W. Indies* **165 C7** 16 50N 62 15W
Guadiamar →, *Spain* **43 J4** 36 55N 6 24W
Guadiana →, *Portugal* **43 H3** 37 14N 7 22W
Guadiana Menor →, *Spain* **43 H7** 37 56N 3 15W
Guadiaro →, *Spain* **43 J5** 36 17N 5 17W
Guadiato →, *Spain* **43 H5** 37 48N 5 5W
Guadiela →, *Spain* **40 E2** 40 22N 2 49W
Guadix, *Spain* **43 H7** 37 18N 3 11W
Guafo, Boca del, *Chile* **176 B2** 43 35 S 74 0W
Guafo, I., *Chile* **176 B2** 43 35 S 74 50W
Guainía □, *Colombia* **168 C4** 2 30N 69 0W
Guainía →, *Colombia* **168 C4** 2 1N 67 7W
Guaíra, *Brazil* **175 A5** 24 5 S 54 10W
Guaitecas, Is., *Chile* **176 B2** 44 0 S 74 30W
Guajará-Mirim, *Brazil* **173 C4** 10 50 S 65 20W
Guajira □, *Colombia* **168 A3** 11 30N 72 30W
Guajira, Pen. de la, *Colombia* **168 A3** 12 0N 72 0W
Gualaceo, *Ecuador* **168 D2** 2 54 S 78 47W
Gualán, *Guatemala* **164 C2** 15 8N 89 22W
Gualdo Tadino, *Italy* **45 E9** 43 14N 12 47 E
Gualeguay, *Argentina* **174 C4** 33 10 S 59 14W
Gualeguaychú, *Argentina* **174 C4** 33 3 S 59 31W
Gualicho, Salina, *Argentina* **176 B3** 40 25 S 65 20W
Gualjaina, *Argentina* **176 B2** 42 45 S 70 30W
Guam ■, *Pac. Oc.* **133 R15** 13 27N 144 45 E
Guamá, *Brazil* **170 B2** 1 37 S 47 29W
Guamá →, *Brazil* **170 B2** 1 29 S 48 30W
Guamblin, I., *Chile* **176 B2** 44 50 S 75 0W
Guaminí, *Argentina* **174 D3** 37 1 S 62 28W
Guamote, *Ecuador* **168 D2** 1 56 S 78 43W
Guampí, Sierra de, *Venezuela* **169 B4** 6 0N 65 35W
Guamúchil, *Mexico* **162 B3** 25 25N 108 3W
Guan Xian, *China* **76 B4** 31 2N 103 38 E
Guanabacoa, *Cuba* **164 B3** 23 8N 82 18W
Guanacaste, Cordillera del, *Costa Rica* **164 D2** 10 40N 85 4W
Guanaceví, *Mexico* **162 B3** 25 40N 106 0W
Guanahani = San Salvador, *Bahamas* **165 B5** 24 0N 74 40W
Guanajay, *Cuba* **164 B3** 22 56N 82 42W
Guanajuato, *Mexico* **162 C4** 21 0N 101 20W
Guanajuato □, *Mexico* **162 C4** 20 40N 101 20W
Guanambi, *Brazil* **171 D3** 14 13 S 42 47W
Guanare, *Venezuela* **168 B4** 8 42N 69 12W
Guanare →, *Venezuela* **168 B4** 8 13N 67 46W
Guandacol, *Argentina* **174 B2** 29 30 S 68 40W
Guane, *Cuba* **164 B3** 22 10N 84 7W
Guang'an, *China* **76 B6** 30 28N 106 35 E
Guangchang, *China* **77 D11** 26 50N 116 21 E
Guangde, *China* **77 B12** 30 54N 119 25 E
Guangdong □, *China* **77 F9** 23 0N 113 0 E
Guangfeng, *China* **77 C12** 28 20N 118 15 E
Guanghan, *China* **76 B5** 30 58N 104 17 E
Guanghua, *China* **77 A8** 32 22N 111 38 E
Guangji, *China* **77 C10** 29 52N 115 30 E
Guangling, *China* **74 E8** 39 47N 114 22 E
Guangning, *China* **77 F9** 23 40N 112 22 E
Guangrao, *China* **75 F10** 37 5N 118 25 E
Guangshun, *China* **76 D6** 26 8N 106 21 E
Guanguan Peninsula, *Phil.* **81 H6** 6 50N 126 10 E
Guangwu, *China* **74 F3** 37 48N 105 57 E
Guangxi Zhuangzu Zizhiqu □, *China* **76 F7** 24 0N 109 0 E
Guangyuan, *China* **76 A5** 32 26N 105 51 E
Guangze, *China* **77 D11** 27 30N 117 12 E
Guangzhou, *China* **77 F9** 23 5N 113 10 E
Guanhães, *Brazil* **171 E3** 18 47 S 42 57W
Guanipa →, *Venezuela* **169 B5** 9 56N 62 26W
Guanling, *China* **76 E5** 25 56N 105 35 E
Guannan, *China* **75 G10** 34 8N 119 21 E
Guanta, *Venezuela* **169 A5** 10 14N 64 36W
Guantánamo, *Cuba* **165 B4** 20 10N 75 14W
Guantao, *China* **74 F8** 36 42N 115 25 E
Guanyang, *China* **77 E8** 25 30N 111 8 E
Guanyun, *China* **75 G10** 34 20N 119 18 E
Guapí, *Colombia* **168 C2** 2 36N 77 54W
Guápiles, *Costa Rica* **164 D3** 10 10N 83 46W
Guaporé →, *Brazil* **173 C4** 11 55 S 65 4W
Guaqui, *Bolivia* **172 D4** 16 41 S 68 54W
Guara, Sierra de, *Spain* **40 C4** 42 19N 0 15W
Guarabira, *Brazil* **170 C4** 6 51 S 35 29W
Guaranda, *Ecuador* **168 D2** 1 36 S 79 0W
Guarapari, *Brazil* **171 F3** 20 40 S 40 30W
Guarapuava, *Brazil* **171 G1** 25 20 S 51 30W
Guaratinguetá, *Brazil* **175 A6** 22 49 S 45 9W
Guaratuba, *Brazil* **175 B6** 25 53 S 48 38W
Guarda, *Portugal* **42 E3** 40 32N 7 20W
Guarda □, *Portugal* **42 E3** 40 40N 7 20W
Guardafui, C. = Asir, Ras, *Somali Rep.* **120 B4** 11 55N 51 10 E
Guardamar del Segura, *Spain* **41 G4** 38 5N 0 39W
Guardavalle, *Italy* **47 D9** 38 30N 16 30 E
Guárdia Sanframondi, *Italy* **47 A7** 41 15N 14 36 E
Guardiagrele, *Italy* **45 F11** 42 11N 14 13 E
Guardo, *Spain* **42 C6** 42 47N 4 50W
Guareña, *Spain* **43 G4** 38 51N 6 6W
Guareña →, *Spain* **42 D5** 41 29N 5 23W
Guaria □, *Paraguay* **174 B4** 25 45 S 56 30W
Guárico □, *Venezuela* **168 B4** 8 40N 66 35W
Guarrojo →, *Colombia* **168 C3** 4 6N 70 42W
Guarujá, *Brazil* **175 A6** 24 2 S 46 25W
Guarus, *Brazil* **171 F3** 21 44 S 41 20W
Guasave, *Mexico* **162 B3** 25 34N 108 27W
Guascama, Pta., *Colombia* **168 C2** 2 32N 78 24W
Guasdualito, *Venezuela* **168 B3** 7 15N 70 44W
Guasipati, *Venezuela* **169 B5** 7 28N 61 54W
Guasopa, *Papua N. G.* **132 E7** 9 12 S 152 56 E
Guastalla, *Italy* **44 D7** 44 55N 10 39 E
Guatemala, *Guatemala* **164 D1** 14 40N 90 22W
Guatemala ■, *Cent. Amer.* **164 C1** 15 40N 90 30W
Guatire, *Venezuela* **168 A4** 10 28N 66 32W
Guaviare □, *Colombia* **168 C3** 2 0N 72 30W
Guaviare →, *Colombia* **168 C4** 4 3N 67 44W
Guaxupé, *Brazil* **175 A6** 21 10 S 47 5W
Guayabero →, *Colombia* **168 C3** 2 36N 72 47W
Guayama, *Puerto Rico* **165 C6** 17 59N 66 7W
Guayaneco, Arch., *Chile* **176 C1** 47 45 S 75 10W
Guayaquil, *Ecuador* **168 D2** 2 15 S 79 52W
Guayaquil, G. de, *Ecuador* **168 D1** 3 10 S 81 0W
Guayaramerín, *Bolivia* **173 C4** 10 48 S 65 23W
Guayas →, *Ecuador* **168 D2** 2 36 S 79 52W
Guaymas, *Mexico* **162 B2** 27 59N 110 54W
Guazhou, *China* **77 A12** 32 17N 119 21 E
Guba, *Dem. Rep. of the Congo* **119 E2** 10 38 S 26 27 E
Gubakha, *Russia* **62 B6** 58 52N 57 36 E
Gûbâl, Madîq, *Egypt* **106 B3** 27 30N 34 0 E
Gubam, *Papua N. G.* **132 E1** 8 39 S 141 53 E
Gubat, *Phil.* **80 E5** 12 55N 124 7 E
Gúbbio, *Italy* **45 E9** 43 21N 12 35 E
Guben, *Germany* **30 D10** 51 57N 14 42 E
Gubin, *Poland* **55 G1** 51 57N 14 43 E
Gubio, *Nigeria* **113 C7** 12 30N 12 42 E
Gubkin, *Russia* **59 G9** 51 17N 37 32 E
Guča, *Serbia, Yug.* **50 C4** 43 46N 20 15 E
Gudå, *Norway* **18 A8** 63 27N 11 36 E
Gudalur, *India* **95 J3** 11 30N 76 29 E
Gudata = Guadauta, *Georgia* **61 J5** 43 7N 40 32 E
Gudbrandsdalen, *Norway* **15 F14** 61 33N 10 10 E
Gudenå →, *Denmark* **17 H4** 56 29N 10 13 E
Gudermes, *Russia* **61 J8** 43 24N 46 5 E
Gudhjem, *Denmark* **17 J8** 55 12N 14 58 E
Gudivada, *India* **95 F5** 16 30N 81 3 E
Gudiyattam, *India* **95 H4** 12 57N 78 55 E
Gudur, *India* **95 G4** 14 12N 79 55 E
Gudvangen, *Norway* **18 D3** 60 52N 6 49 E
Guebwiller, *France* **27 E14** 47 55N 7 12 E
Guecho = Getxo, *Spain* **40 B2** 43 21N 2 59W
Guékédou, *Guinea* **112 D2** 8 40N 10 5W
Guelma, *Algeria* **111 A6** 36 25N 7 29 E
Guelph, *Canada* **140 D3** 43 35N 80 20W
Guemar, *Algeria* **111 B6** 33 30N 6 49 E
Guéméné-Penfao, *France* **26 E5** 47 38N 1 50W
Guéméné-sur-Scorff, *France* **26 D3** 48 4N 3 13W
Guéné, *Benin* **113 C5** 11 44N 3 16 E
Güepi, *Peru* **168 D2** 0 9 S 75 10W
Guer, *France* **26 E4** 47 54N 2 8W
Güer Aike, *Argentina* **176 D3** 51 39 S 69 35W
Guera Pk., *Chad* **109 F3** 11 55N 18 12 E
Guérande, *France* **26 E4** 47 20N 2 26W
Guercif, *Morocco* **111 B4** 34 14N 3 21W
Guéréda, *Chad* **109 F4** 14 31N 22 5 E
Guéret, *France* **27 F8** 46 11N 1 51 E
Guérigny, *France* **27 E10** 47 6N 3 10 E
Guerneville, *U.S.A.* **160 G4** 38 30N 123 0W
Guernica = Gernika-Lumo, *Spain* **40 B2** 43 19N 2 40W
Guernsey, *U.K.* **21 H5** 49 26N 2 35W
Guernsey, *U.S.A.* **154 D2** 42 19N 104 45W
Guerrara, *Oasis, Algeria* **111 B5** 32 51N 4 22 E
Guerrara, *Saoura, Algeria* **111 C4** 28 5N 0 8W
Guerrero □, *Mexico* **163 D5** 17 30N 100 0W
Guerzim, *Algeria* **111 C4** 29 39N 1 40W
Gueugnon, *France* **27 F11** 46 36N 4 4 E
Gueydan, *U.S.A.* **155 K8** 30 2N 92 31W
Gufuðalur, *Iceland* **11 B4** 65 34N 22 25W
Gügher, *Iran* **97 D8** 29 28N 56 27 E
Guglionesi, *Italy* **45 G11** 41 55N 14 55 E
Gui Jiang →, *China* **77 F8** 23 30N 111 15 E
Gui Xian, *China* **76 F7** 23 8N 109 35 E
Guia, *Canary Is.* **39 F4** 28 8N 15 38W
Guia de Isora, *Canary Is.* **39 F3** 28 12N 16 46W
Guia Lopes da Laguna, *Brazil* **175 A4** 21 26 S 56 7W
Guiana, *S. Amer.* **166 C4** 5 10N 60 40W
Guichi, *China* **77 B11** 30 39N 117 27 E
Guider, *Cameroon* **113 D7** 9 56N 13 57 E
Guidimouni, *Niger* **109 F1** 13 42N 9 31 E
Guiding, *China* **76 D6** 26 34N 107 11 E
Guidong, *China* **77 D9** 26 7N 113 57 E
Guidónia-Montecélio, *Italy* **45 F9** 42 1N 12 45 E
Guiglo, *Ivory C.* **112 D3** 6 45N 7 30W
Guijá, *Mozam.* **117 C5** 24 27 S 33 0 E
Guijuelo, *Spain* **42 E5** 40 33N 5 40W
Guildford, *U.K.* **21 F7** 51 14N 0 34W
Guilford, *U.S.A.* **141 C6** 45 10N 69 23W
Guilin, *China* **77 E8** 25 18N 110 15 E
Guillaumes, *France* **29 D10** 44 5N 6 52 E
Guillestre, *France* **29 D10** 44 39N 6 40 E
Guilvinec, *France* **26 E2** 47 48N 4 17W
Güímar, *Canary Is.* **39 F3** 28 18N 16 24W
Guimarães, *Brazil* **170 B3** 2 9 S 44 42W
Guimarães, *Portugal* **42 D2** 41 28N 8 24W
Guimaras, *Phil.* **81 F4** 10 35N 122 37 E
Guimba, *Phil.* **80 D3** 15 40N 120 46 E
Guinda, *U.S.A.* **160 G4** 38 50N 122 12W
Guindulman, *Phil.* **81 G5** 9 46N 124 29 E
Guinea, *Africa* **104 F4** 8 0N 8 0 E
Guinea ■, *W. Afr.* **112 C2** 10 20N 11 30W
Guinea, Gulf of, *Atl. Oc.* **113 E5** 3 0N 2 30 E
Guinea-Bissau ■, *Africa* **112 C2** 12 0N 15 0W
Güines, *Cuba* **164 B3** 22 50N 82 0W
Guingamp, *France* **26 D3** 48 34N 3 10W
Guinobatan, *Phil.* **80 E4** 13 11N 123 36 E
Guiom, *Phil.* **81 F4** 11 59N 123 44 E
Guipavas, *France* **26 D2** 48 26N 4 29W
Guiping, *China* **77 F8** 23 21N 110 2 E
Guipúzcoa □, *Spain* **40 B2** 43 12N 2 15W
Guir, O. →, *Algeria* **111 B4** 31 29N 2 17W

H

Ḥalvān, *Iran* **97 C8** 33 57N 56 15 E
Ham, *France* **27 C10** 49 45N 3 4 E
Ham Tan, *Vietnam* **87 G6** 10 40N 107 45 E
Ham Yen, *Vietnam* **86 A5** 22 4N 105 3 E
Hamab, *Namibia* **116 D2** 28 7 S 19 16 E
Hamad, *Sudan* **107 D3** 15 20N 33 32 E
Hamada, *Japan* **70 C4** 34 56N 132 4 E
Hamadān, *Iran* **97 C6** 34 52N 48 32 E
Hamadān □, *Iran* **97 C6** 35 0N 49 0 E
Hamadia, *Algeria* **111 A5** 35 28N 1 57 E
Hamāh, *Syria* **100 E7** 35 5N 36 40 E
Hamakita, *Japan* **71 C9** 34 45N 137 47 E
Hamamatsu, *Japan* **71 C9** 34 45N 137 45 E
Hamar, *Norway* **15 F14** 60 48N 11 7 E
Hamâta, Gebel, *Egypt* **106 C4** 24 17N 35 0 E
Hamber Prov. Park, *Canada* **142 C5** 52 20N 118 0W
Hamburg, *Germany* **30 B5** 53 33N 9 59 E
Hamburg, *Ark., U.S.A.* **155 J9** 33 14N 91 48W
Hamburg, *Iowa, U.S.A.* **154 E7** 40 36N 95 39W
Hamburg, *N.Y., U.S.A.* **150 D6** 42 43N 78 50W
Hamburg, *Pa., U.S.A.* **151 F9** 40 33N 75 59W
Hamburg □, *Germany* **30 B5** 53 30N 10 0 E
Ḥamḍ, W. al →, *Si. Arabia* **96 E3** 24 55N 36 20 E
Ḥamḍah, *Si. Arabia* **98 C3** 19 2N 43 36 E
Ḥamdānah, *Si. Arabia* **98 C3** 19 59N 40 34 E
Hamden, *U.S.A.* **151 E12** 41 23N 72 54W
Hamdibey, *Turkey* **49 B9** 39 50N 27 15 E
Häme, *Finland* **15 F20** 61 38N 25 10 E
Hämeenlinna, *Finland* **15 F21** 61 0N 24 28 E
Hamélé, *Ghana* **112 C4** 10 56N 2 45W
Hamelin Pool, *Australia* **125 E1** 26 22 S 114 20 E
Hameln, *Germany* **30 C5** 52 6N 9 21 E
Hamer Koke, *Ethiopia* **107 F4** 5 15N 36 45 E
Hamerkaz □, *Israel* **103 C3** 32 15N 34 55 E
Hamersley Ra., *Australia* **124 D2** 22 0 S 117 45 E
Hamhung, *N. Korea* **75 E14** 39 54N 127 30 E
Hami, *China* **72 B4** 42 55N 93 25 E
Hamilton, *Australia* **128 D5** 37 45 S 142 2 E
Hamilton, *Canada* **140 D4** 43 15N 79 50W
Hamilton, *N.Z.* **130 D4** 37 47 S 175 19 E
Hamilton, *U.K.* **22 F4** 55 46N 4 2W
Hamilton, *Ga., U.S.A.* **152 C5** 32 45N 84 53W
Hamilton, *Ill., U.S.A.* **156 D5** 40 24N 91 21W
Hamilton, *Ind., U.S.A.* **157 C12** 41 33N 84 56W
Hamilton, *Mo., U.S.A.* **154 F8** 39 45N 93 59W
Hamilton, *Mo., U.S.A.* **156 E3** 39 45N 94 0W
Hamilton, *Mont., U.S.A.* **158 C6** 46 15N 114 10W
Hamilton, *N.Y., U.S.A.* **151 D9** 42 50N 75 33W
Hamilton, *Ohio, U.S.A.* **157 E12** 39 24N 84 34W
Hamilton, *Tex., U.S.A.* **155 K5** 31 42N 98 7W
Hamilton →, *Australia* **126 C2** 23 30 S 139 47 E
Hamilton City, *U.S.A.* **160 F4** 39 45N 122 1W
Hamilton Hotel, *Australia* **126 C3** 22 45 S 140 40 E
Hamilton Inlet, *Canada* **141 B8** 54 0N 57 30W
Hamina, *Finland* **15 F22** 60 34N 27 12 E
Hamiota, *Canada* **143 C8** 50 11N 100 38W
Hamitabat, *Turkey* **51 E11** 41 30N 27 17 E
Hamlet, *U.S.A.* **149 H6** 34 53N 79 42W
Hamley Bridge, *Australia* **128 C3** 34 17 S 138 35 E
Hamlin = Hameln, *Germany* **30 C5** 52 6N 9 21 E
Hamlin, *N.Y., U.S.A.* **150 C7** 43 17N 77 55W
Hamlin, *Tex., U.S.A.* **155 J4** 32 53N 100 8W
Hamm, *Germany* **30 D3** 51 40N 7 50 E
Hammam Bouhadjar, *Algeria* **111 A4** 35 23N 0 58W
Hammamet, *Tunisia* **108 A2** 36 24N 10 38 E
Hammamet, G. de, *Tunisia* **108 A2** 36 10N 10 48 E
Hammarstrand, *Sweden* **16 A10** 63 7N 16 20 E
Hammelburg, *Germany* **31 E5** 50 6N 9 53 E
Hammeren, *Denmark* **17 J8** 55 18N 14 47 E
Hammerfest, *Norway* **14 A20** 70 39N 23 41 E
Hammerum, *Denmark* **17 H3** 56 8N 9 3 E
Humminkeln, *Germany* **30 D2** 51 43N 6 35 E
Hammond, *Ill., U.S.A.* **157 E8** 39 48N 88 36W
Hammond, *Ind., U.S.A.* **157 C9** 41 38N 87 30W
Hammond, *La., U.S.A.* **155 K9** 30 30N 90 28W
Hammonton, *U.S.A.* **148 F8** 39 39N 74 48W
Hamneda, *Sweden* **17 H7** 56 41N 13 51 E
Hamoyet, Jebel, *Sudan* **106 D4** 17 33N 38 2 E
Hampden, *N.Z.* **131 F5** 45 18 S 170 50 E
Hampshire □, *U.K.* **21 F6** 51 7N 1 23W
Hampshire Downs, *U.K.* **21 F6** 51 15N 1 10W
Hampton, *Ark., U.S.A.* **155 J8** 33 32N 92 28W
Hampton, *Fla., U.S.A.* **152 F7** 29 52N 82 8W
Hampton, *Ga., U.S.A.* **152 B5** 33 23N 84 17W
Hampton, *Iowa, U.S.A.* **156 B3** 42 45N 93 13W
Hampton, *N.H., U.S.A.* **151 D14** 42 57N 70 50W
Hampton, *S.C., U.S.A.* **152 C8** 32 52N 81 7W
Hampton, *Va., U.S.A.* **148 G7** 37 2N 76 21W
Hampton Springs, *U.S.A.* **152 E6** 30 5N 83 40W
Hampton Tableland, *Australia* **125 F4** 32 0 S 127 0 E
Hamra, *Sweden* **16 C8** 61 39N 14 59 E
Ḥamrā', *Yemen* **98 D3** 15 3N 43 0 E
Hamrat esh Sheykh, *Sudan* **107 E2** 14 38N 27 55 E
Hamtik, *Phil.* **81 F3** 10 42N 121 59 E
Hamur, *Turkey* **101 C10** 39 37N 43 3 E
Hamyang, *S. Korea* **75 G14** 35 32N 127 42 E
Han Jiang →, *China* **77 F11** 23 25N 116 40 E
Han Shui →, *China* **77 B10** 30 35N 114 18 E
Hana, *Hawaii, U.S.A.* **145 C6** 20 45N 155 59W
Hana, *Hawaii, U.S.A.* **146 H17** 20 45N 155 59W
Hanahan, *U.S.A.* **152 C10** 32 55N 80 0W
Hanak, *Si. Arabia* **96 E3** 25 32N 37 0 E
Hanalei, *U.S.A.* **145 A2** 22 12N 159 30W
Hanamaki, *Japan* **68 E10** 39 23N 141 7 E
Hanamaulu, *U.S.A.* **145 B2** 21 59N 159 22W
Hanang, *Tanzania* **118 C4** 4 30 S 35 25 E
Hanapepe, *U.S.A.* **145 B2** 21 55N 159 35W
Hanau, *Germany* **31 E4** 50 7N 8 56 E
Hanauma B., *U.S.A.* **145 K14** 21 15N 157 40W
Hanbogd, *Mongolia* **74 C4** 43 11N 107 10 E
Hançalar, *Turkey* **49 C11** 38 8N 29 24 E
Hânceşti, *Moldova* **53 D13** 46 50N 28 36 E
Hancheng, *China* **74 G6** 35 31N 110 25 E
Hanchuan, *China* **77 B9** 30 40N 113 50 E
Hancock, *Mich., U.S.A.* **154 B10** 47 8N 88 35W
Hancock, *Minn., U.S.A.* **154 C7** 45 30N 95 48W
Hancock, *N.Y., U.S.A.* **151 E9** 41 57N 75 17W
Handa, *Japan* **71 C8** 34 53N 136 55 E
Handa, *Somali Rep.* **120 B4** 10 37N 51 2 E
Handan, *China* **74 F8** 36 35N 114 28 E
Handeni, *Tanzania* **118 D4** 5 25 S 38 2 E
Handeni □, *Tanzania* **118 D4** 5 30 S 38 0 E
Handlová, *Slovak Rep.* **35 C11** 48 45N 18 35 E
Handub, *Sudan* **106 D4** 19 15N 37 16 E
Handwara, *India* **93 B6** 34 21N 74 20 E
Hanegev, *Israel* **103 E4** 30 50N 35 0 E
Haney, *Canada* **142 D4** 49 12N 122 40W
Hanford, *U.S.A.* **160 J7** 36 20N 119 39W
Hang Chat, *Thailand* **86 C2** 18 20N 99 21 E
Hang Dong, *Thailand* **86 C2** 18 41N 98 55 E
Hangang →, *S. Korea* **75 F14** 37 50N 126 30 E
Hangayn Nuruu, *Mongolia* **72 B4** 47 30N 99 0 E
Hangchou = Hangzhou, *China* **77 B13** 30 18N 120 11 E
Hanggin Houqi, *China* **74 D4** 40 58N 107 4 E
Hanggin Qi, *China* **74 E5** 39 52N 108 50 E
Hangu, *China* **75 E9** 39 18N 117 53 E
Hangu, *Pakistan* **91 B3** 33 30N 71 10 E
Hangzhou, *China* **77 B13** 30 18N 120 11 E
Hangzhou Wan, *China* **77 B13** 30 15N 120 45 E
Hanhongor, *Mongolia* **74 C3** 43 55N 104 28 E
Ḥanīdh, *Si. Arabia* **97 E6** 26 35N 48 38 E
Ḥanīsh, *Yemen* **98 D3** 13 45N 42 46 E
Haniska, *Slovak Rep.* **35 C14** 48 37N 21 15 E
Hanjiang, *China* **77 E12** 25 26N 119 6 E
Hankinson, *U.S.A.* **154 B6** 46 4N 96 54W
Hanko, *Finland* **15 G20** 59 50N 22 57 E
Hankou, *China* **77 B10** 30 35N 114 30 E
Hanksville, *U.S.A.* **159 G8** 38 22N 110 43W
Hanle, *India* **93 C8** 32 42N 79 4 E
Hanmer Springs, *N.Z.* **131 C7** 42 32 S 172 50 E
Hann →, *Australia* **124 C4** 17 26 S 126 17 E
Hann, Mt., *Australia* **124 C4** 15 45 S 126 0 E
Hanna, *Canada* **142 C6** 51 40N 111 54W
Hannaford, *U.S.A.* **154 B5** 47 19N 98 11W
Hannah, *U.S.A.* **154 A5** 48 58N 98 42W
Hannah B., *Canada* **140 B4** 51 40N 80 0W
Hannahs Bridge, *Australia* **129 A8** 31 55 S 149 41 E
Hannibal, *U.S.A.* **156 E5** 39 42N 91 22W
Hannik, *Sudan* **106 D3** 18 12N 32 20 E
Hannover, *Germany* **30 C5** 52 22N 9 46 E
Hanö, *Sweden* **17 H8** 56 1N 14 50 E
Hanöbukten, *Sweden* **17 J8** 55 35N 14 30 E
Hanoi, *Vietnam* **72 D5** 21 5N 105 55 E
Hanover = Hannover, *Germany* **30 C5** 52 22N 9 46 E
Hanover, *Canada* **150 B3** 44 9N 81 2W
Hanover, *S. Africa* **116 E3** 31 4 S 24 29 E
Hanover, *Ind., U.S.A.* **157 F11** 38 43N 85 28W
Hanover, *N.H., U.S.A.* **151 C12** 43 42N 72 17W
Hanover, *Ohio, U.S.A.* **150 F2** 40 4N 82 16W
Hanover, *Pa., U.S.A.* **148 F7** 39 48N 76 59W
Hanover, I., *Chile* **176 D2** 51 0 S 74 50W
Hanpan, C., *Papua N. G.* **132 C8** 5 0 S 154 35 E
Hans Meyer Ra., *Papua N. G.* **132 C7** 4 20 S 152 55 E
Hanshou, *China* **77 C8** 28 56N 111 50 E
Hansi, *India* **92 E6** 29 10N 75 57 E
Hanson, *U.S.A.* **152 E6** 30 34N 83 21W
Hanson, L., *Australia* **128 A2** 31 0 S 136 15 E
Hanstholm, *Denmark* **17 G2** 57 7N 8 36 E
Hantsavichy, *Belarus* **59 F4** 52 49N 26 30 E
Hanyang, *China* **77 B10** 30 35N 114 2 E
Hanyin, *China* **76 A7** 32 54N 108 28 E
Hanyū, *Japan* **71 A11** 36 10N 139 32 E
Hanyuan, *China* **76 C4** 29 21N 102 40 E
Hanzhong, *China* **74 H4** 33 10N 107 1 E
Hanzhuang, *China* **75 G9** 34 33N 117 23 E
Haora, *India* **93 H13** 22 37N 88 20 E
Haoxue, *China* **77 B9** 30 3N 112 24 E
Haparanda, *Sweden* **14 D21** 65 52N 24 8 E
Hapeville, *U.S.A.* **152 B5** 33 40N 84 25W
Happy, *U.S.A.* **155 H4** 34 45N 101 52W
Happy Camp, *U.S.A.* **158 F2** 41 48N 123 23W
Happy Valley-Goose Bay, *Canada* **141 B7** 53 15N 60 20W
Hapsu, *N. Korea* **75 D15** 41 13N 128 51 E
Hapur, *India* **92 E7** 28 45N 77 45 E
Ḥaql, *Si. Arabia* **103 F3** 29 10N 34 58 E
Haquira, *Peru* **172 C3** 14 14 S 72 12W
Har, *Indonesia* **83 C4** 5 16 S 133 14 E
Har-Ayrag, *Mongolia* **74 B5** 45 47N 109 16 E
Har Hu, *China* **72 C4** 38 20N 97 38 E
Har Us Nuur, *Mongolia* **72 B4** 48 0N 92 0 E
Har Yehuda, *Israel* **103 D3** 31 35N 34 57 E
Ḥaraḍ, *Si. Arabia* **99 A5** 24 22N 49 0 E
Ḥaraḍ, *Yemen* **98 C3** 16 26N 43 5 E
Haradok, *Belarus* **58 E6** 55 30N 30 3 E
Häradsbäck, *Sweden* **17 H8** 56 32N 14 26 E
Haralson, *U.S.A.* **152 B5** 33 14N 84 34W
Haranomachi, *Japan* **68 F10** 37 38N 140 58 E
Harardera, *Somali Rep.* **120 D3** 4 33N 47 38 E
Harare, *Zimbabwe* **119 F3** 17 43 S 31 2 E
Ḥarāsīs, Jiddat al, *Oman* **99 C7** 19 30N 56 0 E
Harat, *Eritrea* **107 D4** 16 5N 39 26 E
Haraz, *Chad* **109 F3** 14 20N 19 12 E
Harazé, *Chad* **109 G4** 9 57N 20 48 E
Harbin, *China* **75 B14** 45 48N 126 40 E
Harbiye, *Turkey* **100 D7** 36 10N 36 8 E
Harbo, *Sweden* **16 D11** 60 7N 17 12 E
Harbøør, *Denmark* **17 H2** 56 38N 8 10 E
Harbor Beach, *U.S.A.* **148 D4** 43 51N 82 39W
Harbor Springs, *U.S.A.* **148 C3** 45 26N 85 0W
Harbour Breton, *Canada* **141 C8** 47 29N 55 50W
Harbour Grace, *Canada* **141 C9** 47 40N 53 22W
Harburg, *Germany* **30 B5** 53 27N 9 58 E
Hårby, *Denmark* **17 J4** 55 13N 10 7 E
Harda, *India* **92 H7** 22 27N 77 5 E
Hardangerfjorden, *Norway* **15 F12** 60 5N 6 0 E
Hardangerjøkulen, *Norway* **18 D4** 60 30N 7 27 E
Hardangervidda, *Norway* **15 F12** 60 7N 7 20 E
Hardap Dam, *Namibia* **116 C2** 24 32 S 17 50 E
Hardeeville, *U.S.A.* **152 C8** 32 17N 81 5W
Harden, *Australia* **129 C8** 34 32 S 148 24 E
Hardenberg, *Neths.* **24 B6** 52 34N 6 37 E
Harderwijk, *Neths.* **24 B5** 52 21N 5 38 E
Hardey →, *Australia* **124 D2** 22 45 S 116 8 E
Hardin, *Ill., U.S.A.* **156 E6** 39 10N 90 37W
Hardin, *Mont., U.S.A.* **158 D10** 45 44N 107 37W
Harding, *S. Africa* **117 E4** 30 35 S 29 55 E
Harding, L., *U.S.A.* **152 C4** 32 40N 85 5W
Harding Ra., *Australia* **124 C3** 16 17 S 124 55 E
Hardinsburg, *U.S.A.* **157 G10** 37 47N 86 28W
Hardisty, *Canada* **142 C6** 52 40N 111 18W
Hardman, *U.S.A.* **158 D4** 45 10N 119 41W
Hardoi, *India* **93 F9** 27 26N 80 6 E
Hardwar = Haridwar, *India* **92 E8** 29 58N 78 9 E
Hardwick, *Ga., U.S.A.* **152 B6** 33 4N 83 14W
Hardwick, *Vt., U.S.A.* **151 B12** 44 30N 72 22W
Hardy, *U.S.A.* **155 G9** 36 19N 91 29W
Hardy, Pen., *Chile* **176 E3** 55 30 S 68 20W
Hare B., *Canada* **141 B8** 51 15N 55 45W
Hareid, *Norway* **15 E12** 62 22N 6 1 E
Haren, *Germany* **30 C3** 52 47N 7 13 E
Harer, *Ethiopia* **107 F5** 9 20N 42 8 E
Harerge □, *Ethiopia* **107 F5** 7 12N 42 0 E
Harestua, *Norway* **18 D7** 60 11N 10 44 E
Hareto, *Ethiopia* **107 F4** 9 23N 37 6 E
Harfleur, *France* **26 C7** 49 30N 0 10 E
Hargeisa, *Somali Rep.* **120 C2** 9 30N 44 2 E
Harghita □, *Romania* **53 D10** 46 30N 25 30 E
Harghita, Munţii, *Romania* **53 D10** 46 25N 25 35 E
Hargshamn, *Sweden* **16 D12** 60 12N 18 30 E
Hari →, *Indonesia* **84 C2** 1 16 S 104 5 E
Haria, *Canary Is.* **39 E6** 29 8N 13 32W
Ḥarīb, *Yemen* **98 D4** 14 56N 45 30 E
Haricha, Hamada el, *Mali* **110 D4** 22 40N 3 15W
Haridwar, *India* **92 E8** 29 58N 78 9 E
Harihar, *India* **95 G2** 14 32N 75 44 E
Harihari, *N.Z.* **131 D5** 43 9 S 170 33 E
Harima-Nada, *Japan* **70 C6** 34 30N 134 35 E
Haringhata →, *Bangla.* **90 E2** 22 0N 89 58 E
Haripad, *India* **95 K3** 9 14N 76 28 E
Harīrūd →, *Asia* **97 B9** 37 24N 60 38 E
Härjedalen, *Sweden* **16 B7** 62 22N 13 5 E
Harlan, *Iowa, U.S.A.* **154 E7** 41 39N 95 19W
Harlan, *Ky., U.S.A.* **149 G4** 36 51N 83 19W
Hârlău, *Romania* **53 C11** 47 23N 26 55 E
Harlech, *U.K.* **20 E3** 52 52N 4 6W
Harlem, *Ga., U.S.A.* **152 B7** 33 25N 82 19W
Harlem, *Mont., U.S.A.* **158 B9** 48 32N 108 47W
Hårlev, *Denmark* **17 J6** 55 21N 12 14 E
Harleyville, *U.S.A.* **152 B9** 33 13N 80 27W
Harlingen, *Neths.* **24 A5** 53 11N 5 25 E
Harlingen, *U.S.A.* **155 M6** 26 12N 97 42W
Harlow, *U.K.* **21 F8** 51 46N 0 8 E
Harlowton, *U.S.A.* **158 C9** 46 26N 109 50W
Harmancık, *Turkey* **49 B11** 39 41N 29 9 E
Harmånger, *Sweden* **16 C11** 61 55N 17 20 E
Harmil, *Eritrea* **107 D5** 16 30N 40 10 E
Harney, L., *U.S.A.* **153 G8** 28 45N 81 3W
Harney Basin, *U.S.A.* **158 E4** 43 30N 119 0W
Harney L., *U.S.A.* **158 E4** 43 14N 119 8W
Harney Peak, *U.S.A.* **154 D3** 43 52N 103 32W
Härnön, *Sweden* **16 B12** 62 36N 18 0 E
Härnösand, *Sweden* **16 B11** 62 38N 17 55 E
Haro, *Spain* **40 C2** 42 35N 2 55W
Harold, *U.S.A.* **153 E3** 30 40N 86 53W
Haroldswick, *U.K.* **22 A8** 60 48N 0 50W
Harp L., *Canada* **141 A7** 55 5N 61 50W
Harpanahalli, *India* **95 G3** 14 47N 76 2 E
Harper, *Liberia* **112 E3** 4 25N 7 43W
Harper Pass, *N.Z.* **131 C6** 42 42 S 171 55 E
Harplinge, *Sweden* **17 H6** 56 45N 12 45 E
Harrand, *Pakistan* **92 E4** 29 28N 70 3 E
Harriman, *U.S.A.* **149 H3** 35 56N 84 33W
Harrington Harbour, *Canada* **141 B8** 50 31N 59 30W
Harris, *U.K.* **22 D2** 57 50N 6 55W
Harris, *U.S.A.* **153 G8** 28 47N 81 49W
Harris, Sd. of, *U.K.* **22 D1** 57 44N 7 6W
Harris L., *Australia* **127 E2** 31 10 S 135 10 E
Harris Mts., *N.Z.* **131 E3** 44 49 S 168 49 E
Harrisburg, *Ill., U.S.A.* **155 G10** 37 44N 88 32W
Harrisburg, *Nebr., U.S.A.* **154 E3** 41 33N 103 44W
Harrisburg, *Oreg., U.S.A.* **158 D2** 44 16N 123 10W
Harrisburg, *Pa., U.S.A.* **150 F8** 40 16N 76 53W
Harrismith, *S. Africa* **117 D4** 28 15 S 29 8 E
Harrison, *Ark., U.S.A.* **155 G8** 36 14N 93 7W
Harrison, *Ga., U.S.A.* **152 C7** 32 50N 82 43W
Harrison, *Idaho, U.S.A.* **158 C5** 47 27N 116 47W
Harrison, *Nebr., U.S.A.* **154 D3** 42 41N 103 53W
Harrison, C., *Canada* **141 B8** 54 55N 57 55W
Harrison L., *Canada* **142 D4** 49 33N 121 50W
Harrisonburg, *U.S.A.* **148 F6** 38 27N 78 52W
Harrisonville, *U.S.A.* **156 F2** 38 39N 94 21W
Harriston, *Canada* **140 D3** 43 57N 80 53W
Harrisville, *U.S.A.* **150 B1** 44 39N 83 17W
Harrodsburg, *Ind., U.S.A.* **157 E10** 39 1N 86 33W
Harrodsburg, *Ky., U.S.A.* **157 G12** 37 46N 84 51W
Harrogate, *U.K.* **20 C6** 54 0N 1 33W
Harrow, *U.K.* **21 F7** 51 35N 0 21W
Harry S. Truman Reservoir, *U.S.A.* **156 F3** 38 16N 93 24W
Harsefeld, *Germany* **30 B5** 53 27N 9 30 E
Harsewinkel, *Germany* **30 D4** 51 56N 8 14 E
Harsīn, *Iran* **101 E12** 34 18N 47 33 E
Hârşova, *Romania* **53 F12** 44 40N 27 59 E
Harstad, *Norway* **14 B17** 68 48N 16 30 E
Hart, *U.S.A.* **148 D2** 43 42N 86 22W
Hart, L., *Australia* **128 A2** 31 10 S 136 25 E
Hartbees →, *S. Africa* **116 D3** 28 45 S 20 32 E
Hartberg, *Austria* **34 D8** 47 17N 15 58 E
Hårteigen, *Norway* **18 D4** 60 11N 7 3 E
Hartford, *Ala., U.S.A.* **152 D4** 31 6N 85 42W
Hartford, *Conn., U.S.A.* **151 E12** 41 46N 72 41W
Hartford, *Ga., U.S.A.* **152 C6** 32 17N 83 28W
Hartford, *Ky., U.S.A.* **148 G2** 37 27N 86 55W
Hartford, *Mich., U.S.A.* **157 B10** 42 13N 86 10W
Hartford, *S. Dak., U.S.A.* **154 D6** 43 38N 96 57W
Hartford, *Wis., U.S.A.* **154 D10** 43 19N 88 22W
Hartford City, *U.S.A.* **157 D11** 40 27N 85 22W
Hartland, *Canada* **141 C6** 46 20N 67 32W
Hartland, *U.S.A.* **157 A8** 43 6N 88 21W
Hartland Pt., *U.K.* **21 F3** 51 1N 4 32W
Hartlepool, *U.K.* **20 C6** 54 42N 1 13W
Hartlepool □, *U.K.* **20 C6** 54 42N 1 17W
Hartley Bay, *Canada* **142 C3** 53 25N 129 15W
Hartmannberge, *Namibia* **116 B1** 17 0 S 13 0 E
Hartney, *Canada* **143 D8** 49 30N 100 35W
Hârtop, *Moldova* **53 D13** 46 39N 28 40 E
Harts →, *S. Africa* **116 D3** 28 24 S 24 17 E
Hartselle, *U.S.A.* **149 H2** 34 27N 86 56W
Hartshorne, *U.S.A.* **155 H7** 34 51N 95 34W
Hartsville, *U.S.A.* **149 H5** 34 23N 80 4W
Hartwell, *U.S.A.* **149 H4** 34 21N 82 56W
Harunabad, *Pakistan* **92 E5** 29 35N 73 8 E
Harur, *India* **95 H4** 12 3N 78 29 E
Harvand, *Iran* **97 D7** 28 25N 55 43 E
Harvard, *U.S.A.* **157 B8** 42 25N 88 37W
Harvey, *Australia* **125 F2** 33 5 S 115 54 E
Harvey, *Ill., U.S.A.* **157 C9** 41 36N 87 50W
Harvey, *N. Dak., U.S.A.* **154 B5** 47 47N 99 56W
Harwich, *U.K.* **21 F9** 51 56N 1 17 E
Haryana □, *India* **92 E7** 29 0N 76 10 E
Haryn →, *Belarus* **59 F4** 52 7N 27 17 E
Harz, *Germany* **30 D6** 51 38N 10 44 E
Harzgerode, *Germany* **30 D7** 51 38N 11 8 E
Hasaheisa, *Sudan* **107 E3** 14 44N 33 20 E
Hasan Kīādeh, *Iran* **97 B6** 37 24N 49 58 E
Ḥasanābād, *Iran* **97 C7** 32 8N 52 44 E
Hasanpur, *India* **92 E8** 28 43N 78 17 E
Haselünne, *Germany* **30 C3** 52 40N 7 29 E
Hashima, *Japan* **71 B8** 35 20N 136 40 E
Hashimoto, *Japan* **71 C7** 34 19N 135 37 E
Hashtjerd, *Iran* **97 C6** 35 52N 50 40 E
Ḥāsik, *Oman* **99 C6** 17 22N 55 17 E
Haskell, *Okla., U.S.A.* **155 H7** 35 50N 95 40W
Haskell, *Tex., U.S.A.* **155 J5** 33 10N 99 44W
Hasköy, *Turkey* **51 E10** 41 38N 26 52 E
Haslach, *Germany* **31 G4** 48 16N 8 5 E
Hasle, *Denmark* **17 J8** 55 11N 14 44 E
Haslemere, *U.K.* **21 F7** 51 5N 0 43W
Haslev, *Denmark* **17 J5** 55 18N 11 57 E
Hasparren, *France* **28 E2** 43 24N 1 18W
Hassa, *Turkey* **100 D7** 36 48N 36 29 E
Hassela, *Sweden* **16 B10** 62 7N 16 42 E
Hasselt, *Belgium* **24 D5** 50 56N 5 21 E
Hassene, Adrar, *Algeria* **111 D5** 21 0N 4 0 E
Hassfurt, *Germany* **31 E6** 50 2N 10 30 E
Hassi bou Khelala, *Algeria* **111 B4** 30 17N 0 18W
Hassi Djafou, *Algeria* **111 B5** 30 55N 3 35 E
Hassi el Abiod, *Algeria* **111 B5** 31 47N 3 37 E
Hassi el Biod, *Algeria* **111 C6** 28 30N 6 0 E
Hassi el Hadjar, *Algeria* **111 B5** 31 28N 4 45 E
Hassi Imoulaye, *Algeria* **111 C6** 29 54N 9 10 E
Hassi Inifel, *Algeria* **111 C5** 29 50N 3 41 E
Hassi Messaoud, *Algeria* **111 B6** 31 51N 6 1 E
Hassi Tartrat, *Algeria* **111 B6** 30 5N 6 28 E
Hassi Zerzour, *Morocco* **110 B4** 30 51N 3 56W
Hässleholm, *Sweden* **17 H7** 56 10N 13 46 E
Hassloch, *Germany* **31 F4** 49 22N 8 16 E
Hästholmen, *Sweden* **17 F8** 58 17N 14 38 E
Hastings, *Australia* **129 E6** 38 18 S 145 12 E
Hastings, *N.Z.* **130 F5** 39 39 S 176 52 E
Hastings, *U.K.* **21 G8** 50 51N 0 35 E
Hastings, *Fla., U.S.A.* **152 F8** 29 43N 81 31W
Hastings, *Mich., U.S.A.* **157 B11** 42 39N 85 17W
Hastings, *Minn., U.S.A.* **154 C8** 44 44N 92 51W
Hastings, *Nebr., U.S.A.* **154 E5** 40 35N 98 23W
Hastings Ra., *Australia* **129 A10** 31 15 S 152 14 E
Hästveda, *Sweden* **17 H7** 56 17N 13 55 E
Hat Yai, *Thailand* **87 J3** 7 1N 100 27 E
Hatanbulag, *Mongolia* **74 C5** 43 8N 109 5 E
Hatano, *Japan* **71 B11** 35 22N 139 14 E
Hatay = Antalya, *Turkey* **100 D4** 36 52N 30 45 E
Hatch, *U.S.A.* **159 K10** 32 40N 107 9W
Hatches Creek, *Australia* **126 C2** 20 56 S 135 12 E
Hatchet L., *Canada* **143 B8** 58 36N 103 40W
Hatchineha, L., *U.S.A.* **153 G8** 28 2N 81 25W
Haţeg, *Romania* **52 E7** 45 36N 22 55 E
Hateruma-Shima, *Japan* **69 M1** 24 3N 123 47 E
Hatfield P.O., *Australia* **128 B5** 33 54 S 143 49 E
Hatgal, *Mongolia* **72 A5** 50 26N 100 9 E
Hathras, *India* **92 F8** 27 36N 78 6 E
Hato de Corozal, *Colombia* **168 B3** 6 11N 71 45W
Hato Mayor, *Dom. Rep.* **165 C6** 18 46N 69 15W
Hattah, *Australia* **128 C5** 34 48 S 142 17 E
Hatteras, C., *U.S.A.* **149 H8** 35 14N 75 32W
Hattiesburg, *U.S.A.* **155 K10** 31 20N 89 17W
Hatvan, *Hungary* **52 C4** 47 40N 19 45 E
Hau Duc, *Vietnam* **86 E7** 15 20N 108 13 E
Haubstadt, *U.S.A.* **157 F9** 38 12N 87 34W
Haugastøl, *Norway* **18 D4** 60 30N 7 50 E
Hauge, *Norway* **18 F3** 58 20N 6 15 E
Haugesund, *Norway* **15 G11** 59 23N 5 13 E
Hauhungaroa Ra., *N.Z.* **130 E4** 38 42 S 175 40 E
Haukeligrend, *Norway* **18 E4** 59 44N 7 33 E
Haukipudas, *Finland* **14 D21** 65 12N 25 20 E
Haultain →, *Canada* **143 B7** 55 51N 106 46W
Haungpa, *Burma* **90 C6** 25 29N 96 7 E
Hauraki G., *N.Z.* **130 C4** 36 35 S 175 5 E
Hausruck, *Austria* **34 C6** 48 6N 13 30 E
Haut Atlas, *Morocco* **110 B4** 32 30N 5 0W
Haut-Rhin □, *France* **27 E14** 48 0N 7 15 E
Haut-Zaïre □, *Dem. Rep. of the Congo* **118 B2** 2 20N 26 0 E
Haute-Corse □, *France* **29 F13** 42 30N 9 30 E
Haute-Garonne □, *France* **28 E5** 43 30N 1 30 E
Haute-Loire □, *France* **28 C7** 45 5N 3 50 E
Haute-Marne □, *France* **27 D12** 48 10N 5 20 E
Haute-Normandie □, *France* **26 C7** 49 20N 1 0 E
Haute-Saône □, *France* **27 E13** 47 45N 6 10 E
Haute-Savoie □, *France* **29 C10** 46 0N 6 20 E
Haute-Vienne □, *France* **28 C5** 45 50N 1 10 E
Hauterive, *Canada* **141 C6** 49 10N 68 16W
Hautes-Alpes □, *France* **29 D10** 44 42N 6 20 E
Hautes Fagnes = Hohe Venn, *Belgium* **24 D6** 50 30N 6 5 E
Hautes-Pyrénées □, *France* **28 F4** 43 0N 0 10 E
Hauteville-Lompnès, *France* **29 C9** 45 58N 5 36 E
Hautmont, *France* **27 B10** 50 15N 3 55 E
Hauts-de-Seine □, *France* **27 D9** 48 52N 2 15 E
Hauts Plateaux, *Algeria* **111 B5** 35 0N 1 0 E
Hauula, *U.S.A.* **145 J14** 21 37N 157 55W
Hauzenberg, *Germany* **31 G9** 48 40N 13 37 E
Havana = La Habana, *Cuba* **164 B3** 23 8N 82 22W
Havana, *Fla., U.S.A.* **152 E5** 30 37N 84 25W
Havana, *Ill., U.S.A.* **156 D6** 40 18N 90 4W
Havant, *U.K.* **21 G7** 50 51N 0 58W
Havârna, *Romania* **53 B11** 48 4N 26 43 E
Havasu, L., *U.S.A.* **161 L12** 34 18N 114 28W
Havdhem, *Sweden* **17 G12** 57 10N 18 20 E
Havel →, *Germany* **30 C8** 52 50N 12 3 E
Havelian, *Pakistan* **92 B5** 34 2N 73 10 E
Havelock, *N.B., Canada* **141 C6** 46 2N 65 24W
Havelock, *Ont., Canada* **140 D4** 44 26N 77 53W
Havelock, *N.Z.* **131 B8** 41 17 S 173 48 E
Havelock North, *N.Z.* **130 F5** 39 40 S 176 53 E
Haverfordwest, *U.K.* **21 F3** 51 48N 4 58W
Haverhill, *Fla., U.S.A.* **153 J9** 26 42N 80 7W
Haverhill, *Mass., U.S.A.* **151 D13** 42 47N 71 5W
Haveri, *India* **95 G2** 14 53N 75 24 E
Haverstraw, *U.S.A.* **151 E11** 41 12N 73 58W
Håverud, *Sweden* **17 F6** 58 50N 12 28 E
Havířov, *Czech.* **35 B11** 49 46N 18 20 E
Havlíčkův Brod, *Czech Rep.* **34 B8** 49 36N 15 33 E
Havneby, *Denmark* **17 J2** 55 5N 8 34 E
Havran, *Turkey* **49 B9** 39 33N 27 6 E
Havre, *U.S.A.* **158 B9** 48 33N 109 41W
Havre-Aubert, *Canada* **141 C7** 47 12N 61 56W
Havre-St.-Pierre, *Canada* **141 B7** 50 18N 63 33W
Havsa, *Turkey* **51 E10** 41 31N 26 48 E
Havza, *Turkey* **100 B6** 41 0N 35 35 E
Haw →, *U.S.A.* **149 H6** 35 36N 79 3W
Hawaii, *U.S.A.* **145 D6** 19 30N 155 30W
Hawaii □, *U.S.A.* **146 H16** 19 30N 156 30W
Hawaii I., *Pac. Oc.* **146 J17** 20 0N 155 0W
Hawaii Volcanoes National Park, *U.S.A.* **145 D6** 19 23N 155 17W
Hawaiian Is., *Pac. Oc.* **135 E12** 20 30N 156 0W
Hawaiian Ridge, *Pac. Oc.* **135 E11** 24 0N 165 0W
Hawarden, *Canada* **143 C7** 51 25N 106 36W
Hawarden, *U.S.A.* **154 D6** 43 0N 96 29W
Hawea, L., *N.Z.* **131 E4** 44 28 S 169 19 E
Hawea Flat, *N.Z.* **131 E4** 44 40 S 169 19 E
Hawera, *N.Z.* **130 F3** 39 35 S 174 19 E
Hawesville, *U.S.A.* **157 G10** 37 54N 86 45W
Hawi, *U.S.A.* **145 C6** 20 14N 155 50W
Hawick, *U.K.* **22 F6** 55 26N 2 47W

Hillerstorp 223

Hawk Junction, *Canada* **140 C3** 48 5N 84 38W
Hawk Point, *U.S.A.* **156 F5** 38 58N 91 8W
Hawkdun Ra., *N.Z.* **131 E5** 44 53 S 170 5 E
Hawke B., *N.Z.* **130 F6** 39 25 S 177 20 E
Hawker, *Australia* **128 A3** 31 59 S 138 22 E
Hawke's Bay □, *N.Z.* **130 F5** 39 45 S 176 35 E
Hawkesbury, *Canada* **140 C5** 45 37N 74 37W
Hawkesbury I., *Canada* **142 C3** 53 37N 129 3W
Hawkesbury Pt., *Australia* **126 A1** 11 55 S 134 5 E
Hawkinsville, *U.S.A.* **152 C6** 32 17N 83 28W
Hawkwood, *Australia* **127 D5** 25 45 S 150 50 E
Hawley, *U.S.A.* **154 B6** 46 53N 96 19W
Hawrān, *Syria* **96 C3** 32 45N 36 15 E
Ḩawrān, W. →, *Iraq* **101 F10** 33 58N 42 34 E
Hawsh Mūssá, *Lebanon* **103 B4** 33 45N 35 55 E
Hawthorne, *Fla., U.S.A.* **153 F7** 29 36N 82 5W
Hawthorne, *Nev., U.S.A.* **158 G4** 38 32N 118 38W
Hawzen, *Ethiopia* **107 E4** 13 58N 39 28 E
Haxtun, *U.S.A.* **154 E3** 40 39N 102 38W
Hay, *Australia* **129 C6** 34 30 S 144 51 E
Hay →, *Australia* **126 C2** 24 50 S 138 0 E
Hay →, *Canada* **142 A5** 60 50N 116 26W
Hay, C., *Australia* **124 B4** 14 5 S 129 29 E
Hay L., *Canada* **142 B5** 58 50N 118 50W
Hay Lakes, *Canada* **142 C6** 53 12N 113 2W
Hay-on-Wye, *U.K.* **21 E4** 52 5N 3 8W
Hay River, *Canada* **142 A5** 60 51N 115 44W
Hay Springs, *U.S.A.* **154 D3** 42 41N 102 41W
Haya = Tehoru, *Indonesia* **83 B3** 3 19 S 129 37 E
Hayachine-San, *Japan* **68 E10** 39 34N 141 29 E
Hayange, *France* **27 C13** 49 20N 6 2 E
Hayato, *Japan* **70 F2** 31 40N 130 43 E
Haydarlı, *Turkey* **49 C12** 38 16N 30 23 E
Hayden, *Ariz., U.S.A.* **159 K8** 33 0N 110 47W
Hayden, *Colo., U.S.A.* **158 F10** 40 30N 107 16W
Haydon, *Australia* **126 B3** 18 0 S 141 30 E
Hayes, *U.S.A.* **154 C4** 44 23N 101 1W
Hayes →, *Canada* **143 B10** 57 3N 92 12W
Hayes, Mt., *U.S.A.* **144 E11** 63 37N 146 43W
Hayjān, *Yemen* **98 C4** 16 40N 44 5 E
Hayle, *U.K.* **21 G2** 50 11N 5 26W
Hayling I., *U.K.* **21 G7** 50 48N 0 59W
Haylow, *U.S.A.* **152 E7** 30 50N 82 54W
Haymā', *Oman* **99 C7** 19 56N 56 19 E
Haymana, *Turkey* **100 C5** 39 26N 32 31 E
Haynan, *Yemen* **99 D5** 15 50N 48 18 E
Haynesville, *U.S.A.* **155 J8** 32 58N 93 8W
Hayneville, *U.S.A.* **152 C6** 32 23N 83 37W
Hayrabolu, *Turkey* **51 E11** 41 12N 27 5 E
Ḩayrān, *Yemen* **98 C3** 16 8N 43 5 E
Hays, *Canada* **142 C6** 50 6N 111 48W
Hays, *U.S.A.* **154 F5** 38 53N 99 20W
Ḩays, *Yemen* **98 D3** 13 56N 43 29 E
Haysville, *U.S.A.* **157 F10** 38 28N 86 55W
Haysyn, *Ukraine* **59 H5** 48 57N 29 25 E
Hayvoron, *Ukraine* **59 H5** 48 22N 29 52 E
Hayward, *Calif., U.S.A.* **160 H4** 37 40N 122 5W
Hayward, *Wis., U.S.A.* **154 B9** 46 1N 91 29W
Haywards Heath, *U.K.* **21 G7** 51 0N 0 5W
Ḩayy, *Oman* **99 B7** 20 46N 58 18 E
Hazafon □, *Israel* **103 C4** 32 40N 35 20 E
Hazarām, Kūh-e, *Iran* **97 D8** 29 30N 57 18 E
Hazard, *U.S.A.* **148 G4** 37 15N 83 12W
Hazaribag, *India* **93 H11** 23 58N 85 26 E
Hazaribag Road, *India* **93 G11** 24 12N 85 57 E
Hazebrouck, *France* **27 B9** 50 42N 2 31 E
Hazelton, *Canada* **142 B3** 55 20N 127 42W
Hazelton, *U.S.A.* **154 B4** 46 29N 100 17W
Hazen, *N. Dak., U.S.A.* **154 B4** 47 18N 101 38W
Hazen, *Nev., U.S.A.* **158 G4** 39 34N 119 3W
Hazlehurst, *Ga., U.S.A.* **152 D7** 31 52N 82 36W
Hazlehurst, *Miss., U.S.A.* **155 K9** 31 52N 90 24W
Hazleton, *Ind., U.S.A.* **157 F9** 38 29N 87 33W
Hazleton, *Pa., U.S.A.* **151 F9** 40 57N 75 59W
Hazlett, L., *Australia* **124 D4** 21 30 S 128 48 E
He Xian, *Anhui, China* **77 B12** 31 45N 118 20 E
He Xian,
Guangxi Zhuangzu, China **77 E8** 24 27N 111 30 E
Head of Bight, *Australia* **125 F5** 31 30 S 131 25 E
Headland, *U.S.A.* **152 D4** 31 21N 85 21W
Headlands, *Zimbabwe* **119 F3** 18 15 S 32 2 E
Healdsburg, *U.S.A.* **160 G4** 38 37N 122 52W
Healdton, *U.S.A.* **155 H6** 34 14N 97 29W
Healesville, *Australia* **129 D6** 37 35 S 145 30 E
Healy, *U.S.A.* **144 E10** 63 52N 148 58W
Heard I., *Ind. Oc.* **121 K6** 53 0 S 74 0 E
Hearne, *U.S.A.* **155 K6** 30 53N 96 36W
Hearne B., *Canada* **143 A9** 60 10N 99 10W
Hearne L., *Canada* **142 A6** 62 20N 113 10W
Hearst, *Canada* **140 C3** 49 40N 83 41W
Heart →, *U.S.A.* **154 B4** 46 46N 100 50W
Heart's Content, *Canada* **141 C9** 47 54N 53 27W
Heath →, *Bolivia* **172 C4** 12 31 S 68 38W
Heath Mts., *N.Z.* **131 F2** 45 39 S 167 9 E
Heath Pt., *Canada* **141 C7** 49 8N 61 40W
Heath Steele, *Canada* **141 C6** 47 17N 66 5W
Heathcote, *Australia* **129 D6** 36 56 S 144 45 E
Heavener, *U.S.A.* **155 H7** 34 53N 94 36W
Hebbronville, *U.S.A.* **155 M5** 27 18N 98 41W
Hebei □, *China* **74 E9** 39 0N 116 0 E
Hebel, *Australia* **127 D4** 28 58 S 147 47 E
Heber, *U.S.A.* **161 N11** 32 44N 115 32W
Heber Springs, *U.S.A.* **155 H9** 35 30N 92 2W
Hebert, *Canada* **143 C7** 50 30N 107 10W
Hebgen L., *U.S.A.* **158 D8** 44 52N 111 20W
Hebi, *China* **74 G8** 35 57N 114 7 E
Hebrides, *U.K.* **12 D4** 57 30N 7 0W
Hebron = Al Khalīl,
West Bank **103 D4** 31 32N 35 6 E
Hebron, *Canada* **139 C13** 58 5N 62 30W
Hebron, *N. Dak., U.S.A.* **154 B3** 46 54N 102 3W
Hebron, *Nebr., U.S.A.* **154 E6** 40 10N 97 35W
Heby, *Sweden* **16 E10** 59 56N 16 53 E
Hecate Str., *Canada* **142 C2** 53 10N 130 30W
Hechi, *China* **76 E7** 24 40N 108 2 E
Hechingen, *Germany* **31 G4** 48 21N 8 57 E
Hechuan, *China* **76 B6** 30 2N 106 12 E
Hecla, *U.S.A.* **154 C5** 45 53N 98 9W
Hecla I., *Canada* **143 C9** 51 10N 96 43W
Hedal, *Norway* **18 D6** 60 37N 9 41 E
Heddal, *Norway* **18 E6** 59 36N 9 9 E
Hédé, *France* **26 D5** 48 18N 1 49W
Hede, *Sweden* **16 B7** 62 23N 13 30 E
Hedemora, *Sweden* **16 D9** 60 18N 15 58 E
Hedensted, *Denmark* **17 J3** 55 46N 9 42 E
Hedesunda, *Sweden* **16 D10** 60 24N 17 0 E
Hedgehope, *N.Z.* **131 G3** 46 12 S 168 34 E
Hedley, *U.S.A.* **155 H4** 34 52N 100 39W
Hedmark □, *Norway* **18 C8** 61 17N 11 40 E
Hedrick, *U.S.A.* **156 C4** 41 11N 92 19W
Heerde, *Neths.* **24 B6** 52 24N 6 2 E
Heerenveen, *Neths.* **24 B5** 52 57N 5 55 E
Heerhugowaard, *Neths.* **24 B4** 52 40N 4 51 E
Heerlen, *Neths.* **24 D5** 50 55N 5 58 E
Ḩefa, *Israel* **103 C4** 32 46N 35 0 E
Ḩefa □, *Israel* **103 C4** 32 40N 35 0 E
Hefei, *China* **77 B11** 31 52N 117 18 E
Heflin, *U.S.A.* **152 B4** 33 39N 85 35W
Hegang, *China* **73 B8** 47 20N 130 19 E
Heggenes, *Norway* **18 C6** 61 9N 9 4 E
Hegra, *Norway* **18 A8** 63 27N 11 8 E
Heichengzhen, *China* **74 F4** 36 24N 106 3 E
Heidal, *Norway* **18 C6** 61 45N 9 19 E
Heide, *Germany* **30 A5** 54 11N 9 6 E
Heidelberg, *Germany* **31 F4** 49 24N 8 42 E
Heidelberg, *S. Africa* **116 E3** 34 6 S 20 59 E
Heidenau, *Germany* **30 E9** 50 57N 13 52 E
Heidenheim, *Germany* **31 G6** 48 41N 10 9 E
Heigun-To, *Japan* **70 D4** 33 47N 132 14 E
Heijing, *China* **76 E3** 25 22N 101 44 E
Heilbad Heiligenstadt,
Germany **30 D6** 51 22N 10 8 E
Heilbron, *S. Africa* **117 D4** 27 16 S 27 59 E
Heilbronn, *Germany* **31 F5** 49 9N 9 13 E
Heiligenblut, *Austria* **34 D5** 47 2N 12 51 E
Heiligenhafen, *Germany* **30 A6** 54 22N 10 59 E
Heilongjiang □, *China* **75 A14** 48 0N 126 0 E
Heilprin Land, *Greenland* **10 A7** 82 5N 33 0W
Heilunkiang =
Heilongjiang □, *China* **75 A14** 48 0N 126 0 E
Heim, *Norway* **18 A6** 63 26N 9 5 E
Heimaey, *Iceland* **11 D6** 63 26N 20 17W
Heimdal, *Norway* **18 A7** 63 21N 10 22 E
Heinola, *Finland* **15 F22** 61 13N 26 2 E
Heinsberg, *Germany* **30 D2** 51 3N 6 5 E
Heinsun, *Burma* **90 C5** 25 52N 95 35 E
Heirnkut, *Burma* **90 C5** 25 14N 94 44 E
Heishan, *China* **75 D12** 41 40N 122 5 E
Heishui, *Liaoning, China* **75 C10** 42 8N 119 30 E
Heishui, *Sichuan, China* **76 A4** 32 4N 103 2 E
Hejaz = Ḩijāz □, *Si. Arabia* **98 A2** 24 0N 40 0 E
Hejian, *China* **74 E9** 38 25N 116 5 E
Hejiang, *China* **76 C5** 28 43N 105 46 E
Hejin, *China* **74 G6** 35 35N 110 42 E
Hekımhan, *Turkey* **100 C7** 38 50N 37 55 E
Hekinan, *Japan* **71 C9** 34 52N 137 0 E
Hekla, *Iceland* **11 D7** 63 56N 19 35W
Hekou, *Gansu, China* **74 F2** 36 10N 103 28 E
Hekou, *Guangdong, China* **77 F9** 23 13N 112 45 E
Hekou, *Yunnan, China* **72 D5** 22 30N 103 59 E
Hel, *Poland* **54 D5** 54 37N 18 47 E
Helagsfjället, *Sweden* **16 B6** 62 54N 12 25 E
Helan Shan, *China* **74 E3** 38 30N 105 55 E
Helechosa, *Spain* **43 F6** 39 22N 4 53W
Helemano →, *U.S.A.* **145 J13** 21 35N 158 7W
Helena, *Ark., U.S.A.* **155 H9** 34 32N 90 36W
Helena, *Ga., U.S.A.* **152 C7** 32 5N 82 55W
Helena, *Mont., U.S.A.* **158 C7** 46 36N 112 2W
Helendale, *U.S.A.* **161 L9** 34 44N 117 19W
Helensburgh, *Australia* **129 C9** 34 11 S 151 1 E
Helensburgh, *U.K.* **22 E4** 56 1N 4 43W
Helensville, *N.Z.* **130 C3** 36 41 S 174 29 E
Helgasjön, *Sweden* **17 H8** 56 55N 14 50 E
Helgeland, *Norway* **14 C15** 66 7N 13 29 E
Helgoland, *Germany* **30 A3** 54 10N 7 53 E
Heligoland = Helgoland,
Germany **30 A3** 54 10N 7 53 E
Heligoland B. = Deutsche
Bucht, *Germany* **30 A4** 54 15N 8 0 E
Heliopolis, *Egypt* **106 H7** 30 6N 31 17 E
Hell, *Norway* **18 A7** 63 26N 10 54 E
Hella, *Iceland* **11 D6** 63 50N 20 24W
Helleland, *Norway* **18 F3** 58 33N 6 7 E
Hellesylt, *Norway* **18 B3** 62 6N 6 51 E
Hellevoetsluis, *Neths.* **24 C4** 51 50N 4 8 E
Hellín, *Spain* **41 G3** 38 31N 1 40W
Hellissandur, *Iceland* **11 C3** 64 55N 23 54W
Hellnar, *Iceland* **11 C3** 64 45N 23 39W
Hellvik, *Norway* **18 F2** 58 29N 5 52 E
Helmand □, *Afghan.* **91 C2** 31 20N 64 0 E
Helmand →, *Afghan.* **91 C1** 31 12N 61 34 E
Helme →, *Germany* **30 D7** 51 20N 11 21 E
Helmond, *Neths.* **24 C5** 51 29N 5 41 E
Helmsdale, *U.K.* **22 C5** 58 7N 3 39W
Helmsdale →, *U.K.* **22 C5** 58 7N 3 40W
Helmstedt, *Germany* **30 C7** 52 12N 11 0 E
Helong, *China* **75 C15** 42 40N 129 0 E
Helper, *U.S.A.* **158 G8** 39 41N 110 51W
Helsingborg, *Sweden* **17 H6** 56 3N 12 42 E
Helsinge, *Denmark* **17 H6** 56 2N 12 12 E
Helsingfors = Helsinki,
Finland **15 F21** 60 15N 25 3 E
Helsingør, *Denmark* **17 H6** 56 2N 12 35 E
Helsinki, *Finland* **15 F21** 60 15N 25 3 E
Helska, Mierzeja, *Poland* **54 D5** 54 45N 18 40 E
Helston, *U.K.* **21 G2** 50 6N 5 17W
Helvellyn, *U.K.* **20 C4** 54 32N 3 1W
Helwân, *Egypt* **106 J7** 29 50N 31 20 E
Hemavati →, *India* **95 H3** 12 30N 76 20 E
Hemel Hempstead, *U.K.* **21 F7** 51 44N 0 28W
Hemet, *U.S.A.* **161 M10** 33 45N 116 58W
Hemingford, *U.S.A.* **154 D3** 42 19N 103 4W
Hemphill, *U.S.A.* **155 K8** 31 20N 93 51W
Hempstead, *U.S.A.* **155 K6** 30 6N 96 5W
Hemse, *Sweden* **17 G12** 57 15N 18 22 E
Hemsedal, *Norway* **18 D5** 60 53N 8 30 E
Hemsön, *Sweden* **16 B12** 62 42N 18 5 E
Hen, *Norway* **18 D7** 60 13N 10 14 E
Hen & Chickens Is., *N.Z.* **130 B3** 35 58 S 174 45 E
Henån, *Sweden* **17 F5** 58 14N 11 40 E
Henan □, *China* **74 H8** 34 0N 114 0 E
Henares →, *Spain* **42 E7** 40 24N 3 30W
Henashi-Misaki, *Japan* **68 D9** 40 37N 139 51 E
Hendaye, *France* **28 E2** 43 23N 1 47W
Hendek, *Turkey* **100 B4** 40 48N 30 44 E
Henderson, *Argentina* **174 D3** 36 18 S 61 43W
Henderson, *Ga., U.S.A.* **152 C6** 32 21N 83 47W
Henderson, *Ky., U.S.A.* **157 G9** 37 50N 87 35W
Henderson, *N.C., U.S.A.* **149 G6** 36 20N 78 25W
Henderson, *Nev., U.S.A.* **161 J12** 36 2N 114 59W
Henderson, *Tenn., U.S.A.* **149 H1** 35 26N 88 38W
Henderson, *Tex., U.S.A.* **155 J7** 32 9N 94 48W
Hendersonville, *N.C.,*
U.S.A. **149 H4** 35 19N 82 28W
Hendersonville, *S.C., U.S.A.* **152 C9** 32 48N 80 43W
Hendījān, *Iran* **97 D6** 30 14N 49 43 E
Hendon, *Australia* **127 D5** 28 5 S 151 50 E
Heng Xian, *China* **76 F7** 22 40N 109 17 E
Hengcheng, *China* **74 E4** 38 18N 106 28 E
Hengdaohezi, *China* **75 B15** 44 52N 129 0 E
Hengelo, *Neths.* **24 B6** 52 16N 6 48 E
Hengfeng, *China* **77 C10** 28 12N 115 48 E
Hengshan, *Hunan, China* **77 D9** 27 16N 112 45 E
Hengshan, *Shaanxi, China* **74 F5** 37 58N 109 5 E
Hengshui, *China* **74 F8** 37 41N 115 40 E
Hengyang, *Hunan, China* **77 D9** 26 52N 112 33 E
Hengyang, *Hunan, China* **77 D9** 26 59N 112 22 E
Henichesk, *Ukraine* **59 J8** 46 12N 34 50 E
Hénin-Beaumont, *France* **27 B9** 50 25N 2 58 E
Henlopen, C., *U.S.A.* **148 F8** 38 48N 75 6W
Hennan, *Sweden* **16 B9** 62 2N 15 54 E
Hennebont, *France* **26 E3** 47 49N 3 19W
Hennenman, *S. Africa* **116 D4** 27 59 S 27 1 E
Hennepin, *U.S.A.* **156 C7** 41 15N 89 21W
Hennessey, *U.S.A.* **155 G6** 36 6N 97 54W
Hennigsdorf, *Germany* **30 C9** 52 38N 13 12 E
Henrietta, *U.S.A.* **155 J5** 33 49N 98 12W
Henrietta, Ostrov =
Genriyetty, Ostrov, *Russia* **65 B16** 77 6N 156 30 E
Henrietta Maria, C., *Canada* **140 A3** 55 9N 82 20W
Henry, *U.S.A.* **156 C7** 41 7N 89 22W
Henryetta, *U.S.A.* **155 H7** 35 27N 95 59W
Hensall, *Canada* **150 C3** 43 26N 81 30W
Henstedt-Ulzburg, *Germany* **30 B6** 53 47N 10 0 E
Hentiyn Nuruu, *Mongolia* **73 B5** 48 30N 108 30 E
Henty, *Australia* **127 F4** 35 30 S 147 0 E
Henzada, *Burma* **90 G5** 17 38N 95 26 E
Hephaestia, *Greece* **49 B7** 39 55N 25 14 E
Hephzibah, *U.S.A.* **152 B7** 33 19N 82 6W
Heping, *China* **77 E10** 24 29N 115 0 E
Heppner, *U.S.A.* **158 D4** 45 21N 119 33W
Hepu, *China* **76 G7** 21 40N 109 12 E
Hepworth, *Canada* **150 B3** 44 37N 81 9W
Heqing, *China* **76 D3** 26 37N 100 11 E
Hequ, *China* **74 E6** 39 20N 111 15 E
Héraðsflói, *Iceland* **11 B12** 65 42N 14 12W
Héraðsvötn →, *Iceland* **11 B7** 65 45N 19 25W
Heradsbygd, *Norway* **18 D8** 60 49N 11 39 E
Herald Cays, *Australia* **126 B4** 16 58 S 149 9 E
Herand, *Norway* **18 D3** 60 20N 6 22 E
Herāt, *Afghan.* **91 B1** 34 20N 62 7 E
Herāt □, *Afghan.* **91 B1** 35 0N 62 0 E
Hérault □, *France* **28 E7** 43 34N 3 15 E
Hérault →, *France* **28 E7** 43 17N 3 26 E
Herbault, *France* **26 E8** 47 36N 1 8 E
Herbert →, *Australia* **126 B4** 18 31 S 146 17 E
Herbert Downs, *Australia* **126 C2** 23 7 S 139 9 E
Herberton, *Australia* **126 B4** 17 20 S 145 25 E
Herbertsville, *N.Z.* **130 G5** 40 30 S 176 33 E
Herbignac, *France* **26 E4** 47 27N 2 18W
Herborn, *Germany* **30 E4** 50 40N 8 19 E
Herby, *Poland* **55 H5** 50 45N 18 50 E
Herceg-Novi,
Montenegro, Yug. **50 D2** 42 30N 18 33 E
Herculaneum, *U.S.A.* **156 F6** 38 16N 90 23W
Herðubreið, *Iceland* **11 B10** 65 11N 16 21W
Hereford, *U.K.* **21 E5** 52 4N 2 43W
Hereford, *U.S.A.* **155 H3** 34 49N 102 24W
Hereford and Worcester □,
U.K. **21 E5** 52 10N 2 30W
Herefordshire □, *U.K.* **21 E5** 52 8N 2 40W
Herefoss, *Norway* **18 F5** 58 32N 8 23 E
Herehogna, *Sweden* **18 C9** 61 44N 12 8 E
Hereke, *Turkey* **51 F13** 40 47N 29 38 E
Herekino, *N.Z.* **130 B2** 35 18 S 173 11 E
Herencia, *Spain* **43 F7** 39 21N 3 22W
Herentals, *Belgium* **24 C4** 51 12N 4 51 E
Herford, *Germany* **30 C4** 52 7N 8 39 E
Héricourt, *France* **27 E13** 47 32N 6 45 E
Herington, *U.S.A.* **154 F6** 38 40N 96 57W
Herisau, *Switz.* **33 B8** 47 22N 9 17 E
Hérisson, *France* **27 F9** 46 32N 2 42 E
Herkimer, *U.S.A.* **151 D10** 43 0N 74 59W
Herlong, *U.S.A.* **160 E6** 40 8N 120 8W
Herm, *U.K.* **21 H5** 49 30N 2 28W
Herman, *U.S.A.* **154 C6** 45 49N 96 9W
Hermann, *U.S.A.* **154 F9** 38 42N 91 27W
Hermannsburg, *Germany* **30 C6** 52 50N 10 5 E
Hermannsburg Mission,
Australia **124 D5** 23 57 S 132 45 E
Hermansverk, *Norway* **18 C3** 61 11N 6 52 E
Hermanus, *S. Africa* **116 E2** 34 27 S 19 12 E
Herment, *France* **28 C6** 45 45N 2 24 E
Hermidale, *Australia* **129 A7** 31 30 S 146 42 E
Hermiston, *U.S.A.* **158 D4** 45 51N 119 17W
Hermitage, *N.Z.* **131 D5** 43 44 S 170 5 E
Hermitage, *U.S.A.* **156 G3** 37 56N 93 19W
Hermite, I., *Chile* **176 E3** 55 50 S 68 0W
Hermon, Mt. = Ash Shaykh,
J., *Lebanon* **103 B4** 33 25N 35 50 E
Hermosillo, *Mexico* **162 B2** 29 10N 111 0W
Hernád →, *Hungary* **52 C6** 47 56N 21 8 E
Hernandarias, *Paraguay* **175 B5** 25 20 S 54 40W
Hernandez, *U.S.A.* **160 J6** 36 24N 120 46W
Hernando, *Argentina* **174 C3** 32 28 S 63 40W
Hernando, *Fla., U.S.A.* **153 G7** 28 54N 82 23W
Hernando, *Miss., U.S.A.* **155 H10** 34 50N 90 0W
Hernani, *Spain* **40 B3** 43 16N 1 58W
Herne, *Germany* **30 D3** 51 32N 7 14 E
Herne Bay, *U.K.* **21 F9** 51 21N 1 8 E
Herning, *Denmark* **17 H2** 56 8N 8 58 E
Herod, *U.S.A.* **152 D5** 31 42N 84 26W
Heroica = Caborca, *Mexico* **162 A2** 30 40N 112 10W
Heroica Nogales = Nogales,
Mexico **162 A2** 31 20N 110 56W
Heron Bay, *Canada* **140 C2** 48 40N 86 25W
Herradura, Pta. de la,
Canary Is. **39 F5** 28 26N 14 8W
Herre, *Norway* **18 E6** 59 6N 9 34 E
Herreid, *U.S.A.* **154 C4** 45 50N 100 4W
Herrenberg, *Germany* **31 G4** 48 35N 8 52 E
Herrera, *Spain* **43 H6** 37 26N 4 55W
Herrera de Alcántara, *Spain* **43 F3** 39 39N 7 25W
Herrera de Pisuerga, *Spain* **42 C6** 42 35N 4 20W
Herrera del Duque, *Spain* **43 F5** 39 10N 5 3W
Herrestad, *Sweden* **17 F5** 58 21N 11 50 E
Herrick, *Australia* **126 G4** 41 5 S 147 55 E
Herrin, *U.S.A.* **155 G10** 37 48N 89 2W
Herrljunga, *Sweden* **17 F7** 58 5N 13 1 E
Hersbruck, *Germany* **31 F7** 49 30N 11 26 E
Hersonissos, *Greece* **38 D7** 35 18N 25 22 E
Herstal, *Belgium* **24 D5** 50 40N 5 38 E
Hertford, *U.K.* **21 F7** 51 48N 0 4W
Hertfordshire □, *U.K.* **21 F7** 51 51N 0 5W
's-Hertogenbosch, *Neths.* **24 C5** 51 42N 5 17 E
Hertzogville, *S. Africa* **116 D4** 28 9 S 25 30 E
Hervás, *Spain* **42 E5** 40 16N 5 52W
Herzberg, *Brandenburg,*
Germany **30 D9** 51 41N 13 14 E
Herzberg, *Niedersachsen,*
Germany **30 D6** 51 38N 10 20 E
Herzliyya, *Israel* **103 C3** 32 10N 34 50 E
Herzogenbuchsee, *Switz.* **32 B5** 47 11N 7 42 E
Herzogenburg, *Austria* **34 C8** 48 17N 15 41 E
Ḩeşār, *Fārs, Iran* **97 D6** 29 52N 50 16 E
Ḩeşār, *Markazī, Iran* **97 C6** 35 50N 49 12 E
Hesdin, *France* **27 B9** 50 21N 2 2 E
Heshui, *China* **74 G5** 36 0N 108 0 E
Heshun, *China* **74 F7** 37 22N 113 32 E
Heskestad, *Norway* **18 F3** 58 28N 6 22 E
Hesperia, *U.S.A.* **161 L9** 34 25N 117 18W
Hessdalen, *Norway* **18 B8** 62 48N 11 10 E
Hesse = Hessen □, *Germany* **30 E4** 50 30N 9 0 E
Hessen □, *Germany* **30 E4** 50 30N 9 0 E
Hesteyri, *Iceland* **11 A4** 66 20N 22 53W
Hestra, *Sweden* **17 G7** 57 26N 13 35 E
Hetch Hetchy Aqueduct,
U.S.A. **160 H5** 37 29N 122 19W
Hettinger, *U.S.A.* **154 C3** 46 0N 102 42W
Hettstedt, *Germany* **30 D7** 51 39N 11 31 E
Heves, *Hungary* **52 C5** 47 36N 20 17 E
Heves □, *Hungary* **52 C5** 47 50N 20 0 E
Hexham, *U.K.* **20 C5** 54 58N 2 4W
Hexi, *Yunnan, China* **76 E4** 24 9N 102 38 E
Hexi, *Zhejiang, China* **77 D12** 27 58N 119 38 E
Hexigten Qi, *China* **75 C9** 43 18N 117 30 E
Ḩeydarābād, *Iran* **97 D7** 30 33N 55 38 E
Heyfield, *Australia* **129 D7** 37 59 S 146 47 E
Heysham, *U.K.* **20 C5** 54 3N 2 53W
Heyuan, *China* **77 F10** 23 39N 114 40 E
Heywood, *Australia* **128 E4** 38 8 S 141 37 E
Heze, *China* **74 G8** 35 14N 115 20 E
Hezhang, *China* **76 D5** 27 8N 104 41 E
Hi-no-Misaki, *Japan* **70 B4** 35 26N 132 38 E
Hi Vista, *U.S.A.* **161 L9** 34 45N 117 46W
Hialeah, *U.S.A.* **149 N5** 25 50N 80 17W
Hiawatha, *Kans., U.S.A.* **154 F7** 39 51N 95 32W
Hiawatha, *Utah, U.S.A.* **158 G8** 39 29N 111 1W
Hibbing, *U.S.A.* **154 B8** 47 25N 92 56W
Hibbs B., *Australia* **126 G4** 42 35 S 145 15 E
Hibernia Reef, *Australia* **124 B3** 12 0 S 123 23 E
Hibiki-Nada, *Japan* **70 D2** 34 0N 130 0 E
Hickory, *U.S.A.* **149 H5** 35 44N 81 21W
Hicks, Pt., *Australia* **129 D8** 37 49 S 149 17 E
Hicks Bay, *N.Z.* **130 D7** 37 34 S 178 21 E
Hicksville, *N.Y., U.S.A.* **151 F11** 40 46N 73 32W
Hicksville, *Ohio, U.S.A.* **157 C12** 41 18N 84 46W
Hida, *Romania* **53 C8** 47 10N 23 19 E
Hida-Gawa →, *Japan* **71 B9** 35 26N 137 3 E
Hida-Sammyaku, *Japan* **71 A9** 36 30N 137 40 E
Hida-Sanchi, *Japan* **71 A9** 36 10N 137 0 E
Hidaka, *Japan* **70 B6** 35 30N 134 44 E
Hidaka-Sammyaku, *Japan* **68 C11** 42 35N 142 45 E
Hidalgo, *Mexico* **163 C5** 24 15N 99 26W
Hidalgo, *U.S.A.* **157 E8** 39 9N 88 9W
Hidalgo □, *Mexico* **163 C5** 20 30N 99 10W
Hidalgo, Presa M., *Mexico* **162 B3** 26 30N 108 35W
Hidalgo, Pta. del, *Canary Is.* **39 F3** 28 33N 16 19W
Hidalgo del Parral, *Mexico* **162 B3** 26 58N 105 40W
Hiddensee, *Germany* **30 A9** 54 32N 13 6 E
Hidrolândia, *Brazil* **171 E2** 17 0 S 49 15W
Hieflau, *Austria* **34 D7** 47 36N 14 46 E
Hiendelaencina, *Spain* **40 D2** 41 5N 3 0W
Hienghène, *N. Cal.* **133 T18** 20 41 S 164 56 E
Hierro, *Canary Is.* **39 G1** 27 44N 18 0W
Higashi-matsuyama, *Japan* **71 A11** 36 2N 139 25 E
Higashiajima-San, *Japan* **68 F10** 37 40N 140 10 E
Higashiōsaka, *Japan* **71 C7** 34 40N 135 37 E
Higasi-Suidō, *Japan* **70 D1** 34 0N 129 30 E
Higbee, *U.S.A.* **156 E4** 39 19N 92 31W
Higgins, *U.S.A.* **155 G4** 36 7N 100 2W
Higgins Corner, *U.S.A.* **160 F5** 39 2N 121 5W
Higginsville, *Australia* **125 F3** 31 42 S 121 38 E
Higginsville, *U.S.A.* **156 E3** 39 4N 93 43W
High Atlas = Haut Atlas,
Morocco **110 B4** 32 30N 5 0W
High I., *Canada* **141 A7** 56 40N 61 10W
High Island, *U.S.A.* **155 L7** 29 34N 94 24W
High Level, *Canada* **142 B5** 58 31N 117 8W
High Point, *U.S.A.* **149 H6** 35 57N 80 0W
High Prairie, *Canada* **142 B5** 55 30N 116 30W
High River, *Canada* **142 C6** 50 30N 113 50W
High Springs, *U.S.A.* **152 F7** 29 50N 82 36W
High Tatra = Tatry,
Slovak Rep. **35 B13** 49 20N 20 0 E
High Veld, *Africa* **104 J6** 27 0 S 27 0 E
High Wycombe, *U.K.* **21 F7** 51 37N 0 45W
Highbank, *N.Z.* **131 D6** 43 37 S 171 45 E
Highbury, *Australia* **126 B3** 16 25 S 143 9 E
Highland, *Ill., U.S.A.* **156 F7** 38 44N 89 41W
Highland, *Ind., U.S.A.* **157 C9** 41 33N 87 28W
Highland, *Wis., U.S.A.* **156 A6** 43 5N 90 22W
Highland □, *U.K.* **22 D4** 57 17N 4 21W
Highland City, *U.S.A.* **153 H8** 27 58N 81 53W
Highland Home, *U.S.A.* **152 D3** 31 57N 86 19W
Highland Mills, *U.S.A.* **152 B5** 33 17N 84 17W
Highland Park, *U.S.A.* **157 B9** 42 11N 87 48W
Highland View, *U.S.A.* **152 F4** 29 50N 85 19W
Highmore, *U.S.A.* **154 C5** 44 31N 99 27W
Highrock L., *Canada* **143 B7** 57 5N 105 32W
Higüey, *Dom. Rep.* **165 C6** 18 37N 68 42W
Hihya, *Egypt* **106 H7** 30 40N 31 36 E
Hiiumaa, *Estonia* **15 G20** 58 50N 22 45 E
Híjar, *Spain* **40 D4** 41 10N 0 27W
Ḩijāz □, *Si. Arabia* **98 A2** 24 0N 40 0 E
Hijāz, Jabal al, *Si. Arabia* **98 C3** 19 45N 41 55 E
Hiji, *Japan* **70 D3** 33 22N 131 32 E
Hijo = Tagum, *Phil.* **81 H5** 7 33N 125 53 E
Hikari, *Japan* **70 D3** 33 58N 131 58 E
Hiketa, *Japan* **70 C6** 34 13N 134 24 E
Hiko, *U.S.A.* **160 H11** 37 32N 115 14W
Hikone, *Japan* **71 B8** 35 15N 136 10 E
Hikurangi, *N.Z.* **130 B3** 35 36 S 174 17 E
Hikurangi, Mt., *N.Z.* **130 E5** 38 21 S 176 52 E
Hiland Park, *U.S.A.* **152 E4** 30 12N 85 33W
Hilawng, *Burma* **90 E4** 21 23N 93 48 E
Hilda, *U.S.A.* **152 B9** 33 16N 81 15W
Hildburghausen, *Germany* **30 E6** 50 25N 10 42 E
Hildesheim, *Germany* **30 C5** 52 9N 9 56 E
Hill →, *Australia* **125 F2** 30 23 S 115 3 E
Hill City, *Idaho, U.S.A.* **158 E6** 43 18N 115 3W
Hill City, *Kans., U.S.A.* **154 F5** 39 22N 99 51W
Hill City, *Minn., U.S.A.* **154 B8** 46 59N 93 36W
Hill City, *S. Dak., U.S.A.* **154 D3** 43 56N 103 35W
Hill End, *Australia* **129 E7** 38 1 S 146 9 E
Hill Island L., *Canada* **143 A7** 60 30N 109 50W
Hillared, *Sweden* **17 G7** 57 37N 13 10 E
Hillcrest Center, *U.S.A.* **161 K8** 35 23N 118 57W
Hillegom, *Neths.* **24 B4** 52 18N 4 35 E
Hillerød, *Denmark* **17 J6** 55 56N 12 19 E
Hillerstorp, *Sweden* **17 G7** 57 20N 13 52 E

Ituiutaba 227

Ingul → = Inhul →, *Ukraine* **59 J7** 46 50N 32 0 E
Ingulec = Inhulec, *Ukraine* **59 J7** 47 42N 33 14 E
Ingulets → = Inhulets →, *Ukraine* **59 J7** 46 46N 32 47 E
Inguri → = Enguri →, *Georgia* **61 J5** 42 27N 41 38 E
Ingushetia □, *Russia* **61 J7** 43 20N 44 50 E
Ingwavuma, *S. Africa* **117 D5** 27 9 S 31 59 E
Inhaca, I., *Mozam.* **117 D5** 26 1 S 32 57 E
Inhafenga, *Mozam.* **117 C5** 20 36 S 33 53 E
Inhambane, *Mozam.* **117 C6** 23 54 S 35 30 E
Inhambane □, *Mozam.* **117 C5** 22 30 S 34 20 E
Inhambupe, *Brazil* **171 D4** 11 47 S 38 21W
Inhaminga, *Mozam.* **119 F4** 18 26 S 35 0 E
Inharrime, *Mozam.* **117 C6** 24 30 S 35 0 E
Inharrime →, *Mozam.* **117 C6** 24 30 S 35 0 E
Inhisar, *Turkey* **49 A12** 40 3N 30 23 E
Inhul →, *Ukraine* **59 J7** 46 50N 32 0 E
Inhulec, *Ukraine* **59 J7** 47 42N 33 14 E
Inhulets →, *Ukraine* **59 J7** 46 46N 32 47 E
Inhuma, *Brazil* **170 C3** 6 40 S 41 42W
Inhumas, *Brazil* **171 E2** 16 22 S 49 30W
Iniesta, *Spain* **41 F3** 39 27N 1 45W
Ining = Yining, *China* **64 E9** 43 58N 81 10 E
Inini □, *Fr. Guiana* **169 C7** 4 0N 53 0W
Inírida →, *Colombia* **168 C4** 3 55N 67 52W
Inishbofin, *Ireland* **23 C1** 53 37N 10 13W
Inisheer, *Ireland* **23 C2** 53 3N 9 32W
Inishfree B., *Ireland* **23 A3** 55 4N 8 23W
Inishkea North, *Ireland* **23 B1** 54 9N 10 11W
Inishkea South, *Ireland* **23 B1** 54 7N 10 12W
Inishmaan, *Ireland* **23 C2** 53 5N 9 35W
Inishmore, *Ireland* **23 C2** 53 8N 9 45W
Inishowen Pen., *Ireland* **23 A4** 55 14N 7 15W
Inishshark, *Ireland* **23 C1** 53 37N 10 16W
Inishturk, *Ireland* **23 C1** 53 42N 10 7W
Inishvickillane, *Ireland* **23 D1** 52 3N 10 37W
Injune, *Australia* **127 D4** 25 53 S 148 32 E
Inklin, *Canada* **142 B2** 58 56N 133 5W
Inklin →, *Canada* **142 B2** 58 50N 133 10W
Inkom, *U.S.A.* **158 E7** 42 48N 112 15W
Inle L., *Burma* **90 E6** 20 30N 96 58 E
Inn →, *Austria* **34 C6** 48 35N 13 28 E
Innamincka, *Australia* **127 D3** 27 44 S 140 46 E
Innbygda, *Norway* **18 C9** 61 19N 12 17 E
Inner Hebrides, *U.K.* **22 E2** 57 0N 6 30W
Inner Mongolia = Nei Monggol Zizhiqu □, *China* **74 D7** 42 0N 112 0 E
Inner Sound, *U.K.* **22 D3** 57 30N 5 55W
Innerkip, *Canada* **150 C4** 43 13N 80 42W
Innerkirchen, *Switz.* **32 C6** 46 43N 8 14 E
Innetalling I., *Canada* **140 A4** 56 0N 79 0W
Innisfail, *Australia* **126 B4** 17 33 S 146 5 E
Innisfail, *Canada* **142 C6** 52 0N 113 57W
In'no-shima, *Japan* **70 C5** 34 19N 133 10 E
Innsbruck, *Austria* **34 D4** 47 16N 11 23 E
Innviertel, *Austria* **34 C6** 48 15N 13 15 E
Innvik, *Norway* **18 C3** 61 51N 6 37 E
Inny →, *Ireland* **23 C4** 53 30N 7 50W
Ino, *Japan* **70 D5** 33 33N 133 26 E
Inocência, *Brazil* **171 E1** 19 47 S 51 48W
Inongo, *Dem. Rep. of the Congo* **114 C3** 1 55 S 18 30 E
Inoni, *Congo* **114 C3** 3 4 S 15 39 E
İnönü, *Turkey* **49 B12** 39 48N 30 9 E
Inoucdjouac = Inukjuak, *Canada* **139 C12** 58 25N 78 15W
Inowrocław, *Poland* **55 F5** 52 50N 18 12 E
Inpundong, *N. Korea* **75 D14** 41 25N 126 34 E
Inquisivi, *Bolivia* **172 D4** 16 50 S 67 10W
Ins, *Switz.* **32 B4** 47 1N 7 7 E
Inscription, C., *Australia* **125 E1** 25 29 S 112 59 E
Insein, *Burma* **90 G6** 16 50N 96 5 E
Insjön, *Sweden* **16 D9** 60 41N 15 6 E
Ińsko, *Poland* **54 E2** 53 25N 15 32 E
Însurăţei, *Romania* **53 F12** 44 50N 27 40 E
Inta, *Russia* **56 A11** 66 5N 60 8 E
Intendente Alvear, *Argentina* **174 D3** 35 12 S 63 32W
İntepe, *Turkey* **49 A8** 40 1N 26 20 E
Intercession City, *U.S.A.* **153 G8** 28 16N 81 31W
Interior, *U.S.A.* **154 D4** 43 44N 101 59W
Interlachen, *U.S.A.* **153 F8** 29 37N 81 53W
Interlaken, *Switz.* **32 C5** 46 41N 7 50 E
International Falls, *U.S.A.* **154 A8** 48 36N 93 25W
Intiyaco, *Argentina* **174 B3** 28 43 S 60 5W
Întorsura Buzăului, *Romania* **53 E11** 45 41N 26 2 E
Intragna, *Switz.* **33 D7** 46 11N 8 42 E
Intutu, *Peru* **168 D3** 3 32 S 74 48W
Inubō-Zaki, *Japan* **71 B12** 35 42N 140 52 E
Inukjuak, *Canada* **139 C12** 58 25N 78 15W
Inútil, B., *Chile* **176 D2** 53 30 S 70 15W
Inuvik, *Canada* **138 B6** 68 16N 133 40W
Inuyama, *Japan* **71 B8** 35 23N 136 56 E
Inveraray, *U.K.* **22 E3** 56 14N 5 5W
Inverbervie, *U.K.* **22 E6** 56 51N 2 17W
Invercargill, *N.Z.* **131 G3** 46 24 S 168 24 E
Inverclyde □, *U.K.* **22 F4** 55 55N 4 49W
Inverell, *Australia* **127 D5** 29 45 S 151 8 E
Invergordon, *U.K.* **22 D4** 57 41N 4 10W
Inverleigh, *Australia* **128 E6** 38 6 S 144 3 E
Invermere, *Canada* **142 C5** 50 30N 116 2W
Inverness, *Canada* **141 C7** 46 15N 61 19W
Inverness, *U.K.* **22 D4** 57 29N 4 13W
Inverness, *Ala., U.S.A.* **152 C4** 32 1N 85 45W
Inverness, *Fla., U.S.A.* **149 L4** 28 50N 82 20W
Inverurie, *U.K.* **22 D6** 57 17N 2 23W
Inverway, *Australia* **124 C4** 17 50 S 129 38 E
Investigator Group, *Australia* **127 E1** 34 45 S 134 20 E
Investigator Str., *Australia* **128 C2** 35 30 S 137 0 E
Inya, *Russia* **64 D9** 50 28N 86 37 E
Inyanga, *Zimbabwe* **119 F3** 18 12 S 32 40 E
Inyangani, *Zimbabwe* **119 F3** 18 5 S 32 50 E
Inyantue, *Zimbabwe* **119 F2** 18 30 S 26 40 E
Inyo Mts., *U.S.A.* **159 H5** 36 40N 118 0W
Inyokern, *U.S.A.* **161 K9** 35 39N 117 49W
Inywa, *Burma* **90 D6** 23 56N 96 17 E
Inza, *Russia* **60 D8** 53 55N 46 25 E
Inzer, *Russia* **62 D6** 54 14N 57 34 E
Inzhavino, *Russia* **60 D6** 52 22N 42 30 E
Iō-Jima, *Japan* **69 J5** 30 48N 130 18 E
Ioánnina, *Greece* **48 B2** 39 42N 20 47 E
Ioánnina □, *Greece* **48 B2** 39 39N 20 57 E
Iola, *U.S.A.* **155 G7** 37 55N 95 24W
Ioma, *Papua N. G.* **132 E4** 8 19 S 147 52 E
Ion Corvin, *Romania* **53 F12** 44 7N 27 50 E
Iona, *U.K.* **22 E2** 56 20N 6 25W
Ione, *Calif., U.S.A.* **160 G6** 38 21N 120 56W
Ione, *Wash., U.S.A.* **158 B5** 48 45N 117 25W
Ionia, *U.S.A.* **157 B11** 42 59N 85 4W
Ionian Sea, *Medit. S.* **12 H9** 37 30N 17 30 E
Iónioi Nísoi □, *Greece* **48 C2** 38 40N 20 0 E
Íos, *Greece* **49 E7** 36 41N 25 20 E
Iowa □, *U.S.A.* **154 D8** 42 18N 93 30W
Iowa →, *U.S.A.* **156 C5** 41 10N 91 1W
Iowa City, *U.S.A.* **156 C5** 41 40N 91 32W
Iowa Falls, *U.S.A.* **156 B3** 42 31N 93 16W
Ipala, *Tanzania* **118 C3** 4 30 S 32 52 E
Ipameri, *Brazil* **171 E2** 17 44 S 48 9W
Iparía, *Peru* **172 B3** 9 17 S 74 29W
Ipáti, *Greece* **48 C4** 38 52N 22 14 E
Ipatinga, *Brazil* **171 E3** 19 32 S 42 30W
Ipatovo, *Russia* **61 H6** 45 45N 42 50 E
Ipel' →, *Europe* **35 D11** 47 48N 18 53 E
Ipiales, *Colombia* **168 C2** 0 50N 77 37W
Ipiaú, *Brazil* **171 D4** 14 8 S 39 44W
Ipil, *Phil.* **81 H4** 7 47N 122 35 E
Ipin = Yibin, *China* **76 C5** 28 45N 104 32 E
Ipirá, *Brazil* **171 D4** 12 10 S 39 44W
Ipiranga, *Brazil* **168 D4** 3 13 S 65 57W
Ípiros □, *Greece* **48 B2** 39 30N 20 30 E
Ipixuna, *Brazil* **172 B3** 7 0 S 71 40W
Ipixuna →, *Amazonas, Brazil* **172 B3** 7 11 S 71 51W
Ipixuna →, *Amazonas, Brazil* **173 B5** 5 45 S 63 2W
Ipoh, *Malaysia* **87 K3** 4 35N 101 5 E
Iporá, *Brazil* **171 D1** 11 23 S 50 40W
Ippy, *C.A.R.* **114 A4** 6 5N 21 7 E
Ipsala, *Turkey* **51 F10** 40 55N 26 23 E
Ipsárion Óros, *Greece* **51 F8** 40 40N 24 40 E
Ipswich, *Australia* **127 D5** 27 35 S 152 40 E
Ipswich, *U.K.* **21 E9** 52 4N 1 10 E
Ipswich, *Mass., U.S.A.* **151 D14** 42 41N 70 50W
Ipswich, *S. Dak., U.S.A.* **154 C5** 45 27N 99 2W
Ipu, *Brazil* **170 B3** 4 23 S 40 44W
Ipueiras, *Brazil* **170 B3** 4 33 S 40 43W
Ipupiara, *Brazil* **171 D3** 11 49 S 42 37W
Iqaluit, *Canada* **139 B13** 63 44N 68 31W
Iquique, *Chile* **172 E3** 20 19 S 70 5W
Iquitos, *Peru* **168 D3** 3 45 S 73 10W
Irabu-Jima, *Japan* **69 M2** 24 50N 125 10 E
Iracoubo, *Fr. Guiana* **169 B7** 5 30N 53 10W
Īrafshān, *Iran* **97 E9** 26 42N 61 56 E
Irahuan, *Phil.* **81 G2** 9 48N 118 41 E
Iráklia, *Kikládhes, Greece* **49 E7** 36 50N 25 28 E
Iráklia, *Sérrai, Greece* **50 E7** 41 10N 23 16 E
Iráklion, *Greece* **38 D7** 35 20N 25 12 E
Iráklion □, *Greece* **38 D7** 35 10N 25 10 E
Irako-Zaki, *Japan* **71 C9** 34 35N 137 1 E
Irala, *Paraguay* **175 B5** 25 55 S 54 35W
Iramba □, *Tanzania* **118 C3** 4 30 S 34 30 E
Iran ■, *Asia* **97 C7** 33 0N 53 0 E
Iran, Gunung-Gunung, *Malaysia* **85 B4** 2 20N 114 50 E
Iran, Plateau of, *Asia* **66 F9** 32 0N 55 0 E
Iran Ra. = Iran, Gunung-Gunung, *Malaysia* **85 B4** 2 20N 114 50 E
Iranamadu Tank, *Sri Lanka* **95 K5** 9 23N 80 29 E
Īrānshahr, *Iran* **97 E9** 27 15N 60 40 E
Irapa, *Venezuela* **169 A5** 10 34N 62 35W
Irapuato, *Mexico* **162 C4** 20 40N 101 30W
Iraq ■, *Asia* **101 F10** 33 0N 44 0 E
Irarrar, O. →, *Mali* **111 E5** 20 0N 1 30 E
Irati, *Brazil* **175 B5** 25 25 S 50 38W
Irbes saurums, *Latvia* **54 A9** 57 45N 22 5 E
Irbid, *Jordan* **103 C4** 32 35N 35 48 E
Irbid □, *Jordan* **103 C5** 32 15N 36 35 E
Irbit, *Russia* **62 C9** 57 41N 63 3 E
Irebu, *Dem. Rep. of the Congo* **114 C3** 0 40 S 17 46 E
Irecê, *Brazil* **170 D3** 11 18 S 41 52W
Iregua →, *Spain* **40 C7** 42 27N 2 24 E
Ireland ■, *Europe* **23 C4** 53 50N 7 52W
Irele, *Nigeria* **113 D6** 7 40N 5 40 E
Iremel, Gora, *Russia* **62 D7** 54 33N 58 50 E
Ireng →, *Brazil* **169 C6** 3 33N 59 51W
Irgiz, Bolshaya →, *Russia* **60 D9** 52 10N 49 10 E
Irhârharene, *Algeria* **111 C6** 27 37N 7 30 E
Irharrhar, O. →, *Algeria* **111 C6** 28 3N 6 15 E
Irherm, *Morocco* **110 B3** 30 7N 8 18W
Irhil Mgoun, *Morocco* **110 B3** 31 30N 6 28W
Irhyangdong, *N. Korea* **75 D15** 41 15N 129 30 E
Iri, *S. Korea* **75 G14** 35 59N 127 0 E
Irian Jaya □, *Indonesia* **83 B5** 4 0 S 137 0 E
Iriba, *Chad* **109 E4** 15 7N 22 15 E
Irid, Mt., *Phil.* **80 D3** 14 47N 121 19 E
Irié, *Guinea* **112 D3** 8 15N 9 10W
Iriga, *Phil.* **80 E4** 13 25N 123 25 E
Iriklinskiy, *Russia* **62 F7** 51 39N 58 38 E
Iriklinskoye Vdkhr., *Russia* **62 F7** 52 0N 59 0 E
Iringa, *Tanzania* **118 D4** 7 48 S 35 43 E
Iringa □, *Tanzania* **118 D4** 7 48 S 35 43 E
Irinjalakuda, *India* **95 J3** 10 21N 76 14 E
Iriomote-Jima, *Japan* **69 M1** 24 19N 123 48 E
Iriona, *Honduras* **164 C2** 15 57N 85 11W
Iriri →, *Brazil* **169 D7** 3 52 S 52 37W
Iriri Novo →, *Brazil* **173 B7** 8 46 S 53 22W
Irish Republic ■, *Europe* **23 C3** 53 0N 8 0W
Irish Sea, *U.K.* **20 D3** 53 38N 4 48W
Irkeshtam, *Kyrgyzstan* **63 D6** 39 41N 73 55 E
Irkutsk, *Russia* **65 D11** 52 18N 104 20 E
Irlığanlı, *Turkey* **49 D11** 37 53N 29 12 E
Irma, *Canada* **143 C6** 52 55N 111 14W
Irō-Zaki, *Japan* **71 C10** 34 36N 138 51 E
Iroise, Mer d', *France* **26 D2** 48 15N 4 45W
Iron Baron, *Australia* **128 B2** 32 58 S 137 11 E
Iron City, *U.S.A.* **152 D5** 31 1N 84 49W
Iron Gate = Portile de Fier, *Europe* **52 F7** 44 44N 22 30 E
Iron Knob, *Australia* **128 B2** 32 46 S 137 8 E
Iron Mountain, *U.S.A.* **148 C1** 45 49N 88 4W
Iron Ra., *Australia* **126 A3** 12 46 S 143 16 E
Iron River, *U.S.A.* **154 B10** 46 6N 88 39W
Irondequoit, *U.S.A.* **150 C7** 43 13N 77 35W
Ironstone Kopje, *Botswana* **116 D3** 25 17 S 24 5 E
Ironton, *Mo., U.S.A.* **155 G9** 37 36N 90 38W
Ironton, *Ohio, U.S.A.* **148 F4** 38 32N 82 41W
Ironwood, *U.S.A.* **154 B9** 46 27N 90 9W
Iroquois →, *U.S.A.* **157 D9** 41 5N 87 49W
Iroquois Falls, *Canada* **140 C3** 48 46N 80 41W
Irosin, *Phil.* **80 E5** 12 42N 124 2 E
Irpin, *Ukraine* **59 G6** 50 30N 30 15 E
Irrara Cr. →, *Australia* **127 D4** 29 35 S 145 31 E
Irrawaddy □, *Burma* **90 G5** 17 0N 95 0 E
Irrawaddy →, *Burma* **90 G5** 15 50N 95 6 E
Irsina, *Italy* **47 B9** 40 45N 16 14 E
Irtysh →, *Russia* **64 C7** 61 4N 68 52 E
Irumu, *Dem. Rep. of the Congo* **118 B2** 1 32N 29 53 E
Irún, *Spain* **40 B3** 43 20N 1 52W
Irunea = Pamplona, *Spain* **40 C3** 42 48N 1 38W
Irurzun, *Spain* **40 C3** 42 55N 1 50W
Irvine, *Canada* **143 D6** 49 57N 110 16W
Irvine, *U.K.* **22 F4** 55 37N 4 41W
Irvine, *Calif., U.S.A.* **161 M9** 33 41N 117 46W
Irvine, *Ky., U.S.A.* **157 G13** 37 42N 83 58W
Irvinestown, *U.K.* **23 B4** 54 28N 7 39W
Irving, *U.S.A.* **155 J6** 32 49N 96 56W
Irvington, *U.S.A.* **157 G10** 37 53N 86 17W
Irvona, *U.S.A.* **150 F6** 40 46N 78 33W
Irwin →, *Australia* **125 E1** 29 15 S 114 54 E
Irwinton, *U.S.A.* **152 C6** 32 49N 83 10W
Irwinville, *U.S.A.* **152 D6** 31 39N 83 23W
Irymple, *Australia* **128 C5** 34 14 S 142 8 E
Is-sur-Tille, *France* **27 E12** 47 30N 5 8 E
Isa, *Nigeria* **113 C6** 13 14N 6 24 E
Isaac →, *Australia* **126 C4** 22 55 S 149 20 E
Isabel, *U.S.A.* **154 C4** 45 24N 101 26W
Isabela, *Phil.* **81 F4** 10 12N 122 59 E
Isabela □, *Phil.* **80 C4** 17 0N 122 0 E
Isabela, I., *Mexico* **162 C3** 21 51N 105 55W
Isabelia, Cord., *Nic.* **164 D2** 13 30N 85 25W
Isabella, *Phil.* **81 H4** 6 40N 122 10 E
Isabella Ra., *Australia* **124 D3** 21 0 S 121 4 E
Isaccea, *Romania* **53 E13** 45 16N 28 28 E
Ísafjarðardjúp, *Iceland* **11 A3** 66 10N 23 0W
Ísafjarðarsýsla □, *Iceland* **11 B3** 66 0N 23 0W
Ísafjörður, *Iceland* **11 A3** 66 5N 23 9W
Isagarh, *India* **92 G7** 24 48N 77 51 E
Isahaya, *Japan* **69 H5** 32 52N 130 2 E
Isaka, *Tanzania* **118 C3** 3 56 S 32 59 E
Isakly, *Russia* **60 C10** 54 8N 51 32 E
Işalniţa, *Romania* **53 F8** 44 24N 23 44 E
Isana = Içana →, *Brazil* **168 C4** 0 26N 67 19W
Isangi, *Dem. Rep. of the Congo* **114 B4** 0 52N 24 10 E
Isar →, *Germany* **31 G8** 48 48N 12 57 E
Isarco →, *Italy* **45 B8** 46 27N 11 18 E
Ísari, *Greece* **48 D3** 37 22N 22 0 E
Íscar, *Spain* **42 D6** 41 22N 4 32W
Iscayachi, *Bolivia* **173 E4** 21 31 S 65 3W
İscehisar, *Turkey* **49 C12** 38 51N 30 45 E
Íschia, *Italy* **46 B6** 40 44N 13 57 E
Iscuandé, *Colombia* **168 C2** 2 28N 77 59W
Isdell →, *Australia* **124 C3** 16 27 S 124 51 E
Ise, *Japan* **71 C8** 34 25N 136 45 E
Ise-Heiya, *Japan* **71 C8** 34 40N 136 30 E
Ise-Wan, *Japan* **71 C8** 34 43N 136 43 E
Isefjord, *Denmark* **17 J5** 55 53N 11 50 E
Isel →, *Austria* **34 E5** 46 50N 12 47 E
Iseltwald, *Switz.* **32 C5** 46 43N 7 58 E
Isenthal, *Switz.* **33 C7** 46 55N 8 34 E
Iseo, *Italy* **44 C7** 45 39N 10 3 E
Iseo, L. d', *Italy* **44 C7** 45 43N 10 4 E
Iseramagazi, *Tanzania* **118 C3** 4 37 S 32 10 E
Isère □, *France* **29 C9** 45 15N 5 40 E
Isère →, *France* **29 D8** 44 59N 4 51 E
Iserlohn, *Germany* **30 D3** 51 22N 7 41 E
Isérnia, *Italy* **47 A7** 41 36N 14 14 E
Isesaki, *Japan* **71 A11** 36 19N 139 12 E
Iseyin, *Nigeria* **113 D5** 8 0N 3 36 E
Isfara, *Tajikistan* **63 C5** 40 7N 70 38 E
Isfjorden, *Norway* **18 B4** 62 35N 7 49 E
Isherton, *Guyana* **169 C6** 2 20N 59 25W
Ishigaki-Shima, *Japan* **69 M2** 24 20N 124 10 E
Ishikari-Gawa →, *Japan* **68 C10** 43 15N 141 23 E
Ishikari-Sammyaku, *Japan* **68 C11** 43 30N 143 0 E
Ishikari-Wan, *Japan* **68 C10** 43 25N 141 1 E
Ishikawa □, *Japan* **71 A8** 36 30N 136 30 E
Ishim, *Russia* **64 D7** 56 10N 69 30 E
Ishim →, *Russia* **64 D8** 57 45N 71 10 E
Ishimbay, *Russia* **62 E6** 53 28N 56 2 E
Ishinomaki, *Japan* **68 E10** 38 32N 141 20 E
Ishioka, *Japan* **71 A12** 36 11N 140 16 E
Ishizuchi-Yama, *Japan* **70 D5** 33 45N 133 6 E
Ishkashim = Eshkāshem, *Tajikistan* **63 E5** 36 44N 71 37 E
Ishkuman, *Pakistan* **93 A5** 36 30N 73 50 E
Ishmi, *Albania* **50 E3** 41 33N 19 34 E
Ishpeming, *U.S.A.* **148 B2** 46 29N 87 40W
Ishurdi, *Bangla.* **90 C2** 24 9N 89 3 E
Isigny-sur-Mer, *France* **26 C5** 49 19N 1 6W
Isıklar Dağı, *Turkey* **51 F11** 40 45N 27 15 E
Işıklı, *Turkey* **49 C11** 38 19N 29 51 E
Isil Kul, *Russia* **64 D8** 54 55N 71 16 E
Ísili, *Italy* **46 C2** 39 44N 9 6 E
Isiolo, *Kenya* **118 B4** 0 24N 37 33 E
Isiolo □, *Kenya* **118 B4** 2 30N 37 30 E
Isiro, *Dem. Rep. of the Congo* **118 B2** 2 53N 27 40 E
Isisford, *Australia* **126 C3** 24 15 S 144 21 E
Iskandar, *Uzbekistan* **63 C4** 41 36N 69 41 E
İskenderun, *Turkey* **100 D7** 36 32N 36 10 E
İskenderun Körfezi, *Turkey* **100 D6** 36 40N 35 50 E
Iski-Naukat, *Kyrgyzstan* **63 C6** 40 16N 72 36 E
İskilip, *Turkey* **100 B6** 40 45N 34 29 E
Iskŭr →, *Bulgaria* **51 C8** 43 45N 24 25 E
Iskŭr, Yazovir, *Bulgaria* **50 D7** 42 23N 23 30 E
Iskut →, *Canada* **142 B2** 56 45N 131 49W
Isla →, *U.K.* **22 E5** 56 32N 3 20W
Isla Cristina, *Spain* **43 H3** 37 13N 7 17W
Isla Vista, *U.S.A.* **161 L7** 34 25N 119 53W
Islâhiye, *Turkey* **100 D7** 37 0N 36 35 E
Islamabad, *Pakistan* **91 B4** 33 40N 73 10 E
Islamkot, *Pakistan* **92 G4** 24 42N 70 13 E
Islamorada, *U.S.A.* **153 L9** 24 56N 80 37W
Islampur, *India* **94 F2** 17 2N 74 20 E
Island →, *Canada* **142 A4** 60 25N 121 12W
Island Bay, *Phil.* **81 G2** 9 6N 118 10 E
Island Falls, *Canada* **140 C3** 49 35N 81 20W
Island Falls, *U.S.A.* **141 C6** 46 1N 68 16W
Island L., *Canada* **143 C10** 53 47N 94 25W
Island Lagoon, *Australia* **128 A2** 31 30 S 136 40 E
Island Pond, *U.S.A.* **151 B13** 44 49N 71 53W
Islands, B. of, *Canada* **141 C8** 49 11N 58 15W
Islands, B. of, *N.Z.* **130 B3** 35 15 S 174 6 E
Islay, *U.K.* **22 F2** 55 46N 6 10W
Isle →, *France* **28 D3** 44 55N 0 15W
Isle aux Morts, *Canada* **141 C8** 47 35N 59 0W
Isle of Hope, *U.S.A.* **152 D8** 31 58N 81 5W
Isle of Wight □, *U.K.* **21 G6** 50 41N 1 17W
Isle Royale, *U.S.A.* **154 B10** 48 0N 88 54W
Isleta, *U.S.A.* **159 J10** 34 55N 106 42W
Isleton, *U.S.A.* **160 G5** 38 10N 121 37W
Ismail = Izmayil, *Ukraine* **59 K5** 45 22N 28 46 E
Ismâ'ilîya, *Egypt* **106 H8** 30 37N 32 18 E
Ismaning, *Germany* **31 G7** 48 13N 11 40 E
Ismay, *U.S.A.* **154 B2** 46 30N 104 48W
Isna, *Egypt* **106 B3** 25 17N 32 30 E
Isogstalo, *India* **93 B8** 34 15N 78 46 E
Ísola del Liri, *Italy* **45 G10** 41 41N 13 34 E
Ísola della Scala, *Italy* **44 C7** 45 16N 11 0 E
Ísola di Capo Rizzuto, *Italy* **47 D10** 38 58N 17 6 E
Isparta, *Turkey* **57 G5** 37 47N 30 30 E
Isperikh, *Bulgaria* **51 C10** 43 43N 26 50 E
Íspica, *Italy* **47 F7** 36 47N 14 55 E
Israel ■, *Asia* **103 D3** 32 0N 34 50 E
Issano, *Guyana* **169 B6** 5 49N 59 26W
Issia, *Ivory C.* **112 D3** 6 33N 6 33W
Issoire, *France* **28 C7** 45 32N 3 15 E
Issoudun, *France* **27 F8** 46 57N 1 59 E
Issyk-Kul = Ysyk-Köl, *Kyrgyzstan* **63 B8** 42 26N 76 12 E
Issyk-Kul, Ozero = Ysyk-Köl, Ozero, *Kyrgyzstan* **63 B8** 42 25N 77 15 E
Ist, *Croatia* **45 D11** 44 17N 14 47 E
Istaihah, *U.A.E.* **97 F7** 23 19N 54 4 E
Istállós-kő, *Hungary* **52 B5** 48 4N 20 26 E
İstanbul, *Turkey* **51 E12** 41 0N 29 0 E
İstanbul □, *Turkey* **51 E12** 41 0N 29 0 E
İstanbul Boğazı, *Turkey* **51 E13** 41 10N 29 10 E
Isteren, *Norway* **18 B8** 61 58N 11 47 E
Istiaía, *Greece* **48 C5** 38 57N 23 9 E
Istmina, *Colombia* **168 B2** 5 10N 76 39W
Isto, Mt., *U.S.A.* **144 B12** 69 12N 143 48W
Istok, *Serbia, Yug.* **50 D4** 42 45N 20 24 E
Istokpoga, L., *U.S.A.* **149 M5** 27 23N 81 17W
Istra, *Croatia* **45 C10** 45 10N 14 0 E
Istres, *France* **29 E8** 43 31N 4 59 E
Istria = Istra, *Croatia* **45 C10** 45 10N 14 0 E
Isulan, *Phil.* **81 H5** 6 30N 124 29 E
Itá, *Paraguay* **174 B4** 25 29 S 57 21W
'Itāb, *Yemen* **99 D5** 15 20N 51 29 E
Itabaiana, *Paraíba, Brazil* **170 C4** 7 18 S 35 19W
Itabaiana, *Sergipe, Brazil* **170 D4** 10 41 S 37 37W
Itabaianinha, *Brazil* **170 D4** 11 16 S 37 47W
Itaberaba, *Brazil* **171 D3** 12 32 S 40 18W
Itaberaí, *Brazil* **171 E2** 16 2 S 49 48W
Itabira, *Brazil* **171 E3** 19 37 S 43 13W
Itabirito, *Brazil* **171 F3** 20 15 S 43 48W
Itaboca, *Brazil* **169 D5** 4 50 S 62 40W
Itabuna, *Brazil* **171 D4** 14 48 S 39 16W
Itacajá, *Brazil* **170 C2** 8 19 S 47 46W
Itacaunas →, *Brazil* **170 C2** 5 21 S 49 8W
Itacoatiara, *Brazil* **169 D6** 3 8 S 58 25W
Itacuaí →, *Brazil* **172 A3** 4 20 S 70 12W
Itaguaçu, *Brazil* **171 E3** 19 48 S 40 51W
Itaguari →, *Brazil* **171 D3** 14 11 S 44 40W
Itaguatins, *Brazil* **170 C2** 5 47 S 47 29W
Itaim →, *Brazil* **170 C3** 7 2 S 42 2W
Itainópolis, *Brazil* **170 C3** 7 24 S 41 31W
Itaipú, Reprêsa de, *Brazil* **175 B5** 25 30 S 54 30W
Itaituba, *Brazil* **169 D6** 4 10 S 55 50W
Itajaí, *Brazil* **175 B6** 27 50 S 48 39W
Itajubá, *Brazil* **171 F2** 22 24 S 45 30W
Itajuípe, *Brazil* **171 D4** 14 41 S 39 22W
Itaka, *Tanzania* **119 D3** 8 50 S 32 49 E
Itako, *Japan* **71 B12** 35 56N 140 33 E
Italy ■, *Europe* **13 G8** 42 0N 13 0 E
Itamataré, *Brazil* **170 B2** 2 16 S 46 24W
Itambacuri, *Brazil* **171 E3** 18 1 S 41 42W
Itambé, *Brazil* **171 E3** 15 15 S 40 37W
Itampolo, *Madag.* **117 C7** 24 41 S 43 57 E
Itanhauã →, *Brazil* **169 D5** 4 45 S 63 48W
Itanhém, *Brazil* **171 E3** 17 9 S 40 20W
Itano, *Japan* **70 C6** 34 7N 134 28 E
Itapaci, *Brazil* **171 D2** 14 57 S 49 34W
Itapagé, *Brazil* **170 B4** 3 41 S 39 34W
Itaparica, I. de, *Brazil* **171 D4** 12 54 S 38 42W
Itapebi, *Brazil* **171 E4** 15 56 S 39 32W
Itapecuru-Mirim, *Brazil* **170 B3** 3 24 S 44 20W
Itaperuna, *Brazil* **171 F3** 21 10 S 41 54W
Itapetinga, *Brazil* **171 E3** 15 15 S 40 15W
Itapetininga, *Brazil* **175 A6** 23 36 S 48 7W
Itapeva, *Brazil* **175 A6** 23 59 S 48 59W
Itapicuru →, *Bahia, Brazil* **170 D4** 11 47 S 37 32W
Itapicuru →, *Maranhão, Brazil* **170 B3** 2 52 S 44 12W
Itapinima, *Brazil* **173 B5** 5 25 S 60 44W
Itapipoca, *Brazil* **170 B4** 3 30 S 39 35W
Itapiranga, *Brazil* **169 D6** 2 45 S 58 1W
Itapiúna, *Brazil* **170 B4** 4 33 S 38 57W
Itaporanga, *Brazil* **170 C4** 7 18 S 38 0W
Itapuá □, *Paraguay* **175 B4** 26 40 S 55 40W
Itapuranga, *Brazil* **171 E2** 15 40 S 49 59W
Itaquari, *Brazil* **171 F3** 20 20 S 40 25W
Itaquatiara, *Brazil* **169 D6** 2 58 S 58 30W
Itaquí, *Brazil* **174 B4** 29 8 S 56 30W
Itararé, *Brazil* **175 A6** 24 6 S 49 23W
Itarsi, *India* **92 H7** 22 36N 77 51 E
Itarumã, *Brazil* **171 E1** 18 42 S 51 25W
Itatí, *Argentina* **174 B4** 27 16 S 58 15W
Itatira, *Brazil* **170 B4** 4 30 S 39 37W
Itatuba, *Brazil* **173 B5** 5 46 S 63 20W
Itatupa, *Brazil* **169 D7** 0 37 S 51 12W
Itaueira, *Brazil* **170 C3** 7 36 S 43 2W
Itaueira →, *Brazil* **170 C3** 6 41 S 42 55W
Itaúna, *Brazil* **171 F3** 20 4 S 44 34W
Itbayat, *Phil.* **80 A3** 20 47N 121 51 E
Itbayat I., *Phil.* **80 A3** 20 46N 121 50 E
Itchen →, *U.K.* **21 G6** 50 55N 1 22W
Ite, *Peru* **172 D3** 17 55 S 70 57W
Itéa, *Greece* **48 C4** 38 25N 22 25 E
Itezhi Tezhi, L., *Zambia* **119 F2** 15 30 S 25 30 E
Ithaca = Itháki, *Greece* **48 C2** 38 25N 20 40 E
Ithaca, *U.S.A.* **151 D8** 42 27N 76 30W
Itháki, *Greece* **48 C2** 38 25N 20 40 E
Itinga, *Brazil* **171 E3** 16 36 S 41 47W
Itiquira, *Brazil* **173 D7** 17 12 S 54 7W
Itiquira →, *Brazil* **173 D6** 17 18 S 56 44W
Itiruçu, *Brazil* **171 D3** 13 31 S 40 9W
Itiúba, *Brazil* **170 D4** 10 43 S 39 51W
Itkilik →, *U.S.A.* **144 A10** 70 9N 150 56W
Ito, *Japan* **71 C11** 34 58N 139 5 E
Itoigawa, *Japan* **69 F8** 37 2N 137 51 E
Iton →, *France* **26 C8** 49 9N 1 12 E
Itonamas →, *Bolivia* **173 C5** 12 28 S 64 24W
Itri, *Italy* **46 A6** 41 17N 13 32 E
Itsa, *Egypt* **106 J7** 29 15N 30 47 E
Itsukaichi, *Japan* **70 C4** 34 22N 132 22 E
Itsuki, *Japan* **69 H5** 32 24N 130 50 E
Íttiri, *Italy* **46 B1** 40 36N 8 34 E
Ittoqqortoormiit = Scoresbysund, *Greenland* **10 C8** 70 20N 23 0W
Itu, *Brazil* **175 A6** 23 17 S 47 15W
Itu, *Nigeria* **113 D6** 5 10N 7 58 E
Ituaçu, *Brazil* **171 D3** 13 50 S 41 18W
Ituango, *Colombia* **168 B2** 7 4N 75 45W
Ituiutaba, *Brazil* **171 E2** 19 0 S 49 25W

J

Jelenia Góra, Poland 55 H2 50 50N 15 45 E
Jelenia Góra □, Poland 55 G2 51 0N 15 30 E
Jelgava, Latvia 15 H20 56 41N 23 49 E
Jelgava □, Latvia 54 B10 56 35N 23 45 E
Jelica, Serbia, Yug. 50 C4 43 50N 20 17 E
Jelli, Sudan 107 F3 5 25N 31 45 E
Jellicoe, Canada 140 C2 49 40N 87 30W
Jelšava, Slovak Rep. 35 C13 48 37N 20 15 E
Jemaja, Indonesia 87 L5 3 5N 105 45 E
Jemaluang, Malaysia 87 L4 2 16N 103 52 E
Jember, Indonesia 85 D4 8 11 S 113 41 E
Jena, Germany 30 E7 50 54N 11 35 E
Jena, Fla., U.S.A. 153 F6 29 40N 83 22W
Jena, La., U.S.A. 155 K8 31 41N 92 8W
Jenbach, Austria 34 D4 47 24N 11 47 E
Jendouba, Tunisia 108 A1 36 29N 8 47 E
Jenkins, U.S.A. 148 G4 37 10N 82 38W
Jenner, U.S.A. 160 G3 38 27N 123 7W
Jennings, Fla., U.S.A. 152 E6 30 36N 83 6W
Jennings, La., U.S.A. 155 K8 30 13N 92 40W
Jennings, Mo., U.S.A. 156 F6 38 43N 90 16W
Jennings →, Canada 142 B2 59 38N 132 5W
Jensen Beach, U.S.A. 153 H9 27 15N 80 14W
Jepara, Indonesia 85 D3 7 40 S 109 14 E
Jeparit, Australia 128 D5 36 8 S 142 1 E
Jequié, Brazil 171 D3 13 51 S 40 5W
Jequitaí →, Brazil 171 E3 17 4 S 44 50W
Jequitinhonha, Brazil 171 E3 16 30 S 41 0W
Jequitinhonha →, Brazil 171 E4 15 51 S 38 53W
Jerada, Morocco 111 B4 34 17N 2 10W
Jerantut, Malaysia 87 L4 3 56N 102 22 E
Jérémie, Haiti 165 C5 18 40N 74 10W
Jeremoabo, Brazil 170 D4 10 4 S 38 21W
Jerez, Punta, Mexico 163 C5 22 58N 97 40W
Jerez de García Salinas, Mexico 162 C4 22 39N 103 0W
Jerez de la Frontera, Spain 43 J4 36 41N 6 7W
Jerez de los Caballeros, Spain 43 G4 38 20N 6 45W
Jericho = El Arīḩā, West Bank 103 D4 31 52N 35 27 E
Jericho, Australia 126 C4 23 38 S 146 6 E
Jerichow, Germany 30 C8 52 30N 12 1 E
Jerico Springs, U.S.A. 156 G2 37 37N 94 1W
Jerilderie, Australia 129 C6 35 20 S 145 41 E
Jermyn, U.S.A. 151 E9 41 31N 75 31W
Jerome, U.S.A. 159 J8 34 45N 112 7W
Jersey, U.K. 21 H5 49 11N 2 7W
Jersey, U.S.A. 152 B6 33 43N 83 47W
Jersey City, U.S.A. 151 F10 40 44N 74 4W
Jersey Shore, U.S.A. 150 E7 41 12N 77 15W
Jerseyville, U.S.A. 156 E6 39 7N 90 20W
Jerusalem, Israel 103 D4 31 47N 35 10 E
Jerusalem, U.S.A. 152 E8 30 58N 81 50W
Jervis B., Australia 129 C9 35 8 S 150 46 E
Jerzu, Italy 46 C2 39 47N 9 31 E
Jesenice, Slovenia 45 B11 46 28N 14 3 E
Jeseník, Czech Rep. 35 A10 50 14N 17 8 E
Jesenké, Slovak Rep. 35 C13 48 20N 20 10 E
Jesselton = Kota Kinabalu, Malaysia 85 A5 6 0N 116 4 E
Jessheim, Norway 18 D8 60 9N 11 10 E
Jessnitz, Germany 30 D8 51 40N 12 18 E
Jessore, Bangla. 90 D2 23 10N 89 10 E
Jessup L., U.S.A. 153 G8 28 43N 81 14W
Jesup, Ga., U.S.A. 152 D8 31 36N 81 53W
Jesup, Iowa, U.S.A. 156 B4 42 29N 92 4W
Jesús, Peru 172 B2 7 15 S 78 25W
Jesús Carranza, Mexico 163 D5 17 28N 95 1W
Jesús María, Argentina 174 C3 30 59 S 64 5W
Jetafe, Phil. 81 F5 10 9N 124 9 E
Jetmore, U.S.A. 155 F5 38 4N 99 54W
Jetpur, India 92 J4 21 45N 70 10 E
Jeumont, France 27 B11 50 18N 4 6 E
Jevnaker, Norway 15 F14 60 15N 10 26 E
Jewell, U.S.A. 156 B3 42 20N 93 39W
Jewett, Ohio, U.S.A. 150 F3 40 22N 81 2W
Jewett, Tex., U.S.A. 155 K6 31 22N 96 9W
Jewett City, U.S.A. 151 E13 41 36N 72 0W
Jeyḩūnābād, Iran 97 C6 34 58N 48 59 E
Jeypore, India 94 E6 18 50N 82 38 E
Jeziorak, Jezioro, Poland 54 E6 53 40N 19 35 E
Jeziorany, Poland 54 E7 53 58N 20 46 E
Jeziorka →, Poland 55 F8 52 8N 21 9 E
Jhajjar, India 92 E7 28 37N 76 42 E
Jhal Jhao, Pakistan 91 D2 26 20N 65 35 E
Jhalakati, Bangla. 90 D3 22 39N 90 12 E
Jhalawar, India 92 G7 24 40N 76 10 E
Jhang Maghiana, Pakistan 91 C4 31 15N 72 22 E
Jhansi, India 93 G8 25 30N 78 36 E
Jharia, India 93 H12 23 45N 86 26 E
Jharsuguda, India 94 D7 21 56N 84 5 E
Jhelum, Pakistan 91 B4 33 0N 73 45 E
Jhelum →, Pakistan 92 D5 31 20N 72 10 E
Jhunjhunu, India 92 E6 28 10N 75 30 E
Ji Xian, Hebei, China 74 F8 37 35N 115 30 E
Ji Xian, Henan, China 74 G8 35 22N 114 5 E
Ji Xian, Shanxi, China 74 F6 36 7N 110 40 E
Jia Xian, Henan, China 74 H7 33 59N 113 12 E
Jia Xian, Shaanxi, China 74 E6 38 12N 110 28 E
Jiading, China 77 B13 31 22N 121 15 E
Jiahe, China 77 E9 25 38N 112 19 E
Jialing Jiang →, China 76 C6 29 30N 106 20 E
Jiamusi, China 73 B8 46 40N 130 26 E
Ji'an, Jiangxi, China 77 D10 27 6N 114 59 E
Ji'an, Jilin, China 75 D14 41 5N 126 10 E
Jianchang, China 75 D11 40 55N 120 35 E
Jianchangying, China 75 D10 40 10N 118 50 E
Jianchuan, China 76 D2 26 38N 99 55 E
Jiande, China 77 C12 29 23N 119 15 E
Jiangbei, China 76 C6 29 40N 106 34 E
Jiangcheng, China 76 F3 22 36N 101 52 E
Jiangdi, China 76 D4 26 57N 103 37 E
Jiange, China 76 A5 32 4N 105 32 E
Jiangjin, China 76 C6 29 14N 106 14 E
Jiangkou, China 76 D7 27 40N 108 49 E
Jiangle, China 77 D11 26 42N 117 23 E
Jiangling, China 77 B9 30 25N 112 12 E
Jiangmen, China 77 F9 22 32N 113 0 E
Jiangshan, China 77 C12 28 40N 118 37 E
Jiangsu □, China 75 H11 33 0N 120 0 E
Jiangxi □, China 77 D11 27 30N 116 0 E
Jiangyin, China 77 B13 31 54N 120 17 E
Jiangyong, China 77 E8 25 20N 111 22 E
Jiangyou, China 76 B5 31 44N 104 43 E
Jianhe, China 76 D7 26 37N 108 31 E
Jianli, China 77 C9 29 46N 112 56 E
Jianning, China 77 D11 26 50N 116 50 E
Jian'ou, China 77 D12 27 3N 118 17 E
Jianshi, China 76 B7 30 37N 109 38 E
Jianshui, China 76 F4 23 36N 102 43 E
Jianyang, Fujian, China 77 D12 27 20N 118 5 E
Jianyang, Sichuan, China 76 B5 30 24N 104 33 E
Jiao Xian, China 75 F11 36 18N 120 1 E
Jiaohe, Hebei, China 74 E9 38 2N 116 20 E
Jiaohe, Jilin, China 75 C14 43 40N 127 22 E
Jiaoling, China 77 E11 24 41N 116 12 E
Jiaozhou Wan, China 75 F11 36 5N 120 10 E
Jiaozuo, China 74 G7 35 16N 113 12 E
Jiashan, China 77 A11 32 46N 117 59 E
Jiawang, China 75 G9 34 28N 117 26 E
Jiaxiang, China 74 G9 35 25N 116 20 E
Jiaxing, China 77 B13 30 49N 120 45 E
Jiayi = Chiai, Taiwan 77 F13 23 29N 120 25 E
Jiayu, China 77 C9 29 55N 113 55 E
Jibão, Serra do, Brazil 171 D3 14 48 S 45 0W
Jibiya, Nigeria 113 C6 13 5N 7 12 E
Jibou, Romania 53 C8 47 15N 23 17 E
Jibuti = Djibouti ■, Africa 107 E5 12 0N 43 0 E
Jicarón, I., Panama 164 E3 7 10N 81 50W
Jičín, Czech Rep. 34 A8 50 25N 15 28 E
Jiddah, Si. Arabia 98 B2 21 29N 39 10 E
Jieshou, China 74 H8 33 18N 115 22 E
Jiexiu, China 74 F6 37 2N 111 55 E
Jieyang, China 77 F11 23 35N 116 21 E
Jigawa □, Nigeria 113 C6 12 0N 9 45 E
Jiggalong, Australia 124 D3 23 21 S 120 47 E
Jihlava, Czech Rep. 34 B8 49 28N 15 35 E
Jihlava →, Czech Rep. 35 C9 48 55N 16 36 E
Jihočeský □, Czech Rep. 34 B7 49 8N 14 35 E
Jihomoravský □, Czech Rep. 35 B9 49 5N 16 30 E
Jijel, Algeria 111 A6 36 52N 5 50 E
Jijiga, Ethiopia 120 C2 9 20N 42 50 E
Jikamshi, Nigeria 113 C6 12 12N 7 45 E
Jilin, China 75 C14 43 44N 126 30 E
Jilin □, China 75 C14 44 0N 127 0 E
Jiloca →, Spain 40 D3 41 21N 1 39W
Jilong = Chilung, Taiwan 77 E13 25 3N 121 45 E
Jima, Ethiopia 107 F4 7 40N 36 47 E
Jimbolia, Romania 52 E5 45 47N 20 43 E
Jimena de la Frontera, Spain 43 J5 36 27N 5 24W
Jimenbuen, Australia 129 D8 36 42 S 148 53 E
Jiménez, Mexico 162 B4 27 10N 104 54W
Jimenez, Phil. 81 G4 8 20N 123 50 E
Jimo, China 75 F11 36 23N 120 30 E
Jin Jiang →, China 77 C10 28 24N 115 48 E
Jin Xian, Hebei, China 74 E8 38 2N 115 2 E
Jin Xian, Liaoning, China 75 E11 38 55N 121 42 E
Jinan, China 74 F9 36 38N 117 1 E
Jincheng, China 74 G7 35 29N 112 50 E
Jinchuan, China 76 B4 31 30N 102 3 E
Jind, India 92 E7 29 19N 76 22 E
Jindabyne, Australia 129 D8 36 25 S 148 35 E
Jindřichův Hradec, Czech Rep. 34 B8 49 10N 15 2 E
Jing He →, China 74 G5 34 27N 109 4 E
Jing Shan, China 77 B8 31 20N 111 35 E
Jing Xian, Anhui, China 77 B12 30 38N 118 25 E
Jing Xian, Hunan, China 76 D7 26 33N 109 40 E
Jing'an, China 77 C10 28 50N 115 17 E
Jingbian, China 74 F5 37 20N 108 30 E
Jingchuan, China 74 G4 35 20N 107 20 E
Jingde, China 77 B12 30 15N 118 27 E
Jingdezhen, China 77 C11 29 20N 117 11 E
Jingdong, China 76 E3 24 23N 100 47 E
Jinggu, China 76 F3 23 35N 100 41 E
Jinghai, China 74 E9 38 55N 116 55 E
Jinghong, China 76 G3 22 0N 100 45 E
Jingjiang, China 77 A13 32 2N 120 16 E
Jingle, China 74 E6 38 20N 111 55 E
Jingmen, China 77 B9 31 0N 112 10 E
Jingning, China 74 G3 35 30N 105 43 E
Jingpo Hu, China 75 C15 43 55N 128 55 E
Jingshan, China 77 B9 31 1N 113 7 E
Jingtai, China 74 F3 37 10N 104 6 E
Jingxi, China 76 F6 23 8N 106 27 E
Jingxing, China 74 E8 38 2N 114 8 E
Jingyang, China 74 G5 34 30N 108 50 E
Jingyu, China 75 C14 42 25N 126 45 E
Jingyuan, China 74 F3 36 30N 104 40 E
Jingziguan, China 74 H6 33 15N 111 0 E
Jinhua, China 77 C12 29 8N 119 38 E
Jining, Nei Mongol Zizhiqu, China 74 D7 41 5N 113 0 E
Jining, Shandong, China 74 G9 35 22N 116 34 E
Jinja, Uganda 118 B3 0 25N 33 12 E
Jinjang, Malaysia 87 L3 3 13N 101 39 E
Jinji, China 74 F4 37 58N 106 8 E
Jinjiang, Fujian, China 77 E12 24 43N 118 33 E
Jinjiang, Yunnan, China 76 D3 26 14N 100 34 E
Jinjie, China 76 F6 23 15N 107 18 E
Jinjini, Ghana 112 D4 7 26N 3 42W
Jinkou, China 77 B10 30 20N 114 8 E
Jinmen Dao, China 77 E12 24 25N 118 25 E
Jinning, China 76 E4 24 38N 102 38 E
Jinotega, Nic. 164 D2 13 6N 85 59W
Jinotepe, Nic. 164 D2 11 50N 86 10W
Jinping, Guizhou, China 76 D7 26 41N 109 10 E
Jinping, Yunnan, China 76 F4 22 45N 103 18 E
Jinsha, China 76 D6 27 29N 106 12 E
Jinsha Jiang →, China 76 C5 28 50N 104 36 E
Jinshan, China 77 B13 30 54N 121 10 E
Jinshi, China 77 C8 29 40N 111 50 E
Jintan, China 77 B12 31 42N 119 36 E
Jintotolo Channel, Phil. 81 F4 11 48N 123 5 E
Jinxi, Jiangxi, China 77 D11 27 56N 116 45 E
Jinxi, Liaoning, China 75 D11 40 52N 120 50 E
Jinxian, China 77 C11 28 26N 116 17 E
Jinxiang, China 74 G9 35 5N 116 22 E
Jinyun, China 77 C13 28 35N 120 5 E
Jinzhai, China 77 B10 31 40N 115 53 E
Jinzhou, China 75 D11 41 5N 121 3 E
Jiparaná →, Brazil 173 B5 8 3 S 62 52W
Jipijapa, Ecuador 168 D1 1 0 S 80 40W
Jiquilpan, Mexico 162 D4 19 57N 102 42W
Jirwān, Si. Arabia 99 B5 23 27N 50 53 E
Jishan, China 74 G6 35 34N 110 58 E
Jishou, China 76 C7 28 21N 109 43 E
Jishui, China 77 D10 27 12N 115 8 E
Jisr ash Shughūr, Syria 100 E7 35 49N 36 18 E
Jitarning, Australia 125 F2 32 48 S 117 57 E
Jitra, Malaysia 87 J3 6 16N 100 25 E
Jiu →, Romania 53 G8 43 47N 23 48 E
Jiudengkou, China 74 E4 39 56N 106 40 E
Jiujiang, Guangdong, China 77 F9 22 50N 113 0 E
Jiujiang, Jiangxi, China 77 C10 29 42N 115 58 E
Jiuling Shan, China 77 C10 28 40N 114 40 E
Jiulong, China 76 C3 28 57N 101 31 E
Jiutai, China 75 B13 44 10N 125 50 E
Jiuxiangcheng, China 74 H8 33 12N 114 50 E
Jiuxincheng, China 74 E8 39 17N 115 59 E
Jiuyuhang, China 77 B12 30 18N 119 56 E
Jixi, Anhui, China 77 B12 30 5N 118 34 E
Jixi, Heilongjiang, China 75 B16 45 20N 130 50 E
Jiyang, China 75 F9 37 0N 117 12 E
Jiz', W. →, Yemen 99 C6 16 12N 52 14 E
Jīzān, Si. Arabia 98 C3 17 0N 42 20 E
Jize, China 74 F8 36 54N 114 56 E
Jizera →, Czech Rep. 34 A7 50 10N 14 43 E
Jizō-Zaki, Japan 70 B5 35 34N 133 20 E
Jizzakh, Uzbekistan 63 C3 40 6N 67 50 E
Joaçaba, Brazil 175 B5 27 5 S 51 31W
Joaíma, Brazil 171 E3 16 39 S 41 2W
João, Brazil 170 B1 2 46 S 50 59W
João Amaro, Brazil 171 D3 12 46 S 40 22W
João Câmara, Brazil 170 C4 5 32 S 35 48W
João Pessoa, Brazil 170 C5 7 10 S 34 52W
João Pinheiro, Brazil 171 E2 17 45 S 46 10W
Joaquim Távora, Brazil 171 F2 23 30 S 49 58W
Joaquín V. González, Argentina 174 B3 25 10 S 64 0W
Jobourg, Nez de, France 26 C5 49 41N 1 57W
Jódar, Spain 43 H7 37 50N 3 21W
Jodhpur, India 92 F5 26 23N 73 8 E
Joensuu, Finland 56 B4 62 37N 29 49 E
Jœuf, France 27 C12 49 12N 6 0 E
Jofane, Mozam. 117 C5 21 15 S 34 18 E
Jõgeva, Estonia 15 G22 58 45N 26 24 E
Joggins, Canada 141 C7 45 42N 64 27W
Jogjakarta = Yogyakarta, Indonesia 85 D4 7 49 S 110 22 E
Jōhana, Japan 71 A8 36 30N 136 57 E
Johannesburg, S. Africa 117 D4 26 10 S 28 2 E
Johannesburg, U.S.A. 161 K9 35 22N 117 38W
Johansfors, Sweden 17 H9 56 42N 15 32 E
Jöhen, Japan 70 E4 32 58N 132 32 E
John Day, U.S.A. 158 D4 44 25N 118 57W
John Day →, U.S.A. 158 D3 45 44N 120 39W
John H. Kerr Reservoir, U.S.A. 149 G6 36 36N 78 18W
John o' Groats, U.K. 22 C5 58 38N 3 4W
Johnnie, U.S.A. 161 J10 36 25N 116 5W
Johns I., U.S.A. 152 C9 32 40N 80 10W
Johns Island, U.S.A. 152 C9 32 47N 80 7W
John's Ra., Australia 126 C1 21 55 S 133 23 E
Johnson, U.S.A. 155 G4 37 34N 101 45W
Johnson City, Ill., U.S.A. 156 G8 37 49N 88 56W
Johnson City, N.Y., U.S.A. 151 D9 42 7N 75 58W
Johnson City, Tenn., U.S.A. 149 G4 36 19N 82 21W
Johnson City, Tex., U.S.A. 155 K5 30 17N 98 25W
Johnsonburg, U.S.A. 150 E6 41 29N 78 41W
Johnsondale, U.S.A. 161 K8 35 58N 118 32W
Johnson's Crossing, Canada 142 A2 60 29N 133 18W
Johnsonville, N.Z. 130 H3 41 13 S 174 48 E
Johnston, U.S.A. 152 B9 33 50N 81 48W
Johnston, L., Australia 125 F3 32 25 S 120 30 E
Johnston Falls = Mambilima Falls, Zambia 119 E2 10 31 S 28 45 E
Johnston I., Pac. Oc. 135 F11 17 10N 169 8W
Johnstone Str., Canada 142 C3 50 28N 126 0W
Johnstown, N.Y., U.S.A. 151 C10 43 0N 74 22W
Johnstown, Pa., U.S.A. 150 F6 40 20N 78 55W
Johor □, Malaysia 84 B2 2 5N 103 20 E
Johor Baharu, Malaysia 87 M4 1 28N 103 46 E
Jõhvi, Estonia 15 G22 59 22N 27 27 E
Joigny, France 27 E10 47 58N 3 20 E
Joinville, Brazil 175 B6 26 15 S 48 55W
Joinville, France 27 D12 48 27N 5 10 E
Joinville I., Antarctica 7 C18 65 0 S 55 30W
Jojutla, Mexico 163 D5 18 37N 99 11W
Jokkmokk, Sweden 14 C18 66 35N 19 50 E
Jökulheimar, Iceland 11 C8 64 20N 18 18W
Jökulsá á Bru →, Iceland 11 B12 65 40N 14 16W
Jökulsá á Fjöllum →, Iceland 11 A10 66 10N 16 30W
Jolfā, Āzarbājān-e Sharqī, Iran 101 C11 38 57N 45 38 E
Jolfā, Eşfahan, Iran 97 C6 32 58N 51 37 E
Joliet, U.S.A. 157 C8 41 32N 88 5W
Joliette, Canada 140 C5 46 3N 73 24W
Jolo, Phil. 81 H3 6 0N 121 0 E
Jolo Group, Phil. 81 J3 6 0N 121 9 E
Jolon, U.S.A. 160 K5 35 58N 121 9W
Jølstravatnet, Norway 18 C3 61 32N 6 23 E
Jomalig, Phil. 80 D4 14 42N 122 22 E
Jombang, Indonesia 85 D4 7 33 S 112 14 E
Jomda, China 76 B2 31 28N 98 12 E
Jome, Indonesia 82 B3 1 16 S 127 30 E
Jonava, Lithuania 15 J21 55 8N 24 12 E
Jondal, Norway 18 D3 60 16N 6 15 E
Jones, Phil. 80 C3 16 33N 121 42 E
Jones Sound, Canada 6 B3 76 0N 85 0W
Jonesboro, Ark., U.S.A. 155 H9 35 50N 90 42W
Jonesboro, Ga., U.S.A. 152 B5 33 31N 84 22W
Jonesboro, Ill., U.S.A. 155 G10 37 27N 89 16W
Jonesboro, La., U.S.A. 155 J8 32 15N 92 43W
Jonesburg, U.S.A. 156 F5 38 51N 91 18W
Jonesport, U.S.A. 141 D6 44 32N 67 37W
Jonesville, Ind., U.S.A. 157 E11 39 5N 85 54W
Jonesville, Mich., U.S.A. 157 C12 41 59N 84 40W
Jonglei, Sudan 107 F3 6 25N 30 50 E
Jonglei □, Sudan 107 F3 7 30N 32 30 E
Joniškis, Lithuania 15 H20 56 13N 23 35 E
Jönköping, Sweden 17 G8 57 45N 14 8 E
Jönköpings län □, Sweden 17 G8 57 30N 14 30 E
Jonquière, Canada 141 C5 48 27N 71 14W
Jonsered, Sweden 17 G6 57 45N 12 10 E
Jonzac, France 28 C3 45 27N 0 28W
Joplin, U.S.A. 155 G7 37 6N 94 31W
Jordan, U.S.A. 158 C10 47 19N 106 55W
Jordan ■, Asia 103 E5 31 0N 36 0 E
Jordan →, Asia 103 D4 31 48N 35 32 E
Jordan Valley, U.S.A. 158 E5 42 59N 117 3W
Jordânia, Brazil 171 E3 15 55 S 40 11W
Jordanów, Poland 55 J6 49 41N 19 49 E
Jordet, Norway 18 C9 61 25N 12 8 E
Jorge, C., Chile 176 D1 51 40 S 75 35W
Jorgen Brønlund Fjord, Greenland 10 A8 82 30N 29 0W
Jorhat, India 90 B5 26 45N 94 12 E
Jorm, Afghan. 91 A3 36 50N 70 52 E
Jörn, Sweden 14 D19 65 4N 20 1 E
Jorong, Indonesia 85 C4 3 58 S 114 56 E
Jørpeland, Norway 15 G11 59 3N 6 1 E
Jorquera →, Chile 174 B2 28 3 S 69 58W
Jos, Nigeria 113 D6 9 53N 8 51 E
Jošanička Banja, Serbia, Yug. 50 C4 43 24N 20 47 E
Jose Abad Santos, Phil. 81 J5 5 55N 125 39 E
José Batlle y Ordóñez, Uruguay 175 C4 33 20 S 55 10W
José de San Martín, Argentina 176 B2 44 4 S 70 26W
Jose Panganiban, Phil. 80 D4 14 17N 122 41 E
Joseni, Romania 53 D10 46 42N 25 29 E
Joseph, U.S.A. 158 D5 45 21N 117 14W
Joseph, L., Nfld., Canada 141 B6 52 45N 65 18W
Joseph, L., Ont., Canada 150 A5 45 10N 79 44W
Joseph Bonaparte G., Australia 124 B4 14 35 S 128 50 E
Joseph City, U.S.A. 159 J8 34 57N 110 20W
Joshua Tree, U.S.A. 161 L10 34 8N 116 19W
Joshua Tree National Monument, U.S.A. 161 M10 33 55N 116 0W
Josselin, France 26 E4 47 57N 2 33W
Jostedal, Norway 18 C4 61 35N 7 15 E
Jostedalsbreen, Norway 15 F12 61 40N 6 59 E
Jotunheimen, Norway 15 F13 61 35N 8 25 E
Joué-lès-Tours, France 26 E7 47 21N 0 40 E
Jourdanton, U.S.A. 155 L5 28 55N 98 33W
Joussard, Canada 142 B5 55 22N 115 50W
Joutseno, Finland 58 B5 61 7N 28 31 E
Jovellanos, Cuba 164 B3 22 40N 81 10W
Jovellar, Phil. 80 E4 13 4N 123 36 E
Jowai, India 90 C4 25 26N 92 12 E
Jowzjān □, Afghan. 91 A2 36 10N 66 0 E
Joyeuse, France 29 D8 44 29N 4 16 E
Józefów, Warszawa, Poland 55 F8 52 10N 21 11 E
Józefów, Zamość, Poland 55 H10 50 28N 23 2 E
Ju Xian, China 75 F10 36 35N 118 20 E
Juan Aldama, Mexico 162 C4 24 20N 103 23W
Juan Bautista Alberdi, Argentina 174 C3 34 26 S 61 48W
Juan de Fuca Str., Canada 160 B3 48 15N 124 0W
Juan de Nova, Ind. Oc. 117 B7 17 3 S 43 45 E
Juan Fernández, Arch. de, Pac. Oc. 135 L20 33 50 S 80 0W
Juan José Castelli, Argentina 174 B3 25 27 S 60 57W
Juan L. Lacaze, Uruguay 174 C4 34 26 S 57 25W
Juanjuí, Peru 172 B2 7 10 S 76 45W
Juankoski, Finland 14 E23 63 3N 28 19 E
Juárez, Argentina 174 D4 37 40 S 59 43W
Juárez, Mexico 161 N11 32 20N 115 57W
Juárez, Sierra de, Mexico 162 A1 32 0N 116 0W
Juatinga, Ponta de, Brazil 171 F3 23 17 S 44 30W
Juàzeiro, Brazil 170 C3 9 30 S 40 30W
Juàzeiro do Norte, Brazil 170 C4 7 10 S 39 18W
Jûbâ, Sudan 107 G3 4 50N 31 35 E
Jubay, Phil. 81 F5 11 33N 124 18 E
Jubayl, Lebanon 103 A4 34 5N 35 39 E
Jubbah, Si. Arabia 96 D4 28 2N 40 56 E
Jubbulpore = Jabalpur, India 93 H8 23 9N 79 58 E
Jübek, Germany 30 A5 54 33N 9 22 E
Jubga, Russia 61 H4 44 19N 38 48 E
Jubilee L., Australia 125 E4 29 0 S 126 50 E
Juby, C., Morocco 110 C2 28 0N 12 59W
Júcar = Xúquer →, Spain 41 F4 39 5N 0 10W
Júcaro, Cuba 164 B4 21 37N 78 51W
Juchitán, Mexico 163 D5 16 27N 95 5W
Judaberg, Norway 18 E2 59 10N 5 51 E
Judaea = Har Yehuda, Israel 103 D3 31 35N 34 57 E
Judenburg, Austria 34 D7 47 12N 14 38 E
Judith →, U.S.A. 158 C9 47 44N 109 39W
Judith, Pt., U.S.A. 151 E13 41 22N 71 29W
Judith Gap, U.S.A. 158 C9 46 41N 109 45W
Juelsminde, Denmark 17 J4 55 43N 10 1 E
Jufari →, Brazil 169 D5 1 13 S 62 0W
Jugoslavia = Yugoslavia ■, Europe 50 C4 43 20N 20 0 E
Juigalpa, Nic. 164 D2 12 6N 85 26W
Juillac, France 28 C5 45 20N 1 19 E
Juist, Germany 30 B2 53 40N 6 59 E
Juiz de Fora, Brazil 171 F3 21 43 S 43 19W
Jujuy □, Argentina 174 A2 23 20 S 65 40W
Julesburg, U.S.A. 154 E3 40 59N 102 16W
Juli, Peru 172 D4 16 10 S 69 25W
Julia Cr. →, Australia 126 C3 20 0 S 141 11 E
Julia Creek, Australia 126 C3 20 39 S 141 44 E
Juliaca, Peru 172 D3 15 25 S 70 10W
Julian, U.S.A. 161 M10 33 4N 116 38W
Julian Alps = Julijske Alpe, Slovenia 45 B11 46 15N 14 1 E
Julianatop, Surinam 169 C6 3 40N 56 30W
Julianehåb, Greenland 10 E6 60 43N 46 0W
Jülich, Germany 30 E2 50 55N 6 22 E
Julierpass, Switz. 33 D9 46 28N 9 32 E
Juliette, L., U.S.A. 152 B6 33 2N 83 50W
Julijske Alpe, Slovenia 45 B11 46 15N 14 1 E
Julimes, Mexico 162 B3 28 25N 105 27W
Jullundur, India 92 D6 31 20N 75 40 E
Julu, China 74 F8 37 15N 115 2 E
Jumbo, Zimbabwe 119 F3 17 30 S 30 58 E
Jumbo Pk., U.S.A. 161 J12 36 12N 114 11W
Jumentos Cays, Bahamas 165 B4 23 0N 75 40W
Jumilla, Spain 41 G3 38 28N 1 19W
Jumla, Nepal 93 E10 29 15N 82 13 E
Jumna = Yamuna →, India 93 G9 25 30N 81 53 E
Junagadh, India 92 J4 21 30N 70 30 E
Junaynah, Si. Arabia 98 B4 22 33N 46 18 E
Junction, Tex., U.S.A. 155 K5 30 29N 99 46W
Junction, Utah, U.S.A. 159 G7 38 14N 112 13W
Junction B., Australia 126 A1 11 52 S 133 55 E
Junction City, Ga., U.S.A. 152 C5 32 36N 84 28W
Junction City, Kans., U.S.A. 154 F6 39 2N 96 50W
Junction City, Oreg., U.S.A. 158 D2 44 13N 123 12W
Junction Pt., Australia 126 A1 11 45 S 133 50 E
Jundah, Australia 126 C3 24 46 S 143 2 E
Jundiaí, Brazil 175 A6 24 30 S 47 0W
Juneau, U.S.A. 138 C6 58 18N 134 25W
Junee, Australia 129 C7 34 53 S 147 35 E
Jungfrau, Switz. 32 C5 46 32N 7 58 E
Junggar Pendi, China 72 B3 44 30N 86 0 E
Jungsdalshytta, Norway 18 D4 60 49N 7 55 E
Jungshahi, Pakistan 92 G2 24 52N 67 44 E
Juniata →, U.S.A. 150 F7 40 30N 77 40W
Junín, Argentina 174 C3 34 33 S 60 57W
Junín, Peru 172 C2 11 12 S 76 0W
Junín □, Peru 172 C3 11 30 S 75 0W
Junín de los Andes, Argentina 176 A2 39 45 S 71 0W
Juniper, U.S.A. 152 C5 32 32N 84 36W
Jūniyah, Lebanon 103 B4 33 59N 35 38 E
Junnar, India 94 E1 19 12N 73 58 E
Juno Beach, U.S.A. 153 J9 26 52N 80 3W
Juntura, U.S.A. 158 E4 43 45N 118 5W
Juparanã, L., Brazil 171 E3 19 16 S 40 8W
Jupiter, U.S.A. 153 J9 26 57N 80 6W
Jupiter →, Canada 141 C7 49 29N 63 37W
Juquiá, Brazil 171 F2 24 19 S 47 38W
Jur, Nahr el →, Sudan 107 F2 8 45N 29 15 E
Jura = Jura, Mts. du, Europe 27 F13 46 40N 6 5 E

K

Karoi 231

Kamensk-Shakhtinskiy, Russia 61 F5 48 23N 40 20 E
Kamensk Uralskiy, Russia 62 C9 56 25N 62 2 E
Kamenskiy, Russia 60 E7 50 48N 45 25 E
Kamenskoye, Russia 65 C17 62 45N 165 30 E
Kamenyak, Bulgaria 51 C10 43 24N 26 57 E
Kamenz, Germany 30 D10 51 15N 14 5 E
Kameoka, Japan 71 C7 35 0N 135 35 E
Kameyama, Japan 71 C8 34 51N 136 27 E
Kami, Albania 50 D4 42 17N 20 18 E
Kami-Jima, Japan 70 E2 32 27N 130 20 E
Kami-koshiki-Jima, Japan 70 F1 31 50N 129 52 E
Kamiah, U.S.A. 158 C5 46 14N 116 2W
Kamień Krajeński, Poland 54 E4 53 32N 17 32 E
Kamień Pomorski, Poland 54 E1 53 57N 14 43 E
Kamienna →, Poland 55 G8 51 6N 21 47 E
Kamienna Góra, Poland 55 H3 50 47N 16 2 E
Kamieńsk, Poland 55 G6 51 12N 19 29 E
Kamieskroon, S. Africa 116 E2 30 9 S 17 56 E
Kamiita, Japan 70 C6 34 6N 134 22 E
Kamilukuak, L., Canada 143 A8 62 22N 101 40W
Kamin-Kashyrskyy, Ukraine 59 G3 51 39N 24 56 E
Kamina, Dem. Rep. of the Congo 115 D5 8 45 S 25 0 E
Kaminak L., Canada 143 A10 62 10N 95 0W
Kaminka, Ukraine 59 H7 49 3N 32 6 E
Kaminoyama, Japan 68 E10 38 9N 140 17 E
Kamioka, Japan 71 A9 36 25N 137 15 E
Kamiros, Greece 38 C9 36 20N 27 56 E
Kamishak Bay, U.S.A. 144 G9 59 15N 153 45W
Kamituga, Dem. Rep. of the Congo 118 C2 3 2 S 28 10 E
Kamloops, Canada 142 C4 50 40N 120 20W
Kamnik, Slovenia 45 B11 46 14N 14 37 E
Kamo, Armenia 61 K7 40 21N 45 7 E
Kamo, Japan 68 F9 37 39N 139 3 E
Kamo, N.Z. 130 B3 35 42 S 174 20 E
Kamoa Mts., Guyana 169 C6 1 30N 59 0W
Kamogawa, Japan 71 B12 35 5N 140 5 E
Kamoke, Pakistan 92 C6 32 4N 74 4 E
Kamooloa, U.S.A. 145 J13 21 34N 158 7W
Kamp →, Austria 34 C8 48 23N 15 42 E
Kampala, Uganda 118 B3 0 20N 32 30 E
Kampang Chhnang, Cambodia 87 F5 12 20N 104 35 E
Kampar, Malaysia 87 K3 4 18N 101 9 E
Kampar →, Indonesia 84 B2 0 30N 103 8 E
Kampen, Neths. 24 B5 52 33N 5 53 E
Kamphaeng Phet, Thailand 86 D2 16 28N 99 30 E
Kampolombo, L., Zambia 119 E2 11 37 S 29 42 E
Kampong Saom, Cambodia 87 G4 10 38N 103 30 E
Kampong Saom, Chaak, Cambodia 87 G4 10 50N 103 32 E
Kampong To, Thailand 87 J3 6 3N 101 13 E
Kampot, Cambodia 87 G5 10 36N 104 10 E
Kampsville, U.S.A. 156 E6 39 18N 90 37W
Kamptee, India 94 D4 21 9N 79 19 E
Kampti, Burkina Faso 112 C4 10 7N 3 25W
Kampuchea = Cambodia ■, Asia 86 F5 12 15N 105 0 E
Kampung →, Indonesia 83 C5 5 44 S 138 24 E
Kampung Air Putih, Malaysia 87 K4 4 15N 103 10 E
Kampung Jerangau, Malaysia 87 K4 4 50N 103 10 E
Kampung Raja, Malaysia 87 K4 5 45N 102 35 E
Kampungbaru = Tolitoli, Indonesia 82 A2 1 5N 120 50 E
Kamrau, Teluk, Indonesia 83 B4 3 30 S 133 36 E
Kamsack, Canada 143 C8 51 34N 101 54W
Kamskoye Ustye, Russia 60 C9 55 10N 49 20 E
Kamskoye Vdkhr., Russia 62 B6 58 41N 56 7 E
Kamuchawie L., Canada 143 B8 56 18N 101 59W
Kamuela, U.S.A. 145 C6 20 1N 155 41W
Kamui-Misaki, Japan 68 C10 43 20N 140 21 E
Kamyanets-Podilskyy, Ukraine 59 H4 48 45N 26 40 E
Kamyanka-Buzka, Ukraine 59 G3 50 8N 24 16 E
Kamyanka-Dniprovska, Ukraine 59 J8 47 29N 34 28 E
Kāmyārān, Iran 101 E12 34 47N 46 56 E
Kamyshin, Russia 60 E7 50 10N 45 24 E
Kamyshlov, Russia 62 C9 56 50N 62 43 E
Kamyzyak, Russia 61 G9 46 4N 48 10 E
Kan, Burma 90 D5 22 25N 94 5 E
Kanaaupscow, Canada 140 B4 54 2N 76 30W
Kanaaupscow →, Canada 139 C12 53 39N 77 9W
Kanab, U.S.A. 159 H7 37 3N 112 32W
Kanab →, U.S.A. 159 H7 36 24N 112 38W
Kanagawa □, Japan 71 B11 35 20N 139 20 E
Kanagi, Japan 68 D10 40 54N 140 27 E
Kanairiktok →, Canada 141 A7 55 2N 60 18W
K'anak, Greenland 10 B4 77 30N 69 0W
Kanakapura, India 95 H3 12 33N 77 28 E
Kanália, Greece 48 B4 39 30N 22 53 E
Kananga, Dem. Rep. of the Congo 115 D4 5 55 S 22 18 E
Kanarraville, U.S.A. 159 H7 37 32N 113 11W
Kanash, Russia 60 C8 55 30N 47 32 E
Kanaskat, U.S.A. 160 C5 47 19N 121 54W
Kanastraíon, Ákra = Paliouríon, Ákra, Greece 50 G7 39 57N 23 45 E
Kanawha →, U.S.A. 148 F4 38 50N 82 9W
Kanazawa, Japan 71 A8 36 30N 136 38 E
Kanbalu, Burma 90 D5 23 12N 95 31 E
Kanchanaburi, Thailand 86 E2 14 2N 99 31 E
Kanchenjunga, Nepal 93 F13 27 50N 88 10 E
Kanchipuram, India 95 H4 12 52N 79 45 E
Kańczuga, Poland 55 J9 49 59N 22 25 E
Kanda Kanda, Dem. Rep. of the Congo 115 D4 6 52 S 23 48 E
Kandahar = Qandahār, Afghan. 91 C2 31 32N 65 30 E
Kandalaksha, Russia 56 A5 67 9N 32 30 E
Kandalakshkiy Zaliv, Russia 56 A6 66 0N 35 0 E
Kandangan, Indonesia 85 C5 2 50 S 115 20 E
Kandanos, Greece 38 D5 35 19N 23 44 E
Kandava, Latvia 54 A9 57 2N 22 46 E
Kandavu, Fiji 133 B2 19 0 S 178 15 E
Kandavu Passage, Fiji 133 B2 18 45 S 178 0 E
Kandep, Papua N. G. 132 C2 5 54 S 143 32 E
Kander →, Switz. 32 C5 46 33N 7 38 E
Kandersteg, Switz. 32 D5 46 28N 7 40 E
Kandhíla, Greece 48 D4 37 46N 22 22 E
Kandhkot, Pakistan 92 E3 28 16N 69 8 E
Kandhla, India 92 E7 29 18N 77 19 E
Kandi, Benin 113 C5 11 7N 2 55 E
Kandi, India 93 H13 23 58N 88 5 E
Kandıra, Turkey 100 B4 41 4N 30 9 E
Kandla, India 92 H4 23 0N 70 10 E
Kandos, Australia 129 B8 32 45 S 149 58 E
Kandrian, Papua N. G. 132 D5 6 14 S 149 37 E
Kandy, Sri Lanka 95 L5 7 18N 80 43 E
Kane, U.S.A. 150 E6 41 40N 78 49W
Kane Basin, Greenland 10 B4 79 1N 70 0W
Kaneilio Pt., U.S.A. 145 K13 21 27N 158 12W
Kaneohe B., U.S.A. 145 K14 21 30N 157 50W
Kanevskaya, Russia 61 G4 46 3N 38 57 E
Kanfanar, Croatia 45 C10 45 7N 13 50 E
Kangaba, Mali 112 C3 11 56N 8 25W
Kangal, Turkey 100 C7 39 14N 37 23 E
Kangâmiut, Greenland 10 D5 65 50N 53 20W
Kangān, Fārs, Iran 97 E7 27 50N 52 3 E
Kangān, Hormozgān, Iran 97 E8 25 48N 57 28 E
Kangar, Malaysia 87 J3 6 27N 100 12 E
Kangaroo I., Australia 128 C2 35 45 S 137 0 E
Kangasala, Finland 15 F21 61 28N 24 4 E
Kangāvar, Iran 97 C6 34 40N 48 0 E
Kangding, China 76 B3 30 2N 101 57 E
Kangdong, N. Korea 75 E14 39 9N 126 5 E
Kangean, Kepulauan, Indonesia 85 D5 6 55 S 115 23 E
Kangean Is. = Kangean, Kepulauan, Indonesia 85 D5 6 55 S 115 23 E
Kangerdlugssuak, Greenland 10 D7 68 10N 32 20W
Kangerluarsoruseq = Færingehavn, Greenland 10 E5 63 45N 51 27W
Kanggye, N. Korea 75 D14 41 0N 126 35 E
Kanggyŏng, S. Korea 75 F14 36 10N 127 0 E
Kanghwa, S. Korea 75 F14 37 45N 126 30 E
Kangilinnguit = Grønnedal, Greenland 10 E6 61 20N 47 57W
Kangiqsualujjuaq, Canada 139 C13 58 30N 65 59W
Kangiqsujuaq, Canada 139 B12 61 30N 72 0W
Kangirsuk, Canada 139 C13 60 0N 70 0W
Kangnŭng, S. Korea 75 F15 37 45N 128 54 E
Kango, Gabon 114 B2 0 11N 10 5 E
Kangoya, Dem. Rep. of the Congo 115 D4 9 55 S 22 48 E
Kangping, China 75 C12 42 43N 123 18 E
Kangpokpi, India 90 C4 25 8N 93 58 E
Kangyidaung, Burma 90 G5 16 56N 94 54 E
Kanhangad, India 95 H2 12 21N 74 58 E
Kanheri, India 94 E1 19 13N 72 50 E
Kani, Ivory C. 112 D3 8 29N 6 36W
Kaniama, Dem. Rep. of the Congo 115 D4 7 30 S 24 12 E
Kaniapiskau →, Canada 141 A6 56 40N 69 30W
Kaniapiskau L., Canada 141 B6 54 10N 69 55W
Kanibadam, Tajikistan 63 C5 40 17N 70 25 E
Kaniere, L., N.Z. 131 C6 42 50 S 171 10 E
Kanin, Poluostrov, Russia 56 A8 68 0N 45 0 E
Kanin Nos, Mys, Russia 56 A7 68 39N 43 32 E
Kanin Pen. = Kanin, Poluostrov, Russia 56 A8 68 0N 45 0 E
Kanina, Albania 50 F3 40 23N 19 30 E
Kaniva, Australia 128 D4 36 22 S 141 18 E
Kanjiža, Serbia, Yug. 52 D5 46 3N 20 4 E
Kanjut Sar, Pakistan 93 A6 36 7N 75 25 E
Kankaanpää, Finland 15 F20 61 44N 22 50 E
Kankakee, U.S.A. 157 C9 41 7N 87 52W
Kankakee →, U.S.A. 157 C8 41 23N 88 15W
Kankan, Guinea 112 C3 10 23N 9 15W
Kankendy = Xankändi, Azerbaijan 101 C12 39 52N 46 49 E
Kanker, India 94 D5 20 10N 81 40 E
Kanmuri-Yama, Japan 70 C4 34 30N 132 4 E
Kannabe, Japan 70 C5 34 32N 133 23 E
Kannapolis, U.S.A. 149 H5 35 30N 80 37W
Kannauj, India 93 F8 27 3N 79 56 E
Kano, Nigeria 113 C6 12 2N 8 30 E
Kano □, Nigeria 113 C6 11 30N 8 30 E
Kan'onji, Japan 70 C5 34 7N 133 39 E
Kanoroba, Ivory C. 112 D3 9 7N 6 8W
Kanowha, U.S.A. 156 B3 42 57N 93 47W
Kanowit, Malaysia 85 B4 2 14N 112 20 E
Kanowna, Australia 125 F3 30 32 S 121 31 E
Kanoya, Japan 70 F2 31 25N 130 50 E
Kanpetlet, Burma 90 E4 21 10N 93 59 E
Kanpur, India 93 F9 26 28N 80 20 E
Kansas, U.S.A. 157 E9 39 33N 87 56W
Kansas □, U.S.A. 154 F6 38 30N 99 0W
Kansas →, U.S.A. 154 F7 39 7N 94 37W
Kansas City, Kans., U.S.A. 156 E2 39 7N 94 38W
Kansas City, Mo., U.S.A. 156 E2 39 6N 94 35W
Kansenia, Dem. Rep. of the Congo 119 E2 10 20 S 26 0 E
Kansk, Russia 65 D10 56 20N 95 37 E
Kansŏng, S. Korea 75 E15 38 24N 128 30 E
Kansu = Gansu □, China 74 G3 36 0N 104 0 E
Kant, Kyrgyzstan 63 B7 42 53N 74 51 E
Kantché, Niger 109 F1 13 31N 8 30 E
Kanté, Togo 113 D5 9 57N 1 3 E
Kantemirovka, Russia 59 H10 49 43N 39 55 E
Kantharalak, Thailand 86 E5 14 39N 104 39 E
Kantishna →, U.S.A. 144 D10 64 45N 149 58W
Kantō □, Japan 71 A11 36 15N 139 30 E
Kantō-Heiya, Japan 71 B11 36 0N 139 30 E
Kantō-Sanchi, Japan 71 B10 35 59N 138 50 E
Kantu-long, Burma 90 F6 19 57N 97 36 E
Kanturk, Ireland 23 D3 52 11N 8 54W
Kanuma, Japan 71 A11 36 34N 139 42 E
Kanus, Namibia 116 D2 27 50 S 18 39 E
Kanye, Botswana 116 C4 24 55 S 25 28 E
Kanzenze, Dem. Rep. of the Congo 115 E5 10 30 S 25 12 E
Kanzi, Ras, Tanzania 118 D4 7 1 S 39 33 E
Kao, Fiji 133 P13 19 40 S 175 1W
Kaohsiung, Taiwan 77 F13 22 35N 120 16 E
Kaokoveld, Namibia 116 B1 19 15 S 14 30 E
Kaolack, Senegal 112 C1 14 5N 16 8W
Kaoshan, China 75 B13 44 38N 124 50 E
Kaouar, Niger 109 E2 19 5N 12 52 E
Kapadvanj, India 92 H5 23 5N 73 0 E
Kapagere, Papua N. G. 132 E4 9 46 S 147 42 E
Kapahulu, U.S.A. 145 K14 21 16N 157 49W
Kapaklı, Turkey 51 E11 41 19N 27 59 E
Kapan, Armenia 101 C12 39 18N 46 27 E
Kapanga, Dem. Rep. of the Congo 115 D4 8 30 S 22 40 E
Kapapa I., U.S.A. 145 K14 21 29N 157 48W
Kapatagan, Phil. 81 H4 7 52N 123 44 E
Kapchagai = Qapshaghay, Kazakstan 63 B8 43 51N 77 14 E
Kapchagaiskoye Vdkhr. = Qapshaghay Bögeni, Kazakstan 63 B8 43 45N 77 50 E
Kapela = Velika Kapela, Croatia 45 C12 45 10N 15 5 E
Kapéllo, Ákra, Greece 48 E5 36 9N 23 3 E
Kapema, Dem. Rep. of the Congo 119 E2 10 45 S 28 22 E
Kapfenberg, Austria 34 D8 47 26N 15 18 E
Kapı Dağı, Turkey 51 F11 40 28N 27 50 E
Kapia, Dem. Rep. of the Congo 115 C3 4 17 S 19 46 E
Kapiri Mposhi, Zambia 119 E2 13 59 S 28 43 E
Kāpīsā □, Afghan. 91 B3 35 0N 69 20 E
Kapiskau →, Canada 140 B3 52 47N 81 55W
Kapit, Malaysia 85 B4 2 0N 112 55 E
Kapiti I., N.Z. 130 G3 40 50 S 174 56 E
Kapka, Massif du, Chad 109 E4 15 7N 21 45 E
Kaplice, Czech Rep. 34 C7 48 42N 14 30 E
Kapoe, Thailand 87 H2 9 34N 98 32 E
Kapoeta, Sudan 107 G3 4 50N 33 35 E
Kápolnásnyék, Hungary 52 C3 47 16N 18 41 E
Kaponga, N.Z. 130 F3 39 29 S 174 9 E
Kapos →, Hungary 52 D3 46 44N 18 30 E
Kaposvár, Hungary 52 D2 46 25N 17 47 E
Kapowsin, U.S.A. 160 D4 46 59N 122 13W
Kapp, Norway 18 D7 60 43N 10 52 E
Kappeln, Germany 30 A5 54 40N 9 55 E
Kappelshamn, Sweden 17 G12 57 52N 18 47 E
Kapps, Namibia 116 C2 22 32 S 17 18 E
Kaprije, Croatia 45 E12 43 42N 15 43 E
Kapsan, N. Korea 75 D15 41 4N 128 19 E
Kapsukas = Marijampolė, Lithuania 15 J20 54 33N 23 19 E
Kapuas, Indonesia 85 C4 3 10 S 114 5 E
Kapuas →, Indonesia 85 C3 0 25 S 109 20 E
Kapuas Hulu, Pegunungan, Malaysia 85 B4 1 30N 113 30 E
Kapuas Hulu Ra. = Kapuas Hulu, Pegunungan, Malaysia 85 B4 1 30N 113 30 E
Kapulo, Dem. Rep. of the Congo 119 D2 8 18 S 29 15 E
Kapunda, Australia 128 C3 34 20 S 138 56 E
Kapuni, N.Z. 130 F3 39 29 S 174 8 E
Kapurthala, India 92 D6 31 23N 75 25 E
Kapuskasing, Canada 140 C3 49 25N 82 30W
Kapuskasing →, Canada 140 C3 49 49N 82 0W
Kapustin Yar, Russia 61 F7 48 37N 45 40 E
Kaputar, Australia 127 E5 30 15 S 150 10 E
Kaputir, Kenya 118 B4 2 5N 35 28 E
Kapuvár, Hungary 52 C2 47 36N 17 1 E
Kara, Russia 64 C7 69 10N 65 0 E
Kara Ada, Turkey 49 E9 36 58N 27 28 E
Kara-Balta, Kyrgyzstan 63 B6 42 50N 73 49 E
Kara Bogaz Gol, Zaliv = Garabogazköl Aylagy, Turkmenistan 57 F9 41 0N 53 30 E
Kara Burun, Turkey 49 E9 36 32N 27 58 E
Kara Kalpak Republic □ = Karakalpakstan □, Uzbekistan 64 E6 43 0N 58 0 E
Kara Kum, Turkmenistan 64 F7 39 30N 60 0 E
Kara-Saki, Japan 70 C1 34 41N 129 30 E
Kara Sea, Russia 64 B8 75 0N 70 0 E
Kara Su, Kyrgyzstan 63 C6 40 44N 72 53 E
Karaadilli, Turkey 49 C12 38 18N 30 37 E
Karabash, Russia 62 D8 55 29N 60 14 E
Karabekaul = Garabekewül, Turkmenistan 63 D2 38 30N 64 8 E
Karabiğa, Turkey 51 F11 40 23N 27 17 E
Karabük, Turkey 100 B5 41 12N 32 37 E
Karabulak, Kazakstan 63 A9 44 54N 78 30 E
Karaburun, Turkey 49 C8 38 41N 26 28 E
Karaburuni, Albania 50 F3 40 25N 19 20 E
Karabutak = Qarabutaq, Kazakstan 62 G8 49 59N 60 14 E
Karacabey, Turkey 51 F12 40 12N 28 21 E
Karacakılavuz, Turkey 51 E11 41 8N 27 21 E
Karacaköy, Turkey 51 E12 41 24N 28 22 E
Karacasu, Turkey 49 D10 37 43N 28 35 E
Karachala = Qaraçala, Azerbaijan 61 L9 39 45N 48 53 E
Karachayevsk, Russia 61 J5 43 50N 41 55 E
Karachev, Russia 59 F8 53 10N 35 5 E
Karachey-Cherkessia □, Russia 61 J5 43 40N 41 30 E
Karachi, Pakistan 91 D2 24 53N 67 0 E
Karad, India 94 F2 17 15N 74 10 E
Karadirek, Turkey 49 C12 38 34N 30 11 E
Karaga, Ghana 113 D4 9 58N 0 28W
Karaganda = Qaraghandy, Kazakstan 64 E8 49 50N 73 10 E
Karagayly, Kazakstan 64 E8 49 26N 76 0 E
Karaginskiy, Ostrov, Russia 65 D17 58 45N 164 0 E
Karagiye, Vpadina, Kazakstan 57 F9 43 27N 51 45 E
Karagiye Depression = Karagiye, Vpadina, Kazakstan 57 F9 43 27N 51 45 E
Karagüney Dağları, Turkey 100 B6 40 30N 34 40 E
Karagwe □, Tanzania 118 C3 2 0 S 31 0 E
Karahallı, Turkey 49 C11 38 21N 29 33 E
Karaikal, India 95 J4 10 59N 79 50 E
Karaikkudi, India 95 J4 10 5N 78 45 E
Karaisali, Turkey 100 D6 37 16N 35 2 E
Karaitivu I., Sri Lanka 95 K4 9 45N 79 52 E
Karaj, Iran 97 C6 35 48N 51 0 E
Karak, Malaysia 87 L4 3 25N 102 2 E
Karakalpakstan □, Uzbekistan 64 E6 43 0N 58 0 E
Karakitang, Indonesia 82 A3 3 14N 125 28 E
Karaklis = Vanadzor, Armenia 61 K7 40 48N 44 30 E
Karakoçan, Turkey 101 C9 38 57N 40 2 E
Karakoram Pass, Pakistan 93 B7 35 33N 77 50 E
Karakoram Ra., Pakistan 93 B7 35 30N 77 0 E
Karakul, Tajikistan 63 D6 39 2N 73 33 E
Karakul, Uzbekistan 63 D1 39 22N 63 50 E
Karakuldzha, Kyrgyzstan 63 C6 40 39N 73 26 E
Karakurt, Turkey 101 B10 40 10N 42 37 E
Karal, Chad 109 F2 12 50N 14 46 E
Karalon, Russia 65 D12 57 5N 115 50 E
Karaman, Balıkesir, Turkey 49 B9 39 39N 28 0 E
Karaman, Konya, Turkey 100 D5 37 14N 33 13 E
Karamanlı, Turkey 49 D11 37 23N 29 47 E
Karamay, China 72 B3 45 30N 84 58 E
Karambu, Indonesia 85 C5 3 53 S 116 6 E
Karamea, N.Z. 131 B7 41 14 S 172 6 E
Karamea →, N.Z. 131 B7 41 13 S 172 26 E
Karamea Bight, N.Z. 131 B6 41 22 S 171 40 E
Karamet Niyaz = Garamätnyyaz, Turkmenistan 63 E2 37 45N 64 34 E
Karamsad, India 92 H5 22 35N 72 50 E
Karamürsel, Turkey 51 F13 40 41N 29 36 E
Karand, Iran 101 E12 34 16N 46 15 E
Karanganyar, Indonesia 85 D3 7 38 S 109 37 E
Karanja, India 94 D3 20 29N 77 31 E
Karaova, Turkey 49 D9 37 5N 27 40 E
Karapınar, Turkey 100 D5 37 41N 33 30 E
Karapiro, N.Z. 130 D4 37 53 S 175 32 E
Karasburg, Namibia 116 D2 28 0 S 18 44 E
Karasino, Russia 64 C9 66 50N 86 50 E
Karasjok, Norway 14 B21 69 27N 25 30 E
Karasu, Turkey 100 B4 41 4N 30 46 E
Karasu →, Turkey 49 E12 36 18N 30 10 E
Karasuk, Russia 64 D8 53 44N 78 2 E
Karasuyama, Japan 71 A12 36 39N 140 9 E
Karataş, Adana, Turkey 100 D6 36 36N 35 21 E
Karataş, Manisa, Turkey 49 C10 38 35N 28 16 E
Karataş Burnu, Turkey 100 D6 36 31N 35 24 E
Karatau = Qarataū, Kazakstan 63 B5 43 10N 70 28 E
Karatau, Khrebet, Kazakstan 63 B4 43 30N 69 30 E
Karativu, Sri Lanka 95 K4 8 22N 79 47 E
Karatobe, Kazakstan 62 G4 49 44N 53 30 E
Karatoprak, Turkey 49 D9 37 2N 27 15 E
Karatoya →, India 90 C2 24 7N 89 36 E
Karaturuk, Kazakstan 63 B8 43 35N 77 50 E
Karaul-Bazar, Uzbekistan 63 D2 39 30N 64 48 E
Karauli, India 92 F7 26 30N 77 4 E
Karavastasë, L. e, Albania 50 F3 40 55N 19 30 E
Karávi, Greece 48 E5 36 49N 23 37 E
Karavostasi, Cyprus 38 D11 35 8N 32 50 E
Karawa, Dem. Rep. of the Congo 114 B4 3 18N 20 17 E
Karawang, Indonesia 85 D3 6 30 S 107 15 E
Karawanken, Europe 34 E7 46 30N 14 40 E
Karayazı, Turkey 101 C10 39 41N 42 9 E
Karazhal, Kazakstan 64 E8 48 2N 70 49 E
Karbalā, Iraq 101 F11 32 36N 44 3 E
Kårböle, Sweden 16 C9 61 59N 15 22 E
Karcag, Hungary 52 C5 47 19N 20 57 E
Karcha →, Pakistan 93 B7 34 45N 76 10 E
Karczew, Poland 55 F8 52 5N 21 15 E
Kardam, Bulgaria 51 C12 43 45N 28 6 E
Kardeljovo = Ploče, Croatia 45 E14 43 4N 17 26 E
Kardhámila, Greece 49 C8 38 35N 26 5 E
Kardhamíli, Greece 48 E4 36 53N 22 13 E
Kardhítsa, Greece 48 B3 39 23N 21 54 E
Kardhítsa □, Greece 48 B3 39 15N 21 50 E
Kärdla, Estonia 15 G20 58 50N 22 40 E
Kareeberge, S. Africa 116 E3 30 59 S 21 50 E
Kareima, Sudan 106 D3 18 30N 31 49 E
Karelia □, Russia 56 A5 65 30N 32 30 E
Karelian Republic □ = Karelia □, Russia 56 A5 65 30N 32 30 E
Karema, Papua N. G. 132 E4 9 12 S 147 18 E
Kārevāndar, Iran 97 E9 27 53N 60 44 E
Kargasok, Russia 64 D9 59 3N 80 53 E
Kargat, Russia 64 D9 55 10N 80 15 E
Kargı, Turkey 100 B6 41 11N 34 30 E
Kargil, India 93 B7 34 32N 76 12 E
Kargopol, Russia 58 B10 61 30N 38 58 E
Kargowa, Poland 55 F2 52 5N 15 51 E
Karguéri, Niger 109 F2 13 27N 10 30 E
Kariá, Greece 48 C2 38 45N 20 39 E
Karia ba Mohammed, Morocco 110 B3 34 22N 5 12W
Kariaí, Greece 51 F8 40 14N 24 19 E
Karīān, Iran 97 E8 26 57N 57 14 E
Kariba, Zimbabwe 119 F2 16 28 S 28 50 E
Kariba, L., Zimbabwe 119 F2 16 40 S 28 25 E
Kariba Dam, Zimbabwe 119 F2 16 30 S 28 35 E
Kariba Gorge, Zambia 119 F2 16 30 S 28 50 E
Karibib, Namibia 116 C2 22 0 S 15 56 E
Karikari, C., N.Z. 130 A2 34 46 S 173 24 E
Karimata, Kepulauan, Indonesia 85 C3 1 25 S 109 0 E
Karimata, Selat, Indonesia 85 C3 2 0 S 108 40 E
Karimata Is. = Karimata, Kepulauan, Indonesia 85 C3 1 25 S 109 0 E
Karimnagar, India 94 E4 18 26N 79 10 E
Karimunjawa, Kepulauan, Indonesia 85 D4 5 50 S 110 30 E
Karin, Somali Rep. 120 B3 10 50N 45 52 E
Káristos, Greece 48 C6 38 1N 24 29 E
Karīt, Iran 97 C8 33 29N 56 55 E
Kariya, Japan 71 C9 34 58N 137 1 E
Karjala, Finland 58 A5 62 0N 30 25 E
Karkal, India 95 H2 13 15N 74 56 E
Karkar I., Papua N. G. 132 C4 4 40 S 146 0 E
Karkaralinsk = Qarqaraly, Kazakstan 64 E8 49 26N 75 30 E
Karkinitska Zatoka, Ukraine 59 K7 45 56N 33 0 E
Karkinitskiy Zaliv = Karkinitska Zatoka, Ukraine 59 K7 45 56N 33 0 E
Karkur Tohl, Egypt 106 C2 22 5N 25 5 E
Karl Liebknecht, Russia 59 G8 51 40N 35 35 E
Karl-Marx-Stadt = Chemnitz, Germany 30 E8 50 51N 12 54 E
Karlholmsbruk, Sweden 16 D11 60 31N 17 37 E
Karlino, Poland 54 D2 54 3N 15 53 E
Karlivka, Ukraine 59 H8 49 29N 35 8 E
Karlobag, Croatia 45 D12 44 32N 15 5 E
Karlovac, Croatia 45 C12 45 31N 15 36 E
Karlovka = Karlivka, Ukraine 59 H8 49 29N 35 8 E
Karlovo, Bulgaria 51 D8 42 38N 24 47 E
Karlovy Vary, Czech Rep. 34 A5 50 13N 12 51 E
Karlsbad = Karlovy Vary, Czech Rep. 34 A5 50 13N 12 51 E
Karlsborg, Sweden 17 F8 58 33N 14 33 E
Karlshamn, Sweden 17 H8 56 10N 14 51 E
Karlshus, Norway 18 E7 59 21N 10 52 E
Karlskoga, Sweden 16 E8 59 28N 14 33 E
Karlskrona, Sweden 17 H9 56 10N 15 35 E
Karlsruhe, Germany 31 F4 49 0N 8 23 E
Karlstad, Sweden 16 E7 59 23N 13 30 E
Karlstad, U.S.A. 154 A6 48 35N 96 31W
Karlstadt, Germany 31 F5 49 57N 9 47 E
Karluk, U.S.A. 144 H9 57 34N 154 28W
Karluk Indian Reservation, U.S.A. 144 H9 57 35N 154 20W
Karmėlava, Lithuania 54 D11 54 58N 24 4 E
Karmøy, Norway 18 E2 59 15N 5 15 E
Karnal, India 92 E7 29 42N 77 2 E
Karnali →, Nepal 93 E9 28 45N 81 16 E
Karnaphuli Res., Bangla. 90 D4 22 40N 92 20 E
Karnataka □, India 95 H3 13 15N 77 0 E
Karnes City, U.S.A. 155 L6 28 53N 97 54W
Karnische Alpen, Europe 34 E6 46 36N 13 0 E
Karnobat, Bulgaria 51 D10 42 39N 26 59 E
Kärnten □, Austria 34 E6 46 52N 13 30 E
Karo, Mali 112 C4 12 16N 3 18W
Karoi, Zimbabwe 119 F2 16 48 S 29 45 E

236 Korong Vale

Korong Vale, Australia 128 D5 36 22 S 143 45 E
Koróni, Greece 48 E3 36 48N 21 57 E
Korónia, Limni, Greece 50 F7 40 47N 23 10 E
Koronís, Greece 49 D7 37 12N 25 35 E
Koronowo, Poland 55 E4 53 19N 17 55 E
Körös →, Hungary 52 D5 46 43N 20 12 E
Köröstarcsa, Hungary 52 D6 46 53N 21 3 E
Korosten, Ukraine 59 G5 50 54N 28 36 E
Korostyshev, Ukraine 59 G5 50 19N 29 4 E
Korotoyak, Russia 59 G10 51 1N 39 2 E
Korraraika, Helodranon' i, Madag. 117 B7 17 45 S 43 57 E
Korsakov, Russia 65 E15 46 36N 142 42 E
Korsberga, Sweden 17 G9 57 19N 15 7 E
Korshunovo, Russia 65 D12 58 37N 110 10 E
Korsør, Denmark 17 J5 55 20N 11 9 E
Korsun Shevchenkovskiy, Ukraine 59 H6 49 26N 31 16 E
Korsze, Poland 54 D8 54 11N 21 9 E
Korti, Sudan 106 D3 18 6N 31 33 E
Kortrijk, Belgium 24 D3 50 50N 3 17 E
Korucu, Turkey 49 B9 39 28N 27 22 E
Korumburra, Australia 129 E6 38 26 S 145 50 E
Korwai, India 92 G8 24 7N 78 5 E
Koryakskoye Nagorye, Russia 65 C18 61 0N 171 0 E
Koryŏng, S. Korea 75 G15 35 44N 128 15 E
Koryukovka, Ukraine 59 G7 51 46N 32 16 E
Kos, Greece 49 E9 36 53N 27 18 E
Kosa, Ethiopia 107 F4 7 50N 36 50 E
Kosa, Russia 62 B5 59 56N 55 0 E
Kosa →, Russia 62 A5 60 11N 55 10 E
Kosaya Gora, Russia 58 E9 54 10N 37 30 E
Koschagyl, Kazakstan 57 E9 46 40N 54 0 E
Kościan, Poland 55 F3 52 5N 16 40 E
Kościerzyna, Poland 54 D4 54 8N 17 59 E
Kosciusko, U.S.A. 155 J10 33 4N 89 35W
Kosciusko I., U.S.A. 142 B2 56 0N 133 40W
Kosciuszko, Mt., Australia 129 D8 36 27 S 148 16 E
Kösély →, Hungary 52 C6 47 25N 21 5 E
Kosgi, India 94 F3 16 58N 77 43 E
Kosha, Sudan 106 C3 20 50N 30 30 E
Koshava, Bulgaria 50 B7 44 4N 23 2 E
Koshigaya, Japan 71 B11 35 54N 139 48 E
K'oshih = Kashi, China 72 C2 39 30N 76 2 E
Koshiki-Rettō, Japan 70 F1 31 45N 129 49 E
Koshkonong L., U.S.A. 157 B8 42 52N 88 58W
Kōshoku, Japan 71 A10 36 38N 138 6 E
Koshtëbë, Kyrgyzstan 63 C7 41 5N 74 15 E
Kosi, India 92 F7 27 48N 77 29 E
Košice, Slovak Rep. 35 C14 48 42N 21 15 E
Košický □, Slovak Rep. 35 C14 48 45N 21 0 E
Kosjerić, Serbia, Yug. 50 B3 44 0N 19 55 E
Köşk, Turkey 49 D10 37 50N 28 3 E
Koskhinoú, Greece 38 C10 36 23N 28 13 E
Koslan, Russia 56 B8 63 34N 49 14 E
Kosŏng, N. Korea 75 E15 38 40N 128 22 E
Kosovo □, Serbia, Yug. 50 D4 42 30N 21 0 E
Kosovska Mitrovica, Serbia, Yug. 50 D4 42 54N 20 52 E
Kossou, L. de, Ivory C. 112 D3 6 59N 5 31W
Kosta, Sweden 17 H9 56 50N 15 24 E
Kostajnica, Croatia 45 C13 45 17N 16 30 E
Kostamuksa, Russia 56 B5 62 34N 32 44 E
Kostanjevica, Slovenia 45 C12 45 51N 15 27 E
Kostenets, Bulgaria 50 D7 42 15N 23 52 E
Koster, S. Africa 116 D4 25 52 S 26 54 E
Kôstî, Sudan 107 E3 13 8N 32 43 E
Kostinbrod, Bulgaria 50 D7 42 49N 23 13 E
Kostolac, Serbia, Yug. 50 B5 44 37N 21 15 E
Kostopil, Ukraine 59 G4 50 51N 26 22 E
Kostroma, Russia 58 D11 57 50N 40 58 E
Kostromskoye Vdkhr., Russia 58 D11 57 52N 40 49 E
Kostrzyn, Gorzow Wlkp., Poland 55 F1 52 35N 14 39 E
Kostrzyn, Poznań, Poland 55 F4 52 24N 17 14 E
Kostyantynivka, Ukraine 59 H9 48 32N 37 39 E
Kostyukovichi = Kastsyukovichy, Belarus 58 F7 53 20N 32 4 E
Koszalin, Poland 54 D3 54 11N 16 8 E
Koszalin □, Poland 54 E3 53 40N 16 10 E
Kőszeg, Hungary 52 C1 47 23N 16 33 E
Kot Addu, Pakistan 91 C3 30 30N 71 0 E
Kot Moman, Pakistan 92 C5 32 13N 73 0 E
Kota, India 92 G6 25 14N 75 49 E
Kota Baharu, Malaysia 87 J4 6 7N 102 14 E
Kota Belud, Malaysia 85 A5 6 21N 116 26 E
Kota Kinabalu, Malaysia 85 A5 6 0N 116 4 E
Kota Kubu Baharu, Malaysia 87 L3 3 34N 101 39 E
Kota Tinggi, Malaysia 87 M4 1 44N 103 53 E
Kotaagung, Indonesia 84 D2 5 38 S 104 29 E
Kotabaru, Indonesia 85 C5 3 20 S 116 20 E
Kotabumi, Indonesia 84 C2 4 49 S 104 54 E
Kotagede, Indonesia 85 D4 7 54 S 110 26 E
Kotamobagu, Indonesia 82 A2 0 57N 124 31 E
Kotaneelee →, Canada 142 A4 60 11N 123 42W
Kotawaringin, Indonesia 85 C4 2 28 S 111 27 E
Kotchandpur, Bangla. 90 D2 23 24N 89 1 E
Kotcho L., Canada 142 B4 59 7N 121 12W
Kotel, Bulgaria 51 D10 42 52N 26 26 E
Kotelnich, Russia 60 A9 58 22N 48 24 E
Kotelnikovo, Russia 61 G6 47 38N 43 8 E
Kotelnyy, Ostrov, Russia 65 B14 75 10N 139 0 E
Kothagudam, India 94 F5 17 30N 80 40 E
Kothapet, India 94 E4 19 21N 79 28 E
Köthen, Germany 30 D7 51 45N 11 59 E
Kothi, India 93 G9 24 45N 80 40 E
Kotiro, Pakistan 92 F2 26 17N 67 13 E
Kotka, Finland 15 F22 60 28N 26 58 E
Kotlas, Russia 56 B8 61 17N 46 43 E
Kotlenska Planina, Bulgaria 51 D10 42 56N 26 30 E
Kotli, Pakistan 92 C5 33 30N 73 55 E
Kotlik, U.S.A. 144 E7 63 2N 163 33W
Kotmul, Pakistan 93 B6 35 32N 75 10 E
Kotohira, Japan 70 C5 34 11N 133 49 E
Kotonkoro, Nigeria 113 C6 11 3N 5 58 E
Kotor, Montenegro, Yug. 50 D2 42 25N 18 47 E
Kotor Varoš, Bos.-H. 52 F2 44 38N 17 22 E
Kotoriba, Croatia 45 B13 46 23N 16 48 E
Kotovo, Russia 60 E7 50 22N 44 45 E
Kotovsk, Russia 60 D5 52 36N 41 32 E
Kotovsk, Ukraine 59 J5 47 45N 29 35 E
Kotputli, India 92 F7 27 43N 76 12 E
Kotri, Pakistan 91 D3 25 22N 68 22 E
Kotri →, India 94 E5 19 15N 80 35 E
Kótronas, Greece 48 E4 36 38N 22 29 E
Kötschach-Mauthen, Austria 34 E6 46 41N 13 1 E
Kottayam, India 95 K3 9 35N 76 33 E
Kottur, India 95 J3 10 34N 76 56 E
Kotuy →, Russia 65 B11 71 54N 102 6 E
Kotzebue, U.S.A. 138 B3 66 53N 162 39W
Kouango, C.A.R. 114 B4 5 0N 20 10 E
Koudougou, Burkina Faso 112 C4 12 10N 2 20W
Koufonísi, Greece 38 E8 34 56N 26 8 E
Koufonísia, Greece 49 E7 36 57N 25 35 E
Kougaberge, S. Africa 116 E3 33 48 S 23 50 E
Kouibli, Ivory C. 112 D3 7 15N 7 14W
Kouilou →, Congo 115 C2 4 10 S 12 5 E
Kouki, C.A.R. 114 A3 7 22N 17 3 E
Koula Moutou, Gabon 114 C2 1 15 S 12 25 E
Koulen = Kulen, Cambodia 86 F5 13 50N 104 40 E
Koulikoro, Mali 112 C3 12 40N 7 50W
Kouloúra, Greece 38 A3 39 42N 19 54 E
Koúm-bournoú, Ákra, Greece 38 C10 36 15N 28 11 E
Koumac, N. Cal. 133 T18 20 33 S 164 17 E
Koumala, Australia 126 C4 21 38 S 149 15 E
Koumankou, Mali 112 C3 11 58N 6 6W
Koumbia, Burkina Faso 112 C4 11 10N 3 50W
Koumbia, Guinea 112 C2 11 48N 13 29W
Koumboum, Guinea 112 C2 10 25N 13 0W
Koumpenntoum, Senegal 112 C2 13 59N 14 34W
Koumra, Chad 109 G3 8 50N 17 35 E
Koundara, Guinea 112 C2 12 29N 13 18W
Koundé, C.A.R. 114 A2 6 7N 14 38 E
Kounradskiy, Kazakstan 64 E8 46 59N 75 0 E
Kountze, U.S.A. 155 K7 30 22N 94 19W
Koupéla, Burkina Faso 113 C4 12 11N 0 21W
Kouris →, Cyprus 38 E11 34 38N 32 54 E
Kourizo, Passe de, Chad 108 D3 22 28N 15 27 E
Kourou, Fr. Guiana 169 B7 5 9N 52 39W
Kouroussa, Guinea 112 C3 10 45N 9 45W
Koussané, Mali 112 C2 14 53N 11 14W
Kousseri, Cameroon 109 F2 12 0N 14 55 E
Koutiala, Mali 112 C3 12 25N 5 23W
Kouto, Ivory C. 112 D3 9 53N 6 25W
Kouts, U.S.A. 157 C9 41 19N 87 2W
Kouvé, Togo 113 D5 6 25N 1 25 E
Kouvola, Finland 15 F22 60 52N 26 43 E
Kovačica, Serbia, Yug. 52 E5 45 5N 20 38 E
Kovdor, Russia 56 A5 67 34N 30 24 E
Kovel, Ukraine 59 G3 51 11N 24 38 E
Kovilpatti, India 95 K3 9 10N 77 50 E
Kovin, Serbia, Yug. 52 F5 44 44N 20 59 E
Kovrov, Russia 60 B5 56 25N 41 25 E
Kovur, Andhra Pradesh, India 94 F5 17 3N 81 39 E
Kovur, Andhra Pradesh, India 95 G5 14 30N 80 1 E
Kowal, Poland 55 F6 52 32N 19 7 E
Kowalewo Pomorskie, Poland 55 E5 53 10N 18 52 E
Kowanyama, Australia 126 B3 15 29 S 141 44 E
Kowghān, Afghan. 91 B1 34 12N 63 2 E
Kowkash, Canada 140 B2 50 20N 87 12W
Kowloon, H.K. 77 F10 22 20N 114 15 E
Kowŏn, N. Korea 75 E14 39 26N 127 14 E
Kōyama, Japan 70 F2 31 20N 130 56 E
Köyceğiz, Turkey 49 E10 36 57N 28 40 E
Köyceğiz Gölü, Turkey 49 E10 36 56N 28 42 E
Koytash, Uzbekistan 63 C3 40 11N 67 19 E
Koyukuk, U.S.A. 144 D8 64 53N 157 42W
Koyulhisar, Turkey 100 B7 40 20N 37 52 E
Koyunyeri, Turkey 51 F10 40 50N 26 26 E
Koza, Japan 69 L3 26 19N 127 46 E
Kozak, Turkey 49 B9 39 15N 27 6 E
Kozan, Turkey 100 D6 37 26N 35 50 E
Kozáni, Greece 50 F5 40 19N 21 47 E
Kozáni □, Greece 50 F5 40 18N 21 45 E
Kozara, Bos.-H. 45 D14 45 0N 17 0 E
Kozarac, Bos.-H. 45 D13 44 58N 16 48 E
Kozelets, Ukraine 59 G6 50 55N 31 7 E
Kozelsk, Russia 58 E8 54 2N 35 48 E
Kozhikode = Calicut, India 95 J2 11 15N 75 43 E
Kozhva, Russia 56 A10 65 10N 57 0 E
Koziegłowy, Poland 55 H6 50 37N 19 8 E
Kozienice, Poland 55 G8 51 35N 21 34 E
Kozje, Slovenia 45 B12 46 5N 15 35 E
Kozloduy, Bulgaria 50 C7 43 45N 23 42 E
Kozlovets, Bulgaria 51 C9 43 30N 25 20 E
Kozlovka, Russia 60 C9 55 52N 48 14 E
Kozlu, Turkey 100 B4 41 26N 31 45 E
Kozluk, Turkey 101 C9 38 11N 41 31 E
Koźmin, Poland 55 G4 51 48N 17 27 E
Kozmodemyansk, Russia 60 B8 56 20N 46 36 E
Kōzu-Shima, Japan 71 C11 34 13N 139 10 E
Kożuchów, Poland 55 G2 51 45N 15 31 E
Kozyatyn, Ukraine 59 H5 49 45N 28 50 E
Kpabia, Ghana 113 D4 9 10N 0 20W
Kpalimé, Togo 113 D5 6 57N 0 44 E
Kpandae, Ghana 113 D4 8 30N 0 2W
Kpessi, Togo 113 D5 8 4N 1 16 E
Kra, Isthmus of = Kra, Kho Khot, Thailand 87 G2 10 15N 99 30 E
Kra, Kho Khot, Thailand 87 G2 10 15N 99 30 E
Kra Buri, Thailand 87 G2 10 22N 98 46 E
Krabi, Thailand 87 H2 8 4N 98 55 E
Kracheh, Cambodia 86 F6 12 32N 106 10 E
Kragan, Indonesia 85 D4 6 43 S 111 38 E
Kragerø, Norway 15 G13 58 52N 9 25 E
Kragujevac, Serbia, Yug. 50 B4 44 2N 20 56 E
Krajenka, Poland 55 E3 53 18N 16 59 E
Krajina, Bos.-H. 45 D13 44 45N 16 35 E
Krakatau = Rakata, Pulau, Indonesia 84 D3 6 10 S 105 20 E
Krakor, Cambodia 86 F5 12 32N 104 12 E
Kraków, Poland 55 H6 50 4N 19 57 E
Kraków □, Poland 55 H6 50 0N 20 0 E
Kraksaan, Indonesia 85 D4 7 43 S 113 23 E
Kralanh, Cambodia 86 F4 13 35N 103 25 E
Králíky, Czech Rep. 35 A9 50 6N 16 45 E
Kraljevo, Serbia, Yug. 50 C4 43 44N 20 41 E
Královský Chlmec, Slovak Rep. 35 C14 48 27N 22 0 E
Kralupy nad Vltavou, Czech Rep. 34 A7 50 13N 14 20 E
Kramatorsk, Ukraine 59 H9 48 50N 37 30 E
Kramfors, Sweden 16 B11 62 55N 17 48 E
Kramis, C., Algeria 111 A5 36 26N 0 45 E
Kraniá, Greece 50 G5 39 53N 21 18 E
Kranía Elassónas, Greece 48 B4 39 57N 22 2 E
Kranídhion, Greece 48 D5 37 20N 23 10 E
Kranj, Slovenia 45 B11 46 16N 14 22 E
Kranjska Gora, Slovenia 45 B10 46 29N 13 48 E
Krankskop, S. Africa 117 D5 28 0 S 30 47 E
Krapina, Croatia 45 B12 46 10N 15 52 E
Krapina →, Croatia 45 C12 45 50N 15 50 E
Krapkowice, Poland 55 H4 50 29N 17 56 E
Kras, Croatia 45 C10 45 35N 14 0 E
Krasavino, Russia 56 B8 60 58N 46 29 E
Krashyy Klyuch, Russia 62 D6 55 23N 56 39 E
Kraskino, Russia 68 C5 42 44N 130 48 E
Krāslava, Latvia 58 E4 55 54N 27 10 E
Kraslice, Czech Rep. 34 A5 50 19N 12 31 E
Krasnaya Gorbatka, Russia 60 C5 55 52N 41 45 E
Krasnaya Polyana, Russia 61 J5 43 40N 40 13 E
Kraśnik, Poland 55 H9 50 55N 22 15 E
Krasnoarmeisk, Ukraine 59 H9 48 18N 37 11 E
Krasnoarmeysk, Russia 60 E7 51 0N 45 42 E
Krasnoarmeyskiy, Russia 61 G6 47 0N 42 12 E
Krasnobrod, Poland 55 H10 50 33N 23 12 E
Krasnodar, Russia 61 H4 45 5N 39 0 E
Krasnodon, Ukraine 59 H10 48 17N 39 44 E
Krasnogorskiy, Russia 60 B9 56 10N 48 28 E
Krasnograd = Krasnohrad, Ukraine 59 H8 49 27N 35 27 E
Krasnogvardeysk = Bulunghur, Uzbekistan 63 D3 39 46N 67 16 E
Krasnogvardeyskoye, Russia 61 H5 45 52N 41 33 E
Krasnogvardyesk, Ukraine 59 K8 45 32N 34 16 E
Krasnohrad, Ukraine 59 H8 49 27N 35 27 E
Krasnokamsk, Russia 62 B5 58 4N 55 48 E
Krasnokutsk, Ukraine 59 G8 50 10N 34 50 E
Krasnolesnyy, Russia 59 G10 51 53N 39 35 E
Krasnoperekopsk, Ukraine 59 J7 46 0N 33 54 E
Krasnorechenskiy, Russia 68 B7 44 41N 135 14 E
Krasnoselkup, Russia 64 C9 65 20N 82 10 E
Krasnoslobodsk, Mordvinia, Russia 60 C6 54 25N 43 45 E
Krasnoslobodsk, Volgograd, Russia 61 F7 48 42N 44 33 E
Krasnoturinsk, Russia 62 B8 59 46N 60 12 E
Krasnoufimsk, Russia 62 C6 56 36N 57 38 E
Krasnouralsk, Russia 62 B8 58 21N 60 3 E
Krasnousolskiy, Russia 62 E6 53 54N 56 27 E
Krasnovishersk, Russia 62 A6 60 23N 57 3 E
Krasnovodsk = Türkmenbashi, Turkmenistan 57 G9 40 5N 53 5 E
Krasnoyarsk, Russia 65 D10 56 8N 93 0 E
Krasnoyarskiy, Russia 62 F7 51 58N 59 55 E
Krasnoye = Krasnyy, Russia 58 E6 54 25N 31 30 E
Krasnoye, Russia 62 B1 59 15N 47 40 E
Krasnozavodsk, Russia 58 D10 56 27N 38 25 E
Krasny Sulin, Russia 59 J11 47 52N 40 8 E
Krasnystaw, Poland 55 H10 50 57N 23 5 E
Krasnyy, Russia 58 E6 54 25N 31 30 E
Krasnyy Kholm, Orenburg, Russia 62 F5 51 35N 54 9 E
Krasnyy Kholm, Tver, Russia 58 C9 58 10N 37 10 E
Krasnyy Kut, Russia 60 E8 50 50N 47 0 E
Krasnyy Liman, Ukraine 59 H9 48 58N 37 50 E
Krasnyy Luch, Ukraine 59 H10 48 13N 39 0 E
Krasnyy Profintern, Russia 58 D11 57 45N 40 27 E
Krasnyy Yar, Astrakhan, Russia 61 G9 46 43N 48 23 E
Krasnyy Yar, Samara, Russia 60 D10 53 30N 50 22 E
Krasnyy Yar, Volgograd, Russia 60 E7 50 42N 44 45 E
Krasnyye Baki, Russia 60 B7 57 8N 45 10 E
Krasnyyoskolske Vdskh., Ukraine 59 H9 49 27N 37 40 E
Kraszna →, Hungary 52 B7 48 4N 22 20 E
Kratie = Kracheh, Cambodia 86 F6 12 32N 106 10 E
Kratke Ra., Papua N. G. 132 D4 6 45 S 146 0 E
Kratovo, Macedonia 50 D6 42 6N 22 10 E
Krau, Indonesia 83 B6 3 19 S 140 5 E
Kraulshavn, Greenland 10 C5 74 8N 57 3W
Kravanh, Chuor Phnum, Cambodia 87 G4 12 0N 103 32 E
Krefeld, Germany 30 D2 51 20N 6 33 E
Krémaston, Límni, Greece 48 C3 38 52N 21 30 E
Kremen, Croatia 45 D12 44 28N 15 53 E
Kremenchug = Kremenchuk, Ukraine 59 H7 49 5N 33 25 E
Kremenchuk, Ukraine 59 H7 49 5N 33 25 E
Kremenchuksk Vdskh., Ukraine 59 H7 49 20N 32 30 E
Kremenets, Ukraine 59 G3 50 8N 25 43 E
Kremennaya, Ukraine 59 H10 49 1N 38 10 E
Kremges = Svitlovodsk, Ukraine 59 H7 49 2N 33 13 E
Kremmen, Germany 30 C9 52 45N 13 1 E
Kremmling, U.S.A. 158 F10 40 4N 106 24W
Kremnica, Slovak Rep. 35 C11 48 45N 18 50 E
Krems, Austria 34 C8 48 25N 15 36 E
Kremsmünster, Austria 34 C7 48 3N 14 8 E
Kretinga, Lithuania 15 J19 55 53N 21 15 E
Krettamia, Algeria 110 C4 28 47N 3 27W
Krettsy, Russia 58 C7 58 15N 32 30 E
Kreuzberg, Germany 31 E5 50 22N 9 58 E
Kreuzlingen, Switz. 33 A8 47 38N 9 10 E
Kreuztal, Germany 30 E4 50 57N 8 0 E
Kría Vrísi, Greece 50 F6 40 41N 22 18 E
Kribi, Cameroon 113 E6 2 57N 9 56 E
Krichem, Bulgaria 51 D8 42 8N 24 28 E
Krichev = Krychaw, Belarus 58 F6 53 40N 31 41 E
Krim, Slovenia 45 C11 45 53N 14 30 E
Kriós, Ákra, Greece 38 D5 35 13N 23 34 E
Krishna →, India 95 G5 15 57N 80 59 E
Krishnagiri, India 95 H4 12 32N 78 16 E
Krishnanagar, India 93 H13 23 24N 88 33 E
Krishnaraja Sagara, India 95 H3 12 20N 76 30 E
Kristdala, Sweden 17 G10 57 24N 16 13 E
Kristiansand, Norway 15 G13 58 8N 8 1 E
Kristianstad, Sweden 17 H8 56 2N 14 9 E
Kristianstads län □, Sweden 17 H7 56 15N 14 0 E
Kristiansund, Norway 14 E12 63 7N 7 45 E
Kristiinankaupunki, Finland 15 E19 62 16N 21 21 E
Kristinehamn, Sweden 16 E8 59 18N 14 7 E
Kristinestad = Kristiinankaupunki, Finland 15 E19 62 16N 21 21 E
Kríti, Greece 38 D7 35 15N 25 0 E
Kritsá, Greece 38 D7 35 10N 25 41 E
Kriva →, Macedonia 50 D5 42 5N 21 47 E
Kriva Palanka, Macedonia 50 D6 42 11N 22 19 E
Krivaja →, Bos.-H. 52 F3 44 27N 18 9 E
Krivelj, Serbia, Yug. 50 B6 44 10N 22 5 E
Krivoy Rog = Kryvyy Rih, Ukraine 59 J7 47 51N 33 20 E
Križevci, Croatia 45 B13 46 3N 16 32 E
Krk, Croatia 45 C11 45 8N 14 40 E
Krka →, Slovenia 45 C12 45 50N 15 30 E
Krkonoše, Czech Rep. 34 A8 50 50N 15 35 E
Krnov, Czech Rep. 35 A10 50 5N 17 40 E
Krobia, Poland 55 G3 51 47N 16 59 E
Krøderen, Norway 18 D6 60 9N 9 49 E
Krokeaí, Greece 48 E4 36 53N 22 32 E
Krokek, Sweden 17 F10 58 40N 16 24 E
Krokodil →, Mozam. 117 D5 25 14 S 32 18 E
Krokom, Sweden 16 A8 63 20N 14 30 E
Krokowa, Poland 54 D5 54 47N 18 9 E
Króksfjarðarnes, Iceland 11 B5 65 27N 21 56W
Krokstadelva, Norway 18 E7 59 56N 10 1 E
Krolevets, Ukraine 59 G7 51 35N 33 20 E
Kroměříž, Czech Rep. 35 B10 49 18N 17 21 E
Krompachy, Slovak Rep. 35 C13 48 54N 20 52 E
Kromy, Russia 59 F8 52 48N 35 48 E
Kronach, Germany 31 E7 50 14N 11 19 E
Kronobergs län □, Sweden 17 H8 56 45N 14 30 E
Kronprins Christian Land, Greenland 10 A8 80 30N 22 0W
Kronprins Olav Kyst, Antarctica 7 C5 69 0 S 42 0 E
Kronshtadt, Russia 58 C5 59 57N 29 51 E
Kroonstad, S. Africa 116 D4 27 43 S 27 19 E
Kröpelin, Germany 30 A7 54 4N 11 47 E
Kropotkin, Russia 61 H5 45 28N 40 28 E
Kropp, Germany 30 A5 54 24N 9 31 E
Krosna, Lithuania 54 D10 54 23N 23 33 E
Krośniewice, Poland 55 F6 52 15N 19 11 E
Krosno, Poland 55 J8 49 42N 21 46 E
Krosno □, Poland 55 J9 49 35N 22 0 E
Krosno Odrzańskie, Poland 55 F2 52 3N 15 7 E
Krotoszyn, Poland 55 G4 51 42N 17 23 E
Krotovka, Russia 60 D10 53 18N 51 10 E
Kroussón, Greece 38 D6 35 13N 24 59 E
Krraba, Albania 50 E3 41 13N 20 0 E
Krško, Slovenia 45 C12 45 57N 15 30 E
Krstača, Serbia, Yug. 50 D4 42 57N 20 8 E
Kruger Nat. Park, S. Africa 117 C5 23 30 S 31 40 E
Krugersdorp, S. Africa 117 D4 26 5 S 27 46 E
Kruisfontein, S. Africa 116 E3 33 59 S 24 43 E
Kruja, Albania 50 E3 41 32N 19 46 E
Krulevshchina = Krulyewshchyna, Belarus 58 E4 55 5N 27 45 E
Krulyewshchyna, Belarus 58 E4 55 5N 27 45 E
Kruma, Albania 50 D4 42 14N 20 28 E
Krumbach, Germany 31 G6 48 13N 10 22 E
Krumovgrad, Bulgaria 51 E9 41 29N 25 38 E
Krung Thep = Bangkok, Thailand 86 F3 13 45N 100 35 E
Krupanj, Serbia, Yug. 50 B3 44 25N 19 22 E
Krupina, Slovak Rep. 35 C12 48 22N 19 5 E
Krupinica →, Slovak Rep. 35 C11 48 4N 18 55 E
Krupki, Belarus 58 E5 54 19N 29 8 E
Krusenstern, C., U.S.A. 144 C7 67 8N 163 45W
Kruševac, Serbia, Yug. 50 C5 43 35N 21 28 E
Kruševo, Macedonia 50 E5 41 23N 21 19 E
Kruszwica, Poland 55 F5 52 40N 18 20 E
Kruzof I., U.S.A. 142 B1 57 10N 135 40W
Krychaw, Belarus 58 F6 53 40N 31 41 E
Krymsk, Russia 59 K10 44 50N 38 0 E
Krymskiy Poluostrov = Krymskyy Pivostriv, Ukraine 59 K8 45 0N 34 0 E
Krymskyy Pivostriv, Ukraine 59 K8 45 0N 34 0 E
Krynica, Poland 55 J7 49 25N 20 57 E
Krynica Morska, Poland 54 D6 54 23N 19 28 E
Krynki, Poland 55 E10 53 17N 23 43 E
Kryvyy Rih, Ukraine 59 J7 47 51N 33 20 E
Krzepice, Poland 55 H5 50 58N 18 50 E
Krzeszów, Poland 55 H9 50 24N 22 21 E
Krzna →, Poland 55 F10 52 8N 23 32 E
Krzywiń, Poland 55 G3 51 58N 16 50 E
Krzyż Wielkopolski, Poland 55 F2 52 52N 16 0 E
Ksabi, Morocco 110 B4 32 51N 4 13W
Ksar Chellala, Algeria 111 A5 35 13N 2 19 E
Ksar el Boukhari, Algeria 111 A5 35 51N 2 52 E
Ksar el Kebir, Morocco 110 B3 35 0N 6 0W
Ksar es Souk = Ar Rachidiya, Morocco 110 B4 31 58N 4 20W
Ksar Rhilane, Tunisia 108 B1 33 0N 9 39 E
Książ Wielkopolski, Poland 55 F4 52 4N 17 14 E
Ksour, Mts. des, Algeria 111 B4 32 45N 0 30W
Kstovo, Russia 60 B7 56 12N 44 13 E
Ku Tree Reservoir, U.S.A. 145 K14 21 30N 157 59W
Kuala, Indonesia 84 B3 2 55N 105 47 E
Kuala Belait, Malaysia 85 B4 4 35N 114 11 E
Kuala Berang, Malaysia 87 K4 5 5N 103 1 E
Kuala Dungun = Dungun, Malaysia 87 K4 4 45N 103 25 E
Kuala Kangsar, Malaysia 87 K3 4 46N 100 56 E
Kuala Kelawang, Malaysia 87 L4 2 56N 102 5 E
Kuala Kerai, Malaysia 87 K4 5 30N 102 12 E
Kuala Lipis, Malaysia 87 K4 4 10N 102 3 E
Kuala Lumpur, Malaysia 87 L3 3 9N 101 41 E
Kuala Nerang, Malaysia 87 J3 6 16N 100 37 E
Kuala Pilah, Malaysia 87 L4 2 45N 102 15 E
Kuala Rompin, Malaysia 87 L4 2 49N 103 29 E
Kuala Selangor, Malaysia 87 L3 3 20N 101 15 E
Kuala Sepetang, Malaysia 87 K3 4 50N 100 38 E
Kuala Terengganu, Malaysia 87 K4 5 20N 103 8 E
Kualajelai, Indonesia 85 C4 2 58 S 110 46 E
Kualakapuas, Indonesia 85 C4 2 55 S 114 20 E
Kualakurun, Indonesia 85 C4 1 10 S 113 50 E
Kualapembuang, Indonesia 85 C4 3 14 S 112 38 E
Kualapuu, U.S.A. 145 B4 21 10N 157 2W
Kualasimpang, Indonesia 84 B1 4 17N 98 3 E
Kualoa Pt., U.S.A. 145 J14 21 31N 157 50W
Kuamat, Malaysia 85 A5 5 13N 117 30 E
Kuancheng, China 75 D10 40 37N 118 30 E
Kuandang, Indonesia 82 A2 0 56N 123 1 E
Kuandian, China 75 D13 40 45N 124 45 E
Kuangchou = Guangzhou, China 77 F9 23 5N 113 10 E
Kuantan, Malaysia 87 L4 3 49N 103 20 E
Kuapa Pond, U.S.A. 145 K14 21 17N 157 43W
Kuba = Quba, Azerbaijan 61 K9 41 21N 48 32 E
Kuban →, Russia 59 K9 45 20N 37 30 E
Kubenskoye, Ozero, Russia 58 C10 59 40N 39 25 E
Kubokawa, Japan 70 D5 33 12N 133 8 E
Kubor, Mt., Papua N. G. 132 D3 6 10 S 144 44 E
Kubrat, Bulgaria 51 C10 43 49N 26 31 E
Kučevo, Serbia, Yug. 50 B5 44 30N 21 40 E
Kucha Gompa, India 93 B7 34 25N 76 56 E
Kuchaman, India 92 F6 27 13N 74 47 E
Kuchenspitze, Austria 34 D3 47 7N 10 12 E
Kuching, Malaysia 85 B4 1 33N 110 25 E
Kuchino-eruba-Jima, Japan 69 J5 30 28N 130 12 E
Kuchino-Shima, Japan 69 K4 29 57N 129 55 E
Kuchinotsu, Japan 70 E2 32 36N 130 11 E
Kuchl, Austria 34 D6 47 37N 13 9 E
Kucing = Kuching, Malaysia 85 B4 1 33N 110 25 E
Kuçovë, Albania 50 F3 40 47N 19 57 E
Küçükbahçe, Turkey 49 C8 38 33N 26 24 E
Küçükköy, Turkey 49 B8 39 16N 26 42 E
Küçükkuyu, Turkey 49 B8 39 32N 26 36 E

La Cocha 237

Küçükmenderes →, *Turkey* 49 D9 37 57N 27 16 E
Kud →, *Pakistan* 92 F2 26 5N 66 20 E
Kudalier →, *India* 94 E4 18 35N 79 48 E
Kudamatsu, *Japan* 70 D3 34 0N 131 52 E
Kudara, *Tajikistan* 63 D6 38 25N 72 39 E
Kudat, *Malaysia* 85 A5 6 55N 116 55 E
Kudayd, *Si. Arabia* 98 C3 19 21N 41 48 E
Kudirkos Naumiestis, *Lithuania* 54 D9 54 46N 22 53 E
Kudowa-Zdrój, *Poland* 55 H3 50 27N 16 15 E
Kudremukh, *India* 95 H2 13 15N 75 20 E
Kudus, *Indonesia* 85 D4 6 48 S 110 51 E
Kudymkar, *Russia* 62 B5 59 1N 54 39 E
Kueiyang = Guiyang, *China* 76 D6 26 32N 106 40 E
Kufra Oasis = Al Kufrah, *Libya* 108 D4 24 17N 23 15 E
Kufstein, *Austria* 34 D5 47 35N 12 11 E
Kugluktuk, *Canada* 138 B8 67 50N 115 5W
Kugong I., *Canada* 140 A4 56 18N 79 50W
Kūh-e-Hazārān, *Iran* 97 D8 29 35N 57 20 E
Kūhak, *Iran* 91 D1 27 12N 63 10 E
Kūhbonān, *Iran* 97 D8 31 23N 56 19 E
Kūhestak, *Iran* 97 E8 26 47N 57 2 E
Kūhestān, *Afghan.* 91 B1 34 39N 61 12 E
Kūhīn, *Iran* 97 C6 35 13N 48 25 E
Kūhīrī, *Iran* 97 E9 26 55N 61 2 E
Kuhnsdorf, *Austria* 34 E7 46 37N 14 38 E
Kūhpāyeh, *Eşfahan, Iran* 97 C7 32 44N 52 20 E
Kūhpāyeh, *Kermān, Iran* 97 D8 30 35N 57 15 E
Kui Buri, *Thailand* 87 F2 12 3N 99 52 E
Kuito, *Angola* 115 E3 12 22 S 16 55 E
Kujang, *N. Korea* 75 E14 39 57N 126 1 E
Kuji, *Japan* 68 D10 40 11N 141 46 E
Kujū-San, *Japan* 70 D3 33 5N 131 15 E
Kujukuri-Heiya, *Japan* 71 B12 35 45N 140 30 E
Kukavica, *Serbia, Yug.* 50 D5 42 48N 21 57 E
Kukawa, *Nigeria* 113 C7 12 58N 13 27 E
Kukerin, *Australia* 125 F2 33 13 S 118 0 E
Kukës, *Albania* 50 D4 42 5N 20 27 E
Kukmor, *Russia* 60 B10 56 11N 50 54 E
Kukuihaele, *U.S.A.* 145 C6 20 5N 155 35W
Kukup, *Malaysia* 87 M4 1 20N 103 27 E
Kukvidze, *Russia* 60 E6 50 40N 43 0 E
Kula, *Bulgaria* 50 C6 43 52N 22 36 E
Kula, *Serbia, Yug.* 52 E4 45 37N 19 32 E
Kula, *Turkey* 49 C10 38 32N 28 40 E
Kula Gulf, *Solomon Is.* 133 M9 8 5 S 157 18 E
Kulai, *Malaysia* 87 M4 1 44N 103 35 E
Kulal, Mt., *Kenya* 118 B4 2 42N 36 57 E
Kulaly, Ostrov, *Kazakstan* 61 H10 45 0N 50 0 E
Kulanak, *Kyrgyzstan* 63 C7 41 22N 75 30 E
Kulasekarappattinam, *India* 95 K4 8 20N 78 5 E
Kulautuva, *Lithuania* 54 D10 54 56N 23 36 E
Kuldīga, *Latvia* 15 H19 56 58N 21 59 E
Kuldīga □, *Latvia* 54 B8 56 55N 22 0 E
Kuldja = Yining, *China* 64 E9 43 58N 81 10 E
Kuldu, *Sudan* 107 E2 12 50N 28 30 E
Kulebaki, *Russia* 60 C6 55 22N 42 25 E
Kulen, *Cambodia* 86 F5 13 50N 104 40 E
Kulen Vakuf, *Bos.-H.* 45 D13 44 35N 16 2 E
Kulgam, *India* 93 C6 33 36N 75 2 E
Kulim, *Malaysia* 87 K3 5 22N 100 34 E
Kulin, *Australia* 125 F2 32 40 S 118 2 E
Kulja, *Australia* 125 F2 30 28 S 117 18 E
Kullen, *Sweden* 17 H6 56 18N 12 26 E
Kulm, *U.S.A.* 154 B5 46 18N 98 57W
Kulmbach, *Germany* 31 E7 50 6N 11 26 E
Kŭlob, *Tajikistan* 63 E4 37 55N 69 50 E
Kulp, *Turkey* 101 C9 38 29N 41 2 E
Kulsary, *Kazakstan* 57 E9 46 59N 54 1 E
Kulti, *India* 93 H12 23 43N 86 50 E
Kulu, *Turkey* 100 C5 39 5N 33 4 E
Kulumbura, *Australia* 124 B4 13 55 S 126 35 E
Kulunda, *Russia* 64 D8 52 35N 78 57 E
Kulungar, *Afghan.* 92 C3 34 0N 69 2 E
Kūlvand, *Iran* 97 D7 31 21N 54 35 E
Kulwin, *Australia* 128 C5 35 0 S 142 42 E
Kulyab = Kŭlob, *Tajikistan* 63 E4 37 55N 69 50 E
Kuma, *Japan* 70 D4 33 39N 132 54 E
Kuma →, *Russia* 61 H8 44 55N 47 0 E
Kumafşarı, *Turkey* 49 D11 37 19N 29 32 E
Kumaganum, *Nigeria* 113 C7 13 8N 10 38 E
Kumagaya, *Japan* 71 A11 36 9N 139 22 E
Kumai, *Indonesia* 85 C4 2 44 S 111 43 E
Kumak, *Russia* 62 F8 51 10N 60 8 E
Kumalar Dağı, *Turkey* 49 C12 38 15N 30 20 E
Kumamba, Kepulauan, *Indonesia* 83 B5 1 36 S 138 45 E
Kumamoto, *Japan* 70 E2 32 45N 130 45 E
Kumamoto □, *Japan* 70 E2 32 55N 130 55 E
Kumano, *Japan* 71 D8 33 54N 136 5 E
Kumano-Nada, *Japan* 71 D8 33 47N 136 20 E
Kumanovo, *Macedonia* 50 D5 42 9N 21 42 E
Kumara, *N.Z.* 131 C6 42 37 S 171 12 E
Kumarkhali, *Bangla.* 90 D2 23 51N 89 15 E
Kumarl, *Australia* 125 F3 32 47 S 121 33 E
Kumasi, *Ghana* 112 D4 6 41N 1 38W
Kumayri = Gyumri, *Armenia* 61 K6 40 47N 43 50 E
Kumba, *Cameroon* 113 E6 4 36N 9 24 E
Kumbağ, *Turkey* 51 F11 40 51N 27 27 E
Kumbakonam, *India* 95 J4 10 58N 79 25 E
Kumbarilla, *Australia* 127 D5 27 15 S 150 55 E
Kumbo, *Cameroon* 113 D7 6 15N 10 36 E
Kumbukkan Oya →, *Sri Lanka* 95 L5 6 35N 81 40 E
Kŭmchŏn, *N. Korea* 75 E14 38 10N 126 29 E
Kumdah, *Si. Arabia* 98 B4 20 23N 45 5 E
Kumdok, *India* 93 C8 33 32N 78 10 E
Kume-Shima, *Japan* 69 L3 26 20N 126 47 E
Kumeny, *Russia* 60 A9 58 10N 49 47 E
Kumertau, *Russia* 62 E5 52 45N 55 57 E
Kŭmhwa, *S. Korea* 75 E14 38 17N 127 28 E
Kumi, *Uganda* 118 B3 1 30N 33 58 E
Kumkale, *Turkey* 51 G10 39 59N 26 11 E
Kumla, *Sweden* 16 E9 59 8N 15 10 E
Kumluca, *Turkey* 49 E12 36 22N 30 20 E
Kummerower See, *Germany* 30 B8 53 49N 12 51 E
Kumo, *Nigeria* 113 C7 10 1N 11 12 E
Kumon Bum, *Burma* 90 B6 26 30N 97 15 E
Kumotori-Yama, *Japan* 71 B10 35 51N 138 57 E
Kumta, *India* 95 G2 14 29N 74 25 E
Kumusi →, *Papua N. G.* 132 E5 8 16 S 148 13 E
Kumylzhenskaya, *Russia* 60 F6 49 51N 42 38 E
Kunágota, *Hungary* 52 D6 46 26N 21 3 E
Kunak, *Malaysia* 85 B5 4 41N 118 15 E
Kunama, *Australia* 129 C8 35 35 S 148 4 E
Kunashir, Ostrov, *Russia* 65 E15 44 0N 146 0 E
Kunda, *Estonia* 15 G22 59 30N 26 34 E
Kundiawa, *Papua N. G.* 132 D3 6 2 S 145 1 E
Kundla, *India* 92 J4 21 21N 71 25 E
Kundur, *Indonesia* 84 C3 3 8 S 107 48 E
Kungala, *Australia* 127 D5 29 58 S 153 7 E
Kungälv, *Sweden* 17 G5 57 53N 11 59 E
Kungey Alatau, Khrebet, *Kyrgyzstan* 63 B8 42 50N 77 0 E
Kunghit I., *Canada* 142 C2 52 6N 131 3W
Kungrad = Qŭnghirot, *Uzbekistan* 64 E6 43 6N 58 54 E
Kungsängen, *Sweden* 16 E11 59 29N 17 45 E
Kungsbacka, *Sweden* 17 G6 57 30N 12 5 E
Kungsgården, *Sweden* 16 D10 60 37N 16 35 E
Kungshamn, *Sweden* 17 F5 58 22N 11 15 E
Kungsör, *Sweden* 16 E10 59 25N 16 5 E
Kungu, *Dem. Rep. of the Congo* 114 B3 2 47N 19 12 E
Kungur, *Russia* 62 C6 57 25N 56 57 E
Kungurri, *Australia* 126 C4 21 3 S 148 46 E
Kungyangon, *Burma* 90 G6 16 27N 96 20 E
Kunhar →, *Pakistan* 93 B5 34 20N 73 30 E
Kunhegyes, *Hungary* 52 C5 47 22N 20 36 E
Kunia, *U.S.A.* 145 K13 21 28N 158 4W
Kunimi-Dake, *Japan* 70 E3 32 33N 131 1 E
Kuningan, *Indonesia* 85 D3 6 59 S 108 29 E
Kunisaki, *Japan* 70 D3 33 33N 131 45 E
Kunlara, *Australia* 128 C3 34 54 S 139 55 E
Kunlong, *Burma* 90 D7 23 20N 98 50 E
Kunlun Shan, *Asia* 72 C3 36 0N 86 30 E
Kunmadaras, *Hungary* 52 C5 47 28N 20 45 E
Kunming, *China* 76 E4 25 1N 102 41 E
Kunnamkulam, *India* 95 J3 10 38N 76 7 E
Kunów, *Poland* 55 H8 50 57N 21 17 E
Kunsan, *S. Korea* 75 G14 35 59N 126 45 E
Kunshan, *China* 77 B13 31 22N 120 58 E
Kunszentmárton, *Hungary* 52 D5 46 50N 20 20 E
Kunszentmiklós, *Hungary* 52 C4 47 2N 19 8 E
Kununurra, *Australia* 124 C4 15 40 S 128 50 E
Kunwarara, *Australia* 126 C5 22 55 S 150 9 E
Kunya-Urgench = Köneürgench, *Turkmenistan* 64 E6 42 19N 59 10 E
Künzelsau, *Germany* 31 F5 49 17N 9 42 E
Kuopio, *Finland* 14 E22 62 53N 27 35 E
Kupa →, *Croatia* 45 C13 45 28N 16 24 E
Kupang, *Indonesia* 82 D2 10 19 S 123 39 E
Kupiano, *Papua N. G.* 132 F5 10 4 S 148 14 E
Kupres, *Bos.-H.* 52 G2 43 56N 17 15 E
Kupyansk, *Ukraine* 59 H9 49 52N 37 35 E
Kupyansk-Uzlovoi, *Ukraine* 59 H9 49 40N 37 43 E
Kuqa, *China* 72 B3 41 35N 82 30 E
Kür →, *Azerbaijan* 101 C13 39 29N 49 15 E
Kur →, *Bhutan* 90 B3 26 50N 91 0 E
Kura = Kür →, *Azerbaijan* 101 C13 39 29N 49 15 E
Kurahashi-Jima, *Japan* 70 C4 34 8N 132 31 E
Kuranda, *Australia* 126 B4 16 48 S 145 35 E
Kurashiki, *Japan* 70 C5 34 40N 133 50 E
Kurayoshi, *Japan* 70 B5 35 26N 133 50 E
Kürdämir, *Azerbaijan* 61 K9 40 25N 48 3 E
Kurday, *Kazakstan* 63 B7 43 21N 74 59 E
Kurdistan, *Asia* 101 D10 37 20N 43 30 E
Kurduvadi, *India* 94 E2 18 8N 75 29 E
Kŭrdzhali, *Bulgaria* 51 E9 41 38N 25 21 E
Kure, *Japan* 70 C4 34 14N 132 32 E
Küre, *Turkey* 100 B5 41 48N 33 43 E
Küre Dağları, *Turkey* 100 B6 41 50N 34 10 E
Kure I., *U.S.A.* 145 F8 28 25N 178 25W
Kuressaare, *Estonia* 15 G20 58 15N 22 30 E
Kurgan, *Russia* 64 D7 55 26N 65 18 E
Kurgan-Tyube = Qŭrghonteppa, *Tajikistan* 63 E4 37 50N 68 47 E
Kurganinsk, *Russia* 61 H5 44 54N 40 34 E
Kurgannaya = Kurganinsk, *Russia* 61 H5 44 54N 40 34 E
Kuria Maria Is. = Khurīyā Murīyā, Jazā 'ir, *Oman* 99 C6 17 30N 55 58 E
Kuria Muria B., *Oman* 99 C6 17 40N 55 45 E
Kurichchi, *India* 95 J3 11 36N 77 35 E
Kuridala, *Australia* 126 C3 21 16 S 140 29 E
Kurigram, *Bangla.* 90 C2 25 49N 89 39 E
Kurihashi, *Japan* 71 A11 36 8N 139 42 E
Kurikka, *Finland* 15 E20 62 36N 22 24 E
Kuril Is. = Kurilskiye Ostrova, *Russia* 65 E16 45 0N 150 0 E
Kuril Trench, *Pac. Oc.* 134 C7 44 0N 153 0 E
Kurilsk, *Russia* 65 E15 45 14N 147 53 E
Kurilskiye Ostrova, *Russia* 65 E16 45 0N 150 0 E
Kurino, *Japan* 70 F2 31 57N 130 43 E
Kurinskaya Kosa, *Azerbaijan* 101 C13 39 3N 49 13 E
Kurkur, *Egypt* 106 C3 23 50N 32 0 E
Kurla, *India* 94 E1 19 5N 72 52 E
Kurlovskiy, *Russia* 60 C5 55 25N 40 40 E
Kurmuk, *Sudan* 107 E3 10 33N 34 21 E
Kurnool, *India* 95 G4 15 45N 78 0 E
Kuro-Shima, *Kagoshima, Japan* 69 J4 30 50N 129 57 E
Kuro-Shima, *Okinawa, Japan* 69 M2 24 14N 124 1 E
Kurobe-Gawe →, *Japan* 71 A9 36 55N 137 25 E
Kurogi, *Japan* 70 D2 33 12N 130 40 E
Kurow, *N.Z.* 131 E5 44 44 S 170 29 E
Kurów, *Poland* 55 G9 51 23N 22 12 E
Kurrajong, *Australia* 129 B9 33 33 S 150 42 E
Kurram →, *Pakistan* 91 B3 32 36N 71 20 E
Kurri Kurri, *Australia* 129 B9 32 50 S 151 28 E
Kursavka, *Russia* 61 H6 44 29N 42 32 E
Kuršėnai, *Lithuania* 54 B9 56 1N 22 58 E
Kurshskiy Zaliv, *Russia* 15 J19 55 9N 21 6 E
Kursk, *Russia* 59 G9 51 42N 36 11 E
Kuršumlija, *Serbia, Yug.* 50 C5 43 9N 21 19 E
Kuršumlijska Banja, *Serbia, Yug.* 50 C5 43 3N 21 11 E
Kurşunlu, *Bursa, Turkey* 51 F13 40 3N 29 40 E
Kurşunlu, *Çankırı, Turkey* 100 B5 40 51N 33 15 E
Kurtalan, *Turkey* 101 D9 37 56N 41 44 E
Kurtbey, *Turkey* 51 E10 41 9N 26 35 E
Kurtistown, *U.S.A.* 145 D6 19 36N 155 4W
Kurty →, *Kazakstan* 63 A8 44 16N 76 42 E
Kuru, Bahr el →, *Sudan* 107 F2 8 10N 26 50 E
Kurucaşile, *Turkey* 100 B5 41 49N 32 42 E
Kuruktag, *China* 72 B3 41 0N 89 0 E
Kuruman, *S. Africa* 116 D3 27 28 S 23 28 E
Kuruman →, *S. Africa* 116 D3 26 56 S 20 39 E
Kurume, *Japan* 70 D2 33 15N 130 30 E
Kurunegala, *Sri Lanka* 95 L5 7 30N 80 23 E
Kurupukari, *Guyana* 169 C6 4 43N 58 37W
Kus Gölü, *Turkey* 51 F11 40 10N 27 55 E
Kusa, *Russia* 62 D7 55 20N 59 29 E
Kuşadası, *Turkey* 100 D2 37 52N 27 15 E
Kuşadası Körfezi, *Turkey* 49 D8 37 56N 27 0 E
Kusatsu, *Gumma, Japan* 71 A10 36 37N 138 36 E
Kusatsu, *Shiga, Japan* 71 C7 34 58N 135 57 E
Kusawa L., *Canada* 142 A1 60 20N 136 13W
Kusel, *Germany* 31 F3 49 32N 7 24 E
Kushchevskaya, *Russia* 61 G4 46 33N 39 35 E
Kushikino, *Japan* 70 F2 31 44N 130 16 E
Kushima, *Japan* 70 F3 31 29N 131 14 E
Kushimoto, *Japan* 71 D7 33 28N 135 47 E
Kushiro, *Japan* 68 C12 43 0N 144 25 E
Kushiro →, *Japan* 68 C12 42 59N 144 23 E
Kūshk, *Iran* 97 D8 28 46N 56 51 E
Kushka = Gushgy, *Turkmenistan* 64 F7 35 20N 62 18 E
Kūshkī, *Īlām, Iran* 96 C5 33 31N 47 13 E
Kūshkī, *Khorāsān, Iran* 97 B8 37 2N 57 26 E
Kūshkū, *Iran* 97 E7 27 19N 53 28 E
Kushnarenkovo, *Russia* 62 D5 55 6N 55 22 E
Kushol, *India* 93 C7 33 40N 76 36 E
Kushrabat, *Uzbekistan* 63 C3 40 18N 66 32 E
Kushtia, *Bangla.* 90 D2 23 55N 89 5 E
Kushum →, *Kazakstan* 60 F10 49 20N 50 30 E
Kushva, *Russia* 62 B7 58 18N 59 45 E
Kuskokwim B., *U.S.A.* 138 C3 59 45N 162 25W
Küsnacht, *Switz.* 33 B7 47 19N 8 35 E
Kussharo-Ko, *Japan* 68 C12 43 38N 144 21 E
Küssnacht, *Switz.* 33 B6 47 5N 8 26 E
Kustanay = Qostanay, *Kazakstan* 64 D7 53 10N 63 35 E
Kusu, *Japan* 70 D3 33 16N 131 9 E
Kut, Ko, *Thailand* 87 G4 11 40N 102 35 E
Kutacane, *Indonesia* 84 B1 3 50N 97 50 E
Kütahya, *Turkey* 57 G5 39 30N 30 2 E
Kütahya □, *Turkey* 49 B11 39 10N 29 30 E
Kutaisi, *Georgia* 61 J6 42 19N 42 40 E
Kutaraja = Banda Aceh, *Indonesia* 84 A1 5 35N 95 20 E
Kutch, Gulf of = Kachchh, Gulf of, *India* 92 H3 22 50N 69 15 E
Kutch, Rann of = Kachchh, Rann of, *India* 92 H4 24 0N 70 0 E
K'utdligssat, *Greenland* 10 C5 70 3N 53 0W
Kutina, *Croatia* 45 C13 45 29N 16 48 E
Kutiyana, *India* 92 J4 21 36N 70 2 E
Kutjevo, *Croatia* 52 E2 45 23N 17 55 E
Kutkai, *Burma* 90 D6 23 27N 97 56 E
Kutkashen, *Azerbaijan* 61 K8 40 58N 47 47 E
Kutná Hora, *Czech Rep.* 34 B8 49 57N 15 16 E
Kutno, *Poland* 55 F6 52 15N 19 23 E
Kuttabul, *Australia* 126 C4 21 5 S 148 48 E
Kutu, *Dem. Rep. of the Congo* 114 C3 2 40 S 18 11 E
Kutum, *Sudan* 107 E1 14 10N 24 40 E
Kúty, *Slovak Rep.* 35 C10 48 40N 17 3 E
Kuujjuaq, *Canada* 139 C13 58 6N 68 15W
Kuŭp-tong, *N. Korea* 75 D14 40 45N 126 1 E
Kuusamo, *Finland* 14 D23 65 57N 29 8 E
Kuusankoski, *Finland* 15 F22 60 55N 26 38 E
Kuvandyk, *Russia* 62 F6 51 28N 57 21 E
Kuvango, *Angola* 115 E3 14 28 S 16 20 E
Kuvasay, *Uzbekistan* 63 C5 40 18N 71 59 E
Kuvshinovo, *Russia* 58 D8 57 2N 34 11 E
Kuwait = Al Kuwayt, *Kuwait* 96 D5 29 30N 48 0 E
Kuwait ■, *Asia* 96 D5 29 30N 47 30 E
Kuwana, *Japan* 71 B8 35 5N 136 43 E
Kuybyshev = Samara, *Russia* 60 D10 53 8N 50 6 E
Kuybyshev, *Russia* 64 D8 55 27N 78 19 E
Kuybyshevo, *Ukraine* 59 J9 47 25N 36 40 E
Kuybyshevo, *Uzbekistan* 63 C5 40 20N 71 15 E
Kuybyshevskiy, *Tajikistan* 63 E4 37 52N 68 44 E
Kuybyshevskoye Vdkhr., *Russia* 60 C9 55 2N 49 30 E
Kuye He →, *China* 74 E6 38 23N 110 46 E
Kūyeh, *Iran* 96 B5 38 45N 47 57 E
Kuylyuk, *Uzbekistan* 63 C4 41 14N 69 17 E
Kūysanjaq, *Iraq* 101 D11 36 5N 44 38 E
Kuyto, Ozero, *Russia* 56 B5 65 6N 31 20 E
Kuyucak, *Turkey* 49 D10 37 55N 28 28 E
Kuyumba, *Russia* 65 C10 60 58N 96 59 E
Kuzey Anadolu Dağları, *Turkey* 100 B7 41 30N 35 0 E
Kuzhitturai, *India* 95 K3 8 18N 77 11 E
Kuzino, *Russia* 62 C7 57 1N 59 27 E
Kuzitrin →, *U.S.A.* 144 D6 65 10N 165 25W
Kuzmin, *Serbia, Yug.* 52 E4 45 2N 19 25 E
Kuznetsk, *Russia* 60 D8 53 12N 46 40 E
Kuzomen, *Russia* 56 A6 66 22N 36 50 E
Kvænangen, *Norway* 14 A19 70 5N 21 15 E
Kværndrup, *Denmark* 17 J4 55 10N 10 31 E
Kvaløy, *Norway* 14 B18 69 40N 18 30 E
Kvam, *Norway* 18 C6 61 40N 9 42 E
Kvänum, *Sweden* 17 F7 58 18N 13 11 E
Kvareli = Qvareli, *Georgia* 61 K7 41 57N 45 47 E
Kvarner, *Croatia* 45 D11 44 50N 14 10 E
Kvarnerič, *Croatia* 45 D11 44 43N 14 37 E
Kvås, *Norway* 18 F4 58 16N 7 14 E
Kvernaland, *Norway* 18 F2 58 47N 5 45 E
Kvichak B., *U.S.A.* 144 G8 58 48N 157 30W
Kvicksund, *Sweden* 16 E10 59 27N 16 19 E
Kvikne, *Norway* 18 B7 62 35N 10 16 E
Kvillsfors, *Sweden* 17 G9 57 24N 15 29 E
Kvina →, *Norway* 18 F3 58 19N 6 55 E
Kvinlog, *Norway* 18 F3 58 31N 6 55 E
Kvismare kanal, *Sweden* 16 E9 59 11N 15 35 E
Kvissleby, *Sweden* 16 B11 62 18N 17 22 E
Kviteseid, *Norway* 18 E5 59 24N 8 29 E
Kwabhaca, *S. Africa* 117 E4 30 51 S 29 0 E
Kwadacha →, *Canada* 142 B3 57 28N 125 38W
Kwakhanai, *Botswana* 116 C3 21 39 S 21 16 E
Kwakoegron, *Surinam* 169 B6 5 12N 55 25W
Kwale, *Kenya* 118 C4 4 15 S 39 31 E
Kwale, *Nigeria* 113 D6 5 46N 6 26 E
Kwale □, *Kenya* 118 C4 4 15 S 39 10 E
KwaMashu, *S. Africa* 117 D5 29 45 S 30 58 E
Kwamouth, *Dem. Rep. of the Congo* 114 C3 3 9 S 16 12 E
Kwando →, *Africa* 115 F4 18 27 S 23 32 E
Kwangdaeri, *N. Korea* 75 D14 40 31N 127 32 E
Kwangju, *S. Korea* 75 G14 35 9N 126 54 E
Kwango →, *Dem. Rep. of the Congo* 114 C3 3 14 S 17 22 E
Kwangsi-Chuang = Guangxi Zhuangzu Zizhiqu □, *China* 76 F7 24 0N 109 0 E
Kwangtung = Guangdong □, *China* 77 F9 23 0N 113 0 E
Kwara □, *Nigeria* 113 D6 8 45N 4 30 E
Kwataboahegan →, *Canada* 140 B3 51 9N 80 50W
Kwatisore, *Indonesia* 83 B4 3 18 S 134 50 E
KwaZulu Natal □, *S. Africa* 117 D5 29 0 S 30 0 E
Kweichow = Guizhou □, *China* 76 D6 27 0N 107 0 E
Kwekwe, *Zimbabwe* 119 F2 18 58 S 29 48 E
Kwidzyn, *Poland* 54 E5 53 44N 18 55 E
Kwikila, *Papua N. G.* 132 E4 9 49 S 147 38 E
Kwimba □, *Tanzania* 118 C3 3 0 S 33 0 E
Kwinana New Town, *Australia* 125 F2 32 15 S 115 47 E
Kwisa →, *Poland* 55 G2 51 34N 15 24 E
Kwoka, *Indonesia* 83 B4 0 31 S 132 27 E
Kya-in-Seikkyi, *Burma* 90 G7 16 2N 98 8 E
Kyabé, *Chad* 109 G3 9 30N 19 0 E
Kyabra Cr. →, *Australia* 127 D3 25 36 S 142 55 E
Kyabram, *Australia* 127 F4 36 19 S 145 4 E
Kyaiklat, *Burma* 90 G5 16 25N 95 40 E
Kyaikmaraw, *Burma* 90 G6 16 23N 97 44 E
Kyaikthin, *Burma* 90 D5 23 32N 95 40 E
Kyaikto, *Burma* 86 D1 17 20N 97 3 E
Kyakhta, *Russia* 65 D11 50 30N 106 25 E
Kyancutta, *Australia* 127 E2 33 8 S 135 33 E
Kyangin, *Burma* 90 F5 18 20N 95 20 E
Kyaukhnyat, *Burma* 90 F6 18 15N 97 31 E
Kyaukse, *Burma* 90 E6 21 36N 96 10 E
Kyauktaw, *Burma* 90 E4 20 51N 92 59 E
Kyawkku, *Burma* 90 E6 21 48N 96 56 E
Kybartai, *Lithuania* 54 D9 54 39N 22 45 E
Kyburz, *U.S.A.* 160 G6 38 47N 120 18W
Kybybolite, *Australia* 128 D4 36 53 S 140 55 E
Kyeintali, *Burma* 90 G5 18 0N 94 29 E
Kyenjojo, *Uganda* 118 B3 0 40N 30 37 E
Kyidaunggan, *Burma* 90 F6 19 53N 96 12 E
Kyjov, *Czech Rep.* 35 B10 49 1N 17 7 E
Kyle Dam, *Zimbabwe* 119 G3 20 15 S 31 0 E
Kyle of Lochalsh, *U.K.* 22 D3 57 17N 5 44W
Kyll →, *Germany* 31 F2 49 48N 6 41 E
Kyllburg, *Germany* 31 E2 50 2N 6 34 E
Kymijoki →, *Finland* 15 F22 60 30N 26 55 E
Kyneton, *Australia* 128 D6 37 10 S 144 29 E
Kynuna, *Australia* 126 C3 21 37 S 141 55 E
Kyō-ga-Saki, *Japan* 71 B7 35 45N 135 15 E
Kyoga, L., *Uganda* 118 B3 1 35N 33 0 E
Kyogle, *Australia* 127 D5 28 40 S 153 0 E
Kyongju, *S. Korea* 75 G15 35 51N 129 14 E
Kyŏngsŏng, *N. Korea* 75 D15 41 35N 129 36 E
Kyōto, *Japan* 71 C7 35 0N 135 45 E
Kyōto □, *Japan* 71 B7 35 15N 135 45 E
Kyparissovouno, *Cyprus* 38 D12 35 19N 33 10 E
Kyperounda, *Cyprus* 38 E11 34 56N 32 58 E
Kyrenia, *Cyprus* 38 D12 35 20N 33 20 E
Kyrgyzstan ■, *Asia* 63 C7 42 0N 75 0 E
Kyritz, *Germany* 30 C8 52 56N 12 24 E
Kyrkhult, *Sweden* 17 H8 56 22N 14 34 E
Kyrksæterøra, *Norway* 18 A6 63 18N 9 5 E
Kyrönjoki →, *Finland* 14 E19 63 14N 21 45 E
Kyshtym, *Russia* 62 D8 55 42N 60 34 E
Kystatyam, *Russia* 65 C13 67 20N 123 10 E
Kysucké Nové Mesto, *Slovak Rep.* 35 B11 49 18N 18 47 E
Kythréa, *Cyprus* 38 D12 35 15N 33 29 E
Kytlym, *Russia* 62 B7 59 30N 59 12 E
Kyu-hkok, *Burma* 90 C7 24 4N 98 4 E
Kyunhla, *Burma* 90 D5 23 25N 95 15 E
Kyuquot, *Canada* 142 C3 50 3N 127 25W
Kyurdamir = Kürdämir, *Azerbaijan* 61 K9 40 25N 48 3 E
Kyūshū, *Japan* 70 E3 33 0N 131 0 E
Kyūshū □, *Japan* 70 E3 33 0N 131 0 E
Kyūshū-Sanchi, *Japan* 70 E3 32 35N 131 17 E
Kyustendil, *Bulgaria* 50 D6 42 16N 22 41 E
Kyusyur, *Russia* 65 B13 70 19N 127 30 E
Kywong, *Australia* 129 C7 34 58 S 146 44 E
Kyyiv, *Ukraine* 59 G6 50 30N 30 28 E
Kyyivske Vdskh., *Ukraine* 59 G6 51 0N 30 25 E
Kyzyl, *Russia* 65 D10 51 50N 94 30 E
Kyzyl Kum, *Uzbekistan* 63 B2 42 30N 65 0 E
Kyzyl-Kyya, *Kyrgyzstan* 63 C6 40 16N 72 8 E
Kyzylsu →, *Kyrgyzstan* 63 D6 38 50N 70 0 E
Kzyl-Orda = Qyzylorda, *Kazakstan* 63 A2 44 48N 65 28 E

L

La Albuera, *Spain* 43 G4 38 45N 6 49W
La Alcarria, *Spain* 40 E2 40 31N 2 45W
La Almarcha, *Spain* 40 F2 39 41N 2 24W
La Almunia de Doña Godina, *Spain* 40 D3 41 29N 1 23W
La Asunción, *Venezuela* 169 A5 11 2N 63 53W
La Banda, *Argentina* 174 B3 27 45 S 64 10W
La Bañeza, *Spain* 42 C5 42 17N 5 54W
La Barca, *Mexico* 162 C4 20 20N 102 40W
La Barge, *U.S.A.* 158 E8 42 16N 110 12W
La Bastide-Puylaurent, *France* 28 D7 44 35N 3 55 E
La Baule-Escoubiac, *France* 26 E4 47 17N 2 24W
La Belle, *Fla., U.S.A.* 149 M5 26 46N 81 26W
La Belle, *Mo., U.S.A.* 156 D5 40 7N 91 55W
La Biche →, *Canada* 142 B4 59 57N 123 50W
La Bisbal d'Empordà, *Spain* 40 D8 41 58N 3 2 E
La Blanquilla, *Venezuela* 169 A5 11 51N 64 37W
La Bomba, *Mexico* 162 A1 31 53N 115 2W
La Bresse, *France* 27 D13 48 2N 6 53 E
La Bureba, *Spain* 42 C7 42 36N 3 24W
La Cal →, *Bolivia* 173 D6 17 27 S 58 15W
La Calera, *Chile* 174 C1 32 50 S 71 10W
La Campiña, *Spain* 43 H6 37 45N 4 45W
La Canal, *Spain* 39 C7 38 51N 1 23 E
La Cañiza = A Cañiza, *Spain* 42 C2 42 13N 8 16W
La Canourgue, *France* 28 D7 44 26N 3 13 E
La Capelle, *France* 27 C10 49 59N 3 50 E
La Carlota, *Argentina* 174 C3 33 30 S 63 20W
La Carlota, *Phil.* 81 F4 10 25N 122 55 E
La Carlota, *Spain* 43 H6 37 40N 4 56W
La Carolina, *Spain* 43 G7 38 17N 3 38W
La Castellana, *Phil.* 81 F4 10 20N 123 3 E
La Cavalerie, *France* 28 D7 44 1N 3 10 E
La Ceiba, *Honduras* 164 C2 15 40N 86 50W
La Chaise-Dieu, *France* 28 C7 45 18N 3 42 E
La Chapelle d'Angillon, *France* 27 E9 47 21N 2 25 E
La Chapelle-St-Luc, *France* 27 D11 48 20N 4 3 E
La Chapelle-sur-Erdre, *France* 26 E5 47 18N 1 34W
La Charité-sur-Loire, *France* 27 E10 47 10N 3 1 E
La Chartre-sur-le-Loir, *France* 26 E7 47 44N 0 34 E
La Châtaigneraie, *France* 28 B3 46 39N 0 44W
La Châtre, *France* 27 F9 46 35N 2 0 E
La Chaux-de-Fonds, *Switz.* 32 B3 47 7N 6 50 E
La Chorrera, *Colombia* 168 D3 0 44 S 73 1W
La Ciotat, *France* 29 E9 43 10N 5 37 E
La Clayette, *France* 27 F11 46 17N 4 19 E
La Cocha, *Argentina* 174 B2 27 50 S 65 40W

238 La Concepción

La Concepción = Ri-Aba, *Eq. Guin.* **113 E6** 3 28N 8 40 E
La Concepción, *Venezuela* **168 A3** 10 30N 71 50W
La Concordia, *Mexico* **163 D6** 16 8N 92 38W
La Conner, *U.S.A.* **158 B2** 48 23N 122 30W
La Coruña = A Coruña, *Spain* **42 B2** 43 20N 8 25W
La Coruña □, *Spain* **42 B2** 43 10N 8 30W
La Côte, *Switz.* **32 D2** 46 25N 6 15 E
La Côte-St-André, *France* **29 C9** 45 24N 5 15 E
La Courtine-le-Trucq, *France* **28 C6** 45 41N 2 15 E
La Crau, *Bouches-du-Rhône, France* **29 E8** 43 32N 4 40 E
La Crau, *Var, France* **29 E10** 43 9N 6 4 E
La Crete, *Canada* **142 B5** 58 11N 116 24W
La Crosse, *Fla., U.S.A.* **152 F7** 29 51N 82 24W
La Crosse, *Kans., U.S.A.* **154 F5** 38 32N 99 18W
La Crosse, *Wis., U.S.A.* **154 D9** 43 48N 91 15W
La Cruz, *Costa Rica* **164 D2** 11 4N 85 39W
La Cruz, *Mexico* **162 C3** 23 55N 106 54W
La Dorada, *Colombia* **168 B3** 5 30N 74 40W
La Ensenada, *Chile* **176 B2** 41 12 S 72 33W
La Escondida, *Mexico* **162 C5** 24 6N 99 55W
La Esmeralda, *Paraguay* **174 A3** 22 16 S 62 33W
La Esperanza, *Argentina* **176 B3** 40 26 S 68 32W
La Esperanza, *Cuba* **164 B3** 22 46N 83 44W
La Esperanza, *Honduras* **164 D2** 14 15N 88 10W
La Estrada = A Estrada, *Spain* **42 C2** 42 43N 8 27W
La Faouët, *France* **26 D3** 48 2N 3 30W
La Fayette, *U.S.A.* **149 H3** 34 42N 85 17W
La Fé, *Cuba* **164 B3** 22 2N 84 15W
La Fère, *France* **27 C10** 49 39N 3 21 E
La Ferté-Bernard, *France* **26 D7** 48 10N 0 40 E
La Ferté-Gaucher, *France* **27 D10** 48 47N 3 19 E
La Ferté-Macé, *France* **26 D6** 48 35N 0 22W
La Ferté-St-Aubin, *France* **27 E8** 47 42N 1 57 E
La Ferté-sous-Jouarre, *France* **27 D10** 48 56N 3 8 E
La Ferté-Vidame, *France* **26 D7** 48 37N 0 53 E
La Flèche, *France* **26 E6** 47 42N 0 4W
La Foa, *N. Cal.* **133 U19** 21 43 S 165 50 E
La Follette, *U.S.A.* **149 G3** 36 23N 84 7W
La Fontaine, *U.S.A.* **157 D11** 40 40N 85 43W
La Fregeneda, *Spain* **42 E4** 40 58N 6 54W
La Fría, *Venezuela* **168 B3** 8 13N 72 15W
La Fuente de San Esteban, *Spain* **42 E4** 40 49N 6 15W
La Gacilly, *France* **26 E4** 47 45N 2 8W
La Gineta, *Spain* **41 F2** 39 8N 2 1W
La Gloria, *Colombia* **168 B3** 8 37N 73 48W
La Gran Sabana, *Venezuela* **169 B5** 5 30N 61 30W
La Grand-Combe, *France* **29 D8** 44 13N 4 2 E
La Grande, *U.S.A.* **158 D4** 45 20N 118 5W
La Grande-Motte, *France* **29 E8** 43 23N 4 5 E
La Grange, *Calif., U.S.A.* **160 H6** 37 42N 120 27W
La Grange, *Ga., U.S.A.* **152 B4** 33 2N 85 2W
La Grange, *Ky., U.S.A.* **148 F3** 38 25N 85 23W
La Grange, *Mo., U.S.A.* **156 D5** 40 3N 91 35W
La Grange, *Tex., U.S.A.* **155 L6** 29 54N 96 52W
La Grave, *France* **29 C10** 45 3N 6 18 E
La Grita, *Venezuela* **168 B3** 8 8N 71 59W
La Guaira, *Venezuela* **168 A4** 10 36N 66 56W
La Guardia = A Guarda, *Spain* **42 D2** 41 56N 8 52W
La Gudiña = A Gudiña, *Spain* **42 C3** 42 4N 7 8W
La Güera, *Mauritania* **110 D1** 20 51N 17 0W
La Guerche-de-Bretagne, *France* **26 E5** 47 57N 1 16W
La Guerche-sur-l'Aubois, *France* **27 F9** 46 58N 2 56 E
La Habana, *Cuba* **164 B3** 23 8N 82 22W
La Harpe, *U.S.A.* **156 D6** 40 35N 90 58W
La Haye-du-Puits, *France* **26 C5** 49 17N 1 33W
La Horqueta, *Venezuela* **169 B5** 7 55N 60 20W
La Horra, *Spain* **42 D7** 41 44N 3 53W
La Independencia, *Mexico* **163 D6** 16 31N 91 47W
La Isabela, *Dom. Rep.* **165 C5** 19 58N 71 2W
La Jara, *U.S.A.* **159 H11** 37 16N 105 58W
La Jonquera, *Spain* **40 C7** 42 25N 2 53 E
La Joya, *Peru* **172 D3** 16 43 S 71 52W
La Junta, *U.S.A.* **155 F3** 37 59N 103 33W
La Laguna, *Canary Is.* **39 F3** 28 28N 16 18W
La Libertad, *Guatemala* **164 C1** 16 47N 90 7W
La Libertad, *Mexico* **162 B2** 29 55N 112 41W
La Libertad □, *Peru* **172 B2** 8 0 S 78 30W
La Ligua, *Chile* **174 C1** 32 30 S 71 16W
La Línea de la Concepción, *Spain* **43 J5** 36 15N 5 23W
La Loche, *Canada* **143 B7** 56 29N 109 26W
La Londe-les-Maures, *France* **29 E10** 43 8N 6 14 E
La Lora, *Spain* **42 C7** 42 45N 4 0W
La Loupe, *France* **26 D8** 48 29N 1 0 E
La Louvière, *Belgium* **24 D4** 50 27N 4 10 E
La Machine, *France* **27 F10** 46 54N 3 27 E
La Maddalena, *Italy* **46 A2** 41 13N 9 24 E
La Malbaie, *Canada* **141 C5** 47 40N 70 10W
La Mancha, *Spain* **41 F2** 39 10N 2 54W
La Mariña, *Spain* **42 B3** 43 30N 7 40W
La Mesa, *Calif., U.S.A.* **161 N9** 32 46N 117 3W
La Mesa, *N. Mex., U.S.A.* **159 K10** 32 7N 106 42W
La Misión, *Mexico* **162 A1** 32 5N 116 50W
La Moille, *U.S.A.* **156 C7** 41 32N 89 17W
La Moine →, *U.S.A.* **156 E6** 39 59N 90 31W
La Monte, *U.S.A.* **156 F3** 38 46N 93 26W
La Mothe-Achard, *France* **26 F5** 46 37N 1 40W
La Motte, *France* **29 D10** 44 20N 6 3 E
La Motte-Chalançon, *France* **29 D9** 44 30N 5 21 E
La Motte-Servolex, *France* **29 C9** 45 35N 5 53 E
La Moure, *U.S.A.* **154 B5** 46 21N 98 18W
La Muela, *Spain* **40 D3** 41 36N 1 7W
La Mure, *France* **29 D9** 44 55N 5 48 E
La Negra, *Chile* **174 A1** 23 46 S 70 18W
La Neuveville, *Switz.* **32 B4** 47 4N 7 6 E
La Oliva, *Canary Is.* **39 F6** 28 36N 13 57W
La Oraya, *Peru* **172 C2** 11 32 S 75 54W
La Orotava, *Canary Is.* **39 F3** 28 22N 16 31W
La Pacaudière, *France* **27 F10** 46 11N 3 52 E
La Palma, *Canary Is.* **39 F2** 28 40N 17 50W
La Palma, *Panama* **164 E4** 8 15N 78 0W
La Palma del Condado, *Spain* **43 H4** 37 21N 6 38W
La Paloma, *Chile* **174 C1** 30 35 S 71 0W
La Pampa □, *Argentina* **174 D2** 36 50 S 66 0W
La Paragua, *Venezuela* **169 B5** 6 50N 63 20W
La Paz, *Entre Ríos, Argentina* **174 C4** 30 50 S 59 45W
La Paz, *San Luis, Argentina* **174 C2** 33 30 S 67 20W
La Paz, *Bolivia* **172 D4** 16 20 S 68 10W
La Paz, *Honduras* **164 D2** 14 20N 87 47W
La Paz, *Mexico* **162 C2** 24 10N 110 20W
La Paz, *Phil.* **80 D3** 15 26N 120 45 E
La Paz □, *Bolivia* **172 D4** 15 30 S 68 0W
La Paz Centro, *Nic.* **164 D2** 12 20N 86 41W
La Pedrera, *Colombia* **168 D4** 1 18 S 69 43W
La Perouse Str., *Asia* **68 B11** 45 40N 142 0 E
La Pesca, *Mexico* **163 C5** 23 46N 97 47W
La Piedad, *Mexico* **162 C4** 20 20N 102 1W
La Pine, *U.S.A.* **158 E3** 43 40N 121 30W
La Plant, *U.S.A.* **154 C4** 45 9N 100 39W
La Plata, *Argentina* **174 D4** 35 0 S 57 55W
La Plata, *Colombia* **168 C2** 2 23N 75 53W
La Plata, *U.S.A.* **156 D4** 40 2N 92 29W
La Plata, L., *Argentina* **176 B2** 44 55 S 71 50W
La Pobla de Lillet, *Spain* **40 C6** 42 16N 1 59 E
La Pola de Gordón, *Spain* **42 C5** 42 51N 5 41W
La Porta, *France* **29 F13** 42 25N 9 21 E
La Porte, *U.S.A.* **157 C10** 41 36N 86 43W
La Porte City, *U.S.A.* **156 B4** 42 19N 92 12W
La Presanella, *Italy* **44 B7** 46 13N 10 40 E
La Puebla = Sa Pobla, *Spain* **40 F8** 39 46N 3 1 E
La Puebla de Cazalla, *Spain* **43 H5** 37 10N 5 20W
La Puebla de los Infantes, *Spain* **43 H5** 37 47N 5 24W
La Puebla de Montalbán, *Spain* **42 F6** 39 52N 4 22W
La Puebla del Río, *Spain* **43 H4** 37 16N 6 3W
La Puerta de Segura, *Spain* **43 G8** 38 22N 2 45W
La Punt, *Switz.* **33 C9** 46 35N 9 56 E
La Purísima, *Mexico* **162 B2** 26 10N 112 4W
La Push, *U.S.A.* **160 C2** 47 55N 124 38W
La Quiaca, *Argentina* **174 A2** 22 5 S 65 35W
La Reine, *Canada* **140 C4** 48 50N 79 30W
La Réole, *France* **28 D3** 44 35N 0 1W
La Restinga, *Canary Is.* **39 G2** 27 38N 17 59W
La Rioja, *Argentina* **174 B2** 29 20 S 67 0W
La Rioja □, *Argentina* **174 B2** 29 30 S 67 0W
La Rioja □, *Spain* **40 C2** 42 20N 2 20W
La Robla, *Spain* **42 C5** 42 50N 5 41W
La Roche, *Switz.* **32 C4** 46 42N 7 7 E
La Roche-Bernard, *France* **26 E4** 47 31N 2 19W
La Roche-Canillac, *France* **28 C5** 45 12N 1 57 E
La Roche-en-Ardenne, *Belgium* **24 D5** 50 11N 5 35 E
La Roche-sur-Foron, *France* **27 F13** 46 4N 6 19 E
La Roche-sur-Yon, *France* **26 F5** 46 40N 1 25W
La Rochefoucauld, *France* **28 C4** 45 44N 0 24 E
La Rochelle, *France* **28 B2** 46 10N 1 9W
La Roda, *Spain* **41 F2** 39 13N 2 15W
La Roda de Andalucía, *Spain* **43 H6** 37 12N 4 46W
La Romana, *Dom. Rep.* **165 C6** 18 27N 68 57W
La Ronge, *Canada* **143 B7** 55 5N 105 20W
La Rue, *U.S.A.* **157 D13** 40 35N 83 23W
La Rumorosa, *Mexico* **161 N10** 32 33N 116 4W
La Sabina, *Spain* **39 C7** 38 44N 1 25 E
La Sagra, *Spain* **41 H2** 37 57N 2 35W
La Salle, *U.S.A.* **156 C7** 41 20N 89 6W
La Sanabria, *Spain* **42 C4** 42 0N 6 30W
La Santa, *Canary Is.* **39 E6** 29 5N 13 40W
La Sarraz, *Switz.* **32 C3** 46 38N 6 32 E
La Sarre, *Canada* **140 C4** 48 45N 79 15W
La Scie, *Canada* **141 C8** 49 57N 55 36W
La Selva, *Spain* **40 C7** 42 0N 2 45 E
La Selva Beach, *U.S.A.* **160 J5** 36 56N 121 51W
La Selva del Camp, *Spain* **40 D6** 41 13N 1 8 E
La Serena, *Chile* **174 B1** 29 55 S 71 10W
La Serena, *Spain* **43 G5** 38 45N 5 40W
La Seu d'Urgell, *Spain* **40 C6** 42 22N 1 23 E
La Seyne-sur-Mer, *France* **29 E9** 43 7N 5 52 E
La Sila, *Italy* **47 C9** 39 15N 16 35 E
La Solana, *Spain* **43 G7** 38 59N 3 14W
La Souterraine, *France* **27 F8** 46 15N 1 30 E
La Spézia, *Italy* **44 D6** 44 7N 9 50 E
La Suze-sur-Sarthe, *France* **26 E7** 47 53N 0 2 E
La Tagua, *Colombia* **168 C3** 0 3N 74 40W
La Teste, *France* **28 D2** 44 37N 1 8W
La Tortuga, *Venezuela* **165 D6** 11 0N 65 22W
La Tour-du-Pin, *France* **29 C9** 45 33N 5 27 E
La Tranche-sur-Mer, *France* **26 F5** 46 20N 1 27W
La Tremblade, *France* **28 C2** 45 46N 1 8W
La Trinidad, *Phil.* **80 C3** 16 28N 120 35 E
La Tuque, *Canada* **140 C5** 47 30N 72 50W
La Unión, *Chile* **176 B2** 40 10 S 73 0W
La Unión, *Colombia* **168 C2** 1 35N 77 5W
La Unión, *El Salv.* **164 D2** 13 20N 87 50W
La Unión, *Mexico* **162 D4** 17 58N 101 49W
La Unión, *Peru* **172 B2** 9 43 S 76 45W
La Unión, *Spain* **41 H4** 37 38N 0 53W
La Union □, *Phil.* **80 C3** 16 30N 120 25 E
La Urbana, *Venezuela* **168 B4** 7 8N 66 56W
La Vall d'Uixó, *Spain* **40 F4** 39 49N 0 15W
La Vecilla de Curveño, *Spain* **42 C5** 42 51N 5 27W
La Vega, *Dom. Rep.* **165 C5** 19 20N 70 30W
La Vega, *Peru* **172 C2** 10 41 S 77 44W
La Vela, *Venezuela* **168 A4** 11 27N 69 34W
La Veleta, *Spain* **43 H7** 37 1N 3 22W
La Venta, *Mexico* **163 D6** 18 8N 94 3W
La Ventura, *Mexico* **162 C4** 24 38N 100 54W
La Venturosa, *Colombia* **168 B4** 6 8N 68 48W
La Victoria, *Venezuela* **168 A4** 10 14N 67 20W
La Voulte-sur-Rhône, *France* **29 D8** 44 48N 4 46 E
Laa an der Thaya, *Austria* **35 C9** 48 43N 16 23 E
Laaber, Grosse →, *Germany* **31 G8** 48 55N 12 30 E
Laage, *Germany* **30 B8** 53 55N 12 21 E
Laatzen, *Germany* **30 C5** 52 19N 9 48 E
Laba →, *Russia* **61 H4** 45 11N 39 42 E
Laban, *Burma* **90 C6** 25 52N 96 40 E
Labason, *Phil.* **81 G4** 8 4N 122 31 E
Labastide-Murat, *France* **28 D5** 44 39N 1 33 E
Labastide-Rouairoux, *France* **28 E6** 43 28N 2 39 E
Labbézenga, *Mali* **113 B5** 15 2N 0 48 E
Labdah = Leptis Magna, *Libya* **108 B2** 32 40N 14 12 E
Labe = Elbe →, *Europe* **30 B4** 53 50N 9 0 E
Labé, *Guinea* **112 C2** 11 24N 12 16W
Laberge, L., *Canada* **142 A1** 61 11N 135 12W
Labian, Tanjong, *Malaysia* **85 A5** 5 9N 119 13 E
Labig Pt., *Phil.* **80 B4** 18 25N 122 25 E
Labin, *Croatia* **45 C11** 45 5N 14 8 E
Labinsk, *Russia* **61 H5** 44 40N 40 48 E
Labis, *Malaysia* **87 L4** 2 22N 103 2 E
Łabiszyn, *Poland* **55 F4** 52 57N 17 54 E
Labo, *Phil.* **80 D4** 14 9N 122 51 E
Laboe, *Germany* **30 A6** 54 24N 10 13 E
Laboka, *Gabon* **114 B2** 0 19N 11 32 E
Laborec →, *Slovak Rep.* **35 C14** 48 37N 21 58 E
Labouheyre, *France* **28 D3** 44 13N 0 55W
Laboulaye, *Argentina* **174 C3** 34 10 S 63 30W
Labrador, *Canada* **141 B7** 53 20N 61 0W
Labrador City, *Canada* **141 B6** 52 57N 66 55W
Lábrea, *Brazil* **173 B5** 7 15 S 64 51W
Labrède, *France* **28 D3** 44 41N 0 32W
Labruguière, *France* **28 E6** 43 31N 2 16 E
Labuan, *Malaysia* **85 A5** 5 20N 115 14 E
Labuan, Pulau, *Malaysia* **85 A5** 5 21N 115 13 E
Labuha, *Indonesia* **82 B3** 0 30 S 127 30 E
Labuhan, *Indonesia* **84 D3** 6 22 S 105 50 E
Labuhanbajo, *Indonesia* **82 C2** 8 28 S 120 1 E
Labuk, Telok, *Malaysia* **85 A5** 6 10N 117 50 E
Labutta, *Burma* **90 G5** 16 9N 94 46 E
Labyrinth, L., *Australia* **127 E2** 30 40 S 135 11 E
Labytnangi, *Russia* **56 A12** 66 39N 66 21 E
Laç, *Albania* **50 E3** 41 38N 19 43 E
Lac Allard, *Canada* **141 B7** 50 33N 63 24W
Lac Bouchette, *Canada* **141 C5** 48 16N 72 11W
Lac du Flambeau, *U.S.A.* **154 B10** 45 58N 89 53W
Lac Édouard, *Canada* **140 C5** 47 40N 72 16W
Lac La Biche, *Canada* **142 C6** 54 45N 111 58W
Lac la Martre = Wha Ti, *Canada* **138 B8** 63 8N 117 16W
Lac-Mégantic, *Canada* **141 C5** 45 35N 70 53W
Lac Seul, Res., *Canada* **140 B1** 50 25N 92 30W
Lac Thien, *Vietnam* **86 F7** 12 25N 108 11 E
Lacanau, *France* **28 D2** 44 58N 1 5W
Lacanau, Étang de, *France* **28 D2** 44 58N 1 7W
Lacantúm →, *Mexico* **163 D6** 16 36N 90 40W
Lacara →, *Spain* **43 G4** 38 55N 6 25W
Lacaune, *France* **28 E6** 43 43N 2 40 E
Lacaune, Mts. de, *France* **28 E6** 43 43N 2 50 E
Laccadive Is. = Lakshadweep Is., *Ind. Oc.* **121 C6** 10 0N 72 30 E
Lacepede B., *Australia* **128 D3** 36 40 S 139 40 E
Lacepede Is., *Australia* **124 C3** 16 55 S 122 0 E
Lacerdónia, *Mozam.* **119 F4** 18 3 S 35 35 E
Lacey, *U.S.A.* **160 C4** 47 7N 122 49W
Lachay, Pta., *Peru* **172 C2** 11 17 S 77 44W
Lachen, *India* **90 B2** 27 46N 88 36 E
Lachen, *Switz.* **33 B7** 47 12N 8 51 E
Lachhmangarh, *India* **92 F6** 27 50N 75 4 E
Lachi, *Pakistan* **92 C4** 33 25N 71 20 E
Lachine, *Canada* **140 C5** 45 30N 73 40W
Lachlan →, *Australia* **128 C5** 34 22 S 143 55 E
Lachute, *Canada* **140 C5** 45 39N 74 21W
Lackawanna, *U.S.A.* **150 D6** 42 50N 78 50W
Lacolle, *Canada* **151 A11** 45 5N 73 22W
Lacombe, *Canada* **142 C6** 52 30N 113 44W
Lacon, *U.S.A.* **156 C7** 41 2N 89 24W
Lacona, *Iowa, U.S.A.* **156 C3** 41 12N 93 23W
Lacona, *N.Y., U.S.A.* **151 C8** 43 39N 76 10W
Láconi, *Italy* **46 C2** 39 54N 9 4 E
Laconia, *U.S.A.* **151 C13** 43 32N 71 28W
Lacoochee, *U.S.A.* **153 G7** 28 28N 82 11W
Lacq, *France* **28 E3** 43 25N 0 35W
Lacrosse, *U.S.A.* **158 C5** 46 51N 117 58W
Lacub, *Phil.* **80 C3** 17 40N 120 53 E
Ladakh Ra., *India* **93 C8** 34 0N 78 0 E
Ladário, *Brazil* **173 D6** 19 1 S 57 35W
Ladd, *U.S.A.* **156 C7** 41 23N 89 13W
Laddonia, *U.S.A.* **156 E5** 39 15N 91 39W
Lądek-Zdrój, *Poland* **55 H3** 50 21N 16 53 E
Ládhon →, *Greece* **48 D3** 37 40N 21 50 E
Ladik, *Turkey* **100 B6** 40 57N 35 58 E
Ladismith, *S. Africa* **116 E3** 33 28 S 21 15 E
Ladíspoli, *Italy* **45 G9** 41 56N 12 5 E
Lādīz, *Iran* **97 D9** 28 55N 61 15 E
Ladnun, *India* **92 F6** 27 38N 74 25 E
Ladoga, L. = Ladozhskoye Ozero, *Russia* **58 B6** 61 15N 30 30 E
Ladozhskoye Ozero, *Russia* **58 B6** 61 15N 30 30 E
Ladrillero, G., *Chile* **176 C1** 49 20 S 75 35W
Ladson, *U.S.A.* **152 C9** 32 59N 80 6W
Lady Grey, *S. Africa* **116 E4** 30 43 S 27 13 E
Lady Lake, *U.S.A.* **153 G8** 28 55N 81 55W
Ladybrand, *S. Africa* **116 D4** 29 9 S 27 29 E
Ladysmith, *Canada* **142 D4** 49 0N 123 49W
Ladysmith, *S. Africa* **117 D4** 28 32 S 29 46 E
Ladysmith, *U.S.A.* **154 C9** 45 28N 91 12W
Lae, *Papua N. G.* **132 D4** 6 40 S 147 2 E
Laem Ngop, *Thailand* **87 F4** 12 10N 102 26 E
Laem Pho, *Thailand* **87 J3** 6 55N 101 19 E
Lærdalsøyri, *Norway* **18 C4** 61 6N 7 28 E
Læsø, *Denmark* **17 G5** 57 15N 11 5 E
Læsø Rende, *Denmark* **17 G4** 57 20N 10 45 E
Lafayette, *Ala., U.S.A.* **152 C4** 32 54N 85 24W
Lafayette, *Colo., U.S.A.* **154 F2** 39 58N 105 12W
Lafayette, *Ind., U.S.A.* **157 D10** 40 25N 86 54W
Lafayette, *La., U.S.A.* **155 K9** 30 14N 92 1W
Lafayette, *Tenn., U.S.A.* **149 G3** 36 31N 86 2W
Laferte →, *Canada* **142 A5** 61 53N 117 44W
Lafia, *Nigeria* **113 D6** 8 30N 8 34 E
Lafiagi, *Nigeria* **113 D6** 8 52N 5 20 E
Lafleche, *Canada* **143 D7** 49 45N 106 40W
Lafon, *Sudan* **107 F3** 5 5N 32 29 E
Lagaip →, *Papua N. G.* **132 C2** 5 4 S 142 52 E
Lagan, *Sweden* **17 H7** 56 56N 13 58 E
Lagan →, *Sweden* **17 H6** 56 30N 12 58 E
Lagan →, *U.K.* **23 B6** 54 36N 5 55W
Lagangilang, *Phil.* **80 C3** 17 37N 120 44 E
Lagarfljót →, *Iceland* **11 B12** 65 40N 14 18W
Lagarto, *Brazil* **170 D4** 10 54 S 37 41W
Lagawe, *Phil.* **80 C3** 16 49N 121 6 E
Lage, *Germany* **30 D4** 51 59N 8 48 E
Lågen →, *Oppland, Norway* **15 F14** 61 8N 10 25 E
Lågen →, *Vestfold, Norway* **15 G14** 59 3N 10 3 E
Lägerdorf, *Germany* **30 B5** 53 53N 9 34 E
Laghman □, *Afghan.* **91 B3** 34 20N 70 0 E
Laghouat, *Algeria* **111 B5** 33 50N 2 59 E
Lagnieu, *France* **29 C9** 45 55N 5 20 E
Lagny-sur-Marne, *France* **27 D9** 48 52N 2 44 E
Lago, *Italy* **47 C9** 39 10N 16 9 E
Lago Posadas, *Argentina* **176 C2** 47 30 S 71 40W
Lago Ranco, *Chile* **176 B2** 40 19 S 72 30W
Lagôa, *Portugal* **43 H2** 37 8N 8 27W
Lagoaça, *Portugal* **42 D4** 41 11N 6 44W
Lagodekhi, *Georgia* **61 K8** 41 50N 46 22 E
Lagónegro, *Italy* **47 B8** 40 8N 15 45 E
Lagonoy G., *Phil.* **80 E4** 13 50N 123 50 E
Lagos, *Nigeria* **113 D5** 6 25N 3 27 E
Lagos, *Portugal* **43 H2** 37 5N 8 41W
Lagos de Moreno, *Mexico* **162 C4** 21 21N 101 55W
Lagrange, *Australia* **124 C3** 18 45 S 121 43 E
Lagrange, *U.S.A.* **157 C11** 41 39N 85 25W
Lagrange B., *Australia* **124 C3** 18 38 S 121 42 E
Laguardia, *Spain* **40 C2** 42 33N 2 35W
Laguépie, *France* **28 D5** 44 8N 1 57 E
Laguna, *Brazil* **175 B6** 28 30 S 48 50W
Laguna, *U.S.A.* **159 J10** 35 2N 107 25W
Laguna □, *Phil.* **80 D3** 14 10N 121 20 E
Laguna Beach, *U.S.A.* **161 M9** 33 33N 117 47W
Laguna de Duera, *Spain* **42 D6** 41 35N 4 43W
Laguna Limpia, *Argentina* **174 B4** 26 32 S 59 45W
Laguna Madre, *U.S.A.* **163 B5** 27 0N 97 20W
Lagunas, *Chile* **174 A2** 21 0 S 69 45W
Lagunas, *Peru* **172 B2** 5 10 S 75 35W
Lagunillas, *Bolivia* **173 D5** 19 38 S 63 43W
Lahad Datu, *Malaysia* **85 A5** 5 0N 118 20 E
Lahad Datu, Teluk, *Malaysia* **85 B5** 4 50N 118 20 E
Lahan Sai, *Thailand* **86 E4** 14 25N 102 52 E
Lahanam, *Laos* **86 D5** 16 16N 105 16 E
Laharpur, *India* **93 F9** 27 43N 80 56 E
Lahat, *Indonesia* **84 C2** 3 45 S 103 30 E
Lahe, *Burma* **90 B5** 26 20N 95 26 E
Lahewa, *Indonesia* **84 B1** 1 22N 97 12 E
Lahiang Lahiang, *Phil.* **81 H3** 6 1N 121 23 E
Laḥij, *Yemen* **98 D4** 13 4N 44 53 E
Lāhījān, *Iran* **97 B6** 37 10N 50 6 E
Lahilahi Pt., *U.S.A.* **145 K13** 21 28N 158 13W
Lahn →, *Germany* **31 E3** 50 19N 7 37 E
Lahnstein, *Germany* **31 E3** 50 19N 7 37 E
Laholm, *Sweden* **17 H7** 56 30N 13 2 E
Laholmsbukten, *Sweden* **17 H6** 56 30N 12 45 E
Lahontan Reservoir, *U.S.A.* **158 G4** 39 28N 119 4W
Lahore, *Pakistan* **91 C4** 31 32N 74 22 E
Lahpongsel, *Burma* **90 B7** 27 7N 98 25 E
Lahr, *Germany* **31 G3** 48 20N 7 53 E
Lahti, *Finland* **15 F21** 60 58N 25 40 E
Lahtis = Lahti, *Finland* **15 F21** 60 58N 25 40 E
Laï, *Chad* **109 G3** 9 25N 16 18 E
Lai Chau, *Vietnam* **86 A4** 22 5N 103 3 E
Lai-hka, *Burma* **90 E6** 21 16N 97 40 E
Laiagam, *Papua N. G.* **132 C2** 5 33 S 143 30 E
Lai'an, *China* **77 A12** 32 28N 118 30 E
Laibin, *China* **76 F7** 23 42N 109 14 E
Laidley, *Australia* **127 D5** 27 39 S 152 20 E
Laie, *U.S.A.* **145 J14** 21 39N 157 56W
Laifeng, *China* **76 C7** 29 27N 109 20 E
L'Aigle, *France* **26 D7** 48 46N 0 38 E
Laignes, *France* **27 E11** 47 50N 4 20 E
L'Aiguillon-sur-Mer, *France* **28 B2** 46 20N 1 18W
Laikipia □, *Kenya* **118 B4** 0 30N 36 30 E
Laila = Laylá, *Si. Arabia* **98 B4** 22 10N 46 40 E
Laingsburg, *S. Africa* **116 E3** 33 9 S 20 52 E
Lainio älv →, *Sweden* **14 C20** 67 35N 22 40 E
Lairg, *U.K.* **22 C4** 58 2N 4 24W
Lais, *Phil.* **81 H5** 6 20N 125 39 E
Laishui, *China* **74 E8** 39 23N 115 45 E
Laissac, *France* **28 D6** 44 23N 2 50 E
Láives, *Italy* **45 B8** 46 26N 11 20 E
Laiwu, *China* **75 F9** 36 15N 117 40 E
Laixi, *China* **75 F11** 36 50N 120 31 E
Laiyang, *China* **75 F11** 36 59N 120 45 E
Laiyuan, *China* **74 E8** 39 20N 114 40 E
Laizhou Wan, *China* **75 F10** 37 30N 119 30 E
Laja →, *Mexico* **162 C4** 20 55N 100 46W
Lajere, *Nigeria* **113 C7** 12 10N 11 25 E
Lajes, *Rio Grande do N., Brazil* **170 C4** 5 41 S 36 14W
Lajes, *Sta. Catarina, Brazil* **175 B5** 27 48 S 50 20W
Lajinha, *Brazil* **171 F3** 20 9 S 41 37W
Lajkovac, *Serbia, Yug.* **50 B4** 44 27N 20 14 E
Lajosmizse, *Hungary* **52 C4** 47 3N 19 32 E
Lak Sao, *Laos* **86 C5** 18 11N 104 59 E
Lakaband, *Pakistan* **92 D3** 31 2N 69 15 E
Lakatoro, *Vanuatu* **133 F5** 16 0 S 167 0 E
Lake Alfred, *U.S.A.* **153 G8** 28 6N 81 44W
Lake Alpine, *U.S.A.* **160 G7** 38 29N 120 0W
Lake Andes, *U.S.A.* **154 D5** 43 9N 98 32W
Lake Anse, *U.S.A.* **148 B1** 46 42N 88 25W
Lake Arthur, *U.S.A.* **155 K8** 30 5N 92 41W
Lake Bird, *U.S.A.* **152 E6** 30 14N 83 37W
Lake Butler, *U.S.A.* **152 E7** 30 1N 82 21W
Lake Cargelligo, *Australia* **129 B7** 33 15 S 146 22 E
Lake Charles, *U.S.A.* **155 K8** 30 14N 93 13W
Lake City, *Colo., U.S.A.* **159 G10** 38 2N 107 19W
Lake City, *Fla., U.S.A.* **152 E7** 30 11N 82 38W
Lake City, *Iowa, U.S.A.* **156 B2** 42 16N 94 44W
Lake City, *Mich., U.S.A.* **148 C3** 44 20N 85 13W
Lake City, *Minn., U.S.A.* **154 C8** 44 27N 92 16W
Lake City, *Pa., U.S.A.* **150 D4** 42 1N 80 21W
Lake City, *S.C., U.S.A.* **152 B10** 33 52N 79 45W
Lake Clarke Shores, *U.S.A.* **153 J9** 26 39N 80 5W
Lake Coleridge, *N.Z.* **131 D6** 43 17 S 171 30 E
Lake Forest, *U.S.A.* **157 B9** 42 15N 87 50W
Lake Geneva, *U.S.A.* **157 B8** 42 36N 88 26W
Lake George, *U.S.A.* **151 C11** 43 26N 73 43W
Lake Grace, *Australia* **125 F2** 33 7 S 118 28 E
Lake Harbor, *U.S.A.* **153 J9** 26 42N 80 48W
Lake Harbour = Kimmirut, *Canada* **139 B13** 62 50N 69 50W
Lake Havasu City, *U.S.A.* **161 L12** 34 27N 114 22W
Lake Helen, *U.S.A.* **153 G8** 28 59N 81 14W
Lake Hughes, *U.S.A.* **161 L8** 34 41N 118 26W
Lake Isabella, *U.S.A.* **161 K8** 35 38N 118 28W
Lake King, *Australia* **125 F2** 33 5 S 119 45 E
Lake Lenore, *Canada* **143 C8** 52 24N 104 59W
Lake Louise, *Canada* **142 C5** 51 30N 116 10W
Lake Mead National Recreation Area, *U.S.A.* **161 K12** 36 15N 114 30W
Lake Michigan Beach, *U.S.A.* **157 B10** 42 13N 86 25W
Lake Mills, *Iowa, U.S.A.* **154 D8** 43 25N 93 32W
Lake Mills, *Wis., U.S.A.* **157 A8** 43 5N 88 55W
Lake Monroe, *U.S.A.* **153 G8** 28 50N 81 19W
Lake Murray, *Papua N. G.* **132 D1** 6 48 S 141 29 E
Lake Nash, *Australia* **126 C2** 20 57 S 138 0 E
Lake Odessa, *U.S.A.* **157 B11** 42 47N 85 8W
Lake Orion, *U.S.A.* **157 B13** 42 47N 83 14W
Lake Park, *Fla., U.S.A.* **153 J9** 26 48N 80 3W
Lake Park, *Ga., U.S.A.* **152 E6** 30 41N 83 11W
Lake Placid, *U.S.A.* **153 H8** 27 18N 81 22W
Lake Providence, *U.S.A.* **155 J9** 32 48N 91 10W
Lake Pukaki, *N.Z.* **131 E5** 44 11 S 170 8 E
Lake River, *Canada* **140 B3** 54 30N 82 31W
Lake Superior Prov. Park, *Canada* **140 C3** 47 45N 84 45W
Lake Tekapo, *N.Z.* **131 E5** 44 0 S 170 30 E
Lake Villa, *U.S.A.* **157 B8** 42 25N 88 5W
Lake Village, *U.S.A.* **155 J9** 33 20N 91 17W
Lake Wales, *U.S.A.* **149 M5** 27 54N 81 35W
Lake Worth, *U.S.A.* **149 M5** 26 37N 80 3W
Lakefield, *Canada* **140 D4** 44 25N 78 16W
Lakeland, *Australia* **126 B3** 15 49 S 144 57 E
Lakeland, *Fla., U.S.A.* **149 M5** 28 3N 81 57W
Lakeland, *Ga., U.S.A.* **152 D6** 31 2N 83 4W
Lakemba, *Fiji* **133 B3** 18 13 S 178 47W
Lakeport, *U.S.A.* **160 F4** 39 3N 122 55W
Lakes Entrance, *Australia* **129 D8** 37 50 S 148 0 E
Lakeside, *Ariz., U.S.A.* **159 J9** 34 9N 109 58W
Lakeside, *Calif., U.S.A.* **161 N10** 32 52N 116 55W
Lakeside, *Nebr., U.S.A.* **154 D3** 42 3N 102 26W
Lakeview, *U.S.A.* **158 E3** 42 11N 120 21W
Lakewood, *Colo., U.S.A.* **154 F2** 39 44N 105 5W

Lawrenceburg 239

Lakewood, *N.J., U.S.A.* **151 F10** 40 6N 74 13W
Lakewood, *Ohio, U.S.A.* **150 E3** 41 29N 81 48W
Lakewood Center, *U.S.A.* **160 C4** 47 11N 122 32W
Lakhaniá, *Greece* **38 D9** 35 58N 27 54 E
Lakhipur, *Assam, India* **90 C4** 24 48N 93 0 E
Lakhipur, *Assam, India* **90 B3** 26 2N 90 18 E
Lakhonpheng, *Laos* **86 E5** 15 54N 105 34 E
Lakhpat, *India* **92 H3** 23 48N 68 47 E
Läki, *Azerbaijan* **61 K8** 40 34N 47 22 E
Laki, *Iceland* **11 C8** 64 4N 18 14W
Lakin, *U.S.A.* **155 G4** 37 57N 101 15W
Lakitusaki →, *Canada* **140 B3** 54 21N 82 25W
Lákkoi, *Greece* **38 D5** 35 24N 23 57 E
Lakonía □, *Greece* **48 E4** 36 55N 22 30 E
Lakonikós Kólpos, *Greece* **48 E4** 36 40N 22 40 E
Lakor, *Indonesia* **82 C3** 8 15 S 128 17 E
Lakota, *Ivory C.* **112 D3** 5 50N 5 30W
Lakota, *U.S.A.* **154 A5** 48 2N 98 21W
Laksefjorden, *Norway* **14 A22** 70 45N 26 50 E
Lakselv, *Norway* **14 A21** 70 2N 25 0 E
Lakshadweep Is., *Ind. Oc.* **121 C6** 10 0N 72 30 E
Laksham, *Bangla.* **90 D3** 23 14N 91 8 E
Lakshmeshwar, *India* **95 G2** 15 9N 75 28 E
Lakshmikantapur, *India* **93 H13** 22 5N 88 20 E
Lakshmipur, *Bangla.* **90 D3** 22 58N 90 50 E
Lakuramau, *Papua N. G.* **132 B6** 2 54 S 151 15 E
Lal-lo, *Phil.* **80 B3** 18 12N 121 40 E
Lala, *Phil.* **81 H4** 7 59N 123 46 E
Lala Musa, *Pakistan* **92 C5** 32 40N 73 57 E
Lalago, *Tanzania* **118 C3** 3 28 S 33 58 E
Lalapanzi, *Zimbabwe* **119 F3** 19 20 S 30 15 E
Lalapaşa, *Turkey* **51 E10** 41 49N 26 44 E
Lalbenque, *France* **28 D5** 44 19N 1 34 E
L'Albufera, *Spain* **41 F4** 39 20N 0 27W
Lalganj, *India* **93 G11** 25 52N 85 13 E
Lalibela, *Ethiopia* **107 E4** 12 2N 39 2 E
Lalin, *China* **75 B14** 45 12N 127 0 E
Lalín, *Spain* **42 C2** 42 40N 8 5W
Lalin He →, *China* **75 B13** 45 32N 125 40 E
Lalinde, *France* **28 D4** 44 50N 0 44 E
Lalitpur, *India* **93 G8** 24 42N 78 28 E
Lalm, *Norway* **18 C6** 61 50N 9 17 E
Lam, *Vietnam* **86 B6** 21 21N 106 31 E
Lam Pao Res., *Thailand* **86 D4** 16 50N 103 15 E
Lama Kara, *Togo* **113 D5** 9 30N 1 15 E
Lamag, *Malaysia* **85 A5** 5 29N 117 49 E
Lamaipum, *Burma* **90 C6** 25 40N 97 57 E
Lamap, *Vanuatu* **133 F5** 16 26 S 167 43 E
Lamar, *Colo., U.S.A.* **154 F3** 38 5N 102 37W
Lamar, *Mo., U.S.A.* **155 G7** 37 30N 94 16W
Lamarque, *Argentina* **176 A3** 39 24 S 65 40W
Lamas, *Peru* **172 B2** 6 28 S 76 31W
Lamastre, *France* **29 D8** 44 59N 4 35 E
Lambach, *Austria* **34 C6** 48 6N 13 51 E
Lamballe, *France* **26 D4** 48 29N 2 31W
Lambaréné, *Gabon* **114 C2** 0 41 S 10 12 E
Lambasa, *Fiji* **133 A2** 16 30 S 179 10 E
Lambay I., *Ireland* **23 C5** 53 29N 6 1W
Lambayeque □, *Peru* **172 B2** 6 45 S 80 0W
Lambert, *U.S.A.* **154 B2** 47 41N 104 37W
Lambert, C., *Papua N. G.* **132 C6** 4 11 S 151 31 E
Lambert Glacier, *Antarctica* **7 D6** 71 0 S 70 0 E
Lamberts Bay, *S. Africa* **116 E2** 32 5 S 18 17 E
Lambesc, *France* **29 E9** 43 39N 5 16 E
Lambi Kyun, *Burma* **87 G2** 10 50N 98 20 E
Lámbia, *Greece* **48 D3** 37 52N 21 53 E
Lambon, *Papua N. G.* **132 C7** 4 45 S 152 48 E
Lambro →, *Italy* **44 C6** 45 8N 9 32 E
Lambunao, *Phil.* **81 F4** 11 3N 122 29 E
Lame, *Nigeria* **113 C6** 10 30N 9 20 E
Lame Deer, *U.S.A.* **158 D10** 45 37N 106 40W
Lamego, *Portugal* **42 D3** 41 5N 7 52W
Lamèque, *Canada* **141 C7** 47 45N 64 38W
Lameroo, *Australia* **128 C4** 35 19 S 140 33 E
Lamesa, *U.S.A.* **155 J4** 32 44N 101 58W
Lamía, *Greece* **48 C4** 38 55N 22 26 E
Lamitan, *Phil.* **81 H4** 6 39N 122 8 E
Lammermuir Hills, *U.K.* **22 F6** 55 50N 2 40W
Lammhult, *Sweden* **17 G8** 57 10N 14 35 E
Lamon B., *Phil.* **80 D4** 14 30N 122 20 E
Lamongan, *Indonesia* **85 D4** 7 7 S 112 25 E
Lamoni, *U.S.A.* **156 D3** 40 37N 93 56W
Lamont, *Canada* **142 C6** 53 46N 112 50W
Lamont, *Calif., U.S.A.* **161 K8** 35 15N 118 55W
Lamont, *Fla., U.S.A.* **152 E6** 30 23N 83 49W
Lamont, *Iowa, U.S.A.* **156 B5** 42 35N 91 40W
Lamotte-Beuvron, *France* **27 E9** 47 36N 2 2 E
Lampa, *Peru* **172 D3** 15 22 S 70 22W
Lampang, *Thailand* **86 C2** 18 16N 99 32 E
Lampasas, *U.S.A.* **155 K5** 31 4N 98 11W
Lampazos de Naranjo, *Mexico* **162 B4** 27 2N 100 32W
Lampertheim, *Germany* **31 F4** 49 35N 8 27 E
Lampeter, *U.K.* **21 E3** 52 7N 4 4W
Lampione, *Medit. S.* **108 A2** 35 33N 12 20 E
Lampman, *Canada* **143 D8** 49 25N 102 50W
Lamprechtshausen, *Austria* **34 D5** 48 0N 12 58 E
Lamprey, *Canada* **143 B10** 58 33N 94 8W
Lampung □, *Indonesia* **84 D2** 5 30 S 104 30 E
Lamu, *Burma* **90 F5** 19 14N 94 10 E
Lamu, *Kenya* **118 C5** 2 16 S 40 55 E
Lamu □, *Kenya* **118 C5** 2 0 S 40 45 E
Lamud, *Peru* **172 B2** 6 10 S 77 57W
Lamut, *Phil.* **80 C3** 16 39N 121 14 E
Lamy, *U.S.A.* **159 J11** 35 29N 105 53W
Lan Xian, *China* **74 E6** 38 15N 111 35 E
Lanai, *Hawaii, U.S.A.* **145 C5** 20 50N 156 55W
Lanai, *Hawaii, U.S.A.* **146 H16** 20 50N 156 55W
Lanaihale, *U.S.A.* **145 C5** 20 49N 156 53W
Lanak La, *India* **93 B8** 34 27N 79 32 E
Lanak'o Shank'ou = Lanak La, *India* **93 B8** 34 27N 79 32 E
Lanao, L., *Phil.* **81 H5** 7 52N 124 15 E
Lanao del Norte □, *Phil.* **81 H5** 8 0N 124 10 E
Lanao del Sur □, *Phil.* **81 H5** 7 40N 124 15 E
Lanark, *Canada* **151 A8** 45 1N 76 22W
Lanark, *U.K.* **22 F5** 55 40N 3 47W
Lanark Village, *U.S.A.* **152 F5** 29 53N 84 36W
Lancang, *China* **76 F2** 22 36N 99 58 E
Lancang Jiang →, *China* **76 G3** 21 40N 101 10 E
Lancashire □, *U.K.* **20 D5** 53 50N 2 48W
Lancaster, *Canada* **151 A10** 45 10N 74 30W
Lancaster, *U.K.* **20 C5** 54 3N 2 48W
Lancaster, *Calif., U.S.A.* **161 L8** 34 42N 118 8W
Lancaster, *Ky., U.S.A.* **148 G3** 37 37N 84 35W
Lancaster, *Mo., U.S.A.* **156 D4** 40 31N 92 32W
Lancaster, *N.H., U.S.A.* **151 B13** 44 29N 71 34W
Lancaster, *N.Y., U.S.A.* **150 D6** 42 54N 78 40W
Lancaster, *Pa., U.S.A.* **151 F8** 40 2N 76 19W
Lancaster, *S.C., U.S.A.* **149 H5** 34 43N 80 46W
Lancaster, *Wis., U.S.A.* **156 B6** 42 51N 90 43W
Lancaster Sd., *Canada* **139 A11** 74 13N 84 0W
Lancer, *Canada* **143 C7** 50 48N 108 53W
Lanchow = Lanzhou, *China* **74 F2** 36 1N 103 52 E
Lanciano, *Italy* **45 F11** 42 14N 14 23 E
Lanco, *Chile* **176 A2** 39 24 S 72 46W
Lancones, *Peru* **172 A1** 4 30 S 80 30W
Lancun, *China* **75 F11** 36 25N 120 10 E
Łańcut, *Poland* **55 H9** 50 10N 22 13 E
Lancy, *Switz.* **32 D2** 46 12N 6 8 E
Landau, *Bayern, Germany* **31 G8** 48 40N 12 41 E
Landau, *Rhld-Pfz., Germany* **31 F4** 49 12N 8 6 E
Landay, *Afghan.* **91 C1** 30 31N 63 47 E
Landeck, *Austria* **34 D3** 47 9N 10 34 E
Lander, *U.S.A.* **158 E9** 42 50N 108 44W
Lander →, *Australia* **124 D5** 22 0 S 132 0 E
Landerneau, *France* **26 D2** 48 28N 4 17W
Landeryd, *Sweden* **17 G7** 57 7N 13 15 E
Landes, *France* **28 D2** 44 0N 1 0W
Landes □, *France* **28 E3** 43 57N 0 48W
Landete, *Spain* **40 F3** 39 56N 1 25W
Landi Kotal, *Pakistan* **91 B3** 34 7N 71 6 E
Landivisiau, *France* **26 D2** 48 31N 4 6W
Landor, *Australia* **125 E2** 25 10 S 116 54 E
Landquart, *Switz.* **33 C9** 46 58N 9 32 E
Landquart →, *Switz.* **33 C9** 46 50N 9 47 E
Landrecies, *France* **27 B10** 50 7N 3 40 E
Land's End, *U.K.* **21 G2** 50 4N 5 44W
Landsberg, *Germany* **31 G6** 48 2N 10 53 E
Landsborough Cr. →, *Australia* **126 C3** 22 28 S 144 35 E
Landsbro, *Sweden* **17 G8** 57 24N 14 56 E
Landshut, *Germany* **31 G8** 48 34N 12 8 E
Landskrona, *Sweden* **17 J6** 55 53N 12 50 E
Landstuhl, *Germany* **31 F3** 49 24N 7 33 E
Landvetter, *Sweden* **17 G6** 57 41N 12 17 E
Lane, *U.S.A.* **152 B10** 33 32N 79 53W
Lanesboro, *U.S.A.* **151 E9** 41 57N 75 34W
Lanester, *France* **26 E3** 47 46N 3 22W
Lanett, *U.S.A.* **152 C4** 32 52N 85 12W
Lang Bay, *Canada* **142 D4** 49 45N 124 21W
Lang Qua, *Vietnam* **86 A5** 22 16N 104 27 E
Lang Shan, *China* **74 D4** 41 0N 106 30 E
Lang Son, *Vietnam* **86 B6** 21 52N 106 42 E
Lang Suan, *Thailand* **87 H2** 9 57N 99 4 E
Langå, *Denmark* **17 H3** 56 23N 9 54 E
Langádhas, *Greece* **50 F7** 40 46N 23 6 E
Langádhia, *Greece* **48 D4** 37 43N 22 1 E
Långan →, *Sweden* **16 A8** 63 19N 14 44 E
Langanes, *Iceland* **11 A12** 66 20N 14 53W
Langar, *Iran* **97 C9** 35 23N 60 25 E
Langara I., *Canada* **142 C2** 54 14N 133 1W
Långås, *Sweden* **17 H6** 56 58N 12 26 E
Langatabbetje, *Surinam* **169 C7** 4 59N 54 28W
Langdai, *China* **76 D5** 26 6N 105 21 E
Langdon, *U.S.A.* **154 A5** 48 45N 98 22W
Långe Jan = Ölands södra udde, *Sweden* **17 H10** 56 12N 16 23 E
Langeac, *France* **28 C7** 45 7N 3 29 E
Langeais, *France* **26 E7** 47 20N 0 24 E
Langeb Baraka →, *Sudan* **106 D4** 17 28N 36 50 E
Langeberg, *S. Africa* **116 E3** 33 55 S 21 0 E
Langeberge, *S. Africa* **116 D3** 28 15 S 22 33 E
Langeland, *Denmark* **17 K4** 54 56N 10 48 E
Langelands Bælt, *Denmark* **17 K4** 54 50N 10 55 E
Langen, *Hessen, Germany* **31 F4** 49 59N 8 40 E
Langen, *Niedersachsen, Germany* **30 B4** 53 36N 8 36 E
Langenburg, *Canada* **143 C8** 50 51N 101 43W
Langeneß, *Germany* **30 A4** 54 38N 8 36 E
Langenlois, *Austria* **34 C8** 48 29N 15 40 E
Langenthal, *Switz.* **32 B5** 47 13N 7 47 E
Langeoog, *Germany* **30 B3** 53 45N 7 32 E
Langeskov, *Denmark* **17 J4** 55 22N 10 35 E
Langesund, *Norway* **18 F6** 59 0N 9 45 E
Langevåg, *Norway* **18 B3** 62 26N 6 13 E
Länghem, *Sweden* **17 G7** 57 36N 13 14 E
Langhirano, *Italy* **44 D7** 44 37N 10 16 E
Langholm, *U.K.* **22 F5** 55 9N 3 0W
Langidoon, *Australia* **128 A5** 31 36 S 142 2 E
Langjökull, *Iceland* **11 C6** 64 39N 20 12W
Langkawi, Pulau, *Malaysia* **87 J2** 6 25N 99 45 E
Langklip, *S. Africa* **116 D3** 28 12 S 20 20 E
Langkon, *Malaysia* **85 A5** 6 30N 116 40 E
Langlade, *St- P. & M.* **141 C8** 46 50N 56 20W
Langlois, *U.S.A.* **158 E1** 42 56N 124 27W
Langnau, *Switz.* **32 C5** 46 56N 7 47 E
Langogne, *France* **28 D7** 44 43N 3 50 E
Langon, *France* **28 D3** 44 33N 0 16W
Langøya, *Norway* **14 B16** 68 45N 14 50 E
Langreo, *Spain* **42 B5** 43 18N 5 40W
Langres, *France* **27 E12** 47 52N 5 20 E
Langres, Plateau de, *France* **27 E12** 47 45N 5 3 E
Langsa, *Indonesia* **84 B1** 4 30N 97 57 E
Långsele, *Sweden* **16 A11** 63 12N 17 4 E
Långshyttan, *Sweden* **16 D10** 60 27N 16 2 E
Langtao, *Burma* **90 B6** 27 15N 97 34 E
Langting, *India* **90 C4** 25 31N 93 7 E
Langtry, *U.S.A.* **155 L4** 29 49N 101 34W
Langu, *Thailand* **87 J2** 6 53N 99 47 E
Languedoc, *France* **28 E7** 43 58N 3 55 E
Languedoc-Roussillon □, *France* **28 E6** 43 25N 3 0 E
Langwies, *Switz.* **33 C9** 46 50N 9 44 E
Langxi, *China* **77 B12** 31 10N 119 12 E
Langxiangzhen, *China* **74 E9** 39 43N 116 8 E
Langzhong, *China* **76 B5** 31 38N 105 58 E
Lanigan, *Canada* **143 C7** 51 51N 105 2W
Lankao, *China* **74 G8** 34 48N 114 50 E
Länkäran, *Azerbaijan* **101 C13** 38 48N 48 52 E
Lanmeur, *France* **26 D3** 48 39N 3 43W
Lannemezan, *France* **28 E4** 43 8N 0 23 E
Lannilis, *France* **26 D2** 48 35N 4 32W
Lannion, *France* **26 D3** 48 46N 3 29W
L'Annonciation, *Canada* **140 C5** 46 25N 74 55W
Lanouaille, *France* **28 C5** 45 24N 1 9 E
Lanping, *China* **76 D2** 26 28N 99 15 E
Lansdale, *U.S.A.* **151 F9** 40 14N 75 17W
Lansdowne, *Australia* **129 A10** 31 48 S 152 30 E
Lansdowne, *Canada* **151 B8** 44 24N 76 1W
Lansdowne House, *Canada* **140 B2** 52 14N 87 53W
L'Anse, *U.S.A.* **140 C2** 46 45N 88 27W
L'Anse au Loup, *Canada* **141 B8** 51 32N 56 50W
Lansford, *U.S.A.* **151 F9** 40 50N 75 53W
Lanshan, *China* **77 E9** 25 24N 112 10 E
Lansing, *U.S.A.* **157 B12** 42 44N 84 33W
Lanslebourg-Mont-Cenis, *France* **29 C10** 45 17N 6 52 E
Lanta Yai, Ko, *Thailand* **87 J2** 7 35N 99 3 E
Lantana, *U.S.A.* **153 J9** 26 35N 80 3W
Lantian, *China* **74 G5** 34 11N 109 20 E
Lanus, *Argentina* **174 C4** 34 44 S 58 27W
Lanusei, *Italy* **46 C2** 39 52N 9 34 E
Lanuza, *Phil.* **81 G6** 9 14N 126 4 E
Lanxi, *China* **77 C12** 29 13N 119 28 E
Lanzarote, *Canary Is.* **39 F6** 29 0N 13 40W
Lanzhou, *China* **74 F2** 36 1N 103 52 E
Lanzo Torinese, *Italy* **44 C4** 45 16N 7 28 E
Lao →, *Italy* **47 C8** 39 47N 15 48 E
Lao Bao, *Laos* **86 D6** 16 35N 106 30 E
Lao Cai, *Vietnam* **86 A4** 22 30N 103 57 E
Laoag, *Phil.* **80 B3** 18 7N 120 34 E
Laoang, *Phil.* **80 E5** 12 32N 125 8 E
Laoha He →, *China* **75 C11** 43 25N 120 35 E
Laois □, *Ireland* **23 D4** 52 57N 7 36W
Laon, *France* **27 C10** 49 33N 3 35 E
Laona, *U.S.A.* **148 C1** 45 34N 88 40W
Laos ■, *Asia* **86 D5** 17 45N 105 0 E
Lapa, *Brazil* **175 B6** 25 46 S 49 44W
Lapalisse, *France* **27 F10** 46 15N 3 38 E
Lapeer, *U.S.A.* **157 A13** 43 3N 83 19W
Lapeyrade, *France* **28 D3** 44 4N 0 3W
Lapithos, *Cyprus* **38 D12** 35 21N 33 11 E
Lapland = Lappland, *Europe* **14 B21** 68 7N 24 0 E
Lapog, *Phil.* **80 C3** 17 45N 120 27 E
Laporte, *U.S.A.* **151 E8** 41 25N 76 30W
Lapovo, *Serbia, Yug.* **50 B5** 44 10N 21 7 E
Lappeenranta, *Finland* **15 F23** 61 3N 28 12 E
Lappland, *Europe* **14 B21** 68 7N 24 0 E
Laprida, *Argentina* **174 D3** 37 34 S 60 45W
Lapseki, *Turkey* **100 B2** 40 20N 26 41 E
Laptev Sea, *Russia* **65 B13** 76 0N 125 0 E
Lapu-Lapu, *Phil.* **81 F4** 10 20N 123 55 E
Lapua, *Finland* **14 E20** 62 58N 23 0 E
Lăpuş →, *Romania* **53 C8** 47 25N 23 40 E
Lapuş, Munţii, *Romania* **53 C8** 47 20N 23 50 E
Lăpuşna, *Moldova* **53 D13** 46 53N 28 25 E
Łapy, *Poland* **55 F9** 52 59N 22 52 E
L'Aquila, *Italy* **45 F10** 42 22N 13 22 E
Lār, *Āzarbājān-e Sharqī, Iran* **96 B5** 38 30N 47 52 E
Lār, *Fārs, Iran* **97 E7** 27 40N 54 14 E
Lara, *Australia* **128 E6** 38 2 S 144 26 E
Lara, *Phil.* **81 G1** 8 48N 117 52 E
Lara □, *Venezuela* **168 A4** 10 10N 69 50W
Larabanga, *Ghana* **112 D4** 9 16N 1 56W
Larache, *Morocco* **110 A3** 35 10N 6 5W
Laragne-Montéglin, *France* **29 D9** 44 18N 5 49 E
Laramie, *U.S.A.* **154 E2** 41 19N 105 35W
Laramie Mts., *U.S.A.* **154 E2** 42 0N 105 30W
Laranjeiras, *Brazil* **170 D4** 10 48 S 37 10W
Laranjeiras do Sul, *Brazil* **175 B5** 25 23 S 52 23W
Larantuka, *Indonesia* **82 C2** 8 21 S 122 55 E
Larap, *Phil.* **80 D4** 14 18N 122 39 E
Larat, *Indonesia* **83 C4** 7 0 S 132 0 E
L'Arbresle, *France* **29 C8** 45 50N 4 36 E
Lärbro, *Sweden* **17 G12** 57 47N 18 50 E
Lårdal, *Norway* **18 E5** 59 25N 8 10 E
Larde, *Mozam.* **119 F4** 16 28 S 39 43 E
Larder Lake, *Canada* **140 C4** 48 5N 79 40W
Lardhos, Ákra = Líndhos, Ákra, *Greece* **38 C10** 36 4N 28 10 E
Lardhos, Órmos, *Greece* **38 C10** 36 4N 28 2 E
Laredo, *Spain* **42 B7** 43 26N 3 28W
Laredo, *U.S.A.* **155 M5** 27 30N 99 30W
Laredo Sd., *Canada* **142 C3** 52 30N 128 53W
Larena, *Phil.* **81 G4** 9 15N 123 35 E
Largentière, *France* **29 D8** 44 34N 4 18 E
L'Argentière-la-Bessée, *France* **29 D10** 44 47N 6 33 E
Largo, *U.S.A.* **149 M4** 27 55N 82 47W
Largo Key, *U.S.A.* **153 K9** 25 15N 80 15W
Largs, *U.K.* **22 F4** 55 47N 4 52W
Lari, *Italy* **44 E7** 43 34N 10 35 E
Lariang, *Indonesia* **82 B1** 1 26 S 119 17 E
Larimore, *U.S.A.* **154 B6** 47 54N 97 38W
Lārīn, *Iran* **97 C7** 35 55N 52 19 E
Larino, *Italy* **45 G11** 41 48N 14 54 E
Lárisa, *Greece* **48 B4** 39 36N 22 27 E
Lárisa □, *Greece* **48 B4** 39 39N 22 24 E
Larkana, *Pakistan* **91 D3** 27 32N 68 18 E
Larnaca, *Cyprus* **38 E12** 34 55N 33 38 E
Larnaca Bay, *Cyprus* **38 E12** 34 53N 33 45 E
Larne, *U.K.* **23 B6** 54 51N 5 51W
Larned, *U.S.A.* **154 F5** 38 11N 99 6W
Laroquebrou, *France* **28 D6** 44 58N 2 12 E
Larrimah, *Australia* **124 C5** 15 35 S 133 12 E
Larsen Bay, *U.S.A.* **144 H9** 57 32N 153 59W
Larsen Ice Shelf, *Antarctica* **7 C17** 67 0 S 62 0W
Laruns, *France* **28 F3** 43 0N 0 26W
Larvik, *Norway* **15 G14** 59 4N 10 2 E
Larzac, Causse du, *France* **28 E7** 43 50N 3 17 E
Las Alpujarras, *Spain* **41 J1** 36 55N 3 20W
Las Animas, *U.S.A.* **154 F3** 38 4N 103 13W
Las Anod, *Somali Rep.* **120 C3** 8 26N 47 19 E
Las Arenas, *Spain* **42 B6** 43 17N 4 50W
Las Brenãs, *Argentina* **174 B3** 27 5 S 61 7W
Las Cabezas de San Juan, *Spain* **43 J5** 36 57N 5 58W
Las Chimeneas, *Mexico* **161 N10** 32 8N 116 5W
Las Coloradas, *Argentina* **176 A2** 39 34 S 70 36W
Las Cruces, *U.S.A.* **159 K10** 32 19N 106 47W
Las Flores, *Argentina* **174 D4** 36 10 S 59 7W
Las Heras, *Argentina* **174 C2** 32 51 S 68 49W
Las Horquetas, *Argentina* **176 C2** 48 14 S 71 11W
Las Khoreh, *Somali Rep.* **120 B3** 11 10N 48 20 E
Las Lajas, *Argentina* **176 A2** 38 30 S 70 25W
Las Lomas, *Peru* **172 A1** 4 40 S 80 10W
Las Lomitas, *Argentina* **174 A3** 24 43 S 60 35W
Las Marismas, *Spain* **43 H4** 37 5N 6 20W
Las Mercedes, *Venezuela* **168 B4** 9 7N 66 24W
Las Minas, *Spain* **41 G3** 38 20N 1 41W
Las Navas de la Concepción, *Spain* **43 H5** 37 56N 5 30W
Las Navas del Marqués, *Spain* **42 E6** 40 36N 4 20W
Las Palmas, *Argentina* **174 B4** 27 8 S 58 45W
Las Palmas, *Canary Is.* **39 F4** 28 7N 15 26W
Las Palmas →, *Mexico* **161 N10** 32 26N 116 54W
Las Pedroñas, *Spain* **41 F2** 39 26N 2 40W
Las Piedras, *Uruguay* **175 C4** 34 44 S 56 14W
Las Pipinas, *Argentina* **174 D4** 35 30 S 57 19W
Las Plumas, *Argentina* **176 B3** 43 40 S 67 15W
Las Rosas, *Argentina* **174 C3** 32 30 S 61 35W
Las Rozas, *Spain* **42 E7** 40 29N 3 52W
Las Tablas, *Panama* **164 E3** 7 49N 80 14W
Las Termas, *Argentina* **174 B3** 27 29 S 64 52W
Las Truchas, *Mexico* **162 D4** 17 57N 102 13W
Las Varillas, *Argentina* **174 C3** 31 50 S 62 50W
Las Vegas, *N. Mex., U.S.A.* **159 J11** 35 36N 105 13W
Las Vegas, *Nev., U.S.A.* **161 J11** 36 10N 115 9W
Lasarte, *Spain* **40 B2** 43 16N 2 1W
Lascano, *Uruguay* **175 C5** 33 35 S 54 12W
Lashburn, *Canada* **143 C7** 53 10N 109 40W
Lashio, *Burma* **90 D6** 22 56N 97 45 E
Lashkar, *India* **92 F8** 26 10N 78 10 E
Lashkar Gāh, *Afghan.* **91 C2** 31 35N 64 21 E
Łasin, *Poland* **54 E6** 53 30N 19 2 E
Lasíthi, *Greece* **38 D7** 35 11N 25 31 E
Lasíthi □, *Greece* **38 D7** 35 5N 25 50 E
Lask, *Poland* **55 G6** 51 34N 19 8 E
Łaskarzew, *Poland* **55 G8** 51 48N 21 36 E
Laško, *Slovenia* **45 B12** 46 10N 15 16 E
Lassance, *Brazil* **171 E3** 17 54 S 44 34W
Lassay-les-Châteaux, *France* **26 D6** 48 27N 0 30W
Lassen Pk., *U.S.A.* **158 F3** 40 29N 121 31W
Last Mountain L., *Canada* **143 C7** 51 5N 105 14W
Lastchance Cr. →, *U.S.A.* **160 E5** 40 2N 121 15W
Lastoursville, *Gabon* **114 C2** 0 55 S 12 38 E
Lastovo, *Croatia* **45 F13** 42 46N 16 55 E
Lastovski Kanal, *Croatia* **45 F14** 42 50N 17 0 E
Lat Yao, *Thailand* **86 E2** 15 45N 99 48 E
Latacunga, *Ecuador* **168 D2** 0 50 S 78 35W
Latakia = Al Lādhiqīyah, *Syria* **100 E6** 35 30N 35 45 E
Latchford, *Canada* **140 C4** 47 20N 79 50W
Late, *Tonga* **133 P13** 18 48 S 174 39W
Laterza, *Italy* **47 B9** 40 37N 16 48 E
Latham, *Australia* **125 E2** 29 44 S 116 20 E
Lathen, *Germany* **30 C3** 52 52N 7 19 E
Lathrop, *U.S.A.* **156 E2** 39 33N 94 20W
Lathrop Wells, *U.S.A.* **161 J10** 36 39N 116 24W
Latiano, *Italy* **47 B10** 40 33N 17 43 E
Latina, *Italy* **46 A5** 41 28N 12 52 E
Latisana, *Italy* **45 C10** 45 47N 13 0 E
Latium = Lazio □, *Italy* **45 F9** 42 10N 12 30 E
Laton, *U.S.A.* **160 J7** 36 26N 119 41W
Latorytsya →, *Slovak Rep.* **35 C14** 48 28N 21 50 E
Latouche Treville, C., *Australia* **124 C3** 18 27 S 121 49 E
Látrar, *Iceland* **11 A3** 66 24N 23 2W
Latrobe, *Australia* **126 G4** 41 14 S 146 30 E
Latrobe, *U.S.A.* **150 F5** 40 19N 79 23W
Latrónico, *Italy* **47 B9** 40 5N 16 1 E
Latur, *India* **94 E3** 18 25N 76 40 E
Latvia ■, *Europe* **15 H20** 56 50N 24 0 E
Lau Group, *Fiji* **133 A3** 17 0 S 178 30W
Lauca →, *Bolivia* **172 D4** 19 9 S 68 10W
Lauchhammer, *Germany* **30 D9** 51 29N 13 47 E
Lauda-Königshofen, *Germany* **31 F5** 49 33N 9 42 E
Laudal, *Norway* **18 F4** 58 15N 7 30 E
Lauenburg, *Germany* **30 B6** 53 22N 10 32 E
Lauf, *Germany* **31 F7** 49 30N 11 16 E
Laufás, *Iceland* **11 B8** 65 53N 18 4W
Läufelfingen, *Switz.* **32 B5** 47 24N 7 52 E
Laufen, *Switz.* **32 B5** 47 25N 7 30 E
Laugarbakki, *Iceland* **11 B6** 65 20N 20 55W
Laugarvatn, *Iceland* **11 C6** 64 13N 20 44W
Laujar de Andarax, *Spain* **41 H2** 37 0N 2 54W
Laukaa, *Finland* **15 E21** 62 24N 25 56 E
Launceston, *Australia* **126 G4** 41 24 S 147 8 E
Launceston, *U.K.* **21 G3** 50 38N 4 22W
Laune →, *Ireland* **23 D2** 52 7N 9 47W
Launglon Bok, *Burma* **86 F1** 13 50N 97 54 E
Laupheim, *Germany* **31 G5** 48 14N 9 52 E
Laur, *Phil.* **80 D3** 15 35N 121 11 E
Laura, *Queens., Australia* **126 B3** 15 32 S 144 32 E
Laura, *S. Austral., Australia* **128 B3** 33 10 S 138 18 E
Laureana di Borrello, *Italy* **47 D9** 38 30N 16 5 E
Laurel, *Fla., U.S.A.* **153 H7** 27 8N 82 27W
Laurel, *Ind., U.S.A.* **157 E11** 39 31N 85 11W
Laurel, *Miss., U.S.A.* **155 K10** 31 41N 89 8W
Laurel, *Mont., U.S.A.* **158 D9** 45 40N 108 46W
Laurel Bay, *U.S.A.* **152 C9** 32 27N 80 47W
Laurencekirk, *U.K.* **22 E6** 56 50N 2 28W
Laurens, *U.S.A.* **149 H4** 34 30N 82 1W
Laurentian Plateau, *Canada* **141 B6** 52 0N 70 0W
Laurentides, Parc Prov. des, *Canada* **141 C5** 47 45N 71 15W
Lauria, *Italy* **47 B8** 40 2N 15 50 E
Laurie L., *Canada* **143 B8** 56 35N 101 57W
Laurinburg, *U.S.A.* **149 H6** 34 47N 79 28W
Laurium, *U.S.A.* **148 B1** 47 14N 88 27W
Lausanne, *Switz.* **32 C3** 46 32N 6 38 E
Laut, *Indonesia* **85 B3** 4 45N 108 0 E
Laut, Pulau, *Indonesia* **85 C5** 3 40 S 116 10 E
Laut Kecil, Kepulauan, *Indonesia* **85 C5** 4 45 S 115 40 E
Lautaro, *Chile* **176 A2** 38 31 S 72 27W
Lauterbach, *Germany* **30 E5** 50 37N 9 24 E
Lauterbrunnen, *Switz.* **32 C5** 46 36N 7 55 E
Lauterecken, *Germany* **31 F3** 49 38N 7 35 E
Lautoka, *Fiji* **133 A1** 17 37 S 177 27 E
Lauzès, *France* **28 D5** 44 34N 1 35 E
Lauzon, *Canada* **141 C5** 46 48N 71 10W
Lava Hot Springs, *U.S.A.* **158 E7** 42 37N 112 1W
Lavagh More, *Ireland* **23 B3** 54 46N 8 6W
Lavagna, *Italy* **44 D6** 44 18N 9 20 E
Laval, *France* **26 D6** 48 4N 0 48W
Lavalle, *Argentina* **174 B2** 28 15 S 65 15W
Lávara, *Greece* **51 E10** 41 19N 26 22 E
Lavardac, *France* **28 D4** 44 12N 0 20 E
Lavaur, *France* **28 E5** 43 40N 1 49 E
Lavaux, *Switz.* **32 D3** 46 30N 6 45 E
Lavelanet, *France* **28 F5** 42 57N 1 51 E
Lavello, *Italy* **47 A8** 41 3N 15 48 E
Laverne, *U.S.A.* **155 G5** 36 43N 99 54W
Lavers Hill, *Australia* **128 E5** 38 40 S 143 25 E
Laverton, *Australia* **125 E3** 28 44 S 122 29 E
Lavik, *Norway* **18 C2** 61 6N 5 25 E
Lavis, *Italy* **44 B8** 46 8N 11 7 E
Lávkos, *Greece* **48 B5** 39 9N 23 14 E
Lavos, *Portugal* **42 E2** 40 6N 8 49W
Lavradio, *Portugal* **43 G1** 38 40N 9 3W
Lavras, *Brazil* **171 F3** 21 20 S 45 0W
Lavre, *Portugal* **43 G2** 38 46N 8 22W
Lávrion, *Greece* **48 D6** 37 40N 24 4 E
Lávris, *Greece* **38 D6** 35 25N 24 40 E
Lavumisa, *Swaziland* **117 D5** 27 20 S 31 55 E
Lawa, *Phil.* **81 H5** 6 12N 125 41 E
Lawa-an, *Phil.* **81 F5** 11 51N 125 5 E
Lawas, *Malaysia* **85 B5** 4 55N 115 25 E
Lawdar, *Yemen* **98 D4** 13 53N 45 52 E
Lawele, *Indonesia* **82 C2** 5 16 S 123 3 E
Lawksawk, *Burma* **90 E6** 21 15N 96 52 E
Lawn Hill, *Australia* **126 B2** 18 36 S 138 33 E
Lawqah, *Si. Arabia* **96 D4** 29 49N 42 45 E
Lawra, *Ghana* **112 C4** 10 39N 2 51W
Lawrence, *N.Z.* **131 F4** 45 55 S 169 41 E
Lawrence, *Ind., U.S.A.* **157 E10** 39 50N 86 2W
Lawrence, *Kans., U.S.A.* **154 F7** 38 58N 95 14W
Lawrence, *Mass., U.S.A.* **151 D13** 42 43N 71 10W
Lawrenceburg, *Ind., U.S.A.* **157 E12** 39 6N 84 52W
Lawrenceburg, *Ky., U.S.A.* **157 F12** 38 2N 84 54W

240 Lawrenceburg

Lawrenceburg, *Tenn., U.S.A.* **149 H2** 35 14N 87 20W
Lawrenceville, *Ga., U.S.A.* **152 B6** 33 57N 83 59W
Lawrenceville, *Ill., U.S.A.* **157 F9** 38 44N 87 41W
Laws, *U.S.A.* **160 H8** 37 24N 118 20W
Lawson, *U.S.A.* **156 E2** 39 26N 94 12W
Lawtey, *U.S.A.* **152 E7** 30 3N 82 5W
Lawton, *Mich., U.S.A.* **157 B11** 42 10N 85 50W
Lawton, *Okla., U.S.A.* **155 H5** 34 37N 98 25W
Lawu, *Indonesia* **85 D4** 7 40S 111 13 E
Laxå, *Sweden* **17 F8** 58 59N 14 37 E
Laxamýri, *Iceland* **11 B9** 65 58N 17 24W
Laxford, L., *U.K.* **22 C3** 58 24N 5 6W
Laxou, *France* **27 D13** 48 41N 6 8 E
Lay →, *France* **28 B2** 46 18N 1 17W
Layht, Ra's, *Yemen* **99 D6** 12 38N 53 25 E
Laylá, *Si. Arabia* **98 B4** 22 10N 46 40 E
Laylān, *Iraq* **96 C5** 35 18N 44 31 E
Layon →, *France* **26 E6** 47 20N 0 45W
Laysan I., *Pac. Oc.* **135 E11** 25 30N 167 0W
Laysan I., *U.S.A.* **145 F9** 25 50N 171 50W
Layton, *U.S.A.* **153 L9** 24 50N 80 47W
Laytonville, *U.S.A.* **158 G2** 39 41N 123 29W
Laza, *Burma* **90 B6** 26 30N 97 38 E
Lazarevac, *Serbia, Yug.* **50 B4** 44 23N 20 17 E
Lazarevskoye, *Russia* **61 J4** 43 55N 39 21 E
Lazdijai, *Lithuania* **54 D10** 54 14N 23 31 E
Lazi, *Phil.* **81 G4** 9 8N 123 38 E
Lazio □, *Italy* **45 F9** 42 10N 12 30 E
Lazo, *Moldova* **53 C13** 47 33N 28 2 E
Lazo, *Russia* **68 C6** 43 25N 133 55 E
Le Beausset, *France* **29 E9** 43 12N 5 48 E
Le Blanc, *France* **28 B5** 46 37N 1 3 E
Le Bleymard, *France* **28 D7** 44 30N 3 42 E
Le Bourgneuf-la-Fôret, *France* **26 D6** 48 10N 0 59W
Le Brassus, *Switz.* **32 C2** 46 35N 6 13 E
Le Bugue, *France* **28 D4** 44 55N 0 56 E
Le Canourgue = La Canourgue, *France* **28 D7** 44 26N 3 13 E
Le Cateau Cambrésis, *France* **27 B10** 50 7N 3 32 E
Le Caylar, *France* **28 E7** 43 51N 3 19 E
Le Chambon-Feugerolles, *France* **29 C8** 45 24N 4 19 E
Le Châtelard, *Switz.* **32 D3** 46 4N 6 57 E
Le Châtelet, *France* **27 F9** 46 38N 2 16 E
Le Chesne, *France* **27 C11** 49 30N 4 45 E
Le Cheylard, *France* **29 D8** 44 55N 4 25 E
Le Claire, *U.S.A.* **156 C6** 41 36N 90 21W
Le Conquet, *France* **26 D2** 48 21N 4 46W
Le Creusot, *France* **27 F11** 46 48N 4 24 E
Le Croisic, *France* **26 E4** 47 18N 2 30W
Le Donjon, *France* **27 F10** 46 22N 3 48 E
Le Dorat, *France* **28 B5** 46 14N 1 5 E
Le François, *Martinique* **165 D7** 14 38N 60 57W
Le Grand-Lucé, *France* **26 E7** 47 52N 0 28 E
Le Grand-Pressigny, *France* **26 F7** 46 55N 0 48 E
Le Grand-Quevilly, *France* **26 C8** 49 24N 1 3 E
Le Havre, *France* **26 C7** 49 30N 0 5 E
Le Lavandou, *France* **29 E10** 43 8N 6 22 E
Le Lion-d'Angers, *France* **26 E6** 47 37N 0 43W
Le Locle, *Switz.* **32 B3** 47 3N 6 44 E
Le Louroux-Béconnais, *France* **26 E6** 47 30N 0 55W
Le Luc, *France* **29 E10** 43 23N 6 21 E
Le Lude, *France* **26 E7** 47 39N 0 9 E
Le Maire, Estr. de, *Argentina* **176 D4** 54 50S 65 0W
Le Mans, *France* **26 E7** 48 0N 0 10 E
Le Mars, *U.S.A.* **154 D6** 42 47N 96 10W
Le Mayet-de-Montagne, *France* **27 F10** 46 4N 3 40 E
Le Mêle-sur-Sarthe, *France* **26 D7** 48 31N 0 22 E
Le Monastier-sur-Gazeille, *France* **28 D7** 44 57N 3 59 E
Le Monêtier-les-Bains, *France* **29 D10** 44 58N 6 30 E
Le Mont-Dore, *France* **28 C6** 45 35N 2 49 E
Le Mont-St-Michel, *France* **26 D5** 48 40N 1 30W
Le Moule, *Guadeloupe* **165 C7** 16 20N 61 22W
Le Muy, *France* **29 E10** 43 28N 6 34 E
Le Palais, *France* **26 E3** 47 20N 3 10W
Le Perthus, *France* **28 F6** 42 30N 2 53 E
Le Pont, *Switz.* **32 C2** 46 41N 6 20 E
Le Puy-en-Velay, *France* **28 C7** 45 3N 3 52 E
Le Roy, *Ill., U.S.A.* **157 D8** 40 21N 88 46W
Le Roy, *Kans., U.S.A.* **155 F7** 38 5N 95 38W
Le Sentier, *Switz.* **32 C2** 46 37N 6 15 E
Le Sueur, *U.S.A.* **154 C8** 44 28N 93 55W
Le Teil, *France* **29 D8** 44 33N 4 40 E
Le Teilleul, *France* **26 D6** 48 32N 0 53W
Le Theil, *France* **26 D7** 48 16N 0 42 E
Le Thillot, *France* **27 E13** 47 53N 6 46 E
Le Thuy, *Vietnam* **86 D6** 17 14N 106 49 E
Le Touquet-Paris-Plage, *France* **27 B8** 50 30N 1 36 E
Le Tréport, *France* **26 B8** 50 3N 1 20 E
Le Val-d'Ajol, *France* **27 E13** 47 55N 6 30 E
Le Verdon-sur-Mer, *France* **28 C2** 45 33N 1 4W
Le Vigan, *France* **28 E7** 43 59N 3 36 E
Lea →, *U.K.* **21 F8** 51 31N 0 1 E
Leach, *Cambodia* **87 F4** 12 21N 103 46 E
Lead, *U.S.A.* **154 C3** 44 21N 103 46W
Leader, *Canada* **143 C7** 50 50N 109 30W
Leadville, *U.S.A.* **159 G10** 39 15N 106 18W
Leaf →, *U.S.A.* **155 K10** 30 59N 88 44W
Leakey, *U.S.A.* **155 L5** 29 44N 99 46W
Lealui, *Zambia* **115 F4** 15 10S 23 2 E
Leamington, *Canada* **140 D3** 42 3N 82 36W
Leamington, *N.Z.* **130 D4** 37 55S 175 30 E
Leamington, *U.S.A.* **158 G7** 39 32N 112 17W
Leamington Spa = Royal Leamington Spa, *U.K.* **21 E6** 52 18N 1 31W
Le'an, *China* **77 D10** 27 22N 115 48 E
Leandro Norte Alem, *Argentina* **175 B4** 27 34S 55 15W
Leane, L., *Ireland* **23 D2** 52 2N 9 32W
Learmonth, *Australia* **124 D1** 22 13S 114 10 E
Leary, *U.S.A.* **152 D5** 31 29N 84 31W
Leask, *Canada* **143 C7** 53 5N 106 45W
Leatherhead, *U.K.* **21 F7** 51 18N 0 20W
Leavenworth, *Ind., U.S.A.* **157 F10** 38 12N 86 21W
Leavenworth, *Kans., U.S.A.* **154 F7** 39 19N 94 55W
Leavenworth, *Wash., U.S.A.* **158 C3** 47 36N 120 40W
Leawood, *U.S.A.* **156 F2** 38 57N 94 37W
Łeba, *Poland* **54 D4** 54 45N 17 32 E
Łeba →, *Poland* **54 D4** 54 46N 17 33 E
Lebach, *Germany* **31 F2** 49 25N 6 54 E
Lebak, *Phil.* **81 H5** 6 32N 124 5 E
Lebam, *U.S.A.* **160 D3** 46 34N 123 33W
Lebane, *Serbia, Yug.* **50 D5** 42 56N 21 44 E
Lebango, *Congo* **114 B2** 0 39N 14 21 E
Lebanon, *Ill., U.S.A.* **156 F7** 38 38N 89 49W
Lebanon, *Ind., U.S.A.* **157 D10** 40 3N 86 28W
Lebanon, *Kans., U.S.A.* **154 F5** 39 49N 98 33W
Lebanon, *Ky., U.S.A.* **148 G3** 37 34N 85 15W
Lebanon, *Mo., U.S.A.* **155 G8** 37 41N 92 40W
Lebanon, *Ohio, U.S.A.* **157 E12** 39 26N 84 13W
Lebanon, *Oreg., U.S.A.* **158 D2** 44 32N 122 55W
Lebanon, *Pa., U.S.A.* **151 F8** 40 20N 76 26W
Lebanon, *Tenn., U.S.A.* **149 G2** 36 12N 86 18W
Lebanon ■, *Asia* **103 B5** 34 0N 36 0 E
Lebanon Junction, *U.S.A.* **157 G11** 37 50N 85 44W
Lebec, *U.S.A.* **161 L8** 34 50N 118 52W
Lebedyan, *Russia* **59 F10** 53 0N 39 10 E
Lebedyn, *Ukraine* **59 G8** 50 35N 34 30 E
Lebomboberge, *S. Africa* **117 C5** 24 30S 32 0 E
Lębork, *Poland* **54 D4** 54 33N 17 46 E
Lebrija, *Spain* **43 J4** 36 53N 6 5W
Łebsko, Jezioro, *Poland* **54 D4** 54 40N 17 25 E
Lebu, *Chile* **174 D1** 37 40S 73 47W
Leca da Palmeira, *Portugal* **42 D2** 41 12N 8 42W
Lecce, *Italy* **47 B11** 40 23N 18 11 E
Lecco, *Italy* **44 C6** 45 51N 9 23 E
Lecco, L. di, *Italy* **44 C6** 45 55N 9 19 E
Lécera, *Spain* **40 D4** 41 13N 0 43W
Lech, *Austria* **34 D3** 47 13N 10 9 E
Lech →, *Germany* **31 G6** 48 43N 10 56 E
Lechang, *China* **77 E9** 25 10N 113 20 E
Lechtaler Alpen, *Austria* **34 D3** 47 15N 10 30 E
Lectoure, *France* **28 E4** 43 56N 0 38 E
Łęczna, *Poland* **55 G9** 51 18N 22 53 E
Łęczyca, *Poland* **55 F6** 52 5N 19 15 E
Ledang, Gunong, *Malaysia* **84 B2** 2 22N 102 37 E
Ledesma, *Spain* **42 D5** 41 6N 5 59W
Ledong, *China* **86 C7** 18 41N 109 5 E
Leduc, *Canada* **142 C6** 53 15N 113 30W
Lee, *Fla., U.S.A.* **152 E6** 30 25N 83 18W
Lee, *Mass., U.S.A.* **151 D11** 42 19N 73 15W
Lee →, *Ireland* **23 E3** 51 53N 8 56W
Lee Vining, *U.S.A.* **160 H7** 37 58N 119 7W
Leech L., *U.S.A.* **154 B7** 47 10N 94 24W
Leedey, *U.S.A.* **155 H5** 35 52N 99 21W
Leeds, *U.K.* **20 D6** 53 48N 1 33W
Leeds, *U.S.A.* **149 J2** 33 33N 86 33W
Leek, *Neths.* **24 A6** 53 10N 6 24 E
Leek, *U.K.* **20 D5** 53 7N 2 1W
Leer, *Germany* **30 B3** 53 13N 7 26 E
Lee's Summit, *U.S.A.* **156 F2** 38 55N 94 23W
Leesburg, *Fla., U.S.A.* **149 L5** 28 49N 81 53W
Leesburg, *Ga., U.S.A.* **152 D5** 31 44N 84 10W
Leesburg, *Ohio, U.S.A.* **157 E13** 39 21N 83 33W
Leeston, *N.Z.* **131 D7** 43 45S 172 19 E
Leesville, *U.S.A.* **155 K8** 31 9N 93 16W
Leeton, *Australia* **129 C7** 34 33S 146 23 E
Leetonia, *U.S.A.* **150 F4** 40 53N 80 45W
Leeu Gamka, *S. Africa* **116 E3** 32 47S 21 59 E
Leeuwarden, *Neths.* **24 A5** 53 15N 5 48 E
Leeuwin, C., *Australia* **125 F2** 34 20S 115 9 E
Leeward Is., *Atl. Oc.* **165 C7** 16 30N 63 30W
Léfini, *Congo* **114 C3** 2 55S 15 39 E
Lefka, *Cyprus* **38 D11** 35 6N 32 51 E
Lefkoniko, *Cyprus* **38 D12** 35 18N 33 44 E
Lefors, *U.S.A.* **155 H4** 35 26N 100 48W
Lefroy, L., *Australia* **125 F3** 31 21S 121 40 E
Łeg →, *Poland* **55 H8** 50 42N 21 50 E
Legal, *Canada* **142 C6** 53 55N 113 35W
Leganés, *Spain* **42 E7** 40 19N 3 45W
Legazpi, *Phil.* **80 E4** 13 10N 123 45 E
Lège, *France* **28 D2** 44 48N 1 9W
Legendre I., *Australia* **124 D2** 20 22S 116 55 E
Leghorn = Livorno, *Italy* **44 E7** 43 33N 10 19 E
Legionowo, *Poland* **55 F7** 52 25N 20 50 E
Legnago, *Italy* **45 C8** 45 11N 11 18 E
Legnano, *Italy* **44 C5** 45 36N 8 54 E
Legnica, *Poland* **55 G3** 51 12N 16 10 E
Legnica □, *Poland* **55 G3** 51 30N 16 0 E
Legrad, *Croatia* **45 B13** 46 17N 16 51 E
Legume, *Australia* **127 D5** 28 20S 152 19 E
Leh, *India* **93 B7** 34 9N 77 35 E
Lehi, *U.S.A.* **158 F8** 40 24N 111 51W
Lehigh Acres, *U.S.A.* **153 J8** 26 36N 81 39W
Lehighton, *U.S.A.* **151 F9** 40 50N 75 43W
Lehliu, *Romania* **53 F11** 44 29N 26 50 E
Lehrte, *Germany* **30 C5** 52 22N 9 58 E
Lehututu, *Botswana* **116 C3** 23 54S 21 55 E
Lei Shui →, *China* **77 D9** 26 55N 112 35 E
Leiah, *Pakistan* **91 C3** 30 58N 70 58 E
Leibnitz, *Austria* **34 E8** 46 47N 15 34 E
Leibo, *China* **76 C4** 28 11N 103 34 E
Leicester, *U.K.* **21 E6** 52 38N 1 8W
Leicester City □, *U.K.* **21 E6** 52 38N 1 8W
Leicestershire □, *U.K.* **21 E6** 52 41N 1 17W
Leichhardt →, *Australia* **126 B2** 17 35S 139 48 E
Leichhardt Ra., *Australia* **126 C4** 20 46S 147 40 E
Leiden, *Neths.* **24 B4** 52 9N 4 30 E
Leie →, *Belgium* **24 C3** 51 2N 3 45 E
Leifers = Láives, *Italy* **45 B8** 46 26N 11 20 E
Leigh →, *Australia* **128 E6** 38 18S 144 30 E
Leikanger, *Sogn og Fjordane, Norway* **18 B2** 62 8N 5 18 E
Leikanger, *Sogn og Fjordane, Norway* **18 C3** 61 10N 6 51 E
Leikong, *Norway* **18 B2** 62 15N 5 47 E
Leiktho, *Burma* **90 F6** 19 13N 96 35 E
Leimen, *Germany* **31 F4** 49 21N 8 41 E
Leine →, *Germany* **30 C5** 52 43N 9 36 E
Leinefelde, *Germany* **30 D6** 51 23N 10 19 E
Leinster, *Australia* **125 E3** 27 51S 120 36 E
Leinster □, *Ireland* **23 C4** 53 3N 7 8W
Leinster, Mt., *Ireland* **23 D5** 52 37N 6 46W
Leipalingis, *Lithuania* **54 D10** 54 5N 23 51 E
Leipzig, *Germany* **30 D8** 51 18N 12 22 E
Leira, *Norway* **18 D6** 60 58N 9 17 E
Leiria, *Portugal* **42 F2** 39 46N 8 53W
Leiria □, *Portugal* **42 F2** 39 46N 8 53W
Leirvassbu, *Norway* **18 C5** 61 33N 8 13 E
Leirvik, *Norway* **15 G11** 59 47N 5 28 E
Leisler, Mt., *Australia* **124 D4** 23 23S 129 20 E
Leisure City, *U.S.A.* **153 K9** 25 31N 80 26W
Leith, *U.K.* **22 F5** 55 59N 3 11W
Leith Hill, *U.K.* **21 F7** 51 11N 0 22W
Leitha →, *Europe* **35 D10** 47 50N 17 15 E
Leitrim, *Ireland* **23 B3** 54 0N 8 5W
Leitrim □, *Ireland* **23 B4** 54 8N 8 0W
Leitza, *Spain* **40 B3** 43 5N 1 55W
Leiyang, *China* **77 D9** 26 27N 112 45 E
Leizhou Bandao, *China* **73 D6** 21 0N 110 0 E
Leizhou Wan, *China* **77 G8** 20 50N 110 20 E
Lek →, *Neths.* **24 C4** 51 54N 4 35 E
Leka, *Norway* **14 D14** 65 5N 11 35 E
Lekáni, *Greece* **51 E8** 41 10N 24 35 E
Lekbibaj, *Albania* **50 D3** 42 17N 19 56 E
Lekeitio, *Spain* **40 B2** 43 20N 2 32W
Lekhainá, *Greece* **48 D3** 37 57N 21 16 E
Leksand, *Sweden* **16 D9** 60 44N 15 1 E
Leksula, *Indonesia* **82 B3** 3 46S 126 31 E
Lékva Óros, *Greece* **38 D6** 35 18N 24 3 E
Leland, *U.S.A.* **155 J9** 33 24N 90 54W
Leland Lakes, *Canada* **143 B6** 60 0N 110 59W
Lelång, *Sweden* **16 E6** 59 10N 12 5 E
Leleiwi Pt., *U.S.A.* **145 D7** 19 44N 155 0W
Leleque, *Argentina* **176 B2** 42 28S 71 0W
Lelu, *Burma* **90 F5** 19 4N 95 30 E
Lelystad, *Neths.* **24 B5** 52 30N 5 25 E
Lem, *Denmark* **17 H2** 56 1N 8 24 E
Lema, *Nigeria* **113 C5** 12 58N 4 13 E
Léman, L., *Europe* **32 D3** 46 26N 6 30 E
Lemera, *Dem. Rep. of the Congo* **118 C2** 3 0S 28 55 E
Lemery, *Phil.* **80 E3** 13 51N 120 56 E
Lemeta, *U.S.A.* **144 D11** 64 52N 147 44W
Lemfu, *Dem. Rep. of the Congo* **115 D3** 5 18S 15 13 E
Lemhi Ra., *U.S.A.* **158 D7** 44 30N 113 30W
Lemmer, *Neths.* **24 B5** 52 51N 5 43 E
Lemmon, *U.S.A.* **154 C3** 45 57N 102 10W
Lemon Grove, *U.S.A.* **161 N9** 32 45N 117 2W
Lemoore, *U.S.A.* **160 J7** 36 18N 119 46W
Lempdes, *France* **28 C7** 45 22N 3 17 E
Lemsid, *W. Sahara* **110 C2** 26 33N 13 51W
Lemvig, *Denmark* **17 H2** 56 33N 8 20 E
Lemyethna, *Burma* **90 G5** 17 36N 95 9 E
Lena →, *Russia* **65 B13** 72 52N 126 40 E
Lenart, *Slovenia* **45 B12** 46 36N 15 48 E
Lenartovce, *Slovak Rep.* **35 C13** 48 18N 20 19 E
Lencloître, *France* **26 F7** 46 50N 0 20 E
Lençóis, *Brazil* **171 D3** 12 35S 41 24W
Léndas, *Greece* **38 E6** 34 56N 24 56 E
Lendava, *Slovenia* **45 B13** 46 35N 16 25 E
Lendeh, *Iran* **97 D6** 30 58N 50 25 E
Lendinara, *Italy* **45 C8** 45 5N 11 36 E
Lenger, *Kazakstan* **63 B4** 42 12N 69 54 E
Lengerich, *Germany* **30 C3** 52 11N 7 52 E
Lenggong, *Malaysia* **87 K3** 5 6N 100 58 E
Lenggries, *Germany* **31 H7** 47 41N 11 35 E
Lengua de Vaca, Pta., *Chile* **174 C1** 30 14S 71 38W
Lengyeltóti, *Hungary* **52 D2** 46 40N 17 40 E
Lenhovda, *Sweden* **17 G9** 57 0N 15 16 E
Lenina, Kanal →, *Russia* **61 J7** 43 44N 45 17 E
Lenina, Pik, *Kyrgyzstan* **63 D6** 39 20N 72 55 E
Leninabad = Khudzhand, *Tajikistan* **63 C4** 40 17N 69 37 E
Leninakan = Gyumri, *Armenia* **61 K6** 40 47N 43 50 E
Leningrad = Sankt-Peterburg, *Russia* **58 C6** 59 55N 30 20 E
Lenino, *Ukraine* **59 K8** 45 17N 35 46 E
Leninogorsk, *Kazakstan* **64 D9** 50 20N 83 30 E
Leninogorsk, *Russia* **62 D4** 54 36N 52 30 E
Leninpol, *Kyrgyzstan* **63 B5** 42 29N 71 55 E
Leninsk, *Russia* **61 F7** 48 40N 45 15 E
Leninsk, *Uzbekistan* **63 C6** 40 38N 72 15 E
Leninsk-Kuznetskiy, *Russia* **64 D9** 54 44N 86 10 E
Leninskoye, *Aqtöbe, Kazakstan* **62 F6** 50 44N 57 53 E
Leninskoye, *Shymkent, Kazakstan* **63 C4** 41 45N 69 23 E
Leninskoye, *Russia* **60 A8** 58 23N 47 3 E
Lenk, *Switz.* **32 D4** 46 27N 7 28 E
Lenkoran = Länkäran, *Azerbaijan* **101 C13** 38 48N 48 52 E
Lenmalu, *Indonesia* **83 B4** 1 45S 130 15 E
Lenne →, *Germany* **30 D3** 51 25N 7 29 E
Lennestadt, *Germany* **30 D4** 51 8N 8 2 E
Lennox, I., *Chile* **176 E3** 55 18S 66 50W
Lennoxville, *Canada* **151 A13** 45 22N 71 51W
Leno, *Italy* **44 C7** 45 22N 10 13 E
Lenoir, *U.S.A.* **149 H5** 35 55N 81 32W
Lenoir City, *U.S.A.* **149 H3** 35 48N 84 16W
Lenora, *U.S.A.* **154 F4** 39 37N 100 0W
Lenore L., *Canada* **143 C8** 52 30N 104 59W
Lenox, *Ga., U.S.A.* **152 D6** 31 16N 83 28W
Lenox, *Iowa, U.S.A.* **156 D2** 40 53N 94 34W
Lenox, *Mass., U.S.A.* **151 D11** 42 22N 73 17W
Lens, *France* **27 B9** 50 26N 2 50 E
Lensahn, *Germany* **30 A6** 54 13N 10 53 E
Lensk, *Russia* **65 C12** 60 48N 114 55 E
Lensvik, *Norway* **18 A6** 63 31N 9 48 E
Lentekhi, *Georgia* **61 J6** 42 47N 42 45 E
Lenti, *Hungary* **52 D1** 46 37N 16 33 E
Lentini, *Italy* **47 E8** 37 17N 15 0 E
Lenwood, *U.S.A.* **161 L9** 34 53N 117 7W
Lenzburg, *Switz.* **32 B6** 47 23N 8 11 E
Lenzen, *Germany* **30 B7** 53 5N 11 29 E
Lenzerheide, *Switz.* **33 C9** 46 44N 9 34 E
Léo, *Burkina Faso* **112 C4** 11 3N 2 2W
Leoben, *Austria* **34 D8** 47 22N 15 5 E
Leodhas = Lewis, *U.K.* **22 C2** 58 9N 6 40W
Leola, *U.S.A.* **154 C5** 45 43N 98 56W
Leominster, *U.K.* **21 E5** 52 14N 2 43W
Leominster, *U.S.A.* **151 D13** 42 32N 71 46W
Léon, *France* **28 E2** 43 53N 1 18W
León, *Mexico* **162 C4** 21 7N 101 40W
León, *Nic.* **164 D2** 12 20N 86 51W
León, *Spain* **42 C5** 42 38N 5 34W
Leon, *U.S.A.* **156 D3** 40 44N 93 45W
León □, *Spain* **42 C5** 42 40N 5 55W
León, Montes de, *Spain* **42 C4** 42 30N 6 18W
Leonardtown, *U.S.A.* **148 F7** 38 17N 76 38W
Leonberg, *Germany* **31 G5** 48 48N 9 1 E
Leonding, *Austria* **34 C7** 48 16N 14 15 E
Leone, Mte., *Switz.* **32 D6** 46 15N 8 5 E
Leonessa, *Italy* **45 F9** 42 34N 12 58 E
Leonforte, *Italy* **47 E7** 37 38N 14 23 E
Leongatha, *Australia* **129 E6** 38 30S 145 58 E
Leonia, *U.S.A.* **152 E3** 30 55N 86 1W
Leonídhion, *Greece* **48 D4** 37 9N 22 52 E
Leonora, *Australia* **125 E3** 28 49S 121 19 E
Leonora Downs, *Australia* **128 B5** 32 29S 142 5 E
Léopold II, Lac = Mai-Ndombe, L., *Dem. Rep. of the Congo* **114 C3** 2 0S 18 20 E
Leopoldina, *Brazil* **171 F3** 21 28S 42 40W
Leopoldo Bulhões, *Brazil* **171 E2** 16 37S 48 46W
Leopoldsburg, *Belgium* **24 C5** 51 7N 5 13 E
Léopoldville = Kinshasa, *Dem. Rep. of the Congo* **115 C3** 4 20S 15 15 E
Leoti, *U.S.A.* **154 F4** 38 29N 101 21W
Leova, *Moldova* **53 D13** 46 28N 28 15 E
Leoville, *Canada* **143 C7** 53 39N 107 33W
Lépa, L. do, *Angola* **115 F3** 17 0S 19 0 E
Lepe, *Spain* **43 H3** 37 15N 7 12W
Lepel = Lyepyel, *Belarus* **58 E5** 54 50N 28 40 E
Lepenoú, *Greece* **48 C3** 38 42N 21 17 E
Leping, *China* **77 C11** 28 47N 117 7 E
Lepontine, Alpi, *Italy* **33 D6** 46 22N 8 27 E
Leppävirta, *Finland* **15 E22** 62 29N 27 46 E
Lepsény, *Hungary* **52 D3** 47 0N 18 15 E
Leptis Magna, *Libya* **108 B2** 32 40N 14 12 E
Lequeitio = Lekeitio, *Spain* **40 B2** 43 20N 2 32W
Lercara Friddi, *Italy* **46 E6** 37 45N 13 36 E
Lerdo, *Mexico* **162 B4** 25 32N 103 32W
Léré, *C.A.R.* **114 A3** 6 46N 17 25 E
Léré, *Chad* **109 G2** 9 39N 14 13 E
Lere, *Nigeria* **113 D6** 9 43N 9 18 E
Leribe, *Lesotho* **117 D4** 28 51S 28 3 E
Lérici, *Italy* **44 D6** 44 4N 9 55 E
Lérida = Lleida, *Spain* **40 D5** 41 37N 0 39 E
Lérins, Îs. de, *France* **29 E11** 43 31N 7 3 E
Lerma, *Spain* **42 C7** 42 0N 3 47W
Léros, *Greece* **49 D8** 37 10N 26 50 E
Lérouville, *France* **27 D12** 48 44N 5 30 E
Lerum, *Sweden* **17 G6** 57 46N 12 16 E
Lerwick, *U.K.* **22 A7** 60 9N 1 9W
Leş, *Romania* **52 D6** 46 58N 21 50 E
Les Abrets, *France* **29 C9** 45 32N 5 35 E
Les Andelys, *France* **26 C8** 49 15N 1 25 E
Les Bois, *Switz.* **32 B3** 47 11N 6 50 E
Les Borges Blanques, *Spain* **40 D5** 41 31N 0 52 E
Les Cayes, *Haiti* **165 C5** 18 15N 73 46W
Les Diablerets, *Switz.* **32 D4** 46 22N 7 10 E
Les Essarts, *France* **26 F5** 46 47N 1 12W
Les Étroits, *Canada* **141 C6** 47 24N 68 54W
Les Herbiers, *France* **26 F5** 46 52N 1 1W
Les Minquiers, Plateau des, *Chan. Is.* **26 D4** 48 58N 2 8W
Les Pieux, *France* **26 C5** 49 30N 1 48W
Les Ponts-de-Cé, *France* **26 E6** 47 25N 0 30W
Les Riceys, *France* **27 E11** 47 59N 4 22 E
Les Sables-d'Olonne, *France* **28 B2** 46 30N 1 45W
Les Vans, *France* **29 D8** 44 25N 4 7 E
Les Verrières, *Switz.* **32 C2** 46 55N 6 28 E
Lesbos = Lésvos, *Greece* **49 B8** 39 10N 26 20 E
L'Escala, *Spain* **40 C8** 42 7N 3 8 E
Lèshan, *China* **76 C4** 29 33N 103 41 E
Leshukonskoye, *Russia* **56 B8** 64 54N 45 46 E
Lésina, *Italy* **45 G12** 41 52N 15 21 E
Lésina, L. di, *Italy* **45 G12** 41 53N 15 26 E
Lesjaskog, *Norway* **18 B5** 62 14N 8 22 E
Lesjaverk, *Norway* **18 B5** 62 12N 8 34 E
Lesjöfors, *Sweden* **16 E8** 59 58N 14 11 E
Lesko, *Poland* **55 J9** 49 30N 22 23 E
Leskov I., *Antarctica* **7 B1** 56 0S 28 0W
Leskovac, *Serbia, Yug.* **50 C5** 43 0N 21 58 E
Leskoviku, *Albania* **50 F4** 40 10N 20 34 E
Leslie, *Ark., U.S.A.* **155 H8** 35 50N 92 34W
Leslie, *Ga., U.S.A.* **152 D5** 31 57N 84 5W
Leslie, *Mich., U.S.A.* **157 B12** 42 27N 84 26W
Leśna, *Poland* **55 G2** 51 1N 15 15 E
Lesneven, *France* **26 D2** 48 35N 4 20W
Leśnica, *Poland* **55 H5** 50 26N 18 11 E
Lešnica, *Serbia, Yug.* **50 B3** 44 39N 19 20 E
Lesnoy, *Russia* **62 B4** 59 47N 52 9 E
Lesnoye, *Russia* **58 C8** 58 15N 35 18 E
Lesopilnoye, *Russia* **68 A7** 46 44N 134 20 E
Lesotho ■, *Africa* **117 D4** 29 40S 28 0 E
Lesozavodsk, *Russia* **68 B6** 45 30N 133 29 E
Lesparre-Médoc, *France* **28 C3** 45 18N 0 57W
Lessay, *France* **26 C5** 49 14N 1 30W
Lesse →, *Belgium* **24 D4** 50 15N 4 54 E
Lessebo, *Sweden* **17 H9** 56 45N 15 16 E
Lesser Antilles, *W. Indies* **165 D7** 15 0N 61 0W
Lesser Slave L., *Canada* **142 B5** 55 30N 115 25W
Lesser Sunda Is., *Indonesia* **85 D5** 7 0S 120 0 E
Lessines, *Belgium* **24 D3** 50 42N 3 50 E
Lester, *U.S.A.* **160 C5** 47 12N 121 29W
Lestock, *Canada* **143 C8** 51 19N 103 59W
Lesueur I., *Australia* **124 B4** 13 50S 127 17 E
Lésvos, *Greece* **49 B8** 39 10N 26 20 E
Leszno, *Poland* **55 G3** 51 50N 16 30 E
Leszno □, *Poland* **55 G3** 51 45N 16 30 E
Letälven, *Sweden* **16 E8** 59 5N 14 20 E
Létavértes, *Hungary* **52 C6** 47 23N 21 55 E
Letchworth, *U.K.* **21 F7** 51 59N 0 13W
Letea, Ostrov, *Romania* **53 E14** 45 18N 29 20 E
Lethbridge, *Canada* **142 D6** 49 45N 112 45W
Lethem, *Guyana* **169 C6** 3 20N 59 50W
Lethero, *Australia* **128 B5** 33 33S 142 30 E
Leti, Kepulauan, *Indonesia* **82 C3** 8 10S 128 0 E
Leti Is. = Leti, Kepulauan, *Indonesia* **82 C3** 8 10S 128 0 E
Letiahau →, *Botswana* **116 C3** 21 16S 24 0 E
Leticia, *Colombia* **168 D4** 4 9S 70 0W
Leting, *China* **75 E10** 39 23N 118 55 E
Letjiesbos, *S. Africa* **116 E3** 32 34S 22 16 E
Letlhakeng, *Botswana* **116 C3** 24 0S 24 59 E
Letpadan, *Burma* **90 G5** 17 45N 95 45 E
Letpan, *Burma* **90 F5** 19 28N 94 10 E
Letsôk-aw Kyun, *Burma* **87 G2** 11 30N 98 25 E
Letterkenny, *Ireland* **23 B4** 54 57N 7 45W
Leu, *Romania* **53 F9** 44 10N 24 0 E
Léua, *Angola* **115 E4** 11 34S 20 32 E
Leucadia, *U.S.A.* **161 M9** 33 4N 117 18W
Leucate, *France* **28 F7** 42 56N 3 3 E
Leucate, Étang de, *France* **28 F7** 42 50N 3 0 E
Leuk, *Switz.* **32 D5** 46 19N 7 37 E
Leukerbad, *Switz.* **32 D5** 46 24N 7 36 E
Leuşeni, *Moldova* **53 D13** 46 49N 28 12 E
Leuser, G., *Indonesia* **84 B1** 3 46N 97 12 E
Leutkirch, *Germany* **31 H6** 47 49N 10 1 E
Leuven, *Belgium* **24 D4** 50 52N 4 42 E
Leuze-en-Hainaut, *Belgium* **24 D3** 50 36N 3 37 E
Lev Tolstoy, *Russia* **58 F10** 53 13N 39 29 E
Levádhia, *Greece* **48 C4** 38 27N 22 54 E
Levan, *U.S.A.* **158 G8** 39 33N 111 52W
Levanger, *Norway* **14 E14** 63 45N 11 19 E
Levani, *Albania* **50 F3** 40 40N 19 28 E
Levant, Î. du, *France* **29 E10** 43 3N 6 28 E
Lévanto, *Italy* **44 D6** 44 10N 9 38 E
Lévanzo, *Italy* **46 D5** 38 0N 12 20 E
Leveld, *Norway* **18 D5** 60 44N 8 33 E
Levelland, *U.S.A.* **155 J3** 33 35N 102 23W
Levelock, *U.S.A.* **144 G8** 59 7N 156 51W
Leven, *U.K.* **22 E6** 56 12N 3 0W
Leven, L., *U.K.* **22 E5** 56 12N 3 22W
Leven, Toraka, *Madag.* **117 A8** 12 30S 47 45 E
Leveque C., *Australia* **124 C3** 16 20S 123 0 E
Leverano, *Italy* **47 B10** 40 17N 18 0 E
Leverkusen, *Germany* **30 D3** 51 3N 6 59 E
Leverville, *Dem. Rep. of the Congo* **115 C3** 4 50S 18 44 E
Levice, *Slovak Rep.* **35 C11** 48 13N 18 35 E
Levico Terme, *Italy* **45 C8** 46 0N 11 18 E

242 Litang

Litang, *Malaysia* **85 A5** 5 27N 118 31 E
Litang Qu →, *China* **76 C3** 28 4N 101 32 E
Litani →, *Lebanon* **103 B4** 33 20N 35 15 E
Litchfield, *Australia* **128 D5** 36 18 S 142 52 E
Litchfield, *Calif., U.S.A.* **160 E6** 40 24N 120 23W
Litchfield, *Conn., U.S.A.* **151 E11** 41 45N 73 11W
Litchfield, *Ill., U.S.A.* **156 E7** 39 11N 89 39W
Litchfield, *Minn., U.S.A.* **154 C7** 45 8N 94 32W
Liteni, *Romania* **53 C11** 47 32N 26 32 E
Lithgow, *Australia* **129 B9** 33 25 S 150 8 E
Líthinon, Ákra, *Greece* **38 E6** 34 55N 24 44 E
Lithuania ■, *Europe* **15 J20** 55 30N 24 0 E
Litija, *Slovenia* **45 B11** 46 3N 14 50 E
Litókhoron, *Greece* **50 F6** 40 8N 22 34 E
Litoměřice, *Czech Rep.* **34 A7** 50 33N 14 10 E
Litomyšl, *Czech Rep.* **35 B9** 49 52N 16 20 E
Litschau, *Austria* **34 C8** 48 58N 15 4 E
Little Abaco I., *Bahamas* **164 A4** 26 50N 77 30W
Little Aden, *Yemen* **98 D4** 12 45N 44 52 E
Little Barrier I., *N.Z.* **130 C4** 36 12 S 175 8 E
Little Belt Mts., *U.S.A.* **158 C8** 46 40N 110 45W
Little Blue →, *U.S.A.* **154 F6** 39 42N 96 41W
Little Cadotte →, *Canada* **142 B5** 56 41N 117 6W
Little Cayman, I., *Cayman Is.* **164 C3** 19 41N 80 3W
Little Churchill →, *Canada* **143 B9** 57 30N 95 22W
Little Colorado →, *U.S.A.* **159 H8** 36 12N 111 48W
Little Current, *Canada* **140 C3** 45 55N 82 0W
Little Current →, *Canada* **140 B3** 50 57N 84 36W
Little Diomede I., *U.S.A.* **144 D5** 65 45N 168 56W
Little Falls, *Minn., U.S.A.* **154 C7** 45 59N 94 22W
Little Falls, *N.Y., U.S.A.* **151 C10** 43 3N 74 51W
Little Fork →, *U.S.A.* **154 A8** 48 31N 93 35W
Little Grand Rapids, *Canada* **143 C9** 52 0N 95 29W
Little Haw Cr. →, *U.S.A.* **153 F8** 29 23N 81 24W
Little Humboldt →, *U.S.A.* **158 F5** 41 1N 117 43W
Little Inagua I., *Bahamas* **165 B5** 21 40N 73 50W
Little Karoo, *S. Africa* **116 E3** 33 45 S 21 0 E
Little Lake, *U.S.A.* **161 K9** 35 56N 117 55W
Little Laut Is. = Laut Kecil, Kepulauan, *Indonesia* **85 C5** 4 45 S 115 40 E
Little Minch, *U.K.* **22 D2** 57 35N 6 45W
Little Missouri →, *U.S.A.* **154 B3** 47 36N 102 25W
Little Ouse →, *U.K.* **21 E9** 52 22N 1 12 E
Little Rann, *India* **92 H4** 23 25N 71 25 E
Little Red →, *U.S.A.* **155 H9** 35 11N 91 27W
Little River, *N.Z.* **131 D7** 43 45 S 172 49 E
Little Rock, *U.S.A.* **155 H8** 34 45N 92 17W
Little Ruaha →, *Tanzania* **118 D4** 7 57 S 37 53 E
Little Sable Pt., *U.S.A.* **148 D2** 43 38N 86 33W
Little Sioux →, *U.S.A.* **154 E6** 41 48N 96 4W
Little Sitkin I., *U.S.A.* **144 L2** 51 57N 178 31 E
Little Smoky →, *Canada* **142 C5** 54 44N 117 11W
Little Snake →, *U.S.A.* **158 F9** 40 27N 108 26W
Little Tallapoosa →, *U.S.A.* **152 B4** 33 18N 85 34W
Little Valley, *U.S.A.* **150 D6** 42 15N 78 48W
Little Wabash →, *U.S.A.* **157 G8** 37 55N 88 5W
Little York, *U.S.A.* **156 C6** 41 1N 90 45W
Littlefield, *U.S.A.* **155 J3** 33 55N 102 20W
Littlefork, *U.S.A.* **154 A8** 48 24N 93 34W
Littlehampton, *U.K.* **21 G7** 50 49N 0 32W
Littleton, *U.S.A.* **151 B13** 44 18N 71 46W
Litvinov, *Czech Rep.* **34 A6** 50 36N 13 37 E
Liu He →, *China* **75 D11** 40 55N 121 35 E
Liu Jiang →, *China* **76 F7** 23 55N 109 30 E
Liuba, *China* **74 H4** 33 38N 106 55 E
Liucheng, *China* **76 E7** 24 38N 109 14 E
Liugou, *China* **75 D10** 40 57N 118 15 E
Liuhe, *China* **75 C13** 42 17N 125 43 E
Liuheng Dao, *China* **77 C14** 29 40N 122 5 E
Liukang Tenggaja = Sabalana, Kepulauan, *Indonesia* **82 C1** 6 45 S 118 50 E
Liuli, *Tanzania* **119 E3** 11 3 S 34 38 E
Liuwa Plain, *Zambia* **115 E4** 14 20 S 22 30 E
Liuyang, *China* **77 C9** 28 10N 113 37 E
Liuzhou, *China* **76 E7** 24 22N 109 22 E
Liuzhuang, *China* **75 H11** 33 12N 120 18 E
Livada, *Romania* **52 C8** 47 52N 23 5 E
Livadherón, *Greece* **50 F5** 40 2N 21 57 E
Livadhia, *Cyprus* **38 E12** 34 57N 33 38 E
Livádhion, *Greece* **48 A4** 40 8N 22 9 E
Livarot, *France* **26 D7** 48 58N 0 9 E
Live Oak, *Calif., U.S.A.* **160 F5** 39 17N 121 40W
Live Oak, *Fla., U.S.A.* **153 E7** 30 18N 82 59W
Livengood, *U.S.A.* **144 D10** 65 32N 148 33W
Liveras, *Cyprus* **38 D11** 35 23N 32 57 E
Liveringa, *Australia* **124 C3** 18 3 S 124 10 E
Livermore, *U.S.A.* **160 H5** 37 41N 121 47W
Livermore, Mt., *U.S.A.* **155 K2** 30 38N 104 11W
Liverpool, *Australia* **129 B9** 33 54 S 150 58 E
Liverpool, *Canada* **141 D7** 44 5N 64 41W
Liverpool, *U.K.* **20 D4** 53 25N 3 0W
Liverpool Bay, *U.K.* **20 D4** 53 30N 3 20W
Liverpool Plains, *Australia* **129 A9** 31 15 S 150 15 E
Liverpool Ra., *Australia* **129 A9** 31 50 S 150 30 E
Livigno, *Italy* **44 B7** 46 35N 10 10 E
Livingston, *Guatemala* **164 C2** 15 50N 88 50W
Livingston, *U.K.* **22 F5** 55 54N 3 30W
Livingston, *Calif., U.S.A.* **160 H6** 37 23N 120 43W
Livingston, *Mont., U.S.A.* **158 D8** 45 40N 110 34W
Livingston, *Tex., U.S.A.* **155 K7** 30 43N 94 56W
Livingston, *Wis., U.S.A.* **156 B6** 42 54N 90 26W
Livingstone, *Zambia* **119 F2** 17 46 S 25 52 E
Livingstone Mts., *N.Z.* **131 F3** 45 15 S 168 9 E
Livingstone Mts., *Tanzania* **119 D3** 9 40 S 34 20 E
Livingstonia, *Malawi* **119 E3** 10 38 S 34 5 E
Livno, *Bos.-H.* **52 G2** 43 50N 17 1 E
Livny, *Russia* **59 F9** 52 30N 37 30 E
Livonia, *U.S.A.* **157 B13** 42 23N 83 23W
Livorno, *Italy* **44 E7** 43 33N 10 19 E
Livramento, *Brazil* **175 C4** 30 55 S 55 30W
Livramento do Brumado, *Brazil* **171 D3** 13 39 S 41 50W
Livron-sur-Drôme, *France* **29 D8** 44 46N 4 51 E
Liwale, *Tanzania* **119 D4** 9 48 S 37 58 E
Liwale □, *Tanzania* **119 D4** 9 0 S 38 0 E
Liwiec →, *Poland* **55 F8** 52 36N 21 34 E
Lixi, *China* **76 D3** 26 23N 101 59 E
Lixoúrion, *Greece* **48 C2** 38 14N 20 24 E
Liyang, *China* **77 B12** 31 26N 119 28 E
Lizard I., *Australia* **126 A4** 14 42 S 145 30 E
Lizard Pt., *U.K.* **21 H2** 49 57N 5 13W
Lizarda, *Brazil* **170 C2** 9 36 S 46 41W
Lizella, *U.S.A.* **152 C6** 32 48N 83 49W
Lizzano, *Italy* **47 B10** 40 23N 17 27 E
Ljig, *Serbia, Yug.* **50 B4** 44 13N 20 18 E
Ljøra →, *Norway* **18 C9** 61 24N 12 35 E
Ljørdal, *Norway* **18 C9** 61 23N 12 41 E
Ljosland, *Norway* **18 F4** 58 47N 7 22 E
Ljubija, *Bos.-H.* **45 D13** 44 55N 16 35 E
Ljubinje, *Bos.-H.* **50 D2** 42 58N 18 5 E
Ljubljana, *Slovenia* **45 B11** 46 4N 14 33 E
Ljubno, *Slovenia* **45 B11** 46 25N 14 46 E
Ljubovija, *Serbia, Yug.* **50 B3** 44 11N 19 22 E
Ljugarn, *Sweden* **17 G12** 57 20N 18 43 E
Ljung, *Sweden* **17 G7** 57 59N 13 3 E
Ljungan →, *Sweden* **16 B11** 62 18N 17 23 E
Ljungaverk, *Sweden* **16 B10** 62 30N 16 5 E
Ljungby, *Sweden* **15 H15** 56 49N 13 55 E
Ljungbyholm, *Sweden* **17 H10** 56 39N 16 5 E
Ljungdalen, *Sweden* **16 B6** 62 51N 12 47 E
Ljungsbro, *Sweden* **17 F9** 58 31N 15 30 E
Ljungskile, *Sweden* **17 F5** 58 14N 11 55 E
Ljusdal, *Sweden* **16 C10** 61 46N 16 3 E
Ljusfallshammar, *Sweden* **17 F9** 58 48N 15 30 E
Ljusnan →, *Sweden* **16 C11** 61 12N 17 8 E
Ljusne, *Sweden* **16 C11** 61 13N 17 7 E
Ljutomer, *Slovenia* **45 B13** 46 31N 16 11 E
Llagostera, *Spain* **40 D7** 41 50N 2 54 E
Llamellín, *Peru* **172 B2** 9 0 S 76 54W
Llancanelo, Salina, *Argentina* **174 D2** 35 40 S 69 8W
Llandeilo, *U.K.* **21 F4** 51 53N 3 59W
Llandovery, *U.K.* **21 F4** 51 59N 3 48W
Llandrindod Wells, *U.K.* **21 E4** 52 14N 3 22W
Llandudno, *U.K.* **20 D4** 53 19N 3 50W
Llanelli, *U.K.* **21 F3** 51 41N 4 10W
Llanes, *Spain* **42 B6** 43 25N 4 50W
Llangollen, *U.K.* **20 E4** 52 58N 3 11W
Llanidloes, *U.K.* **21 E4** 52 27N 3 31W
Llano, *U.S.A.* **155 K5** 30 45N 98 41W
Llano →, *U.S.A.* **155 K5** 30 39N 98 26W
Llano Estacado, *U.S.A.* **155 J3** 33 30N 103 0W
Llanos, *S. Amer.* **166 C3** 5 0N 71 35W
Llanquihue □, *Chile* **176 B2** 41 30 S 73 0W
Llanquihue, L., *Chile* **176 B1** 41 10 S 75 50W
Llanwrtyd Wells, *U.K.* **21 E4** 52 7N 3 38W
Llebetx, C., *Spain* **39 B9** 39 33N 2 18 E
Lleida, *Spain* **40 D5** 41 37N 0 39 E
Lleida □, *Spain* **40 C6** 42 6N 1 0 E
Llentrisca, C., *Spain* **39 C7** 38 52N 1 15 E
Llera, *Mexico* **163 C5** 23 19N 99 1W
Llerena, *Spain* **43 G5** 38 17N 6 0W
Lleyn Peninsula, *U.K.* **20 E3** 52 51N 4 36W
Llica, *Bolivia* **172 D4** 19 52 S 68 16W
Llico, *Chile* **174 C1** 34 46 S 72 5W
Lliria, *Spain* **41 F4** 39 37N 0 35W
Llobregat →, *Spain* **40 D7** 41 19N 2 9 E
Llodio, *Spain* **40 B2** 43 9N 2 58W
Llorente, *Phil.* **81 F5** 11 25N 125 33 E
Lloret de Mar, *Spain* **40 D7** 41 41N 2 53 E
Lloyd B., *Australia* **126 A3** 12 45 S 143 27 E
Lloyd L., *Canada* **143 B7** 57 22N 108 57W
Lloydminster, *Canada* **143 C7** 53 17N 110 0W
Llucena del Cid, *Spain* **40 E4** 40 9N 0 17W
Llucmajor, *Spain* **39 B9** 39 29N 2 53 E
Llullaillaco, Volcán, *S. Amer.* **174 A2** 24 43 S 68 30W
Lo →, *Vietnam* **86 B5** 21 18N 105 25 E
Loa, *U.S.A.* **159 G8** 38 24N 111 39W
Loa →, *Chile* **174 A1** 21 26 S 70 41W
Loano, *Italy* **44 D5** 44 8N 8 15 E
Loay, *Phil.* **81 G5** 9 36N 124 1 E
Lobatse, *Botswana* **116 D4** 25 12 S 25 40 E
Löbau, *Germany* **30 D10** 51 5N 14 40 E
Lobaye →, *C.A.R.* **114 B3** 3 41N 18 35 E
Lobenstein, *Germany* **30 E7** 50 25N 11 39 E
Lobería, *Argentina* **174 D4** 38 10 S 58 40W
Löberöd, *Sweden* **17 J7** 55 47N 13 31 E
Łobez, *Poland* **54 E2** 53 38N 15 39 E
Lobito, *Angola* **115 E2** 12 18 S 13 35 E
Lobo, *Phil.* **80 E3** 13 39N 121 13 E
Lobos, *Argentina* **174 D4** 35 10 S 59 0W
Lobos, I., *Mexico* **162 B2** 27 15N 110 30W
Lobos, I. de, *Canary Is.* **39 F6** 28 45N 13 50W
Lobos de Tierra, I., *Peru* **172 B1** 6 27 S 80 52W
Lobva, *Russia* **62 B8** 59 10N 60 30 E
Lobva →, *Russia* **62 B8** 59 8N 60 48 E
Łobżenica, *Poland* **55 E4** 53 18N 17 15 E
Loc Binh, *Vietnam* **86 B6** 21 46N 106 54 E
Loc Ninh, *Vietnam* **87 G6** 11 50N 106 34 E
Locarno, *Switz.* **33 D7** 46 10N 8 47 E
Loch Baghasdail = Lochboisdale, *U.K.* **22 D1** 57 9N 7 20W
Loch Garman = Wexford, *Ireland* **23 D5** 52 20N 6 28W
Loch Nam Madadh = Lochmaddy, *U.K.* **22 D1** 57 36N 7 10W
Lochaber, *U.K.* **22 E3** 56 59N 5 1W
Locharbriggs, *U.K.* **22 F5** 55 7N 3 35W
Lochboisdale, *U.K.* **22 D1** 57 9N 7 20W
Lochem, *Neths.* **24 B6** 52 9N 6 26 E
Loches, *France* **26 E7** 47 7N 1 0 E
Lochgilphead, *U.K.* **22 E3** 56 2N 5 26W
Lochinver, *U.K.* **22 C3** 58 9N 5 14W
Lochloose L., *U.S.A.* **153 F7** 29 30N 82 7W
Lochmaddy, *U.K.* **22 D1** 57 36N 7 10W
Lochnagar, *Australia* **126 C4** 23 33 S 145 38 E
Lochnagar, *U.K.* **22 E5** 56 57N 3 15W
Łochów, *Poland* **55 F8** 52 33N 21 42 E
Lochy, L., *U.K.* **22 E4** 57 0N 4 53W
Lock, *Australia* **127 E2** 33 34 S 135 46 E
Lock Haven, *U.S.A.* **150 E7** 41 8N 77 28W
Lockeford, *U.S.A.* **160 G5** 38 10N 121 9W
Lockeport, *Canada* **141 D6** 43 47N 65 4W
Lockerbie, *U.K.* **22 F5** 55 7N 3 21W
Lockhart, *Australia* **129 C7** 35 14 S 146 40 E
Lockhart, *U.S.A.* **155 L6** 29 53N 97 40W
Lockhart, L., *Australia* **125 F2** 33 15 S 119 3 E
Lockington, *Australia* **128 D6** 36 16 S 144 34 E
Lockney, *U.S.A.* **155 H4** 34 7N 101 27W
Lockport, *Ill., U.S.A.* **157 C8** 41 35N 88 3W
Lockport, *N.Y., U.S.A.* **150 C6** 43 10N 78 42W
Locminé, *France* **26 E4** 47 54N 2 51W
Locri, *Italy* **47 D9** 38 14N 16 16 E
Locronan, *France* **26 D2** 48 7N 4 15W
Locust Cr. →, *U.S.A.* **156 E3** 39 40N 93 17W
Locust Grove, *U.S.A.* **152 B5** 33 21N 84 7W
Lod, *Israel* **103 D3** 31 57N 34 54 E
Lodalskåpa, *Norway* **18 C4** 61 47N 7 13 E
Lodeinoye Pole, *Russia* **58 B7** 60 44N 33 33 E
Lodève, *France* **28 E7** 43 44N 3 19 E
Lodge Grass, *U.S.A.* **158 D10** 45 19N 107 22W
Lodgepole, *U.S.A.* **154 E3** 41 9N 102 38W
Lodgepole Cr. →, *U.S.A.* **154 E2** 41 20N 104 30W
Lodhran, *Pakistan* **92 E4** 29 32N 71 30 E
Lodi, *Italy* **44 C6** 45 19N 9 30 E
Lodi, *U.S.A.* **160 G5** 38 8N 121 16W
Lodja, *Dem. Rep. of the Congo* **114 C4** 3 30 S 23 23 E
Lodosa, *Spain* **40 C2** 42 25N 2 4W
Lödöse, *Sweden* **17 F6** 58 2N 12 9 E
Lodwar, *Kenya* **118 B4** 3 10N 35 40 E
Łódź, *Poland* **55 G6** 51 45N 19 27 E
Łódź □, *Poland* **55 G6** 51 45N 19 27 E
Loei, *Thailand* **86 D3** 17 29N 101 35 E
Loengo, *Dem. Rep. of the Congo* **115 C5** 4 48 S 26 30 E
Loeriesfontein, *S. Africa* **116 E2** 31 0 S 19 26 E
Lofer, *Austria* **34 D5** 47 35N 12 41 E
Lofoten, *Norway* **14 B15** 68 30N 14 0 E
Lofsdalen, *Sweden* **16 B7** 62 10N 13 20 E
Lofsen →, *Sweden* **16 B7** 62 7N 13 57 E
Loftahammar, *Sweden* **17 G10** 57 54N 16 41 E
Logan, *Kans., U.S.A.* **154 F5** 39 40N 99 34W
Logan, *Ohio, U.S.A.* **148 F4** 39 32N 82 25W
Logan, *Utah, U.S.A.* **158 F8** 41 44N 111 50W
Logan, *W. Va., U.S.A.* **148 G5** 37 51N 81 59W
Logan, Mt., *Canada* **138 B5** 60 31N 140 22W
Logan Martin Reservoir, *U.S.A.* **152 B3** 33 26N 86 20W
Logan Pass, *U.S.A.* **142 D6** 48 41N 113 44W
Logandale, *U.S.A.* **161 J12** 36 36N 114 29W
Logansport, *Ind., U.S.A.* **157 D10** 40 45N 86 22W
Logansport, *La., U.S.A.* **155 K8** 31 58N 94 0W
Loganville, *U.S.A.* **152 B6** 33 50N 83 54W
Logo, *Sudan* **107 F3** 5 20N 30 18 E
Logone →, *Chad* **109 F3** 12 6N 15 2 E
Logroño, *Spain* **40 C2** 42 28N 2 27W
Logrosán, *Spain* **43 F5** 39 20N 5 32W
Løgstør, *Denmark* **17 H3** 56 58N 9 14 E
Løgumkloster, *Denmark* **17 J2** 55 4N 8 57 E
Loh, *Vanuatu* **133 C4** 13 21 S 166 38 E
Lohals, *Denmark* **17 J4** 55 8N 10 55 E
Lohardaga, *India* **93 H11** 23 27N 84 45 E
Lohja, *Finland* **15 F21** 60 12N 24 5 E
Löhne, *Germany* **30 C4** 52 11N 8 40 E
Lohr, *Germany* **31 F5** 49 59N 9 35 E
Lohrville, *U.S.A.* **156 B2** 42 17N 94 33W
Loi-kaw, *Burma* **90 F6** 19 40N 97 17 E
Loimaa, *Finland* **15 F20** 60 50N 23 5 E
Loir →, *France* **26 E6** 47 33N 0 32W
Loir-et-Cher □, *France* **26 E8** 47 40N 1 20 E
Loire □, *France* **29 C8** 45 40N 4 5 E
Loire →, *France* **26 E4** 47 16N 2 10W
Loire-Atlantique □, *France* **26 E5** 47 25N 1 40W
Loiret □, *France* **27 E9** 47 55N 2 30 E
Loitz, *Germany* **30 B9** 53 58N 13 8 E
Loja, *Ecuador* **172 A2** 3 59 S 79 16W
Loja, *Spain* **43 H6** 37 10N 4 10W
Loja □, *Ecuador* **168 D2** 4 0 S 79 13W
Loji = Kawasi, *Indonesia* **82 B3** 1 38 S 127 28 E
Løjt Kirkeby, *Denmark* **17 J3** 55 7N 9 26 E
Loka, *Sudan* **107 G3** 4 13N 31 0 E
Lokandu, *Dem. Rep. of the Congo* **114 C5** 2 30 S 25 45 E
Løken, *Norway* **18 E8** 59 48N 11 29 E
Lokeren, *Belgium* **24 C3** 51 6N 3 59 E
Lokhvitsa, *Ukraine* **59 G7** 50 25N 33 18 E
Lokichokio, *Kenya* **118 B3** 4 19N 34 13 E
Lokitaung, *Kenya* **118 B4** 4 12N 35 48 E
Lokkan tekojärvi, *Finland* **14 C22** 67 55N 27 35 E
Løkken, *Denmark* **17 G3** 57 22N 9 41 E
Løkken, *Norway* **18 A6** 63 9N 9 41 E
Loknya, *Russia* **58 D6** 56 49N 30 4 E
Lokoja, *Nigeria* **113 D6** 7 47N 6 45 E
Lokolama, *Dem. Rep. of the Congo* **114 C3** 2 35 S 19 50 E
Lokot, *Russia* **59 F8** 52 34N 34 36 E
Lokuru, *Solomon Is.* **133 M9** 8 20 S 157 0 E
Lol →, *Sudan* **107 F2** 9 13N 26 30 E
Lola, *Guinea* **112 D3** 7 52N 8 29W
Lola, Mt., *U.S.A.* **160 F6** 39 26N 120 22W
Lolibai, Gebel, *Sudan* **107 G3** 3 50N 33 0 E
Lolimi, *Sudan* **107 G3** 4 35N 34 0 E
Loliondo, *Tanzania* **118 C4** 2 2 S 35 39 E
Lolland, *Denmark* **17 K5** 54 45N 11 30 E
Lollar, *Germany* **30 E4** 50 37N 8 43 E
Lolo, *U.S.A.* **158 C6** 46 45N 114 5W
Lolodorf, *Cameroon* **113 E7** 3 16N 10 49 E
Lolowai, *Vanuatu* **133 E6** 15 18 S 168 0 E
Lom, *Bulgaria* **50 C7** 43 48N 23 12 E
Lom, *Norway* **18 C5** 61 50N 8 34 E
Lom →, *Bulgaria* **50 C7** 43 45N 23 15 E
Lom Kao, *Thailand* **86 D3** 16 53N 101 14 E
Lom Sak, *Thailand* **86 D3** 16 47N 101 15 E
Loma, *U.S.A.* **158 C8** 47 56N 110 30W
Loma Linda, *U.S.A.* **161 L9** 34 3N 117 16W
Lomaloma, *Fiji* **133 A3** 17 17 S 178 59W
Lomami →, *Dem. Rep. of the Congo* **114 B4** 0 46N 24 16 E
Lomas de Zamóra, *Argentina* **174 C4** 34 45 S 58 25W
Lombadina, *Australia* **124 C3** 16 31 S 122 54 E
Lombard, *U.S.A.* **157 C8** 41 53N 88 1W
Lombárdia □, *Italy* **44 C6** 45 40N 9 30 E
Lombardy = Lombárdia □, *Italy* **44 C6** 45 40N 9 30 E
Lombe, *Angola* **115 D3** 9 27 S 16 13 E
Lombez, *France* **28 E4** 43 29N 0 55 E
Lomblen, *Indonesia* **82 C2** 8 30 S 123 32 E
Lombok, *Indonesia* **85 D5** 8 45 S 116 30 E
Lomé, *Togo* **113 D5** 6 9N 1 20 E
Lomela, *Dem. Rep. of the Congo* **114 C4** 2 19 S 23 15 E
Lomela →, *Dem. Rep. of the Congo* **114 C4** 0 15 S 20 40 E
Lometa, *U.S.A.* **155 K5** 31 13N 98 24W
Lomianki, *Poland* **55 F7** 52 21N 20 54 E
Lomié, *Cameroon* **114 B2** 3 13N 13 38 E
Lomma, *Sweden* **17 J7** 55 43N 13 6 E
Lommel, *Belgium* **24 C5** 51 14N 5 19 E
Lomond, *Canada* **142 C6** 50 24N 112 36W
Lomond, L., *U.K.* **22 E4** 56 8N 4 38W
Lomphat, *Cambodia* **86 F6** 13 30N 106 59 E
Lompobatang, *Indonesia* **82 C1** 5 24 S 119 56 E
Lompoc, *U.S.A.* **161 L6** 34 38N 120 28W
Lomseggi, *Norway* **18 C5** 61 49N 8 21 E
Łomża, *Poland* **55 E9** 53 10N 22 2 E
Łomża □, *Poland* **55 E9** 53 0N 22 30 E
Lonavale, *India* **94 E1** 18 46N 73 29 E
Loncoche, *Chile* **176 A2** 39 20 S 72 50W
Loncopuè, *Argentina* **176 A2** 38 4 S 70 37W
Londa, *India* **95 G2** 15 30N 74 30 E
Londiani, *Kenya* **118 C4** 0 10 S 35 33 E
Londinières, *France* **26 C8** 49 50N 1 25 E
London, *Canada* **140 D3** 42 59N 81 15W
London, *U.K.* **21 F7** 51 30N 0 3W
London, *Ky., U.S.A.* **148 G3** 37 8N 84 5W
London, *Ohio, U.S.A.* **157 E13** 39 53N 83 27W
London, Greater □, *U.K.* **21 F7** 51 36N 0 5W
London Mills, *U.S.A.* **156 D6** 40 43N 90 11W
Londonderry, *U.K.* **23 B4** 55 0N 7 20W
Londonderry □, *U.K.* **23 B4** 55 0N 7 20W
Londonderry, C., *Australia* **124 B4** 13 45 S 126 55 E
Londonderry, I., *Chile* **176 E2** 55 0 S 71 0W
Londrina, *Brazil* **175 A5** 23 18 S 51 10W
Londuimbale, *Angola* **115 E3** 12 15 S 15 19 E
Lone Pine, *U.S.A.* **160 J8** 36 36N 118 4W
Long Akah, *Malaysia* **85 B4** 3 19N 114 47 E
Long Beach, *Calif., U.S.A.* **161 M8** 33 47N 118 11W
Long Beach, *N.Y., U.S.A.* **151 F11** 40 35N 73 39W
Long Beach, *Wash., U.S.A.* **160 D2** 46 21N 124 3W
Long Branch, *U.S.A.* **151 F11** 40 18N 74 0W
Long Creek, *U.S.A.* **158 D4** 44 43N 119 6W
Long Eaton, *U.K.* **20 E6** 52 53N 1 15W
Long I., *Australia* **126 C4** 22 8 S 149 53 E
Long I., *Bahamas* **165 B4** 23 20N 75 10W
Long I., *Ireland* **23 E2** 51 30N 9 34W
Long I., *Papua N. G.* **132 C4** 5 20 S 147 5 E
Long I., *U.S.A.* **151 F11** 40 45N 73 30W
Long Island Sd., *U.S.A.* **151 E12** 41 10N 73 0W
Long L., *Canada* **140 C2** 49 30N 86 50W
Long Lake, *U.S.A.* **151 C10** 43 58N 74 25W
Long Pine, *U.S.A.* **154 D5** 42 32N 99 42W
Long Point B., *Canada* **150 D4** 42 40N 80 10W
Long Pt., *Nfld., Canada* **141 C8** 48 47N 58 46W
Long Pt., *Ont., Canada* **150 D4** 42 35N 80 2W
Long Pt., *N.Z.* **131 G4** 46 34 S 169 36 E
Long Range Mts., *Canada* **141 C8** 49 30N 57 30W
Long Reef, *Australia* **124 B4** 14 1 S 125 48 E
Long Str. = Longa, Proliv, *Russia* **6 C16** 70 0N 175 0 E
Long Thanh, *Vietnam* **87 G6** 10 47N 106 57 E
Long Xian, *China* **74 G4** 34 55N 106 55 E
Long Xuyen, *Vietnam* **87 G5** 10 19N 105 28 E
Longa, *Angola* **115 E3** 14 42 S 18 32 E
Longá, *Greece* **48 E3** 36 53N 21 55 E
Longa, Proliv, *Russia* **6 C16** 70 0N 175 0 E
Long'an, *China* **76 F6** 23 10N 107 40 E
Longarone, *Italy* **45 B9** 46 16N 12 18 E
Longbenton, *U.K.* **20 B6** 55 1N 1 31W
Longboat Key, *U.S.A.* **153 H7** 27 23N 82 39W
Longburn, *N.Z.* **130 G4** 40 23 S 175 35 E
Longchang, *China* **76 C5** 29 18N 105 15 E
Longchi, *China* **76 C4** 29 25N 103 24 E
Longchuan, *Guangdong, China* **77 E10** 24 5N 115 17 E
Longchuan, *Yunnan, China* **76 E1** 24 23N 97 58 E
Longde, *China* **74 G4** 35 30N 106 20 E
Longeau, *France* **27 E12** 47 47N 5 20 E
Longford, *Australia* **126 G4** 41 32 S 147 3 E
Longford, *Ireland* **23 C4** 53 43N 7 49W
Longford □, *Ireland* **23 C4** 53 42N 7 45W
Longguan, *China* **74 D8** 40 45N 115 30 E
Longhua, *China* **75 D9** 41 18N 117 45 E
Longhui, *China* **77 D8** 27 7N 111 2 E
Longido, *Tanzania* **118 C4** 2 43 S 36 42 E
Longiram, *Indonesia* **85 C5** 0 5 S 115 45 E
Longkou, *Jiangxi, China* **77 D10** 26 8N 115 10 E
Longkou, *Shandong, China* **75 F11** 37 40N 120 18 E
Longlac, *Canada* **140 C2** 49 45N 86 25W
Longli, *China* **76 D6** 26 25N 106 58 E
Longlin, *China* **76 E5** 24 47N 105 20 E
Longling, *China* **76 E2** 24 37N 98 39 E
Longmen, *China* **77 F10** 23 40N 114 18 E
Longming, *China* **76 F6** 22 59N 107 7 E
Longmont, *U.S.A.* **154 E2** 40 10N 105 6W
Longnan, *China* **77 E10** 24 55N 114 47 E
Longnawan, *Indonesia* **85 B4** 1 51N 114 55 E
Longobucco, *Italy* **47 C9** 39 27N 16 37 E
Longquan, *China* **77 C12** 28 7N 119 10 E
Longreach, *Australia* **126 C3** 23 28 S 144 14 E
Longshan, *China* **76 C7** 29 29N 109 25 E
Longsheng, *China* **77 E8** 25 48N 110 0 E
Longton, *Australia* **126 C4** 20 58 S 145 55 E
Longué-Jumelles, *France* **26 E6** 47 22N 0 8W
Longueau, *France* **27 C9** 49 52N 2 21 E
Longueuil, *Canada* **151 A11** 45 32N 73 28W
Longuyon, *France* **27 C12** 49 27N 5 35 E
Longvic, *France* **27 E12** 47 17N 5 4 E
Longview, *Canada* **142 C6** 50 32N 114 10W
Longview, *Tex., U.S.A.* **155 J7** 32 30N 94 44W
Longview, *Wash., U.S.A.* **160 D4** 46 8N 122 57W
Longwood, *U.S.A.* **153 G8** 28 42N 81 21W
Longwy, *France* **27 C12** 49 30N 5 46 E
Longxi, *China* **74 G3** 34 53N 104 40 E
Longyou, *China* **77 C12** 29 1N 119 8 E
Longzhou, *China* **76 F6** 22 22N 106 50 E
Lonigo, *Italy* **45 C8** 45 23N 11 23 E
Löningen, *Germany* **30 C3** 52 44N 7 46 E
Lonja →, *Croatia* **45 C13** 45 22N 16 40 E
Lonkin, *Burma* **90 C6** 25 39N 96 22 E
Lonoke, *U.S.A.* **155 H9** 34 47N 91 54W
Lonquimay, *Chile* **176 A2** 38 26 S 71 14W
Lons-le-Saunier, *France* **27 F12** 46 40N 5 31 E
Lönsboda, *Sweden* **17 H8** 56 24N 14 20 E
Lønset, *Norway* **18 B4** 62 46N 7 24 E
Looe, *U.K.* **21 G3** 50 22N 4 28W
Loogootee, *U.S.A.* **157 F10** 38 41N 86 55W
Lookout, C., *Canada* **140 A3** 55 18N 83 56W
Lookout, C., *U.S.A.* **149 H7** 34 35N 76 32W
Loolmalasin, *Tanzania* **118 C4** 3 0 S 35 53 E
Loon →, *Alta., Canada* **142 B5** 57 8N 115 3W
Loon →, *Man., Canada* **143 B8** 55 53N 101 59W
Loon Lake, *Canada* **143 C7** 54 2N 109 10W
Loongana, *Australia* **125 F4** 30 52 S 127 5 E
Loop Hd., *Ireland* **23 D2** 52 34N 9 56W
Lop Buri, *Thailand* **86 E3** 14 48N 100 37 E
Lop Nor = Lop Nur, *China* **72 B4** 40 20N 90 10 E
Lop Nur, *China* **72 B4** 40 20N 90 10 E
Lopare, *Bos.-H.* **52 F3** 44 39N 18 46 E
Lopatin, *Russia* **61 J8** 43 50N 47 35 E
Lopatina, Gora, *Russia* **65 D15** 50 47N 143 10 E
Lopaye, *Sudan* **107 F3** 6 37N 33 40 E
Lopevi, *Vanuatu* **133 F6** 16 30 S 168 21 E
Lopez, *Phil.* **80 E4** 13 53N 122 15 E
Lopez, C., *Gabon* **114 C1** 0 47 S 8 40 E
Lopez I., *Gabon* **114 C1** 0 50 S 8 47 E
Lopphavet, *Norway* **14 A19** 70 27N 21 15 E
Lora, *Norway* **18 B5** 62 8N 8 45 E
Lora →, *Afghan.* **91 C2** 31 35N 65 50 E
Lora →, *Norway* **18 B5** 62 8N 8 42 E
Lora, Hamun-i-, *Pakistan* **91 C2** 29 38N 64 58 E
Lora Cr. →, *Australia* **127 D2** 28 10 S 135 22 E
Lora del Río, *Spain* **43 H5** 37 39N 5 33W
Lorain, *U.S.A.* **150 E2** 41 28N 82 11W
Loraine, *U.S.A.* **156 D5** 40 9N 91 13W
Loralai, *Pakistan* **91 C3** 30 20N 68 41 E
Lorca, *Spain* **41 H3** 37 41N 1 42W
Lord Howe I., *Pac. Oc.* **134 L7** 31 33 S 159 6 E
Lord Howe Ridge, *Pac. Oc.* **134 L8** 30 0 S 162 30 E
Lordsburg, *U.S.A.* **159 K9** 32 21N 108 43W
Lorengau, *Papua N. G.* **132 B4** 2 1 S 147 15 E
Loreto, *Bolivia* **173 D5** 15 13 S 64 40W

Lusaka 243

Loreto, *Brazil* **170 C2** 7 5 S 45 10W
Loreto, *Italy* **45 E10** 43 26N 13 36 E
Loreto, *Mexico* **162 B2** 26 1N 111 21W
Loreto □, *Peru* **168 E3** 5 0 S 75 0W
Lorgues, *France* **29 E10** 43 28N 6 22 E
Lorica, *Colombia* **168 B2** 9 14N 75 49W
Lorient, *France* **26 E3** 47 45N 3 23W
Lorimor, *U.S.A.* **156 C2** 41 8N 94 3W
Lőrinci, *Hungary* **52 C4** 47 44N 19 41 E
Lorn, *U.K.* **22 E3** 56 26N 5 10W
Lorn, Firth of, *U.K.* **22 E3** 56 20N 5 40W
Lorne, *Australia* **128 E5** 38 33 S 143 59 E
Lorovouno, *Cyprus* **38 D11** 35 8N 32 36 E
Lörrach, *Germany* **31 H3** 47 36N 7 40 E
Lorraine □, *France* **27 D13** 48 53N 6 0 E
Lorrainville, *Canada* **140 C4** 47 21N 79 23W
Los, *Sweden* **16 C9** 61 45N 15 10 E
Los, Îles de, *Guinea* **112 D2** 9 30N 13 50W
Los Alamos, *Calif., U.S.A.* **161 L6** 34 44N 120 17W
Los Alamos, *N. Mex., U.S.A.* **159 J10** 35 53N 106 19W
Los Altos, *U.S.A.* **160 H4** 37 23N 122 7W
Los Andes, *Chile* **174 C1** 32 50 S 70 40W
Los Angeles, *Chile* **174 D1** 37 28 S 72 23W
Los Angeles, *U.S.A.* **161 M8** 34 4N 118 15W
Los Angeles Aqueduct, *U.S.A.* **161 K9** 35 22N 118 5W
Los Antiguos, *Argentina* **176 C2** 46 35 S 71 40W
Los Banos, *U.S.A.* **160 H6** 37 4N 120 51W
Los Barrios, *Spain* **43 J5** 36 11N 5 30W
Los Blancos, *Argentina* **174 A3** 23 40 S 62 30W
Los Corrales de Buelna, *Spain* **42 B6** 43 16N 4 4W
Los Cristianos, *Canary Is.* **39 F3** 28 3N 16 42W
Los Gallardos, *Spain* **41 H3** 37 10N 1 57W
Los Gatos, *U.S.A.* **160 H5** 37 14N 121 59W
Los Hermanos, *Venezuela* **165 D7** 11 45N 64 25W
Los Islotes, *Canary Is.* **39 E6** 29 4N 13 44W
Los Lagos, *Chile* **176 A2** 39 51 S 72 50W
Los Llanos de Aridane, *Canary Is.* **39 F2** 28 38N 17 54W
Los Lomas, *Peru* **172 A1** 4 40 S 80 10W
Los Lunas, *U.S.A.* **159 J10** 34 48N 106 44W
Los Menucos, *Argentina* **176 B3** 40 50 S 68 10W
Los Mochis, *Mexico* **162 B3** 25 45N 108 57W
Los Monegros, *Spain* **40 D4** 41 29N 0 13W
Los Monos, *Argentina* **176 C3** 46 1 S 69 36W
Los Nietos, *Spain* **41 H4** 37 39N 0 47W
Los Olivos, *U.S.A.* **161 L6** 34 40N 120 7W
Los Palacios, *Cuba* **164 B3** 22 35N 83 15W
Los Palacios y Villafranca, *Spain* **43 H5** 37 10N 5 55W
Los Reyes, *Mexico* **162 D4** 19 34N 102 30W
Los Ríos □, *Ecuador* **168 D2** 1 30 S 79 25W
Los Roques, *Venezuela* **168 A4** 11 50N 66 45W
Los Santos de Maimona, *Spain* **43 G4** 38 27N 6 22W
Los Teques, *Venezuela* **168 A4** 10 21N 67 2W
Los Testigos, *Venezuela* **169 A5** 11 23N 63 6W
Los Vilos, *Chile* **174 C1** 32 10 S 71 30W
Los Yébenes, *Spain* **43 F7** 39 36N 3 55W
Losada →, *Colombia* **168 C3** 2 12N 73 55W
Łosice, *Poland* **55 F9** 52 13N 22 43 E
Løsning, *Denmark* **17 J3** 55 48N 9 42 E
Lossiemouth, *U.K.* **22 D5** 57 42N 3 17W
Lostwithiel, *U.K.* **21 G3** 50 24N 4 41W
Losuia, *Papua N. G.* **132 E6** 8 30 S 151 4 E
Lot □, *France* **28 D5** 44 39N 1 40 E
Lot →, *France* **28 D4** 44 18N 0 20 E
Lot-et-Garonne □, *France* **28 D4** 44 22N 0 30 E
Lota, *Chile* **174 D1** 37 5 S 73 10W
Løten, *Norway* **18 D8** 60 51N 11 21 E
Loṭfābād, *Iran* **97 B8** 37 32N 59 20 E
Lothair, *S. Africa* **117 D5** 26 22 S 30 27 E
Lotofaga, *W. Samoa* **133 X24** 14 1 S 171 30W
Lotorp, *Sweden* **17 F9** 58 44N 15 50 E
Lötschbergtunnel, *Switz.* **32 D5** 46 26N 7 43 E
Löttorp, *Sweden* **17 G11** 57 10N 17 0 E
Lotung, *Taiwan* **77 E13** 24 41N 121 46 E
Lotzwil, *Switz.* **32 B5** 47 12N 7 48 E
Loubomo, *Congo* **114 C2** 4 9 S 12 47 E
Loudéac, *France* **26 D4** 48 11N 2 47W
Loudi, *China* **77 D8** 27 42N 111 59 E
Loudima, *Congo* **114 C2** 4 6 S 13 5 E
Loudon, *U.S.A.* **149 H3** 35 45N 84 20W
Loudonville, *U.S.A.* **152 F2** 40 38N 82 14W
Loudun, *France* **26 E7** 47 3N 0 5 E
Loue →, *France* **27 E12** 47 1N 5 28 E
Louga, *Senegal* **112 B1** 15 45N 16 5W
Loughborough, *U.K.* **20 E6** 52 47N 1 11W
Loughman, *U.S.A.* **153 G8** 28 14N 81 34W
Loughrea, *Ireland* **23 C3** 53 12N 8 33W
Loughros More B., *Ireland* **23 B3** 54 48N 8 32W
Louhans, *France* **27 F12** 46 38N 5 12 E
Louis Trichardt, *S. Africa* **117 C4** 23 1 S 29 43 E
Louis XIV, Pte., *Canada* **140 B4** 54 37N 79 45W
Louisa, *U.S.A.* **148 F4** 38 7N 82 36W
Louisbourg, *Canada* **141 C8** 45 55N 60 0W
Louisburg, *U.S.A.* **156 F2** 38 37N 94 41W
Louise I., *Canada* **142 C2** 52 55N 131 50W
Louiseville, *Canada* **140 C5** 46 20N 72 56W
Louisiade Arch., *Papua N. G.* **132 F7** 11 10 S 153 0 E
Louisiana, *U.S.A.* **156 E6** 39 27N 91 3W
Louisiana □, *U.S.A.* **155 K9** 30 50N 92 0W
Louisville, *Ala., U.S.A.* **152 D4** 31 47N 85 33W
Louisville, *Ga., U.S.A.* **152 B7** 33 0N 82 25W
Louisville, *Ky., U.S.A.* **157 F11** 38 15N 85 46W
Louisville, *Miss., U.S.A.* **155 J10** 33 7N 89 3W
Loukouo, *Congo* **114 C2** 3 38 S 14 39 E
Loulay, *France* **28 B3** 46 3N 0 30W
Loulé, *Portugal* **43 H3** 37 9N 8 0W
Louny, *Czech Rep.* **34 A6** 50 20N 13 48 E
Loup City, *U.S.A.* **154 E5** 41 17N 98 58W
Lourdes, *France* **28 E3** 43 6N 0 3W
Lourdes-du-Blanc-Sablon, *Canada* **141 B8** 51 24N 57 12W
Lourenço, *Brazil* **169 C7** 2 30N 51 40W
Lourenço-Marques = Maputo, *Mozam.* **117 D5** 25 58 S 32 32 E
Lourinhã, *Portugal* **43 F1** 39 14N 9 17W
Lousã, *Portugal* **42 E2** 40 7N 8 14W
Louth, *Australia* **129 A6** 30 30 S 145 8 E
Louth, *Ireland* **23 C5** 53 58N 6 32W
Louth, *U.K.* **20 D7** 53 22N 0 1W
Louth □, *Ireland* **23 C5** 53 56N 6 34W
Loutrá Aidhipsoú, *Greece* **48 C5** 38 54N 23 2 E
Loutráki, *Greece* **48 D4** 37 58N 22 57 E
Louvain = Leuven, *Belgium* **24 D4** 50 52N 4 42 E
Louvale, *U.S.A.* **152 C5** 32 10N 84 50W
Louviers, *France* **26 C8** 49 12N 1 10 E
Louwsburg, *S. Africa* **117 D5** 27 37 S 31 7 E
Lovat →, *Russia* **58 C6** 58 14N 31 28 E
Lovćen, *Montenegro, Yug.* **50 D2** 42 23N 18 51 E
Love, *Canada* **143 C8** 53 29N 104 10W
Lovech, *Bulgaria* **51 C8** 43 8N 24 42 E
Lovech □, *Bulgaria* **51 C8** 43 15N 24 45 E
Loveland, *Colo., U.S.A.* **154 E2** 40 24N 105 5W
Loveland, *Ohio, U.S.A.* **157 E12** 39 16N 84 16W
Lovell, *U.S.A.* **158 D9** 44 50N 108 24W
Lovelock, *U.S.A.* **158 F4** 40 11N 118 28W
Lóvere, *Italy* **44 C7** 45 49N 10 4 E
Loves Park, *U.S.A.* **156 B7** 42 19N 89 3W
Lövestad, *Sweden* **17 J7** 55 40N 13 54 E
Lovett, *U.S.A.* **152 C7** 32 38N 82 46W
Loviisa, *Finland* **15 F22** 60 28N 26 12 E
Lovilia, *U.S.A.* **156 C4** 41 8N 92 55W
Loving, *U.S.A.* **155 J2** 32 17N 104 6W
Lovington, *Ill., U.S.A.* **157 E8** 39 43N 88 38W
Lovington, *N. Mex., U.S.A.* **155 J3** 32 57N 103 21W
Lovisa = Loviisa, *Finland* **15 F22** 60 28N 26 12 E
Lovosice, *Czech Rep.* **34 A7** 50 30N 14 2 E
Lovran, *Croatia* **45 C11** 45 18N 14 15 E
Lovrin, *Romania* **52 E5** 45 58N 20 48 E
Lövstabruk, *Sweden* **16 D11** 60 25N 17 53 E
Lövstabukten, *Sweden* **16 D11** 60 35N 17 45 E
Low Pt., *Australia* **125 F4** 32 25 S 127 25 E
Low Tatra = Nízké Tatry, *Slovak Rep.* **35 C12** 48 55N 19 30 E
Lowa, *Dem. Rep. of the Congo* **118 C2** 1 25 S 25 47 E
Lowa →, *Dem. Rep. of the Congo* **118 C2** 1 24 S 25 51 E
Lowden, *U.S.A.* **156 C6** 41 52N 90 56W
Lowell, *Fla., U.S.A.* **153 F7** 29 20N 82 12W
Lowell, *Ind., U.S.A.* **157 C9** 41 18N 87 25W
Lowell, *Mass., U.S.A.* **151 D13** 42 38N 71 19W
Lower Arrow L., *Canada* **142 D5** 49 40N 118 5W
Lower Austria = Niederösterreich □, *Austria* **34 C8** 48 25N 15 40 E
Lower California = Baja California, *Mexico* **162 A1** 31 10N 115 12W
Lower Hutt, *N.Z.* **130 H3** 41 10 S 174 55 E
Lower Kalskag, *U.S.A.* **144 F7** 61 31N 160 22W
Lower L., *U.S.A.* **158 F3** 41 16N 120 2W
Lower Lake, *U.S.A.* **160 G4** 38 55N 122 37W
Lower Paia, *U.S.A.* **145 C5** 20 55N 156 23W
Lower Post, *Canada* **142 B3** 59 58N 128 30W
Lower Red L., *U.S.A.* **154 B7** 47 58N 95 0W
Lower Saxony = Niedersachsen □, *Germany* **30 C4** 52 50N 9 0 E
Lower Tunguska = Tunguska, Nizhnyaya →, *Russia* **65 C9** 65 48N 88 4 E
Lowestoft, *U.K.* **21 E9** 52 29N 1 45 E
Łowicz, *Poland* **55 F6** 52 6N 19 55 E
Lowry City, *U.S.A.* **156 F3** 38 8N 93 44W
Lowville, *U.S.A.* **151 C9** 43 47N 75 29W
Loxton, *Australia* **128 C4** 34 28 S 140 31 E
Loxton, *S. Africa* **116 E3** 31 30 S 22 22 E
Loyalton, *U.S.A.* **160 F6** 39 41N 120 14W
Loyalty Is. = Loyauté, Is., *N. Cal.* **133 K4** 20 50 S 166 30 E
Loyang = Luoyang, *China* **74 G7** 34 40N 112 26 E
Loyauté, Is., *N. Cal.* **133 K4** 20 50 S 166 30 E
Loyev = Loyew, *Belarus* **59 G6** 51 56N 30 46 E
Loyew, *Belarus* **59 G6** 51 56N 30 46 E
Loyoro, *Uganda* **118 B3** 3 22N 34 14 E
Lož, *Slovenia* **45 C11** 45 43N 14 30 E
Lozère □, *France* **28 D7** 44 35N 3 30 E
Loznica, *Serbia, Yug.* **50 B3** 44 32N 19 12 E
Lozova, *Ukraine* **59 H9** 49 0N 36 20 E
Lozva →, *Russia* **62 B9** 59 36N 62 20 E
Lua Makiki, *U.S.A.* **145 C5** 20 33N 156 37W
Luachimo, *Angola* **115 D4** 7 23 S 20 48 E
Luacono, *Angola* **115 E4** 11 15 S 21 37 E
Lualaba →, *Dem. Rep. of the Congo* **114 B5** 0 26N 25 20 E
Luampa, *Zambia* **119 F1** 15 4 S 24 20 E
Lu'an, *China* **77 B11** 31 45N 116 29 E
Luan Chau, *Vietnam* **86 B4** 21 38N 103 24 E
Luan He →, *China* **75 E10** 39 20N 119 5 E
Luan Xian, *China* **75 E10** 39 40N 118 40 E
Luancheng, *Guangxi Zhuangzu, China* **76 F7** 22 48N 108 55 E
Luancheng, *Hebei, China* **74 F8** 37 53N 114 40 E
Luanco, *Spain* **42 B5** 43 37N 5 48W
Luanda, *Angola* **115 D2** 8 50 S 13 15 E
Luang, Thale, *Thailand* **87 J3** 7 30N 100 15 E
Luang Prabang, *Laos* **86 C4** 19 52N 102 10 E
Luangwa, *Zambia* **119 F3** 15 35 S 30 16 E
Luangwa →, *Zambia* **119 E3** 14 25 S 30 25 E
Luangwa Valley, *Zambia* **119 E3** 13 30 S 31 30 E
Luanne, *China* **75 D9** 40 55N 117 40 E
Luanping, *China* **75 D9** 40 53N 117 23 E
Luanshya, *Zambia* **119 E2** 13 3 S 28 28 E
Luapula □, *Zambia* **119 E2** 11 0 S 29 0 E
Luapula →, *Africa* **119 D2** 9 26 S 28 33 E
Luarca, *Spain* **42 B4** 43 32N 6 32W
Luashi, *Dem. Rep. of the Congo* **115 E4** 10 50 S 23 36 E
Luau, *Angola* **115 E4** 10 40 S 22 10 E
Luba, *Phil.* **80 C3** 17 19N 120 42 E
Lubaczów, *Poland* **55 H10** 50 10N 23 8 E
Lubalo, *Angola* **115 D3** 9 10 S 19 15 E
Luban, *Phil.* **81 H6** 6 26N 126 13 E
Lubań, *Poland* **55 G2** 51 5N 15 15 E
Lubana, Ozero = Lubānas Ezers, *Latvia* **15 H22** 56 45N 27 0 E
Lubānas Ezers, *Latvia* **15 H22** 56 45N 27 0 E
Lubang, *Phil.* **80 E3** 13 52N 120 7 E
Lubang Is., *Phil.* **80 E3** 13 50N 120 12 E
Lubango, *Angola* **115 E2** 14 55 S 13 30 E
Lubao, *Phil.* **80 D3** 14 56N 120 36 E
Lubartów, *Poland* **55 G9** 51 28N 22 42 E
Lubawa, *Poland* **54 E6** 53 30N 19 48 E
Lübbecke, *Germany* **30 C4** 52 18N 8 37 E
Lübben, *Germany* **30 D9** 51 56N 13 54 E
Lübbenau, *Germany* **30 D9** 51 52N 13 57 E
Lubbock, *U.S.A.* **155 J4** 33 35N 101 51W
Lübeck, *Germany* **30 B6** 53 52N 10 40 E
Lübecker Bucht, *Germany* **30 A6** 54 3N 10 54 E
Lubefu, *Dem. Rep. of the Congo* **115 C4** 4 47 S 24 27 E
Lubefu →, *Dem. Rep. of the Congo* **115 C4** 4 10 S 23 0 E
Lubero = Luofu, *Dem. Rep. of the Congo* **118 C2** 0 10 S 29 15 E
Lubersac, *France* **28 C5** 45 26N 1 23 E
Lubicon L., *Canada* **142 B5** 56 23N 115 56W
Lubień Kujawski, *Poland* **55 F6** 52 23N 19 9 E
Lubin, *Poland* **55 G3** 51 24N 16 11 E
Lublin, *Poland* **55 G9** 51 12N 22 38 E
Lublin □, *Poland* **55 G9** 51 25N 22 30 E
Lubliniec, *Poland* **55 H5** 50 43N 18 45 E
Lubnān, J., *Lebanon* **103 B4** 33 50N 35 45 E
Lubniewice, *Poland* **55 F2** 52 31N 15 15 E
Lubny, *Ukraine* **59 G7** 50 3N 32 58 E
Lubomierz, *Poland* **55 G2** 51 1N 15 31 E
Luboń, *Poland* **55 F3** 52 21N 16 51 E
Lubongola, *Dem. Rep. of the Congo* **118 C2** 2 35 S 27 50 E
L'ubotín, *Slovak Rep.* **35 B13** 49 17N 20 53 E
Lubraniec, *Poland* **55 F5** 52 33N 18 50 E
Lubsko, *Poland* **55 G1** 51 45N 14 57 E
Lübtheen, *Germany* **30 B7** 53 18N 11 5 E
Lubuagan, *Phil.* **80 C3** 17 21N 121 10 E
Lubudi →, *Dem. Rep. of the Congo* **115 D5** 9 0 S 25 35 E
Lubuklinggau, *Indonesia* **84 C2** 3 15 S 102 55 E
Lubuksikaping, *Indonesia* **84 B2** 0 10N 100 15 E
Lubumbashi, *Dem. Rep. of the Congo* **119 E2** 11 40 S 27 28 E
Lubunda, *Dem. Rep. of the Congo* **115 D5** 5 12 S 26 41 E
Lubungu, *Zambia* **119 E2** 14 35 S 26 24 E
Lubutu, *Dem. Rep. of the Congo* **118 C2** 0 45 S 26 30 E
Luc An Chau, *Vietnam* **86 A5** 22 6N 104 43 E
Luc-en-Diois, *France* **29 D9** 44 36N 5 28 E
Lucala, *Angola* **115 D3** 9 7 S 15 58 E
Lucan, *Canada* **150 C3** 43 11N 81 24W
Lucapa, *Angola* **115 D4** 8 25 S 20 45 E
Lucban, *Phil.* **80 D3** 14 6N 121 33 E
Lucca, *Italy* **44 E7** 43 50N 10 29 E
Lucé, *France* **26 D8** 48 26N 1 27 E
Luce Bay, *U.K.* **22 G4** 54 45N 4 48W
Lucea, *Jamaica* **164 C4** 18 25N 78 10W
Lucedale, *U.S.A.* **149 K1** 30 56N 88 35W
Lucena, *Phil.* **80 E3** 13 56N 121 37 E
Lucena, *Spain* **43 H6** 37 27N 4 31W
Lučenec, *Slovak Rep.* **35 C12** 48 18N 19 42 E
Lucens, *Switz.* **32 C3** 46 43N 6 51 E
Lucera, *Italy* **47 A8** 41 30N 15 20 E
Lucerne = Luzern, *Switz.* **33 B6** 47 3N 8 18 E
Lucerne, *U.S.A.* **160 F4** 39 6N 122 48W
Lucerne Valley, *U.S.A.* **161 L10** 34 27N 116 57W
Lucero, *Mexico* **162 A3** 30 49N 106 30W
Luchena →, *Spain* **41 H3** 37 44N 1 50W
Lucheng, *China* **74 F7** 36 20N 113 11 E
Lucheringo →, *Mozam.* **119 E4** 11 43 S 36 17 E
Luchiang, *Taiwan* **77 E13** 24 1N 120 22 E
Lüchow, *Germany* **30 C7** 52 58N 11 8 E
Luchuan, *China* **77 F8** 22 21N 110 12 E
Lucie →, *Surinam* **169 C6** 3 35N 57 38W
Lucira, *Angola* **115 E2** 14 0 S 12 35 E
Luckau, *Germany* **30 D9** 51 50N 13 42 E
Luckenwalde, *Germany* **30 C9** 52 5N 13 10 E
Luckey, *U.S.A.* **157 C13** 41 27N 83 29W
Lucknow, *India* **93 F9** 26 50N 81 0 E
Luçon, *France* **28 B2** 46 28N 1 10W
Lucusse, *Angola* **115 E4** 12 32 S 20 48 E
Lüda = Dalian, *China* **75 E11** 38 50N 121 40 E
Luda Kamchiya →, *Bulgaria* **51 C11** 43 3N 27 29 E
Ludbreg, *Croatia* **45 B13** 46 15N 16 38 E
Lüdenscheid, *Germany* **30 D3** 51 13N 7 37 E
Lüderitz, *Namibia* **116 D2** 26 41 S 15 8 E
Ludewe □, *Tanzania* **119 E3** 10 0 S 34 50 E
Ludhiana, *India* **92 D6** 30 57N 75 56 E
Ludian, *China* **76 D4** 27 10N 103 33 E
Luding Qiao, *China* **76 C4** 29 53N 102 12 E
Lüdinghausen, *Germany* **30 D3** 51 46N 7 27 E
Ludington, *U.S.A.* **148 D2** 43 57N 86 27W
Ludlow, *U.K.* **21 E5** 52 22N 2 42W
Ludlow, *Calif., U.S.A.* **161 L10** 34 43N 116 10W
Ludlow, *Vt., U.S.A.* **151 C12** 43 24N 72 42W
Ludowici, *U.S.A.* **152 D8** 31 43N 81 45W
Ludus, *Romania* **53 D9** 46 29N 24 5 E
Ludvika, *Sweden* **16 D9** 60 8N 15 14 E
Ludwigsburg, *Germany* **31 G5** 48 53N 9 11 E
Ludwigsfelde, *Germany* **30 C9** 52 17N 13 17 E
Ludwigshafen, *Germany* **31 F4** 49 29N 8 26 E
Ludwigslust, *Germany* **30 B7** 53 19N 11 30 E
Ludza, *Latvia* **58 D4** 56 32N 27 43 E
Lue, *Australia* **129 B8** 32 38 S 149 50 E
Luebo, *Dem. Rep. of the Congo* **115 D4** 5 21 S 21 23 E
Lueki, *Dem. Rep. of the Congo* **114 C5** 3 20 S 25 48 E
Luena, *Angola* **115 E3** 12 13 S 19 51 E
Luena, *Dem. Rep. of the Congo* **119 D2** 9 28 S 25 43 E
Luena, *Zambia* **119 E3** 10 40 S 30 25 E
Luepa, *Venezuela* **169 B5** 5 43N 61 31W
Lüeyang, *China* **74 H4** 33 22N 106 10 E
Lufeng, *Guangdong, China* **77 F10** 22 57N 115 38 E
Lufeng, *Yunnan, China* **76 E4** 25 0N 102 5 E
Lufico, *Angola* **115 D2** 6 24 S 13 23 E
Lufira →, *Dem. Rep. of the Congo* **119 D2** 9 30 S 27 0 E
Lufkin, *U.S.A.* **155 K7** 31 21N 94 44W
Lufupa, *Dem. Rep. of the Congo* **115 E4** 10 37 S 24 56 E
Luga, *Russia* **58 C5** 58 40N 29 55 E
Luga →, *Russia* **58 C5** 59 40N 28 18 E
Lugano, *Switz.* **33 E7** 46 1N 8 57 E
Lugano, L. di, *Switz.* **33 E8** 46 0N 9 0 E
Lugansk = Luhansk, *Ukraine* **59 H10** 48 38N 39 15 E
Lugard's Falls, *Kenya* **118 C4** 3 6 S 38 41 E
Lugela, *Mozam.* **119 F4** 16 25 S 36 43 E
Lugenda →, *Mozam.* **119 E4** 11 25 S 38 33 E
Lugh Ganana, *Somali Rep.* **120 D2** 3 48N 42 34 E
Lugnaquilla, *Ireland* **23 D5** 52 58N 6 28W
Lugo, *Italy* **45 D8** 44 25N 11 54 E
Lugo, *Spain* **42 B3** 43 2N 7 35W
Lugo □, *Spain* **42 C3** 43 0N 7 30W
Lugoj, *Romania* **52 E6** 45 42N 21 57 E
Lugovoy, *Kazakstan* **63 B6** 42 55N 72 43 E
Luhansk, *Ukraine* **59 H10** 48 38N 39 15 E
Luhe, *China* **77 A12** 32 19N 118 50 E
Luhe →, *Germany* **30 B6** 53 23N 10 13 E
Luhuo, *China* **76 B3** 31 21N 100 48 E
Luiana, *Angola* **115 F4** 17 25 S 22 59 E
Luimneach = Limerick, *Ireland* **23 D3** 52 40N 8 37W
Luing, *U.K.* **22 E3** 56 14N 5 39W
Luino, *Italy* **44 C5** 45 59N 8 44 E
Luís Correia, *Brazil* **170 B3** 3 0 S 41 35W
Luís Gonçalves, *Brazil* **170 C1** 5 37 S 50 25W
Luitpold Coast, *Antarctica* **7 D1** 78 30 S 32 0W
Luiza, *Dem. Rep. of the Congo* **115 D4** 7 40 S 22 30 E
Luizi, *Dem. Rep. of the Congo* **118 D2** 6 0 S 27 25 E
Luján, *Argentina* **174 C4** 34 45 S 59 5W
Lujiang, *China* **77 B11** 31 20N 117 15 E
Lukala, *Dem. Rep. of the Congo* **115 D2** 5 31 S 14 32 E
Lukanga Swamp, *Zambia* **119 E2** 14 30 S 27 40 E
Lukavac, *Bos.-H.* **52 F3** 44 33N 18 32 E
Lukenie →, *Dem. Rep. of the Congo* **114 C3** 3 0 S 18 50 E
Lukhisaral, *India* **93 G12** 25 11N 86 5 E
Lŭki, *Bulgaria* **51 E8** 41 50N 24 43 E
Lukk, *Libya* **108 B4** 32 1N 24 46 E
Lukolela, *Equateur, Dem. Rep. of the Congo* **114 C3** 1 10 S 17 12 E
Lukolela, *Kasai Or., Dem. Rep. of the Congo* **115 D4** 5 23 S 24 32 E
Lukosi, *Zimbabwe* **119 F2** 18 30 S 26 30 E
Lukovë, *Albania* **50 G3** 39 59N 19 54 E
Lukovit, *Bulgaria* **51 C8** 43 13N 24 11 E
Łuków, *Poland* **55 G9** 51 55N 22 23 E
Lukoyanov, *Russia* **60 C7** 55 2N 44 29 E
Luksefjell, *Norway* **18 E6** 59 23N 9 34 E
Lule älv →, *Sweden* **14 D19** 65 35N 22 10 E
Luleå, *Sweden* **14 D20** 65 35N 22 10 E
Lüleburgaz, *Turkey* **51 E11** 41 23N 27 22 E
Luliang, *China* **76 E4** 25 0N 103 40 E
Luling, *U.S.A.* **155 L6** 29 41N 97 39W
Lulong, *China* **75 E10** 39 53N 118 51 E
Lulonga →, *Dem. Rep. of the Congo* **114 B3** 1 0N 18 10 E
Lulu, *U.S.A.* **152 E7** 30 7N 82 29W
Lulua →, *Dem. Rep. of the Congo* **115 C4** 4 30 S 20 30 E
Luluabourg = Kananga, *Dem. Rep. of the Congo* **115 D4** 5 55 S 22 18 E
Lumai, *Angola* **115 E4** 13 13 S 21 25 E
Lumajang, *Indonesia* **85 D4** 8 8 S 113 13 E
Lumaku, Gunong, *Malaysia* **85 B5** 4 52N 115 38 E
Lumbala Kaquengue, *Angola* **115 E4** 12 39 S 22 34 E
Lumbala N'guimbo, *Angola* **115 E4** 14 18 S 21 18 E
Lumber City, *U.S.A.* **152 D7** 31 56N 82 41W
Lumberton, *Miss., U.S.A.* **155 K10** 31 0N 89 27W
Lumberton, *N.C., U.S.A.* **149 H6** 34 37N 79 0W
Lumberton, *N. Mex., U.S.A.* **159 H10** 36 56N 106 56W
Lumbwa, *Kenya* **118 C4** 0 12 S 35 28 E
Lumding, *India* **90 C4** 25 46N 93 10 E
Lumi, *Papua N. G.* **132 B2** 3 30 S 142 2 E
Lumpkin, *U.S.A.* **152 C5** 32 3N 84 48W
Lumsden, *N.Z.* **131 F3** 45 44 S 168 27 E
Lumut, *Malaysia* **87 K3** 4 13N 100 37 E
Lumut, Tanjung, *Indonesia* **84 C3** 3 50 S 105 58 E
Luna, *Luzon, Phil.* **80 B3** 18 18N 121 21 E
Luna, *Luzon, Phil.* **80 C3** 16 51N 120 23 E
Lunan, *China* **76 E4** 24 40N 103 18 E
Lunavada, *India* **92 H5** 23 8N 73 37 E
Lunca, *Romania* **53 C10** 47 22N 25 1 E
Lunca Corbului, *Romania* **53 F9** 44 42N 24 45 E
Lund, *Sweden* **17 J7** 55 44N 13 12 E
Lund, *U.S.A.* **158 G6** 38 52N 115 0W
Lunda Norte □, *Angola* **115 D4** 8 0 S 20 0 E
Lunda Sul □, *Angola* **115 E4** 10 0 S 20 0 E
Lundamo, *Norway* **18 A7** 63 9N 10 19 E
Lundarbrekka, *Iceland* **11 B9** 65 27N 17 24W
Lundazi, *Zambia* **119 E3** 12 20 S 33 7 E
Lunderskov, *Denmark* **17 J3** 55 29N 9 19 E
Lundi →, *Zimbabwe* **119 G3** 21 43 S 32 34 E
Lundu, *Malaysia* **85 B3** 1 40N 109 50 E
Lundy, *U.K.* **21 F3** 51 10N 4 41W
Lune →, *U.K.* **20 C5** 54 0N 2 51W
Lüneburg, *Germany* **30 B6** 53 15N 10 24 E
Lüneburg Heath = Lüneburger Heide, *Germany* **30 B6** 53 10N 10 12 E
Lüneburger Heide, *Germany* **30 B6** 53 10N 10 12 E
Lunel, *France* **29 E8** 43 39N 4 9 E
Lünen, *Germany* **30 D3** 51 37N 7 32 E
Lunenburg, *Canada* **141 D7** 44 22N 64 18W
Lunéville, *France* **27 D13** 48 36N 6 30 E
Lunga →, *Zambia* **119 E2** 14 34 S 26 25 E
Lungern, *Switz.* **32 C6** 46 48N 8 10 E
Lungi Airport, *S. Leone* **112 D2** 8 40N 13 17W
Lunglei, *India* **90 D4** 22 55N 92 45 E
Lungngo, *Burma* **90 E4** 21 57N 93 36 E
Luni, *India* **92 G5** 26 0N 73 6 E
Luni →, *India* **92 G4** 24 41N 71 14 E
Luninets = Luninyets, *Belarus* **59 F4** 52 15N 26 50 E
Luning, *U.S.A.* **158 G4** 38 30N 118 11W
Lunino, *Russia* **60 D7** 53 35N 45 18 E
Luninyets, *Belarus* **59 F4** 52 15N 26 50 E
Lunner, *Norway* **18 D7** 60 19N 10 35 E
Lunsemfwa →, *Zambia* **119 E3** 14 54 S 30 12 E
Lunsemfwa Falls, *Zambia* **119 E2** 14 30 S 29 6 E
Luo He →, *China* **74 G6** 34 35N 110 20 E
Luocheng, *China* **76 E7** 24 48N 108 53 E
Luochuan, *China* **74 G5** 35 45N 109 26 E
Luoci, *China* **76 E4** 25 19N 102 18 E
Luodian, *China* **76 E6** 25 24N 106 43 E
Luoding, *China* **77 F8** 22 45N 111 40 E
Luofu, *Dem. Rep. of the Congo* **118 C2** 0 10 S 29 15 E
Luohe, *China* **74 H8** 33 32N 114 2 E
Luojiang, *China* **76 B5** 31 18N 104 33 E
Luonan, *China* **74 G6** 34 5N 110 10 E
Luoning, *China* **74 G6** 34 35N 111 40 E
Luoshan, *China* **77 A10** 32 13N 114 30 E
Luotian, *China* **77 B10** 30 46N 115 22 E
Luoyang, *China* **74 G7** 34 40N 112 26 E
Luoyuan, *China* **77 D12** 26 28N 119 30 E
Luozi, *Dem. Rep. of the Congo* **115 C2** 4 54 S 14 0 E
Luozigou, *China* **75 C16** 43 42N 130 18 E
Lupanshui, *China* **76 D5** 26 38N 104 48 E
Lupeni, *Romania* **53 E8** 45 21N 23 13 E
Lupilichi, *Mozam.* **119 E4** 11 47 S 35 13 E
Lupire, *Angola* **115 E3** 14 36 S 19 29 E
Łupków, *Poland* **55 J9** 49 15N 22 4 E
Lupoing, *China* **76 E5** 24 53N 104 21 E
Lupon, *Phil.* **81 H6** 6 54N 126 0 E
Luquan, *China* **76 E4** 25 35N 102 25 E
Luque, *Paraguay* **174 B4** 25 19 S 57 25W
Lūras, *Italy* **46 B2** 40 56N 9 10 E
Luray, *S.C., U.S.A.* **152 C8** 32 49N 81 14W
Luray, *Va., U.S.A.* **148 F6** 38 40N 78 28W
Lure, *France* **27 E13** 47 40N 6 30 E
Luremo, *Angola* **115 D3** 8 30 S 17 50 E
Lurgan, *U.K.* **23 B5** 54 28N 6 19W
Luribay, *Bolivia* **172 D4** 17 6 S 67 39W
Lurin, *Peru* **172 C2** 12 17 S 76 52W
Lusaka, *Zambia* **119 F2** 15 28 S 28 16 E

Maltepe 245

Mafeking = Mafikeng, *S. Africa* **116 D4** 25 50 S 25 38 E
Mafeking, *Canada* **143 C8** 52 40N 101 10W
Maféré, *Ivory C.* **112 D4** 5 30N 3 2W
Mafeteng, *Lesotho* **116 D4** 29 51 S 27 15 E
Maffra, *Australia* **129 D7** 37 53 S 146 58 E
Mafia I., *Tanzania* **118 D4** 7 45 S 39 50 E
Mafikeng, *S. Africa* **116 D4** 25 50 S 25 38 E
Mafra, *Brazil* **175 B6** 26 10 S 49 55W
Mafra, *Portugal* **43 G1** 38 55N 9 20W
Mafungabusi Plateau, *Zimbabwe* **119 F2** 18 30 S 29 8 E
Magadan, *Russia* **65 D16** 59 38N 150 50 E
Magadi, *Kenya* **118 C4** 1 54 S 36 19 E
Magadi, L., *Kenya* **118 C4** 1 54 S 36 19 E
Magaliesburg, *S. Africa* **117 D4** 26 0 S 27 32 E
Magallanes, *Phil.* **80 E4** 12 50N 123 50 E
Magallanes □, *Chile* **176 D2** 52 0 S 72 0W
Magallanes, Estrecho de, *Chile* **176 D2** 52 30 S 75 0W
Magaluf, *Spain* **41 F7** 39 29N 2 32 E
Magangué, *Colombia* **168 B3** 9 14N 74 45W
Maganoy, *Phil.* **81 H5** 6 51N 124 31 E
Magaria, *Niger* **109 F1** 13 4N 9 5 E
Magburaka, *S. Leone* **112 D2** 8 47N 12 0W
Magdalen Is. = Madeleine, Is. de la, *Canada* **141 C7** 47 30N 61 40W
Magdalena, *Argentina* **174 D4** 35 5 S 57 30W
Magdalena, *Bolivia* **173 C5** 13 13 S 63 57W
Magdalena, *Mexico* **162 A2** 30 50N 112 0W
Magdalena, *U.S.A.* **159 J10** 34 7N 107 15W
Magdalena □, *Colombia* **168 B3** 10 0N 74 0W
Magdalena →, *Colombia* **168 A3** 11 6N 74 51W
Magdalena →, *Mexico* **162 A2** 30 40N 112 25W
Magdalena, B., *Mexico* **162 C2** 24 30N 112 10W
Magdalena, I., *Chile* **176 B2** 44 40 S 73 0W
Magdalena, Llano de la, *Mexico* **162 C2** 25 0N 111 30W
Magdeburg, *Germany* **30 C7** 52 7N 11 38 E
Magdelaine Cays, *Australia* **126 B5** 16 33 S 150 18 E
Magdub, *Sudan* **107 E2** 13 42N 25 5 E
Magee, *U.S.A.* **155 K10** 31 52N 89 44W
Magelang, *Indonesia* **85 D4** 7 29 S 110 13 E
Magellan's Str. = Magallanes, Estrecho de, *Chile* **176 D2** 52 30 S 75 0W
Magenta, *Australia* **128 B5** 33 51 S 143 34 E
Magenta, *Italy* **44 C5** 45 28N 8 53 E
Magenta, L., *Australia* **125 F2** 33 30 S 119 2 E
Magerøya, *Norway* **14 A21** 71 3N 25 40 E
Maggea, *Australia* **128 C4** 34 28 S 140 2 E
Maggia, *Switz.* **33 D7** 46 15N 8 42 E
Maggia →, *Switz.* **33 D7** 46 18N 8 36 E
Maggiorasca, Mte., *Italy* **44 D6** 44 33N 9 29 E
Maggiore, L., *Italy* **44 C5** 45 57N 8 39 E
Maghama, *Mauritania* **112 B2** 15 32N 12 57W
Magherafelt, *U.K.* **23 B5** 54 45N 6 37W
Maghnia, *Algeria* **111 B4** 34 50N 1 43W
Magione, *Italy* **45 E9** 43 8N 12 12 E
Magistralnyy, *Russia* **65 D11** 56 16N 107 36 E
Maglaj, *Bos.-H.* **52 F3** 44 33N 18 7 E
Magliano in Toscana, *Italy* **45 F8** 42 36N 11 17 E
Máglie, *Italy* **47 B11** 40 7N 18 18 E
Magnac-Laval, *France* **28 B5** 46 13N 1 11 E
Magnetic Pole (North) = North Magnetic Pole, *Canada* **6 B2** 77 58N 102 8W
Magnetic Pole (South) = South Magnetic Pole, *Antarctica* **7 C9** 64 8 S 138 8 E
Magnisía □, *Greece* **48 B5** 39 15N 23 0 E
Magnitogorsk, *Russia* **62 E7** 53 27N 59 4 E
Magnolia, *Ark., U.S.A.* **155 J8** 33 16N 93 14W
Magnolia, *Miss., U.S.A.* **155 K9** 31 9N 90 28W
Magnor, *Norway* **18 E9** 59 56N 12 15 E
Magny-en-Vexin, *France* **27 C8** 49 9N 1 47 E
Magog, *Canada* **141 C5** 45 18N 72 9W
Magoro, *Uganda* **118 B3** 1 45N 34 12 E
Magosa = Famagusta, *Cyprus* **38 D12** 35 8N 33 55 E
Magouládhes, *Greece* **38 A3** 39 45N 19 42 E
Magoye, *Zambia* **119 F2** 16 1 S 27 30 E
Magpie L., *Canada* **141 B7** 51 0N 64 41W
Magrath, *Canada* **142 D6** 49 25N 112 50W
Magre →, *Spain* **41 F4** 39 11N 0 25W
Magrur, Wadi →, *Sudan* **107 D2** 16 5N 26 30 E
Magsingal, *Phil.* **80 C3** 17 41N 120 25 E
Magu □, *Tanzania* **118 C3** 2 31 S 33 28 E
Maguan, *China* **76 F5** 23 0N 104 21 E
Maguarinho, C., *Brazil* **170 B2** 0 15 S 48 30W
Maguindanao □, *Phil.* **81 H5** 7 5N 124 0 E
Maġusa = Famagusta, *Cyprus* **38 D12** 35 8N 33 55 E
Maguse L., *Canada* **143 A9** 61 40N 95 10W
Maguse Pt., *Canada* **143 A10** 61 20N 93 50W
Magwe, *Burma* **90 E5** 20 10N 95 0 E
Maha Sarakham, *Thailand* **86 D4** 16 12N 103 16 E
Mahābād, *Iran* **101 D11** 36 50N 45 45 E
Mahabaleshwar, *India* **94 F1** 17 58N 73 43 E
Mahabharat Lekh, *Nepal* **93 E10** 28 30N 82 0 E
Mahabo, *Madag.* **117 C7** 20 23 S 44 40 E
Mahad, *India* **94 E1** 18 6N 73 29 E
Mahaddei Uen, *Somali Rep.* **120 D3** 2 58N 45 32 E
Mahadeo Hills, *India* **92 H8** 22 20N 78 30 E
Mahadeopur, *India* **94 E5** 18 48N 80 0 E
Mahagi, *Dem. Rep. of the Congo* **118 B3** 2 20N 31 0 E
Mahaicony, *Guyana* **169 B6** 6 36N 57 48W
Mahajamba →, *Madag.* **117 B8** 15 33 S 47 8 E
Mahajamba, Helodranon' i, *Madag.* **117 B8** 15 24 S 47 5 E
Mahajan, *India* **92 E5** 28 48N 73 56 E
Mahajanga, *Madag.* **117 B8** 15 40 S 46 25 E
Mahajanga □, *Madag.* **117 B8** 17 0 S 47 0 E
Mahajilo →, *Madag.* **117 B8** 19 42 S 45 22 E
Mahakam →, *Indonesia* **85 C5** 0 35 S 117 17 E
Mahalapye, *Botswana* **116 C4** 23 1 S 26 51 E
Maḥallāt, *Iran* **97 C6** 33 55N 50 30 E
Māhān, *Iran* **97 D8** 30 5N 57 18 E
Mahanadi →, *India* **94 D8** 20 20N 86 25 E
Mahanoro, *Madag.* **117 B8** 19 54 S 48 48 E
Mahanoy City, *U.S.A.* **151 F8** 40 49N 76 9W
Mahaplag, *Phil.* **81 F5** 10 35N 124 57 E
Maharashtra □, *India* **94 D2** 20 30N 75 30 E
Maharès, *Tunisia* **108 B2** 34 32N 10 29 E
Mahari Mts., *Tanzania* **118 D3** 6 20 S 30 0 E
Mahasham, W. →, *Egypt* **103 E3** 30 15N 34 10 E
Mahasolo, *Madag.* **117 B8** 19 7 S 46 22 E
Mahattat ash Shīdīyah, *Jordan* **103 F4** 29 55N 35 55 E
Mahattat 'Unayzah, *Jordan* **103 E4** 30 30N 35 47 E
Mahaweli →, *Sri Lanka* **95 K5** 8 30N 81 15 E
Mahaxay, *Laos* **86 D5** 17 22N 105 12 E
Mahbes, *W. Sahara* **110 C3** 27 10N 9 50W
Mahbubabad, *India* **94 F5** 17 42N 80 2 E
Mahbubnagar, *India* **94 F3** 16 45N 77 59 E
Maḥḍah, *Oman* **97 E7** 24 24N 55 59 E
Mahdia, *Guyana* **169 B6** 5 13N 59 8W
Mahdia, *Tunisia* **108 A2** 35 28N 11 0 E
Mahe, *Jammu & Kashmir, India* **93 C8** 33 10N 78 32 E
Mahé, *Pondicherry, India* **95 J2** 11 42N 75 34 E
Mahé, *Seychelles* **121 E4** 5 0 S 55 30 E
Mahendra Giri, *India* **95 K3** 8 20N 77 30 E
Mahendraganj, *India* **90 C2** 25 20N 89 45 E
Mahenge, *Tanzania* **119 D4** 8 45 S 36 41 E
Maheno, *N.Z.* **131 F5** 45 10 S 170 50 E
Mahesana, *India* **92 H5** 23 39N 72 26 E
Mahia Pen., *N.Z.* **130 F6** 39 9 S 177 55 E
Mahilyow, *Belarus* **58 F6** 53 55N 30 18 E
Mahirija, *Morocco* **111 B4** 34 0N 3 16W
Mahlaing, *Burma* **90 E5** 21 6N 95 39 E
Mahmiya, *Sudan* **107 D3** 17 12N 33 43 E
Mahmud Kot, *Pakistan* **92 D4** 30 16N 71 0 E
Mahmudia, *Romania* **53 E14** 45 5N 29 5 E
Mahmudiye, *Turkey* **49 B12** 39 48N 30 15 E
Mahmutbey, *Turkey* **51 E12** 41 3N 28 49 E
Mahnomen, *U.S.A.* **154 B7** 47 19N 95 58W
Mahoba, *India* **93 G8** 25 15N 79 55 E
Mahomet, *U.S.A.* **157 D8** 40 12N 88 24W
Mahón = Maó, *Spain* **39 B11** 39 53N 4 16 E
Mahone Bay, *Canada* **141 D7** 44 30N 64 20W
Mahuta, *Nigeria* **113 C5** 11 32N 4 58 E
Mahya Daği, *Turkey* **51 E11** 41 47N 27 36 E
Mai-Ndombe, L., *Dem. Rep. of the Congo* **114 C3** 2 0 S 18 20 E
Mai-Sai, *Thailand* **86 B2** 20 20N 99 55 E
Maia, *Portugal* **42 D2** 41 14N 8 37W
Maia, *Spain* **40 B3** 43 12N 1 29W
Maials, *Spain* **40 D5** 41 22N 0 30 E
Maibara, *Japan* **71 B8** 35 19N 136 17 E
Maicao, *Colombia* **168 A3** 11 23N 72 13W
Maîche, *France* **27 E13** 47 16N 6 48 E
Maici →, *Brazil* **173 B5** 6 30 S 61 43W
Maicurú →, *Brazil* **169 D7** 2 14 S 54 17W
Máida, *Italy* **47 D9** 38 51N 16 22 E
Maidan Khula, *Afghan.* **92 C3** 33 36N 69 50 E
Maidenhead, *U.K.* **21 F7** 51 31N 0 42W
Maidi, *Yemen* **107 D5** 16 20N 42 45 E
Maidstone, *Canada* **143 C7** 53 5N 109 20W
Maidstone, *U.K.* **21 F8** 51 16N 0 32 E
Maiduguri, *Nigeria* **113 C7** 12 0N 13 20 E
Măieruş, *Romania* **53 E10** 45 53N 25 31 E
Maignelay Montigny, *France* **27 C9** 49 32N 2 30 E
Maigo, *Phil.* **81 G4** 8 10N 123 57 E
Maigualida, Sierra, *Venezuela* **169 B4** 5 30N 65 10W
Maigudo, *Ethiopia* **107 F4** 7 30N 37 8 E
Maijdi, *Bangla.* **90 D3** 22 48N 91 10 E
Maikala Ra., *India* **94 D5** 22 0N 81 0 E
Maikoor, *Indonesia* **83 C4** 6 8 S 134 6 E
Maili, *U.S.A.* **145 K13** 21 25N 158 11W
Maili Pt, *U.S.A.* **145 K13** 21 24N 158 11W
Maillezais, *France* **28 B3** 46 22N 0 45W
Mailsi, *Pakistan* **92 E5** 29 48N 72 15 E
Maimbung, *Phil.* **81 J3** 5 56N 121 2 E
Main →, *Germany* **31 F4** 50 0N 8 18 E
Main →, *U.K.* **23 B5** 54 48N 6 18W
Main Centre, *Canada* **143 C7** 50 35N 107 21W
Mainburg, *Germany* **31 G7** 48 38N 11 47 E
Maine, *France* **26 D6** 48 20N 0 15W
Maine □, *U.S.A.* **141 C6** 45 20N 69 0W
Maine →, *Ireland* **23 D2** 52 9N 9 45W
Maine-et-Loire □, *France* **26 E6** 47 31N 0 30W
Maïne-Soroa, *Niger* **113 C7** 13 13N 12 2 E
Maingkwan, *Burma* **90 B6** 26 15N 96 37 E
Mainit, *Phil.* **81 G5** 9 32N 125 32 E
Mainit, L., *Phil.* **81 G5** 9 31N 125 30 E
Mainkaing, *Burma* **90 C5** 24 48N 95 16 E
Mainland, *Orkney, U.K.* **22 C5** 58 59N 3 8W
Mainland, *Shet., U.K.* **22 A7** 60 15N 1 22W
Mainpuri, *India* **93 F8** 27 18N 79 4 E
Maintal, *Germany* **31 E4** 50 7N 8 52 E
Maintenon, *France* **27 D8** 48 35N 1 35 E
Maintirano, *Madag.* **117 B7** 18 3 S 44 1 E
Mainz, *Germany* **31 E4** 50 1N 8 14 E
Maipú, *Argentina* **174 D4** 36 52 S 57 50W
Maiquetía, *Venezuela* **168 A4** 10 36N 66 57W
Máira →, *Italy* **44 D4** 44 49N 7 38 E
Mairabari, *India* **90 B4** 26 30N 92 22 E
Mairipotaba, *Brazil* **171 E2** 17 18 S 49 28W
Maisí, *Cuba* **165 B5** 20 17N 74 9W
Maisi, Pta. de, *Cuba* **165 B5** 20 10N 74 10W
Maitland, *N.S.W., Australia* **129 B9** 32 33 S 151 36 E
Maitland, *S. Austral., Australia* **128 C2** 34 23 S 137 40 E
Maitland →, *Canada* **150 C3** 43 45N 81 43W
Maitland, Banjaran, *Malaysia* **85 B5** 4 55N 116 37 E
Maiyema, *Nigeria* **113 C5** 12 5N 4 25 E
Maiyuan, *China* **77 E11** 25 34N 117 28 E
Maiz, Is. del, *Nic.* **164 D3** 12 15N 83 4W
Maizuru, *Japan* **71 B7** 35 25N 135 22 E
Majagual, *Colombia* **168 B3** 8 33N 74 38W
Majalengka, *Indonesia* **85 D3** 6 50 S 108 13 E
Majari →, *Brazil* **169 C5** 3 29N 60 58W
Majene, *Indonesia* **82 B1** 3 38 S 118 57 E
Majes →, *Peru* **172 D3** 16 40 S 72 44W
Majevica, *Bos.-H.* **52 F3** 44 45N 18 50 E
Maji, *Ethiopia* **107 F4** 6 12N 35 30 E
Majiang, *China* **76 D6** 26 28N 107 32 E
Major, *Canada* **143 C7** 51 52N 109 37W
Majorca = Mallorca, *Spain* **39 B10** 39 30N 3 0 E
Majors Creek, *Australia* **129 C8** 35 33 S 149 45 E
Majuriã, *Brazil* **173 B5** 7 30 S 64 55W
Maka, *Senegal* **112 C2** 13 40N 14 10W
Makaha, *U.S.A.* **145 K13** 21 29N 158 13W
Makahoa Pt., *U.S.A.* **145 J14** 21 41N 157 56W
Makahuena Pt., *U.S.A.* **145 B2** 21 52N 159 27W
Makak, *Cameroon* **113 E7** 3 36N 11 0 E
Makakilo City, *U.S.A.* **145 K13** 21 22N 158 5W
Makakou, *Gabon* **114 C2** 0 11 S 12 12 E
Makale, *Indonesia* **82 B1** 3 6 S 119 51 E
Makamba, *Burundi* **118 C2** 4 8 S 29 49 E
Makapuu Pt., *U.S.A.* **145 K14** 21 19N 157 39W
Makarewa, *N.Z.* **131 G3** 46 20 S 168 21 E
Makarikari = Makgadikgadi Salt Pans, *Botswana* **116 C4** 20 40 S 25 45 E
Makarovo, *Russia* **65 D11** 57 40N 107 45 E
Makarska, *Croatia* **45 E14** 43 20N 17 2 E
Makaryev, *Russia* **60 B6** 57 52N 43 50 E
Makasar = Ujung Pandang, *Indonesia* **82 C1** 5 10 S 119 20 E
Makasar, Selat, *Indonesia* **82 B1** 1 0 S 118 20 E
Makasar, Str. of = Makasar, Selat, *Indonesia* **82 B1** 1 0 S 118 20 E
Makat, *Kazakstan* **57 E9** 47 39N 53 19 E
Makawao, *U.S.A.* **145 C5** 20 52N 156 17W
Makedonija = Macedonia ■, *Europe* **50 E5** 41 53N 21 40 E
Makena, *U.S.A.* **146 H16** 20 39N 156 27W
Makeni, *S. Leone* **112 D2** 8 55N 12 5W
Makeyevka = Makiyivka, *Ukraine* **59 H9** 48 0N 38 0 E
Makgadikgadi Salt Pans, *Botswana* **116 C4** 20 40 S 25 45 E
Makhachkala, *Russia* **61 J8** 43 0N 47 30 E
Makharadze = Ozurgeti, *Georgia* **61 K5** 41 55N 42 2 E
Makhmūr, *Iraq* **101 E10** 35 46N 43 35 E
Makhyah, W. →, *Yemen* **99 C5** 17 40N 49 1 E
Makian, *Indonesia* **82 A3** 0 20N 127 20 E
Makindu, *Kenya* **118 C4** 2 18 S 37 50 E
Makinsk, *Kazakstan* **64 D8** 52 37N 70 26 E
Makiyivka, *Ukraine* **59 H9** 48 0N 38 0 E
Makkah, *Si. Arabia* **98 B2** 21 30N 39 54 E
Makkovik, *Canada* **141 A8** 55 10N 59 10W
Makó, *Hungary* **52 D5** 46 14N 20 33 E
Makok, *Gabon* **114 C1** 0 1 S 9 35 E
Makokou, *Gabon* **114 B2** 0 40N 12 50 E
Makongo, *Dem. Rep. of the Congo* **118 B2** 3 25N 26 17 E
Makoro, *Dem. Rep. of the Congo* **118 B2** 3 10N 29 59 E
Makoua, *Congo* **114 C3** 0 5 S 15 50 E
Maków Mazowiecki, *Poland* **55 F8** 52 52N 21 6 E
Maków Podhalański, *Poland* **55 J6** 49 43N 19 45 E
Makrá, *Greece* **49 E7** 36 15N 25 54 E
Makran, *Asia* **91 D1** 26 13N 61 30 E
Makran Coast Range, *Pakistan* **91 D2** 25 40N 64 0 E
Makrana, *India* **92 F6** 27 2N 74 46 E
Mákri, *Greece* **51 F9** 40 52N 25 40 E
Makriyialos, *Greece* **38 D7** 35 2N 25 59 E
Maktar, *Tunisia* **108 A1** 35 48N 9 12 E
Mākū, *Iran* **101 C11** 39 15N 44 31 E
Makum, *India* **90 B5** 27 30N 95 23 E
Makumbi, *Dem. Rep. of the Congo* **115 D4** 5 50 S 20 43 E
Makunda, *Botswana* **116 C3** 22 30 S 20 7 E
Makurazaki, *Japan* **70 F2** 31 15N 130 20 E
Makurdi, *Nigeria* **113 D6** 7 43N 8 35 E
Makushin Volcano, *U.S.A.* **144 K6** 53 53N 166 55W
Makūyeh, *Iran* **97 D7** 28 7N 53 9 E
Makwassie, *S. Africa* **116 D4** 27 17 S 26 0 E
Mal, *India* **90 B2** 26 51N 88 45 E
Mal B., *Ireland* **23 D2** 52 50N 9 30W
Mal i Gjalicës së Lumës, *Albania* **50 D4** 42 2N 20 25 E
Mal i Gribës, *Albania* **50 F3** 40 17N 19 45 E
Mal i Nemërçkës, *Albania* **50 F4** 40 15N 20 15 E
Mala, *Peru* **172 C2** 12 40 S 76 38W
Mala, Pta., *Panama* **164 E3** 7 28N 80 2W
Mala Belozërka, *Ukraine* **59 J8** 47 12N 34 56 E
Mala Kapela, *Croatia* **45 D12** 44 45N 15 30 E
Mała Panew →, *Poland* **55 H4** 50 43N 17 54 E
Mala Vyska, *Ukraine* **59 H6** 48 39N 31 36 E
Malabang, *Phil.* **81 H5** 7 36N 124 3 E
Malabar, *U.S.A.* **153 H9** 28 0N 80 34W
Malabar Coast, *India* **95 J2** 11 0N 75 0 E
Malabo = Rey Malabo, *Eq. Guin.* **113 E6** 3 45N 8 50 E
Malabon, *Phil.* **80 D3** 14 21N 121 0 E
Malabrigo Pt., *Phil.* **80 E3** 13 36N 121 15 E
Malabungan, *Phil.* **81 G1** 9 3N 117 38 E
Malacca, Str. of, *Indonesia* **87 L3** 3 0N 101 0 E
Malacky, *Slovak Rep.* **35 C10** 48 27N 17 0 E
Malad City, *U.S.A.* **158 E7** 42 12N 112 15W
Maladeta, *Spain* **40 C5** 42 39N 0 39 E
Maladzyechna, *Belarus* **58 E4** 54 20N 26 50 E
Malae Pt., *U.S.A.* **145 C6** 20 7N 155 53W
Málaga, *Colombia* **168 B3** 6 42N 72 44W
Málaga, *Spain* **43 J6** 36 43N 4 23W
Malaga, *U.S.A.* **155 J2** 32 14N 104 4W
Málaga □, *Spain* **43 J6** 36 38N 4 58W
Malagarasi, *Tanzania* **118 D3** 5 5 S 30 50 E
Malagarasi →, *Tanzania* **118 D2** 5 12 S 29 47 E
Malagón, *Spain* **43 F7** 39 11N 3 52W
Malagón →, *Spain* **43 H3** 37 35N 7 29W
Malahide, *Ireland* **23 C5** 53 26N 6 9W
Malaimbandy, *Madag.* **117 C8** 20 20 S 45 36 E
Malaita, *Pac. Oc.* **133 M11** 9 0 S 161 0 E
Malakâl, *Sudan* **107 F3** 9 33N 31 40 E
Malakand, *Pakistan* **91 B3** 34 40N 71 55 E
Malakoff, *U.S.A.* **155 J7** 32 10N 96 1W
Malakula, *Vanuatu* **133 F5** 16 15 S 167 30 E
Malalag, *Phil.* **81 H5** 6 36N 125 24 E
Malam, *Chad* **109 F4** 11 27N 20 59 E
Malang, *Indonesia* **85 D4** 7 59 S 112 45 E
Malangas, *Phil.* **81 H4** 7 37N 123 1 E
Malange □, *Angola* **115 D3** 9 30 S 16 0 E
Malangen, *Norway* **14 B18** 69 24N 18 37 E
Malanje, *Angola* **115 D3** 9 36 S 16 17 E
Mälaren, *Sweden* **16 E11** 59 30N 17 10 E
Malargüe, *Argentina* **174 D2** 35 32 S 69 30W
Malartic, *Canada* **140 C4** 48 9N 78 9W
Malaryta, *Belarus* **59 G3** 51 50N 24 3 E
Malaspina Glacier, *U.S.A.* **144 G12** 59 50N 140 30W
Malatya, *Turkey* **101 C8** 38 25N 38 20 E
Malawali, *Malaysia* **85 A5** 7 3N 117 18 E
Malawi ■, *Africa* **119 E3** 11 55 S 34 0 E
Malawi, L. = Nyasa, L., *Africa* **119 E3** 12 30 S 34 30 E
Malay, *Phil.* **81 F3** 11 54N 121 55 E
Malay Pen., *Asia* **87 J3** 7 25N 100 0 E
Malaya Belozërka = Mala Belozërka, *Ukraine* **59 J8** 47 12N 34 56 E
Malaya Vishera, *Russia* **58 C7** 58 55N 32 25 E
Malaya Viska = Mala Vyska, *Ukraine* **59 H6** 48 39N 31 36 E
Malaybalay, *Phil.* **81 G5** 8 5N 125 7 E
Malāyer, *Iran* **97 C6** 34 19N 48 51 E
Malaysia ■, *Asia* **84 B4** 5 0N 110 0 E
Malazgirt, *Turkey* **101 C10** 39 10N 42 33 E
Malbon, *Australia* **126 C3** 21 5 S 140 17 E
Malbooma, *Australia* **127 E1** 30 41 S 134 11 E
Malbork, *Poland* **54 D6** 54 3N 19 1 E
Malca Dube, *Ethiopia* **120 C2** 6 47N 42 4 E
Malcésine, *Italy* **44 C7** 45 46N 10 48 E
Malchin, *Germany* **30 B8** 53 44N 12 44 E
Malchow, *Germany* **30 B8** 53 28N 12 25 E
Malcolm, *Australia* **125 E3** 28 51 S 121 25 E
Malcolm, Pt., *Australia* **125 F3** 33 48 S 123 45 E
Malczyce, *Poland* **55 G3** 51 14N 16 29 E
Maldegem, *Belgium* **24 C3** 51 14N 3 26 E
Malden, *Mass., U.S.A.* **151 D13** 42 26N 71 4W
Malden, *Mo., U.S.A.* **155 G10** 36 34N 89 57W
Malden I., *Kiribati* **135 H12** 4 3 S 155 1W
Maldives ■, *Ind. Oc.* **121 D6** 5 0N 73 0 E
Maldon, *Australia* **128 D6** 37 0 S 144 6 E
Maldonado, *Uruguay* **175 C5** 34 59 S 55 0W
Maldonado, Punta, *Mexico* **163 D5** 16 19N 98 35W
Malè, *Italy* **44 B7** 46 21N 10 55 E
Malé Karpaty, *Slovak Rep.* **35 C10** 48 30N 17 20 E
Maléa, Ákra, *Greece* **48 E5** 36 28N 23 7 E
Malebo, Pool, *Africa* **115 C3** 4 17 S 15 20 E
Malegaon, *India* **94 D2** 20 30N 74 38 E
Malei, *Mozam.* **119 F4** 17 12 S 36 58 E
Malek Kandī, *Iran* **101 D12** 37 9N 46 6 E
Malela, *Bas Zaïre, Dem. Rep. of the Congo* **115 D2** 5 59 S 12 37 E
Malela, *Kivu, Dem. Rep. of the Congo* **115 C5** 4 22 S 26 8 E
Malema, *Mozam.* **119 E4** 14 57 S 37 20 E
Máleme, *Greece* **38 D5** 35 31N 23 49 E
Målerås, *Sweden* **17 H9** 56 54N 15 34 E
Malerkotla, *India* **92 D6** 30 32N 75 58 E
Máles, *Greece* **38 D7** 35 6N 25 35 E
Malesherbes, *France* **27 D9** 48 15N 2 24 E
Maleshevska Planina, *Europe* **50 E7** 41 38N 23 7 E
Malesína, *Greece* **48 C5** 38 37N 23 14 E
Malestroit, *France* **26 E4** 47 49N 2 25W
Malfa, *Italy* **47 D7** 38 35N 14 50 E
Malgobek, *Russia* **61 J7** 43 30N 44 34 E
Malgomaj, *Sweden* **14 D17** 64 40N 16 30 E
Malgrat = Malgrat de Mar, *Spain* **40 D7** 41 39N 2 46 E
Malgrat de Mar, *Spain* **40 D7** 41 39N 2 46 E
Malha, *Sudan* **107 D2** 15 8N 25 10 E
Malheur →, *U.S.A.* **158 D5** 44 4N 116 59W
Malheur L., *U.S.A.* **158 E4** 43 20N 118 48W
Mali, *Guinea* **112 C2** 12 10N 12 20W
Mali ■, *Africa* **112 B4** 17 0N 3 0W
Mali Hka →, *Burma* **90 C6** 25 42N 97 30 E
Mali Kanal, *Serbia, Yug.* **52 E4** 45 36N 19 24 E
Mali Kyun, *Burma* **86 F2** 13 0N 98 20 E
Malibu, *U.S.A.* **161 L8** 34 2N 118 41W
Maligaya, *Phil.* **80 E3** 12 59N 121 30 E
Maliku, *Indonesia* **82 B2** 0 39 S 123 16 E
Malili, *Indonesia* **82 B2** 2 42 S 121 6 E
Målilla, *Sweden* **17 G9** 57 23N 15 48 E
Malimba, Mts., *Dem. Rep. of the Congo* **118 D2** 7 30 S 29 30 E
Malin Hd., *Ireland* **23 A4** 55 23N 7 23W
Malin Pen., *Ireland* **23 A4** 55 20N 7 17W
Malindang, Mt., *Phil.* **81 G4** 8 13N 123 38 E
Malindi, *Kenya* **118 C5** 3 12 S 40 5 E
Malines = Mechelen, *Belgium* **24 C4** 51 2N 4 29 E
Malino, *Indonesia* **82 A2** 1 0N 121 0 E
Malinyi, *Tanzania* **119 D4** 8 56 S 36 0 E
Malipo, *China* **76 F5** 23 7N 104 42 E
Maliqi, *Albania* **50 F4** 40 45N 20 48 E
Malita, *Phil.* **81 H5** 6 19N 125 39 E
Maljenik, *Serbia, Yug.* **50 C5** 43 54N 21 43 E
Malkapur, *Maharashtra, India* **94 F3** 16 57N 76 17 E
Malkapur, *Maharashtra, India* **94 D1** 20 53N 73 58 E
Malkara, *Turkey* **51 F10** 40 53N 26 53 E
Małkinia Górna, *Poland* **55 F9** 52 42N 22 5 E
Malko Tŭrnovo, *Bulgaria* **51 E11** 41 59N 27 31 E
Mallacoota, *Australia* **129 D8** 37 40 S 149 40 E
Mallacoota Inlet, *Australia* **129 D8** 37 34 S 149 40 E
Mallaig, *U.K.* **22 D3** 57 0N 5 50W
Mallala, *Australia* **128 C3** 34 26 S 138 30 E
Mallard, *U.S.A.* **156 B2** 42 56N 94 41W
Mallawan, *India* **93 F9** 27 4N 80 12 E
Mallawi, *Egypt* **106 B3** 27 44N 30 44 E
Malleco □, *Chile* **176 A2** 38 10 S 72 20W
Mallemort, *France* **29 E9** 43 43N 5 11 E
Mallemukfjeld, *Greenland* **10 A9** 80 15N 16 15W
Málles Venosta, *Italy* **44 B7** 46 41N 10 32 E
Mállia, *Greece* **38 D7** 35 17N 25 32 E
Mallicolo = Malakula, *Vanuatu* **133 F5** 16 15 S 167 30 E
Mallig, *Phil.* **80 C3** 17 8N 121 42 E
Mallión, Kólpos, *Greece* **38 D7** 35 19N 25 27 E
Mallorca, *Spain* **39 B10** 39 30N 3 0 E
Mallorytown, *Canada* **151 B9** 44 29N 75 53W
Mallow, *Ireland* **23 D3** 52 8N 8 39W
Malmbäck, *Sweden* **17 G8** 57 34N 14 28 E
Malmberget, *Sweden* **14 C19** 67 11N 20 40 E
Malmédy, *Belgium* **24 D6** 50 25N 6 2 E
Malmesbury, *S. Africa* **116 E2** 33 28 S 18 41 E
Malmköping, *Sweden* **16 E10** 59 8N 16 44 E
Malmö, *Sweden* **17 J6** 55 36N 12 59 E
Malmöhus län □, *Sweden* **17 J7** 55 45N 13 30 E
Malmslätt, *Sweden* **17 F9** 58 27N 15 33 E
Malmyzh, *Russia* **60 B10** 56 31N 50 41 E
Malnaş, *Romania* **53 D10** 46 2N 25 49 E
Malo, *Vanuatu* **133 E5** 15 40 S 167 11 E
Malo Konare, *Bulgaria* **51 D8** 42 12N 24 24 E
Maloarkhangelsk, *Russia* **59 F9** 52 28N 36 30 E
Maloca, *Brazil* **169 C6** 0 43N 55 57W
Maloja, *Switz.* **33 D9** 46 25N 9 35 E
Maloja, P., *Switz.* **33 D9** 46 23N 9 42 E
Malolos, *Phil.* **80 D3** 14 50N 120 49 E
Malombe L., *Malawi* **119 E4** 14 40 S 35 15 E
Małomice, *Poland* **55 G2** 51 34N 15 29 E
Malomir, *Bulgaria* **51 D10** 42 16N 26 32 E
Malone, *Fla., U.S.A.* **152 E4** 30 57N 85 10W
Malone, *N.Y., U.S.A.* **151 B10** 44 51N 74 18W
Malong, *China* **76 E4** 25 24N 103 34 E
Malonga, *Dem. Rep. of the Congo* **115 E4** 10 24 S 23 10 E
Malorad, *Bulgaria* **50 C7** 43 28N 23 41 E
Måløy, *Norway* **15 F11** 61 57N 5 6 E
Maloyaroslovets, *Russia* **58 E9** 55 2N 36 20 E
Malozemelskaya Tundra, *Russia* **56 A9** 67 0N 50 0 E
Malpartida, *Spain* **43 F4** 39 26N 6 30W
Malpaso, *Canary Is.* **39 G1** 27 43N 18 3W
Malpica = Malpica de Bergantiños, *Spain* **42 B2** 43 19N 8 50W
Malpica de Bergantiños, *Spain* **42 B2** 43 19N 8 50W
Malprabha →, *India* **95 F3** 16 20N 76 5 E
Mals = Málles Venosta, *Italy* **44 B7** 46 41N 10 32 E
Malta, *Brazil* **170 C4** 6 54 S 37 31W
Malta, *Idaho, U.S.A.* **158 E7** 42 18N 113 22W
Malta, *Mont., U.S.A.* **158 B10** 48 21N 107 52W
Malta ■, *Europe* **38 D2** 35 50N 14 30 E
Maltahöhe, *Namibia* **116 C2** 24 55 S 17 0 E
Maltepe, *Turkey* **51 F13** 40 55N 29 8 E

246 Malters

Malters, *Switz.* **32 B6** 47 3N 8 11 E
Malton, *Canada* **150 C5** 43 42N 79 38W
Malton, *U.K.* **20 C7** 54 8N 0 49W
Malu'a, *Solomon Is.* **133 M11** 8 0 S 160 0 E
Maluku, *Indonesia* **82 B3** 1 0 S 127 0 E
Maluku □, *Indonesia* **82 B3** 3 0 S 128 0 E
Maluku Sea = Molucca Sea, *Indonesia* **82 A3** 2 0 S 124 0 E
Malumfashi, *Nigeria* **113 C6** 11 48N 7 39 E
Malung, *Sweden* **16 D7** 60 42N 13 44 E
Malungsfors, *Sweden* **16 D7** 60 44N 13 33 E
Malungun, *Phil.* **81 H5** 6 16N 125 14 E
Maluso, *Phil.* **81 H3** 6 33N 121 53 E
Malvalli, *India* **95 H3** 12 28N 77 8 E
Malvan, *India* **95 F1** 16 2N 73 30 E
Malvern, *U.S.A.* **155 H8** 34 22N 92 49W
Malvern Hills, *U.K.* **21 E5** 52 0N 2 19W
Malvik, *Norway* **18 A7** 63 25N 10 40 E
Malvinas, Is. = Falkland Is. □, *Atl. Oc.* **176 D5** 51 30 S 59 0W
Malý Dunaj →, *Slovak Rep.* **35 D11** 47 45N 18 9 E
Malya, *Tanzania* **118 C3** 3 5 S 33 38 E
Malybay, *Kazakstan* **63 B9** 43 30N 78 25 E
Malyn, *Ukraine* **59 G5** 50 46N 29 3 E
Malyy Lyakhovskiy, Ostrov, *Russia* **65 B15** 74 7N 140 36 E
Mama, *Russia* **65 D12** 58 18N 112 54 E
Mamadysh, *Russia* **60 C10** 55 44N 51 23 E
Mamaku, *N.Z.* **130 E5** 38 5 S 176 8 E
Mamala B., *U.S.A.* **145 K14** 21 15N 157 55W
Mamanguape, *Brazil* **170 C4** 6 50 S 35 4W
Mamasa, *Indonesia* **82 B1** 2 55 S 119 20 E
Mambajao, *Phil.* **81 G5** 9 15N 124 43 E
Mambasa, *Dem. Rep. of the Congo* **118 B2** 1 22N 29 3 E
Mamberamo →, *Indonesia* **83 B5** 2 0 S 137 50 E
Mambilima Falls, *Zambia* **119 E2** 10 31 S 28 45 E
Mambirima, *Dem. Rep. of the Congo* **119 E2** 11 25 S 27 33 E
Mambo, *Tanzania* **118 C4** 4 52 S 38 22 E
Mambrui, *Kenya* **118 C5** 3 5 S 40 5 E
Mamburao, *Phil.* **80 E3** 13 13N 120 39 E
Mameigwess L., *Canada* **140 B2** 52 35N 87 50W
Mamers, *France* **26 D7** 48 21N 0 22 E
Mamfé, *Cameroon* **113 D6** 5 50N 9 15 E
Māmī, Ra's, *Yemen* **99 D6** 12 32N 54 30 E
Mamiña, *Chile* **172 E4** 20 5 S 69 14W
Mammoth, *U.S.A.* **159 K8** 32 43N 110 39W
Mamoré →, *Bolivia* **173 C4** 10 23 S 65 53W
Mamou, *Guinea* **112 C2** 10 15N 12 0W
Mamparang Mts., *Phil.* **80 C3** 16 21N 121 28 E
Mampatá, *Guinea-Biss.* **112 C2** 11 54N 14 53W
Mampong, *Ghana* **113 D4** 7 6N 1 26W
Mamry, Jezioro, *Poland* **54 D8** 54 5N 21 50 E
Mamuil Malal, Paso, *S. Amer.* **176 A2** 39 35 S 71 28W
Mamuju, *Indonesia* **82 B1** 2 41 S 118 50 E
Ma'mūl, *Oman* **99 C6** 18 8N 55 16 E
Mamuras, *Albania* **50 E3** 41 34N 19 41 E
Man, *Ivory C.* **112 D3** 7 30N 7 40W
Man →, *India* **94 F2** 17 31N 75 32 E
Man, I. of, *U.K.* **20 C3** 54 15N 4 30W
Man Na, *Burma* **90 D6** 23 27N 97 19 E
Man Tun, *Burma* **90 D7** 23 52N 98 38 E
Mana, *Fr. Guiana* **169 B7** 5 45N 53 55W
Mana →, *Fr. Guiana* **169 B7** 5 45N 53 55W
Manaar, G. of = Mannar, G. of, *Asia* **95 K4** 8 30N 79 0 E
Manabí □, *Ecuador* **168 D1** 0 40 S 80 5W
Manacacías →, *Colombia* **168 C3** 4 23N 72 4W
Manacapuru, *Brazil* **169 D5** 3 16 S 60 37W
Manacapuru →, *Brazil* **169 D5** 3 18 S 60 37W
Manacor, *Spain* **39 B10** 39 34N 3 13 E
Manado, *Indonesia* **82 A2** 1 29N 124 51 E
Managua, *Nic.* **164 D2** 12 6N 86 20W
Managua, L., *Nic.* **164 D2** 12 20N 86 30W
Manaia, *N.Z.* **130 F3** 39 33 S 174 8 E
Manakara, *Madag.* **117 C8** 22 8 S 48 1 E
Manakau Mt., *N.Z.* **131 C8** 42 15 S 173 42 E
Manākhah, *Yemen* **98 D3** 15 5N 43 44 E
Manakino, *N.Z.* **130 E4** 38 22 S 175 47 E
Manam I., *Papua N. G.* **132 C3** 4 5 S 145 0 E
Manama = Al Manāmah, *Bahrain* **97 E6** 26 10N 50 30 E
Manambao →, *Madag.* **117 B7** 17 35 S 44 0 E
Manambato, *Madag.* **117 A8** 13 43 S 49 7 E
Manambolo →, *Madag.* **117 B7** 19 18 S 44 22 E
Manambolosy, *Madag.* **117 B8** 16 2 S 49 40 E
Manana I., *U.S.A.* **145 K14** 21 20N 157 40W
Mananara, *Madag.* **117 B8** 16 10 S 49 46 E
Mananara →, *Madag.* **117 C8** 23 21 S 47 42 E
Mananjary, *Madag.* **117 C8** 21 13 S 48 20 E
Manantenina, *Madag.* **117 C8** 24 17 S 47 19 E
Manaos = Manaus, *Brazil* **169 D6** 3 0 S 60 0W
Manapala, *Phil.* **81 F4** 10 58N 123 5 E
Manapire →, *Venezuela* **168 B4** 7 42N 66 7W
Manapouri, *N.Z.* **131 F2** 45 34 S 167 39 E
Manapouri, L., *N.Z.* **131 F2** 45 32 S 167 32 E
Manar →, *India* **94 E3** 18 50N 77 20 E
Manār, Jabal, *Yemen* **98 D4** 14 2N 44 17 E
Manas, *China* **72 B3** 44 17N 85 56 E
Manas, *Somali Rep.* **120 D2** 2 57N 43 28 E
Manas, Gora, *Kyrgyzstan* **63 B5** 42 22N 71 2 E
Manaslu, *Nepal* **93 E11** 28 33N 84 33 E
Manasquan, *U.S.A.* **151 F10** 40 8N 74 3W
Manassa, *U.S.A.* **159 H11** 37 11N 105 56W
Manatuto, *Indonesia* **82 C3** 8 30 S 126 1 E
Manaung, *Burma* **90 F4** 18 45N 93 40 E
Manaus, *Brazil* **169 D6** 3 0 S 60 0W
Manavgat, *Turkey* **100 D4** 36 47N 31 26 E
Manawan L., *Canada* **143 B8** 55 24N 103 14W
Manawatu →, *N.Z.* **130 G4** 40 28 S 175 12 E
Manay, *Phil.* **81 H6** 7 17N 126 33 E
Manbij, *Syria* **100 D7** 36 31N 37 57 E
Mancelona, *U.S.A.* **148 C3** 44 54N 85 4W
Mancha Real, *Spain* **43 H7** 37 48N 3 39W
Manche □, *France* **26 C5** 49 10N 1 20W
Manchester, *U.K.* **20 D5** 53 29N 2 12W
Manchester, *Calif., U.S.A.* **160 G3** 38 58N 123 41W
Manchester, *Conn., U.S.A.* **151 E12** 41 47N 72 31W
Manchester, *Ga., U.S.A.* **152 C5** 32 51N 84 37W
Manchester, *Iowa, U.S.A.* **156 B5** 42 29N 91 27W
Manchester, *Ky., U.S.A.* **148 G4** 37 9N 83 46W
Manchester, *Mich., U.S.A.* **157 B12** 42 9N 84 2W
Manchester, *N.H., U.S.A.* **151 D13** 42 59N 71 28W
Manchester, *N.Y., U.S.A.* **150 D7** 42 56N 77 16W
Manchester, *Vt., U.S.A.* **151 C11** 43 10N 73 5W
Manchester L., *Canada* **143 A7** 61 28N 107 29W
Manchuria = Dongbei, *China* **75 D13** 45 0N 125 0 E
Manchurian Plain, *China* **66 E16** 47 0N 124 0 E
Manciano, *Italy* **45 F8** 42 35N 11 31 E
Mancifa, *Ethiopia* **107 F5** 6 53N 41 50 E
Mancora, Pta., *Peru* **172 A1** 4 9 S 81 1W
Mand →, *Iran* **97 D7** 28 20N 52 30 E
Manda, *Chunya, Tanzania* **118 D3** 6 51 S 32 29 E
Manda, *Ludewe, Tanzania* **119 E3** 10 30 S 34 40 E
Mandabé, *Madag.* **117 C7** 21 0 S 44 55 E
Mandaguari, *Brazil* **175 A5** 23 32 S 51 42W
Mandah, *Mongolia* **74 B5** 44 27N 108 2 E
Mandal, *Norway* **15 G12** 58 2N 7 25 E
Mandalay, *Burma* **90 E6** 22 0N 96 4 E
Mandale = Mandalay, *Burma* **90 E6** 22 0N 96 4 E
Mandalgovi, *Mongolia* **74 B4** 45 45N 106 10 E
Mandalī, *Iraq* **101 F11** 33 43N 45 28 E
Mandan, *U.S.A.* **154 B4** 46 50N 100 54W
Mandaon, *Phil.* **80 E4** 12 13N 123 17 E
Mandapeta, *India* **94 F5** 16 47N 81 56 E
Mandar, Teluk, *Indonesia* **82 B1** 3 35 S 119 15 E
Mándas, *Italy* **46 C2** 39 40N 9 8 E
Mandaue, *Phil.* **81 F4** 10 20N 123 56 E
Mandayar, *Phil.* **81 H6** 7 34N 126 14 E
Mandelieu-la-Napoule, *France* **29 E10** 43 34N 6 57 E
Mandera, *Kenya* **118 B5** 3 55N 41 53 E
Mandera □, *Kenya* **118 B5** 3 30N 41 0 E
Mandi, *India* **92 D7** 31 39N 76 58 E
Mandimba, *Mozam.* **119 E4** 14 20 S 35 40 E
Mandioli, *Indonesia* **82 B3** 0 40 S 127 20 E
Mandioré, L., *S. Amer.* **173 D6** 18 8 S 57 33W
Mandji I. = Lopez I., *Gabon* **114 C1** 0 50 S 8 47 E
Mandla, *India* **93 H9** 22 39N 80 30 E
Mandø, *Denmark* **17 J2** 55 18N 8 33 E
Mandoto, *Madag.* **117 B8** 19 34 S 46 17 E
Mandoúdhion, *Greece* **48 C5** 38 48N 23 29 E
Mándra, *Greece* **48 C5** 38 4N 23 30 E
Mandra, *Pakistan* **92 C5** 33 23N 73 12 E
Mandrákhi, *Greece* **49 E9** 36 36N 27 11 E
Mandrare →, *Madag.* **117 D8** 25 10 S 46 30 E
Mandritsara, *Madag.* **117 B8** 15 50 S 48 49 E
Mandsaur, *India* **92 G6** 24 3N 75 8 E
Mandurah, *Australia* **125 F2** 32 36 S 115 48 E
Mandúria, *Italy* **47 B10** 40 24N 17 38 E
Mandvi, *India* **92 H3** 22 51N 69 22 E
Mandya, *India* **95 H3** 12 30N 77 0 E
Mandzai, *Pakistan* **92 D2** 30 55N 67 6 E
Mané, *Burkina Faso* **113 C4** 12 59N 1 21W
Maneh, *Iran* **97 B8** 37 39N 57 7 E
Manengouba, Mts., *Cameroon* **113 E6** 5 0N 9 50 E
Maner →, *India* **94 E4** 18 30N 79 40 E
Manérbio, *Italy* **44 C7** 45 21N 10 8 E
Maneroo, *Australia* **126 C3** 23 22 S 143 53 E
Maneroo Cr. →, *Australia* **126 C3** 23 21 S 143 53 E
Manfalût, *Egypt* **106 B3** 27 20N 30 52 E
Manfred, *Australia* **128 B5** 33 19 S 143 45 E
Manfredónia, *Italy* **45 G12** 41 38N 15 55 E
Manfredónia, G. di, *Italy* **45 G13** 41 35N 16 5 E
Manga, *Brazil* **171 D3** 14 46 S 43 56W
Manga, *Burkina Faso* **113 C4** 11 40N 1 4W
Manga, *Niger* **109 F2** 15 0N 14 0 E
Mangabeiras, Chapada das, *Brazil* **170 D2** 10 0 S 46 30W
Mangal, *Phil.* **81 H3** 6 25N 121 58 E
Mangalagiri, *India* **95 F5** 16 26N 80 36 E
Mangaldai, *India* **90 B4** 26 26N 92 2 E
Mangalia, *Romania* **53 G13** 43 50N 28 35 E
Mangalore, *Australia* **129 D6** 36 56 S 145 10 E
Mangalore, *India* **95 H2** 12 55N 74 47 E
Mangaon, *India* **94 E1** 18 15N 73 20 E
Mangaweka, *N.Z.* **130 F4** 39 48 S 175 47 E
Mangaweka, Mt., *N.Z.* **130 F5** 39 49 S 176 5 E
Mange, *Dem. Rep. of the Congo* **114 B4** 0 54N 20 30 E
Manger, *Norway* **18 D2** 60 38N 5 3 E
Manggar, *Indonesia* **85 C3** 2 50 S 108 10 E
Manggawitu, *Indonesia* **83 B4** 4 8 S 133 32 E
Mangin Range, *Burma* **90 C5** 24 15N 95 45 E
Mangkalihat, Tanjung, *Indonesia* **85 B5** 1 2N 118 59 E
Mangla Dam, *Pakistan* **93 C5** 33 9N 73 44 E
Manglares, C., *Colombia* **168 C2** 1 36N 79 2W
Manglaur, *India* **92 E7** 29 44N 77 49 E
Mangnai, *China* **72 C4** 37 52N 91 43 E
Mango, *Togo* **113 C5** 10 20N 0 30 E
Mangoche, *Malawi* **119 E4** 14 25 S 35 16 E
Mangoky →, *Madag.* **117 C7** 21 29 S 43 41 E
Mangole, *Indonesia* **82 B3** 1 50 S 125 55 E
Mangombe, *Dem. Rep. of the Congo* **118 C2** 1 20 S 26 48 E
Mangonia Park, *U.S.A.* **153 J9** 26 45N 80 4W
Mangonui, *N.Z.* **130 B2** 35 1 S 173 32 E
Mangualde, *Portugal* **42 E3** 40 38N 7 48W
Mangueigne, *Chad* **109 F4** 10 30N 21 15 E
Mangueira, L. da, *Brazil* **175 C5** 33 0 S 52 50W
Manguéni, Hamada, *Niger* **108 D2** 22 35N 12 40 E
Mangum, *U.S.A.* **155 H5** 34 53N 99 30W
Mangyshlak Poluostrov, *Kazakstan* **64 E6** 44 30N 52 30 E
Manhattan, *U.S.A.* **154 F6** 39 11N 96 35W
Manhatten, *U.S.A.* **157 C9** 41 26N 87 59W
Manhiça, *Mozam.* **117 D5** 25 23 S 32 49 E
Manhuaçu, *Brazil* **171 F3** 20 15 S 42 2W
Manhumirim, *Brazil* **171 F3** 20 22 S 41 57W
Maní, *Colombia* **168 C3** 4 49N 72 17W
Mania →, *Madag.* **117 B8** 19 42 S 45 22 E
Maniago, *Italy* **45 B9** 46 10N 12 43 E
Manica, *Mozam.* **117 B5** 18 58 S 32 59 E
Manica e Sofala □, *Mozam.* **117 B5** 19 10 S 33 45 E
Manicaland □, *Zimbabwe* **119 F3** 19 0 S 32 30 E
Manicoré, *Brazil* **173 B5** 5 48 S 61 16W
Manicoré →, *Brazil* **173 B5** 5 51 S 61 19W
Manicouagan →, *Canada* **141 C6** 49 30N 68 30W
Manīfah, *Si. Arabia* **97 E6** 27 44N 49 0 E
Manifold, *Australia* **126 C5** 22 41 S 150 40 E
Manifold, C., *Australia* **126 C5** 22 41 S 150 50 E
Maniganggo, *China* **76 B2** 31 56N 99 10 E
Manigotagan, *Canada* **143 C9** 51 6N 96 18W
Manihiki, *Cook Is.* **135 J11** 10 24 S 161 1W
Maniitsoq = Sukkertoppen, *Greenland* **10 D5** 65 26N 52 55W
Manika, Plateau de la, *Dem. Rep. of the Congo* **119 E2** 10 0 S 25 5 E
Manikganj, *Bangla.* **90 D3** 23 52N 90 0 E
Manila, *Phil.* **80 D3** 14 40N 121 3 E
Manila, *U.S.A.* **158 F9** 40 59N 109 43W
Manila B., *Phil.* **80 D3** 14 40N 120 35 E
Manilla, *Australia* **129 A9** 30 45 S 150 43 E
Manimpé, *Mali* **112 C3** 14 11N 5 28W
Maningrida, *Australia* **126 A1** 12 3 S 134 13 E
Manipur □, *India* **90 C5** 25 0N 94 0 E
Manipur →, *Burma* **90 D5** 23 45N 94 20 E
Manisa, *Turkey* **57 G4** 38 38N 27 30 E
Manisa □, *Turkey* **49 C9** 38 40N 28 0 E
Manistee, *U.S.A.* **148 C2** 44 15N 86 19W
Manistee →, *U.S.A.* **148 C2** 44 15N 86 21W
Manistique, *U.S.A.* **148 C2** 45 57N 86 15W
Manito, *U.S.A.* **156 D7** 40 26N 89 47W
Manito L., *Canada* **143 C7** 52 43N 109 43W
Manitoba □, *Canada* **143 B9** 55 30N 97 0W
Manitoba, L., *Canada* **143 C9** 51 0N 98 45W
Manitou, *Canada* **143 D9** 49 15N 98 32W
Manitou Beach, *U.S.A.* **157 C12** 41 58N 84 19W
Manitou I., *U.S.A.* **140 C2** 47 25N 87 37W
Manitou Is., *U.S.A.* **148 C3** 45 8N 86 0W
Manitou L., *Canada* **141 B6** 50 55N 65 17W
Manitou Springs, *U.S.A.* **154 F2** 38 52N 104 55W
Manitoulin I., *Canada* **140 C3** 45 40N 82 30W
Manitowaning, *Canada* **140 C3** 45 46N 81 49W
Manitowoc, *U.S.A.* **148 C2** 44 5N 87 40W
Manitsauá-Missu →, *Brazil* **173 C7** 10 58 S 53 20W
Manītsoé, *Greenland* **10 D5** 65 26N 52 55W
Manizales, *Colombia* **168 B2** 5 5N 75 32W
Manja, *Madag.* **117 C7** 21 26 S 44 20 E
Manjacaze, *Mozam.* **117 C5** 24 45 S 34 0 E
Manjakandriana, *Madag.* **117 B8** 18 55 S 47 47 E
Manjeri, *India* **95 J3** 11 7N 76 11 E
Manjhand, *Pakistan* **91 D3** 25 50N 68 10 E
Manjil, *Iran* **97 B6** 36 46N 49 30 E
Manjimup, *Australia* **125 F2** 34 15 S 116 6 E
Manjra →, *India* **94 E3** 18 49N 77 52 E
Mankato, *Kans., U.S.A.* **154 F5** 39 47N 98 13W
Mankato, *Minn., U.S.A.* **154 C8** 44 10N 94 0W
Mankayan, *Phil.* **80 C3** 16 52N 120 47 E
Mankayane, *Swaziland* **117 D5** 26 40 S 31 4 E
Mankono, *Ivory C.* **112 D3** 8 1N 6 10W
Mankota, *Canada* **143 D7** 49 25N 107 5W
Manlay, *Mongolia* **74 B4** 44 9N 107 0 E
Manley Hot Springs, *U.S.A.* **144 D10** 65 0N 150 38W
Manlleu, *Spain* **40 C7** 42 2N 2 17 E
Manly, *Australia* **129 B9** 33 48 S 151 17 E
Manmad, *India* **94 D2** 20 18N 74 28 E
Mann Ranges, *Australia* **125 E5** 26 6 S 130 5 E
Manna, *Indonesia* **84 C2** 4 25 S 102 55 E
Mannahill, *Australia* **128 B4** 32 25 S 140 0 E
Mannar, *Sri Lanka* **95 K4** 9 1N 79 54 E
Mannar, G. of, *Asia* **95 K4** 8 30N 79 0 E
Mannar I., *Sri Lanka* **95 K4** 9 5N 79 45 E
Mannargudi, *India* **95 J4** 10 45N 79 51 E
Mannduque □, *Phil.* **80 E4** 13 18N 122 0 E
Männedorf, *Switz.* **33 B7** 47 15N 8 43 E
Mannheim, *Germany* **31 F4** 49 29N 8 29 E
Manning, *Canada* **142 B5** 56 53N 117 39W
Manning, *Oreg., U.S.A.* **160 E3** 45 45N 123 13W
Manning, *S.C., U.S.A.* **152 B9** 33 42N 80 13W
Manning Prov. Park, *Canada* **142 D4** 49 5N 120 45W
Manning Str., *Solomon Is.* **133 L10** 7 30 S 158 0 E
Mannington, *U.S.A.* **148 F5** 39 32N 80 21W
Mannu →, *Italy* **46 C2** 39 16N 9 0 E
Mannu, C., *Italy* **46 B1** 40 3N 8 21 E
Mannum, *Australia* **128 C3** 34 50 S 139 20 E
Mano, *S. Leone* **112 D2** 8 3N 12 2W
Manoa, *Bolivia* **173 B4** 9 40 S 65 27W
Manokotak, *U.S.A.* **144 G8** 58 58N 159 3W
Manokwari, *Indonesia* **83 B4** 0 54 S 134 0 E
Manolás, *Greece* **48 C3** 38 4N 21 21 E
Manolo Fortich, *Phil.* **81 G5** 8 28N 124 50 E
Manombo, *Madag.* **117 C7** 22 57 S 43 28 E
Manono, *Dem. Rep. of the Congo* **118 D2** 7 15 S 27 25 E
Manoppello, *Italy* **45 F11** 42 15N 14 3 E
Manosque, *France* **29 E9** 43 49N 5 47 E
Manouane, L., *Canada* **141 B5** 50 45N 70 45W
Manouro, Pt., *Vanuatu* **133 G6** 17 41 S 168 36 E
Manpojin, *N. Korea* **75 D14** 41 6N 126 24 E
Manresa, *Spain* **40 D6** 41 48N 1 50 E
Mansa, *Gujarat, India* **92 H5** 23 27N 72 45 E
Mansa, *Punjab, India* **92 E6** 30 0N 75 27 E
Mansa, *Zambia* **119 E2** 11 13 S 28 55 E
Mansalay, *Phil.* **80 E3** 12 31N 121 26 E
Månsåsen, *Sweden* **16 A8** 63 5N 14 18 E
Mansehra, *Pakistan* **92 B5** 34 20N 73 15 E
Mansel I., *Canada* **139 B12** 62 0N 80 0W
Mansfield, *Australia* **129 D7** 37 4 S 146 6 E
Mansfield, *U.K.* **20 D6** 53 9N 1 11W
Mansfield, *Ga., U.S.A.* **152 B6** 33 31N 83 44W
Mansfield, *La., U.S.A.* **155 J8** 32 2N 93 43W
Mansfield, *Mass., U.S.A.* **151 D13** 42 2N 71 13W
Mansfield, *Ohio, U.S.A.* **150 F2** 40 45N 82 31W
Mansfield, *Pa., U.S.A.* **150 E7** 41 48N 77 5W
Mansfield, *Wash., U.S.A.* **158 C4** 47 49N 119 38W
Mansi, *Burma* **90 C5** 24 48N 95 52 E
Mansidão, *Brazil* **170 D3** 10 43 S 44 2W
Mansilla de las Mulas, *Spain* **42 C5** 42 30N 5 25W
Mansle, *France* **28 C4** 45 52N 0 12 E
Manso →, *Brazil* **171 D2** 13 50 S 47 0W
Mansoa, *Guinea-Biss.* **112 C1** 12 0N 15 20W
Manson, *U.S.A.* **156 B2** 42 32N 94 32W
Manson Creek, *Canada* **142 B4** 55 37N 124 32W
Mansoura, *Algeria* **111 A5** 36 1N 4 31 E
Manta, *Ecuador* **168 D1** 1 0 S 80 40W
Manta, B. de, *Ecuador* **168 D1** 0 54 S 80 44W
Mantalingajan, Mt., *Phil.* **81 G1** 8 55N 117 45 E
Mantare, *Tanzania* **118 C3** 2 42 S 33 13 E
Manteca, *U.S.A.* **160 H5** 37 48N 121 13W
Mantecal, *Venezuela* **168 B4** 7 34N 69 17W
Mantena, *Brazil* **171 E3** 18 47 S 40 59W
Manteno, *U.S.A.* **157 C9** 41 15N 87 50W
Manteo, *U.S.A.* **149 H8** 35 55N 75 40W
Mantes-la-Jolie, *France* **27 D8** 48 58N 1 41 E
Manthani, *India* **94 E4** 18 40N 79 35 E
Manti, *U.S.A.* **158 G8** 39 16N 111 38W
Mantiqueira, Serra da, *Brazil* **171 F3** 22 0 S 44 0W
Manton, *U.S.A.* **148 C3** 44 25N 85 24W
Mantorp, *Sweden* **17 F9** 58 21N 15 20 E
Mántova, *Italy* **44 C7** 45 9N 10 48 E
Mänttä, *Finland* **15 E21** 62 0N 24 40 E
Mantua = Mántova, *Italy* **44 C7** 45 9N 10 48 E
Mantung, *Australia* **128 C4** 34 35 S 140 3 E
Manturovo, *Russia* **60 A7** 58 23N 44 45 E
Manu, *Peru* **172 C3** 12 10 S 70 51W
Manu →, *Peru* **172 C3** 12 16 S 70 55W
Manua Is., *Amer. Samoa* **133 X25** 14 13 S 169 35W
Manuae, *Cook Is.* **135 J12** 19 30 S 159 0W
Manuel Alves →, *Brazil* **171 D2** 11 19 S 48 28W
Manuel Alves Grande →, *Brazil* **170 C2** 7 27 S 47 35W
Manuel Urbano, *Brazil* **172 B4** 8 53 S 69 18W
Manui, *Indonesia* **82 B2** 3 35 S 123 5 E
Manukau Harbour, *N.Z.* **130 D3** 37 3 S 174 45 E
Manunui, *N.Z.* **130 E4** 38 54 S 175 21 E
Manurewa, *N.Z.* **130 D3** 37 1 S 174 54 E
Manuripi →, *Bolivia* **172 C4** 11 6 S 67 36W
Manus I., *Papua N. G.* **132 B4** 2 0 S 147 0 E
Manvi, *India* **95 G3** 15 57N 76 59 E
Manville, *U.S.A.* **154 D2** 42 47N 104 37W
Manwath, *India* **94 E3** 19 19N 76 32 E
Many, *U.S.A.* **155 K8** 31 34N 93 29W
Manyara, L., *Tanzania* **118 C4** 3 40 S 35 50 E
Manyas, *Turkey* **51 F11** 40 2N 27 59 E
Manych →, *Russia* **61 G5** 47 13N 40 40 E
Manych-Gudilo, Ozero, *Russia* **61 G6** 46 24N 42 38 E
Manyonga →, *Tanzania* **118 C3** 4 10 S 34 15 E
Manyoni, *Tanzania* **118 D3** 5 45 S 34 55 E
Manyoni □, *Tanzania* **118 D3** 6 30 S 34 30 E
Manzai, *Pakistan* **91 B3** 32 12N 70 15 E
Manzala, Bahra el, *Egypt* **106 H7** 31 10N 31 56 E
Manzanares, *Spain* **43 F7** 39 2N 3 22W
Manzaneda, *Spain* **42 C3** 42 12N 7 15W
Manzanillo, *Cuba* **164 B4** 20 20N 77 31W
Manzanillo, *Mexico* **162 D4** 19 0N 104 20W
Manzanillo, Pta., *Panama* **164 E4** 9 30N 79 40W
Manzano Mts., *U.S.A.* **159 J10** 34 40N 106 20W
Manzarīyeh, *Iran* **97 C6** 34 53N 50 50 E
Manzhouli, *China* **73 B6** 49 35N 117 25 E
Manzini, *Swaziland* **117 D5** 26 30 S 31 25 E
Mao, *Chad* **109 F3** 14 4N 15 19 E
Maó, *Spain* **39 B11** 39 53N 4 16 E
Maoke, Pegunungan, *Indonesia* **83 B5** 3 40 S 137 30 E
Maolin, *China* **75 C12** 43 58N 123 30 E
Maoming, *China* **77 G8** 21 50N 110 54 E
Maowen, *China* **76 B4** 31 41N 103 49 E
Maoxing, *China* **75 B13** 45 28N 124 40 E
Mapam Yumco, *China* **72 C3** 30 45N 81 28 E
Mapastepec, *Mexico* **163 D6** 15 26N 92 54W
Mapia, Kepulauan, *Indonesia* **83 A4** 0 50N 134 20 E
Mapimí, *Mexico* **162 B4** 25 50N 103 50W
Mapimí, Bolsón de, *Mexico* **162 B4** 27 30N 104 15W
Maping, *China* **77 B9** 31 34N 113 32 E
Mapinga, *Tanzania* **118 D4** 6 40 S 39 12 E
Mapinhane, *Mozam.* **117 C6** 22 20 S 35 0 E
Mapire, *Venezuela* **169 B5** 7 45N 64 42W
Maple →, *U.S.A.* **157 B12** 42 59N 84 57W
Maple Creek, *Canada* **143 D7** 49 55N 109 29W
Maple Valley, *U.S.A.* **160 C4** 47 25N 122 3W
Mapleton, *U.S.A.* **158 D2** 44 2N 123 52W
Mapourika, L., *N.Z.* **131 D5** 43 16 S 170 12 E
Maprik, *Papua N. G.* **132 B2** 3 44 S 143 2 E
Mapuca, *India* **95 G1** 15 36N 73 46 E
Mapuera →, *Brazil* **169 D6** 1 5 S 57 2W
Maputing Baybay, *Phil.* **80 E4** 12 45N 123 20 E
Maputo, *Mozam.* **117 D5** 25 58 S 32 32 E
Maputo, B. de, *Mozam.* **117 D5** 25 50 S 32 45 E
Maqiaohe, *China* **75 B16** 44 40N 130 30 E
Maqnā, *Si. Arabia* **96 D2** 28 25N 34 50 E
Maqran, W. →, *Si. Arabia* **98 B4** 20 55N 47 12 E
Maqteïr, *Mauritania* **110 D2** 21 50N 11 40W
Maqueda, *Spain* **42 E6** 40 4N 4 22W
Maqueda Channel, *Phil.* **80 E5** 13 42N 124 1 E
Maquela do Zombo, *Angola* **115 D3** 6 0 S 15 15 E
Maquinchao, *Argentina* **176 B3** 41 15 S 68 50W
Maquoketa, *U.S.A.* **156 B6** 42 4N 90 40W
Mar, Serra do, *Brazil* **175 B6** 25 30 S 49 0W
Mar Chiquita, L., *Argentina* **174 C3** 30 40 S 62 50W
Mar del Plata, *Argentina* **174 D4** 38 0 S 57 30W
Mar Menor, *Spain* **41 H4** 37 40N 0 45W
Mara, *Guyana* **169 B6** 6 0N 57 36W
Mara, *India* **90 A5** 28 11N 94 14 E
Mara, *Tanzania* **118 C3** 1 30 S 34 32 E
Mara □, *Tanzania* **118 C3** 1 45 S 34 20 E
Maraã, *Brazil* **168 D4** 1 52 S 65 25W
Marabá, *Brazil* **170 C2** 5 20 S 49 5W
Maracá, I. de, *Brazil* **169 C7** 2 10N 50 30W
Maracaibo, *Venezuela* **168 A3** 10 40N 71 37W
Maracaibo, L. de, *Venezuela* **168 B3** 9 40N 71 30W
Maracaju, *Brazil* **175 A4** 21 38 S 55 9W
Maracajú, Serra de, *Brazil* **173 E6** 23 57 S 55 1W
Maracanã, *Brazil* **170 B2** 0 46 S 47 27W
Maracás, *Brazil* **171 D3** 13 26 S 40 18W
Maracay, *Venezuela* **168 A4** 10 15N 67 28W
Maracena, *Spain* **43 H7** 37 12N 3 38W
Marādah, *Libya* **108 C3** 29 15N 19 15 E
Maradi, *Niger* **113 C6** 13 29N 7 20 E
Maradun, *Nigeria* **113 C6** 12 35N 6 18 E
Marāgheh, *Iran* **101 D12** 37 30N 46 12 E
Maragogipe, *Brazil* **171 D4** 12 46 S 38 55W
Maragondon, *Phil.* **80 D3** 14 16N 120 44 E
Marāh, *Si. Arabia* **96 E5** 25 0N 45 35 E
Marajó, B. de, *Brazil* **170 B2** 1 0 S 48 30W
Marajó, I. de, *Brazil* **170 B2** 1 0 S 49 30W
Marākand, *Iran* **96 B5** 38 51N 45 16 E
Maralal, *Kenya* **118 B4** 1 0N 36 38 E
Maralinga, *Australia* **125 F5** 30 13 S 131 32 E
Marama, *Australia* **128 C4** 35 10 S 140 10 E
Maramaraereğlisi, *Turkey* **51 F11** 40 57N 27 57 E
Maramasike, *Solomon Is.* **133 M11** 9 30 S 161 25 E
Marampa, *S. Leone* **112 D2** 8 45N 12 28W
Maramureş □, *Romania* **53 C9** 47 45N 24 0 E
Maran, *Malaysia* **87 L4** 3 35N 102 45 E
Marana, *U.S.A.* **159 K8** 32 27N 111 13W
Maranboy, *Australia* **124 B5** 14 40 S 132 39 E
Maranchón, *Spain* **40 D2** 41 6N 2 15W
Marand, *Iran* **101 C11** 38 30N 45 45 E
Marang, *Malaysia* **87 K4** 5 12N 103 13 E
Maranguape, *Brazil* **170 B4** 3 55 S 38 50W
Maranhão = São Luís, *Brazil* **170 B3** 2 39 S 44 15W
Maranhão □, *Brazil* **170 C2** 5 0 S 46 0W
Marano, L. di, *Italy* **45 C10** 45 43N 13 10 E
Maranoa →, *Australia* **127 D4** 27 50 S 148 37 E
Marañón →, *Peru* **172 A3** 4 30 S 73 35W
Marão, *Mozam.* **117 C5** 24 18 S 34 2 E
Marapi →, *Brazil* **169 C6** 0 37N 55 58W
Marari, *Brazil* **172 B4** 5 43 S 67 47W
Maraş = Kahramanmaraş, *Turkey* **100 D7** 37 37N 36 53 E
Mărăşeşti, *Romania* **53 E12** 45 52N 27 14 E
Maratea, *Italy* **47 C8** 39 59N 15 43 E
Marateca, *Portugal* **43 G2** 38 34N 8 40W
Marathasa □, *Cyprus* **38 E11** 34 59N 32 51 E
Marathókambos, *Greece* **49 D8** 37 43N 26 42 E
Marathon, *Australia* **126 C3** 20 51 S 143 32 E
Marathon, *Canada* **140 C2** 48 44N 86 23W
Marathón, *Greece* **48 C5** 38 11N 23 58 E
Marathon, *Fla., U.S.A.* **153 L8** 24 43N 81 5W
Marathon, *Iowa, U.S.A.* **156 B2** 42 52N 94 59W
Marathon, *N.Y., U.S.A.* **151 D8** 42 27N 76 2W
Marathon, *Tex., U.S.A.* **155 K3** 30 12N 103 15W
Marathóvouno, *Cyprus* **38 D12** 35 13N 33 37 E
Maratua, *Indonesia* **85 B5** 2 10N 118 35 E
Maraú, *Brazil* **171 D4** 14 6 S 39 0W
Maravatío, *Mexico* **162 D4** 19 51N 100 25W

Middelburg 249

Mehr Jān, *Iran* **97 C7** 33 50N 55 6 E
Mehrābād, *Iran* **101 D12** 36 53N 47 55 E
Mehrān, *Iran* **101 F12** 33 7N 46 10 E
Mehrīz, *Iran* **97 D7** 31 35N 54 28 E
Mehun-sur-Yèvre, *France* **27 E9** 47 10N 2 13 E
Mei Jiang →, *China* **77 E11** 24 25N 116 35 E
Mei Xian, *Guangdong, China* **77 E11** 24 16N 116 6 E
Mei Xian, *Shaanxi, China* **74 G4** 34 18N 107 55 E
Meia Ponte →, *Brazil* **171 E2** 18 32 S 49 36W
Meicheng, *China* **77 C12** 29 29N 119 16 E
Meichengzhen, *China* **77 C8** 28 9N 111 40 E
Meichuan, *China* **77 B10** 30 8N 115 31 E
Meiganga, *Cameroon* **114 A2** 6 30N 14 25 E
Meigs, *U.S.A.* **152 D5** 31 4N 84 6W
Meiktila, *Burma* **90 E5** 20 53N 95 54 E
Meilen, *Switz.* **33 B7** 47 16N 8 39 E
Meinerzhagen, *Germany* **30 D3** 51 6N 7 38 E
Meiningen, *Germany* **30 E6** 50 34N 10 25 E
Meio →, *Brazil* **171 D3** 13 36 S 44 7W
Meira, Serra de, *Spain* **42 B3** 43 15N 7 15W
Meiringen, *Switz.* **32 C6** 46 43N 8 12 E
Meishan, *China* **76 B4** 30 3N 103 23 E
Meissen, *Germany* **30 D9** 51 9N 13 29 E
Meissner, *Germany* **30 D5** 51 14N 9 50 E
Meitan, *China* **76 D6** 27 45N 107 29 E
Mejillones, *Chile* **174 A1** 23 10 S 70 30W
Meka, *Australia* **125 E2** 27 25 S 116 48 E
Mékambo, *Gabon* **114 B2** 1 2N 13 50 E
Mekdela, *Ethiopia* **107 E4** 11 24N 39 10 E
Mekele, *Ethiopia* **107 E4** 13 33N 39 30 E
Mekhtar, *Pakistan* **91 C3** 30 30N 69 15 E
Meknès, *Morocco* **110 B3** 33 57N 5 33W
Meko, *Nigeria* **113 D5** 7 27N 2 52 E
Mekong →, *Asia* **87 H6** 9 30N 106 15 E
Mekongga, *Indonesia* **85 E6** 3 39 S 121 15 E
Mekoryuk, *U.S.A.* **144 F6** 60 23N 166 11W
Mekvari = Kür →, *Azerbaijan* **101 C13** 39 29N 49 15 E
Mel, *Italy* **45 B9** 46 4N 12 5 E
Melagiri Hills, *India* **95 H3** 12 20N 77 30 E
Melah, Sebkhet el, *Algeria* **111 C4** 29 20N 1 30W
Melaka, *Malaysia* **87 L4** 2 15N 102 15 E
Melaka □, *Malaysia* **84 B2** 2 15N 102 15 E
Mélambes, *Greece* **38 D6** 35 8N 24 40 E
Melanesia, *Pac. Oc.* **134 H7** 4 0 S 155 0 E
Melapalaiyam, *India* **95 K3** 8 39N 77 44 E
Melawi →, *Indonesia* **85 B4** 0 5N 111 29 E
Melbourne, *Australia* **129 D6** 37 50 S 145 0 E
Melbourne, *Fla., U.S.A.* **149 L5** 28 5N 80 37W
Melbourne, *Iowa, U.S.A.* **156 C3** 41 57N 93 6W
Melcher, *U.S.A.* **156 C3** 41 14N 93 15W
Melchor Múzquiz, *Mexico* **162 B4** 27 50N 101 30W
Melchor Ocampo, *Mexico* **162 C4** 24 52N 101 40W
Meldal, *Norway* **18 A6** 63 3N 9 44 E
Méldola, *Italy* **45 D9** 44 7N 12 5 E
Meldorf, *Germany* **30 A5** 54 5N 9 5 E
Meleden, *Somali Rep.* **120 B3** 10 25N 49 51 E
Melegnano, *Italy* **44 C6** 45 21N 9 19 E
Melenci, *Serbia, Yug.* **52 E5** 45 32N 20 20 E
Melenki, *Russia* **60 C5** 55 20N 41 37 E
Meleuz, *Russia* **62 E6** 52 58N 55 55 E
Mélèzes →, *Canada* **139 C12** 57 30N 71 0W
Melfi, *Chad* **109 F3** 11 0N 17 59 E
Melfi, *Italy* **47 B8** 41 0N 15 39 E
Melfort, *Canada* **143 C8** 52 50N 104 37W
Melfort, *Zimbabwe* **119 F3** 18 0 S 31 25 E
Melgaço, *Portugal* **42 C2** 42 7N 8 15W
Melgar de Fernamental, *Spain* **42 C6** 42 27N 4 17W
Melhus, *Norway* **14 E14** 63 17N 10 18 E
Melide, *Spain* **42 C2** 42 55N 8 1W
Melide, *Switz.* **33 E7** 45 57N 8 57 E
Meligalá, *Greece* **48 D3** 37 15N 21 59 E
Melilla, *N. Afr.* **111 A4** 35 21N 2 57W
Melilli, *Italy* **47 E8** 37 11N 15 7 E
Melipilla, *Chile* **174 C1** 33 42 S 71 15W
Mélissa, Ákra, *Greece* **38 D6** 35 6N 24 33 E
Mélissa Óros, *Greece* **49 D8** 37 32N 26 4 E
Melita, *Canada* **143 D8** 49 15N 101 0W
Mélito di Porto Salvo, *Italy* **47 E8** 37 55N 15 47 E
Melitopol, *Ukraine* **59 J8** 46 50N 35 22 E
Melk, *Austria* **34 C8** 48 13N 15 20 E
Mellan Fryken, *Sweden* **16 E7** 59 45N 13 10 E
Mellansel, *Sweden* **14 E18** 63 25N 18 17 E
Mellbystrand, *Sweden* **17 H6** 56 30N 12 56 E
Melle, *France* **28 B3** 46 14N 0 10W
Melle, *Germany* **30 C4** 52 12N 8 20 E
Mellègue, O. →, *Tunisia* **108 A1** 36 36N 9 14 E
Mellen, *U.S.A.* **154 B9** 46 20N 90 40W
Mellerud, *Sweden* **17 F6** 58 41N 12 28 E
Mellette, *U.S.A.* **154 C5** 45 9N 98 30W
Mellid = Melide, *Spain* **42 C2** 42 55N 8 1W
Mellieha, *Malta* **38 D1** 35 57N 14 21 E
Mellit, *Sudan* **107 E2** 14 7N 25 34 E
Mellizo Sur, Cerro, *Chile* **176 C2** 48 33 S 73 10W
Mellrichstadt, *Germany* **30 E6** 50 25N 10 17 E
Melnik, *Bulgaria* **50 E7** 41 30N 23 25 E
Mělník, *Czech Rep.* **34 A7** 50 22N 14 23 E
Melo, *Uruguay* **175 C5** 32 20 S 54 10W
Melolo, *Indonesia* **82 C2** 9 53 S 120 40 E
Melouprey, *Cambodia* **86 F5** 13 48N 105 16 E
Melrhir, Chott, *Algeria* **111 B6** 34 13N 6 30 E
Melrose, *N.S.W., Australia* **129 B7** 32 42 S 146 57 E
Melrose, *W. Austral., Australia* **125 E3** 27 50 S 121 15 E
Melrose, *U.K.* **22 F6** 55 36N 2 43W
Melrose, *Iowa, U.S.A.* **156 D3** 40 59N 93 3W
Melrose, *N. Mex., U.S.A.* **155 H3** 34 26N 103 38W
Mels, *Switz.* **33 B8** 47 3N 9 25 E
Melstone, *U.S.A.* **158 C10** 46 36N 107 52W
Melsungen, *Germany* **30 D5** 51 7N 9 32 E
Melton Mowbray, *U.K.* **20 E7** 52 47N 0 54W
Melun, *France* **27 D9** 48 32N 2 39 E
Melur, *India* **95 J4** 10 2N 78 23 E
Melut, *Sudan* **107 E3** 10 30N 32 13 E
Melville, *Canada* **143 C8** 50 55N 102 50W
Melville, C., *Australia* **126 A3** 14 11 S 144 30 E
Melville, L., *Canada* **141 B8** 53 30N 60 0W
Melville B., *Australia* **126 A2** 12 0 S 136 45 E
Melville Bugt, *Greenland* **10 B4** 75 30N 63 0W
Melville I., *Australia* **124 B5** 11 30 S 131 0 E
Melville I., *Canada* **6 B2** 75 30N 112 0W
Melville Pen., *Canada* **139 B11** 68 0N 84 0W
Melvin →, *Canada* **142 B5** 59 11N 117 31W
Mélykút, *Hungary* **52 D4** 46 11N 19 25 E
Memaliaj, *Albania* **50 F3** 40 25N 19 56 E
Memba, *Mozam.* **119 E5** 14 11 S 40 30 E
Memboro, *Indonesia* **82 C1** 9 30 S 119 30 E
Membrilla, *Spain* **43 G7** 38 59N 3 21W
Memel = Klaipėda, *Lithuania* **15 J19** 55 43N 21 10 E
Memel, *S. Africa* **117 D4** 27 38 S 29 36 E
Memmingen, *Germany* **31 H6** 47 58N 10 10 E
Mempawah, *Indonesia* **85 B3** 0 30N 109 5 E
Memphis, *Fla., U.S.A.* **153 H7** 27 32N 82 34W
Memphis, *Mo., U.S.A.* **156 D4** 40 28N 92 10W
Memphis, *Tenn., U.S.A.* **155 H10** 35 8N 90 3W
Memphis, *Tex., U.S.A.* **155 H4** 34 44N 100 33W
Mena, *Ukraine* **59 G7** 51 31N 32 13 E
Mena, *U.S.A.* **155 H7** 34 35N 94 15W
Mena →, *Ethiopia* **107 F5** 5 40N 40 50 E
Menai Strait, *U.K.* **20 D3** 53 11N 4 13W
Ménaka, *Mali* **113 B5** 15 59N 2 18 E
Menamurtee, *Australia* **128 A5** 31 25 S 143 11 E
Menan = Chao Phraya →, *Thailand* **86 F3** 13 32N 100 36 E
Menarandra →, *Madag.* **117 D7** 25 17 S 44 30 E
Menard, *U.S.A.* **155 K5** 30 55N 99 47W
Menasha, *U.S.A.* **148 C1** 44 13N 88 26W
Menate, *Indonesia* **85 C4** 0 12 S 113 3 E
Mendawai →, *Indonesia* **85 C4** 3 30 S 113 0 E
Mende, *France* **28 D7** 44 31N 3 30 E
Mendebo, *Ethiopia* **107 F4** 7 0N 39 22 E
Menden, *Germany* **30 D3** 51 26N 7 47 E
Menderes, *Turkey* **49 C9** 38 14N 27 8 E
Mendez, *Mexico* **163 B5** 25 7N 98 34W
Mendez Nunez, *Phil.* **80 D3** 14 8N 120 54 E
Mendhar, *India* **93 C6** 33 35N 74 10 E
Mendi, *Ethiopia* **107 F4** 9 47N 35 4 E
Mendi, *Papua N. G.* **132 D2** 6 11 S 143 39 E
Mendip Hills, *U.K.* **21 F5** 51 17N 2 40W
Mendocino, *U.S.A.* **158 G2** 39 19N 123 48W
Mendocino, C., *U.S.A.* **158 F1** 40 26N 124 25W
Mendon, *U.S.A.* **157 C11** 42 0N 85 27W
Mendota, *Calif., U.S.A.* **160 J6** 36 45N 120 23W
Mendota, *Ill., U.S.A.* **156 C7** 41 33N 89 7W
Mendoza, *Argentina* **174 C2** 32 50 S 68 52W
Mendoza □, *Argentina* **174 C2** 33 0 S 69 0W
Mendrisio, *Switz.* **33 E7** 45 52N 8 59 E
Mene Grande, *Venezuela* **168 B3** 9 49N 70 56W
Menemen, *Turkey* **49 C9** 38 34N 27 3 E
Menen, *Belgium* **24 D3** 50 47N 3 7 E
Menéndez, L., *Argentina* **176 B2** 42 40 S 71 51W
Menfi, *Italy* **46 E5** 37 36N 12 58 E
Mengcheng, *China* **77 A11** 33 18N 116 31 E
Mengdingjie, *China* **76 F2** 23 31N 98 58 E
Mengeš, *Slovenia* **45 B11** 46 10N 14 35 E
Menggala, *Indonesia* **84 C3** 4 30 S 105 15 E
Menghai, *China* **76 G3** 21 49N 100 55 E
Mengíbar, *Spain* **43 H7** 37 58N 3 48W
Mengjin, *China* **74 G7** 34 55N 112 45 E
Mengla, *China* **76 G3** 21 20N 101 25 E
Menglian, *China* **76 F2** 22 21N 99 27 E
Mengoub, *Algeria* **110 C3** 29 49N 5 26W
Mengshan, *China* **77 E8** 24 14N 110 55 E
Mengyin, *China* **75 G9** 35 40N 117 58 E
Mengzhe, *China* **76 F3** 22 2N 100 15 E
Mengzi, *China* **76 F4** 23 20N 103 22 E
Menihek L., *Canada* **141 B6** 54 0N 67 0W
Menin = Menen, *Belgium* **24 D3** 50 47N 3 7 E
Menindee, *Australia* **128 B5** 32 20 S 142 25 E
Menindee L., *Australia* **128 B5** 32 20 S 142 25 E
Meningie, *Australia* **128 C3** 35 50 S 139 18 E
Menlo Park, *U.S.A.* **160 H4** 37 27N 122 12W
Menominee, *U.S.A.* **148 C2** 45 6N 87 37W
Menominee →, *U.S.A.* **148 C2** 45 6N 87 36W
Menomonee Falls, *U.S.A.* **157 A8** 43 11N 88 7W
Menomonie, *U.S.A.* **154 C9** 44 53N 91 55W
Menongue, *Angola* **115 E3** 14 48 S 17 52 E
Menorca, *Spain* **39 B11** 40 0N 4 0 E
Mentakab, *Malaysia* **87 L4** 3 29N 102 21 E
Mentasta Lake, *U.S.A.* **144 E12** 62 55N 143 45W
Mentawai, Kepulauan, *Indonesia* **84 C1** 2 0 S 99 0 E
Menton, *France* **29 E11** 43 50N 7 29 E
Mentone, *U.S.A.* **157 C10** 41 10N 86 2W
Mentor, *U.S.A.* **150 E3** 41 40N 81 21W
Menyamya, *Papua N. G.* **132 D3** 7 10 S 145 59 E
Menzel-Bourguiba, *Tunisia* **108 A1** 37 9N 9 49 E
Menzel Chaker, *Tunisia* **108 B1** 35 0N 10 26 E
Menzel-Temime, *Tunisia* **108 A2** 36 46N 11 0 E
Menzelinsk, *Russia* **62 D4** 55 47N 53 11 E
Menzies, *Australia* **125 E3** 29 40 S 121 2 E
Me'ona, *Israel* **103 B4** 33 1N 35 15 E
Meoqui, *Mexico* **162 B3** 28 17N 105 29W
Mepaco, *Mozam.* **119 F3** 15 57 S 30 48 E
Meppel, *Neths.* **24 B6** 52 42N 6 12 E
Meppen, *Germany* **30 C3** 52 42N 7 17 E
Mequinenza, *Spain* **40 D5** 41 22N 0 17 E
Mequinenza, Embalse de, *Spain* **40 D5** 41 25N 0 15 E
Mequon, *U.S.A.* **157 A9** 43 14N 87 59W
Mer, *France* **26 E8** 47 42N 1 30 E
Mer Rouge, *U.S.A.* **155 J9** 32 47N 91 48W
Mera Lava, *Vanuatu* **133 D6** 14 25 S 168 3 E
Merabéllou, Kólpos, *Greece* **38 D7** 35 10N 25 50 E
Merai, *Papua N. G.* **132 C7** 4 52 S 152 19 E
Meråker, *Norway* **18 A8** 63 25N 11 46 E
Meramangye, L., *Australia* **125 E5** 28 25 S 132 13 E
Meramec →, *U.S.A.* **156 F6** 38 24N 90 21W
Meran = Merano, *Italy* **45 B8** 46 40N 11 9 E
Merano, *Italy* **45 B8** 46 40N 11 9 E
Merate, *Italy* **44 C6** 45 42N 9 25 E
Merauke, *Indonesia* **83 C6** 8 29 S 140 24 E
Merbabu, *Indonesia* **85 D4** 7 30 S 110 40 E
Merbein, *Australia* **128 C5** 34 10 S 142 2 E
Merca, *Somali Rep.* **120 D2** 1 48N 44 50 E
Mercato Saraceno, *Italy* **45 E9** 43 57N 12 12 E
Merced, *U.S.A.* **160 H6** 37 18N 120 29W
Merced Pk., *U.S.A.* **160 H7** 37 36N 119 24W
Mercedes, *Buenos Aires, Argentina* **174 C4** 34 40 S 59 30W
Mercedes, *Corrientes, Argentina* **174 B4** 29 10 S 58 5W
Mercedes, *San Luis, Argentina* **174 C2** 33 40 S 65 21W
Mercedes, *Camarines N., Phil.* **80 D4** 14 7N 123 1 E
Mercedes, *Leyte, Phil.* **81 F5** 10 41N 124 24 E
Mercedes, *Zamboanga del S., Phil.* **81 H4** 6 57N 122 9 E
Mercedes, *Uruguay* **174 C4** 33 12 S 58 0W
Merceditas, *Chile* **174 B1** 28 20 S 70 35W
Mercer, *N.Z.* **130 D4** 37 16 S 175 5 E
Mercer, *Mo., U.S.A.* **156 D3** 40 31N 93 32W
Mercer, *Pa., U.S.A.* **150 E4** 41 14N 80 15W
Mercier, *Bolivia* **172 C4** 10 42 S 68 5W
Mercury, *U.S.A.* **161 J11** 36 40N 115 58W
Mercury B., *N.Z.* **130 C4** 36 48 S 175 45 E
Mercury Is., *N.Z.* **130 C4** 36 37 S 175 52 E
Mercy C., *Canada* **139 B13** 65 0N 63 30W
Merdrignac, *France* **26 D4** 48 11N 2 27W
Mere, *U.K.* **21 F5** 51 6N 2 16W
Meredith, C., *Falk. Is.* **176 D4** 52 15 S 60 40W
Meredith, L., *U.S.A.* **155 H4** 35 43N 101 33W
Meredosia, *U.S.A.* **156 E6** 39 50N 90 34W
Merefa, *Ukraine* **59 H9** 49 48N 36 3 E
Meregh, *Somali Rep.* **120 D3** 3 46N 47 18 E
Merei, *Romania* **53 E11** 45 7N 26 43 E
Merga = Nukheila, *Sudan* **106 D2** 19 1N 26 21 E
Mergui, *Burma* **86 F2** 12 26N 98 34 E
Mergui Arch. = Myeik Kyunzu, *Burma* **87 G1** 11 30N 97 30 E
Meribah, *Australia* **128 C4** 34 43 S 140 51 E
Meriç, *Turkey* **51 E10** 41 11N 26 25 E
Meriç →, *Turkey* **51 F10** 40 52N 26 12 E
Mérida, *Mexico* **163 C7** 20 58N 89 37W
Mérida, *Phil.* **81 F5** 10 55N 124 32 E
Mérida, *Spain* **43 G4** 38 55N 6 25W
Mérida, *Venezuela* **168 B3** 8 24N 71 8W
Mérida □, *Venezuela* **168 B3** 8 30N 71 10W
Mérida, Cord. de, *Venezuela* **168 B3** 9 0N 71 0W
Meriden, *U.K.* **21 E6** 52 26N 1 38W
Meriden, *U.S.A.* **151 E12** 41 32N 72 48W
Meridian, *Calif., U.S.A.* **160 F5** 39 9N 121 55W
Meridian, *Ga., U.S.A.* **152 D8** 31 27N 81 23W
Meridian, *Idaho, U.S.A.* **158 E5** 43 37N 116 24W
Meridian, *Miss., U.S.A.* **149 J1** 32 22N 88 42W
Meridian, *Tex., U.S.A.* **155 K6** 31 56N 97 39W
Mérignac, *France* **28 D3** 44 51N 0 39W
Mering, *Germany* **31 G6** 48 16N 10 59 E
Meriruma, *Brazil* **169 C7** 1 15N 54 50W
Merke, *Kazakstan* **63 B6** 42 52N 73 11 E
Merkel, *U.S.A.* **155 J5** 32 28N 100 1W
Mermaid Reef, *Australia* **124 C2** 17 6 S 119 36 E
Merowe, *Sudan* **106 D3** 18 29N 31 46 E
Merredin, *Australia* **125 F2** 31 28 S 118 18 E
Merrick, *U.K.* **22 F4** 55 8N 4 28W
Merrickville, *Canada* **151 B9** 44 55N 75 50W
Merrill, *Oreg., U.S.A.* **158 E3** 42 1N 121 36W
Merrill, *Wis., U.S.A.* **154 C10** 45 11N 89 41W
Merrillville, *Ga., U.S.A.* **152 E6** 30 57N 83 53W
Merrillville, *Ind., U.S.A.* **157 C9** 41 29N 87 20W
Merriman, *U.S.A.* **154 D4** 42 55N 101 42W
Merritt, *Canada* **142 C4** 50 10N 120 45W
Merritt Island, *U.S.A.* **153 G9** 28 21N 80 42W
Merriwa, *Australia* **129 B9** 32 6 S 150 22 E
Merriwagga, *Australia* **129 B6** 33 47 S 145 43 E
Merry I., *Canada* **140 A4** 55 29N 77 31W
Merrygoen, *Australia* **129 A8** 31 51 S 149 12 E
Merryville, *U.S.A.* **155 K8** 30 45N 93 33W
Mersa Fatma, *Eritrea* **107 E5** 14 57N 40 17 E
Mersch, *Lux.* **24 E6** 49 44N 6 7 E
Merse →, *Italy* **45 E8** 43 5N 11 22 E
Mersea I., *U.K.* **21 F8** 51 47N 0 58 E
Merseburg, *Germany* **30 D7** 51 22N 11 59 E
Mersey →, *U.K.* **20 D4** 53 25N 3 1W
Merseyside □, *U.K.* **20 D4** 53 31N 3 2W
Mershon, *U.S.A.* **152 D7** 31 28N 82 15W
Mersin, *Turkey* **100 D6** 36 51N 34 36 E
Mersing, *Malaysia* **87 L4** 2 25N 103 50 E
Merta, *India* **92 F6** 26 39N 74 4 E
Merthyr Tydfil, *U.K.* **21 F4** 51 45N 3 22W
Merthyr Tydfil □, *U.K.* **21 F4** 51 46N 3 21W
Mértola, *Portugal* **43 H3** 37 40N 7 40W
Mertzon, *U.S.A.* **155 K4** 31 16N 100 49W
Méru, *France* **27 C9** 49 13N 2 8 E
Meru, *Kenya* **118 B4** 0 3N 37 40 E
Meru, *Tanzania* **118 C4** 3 15 S 36 46 E
Meru □, *Kenya* **118 B4** 0 3N 37 46 E
Merville, *France* **27 B9** 50 38N 2 38 E
Méry-sur-Seine, *France* **27 D10** 48 31N 3 54 E
Merzifon, *Turkey* **100 B6** 40 53N 35 32 E
Merzig, *Germany* **31 F2** 49 26N 6 38 E
Merzouga, Erg Tin, *Algeria* **111 D7** 24 0N 11 4 E
Mesa, *U.S.A.* **159 K8** 33 25N 111 50W
Mesach Mellet, *Libya* **108 D2** 24 30N 11 30 E
Mesagne, *Italy* **47 B10** 40 34N 17 48 E
Mesanagrós, *Greece* **38 C9** 36 1N 27 49 E
Mesaoría □, *Cyprus* **38 D12** 35 12N 33 14 E
Mesarás, Kólpos, *Greece* **38 D6** 35 6N 24 47 E
Meschede, *Germany* **30 D4** 51 20N 8 18 E
Mescit, *Turkey* **101 B9** 40 21N 41 11 E
Mesfinto, *Ethiopia* **107 E4** 13 20N 37 22 E
Mesgouez, L., *Canada* **140 B5** 51 20N 75 0W
Meshchovsk, *Russia* **58 E8** 54 22N 35 17 E
Meshed = Mashhad, *Iran* **97 B8** 36 20N 59 35 E
Meshoppen, *U.S.A.* **151 E8** 41 36N 76 3W
Meshra er Req, *Sudan* **107 F2** 8 25N 29 18 E
Mesick, *U.S.A.* **148 C3** 44 24N 85 43W
Mesilinka →, *Canada* **142 B4** 56 6N 124 30W
Mesilla, *U.S.A.* **159 K10** 32 16N 106 48W
Meslay-du-Maine, *France* **26 E6** 47 58N 0 33W
Mesocco, *Switz.* **33 D8** 46 23N 9 12 E
Mesolóngion, *Greece* **48 C3** 38 21N 21 28 E
Mesopotamia = Al Jazirah, *Iraq* **101 E10** 33 30N 44 0 E
Mesopótamon, *Greece* **48 B2** 39 14N 20 32 E
Mesoraca, *Italy* **47 C9** 39 5N 16 48 E
Mésou Volímais = Volímai, *Greece* **48 D2** 37 53N 20 35 E
Mesquite, *U.S.A.* **159 H6** 36 47N 114 6W
Mess Cr. →, *Canada* **142 B2** 57 55N 131 14W
Messac, *France* **26 E5** 47 49N 1 50W
Messad, *Algeria* **111 B5** 34 8N 3 30 E
Messalo →, *Mozam.* **119 E4** 12 25 S 39 15 E
Méssaména, *Cameroon* **113 E7** 3 48N 12 49 E
Messeue, *Greece* **48 D3** 37 12N 21 58 E
Messier, Canal, *Chile* **176 C2** 48 20 S 74 33W
Messina, *Italy* **47 D8** 38 11N 15 34 E
Messina, *S. Africa* **117 C5** 22 20 S 30 5 E
Messina, Str. di, *Italy* **47 D8** 38 15N 15 35 E
Messíni, *Greece* **48 D4** 37 4N 22 1 E
Messínia □, *Greece* **48 D3** 37 10N 22 0 E
Messiniakós Kólpos, *Greece* **48 E4** 36 45N 22 5 E
Messkirch, *Germany* **31 H5** 47 59N 9 7 E
Messonghi, *Greece* **38 B3** 39 29N 19 56 E
Mesta →, *Bulgaria* **50 E7** 40 54N 24 49 E
Mestá, Ákra, *Greece* **49 C7** 38 16N 25 53 E
Mestanza, *Spain* **43 G6** 38 35N 4 4W
Mestersvig, *Greenland* **10 C6** 72 10N 23 40W
Mestre, *Italy* **45 C9** 45 29N 12 15 E
Mestre, Espigão, *Brazil* **171 D2** 12 30 S 46 10W
Mesudiye, *Turkey* **100 B7** 40 28N 37 46 E
Meta, *U.S.A.* **156 F4** 38 19N 92 10W
Meta □, *Colombia* **168 C3** 3 30N 73 0W
Meta →, *S. Amer.* **168 B4** 6 12N 67 28W
Metairie, *U.S.A.* **155 L9** 29 58N 90 10W
Metalici, Munţii, *Romania* **52 D7** 46 15N 22 50 E
Metaline Falls, *U.S.A.* **158 B5** 48 52N 117 22W
Metallifere, Colline, *Italy* **44 E8** 43 10N 11 0 E
Metamora, *U.S.A.* **156 D7** 40 47N 89 22W
Metán, *Argentina* **174 B3** 25 30 S 65 0W
Metangula, *Mozam.* **119 E3** 12 40 S 34 50 E
Metauro →, *Italy* **45 E10** 43 50N 13 3 E
Metcalf, *U.S.A.* **152 E6** 30 43N 83 59W
Metema, *Ethiopia* **107 E4** 12 56N 36 13 E
Metengobalame, *Mozam.* **119 E3** 14 49 S 34 30 E
Méthana, *Greece* **48 D5** 37 34N 23 23 E
Methóni, *Greece* **48 E3** 36 49N 21 42 E
Methven, *N.Z.* **131 D6** 43 38 S 171 40 E
Methy L., *Canada* **143 B7** 56 28N 109 30W
Metil, *Mozam.* **119 F4** 16 24 S 39 0 E
Metkovets, *Bulgaria* **50 C7** 43 37N 23 10 E
Metković, *Croatia* **45 E14** 43 6N 17 39 E
Metlakatla, *U.S.A.* **142 B2** 55 8N 131 35W
Metlaoui, *Tunisia* **108 B1** 34 24N 8 24 E
Metlika, *Slovenia* **45 C12** 45 40N 15 20 E
Metro, *Indonesia* **84 D3** 5 5 S 105 20 E
Metropolis, *U.S.A.* **155 G10** 37 9N 88 44W
Métsovon, *Greece* **48 B3** 39 48N 21 12 E
Metter, *U.S.A.* **152 C7** 32 24N 82 3W
Mettuppalaiyam, *India* **95 J3** 11 18N 76 59 E
Mettur, *India* **95 J3** 11 48N 77 47 E
Metz, *France* **27 C13** 49 8N 6 10 E
Metzingen, *Germany* **31 G5** 48 31N 9 17 E
Meulaboh, *Indonesia* **84 B1** 4 11N 96 3 E
Meung-sur-Loire, *France* **27 E8** 47 50N 1 40 E
Meureudu, *Indonesia* **84 A1** 5 19N 96 10 E
Meurthe →, *France* **27 D13** 48 47N 6 9 E
Meurthe-et-Moselle □, *France* **27 D13** 48 52N 6 0 E
Meuse □, *France* **27 C12** 49 8N 5 25 E
Meuse →, *Europe* **24 D5** 50 45N 5 41 E
Meuselwitz, *Germany* **30 D8** 51 2N 12 17 E
Meutapok, Mt., *Malaysia* **85 A5** 5 40N 117 0 E
Mexia, *U.S.A.* **155 K6** 31 41N 96 29W
Mexiana, I., *Brazil* **170 B2** 0 0 49 30W
Mexicali, *Mexico* **162 A1** 32 40N 115 30W
Mexican Plateau, *Mexico* **136 G9** 25 0N 104 0W
México, *Mexico* **163 D5** 19 20N 99 10W
Mexico, *Maine, U.S.A.* **151 B14** 44 34N 70 33W
Mexico, *Mo., U.S.A.* **156 E5** 39 10N 91 53W
México □, *Mexico* **162 D5** 19 20N 99 10W
Mexico ■, *Cent. Amer.* **162 C4** 25 0N 105 0W
Mexico, G. of, *Cent. Amer.* **163 C7** 25 0N 90 0W
Mexico Beach, *U.S.A.* **152 F4** 29 57N 85 25W
Meyenburg, *Germany* **30 B8** 53 18N 12 14 E
Meyers Chuck, *U.S.A.* **144 J14** 55 45N 132 15W
Meymac, *France* **28 C6** 45 32N 2 10 E
Meymaneh, *Afghan.* **91 B2** 35 53N 64 38 E
Meyrueis, *France* **28 D7** 44 12N 3 27 E
Meyssac, *France* **28 C5** 45 3N 1 40 E
Meyzieu, *France* **29 C8** 45 46N 4 59 E
Mezdra, *Bulgaria* **50 C7** 43 12N 23 42 E
Mèze, *France* **28 E7** 43 27N 3 36 E
Mezen, *Russia* **56 A7** 65 50N 44 20 E
Mezen →, *Russia* **56 A7** 65 44N 44 22 E
Mézenc, Mt., *France* **29 D8** 44 54N 4 11 E
Mezeș, Munţii, *Romania* **52 C8** 47 5N 23 5 E
Mezha →, *Russia* **58 E6** 55 44N 31 33 E
Mézidon-Canon, *France* **26 C6** 49 5N 0 1W
Mézières-en-Brenne, *France* **28 B5** 46 49N 1 13 E
Mézilhac, *France* **29 D8** 44 49N 4 21 E
Mézin, *France* **28 D4** 44 4N 0 16 E
Mezöberény, *Hungary* **52 D6** 46 49N 21 3 E
Mezöfalva, *Hungary* **52 D3** 46 55N 18 49 E
Mezöhegyes, *Hungary* **52 D5** 46 19N 20 49 E
Mezökövácsháza, *Hungary* **52 D5** 46 25N 20 57 E
Mezökövesd, *Hungary* **52 C5** 47 49N 20 35 E
Mézos, *France* **28 D2** 44 5N 1 10W
Mezötúr, *Hungary* **52 C5** 47 1N 20 41 E
Mezquital, *Mexico* **162 C4** 23 29N 104 23W
Mezzolombardo, *Italy* **44 B8** 46 13N 11 5 E
Mgeta, *Tanzania* **119 D4** 8 22 S 36 6 E
Mglin, *Russia* **59 F7** 53 2N 32 50 E
Mhlaba Hills, *Zimbabwe* **119 F3** 18 30 S 30 30 E
Mhow, *India* **92 H6** 22 33N 75 50 E
Mi-Shima, *Japan* **70 C3** 34 46N 131 9 E
Miahuatlán, *Mexico* **163 D5** 16 21N 96 36W
Miajadas, *Spain* **43 F5** 39 9N 5 54W
Miallo, *Australia* **126 B4** 16 28 S 145 22 E
Miami, *Ariz., U.S.A.* **159 K8** 33 24N 110 52W
Miami, *Fla., U.S.A.* **149 N5** 25 47N 80 11W
Miami, *Tex., U.S.A.* **155 H4** 35 42N 100 38W
Miami →, *U.S.A.* **148 F3** 39 20N 84 40W
Miami Beach, *U.S.A.* **149 N5** 25 47N 80 8W
Miami Canal, *U.S.A.* **153 J9** 26 30N 80 45W
Miami Shores, *U.S.A.* **153 K9** 25 52N 80 12W
Miami Springs, *U.S.A.* **153 K9** 25 49N 80 17W
Miamisburg, *U.S.A.* **157 E12** 39 38N 84 17W
Mian Xian, *China* **74 H4** 33 10N 106 32 E
Mianchi, *China* **74 G6** 34 48N 111 48 E
Mīāndowāb, *Iran* **101 D12** 37 0N 46 5 E
Miandrivazo, *Madag.* **117 B8** 19 31 S 45 29 E
Mīāneh, *Iran* **101 D12** 37 30N 47 40 E
Mianning, *China* **76 C4** 28 32N 102 9 E
Mianwali, *Pakistan* **91 B3** 32 38N 71 28 E
Mianyang, *Hubei, China* **77 B9** 30 25N 113 25 E
Mianyang, *Sichuan, China* **76 B5** 31 22N 104 47 E
Mianzhu, *China* **76 B5** 31 22N 104 7 E
Miaoli, *Taiwan* **77 E13** 24 37N 120 49 E
Miarinarivo, *Madag.* **117 B8** 18 57 S 46 55 E
Miass, *Russia* **62 D8** 54 59N 60 6 E
Miasteczko Krajeńskie, *Poland* **55 E4** 53 7N 17 1 E
Miastko, *Poland* **54 E3** 54 0N 16 58 E
Micanopy, *U.S.A.* **153 F7** 29 30N 82 17W
Micăsasa, *Romania* **53 D9** 46 7N 24 7 E
Miccasukee, L., *U.S.A.* **152 E6** 30 33N 83 53W
Micco, *U.S.A.* **153 H9** 27 53N 80 30W
Miccosukee, *U.S.A.* **152 E5** 30 36N 84 3W
Michael, Mt., *Papua N. G.* **132 D3** 6 27 S 145 22 E
Michalovce, *Slovak Rep.* **35 C14** 48 47N 21 58 E
Michigan □, *U.S.A.* **148 C3** 44 0N 85 0W
Michigan, L., *U.S.A.* **148 D2** 44 0N 87 0W
Michigan Center, *U.S.A.* **157 B12** 42 14N 84 20W
Michigan City, *U.S.A.* **157 C10** 41 43N 86 54W
Michikamau L., *Canada* **141 B7** 54 20N 63 10W
Michipicoten, *Canada* **140 C3** 47 55N 84 55W
Michipicoten I., *Canada* **140 C2** 47 40N 85 40W
Michoacan □, *Mexico* **162 D4** 19 0N 102 0W
Michurin, *Bulgaria* **51 D11** 42 9N 27 51 E
Michurinsk, *Russia* **60 D5** 52 58N 40 27 E
Miclere, *Australia* **126 C4** 22 34 S 147 32 E
Mico, Pta., *Nic.* **164 D3** 12 0N 83 30W
Micronesia, Federated States of ■, *Pac. Oc.* **134 G7** 9 0N 150 0 E
Mid-Indian Ridge, *Ind. Oc.* **121 H6** 40 0 S 75 0 E
Midai, *Indonesia* **85 B3** 3 0N 107 47 E
Midale, *Canada* **143 D8** 49 25N 103 20W
Middelburg, *Neths.* **24 C3** 51 30N 3 36 E
Middelburg, *Eastern Cape, S. Africa* **116 E4** 31 30 S 25 0 E

250 Middelburg

Middelburg, Mpumalanga, S. Africa 117 D4 25 49 S 29 28 E
Middelfart, Denmark 17 J3 55 30N 9 43 E
Middelwit, S. Africa 116 C4 24 51 S 27 3 E
Middle →, U.S.A. 156 C3 41 26N 93 30W
Middle Alkali L., U.S.A. 158 F3 41 27N 120 5W
Middle Fork Feather →, U.S.A. 160 F5 38 33N 121 30W
Middle I., Australia 125 F3 34 6 S 123 11 E
Middle Loup →, U.S.A. 154 E5 41 17N 98 24W
Middle Raccoon →, U.S.A. 156 C3 41 35N 93 35W
Middleboro, U.S.A. 151 E14 41 54N 70 55W
Middleburg, Fla., U.S.A. 152 E8 30 4N 81 52W
Middleburg, N.Y., U.S.A. 151 D10 42 36N 74 20W
Middleburg, Pa., U.S.A. 150 F7 40 47N 77 3W
Middlebury, Ind., U.S.A. 157 C11 41 41N 85 42W
Middlebury, Vt., U.S.A. 151 B11 44 1N 73 10W
Middlemarch, N.Z. 131 F5 45 30 S 170 9 E
Middleport, U.S.A. 148 F4 39 0N 82 3W
Middlesboro, U.S.A. 149 G4 36 36N 83 43W
Middlesbrough, U.K. 20 C6 54 35N 1 13W
Middlesbrough □, U.K. 20 C6 54 28N 1 13W
Middlesex, Belize 164 C2 17 2N 88 31W
Middlesex, U.S.A. 151 F10 40 36N 74 30W
Middleton, Australia 126 C3 22 22 S 141 32 E
Middleton, Canada 141 D6 44 57N 65 4W
Middleton, U.S.A. 156 A7 43 6N 89 30W
Middletown, U.K. 23 B5 54 17N 6 51W
Middletown, Calif., U.S.A. 160 G4 38 45N 122 37W
Middletown, Conn., U.S.A. 151 E12 41 34N 72 39W
Middletown, N.Y., U.S.A. 151 E10 41 27N 74 25W
Middletown, Ohio, U.S.A. 157 E12 39 31N 84 24W
Middletown, Pa., U.S.A. 151 F8 40 12N 76 44W
Middleville, U.S.A. 157 B11 42 43N 85 28W
Midelt, Morocco 110 B4 32 46N 4 44W
Miðgarðar, Iceland 11 A9 66 32N 18 0W
Midhirst, N.Z. 130 F3 39 17 S 174 18 E
Miðsandur, Iceland 11 C5 64 24N 21 28W
Midhurst, U.K. 21 G7 50 59N 0 44W
Midi, Canal du →, France 28 E5 43 45N 1 21 E
Midi d'Ossau, Pic du, France 28 F3 42 50N 0 26W
Midi-Pyrénées □, France 28 E5 43 55N 1 45 E
Midland, Canada 140 D4 44 45N 79 50W
Midland, Calif., U.S.A. 161 M12 33 52N 114 48W
Midland, Mich., U.S.A. 148 D3 43 37N 84 14W
Midland, Pa., U.S.A. 150 F4 40 39N 80 27W
Midland, Tex., U.S.A. 155 K3 32 0N 102 3W
Midlands □, Zimbabwe 119 F2 19 40 S 29 0 E
Midleton, Ireland 23 E3 51 55N 8 10W
Midlothian, U.S.A. 155 J6 32 30N 97 0W
Midlothian □, U.K. 22 F5 55 51N 3 5W
Midongy, Tangorombohitr' i, Madag. 117 C8 23 30 S 47 0 E
Midongy Atsimo, Madag. 117 C8 23 35 S 47 1 E
Midou →, France 28 E3 43 54N 0 30W
Midouze →, France 28 E3 43 48N 0 51W
Midsayap, Phil. 81 H5 7 12N 124 32 E
Midsund, Norway 18 B3 62 41N 6 40 E
Midtgulen, Norway 18 C2 61 44N 5 11 E
Midu, China 76 E3 25 18N 100 30 E
Midville, U.S.A. 152 C7 32 49N 82 14W
Midway, Ala., U.S.A. 152 C4 32 5N 85 31W
Midway, Fla., U.S.A. 152 E5 30 30N 84 27W
Midway Is., Pac. Oc. 134 E10 28 13N 177 22W
Midway Wells, U.S.A. 161 N11 32 41N 115 7W
Midwest, U.S.A. 147 B9 42 0N 90 0W
Midwest, Wyo., U.S.A. 158 E10 43 25N 106 16W
Midwest City, U.S.A. 155 H6 35 27N 97 24W
Midyat, Turkey 101 D9 37 25N 41 23 E
Midzŏr, Bulgaria 50 C6 43 24N 22 40 E
Mie □, Japan 71 C8 34 30N 136 10 E
Miechów, Poland 55 H7 50 21N 20 5 E
Miedwie, Jezioro, Poland 55 E1 53 17N 14 54 E
Międzybórz, Poland 55 G4 51 25N 17 34 E
Międzychód, Poland 55 F2 52 35N 15 53 E
Międzylesie, Poland 55 H3 50 8N 16 40 E
Międzyrzec Podlaski, Poland 55 G9 51 58N 22 45 E
Międzyrzecz, Poland 55 F2 52 26N 15 35 E
Międzyzdroje, Poland 54 E1 53 56N 14 26 E
Miejska Górka, Poland 55 G3 51 39N 16 58 E
Miélan, France 28 E4 43 27N 0 19 E
Mielec, Poland 55 H8 50 15N 21 25 E
Mienga, Angola 115 F3 17 12 S 19 48 E
Miercurea-Ciuc, Romania 53 D10 46 21N 25 48 E
Miercurea Sibiului, Romania 53 E8 45 53N 23 48 E
Mieres, Spain 42 B5 43 18N 5 48W
Mieroszów, Poland 55 H3 50 40N 16 11 E
Mieso, Ethiopia 107 F5 9 15N 40 43 E
Mieszkowice, Poland 55 F1 52 47N 14 30 E
Mifflintown, U.S.A. 150 F7 40 34N 77 24W
Mifraz Hefa, Israel 103 C4 32 52N 35 0 E
Migennes, France 27 E10 47 58N 3 31 E
Migliarino, Italy 45 D8 44 46N 11 56 E
Miguel Alemán, Presa, Mexico 163 D5 18 15N 96 40W
Miguel Alves, Brazil 170 B3 4 11 S 42 55W
Miguel Calmon, Brazil 170 D3 11 26 S 40 36W
Miguelturra, Spain 43 G7 38 58N 3 53W
Mihăileni, Romania 53 C11 47 58N 26 9 E
Mihăileşti, Romania 53 F10 44 20N 25 54 E
Mihailovca, Moldova 53 D13 46 33N 28 56 E
Mihalgazi, Turkey 49 A12 40 2N 30 34 E
Mihaliççık, Turkey 100 C4 39 53N 31 30 E
Mihara, Japan 70 C5 34 24N 133 5 E
Mihara-Yama, Japan 71 C11 34 43N 139 23 E
Miheşu de Cîmpie, Romania 53 D9 46 41N 24 9 E
Mijas, Spain 43 J6 36 36N 4 40W
Mikese, Tanzania 118 D4 6 48 S 37 55 E
Mikha-Tskhakaya = Senaki, Georgia 61 J6 42 15N 42 7 E
Mikhailovka = Mykhaylivka, Ukraine 59 J8 47 12N 35 15 E
Mikhaylov, Russia 58 E10 54 14N 39 0 E
Mikhaylovgrad = Montana, Bulgaria 50 C7 43 27N 23 16 E
Mikhaylovka, Russia 60 E6 50 3N 43 5 E
Mikhaylovski, Russia 62 C7 56 27N 59 7 E
Mikhnevo, Russia 58 E9 55 4N 37 59 E
Miki, Hyōgo, Japan 70 C6 34 48N 134 59 E
Miki, Kagawa, Japan 70 C6 34 12N 134 7 E
Mikínai, Greece 48 D4 37 43N 22 46 E
Mikkeli, Finland 15 F22 61 43N 27 15 E
Mikkwa →, Canada 142 B6 58 25N 114 46W
Mikniya, Sudan 107 D3 17 0N 33 45 E
Mikołajki, Poland 54 E8 53 49N 21 37 E
Míkonos, Greece 49 D7 37 26N 25 20 E
Mikrí Préspa, Límni, Greece 50 F5 40 47N 21 3 E
Mikrón Dhérion, Greece 51 E10 41 19N 26 6 E
Mikstat, Poland 55 G4 51 32N 17 59 E
Mikulov, Czech Rep. 35 C9 48 48N 16 39 E
Mikumi, Tanzania 118 D4 7 26 S 37 0 E
Mikun, Russia 56 B9 62 20N 50 0 E
Mikuni, Japan 71 A8 36 13N 136 9 E
Mikuni-Tōge, Japan 71 A10 36 50N 138 50 E
Mikura-Jima, Japan 71 D11 33 52N 139 36 E
Milaca, U.S.A. 154 C8 45 45N 93 39W
Milagro, Ecuador 168 D2 2 11 S 79 36W
Milagros, Phil. 80 E4 12 13N 123 30 E
Milan = Milano, Italy 44 C6 45 28N 9 12 E
Milan, Ga., U.S.A. 152 C6 32 1N 83 4W
Milan, Ill., U.S.A. 156 C6 41 27N 90 34W
Milan, Mich., U.S.A. 157 B13 42 5N 83 41W
Milan, Mo., U.S.A. 156 D3 40 12N 93 7W
Milan, Tenn., U.S.A. 149 H1 35 55N 88 46W
Miland, Norway 18 E5 59 54N 8 45 E
Milang, S. Austral., Australia 127 E2 32 2 S 139 10 E
Milang, S. Austral., Australia 128 C3 35 24 S 138 58 E
Milange, Mozam. 119 F4 16 3 S 35 45 E
Milano, Italy 44 C6 45 28N 9 12 E
Milâs, Turkey 57 G4 37 20N 27 50 E
Milatos, Greece 38 D7 35 18N 25 34 E
Milazzo, Italy 47 D8 38 13N 15 15 E
Milbank, U.S.A. 154 C6 45 13N 96 38W
Milden, Canada 143 C7 51 29N 107 32W
Mildenhall, U.K. 21 E8 52 21N 0 32 E
Mildmay, Canada 150 B3 44 3N 81 7W
Mildura, Australia 128 C5 34 13 S 142 9 E
Mile, China 76 E4 24 28N 103 20 E
Miléai, Greece 48 B5 39 20N 23 9 E
Miles, Australia 127 D5 26 40 S 150 9 E
Miles, U.S.A. 155 K4 31 36N 100 11W
Miles City, U.S.A. 154 B2 46 25N 105 51W
Mileşti, Moldova 53 C13 47 13N 28 3 E
Milestone, Canada 143 D8 49 59N 104 31W
Mileto, Italy 47 D9 38 36N 16 4 E
Miletto, Mte., Italy 47 A7 41 27N 14 22 E
Miletus, Turkey 49 D9 37 30N 27 18 E
Mileura, Australia 125 E2 26 22 S 117 20 E
Milevsko, Czech Rep. 34 B7 49 27N 14 21 E
Milford, Calif., U.S.A. 160 E6 40 10N 120 22W
Milford, Conn., U.S.A. 151 E11 41 14N 73 3W
Milford, Del., U.S.A. 148 F8 38 55N 75 26W
Milford, Ga., U.S.A. 152 D5 31 23N 84 33W
Milford, Ill., U.S.A. 157 D9 40 38N 87 42W
Milford, Mass., U.S.A. 151 D13 42 8N 71 31W
Milford, Mich., U.S.A. 157 B13 42 35N 83 36W
Milford, Pa., U.S.A. 151 E10 41 19N 74 48W
Milford, Utah, U.S.A. 159 G7 38 24N 113 1W
Milford Haven, U.K. 21 F2 51 42N 5 7W
Milford Sd., N.Z. 131 E2 44 41 S 167 47 E
Milgun, Australia 125 D2 24 56 S 118 18 E
Milḩ, Baḩr al, Iraq 101 F10 32 40N 43 35 E
Miliana, Aïn Salah, Algeria 111 C5 27 20N 2 32 E
Miliana, Médéa, Algeria 111 A5 36 20N 2 15 E
Milicz, Poland 55 G4 51 31N 17 19 E
Mililani Town, U.S.A. 145 K13 21 28N 158 1W
Miling, Australia 125 F2 30 30 S 116 17 E
Militello in Val di Catánia, Italy 47 E7 37 16N 14 48 E
Milk →, U.S.A. 158 B10 48 4N 106 19W
Milk, Wadi el →, Sudan 106 D3 17 55N 30 20 E
Milk River, Canada 142 D6 49 10N 112 5W
Mill City, U.S.A. 158 D2 44 45N 122 29W
Mill I., Antarctica 7 C8 66 0 S 101 30 E
Mill Shoals, U.S.A. 157 F8 38 15N 88 21W
Mill Valley, U.S.A. 160 H4 37 54N 122 32W
Millárs →, Spain 40 F4 39 55N 0 1W
Millau, France 28 D7 44 8N 3 4 E
Millbridge, Canada 150 B7 44 41N 77 36W
Millbrook, Canada 150 B6 44 10N 78 29W
Mille Lacs, L. des, Canada 140 C1 48 45N 90 35W
Mille Lacs L., U.S.A. 154 B8 46 15N 93 39W
Milledgeville, Ga., U.S.A. 152 B6 33 5N 83 14W
Milledgeville, Ill., U.S.A. 156 C7 41 58N 89 46W
Millen, U.S.A. 152 C8 32 48N 81 57W
Miller, U.S.A. 154 C5 44 31N 98 59W
Millerovo, Russia 61 F5 48 57N 40 28 E
Miller's Flat, N.Z. 131 F4 45 39 S 169 23 E
Millersburg, Ind., U.S.A. 157 C11 41 32N 85 42W
Millersburg, Ohio, U.S.A. 150 F3 40 33N 81 55W
Millersburg, Pa., U.S.A. 150 F8 40 32N 76 58W
Millerton, N.Z. 131 B6 41 39 S 171 54 E
Millerton, U.S.A. 151 E11 41 57N 73 31W
Millerton L., U.S.A. 160 J7 37 1N 119 41W
Millevaches, Plateau de, France 28 C6 45 45N 2 0 E
Millicent, Australia 128 D4 37 34 S 140 21 E
Milligan, U.S.A. 153 E3 30 45N 86 38W
Millinocket, U.S.A. 141 C6 45 39N 68 43W
Millmerran, Australia 127 D5 27 53 S 151 16 E
Millom, U.K. 20 C4 54 13N 3 16W
Mills L., Canada 142 A5 61 30N 118 20W
Millsboro, U.S.A. 150 G5 40 0N 80 0W
Milltown Malbay, Ireland 23 D2 52 52N 9 24W
Millville, U.S.A. 148 F8 39 24N 75 2W
Millwood, U.S.A. 152 D7 31 16N 82 40W
Millwood L., U.S.A. 155 J8 33 42N 93 58W
Milna, Croatia 45 E13 43 20N 16 28 E
Milne →, Australia 126 C2 21 10 S 137 33 E
Milne Land, Greenland 10 C8 70 40N 26 30W
Milnor, U.S.A. 154 B6 46 16N 97 27W
Milo, Canada 142 C6 50 34N 112 53W
Mílos, Greece 48 E6 36 44N 24 25 E
Miłosław, Poland 55 F4 52 12N 17 32 E
Milot, Albania 50 E3 41 41N 19 43 E
Milparinka P.O., Australia 127 D3 29 46 S 141 57 E
Milroy, U.S.A. 157 E11 39 30N 85 28W
Miltenberg, Germany 31 F5 49 41N 9 16 E
Milton, Canada 150 C5 43 31N 79 53W
Milton, N.Z. 131 G4 46 7 S 169 59 E
Milton, Calif., U.S.A. 160 G6 38 3N 120 51W
Milton, Fla., U.S.A. 149 K2 30 38N 87 3W
Milton, Iowa, U.S.A. 156 D4 40 41N 92 10W
Milton, Pa., U.S.A. 150 F8 41 1N 76 51W
Milton, Wis., U.S.A. 157 B8 42 47N 88 56W
Milton-Freewater, U.S.A. 158 D4 45 56N 118 23W
Milton Keynes, U.K. 21 E7 52 1N 0 44W
Milton Keynes □, U.K. 21 E7 52 1N 0 44W
Miltou, Chad 109 F3 10 14N 17 26 E
Milverton, Canada 150 C4 43 34N 80 55W
Milwaukee, U.S.A. 157 A9 43 2N 87 55W
Milwaukee Deep, Atl. Oc. 8 G2 19 50N 68 0W
Milwaukie, U.S.A. 160 E4 45 27N 122 38W
Mim, Ghana 112 D4 6 57N 2 33W
Mimizan, France 28 D2 44 12N 1 13W
Mimoň, Czech Rep. 34 A7 50 38N 14 43 E
Mimongo, Gabon 114 C2 1 11 S 11 36 E
Mimoso, Brazil 171 E2 15 10 S 48 5W
Mims, U.S.A. 153 G9 28 40N 80 51W
Min Jiang →, Fujian, China 77 E12 26 0N 119 35 E
Min Jiang →, Sichuan, China 76 C5 28 45N 104 40 E
Min-Kush, Kyrgyzstan 63 C7 41 40N 74 28 E
Min Xian, China 74 G3 34 25N 104 5 E
Mina, U.S.A. 159 G4 38 24N 118 7W
Mina Pirquitas, Argentina 174 A2 22 40 S 66 30W
Mīnā Su'ud, Si. Arabia 97 D6 28 45N 48 28 E
Mīnā'al Aḩmadī, Kuwait 97 D6 29 5N 48 10 E
Mīnāb, Iran 97 E8 27 10N 57 1 E
Minago →, Canada 143 C9 54 33N 98 59W
Minakami, Japan 71 A10 36 49N 138 59 E
Minaki, Canada 143 D10 49 59N 94 40W
Minakuchi, Japan 71 C8 34 58N 136 10 E
Minamata, Japan 70 E2 32 10N 130 30 E
Minami-Tori-Shima, Pac. Oc. 134 E7 24 0N 153 45 E
Minas, Uruguay 175 C4 34 20 S 55 10W
Minas, Sierra de las, Guatemala 164 C2 15 9N 89 31W
Minas Basin, Canada 141 C7 45 20N 64 12W
Minas de Rio Tinto = Minas de Riotinto, Spain 43 H4 37 42N 6 35W
Minas de Riotinto, Spain 43 H4 37 42N 6 35W
Minas Gerais □, Brazil 171 E2 18 50 S 46 0W
Minas Novas, Brazil 171 E3 17 15 S 42 36W
Minatitlán, Mexico 163 D6 17 59N 94 31W
Minbu, Burma 90 E5 20 10N 94 52 E
Minbya, Burma 90 E4 20 22N 93 16 E
Mincio →, Italy 44 C7 45 4N 10 59 E
Minčol, Slovak Rep. 35 B13 49 15N 20 58 E
Mindanao, Phil. 81 H5 8 0N 125 0 E
Mindanao Sea = Bohol Sea, Phil. 81 G5 9 0N 124 0 E
Mindanao Trench, Pac. Oc. 80 E5 12 0N 126 6 E
Mindel →, Germany 31 G6 48 31N 10 23 E
Mindelheim, Germany 31 G6 48 2N 10 29 E
Minden, Canada 150 B6 44 55N 78 43W
Minden, Germany 30 C4 52 17N 8 55 E
Minden, La., U.S.A. 155 J8 32 37N 93 17W
Minden, Nev., U.S.A. 160 G7 38 57N 119 46W
Mindiptana, Indonesia 83 C6 5 55 S 140 22 E
Mindon, Burma 90 F5 19 21N 94 44 E
Mindoro, Phil. 80 E3 13 0N 121 0 E
Mindoro Occidental □, Phil. 80 E3 13 0N 120 55 E
Mindoro Oriental □, Phil. 80 E3 13 0N 121 5 E
Mindoro Str., Phil. 80 E3 12 30N 120 30 E
Mindouli, Congo 115 C2 4 12 S 14 28 E
Mine, Japan 70 C3 34 12N 131 7 E
Minehead, U.K. 21 F4 51 12N 3 29W
Mineiros, Brazil 173 D7 17 34 S 52 34W
Mineola, U.S.A. 155 J7 32 40N 95 29W
Mineral King, U.S.A. 160 J8 36 27N 118 36W
Mineral Point, U.S.A. 156 B6 42 52N 90 11W
Mineral Wells, U.S.A. 155 J5 32 48N 98 7W
Mineralnyye Vody, Russia 61 H6 44 15N 43 8 E
Minersville, Pa., U.S.A. 151 F8 40 41N 76 16W
Minersville, Utah, U.S.A. 159 G7 38 13N 112 56W
Minerva, U.S.A. 150 F3 40 44N 81 6W
Minervino Murge, Italy 47 A9 41 5N 16 5 E
Minetto, U.S.A. 151 C8 43 24N 76 28W
Mingäçevir, Azerbaijan 61 K8 40 45N 47 0 E
Mingäçevir Su Anbarı, Azerbaijan 61 K8 40 57N 46 50 E
Mingan, Canada 141 B7 50 20N 64 0W
Mingary, Australia 128 B4 32 8 S 140 45 E
Mingechaur = Mingäçevir, Azerbaijan 61 K8 40 45N 47 0 E
Mingechaurskoye Vdkhr. = Mingäçevir Su Anbarı, Azerbaijan 61 K8 40 57N 46 50 E
Mingela, Australia 126 B4 19 52 S 146 38 E
Mingenew, Australia 125 E2 29 12 S 115 21 E
Mingera Cr. →, Australia 126 C2 20 38 S 137 45 E
Minggang, China 77 A10 32 24N 114 3 E
Mingin, Burma 90 D5 22 50N 94 30 E
Mingir, Moldova 53 D13 46 38N 28 20 E
Minglanilla, Spain 41 F3 39 34N 1 38W
Minglun, China 76 E7 25 10N 108 21 E
Mingorria, Spain 42 E6 40 45N 4 40W
Mingt'iehkaitafan = Mintaka Pass, Pakistan 93 A6 37 0N 74 58 E
Mingxi, China 77 D11 26 18N 117 12 E
Mingyuegue, China 75 C15 43 2N 128 50 E
Minho = Miño →, Spain 42 D2 41 52N 8 40W
Minhou, China 77 E12 26 0N 119 15 E
Minićevo, Serbia, Yug. 52 G7 43 42N 22 18 E
Minidoka, U.S.A. 158 E7 42 45N 113 29W
Minier, U.S.A. 156 D7 40 26N 89 19W
Minigwal, L., Australia 125 E3 29 31 S 123 14 E
Minilya, Australia 125 D1 23 55 S 114 0 E
Minilya →, Australia 125 D1 23 45 S 114 0 E
Mininera, Australia 128 D5 37 37 S 142 58 E
Minipi, L., Canada 141 B7 52 25N 60 45W
Minj, Papua N. G. 132 C3 5 54 S 144 37 E
Mink L., Canada 142 A5 61 54N 117 40W
Minlaton, Australia 128 C2 34 45 S 137 35 E
Minna, Nigeria 113 D6 9 37N 6 30 E
Minneapolis, Kans., U.S.A. 154 F6 39 8N 97 42W
Minneapolis, Minn., U.S.A. 154 C8 44 59N 93 16W
Minnedosa, Canada 143 C9 50 14N 99 50W
Minnesota □, U.S.A. 154 B8 46 0N 94 15W
Minnesund, Norway 18 D8 60 23N 11 14 E
Minnie Creek, Australia 125 D2 24 3 S 115 42 E
Minnipa, Australia 127 E2 32 51 S 135 9 E
Minnitaki L., Canada 140 C1 49 57N 92 10W
Mino, Japan 71 B8 35 32N 136 55 E
Miño, Spain 42 B2 43 21N 8 12W
Miño →, Spain 42 D2 41 52N 8 40W
Mino-Kamo, Japan 71 B9 35 23N 137 2 E
Mino-Mikawa-Kōgen, Japan 71 B9 35 10N 137 23 E
Minoa, Greece 49 F7 35 6N 25 45 E
Minobu, Japan 71 B10 35 22N 138 26 E
Minobu-Sanchi, Japan 71 B10 35 14N 138 20 E
Minonk, U.S.A. 156 D7 40 54N 89 2W
Minooka, U.S.A. 157 C8 41 27N 88 16W
Minorca = Menorca, Spain 39 B11 40 0N 4 0 E
Minore, Australia 129 B8 32 14 S 148 27 E
Minot, U.S.A. 154 A4 48 14N 101 18W
Minqin, China 74 E2 38 38N 103 20 E
Minqing, China 77 D12 26 15N 118 50 E
Minsen, Germany 30 B3 53 41N 7 58 E
Minsk, Belarus 58 F4 53 52N 27 30 E
Mińsk Mazowiecki, Poland 55 F8 52 10N 21 33 E
Minster, U.S.A. 157 D12 40 24N 84 23W
Mintaka Pass, Pakistan 93 A6 37 0N 74 58 E
Minthami, Burma 90 D5 23 55N 94 16 E
Minton, Canada 143 D8 49 10N 104 35W
Mintoum, Gabon 114 B2 0 27N 12 16 E
Minturn, U.S.A. 158 G10 39 35N 106 26W
Minturno, Italy 46 A6 41 15N 13 45 E
Minûf, Egypt 106 H7 30 26N 30 52 E
Minusinsk, Russia 65 D10 53 43N 91 20 E
Minutang, India 90 A6 28 15N 96 30 E
Minvoul, Gabon 114 B2 2 9N 12 8 E
Minwakh, Yemen 99 C5 16 48N 48 6 E
Minya el Qamh, Egypt 106 H7 30 31N 31 21 E
Minyar, Russia 62 D6 55 4N 57 33 E
Minyip, Australia 128 D5 36 29 S 142 36 E
Mionica, Bos.-H. 52 F3 44 51N 18 29 E
Mionica, Serbia, Yug. 50 B4 44 14N 20 6 E
Mir, Niger 109 F2 14 5N 11 59 E
Mīr Kūh, Iran 97 E8 26 22N 58 55 E
Mīr Shahdād, Iran 97 E8 26 15N 58 29 E
Mira, Italy 45 C9 45 26N 12 8 E
Mira, Portugal 42 E2 40 26N 8 44W
Mira →, Colombia 168 C2 1 36N 79 1W
Mira →, Portugal 43 H2 37 43N 8 47W
Mira por vos Cay, Bahamas 165 B5 22 9N 74 30W
Mīrābād, Afghan. 91 C1 30 25N 61 50 E
Mirabella Eclano, Italy 47 A7 41 2N 14 59 E
Miracema do Norte, Brazil 170 C2 9 33 S 48 24W
Mirador, Brazil 170 C3 6 22 S 44 22W
Miraflores, Colombia 168 C3 1 25N 72 13W
Miraj, India 94 F2 16 50N 74 45 E
Miram Shah, Pakistan 91 B3 33 0N 70 2 E
Miramar, Argentina 174 D4 38 15 S 57 50W
Miramar, Mozam. 117 C6 23 50 S 35 35 E
Miramar, U.S.A. 153 K9 25 59N 80 15W
Miramas, France 29 E8 43 33N 4 59 E
Mirambeau, France 28 C3 45 23N 0 35W
Miramichi B., Canada 141 C7 47 15N 65 0W
Miramont-de-Guyenne, France 28 D4 44 37N 0 21 E
Miranda, Brazil 173 E6 20 10 S 56 15W
Miranda □, Venezuela 168 A4 10 15N 66 25W
Miranda →, Brazil 173 D6 19 25 S 57 20W
Miranda de Ebro, Spain 40 C2 42 41N 2 57W
Miranda do Corvo, Spain 42 E2 40 6N 8 20W
Miranda do Douro, Portugal 42 D4 41 30N 6 16W
Mirande, France 28 E4 43 31N 0 25 E
Mirandela, Portugal 42 D3 41 32N 7 10W
Mirando City, U.S.A. 155 M5 27 26N 99 0W
Mirándola, Italy 44 D8 44 53N 11 4 E
Mirandópolis, Brazil 175 A5 21 9 S 51 6W
Mirango, Malawi 119 E3 13 32 S 34 58 E
Mirani, Australia 126 C4 21 9 S 148 53 E
Mirano, Italy 45 C9 45 30N 12 7 E
Miras, Albania 50 F4 40 30N 20 56 E
Mirassol, Brazil 175 A6 20 46 S 49 28W
Mirbāţ, Oman 99 C6 17 0N 54 45 E
Mirboo North, Australia 129 E7 38 24 S 146 10 E
Mirear, Egypt 106 C4 23 15N 35 41 E
Mirebeau, Côte-d'Or, France 27 E12 47 25N 5 20 E
Mirebeau, Vienne, France 26 F7 46 49N 0 10 E
Mirecourt, France 27 D13 48 20N 6 10 E
Mirgorod = Myrhorod, Ukraine 59 H7 49 58N 33 37 E
Miri, Malaysia 85 B4 4 23N 113 59 E
Miriam Vale, Australia 126 C5 24 20 S 151 33 E
Miribel, France 27 G11 45 50N 4 57 E
Mirim, L., S. Amer. 175 C5 32 45 S 52 50W
Mirimire, Venezuela 168 A4 11 10N 68 43W
Miriti, Brazil 173 B6 6 15 S 59 0W
Mirnyy, Russia 65 C12 62 33N 113 53 E
Miroč, Serbia, Yug. 50 B6 44 32N 22 16 E
Mirond L., Canada 143 B8 55 6N 102 47W
Mirosławiec, Poland 54 E3 53 20N 16 5 E
Mirpur, Pakistan 91 B4 33 32N 73 56 E
Mirpur Bibiwari, Pakistan 92 E2 28 33N 67 44 E
Mirpur Khas, Pakistan 91 D3 25 30N 69 0 E
Mirpur Sakro, Pakistan 92 G2 24 33N 67 41 E
Mirria, Niger 109 F1 13 43N 9 7 E
Mirror, Canada 142 C6 52 30N 113 7W
Mirsk, Poland 55 H2 50 58N 15 23 E
Miryang, S. Korea 75 G15 35 31N 128 44 E
Mirzaani, Georgia 61 K8 41 24N 46 5 E
Mirzapur, India 93 G10 25 10N 82 34 E
Mirzapur-cum-Vindhyachal = Mirzapur, India 93 G10 25 10N 82 34 E
Misamis Occidental □, Phil. 81 G4 8 20N 123 42 E
Misamis Oriental □, Phil. 81 G5 8 45N 125 0 E
Misantla, Mexico 163 D5 19 56N 96 50W
Misawa, Japan 68 D10 40 41N 141 24 E
Miscou I., Canada 141 C7 47 57N 64 31W
Mish'āb, Ra'as al, Si. Arabia 97 D6 28 15N 48 43 E
Mishagua →, Peru 172 C3 11 12 S 72 58W
Mishan, China 73 B8 45 37N 131 48 E
Mishawaka, U.S.A. 157 C10 41 40N 86 11W
Mishbih, Gebel, Egypt 106 C3 22 38N 34 44 E
Mishima, Japan 71 B10 35 10N 138 52 E
Mishmi Hills, India 90 A6 29 0N 96 0 E
Misima I., Papua N. G. 132 F7 10 40 S 152 45 E
Misión, Mexico 161 N10 32 6N 116 53W
Misión Fagnano, Argentina 176 D3 54 32 S 67 17W
Misiones □, Argentina 175 B5 27 0 S 55 0W
Misiones □, Paraguay 174 B4 27 0 S 56 0W
Miskah, Si. Arabia 96 E4 24 49N 42 56 E
Miskitos, Cayos, Nic. 164 D3 14 26N 82 50W
Miskolc, Hungary 52 B5 48 7N 20 50 E
Misoke, Dem. Rep. of the Congo 118 C2 0 42 S 28 2 E
Misool, Indonesia 83 B4 1 52 S 130 10 E
Misrātah, Libya 108 B3 32 24N 15 3 E
Misrātah □, Libya 108 C3 33 30N 15 0 E
Missanabie, Canada 140 C3 48 20N 84 6W
Missão Velha, Brazil 170 C4 7 15 S 39 10W
Missinaibi →, Canada 140 B3 50 43N 81 29W
Missinaibi L., Canada 140 C3 48 23N 83 40W
Mission, S. Dak., U.S.A. 154 D4 43 18N 100 39W
Mission, Tex., U.S.A. 155 M5 26 13N 98 20W
Mission City, Canada 142 D4 49 10N 122 15W
Mission Viejo, U.S.A. 161 M9 33 36N 117 40W
Missisa L., Canada 140 B2 52 20N 85 7W
Missisagi →, Canada 140 C3 46 15N 83 9W
Mississinewa Res., U.S.A. 157 D10 40 46N 86 3W
Mississippi □, U.S.A. 155 J10 33 0N 90 0W
Mississippi →, U.S.A. 155 L10 29 9N 89 15W
Mississippi L., Canada 151 A8 45 5N 76 10W
Mississippi River Delta, U.S.A. 155 L9 29 10N 89 15W
Mississippi Sd., U.S.A. 155 K10 30 20N 89 0W
Missoula, U.S.A. 158 C7 46 52N 114 1W
Missour, Morocco 110 B4 33 3N 4 0W
Missouri □, U.S.A. 154 F8 38 25N 92 30W
Missouri →, U.S.A. 154 F9 38 49N 90 7W
Missouri Valley, U.S.A. 154 E7 41 34N 95 53W
Mist, U.S.A. 160 E3 45 59N 123 15W
Mistake B., Canada 143 A10 62 8N 93 0W
Mistassini →, Canada 141 C5 48 42N 72 20W
Mistassini L., Canada 140 B5 51 0N 73 30W
Mistastin L., Canada 141 A7 55 57N 63 20W
Mistatim, Canada 143 C8 52 52N 103 22W
Mistelbach, Austria 35 C9 48 34N 16 34 E
Misterbianco, Italy 47 E8 37 31N 15 1 E
Mistretta, Italy 47 E7 37 56N 14 22 E

252 Montagnana

Montagnana, *Italy* **45 C8** 45 14N 11 28 E
Montagu, *S. Africa* **116 E3** 33 45 S 20 8 E
Montagu I., *Antarctica* **7 B1** 58 25 S 26 20W
Montague, *Canada* **141 C7** 46 10N 62 39W
Montague, *U.S.A.* **158 F2** 41 44N 122 32W
Montague, I., *Mexico* **162 A2** 31 40N 114 56W
Montague Ra., *Australia* **125 E2** 27 15 S 119 30 E
Montague Sd., *Australia* **124 B4** 14 28 S 125 20 E
Montaigu, *France* **26 F5** 46 59N 1 18W
Montalbán, *Spain* **40 E4** 40 50N 0 45W
Montalbano Iónico, *Italy* **47 B9** 40 17N 16 34 E
Montalbo, *Spain* **40 F2** 39 53N 2 42W
Montalcino, *Italy* **45 E8** 43 3N 11 29 E
Montalegre, *Portugal* **42 D3** 41 49N 7 47W
Montalto, *Italy* **47 D8** 38 10N 15 55 E
Montalto di Castro, *Italy* **45 F8** 42 21N 11 37 E
Montalto Uffugo, *Italy* **47 C9** 39 24N 16 9 E
Montalvo, *U.S.A.* **161 L7** 34 15N 119 12W
Montamarta, *Spain* **42 D5** 41 39N 5 49W
Montana, *Bulgaria* **50 C7** 43 27N 23 16 E
Montaña, *Peru* **172 B3** 6 0 S 73 0W
Montana, *Switz.* **32 D4** 46 19N 7 29 E
Montana □, *Bulgaria* **50 C7** 43 30N 23 20 E
Montana □, *U.S.A.* **158 C9** 47 0N 110 0W
Montaña Clara, I., *Canary Is.* **39 E6** 29 17N 13 33W
Montánchez, *Spain* **43 F4** 39 15N 6 8W
Montañita, *Colombia* **168 C2** 1 22N 75 28W
Montargil, *Portugal* **43 F2** 39 5N 8 10W
Montargis, *France* **27 E9** 47 59N 2 43 E
Montauban, *France* **28 D5** 44 2N 1 21 E
Montauk, *U.S.A.* **151 E13** 41 3N 71 57W
Montauk Pt., *U.S.A.* **151 E13** 41 4N 71 52W
Montbard, *France* **27 E11** 47 38N 4 20 E
Montbarrey, *France* **27 E12** 47 1N 5 39 E
Montbéliard, *France* **27 E13** 47 31N 6 48 E
Montblanc, *Spain* **40 D6** 41 23N 1 4 E
Montbrison, *France* **29 C8** 45 36N 4 3 E
Montcalm, Pic de, *France* **28 F5** 42 40N 1 25 E
Montceau-les-Mines, *France* **27 F11** 46 40N 4 23 E
Montcenis, *France* **29 B8** 46 47N 4 23 E
Montclair, *U.S.A.* **151 F10** 40 49N 74 13W
Montcornet, *France* **27 C11** 49 40N 4 1 E
Montcuq, *France* **28 D5** 44 21N 1 13 E
Montdidier, *France* **27 C9** 49 38N 2 35 E
Monte Albán, *Mexico* **163 D5** 17 2N 96 45W
Monte Alegre, *Brazil* **169 D7** 2 0 S 54 0W
Monte Alegre de Goiás, *Brazil* **171 D2** 13 14 S 47 10W
Monte Alegre de Minas, *Brazil* **171 E2** 18 52 S 48 52W
Monte Azul, *Brazil* **171 E3** 15 9 S 42 53W
Monte Bello Is., *Australia* **124 D2** 20 30 S 115 45 E
Monte-Carlo, *Monaco* **29 E11** 43 46N 7 23 E
Monte Carmelo, *Brazil* **171 E2** 18 43 S 47 29W
Monte Caseros, *Argentina* **174 C4** 30 10 S 57 50W
Monte Comán, *Argentina* **174 C2** 34 40 S 67 53W
Monte Cristi, *Dom. Rep.* **165 C5** 19 52N 71 39W
Monte Dinero, *Argentina* **176 D3** 52 18 S 68 33W
Monte Lindo →, *Paraguay* **174 A4** 23 56 S 57 12W
Monte Quemado, *Argentina* **174 B3** 25 53 S 62 41W
Monte Redondo, *Portugal* **42 F2** 39 53N 8 50W
Monte Rio, *U.S.A.* **160 G4** 38 28N 123 0W
Monte San Giovanni Campano, *Italy* **46 A6** 41 38N 13 31 E
Monte San Savino, *Italy* **45 E8** 43 20N 11 43 E
Monte Sant' Ángelo, *Italy* **45 G12** 41 42N 15 59 E
Monte Santu, C. di, *Italy* **46 B2** 40 5N 9 44 E
Monte Vista, *U.S.A.* **159 H10** 37 35N 106 9W
Monteagudo, *Argentina* **175 B5** 27 14 S 54 8W
Monteagudo, *Bolivia* **173 D5** 19 49 S 63 59W
Montealegre del Castillo, *Spain* **41 G3** 38 48N 1 17W
Montebello, *Canada* **140 C5** 45 40N 74 55W
Montebello Iónico, *Italy* **47 E8** 37 59N 15 45 E
Montebelluna, *Italy* **45 C9** 45 47N 12 3 E
Montebourg, *France* **26 C5** 49 30N 1 20W
Montecastrilli, *Italy* **45 F9** 42 39N 12 29 E
Montecatini Terme, *Italy* **44 E7** 43 53N 10 46 E
Montecito, *U.S.A.* **161 L7** 34 26N 119 40W
Montecristi, *Ecuador* **168 D1** 1 0 S 80 40W
Montecristo, *Italy* **44 F7** 42 20N 10 19 E
Montefalco, *Italy* **45 F9** 42 54N 12 39 E
Montefiascone, *Italy* **45 F9** 42 32N 12 2 E
Montefrío, *Spain* **43 H7** 37 20N 4 0W
Montegiórgio, *Italy* **45 E10** 43 6N 13 33 E
Montego Bay, *Jamaica* **164 C4** 18 30N 78 0W
Montehermoso, *Spain* **42 E4** 40 5N 6 21W
Monteiro, *Brazil* **170 C4** 7 48 S 37 2W
Monteith, *Australia* **128 C3** 35 11 S 139 23 E
Montejicar, *Spain* **43 H7** 37 33N 3 30W
Montejinnie, *Australia* **124 C5** 16 40 S 131 38 E
Montelíbano, *Colombia* **168 B2** 8 5N 75 29W
Montélimar, *France* **29 D8** 44 33N 4 45 E
Montella, *Italy* **47 B8** 40 51N 15 1 E
Montellano, *Spain* **43 J5** 36 59N 5 29W
Montello, *U.S.A.* **154 D10** 43 48N 89 20W
Montemor-o-Novo, *Portugal* **43 G2** 38 40N 8 12W
Montemor-o-Velho, *Portugal* **42 E2** 40 11N 8 40W
Montemorelos, *Mexico* **163 B5** 25 11N 99 42W
Montendre, *France* **28 C3** 45 16N 0 26W
Montenegro, *Brazil* **175 B5** 29 39 S 51 29W
Montenegro □, *Yugoslavia* **50 D3** 42 40N 19 20 E
Montenero di Bisáccia, *Italy* **45 G11** 41 57N 14 47 E
Montepuez, *Mozam.* **119 E4** 13 8 S 38 59 E
Montepuez →, *Mozam.* **119 E5** 12 32 S 40 27 E
Montepulciano, *Italy* **45 E8** 43 5N 11 47 E
Montereale, *Italy* **45 F10** 42 31N 13 13 E
Montereau-Faut-Yonne, *France* **27 D9** 48 22N 2 57 E
Monterey, *Calif., U.S.A.* **160 J5** 36 37N 121 55W
Monterey, *Ind., U.S.A.* **157 C10** 41 11N 86 30W
Monterey B., *U.S.A.* **160 J5** 36 45N 122 0W
Montería, *Colombia* **168 B2** 8 46N 75 53W
Montero, *Bolivia* **173 D5** 17 20 S 63 15W
Monteros, *Argentina* **174 B2** 27 11 S 65 30W
Monterotondo, *Italy* **45 F9** 42 3N 12 37 E
Monterrey, *Mexico* **162 B4** 25 40N 100 30W
Montes Altos, *Brazil* **170 C2** 5 50 S 47 4W
Montes Claros, *Brazil* **171 E3** 16 30 S 43 50W
Montesano, *U.S.A.* **160 D3** 46 59N 123 36W
Montesano sulla Marcellana, *Italy* **47 B8** 40 16N 15 42 E
Montesárchio, *Italy* **47 A7** 41 4N 14 38 E
Montescaglioso, *Italy* **47 B9** 40 33N 16 40 E
Montesilvano, *Italy* **45 F11** 42 29N 14 8 E
Montevarchi, *Italy* **45 E8** 43 31N 11 34 E
Montevideo, *Uruguay* **175 C4** 34 50 S 56 11W
Montevideo, *U.S.A.* **154 C7** 44 57N 95 43W
Montezuma, *Ga., U.S.A.* **152 C5** 32 18N 84 2W
Montezuma, *Ind., U.S.A.* **157 E9** 39 48N 87 22W
Montezuma, *Iowa, U.S.A.* **156 C4** 41 35N 92 32W
Montfaucon, *France* **26 E5** 47 6N 1 7W
Montfaucon-d'Argonne, *France* **27 C12** 49 16N 5 8 E
Montfaucon-en-Velay, *France* **29 C8** 45 11N 4 20 E
Montfort, *France* **26 D5** 48 9N 1 58W
Montfort-le-Gesnois, *France* **26 D7** 48 3N 0 25 E
Montgenèvre, *France* **29 D10** 44 56N 6 43 E
Montgomery = Sahiwal, *Pakistan* **91 C4** 30 45N 73 8 E
Montgomery, *U.K.* **21 E4** 52 34N 3 8W
Montgomery, *Ala., U.S.A.* **152 C3** 32 23N 86 19W
Montgomery, *Ga., U.S.A.* **152 D8** 31 57N 81 7W
Montgomery, *Ill., U.S.A.* **157 C8** 41 44N 88 21W
Montgomery, *W. Va., U.S.A.* **148 F5** 38 11N 81 19W
Montgomery City, *U.S.A.* **156 F5** 38 59N 91 30W
Montguyon, *France* **28 C3** 45 12N 0 12W
Monthermé, *France* **27 C11** 49 52N 4 42 E
Monthey, *Switz.* **32 D3** 46 15N 6 56 E
Monthois, *France* **27 C11** 49 19N 4 43 E
Monti, *Italy* **46 B2** 40 49N 9 19 E
Monticelli d'Ongina, *Italy* **44 C6** 45 5N 9 56 E
Monticello, *Ark., U.S.A.* **155 J9** 33 38N 91 47W
Monticello, *Fla., U.S.A.* **152 E6** 30 33N 83 52W
Monticello, *Ga., U.S.A.* **152 B6** 33 18N 83 40W
Monticello, *Ill., U.S.A.* **157 D8** 40 1N 88 34W
Monticello, *Ind., U.S.A.* **157 D10** 40 45N 86 46W
Monticello, *Iowa, U.S.A.* **156 B5** 42 15N 91 12W
Monticello, *Ky., U.S.A.* **149 G3** 36 50N 84 51W
Monticello, *Minn., U.S.A.* **154 C8** 45 18N 93 48W
Monticello, *Miss., U.S.A.* **155 K9** 31 33N 90 7W
Monticello, *Mo., U.S.A.* **156 D5** 40 7N 91 43W
Monticello, *N.Y., U.S.A.* **151 E10** 41 39N 74 42W
Monticello, *Utah, U.S.A.* **159 H9** 37 52N 109 21W
Montichiari, *Italy* **44 C7** 45 25N 10 23 E
Montier-en-Der, *France* **27 D11** 48 30N 4 45 E
Montignac, *France* **28 C5** 45 4N 1 10 E
Montigny-les-Metz, *France* **27 C13** 49 7N 6 10 E
Montigny-sur-Aube, *France* **27 E11** 47 57N 4 45 E
Montijo, *Portugal* **43 G2** 38 41N 8 54W
Montijo, *Spain* **43 G4** 38 52N 6 39W
Montilla, *Spain* **43 H6** 37 36N 4 40W
Montivilliers, *France* **26 C7** 49 33N 0 12 E
Montluçon, *France* **27 F9** 46 22N 2 36 E
Montmagny, *Canada* **141 C5** 46 58N 70 34W
Montmarault, *France* **27 F9** 46 19N 2 57 E
Montmartre, *Canada* **143 C8** 50 14N 103 27W
Montmédy, *France* **27 C12** 49 30N 5 20 E
Montmélian, *France* **29 C10** 45 30N 6 4 E
Montmirail, *France* **27 D10** 48 51N 3 30 E
Montmoreau-St-Cybard, *France* **28 C4** 45 23N 0 8 E
Montmorency, *Canada* **141 C5** 46 53N 71 11W
Montmorillon, *France* **28 B4** 46 26N 0 50 E
Montmort-Lucy, *France* **27 D10** 48 55N 3 49 E
Monto, *Australia* **126 C5** 24 52 S 151 6 E
Montoire-sur-le-Loir, *France* **26 E7** 47 45N 0 52 E
Montório al Vomano, *Italy* **45 F10** 42 35N 13 38 E
Montoro, *Spain* **43 G6** 38 1N 4 27W
Montour Falls, *U.S.A.* **150 D8** 42 21N 76 51W
Montpelier, *Idaho, U.S.A.* **158 E8** 42 19N 111 18W
Montpelier, *Ind., U.S.A.* **157 D11** 40 33N 85 17W
Montpelier, *Ohio, U.S.A.* **157 C12** 41 35N 84 37W
Montpelier, *Vt., U.S.A.* **151 B12** 44 16N 72 35W
Montpellier, *France* **28 E7** 43 37N 3 52 E
Montpezat-de-Quercy, *France* **28 D5** 44 15N 1 30 E
Montpon-Ménestérol, *France* **28 D4** 45 0N 0 11 E
Montréal, *Canada* **140 C5** 45 31N 73 34W
Montréal, *Aude, France* **28 E6** 43 13N 2 8 E
Montréal, *Gers, France* **28 E4** 43 56N 0 11 E
Montreal L., *Canada* **143 C7** 54 20N 105 45W
Montreal Lake, *Canada* **143 C7** 54 3N 105 46W
Montredon-Labessonnié, *France* **28 E6** 43 45N 2 18 E
Montrésor, *France* **26 E8** 47 10N 1 10 E
Montret, *France* **27 F12** 46 40N 5 7 E
Montreuil, *Pas-de-Calais, France* **27 B8** 50 27N 1 45 E
Montreuil, *Seine-St.-Denis, France* **27 D9** 48 51N 2 27 E
Montreuil-Bellay, *France* **26 E6** 47 8N 0 9W
Montreux, *Switz.* **32 D3** 46 26N 6 55 E
Montrevel-en-Bresse, *France* **27 F12** 46 21N 5 8 E
Montrichard, *France* **26 E8** 47 20N 1 10 E
Montrose, *U.K.* **22 E6** 56 44N 2 27W
Montrose, *Colo., U.S.A.* **159 G10** 38 29N 107 53W
Montrose, *Pa., U.S.A.* **151 E9** 41 50N 75 53W
Montrose, L., *U.S.A.* **156 F3** 38 18N 93 50W
Monts, Pte. des, *Canada* **141 C6** 49 20N 67 12W
Montsalvy, *France* **28 D6** 44 41N 2 30 E
Montsant, Serra de, *Spain* **40 D6** 41 17N 1 0 E
Montsauche-les-Settons, *France* **27 E11** 47 13N 4 2 E
Montsech, Serra del, *Spain* **40 C5** 42 5N 0 45 E
Montserrat, *Spain* **40 D6** 41 36N 1 49 E
Montserrat ■, *W. Indies* **165 C7** 16 40N 62 10W
Montuenga, *Spain* **42 D6** 41 3N 4 38W
Montuiri, *Spain* **39 B9** 39 34N 2 59 E
Monveda, *Dem. Rep. of the Congo* **114 B4** 2 52N 21 30 E
Monyo, *Burma* **90 G5** 17 59N 95 30 E
Monywa, *Burma* **90 D5** 22 7N 95 11 E
Monza, *Italy* **44 C6** 45 35N 9 16 E
Monze, *Zambia* **119 F2** 16 17 S 27 29 E
Monze, C., *Pakistan* **91 D2** 24 47N 66 37 E
Monzón, *Spain* **40 D5** 41 52N 0 10 E
Mooi River, *S. Africa* **117 D4** 29 13 S 29 50 E
Mo'oka, *Japan* **71 A12** 36 26N 140 1 E
Moolawatana, *Australia* **127 D2** 29 55 S 139 45 E
Mooleulooloo, *Australia* **128 A4** 31 36 S 140 32 E
Mooliabeenee, *Australia* **125 F2** 31 20 S 116 2 E
Mooloogool, *Australia* **125 E2** 26 2 S 119 5 E
Moomin Cr. →, *Australia* **127 D4** 29 44 S 149 20 E
Moonah →, *Australia* **126 C2** 22 3 S 138 33 E
Moonbeam, *Canada* **140 C3** 49 20N 82 10W
Moonda, L., *Australia* **126 D3** 25 52 S 140 25 E
Moonie, *Australia* **127 D5** 27 46 S 150 20 E
Moonie →, *Australia* **127 D4** 29 19 S 148 43 E
Moonta, *Australia* **128 C2** 34 6 S 137 32 E
Moora, *Australia* **125 F2** 30 37 S 115 58 E
Mooraberree, *Australia* **126 D3** 25 13 S 140 54 E
Moorarie, *Australia* **125 E2** 25 56 S 117 35 E
Moorcroft, *U.S.A.* **154 C2** 44 16N 104 57W
Moore →, *Australia* **125 F2** 31 22 S 115 30 E
Moore, L., *Australia* **125 E2** 29 50 S 117 35 E
Moore Haven, *U.S.A.* **153 J8** 26 50N 81 6W
Moore Reefs, *Australia* **126 B4** 16 0 S 149 5 E
Moorefield, *U.S.A.* **148 F6** 39 5N 78 59W
Moores Res., *U.S.A.* **151 B13** 44 45N 71 50W
Mooresville, *Ind., U.S.A.* **157 E10** 39 37N 86 22W
Mooresville, *N.C., U.S.A.* **149 H5** 35 35N 80 48W
Moorfoot Hills, *U.K.* **22 F5** 55 44N 3 8W
Moorhead, *U.S.A.* **154 B6** 46 53N 96 45W
Moorland, *Australia* **129 A10** 31 46 S 152 38 E
Moormerland, *Germany* **30 B3** 53 20N 7 20 E
Mooroopna, *Australia* **129 D6** 36 25 S 145 22 E
Moorpark, *U.S.A.* **161 L8** 34 17N 118 53W
Moorreesburg, *S. Africa* **116 E2** 33 6 S 18 38 E
Moosburg, *Germany* **31 G7** 48 27N 11 56 E
Moose →, *Canada* **140 B3** 51 20N 80 25W
Moose Factory, *Canada* **140 B3** 51 16N 80 32W
Moose I., *Canada* **143 C9** 51 42N 97 10W
Moose Jaw, *Canada* **143 C7** 50 24N 105 30W
Moose Jaw →, *Canada* **143 C7** 50 34N 105 18W
Moose Lake, *Canada* **143 C8** 53 43N 100 20W
Moose Lake, *U.S.A.* **154 B8** 46 27N 92 46W
Moose Mountain Cr. →, *Canada* **143 D8** 49 13N 102 12W
Moose Mountain Prov. Park, *Canada* **143 D8** 49 48N 102 25W
Moose Pass, *U.S.A.* **144 F10** 60 29N 149 22W
Moose River, *Canada* **140 B3** 50 48N 81 17W
Moosehead L., *U.S.A.* **141 C6** 45 38N 69 40W
Moosomin, *Canada* **143 C8** 50 9N 101 40W
Moosonee, *Canada* **140 B3** 51 17N 80 39W
Moosup, *U.S.A.* **151 E13** 41 43N 71 53W
Mopeia Velha, *Mozam.* **119 F4** 17 30 S 35 40 E
Mopipi, *Botswana* **116 C3** 21 6 S 24 55 E
Mopoi, *C.A.R.* **114 A5** 5 6N 26 54 E
Mopti, *Mali* **112 C4** 14 30N 4 0W
Moqatta, *Sudan* **107 E4** 14 38N 35 50 E
Moquegua, *Peru* **172 D3** 17 15 S 70 46W
Moquegua □, *Peru* **172 D3** 16 50 S 70 55W
Mór, *Hungary* **52 C3** 47 25N 18 12 E
Móra, *Portugal* **43 G2** 38 55N 8 10W
Mora, *Spain* **43 F7** 39 41N 3 46W
Mora, *Sweden* **16 C8** 61 2N 14 38 E
Mora, *Ga., U.S.A.* **152 D7** 31 25N 82 57W
Mora, *Minn., U.S.A.* **154 C8** 45 53N 93 18W
Mora, *N. Mex., U.S.A.* **159 J11** 35 58N 105 20W
Mora de Ebro = Mòra d'Ebre, *Spain* **40 D5** 41 6N 0 38 E
Mora de Rubielos, *Spain* **40 E4** 40 15N 0 45W
Mòra d'Ebre, *Spain* **40 D5** 41 6N 0 38 E
Mòra la Nova, *Spain* **40 D5** 41 7N 0 39 E
Morača →, *Montenegro, Yug.* **50 D3** 42 20N 19 9 E
Morada Nova, *Brazil* **170 C4** 5 7 S 38 23W
Morada Nova de Minas, *Brazil* **171 E2** 18 37 S 45 22W
Moradabad, *India* **93 E8** 28 50N 78 50 E
Morafenobe, *Madag.* **117 B7** 17 50 S 44 53 E
Morąg, *Poland* **54 E6** 53 55N 19 56 E
Moral de Calatrava, *Spain* **43 G7** 38 51N 3 33W
Moraleja, *Spain* **42 E4** 40 6N 6 43W
Morales, *Colombia* **168 C2** 2 45N 76 38W
Moramanga, *Madag.* **117 B8** 18 56 S 48 12 E
Moran, *Kans., U.S.A.* **155 G7** 37 55N 95 10W
Moran, *Wyo., U.S.A.* **158 E8** 43 53N 110 37W
Moranbah, *Australia* **126 C4** 22 1 S 148 6 E
Morano Cálabro, *Italy* **47 C9** 39 50N 16 8 E
Morant Cays, *Jamaica* **164 C4** 17 22N 76 0W
Morant Pt., *Jamaica* **164 C4** 17 55N 76 12W
Morar, L., *U.K.* **22 E3** 56 57N 5 40W
Moratalla, *Spain* **41 G3** 38 14N 1 49W
Moratuwa, *Sri Lanka* **95 L4** 6 45N 79 55 E
Morava →, *Slovak Rep.* **35 C9** 48 10N 16 59 E
Moravia, *U.S.A.* **156 D4** 40 53N 92 49W
Moravian Hts. = Českomoravská Vrchovina, *Czech Rep.* **34 B8** 49 30N 15 40 E
Moravica →, *Serbia, Yug.* **50 C4** 43 52N 20 8 E
Moraviţa, *Romania* **52 E6** 45 17N 21 14 E
Moravská Třebová, *Czech Rep.* **35 B9** 49 45N 16 40 E
Moravské Budějovice, *Czech Rep.* **34 B8** 49 4N 15 49 E
Morawa, *Australia* **125 E2** 29 13 S 116 0 E
Morawhanna, *Guyana* **169 B6** 8 30N 59 40W
Moray □, *U.K.* **22 D5** 57 31N 3 18W
Moray Firth, *U.K.* **22 D5** 57 40N 3 52W
Morbach, *Germany* **31 F3** 49 48N 7 7 E
Morbegno, *Italy* **44 B6** 46 8N 9 34 E
Morbi, *India* **92 H4** 22 50N 70 42 E
Morbihan □, *France* **26 E4** 47 55N 2 50W
Mörbylånga, *Sweden* **17 H10** 56 32N 16 22 E
Morcenx, *France* **28 D3** 44 3N 0 55W
Morcone, *Italy* **47 A7** 41 20N 14 40 E
Mordelles, *France* **26 D5** 48 5N 1 52W
Morden, *Canada* **143 D9** 49 15N 98 10W
Mordoğan, *Turkey* **49 C8** 38 30N 26 37 E
Mordovian Republic □ = Mordvinia □, *Russia* **60 C7** 54 20N 44 30 E
Mordovo, *Russia* **60 D5** 52 6N 40 50 E
Mordvinia □, *Russia* **60 C7** 54 20N 44 30 E
Mordy, *Poland* **55 F9** 52 13N 22 31 E
Møre og Romsdal □, *Norway* **18 B4** 62 30N 8 0 E
Morea, *Australia* **128 D4** 36 45 S 141 18 E
Morea, *Greece* **12 H10** 37 45N 22 10 E
Moreau →, *U.S.A.* **154 C4** 45 18N 100 43W
Morecambe, *U.K.* **20 C5** 54 5N 2 52W
Morecambe B., *U.K.* **20 C5** 54 7N 3 0W
Moree, *Australia* **127 D4** 29 28 S 149 54 E
Morehead, *Papua N. G.* **132 E1** 8 41 S 141 41 E
Morehead, *U.S.A.* **157 F13** 38 11N 83 26W
Morehead City, *U.S.A.* **149 H7** 34 43N 76 43W
Moreland, *U.S.A.* **152 B5** 33 17N 84 46W
Morelia, *Mexico* **162 D4** 19 42N 101 7W
Morella, *Australia* **126 C3** 23 0 S 143 52 E
Morella, *Spain* **40 E4** 40 35N 0 5W
Morelos, *Mexico* **162 B3** 26 42N 107 40W
Morelos □, *Mexico* **163 D5** 18 40N 99 10W
Morena, Sierra, *Spain* **43 G7** 38 20N 4 0W
Morenci, *Ariz., U.S.A.* **159 K9** 33 5N 109 22W
Morenci, *Mich., U.S.A.* **157 C12** 41 43N 84 13W
Moreni, *Romania* **53 F10** 44 59N 25 36 E
Moreno Valley, *U.S.A.* **161 M10** 33 56N 116 58W
Morero, *Bolivia* **173 C4** 11 9 S 66 15W
Moreru →, *Brazil* **173 C6** 10 10 S 59 15W
Moresby I., *Canada* **142 C2** 52 30N 131 40W
Morestel, *France* **29 C9** 45 40N 5 28 E
Moreton, *Australia* **126 A3** 12 22 S 142 40 E
Moreton I., *Australia* **127 D5** 27 10 S 153 25 E
Moreuil, *France* **27 C9** 49 46N 2 30 E
Morey, *Spain* **39 B10** 39 44N 3 20 E
Morez, *France* **27 F13** 46 31N 6 2 E
Morgan, *Australia* **128 C3** 34 2 S 139 35 E
Morgan, *Ga., U.S.A.* **152 D5** 31 32N 84 36W
Morgan, *Utah, U.S.A.* **158 F8** 41 2N 111 41W
Morgan City, *U.S.A.* **155 L9** 29 42N 91 12W
Morgan Hill, *U.S.A.* **160 H5** 37 8N 121 39W
Morgan Vale, *Australia* **128 B4** 33 10 S 140 32 E
Morganfield, *U.S.A.* **148 G2** 37 41N 87 55W
Morganton, *U.S.A.* **149 H5** 35 45N 81 41W
Morgantown, *Ind., U.S.A.* **157 E10** 39 22N 86 16W
Morgantown, *W. Va., U.S.A.* **148 F6** 39 38N 79 57W
Morgenzon, *S. Africa* **117 D4** 26 45 S 29 36 E
Morges, *Switz.* **32 C2** 46 31N 6 29 E
Morghak, *Iran* **97 D8** 29 7N 57 54 E
Morgongåva, *Sweden* **16 E10** 59 57N 16 58 E
Morhange, *France* **27 D13** 48 55N 6 38 E
Mori, *Italy* **44 C7** 45 51N 10 59 E
Morice L., *Canada* **142 C3** 53 50N 127 40W
Morichal, *Colombia* **168 C3** 2 10N 70 34W
Morichal Largo →, *Venezuela* **169 B5** 9 27N 62 25W
Moriguchi, *Japan* **71 C7** 34 44N 135 34 E
Moriki, *Nigeria* **113 C6** 12 52N 6 30 E
Morinville, *Canada* **142 C6** 53 49N 113 41W
Morioka, *Japan* **68 E10** 39 45N 141 8 E
Moris, *Mexico* **162 B3** 28 8N 108 32W
Morisset, *Australia* **129 B9** 33 6 S 151 30 E
Morlaàs, *France* **28 E3** 43 21N 0 18W
Morlaix, *France* **26 D3** 48 36N 3 52W
Mörlunda, *Sweden* **17 G9** 57 19N 15 52 E
Mormanno, *Italy* **47 C8** 39 53N 15 59 E
Mormant, *France* **27 D9** 48 37N 2 52 E
Mornington, *Vic., Australia* **129 E6** 38 15 S 145 5 E
Mornington, *W. Austral., Australia* **124 C4** 17 31 S 126 6 E
Mornington, I., *Chile* **176 C1** 49 50 S 75 30W
Mornington I., *Australia* **126 B2** 16 30 S 139 30 E
Mórnos →, *Greece* **48 C3** 38 25N 21 50 E
Moro, *Sudan* **107 E3** 10 50N 30 9 E
Moro G., *Phil.* **81 H4** 6 30N 123 0 E
Morobe, *Papua N. G.* **132 D4** 7 49 S 147 38 E
Morocco, *U.S.A.* **157 D9** 40 57N 87 27W
Morocco ■, *N. Afr.* **110 B3** 32 0N 5 50W
Morococha, *Peru* **172 C2** 11 40 S 76 5W
Morogoro, *Tanzania* **118 D4** 6 50 S 37 40 E
Morogoro □, *Tanzania* **118 D4** 8 0 S 37 0 E
Moroleón, *Mexico* **162 C4** 20 8N 101 32W
Morombe, *Madag.* **117 C7** 21 45 S 43 22 E
Moron, *Argentina* **174 C4** 34 39 S 58 37W
Morón, *Cuba* **164 B4** 22 8N 78 39W
Morón de Almazán, *Spain* **40 D2** 41 29N 2 27W
Morón de la Frontera, *Spain* **43 H5** 37 6N 5 28W
Morona →, *Peru* **168 D2** 4 45 S 77 10W
Morona-Santiago □, *Ecuador* **168 D2** 2 30 S 78 0W
Morondava, *Madag.* **117 C7** 20 17 S 44 17 E
Morondo, *Ivory C.* **112 D3** 8 57N 6 47W
Morong, *Phil.* **80 D3** 14 41N 120 16 E
Morongo Valley, *U.S.A.* **161 L10** 34 3N 116 37W
Moronou, *Ivory C.* **112 D4** 6 16N 4 59W
Morotai, *Indonesia* **82 A3** 2 10N 128 30 E
Moroto, *Uganda* **118 B3** 2 28N 34 42 E
Moroto Summit, *Kenya* **118 B3** 2 30N 34 43 E
Morozov, *Bulgaria* **51 D9** 42 30N 25 10 E
Morozovsk, *Russia* **61 F5** 48 25N 41 50 E
Morpeth, *U.K.* **20 B6** 55 10N 1 41W
Morphou, *Cyprus* **38 D11** 35 12N 32 59 E
Morphou Bay, *Cyprus* **38 D11** 35 15N 32 50 E
Morrelganj, *Bangla.* **90 D2** 22 28N 89 51 E
Morrilton, *U.S.A.* **155 H8** 35 9N 92 44W
Morrinhos, *Ceara, Brazil* **170 B3** 3 14 S 40 7W
Morrinhos, *Minas Gerais, Brazil* **171 E2** 17 45 S 49 10W
Morrinsville, *N.Z.* **130 D4** 37 40 S 175 32 E
Morris, *Canada* **143 D9** 49 25N 97 22W
Morris, *Ga., U.S.A.* **152 D5** 31 48N 84 57W
Morris, *Ill., U.S.A.* **157 C8** 41 22N 88 26W
Morris, *Minn., U.S.A.* **154 C7** 45 35N 95 55W
Morris, Mt., *Australia* **125 E5** 26 9 S 131 4 E
Morris Jesup, Kap, *Greenland* **10 A7** 83 40N 34 0W
Morrisburg, *Canada* **140 D4** 44 55N 75 7W
Morrison, *U.S.A.* **156 C7** 41 49N 89 58W
Morrisonville, *U.S.A.* **156 E7** 39 25N 89 27W
Morristown, *Ariz., U.S.A.* **159 K7** 33 51N 112 37W
Morristown, *Ind., U.S.A.* **157 E11** 39 40N 85 42W
Morristown, *N.J., U.S.A.* **151 F10** 40 48N 74 29W
Morristown, *S. Dak., U.S.A.* **154 C4** 45 56N 101 43W
Morristown, *Tenn., U.S.A.* **149 G4** 36 13N 83 18W
Morro, Pta., *Chile* **174 B1** 27 6 S 71 0W
Morro Bay, *U.S.A.* **160 K6** 35 22N 120 51W
Morro del Jable, *Canary Is.* **39 F5** 28 3N 14 23W
Morro do Chapéu, *Brazil* **171 D3** 11 33 S 41 9W
Morro Jable, Pta. de, *Canary Is.* **39 F5** 28 2N 14 20W
Morros, *Brazil* **170 B3** 2 52 S 44 3W
Morrosquillo, G. de, *Colombia* **164 E4** 9 35N 75 40W
Mörrum, *Sweden* **17 H8** 56 12N 14 45 E
Morrumbene, *Mozam.* **117 C6** 23 31 S 35 16 E
Mörrumsån →, *Sweden* **17 H8** 56 10N 14 45 E
Mors, *Denmark* **17 H2** 56 50N 8 45 E
Morshansk, *Russia* **60 D5** 53 28N 41 50 E
Mörsil, *Sweden* **16 A7** 63 19N 13 40 E
Mortagne →, *France* **27 D13** 48 33N 6 27 E
Mortagne-au-Perche, *France* **26 D7** 48 31N 0 33 E
Mortagne-sur-Gironde, *France* **28 C3** 45 28N 0 47W
Mortagne-sur-Sèvre, *France* **26 F6** 47 0N 0 59W
Mortain, *France* **26 D6** 48 40N 0 57W
Mortara, *Italy* **44 C5** 45 15N 8 44 E
Mortcha, *Chad* **109 E4** 16 0N 21 10 E
Morteau, *France* **27 E13** 47 3N 6 35 E
Morteros, *Argentina* **174 C3** 30 50 S 62 0W
Mortes, R. das →, *Brazil* **171 D1** 11 45 S 50 44W
Mortlake, *Australia* **128 E5** 38 5 S 142 50 E
Morton, *Ill., U.S.A.* **156 D7** 40 37N 89 28W
Morton, *Tex., U.S.A.* **155 J3** 33 44N 102 46W
Morton, *Wash., U.S.A.* **160 D4** 46 34N 122 17W
Morundah, *Australia* **129 C7** 34 57 S 146 19 E
Moruya, *Australia* **129 C9** 35 58 S 150 3 E
Morvan, *France* **27 E11** 47 5N 4 3 E
Morven, *Australia* **127 D4** 26 22 S 147 5 E
Morven, *N.Z.* **131 E6** 44 50 S 171 6 E
Morven, *U.S.A.* **152 E6** 30 57N 83 30W
Morvern, *U.K.* **22 E3** 56 38N 5 44W
Morwell, *Australia* **129 E7** 38 10 S 146 22 E
Moryń, *Poland* **55 F1** 52 51N 14 22 E
Morzhovets, Ostrov, *Russia* **56 A7** 66 44N 42 35 E
Morzine, *France* **27 F13** 46 11N 6 42 E
Mosalsk, *Russia* **58 E8** 54 30N 34 55 E
Mosbach, *Germany* **31 F5** 49 21N 9 9 E

Munchŏn 253

Mosby, *Norway* **18 F4** 58 12N 7 55 E
Moščenice, *Croatia* **45 C11** 45 17N 14 16 E
Mosciano Sant' Ángelo, *Italy* **45 F10** 42 42N 13 52 E
Moscos Is. = Maungmagan Is., *Burma* **86 F1** 14 0N 97 30 E
Moscow = Moskva, *Russia* **58 E9** 55 45N 37 35 E
Moscow, *U.S.A.* **158 C5** 46 44N 117 0W
Mosel →, *Europe* **27 B14** 50 22N 7 36 E
Moselle = Mosel →, *Europe* **27 B14** 50 22N 7 36 E
Moselle □, *France* **27 D13** 48 59N 6 33 E
Moses Lake, *U.S.A.* **158 C4** 47 8N 119 17W
Mosgiel, *N.Z.* **131 F5** 45 53 S 170 21 E
Moshi, *Tanzania* **118 C4** 3 22 S 37 18 E
Moshi □, *Tanzania* **118 C4** 3 22 S 37 18 E
Moshupa, *Botswana* **116 C4** 24 46 S 25 29 E
Mosina, *Poland* **55 F3** 52 15N 16 50 E
Mosjøen, *Norway* **14 D15** 65 51N 13 12 E
Moskenesøya, *Norway* **14 C15** 67 58N 13 0 E
Moskenstraumen, *Norway* **14 C15** 67 47N 12 45 E
Moskog, *Norway* **18 C2** 61 26N 6 0 E
Moskva, *Russia* **58 E9** 55 45N 37 35 E
Moskva →, *Russia* **58 E10** 55 5N 38 51 E
Moslavačka Gora, *Croatia* **45 C13** 45 40N 16 37 E
Moso, *Vanuatu* **133 G6** 17 30 S 168 15 E
Mosomane, *Botswana* **116 C4** 24 2 S 26 19 E
Moson-magyaróvár, *Hungary* **52 C2** 47 52N 17 18 E
Mošorin, *Serbia, Yug.* **52 E5** 45 19N 20 4 E
Mospino, *Ukraine* **59 J9** 47 52N 38 0 E
Mosquera, *Colombia* **168 C2** 2 35N 78 24W
Mosquero, *U.S.A.* **155 H3** 35 47N 103 58W
Mosqueruela, *Spain* **40 E4** 40 21N 0 27W
Mosquitia, *Honduras* **164 C3** 15 20N 84 10W
Mosquitos, G. de los, *Panama* **164 E3** 9 15N 81 10W
Moss, *Norway* **15 G14** 59 27N 10 40 E
Moss Vale, *Australia* **129 C9** 34 32 S 150 25 E
Mossaka, *Congo* **114 C3** 1 15 S 16 45 E
Mossâmedes, *Brazil* **171 E1** 16 7 S 50 11W
Mossbank, *Canada* **143 D7** 49 56N 105 56W
Mossburn, *N.Z.* **131 F3** 45 41 S 168 15 E
Mosselbaai, *S. Africa* **116 E3** 34 11 S 22 8 E
Mossendjo, *Congo* **114 C2** 2 55 S 12 42 E
Mosses, Col des, *Switz.* **32 D4** 46 25N 7 7 E
Mossgiel, *Australia* **128 B6** 33 15 S 144 5 E
Mossingen, *Germany* **31 G5** 48 24N 9 4 E
Mossman, *Australia* **126 B4** 16 21 S 145 15 E
Mossoró, *Brazil* **170 C4** 5 10 S 37 15W
Mossuril, *Mozam.* **119 E5** 14 58 S 40 42 E
Mossy →, *Canada* **143 C8** 54 5N 102 58W
Mossy Head, *U.S.A.* **153 E3** 30 45N 86 19W
Most, *Czech Rep.* **34 A6** 50 31N 13 38 E
Mosta, *Malta* **38 D1** 35 54N 14 24 E
Moştafáābād, *Iran* **97 C7** 33 39N 54 53 E
Mostaganem, *Algeria* **111 A5** 35 54N 0 5 E
Mostar, *Bos.-H.* **52 G2** 43 22N 17 50 E
Mostardas, *Brazil* **175 C5** 31 2 S 50 51W
Mostefa, Rass, *Tunisia* **108 A2** 36 55N 11 3 E
Mosterhamn, *Norway* **18 E2** 59 42N 5 21 E
Mostiska = Mostyska, *Ukraine* **59 H2** 49 48N 23 4 E
Móstoles, *Spain* **42 E7** 40 19N 3 53W
Mosty = Masty, *Belarus* **58 F3** 53 27N 24 38 E
Mostyska, *Ukraine* **59 H2** 49 48N 23 4 E
Mosul = Al Mawşil, *Iraq* **101 D10** 36 15N 43 5 E
Mosulpo, *S. Korea* **75 H14** 33 20N 126 17 E
Møsvatnet, *Norway* **18 E5** 59 52N 8 5 E
Mota, *Vanuatu* **133 C5** 13 49 S 167 42 E
Mota del Cuervo, *Spain* **41 F2** 39 30N 2 52W
Mota del Marqués, *Spain* **42 D5** 41 38N 5 11W
Mota Lava, *Vanuatu* **133 C5** 13 40 S 167 40 E
Motagua →, *Guatemala* **164 C2** 15 44N 88 14W
Motala, *Sweden* **17 F9** 58 32N 15 1 E
Moţca, *Romania* **53 C11** 47 15N 26 37 E
Motegi, *Japan* **71 A12** 36 32N 140 11 E
Motherwell, *U.K.* **22 F5** 55 47N 3 58W
Motihari, *India* **93 F11** 26 30N 84 55 E
Motilla del Palancar, *Spain* **41 F3** 39 34N 1 55W
Motiti I., *N.Z.* **130 D5** 37 38 S 176 25 E
Motnik, *Slovenia* **45 B11** 46 14N 14 54 E
Motocurunya, *Venezuela* **169 C5** 4 24N 64 5W
Motovun, *Croatia* **45 C10** 45 20N 13 50 E
Motozintla de Mendoza, *Mexico* **163 D6** 15 21N 92 14W
Motril, *Spain* **43 J7** 36 31N 3 37W
Motru, *Romania* **52 F7** 44 48N 22 59 E
Motru →, *Romania* **53 F8** 44 32N 23 31 E
Mott, *U.S.A.* **154 B3** 46 23N 102 20W
Mòttola, *Italy* **47 B10** 40 38N 17 2 E
Motu, *N.Z.* **130 E6** 38 18 S 177 40 E
Motu →, *N.Z.* **130 D6** 37 51 S 177 35 E
Motueka, *N.Z.* **131 B8** 41 7 S 173 1 E
Motueka →, *N.Z.* **131 B8** 41 5 S 173 1 E
Motul, *Mexico* **163 C7** 21 0N 89 20W
Motupena Pt., *Papua N. G.* **132 D8** 6 30 S 155 10 E
Mouanda, *Gabon* **114 C2** 1 28 S 13 7 E
Mouchalagane →, *Canada* **141 B6** 50 56N 68 41W
Moúdhros, *Greece* **49 B7** 39 50N 25 18 E
Mouding, *China* **76 E3** 25 20N 101 28 E
Moudjeria, *Mauritania* **112 B2** 17 50N 12 28W
Moudon, *Switz.* **32 C3** 46 40N 6 49 E
Mougoundou, *Congo* **114 C2** 2 40 S 12 41 E
Mouila, *Gabon* **114 C2** 1 50 S 11 0 E
Mouka, *C.A.R.* **114 A4** 7 16N 21 52 E
Moulamein, *Australia* **128 C6** 35 3 S 144 1 E
Moulianá, *Greece* **38 D7** 35 10N 25 59 E
Moulins, *France* **27 F10** 46 35N 3 19 E
Moulmein, *Burma* **90 G6** 16 30N 97 40 E
Moulmeingyun, *Burma* **90 G5** 16 23N 95 16 E
Moulouya, O. →, *Morocco* **111 A4** 35 5N 2 25W
Moulton, *Iowa, U.S.A.* **156 D4** 40 41N 92 41W
Moulton, *Tex., U.S.A.* **155 L6** 29 35N 97 9W
Moultrie, *U.S.A.* **152 D6** 31 11N 83 47W
Moultrie, L., *U.S.A.* **152 B9** 33 20N 80 5W
Mound City, *Mo., U.S.A.* **154 E7** 40 7N 95 14W
Mound City, *S. Dak., U.S.A.* **154 C4** 45 44N 100 4W
Moúnda, Ákra, *Greece* **48 C2** 38 5N 20 45 E
Moundou, *Chad* **109 G3** 8 40N 16 10 E
Moundsville, *U.S.A.* **150 G4** 39 55N 80 44W
Mounembé, *Congo* **114 C2** 3 20 S 12 32 E
Moung, *Cambodia* **86 F4** 12 46N 103 27 E
Moungoudi, *Congo* **114 C2** 2 45 S 11 46 E
Mount Airy, *U.S.A.* **149 G5** 36 31N 80 37W
Mount Albert, *Canada* **150 B5** 44 8N 79 19W
Mount Amherst, *Australia* **124 C4** 18 24 S 126 58 E
Mount Angel, *U.S.A.* **158 D2** 45 4N 122 48W
Mount Augustus, *Australia* **124 D2** 24 20 S 116 56 E
Mount Ayr, *U.S.A.* **156 D2** 40 43N 94 14W
Mount Barker, *S. Austral., Australia* **128 C3** 35 5 S 138 52 E
Mount Barker, *W. Austral., Australia* **125 F2** 34 38 S 117 40 E
Mount Beauty, *Australia* **129 D7** 36 47 S 147 10 E
Mount Carmel, *U.S.A.* **157 F9** 38 25N 87 46W
Mount Carroll, *U.S.A.* **156 B7** 42 6N 89 59W
Mount Clemens, *U.S.A.* **140 D3** 42 35N 82 53W
Mount Coolon, *Australia* **126 C4** 21 25 S 147 25 E
Mount Darwin, *Zimbabwe* **119 F3** 16 47 S 31 38 E
Mount Desert I., *U.S.A.* **141 D6** 44 21N 68 20W
Mount Dora, *U.S.A.* **149 L5** 28 48N 81 38W
Mount Douglas, *Australia* **126 C4** 21 35 S 146 50 E
Mount Eba, *Australia* **127 E2** 30 11 S 135 40 E
Mount Eden, *U.S.A.* **157 F11** 38 3N 85 9W
Mount Edgecumbe, *U.S.A.* **142 B1** 57 3N 135 21W
Mount Elizabeth, *Australia* **124 C4** 16 0 S 125 50 E
Mount Fletcher, *S. Africa* **117 E4** 30 40 S 28 30 E
Mount Forest, *Canada* **140 D3** 43 59N 80 43W
Mount Gambier, *Australia* **128 D4** 37 50 S 140 46 E
Mount Garnet, *Australia* **126 B4** 17 37 S 145 6 E
Mount Hagen, *Papua N. G.* **132 C3** 5 52 S 144 16 E
Mount Hope, *N.S.W., Australia* **129 B6** 32 51 S 145 51 E
Mount Hope, *S. Austral., Australia* **127 E2** 34 7 S 135 23 E
Mount Hope, *U.S.A.* **148 G5** 37 54N 81 10W
Mount Horeb, *U.S.A.* **156 B7** 43 1N 89 44W
Mount Howitt, *Australia* **127 D3** 26 31 S 142 16 E
Mount Isa, *Australia* **126 C2** 20 42 S 139 26 E
Mount Keith, *Australia* **125 E3** 27 15 S 120 30 E
Mount Laguna, *U.S.A.* **161 N10** 32 52N 116 25W
Mount Larcom, *Australia* **126 C5** 23 48 S 150 59 E
Mount Lofty Ra., *Australia* **128 C3** 34 35 S 139 5 E
Mount Magnet, *Australia* **125 E2** 28 2 S 117 47 E
Mount Manara, *Australia* **128 B5** 32 29 S 143 58 E
Mount Margaret, *Australia* **127 D3** 26 54 S 143 21 E
Mount Maunganui, *N.Z.* **130 D5** 37 40 S 176 14 E
Mount Molloy, *Australia* **126 B4** 16 42 S 145 20 E
Mount Monger, *Australia* **125 F3** 31 0 S 122 0 E
Mount Morgan, *Australia* **126 C5** 23 40 S 150 25 E
Mount Morris, *U.S.A.* **150 D7** 42 44N 77 52W
Mount Mulligan, *Australia* **126 B3** 16 45 S 144 47 E
Mount Narryer, *Australia* **125 E2** 26 30 S 115 55 E
Mount Olive, *U.S.A.* **156 E7** 39 4N 89 44W
Mount Olivet, *U.S.A.* **157 F12** 38 32N 84 2W
Mount Orab, *U.S.A.* **157 E13** 39 2N 83 55W
Mount Oxide Mine, *Australia* **126 B2** 19 30 S 139 29 E
Mount Pearl, *Canada* **141 C9** 47 31N 52 47W
Mount Perry, *Australia* **127 D5** 25 13 S 151 42 E
Mount Phillips, *Australia* **124 D2** 24 25 S 116 15 E
Mount Pleasant, *Iowa, U.S.A.* **156 D5** 40 58N 91 33W
Mount Pleasant, *Mich., U.S.A.* **148 D3** 43 36N 84 46W
Mount Pleasant, *Pa., U.S.A.* **150 F5** 40 9N 79 33W
Mount Pleasant, *S.C., U.S.A.* **152 C10** 32 47N 79 52W
Mount Pleasant, *Tenn., U.S.A.* **149 H2** 35 32N 87 12W
Mount Pleasant, *Tex., U.S.A.* **155 J7** 33 9N 94 58W
Mount Pleasant, *Utah, U.S.A.* **158 G8** 39 33N 111 27W
Mount Pocono, *U.S.A.* **151 E9** 41 7N 75 22W
Mount Pulaski, *U.S.A.* **156 D7** 40 1N 89 17W
Mount Rainier National Park, *U.S.A.* **160 D5** 46 55N 121 50W
Mount Revelstoke Nat. Park, *Canada* **142 C5** 51 5N 118 30W
Mount Robson Prov. Park, *Canada* **142 C5** 53 0N 119 0W
Mount Roskill, *N.Z.* **130 C3** 36 55 S 174 45 E
Mount Sandiman, *Australia* **125 D2** 24 25 S 115 30 E
Mount Shasta, *U.S.A.* **158 F2** 41 19N 122 19W
Mount Signal, *U.S.A.* **161 N11** 32 39N 115 37W
Mount Somers, *N.Z.* **131 D6** 43 45 S 171 27 E
Mount Sterling, *Ill., U.S.A.* **156 E6** 39 59N 90 45W
Mount Sterling, *Ky., U.S.A.* **157 F13** 38 4N 83 56W
Mount Sterling, *Ohio, U.S.A.* **157 E13** 39 43N 83 16W
Mount Surprise, *Australia* **126 B3** 18 10 S 144 17 E
Mount Union, *U.S.A.* **150 F7** 40 23N 77 53W
Mount Vernon, *Australia* **124 D2** 24 9 S 118 2 E
Mount Vernon, *Ga., U.S.A.* **152 C7** 32 11N 82 36W
Mount Vernon, *Ind., U.S.A.* **154 F10** 38 17N 88 57W
Mount Vernon, *Ind., U.S.A.* **157 F8** 37 56N 87 54W
Mount Vernon, *Iowa, U.S.A.* **156 C5** 41 55N 91 23W
Mount Vernon, *N.Y., U.S.A.* **151 F11** 40 55N 73 50W
Mount Vernon, *Ohio, U.S.A.* **150 F2** 40 23N 82 29W
Mount Vernon, *Wash., U.S.A.* **160 B4** 48 25N 122 20W
Mount Victor, *Australia* **128 B3** 32 11 S 139 44 E
Mount Washington, *U.S.A.* **157 F11** 38 3N 85 33W
Mount Wellington, *N.Z.* **130 C3** 36 55 S 174 52 E
Mount Zion, *U.S.A.* **157 E8** 39 46N 88 53W
Mountain □, *Phil.* **80 C3** 17 20N 121 10 E
Mountain Ash, *U.K.* **21 F4** 51 40N 3 23W
Mountain Center, *U.S.A.* **161 M10** 33 42N 116 44W
Mountain City, *Nev., U.S.A.* **158 F6** 41 50N 115 58W
Mountain City, *Tenn., U.S.A.* **149 G5** 36 29N 81 48W
Mountain Grove, *U.S.A.* **155 G8** 37 8N 92 16W
Mountain Home, *Ark., U.S.A.* **155 G8** 36 20N 92 23W
Mountain Home, *Idaho, U.S.A.* **158 E6** 43 8N 115 41W
Mountain Iron, *U.S.A.* **154 B8** 47 32N 92 37W
Mountain Park, *Canada* **142 C5** 52 50N 117 15W
Mountain Pass, *U.S.A.* **161 K11** 35 29N 115 35W
Mountain View, *Ark., U.S.A.* **155 H8** 35 52N 92 7W
Mountain View, *Calif., U.S.A.* **160 H4** 37 23N 122 5W
Mountain View, *Hawaii, U.S.A.* **145 D6** 19 33N 155 7W
Mountainair, *U.S.A.* **159 J10** 34 31N 106 15W
Mountmellick, *Ireland* **23 C4** 53 7N 7 20W
Mountrath, *Ireland* **23 D4** 53 0N 7 28W
Moura, *Australia* **126 C4** 24 35 S 149 58 E
Moura, *Brazil* **169 D5** 1 32 S 61 38W
Moura, *Portugal* **43 G3** 38 7N 7 30W
Mourão, *Portugal* **43 G3** 38 22N 7 22W
Mourdi, Dépression du, *Chad* **109 E4** 18 10N 23 0 E
Mourdiah, *Mali* **112 C3** 14 35N 7 25W
Mourenx-Ville-Nouvelle, *France* **28 E3** 43 22N 0 38W
Mouri, *Ghana* **113 D4** 5 6N 1 14W
Mourilyan, *Australia* **126 B4** 17 35 S 146 3 E
Mourmelon-le-Grand, *France* **27 C11** 49 8N 4 22 E
Mourne →, *U.K.* **23 B4** 54 52N 7 26W
Mourne Mts., *U.K.* **23 B5** 54 10N 6 0W
Mourniaí, *Greece* **38 D6** 35 29N 24 1 E
Mournies = Mourniaí, *Greece* **38 D6** 35 29N 24 1 E
Mouscron, *Belgium* **24 D3** 50 45N 3 12 E
Moussoro, *Chad* **109 F3** 13 41N 16 35 E
Mouthe, *France* **27 F13** 46 44N 6 12 E
Moutier, *Switz.* **32 B4** 47 16N 7 21 E
Moûtiers, *France* **29 C10** 45 29N 6 32 E
Moutohara, *N.Z.* **130 E6** 38 27 S 177 32 E
Moutong, *Indonesia* **82 A2** 0 28N 121 13 E
Mouy, *France* **27 C9** 49 18N 2 20 E
Mouzáki, *Greece* **48 B3** 39 25N 21 37 E
Mouzon, *France* **27 C12** 49 36N 5 3 E
Movas, *Mexico* **162 B3** 28 10N 109 25W
Moville, *Ireland* **23 A4** 55 11N 7 3W
Moweaqua, *U.S.A.* **156 E7** 39 38N 89 1W
Moxico □, *Angola* **115 E4** 12 0 S 20 30 E
Moxotó →, *Brazil* **170 C4** 9 19 S 38 14W
Moy →, *Ireland* **23 B2** 54 8N 9 8W
Moyale, *Kenya* **107 G4** 3 30N 39 0 E
Moyamba, *S. Leone* **112 D2** 8 4N 12 30W
Moyen Atlas, *Morocco* **110 B4** 33 0N 5 0W
Moyo, *Indonesia* **82 C1** 8 10 S 117 40 E
Moyobamba, *Peru* **172 B2** 6 0 S 77 0W
Moyyero →, *Russia* **65 C11** 68 44N 103 42 E
Moyynty, *Kazakstan* **64 E8** 47 10N 73 18 E
Mozambique = Moçambique, *Mozam.* **119 F5** 15 3 S 40 42 E
Mozambique ■, *Africa* **119 F4** 19 0 S 35 0 E
Mozambique Chan., *Africa* **117 B7** 17 30 S 42 30 E
Mozdok, *Russia* **61 J7** 43 45N 44 48 E
Mozdūrān, *Iran* **97 B9** 36 9N 60 35 E
Mozhaysk, *Russia* **58 E9** 55 30N 36 2 E
Mozhga, *Russia* **60 B11** 56 26N 52 15 E
Mozhnābād, *Iran* **97 C9** 34 7N 60 6 E
Mozirje, *Slovenia* **45 B11** 46 22N 14 58 E
Mozyr = Mazyr, *Belarus* **59 F5** 51 59N 29 15 E
Mpanda, *Tanzania* **118 D3** 6 23 S 31 1 E
Mpanda □, *Tanzania* **118 D3** 6 23 S 31 40 E
Mpésoba, *Mali* **112 C3** 12 31N 5 39W
Mpika, *Zambia* **119 E3** 11 51 S 31 25 E
Mpulungu, *Zambia* **119 D3** 8 51 S 31 5 E
Mpumalanga, *S. Africa* **117 D5** 29 50 S 30 33 E
Mpumalanga □, *S. Africa* **117 B5** 26 0 S 30 0 E
Mpwapwa, *Tanzania* **118 D4** 6 23 S 36 30 E
Mpwapwa □, *Tanzania* **118 D4** 6 30 S 36 20 E
Mqinvartsveri = Kazbek, *Russia* **61 J7** 42 42N 44 30 E
Mrągowo, *Poland* **54 E8** 53 52N 21 18 E
Mramor, *Serbia, Yug.* **50 C5** 43 20N 21 45 E
Mrimina, *Morocco* **110 C3** 29 50N 7 9W
Mrkonjić Grad, *Bos.-H.* **52 F2** 44 26N 17 4 E
Mrkopalj, *Croatia* **45 C11** 45 21N 14 52 E
Mrocza, *Poland* **55 E4** 53 16N 17 35 E
Msab, Oued en →, *Algeria* **111 B6** 32 15N 5 0 E
Msaken, *Tunisia* **108 A2** 35 49N 10 33 E
Msambansovu, *Zimbabwe* **119 F3** 15 50 S 30 3 E
M'sila, *Algeria* **111 A5** 35 46N 4 30 E
Msoro, *Zambia* **119 E3** 13 35 S 31 50 E
Msta →, *Russia* **58 C6** 58 25N 31 20 E
Mstislavl = Mstsislaw, *Belarus* **58 E6** 54 0N 31 50 E
Mstsislaw, *Belarus* **58 E6** 54 0N 31 50 E
Mszana Dolna, *Poland* **55 J7** 49 41N 20 5 E
Mszczonów, *Poland* **55 G7** 51 58N 20 33 E
Mtama, *Tanzania* **119 E4** 10 17 S 39 21 E
Mtilikwe →, *Zimbabwe* **119 G3** 21 9 S 31 30 E
Mtsensk, *Russia* **58 F9** 53 17N 36 36 E
Mtskheta, *Georgia* **61 K7** 41 52N 44 45 E
Mtubatuba, *S. Africa* **117 D5** 28 30 S 32 8 E
Mtwara-Mikindani, *Tanzania* **119 E5** 10 20 S 40 20 E
Mu →, *Burma* **90 E5** 21 56N 95 38 E
Mu Gia, Deo, *Vietnam* **86 D5** 17 40N 105 47 E
Mu Us Shamo, *China* **74 E5** 39 0N 109 0 E
Muacandalo, *Angola* **115 E3** 10 2 S 19 40 E
Muaná, *Brazil* **170 B2** 1 25 S 49 15W
Muanda, *Dem. Rep. of the Congo* **115 D2** 6 0 S 12 20 E
Muang Chiang Rai, *Thailand* **86 C2** 19 52N 99 50 E
Muang Khong, *Laos* **86 E5** 14 7N 105 51 E
Muang Lamphun, *Thailand* **86 C2** 18 40N 99 2 E
Muang Pak Beng, *Laos* **86 C3** 19 54N 101 8 E
Muar, *Malaysia* **87 L4** 2 3N 102 34 E
Muarabungo, *Indonesia* **84 C2** 1 28 S 102 52 E
Muaraenim, *Indonesia* **84 C2** 3 40 S 103 50 E
Muarajuloi, *Indonesia* **85 C4** 0 12 S 114 3 E
Muarakaman, *Indonesia* **85 C5** 0 2 S 116 45 E
Muaratebo, *Indonesia* **84 C2** 1 30 S 102 26 E
Muaratembesi, *Indonesia* **84 C2** 1 42 S 103 8 E
Muaratewe, *Indonesia* **85 C4** 0 58 S 114 52 E
Mubarakpur, *India* **93 F10** 26 6N 83 18 E
Mubarraz = Al Mubarraz, *Si. Arabia* **97 E6** 25 30N 49 40 E
Mubende, *Uganda* **118 B3** 0 33N 31 22 E
Mubi, *Nigeria* **113 C7** 10 18N 13 16 E
Mubur, P., *Indonesia* **87 L6** 3 20N 106 12 E
Mucajaí →, *Brazil* **169 C5** 2 25N 60 52W
Mucajaí, Serra do, *Brazil* **169 C5** 2 23N 61 10W
Mucari, *Angola* **115 D3** 9 30 S 16 54 E
Muchachos, Roque de los, *Canary Is.* **39 F2** 28 44N 17 52W
Mücheln, *Germany* **30 D7** 51 17N 11 47 E
Muchinga Mts., *Zambia* **119 E3** 11 30 S 31 30 E
Muchkapskiy, *Russia* **60 E6** 51 52N 42 28 E
Muck, *U.K.* **22 E2** 56 50N 6 15W
Muckadilla, *Australia* **127 D4** 26 35 S 148 23 E
Muckalee Cr. →, *U.S.A.* **152 D5** 31 38N 84 9W
Muco →, *Colombia* **168 C3** 4 15N 70 21W
Mucoma, *Angola* **115 F2** 15 18 S 13 39 E
Muconda, *Angola* **115 E4** 10 31 S 21 15 E
Mucuim →, *Brazil* **173 B5** 6 33 S 64 18W
Mucur, *Turkey* **100 C6** 39 3N 34 22 E
Mucura, *Brazil* **169 D5** 2 31 S 62 43W
Mucuri, *Brazil* **171 E4** 18 0 S 39 36W
Mucurici, *Brazil* **171 E3** 18 6 S 40 31W
Mucusso, *Angola* **115 F4** 18 1 S 21 25 E
Muda, *Canary Is.* **39 F6** 28 34N 13 57W
Mudan Jiang →, *China* **75 A15** 46 20N 129 30 E
Mudanjiang, *China* **75 B15** 44 38N 129 30 E
Mudanya, *Turkey* **51 F12** 40 25N 28 50 E
Muddy Cr. →, *U.S.A.* **159 H8** 38 24N 110 42W
Mudgee, *Australia* **129 B8** 32 32 S 149 31 E
Mudjatik →, *Canada* **143 B7** 56 1N 107 36W
Mudon, *Burma* **90 G6** 16 15N 97 44 E
Mudugh, *Somali Rep.* **120 C3** 7 0N 47 30 E
Mudurnu, *Turkey* **100 B4** 40 27N 31 12 E
Muecate, *Mozam.* **119 E4** 14 55 S 39 40 E
Mueda, *Mozam.* **119 E4** 11 36 S 39 28 E
Mueller Ra., *Australia* **124 C4** 18 18 S 126 46 E
Muende, *Mozam.* **119 E3** 14 28 S 33 0 E
Muerto, Mar, *Mexico* **163 D6** 16 10N 94 10W
Mufindi □, *Tanzania* **119 D4** 8 30 S 35 20 E
Mufu Shan, *China* **77 C10** 29 20N 114 30 E
Mufulira, *Zambia* **119 E2** 12 32 S 28 15 E
Mufumbiro Range, *Africa* **118 C2** 1 25 S 29 30 E
Mugardos, *Spain* **42 B2** 43 27N 8 15W
Muge →, *Portugal* **43 F2** 39 8N 8 44W
Múggia, *Italy* **45 C10** 45 36N 13 46 E
Mughayrā', *Si. Arabia* **96 D3** 29 17N 37 41 E
Mugi, *Japan* **70 D6** 33 40N 134 25 E
Mugia = Muxía, *Spain* **42 B1** 43 3N 9 10W
Mugila, Mts., *Dem. Rep. of the Congo* **118 D2** 7 0 S 28 50 E
Muğla, *Turkey* **57 G4** 37 15N 28 22 E
Muğla □, *Turkey* **49 D10** 37 15N 28 0 E
Mŭglizh, *Bulgaria* **51 D9** 42 37N 25 32 E
Mugu, *Nepal* **93 E10** 29 45N 82 30 E
Muhammad, Râs, *Egypt* **106 B3** 27 44N 34 16 E
Muhammad Qol, *Sudan* **106 C4** 20 53N 37 9 E
Muhammadabad, *India* **93 F10** 26 4N 83 25 E
Muḩayriqah, *Si. Arabia* **98 B4** 23 59N 45 4 E
Muhesi →, *Tanzania* **118 D4** 7 0 S 35 20 E
Muheza □, *Tanzania* **118 D4** 5 0 S 39 0 E
Mühlacker, *Germany* **31 G4** 48 57N 8 51 E
Mühldorf, *Germany* **31 G8** 48 14N 12 32 E
Mühlhausen, *Germany* **30 D6** 51 12N 10 27 E
Mühlig Hofmann fjell, *Antarctica* **7 D3** 72 30 S 5 0 E
Mühlviertel, *Austria* **34 C7** 48 30N 14 10 E
Muhos, *Finland* **14 D22** 64 47N 25 59 E
Muhu, *Estonia* **15 G20** 58 36N 23 11 E
Muhutwe, *Tanzania* **118 C3** 1 35 S 31 45 E
Muikamachi, *Japan* **69 F9** 37 15N 138 50 E
Muine Bheag, *Ireland* **23 D5** 52 42N 6 58W
Muir, L., *Australia* **125 F2** 34 30 S 116 40 E
Mukacheve, *Ukraine* **59 H2** 48 27N 22 45 E
Mukachevo = Mukacheve, *Ukraine* **59 H2** 48 27N 22 45 E
Mukah, *Malaysia* **85 B4** 2 55N 112 5 E
Mukawwa, Geziret, *Egypt* **106 C4** 23 55N 35 53 E
Mukdahan, *Thailand* **86 D5** 16 32N 104 43 E
Mukden = Shenyang, *China* **75 D12** 41 48N 123 27 E
Mukhtolovo, *Russia* **60 C6** 55 29N 43 15 E
Mukhtuya = Lensk, *Russia* **65 C12** 60 48N 114 55 E
Mukinbudin, *Australia* **125 F2** 30 55 S 118 5 E
Mukishi, *Dem. Rep. of the Congo* **115 D4** 8 30 S 24 44 E
Mukomuko, *Indonesia* **84 C2** 2 30 S 101 10 E
Mukomwenze, *Dem. Rep. of the Congo* **118 D2** 6 49 S 27 15 E
Mukry, *Turkmenistan* **63 E2** 37 54N 65 12 E
Muktsar, *India* **92 D6** 30 30N 74 30 E
Mukur, *Afghan.* **92 C2** 32 50N 67 42 E
Mukutawa →, *Canada* **143 C9** 53 10N 97 24W
Mukwela, *Zambia* **119 F2** 17 0 S 26 40 E
Mukwonago, *U.S.A.* **157 B8** 42 52N 88 20W
Mula, *Spain* **41 G3** 38 3N 1 33W
Mula →, *India* **94 E2** 18 34N 74 21 E
Mulanay, *Phil.* **80 E4** 13 31N 122 24 E
Mulange, *Dem. Rep. of the Congo* **118 C2** 3 40 S 27 10 E
Mulanje, *Malawi* **119 F4** 16 2 S 35 33 E
Mulberry, *U.S.A.* **153 H8** 27 54N 81 59W
Mulberry Grove, *U.S.A.* **156 F7** 38 56N 89 16W
Mulchatna →, *U.S.A.* **144 G8** 59 40N 157 7W
Mulchén, *Chile* **174 D1** 37 45 S 72 20W
Mulde →, *Germany* **30 D8** 51 53N 12 15 E
Muldraugh, *U.S.A.* **157 G11** 37 56N 85 59W
Mule Creek, *U.S.A.* **154 D2** 43 19N 104 8W
Muleba, *Tanzania* **118 C3** 1 50 S 31 37 E
Muleba □, *Tanzania* **118 C3** 2 0 S 31 30 E
Mulegns, *Switz.* **33 C9** 46 32N 9 38 E
Muleshoe, *U.S.A.* **155 H3** 34 13N 102 43W
Mulga Valley, *Australia* **128 A4** 31 8 S 141 3 E
Mulgathing, *Australia* **127 E1** 30 15 S 134 8 E
Mulgrave, *Canada* **141 C7** 45 38N 61 31W
Mulgrave I., *Papua N. G.* **132 F2** 10 5 S 142 10 E
Mulhacén, *Spain* **43 H7** 37 4N 3 20W
Mülheim, *Germany* **30 D2** 51 25N 6 54 E
Mulhouse, *France* **27 E14** 47 40N 7 20 E
Muli, *China* **76 D3** 27 52N 101 8 E
Mulifanua, *W. Samoa* **133 W24** 13 50 S 171 59W
Muling, *China* **75 B16** 44 35N 130 10 E
Mull, *U.K.* **22 E3** 56 25N 5 56W
Mull, Sound of, *U.K.* **22 E3** 56 30N 5 50W
Mullaittivu, *Sri Lanka* **95 K5** 9 15N 80 49 E
Mullen, *U.S.A.* **154 D4** 42 3N 101 1W
Mullengudgery, *Australia* **129 A7** 31 43 S 147 23 E
Mullens, *U.S.A.* **148 G5** 37 35N 81 23W
Muller, Pegunungan, *Indonesia* **85 B4** 0 30N 113 30 E
Mullet Pen., *Ireland* **23 B1** 54 13N 10 2W
Mullewa, *Australia* **125 E2** 28 29 S 115 30 E
Müllheim, *Germany* **31 H3** 47 47N 7 36 E
Mulligan →, *Australia* **126 D2** 25 0 S 139 0 E
Mullin, *U.S.A.* **155 K5** 31 33N 98 40W
Mullingar, *Ireland* **23 C4** 53 31N 7 21W
Mullins, *U.S.A.* **149 H6** 34 12N 79 15W
Mullsjö, *Sweden* **17 G7** 57 56N 13 55 E
Mullumbimby, *Australia* **127 D5** 28 30 S 153 30 E
Mulobezi, *Zambia* **119 F2** 16 45 S 25 7 E
Mulroy B., *Ireland* **23 A4** 55 15N 7 46W
Mulshi L., *India* **94 E1** 18 30N 73 48 E
Multai, *India* **94 D4** 21 50N 78 21 E
Multan, *Pakistan* **91 C3** 30 15N 71 36 E
Mulu, Gunong, *Malaysia* **85 B4** 4 3N 114 56 E
Mulumbe, Mts., *Dem. Rep. of the Congo* **119 D2** 8 40 S 27 30 E
Mulungushi Dam, *Zambia* **119 E2** 14 48 S 28 48 E
Mulvane, *U.S.A.* **155 G6** 37 29N 97 15W
Mulwad, *Sudan* **106 D3** 18 45N 30 39 E
Mulwala, *Australia* **129 C7** 35 59 S 146 0 E
Mumbai, *India* **94 E1** 18 55N 72 50 E
Mumbondo, *Angola* **115 E2** 10 9 S 14 15 E
Mumbwa, *Zambia* **119 F2** 15 0 S 27 0 E
Mumeng, *Papua N. G.* **132 D4** 7 1 S 146 37 E
Mumra, *Russia* **61 H8** 45 45N 47 41 E
Mun →, *Thailand* **86 E5** 15 19N 105 30 E
Muna, *Indonesia* **82 C2** 5 0 S 122 30 E
Munamagi, *Estonia* **15 H22** 57 43N 27 4 E
Münchberg, *Germany* **31 E7** 50 11N 11 47 E
Müncheberg, *Germany* **30 C10** 52 30N 14 9 E
München, *Germany* **31 G7** 48 8N 11 34 E
Munchen-Gladbach = Mönchengladbach, *Germany* **30 D2** 51 11N 6 27 E
Muncho Lake, *Canada* **142 B3** 59 0N 125 50W
Munchŏn, *N. Korea* **75 E14** 39 14N 127 19 E

254 Münchwilen

Münchwilen, *Switz.* **33 B7** 47 28N 8 59 E
Muncie, *U.S.A.* **157 D11** 40 12N 85 23W
Muncoonie, L., *Australia* **126 D2** 25 12 S 138 40 E
Munda, *Solomon Is.* **133 M9** 8 20 S 157 16 E
Mundakayam, *India* **95 K3** 9 30N 76 50 E
Mundala, *Indonesia* **83 B6** 4 30 S 141 0 E
Mundare, *Canada* **142 C6** 53 35N 112 20W
Munday, *U.S.A.* **155 J5** 33 27N 99 38W
Münden, *Germany* **30 D5** 51 25N 9 38 E
Mundiwindi, *Australia* **124 D3** 23 47 S 120 9 E
Mundo →, *Spain* **41 G2** 38 30N 2 15W
Mundo Novo, *Brazil* **171 D3** 11 50 S 40 29W
Mundra, *India* **92 H3** 22 54N 69 48 E
Mundrabilla, *Australia* **125 F4** 31 52 S 127 51 E
Munducurus, *Brazil* **169 D6** 4 47 S 58 16W
Munenga, *Angola* **115 E2** 10 2 S 14 41 E
Munera, *Spain* **41 F2** 39 2N 2 29W
Muneru →, *India* **95 F5** 16 45N 80 3 E
Mungallala, *Australia* **127 D4** 26 28 S 147 34 E
Mungallala Cr. →, *Australia* **127 D4** 28 53 S 147 5 E
Mungana, *Australia* **126 B3** 17 8 S 144 27 E
Mungaoli, *India* **92 G8** 24 24N 78 7 E
Mungari, *Mozam.* **119 F3** 17 12 S 33 30 E
Mungbere,
Dem. Rep. of the Congo **118 B2** 2 36N 28 28 E
Munger, *India* **93 G12** 25 23N 86 30 E
Mungindi, *Australia* **127 D4** 28 58 S 149 1 E
Munhango, *Angola* **115 E3** 12 10 S 18 38 E
Munich = München,
Germany **31 G7** 48 8N 11 34 E
Munising, *U.S.A.* **148 B2** 46 25N 86 40W
Munka-Ljungby, *Sweden* **17 H6** 56 16N 12 58 E
Munkebo, *Denmark* **17 J4** 55 27N 10 34 E
Munkedal, *Sweden* **17 F5** 58 28N 11 40 E
Munkfors, *Sweden* **16 E7** 59 47N 13 30 E
Munku-Sardyk, *Russia* **65 D11** 51 45N 100 20 E
Münnerstadt, *Germany* **31 E6** 50 14N 10 12 E
Munoz, *Phil.* **80 D3** 15 43N 120 54 E
Muñoz Gamero, Pen., *Chile* **176 D2** 52 30 S 73 5W
Munro, *Australia* **129 D7** 37 56 S 147 11 E
Munroe L., *Canada* **143 B9** 59 13N 98 35W
Munsan, *S. Korea* **75 F14** 37 51N 126 48 E
Munshiganj, *Bangla.* **90 D3** 23 33N 90 32 E
Münsingen, *Switz.* **32 C5** 46 52N 7 32 E
Munson, *U.S.A.* **153 E3** 30 52N 86 52W
Munster, *France* **27 D14** 48 2N 7 8 E
Munster, *Niedersachsen,*
Germany **30 C6** 52 58N 10 5 E
Münster,
Nordrhein-Westfalen,
Germany **30 D3** 51 58N 7 37 E
Münster, *Switz.* **33 D6** 46 29N 8 17 E
Munster □, *Ireland* **23 D3** 52 18N 8 44W
Muntadgin, *Australia* **125 F2** 31 45 S 118 33 E
Muntele Mare, Vf., *Romania* **53 D8** 46 30N 23 12 E
Muntok, *Indonesia* **84 C3** 2 5 S 105 10 E
Munyama, *Zambia* **119 F2** 16 5 S 28 31 E
Munzur Dağları, *Turkey* **101 C8** 39 30N 39 10 E
Muong Beng, *Laos* **86 B3** 20 23N 101 46 E
Muong Boum, *Vietnam* **86 A4** 22 24N 102 49 E
Muong Et, *Laos* **86 B5** 20 49N 104 1 E
Muong Hai, *Laos* **86 B3** 21 3N 101 49 E
Muong Hiem, *Laos* **86 B4** 20 5N 103 22 E
Muong Houn, *Laos* **86 B3** 20 8N 101 23 E
Muong Hung, *Vietnam* **86 B4** 20 56N 103 53 E
Muong Kau, *Laos* **86 E5** 15 6N 105 47 E
Muong Khao, *Laos* **86 C4** 19 38N 103 32 E
Muong Khoua, *Laos* **86 B4** 21 5N 102 31 E
Muong Liep, *Laos* **86 C3** 18 29N 101 40 E
Muong May, *Laos* **86 E6** 14 49N 106 56 E
Muong Ngeun, *Laos* **86 B3** 20 36N 101 3 E
Muong Ngoi, *Laos* **86 B4** 20 43N 102 41 E
Muong Nhie, *Vietnam* **86 A4** 22 12N 102 28 E
Muong Nong, *Laos* **86 D6** 16 22N 106 30 E
Muong Ou Tay, *Laos* **86 A3** 22 7N 101 48 E
Muong Oua, *Laos* **86 C3** 18 18N 101 20 E
Muong Peun, *Laos* **86 B4** 20 13N 103 52 E
Muong Phalane, *Laos* **86 D5** 16 39N 105 34 E
Muong Phieng, *Laos* **86 C3** 19 6N 101 32 E
Muong Phine, *Laos* **86 D6** 16 32N 106 2 E
Muong Sai, *Laos* **86 B3** 20 42N 101 59 E
Muong Saiapoun, *Laos* **86 C3** 18 24N 101 31 E
Muong Sen, *Vietnam* **86 C5** 19 24N 104 8 E
Muong Sing, *Laos* **86 B3** 21 11N 101 9 E
Muong Son, *Laos* **86 B4** 20 27N 103 19 E
Muong Soui, *Laos* **86 C4** 19 33N 102 52 E
Muong Va, *Laos* **86 B4** 21 53N 102 19 E
Muong Xia, *Vietnam* **86 B5** 20 19N 104 50 E
Muonio, *Finland* **14 C20** 67 57N 23 40 E
Muonionjoki →, *Finland* **14 C20** 67 11N 23 34 E
Muotathal, *Switz.* **33 C7** 46 58N 8 46 E
Mupa, *Angola* **115 F3** 16 5 S 15 50 E
Muping, *China* **75 F11** 37 22N 121 36 E
Muqaddam, Wadi →, *Sudan* **106 D3** 18 4N 31 30 E
Muqdisho, *Somali Rep.* **120 D3** 2 2N 45 25 E
Muqshin, W. →, *Oman* **99 C6** 19 44N 55 14 E
Muquequete, *Angola* **115 E2** 14 50 S 14 16 E
Mur →, *Austria* **35 E9** 46 18N 16 52 E
Mur-de-Bretagne, *France* **26 D4** 48 12N 3 0W
Muradiye, *Manisa, Turkey* **49 C9** 38 39N 27 21 E
Muradiye, *Van, Turkey* **101 C10** 39 0N 43 44 E
Murakami, *Japan* **68 E9** 38 14N 139 29 E
Murallón, Cerro, *Chile* **176 C2** 49 48 S 73 30W
Muralto, *Switz.* **33 D7** 46 11N 8 49 E
Muranda, *Rwanda* **118 C2** 1 52 S 29 20 E
Murang'a, *Kenya* **118 C4** 0 45 S 37 9 E
Murashi, *Russia* **62 B2** 59 30N 49 0 E
Murat, *France* **28 C6** 45 7N 2 53 E
Murat →, *Turkey* **101 C9** 38 46N 40 0 E
Murat Dağı, *Turkey* **49 C11** 38 55N 29 43 E
Muratlı, *Turkey* **51 E11** 41 10N 27 29 E
Murato, *France* **29 F13** 42 35N 9 20 E
Murau, *Austria* **34 D7** 47 6N 14 10 E
Muravera, *Italy* **46 C2** 39 25N 9 34 E
Murayama, *Japan* **68 E10** 38 30N 140 25 E
Murban, *U.A.E.* **97 F7** 23 50N 53 45 E
Murça, *Portugal* **42 D3** 41 24N 7 28W
Murchison, *N.Z.* **131 B7** 41 49 S 172 21 E
Murchison →, *Australia* **125 E1** 27 45 S 114 0 E
Murchison, Mt., *Antarctica* **7 D11** 73 0 S 168 0 E
Murchison Falls, *Uganda* **118 B3** 2 15N 31 30 E
Murchison House, *Australia* **125 E1** 27 39 S 114 14 E
Murchison Mt., *N.Z.* **131 D6** 43 0 S 171 22 E
Murchison Mts., *N.Z.* **131 F2** 45 13 S 167 23 E
Murchison Ra., *Australia* **126 C1** 20 0 S 134 10 E
Murchison Rapids, *Malawi* **119 F3** 15 55 S 34 35 E
Murcia, *Spain* **41 G3** 38 5N 1 10W
Murcia □, *Spain* **41 H3** 37 50N 1 30W
Murdo, *U.S.A.* **154 D4** 43 53N 100 43W
Murdoch Pt., *Australia* **126 A3** 14 37 S 144 55 E
Mürefte, *Turkey* **51 F11** 40 40N 27 14 E
Mureş □, *Romania* **53 D9** 46 45N 24 40 E
Mureş →, *Romania* **52 D5** 46 15N 20 13 E
Mureşul = Mureş →,
Romania **52 D5** 46 15N 20 13 E
Muret, *France* **28 E5** 43 30N 1 20 E
Murfreesboro, *U.S.A.* **149 H2** 35 51N 86 24W
Murg, *Switz.* **33 B8** 47 6N 9 13 E
Murgab = Murghob,
Tajikistan **63 D7** 38 10N 74 2 E
Murgeni, *Romania* **53 D13** 46 12N 28 1 E
Murgenthal, *Switz.* **32 B5** 47 16N 7 50 E
Murghob, *Tajikistan* **63 D7** 38 10N 74 2 E
Murgon, *Australia* **127 D5** 26 15 S 151 54 E
Murgoo, *Australia* **125 E2** 27 24 S 116 28 E
Muri, *Switz.* **33 B6** 47 17N 8 21 E
Muria, *Indonesia* **85 D4** 6 36 S 110 53 E
Muriaé, *Brazil* **171 F3** 21 8 S 42 23W
Murias de Paredes, *Spain* **42 C4** 42 52N 6 11W
Murici, *Brazil* **170 C4** 9 19 S 35 56W
Muriége, *Angola* **115 D4** 9 58 S 21 11 E
Muriel Mine, *Zimbabwe* **119 F3** 17 14 S 30 40 E
Murila, *Angola* **115 E4** 10 44 S 20 20 E
Müritz, *Germany* **30 B8** 53 25N 12 42 E
Murka, *Kenya* **118 C4** 3 27 S 38 0 E
Murmansk, *Russia* **56 A5** 68 57N 33 10 E
Murnau, *Germany* **31 H7** 47 40N 11 12 E
Muro, *France* **29 F12** 42 34N 8 54 E
Muro, *Spain* **39 B10** 39 44N 3 3 E
Muro, C. de, *France* **29 G12** 41 44N 8 37 E
Muro de Alcoy, *Spain* **41 G4** 38 46N 0 26W
Muro Lucano, *Italy* **47 B8** 40 45N 15 29 E
Murom, *Russia* **60 C6** 55 35N 42 3 E
Muroran, *Japan* **68 C10** 42 25N 141 0 E
Muros, *Spain* **42 C1** 42 45N 9 5W
Muros y de Noya, Ría de,
Spain **42 C1** 42 45N 9 0W
Muroto, *Japan* **70 D6** 33 18N 134 9 E
Muroto-Misaki, *Japan* **70 D6** 33 15N 134 10 E
Murowana Goślina, *Poland* **55 F3** 52 35N 17 0 E
Murphy, *U.S.A.* **158 E5** 43 13N 116 33W
Murphys, *U.S.A.* **160 G6** 38 8N 120 28W
Murphysboro, *U.S.A.* **155 G10** 37 46N 89 20W
Murrat, *Sudan* **106 D2** 18 51N 29 33 E
Murray, *Iowa, U.S.A.* **156 C3** 41 3N 93 57W
Murray, *Ky., U.S.A.* **149 G1** 36 37N 88 19W
Murray, *Utah, U.S.A.* **158 F8** 40 40N 111 53W
Murray →, *Australia* **128 C3** 35 20 S 139 22 E
Murray →, *Canada* **142 B4** 56 11N 120 45W
Murray, L., *Papua N. G.* **132 D1** 7 0 S 141 35 E
Murray, L., *U.S.A.* **152 A8** 34 3N 81 13W
Murray Bridge, *Australia* **128 C3** 35 6 S 139 14 E
Murray Downs, *Australia* **126 C1** 21 4 S 134 40 E
Murray Harbour, *Canada* **141 C7** 46 0N 62 28W
Murraysburg, *S. Africa* **116 E3** 31 58 S 23 47 E
Murrayville, *U.S.A.* **156 E6** 39 35N 90 15W
Murree, *Pakistan* **92 C5** 33 56N 73 28 E
Murrieta, *U.S.A.* **161 M9** 33 33N 117 13W
Murrin Murrin, *Australia* **125 E3** 28 58 S 121 33 E
Murro di Porco, Capo, *Italy* **47 F8** 37 0N 15 20 E
Murrumbidgee →, *Australia* **128 C5** 34 43 S 143 12 E
Murrumburrah, *Australia* **129 C8** 34 32 S 148 22 E
Murrurundi, *Australia* **129 A9** 31 42 S 150 51 E
Murshid, *Sudan* **106 C3** 21 40N 31 10 E
Murshidabad, *India* **93 G13** 24 11N 88 19 E
Murska Sobota, *Slovenia* **45 B13** 46 39N 16 12 E
Murtazapur, *India* **94 D3** 20 40N 77 25 E
Murten, *Switz.* **32 C4** 46 56N 7 4 E
Murtensee, *Switz.* **32 C4** 46 56N 7 7 E
Murtle L., *Canada* **142 C5** 52 8N 119 38W
Murtoa, *Australia* **128 D5** 36 35 S 142 28 E
Murtosa, *Portugal* **42 E2** 40 44N 8 40W
Muru →, *Brazil* **172 B3** 8 9 S 70 45W
Murungu, *Tanzania* **118 C3** 4 12 S 31 10 E
Murupara, *N.Z.* **130 E5** 38 28 S 176 42 E
Murwara, *India* **93 H9** 23 46N 80 28 E
Murwillumbah, *Australia* **127 D5** 28 18 S 153 27 E
Mürz →, *Austria* **34 D8** 47 30N 15 25 E
Mürzzuschlag, *Austria* **34 D8** 47 36N 15 41 E
Muş, *Turkey* **101 C9** 38 45N 41 30 E
Musa,
Dem. Rep. of the Congo **114 B3** 2 40N 19 18 E
Musa →, *Papua N. G.* **132 E5** 9 3 S 148 55 E
Mûsa, Gebel, *Egypt* **106 J8** 28 33N 33 59 E
Musa Khel, *Pakistan* **91 C3** 30 59N 69 52 E
Mûsá Qal'eh, *Afghan.* **91 B2** 32 20N 64 50 E
Musala, *Bulgaria* **50 D7** 42 13N 23 37 E
Musala, *Indonesia* **84 B1** 1 41N 98 28 E
Musan, *N. Korea* **75 C15** 42 12N 129 12 E
Musangu,
Dem. Rep. of the Congo **115 E4** 10 28 S 23 55 E
Musasa, *Tanzania* **118 C3** 3 25 S 31 30 E
Musashino, *Japan* **71 B11** 35 42N 139 34 E
Musay'īd, *Qatar* **97 E6** 25 0N 51 33 E
Musaymīr, *Yemen* **98 D4** 13 27N 44 37 E
Muscat = Masqat, *Oman* **99 B7** 23 37N 58 36 E
Muscat & Oman = Oman ■,
Asia **99 B7** 23 0N 58 0 E
Muscatine, *U.S.A.* **156 C5** 41 25N 91 3W
Muscoda, *U.S.A.* **156 A6** 43 11N 90 27W
Musella, *U.S.A.* **152 C5** 32 48N 84 2W
Musgrave, *Australia* **126 A3** 14 47 S 143 30 E
Musgrave Ranges, *Australia* **125 E5** 26 0 S 132 0 E
Mushie,
Dem. Rep. of the Congo **114 C3** 2 56 S 16 55 E
Mushin, *Nigeria* **113 D5** 6 32N 3 21 E
Musi →, *India* **94 F4** 16 41N 79 40 E
Musi →, *Indonesia* **84 C2** 2 20 S 104 56 E
Muskeg →, *Canada* **142 A4** 60 20N 123 20W
Muskegon, *U.S.A.* **157 A10** 43 14N 86 16W
Muskegon →, *U.S.A.* **148 D2** 43 14N 86 21W
Muskegon Heights, *U.S.A.* **157 A10** 43 12N 86 16W
Muskogee, *U.S.A.* **155 H7** 35 45N 95 22W
Muskwa →, *Canada* **142 B4** 58 47N 122 48W
Muslīmiyah, *Syria* **96 B3** 36 19N 37 12 E
Musmar, *Sudan* **106 D4** 18 13N 35 40 E
Musofu, *Zambia* **119 E2** 13 30 S 29 0 E
Musoma, *Tanzania* **118 C3** 1 30 S 33 48 E
Musoma □, *Tanzania* **118 C3** 1 50 S 34 30 E
Musquaro, L., *Canada* **141 B7** 50 38N 61 5W
Musquodoboit Harbour,
Canada **141 D7** 44 50N 63 9W
Mussau I., *Papua N. G.* **132 A5** 1 30 S 149 40 E
Musselburgh, *U.K.* **22 F5** 55 57N 3 2W
Musselshell →, *U.S.A.* **158 C10** 47 21N 107 57W
Mussende, *Angola* **115 E3** 10 32 S 16 5 E
Mussidan, *France* **28 C4** 45 2N 0 22 E
Mussolo, *Angola* **115 D3** 9 59 S 17 19 E
Mussomeli, *Italy* **46 E6** 37 35N 13 45 E
Mussoorie, *India* **92 D8** 30 27N 78 6 E
Mussuco, *Angola* **115 F3** 17 2 S 19 3 E
Mustafakemalpaşa, *Turkey* **51 F12** 40 2N 28 24 E
Mustahil, *Ethiopia* **120 C2** 5 16N 44 45 E
Mustang, *Nepal* **93 E10** 29 10N 83 55 E
Musters, L., *Argentina* **176 C3** 45 20 S 69 25W
Musudan, *N. Korea* **75 D15** 40 50N 129 43 E
Muswellbrook, *Australia* **129 B9** 32 16 S 150 56 E
Muszyna, *Poland* **55 J7** 49 22N 20 55 E
Mût, *Egypt* **106 B2** 25 28N 28 58 E
Mut, *Turkey* **100 D5** 36 40N 33 28 E
Mutanda, *Mozam.* **117 C5** 21 0 S 33 34 E
Mutanda, *Zambia* **119 E2** 12 24 S 26 13 E
Mutare, *Zimbabwe* **119 F3** 18 58 S 32 38 E
Mu'tariḍah, Al 'Urūq al,
Si. Arabia **99 B6** 21 15N 54 0 E
Muting, *Indonesia* **83 C6** 7 23 S 140 20 E
Mutooroo, *Australia* **128 B4** 32 26 S 140 55 E
Mutoray, *Russia* **65 C11** 60 56N 101 0 E
Mutoto,
Dem. Rep. of the Congo **115 D4** 5 42 S 22 42 E
Mutshatsha,
Dem. Rep. of the Congo **115 E4** 10 35 S 24 20 E
Mutsu, *Japan* **68 D10** 41 5N 140 55 E
Mutsu-Wan, *Japan* **68 D10** 41 5N 140 55 E
Muttaburra, *Australia* **126 C3** 22 38 S 144 29 E
Muttalip, *Turkey* **49 B12** 39 50N 30 32 E
Muttama, *Australia* **129 C8** 34 46 S 148 8 E
Mutton I., *Ireland* **23 D2** 52 49N 9 32W
Mutuáli, *Mozam.* **119 E4** 14 55 S 37 0 E
Mutunópolis, *Brazil* **171 D2** 13 40 S 49 15W
Muvatupusha, *India* **95 K3** 9 53N 76 35 E
Muweilih, *Egypt* **103 E3** 30 42N 34 19 E
Muxía, *Spain* **42 B1** 43 3N 9 10W
Muxima, *Angola* **115 D2** 9 33 S 13 58 E
Muy Muy, *Nic.* **164 D2** 12 39N 85 36W
Muyinga, *Burundi* **118 C3** 3 14 S 30 33 E
Muynak, *Uzbekistan* **64 E6** 43 44N 59 10 E
Muyunkum, Peski,
Kazakstan **63 A5** 44 12N 71 0 E
Muzaffarabad, *Pakistan* **93 B5** 34 25N 73 30 E
Muzaffargarh, *Pakistan* **91 C3** 30 5N 71 14 E
Muzaffarnagar, *India* **92 E7** 29 26N 77 40 E
Muzaffarpur, *India* **93 F11** 26 7N 85 23 E
Muzeze, *Angola* **115 F3** 15 3 S 17 43 E
Muzhi, *Russia* **64 C7** 65 25N 64 40 E
Muzillac, *France* **26 E4** 47 35N 2 30W
Muzkol, Khrebet, *Tajikistan* **63 D6** 38 22N 73 20 E
Muzon, C., *U.S.A.* **142 C2** 54 40N 132 42W
Mvadhi-Ousyé, *Gabon* **114 B2** 1 13N 13 12 E
Mvam, *Gabon* **114 C1** 0 13 S 9 39 E
Mvôlô, *Sudan* **107 F2** 6 2N 29 53 E
Mvuma, *Zimbabwe* **119 F3** 19 16 S 30 30 E
Mvurwi, *Zimbabwe* **119 F3** 17 0 S 30 57 E
Mwadui, *Tanzania* **118 C3** 3 26 S 33 32 E
Mwambo, *Tanzania* **119 E5** 10 30 S 40 22 E
Mwandi, *Zambia* **119 F1** 17 30 S 24 51 E
Mwanza,
Dem. Rep. of the Congo **115 D5** 7 55 S 26 43 E
Mwanza, *Tanzania* **118 C3** 2 30 S 32 58 E
Mwanza, *Zambia* **119 F1** 16 58 S 24 28 E
Mwanza □, *Tanzania* **118 C3** 2 0 S 33 0 E
Mwaya, *Tanzania* **119 D3** 9 32 S 33 55 E
Mweelrea, *Ireland* **23 C2** 53 39N 9 49W
Mweka,
Dem. Rep. of the Congo **115 C4** 4 50 S 21 34 E
Mwendjila,
Dem. Rep. of the Congo **115 D3** 7 12 S 18 51 E
Mwene-Ditu,
Dem. Rep. of the Congo **115 D4** 6 35 S 22 27 E
Mwenezi, *Zimbabwe* **119 G3** 21 15 S 30 48 E
Mwenezi →, *Mozam.* **119 G3** 22 40 S 31 50 E
Mwenga,
Dem. Rep. of the Congo **118 C2** 3 1 S 28 28 E
Mweru, L., *Zambia* **119 D2** 9 0 S 28 40 E
Mweza Range, *Zimbabwe* **119 G3** 21 0 S 30 0 E
Mwilambwe,
Dem. Rep. of the Congo **115 D5** 8 7 S 25 5 E
Mwimbi, *Tanzania* **119 D3** 8 38 S 31 39 E
Mwinilunga, *Zambia* **119 E1** 11 43 S 24 25 E
My Tho, *Vietnam* **87 G6** 10 29N 106 23 E
Mya, O. →, *Algeria* **111 B5** 30 46N 4 54 E
Myajlar, *India* **92 F4** 26 15N 70 20 E
Myakka →, *U.S.A.* **153 J7** 26 56N 82 11W
Myanaung, *Burma* **90 F5** 18 18N 95 22 E
Myanmar = Burma ■, *Asia* **90 E6** 21 0N 96 30 E
Myaungmya, *Burma* **90 G5** 16 30N 94 40 E
Myeik Kyunzu, *Burma* **87 G1** 11 30N 97 30 E
Myerstown, *U.S.A.* **151 F8** 40 22N 76 19W
Myingyan, *Burma* **90 E5** 21 30N 95 20 E
Myitkyina, *Burma* **90 C6** 25 24N 97 26 E
Myittha →, *Burma* **90 D5** 23 12N 94 17 E
Myjava, *Slovak Rep.* **35 C10** 48 41N 17 37 E
Mykhaylivka, *Ukraine* **59 J8** 47 12N 35 15 E
Mykines, *Færoe Is.* **14 E9** 62 7N 7 35W
Myking, *Norway* **18 D2** 60 41N 5 19 E
Mykolayiv, *Ukraine* **59 J7** 46 58N 32 0 E
Mylius Erichsen Land,
Greenland **10 A8** 81 30N 27 0W
Mymensingh, *Bangla.* **90 C3** 24 45N 90 24 E
Mynydd Du, *U.K.* **21 F4** 51 52N 3 50W
Mynzhilgi, Gora, *Kazakstan* **63 B4** 43 48N 68 51 E
Myrasýsla □, *Iceland* **11 C5** 64 45N 21 30W
Mýrdalsjökull, *Iceland* **11 D7** 63 40N 19 6W
Myrhorod, *Ukraine* **59 H7** 49 58N 33 37 E
Mýri, *Iceland* **11 B9** 65 23N 17 23W
Myroodah, *Australia* **124 C3** 18 7 S 124 16 E
Myrtle Beach, *U.S.A.* **149 J6** 33 42N 78 53W
Myrtle Creek, *U.S.A.* **158 E2** 43 1N 123 17W
Myrtle Grove, *U.S.A.* **153 E2** 30 23N 87 17W
Myrtle Point, *U.S.A.* **158 E1** 43 4N 124 8W
Myrtleford, *Australia* **129 D7** 36 34 S 146 44 E
Myrtou, *Cyprus* **38 D12** 35 18N 33 4 E
Mysen, *Norway* **18 E8** 59 33N 11 20 E
Mysia, *Turkey* **51 G11** 39 50N 27 0 E
Myślenice, *Poland* **55 J6** 49 51N 19 57 E
Myślibórz, *Poland* **55 F1** 52 55N 14 50 E
Mysłowice, *Poland* **55 H6** 50 15N 19 12 E
Mysore = Karnataka □,
India **95 H3** 13 15N 77 0 E
Mysore, *India* **95 H3** 12 17N 76 41 E
Mystic, *Conn., U.S.A.* **151 E13** 41 21N 71 58W
Mystic, *Iowa, U.S.A.* **156 D4** 40 47N 92 57W
Myszków, *Poland* **55 H6** 50 45N 19 22 E
Myszyniec, *Poland* **54 E8** 53 23N 21 21 E
Mythen, *Switz.* **33 B7** 47 2N 8 42 E
Mytishchi, *Russia* **58 E9** 55 50N 37 50 E
Myton, *U.S.A.* **158 F8** 40 12N 110 4W
Mývatn, *Iceland* **11 B10** 65 36N 17 0W
Mže →, *Czech Rep.* **34 B6** 49 46N 13 24 E
Mzimba, *Malawi* **119 E3** 11 55 S 33 39 E
Mzimkulu →, *S. Africa* **117 E5** 30 44 S 30 28 E
Mzimvubu →, *S. Africa* **117 E4** 31 38 S 29 33 E
Mzuzu, *Malawi* **119 E3** 11 30 S 33 55 E

N

N' Dioum, *Senegal* **112 B2** 16 31N 14 39W
Na Hearadh = Harris, *U.K.* **22 D2** 57 50N 6 55W
Na-lang, *Burma* **90 D6** 22 42N 97 33 E
Na Noi, *Thailand* **86 C3** 18 19N 100 43 E
Na Phao, *Laos* **86 D5** 17 35N 105 44 E
Na Sam, *Vietnam* **86 A6** 22 3N 106 37 E
Na San, *Vietnam* **86 B5** 21 12N 104 2 E
Naab →, *Germany* **31 F8** 49 1N 12 2 E
Na'am, *Sudan* **107 F2** 9 42N 28 27 E
Naantali, *Finland* **15 F19** 60 29N 22 2 E
Naas, *Ireland* **23 C5** 53 12N 6 40W
Nababiep, *S. Africa* **116 D2** 29 36 S 17 46 E
Nabadwip = Navadwip,
India **93 H13** 23 34N 88 20 E
Nabari, *Japan* **71 C8** 34 37N 136 5 E
Nabawa, *Australia* **125 E1** 28 30 S 114 48 E
Nabberu, L., *Australia* **125 E3** 25 50 S 120 30 E
Nabburg, *Germany* **31 F8** 49 27N 12 11 E
Naberezhnyye Chelny, *Russia* **60 C11** 55 42N 52 19 E
Nabeul, *Tunisia* **108 A2** 36 30N 10 44 E
Nabha, *India* **92 D7** 30 26N 76 14 E
Nabīd, *Iran* **97 D8** 29 40N 57 38 E
Nabire, *Indonesia* **83 B5** 3 15 S 135 26 E
Nabisar, *Pakistan* **92 G3** 25 8N 69 40 E
Nabisipi →, *Canada* **141 B7** 50 14N 62 13W
Nabiswera, *Uganda* **118 B3** 1 27N 32 15 E
Nablus = Nābulus,
West Bank **103 C4** 32 14N 35 15 E
Naboomspruit, *S. Africa* **117 C4** 24 32 S 28 40 E
Nabua, *Phil.* **80 E4** 13 24N 123 22 E
Nābulus, *West Bank* **103 C4** 32 14N 35 15 E
Nabunturan, *Phil.* **81 H5** 7 35N 125 58 E
Nacala, *Mozam.* **119 E5** 14 31 S 40 34 E
Nacala-Velha, *Mozam.* **119 E5** 14 32 S 40 34 E
Nacaome, *Honduras* **164 D2** 13 31N 87 30W
Nacaroa, *Mozam.* **119 E4** 14 22 S 39 56 E
Naches, *U.S.A.* **158 C3** 46 44N 120 42W
Naches →, *U.S.A.* **160 D6** 46 38N 120 31W
Nachikatsuura, *Japan* **71 D7** 33 33N 135 58 E
Nachingwea, *Tanzania* **119 E4** 10 23 S 38 49 E
Nachingwea □, *Tanzania* **119 E4** 10 30 S 38 30 E
Nachna, *India* **92 F4** 27 34N 71 41 E
Náchod, *Czech Rep.* **34 A9** 50 25N 16 8 E
Nacimiento Reservoir,
U.S.A. **160 K6** 35 46N 120 53W
Nackara, *Australia* **128 B3** 32 48 S 139 12 E
Naco, *Mexico* **162 A3** 31 20N 109 56W
Naco, *U.S.A.* **159 L9** 31 20N 109 57W
Nacogdoches, *U.S.A.* **155 K7** 31 36N 94 39W
Nácori Chico, *Mexico* **162 B3** 29 39N 109 1W
Nacozari, *Mexico* **162 A3** 30 24N 109 39W
Nadi, *Sudan* **106 D3** 18 40N 33 41 E
Nadiad, *India* **92 H5** 22 41N 72 56 E
Nădlac, *Romania* **52 D5** 46 10N 20 50 E
Nador, *Morocco* **111 A4** 35 14N 2 58W
Nadur, *Malta* **38 C1** 36 2N 14 17 E
Nadūshan, *Iran* **97 C7** 32 2N 53 35 E
Nadvirna, *Ukraine* **59 H3** 48 37N 24 30 E
Nadvoitsy, *Russia* **56 B5** 63 52N 34 14 E
Nadvornaya = Nadvirna,
Ukraine **59 H3** 48 37N 24 30 E
Nadym, *Russia* **64 C8** 65 35N 72 42 E
Nadym →, *Russia* **64 C8** 66 12N 72 0 E
Nærbø, *Norway* **15 G11** 58 40N 5 39 E
Næstved, *Denmark* **17 J5** 55 13N 11 44 E
Nafada, *Nigeria* **113 C7** 11 8N 11 20 E
Näfels, *Switz.* **33 B8** 47 6N 9 4 E
Naftshahr, *Iran* **101 E11** 34 0N 45 30 E
Nafud Desert = An Nafūd,
Si. Arabia **96 D4** 28 15N 41 0 E
Nafūsah, Jabal, *Libya* **108 B2** 32 12N 12 30 E
Nag Hammâdi, *Egypt* **106 B3** 26 2N 32 18 E
Naga, *Cebu, Phil.* **81 F4** 10 13N 123 45 E
Naga, *Luzon, Phil.* **80 E4** 13 38N 123 15 E
Naga, *Zamboanga del S.,*
Phil. **81 H4** 7 46N 122 45 E
Naga, Kreb en, *Africa* **110 D3** 24 12N 6 0W
Naga-Shima, *Kagoshima,*
Japan **70 E2** 32 10N 130 9 E
Naga-Shima, *Yamaguchi,*
Japan **70 D4** 33 49N 132 5 E
Nagagami →, *Canada* **140 C3** 49 40N 84 40W
Nagahama, *Ehime, Japan* **70 D4** 33 36N 132 29 E
Nagahama, *Shiga, Japan* **71 B8** 35 23N 136 16 E
Nagai, *Japan* **68 E10** 38 6N 140 2 E
Nagai I., *U.S.A.* **144 J8** 55 5N 160 0W
Nagaland □, *India* **90 C5** 26 0N 94 30 E
Nagambie, *Australia* **129 D6** 36 47 S 145 10 E
Nagano, *Japan* **71 A10** 36 40N 138 10 E
Nagano □, *Japan* **71 A10** 36 15N 138 0 E
Nagaoka, *Japan* **69 F9** 37 27N 138 51 E
Nagappattinam, *India* **95 J4** 10 46N 79 51 E
Nagar Parkar, *Pakistan* **92 G4** 24 28N 70 46 E
Nagara →, *Japan* **71 B8** 35 1N 136 43 E
Nagari Hills, *India* **95 H4** 13 3N 79 45 E
Nagarjuna Sagar, *India* **95 F4** 16 35N 79 17 E
Nagasaki, *Japan* **70 E1** 32 47N 129 50 E
Nagasaki □, *Japan* **70 E1** 32 50N 129 40 E
Nagato, *Japan* **70 C3** 34 19N 131 5 E
Nagaur, *India* **92 F5** 27 15N 73 45 E
Nagbhir, *India* **94 D4** 20 34N 79 55 E
Nagercoil, *India* **95 K3** 8 12N 77 26 E
Nagina, *India* **93 E8** 29 30N 78 30 E
Nagīneh, *Iran* **97 C8** 34 20N 57 15 E
Nagir, *Pakistan* **93 A6** 36 12N 74 42 E
Naglarby, *Sweden* **16 D9** 60 25N 15 34 E
Nagold, *Germany* **31 G4** 48 32N 8 43 E
Nagold →, *Germany* **31 G4** 48 52N 8 42 E
Nagoorin, *Australia* **126 C5** 24 17 S 151 15 E
Nagorno-Karabakh,
Azerbaijan **101 C12** 39 55N 46 45 E
Nagornyy, *Russia* **65 D13** 55 58N 124 57 E
Nagorsk, *Russia* **62 B3** 59 18N 50 48 E
Nagoya, *Japan* **71 B8** 35 10N 136 50 E
Nagpur, *India* **94 D4** 21 8N 79 10 E
Nagua, *Dom. Rep.* **165 C6** 19 23N 69 50W
Nagyatád, *Hungary* **52 D2** 46 14N 17 22 E
Nagyecsed, *Hungary* **52 C7** 47 53N 22 24 E
Nagykálló, *Hungary* **52 C6** 47 53N 21 51 E
Nagykanizsa, *Hungary* **52 D2** 46 28N 17 0 E
Nagykáta, *Hungary* **52 C4** 47 25N 19 45 E
Nagykőrös, *Hungary* **52 C4** 47 5N 19 48 E
Naha, *Japan* **69 L3** 26 13N 127 42 E
Nahanni Butte, *Canada* **142 A4** 61 2N 123 31W
Nahanni Nat. Park, *Canada* **142 A4** 61 15N 125 0W
Nahariyya, *Israel* **100 F6** 33 1N 35 5 E
Nahāvand, *Iran* **97 C6** 34 10N 48 22 E
Nahe →, *Germany* **31 F3** 49 58N 7 54 E

Ndali 255

Nahîya, W. →, *Egypt* **106 J7** 28 55N 31 0 E
Nahlin, *Canada* **142 B2** 58 55N 131 38W
Nahuel Huapi, L., *Argentina* **176 B2** 41 0 S 71 32W
Nahunta, *U.S.A.* **152 D8** 31 12N 81 59W
Naicá, *Mexico* **162 B3** 27 53N 105 31W
Naicam, *Canada* **143 C8** 52 30N 104 30W
Naila, *Germany* **31 E7** 50 19N 11 42 E
Nain, *Canada* **141 A7** 56 34N 61 40W
Nā'īn, *Iran* **97 C7** 32 54N 53 0 E
Naini Tal, *India* **93 E8** 29 30N 79 30 E
Naintré, *France* **26 F7** 46 46N 0 29 E
Naipu, *Romania* **53 F10** 44 12N 25 47 E
Nairn, *U.K.* **22 D5** 57 35N 3 53W
Nairobi, *Kenya* **118 C4** 1 17 S 36 48 E
Naissaar, *Estonia* **15 G21** 59 34N 24 29 E
Naivasha, *Kenya* **118 C4** 0 40 S 36 30 E
Naivasha, L., *Kenya* **118 C4** 0 48 S 36 20 E
Najac, *France* **28 D5** 44 14N 1 58 E
Najafābād, *Iran* **97 C6** 32 40N 51 15 E
Nájera, *Spain* **40 C2** 42 26N 2 48W
Najerilla →, *Spain* **40 C2** 42 32N 2 48W
Najibabad, *India* **92 E8** 29 40N 78 20 E
Najin, *N. Korea* **75 C16** 42 12N 130 15 E
Najmah, *Si. Arabia* **97 E6** 26 42N 50 6 E
Naju, *S. Korea* **75 G14** 35 3N 126 43 E
Naka →, *Japan* **71 A12** 36 20N 140 36 E
Nakadōri-Shima, *Japan* **69 H4** 32 57N 129 4 E
Nakalagba,
Dem. Rep. of the Congo **118 B2** 2 50N 27 58 E
Nakama, *Japan* **70 D2** 33 56N 130 43 E
Nakaminato, *Japan* **71 A12** 36 21N 140 36 E
Nakamura, *Japan* **70 E4** 32 59N 132 56 E
Nakanai Mts., *Papua N. G.* **132 C6** 5 40 S 151 0 E
Nakano, *Japan* **71 A10** 36 45N 138 22 E
Nakano-Shima, *Japan* **69 K4** 29 51N 129 52 E
Nakanojō, *Japan* **71 A10** 36 35N 138 51 E
Nakashibetsu, *Japan* **68 C12** 43 33N 144 59 E
Nakatsu, *Japan* **70 D3** 33 34N 131 15 E
Nakatsugawa, *Japan* **71 B9** 35 29N 137 30 E
Nakfa, *Eritrea* **107 D4** 16 40N 38 32 E
Nakhichevan = Naxçıvan,
Azerbaijan **101 C11** 39 12N 45 15 E
Nakhichevan Republic □ =
Naxçıvan □, *Azerbaijan* **101 C11** 39 25N 45 26 E
Nakhl, *Egypt* **103 F2** 29 55N 33 43 E
Nakhl-e Taqī, *Iran* **97 E7** 27 28N 52 36 E
Nakhodka, *Russia* **68 C6** 42 53N 132 54 E
Nakhon Nayok, *Thailand* **86 E3** 14 12N 101 13 E
Nakhon Pathom, *Thailand* **86 F3** 13 49N 100 3 E
Nakhon Phanom, *Thailand* **86 D5** 17 23N 104 43 E
Nakhon Ratchasima,
Thailand **86 E4** 14 59N 102 12 E
Nakhon Sawan, *Thailand* **86 E3** 15 35N 100 10 E
Nakhon Si Thammarat,
Thailand **87 H3** 8 29N 100 0 E
Nakhon Thai, *Thailand* **86 D3** 17 5N 100 44 E
Nakina, *B.C., Canada* **142 B2** 59 12N 132 52W
Nakina, *Ont., Canada* **140 B2** 50 10N 86 40W
Nakło nad Notecią, *Poland* **55 E4** 53 9N 17 38 E
Nakodar, *India* **92 D6** 31 8N 75 31 E
Nakskov, *Denmark* **17 K5** 54 50N 11 8 E
Naktong →, *S. Korea* **75 G15** 35 7N 128 57 E
Nakuru, *Kenya* **118 C4** 0 15 S 36 4 E
Nakuru □, *Kenya* **118 C4** 0 15 S 35 5 E
Nakuru, L., *Kenya* **118 C4** 0 23 S 36 5 E
Nakusp, *Canada* **142 C5** 50 20N 117 45W
Nal →, *Pakistan* **91 D2** 25 20N 65 30 E
Nalchik, *Russia* **61 J6** 43 30N 43 33 E
Nałęczów, *Poland* **55 G9** 51 17N 22 9 E
Nalerigu, *Ghana* **113 C4** 10 35N 0 25W
Nalgonda, *India* **94 F4** 17 6N 79 15 E
Nalhati, *India* **93 G12** 24 17N 87 52 E
Nallamalai Hills, *India* **95 G4** 15 30N 78 50 E
Nallıhan, *Turkey* **100 B4** 40 11N 31 20 E
Nalón →, *Spain* **42 B4** 43 32N 6 4W
Nālūt, *Libya* **108 B2** 31 54N 11 0 E
Nam Can, *Vietnam* **87 H5** 8 46N 104 59 E
Nam Co, *China* **72 C4** 30 30N 90 45 E
Nam Dinh, *Vietnam* **86 B6** 20 25N 106 5 E
Nam Du, Hon, *Vietnam* **87 H5** 9 41N 104 21 E
Nam Ngum Dam, *Laos* **86 C4** 18 35N 102 34 E
Nam-Phan = Cochin China,
Vietnam **87 G6** 10 30N 106 0 E
Nam Phong, *Thailand* **86 D4** 16 42N 102 52 E
Nam Tha, *Laos* **86 B3** 20 58N 101 30 E
Nam Tok, *Thailand* **86 E2** 14 21N 99 4 E
Namachire, *Angola* **115 E4** 11 26 S 22 43 E
Namacunde, *Angola* **115 F3** 17 18 S 15 50 E
Namacurra, *Mozam.* **117 B6** 17 30 S 36 50 E
Namak, Daryācheh-ye, *Iran* **97 C7** 34 30N 52 0 E
Namak, Kavir-e, *Iran* **97 C8** 34 30N 57 30 E
Namakkal, *India* **95 J4** 11 13N 78 13 E
Namaland, *Namibia* **116 C2** 26 0 S 17 0 E
Namangan, *Uzbekistan* **63 C5** 41 0N 71 40 E
Namapa, *Mozam.* **119 E4** 13 43 S 39 50 E
Namaqualand, *S. Africa* **116 E2** 30 0 S 17 25 E
Namasagali, *Uganda* **118 B3** 1 2N 33 0 E
Namatanai, *Papua N. G.* **132 B7** 3 40 S 152 29 E
Namber, *Indonesia* **83 B4** 1 2 S 134 49 E
Nambour, *Australia* **127 D5** 26 32 S 152 58 E
Nambouwalu, *Fiji* **133 A2** 17 0 S 178 45 E
Nambucca Heads, *Australia* **129 A10** 30 37 S 153 0 E
Namcha Barwa, *China* **72 D4** 29 40N 95 10 E
Namche Bazar, *Nepal* **93 F12** 27 51N 86 47 E
Namchonjŏm, *N. Korea* **75 E14** 38 15N 126 26 E
Namecunda, *Mozam.* **119 E4** 14 54 S 37 37 E
Nameh, *Indonesia* **85 B5** 2 34N 116 21 E
Nameponda, *Mozam.* **119 F4** 15 50 S 39 50 E
Namerikawa, *Japan* **71 A9** 36 46N 137 20 E
Náměšť nad Oslavou,
Czech Rep. **35 B9** 49 12N 16 10 E
Námestovo, *Slovak Rep.* **35 B12** 49 24N 19 25 E
Nametil, *Mozam.* **119 F4** 15 40 S 39 21 E
Namew L., *Canada* **143 C8** 54 14N 101 56W
Namhsan, *Burma* **90 D6** 22 48N 97 2 E
Namib Desert =
Namibwoestyn, *Namibia* **116 C2** 22 30 S 15 0 E
Namibe, *Angola* **115 F2** 15 7 S 12 11 E
Namibe □, *Angola* **115 F2** 16 35 S 12 30 E
Namibia ■, *Africa* **116 C2** 22 0 S 18 9 E
Namibwoestyn, *Namibia* **116 C2** 22 30 S 15 0 E
Namīn, *Iran* **101 C13** 38 25N 48 30 E
Namkhan, *Burma* **90 D6** 23 50N 97 41 E
Namlea, *Indonesia* **82 B3** 3 18 S 127 5 E
Namoi →, *Australia* **129 A8** 30 12 S 149 30 E
Namous, O. en →, *Algeria* **111 B4** 31 0N 0 15W
Nampa, *U.S.A.* **158 E5** 43 34N 116 34W
Nampo, *N. Korea* **75 E13** 38 52N 125 10 E
Nampō-Shotō, *Japan* **69 J10** 32 0N 140 0 E
Nampula, *Mozam.* **119 F4** 15 6 S 39 15 E
Namrole, *Indonesia* **82 B3** 3 46 S 126 46 E
Namsen →, *Norway* **14 D14** 64 28N 11 37 E
Namsos, *Norway* **14 D14** 64 29N 11 30 E
Namtsy, *Russia* **65 C13** 62 43N 129 37 E
Namtu, *Burma* **90 D6** 23 5N 97 28 E
Namtumbo, *Tanzania* **119 E4** 10 30 S 36 4 E
Namu, *Canada* **142 C3** 51 52N 127 50W
Namuac, *Phil.* **80 B3** 18 37N 121 10 E
Namur, *Belgium* **24 D4** 50 27N 4 52 E
Namur □, *Belgium* **24 D4** 50 17N 5 0 E
Namutoni, *Namibia* **116 B2** 18 49 S 16 55 E
Namwala, *Zambia* **119 F2** 15 44 S 26 30 E
Namwŏn, *S. Korea* **75 G14** 35 23N 127 23 E
Namysłów, *Poland* **55 G4** 51 6N 17 42 E
Nan, *Thailand* **86 C3** 18 48N 100 46 E
Nan →, *Thailand* **86 E3** 15 42N 100 9 E
Nan Xian, *China* **77 C9** 29 20N 112 22 E
Nana, *Romania* **53 F11** 44 17N 26 34 E
Nanaimo, *Canada* **142 D4** 49 10N 124 0W
Nanakuli, *U.S.A.* **145 K13** 21 24N 158 9W
Nanam, *N. Korea* **75 D15** 41 44N 129 40 E
Nanan, *China* **77 E12** 24 59N 118 21 E
Nanango, *Australia* **127 D5** 26 40 S 152 0 E
Nan'ao, *China* **77 F11** 23 28N 117 5 E
Nanao, *Japan* **69 F8** 37 0N 137 0 E
Nanbu, *China* **76 B6** 31 18N 106 3 E
Nanchang, *China* **77 C10** 28 42N 115 55 E
Nancheng, *China* **77 D11** 27 33N 116 35 E
Nanching = Nanjing, *China* **77 A12** 32 2N 118 47 E
Nanchong, *China* **76 B6** 30 43N 106 2 E
Nanchuan, *China* **76 C6** 29 9N 107 6 E
Nancy, *France* **27 D13** 48 42N 6 12 E
Nanda Devi, *India* **93 D8** 30 23N 79 59 E
Nandan, *China* **76 E6** 24 58N 107 29 E
Nandan, *Japan* **70 C6** 34 10N 134 42 E
Nanded, *India* **94 E3** 19 10N 77 20 E
Nandewar Ra., *Australia* **127 E5** 30 15 S 150 35 E
Nandi □, *Kenya* **118 B4** 0 15N 35 0 E
Nandikotkur, *India* **95 G4** 15 52N 78 18 E
Nandura, *India* **94 D3** 20 52N 76 25 E
Nandurbar, *India* **94 D2** 21 20N 74 15 E
Nandyal, *India* **95 G4** 15 30N 78 30 E
Nanfeng, *Guangdong, China* **77 F8** 23 45N 111 47 E
Nanfeng, *Jiangxi, China* **77 D11** 27 12N 116 28 E
Nanga, *Australia* **125 E1** 26 7 S 113 45 E
Nanga-Eboko, *Cameroon* **113 E7** 4 41N 12 22 E
Nanga Parbat, *Pakistan* **93 B6** 35 10N 74 35 E
Nangade, *Mozam.* **119 E4** 11 5 S 39 36 E
Nangapinoh, *Indonesia* **85 C4** 0 20 S 111 44 E
Nangarhár □, *Afghan.* **91 B3** 34 20N 70 0 E
Nangatayap, *Indonesia* **85 C4** 1 32 S 110 34 E
Nangeya Mts., *Uganda* **118 B3** 3 30N 33 30 E
Nangis, *France* **27 D10** 48 33N 3 1 E
Nangong, *China* **74 F8** 37 23N 115 22 E
Nangwarry, *Australia* **128 D4** 37 33 S 140 48 E
Nanhua, *China* **76 E3** 25 13N 101 21 E
Nanhuang, *China* **75 F11** 36 58N 121 48 E
Nanhui, *China* **77 B13** 31 5N 121 44 E
Nanjangud, *India* **95 H3** 12 6N 76 43 E
Nanjeko, *Zambia* **119 F1** 15 31 S 23 30 E
Nanji Shan, *China* **77 D13** 27 27N 121 4 E
Nanjian, *China* **76 E3** 25 2N 100 25 E
Nanjiang, *China* **76 A6** 32 28N 106 51 E
Nanjing, *Fujian, China* **77 E11** 24 25N 117 20 E
Nanjing, *Jiangsu, China* **77 A12** 32 2N 118 47 E
Nanjirinji, *Tanzania* **119 D4** 9 41 S 39 5 E
Nankana Sahib, *Pakistan* **92 D5** 31 27N 73 38 E
Nankang, *China* **77 E10** 25 40N 114 45 E
Nanking = Nanjing, *China* **77 A12** 32 2N 118 47 E
Nankoku, *Japan* **70 D5** 33 39N 133 44 E
Nanling, *China* **77 B12** 30 55N 118 20 E
Nanning, *China* **76 F7** 22 48N 108 20 E
Nannup, *Australia* **125 F2** 33 59 S 115 48 E
Nanortalik, *Greenland* **10 E6** 60 10N 45 17W
Nanpan Jiang →, *China* **76 E6** 25 10N 106 5 E
Nanpara, *India* **93 F9** 27 52N 81 33 E
Nanpi, *China* **74 E9** 38 2N 116 45 E
Nanping, *Fujian, China* **77 D12** 26 38N 118 10 E
Nanping, *Henan, China* **77 C9** 29 55N 112 3 E
Nanri Dao, *China* **77 E12** 25 15N 119 25 E
Nanripe, *Mozam.* **119 E4** 13 52 S 38 52 E
Nansei-Shotō = Ryūkyū-rettō, *Japan* **69 M3** 26 0N 126 0 E
Nansen Land, *Greenland* **10 A6** 83 0N 43 0W
Nansen Sd., *Canada* **6 A3** 81 0N 91 0W
Nansio, *Tanzania* **118 C3** 2 3 S 33 4 E
Nant, *France* **28 D7** 44 1N 3 18 E
Nanterre, *France* **27 D9** 48 53N 2 13 E
Nantes, *France* **26 E5** 47 12N 1 33W
Nantiat, *France* **28 B5** 46 1N 1 11 E
Nanticoke, *U.S.A.* **151 E8** 41 12N 76 0W
Nanton, *Canada* **142 C6** 50 21N 113 46W
Nantong, *China* **77 A13** 32 1N 120 52 E
Nantua, *France* **27 F12** 46 10N 5 35 E
Nantucket I., *U.S.A.* **136 E12** 41 16N 70 5W
Nantwich, *U.K.* **20 D5** 53 4N 2 31W
Nanuku Passage, *Fiji* **133 A3** 16 45 S 179 15W
Nanuque, *Brazil* **171 E3** 17 50 S 40 21W
Nanutarra, *Australia* **124 D2** 22 32 S 115 30 E
Nanxiong, *China* **77 E10** 25 6N 114 15 E
Nanyang, *China* **74 H7** 33 11N 112 30 E
Nanyi Hu, *China* **77 B12** 31 5N 119 0 E
Nan'yō, *Japan* **70 C3** 34 3N 131 49 E
Nanyuan, *China* **74 E9** 39 44N 116 22 E
Nanyuki, *Kenya* **118 B4** 0 2N 37 4 E
Nanzhang, *China* **77 B8** 31 45N 111 50 E
Nao, C. de la, *Spain* **41 G5** 38 44N 0 14 E
Naococane L., *Canada* **141 B5** 52 50N 70 45W
Naoetsu, *Japan* **69 F9** 37 12N 138 10 E
Naogaon, *Bangla.* **90 C2** 24 52N 88 52 E
Náousa, *Imathía, Greece* **50 F6** 40 42N 22 9 E
Náousa, *Kikládhes, Greece* **49 D7** 37 7N 25 14 E
Naozhou Dao, *China* **77 G8** 20 55N 110 20 E
Napa, *U.S.A.* **160 G4** 38 18N 122 17W
Napa →, *U.S.A.* **160 G4** 38 10N 122 19W
Napakiak, *U.S.A.* **144 F7** 60 42N 161 57W
Napamute, *U.S.A.* **144 F8** 61 33N 158 42W
Napanee, *Canada* **140 D4** 44 15N 77 0W
Napanoch, *U.S.A.* **151 E10** 41 44N 74 22W
Napaskiak, *U.S.A.* **144 F7** 60 43N 161 55W
Nape, *Laos* **86 C5** 18 18N 105 6 E
Nape Pass = Keo Neua,
Deo, *Vietnam* **86 C5** 18 23N 105 10 E
Naperville, *U.S.A.* **157 C8** 41 46N 88 9W
Napf, *Switz.* **32 B5** 47 1N 7 56 E
Napier, *N.Z.* **130 F5** 39 30 S 176 56 E
Napier Broome B., *Australia* **124 B4** 14 2 S 126 37 E
Napier Downs, *Australia* **124 C3** 17 11 S 124 36 E
Napier Pen., *Australia* **126 A2** 12 4 S 135 43 E
Naples = Nápoli, *Italy* **47 B7** 40 50N 14 15 E
Naples, *U.S.A.* **149 M5** 26 8N 81 48W
Naples Park, *U.S.A.* **153 J8** 26 17N 81 46W
Napo, *China* **76 F5** 23 22N 105 50 E
Napo □, *Ecuador* **168 D2** 0 30 S 77 0W
Napo →, *Peru* **168 D3** 3 20 S 72 40W
Napoleon, *N. Dak., U.S.A.* **154 B5** 46 30N 99 46W
Napoleon, *Ohio, U.S.A.* **157 C12** 41 23N 84 8W
Nápoli, *Italy* **47 B7** 40 50N 14 15 E
Nápoli, G. di, *Italy* **47 B7** 40 43N 14 10 E
Napopo,
Dem. Rep. of the Congo **118 B2** 4 15N 28 0 E
Nappa Merrie, *Australia* **127 D3** 27 36 S 141 7 E
Nappanee, *U.S.A.* **157 C11** 41 27N 86 0W
Naqâda, *Egypt* **106 B3** 25 53N 32 42 E
Naqadeh, *Iran* **101 D11** 36 57N 45 23 E
Naqqāsh, *Iran* **97 C6** 35 40N 49 6 E
Nara, *Japan* **71 C7** 34 40N 135 49 E
Nara, *Mali* **112 B3** 15 10N 7 20W
Nara □, *Japan* **71 C8** 34 30N 136 0 E
Nara Canal, *Pakistan* **92 G3** 24 30N 69 20 E
Nara Visa, *U.S.A.* **155 H3** 35 37N 103 6W
Naracoorte, *Australia* **128 D4** 36 58 S 140 45 E
Naradhan, *Australia* **129 B7** 33 34 S 146 17 E
Narasapur, *India* **95 F5** 16 26N 81 40 E
Narasaropet, *India* **95 F5** 16 14N 80 4 E
Narathiwat, *Thailand* **87 J3** 6 30N 101 48 E
Narayanganj, *Bangla.* **90 D3** 23 40N 90 33 E
Narayanpet, *India* **94 F3** 16 45N 77 30 E
Narbonne, *France* **28 E7** 43 11N 3 0 E
Narbuvollen, *Norway* **18 B8** 62 21N 11 27 E
Narcea →, *Spain* **42 B4** 43 33N 6 44W
Nardīn, *Iran* **97 B7** 37 3N 55 59 E
Nardò, *Italy* **47 B11** 40 11N 18 2 E
Narembeen, *Australia* **125 F2** 32 7 S 118 24 E
Nares Str., *Arctic* **10 B3** 80 0N 70 0W
Naretha, *Australia* **125 F3** 31 0 S 124 45 E
Narew →, *Poland* **55 F7** 52 26N 20 41 E
Nari →, *Pakistan* **92 F2** 28 0N 67 40 E
Narindra, Helodranon' i,
Madag. **117 A8** 14 55 S 47 30 E
Narino □, *Colombia* **168 C2** 1 30N 78 0W
Narita, *Japan* **71 B12** 35 47N 140 19 E
Närke, *Sweden* **16 E8** 59 10N 15 0 E
Narmada →, *India* **92 J5** 21 38N 72 36 E
Narman, *Turkey* **101 B9** 40 26N 41 57 E
Narmland, *Sweden* **15 F15** 60 0N 13 30 E
Narnaul, *India* **92 E7** 28 5N 76 11 E
Narni, *Italy* **45 F9** 42 30N 12 30 E
Naro, *Ghana* **112 C4** 10 22N 2 27W
Naro Fominsk, *Russia* **58 E9** 55 23N 36 43 E
Narodnaya, *Russia* **56 A10** 65 5N 59 58 E
Narok, *Kenya* **118 C4** 1 55 S 35 52 E
Narok □, *Kenya* **118 C4** 1 20 S 36 30 E
Narón, *Spain* **42 B2** 43 32N 8 9W
Narooma, *Australia* **129 D9** 36 14 S 150 4 E
Narowal, *Pakistan* **91 B4** 32 6N 74 52 E
Narrabri, *Australia* **127 E4** 30 19 S 149 46 E
Narran →, *Australia* **127 D4** 28 37 S 148 12 E
Narrandera, *Australia* **129 C7** 34 42 S 146 31 E
Narraway →, *Canada* **142 B5** 55 44N 119 55W
Narrogin, *Australia* **125 F2** 32 58 S 117 14 E
Narromine, *Australia* **129 B8** 32 12 S 148 12 E
Narsampet, *India* **94 F4** 17 57N 79 58 E
Narsaq, *Greenland* **10 E6** 60 57N 46 4W
Narsimhapur, *India* **93 H8** 22 54N 79 14 E
Nartes, L. e, *Albania* **50 F3** 40 32N 19 25 E
Nartkala, *Russia* **61 J6** 43 33N 43 51 E
Naruto, *Japan* **70 C6** 34 11N 134 37 E
Narutō, *Shikoku, Japan* **71 B12** 35 36N 140 25 E
Naruto-Kaikyō, *Japan* **70 C6** 34 14N 134 39 E
Narva, *Estonia* **58 C5** 59 23N 28 12 E
Narva →, *Russia* **15 G22** 59 27N 28 2 E
Narvacan, *Phil.* **80 C3** 17 25N 120 28 E
Narvik, *Norway* **14 B17** 68 28N 17 26 E
Narvskoye Vdkhr., *Russia* **58 C5** 59 18N 28 14 E
Narwana, *India* **92 E7** 29 39N 76 6 E
Naryan-Mar, *Russia* **56 A9** 67 42N 53 12 E
Naryilco, *Australia* **127 D3** 28 37 S 141 53 E
Narym, *Russia* **64 D9** 59 0N 81 30 E
Naryn, *Kyrgyzstan* **63 C7** 41 26N 75 58 E
Naryn →, *Uzbekistan* **63 C5** 40 52N 71 36 E
Nasa, *Norway* **14 C16** 66 29N 15 23 E
Nasarawa, *Nigeria* **113 D6** 8 32N 7 41 E
Năsăud, *Romania* **53 C9** 47 19N 24 29 E
Nasawa, *Vanuatu* **133 E6** 15 0 S 168 0 E
Naseby, *N.Z.* **131 F5** 45 1 S 170 10 E
Naselle, *U.S.A.* **160 D3** 46 22N 123 49W
Naser, Buheirat en, *Egypt* **106 C3** 23 0N 32 30 E
Nashua, *Iowa, U.S.A.* **156 B4** 42 57N 92 32W
Nashua, *Mont., U.S.A.* **158 B10** 48 8N 106 22W
Nashua, *N.H., U.S.A.* **151 D13** 42 45N 71 28W
Nashville, *Ark., U.S.A.* **155 J8** 33 57N 93 51W
Nashville, *Ga., U.S.A.* **152 D6** 31 12N 83 15W
Nashville, *Ill., U.S.A.* **156 F7** 38 21N 89 23W
Nashville, *Ind., U.S.A.* **157 E10** 39 12N 86 15W
Nashville, *Mich., U.S.A.* **157 B11** 42 36N 85 5W
Nashville, *Tenn., U.S.A.* **149 G2** 36 10N 86 47W
Našice, *Croatia* **52 E3** 45 32N 18 4 E
Nasielsk, *Poland* **55 F7** 52 35N 20 50 E
Nasik, *India* **94 E1** 19 58N 73 50 E
Nasipit, *Phil.* **81 G5** 8 57N 125 19 E
Nasirabad, *India* **92 F6** 26 15N 74 45 E
Naskaupi →, *Canada* **141 B7** 53 47N 60 51W
Naso, *Italy* **47 D7** 38 7N 14 47 E
Naso Pt., *Phil.* **81 F3** 10 25N 121 57 E
Naṣriān-e Pā'īn, *Iran* **96 C5** 32 52N 46 52 E
Nass →, *Canada* **142 C3** 55 0N 129 40W
Nassau, *Bahamas* **164 A4** 25 5N 77 20W
Nassau, *U.S.A.* **151 D11** 42 31N 73 37W
Nassau, B., *Chile* **176 E3** 55 20 S 68 0W
Nasser, L. = Naser, Buheirat
en, *Egypt* **106 C3** 23 0N 32 30 E
Nasser City = Kôm Ombo,
Egypt **106 C3** 24 25N 32 52 E
Nassian, *Ivory C.* **112 D4** 8 28N 3 28W
Nässjö, *Sweden* **17 G8** 57 39N 14 42 E
Nasugbu, *Phil.* **80 D3** 14 5N 120 38 E
Näsum, *Sweden* **17 H8** 56 10N 14 29 E
Näsviken, *Sweden* **16 C10** 61 46N 16 52 E
Nat Kyizin, *Burma* **86 E1** 14 57N 97 59 E
Nata, *Botswana* **116 C4** 20 12 S 26 12 E
Natagaima, *Colombia* **168 C2** 3 37N 75 6W
Natal, *Brazil* **170 C4** 5 47 S 35 13W
Natal, *Canada* **142 D6** 49 43N 114 51W
Natal, *Indonesia* **84 B1** 0 35N 99 7 E
Natalinci, *Serbia, Yug.* **52 F5** 44 15N 20 49 E
Naṭanz, *Iran* **97 C6** 33 30N 51 55 E
Natashquan, *Canada* **141 B7** 50 14N 61 46W
Natashquan →, *Canada* **141 B7** 50 7N 61 50W
Natchez, *U.S.A.* **155 K9** 31 34N 91 24W
Natchitoches, *U.S.A.* **155 K8** 31 46N 93 5W
Naters, *Switz.* **32 D5** 46 19N 7 58 E
Natewa B., *Fiji* **133 A2** 16 35 S 179 40 E
Nathalia, *Australia* **129 D6** 36 1 S 145 13 E
Nathdwara, *India* **92 G5** 24 55N 73 50 E
Nati, Pta., *Spain* **39 A10** 40 3N 3 50 E
Natimuk, *Australia* **128 D5** 36 42 S 142 0 E
Nation →, *Canada* **142 B4** 55 30N 123 32W
National City, *U.S.A.* **161 N9** 32 41N 117 6W
Natitingou, *Benin* **113 C5** 10 20N 1 26 E
Natividad, I., *Mexico* **162 B1** 27 50N 115 10W
Natogyi, *Burma* **90 E5** 21 25N 95 39 E
Natoma, *U.S.A.* **154 F5** 39 11N 99 2W
Natonin, *Phil.* **80 C3** 17 6N 121 18 E
Natron, L., *Tanzania* **118 C4** 2 20 S 36 0 E
Natrona Heights, *U.S.A.* **150 F5** 40 37N 79 44W
Natrûn, W. el →, *Egypt* **106 H7** 30 25N 30 13 E
Nättraby, *Sweden* **17 H9** 56 13N 15 31 E
Natuna Besar, Kepulauan,
Indonesia **87 L7** 4 0N 108 15 E
Natuna Is. = Natuna Besar,
Kepulauan, *Indonesia* **87 L7** 4 0N 108 15 E
Natuna Selatan, Kepulauan,
Indonesia **85 B3** 2 45N 109 0 E
Natural Bridge, *U.S.A.* **151 B9** 44 5N 75 30W
Naturaliste, C., *Australia* **126 G4** 40 50 S 148 15 E
Natya, *Australia* **128 C5** 34 57 S 143 13 E
Nau, *Tajikistan* **63 C4** 40 9N 69 22 E
Nau Qala, *Afghan.* **92 B3** 34 5N 68 5 E
Naubinway, *U.S.A.* **140 C2** 46 6N 85 27W
Naucelle, *France* **28 D6** 44 13N 2 20 E
Nauders, *Austria* **34 E3** 46 54N 10 30 E
Nauen, *Germany* **30 C8** 52 36N 12 52 E
Naugatuck, *U.S.A.* **151 E11** 41 30N 73 3W
Naujan, *Phil.* **80 E3** 13 20N 121 18 E
Naujoji Akmenė, *Lithuania* **54 B9** 56 19N 22 54 E
Naumburg, *Germany* **30 D7** 51 9N 11 47 E
Nā'ūr at Tunayb, *Jordan* **103 D4** 31 48N 35 57 E
Nauru ■, *Pac. Oc.* **134 H8** 1 0 S 166 0 E
Naushahra = Nowshera,
Pakistan **91 B4** 34 0N 72 0 E
Nausori, *Fiji* **133 B2** 18 2 S 178 32 E
Naustdal, *Norway* **18 C2** 61 31N 5 43 E
Nauta, *Peru* **168 D3** 4 31 S 73 35W
Nauteyri, *Iceland* **11 B4** 65 55N 22 21W
Nautla, *Mexico* **163 C5** 20 20N 96 50W
Nauvoo, *U.S.A.* **156 D5** 40 33N 91 23W
Nava, *Mexico* **162 B4** 28 25N 100 46W
Nava, *Spain* **42 B5** 43 21N 5 31W
Nava del Rey, *Spain* **42 D5** 41 22N 5 6W
Navadwip, *India* **93 H13** 23 34N 88 20 E
Navahermosa, *Spain* **43 F6** 39 41N 4 28W
Navahrudak, *Belarus* **58 F3** 53 40N 25 50 E
Navajo Reservoir, *U.S.A.* **159 H10** 36 48N 107 36W
Naval, *Phil.* **81 F5** 11 34N 124 23 E
Navalcarnero, *Spain* **42 E6** 40 17N 4 5W
Navalmoral de la Mata,
Spain **42 F5** 39 52N 5 33W
Navalvillar de Pela, *Spain* **43 F5** 39 9N 5 24W
Navan = An Uaimh, *Ireland* **23 C5** 53 39N 6 41W
Navapolatsk, *Belarus* **58 E5** 55 32N 28 37 E
Navarino, I., *Chile* **176 E3** 55 0 S 67 40W
Navarra □, *Spain* **40 C3** 42 40N 1 40W
Navarre, *Fla., U.S.A.* **153 E3** 30 24N 86 52W
Navarre, *Ohio, U.S.A.* **150 F3** 40 43N 81 31W
Navarro →, *U.S.A.* **160 F3** 39 11N 123 45W
Navas de San Juan, *Spain* **43 G7** 38 30N 3 19W
Navasota, *U.S.A.* **155 K6** 30 23N 96 5W
Navassa, *W. Indies* **165 C5** 18 30N 75 0W
Nävekvarn, *Sweden* **17 F10** 58 38N 16 49 E
Naver →, *U.K.* **22 C4** 58 32N 4 14W
Navia, *Spain* **42 B4** 43 35N 6 42W
Navia →, *Spain* **42 B4** 43 15N 6 50W
Navia de Suarna, *Spain* **42 C3** 42 58N 7 3W
Navidad, *Chile* **174 C1** 33 57 S 71 50W
Navlya, *Russia* **59 F8** 52 53N 34 30 E
Năvodari, *Romania* **53 F13** 44 19N 28 36 E
Navoi = Nawoiy, *Uzbekistan* **63 C2** 40 9N 65 22 E
Navojoa, *Mexico* **162 B3** 27 0N 109 30W
Navolato, *Mexico* **162 C3** 24 47N 107 42W
Návpaktos, *Greece* **48 C3** 38 24N 21 50 E
Návplion, *Greece* **48 D4** 37 33N 22 50 E
Navrongo, *Ghana* **113 C4** 10 51N 1 3W
Navsari, *India* **94 D1** 20 57N 72 59 E
Nawa Kot, *Pakistan* **92 E4** 28 21N 71 24 E
Nawabganj, *Bangla.* **90 C2** 24 35N 88 14 E
Nawabganj, *Ut. P., India* **93 F9** 26 56N 81 14 E
Nawabganj, *Ut. P., India* **93 E8** 28 32N 79 40 E
Nawabshah, *Pakistan* **91 D3** 26 15N 68 25 E
Nawada, *India* **93 G11** 24 50N 85 33 E
Nāwah, *Afghan.* **91 B2** 32 19N 67 53 E
Nawakot, *Nepal* **93 F11** 27 55N 85 10 E
Nawalgarh, *India* **92 F6** 27 50N 75 15 E
Nawanshahr, *India* **93 C6** 32 33N 74 48 E
Nawapara, *India* **94 D6** 20 46N 82 33 E
Nawāsīf, Harrat, *Si. Arabia* **98 B3** 21 20N 42 10 E
Nawi, *Sudan* **106 D3** 18 32N 30 50 E
Nawng Hpa, *Burma* **90 D7** 22 30N 98 30 E
Nawoiy, *Uzbekistan* **63 C2** 40 9N 65 22 E
Nawş, Ra's, *Oman* **99 C6** 17 15N 55 16 E
Naxçıvan, *Azerbaijan* **101 C11** 39 12N 45 15 E
Naxçıvan □, *Azerbaijan* **101 C11** 39 25N 45 26 E
Náxos, *Greece* **49 D7** 37 8N 25 25 E
Nay, *France* **28 E3** 43 10N 0 18W
Nāy Band, *Iran* **97 E7** 27 20N 52 40 E
Naya →, *Colombia* **168 C2** 3 13N 77 22W
Nayakhan, *Russia* **65 C16** 61 56N 159 0 E
Nayarit □, *Mexico* **162 C4** 22 0N 105 0W
Nayé, *Senegal* **112 C2** 14 28N 12 12W
Nayong, *China* **76 D5** 26 50N 105 20 E
Nayoro, *Japan* **68 B11** 44 21N 142 28 E
Nayyāl, W. →, *Si. Arabia* **96 D3** 28 35N 39 4 E
Nazaré, *Bahia, Brazil* **171 D4** 13 2 S 39 0W
Nazaré, *Pará, Brazil* **173 B7** 6 25 S 52 29W
Nazaré, *Tocantins, Brazil* **170 C2** 6 23 S 47 40W
Nazaré, *Portugal* **43 F1** 39 36N 9 4W
Nazareth = Nazerat, *Israel* **103 C4** 32 42N 35 17 E
Nazas, *Mexico* **162 B4** 25 10N 104 6W
Nazas →, *Mexico* **162 B4** 25 35N 103 25W
Nazca, *Peru* **172 C2** 14 53 S 74 54W
Naze, The, *U.K.* **21 F9** 51 53N 1 18 E
Nazerat, *Israel* **103 C4** 32 42N 35 17 E
Nāzīk, *Iran* **101 C11** 39 1N 45 4 E
Nazilli, *Turkey* **49 D10** 37 55N 28 15 E
Nazir Hat, *Bangla.* **90 D3** 22 35N 91 49 E
Nazko, *Canada* **142 C4** 53 1N 123 37W
Nazko →, *Canada* **142 C4** 53 7N 123 34W
Nazret, *Ethiopia* **107 F4** 8 32N 39 22 E
Nazwá, *Oman* **99 B7** 22 56N 57 32 E
Nchanga, *Zambia* **119 E2** 12 30 S 27 49 E
Ncheu, *Malawi* **119 E3** 14 50 S 34 47 E
Ndala, *Tanzania* **118 C3** 4 45 S 33 15 E
Ndalatando, *Angola* **115 D2** 9 12 S 14 48 E
Ndali, *Benin* **113 D5** 9 50N 2 46 E

Nordfjordeid 257

Newton, *Mass., U.S.A.* **151 D13** 42 21N 71 12W
Newton, *Miss., U.S.A.* **155 J10** 32 19N 89 10W
Newton, *N.C., U.S.A.* **149 H5** 35 40N 81 13W
Newton, *N.J., U.S.A.* **151 E10** 41 3N 74 45W
Newton, *Tex., U.S.A.* **155 K8** 30 51N 93 46W
Newton Abbot, *U.K.* **21 G4** 50 32N 3 37W
Newton Aycliffe, *U.K.* **20 C6** 54 37N 1 34W
Newton Boyd, *Australia* **127 D5** 29 45 S 152 16 E
Newton Stewart, *U.K.* **22 G4** 54 57N 4 30W
Newtonmore, *U.K.* **22 D4** 57 4N 4 8W
Newtown, *U.K.* **21 E4** 52 31N 3 19W
Newtown, *U.S.A.* **156 D3** 40 22N 93 20W
Newtownabbey, *U.K.* **23 B6** 54 40N 5 56W
Newtownards, *U.K.* **23 B6** 54 36N 5 42W
Newtownbarry = Bunclody, *Ireland* **23 D5** 52 39N 6 40W
Newtownstewart, *U.K.* **23 B4** 54 43N 7 23W
Newville, *U.S.A.* **150 F7** 40 10N 77 24W
Nexon, *France* **28 C5** 45 41N 1 11 E
Neya, *Russia* **60 A6** 58 21N 43 49 E
Neyrīz, *Iran* **97 D7** 29 15N 54 19 E
Neyshābūr, *Iran* **97 B8** 36 10N 58 50 E
Neyyattinkara, *India* **95 K3** 8 26N 77 5 E
Nezhin = Nizhyn, *Ukraine* **59 G6** 51 5N 31 55 E
Nezperce, *U.S.A.* **158 C5** 46 14N 116 14W
Ngabang, *Indonesia* **85 B3** 0 23N 109 55 E
Ngabordamlu, Tanjung, *Indonesia* **83 C4** 6 56 S 134 11 E
N'Gage, *Angola* **115 D3** 7 46 S 15 15 E
Ngaiphaipi, *Burma* **90 D4** 22 14N 93 15 E
Ngambé, *Cameroon* **113 D7** 5 48N 11 29 E
Ngami Depression, *Botswana* **116 C3** 20 30 S 22 46 E
Ngamo, *Zimbabwe* **119 F2** 19 3 S 27 32 E
Nganjuk, *Indonesia* **85 D4** 7 32 S 111 55 E
Ngao, *Thailand* **86 C2** 18 46N 99 59 E
Ngaoundéré, *Cameroon* **114 A2** 7 15N 13 35 E
Ngapara, *N.Z.* **131 E5** 44 57 S 170 46 E
Ngara, *Tanzania* **118 C3** 2 29 S 30 40 E
Ngara □, *Tanzania* **118 C3** 2 29 S 30 40 E
Ngaruawahia, *N.Z.* **130 D4** 37 42 S 175 11 E
Ngaruroro →, *N.Z.* **130 F5** 39 34 S 176 55 E
Ngatapa, *N.Z.* **130 E6** 38 32 S 177 45 E
Ngathainggyaung, *Burma* **90 G5** 17 24N 95 5 E
Ngauruhoe, Mt., *N.Z.* **130 F4** 39 13 S 175 45 E
Ngawi, *Indonesia* **85 D4** 7 24 S 111 26 E
Nggela, *Solomon Is.* **133 M11** 9 5 S 160 15 E
Nghia Lo, *Vietnam* **86 B5** 21 33N 104 28 E
Ngidinga, *Dem. Rep. of the Congo* **115 D3** 5 37 S 15 17 E
Ngo, *Congo* **114 C3** 2 29 S 15 45 E
N'Gola, *Angola* **115 E2** 14 10 S 14 30 E
Ngoma, *Malawi* **119 E3** 13 8 S 33 45 E
Ngomahura, *Zimbabwe* **119 G3** 20 26 S 30 43 E
Ngomba, *Tanzania* **119 D3** 8 20 S 32 53 E
Ngongotaha, *N.Z.* **130 E5** 38 5 S 176 12 E
Ngop, *Sudan* **107 F3** 6 17N 30 9 E
Ngoring Hu, *China* **72 C4** 34 55N 97 5 E
Ngorkou, *Mali* **112 B4** 15 40N 3 41W
Ngorongoro, *Tanzania* **118 C4** 3 11 S 35 32 E
Ngouri, *Chad* **109 F3** 13 38N 15 22 E
Ngourti, *Niger* **109 E2** 15 19N 13 12 E
Ngozi, *Burundi* **118 C2** 2 54 S 29 50 E
Ngudu, *Tanzania* **118 C3** 2 58 S 33 25 E
Nguigmi, *Niger* **109 F2** 14 20N 13 20 E
Ngukurr, *Australia* **126 A1** 14 44 S 134 44 E
Ngunga, *Tanzania* **118 C3** 3 37 S 33 37 E
Nguru, *Nigeria* **113 C7** 12 56N 10 29 E
Nguru Mts., *Tanzania* **118 D4** 6 0 S 37 30 E
Nguyen Binh, *Vietnam* **86 A5** 22 39N 105 56 E
Nha Trang, *Vietnam* **87 F7** 12 16N 109 10 E
Nhacoongo, *Mozam.* **117 C6** 24 18 S 35 14 E
Nhamaabué, *Mozam.* **119 F4** 17 25 S 35 5 E
Nhambiquara, *Brazil* **173 C6** 12 50 S 59 49W
Nhamundá, *Brazil* **169 D6** 2 14 S 56 43W
Nhamundá →, *Brazil* **169 D6** 2 12 S 56 41W
Nhangutazi, L., *Mozam.* **117 C5** 24 0 S 34 30 E
Nhecolândia, *Brazil* **173 D6** 19 17 S 56 58W
Nhill, *Australia* **128 D4** 36 18 S 141 40 E
Nho Quan, *Vietnam* **86 B5** 20 18N 105 45 E
Nhulunbuy, *Australia* **126 A2** 12 10 S 137 20 E
Nhundo, *Angola* **115 E4** 14 25 S 21 23 E
Nia-nia, *Dem. Rep. of the Congo* **118 B2** 1 30N 27 40 E
Niafounké, *Mali* **112 B4** 16 0N 4 5W
Niagara, *U.S.A.* **148 C2** 45 45N 88 0W
Niagara Falls, *Canada* **140 D4** 43 7N 79 5W
Niagara Falls, *U.S.A.* **150 C6** 43 5N 79 4W
Niagara-on-the-Lake, *Canada* **150 C5** 43 15N 79 4W
Niah, *Malaysia* **85 B4** 3 58N 113 46 E
Niamey, *Niger* **113 C5** 13 27N 2 6 E
Nianforando, *Guinea* **112 D2** 9 37N 10 36W
Niangara, *Dem. Rep. of the Congo* **118 B2** 3 42N 27 50 E
Niangua →, *U.S.A.* **156 G4** 38 0N 92 48W
Nias, *Indonesia* **84 B1** 1 0N 97 30 E
Niassa □, *Mozam.* **119 E4** 13 30 S 36 0 E
Nibāk, *Si. Arabia* **99 A5** 24 25N 50 50 E
Nibe, *Denmark* **17 H3** 56 59N 9 38 E
Nicaragua ■, *Cent. Amer.* **164 D2** 11 40N 85 30W
Nicaragua, L. de, *Nic.* **164 D2** 12 0N 85 30W
Nicastro, *Italy* **47 D9** 38 59N 16 19 E
Nice, *France* **29 E11** 43 42N 7 14 E
Niceville, *U.S.A.* **149 K2** 30 31N 86 30W
Nichinan, *Japan* **70 F3** 31 38N 131 23 E
Nicholás, Canal, *W. Indies* **164 B3** 23 30N 80 5W
Nicholasville, *U.S.A.* **157 G12** 37 53N 84 34W
Nicholls, *U.S.A.* **152 D7** 31 31N 82 38W
Nichols, *U.S.A.* **151 D8** 42 1N 76 22W
Nicholson, *Australia* **124 C4** 18 2 S 128 54 E
Nicholson, *Ga., U.S.A.* **152 A6** 34 7N 83 26W
Nicholson, *Pa., U.S.A.* **151 E9** 41 37N 75 47W
Nicholson →, *Australia* **126 B2** 17 31 S 139 36 E
Nicholson Ra., *Australia* **125 E2** 27 15 S 116 45 E
Nickerie □, *Surinam* **169 C6** 4 0N 57 0W
Nickerie →, *Surinam* **169 B6** 5 58N 57 0W
Nicobar Is., *Ind. Oc.* **121 D8** 9 0N 93 0 E
Nicoclí, *Colombia* **168 B2** 8 26N 76 48W
Nicola, *Canada* **142 C4** 50 12N 120 40W
Nicolet, *Canada* **140 C5** 46 17N 72 35W
Nicolls Town, *Bahamas* **164 A4** 25 8N 78 0W
Nicopolis, *Greece* **48 B2** 39 2N 20 37 E
Nicosia, *Cyprus* **38 D12** 35 10N 33 25 E
Nicosia, *Italy* **47 E7** 37 45N 14 24 E
Nicótera, *Italy* **47 D8** 38 33N 15 56 E
Nicoya, *Costa Rica* **164 D2** 10 9N 85 27W
Nicoya, G. de, *Costa Rica* **164 E3** 10 0N 85 0W
Nicoya, Pen. de, *Costa Rica* **164 E2** 9 45N 85 40W
Nidau, *Switz.* **32 B4** 47 7N 7 15 E
Nidd →, *U.K.* **20 D6** 53 59N 1 23W
Nidda, *Germany* **31 E5** 50 23N 9 1 E
Nidda →, *Germany* **31 E4** 50 17N 8 48 E
Nidwalden □, *Switz.* **33 C6** 46 50N 8 25 E
Nidzica, *Poland* **55 E7** 53 25N 20 28 E
Niebüll, *Germany* **30 A4** 54 46N 8 48 E
Nied →, *Germany* **27 C13** 49 23N 6 40 E
Niederaula, *Germany* **30 E5** 50 47N 9 36 E
Niederbayern □, *Germany* **31 G8** 48 40N 12 50 E
Niederbipp, *Switz.* **32 B5** 47 16N 7 42 E
Niederbronn-les-Bains, *France* **27 D14** 48 57N 7 39 E
Niedere Tauern, *Austria* **34 D7** 47 20N 14 0 E
Niederlausitz, *Germany* **30 D9** 51 42N 13 59 E
Niederösterreich □, *Austria* **34 C8** 48 25N 15 40 E
Niedersachsen □, *Germany* **30 C4** 52 50N 9 0 E
Niefang, *Eq. Guin.* **114 B2** 1 50N 10 14 E
Niekerkshoop, *S. Africa* **116 D3** 29 19 S 22 51 E
Niellé, *Ivory C.* **112 C3** 10 5N 5 38W
Niem, *C.A.R.* **114 A3** 6 12N 15 14 E
Niemba, *Dem. Rep. of the Congo* **118 D2** 5 58 S 28 24 E
Niemen = Neman →, *Lithuania* **15 J20** 55 25N 21 10 E
Niemodlin, *Poland* **55 H4** 50 38N 17 38 E
Niemur, *Australia* **128 C6** 35 17 S 144 9 E
Nienburg, *Germany* **30 C5** 52 39N 9 13 E
Niepołomice, *Poland* **55 H7** 50 3N 20 13 E
Niers →, *Germany* **30 D1** 51 43N 5 57 E
Niesen, *Switz.* **32 C5** 46 38N 7 39 E
Niesky, *Germany* **30 D10** 51 17N 14 49 E
Nieszawa, *Poland* **55 F5** 52 52N 18 50 E
Nieu Bethesda, *S. Africa* **116 E3** 31 51 S 24 34 E
Nieuw Amsterdam, *Surinam* **169 B6** 5 53N 55 5W
Nieuw Nickerie, *Surinam* **169 B6** 6 0N 56 59W
Nieuwoudtville, *S. Africa* **116 E2** 31 23 S 19 7 E
Nieuwpoort, *Belgium* **24 C2** 51 8N 2 45 E
Nieves, Pico de las, *Canary Is.* **39 G4** 27 57N 15 35W
Nièvre □, *France* **27 E10** 47 10N 3 40 E
Nigata, *Japan* **70 C4** 34 13N 132 39 E
Niğde, *Turkey* **100 D6** 37 58N 34 40 E
Nigel, *S. Africa* **117 D4** 26 27 S 28 25 E
Niger □, *Nigeria* **113 D6** 10 0N 5 30 E
Niger ■, *W. Afr.* **109 E2** 17 30N 10 0 E
Niger →, *W. Afr.* **113 D6** 5 33N 6 33 E
Nigeria ■, *W. Afr.* **113 D6** 8 30N 8 0 E
Nightcaps, *N.Z.* **131 F3** 45 57 S 168 2 E
Nightmute, *U.S.A.* **144 F6** 60 29N 164 44W
Nigríta, *Greece* **50 F7** 40 56N 23 29 E
Nihoa, *U.S.A.* **145 G11** 23 6N 161 58W
Nihtaur, *India* **93 E8** 29 20N 78 23 E
Nii-Jima, *Japan* **71 C11** 34 20N 139 15 E
Niigata, *Japan* **68 F9** 37 58N 139 0 E
Niigata □, *Japan* **69 F9** 37 15N 138 45 E
Niihama, *Japan* **70 D5** 33 55N 133 16 E
Niihau, *U.S.A.* **146 H14** 21 54N 160 9W
Niimi, *Japan* **70 C5** 34 59N 133 28 E
Niitsu, *Japan* **68 F9** 37 48N 139 7 E
Nijar, *Spain* **41 J2** 36 53N 2 15W
Nijil, *Jordan* **103 E4** 30 32N 35 33 E
Nijkerk, *Neths.* **24 B5** 52 13N 5 30 E
Nijmegen, *Neths.* **24 C5** 51 50N 5 52 E
Nijverdal, *Neths.* **24 B6** 52 22N 6 28 E
Nīk Pey, *Iran* **97 B6** 36 50N 48 10 E
Nike, *Nigeria* **113 D6** 6 26N 7 29 E
Nikiniki, *Indonesia* **82 C2** 9 49 S 124 30 E
Nikísiani, *Greece* **51 F8** 40 57N 24 9 E
Nikítas, *Greece* **50 F7** 40 13N 23 43 E
Nikki, *Benin* **113 D5** 9 58N 3 12 E
Nikkō, *Japan* **71 A11** 36 45N 139 35 E
Nikolai, *U.S.A.* **144 E9** 62 58N 154 10W
Nikolayev = Mykolayiv, *Ukraine* **59 J7** 46 58N 32 0 E
Nikolayevsk, *Russia* **60 E7** 50 0N 45 35 E
Nikolayevsk-na-Amur, *Russia* **65 D15** 53 8N 140 44 E
Nikolsk, *Russia* **60 D8** 53 49N 46 4 E
Nikolskoye, *Russia* **65 D17** 55 12N 166 0 E
Nikopol, *Bulgaria* **51 C8** 43 43N 24 54 E
Nikopol, *Ukraine* **59 J8** 47 35N 34 25 E
Niksar, *Turkey* **100 B7** 40 31N 37 2 E
Nīkshahr, *Iran* **97 E9** 26 15N 60 10 E
Nikšić, *Montenegro, Yug.* **50 D2** 42 50N 18 57 E
Nîl, Nahr en →, *Africa* **106 H7** 30 10N 31 6 E
Nîl el Abyad →, *Sudan* **107 D3** 15 38N 32 31 E
Nîl el Azraq →, *Sudan* **107 D3** 15 38N 32 31 E
Niland, *U.S.A.* **161 M11** 33 14N 115 31W
Nile = Nîl, Nahr en →, *Africa* **106 H7** 30 10N 31 6 E
Nile Delta, *Egypt* **106 H7** 31 40N 31 0 E
Niles, *U.S.A.* **150 E4** 41 11N 80 46W
Nilgiri Hills, *India* **95 J3** 11 30N 76 30 E
Nilo Peçanha, *Brazil* **171 D4** 13 37 S 39 6W
Nilpena, *Australia* **128 A3** 30 58 S 138 20 E
Nilsebu, *Norway* **18 E3** 59 11N 6 39 E
Nilüfer →, *Turkey* **51 F12** 40 18N 28 27 E
Nimach, *India* **92 G6** 24 30N 74 56 E
Nimbahera, *India* **92 G6** 24 37N 74 45 E
Nîmes, *France* **29 E8** 43 50N 4 23 E
Nimfaíon, Ákra = Pínnes, Ákra, *Greece* **51 F8** 40 5N 24 20 E
Nimmitabel, *Australia* **129 D8** 36 29 S 149 15 E
Nimule, *Sudan* **107 G3** 3 32N 32 3 E
Nin, *Croatia* **45 D12** 44 16N 15 12 E
Nīnawá, *Iraq* **101 D10** 36 25N 43 10 E
Ninda, *Angola* **115 E4** 14 47 S 21 24 E
Nindigully, *Australia* **127 D4** 28 21 S 148 50 E
Ninemile, *U.S.A.* **142 B2** 56 0N 130 7W
Ninety Mile Beach, *N.Z.* **130 A2** 34 48 S 173 0 E
Ninety Mile Beach, The, *Australia* **129 E7** 38 15 S 147 24 E
Ninety Six, *U.S.A.* **152 A7** 34 11N 82 1W
Nineveh = Nīnawá, *Iraq* **101 D10** 36 25N 43 10 E
Ning Xian, *China* **74 G4** 35 30N 107 58 E
Ningaloo, *Australia* **124 D1** 22 41 S 113 41 E
Ning'an, *China* **75 B15** 44 22N 129 20 E
Ningbo, *China* **77 C13** 29 51N 121 28 E
Ningcheng, *China* **75 D10** 41 32N 119 53 E
Ningde, *China* **77 D12** 26 38N 119 23 E
Ningdu, *China* **77 D10** 26 25N 115 59 E
Ninggang, *China* **77 D9** 26 42N 113 55 E
Ningguo, *China* **77 B12** 30 35N 119 0 E
Ninghai, *China* **77 C13** 29 15N 121 27 E
Ninghua, *China* **77 D11** 26 14N 116 45 E
Ningjin, *China* **74 F8** 37 35N 114 57 E
Ningjing Shan, *China* **76 C2** 30 0N 98 20 E
Ninglang, *China* **76 D3** 27 20N 100 55 E
Ningling, *China* **74 G8** 34 25N 115 22 E
Ningming, *China* **76 F6** 22 8N 107 4 E
Ningnan, *China* **76 D4** 27 5N 102 36 E
Ningpo = Ningbo, *China* **77 C13** 29 51N 121 28 E
Ningqiang, *China* **74 H4** 32 47N 106 15 E
Ningshan, *China* **74 H5** 33 21N 108 21 E
Ningsia Hui A.R. = Ningxia Huizu Zizhiqu □, *China* **74 F4** 38 0N 106 0 E
Ningwu, *China* **74 E7** 39 0N 112 18 E
Ningxia Huizu Zizhiqu □, *China* **74 F4** 38 0N 106 0 E
Ningxiang, *China* **77 C9** 28 15N 112 30 E
Ningyang, *China* **74 G9** 35 47N 116 45 E
Ningyuan, *China* **77 E8** 25 37N 111 57 E
Ninh Binh, *Vietnam* **86 B5** 20 15N 105 55 E
Ninh Giang, *Vietnam* **86 B6** 20 44N 106 24 E
Ninh Hoa, *Vietnam* **86 F7** 12 30N 109 7 E
Ninh Ma, *Vietnam* **86 F7** 12 48N 109 21 E
Ninini Pt., *U.S.A.* **145 B2** 21 58N 159 20W
Ninove, *Belgium* **24 D4** 50 51N 4 2 E
Nioaque, *Brazil* **175 A4** 21 5 S 55 50W
Niobrara, *U.S.A.* **154 D6** 42 45N 98 2W
Niobrara →, *U.S.A.* **154 D6** 42 46N 98 3W
Nioki, *Dem. Rep. of the Congo* **114 C3** 2 47 S 17 40 E
Niono, *Mali* **112 C3** 14 15N 6 0W
Nioro du Rip, *Senegal* **112 C1** 13 40N 15 50W
Nioro du Sahel, *Mali* **112 B3** 15 15N 9 30W
Niort, *France* **28 B3** 46 19N 0 29W
Nipa, *Papua N. G.* **132 D2** 6 9 S 143 29 E
Nipani, *India* **95 F2** 16 20N 74 25 E
Nipawin, *Canada* **143 C8** 53 20N 104 0W
Nipawin Prov. Park, *Canada* **143 C8** 54 0N 104 37W
Nipfjället, *Sweden* **16 C6** 61 59N 12 50 E
Nipigon, *Canada* **140 C2** 49 0N 88 17W
Nipigon, L., *Canada* **140 C2** 49 50N 88 30W
Nipin →, *Canada* **143 B7** 55 46N 108 35W
Nipishish L., *Canada* **141 B7** 54 12N 60 45W
Nipissing L., *Canada* **140 C4** 46 20N 80 0W
Nipomo, *U.S.A.* **161 K6** 35 3N 120 29W
Nipton, *U.S.A.* **161 K11** 35 28N 115 16W
Niquelândia, *Brazil* **171 D2** 14 33 S 48 23W
Nīr, *Iran* **101 C12** 38 2N 47 59 E
Nira →, *India* **94 F2** 17 58N 75 8 E
Nirasaki, *Japan* **71 B10** 35 42N 138 27 E
Nirmal, *India* **94 E4** 19 3N 78 20 E
Nirmali, *India* **93 F12** 26 20N 86 35 E
Niš, *Serbia, Yug.* **50 C5** 43 19N 21 58 E
Nisa, *Portugal* **43 F3** 39 30N 7 41W
Nişāb, *Si. Arabia* **96 D5** 29 11N 44 43 E
Nişāb, *Yemen* **98 D4** 14 25N 46 29 E
Nišava →, *Serbia, Yug.* **50 C5** 43 20N 21 46 E
Niscemi, *Italy* **47 E7** 37 9N 14 23 E
Nishi-Sonogi-Hantō, *Japan* **70 E1** 32 55N 129 45 E
Nishinomiya, *Japan* **71 C7** 34 45N 135 20 E
Nishino'omote, *Japan* **69 J5** 30 43N 130 59 E
Nishio, *Japan* **71 C9** 34 52N 137 3 E
Nishiwaki, *Japan* **70 C6** 34 59N 134 58 E
Nísíros, *Greece* **49 E9** 36 35N 27 12 E
Niška Banja, *Serbia, Yug.* **50 C6** 43 18N 22 1 E
Niskibi →, *Canada* **140 A2** 56 29N 88 9W
Nisko, *Poland* **55 H9** 50 35N 22 7 E
Nisporeni, *Moldova* **53 C13** 47 4N 28 10 E
Nisqually →, *U.S.A.* **160 C4** 47 6N 122 42W
Nissáki, *Greece* **38 A3** 39 43N 19 52 E
Nissan →, *Sweden* **17 H6** 56 40N 12 51 E
Nissedal, *Norway* **18 E5** 59 10N 8 30 E
Nisser, *Norway* **18 E5** 59 7N 8 28 E
Nissum Bredning, *Denmark* **17 H2** 56 40N 8 20 E
Nissum Fjord, *Denmark* **17 H2** 56 20N 8 11 E
Nistru = Dnister →, *Europe* **59 J6** 46 18N 30 17 E
Nisutlin →, *Canada* **142 A2** 60 14N 132 34W
Nitchequon, *Canada* **141 B5** 53 10N 70 58W
Niterói, *Brazil* **171 F3** 22 52 S 43 0W
Nith →, *U.K.* **22 F5** 55 14N 3 33W
Nitra, *Slovak Rep.* **35 C11** 48 19N 18 4 E
Nitra →, *Slovak Rep.* **35 D11** 47 46N 18 10 E
Nitriansky □, *Slovak Rep.* **35 C11** 48 10N 18 30 E
Nittenau, *Germany* **31 F8** 49 12N 12 16 E
Niu, *U.S.A.* **145 K14** 21 19N 157 44W
Niuafo'ou, *Tonga* **123 D15** 15 30 S 175 58W
Niue, *Cook Is.* **135 J11** 19 2 S 169 54W
Niulan Jiang →, *China* **76 D4** 27 30N 103 5 E
Niut, *Indonesia* **85 B4** 0 55N 110 6 E
Niutou Shan, *China* **77 C13** 29 5N 121 59 E
Niuzhuang, *China* **75 D12** 40 58N 122 28 E
Nivala, *Finland* **14 E21** 63 56N 24 57 E
Nivelles, *Belgium* **24 D4** 50 35N 4 20 E
Nivernais, *France* **27 E10** 47 15N 3 30 E
Nixon, *U.S.A.* **155 L6** 29 16N 97 46W
Nizam Sagar, *India* **94 E3** 18 10N 77 58 E
Nizamabad, *India* **94 E4** 18 45N 78 7 E
Nizamghat, *India* **90 A5** 28 20N 95 45 E
Nizhne Kolymsk, *Russia* **65 C17** 68 34N 160 55 E
Nizhnegorskiy = Nyzhnohirskyy, *Ukraine* **59 K8** 45 27N 34 38 E
Nizhnekamsk, *Russia* **60 C10** 55 38N 51 49 E
Nizhnekamskoye Vdkhr., *Russia* **62 D4** 55 56N 52 56 E
Nizhneudinsk, *Russia* **65 D10** 54 54N 99 3 E
Nizhnevartovsk, *Russia* **64 C8** 60 56N 76 38 E
Nizhniy Chir, *Russia* **61 F6** 48 22N 43 5 E
Nizhniy Lomov, *Russia* **60 D6** 53 34N 43 38 E
Nizhniy Novgorod, *Russia* **60 B7** 56 20N 44 0 E
Nizhniy Tagil, *Russia* **62 C7** 57 55N 59 57 E
Nizhniye Sergi, *Russia* **62 C7** 56 40N 59 18 E
Nizhnyaya Salda, *Russia* **62 B8** 58 8N 60 42 E
Nizhyn, *Ukraine* **59 G6** 51 5N 31 55 E
Nizina Mazowiecka, *Poland* **55 F8** 52 30N 21 0 E
Nizip, *Turkey* **100 D7** 37 5N 37 50 E
Nízké Tatry, *Slovak Rep.* **35 C12** 48 55N 19 30 E
Nízký Jeseník, *Czech Rep.* **35 B10** 49 50N 17 30 E
Nizza Monferrato, *Italy* **44 D5** 44 46N 8 21 E
Njakwa, *Malawi* **119 E3** 11 1 S 33 56 E
Njanji, *Zambia* **119 E3** 14 25 S 31 46 E
Njarðvíkur, *Iceland* **11 D4** 63 59N 22 32W
Njegoš, *Montenegro, Yug.* **50 D2** 42 53N 18 45 E
Njinjo, *Tanzania* **119 D4** 8 48 S 38 54 E
Njombe, *Tanzania* **119 D3** 9 20 S 34 50 E
Njombe □, *Tanzania* **119 D3** 9 20 S 34 49 E
Njombe →, *Tanzania* **118 D4** 6 56 S 35 6 E
Njurundabommen, *Sweden* **16 B11** 62 15N 17 25 E
Nkambe, *Cameroon* **113 D7** 6 35N 10 40 E
Nkana, *Zambia* **119 E2** 12 50 S 28 8 E
Nkawkaw, *Ghana* **113 D4** 6 36N 0 49W
Nkayi, *Zimbabwe* **119 F2** 19 41 S 29 20 E
Nkhotakota, *Malawi* **119 E3** 12 56 S 34 15 E
Nkolabona, *Gabon* **114 B2** 1 14N 11 43 E
Nkone, *Dem. Rep. of the Congo* **114 C4** 1 2 S 22 20 E
Nkongsamba, *Cameroon* **113 E6** 4 55N 9 55 E
Nkunga, *Dem. Rep. of the Congo* **115 C3** 4 41 S 18 34 E
Nkurenkuru, *Namibia* **116 B2** 17 42 S 18 32 E
Nkwanta, *Ghana* **112 D4** 6 10N 2 10W
Noakhali = Maijdi, *Bangla.* **90 D3** 22 48N 91 10 E
Nobel, *Canada* **150 A4** 45 25N 80 6W
Nobeoka, *Japan* **70 E3** 32 36N 131 41 E
Nōbi-Heiya, *Japan* **71 B8** 35 15N 136 45 E
Noble, *U.S.A.* **157 F8** 38 42N 88 14W
Noblejas, *Spain* **42 F7** 39 58N 3 26W
Noblesville, *U.S.A.* **157 D11** 40 3N 86 1W
Nocatee, *U.S.A.* **153 H8** 27 10N 81 53W
Noce →, *Italy* **44 B8** 46 9N 11 4 E
Nocera Inferiore, *Italy* **47 B7** 40 44N 14 38 E
Nocera Umbra, *Italy* **45 E9** 43 5N 12 47 E
Noci, *Italy* **47 B10** 40 48N 17 7 E
Nockatunga, *Australia* **127 D3** 27 42 S 142 42 E
Nocona, *U.S.A.* **155 J6** 33 47N 97 44W
Nocrich, *Romania* **53 E9** 45 55N 24 26 E
Noda, *Japan* **71 B11** 35 56N 139 52 E
Nodeland, *Norway* **18 F4** 58 8N 7 51 E
Noel, *U.S.A.* **155 G7** 36 33N 94 29W
Nogal Valley, *Somali Rep.* **120 C3** 8 35N 48 35 E
Nogales, *Mexico* **162 A2** 31 20N 110 56W
Nogales, *U.S.A.* **159 L8** 31 20N 110 56W
Nogaro, *France* **28 E3** 43 45N 0 2W
Nogat →, *Poland* **54 D6** 54 17N 19 17 E
Nōgata, *Japan* **70 D2** 33 48N 130 44 E
Nogent, *France* **27 D12** 48 1N 5 20 E
Nogent-le-Rotrou, *France* **26 D7** 48 20N 0 50 E
Nogent-sur-Seine, *France* **27 D10** 48 30N 3 30 E
Noggerup, *Australia* **125 F2** 33 32 S 116 5 E
Noginsk, *Moskva, Russia* **58 E10** 55 50N 38 25 E
Noginsk, *Tunguska, Russia* **65 C10** 64 30N 90 50 E
Nogoa →, *Australia* **126 C4** 23 40 S 147 55 E
Nogoyá, *Argentina* **174 C4** 32 24 S 59 48W
Nógrád □, *Hungary* **52 C4** 48 0N 19 30 E
Noguera Pallaresa →, *Spain* **40 D5** 41 55N 0 55 E
Noguera Ribagorzana →, *Spain* **40 D5** 41 40N 0 43 E
Nohar, *India* **92 E6** 29 11N 74 49 E
Nohfelden, *Germany* **31 F3** 49 35N 7 7 E
Nohili Pt., *U.S.A.* **145 A2** 22 4N 159 47W
Noia, *Spain* **42 C2** 42 48N 8 53W
Noing, *Phil.* **81 J5** 5 40N 125 28 E
Noire, Montagne, *France* **28 E6** 43 28N 2 18 E
Noire, Mt., *France* **26 D3** 48 11N 3 40W
Noirétable, *France* **28 C7** 45 48N 3 46 E
Noirmoutier, Î. de, *France* **26 F4** 46 58N 2 10W
Noirmoutier-en-l'Île, *France* **26 F4** 47 0N 2 14W
Nojane, *Botswana* **116 C3** 23 15 S 20 14 E
Nojima-Zaki, *Japan* **71 C11** 34 54N 139 53 E
Nok Kundi, *Pakistan* **91 C1** 28 50N 62 45 E
Nokaneng, *Botswana* **116 B3** 19 40 S 22 17 E
Nokia, *Finland* **15 F20** 61 30N 23 30 E
Nokomis, *Canada* **143 C8** 51 35N 105 0W
Nokomis, *Fla., U.S.A.* **153 H7** 27 7N 82 27W
Nokomis, *Ill., U.S.A.* **156 E7** 39 18N 89 18W
Nokomis L., *Canada* **143 B8** 57 0N 103 0W
Nokou, *Chad* **109 F2** 14 35N 14 47 E
Nol, *Sweden* **17 G6** 57 56N 12 5 E
Nola, *C.A.R.* **114 B3** 3 35N 16 4 E
Nola, *Italy* **47 B7** 40 55N 14 33 E
Nolay, *France* **27 F11** 46 58N 4 35 E
Noli, C. di, *Italy* **44 D5** 44 12N 8 25 E
Nolinsk, *Russia* **60 B9** 57 28N 49 57 E
Noma, *U.S.A.* **152 E4** 30 59N 85 37W
Noma Omuramba →, *Namibia* **116 B3** 18 52 S 20 53 E
Noma-Saki, *Japan* **70 F2** 31 25N 130 7 E
Nomad, *Papua N. G.* **132 D2** 6 19 S 142 13 E
Noman L., *Canada* **143 A7** 62 15N 108 55W
Nombre de Dios, *Panama* **164 E4** 9 34N 79 28W
Nome, *U.S.A.* **138 B3** 64 30N 165 25W
Nomo-Zaki, *Japan* **70 E1** 32 35N 129 44 E
Nomuka, *Tonga* **133 Q13** 20 17 S 174 48W
Nomuka Group, *Tonga* **133 Q13** 20 20 S 174 48W
Nonacho L., *Canada* **143 A7** 61 42N 109 40W
Nonancourt, *France* **26 D8** 48 47N 1 11 E
Nonda, *Australia* **126 C3** 20 40 S 142 28 E
Nondalton, *U.S.A.* **144 G9** 60 0N 154 51W
None, *Italy* **44 D4** 44 56N 7 32 E
Nong Chang, *Thailand* **86 E2** 15 23N 99 51 E
Nong Het, *Laos* **86 C4** 19 29N 103 59 E
Nong Khai, *Thailand* **86 D4** 17 50N 102 46 E
Nong'an, *China* **75 B13** 44 25N 125 5 E
Nongoma, *S. Africa* **117 D5** 27 58 S 31 35 E
Nonoava, *Mexico* **162 B3** 27 28N 106 44W
Nonoc I., *Phil.* **81 G5** 9 51N 125 37 E
Nonthaburi, *Thailand* **86 F3** 13 51N 100 34 E
Nontron, *France* **28 C4** 45 31N 0 40 E
Nonza, *France* **29 F13** 42 47N 9 21 E
Noonamah, *Australia* **124 B5** 12 40 S 131 4 E
Noonan, *U.S.A.* **154 A3** 48 54N 103 1W
Noondoo, *Australia* **127 D4** 28 35 S 148 30 E
Noonkanbah, *Australia* **124 C3** 18 30 S 124 50 E
Noord Brabant □, *Neths.* **24 C5** 51 40N 5 0 E
Noord Holland □, *Neths.* **24 B4** 52 30N 4 45 E
Noordbeveland, *Neths.* **24 C3** 51 35N 3 50 E
Noordoostpolder, *Neths.* **24 B5** 52 45N 5 45 E
Noordwijk, *Neths.* **24 B4** 52 14N 4 26 E
Noorvik Indian Reservation, *U.S.A.* **144 C7** 66 50N 161 5W
Nootka, *Canada* **142 D3** 49 38N 126 38W
Nootka I., *Canada* **142 D3** 49 32N 126 42W
Nóqui, *Angola* **115 D2** 5 55 S 13 30 E
Nora, *Eritrea* **107 D5** 16 6N 40 4 E
Nora, *Sweden* **16 E9** 59 32N 15 2 E
Nora Springs, *U.S.A.* **156 A4** 43 9N 93 1W
Noranda = Rouyn-Noranda, *Canada* **140 C4** 48 20N 79 0W
Norberg, *Sweden* **16 D9** 60 4N 15 56 E
Norborne, *U.S.A.* **156 E3** 39 18N 93 40W
Nórcia, *Italy* **45 F10** 42 48N 13 5 E
Norco, *U.S.A.* **161 M9** 33 56N 117 33W
Norcross, *U.S.A.* **152 B5** 33 56N 84 13W
Nord, *Greenland* **10 A9** 81 40N 17 20W
Nord □, *France* **27 B10** 50 15N 3 30 E
Nord-Ostsee-Kanal →, *Germany* **30 A5** 54 12N 9 32 E
Nord-Pas-de-Calais □, *France* **27 B9** 50 30N 2 50 E
Nordagutu, *Norway* **18 E6** 59 25N 9 20 E
Nordaustlandet, *Svalbard* **6 B9** 79 14N 23 0 E
Nordberg, *Norway* **18 C5** 61 54N 8 10 E
Nordborg, *Denmark* **17 J3** 55 5N 9 50 E
Nordby, *Denmark* **17 J2** 55 27N 8 24 E
Norddeich, *Germany* **30 B3** 53 36N 7 9 E
Nordegg, *Canada* **142 C5** 52 29N 116 5W
Norden, *Germany* **30 B3** 53 35N 7 12 E
Nordenham, *Germany* **30 B4** 53 30N 8 30 E
Norderhov, *Norway* **18 D7** 60 7N 10 17 E
Norderney, *Germany* **30 B3** 53 42N 7 9 E
Norderstedt, *Germany* **30 B6** 53 42N 10 1 E
Nordfjord, *Norway* **15 F11** 61 55N 5 30 E
Nordfjordeid, *Norway* **18 C2** 61 54N 6 0 E

258 Nordfriesische Inseln

Nordfriesische Inseln, *Germany* **30 A4** 54 40N 8 20 E
Nordgrønland □, *Greenland* **10 B5** 80 0N 55 0W
Nordhausen, *Germany* **30 D6** 51 30N 10 47 E
Nordhorn, *Germany* **30 C3** 52 26N 7 4 E
Norðoyar, *Færoe Is.* **14 E9** 62 17N 6 35W
Norðtunga, *Iceland* **11 C5** 64 44N 21 24W
Norður-múlasýla □, *Iceland* **11 B11** 65 30N 15 0W
Norður-þingeyjarsýsla, *Iceland* **11 A10** 66 0N 16 0W
Norðurbotn, *Iceland* **11 B3** 65 36N 23 46W
Norðurfjörður, *Iceland* **11 A5** 66 3N 21 34W
Nordingrå, *Sweden* **16 B12** 62 56N 18 17 E
Nordjyllands Amtskommune □, *Denmark* **17 G4** 57 20N 10 0 E
Nordkapp, *Norway* **14 A21** 71 10N 25 50 E
Nordkapp, *Svalbard* **6 A9** 80 31N 20 0 E
Nordkinn = Kinnarodden, *Norway* **12 A11** 71 8N 27 40 E
Nordkinn-halvøya, *Norway* **14 A22** 70 55N 27 40 E
Nördlingen, *Germany* **31 G6** 48 48N 10 30 E
Nordmøre, *Norway* **18 B4** 63 6N 8 15 E
Nordostrundingen, *Greenland* **10 A9** 81 22N 11 40W
Nordøyane, *Norway* **18 B3** 62 40N 6 0 E
Nordre Aputitê, *Greenland* **10 D7** 67 55N 32 0W
Nordre Osen, *Norway* **18 C8** 61 18N 11 45 E
Nordre Strømfjord, *Greenland* **10 D5** 67 10N 53 35W
Nordrhein-Westfalen □, *Germany* **30 D3** 51 45N 7 30 E
Nordsinni, *Norway* **18 D6** 60 50N 9 57 E
Nørdstedalsseter, *Norway* **18 C4** 61 40N 7 48 E
Nordstrand, *Germany* **30 A4** 54 30N 8 52 E
Nordvik, *Russia* **65 B12** 74 2N 111 32 E
Nore →, *Ireland* **23 D4** 52 25N 6 58W
Norefjell, *Norway* **18 D6** 60 16N 9 29 E
Norembega, *Canada* **140 C3** 48 59N 80 43W
Noresund, *Norway* **18 D6** 60 11N 9 37 E
Norfolk, *Nebr., U.S.A.* **154 D6** 42 2N 97 25W
Norfolk, *Va., U.S.A.* **148 G7** 36 51N 76 17W
Norfolk □, *U.K.* **21 E8** 52 39N 0 54 E
Norfolk I., *Pac. Oc.* **134 K8** 28 58 S 168 3 E
Norfork Res., *U.S.A.* **155 G8** 36 13N 92 15W
Norheimsund, *Norway* **18 D3** 60 23N 6 6 E
Norilsk, *Russia* **65 C9** 69 20N 88 6 E
Norley, *Australia* **127 D3** 27 45 S 143 48 E
Norma, Mt., *Australia* **126 C3** 20 55 S 140 42 E
Normal, *U.S.A.* **156 D8** 40 31N 88 59W
Norman, *U.S.A.* **155 H6** 35 13N 97 26W
Norman →, *Australia* **126 B3** 19 18 S 141 51 E
Norman Park, *U.S.A.* **152 D6** 31 16N 83 41W
Norman Wells, *Canada* **138 B7** 65 17N 126 51W
Normanby, *N.Z.* **130 F3** 39 32 S 174 18 E
Normanby →, *Australia* **126 A3** 14 23 S 144 10 E
Normanby I., *Papua N. G.* **132 F6** 10 5 S 151 5 E
Normandin, *Canada* **140 C5** 48 49N 72 31W
Normanhurst, Mt., *Australia* **125 E3** 25 4 S 122 30 E
Normanton, *Australia* **126 B3** 17 40 S 141 10 E
Normanville, *Australia* **128 C3** 35 27 S 138 18 E
Norquay, *Canada* **143 C8** 51 53N 102 5W
Norquinco, *Argentina* **176 B2** 41 51 S 70 55W
Norra Dellen, *Sweden* **16 C10** 61 53N 16 43 E
Norra Ulvön, *Sweden* **16 A12** 63 3N 18 40 E
Norrahammar, *Sweden* **17 G8** 57 43N 14 7 E
Norrbotten □, *Sweden* **14 C19** 66 30N 22 30 E
Nørre Åby, *Denmark* **17 J3** 55 27N 9 52 E
Nørre Alslev, *Denmark* **17 K5** 54 54N 11 52 E
Nørresundby, *Denmark* **17 G3** 57 5N 9 52 E
Norrhult, *Sweden* **17 G9** 57 7N 15 10 E
Norris, *U.S.A.* **158 D8** 45 34N 111 41W
Norris City, *U.S.A.* **157 G8** 37 59N 88 20W
Norristown, *Ga., U.S.A.* **152 C7** 32 30N 82 30W
Norristown, *Pa., U.S.A.* **151 F9** 40 7N 75 21W
Norrköping, *Sweden* **17 F10** 58 37N 16 11 E
Norrland, *Sweden* **15 E16** 62 15N 15 45 E
Norrsundet, *Sweden* **16 D11** 60 56N 17 8 E
Norrtälje, *Sweden* **16 E12** 59 46N 18 42 E
Norseman, *Australia* **125 F3** 32 8 S 121 43 E
Norsewood, *N.Z.* **130 G5** 40 3 S 176 13 E
Norsjø, *Norway* **18 E6** 59 18N 9 20 E
Norsk, *Russia* **65 D14** 52 30N 130 5 E
Norsup, *Vanuatu* **133 F5** 16 3 S 167 24 E
Norte, Pta., *Argentina* **176 B4** 42 5 S 63 46W
Norte, Pta. del, *Canary Is.* **39 G2** 27 51N 17 57W
Norte de Santander □, *Colombia* **168 B3** 8 0N 73 0W
Nortelândia, *Brazil* **173 C6** 14 25 S 56 48W
North, *U.S.A.* **152 B9** 33 37N 81 6W
North Adams, *U.S.A.* **151 D11** 42 42N 73 7W
North Atlanta, *U.S.A.* **152 B5** 33 52N 84 21W
North Atlantic Ocean, *Atl. Oc.* **8 F4** 30 0N 50 0W
North Augusta, *U.S.A.* **152 B9** 33 30N 81 59W
North Ayrshire □, *U.K.* **22 F4** 55 45N 4 44W
North Baltimore, *U.S.A.* **157 C13** 41 11N 83 41W
North Battleford, *Canada* **143 C7** 52 50N 108 17W
North Bay, *Canada* **140 C4** 46 20N 79 30W
North Belcher Is., *Canada* **140 A4** 56 50N 79 50W
North Bend, *Canada* **142 D4** 49 50N 121 27W
North Bend, *Oreg., U.S.A.* **158 E1** 43 24N 124 14W
North Bend, *Pa., U.S.A.* **150 E7** 41 20N 77 42W
North Bend, *Wash., U.S.A.* **160 C5** 47 30N 121 47W
North Berwick, *U.K.* **22 E6** 56 4N 2 42W
North Berwick, *U.S.A.* **151 C14** 43 18N 70 44W
North C., *Canada* **141 C7** 47 2N 60 20W
North C., *N.Z.* **130 A2** 34 23 S 173 4 E
North C., *Papua N. G.* **132 B6** 2 32 S 150 50 E
North Canadian →, *U.S.A.* **155 H7** 35 16N 95 31W
North Cape = Nordkapp, *Norway* **14 A21** 71 10N 25 50 E
North Cape = Nordkapp, *Svalbard* **6 A9** 80 31N 20 0 E
North Caribou L., *Canada* **140 B1** 52 50N 90 40W
North Carolina □, *U.S.A.* **149 H6** 35 30N 80 0W
North Channel, *Canada* **140 C3** 46 0N 83 0W
North Channel, *U.K.* **22 F3** 55 13N 5 52W
North Charleston, *U.S.A.* **152 C10** 32 53N 79 58W
North Chicago, *U.S.A.* **157 B9** 42 19N 87 51W
North College Hill, *U.S.A.* **157 E12** 39 13N 84 33W
North Cotabato □, *Phil.* **81 H5** 7 10N 125 0 E
North Dakota □, *U.S.A.* **154 B5** 47 30N 100 15W
North Dandalup, *Australia* **125 F2** 32 30 S 115 57 E
North Downs, *U.K.* **21 F8** 51 19N 0 21 E
North Druid Hills, *U.S.A.* **152 B5** 33 49N 84 19W
North East, *U.S.A.* **150 D5** 42 13N 79 50W
North East Frontier Agency = Arunachal Pradesh □, *India* **90 B5** 28 0N 95 0 E
North East Lincolnshire □, *U.K.* **20 D7** 53 34N 0 2W
North East Providence Chan., *W. Indies* **164 A4** 26 0N 76 0W
North Eastern □, *Kenya* **118 B5** 1 30N 40 0 E
North English, *U.S.A.* **156 C4** 41 31N 92 5W
North Esk →, *U.K.* **22 E6** 56 46N 2 24W
North European Plain, *Europe* **12 E10** 55 0N 25 0 E
North Fabius →, *U.S.A.* **156 E5** 39 54N 91 30W
North Foreland, *U.K.* **21 F9** 51 22N 1 28 E
North Fork, *U.S.A.* **160 H7** 37 14N 119 21W
North Fork, Salt →, *U.S.A.* **156 E5** 39 26N 91 53W
North Fork American →, *U.S.A.* **160 G5** 38 57N 120 59W
North Fork Edisto →, *U.S.A.* **152 B9** 33 16N 80 54W
North Fork Feather →, *U.S.A.* **160 F5** 38 33N 121 30W
North Fort Myers, *U.S.A.* **153 J8** 26 41N 81 53W
North Frisian Is. = Nordfriesische Inseln, *Germany* **30 A4** 54 40N 8 20 E
North Henik L., *Canada* **143 A9** 61 45N 97 40W
North Highlands, *U.S.A.* **160 G5** 38 40N 121 23W
North Horr, *Kenya* **118 B4** 3 20N 37 8 E
North I., *Kenya* **118 B4** 4 5N 36 5 E
North I., *N.Z.* **130 E4** 38 0 S 175 0 E
North Judson, *U.S.A.* **157 C10** 41 13N 86 46W
North Kingsville, *U.S.A.* **150 E4** 41 54N 80 42W
North Knife →, *Canada* **143 B10** 58 53N 94 45W
North Koel →, *India* **93 G10** 24 45N 83 50 E
North Korea ■, *Asia* **75 E14** 40 0N 127 0 E
North Lakhimpur, *India* **90 B5** 27 14N 94 7 E
North Lanarkshire □, *U.K.* **22 F5** 55 52N 3 56W
North Las Vegas, *U.S.A.* **161 J11** 36 12N 115 7W
North Liberty, *U.S.A.* **157 C10** 41 32N 86 26W
North Lincolnshire □, *U.K.* **20 D7** 53 36N 0 30W
North Little Rock, *U.S.A.* **155 H8** 34 45N 92 16W
North Loup →, *U.S.A.* **154 E5** 41 17N 98 24W
North Magnetic Pole, *Canada* **6 B2** 77 58N 102 8W
North Manchester, *U.S.A.* **157 D11** 41 0N 85 46W
North Miami, *U.S.A.* **153 K9** 25 54N 80 11W
North Miami Beach, *U.S.A.* **153 K9** 25 56N 80 10W
North Minch, *U.K.* **22 C3** 58 5N 5 55W
North Nahanni →, *Canada* **142 A4** 62 15N 123 20W
North Naples, *U.S.A.* **153 J8** 26 12N 81 48W
North New River Canal, *U.S.A.* **153 J9** 26 30N 80 30W
North Olmsted, *U.S.A.* **150 E3** 41 25N 81 56W
North Ossetia □, *Russia* **61 J7** 43 30N 44 30 E
North Pagai, I. = Pagai Utara, Pulau, *Indonesia* **84 C2** 2 35 S 100 0 E
North Palisade, *U.S.A.* **160 H8** 37 6N 118 31W
North Platte, *U.S.A.* **154 E4** 41 8N 100 46W
North Platte →, *U.S.A.* **154 E4** 41 7N 100 42W
North Pole, *Arctic* **6 A** 90 0N 0 0 E
North Pole, *U.S.A.* **144 D11** 64 45N 147 21W
North Portal, *Canada* **143 D8** 49 0N 102 33W
North Powder, *U.S.A.* **158 D5** 45 2N 117 55W
North Pt., *Canada* **141 C7** 47 5N 64 0W
North Pt., *Vanuatu* **133 D6** 14 56 S 168 6 E
North Rhine Westphalia □ = Nordrhein-Westfalen □, *Germany* **30 D3** 51 45N 7 30 E
North Ronaldsay, *U.K.* **22 B6** 59 22N 2 26W
North Saskatchewan →, *Canada* **143 C7** 53 15N 105 5W
North Sea, *Europe* **12 D6** 56 0N 4 0 E
North Somerset □, *U.K.* **21 F5** 51 24N 2 45W
North Sporades = Vóriai Sporádhes, *Greece* **48 B5** 39 15N 23 30 E
North Sydney, *Canada* **141 C7** 46 12N 60 15W
North Taranaki Bight, *N.Z.* **130 E3** 38 50 S 174 15 E
North Thompson →, *Canada* **142 C4** 50 40N 120 20W
North Tonawanda, *U.S.A.* **150 C6** 43 2N 78 53W
North Troy, *U.S.A.* **151 B12** 45 0N 72 24W
North Truchas Pk., *U.S.A.* **159 J11** 36 0N 105 30W
North Twin I., *Canada* **140 B4** 53 20N 80 0W
North Tyne →, *U.K.* **20 B5** 55 0N 2 8W
North Uist, *U.K.* **22 D1** 57 40N 7 15W
North Vancouver, *Canada* **142 D4** 49 25N 123 3W
North Vernon, *U.S.A.* **157 F11** 39 0N 85 38W
North Wabasca L., *Canada* **142 B6** 56 0N 113 55W
North Walsham, *U.K.* **20 E9** 52 50N 1 22 E
North Webster, *U.S.A.* **157 C11** 41 25N 85 48W
North-West □, *S. Africa* **116 D4** 27 0 S 25 0 E
North West C., *Australia* **124 D1** 21 45 S 114 9 E
North West Christmas I. Ridge, *Pac. Oc.* **135 G11** 6 30N 165 0W
North West Frontier □, *Pakistan* **91 B3** 34 0N 72 0 E
North West Highlands, *U.K.* **22 D4** 57 33N 4 58W
North West Providence Channel, *W. Indies* **164 A4** 26 0N 78 0W
North West River, *Canada* **141 B7** 53 30N 60 10W
North Western □, *Zambia* **119 E2** 13 30 S 25 30 E
North York Moors, *U.K.* **20 C7** 54 23N 0 53W
North Yorkshire □, *U.K.* **20 C6** 54 15N 1 25W
Northallerton, *U.K.* **20 C6** 54 20N 1 26W
Northam, *S. Africa* **116 C4** 24 56 S 27 18 E
Northampton, *Australia* **125 E1** 28 27 S 114 33 E
Northampton, *U.K.* **21 E7** 52 15N 0 53W
Northampton, *Mass., U.S.A.* **151 D12** 42 19N 72 38W
Northampton, *Pa., U.S.A.* **151 F9** 40 41N 75 30W
Northampton Downs, *Australia* **126 C4** 24 35 S 145 48 E
Northamptonshire □, *U.K.* **21 E7** 52 16N 0 55W
Northbridge, *U.S.A.* **151 D13** 42 9N 71 39W
Northcliffe, *Australia* **125 F2** 34 39 S 116 7 E
Northeast C., *U.S.A.* **144 E5** 63 18N 168 42W
Northeim, *Germany* **30 D6** 51 42N 10 0 E
Northern □, *Malawi* **119 E3** 11 0 S 34 0 E
Northern □, *Uganda* **118 B3** 3 5N 32 30 E
Northern □, *Zambia* **119 E3** 10 30 S 31 0 E
Northern Cape □, *S. Africa* **116 D3** 30 0 S 20 0 E
Northern Circars, *India* **94 F6** 17 30N 82 30 E
Northern Indian L., *Canada* **143 B9** 57 20N 97 20W
Northern Ireland □, *U.K.* **23 B5** 54 45N 7 0W
Northern Light, L., *Canada* **140 C1** 48 15N 90 39W
Northern Marianas ■, *Pac. Oc.* **134 F6** 17 0N 145 0 E
Northern Province □, *S. Leone* **112 D2** 9 15N 11 30W
Northern Samar □, *Phil.* **80 E5** 12 30N 124 40 E
Northern Territory □, *Australia* **124 D5** 20 0 S 133 0 E
Northern Transvaal □, *S. Africa* **117 C4** 24 0 S 29 0 E
Northfield, *U.S.A.* **154 C8** 44 27N 93 9W
Northland □, *N.Z.* **130 B2** 35 30 S 173 30 E
Northome, *U.S.A.* **154 B7** 47 52N 94 17W
Northport, *Ala., U.S.A.* **149 J2** 33 14N 87 35W
Northport, *Mich., U.S.A.* **148 C3** 45 8N 85 37W
Northport, *Wash., U.S.A.* **158 B5** 48 55N 117 48W
Northumberland □, *U.K.* **20 B6** 55 12N 2 0W
Northumberland, C., *Australia* **128 E4** 38 5 S 140 40 E
Northumberland Is., *Australia* **126 C4** 21 30 S 149 50 E
Northumberland Str., *Canada* **141 C7** 46 20N 64 0W
Northwest Territories □, *Canada* **138 B9** 67 0N 110 0W
Northwood, *Iowa, U.S.A.* **154 D8** 43 27N 93 13W
Northwood, *N. Dak., U.S.A.* **154 B6** 47 44N 97 34W
Norton, *U.S.A.* **154 F5** 39 50N 99 53W
Norton, *Zimbabwe* **119 F3** 17 52 S 30 40 E
Norton Sd., *U.S.A.* **138 B3** 63 50N 164 0W
Norton Shores, *U.S.A.* **157 A10** 43 8N 86 15W
Nortorf, *Germany* **30 A5** 54 10N 9 50 E
Norwalk, *Calif., U.S.A.* **161 M8** 33 54N 118 5W
Norwalk, *Conn., U.S.A.* **151 E11** 41 7N 73 22W
Norwalk, *Ohio, U.S.A.* **150 E2** 41 15N 82 37W
Norway, *Mich., U.S.A.* **148 C2** 45 47N 87 55W
Norway, *S.C., U.S.A.* **152 B9** 33 27N 81 7W
Norway ■, *Europe* **14 E14** 63 0N 11 0 E
Norway House, *Canada* **143 C9** 53 59N 97 50W
Norwegian Sea, *Atl. Oc.* **8 B9** 66 0N 1 0 E
Norwich, *Canada* **150 D4** 42 59N 80 36W
Norwich, *U.K.* **21 E9** 52 38N 1 18 E
Norwich, *Conn., U.S.A.* **151 E12** 41 31N 72 5W
Norwich, *N.Y., U.S.A.* **151 D9** 42 32N 75 32W
Norwood, *Canada* **150 B7** 44 23N 77 59W
Norwood, *U.S.A.* **157 E12** 39 10N 84 27W
Noshiro, *Japan* **68 D10** 40 12N 140 0 E
Nosivka, *Ukraine* **59 G6** 50 50N 31 37 E
Nosovka = Nosivka, *Ukraine* **59 G6** 50 50N 31 37 E
Noss Hd., *U.K.* **22 C5** 58 28N 3 3W
Nossa Senhora da Glória, *Brazil* **170 D4** 10 14 S 37 25W
Nossa Senhora das Dores, *Brazil* **170 D4** 10 29 S 37 13W
Nossa Senhora do Livramento, *Brazil* **173 D6** 15 48 S 56 22W
Nossebro, *Sweden* **17 F6** 58 12N 12 43 E
Nossob →, *S. Africa* **116 D3** 26 55 S 20 45 E
Nosy Boraha, *Madag.* **117 B8** 16 50 S 49 55 E
Nosy Varika, *Madag.* **117 C8** 20 35 S 48 32 E
Notasulga, *U.S.A.* **152 C4** 32 34N 85 41W
Noteć →, *Poland* **55 F2** 52 44N 15 26 E
Notigi Dam, *Canada* **143 B9** 56 40N 99 10W
Notikewin →, *Canada* **142 B5** 57 2N 117 38W
Notios Evvoïkos Kólpos, *Greece* **48 C5** 38 20N 24 0 E
Noto, *Italy* **47 F8** 36 53N 15 4 E
Noto, G. di, *Italy* **47 F8** 36 50N 15 12 E
Notodden, *Norway* **15 G13** 59 35N 9 17 E
Notre-Dame, *Canada* **141 C7** 46 18N 64 46W
Notre Dame B., *Canada* **141 C8** 49 45N 55 30W
Notre Dame de Koartac = Quaqtaq, *Canada* **139 B13** 60 55N 69 40W
Notre Dame d'Ivugivic = Ivujivik, *Canada* **139 B12** 62 24N 77 55W
Notsé, *Togo* **113 D5** 7 0N 1 17 E
Nottaway →, *Canada* **140 B4** 51 22N 78 55W
Nottingham, *U.K.* **20 E6** 52 58N 1 10W
Nottingham, City of □, *U.K.* **20 E6** 52 58N 1 10W
Nottinghamshire □, *U.K.* **20 D6** 53 10N 1 3W
Nottoway →, *U.S.A.* **148 G7** 36 33N 76 55W
Notwane →, *Botswana* **116 C4** 23 35 S 26 58 E
Nouâdhibou, *Mauritania* **110 D1** 20 54N 17 0W
Nouâdhibou, Ras, *Mauritania* **110 D1** 20 50N 17 0W
Nouakchott, *Mauritania* **112 B1** 18 9N 15 58W
Nouméa, *N. Cal.* **134 K8** 22 17 S 166 30 E
Noupoort, *S. Africa* **116 E3** 31 10 S 24 57 E
Nouveau Comptoir = Wemindji, *Canada* **140 B4** 53 0N 78 49W
Nouvelle-Calédonie = New Caledonia ■, *Pac. Oc.* **133 U19** 21 0 S 165 0 E
Nouzonville, *France* **27 C11** 49 48N 4 44 E
Nová Baňa, *Slovak Rep.* **35 C11** 48 28N 18 39 E
Nová Bystřice, *Czech Rep.* **34 B8** 49 2N 15 8 E
Nova Casa Nova, *Brazil* **170 C3** 9 25 S 41 5W
Nova Cruz, *Brazil* **170 C4** 6 28 S 35 25W
Nova Era, *Brazil* **171 E3** 19 45 S 43 3W
Nova Esperança, *Brazil* **175 A5** 23 8 S 52 24W
Nova Friburgo, *Brazil* **171 F3** 22 16 S 42 30W
Nova Gaia = Cambundi-Catembo, *Angola* **115 E3** 10 10 S 17 35 E
Nova Gorica, *Slovenia* **45 C10** 45 7N 13 39 E
Nova Gradiška, *Croatia* **52 E2** 45 17N 17 28 E
Nova Granada, *Brazil* **171 F2** 20 30 S 49 20W
Nova Iguaçu, *Brazil* **171 F3** 22 45 S 43 28W
Nova Iorque, *Brazil* **170 C3** 7 0 S 44 5W
Nova Kakhovka, *Ukraine* **59 J7** 46 42N 33 27 E
Nova Lamego, *Guinea-Biss.* **112 C2** 12 19N 14 11W
Nova Lima, *Brazil* **175 A7** 19 59 S 43 51W
Nova Lisboa = Huambo, *Angola* **115 E3** 12 42 S 15 54 E
Nova Lusitânia, *Mozam.* **119 F3** 19 50 S 34 34 E
Nova Mambone, *Mozam.* **117 C6** 21 0 S 35 3 E
Nova Odesa, *Ukraine* **59 J6** 47 19N 31 47 E
Nová Paka, *Czech Rep.* **34 A8** 50 29N 15 30 E
Nova Pavova, *Serbia, Yug.* **52 F5** 44 56N 20 14 E
Nova Ponte, *Brazil* **171 E2** 19 8 S 47 41W
Nova Scotia □, *Canada* **141 C7** 45 10N 63 0W
Nova Siri, *Italy* **47 B9** 40 10N 16 35 E
Nova Sofala, *Mozam.* **117 C5** 20 7 S 34 42 E
Nova Varoš, *Serbia, Yug.* **50 C3** 43 29N 19 48 E
Nova Venécia, *Brazil* **171 E3** 18 45 S 40 24W
Nova Vida, *Brazil* **173 C5** 10 11 S 62 47W
Nova Zagora, *Bulgaria* **51 D10** 42 32N 26 1 E
Novaci, *Macedonia* **50 E5** 41 5N 21 29 E
Novaci, *Romania* **53 E8** 45 10N 23 42 E
Novaféltria, *Italy* **45 E9** 43 53N 12 17 E
Novaleksandrovskaya = Novoaleksandrovsk, *Russia* **61 H5** 45 29N 41 17 E
Novannenskiy = Novoannenskiy, *Russia* **60 E6** 50 32N 42 39 E
Novara, *Italy* **44 C5** 45 28N 8 38 E
Novato, *U.S.A.* **160 G4** 38 6N 122 35W
Novaya Kakhovka = Nova Kakhovka, *Ukraine* **59 J7** 46 42N 33 27 E
Novaya Kazanka, *Kazakstan* **61 F9** 48 56N 49 36 E
Novaya Ladoga, *Russia* **58 B7** 60 7N 32 16 E
Novaya Lyalya, *Russia* **62 B8** 59 4N 60 45 E
Novaya Sibir, Ostrov, *Russia* **65 B16** 75 10N 150 0 E
Novaya Zemlya, *Russia* **64 B6** 75 0N 56 0 E
Nové Město, *Slovak Rep.* **35 C10** 48 45N 17 50 E
Nové Město na Moravě, *Czech Rep.* **34 B9** 49 34N 16 5 E
Nové Město nad Metují, *Czech Rep.* **35 A9** 50 20N 16 10 E
Nové Zámky, *Slovak Rep.* **35 C11** 48 2N 18 8 E
Novelda, *Spain* **41 G4** 38 24N 0 45W
Novellara, *Italy* **44 D7** 44 51N 10 44 E
Novelty, *U.S.A.* **156 D4** 40 1N 92 12W
Noventa Vicentina, *Italy* **45 C8** 45 17N 11 32 E
Novgorod, *Russia* **58 C6** 58 30N 31 25 E
Novgorod-Severskiy = Novhorod-Siverskyy, *Ukraine* **59 G7** 52 2N 33 10 E
Novhorod-Siverskyy, *Ukraine* **59 G7** 52 2N 33 10 E
Novi Bečej, *Serbia, Yug.* **52 E5** 45 36N 20 10 E
Novi Iskar, *Bulgaria* **50 D7** 42 48N 23 21 E
Novi Kneževac, *Serbia, Yug.* **52 D5** 46 4N 20 8 E
Novi Lígure, *Italy* **44 D5** 44 46N 8 47 E
Novi Pazar, *Bulgaria* **51 C11** 43 25N 27 15 E
Novi Pazar, *Serbia, Yug.* **50 C4** 43 12N 20 28 E
Novi Sad, *Serbia, Yug.* **52 E4** 45 18N 19 52 E
Novi Slankamen, *Serbia, Yug.* **52 E5** 45 8N 20 15 E
Novi Travnik, *Bos.-H.* **52 F2** 44 10N 17 40 E
Novi Vinodolski, *Croatia* **45 C11** 45 10N 14 48 E
Novigrad, *Istra, Croatia* **45 C10** 45 19N 13 33 E
Novigrad, *Zadar, Croatia* **45 D12** 44 10N 15 32 E
Novigradsko More, *Croatia* **45 D12** 44 12N 15 33 E
Novinger, *U.S.A.* **156 D4** 40 14N 92 43W
Novo Acôrdo, *Brazil* **170 D2** 10 10 S 46 48W
Novo Aripuanã, *Brazil* **169 E5** 5 8 S 60 22W
Nôvo Cruzeiro, *Brazil* **171 E3** 17 29 S 41 53W
Nôvo Hamburgo, *Brazil* **175 B5** 29 37 S 51 7W
Novo Horizonte, *Brazil* **171 F2** 21 25 S 49 10W
Novo Mesto, *Slovenia* **45 C12** 45 47N 15 12 E
Novo Miloševo, *Serbia, Yug.* **52 E5** 45 42N 20 20 E
Novo Remanso, *Brazil* **170 C3** 9 41 S 42 4W
Novo-Sergiyevskiy, *Russia* **62 E4** 52 5N 53 38 E
Novoaleksandrovsk, *Russia* **61 H5** 45 29N 41 17 E
Novoalekseyevka, *Kazakstan* **62 F5** 50 8N 55 39 E
Novoannenskiy, *Russia* **60 E6** 50 32N 42 39 E
Novoataysk, *Russia* **64 D9** 53 30N 84 0 E
Novoazovsk, *Ukraine* **59 J10** 47 15N 38 4 E
Novocheboksarsk, *Russia* **60 B8** 56 5N 47 27 E
Novocherkassk, *Russia* **61 G5** 47 27N 40 15 E
Novodevichye, *Russia* **60 D9** 53 37N 48 50 E
Novogrudok = Navahrudak, *Belarus* **58 F3** 53 40N 25 50 E
Novohrad-Volynskyy, *Ukraine* **59 G4** 50 34N 27 35 E
Novokachalinsk, *Russia* **68 B6** 45 5N 132 0 E
Novokazalinsk = Zhangaqazaly, *Kazakstan* **64 E7** 45 48N 62 6 E
Novokhopersk, *Russia* **60 E5** 51 5N 41 39 E
Novokuybyshevsk, *Russia* **60 D9** 53 7N 49 58 E
Novokuznetsk, *Russia* **64 D9** 53 45N 87 10 E
Novomirgorod, *Ukraine* **59 H6** 48 45N 31 33 E
Novomoskovsk, *Russia* **58 E10** 54 5N 38 15 E
Novomoskovsk, *Ukraine* **59 H8** 48 33N 35 17 E
Novoorsk, *Russia* **62 F7** 51 21N 59 2 E
Novopolotsk = Navapolatsk, *Belarus* **58 E5** 55 32N 28 37 E
Novorossiysk, *Russia* **59 K9** 44 43N 37 46 E
Novorossiyskoye, *Kazakstan* **62 F7** 50 13N 58 0 E
Novorybnoye, *Russia* **65 B11** 72 50N 105 50 E
Novorzhev, *Russia* **58 D5** 57 3N 29 25 E
Novosej, *Albania* **50 E4** 41 56N 20 35 E
Novoselytsya, *Ukraine* **59 H4** 48 14N 26 15 E
Novoshakhtinsk, *Russia* **59 J10** 47 46N 39 58 E
Novosibirsk, *Russia* **64 D9** 55 0N 83 5 E
Novosibirskiye Ostrova, *Russia* **65 B15** 75 0N 142 0 E
Novosil, *Russia* **59 F9** 52 59N 37 2 E
Novosineglazovskiy, *Russia* **62 D8** 55 2N 61 21 E
Novosokolniki, *Russia* **58 D6** 56 20N 30 2 E
Novotitarovskaya, *Russia* **61 H4** 45 17N 39 2 E
Novotroitsk, *Russia* **62 F7** 51 10N 58 15 E
Novotroitskoye, *Kazakstan* **63 B6** 43 42N 73 46 E
Novoukrayinka, *Ukraine* **59 H6** 48 25N 31 30 E
Novouljanovsk, *Russia* **60 C9** 54 8N 48 24 E
Novouzensk, *Russia* **60 E9** 50 32N 48 17 E
Novovolynsk, *Ukraine* **59 G3** 50 45N 24 4 E
Novovoronezhskiy, *Russia* **59 G10** 51 19N 39 13 E
Novovyatsk, *Russia* **62 B2** 58 24N 49 45 E
Novozybkov, *Russia* **59 F6** 52 30N 32 0 E
Novska, *Croatia* **45 C14** 45 19N 17 0 E
Nový Bor, *Czech Rep.* **34 A7** 50 46N 14 35 E
Novy Bug = Novyy Buh, *Ukraine* **59 J7** 47 34N 32 29 E
Nový Bydžov, *Czech Rep.* **34 A8** 50 14N 15 29 E
Novy Dwór Mazowiecki, *Poland* **55 F7** 52 26N 20 44 E
Nový Jičín, *Czech Rep.* **35 B11** 49 30N 18 2 E
Novyy Afon, *Georgia* **61 J5** 43 7N 40 50 E
Novyy Buh, *Ukraine* **59 J7** 47 34N 32 29 E
Novyy Oskol, *Russia* **59 G9** 50 44N 37 55 E
Novyy Port, *Russia* **64 C8** 67 40N 72 30 E
Now Shahr, *Iran* **97 B6** 36 40N 51 30 E
Nowa Deba, *Poland* **55 H8** 50 26N 21 41 E
Nowa Nowa, *Australia* **129 D8** 37 44 S 148 3 E
Nowa Ruda, *Poland* **55 H3** 50 35N 16 30 E
Nowa Sarzyna, *Poland* **55 H9** 50 21N 22 21 E
Nowa Sól, *Poland* **55 G2** 51 48N 15 44 E
Nowbarān, *Iran* **97 C6** 35 8N 49 42 E
Nowe, *Poland* **54 E5** 53 41N 18 44 E
Nowe Miasteczko, *Poland* **55 G2** 51 42N 15 42 E
Nowe Miasto, *Poland* **55 G7** 51 38N 20 34 E
Nowe Miasto Lubawskie, *Poland* **54 E6** 53 27N 19 33 E
Nowe Skalmierzyce, *Poland* **55 G4** 51 43N 18 0 E
Nowe Warpno, *Poland* **54 E1** 53 42N 14 18 E
Nowendoc, *Australia* **129 A9** 31 32 S 151 44 E
Nowghāb, *Iran* **97 C8** 33 53N 59 4 E
Nowgong, *India* **90 B4** 26 20N 92 50 E
Nowingi, *Australia* **128 C5** 34 33 S 142 15 E
Nowogard, *Poland* **54 E2** 53 41N 15 10 E
Nowogród, *Poland* **55 E8** 53 14N 21 53 E
Nowogród Bobrzanski, *Poland* **55 G2** 51 48N 15 15 E
Nowogrodziec, *Poland* **55 G2** 51 12N 15 24 E
Nowra, *Australia* **129 C9** 34 53 S 150 35 E
Nowshera, *Pakistan* **91 B4** 34 0N 72 0 E
Nowy Dwór Gdański, *Poland* **54 D6** 54 13N 19 7 E
Nowy Sącz, *Poland* **55 J7** 49 40N 20 41 E
Nowy Sącz □, *Poland* **55 J7** 49 30N 20 30 E
Nowy Staw, *Poland* **54 D6** 54 13N 19 2 E
Nowy Targ, *Poland* **55 J7** 49 29N 20 2 E
Nowy Tomyśl, *Poland* **55 F3** 52 19N 16 10 E
Nowy Wiśnicz, *Poland* **55 J7** 49 55N 20 28 E
Noxen, *U.S.A.* **151 E8** 41 25N 76 4W
Noxon, *U.S.A.* **158 C6** 48 0N 115 43W
Noyant, *France* **26 E7** 47 30N 0 6 E
Noyers, *France* **27 E10** 47 40 N 4 0 E

Ohau, L. 259

Noyes I., *U.S.A.* **142 B2** 55 30N 133 40W
Noyon, *France* **27 C9** 49 34N 2 59 E
Noyon, *Mongolia* **74 C2** 43 2N 102 4 E
Nozay, *France* **26 E5** 47 34N 1 38W
Nsa, O. en →, *Algeria* **111 B6** 32 28N 5 24 E
Nsa, Plateau de, *Congo* **114 C3** 2 26 S 15 20 E
Nsah, *Congo* **114 C3** 2 22 S 15 19 E
Nsanje, *Malawi* **119 F4** 16 55 S 35 12 E
Nsawam, *Ghana* **113 D4** 5 50N 0 24W
Nsomba, *Zambia* **119 E2** 10 45 S 29 51 E
Nsopzup, *Burma* **90 C6** 25 51N 97 30 E
Nsukka, *Nigeria* **113 D6** 6 51N 7 29 E
Ntoum, *Gabon* **114 B1** 0 22N 9 47 E
Nu Jiang →, *China* **76 C1** 29 58N 97 25 E
Nu Shan, *China* **76 E2** 26 0N 99 20 E
Nuba Mts. = Nubah, Jibalan, *Sudan* **107 E3** 12 0N 31 0 E
Nubah, Jibalan, *Sudan* **107 E3** 12 0N 31 0 E
Nubia, *Africa* **104 D7** 21 0N 32 0 E
Nubian Desert = Nûbîya, Es Sahrâ en, *Sudan* **106 C3** 21 30N 33 30 E
Nûbîya, Es Sahrâ en, *Sudan* **106 C3** 21 30N 33 30 E
Ñuble □, *Chile* **174 D1** 37 0 S 72 0W
Nubledo, *Spain* **42 B5** 43 31N 5 52W
Nuboai, *Indonesia* **83 B5** 2 10 S 136 30 E
Nubra →, *India* **93 B7** 34 35N 77 35 E
Nucet, *Romania* **52 D7** 46 28N 22 35 E
Nueces →, *U.S.A.* **155 M6** 27 51N 97 30W
Nueltin L., *Canada* **143 A9** 60 30N 99 30W
Nueva, I., *Chile* **176 E3** 55 13 S 66 30W
Nueva Antioquia, *Colombia* **168 B4** 6 5N 69 26W
Nueva Asunción □, *Paraguay* **174 A3** 21 0 S 61 0W
Nueva Carteya, *Spain* **43 H6** 37 35N 4 28W
Nueva Ecija □, *Phil.* **80 D3** 15 35N 121 0 E
Nueva Esparta □, *Venezuela* **169 A5** 11 0N 64 0W
Nueva Gerona, *Cuba* **164 B3** 21 53N 82 49W
Nueva Imperial, *Chile* **176 A2** 38 45 S 72 58W
Nueva Palmira, *Uruguay* **174 C4** 33 52 S 58 20W
Nueva Rosita, *Mexico* **162 B4** 28 0N 101 11W
Nueva San Salvador, *El Salv.* **164 D2** 13 40N 89 18W
Nueva Tabarca, *Spain* **41 G4** 38 17N 0 30W
Nueva Vizcaya □, *Phil.* **80 C3** 16 20N 121 20 E
Nuéve de Julio, *Argentina* **174 D3** 35 30 S 61 0W
Nuevitas, *Cuba* **164 B4** 21 30N 77 20W
Nuevo, G., *Argentina* **176 B4** 43 0 S 64 30W
Nuevo Guerrero, *Mexico* **163 B5** 26 34N 99 15W
Nuevo Laredo, *Mexico* **163 B5** 27 30N 99 30W
Nuevo León □, *Mexico* **162 C5** 25 0N 100 0W
Nuevo Mundo, Cerro, *Bolivia* **172 E4** 21 55 S 66 53W
Nuevo Rocafuerte, *Ecuador* **168 D2** 0 55 S 75 27W
Nugget Pt., *N.Z.* **131 G4** 46 27 S 169 50 E
Nugrus, Gebel, *Egypt* **106 C3** 24 47N 34 35 E
Nuhaka, *N.Z.* **130 F6** 39 3 S 177 45 E
Nuits-St-Georges, *France* **27 E11** 47 10N 4 56 E
Nûk, *Greenland* **10 E5** 64 10N 51 46W
Nukey Bluff, *Australia* **127 E2** 32 26 S 135 29 E
Nukheila, *Sudan* **106 D2** 19 1N 26 21 E
Nukhuyb, *Iraq* **101 F10** 32 4N 42 3 E
Nuku'alofa, *Tonga* **133 Q14** 21 10 S 174 0W
Nukus, *Uzbekistan* **64 E6** 42 27N 59 41 E
Nules, *Spain* **40 F4** 39 51N 0 9W
Nullagine →, *Australia* **124 D3** 21 20 S 120 20 E
Nullarbor, *Australia* **125 F5** 31 28 S 130 55 E
Nullarbor Plain, *Australia* **125 F4** 31 10 S 129 0 E
Numalla, L., *Australia* **127 D3** 28 43 S 144 20 E
Numan, *Nigeria* **113 D7** 9 29N 12 3 E
Numata, *Japan* **71 A11** 36 45N 139 4 E
Numatinna →, *Sudan* **107 F2** 7 38N 27 20 E
Numazu, *Japan* **71 B10** 35 7N 138 51 E
Numbulwar, *Australia* **126 A2** 14 15 S 135 45 E
Numedal, *Norway* **18 D6** 60 6N 9 6 E
Numfoor, *Indonesia* **83 B4** 1 0 S 134 50 E
Numurkah, *Australia* **129 D6** 36 5 S 145 26 E
Nunaksaluk I., *Canada* **141 A7** 55 49N 60 20W
Nunavut □, *Canada* **139 B11** 66 0N 85 0W
Nungo, *Mozam.* **119 E4** 13 23 S 37 43 E
Nungwe, *Tanzania* **118 C3** 2 48 S 32 2 E
Nunivak I., *U.S.A.* **138 C3** 60 10N 166 30W
Nunkun, *India* **93 C7** 33 57N 76 2 E
Nùoro, *Italy* **46 B2** 40 20N 9 20 E
Núpur, *Iceland* **11 B3** 65 56N 23 36W
Nuqayy, Jabal, *Libya* **108 D3** 23 11N 19 30 E
Nuqûb, *Yemen* **98 D4** 14 59N 45 48 E
Nuquí, *Colombia* **168 B2** 5 42N 77 17W
Nūrābād, *Iran* **97 E8** 27 47N 57 12 E
Nurata, *Uzbekistan* **63 C2** 40 33N 65 41 E
Nuratau, Khrebet, *Uzbekistan* **63 C3** 40 40N 66 30 E
Nure →, *Italy* **44 C6** 45 3N 9 49 E
Nuremberg = Nürnberg, *Germany* **31 F7** 49 27N 11 3 E
Nūrestān, *Afghan.* **91 B3** 35 30N 70 45 E
Nuri, *Mexico* **162 B3** 28 2N 109 22W
Nurina, *Australia* **125 F4** 30 56 S 126 33 E
Nuriootpa, *Australia* **128 C3** 34 27 S 139 0 E
Nurlat, *Russia* **60 C10** 54 29N 50 45 E
Nurmes, *Finland* **14 E23** 63 33N 29 10 E
Nürnberg, *Germany* **31 F7** 49 27N 11 3 E
Nurra, La, *Italy* **46 B1** 40 45N 8 15 E
Nurran, L. = Terewah, L., *Australia* **127 D4** 29 52 S 147 35 E
Nurrari Lakes, *Australia* **125 E5** 29 1 S 130 5 E
Nurri, *Italy* **46 C2** 39 43N 9 14 E
Nürtingen, *Germany* **31 G5** 48 37N 9 19 E
Nurzec →, *Poland* **55 F9** 52 37N 22 25 E
Nus, *Italy* **44 C4** 45 45N 7 28 E
Nusa Barung, *Indonesia* **85 D4** 8 30 S 113 30 E
Nusa Kambangan, *Indonesia* **85 D3** 7 40 S 108 10 E
Nusa Tenggara Barat □, *Indonesia* **85 D5** 8 50 S 117 30 E
Nusa Tenggara Timur □, *Indonesia* **82 C2** 9 30 S 122 0 E
Nusaybin, *Turkey* **101 D9** 37 3N 41 10 E
Nushki, *Pakistan* **91 C2** 29 35N 66 0 E
Nutwood Downs, *Australia* **126 B1** 15 49 S 134 10 E
Nuuk, *Greenland* **10 E5** 64 10N 51 35W
Nuussauq = Kraulshavn, *Greenland* **10 C5** 74 8N 57 3W
Nuwakot, *Nepal* **93 E10** 28 10N 83 55 E
Nuwara Eliya, *Sri Lanka* **95 L5** 6 58N 80 48 E
Nuweiba', *Egypt* **106 B3** 28 59N 34 39 E
Nuweveldberge, *S. Africa* **116 E3** 32 10 S 21 45 E
Nuyts, C., *Australia* **125 F5** 32 2 S 132 21 E
Nuyts Arch., *Australia* **127 E1** 32 35 S 133 20 E
Nuzvid, *India* **94 F5** 16 47N 80 53 E
Nxau-Nxau, *Botswana* **116 B3** 18 57 S 21 4 E
Nyaake, *Liberia* **112 E3** 4 52N 7 37W
Nyack, *U.S.A.* **151 E11** 41 5N 73 55W
Nyah West, *Australia* **128 C5** 35 16 S 143 21 E
Nyahanga, *Tanzania* **118 C3** 2 20 S 33 37 E
Nyahua, *Tanzania* **118 D3** 5 25 S 33 23 E
Nyahururu, *Kenya* **118 B4** 0 2N 36 27 E
Nyainqentanglha Shan, *China* **72 D4** 30 0N 90 0 E
Nyakanazi, *Tanzania* **118 C3** 3 2 S 31 10 E
Nyakrom, *Ghana* **113 D4** 5 40N 0 50W
Nyâlâ, *Sudan* **107 E1** 12 2N 24 58 E
Nyamandhlovu, *Zimbabwe* **119 F2** 19 55 S 28 16 E
Nyambiti, *Tanzania* **118 C3** 2 48 S 33 27 E
Nyamwaga, *Tanzania* **118 C3** 1 27 S 34 33 E
Nyandekwa, *Tanzania* **118 C3** 3 57 S 32 32 E
Nyanding →, *Sudan* **107 F3** 8 40N 32 41 E
Nyandoma, *Russia* **58 B11** 61 40N 40 12 E
Nyanga →, *Gabon* **114 C2** 2 58 S 10 15 E
Nyangana, *Namibia* **116 B3** 18 0 S 20 40 E
Nyanguge, *Tanzania* **118 C3** 2 30 S 33 12 E
Nyankpala, *Ghana* **113 D4** 9 21N 0 58W
Nyanza, *Rwanda* **118 C2** 2 20 S 29 42 E
Nyanza □, *Kenya* **118 C3** 0 10 S 34 15 E
Nyanza-Lac, *Burundi* **118 C2** 4 21 S 29 36 E
Nyarling →, *Canada* **142 A6** 60 41N 113 23W
Nyasa, L., *Africa* **119 E3** 12 30 S 34 30 E
Nyasvizh, *Belarus* **59 F4** 53 14N 26 38 E
Nyaunglebin, *Burma* **90 G6** 17 52N 96 42 E
Nyazepetrovsk, *Russia* **62 C7** 56 3N 59 36 E
Nyazura, *Zimbabwe* **119 F3** 18 40 S 32 16 E
Nyazwidzi →, *Zimbabwe* **119 G3** 20 0 S 31 17 E
Nybergsund, *Norway* **18 C9** 61 15N 12 19 E
Nyborg, *Denmark* **17 J4** 55 18N 10 47 E
Nybro, *Sweden* **17 H9** 56 44N 15 55 E
Nyda, *Russia* **64 C8** 66 40N 72 58 E
Nyeboe Land, *Greenland* **10 A5** 82 0N 57 0W
Nyeri, *Kenya* **118 C4** 0 23 S 36 56 E
Nyerol, *Sudan* **107 F3** 8 41N 32 1 E
Nyhammar, *Sweden* **16 D8** 60 17N 14 58 E
Nyiel, *Sudan* **107 F3** 6 9N 31 13 E
Nyinahin, *Ghana* **112 D4** 6 43N 2 3W
Nyíradony, *Hungary* **52 C6** 47 41N 21 55 E
Nyírbátor, *Hungary* **52 C7** 47 49N 22 9 E
Nyíregyháza, *Hungary* **52 C6** 47 58N 21 47 E
Nykirke, *Norway* **18 D7** 60 54N 10 19 E
Nykøbing, *Storstrøm, Denmark* **17 K5** 54 56N 11 52 E
Nykøbing, *Vestsjælland, Denmark* **17 J5** 55 55N 11 40 E
Nykøbing, *Viborg, Denmark* **17 H2** 56 48N 8 51 E
Nyköping, *Sweden* **17 F11** 58 45N 17 1 E
Nykroppa, *Sweden* **16 E8** 59 37N 14 18 E
Nykvarn, *Sweden* **16 E11** 59 11N 17 25 E
Nyland, *Sweden* **16 A11** 63 1N 17 45 E
Nylstroom, *S. Africa* **117 C4** 24 42 S 28 22 E
Nymagee, *Australia* **129 B7** 32 7 S 146 20 E
Nymburk, *Czech Rep.* **34 A8** 50 10N 15 1 E
Nynäshamn, *Sweden* **17 F11** 58 54N 17 57 E
Nyngan, *Australia* **127 E4** 31 30 S 147 8 E
Nyoman = Neman →, *Lithuania* **15 J20** 55 25N 21 10 E
Nyon, *Switz.* **32 D2** 46 23N 6 14 E
Nyong →, *Cameroon* **113 E6** 3 17N 9 54 E
Nyons, *France* **29 D9** 44 22N 5 10 E
Nyora, *Australia* **129 E6** 38 20 S 145 41 E
Nyou, *Burkina Faso* **113 C4** 12 42N 2 1W
Nýrsko, *Czech Rep.* **34 B6** 49 18N 13 9 E
Nysa, *Poland* **55 H4** 50 30N 17 22 E
Nysa →, *Europe* **30 C10** 52 4N 14 46 E
Nysa Kłodzka →, *Poland* **55 H4** 50 49N 17 40 E
Nysäter, *Sweden* **16 E6** 59 17N 12 47 E
Nyseter, *Norway* **18 B5** 62 2N 8 20 E
Nyssa, *U.S.A.* **158 E5** 43 53N 117 0W
Nysted, *Denmark* **17 K5** 54 40N 11 44 E
Nytva, *Russia* **62 C5** 57 56N 55 20 E
Nyûgawa, *Japan* **70 D5** 33 56N 133 5 E
Nyunzu, *Dem. Rep. of the Congo* **118 D2** 5 57 S 27 58 E
Nyurba, *Russia* **65 C12** 63 17N 118 28 E
Nyzhnohirskyy, *Ukraine* **59 K8** 45 27N 34 38 E
Nzega, *Tanzania* **118 C3** 4 10 S 33 12 E
Nzega □, *Tanzania* **118 C3** 4 10 S 33 10 E
N'zérékoré, *Guinea* **112 D3** 7 49N 8 48W
Nzeto, *Angola* **115 D2** 7 10 S 12 52 E
Nzilo, Chutes de, *Dem. Rep. of the Congo* **115 E5** 10 18 S 25 27 E
Nzubuka, *Tanzania* **118 C3** 4 45 S 32 50 E

O

O Barco, *Spain* **42 C4** 42 23N 6 58W
O Carballiño, *Spain* **42 C2** 42 26N 8 5W
O Corgo, *Spain* **42 C3** 42 56N 7 25W
O Pino, *Spain* **42 C2** 42 56N 8 20W
O Porriño, *Spain* **42 C2** 42 10N 8 37W
Ō-Shima, *Fukuoka, Japan* **70 D2** 33 54N 130 25 E
Ō-Shima, *Nagasaki, Japan* **70 C1** 34 29N 129 33 E
Ō-Shima, *Shizuoka, Japan* **71 C11** 34 44N 139 24 E
Oa, Mull of, *U.K.* **22 F2** 55 35N 6 20W
Oacoma, *U.S.A.* **154 D5** 43 48N 99 24W
Oahe, L., *U.S.A.* **154 C4** 44 27N 100 24W
Oahe Dam, *U.S.A.* **154 C4** 44 27N 100 24W
Oahu, *U.S.A.* **146 H16** 21 28N 157 58W
Oak Creek, *Colo., U.S.A.* **158 F10** 40 16N 106 57W
Oak Creek, *Wis., U.S.A.* **157 B9** 42 52N 87 55W
Oak Harbor, *U.S.A.* **160 B4** 48 18N 122 39W
Oak Hill, *Fla., U.S.A.* **153 G9** 28 52N 80 51W
Oak Hill, *W. Va., U.S.A.* **148 G5** 37 59N 81 9W
Oak Lawn, *U.S.A.* **157 C9** 41 43N 87 44W
Oak Park, *Ga., U.S.A.* **152 C7** 32 22N 82 19W
Oak Park, *Ill., U.S.A.* **157 C9** 41 53N 87 47W
Oak Ridge, *U.S.A.* **149 G3** 36 1N 84 16W
Oak View, *U.S.A.* **161 L7** 34 24N 119 18W
Oakan-Dake, *Japan* **68 C12** 43 27N 144 10 E
Oakbank, *Australia* **128 B4** 33 4 S 140 33 E
Oakdale, *Calif., U.S.A.* **160 H6** 37 46N 120 51W
Oakdale, *La., U.S.A.* **155 K8** 30 49N 92 40W
Oakes, *U.S.A.* **154 B5** 46 8N 98 6W
Oakesdale, *U.S.A.* **158 C5** 47 8N 117 15W
Oakey, *Australia* **127 D5** 27 25 S 151 43 E
Oakfield, *U.S.A.* **152 D6** 31 47N 83 58W
Oakford, *U.S.A.* **156 D7** 40 6N 89 58W
Oakham, *U.K.* **21 E7** 52 40N 0 43W
Oakhurst, *U.S.A.* **160 H7** 37 19N 119 40W
Oakland, *Calif., U.S.A.* **160 H4** 37 49N 122 16W
Oakland, *Ill., U.S.A.* **157 E8** 39 39N 88 2W
Oakland, *Oreg., U.S.A.* **158 E2** 43 25N 123 18W
Oakland City, *U.S.A.* **157 F9** 38 20N 87 21W
Oakland Park, *U.S.A.* **153 J9** 26 10N 80 8W
Oaklands, *Australia* **129 C7** 35 34 S 146 10 E
Oakley, *Idaho, U.S.A.* **158 E7** 42 15N 113 53W
Oakley, *Kans., U.S.A.* **154 F4** 39 8N 100 51W
Oakley Creek, *Australia* **129 A8** 31 37 S 149 46 E
Oakover →, *Australia* **124 D3** 21 0 S 120 40 E
Oakridge, *U.S.A.* **158 E2** 43 45N 122 28W
Oaktown, *U.S.A.* **157 F9** 38 52N 87 27W
Oakville, *U.S.A.* **160 D3** 46 51N 123 14W
Oakwood, *U.S.A.* **157 C12** 41 6N 84 23W
Oamaru, *N.Z.* **131 F5** 45 5 S 170 59 E
Ōamishirasato, *Japan* **71 B12** 35 31N 140 18 E
Oancea, *Romania* **53 E12** 45 21N 27 42 E
Oarai, *Japan* **71 A12** 36 21N 140 34 E
Oasis, *Calif., U.S.A.* **161 M10** 33 28N 116 6W
Oasis, *Nev., U.S.A.* **160 H9** 37 29N 117 55W
Oates Land, *Antarctica* **7 C11** 69 0 S 160 0 E
Oatman, *U.S.A.* **161 K12** 35 1N 114 19W
Oaxaca, *Mexico* **163 D5** 17 2N 96 40W
Oaxaca □, *Mexico* **163 D5** 17 0N 97 0W
Ob →, *Russia* **64 C7** 66 45N 69 30 E
Oba, *Canada* **140 C3** 49 4N 84 7W
Obala, *Cameroon* **113 E7** 4 9N 11 32 E
Obama, *Fukui, Japan* **71 B7** 35 30N 135 45 E
Obama, *Nagasaki, Japan* **70 E2** 32 43N 130 13 E
Oban, *U.K.* **22 E3** 56 25N 5 29W
Obbia, *Somali Rep.* **120 C3** 5 25N 48 30 E
Obed, *Canada* **142 C5** 53 30N 117 10W
Ober-Aagau, *Switz.* **32 B5** 47 10N 7 45 E
Obera, *Argentina* **175 B4** 27 21 S 55 2W
Oberalppass, *Switz.* **33 C7** 46 39N 8 35 E
Oberalpstock, *Switz.* **33 C7** 46 45N 8 47 E
Oberammergau, *Germany* **31 H7** 47 36N 11 4 E
Oberasbach, *Germany* **31 F6** 49 25N 10 57 E
Oberbayern □, *Germany* **31 G7** 48 5N 11 50 E
Oberdrauburg, *Austria* **34 E5** 46 44N 12 58 E
Oberengadin, *Switz.* **33 C9** 46 35N 9 55 E
Oberentfelden, *Switz.* **32 B6** 47 21N 8 2 E
Oberfranken □, *Germany* **31 E7** 50 10N 11 20 E
Oberhausen, *Germany* **30 D2** 51 28N 6 51 E
Oberkirch, *Germany* **31 G4** 48 31N 8 4 E
Oberland, *Switz.* **32 C5** 46 35N 7 38 E
Oberlausitz, *Germany* **30 D10** 51 16N 14 18 E
Oberlin, *Kans., U.S.A.* **154 F4** 39 49N 100 32W
Oberlin, *La., U.S.A.* **155 K8** 30 37N 92 46W
Oberlin, *Ohio, U.S.A.* **150 E2** 41 18N 82 13W
Obernai, *France* **27 D14** 48 28N 7 30 E
Oberndorf, *Germany* **31 G4** 48 17N 8 34 E
Oberon, *Australia* **129 B8** 33 45 S 149 52 E
Oberösterreich □, *Austria* **34 C7** 48 10N 14 0 E
Oberpfalz □, *Germany* **31 F8** 49 20N 12 10 E
Oberpfälzer Wald, *Germany* **31 F8** 49 30N 12 30 E
Obersiggenthal, *Switz.* **33 B6** 47 29N 8 18 E
Oberstdorf, *Germany* **31 H6** 47 24N 10 15 E
Oberting, *Gabon* **114 C1** 0 22 S 9 46 E
Oberursel, *Germany* **31 E4** 50 11N 8 35 E
Oberwart, *Austria* **35 D9** 47 17N 16 12 E
Oberwil, *Switz.* **32 A5** 47 32N 7 33 E
Obi, Kepulauan, *Indonesia* **82 B3** 1 23 S 127 45 E
Obi Is. = Obi, Kepulauan, *Indonesia* **82 B3** 1 23 S 127 45 E
Obiaruku, *Nigeria* **113 D6** 5 51N 6 9 E
Óbidos, *Brazil* **169 D6** 1 50 S 55 30W
Óbidos, *Portugal* **43 F1** 39 19N 9 10W
Obihiro, *Japan* **68 C11** 42 56N 143 12 E
Obilatu, *Indonesia* **82 B3** 1 25 S 127 20 E
Obilnoye, *Russia* **61 G7** 47 32N 44 30 E
Obing, *Germany* **31 G8** 48 0N 12 24 E
Objat, *France* **28 C5** 45 16N 1 24 E
Oblong, *U.S.A.* **157 F9** 39 0N 87 55W
Obluchye, *Russia* **65 E14** 49 1N 131 4 E
Obninsk, *Russia* **58 E9** 55 8N 36 37 E
Obo, *C.A.R.* **114 A5** 5 20N 26 32 E
Obo, *Ethiopia* **107 G4** 3 46N 38 52 E
Oboa, Mt., *Uganda* **118 B3** 1 45N 34 45 E
Obock, *Djibouti* **107 E5** 12 0N 43 20 E
Oborniki, *Poland* **55 F3** 52 39N 16 50 E
Oborniki Śląskie, *Poland* **55 G3** 51 17N 16 53 E
Obouya, *Congo* **114 C3** 0 56 S 15 43 E
Oboyan, *Russia* **59 G9** 51 15N 36 21 E
Obozerskaya = Obozerskiy, *Russia* **64 C5** 63 34N 40 21 E
Obozerskiy, *Russia* **64 C5** 63 34N 40 21 E
Obrenovac, *Serbia, Yug.* **50 B4** 44 40N 20 11 E
O'Brien, *U.S.A.* **152 E7** 30 2N 82 57W
Obrovac, *Croatia* **45 D12** 44 11N 15 41 E
Obruk, *Turkey* **100 C5** 38 7N 33 12 E
Obrzycko, *Poland* **55 F3** 52 42N 16 32 E
Observatory Inlet, *Canada* **142 B3** 55 10N 129 54W
Obshchi Syrt, *Russia* **62 E4** 52 0N 53 0 E
Obskaya Guba, *Russia* **64 C8** 69 0N 73 0 E
Obuasi, *Ghana* **113 D4** 6 17N 1 40W
Obubra, *Nigeria* **113 D6** 6 8N 8 20 E
Obwalden □, *Switz.* **32 C6** 46 55N 8 15 E
Obzor, *Bulgaria* **51 D11** 42 50N 27 52 E
Ocala, *U.S.A.* **149 L4** 29 11N 82 8W
Ocamo →, *Venezuela* **169 C4** 2 48N 65 14W
Ocampo, *Mexico* **162 B3** 28 9N 108 24W
Ocaña, *Colombia* **168 B3** 8 15N 73 20W
Ocaña, *Spain* **42 F7** 39 55N 3 30W
Ocanomowoc, *U.S.A.* **154 D10** 43 7N 88 30W
Ocate, *U.S.A.* **155 G2** 36 11N 105 3W
Occidental, Cordillera, *Colombia* **168 C2** 5 0N 76 0W
Occidental, Cordillera, *Peru* **172 C3** 14 0 S 74 0W
Ocean City, *N.J., U.S.A.* **148 F8** 39 17N 74 35W
Ocean City, *Wash., U.S.A.* **160 C2** 47 4N 124 10W
Ocean I. = Banaba, *Kiribati* **134 H8** 0 45 S 169 50 E
Ocean Park, *U.S.A.* **160 D2** 46 30N 124 3W
Oceano, *U.S.A.* **161 K6** 35 6N 120 37W
Oceanport, *U.S.A.* **151 F10** 40 19N 74 3W
Oceanside, *U.S.A.* **161 M9** 33 12N 117 23W
Ochagavía, *Spain* **40 C3** 42 55N 1 5W
Ochakiv, *Ukraine* **59 J6** 46 37N 31 33 E
Ochamchira, *Georgia* **61 J5** 42 46N 41 32 E
Ocher, *Russia* **62 C5** 57 53N 54 42 E
Ochiai, *Japan* **70 B5** 35 1N 133 45 E
Ochil Hills, *U.K.* **22 E5** 56 14N 3 40W
Ochlocknee, *U.S.A.* **152 E5** 30 58N 84 3W
Ochlockonee →, *U.S.A.* **152 F5** 29 59N 84 26W
Ochopee, *U.S.A.* **153 K8** 25 54N 81 18W
Ochre River, *Canada* **143 C9** 51 4N 99 47W
Ochsenfurt, *Germany* **31 F6** 49 40N 10 4 E
Ochsenhausen, *Germany* **31 G5** 48 4N 9 57 E
Ocilla, *U.S.A.* **152 D6** 31 36N 83 15W
Ockelbo, *Sweden* **16 D10** 60 54N 16 45 E
Ocmulgee →, *U.S.A.* **152 D7** 31 58N 82 33W
Ocna Mureș, *Romania* **53 D8** 46 23N 23 55 E
Ocna Sibiului, *Romania* **53 E9** 45 52N 24 2 E
Ocnele Mari, *Romania* **53 E9** 45 8N 24 18 E
Ocnița, *Moldova* **53 B12** 48 25N 27 30 E
Ocoee, *U.S.A.* **153 G8** 28 34N 81 33W
Ocoña, *Peru* **172 D3** 16 26 S 73 8W
Ocoña →, *Peru* **172 D3** 16 28 S 73 8W
Oconee →, *U.S.A.* **152 D7** 31 58N 82 33W
Oconee, L., *U.S.A.* **152 B6** 33 28N 83 15W
Oconee National Forest, *U.S.A.* **152 B6** 33 15N 83 45W
Oconomowoc, *U.S.A.* **157 A8** 43 7N 88 30W
Oconto, *U.S.A.* **148 C2** 44 53N 87 52W
Oconto Falls, *U.S.A.* **148 C1** 44 52N 88 9W
Ocosingo, *Mexico* **163 D6** 17 10N 92 15W
Ocotal, *Nic.* **164 D2** 13 41N 86 31W
Ocotlán, *Mexico* **162 C4** 20 21N 102 42W
Ocreza →, *Portugal* **43 F3** 39 32N 7 50W
Ócsa, *Hungary* **52 C4** 47 17N 19 15 E
Octave, *U.S.A.* **159 J7** 34 10N 112 43W
Octeville, *France* **26 C5** 49 38N 1 40W
Ocumare del Tuy, *Venezuela* **168 A4** 10 7N 66 46W
Ocuri, *Bolivia* **173 D4** 18 45 S 65 50W
Oda, *Ghana* **113 D4** 5 50N 0 51W
Oda, *Ehime, Japan* **70 D4** 33 36N 132 53 E
Ōda, *Shimane, Japan* **70 B4** 35 11N 132 30 E
Oda, J., *Sudan* **106 C4** 20 21N 36 39 E
Ódáðahraun, *Iceland* **11 B9** 65 5N 17 0W
Ödåkra, *Sweden* **17 H6** 56 7N 12 45 E
Odate, *Japan* **68 D10** 40 16N 140 34 E
Odawara, *Japan* **71 B11** 35 20N 139 6 E
Odda, *Norway* **15 F12** 60 3N 6 35 E
Odder, *Denmark* **17 J4** 55 58N 10 10 E
Oddi, *Iceland* **11 D6** 63 46N 20 24W
Oddur, *Somali Rep.* **120 D2** 4 11N 43 52 E
Odei →, *Canada* **143 B9** 56 6N 96 54W
Odell, *U.S.A.* **157 D8** 41 0N 88 31W
Odemira, *Portugal* **43 H2** 37 35N 8 40W
Ödemiș, *Turkey* **49 C9** 38 15N 28 0 E
Odendaalsrus, *S. Africa* **116 D4** 27 48 S 26 45 E
Odensbacken, *Sweden* **16 E9** 59 10N 15 32 E
Odense, *Denmark* **17 J4** 55 22N 10 23 E
Odenwald, *Germany* **31 F5** 49 35N 9 0 E
Oder →, *Europe* **30 B10** 53 33N 14 38 E
Oder-Havel Kanal, *Germany* **30 C10** 52 52N 14 2 E
Oderzo, *Italy* **45 C9** 45 47N 12 29 E
Odesa, *Ukraine* **59 J6** 46 30N 30 45 E
Ödeshög, *Sweden* **17 F8** 58 14N 14 39 E
Odessa = Odesa, *Ukraine* **59 J6** 46 30N 30 45 E
Odessa, *Canada* **151 B8** 44 17N 76 43W
Odessa, *Mo., U.S.A.* **156 F3** 39 0N 93 57W
Odessa, *Tex., U.S.A.* **155 K3** 31 52N 102 23W
Odessa, *Wash., U.S.A.* **158 C4** 47 20N 118 41W
Odiakwe, *Botswana* **116 C4** 20 12 S 25 17 E
Odiel →, *Spain* **43 H4** 37 10N 6 55W
Odienné, *Ivory C.* **112 D3** 9 30N 7 34W
Odintsovo, *Russia* **58 E9** 55 39N 37 15 E
Odiongan, *Phil.* **80 E3** 12 24N 121 59 E
Odobești, *Romania* **53 E12** 45 43N 27 4 E
Odolanów, *Poland* **55 G4** 51 34N 17 40 E
O'Donnell, *Phil.* **80 D3** 15 21N 120 27 E
O'Donnell, *U.S.A.* **155 J4** 32 58N 101 50W
Odorheiu Secuiesc, *Romania* **53 D10** 46 21N 25 21 E
Odoyevo, *Russia* **58 F9** 53 56N 36 42 E
Odra = Oder →, *Europe* **30 B10** 53 33N 14 38 E
Odra →, *Europe* **54 E1** 53 33N 14 38 E
Odra →, *Spain* **42 C6** 42 14N 4 17W
Odum, *U.S.A.* **152 D7** 31 40N 82 2W
Odweina, *Somali Rep.* **120 C3** 9 25N 45 4 E
Odžaci, *Serbia, Yug.* **52 E4** 45 30N 19 17 E
Odžak, *Bos.-H.* **52 E3** 45 3N 18 18 E
Odzi, *Zimbabwe* **117 B5** 19 0 S 32 20 E
Oebisfelde, *Germany* **30 C6** 52 27N 10 57 E
Oeiras, *Brazil* **170 C3** 7 0 S 42 8W
Oeiras, *Portugal* **43 G1** 38 41N 9 18W
Oelrichs, *U.S.A.* **154 D3** 43 11N 103 14W
Oelsnitz, *Germany* **31 E8** 50 24N 12 10 E
Oelwein, *U.S.A.* **154 D9** 42 41N 91 55W
Oenpelli, *Australia* **124 B5** 12 20 S 133 4 E
Oetz, *Austria* **34 D3** 47 13N 10 53 E
Of, *Turkey* **101 B9** 40 59N 40 23 E
O'Fallon, *U.S.A.* **156 F6** 38 49N 90 42W
Ofanto →, *Italy* **47 A9** 41 22N 16 13 E
Offa, *Nigeria* **113 D5** 8 13N 4 42 E
Offaly □, *Ireland* **23 C4** 53 15N 7 30W
Offenbach, *Germany* **31 E4** 50 6N 8 44 E
Offenburg, *Germany* **31 G3** 48 28N 7 56 E
Offida, *Italy* **45 F10** 42 56N 13 41 E
Ofidhousa, *Greece* **49 E8** 36 33N 26 8 E
Ofotfjorden, *Norway* **14 B17** 68 27N 17 0 E
Ofte, *Norway* **18 E5** 59 30N 8 12 E
Ofu, *Amer. Samoa* **133 X25** 14 11 S 169 41W
Ōfunato, *Japan* **68 E10** 39 4N 141 43 E
Oga, *Japan* **68 E9** 39 55N 139 50 E
Oga-Hantō, *Japan* **68 E9** 39 58N 139 47 E
Ogaden, *Ethiopia* **120 C3** 7 30N 45 30 E
Ogahalla, *Canada* **140 B2** 50 6N 85 51W
Ōgaki, *Japan* **71 B8** 35 21N 136 37 E
Ogallala, *U.S.A.* **154 E4** 41 8N 101 43W
Ogan →, *Indonesia* **84 C2** 3 1 S 104 44 E
Ogasawara Gunto, *Pac. Oc.* **134 E6** 27 0N 142 0 E
Ogbomosho, *Nigeria* **113 D5** 8 1N 4 11 E
Ogden, *Iowa, U.S.A.* **156 B3** 42 2N 94 2W
Ogden, *Utah, U.S.A.* **158 F7** 41 13N 111 58W
Ogdensburg, *U.S.A.* **151 B9** 44 42N 75 30W
Ogeechee →, *U.S.A.* **152 D8** 31 50N 81 3W
Ogilby, *U.S.A.* **161 N12** 32 49N 114 50W
Oglesby, *U.S.A.* **156 C7** 41 18N 89 4W
Oglethorpe, *U.S.A.* **152 C5** 32 18N 84 4W
Oglio →, *Italy* **44 C7** 45 2N 10 39 E
Ogmore, *Australia* **126 C4** 22 37 S 149 35 E
Ognon →, *France* **27 E12** 47 16N 5 28 E
Ogoamas, *Indonesia* **82 A2** 0 50N 120 5 E
Ogoja, *Nigeria* **113 D6** 6 38N 8 39 E
Ogoki →, *Canada* **140 B2** 51 38N 85 57W
Ogoki L., *Canada* **140 B2** 50 50N 87 10W
Ogoki Res., *Canada* **140 B2** 50 45N 88 15W
Ogooué →, *Gabon* **114 C1** 1 0 S 9 0 E
Ōgori, *Japan* **70 C3** 34 6N 131 24 E
Ogosta →, *Bulgaria* **50 C7** 43 48N 23 55 E
Ogowe = Ogooué →, *Gabon* **114 C1** 1 0 S 9 0 E
Ogr = Sharafa, *Sudan* **107 E2** 11 59N 27 7 E
Ograzden, *Macedonia* **50 E6** 41 30N 22 50 E
Ogre, *Latvia* **15 H21** 56 49N 24 36 E
Ogrein, *Sudan* **106 D3** 17 55N 34 50 E
Ogulin, *Croatia* **45 C12** 45 16N 15 16 E
Ogun □, *Nigeria* **113 D5** 7 0N 3 0 E
Oguni, *Japan* **70 D3** 33 11N 131 8 E
Ögur, *Iceland* **11 A4** 66 2N 22 44W
Oguta, *Nigeria* **113 D6** 5 44N 6 44 E
Ogwashi-Uku, *Nigeria* **113 D6** 6 15N 6 30 E
Ogwe, *Nigeria* **113 E6** 5 0N 7 14 E
Ohai, *N.Z.* **131 F3** 45 55 S 168 0 E
Ohakune, *N.Z.* **130 F4** 39 24 S 175 24 E
Ohanet, *Algeria* **111 C6** 28 44N 8 46 E
Ōhara, *Japan* **71 B12** 35 15N 140 23 E
Ohata, *Japan* **68 D10** 41 24N 141 10 E
Ohatchee, *U.S.A.* **152 B4** 33 47N 86 0W
Ohau, L., *N.Z.* **131 E4** 44 15 S 169 53 E

260 Ohaupo

Ohaupo, *N.Z.* **130 D4** 37 56 S 175 20 E
Ohio □, *U.S.A.* **148 E3** 40 15N 82 45W
Ohio →, *U.S.A.* **148 G1** 36 59N 89 8W
Ohio City, *U.S.A.* **157 D12** 40 46N 84 37W
Ohiwa Harbour, *N.Z.* **130 D6** 37 59 S 177 10 E
Ohře →, *Czech Rep.* **34 A7** 50 30N 14 10 E
Ohre →, *Germany* **30 C7** 52 18N 11 46 E
Ohrid, *Macedonia* **50 E4** 41 8N 20 52 E
Ohridsko Jezero, *Macedonia* **50 E4** 41 8N 20 52 E
Ohrigstad, *S. Africa* **117 C5** 24 39 S 30 36 E
Öhringen, *Germany* **31 F5** 49 12N 9 31 E
Ohura, *N.Z.* **130 E3** 38 51 S 174 59 E
Oiapoque →, *Brazil* **169 C7** 4 8N 51 40W
Oikou, *China* **75 E9** 38 35N 117 42 E
Oil City, *U.S.A.* **150 E5** 41 26N 79 42W
Oildale, *U.S.A.* **161 K7** 35 25N 119 1W
Oinousa, *Greece* **49 C8** 38 33N 26 14 E
Oise □, *France* **27 C9** 49 28N 2 30 E
Oise →, *France* **27 C9** 49 0N 2 4 E
Ōita, *Japan* **70 D3** 33 14N 131 36 E
Ōita □, *Japan* **70 D3** 33 15N 131 30 E
Oiticica, *Brazil* **170 C3** 5 3 S 41 5W
Ojai, *U.S.A.* **161 L7** 34 27N 119 15W
Ojinaga, *Mexico* **162 B4** 29 34N 104 25W
Ojiya, *Japan* **69 F9** 37 18N 138 48 E
Ojos del Salado, Cerro, *Argentina* **174 B2** 27 0 S 68 40W
Oka →, *Russia* **60 B7** 56 20N 43 59 E
Okaba, *Indonesia* **83 C5** 8 6 S 139 42 E
Okahandja, *Namibia* **116 C2** 22 0 S 16 59 E
Okahukura, *N.Z.* **130 E4** 38 48 S 175 14 E
Okaihau, *N.Z.* **130 B2** 35 19 S 173 47 E
Okanagan L., *Canada* **142 D5** 50 0N 119 30W
Okandja, *Gabon* **114 C2** 0 35 S 13 45 E
Okanogan, *U.S.A.* **158 B4** 48 22N 119 35W
Okanogan →, *U.S.A.* **158 B4** 48 6N 119 44W
Okány, *Hungary* **52 D6** 46 52N 21 21 E
Okapa, *Papua N. G.* **132 D3** 6 38 S 145 39 E
Okaputa, *Namibia* **116 C2** 20 5 S 17 0 E
Okara, *Pakistan* **91 C4** 30 50N 73 31 E
Okarito, *N.Z.* **131 D5** 43 15 S 170 9 E
Okato, *N.Z.* **130 F2** 39 12 S 173 53 E
Okaukuejo, *Namibia* **116 B2** 19 10 S 16 0 E
Okavango Swamps, *Botswana* **116 B3** 18 45 S 22 45 E
Okawa, *Japan* **70 D2** 33 9N 130 21 E
Okawville, *U.S.A.* **156 F7** 38 26N 89 33W
Okaya, *Japan* **71 A10** 36 5N 138 10 E
Okayama, *Japan* **70 C5** 34 40N 133 54 E
Okayama □, *Japan* **70 C5** 35 0N 133 50 E
Okazaki, *Japan* **71 C9** 34 57N 137 10 E
Oke-Iho, *Nigeria* **113 D5** 8 1N 3 18 E
Okeechobee, *U.S.A.* **149 M5** 27 15N 80 50W
Okeechobee, L., *U.S.A.* **149 M5** 27 0N 80 50W
Okefenokee Swamp, *U.S.A.* **152 E7** 30 40N 82 20W
Okehampton, *U.K.* **21 G4** 50 44N 4 0W
Okene, *Nigeria* **113 D6** 7 32N 6 11 E
Oker →, *Germany* **30 C6** 52 32N 10 22 E
Okha, *Russia* **65 D15** 53 40N 143 0 E
Ókhi Óros, *Greece* **48 C6** 38 5N 24 25 E
Okhotsk, *Russia* **65 D15** 59 20N 143 10 E
Okhotsk, Sea of, *Asia* **65 D15** 55 0N 145 0 E
Okhotskiy Perevoz, *Russia* **65 C14** 61 52N 135 35 E
Okhtyrka, *Ukraine* **59 G8** 50 25N 35 0 E
Oki-no-Shima, *Japan* **70 E4** 32 44N 132 33 E
Oki-Shotō, *Japan* **70 A5** 36 5N 133 15 E
Okiep, *S. Africa* **116 D2** 29 39 S 17 53 E
Okigwi, *Nigeria* **113 D6** 5 52N 7 20 E
Okija, *Nigeria* **113 D6** 5 54N 6 55 E
Okinawa □, *Japan* **69 L4** 26 40N 128 0 E
Okinawa-Guntō, *Japan* **69 L4** 26 40N 128 0 E
Okinawa-Jima, *Japan* **69 L4** 26 32N 128 0 E
Okino-erabu-Shima, *Japan* **69 L4** 27 21N 128 33 E
Okitipupa, *Nigeria* **113 D5** 6 31N 4 50 E
Oklahoma □, *U.S.A.* **155 H6** 35 20N 97 30W
Oklahoma City, *U.S.A.* **155 H6** 35 30N 97 30W
Oklawaha →, *U.S.A.* **153 F8** 29 28N 81 41W
Oklawaha, L., *U.S.A.* **153 F8** 29 30N 81 45W
Okmulgee, *U.S.A.* **155 H7** 35 37N 95 58W
Oknitsa = Ocniţa, *Moldova* **53 B12** 48 25N 27 30 E
Okolo, *Uganda* **118 B3** 2 37N 31 8 E
Okolona, *Ky., U.S.A.* **157 F11** 38 8N 85 41W
Okolona, *Miss., U.S.A.* **155 J10** 34 0N 88 45W
Okonek, *Poland* **54 E3** 53 32N 16 51 E
Okrika, *Nigeria* **113 E6** 4 40N 7 10 E
Øksendal, *Norway* **18 B5** 62 42N 8 27 E
Oksibil, *Indonesia* **83 B6** 4 59 S 140 35 E
Øksna, *Norway* **18 D8** 60 58N 11 27 E
Oksovskiy, *Russia* **56 B6** 62 33N 39 57 E
Oktabrsk = Oktyabrsk, *Kazakstan* **57 E10** 49 28N 57 25 E
Oktyabr, *Kazakstan* **63 B8** 43 41N 77 12 E
Oktyabrsk, *Kazakstan* **57 E10** 49 28N 57 25 E
Oktyabrsk, *Russia* **60 D9** 53 11N 48 40 E
Oktyabrskiy = Aktsyabrski, *Belarus* **59 F5** 52 38N 28 53 E
Oktyabrskiy, *Bashkortostan, Russia* **62 D4** 54 28N 53 28 E
Oktyabrskiy, *Perm, Russia* **62 C6** 56 31N 57 12 E
Oktyabrskiy, *Rostov, Russia* **61 G5** 47 30N 40 4 E
Oktyabrskoy Revolyutsii, Ostrov, *Russia* **65 B10** 79 30N 97 0 E
Oktyabrskoye = Zhovtneve, *Ukraine* **59 J7** 46 54N 32 3 E
Oktyabrskoye, *Russia* **62 D9** 54 26N 62 44 E
Ōkuchi, *Japan* **70 E2** 32 4N 130 37 E
Okulovka, *Russia* **58 C7** 58 25N 33 19 E
Okuru, *N.Z.* **131 D3** 43 55 S 168 55 E
Okushiri-Tō, *Japan* **68 C9** 42 15N 139 30 E
Okuta, *Nigeria* **113 D5** 9 14N 3 12 E
Okwa →, *Botswana* **116 C3** 22 30 S 23 0 E
Ola, *U.S.A.* **155 H8** 35 2N 93 13W
Ólafsfjörður, *Iceland* **11 A8** 66 4N 18 39W
Ólafsvík, *Iceland* **11 C3** 64 53N 23 43W
Olaine, *Latvia* **54 B10** 56 48N 23 59 E
Olancha, *U.S.A.* **161 J8** 36 17N 118 1W
Olancha Pk., *U.S.A.* **161 J8** 36 15N 118 7W
Olanchito, *Honduras* **164 C2** 15 30N 86 30W
Öland, *Sweden* **17 H10** 56 45N 16 38 E
Ölands norra udde, *Sweden* **17 G11** 57 22N 17 5 E
Ölands södra udde, *Sweden* **17 H10** 56 12N 16 23 E
Olanta, *U.S.A.* **152 B10** 33 56N 79 56W
Olar, *U.S.A.* **152 B9** 33 11N 81 11W
Olargues, *France* **28 E6** 43 34N 2 53 E
Olary, *Australia* **128 B4** 32 18 S 140 19 E
Olascoaga, *Argentina* **174 D3** 35 15 S 60 39W
Olathe, *U.S.A.* **154 F7** 38 53N 94 49W
Olavarría, *Argentina* **174 D3** 36 55 S 60 20W
Oława, *Poland* **55 H4** 50 57N 17 20 E
Olbernhau, *Germany* **30 E9** 50 40N 13 19 E
Ólbia, *Italy* **46 B2** 40 55N 9 31 E
Ólbia, G. di, *Italy* **46 B2** 40 55N 9 39 E
Olching, *Germany* **31 G7** 48 12N 11 21 E
Old Bahama Chan. = Bahama, Canal Viejo de, *W. Indies* **164 B4** 22 10N 77 30W
Old Baldy Pk. = San Antonio, Mt., *U.S.A.* **161 L9** 34 17N 117 38W
Old Cork, *Australia* **126 C3** 22 57 S 141 52 E
Old Crow, *Canada* **138 B6** 67 30N 139 55W
Old Dale, *U.S.A.* **161 L11** 34 8N 115 47W
Old Dongola, *Sudan* **106 D3** 18 11N 30 44 E
Old Forge, *N.Y., U.S.A.* **151 C10** 43 43N 74 58W
Old Forge, *Pa., U.S.A.* **151 E9** 41 22N 75 45W
Old Fort →, *Canada* **143 B6** 58 36N 110 24W
Old Shinyanga, *Tanzania* **118 C3** 3 33 S 33 27 E
Old Speck Mt., *U.S.A.* **151 B14** 44 34N 70 57W
Old Town, *Fla., U.S.A.* **153 F7** 29 36N 82 59W
Old Town, *Maine, U.S.A.* **141 D6** 44 56N 68 39W
Old Wives L., *Canada* **143 C7** 50 5N 106 0W
Oldbury, *U.K.* **21 F5** 51 38N 2 33W
Oldcastle, *Ireland* **23 C4** 53 46N 7 10W
Oldeani, *Tanzania* **118 C4** 3 22 S 35 35 E
Olden, *Norway* **18 C3** 61 49N 6 49 E
Oldenburg, *Niedersachsen, Germany* **30 B4** 53 9N 8 13 E
Oldenburg, *Schleswig-Holstein, Germany* **30 A6** 54 17N 10 52 E
Oldenzaal, *Neths.* **24 B6** 52 19N 6 53 E
Oldham, *U.K.* **20 D5** 53 33N 2 7W
Oldman →, *Canada* **142 D6** 49 57N 111 42W
Oldmeldrum, *U.K.* **22 D6** 57 20N 2 19W
Olds, *Canada* **142 C6** 51 50N 114 10W
Oldsmar, *U.S.A.* **153 G7** 28 2N 82 40W
Ole Rømer Land, *Greenland* **10 C8** 74 10N 24 30W
Olean, *U.S.A.* **150 D6** 42 5N 78 26W
Olecko, *Poland* **54 D9** 54 2N 22 31 E
Oléggio, *Italy* **44 C5** 45 36N 8 38 E
Oleiros, *Portugal* **42 F3** 39 56N 7 56W
Oleiros, *Spain* **42 B2** 43 20N 8 19W
Olekma →, *Russia* **65 C13** 60 22N 120 42 E
Olekminsk, *Russia* **65 C13** 60 25N 120 30 E
Oleksandriya, *Kirovohrad, Ukraine* **59 H7** 48 42N 33 3 E
Oleksandriya, *Rivne, Ukraine* **59 G4** 50 37N 26 19 E
Oleksandrovka, *Ukraine* **59 H7** 48 55N 32 20 E
Olema, *U.S.A.* **160 G4** 38 3N 122 47W
Ølen, *Norway* **18 E2** 59 36N 5 48 E
Olenegorsk, *Russia* **56 A5** 68 9N 33 18 E
Olenek, *Russia* **65 C12** 68 28N 112 18 E
Olenek →, *Russia* **65 B13** 73 0N 120 10 E
Olenino, *Russia* **58 D7** 56 15N 33 30 E
Oléron, Î. d', *France* **28 C2** 45 55N 1 15W
Oleśnica, *Poland* **55 G4** 51 13N 17 22 E
Olesno, *Poland* **55 H5** 50 51N 18 26 E
Olevsk, *Ukraine* **59 G4** 51 12N 27 39 E
Olga, *Russia* **65 E14** 43 50N 135 14 E
Olga, L., *Canada* **140 C4** 49 47N 77 15W
Olga, Mt., *Australia* **125 E5** 25 20 S 130 50 E
Ølgod, *Denmark* **17 J2** 55 49N 8 36 E
Olhão, *Portugal* **43 H3** 37 3N 7 48W
Olib, *Croatia* **45 D11** 44 23N 14 44 E
Oliena, *Italy* **46 B2** 40 16N 9 24 E
Oliete, *Spain* **40 D4** 41 1N 0 41W
Olifants →, *Africa* **117 C5** 23 57 S 31 58 E
Olifantshoek, *S. Africa* **116 D3** 27 57 S 22 42 E
Ólimbos, *Greece* **49 F9** 35 44N 27 11 E
Ólimbos, Óros, *Greece* **50 F6** 40 6N 22 23 E
Olímpia, *Brazil* **175 A6** 20 44 S 48 54W
Olin, *U.S.A.* **156 C5** 42 0N 91 9W
Olinda, *Brazil* **170 C5** 8 1 S 34 51W
Olindiná, *Brazil* **170 D4** 11 22 S 38 21W
Olite, *Spain* **40 C3** 42 29N 1 40W
Oliva, *Argentina* **174 C3** 32 0 S 63 38W
Oliva, *Spain* **41 G4** 38 58N 0 9W
Oliva, Punta del, *Spain* **42 B5** 43 37N 5 28W
Oliva de la Frontera, *Spain* **43 G4** 38 17N 6 54W
Olivares, *Spain* **40 F2** 39 46N 2 20W
Olive Hill, *U.S.A.* **157 F13** 38 18N 83 13W
Olivehurst, *U.S.A.* **160 F5** 39 6N 121 34W
Oliveira, *Brazil* **171 F3** 20 39 S 44 50W
Oliveira de Azeméis, *Portugal* **42 E2** 40 49N 8 29W
Oliveira do Douro, *Portugal* **42 D2** 41 5N 8 2W
Oliveira dos Brejinhos, *Brazil* **171 D3** 12 19 S 42 54W
Olivenza, *Spain* **43 G3** 38 41N 7 9W
Oliver, *Canada* **142 D5** 49 13N 119 37W
Oliver, *U.S.A.* **152 C8** 32 31N 81 32W
Oliver L., *Canada* **143 B8** 56 56N 103 22W
Olivet, *France* **27 E8** 47 51N 1 55 E
Olivine Ra., *N.Z.* **131 E3** 44 15 S 168 30 E
Olivone, *Switz.* **33 C7** 46 32N 8 57 E
Olkhovka, *Russia* **60 F7** 49 48N 44 32 E
Olkusz, *Poland* **55 H6** 50 18N 19 33 E
Ollagüe, *Chile* **174 A2** 21 15 S 68 10W
Olmaliq, *Uzbekistan* **63 C4** 40 50N 69 35 E
Olmedo, *Spain* **42 D6** 41 20N 4 43W
Olmeto, *France* **29 G12** 41 43N 8 55 E
Olmos, *Peru* **172 B2** 5 59 S 79 46W
Olney, *Ill., U.S.A.* **157 F8** 38 44N 88 5W
Olney, *Tex., U.S.A.* **155 J5** 33 22N 98 45W
Olofström, *Sweden* **17 H8** 56 17N 14 32 E
Oloma, *Cameroon* **113 E7** 3 29N 11 19 E
Olomane →, *Canada* **141 B7** 50 14N 60 37W
Olombo, *Congo* **114 C3** 1 18 S 15 53 E
Olomouc, *Czech Rep.* **35 B10** 49 38N 17 12 E
Olonets, *Russia* **58 B7** 61 0N 32 54 E
Olongapo, *Phil.* **80 D3** 14 50N 120 18 E
Olonne-sur-Mer, *France* **28 B2** 46 32N 1 47W
Oloron, Gave d' →, *France* **28 E2** 43 33N 1 5W
Oloron-Ste-Marie, *France* **28 E3** 43 11N 0 38W
Olot, *Spain* **40 C7** 42 11N 2 30 E
Olovo, *Bos.-H.* **52 F3** 44 8N 18 35 E
Olovyannaya, *Russia* **65 D12** 50 58N 115 35 E
Olowalu, *U.S.A.* **145 C5** 20 49N 156 38W
Oloy →, *Russia* **65 C16** 66 29N 159 29 E
Olsberg, *Germany* **30 D4** 51 21N 8 31 E
Olshammar, *Sweden* **17 F8** 58 45N 14 48 E
Olshanka, *Ukraine* **59 H6** 48 16N 30 58 E
Olshany, *Ukraine* **59 G8** 50 3N 35 53 E
Olsztyn, *Poland* **54 E7** 53 48N 20 29 E
Olsztyn □, *Poland* **54 E7** 53 30N 20 30 E
Olsztynek, *Poland* **54 E7** 53 34N 20 19 E
Olt □, *Romania* **53 F9** 44 20N 24 30 E
Olt →, *Romania* **53 G9** 43 43N 24 51 E
Oltedal, *Norway* **18 F3** 58 49N 6 2 E
Olten, *Switz.* **32 B5** 47 21N 7 53 E
Olteniţa, *Romania* **53 F11** 44 7N 26 42 E
Olton, *U.S.A.* **155 H3** 34 11N 102 8W
Oltu, *Turkey* **101 B9** 40 35N 41 58 E
Olula del Río, *Spain* **41 H2** 37 21N 2 18W
Olur, *Turkey* **101 B10** 40 49N 42 8 E
Olustee, *U.S.A.* **152 E7** 30 12N 82 26W
Olutanga, *Phil.* **81 H4** 7 26N 122 54 E
Olutanga I., *Phil.* **81 H4** 7 22N 122 52 E
Olvega, *Spain* **40 D2** 41 47N 2 0W
Olvera, *Spain* **43 J5** 36 55N 5 18W
Olymbos, *Cyprus* **38 D12** 35 21N 33 45 E
Olympia, *Greece* **48 D3** 37 39N 21 39 E
Olympia, *U.S.A.* **160 D4** 47 3N 122 53W
Olympic Mts., *U.S.A.* **160 C3** 47 55N 123 45W
Olympic Nat. Park, *U.S.A.* **160 C3** 47 48N 123 30W
Olympus, *Cyprus* **38 E11** 34 56N 32 52 E
Olympus, Mt. = Ólimbos, Óros, *Greece* **50 F6** 40 6N 22 23 E
Olympus, Mt. = Uludağ, *Turkey* **51 F13** 40 4N 29 13 E
Olympus, Mt., *U.S.A.* **160 C3** 47 48N 123 43W
Olyphant, *U.S.A.* **151 E9** 41 27N 75 36W
Om →, *Russia* **64 D8** 54 59N 73 22 E
Om Hajer, *Eritrea* **107 E4** 14 20N 36 41 E
Om Koi, *Thailand* **86 D2** 17 48N 98 22 E
Ōma, *Japan* **68 D10** 41 45N 141 5 E
Ōmachi, *Japan* **71 A9** 36 30N 137 50 E
Omae-Zaki, *Japan* **71 C10** 34 36N 138 14 E
Ōmagari, *Japan* **68 E10** 39 27N 140 29 E
Omagh, *U.K.* **23 B4** 54 36N 7 19W
Omagh □, *U.K.* **23 B4** 54 35N 7 15W
Omaha, *U.S.A.* **154 E7** 41 17N 95 58W
Omak, *U.S.A.* **158 B4** 48 25N 119 31W
Omalos, *Greece* **38 D5** 35 19N 23 55 E
Oman ■, *Asia* **99 B7** 23 0N 58 0 E
Oman, G. of, *Asia* **97 E8** 24 30N 58 30 E
Omapere, *N.Z.* **130 B2** 35 37 S 173 25 E
Omar Combon, *Somali Rep.* **120 D3** 3 10N 45 47 E
Omaruru, *Namibia* **116 C2** 21 26 S 16 0 E
Omaruru →, *Namibia* **116 C1** 22 7 S 14 15 E
Omate, *Peru* **172 D3** 16 45 S 71 0W
Ombai, Selat, *Indonesia* **82 C2** 8 30 S 124 50 E
Omboué, *Gabon* **114 C1** 1 35 S 9 15 E
Ombrone →, *Italy* **44 F8** 42 42N 11 5 E
Omchi, *Chad* **109 D3** 21 22N 17 53 E
Omdurmân, *Sudan* **107 D3** 15 40N 32 28 E
Ōme, *Japan* **71 B11** 35 47N 139 15 E
Omega, *U.S.A.* **152 D6** 31 21N 83 36W
Omegna, *Italy* **44 C5** 45 53N 8 24 E
Omeonga, *Dem. Rep. of the Congo* **114 C4** 3 40 S 24 22 E
Ometepe, I. de, *Nic.* **164 D2** 11 32N 85 35W
Ometepec, *Mexico* **163 D5** 16 39N 98 23W
Ōmi-Shima, *Ehime, Japan* **70 C5** 34 15N 133 0 E
Ōmi-Shima, *Yamaguchi, Japan* **70 C3** 34 25N 131 9 E
Omihachiman, *Japan* **71 B8** 35 7N 136 3 E
Ominato, *Japan* **68 D10** 41 17N 141 10 E
Omineca →, *Canada* **142 B4** 56 3N 124 16W
Omiš, *Croatia* **45 E13** 43 28N 16 40 E
Omišalj, *Croatia* **45 C11** 45 13N 14 32 E
Omitara, *Namibia* **116 C2** 22 16 S 18 2 E
Ōmiya, *Japan* **71 B11** 35 54N 139 38 E
Ommaney, C., *U.S.A.* **144 H14** 56 10N 134 40W
Omme Å →, *Denmark* **17 J2** 55 56N 8 32 E
Ommen, *Neths.* **24 B6** 52 31N 6 26 E
Ömnögovĭ □, *Mongolia* **74 C3** 43 15N 104 0 E
Omo →, *Ethiopia* **107 F4** 6 25N 36 10 E
Omodeo, L., *Italy* **46 B1** 40 8N 8 56 E
Omodhos, *Cyprus* **38 E11** 34 51N 32 48 E
Omolon →, *Russia* **65 C16** 68 42N 158 36 E
Omono-Gawa →, *Japan* **68 E10** 39 46N 140 3 E
Omsk, *Russia* **64 D8** 55 0N 73 12 E
Omsukchan, *Russia* **65 C16** 62 32N 155 48 E
Ōmu, *Japan* **68 B11** 44 34N 142 58 E
Omul, Vf., *Romania* **53 E10** 45 27N 25 29 E
Omulew →, *Poland* **55 E8** 53 5N 21 33 E
Ōmura, *Japan* **70 E1** 32 56N 129 57 E
Ōmura-Wan, *Japan* **70 E1** 32 57N 129 52 E
Omurtag, *Bulgaria* **51 C10** 43 8N 26 26 E
Ōmuta, *Japan* **70 D2** 33 5N 130 26 E
Omutninsk, *Russia* **62 B4** 58 45N 52 4 E
On-Take, *Japan* **70 F2** 31 35N 130 39 E
Oña, *Spain* **42 C7** 42 43N 3 25W
Ona, *U.S.A.* **153 H8** 27 29N 81 55W
Onaga, *U.S.A.* **154 F6** 39 29N 96 10W
Onalaska, *U.S.A.* **154 D9** 43 53N 91 14W
Onamia, *U.S.A.* **154 B8** 46 4N 93 40W
Onancock, *U.S.A.* **148 G8** 37 43N 75 45W
Onang, *Indonesia* **82 B1** 3 2 S 118 49 E
Onaping L., *Canada* **140 C3** 47 3N 81 30W
Onarga, *U.S.A.* **157 D8** 40 43N 88 1W
Onarhā, *Afghan.* **91 B3** 35 30N 71 0 E
Oñati, *Spain* **40 B2** 43 3N 2 25W
Onavas, *Mexico* **162 B3** 28 28N 109 30W
Onawa, *U.S.A.* **154 D6** 42 2N 96 6W
Onaway, *U.S.A.* **148 C3** 45 21N 84 14W
Oncócua, *Angola* **115 F2** 16 30 S 13 25 E
Onda, *Spain* **40 F4** 39 55N 0 17W
Ondaejin, *N. Korea* **75 D15** 41 34N 129 40 E
Ondangua, *Namibia* **116 B2** 17 57 S 16 4 E
Ondarroa, *Spain* **40 B2** 43 19N 2 25W
Ondas →, *Brazil* **171 D3** 12 8 S 44 55W
Ondava →, *Slovak Rep.* **35 C14** 48 27N 21 48 E
Ondjiva, *Angola* **115 F3** 16 48 S 15 50 E
Ondo, *Japan* **70 C4** 34 11N 132 32 E
Ondo, *Nigeria* **113 D5** 7 4N 4 47 E
Ondo □, *Nigeria* **113 D6** 7 0N 5 0 E
Öndörshil, *Mongolia* **74 B5** 45 13N 108 5 E
Öndverðarnes, *Iceland* **11 C2** 64 52N 24 0W
Oneco, *U.S.A.* **153 H7** 27 25N 82 31W
Onega, *Russia* **56 B6** 64 0N 38 10 E
Onega →, *Russia* **56 B6** 63 58N 38 2 E
Onega, G. of = Onezhskaya Guba, *Russia* **56 B6** 64 24N 36 38 E
Onega, L. = Onezhskoye Ozero, *Russia* **58 B8** 61 44N 35 22 E
Onehunga, *N.Z.* **130 C3** 36 55 S 174 48 E
Oneida, *Ill., U.S.A.* **156 C6** 41 4N 90 13W
Oneida, *N.Y., U.S.A.* **151 C9** 43 6N 75 39W
Oneida L., *U.S.A.* **151 C9** 43 12N 75 54W
O'Neill, *U.S.A.* **154 D5** 42 27N 98 39W
Onekotan, Ostrov, *Russia* **65 E16** 49 25N 154 45 E
Onema, *Dem. Rep. of the Congo* **115 C4** 4 35 S 24 30 E
Oneonta, *Ala., U.S.A.* **149 J2** 33 57N 86 28W
Oneonta, *N.Y., U.S.A.* **151 D9** 42 27N 75 4W
Onerahi, *N.Z.* **130 B3** 35 45 S 174 22 E
Oneşti, *Romania* **53 D11** 46 17N 26 47 E
Onezhskaya Guba, *Russia* **56 B6** 64 24N 36 38 E
Onezhskoye Ozero, *Russia* **58 B8** 61 44N 35 22 E
Ongarue, *N.Z.* **130 E4** 38 42 S 175 19 E
Ongea Levu, *Fiji* **133 B3** 19 8 S 178 24W
Ongerup, *Australia* **125 F2** 33 58 S 118 28 E
Ongjin, *N. Korea* **75 F13** 37 56N 125 21 E
Ongkharak, *Thailand* **86 E3** 14 8N 101 1 E
Ongniud Qi, *China* **75 C10** 43 0N 118 38 E
Ongoka, *Dem. Rep. of the Congo* **118 C2** 1 20 S 26 0 E
Ongole, *India* **95 G5** 15 33N 80 2 E
Ongon, *Mongolia* **74 B7** 45 41N 113 5 E
Oni, *Georgia* **61 J6** 42 33N 43 26 E
Onida, *U.S.A.* **154 C4** 44 42N 100 4W
Onilahy →, *Madag.* **117 C7** 23 34 S 43 45 E
Onitsha, *Nigeria* **113 D6** 6 6N 6 42 E
Onmaka, *Burma* **90 D6** 22 17N 96 41 E
Ono, *Fiji* **133 B2** 18 55 S 178 29 E
Ono, *Fukui, Japan* **71 B8** 35 59N 136 29 E
Ono, *Hyōgo, Japan* **70 C6** 34 51N 134 56 E
Onoda, *Japan* **70 C3** 34 2N 131 25 E
Onoke, L., *N.Z.* **130 H4** 41 22 S 175 8 E
Onomichi, *Japan* **70 C5** 34 25N 133 12 E
Onpyŏng-ni, *S. Korea* **75 H14** 33 25N 126 55 E
Ons, I. de, *Spain* **42 C2** 42 23N 8 55W
Onslow, *Australia* **124 D2** 21 40 S 115 12 E
Onslow B., *U.S.A.* **149 H7** 34 20N 77 15W
Ontake-San, *Japan* **71 B9** 35 53N 137 29 E
Ontario, *Calif., U.S.A.* **161 L9** 34 4N 117 39W
Ontario, *Oreg., U.S.A.* **158 D5** 44 2N 116 58W
Ontario □, *Canada* **140 B2** 48 0N 83 0W
Ontario, L., *U.S.A.* **140 D4** 43 20N 78 0W
Ontinyent, *Spain* **41 G4** 38 50N 0 35W
Ontonagon, *U.S.A.* **154 B10** 46 52N 89 19W
Ontur, *Spain* **41 G3** 38 38N 1 29W
Onyx, *U.S.A.* **161 K8** 35 41N 118 14W
Oodnadatta, *Australia* **127 D2** 27 33 S 135 30 E
Ooldea, *Australia* **125 F5** 30 27 S 131 50 E
Oombulgurri, *Australia* **124 C4** 15 15 S 127 45 E
Oona River, *Canada* **142 C2** 53 57N 130 16W
Oorindi, *Australia* **126 C3** 20 40 S 141 1 E
Oost-Vlaanderen □, *Belgium* **24 C3** 51 5N 3 50 E
Oostende, *Belgium* **24 C2** 51 15N 2 54 E
Oosterhout, *Neths.* **24 C4** 51 39N 4 47 E
Oosterschelde, *Neths.* **24 C4** 51 33N 4 0 E
Oosterwolde, *Neths.* **24 B6** 53 0N 6 17 E
Ootacamund = Udagamandalam, *India* **95 J3** 11 30N 76 44 E
Ootha, *Australia* **129 B7** 33 6 S 147 29 E
Ootsa L., *Canada* **142 C3** 53 50N 126 2W
Opaka, *Bulgaria* **51 C10** 43 28N 26 10 E
Opala, *Dem. Rep. of the Congo* **114 C4** 0 40 S 24 20 E
Opalenica, *Poland* **55 F3** 52 18N 16 24 E
Opan, *Bulgaria* **51 D9** 42 13N 25 41 E
Opanake, *Sri Lanka* **95 L5** 6 35N 80 40 E
Opapa, *N.Z.* **130 F5** 39 47 S 176 42 E
Opasatika, *Canada* **140 C3** 49 30N 82 50W
Opasquia, *Canada* **143 C10** 53 16N 93 34W
Opatija, *Croatia* **45 C11** 45 21N 14 17 E
Opatów, *Poland* **55 H8** 50 50N 21 27 E
Opava, *Czech Rep.* **35 B10** 49 57N 17 58 E
Opelika, *U.S.A.* **152 C4** 32 39N 85 23W
Opelousas, *U.S.A.* **155 K8** 30 32N 92 5W
Opémisca, L., *Canada* **140 C5** 49 56N 74 52W
Open Bay Is., *N.Z.* **131 D3** 43 51 S 168 51 E
Opheim, *U.S.A.* **158 B10** 48 51N 106 24W
Ophthalmia Ra., *Australia* **124 D2** 23 15 S 119 30 E
Opi, *Nigeria* **113 D6** 6 36N 7 28 E
Opihikao, *U.S.A.* **145 D7** 19 26N 154 53W
Opinaca →, *Canada* **140 B4** 52 15N 78 2W
Opinaca L., *Canada* **140 B4** 52 39N 76 20W
Opiskotish, L., *Canada* **141 B6** 53 10N 67 50W
Opobo, *Nigeria* **113 E6** 4 35N 7 34 E
Opochka, *Russia* **58 D5** 56 42N 28 45 E
Opoczno, *Poland* **55 G7** 51 22N 20 18 E
Opol, *Phil.* **81 G5** 8 31N 124 34 E
Opole, *Poland* **55 H4** 50 42N 17 58 E
Opole □, *Poland* **55 H4** 50 40N 17 56 E
Opole Lubelskie, *Poland* **55 G8** 51 9N 21 58 E
Opon = Lapu-Lapu, *Phil.* **81 F4** 10 20N 123 55 E
Oporto = Porto, *Portugal* **42 D2** 41 8N 8 40W
Opotiki, *N.Z.* **130 E6** 38 1 S 177 19 E
Opp, *U.S.A.* **152 D3** 31 17N 86 16W
Oppdal, *Norway* **15 E13** 62 35N 9 41 E
Óppido Mamertina, *Italy* **47 D8** 38 16N 15 59 E
Oppland □, *Norway* **18 C6** 61 15N 9 40 E
Oprişor, *Romania* **52 F8** 44 17N 23 5 E
Oprtalj, *Croatia* **45 C10** 45 23N 13 50 E
Opua, *N.Z.* **130 B3** 35 19 S 174 9 E
Opunake, *N.Z.* **130 F2** 39 26 S 173 52 E
Opuzen, *Croatia* **45 E14** 43 1N 17 34 E
Oquawka, *U.S.A.* **156 D6** 40 56N 90 57W
Ora, *Cyprus* **38 E12** 34 51N 33 12 E
Ora Banda, *Australia* **125 F3** 30 20 S 121 0 E
Oracle, *U.S.A.* **159 K8** 32 37N 110 46W
Oradea, *Romania* **52 C6** 47 2N 21 58 E
Öræfajökull, *Iceland* **11 C10** 64 2N 16 39W
Orahovac, *Serbia, Yug.* **50 D4** 42 24N 20 40 E
Orahovica, *Croatia* **52 E2** 45 35N 17 52 E
Orai, *India* **93 G8** 25 58N 79 30 E
Oraison, *France* **29 E9** 43 55N 5 55 E
Oral = Zhayyq →, *Kazakstan* **57 E9** 47 0N 51 48 E
Oral, *Kazakstan* **60 E10** 51 20N 51 20 E
Oran, *Algeria* **111 A4** 35 45N 0 39W
Orán, *Argentina* **174 A3** 23 10 S 64 20W
Orange, *Australia* **129 B8** 33 15 S 149 7 E
Orange, *France* **29 D8** 44 8N 4 47 E
Orange, *Calif., U.S.A.* **161 M9** 33 47N 117 51W
Orange, *Mass., U.S.A.* **151 D12** 42 35N 72 19W
Orange, *Tex., U.S.A.* **155 K8** 30 6N 93 44W
Orange, *Va., U.S.A.* **148 F6** 38 15N 78 7W
Orange →, *S. Africa* **116 D2** 28 41 S 16 28 E
Orange, C., *Brazil* **169 C7** 4 20N 51 30W
Orange City, *U.S.A.* **153 G8** 28 57N 81 18W
Orange Cove, *U.S.A.* **160 J7** 36 38N 119 19W
Orange Free State □ = Free State □, *S. Africa* **116 D4** 28 30 S 27 0 E
Orange Grove, *U.S.A.* **155 M6** 27 58N 97 56W
Orange L., *U.S.A.* **153 F7** 29 25N 82 13W
Orange Park, *U.S.A.* **152 E8** 30 10N 81 42W
Orange Walk, *Belize* **163 D7** 18 6N 88 33W
Orangeburg, *U.S.A.* **152 B9** 33 30N 80 52W
Orangeville, *Canada* **140 D3** 43 55N 80 5W
Orangeville, *U.S.A.* **156 B7** 42 28N 89 39W
Orani, *Phil.* **80 D3** 14 49N 120 32 E
Oranienburg, *Germany* **30 C9** 52 45N 13 14 E
Oranje = Orange →, *S. Africa* **116 D2** 28 41 S 16 28 E
Oranje Vrystaat □ = Free State □, *S. Africa* **116 D4** 28 30 S 27 0 E
Oranjemund, *Namibia* **116 D2** 28 38 S 16 29 E
Oranjerivier, *S. Africa* **116 D3** 29 40 S 24 12 E
Oras, *Phil.* **80 E5** 12 9N 125 28 E
Orašje, *Bos.-H.* **52 E3** 45 1N 18 42 E

Patrocínio 263

Pamanukan, *Indonesia* . . . . . **85 D3** 6 16 S 107 49 E
Pamban I., *India* . . . . . **95 K4** 9 15N 79 20 E
Pambuhan, *Phil.* . . . . . **80 E4** 13 59N 123 5 E
Pamekasan, *Indonesia* . . . . . **85 D4** 7 10 S 113 28 E
Pamiers, *France* . . . . . **28 E5** 43 7N 1 39 E
Pamir, *Tajikistan* . . . . . **63 E6** 37 40N 73 0 E
Pamir →, *Tajikistan* . . . . . **63 E6** 37 1N 72 41 E
Pâmiut, *Greenland* **10 E6** 62 0N 49 50W
Pamlico →, *U.S.A.* **149 H7** 35 20N 76 28W
Pamlico Sd., *U.S.A.* **149 H8** 35 20N 76 0W
Pampa, *U.S.A.* **155 H4** 35 32N 100 58W
Pampa de Agma, *Argentina* **176 B3** 43 45 S 69 40W
Pampa de las Salinas, *Argentina* . . . . . **174 C2** 32 1 S 66 58W
Pampa Grande, *Bolivia* . . . . . **173 D5** 18 5 S 64 6W
Pampa Hermosa, *Peru* . . . . . **172 B2** 7 7 S 75 4W
Pampanga □, *Phil.* . . . . . **80 D3** 15 4N 120 40 E
Pampanua, *Indonesia* . . . . . **82 B2** 4 16 S 120 8 E
Pampas, *Argentina* . . . . . **174 D3** 35 0 S 63 0W
Pampas, *Peru* . . . . . **172 C3** 12 20 S 74 50W
Pampas →, *Peru* . . . . . **172 C3** 13 24 S 73 12W
Pamphylia, *Turkey* . . . . . **100 D4** 37 0N 31 20 E
Pamplona, *Colombia* **168 B3** 7 23N 72 39W
Pamplona, *Phil.* . . . . . **80 B3** 18 31N 121 20 E
Pamplona, *Spain* **40 C3** 42 48N 1 38W
Pampoenpoort, *S. Africa* . . **116 E3** 31 3 S 22 40 E
Pamukçu, *Turkey* . . . . . **49 B9** 39 30N 27 54 E
Pamukkale, *Turkey* . . . . . **49 D11** 37 55N 29 8 E
Pan Xian, *China* . . . . . **76 E5** 25 46N 104 38 E
Pana, *U.S.A.* **156 E7** 39 23N 89 5W
Panabo, *Phil.* . . . . . **81 H5** 7 19N 125 42 E
Panaca, *U.S.A.* **159 H6** 37 47N 114 23W
Panacea, *U.S.A.* **152 E5** 30 2N 84 23W
Panagyurishte, *Bulgaria* . . . . . **51 D8** 42 30N 24 15 E
Panaitan, *Indonesia* . . . . . **84 D3** 6 36 S 105 12 E
Panaji, *India* . . . . . **95 G1** 15 25N 73 50 E
Panamá, *Panama* **164 E4** 9 0N 79 25W
Panama ■, *Cent. Amer.* **164 E4** 8 48N 79 55W
Panamá, G. de, *Panama* . . . **164 E4** 8 4N 79 20W
Panama Canal, *Panama* **164 E4** 9 10N 79 37W
Panama City, *U.S.A.* **152 E4** 30 10N 85 40W
Panama City Beach, *U.S.A.* **152 E4** 30 11N 85 48W
Panamint Range, *U.S.A.* . . . **161 J9** 36 20N 117 20W
Panamint Springs, *U.S.A.* . . . **161 J9** 36 20N 117 28W
Panão, *Peru* . . . . . **172 B2** 9 55 S 75 55W
Panaon I., *Phil.* . . . . . **81 F5** 10 3N 125 13 E
Panare, *Thailand* . . . . . **87 J3** 6 51N 101 30 E
Panarea, *Italy* . . . . . **47 D8** 38 38N 15 4 E
Panaro →, *Italy* . . . . . **45 D8** 44 55N 11 25 E
Panarukan, *Indonesia* . . . . . **85 D4** 7 42 S 113 56 E
Panay, *Phil.* . . . . . **81 F4** 11 10N 122 30 E
Panay, G., *Phil.* . . . . . **81 F4** 11 0N 122 30 E
Pancake Range, *U.S.A.* **159 G6** 38 30N 115 50W
Pančevo, *Serbia, Yug.* . . . . . **52 F5** 44 52N 20 41 E
Panciu, *Romania* . . . . . **53 E12** 45 54N 27 8 E
Pancol, *Phil.* . . . . . **81 F2** 10 52N 119 25 E
Pancorbo, Desfiladero, *Spain* **42 C7** 42 32N 3 5W
Pâncota, *Romania* . . . . . **52 D6** 46 20N 21 45 E
Pandan, *Antique, Phil.* . . . . . **81 F4** 11 45N 122 10 E
Pandan, *Catanduanes, Phil.* **80 D5** 14 3N 124 10 E
Pandan Bay, *Phil.* . . . . . **81 F4** 11 43N 122 0 E
Pandegelang, *Indonesia* . . . . . **84 D3** 6 25 S 106 5 E
Pandharpur, *India* . . . . . **94 F2** 17 41N 75 20 E
Pandhurna, *India* . . . . . **94 D4** 21 36N 78 35 E
Pando, *Uruguay* . . . . . **175 C4** 34 44 S 56 0W
Pando □, *Bolivia* . . . . . **172 C4** 11 20 S 67 40W
Pando, L. = Hope, L., *Australia* . . . . . **127 D2** 28 24 S 139 18 E
Pandokrátor, *Greece* . . . . . **38 A3** 39 45N 19 50 E
Pandora, *Costa Rica* **164 E3** 9 43N 83 3W
Pandrup, *Denmark* . . . . . **17 G3** 57 14N 9 40 E
Pandu, *Dem. Rep. of the Congo* . **114 B3** 4 59N 19 16 E
Panevėžys, *Lithuania* . . . . . **15 J21** 55 42N 24 25 E
Panfilov, *Kazakstan* . . . . . **64 E9** 44 10N 80 0 E
Panfilovo, *Russia* . . . . . **60 E6** 50 25N 42 46 E
Panga, *Dem. Rep. of the Congo* . **118 B2** 1 52N 26 18 E
Pangaíon Óros, *Greece* . . . . . **51 F8** 40 50N 24 0 E
Pangala, *Congo* . . . . . **114 C2** 4 1 S 13 52 E
Pangalanes, Canal des, *Madag.* . . . . . **117 C8** 22 48 S 47 50 E
Pangani, *Tanzania* . . . . . **118 D4** 5 25 S 38 58 E
Pangani □, *Tanzania* . . . . . **118 D4** 5 25 S 39 0 E
Pangani →, *Tanzania* . . . . . **118 D4** 5 26 S 38 58 E
Panganiban, *Phil.* . . . . . **80 E5** 13 55N 124 18 E
Panganuran, *Phil.* . . . . . **81 G4** 8 2N 122 22 E
Pangasinan □, *Phil.* . . . . . **80 D3** 15 55N 120 20 E
Pangfou = Bengbu, *China* . **75 H9** 32 58N 117 20 E
Pangil, *Dem. Rep. of the Congo* . **118 C2** 3 10 S 26 35 E
Pangkah, Tanjung, *Indonesia* **85 D4** 6 51 S 112 33 E
Pangkai, *Burma* . . . . . **90 D7** 22 40N 98 40 E
Pangkajene, *Indonesia* . . . . . **82 B1** 4 46 S 119 34 E
Pangkalanbrandan, *Indonesia* **84 B1** 4 1N 98 20 E
Pangkalanbuun, *Indonesia* . . **85 C4** 2 41 S 111 37 E
Pangkalansusu, *Indonesia* . . **84 B1** 4 2N 98 13 E
Pangkalpinang, *Indonesia* . . **84 C3** 2 0 S 106 0 E
Pangkoh, *Indonesia* . . . . . **85 C4** 3 5 S 114 8 E
Panglao, *Phil.* . . . . . **81 G4** 9 35N 123 45 E
Panglao I., *Phil.* . . . . . **81 G4** 9 35N 123 48 E
Pangnirtung, *Canada* **139 B13** 66 8N 65 54W
Pangrango, *Indonesia* . . . . . **84 D3** 6 46 S 107 1 E
Pangsau Pass, *Burma* . . . . . **90 B6** 27 15N 96 10 E
Pangtara, *Burma* . . . . . **90 E6** 20 57N 96 40 E
Panguipulli, *Chile* . . . . . **176 A2** 39 38 S 72 20W
Panguitch, *U.S.A.* **159 H7** 37 50N 112 26W
Pangutaran, *Phil.* . . . . . **81 H3** 6 18N 120 35 E
Pangutaran Group, *Phil.* . . . **81 H3** 6 18N 120 34 E
Panhandle, *U.S.A.* **155 H4** 35 21N 101 23W
Pani Mines, *India* . . . . . **92 H5** 22 29N 73 50 E
Pania-Mutombo, *Dem. Rep. of the Congo* . **115 D4** 5 11 S 23 51 E
Paniau, *U.S.A.* **145 B1** 21 56N 160 5W
Panié, Mt., *N. Cal.* . . . . . **133 T18** 20 36 S 164 46 E
Panipat, *India* . . . . . **92 E7** 29 25N 77 2 E
Panitan, *Phil.* . . . . . **81 F4** 11 28N 122 46 E
Panjal Range, *India* . . . . . **92 C7** 32 30N 76 50 E
Panjgur, *Pakistan* . . . . . **91 D2** 27 0N 64 5 E
Panji Poyon, *Tajikistan* . . . . . **63 E4** 37 12N 68 35 E
Panjim = Panaji, *India* . . . . . **95 G1** 15 25N 73 50 E
Panjwai, *Afghan.* . . . . . **92 D1** 31 26N 65 27 E
Pankshin, *Nigeria* . . . . . **113 D6** 9 16N 9 25 E
Panmunjŏm, *N. Korea* . . . . . **75 F14** 37 59N 126 38 E
Panna, *India* . . . . . **93 G9** 24 40N 80 15 E
Panna Hills, *India* . . . . . **93 G9** 24 40N 81 15 E
Pano Lefkara, *Cyprus* . . . . . **38 E12** 34 53N 33 20 E
Pano Panayia, *Cyprus* . . . . . **38 E11** 34 55N 32 38 E
Panora, *U.S.A.* **156 C2** 41 42N 94 22W
Panorama, *Brazil* . . . . . **175 A5** 21 21 S 51 51W
Pánormon, *Greece* . . . . . **38 D6** 35 25N 24 41 E
Panruti, *India* . . . . . **95 J4** 11 46N 79 35 E
Panshan, *China* . . . . . **75 D12** 41 3N 122 2 E
Panshi, *China* . . . . . **75 C14** 42 58N 126 5 E
Pantao, *Phil.* . . . . . **80 E4** 13 12N 123 20 E
Pantar, *Indonesia* . . . . . **82 C2** 8 28 S 124 10 E
Pante Macassar, *Indonesia* . **82 C2** 9 30 S 123 58 E
Pantelleria, *Italy* . . . . . **46 F4** 36 50N 11 57 E
Pantha, *Burma* . . . . . **90 D5** 23 55N 94 35 E
Pantin Sakan, *Burma* . . . . . **90 F6** 18 38N 97 33 E
Pantón, *Spain* **42 C3** 42 31N 7 37W
Pánuco, *Mexico* **163 C5** 22 0N 98 15W
Panukulan, *Phil.* . . . . . **80 D3** 14 56N 121 49 E
Panyam, *Nigeria* . . . . . **113 D6** 9 27N 9 8 E
Panyu, *China* . . . . . **77 F9** 22 51N 113 20 E
Pao →, *Anzoátegui, Venezuela* **169 B5** 8 6N 64 17W
Pao →, *Apure, Venezuela* . **168 B4** 8 33N 68 1W
Páola, *Italy* . . . . . **47 C9** 39 21N 16 2 E
Paola, *Malta* . . . . . **38 D2** 35 52N 14 30 E
Paola, *U.S.A.* **154 F7** 38 35N 94 53W
Paoli, *U.S.A.* **157 F10** 38 33N 86 28W
Paonia, *U.S.A.* **159 G10** 38 52N 107 36W
Paoting = Baoding, *China* . **74 E8** 38 50N 115 28 E
Paot'ou = Baotou, *China* . . **74 D6** 40 32N 110 2 E
Paoua, *C.A.R.* . . . . . **114 A3** 7 9N 16 20 E
Pápa, *Hungary* . . . . . **52 C2** 47 22N 17 30 E
Papa, *U.S.A.* **145 D6** 19 13N 155 52W
Papa Stour, *U.K.* **22 A7** 60 20N 1 42W
Papa Westray, *U.K.* **22 B6** 59 20N 2 55W
Papaaloa, *U.S.A.* **145 D6** 19 59N 155 13W
Papagayo →, *Mexico* **163 D5** 16 36N 99 43W
Papagayo, G. de, *Costa Rica* **164 D2** 10 30N 85 50W
Papagni →, *India* . . . . . **95 G3** 15 35N 77 45 E
Papakura, *N.Z.* . . . . . **130 D3** 37 4 S 174 59 E
Papantla, *Mexico* **163 C5** 20 30N 97 30W
Paparoa, *N.Z.* . . . . . **130 C3** 36 6 S 174 16 E
Paparoa Nat. Park, *N.Z.* . . . **131 C6** 42 7 S 171 26 E
Paparoa Ra., *N.Z.* . . . . . **131 C6** 42 5 S 171 35 E
Pápas, Ákra, *Greece* . . . . . **48 C3** 38 13N 21 20 E
Papatoetoe, *N.Z.* . . . . . **130 C3** 36 59 S 174 51 E
Papawai Pt., *U.S.A.* **145 C5** 20 47N 156 32W
Papeete, *Tahiti* . . . . . **133 S16** 17 32 S 149 34W
Papenburg, *Germany* . . . . . **30 B3** 53 5N 7 23 E
Paphlagonia, *Turkey* . . . . . **100 B5** 41 30N 33 0 E
Paphos, *Cyprus* . . . . . **38 E11** 34 46N 32 25 E
Papien Chiang = Da →, *Vietnam* . . . . . **86 B5** 21 15N 105 20 E
Papigochic →, *Mexico* **162 B3** 29 9N 109 40W
Paposo, *Chile* . . . . . **174 B1** 25 0 S 70 30W
Papoutsa, *Cyprus* . . . . . **38 E12** 34 54N 33 4 E
Papua, G. of, *Papua N. G.* . **132 E3** 9 0 S 144 50 E
Papua New Guinea ■, *Oceania* . . . . . **132 E3** 8 0 S 145 0 E
Papudo, *Chile* . . . . . **174 C1** 32 29 S 71 27W
Papuk, *Croatia* . . . . . **52 E2** 45 30N 17 30 E
Papun, *Burma* . . . . . **90 F6** 18 2N 97 30 E
Papunya, *Australia* . . . . . **124 D5** 23 15 S 131 54 E
Pará = Belém, *Brazil* . . . . . **170 B2** 1 20 S 48 30W
Pará □, *Brazil* . . . . . **173 A7** 3 20 S 52 0W
Pará □, *Surinam* **169 B6** 5 20N 55 5W
Paraburdoo, *Australia* . . . . . **124 D2** 23 14 S 117 32 E
Paracale, *Phil.* . . . . . **80 D4** 14 17N 122 48 E
Paracas, Pen., *Peru* . . . . . **172 C2** 13 53 S 76 20W
Paracatu, *Brazil* . . . . . **171 E2** 17 10 S 46 50W
Paracatu →, *Brazil* . . . . . **171 E2** 16 30 S 45 4W
Parachilna, *Australia* . . . . . **128 A3** 31 10 S 138 21 E
Parachinar, *Pakistan* . . . . . **91 B3** 33 55N 70 5 E
Paraćin, *Serbia, Yug.* . . . . . **50 C5** 43 54N 21 27 E
Paracuru, *Brazil* . . . . . **170 B4** 3 24 S 39 4W
Parada, Punta, *Peru* . . . . . **172 D2** 15 22 S 75 11W
Paradas, *Spain* **43 H5** 37 18N 5 29W
Paradela, *Spain* **42 C3** 42 44N 7 37W
Paradhísi, *Greece* . . . . . **38 C10** 36 18N 28 7 E
Paradip, *India* . . . . . **94 D8** 20 15N 86 35 E
Paradise, *Calif., U.S.A.* **160 F5** 39 46N 121 37W
Paradise, *Mont., U.S.A.* **158 C6** 47 23N 114 48W
Paradise, *Nev., U.S.A.* **161 J11** 36 9N 115 10W
Paradise →, *Canada* **141 B8** 53 27N 57 19W
Paradise Valley, *U.S.A.* **158 F5** 41 30N 117 32W
Parado, *Indonesia* . . . . . **85 D5** 8 42 S 118 30 E
Paragould, *U.S.A.* **155 G9** 36 3N 90 29W
Paraguá →, *Bolivia* . . . . . **173 C5** 13 34 S 61 53W
Paragua →, *Venezuela* **169 B5** 6 55N 62 55W
Paraguaçu →, *Brazil* . . . . . **171 D4** 12 45 S 38 54W
Paraguaçu Paulista, *Brazil* . . **175 A5** 22 22 S 50 35W
Paraguaipoa, *Venezuela* **168 A3** 11 21N 71 57W
Paraguaná, Pen. de, *Venezuela* **168 A4** 12 0N 70 0W
Paraguarí, *Paraguay* . . . . . **174 B4** 25 36 S 57 0W
Paraguarí □, *Paraguay* . . . . . **174 B4** 26 0 S 57 10W
Paraguay ■, *S. Amer.* . . . . . **174 A4** 23 0 S 57 0W
Paraguay →, *Paraguay* . . . . . **174 B4** 27 18 S 58 38W
Paraíba = João Pessoa, *Brazil* . . . . . **170 C5** 7 10 S 34 52W
Paraíba □, *Brazil* . . . . . **170 C4** 7 0 S 36 0W
Paraíba do Sul →, *Brazil* . . **171 F3** 21 37 S 41 3W
Parainen, *Finland* . . . . . **15 F20** 60 18N 22 18 E
Paraiso, *Mexico* **163 D6** 18 24N 93 14W
Parak, *Iran* . . . . . **97 E7** 27 38N 52 25 E
Parakhino Paddubye, *Russia* **58 C7** 58 26N 33 10 E
Parakou, *Benin* . . . . . **113 D5** 9 25N 2 40 E
Parakylia, *Australia* . . . . . **128 A2** 30 24 S 136 25 E
Paralimni, *Cyprus* . . . . . **38 D12** 35 2N 33 58 E
Parálion-Astrous, *Greece* . . . **48 D4** 37 25N 22 45 E
Paramakkudi, *India* . . . . . **95 K4** 9 31N 78 39 E
Paramaribo, *Surinam* **169 B6** 5 50N 55 10W
Parambu, *Brazil* . . . . . **170 C3** 6 13 S 40 43W
Paramillo, Nudo del, *Colombia* **168 B2** 7 4N 75 55W
Paramirim, *Brazil* . . . . . **171 D3** 13 26 S 42 15W
Paramirim →, *Brazil* . . . . . **171 D3** 11 34 S 43 18W
Paramithiá, *Greece* . . . . . **48 B2** 39 30N 20 35 E
Paramushir, Ostrov, *Russia* . **65 D16** 50 24N 156 0 E
Paran →, *Israel* . . . . . **103 E4** 30 20N 35 10 E
Paraná, *Argentina* . . . . . **174 C3** 31 45 S 60 30W
Paranã, *Brazil* . . . . . **171 D2** 12 30 S 47 48W
Paraná □, *Brazil* . . . . . **175 A5** 24 30 S 51 0W
Paraná →, *Argentina* . . . . . **174 C4** 33 43 S 59 15W
Paranaguá, *Brazil* . . . . . **175 B6** 25 30 S 48 30W
Paranaíba →, *Brazil* . . . . . **171 F1** 20 6 S 51 4W
Paranapanema →, *Brazil* . . **175 A5** 22 40 S 53 9W
Paranapiacaba, Serra do, *Brazil* . . . . . **175 A6** 24 31 S 48 35W
Paranavaí, *Brazil* . . . . . **175 A5** 23 4 S 52 56W
Parang, *Jolo, Phil.* . . . . . **81 J3** 5 55N 120 54 E
Parang, *Mindanao, Phil.* . . . . . **81 H5** 7 23N 124 16 E
Parangaba, *Brazil* . . . . . **170 B4** 3 45 S 38 33W
Parangippettai, *India* . . . . . **95 J4** 11 30N 79 38 E
Parângul Mare, Vf., *Romania* . . . . . **53 E8** 45 20N 23 37 E
Paraparauma, *N.Z.* . . . . . **130 G4** 40 57 S 175 3 E
Parapóla, *Greece* . . . . . **48 E5** 36 55N 23 27 E
Paraspóri, Ákra, *Greece* . . . **49 F9** 35 55N 27 15 E
Paratinga, *Brazil* . . . . . **171 D3** 12 40 S 43 10W
Paratoo, *Australia* . . . . . **128 B3** 32 42 S 139 20 E
Parattah, *Australia* . . . . . **126 G4** 42 22 S 147 23 E
Paraúna, *Brazil* . . . . . **171 E1** 16 55 S 50 26W
Paray-le-Monial, *France* . . . . **27 F11** 46 27N 4 7 E
Parbati →, *India* . . . . . **92 G7** 25 50N 76 30 E
Parbatipur, *Bangla.* . . . . . **90 C2** 25 39N 88 55 E
Parbhani, *India* . . . . . **94 E3** 19 8N 76 52 E
Parchim, *Germany* . . . . . **30 B7** 53 26N 11 52 E
Parczew, *Poland* . . . . . **55 G9** 51 40N 22 52 E
Pardes Hanna-Karkur, *Israel* **103 C3** 32 28N 34 57 E
Pardilla, *Spain* **42 D7** 41 33N 3 43W
Pardo →, *Bahia, Brazil* . . . **171 E4** 15 40 S 39 0W
Pardo →, *Mato Grosso, Brazil* . . . . . **175 A5** 21 46 S 52 9W
Pardo →, *Minas Gerais, Brazil* . . . . . **171 E3** 15 48 S 44 48W
Pardo →, *São Paulo, Brazil* **171 F2** 20 10 S 48 38W
Pardubice, *Czech Rep.* . . . . . **34 A8** 50 3N 15 45 E
Pare, *Indonesia* . . . . . **85 D4** 7 43 S 112 12 E
Pare □, *Tanzania* . . . . . **118 C4** 4 10 S 38 0 E
Pare Mts., *Tanzania* . . . . . **118 C4** 4 0 S 37 45 E
Parecis, Serra dos, *Brazil* . . **173 C6** 13 0 S 60 0W
Paredes de Nava, *Spain* **42 C6** 42 9N 4 42W
Pareh, *Iran* . . . . . **96 B5** 38 52N 45 42 E
Parelhas, *Brazil* . . . . . **170 C4** 6 41 S 36 39W
Paren, *Russia* . . . . . **65 C17** 62 30N 163 15 E
Parengarenga Harbour, *N.Z.* **130 A2** 34 31 S 173 0 E
Parent, *Canada* **140 C5** 47 55N 74 35W
Parent, L., *Canada* **140 C4** 48 31N 77 1W
Parentis-en-Born, *France* . . . **28 D2** 44 21N 1 4W
Parepare, *Indonesia* . . . . . **82 B1** 4 0 S 119 40 E
Parfino, *Russia* . . . . . **58 D6** 57 59N 31 34 E
Párga, *Greece* . . . . . **48 B2** 39 15N 20 29 E
Pargo, Pta. do, *Madeira* **39 D2** 32 49N 17 17W
Parguba, *Russia* . . . . . **56 B5** 62 20N 34 27 E
Paria, G. de, *Venezuela* . . . **169 A5** 10 20N 62 0W
Paria, Pen. de, *Venezuela* . . **169 A5** 10 50N 62 30W
Pariaguán, *Venezuela* **169 B5** 8 51N 64 34W
Pariaman, *Indonesia* . . . . . **84 C2** 0 47 S 100 11 E
Paricatuba, *Brazil* . . . . . **169 D5** 4 26 S 61 53W
Paricutín, Cerro, *Mexico* . . . **162 D4** 19 28N 102 15W
Parigi, *Java, Indonesia* . . . . . **85 D3** 7 42 S 108 29 E
Parigi, *Sulawesi, Indonesia* . **82 B2** 0 50 S 120 5 E
Parika, *Guyana* **169 B6** 6 50N 58 20W
Parikkala, *Finland* . . . . . **58 B5** 61 33N 29 31 E
Parima, Serra, *Brazil* **169 C5** 2 30N 64 0W
Parinari, *Peru* . . . . . **172 A3** 4 35 S 74 25W
Pariñas, Pta., *S. Amer.* . . . . . **166 D2** 4 30 S 82 0W
Parincea, *Romania* . . . . . **53 D12** 46 27N 27 9 E
Parintins, *Brazil* . . . . . **169 D6** 2 40 S 56 50W
Pariparit Kyun, *Burma* . . . . . **90 G4** 14 55N 93 45 E
Paris, *Canada* **140 D3** 43 12N 80 25W
Paris, *France* . . . . . **27 D9** 48 50N 2 20 E
Paris, *Idaho, U.S.A.* **158 E8** 42 14N 111 24W
Paris, *Ill., U.S.A.* **157 E9** 39 36N 87 42W
Paris, *Ky., U.S.A.* **157 F12** 38 13N 84 15W
Paris, *Mo., U.S.A.* **156 E5** 39 29N 92 0W
Paris, *Tenn., U.S.A.* **149 G1** 36 18N 88 19W
Paris, *Tex., U.S.A.* **155 J7** 33 40N 95 33W
Paris, Ville de □, *France* . . . **27 D9** 48 50N 2 20 E
Parish, *U.S.A.* **151 C8** 43 25N 76 8W
Pariti, *Indonesia* . . . . . **82 D2** 10 15 S 123 45 E
Park, *U.S.A.* **160 B4** 48 45N 122 18W
Park City, *U.S.A.* **158 F8** 40 39N 111 30W
Park Falls, *U.S.A.* **154 C9** 45 56N 90 27W
Park Forest, *U.S.A.* **157 C9** 41 29N 87 40W
Park Range, *U.S.A.* **158 G10** 40 0N 106 30W
Park Rapids, *U.S.A.* **154 B7** 46 55N 95 4W
Park Ridge, *U.S.A.* **157 B9** 42 2N 87 51W
Park River, *U.S.A.* **154 A6** 48 24N 97 45W
Park Rynie, *S. Africa* . . . . . **117 E5** 30 25 S 30 45 E
Parkā Bandar, *Iran* . . . . . **97 E8** 25 55N 59 35 E
Parkano, *Finland* . . . . . **15 E20** 62 1N 23 0 E
Parkent, *Uzbekistan* . . . . . **63 C4** 41 18N 69 40 E
Parker, *Ariz., U.S.A.* **161 L12** 34 9N 114 17W
Parker, *S. Dak., U.S.A.* **154 D6** 43 24N 97 8W
Parker Dam, *U.S.A.* **161 L12** 34 18N 114 8W
Parkersburg, *Iowa, U.S.A.* . **156 B4** 42 35N 92 47W
Parkersburg, *W. Va., U.S.A.* **148 F5** 39 16N 81 34W
Parkerview, *Canada* **143 C8** 51 21N 103 18W
Parkes, *Australia* . . . . . **129 B8** 33 9 S 148 11 E
Parkfield, *U.S.A.* **160 K6** 35 54N 120 26W
Parkhar, *Tajikistan* . . . . . **63 E4** 37 30N 69 34 E
Parkland, *U.S.A.* **160 C4** 47 9N 122 26W
Parkside, *Canada* **143 C7** 53 10N 106 33W
Parkston, *U.S.A.* **154 D6** 43 24N 97 59W
Parksville, *Canada* **142 D4** 49 20N 124 21W
Parksville, *U.S.A.* **152 B7** 33 47N 82 13W
Parla, *Spain* **42 E7** 40 14N 3 46W
Parlakimidi, *India* . . . . . **94 E7** 18 45N 84 5 E
Parli, *India* . . . . . **94 E3** 18 50N 76 35 E
Pârliţa, *Moldova* . . . . . **53 C12** 47 19N 27 52 E
Parma, *Italy* . . . . . **44 D7** 44 48N 10 20 E
Parma, *Idaho, U.S.A.* **158 E5** 43 47N 116 57W
Parma, *Ohio, U.S.A.* **150 E3** 41 23N 81 43W
Parma →, *Italy* . . . . . **44 D7** 44 56N 10 26 E
Parnaguá, *Brazil* . . . . . **170 D3** 10 10 S 44 38W
Parnaíba, *Piauí, Brazil* . . . . **170 B3** 2 54 S 41 47W
Parnaíba, *São Paulo, Brazil* **173 D7** 19 34 S 51 14W
Parnaíba →, *Brazil* . . . . . **170 B3** 3 0 S 41 50W
Parnamirim, *Brazil* . . . . . **170 C4** 8 5 S 39 34W
Parnarama, *Brazil* . . . . . **170 C3** 5 31 S 43 6W
Parnassós, *Greece* . . . . . **48 C4** 38 35N 22 30 E
Parnassus, *N.Z.* . . . . . **131 C8** 42 42 S 173 23 E
Párnis, *Greece* . . . . . **48 C5** 38 14N 23 45 E
Párnon Óros, *Greece* . . . . . **48 D4** 37 15N 22 45 E
Pärnu, *Estonia* . . . . . **15 G21** 58 28N 24 33 E
Parola, *India* . . . . . **94 D2** 20 47N 75 7 E
Paroo →, *Australia* . . . . . **127 E3** 31 28 S 143 32 E
Páros, *Greece* . . . . . **49 D7** 37 5N 25 12 E
Parowan, *U.S.A.* **159 H7** 37 51N 112 50W
Parpaillon, *France* . . . . . **29 D10** 44 30N 6 40 E
Parral, *Chile* . . . . . **174 D1** 36 10 S 71 52W
Parramatta, *Australia* . . . . . **129 B9** 33 48 S 151 1 E
Parras, *Mexico* **162 B4** 25 30N 102 20W
Parrett →, *U.K.* **21 F4** 51 12N 3 1W
Parris I., *U.S.A.* **152 C9** 32 20N 80 41W
Parrish, *U.S.A.* **153 H7** 27 35N 82 26W
Parrott, *U.S.A.* **152 D5** 31 54N 84 31W
Parrsboro, *Canada* **141 C7** 45 30N 64 25W
Parry Is., *Canada* **6 B2** 77 0N 110 0W
Parry Sound, *Canada* **140 C4** 45 20N 80 0W
Parsberg, *Germany* . . . . . **31 F7** 49 10N 11 43 E
Parsęta →, *Poland* . . . . . **54 D2** 54 11N 15 34 E
Parshall, *U.S.A.* **154 B3** 47 57N 102 8W
Parsnip →, *Canada* **142 B4** 55 10N 123 2W
Parsons, *U.S.A.* **155 G7** 37 20N 95 16W
Parsons Ra., *Australia* . . . . **126 A2** 13 30 S 135 15 E
Partabpur, *India* . . . . . **94 E5** 20 0N 80 42 E
Partanna, *Italy* . . . . . **46 E5** 37 43N 12 53 E
Parthenay, *France* **26 F6** 46 38N 0 16W
Partinico, *Italy* . . . . . **46 D6** 38 3N 13 7 E
Partizánske, *Slovak Rep.* . . . **35 C11** 48 38N 18 23 E
Partur, *India* . . . . . **94 E3** 19 40N 76 14 E
Paru →, *Brazil* . . . . . **169 D7** 1 33 S 52 38W
Parú →, *Venezuela* **168 C4** 4 20N 66 27W
Paru de Oeste →, *Brazil* . . **169 C6** 1 30N 56 0W
Parubcan, *Phil.* . . . . . **80 E4** 13 43N 123 45 E
Parucito →, *Venezuela* **168 B4** 5 18N 65 59W
Parur, *India* . . . . . **95 J3** 10 13N 76 14 E
Paruro, *Peru* . . . . . **172 C3** 13 45 S 71 50W
Parvān □, *Afghan.* . . . . . **91 B3** 35 0N 69 0 E
Parvatipuram, *India* . . . . . **94 E6** 18 50N 83 25 E
Påryd, *Sweden* . . . . . **17 H9** 56 34N 15 55 E
Parys, *S. Africa* . . . . . **116 D4** 26 52 S 27 29 E
Pas-de-Calais □, *France* . . . . **27 B9** 50 30N 2 10 E
Pasada, *Spain* **42 B5** 43 23N 5 40W
Pasadena, *Calif., U.S.A.* **161 L8** 34 9N 118 9W
Pasadena, *Tex., U.S.A.* **155 L7** 29 43N 95 13W
Pasaje, *Ecuador* . . . . . **168 D2** 3 23 S 79 50W
Pasaje →, *Argentina* . . . . . **174 B3** 25 39 S 63 56W
Paşalimanı, *Turkey* . . . . . **51 F11** 40 29N 27 36 E
Pasay, *Phil.* . . . . . **80 D3** 14 33N 121 0 E
Pascagoula, *U.S.A.* **155 K10** 30 21N 88 33W
Pascagoula →, *U.S.A.* **155 K10** 30 23N 88 37W
Paşcani, *Romania* . . . . . **53 C11** 47 14N 26 45 E
Pasco, *U.S.A.* **158 C4** 46 14N 119 6W
Pasco □, *Peru* . . . . . **172 C3** 10 40 S 75 0W
Pasco, Cerro de, *Peru* . . . . . **172 C2** 10 45 S 76 10W
Pascua, I. de, *Pac. Oc.* . . . . . **135 K17** 27 0 S 109 0W
Pasewalk, *Germany* . . . . . **30 B9** 53 30N 13 58 E
Pasfield L., *Canada* **143 B7** 58 24N 105 20W
Pasha →, *Russia* . . . . . **58 B7** 60 29N 32 55 E
Pashiwari, *Pakistan* . . . . . **93 B6** 34 40N 75 10 E
Pashiya, *Russia* . . . . . **62 B7** 58 33N 58 26 E
Pashmakli = Smolyan, *Bulgaria* . . . . . **51 E8** 41 36N 24 38 E
Pasighat, *India* . . . . . **90 A5** 28 4N 95 21 E
Pasinler, *Turkey* . . . . . **101 C9** 39 59N 41 41 E
Pasir Mas, *Malaysia* . . . . . **87 J4** 6 2N 102 8 E
Pasirian, *Indonesia* . . . . . **85 D4** 8 13 S 113 8 E
Pasirkuning, *Indonesia* . . . . . **84 C2** 0 30 S 104 33 E
Påskallavik, *Sweden* . . . . . **17 G10** 57 10N 16 26 E
Paskūh, *Iran* . . . . . **97 E9** 27 34N 61 39 E
Pasłęk, *Poland* . . . . . **54 D6** 54 3N 19 41 E
Pasłęka →, *Poland* . . . . . **54 D6** 54 26N 19 46 E
Pasley, C., *Australia* . . . . . **125 F3** 33 52 S 123 35 E
Pašman, *Croatia* . . . . . **45 E12** 43 58N 15 20 E
Pasmore →, *Australia* . . . . . **128 A3** 31 5 S 139 49 E
Pasni, *Pakistan* . . . . . **91 D1** 25 15N 63 27 E
Paso Cantinela, *Mexico* **161 N11** 32 33N 115 47W
Paso de Indios, *Argentina* . . **176 B3** 43 55 S 69 0W
Paso de los Libres, *Argentina* **174 B4** 29 44 S 57 10W
Paso de los Toros, *Uruguay* **174 C4** 32 45 S 56 30W
Paso Flores, *Argentina* . . . . . **176 B2** 40 35 S 70 38W
Paso Robles, *U.S.A.* **159 J3** 35 38N 120 41W
Pasorapa, *Bolivia* . . . . . **173 D5** 18 16 S 64 37W
Paspébiac, *Canada* **141 C6** 48 3N 65 17W
Pasrur, *Pakistan* . . . . . **92 C6** 32 16N 74 43 E
Passage West, *Ireland* **23 E3** 51 52N 8 21W
Passaic, *U.S.A.* **151 F10** 40 51N 74 7W
Passau, *Germany* . . . . . **31 G9** 48 34N 13 28 E
Passero, C., *Italy* . . . . . **47 F8** 36 41N 15 10 E
Passi, *Phil.* . . . . . **81 F4** 11 6N 122 38 E
Passo Fundo, *Brazil* . . . . . **175 B5** 28 10 S 52 20W
Passos, *Brazil* . . . . . **171 F2** 20 45 S 46 37W
Passow, *Germany* . . . . . **30 B10** 53 8N 14 6 E
Passwang, *Switz.* . . . . . **32 B5** 47 22N 7 41 E
Passy, *France* . . . . . **29 C10** 45 55N 6 41 E
Pastavy, *Belarus* . . . . . **15 J22** 55 4N 26 50 E
Pastaza □, *Ecuador* . . . . . **168 D2** 2 0 S 77 0W
Pastaza →, *Peru* . . . . . **168 D2** 4 50 S 76 52W
Pasto, *Colombia* **168 C2** 1 13N 77 17W
Pastol B., *U.S.A.* **144 E7** 63 7N 163 15W
Pastos Bons, *Brazil* . . . . . **170 C3** 6 36 S 44 5W
Pastrana, *Spain* **40 E2** 40 27N 2 53W
Pasuquin, *Phil.* . . . . . **80 B3** 18 20N 120 37 E
Pasuruan, *Indonesia* . . . . . **85 D4** 7 40 S 112 44 E
Pasym, *Poland* . . . . . **54 E7** 53 48N 20 49 E
Pásztó, *Hungary* . . . . . **52 C4** 47 52N 19 43 E
Patagonia, *Argentina* . . . . . **176 C3** 45 0 S 69 0W
Patagonia, *U.S.A.* **159 L8** 31 33N 110 45W
Patambar, *Iran* . . . . . **97 D9** 29 45N 60 17 E
Patan, *Gujarat, India* . . . . . **94 F1** 17 22N 73 57 E
Patan, *Maharashtra, India* . . **92 H5** 23 54N 72 14 E
Patani, *Indonesia* . . . . . **82 A3** 0 20N 128 50 E
Pătârlagele, *Romania* . . . . . **53 E11** 45 19N 26 21 E
Pataudi, *India* . . . . . **92 E7** 28 18N 76 48 E
Patchewollock, *Australia* . . . **128 C5** 35 22 S 142 12 E
Patchogue, *U.S.A.* **151 F11** 40 46N 73 1W
Patea, *N.Z.* . . . . . **130 F3** 39 45 S 174 30 E
Pategi, *Nigeria* . . . . . **113 D6** 8 50N 5 45 E
Patensie, *S. Africa* . . . . . **116 E3** 33 46 S 24 49 E
Paterna, *Spain* **41 F4** 39 30N 0 26W
Paternion, *Austria* . . . . . **34 E6** 46 43N 13 38 E
Paternò, *Italy* . . . . . **47 E7** 37 34N 14 54 E
Pateros, *U.S.A.* **158 B4** 48 3N 119 54W
Paterson, *Australia* . . . . . **129 B9** 32 35 S 151 36 E
Paterson, *U.S.A.* **151 F10** 40 55N 74 11W
Paterson Inlet, *N.Z.* . . . . . **131 G3** 46 56 S 168 12 E
Paterson Ra., *Australia* . . . . **124 D3** 21 45 S 122 10 E
Pathankot, *India* . . . . . **92 C6** 32 18N 75 45 E
Patharghata, *Bangla.* . . . . . **90 D2** 22 2N 89 58 E
Pathfinder Reservoir, *U.S.A.* **158 E10** 42 28N 106 51W
Pathiu, *Thailand* . . . . . **87 G2** 10 42N 99 19 E
Pathum Thani, *Thailand* . . . . **86 E3** 14 1N 100 32 E
Pati, *Indonesia* . . . . . **85 D4** 6 45 S 111 1 E
Pati Pt., *Guam* . . . . . **133 R15** 13 40N 144 50 E
Patía, *Colombia* **168 C2** 2 4N 77 4W
Patía →, *Colombia* **168 C2** 2 13N 78 40W
Patiala, *India* . . . . . **92 D7** 30 23N 76 26 E
Patine Kouka, *Senegal* **112 C2** 12 45N 13 45W
Patitírion, *Greece* . . . . . **48 B5** 39 8N 23 50 E
Pativilca, *Peru* . . . . . **172 C2** 10 42 S 77 48W
Patkai Bum, *India* . . . . . **90 B5** 27 0N 95 30 E
Pátmos, *Greece* . . . . . **49 D8** 37 21N 26 36 E
Patna, *India* . . . . . **93 G11** 25 35N 85 12 E
Patnongon, *Phil.* . . . . . **81 F4** 10 55N 122 0 E
Patnos, *Turkey* . . . . . **101 C10** 39 13N 42 51 E
Patonga, *Uganda* . . . . . **118 B3** 2 45N 33 15 E
Patos, *Brazil* . . . . . **170 C4** 6 55 S 37 16W
Patos, L. dos, *Brazil* . . . . . **175 C5** 31 20 S 51 0W
Patos de Minas, *Brazil* . . . . **171 E2** 18 35 S 46 32W
Patosi, *Albania* . . . . . **50 F3** 40 42N 19 38 E
Patquía, *Argentina* . . . . . **174 C2** 30 2 S 66 55W
Pátrai, *Greece* . . . . . **48 C3** 38 14N 21 47 E
Pátraikós Kólpos, *Greece* . . . **48 C3** 38 17N 21 30 E
Patras = Pátrai, *Greece* . . . . **48 C3** 38 14N 21 47 E
Patricio Lynch, I., *Chile* . . . **176 C1** 48 35 S 75 30W
Patrocínio, *Brazil* . . . . . **171 E2** 18 57 S 47 0W

Piombino, Canale di 265

Pervomayskiy, *Russia* **62 F5** 51 32N 55 2 E
Pervouralsk, *Russia* **62 C7** 56 59N 59 59 E
Pes, Pta. del, *Spain* **39 C7** 38 46N 1 26 E
Pésaro, *Italy* **45 E9** 43 54N 12 55 E
Pescara, *Italy* **45 F11** 42 28N 14 13 E
Pescara →, *Italy* **45 F11** 42 28N 14 13 E
Peschanokopskoye, *Russia* **61 G5** 46 14N 41 4 E
Péscia, *Italy* **44 E7** 43 54N 10 41 E
Pescina, *Italy* **45 F10** 42 2N 13 39 E
Peseux, *Switz.* **32 C3** 46 59N 6 53 E
Peshawar, *Pakistan* **91 B3** 34 2N 71 37 E
Peshkopi, *Albania* **50 E4** 41 41N 20 25 E
Peshtera, *Bulgaria* **51 D8** 42 2N 24 18 E
Peshtigo, *U.S.A.* **148 C2** 45 4N 87 46W
Peski, *Russia* **60 E6** 51 14N 42 29 E
Peskovka, *Russia* **62 B4** 59 4N 52 22 E
Pêso da Régua, *Portugal* **42 D3** 41 10N 7 47W
Pesqueira, *Brazil* **170 C4** 8 20 S 36 42W
Pessac, *France* **28 D3** 44 48N 0 37W
Pest □, *Hungary* **52 C4** 47 29N 19 5 E
Pestovo, *Russia* **58 C8** 58 33N 35 42 E
Pestravka, *Russia* **60 D9** 52 28N 49 57 E
Péta, *Greece* **48 B3** 39 10N 21 2 E
Petah Tiqwa, *Israel* **103 C3** 32 6N 34 53 E
Petalídhion, *Greece* **48 E3** 36 57N 21 55 E
Petaling Jaya, *Malaysia* **87 L3** 3 4N 101 42 E
Petaloudhes, *Greece* **38 C10** 36 18N 28 5 E
Petaluma, *U.S.A.* **160 G4** 38 14N 122 39W
Pétange, *Lux.* **24 E5** 49 33N 5 55 E
Petatlán, *Mexico* **162 D4** 17 31N 101 16W
Petauke, *Zambia* **119 E3** 14 14 S 31 20 E
Petawawa, *Canada* **140 C4** 45 54N 77 17W
Petén Itzá, L., *Guatemala* **164 C2** 16 58N 89 50W
Peter I.s Øy, *Antarctica* **7 C16** 69 0 S 91 0W
Peter Pond L., *Canada* **143 B7** 55 55N 108 44W
Peterbell, *Canada* **140 C3** 48 36N 83 21W
Peterborough, *Australia* **128 B3** 32 58 S 138 51 E
Peterborough, *Canada* **150 B6** 44 20N 78 20W
Peterborough, *U.K.* **21 E7** 52 35N 0 15W
Peterborough, *U.S.A.* **151 D13** 42 53N 71 57W
Peterborough □, *U.K.* **21 E7** 52 35N 0 15W
Peterculter, *U.K.* **22 D6** 57 6N 2 16W
Peterhead, *U.K.* **22 D7** 57 31N 1 48W
Peterlee, *U.K.* **20 C6** 54 47N 1 20W
Petermann Bjerg, *Greenland* **10 C8** 73 7N 28 25W
Petermann Gletscher, *Greenland* **10 A4** 80 30N 60 0W
Peter's Mine, *Guyana* **169 B6** 6 14N 59 20W
Petersburg, *Alaska, U.S.A.* **142 B2** 56 48N 132 58W
Petersburg, *Ill., U.S.A.* **156 D7** 40 1N 89 51W
Petersburg, *Ind., U.S.A.* **157 F9** 38 30N 87 17W
Petersburg, *Va., U.S.A.* **148 G7** 37 14N 77 24W
Petersburg, *W. Va., U.S.A.* **148 F6** 39 1N 79 5W
Petersfield, *U.K.* **21 F7** 51 1N 0 56W
Petershagen, *Germany* **30 C4** 52 23N 8 58 E
Petford, *Australia* **126 B3** 17 20 S 144 58 E
Petília Policastro, *Italy* **47 C9** 39 7N 16 48 E
Petit Bois I., *U.S.A.* **149 K1** 30 12N 88 26W
Petit-Cap, *Canada* **141 C7** 49 3N 64 30W
Petit Goâve, *Haiti* **165 C5** 18 27N 72 51W
Petit Lac Manicouagan, *Canada* **141 B6** 51 25N 67 40W
Petit Saint Bernard, Col du, *Italy* **29 C10** 45 40N 6 52 E
Petitcodiac, *Canada* **141 C6** 45 57N 65 11W
Petite Baleine →, *Canada* **140 A4** 56 0N 76 45W
Petite Saguenay, *Canada* **141 C5** 48 15N 70 4W
Petitsikapau, L., *Canada* **141 B6** 54 37N 66 25W
Petlad, *India* **92 H5** 22 30N 72 45 E
Peto, *Mexico* **163 C7** 20 10N 88 53W
Petone, *N.Z.* **130 H3** 41 13 S 174 53 E
Petoskey, *U.S.A.* **148 C3** 45 22N 84 57W
Petra, *Jordan* **103 E4** 30 20N 35 22 E
Petra, *Spain* **39 B10** 39 37N 3 6 E
Petra, Ostrova, *Russia* **6 B13** 76 15N 118 30 E
Petra Velikogo, Zaliv, *Russia* **68 C6** 42 40N 132 0 E
Petrel = Petrer, *Spain* **41 G4** 38 30N 0 46W
Petrella, Monte, *Italy* **46 A6** 41 18N 13 40 E
Petrer, *Spain* **41 G4** 38 30N 0 46W
Petreto-Bicchisano, *France* **29 G12** 41 47N 8 58 E
Petrey, *U.S.A.* **152 D3** 31 51N 86 13W
Petrich, *Bulgaria* **50 E7** 41 24N 23 13 E
Petrijanec, *Croatia* **45 B13** 46 23N 16 17 E
Petrikov = Pyetrikaw, *Belarus* **59 F5** 52 11N 28 29 E
Petrila, *Romania* **53 E8** 45 29N 23 29 E
Petrinja, *Croatia* **45 C13** 45 28N 16 18 E
Petrodvorets, *Russia* **58 C5** 59 52N 29 54 E
Petrograd = Sankt-Peterburg, *Russia* **58 C6** 59 55N 30 20 E
Petrolândia, *Brazil* **170 C4** 9 5 S 38 20W
Petrolia, *Canada* **140 D3** 42 54N 82 9W
Petrolina, *Brazil* **170 C3** 9 24 S 40 30W
Petropavl, *Kazakstan* **64 D7** 54 53N 69 13 E
Petropavlovsk = Petropavl, *Kazakstan* **64 D7** 54 53N 69 13 E
Petropavlovsk-Kamchatskiy, *Russia* **65 D16** 53 3N 158 43 E
Petropavlovskiy = Akhtubinsk, *Russia* **61 F8** 48 13N 46 7 E
Petrópolis, *Brazil* **171 F3** 22 33 S 43 9W
Petroşani, *Romania* **53 E8** 45 28N 23 20 E
Petrova Gora, *Croatia* **45 C12** 45 15N 15 45 E
Petrovac, *Montenegro, Yug.* **50 D2** 42 13N 18 57 E
Petrovac, *Serbia, Yug.* **50 B5** 44 22N 21 26 E
Petrovaradin, *Serbia, Yug.* **52 E4** 45 16N 19 55 E
Petrovsk, *Russia* **60 D7** 52 22N 45 19 E
Petrovsk-Zabaykalskiy, *Russia* **65 D11** 51 20N 108 55 E
Petrovskaya, *Russia* **59 K9** 45 25N 37 58 E
Petrovskoye = Svetlograd, *Russia* **61 H6** 45 25N 42 58 E
Petrovskoye, *Russia* **62 E6** 53 37N 56 23 E
Petrozavodsk, *Russia* **58 B8** 61 41N 34 20 E
Petrus Steyn, *S. Africa* **117 D4** 27 38 S 28 8 E
Petrusburg, *S. Africa* **116 D4** 29 4 S 25 26 E
Pettitts, *Australia* **129 C8** 34 56 S 148 10 E
Petzeck, *Austria* **34 E5** 46 57N 12 48 E
Peumo, *Chile* **174 C1** 34 21 S 71 12W
Peureulak, *Indonesia* **84 B1** 4 48N 97 45 E
Peusangan →, *Indonesia* **84 A1** 5 16N 96 51 E
Pevek, *Russia* **65 C18** 69 41N 171 19 E
Peveragno, *Italy* **44 D4** 44 20N 7 37 E
Peyrehorade, *France* **28 E2** 43 34N 1 7W
Peyruis, *France* **29 D9** 44 1N 5 56 E
Pézenas, *France* **28 E7** 43 28N 3 24 E
Pezinok, *Slovak Rep.* **35 C10** 48 17N 17 17 E
Pfaffenhofen, *Germany* **31 G7** 48 31N 11 31 E
Pfäffikon, *Switz.* **33 B7** 47 13N 8 46 E
Pfarrkirchen, *Germany* **31 G8** 48 25N 12 56 E
Pfeffenhausen, *Germany* **31 G7** 48 39N 11 58 E
Pforzheim, *Germany* **31 G4** 48 52N 8 41 E
Pfullendorf, *Germany* **31 H5** 47 55N 9 15 E
Pfungstadt, *Germany* **31 F4** 49 48N 8 35 E
Phaistós, *Greece* **38 D6** 35 2N 24 50 E
Phala, *Botswana* **116 C4** 23 45 S 26 50 E
Phalera = Phulera, *India* **92 F6** 26 52N 75 16 E
Phalodi, *India* **92 F5** 27 12N 72 24 E
Phalsbourg, *France* **27 D14** 48 46N 7 15 E
Phan, *Thailand* **86 C2** 19 28N 99 43 E
Phan Rang, *Vietnam* **87 G7** 11 34N 109 0 E
Phan Ri = Hoa Da, *Vietnam* **87 G7** 11 16N 108 40 E
Phan Thiêt, *Vietnam* **87 G7** 11 1N 108 9 E
Phanae, *Greece* **49 C7** 38 8N 25 57 E
Phanat Nikhom, *Thailand* **86 F3** 13 27N 101 11 E
Phangan, Ko, *Thailand* **87 H3** 9 45N 100 0 E
Phangnga, *Thailand* **87 H2** 8 28N 98 30 E
Phanh Bho Ho Chi Minh, *Vietnam* **87 G6** 10 58N 106 40 E
Phanom Sarakham, *Thailand* **86 F3** 13 45N 101 21 E
Pharenda, *India* **93 F10** 27 5N 83 17 E
Phatthalung, *Thailand* **87 J3** 7 39N 100 6 E
Phayao, *Thailand* **86 C2** 19 11N 99 55 E
Phelps, *N.Y., U.S.A.* **150 D7** 42 58N 77 3W
Phelps, *Wis., U.S.A.* **154 B10** 46 4N 89 5W
Phelps L., *Canada* **143 B8** 59 15N 103 15W
Phenix City, *U.S.A.* **149 J3** 32 28N 85 0W
Phet Buri, *Thailand* **86 F2** 13 1N 99 55 E
Phetchabun, *Thailand* **86 D3** 16 25N 101 8 E
Phetchabun, Thiu Khao, *Thailand* **86 E3** 16 0N 101 20 E
Phetchaburi = Phet Buri, *Thailand* **86 F2** 13 1N 99 55 E
Phi Phi, Ko, *Thailand* **87 J2** 7 45N 98 46 E
Phiafay, *Laos* **86 E6** 14 48N 106 0 E
Phibun Mangsahan, *Thailand* **86 E5** 15 14N 105 14 E
Phichai, *Thailand* **86 D3** 17 22N 100 10 E
Phichit, *Thailand* **86 D3** 16 26N 100 22 E
Philadelphia, *Miss., U.S.A.* **155 J10** 32 46N 89 7W
Philadelphia, *N.Y., U.S.A.* **151 B9** 44 9N 75 43W
Philadelphia, *Pa., U.S.A.* **151 G9** 39 57N 75 10W
Philip, *U.S.A.* **154 C4** 44 2N 101 40W
Philip Smith Mts., *U.S.A.* **144 C11** 68 0N 148 0W
Philippeville, *Belgium* **24 D4** 50 12N 4 33 E
Philippi, *Greece* **51 E8** 41 1N 24 16 E
Philippi L., *Australia* **126 C2** 24 20 S 138 55 E
Philippines ■, *Asia* **80 E4** 12 0N 123 0 E
Philippolis, *S. Africa* **116 E4** 30 15 S 25 16 E
Philippopolis = Plovdiv, *Bulgaria* **51 D8** 42 8N 24 44 E
Philipsburg, *Mont., U.S.A.* **158 C7** 46 20N 113 18W
Philipsburg, *Pa., U.S.A.* **150 F6** 40 54N 78 13W
Philipstown = Daingean, *Ireland* **23 C4** 53 18N 7 17W
Philipstown, *S. Africa* **116 E3** 30 28 S 24 30 E
Phillip I., *Australia* **129 E6** 38 30 S 145 12 E
Phillips, *Tex., U.S.A.* **155 H4** 35 42N 101 22W
Phillips, *Wis., U.S.A.* **154 C9** 45 42N 90 24W
Phillipsburg, *Ga., U.S.A.* **152 D6** 31 25N 83 30W
Phillipsburg, *Kans., U.S.A.* **154 F5** 39 45N 99 19W
Phillipsburg, *N.J., U.S.A.* **151 F9** 40 42N 75 12W
Phillott, *Australia* **127 D4** 27 53 S 145 50 E
Philmont, *U.S.A.* **151 D11** 42 15N 73 39W
Philomath, *Ga., U.S.A.* **152 B7** 33 44N 82 59W
Philomath, *Oreg., U.S.A.* **158 D2** 44 32N 123 22W
Phimai, *Thailand* **86 E4** 15 13N 102 30 E
Phitsanulok, *Thailand* **86 D3** 16 50N 100 12 E
Phnom Dang, *Cambodia* **86 F5** 14 20N 104 0 E
Phnom Penh, *Cambodia* **87 G5** 11 33N 104 55 E
Phnum Penh = Phnom Penh, *Cambodia* **87 G5** 11 33N 104 55 E
Phoenix, *Ariz., U.S.A.* **159 K7** 33 27N 112 4W
Phoenix, *N.Y., U.S.A.* **151 C8** 43 14N 76 18W
Phoenix Is., *Kiribati* **134 H10** 3 30 S 172 0W
Phoenixville, *U.S.A.* **151 F9** 40 8N 75 31W
Phon, *Thailand* **86 E4** 15 49N 102 36 E
Phon Tiou, *Laos* **86 D5** 17 53N 104 37 E
Phong →, *Thailand* **86 D4** 16 23N 102 56 E
Phong Saly, *Laos* **86 B4** 21 42N 102 9 E
Phong Tho, *Vietnam* **86 A4** 22 32N 103 21 E
Phonhong, *Laos* **86 C4** 18 30N 102 25 E
Phonum, *Thailand* **87 H2** 8 49N 98 48 E
Phosphate Hill, *Australia* **126 C2** 21 53 S 139 58 E
Photharam, *Thailand* **86 F2** 13 41N 99 51 E
Phra Nakhon Si Ayutthaya, *Thailand* **86 E3** 14 25N 100 30 E
Phra Thong, Ko, *Thailand* **87 H2** 9 5N 98 17 E
Phrae, *Thailand* **86 C3** 18 7N 100 9 E
Phrom Phiram, *Thailand* **86 D3** 17 2N 100 12 E
Phrygia, *Turkey* **100 C4** 38 40N 30 0 E
Phu Dien, *Vietnam* **86 C5** 18 58N 105 31 E
Phu Loi, *Laos* **86 B4** 20 14N 103 14 E
Phu Ly, *Vietnam* **86 B5** 20 35N 105 50 E
Phu Quoc, Dao, *Vietnam* **87 G4** 10 20N 104 0 E
Phu Tho, *Vietnam* **86 B5** 21 24N 105 13 E
Phuc Yen, *Vietnam* **86 B5** 21 16N 105 45 E
Phuket, *Thailand* **87 J2** 7 52N 98 22 E
Phuket, Ko, *Thailand* **87 J2** 8 0N 98 22 E
Phulbari, *India* **90 C3** 25 55N 90 2 E
Phulera, *India* **92 F6** 26 52N 75 16 E
Phun Phin, *Thailand* **87 H2** 9 7N 99 12 E
Piacá, *Brazil* **170 C2** 7 42 S 47 18W
Piacenza, *Italy* **44 C6** 45 1N 9 40 E
Piaçubaçu, *Brazil* **170 D4** 10 24 S 36 25W
Piako →, *N.Z.* **130 D4** 37 12 S 175 30 E
Pialba, *Australia* **127 D5** 25 20 S 152 45 E
Pian Cr. →, *Australia* **127 E4** 30 2 S 148 12 E
Piana, *France* **29 F12** 42 15N 8 34 E
Pianella, *Italy* **45 F11** 42 24N 14 2 E
Piangil, *Australia* **128 C5** 35 5 S 143 20 E
Pianosa, *Puglia, Italy* **45 F12** 42 12N 15 44 E
Pianosa, *Toscana, Italy* **44 F7** 42 35N 10 5 E
Piapot, *Canada* **143 D7** 49 59N 109 8W
Pias, *Portugal* **43 G3** 38 1N 7 29W
Piaseczno, *Poland* **55 F8** 52 5N 21 2 E
Piaski, *Poland* **55 G9** 51 8N 22 52 E
Piastów, *Poland* **55 F7** 52 12N 20 48 E
Piatã, *Brazil* **171 D3** 13 9 S 41 48W
Piatra, *Romania* **53 G10** 43 51N 25 9 E
Piatra Neamţ, *Romania* **53 D11** 46 56N 26 21 E
Piatra Olt, *Romania* **53 F9** 44 22N 24 16 E
Piauí □, *Brazil* **170 C3** 7 0 S 43 0W
Piaui →, *Brazil* **170 C3** 6 38 S 42 42W
Piave →, *Italy* **45 C9** 45 32N 12 44 E
Piazza Ármerina, *Italy* **47 E7** 37 21N 14 20 E
Pibor →, *Sudan* **107 F3** 7 35N 33 0 E
Pibor Post, *Sudan* **107 F3** 6 47N 33 3 E
Pica, *Chile* **172 E4** 20 35 S 69 25W
Picardie, *France* **27 C10** 49 50N 3 0 E
Picardie, Plaine de, *France* **27 C9** 50 0N 2 0 E
Picardy = Picardie, *France* **27 C10** 49 50N 3 0 E
Picayune, *U.S.A.* **155 K10** 30 32N 89 41W
Picerno, *Italy* **47 B8** 40 38N 15 38 E
Pichilemu, *Chile* **174 C1** 34 22 S 72 0W
Pichincha, □, *Ecuador* **168 D2** 0 10 S 78 40W
Pickerel L., *Canada* **140 C1** 48 40N 91 25W
Pickering, *U.K.* **20 C7** 54 15N 0 46W
Pickering, Vale of, *U.K.* **20 C7** 54 14N 0 45W
Pickle Lake, *Canada* **140 B1** 51 30N 90 12W
Pico Truncado, *Argentina* **176 C3** 46 40 S 68 0W
Picos, *Brazil* **170 C3** 7 5 S 41 28W
Picota, *Peru* **172 B2** 6 54 S 76 24W
Picton, *Australia* **129 C9** 34 12 S 150 34 E
Picton, *Canada* **140 D4** 44 1N 77 9W
Picton, *N.Z.* **131 B9** 41 18 S 174 3 E
Picton, I., *Chile* **176 E3** 55 2 S 66 57W
Pictou, *Canada* **141 C7** 45 41N 62 42W
Picture Butte, *Canada* **142 D6** 49 55N 112 45W
Picuí, *Brazil* **170 C4** 6 31 S 36 21W
Picún Leufú, *Argentina* **176 A3** 39 30 S 69 5W
Pidurutalagala, *Sri Lanka* **95 L5** 7 10N 80 50 E
Piechowice, *Poland* **55 H2** 50 51N 15 36 E
Piedecuesta, *Colombia* **168 B3** 6 59N 73 3W
Piedmont = Piemonte □, *Italy* **44 D5** 45 0N 8 0 E
Piedmont, *U.S.A.* **152 B4** 33 55N 85 37W
Piedmont Plateau, *U.S.A.* **149 J5** 34 0N 81 30W
Piedmonte Matese, *Italy* **47 A7** 41 22N 14 22 E
Piedra →, *Spain* **40 D3** 41 18N 1 47W
Piedra del Anguila, *Argentina* **176 B2** 40 2 S 70 4W
Piedra Lais, *Venezuela* **168 C4** 3 10N 65 50W
Piedrabuena, *Spain* **43 G6** 39 0N 4 10W
Piedrahita, *Spain* **42 E5** 40 28N 5 23W
Piedralaves, *Spain* **42 E6** 40 19N 4 42W
Piedras, R. de las →, *Peru* **172 C4** 12 30 S 69 15W
Piedras Blancas, *Spain* **42 B5** 43 33N 5 58W
Piedras Negras, *Mexico* **162 B4** 28 42N 100 31W
Piedras Pt., *Phil.* **81 F2** 10 11N 118 48 E
Piekary Śląskie, *Poland* **55 H5** 50 24N 18 57 E
Pieksämäki, *Finland* **15 E22** 62 18N 27 10 E
Piemonte □, *Italy* **44 D5** 45 0N 8 0 E
Pieniężno, *Poland* **54 D7** 54 14N 20 8 E
Pieńsk, *Poland* **55 G2** 51 16N 15 2 E
Pier Millan, *Australia* **128 C5** 35 14 S 142 40 E
Pierce, *U.S.A.* **158 C6** 46 30N 115 48W
Piercefield, *U.S.A.* **151 B10** 44 13N 74 35W
Piería □, *Greece* **50 F6** 40 13N 22 25 E
Pierre, *U.S.A.* **154 C4** 44 22N 100 21W
Pierre-Buffière, *France* **28 C5** 45 41N 1 22 E
Pierre-de-Bresse, *France* **27 F12** 46 54N 5 13 E
Pierrefontaine-les-Varans, *France* **27 E13** 47 14N 6 32 E
Pierrefort, *France* **28 D6** 44 55N 2 50 E
Pierrelatte, *France* **29 D8** 44 23N 4 43 E
Pierson, *U.S.A.* **153 F8** 29 14N 81 28W
Piešťany, *Slovak Rep.* **35 C10** 48 38N 17 55 E
Piesting →, *Austria* **35 C9** 48 6N 16 40 E
Pieszyce, *Poland* **55 H3** 50 43N 16 33 E
Piet Retief, *S. Africa* **117 D5** 27 1 S 30 50 E
Pietarsaari, *Finland* **14 E20** 63 40N 22 43 E
Pietermaritzburg, *S. Africa* **117 D5** 29 35 S 30 25 E
Pietersburg, *S. Africa* **117 C4** 23 54 S 29 25 E
Pietragalla, *Italy* **47 B8** 40 45N 15 53 E
Pietrasanta, *Italy* **44 E7** 43 57N 10 14 E
Pietroşiţa, *Romania* **53 E10** 45 11N 25 26 E
Pietrosul, Vf., *Maramureş, Romania* **53 C9** 47 35N 24 43 E
Pietrosul, Vf., *Suceava, Romania* **53 C10** 47 12N 25 18 E
Pieve di Cadore, *Italy* **45 B9** 46 26N 12 22 E
Pieve di Teco, *Italy* **44 D4** 44 3N 7 56 E
Pievepélago, *Italy* **44 D7** 44 12N 10 37 E
Pigadhítsa, *Greece* **50 G5** 39 59N 21 23 E
Pigeon, *U.S.A.* **148 D4** 43 50N 83 16W
Pigeon I., *India* **95 G2** 14 2N 74 20 E
Piggott, *U.S.A.* **155 G9** 36 23N 90 11W
Pigna, *Italy* **44 E4** 43 56N 7 40 E
Pigüe, *Argentina* **174 D3** 37 36 S 62 25W
Pihani, *India* **93 F9** 27 36N 80 15 E
Pihlajavesi, *Finland* **15 F23** 61 45N 28 45 E
Pikalevo, *Russia* **58 C8** 59 37N 34 9 E
Pike Road, *U.S.A.* **152 C3** 32 17N 86 6W
Pikes Peak, *U.S.A.* **154 F2** 38 50N 105 3W
Piketberg, *S. Africa* **116 E2** 32 55 S 18 40 E
Pikeville, *U.S.A.* **148 G4** 37 29N 82 31W
Pikou, *China* **75 E12** 39 18N 122 22 E
Pikwitonei, *Canada* **143 B9** 55 35N 97 9W
Piła, *Poland* **55 E3** 53 10N 16 48 E
Pila, *Spain* **41 G3** 38 16N 1 11W
Piła □, *Poland* **55 E3** 53 0N 17 0 E
Pilaía, *Greece* **50 F6** 40 32N 22 59 E
Pilani, *India* **92 E6** 28 22N 75 33 E
Pilar, *Brazil* **170 C4** 9 36 S 35 56W
Pilar, *Paraguay* **174 B4** 26 50 S 58 20W
Pilar de la Horadada, *Spain* **41 H4** 37 52N 0 47W
Pilas Group, *Phil.* **81 H3** 6 45N 121 35 E
Pilawa, *Poland* **55 G8** 51 57N 21 32 E
Pilaya →, *Bolivia* **173 E5** 20 55 S 64 4W
Pilcomayo →, *Paraguay* **174 B4** 25 21 S 57 42W
Pilgrimstad, *Sweden* **16 B9** 62 57N 15 2 E
Pílí, *Greece* **49 E9** 36 50N 27 15 E
Pili, *Phil.* **80 E4** 13 33N 123 19 E
Pilibhit, *India* **93 E8** 28 40N 79 50 E
Pilica →, *Poland* **55 G8** 51 52N 21 17 E
Pilion, *Greece* **48 B5** 39 27N 23 3 E
Pilis, *Hungary* **52 C4** 47 17N 19 35 E
Pilisvörösvár, *Hungary* **52 C3** 47 38N 18 56 E
Pilkhawa, *India* **92 E7** 28 43N 77 42 E
Pillaro, *Ecuador* **168 D2** 1 10 S 78 32W
Pílos, *Greece* **48 E3** 36 55N 21 42 E
Pilot Grove, *U.S.A.* **156 F4** 38 53N 92 55W
Pilot Mound, *Canada* **143 D9** 49 15N 98 54W
Pilot Point, *Alaska, U.S.A.* **144 H8** 57 34N 157 35W
Pilot Point, *Tex., U.S.A.* **155 J6** 33 24N 96 58W
Pilot Rock, *U.S.A.* **158 D4** 45 29N 118 50W
Pilot Station, *U.S.A.* **144 F7** 61 56N 162 53W
Pilsen = Plzeň, *Czech Rep.* **34 B6** 49 45N 13 22 E
Pilštanj, *Slovenia* **45 B12** 46 8N 15 39 E
Piltene, *Latvia* **54 A8** 57 13N 21 40 E
Pilzno, *Poland* **55 J8** 49 58N 21 16 E
Pima, *U.S.A.* **159 K9** 32 54N 109 50W
Pimba, *Australia* **128 A2** 31 18 S 136 46 E
Pimenta Bueno, *Brazil* **173 C5** 11 35 S 61 10W
Pimentel, *Peru* **172 B2** 6 45 S 79 55W
Pina de Ebro, *Spain* **40 D4** 41 29N 0 33W
Pinamalayan, *Phil.* **80 E3** 13 2N 121 29 E
Pinang, *Malaysia* **87 K3** 5 25N 100 15 E
Pinang □, *Malaysia* **84 A2** 5 20N 100 20 E
Pinar, C. del, *Spain* **39 B10** 39 53N 3 12 E
Pinar del Río, *Cuba* **164 B3** 22 26N 83 40W
Pınarbaşı, *Çanakkale, Turkey* **49 B8** 39 53N 26 15 E
Pınarbaşı, *Kayseri, Turkey* **100 C7** 38 43N 36 23 E
Pınarhisar, *Turkey* **51 E11** 41 37N 27 30 E
Pinatubo, *Phil.* **80 D3** 15 8N 120 21 E
Pincehely, *Hungary* **52 D3** 46 41N 18 27 E
Pinchang, *China* **76 B6** 31 36N 107 3 E
Pincher Creek, *Canada* **142 D6** 49 30N 113 57W
Pinchi L., *Canada* **142 C4** 54 38N 124 30W
Pinckard, *U.S.A.* **152 D4** 31 19N 85 33W
Pinckneyville, *U.S.A.* **156 F7** 38 5N 89 23W
Pińczów, *Poland* **55 H7** 50 32N 20 32 E
Pind Dadan Khan, *Pakistan* **92 C5** 32 36N 73 7 E
Pindar, *Australia* **125 E2** 28 30 S 115 47 E
Pindaré →, *Brazil* **170 B3** 3 17 S 44 47W
Pindaré Mirim, *Brazil* **170 B2** 3 37 S 45 21W
Pindi Gheb, *Pakistan* **92 C5** 33 14N 72 21 E
Pindiga, *Nigeria* **113 D7** 9 58N 10 53 E
Pindobal, *Brazil* **170 B2** 3 16 S 48 25W
Pindos Óros, *Greece* **48 B3** 40 0N 21 0 E
Pindus Mts. = Pindos Óros, *Greece* **48 B3** 40 0N 21 0 E
Pine, *U.S.A.* **159 J8** 34 23N 111 27W
Pine →, *Canada* **143 B7** 58 50N 105 38W
Pine, C., *Canada* **141 C9** 46 37N 53 32W
Pine Bluff, *U.S.A.* **155 H9** 34 13N 92 1W
Pine City, *U.S.A.* **154 C8** 45 50N 92 59W
Pine Falls, *Canada* **143 C9** 50 34N 96 11W
Pine Flat L., *U.S.A.* **160 J7** 36 50N 119 20W
Pine Hill, *U.S.A.* **153 G8** 28 32N 81 28W
Pine Is., *U.S.A.* **153 J7** 26 36N 82 7W
Pine Level, *U.S.A.* **152 C3** 32 4N 86 4W
Pine Mountain, *U.S.A.* **152 C5** 32 52N 84 51W
Pine Pass, *Canada* **142 B4** 55 25N 122 42W
Pine Point, *Canada* **142 A6** 60 50N 114 28W
Pine Ridge, *Australia* **129 A9** 31 30 S 150 28 E
Pine Ridge, *U.S.A.* **154 D3** 43 2N 102 33W
Pine River, *Canada* **143 C8** 51 45N 100 30W
Pine River, *U.S.A.* **154 B7** 46 43N 94 24W
Pine Valley, *U.S.A.* **161 N10** 32 50N 116 32W
Pinecrest, *U.S.A.* **160 G6** 38 12N 120 1W
Pineda de Mar, *Spain* **40 D7** 41 37N 2 42 E
Pinedale, *U.S.A.* **160 J7** 36 50N 119 48W
Pinega →, *Russia* **56 B8** 64 30N 44 19 E
Pinehill, *Australia* **126 C4** 23 38 S 146 57 E
Pinehurst, *U.S.A.* **152 C6** 32 12N 83 46W
Pinellas Park, *U.S.A.* **153 H7** 27 50N 82 43W
Pinerolo, *Italy* **44 D4** 44 53N 7 21 E
Pineto, *Italy* **45 F11** 42 36N 14 4 E
Pinetop, *U.S.A.* **159 J9** 34 8N 109 56W
Pinetown, *S. Africa* **117 D5** 29 48 S 30 54 E
Pinetree, *U.S.A.* **158 E11** 43 42N 105 52W
Pinetta, *U.S.A.* **152 E6** 30 36N 83 21W
Pineview, *U.S.A.* **152 C6** 32 7N 83 30W
Pineville, *Ky., U.S.A.* **149 G4** 36 46N 83 42W
Pineville, *La., U.S.A.* **155 K8** 31 19N 92 26W
Pineville, *S.C., U.S.A.* **152 B9** 33 26N 80 1W
Pinewood, *U.S.A.* **152 B9** 33 44N 80 27W
Piney, *France* **27 D11** 48 22N 4 21 E
Ping →, *Thailand* **86 E3** 15 42N 100 9 E
Pingaring, *Australia* **125 F2** 32 40 S 118 32 E
Pingba, *China* **76 D6** 26 23N 106 12 E
Pingchuan, *China* **76 D3** 27 35N 101 55 E
Pingding, *China* **74 F7** 37 47N 113 38 E
Pingdingshan, *China* **74 H7** 33 43N 113 27 E
Pingdong, *Taiwan* **77 F13** 22 39N 120 30 E
Pingdu, *China* **75 F10** 36 42N 119 59 E
Pingelly, *Australia* **125 F2** 32 32 S 117 5 E
Pingguo, *China* **76 F6** 23 19N 107 36 E
Pinghe, *China* **77 E11** 24 17N 117 21 E
Pinghu, *China* **77 B13** 30 40N 121 2 E
Pingjiang, *China* **77 C9** 28 45N 113 36 E
Pingle, *China* **77 E8** 24 40N 110 40 E
Pingli, *China* **76 A7** 32 27N 109 22 E
Pingliang, *China* **74 G4** 35 35N 106 31 E
Pinglu, *China* **74 E7** 39 31N 112 30 E
Pingluo, *China* **74 E4** 38 52N 106 30 E
Pingnan, *Fujian, China* **77 D12** 26 55N 119 0 E
Pingnan, *Guangxi Zhuangzu, China* **77 F8** 23 33N 110 22 E
Pingquan, *China* **75 D10** 41 1N 118 37 E
Pingrup, *Australia* **125 F2** 33 32 S 118 29 E
Pingtan, *China* **77 E12** 25 31N 119 47 E
Pingtang, *China* **76 E6** 25 49N 107 17 E
P'ingtung, *Taiwan* **77 F13** 22 38N 120 30 E
Pingwu, *China* **74 H3** 32 25N 104 30 E
Pingxiang, *Guangxi Zhuangzu, China* **76 F6** 22 6N 106 46 E
Pingxiang, *Jiangxi, China* **77 D9** 27 43N 113 48 E
Pingyao, *China* **74 F7** 37 12N 112 10 E
Pingyi, *China* **75 G9** 35 30N 117 35 E
Pingyin, *China* **74 F9** 36 20N 116 25 E
Pingyuan, *Guangdong, China* **77 E10** 24 37N 115 57 E
Pingyuan, *Shandong, China* **74 F9** 37 10N 116 22 E
Pingyuanjie, *China* **76 F4** 23 45N 103 48 E
Pinhal, *Brazil* **175 A6** 22 10 S 46 46W
Pinhal Novo, *Portugal* **43 G2** 38 38N 8 55W
Pinheiro, *Brazil* **170 B2** 2 31 S 45 5W
Pinhel, *Portugal* **42 E3** 40 50N 7 1W
Pinhuá →, *Brazil* **173 B5** 6 21 S 65 0W
Pini, *Indonesia* **84 B1** 0 10N 98 40 E
Piniós →, *Ilía, Greece* **48 D3** 37 48N 21 20 E
Piniós →, *Trikkala, Greece* **48 B4** 39 55N 22 41 E
Pinjarra, *Australia* **125 F2** 32 37 S 115 52 E
Pink →, *Canada* **143 B8** 56 50N 103 50W
Pinkafeld, *Austria* **35 D9** 47 22N 16 9 E
Pinlebu, *Burma* **90 C5** 24 5N 95 22 E
Pinnacles, *Australia* **125 E3** 28 12 S 120 26 E
Pinnacles, *U.S.A.* **160 J5** 36 33N 121 19W
Pinnaroo, *Australia* **128 C4** 35 17 S 140 53 E
Pinneberg, *Germany* **30 B5** 53 39N 9 48 E
Pínnes, Akra, *Greece* **51 F8** 40 5N 24 20 E
Pino Hachado, Paso, *S. Amer.* **176 A2** 38 39 S 70 54W
Pinon Hills, *U.S.A.* **161 L9** 34 26N 117 39W
Pinos, *Mexico* **162 C4** 22 20N 101 40W
Pinos, Mt., *U.S.A.* **161 L7** 34 49N 119 8W
Pinos Pt., *U.S.A.* **159 H3** 36 38N 121 57W
Pinos Puente, *Spain* **43 H7** 37 15N 3 45W
Pinotepa Nacional, *Mexico* **163 D5** 16 19N 98 3W
Pinrang, *Indonesia* **82 B1** 3 46 S 119 41 E
Pins, I. des, *N. Cal.* **133 V21** 22 37 S 167 30 E
Pinsk, *Belarus* **59 F4** 52 10N 26 1 E
Pintados, *Chile* **172 E4** 20 35 S 69 40W
Pintumba, *Australia* **125 F5** 31 30 S 132 12 E
Pintuyan, *Phil.* **81 G5** 9 57N 125 15 E
Pinukpuk, *Phil.* **80 C3** 17 35N 121 22 E
Pinyang, *China* **77 D13** 27 42N 120 31 E
Pinyug, *Russia* **62 A1** 60 5N 48 0 E
Pio V. Corpuz, *Phil.* **81 F5** 11 55N 124 2 E
Pio XII, *Brazil* **170 B2** 3 53 S 45 17W
Pioche, *U.S.A.* **159 H6** 37 56N 114 27W
Pioduran, *Phil.* **80 E4** 13 2N 123 25 E
Piombino, *Italy* **44 F7** 42 55N 10 32 E
Piombino, Canale di, *Italy* **44 F7** 42 53N 10 30 E

266 Pioner, Ostrov

Pioner, Ostrov, Russia 65 B10 79 50N 92 0 E
Pionki, Poland 55 G8 51 29N 21 28 E
Piorini →, Brazil 169 D5 3 23 S 63 30W
Piorini, L., Brazil 169 D5 3 15 S 62 35W
Piotrków Trybunalski, Poland 55 G6 51 23N 19 43 E
Piotrków Trybunalski □, Poland 55 G6 51 30N 19 45 E
Piove di Sacco, Italy 45 C9 45 18N 12 2 E
Pīp, Iran 97 E9 26 45N 60 10 E
Pipar, India 92 F5 26 25N 73 31 E
Piparia, India 92 H8 22 45N 78 23 E
Pipéri, Greece 48 B6 39 20N 24 19 E
Pipestone, U.S.A. 154 D6 44 0N 96 19W
Pipestone →, Canada 140 B2 52 53N 89 23W
Pipestone Cr. →, Canada 143 D8 49 38N 100 15W
Pipiriki, N.Z. 130 F4 39 28 S 175 5 E
Pipmuacan, Rés., Canada 141 C5 49 45N 70 30W
Pippingarra, Australia 124 D2 20 27 S 118 42 E
Pipriac, France 26 E5 47 59N 1 58W
Piqua, U.S.A. 157 D12 40 9N 84 15W
Piquet Carneiro, Brazil 170 C4 5 48 S 39 25W
Piquiri →, Brazil 175 A5 24 3 S 54 14W
Pīr Sohrāb, Iran 97 E9 25 44N 60 54 E
Piracanjuba, Brazil 171 E2 17 18 S 49 1W
Piracicaba, Brazil 175 A6 22 45 S 47 40W
Piracuruca, Brazil 170 B3 3 50 S 41 50W
Piræus = Piraiévs, Greece 48 D5 37 57N 23 42 E
Piraiévs, Greece 48 D5 37 57N 23 42 E
Pirajuí, Brazil 175 A6 21 59 S 49 29W
Piran, Slovenia 45 C10 45 31N 13 33 E
Pirané, Argentina 174 B4 25 42 S 59 6W
Piranhas, Brazil 170 C4 9 27 S 37 46W
Pirano = Piran, Slovenia 45 C10 45 31N 13 33 E
Pirapemas, Brazil 170 B3 3 43 S 44 14W
Pirapora, Brazil 171 E3 17 20 S 44 56W
Piray →, Bolivia 173 D5 16 32 S 63 45W
Pirdop, Bulgaria 51 D8 42 40N 24 10 E
Pires do Rio, Brazil 171 E2 17 18 S 48 17W
Pirganj, Bangla. 90 C2 25 51N 88 24 E
Pírgos, Ilía, Greece 48 D3 37 40N 21 27 E
Pírgos, Kríti, Greece 49 F7 35 0N 25 9 E
Pirgovo, Bulgaria 51 C9 43 44N 25 43 E
Piribebuy, Paraguay 174 B4 25 26 S 57 2W
Pirin Planina, Bulgaria 50 E7 41 40N 23 30 E
Píríneos = Pyrénées, Europe 28 F4 42 45N 0 18 E
Piripiri, Brazil 170 B3 4 15 S 41 46W
Piritu, Venezuela 168 B4 9 23N 69 12W
Pirmasens, Germany 31 F3 49 12N 7 36 E
Pirna, Germany 30 E9 50 57N 13 56 E
Pirojpur, Bangla. 90 D3 22 35N 90 1 E
Pirot, Serbia, Yug. 50 C6 43 9N 22 33 E
Piru, Indonesia 82 B3 3 4 S 128 12 E
Piru, U.S.A. 161 L8 34 25N 118 48W
Piryatin = Pyryatyn, Ukraine 59 G7 50 15N 32 25 E
Piryí, Greece 49 C7 38 13N 25 59 E
Pisa, Italy 44 E7 43 43N 10 23 E
Pisa →, Poland 55 E8 53 14N 21 52 E
Pisa Ra., N.Z. 131 E4 44 52 S 169 12 E
Pisac, Peru 172 C3 13 25 S 71 50W
Pisagne, Italy 44 C7 45 45N 10 10 E
Pisagua, Chile 172 D3 19 40 S 70 15W
Pisarovina, Croatia 45 C12 45 35N 15 50 E
Pisau, Tanjong, Malaysia 85 A5 6 4N 117 59 E
Pisco, Peru 172 C2 13 50 S 76 12W
Piscu, Romania 53 E12 45 30N 27 43 E
Písek, Czech Rep. 34 B7 49 19N 14 10 E
Pishan, China 72 C2 37 30N 78 33 E
Pishin Lora →, Pakistan 92 E1 29 9N 64 5 E
Pisidia, Turkey 100 D4 37 30N 31 40 E
Pising, Indonesia 82 C2 5 8 S 121 53 E
Pismo Beach, U.S.A. 161 K6 35 9N 120 38W
Pissis, Cerro, Argentina 174 B2 27 45 S 68 48W
Pissos, France 28 D3 44 19N 0 49W
Pissouri, Cyprus 38 E11 34 40N 32 42 E
Pisticci, Italy 47 B9 40 23N 16 33 E
Pistóia, Italy 44 E7 43 55N 10 54 E
Pistol B., Canada 143 A10 62 25N 92 37W
Pisuerga →, Spain 42 D6 41 33N 4 52W
Pisz, Poland 54 E8 53 38N 21 49 E
Pitalito, Colombia 168 C2 1 51N 76 2W
Pitanga, Brazil 171 F1 24 46 S 51 44W
Pitangui, Brazil 171 E3 19 40 S 44 54W
Pitarpunga, L., Australia 128 C5 34 24 S 143 30 E
Pitcairn I., Pac. Oc. 135 K14 25 5 S 130 5W
Pite älv →, Sweden 14 D19 65 20N 21 25 E
Piteå, Sweden 14 D19 65 20N 21 25 E
Piterka, Russia 60 E8 50 41N 47 29 E
Pitești, Romania 53 F9 44 52N 24 54 E
Pithapuram, India 94 F6 17 10N 82 15 E
Pithara, Australia 125 F2 30 20 S 116 35 E
Píthion, Greece 51 E10 41 24N 26 40 E
Pithiviers, France 27 D9 48 10N 2 13 E
Pitigliano, Italy 45 F8 42 38N 11 40 E
Pitkyaranta, Russia 58 B6 61 34N 31 37 E
Pitlochry, U.K. 22 E5 56 42N 3 44W
Pitoco, Phil. 81 F5 10 8N 124 33 E
Pitrufquén, Chile 176 A2 38 59 S 72 39W
Pitsilia □, Cyprus 38 E12 34 55N 33 0 E
Pitt I., Canada 142 C3 53 30N 129 50W
Pittsburg, Kans., U.S.A. 155 G7 37 25N 94 42W
Pittsburg, Tex., U.S.A. 155 J7 33 0N 94 59W
Pittsburgh, U.S.A. 150 F5 40 26N 80 1W
Pittsfield, Ill., U.S.A. 156 E6 39 36N 90 49W
Pittsfield, Mass., U.S.A. 151 D11 42 27N 73 15W
Pittsfield, N.H., U.S.A. 151 C13 43 18N 71 20W
Pittston, U.S.A. 151 E9 41 19N 75 47W
Pittsview, U.S.A. 152 C4 32 11N 85 10W
Pittsworth, Australia 127 D5 27 41 S 151 37 E
Pituri →, Australia 126 C2 22 35 S 138 30 E
Piuí, Brazil 171 F2 20 28 S 45 58W
Pium, Brazil 170 D2 10 27 S 49 11W
Piura, Peru 172 B1 5 15 S 80 38W
Piura □, Peru 172 B2 5 10 S 80 0W
Piva →, Montenegro, Yug. 50 C2 43 20N 18 50 E
Pivijay, Colombia 168 A3 10 28N 74 37W
Piwniczna, Poland 55 J7 49 27N 20 42 E
Pixley, U.S.A. 160 K7 35 58N 119 18W
Piyai, Greece 48 B3 39 17N 21 25 E
Pizarra, Spain 43 J6 36 36N 4 42W
Pizarro, Colombia 168 C2 4 58N 77 22W
Pizol, Switz. 33 C8 46 57N 9 23 E
Pizzo, Italy 47 D9 38 44N 16 10 E
Placentia, Canada 141 C9 47 20N 54 0W
Placentia B., Canada 141 C9 47 0N 54 40W
Placer, Phil. 81 F4 11 52N 123 55 E
Placerville, U.S.A. 160 G6 38 44N 120 48W
Placetas, Cuba 164 B4 22 15N 79 44W
Placid, L., U.S.A. 153 H8 27 15N 81 22W
Plačkovica, Macedonia 50 E6 41 45N 22 30 E
Plaffeien, Switz. 32 C4 46 45N 7 17 E
Plain Dealing, U.S.A. 155 J8 32 54N 93 42W
Plainfield, Ill., U.S.A. 157 C8 41 37N 88 12W
Plainfield, N.J., U.S.A. 151 F10 40 37N 74 25W
Plains, Ga., U.S.A. 152 C5 32 2N 84 24W
Plains, Kans., U.S.A. 155 G4 37 16N 100 35W
Plains, Mont., U.S.A. 158 C6 47 28N 114 53W
Plains, Tex., U.S.A. 155 J3 33 11N 102 50W
Plainview, Nebr., U.S.A. 154 D6 42 21N 97 47W
Plainview, Tex., U.S.A. 155 H4 34 11N 101 43W
Plainville, U.S.A. 154 F5 39 14N 99 18W
Plainwell, U.S.A. 148 D3 42 27N 85 38W
Plaisance, France 28 E4 43 36N 0 3 E
Pláka, Greece 49 B7 40 0N 25 24 E
Pláka, Ákra, Greece 38 D8 35 11N 26 19 E
Plakenska Planina, Macedonia 50 E5 41 14N 21 2 E
Planá, Czech Rep. 34 B5 49 50N 12 44 E
Plana Cays, Bahamas 165 B5 22 38N 73 30W
Planada, U.S.A. 160 H6 37 16N 120 19W
Plancoët, France 26 D4 48 32N 2 13W
Plandište, Serbia, Yug. 52 E6 45 16N 21 10 E
Planeta Rica, Colombia 168 B2 8 25N 75 36W
Plankinton, U.S.A. 154 D5 43 43N 98 29W
Plano, U.S.A. 155 J6 33 1N 96 42W
Plant City, U.S.A. 149 M4 28 1N 82 7W
Plantation, U.S.A. 153 J9 26 8N 80 15W
Plaquemine, U.S.A. 155 K9 30 17N 91 14W
Plaridel, Phil. 81 G4 8 37N 123 43 E
Plasencia, Spain 42 E4 40 3N 6 8W
Plaški, Croatia 45 C12 45 4N 15 22 E
Plassen, Norway 18 C9 61 9N 12 30 E
Plast, Russia 62 D8 54 22N 60 50 E
Plaster City, U.S.A. 161 N11 32 47N 115 51W
Plaster Rock, Canada 141 C6 46 53N 67 22W
Plastun, Russia 68 B8 44 45N 136 19 E
Plasy, Czech Rep. 34 B6 49 56N 13 24 E
Plata, Río de la, S. Amer. 174 C4 34 45 S 57 30W
Plátani →, Italy 46 E6 37 23N 13 16 E
Plátanos, Greece 38 D5 35 28N 23 33 E
Plateau □, Nigeria 113 D6 8 0N 8 30 E
Plateau du Coteau du Missouri, U.S.A. 154 B4 47 9N 101 5W
Platí, Ákra, Greece 51 F8 40 27N 24 0 E
Plato, Colombia 168 B3 9 47N 74 47W
Platta, Piz, Switz. 33 D9 46 28N 9 35 E
Platte, U.S.A. 154 D5 43 23N 98 51W
Platte →, Mo., U.S.A. 156 E2 39 16N 94 50W
Platte →, Nebr., U.S.A. 136 E10 41 4N 95 53W
Platte City, U.S.A. 156 E2 39 22N 94 47W
Platteville, Colo., U.S.A. 154 E2 40 13N 104 49W
Platteville, Wis., U.S.A. 156 B6 42 44N 90 29W
Plattling, Germany 31 G8 48 46N 12 53 E
Plattsburg, U.S.A. 156 E2 39 34N 94 27W
Plattsburgh, U.S.A. 151 B11 44 42N 73 28W
Plattsmouth, U.S.A. 154 E7 41 1N 95 53W
Plau, Germany 30 B8 53 27N 12 15 E
Plauen, Germany 30 E8 50 30N 12 8 E
Plauer See, Germany 30 B8 53 28N 12 17 E
Plav, Montenegro, Yug. 50 D3 42 38N 19 57 E
Plavinas, Latvia 15 H21 56 35N 25 46 E
Plavnica, Montenegro, Yug. 50 D3 42 20N 19 13 E
Plavno, Croatia 45 D13 44 9N 16 10 E
Plavsk, Russia 58 F9 53 40N 37 18 E
Playa Blanca, Canary Is. 39 F6 28 55N 13 37W
Playa Blanca Sur, Canary Is. 39 F6 28 51N 13 50W
Playa de las Americas, Canary Is. 39 F3 28 5N 16 43W
Playa de Mogán, Canary Is. 39 G4 27 48N 15 47W
Playa del Inglés, Canary Is. 39 G4 27 45N 15 33W
Playa Esmerelda, Canary Is. 39 F5 28 8N 14 16W
Playgreen L., Canada 143 C9 54 0N 98 15W
Pleasant Bay, Canada 141 C7 46 51N 60 48W
Pleasant Hill, Calif., U.S.A. 160 H4 37 57N 122 4W
Pleasant Hill, Ill., U.S.A. 156 E6 39 27N 90 52W
Pleasant Hill, Mo., U.S.A. 156 F2 38 47N 94 16W
Pleasant Hills, Australia 129 C7 35 28 S 146 50 E
Pleasant Pt., N.Z. 131 E6 44 16 S 171 9 E
Pleasanton, U.S.A. 155 L5 28 58N 98 29W
Pleasantville, Iowa, U.S.A. 156 C3 41 23N 93 18W
Pleasantville, N.J., U.S.A. 148 F8 39 24N 74 32W
Pleasure Ridge Park, U.S.A. 157 F11 38 9N 85 50W
Pléaux, France 28 C6 45 8N 2 13 E
Plei Ku, Vietnam 86 F7 13 57N 108 0 E
Plélan-le-Grand, France 26 D4 48 0N 2 7W
Pléneuf-Val-André, France 26 D4 48 35N 2 32W
Plenița, Romania 53 F8 44 14N 23 10 E
Plenty →, Australia 126 C2 23 25 S 136 31 E
Plenty, B. of, N.Z. 130 D6 37 45 S 177 0 E
Plentywood, U.S.A. 154 A2 48 47N 104 34W
Plérin, France 26 D4 48 32N 2 46W
Plesetsk, Russia 56 B7 62 43N 40 20 E
Plessisville, Canada 141 C5 46 14N 71 47W
Plestin-les-Grèves, France 26 D3 48 40N 3 39W
Pleszew, Poland 55 G4 51 53N 17 47 E
Pleternica, Croatia 52 E2 45 17N 17 48 E
Pletipi L., Canada 141 B5 51 44N 70 6W
Pleven, Bulgaria 51 C8 43 26N 24 37 E
Plevlja, Montenegro, Yug. 50 C3 43 21N 19 21 E
Plješevica, Croatia 45 D12 44 45N 15 45 E
Ploaghe, Italy 46 B1 40 40N 8 45 E
Plo'ana, Slovak Rep. 35 C12 48 38N 19 29 E
Ploče, Croatia 45 E14 43 4N 17 26 E
Płock, Poland 55 F6 52 32N 19 40 E
Płock □, Poland 55 F6 52 30N 19 45 E
Plöckenpass, Italy 45 B9 46 37N 12 57 E
Plöckenstein, Germany 31 G9 48 46N 13 51 E
Ploemeur, France 26 E3 47 44N 3 26W
Ploërmel, France 26 E4 47 55N 2 26W
Ploiești, Romania 53 F11 44 57N 26 5 E
Plomárion, Greece 49 C8 38 59N 26 22 E
Plombières-les-Bains, France 27 E13 47 58N 6 27 E
Plomin, Croatia 45 C11 45 8N 14 10 E
Plön, Germany 30 A6 54 9N 10 24 E
Plonge, Lac la, Canada 143 B7 55 8N 107 20W
Płońsk, Poland 55 F7 52 37N 20 21 E
Plopeni, Romania 53 E10 45 4N 25 59 E
Plopișului, Munții, Romania 52 C7 47 5N 22 30 E
Płoty, Poland 54 E2 53 48N 15 18 E
Plouaret, France 26 D3 48 37N 3 28W
Plouay, France 26 E3 47 55N 3 21W
Ploučnice →, Czech Rep. 34 A7 50 46N 14 13 E
Ploudalmézeau, France 26 D2 48 34N 4 41W
Plouescat, France 26 D2 48 39N 4 10W
Plougasnou, France 26 D3 48 42N 3 49W
Plougastel-Daoulas, France 26 D2 48 22N 4 17W
Plouguerneau, France 26 D2 48 36N 4 30W
Plouha, France 26 D4 48 41N 2 57W
Plouhinec, France 26 E2 48 0N 4 29W
Plovdiv, Bulgaria 51 D8 42 8N 24 44 E
Plovdiv □, Bulgaria 51 D8 42 15N 24 30 E
Plum, U.S.A. 150 F5 40 29N 79 47W
Plum I., U.S.A. 151 E12 41 11N 72 12W
Plumas, U.S.A. 160 F7 39 45N 119 4W
Plummer, U.S.A. 158 C5 47 20N 116 53W
Plumtree, Zimbabwe 119 G2 20 27 S 27 55 E
Plungė, Lithuania 15 J19 55 53N 21 59 E
Pluvigner, France 26 E3 47 46N 3 1W
Plymouth, U.K. 21 G3 50 22N 4 10W
Plymouth, Calif., U.S.A. 160 G6 38 29N 120 51W
Plymouth, Ill., U.S.A. 156 D6 40 18N 90 58W
Plymouth, Ind., U.S.A. 157 C10 41 21N 86 19W
Plymouth, Mass., U.S.A. 151 E14 41 57N 70 40W
Plymouth, N.C., U.S.A. 149 H7 35 52N 76 43W
Plymouth, N.H., U.S.A. 151 C13 43 46N 71 41W
Plymouth, Pa., U.S.A. 151 E9 41 14N 75 57W
Plymouth, Wis., U.S.A. 148 D2 43 45N 87 59W
Plynlimon = Pumlumon Fawr, U.K. 21 E4 52 28N 3 46W
Plyusa, Russia 58 C5 58 28N 29 27 E
Plyusa →, Russia 58 C5 59 4N 28 6 E
Plyussa = Plyusa, Russia 58 C5 58 28N 29 27 E
Plyussa → = Plyusa →, Russia 58 C5 59 4N 28 6 E
Plzeň, Czech Rep. 34 B6 49 45N 13 22 E
Pniewy, Poland 55 F3 52 31N 16 16 E
Pô, Burkina Faso 113 C4 11 14N 1 5W
Po →, Italy 45 D9 44 57N 12 4 E
Po, Foci del, Italy 45 D9 44 55N 12 30 E
Po Hai = Bo Hai, China 75 E10 39 0N 119 0 E
Pobé, Benin 113 D5 7 0N 2 56 E
Pobeda, Russia 65 C15 65 12N 146 12 E
Pobedy, Pik, Kyrgyzstan 64 E8 42 0N 79 58 E
Pobiedziska, Poland 55 F4 52 29N 17 11 E
Pobla de Segur, Spain 40 C5 42 15N 0 58 E
Pobladura del Valle, Spain 42 C5 42 6N 5 44W
Pobra de Trives, Spain 42 C3 42 20N 7 10W
Pocahontas, Ark., U.S.A. 155 G9 36 16N 90 58W
Pocahontas, Ill., U.S.A. 156 F7 38 50N 89 33W
Pocahontas, Iowa, U.S.A. 156 B2 42 44N 94 40W
Pocatello, U.S.A. 158 E7 42 52N 112 27W
Počátky, Czech Rep. 34 B8 49 15N 15 14 E
Pochep, Russia 59 F7 52 58N 33 29 E
Pochinki, Russia 60 C7 54 41N 44 59 E
Pochinok, Russia 58 E7 54 28N 32 29 E
Pöchlarn, Austria 34 C8 48 12N 15 12 E
Pochutla, Mexico 163 D5 15 50N 96 31W
Poci, Venezuela 169 B5 5 57N 61 29W
Pocinhos, Brazil 170 C4 7 4 S 36 3W
Pocito Casas, Mexico 162 B2 28 32N 111 6W
Pocking, Germany 31 G9 48 23N 13 19 E
Poções, Brazil 171 D3 14 31 S 40 21W
Pocomoke City, U.S.A. 148 F8 38 5N 75 34W
Poconé, Brazil 173 D6 16 15 S 56 37W
Poços de Caldas, Brazil 175 A6 21 50 S 46 33W
Podbořany, Czech Rep. 34 A6 50 14N 13 25 E
Poddębice, Poland 55 G5 51 54N 18 58 E
Poděbrady, Czech Rep. 34 A8 50 9N 15 8 E
Podensac, France 28 D3 44 40N 0 22W
Podenzano, Italy 44 D6 44 57N 9 41 E
Podgorač, Croatia 52 E3 45 27N 18 13 E
Podgorica, Montenegro, Yug. 50 D3 42 30N 19 19 E
Podgorie, Albania 50 F4 40 49N 20 48 E
Podilska Vysochyna, Ukraine 59 H4 49 0N 28 0 E
Podkova, Bulgaria 51 E9 41 24N 25 24 E
Podlapača, Croatia 45 D12 44 37N 15 47 E
Podoleni, Romania 53 D11 46 46N 26 39 E
Podolínec, Slovak Rep. 35 B13 49 16N 20 31 E
Podolsk, Russia 58 E9 55 25N 37 30 E
Podor, Senegal 112 B1 16 40N 15 2W
Podporozhy, Russia 58 B8 60 55N 34 2 E
Podu Iloaiei, Romania 53 C12 47 13N 27 16 E
Podu Turcului, Romania 53 D12 46 11N 27 25 E
Podujevo, Serbia, Yug. 50 D5 42 54N 21 10 E
Poel, Germany 30 B6 54 0N 11 25 E
Pofadder, S. Africa 116 D2 29 10 S 19 22 E
Pogamasing, Canada 140 C3 46 55N 81 50W
Poggiardo, Italy 47 B11 40 3N 18 23 E
Poggibonsi, Italy 44 E8 43 28N 11 9 E
Pòggio Mirteto, Italy 45 F9 42 16N 12 41 E
Pogoanele, Romania 53 F12 44 55N 27 0 E
Pogorzela, Poland 55 G4 51 50N 17 12 E
Pogoso, Dem. Rep. of the Congo 115 D3 6 46 S 17 12 E
Pogradeci, Albania 50 F4 40 54N 20 37 E
Pogranitšnyi, Russia 68 B5 44 25N 131 24 E
Poh, Indonesia 82 B2 0 46 S 122 51 E
Pohang, S. Korea 75 F15 36 1N 129 23 E
Pohjanmaa, Finland 14 E20 62 58N 22 50 E
Pohnpei, Pac. Oc. 134 G7 6 55N 158 10 E
Pohorelá, Slovak Rep. 35 C13 48 50N 20 2 E
Pohořelice, Czech Rep. 35 C9 48 59N 16 31 E
Pohorje, Slovenia 45 B12 46 30N 15 20 E
Pohue B., U.S.A. 145 E6 19 0N 155 48W
Poiana Mare, Romania 52 G8 43 57N 23 5 E
Poiana Ruscăi, Munții, Romania 52 E7 45 45N 22 25 E
Poiana Stampei, Romania 53 C10 47 19N 25 8 E
Poindimié, N. Cal. 133 T19 20 56 S 165 20 E
Poinsett, C., Antarctica 7 C8 65 42 S 113 18 E
Point Edward, Canada 140 D3 43 0N 82 30W
Point Pass, Australia 128 C3 34 5 S 139 5 E
Point Pedro, Sri Lanka 95 K5 9 50N 80 15 E
Point Pleasant, N.J., U.S.A. 151 F10 40 5N 74 4W
Point Pleasant, W. Va., U.S.A. 148 F4 38 51N 82 8W
Pointe-à-la Hache, U.S.A. 155 L10 29 35N 89 55W
Pointe-à-Pitre, Guadeloupe 165 C7 16 10N 61 30W
Pointe-Noire, Congo 115 C2 4 48 S 11 53 E
Poio, Spain 42 C2 42 28N 8 41W
Poirino, Italy 44 D4 44 56N 7 48 E
Poisonbush Ra., Australia 124 D3 22 30 S 121 30 E
Poitiers, France 26 F7 46 35N 0 20 E
Poitou, France 28 B3 46 40N 0 10W
Poitou-Charentes □, France 28 B4 46 10N 0 15 E
Poix-de-Picardie, France 27 C8 49 47N 1 58 E
Poix-Terron, France 27 C11 49 38N 4 38 E
Pojoaque Valley, U.S.A. 159 J11 35 54N 106 1W
Pokai B., U.S.A. 145 K13 21 27N 158 12W
Pokataroo, Australia 127 D4 29 30 S 148 36 E
Pokhara, Nepal 93 E10 28 14N 83 58 E
Pokhvistnevo, Russia 60 D11 53 36N 52 0 E
Poko, Dem. Rep. of the Congo 118 B2 3 7N 26 52 E
Poko, Sudan 107 F3 5 41N 31 55 E
Pokrov, Russia 58 E10 55 55N 39 7 E
Pokrovka, Kyrgyzstan 63 B9 42 20N 78 0 E
Pokrovsk = Engels, Russia 60 E8 51 28N 46 6 E
Pokrovsk, Russia 65 C13 61 29N 129 0 E
Pokrovsk-Uralskiy, Russia 62 A7 60 10N 59 49 E
Pokrovskoye, Russia 59 J10 47 25N 38 54 E
Pola = Pula, Croatia 45 D10 44 54N 13 57 E
Pola, Russia 58 D7 57 55N 32 0 E
Pola de Allande, Spain 42 B4 43 16N 6 37W
Pola de Lena, Spain 42 B5 43 10N 5 49W
Pola de Siero, Spain 42 B5 43 24N 5 39W
Pola de Somiedo, Spain 42 B4 43 5N 6 15W
Polacca, U.S.A. 159 J8 35 50N 110 23W
Polan, Iran 97 E9 25 30N 61 10 E
Poland ■, Europe 55 G3 52 0N 20 0 E
Polanica-Zdrój, Poland 55 H3 50 24N 16 32 E
Połaniec, Poland 55 H8 50 26N 21 17 E
Polanów, Poland 54 D3 54 7N 16 41 E
Polatlı, Turkey 100 C5 39 36N 32 9 E
Polatsk, Belarus 58 E5 55 30N 28 50 E
Polcura, Chile 174 D1 37 17 S 71 43W
Połczyn-Zdrój, Poland 54 E3 53 47N 16 5 E
Polessk, Russia 15 J19 54 50N 21 8 E
Polesye = Pripet Marshes, Europe 59 F5 52 10N 28 10 E
Polevskoy, Russia 62 C8 56 26N 60 11 E
Polgar, Hungary 52 C6 47 54N 21 6 E
Pŏlgyo-ri, S. Korea 75 G14 34 51N 127 21 E
Poli, Cameroon 114 A2 8 34N 13 15 E
Políaigos, Greece 48 E6 36 45N 24 38 E
Policastro, G. di, Italy 47 C8 40 0N 15 35 E
Police, Poland 54 E1 53 33N 14 33 E
Polička, Czech Rep. 35 B9 49 43N 16 15 E
Policoro, Italy 47 B9 40 13N 16 41 E
Polignano a Mare, Italy 47 A9 41 0N 17 13 E
Poligny, France 27 F12 46 50N 5 42 E
Políkhnitas, Greece 49 B8 39 6N 26 10 E
Polillo, Phil. 80 D3 14 43N 121 56 E
Polillo Is., Phil. 80 D4 14 56N 122 0 E
Polillo Strait, Phil. 80 D3 14 44N 121 51 E
Polis, Cyprus 38 D11 35 2N 32 26 E
Polístena, Italy 47 D9 38 24N 16 4 E
Políyiros, Greece 50 F7 40 23N 23 25 E
Polk, U.S.A. 150 E5 41 22N 79 56W
Polkowice, Poland 55 G3 51 29N 16 3 E
Polla, Italy 47 B8 40 31N 15 29 E
Pollachi, India 95 J3 10 35N 77 0 E
Pollença, Spain 39 B10 39 54N 3 1 E
Pollença, B. de, Spain 39 B10 39 53N 3 8 E
Pollfoss, Norway 18 C4 61 58N 7 54 E
Póllica, Italy 47 B8 40 11N 15 3 E
Pollino, Mte., Italy 47 C9 39 55N 16 11 E
Pollock, U.S.A. 154 C4 45 55N 100 17W
Polna, Russia 58 C5 58 31N 28 5 E
Polnovat, Russia 64 C7 63 50N 65 54 E
Polo, Ill., U.S.A. 156 C7 41 59N 89 35W
Polo, Mo., U.S.A. 156 E2 39 33N 94 3W
Pology, Ukraine 59 J9 47 29N 36 15 E
Polonne, Ukraine 59 G4 50 6N 27 30 E
Polonnoye = Polonne, Ukraine 59 G4 50 6N 27 30 E
Polski Trŭmbesh, Bulgaria 51 C9 43 20N 25 38 E
Polsko Kosovo, Bulgaria 51 C9 43 23N 25 38 E
Polson, U.S.A. 158 C6 47 41N 114 9W
Poltár, Slovak Rep. 35 C12 48 26N 19 48 E
Poltava, Ukraine 59 H8 49 35N 34 35 E
Põltsamaa, Estonia 15 G21 58 41N 25 58 E
Polunochnoye, Russia 56 B11 60 52N 60 25 E
Polur, India 95 H4 12 32N 79 11 E
Põlva, Estonia 15 G22 58 3N 27 3 E
Polyarny, Russia 56 A5 69 8N 33 20 E
Polynesia, Pac. Oc. 135 J11 10 0 S 162 0W
Polynésie française = French Polynesia ■, Pac. Oc. 135 K13 20 0 S 145 0W
Pomarance, Italy 44 E7 43 18N 10 52 E
Pomaro, Mexico 162 D4 18 20N 103 18W
Pombal, Brazil 170 C4 6 45 S 37 50W
Pombal, Portugal 42 F2 39 55N 8 40W
Pómbia, Greece 38 E6 35 0N 24 51 E
Pomeroy, Ohio, U.S.A. 148 F4 39 2N 82 2W
Pomeroy, Wash., U.S.A. 158 C5 46 28N 117 36W
Pomézia, Italy 46 A5 41 40N 12 30 E
Pomichna, Ukraine 59 H6 48 13N 31 36 E
Pomio, Papua N. G. 132 C6 5 32 S 151 33 E
Pomme de Terre L., U.S.A. 156 G3 37 54N 93 19W
Pomona, U.S.A. 161 L9 34 4N 117 45W
Pomona Park, U.S.A. 153 F8 29 30N 81 36W
Pomorie, Bulgaria 51 D11 42 32N 27 41 E
Pomorskie, Pojezierze, Poland 54 E3 53 40N 16 37 E
Pomos, Cyprus 38 D11 35 9N 32 33 E
Pomos, C., Cyprus 38 D11 35 10N 32 33 E
Pompano Beach, U.S.A. 149 M5 26 14N 80 8W
Pompei, Italy 47 B7 40 45N 14 30 E
Pompey, France 27 D13 48 46N 6 6 E
Pompeys Pillar, U.S.A. 158 D10 45 59N 107 57W
Ponape = Pohnpei, Pac. Oc. 134 G7 6 55N 158 10 E
Ponask, L., Canada 140 B1 54 0N 92 41W
Ponass L., Canada 143 C8 52 16N 103 58W
Ponca, U.S.A. 154 D6 42 34N 96 43W
Ponca City, U.S.A. 155 G6 36 42N 97 5W
Ponce, Puerto Rico 165 C6 18 1N 66 37W
Ponce de Leon, U.S.A. 152 E4 30 44N 85 56W
Ponce de Leon B., U.S.A. 153 K8 25 15N 81 10W
Ponchatoula, U.S.A. 155 K9 30 26N 90 26W
Poncheville, L., Canada 140 B4 50 10N 76 55W
Pond, U.S.A. 161 K7 35 43N 119 20W
Pond Inlet, Canada 139 A12 72 40N 77 0W
Pondicherry, India 95 J4 11 59N 79 50 E
Pondooma, Australia 128 B2 33 29 S 136 59 E
Ponds, I. of, Canada 141 B8 53 27N 55 52W
Ponérihouen, N. Cal. 133 U19 21 5 S 165 24 E
Ponferrada, Spain 42 C4 42 32N 6 35W
Pongo, Wadi →, Sudan 107 F2 8 42N 27 40 E
Poniatowa, Poland 55 G9 51 11N 22 3 E
Poniec, Poland 55 G3 51 48N 16 50 E
Ponikva, Slovenia 45 B12 46 16N 15 26 E
Ponnaiyar →, India 95 J4 11 50N 79 45 E
Ponnani, India 95 J2 10 45N 75 59 E
Ponneri, India 95 H5 13 20N 80 15 E
Ponnuru, India 95 F5 16 5N 80 34 E
Ponoka, Canada 142 C6 52 42N 113 40W
Ponomarevka, Russia 62 E5 53 19N 54 8 E
Ponorogo, Indonesia 85 D4 7 52 S 111 27 E
Ponot, Phil. 81 G4 8 25N 123 0 E
Ponoy, Russia 56 A7 67 0N 41 13 E
Ponoy →, Russia 56 A7 66 59N 41 17 E
Pons = Ponts, Spain 40 D6 41 55N 1 12 E
Pons, France 28 C3 45 35N 0 34W
Ponsul →, Portugal 42 F3 39 40N 7 31W
Pont-à-Mousson, France 27 D13 48 54N 6 1 E
Pont-Audemer, France 26 C7 49 21N 0 30 E
Pont-Aven, France 26 E3 47 51N 3 47W
Pont Canavese, Italy 44 C4 45 25N 7 36 E
Pont-d'Ain, France 27 F12 46 3N 5 21 E
Pont-de-Roide, France 27 E13 47 23N 6 45 E
Pont-de-Salars, France 28 D6 44 18N 2 44 E
Pont-de-Vaux, France 27 F11 46 26N 4 56 E
Pont-de-Veyle, France 27 F11 46 17N 4 53 E

268 Powell, L.

Powell, L., *U.S.A.* **159 H8** 36 57N 111 29W
Powell River, *Canada* **142 D4** 49 50N 124 35W
Powelton, *U.S.A.* **152 B7** 33 26N 82 52W
Powers, *Mich., U.S.A.* **148 C2** 45 41N 87 32W
Powers, *Oreg., U.S.A.* **158 E1** 42 53N 124 4W
Powers Lake, *U.S.A.* **154 A3** 48 34N 102 39W
Powys □, *U.K.* **21 E4** 52 20N 3 20W
Poxoreu, *Brazil* **173 D7** 15 50 S 54 23W
Poya, *N. Cal.* **133 U19** 21 19 S 165 7 E
Poyang Hu, *China* **77 C11** 29 5N 116 20 E
Poyarkovo, *Russia* **65 E13** 49 36N 128 41 E
Poysdorf, *Austria* **35 C9** 48 40N 16 37 E
Poza de la Sal, *Spain* **42 C7** 42 35N 3 31W
Poza Rica, *Mexico* **163 C5** 20 33N 97 27W
Pozanti, *Turkey* **100 D6** 37 25N 34 50 E
Požarevac, *Serbia, Yug.* **50 B5** 44 35N 21 18 E
Pozazal, Puerto, *Spain* **42 C6** 42 56N 4 10W
Požega, *Croatia* **52 E2** 45 20N 17 40 E
Požega, *Serbia, Yug.* **50 C4** 43 53N 20 2 E
Pozhva, *Russia* **62 B6** 59 5N 56 5 E
Poznań, *Poland* **55 F3** 52 25N 16 55 E
Poznań □, *Poland* **55 F3** 52 30N 17 0 E
Pozo, *U.S.A.* **161 K6** 35 20N 120 24W
Pozo Alcón, *Spain* **43 H8** 37 42N 2 56W
Pozo Almonte, *Chile* **172 E4** 20 10 S 69 50W
Pozo Colorado, *Paraguay* **174 A4** 23 30 S 58 45W
Pozo del Dátil, *Mexico* **162 B2** 30 0N 112 15W
Pozoblanco, *Spain* **43 G6** 38 23N 4 51W
Pozorrubio, *Phil.* **80 C3** 16 7N 120 33 E
Pozuzo, *Peru* **172 C2** 10 5 S 75 35W
Pozzallo, *Italy* **47 F7** 36 43N 14 51 E
Pozzomaggiore, *Italy* **46 B1** 40 24N 8 39 E
Pozzuoli, *Italy* **47 B7** 40 49N 14 7 E
Pra →, *Ghana* **113 D4** 5 1N 1 37W
Prabuty, *Poland* **54 E6** 53 47N 19 15 E
Prača, *Bos.-H.* **52 G3** 43 47N 18 43 E
Prachatice, *Czech Rep.* **34 B6** 49 1N 14 0 E
Prachin Buri, *Thailand* **86 F3** 14 0N 101 25 E
Prachuap Khiri Khan, *Thailand* **87 G2** 11 49N 99 48 E
Pradelles, *France* **28 D7** 44 46N 3 52 E
Pradera, *Colombia* **168 C2** 3 25N 76 15W
Prades, *France* **28 F6** 42 38N 2 23 E
Prado, *Brazil* **171 E4** 17 20 S 39 13W
Prado del Rey, *Spain* **43 J5** 36 48N 5 33W
Præstø, *Denmark* **17 J6** 55 8N 12 2 E
Pragersko, *Slovenia* **45 B12** 46 27N 15 42 E
Prague = Praha, *Czech Rep.* **34 A7** 50 5N 14 22 E
Praha, *Czech Rep.* **34 A7** 50 5N 14 22 E
Prahecq, *France* **28 B3** 46 19N 0 26W
Prahita →, *India* **94 E4** 19 0N 79 55 E
Prahova □, *Romania* **53 E10** 45 10N 26 0 E
Prahova →, *Romania* **53 F10** 44 50N 25 50 E
Prahovo, *Serbia, Yug.* **50 B6** 44 18N 22 39 E
Praia, *C. Verde Is.* **8 G6** 14 55N 23 30W
Práia a Mare, *Italy* **47 C8** 39 50N 15 45 E
Praid, *Romania* **53 D10** 46 32N 25 10 E
Prainha, *Amazonas, Brazil* **173 B5** 7 10 S 60 30W
Prainha, *Pará, Brazil* **169 D7** 1 45 S 53 30W
Prairie, *Australia* **126 C3** 20 50 S 144 35 E
Prairie →, *U.S.A.* **155 H5** 34 30N 99 23W
Prairie City, *U.S.A.* **158 D4** 44 28N 118 43W
Prairie du Chien, *U.S.A.* **156 A5** 43 3N 91 9W
Prairie du Rocher, *U.S.A.* **156 F6** 38 5N 90 6W
Pramánda, *Greece* **48 B3** 39 32N 21 8 E
Pran Buri, *Thailand* **86 F2** 12 23N 99 55 E
Prándarjökull, *Iceland* **11 C12** 64 40N 14 55W
Prang, *Ghana* **113 D4** 8 1N 0 56W
Prasonísi, Ákra, *Greece* **38 D9** 35 42N 27 46 E
Prästmon, *Sweden* **16 A11** 63 5N 17 45 E
Praszka, *Poland* **55 G5** 51 5N 18 31 E
Prata, *Brazil* **171 E2** 19 25 S 48 54W
Pratapgarh, *India* **92 G6** 24 2N 74 40 E
Prätigau, *Switz.* **33 C9** 46 56N 9 44 E
Prato, *Italy* **44 E8** 43 53N 11 6 E
Prátola Peligna, *Italy* **45 F10** 42 6N 13 52 E
Prats-de-Mollo-la-Preste, *France* **28 F6** 42 25N 2 27 E
Pratt, *U.S.A.* **155 G5** 37 39N 98 44W
Pratteln, *Switz.* **32 A5** 47 31N 7 41 E
Prattville, *U.S.A.* **149 J2** 32 28N 86 29W
Pravara →, *India* **94 E2** 19 35N 74 45 E
Pravdinsk, *Russia* **60 B6** 56 29N 43 28 E
Pravets, *Bulgaria* **50 D7** 42 53N 23 55 E
Pravia, *Spain* **42 B4** 43 30N 6 12W
Praya, *Indonesia* **85 D5** 8 39 S 116 17 E
Pré-en-Pail, *France* **26 D6** 48 28N 0 12W
Precordillera, *Argentina* **174 C2** 30 0 S 69 1W
Predáppio, *Italy* **45 D8** 44 6N 11 59 E
Predazzo, *Italy* **45 B8** 46 19N 11 36 E
Predeal, *Romania* **53 E10** 45 30N 25 34 E
Predejane, *Serbia, Yug.* **50 D6** 42 51N 22 9 E
Preeceville, *Canada* **143 C8** 51 57N 102 40W
Preetz, *Germany* **30 A6** 54 14N 10 18 E
Pregonero, *Venezuela* **168 B3** 8 1N 71 46W
Pregrada, *Croatia* **45 B12** 46 11N 15 45 E
Preiļi, *Latvia* **15 H22** 56 18N 26 43 E
Preko, *Croatia* **45 D12** 44 7N 15 10 E
Prelate, *Canada* **143 C7** 50 51N 109 24W
Prelog, *Croatia* **45 B13** 46 18N 16 32 E
Prémery, *France* **27 E10** 47 10N 3 18 E
Premià de Mar, *Spain* **40 D7** 41 29N 2 22 E
Premier, *Canada* **142 B3** 56 4N 129 56W
Premont, *U.S.A.* **155 M5** 27 22N 98 7W
Premuda, *Croatia* **45 D11** 44 20N 14 36 E
Prenjasi, *Albania* **50 E4** 41 4N 20 32 E
Prentice, *U.S.A.* **154 C9** 45 33N 90 17W
Prenzlau, *Germany* **30 B9** 53 19N 13 50 E
Preobrazheniye, *Russia* **68 C6** 42 54N 133 54 E
Přerov, *Czech Rep.* **35 B10** 49 28N 17 27 E
Prescott, *Canada* **140 D4** 44 45N 75 30W
Prescott, *Ariz., U.S.A.* **159 J7** 34 33N 112 28W
Prescott, *Ark., U.S.A.* **155 J8** 33 48N 93 23W
Preservation Inlet, *N.Z.* **131 G1** 46 8 S 166 35 E
Preševo, *Serbia, Yug.* **50 D5** 42 19N 21 39 E
Presho, *U.S.A.* **154 D4** 43 54N 100 3W
Presicce, *Italy* **47 C11** 39 54N 18 16 E
Presidencia de la Plaza, *Argentina* **174 B4** 27 0 S 59 50W
Presidencia Roque Saenz Peña, *Argentina* **174 B3** 26 45 S 60 30W
Presidente Epitácio, *Brazil* **171 F1** 21 56 S 52 6W
Presidente Hayes □, *Paraguay* **174 A4** 24 0 S 59 0W
Presidente Hermes, *Brazil* **173 C5** 11 17 S 61 55W
Presidente Prudente, *Brazil* **175 A5** 22 5 S 51 25W
Presidente Roxas, *Phil.* **81 F4** 11 26N 122 56 E
Presidio, *Mexico* **162 B4** 29 29N 104 23W
Presidio, *U.S.A.* **155 L2** 29 34N 104 22W
Preslav, *Bulgaria* **51 C10** 43 10N 26 52 E
Preslavska Planina, *Bulgaria* **51 C10** 43 10N 26 45 E
Prešov, *Slovak Rep.* **35 B14** 49 0N 21 15 E
Prešovský □, *Slovak Rep.* **35 B13** 49 10N 21 0 E
Prespa, *Bulgaria* **51 E8** 41 44N 24 55 E
Prespa, L. = Prespansko Jezero, *Macedonia* **50 F5** 40 55N 21 0 E
Prespansko Jezero, *Macedonia* **50 F5** 40 55N 21 0 E
Presque Isle, *U.S.A.* **141 C6** 46 41N 68 1W
Prestatyn, *U.K.* **20 D4** 53 20N 3 24W
Prestea, *Ghana* **112 D4** 5 22N 2 7W
Presteigne, *U.K.* **21 E5** 52 17N 3 0W
Přeštice, *Czech Rep.* **34 B6** 49 34N 13 20 E
Presto, *Bolivia* **173 D5** 18 55 S 64 56W
Preston, *Canada* **150 C4** 43 23N 80 21W
Preston, *U.K.* **20 D5** 53 46N 2 42W
Preston, *Ga., U.S.A.* **152 C5** 32 4N 84 32W
Preston, *Idaho, U.S.A.* **158 E8** 42 6N 111 53W
Preston, *Iowa, U.S.A.* **156 B6** 42 3N 90 24W
Preston, *Minn., U.S.A.* **154 D8** 43 40N 92 5W
Preston, *Nev., U.S.A.* **158 G6** 38 55N 115 4W
Preston, C., *Australia* **124 D2** 20 51 S 116 12 E
Prestranda, *Norway* **18 E6** 59 6N 9 4 E
Prestwick, *U.K.* **22 F4** 55 29N 4 37W
Prêto →, *Amazonas, Brazil* **169 D5** 0 8 S 64 6W
Prêto →, *Bahia, Brazil* **170 D3** 11 21 S 43 52W
Prêto do Igapó-Açu →, *Brazil* **169 D6** 4 26 S 59 48W
Pretoria, *S. Africa* **117 D4** 25 44 S 28 12 E
Preuilly-sur-Claise, *France* **26 F7** 46 51N 0 56 E
Préveza, *Greece* **48 C2** 38 57N 20 47 E
Préveza □, *Greece* **48 B2** 39 10N 20 40 E
Prey Veng, *Cambodia* **87 G5** 11 35N 105 29 E
Priazovskoye, *Ukraine* **59 J8** 46 44N 35 40 E
Pribilof Is., *U.S.A.* **144 H5** 57 0N 170 0W
Priboj, *Serbia, Yug.* **50 C3** 43 35N 19 32 E
Příbram, *Czech Rep.* **34 B7** 49 41N 14 2 E
Price, *U.S.A.* **158 G8** 39 36N 110 49W
Price I., *Canada* **142 C3** 52 23N 128 41W
Prichard, *U.S.A.* **149 K1** 30 44N 88 5W
Priego, *Spain* **40 E2** 40 26N 2 21W
Priego de Córdoba, *Spain* **43 H6** 37 27N 4 12W
Priekule, *Latvia* **15 H19** 56 26N 21 35 E
Prien, *Germany* **31 H8** 47 51N 12 20 E
Prienai, *Lithuania* **54 D10** 54 38N 23 57 E
Prieska, *S. Africa* **116 D3** 29 40 S 22 42 E
Priest →, *U.S.A.* **158 B5** 48 12N 116 54W
Priest L., *U.S.A.* **158 B5** 48 35N 116 52W
Priest Valley, *U.S.A.* **160 J6** 36 10N 120 39W
Priestly, *Canada* **142 C3** 54 8N 125 20W
Prieto Diaz, *Phil.* **80 E5** 13 2N 124 12 E
Prievidza, *Slovak Rep.* **35 C11** 48 46N 18 36 E
Prignitz, *Germany* **30 B7** 53 6N 11 45 E
Prijedor, *Bos.-H.* **45 D13** 44 58N 16 41 E
Prijepolje, *Serbia, Yug.* **50 C3** 43 27N 19 40 E
Prikaspiyskaya Nizmennost = Caspian Depression, *Eurasia* **61 G9** 47 0N 48 0 E
Prikubanskaya Nizmennost, *Russia* **59 K10** 45 39N 38 33 E
Prilep, *Macedonia* **50 E5** 41 21N 21 32 E
Priluki = Pryluky, *Ukraine* **59 G7** 50 30N 32 24 E
Prime Seal I., *Australia* **126 G4** 40 3 S 147 43 E
Primeira Cruz, *Brazil* **170 B3** 2 30 S 43 26W
Primorsk, *Russia* **58 B5** 60 22N 28 37 E
Primorsko, *Bulgaria* **51 D11** 42 15N 27 44 E
Primorsko-Akhtarsk, *Russia* **59 J10** 46 2N 38 10 E
Primorskoye, *Ukraine* **59 J9** 46 48N 36 20 E
Primrose L., *Canada* **143 C7** 54 55N 109 45W
Prince Albert, *Canada* **143 C7** 53 15N 105 50W
Prince Albert, *S. Africa* **116 E3** 33 12 S 22 2 E
Prince Albert Mts., *Antarctica* **7 D11** 76 0 S 161 30 E
Prince Albert Nat. Park, *Canada* **143 C7** 54 0N 106 25W
Prince Albert Pen., *Canada* **138 A8** 72 30N 116 0W
Prince Albert Sd., *Canada* **138 A8** 70 25N 115 0W
Prince Alfred, C., *Canada* **6 B1** 74 20N 124 40W
Prince Charles I., *Canada* **139 B12** 67 47N 76 12W
Prince Charles Mts., *Antarctica* **7 D6** 72 0 S 67 0 E
Prince Edward I. □, *Canada* **141 C7** 46 20N 63 20W
Prince Edward Is., *Ind. Oc.* **121 J2** 46 35 S 38 0 E
Prince George, *Canada* **142 C4** 53 55N 122 50W
Prince of Wales, C., *U.S.A.* **136 C3** 65 36N 168 5W
Prince of Wales I., *Australia* **126 A3** 10 40 S 142 10 E
Prince of Wales I., *Canada* **138 A10** 73 0N 99 0W
Prince of Wales I., *U.S.A.* **142 B2** 55 47N 132 50W
Prince Patrick I., *Canada* **6 B2** 77 0N 120 0W
Prince Regent Inlet, *Canada* **6 B3** 73 0N 90 0W
Prince Rupert, *Canada* **142 C2** 54 20N 130 20W
Princesa Isabel, *Brazil* **170 C4** 7 44 S 38 0W
Princess Charlotte B., *Australia* **126 A3** 14 25 S 144 0 E
Princess May Ranges, *Australia* **124 C4** 15 30 S 125 30 E
Princess Royal I., *Canada* **142 C3** 53 0N 128 40W
Princeton, *Canada* **142 D4** 49 27N 120 30W
Princeton, *Calif., U.S.A.* **160 F4** 39 24N 122 1W
Princeton, *Ill., U.S.A.* **156 C7** 41 23N 89 28W
Princeton, *Ind., U.S.A.* **157 F9** 38 21N 87 34W
Princeton, *Ky., U.S.A.* **148 G2** 37 7N 87 53W
Princeton, *Mo., U.S.A.* **156 D3** 40 24N 93 35W
Princeton, *N.J., U.S.A.* **151 F10** 40 21N 74 39W
Princeton, *W. Va., U.S.A.* **148 G5** 37 22N 81 6W
Princeville, *U.S.A.* **156 D7** 40 56N 89 46W
Principe, I. de, *Atl. Oc.* **114 B1** 1 37N 7 27 E
Principe Chan., *Canada* **142 C3** 53 28N 130 0W
Principe da Beira, *Brazil* **173 C5** 12 20 S 64 30W
Prineville, *U.S.A.* **158 D3** 44 18N 120 51W
Prins Christian Sund, *Greenland* **10 E6** 60 4N 43 10W
Prins Harald Kyst, *Antarctica* **7 D4** 70 0 S 35 1 E
Prinsesse Astrid Kyst, *Antarctica* **7 D3** 70 45 S 12 30 E
Prinsesse Ragnhild Kyst, *Antarctica* **7 D4** 70 15 S 27 30 E
Prinzapolca, *Nic.* **164 D3** 13 20N 83 35W
Prior, C., *Spain* **42 B2** 43 34N 8 17W
Priozersk, *Russia* **58 B6** 61 2N 30 7 E
Pripet → = Prypyat →, *Europe* **59 G6** 51 20N 30 15 E
Pripet Marshes, *Europe* **59 F5** 52 10N 28 10 E
Pripyat Marshes = Pripet Marshes, *Europe* **59 F5** 52 10N 28 10 E
Pripyats = Prypyat →, *Europe* **59 G6** 51 20N 30 15 E
Prislop, Pasul, *Romania* **53 C9** 47 37N 24 48 E
Pristen, *Russia* **59 G9** 51 15N 36 44 E
Priština, *Serbia, Yug.* **50 D5** 42 40N 21 13 E
Pritzwalk, *Germany* **30 B8** 53 9N 12 10 E
Privas, *France* **29 D8** 44 45N 4 37 E
Priverno, *Italy* **46 A6** 41 28N 13 11 E
Privolzhsk, *Russia* **60 B5** 57 23N 41 16 E
Privolzhskaya Vozvyshennost, *Russia* **60 E7** 51 0N 46 0 E
Privolzhskiy, *Russia* **60 E8** 51 25N 46 3 E
Privolzhye, *Russia* **60 D9** 52 52N 48 33 E
Priyutnoye, *Russia* **61 G6** 46 12N 43 40 E
Priyutovo, *Russia* **62 E4** 53 55N 53 59 E
Prizren, *Serbia, Yug.* **50 D4** 42 13N 20 45 E
Prizzi, *Italy* **46 E6** 37 43N 13 26 E
Prnjavor, *Bos.-H.* **52 F2** 44 52N 17 43 E
Probolinggo, *Indonesia* **85 D4** 7 46 S 113 13 E
Prochowice, *Poland* **55 G3** 51 17N 16 20 E
Proddatur, *India* **95 G4** 14 45N 78 30 E
Prodhromos, *Cyprus* **38 E11** 34 57N 32 50 E
Proença-a-Nova, *Portugal* **42 F3** 39 45N 7 54W
Profítis Ilías, *Greece* **38 C9** 36 17N 27 56 E
Profondeville, *Belgium* **24 D4** 50 23N 4 52 E
Progreso, *Mexico* **163 C7** 21 20N 89 40W
Prokhladnyy, *Russia* **61 J7** 43 50N 44 2 E
Prokletije, *Albania* **50 D3** 42 30N 19 45 E
Prokopyevsk, *Russia* **64 D9** 54 0N 86 45 E
Prokuplje, *Serbia, Yug.* **50 C5** 43 16N 21 36 E
Proletarsk, *Russia* **61 G5** 46 42N 41 50 E
Prome = Pyè, *Burma* **90 F5** 18 49N 95 13 E
Prophet →, *Canada* **142 B4** 58 48N 122 40W
Prophetstown, *U.S.A.* **156 C7** 41 40N 89 56W
Propriá, *Brazil* **170 D4** 10 13 S 36 51W
Propriano, *France* **29 G12** 41 41N 8 52 E
Proserpine, *Australia* **126 C4** 20 21 S 148 36 E
Prosna →, *Poland* **55 F4** 52 6N 17 44 E
Prosperidad, *Phil.* **81 G5** 8 34N 125 52 E
Prosser, *U.S.A.* **158 C4** 46 12N 119 46W
Prostějov, *Czech Rep.* **35 B10** 49 30N 17 9 E
Prostki, *Poland* **54 E9** 53 42N 22 25 E
Proston, *Australia* **127 D5** 26 8 S 151 32 E
Proszowice, *Poland* **55 H7** 50 13N 20 16 E
Protection, *U.S.A.* **155 G5** 37 12N 99 29W
Próti, *Greece* **48 D3** 37 5N 21 32 E
Provadiya, *Bulgaria* **51 C11** 43 12N 27 30 E
Provence, *France* **29 E9** 43 40N 5 46 E
Provence-Alpes-Côte d'Azur □, *France* **29 D10** 44 0N 6 15 E
Providence, *Ky., U.S.A.* **148 G2** 37 24N 87 46W
Providence, *R.I., U.S.A.* **151 E13** 41 49N 71 24W
Providence Bay, *Canada* **140 C3** 45 41N 82 15W
Providence C., *N.Z.* **131 F1** 45 59 S 166 29 E
Providence Mts., *U.S.A.* **159 J6** 35 10N 115 15W
Providencia, *Ecuador* **168 D2** 0 28 S 76 28W
Providencia, I. de, *Colombia* **164 D3** 13 25N 81 26W
Provideniya, *Russia* **65 C19** 64 23N 173 18W
Provins, *France* **27 D10** 48 33N 3 15 E
Provo, *U.S.A.* **158 F8** 40 14N 111 39W
Provost, *Canada* **143 C6** 52 25N 110 20W
Prozor, *Bos.-H.* **52 G2** 43 50N 17 34 E
Prudentópolis, *Brazil* **171 G1** 25 12 S 50 57W
Prud'homme, *Canada* **143 C7** 52 20N 105 54W
Prudnik, *Poland* **55 H4** 50 20N 17 38 E
Prüm, *Germany* **31 E2** 50 12N 6 25 E
Prundu, *Romania* **53 F11** 44 6N 26 14 E
Pruszcz Gdański, *Poland* **54 D5** 54 17N 18 40 E
Pruszków, *Poland* **55 F7** 52 9N 20 49 E
Prut →, *Romania* **53 E13** 45 28N 28 10 E
Pruzhany, *Belarus* **59 F3** 52 33N 24 28 E
Prvič, *Croatia* **45 D11** 44 55N 14 47 E
Prydz B., *Antarctica* **7 C6** 69 0 S 74 0 E
Pryluky, *Ukraine* **59 G7** 50 30N 32 24 E
Pryor, *U.S.A.* **155 G7** 36 19N 95 19W
Prypyat →, *Europe* **59 G6** 51 20N 30 15 E
Przasnysz, *Poland* **55 E7** 53 2N 20 54 E
Przedbórz, *Poland* **55 G6** 51 6N 19 53 E
Przedecz, *Poland* **55 F5** 52 20N 18 53 E
Przemków, *Poland* **55 G2** 51 31N 15 48 E
Przemyśl, *Poland* **55 J9** 49 50N 22 45 E
Przemyśl □, *Poland* **55 J9** 50 0N 22 45 E
Przeworsk, *Poland* **55 H9** 50 6N 22 32 E
Przewóz, *Poland* **55 G1** 51 28N 14 57 E
Przhevalsk, *Kyrgyzstan* **63 B9** 42 30N 78 20 E
Przysucha, *Poland* **55 G7** 51 22N 20 38 E
Psakhná, *Greece* **48 C5** 38 34N 23 35 E
Psará, *Greece* **49 C7** 38 37N 25 38 E
Psathoúra, *Greece* **48 B6** 39 30N 24 12 E
Psel →, *Ukraine* **59 H7** 49 10N 33 37 E
Pserimos, *Greece* **49 E9** 36 56N 27 8 E
Psíra, *Greece* **38 D7** 35 12N 25 52 E
Pskem →, *Uzbekistan* **63 C5** 41 38N 70 1 E
Pskemskiy Khrebet, *Uzbekistan* **63 C5** 42 0N 70 45 E
Pskent, *Uzbekistan* **63 C4** 40 54N 69 20 E
Pskov, *Russia* **58 D5** 57 50N 28 25 E
Pskovskoye, Ozero, *Russia* **15 H22** 58 0N 27 58 E
Psunj, *Croatia* **52 E2** 45 25N 17 19 E
Pteleón, *Greece* **48 B4** 39 3N 22 57 E
Ptich = Ptsich →, *Belarus* **59 F5** 52 9N 28 52 E
Ptolemaís, *Greece* **50 F5** 40 30N 21 43 E
Ptsich →, *Belarus* **59 F5** 52 9N 28 52 E
Ptuj, *Slovenia* **45 B12** 46 28N 15 50 E
Ptujska Gora, *Slovenia* **45 B12** 46 23N 15 47 E
Pu Xian, *China* **74 F6** 36 24N 111 6 E
Pua, *Thailand* **86 C3** 19 11N 100 55 E
Puaena Pt., *U.S.A.* **145 J13** 21 36N 158 6W
Puán, *Argentina* **174 D3** 37 30 S 62 45W
Pu'an, *China* **76 E5** 25 46N 104 57 E
Puan, *S. Korea* **75 G14** 35 44N 126 44 E
Pu'apu'a, *W. Samoa* **133 W23** 13 34 S 172 9W
Pubei, *China* **76 F7** 22 16N 109 31 E
Pucacuro →, *Peru* **168 D3** 3 20 S 74 58W
Pucallpa, *Peru* **172 B3** 8 25 S 74 30W
Pucará, *Bolivia* **173 D5** 18 43 S 64 11W
Pucará, *Peru* **172 D3** 15 5 S 70 24W
Pucarani, *Bolivia* **172 D4** 16 23 S 68 30W
Pucheng, *China* **77 D12** 27 59N 118 31 E
Pucheni, *Romania* **53 E10** 45 12N 25 17 E
Puchheim, *Germany* **31 G7** 48 9N 11 21 E
Púchov, *Slovak Rep.* **35 B11** 49 8N 18 20 E
Pucio Pt., *Phil.* **81 F3** 11 46N 121 51 E
Pucioasa, *Romania* **53 E10** 45 5N 25 25 E
Pučišća, *Croatia* **45 E13** 43 22N 16 43 E
Puck, *Poland* **54 D5** 54 45N 18 23 E
Pucka, Zatoka, *Poland* **54 D5** 54 30N 18 40 E
Puçol, *Spain* **41 F4** 39 37N 0 18W
Pudasjärvi, *Finland* **14 D22** 65 23N 26 53 E
Puding, *China* **76 D5** 26 18N 105 44 E
Pudozh, *Russia* **58 B9** 61 48N 36 32 E
Pudtol, *Phil.* **80 B3** 18 13N 121 22 E
Pudukkottai, *India* **95 J4** 10 28N 78 47 E
Puebla, *Mexico* **163 D5** 19 3N 98 12W
Puebla □, *Mexico* **163 D5** 18 30N 98 0W
Puebla de Alcocer, *Spain* **43 G5** 38 59N 5 14W
Puebla de Don Fadrique, *Spain* **41 H2** 37 58N 2 25W
Puebla de Don Rodrigo, *Spain* **43 F6** 39 5N 4 37W
Puebla de Guzmán, *Spain* **43 H3** 37 37N 7 15W
Puebla de la Calzada, *Spain* **43 G4** 38 54N 6 37W
Puebla de Sanabria, *Spain* **42 C4** 42 4N 6 38W
Puebla de Trives = Pobra de Trives, *Spain* **42 C3** 42 20N 7 10W
Pueblo, *U.S.A.* **154 F2** 38 16N 104 37W
Pueblo Hundido, *Chile* **174 B1** 26 20 S 70 5W
Pueblo Nuevo, *Venezuela* **168 B3** 8 26N 71 26W
Puelches, *Argentina* **174 D2** 38 5 S 65 51W
Puelén, *Argentina* **174 D2** 37 32 S 67 38W
Puente Alto, *Chile* **174 C1** 33 32 S 70 35W
Puente-Genil, *Spain* **43 H6** 37 22N 4 47W
Puente la Reina, *Spain* **40 C3** 42 40N 1 49W
Puenteareas = Ponteareas, *Spain* **42 C2** 42 10N 8 28W
Puentedeume = Pontedeume, *Spain* **42 B2** 43 24N 8 10W
Puentes de Garcia Rodriguez = As Pontes de García Rodríquez, *Spain* **42 B3** 43 27N 7 50W
Pueo Pt., *U.S.A.* **145 B1** 21 54N 160 4W
Pu'er, *China* **76 F3** 23 0N 101 15 E
Puerco →, *U.S.A.* **159 J10** 34 22N 107 50W
Puerta Galera, *Phil.* **80 E3** 13 30N 120 57 E
Puerto, *Canary Is.* **39 F2** 28 5N 17 20W
Puerto Acosta, *Bolivia* **172 D4** 15 32 S 69 15W
Puerto Aisén, *Chile* **176 C2** 45 27 S 73 0W
Puerto Ángel, *Mexico* **163 D5** 15 40N 96 29W
Puerto Arista, *Mexico* **163 D6** 15 56N 93 48W
Puerto Armuelles, *Panama* **164 E3** 8 20N 82 51W
Puerto Ayacucho, *Venezuela* **168 B4** 5 40N 67 35W
Puerto Barrios, *Guatemala* **164 C2** 15 40N 88 32W
Puerto Bermejo, *Argentina* **174 B4** 26 55 S 58 34W
Puerto Bermúdez, *Peru* **172 C3** 10 20 S 74 58W
Puerto Bolívar, *Ecuador* **168 D2** 3 19 S 79 55W
Puerto Cabello, *Venezuela* **168 A4** 10 28N 68 1W
Puerto Cabezas, *Nic.* **164 D3** 14 0N 83 30W
Puerto Cabo Gracias á Dios, *Nic.* **164 D3** 15 0N 83 10W
Puerto Capaz = Jebba, *Morocco* **110 A4** 35 11N 4 43W
Puerto Carreño, *Colombia* **168 B4** 6 12N 67 22W
Puerto Castilla, *Honduras* **164 C2** 16 0N 86 0W
Puerto Chicama, *Peru* **172 B2** 7 45 S 79 20W
Puerto Coig, *Argentina* **176 D3** 50 54 S 69 15W
Puerto Cortés, *Costa Rica* **164 E3** 8 55N 84 0W
Puerto Cortés, *Honduras* **164 C2** 15 51N 88 0W
Puerto Cumarebo, *Venezuela* **168 A4** 11 29N 69 30W
Puerto de Alcudia, *Spain* **39 B10** 39 50N 3 7 E
Puerto de Andraitx, *Spain* **39 B9** 39 32N 2 23 E
Puerto de Cabrera, *Spain* **39 B9** 39 8N 2 56 E
Puerto de Gran Tarajal, *Canary Is.* **39 F5** 28 13N 14 1W
Puerto de la Cruz, *Canary Is.* **39 F3** 28 24N 16 32W
Puerto de Mazarrón, *Spain* **41 H3** 37 34N 1 15W
Puerto de Pozo Negro, *Canary Is.* **39 F6** 28 19N 13 55W
Puerto de Sóller, *Spain* **39 B9** 39 48N 2 42 E
Puerto de Somosierra, *Spain* **42 D7** 41 9N 3 35W
Puerto del Carmen, *Canary Is.* **39 F6** 28 55N 13 38W
Puerto del Rosario, *Canary Is.* **39 F6** 28 30N 13 52W
Puerto Deseado, *Argentina* **176 C3** 47 55 S 66 0W
Puerto Guaraní, *Paraguay* **173 E6** 21 18 S 57 55W
Puerto Heath, *Bolivia* **172 C4** 12 34 S 68 39W
Puerto Huitoto, *Colombia* **168 C3** 0 18N 74 3W
Puerto Inca, *Peru* **172 B3** 9 22 S 74 54W
Puerto Juárez, *Mexico* **163 C7** 21 11N 86 49W
Puerto La Cruz, *Venezuela* **169 A5** 10 13N 64 38W
Puerto Leguízamo, *Colombia* **168 D3** 0 12 S 74 46W
Puerto Limón, *Colombia* **168 C3** 3 23N 73 30W
Puerto Lobos, *Argentina* **176 B3** 42 0 S 65 3W
Puerto López, *Colombia* **168 C3** 4 5N 72 58W
Puerto Lumbreras, *Spain* **41 H3** 37 34N 1 48W
Puerto Madryn, *Argentina* **176 B3** 42 48 S 65 4W
Puerto Maldonado, *Peru* **172 C4** 12 30 S 69 10W
Puerto Manotí, *Cuba* **164 B4** 21 22N 76 50W
Puerto Mazarrón = Puerto de Mazarrón, *Spain* **41 H3** 37 34N 1 15W
Puerto Mercedes, *Colombia* **168 C3** 1 11N 72 53W
Puerto Miraña, *Colombia* **168 D3** 1 20 S 70 19W
Puerto Montt, *Chile* **176 B2** 41 28 S 73 0W
Puerto Morelos, *Mexico* **163 C7** 20 49N 86 52W
Puerto Nariño, *Colombia* **168 C4** 4 56N 67 48W
Puerto Natales, *Chile* **176 D2** 51 45 S 72 15W
Puerto Nuevo, *Colombia* **168 B4** 5 53N 69 56W
Puerto Nutrias, *Venezuela* **168 B4** 8 5N 69 18W
Puerto Ordaz, *Venezuela* **169 B5** 8 16N 62 44W
Puerto Padre, *Cuba* **164 B4** 21 13N 76 35W
Puerto Páez, *Venezuela* **168 B4** 6 13N 67 28W
Puerto Peñasco, *Mexico* **162 A2** 31 20N 113 33W
Puerto Pinasco, *Paraguay* **174 A4** 22 36 S 57 50W
Puerto Pirámides, *Argentina* **176 B4** 42 35 S 64 20W
Puerto Plata, *Dom. Rep.* **165 C5** 19 48N 70 45W
Puerto Pollensa, *Spain* **39 B10** 39 54N 3 4 E
Puerto Portillo, *Peru* **172 B3** 9 45 S 72 42W
Puerto Princesa, *Phil.* **81 G2** 9 46N 118 45 E
Puerto Quellón, *Chile* **176 B2** 43 7 S 73 37W
Puerto Quepos, *Costa Rica* **164 E3** 9 29N 84 6W
Puerto Real, *Spain* **43 J4** 36 33N 6 12W
Puerto Rico, *Bolivia* **172 C4** 11 5 S 67 38W
Puerto Rico, *Canary Is.* **39 G4** 27 47N 15 42W
Puerto Rico ■, *W. Indies* **165 C6** 18 15N 66 45W
Puerto Rico Trench, *Atl. Oc.* **165 C6** 19 50N 66 0W
Puerto Saavedra, *Chile* **176 A2** 38 47 S 73 24W
Puerto Sastre, *Paraguay* **174 A4** 22 2 S 57 55W
Puerto Serrano, *Spain* **43 J5** 36 56N 5 33W
Puerto Siles, *Bolivia* **173 C4** 12 48 S 65 5W
Puerto Suárez, *Bolivia* **173 D6** 18 58 S 57 52W
Puerto Tejada, *Colombia* **168 C2** 3 14N 76 24W
Puerto Umbría, *Colombia* **168 C2** 0 52N 76 33W
Puerto Vallarta, *Mexico* **162 C3** 20 36N 105 15W
Puerto Varas, *Chile* **176 B2** 41 19 S 72 59W
Puerto Villazón, *Bolivia* **173 C5** 13 32 S 61 57W
Puerto Wilches, *Colombia* **168 B3** 7 21N 73 54W
Puertollano, *Spain* **43 G6** 38 43N 4 7W
Puesto Cunambo, *Peru* **168 D2** 2 10 S 76 0W
Pueyrredón, L., *Argentina* **176 C2** 47 20 S 72 0W
Puffin I., *Ireland* **23 E1** 51 50N 10 24W
Pugachev, *Russia* **60 D9** 52 0N 48 49 E
Puge, *China* **76 D4** 27 20N 102 31 E
Puge, *Tanzania* **118 C3** 4 45 S 33 11 E
Puget Sound, *U.S.A.* **158 C2** 47 50N 122 30W
Puget-Théniers, *France* **29 E10** 43 58N 6 53 E
Púglia □, *Italy* **47 A9** 41 15N 16 15 E
Pugo, *Phil.* **80 C3** 16 8N 121 31 E
Pugŏdong, *N. Korea* **75 C16** 42 5N 130 0 E
Pugu, *Tanzania* **118 D4** 6 55 S 39 4 E

Quilpué 269

Pūgūnzī, *Iran* 97 E8 25 49N 59 10 E
Puha, *N.Z.* 130 E6 38 30 S 177 50 E
Pui, *Romania* 52 E8 45 30N 23 4 E
Puica, *Peru* 172 D3 15 0 S 72 33W
Puieşti, *Romania* 53 D12 46 25N 27 33 E
Puig Major, *Spain* 39 B9 39 48N 2 47 E
Puigcerdà, *Spain* 40 C6 42 24N 1 50 E
Puigmal, *Spain* 40 C7 42 23N 2 7 E
Puigpuñent, *Spain* 39 B9 39 38N 2 32 E
Puisaye, Collines de la, *France* 27 E10 47 37N 3 20 E
Puiseaux, *France* 27 D9 48 11N 2 30 E
Pujilí, *Ecuador* 168 D2 0 57 S 78 41W
Pujols, *France* 28 D3 44 48N 0 2W
Pujon-chosuji, *N. Korea* 75 D14 40 35N 127 35 E
Puka, *Albania* 50 D3 42 2N 19 53 E
Pukaki L., *N.Z.* 131 E5 44 4 S 170 1 E
Pukalani, *U.S.A.* 145 C5 20 51N 156 20W
Pukapuka, *Cook Is.* 135 J11 10 53 S 165 49W
Pukatawagan, *Canada* 143 B8 55 45N 101 20W
Pukchin, *N. Korea* 75 D13 40 12N 125 45 E
Pukchŏng, *N. Korea* 75 D15 40 14N 128 10 E
Pukearuhe, *N.Z.* 130 E3 38 55 S 174 31 E
Pukekohe, *N.Z.* 130 D3 37 12 S 174 55 E
Puketeraki Ra., *N.Z.* 131 C7 42 58 S 172 13 E
Puketoi Ra., *N.Z.* 130 G5 40 30 S 176 5 E
Pukeuri, *N.Z.* 131 F6 45 4 S 171 2 E
Pukou, *China* 77 A12 32 7N 118 38 E
Pula, *Croatia* 45 D10 44 54N 13 57 E
Pula, *Italy* 46 C1 39 1N 9 0 E
Pulacayo, *Bolivia* 172 E4 20 25 S 66 41W
Pulaski, *N.Y., U.S.A.* 151 C8 43 34N 76 8W
Pulaski, *Tenn., U.S.A.* 149 H2 35 12N 87 2W
Pulaski, *Va., U.S.A.* 148 G5 37 3N 80 47W
Puławy, *Poland* 55 G8 51 23N 21 59 E
Pulga, *U.S.A.* 160 F5 39 48N 121 29W
Pulgaon, *India* 94 D4 20 44N 78 21 E
Pulicat, L., *India* 95 H5 13 40N 80 15 E
Puliyangudi, *India* 95 K3 9 11N 77 24 E
Pullabooka, *Australia* 129 B7 33 44 S 147 46 E
Pullman, *U.S.A.* 158 C5 46 44N 117 10W
Pulog, *Phil.* 80 C3 16 40N 120 50 E
Púlpito do Sul, *Angola* 115 F2 15 46 S 12 0 E
Pułtusk, *Poland* 55 F8 52 43N 21 6 E
Pülümür, *Turkey* 101 C8 39 30N 39 51 E
Pulupandan, *Phil.* 81 F4 10 31N 122 48 E
Pumlumon Fawr, *U.K.* 21 E4 52 28N 3 46W
Puna, *Bolivia* 173 D4 19 45 S 65 28W
Puná, I., *Ecuador* 168 D1 2 55 S 80 5W
Punakha, *Bhutan* 90 B2 27 42N 89 52 E
Punalur, *India* 95 K3 9 0N 76 56 E
Punaluu, *U.S.A.* 145 J14 21 35N 157 53W
Punasar, *India* 92 F5 27 6N 73 6 E
Punata, *Bolivia* 173 D4 17 32 S 65 50W
Punch, *India* 93 C6 33 48N 74 4 E
Pune, *India* 94 E1 18 29N 73 57 E
Pungsan, *N. Korea* 75 D15 40 50N 128 9 E
Pungue, Ponte de, *Mozam.* 119 F3 19 0 S 34 0 E
Puning, *China* 77 F11 23 20N 116 12 E
Punjab □, *India* 92 D7 31 0N 76 0 E
Punjab □, *Pakistan* 91 C4 32 0N 74 30 E
Puno, *Peru* 172 D3 15 55 S 70 3W
Punta Alta, *Argentina* 176 A4 38 53 S 62 4W
Punta Arenas, *Chile* 176 D2 53 10 S 71 0W
Punta Cardón, *Venezuela* 168 A3 11 38N 70 14W
Punta Coles, *Peru* 172 D3 17 43 S 71 23W
Punta de Bombón, *Peru* 172 D3 17 10 S 71 48W
Punta de Díaz, *Chile* 174 B1 28 0 S 70 45W
Punta Delgado, *Argentina* 176 B4 42 43 S 63 38W
Punta Gorda, *Belize* 163 D7 16 10N 88 45W
Punta Gorda, *U.S.A.* 149 M5 26 56N 82 3W
Punta Prieta, *Mexico* 162 B2 28 58N 114 17W
Punta Prima, *Spain* 39 B11 39 48N 4 16 E
Punta Rassa, *U.S.A.* 153 J8 26 26N 81 59W
Punta Umbría, *Spain* 43 H4 37 10N 6 56W
Puntabie, *Australia* 127 E1 32 12 S 134 13 E
Puntarenas, *Costa Rica* 164 E3 10 0N 84 50W
Punto Fijo, *Venezuela* 168 A3 11 50N 70 13W
Punxsatawney, *U.S.A.* 150 F6 40 57N 78 59W
Puolo Pt., *U.S.A.* 145 B2 21 54N 159 36W
Puqi, *China* 77 C9 29 40N 113 50 E
Puquio, *Peru* 172 C3 14 45 S 74 10W
Pur →, *Russia* 64 C8 67 31N 77 55 E
Purace, Vol., *Colombia* 168 C2 2 21N 76 23W
Puračić, *Bos.-H.* 52 F3 44 33N 18 28 E
Puralia = Puruliya, *India* 93 H12 23 17N 86 24 E
Purari →, *Papua N. G.* 132 D3 7 49 S 145 0 E
Purbeck, Isle of, *U.K.* 21 G6 50 39N 1 59W
Purcell, *U.S.A.* 155 H6 35 1N 97 22W
Puri, *India* 94 E7 19 50N 85 58 E
Purificación, *Colombia* 168 C3 3 51N 74 55W
Purmerend, *Neths.* 24 B4 52 32N 4 58 E
Purna →, *India* 94 E3 19 6N 77 2 E
Purnia, *India* 93 G12 25 45N 87 31 E
Pursat = Pouthisat, *Cambodia* 86 F4 12 34N 103 50 E
Purukcahu, *Indonesia* 85 C4 0 35 S 114 35 E
Puruliya, *India* 93 H12 23 17N 86 24 E
Purus →, *Brazil* 169 D5 3 42 S 61 28W
Puruvesi, *Finland* 58 B5 61 50N 29 30 E
Pŭrvomay, *Bulgaria* 51 D9 42 8N 25 17 E
Purwakarta, *Indonesia* 85 D3 6 35 S 107 29 E
Purwodadi, *Jawa, Indonesia* 85 D4 7 7 S 110 55 E
Purwodadi, *Jawa, Indonesia* 85 D4 7 51 S 110 0 E
Purwokerto, *Indonesia* 85 D3 7 25 S 109 14 E
Purworejo, *Indonesia* 85 D4 7 43 S 110 2 E
Puryŏng, *N. Korea* 75 C15 42 5N 129 43 E
Pus →, *India* 94 E3 19 55N 77 55 E
Pusad, *India* 94 E3 19 56N 77 36 E
Pusan, *S. Korea* 75 G15 35 5N 129 0 E
Pushkin, *Russia* 58 C6 59 45N 30 25 E
Pushkino, *Moskva, Russia* 58 D9 56 2N 37 49 E
Pushkino, *Saratov, Russia* 60 E8 51 16N 47 0 E
Püspökladány, *Hungary* 52 C6 47 19N 21 6 E
Pustoshka, *Russia* 58 D5 56 20N 29 30 E
Puszczykowo, *Poland* 55 F3 52 18N 16 49 E
Putahow L., *Canada* 143 B8 59 54N 100 40W
Putao, *Burma* 90 B6 27 28N 97 30 E
Putaruru, *N.Z.* 130 E4 38 2 S 175 50 E
Putbus, *Germany* 30 A9 54 22N 13 28 E
Putian, *China* 77 E12 25 23N 119 0 E
Putignano, *Italy* 47 B10 40 51N 17 7 E
Putina, *Peru* 172 C4 14 55 S 69 55W
Puting, Tanjung, *Indonesia* 85 C4 3 31 S 111 46 E
Putlitz, *Germany* 30 B8 53 15N 12 2 E
Putna, *Romania* 53 C10 47 50N 25 33 E
Putna →, *Romania* 53 E12 45 42N 27 26 E
Putnam, *U.S.A.* 151 E13 41 55N 71 55W
Putney, *U.S.A.* 152 D5 31 29N 84 8W
Putnok, *Hungary* 52 B5 48 18N 20 26 E
Putorana, Gory, *Russia* 65 C10 69 0N 95 0 E
Putorino, *N.Z.* 130 F5 39 4 S 176 58 E
Putre, *Chile* 172 D4 18 12 S 69 35W
Puttalam Lagoon, *Sri Lanka* 95 K4 8 15N 79 45 E
Puttgarden, *Germany* 30 A7 54 30N 11 10 E
Püttlingen, *Germany* 31 F2 49 17N 6 53 E
Puttur, *India* 95 H2 12 46N 75 12 E
Putty, *Australia* 129 B9 32 57 S 150 42 E
Putumayo →, *S. Amer.* 168 D4 3 7 S 67 58W
Putuo, *China* 77 C14 29 56N 122 20 E
Putussibau, *Indonesia* 85 B4 0 50N 112 56 E
Pututahi, *N.Z.* 130 E6 38 39 S 177 53 E
Puu Kaaumakua, *U.S.A.* 145 K14 21 30N 157 54W
Puu Keahiakahoe, *U.S.A.* 145 K14 21 23N 157 49W
Puu o Keokeo, *U.S.A.* 145 D6 19 13N 155 44W
Puuanahulu, *U.S.A.* 145 D6 19 49N 155 51W
Puukolii, *U.S.A.* 145 C5 20 56N 156 41W
Puunene, *U.S.A.* 145 C5 20 53N 156 23W
Puuwai, *U.S.A.* 145 B1 21 54N 160 12W
Puy-de-Dôme, *France* 28 C6 45 46N 2 57 E
Puy-de-Dôme □, *France* 28 C7 45 40N 3 5 E
Puy-l'Évêque, *France* 28 D5 44 31N 1 9 E
Puyallup, *U.S.A.* 160 C4 47 12N 122 18W
Puyang, *China* 74 G8 35 40N 115 1 E
Puyehue, *Chile* 176 B2 40 40 S 72 37W
Puylaurens, *France* 28 E6 43 35N 2 0 E
Puyo, *Ecuador* 168 D2 1 28 S 77 59W
Puysegur Pt., *N.Z.* 131 G1 46 9 S 166 37 E
Pūzeh Rīg, *Iran* 97 E8 27 20N 58 40 E
Pwani □, *Tanzania* 118 D4 7 0 S 39 0 E
Pweto, *Dem. Rep. of the Congo* 119 D2 8 25 S 28 51 E
Pwinbyu, *Burma* 90 E5 20 23N 94 40 E
Pwllheli, *U.K.* 20 E3 52 53N 4 25W
Pya-ozero, *Russia* 56 A5 66 5N 30 58 E
Pyana →, *Russia* 60 C8 55 43N 46 1 E
Pyandzh, *Tajikistan* 63 E4 37 14N 69 6 E
Pyandzh →, *Afghan.* 91 A2 37 15N 67 15 E
Pyandzh →, *Tajikistan* 63 E4 37 6N 68 20 E
Pyapon, *Burma* 90 G5 16 20N 95 40 E
Pyasina →, *Russia* 65 B9 73 30N 87 0 E
Pyatigorsk, *Russia* 61 H6 44 2N 43 6 E
Pyatykhatky, *Ukraine* 59 H7 48 28N 33 38 E
Pyaye, *Burma* 90 F5 19 12N 95 10 E
Pydna, *Greece* 50 F6 40 20N 22 34 E
Pyè, *Burma* 90 F5 18 49N 95 13 E
Pyetrikaw, *Belarus* 59 F5 52 11N 28 29 E
Pyhäjoki, *Finland* 14 D21 64 28N 24 14 E
Pyinbauk, *Burma* 90 F5 19 10N 95 12 E
Pyinmana, *Burma* 90 F6 19 45N 96 12 E
Pyla, C., *Cyprus* 38 E12 34 56N 33 51 E
Pyŏktong, *N. Korea* 75 D13 40 50N 125 50 E
Pyŏnggang, *N. Korea* 75 E14 38 24N 127 17 E
Pyŏngtaek, *S. Korea* 75 F14 37 1N 127 4 E
P'yŏngyang, *N. Korea* 75 E13 39 0N 125 30 E
Pyote, *U.S.A.* 155 K3 31 32N 103 8W
Pyramid L., *U.S.A.* 158 G4 40 1N 119 35W
Pyramid Pk., *U.S.A.* 161 J10 36 25N 116 37W
Pyramids, *Egypt* 106 J7 29 58N 31 9 E
Pyrénées, *Europe* 28 F4 42 45N 0 18 E
Pyrénées-Atlantiques □, *France* 28 E3 43 10N 0 50W
Pyrénées-Orientales □, *France* 28 F6 42 35N 2 26 E
Pyryatyn, *Ukraine* 59 G7 50 15N 32 25 E
Pyrzyce, *Poland* 55 E1 53 10N 14 55 E
Pyskowice, *Poland* 55 H5 50 24N 18 38 E
Pytalovo, *Russia* 58 D4 57 5N 27 55 E
Pyttegga, *Norway* 18 B4 62 13N 7 42 E
Pyu, *Burma* 90 F6 18 30N 96 28 E
Pyzdry, *Poland* 55 F4 52 11N 17 42 E

Q

Qaanaaq = Thule, *Greenland* 10 B4 77 40N 69 0W
Qabirri →, *Azerbaijan* 61 K8 41 3N 46 17 E
Qabr Hūd, *Yemen* 99 C5 16 9N 49 34 E
Qachasnek, *S. Africa* 117 E4 30 6 S 28 42 E
Qādib, *Yemen* 99 D6 12 37N 53 57 E
Qa'el Jafr, *Jordan* 103 E5 30 20N 36 25 E
Qa'emābād, *Iran* 97 D9 31 44N 60 2 E
Qā'emshahr, *Iran* 97 B7 36 30N 52 53 E
Qagan Nur, *China* 74 C8 43 30N 114 55 E
Qahar Youyi Zhongqi, *China* 74 D7 41 12N 112 40 E
Qahremānshahr = Bākhtarān, *Iran* 101 E12 34 23N 47 0 E
Qaidam Pendi, *China* 72 C4 37 0N 95 0 E
Qajarīyeh, *Iran* 97 D6 31 1N 48 22 E
Qala, Ras il, *Malta* 38 C1 36 1N 14 20 E
Qala-i-Jadid, *Afghan.* 92 D2 31 1N 66 25 E
Qala Yangi, *Afghan.* 92 B2 34 20N 66 30 E
Qalācheh, *Afghan.* 91 B2 35 30N 67 43 E
Qalansīyah, *Yemen* 99 D6 12 41N 53 29 E
Qalāt, *Afghan.* 91 B2 32 15N 66 58 E
Qal'at al Akhḍar, *Si. Arabia* 96 E3 28 0N 37 10 E
Qal'at Bīshah, *Si. Arabia* 98 C3 20 0N 42 36 E
Qal'at Dīzah, *Iraq* 101 D11 36 11N 45 7 E
Qal'at Sukkar, *Iraq* 101 G12 31 51N 46 5 E
Qal'eh Darreh, *Iran* 96 B5 38 47N 47 2 E
Qal'eh-ye Best, *Afghan.* 91 C2 31 30N 64 21 E
Qal'eh-ye Now, *Afghan.* 91 B1 35 0N 63 5 E
Qal'eh-ye Panjeh, *Afghan.* 91 A4 37 0N 72 35 E
Qal'eh-ye Sarkari, *Afghan.* 91 B2 35 54N 67 17 E
Qal'eh-ye Valī, *Afghan.* 91 B1 35 46N 63 45 E
Qalyûb, *Egypt* 106 H7 30 12N 31 11 E
Qamar, Ghubbat al, *Yemen* 99 C6 16 20N 52 30 E
Qamar, Jabal al, *Oman* 99 C6 16 48N 53 15 E
Qamdo, *China* 76 B1 31 15N 97 6 E
Qamruddin Karez, *Pakistan* 91 C3 31 45N 68 20 E
Qandahār, *Afghan.* 91 C2 31 32N 65 30 E
Qandahār □, *Afghan.* 91 C2 31 0N 65 0 E
Qapān, *Iran* 97 B7 37 40N 55 47 E
Qapshaghay, *Kazakstan* 63 B8 43 51N 77 14 E
Qapshaghay Bögeni, *Kazakstan* 63 B8 43 45N 77 50 E
Qaqortoq = Julianehåb, *Greenland* 10 E6 60 43N 46 0W
Qâra, *Egypt* 106 B2 29 38N 26 30 E
Qarā', Jabal al, *Oman* 99 C6 17 15N 54 15 E
Qara Qash →, *India* 93 B8 35 0N 78 30 E
Qarabutaq, *Kazakstan* 62 G8 49 59N 60 14 E
Qaraçala, *Azerbaijan* 61 L9 39 45N 48 53 E
Qaraghandy, *Kazakstan* 64 E8 49 50N 73 10 E
Qārah, *Si. Arabia* 96 D4 29 55N 40 3 E
Qarataū, *Kazakstan* 63 B5 43 10N 70 28 E
Qarāvol, *Afghan.* 91 A3 37 14N 68 46 E
Qardud, *Sudan* 107 E2 10 20N 29 56 E
Qareh →, *Iran* 101 C12 39 25N 47 22 E
Qareh Tekān, *Iran* 97 B6 36 38N 49 29 E
Qarqan He →, *China* 72 C3 39 30N 88 30 E
Qarqaraly, *Kazakstan* 64 E8 49 26N 75 30 E
Qarrasa, *Sudan* 107 E3 14 38N 32 5 E
Qarshi, *Uzbekistan* 63 D2 38 53N 65 48 E
Qartabā, *Lebanon* 103 A4 34 4N 35 50 E
Qaryat al Gharab, *Iraq* 96 D5 31 27N 44 48 E
Qaryat al 'Ulyā, *Si. Arabia* 96 E5 27 33N 47 42 E
Qasigiannguit = Christianshåb, *Greenland* 10 D5 68 50N 51 18W
Qaskeleng, *Kazakstan* 63 B8 43 20N 76 35 E
Qaşr 'Amra, *Jordan* 96 D3 31 48N 36 35 E
Qaşr Bū Hadi, *Libya* 108 B3 31 1N 16 45 E
Qaşr-e Qand, *Iran* 97 E9 26 15N 60 45 E
Qasr Farâfra, *Egypt* 106 B2 27 0N 28 1 E
Qat Lesh, *Afghan.* 91 B2 34 40N 66 18 E
Qa'ṭabah, *Yemen* 98 D4 13 51N 44 42 E
Qatanā, *Syria* 103 B5 33 26N 36 4 E
Qaṭānan, Ra's, *Yemen* 99 D6 12 21N 53 33 E
Qatar ■, *Asia* 97 E6 25 30N 51 15 E
Qatlīsh, *Iran* 97 B8 37 50N 57 19 E
Qattâra, *Egypt* 106 A2 30 12N 27 3 E
Qattâra, Munkhafed el, *Egypt* 106 B2 29 30N 27 30 E
Qattâra Depression = Qattâra, Munkhafed el, *Egypt* 106 B2 29 30N 27 30 E
Qawām al Ḩamzah, *Iraq* 96 D5 31 43N 44 58 E
Qāyen, *Iran* 97 C8 33 40N 59 10 E
Qazaqstan = Kazakstan ■, *Asia* 64 E8 50 0N 70 0 E
Qazimämmäd, *Azerbaijan* 61 K9 40 3N 49 0 E
Qazvin, *Iran* 97 B6 36 15N 50 0 E
Qena, *Egypt* 106 B3 26 10N 32 43 E
Qena, W. →, *Egypt* 106 B3 26 12N 32 44 E
Qeqertarsuaq = Disko, *Greenland* 10 D5 69 45N 53 30W
Qeqertarsuaq = Godhavn, *Greenland* 10 D5 69 15N 53 38W
Qeqertarsuatsiaat = Fiskenæsset, *Greenland* 10 E5 63 5N 50 45W
Qeshlāq, *Iran* 101 E12 34 55N 46 28 E
Qeshm, *Iran* 97 E8 26 55N 56 10 E
Qezi'ot, *Israel* 103 E3 30 52N 34 26 E
Qi Xian, *China* 74 G8 34 40N 114 48 E
Qian Gorlos, *China* 75 B13 45 5N 124 42 E
Qian Xian, *China* 74 G5 34 31N 108 15 E
Qiancheng, *China* 76 D7 27 12N 109 50 E
Qianjiang, *Guangxi Zhuangzu, China* 76 F7 23 38N 108 58 E
Qianjiang, *Hubei, China* 77 B9 30 24N 112 55 E
Qianjiang, *Sichuan, China* 76 C7 29 33N 108 47 E
Qianshan, *China* 77 B11 30 37N 116 35 E
Qianwei, *China* 76 C4 29 13N 103 56 E
Qianxi, *China* 76 D6 27 3N 106 3 E
Qianyang, *Hunan, China* 76 D8 27 18N 110 10 E
Qianyang, *Shaanxi, China* 74 G4 34 40N 107 8 E
Qianyang, *Zhejiang, China* 77 B12 30 11N 119 25 E
Qiaojia, *China* 76 D4 26 56N 102 58 E
Qibā', *Si. Arabia* 96 E5 27 24N 44 20 E
Qichun, *China* 77 B10 30 18N 115 25 E
Qidong, *Hunan, China* 77 D9 26 49N 112 7 E
Qidong, *Jiangsu, China* 77 B13 31 48N 121 38 E
Qijiang, *China* 76 C6 28 57N 106 35 E
Qila Saifullāh, *Pakistan* 91 C3 30 45N 68 17 E
Qilian Shan, *China* 72 C4 38 30N 96 0 E
Qimen, *China* 77 C11 29 50N 117 42 E
Qin He →, *China* 74 G7 35 1N 113 22 E
Qin Jiang →, *China* 77 D10 26 15N 115 55 E
Qin Ling = Qinling Shandi, *China* 74 H5 33 50N 108 10 E
Qināb, W. →, *Yemen* 99 C5 17 55N 49 59 E
Qin'an, *China* 74 G3 34 48N 105 40 E
Qing Xian, *China* 74 E9 38 35N 116 45 E
Qingcheng, *China* 75 F9 37 15N 117 40 E
Qingdao, *China* 75 F11 36 5N 120 20 E
Qingfeng, *China* 74 G8 35 52N 115 8 E
Qinghai □, *China* 72 C4 36 0N 98 0 E
Qinghai Hu, *China* 72 C5 36 40N 100 10 E
Qinghecheng, *China* 75 D13 41 15N 124 30 E
Qinghemen, *China* 75 D11 41 48N 121 25 E
Qingjian, *China* 74 F6 37 8N 110 8 E
Qingjiang, *Jiangsu, China* 75 H10 33 30N 119 2 E
Qingjiang, *Jiangxi, China* 77 C10 28 4N 115 29 E
Qingliu, *China* 77 D11 26 11N 116 48 E
Qinglong, *China* 76 E5 25 49N 105 12 E
Qingping, *China* 76 D6 26 39N 107 47 E
Qingpu, *China* 77 B13 31 10N 121 6 E
Qingshui, *China* 74 G4 34 48N 106 8 E
Qingshuihe, *China* 74 E6 39 55N 111 35 E
Qingtian, *China* 77 C13 28 12N 120 15 E
Qingtongxia Shuiku, *China* 74 F3 37 50N 105 58 E
Qingxi, *China* 76 D7 27 8N 108 43 E
Qingxu, *China* 74 F7 37 34N 112 22 E
Qingyang, *Anhui, China* 77 B11 30 38N 117 50 E
Qingyang, *Gansu, China* 74 F4 36 2N 107 55 E
Qingyi Jiang →, *China* 76 C4 29 32N 103 44 E
Qingyuan, *Guangdong, China* 77 F9 23 40N 112 59 E
Qingyuan, *Liaoning, China* 75 C13 42 10N 124 55 E
Qingyuan, *Zhejiang, China* 77 D12 27 36N 119 3 E
Qingyun, *China* 75 F9 37 45N 117 20 E
Qingzhen, *China* 76 D6 26 31N 106 25 E
Qinhuangdao, *China* 75 E10 39 56N 119 30 E
Qinling Shandi, *China* 74 H5 33 50N 108 10 E
Qinshui, *China* 74 G7 35 40N 112 8 E
Qinyang, *China* 74 G7 35 7N 112 57 E
Qinyuan, *China* 74 F7 36 29N 112 20 E
Qinzhou, *China* 76 G7 21 58N 108 38 E
Qionghai, *China* 86 C8 19 15N 110 26 E
Qionglai, *China* 76 B4 30 25N 103 31 E
Qionglai Shan, *China* 76 B4 31 0N 102 30 E
Qiongzhou Haixia, *China* 86 B8 20 10N 110 15 E
Qiqihar, *China* 65 E13 47 26N 124 0 E
Qiraîya, W. →, *Egypt* 103 E3 30 27N 34 0 E
Qiryat Ata, *Israel* 103 C4 32 47N 35 6 E
Qiryat Gat, *Israel* 103 D3 31 32N 34 46 E
Qiryat Mal'akhi, *Israel* 103 D3 31 44N 34 44 E
Qiryat Shemona, *Israel* 103 B4 33 13N 35 35 E
Qiryat Yam, *Israel* 103 C4 32 51N 35 4 E
Qishan, *China* 74 G4 34 25N 107 38 E
Qishn, *Yemen* 99 D5 15 26N 51 40 E
Qitai, *China* 72 B3 44 2N 89 35 E
Qitbīt, W. →, *Oman* 99 C6 19 15N 54 23 E
Qiubei, *China* 76 E5 24 2N 104 12 E
Qixia, *China* 75 F11 37 17N 120 52 E
Qiyang, *China* 77 D8 26 35N 111 50 E
Qızılağac Körfäzi, *Azerbaijan* 101 C13 39 9N 49 0 E
Qojūr, *Iran* 96 B5 36 12N 47 55 E
Qom, *Iran* 97 C6 34 40N 51 0 E
Qomsheh, *Iran* 97 D6 32 0N 51 55 E
Qondūz, *Afghan.* 91 A3 36 50N 68 50 E
Qondūz □, *Afghan.* 91 A3 36 50N 68 50 E
Qorveh, *Iran* 101 E12 35 10N 47 48 E
Qostanay, *Kazakstan* 64 D7 53 10N 63 35 E
Qoţūr, *Iran* 101 C11 38 28N 44 25 E
Qu Jiang →, *China* 76 B6 30 1N 106 24 E
Qu Xian, *Sichuan, China* 76 B6 30 48N 106 58 E
Qu Xian, *Zhejiang, China* 77 C12 28 57N 118 54 E
Quairading, *Australia* 125 F2 32 0 S 117 21 E
Quakenbrück, *Germany* 30 C3 52 41N 7 57 E
Quakertown, *U.S.A.* 151 F9 40 26N 75 21W
Qualeup, *Australia* 125 F2 33 48 S 116 48 E
Quambatook, *Australia* 128 C5 35 49 S 143 34 E
Quambone, *Australia* 129 A7 30 57 S 147 53 E
Quamby, *Australia* 126 C3 20 22 S 140 17 E
Quan Long = Ca Mau, *Vietnam* 87 H5 9 7N 105 8 E
Quanah, *U.S.A.* 155 H5 34 18N 99 44W
Quandialla, *Australia* 129 C7 34 1 S 147 47 E
Quang Ngai, *Vietnam* 86 E7 15 13N 108 58 E
Quang Tri, *Vietnam* 86 D6 16 45N 107 13 E
Quang Yen, *Vietnam* 86 B6 20 56N 106 52 E
Quannan, *China* 77 E10 24 45N 114 33 E
Quantock Hills, *U.K.* 21 F4 51 8N 3 10W
Quanzhou, *Fujian, China* 77 E12 24 55N 118 34 E
Quanzhou, *Guangxi Zhuangzu, China* 77 E8 25 57N 111 5 E
Quaqtaq, *Canada* 139 B13 60 55N 69 40W
Quaraí, *Brazil* 174 C4 30 15 S 56 20W
Quarré-les-Tombes, *France* 27 E11 47 21N 4 0 E
Quarteira, *Portugal* 43 H2 37 4N 8 6W
Quartu Sant'Èlena, *Italy* 46 C2 39 15N 9 10 E
Quartzsite, *U.S.A.* 161 M12 33 40N 114 13W
Quatsino, *Canada* 142 C3 50 30N 127 40W
Quatsino Sd., *Canada* 142 C3 50 25N 127 58W
Quba, *Azerbaijan* 61 K9 41 21N 48 32 E
Qūchān, *Iran* 97 B8 37 10N 58 27 E
Queanbeyan, *Australia* 129 C8 35 17 S 149 14 E
Québec, *Canada* 141 C5 46 52N 71 13W
Québec □, *Canada* 141 C6 48 0N 74 0W
Quedlinburg, *Germany* 30 D7 51 47N 11 8 E
Queen Alexandra Ra., *Antarctica* 7 E11 85 0 S 170 0 E
Queen Charlotte, *Canada* 142 C2 53 15N 132 2W
Queen Charlotte Bay, *Falk. Is.* 176 D4 51 50 S 60 40W
Queen Charlotte Is., *Canada* 142 C2 53 20N 132 10W
Queen Charlotte Sd., *Canada* 142 C3 51 0N 128 0W
Queen Charlotte Sd., *N.Z.* 131 B9 41 10 S 174 15 E
Queen City, *U.S.A.* 156 D4 40 25N 92 34W
Queen Elizabeth Is., *Canada* 136 B10 76 0N 95 0W
Queen Elizabeth Nat. Park, *Uganda* 118 C3 0 0 30 0 E
Queen Mary Land, *Antarctica* 7 D7 70 0 S 95 0 E
Queen Maud G., *Canada* 138 B9 68 15N 102 30W
Queen Maud Land, *Antarctica* 7 D3 72 30 S 12 0 E
Queen Maud Mts., *Antarctica* 7 E13 86 0 S 160 0W
Queens Chan., *Australia* 124 C4 15 0 S 129 30 E
Queenscliff, *Australia* 127 F3 38 16 S 144 39 E
Queensland, *U.S.A.* 152 D6 31 46N 83 14W
Queensland □, *Australia* 126 C3 22 0 S 142 0 E
Queenstown, *Australia* 126 G4 42 4 S 145 35 E
Queenstown, *N.Z.* 131 F3 45 1 S 168 40 E
Queenstown, *S. Africa* 116 E4 31 52 S 26 52 E
Queets, *U.S.A.* 160 C2 47 32N 124 20W
Queguay Grande →, *Uruguay* 174 C4 32 9 S 58 9W
Queimadas, *Brazil* 170 D4 11 0 S 39 38W
Queiros, C., *Vanuatu* 133 D5 14 55 S 167 1 E
Quela, *Angola* 115 D3 9 10 S 16 56 E
Quelimane, *Mozam.* 119 F4 17 53 S 36 58 E
Quelpart = Cheju Do, *S. Korea* 75 H14 33 29N 126 34 E
Queluz, *Portugal* 43 G1 38 45N 9 15W
Quemado, *N. Mex., U.S.A.* 159 J9 34 20N 108 30W
Quemado, *Tex., U.S.A.* 155 L4 28 58N 100 35W
Quemoy = Chinmen, *Taiwan* 77 E13 24 26N 118 19 E
Quemú-Quemú, *Argentina* 174 D3 36 3 S 63 36W
Quequén, *Argentina* 174 D4 38 30 S 58 30W
Querco, *Peru* 172 C3 13 50 S 74 52W
Querétaro, *Mexico* 162 C4 20 36N 100 23W
Querétaro □, *Mexico* 162 C5 20 30N 100 0W
Querfurt, *Germany* 30 D7 51 23N 11 35 E
Quérigut, *France* 28 F6 42 42N 2 6 E
Querqueville, *France* 26 C5 49 40N 1 42W
Quesada, *Spain* 43 H7 37 51N 3 4W
Queshan, *China* 74 H8 32 55N 114 2 E
Quesnel, *Canada* 142 C4 53 0N 122 30W
Quesnel →, *Canada* 142 C4 52 58N 122 29W
Quesnel L., *Canada* 142 C4 52 30N 121 20W
Questa, *U.S.A.* 159 H11 36 42N 105 36W
Questembert, *France* 26 E4 47 40N 2 28W
Quetena, *Bolivia* 172 E4 22 10 S 67 25W
Quetico Prov. Park, *Canada* 140 C1 48 30N 91 45W
Quetrequile, *Argentina* 176 B3 41 33 S 69 22W
Quetta, *Pakistan* 91 C2 30 15N 66 55 E
Quevedo, *Ecuador* 168 D2 1 2 S 79 29W
Quezaltenango, *Guatemala* 164 D1 14 50N 91 30W
Quezon □, *Phil.* 80 D3 14 40N 121 30 E
Quezon City, *Phil.* 80 D3 14 38N 121 0 E
Qufār, *Si. Arabia* 96 E4 27 26N 41 37 E
Qui Nhon, *Vietnam* 86 F7 13 40N 109 13 E
Quibala, *Angola* 115 E2 10 46 S 14 59 E
Quibaxe, *Angola* 115 D2 8 24 S 14 27 E
Quibdo, *Colombia* 168 B2 5 42N 76 40W
Quiberon, *France* 26 E3 47 29N 3 9W
Quiberon, Presqu'île de, *France* 26 E3 47 30N 3 8W
Quíbor, *Venezuela* 168 B4 9 56N 69 37W
Quick, *Canada* 142 C3 54 36N 126 54W
Quickborn, *Germany* 30 B5 53 42N 9 52 E
Quiet L., *Canada* 142 A2 61 5N 133 5W
Quiindy, *Paraguay* 174 B4 25 58 S 57 14W
Quila, *Mexico* 162 C3 24 23N 107 13W
Quilán, C., *Chile* 176 B2 43 15 S 74 30W
Quilcene, *U.S.A.* 160 C4 47 49N 122 53W
Quilengues, *Angola* 115 E2 14 12 S 14 12 E
Quilimarí, *Chile* 174 C1 32 5 S 71 30W
Quilino, *Argentina* 174 C3 30 14 S 64 29W
Quillabamba, *Peru* 172 C3 12 50 S 72 50W
Quillacollo, *Bolivia* 172 D4 17 26 S 66 17W
Quillagua, *Chile* 174 A2 21 40 S 69 40W
Quillaicillo, *Chile* 174 C1 31 17 S 71 40W
Quillan, *France* 28 F6 42 53N 2 10 E
Quillota, *Chile* 174 C1 32 54 S 71 16W
Quilmes, *Argentina* 174 C4 34 43 S 58 15W
Quilon, *India* 95 K3 8 50N 76 38 E
Quilpie, *Australia* 127 D3 26 35 S 144 11 E
Quilpué, *Chile* 174 C1 33 5 S 71 33W

270 Quilua

Quilua, *Mozam.* **119 F4** 16 17 S 39 54 E
Quimbele, *Angola* **115 D3** 6 17 S 16 41 E
Quimbonge, *Angola* **115 D3** 8 36 S 18 30 E
Quime, *Bolivia* **172 D4** 17 2 S 67 15W
Quimilí, *Argentina* **174 B3** 27 40 S 62 30W
Quimper, *France* **26 E2** 48 0N 4 9W
Quimperlé, *France* **26 E3** 47 53N 3 33W
Quinault →, *U.S.A.* **160 C2** 47 21N 124 18W
Quincemil, *Peru* **172 C3** 13 15 S 70 40W
Quincy, *Calif., U.S.A.* **160 F6** 39 56N 120 57W
Quincy, *Fla., U.S.A.* **152 E5** 30 35N 84 34W
Quincy, *Ill., U.S.A.* **154 F9** 39 56N 91 23W
Quincy, *Mass., U.S.A.* **151 D14** 42 15N 71 0W
Quincy, *Wash., U.S.A.* **158 C4** 47 22N 119 56W
Quines, *Argentina* **174 C2** 32 13 S 65 48W
Quinga, *Mozam.* **119 F5** 15 49 S 40 15 E
Quingey, *France* **27 E12** 47 7N 5 52 E
Quiniluban Group, *Phil.* **81 F3** 11 27N 120 48 E
Quintana de la Serena, *Spain* **43 G5** 38 45N 5 40W
Quintana Roo □, *Mexico* **163 D7** 19 0N 88 0W
Quintanar de la Orden, *Spain* **43 F7** 39 36N 3 5W
Quintanar de la Sierra, *Spain* **40 D2** 41 57N 2 55W
Quintanar del Rey, *Spain* **41 F3** 39 21N 1 56W
Quintero, *Chile* **174 C1** 32 45 S 71 30W
Quintin, *France* **26 D4** 48 26N 2 56W
Quinto, *Spain* **40 D4** 41 25N 0 32W
Quinyambie, *Australia* **127 E3** 30 15 S 141 0 E
Quípar →, *Spain* **41 G3** 38 15N 1 40W
Quipungo, *Angola* **115 E2** 14 37 S 14 40 E
Quirihue, *Chile* **174 D1** 36 15 S 72 35W
Quirimbo, *Angola* **115 E2** 10 36 S 14 12 E
Quirindi, *Australia* **129 A9** 31 28 S 150 40 E
Quirino □, *Phil.* **80 C3** 16 15N 121 40 E
Quiroga, *Spain* **42 C3** 42 28N 7 18W
Quiruvilca, *Peru* **172 B2** 8 1 S 78 19W
Quissac, *France* **29 E8** 43 55N 4 0 E
Quissanga, *Mozam.* **119 E5** 12 24 S 40 28 E
Quitapa, *Angola* **115 E3** 10 20 S 18 19 E
Quitilipi, *Argentina* **174 B3** 26 50 S 60 13W
Quitman, *Ga., U.S.A.* **153 E4** 30 47N 83 34W
Quitman, *Miss., U.S.A.* **149 J1** 32 2N 88 44W
Quitman, *Tex., U.S.A.* **155 J7** 32 48N 95 27W
Quito, *Ecuador* **168 D2** 0 15 S 78 35W
Quixadá, *Brazil* **170 B4** 4 55 S 39 0W
Quixaxe, *Mozam.* **119 F5** 15 17 S 40 4 E
Quixeramobim, *Brazil* **170 C4** 5 12 S 39 17W
Quixinge, *Angola* **115 D2** 9 52 S 14 23 E
Quizenga, *Angola* **115 D3** 9 21 S 15 28 E
Qujing, *China* **76 E4** 25 32N 103 41 E
Qul'ân, Jazâ'ir, *Egypt* **106 C4** 24 22N 35 31 E
Qumbu, *S. Africa* **117 E4** 31 10 S 28 48 E
Quneitra, *Syria* **103 B4** 33 7N 35 48 E
Qunfudh, *Yemen* **99 C5** 16 39N 49 33 E
Qŭnghirot, *Uzbekistan* **64 E6** 43 6N 58 54 E
Quoin I., *Australia* **124 B4** 14 54 S 129 32 E
Quoin Pt., *S. Africa* **116 E2** 34 46 S 19 37 E
Quondong, *Australia* **128 B3** 33 6 S 140 18 E
Quorn, *Australia* **128 B3** 32 25 S 138 5 E
Qŭqon, *Uzbekistan* **63 C5** 40 30N 70 57 E
Qurein, *Sudan* **107 E3** 13 30N 34 50 E
Qŭrghonteppa, *Tajikistan* **63 E4** 37 50N 68 47 E
Qurnat as Sawdā', *Lebanon* **103 A5** 34 18N 36 6 E
Qûs, *Egypt* **106 B3** 25 55N 32 50 E
Qusar, *Azerbaijan* **61 K9** 41 25N 48 26 E
Qusaybah, *Iraq* **101 E9** 34 24N 40 59 E
Quşay'ir, *Yemen* **99 D5** 14 55N 50 20 E
Quseir, *Egypt* **106 B3** 26 7N 34 16 E
Qūshchī, *Iran* **101 D11** 37 59N 45 3 E
Quthing, *Lesotho* **117 E4** 30 25 S 27 36 E
Qūţīābād, *Iran* **97 C6** 35 47N 48 30 E
Quwo, *China* **74 G6** 35 38N 111 25 E
Quyang, *China* **74 E8** 38 35N 114 40 E
Quynh Nhai, *Vietnam* **86 B4** 21 49N 103 33 E
Quzi, *China* **74 F4** 36 20N 107 20 E
Qvareli, *Georgia* **61 K7** 41 57N 45 47 E
Qytet Stalin = Kuçovë, *Albania* **50 F3** 40 47N 19 57 E
Qyzylorda, *Kazakstan* **63 A2** 44 48N 65 28 E

R

Ra, Ko, *Thailand* **87 H2** 9 13N 98 16 E
Raab, *Austria* **34 C6** 48 21N 13 39 E
Raahe, *Finland* **14 D21** 64 40N 24 28 E
Raalte, *Neths.* **24 B6** 52 23N 6 16 E
Raasay, *U.K.* **22 D2** 57 25N 6 4W
Raasay, Sd. of, *U.K.* **22 D2** 57 30N 6 8W
Rab, *Croatia* **45 D11** 44 45N 14 45 E
Raba, *Indonesia* **82 C1** 8 36 S 118 55 E
Rába →, *Hungary* **52 C2** 47 38N 17 38 E
Raba →, *Poland* **55 H7** 50 8N 20 30 E
Rabaçal →, *Portugal* **42 D3** 41 30N 7 12W
Rabah, *Nigeria* **113 C6** 13 5N 5 30 E
Rabai, *Kenya* **118 C4** 3 50 S 39 31 E
Rabaraba, *Papua N. G.* **132 E5** 9 58 S 149 49 E
Rabastens, *France* **28 E5** 43 50N 1 43 E
Rabastens-de-Bigorre, *France* **28 E4** 43 23N 0 9 E
Rabat, *Malta* **38 D1** 35 53N 14 25 E
Rabat, *Morocco* **110 B3** 34 2N 6 48W
Rabaul, *Papua N. G.* **132 C7** 4 24 S 152 18 E
Rabbit →, *Canada* **142 B3** 59 41N 127 12W
Rabbit Lake, *Canada* **143 C7** 53 8N 107 46W
Rabbitskin →, *Canada* **142 A4** 61 47N 120 42W
Rābigh, *Si. Arabia* **98 B2** 22 50N 39 5 E
Rabka, *Poland* **55 J6** 49 37N 19 59 E
Râbniţa, *Moldova* **53 C14** 47 45N 29 0 E
Rābor, *Iran* **97 D8** 29 17N 56 55 E
Rača, *Serbia, Yug.* **50 B4** 44 14N 21 0 E
Răcăciuni, *Romania* **53 D11** 46 20N 26 59 E
Răcăşdia, *Romania* **52 F6** 44 59N 21 36 E
Racconigi, *Italy* **44 D4** 44 46N 7 46 E
Raccoon →, *U.S.A.* **156 C3** 41 35N 93 37W
Raccoon Cr. →, *U.S.A.* **157 E9** 39 47N 87 23W
Race, C., *Canada* **141 C9** 46 40N 53 5W
Rach Gia, *Vietnam* **87 G5** 10 5N 105 5 E
Raciąż, *Poland* **55 F7** 52 46N 20 10 E
Racibórz, *Poland* **55 H5** 50 7N 18 18 E
Racine, *U.S.A.* **157 B9** 42 41N 87 51W
Rackerby, *U.S.A.* **160 F5** 39 26N 121 22W
Radama, Nosy, *Madag.* **117 A8** 14 0 S 47 47 E
Radama, Saikanosy, *Madag.* **117 A8** 14 16 S 47 53 E
Radan, *Serbia, Yug.* **50 D5** 42 59N 21 29 E
Rădăuţi, *Romania* **53 C10** 47 50N 25 59 E
Rădăuţi-Prut, *Romania* **53 B11** 48 14N 26 48 E
Radbuza →, *Czech Rep.* **34 B6** 49 45N 13 22 E
Radcliff, *U.S.A.* **157 G11** 37 51N 85 57W
Radebeul, *Germany* **30 D9** 51 6N 13 39 E
Radeburg, *Germany* **30 D9** 51 7N 13 53 E
Radeče, *Slovenia* **45 B12** 46 5N 15 14 E
Radekhiv, *Ukraine* **59 G3** 50 25N 24 32 E
Radekhov = Radekhiv, *Ukraine* **59 G3** 50 25N 24 32 E
Radenthein, *Austria* **34 E6** 46 48N 13 43 E
Radew →, *Poland* **54 D2** 54 2N 15 52 E
Radford, *U.S.A.* **148 G5** 37 8N 80 34W
Radhanpur, *India* **92 H4** 23 50N 71 38 E
Radika →, *Macedonia* **50 E4** 41 38N 20 37 E
Radisson, *Canada* **143 C7** 52 30N 107 20W
Radium Hot Springs, *Canada* **142 C5** 50 35N 116 2W
Radlje ob Dravi, *Slovenia* **45 B12** 46 38N 15 13 E
Radnevo, *Bulgaria* **51 D9** 42 17N 25 58 E
Radnice, *Czech Rep.* **34 B6** 49 51N 13 35 E
Radnor Forest, *U.K.* **21 E4** 52 17N 3 10W
Radolfzell, *Germany* **31 H4** 47 39N 8 59 E
Radom, *Poland* **55 G8** 51 23N 21 12 E
Radom □, *Poland* **55 G8** 51 30N 21 0 E
Radomir, *Bulgaria* **50 D6** 42 37N 22 59 E
Radomka →, *Poland* **55 G8** 51 43N 21 28 E
Radomsko, *Poland* **55 G6** 51 5N 19 28 E
Radomyshl, *Ukraine* **59 G5** 50 30N 29 12 E
Radomyśl Wielki, *Poland* **55 H8** 50 14N 21 15 E
Radoszyce, *Poland* **55 G7** 51 4N 20 15 E
Radoviš, *Macedonia* **50 E6** 41 38N 22 28 E
Radovljica, *Slovenia* **45 B11** 46 22N 14 12 E
Radstadt, *Austria* **34 D6** 47 24N 13 28 E
Radstock, C., *Australia* **127 E1** 33 12 S 134 20 E
Răducăneni, *Romania* **53 D12** 46 58N 27 54 E
Raduša, *Serbia, Yug.* **50 D5** 42 7N 21 15 E
Radviliškis, *Lithuania* **15 J20** 55 49N 23 33 E
Radville, *Canada* **143 D8** 49 30N 104 15W
Radymno, *Poland* **55 J9** 49 59N 22 52 E
Radzanów, *Poland* **55 F7** 52 56N 20 8 E
Radziejów, *Poland* **55 F5** 52 40N 18 30 E
Radzyń Chełmiński, *Poland* **54 E5** 53 23N 18 55 E
Radzyń Podlaski, *Poland* **55 G9** 51 47N 22 37 E
Rae, *Canada* **142 A5** 62 50N 116 3W
Rae Bareli, *India* **93 F9** 26 18N 81 20 E
Rae Isthmus, *Canada* **139 B11** 66 40N 87 30W
Raeren, *Belgium* **24 D6** 50 41N 6 7 E
Raeside, L., *Australia* **125 E3** 29 20 S 122 0 E
Raetihi, *N.Z.* **130 F4** 39 25 S 175 17 E
Rafaela, *Argentina* **174 C3** 31 10 S 61 30W
Rafah, *Gaza Strip* **103 D3** 31 18N 34 14 E
Rafai, *C.A.R.* **114 B4** 4 59N 23 58 E
Raffadali, *Italy* **46 E6** 37 24N 13 32 E
Rafḩā, *Si. Arabia* **96 D4** 29 35N 43 35 E
Rafsanjān, *Iran* **97 D8** 30 30N 56 5 E
Raft Pt., *Australia* **124 C3** 16 4 S 124 26 E
Ragachow, *Belarus* **59 F6** 53 8N 30 5 E
Ragag, *Sudan* **107 E1** 10 59N 24 40 E
Ragang, Mt., *Phil.* **81 H5** 7 43N 124 32 E
Ragay, *Phil.* **80 E4** 13 49N 122 47 E
Ragay G., *Phil.* **80 E4** 13 30N 122 45 E
Ragged, Mt., *Australia* **125 F3** 33 27 S 123 25 E
Raglan, *Australia* **126 C5** 23 42 S 150 49 E
Raglan, *N.Z.* **130 D3** 37 55 S 174 55 E
Raglan Harbour, *N.Z.* **130 D3** 37 47 S 174 50 E
Ragland, *U.S.A.* **152 B3** 33 45N 86 9W
Ragusa, *Italy* **47 F7** 36 55N 14 44 E
Raha, *Indonesia* **82 B2** 4 55 S 123 0 E
Rahad, Nahr ed →, *Sudan* **107 E3** 14 28N 33 31 E
Rahad al Bardī, *Sudan* **109 F4** 11 20N 23 40 E
Rahaeng = Tak, *Thailand* **86 D2** 16 52N 99 8 E
Rahden, *Germany* **30 C4** 52 26N 8 36 E
Raheita, *Eritrea* **107 E5** 12 46N 43 4 E
Raḩīmah, *Si. Arabia* **97 E6** 26 42N 50 4 E
Rahimyar Khan, *Pakistan* **91 C3** 28 30N 70 25 E
Rāhjerd, *Iran* **97 C6** 34 22N 50 22 E
Råholt, *Norway* **18 D8** 60 16N 11 11 E
Rahotu, *N.Z.* **130 F2** 39 20 S 173 49 E
Raichur, *India* **95 F3** 16 10N 77 20 E
Raiford, *U.S.A.* **152 E7** 30 4N 82 14W
Raiganj, *India* **93 G13** 25 37N 88 10 E
Raigarh, *India* **94 D6** 21 56N 83 25 E
Raighar, *India* **94 E6** 19 51N 82 6 E
Raijua, *Indonesia* **82 D2** 10 37 S 121 36 E
Railton, *Australia* **126 G4** 41 25 S 146 28 E
Rainbow, *Australia* **128 C5** 35 55 S 142 0 E
Rainbow City, *U.S.A.* **152 B3** 33 57N 86 5W
Rainbow Lake, *Canada* **142 B5** 58 30N 119 23W
Rainier, *U.S.A.* **160 D4** 46 53N 122 41W
Rainier, Mt., *U.S.A.* **160 D5** 46 52N 121 46W
Rainy L., *Canada* **143 D10** 48 42N 93 10W
Rainy River, *Canada* **143 D10** 48 43N 94 29W
Raippaluoto, *Finland* **14 E19** 63 13N 21 14 E
Raipur, *India* **94 D5** 21 17N 81 45 E
Raisio, *Finland* **15 F20** 60 28N 22 11 E
Raj Nandgaon, *India* **94 D5** 21 5N 81 5 E
Raja, Ujung, *Indonesia* **84 B1** 3 40N 96 25 E
Raja Ampat, Kepulauan, *Indonesia* **83 B4** 0 30 S 130 0 E
Rajahmundry, *India* **94 F5** 17 1N 81 48 E
Rajang →, *Malaysia* **85 B4** 2 30N 112 0 E
Rajapalaiyam, *India* **95 K3** 9 25N 77 35 E
Rajasthan □, *India* **92 F5** 26 45N 73 30 E
Rajasthan Canal, *India* **92 F5** 28 0N 72 0 E
Rajauri, *India* **93 C6** 33 25N 74 21 E
Rajbari, *Bangla.* **90 D2** 23 47N 89 41 E
Rajgarh, *Mad. P., India* **92 G7** 24 2N 76 45 E
Rajgarh, *Raj., India* **92 E6** 28 40N 75 25 E
Rajgród, *Poland* **54 E9** 53 42N 22 42 E
Rajkot, *India* **92 H4** 22 15N 70 56 E
Rajmahal Hills, *India* **93 G12** 24 30N 87 30 E
Rajpipla, *India* **94 D1** 21 50N 73 30 E
Rajpura, *India* **92 D7** 30 25N 76 32 E
Rajshahi, *Bangla.* **90 C2** 24 22N 88 39 E
Rajshahi □, *Bangla.* **93 G13** 25 0N 89 0 E
Rakaia, *N.Z.* **131 D7** 43 45 S 172 1 E
Rakaia →, *N.Z.* **131 D7** 43 36 S 172 15 E
Rakan, Ra's, *Qatar* **97 E6** 26 10N 51 20 E
Rakaposhi, *Pakistan* **93 A6** 36 10N 74 25 E
Rakata, Pulau, *Indonesia* **84 D3** 6 10 S 105 20 E
Rakhawt, W. →, *Yemen* **99 C5** 18 16N 51 50 E
Rakhiv, *Ukraine* **59 H3** 48 3N 24 12 E
Rakhneh-ye Jamshīdī, *Afghan.* **91 B1** 34 22N 62 19 E
Rakhni, *Pakistan* **92 D3** 30 4N 69 56 E
Rakhyūt, *Oman* **99 C6** 16 44N 53 20 E
Rakitnoye, *Russia* **68 B7** 45 36N 134 17 E
Rakitovo, *Bulgaria* **51 E8** 41 59N 24 5 E
Rakkestad, *Norway* **18 E8** 59 25N 11 21 E
Rakoniewice, *Poland* **55 F3** 52 10N 16 16 E
Rakops, *Botswana* **116 C3** 21 1 S 24 28 E
Rakovica, *Croatia* **45 D12** 44 59N 15 38 E
Rakovník, *Czech Rep.* **34 A6** 50 6N 13 42 E
Rakovski, *Bulgaria* **51 D8** 42 21N 24 57 E
Rakvere, *Estonia* **15 G22** 59 20N 26 25 E
Raleigh, *Fla., U.S.A.* **153 F7** 29 25N 82 32W
Raleigh, *N.C., U.S.A.* **149 H6** 35 47N 78 39W
Raleigh B., *U.S.A.* **149 H7** 34 50N 76 15W
Ralja, *Serbia, Yug.* **50 B4** 44 33N 20 34 E
Ralls, *U.S.A.* **155 J4** 33 41N 101 24W
Ram →, *Canada* **142 A4** 62 1N 123 41W
Rām Allāh, *West Bank* **103 D4** 31 55N 35 10 E
Ram Hd., *Australia* **129 D8** 37 47 S 149 30 E
Rama, *Nic.* **164 D3** 12 9N 84 15W
Ramacca, *Italy* **47 E7** 37 23N 14 42 E
Ramachandrapuram, *India* **94 F6** 16 50N 82 4 E
Ramales de la Victoria, *Spain* **42 B7** 43 15N 3 28W
Ramalho, Serra do, *Brazil* **171 D3** 13 45 S 44 0W
Raman, *Thailand* **87 J3** 6 29N 101 18 E
Ramanathapuram, *India* **95 K4** 9 25N 78 55 E
Ramanetaka, B. de, *Madag.* **117 A8** 14 13 S 47 52 E
Ramas C., *India* **95 G1** 15 5N 73 55 E
Ramat Gan, *Israel* **103 C3** 32 4N 34 48 E
Ramatlhabama, *S. Africa* **116 D4** 25 37 S 25 33 E
Ramban, *India* **93 C6** 33 14N 75 12 E
Rambervillers, *France* **27 D13** 48 20N 6 38 E
Rambi, *Fiji* **133 A3** 16 30 S 179 59W
Rambipuji, *Indonesia* **85 D4** 8 12 S 113 37 E
Rambouillet, *France* **27 D8** 48 39N 1 50 E
Ramdurg, *India* **95 G2** 15 58N 75 22 E
Ramea, *Canada* **141 C8** 47 31N 57 23W
Ramechhap, *Nepal* **93 F12** 27 25N 86 10 E
Ramelau, *Indonesia* **82 C3** 8 55 S 126 22 E
Ramenskoye, *Russia* **58 E10** 55 32N 38 15 E
Ramer, *U.S.A.* **152 C3** 32 3N 86 13W
Ramgarh, *Bihar, India* **93 H11** 23 40N 85 35 E
Ramgarh, *Raj., India* **92 F6** 27 16N 75 14 E
Ramgarh, *Raj., India* **92 F4** 27 30N 70 36 E
Rāmhormoz, *Iran* **97 D6** 31 15N 49 35 E
Ramīān, *Iran* **97 B7** 37 3N 55 16 E
Ramingining, *Australia* **126 A2** 12 19 S 135 3 E
Ramla, *Israel* **103 D3** 31 55N 34 52 E
Ramlat Zalṭan, *Libya* **108 C3** 28 30N 19 30 E
Ramlu, *Eritrea* **107 E5** 13 32N 41 40 E
Râmna →, *Romania* **53 E12** 45 36N 27 3 E
Ramnad = Ramanathapuram, *India* **95 K4** 9 25N 78 55 E
Ramnagar, *India* **93 C6** 32 47N 75 18 E
Ramnäs, *Sweden* **16 E10** 59 46N 16 12 E
Râmnicu Sărat, *Romania* **53 E12** 45 26N 27 3 E
Râmnicu Vâlcea, *Romania* **53 E9** 45 9N 24 21 E
Ramon, *Russia* **59 G10** 51 55N 39 21 E
Ramona, *U.S.A.* **161 M10** 33 2N 116 52W
Ramonville-St-Agne, *France* **28 E5** 43 33N 1 28 E
Ramore, *Canada* **140 C3** 48 30N 80 25W
Ramotswa, *Botswana* **116 C4** 24 50 S 25 52 E
Rampur, *H.P., India* **92 D7** 31 26N 77 43 E
Rampur, *Mad. P., India* **92 H5** 23 25N 73 53 E
Rampur, *Orissa, India* **94 D6** 21 48N 83 58 E
Rampur, *Ut. P., India* **93 E8** 28 50N 79 5 E
Rampur Hat, *India* **93 G12** 24 10N 87 50 E
Rampura, *India* **92 G6** 24 30N 75 27 E
Rāmsar, *Iran* **97 B6** 36 53N 50 41 E
Ramsey, *Canada* **140 C3** 47 25N 82 20W
Ramsey, *U.K.* **20 C3** 54 20N 4 22W
Ramsey, *U.S.A.* **156 E7** 39 8N 89 7W
Ramsgate, *U.K.* **21 F9** 51 20N 1 25 E
Ramshai, *India* **90 B2** 26 44N 88 51 E
Ramsjö, *Sweden* **16 B9** 62 11N 15 37 E
Ramstein, *Germany* **31 F3** 49 27N 7 32 E
Ramtek, *India* **94 D4** 21 20N 79 15 E
Ramu →, *Papua N. G.* **132 C3** 4 0 S 144 41 E
Ramvik, *Sweden* **16 B11** 62 49N 17 51 E
Ranaghat, *India* **93 H13** 23 15N 88 35 E
Ranahu, *Pakistan* **92 G3** 25 55N 69 45 E
Rancagua, *Chile* **174 C1** 34 10 S 70 50W
Rance →, *France* **26 D5** 48 34N 1 59W
Rancharia, *Brazil* **171 F1** 22 15 S 50 55W
Rancheria →, *Canada* **142 A3** 60 13N 129 7W
Ranchester, *U.S.A.* **158 D10** 44 54N 107 10W
Ranchi, *India* **93 H11** 23 19N 85 27 E
Rancho Cucamonga, *U.S.A.* **161 L9** 34 10N 117 30W
Ranco, L., *Chile* **176 B2** 40 15 S 72 25W
Rand, *Australia* **129 C7** 35 33 S 146 32 E
Randaberg, *Norway* **18 E2** 59 1N 5 36 E
Randabygd, *Norway* **18 C3** 61 51N 6 20 E
Randalstown, *U.K.* **23 B5** 54 45N 6 19W
Randan, *France* **27 F10** 46 2N 3 21 E
Randazzo, *Italy* **47 E7** 37 53N 14 57 E
Randers, *Denmark* **17 H4** 56 29N 10 1 E
Randers Fjord, *Denmark* **17 H4** 56 37N 10 20 E
Randfontein, *S. Africa* **117 D4** 26 8 S 27 45 E
Randle, *U.S.A.* **160 D5** 46 32N 121 57W
Randolph, *Mass., U.S.A.* **151 D13** 42 10N 71 2W
Randolph, *N.Y., U.S.A.* **150 D6** 42 10N 78 59W
Randolph, *Utah, U.S.A.* **158 F8** 41 40N 111 11W
Randolph, *Vt., U.S.A.* **151 C12** 43 55N 72 40W
Randsfjorden, *Norway* **18 D7** 60 25N 10 24 E
Randsverk, *Norway* **18 C6** 61 44N 9 3 E
Råne älv →, *Sweden* **14 D20** 65 50N 22 20 E
Ranfurly, *N.Z.* **131 F5** 45 7 S 170 6 E
Rangae, *Thailand* **87 J3** 6 19N 101 44 E
Rangamati, *Bangla.* **90 D4** 22 38N 92 12 E
Rangárvallasýsla □, *Iceland* **11 D7** 63 55N 20 0W
Rangataua, *N.Z.* **130 F4** 39 26 S 175 28 E
Rangaunu B., *N.Z.* **130 A2** 34 51 S 173 15 E
Rangeley, *U.S.A.* **151 B14** 44 58N 70 39W
Rangely, *U.S.A.* **158 F9** 40 5N 108 48W
Ranger, *U.S.A.* **155 J5** 32 28N 98 41W
Ranga, *India* **90 B3** 26 28N 91 38 E
Rangiora, *N.Z.* **131 D7** 43 19 S 172 36 E
Rangitaiki, *N.Z.* **130 E5** 38 52 S 176 24 E
Rangitaiki →, *N.Z.* **130 D5** 37 54 S 176 49 E
Rangitata →, *N.Z.* **131 D6** 43 45 S 171 15 E
Rangitikei →, *N.Z.* **130 G4** 40 17 S 175 15 E
Rangitoto Ra., *N.Z.* **130 E4** 38 25 S 175 35 E
Rangkasbitung, *Indonesia* **84 D3** 6 21 S 106 15 E
Rangoon, *Burma* **90 G6** 16 45N 96 20 E
Rangpur, *Bangla.* **90 C2** 25 42N 89 22 E
Rangsang, *Indonesia* **84 B2** 1 20N 103 30 E
Rangsit, *Thailand* **86 F3** 13 59N 100 37 E
Ranheim, *Norway* **18 A7** 63 26N 10 32 E
Ranibennur, *India* **95 G2** 14 35N 75 30 E
Raniganj, *India* **93 H12** 23 40N 87 5 E
Ranippettai, *India* **95 H4** 12 56N 79 23 E
Rāniyah, *Iraq* **96 B5** 36 15N 44 53 E
Ranken →, *Australia* **126 C2** 20 31 S 137 36 E
Rankin, *Ill., U.S.A.* **157 D9** 40 28N 87 54W
Rankin, *Tex., U.S.A.* **155 K4** 31 13N 101 56W
Rankin Inlet, *Canada* **138 B10** 62 30N 93 0W
Rankins Springs, *Australia* **129 B7** 33 49 S 146 14 E
Rankweil, *Austria* **34 D2** 47 17N 9 39 E
Rannoch, L., *U.K.* **22 E4** 56 41N 4 20W
Rannoch Moor, *U.K.* **22 E4** 56 38N 4 48W
Ranobe, Helodranon' i, *Madag.* **117 C7** 23 3 S 43 33 E
Ranohira, *Madag.* **117 C8** 22 29 S 45 24 E
Ranomafana, *Toamasina, Madag.* **117 B8** 18 57 S 48 50 E
Ranomafana, *Toliara, Madag.* **117 C8** 24 34 S 47 0 E
Ranong, *Thailand* **87 H2** 9 56N 98 40 E
Rānsa, *Iran* **97 C6** 33 39N 48 18 E
Ransiki, *Indonesia* **83 B4** 1 30 S 134 10 E
Ransom, *U.S.A.* **157 C8** 41 9N 88 39W
Rantau, *Indonesia* **85 C5** 2 56 S 115 9 E
Rantauprapat, *Indonesia* **84 B1** 2 15N 99 50 E
Rantemario, *Indonesia* **82 B1** 3 15 S 119 57 E
Rantoul, *U.S.A.* **157 D8** 40 19N 88 9W
Ranum, *Denmark* **17 H3** 56 54N 9 14 E
Ranyah, W. →, *Si. Arabia* **98 B3** 21 18N 43 20 E
Raon l'Étape, *France* **27 D13** 48 24N 6 50 E
Raoui, Erg er, *Algeria* **111 C4** 29 0N 2 0W
Raoyang, *China* **74 E8** 38 15N 115 45 E
Rapa, *Pac. Oc.* **135 K13** 27 35 S 144 20W
Rapallo, *Italy* **44 D6** 44 21N 9 14 E
Rāpch, *Iran* **97 E8** 25 40N 59 15 E
Rapid →, *Canada* **142 B3** 59 15N 129 5W
Rapid City, *U.S.A.* **154 D3** 44 5N 103 14W
Rapid River, *U.S.A.* **148 C2** 45 55N 86 58W
Rapides des Joachims, *Canada* **140 C4** 46 13N 77 43W
Rapla, *Estonia* **15 G21** 59 1N 24 52 E
Rapperswil, *Switz.* **33 B7** 47 14N 8 45 E
Rapu Rapu I., *Phil.* **80 E5** 13 12N 124 9 E
Rarotonga, *Cook Is.* **135 K12** 21 30 S 160 0W
Ra's al 'Ayn, *Syria* **101 D9** 36 45N 40 12 E
Ra's al Khaymah, *U.A.E.* **97 E8** 25 50N 56 5 E
Ra's al-Unuf, *Libya* **108 B3** 30 46N 18 11 E
Ra's an Naqb, *Jordan* **103 F4** 30 0N 35 29 E
Ras Dashen, *Ethiopia* **107 E4** 13 8N 38 26 E
Ras el Ma, *Algeria* **111 B4** 34 26N 0 50W
Râs Ghârib, *Egypt* **106 J8** 28 6N 33 18 E
Ras Mallap, *Egypt* **106 J8** 29 18N 32 50 E
Râs Timirist, *Mauritania* **112 B1** 19 21N 16 30W
Rasa, Punta, *Argentina* **176 B4** 40 50 S 62 15W
Rasca, Pta. de la, *Canary Is.* **39 G3** 27 59N 16 41W
Răşcani, *Moldova* **53 C12** 47 58N 27 33 E
Raseiniai, *Lithuania* **15 J20** 55 25N 23 5 E
Rashad, *Sudan* **107 E3** 11 55N 31 0 E
Rashîd, *Egypt* **106 H7** 31 21N 30 22 E
Rashîd, Masabb, *Egypt* **106 H7** 31 22N 30 17 E
Rasht, *Iran* **97 B6** 37 20N 49 40 E
Rasi Salai, *Thailand* **86 E5** 15 20N 104 9 E
Rasipuram, *India* **95 J4** 11 30N 78 15 E
Raška, *Serbia, Yug.* **50 C4** 43 19N 20 39 E
Râsnov, *Romania* **53 E10** 45 35N 25 27 E
Rason L., *Australia* **125 E3** 28 45 S 124 25 E
Raşova, *Romania* **53 F12** 44 15N 27 55 E
Rasovo, *Bulgaria* **50 C7** 43 42N 23 17 E
Rasra, *India* **93 G10** 25 50N 83 50 E
Rass el Oued, *Algeria* **111 A6** 35 57N 5 2 E
Rasskazovo, *Russia* **60 D5** 52 35N 41 50 E
Rast, *Romania* **53 G8** 43 53N 23 16 E
Rastatt, *Germany* **31 G4** 48 50N 8 11 E
Rastede, *Germany* **30 B4** 53 15N 8 12 E
Răstoliţa, *Romania* **53 D9** 46 59N 24 58 E
Raszków, *Poland* **55 G4** 51 43N 17 40 E
Rat Buri, *Thailand* **86 F2** 13 30N 99 54 E
Rat Islands, *U.S.A.* **138 C1** 52 0N 178 0 E
Rat River, *Canada* **142 A6** 61 7N 112 36W
Ratangarh, *India* **92 E6** 28 5N 74 35 E
Rätansbyn, *Sweden* **16 B8** 62 29N 14 33 E
Raţāwī, *Iraq* **96 D5** 30 38N 47 13 E
Rath, *India* **93 G8** 25 36N 79 37 E
Rath Luirc, *Ireland* **23 D3** 52 21N 8 40W
Rathbun Res., *U.S.A.* **156 D4** 40 49N 92 53W
Rathdrum, *Ireland* **23 D5** 52 56N 6 14W
Rathedaung, *Burma* **90 E4** 20 29N 92 45 E
Rathenow, *Germany* **30 C8** 52 37N 12 19 E
Rathkeale, *Ireland* **23 D3** 52 32N 8 56W
Rathlin I., *U.K.* **23 A5** 55 18N 6 14W
Rathmelton, *Ireland* **23 A4** 55 2N 7 38W
Ratibor = Racibórz, *Poland* **55 H5** 50 7N 18 18 E
Rätikon, *Austria* **34 D2** 47 0N 9 55 E
Ratingen, *Germany* **30 D2** 51 18N 6 52 E
Ratlam, *India* **92 H6** 23 20N 75 0 E
Ratnagiri, *India* **94 F1** 16 57N 73 18 E
Ratnapura, *Sri Lanka* **95 L5** 6 40N 80 20 E
Raton, *U.S.A.* **155 G2** 36 54N 104 24W
Rattaphum, *Thailand* **87 J3** 7 8N 100 16 E
Ratten, *Austria* **34 D8** 47 28N 15 44 E
Rattray Hd., *U.K.* **22 D7** 57 38N 1 50W
Rättvik, *Sweden* **16 D9** 60 52N 15 7 E
Ratz, Mt., *Canada* **142 B2** 57 23N 132 12W
Ratzeburg, *Germany* **30 B6** 53 40N 10 46 E
Raub, *Malaysia* **87 L3** 3 47N 101 52 E
Rauch, *Argentina* **174 D4** 36 45 S 59 5W
Raudeberg, *Norway* **18 C2** 61 59N 5 7 E
Raufarhöfn, *Iceland* **11 A11** 66 27N 15 57W
Raufoss, *Norway* **15 F14** 60 44N 10 37 E
Rauhellern, *Norway* **18 D4** 60 15N 7 50 E
Raukumara Ra., *N.Z.* **130 E6** 38 5 S 177 55 E
Raul Soares, *Brazil* **171 F3** 20 5 S 42 22W
Rauma, *Finland* **15 F19** 61 10N 21 30 E
Rauma →, *Norway* **18 B4** 62 34N 7 43 E
Raurkela, *India* **93 H11** 22 14N 84 50 E
Rausu-Dake, *Japan* **68 B12** 44 4N 145 7 E
Răut →, *Moldova* **53 C14** 47 15N 29 9 E
Rava-Ruska, *Poland* **59 G2** 50 15N 23 42 E
Rava Russkaya = Rava-Ruska, *Poland* **59 G2** 50 15N 23 42 E
Ravānsar, *Iran* **101 E12** 34 43N 46 40 E
Ravanusa, *Italy* **46 E6** 37 16N 13 58 E
Rāvar, *Iran* **97 D8** 31 20N 56 51 E
Ravena, *U.S.A.* **151 D11** 42 28N 73 49W
Ravenel, *U.S.A.* **152 C9** 32 46N 80 15W
Ravenna, *Italy* **45 D9** 44 25N 12 12 E
Ravenna, *Ky., U.S.A.* **157 G13** 37 42N 83 55W
Ravenna, *Nebr., U.S.A.* **154 E5** 41 1N 98 55W
Ravenna, *Ohio, U.S.A.* **150 E3** 41 9N 81 15W
Ravensburg, *Germany* **31 H5** 47 46N 9 36 E
Ravenshoe, *Australia* **126 B4** 17 37 S 145 29 E
Ravensthorpe, *Australia* **125 F3** 33 35 S 120 2 E
Ravenswood, *Australia* **126 C4** 20 6 S 146 54 E
Ravenswood, *U.S.A.* **148 F5** 38 57N 81 46W
Ravensworth, *Australia* **129 B9** 32 26 S 151 4 E
Ravenwood, *U.S.A.* **156 D2** 40 22N 94 41W
Ravi →, *Pakistan* **92 D4** 30 35N 71 49 E
Ravna Gora, *Croatia* **45 C11** 45 24N 14 50 E
Ravna Reka, *Serbia, Yug.* **50 B5** 44 1N 21 35 E
Ravne na Koroškem, *Slovenia* **45 B11** 46 36N 14 59 E
Rawa Mazowiecka, *Poland* **55 G7** 51 46N 20 12 E
Rawalpindi, *Pakistan* **91 B4** 33 38N 73 8 E
Rawāndūz, *Iraq* **101 D11** 36 40N 44 30 E
Rawang, *Malaysia* **87 L3** 3 20N 101 35 E

Ridgewood 271

Rawdon, Canada **140 C5** 46 3N 73 40W
Rawene, N.Z. **130 B2** 35 25 S 173 32 E
Rawicz, Poland **55 G3** 51 36N 16 52 E
Rawka →, Poland **55 F7** 52 9N 20 8 E
Rawlinna, Australia **125 F4** 30 58 S 125 28 E
Rawlins, U.S.A. **158 F10** 41 47N 107 14W
Rawlinson Ra., Australia **125 D4** 24 40 S 128 30 E
Rawson, Argentina **176 B3** 43 15 S 65 5W
Ray, U.S.A. **154 A3** 48 21N 103 10W
Ray, C., Canada **141 C8** 47 33N 59 15W
Ray City, U.S.A. **152 D6** 31 5N 83 11W
Rayachoti, India **95 G4** 14 4N 78 50 E
Rayadurg, India **95 G3** 14 40N 76 50 E
Rayagada, India **94 E6** 19 15N 83 20 E
Raychikhinsk, Russia **65 E13** 49 46N 129 25 E
Rāyen, Iran **97 D8** 29 34N 57 26 E
Rayevskiy, Russia **62 D5** 54 4N 56 56 E
Rayle, U.S.A. **152 B7** 33 48N 82 54W
Rayleigh, U.K. **21 F8** 51 36N 0 37 E
Raymond, Canada **142 D6** 49 30N 112 35W
Raymond, Calif., U.S.A. **160 H7** 37 13N 119 54W
Raymond, Ga., U.S.A. **152 B5** 33 20N 84 43W
Raymond, Ill., U.S.A. **156 E7** 39 19N 89 34W
Raymond, Wash., U.S.A. **160 D3** 46 41N 123 44W
Raymond Terrace, Australia **129 B9** 32 45 S 151 44 E
Raymondville, U.S.A. **155 M6** 26 29N 97 47W
Raymore, Canada **143 C8** 51 25N 104 31W
Rayne, U.S.A. **155 K8** 30 14N 92 16W
Rayón, Mexico **162 B2** 29 43N 110 35W
Rayong, Thailand **86 F3** 12 40N 101 20 E
Raytown, U.S.A. **156 E2** 39 1N 94 28W
Rayville, U.S.A. **155 J9** 32 29N 91 46W
Raz, Pte. du, France **26 D2** 48 2N 4 47W
Razan, Iran **97 C6** 35 23N 49 2 E
Ražana, Serbia, Yug. **50 B3** 44 6N 19 55 E
Ražanj, Serbia, Yug. **50 C5** 43 40N 21 31 E
Razdelna, Bulgaria **51 C11** 43 13N 27 41 E
Razdel'naya = Rozdilna, Ukraine **59 J6** 46 50N 30 2 E
Razdolnoye, Russia **68 C5** 43 30N 131 52 E
Razdolnoye, Ukraine **59 K7** 45 46N 33 29 E
Razeh, Iran **97 C6** 32 47N 48 9 E
Razgrad, Bulgaria **51 C10** 43 33N 26 34 E
Razim, Lacul, Romania **53 F14** 44 50N 29 0 E
Razlog, Bulgaria **50 E7** 41 53N 23 28 E
Razmak, Pakistan **91 B3** 32 45N 69 50 E
Razole, India **95 F5** 16 36N 81 48 E
Ré, I. de, France **28 B2** 46 12N 1 30W
Reading, U.K. **21 F7** 51 27N 0 58W
Reading, Mich., U.S.A. **157 C12** 41 50N 84 45W
Reading, Ohio, U.S.A. **157 E12** 39 13N 84 26W
Reading, Pa., U.S.A. **151 F9** 40 20N 75 56W
Reading □, U.K. **21 F7** 51 27N 0 58W
Real, Cordillera, Bolivia **172 D4** 17 0 S 67 10W
Realicó, Argentina **174 D3** 35 0 S 64 15W
Réalmont, France **28 E6** 43 48N 2 10 E
Ream, Cambodia **87 G4** 10 34N 103 39 E
Reata, Mexico **162 B4** 26 8N 101 5W
Reay Forest, U.K. **22 C4** 58 22N 4 55W
Rebais, France **27 D10** 48 50N 3 10 E
Rebecca, L., Australia **125 F3** 30 0 S 122 15 E
Rebi, Indonesia **83 C4** 6 23 S 134 7 E
Rebiana, Libya **108 D4** 24 12N 22 10 E
Rebun-Tō, Japan **68 B10** 45 23N 141 2 E
Recanati, Italy **45 E10** 43 24N 13 32 E
Recaş, Romania **52 E6** 45 46N 21 30 E
Recco, Italy **44 D6** 44 22N 9 9 E
Recherche, Arch. of the, Australia **125 F3** 34 15 S 122 50 E
Rechytsa, Belarus **59 F6** 52 21N 30 24 E
Recife, Brazil **170 C5** 8 0 S 35 0W
Recklinghausen, Germany **30 D3** 51 37N 7 12 E
Reconquista, Argentina **174 B4** 29 10 S 59 45W
Recreio, Brazil **173 B6** 8 0 S 58 25W
Recreo, Argentina **174 B2** 29 25 S 65 10W
Recuay, Peru **172 B2** 9 43 S 77 28W
Recz, Poland **55 E2** 53 16N 15 31 E
Red →, La., U.S.A. **155 K9** 31 1N 91 45W
Red →, N. Dak., U.S.A. **154 A6** 49 0N 97 15W
Red Bank, U.S.A. **151 F10** 40 21N 74 5W
Red Bay, Canada **141 B8** 51 44N 56 25W
Red Bluff, U.S.A. **158 F2** 40 11N 122 15W
Red Bluff L., U.S.A. **155 K3** 31 54N 103 55W
Red Bud, U.S.A. **156 F7** 38 13N 89 59W
Red Cliffs, Australia **128 C5** 34 19 S 142 11 E
Red Cloud, U.S.A. **154 E5** 40 5N 98 32W
Red Deer, Canada **142 C6** 52 20N 113 50W
Red Deer →, Alta., Canada **143 C7** 50 58N 110 0W
Red Deer →, Man., Canada **143 C8** 52 53N 101 1W
Red Deer L., Canada **143 C8** 52 55N 101 20W
Red Devil, U.S.A. **144 F8** 61 46N 157 19W
Red Head, U.S.A. **152 E4** 30 29N 85 51W
Red Indian L., Canada **141 C8** 48 35N 57 0W
Red Lake, Canada **143 C10** 51 3N 93 49W
Red Lake Falls, U.S.A. **154 B6** 47 53N 96 16W
Red Lodge, U.S.A. **158 D9** 45 11N 109 15W
Red Mountain, U.S.A. **161 K9** 35 37N 117 38W
Red Oak, U.S.A. **154 E7** 41 1N 95 14W
Red Rock, Canada **140 C2** 48 55N 88 15W
Red Rock, L., U.S.A. **156 C3** 41 22N 92 59W
Red Rocks Pt., Australia **125 F4** 32 13 S 127 32 E
Red Sea, Asia **102 C2** 25 0N 36 0 E
Red Slate Mt., U.S.A. **160 H8** 37 31N 118 52W
Red Sucker L., Canada **143 C10** 54 9N 93 40W
Red Tower Pass = Turnu Roşu, P., Romania **53 E9** 45 33N 24 17 E
Red Wing, U.S.A. **154 C8** 44 34N 92 31W
Reda, Poland **54 D5** 54 40N 18 19 E
Redange, Lux. **24 E5** 49 46N 5 52 E
Redbay, U.S.A. **152 E4** 30 35N 85 57W
Redcar, U.K. **20 C6** 54 37N 1 4W
Redcar & Cleveland □, U.K. **20 C7** 54 29N 1 0W
Redcliff, Canada **143 C6** 50 10N 110 50W
Redcliffe, Australia **127 D5** 27 12 S 153 0 E
Redcliffe, Mt., Australia **125 E3** 28 30 S 121 30 E
Reddersburg, S. Africa **116 D4** 29 41 S 26 10 E
Reddick, U.S.A. **153 F7** 29 22N 82 12W
Redding, U.S.A. **158 F2** 40 35N 122 24W
Redditch, U.K. **21 E6** 52 18N 1 55W
Redenção, Brazil **170 B4** 4 13 S 38 43W
Redfield, U.S.A. **154 C5** 44 53N 98 31W
Redkey, U.S.A. **157 D11** 40 21N 85 9W
Redkino, Russia **58 D9** 56 39N 36 16 E
Redknife →, Canada **142 A5** 61 14N 119 22W
Redlands, U.S.A. **161 M9** 34 4N 117 11W
Redmond, Australia **125 F2** 34 55 S 117 40 E
Redmond, Oreg., U.S.A. **158 D3** 44 17N 121 11W
Redmond, Wash., U.S.A. **160 C4** 47 41N 122 7W
Redon, France **26 E4** 47 40N 2 6W
Redonda, Antigua **165 C7** 16 58N 62 19W
Redondela, Spain **42 C2** 42 15N 8 38W
Redondo, Portugal **43 G3** 38 39N 7 37W
Redondo Beach, U.S.A. **161 M8** 33 50N 118 23W
Redoubt Volcano, U.S.A. **144 F9** 60 29N 152 45W
Redrock Pt., Canada **142 A5** 62 11N 115 2W
Redruth, U.K. **21 G2** 50 14N 5 14W
Redvers, Canada **143 D8** 49 35N 101 40W
Redwater, Canada **142 C6** 53 55N 113 6W
Redwood, U.S.A. **151 B9** 44 18N 75 48W
Redwood City, U.S.A. **160 H4** 37 30N 122 15W
Redwood Falls, U.S.A. **154 C7** 44 32N 95 7W
Ree, L., Ireland **23 C3** 53 35N 8 0W
Reed, L., Canada **143 C8** 54 38N 100 30W
Reed City, U.S.A. **148 D3** 43 53N 85 31W
Reeder, U.S.A. **154 B3** 46 7N 102 57W
Reedley, U.S.A. **160 J7** 36 36N 119 27W
Reedsburg, U.S.A. **154 D9** 43 32N 90 0W
Reedsport, U.S.A. **158 E1** 43 42N 124 6W
Reedy Creek, Australia **128 D4** 36 58 S 140 2 E
Reefton, Australia **129 C7** 34 15 S 147 27 E
Reefton, N.Z. **131 C6** 42 6 S 171 51 E
Rees, Germany **30 D2** 51 46N 6 24 E
Refahiye, Turkey **101 C8** 39 54N 38 47 E
Reftele, Sweden **17 G7** 57 11N 13 35 E
Refugio, U.S.A. **155 L6** 28 18N 97 17W
Rega →, Poland **54 D2** 54 10N 15 18 E
Regalbuto, Italy **47 E7** 37 39N 14 38 E
Regen, Germany **31 G9** 48 58N 13 8 E
Regen →, Germany **31 F8** 49 1N 12 6 E
Regeneração, Brazil **170 C3** 6 15 S 42 41W
Regensburg, Germany **31 F8** 49 1N 12 6 E
Regensdorf, Switz. **33 B6** 47 26N 8 28 E
Regenstauf, Germany **31 F8** 49 7N 12 8 E
Reggello, Italy **45 E8** 43 41N 11 32 E
Réggio di Calábria, Italy **47 D8** 38 6N 15 39 E
Réggio nell'Emília, Italy **44 D7** 44 43N 10 36 E
Reghin, Romania **53 D9** 46 46N 24 42 E
Regina, Canada **143 C8** 50 27N 104 35W
Régina, Fr. Guiana **169 C7** 4 19N 52 8W
Register, U.S.A. **152 C8** 32 22N 81 53W
Registro, Brazil **175 A6** 24 29 S 47 49W
Reguengos de Monsaraz, Portugal **43 G3** 38 25N 7 32W
Rehar →, India **93 H10** 23 55N 82 40 E
Rehoboth, Namibia **116 C2** 23 15 S 17 4 E
Rehovot, Israel **103 D3** 31 54N 34 48 E
Rei-Bouba, Cameroon **114 A2** 8 40N 14 15 E
Reichenbach, Germany **30 E8** 50 37N 12 17 E
Reichenbach, Switz. **32 C5** 46 38N 7 42 E
Reid, Australia **125 F4** 30 49 S 128 26 E
Reid River, Australia **126 B4** 19 40 S 146 48 E
Reiden, Switz. **32 B5** 47 14N 7 59 E
Reidsville, Ga., U.S.A. **152 C7** 32 6N 82 7W
Reidsville, N.C., U.S.A. **149 G6** 36 21N 79 40W
Reigate, U.K. **21 F7** 51 14N 0 12W
Reillo, Spain **40 F3** 39 54N 1 53W
Reims, France **27 C11** 49 15N 4 1 E
Reina Adelaida, Arch., Chile **176 D2** 52 20 S 74 0W
Reinach, Aargau, Switz. **32 B6** 47 14N 8 11 E
Reinach, Basel, Switz. **32 B5** 47 29N 7 35 E
Reinbeck, U.S.A. **156 B4** 42 19N 92 36W
Reinbek, Germany **30 B6** 53 30N 10 16 E
Reindeer →, Canada **143 B8** 55 36N 103 11W
Reindeer I., Canada **143 C9** 52 30N 98 0W
Reindeer L., Canada **143 B8** 57 15N 102 15W
Reinga, C., N.Z. **130 A1** 34 25 S 172 43 E
Reinosa, Spain **42 B6** 43 2N 4 15W
Reinsvoll, Norway **18 D7** 60 40N 10 38 E
Reitan, Norway **18 B8** 62 49N 11 22 E
Reitz, S. Africa **117 D4** 27 48 S 28 29 E
Reivilo, S. Africa **116 D3** 27 36 S 24 8 E
Rejmyre, Sweden **17 F9** 58 50N 15 55 E
Rejowiec Fabryczny, Poland **55 G10** 51 5N 23 17 E
Reka →, Slovenia **45 C11** 45 40N 14 0 E
Rekovac, Serbia, Yug. **50 C5** 43 51N 21 3 E
Reliance, Canada **143 A7** 63 0N 109 20W
Remad, Oued →, Algeria **111 B4** 33 28N 1 20W
Rémalard, France **26 D7** 48 26N 0 47 E
Remarkable, Mt., Australia **128 B3** 32 48 S 138 10 E
Rembang, Indonesia **85 D4** 6 42 S 111 21 E
Rembau, Malaysia **84 B2** 2 35N 102 6 E
Rembert, U.S.A. **152 A9** 34 6N 80 32W
Remchi, Algeria **111 A4** 35 2N 1 26W
Remedios, Colombia **168 B3** 7 2N 74 41W
Remedios, Panama **164 E3** 8 15N 81 50W
Remeshk, Iran **97 E8** 26 55N 58 50 E
Remetea, Romania **53 D10** 46 45N 25 29 E
Remich, Lux. **24 E6** 49 32N 6 22 E
Remington, U.S.A. **157 D9** 40 46N 87 9W
Rémire, Fr. Guiana **169 C7** 4 53N 52 17W
Remiremont, France **27 D13** 48 2N 6 36 E
Remo, Ethiopia **107 F5** 6 48N 41 20 E
Remontnoye, Russia **61 G6** 46 34N 43 37 E
Remoulins, France **29 E8** 43 55N 4 35 E
Remscheid, Germany **30 D3** 51 11N 7 12 E
Ren Xian, China **74 F8** 37 8N 114 40 E
Rena, Norway **18 C8** 61 8N 11 20 E
Rena →, Norway **18 C8** 61 8N 11 23 E
Renascença, Brazil **168 D4** 3 50 S 66 21W
Rend Lake, U.S.A. **156 F8** 38 2N 88 58W
Rende, Italy **47 C9** 39 20N 16 11 E
Rendína, Greece **48 B3** 39 4N 21 58 E
Rendova, Solomon Is. **133 M9** 8 33 S 157 17 E
Rendsburg, Germany **30 A5** 54 17N 9 39 E
Renfrew, Canada **140 C4** 45 30N 76 40W
Renfrewshire □, U.K. **22 F4** 55 49N 4 38W
Renfroe, U.S.A. **152 C5** 32 14N 84 43W
Rengat, Indonesia **84 C2** 0 30 S 102 45 E
Rengo, Chile **174 C1** 34 24 S 70 50W
Renhua, China **77 E9** 25 5N 113 40 E
Renhuai, China **76 D6** 27 48N 106 24 E
Reni, Ukraine **59 K5** 45 28N 28 15 E
Renigunta, India **95 H4** 13 38N 79 30 E
Renk, Sudan **107 E3** 11 50N 32 50 E
Renland, Greenland **10 C8** 71 10N 26 30W
Renmark, Australia **128 C4** 34 11 S 140 43 E
Rennebu, Norway **18 B6** 62 52N 9 49 E
Rennell, Solomon Is. **133 N11** 11 40 S 160 10 E
Rennell Sd., Canada **142 C2** 53 23N 132 35W
Renner Springs T.O., Australia **126 B1** 18 20 S 133 47 E
Rennes, France **26 D5** 48 7N 1 41W
Reno, U.S.A. **160 F7** 39 31N 119 48W
Reno →, Italy **45 D9** 44 38N 12 16 E
Renovo, U.S.A. **150 E7** 41 20N 77 45W
Renqiu, China **74 E9** 38 43N 116 5 E
Rens, Denmark **17 K3** 54 54N 9 5 E
Rensselaer, Ind., U.S.A. **157 D9** 40 57N 87 9W
Rensselaer, N.Y., U.S.A. **151 D11** 42 38N 73 45W
Rentería, Spain **40 B3** 43 19N 1 54W
Renton, U.S.A. **160 C4** 47 29N 122 12W
Rentz, U.S.A. **152 C7** 32 25N 82 59W
Renwick, N.Z. **131 B8** 41 30 S 173 51 E
Réo, Burkina Faso **112 C4** 12 28N 2 35W
Reocín, Spain **42 B6** 43 21N 4 5W
Reotipur, India **93 G10** 25 33N 83 45 E
Repalle, India **95 F5** 16 2N 80 45 E
Répcelak, Hungary **52 C2** 47 24N 17 1 E
Republic, Mich., U.S.A. **148 B2** 46 25N 87 59W
Republic, Wash., U.S.A. **158 B4** 48 39N 118 44W
Republican →, U.S.A. **154 F6** 39 4N 96 48W
Republican City, U.S.A. **154 E5** 40 6N 99 13W
Republiek, Surinam **169 B6** 5 30N 55 13W
Repulse Bay, Canada **139 B11** 66 30N 86 30W
Requena, Peru **172 B3** 5 5 S 73 52W
Requena, Spain **41 F3** 39 30N 1 4W
Réquista, France **28 D6** 44 1N 2 32 E
Reşadiye = Datça, Turkey **49 E9** 36 46N 27 40 E
Reşadiye, Turkey **100 B7** 40 23N 37 20 E
Reşadiye Yarımadası, Turkey **49 E9** 36 40N 27 45 E
Resavica, Serbia, Yug. **50 B5** 44 4N 21 31 E
Resen, Macedonia **50 E5** 41 5N 21 0 E
Reserve, Canada **143 C8** 52 28N 102 39W
Reserve, U.S.A. **159 K9** 33 43N 108 45W
Resht = Rasht, Iran **97 B6** 37 20N 49 40 E
Resistencia, Argentina **174 B4** 27 30 S 59 0W
Reşiţa, Romania **52 E6** 45 18N 21 53 E
Resko, Poland **54 E2** 53 47N 15 25 E
Resolution I., Canada **139 B13** 61 30N 65 0W
Resolution I., N.Z. **131 F1** 45 40 S 166 40 E
Resplandes, Brazil **170 C2** 6 17 S 45 13W
Resplendor, Brazil **171 E3** 19 20 S 41 15W
Ressano Garcia, Mozam. **117 D5** 25 25 S 32 0 E
Reston, Canada **143 D8** 49 33N 101 6W
Reszel, Poland **54 D8** 54 4N 21 10 E
Retalhuleu, Guatemala **164 D1** 14 33N 91 46W
Retenue, L. de, Dem. Rep. of the Congo **119 E2** 11 0 S 27 0 E
Retezat, Munţii, Romania **52 E8** 45 25N 23 0 E
Retford, U.K. **20 D7** 53 19N 0 56W
Rethel, France **27 C11** 49 30N 4 20 E
Rethem, Germany **30 C5** 52 47N 9 22 E
Réthímnon, Greece **38 D6** 35 18N 24 30 E
Réthímnon □, Greece **38 D6** 35 23N 24 28 E
Retiche, Alpi, Switz. **33 D10** 46 30N 10 0 E
Retiers, France **26 E5** 47 55N 1 23W
Retortillo, Spain **42 E4** 40 48N 6 21W
Retournac, France **29 C8** 45 12N 4 2 E
Rétság, Hungary **52 C4** 47 58N 19 10 E
Réunion ■, Ind. Oc. **121 G4** 21 0 S 56 0 E
Reus, Spain **40 D6** 41 10N 1 5 E
Reuss →, Switz. **33 B6** 47 16N 8 24 E
Reuterstadt Stavenhagen, Germany **30 B8** 53 42N 12 54 E
Reutlingen, Germany **31 G5** 48 29N 9 12 E
Reutte, Austria **34 D3** 47 29N 10 42 E
Reval = Tallinn, Estonia **15 G21** 59 22N 24 48 E
Revda, Russia **62 C7** 56 48N 59 57 E
Revel, France **28 E6** 43 28N 2 1 E
Revelganj, India **93 G11** 25 50N 84 40 E
Revelstoke, Canada **142 C5** 51 0N 118 10W
Reventazón, Peru **172 B1** 6 10 S 80 58W
Revigny-sur-Ornain, France **27 D11** 48 49N 4 59 E
Revilla Gigedo, Is., Pac. Oc. **135 F16** 18 40N 112 0W
Revillagigedo I., U.S.A. **142 B2** 55 50N 131 20W
Revin, France **27 C11** 49 55N 4 39 E
Revolyutsii, Pik, Tajikistan **63 D6** 38 31N 72 21 E
Revúca, Slovak Rep. **35 C13** 48 41N 20 7 E
Revuè →, Mozam. **119 F3** 19 50 S 34 0 E
Rewa, India **93 G9** 24 33N 81 25 E
Rewa →, Guyana **169 C6** 3 19N 58 42W
Rewari, India **92 E7** 28 15N 76 40 E
Rexburg, U.S.A. **158 E8** 43 49N 111 47W
Rey, Iran **97 C6** 35 35N 51 25 E
Rey, I. del, Panama **164 E4** 8 20N 78 30W
Rey, Rio del →, Nigeria **113 E6** 4 30N 8 48 E
Rey Malabo, Eq. Guin. **113 E6** 3 45N 8 50 E
Reyðarfjörður, Iceland **11 B12** 65 2N 14 13W
Reyes, Bolivia **172 C4** 14 19 S 67 23W
Reyes, Pt., U.S.A. **160 H3** 38 0N 123 0W
Reyhanlı, Turkey **100 D7** 36 16N 36 35 E
Reykhólar, Iceland **11 B4** 65 27N 22 12W
Reykholt, Iceland **11 C5** 64 40N 21 18W
Reykjahlið, Iceland **11 B10** 65 40N 16 55W
Reykjanes, Iceland **11 D4** 63 48N 22 40W
Reykjavík, Iceland **11 C5** 64 10N 21 57W
Reynolds, Canada **143 D9** 49 40N 95 55W
Reynolds, Ga., U.S.A. **152 C5** 32 33N 84 6W
Reynolds, Ill., U.S.A. **156 C6** 41 20N 90 40W
Reynolds Ra., Australia **124 D5** 22 30 S 133 0 E
Reynoldsville, Ga., U.S.A. **152 E5** 30 51N 84 47W
Reynoldsville, Pa., U.S.A. **150 E6** 41 5N 78 58W
Reynosa, Mexico **163 B5** 26 5N 98 18W
Rēzekne, Latvia **15 H22** 56 30N 27 17 E
Rezh, Russia **62 C8** 57 23N 61 24 E
Rezina, Moldova **53 C13** 47 45N 28 58 E
Rezovo, Bulgaria **51 D12** 42 0N 28 0 E
Rezvān, Iran **97 E8** 27 34N 56 6 E
Rgotina, Serbia, Yug. **50 B6** 44 1N 22 17 E
Rhamnus, Greece **48 C6** 38 12N 24 3 E
Rharis, O. →, Algeria **111 C6** 26 0N 5 4 E
Rhayader, U.K. **21 E4** 52 18N 3 29W
Rheda-Wiedenbrück, Germany **30 D4** 51 50N 8 20 E
Rhede, Germany **30 D2** 51 50N 6 42 E
Rhein, Canada **143 C8** 51 25N 102 15W
Rhein →, Europe **24 C6** 51 52N 6 2 E
Rhein-Main-Donau-Kanal, Germany **31 F7** 49 1N 11 27 E
Rheinbach, Germany **30 E2** 50 38N 6 57 E
Rheine, Germany **30 C3** 52 17N 7 26 E
Rheineck, Switz. **33 B9** 47 28N 9 31 E
Rheinfelden, Germany **31 H3** 47 33N 7 47 E
Rheinfelden, Switz. **32 A5** 47 32N 7 47 E
Rheinhessen-Pfalz □, Germany **31 F3** 49 20N 8 0 E
Rheinland-Pfalz □, Germany **31 E2** 50 0N 7 0 E
Rheinsberg, Germany **30 B8** 53 6N 12 54 E
Rheinwaldhorn, Switz. **33 D8** 46 30N 9 3 E
Rheriss, Oued →, Morocco **110 B4** 30 50N 4 34W
Rhin = Rhein →, Europe **24 C6** 51 52N 6 2 E
Rhine = Rhein →, Europe **24 C6** 51 52N 6 2 E
Rhine, U.S.A. **152 D6** 31 59N 83 12W
Rhineland-Palatinate □ = Rheinland-Pfalz □, Germany **31 E2** 50 0N 7 0 E
Rhinelander, U.S.A. **154 C10** 45 38N 89 25W
Rhinns Pt., U.K. **22 F2** 55 40N 6 29W
Rhino Camp, Uganda **118 B3** 3 0N 31 22 E
Rhir, Cap, Morocco **110 B3** 30 38N 9 54W
Rho, Italy **44 C6** 45 32N 9 2 E
Rhode Island □, U.S.A. **151 E13** 41 40N 71 30W
Rhodes = Ródhos, Greece **38 C10** 36 15N 28 10 E
Rhodesia = Zimbabwe ■, Africa **119 F3** 19 0 S 30 0 E
Rhodope Mts. = Rhodopi Planina, Bulgaria **51 E8** 41 40N 24 20 E
Rhodopi Planina, Bulgaria **51 E8** 41 40N 24 20 E
Rhön = Rhön, Germany **30 E5** 50 24N 9 58 E
Rhön, Germany **30 E5** 50 24N 9 58 E
Rhondda, U.K. **21 F4** 51 39N 3 31W
Rhondda Cynon Taff □, U.K. **21 F4** 51 42N 3 27W
Rhône □, France **29 C8** 45 54N 4 35 E
Rhône →, France **29 E8** 43 28N 4 42 E
Rhône-Alpes □, France **29 C9** 45 40N 6 0 E
Rhum, U.K. **22 E2** 57 0N 6 20W
Rhyl, U.K. **20 D4** 53 20N 3 29W
Ri-Aba, Eq. Guin. **113 E6** 3 28N 8 40 E
Riachão, Brazil **170 C2** 7 20 S 46 37W
Riacho de Santana, Brazil **171 D3** 13 37 S 42 57W
Rialma, Brazil **171 E2** 15 18 S 49 34W
Riang, India **90 B4** 27 31N 92 56 E
Riaño, Spain **42 C6** 42 59N 4 59W
Rians, France **29 E9** 43 37N 5 44 E
Riansáres →, Spain **43 F7** 39 32N 3 18W
Riasi, India **93 C6** 33 10N 74 50 E
Riau □, Indonesia **84 B2** 0 0 102 35 E
Riau, Kepulauan, Indonesia **84 B2** 0 30N 104 20 E
Riau Arch. = Riau, Kepulauan, Indonesia **84 B2** 0 30N 104 20 E
Riaza, Spain **42 D7** 41 18N 3 30W
Riaza →, Spain **42 D7** 41 42N 3 55W
Riba de Saelices, Spain **40 E2** 40 55N 2 17W
Riba-Roja de Turia, Spain **41 F4** 39 33N 0 34W
Ribadavia, Spain **42 C2** 42 17N 8 8W
Ribadeo, Spain **42 B3** 43 35N 7 5W
Ribadesella, Spain **42 B5** 43 30N 5 7W
Ribamar, Brazil **170 B3** 2 33 S 44 3W
Ribas = Ribes de Freser, Spain **40 C7** 42 19N 2 15 E
Ribas do Rio Pardo, Brazil **173 E7** 20 27 S 53 46W
Ribāţ, Yemen **98 D4** 14 18N 44 15 E
Ribble →, U.K. **20 D5** 53 52N 2 25W
Ribe, Denmark **17 J2** 55 19N 8 44 E
Ribe Amtskommune □, Denmark **17 J2** 55 35N 8 45 E
Ribeauvillé, France **27 D14** 48 10N 7 20 E
Ribécourt-Dreslincourt, France **27 C9** 49 30N 2 55 E
Ribeira = Santa Uxía, Spain **42 C2** 42 36N 8 58W
Ribeira Brava, Madeira **39 D2** 32 41N 17 4W
Ribeira do Pombal, Brazil **170 D4** 10 50 S 38 32W
Ribeirão Prêto, Brazil **175 A6** 21 10 S 47 50W
Ribeiro Gonçalves, Brazil **170 C2** 7 32 S 45 14W
Ribemont, France **27 C10** 49 47N 3 27 E
Ribera, Italy **46 E6** 37 30N 13 16 E
Ribérac, France **28 C4** 45 15N 0 20 E
Riberalta, Bolivia **173 C4** 11 0 S 66 0W
Ribes de Freser, Spain **40 C7** 42 19N 2 15 E
Ribnica, Slovenia **45 C11** 45 45N 14 45 E
Ribnitz-Damgarten, Germany **30 A8** 54 15N 12 27 E
Ričany, Czech Rep. **34 B7** 50 0N 14 40 E
Riccarton, N.Z. **131 D7** 43 32 S 172 37 E
Ríccia, Italy **47 A7** 41 30N 14 50 E
Riccione, Italy **45 E9** 43 59N 12 39 E
Rice, U.S.A. **161 L12** 34 5N 114 51W
Rice L., Canada **150 B6** 44 12N 78 10W
Rice Lake, U.S.A. **154 C9** 45 30N 91 44W
Riceboro, U.S.A. **152 D8** 31 44N 81 26W
Rich, Morocco **110 B4** 32 16N 4 30W
Rich Hill, U.S.A. **155 F7** 38 6N 94 22W
Richards Bay, S. Africa **117 D5** 28 48 S 32 6 E
Richards L., Canada **143 B7** 59 10N 107 10W
Richardson →, Canada **143 B6** 58 25N 111 14W
Richardson Mts., N.Z. **131 E3** 44 49 S 168 34 E
Richardson Springs, U.S.A. **160 F5** 39 51N 121 46W
Richardton, U.S.A. **154 B3** 46 53N 102 19W
Riche, C., Australia **125 F2** 34 36 S 118 47 E
Richelieu, France **26 E7** 47 1N 0 20 E
Richey, U.S.A. **154 B2** 47 39N 105 4W
Richfield, Idaho, U.S.A. **158 E6** 43 3N 114 9W
Richfield, Utah, U.S.A. **159 G8** 38 46N 112 5W
Richford, U.S.A. **151 B12** 45 0N 72 40W
Richibucto, Canada **141 C7** 46 42N 64 54W
Richland, Ga., U.S.A. **152 C5** 32 5N 84 40W
Richland, Iowa, U.S.A. **156 C5** 41 13N 92 0W
Richland, Mo., U.S.A. **156 G4** 37 51N 92 26W
Richland, Oreg., U.S.A. **158 D5** 44 46N 117 10W
Richland, Wash., U.S.A. **158 C4** 46 17N 119 18W
Richland Center, U.S.A. **154 D9** 43 21N 90 23W
Richlands, U.S.A. **148 G5** 37 6N 81 48W
Richmond, N.S.W., Australia **129 B9** 33 35 S 150 42 E
Richmond, Queens., Australia **126 C3** 20 43 S 143 8 E
Richmond, N.Z. **131 B8** 41 20 S 173 12 E
Richmond, U.K. **20 C6** 54 25N 1 43W
Richmond, Calif., U.S.A. **160 H4** 37 56N 122 21W
Richmond, Ind., U.S.A. **157 E12** 39 50N 84 53W
Richmond, Ky., U.S.A. **157 G12** 37 45N 84 18W
Richmond, Mich., U.S.A. **150 D2** 42 49N 82 45W
Richmond, Mo., U.S.A. **154 F8** 39 17N 93 58W
Richmond, Tex., U.S.A. **155 L7** 29 35N 95 46W
Richmond, Utah, U.S.A. **158 F8** 41 56N 111 48W
Richmond, Va., U.S.A. **148 G7** 37 33N 77 27W
Richmond, Mt., N.Z. **131 B8** 41 32 S 173 22 E
Richmond Heights, U.S.A. **153 K9** 25 38N 80 23W
Richmond Hill, U.S.A. **152 D8** 31 56N 81 18W
Richmond Ra., Australia **127 D5** 29 0 S 152 45 E
Richmond Ra., N.Z. **131 B8** 41 32 S 173 22 E
Richterswil, Switz. **33 B7** 47 13N 8 43 E
Richton, U.S.A. **149 K1** 31 16N 88 56W
Richwood, Ohio, U.S.A. **157 D13** 40 26N 83 18W
Richwood, W. Va., U.S.A. **148 F5** 38 14N 80 32W
Ricla, Spain **40 D3** 41 31N 1 24W
Ricupe, Angola **115 E4** 14 37 S 21 25 E
Ridā', Yemen **98 D4** 14 25N 44 50 E
Ridder = Leninogorsk, Kazakstan **64 D9** 50 20N 83 30 E
Riddes, Switz. **32 D4** 46 11N 7 14 E
Ridge Farm, U.S.A. **157 E9** 39 54N 87 39W
Ridge Spring, U.S.A. **152 B9** 33 51N 81 40W
Ridgecrest, U.S.A. **161 K9** 35 38N 117 40W
Ridgedale, Canada **143 C8** 53 0N 104 10W
Ridgefield, U.S.A. **160 E4** 45 49N 122 45W
Ridgeland, U.S.A. **152 C9** 32 29N 80 59W
Ridgelands, Australia **126 C5** 23 16 S 150 17 E
Ridgetown, Canada **140 D3** 42 26N 81 52W
Ridgeville, Ind., U.S.A. **157 D11** 40 18N 85 2W
Ridgeville, S.C., U.S.A. **152 B9** 33 6N 80 19W
Ridgewood, U.S.A. **151 F10** 40 59N 74 7W

Rzhev 273

Rønne, *Denmark* 17 J8 55 6N 14 43 E
Ronne Ice Shelf, *Antarctica* 7 D18 78 0 S 60 0W
Ronneby, *Sweden* 17 H9 56 12N 15 17 E
Ronnebyån →, *Sweden* 17 H9 56 11N 15 18 E
Rönneshytta, *Sweden* 17 F9 58 56N 15 2 E
Ronsard, C., *Australia* 125 D1 24 46 S 113 10 E
Ronse, *Belgium* 24 D3 50 45N 3 35 E
Ronuro →, *Brazil* 173 C7 11 56 S 53 33W
Roodepoort, *S. Africa* 117 D4 26 11 S 27 54 E
Roodhouse, *U.S.A.* 156 E6 39 29N 90 24W
Roof Butte, *U.S.A.* 159 H9 36 28N 109 5W
Roopville, *U.S.A.* 152 B4 33 27N 85 8W
Roorkee, *India* 92 E7 29 52N 77 59 E
Roosendaal, *Neths.* 24 C4 51 32N 4 29 E
Roosevelt, *Minn., U.S.A.* 154 A7 48 48N 95 6W
Roosevelt, *Utah, U.S.A.* 158 F8 40 18N 109 59W
Roosevelt →, *Brazil* 173 B5 7 35 S 60 20W
Roosevelt, Mt., *Canada* 142 B3 58 26N 125 20W
Roosevelt I., *Antarctica* 7 D12 79 30 S 162 0W
Roosevelt Res., *U.S.A.* 159 K8 33 46N 111 0W
Ropczyce, *Poland* 55 H8 50 4N 21 38 E
Roper →, *Australia* 126 A2 14 43 S 135 27 E
Ropesville, *U.S.A.* 155 J3 33 26N 102 9W
Roque Pérez, *Argentina* 174 D4 35 25 S 59 24W
Roquefort, *France* 28 D3 44 2N 0 20W
Roquemaure, *France* 29 D8 44 3N 4 48 E
Roquetas de Mar, *Spain* 41 J2 36 46N 2 36W
Roquetes, *Spain* 40 E5 40 50N 0 30 E
Roquevaire, *France* 29 E9 43 20N 5 36 E
Roraima □, *Brazil* 169 C5 2 0N 61 30W
Roraima, Mt., *Venezuela* 169 B5 5 10N 60 40W
Rorketon, *Canada* 143 C9 51 24N 99 35W
Røros, *Norway* 15 E14 62 35N 11 23 E
Rorschach, *Switz.* 33 B8 47 28N 9 28 E
Rosa, *Zambia* 119 D3 9 33 S 31 15 E
Rosa, C., *Algeria* 111 A6 37 0N 8 16 E
Rosa, Monte, *Europe* 32 E5 45 57N 7 53 E
Rosa, Monte, *Italy* 44 C4 45 55N 7 53 E
Rosal de la Frontera, *Spain* 43 H3 37 59N 7 13W
Rosales, *Phil.* 80 D3 15 54N 120 38 E
Rosalia, *U.S.A.* 158 C5 47 14N 117 22W
Rosamond, *U.S.A.* 161 L8 34 52N 118 10W
Rosans, *France* 29 D9 44 24N 5 29 E
Rosario, *Argentina* 174 C3 33 0 S 60 40W
Rosário, *Brazil* 170 B3 3 0 S 44 15W
Rosario, *Baja Calif., Mexico* 162 B1 30 0N 115 50W
Rosario, *Sinaloa, Mexico* 162 C3 23 0N 105 52W
Rosario, *Paraguay* 174 A4 24 30 S 57 35W
Rosario, *Phil.* 81 G5 8 24N 125 59 E
Rosario, Villa del, *Venezuela* 168 A3 10 19N 72 19W
Rosario de la Frontera, *Argentina* 174 B3 25 50 S 65 0W
Rosario de Lerma, *Argentina* 174 A2 24 59 S 65 35W
Rosario del Tala, *Argentina* 174 C4 32 20 S 59 10W
Rosário do Sul, *Brazil* 175 C5 30 15 S 54 55W
Rosário Oeste, *Brazil* 173 C6 14 50 S 56 25W
Rosarito, *Mexico* 161 N9 32 18N 117 4W
Rosarno, *Italy* 47 D8 38 29N 15 58 E
Rosas = Roses, *Spain* 40 C8 42 19N 3 10 E
Roscoe, *Miss., U.S.A.* 156 G3 37 58N 93 48W
Roscoe, *S. Dak., U.S.A.* 154 C5 45 27N 99 20W
Roscoff, *France* 26 D3 48 44N 3 58W
Roscommon, *Ireland* 23 C3 53 38N 8 11W
Roscommon, *U.S.A.* 148 C3 44 30N 84 35W
Roscommon □, *Ireland* 23 C3 53 49N 8 23W
Roscrea, *Ireland* 23 D4 52 57N 7 49W
Rose →, *Australia* 126 A2 14 16 S 135 45 E
Rose Blanche, *Canada* 141 C8 47 38N 58 45W
Rose Harbour, *Canada* 142 C2 52 15N 131 10W
Rose Pt., *Canada* 142 C2 54 11N 131 39W
Rose Valley, *Canada* 143 C8 52 19N 103 49W
Roseau, *Domin.* 165 C7 15 20N 61 24W
Roseau, *U.S.A.* 154 A7 48 51N 95 46W
Rosebery, *Australia* 126 G4 41 46 S 145 33 E
Rosebud, *U.S.A.* 155 K6 31 4N 96 59W
Roseburg, *U.S.A.* 158 E2 43 13N 123 20W
Rosedale, *Australia* 126 C5 24 38 S 151 53 E
Rosedale, *U.S.A.* 155 J9 33 51N 91 2W
Roseland, *U.S.A.* 160 G4 38 25N 122 43W
Rosemary, *Canada* 142 C6 50 46N 112 5W
Rosenberg, *U.S.A.* 155 L7 29 34N 95 49W
Rosendaël, *France* 27 A9 51 3N 2 24 E
Rosendal, *Norway* 18 E3 59 59N 6 0 E
Rosendale, *U.S.A.* 156 D2 40 4N 94 51W
Rosenheim, *Germany* 31 H8 47 51N 12 7 E
Roses, *Spain* 40 C8 42 19N 3 10 E
Roses, G. de, *Spain* 40 C8 42 10N 3 15 E
Roseto degli Abruzzi, *Italy* 45 F11 42 41N 14 1 E
Rosetown, *Canada* 143 C7 51 35N 107 59W
Rosetta = Rashîd, *Egypt* 106 H7 31 21N 30 22 E
Roseville, *Calif., U.S.A.* 160 G5 38 45N 121 17W
Roseville, *Ill., U.S.A.* 156 D6 40 44N 90 40W
Roseville, *Mich., U.S.A.* 157 B14 42 30N 82 56W
Rosewood, *N.S.W., Australia* 129 C7 35 38 S 147 52 E
Rosewood, *N. Terr., Australia* 124 C4 16 28 S 128 58 E
Rosewood, *Queens., Australia* 127 D5 27 38 S 152 36 E
Roshkhvār, *Iran* 97 C8 34 58N 59 37 E
Rosier, *U.S.A.* 152 C7 32 59N 82 15W
Rosières-en-Santerre, *France* 27 C9 49 49N 2 42 E
Rosignano Maríttimo, *Italy* 44 E7 43 24N 10 28 E
Rosignol, *Guyana* 169 B6 6 15N 57 30W
Roşiori de Vede, *Romania* 53 F10 44 9N 25 0 E
Rositsa, *Bulgaria* 51 C11 43 57N 27 57 E
Rositsa →, *Bulgaria* 51 C9 43 10N 25 30 E
Roskilde, *Denmark* 17 J6 55 38N 12 3 E
Roskilde Amtskommune □, *Denmark* 17 J6 55 35N 12 5 E
Roskovec, *Albania* 50 F3 40 44N 19 43 E
Roslavl, *Russia* 58 F7 53 57N 32 55 E
Roslyn, *Australia* 129 C8 34 29 S 149 37 E
Rosmaninhal, *Portugal* 42 F3 39 44N 7 5W
Rosmead, *S. Africa* 116 E4 31 29 S 25 8 E
Røsnæs, *Denmark* 17 J4 55 44N 10 55 E
Rosolini, *Italy* 47 F7 36 49N 14 57 E
Rosporden, *France* 26 E3 47 57N 3 50W
Ross, *Australia* 126 G4 42 2 S 147 30 E
Ross, *N.Z.* 131 C5 42 53 S 170 49 E
Ross I., *Antarctica* 7 D11 77 30 S 168 0 E
Ross Ice Shelf, *Antarctica* 7 E12 80 0 S 180 0 E
Ross L., *U.S.A.* 158 B3 48 44N 121 4W
Ross-on-Wye, *U.K.* 21 F5 51 54N 2 34W
Ross Sea, *Antarctica* 7 D11 74 0 S 178 0 E
Rossa, *Switz.* 33 D8 46 23N 9 8 E
Rossall Pt., *U.K.* 20 D4 53 55N 3 3W
Rossan Pt., *Ireland* 23 B3 54 42N 8 47W
Rossano, *Italy* 47 C9 39 36N 16 39 E
Rossburn, *Canada* 143 C8 50 40N 100 49W
Rosseau, *Canada* 150 A5 45 16N 79 39W
Rossel, C., *Vanuatu* 133 K4 20 23 S 166 36 E
Rosses, The, *Ireland* 23 A3 55 2N 8 20W
Rossford, *U.S.A.* 157 C13 41 36N 83 34W
Rossignol, L., *Canada* 140 B5 52 43N 73 40W
Rossignol Res., *Canada* 141 D6 44 12N 65 10W
Rosskreppfjorden, *Norway* 18 E4 59 4N 7 10 E
Rossland, *Canada* 142 D5 49 6N 117 50W
Rosslare, *Ireland* 23 D5 52 17N 6 24W
Rosslau, *Germany* 30 D8 51 52N 12 15 E
Rosso, *Mauritania* 112 B1 16 40N 15 45W
Rosso, C., *France* 29 F12 42 13N 8 32 E
Rossosh, *Russia* 59 G10 50 15N 39 28 E
Rossport, *Canada* 140 C2 48 50N 87 30W
Røssvatnet, *Norway* 14 D16 65 45N 14 5 E
Rossville, *Australia* 126 B4 15 48 S 145 15 E
Rossville, *U.S.A.* 157 D10 40 25N 86 36W
Røst, *Norway* 14 C15 67 32N 12 0 E
Rostāq, *Afghan.* 93 A3 37 7N 69 49 E
Rosthern, *Canada* 143 C7 52 40N 106 20W
Rostock, *Germany* 30 A8 54 5N 12 8 E
Rostov, *Don, Russia* 59 J10 47 15N 39 45 E
Rostov, *Yaroslavl, Russia* 58 D10 57 14N 39 25 E
Rostrenen, *France* 26 D3 48 14N 3 21W
Roswell, *Ga., U.S.A.* 152 A5 34 2N 84 22W
Roswell, *N. Mex., U.S.A.* 155 J2 33 24N 104 32W
Rota, *Spain* 43 J4 36 37N 6 20W
Rotan, *U.S.A.* 155 J4 32 51N 100 28W
Rotenburg, *Hessen, Germany* 30 E5 50 59N 9 44 E
Rotenburg, *Niedersachsen, Germany* 30 B5 53 6N 9 25 E
Roth, *Germany* 31 F7 49 15N 11 5 E
Rothaargebirge, *Germany* 30 D4 51 2N 8 13 E
Rothenburg, *Switz.* 33 B6 47 6N 8 16 E
Rothenburg ob der Tauber, *Germany* 31 F6 49 23N 10 11 E
Rother →, *U.K.* 21 G8 50 59N 0 45 E
Rotherham, *U.K.* 20 D6 53 26N 1 20W
Rothes, *U.K.* 22 D5 57 32N 3 13W
Rothesay, *Canada* 141 C6 45 23N 66 0W
Rothesay, *U.K.* 22 F3 55 50N 5 3W
Rothrist, *Switz.* 32 B5 47 18N 7 54 E
Roti, *Indonesia* 82 D2 10 50 S 123 0 E
Rotja, Pta., *Spain* 41 G6 38 38N 1 35 E
Rotnes, *Norway* 18 D7 60 3N 10 51 E
Roto, *Australia* 129 B6 33 0 S 145 30 E
Roto Aira L., *N.Z.* 130 F4 39 3 S 175 45 E
Rotoehu L., *N.Z.* 130 E5 38 1 S 176 32 E
Rotoiti, L., *N.Z.* 130 E5 38 2 S 176 26 E
Rotoiti L., *N.Z.* 131 B7 41 51 S 172 49 E
Rotoma L., *N.Z.* 130 E5 38 2 S 176 35 E
Rotoroa, L., *N.Z.* 131 B7 41 55 S 172 39 E
Rotorua, *N.Z.* 130 E5 38 9 S 176 16 E
Rotorua, L., *N.Z.* 130 E5 38 5 S 176 18 E
Rott →, *Germany* 31 G9 48 27N 13 25 E
Rotten →, *Switz.* 32 D5 46 18N 7 36 E
Rottenburg, *Germany* 31 G4 48 28N 8 55 E
Rottenmann, *Austria* 34 D7 47 31N 14 22 E
Rotterdam, *Neths.* 24 C4 51 55N 4 30 E
Rottne, *Sweden* 17 G8 57 1N 14 54 E
Rottnest I., *Australia* 125 F2 32 0 S 115 27 E
Rottumeroog, *Neths.* 24 A6 53 33N 6 34 E
Rottweil, *Germany* 31 G4 48 9N 8 37 E
Rotuma, *Fiji* 134 J9 12 25 S 177 5 E
Roubaix, *France* 27 B10 50 40N 3 10 E
Roudnice nad Labem, *Czech Rep.* 34 A7 50 25N 14 15 E
Rouen, *France* 26 C8 49 27N 1 4 E
Rouergue, *France* 28 D5 44 15N 2 0 E
Rough Ridge, *N.Z.* 131 F4 45 10 S 169 55 E
Rouillac, *France* 28 C3 45 47N 0 4W
Rouleau, *Canada* 143 C8 50 10N 104 56W
Round Mountain, *U.S.A.* 158 G5 38 43N 117 4W
Round Mt., *Australia* 127 E5 30 26 S 152 16 E
Round Oak, *U.S.A.* 152 B6 33 7N 83 37W
Roundup, *U.S.A.* 158 C9 46 27N 108 33W
Roura, *Fr. Guiana* 169 C7 4 44N 52 20W
Rousay, *U.K.* 22 B5 59 10N 3 2W
Rouses Point, *U.S.A.* 151 B11 44 59N 73 22W
Roussillon, *Isère, France* 29 C8 45 24N 4 49 E
Roussillon, *Pyrénées-Or., France* 28 F6 42 30N 2 35 E
Roussin, C., *N. Cal.* 133 U21 21 20 S 167 59 E
Rouxville, *S. Africa* 116 E4 30 25 S 26 50 E
Rouyn-Noranda, *Canada* 140 C4 48 20N 79 0W
Rovaniemi, *Finland* 14 C21 66 29N 25 41 E
Rovato, *Italy* 44 C7 45 34N 10 0 E
Rovenki, *Ukraine* 59 H10 48 5N 39 21 E
Rovereto, *Italy* 44 C8 45 53N 11 3 E
Roverud, *Norway* 18 D9 60 15N 12 3 E
Rovigo, *Italy* 45 C8 45 4N 11 47 E
Rovinj, *Croatia* 45 C10 45 5N 13 40 E
Rovira, *Colombia* 168 C2 4 15N 75 20W
Rovno = Rivne, *Ukraine* 59 G4 50 40N 26 10 E
Rovnoye, *Russia* 60 E8 50 52N 46 3 E
Rovuma = Ruvuma →, *Tanzania* 119 E5 10 29 S 40 28 E
Row'ān, *Iran* 97 C6 35 8N 48 51 E
Rowena, *Australia* 129 A8 29 48 S 148 55 E
Rowes, *Australia* 129 D8 37 0 S 149 6 E
Rowesville, *U.S.A.* 152 B9 33 22N 80 50W
Rowley Shoals, *Australia* 124 C2 17 30 S 119 0 E
Roxa, *Guinea-Biss.* 112 C1 11 15N 15 45W
Roxas = Barbacan, *Phil.* 81 F2 10 20N 119 21 E
Roxas, *Capiz, Phil.* 81 F4 11 36N 122 49 E
Roxas, *Isabela, Phil.* 80 C3 17 8N 121 36 E
Roxas, *Mindoro, Phil.* 80 E3 12 35N 121 31 E
Roxboro, *U.S.A.* 149 G6 36 24N 78 59W
Roxborough Downs, *Australia* 126 C2 22 30 S 138 45 E
Roxburgh, *N.Z.* 131 F4 45 33 S 169 19 E
Roxen, *Sweden* 17 F9 58 30N 15 40 E
Roy, *Fla., U.S.A.* 153 F8 29 37N 81 29W
Roy, *Mont., U.S.A.* 158 C9 47 20N 108 58W
Roy, *N. Mex., U.S.A.* 155 H2 35 57N 104 12W
Roy Hill, *Australia* 124 D2 22 37 S 119 58 E
Royal Canal, *Ireland* 23 C4 53 30N 7 13W
Royal Center, *U.S.A.* 157 D10 40 52N 86 30W
Royal Leamington Spa, *U.K.* 21 E6 52 18N 1 31W
Royal Tunbridge Wells, *U.K.* 21 F8 51 7N 0 16 E
Royalla, *Australia* 129 C8 35 30 S 149 9 E
Royan, *France* 28 C2 45 37N 1 2W
Roye, *France* 27 C9 49 42N 2 48 E
Royston, *U.K.* 21 E7 52 3N 0 0W
Rožaj, *Montenegro, Yug.* 50 D4 42 50N 20 11 E
Różan, *Poland* 55 F8 52 52N 21 25 E
Rozay-en-Brie, *France* 27 D9 48 41N 2 58 E
Rozdilna, *Ukraine* 59 J6 46 50N 30 2 E
Rozhyshche, *Ukraine* 59 G3 50 54N 25 15 E
Rožmitál pod Třemšínem, *Czech Rep.* 34 B6 49 36N 13 53 E
Rožňava, *Slovak Rep.* 35 C13 48 37N 20 35 E
Rozogi, *Poland* 54 E8 53 28N 21 19 E
Rozoy-sur-Serre, *France* 27 C11 49 40N 4 8 E
Rozzano, *Italy* 44 C6 45 22N 9 10 E
Rrësheni, *Albania* 50 E3 41 47N 19 49 E
Rrogozhino, *Albania* 50 E3 41 4N 19 50 E
Rtanj, *Serbia, Yug.* 50 C5 43 45N 21 50 E
Rtishchevo, *Russia* 60 D6 52 18N 43 46 E
Rúa = A Rúa, *Spain* 42 C3 42 24N 7 6W
Ruacaná, *Angola* 115 F2 17 20 S 14 12 E
Ruahine Ra., *N.Z.* 130 F5 39 55 S 176 2 E
Ruamahanga →, *N.Z.* 130 H4 41 24 S 175 8 E
Ruapehu, *N.Z.* 130 F4 39 17 S 175 35 E
Ruapuke I., *N.Z.* 131 G3 46 46 S 168 31 E
Ruâq, W. →, *Egypt* 103 F2 30 0N 33 49 E
Ruatoria, *N.Z.* 130 D7 37 55 S 178 20 E
Ruawai, *N.Z.* 130 C3 36 8 S 173 59 E
Rub' al Khālī, *Si. Arabia* 99 C5 18 0N 48 0 E
Rubeho Mts., *Tanzania* 118 D4 6 50 S 36 25 E
Rubezhnoye = Rubizhne, *Ukraine* 59 H10 49 6N 38 25 E
Rubh a' Mhail, *U.K.* 22 F2 55 56N 6 8W
Rubha Hunish, *U.K.* 22 D2 57 42N 6 20W
Rubha Robhanais = Lewis, Butt of, *U.K.* 22 C2 58 31N 6 16W
Rubí, *Spain* 40 D7 41 29N 2 2 E
Rubiataba, *Brazil* 171 E2 15 8 S 49 48W
Rubicon →, *U.S.A.* 160 G5 38 53N 121 4W
Rubicone →, *Italy* 45 D9 44 8N 12 28 E
Rubik, *Albania* 50 E3 41 46N 19 47 E
Rubinéia, *Brazil* 171 F1 20 13 S 51 2W
Rubino, *Ivory C.* 112 D4 6 4N 4 18W
Rubio, *Venezuela* 168 B3 7 43N 72 22W
Rubizhne, *Ukraine* 59 H10 49 6N 38 25 E
Rubtsovsk, *Russia* 64 D9 51 30N 81 10 E
Ruby L., *U.S.A.* 158 F6 40 10N 115 28W
Ruby Mts., *U.S.A.* 158 F6 40 30N 115 20W
Rucheng, *China* 77 E9 25 33N 113 38 E
Ruciane-Nida, *Poland* 54 E8 53 40N 21 32 E
Rūd Sar, *Iran* 97 B6 37 8N 50 18 E
Ruda, *Sweden* 17 G10 57 6N 16 7 E
Ruda Śląska, *Poland* 55 H5 50 16N 18 50 E
Rudall, *Australia* 128 B2 33 43 S 136 17 E
Rudall →, *Australia* 124 D3 22 34 S 122 13 E
Rūdbār, *Afghan.* 91 C1 30 9N 62 36 E
Rüdersdorf, *Germany* 30 C9 52 27N 13 47 E
Rudewa, *Tanzania* 119 E3 10 7 S 34 40 E
Rudkøbing, *Denmark* 17 K4 54 56N 10 41 E
Rudna, *Poland* 55 G3 51 30N 16 17 E
Rudnichnyy, *Russia* 56 C9 59 38N 52 26 E
Rudnik, *Bulgaria* 51 D11 42 36N 27 30 E
Rudnik, *Poland* 55 H9 50 26N 22 15 E
Rudnik, *Serbia, Yug.* 50 B4 44 7N 20 35 E
Rudnya, *Russia* 58 E6 54 55N 31 7 E
Rudnyy, *Kazakstan* 62 E9 52 57N 63 7 E
Rudo, *Bos.-H.* 52 G4 43 41N 19 23 E
Rudolfa, Ostrov, *Russia* 64 A6 81 45N 58 30 E
Rudolstadt, *Germany* 30 E7 50 44N 11 19 E
Rudong, *China* 77 A13 32 20N 121 12 E
Rudozem, *Bulgaria* 51 E8 41 29N 24 51 E
Rudyard, *U.S.A.* 148 B3 46 14N 84 36W
Rue, *France* 27 B8 50 15N 1 40 E
Rufa'a, *Sudan* 107 E3 14 44N 33 22 E
Rufflin, *U.S.A.* 152 B9 33 0N 80 49W
Rufiji □, *Tanzania* 118 D4 8 0 S 38 30 E
Rufiji →, *Tanzania* 118 D4 7 50 S 39 15 E
Rufino, *Argentina* 174 C3 34 20 S 62 50W
Rufisque, *Senegal* 112 C1 14 40N 17 15W
Rufunsa, *Zambia* 119 F2 15 4 S 29 34 E
Rugao, *China* 77 A13 32 23N 120 31 E
Rugby, *U.K.* 21 E6 52 23N 1 16W
Rugby, *U.S.A.* 154 A5 48 22N 100 0W
Rügen, *Germany* 30 A9 54 22N 13 24 E
Rugles, *France* 26 D7 48 50N 0 40 E
Ruhea, *Bangla.* 90 B2 26 10N 88 25 E
Ruhengeri, *Rwanda* 118 C2 1 30 S 29 36 E
Ruhla, *Germany* 30 E6 50 54N 10 23 E
Ruhland, *Germany* 30 D9 51 27N 13 51 E
Ruhnu, *Estonia* 15 H20 57 48N 23 15 E
Ruhr →, *Germany* 30 D2 51 27N 6 43 E
Ruhuhu →, *Tanzania* 119 E3 10 31 S 34 34 E
Rui Barbosa, *Brazil* 171 D3 12 18 S 40 27W
Rui'an, *China* 77 D13 27 47N 120 40 E
Ruichang, *China* 77 C10 29 40N 115 39 E
Ruidosa, *U.S.A.* 155 L2 29 59N 104 41W
Ruidoso, *U.S.A.* 159 K11 33 20N 105 41W
Ruili, *China* 76 E1 24 1N 97 43 E
Ruivo, Pico, *Madeira* 39 D3 32 45N 16 56W
Ruj, *Bulgaria* 50 D6 42 52N 22 34 E
Rujen, *Macedonia* 50 D6 42 9N 22 30 E
Rujm Tal'at al Jamā'ah, *Jordan* 103 E4 30 24N 35 30 E
Ruk, *Pakistan* 92 F3 27 50N 68 42 E
Rukwa □, *Tanzania* 118 D3 7 0 S 31 30 E
Rukwa, L., *Tanzania* 118 D3 8 0 S 32 20 E
Rulhieres, C., *Australia* 124 B4 13 56 S 127 22 E
Rum = Rhum, *U.K.* 22 E2 57 0N 6 20W
Rum Cay, *Bahamas* 165 B5 23 40N 74 58W
Rum Jungle, *Australia* 124 B5 13 0 S 130 59 E
Ruma, *Serbia, Yug.* 52 E4 45 0N 19 50 E
Rumādah, *Yemen* 98 D3 13 34N 43 52 E
Rumāḩ, *Si. Arabia* 96 E5 25 29N 47 10 E
Rumania = Romania ■, *Europe* 53 D10 46 0N 25 0 E
Rumaylah, *Iraq* 96 D5 30 47N 47 37 E
Rumaylah, 'Urūq ar, *Si. Arabia* 98 B4 21 0N 47 30 E
Rumbalara, *Australia* 126 D1 25 20 S 134 29 E
Rumbêk, *Sudan* 107 F2 6 54N 29 37 E
Rumburk, *Czech Rep.* 34 A7 50 57N 14 32 E
Rumford, *U.S.A.* 151 B14 44 33N 70 33W
Rumia, *Poland* 54 D5 54 37N 18 25 E
Rumilly, *France* 29 C9 45 53N 5 56 E
Rumoi, *Japan* 68 C10 43 56N 141 39 E
Rumonge, *Burundi* 118 C2 3 59 S 29 26 E
Rumsey, *Canada* 142 C6 51 51N 112 48W
Rumula, *Australia* 126 B4 16 35 S 145 20 E
Rumuruti, *Kenya* 118 B4 0 17N 36 32 E
Runan, *China* 74 H8 33 0N 114 30 E
Runanga, *N.Z.* 131 C6 42 25 S 171 15 E
Runaway, C., *N.Z.* 130 D6 37 32 S 177 59 E
Runcorn, *U.K.* 20 D5 53 21N 2 44W
Rungwa, *Tanzania* 118 D3 6 55 S 33 32 E
Rungwa →, *Tanzania* 118 D3 7 36 S 31 50 E
Rungwe, *Tanzania* 119 D3 9 11 S 33 32 E
Rungwe □, *Tanzania* 119 D3 9 25 S 33 32 E
Runka, *Nigeria* 113 C6 12 28N 7 20 E
Runn, *Sweden* 16 D9 60 30N 15 40 E
Running Ra., *Australia* 124 D3 23 31 S 123 6 E
Ruokolahti, *Finland* 58 B5 61 17N 28 50 E
Ruoqiang, *China* 72 C3 38 55N 88 10 E
Rupa, *India* 90 B4 27 15N 92 21 E
Rupar, *India* 92 D7 31 2N 76 38 E
Rupat, *Indonesia* 84 B2 1 45N 101 40 E
Rupea, *Romania* 53 D10 46 2N 25 13 E
Rupert →, *Canada* 140 B4 51 29N 78 45W
Rupert House = Waskaganish, *Canada* 140 B4 51 30N 78 40W
Rupsa, *Bangla.* 90 E2 21 44N 89 30 E
Rupununi →, *Guyana* 169 C6 4 3N 58 35W
Rur →, *Germany* 30 D1 51 11N 5 59 E
Rurrenabaque, *Bolivia* 172 C4 14 30 S 67 32W
Rus →, *Spain* 41 F2 39 30N 2 30W
Rusambo, *Zimbabwe* 119 F3 16 30 S 32 4 E
Rusape, *Zimbabwe* 119 F3 18 35 S 32 8 E
Ruschuk = Ruse, *Bulgaria* 51 C9 43 48N 25 59 E
Ruse, *Bulgaria* 51 C9 43 48N 25 59 E
Ruse □, *Bulgaria* 51 C10 43 35N 26 20 E
Ruşeţu, *Romania* 53 F12 44 57N 27 14 E
Rush, *Ireland* 23 C5 53 31N 6 6W
Rushan, *China* 75 F11 36 56N 121 30 E
Rushden, *U.K.* 21 E7 52 18N 0 35W
Rushford, *U.S.A.* 154 D9 43 49N 91 46W
Rushville, *Ill., U.S.A.* 156 D6 40 7N 90 34W
Rushville, *Ind., U.S.A.* 157 E11 39 37N 85 27W
Rushville, *Nebr., U.S.A.* 154 D3 42 43N 102 28W
Rushworth, *Australia* 129 D6 36 32 S 145 1 E
Ruskin, *U.S.A.* 153 H7 27 43N 82 26W
Russas, *Brazil* 170 B4 4 55 S 37 50W
Russell, *Canada* 143 C8 50 50N 101 20W
Russell, *N.Z.* 130 B3 35 16 S 174 10 E
Russell, *Fla., U.S.A.* 152 E8 30 3N 81 45W
Russell, *Kans., U.S.A.* 154 F5 38 54N 98 52W
Russell Is., *Solomon Is.* 133 M10 9 4 S 159 12 E
Russell L., *Man., Canada* 143 B8 56 15N 101 30W
Russell L., *N.W.T., Canada* 142 A5 63 5N 115 44W
Russellkonda, *India* 94 E7 19 57N 84 42 E
Russellville, *Ala., U.S.A.* 149 H2 34 30N 87 44W
Russellville, *Ark., U.S.A.* 155 H8 35 17N 93 8W
Russellville, *Ky., U.S.A.* 149 G2 36 51N 86 53W
Rüsselsheim, *Germany* 31 F4 49 59N 8 23 E
Russi, *Italy* 45 D9 44 22N 12 2 E
Russia ■, *Eurasia* 65 C11 62 0N 105 0 E
Russian →, *U.S.A.* 160 G3 38 27N 123 8W
Russiaville, *U.S.A.* 157 D10 40 25N 86 16W
Russkoye Ustie, *Russia* 6 B15 71 0N 149 0 E
Rust, *Austria* 35 D9 47 49N 16 42 E
Rustam, *Pakistan* 92 B5 34 25N 72 13 E
Rustam Shahr, *Pakistan* 92 F2 26 58N 66 6 E
Rustavi, *Georgia* 61 K7 41 30N 45 0 E
Rustenburg, *S. Africa* 116 D4 25 41 S 27 14 E
Ruston, *U.S.A.* 155 J8 32 32N 92 38W
Ruswil, *Switz.* 32 B6 47 5N 8 8 E
Rutana, *Burundi* 118 C3 3 55 S 30 0 E
Rute, *Spain* 43 H6 37 19N 4 23W
Ruteng, *Indonesia* 82 C2 8 35 S 120 30 E
Ruth, *Mich., U.S.A.* 150 C2 43 42N 82 45W
Ruth, *Nev., U.S.A.* 158 G6 39 17N 114 59W
Rutherford, *U.S.A.* 160 G4 38 26N 122 24W
Rutherglen, *Australia* 129 D7 36 5 S 146 29 E
Rüti, *Switz.* 33 B7 47 16N 8 51 E
Rutland □, *U.K.* 21 E7 52 38N 0 40W
Rutland Plains, *Australia* 126 B3 15 38 S 141 43 E
Rutland Water, *U.K.* 21 E7 52 39N 0 38W
Rutledal, *Norway* 18 C2 61 4N 5 10 E
Rutledge, *U.S.A.* 152 B6 33 38N 83 37W
Rutledge →, *Canada* 143 A6 61 4N 112 0W
Rutledge L., *Canada* 143 A6 61 33N 110 47W
Rutqa, W. →, *Syria* 101 E9 34 30N 41 3 E
Rutshuru, *Dem. Rep. of the Congo* 118 C2 1 13 S 29 25 E
Ruvo di Púglia, *Italy* 47 A9 41 7N 16 29 E
Ruvu, *Tanzania* 118 D4 6 49 S 38 43 E
Ruvu →, *Tanzania* 118 D4 6 23 S 38 52 E
Ruvuma □, *Tanzania* 119 E4 10 20 S 36 0 E
Ruvuma →, *Tanzania* 119 E5 10 29 S 40 28 E
Ruwais, *U.A.E.* 97 E7 24 5N 52 50 E
Ruwenzori, *Africa* 118 B2 0 30N 29 55 E
Ruyigi, *Burundi* 118 C3 3 29 S 30 15 E
Ruyuan, *China* 77 E9 24 46N 113 16 E
Ruzayevka, *Russia* 60 C7 54 4N 45 0 E
Ruzhevo Konare, *Bulgaria* 51 D8 42 23N 24 46 E
Ružomberok, *Slovak Rep.* 35 B12 49 3N 19 17 E
Rwanda ■, *Africa* 118 C3 2 0 S 30 0 E
Ryakhovo, *Bulgaria* 51 C10 43 58N 26 18 E
Ryan, L., *U.K.* 22 G3 55 0N 5 2W
Ryazan, *Russia* 58 E10 54 40N 39 40 E
Ryazhsk, *Russia* 58 F11 53 45N 40 3 E
Rybache = Rybachye, *Kazakstan* 64 E9 46 40N 81 20 E
Rybachiy Poluostrov, *Russia* 56 A5 69 43N 32 0 E
Rybachye = Ysyk-Köl, *Kyrgyzstan* 63 B8 42 26N 76 12 E
Rybachye, *Kazakstan* 64 E9 46 40N 81 20 E
Rybinsk, *Russia* 58 C10 58 5N 38 50 E
Rybinskoye Vdkhr., *Russia* 58 C10 58 30N 38 25 E
Rybnik, *Poland* 55 H5 50 6N 18 32 E
Rybnitsa = Râbniţa, *Moldova* 53 C14 47 45N 29 0 E
Rybnoye, *Russia* 58 E10 54 45N 39 30 E
Rychnov nad Kněžnou, *Czech Rep.* 35 A9 50 10N 16 17 E
Rychwał, *Poland* 55 F5 52 4N 18 10 E
Ryd, *Sweden* 17 H8 56 27N 14 42 E
Rydaholm, *Sweden* 17 H8 56 59N 14 18 E
Ryde, *U.K.* 21 G6 50 43N 1 9W
Ryderwood, *U.S.A.* 160 D3 46 23N 123 3W
Rydzyna, *Poland* 55 G3 51 47N 16 39 E
Rye, *U.K.* 21 G8 50 57N 0 45 E
Rye →, *U.K.* 20 C7 54 11N 0 44W
Rye Bay, *U.K.* 21 G8 50 52N 0 49 E
Rye Patch Reservoir, *U.S.A.* 158 F4 40 28N 118 19W
Ryegate, *U.S.A.* 158 C9 46 18N 109 15W
Ryfylke, *Norway* 18 E3 59 25N 6 25 E
Rykene, *Norway* 18 F5 58 24N 8 37 E
Ryki, *Poland* 55 G8 51 38N 21 56 E
Rylsk, *Russia* 59 G8 51 36N 34 43 E
Rylstone, *Australia* 129 B8 32 46 S 149 58 E
Rymanów, *Poland* 55 J8 49 35N 21 51 E
Ryn, *Poland* 54 E8 53 57N 21 34 E
Ryn Peski, *Kazakstan* 61 G9 47 30N 49 0 E
Ryōhaku-Sanchi, *Japan* 71 A8 36 9N 136 49 E
Ryōthu, *Japan* 68 E9 38 5N 138 26 E
Rypin, *Poland* 55 E6 53 3N 19 25 E
Ryssby, *Sweden* 17 H8 56 52N 14 10 E
Ryūgasaki, *Japan* 71 B12 35 54N 140 11 E
Ryūkyū Is. = Ryūkyū-rettō, *Japan* 69 M3 26 0N 126 0 E
Ryūkyū-rettō, *Japan* 69 M3 26 0N 126 0 E
Rzepin, *Poland* 55 F1 52 20N 14 49 E
Rzeszów, *Poland* 55 H8 50 5N 21 58 E
Rzeszów □, *Poland* 55 J8 50 0N 22 0 E
Rzhev, *Russia* 58 D8 56 20N 34 20 E

S

Samer 275

St. Magnus B., *U.K.* **22 A7** 60 25N 1 35W
St-Maixent-l'École, *France* **28 B3** 46 24N 0 12W
St-Malo, *France* **26 D4** 48 39N 2 1W
St-Malo, G. de, *France* **26 D4** 48 50N 2 30W
St-Mandrier-sur-Mer, *France* **29 E9** 43 4N 5 57 E
St-Marc, *Haiti* **165 C5** 19 10N 72 41W
St-Marcellin, *France* **29 C9** 45 9N 5 20 E
St-Marcouf, Îs., *France* **26 C5** 49 30N 1 10W
St. Maries, *U.S.A.* **158 C5** 47 19N 116 35W
St. Marks, *U.S.A.* **152 E5** 30 9N 84 12W
St. Marks →, *U.S.A.* **152 E5** 30 8N 84 12W
St. Martin, *W. Indies* **165 C7** 18 0N 63 0W
St. Martin, L., *Canada* **143 C9** 51 40N 98 30W
St-Martin-de-Crau, *France* **29 E8** 43 38N 4 48 E
St-Martin-de-Ré, *France* **28 B2** 46 12N 1 21W
St-Martin-d'Hères, *France* **29 C9** 45 9N 5 45 E
St-Martin-Vésubie, *France* **29 D11** 44 4N 7 15 E
St. Martins, *Canada* **141 C6** 45 22N 65 34W
St. Martinville, *U.S.A.* **155 K9** 30 7N 91 50W
St-Martory, *France* **28 E4** 43 9N 0 56 E
St. Mary, Mt., *Papua N. G.* **132 E4** 8 8 S 147 2 E
St. Mary Is., *India* **95 H2** 13 20N 74 35 E
St. Mary Pk., *Australia* **128 A3** 31 32 S 138 34 E
St. Marys, *Australia* **126 G4** 41 35 S 148 11 E
St. Marys, *Canada* **150 C3** 43 20N 81 10W
St. Mary's, *Corn., U.K.* **21 H1** 49 55N 6 18W
St. Mary's, *Orkney, U.K.* **22 C6** 58 54N 2 54W
St. Mary's, *Alaska, U.S.A.* **144 E7** 62 4N 163 10W
St. Marys, *Ga., U.S.A.* **152 E8** 30 44N 81 33W
St. Marys, *Mo., U.S.A.* **156 G7** 37 53N 89 57W
St. Marys, *Pa., U.S.A.* **150 E6** 41 26N 78 34W
St. Marys →, *U.S.A.* **152 E8** 30 43N 81 27W
St. Mary's, C., *Canada* **141 C9** 46 50N 54 12W
St. Mary's B., *Canada* **141 C9** 46 50N 53 50W
St. Marys Bay, *Canada* **141 D6** 44 25N 66 10W
St-Mathieu, Pte., *France* **26 D2** 48 20N 4 45W
St. Matthew I., *U.S.A.* **144 F4** 60 24N 172 42W
St. Matthews, *Ky., U.S.A.* **157 F11** 38 15N 85 39W
St. Matthews, *S.C., U.S.A.* **152 B9** 33 40N 80 46W
St. Matthews, I. = Zadetkyi Kyun, *Burma* **87 H2** 10 0N 98 25 E
St. Matthias Group, *Papua N. G.* **132 A6** 1 30 S 150 0 E
St. Maurice, *Switz.* **32 D4** 46 13N 7 0 E
St-Maurice →, *Canada* **140 C5** 46 21N 72 31W
St-Maximin-la-Ste-Baume, *France* **29 E9** 43 27N 5 52 E
St-Médard-en-Jalles, *France* **28 D3** 44 53N 0 43W
St-Méen-le-Grand, *France* **26 D4** 48 11N 2 12W
St. Meinrad, *U.S.A.* **157 F10** 38 10N 86 49W
St-Mihiel, *France* **27 D12** 48 54N 5 32 E
St. Moritz, *Switz.* **31 J5** 46 30N 9 51 E
St-Nazaire, *France* **26 E4** 47 17N 2 12W
St. Neots, *U.K.* **21 E7** 52 14N 0 15W
St-Nicolas-de-Port, *France* **27 D13** 48 38N 6 18 E
St-Niklaas, *Belgium* **24 C4** 51 10N 4 8 E
St. Niklaus, *Switz.* **32 D5** 46 10N 7 49 E
St-Omer, *France* **27 B9** 50 45N 2 15 E
St-Pacome, *Canada* **141 C6** 47 24N 69 58W
St-Palais-sur-Mer, *France* **28 C2** 45 38N 1 5W
St-Pamphile, *Canada* **141 C6** 46 58N 69 48W
St-Pardoux-la-Rivière, *France* **28 C4** 45 29N 0 45 E
St. Paris, *U.S.A.* **157 D13** 40 8N 83 58W
St. Pascal, *Canada* **141 C6** 47 32N 69 48W
St. Paul, *Canada* **142 C6** 54 0N 111 17W
St-Paul, *France* **29 D10** 44 31N 6 45 E
St. Paul, *Alaska, U.S.A.* **144 H5** 57 7N 170 17W
St. Paul, *Ind., U.S.A.* **157 E11** 39 26N 85 38W
St. Paul, *Minn., U.S.A.* **154 C8** 44 57N 93 6W
St. Paul, *Nebr., U.S.A.* **154 E5** 41 13N 98 27W
St. Paul, I., *Ind. Oc.* **121 H6** 38 55 S 77 34 E
St-Paul-de-Fenouillet, *France* **28 F6** 42 48N 2 30 E
St. Paul I., *Canada* **141 C7** 47 12N 60 9W
St. Paul I., *U.S.A.* **144 H5** 57 10N 170 15W
St-Paul-lès-Dax, *France* **28 E2** 43 44N 1 3W
St-Péray, *France* **29 D8** 44 57N 4 50 E
St. Peter, *U.S.A.* **154 C8** 44 20N 93 57W
St-Peter-Ording, *Germany* **30 A4** 54 20N 8 36 E
St. Peter Port, *U.K.* **21 H5** 49 26N 2 33W
St. Peters, *N.S., Canada* **141 C7** 45 40N 60 53W
St. Peters, *P.E.I., Canada* **141 C7** 46 25N 62 35W
St. Petersburg = Sankt-Peterburg, *Russia* **58 C6** 59 55N 30 20 E
St. Petersburg, *U.S.A.* **149 M4** 27 46N 82 39W
St. Petersburg Beach, *U.S.A.* **153 H7** 27 45N 82 45W
St-Philbert-de-Grand-Lieu, *France* **26 E5** 47 2N 1 39W
St-Pierre, *St- P. & M.* **141 C8** 46 46N 56 12W
St. Pierre, *Seychelles* **121 E3** 9 20 S 46 0 E
St-Pierre, L., *Canada* **140 C5** 46 12N 72 52W
St-Pierre-d'Oléron, *France* **28 C2** 45 57N 1 19W
St-Pierre-en-Port, *France* **26 C7** 49 48N 0 30 E
St-Pierre et Miquelon □, *St- P. & M.* **141 C8** 46 55N 56 10W
St-Pierre-le-Moûtier, *France* **27 F10** 46 47N 3 7 E
St-Pierre-sur-Dives, *France* **26 C6** 49 2N 0 1W
St-Pol-de-Léon, *France* **26 D3** 48 41N 4 0W
St-Pol-sur-Mer, *France* **27 A9** 51 1N 2 20 E
St-Pol-sur-Ternoise, *France* **27 B9** 50 23N 2 20 E
St-Pons, *France* **28 E6** 43 30N 2 45 E
St-Pourçain-sur-Sioule, *France* **27 F10** 46 18N 3 18 E
St-Priest, *France* **29 C8** 45 42N 4 57 E
St-Quay-Portrieux, *France* **26 D4** 48 39N 2 51W
St-Quentin, *France* **27 C10** 49 50N 3 16 E
St-Rambert-d'Albon, *France* **29 C8** 45 17N 4 49 E
St-Raphaël, *France* **29 E10** 43 25N 6 46 E
St. Regis, *U.S.A.* **158 C6** 47 18N 115 6W
St-Renan, *France* **26 D2** 48 26N 4 37W
St-Saëns, *France* **26 C8** 49 41N 1 16 E
St-Savin, *France* **28 B4** 46 34N 0 53 E
St-Savinien, *France* **28 C3** 45 53N 0 42W
St. Sebastien, Tanjon' i, *Madag.* **117 A8** 12 26 S 48 44 E
St-Seine-l'Abbaye, *France* **27 E11** 47 26N 4 47 E
St-Sernin-sur-Rance, *France* **28 E6** 43 54N 2 35 E
St-Sever, *France* **28 E3** 43 45N 0 35W
St-Siméon, *Canada* **141 C6** 47 51N 69 54W
St. Simons I., *U.S.A.* **152 D8** 31 12N 81 15W
St. Stephen, *Canada* **141 C6** 45 16N 67 17W
St. Stephen, *U.S.A.* **152 B10** 33 24N 79 55W
St-Sulpice, *France* **28 E5** 43 46N 1 41 E
St-Sulpice-Laurière, *France* **28 B5** 46 3N 1 29 E
St-Sulpice-les-Feuilles, *France* **28 B5** 46 19N 1 21 E
St-Syprien = St-Cyprien, *France* **28 F7** 42 37N 3 2 E
St-Thégonnec, *France* **26 D3** 48 31N 3 57W
St. Thomas, *Canada* **140 D3** 42 45N 81 10W
St. Thomas I., *Virgin Is.* **165 C7** 18 20N 64 55W
St-Tite, *Canada* **140 C5** 46 45N 72 34W
St-Tropez, *France* **29 E10** 43 17N 6 38 E
St. Troud = St. Truiden, *Belgium* **24 D5** 50 48N 5 10 E
St. Truiden, *Belgium* **24 D5** 50 48N 5 10 E
St-Vaast-la-Hougue, *France* **26 C5** 49 35N 1 17W
St-Valery-en-Caux, *France* **26 C7** 49 52N 0 43 E
St-Valéry-sur-Somme, *France* **27 B8** 50 11N 1 38 E
St-Vallier, *France* **27 F11** 46 38N 4 22 E
St-Vallier-de-Thiey, *France* **29 E10** 43 42N 6 51 E
St-Varent, *France* **26 F6** 46 53N 0 13W
St-Vaury, *France* **28 B5** 46 12N 1 46 E
St. Vincent = São Vicente, *C. Verde Is.* **8 G6** 18 0N 26 1W
St. Vincent, *Italy* **44 C4** 45 45N 7 39 E
St. Vincent, *W. Indies* **165 D7** 13 10N 61 10W
St. Vincent, G., *Australia* **128 C3** 35 0 S 138 0 E
St. Vincent & the Grenadines ■, *W. Indies* **165 D7** 13 0N 61 10W
St-Vincent-de-Tyrosse, *France* **28 E2** 43 39N 1 19W
St. Vincent Passage, *W. Indies* **165 D7** 13 30N 61 0W
St-Vith, *Belgium* **24 D6** 50 17N 6 9 E
St-Vivien-de-Médoc, *France* **28 C2** 45 25N 1 2W
St-Yrieix-la-Perche, *France* **28 C5** 45 31N 1 12 E
Ste-Adresse, *France* **26 C7** 49 31N 0 5 E
Ste-Agathe-des-Monts, *Canada* **140 C5** 46 3N 74 17W
Ste-Anne de Beaupré, *Canada* **141 C5** 47 2N 70 58W
Ste-Anne-des-Monts, *Canada* **141 C6** 49 8N 66 30W
Ste-Croix, *Switz.* **32 C3** 46 49N 6 34 E
Ste-Enimie, *France* **28 D7** 44 22N 3 26 E
Ste-Foy-la-Grande, *France* **28 D4** 44 50N 0 13 E
Ste. Genevieve, *U.S.A.* **156 G6** 37 59N 90 2W
Ste-Hermine, *France* **28 B2** 46 32N 1 4W
Ste-Livrade-sur-Lot, *France* **28 D4** 44 24N 0 36 E
Ste-Marguerite →, *Canada* **141 B6** 50 9N 66 36W
Ste-Marie, *Martinique* **165 D7** 14 48N 61 1W
Ste-Marie-aux-Mines, *France* **27 D14** 48 15N 7 12 E
Ste-Marie de la Madeleine, *Canada* **141 C5** 46 26N 71 0W
Ste-Maure-de-Touraine, *France* **26 E7** 47 7N 0 37 E
Ste-Maxime, *France* **29 E10** 43 19N 6 39 E
Ste-Menehould, *France* **27 C11** 49 5N 4 54 E
Ste-Mère-Église, *France* **26 C5** 49 24N 1 19W
Ste-Rose, *Guadeloupe* **165 C7** 16 20N 61 45W
Ste. Rose du Lac, *Canada* **143 C9** 51 4N 99 30W
Ste-Savine, *France* **27 D11** 48 18N 4 3 E
Ste-Sigolène, *France* **29 C8** 45 15N 4 14 E
Saintes, *France* **28 C3** 45 45N 0 37W
Saintes, I. des, *Guadeloupe* **165 C7** 15 50N 61 35W
Stes-Maries-de-la-Mer, *France* **29 E8** 43 26N 4 26 E
Saintfield, *U.K.* **23 B6** 54 28N 5 49W
Saintonge, *France* **28 C3** 45 40N 0 50W
Saipan, *Pac. Oc.* **134 F6** 15 12N 145 45 E
Sairecábur, Cerro, *Bolivia* **174 A2** 22 43 S 67 54W
Saitama □, *Japan* **71 A11** 36 25N 139 30 E
Saiţi, *Moldova* **53 D14** 46 30N 29 24 E
Saito, *Japan* **70 E3** 32 3N 131 24 E
Sajama, *Bolivia* **172 D4** 18 7 S 69 0W
Sajan, *Serbia, Yug.* **52 E5** 45 50N 20 20 E
Sajó →, *Hungary* **52 C6** 47 56N 21 7 E
Sajószentpéter, *Hungary* **52 B5** 48 12N 20 44 E
Sajum, *India* **93 C8** 33 20N 79 0 E
Sak →, *S. Africa* **116 E3** 30 52 S 20 25 E
Sakai, *Japan* **71 C7** 34 30N 135 30 E
Sakaide, *Japan* **70 C5** 34 15N 133 50 E
Sakaiminato, *Japan* **70 B5** 35 38N 133 11 E
Sakākah, *Si. Arabia* **96 D4** 30 0N 40 8 E
Sakakawea, L., *U.S.A.* **154 B4** 47 30N 101 25W
Sakami, L., *Canada* **140 B4** 53 15N 77 0W
Sâkâne, 'Erg i-n, *Mali* **110 D4** 20 30N 1 30W
Sakania, *Dem. Rep. of the Congo* **119 E2** 12 43 S 28 30 E
Sakarya = Adapazarı, *Turkey* **100 B4** 40 48N 30 25 E
Sakarya →, *Turkey* **100 B4** 41 7N 30 39 E
Sakashima-Guntō, *Japan* **69 M2** 24 46N 124 0 E
Sakata, *Japan* **68 E9** 38 55N 139 50 E
Sakchu, *N. Korea* **75 D13** 40 23N 125 2 E
Sakeny →, *Madag.* **117 C8** 20 0 S 45 25 E
Sakété, *Benin* **113 D5** 6 40N 2 45 E
Sakha □, *Russia* **65 C14** 66 0N 130 0 E
Sakhalin, *Russia* **65 D15** 51 0N 143 0 E
Sakhalinskiy Zaliv, *Russia* **65 D15** 54 0N 141 0 E
Sakhi Gopal, *India* **94 F7** 19 58N 85 50 E
Şaki, *Azerbaijan* **61 K8** 41 10N 47 5 E
Šakiai, *Lithuania* **15 J20** 54 59N 23 2 E
Sakmara →, *Russia* **62 F5** 51 46N 55 1 E
Sakon Nakhon, *Thailand* **86 D5** 17 10N 104 9 E
Sakrand, *Pakistan* **92 F3** 26 10N 68 15 E
Sakri, *India* **94 D2** 21 2N 74 20 E
Sakrivier, *S. Africa* **116 E3** 30 54 S 20 28 E
Sakskøbing, *Denmark* **17 K5** 54 49N 11 39 E
Saku, *Japan* **71 A10** 36 17N 138 31 E
Sakuma, *Japan* **71 B9** 35 3N 137 49 E
Sakura, *Japan* **71 B12** 35 43N 140 14 E
Sakurai, *Japan* **71 C7** 34 30N 135 51 E
Saky, *Ukraine* **59 K7** 45 9N 33 34 E
Sal →, *Russia* **61 G5** 47 31N 40 45 E
Šal'a, *Slovak Rep.* **35 C10** 48 10N 17 50 E
Sala, *Sweden* **16 E10** 59 58N 16 35 E
Sala Consilina, *Italy* **47 B8** 40 23N 15 36 E
Sala-y-Gómez, *Pac. Oc.* **135 K17** 26 28 S 105 28W
Salaberry-de-Valleyfield, *Canada* **140 C5** 45 15N 74 8W
Saladas, *Argentina* **174 B4** 28 15 S 58 40W
Saladillo, *Argentina* **174 D4** 35 40 S 59 55W
Salado →, *Buenos Aires, Argentina* **174 D4** 35 44 S 57 22W
Salado →, *La Pampa, Argentina* **176 A3** 37 30 S 67 0W
Salado →, *Río Negro, Argentina* **176 B3** 41 34 S 65 3W
Salado →, *Santa Fe, Argentina* **174 C3** 31 40 S 60 41W
Salado →, *Mexico* **162 B5** 26 52N 99 19W
Salaga, *Ghana* **113 D4** 8 31N 0 31W
Salāh, *Syria* **103 C5** 32 40N 36 45 E
Sălaj □, *Romania* **52 C8** 47 15N 23 0 E
Sálakhos, *Greece* **38 C9** 36 17N 27 57 E
Salala, *Liberia* **112 D2** 6 42N 10 7W
Salala, *Sudan* **106 C4** 21 17N 36 16 E
Salālah, *Oman* **99 C6** 16 56N 53 59 E
Salamanca, *Chile* **174 C1** 31 46 S 70 59W
Salamanca, *Spain* **42 E5** 40 58N 5 39W
Salamanca, *U.S.A.* **150 D6** 42 10N 78 43W
Salamanca □, *Spain* **42 E5** 40 57N 5 40W
Salāmatābād, *Iran* **96 C5** 35 39N 47 50 E
Salamina, *Colombia* **168 B2** 5 25N 75 29W
Salamis, *Cyprus* **38 D12** 35 11N 33 54 E
Salamís, *Greece* **48 D5** 37 56N 23 30 E
Salamonie L., *U.S.A.* **157 D11** 40 46N 85 37W
Salar de Atacama, *Chile* **174 A2** 23 30 S 68 25W
Salar de Uyuni, *Bolivia* **172 E4** 20 30 S 67 45W
Sălard, *Romania* **52 C7** 47 12N 22 3 E
Salas, *Spain* **42 B4** 43 25N 6 15W
Salas de los Infantes, *Spain* **42 C7** 42 2N 3 17W
Salatiga, *Indonesia* **85 D4** 7 19 S 110 30 E
Salavat, *Russia* **62 E5** 53 21N 55 55 E
Salaverry, *Peru* **172 B2** 8 15 S 79 0W
Salawati, *Indonesia* **83 B4** 1 7 S 130 52 E
Salay, *Phil.* **81 G5** 8 52N 124 47 E
Salayar, *Indonesia* **82 C2** 6 7 S 120 30 E
Salazar →, *Spain* **40 C3** 42 40N 1 20W
Salbris, *France* **27 E9** 47 25N 2 3 E
Salcia, *Romania* **53 G9** 43 56N 24 55 E
Sălciua, *Romania* **53 D8** 46 24N 23 26 E
Salcombe, *U.K.* **21 G4** 50 14N 3 47W
Saldaña, *Spain* **42 C6** 42 32N 4 48W
Saldanha, *S. Africa* **116 E2** 33 0 S 17 58 E
Saldanha B., *S. Africa* **116 E2** 33 6 S 18 0 E
Saldus, *Latvia* **15 H20** 56 38N 22 30 E
Saldus □, *Latvia* **54 B9** 56 38N 22 30 E
Sale, *Australia* **129 E7** 38 6 S 147 6 E
Sale, *Italy* **44 D5** 44 59N 8 48 E
Salé, *Morocco* **110 B3** 34 3N 6 48W
Sale, *U.K.* **20 D5** 53 26N 2 19W
Sale City, *U.S.A.* **152 D5** 31 16N 84 1W
Salekhard, *Russia* **56 A12** 66 30N 66 35 E
Salem, *India* **95 J4** 11 40N 78 11 E
Salem, *Ala., U.S.A.* **152 C4** 32 36N 85 14W
Salem, *Fla., U.S.A.* **152 F6** 29 53N 83 25W
Salem, *Ill., U.S.A.* **156 F8** 38 38N 88 57W
Salem, *Ind., U.S.A.* **157 F10** 38 36N 86 6W
Salem, *Mass., U.S.A.* **151 D14** 42 31N 70 53W
Salem, *Mo., U.S.A.* **155 G9** 37 39N 91 32W
Salem, *N.J., U.S.A.* **148 F8** 39 34N 75 28W
Salem, *Ohio, U.S.A.* **150 F4** 40 54N 80 52W
Salem, *Oreg., U.S.A.* **158 D2** 44 56N 123 2W
Salem, *S. Dak., U.S.A.* **154 D6** 43 44N 97 23W
Salem, *Va., U.S.A.* **148 G5** 37 18N 80 3W
Salemi, *Italy* **46 E5** 37 49N 12 48 E
Sälen, *Sweden* **16 C7** 61 15N 13 22 E
Salernes, *France* **29 E10** 43 34N 6 15 E
Salerno, *Italy* **47 B7** 40 41N 14 47 E
Salerno, G. di, *Italy* **47 B7** 40 32N 14 42 E
Salford, *U.K.* **20 D5** 53 30N 2 18W
Salgir →, *Ukraine* **59 K8** 45 38N 35 1 E
Salgótarján, *Hungary* **52 B4** 48 5N 19 47 E
Salgueiro, *Brazil* **170 C4** 8 4 S 39 6W
Salhus, *Norway* **16 D2** 60 30N 5 15 E
Salida, *U.S.A.* **146 C5** 38 32N 106 0W
Salies-de-Béarn, *France* **28 E3** 43 28N 0 56W
Şalif, *Yemen* **98 D3** 15 18N 42 41 E
Salihli, *Turkey* **100 C3** 38 28N 28 8 E
Salihorsk, *Belarus* **55 F4** 52 51N 27 27 E
Salin, *Burma* **90 E5** 20 35N 94 40 E
Salina, *Italy* **47 D7** 38 34N 14 50 E
Salina, *U.S.A.* **154 F6** 38 50N 97 37W
Salina Cruz, *Mexico* **163 D5** 16 10N 95 10W
Salinas, *Brazil* **171 E3** 16 10 S 42 10W
Salinas, *Chile* **174 A2** 23 31 S 69 29W
Salinas, *Ecuador* **168 D1** 2 10 S 80 58W
Salinas, *U.S.A.* **160 J5** 36 40N 121 39W
Salinas →, *Guatemala* **163 D6** 16 28N 90 31W
Salinas →, *U.S.A.* **160 J5** 36 45N 121 48W
Salinas, B. de, *Nic.* **164 D2** 11 4N 85 45W
Salinas, Pampa de las, *Argentina* **174 C2** 31 58 S 66 42W
Salinas Ambargasta, *Argentina* **174 B3** 29 0 S 65 0W
Salinas de Hidalgo, *Mexico* **162 C4** 22 30N 101 40W
Salinas Grandes, *Argentina* **174 C3** 30 0 S 65 0W
Saline →, *Ark., U.S.A.* **155 J8** 33 10N 92 8W
Saline →, *Kans., U.S.A.* **154 F6** 38 52N 97 30W
Salines, *Spain* **39 B10** 39 21N 3 3 E
Salines, C. de ses, *Spain* **39 B10** 39 16N 3 4 E
Salinópolis, *Brazil* **170 B2** 0 40 S 47 20W
Salins-les-Bains, *France* **27 F12** 46 58N 5 52 E
Salir, *Portugal* **43 H2** 37 14N 8 2W
Salisbury = Harare, *Zimbabwe* **119 F3** 17 43 S 31 2 E
Salisbury, *Australia* **128 C3** 34 46 S 138 40 E
Salisbury, *U.K.* **21 F6** 51 4N 1 47W
Salisbury, *Md., U.S.A.* **148 F8** 38 22N 75 36W
Salisbury, *Mo., U.S.A.* **156 E4** 39 25N 92 48W
Salisbury, *N.C., U.S.A.* **149 H5** 35 40N 80 29W
Salisbury Plain, *U.K.* **21 F6** 51 14N 1 55W
Săliște, *Romania* **53 E8** 45 45N 23 56 E
Salitre →, *Brazil* **170 C3** 9 29 S 40 39W
Salka, *Nigeria* **113 C5** 10 20N 4 58 E
Salkehatchie →, *U.S.A.* **152 C9** 32 37N 80 53W
Şalkhad, *Syria* **103 C5** 32 29N 36 43 E
Salla, *Finland* **14 C23** 66 50N 28 49 E
Sallanches, *France* **29 C10** 45 55N 6 38 E
Sallent, *Spain* **40 D6** 41 49N 1 54 E
Salles, *France* **28 D3** 44 33N 0 52W
Salles-Curan, *France* **28 D6** 44 11N 2 48 E
Salley, *U.S.A.* **152 B9** 33 34N 81 18W
Salling, *Denmark* **17 H2** 56 40N 8 55 E
Sallisaw, *U.S.A.* **155 H7** 35 28N 94 47W
Sallom Junction, *Sudan* **106 D4** 19 17N 37 6 E
Salluit, *Canada* **139 B12** 62 14N 75 38W
Salmās, *Iran* **101 C11** 38 11N 44 47 E
Salmerón, *Spain* **40 E2** 40 33N 2 29W
Salmo, *Canada* **142 D5** 49 10N 117 20W
Salmon, *U.S.A.* **158 D7** 45 11N 113 54W
Salmon →, *Canada* **142 C4** 54 3N 122 40W
Salmon →, *U.S.A.* **158 D5** 45 51N 116 47W
Salmon Arm, *Canada* **142 C5** 50 40N 119 15W
Salmon Falls, *U.S.A.* **158 E6** 42 48N 114 59W
Salmon Gums, *Australia* **125 F3** 32 59 S 121 38 E
Salmon Res., *Canada* **141 C8** 48 5N 56 0W
Salmon River Mts., *U.S.A.* **158 D6** 45 0N 114 30W
Salo, *Finland* **15 F20** 60 22N 23 10 E
Salò, *Italy* **44 C7** 45 36N 10 31 E
Salobreña, *Spain* **43 J7** 36 44N 3 35W
Salome, *U.S.A.* **161 M13** 33 47N 113 37W
Salon-de-Provence, *France* **29 E9** 43 39N 5 6 E
Salonica = Thessaloníki, *Greece* **50 F6** 40 38N 22 58 E
Salonta, *Romania* **52 D6** 46 49N 21 42 E
Salor →, *Spain* **43 F3** 39 39N 7 3W
Salou, *Spain* **40 D6** 41 4N 1 8 E
Salou, C. de, *Spain* **40 D6** 41 3N 1 10 E
Salpausselkä, *Finland* **15 F22** 61 0N 27 0 E
Salsacate, *Argentina* **174 C2** 31 20 S 65 5W
Salses, *France* **28 F6** 42 50N 2 55 E
Salsette I., *India* **94 E1** 19 5N 72 50 E
Salsk, *Russia* **61 G5** 46 28N 41 30 E
Salso →, *Italy* **46 E6** 37 6N 13 57 E
Salsomaggiore Terme, *Italy* **44 D6** 44 49N 9 59 E
Salt, *Spain* **40 D7** 41 59N 2 47 E
Salt →, *Canada* **142 B6** 60 0N 112 25W
Salt →, *Ariz., U.S.A.* **159 K7** 33 23N 112 19W
Salt →, *Mo., U.S.A.* **156 E5** 39 28N 91 4W
Salt Creek, *Australia* **128 D3** 36 8 S 139 38 E
Salt Fork Arkansas →, *U.S.A.* **155 G6** 36 36N 97 3W
Salt L., *U.S.A.* **145 K14** 21 21N 157 55W
Salt Lake City, *U.S.A.* **158 F8** 40 45N 111 53W
Salt Range, *Pakistan* **92 C5** 32 30N 72 25 E
Salt Springs, *U.S.A.* **153 F8** 29 21N 81 44W
Salta, *Argentina* **174 A2** 24 57 S 65 25W
Salta □, *Argentina* **174 A2** 24 48 S 65 30W
Saltara, *Italy* **45 E9** 43 45N 12 50 E
Saltash, *U.K.* **21 G3** 50 24N 4 14W
Saltburn by the Sea, *U.K.* **20 C7** 54 35N 0 58W
Saltcoats, *U.K.* **22 F4** 55 38N 4 47W
Saltee Is., *Ireland* **23 D5** 52 7N 6 37W
Salters, *U.S.A.* **152 B10** 33 36N 79 51W
Saltfjellet, *Norway* **14 C16** 66 40N 15 15 E
Saltfjorden, *Norway* **14 C16** 67 15N 14 10 E
Saltholm, *Denmark* **17 J6** 55 38N 12 43 E
Salthólmavík, *Iceland* **11 B5** 65 24N 21 57W
Saltillo, *Mexico* **162 B4** 25 25N 101 0W
Salto, *Argentina* **174 C3** 34 20 S 60 15W
Salto, *Uruguay* **174 C4** 31 27 S 57 50W
Salto da Divisa, *Brazil* **171 E4** 16 0 S 39 57W
Salton City, *U.S.A.* **161 M11** 33 29N 115 51W
Salton Sea, *U.S.A.* **161 M11** 33 15N 115 45W
Saltpond, *Ghana* **113 D4** 5 15N 1 3W
Saltrød, *Norway* **18 F5** 58 30N 8 49 E
Saltsjöbaden, *Sweden* **16 E12** 59 15N 18 20 E
Saltville, *U.S.A.* **148 G5** 36 53N 81 46W
Saluda, *U.S.A.* **152 A8** 34 0N 81 46W
Saluda →, *U.S.A.* **149 J5** 34 1N 81 4W
Salûm, *Egypt* **106 A2** 31 31N 25 7 E
Salûm, Khâlig el, *Egypt* **106 A2** 31 35N 25 24 E
Salur, *India* **94 E6** 18 27N 83 18 E
Salut, Is. du, *Fr. Guiana* **169 B7** 5 15N 52 35W
Saluzzo, *Italy* **44 D4** 44 39N 7 29 E
Salvacion, *Phil.* **81 G2** 9 56N 118 47 E
Salvación, B., *Chile* **176 D1** 50 50 S 75 10W
Salvador, *Brazil* **171 D4** 13 0 S 38 30W
Salvador, *Canada* **143 C7** 52 10N 109 32W
Salvador, L., *U.S.A.* **155 L9** 29 43N 90 15W
Salvaterra, *Brazil* **170 B2** 0 46 S 48 31W
Salvaterra de Magos, *Portugal* **43 F2** 39 1N 8 47W
Salvisa, *U.S.A.* **157 G12** 37 54N 84 51W
Sálvora, I. de, *Spain* **42 C2** 42 30N 8 58W
Salween →, *Burma* **90 G6** 16 31N 97 37 E
Salyan, *Azerbaijan* **101 C13** 39 33N 48 59 E
Salyersville, *U.S.A.* **148 G4** 37 45N 83 4W
Salza →, *Austria* **34 D7** 47 40N 14 43 E
Salzach →, *Austria* **34 C5** 48 12N 12 56 E
Salzburg, *Austria* **34 D6** 47 48N 13 2 E
Salzburg □, *Austria* **34 D6** 47 15N 13 0 E
Salzgitter, *Germany* **30 C6** 52 9N 10 19 E
Salzkotten, *Germany* **30 D4** 51 40N 8 37 E
Salzwedel, *Germany* **30 C7** 52 52N 11 10 E
Sam, *Gabon* **114 B2** 0 58N 11 16 E
Sam Neua, *Laos* **86 B5** 20 29N 104 5 E
Sam Ngao, *Thailand* **86 D2** 17 18N 99 0 E
Sam Rayburn Reservoir, *U.S.A.* **155 K7** 31 4N 94 5W
Sam Son, *Vietnam* **86 C5** 19 44N 105 54 E
Sam Teu, *Laos* **86 C5** 19 59N 104 38 E
Sama de Langreo = Langreo, *Spain* **42 B5** 43 18N 5 40W
Samacimbo, *Angola* **115 E3** 13 33 S 16 59 E
Samagaltay, *Russia* **65 D10** 50 36N 95 3 E
Samā'il, *Oman* **99 B7** 23 40N 57 50 E
Samaipata, *Bolivia* **173 D5** 18 9 S 63 52W
Samal, *Phil.* **81 H5** 7 5N 125 42 E
Samal I., *Phil.* **81 H5** 7 3N 125 44 E
Samales Group, *Phil.* **81 J3** 6 0N 122 0 E
Samalkot, *India* **94 F6** 17 3N 82 13 E
Samâlût, *Egypt* **106 J7** 28 20N 30 42 E
Samana, *India* **92 D7** 30 10N 76 13 E
Samana Cay, *Bahamas* **165 B5** 23 3N 73 45W
Samandağı, *Turkey* **100 D6** 36 5N 35 59 E
Samandıra, *Turkey* **51 F13** 40 59N 29 13 E
Samanga, *Tanzania* **119 D4** 8 20 S 39 13 E
Samangān □, *Afghan.* **91 A3** 36 15N 68 3 E
Samangwa, *Dem. Rep. of the Congo* **115 C4** 4 23 S 24 10 E
Samani, *Japan* **68 C11** 42 7N 142 56 E
Samanli Dağları, *Turkey* **51 F13** 40 32N 29 10 E
Samar, *Phil.* **81 E5** 12 0N 125 0 E
Samar □, *Phil.* **81 F5** 11 50N 125 0 E
Samar Sea, *Phil.* **80 E5** 12 0N 124 15 E
Samara, *Russia* **60 D10** 53 8N 50 6 E
Samara →, *Russia* **60 D10** 53 10N 50 4 E
Samara →, *Ukraine* **59 H8** 48 28N 35 7 E
Samarai, *Papua N. G.* **132 F6** 10 39 S 150 41 E
Samaria = Shōmrōn, *West Bank* **103 C4** 32 15N 35 13 E
Samariá, *Greece* **38 D5** 35 17N 23 58 E
Samarinda, *Indonesia* **85 C5** 0 30 S 117 9 E
Samarkand = Samarqand, *Uzbekistan* **63 D3** 39 40N 66 55 E
Samarqand, *Uzbekistan* **63 D3** 39 40N 66 55 E
Sāmarrā, *Iraq* **101 E10** 34 12N 43 52 E
Samastipur, *India* **93 G11** 25 50N 85 50 E
Samaúma, *Brazil* **173 B5** 7 50 S 60 2W
Şamaxi, *Azerbaijan* **61 K9** 40 38N 48 37 E
Samba, *Dem. Rep. of the Congo* **115 C5** 4 38 S 26 22 E
Samba, *India* **93 C6** 32 32N 75 10 E
Samba Caju, *Angola* **115 D3** 8 46 S 15 24 E
Sambaíba, *Brazil* **170 C2** 7 8 S 45 21W
Sambalpur, *India* **94 D7** 21 28N 84 4 E
Sambar, Tanjung, *Indonesia* **85 C4** 2 59 S 110 19 E
Sambas, *Indonesia* **85 B3** 1 20N 109 20 E
Sambava, *Madag.* **117 A9** 14 16 S 50 10 E
Sambawizi, *Zimbabwe* **119 F2** 18 24 S 26 13 E
Sambhal, *India* **93 E8** 28 35N 78 37 E
Sambhar, *India* **92 F6** 26 52N 75 6 E
Sambiase, *Italy* **47 D9** 38 58N 16 17 E
Sambir, *Ukraine* **59 H2** 49 30N 23 10 E
Sambor, *Cambodia* **86 F6** 12 46N 106 0 E
Sambuca di Sicília, *Italy* **46 E6** 37 39N 13 7 E
Samburu □, *Kenya* **118 B4** 1 10N 37 0 E
Samchŏk, *S. Korea* **75 F15** 37 30N 129 10 E
Samchonpo, *S. Korea* **75 G15** 35 0N 128 6 E
Same, *Tanzania* **118 C4** 4 2 S 37 38 E
Samedan, *Switz.* **33 C9** 46 32N 9 52 E
Samer, *France* **27 B8** 50 38N 1 44 E

276 Samfya

Samfya, *Zambia* **119 E2** 11 22 S 29 31 E
Samḩān, Jabal, *Oman* **99 C6** 17 12N 54 55 E
Sámi, *Greece* **48 C2** 38 15N 20 39 E
Şämkir, *Azerbaijan* **61 K8** 40 50N 46 0 E
Şamlı, *Turkey* **49 B9** 39 48N 27 51 E
Samnah, *Si. Arabia* **96 E3** 25 10N 37 15 E
Samnaun, *Switz.* **33 C10** 46 57N 10 22 E
Samnū, *Libya* **108 C2** 27 15N 14 55 E
Samo Alto, *Chile* **174 C1** 30 22 S 71 0W
Samoan Is., *Pac. Oc.* **133 X24** 14 0 S 171 0W
Samobor, *Croatia* **45 C12** 45 47N 15 44 E
Samoëns, *France* **27 F13** 46 5N 6 45 E
Samokov, *Bulgaria* **50 D7** 42 18N 23 35 E
Samoorombón, B., *Argentina* **174 D4** 36 5 S 57 20W
Šamorín, *Slovak Rep.* **35 C10** 48 2N 17 19 E
Samorogouan, *Burkina Faso* **112 C4** 11 21N 4 57W
Sámos, *Greece* **49 D8** 37 41N 26 56 E
Samoš, *Serbia, Yug.* **52 E5** 45 13N 20 46 E
Samos, *Spain* **42 C3** 42 44N 7 20W
Sámos □, *Greece* **49 D8** 37 45N 26 50 E
Samoset, *U.S.A.* **153 H7** 27 28N 82 33W
Samosir, *Indonesia* **84 B1** 2 55N 98 50 E
Samothráki = Mathráki, *Greece* **38 A3** 39 48N 19 31 E
Samothráki, *Greece* **51 F9** 40 28N 25 28 E
Samoylovka, *Russia* **60 E6** 51 12N 43 43 E
Sampa, *Ghana* **112 D4** 8 0N 2 36W
Sampacho, *Argentina* **174 C3** 33 20 S 64 50W
Sampang, *Indonesia* **85 D4** 7 11 S 113 13 E
Samper de Calanda, *Spain* **40 D4** 41 11N 0 28W
Sampéyre, *Italy* **44 D4** 44 34N 7 11 E
Sampit, *Indonesia* **85 C4** 2 34 S 113 0 E
Sampit →, *Indonesia* **85 C4** 2 44 S 112 54 E
Sampit, Teluk, *Indonesia* **85 C4** 3 5 S 113 3 E
Samrong, *Cambodia* **86 E4** 14 15N 103 30 E
Samrong, *Thailand* **86 E3** 15 10N 100 40 E
Samsø, *Denmark* **17 J4** 55 50N 10 35 E
Samsø Bælt, *Denmark* **17 J4** 55 45N 10 45 E
Samson, *U.S.A.* **152 D3** 31 7N 86 3W
Samsonovo, *Turkmenistan* **63 E2** 37 53N 65 15 E
Samsun, *Turkey* **100 B7** 41 15N 36 22 E
Samtredia, *Georgia* **61 J6** 42 7N 42 24 E
Samui, Ko, *Thailand* **87 H3** 9 30N 100 0 E
Samur →, *Russia* **61 K9** 41 53N 48 32 E
Samurskiy Khrebet, *Russia* **61 K8** 41 55N 47 11 E
Samusole, *Dem. Rep. of the Congo* **115 E4** 10 2 S 24 0 E
Samut Prakan, *Thailand* **86 F3** 13 32N 100 40 E
Samwari, *Pakistan* **92 E2** 28 30N 66 46 E
San, *Mali* **112 C4** 13 15N 4 57W
San →, *Cambodia* **86 F5** 13 32N 105 57 E
San →, *Poland* **55 H8** 50 45N 21 51 E
San Adrián, *Spain* **40 C3** 42 20N 1 56W
San Adrián, C. de, *Spain* **42 B2** 43 21N 8 50W
San Agustín, *Colombia* **168 C2** 1 53N 76 16W
San Agustin, C., *Phil.* **81 H6** 6 20N 126 13 E
San Agustín de Valle Fértil, *Argentina* **174 C2** 30 35 S 67 30W
San Ambrosio, *Pac. Oc.* **135 K20** 26 28 S 79 53W
San Andreas, *U.S.A.* **160 G6** 38 12N 120 41W
San Andres, *Phil.* **80 E4** 13 19N 122 41 E
San Andrés, I. de, *Caribbean* **164 D3** 12 42N 81 46W
San Andrês del Rabanedo, *Spain* **42 C5** 42 37N 5 36W
San Andres Mts., *U.S.A.* **159 K10** 33 0N 106 30W
San Andrés Tuxtla, *Mexico* **163 D5** 18 30N 95 20W
San Angelo, *U.S.A.* **155 K4** 31 28N 100 26W
San Anselmo, *U.S.A.* **160 H4** 37 59N 122 34W
San Antonio, *Belize* **163 D7** 16 15N 89 2W
San Antonio, *Chile* **174 C1** 33 40 S 71 40W
San Antonio, *Phil.* **80 D3** 14 57N 120 5 E
San Antonio, *N. Mex.*, *U.S.A.* **159 K10** 33 55N 106 52W
San Antonio, *Tex.*, *U.S.A.* **155 L5** 29 25N 98 30W
San Antonio, *Venezuela* **168 C4** 3 30N 66 44W
San Antonio →, *U.S.A.* **155 L6** 28 30N 96 54W
San Antonio, C., *Argentina* **174 D4** 36 15 S 56 40W
San Antonio, C., *Cuba* **164 B3** 21 50N 84 57W
San Antonio, C. de, *Spain* **41 G5** 38 48N 0 12 E
San Antonio, Mt., *U.S.A.* **161 L9** 34 17N 117 38W
San Antonio Bay, *Phil.* **81 G1** 8 38N 117 35 E
San Antonio de los Baños, *Cuba* **164 B3** 22 54N 82 31W
San Antonio de los Cobres, *Argentina* **174 A2** 24 10 S 66 17W
San Antonio Oeste, *Argentina* **176 B4** 40 40 S 65 0W
San Arcángelo, *Italy* **47 B9** 40 14N 16 14 E
San Ardo, *U.S.A.* **160 J6** 36 1N 120 54W
San Augustín, *Canary Is.* **39 G4** 27 47N 15 32W
San Augustine, *U.S.A.* **155 K7** 31 30N 94 7W
San Bartolomé, *Canary Is.* **39 F6** 28 59N 13 37W
San Bartolomé de Tirajana, *Canary Is.* **39 G4** 27 54N 15 34W
San Bartolomeo in Galdo, *Italy* **47 A8** 41 24N 15 1 E
San Benedetto del Tronto, *Italy* **45 F10** 42 57N 13 53 E
San Benedetto Po, *Italy* **44 C7** 45 2N 10 55 E
San Benedicto, I., *Mexico* **162 D2** 19 18N 110 49W
San Benito, *U.S.A.* **155 M6** 26 8N 97 38W
San Benito →, *U.S.A.* **160 J5** 36 53N 121 34W
San Benito Mt., *U.S.A.* **160 J6** 36 22N 120 37W
San Bernardino, *U.S.A.* **161 L9** 34 7N 117 19W
San Bernardino, Paso del, *Switz.* **33 D8** 46 28N 9 11 E
San Bernardino Mts., *U.S.A.* **161 L10** 34 10N 116 45W
San Bernardino Str., *Phil.* **80 E5** 13 0N 125 0 E
San Bernardo, *Chile* **174 C1** 33 40 S 70 50W
San Bernardo, I. de, *Colombia* **168 B2** 9 45N 75 50W
San Blas, *Mexico* **162 B3** 26 4N 108 46W
San Blas, Arch. de, *Panama* **164 E4** 9 50N 78 31W
San Blas, C., *U.S.A.* **152 F4** 29 40N 85 21W
San Bonifacio, *Italy* **45 C8** 45 24N 11 16 E
San Borja, *Bolivia* **172 C4** 14 50 S 66 52W
San Buenaventura, *Bolivia* **172 C4** 14 28 S 67 35W
San Buenaventura, *Mexico* **162 B4** 27 5N 101 32W
San Carlos = Butuku-Luba, *Eq. Guin.* **113 E6** 3 29N 8 33 E
San Carlos, *Argentina* **174 C2** 33 50 S 69 0W
San Carlos, *Bolivia* **173 D5** 17 24 S 63 45W
San Carlos, *Chile* **174 D1** 36 10 S 72 0W
San Carlos, *Mexico* **162 B4** 29 0N 100 54W
San Carlos, *Nic.* **164 D3** 11 12N 84 50W
San Carlos, *Negros, Phil.* **81 F4** 10 29N 123 25 E
San Carlos, *Pangasinan, Phil.* **80 D3** 15 55N 120 20 E
San Carlos, *Spain* **39 B8** 39 3N 1 34 E
San Carlos, *Uruguay* **175 C5** 34 46 S 54 58W
San Carlos, *U.S.A.* **159 K8** 33 21N 110 27W
San Carlos, *Amazonas, Venezuela* **168 C4** 1 55N 67 4W
San Carlos, *Cojedes, Venezuela* **168 B4** 9 40N 68 36W
San Carlos de Bariloche, *Argentina* **176 B2** 41 10 S 71 25W
San Carlos de la Ràpita = Sant Carles de la Ràpita, *Spain* **40 E5** 40 37N 0 35 E
San Carlos del Zulia, *Venezuela* **168 B3** 9 1N 71 55W
San Carlos L., *U.S.A.* **159 K8** 33 11N 110 32W
San Cataldo, *Italy* **46 E6** 37 29N 13 59 E
San Celoni = Sant Celoni, *Spain* **40 D7** 41 42N 2 30 E
San Clemente, *Chile* **174 D1** 35 30 S 71 29W
San Clemente, *Spain* **41 F2** 39 24N 2 25W
San Clemente, *U.S.A.* **161 M9** 33 26N 117 37W
San Clemente I., *U.S.A.* **161 N8** 32 53N 118 29W
San Cristóbal, *Argentina* **174 C3** 30 20 S 61 10W
San Cristóbal, *Colombia* **168 D3** 2 18 S 73 2W
San Cristóbal, *Dom. Rep.* **165 C5** 18 25N 70 6W
San Cristóbal, *Mexico* **163 D6** 16 50N 92 33W
San Cristóbal, *Solomon Is.* **133 N11** 10 30 S 161 0 E
San Cristóbal, *Spain* **39 B11** 39 57N 4 3 E
San Cristóbal, *Venezuela* **168 B3** 7 46N 72 14W
San Damiano d'Asti, *Italy* **44 D5** 44 50N 8 4 E
San Daniele del Friuli, *Italy* **45 B10** 46 9N 13 1 E
San Diego, *Calif., U.S.A.* **161 N9** 32 43N 117 9W
San Diego, *Tex., U.S.A.* **155 M5** 27 46N 98 14W
San Diego, C., *Argentina* **176 D3** 54 40 S 65 10W
San Diego de la Unión, *Mexico* **162 C4** 21 28N 100 52W
San Dimitri, Ras, *Malta* **38 C1** 36 4N 14 11 E
San Dionosio, *Phil.* **81 F4** 11 16N 123 6 E
San Donà di Piave, *Italy* **45 C9** 45 38N 12 34 E
San Emilio, *Phil.* **80 C3** 17 14N 120 37 E
San Estanislao, *Paraguay* **174 A4** 24 39 S 56 26W
San Esteban de Gormaz, *Spain* **40 D1** 41 34N 3 13W
San Fabian, *Phil.* **80 C3** 16 5N 120 25 E
San Felice Circeo, *Italy* **46 A6** 41 14N 13 5 E
San Felice sul Panaro, *Italy* **45 D8** 44 50N 11 8 E
San Felipe, *Chile* **174 C1** 32 43 S 70 42W
San Felipe, *Colombia* **168 C4** 1 55N 67 6W
San Felipe, *Mexico* **162 A2** 31 0N 114 52W
San Felipe, *Phil.* **80 D3** 15 4N 120 4 E
San Felipe, *Venezuela* **168 A4** 10 20N 68 44W
San Felipe →, *U.S.A.* **161 M11** 33 12N 115 49W
San Félix, *Pac. Oc.* **135 K20** 26 23 S 80 0W
San Fernando, *Chile* **174 C1** 34 30 S 71 0W
San Fernando, *Mexico* **162 B1** 29 55N 115 10W
San Fernando, *Cebu, Phil.* **81 F4** 10 10N 123 42 E
San Fernando, *La Union, Phil.* **80 C3** 16 40N 120 23 E
San Fernando, *Pampanga, Phil.* **80 D3** 15 5N 120 37 E
San Fernando, *Tablas, Phil.* **80 E4** 12 18N 122 36 E
San Fernando, *Baleares, Spain* **39 C7** 38 42N 1 28 E
San Fernando, *Cádiz, Spain* **43 J4** 36 28N 6 17W
San Fernando, *Trin. & Tob.* **165 D7** 10 20N 61 30W
San Fernando, *U.S.A.* **161 L8** 34 17N 118 26W
San Fernando →, *Mexico* **162 C5** 24 55N 98 10W
San Fernando de Apure, *Venezuela* **168 B4** 7 54N 67 15W
San Fernando de Atabapo, *Venezuela* **168 C4** 4 3N 67 42W
San Fernando di Púglia, *Italy* **47 A9** 41 18N 16 5 E
San Francisco, *Argentina* **174 C3** 31 30 S 62 5W
San Francisco, *Bolivia* **173 D4** 15 16 S 65 31W
San Francisco, *Cebu, Phil.* **81 F5** 10 39N 124 23 E
San Francisco, *Leyte, Phil.* **81 F5** 10 4N 125 9 E
San Francisco, *Mindanao, Phil.* **81 G5** 8 30N 125 56 E
San Francisco, *U.S.A.* **160 H4** 37 47N 122 25W
San Francisco →, *U.S.A.* **159 K9** 32 59N 109 22W
San Francisco, Paso de, *S. Amer.* **174 B2** 27 0 S 68 0W
San Francisco de Macorís, *Dom. Rep.* **165 C5** 19 19N 70 15W
San Francisco del Monte de Oro, *Argentina* **174 C2** 32 36 S 66 8W
San Francisco del Oro, *Mexico* **162 B3** 26 52N 105 50W
San Francisco Javier = Sant Francesc de Formentera, *Spain* **39 C7** 38 42N 1 26 E
San Francisco Solano, Pta., *Colombia* **168 B2** 6 18N 77 29W
San Fratello, *Italy* **47 D7** 38 1N 14 36 E
San Gabriel, *Ecuador* **168 C2** 0 36N 77 49W
San Gavino Monreale, *Italy* **46 C1** 39 33N 8 47 E
San Gil, *Colombia* **168 B3** 6 33N 73 8W
San Gimignano, *Italy* **44 E8** 43 28N 11 2 E
San Giórgio di Nogaro, *Italy* **45 C10** 45 50N 13 13 E
San Giórgio Iónico, *Italy* **47 B10** 40 27N 17 23 E
San Giovanni Bianco, *Italy* **44 C6** 45 52N 9 39 E
San Giovanni in Fiore, *Italy* **47 C9** 39 15N 16 42 E
San Giovanni in Persiceto, *Italy* **45 D8** 44 38N 11 11 E
San Giovanni Rotondo, *Italy* **45 G12** 41 42N 15 44 E
San Giovanni Valdarno, *Italy* **45 E8** 43 34N 11 32 E
San Giuliano Terme, *Italy* **44 E7** 43 46N 10 26 E
San Gorgonio Mt., *U.S.A.* **161 L10** 34 7N 116 51W
San Gottardo, P. del, *Switz.* **33 C7** 46 33N 8 33 E
San Gregorio, *Uruguay* **175 C4** 32 37 S 55 40W
San Gregorio, *U.S.A.* **160 H4** 37 20N 122 23W
San Guiseppe Jato, *Italy* **46 E6** 37 57N 13 11 E
San Ignacio, *Belize* **163 D7** 17 10N 89 0W
San Ignacio, *Bolivia* **173 D5** 16 20 S 60 55W
San Ignacio, *Mexico* **162 B2** 27 27N 113 0W
San Ignacio, *Paraguay* **174 B4** 26 52 S 57 3W
San Ignacio, L., *Mexico* **162 B2** 26 50N 113 11W
San Ildefonso, *Phil.* **80 D3** 15 5N 120 56 E
San Ildefonso, C., *Phil.* **80 C4** 16 0N 122 1 E
San Isidro, *Argentina* **174 C4** 34 29 S 58 31W
San Jacinto, *Colombia* **168 B2** 9 50N 75 8W
San Jacinto, *Phil.* **80 E4** 12 34N 123 44 E
San Jacinto, *U.S.A.* **161 M10** 33 47N 116 57W
San Jaime, *Spain* **39 B11** 39 54N 4 4 E
San Javier, *Misiones, Argentina* **175 B4** 27 55 S 55 5W
San Javier, *Santa Fe, Argentina* **174 C4** 30 40 S 59 55W
San Javier, *Beni, Bolivia* **173 C5** 14 34 S 64 42W
San Javier, *Santa Cruz, Bolivia* **173 D5** 16 18 S 62 30W
San Javier, *Chile* **174 D1** 35 40 S 71 45W
San Javier, *Spain* **41 H4** 37 49N 0 50W
San Jerónimo, Sa. de, *Colombia* **168 B2** 8 0N 75 50W
San Jeronimo Taviche, *Mexico* **163 D5** 16 38N 96 32W
San Joaquín, *Bolivia* **173 C5** 13 4 S 64 49W
San Joaquin, *Phil.* **81 F4** 10 35N 122 8 E
San Joaquin, *U.S.A.* **160 J6** 36 36N 120 11W
San Joaquín, *Venezuela* **168 A4** 10 16N 67 47W
San Joaquín →, *Bolivia* **173 C5** 13 8 S 63 41W
San Joaquin →, *U.S.A.* **160 G5** 38 4N 121 51W
San Joaquin Valley, *U.S.A.* **160 J6** 37 20N 121 0W
San Jordi, *Spain* **39 B9** 39 33N 2 46 E
San Jorge, *Argentina* **174 C3** 31 54 S 61 50W
San Jorge, *Spain* **39 C7** 38 54N 1 24 E
San Jorge, B. de, *Mexico* **162 A2** 31 20N 113 20W
San Jorge, G., *Argentina* **176 C3** 46 0 S 66 0W
San Jorge, G. of, *Argentina* **166 H4** 46 0 S 66 0W
San José = San Josep, *Spain* **39 C7** 38 55N 1 18 E
San José, *Costa Rica* **164 E3** 9 55N 84 2W
San José, *Guatemala* **164 D1** 14 0N 90 50W
San José, *Mexico* **162 C2** 25 0N 110 50W
San Jose, *Luzon, Phil.* **80 D3** 15 45N 120 55 E
San Jose, *Mindoro, Phil.* **80 E3** 12 27N 121 4 E
San Jose, *Calif., U.S.A.* **160 H5** 37 20N 121 53W
San Jose, *Ill., U.S.A.* **156 D7** 40 18N 89 36W
San Jose →, *U.S.A.* **159 J10** 34 25N 106 45W
San José de Chiquitos, *Bolivia* **173 D5** 17 53 S 60 50W
San José de Feliciano, *Argentina* **174 C4** 30 26 S 58 46W
San José de Jáchal, *Argentina* **174 C2** 30 15 S 68 46W
San José de Mayo, *Uruguay* **174 C4** 34 27 S 56 40W
San José de Ocune, *Colombia* **168 C3** 4 15N 70 20W
San José de Uchapiamonas, *Bolivia* **172 C4** 14 13 S 68 5W
San José del Cabo, *Mexico* **162 C3** 23 0N 109 40W
San José del Guaviare, *Colombia* **168 C3** 2 35N 72 38W
San José do Anauá, *Brazil* **169 C5** 0 58N 61 22W
San Josep, *Spain* **39 C7** 38 55N 1 18 E
San Juan, *Argentina* **174 C2** 31 30 S 68 30W
San Juan, *Colombia* **168 B2** 8 46N 76 32W
San Juan, *Mexico* **162 C4** 21 20N 102 50W
San Juan, *Ica, Peru* **172 D2** 15 22 S 75 7W
San Juan, *Puno, Peru* **172 C4** 14 2 S 69 19W
San Juan, *Luzon, Phil.* **80 C3** 16 40N 120 20 E
San Juan, *Mindanao, Phil.* **81 G6** 8 25N 126 20 E
San Juan, *Puerto Rico* **165 C6** 18 28N 66 7W
San Juan □, *Argentina* **174 C2** 31 9 S 69 0W
San Juan →, *Argentina* **174 C2** 32 20 S 67 25W
San Juan →, *Bolivia* **173 E4** 21 2 S 65 19W
San Juan →, *Colombia* **168 C2** 4 3N 77 27W
San Juan →, *Nic.* **164 D3** 10 56N 83 42W
San Juan →, *U.S.A.* **159 H8** 37 16N 110 26W
San Juan →, *Venezuela* **169 A5** 10 14N 62 38W
San Juan, C., *Eq. Guin.* **114 B1** 1 5N 9 20 E
San Juan Bautista = Sant Joan Baptista, *Spain* **39 B8** 39 5N 1 31 E
San Juan Bautista, *Paraguay* **174 B4** 26 37 S 57 6W
San Juan Bautista, *U.S.A.* **160 J5** 36 51N 121 32W
San Juan Bautista Valle Nacional, *Mexico* **163 D5** 17 47N 96 19W
San Juan Capistrano, *U.S.A.* **161 M9** 33 30N 117 40W
San Juan Cr. →, *U.S.A.* **160 J5** 35 40N 120 22W
San Juan de Alicante, *Spain* **41 G4** 38 24N 0 26W
San Juan de Guadalupe, *Mexico* **162 C4** 24 38N 102 44W
San Juan de los Morros, *Venezuela* **168 B4** 9 55N 67 21W
San Juan del César, *Colombia* **168 A3** 10 46N 73 1W
San Juan del Norte, *Nic.* **164 D3** 10 58N 83 40W
San Juan del Norte, B. de, *Nic.* **164 D3** 11 0N 83 40W
San Juan del Río, *Mexico* **163 C5** 20 25N 100 0W
San Juan del Sur, *Nic.* **164 D2** 11 20N 85 51W
San Juan I., *U.S.A.* **160 B3** 48 32N 123 5W
San Juan Mts., *U.S.A.* **159 H10** 37 30N 107 0W
San Julián, *Argentina* **176 C3** 49 15 S 67 45W
San Julian, *Phil.* **81 F5** 11 45N 125 27 E
San Just, Sierra de, *Spain* **40 E4** 40 45N 0 49W
San Justo, *Argentina* **174 C3** 30 47 S 60 30W
San Kamphaeng, *Thailand* **86 C2** 18 45N 99 8 E
San Lázaro, C., *Mexico* **162 C2** 24 50N 112 18W
San Lázaro, Sa., *Mexico* **162 C3** 23 25N 110 0W
San Leandro, *U.S.A.* **160 H4** 37 44N 122 9W
San Leonardo de Yagüe, *Spain* **40 D1** 41 51N 3 5W
San Lorenzo, *Argentina* **174 C3** 32 45 S 60 45W
San Lorenzo, *Beni, Bolivia* **173 D4** 15 22 S 65 48W
San Lorenzo, *Tarija, Bolivia* **173 E5** 21 26 S 64 47W
San Lorenzo, *Ecuador* **168 C2** 1 15N 78 50W
San Lorenzo, *Paraguay* **174 B4** 25 20 S 57 32W
San Lorenzo, *Spain* **39 B10** 39 37N 3 17 E
San Lorenzo, *Venezuela* **168 B3** 9 47N 71 4W
San Lorenzo →, *Mexico* **162 C3** 24 15N 107 24W
San Lorenzo, I., *Mexico* **162 B2** 28 35N 112 50W
San Lorenzo, I., *Peru* **172 C2** 12 7 S 77 15W
San Lorenzo, Mt., *Argentina* **176 C2** 47 40 S 72 20W
San Lorenzo de la Parrilla, *Spain* **40 F2** 39 51N 2 22W
San Lorenzo de Morunys = Sant Llorenç de Morunys, *Spain* **40 C6** 42 8N 1 35 E
San Lucas, *Bolivia* **173 E4** 20 5 S 65 7W
San Lucas, *Baja Calif. S., Mexico* **162 C3** 22 53N 109 54W
San Lucas, *Baja Calif. S., Mexico* **162 B2** 27 10N 112 14W
San Lucas, *U.S.A.* **160 J5** 36 8N 121 1W
San Lucas, C., *Mexico* **162 C3** 22 50N 110 0W
San Lúcido, *Italy* **47 C9** 39 18N 16 3 E
San Luis, *Argentina* **174 C2** 33 20 S 66 20W
San Luis, *Cuba* **164 B3** 22 17N 83 46W
San Luis, *Guatemala* **164 C2** 16 14N 89 27W
San Luis, *U.S.A.* **159 H11** 37 12N 105 25W
San Luis □, *Argentina* **174 C2** 34 0 S 66 0W
San Luis, I., *Mexico* **162 B2** 29 58N 114 26W
San Luis, L. de, *Bolivia* **173 C5** 13 45 S 64 0W
San Luis, Sierra de, *Argentina* **174 C2** 32 30 S 66 10W
San Luis de la Paz, *Mexico* **162 C4** 21 19N 100 32W
San Luis Obispo, *U.S.A.* **161 K6** 35 17N 120 40W
San Luis Potosí, *Mexico* **162 C4** 22 9N 100 59W
San Luis Potosí □, *Mexico* **162 C4** 22 10N 101 0W
San Luis Reservoir, *U.S.A.* **160 H5** 37 4N 121 5W
San Luis Río Colorado, *Mexico* **162 A2** 32 29N 114 58W
San Manuel, *Phil.* **80 C3** 16 4N 120 40 E
San Marco, C., *Italy* **46 C1** 39 51N 8 26 E
San Marco Argentano, *Italy* **47 C9** 39 33N 16 7 E
San Marco in Lámis, *Italy* **45 G12** 41 43N 15 38 E
San Marcos, *Colombia* **168 B2** 8 39N 75 8W
San Marcos, *Guatemala* **164 D1** 14 59N 91 52W
San Marcos, *Mexico* **162 B2** 27 13N 112 6W
San Marcos, *U.S.A.* **155 L6** 29 53N 97 56W
San Marino, *San Marino* **45 E9** 43 55N 12 30 E
San Marino ■, *Europe* **45 E9** 43 56N 12 25 E
San Martín, *Argentina* **174 C2** 33 5 S 68 28W
San Martín, *Colombia* **168 C3** 3 42N 73 42W
San Martin →, *Bolivia* **173 C5** 13 8 S 63 43W
San Martín, L., *Argentina* **176 C2** 48 50 S 72 50W
San Martín de la Vega, *Spain* **42 E7** 40 13N 3 34W
San Martin de los Andes, *Argentina* **176 B2** 40 10 S 71 20W
San Martín de Valdeiglesias, *Spain* **42 E6** 40 21N 4 24W
San Mateo = Sant Mateu, *Spain* **40 E5** 40 28N 0 10 E
San Mateo, *Agusan del N., Phil.* **81 G5** 8 48N 125 33 E
San Mateo, *Isabela, Phil.* **80 C3** 16 54N 121 33 E
San Mateo, *Spain* **39 B7** 39 3N 1 23 E
San Mateo, *U.S.A.* **160 H4** 37 34N 122 19W
San Matías, *Bolivia* **173 D6** 16 25 S 58 20W
San Matías, G., *Argentina* **176 B4** 41 30 S 64 0W
San Miguel = Linapacan, *Phil.* **81 F2** 11 30N 119 52 E
San Miguel = Sant Miquel, *Spain* **39 B7** 39 3N 1 26 E
San Miguel, *El Salv.* **164 D2** 13 30N 88 12W
San Miguel, *Panama* **164 E4** 8 27N 78 55W
San Miguel, *Lanao del N., Phil.* **81 G5** 9 3N 125 59 E
San Miguel, *Lanao del S., Phil.* **81 G5** 8 13N 124 14 E
San Miguel, *U.S.A.* **160 K6** 35 45N 120 42W
San Miguel, *Venezuela* **168 B4** 9 40N 65 11W
San Miguel →, *Bolivia* **173 C5** 13 52 S 63 56W
San Miguel →, *S. Amer.* **168 C2** 0 25N 76 30W
San Miguel de Huachi, *Bolivia* **172 D4** 15 40 S 67 15W
San Miguel de Tucumán, *Argentina* **174 B2** 26 50 S 65 20W
San Miguel del Monte, *Argentina* **174 D4** 35 23 S 58 50W
San Miguel I., *U.S.A.* **161 L6** 34 2N 120 23W
San Miguel Is., *Phil.* **81 H2** 7 45N 118 28 E
San Miniato, *Italy* **44 E7** 43 41N 10 51 E
San Narciso, *Quezon, Phil.* **80 E4** 13 34N 122 34 E
San Narciso, *Zambales, Phil.* **80 D3** 15 2N 120 3 E
San Nicolás, *Canary Is.* **39 G4** 27 58N 15 47W
San Nicolas, *Phil.* **80 B3** 18 10N 120 36 E
San Nicolás de los Arroyas, *Argentina* **174 C3** 33 25 S 60 10W
San Nicolas I., *U.S.A.* **161 M7** 33 15N 119 30W
San Onofre, *Colombia* **168 B2** 9 44N 75 32W
San Onofre, *U.S.A.* **161 M9** 33 22N 117 34W
San Pablo, *Bolivia* **174 A2** 21 43 S 66 38W
San Pablo, *Isabela, Phil.* **80 C3** 17 27N 121 48 E
San Pablo, *Laguna, Phil.* **80 D3** 14 11N 121 31 E
San Páolo di Civitate, *Italy* **45 G12** 41 44N 15 15 E
San Pascual, *Phil.* **80 E4** 13 8N 122 59 E
San Pedro, *Buenos Aires, Argentina* **175 B5** 26 30 S 54 10W
San Pedro, *Jujuy, Argentina* **174 A3** 24 12 S 64 55W
San Pedro, *Colombia* **168 C3** 4 56N 71 53W
San Pedro, *Ivory C.* **112 E3** 4 50N 6 33W
San Pedro, *Mexico* **162 C2** 23 55N 110 17W
San Pedro, *Peru* **172 C3** 14 49 S 74 5W
San Pedro □, *Paraguay* **174 A4** 24 0 S 57 0W
San Pedro →, *Chihuahua, Mexico* **162 B3** 28 20N 106 10W
San Pedro →, *Michoacan, Mexico* **162 D4** 19 23N 103 51W
San Pedro →, *Nayarit, Mexico* **162 C3** 21 45N 105 30W
San Pedro →, *U.S.A.* **159 K8** 32 59N 110 47W
San Pedro, Pta., *Chile* **174 B1** 25 30 S 70 38W
San Pedro, Sierra de, *Spain* **43 F4** 39 18N 6 40W
San Pedro Channel, *U.S.A.* **161 M8** 33 30N 118 25W
San Pedro de Arimena, *Colombia* **168 C3** 4 37N 71 42W
San Pedro de Atacama, *Chile* **174 A2** 22 55 S 68 15W
San Pedro de Jujuy, *Argentina* **174 A3** 24 12 S 64 55W
San Pedro de las Colonias, *Mexico* **162 B4** 25 50N 102 59W
San Pedro de Lloc, *Peru* **172 B2** 7 15 S 79 28W
San Pedro de Macorís, *Dom. Rep.* **165 C6** 18 30N 69 18W
San Pedro del Norte, *Nic.* **164 D3** 13 4N 84 33W
San Pedro del Paraná, *Paraguay* **174 B4** 26 43 S 56 13W
San Pedro del Pinatar, *Spain* **41 H4** 37 50N 0 50W
San Pedro Mártir, Sierra, *Mexico* **162 A1** 31 0N 115 30W
San Pedro Mixtepec, *Mexico* **163 D5** 16 2N 97 7W
San Pedro Ocampo = Melchor Ocampo, *Mexico* **162 C4** 24 52N 101 40W
San Pedro Sula, *Honduras* **164 C2** 15 30N 88 0W
San Pietro, *Italy* **46 C1** 39 8N 8 17 E
San Pietro Vernótico, *Italy* **47 B11** 40 29N 18 0 E
San Quintín, *Mexico* **162 A1** 30 29N 115 57W
San Rafael, *Argentina* **174 C2** 34 40 S 68 21W
San Rafael, *Calif., U.S.A.* **160 H4** 37 58N 122 32W
San Rafael, *N. Mex., U.S.A.* **159 J10** 35 7N 107 53W
San Rafael, *Venezuela* **168 A3** 10 58N 71 46W
San Rafael Mt., *U.S.A.* **161 L7** 34 41N 119 52W
San Rafael Mts., *U.S.A.* **161 L7** 34 40N 119 50W
San Ramón, *Bolivia* **173 C5** 13 17 S 64 43W
San Ramón, *Peru* **172 C2** 11 8 S 75 20W
San Ramón de la Nueva Orán, *Argentina* **174 A3** 23 10 S 64 20W
San Remo, *Italy* **44 E4** 43 49N 7 46 E
San Román, C., *Venezuela* **168 A3** 12 12N 70 0W
San Roque, *Argentina* **174 B4** 28 25 S 58 45W
San Roque, *Phil.* **80 E5** 12 37N 124 52 E
San Roque, *Spain* **43 J5** 36 17N 5 21W
San Rosendo, *Chile* **174 D1** 37 16 S 72 43W
San Saba, *U.S.A.* **155 K5** 31 12N 98 43W
San Salvador, *Bahamas* **165 B5** 24 0N 74 40W
San Salvador, *El Salv.* **164 D2** 13 40N 89 10W
San Salvador, *Spain* **39 B10** 39 27N 3 11 E
San Salvador de Jujuy, *Argentina* **174 A3** 24 10 S 64 48W
San Salvador I., *Bahamas* **165 B5** 24 0N 74 32W
San Salvo, *Italy* **45 F11** 42 3N 14 44 E
San Sebastián = Donostia-San Sebastián, *Spain* **40 B3** 43 17N 1 58W
San Sebastián, *Argentina* **176 D3** 53 10 S 68 30W

Santiago 277

San Sebastián, Venezuela 168 B4 9 57N 67 11W
San Sebastian de la Gomera, Canary Is. 39 F2 28 5N 17 7W
San Serra, Spain 39 B10 39 43N 3 13 E
San Serverino Marche, Italy 45 E10 43 13N 13 10 E
San Severo, Italy 45 G12 41 41N 15 23 E
San Simeon, U.S.A. 160 K5 35 39N 121 11W
San Simon, U.S.A. 159 K9 32 16N 109 14W
San Stéfano di Cadore, Italy 45 B9 46 34N 12 33 E
San Stino di Livenza, Italy 45 C9 45 44N 12 41 E
San Telmo, Mexico 162 A1 30 58N 116 6W
San Telmo, Spain 39 B9 39 35N 2 21 E
San Teodoro, Phil. 80 E3 13 26N 121 1 E
San Tiburcio, Mexico 162 C4 24 8N 101 32W
San Valentin, Mte., Chile 176 C2 46 30 S 73 30W
San Vicente, Phil. 80 B4 18 30N 122 8 E
San Vicente de Alcántara, Spain 43 F3 39 22N 7 8W
San Vicente de la Barquera, Spain 42 B6 43 23N 4 29W
San Vicente del Caguán, Colombia 168 C3 2 7N 74 46W
San Vincente del Raspeig, Spain 41 G4 38 24N 0 31W
San Vincenzo, Italy 44 E7 43 6N 10 32 E
San Vito, Italy 46 C2 39 26N 9 32 E
San Vito, C., Italy 46 D5 38 11N 12 41 E
San Vito al Tagliamento, Italy 45 C9 45 54N 12 52 E
San Vito Chietino, Italy 45 F11 42 18N 14 27 E
San Vito dei Normanni, Italy 47 B10 40 39N 17 42 E
San Yanaro, Colombia 168 C4 2 47N 69 42W
San Ygnacio, U.S.A. 155 M5 27 3N 99 26W
Saña, Peru 172 B2 6 54 S 79 36W
Sana', Yemen 98 D4 15 27N 44 12 E
Sana →, Bos.-H. 45 C13 45 3N 16 23 E
Sanaba, Burkina Faso 112 C4 12 25N 3 47W
Şanâfîr, Si. Arabia 106 B3 27 56N 34 42 E
Sanaga →, Cameroon 113 E6 3 35N 9 38 E
Sanaloa, Presa, Mexico 162 C3 24 50N 107 20W
Sanām, Si. Arabia 98 B4 23 40N 44 45 E
Sanana, Indonesia 82 B3 2 4 S 125 58 E
Sanand, India 92 H5 22 59N 72 25 E
Sanandaj, Iran 101 E12 35 18N 47 1 E
Sanandita, Bolivia 174 A3 21 40 S 63 45W
Sanary-sur-Mer, France 29 E9 43 7N 5 49 E
Sanāw, Yemen 99 C5 17 50N 51 5 E
Sanawad, India 92 H7 22 11N 76 5 E
Sanbe-San, Japan 70 B4 35 6N 132 38 E
Sancellas, Spain 39 B9 39 39N 2 54 E
Sancergues, France 27 E9 47 10N 2 54 E
Sancerre, France 27 E9 47 20N 2 50 E
Sancerrois, Collines du, France 27 E9 47 20N 2 40 E
Sancha He →, China 76 D6 26 48N 106 7 E
Sanchahe, China 75 B14 44 50N 126 2 E
Sánchez, Dom. Rep. 165 C6 19 15N 69 36W
Sanchor, India 92 G4 24 45N 71 55 E
Sanco Pt., Phil. 81 G6 8 15N 126 27 E
Sancoins, France 27 F9 46 47N 2 55 E
Sancti Spíritus, Cuba 164 B4 21 52N 79 33W
Sancy, Puy de, France 28 C6 45 32N 2 50 E
Sand, Norway 18 E3 59 29N 6 16 E
Sand →, S. Africa 117 C5 22 25 S 30 5 E
Sand Cr. →, U.S.A. 157 E11 39 3N 85 51W
Sand I., U.S.A. 145 K14 21 19N 157 53W
Sand Springs, U.S.A. 155 G6 36 9N 96 7W
Sanda, Japan 71 C7 34 53N 135 14 E
Sandakan, Malaysia 85 A5 5 53N 118 4 E
Sandalwood, Australia 128 C4 34 55 S 140 9 E
Sandan = Sambor, Cambodia 86 F6 12 46N 106 0 E
Sandane, Norway 18 C3 61 46N 6 13 E
Sandanski, Bulgaria 50 E7 41 35N 23 16 E
Sandaré, Mali 112 C2 14 40N 10 15W
Sandared, Sweden 17 G6 57 43N 12 47 E
Sandarne, Sweden 16 C11 61 16N 17 9 E
Sanday, U.K. 22 B6 59 16N 2 31W
Sande, Møre og Romsdal, Norway 18 B2 62 15N 5 27 E
Sande, Sogn og Fjordane, Norway 18 C2 61 20N 5 47 E
Sande, Vestfold, Norway 18 E7 59 36N 10 12 E
Sandefjord, Norway 15 G14 59 10N 10 15 E
Sandeid, Norway 18 E2 59 33N 5 52 E
Sanders, Ariz., U.S.A. 159 J9 35 13N 109 20W
Sanders, Ky., U.S.A. 157 F12 38 40N 84 56W
Sanderson, Fla., U.S.A. 152 E7 30 15N 82 16W
Sanderson, Tex., U.S.A. 155 K3 30 9N 102 24W
Sanderston, Australia 128 C3 34 46 S 139 15 E
Sandersville, U.S.A. 152 C7 32 59N 82 48W
Sandfloegga, Norway 18 E4 59 58N 7 10 E
Sandfly L., Canada 143 B7 55 43N 106 6W
Sandgate, Australia 127 D5 27 18 S 153 3 E
Sandgerði, Iceland 11 C4 64 3N 22 42W
Sandhammaren, C., Sweden 17 J8 55 23N 14 14 E
Sandía, Peru 172 C4 14 10 S 69 30W
Sandıklı, Turkey 49 C12 38 28N 30 17 E
Sandnes, Aust-Agder, Norway 18 F4 58 53N 7 45 E
Sandnes, Rogaland, Norway 15 G11 58 50N 5 45 E
Sandnessjøen, Norway 14 C15 66 2N 12 38 E
Sandoa, Dem. Rep. of the Congo 115 D4 9 41 S 23 0 E
Sandomierz, Poland 55 H8 50 40N 21 43 E
Sândominic, Romania 53 D10 46 35N 25 47 E
Sandona, Colombia 168 C2 1 17N 77 28W
Sandongo, Angola 115 F4 15 30 S 21 28 E
Sandoval, U.S.A. 156 F7 38 37N 89 7W
Sandover →, Australia 126 C2 21 43 S 136 32 E
Sandoy, Færoe Is. 14 F9 61 52N 6 46W
Sandpoint, U.S.A. 158 B5 48 17N 116 33W
Sandray, U.K. 22 E1 56 53N 7 31W
Sandringham, U.K. 20 E8 52 51N 0 31 E
Sandspit, Canada 142 C2 53 14N 131 49W
Sandstone, Australia 125 E2 27 59 S 119 16 E
Sandu, China 76 E6 26 0N 107 52 E
Sandusky, Mich., U.S.A. 140 D3 43 25N 82 50W
Sandusky, Ohio, U.S.A. 150 E2 41 27N 82 42W
Sandusky →, U.S.A. 157 C14 41 27N 83 0W
Sandvig, Sweden 17 J8 55 18N 14 47 E
Sandvika, Norway 18 E7 59 54N 10 31 E
Sandviken, Sweden 16 D10 60 38N 16 46 E
Sandwich, U.S.A. 157 C8 41 39N 88 37W
Sandwich, C., Australia 126 B4 18 14 S 146 18 E
Sandwich B., Canada 141 B8 53 40N 57 15W
Sandwich B., Namibia 116 C1 23 25 S 14 20 E
Sandy, Nev., U.S.A. 161 K11 35 49N 115 36W
Sandy, Oreg., U.S.A. 160 E4 45 24N 122 16W
Sandy, Utah, U.S.A. 158 F8 40 35N 111 50W
Sandy Bight, Australia 125 F3 33 50 S 123 20 E
Sandy C., Queens., Australia 126 C5 24 42 S 153 15 E
Sandy C., Tas., Australia 126 G3 41 25 S 144 45 E
Sandy Cay, Bahamas 165 B4 23 13N 75 18W
Sandy Cr. →, U.S.A. 158 F9 41 51N 109 47W
Sandy L., Canada 140 B1 53 2N 93 0W
Sandy Lake, Canada 140 B1 53 0N 93 15W
Sandy Narrows, Canada 143 B8 55 5N 103 4W
Sandy Springs, U.S.A. 152 B5 33 56N 84 23W
Sanford, Fla., U.S.A. 149 L5 28 48N 81 16W
Sanford, Maine, U.S.A. 151 C14 43 27N 70 47W
Sanford, N.C., U.S.A. 149 H6 35 29N 79 10W
Sanford →, Australia 125 E2 27 22 S 115 53 E
Sanford, Mt., U.S.A. 138 B5 62 13N 144 8W
Sang-i-Masha, Afghan. 92 C2 33 8N 67 27 E
Sanga, Mozam. 119 E4 12 22 S 35 21 E
Sanga →, Congo 114 C3 1 5 S 17 0 E
Sangamner, India 94 E2 19 37N 74 15 E
Sangamon →, U.S.A. 156 D6 40 7N 90 20W
Sangar, Afghan. 92 C1 32 56N 65 30 E
Sangar, Russia 65 C13 64 2N 127 31 E
Sangar Sarai, Afghan. 92 B4 34 27N 70 35 E
Sangasangadalam, Indonesia 85 C5 0 36 S 117 13 E
Sangatte, France 27 B8 50 56N 1 44 E
Sangay, Ecuador 168 D2 2 0 S 78 20W
Sange, Dem. Rep. of the Congo 118 D2 6 58 S 28 21 E
Sangeang, Indonesia 82 C1 8 12 S 119 6 E
Sângeorz-Băi, Romania 53 C9 47 22N 24 41 E
Sanger, U.S.A. 160 J7 36 42N 119 33W
Sângera, Moldova 53 D13 46 55N 28 58 E
Sangerhausen, Germany 30 D7 51 28N 11 18 E
Sanggan He →, China 74 E9 38 12N 117 15 E
Sanggau, Indonesia 85 B4 0 5N 110 30 E
Sangihe, Kepulauan, Indonesia 82 A3 3 0N 126 0 E
Sangihe, Pulau, Indonesia 82 A3 3 45N 125 30 E
Sangju, S. Korea 75 F15 36 25N 128 10 E
Sangkapura, Indonesia 85 D4 5 52 S 112 40 E
Sangkhla, Thailand 86 E2 14 57N 98 28 E
Sangli, India 94 F2 16 55N 74 33 E
Sangmélima, Cameroon 113 E7 2 57N 12 1 E
Sangpang Bum, Burma 90 B5 26 30N 95 50 E
Sangre de Cristo Mts., U.S.A. 155 G2 37 0N 105 0W
Sangro →, Italy 45 F11 42 14N 14 32 E
Sangudo, Canada 142 C6 53 50N 114 54W
Sangue →, Brazil 173 C6 11 1 S 58 39W
Sangüesa, Spain 40 C3 42 37N 1 17W
Sanguinaires, Îs., France 29 G12 41 51N 8 36 E
Sangzhi, China 77 C8 29 25N 110 12 E
Sanhala, Ivory C. 112 C3 10 3N 6 51W
Sanibel I., U.S.A. 153 J7 26 26N 82 6W
Sanje, Uganda 118 C3 0 49 S 31 30 E
Sanjiang, China 76 E7 25 48N 109 37 E
Sanjo, Japan 68 F9 37 37N 138 57 E
Sankarankovil, India 95 K3 9 10N 77 35 E
Sankeshwar, India 95 F2 16 23N 74 32 E
Sankosh →, India 90 B2 26 24N 89 47 E
Sankt Andrä, Austria 34 E7 46 46N 14 50 E
Sankt Antönien, Switz. 33 C9 46 58N 9 48 E
Sankt Augustin, Germany 30 E3 50 45N 7 10 E
Sankt Blasien, Germany 31 H4 47 47N 8 7 E
Sankt Gallen, Switz. 33 B8 47 26N 9 22 E
Sankt Gallen □, Switz. 33 B8 47 25N 9 22 E
Sankt Goar, Germany 31 E3 50 12N 7 43 E
Sankt Ingbert, Germany 31 F3 49 16N 7 6 E
Sankt Johann im Pongau, Austria 34 D6 47 22N 13 12 E
Sankt Johann in Tirol, Austria 34 D5 47 30N 12 25 E
Sankt Margrethen, Switz. 33 B9 47 28N 9 37 E
Sankt Moritz, Switz. 33 D9 46 30N 9 50 E
Sankt-Peterburg, Russia 58 C6 59 55N 30 20 E
Sankt Pölten, Austria 34 C8 48 12N 15 38 E
Sankt Ulrich = Ortisei, Italy 45 B8 46 34N 11 40 E
Sankt Valentin, Austria 34 C7 48 11N 14 33 E
Sankt Veit an der Glan, Austria 34 E7 46 47N 14 22 E
Sankt Wendel, Germany 31 F3 49 27N 7 9 E
Sankt Wolfgang, Austria 34 D6 47 43N 13 27 E
Sankuru →, Dem. Rep. of the Congo 115 C4 4 17 S 20 25 E
Sanliurfa, Turkey 101 D8 37 12N 38 50 E
Sanlúcar de Barrameda, Spain 43 J4 36 46N 6 21W
Sanluri, Italy 46 C1 39 34N 8 54 E
Sânmartin, Romania 53 D10 46 19N 25 58 E
Sanmenxia, China 74 G6 34 47N 111 12 E
Sanming, China 77 D11 26 15N 117 40 E
Sannan, Japan 71 B7 35 2N 135 1 E
Sannaspos, S. Africa 116 D4 29 6 S 26 34 E
Sannicandro Gargánico, Italy 45 G12 41 50N 15 34 E
Sânnicolau Mare, Romania 52 D5 46 5N 20 39 E
Sannieshof, S. Africa 116 D4 26 30 S 25 47 E
Sannîn, J., Lebanon 103 B4 33 57N 35 52 E
Sano, Japan 71 A11 36 19N 139 35 E
Sanok, Poland 55 J9 49 35N 22 10 E
Sanquhar, U.K. 22 F5 55 22N 3 54W
Sansanding Dam, Mali 112 C3 13 48N 6 0W
Sansepolcro, Italy 45 E9 43 34N 12 8 E
Sansha, China 77 D13 26 58N 120 12 E
Sanshui, China 77 F9 23 10N 112 56 E
Sanski Most, Bos.-H. 45 D13 44 46N 16 40 E
Sansui, China 76 D7 26 58N 108 39 E
Sant Antoni Abat, Spain 39 C7 38 59N 1 19 E
Sant Boi de Llobregat, Spain 40 D7 41 21N 2 2 E
Sant Carles de la Ràpita, Spain 40 E5 40 37N 0 35 E
Sant Celoni, Spain 40 D7 41 42N 2 30 E
Sant' Egídio alla Vibrata, Italy 45 F10 42 49N 13 42 E
Sant Feliu de Guíxols, Spain 40 D8 41 45N 3 1 E
Sant Feliu de Llobregat, Spain 40 D7 41 23N 2 2 E
Sant Francesc de Formentera, Spain 39 C7 38 42N 1 26 E
Sant Joan Baptista, Spain 39 B8 39 5N 1 31 E
Sant Jordi, G. de, Spain 40 E6 40 53N 1 2 E
Sant Llorenç de Morunys, Spain 40 C6 42 8N 1 35 E
Sant Mateu, Spain 40 E5 40 28N 0 10 E
Sant Miquel, Spain 39 B7 39 3N 1 26 E
Santa, Peru 172 B2 8 59 S 78 40W
Sant' Ágata Militello, Italy 47 D7 38 2N 14 8 E
Santa Ana, Beni, Bolivia 173 C4 13 50 S 65 40W
Santa Ana, Santa Cruz, Bolivia 173 D6 18 43 S 58 44W
Santa Ana, Santa Cruz, Bolivia 173 D5 16 37 S 60 43W
Santa Ana, Ecuador 168 D1 1 16 S 80 20W
Santa Ana, El Salv. 164 D2 14 0N 89 31W
Santa Ana, Mexico 162 A2 30 31N 111 8W
Santa Ana, Phil. 80 B4 18 28N 122 20 E
Santa Ana, U.S.A. 161 M9 33 46N 117 52W
Santa Ana →, Venezuela 168 B3 9 30N 71 57W
Sant' Ángelo Lodigiano, Italy 44 C6 45 14N 9 25 E
Sant'Antíoco, Italy 46 C1 39 4N 8 27 E
Santa Bárbara, Colombia 168 B2 5 53N 75 35W
Santa Barbara, Honduras 164 D2 14 53N 88 14W
Santa Bárbara, Mexico 162 B3 26 48N 105 50W
Santa Barbara, Phil. 81 F4 10 50N 122 32 E
Santa Bárbara, Spain 40 E5 40 42N 0 29 E
Santa Barbara, U.S.A. 161 L7 34 25N 119 42W
Santa Bárbara, Venezuela 168 B3 7 47N 71 10W
Santa Bárbara, Mt., Spain 41 H2 37 23N 2 50W
Santa Barbara Channel, U.S.A. 161 L7 34 15N 120 0W
Santa Barbara I., U.S.A. 161 M7 33 29N 119 2W
Santa Catalina, Colombia 168 A2 10 36N 75 17W
Santa Catalina, Mexico 162 B2 25 40N 110 50W
Santa Catalina, Gulf of, U.S.A. 161 N9 33 10N 117 50W
Santa Catalina I., U.S.A. 161 M8 33 23N 118 25W
Santa Catarina □, Brazil 175 B6 27 25 S 48 30W
Santa Catarina, I. de, Brazil 175 B6 27 30 S 48 40W
Santa Caterina di Pittinuri, Italy 46 B1 40 6N 8 27 E
Santa Caterina Villarmosa, Italy 47 E7 37 35N 14 2 E
Santa Cecília, Brazil 175 B5 26 56 S 50 18W
Santa Clara, Cuba 164 B4 22 20N 80 0W
Santa Clara, Calif., U.S.A. 160 H5 37 21N 121 57W
Santa Clara, Utah, U.S.A. 159 H7 37 8N 113 39W
Santa Clara de Olimar, Uruguay 175 C5 32 50 S 54 54W
Santa Clotilde, Peru 168 D3 2 33 S 73 45W
Santa Coloma de Farners, Spain 40 D7 41 50N 2 39 E
Santa Coloma de Gramenet, Spain 40 D7 41 27N 2 13 E
Santa Comba, Spain 42 B2 43 2N 8 49W
Santa Croce Camerina, Italy 47 F7 36 50N 14 31 E
Santa Croce di Magliano, Italy 45 G11 41 42N 14 59 E
Santa Cruz, Argentina 176 D3 50 0 S 68 32W
Santa Cruz, Bolivia 173 D5 17 43 S 63 10W
Santa Cruz, Brazil 170 C4 6 13 S 36 1W
Santa Cruz, Chile 174 C1 34 38 S 71 27W
Santa Cruz, Costa Rica 164 D2 10 15N 85 35W
Santa Cruz, Madeira 39 D3 32 42N 16 46W
Santa Cruz, Peru 172 B2 5 40 S 75 56W
Santa Cruz, Davao del S., Phil. 81 H5 6 50N 125 25 E
Santa Cruz, Laguna, Phil. 80 D3 14 20N 121 24 E
Santa Cruz, Marinduque, Phil. 80 E4 13 28N 122 2 E
Santa Cruz, U.S.A. 160 J4 36 58N 122 1W
Santa Cruz, Venezuela 169 B5 8 3N 64 27W
Santa Cruz □, Argentina 176 C3 49 0 S 70 0W
Santa Cruz □, Bolivia 173 D5 17 43 S 63 10W
Santa Cruz →, Argentina 176 D3 50 10 S 68 20W
Santa Cruz Cabrália, Brazil 171 E4 16 17 S 39 2W
Santa Cruz de la Palma, Canary Is. 39 F2 28 41N 17 46W
Santa Cruz de Mudela, Spain 43 G7 38 39N 3 28W
Santa Cruz de Tenerife, Canary Is. 39 F3 28 28N 16 15W
Santa Cruz del Norte, Cuba 164 B3 23 9N 81 55W
Santa Cruz del Retamar, Spain 42 E6 40 8N 4 14W
Santa Cruz del Sur, Cuba 164 B4 20 44N 78 0W
Santa Cruz do Rio Pardo, Brazil 175 A6 22 54 S 49 37W
Santa Cruz do Sul, Brazil 175 B5 29 42 S 52 25W
Santa Cruz I., U.S.A. 161 M7 34 1N 119 43W
Santa Cruz Is., Solomon Is. 134 J8 10 30 S 166 0 E
Santa Domingo, Cay, Bahamas 164 B4 21 25N 75 15W
Santa Elena, Argentina 174 C4 30 58 S 59 47W
Santa Elena, Ecuador 168 D1 2 16 S 80 52W
Santa Elena, C., Costa Rica 164 D2 10 54N 85 56W
Sant' Eufémia, G. di, Italy 47 D9 38 51N 16 4 E
Santa Eugenia, Pta., Mexico 162 B1 27 50N 115 5W
Santa Eulàlia des Riu, Spain 39 C8 38 59N 1 32 E
Santa Fe, Argentina 174 C3 31 35 S 60 41W
Santa Fe, Nueva Vizcaya, Phil. 80 C3 16 10N 120 57 E
Santa Fe, Tablas, Phil. 80 E4 12 10N 122 0 E
Santa Fe, Spain 43 H7 37 11N 3 43W
Santa Fe, U.S.A. 159 J11 35 41N 105 57W
Santa Fé □, Argentina 174 C3 31 50 S 60 55W
Santa Fe, L., U.S.A. 152 F7 29 45N 82 5W
Santa Filomena, Brazil 170 C2 9 6 S 45 50W
Santa Fiora, Italy 45 F8 42 50N 11 35 E
Santa Galdana, Spain 39 B10 39 56N 3 58 E
Santa Gertrudis, Spain 39 C7 39 0N 1 26 E
Santa Giustina, Italy 45 B9 46 10N 12 5 E
Santa Helena, Brazil 170 B2 2 14 S 45 18W
Santa Helena de Goiás, Brazil 171 E1 17 53 S 50 35W
Santa Inês, Brazil 171 D4 13 17 S 39 48W
Santa Inés, Baleares, Spain 39 B7 39 3N 1 21 E
Santa Inés, Extremadura, Spain 43 G5 38 32N 5 37W
Santa Inés, I., Chile 176 D2 54 0 S 73 0W
Santa Isabel = Rey Malabo, Eq. Guin. 113 E6 3 45N 8 50 E
Santa Isabel, Argentina 174 D2 36 10 S 66 54W
Santa Isabel, Brazil 171 D1 11 45 S 51 30W
Santa Isabel, Solomon Is. 133 M10 8 0 S 159 0 E
Santa Isabel, Pico, Eq. Guin. 113 E6 3 36N 8 49 E
Santa Isabel do Araguaia, Brazil 170 C2 6 7 S 48 19W
Santa Isabel do Morro, Brazil 171 D1 11 34 S 50 40W
Santa Lucía, Corrientes, Argentina 174 B4 28 58 S 59 5W
Santa Lucía, San Juan, Argentina 174 C2 31 30 S 68 30W
Santa Lucia, Phil. 80 C3 17 7N 120 27 E
Santa Lucía, Spain 41 H4 37 35N 0 58W
Santa Lucía, Uruguay 174 C4 34 27 S 56 24W
Santa Lucia Range, U.S.A. 160 K5 36 0N 121 20W
Santa Magdalena, I., Mexico 162 C2 24 40N 112 15W
Santa Margarita, Argentina 174 D3 38 28 S 61 35W
Santa Margarita, Mexico 162 C2 24 30N 111 50W
Santa Margarita, Spain 39 B10 39 42N 3 6 E
Santa Margarita, U.S.A. 160 K6 35 23N 120 37W
Santa Margarita →, U.S.A. 161 M9 33 13N 117 23W
Santa Margherita, Italy 46 D1 38 58N 8 58 E
Santa Margherita Ligure, Italy 44 D6 44 20N 9 11 E
Santa María, Argentina 174 B2 26 40 S 66 0W
Santa Maria, Brazil 175 B5 29 40 S 53 48W
Santa Maria, Ilocos S., Phil. 80 C3 17 22N 120 29 E
Santa Maria, Isabela, Phil. 80 C3 17 28N 121 45 E
Santa Maria, Spain 39 B9 39 38N 2 47 E
Santa Maria, Switz. 33 C10 46 36N 10 25 E
Santa Maria, U.S.A. 161 L6 34 57N 120 26W
Santa María →, Mexico 162 A3 31 0N 107 14W
Santa María, B. de, Mexico 162 B3 25 10N 108 40W
Santa Maria, C. de, Portugal 43 J3 36 58N 7 53W
Santa Maria Cápua Vétere, Italy 47 A7 41 5N 14 15 E
Santa Maria da Feira, Portugal 42 E2 40 55N 8 35W
Santa Maria da Vitória, Brazil 171 D3 13 24 S 44 12W
Santa María de Ipire, Venezuela 169 B4 8 49N 65 19W
Santa Maria di Léuca, C., Italy 47 C11 39 47N 18 22 E
Santa Maria do Suaçuí, Brazil 171 E3 18 12 S 42 25W
Santa Maria dos Marmelos, Brazil 173 B5 6 7 S 61 51W
Santa María la Real de Nieva, Spain 42 D6 41 4N 4 24W
Santa Marinella, Italy 45 F8 42 2N 11 52 E
Santa Marta, Colombia 168 A3 11 15N 74 13W
Santa Marta, Sierra Nevada de, Colombia 168 A3 10 55N 73 50W
Santa Marta de Tormes, Spain 42 E5 40 57N 5 38W
Santa Marta Grande, C., Brazil 175 B6 28 43 S 48 50W
Santa Marta Ortigueira, Ría de, Spain 42 B3 43 44N 7 45W
Santa Monica, U.S.A. 161 M8 34 1N 118 29W
Santa Olalla, Huelva, Spain 43 H4 37 54N 6 14W
Santa Olalla, Toledo, Spain 42 E6 40 2N 4 25W
Santa Pola, Spain 41 G4 38 13N 0 35W
Santa Ponsa, Spain 39 B9 39 30N 2 28 E
Santa Quitéria, Brazil 170 B3 4 20 S 40 10W
Santa Rita, U.S.A. 159 K10 32 48N 108 4W
Santa Rita, Guarico, Venezuela 168 B4 8 8N 66 16W
Santa Rita, Zulia, Venezuela 168 A3 10 32N 71 32W
Santa Rita do Araquaia, Brazil 173 D7 17 20 S 53 12W
Santa Rosa, La Pampa, Argentina 174 D3 36 40 S 64 17W
Santa Rosa, San Luis, Argentina 174 C2 32 21 S 65 10W
Santa Rosa, Bolivia 172 C4 10 36 S 67 20W
Santa Rosa, Brazil 175 B5 27 52 S 54 29W
Santa Rosa, Colombia 168 C4 3 32N 69 48W
Santa Rosa, Ecuador 168 D2 3 27 S 79 58W
Santa Rosa, Peru 172 C3 14 30 S 70 50W
Santa Rosa, Phil. 80 D3 15 25N 120 57 E
Santa Rosa, Calif., U.S.A. 160 G4 38 26N 122 43W
Santa Rosa, N. Mex., U.S.A. 155 H2 34 57N 104 41W
Santa Rosa, Venezuela 168 C4 1 29N 66 55W
Santa Rosa Beach, U.S.A. 152 E3 30 22N 86 14W
Santa Rosa de Cabal, Colombia 168 C2 4 52N 75 38W
Santa Rosa de Copán, Honduras 164 D2 14 47N 88 46W
Santa Rosa de Osos, Colombia 168 B2 6 39N 75 28W
Santa Rosa de Río Primero, Argentina 174 C3 31 8 S 63 20W
Santa Rosa de Viterbo, Colombia 168 B3 5 53N 72 59W
Santa Rosa del Palmar, Bolivia 173 D5 16 54 S 62 24W
Santa Rosa I., Calif., U.S.A. 161 M6 33 58N 120 6W
Santa Rosa I., Fla., U.S.A. 149 K2 30 20N 86 50W
Santa Rosa Range, U.S.A. 158 F5 41 45N 117 40W
Santa Rosalía, Mexico 162 B2 27 20N 112 20W
Santa Sylvina, Argentina 174 B3 27 50 S 61 10W
Santa Tecla = Nueva San Salvador, El Salv. 164 D2 13 40N 89 18W
Santa Teresa, Argentina 174 C3 33 25 S 60 47W
Santa Teresa, Brazil 171 E3 19 55 S 40 36W
Santa Teresa, Mexico 163 B5 25 17N 97 51W
Santa Teresa, Venezuela 169 C5 4 43N 61 4W
Santa Teresa di Riva, Italy 47 E8 37 57N 15 22 E
Santa Teresa Gallura, Italy 46 A2 41 14N 9 11 E
Santa Uxía, Spain 42 C2 42 36N 8 58W
Santa Vitória, Brazil 171 E1 18 50 S 50 8W
Santa Vitória do Palmar, Brazil 175 C5 33 32 S 53 25W
Santa Ynez →, U.S.A. 161 L6 35 41N 120 36W
Santa Ynez Mts., U.S.A. 161 L6 34 30N 120 0W
Santa Ysabel, U.S.A. 161 M10 33 7N 116 40W
Santadi, Italy 46 C1 39 5N 8 43 E
Santaella, Spain 43 H6 37 34N 4 51W
Santahar, Bangla. 90 C2 24 48N 88 59 E
Santai, China 76 B5 31 5N 104 58 E
Santaluz, Brazil 170 D4 11 15 S 39 22W
Santana, Brazil 171 D3 13 2 S 44 5W
Santana, Madeira 39 D3 32 48N 16 52W
Sântana, Romania 52 D6 46 20N 21 30 E
Santana, Coxilha de, Brazil 175 C4 30 50 S 55 35W
Santana do Ipanema, Brazil 170 C4 9 22 S 37 14W
Santana do Livramento, Brazil 175 C4 30 55 S 55 30W
Santander, Colombia 168 C2 3 1N 76 28W
Santander, Phil. 81 G4 9 25N 123 20 E
Santander, Spain 42 B7 43 27N 3 51W
Santander Jiménez, Mexico 163 C5 24 11N 98 29W
Santanyí, Spain 39 B10 39 20N 3 5 E
Santaquin, U.S.A. 158 G8 39 59N 111 47W
Santarcángelo di Romagna, Italy 45 D9 44 4N 12 26 E
Santarém, Brazil 169 D7 2 25 S 54 42W
Santarém, Portugal 43 F2 39 12N 8 42W
Santarém □, Portugal 43 F2 39 10N 8 40W
Santaren Channel, W. Indies 164 B4 24 0N 79 30W
Santee, U.S.A. 161 N10 32 50N 116 58W
Santéramo in Colle, Italy 47 B9 40 49N 16 45 E
Santerno →, Italy 45 D8 44 34N 11 58 E
Santhià, Italy 44 C5 45 22N 8 10 E
Santiago, Bolivia 173 D6 18 19 S 59 34W
Santiago, Brazil 175 B5 29 11 S 54 52W
Santiago, C. Verde Is. 8 G6 15 10N 23 30W
Santiago, Chile 174 C1 33 24 S 70 40W
Santiago, Panama 164 E3 8 0N 81 0W
Santiago, Peru 172 C2 14 11 S 75 43W
Santiago, Ilocos S., Phil. 80 C3 17 18N 120 27 E
Santiago, Isabela, Phil. 80 C3 16 41N 121 33 E
Santiago □, Chile 174 C1 33 30 S 70 50W

278 Santiago

Santiago →, Mexico 136 G9 25 11N 105 26W
Santiago →, Peru 168 D2 4 27 S 77 38W
Santiago, C., Chile 176 D1 50 46 S 75 27W
Santiago, Punta de, Eq. Guin. 113 E6 3 12N 8 40 E
Santiago, Serranía de, Bolivia 173 D6 18 25 S 59 25W
Santiago de Chuco, Peru 172 B2 8 9 S 78 11W
Santiago de Compostela, Spain 42 C2 42 52N 8 37W
Santiago de Cuba, Cuba 164 C4 20 0N 75 49W
Santiago de los Cabelleros, Dom. Rep. 165 C5 19 30N 70 40W
Santiago del Estero, Argentina 174 B3 27 50 S 64 15W
Santiago del Estero □, Argentina 174 B3 27 40 S 63 15W
Santiago del Teide, Canary Is. 39 F3 28 17N 16 48W
Santiago do Cacém, Portugal 43 G2 38 1N 8 42W
Santiago Ixcuintla, Mexico 162 C3 21 50N 105 11W
Santiago Papasquiaro, Mexico 162 C3 25 0N 105 20W
Santiaguillo, L. de, Mexico 162 C4 24 50N 104 50W
Santillana, Spain 42 B6 43 24N 4 6W
Säntis, Switz. 33 B8 47 15N 9 22 E
Santisteban del Puerto, Spain 43 G7 38 17N 3 15W
Santo, Vanuatu 133 E5 15 27 S 167 10 E
Santo →, Peru 172 B2 8 56 S 78 37W
Santo Amaro, Brazil 171 D4 12 30 S 38 43W
Santo Anastácio, Brazil 175 A5 21 58 S 51 39W
Santo André, Brazil 175 A6 23 39 S 46 29W
Santo Ângelo, Brazil 175 B5 28 15 S 54 15W
Santo Antonio, Brazil 173 D6 15 50 S 56 0W
Santo Antônio de Jesus, Brazil 171 D4 12 58 S 39 16W
Santo Antônio do Içá, Brazil 168 D4 3 5 S 67 57W
Santo Antônio do Leverger, Brazil 173 D6 15 52 S 56 5W
Santo Corazón, Bolivia 173 D6 18 0 S 58 45W
Santo Domingo, Dom. Rep. 165 C6 18 30N 69 59W
Santo Domingo, Baja Calif., Mexico 162 A1 30 43N 116 2W
Santo Domingo, Baja Calif. S., Mexico 162 B2 25 32N 112 2W
Santo Domingo, Nic. 164 D3 12 14N 84 59W
Santo Domingo de la Calzada, Spain 40 C2 42 26N 2 57W
Santo Domingo de los Colorados, Ecuador 168 D2 0 15 S 79 9W
Santo Nino I., Phil. 81 F5 11 55N 124 27 E
Santo Stéfano di Camastro, Italy 47 D7 38 1N 14 22 E
Santo Tirso, Portugal 42 D2 41 21N 8 28W
Santo Tomás, Mexico 162 A1 31 33N 116 24W
Santo Tomás, Peru 172 C3 14 26 S 72 8W
Santo Tomé, Argentina 175 B4 28 40 S 56 5W
Santo Tomé de Guayana = Ciudad Guayana, Venezuela 169 B5 8 0N 62 30W
Santol, Phil. 80 C3 16 47N 120 27 E
Santomera, Spain 41 G3 38 4N 1 3W
Santoña, Spain 42 B7 43 29N 3 27W
Santoríni = Thíra, Greece 49 E7 36 23N 25 27 E
Santos, Brazil 175 A6 24 0 S 46 20W
Santos, Sierra de los, Spain 43 G5 38 7N 5 12W
Santos Dumont, Brazil 171 F3 22 55 S 43 10W
Sanxenxo, Spain 42 C2 42 24N 8 49W
San'yō, Japan 70 C3 34 2N 131 5 E
Sanyuan, China 74 G5 34 35N 108 58 E
Sanyuki-Sammyaku, Japan 70 C6 34 5N 134 0 E
Sanza Pombo, Angola 115 D3 7 18 S 15 56 E
São Anastácio, Brazil 175 A5 22 0 S 51 40W
São Bartolomeu de Messines, Portugal 43 H2 37 15N 8 17W
São Benedito, Brazil 170 B3 4 3 S 40 53W
São Bento, Brazil 170 B3 2 42 S 44 50W
São Bento do Norte, Brazil 170 C4 5 4 S 36 2W
São Bernardo do Campo, Brazil 171 F2 23 45 S 46 34W
São Borja, Brazil 175 B4 28 39 S 56 0W
São Brás de Alportel, Portugal 43 H3 37 8N 7 37W
São Caitano, Brazil 170 C4 8 21 S 36 6W
São Carlos, Brazil 175 A6 22 0 S 47 50W
São Cristóvão, Brazil 170 D4 11 1 S 37 15W
São Domingos, Brazil 171 D2 13 25 S 46 19W
São Domingos do Maranhão, Brazil 170 C3 5 42 S 44 22W
São Félix, Brazil 171 D1 11 36 S 50 39W
São Francisco, Brazil 171 E3 16 0 S 44 50W
São Francisco →, Brazil 170 D4 10 30 S 36 24W
São Francisco do Maranhão, Brazil 170 C3 6 15 S 42 52W
São Francisco do Sul, Brazil 175 B6 26 15 S 48 36W
São Gabriel, Brazil 175 C5 30 20 S 54 20W
São Gabriel da Palha, Brazil 171 E3 18 47 S 40 39W
São Gonçalo, Brazil 171 F3 22 48 S 43 5W
São Gotardo, Brazil 171 E2 19 19 S 46 3W
Sao Hill, Tanzania 119 D4 8 20 S 35 12 E
São João da Boa Vista, Brazil 175 A6 22 0 S 46 52W
São João da Madeira, Portugal 42 E2 40 54N 8 30W
São João da Pesqueira, Portugal 42 D3 41 8N 7 24W
São João da Ponte, Brazil 171 E3 15 56 S 44 1W
São João del Rei, Brazil 171 F3 21 8 S 44 15W
São João do Araguaia, Brazil 170 C2 5 23 S 48 46W
São João do Paraíso, Brazil 171 E3 15 19 S 42 1W
São João do Piauí, Brazil 170 C3 8 21 S 42 15W
São João dos Patos, Brazil 170 C3 6 30 S 43 42W
São Joaquim da Barra, Brazil 171 F2 20 35 S 47 53W
São Jorge, Pta. de, Madeira 39 D3 32 50N 16 53W
São José, B. de, Brazil 170 B3 2 38 S 44 4W
São José da Laje, Brazil 170 C4 9 1 S 36 3W
São José de Mipibu, Brazil 170 C4 6 5 S 35 15W
São José do Peixe, Brazil 170 C3 7 24 S 42 34W
São José do Rio Prêto, Brazil 175 A6 20 50 S 49 20W
São José dos Campos, Brazil 175 A6 23 7 S 45 52W
São Leopoldo, Brazil 175 B5 29 50 S 51 10W
São Lourenço, Brazil 171 F2 22 7 S 45 3W
São Lourenço →, Brazil 173 D6 17 53 S 57 27W
São Lourenço, Pantanal do, Brazil 173 D6 17 30 S 56 20W
São Lourenço, Pta. de, Madeira 39 D3 32 44N 16 39W
São Luís, Brazil 170 B3 2 39 S 44 15W
São Luís do Curu, Brazil 170 B4 3 40 S 39 14W
São Luís Gonzaga, Brazil 175 B5 28 25 S 55 0W
São Marcos →, Brazil 171 E2 18 15 S 47 37W
São Marcos, B. de, Brazil 170 B3 2 0 S 44 0W
São Martinho da Cortiça, Portugal 42 E2 40 18N 8 8W
São Mateus, Brazil 171 E4 18 44 S 39 50W
São Mateus →, Brazil 171 E4 18 35 S 39 44W
São Miguel do Araguaia, Brazil 171 D1 13 19 S 50 13W
São Miguel dos Campos, Brazil 170 C4 9 47 S 36 5W
São Nicolau →, Brazil 170 C3 5 45 S 42 2W
São Paulo, Brazil 175 A6 23 32 S 46 37W
São Paulo □, Brazil 175 A6 22 0 S 49 0W
São Paulo, I., Atl. Oc. 8 H5 0 50N 31 40W
São Paulo de Olivença, Brazil 168 D4 3 27 S 68 48W
São Pedro do Sul, Portugal 42 E2 40 46N 8 4W
São Rafael, Brazil 170 C4 5 47 S 36 55W
São Raimundo das Mangabeiras, Brazil 170 C2 7 1 S 45 29W
São Raimundo Nonato, Brazil 170 C3 9 1 S 42 42W
São Romão, Brazil 171 E2 16 22 S 45 4W
São Roque, Madeira 39 D3 32 46N 16 48W
São Roque, C. de, Brazil 170 C4 5 30 S 35 16W
São Sebastião, I. de, Brazil 175 A6 23 50 S 45 18W
São Sebastião do Paraíso, Brazil 175 A6 20 54 S 46 59W
São Simão, Brazil 171 E1 18 56 S 50 30W
São Teotónio, Portugal 43 H2 37 30N 8 42W
São Tomé, Atl. Oc. 115 G6 0 10N 6 39 E
São Tomé, Brazil 170 C4 5 58 S 36 4W
São Tomé, C. de, Brazil 171 F3 22 0 S 40 59W
São Tomé & Principe ■, Africa 115 G6 0 12N 6 39 E
São Vicente, Brazil 175 A6 23 57 S 46 23W
São Vicente, C. Verde Is. 8 G6 18 0N 26 1W
São Vicente, Madeira 39 D2 32 48N 17 3W
São Vicente, C. de, Portugal 43 H1 37 0N 9 0W
Saona, I., Dom. Rep. 165 C6 18 10N 68 40W
Saône →, France 27 G11 45 44N 4 50 E
Saône-et-Loire □, France 27 F11 46 30N 4 50 E
Saonek, Indonesia 83 B4 0 22 S 130 55 E
Saoura, O. →, Algeria 111 C4 29 0N 0 55W
Sápai, Greece 51 E9 41 2N 25 43 E
Sapanca, Turkey 100 B4 40 41N 30 16 E
Sapão →, Brazil 170 D2 11 1 S 45 32W
Saparua, Indonesia 82 B3 3 33 S 128 40 E
Sapé, Brazil 170 C4 7 6 S 35 13W
Sape, Indonesia 82 C1 8 34 S 118 59 E
Sapele, Nigeria 113 D6 5 50N 5 40 E
Sapelo I., U.S.A. 152 D8 31 25N 81 12W
Sapelo Island, U.S.A. 152 D8 31 23N 81 17W
Sapelo Sound, U.S.A. 152 D8 31 30N 81 10W
Şaphane, Turkey 49 B11 39 1N 29 13 E
Sapiéntza, Greece 48 E3 36 45N 21 43 E
Sapone, Burkina Faso 113 C4 12 3N 1 35W
Saposoa, Peru 172 B2 6 55 S 76 45W
Sapozhok, Russia 60 D5 53 59N 40 41 E
Sappho, U.S.A. 160 B2 48 4N 124 16W
Sapporo, Japan 68 C10 43 0N 141 21 E
Sapri, Italy 47 B8 40 4N 15 38 E
Sapu Grande, Phil. 81 J5 5 55N 125 16 E
Sapudi, Indonesia 85 D4 7 6 S 114 20 E
Sapulpa, U.S.A. 155 H7 35 59N 96 5W
Saqqez, Iran 101 D12 36 15N 46 20 E
Sar Dasht, Iran 97 C6 32 32N 48 52 E
Sar-e Pol, Afghan. 91 A2 36 10N 66 0 E
Sar-e Pol □, Afghan. 91 A2 36 20N 65 50 E
Sar Gachīneh, Iran 97 D6 30 31N 51 31 E
Sar Planina, Macedonia 50 E4 42 0N 21 0 E
Sara, Burkina Faso 112 C4 11 40N 3 53W
Sara, Phil. 81 F4 11 16N 123 1 E
Sara Buri = Saraburi, Thailand 86 E3 14 30N 100 55 E
Sarāb, Iran 101 D12 37 55N 47 40 E
Sarab, Yemen 99 D5 14 51N 48 31 E
Sarabadi, Iraq 96 C5 33 1N 44 48 E
Saraburi, Thailand 86 E3 14 30N 100 55 E
Saragossa = Zaragoza, Spain 40 D4 41 39N 0 53W
Saraguro, Ecuador 168 D2 3 35 S 79 16W
Saraipali, India 94 D6 21 20N 82 59 E
Saraiu, Romania 53 F13 44 43N 28 10 E
Sarajevo, Bos.-H. 52 G3 43 52N 18 26 E
Saraktash, Russia 62 F6 51 47N 56 22 E
Saramacca □, Surinam 169 C6 5 0N 56 0W
Saramacca →, Surinam 169 B6 5 50N 55 55W
Saramati, Burma 90 C5 25 44N 95 2 E
Saran, Gunung, Indonesia 85 C4 0 30 S 111 25 E
Saranac, U.S.A. 157 B11 42 56N 85 13W
Saranac Lake, U.S.A. 151 B10 44 20N 74 8W
Saranda, Tanzania 118 D3 5 45 S 34 59 E
Sarandë, Albania 50 G3 39 52N 19 55 E
Sarandí del Yi, Uruguay 175 C4 33 18 S 55 38W
Sarandí Grande, Uruguay 174 C4 33 44 S 56 20W
Sarangani B., Phil. 81 J5 6 0N 125 13 E
Sarangani Is., Phil. 81 J5 5 25N 125 25 E
Sarangarh, India 94 D6 21 30N 83 5 E
Saransk, Russia 60 C7 54 10N 45 10 E
Sarapul, Russia 62 C4 56 28N 53 48 E
Sarar Plain, Somali Rep. 120 C3 9 25N 46 17 E
Sarasota, U.S.A. 149 M4 27 20N 82 32W
Saratoga, Calif., U.S.A. 160 H4 37 16N 122 2W
Saratoga, Wyo., U.S.A. 158 F10 41 27N 106 49W
Saratoga Springs, U.S.A. 151 C11 43 5N 73 47W
Saratov, Russia 60 E7 51 30N 46 2 E
Saravane, Laos 86 E6 15 43N 106 25 E
Sarawak □, Malaysia 85 B4 2 0N 113 0 E
Saray, Tekirdağ, Turkey 51 E11 41 26N 27 55 E
Saray, Van, Turkey 101 C11 38 38N 44 9 E
Saraya, Senegal 112 C2 12 50N 11 45W
Saraycık, Turkey 49 B11 39 1N 29 49 E
Sarayköy, Turkey 49 D10 37 55N 28 54 E
Saraylar, Turkey 51 F11 40 39N 27 40 E
Sarayönü, Turkey 100 C5 38 16N 32 24 E
Sarbāz, Iran 97 E9 26 38N 61 19 E
Sarbīsheh, Iran 97 C8 32 30N 59 40 E
Sárbogárd, Hungary 52 D3 46 50N 18 40 E
Sarca →, Italy 44 C7 45 52N 10 52 E
Sarcelles, France 27 D9 48 59N 2 23 E
Sardalas, Libya 108 C2 25 50N 10 34 E
Sardarshahr, India 92 E6 28 30N 74 29 E
Sardegna □, Italy 46 B1 40 0N 9 0 E
Sardhana, India 92 E7 29 9N 77 39 E
Sardina, Pta., Canary Is. 39 F4 28 9N 15 44W
Sardinata, Colombia 168 B3 8 5N 72 48W
Sardinia = Sardegna □, Italy 46 B1 40 0N 9 0 E
Sardinia, U.S.A. 157 F13 39 0N 83 49W
Sardis, Turkey 49 C10 38 28N 28 2 E
Sardis, U.S.A. 152 C8 32 58N 81 46W
Sārdūīyeh = Dar Mazār, Iran 97 D8 29 14N 57 20 E
S'Arenal, Spain 39 B9 39 30N 2 45 E
Sarentino, Italy 45 B8 46 38N 11 21 E
Saréyamou, Mali 112 B4 16 7N 3 10W
Sargasso Sea, Atl. Oc. 8 F1 27 0N 72 0W
Sargent, Ga., U.S.A. 152 B5 33 26N 84 52W
Sargent, Nebr., U.S.A. 154 E5 41 39N 99 22W
Sargodha, Pakistan 91 B4 32 10N 72 40 E
Sarh, Chad 109 G3 9 5N 18 23 E
Sarhro, Djebel, Morocco 110 B4 31 6N 5 0W
Sārī, Iran 97 B7 36 30N 53 4 E
Sari d'Orcino, France 29 F12 42 3N 8 49 E
Sária, Greece 49 F9 35 54N 27 17 E
Sarıbeyler, Turkey 49 B9 39 24N 27 35 E
Saricumbe, Angola 115 E3 12 12 S 19 46 E
Sarıgöl, Turkey 100 C3 38 14N 28 41 E
Sarikamiş, Turkey 101 B10 40 22N 42 35 E
Sarıkaya, Turkey 100 C6 39 29N 35 22 E
Sarikei, Malaysia 85 B4 2 8N 111 30 E
Sarıköy, Turkey 51 F11 40 12N 27 37 E
Sarina, Australia 126 C4 21 22 S 149 13 E
Sariñena, Spain 40 D4 41 47N 0 10W
Sarīr Tibasti, Libya 108 D3 22 50N 18 30 E
Sarita, U.S.A. 155 M6 27 13N 97 47W
Sariwōn, N. Korea 75 E13 38 31N 125 46 E
Sariyar Baraji, Turkey 100 B4 40 2N 31 33 E
Sariyer, Turkey 51 E13 41 10N 29 3 E
Sark, U.K. 21 H5 49 25N 2 22W
Sarkad, Hungary 52 D6 46 47N 21 23 E
Şarkışla, Turkey 100 C7 39 21N 36 25 E
Şarköy, Turkey 51 F11 40 36N 27 6 E
Sarlat-la-Canéda, France 28 D5 44 54N 1 13 E
Sarles, U.S.A. 154 A5 48 58N 99 0W
Sărmăşag, Romania 52 C7 47 21N 22 50 E
Sărmaşu, Romania 53 D9 46 45N 24 13 E
Sarmi, Indonesia 83 B5 1 49 S 138 44 E
Sarmiento, Argentina 176 C3 45 35 S 69 5W
Sarmizegetusa, Romania 52 E7 45 31N 22 47 E
Särna, Sweden 16 C7 61 41N 13 8 E
Sarnano, Italy 45 E10 43 2N 13 18 E
Sarnen, Switz. 32 C6 46 53N 8 13 E
Sarnia, Canada 140 D3 42 58N 82 23W
Sarno, Italy 47 B7 40 49N 14 37 E
Sarnthein = Sarentino, Italy 45 B8 46 38N 11 21 E
Sarny, Ukraine 56 D4 51 17N 26 40 E
Särö, Sweden 17 G5 57 31N 11 57 E
Sarolangun, Indonesia 84 C2 2 19 S 102 42 E
Saronikós Kólpos, Greece 48 D5 37 45N 23 45 E
Saronno, Italy 44 C6 45 38N 9 2 E
Saros Körfezi, Turkey 51 F10 40 30N 26 15 E
Sárospatak, Hungary 52 B6 48 18N 21 33 E
Sarpsborg, Norway 15 G14 59 16N 11 7 E
Sarracín, Spain 42 C7 42 15N 3 45W
Sarralbe, France 27 D14 49 0N 7 1 E
Sarrat, Phil. 80 B3 18 9N 120 39 E
Sarre = Saar →, Europe 24 E6 49 41N 6 32 E
Sarre, Italy 44 C4 45 40N 7 15 E
Sarre-Union, France 27 D14 48 57N 7 4 E
Sarrebourg, France 27 D14 48 43N 7 3 E
Sarreguemines, France 27 C14 49 5N 7 4 E
Sarria, Spain 42 C3 42 49N 7 29W
Sarrión, Spain 40 E4 40 9N 0 49W
Sarro, Mali 112 C3 13 40N 5 15W
Sarstedt, Germany 30 C5 52 14N 9 52 E
Sartène, France 29 G12 41 38N 8 58 E
Sarthe □, France 26 D7 48 10N 0 10 E
Sarthe →, France 26 E6 47 33N 0 31W
Sartilly, France 26 D5 48 45N 1 28W
Saruhanlı, Turkey 49 C9 38 44N 27 36 E
Sărulești, Romania 53 F11 44 25N 26 39 E
Sárvár, Hungary 52 C1 47 15N 16 56 E
Sarvestān, Iran 97 D7 29 20N 53 10 E
Särvfjället, Sweden 16 B7 62 42N 13 30 E
Sárviz →, Hungary 52 D3 46 24N 18 41 E
Sary Ozek, Kazakstan 63 A8 44 22N 77 59 E
Sary-Tash, Kyrgyzstan 63 D6 39 44N 73 15 E
Saryagach, Kazakstan 63 C4 41 27N 69 9 E
Sarych, Mys, Ukraine 59 K7 44 25N 33 45 E
Sarykolskiy Khrebet, Asia 63 D7 38 30N 74 30 E
Saryshagan, Kazakstan 64 E8 46 12N 73 38 E
Sarzana, Italy 44 D6 44 7N 9 58 E
Sarzeau, France 26 E4 47 31N 2 48W
Sasabeneh, Ethiopia 120 C2 7 59N 44 43 E
Sasamungga, Solomon Is. 133 L9 7 0 S 156 50 E
Sasaram, India 93 G11 24 57N 84 5 E
Sasayama, Japan 71 B7 35 4N 135 13 E
Sasebo, Japan 70 D1 33 10N 129 43 E
Saser, India 93 B7 34 50N 77 50 E
Saskatchewan □, Canada 143 C7 54 40N 106 0W
Saskatchewan →, Canada 143 C8 53 37N 100 40W
Saskatoon, Canada 143 C7 52 10N 106 38W
Saskylakh, Russia 65 B12 71 55N 114 1 E
Sasolburg, S. Africa 117 D4 26 46 S 27 49 E
Sasovo, Russia 60 C5 54 25N 41 55 E
Sassandra, Ivory C. 112 E3 4 55N 6 8W
Sassandra →, Ivory C. 112 E3 4 58N 6 5W
Sássari, Italy 46 B1 40 43N 8 34 E
Sasser, U.S.A. 152 D5 31 43N 84 21W
Sassnitz, Germany 30 A9 54 29N 13 39 E
Sasso Marconi, Italy 45 D8 44 24N 11 15 E
Sassocorvaro, Italy 45 E9 43 47N 12 30 E
Sassoferrato, Italy 45 E9 43 26N 12 51 E
Sasstown, Liberia 112 E3 4 45N 8 27W
Sassuolo, Italy 44 D7 44 33N 10 47 E
Sástago, Spain 40 D4 41 19N 0 21W
Sasumua Dam, Kenya 118 C4 0 45 S 36 40 E
Sasyk, Ozero, Ukraine 59 K5 45 45N 29 20 E
Sasykkul, Tajikistan 63 E6 37 41N 73 11 E
Sata-Misaki, Japan 70 F2 31 0N 130 40 E
Satadougou, Mali 112 C2 12 25N 11 25W
Satakunta, Finland 15 F20 61 45N 23 0 E
Satanta, U.S.A. 155 G4 37 26N 100 59W
Satara, India 94 F1 17 44N 73 58 E
Sataua, W. Samoa 133 W23 13 28 S 172 40W
Satellite Beach, U.S.A. 153 G9 28 10N 80 36W
Såtenäs, Sweden 17 F6 58 27N 12 41 E
Säter, Sweden 16 D9 60 21N 15 45 E
Satilla →, U.S.A. 152 E8 30 59N 81 29W
Satipo, Peru 172 C3 11 15 S 74 25W
Satka, Russia 62 D7 55 3N 59 1 E
Satkania, Bangla. 90 D4 22 4N 92 3 E
Satkhira, Bangla. 90 D2 22 43N 89 8 E
Satmala Hills, India 94 D2 20 15N 74 40 E
Satna, India 93 G9 24 35N 80 50 E
Šator, Bos.-H. 45 D13 44 11N 16 37 E
Sátoraljaújhely, Hungary 52 B6 48 25N 21 41 E
Satpura Ra., India 92 J7 21 25N 76 10 E
Satrup, Germany 30 A5 54 41N 9 36 E
Satsuma-Hantō, Japan 70 F2 31 25N 130 25 E
Satsuna-Shotō, Japan 69 K5 30 0N 130 0 E
Sattahip, Thailand 86 F3 12 41N 100 54 E
Sattenapalle, India 95 F5 16 25N 80 6 E
Satu Mare, Romania 52 C7 47 46N 22 55 E
Satu Mare □, Romania 52 C7 47 45N 23 0 E
Satui, Indonesia 85 C5 3 50 S 115 27 E
Satun, Thailand 87 J3 6 43N 100 2 E
Satupe'itea, W. Samoa 133 W23 13 45 S 172 18W
Saturnina →, Brazil 173 C6 12 15 S 58 10W
Sauce, Argentina 174 C4 30 5 S 58 46W
Sauceda, Mexico 162 B4 25 55N 101 18W
Saucillo, Mexico 162 B3 28 1N 105 17W
Sauda, Norway 15 G12 59 40N 6 20 E
Saudasjøen, Norway 18 E3 59 38N 6 17 E
Saúde, Brazil 170 D3 10 56 S 40 24W
Sauðarkrókur, Iceland 11 B7 65 45N 19 40W
Saudi Arabia ■, Asia 98 A4 26 0N 44 0 E
Saugatuck, U.S.A. 157 B10 42 40N 86 12W
Saugeen →, Canada 150 B3 44 30N 81 22W
Saugerties, U.S.A. 151 D11 42 5N 73 57W
Saugues, France 28 D7 44 58N 3 32 E
Saujon, France 28 C3 45 41N 0 55W
Sauk Centre, U.S.A. 154 C7 45 44N 94 57W
Sauk City, U.S.A. 156 A7 43 17N 89 43W
Sauk Rapids, U.S.A. 154 C7 45 35N 94 10W
Saül, Fr. Guiana 169 C7 3 37N 53 12W
Sauland, Norway 18 E5 59 37N 8 56 E
Saulgau, Germany 31 G5 48 1N 9 29 E
Saulieu, France 27 E11 47 17N 4 14 E
Sault, France 29 D9 44 6N 5 24 E
Sault Ste. Marie, Canada 140 C3 46 30N 84 20W
Sault Ste. Marie, U.S.A. 148 B3 46 30N 84 21W
Saumlaki, Indonesia 83 C4 7 55 S 131 20 E
Saumur, France 26 E6 47 15N 0 5W
Saunders C., N.Z. 131 F5 45 53 S 170 45 E
Saunders I., Antarctica 7 B1 57 48 S 26 28W
Saunders Point, Australia 125 E4 27 52 S 125 38 E
Saunemin, U.S.A. 157 D8 40 54N 88 24W
Saurbær, Barðastrandarsýsla, Iceland 11 B2 65 29N 24 0W
Saurbær, Borgarfjarðarsýsla, Iceland 11 C5 64 24N 21 35W
Saurbær, Eyjafjarðarsýsla, Iceland 11 B8 65 27N 18 13W
Sauri, Nigeria 113 C6 11 42N 6 44 E
Saurimo, Angola 115 D4 9 40 S 20 12 E
Sausalito, U.S.A. 160 H4 37 51N 122 29W
Sautatá, Colombia 168 B2 7 50N 77 4W
Sauveterre-de-Béarn, France 28 E3 43 24N 0 57W
Sauzé-Vaussais, France 28 B4 46 8N 0 8 E
Savá, Honduras 164 C2 15 32N 86 15W
Sava, Italy 47 B10 40 24N 17 33 E
Sava →, Serbia, Yug. 52 F5 44 50N 20 26 E
Savage, U.S.A. 154 B2 47 27N 104 21W
Savage I. = Niue, Cook Is. 135 J11 19 2 S 169 54W
Savai'i, W. Samoa 133 W23 13 28 S 172 24W
Savalou, Benin 113 D5 7 57N 1 58 E
Savane, Mozam. 119 F4 19 37 S 35 8 E
Savanna, U.S.A. 156 B6 42 5N 90 8W
Savanna-la-Mar, Jamaica 164 C4 18 10N 78 10W
Savannah, Ga., U.S.A. 152 C8 32 5N 81 6W
Savannah, Mo., U.S.A. 156 E2 39 56N 94 50W
Savannah, Tenn., U.S.A. 149 H1 35 14N 88 15W
Savannah →, U.S.A. 152 C9 32 2N 80 53W
Savannah Beach, U.S.A. 152 C9 32 1N 80 51W
Savannakhet, Laos 86 D5 16 30N 104 49 E
Savant L., Canada 140 B1 50 16N 90 44W
Savant Lake, Canada 140 B1 50 14N 90 40W
Savantvadi, India 95 G1 15 55N 73 54 E
Savanur, India 95 G2 14 59N 75 21 E
Săvârşin, Romania 52 D7 46 1N 22 14 E
Savaştepe, Turkey 49 B9 39 22N 27 42 E
Savda, India 94 D2 21 9N 75 56 E
Savé, Benin 113 D5 8 2N 2 29 E
Save →, France 28 E5 43 47N 1 17 E
Save →, Mozam. 117 C5 21 16 S 34 0 E
Sāveh, Iran 97 C6 35 2N 50 20 E
Savelugu, Ghana 113 D4 9 38N 0 54W
Savenay, France 26 E5 47 20N 1 55W
Săveni, Romania 53 C11 47 57N 26 52 E
Saverdun, France 28 E5 43 14N 1 34 E
Saverne, France 27 D14 48 43N 7 20 E
Savièse, Switz. 32 D4 46 17N 7 22 E
Savigliano, Italy 44 D4 44 38N 7 40 E
Savigny-sur-Braye, France 26 E7 47 53N 0 49 E
Sávio →, Italy 45 D9 44 19N 12 20 E
Šavnik, Montenegro, Yug. 50 D3 42 59N 19 10 E
Savo, Finland 14 E22 62 45N 27 30 E
Savo, Solomon Is. 133 M10 9 8 S 159 48 E
Savognin, Switz. 33 C9 46 36N 9 37 E
Savoie □, France 29 C10 45 26N 6 25 E
Savona, Italy 44 D5 44 17N 8 30 E
Savonlinna, Finland 58 B5 61 52N 28 53 E
Savoonga, U.S.A. 144 E5 63 42N 170 29W
Şavşat, Turkey 101 B10 41 15N 42 20 E
Sävsjö, Sweden 17 G8 57 20N 14 40 E
Savusavu, Fiji 133 A2 16 34 S 179 15 E
Savusavu B., Fiji 133 A2 16 45 S 179 15 E
Sawahlunto, Indonesia 84 C2 0 40 S 100 52 E
Sawai, Indonesia 83 B3 3 0 S 129 5 E
Sawai Madhopur, India 92 G7 26 0N 76 25 E
Sawang Daen Din, Thailand 86 D4 17 28N 103 28 E
Sawankhalok, Thailand 86 D2 17 19N 99 50 E
Sawara, Japan 71 B12 35 55N 140 30 E
Sawatch Mts., U.S.A. 159 G10 38 30N 106 30W
Sawdā, Jabal as, Libya 108 C3 28 51N 15 12 E
Sawel Mt., U.K. 23 B4 54 50N 7 2W
Sawfajjin, W. →, Libya 108 B2 31 41N 14 44 E
Sawi, Thailand 87 G2 10 14N 99 5 E
Sawmills, Zimbabwe 119 F2 19 30 S 28 2 E
Şawqirah, Oman 99 C7 18 7N 56 32 E
Sawqirah, Ghubbat, Oman 99 C7 18 35N 57 0 E
Sawu, Indonesia 82 C2 10 35 S 121 50 E
Sawu Sea, Indonesia 82 C2 9 30 S 121 50 E
Saxby →, Australia 126 B3 18 25 S 140 53 E
Saxmundham, U.K. 21 E9 52 13N 1 30 E
Saxon, Switz. 32 D4 46 9N 7 11 E
Saxony, Lower = Niedersachsen □, Germany 30 C4 52 50N 9 0 E
Saxton, U.S.A. 150 F6 40 13N 78 15W
Say, Niger 113 C5 13 8N 2 22 E
Saya, Nigeria 113 D5 9 30N 3 18 E
Sayabec, Canada 141 C6 48 35N 67 41W
Sayaboury, Laos 86 C3 19 15N 101 45 E
Sayán, Peru 172 C2 11 8 S 77 12W
Sayan, Vostochnyy, Russia 65 D10 54 0N 96 0 E
Sayan, Zapadnyy, Russia 65 D10 52 30N 94 0 E
Saydā, Lebanon 103 B4 33 35N 35 25 E
Sayghān, Afghan. 91 B2 35 10N 67 55 E
Sayhan-Ovoo, Mongolia 74 B2 45 27N 103 54 E
Sayhandulaan, Mongolia 74 B5 44 40N 109 1 E
Sayhūt, Yemen 99 D5 15 12N 51 10 E
Saykhin, Kazakstan 61 F8 48 50N 46 47 E
Saylorville L., U.S.A. 156 C3 41 48N 93 46W

282 Silver City

Silver City, *Nev., U.S.A.* ... **158 G4** 39 15N 119 48W
Silver Cr. →, *U.S.A.* **158 E4** 43 16N 119 13W
Silver Creek, *U.S.A.* **150 D5** 42 33N 79 10W
Silver Grove, *U.S.A.* **157 E12** 39 2N 84 24W
Silver L., *Calif., U.S.A.* **160 G6** 38 39N 120 6W
Silver L., *Calif., U.S.A.* **161 K10** 35 21N 116 7W
Silver Lake, *Ind., U.S.A.* .. **157 C11** 41 4N 85 53W
Silver Lake, *Oreg., U.S.A.* . **158 E3** 43 8N 121 3W
Silver Lake, *Wis., U.S.A.* .. **157 B8** 42 33N 88 13W
Silver Springs, *U.S.A.* **153 F7** 29 13N 82 3W
Silver Streams, *S. Africa* ... **116 D3** 28 20 S 23 33 E
Silverdalen, *Sweden* .... **17 G9** 57 32N 15 45 E
Silverton, *Australia* .... **128 A4** 31 52 S 141 10 E
Silverton, *Colo., U.S.A.* ... **159 H10** 37 49N 107 40W
Silverton, *Tex., U.S.A.* **155 H4** 34 28N 101 19W
Silves, *Portugal* **43 H2** 37 11N 8 26W
Silvi Marina, *Italy* .... **45 F11** 42 34N 14 5 E
Silvia, *Colombia* **168 C2** 2 37N 76 21W
Silvies →, *U.S.A.* **158 E4** 43 34N 119 2W
Silvrettahorn, *Switz.* .... **33 C10** 46 50N 10 6 E
Silwa Bahari, *Egypt* .... **106 C3** 24 45N 32 55 E
Silz, *Austria* .... **34 D3** 47 16N 10 56 E
Sim, C., *Morocco* **110 B3** 31 26N 9 51W
Simanggang = Bandar Sri
Aman, *Malaysia* .... **85 B4** 1 15N 111 32 E
Simao, *China* .... **76 F3** 22 47N 101 5 E
Simão Dias, *Brazil* .... **170 D4** 10 44 S 37 49W
Simara I., *Phil.* .... **80 E4** 12 48N 122 3 E
Simard, L., *Canada* **140 C4** 47 40N 78 40W
Şīmareh →, *Iran* .... **101 F12** 33 9N 47 41 E
Simav, *Turkey* .... **49 B10** 39 4N 28 58 E
Simav →, *Turkey* .... **51 F12** 40 23N 28 31 E
Simav Dağları, *Turkey* ..... **49 B10** 39 10N 28 30 E
Simba, *Tanzania* .... **118 C4** 2 10 S 37 36 E
Simbach, *Germany* .... **31 G9** 48 16N 13 2 E
Simbirsk, *Russia* .... **60 C9** 54 20N 48 25 E
Simbo, *Tanzania* .... **118 C2** 4 51 S 29 41 E
Simcoe, *Canada* **140 D3** 42 50N 80 20W
Simcoe, L., *Canada* **140 D4** 44 25N 79 20W
Simeonovgrad, *Bulgaria* .... **51 D9** 42 1N 25 50 E
Simeria, *Romania* .... **52 E8** 45 51N 23 1 E
Simeto →, *Italy* .... **47 E8** 37 24N 15 6 E
Simeulue, *Indonesia* ...... **84 B1** 2 45N 95 45 E
Simferopol, *Ukraine* .... **59 K8** 44 55N 34 3 E
Sími, *Greece* .... **49 E9** 36 35N 27 50 E
Simi Valley, *U.S.A.* **161 L8** 34 16N 118 47W
Simikot, *Nepal* .... **93 E9** 30 0N 81 50 E
Simití, *Colombia* **168 B3** 7 58N 73 57W
Simitli, *Bulgaria* .... **50 E7** 41 52N 23 7 E
Simla, *India* .... **92 D7** 31 2N 77 9 E
Simlångsdalen, *Sweden* ..... **17 H7** 56 43N 13 6 E
Şimleu-Silvaniei, *Romania* .. **52 C7** 47 17N 22 50 E
Simme →, *Switz.* .... **32 C4** 46 38N 7 25 E
Simmern, *Germany* .... **31 F3** 49 58N 7 30 E
Simmie, *Canada* **143 D7** 49 56N 108 6W
Simmler, *U.S.A.* **161 K7** 35 21N 119 59W
Simnas, *Lithuania* .... **54 D10** 54 24N 23 39 E
Simões, *Brazil* .... **170 C3** 7 36 S 40 49W
Simojoki →, *Finland* ...... **14 D21** 65 35N 25 1 E
Simojovel, *Mexico* **163 D6** 17 12N 92 38W
Simonette →, *Canada* **142 B5** 55 9N 118 15W
Simonstown, *S. Africa* ..... **116 E2** 34 14 S 18 26 E
Simontornya, *Hungary* ..... **52 D3** 46 45N 18 33 E
Simpangkiri →, *Indonesia* . **84 B1** 2 50N 97 40 E
Simplício Mendes, *Brazil* ... **170 C3** 7 51 S 41 54W
Simplon, *Switz.* .... **32 D6** 46 12N 8 4 E
Simplon P., *Switz.* .... **32 D6** 46 15N 8 3 E
Simplontunnel, *Switz.* ...... **32 D6** 46 15N 8 7 E
Simpson Desert, *Australia* .. **126 D2** 25 0 S 137 0 E
Simpungdong, *N. Korea* .... **75 D15** 40 56N 129 29 E
Simrishamn, *Sweden* .... **17 J8** 55 33N 14 22 E
Simunjan, *Malaysia* .... **85 B4** 1 25N 110 45 E
Simushir, Ostrov, *Russia* ... **65 E16** 46 50N 152 30 E
Sina →, *India* .... **94 F2** 17 30N 75 55 E
Sinabang, *Indonesia* .... **84 B1** 2 30N 96 24 E
Sinadogo, *Somali Rep.* ..... **120 C3** 5 50N 47 0 E
Sinai = Es Sînâ', *Egypt* .... **106 J8** 29 0N 34 0 E
Sinai, Mt. = Mûsa, Gebel,
Egypt .... **106 J8** 28 33N 33 59 E
Sinai Peninsula, *Egypt* ..... **103 F3** 29 30N 34 0 E
Sinaia, *Romania* .... **53 E10** 45 21N 25 38 E
Sinait, *Phil.* .... **80 C3** 17 52N 120 27 E
Sinako, Mt., *Phil.* .... **81 H5** 7 30N 125 17 E
Sinaloa □, *Mexico* **162 C3** 25 0N 107 30W
Sinaloa de Leyva, *Mexico* .. **162 B3** 25 50N 108 20W
Sinalunga, *Italy* .... **45 E8** 43 12N 11 44 E
Sinan, *China* .... **76 D7** 27 56N 108 13 E
Sînandrei, *Romania* .... **52 E6** 45 52N 21 13 E
Sinarádhes, *Greece* .... **38 A3** 39 34N 19 51 E
Sīnāwan, *Libya* .... **108 B2** 31 0N 10 37 E
Sinbaungwe, *Burma* .... **90 F5** 19 43N 95 10 E
Sinbo, *Burma* .... **90 C6** 24 46N 97 3 E
Sincan, *Turkey* .... **100 B5** 39 58N 32 36 E
Sincanlı, *Turkey* .... **49 C12** 38 45N 30 15 E
Sincé, *Colombia* **168 B2** 9 15N 75 9W
Sincelejo, *Colombia* **168 B2** 9 18N 75 24W
Sinchang, *N. Korea* .... **75 D15** 40 7N 128 28 E
Sinchang-ni, *N. Korea* ..... **75 E14** 39 24N 126 8 E
Sinclair, *U.S.A.* **158 F10** 41 47N 107 7W
Sinclair, L., *U.S.A.* **152 B6** 33 8N 83 12W
Sinclair Mills, *Canada* **142 C4** 54 5N 121 40W
Sinclair's B., *U.K.* **22 C5** 58 31N 3 5W
Sincorá, Serra do, *Brazil* ... **171 D3** 13 30 S 41 0W
Sind, *Pakistan* .... **92 G3** 26 0N 68 30 E
Sind □, *Pakistan* .... **91 D3** 26 0N 69 0 E
Sind →, *India* .... **93 B6** 34 18N 74 45 E
Sind Sagar Doab, *Pakistan* . **92 D4** 32 0N 71 30 E
Sindal, *Denmark* .... **17 G4** 57 28N 10 10 E
Sindangan, *Phil.* .... **81 G4** 8 10N 123 5 E
Sindangan Bay, *Phil.* .... **81 G4** 8 11N 122 50 E
Sindangbarang, *Indonesia* .. **85 D3** 7 27 S 107 1 E
Sinde, *Zambia* .... **119 F2** 17 28 S 25 51 E
Sindelfingen, *Germany* ..... **31 G4** 48 42N 9 0 E
Sındırgı, *Turkey* .... **49 B10** 39 13N 28 10 E
Sinegorskiy, *Russia* .... **61 G5** 47 55N 40 52 E
Sinekli, *Turkey* .... **51 E12** 41 14N 28 12 E
Sinelnikovo = Synelnykove,
Ukraine .... **59 H8** 48 25N 35 30 E
Sines, *Portugal* **43 H2** 37 56N 8 51W
Sines, C. de, *Portugal* **43 H2** 37 58N 8 53W
Sineu, *Spain* .... **39 B10** 39 38N 3 1 E
Sinewit, Mt., *Papua N. G.* . **132 C7** 4 44 S 152 2 E
Sinfra, *Ivory C.* **112 D3** 6 35N 5 56W
Sing Buri, *Thailand* .... **86 E3** 14 53N 100 25 E
Singa, *Sudan* .... **107 E3** 13 10N 33 57 E
Singanallur, *India* .... **95 J3** 11 2N 77 1 E
Singapore ■, *Asia* .... **87 M4** 1 17N 103 51 E
Singapore, Straits of, *Asia* .. **87 M5** 1 15N 104 0 E
Singaraja, *Indonesia* ...... **85 D5** 8 6 S 115 10 E
Singen, *Germany* .... **31 H4** 47 45N 8 50 E
Singida, *Tanzania* .... **118 C3** 4 49 S 34 48 E
Singida □, *Tanzania* .... **118 D3** 6 0 S 34 30 E
Singitikós Kólpos, *Greece* .. **50 F7** 40 6N 24 0 E
Singkaling Hkamti, *Burma* . **90 C5** 26 0N 95 39 E
Singkang, *Indonesia* .... **82 E2** 4 8 S 120 1 E
Singkawang, *Indonesia* ..... **85 B3** 1 0N 108 57 E
Singleton, *Australia* .... **129 B9** 32 33 S 151 0 E
Singleton, Mt., *N. Terr.*,
Australia .... **124 D5** 22 0 S 130 46 E
Singleton, Mt., *W. Austral.*,
Australia .... **125 E2** 29 27 S 117 15 E
Singö, *Sweden* .... **16 D12** 60 12N 18 45 E
Singoli, *India* .... **92 G6** 25 0N 75 22 E
Singora = Songkhla,
Thailand .... **87 J3** 7 13N 100 37 E
Singosan, *N. Korea* .... **75 E14** 38 52N 127 25 E
Singsås, *Norway* .... **18 B7** 62 56N 10 37 E
Sinhung, *N. Korea* .... **75 D14** 40 11N 127 34 E
Sînî □, *Egypt* .... **103 F3** 30 0N 34 0 E
Siniátsikon, Óros, *Greece* .. **50 F5** 40 25N 21 35 E
Siniloan, *Phil.* .... **80 D3** 14 25N 121 27 E
Siniscóla, *Italy* .... **46 B2** 40 34N 9 41 E
Sinj, *Croatia* .... **45 E13** 43 42N 16 39 E
Sinjai, *Indonesia* .... **82 C2** 5 7 S 120 20 E
Sinjajevina,
Montenegro, Yug. .... **50 D3** 42 57N 19 22 E
Sinjār, *Iraq* .... **101 D9** 36 19N 41 52 E
Sinkat, *Sudan* .... **106 D4** 18 55N 36 49 E
Sinkiang Uighur = Xinjiang
Uygur Zizhiqu □, *China* . **72 C3** 42 0N 86 0 E
Sinmak, *N. Korea* .... **75 E14** 38 25N 126 14 E
Sinnai, *Italy* .... **46 C2** 39 18N 9 13 E
Sinnar, *India* .... **94 E2** 19 48N 74 0 E
Sinni →, *Italy* .... **47 B9** 40 8N 16 41 E
Sinnuris, *Egypt* .... **106 J7** 29 26N 30 31 E
Sinoie, Lacul, *Romania* .... **53 F13** 44 35N 28 50 E
Sinop, *Turkey* .... **100 A6** 42 1N 35 11 E
Sinpo, *N. Korea* .... **75 E15** 40 0N 128 13 E
Sins, *Switz.* .... **33 B6** 47 12N 8 24 E
Sinsheim, *Germany* .... **31 F4** 49 15N 8 52 E
Sinsk, *Russia* .... **65 C13** 61 8N 126 48 E
Sint Eustatius, I., *Neth. Ant.* **165 C7** 17 30N 62 59W
Sint Maarten, I., *W. Indies* . **165 C7** 18 4N 63 4W
Sintang, *Indonesia* .... **85 B4** 0 5N 111 35 E
Sinton, *U.S.A.* **155 L6** 28 2N 97 31W
Sintra, *Portugal* **43 G1** 38 47N 9 25W
Sinugif, *Somali Rep.* .... **120 C3** 8 33N 48 59 E
Sinŭiju, *N. Korea* .... **75 D13** 40 5N 124 24 E
Sinyukha →, *Ukraine* ..... **59 H6** 48 3N 30 51 E
Sinzig, *Germany* .... **30 E3** 50 32N 7 14 E
Sio →, *Hungary* .... **52 D3** 46 20N 18 53 E
Siocon, *Phil.* .... **81 H4** 7 40N 122 10 E
Siófok, *Hungary* .... **52 D3** 46 54N 18 3 E
Sioma, *Zambia* .... **116 B3** 16 25 S 23 28 E
Sion, *Switz.* .... **32 D4** 46 14N 7 20 E
Sion Mills, *U.K.* **23 B4** 54 48N 7 29W
Sioux City, *U.S.A.* **154 D6** 42 30N 96 24W
Sioux Falls, *U.S.A.* **154 D6** 43 33N 96 44W
Sioux Lookout, *Canada* **140 B1** 50 10N 91 50W
Sipalay, *Phil.* .... **81 G4** 9 45N 122 24 E
Šipan, *Croatia* .... **50 D1** 42 45N 17 52 E
Sipang, Tanjong, *Malaysia* . **85 B4** 1 48N 110 20 E
Siping, *China* .... **75 C13** 43 8N 124 21 E
Sipiwesk L., *Canada* **143 B9** 55 5N 97 35W
Sipocot, *Phil.* .... **80 E4** 13 46N 122 58 E
Šipovo, *Bos.-H.* .... **52 F2** 44 16N 17 6 E
Sipura, *Indonesia* .... **84 C1** 2 18 S 99 40 E
Siquia →, *Nic.* **164 D3** 12 10N 84 20W
Siquijor, *Phil.* .... **81 G4** 9 12N 123 35 E
Siquijor □, *Phil.* .... **81 G4** 9 11N 123 35 E
Siquirres, *Costa Rica* **164 D3** 10 6N 83 30W
Siquisique, *Venezuela* **168 A4** 10 34N 69 42W
Sir Edward Pellew Group,
Australia .... **126 B2** 15 40 S 137 10 E
Sir Graham Moore Is.,
Australia .... **124 B4** 13 53 S 126 34 E
Sira, *India* .... **95 H3** 13 41N 76 49 E
Sira, *Norway* .... **18 F3** 58 24N 6 40 E
Sira →, *Norway* .... **15 G12** 58 23N 6 34 E
Siracusa, *Italy* .... **47 E8** 37 4N 15 17 E
Sirajganj, *Bangla.* .... **93 G13** 24 25N 89 47 E
Sirakoro, *Mali* **112 C3** 12 41N 9 14W
Şiran, *Turkey* .... **101 B8** 40 11N 39 7 E
Sirasso, *Ivory C.* **112 D3** 9 16N 6 6W
Siraway, *Phil.* .... **81 H4** 7 34N 122 8 E
Sirdalsvatnet, *Norway* ..... **18 F3** 58 34N 6 43 E
Sirdān, *Iran* .... **97 B6** 36 39N 49 12 E
Sirdaryo = Syrdarya →,
Kazakstan .... **64 E7** 46 3N 61 0 E
Sirdaryo, *Uzbekistan* ...... **63 C4** 40 50N 68 40 E
Sirer, *Spain* .... **39 C7** 38 56N 1 22 E
Siret, *Romania* .... **53 C11** 47 55N 26 5 E
Siret →, *Romania* .... **53 E12** 45 24N 28 1 E
Sirevåg, *Norway* .... **18 F2** 58 30N 5 48 E
Şiria, *Romania* .... **52 D6** 46 16N 21 38 E
Sirik, Tanjong, *Malaysia* ... **85 B4** 2 47N 111 15 E
Sirino, Mte., *Italy* .... **47 B8** 40 7N 15 50 E
Sirkali = Sirkazhi, *India* ... **95 J4** 11 15N 79 41 E
Sirkazhi, *India* .... **95 J4** 11 15N 79 41 E
Sirmans, *U.S.A.* **152 E6** 30 21N 83 39W
Sírna, *Greece* .... **49 E8** 36 22N 26 42 E
Sirnach, *Switz.* .... **33 B7** 47 28N 8 59 E
Şirnak, *Turkey* .... **101 D10** 37 32N 42 28 E
Sirohi, *India* .... **92 G5** 24 52N 72 53 E
Sironj, *India* .... **92 G7** 24 5N 77 39 E
Síros, *Greece* .... **49 D7** 37 28N 24 57 E
Sirrayn, *Si. Arabia* .... **98 C3** 19 38N 40 36 E
Sirretta Pk., *U.S.A.* **161 K8** 35 56N 118 19W
Sirsa, *India* .... **92 E6** 29 33N 75 4 E
Sirsi, *India* .... **95 G2** 14 40N 74 49 E
Siruela, *Spain* **43 G5** 38 58N 5 3W
Sisak, *Croatia* .... **45 C13** 45 30N 16 21 E
Sisaket, *Thailand* .... **86 E5** 15 8N 104 23 E
Sisante, *Spain* **41 F2** 39 25N 2 12W
Sisargas, Is., *Spain* **42 B2** 43 21N 8 50W
Sishen, *S. Africa* .... **116 D3** 27 47 S 22 59 E
Sishui, *Henan, China* ...... **74 G7** 34 48N 113 15 E
Sishui, *Shandong, China* ... **75 G9** 35 42N 117 18 E
Sisimiut = Holsteinsborg,
Greenland **10 D5** 66 40N 53 30W
Sisipuk L., *Canada* **143 B8** 55 45N 101 50W
Sisophon, *Cambodia* .... **86 F4** 13 38N 102 59 E
Sissach, *Switz.* .... **32 B5** 47 27N 7 48 E
Sisseton, *U.S.A.* **154 C6** 45 40N 97 3W
Sissonne, *France* .... **27 C10** 49 34N 3 51 E
Sīstān va Balūchestān □,
Iran .... **97 E9** 27 0N 62 0 E
Sisteron, *France* .... **29 D9** 44 12N 5 57 E
Sisters, *U.S.A.* **158 D3** 44 18N 121 33W
Sitamarhi, *India* .... **93 F11** 26 37N 85 30 E
Sitapur, *India* .... **93 F9** 27 38N 80 45 E
Siteki, *Swaziland* .... **117 D5** 26 32 S 31 58 E
Sitges, *Spain* .... **40 D6** 41 17N 1 47 E
Sithoniá, *Greece* .... **50 F7** 40 0N 23 45 E
Sitía, *Greece* .... **38 D8** 35 13N 26 6 E
Sítio da Abadia, *Brazil* .... **171 D2** 14 48 S 46 16W
Sitka, *U.S.A.* **138 C6** 57 3N 135 20W
Sitkinak I., *U.S.A.* **144 H9** 56 33N 154 10W
Sitoti, *Botswana* .... **116 C3** 23 15 S 23 40 E
Sitra, *Egypt* .... **106 B2** 28 40N 26 53 E
Sittang Myit →, *Burma* ... **90 G6** 17 20N 96 45 E
Sittard, *Neths.* .... **24 C5** 51 0N 5 52 E
Sittaung, *Burma* .... **90 C5** 24 10N 94 35 E
Sittensen, *Germany* .... **30 B5** 53 17N 9 32 E
Sittingbourne, *U.K.* .... **21 F8** 51 21N 0 45 E
Sittona, *Eritrea* .... **107 E4** 14 25N 37 23 E
Sittwe, *Burma* .... **90 E4** 20 18N 92 45 E
Siuna, *Nic.* **164 D3** 13 37N 84 45W
Siuri, *India* .... **93 H12** 23 50N 87 34 E
Siutghiol, Lacul, *Romania* .. **53 F13** 44 15N 28 35 E
Sivaganga, *India* .... **95 K4** 9 50N 78 28 E
Sivagiri, *India* .... **95 K3** 9 16N 77 26 E
Sivakasi, *India* .... **95 K3** 9 24N 77 47 E
Sivana, *India* .... **92 E8** 28 37N 78 6 E
Sīvand, *Iran* .... **97 D7** 30 5N 52 55 E
Sivas, *Turkey* .... **100 C7** 39 43N 36 58 E
Sivaslı, *Turkey* .... **49 C11** 38 31N 29 42 E
Siverek, *Turkey* .... **101 D8** 37 50N 39 19 E
Sivomaskinskiy, *Russia* ..... **56 A11** 66 40N 62 35 E
Sivrihisar, *Turkey* .... **100 C4** 39 30N 31 35 E
Sîwa, *Egypt* .... **106 B2** 29 11N 25 31 E
Sîwa, El Wâhât es, *Egypt* .. **106 B2** 29 10N 25 30 E
Siwa Oasis, *Egypt* .... **104 D6** 29 10N 25 30 E
Siwalik Range, *Nepal* ...... **93 F10** 28 0N 83 0 E
Siwan, *India* .... **93 F11** 26 13N 84 21 E
Sixmilebridge, *Ireland* **23 D3** 52 44N 8 46W
Siyâl, Jazâ'ir, *Egypt* ...... **106 C4** 22 49N 36 12 E
Siyäzän, *Azerbaijan* .... **61 K9** 41 3N 49 10 E
Siziwang Qi, *China* .... **74 D6** 41 25N 111 40 E
Sjælland, *Denmark* .... **17 J5** 55 30N 11 30 E
Sjællands Odde, *Denmark* .. **17 J5** 55 58N 11 24 E
Sjenica, *Serbia, Yug.* ...... **50 C3** 43 16N 20 0 E
Sjoa, *Norway* .... **18 C6** 61 41N 9 33 E
Sjöbo, *Sweden* .... **17 J7** 55 37N 13 45 E
Sjøholt, *Norway* .... **18 B3** 62 27N 6 52 E
Sjøli, *Norway* .... **18 C8** 61 29N 11 18 E
Sjötofta, *Sweden* .... **17 G7** 57 22N 13 17 E
Sjötorp, *Sweden* .... **17 F8** 58 50N 14 0 E
Sjumen = Shumen, *Bulgaria* **51 C10** 43 18N 26 55 E
Sjuntorp, *Sweden* .... **17 F6** 58 12N 12 13 E
Skåbu, *Norway* .... **18 C6** 61 32N 9 24 E
Skadarsko Jezero,
Montenegro, Yug. .... **50 D3** 42 10N 19 20 E
Skadovsk, *Ukraine* .... **59 J7** 46 17N 32 52 E
Skælskør, *Denmark* .... **17 J5** 55 15N 11 18 E
Skærbæk, *Denmark* .... **17 J2** 55 10N 8 45 E
Skaftafell, *Iceland* **11 C9** 64 1N 17 0W
Skagafjarðarsýsla □, *Iceland* **11 B7** 65 30N 19 0W
Skagafjörður, *Iceland* **11 B7** 65 54N 19 35W
Skagastølstindane, *Norway* . **15 F12** 61 28N 7 52 E
Skagaströnd, *Iceland* **11 B6** 65 50N 20 19W
Skagatá, *Iceland* **11 A6** 66 7N 20 6W
Skagen, *Denmark* .... **17 G4** 57 43N 10 35 E
Skagern, *Sweden* .... **17 F8** 59 0N 14 20 E
Skagerrak, *Denmark* ...... **17 G2** 57 30N 9 0 E
Skagit →, *U.S.A.* **160 B4** 48 23N 122 22W
Skagway, *U.S.A.* **142 B1** 59 28N 135 19W
Skala-Podilska, *Ukraine* .... **59 H4** 48 50N 26 15 E
Skala Podolskaya = Skala-
Podilska, *Ukraine* .... **59 H4** 48 50N 26 15 E
Skalat, *Ukraine* .... **59 H3** 49 23N 25 55 E
Skálavík, *Iceland* **11 A3** 66 11N 23 29W
Skalbmierz, *Poland* .... **55 H7** 50 20N 20 25 E
Skälderviken, *Sweden* ...... **17 H6** 56 22N 12 30 E
Skålevik, *Norway* .... **18 F5** 58 5N 8 1 E
Skálholt, *Iceland* **11 C6** 64 7N 20 32W
Skalica, *Slovak Rep.* ...... **35 C10** 48 50N 17 15 E
Skallingen, *Denmark* ...... **17 J2** 55 32N 8 13 E
Skalni Dol = Kamenyak,
Bulgaria .... **51 C10** 43 24N 26 57 E
Skanderborg, *Denmark* ..... **17 H3** 56 2N 9 55 E
Skåne, *Sweden* .... **17 J7** 55 59N 13 30 E
Skånevik, *Norway* .... **18 E2** 59 43N 5 53 E
Skänninge, *Sweden* .... **17 F9** 58 24N 15 5 E
Skanör med Falsterbo,
Sweden .... **17 J6** 55 24N 12 50 E
Skantzoúra, *Greece* .... **48 B6** 39 5N 24 6 E
Skara, *Sweden* .... **17 F7** 58 25N 13 30 E
Skaraborgs län □, *Sweden* .. **17 F7** 58 20N 13 30 E
Skärblacka, *Sweden* .... **17 F9** 58 35N 15 54 E
Skarð, *Iceland* **11 B4** 65 17N 22 19W
Skardu, *Pakistan* .... **93 B6** 35 20N 75 44 E
Skare, *Norway* .... **18 E3** 59 55N 6 36 E
Skåre, *Sweden* .... **16 E7** 59 26N 13 26 E
Skärhamn, *Sweden* .... **17 G5** 57 59N 11 34 E
Skarnes, *Norway* .... **18 D8** 60 15N 11 41 E
Skarszewy, *Poland* .... **54 D5** 54 4N 18 25 E
Skaryszew, *Poland* .... **55 G8** 51 19N 21 15 E
Skarżysko-Kamienna, *Poland* **55 G7** 51 7N 20 52 E
Skattkärr, *Sweden* .... **16 E7** 59 25N 13 40 E
Skattungbyn, *Sweden* ...... **16 C8** 61 10N 14 56 E
Skawina, *Poland* .... **55 J6** 49 59N 19 50 E
Skebobruk, *Sweden* .... **16 E12** 59 58N 18 36 E
Skeena →, *Canada* **142 C2** 54 9N 130 5W
Skeena Mts., *Canada* **142 B3** 56 40N 128 30W
Skegness, *U.K.* .... **20 D8** 53 9N 0 20 E
Skei, *Norway* .... **18 C3** 61 34N 6 28 E
Skeiðarársandur, *Iceland* ... **11 D9** 63 54N 17 14W
Skeiðflötur, *Iceland* **11 D7** 63 26N 19 11W
Skeldon, *Guyana* **169 B6** 5 55N 57 20W
Skellefte älv →, *Sweden* ... **14 D19** 64 45N 21 10 E
Skellefteå, *Sweden* .... **14 D19** 64 45N 20 50 E
Skelleftehamn, *Sweden* .... **14 D19** 64 40N 21 9 E
Skender Vakuf, *Bos.-H.* .... **52 F2** 44 29N 17 22 E
Skerries, The, *U.K.* **20 D3** 53 25N 4 36W
Skhíza, *Greece* .... **48 E3** 36 41N 21 40 E
Skhoinoúsa, *Greece* .... **49 E7** 36 53N 25 31 E
Ski, *Norway* .... **15 G14** 59 43N 10 52 E
Skíathos, *Greece* .... **48 B5** 39 12N 23 30 E
Skibbereen, *Ireland* **23 E2** 51 33N 9 16W
Skiddaw, *U.K.* **20 C4** 54 39N 3 9W
Skídhra, *Greece* .... **50 F6** 40 46N 22 10 E
Skien, *Norway* .... **15 G13** 59 12N 9 35 E
Skierniewice, *Poland* ...... **55 G7** 51 58N 20 10 E
Skierniewice □, *Poland* .... **55 G7** 51 50N 20 10 E
Skikda, *Algeria* .... **111 A6** 36 50N 6 58 E
Skillet →, *U.S.A.* **157 F8** 38 5N 88 5W
Skillingaryd, *Sweden* ...... **17 G8** 57 27N 14 5 E
Skillinge, *Sweden* .... **17 J8** 55 30N 14 16 E
Skilloura, *Cyprus* .... **38 D12** 35 14N 33 10 E
Skinári, Ákra, *Greece* ...... **48 D2** 37 56N 20 40 E
Skinnastaður, *Iceland* **11 A10** 66 4N 16 27W
Skinnskatteberg, *Sweden* ... **16 E9** 59 50N 15 42 E
Skipton, *Australia* .... **128 D5** 37 39 S 143 40 E
Skipton, *U.K.* **20 D5** 53 58N 2 3W
Skiptvet, *Norway* .... **18 E8** 59 28N 11 11 E
Skirmish Pt., *Australia* ..... **126 A1** 11 59 S 134 17 E
Skiropoúla, *Greece* .... **48 C6** 38 50N 24 21 E
Skíros, *Greece* .... **48 C6** 38 55N 24 34 E
Skivarp, *Sweden* .... **17 J7** 55 26N 13 34 E
Skive, *Denmark* .... **17 H3** 56 33N 9 2 E
Skjærhalden, *Norway* ...... **18 E8** 59 2N 11 2 E
Skjálfandafljót →, *Iceland* . **11 B9** 65 59N 17 25W
Skjálfandi, *Iceland* **11 A9** 66 5N 17 30W
Skjeberg, *Norway* .... **18 E8** 59 12N 11 12 E
Skjern, *Denmark* .... **17 J2** 55 57N 8 30 E
Skjold, *Norway* .... **18 E2** 59 31N 5 34 E
Skjolden, *Norway* .... **18 C4** 61 29N 7 36 E
Skjöldólfsstaðir, *Iceland* **11 B11** 65 19N 15 7W
Skjønhaug, *Norway* .... **18 E8** 59 39N 11 19 E
Skoczów, *Poland* .... **55 J5** 49 49N 18 45 E
Skodje, *Norway* .... **18 B3** 62 30N 6 43 E
Škofja Loka, *Slovenia* ..... **45 B11** 46 9N 14 19 E
Skógarnes, *Iceland* **11 C4** 64 46N 22 34W
Skoghall, *Sweden* .... **16 E7** 59 20N 13 30 E
Skogstorp, *Sweden* .... **16 E10** 59 19N 16 29 E
Skoki, *Poland* .... **55 F4** 52 40N 17 11 E
Skokie, *U.S.A.* **157 B9** 42 3N 87 45W
Skole, *Ukraine* .... **59 H2** 49 3N 23 30 E
Skollenborg, *Norway* ...... **18 E6** 59 38N 9 43 E
Skópelos, *Greece* .... **48 B5** 39 9N 23 47 E
Skopí, *Greece* .... **38 D8** 35 11N 26 2 E
Skopin, *Russia* .... **58 F10** 53 55N 39 32 E
Skopje, *Macedonia* .... **50 D5** 42 1N 21 26 E
Skoppum, *Norway* .... **18 E7** 59 23N 10 26 E
Skórcz, *Poland* .... **54 E5** 53 47N 18 30 E
Skørping, *Denmark* .... **17 H3** 56 50N 9 53 E
Skotfoss, *Norway* .... **18 E6** 59 12N 9 30 E
Skotterud, *Norway* .... **18 D9** 60 0N 12 7 E
Skövde, *Sweden* .... **17 F7** 58 24N 13 50 E
Skovorodino, *Russia* ...... **65 D13** 54 0N 124 0 E
Skowhegan, *U.S.A.* **141 D6** 44 46N 69 43W
Skownan, *Canada* **143 C9** 51 58N 99 35W
Skradin, *Croatia* .... **45 E12** 43 52N 15 53 E
Skrea, *Sweden* .... **17 H6** 56 53N 12 34 E
Skreia, *Norway* .... **18 D7** 60 40N 10 56 E
Skrim, *Norway* .... **18 E6** 59 31N 9 38 E
Skrunda, *Latvia* .... **54 B9** 56 41N 22 1 E
Skrwa →, *Poland* .... **55 F6** 52 35N 19 32 E
Skudeneshavn, *Norway* .... **18 E2** 59 10N 5 10 E
Skull, *Ireland* **23 E2** 51 32N 9 34W
Skultorp, *Sweden* .... **17 F7** 58 24N 13 51 E
Skultuna, *Sweden* .... **16 E10** 59 43N 16 25 E
Skunk →, *U.S.A.* **156 D5** 40 42N 91 7W
Skuodas, *Lithuania* .... **15 H19** 56 16N 21 33 E
Skurup, *Sweden* .... **17 J7** 55 28N 13 30 E
Skutskär, *Sweden* .... **16 D11** 60 37N 17 25 E
Skútustaðir, *Iceland* **11 B9** 65 34N 17 2W
Skvyra, *Ukraine* .... **59 H5** 49 44N 29 40 E
Skwierzyna, *Poland* .... **55 F2** 52 33N 15 30 E
Skye, *U.K.* **22 D2** 57 15N 6 10W
Skykomish, *U.S.A.* **158 C3** 47 42N 121 22W
Skyttorp, *Sweden* .... **16 D11** 60 5N 17 44 E
Slættaratindur, *Færoe Is.* ... **14 E9** 62 18N 7 1W
Slagelse, *Denmark* .... **17 J5** 55 23N 11 19 E
Slamannon, *Australia* ..... **128 B5** 32 1 S 143 41 E
Slamet, *Indonesia* .... **85 D3** 7 16 S 109 8 E
Slaney →, *Ireland* **23 D5** 52 26N 6 33W
Slănic, *Romania* .... **53 E10** 45 14N 25 58 E
Slano, *Croatia* .... **50 D1** 42 48N 17 53 E
Slantsy, *Russia* .... **58 C5** 59 7N 28 5 E
Slaný, *Czech Rep.* .... **34 A7** 50 13N 14 6 E
Slätbaken, *Sweden* .... **17 F10** 58 25N 16 45 E
Slate Is., *Canada* **140 C2** 48 40N 87 0W
Slater, *U.S.A.* **156 E3** 39 13N 93 4W
Slatina, *Croatia* .... **52 E2** 45 42N 17 45 E
Slatina, *Romania* .... **53 F9** 44 28N 24 22 E
Slatina Timiş, *Romania* .... **52 E7** 45 15N 22 17 E
Slaton, *U.S.A.* **155 J4** 33 26N 101 39W
Slave →, *Canada* **142 A6** 61 18N 113 39W
Slave Coast, *W. Afr.* ...... **113 D5** 6 0N 2 30 E
Slave Lake, *Canada* **142 B6** 55 17N 114 43W
Slave Pt., *Canada* **142 A5** 61 11N 115 56W
Slavgorod, *Russia* .... **64 D8** 53 1N 78 37 E
Slavinja, *Serbia, Yug.* ..... **50 C6** 43 9N 22 50 E
Slavkov u Brna, *Czech Rep.* . **35 B9** 49 10N 16 52 E
Slavonija, *Europe* .... **52 E2** 45 20N 17 40 E
Slavonski Brod, *Croatia* .... **52 E3** 45 11N 18 1 E
Slavuta, *Ukraine* .... **59 G4** 50 15N 27 2 E
Slavyanka, *Russia* .... **68 C5** 42 53N 131 21 E
Slavyanovo, *Bulgaria* ...... **51 C8** 43 28N 24 52 E
Slavyansk = Slovyansk,
Ukraine .... **59 H9** 48 55N 37 36 E
Slavyansk-na-Kubani, *Russia* **59 K10** 45 15N 38 11 E
Sława, *Poland* .... **55 G3** 51 52N 16 2 E
Slawharad, *Belarus* .... **58 F6** 53 27N 31 0 E
Sławno, *Poland* .... **54 D3** 54 20N 16 41 E
Sławoborze, *Poland* .... **54 E2** 53 55N 15 42 E
Sleaford, *U.K.* **20 D7** 53 0N 0 24W
Sleaford B., *Australia* ..... **127 E2** 34 55 S 135 45 E
Sleat, Sd. of, *U.K.* **22 D3** 57 5N 5 47W
Sleðbrjótur, *Iceland* **11 B12** 65 34N 14 30W
Sleeper Is., *Canada* **139 C11** 58 30N 81 0W
Sleepy Eye, *U.S.A.* **154 C7** 44 18N 94 43W
Sleetmute, *U.S.A.* **144 F8** 61 42N 157 10W
Sleman, *Indonesia* .... **85 D4** 7 40 S 110 20 E
Ślemon L., *Canada* **142 A5** 63 13N 116 4W
Ślesin, *Poland* .... **55 F5** 52 22N 18 14 E
Slidell, *U.S.A.* **155 K10** 30 17N 89 47W
Sliema, *Malta* .... **38 D2** 35 54N 14 30 E
Slieve Aughty, *Ireland* **23 C3** 53 4N 8 30W
Slieve Bloom, *Ireland* **23 C4** 53 4N 7 40W
Slieve Donard, *U.K.* **23 B6** 54 11N 5 55W
Slieve Gamph, *Ireland* **23 B3** 54 6N 9 0W
Slieve Gullion, *U.K.* **23 B5** 54 7N 6 26W
Slieve Mish, *Ireland* **23 D2** 52 12N 9 50W
Slievenamon, *Ireland* **23 D4** 52 25N 7 34W
Sligeach = Sligo, *Ireland* ... **23 B3** 54 16N 8 28W
Sligo, *Ireland* **23 B3** 54 16N 8 28W
Sligo □, *Ireland* **23 B3** 54 8N 8 42W
Sligo B., *Ireland* **23 B3** 54 18N 8 40W
Slite, *Sweden* .... **17 G12** 57 42N 18 48 E
Sliven, *Bulgaria* .... **51 D10** 42 42N 26 19 E
Slivnitsa, *Bulgaria* .... **50 D7** 42 50N 23 2 E
Sljeme, *Croatia* .... **45 C12** 45 57N 15 58 E
Sloan, *U.S.A.* **161 K11** 35 57N 115 13W
Sloansville, *U.S.A.* **151 D10** 42 45N 74 22W
Slobodskoy, *Russia* .... **62 B3** 58 40N 50 6 E
Slobozia, *Moldova* .... **53 D14** 46 45N 29 42 E
Slobozia, *Argeş, Romania* .. **53 F10** 44 30N 25 14 E
Slobozia, *Ialomiţa, Romania* **53 F12** 44 34N 27 23 E
Slocan, *Canada* **142 D5** 49 48N 117 28W
Slocomb, *U.S.A.* **152 D4** 31 7N 85 36W
Słomniki, *Poland* .... **55 H7** 50 16N 20 4 E

Sorrento 283

Slonim, *Belarus* 59 F3 53 4N 25 19 E
Slough, *U.K.* 21 F7 51 30N 0 36W
Slough □, *U.K.* 21 F7 51 30N 0 36W
Sloughhouse, *U.S.A.* 160 G5 38 26N 121 12W
Slovak Rep. ■, *Europe* 35 C13 48 30N 20 0 E
Slovakia = Slovak Rep. ■, *Europe* 35 C13 48 30N 20 0 E
Slovakian Ore Mts. = Slovenské Rudohorie, *Slovak Rep.* 35 C12 48 45N 20 0 E
Slovenia ■, *Europe* 45 C11 45 58N 14 30 E
Slovenija = Slovenia ■, *Europe* 45 C11 45 58N 14 30 E
Slovenj Gradec, *Slovenia* 45 B12 46 31N 15 5 E
Slovenska Bistrica, *Slovenia* 45 B12 46 24N 15 35 E
Slovenská Republika = Slovak Rep. ■, *Europe* 35 C13 48 30N 20 0 E
Slovenske Konjice, *Slovenia* 45 B12 46 20N 15 28 E
Slovenské Rudohorie, *Slovak Rep.* 35 C12 48 45N 20 0 E
Slovyansk, *Ukraine* 59 H9 48 55N 37 36 E
Słubice, *Poland* 55 F1 52 22N 14 35 E
Sluch →, *Ukraine* 59 G4 51 37N 26 38 E
Sluis, *Neths.* 24 C3 51 18N 3 23 E
Slŭnchev Bryag, *Bulgaria* 51 D11 42 40N 27 41 E
Slunj, *Croatia* 45 C12 45 6N 15 33 E
Słupca, *Poland* 55 F4 52 15N 17 52 E
Słupia →, *Poland* 54 D3 54 35N 16 51 E
Słupsk, *Poland* 54 D4 54 30N 17 3 E
Słupsk □, *Poland* 54 D4 54 15N 17 30 E
Slurry, *S. Africa* 116 D4 25 49 S 25 42 E
Slutsk, *Belarus* 59 F4 53 2N 27 31 E
Slyne Hd., *Ireland* 23 C1 53 25N 10 10W
Slyudyanka, *Russia* 65 D11 51 40N 103 40 E
Småland, *Sweden* 17 G9 57 15N 15 25 E
Smålandsfarvandet, *Denmark* 17 J5 55 10N 11 20 E
Smålandsstenar, *Sweden* 17 G7 57 10N 13 25 E
Small Nggela, *Solomon Is.* 133 M11 9 0 S 160 0 E
Smalltree L., *Canada* 143 A8 61 0N 105 0W
Smallwood Res., *Canada* 141 B7 54 0N 64 0W
Smarhon, *Belarus* 58 E4 54 20N 26 24 E
Smarje, *Slovenia* 45 B12 46 15N 15 34 E
Smarr, *U.S.A.* 152 C6 32 59N 83 53W
Smartt Syndicate Dam, *S. Africa* 116 E3 30 45 S 23 10 E
Smartville, *U.S.A.* 160 F5 39 13N 121 18W
Smeaton, *Canada* 143 C8 53 30N 104 49W
Smedby, *Sweden* 17 H10 56 41N 16 13 E
Smederevo, *Serbia, Yug.* 50 B4 44 40N 20 57 E
Smederevska Palanka, *Serbia, Yug.* 50 B4 44 22N 20 58 E
Smedjebacken, *Sweden* 16 D9 60 8N 15 25 E
Smela = Smila, *Ukraine* 59 H6 49 15N 31 58 E
Smerwick Harbour, *Ireland* 23 D1 52 12N 10 23W
Smethport, *U.S.A.* 150 E6 41 49N 78 27W
Smidovich, *Russia* 65 E14 48 36N 133 49 E
Śmigiel, *Poland* 55 F3 52 1N 16 32 E
Smila, *Ukraine* 59 H6 49 15N 31 58 E
Smiley, *Canada* 143 C7 51 38N 109 29W
Smilyan, *Bulgaria* 51 E8 41 29N 24 46 E
Smith, *Canada* 142 B6 55 10N 114 0W
Smith →, *Canada* 142 B3 59 34N 126 30W
Smith B., *U.S.A.* 144 A9 70 30N 154 20W
Smith Center, *U.S.A.* 154 F5 39 47N 98 47W
Smith Sund, *Greenland* 10 B3 78 30N 74 0W
Smithburne →, *Australia* 126 B3 17 3 S 140 57 E
Smithers, *Canada* 142 C3 54 45N 127 10W
Smithfield, *S. Africa* 117 E4 30 9 S 26 30 E
Smithfield, *N.C., U.S.A.* 149 H6 35 31N 78 21W
Smithfield, *Utah, U.S.A.* 158 F8 41 50N 111 50W
Smiths, *U.S.A.* 152 C4 32 32N 85 6W
Smiths Falls, *Canada* 140 D4 44 55N 76 0W
Smithton, *Australia* 126 G4 40 53 S 145 6 E
Smithtown, *Australia* 129 A10 30 58 S 152 48 E
Smithville, *Canada* 150 C5 43 6N 79 33W
Smithville, *Ga., U.S.A.* 152 D5 31 54N 84 15W
Smithville, *Mo., U.S.A.* 156 E2 39 23N 94 35W
Smithville, *Tex., U.S.A.* 155 K6 30 1N 97 10W
Smoaks, *U.S.A.* 152 B9 33 5N 80 49W
Smoky →, *Canada* 142 B5 56 10N 117 21W
Smoky Bay, *Australia* 127 E1 32 22 S 134 13 E
Smoky Falls, *Canada* 140 B3 50 4N 82 10W
Smoky Hill →, *U.S.A.* 154 F6 39 4N 96 48W
Smoky Lake, *Canada* 142 C6 54 10N 112 30W
Smøla, *Norway* 14 E13 63 23N 8 3 E
Smolensk, *Russia* 58 E7 54 45N 32 5 E
Smolikas, Óros, *Greece* 50 F4 40 9N 20 58 E
Smolník, *Slovak Rep.* 35 C13 48 43N 20 44 E
Smolyan, *Bulgaria* 51 E8 41 36N 24 38 E
Smooth Rock Falls, *Canada* 140 C3 49 17N 81 37W
Smoothstone L., *Canada* 143 C7 54 40N 106 50W
Smorgon = Smarhon, *Belarus* 58 E4 54 20N 26 24 E
Smulţi, *Romania* 53 E12 45 57N 27 44 E
Smyadovo, *Bulgaria* 51 C11 43 2N 27 1 E
Smygehamn, *Sweden* 17 J7 55 21N 13 22 E
Smyrna = İzmir, *Turkey* 57 G4 38 25N 27 8 E
Smyrna, *U.S.A.* 152 B5 33 53N 84 31W
Snæfell, *Iceland* 11 C11 64 48N 15 34W
Snaefell, *U.K.* 20 C3 54 16N 4 27W
Snæfellsjökull, *Iceland* 11 C3 64 49N 23 46W
Snæfellsnessýsla □, *Iceland* 11 C3 65 0N 23 0W
Snake →, *U.S.A.* 158 C4 46 12N 119 2W
Snake I., *Australia* 129 E7 38 47 S 146 33 E
Snake L., *Canada* 143 B7 55 32N 106 35W
Snake Range, *U.S.A.* 158 G6 39 0N 114 20W
Snake River Plain, *U.S.A.* 158 E7 42 50N 114 0W
Snasahögarna, *Sweden* 16 A6 63 13N 12 21 E
Snåsavatnet, *Norway* 14 D14 64 12N 12 0 E
Snedsted, *Denmark* 17 H2 56 55N 8 32 E
Sneek, *Neths.* 24 A5 53 2N 5 40 E
Sneeuberge, *S. Africa* 116 E3 31 46 S 24 20 E
Snejbjerg, *Denmark* 17 H2 56 8N 8 54 E
Snelling, *Calif., U.S.A.* 160 H6 37 31N 120 26W
Snelling, *S.C., U.S.A.* 152 B9 33 15N 81 27W
Snezhnoye, *Ukraine* 59 J10 48 0N 38 58 E
Snežnik, *Slovenia* 45 C11 45 36N 14 35 E
Śniadowo, *Poland* 55 E8 53 2N 22 0 E
Śniardwy, Jezioro, *Poland* 54 E8 53 48N 21 50 E
Śnieżka, *Europe* 34 A8 50 44N 15 44 E
Snigirevka = Snihurivka, *Ukraine* 59 J7 47 2N 32 49 E
Snihurivka, *Ukraine* 59 J7 47 2N 32 49 E
Snillfjord, *Norway* 18 A6 63 24N 9 30 E
Snina, *Slovak Rep.* 35 C15 48 58N 22 9 E
Snizort, L., *U.K.* 22 D2 57 33N 6 28W
Snøhetta, *Norway* 15 E13 62 19N 9 16 E
Snohomish, *U.S.A.* 160 C4 47 55N 122 6W
Snønuten, *Norway* 18 E3 59 31N 6 52 E
Snoul, *Cambodia* 87 F6 12 4N 106 26 E
Snow Hill, *U.S.A.* 148 F8 38 11N 75 24W
Snow Lake, *Canada* 143 C8 54 52N 100 3W
Snow Mt., *U.S.A.* 160 F4 39 23N 122 45W
Snowbird L., *Canada* 143 A8 60 45N 103 0W
Snowdon, *U.K.* 20 D3 53 4N 4 5W
Snowdoun, *U.S.A.* 152 C3 32 15N 86 18W
Snowdrift →, *Canada* 143 A6 62 24N 110 44W
Snowflake, *U.S.A.* 159 J8 34 30N 110 5W
Snowshoe Pk., *U.S.A.* 158 B6 48 13N 115 41W
Snowtown, *Australia* 128 B3 33 46 S 138 14 E
Snowville, *U.S.A.* 158 F7 41 58N 112 43W
Snowy →, *Australia* 129 D8 37 46 S 148 30 E
Snowy Mts., *Australia* 129 D8 36 30 S 148 20 E
Snug Corner, *Bahamas* 165 B5 22 33N 73 52W
Snyatyn, *Ukraine* 59 H3 48 27N 25 38 E
Snyder, *Okla., U.S.A.* 155 H5 34 40N 98 57W
Snyder, *Tex., U.S.A.* 155 J4 32 44N 100 55W
Soacha, *Colombia* 168 C3 4 35N 74 13W
Soahanina, *Madag.* 117 B7 18 42 S 44 13 E
Soalala, *Madag.* 117 B8 16 6 S 45 20 E
Soan →, *Pakistan* 92 C4 33 1N 71 44 E
Soanierana-Ivongo, *Madag.* 117 B8 16 55 S 49 35 E
Soap Lake, *U.S.A.* 158 C4 47 23N 119 29W
Şoarş, *Romania* 53 E9 45 56N 24 55 E
Sobat, Nahr →, *Sudan* 107 F3 9 22N 31 33 E
Soběslav, *Czech Rep.* 34 B7 49 16N 14 45 E
Sobhapur, *India* 92 H8 22 47N 78 17 E
Sobinka, *Russia* 58 E11 56 0N 40 0 E
Sobo-Yama, *Japan* 70 E3 32 51N 131 22 E
Sobótka, *Poland* 55 H3 50 54N 16 44 E
Sobra, *Croatia* 45 F14 42 44N 17 34 E
Sobradinho, Reprêsa de, *Brazil* 170 C3 9 30 S 42 0 E
Sobral, *Brazil* 170 B3 3 50 S 40 20W
Sobrance, *Slovak Rep.* 35 C15 48 45N 22 11 E
Sobreira Formosa, *Portugal* 42 F3 39 46N 7 51W
Soc Giang, *Vietnam* 86 A6 22 54N 106 1 E
Soc Trang, *Vietnam* 87 H5 9 37N 105 50 E
Soča →, *Europe* 34 E6 46 20N 13 40 E
Sochaczew, *Poland* 55 F7 52 15N 20 13 E
Soch'e = Shache, *China* 72 C2 38 20N 77 10 E
Sochi, *Russia* 61 J4 43 35N 39 40 E
Social Circle, *U.S.A.* 152 B6 33 39N 83 43W
Société, Is. de la, *Pac. Oc.* 135 J12 17 0 S 151 0W
Society Hill, *U.S.A.* 152 C4 32 26N 85 27W
Society Is. = Société, Is. de la, *Pac. Oc.* 135 J12 17 0 S 151 0W
Socompa, Portezuelo de, *Chile* 174 A2 24 27 S 68 18W
Socorro, *Colombia* 168 B3 6 29N 73 16W
Socorro, *Phil.* 81 G5 9 37N 125 58 E
Socorro, *U.S.A.* 159 J10 34 4N 106 54W
Socorro, I., *Mexico* 162 D2 18 45N 110 58W
Socotra, *Ind. Oc.* 99 D6 12 30N 54 0 E
Socovos, *Spain* 41 G3 38 20N 1 58W
Socuéllamos, *Spain* 41 F2 39 16N 2 47W
Soda L., *U.S.A.* 159 J5 35 10N 116 4W
Soda Plains, *India* 93 B8 35 30N 79 0 E
Soda Springs, *U.S.A.* 158 E8 42 39N 111 36W
Sodankylä, *Finland* 14 C22 67 29N 26 40 E
Söderala, *Sweden* 16 C10 61 17N 16 55 E
Söderbärke, *Sweden* 16 D9 60 5N 15 33 E
Söderfors, *Sweden* 16 D11 60 23N 17 25 E
Söderhamn, *Sweden* 16 C11 61 18N 17 10 E
Söderköping, *Sweden* 17 F10 58 31N 16 20 E
Södermanland, *Sweden* 15 G17 58 56N 16 55 E
Södermanlands län □, *Sweden* 16 E10 59 10N 16 30 E
Södertälje, *Sweden* 16 E11 59 12N 17 39 E
Sodiri, *Sudan* 107 E2 14 27N 29 0 E
Sodo, *Ethiopia* 107 F4 7 0N 37 41 E
Södra Dellen, *Sweden* 16 C10 61 48N 16 43 E
Södra Finnskoga, *Sweden* 16 D6 60 42N 12 34 E
Södra Sandby, *Sweden* 17 J7 55 43N 13 21 E
Södra Ulvön, *Sweden* 16 B12 62 59N 18 38 E
Södra Vi, *Sweden* 17 G9 57 45N 15 45 E
Sodražica, *Slovenia* 45 C11 45 45N 14 39 E
Sodus, *U.S.A.* 150 C7 43 14N 77 4W
Soe, *Indonesia* 82 C2 9 52 S 124 17 E
Soekmekaar, *S. Africa* 117 C4 23 30 S 29 55 E
Soest, *Germany* 30 D4 51 34N 8 7 E
Soest, *Neths.* 24 B5 52 9N 5 19 E
Sofádhes, *Greece* 48 B4 39 20N 22 4 E
Sofara, *Mali* 112 C4 13 59N 4 9W
Sofia = Sofiya, *Bulgaria* 50 D7 42 45N 23 20 E
Sofia →, *Madag.* 117 B8 15 27 S 47 23 E
Sofievka, *Ukraine* 59 H7 48 6N 33 55 E
Sofikón, *Greece* 48 D5 37 47N 23 3 E
Sofiya, *Bulgaria* 50 D7 42 45N 23 20 E
Sofiya □, *Bulgaria* 50 D7 42 15N 23 20 E
Sōfu-Gan, *Japan* 69 K10 29 49N 140 21 E
Sogakofe, *Ghana* 113 D5 6 2N 0 39 E
Sogamoso, *Colombia* 168 B3 5 43N 72 56W
Soğār, *Iran* 97 E8 25 53N 58 6 E
Sögel, *Germany* 30 C3 52 50N 7 31 E
Sogeri, *Papua N. G.* 132 E4 9 26 S 147 35 E
Sogn og Fjordane □, *Norway* 18 C3 61 40N 6 45 E
Sogndalsfjøra, *Norway* 15 F12 61 14N 7 5 E
Søgne, *Norway* 15 G12 58 5N 7 48 E
Sognefjorden, *Norway* 15 F11 61 10N 5 50 E
Söğüt, *Bilecik, Turkey* 49 A12 40 2N 30 11 E
Söğüt, *Burdur, Turkey* 49 D11 37 2N 29 50 E
Söğüt Dağı, *Turkey* 49 D11 37 50N 29 55 E
Söğütköy, *Turkey* 49 E10 36 40N 28 8 E
Sŏgwi-po, *S. Korea* 75 H14 33 13N 126 34 E
Soh, *Iran* 97 C6 33 26N 51 27 E
Sohâg, *Egypt* 106 B3 26 33N 31 43 E
Sohano, *Papua N. G.* 132 C8 5 22 S 154 37 E
Sŏhori, *N. Korea* 75 D15 40 7N 128 23 E
Soignies, *Belgium* 24 D4 50 35N 4 5 E
Soira, *Eritrea* 107 E4 14 45N 39 30 E
Soissons, *France* 27 C10 49 25N 3 19 E
Sōja, *Japan* 70 C5 34 40N 133 45 E
Sojat, *India* 92 G5 25 55N 73 45 E
Sok →, *Russia* 60 D10 53 24N 50 8 E
Sokal, *Ukraine* 59 G3 50 31N 24 15 E
Söke, *Turkey* 49 D9 37 48N 27 28 E
Sokelo, *Dem. Rep. of the Congo* 115 D4 9 55 S 24 36 E
Sokhós, *Greece* 50 F7 40 48N 23 22 E
Sokhumi, *Georgia* 61 J5 43 0N 41 0 E
Sokki, Oued In →, *Algeria* 111 C5 29 30N 3 42 E
Sokna, *Norway* 18 D6 60 16N 9 58 E
Soknedal, *Norway* 18 B7 62 57N 10 13 E
Soko Banja, *Serbia, Yug.* 50 C5 43 40N 21 51 E
Sokodé, *Togo* 113 D5 9 0N 1 11 E
Sokol, *Russia* 58 C11 59 30N 40 5 E
Sokolac, *Bos.-H.* 52 G3 43 56N 18 48 E
Sokółka, *Poland* 54 E10 53 25N 23 30 E
Sokolo, *Mali* 112 C3 14 53N 6 8W
Sokolov, *Czech Rep.* 34 A5 50 12N 12 40 E
Sokołów Małpolski, *Poland* 55 H9 50 12N 22 7 E
Sokołów Podlaski, *Poland* 55 F9 52 25N 22 15 E
Sokoły, *Poland* 55 F9 52 59N 22 42 E
Sokoto, *Nigeria* 113 C6 13 2N 5 16 E
Sokoto □, *Nigeria* 113 C6 12 30N 6 0 E
Sokoto →, *Nigeria* 113 C5 11 20N 4 10 E
Sokuluk, *Kyrgyzstan* 63 B7 42 52N 74 18 E
Sol Iletsk, *Russia* 62 F5 51 10N 55 0 E
Sola, *Norway* 18 F2 58 53N 5 36 E
Sola →, *Poland* 55 H6 50 4N 19 15 E
Solai, *Kenya* 118 B4 0 2N 36 12 E
Solana, *Phil.* 80 C3 17 39N 121 41 E
Solander I., *N.Z.* 131 G1 46 34 S 166 54 E
Solano, *Phil.* 80 C3 16 31N 121 15 E
Solapur, *India* 94 F2 17 43N 75 56 E
Solca, *Romania* 53 C10 47 40N 25 50 E
Solda Gölü, *Turkey* 49 D11 37 33N 29 42 E
Soldăneşti, *Moldova* 53 C13 47 49N 28 48 E
Soldotna, *U.S.A.* 144 F10 60 29N 151 3W
Soléa □, *Cyprus* 38 D12 35 5N 33 4 E
Solec Kujawski, *Poland* 55 E5 53 5N 18 14 E
Soledad, *Colombia* 168 A3 10 55N 74 46W
Soledad, *U.S.A.* 160 J5 36 26N 121 20W
Soledad, *Venezuela* 169 B5 8 10N 63 34W
Sølen, *Norway* 18 C8 61 53N 11 31 E
Solent, The, *U.K.* 21 G6 50 45N 1 25W
Solenzara, *France* 29 G13 41 53N 9 23 E
Solesmes, *France* 27 B10 50 10N 3 30 E
Solfonn, *Norway* 15 F12 60 2N 6 57 E
Solhan, *Turkey* 101 C9 38 57N 41 3 E
Sølheim, *Norway* 18 D2 60 53N 5 27 E
Soligalich, *Russia* 56 C7 59 5N 42 10 E
Soligorsk = Salihorsk, *Belarus* 59 F4 52 51N 27 27 E
Solihull, *U.K.* 21 E6 52 26N 1 47W
Solikamsk, *Russia* 62 B6 59 38N 56 50 E
Solila, *Madag.* 117 C8 21 25 S 46 37 E
Solimões = Amazonas →, *S. Amer.* 169 D8 0 5 S 50 0W
Solin, *Croatia* 45 E13 43 33N 16 30 E
Solingen, *Germany* 30 D3 51 10N 7 5 E
Sollebrunn, *Sweden* 17 F6 58 8N 12 32 E
Sollefteå, *Sweden* 16 A11 63 12N 17 20 E
Sollentuna, *Sweden* 16 E11 59 26N 17 56 E
Sóller, *Spain* 39 B9 39 46N 2 43 E
Sollerön, *Sweden* 16 D8 60 55N 14 37 E
Solling, *Germany* 30 D5 51 42N 9 38 E
Solnechnogorsk, *Russia* 58 D9 56 10N 36 57 E
Solofra, *Italy* 47 B7 40 50N 14 51 E
Sologne, *France* 27 E8 47 40N 1 45 E
Solok, *Indonesia* 84 C2 0 45 S 100 40 E
Sololá, *Guatemala* 164 D1 14 49N 91 10W
Solomon, N. Fork →, *U.S.A.* 154 F5 39 29N 98 26W
Solomon, S. Fork →, *U.S.A.* 154 F5 39 25N 99 12W
Solomon Is. ■, *Pac. Oc.* 133 L8 6 0 S 155 0 E
Solomon Sea, *Papua N. G.* 132 D6 7 0 S 150 0 E
Solon, *China* 73 B7 46 32N 121 10 E
Solon Springs, *U.S.A.* 154 B9 46 22N 91 49W
Solonópole, *Brazil* 170 C4 5 44 S 39 1W
Solor, *Indonesia* 82 C2 8 27 S 123 0 E
Solotcha, *Russia* 58 E10 54 48N 39 53 E
Solothurn, *Switz.* 32 B5 47 13N 7 32 E
Solothurn □, *Switz.* 32 B5 47 18N 7 40 E
Solotobe, *Kazakstan* 63 A3 44 37N 66 3 E
Solsona, *Spain* 40 C6 42 0N 1 31 E
Solsvik, *Norway* 18 D1 60 26N 4 58 E
Solt, *Hungary* 52 D4 46 45N 19 1 E
Šolta, *Croatia* 45 E13 43 24N 16 15 E
Soltānābād, *Khorāsān, Iran* 97 C8 34 13N 59 58 E
Soltānābād, *Khorāsān, Iran* 97 B8 36 29N 58 5 E
Soltānābād, *Markazī, Iran* 97 C6 35 31N 51 10 E
Soltau, *Germany* 30 C5 52 59N 9 50 E
Soltsy, *Russia* 58 C6 58 10N 30 30 E
Solund, *Norway* 18 C1 61 5N 4 50 E
Solunska Glava, *Macedonia* 50 E5 41 44N 21 31 E
Solvang, *U.S.A.* 161 L6 34 36N 120 8W
Solvay, *U.S.A.* 151 C8 43 3N 76 13W
Sölvesborg, *Sweden* 17 H8 56 5N 14 35 E
Solvychegodsk, *Russia* 56 B8 61 21N 46 56 E
Solway Firth, *U.K.* 20 C4 54 49N 3 35W
Solwezi, *Zambia* 119 E2 12 11 S 26 21 E
Sōma, *Japan* 68 F10 37 40N 140 50 E
Soma, *Turkey* 49 B9 39 10N 27 35 E
Somali Pen., *Africa* 104 F8 7 0N 46 0 E
Somali Rep. ■, *Africa* 120 C3 7 0N 47 0 E
Somalia = Somali Rep. ■, *Africa* 120 C3 7 0N 47 0 E
Sombe Dzong, *Bhutan* 90 B2 27 13N 89 8 E
Sombernon, *France* 27 E11 47 20N 4 40 E
Sombor, *Serbia, Yug.* 52 E4 45 46N 19 9 E
Sombra, *Canada* 150 D2 42 43N 82 29W
Sombrerete, *Mexico* 162 C4 23 40N 103 40W
Sombrero, *Anguilla* 165 C7 18 37N 63 30W
Şomcuta Mare, *Romania* 53 C8 47 31N 23 28 E
Somers, *U.S.A.* 158 B6 48 5N 114 13W
Somerset, *Canada* 143 D9 49 25N 98 39W
Somerset, *Colo., U.S.A.* 159 G10 38 56N 107 28W
Somerset, *Ky., U.S.A.* 148 G3 37 5N 84 36W
Somerset, *Mass., U.S.A.* 151 E13 41 47N 71 8W
Somerset, *Pa., U.S.A.* 150 F5 40 1N 79 5W
Somerset □, *U.K.* 21 F5 51 9N 3 0W
Somerset East, *S. Africa* 116 E4 32 42 S 25 35 E
Somerset I., *Canada* 138 A10 73 30N 93 0W
Somerset West, *S. Africa* 116 E2 34 8 S 18 50 E
Somerton, *U.S.A.* 159 K6 32 36N 114 43W
Somerville, *U.S.A.* 151 F10 40 35N 74 38W
Someş →, *Romania* 52 C7 47 49N 22 43 E
Someşul Mare →, *Romania* 53 C8 47 9N 23 55 E
Sommariva, *Australia* 127 D4 26 24 S 146 36 E
Somme □, *France* 27 C9 49 57N 2 20 E
Somme →, *France* 27 B8 50 11N 1 38 E
Somme, B. de la, *France* 26 B8 50 14N 1 33 E
Sommen, *Jönköping, Sweden* 17 F8 58 12N 14 58 E
Sommen, *Östergötland, Sweden* 17 F9 58 0N 15 15 E
Sommepy-Tahure, *France* 27 C11 49 15N 4 31 E
Sömmerda, *Germany* 30 D7 51 9N 11 7 E
Sommesous, *France* 27 D11 48 44N 4 12 E
Sommières, *France* 29 E8 43 47N 4 6 E
Somogy □, *Hungary* 52 D2 46 19N 17 30 E
Somogyszob, *Hungary* 52 D2 46 18N 17 20 E
Somosomo Str., *Fiji* 133 A3 16 0 S 180 0 E
Somoto, *Nic.* 164 D2 13 28N 86 37W
Sompolno, *Poland* 55 F5 52 26N 18 30 E
Somport, Puerto de, *Spain* 40 C4 42 48N 0 31W
Somuncurá, Meseta de, *Argentina* 176 B3 41 30 S 67 0W
Son, *Norway* 18 E7 59 32N 10 42 E
Son Ha, *Vietnam* 86 E7 15 3N 108 34 E
Son Hoa, *Vietnam* 86 F7 13 2N 108 58 E
Son La, *Vietnam* 86 B4 21 20N 103 50 E
Son Servera, *Spain* 40 F8 39 37N 3 21 E
Son Tay, *Vietnam* 86 B5 21 8N 105 30 E
Soná, *Panama* 164 E3 8 0N 81 20W
Sonamarg, *India* 93 B6 34 18N 75 21 E
Sonamukhi, *India* 93 H12 23 18N 87 27 E
Sonamura, *India* 90 D3 23 29N 91 15 E
Sŏnchŏn, *N. Korea* 75 E13 39 48N 124 55 E
Sondags →, *S. Africa* 116 E4 33 44 S 25 51 E
Sóndalo, *Italy* 44 B7 46 20N 10 19 E
Sondar, *India* 93 C6 33 28N 75 56 E
Søndeled, *Norway* 18 F6 58 46N 9 3 E
Sønder Felding, *Denmark* 17 J2 55 57N 8 47 E
Sønder Omme, *Denmark* 17 J2 55 50N 8 54 E
Sønderborg, *Denmark* 17 K3 54 55N 9 49 E
Sønderjyllands Amtskommune □, *Denmark* 17 J3 55 10N 9 10 E
Sondershausen, *Germany* 30 D6 51 22N 10 51 E
Søndre Strømfjord, *Greenland* 10 D6 66 59N 50 40W
Sóndrio, *Italy* 44 B6 46 10N 9 52 E
Sone, *Mozam.* 119 F3 17 23 S 34 55 E
Sonepur, *India* 94 D6 20 55N 83 50 E
Song, *Thailand* 86 C3 18 28N 100 11 E
Song Cau, *Vietnam* 86 F7 13 27N 109 18 E
Song Xian, *China* 74 G7 34 12N 112 8 E
Songavatnet, *Norway* 18 E4 59 52N 7 32 E
Songchŏn, *N. Korea* 75 E14 39 12N 126 15 E
Songea, *Tanzania* 119 E4 10 40 S 35 40 E
Songea □, *Tanzania* 119 E4 10 30 S 36 0 E
Songeons, *France* 27 C8 49 32N 1 50 E
Songhua Hu, *China* 75 C14 43 35N 126 50 E
Songhua Jiang →, *China* 73 B8 47 45N 132 30 E
Songjiang, *China* 77 B13 31 1N 121 12 E
Songjin, *N. Korea* 75 D15 40 40N 129 10 E
Songjŏng-ni, *S. Korea* 75 G14 35 8N 126 47 E
Songkan, *China* 76 C6 28 35N 106 52 E
Songkhla, *Thailand* 87 J3 7 13N 100 37 E
Songming, *China* 76 E4 25 12N 103 2 E
Songnim, *N. Korea* 75 E13 38 45N 125 39 E
Songo, *Angola* 115 D2 7 22 S 14 51 E
Songololo, *Dem. Rep. of the Congo* 115 D2 5 42 S 14 2 E
Songpan, *China* 76 A4 32 40N 103 30 E
Songtao, *China* 76 C7 28 11N 109 10 E
Songwe, *Dem. Rep. of the Congo* 118 C2 3 20 S 26 16 E
Songwe →, *Africa* 119 D3 9 44 S 33 58 E
Songxi, *China* 77 D12 27 31N 118 44 E
Songzi, *China* 77 B8 30 12N 111 45 E
Sonid Youqi, *China* 74 C7 42 45N 112 48 E
Sonipat, *India* 92 E7 29 0N 77 5 E
Sonkel, Ozero, *Kyrgyzstan* 63 C7 41 50N 75 12 E
Sonkovo, *Russia* 58 D9 57 50N 37 5 E
Sonmiani, *Pakistan* 91 D2 25 25N 66 40 E
Sonnino, *Italy* 46 A6 41 25N 13 14 E
Sono →, *Minas Gerais, Brazil* 171 E2 17 2 S 45 32W
Sono →, *Tocantins, Brazil* 170 C2 9 58 S 48 11W
Sonobe, *Japan* 71 B7 35 6N 135 28 E
Sonogno, *Switz.* 33 D7 46 22N 8 47 E
Sonora, *Calif., U.S.A.* 160 H6 37 59N 120 23W
Sonora, *Tex., U.S.A.* 155 K4 30 34N 100 39W
Sonora □, *Mexico* 162 B2 29 0N 111 0W
Sonora →, *Mexico* 162 B2 28 50N 111 33W
Sonora Desert, *U.S.A.* 161 L12 33 40N 114 15W
Sonoyta, *Mexico* 162 A2 31 51N 112 50W
Sonqor, *Iran* 101 E12 34 47N 47 36 E
Sŏnsan, *S. Korea* 75 F15 36 14N 128 17 E
Sonseca, *Spain* 43 F7 39 42N 3 57W
Sonsonate, *El Salv.* 164 D2 13 43N 89 44W
Sonstorp, *Sweden* 17 F9 58 44N 15 38 E
Sonthofen, *Germany* 31 H6 47 30N 10 16 E
Soochow = Suzhou, *China* 77 B13 31 19N 120 38 E
Sop Hao, *Laos* 86 B5 20 33N 104 27 E
Sop Prap, *Thailand* 86 D2 17 53N 99 20 E
Sopachuy, *Bolivia* 173 D5 19 29 S 64 31W
Sopchoppy, *U.S.A.* 152 E5 30 4N 84 29W
Sopelana, *Spain* 40 B2 43 23N 2 58W
Soperton, *U.S.A.* 152 C7 32 23N 82 35W
Sopi, *Indonesia* 82 A3 2 34N 128 28 E
Sopo, Nahr →, *Sudan* 107 F2 8 40N 26 30 E
Sopot, *Bulgaria* 51 D8 42 39N 24 45 E
Sopot, *Poland* 54 D5 54 27N 18 31 E
Sopot, *Serbia, Yug.* 50 B4 44 29N 20 36 E
Sopotnica, *Macedonia* 50 E5 41 18N 21 13 E
Sopron, *Hungary* 52 C1 47 45N 16 32 E
Sop's Arm, *Canada* 141 C8 49 46N 56 56W
Sopur, *India* 93 B6 34 18N 74 27 E
Sør-Rondane, *Antarctica* 7 D4 72 0 S 25 0 E
Sør-Trøndelag □, *Norway* 18 A7 63 0N 11 0 E
Sora, *Italy* 45 G10 41 43N 13 37 E
Sorada, *India* 94 E7 19 45N 84 26 E
Sorah, *Pakistan* 92 F3 27 13N 68 56 E
Soråker, *Sweden* 16 B11 62 30N 17 32 E
Sorano, *Italy* 45 F8 42 41N 11 43 E
Sorata, *Bolivia* 172 D4 15 50 S 68 40W
Sorbas, *Spain* 41 H2 37 6N 2 7W
Sörbygden, *Sweden* 16 B10 62 48N 16 12 E
Sore, *France* 28 D3 44 18N 0 35W
Sorel, *Canada* 140 C5 46 0N 73 10W
Sörenberg, *Switz.* 32 C6 46 50N 8 2 E
Sorento, *U.S.A.* 156 F7 39 1N 89 35W
Soresina, *Italy* 44 C6 45 17N 9 51 E
Sørfjorden, *Norway* 18 D3 60 20N 6 37 E
Sörforsa, *Sweden* 16 C10 61 43N 16 58 E
Sórgono, *Italy* 46 B2 40 1N 9 6 E
Sorgues, *France* 29 D8 44 1N 4 53 E
Sorgun, *Turkey* 100 C6 39 46N 35 11 E
Soria, *Spain* 40 D2 41 43N 2 32W
Soria □, *Spain* 40 D2 41 46N 2 28W
Soriano, *Uruguay* 174 C4 33 24 S 58 19W
Soriano nel Cimino, *Italy* 45 F9 42 25N 12 14 E
Sorkh, Kuh-e, *Iran* 97 C8 35 40N 58 30 E
Sorø, *Denmark* 17 J5 55 26N 11 32 E
Soro, *Guinea* 112 C3 10 9N 9 48W
Soroca, *Moldova* 53 B13 48 8N 28 12 E
Sorocaba, *Brazil* 175 A6 23 31 S 47 27W
Sorochinsk, *Russia* 62 E4 52 26N 53 10 E
Soroki = Soroca, *Moldova* 53 B13 48 8N 28 12 E
Soron, *India* 93 F8 27 55N 78 45 E
Sorong, *Indonesia* 83 B4 0 55 S 131 15 E
Soroní, *Greece* 38 C10 36 21N 28 1 E
Soroti, *Uganda* 118 B3 1 43N 33 35 E
Sørøya, *Norway* 14 A20 70 40N 22 30 E
Sørøyane, *Norway* 18 B2 62 25N 5 32 E
Sørøysundet, *Norway* 14 A20 70 25N 23 0 E
Sorraia →, *Portugal* 43 G2 38 55N 8 53W
Sorrento, *Australia* 127 F3 38 22 S 144 47 E
Sorrento, *Italy* 47 B7 40 37N 14 22 E

284 Sorsele

Sorsele, *Sweden* **14 D17** 65 31N 17 30 E
Sörsjön, *Sweden* **16 C7** 61 24N 13 5 E
Sorso, *Italy* **46 B1** 40 48N 8 34 E
Sorsogon, *Phil.* **80 E5** 13 0N 124 0 E
Sorsogon □, *Phil.* **80 E4** 12 50N 123 55 E
Sortavala, *Russia* **58 B6** 61 42N 30 41 E
Sortino, *Italy* **47 E8** 37 9N 15 2 E
Sortland, *Norway* **14 B16** 68 42N 15 25 E
Sorūbī, *Afghan.* **91 B3** 34 36N 69 43 E
Sørum, *Norway* **18 D7** 60 30N 10 17 E
Sørvika, *Norway* **18 B8** 62 25N 11 54 E
Sorvizhi, *Russia* **60 B9** 57 52N 48 32 E
Sos = Sos del Rey Católico, *Spain* **40 C3** 42 30N 1 13W
Sos del Rey Católico, *Spain* **40 C3** 42 30N 1 13W
Sŏsan, *S. Korea* **75 F14** 36 47N 126 27 E
Soscumica, L., *Canada* **140 B4** 50 15N 77 27W
Sösdala, *Sweden* **17 H7** 56 2N 13 41 E
Sosna →, *Russia* **59 F10** 52 42N 38 55 E
Sosnogorsk, *Russia* **56 B9** 63 37N 53 51 E
Sosnovka, *Kirov, Russia* **60 B10** 56 17N 51 17 E
Sosnovka, *Tambov, Russia* **60 D5** 53 13N 41 24 E
Sosnovyy Bor, *Russia* **58 C5** 59 55N 29 6 E
Sosnowiec, *Poland* **55 H6** 50 20N 19 10 E
Sospel, *France* **29 E11** 43 52N 7 27 E
Šoštanj, *Slovenia* **45 B12** 46 23N 15 4 E
Sŏsura, *N. Korea* **75 C16** 42 16N 130 36 E
Sosva, *Russia* **62 B8** 59 10N 61 50 E
Sosva →, *Russia* **62 B9** 59 21N 62 23 E
Sotaseter, *Norway* **18 C4** 61 49N 7 43 E
Sotkamo, *Finland* **14 D23** 64 8N 28 23 E
Soto del Barco, *Spain* **42 B4** 43 32N 6 4W
Soto la Marina →, *Mexico* **163 C5** 23 40N 97 40W
Soto y Amío, *Spain* **42 C5** 42 46N 5 53W
Sotrondio, *Spain* **42 B5** 43 17N 5 36W
Sotuta, *Mexico* **163 C7** 20 29N 89 43W
Souanké, *Congo* **114 B2** 2 10N 14 3 E
Soúdha, *Greece* **38 D6** 35 29N 24 4 E
Soúdhas, Kólpos, *Greece* **38 D6** 35 25N 24 10 E
Soufflay, *Congo* **114 B2** 2 1N 14 54 E
Souflíon, *Greece* **51 E10** 41 12N 26 18 E
Souillac, *France* **28 D5** 44 53N 1 29 E
Souilly, *France* **27 C12** 49 3N 5 17 E
Souk-Ahras, *Algeria* **111 A6** 36 23N 7 57 E
Souk el Arba du Rharb, *Morocco* **110 B3** 34 43N 5 59W
Soukhouma, *Laos* **86 E5** 14 38N 105 48 E
Sŏul, *S. Korea* **75 F14** 37 31N 126 58 E
Soulac-sur-Mer, *France* **28 C2** 45 30N 1 7W
Soultz-sous-Forêts, *France* **27 D14** 48 57N 7 52 E
Sound, The = Øresund, *Europe* **17 J6** 55 45N 12 45 E
Sound, The, *U.K.* **21 G3** 50 20N 4 10W
Soúnion, Ákra, *Greece* **48 D6** 37 37N 24 1 E
Sour el Ghozlane, *Algeria* **111 A5** 36 10N 3 45 E
Sources, Mt. aux, *Lesotho* **117 D4** 28 45 S 28 50 E
Soure, *Brazil* **170 B2** 0 35 S 48 30W
Soure, *Portugal* **42 E2** 40 4N 8 38W
Souris, *Man., Canada* **143 D8** 49 40N 100 20W
Souris, *P.E.I., Canada* **141 C7** 46 21N 62 15W
Souris →, *Canada* **154 A5** 49 40N 99 34W
Soúrpi, *Greece* **48 B4** 39 6N 22 54 E
Sousa, *Brazil* **170 C4** 6 45 S 38 10W
Sousel, *Brazil* **170 B1** 2 38 S 52 29W
Sousel, *Portugal* **43 G3** 38 57N 7 40W
Souss, O. →, *Morocco* **110 B3** 30 27N 9 31W
Sousse, *Tunisia* **108 A2** 35 50N 10 38 E
Soustons, *France* **28 E2** 43 45N 1 19W
South Africa ■, *Africa* **116 E3** 32 0 S 23 0 E
South Atlantic Ocean **9 L8** 20 0 S 10 0W
South Aulatsivik I., *Canada* **141 A7** 56 45N 61 30W
South Australia □, *Australia* **127 E2** 32 0 S 139 0 E
South Ayrshire □, *U.K.* **22 F4** 55 18N 4 41W
South Baldy, *U.S.A.* **159 J10** 33 59N 107 11W
South Bay, *U.S.A.* **153 J9** 26 40N 80 43W
South Beloit, *U.S.A.* **156 B7** 42 29N 89 2W
South Bend, *Ind., U.S.A.* **157 C10** 41 41N 86 15W
South Bend, *Wash., U.S.A.* **160 D3** 46 40N 123 48W
South Boston, *U.S.A.* **149 G6** 36 42N 78 54W
South Branch, *Canada* **141 C8** 47 55N 59 2W
South Brook, *Canada* **141 C8** 49 26N 56 5W
South Carolina □, *U.S.A.* **149 J5** 34 0N 81 0W
South Charleston, *U.S.A.* **148 F5** 38 22N 81 44W
South China Sea, *Asia* **134 G3** 10 0N 113 0 E
South Congaree, *U.S.A.* **152 B9** 33 53N 81 9W
South Cotabato □, *Phil.* **81 J5** 6 0N 125 0 E
South Dakota □, *U.S.A.* **154 C5** 44 15N 100 0W
South Daytona, *U.S.A.* **153 F8** 29 10N 81 0W
South Downs, *U.K.* **21 G7** 50 52N 0 25W
South East C., *Australia* **126 G4** 43 40 S 146 50 E
South East Is., *Australia* **125 F3** 34 17 S 123 30 E
South Esk →, *U.K.* **22 E6** 56 43N 2 31W
South Foreland, *U.K.* **21 F9** 51 8N 1 24 E
South Fork →, *U.S.A.* **158 C7** 47 54N 113 15W
South Fork, American →, *U.S.A.* **160 G5** 38 45N 121 5W
South Fork, Feather →, *U.S.A.* **160 F5** 39 17N 121 36W
South Fork Edisto →, *U.S.A.* **152 B9** 33 16N 80 54W
South Georgia, *Antarctica* **7 B1** 54 30 S 37 0W
South Gloucestershire □, *U.K.* **21 F5** 51 32N 2 28W
South Grand →, *U.S.A.* **156 F3** 38 17N 93 25W
South Haven, *U.S.A.* **157 B10** 42 24N 86 16W
South Henik, L., *Canada* **143 A9** 61 30N 97 30W
South Honshu Ridge, *Pac. Oc.* **134 E6** 23 0N 143 0 E
South Horr, *Kenya* **118 B4** 2 12N 36 56 E
South I., *Kenya* **118 B4** 2 35N 36 35 E
South I., *N.Z.* **131 E5** 44 0 S 170 0 E
South Invercargill, *N.Z.* **131 G3** 46 26 S 168 23 E
South Knife →, *Canada* **143 B10** 58 55N 94 37W
South Korea ■, *Asia* **75 G15** 36 0N 128 0 E
South Lake Tahoe, *U.S.A.* **160 G6** 38 57N 119 59W
South Lanarkshire □, *U.K.* **22 F5** 55 37N 3 53W
South Loup →, *U.S.A.* **154 E5** 41 4N 98 39W
South Lyon, *U.S.A.* **157 B13** 42 28N 83 39W
South Magnetic Pole, *Antarctica* **7 C9** 64 8 S 138 8 E
South Miami, *U.S.A.* **153 K9** 25 42N 80 18W
South Milwaukee, *U.S.A.* **157 B9** 42 55N 87 52W
South Molton, *U.K.* **21 F4** 51 1N 3 51W
South Nahanni →, *Canada* **142 A4** 61 3N 123 21W
South Natuna Is. = Natuna Selatan, Kepulauan, *Indonesia* **85 B3** 2 45N 109 0 E
South Negril Pt., *Jamaica* **164 C4** 18 14N 78 30W
South Newport, *U.S.A.* **152 D8** 31 38N 81 24W
South Orkney Is., *Antarctica* **7 C18** 63 0 S 45 0W
South Ossetia □, *Georgia* **61 J7** 42 21N 44 2 E
South Pagai, I. = Pagai Selatan, *Indonesia* **84 C2** 3 0 S 100 15 E
South Pass, *U.S.A.* **158 E9** 42 20N 108 58W
South Pekin, *U.S.A.* **156 D7** 40 30N 89 39W
South Pittsburg, *U.S.A.* **149 H3** 35 1N 85 42W
South Platte →, *U.S.A.* **154 E4** 41 7N 100 42W
South Pole, *Antarctica* **7 E** 90 0 S 0 0 E
South Ponte Vedra Beach, *U.S.A.* **152 E8** 30 3N 81 20W
South Porcupine, *Canada* **140 C3** 48 30N 81 12W
South River, *Canada* **140 C4** 45 52N 79 23W
South River, *U.S.A.* **151 F10** 40 27N 74 23W
South Ronaldsay, *U.K.* **22 C6** 58 48N 2 58W
South Sandwich Is., *Antarctica* **9 P6** 57 0 S 27 0W
South Saskatchewan →, *Canada* **143 C7** 53 15N 105 5W
South Seal →, *Canada* **143 B9** 58 48N 98 8W
South Shetland Is., *Antarctica* **7 C18** 62 0 S 59 0W
South Shields, *U.K.* **20 C6** 55 0N 1 25W
South Sioux City, *U.S.A.* **154 D6** 42 28N 96 24W
South Taranaki Bight, *N.Z.* **130 F3** 39 40 S 174 5 E
South Thompson →, *Canada* **142 C4** 50 40N 120 20W
South Twin I., *Canada* **140 B4** 53 7N 79 52W
South Tyne →, *U.K.* **20 C5** 54 59N 2 8W
South Uist, *U.K.* **22 D1** 57 20N 7 15W
South Venice, *U.S.A.* **153 H7** 27 3N 82 25W
South Wayne, *U.S.A.* **156 B7** 42 34N 89 53W
South West Africa = Namibia ■, *Africa* **116 C2** 22 0 S 18 9 E
South West C., *Australia* **126 G4** 43 34 S 146 3 E
South Whitley, *U.S.A.* **157 C11** 41 5N 85 38W
South Yorkshire □, *U.K.* **20 D6** 53 27N 1 36W
Southampton, *Canada* **140 D3** 44 30N 81 25W
Southampton, *U.K.* **21 G6** 50 54N 1 23W
Southampton, *U.S.A.* **151 F12** 40 53N 72 23W
Southampton □, *U.K.* **21 G6** 50 54N 1 23W
Southampton I., *Canada* **139 B11** 64 30N 84 0W
Southbridge, *N.Z.* **131 D7** 43 48 S 172 16 E
Southbridge, *U.S.A.* **151 D12** 42 5N 72 2W
Southend, *Canada* **143 B8** 56 19N 103 22W
Southend-on-Sea, *U.K.* **21 F8** 51 32N 0 44 E
Southend-on-Sea □, *U.K.* **21 F8** 51 32N 0 44 E
Southern □, *Malawi* **119 F4** 15 0 S 35 0 E
Southern □, *S. Leone* **112 D2** 8 0N 12 30W
Southern □, *Uganda* **118 C3** 0 15 S 31 30 E
Southern □, *Zambia* **119 F2** 16 20 S 26 20 E
Southern Alps, *N.Z.* **131 D5** 43 41 S 170 11 E
Southern Cross, *Australia* **125 F2** 31 12 S 119 15 E
Southern Hills, *Australia* **125 F3** 32 15 S 122 40 E
Southern Indian L., *Canada* **143 B9** 57 10N 98 30W
Southern Leyte □, *Phil.* **81 F5** 10 30N 125 10 E
Southern Ocean, *Antarctica* **7 C6** 62 0 S 60 0 E
Southern Pines, *U.S.A.* **149 H6** 35 11N 79 24W
Southern Uplands, *U.K.* **22 F5** 55 28N 3 52W
Southfield, *U.S.A.* **157 B13** 42 29N 83 17W
Southington, *U.S.A.* **151 E12** 41 36N 72 53W
Southland □, *N.Z.* **131 F3** 45 30 S 168 0 E
Southold, *U.S.A.* **151 E12** 41 4N 72 26W
Southport, *Australia* **127 D5** 27 58 S 153 25 E
Southport, *U.K.* **20 D4** 53 39N 3 0W
Southport, *U.S.A.* **149 J6** 33 55N 78 1W
Southwest C., *N.Z.* **131 H2** 47 17 S 167 28 E
Southwold, *U.K.* **21 E9** 52 20N 1 41 E
Soutpansberg, *S. Africa* **117 C4** 23 0 S 29 30 E
Souvigny, *France* **27 F10** 46 33N 3 10 E
Sovata, *Romania* **53 D10** 46 35N 25 3 E
Soverato, *Italy* **47 D9** 38 41N 16 33 E
Sovetsk, *Kaliningd., Russia* **15 J19** 55 6N 21 50 E
Sovetsk, *Kirov, Russia* **60 B9** 57 38N 48 53 E
Sovetskaya Gavan = Vanino, *Russia* **65 E15** 48 50N 140 5 E
Sovicille, *Italy* **45 E8** 43 17N 11 13 E
Søvik, *Norway* **18 B3** 62 33N 6 17 E
Soweto, *S. Africa* **117 D4** 26 14 S 27 54 E
Sōya-Kaikyō = La Perouse Str., *Asia* **68 B11** 45 40N 142 0 E
Sōya-Misaki, *Japan* **68 B10** 45 30N 141 55 E
Soyaux, *France* **28 C4** 45 39N 0 12 E
Soyo, *Angola* **115 D2** 6 13 S 12 20 E
Sozh →, *Belarus* **59 F6** 51 57N 30 48 E
Sozopol, *Bulgaria* **51 D11** 42 23N 27 42 E
Spa, *Belgium* **24 D5** 50 29N 5 53 E
Spain ■, *Europe* **13 H5** 39 0N 4 0W
Spalding, *Australia* **128 B3** 33 30 S 138 37 E
Spalding, *U.K.* **20 E7** 52 48N 0 9W
Spalding, *U.S.A.* **154 E5** 41 42N 98 22W
Spangler, *U.S.A.* **150 F6** 40 39N 78 48W
Spaniard's Bay, *Canada* **141 C9** 47 38N 53 20W
Spanish, *Canada* **140 C3** 46 12N 82 20W
Spanish Fork, *U.S.A.* **158 F8** 40 7N 111 39W
Spanish Town, *Jamaica* **164 C4** 18 0N 76 57W
Sparks, *Ga., U.S.A.* **152 D6** 31 11N 83 26W
Sparks, *Nev., U.S.A.* **160 F7** 39 32N 119 45W
Sparr, *U.S.A.* **153 F7** 29 20N 82 7W
Sparreholm, *Sweden* **16 E10** 59 4N 16 49 E
Sparta = Spárti, *Greece* **48 D4** 37 5N 22 25 E
Sparta, *Ga., U.S.A.* **152 B7** 33 17N 82 58W
Sparta, *Ill., U.S.A.* **156 F7** 38 8N 89 42W
Sparta, *Mich., U.S.A.* **157 A11** 43 10N 85 42W
Sparta, *Wis., U.S.A.* **154 D9** 43 56N 90 49W
Spartanburg, *U.S.A.* **149 H5** 34 56N 81 57W
Spartansburg, *U.S.A.* **150 E5** 41 49N 79 41W
Spartel, C., *Morocco* **110 A3** 35 47N 5 56W
Spárti, *Greece* **48 D4** 37 5N 22 25 E
Spartivento, C., *Calabria, Italy* **47 E9** 37 55N 16 4 E
Spartivento, C., *Sard., Italy* **46 D1** 38 53N 8 50 E
Spas-Demensk, *Russia* **58 E7** 54 20N 34 0 E
Spas-Klepiki, *Russia* **58 E11** 55 10N 40 10 E
Spassk Dalniy, *Russia* **68 B6** 44 40N 132 48 E
Spassk-Ryazanskiy, *Russia* **58 E11** 54 24N 40 25 E
Spátha, Ákra, *Greece* **38 D5** 35 42N 23 43 E
Spatsizi →, *Canada* **142 B3** 57 42N 128 7W
Spean →, *U.K.* **22 E4** 56 55N 4 59W
Spearfish, *U.S.A.* **154 C3** 44 30N 103 52W
Spearman, *U.S.A.* **155 G4** 36 12N 101 12W
Speed, *Australia* **128 C5** 35 21 S 142 27 E
Speedway, *U.S.A.* **157 E10** 39 47N 86 15W
Speer, *Switz.* **33 B8** 47 12N 9 8 E
Speers, *Canada* **143 C7** 52 43N 107 34W
Speia, *Moldova* **53 D14** 46 59N 29 19 E
Speightstown, *Barbados* **165 D8** 13 15N 59 39W
Speke Gulf, *Tanzania* **118 C3** 2 20 S 32 50 E
Spello, *Italy* **45 F9** 42 59N 12 40 E
Spencer, *Idaho, U.S.A.* **158 D7** 44 22N 112 11W
Spencer, *Ind., U.S.A.* **157 E10** 39 17N 86 46W
Spencer, *Iowa, U.S.A.* **154 D7** 43 9N 95 9W
Spencer, *N.Y., U.S.A.* **151 D8** 42 13N 76 30W
Spencer, *Nebr., U.S.A.* **154 D5** 42 53N 98 42W
Spencer, *W. Va., U.S.A.* **148 F5** 38 48N 81 21W
Spencer, C., *Australia* **128 C2** 35 20 S 136 53 E
Spencer B., *Namibia* **116 D1** 25 30 S 14 47 E
Spencer G., *Australia* **128 C2** 34 0 S 137 20 E
Spencerville, *Canada* **151 B9** 44 51N 75 33W
Spencerville, *U.S.A.* **157 D12** 40 43N 84 21W
Spences Bridge, *Canada* **142 C4** 50 25N 121 20W
Spennymoor, *U.K.* **20 C6** 54 42N 1 36W
Spenser Mts., *N.Z.* **131 C7** 42 15 S 172 45 E
Spentrup, *Denmark* **17 H4** 56 33N 10 2 E
Sperillen, *Norway* **18 D7** 60 28N 10 3 E
Sperkhiós →, *Greece* **48 C4** 38 57N 22 3 E
Sperrin Mts., *U.K.* **23 B5** 54 50N 7 0W
Spessart, *Germany* **31 F5** 49 56N 9 18 E
Spetalen, *Norway* **18 E7** 59 18N 10 45 E
Spétsai, *Greece* **48 D5** 37 15N 23 10 E
Spey →, *U.K.* **22 D5** 57 40N 3 6W
Speyer, *Germany* **31 F4** 49 29N 8 25 E
Spezzano Albanese, *Italy* **47 C9** 39 40N 16 19 E
Spickard, *U.S.A.* **156 D3** 40 14N 93 36W
Spiekeroog, *Germany* **30 B3** 53 46N 7 42 E
Spiez, *Switz.* **32 C5** 46 40N 7 40 E
Spíli, *Greece* **38 D6** 35 13N 24 31 E
Spilimbergo, *Italy* **45 B9** 46 7N 12 54 E
Spin Baldak = Qala-i-Jadid, *Afghan.* **92 D2** 31 1N 66 25 E
Spinalónga, *Greece* **38 D7** 35 18N 25 44 E
Spinazzola, *Italy* **47 B9** 40 58N 16 5 E
Spineni, *Romania* **53 F9** 44 43N 24 37 E
Spirit Lake, *Idaho, U.S.A.* **158 C5** 47 58N 116 52W
Spirit Lake, *Wash., U.S.A.* **160 D4** 46 15N 122 9W
Spirit River, *Canada* **142 B5** 55 45N 118 50W
Spiritwood, *Canada* **143 C7** 53 24N 107 33W
Spišská Nová Ves, *Slovak Rep.* **35 C13** 48 58N 20 34 E
Spišské Podhradie, *Slovak Rep.* **35 B13** 49 0N 20 48 E
Spital, *Austria* **34 D7** 47 42N 14 18 E
Spithead, *U.K.* **21 G6** 50 45N 1 10W
Spittal an der Drau, *Austria* **34 E6** 46 48N 13 31 E
Spitzbergen = Svalbard, *Arctic* **6 B8** 78 0N 17 0 E
Spjelkavik, *Norway* **15 E12** 62 28N 6 22 E
Split, *Croatia* **45 E13** 43 31N 16 26 E
Split L., *Canada* **143 B9** 56 8N 96 15W
Splitski Kanal, *Croatia* **45 E13** 43 31N 16 20 E
Splügen, *Switz.* **33 C8** 46 34N 9 21 E
Splügenpass, *Switz.* **33 D8** 46 30N 9 20 E
Spofford, *U.S.A.* **155 L4** 29 10N 100 25W
Spokane, *U.S.A.* **158 C5** 47 40N 117 24W
Spoleto, *Italy* **45 F9** 42 44N 12 44 E
Sponvika, *Norway* **18 E8** 59 7N 11 15 E
Spoon →, *U.S.A.* **156 D6** 40 19N 90 4W
Spooner, *U.S.A.* **154 C9** 45 50N 91 53W
Sporyy Navolok, Mys, *Russia* **64 B7** 75 50N 68 40 E
Spragge, *Canada* **140 C3** 46 15N 82 40W
Sprague, *U.S.A.* **158 C5** 47 18N 117 59W
Sprague River, *U.S.A.* **158 E3** 42 27N 121 30W
Spray, *U.S.A.* **158 D4** 44 50N 119 48W
Spreča →, *Bos.-H.* **52 F3** 44 44N 18 6 E
Spree →, *Germany* **30 C9** 52 32N 13 13 E
Spreewald, *Germany* **30 D9** 51 58N 13 51 E
Spremberg, *Germany* **30 D10** 51 34N 14 22 E
Sprengisandur, *Iceland* **11 C8** 64 52N 18 7W
Spring City, *U.S.A.* **158 G8** 39 29N 111 30W
Spring Cr. →, *U.S.A.* **158 E5** 30 54N 84 45W
Spring Garden, *U.S.A.* **160 F6** 39 52N 120 47W
Spring Green, *U.S.A.* **156 A6** 43 11N 90 4W
Spring Hill, *Australia* **129 B8** 33 23 S 149 9 E
Spring Hill, *Ala., U.S.A.* **152 D4** 31 42N 85 58W
Spring Hill, *Fla., U.S.A.* **153 G7** 28 27N 82 41W
Spring Mts., *U.S.A.* **159 H6** 36 0N 115 45W
Spring Valley, *Calif., U.S.A.* **161 N10** 32 45N 117 5W
Spring Valley, *Ill., U.S.A.* **156 C7** 41 20N 89 12W
Spring Valley, *Minn., U.S.A.* **154 D8** 43 41N 92 23W
Springbok, *S. Africa* **116 D2** 29 42 S 17 54 E
Springdale, *Canada* **141 C8** 49 30N 56 6W
Springdale, *Ark., U.S.A.* **155 G7** 36 11N 94 8W
Springdale, *Wash., U.S.A.* **158 B5** 48 4N 117 45W
Springe, *Germany* **30 C5** 52 13N 9 33 E
Springer, *U.S.A.* **155 G2** 36 22N 104 36W
Springerville, *U.S.A.* **159 J9** 34 8N 109 17W
Springfield, *Canada* **150 D4** 42 50N 80 56W
Springfield, *N.Z.* **131 D6** 43 19 S 171 56 E
Springfield, *Colo., U.S.A.* **155 G3** 37 24N 102 37W
Springfield, *Fla., U.S.A.* **152 E4** 30 10N 85 37W
Springfield, *Ga., U.S.A.* **152 C8** 32 22N 81 18W
Springfield, *Ill., U.S.A.* **156 E7** 39 48N 89 39W
Springfield, *Ky., U.S.A.* **157 G11** 37 41N 85 13W
Springfield, *Mass., U.S.A.* **151 D12** 42 6N 72 35W
Springfield, *Mo., U.S.A.* **155 G8** 37 13N 93 17W
Springfield, *Ohio, U.S.A.* **157 E13** 39 55N 83 49W
Springfield, *Oreg., U.S.A.* **158 D2** 44 3N 123 1W
Springfield, *S.C., U.S.A.* **152 B9** 33 30N 81 17W
Springfield, *Tenn., U.S.A.* **149 G2** 36 31N 86 53W
Springfield, *Vt., U.S.A.* **151 C12** 43 18N 72 29W
Springfield, L., *U.S.A.* **156 E7** 39 46N 89 36W
Springfontein, *S. Africa* **116 E4** 30 15 S 25 40 E
Springhill, *Canada* **141 C7** 45 40N 64 4W
Springhouse, *Canada* **142 C4** 51 56N 122 7W
Springhurst, *Australia* **129 D7** 36 10 S 146 31 E
Springs, *S. Africa* **117 D4** 26 13 S 28 25 E
Springsure, *Australia* **126 C4** 24 8 S 148 6 E
Springvale, *Queens., Australia* **126 C3** 23 33 S 140 42 E
Springvale, *W. Austral., Australia* **124 C4** 17 48 S 127 41 E
Springvale, *Ga., U.S.A.* **152 D5** 31 50N 84 53W
Springvale, *Maine, U.S.A.* **151 C14** 43 28N 70 48W
Springville, *Calif., U.S.A.* **160 J8** 36 8N 118 49W
Springville, *N.Y., U.S.A.* **150 D6** 42 31N 78 40W
Springville, *Utah, U.S.A.* **158 F8** 40 10N 111 37W
Springwater, *Canada* **143 C7** 51 58N 108 23W
Spruce-Creek, *U.S.A.* **150 F6** 40 36N 78 9W
Spur, *U.S.A.* **155 J4** 33 28N 100 52W
Spurgeon, *U.S.A.* **157 F9** 38 14N 87 15W
Spurn Hd., *U.K.* **20 D8** 53 35N 0 8 E
Spuž, *Montenegro, Yug.* **50 D3** 42 32N 19 10 E
Spuzzum, *Canada* **142 D4** 49 37N 121 23W
Spydeberg, *Norway* **18 E8** 59 36N 11 5 E
Squam L., *U.S.A.* **151 C13** 43 45N 71 32W
Squamish, *Canada* **142 D4** 49 45N 123 10W
Square Islands, *Canada* **141 B8** 52 47N 55 47W
Squillace, G. di, *Italy* **47 D9** 38 45N 16 50 E
Squinzano, *Italy* **47 B11** 40 26N 18 2 E
Squires, Mt., *Australia* **125 E4** 26 14 S 127 28 E
Sragen, *Indonesia* **85 D4** 7 26 S 111 2 E
Srbac, *Bos.-H.* **52 E2** 45 7N 17 30 E
Srbija = Serbia □, *Yugoslavia* **50 C5** 43 30N 21 0 E
Srbobran, *Serbia, Yug.* **52 E4** 45 32N 19 48 E
Sre Ambel, *Cambodia* **87 G4** 11 8N 103 46 E
Sre Khtum, *Cambodia* **87 F6** 12 10N 106 52 E
Sre Umbell = Sre Ambel, *Cambodia* **87 G4** 11 8N 103 46 E
Srebrenica, *Bos.-H.* **52 F4** 44 6N 19 18 E
Sredinny Ra. = Sredinnyy Khrebet, *Russia* **65 D17** 57 0N 160 0 E
Sredinnyy Khrebet, *Russia* **65 D17** 57 0N 160 0 E
Središče, *Slovenia* **45 B13** 46 24N 16 17 E
Sredna Gora, *Bulgaria* **51 D8** 42 40N 24 20 E
Srednekolymsk, *Russia* **65 C16** 67 27N 153 40 E
Sredni Rodopi, *Bulgaria* **51 E8** 41 40N 24 45 E
Srednogorie, *Bulgaria* **51 D8** 42 43N 24 10 E
Śrem, *Poland* **55 F4** 52 6N 17 2 E
Sremska Mitrovica, *Serbia, Yug.* **52 F4** 44 59N 19 38 E
Sremski Karlovci, *Serbia, Yug.* **52 E4** 45 12N 19 56 E
Srepok →, *Cambodia* **86 F6** 13 33N 106 16 E
Sretensk, *Russia* **65 D12** 52 10N 117 40 E
Sri Kalahasti, *India* **95 H4** 13 45N 79 44 E
Sri Lanka ■, *Asia* **95 L5** 7 30N 80 50 E
Sriharikota I., *India* **95 H5** 13 40N 80 20 E
Srikakulam, *India* **94 E6** 18 14N 83 58 E
Srinagar, *India* **93 B6** 34 5N 74 50 E
Sripur, *Bangla.* **90 C3** 24 14N 90 30 E
Srirangam, *India* **95 J4** 10 54N 78 42 E
Srivardhan, *India* **94 E1** 18 4N 73 3 E
Srivilliputtur, *India* **95 K3** 9 31N 77 40 E
Środa Śląska, *Poland* **55 G3** 51 10N 16 36 E
Środa Wielkopolski, *Poland* **55 F4** 52 15N 17 19 E
Srpska Crnja, *Serbia, Yug.* **52 E5** 45 38N 20 44 E
Srpski Itebej, *Serbia, Yug.* **52 E5** 45 35N 20 44 E
Staaten →, *Australia* **126 B3** 16 24 S 141 17 E
Staberhuk, *Germany* **30 A7** 54 23N 11 18 E
Stade, *Germany* **30 B5** 53 35N 9 29 E
Staðarhólskirkja, *Iceland* **11 B5** 65 23N 21 58W
Staðastaður, *Iceland* **11 C3** 64 49N 23 1W
Stadhavet, *Norway* **18 B2** 62 13N 5 0 E
Staður, *Húnavatnssýsla, Iceland* **11 B5** 65 9N 21 3W
Staður, *Ísafjarðarsýsla, Iceland* **11 A4** 66 15N 22 50W
Stadlandet, *Norway* **18 B2** 62 10N 5 10 E
Stadskanaal, *Neths.* **24 A6** 53 4N 6 55 E
Stadtallendorf, *Germany* **30 E5** 50 48N 9 1 E
Stadthagen, *Germany* **30 C5** 52 19N 9 13 E
Stadtlohn, *Germany* **30 D2** 51 59N 6 55 E
Stadtroda, *Germany* **30 E7** 50 52N 11 44 E
Stäfa, *Switz.* **33 B7** 47 14N 8 45 E
Stafafell, *Iceland* **11 C12** 64 25N 14 52W
Staffa, *U.K.* **22 E2** 56 27N 6 21W
Staffanstorp, *Sweden* **17 J7** 55 39N 13 13 E
Stafford, *U.K.* **20 E5** 52 49N 2 7W
Stafford, *U.S.A.* **155 G5** 37 58N 98 36W
Stafford, L., *U.S.A.* **153 F7** 29 20N 82 29W
Stafford Springs, *U.S.A.* **151 E12** 41 57N 72 18W
Staffordshire □, *U.K.* **20 E5** 52 53N 2 10W
Stagnone, *Italy* **46 E5** 37 53N 12 26 E
Staines, *U.K.* **21 F7** 51 26N 0 29W
Stainz, *Austria* **34 E8** 46 53N 15 17 E
Stakhanov, *Ukraine* **59 H10** 48 35N 38 40 E
Stalać, *Serbia, Yug.* **50 C5** 43 43N 21 28 E
Stalden, *Switz.* **32 D5** 46 14N 7 52 E
Stalingrad = Volgograd, *Russia* **61 F7** 48 40N 44 25 E
Staliniri = Tskhinvali, *Georgia* **61 J7** 42 14N 44 1 E
Stalino = Donetsk, *Ukraine* **59 J9** 48 0N 37 45 E
Stalinogorsk = Novomoskovsk, *Russia* **58 E10** 54 5N 38 15 E
Stalis, *Greece* **38 D7** 35 17N 25 25 E
Stallarholmen, *Sweden* **16 E11** 59 22N 17 12 E
Ställdalen, *Sweden* **16 E8** 59 56N 14 56 E
Stalowa Wola, *Poland* **55 H9** 50 34N 22 3 E
Stalybridge, *U.K.* **20 D5** 53 28N 2 3W
Stamford, *Australia* **126 C3** 21 15 S 143 46 E
Stamford, *U.K.* **21 E7** 52 39N 0 29W
Stamford, *Conn., U.S.A.* **151 E11** 41 3N 73 32W
Stamford, *Tex., U.S.A.* **155 J5** 32 57N 99 48W
Stamnes, *Norway* **18 D2** 60 40N 5 45 E
Stamping Ground, *U.S.A.* **157 F12** 38 16N 84 41W
Stamps, *U.S.A.* **155 J8** 33 22N 93 30W
Stanberry, *U.S.A.* **154 E7** 40 13N 94 35W
Stančevo = Kalipetrovo, *Bulgaria* **51 B11** 44 5N 27 14 E
Standerton, *S. Africa* **117 D4** 26 55 S 29 7 E
Standish, *U.S.A.* **148 D4** 43 59N 83 57W
Stanford, *U.S.A.* **158 C8** 47 9N 110 13W
Stånga, *Sweden* **17 G12** 57 17N 18 29 E
Stange, *Norway* **18 D8** 60 43N 11 5 E
Stanger, *S. Africa* **117 D5** 29 27 S 31 14 E
Stangvik, *Norway* **18 B5** 62 55N 8 28 E
Stanhope, *Australia* **129 D6** 36 27 S 144 59 E
Stanhope, *U.S.A.* **156 B3** 42 17N 93 48W
Stanišić, *Serbia, Yug.* **52 E4** 45 56N 19 10 E
Stanislaus →, *U.S.A.* **160 H5** 37 40N 121 14W
Stanislav = Ivano-Frankivsk, *Ukraine* **59 H3** 48 40N 24 40 E
Stanisławów, *Poland* **55 F8** 52 18N 21 33 E
Stanley, *Australia* **126 G4** 40 46 S 145 19 E
Stanley, *N.B., Canada* **141 C6** 46 20N 66 44W
Stanley, *Sask., Canada* **143 B8** 55 24N 104 22W
Stanley, *Falk. Is.* **176 D5** 51 40 S 59 51W
Stanley, *U.K.* **20 C6** 54 53N 1 41W
Stanley, *Idaho, U.S.A.* **158 D6** 44 13N 114 56W
Stanley, *N. Dak., U.S.A.* **154 A3** 48 19N 102 23W
Stanley, *N.Y., U.S.A.* **150 D7** 42 48N 77 6W
Stanley, *Wis., U.S.A.* **154 C9** 44 58N 90 56W
Stanley Res., *India* **95 J3** 11 50N 77 40 E
Stanovoy Khrebet, *Russia* **65 D14** 55 0N 130 0 E
Stanovoy Ra. = Stanovoy Khrebet, *Russia* **65 D14** 55 0N 130 0 E
Stans, *Switz.* **33 C6** 46 58N 8 21 E
Stansmore Ra., *Australia* **124 D4** 21 23 S 128 33 E
Stanthorpe, *Australia* **127 D5** 28 36 S 151 59 E
Stanton, *U.S.A.* **155 J4** 32 8N 101 48W
Stantsiya Karshi, *Uzbekistan* **63 D2** 38 49N 65 47 E
Stanwood, *U.S.A.* **160 B4** 48 15N 122 23W
Stapi, *Iceland* **11 C3** 64 46N 23 38W
Staples, *U.S.A.* **154 B7** 46 21N 94 48W
Stapleton, *U.S.A.* **154 E4** 41 29N 100 31W
Stąporków, *Poland* **55 G7** 51 9N 20 31 E
Star City, *Canada* **143 C8** 52 50N 104 20W
Stará L'ubovňa, *Slovak Rep.* **35 B13** 49 18N 20 42 E
Stara Moravica, *Serbia, Yug.* **52 E4** 45 50N 19 30 E
Stara Pazova, *Serbia, Yug.* **52 F5** 44 58N 20 10 E
Stara Planina, *Bulgaria* **50 C7** 43 15N 23 0 E
Stará Turá, *Slovak Rep.* **35 C10** 48 47N 17 42 E
Stara Zagora, *Bulgaria* **51 D9** 42 26N 25 39 E

Süderbrarup 285

Starachowice, *Poland* **55 G8** 51 3N 21 2 E
Staraya Russa, *Russia* **58 D6** 57 58N 31 23 E
Starbuck I., *Kiribati* **135 H12** 5 37 S 155 55W
Starchiojd, *Romania* **53 E11** 45 19N 26 11 E
Stargard Szczeciński, *Poland* **54 E2** 53 20N 15 0 E
Stårheim, *Norway* **18 C2** 61 56N 5 40 E
Stari Bar, *Montenegro, Yug.* **50 D3** 42 7N 19 10 E
Stari Trg, *Slovenia* **45 C12** 45 29N 15 7 E
Staritsa, *Russia* **58 D8** 56 33N 34 55 E
Starke, *U.S.A.* **152 F7** 29 57N 82 7W
Starkville, *Colo., U.S.A.* **155 G2** 37 8N 104 30W
Starkville, *Miss., U.S.A.* **149 J1** 33 28N 88 49W
Starnberg, *Germany* **31 H7** 48 0N 11 21 E
Starnberger See, *Germany* **31 H7** 47 54N 11 19 E
Starobilsk, *Ukraine* **59 H10** 49 16N 39 0 E
Starodub, *Russia* **59 F7** 52 30N 32 50 E
Starogard Gdański, *Poland* **54 E5** 53 59N 18 30 E
Starokonstantinov = Starokonstyantyniv, *Ukraine* **59 H4** 49 48N 27 10 E
Starokonstyantyniv, *Ukraine* **59 H4** 49 48N 27 10 E
Starominskaya, *Russia* **59 J10** 46 33N 39 0 E
Staroshcherbinovskaya, *Russia* **59 J10** 46 40N 38 53 E
Stars Mill, *U.S.A.* **152 B5** 33 19N 84 31W
Start Pt., *U.K.* **21 G4** 50 13N 3 39W
Stary Sącz, *Poland* **55 J7** 49 33N 20 35 E
Staryy Biryuzyak, *Russia* **61 H8** 44 46N 46 50 E
Staryy Chartoriysk, *Ukraine* **59 G3** 51 15N 25 54 E
Staryy Krym, *Ukraine* **59 K8** 45 3N 35 8 E
Staryy Oskol, *Russia* **59 G9** 51 19N 37 55 E
Stassfurt, *Germany* **30 D7** 51 51N 11 35 E
Staszów, *Poland* **55 H8** 50 33N 21 10 E
State Center, *U.S.A.* **156 B3** 42 1N 93 10W
State College, *U.S.A.* **150 F7** 40 48N 77 52W
Stateline, *U.S.A.* **160 G7** 38 57N 119 56W
Staten, I. = Estados, I. de Los, *Argentina* **176 D4** 54 40 S 64 30W
Staten I., *U.S.A.* **151 F10** 40 35N 74 9W
Statenville, *U.S.A.* **152 E6** 30 42N 83 2W
Statesboro, *U.S.A.* **152 C8** 32 27N 81 47W
Statesville, *U.S.A.* **149 H5** 35 47N 80 53W
Statham, *U.S.A.* **152 B6** 33 58N 83 35W
Stathelle, *Norway* **18 E6** 59 3N 9 41 E
Stauffer, *U.S.A.* **161 L7** 34 45N 119 3W
Staunton, *Ill., U.S.A.* **156 F7** 39 1N 89 47W
Staunton, *Va., U.S.A.* **148 F6** 38 9N 79 4W
Stavanger, *Norway* **15 G11** 58 57N 5 40 E
Staveley, *N.Z.* **131 D6** 43 40 S 171 32 E
Stavelot, *Belgium* **24 D5** 50 23N 5 55 E
Stavern, *Norway* **15 G14** 59 0N 10 1 E
Stavoren, *Neths.* **24 B5** 52 53N 5 22 E
Stavropol, *Russia* **61 H6** 45 5N 42 0 E
Stavros, *Cyprus* **38 D11** 35 1N 32 38 E
Stavrós, *Greece* **38 D6** 35 12N 24 45 E
Stavros, Ákra, *Greece* **38 D6** 35 26N 24 58 E
Stavroúpolis, *Greece* **51 E8** 41 12N 24 45 E
Stawell, *Australia* **128 D5** 37 5 S 142 47 E
Stawell →, *Australia* **126 C3** 20 20 S 142 55 E
Stawiski, *Poland* **54 E9** 53 22N 22 9 E
Stawiszyn, *Poland* **55 G5** 51 56N 18 4 E
Stayner, *Canada* **150 B4** 44 25N 80 5W
Steamboat Springs, *U.S.A.* **158 F10** 40 29N 106 50W
Steane, *Norway* **18 E5** 59 16N 8 33 E
Stebbins, *U.S.A.* **144 E7** 63 31N 162 17W
Stebleva, *Albania* **50 E4** 41 23N 20 33 E
Steckborn, *Switz.* **33 A7** 47 44N 8 59 E
Steele, *Ala., U.S.A.* **152 B3** 33 56N 86 12W
Steele, *N. Dak., U.S.A.* **154 B5** 46 51N 99 55W
Steelton, *U.S.A.* **150 F8** 40 14N 76 50W
Steelville, *U.S.A.* **155 G9** 37 58N 91 22W
Steen River, *Canada* **142 B5** 59 40N 117 12W
Steenkool = Bintuni, *Indonesia* **83 B4** 2 7 S 133 32 E
Steenstrup Gletscher, *Greenland* **10 B5** 75 15N 57 0W
Steenwijk, *Neths.* **24 B6** 52 47N 6 7 E
Steep Pt., *Australia* **125 E1** 26 8 S 113 8 E
Steep Rock, *Canada* **143 C9** 51 30N 98 48W
Ştefan Vodă, *Moldova* **53 D14** 46 27N 29 42 E
Ştefăneşti, *Romania* **53 C12** 47 44N 27 15 E
Stefanie L. = Chew Bahir, *Ethiopia* **107 G4** 4 40N 36 50 E
Stefansson Bay, *Antarctica* **7 C5** 67 20 S 59 8 E
Steffisburg, *Switz.* **32 C5** 46 47N 7 38 E
Stege, *Denmark* **17 K6** 54 59N 12 18 E
Ştei, *Romania* **52 D7** 46 32N 22 27 E
Steiermark □, *Austria* **34 D8** 47 26N 15 0 E
Steigerwald, *Germany* **31 F6** 49 44N 10 26 E
Steilacoom, *U.S.A.* **160 C4** 47 10N 122 36W
Steinbach, *Canada* **143 D9** 49 32N 96 40W
Steinfurt, *Germany* **30 C3** 52 9N 7 20 E
Steinhatchee, *U.S.A.* **152 F6** 29 40N 83 23W
Steinheim, *Germany* **30 D5** 51 51N 9 5 E
Steinhuder Meer, *Germany* **30 C5** 52 29N 9 21 E
Steinkjer, *Norway* **14 D14** 64 1N 11 31 E
Steinkopf, *S. Africa* **116 D2** 29 18 S 17 43 E
Steinshamn, *Norway* **18 B3** 62 47N 6 28 E
Stellarton, *Canada* **141 C7** 45 32N 62 30W
Stellenbosch, *S. Africa* **116 E2** 33 58 S 18 50 E
Stelvio, Paso dello, *Italy* **33 C10** 46 32N 10 27 E
Stenay, *France* **27 C12** 49 29N 5 12 E
Stendal, *Germany* **30 C7** 52 36N 11 53 E
Stende, *Latvia* **54 A9** 57 11N 22 33 E
Stenhamra, *Sweden* **16 E11** 59 20N 17 41 E
Stenstorp, *Sweden* **17 F7** 58 17N 13 45 E
Stenungsund, *Sweden* **17 F5** 58 6N 11 50 E
Steornabhaigh = Stornoway, *U.K.* **22 C2** 58 13N 6 23W
Stepanakert = Xankändi, *Azerbaijan* **101 C12** 39 52N 46 49 E
Stepanavan, *Armenia* **61 K7** 41 1N 44 23 E
Stephen, *U.S.A.* **154 A6** 48 27N 96 53W
Stephens, C., *N.Z.* **131 A8** 40 42 S 173 58 E
Stephens Creek, *Australia* **128 A4** 31 50 S 141 30 E
Stephens I., *Canada* **142 C2** 54 10N 130 45W
Stephens I., *N.Z.* **131 A9** 40 40 S 174 1 E
Stephenville, *Canada* **141 C8** 48 31N 58 35W
Stephenville, *U.S.A.* **155 J5** 32 13N 98 12W
Stepnica, *Poland* **54 E1** 53 38N 14 36 E
Stepnoi = Elista, *Russia* **61 G7** 46 16N 44 14 E
Stepnoye, *Russia* **62 D8** 54 4N 60 26 E
Steppe, *Asia* **66 D9** 50 0N 50 0 E
Stereá Ellas □, *Greece* **48 C4** 38 50N 23 0 E
Sterkstroom, *S. Africa* **116 E4** 31 32 S 26 32 E
Sterling, *Alaska, U.S.A.* **144 F10** 60 32N 150 46W
Sterling, *Colo., U.S.A.* **154 E3** 40 37N 103 13W
Sterling, *Ga., U.S.A.* **152 D8** 31 16N 81 34W
Sterling, *Ill., U.S.A.* **156 C7** 41 48N 89 42W
Sterling, *Kans., U.S.A.* **154 F5** 38 13N 98 12W
Sterling City, *U.S.A.* **155 K4** 31 51N 101 0W
Sterling Heights, *U.S.A.* **157 B13** 42 35N 83 0W
Sterling Run, *U.S.A.* **150 E6** 41 25N 78 12W
Sterlitamak, *Russia* **62 E6** 53 40N 56 0 E
Sternberg, *Germany* **30 B7** 53 42N 11 50 E
Šternberk, *Czech Rep.* **35 B10** 49 45N 17 15 E
Stérnes, *Greece* **38 D6** 35 30N 24 9 E
Sterzing = Vipiteno, *Italy* **45 B8** 46 54N 11 26 E
Stettin = Szczecin, *Poland* **54 E1** 53 27N 14 27 E
Stettiner Haff, *Germany* **30 B10** 53 47N 14 15 E
Stettler, *Canada* **142 C6** 52 19N 112 40W
Steubenville, *U.S.A.* **150 F4** 40 22N 80 37W
Stevenage, *U.K.* **21 F7** 51 55N 0 13W
Stevens Point, *U.S.A.* **154 C10** 44 31N 89 34W
Stevens Pottery, *U.S.A.* **152 C6** 32 57N 83 17W
Stevenson, *U.S.A.* **160 E5** 45 42N 121 53W
Stevenson L., *Canada* **143 C9** 53 55N 96 0W
Stevns Klint, *Denmark* **17 J6** 55 17N 12 28 E
Steward, *U.S.A.* **156 C7** 41 51N 89 1W
Stewardson, *U.S.A.* **157 E8** 39 16N 88 38W
Stewart, *Canada* **142 B3** 55 56N 129 57W
Stewart, *Ga., U.S.A.* **152 B6** 33 25N 83 52W
Stewart, *Nev., U.S.A.* **160 F7** 39 5N 119 46W
Stewart, C., *Australia* **126 A1** 11 57 S 134 56 E
Stewart, I., *Chile* **176 D2** 54 50 S 71 15W
Stewart I., *N.Z.* **131 G2** 46 58 S 167 54 E
Stewarts Point, *U.S.A.* **160 G3** 38 39N 123 24W
Stewartsville, *U.S.A.* **156 E2** 39 45N 94 30W
Stewiacke, *Canada* **141 C7** 45 9N 63 22W
Steynsburg, *S. Africa* **116 E4** 31 15 S 25 49 E
Steyr, *Austria* **34 C7** 48 3N 14 25 E
Steyr →, *Austria* **34 C7** 48 3N 14 25 E
Steytlerville, *S. Africa* **116 E3** 33 17 S 24 19 E
Stia, *Italy* **45 E8** 43 48N 11 42 E
Stigler, *U.S.A.* **155 H7** 35 15N 95 8W
Stigliano, *Italy* **47 B9** 40 24N 16 14 E
Stigtomta, *Sweden* **17 F10** 58 47N 16 48 E
Stikine →, *Canada* **142 B2** 56 40N 132 30W
Stilfontein, *S. Africa* **116 D4** 26 51 S 26 50 E
Stilís, *Greece* **48 C4** 38 55N 22 47 E
Stillmore, *U.S.A.* **152 C7** 32 27N 82 13W
Stillwater, *N.Z.* **131 C6** 42 27 S 171 20 E
Stillwater, *Minn., U.S.A.* **154 C8** 45 3N 92 49W
Stillwater, *N.Y., U.S.A.* **151 D11** 42 55N 73 41W
Stillwater, *Okla., U.S.A.* **155 G6** 36 7N 97 4W
Stillwater Range, *U.S.A.* **158 G4** 39 50N 118 5W
Stillwell, *U.S.A.* **152 C8** 32 23N 81 15W
Stilo, Pta., *Italy* **47 D9** 38 25N 16 35 E
Stilwell, *U.S.A.* **155 H7** 35 49N 94 38W
Štip, *Macedonia* **50 E6** 41 42N 22 10 E
Stíra, *Greece* **48 C6** 38 9N 24 14 E
Stirling, *Australia* **126 B3** 17 12 S 141 35 E
Stirling, *Canada* **142 D6** 49 30N 112 30W
Stirling, *N.Z.* **131 G4** 46 14 S 169 49 E
Stirling, *U.K.* **22 E5** 56 8N 3 57W
Stirling □, *U.K.* **22 E4** 56 12N 4 18W
Stirling Ra., *Australia* **125 F2** 34 23 S 118 0 E
Stittsville, *Canada* **151 A9** 45 15N 75 55W
Stjernøya, *Norway* **14 A20** 70 20N 22 40 E
Stjørdalshalsen, *Norway* **14 E14** 63 29N 10 51 E
Stock Island, *U.S.A.* **153 L8** 24 32N 81 34W
Stockach, *Germany* **31 H5** 47 50N 9 1 E
Stockaryd, *Sweden* **17 G8** 57 19N 14 36 E
Stockbridge, *Ga., U.S.A.* **152 B5** 33 33N 84 14W
Stockbridge, *Mich., U.S.A.* **157 B12** 42 27N 84 11W
Stockerau, *Austria* **35 C9** 48 24N 16 12 E
Stockett, *U.S.A.* **158 C8** 47 21N 111 10W
Stockholm, *Sweden* **16 E12** 59 20N 18 3 E
Stockholms län □, *Sweden* **16 E12** 59 30N 18 20 E
Stockhorn, *Switz.* **32 C5** 46 42N 7 33 E
Stockport, *U.K.* **20 D5** 53 25N 2 9W
Stocksbridge, *U.K.* **20 D6** 53 29N 1 35W
Stockton, *Australia* **129 B9** 32 50 S 151 47 E
Stockton, *Calif., U.S.A.* **160 H5** 37 58N 121 17W
Stockton, *Ill., U.S.A.* **156 B6** 42 21N 90 1W
Stockton, *Kans., U.S.A.* **154 F5** 39 26N 99 16W
Stockton, *Mo., U.S.A.* **155 G8** 37 42N 93 48W
Stockton-on-Tees, *U.K.* **20 C6** 54 35N 1 19W
Stockton-on-Tees □, *U.K.* **20 C6** 54 35N 1 19W
Stoczek Łukowski, *Poland* **55 G8** 51 58N 21 58 E
Stöde, *Sweden* **16 B10** 62 28N 16 35 E
Stoeng Treng, *Cambodia* **86 F5** 13 31N 105 58 E
Stoer, Pt. of, *U.K.* **22 C3** 58 16N 5 23W
Stogovo, *Macedonia* **50 E4** 41 31N 20 38 E
Stoholm, *Denmark* **17 H3** 56 30N 9 8 E
Stoke, *N.Z.* **131 B8** 41 19 S 173 14 E
Stoke-on-Trent, *U.K.* **20 D5** 53 1N 2 11W
Stoke-on-Trent □, *U.K.* **20 D5** 53 1N 2 11W
Stokes Bay, *Canada* **140 D3** 45 0N 81 28W
Stokes Pt., *Australia* **126 G3** 40 10 S 143 56 E
Stokes Ra., *Australia* **124 C5** 15 50 S 130 50 E
Stokkseyri, *Iceland* **11 D5** 63 50N 21 2W
Stokksnes, *Iceland* **11 C12** 64 14N 14 58W
Stokmarknes, *Norway* **14 B16** 68 34N 14 54 E
Stolac, *Bos.-H.* **50 C1** 43 5N 17 59 E
Stolberg, *Germany* **30 E2** 50 47N 6 13 E
Stolbovoy, Ostrov, *Russia* **65 D17** 74 44N 135 14 E
Stolbtsy = Stowbtsy, *Belarus* **58 F4** 53 30N 26 43 E
Stolin, *Belarus* **59 G4** 51 53N 26 50 E
Stöllet, *Sweden* **16 D7** 60 26N 13 15 E
Stolnici, *Romania* **53 F9** 44 31N 24 48 E
Stomíon, *Greece* **38 D5** 35 21N 23 32 E
Ston, *Croatia* **45 F14** 42 51N 17 43 E
Stone, *U.K.* **20 E5** 52 55N 2 9W
Stone Mountain, *U.S.A.* **152 B5** 33 49N 84 10W
Stonehaven, *U.K.* **22 E6** 56 59N 2 12W
Stonehenge, *Australia* **126 C3** 24 22 S 143 17 E
Stonehenge, *U.K.* **21 F6** 51 9N 1 45W
Stonewall, *Canada* **143 C9** 50 10N 97 19W
Stongfjorden, *Norway* **18 C2** 61 26N 5 10 E
Stonington, *U.S.A.* **156 E7** 39 44N 89 12W
Stony L., *Man., Canada* **143 B9** 58 51N 98 40W
Stony L., *Ont., Canada* **150 B6** 44 30N 78 5W
Stony Rapids, *Canada* **143 B7** 59 16N 105 50W
Stony Tunguska = Tunguska, Podkamennaya →, *Russia* **65 C10** 61 50N 90 13 E
Stonyford, *U.S.A.* **160 F4** 39 23N 122 33W
Stopnica, *Poland* **55 H7** 50 27N 20 57 E
Storå, *Sweden* **16 E9** 59 42N 15 6 E
Storå →, *Denmark* **17 H2** 56 20N 8 19 E
Stora Gla, *Sweden* **16 E6** 59 30N 12 30 E
Stora Le, *Sweden* **16 E5** 59 5N 11 55 E
Stora Lulevatten, *Sweden* **14 C18** 67 10N 19 30 E
Stóra-Vatnshorn, *Iceland* **11 B5** 65 4N 21 33W
Storavan, *Sweden* **14 D18** 65 45N 18 10 E
Stord, *Norway* **15 G11** 59 52N 5 23 E
Stordal, *Norway* **18 B4** 62 23N 7 0 E
Store Bælt, *Denmark* **17 J4** 55 20N 11 0 E
Store Creek, *Australia* **129 B8** 32 54 S 149 6 E
Store Heddinge, *Denmark* **17 J6** 55 18N 12 23 E
Store-Jukleeggi, *Norway* **18 C5** 61 3N 8 12 E
Store Koldewey, *Greenland* **10 B9** 76 30N 19 0W
Store Sølnkletten, *Norway* **18 C7** 61 59N 10 16 E
Store Sotra, *Norway* **18 D1** 60 18N 5 4 E
Storebro, *Sweden* **17 G9** 57 35N 15 52 E
Støren, *Norway* **18 A7** 63 3N 10 18 E
Storerikvollen, *Norway* **18 A8** 63 7N 11 58 E
Storfjellseter, *Norway* **18 C7** 61 40N 10 30 E
Storfjorden, *Møre og Romsdal, Norway* **18 B3** 62 8N 6 33 E
Storfjorden, *Møre og Romsdal, Norway* **18 B3** 62 28N 6 35 E
Storfors, *Sweden* **16 E8** 59 32N 14 17 E
Stóridalur, *Iceland* **11 D7** 63 38N 19 57W
Stórinúpur, *Iceland* **11 C6** 64 3N 20 10W
Storli, *Norway* **18 B6** 62 42N 9 5 E
Storlien, *Sweden* **16 A6** 63 20N 12 5 E
Storm B., *Australia* **126 G4** 43 10 S 147 30 E
Storm Lake, *U.S.A.* **154 D7** 42 39N 95 13W
Stormberge, *S. Africa* **116 E4** 31 16 S 26 17 E
Stormsrivier, *S. Africa* **116 E3** 33 59 S 23 52 E
Stornoway, *U.K.* **22 C2** 58 13N 6 23W
Storo, *Italy* **44 C7** 45 51N 10 35 E
Storozhinets = Storozhynets, *Ukraine* **59 H3** 48 14N 25 45 E
Storozhynets, *Ukraine* **59 H3** 48 14N 25 45 E
Storsjøen, *Hedmark, Norway* **18 D8** 60 20N 11 40 E
Storsjøen, *Hedmark, Norway* **18 C8** 61 30N 11 14 E
Storsjön, *Gävleborg, Sweden* **16 D10** 60 35N 16 45 E
Storsjön, *Jämtland, Sweden* **16 B7** 62 48N 13 7 E
Storsjön, *Jämtland, Sweden* **16 A8** 63 9N 14 30 E
Storstrøms Amtskommune □, *Denmark* **17 J5** 54 50N 11 45 E
Storuman, *Sweden* **14 D17** 65 5N 17 10 E
Storuman, sjö, *Sweden* **14 D17** 65 13N 16 50 E
Storvätteshågna, *Sweden* **16 B6** 62 6N 12 30 E
Storvigelen, *Norway* **18 B9** 62 32N 12 2 E
Storvik, *Sweden* **16 D10** 60 35N 16 33 E
Storvreta, *Sweden* **16 E11** 59 58N 17 44 E
Story City, *U.S.A.* **156 B3** 42 11N 93 36W
Stoughton, *Canada* **143 D8** 49 40N 103 0W
Stoughton, *U.S.A.* **156 B8** 42 55N 89 13W
Stour →, *Dorset, U.K.* **21 G6** 50 43N 1 47W
Stour →, *Kent, U.K.* **21 F9** 51 18N 1 22 E
Stour →, *Suffolk, U.K.* **21 F9** 51 57N 1 4 E
Stourbridge, *U.K.* **21 E5** 52 28N 2 8W
Stout, L., *Canada* **143 C10** 52 0N 94 40W
Stovall, *U.S.A.* **152 C5** 32 58N 84 51W
Stove Pipe Wells Village, *U.S.A.* **161 J9** 36 35N 117 11W
Støvring, *Denmark* **17 H3** 56 54N 9 50 E
Stowbtsy, *Belarus* **58 F4** 53 30N 26 43 E
Stowmarket, *U.K.* **21 E9** 52 12N 1 0 E
Strabane, *U.K.* **23 B4** 54 50N 7 27W
Stracin, *Macedonia* **50 D6** 42 13N 22 2 E
Stradella, *Italy* **44 C6** 45 5N 9 18 E
Strahan, *Australia* **126 G4** 42 9 S 145 20 E
Strajitsa, *Bulgaria* **51 C9** 43 14N 25 58 E
Strakonice, *Czech Rep.* **34 B6** 49 15N 13 53 E
Straldzha, *Bulgaria* **51 D10** 42 35N 26 40 E
Stralsund, *Germany* **30 A9** 54 18N 13 4 E
Strand, *Norway* **18 C8** 61 17N 11 17 E
Strand, *S. Africa* **116 E2** 34 9 S 18 48 E
Stranda, *Møre og Romsdal, Norway* **15 E12** 62 19N 6 58 E
Stranda, *Nord-Trøndelag, Norway* **14 E14** 63 33N 10 14 E
Strandarkirkja, *Iceland* **11 D5** 63 50N 21 42W
Strandasýsla □, *Iceland* **11 B5** 65 45N 21 45W
Strandby, *Denmark* **17 G4** 57 30N 10 29 E
Strangford L., *U.K.* **23 B6** 54 30N 5 37W
Strängnäs, *Sweden* **16 E11** 59 23N 17 2 E
Strangsville, *U.S.A.* **150 E3** 41 19N 81 50W
Stranraer, *U.K.* **22 G3** 54 54N 5 1W
Strasbourg, *Canada* **143 C8** 51 4N 104 55W
Strasbourg, *France* **27 D14** 48 35N 7 42 E
Strasburg, *Germany* **30 B9** 53 30N 13 43 E
Strasburg, *U.S.A.* **154 B4** 46 8N 100 10W
Strășeni, *Moldova* **53 C13** 47 8N 28 36 E
Stråssa, *Sweden* **16 E9** 59 44N 15 12 E
Stratford, *N.S.W., Australia* **129 B9** 32 7 S 151 55 E
Stratford, *Vic., Australia* **129 D7** 37 59 S 147 7 E
Stratford, *Canada* **140 D3** 43 23N 81 0W
Stratford, *N.Z.* **130 F3** 39 20 S 174 19 E
Stratford, *Calif., U.S.A.* **160 J7** 36 11N 119 49W
Stratford, *Conn., U.S.A.* **151 E11** 41 12N 73 8W
Stratford, *Tex., U.S.A.* **155 G3** 36 20N 102 4W
Stratford-upon-Avon, *U.K.* **21 E6** 52 12N 1 42W
Strath Spey, *U.K.* **22 D5** 57 9N 3 49W
Strathalbyn, *Australia* **128 C3** 35 13 S 138 53 E
Strathaven, *U.K.* **22 F4** 55 40N 4 5W
Strathcona Prov. Park, *Canada* **142 D3** 49 38N 125 40W
Strathmore, *Australia* **126 B3** 17 50 S 142 35 E
Strathmore, *Canada* **142 C6** 51 5N 113 18W
Strathmore, *U.K.* **22 E5** 56 37N 3 7W
Strathmore, *U.S.A.* **160 J7** 36 9N 119 4W
Strathnaver, *Canada* **142 C4** 53 20N 122 33W
Strathpeffer, *U.K.* **22 D4** 57 35N 4 32W
Strathroy, *Canada* **140 D3** 42 58N 81 38W
Strathy Pt., *U.K.* **22 C4** 58 36N 4 1W
Stratton, *U.S.A.* **154 F3** 39 19N 102 36W
Straubing, *Germany* **31 G8** 48 52N 12 34 E
Straumnes, *Iceland* **11 A3** 66 26N 23 8W
Strausberg, *Germany* **30 C9** 52 35N 13 54 E
Strawberry Point, *U.S.A.* **156 B5** 42 41N 91 32W
Strawberry Reservoir, *U.S.A.* **158 F8** 40 8N 111 9W
Strawn, *U.S.A.* **155 J5** 32 33N 98 30W
Strážnice, *Czech Rep.* **35 C10** 48 54N 17 19 E
Streaky B., *Australia* **127 E1** 32 48 S 134 13 E
Streaky Bay, *Australia* **127 E1** 32 51 S 134 18 E
Streator, *U.S.A.* **154 E10** 41 8N 88 50W
Středočeský □, *Czech Rep.* **34 B7** 49 55N 14 30 E
Streeter, *U.S.A.* **154 B5** 46 39N 99 21W
Streetsville, *Canada* **150 C5** 43 35N 79 42W
Strehaia, *Romania* **53 F8** 44 37N 23 10 E
Strelcha, *Bulgaria* **51 D8** 42 25N 24 19 E
Strelka, *Russia* **65 D10** 58 5N 93 3 E
Streng →, *Cambodia* **86 F4** 13 12N 103 37 E
Stresa, *Italy* **44 C5** 45 52N 8 28 E
Streymoy, *Færoe Is.* **14 E9** 62 8N 7 5W
Strezhevoy, *Russia* **64 C8** 60 42N 77 34 E
Stříbro, *Czech Rep.* **34 A6** 49 44N 13 0 E
Strickland →, *Papua N. G.* **132 D1** 7 35 S 141 36 E
Strimón →, *Greece* **50 F7** 40 46N 23 51 E
Strimonikós Kólpos, *Greece* **50 F7** 40 33N 24 0 E
Stroeder, *Argentina* **176 B4** 40 12 S 62 37W
Strofádhes, *Greece* **48 D3** 37 15N 21 0 E
Stroma, *U.K.* **22 C5** 58 41N 3 7W
Strómboli, *Italy* **47 D8** 38 47N 15 13 E
Stromeferry, *U.K.* **22 D3** 57 21N 5 33W
Strømmen, *Norway* **18 E7** 59 58N 10 59 E
Stromness, *U.K.* **22 C5** 58 58N 3 17W
Strömsbruk, *Sweden* **16 C11** 61 52N 17 18 E
Stromsburg, *U.S.A.* **154 E6** 41 7N 97 36W
Strömsnäsbruk, *Sweden* **17 H7** 56 35N 13 45 E
Strömstad, *Sweden* **17 F5** 58 56N 11 10 E
Strömsund, *Sweden* **14 E16** 63 51N 15 33 E
Stronghurst, *U.S.A.* **156 D6** 40 45N 90 55W
Strongilí, *Greece* **49 E11** 36 6N 29 42 E
Stróngoli, *Italy* **47 C10** 39 16N 17 3 E
Stronie Śląskie, *Poland* **55 H3** 50 18N 16 53 E
Stronsay, *U.K.* **22 B6** 59 7N 2 35W
Stropkov, *Slovak Rep.* **35 B14** 49 13N 21 39 E
Stroud, *U.K.* **21 F5** 51 45N 2 13W
Stroud Road, *Australia* **129 B9** 32 18 S 151 57 E
Stroudsburg, *U.S.A.* **151 F9** 40 59N 75 12W
Stroumbi, *Cyprus* **38 E11** 34 53N 32 29 E
Struer, *Denmark* **17 H2** 56 30N 8 35 E
Struga, *Macedonia* **50 E4** 41 11N 20 44 E
Strugi Krasnyye, *Russia* **58 C5** 58 21N 29 1 E
Strumica, *Macedonia* **50 E6** 41 28N 22 41 E
Strumica →, *Europe* **50 E7** 41 20N 23 22 E
Struthers, *Canada* **140 C2** 48 41N 85 51W
Struthers, *U.S.A.* **150 E4** 41 4N 80 39W
Stryama, *Bulgaria* **51 D8** 42 16N 24 54 E
Stryker, *U.S.A.* **158 B6** 48 41N 114 46W
Stryków, *Poland* **55 G6** 51 55N 19 33 E
Stryn, *Norway* **18 C3** 61 54N 6 43 E
Stryy, *Ukraine* **59 H2** 49 16N 23 48 E
Strzegom, *Poland* **55 H3** 50 58N 16 20 E
Strzelce Krajeńskie, *Poland* **55 F2** 52 52N 15 33 E
Strzelce Opolskie, *Poland* **55 H5** 50 31N 18 18 E
Strzelecki Cr. →, *Australia* **127 D2** 29 37 S 139 59 E
Strzelin, *Poland* **55 H4** 50 46N 17 2 E
Strzelno, *Poland* **55 F5** 52 35N 18 9 E
Strzybnica, *Poland* **55 H5** 50 28N 18 48 E
Strzyżów, *Poland* **55 J8** 49 52N 21 47 E
Stuart, *Fla., U.S.A.* **149 M5** 27 12N 80 15W
Stuart, *Iowa, U.S.A.* **156 C2** 41 30N 94 19W
Stuart, *Nebr., U.S.A.* **154 D5** 42 36N 99 8W
Stuart →, *Canada* **142 C4** 54 0N 123 35W
Stuart Bluff Ra., *Australia* **124 D5** 22 50 S 131 52 E
Stuart I., *U.S.A.* **144 E7** 63 35N 162 30W
Stuart L., *Canada* **142 C4** 54 30N 124 30W
Stuart Mts., *N.Z.* **131 F2** 45 2 S 167 39 E
Stuart Ra., *Australia* **127 D1** 29 10 S 134 56 E
Stubbekøbing, *Denmark* **17 K6** 54 53N 12 9 E
Stuben, *Austria* **34 D3** 47 10N 10 8 E
Studen Kladenets, Yazovir, *Bulgaria* **51 E9** 41 37N 25 30 E
Studenka, *Czech Rep.* **35 B11** 49 44N 18 5 E
Studholme, *N.Z.* **131 E6** 44 42 S 171 9 E
Stugudal, *Norway* **18 B8** 62 53N 11 53 E
Stugun, *Sweden* **16 A9** 63 10N 15 40 E
Stuhr, *Germany* **30 B4** 53 5N 8 46 E
Stull, L., *Canada* **140 B1** 54 24N 92 34W
Stung Treng = Stoeng Treng, *Cambodia* **86 F5** 13 31N 105 58 E
Stupart →, *Canada* **143 B10** 56 0N 93 25W
Stupava, *Slovak Rep.* **35 C10** 48 17N 17 2 E
Stupino, *Russia* **58 E10** 54 57N 38 2 E
Sturgeon B., *Canada* **143 C9** 52 0N 97 50W
Sturgeon Bay, *U.S.A.* **148 C2** 44 50N 87 23W
Sturgeon Falls, *Canada* **140 C4** 46 25N 79 57W
Sturgeon L., *Alta., Canada* **142 B5** 55 6N 117 32W
Sturgeon L., *Ont., Canada* **140 C1** 50 0N 90 45W
Sturgeon L., *Ont., Canada* **150 B6** 44 28N 78 43W
Sturgis, *Mich., U.S.A.* **157 C11** 41 48N 85 25W
Sturgis, *S. Dak., U.S.A.* **154 C3** 44 25N 103 31W
Sturkö, *Sweden* **17 H9** 56 5N 15 42 E
Štúrovo, *Slovak Rep.* **35 D11** 47 48N 18 41 E
Sturt Cr. →, *Australia* **124 C4** 19 8 S 127 50 E
Sturt Creek, *Australia* **124 C4** 19 12 S 128 8 E
Sturts Meadows, *Australia* **128 A4** 31 18 S 141 42 E
Stutterheim, *S. Africa* **116 E4** 32 33 S 27 28 E
Stuttgart, *Germany* **31 G5** 48 48N 9 11 E
Stuttgart, *U.S.A.* **155 H9** 34 30N 91 33W
Stuyvesant, *U.S.A.* **151 D11** 42 23N 73 45W
Stykkishólmur, *Iceland* **11 B4** 65 2N 22 40W
Styria = Steiermark □, *Austria* **34 D8** 47 26N 15 0 E
Styrsö, *Sweden* **17 G5** 57 37N 11 46 E
Su-no-Saki, *Japan* **71 C11** 34 58N 139 45 E
Su Xian, *China* **74 H9** 33 41N 116 59 E
Suai, *Indonesia* **82 C3** 9 21 S 125 17 E
Suakin, *Sudan* **106 D4** 19 8N 37 20 E
Sual, *Phil.* **80 C3** 16 4N 120 5 E
Suan, *N. Korea* **75 E14** 38 42N 126 22 E
Suapure →, *Venezuela* **168 B4** 6 48N 67 1W
Suaqui, *Mexico* **162 B3** 29 12N 109 41W
Suatá →, *Venezuela* **169 B4** 7 52N 65 22W
Subang, *Indonesia* **85 D3** 6 34 S 107 45 E
Subansiri →, *India* **90 B4** 26 48N 93 50 E
Subayhah, *Si. Arabia* **96 D3** 30 2N 38 50 E
Subcetate, *Romania* **52 E8** 45 36N 23 0 E
Subi, *Indonesia* **85 B3** 2 58N 108 50 E
Subiaco, *Italy* **45 G10** 41 56N 13 5 E
Subotica, *Serbia, Yug.* **52 D4** 46 6N 19 39 E
Success, *Canada* **143 C7** 50 28N 108 6W
Suceava, *Romania* **53 C11** 47 38N 26 16 E
Suceava □, *Romania* **53 C10** 47 37N 25 40 E
Suceava →, *Romania* **53 C11** 47 32N 26 32 E
Sucha-Beskidzka, *Poland* **55 J6** 49 44N 19 35 E
Suchań, *Poland* **55 E2** 53 18N 15 18 E
Suchan, *Russia* **68 C6** 43 8N 133 9 E
Suchedniów, *Poland* **55 G7** 51 3N 20 49 E
Suchitoto, *El Salv.* **164 D2** 13 56N 89 0W
Suchou = Suzhou, *China* **77 B13** 31 19N 120 38 E
Süchow = Xuzhou, *China* **75 G9** 34 18N 117 10 E
Suchowola, *Poland* **54 E10** 53 33N 23 3 E
Sucio →, *Colombia* **168 B2** 7 27N 77 7W
Suck →, *Ireland* **23 C3** 53 17N 8 3W
Suckling, Mt., *Papua N. G.* **132 E5** 9 49 S 148 53 E
Sucre, *Bolivia* **173 D4** 19 0 S 65 15W
Sucre, *Colombia* **168 B3** 8 49N 74 44W
Sucre □, *Colombia* **168 B2** 8 50N 75 40W
Sucre □, *Venezuela* **169 A5** 10 25N 63 30W
Sucuaro, *Colombia* **168 C4** 4 34N 68 50W
Sućuraj, *Croatia* **45 E14** 43 10N 17 8 E
Sucuriju, *Brazil* **170 A2** 1 39N 49 57W
Sucuriú →, *Brazil* **173 E7** 20 47 S 51 38W
Sud, Pte., *Canada* **141 C7** 49 3N 62 14W
Sud-Ouest, Pte. du, *Canada* **141 C7** 49 23N 63 36W
Suda →, *Russia* **58 C9** 59 0N 37 40 E
Sudak, *Ukraine* **59 K8** 44 51N 34 57 E
Sudan, *U.S.A.* **155 H3** 34 4N 102 32W
Sudan ■, *Africa* **107 E3** 15 0N 30 0 E
Sudbury, *Canada* **140 C3** 46 30N 81 0W
Sudbury, *U.K.* **21 E8** 52 2N 0 45 E
Sûdd, *Sudan* **107 F3** 8 20N 30 0 E
Suddie, *Guyana* **169 B6** 7 8N 58 29W
Süderbrarup, *Germany* **30 A5** 54 38N 9 45 E

Tamarinda 287

Swan L., Canada 143 C8 52 30N 100 40W
Swan Reach, Australia 128 C3 34 35 S 139 37 E
Swan River, Canada 143 C8 52 10N 101 16W
Swanage, U.K. 21 G6 50 36N 1 58W
Swansea, Australia 129 B9 33 3 S 151 35 E
Swansea, U.K. 21 F4 51 37N 3 57W
Swansea, U.S.A. 152 B9 33 44N 81 6W
Swansea □, U.K. 21 F3 51 38N 4 3W
Swar →, Pakistan 93 B5 34 40N 72 5 E
Swartberge, S. Africa 116 E3 33 20 S 22 0 E
Swartmodder, S. Africa 116 D3 28 1 S 20 32 E
Swartruggens, S. Africa 116 D4 25 39 S 26 42 E
Swarzędz, Poland 55 F4 52 25N 17 4 E
Swastika, Canada 140 C3 48 7N 80 6W
Swatow = Shantou, China 77 F11 23 18N 116 40 E
Swaziland ■, Africa 117 D5 26 30 S 31 30 E
Sweden ■, Europe 15 G16 57 0N 15 0 E
Swedru, Ghana 113 D4 5 32N 0 41W
Sweet Home, U.S.A. 158 D2 44 24N 122 44W
Sweet Springs, U.S.A. 156 F3 38 58N 93 25W
Sweetwater, Nev., U.S.A. 160 G7 38 27N 119 9W
Sweetwater, Tex., U.S.A. 155 J4 32 28N 100 25W
Sweetwater →, U.S.A. 158 E10 42 31N 107 2W
Swellendam, S. Africa 116 E3 34 1 S 20 26 E
Swider →, Poland 55 F8 52 6N 21 14 E
Świdnica, Poland 55 H3 50 50N 16 30 E
Świdnik, Poland 55 G9 51 13N 22 39 E
Świdwin, Poland 54 E2 53 47N 15 49 E
Świebodzice, Poland 55 H3 50 51N 16 20 E
Świebodzin, Poland 55 F2 52 15N 15 31 E
Świecie, Poland 54 E5 53 25N 18 30 E
Świerzawa, Poland 55 G2 51 1N 15 54 E
Świętokrzyskie, Góry, Poland 55 H7 51 0N 20 30 E
Swift Current, Canada 143 C7 50 20N 107 45W
Swiftcurrent →, Canada 143 C7 50 38N 107 44W
Swilly, L., Ireland 23 A4 55 12N 7 33W
Swindle, I., Canada 142 C3 52 30N 128 35W
Swindon, U.K. 21 F6 51 34N 1 46W
Swindon □, U.K. 21 F6 51 34N 1 46W
Swinemünde = Świnoujście, Poland 54 E1 53 54N 14 16 E
Swinford, Ireland 23 C3 53 57N 8 58W
Świnoujście, Poland 54 E1 53 54N 14 16 E
Switzerland ■, Europe 32 D6 46 30N 8 0 E
Swords, Ireland 23 C5 53 28N 6 13W
Swords, U.S.A. 152 B6 33 33N 83 18W
Syasstroy, Russia 58 B7 60 9N 32 33 E
Sycamore, Ga., U.S.A. 152 D6 31 40N 83 38W
Sycamore, S.C., U.S.A. 152 B9 33 2N 81 13W
Sychevka, Russia 58 E8 55 59N 34 16 E
Syców, Poland 55 G4 51 19N 17 40 E
Sydney, Australia 129 B9 33 53 S 151 10 E
Sydney, Canada 141 C7 46 7N 60 7W
Sydney Mines, Canada 141 C7 46 18N 60 15W
Sydprøven, Greenland 10 E6 60 30N 45 35W
Sydra, G. of = Surt, Khalīj, Libya 108 B3 31 40N 18 30 E
Syeverodonetsk, Ukraine 59 H10 48 58N 38 35 E
Syfteland, Norway 18 D2 60 14N 5 27 E
Syke, Germany 30 C4 52 55N 8 50 E
Sykkylven, Norway 18 B3 62 23N 6 35 E
Syktyvkar, Russia 56 B9 61 45N 50 40 E
Sylacauga, U.S.A. 152 B3 33 10N 86 15W
Sylarna, Sweden 14 E15 63 2N 12 13 E
Sylhet, Bangla. 90 C3 24 54N 91 52 E
Sylt, Germany 30 A4 54 54N 8 22 E
Sylte, Norway 18 B4 62 18N 7 17 E
Sylva →, Russia 62 B6 58 0N 56 54 E
Sylvan Lake, Canada 142 C6 52 20N 114 3W
Sylvania, Ga., U.S.A. 152 C8 32 45N 81 38W
Sylvania, Ohio, U.S.A. 157 C13 41 43N 83 42W
Sylvester, U.S.A. 152 D6 31 32N 83 50W
Sym, Russia 64 C9 60 20N 88 18 E
Symón, Mexico 162 C4 24 42N 102 35W
Synelnykove, Ukraine 59 H8 48 25N 35 30 E
Synnfjell, Norway 18 C6 61 5N 9 46 E
Synnott Ra., Australia 124 C4 16 30 S 125 20 E
Syracuse, Ind., U.S.A. 157 C11 41 26N 85 45W
Syracuse, Kans., U.S.A. 155 G4 37 59N 101 45W
Syracuse, N.Y., U.S.A. 151 C8 43 3N 76 9W
Syrdarya = Sirdaryo, Uzbekistan 63 C4 40 50N 68 40 E
Syrdarya →, Kazakstan 64 E7 46 3N 61 0 E
Syria ■, Asia 101 E8 35 0N 38 0 E
Syriam, Burma 90 G6 16 44N 96 19 E
Syrian Desert = Ash Shām, Bādiyat, Asia 66 F7 32 0N 40 0 E
Sysert, Russia 62 C8 56 29N 60 49 E
Sysslebäck, Sweden 16 D6 60 44N 12 52 E
Syvde, Norway 18 B2 62 5N 5 44 E
Syzran, Russia 60 D9 53 12N 48 30 E
Szabolcs-Szatmár-Bereg □, Hungary 52 B6 48 2N 21 45 E
Szadek, Poland 55 G5 51 41N 18 59 E
Szamocin, Poland 55 E4 53 2N 17 7 E
Szamos →, Hungary 52 B7 48 7N 22 20 E
Szamotuły, Poland 55 F3 52 37N 16 33 E
Száraz →, Hungary 52 D6 46 10N 21 15 E
Szarvas, Hungary 52 D6 46 50N 20 38 E
Százhalombatta, Hungary 52 C3 47 20N 18 58 E
Szczawnica, Poland 55 J7 49 26N 20 30 E
Szczebrzeszyn, Poland 55 H9 50 42N 22 59 E
Szczecin, Poland 54 E1 53 27N 14 27 E
Szczecin □, Poland 54 E1 53 25N 14 32 E
Szczecinek, Poland 54 E3 53 43N 16 41 E
Szczeciński, Zalew = Stettiner Haff, Germany 30 B10 53 47N 14 15 E
Szczekociny, Poland 55 H6 50 38N 19 48 E
Szczucin, Poland 55 H8 50 18N 21 4 E
Szczuczyn, Poland 54 E9 53 36N 22 19 E
Szczyrk, Poland 55 J6 49 43N 19 2 E
Szczytna, Poland 55 H3 50 25N 16 28 E
Szczytno, Poland 54 E7 53 33N 21 0 E
Szechwan = Sichuan □, China 76 B5 30 30N 103 0 E
Szécsény, Hungary 52 B4 48 7N 19 30 E
Szeged, Hungary 52 D5 46 16N 20 10 E
Szeghalom, Hungary 52 C6 47 1N 21 10 E
Székesfehérvár, Hungary 52 C3 47 15N 18 25 E
Szekszárd, Hungary 52 D3 46 22N 18 42 E
Szendrő, Hungary 52 B5 48 24N 20 41 E
Szentendre, Hungary 52 C4 47 39N 19 4 E
Szentes, Hungary 52 D5 46 39N 20 21 E
Szentgotthárd, Hungary 52 D1 46 58N 16 19 E
Szentlőrinc, Hungary 52 D3 46 3N 18 1 E
Szerencs, Hungary 52 B6 48 10N 21 12 E
Szigetszentmiklós, Hungary 52 C4 47 21N 19 3 E
Szigetvár, Hungary 52 D2 46 3N 17 46 E
Szikszó, Hungary 52 B5 48 12N 20 56 E
Szklarska Poreba, Poland 55 H2 50 50N 15 33 E
Szkwa →, Poland 55 E8 53 11N 21 43 E
Szlichtyngowa, Poland 55 G3 51 42N 16 15 E
Szob, Hungary 52 C3 47 48N 18 53 E
Szolnok, Hungary 52 C5 47 10N 20 15 E
Szombathely, Hungary 52 C1 47 14N 16 38 E
Szprotawa, Poland 55 G2 51 33N 15 35 E
Sztum, Poland 54 E6 53 55N 19 1 E
Sztutowo, Poland 54 D6 54 20N 19 15 E
Szubin, Poland 55 E4 53 1N 17 45 E
Szydłowiec, Poland 55 G7 51 15N 20 51 E
Szypliszki, Poland 54 D10 54 17N 23 2 E

T

Ta Khli Khok, Thailand 86 E3 15 18N 100 20 E
Ta Lai, Vietnam 87 G6 11 24N 107 23 E
Tab, Hungary 52 D3 46 44N 18 2 E
Tabacal, Argentina 174 A3 23 15 S 64 15W
Tabaco, Phil. 80 E4 13 22N 123 44 E
Tabagné, Ivory C. 112 D4 7 59N 3 4W
Ţābah, Si. Arabia 96 E4 26 55N 42 38 E
Tabajara, Brazil 173 B5 8 56 S 62 8W
Tabalos, Peru 172 B2 6 26 S 76 37W
Tabango, Phil. 81 F5 11 19N 124 22 E
Tabar Is., Papua N. G. 132 B7 2 50 S 152 0 E
Tabarka, Tunisia 108 A1 36 56N 8 46 E
Ţabas, Khorāsān, Iran 97 C9 32 48N 60 12 E
Ţabas, Khorāsān, Iran 97 C8 33 35N 56 55 E
Tabasará, Serranía de, Panama 164 E3 8 35N 81 40W
Tabasco □, Mexico 163 D6 17 45N 93 30W
Tabatinga, Serra da, Brazil 170 D3 10 30 S 44 0W
Tabayin, Burma 90 D5 22 42N 95 20 E
Tabāzīn, Iran 97 D8 31 12N 57 54 E
Tabelbala, Kahal de, Algeria 111 C4 28 47N 2 0W
Taber, Canada 142 D6 49 47N 112 8W
Taberg, Sweden 17 G8 57 40N 14 6 E
Tabi, Angola 115 D2 8 10 S 13 18 E
Tabira, Brazil 170 C4 7 35 S 37 33W
Tablas, Phil. 80 E4 12 25N 122 2 E
Tablas Strait, Phil. 80 E3 12 40N 121 48 E
Table B. = Tafelbaai, S. Africa 116 E2 33 35 S 18 25 E
Table B., Canada 141 B8 53 40N 56 25W
Table Grove, U.S.A. 156 D6 40 20N 90 27W
Table Mt., S. Africa 116 E2 34 0 S 18 22 E
Tableland, Australia 124 C4 17 16 S 126 51 E
Tabletop, Mt., Australia 126 C4 23 24 S 147 11 E
Tabogon, Phil. 81 F5 10 57N 124 2 E
Tábor, Czech Rep. 34 B7 49 25N 14 39 E
Tabora, Tanzania 118 D3 5 2 S 32 50 E
Tabora □, Tanzania 118 D3 5 0 S 33 0 E
Tabou, Ivory C. 112 E3 4 30N 7 20W
Tabrīz, Iran 101 C12 38 7N 46 20 E
Tabuaeran, Pac. Oc. 135 G12 3 51N 159 22W
Tabuelan, Phil. 81 F4 10 49N 123 52 E
Tabuenca, Spain 40 D3 41 42N 1 33W
Tabuk, Phil. 80 C3 17 24N 121 25 E
Tabūk, Si. Arabia 96 D3 28 23N 36 36 E
Tabwemasana, Mt., Vanuatu 133 E4 15 20 S 166 44 E
Täby, Sweden 16 E12 59 28N 18 4 E
Tacámbaro de Codallos, Mexico 162 D4 19 14N 101 28W
Tacheng, China 72 B3 46 40N 82 58 E
Tachia, Taiwan 77 E13 24 25N 120 28 E
Tachibana-Wan, Japan 70 E2 32 45N 130 7 E
Tachikawa, Japan 71 B11 35 42N 139 25 E
Tach'ing Shan = Daqing Shan, China 74 D6 40 40N 111 0 E
Táchira □, Venezuela 168 B3 8 7N 72 15W
Tachov, Czech Rep. 34 B5 49 47N 12 39 E
Tácina →, Italy 47 D9 38 57N 16 55 E
Tacloban, Phil. 81 F5 11 15N 124 58 E
Tacna, Peru 172 D3 18 0 S 70 20W
Tacna □, Peru 172 D3 17 40 S 70 20W
Tacoma, U.S.A. 160 C4 47 14N 122 26W
Tacuarembó, Uruguay 175 C4 31 45 S 56 0W
Tacutu →, Brazil 169 C5 3 1N 60 29W
Tademaït, Plateau du, Algeria 111 C5 28 30N 2 30 E
Tadent, O. →, Algeria 111 D6 22 25N 6 40 E
Tadjerdjeri, O. →, Algeria 111 C6 26 0N 8 0 E
Tadjerouna, Algeria 111 B5 33 31N 2 3 E
Tadjettaret, O. →, Algeria 111 D6 21 20N 7 22 E
Tadjmout, Oasis, Algeria 111 B5 33 52N 2 30 E
Tadjmout, Saoura, Algeria 111 C5 25 37N 3 48 E
Tadjoura, Djibouti 107 E5 11 50N 42 55 E
Tadjoura, Golfe de, Djibouti 107 E5 11 50N 43 0 E
Tadmor, N.Z. 131 B7 41 27 S 172 45 E
Tadotsu, Japan 70 C5 34 16N 133 45 E
Tadoule, L., Canada 143 B9 58 36N 98 20W
Tadoussac, Canada 141 C6 48 11N 69 42W
Tadzhikistan = Tajikistan ■, Asia 63 D5 38 30N 70 0 E
Taechŏn-ni, S. Korea 75 F14 36 21N 126 36 E
Taegu, S. Korea 75 G15 35 50N 128 37 E
Taegwan, N. Korea 75 D13 40 13N 125 12 E
Taejŏn, S. Korea 75 F14 36 20N 127 28 E
Tafalla, Spain 40 C3 42 30N 1 41W
Tafar, Sudan 107 F2 6 52N 28 15 E
Tafassasset, O. →, Algeria 111 D6 22 0N 9 57 E
Tafelbaai, S. Africa 116 E2 33 35 S 18 25 E
Tafelney, C., Morocco 110 B3 31 3N 9 51W
Tafermaar, Indonesia 83 C4 6 47 S 134 10 E
Taffermit, Morocco 110 C3 29 37N 9 15W
Tafí Viejo, Argentina 174 B2 26 43 S 65 17W
Tafīhān, Iran 97 D7 29 25N 52 39 E
Tafiré, Ivory C. 112 D3 9 4N 5 4W
Tafjord, Norway 18 B4 62 14N 7 24 E
Tafnidilt, Morocco 110 C2 28 47N 10 58W
Tafraoute, Morocco 110 C3 29 50N 8 58W
Taft, Iran 97 D7 31 45N 54 14 E
Taft, Phil. 81 F5 11 57N 125 30 E
Taft, Calif., U.S.A. 161 K7 35 8N 119 28W
Taft, Fla., U.S.A. 153 G8 28 26N 81 22W
Taft, Tex., U.S.A. 155 M6 27 59N 97 24W
Taga, W. Samoa 133 W23 13 46 S 172 28W
Taga Dzong, Bhutan 90 B2 27 5N 89 55 E
Tagana-an, Phil. 81 G5 9 42N 125 35 E
Taganrog, Russia 59 J10 47 12N 38 50 E
Taganrogskiy Zaliv, Russia 59 J10 47 0N 38 30 E
Tagânt, Mauritania 112 B2 18 20N 11 0W
Tagap Ga, Burma 90 B6 26 56N 96 13 E
Tagapula I., Phil. 80 E5 12 4N 124 12 E
Tagatay, Phil. 80 D3 14 6N 120 56 E
Tagauayan I., Phil. 81 F3 10 58N 121 13 E
Tagbilaran, Phil. 81 G4 9 39N 123 51 E
Tage, Papua N. G. 132 D2 6 19 S 143 20 E
Tággia, Italy 44 E4 43 52N 7 51 E
Taghzout, Morocco 110 B4 33 30N 4 49W
Tagish, Canada 142 A2 60 19N 134 16W
Tagish L., Canada 142 A2 60 10N 134 20W
Tagkawayan, Phil. 80 E4 13 58N 122 32 E
Tagliacozzo, Italy 45 F10 42 4N 13 14 E
Tagliamento →, Italy 45 C10 45 38N 13 6 E
Táglio di Po, Italy 45 D9 45 0N 12 12 E
Tagna, Colombia 168 D3 2 24 S 70 37W
Tago, Phil. 81 G6 9 2N 126 13 E
Tago, Mt., Phil. 81 G5 8 23N 125 5 E
Tagomago, Spain 39 B8 39 2N 1 39 E
Taguatinga, Brazil 171 D3 12 16 S 42 26W
Tagudin, Phil. 80 C3 16 56N 120 27 E
Tagula, Papua N. G. 132 F7 11 22 S 153 15 E
Tagula I., Papua N. G. 132 F7 11 30 S 153 30 E
Tagum, Phil. 81 H5 7 33N 125 53 E
Tagus = Tejo →, Europe 43 F2 38 40N 9 24W
Tahakopa, N.Z. 131 G4 46 30 S 169 23 E
Tahala, Morocco 110 B4 34 0N 4 28W
Tahan, Gunong, Malaysia 87 K4 4 34N 102 17 E
Tahānah-ye sūr Gol, Afghan. 91 C2 31 43N 67 53 E
Tahara, Japan 71 C9 34 40N 137 16 E
Tahat, Algeria 111 D6 23 18N 5 33 E
Tāherī, Iran 97 E7 27 43N 52 20 E
Tahiti, Pac. Oc. 135 J13 17 37 S 149 27W
Tahoe, L., U.S.A. 160 G6 39 6N 120 2W
Tahoe City, U.S.A. 160 F6 39 10N 120 9W
Taholah, U.S.A. 160 C2 47 21N 124 17W
Tahora, N.Z. 130 F3 39 2 S 174 49 E
Tahoua, Niger 113 C6 14 57N 5 16 E
Tahta, Egypt 106 B3 26 44N 31 32 E
Tahtaköprü, Turkey 51 G13 39 57N 29 39 E
Tahtalı Dağları, Turkey 100 C7 38 20N 36 0 E
Tahuamanu →, Bolivia 172 C4 11 6 S 67 36W
Tahulandang, Indonesia 82 A3 2 27N 125 23 E
Tahuna, Indonesia 82 A3 3 38N 125 30 E
Taï, Ivory C. 112 D3 5 55N 7 30W
Tai Shan, China 75 F9 36 25N 117 20 E
Tai Xian, China 77 A13 32 30N 120 7 E
Tai'an, China 75 F9 36 12N 117 8 E
Taibei = T'aipei, Taiwan 77 E13 25 2N 121 30 E
Taibique, Canary Is. 39 G2 27 42N 17 58W
Taibus Qi, China 74 D8 41 54N 115 22 E
T'aichung, Taiwan 77 E13 24 9N 120 37 E
Taieri →, N.Z. 131 G5 46 3 S 170 12 E
Taiga Madema, Libya 108 D3 23 46N 15 25 E
Taigu, China 74 F7 37 28N 112 30 E
Taihang Shan, China 74 G7 36 0N 113 30 E
Taihape, N.Z. 130 F4 39 41 S 175 48 E
Taihe, Anhui, China 74 H8 33 20N 115 42 E
Taihe, Jiangxi, China 77 D10 26 47N 114 52 E
Taihu, China 77 B11 30 22N 116 20 E
Taijiang, China 76 D7 26 39N 108 21 E
Taikang, China 74 G8 34 5N 114 50 E
Taikkyi, Burma 90 G6 17 20N 96 0 E
Tailem Bend, Australia 128 C3 35 12 S 139 29 E
Tailfingen, Germany 31 G5 48 15N 9 1 E
Taimyr Peninsula = Taymyr, Poluostrov, Russia 65 B11 75 0N 100 0 E
Tain, U.K. 22 D4 57 49N 4 4W
T'ainan, Taiwan 77 F13 23 0N 120 10 E
Taínaron, Ákra, Greece 48 E4 36 22N 22 27 E
Tainggyo, Burma 90 G5 17 49N 94 29 E
Taining, China 77 D11 26 54N 117 9 E
Taiobeiras, Brazil 171 E3 15 49 S 42 14W
T'aipei, Taiwan 77 E13 25 2N 121 30 E
Taiping, China 77 B12 30 15N 118 6 E
Taiping, Malaysia 87 K3 4 51N 100 44 E
Taipingzhen, China 74 H6 33 35N 111 42 E
Taipu, Brazil 170 C4 5 37 S 35 36W
Tairbeart = Tarbert, U.K. 22 D2 57 54N 6 49W
Taisha, Japan 70 B4 35 24N 132 40 E
Taishan, China 77 F9 22 14N 112 41 E
Taishun, China 77 D12 27 30N 119 42 E
Taita □, Kenya 118 C4 4 0 S 38 30 E
Taita Hills, Kenya 118 C4 3 25 S 38 15 E
Taitao, C., Chile 176 C1 45 53 S 75 5W
Taitao, Pen. de, Chile 176 C2 46 30 S 75 0W
T'aitung, Taiwan 77 F13 22 43N 121 4 E
Taivalkoski, Finland 14 D23 65 33N 28 12 E
Taiwan ■, Asia 77 F13 23 30N 121 0 E
Taixing, China 77 A13 32 11N 120 0 E
Taïyetos Óros, Greece 48 D4 37 0N 22 23 E
Taiyiba, Israel 103 C4 32 36N 35 27 E
Taiyuan, China 74 F7 37 52N 112 33 E
Taizhong = T'aichung, Taiwan 77 E13 24 9N 120 37 E
Taizhou, China 77 A12 32 28N 119 55 E
Taizhou Liedao, China 77 C13 28 30N 121 55 E
Ta'izz, Yemen 98 D4 13 35N 44 2 E
Tājābād, Iran 97 D7 30 2N 54 24 E
Tajapuru, Furo do, Brazil 170 B1 1 50 S 50 25W
Tajarḥī, Libya 108 D2 24 21N 14 28 E
Tajikistan ■, Asia 63 D5 38 30N 70 0 E
Tajima, Japan 69 F9 37 12N 139 46 E
Tajimi, Japan 71 B9 35 19N 137 8 E
Tajo = Tejo →, Europe 43 F2 38 40N 9 24W
Tajrīsh, Iran 97 C6 35 48N 51 25 E
Tājūrā, Libya 108 B2 32 51N 13 21 E
Tak, Thailand 86 D2 16 52N 99 8 E
Takāb, Iran 101 D12 36 24N 47 7 E
Takachiho, Japan 70 E3 32 42N 131 18 E
Takada, Japan 69 F9 37 7N 138 15 E
Takahagi, Japan 69 F10 36 43N 140 45 E
Takahashi, Japan 70 C5 34 51N 133 39 E
Takaka, N.Z. 131 A7 40 51 S 172 50 E
Takamatsu, Japan 70 C6 34 20N 134 5 E
Takanabe, Japan 70 E3 32 8N 131 30 E
Takaoka, Japan 71 A9 36 47N 137 0 E
Takapau, N.Z. 130 G5 40 2 S 176 21 E
Takapuna, N.Z. 130 C3 36 47 S 174 47 E
Takasago, Japan 70 C6 34 45N 134 48 E
Takasaki, Japan 71 A11 36 20N 139 0 E
Takase, Japan 70 C5 34 7N 133 48 E
Takatsuki, Japan 71 C7 34 51N 135 37 E
Takaungu, Kenya 118 C4 3 38 S 39 52 E
Takawa, Japan 70 D2 33 38N 130 51 E
Takayama, Japan 71 A9 36 18N 137 11 E
Takayama-Bonchi, Japan 71 B9 36 0N 137 18 E
Take-Shima, Japan 69 J5 30 49N 130 26 E
Takefu, Japan 71 B8 35 50N 136 10 E
Takehara, Japan 70 C4 34 21N 132 55 E
Takengon, Indonesia 84 B1 4 45N 96 50 E
Takeo, Japan 70 D2 33 12N 130 1 E
Tåkern, Sweden 17 F8 58 22N 14 45 E
Tākestān, Iran 97 C6 36 0N 49 40 E
Taketa, Japan 70 E3 32 58N 131 24 E
Takev, Cambodia 87 G5 10 59N 104 47 E
Takh, India 93 C7 33 6N 77 32 E
Takhār □, Afghan. 91 A3 36 40N 70 0 E
Taki, Papua N. G. 132 D8 6 29 S 155 52 E
Takikawa, Japan 68 C10 43 33N 141 54 E
Takla L., Canada 142 B3 55 15N 125 45W
Takla Landing, Canada 142 B3 55 30N 125 50W
Takla Makan = Taklamakan Shamo, China 72 C3 38 0N 83 0 E
Taklamakan Shamo, China 72 C3 38 0N 83 0 E
Takotna, U.S.A. 144 E8 62 59N 156 4W
Taku, Japan 70 D2 33 18N 130 3 E
Taku →, Canada 142 B2 58 30N 133 50W
Takum, Nigeria 113 D6 7 18N 9 36 E
Takuma, Japan 70 C5 34 13N 133 40 E
Takundi, Dem. Rep. of the Congo 115 C3 4 45 S 16 34 E
Takuran, Phil. 81 H4 7 51N 123 34 E
Takutu →, Guyana 169 C5 3 1N 60 29W
Tal Halāl, Iran 97 D7 28 54N 55 1 E
Tala, Uruguay 175 C4 34 21 S 55 46W
Talachyn, Belarus 58 E5 54 25N 29 42 E
Talacogan, Phil. 81 G5 8 32N 125 39 E
Talagante, Chile 174 C1 33 40 S 70 50W
Talaïnt, Morocco 110 C3 29 41N 9 40W
Talak, Niger 113 B6 18 0N 5 0 E
Talakag, Phil. 81 G5 8 16N 124 37 E
Talamanca, Cordillera de, Cent. Amer. 164 E3 9 20N 83 20W
Talant, France 27 E11 47 19N 4 58 E
Talara, Peru 172 A1 4 38 S 81 18W
Talas, Kyrgyzstan 63 B6 42 30N 72 13 E
Talas, Turkey 100 C6 38 41N 35 33 E
Talas →, Kazakstan 63 B5 44 0N 70 20 E
Talasea, Papua N. G. 132 C6 5 20 S 150 2 E
Talasskiy Alatau, Khrebet, Kyrgyzstan 63 B6 42 15N 72 0 E
Talâta, Egypt 103 E1 30 36N 32 20 E
Talata Mafara, Nigeria 113 C6 12 38N 6 4 E
Talaud, Kepulauan, Indonesia 82 A3 4 30N 127 10 E
Talaud Is. = Talaud, Kepulauan, Indonesia 82 A3 4 30N 127 10 E
Talavera de la Reina, Spain 42 F6 39 55N 4 46W
Talavera la Real, Spain 43 G4 38 53N 6 46W
Talawana, Australia 124 D3 22 51 S 121 9 E
Talawgyi, Burma 90 C6 25 4N 97 19 E
Talayan, Phil. 81 H5 6 52N 124 24 E
Talayuela, Spain 42 F5 39 59N 5 36W
Talbert, Sillon de, France 26 D3 48 53N 3 5W
Talbot, C., Australia 124 B4 13 48 S 126 43 E
Talbotton, U.S.A. 152 C5 32 41N 84 32W
Talbragar →, Australia 129 B8 32 12 S 148 37 E
Talca, Chile 174 D1 35 28 S 71 40W
Talca □, Chile 174 D1 35 20 S 71 46W
Talcahuano, Chile 174 D1 36 40 S 73 10W
Talcher, India 94 D7 21 0N 85 18 E
Talcho, Niger 113 C5 14 44N 3 28 E
Taldy Kurgan = Taldyqorghan, Kazakstan 64 E8 45 10N 78 45 E
Taldyqorghan, Kazakstan 64 E8 45 10N 78 45 E
Tālesh, Iran 97 B6 37 58N 48 58 E
Ţālesh, Kūhhā-ye, Iran 97 B6 37 42N 48 55 E
Talgar, Kazakstan 63 B8 43 19N 77 15 E
Talgar, Pik, Kazakstan 63 B8 43 5N 77 20 E
Talguharai, Sudan 106 D4 18 19N 35 56 E
Tali Post, Sudan 107 F3 5 55N 30 44 E
Talibon, Phil. 81 F5 10 9N 124 20 E
Talibong, Ko, Thailand 87 J2 7 15N 99 23 E
Talihina, U.S.A. 155 H7 34 45N 95 3W
Talikota, India 95 F3 16 29N 76 17 E
Talimardzhan = Tallymerjen, Turkmenistan 63 D2 38 23N 65 37 E
Talisay, Phil. 81 F4 10 44N 122 58 E
Talisayan, Phil. 81 G5 9 0N 124 55 E
Talitsa, Russia 62 C9 57 0N 63 43 E
Taliwang, Indonesia 82 C1 8 50 S 116 55 E
Tall 'Afar, Iraq 101 D10 36 22N 42 27 E
Tall Kalakh, Syria 103 A5 34 41N 36 15 E
Talla, Egypt 106 J7 28 5N 30 43 E
Talladega, U.S.A. 152 B3 33 26N 86 6W
Tallahassee, U.S.A. 152 E5 30 27N 84 17W
Tallangatta, Australia 129 D7 36 15 S 147 19 E
Tallapoosa, U.S.A. 152 B4 33 45N 85 17W
Tallapoosa →, U.S.A. 152 C3 32 30N 86 16W
Tallard, France 29 D10 44 28N 6 3 E
Tallarook, Australia 129 D6 37 5 S 145 6 E
Tallassee, U.S.A. 152 C4 32 32N 85 54W
Tallawang, Australia 129 B8 32 12 S 149 28 E
Tällberg, Sweden 16 D9 60 51N 15 2 E
Tallering Pk., Australia 125 E2 28 6 S 115 37 E
Tallinn, Estonia 15 G21 59 22N 24 48 E
Tallulah, U.S.A. 155 J9 32 25N 91 11W
Tallymerjen, Turkmenistan 63 D2 38 23N 65 37 E
Tălmaciu, Romania 53 E9 45 38N 24 19 E
Talmest, Morocco 110 B3 31 48N 9 21W
Talmont-St-Hilaire, France 28 B2 46 27N 1 37W
Talne, Ukraine 59 H6 48 50N 30 44 E
Talnoye = Talne, Ukraine 59 H6 48 50N 30 44 E
Taloda, India 94 D2 21 34N 74 11 E
Talodi, Sudan 107 E3 10 35N 30 22 E
Talomo, Phil. 81 H5 7 3N 125 32 E
Tāloqān, Afghan. 91 A3 36 44N 69 33 E
Talovaya, Russia 60 E5 51 6N 40 45 E
Taloyoak, Canada 138 B10 69 32N 93 32W
Talpa de Allende, Mexico 162 C4 20 23N 104 51W
Talquin, L., U.S.A. 152 E5 30 23N 84 39W
Talsi, Latvia 15 H20 57 10N 22 30 E
Talsi □, Latvia 54 A9 57 20N 22 40 E
Talsinnt, Morocco 111 B4 32 33N 3 27W
Taltal, Chile 174 B1 25 23 S 70 33W
Taltson →, Canada 142 A6 61 24N 112 46W
Talwood, Australia 127 D4 28 29 S 149 29 E
Talyawalka Cr. →, Australia 128 B5 32 28 S 142 22 E
Tam Chau, Vietnam 87 G5 10 48N 105 12 E
Tam Ky, Vietnam 86 E7 15 34N 108 29 E
Tam Quan, Vietnam 86 E7 14 35N 109 3 E
Tama, U.S.A. 156 C4 41 58N 92 35W
Tamala, Australia 125 E1 26 42 S 113 47 E
Tamalameque, Colombia 168 B3 8 52N 73 49W
Tamale, Ghana 113 D4 9 22N 0 50W
Taman, Russia 59 K9 45 14N 36 41 E
Tamana, Japan 70 E2 32 58N 130 32 E
Tamanar, Morocco 110 B3 31 1N 9 46W
Tamano, Japan 70 C5 34 29N 133 59 E
Tamanrasset, Algeria 111 D6 22 50N 5 30 E
Tamanrasset, O. →, Algeria 111 D5 22 0N 2 0 E
Tamanthi, Burma 90 C5 25 19N 95 17 E
Tamaqua, U.S.A. 151 F9 40 48N 75 58W
Tamar →, U.K. 21 G3 50 27N 4 15W
Támara, Colombia 168 B3 5 50N 72 10W
Tamarac, U.S.A. 153 J9 26 12N 80 10W
Tamarang, Australia 129 A9 31 27 S 150 5 E
Tamarinda, Spain 39 B10 39 55N 3 49 E

288 Tamarite de Litera

Tamarite de Litera, *Spain* **40 D5** 41 52N 0 25 E
Tamaroa, *U.S.A.* **156 F7** 38 8N 89 14W
Tamashima, *Japan* **70 C5** 34 32N 133 40 E
Tamási, *Hungary* **52 D3** 46 40N 18 18 E
Tamaské, *Niger* **113 C6** 14 49N 5 43 E
Tamaulipas □, *Mexico* **163 C5** 24 0N 99 0W
Tamaulipas, Sierra de, *Mexico* **163 C5** 23 30N 98 20W
Tamazula, *Mexico* **162 C3** 24 55N 106 58W
Tamazunchale, *Mexico* **163 C5** 21 16N 98 47W
Tamba-Dabatou, *Guinea* **112 C2** 11 50N 10 40W
Tambacounda, *Senegal* **112 C2** 13 45N 13 40W
Tambelan, Kepulauan, *Indonesia* **84 B3** 1 0N 107 30 E
Tambellup, *Australia* **125 F2** 34 4 S 117 37 E
Tambo, *Australia* **126 C4** 24 54 S 146 14 E
Tambo, *Peru* **172 C3** 12 57 S 74 1W
Tambo →, *Peru* **172 C3** 10 42 S 73 47W
Tambo de Mora, *Peru* **172 C2** 13 30 S 76 8W
Tambobamba, *Peru* **172 C3** 13 54 S 72 8W
Tambohorano, *Madag.* **117 B7** 17 30 S 43 58 E
Tambopata →, *Peru* **172 C4** 13 21 S 69 36W
Tambora, *Indonesia* **82 C1** 8 12 S 118 5 E
Tamboritha, Mt., *Australia* **129 D7** 37 31 S 146 40 E
Tambov, *Russia* **60 D5** 52 45N 41 28 E
Tambre →, *Spain* **42 C2** 42 49N 8 53W
Tambuku, *Indonesia* **85 D4** 7 8 S 113 40 E
Tambun Sigumbal, *Phil.* **81 H3** 6 5N 121 47 E
Tamburâ, *Sudan* **107 F2** 5 40N 27 25 E
Tambuyukan, Gunong, *Malaysia* **85 A5** 6 13N 116 39 E
Tâmchekket, *Mauritania* **112 B2** 17 25N 10 40W
Tamdybulak, *Uzbekistan* **63 C2** 41 46N 64 36 E
Tame, *Colombia* **168 B3** 6 28N 71 44W
Tâmega →, *Portugal* **42 D2** 41 5N 8 21W
Tamelelt, *Morocco* **110 B3** 31 50N 7 32W
Tamenglong, *India* **90 C4** 25 0N 93 35 E
Tamerlanovka, *Kazakstan* **63 B4** 42 36N 69 17 E
Tamerza, *Tunisia* **108 B1** 34 23N 7 58 E
Tamiahua, L. de, *Mexico* **163 C5** 21 30N 97 30W
Tamiami Canal, *U.S.A.* **153 K8** 25 50N 81 0W
Tamil Nadu □, *India* **95 J3** 11 0N 77 0 E
Tamis →, *Serbia, Yug.* **52 F5** 44 51N 20 39 E
Tamluk, *India* **93 H12** 22 18N 87 58 E
Tammerfors = Tampere, *Finland* **15 F20** 61 30N 23 50 E
Tammisaari, *Finland* **15 F20** 60 0N 23 26 E
Tämnaren, *Sweden* **16 D11** 60 10N 17 25 E
Tamo Abu, Pegunungan, *Malaysia* **85 B5** 3 10N 115 5 E
Tampa, *U.S.A.* **149 M4** 27 57N 82 27W
Tampa B., *U.S.A.* **149 M4** 27 50N 82 30W
Tampere, *Finland* **15 F20** 61 30N 23 50 E
Tampico, *Mexico* **163 C5** 22 20N 97 50W
Tampico, *U.S.A.* **156 C7** 41 38N 89 47W
Tampin, *Malaysia* **87 L4** 2 28N 102 13 E
Tamrah, *Si. Arabia* **98 B4** 20 24N 45 25 E
Tamri, *Morocco* **110 B3** 30 49N 9 50W
Tamrida = Qādib, *Yemen* **99 D6** 12 37N 53 57 E
Tamsweg, *Austria* **34 D6** 47 7N 13 49 E
Tamuja →, *Spain* **43 F4** 39 38N 6 29W
Tamworth, *Australia* **129 A9** 31 7 S 150 58 E
Tamworth, *U.K.* **21 E6** 52 39N 1 41W
Tamyang, *S. Korea* **75 G14** 35 19N 126 59 E
Tan An, *Vietnam* **87 G6** 10 32N 106 25 E
Tan-Tan, *Morocco* **110 C2** 28 29N 11 1W
Tana →, *Kenya* **118 C5** 2 32 S 40 31 E
Tana →, *Norway* **14 A23** 70 30N 28 14 E
Tana, L., *Ethiopia* **107 E4** 13 5N 37 30 E
Tana River, *Kenya* **118 C4** 2 0 S 39 30 E
Tanabe, *Japan* **71 D7** 33 44N 135 22 E
Tanabi, *Brazil* **171 F2** 20 37 S 49 37W
Tanacross, *U.S.A.* **144 E12** 63 23N 143 21W
Tanafjorden, *Norway* **14 A23** 70 45N 28 25 E
Tanaga, Pta., *Canary Is.* **39 G1** 27 42N 18 10W
Tanaga I., *U.S.A.* **144 L3** 51 48N 177 53W
Tanaga Volcano, *U.S.A.* **144 L3** 51 53N 178 8W
Tanah Merah, *Malaysia* **84 A2** 5 48N 102 9 E
Tanahbala, *Indonesia* **84 C1** 0 30 S 98 30 E
Tanahgrogot, *Indonesia* **85 C5** 1 55 S 116 15 E
Tanahjampea, *Indonesia* **82 C2** 7 10 S 120 35 E
Tanahmasa, *Indonesia* **84 C1** 0 12 S 98 39 E
Tanahmerah, *Indonesia* **83 C6** 6 5 S 140 16 E
Tanakura, *Japan* **69 F10** 37 10N 140 20 E
Tanami, *Australia* **124 C4** 19 59 S 129 43 E
Tanami Desert, *Australia* **124 C5** 18 50 S 132 0 E
Tanana →, *U.S.A.* **138 B4** 65 10N 151 58W
Tananarive = Antananarivo, *Madag.* **117 B8** 18 55 S 47 31 E
Tananger, *Norway* **18 F2** 58 57N 5 37 E
Tanannt, *Morocco* **110 B3** 31 54N 6 56W
Tánaro →, *Italy* **44 D5** 44 55N 8 40 E
Tanauan, *Batangas, Phil.* **80 D3** 14 5N 121 10 E
Tanauan, *Leyte, Phil.* **81 F5** 11 7N 125 1 E
Tanay, *Phil.* **80 D3** 14 30N 121 17 E
Tanba-Sanchi, *Japan* **71 B7** 35 7N 135 48 E
Tanbar, *Australia* **126 D3** 25 51 S 141 55 E
Tancheng, *China* **75 G10** 34 25N 118 20 E
Tanchŏn, *N. Korea* **75 D15** 40 27N 128 54 E
Tanda, *Ut. P., India* **93 F10** 26 33N 82 35 E
Tanda, *Ut. P., India* **93 E8** 28 57N 78 56 E
Tanda, *Ivory C.* **112 D4** 7 48N 3 10W
Tandag, *Phil.* **81 G6** 9 4N 126 9 E
Tandaia, *Tanzania* **119 D3** 9 25 S 34 15 E
Tăndărei, *Romania* **53 F12** 44 39N 27 40 E
Tandaué, *Angola* **115 F3** 16 58 S 18 5 E
Tandil, *Argentina* **174 D4** 37 15 S 59 6W
Tandil, Sa. del, *Argentina* **174 D4** 37 30 S 59 0W
Tandlianwala, *Pakistan* **92 D5** 31 3N 73 9 E
Tando Adam, *Pakistan* **91 D3** 25 45N 68 40 E
Tandou L., *Australia* **128 B5** 32 40 S 142 5 E
Tandragee, *U.K.* **23 B5** 54 21N 6 24W
Tandsjöborg, *Sweden* **16 C8** 61 42N 14 43 E
Tandur, *India* **94 E4** 19 11N 79 30 E
Tane-ga-Shima, *Japan* **69 J5** 30 30N 131 0 E
Taneatua, *N.Z.* **130 E6** 38 4 S 177 1 E
Tanen Tong Dan, *Burma* **86 D2** 16 30N 98 30 E
Tanew →, *Poland* **55 H9** 50 29N 22 16 E
Tanezrouft, *Algeria* **111 D5** 23 9N 0 11 E
Tang, Koh, *Cambodia* **87 G4** 10 16N 103 7 E
Tang Krasang, *Cambodia* **86 F5** 12 34N 105 3 E
Tanga, *Tanzania* **118 D4** 5 5 S 39 2 E
Tanga □, *Tanzania* **118 D4** 5 20 S 38 0 E
Tanga Is., *Papua N. G.* **132 B7** 3 20 S 153 15 E
Tangail, *Bangla.* **90 C2** 24 15N 89 55 E
Tanganyika, L., *Africa* **118 D3** 6 40 S 30 0 E
Tangawan, *Phil.* **81 H4** 7 30N 122 20 E
Tangen, *Norway* **18 D8** 60 37N 11 15 E
Tanger = Tangier, *Morocco* **110 A3** 35 50N 5 49W
Tangerang, *Indonesia* **84 D3** 6 11 S 106 37 E
Tangerhütte, *Germany* **30 C7** 52 26N 11 48 E
Tangermünde, *Germany* **30 C7** 52 33N 11 58 E
Tanggu, *China* **75 E9** 39 2N 117 40 E
Tanggula Shan, *China* **72 C4** 32 40N 92 10 E
Tanghe, *China* **74 H7** 32 47N 112 50 E
Tangier, *Morocco* **110 A3** 35 50N 5 49W
Tangkeleboke, *Indonesia* **82 B2** 3 10 S 121 30 E
Tangorin P.O., *Australia* **126 C3** 21 47 S 144 12 E
Tangshan, *China* **75 E10** 39 38N 118 10 E
Tangtou, *China* **75 G10** 35 28N 118 30 E
Tangub, *Phil.* **81 G4** 8 3N 123 44 E
Tanguiéta, *Benin* **113 C5** 10 35N 1 21 E
Tangxi, *China* **77 C12** 29 3N 119 25 E
Tangyan He →, *China* **76 C7** 28 54N 108 19 E
Tanimbar, Kepulauan, *Indonesia* **83 C4** 7 30 S 131 30 E
Tanimbar Is. = Tanimbar, Kepulauan, *Indonesia* **83 C4** 7 30 S 131 30 E
Taninthari, *Burma* **87 F2** 12 6N 99 3 E
Taniyama, *Japan* **70 F2** 31 31N 130 31 E
Tanjay, *Phil.* **81 G4** 9 30N 123 5 E
Tanjong Malim, *Malaysia* **87 L3** 3 42N 101 31 E
Tanjore = Thanjavur, *India* **95 J4** 10 48N 79 12 E
Tanjung, *Indonesia* **85 C5** 2 10 S 115 25 E
Tanjungbalai, *Indonesia* **84 B1** 2 55N 99 44 E
Tanjungbatu, *Indonesia* **85 B5** 2 23N 118 3 E
Tanjungkarang Telukbetung, *Indonesia* **84 D3** 5 20 S 105 10 E
Tanjungpandan, *Indonesia* **85 C3** 2 43 S 107 38 E
Tanjungpinang, *Indonesia* **84 B2** 1 5N 104 30 E
Tanjungpriok, *Indonesia* **84 D3** 6 8 S 106 55 E
Tanjungredeb, *Indonesia* **85 B5** 2 9N 117 29 E
Tanjungselor, *Indonesia* **85 B5** 2 55N 117 25 E
Tank, *Pakistan* **91 B3** 32 14N 70 25 E
Tanna, *Vanuatu* **133 J7** 19 30 S 169 20 E
Tännäs, *Sweden* **16 B6** 62 26N 12 42 E
Tannis Bugt, *Denmark* **17 G4** 57 40N 10 15 E
Tannu-Ola, *Russia* **65 D10** 51 0N 94 0 E
Tano →, *Ghana* **112 D4** 5 7N 2 56W
Tanon Str., *Phil.* **81 F4** 10 20N 123 30 E
Tanout, *Niger* **109 F1** 14 50N 8 55 E
Tanquinho, *Brazil* **171 D4** 11 58 S 39 6W
Tanshui, *Taiwan* **77 E13** 25 10N 121 28 E
Tanta, *Egypt* **106 H7** 30 45N 30 57 E
Tantoyuca, *Mexico* **163 C5** 21 21N 98 10W
Tantung = Dandong, *China* **75 D13** 40 10N 124 20 E
Tanuku, *India* **94 F5** 16 45N 81 44 E
Tanumshede, *Sweden* **17 F5** 58 42N 11 20 E
Tanunda, *Australia* **128 C3** 34 30 S 139 0 E
Tanur, *India* **95 J2** 11 1N 75 52 E
Tanus, *France* **28 D6** 44 8N 2 19 E
Tanza, *Phil.* **80 D3** 14 41N 120 56 E
Tanzania ■, *Africa* **118 D3** 6 0 S 34 0 E
Tanzawa-Sanchi, *Japan* **71 B11** 35 27N 139 0 E
Tanzilla →, *Canada* **142 B2** 58 8N 130 43W
Tao, Ko, *Thailand* **87 G2** 10 5N 99 52 E
Tao'an, *China* **75 B12** 45 22N 122 40 E
Tao'er He →, *China* **75 B13** 45 45N 124 5 E
Taohua Dao, *China* **77 C14** 29 50N 122 20 E
Taolanaro, *Madag.* **117 D8** 25 2 S 47 0 E
Taole, *China* **74 E4** 38 48N 106 40 E
Taormina, *Italy* **47 E8** 37 51N 15 17 E
Taos, *U.S.A.* **159 H11** 36 24N 105 35W
Taoudenni, *Mali* **110 D4** 22 40N 3 55W
Taoudrart, Adrar, *Algeria* **111 D5** 24 25N 2 24 E
Taounate, *Morocco* **110 B4** 34 25N 4 41W
Taourirt, *Algeria* **111 C5** 26 37N 0 20 E
Taourirt, *Morocco* **111 B4** 34 25N 2 53W
Taouz, *Morocco* **110 B4** 30 53N 4 0W
Taoyuan, *China* **77 C8** 28 55N 111 16 E
T'aoyüan, *Taiwan* **77 E13** 25 0N 121 4 E
Tapa, *Estonia* **15 G21** 59 15N 25 50 E
Tapa Shan = Daba Shan, *China* **76 B7** 32 0N 109 0 E
Tapachula, *Mexico* **163 E6** 14 54N 92 17W
Tapah, *Malaysia* **87 K3** 4 12N 101 15 E
Tapajós →, *Brazil* **169 D7** 2 24 S 54 41W
Tapaktuan, *Indonesia* **84 B1** 3 15N 97 10 E
Tapanahoni →, *Surinam* **169 C7** 4 20N 54 25W
Tapanui, *N.Z.* **131 F4** 45 56 S 169 18 E
Tapauá, *Brazil* **173 B5** 5 40 S 64 20W
Tapauá →, *Brazil* **173 B5** 5 40 S 64 21W
Tapaz, *Phil.* **81 F4** 11 16N 122 32 E
Tapeta, *Liberia* **112 D3** 6 29N 8 52W
Taphan Hin, *Thailand* **86 D3** 16 13N 100 26 E
Tapi →, *India* **94 D1** 21 8N 72 41 E
Tapia de Casariego, *Spain* **42 B4** 43 34N 6 56W
Tapiantana Group, *Phil.* **81 H4** 6 20N 122 0 E
Tapilon, *Phil.* **81 F5** 11 17N 124 2 E
Tapini, *Papua N. G.* **132 E4** 8 19 S 147 0 E
Tapiraí, *Brazil* **171 E2** 19 52 S 46 1W
Tapirapé →, *Brazil* **170 D1** 10 41 S 50 38W
Tapirapecó, Serra, *Venezuela* **169 C5** 1 10N 65 0W
Tapirapuã, *Brazil* **173 C6** 14 51 S 57 45W
Taplan, *Australia* **128 C4** 34 33 S 140 52 E
Tapoeripa, *Surinam* **169 B6** 5 22N 56 34W
Tapolca, *Hungary* **52 D2** 46 53N 17 29 E
Tappahannock, *U.S.A.* **148 G7** 37 56N 76 52W
Tapuaenuku, Mt., *N.Z.* **131 C8** 42 0 S 173 39 E
Tapul = Salvacion, *Phil.* **81 G2** 9 56N 118 47 E
Tapul Group, *Phil.* **81 J3** 5 35N 120 50 E
Tapun, *India* **90 B6** 27 35N 96 22 E
Tapurucuará, *Brazil* **169 D4** 0 24 S 65 2W
Taqīābād, *Iran* **97 C8** 35 33N 59 11 E
Țaqțaq, *Iraq* **101 E11** 35 53N 44 35 E
Taquara, *Brazil* **175 B5** 29 36 S 50 46W
Taquari →, *Brazil* **173 D6** 19 15 S 57 17W
Taquaritinga, *Brazil* **171 F2** 21 24 S 48 30W
Tara, *Australia* **127 D5** 27 17 S 150 31 E
Tara, *Canada* **150 B3** 44 28N 81 9W
Tara, *Japan* **70 D2** 33 2N 130 11 E
Tara, *Russia* **64 D8** 56 55N 74 24 E
Tara, *Zambia* **119 F2** 16 58 S 26 45 E
Tara →, *Montenegro, Yug.* **50 C2** 43 21N 18 51 E
Tara-Dake, *Japan* **70 E2** 32 58N 130 6 E
Taraba □, *Nigeria* **113 D7** 8 0N 10 30 E
Tarabagatay, Khrebet, *Kazakstan* **64 E9** 48 0N 83 0 E
Tarabuco, *Bolivia* **173 D5** 19 10 S 64 57W
Tarābulus, *Lebanon* **103 A4** 34 31N 35 50 E
Tarābulus, *Libya* **108 B2** 32 49N 13 7 E
Taraclia, *Moldova* **53 D14** 46 34N 29 7 E
Taraclia, *Moldova* **53 E13** 45 54N 28 40 E
Taradale, *N.Z.* **130 F5** 39 33 S 176 53 E
Tarahouahout, *Algeria* **111 D6** 22 41N 5 59 E
Tarajalejo, *Canary Is.* **39 F5** 28 12N 14 7W
Tarakan, *Indonesia* **85 B5** 3 20N 117 35 E
Tarakit, Mt., *Kenya* **118 B4** 2 2N 35 10 E
Taralga, *Australia* **129 C8** 34 26 S 149 52 E
Tarama-Jima, *Japan* **69 M2** 24 39N 124 42 E
Taramakau →, *N.Z.* **131 C6** 42 34 S 171 8 E
Taran, Mys, *Russia* **15 J18** 54 56N 19 59 E
Tarana, *Australia* **129 B8** 33 31 S 149 52 E
Taranagar, *India* **92 E6** 28 43N 74 50 E
Taranaki □, *N.Z.* **130 F3** 39 25 S 174 30 E
Tarancón, *Spain* **40 E1** 40 1N 3 1W
Taranga, *India* **92 H5** 23 56N 72 43 E
Taranga Hill, *India* **92 H5** 24 0N 72 40 E
Taransay, *U.K.* **22 D1** 57 54N 7 0W
Táranto, *Italy* **47 B10** 40 28N 17 14 E
Táranto, G. di, *Italy* **47 B10** 40 8N 17 20 E
Tarapacá, *Colombia* **168 D4** 2 56 S 69 46W
Tarapacá □, *Chile* **174 A2** 20 45 S 69 30W
Tarapoto, *Peru* **172 B2** 6 30 S 76 20W
Taraquá, *Brazil* **168 C4** 0 6N 68 28W
Tarare, *France* **29 C8** 45 54N 4 26 E
Tararua Ra., *N.Z.* **130 G4** 40 45 S 175 25 E
Tarascon, *France* **29 E8** 43 48N 4 39 E
Tarascon-sur-Ariège, *France* **28 F5** 42 50N 1 36 E
Tarashcha, *Ukraine* **59 H6** 49 30N 30 31 E
Tarata, *Peru* **172 D3** 17 27 S 70 2W
Tarauacá, *Brazil* **172 B3** 8 6 S 70 48W
Tarauacá →, *Brazil* **172 B4** 6 42 S 69 48W
Taravo →, *France* **29 G12** 41 42N 8 49 E
Tarawera, *N.Z.* **130 F5** 39 2 S 176 36 E
Tarawera L., *N.Z.* **130 E5** 38 13 S 176 27 E
Tarawera Mt., *N.Z.* **130 E5** 38 14 S 176 32 E
Tarazona, *Spain* **40 D3** 41 55N 1 43W
Tarazona de la Mancha, *Spain* **41 F3** 39 16N 1 55W
Tarbat Ness, *U.K.* **22 D5** 57 52N 3 47W
Tarbela Dam, *Pakistan* **92 B5** 34 8N 72 52 E
Tarbert, *Arg. & Bute, U.K.* **22 F3** 55 52N 5 25W
Tarbert, *W. Isles, U.K.* **22 D2** 57 54N 6 49W
Tarbes, *France* **28 E4** 43 15N 0 3 E
Tarboro, *Ga., U.S.A.* **152 D8** 31 1N 81 48W
Tarboro, *N.C., U.S.A.* **149 H7** 35 54N 77 32W
Tarbrax, *Australia* **126 C3** 21 7 S 142 26 E
Tarbū, *Libya* **108 C3** 26 0N 15 5 E
Tărcău, Munții, *Romania* **53 D11** 46 39N 26 7 E
Tarcento, *Italy* **45 B10** 46 13N 13 13 E
Tarcoola, *Australia* **127 E1** 30 44 S 134 36 E
Tarcoon, *Australia* **127 E4** 30 15 S 146 43 E
Tardets-Sorholus, *France* **28 E3** 43 8N 0 52W
Tardoire →, *France* **28 C4** 45 52N 0 14 E
Taree, *Australia* **129 A10** 31 50 S 152 30 E
Tarf, Ras, *Morocco* **110 A3** 35 40N 5 11W
Țarfā, Ra's aț, *Si. Arabia* **98 C3** 17 2N 42 22 E
Tarfa, W. el →, *Egypt* **106 J7** 28 25N 30 50 E
Tarfaya, *Morocco* **110 C2** 27 55N 12 55W
Târgoviște, *Romania* **53 F10** 44 55N 25 27 E
Târgu Bujor, *Romania* **53 E12** 45 52N 27 54 E
Târgu Cărbunești, *Romania* **53 F8** 44 57N 23 31 E
Târgu Frumos, *Romania* **53 C12** 47 12N 27 2 E
Târgu-Jiu, *Romania* **53 E8** 45 5N 23 19 E
Târgu Lăpuș, *Romania* **53 C8** 47 27N 23 52 E
Târgu Mureș, *Romania* **53 D9** 46 31N 24 38 E
Târgu Neamț, *Romania* **53 C11** 47 12N 26 25 E
Târgu Ocna, *Romania* **53 D11** 46 16N 26 39 E
Târgu Secuiesc, *Romania* **53 E11** 46 0N 26 10 E
Targuist, *Morocco* **110 B4** 34 59N 4 14W
Tărgușor, *Romania* **53 F13** 44 27N 28 25 E
Tărhăus, Vf., *Romania* **53 D11** 46 40N 26 8 E
Tarhbalt, *Morocco* **110 B3** 30 39N 5 20W
Tarhit, *Algeria* **111 B4** 30 58N 2 0W
Tarhūnah, *Libya* **108 B2** 32 27N 13 36 E
Tari, *Papua N. G.* **132 C2** 5 54 S 142 59 E
Táriba, *Venezuela* **168 B3** 7 49N 72 13W
Tarifa, *Spain* **43 J5** 36 1N 5 36W
Tarija, *Bolivia* **174 A3** 21 30 S 64 40W
Tarija □, *Bolivia* **174 A3** 21 30 S 63 30W
Tariku →, *Indonesia* **83 B5** 2 55 S 138 26 E
Tarīm, *Yemen* **99 C5** 16 3N 49 0 E
Tarim Basin = Tarim Pendi, *China* **72 B3** 40 0N 84 0 E
Tarim He →, *China* **72 C3** 39 30N 88 30 E
Tarim Pendi, *China* **72 B3** 40 0N 84 0 E
Tarime □, *Tanzania* **118 C3** 1 15 S 34 0 E
Taringo Downs, *Australia* **129 B6** 32 13 S 145 33 E
Taritatu →, *Indonesia* **83 B5** 2 54 S 138 27 E
Tarka →, *S. Africa* **116 E4** 32 10 S 26 0 E
Tarkastad, *S. Africa* **116 E4** 32 0 S 26 16 E
Tarkhankut, Mys, *Ukraine* **59 K7** 45 25N 32 30 E
Tarko Sale, *Russia* **64 C8** 64 55N 77 50 E
Tarkwa, *Ghana* **112 D4** 5 20N 2 0W
Tarlac, *Phil.* **80 D3** 15 29N 120 35 E
Tarlac □, *Phil.* **80 D3** 15 30N 120 30 E
Tarlton Downs, *Australia* **126 C2** 22 40 S 136 45 E
Tarm, *Denmark* **17 J2** 55 56N 8 31 E
Tarma, *Peru* **172 C2** 11 25 S 75 45W
Tarn □, *France* **28 E6** 43 49N 2 8 E
Tarn →, *France* **28 D5** 44 5N 1 6 E
Tarn-et-Garonne □, *France* **28 D5** 44 8N 1 20 E
Tarna →, *Hungary* **52 C4** 47 31N 19 59 E
Târnava Mare →, *Romania* **53 D8** 46 10N 23 43 E
Târnava Mică →, *Romania* **53 D8** 46 9N 23 55 E
Târnăveni, *Romania* **53 D9** 46 19N 24 13 E
Tarnica, *Poland* **55 J9** 49 4N 22 44 E
Tarnobrzeg, *Poland* **55 H8** 50 35N 21 41 E
Tarnobrzeg □, *Poland* **55 H8** 50 40N 22 0 E
Tarnogród, *Poland* **55 H9** 50 22N 22 45 E
Tarnos, *France* **28 E2** 43 32N 1 28W
Târnova, *Moldova* **53 B12** 48 10N 27 40 E
Târnova, *Romania* **52 E6** 45 23N 21 59 E
Tarnów, *Poland* **55 H8** 50 3N 21 0 E
Tarnów □, *Poland* **55 J7** 50 0N 21 0 E
Tarnowskie Góry, *Poland* **55 H5** 50 27N 18 54 E
Tärnsjö, *Sweden* **16 D10** 60 9N 16 56 E
Táro →, *Italy* **44 C7** 45 2N 10 15 E
Taroom, *Australia* **127 D4** 25 36 S 149 48 E
Taroudannt, *Morocco* **110 B3** 30 30N 8 52W
Tarp, *Germany* **30 A5** 54 39N 9 24 E
Tarpon Springs, *U.S.A.* **149 L4** 28 9N 82 45W
Tarquínia, *Italy* **45 F8** 42 15N 11 45 E
Tarragona, *Spain* **40 D6** 41 5N 1 17 E
Tarragona □, *Spain* **40 D6** 41 5N 1 0 E
Tarrasa = Terrassa, *Spain* **40 D7** 41 34N 2 1 E
Tàrrega, *Spain* **40 D6** 41 39N 1 9 E
Tarrytown, *Ga., U.S.A.* **152 C7** 32 19N 82 34W
Tarrytown, *N.Y., U.S.A.* **151 E11** 41 4N 73 52W
Tårs, *Denmark* **17 G4** 57 23N 10 7 E
Tarshiha = Me'ona, *Israel* **103 B4** 33 1N 35 15 E
Tarso Emissi, *Chad* **109 D3** 21 27N 18 36 E
Tarso Ourari, *Chad* **109 D3** 21 27N 17 27 E
Tarsus, *Turkey* **100 D6** 36 58N 34 55 E
Tartagal, *Argentina* **174 A3** 22 30 S 63 50W
Tärtär, *Azerbaijan* **61 K8** 40 20N 46 58 E
Tärtär →, *Azerbaijan* **61 K8** 40 26N 47 20 E
Tartas, *France* **28 E3** 43 50N 0 49W
Tartu, *Estonia* **15 G22** 58 20N 26 44 E
Țarțūs, *Syria* **100 E6** 34 55N 35 55 E
Tarumirim, *Brazil* **171 E3** 19 16 S 41 59W
Tarumizu, *Japan* **70 F2** 31 29N 130 42 E
Tarussa, *Russia* **58 E9** 54 44N 37 10 E
Tarutao, Ko, *Thailand* **87 J2** 6 33N 99 40 E
Tarutung, *Indonesia* **84 B1** 2 0N 98 54 E
Tarvísio, *Italy* **45 B10** 46 30N 13 35 E
Tarz Ulli, *Libya* **108 C2** 25 32N 10 8 E
Tas-Buget, *Kazakstan* **63 A2** 44 46N 65 33 E
Tasahku, *Burma* **90 B6** 27 33N 97 52 E
Tasāwah, *Libya* **108 C2** 26 0N 13 30 E
Taschereau, *Canada* **140 C4** 48 40N 78 40W
Taseko →, *Canada* **142 C4** 52 8N 123 45W
Tasgaon, *India* **94 F2** 17 2N 74 39 E
Tash-Kömür, *Kyrgyzstan* **63 C6** 41 40N 72 10 E
Tash-Kumyr = Tash-Kömür, *Kyrgyzstan* **63 C6** 41 40N 72 10 E
Tashauz = Dashhowuz, *Turkmenistan* **64 E6** 41 49N 59 58 E
Tashi Chho Dzong = Thimphu, *Bhutan* **90 B2** 27 31N 89 45 E
Tashkent = Toshkent, *Uzbekistan* **63 C4** 41 20N 69 10 E
Tashtagol, *Russia* **64 D9** 52 47N 87 53 E
Tasikmalaya, *Indonesia* **85 D3** 7 18 S 108 12 E
Tåsinge, *Denmark* **17 J4** 55 0N 10 35 E
Tåsjön, *Sweden* **14 D16** 64 15N 15 40 E
Taskan, *Russia* **65 C16** 62 59N 150 20 E
Taşköprü, *Turkey* **100 B6** 41 30N 34 15 E
Taşlâc, *Moldova* **53 C14** 47 4N 29 24 E
Tasman →, *N.Z.* **131 D5** 43 48 S 170 8 E
Tasman, Mt., *N.Z.* **131 D5** 43 34 S 170 12 E
Tasman B., *N.Z.* **131 A8** 40 59 S 173 25 E
Tasman Mts., *N.Z.* **131 B7** 41 3 S 172 25 E
Tasman Pen., *Australia* **126 G4** 43 10 S 148 0 E
Tasman Sea, *Pac. Oc.* **134 L8** 36 0 S 160 0 E
Tasmania □, *Australia* **126 G4** 42 0 S 146 30 E
Tășnad, *Romania* **52 C7** 47 30N 22 33 E
Tassil Tin-Rerhoh, *Algeria* **111 D5** 20 5N 3 55 E
Tassili n'Ajjer, *Algeria* **111 C6** 25 47N 8 1 E
Tassili-Oua-n-Ahaggar, *Algeria* **111 D6** 20 41N 5 30 E
Tasty, *Kazakstan* **63 A4** 44 47N 69 7 E
Tasu Sd., *Canada* **142 C2** 52 47N 132 2W
Tata, *Hungary* **52 C3** 47 37N 18 19 E
Tata, *Morocco* **110 C3** 29 46N 7 56W
Tatabánya, *Hungary* **52 C3** 47 32N 18 25 E
Tataouine, *Tunisia* **108 B2** 32 57N 10 29 E
Tatar Republic □ = Tatarstan □, *Russia* **60 C10** 55 30N 51 30 E
Tatarbunary, *Ukraine* **59 K5** 45 50N 29 39 E
Tatarsk, *Russia* **64 D8** 55 14N 76 0 E
Tatarstan □, *Russia* **60 C10** 55 30N 51 30 E
Tatau, *Malaysia* **85 B4** 2 53N 112 51 E
Tatebayashi, *Japan* **71 A11** 36 15N 139 32 E
Tateshina-Yama, *Japan* **71 A10** 36 8N 138 11 E
Tateyama, *Japan* **71 C11** 35 0N 139 50 E
Tathlina L., *Canada* **142 A5** 60 33N 117 39W
Tathlīth, *Si. Arabia* **98 C3** 19 32N 43 30 E
Tathlīth, W. →, *Si. Arabia* **98 B4** 20 35N 44 20 E
Tathra, *Australia* **129 D8** 36 44 S 149 59 E
Tatinnai L., *Canada* **143 A9** 60 55N 97 40W
Tatitlek, *U.S.A.* **144 F11** 60 52N 146 41W
Tatlısu, *Turkey* **51 F11** 40 24N 27 55 E
Tatnam, C., *Canada* **143 B10** 57 16N 91 0W
Tatra = Tatry, *Slovak Rep.* **35 B13** 49 20N 20 0 E
Tatry, *Slovak Rep.* **35 B13** 49 20N 20 0 E
Tatsuno, *Japan* **70 C6** 34 52N 134 33 E
Tatta, *Pakistan* **91 D2** 24 42N 67 55 E
Tatuí, *Brazil* **175 A6** 23 25 S 47 53W
Tatum, *U.S.A.* **155 J3** 33 16N 103 19W
Tatum Cr. →, *U.S.A.* **152 E7** 30 43N 82 32W
Tat'ung = Datong, *China* **74 D7** 40 6N 113 18 E
Tatura, *Australia* **129 D6** 36 29 S 145 16 E
Tatvan, *Turkey* **101 C10** 38 31N 42 15 E
Tau, *Amer. Samoa* **133 X25** 14 15 S 169 30W
Tau, *Norway* **18 E2** 59 3N 5 55 E
Tauá, *Brazil* **170 C3** 6 1 S 40 26W
Taubaté, *Brazil* **175 A6** 23 0 S 45 36W
Tauberbischofsheim, *Germany* **31 F5** 49 37N 9 39 E
Taucha, *Germany* **30 D8** 51 23N 12 29 E
Tauern-tunnel, *Austria* **34 D6** 47 0N 13 12 E
Taufikia, *Sudan* **107 F3** 9 24N 31 37 E
Taulé, *France* **26 D3** 48 37N 3 55W
Taumarunui, *N.Z.* **130 E4** 38 53 S 175 15 E
Taumaturgo, *Brazil* **172 B3** 8 54 S 72 51W
Taung, *S. Africa* **116 D3** 27 33 S 24 47 E
Taungdwingyi, *Burma* **90 E5** 20 1N 95 40 E
Taunggyi, *Burma* **90 E6** 20 50N 97 0 E
Taungtha, *Burma* **90 E5** 21 12N 95 25 E
Taungup, *Burma* **90 F5** 18 51N 94 14 E
Taungup Pass, *Burma* **90 F5** 18 40N 94 45 E
Taunsa Barrage, *Pakistan* **92 D4** 30 42N 70 50 E
Taunton, *U.K.* **21 F4** 51 1N 3 5W
Taunton, *U.S.A.* **151 E13** 41 54N 71 6W
Taunus, *Germany* **31 E4** 50 13N 8 34 E
Taupo, *N.Z.* **130 E5** 38 41 S 176 7 E
Taupo, L., *N.Z.* **130 E4** 38 46 S 175 55 E
Tauragė, *Lithuania* **15 J20** 55 14N 22 16 E
Tauragė □, *Lithuania* **54 C9** 55 15N 22 17 E
Tauranga, *N.Z.* **130 D5** 37 42 S 176 11 E
Tauranga Harb., *N.Z.* **130 D5** 37 30 S 176 5 E
Tauri →, *Papua N. G.* **132 E4** 8 8 S 146 8 E
Taurianova, *Italy* **47 D9** 38 21N 16 1 E
Taurus Mts. = Toros Dağları, *Turkey* **100 D5** 37 0N 32 30 E
Tauste, *Spain* **40 D3** 41 58N 1 18W
Tauz = Tovuz, *Azerbaijan* **61 K7** 41 0N 45 40 E
Tavaar, *Somali Rep.* **120 D3** 3 6N 46 1 E
Tavannes, *Switz.* **32 B4** 47 13N 7 12 E
Tavares, *U.S.A.* **153 G8** 28 48N 81 44W
Tavas, *Turkey* **49 D11** 37 34N 29 4 E
Tavda, *Russia* **64 D7** 58 7N 65 8 E
Tavda →, *Russia* **64 D7** 57 47N 67 18 E
Tavernes de la Valldigna, *Spain* **41 F4** 39 5N 0 13W
Tavernier, *U.S.A.* **153 K9** 25 1N 80 31W
Taveta, *Tanzania* **118 C4** 3 23 S 37 37 E
Taveuni, *Fiji* **133 A3** 16 51 S 179 58W
Taviano, *Italy* **47 C11** 39 59N 18 5 E
Tavignano →, *France* **29 F13** 42 7N 9 33 E
Tavira, *Portugal* **43 H3** 37 8N 7 40W
Tavistock, *Canada* **150 C4** 43 19N 80 50W
Tavistock, *U.K.* **21 G3** 50 33N 4 9W
Tavolara, *Italy* **46 B2** 40 54N 9 42 E
Távora →, *Portugal* **42 D3** 41 8N 7 35W
Tavoy = Dawei, *Burma* **86 E2** 14 2N 98 12 E
Tavşanlı, *Turkey* **49 B11** 39 32N 29 30 E
Taw →, *U.K.* **21 F3** 51 4N 4 4W
Tawas City, *U.S.A.* **148 C4** 44 16N 83 31W
Tawau, *Malaysia* **85 B5** 4 20N 117 55 E
Țawī Șulaym, *Oman* **99 B7** 22 33N 58 40 E
Tawitawi, *Phil.* **81 J2** 5 10N 120 0 E

Thanet, I. of 289

Tawitawi □, Phil. 81 J3 5 0N 120 0 E
Tawitawi Group, Phil. 81 J3 5 10N 120 15 E
Tawngche, Burma 90 B5 26 34N 95 38 E
Tāwurgha', Libya 108 B3 32 1N 15 2 E
Taxila, Pakistan 92 C5 33 42N 72 52 E
Tay →, U.K. 22 E5 56 37N 3 38W
Tay, Firth of, U.K. 22 E5 56 25N 3 8W
Tay, L., Australia 125 F3 32 55 S 120 48 E
Tay, L., U.K. 22 E4 56 32N 4 8W
Tay Ninh, Vietnam 87 G6 11 20N 106 5 E
Tayabamba, Peru 172 B2 8 15 S 77 16W
Tayabas, Phil. 80 D3 14 1N 121 35 E
Tayabas Bay, Phil. 80 E3 13 45N 121 45 E
Taylakova, Russia 64 D8 59 13N 74 0 E
Taylakovy = Taylakova, Russia 64 D8 59 13N 74 0 E
Taylor, Canada 142 B4 56 13N 120 40W
Taylor, Fla., U.S.A. 152 E7 30 26N 82 18W
Taylor, Mich., U.S.A. 157 B13 42 14N 83 16W
Taylor, Nebr., U.S.A. 154 E5 41 46N 99 23W
Taylor, Pa., U.S.A. 151 E9 41 23N 75 43W
Taylor, Tex., U.S.A. 155 K6 30 34N 97 25W
Taylor, Mt., N.Z. 131 D6 43 30 S 171 20 E
Taylor, Mt., U.S.A. 159 J10 35 14N 107 37W
Taylorsville, U.S.A. 157 F11 38 2N 85 21W
Taylorville, U.S.A. 156 E7 39 33N 89 18W
Taymā, Si. Arabia 96 E3 27 35N 38 45 E
Taymyr, Oz., Russia 65 B11 74 20N 102 0 E
Taymyr, Poluostrov, Russia 65 B11 75 0N 100 0 E
Tayog, Phil. 80 C3 16 2N 120 45 E
Tayport, U.K. 22 E6 56 27N 2 52W
Tayshet, Russia 65 D10 55 58N 98 1 E
Taytay, Palawan, Phil. 81 F2 10 45N 119 30 E
Taytay, Rizal, Phil. 80 D3 14 34N 121 8 E
Taytay Bay, Phil. 81 F2 10 55N 119 35 E
Taz →, Russia 64 C8 67 32N 78 40 E
Taza, Morocco 110 B4 34 16N 4 6W
Tāzah Khurmātū, Iraq 101 E11 35 18N 44 20 E
Tazawa-Ko, Japan 68 E10 39 43N 140 40 E
Taze, Burma 90 D5 22 57N 95 24 E
Tazenakht, Morocco 110 B3 30 35N 7 12W
Tazerbo, Libya 108 C4 25 45N 21 0 E
Tazin L., Canada 143 B7 59 44N 108 42W
Tazlina, U.S.A. 144 E11 62 4N 146 27W
Tazoult, Algeria 111 A6 35 29N 6 11 E
Tazovskiy, Russia 64 C8 67 30N 78 44 E
Tbilisi, Georgia 61 K7 41 43N 44 50 E
Tchad = Chad ■, Africa 109 F3 15 0N 17 15 E
Tchad, L., Chad 109 F2 13 30N 14 30 E
Tchaourou, Benin 113 D5 8 58N 2 40 E
Tch'eng-tou = Chengdu, China 76 B5 30 38N 104 2 E
Tchentlo L., Canada 142 B4 55 15N 125 0W
Tchibanga, Gabon 114 C2 2 45 S 11 0 E
Tchien, Liberia 112 D3 5 59N 8 15W
Tchikala-Tcholohanga, Angola 115 E3 12 38 S 16 3 E
Tchin Tabaraden, Niger 113 B6 15 58N 5 56 E
Tchingou, Massif de, N. Cal. 133 T19 20 54 S 165 0 E
Tcholliré, Cameroon 114 A2 8 24N 14 10 E
Tch'ong-k'ing = Chongqing, China 76 C6 29 35N 106 25 E
Tczew, Poland 54 D5 54 8N 18 50 E
Te Anau, N.Z. 131 F2 45 25 S 167 43 E
Te Anau, L., N.Z. 131 F2 45 15 S 167 45 E
Te Araroa, N.Z. 130 D7 37 39 S 178 25 E
Te Aroha, N.Z. 130 D4 37 32 S 175 44 E
Te Awamutu, N.Z. 130 E4 38 1 S 175 20 E
Te Kaha, N.Z. 130 D6 37 44 S 177 52 E
Te Karaka, N.Z. 130 E6 38 26 S 177 53 E
Te Kauwhata, N.Z. 130 D4 37 25 S 175 9 E
Te Kopuru, N.Z. 130 C2 36 2 S 173 56 E
Te Kuiti, N.Z. 130 E4 38 20 S 175 11 E
Te Puke, N.Z. 130 D5 37 46 S 176 22 E
Te Teko, N.Z. 130 E5 38 2 S 176 48 E
Te Waewae B., N.Z. 131 G2 46 13 S 167 33 E
Tea →, Brazil 168 D4 0 30 S 65 9W
Tea Tree, Australia 126 C1 22 5 S 133 22 E
Teaca, Romania 53 D9 46 55N 24 30 E
Teague, U.S.A. 155 K6 31 38N 96 17W
Teano, Italy 47 A7 41 15N 14 4 E
Teapa, Mexico 163 D6 18 35N 92 56W
Teba, Spain 43 J6 36 59N 4 55W
Teberau, Malaysia 84 B2 1 32N 103 45 E
Teberda, Russia 61 J5 43 30N 41 46 E
Tébessa, Algeria 111 A6 35 22N 8 8 E
Tebicuary →, Paraguay 174 B4 26 36 S 58 16W
Tebingtinggi, Indonesia 84 B1 3 20N 99 9 E
Tebintingii, Indonesia 84 B2 1 0N 102 45 E
Tébourba, Tunisia 108 A1 36 49N 9 51 E
Téboursouk, Tunisia 108 A1 36 29N 9 10 E
Tebulos, Georgia 61 J7 42 36N 45 17 E
Tecate, Mexico 161 N10 32 34N 116 38W
Tecer Dağları, Turkey 100 C7 39 27N 37 27 E
Tech →, France 28 F7 42 36N 3 3 E
Techa →, Russia 62 C9 56 13N 62 58 E
Techiman, Ghana 112 D4 7 35N 1 58W
Techirghiol, Romania 53 F13 44 4N 28 32 E
Tecka, Argentina 176 B2 43 29 S 70 48W
Tecomán, Mexico 162 D4 18 55N 103 53W
Tecopa, U.S.A. 161 K10 35 51N 116 13W
Tecoripa, Mexico 162 B3 28 37N 109 57W
Tecuala, Mexico 162 C3 22 23N 105 27W
Tecuci, Romania 53 E12 45 51N 27 27 E
Tecumseh, U.S.A. 157 B13 42 0N 83 57W
Ted, Somali Rep. 120 D2 4 24N 43 55 E
Tedzhen = Tejen, Turkmenistan 64 F7 37 23N 60 31 E
Tees →, U.K. 20 C6 54 37N 1 10W
Tees B., U.K. 20 C6 54 40N 1 9W
Teeswater, Canada 150 C3 43 59N 81 17W
Tefé, Brazil 169 D5 3 25 S 64 50W
Tefé →, Brazil 169 D5 3 35 S 64 47W
Tefenni, Turkey 49 D11 37 18N 29 45 E
Tegal, Indonesia 85 D3 6 52 S 109 8 E
Tegernsee, Germany 31 H7 47 43N 11 46 E
Teggiano, Italy 47 B8 40 23N 15 32 E
Teghra, India 93 G11 25 30N 85 34 E
Tegid, L. = Bala, L., U.K. 20 E4 52 53N 3 37W
Tegina, Nigeria 113 C6 10 5N 6 11 E
Tegua, Vanuatu 133 C4 13 15 S 166 37 E
Tegucigalpa, Honduras 164 D2 14 5N 87 14W
Tehachapi, U.S.A. 161 K8 35 8N 118 27W
Tehachapi Mts., U.S.A. 161 L8 35 0N 118 30W
Tehamiyam, Sudan 106 D4 18 20N 36 32 E
Tehilla, Sudan 106 D4 17 42N 36 6 E
Téhini, Ivory C. 112 D4 9 39N 3 40W
Tehoru, Indonesia 83 B3 3 19 S 129 37 E
Tehrān, Iran 97 C6 35 44N 51 30 E
Tehuacán, Mexico 163 D5 18 30N 97 30W
Tehuantepec, Mexico 163 D5 16 21N 95 13W
Tehuantepec, G. de, Mexico 163 D5 15 50N 95 12W
Tehuantepec, Istmo de, Mexico 163 D6 17 0N 94 30W
Teide, Canary Is. 39 F3 28 15N 16 38W
Teifi →, U.K. 21 E3 52 5N 4 41W
Teigebyen, Norway 18 D8 60 12N 11 0 E
Teign →, U.K. 21 G4 50 32N 3 32W
Teignmouth, U.K. 21 G4 50 33N 3 31W
Teiuş, Romania 53 D8 46 12N 23 40 E
Teixeira, Brazil 170 C4 7 13 S 37 15W
Teixeira Pinto, Guinea-Biss. 112 C1 12 3N 16 0W
Tejen, Turkmenistan 64 F7 37 23N 60 31 E
Tejo →, Europe 43 F2 38 40N 9 24W
Tejon Pass, U.S.A. 161 L8 34 49N 118 53W
Tekamah, U.S.A. 154 E6 41 47N 96 13W
Tekapo →, N.Z. 131 E5 44 13 S 170 21 E
Tekapo, L., N.Z. 131 D5 43 53 S 170 33 E
Tekax, Mexico 163 C7 20 11N 89 18W
Teke, Turkey 51 E13 41 4N 29 39 E
Tekeli, Kazakstan 63 A9 44 50N 79 0 E
Tekeze →, Ethiopia 107 E4 14 20N 35 50 E
Tekija, Serbia, Yug. 50 B6 44 42N 22 26 E
Tekirdağ, Turkey 51 F11 40 58N 27 30 E
Tekirdağ □, Turkey 51 E11 41 0N 27 0 E
Tekirova, Turkey 49 E12 36 30N 30 32 E
Tekkali, India 94 E7 18 37N 84 15 E
Tekke, Turkey 100 B7 40 42N 36 12 E
Tekman, Turkey 101 C9 39 38N 41 29 E
Tekoa, U.S.A. 158 C5 47 14N 117 4W
Tekouiât, O. →, Algeria 111 D5 22 25N 2 35 E
Tel Aviv-Yafo, Israel 103 C3 32 4N 34 48 E
Tel Lakhish, Israel 103 D3 31 34N 34 51 E
Tel Megiddo, Israel 103 C4 32 35N 35 11 E
Tela, Honduras 164 C2 15 40N 87 28W
Télagh, Algeria 111 B4 34 51N 0 32W
Telanaipura = Jambi, Indonesia 84 C2 1 38 S 103 30 E
Telavåg, Norway 18 D1 60 15N 4 58 E
Telavi, Georgia 61 J7 42 0N 45 30 E
Telč, Czech Rep. 34 B8 49 11N 15 28 E
Telciu, Romania 53 C9 47 25N 24 24 E
Telde, Canary Is. 39 G4 27 59N 15 25W
Telefomin, Papua N. G. 132 C1 5 10 S 141 31 E
Telegraph Creek, Canada 142 B2 58 0N 131 10W
Telekhany = Tsyelyakhany, Belarus 59 F3 52 30N 25 46 E
Telemark, Norway 15 G12 59 15N 7 40 E
Telemark □, Norway 18 E5 59 25N 8 30 E
Telén, Argentina 174 D2 36 15 S 65 31W
Telen →, Indonesia 85 C5 0 10 S 117 20 E
Teleneşti, Moldova 53 C13 47 30N 28 22 E
Teleng, Iran 97 E9 25 47N 61 3 E
Teleño, Spain 42 C4 42 23N 6 22W
Teleorman □, Romania 53 G10 44 0N 25 0 E
Teleorman →, Romania 53 G10 43 52N 25 26 E
Teles Pires →, Brazil 173 B6 7 21 S 58 3W
Telescope Pk., U.S.A. 161 J9 36 10N 117 5W
Teletaye, Mali 113 B5 16 31N 1 30 E
Telford, U.K. 21 E5 52 40N 2 27W
Telford and Wrekin □, U.K. 20 E5 52 45N 2 27W
Telfs, Austria 34 D4 47 19N 11 4 E
Telida, U.S.A. 144 E9 63 23N 153 16W
Télimélé, Guinea 112 C2 10 54N 13 2W
Telkwa, Canada 142 C3 54 41N 127 5W
Tell City, U.S.A. 157 G10 37 57N 86 46W
Tellicherry, India 95 J2 11 45N 75 30 E
Telluride, U.S.A. 159 H10 37 56N 107 49W
Telogia, U.S.A. 152 E5 30 21N 84 49W
Telok Datok, Malaysia 84 B2 2 49N 101 31 E
Teloloapán, Mexico 163 D5 18 21N 99 51W
Telpos Iz, Russia 56 B10 63 16N 59 13 E
Telsen, Argentina 176 B3 42 30 S 66 50W
Telšiai, Lithuania 15 H20 55 59N 22 14 E
Telšiai □, Lithuania 54 C9 55 59N 22 15 E
Teltow, Germany 30 C9 52 24N 13 15 E
Teluk Anson = Teluk Intan, Malaysia 87 K3 4 3N 101 0 E
Teluk Betung = Tanjungkarang Telukbetung, Indonesia 84 D3 5 20 S 105 10 E
Teluk Cenderawasih, Indonesia 83 B5 2 30 S 135 20 E
Teluk Intan, Malaysia 87 K3 4 3N 101 0 E
Telukbutun, Indonesia 85 B3 4 13N 108 12 E
Telukdalem, Indonesia 84 B1 0 33N 97 50 E
Tema, Ghana 113 D5 5 41N 0 0 E
Temanggung, Indonesia 85 D4 7 18 S 110 10 E
Temapache, Mexico 163 C5 21 4N 97 38W
Temax, Mexico 163 C7 21 10N 88 50W
Temba, S. Africa 117 D4 25 20 S 28 17 E
Tembe, Dem. Rep. of the Congo 118 C2 0 16 S 28 14 E
Tembesi, Indonesia 84 C2 1 43 S 103 6 E
Tembilahan, Indonesia 84 C2 0 19 S 103 9 E
Temblador, Venezuela 169 B5 8 59N 62 44W
Tembleque, Spain 42 F7 39 41N 3 30W
Temblor Range, U.S.A. 161 K7 35 20N 119 50W
Teme →, U.K. 21 E5 52 11N 2 13W
Temecula, U.S.A. 161 M9 33 30N 117 9W
Temerloh, Malaysia 84 B2 3 27N 102 25 E
Temir, Kazakstan 64 E6 49 1N 57 14 E
Temirtau, Kazakstan 64 D8 50 5N 72 56 E
Temirtau, Russia 64 D9 53 10N 87 30 E
Témiscaming, Canada 140 C4 46 44N 79 5W
Temma, Australia 126 G3 41 12 S 144 48 E
Temnikov, Russia 60 C6 54 40N 43 11 E
Temo →, Italy 46 B1 40 17N 8 28 E
Temora, Australia 129 C7 34 30 S 147 30 E
Temosachic, Mexico 162 B3 28 58N 107 50W
Tempe, U.S.A. 159 K8 33 25N 111 56W
Tempe Downs, Australia 124 D5 24 22 S 132 24 E
Témpio Pausánia, Italy 46 B2 40 54N 9 6 E
Tempiute, U.S.A. 160 H11 37 39N 115 38W
Temple, U.S.A. 155 K6 31 6N 97 21W
Temple B., Australia 126 A3 12 15 S 143 3 E
Temple Terrace, U.S.A. 153 G7 28 2N 82 23W
Templemore, Ireland 23 D4 52 47N 7 51W
Templeton, U.S.A. 160 K6 35 33N 120 42W
Templeton →, Australia 126 C2 21 0 S 138 40 E
Templin, Germany 30 B9 53 7N 13 28 E
Tempoal, Mexico 163 C5 21 31N 98 23W
Temryuk, Russia 59 K9 45 15N 37 24 E
Temska →, Serbia, Yug. 50 C6 43 17N 22 33 E
Temuco, Chile 176 A2 38 45 S 72 40W
Temuka, N.Z. 131 E6 44 14 S 171 17 E
Ten Thousand Is., U.S.A. 153 K8 25 55N 81 45W
Tena, Ecuador 168 D2 0 59 S 77 49W
Tenabo, Mexico 163 C6 20 2N 90 12W
Tenaha, U.S.A. 155 K7 31 57N 94 15W
Tenakee Springs, U.S.A. 144 H14 57 47N 135 13W
Tenali, India 95 F5 16 15N 80 35 E
Tenancingo, Mexico 163 D5 19 0N 99 33W
Tenango, Mexico 163 D5 19 7N 99 33W
Tenasserim = Tanintharí, Burma 87 F2 12 6N 99 3 E
Tenasserim □, Burma 86 F2 14 0N 98 30 E
Tenby, U.K. 21 F3 51 40N 4 42W
Tenda, Colle di, France 29 D11 44 7N 7 36 E
Tendaho, Ethiopia 107 E5 11 48N 40 54 E
Tende, France 29 D11 44 5N 7 35 E
Tendelti, Sudan 107 E3 13 1N 31 55 E
Tendjedi, Adrar, Algeria 111 D6 23 41N 7 32 E
Tendrara, Morocco 111 B4 33 3N 1 58W
Tendre, Mt., Switz. 32 C2 46 35N 6 18 E
Tendrovskaya Kosa, Ukraine 59 J6 46 16N 31 35 E
Teneida, Egypt 106 B2 25 30N 29 19 E
Tenente Marques →, Brazil 173 C6 11 10 S 59 56W
Ténéré, Niger 109 E2 19 0N 10 30 E
Ténéré, Erg du, Niger 109 E2 17 35N 10 55 E
Tenerife, Canary Is. 39 F3 28 15N 16 35W
Tenerife, Pico, Canary Is. 39 G1 27 43N 18 1W
Ténès, Algeria 111 A5 36 31N 1 14 E
Teng Xian, Guangxi Zhuangzu, China 77 F8 23 21N 110 56 E
Teng Xian, Shandong, China 75 G9 35 5N 117 10 E
Tengah □, Indonesia 82 B2 2 0 S 122 0 E
Tengah, Kepulauan, Indonesia 85 D5 7 5 S 118 15 E
Tengchong, China 76 E2 25 0N 98 28 E
Tengchowfu = Penglai, China 75 F11 37 48N 120 42 E
Tenggara □, Indonesia 82 B2 3 0 S 122 0 E
Tenggarong, Indonesia 85 C5 0 24 S 116 58 E
Tenggol, Pulau, Malaysia 87 K4 4 48N 103 41 E
Tengiz, Ozero, Kazakstan 64 D7 50 30N 69 0 E
Tenhult, Sweden 17 G8 57 41N 14 20 E
Tenigerbad, Switz. 33 C7 46 42N 8 57 E
Tenino, U.S.A. 160 D4 46 51N 122 51W
Tenkasi, India 95 K3 8 55N 77 20 E
Tenke, Shaba, Dem. Rep. of the Congo 119 E2 11 22 S 26 40 E
Tenke, Shaba, Dem. Rep. of the Congo 119 E2 10 32 S 26 7 E
Tenkodogo, Burkina Faso 113 C4 11 54N 0 19W
Tenna →, Italy 45 E10 43 14N 13 47 E
Tennant Creek, Australia 126 B1 19 30 S 134 15 E
Tennessee □, U.S.A. 149 H2 36 0N 86 30W
Tennessee →, U.S.A. 148 G1 37 4N 88 34W
Tennille, U.S.A. 152 C7 32 56N 82 48W
Tennsift, Oued →, Morocco 110 B3 32 3N 9 28W
Tennyson, U.S.A. 157 F9 38 5N 87 7W
Teno, Pta. de, Canary Is. 39 F3 28 21N 16 55W
Tenom, Malaysia 85 A5 5 4N 115 57 E
Tenosique, Mexico 163 D6 17 30N 91 24W
Tenri, Japan 71 C7 34 39N 135 49 E
Tenryū, Japan 71 C9 34 52N 137 49 E
Tenryū-Gawa →, Japan 71 B9 35 39N 137 48 E
Tent L., Canada 143 A7 62 25N 107 54W
Tentelomatinan, Indonesia 82 A2 0 56N 121 48 E
Tenterden, U.K. 21 F8 51 4N 0 42 E
Tenterfield, Australia 127 D5 29 0 S 152 0 E
Teo, Spain 42 C2 42 45N 8 30W
Teófilo Otoni, Brazil 171 E3 17 50 S 41 30W
Teotihuacán, Mexico 163 D5 19 44N 98 50W
Tepa, Indonesia 83 C3 7 52 S 129 31 E
Tepalcatepec →, Mexico 162 D4 18 35N 101 59W
Tepecik, Bursa, Turkey 51 F12 40 7N 28 25 E
Tepecik, Kütahya, Turkey 49 B11 39 32N 29 28 E
Tepehuanes, Mexico 162 B3 25 21N 105 44W
Tepelena, Albania 50 F4 40 17N 20 2 E
Tepequem, Serra, Brazil 169 C5 3 45N 61 45W
Tepetongo, Mexico 162 C4 22 28N 103 9W
Tepic, Mexico 162 C4 21 30N 104 54W
Teplá, Czech Rep. 34 B5 49 59N 12 52 E
Teplice, Czech Rep. 34 A6 50 40N 13 48 E
Teploklyuchenka, Kyrgyzstan 63 B9 42 30N 78 30 E
Tepoca, C., Mexico 162 A2 30 20N 112 25W
Tequila, Mexico 162 C4 20 54N 103 47W
Ter →, Spain 40 C8 42 2N 3 12 E
Ter Apel, Neths. 24 B7 52 53N 7 5 E
Téra, Niger 113 C5 14 0N 0 45 E
Tera →, Spain 42 D5 41 54N 5 44W
Teraina, Kiribati 135 G11 4 43N 160 25W
Téramo, Italy 45 F10 42 39N 13 42 E
Terang, Australia 128 E5 38 15 S 142 55 E
Terawhiti, C., N.Z. 130 H3 41 16 S 174 38 E
Terazit, Massif de, Niger 109 D1 20 2N 8 30 E
Tercan, Turkey 101 C9 39 47N 40 23 E
Tercero →, Argentina 174 C3 32 58 S 61 47W
Terdal, India 94 F2 16 33N 75 3 E
Terebovlya, Ukraine 59 H3 49 18N 25 44 E
Teregova, Romania 52 E7 45 10N 22 16 E
Terek →, Russia 61 J8 44 0N 47 30 E
Terek-Say, Kyrgyzstan 63 C5 41 30N 71 11 E
Terengganu □, Malaysia 84 B2 4 55N 103 0 E
Terenos, Brazil 173 E7 20 26 S 54 50W
Tereshka →, Russia 60 E8 51 48N 46 26 E
Teresina, Brazil 170 C3 5 9 S 42 45W
Teresinha, Brazil 169 C7 0 58N 52 2W
Terespol, Poland 55 F10 52 5N 23 37 E
Terewah, L., Australia 127 D4 29 52 S 147 35 E
Terges →, Portugal 43 H3 37 49N 7 41W
Tergnier, France 27 C10 49 40N 3 17 E
Terhazza, Mali 110 D3 23 38N 5 22W
Teridgerie Cr. →, Australia 127 E4 30 25 S 148 50 E
Terifa, Yemen 98 D3 14 24N 43 48 E
Terlizzi, Italy 47 A9 41 8N 16 32 E
Terme, Turkey 100 B7 41 11N 37 0 E
Termez = Termiz, Uzbekistan 63 E3 37 15N 67 15 E
Términi Imerese, Italy 46 E6 37 59N 13 42 E
Términos, L. de, Mexico 163 D6 18 35N 91 30W
Termiz, Uzbekistan 63 E3 37 15N 67 15 E
Térmoli, Italy 45 F12 42 0N 15 0 E
Ternate, Indonesia 82 A3 0 45N 127 25 E
Terneuzen, Neths. 24 C3 51 20N 3 50 E
Terney, Russia 68 B8 45 3N 136 37 E
Terni, Italy 45 F9 42 34N 12 37 E
Ternitz, Austria 34 D9 47 43N 16 2 E
Ternopil, Ukraine 59 H3 49 30N 25 40 E
Ternopol = Ternopil, Ukraine 59 H3 49 30N 25 40 E
Terowie, N.S.W., Australia 127 E4 32 27 S 147 52 E
Terowie, S. Austral., Australia 127 E2 33 8 S 138 55 E
Terpní, Greece 50 F7 40 55N 23 26 E
Terra Bella, U.S.A. 161 K7 35 58N 119 3W
Terrace, Canada 142 C3 54 30N 128 35W
Terrace Bay, Canada 140 C2 48 47N 87 5W
Terracina, Italy 46 A6 41 17N 13 15 E
Terralba, Italy 46 C1 39 43N 8 39 E
Terranova = Ólbia, Italy 46 B2 40 55N 9 31 E
Terrasini, Italy 46 D6 38 10N 13 4 E
Terrassa, Spain 40 D7 41 34N 2 1 E
Terrasson-la-Villedieu, France 28 C5 45 8N 1 18 E
Terre Haute, U.S.A. 157 E9 39 28N 87 25W
Terrebonne B., U.S.A. 155 L9 29 5N 90 35W
Terrecht, Mali 111 D4 20 10N 0 10W
Terrell, U.S.A. 155 J6 32 44N 96 17W
Terrenceville, Canada 141 C9 47 40N 54 44W
Terrick Terrick, Australia 126 C4 24 44 S 145 5 E
Terry, U.S.A. 154 B2 46 47N 105 19W
Terschelling, Neths. 24 A5 53 25N 5 20 E
Terskey Alatau, Khrebet, Kyrgyzstan 63 C8 41 50N 77 0 E
Tersko-Kumskiy Kanal →, Russia 61 H7 44 32N 44 38 E
Tertenía, Italy 46 C2 39 42N 9 34 E
Terter → = Tärtär →, Azerbaijan 61 K8 40 26N 47 20 E
Teruel, Spain 40 E3 40 22N 1 8W
Teruel □, Spain 40 E4 40 48N 1 0W
Tervel, Bulgaria 51 C11 43 45N 27 28 E
Tervola, Finland 14 C21 66 6N 24 49 E
Teryaweyna L., Australia 128 B5 32 18 S 143 22 E
Tešanj, Bos.-H. 52 F3 44 38N 18 1 E
Teseney, Eritrea 107 D4 15 5N 36 42 E
Tesha →, Russia 60 C6 55 38N 42 9 E
Teshekpuk L., U.S.A. 144 A9 70 35N 153 26W
Teshio, Japan 68 B10 44 53N 141 44 E
Teshio-Gawa →, Japan 68 B10 44 53N 141 45 E
Tešica, Serbia, Yug. 50 C5 43 27N 21 45 E
Tesiyn Gol →, Mongolia 72 A4 50 40N 93 20 E
Teslić, Bos.-H. 52 F2 44 37N 17 54 E
Teslin, Canada 142 A2 60 10N 132 43W
Teslin →, Canada 142 A2 61 34N 134 35W
Teslin L., Canada 142 A2 60 15N 132 57W
Tesouro, Brazil 173 D7 16 4 S 53 34W
Tessalit, Mali 113 A5 20 12N 1 0 E
Tessaoua, Niger 109 F1 13 47N 7 56 E
Tessin, Germany 30 A8 54 2N 12 26 E
Tessit, Mali 113 B5 15 13N 0 18 E
Test →, U.K. 21 G6 50 56N 1 29W
Testa del Gargano, Italy 45 G13 41 50N 16 10 E
Tét, Hungary 52 C2 47 30N 17 33 E
Têt →, France 28 F7 42 44N 3 2 E
Tetachuck L., Canada 142 C3 53 18N 125 55W
Tetas, Pta., Chile 174 A1 23 31 S 70 38W
Tete, Mozam. 119 F3 16 13 S 33 33 E
Tete □, Mozam. 119 F3 15 15 S 32 40 E
Teterev →, Ukraine 59 G6 51 1N 30 5 E
Teterow, Germany 30 B8 53 46N 12 34 E
Teteven, Bulgaria 51 D8 42 58N 24 17 E
Tethul →, Canada 142 A6 60 35N 112 12W
Tetiyev, Ukraine 59 H5 49 22N 29 38 E
Tetlin Indian Reservation, U.S.A. 144 E12 63 0N 142 30W
Teton →, U.S.A. 158 C8 47 56N 110 31W
Tétouan, Morocco 110 A3 35 35N 5 21W
Tetovo, Macedonia 50 D4 42 1N 20 59 E
Tetyukhe Pristan, Russia 68 B7 44 22N 135 48 E
Tetyushi, Russia 60 C9 54 55N 48 49 E
Teuco →, Argentina 174 B3 25 35 S 60 11W
Teufen, Switz. 33 B8 47 24N 9 23 E
Teulada, Italy 46 D1 38 58N 8 46 E
Teulon, Canada 143 C9 50 23N 97 16W
Teun, Indonesia 83 C3 6 59 S 129 8 E
Teutoburger Wald, Germany 30 C4 52 5N 8 22 E
Tevere →, Italy 45 G9 41 44N 12 14 E
Teverya, Israel 103 C4 32 47N 35 32 E
Teviot →, U.K. 22 F6 55 29N 2 38W
Tewantin, Australia 127 D5 26 27 S 153 3 E
Tewkesbury, U.K. 21 F5 51 59N 2 9W
Texada I., Canada 142 D4 49 40N 124 25W
Texarkana, Ark., U.S.A. 155 J8 33 26N 94 2W
Texarkana, Tex., U.S.A. 155 J7 33 26N 94 3W
Texas, Australia 127 D5 28 49 S 151 9 E
Texas □, U.S.A. 155 K5 31 40N 98 30W
Texas City, U.S.A. 155 L7 29 24N 94 54W
Texel, Neths. 24 A4 53 5N 4 50 E
Texhoma, U.S.A. 155 G4 36 30N 101 47W
Texline, U.S.A. 155 G3 36 23N 103 2W
Texoma, L., U.S.A. 155 J6 33 50N 96 34W
Teykovo, Russia 58 D11 56 55N 40 30 E
Teyvareh, Afghan. 91 B2 33 30N 64 24 E
Teza →, Russia 60 B5 56 32N 41 53 E
Tezin, Afghan. 92 B3 34 24N 69 30 E
Teziutlán, Mexico 163 D5 19 50N 97 22W
Tezpur, India 90 B4 26 40N 92 45 E
Tezzeron L., Canada 142 C4 54 43N 124 30W
Tha-anne →, Canada 143 A10 60 31N 94 37W
Tha Deua, Laos 86 D4 17 57N 102 53 E
Tha Deua, Laos 86 C3 19 26N 101 50 E
Tha Pla, Thailand 86 D3 17 48N 100 32 E
Tha Rua, Thailand 86 E3 14 34N 100 44 E
Tha Sala, Thailand 87 H2 8 40N 99 56 E
Tha Song Yang, Thailand 86 D1 17 34N 97 55 E
Thaba Putsoa, Lesotho 117 D4 29 45 S 28 0 E
Thabana Ntlenyana, Lesotho 117 D4 29 30 S 29 16 E
Thabazimbi, S. Africa 117 C4 24 40 S 27 21 E
Thabeikkyin, Burma 90 D5 22 53N 95 59 E
Thai Binh, Vietnam 86 B6 20 35N 106 1 E
Thai Muang, Thailand 87 H2 8 24N 98 16 E
Thai Nguyen, Vietnam 86 B5 21 35N 105 55 E
Thailand ■, Asia 86 E4 16 0N 102 0 E
Thailand, G. of, Asia 87 G3 11 30N 101 0 E
Thakhek, Laos 86 D5 17 25N 104 45 E
Thakurgaon, Bangla. 90 B2 26 2N 88 34 E
Thal, Pakistan 91 B3 33 28N 70 33 E
Thal Desert, Pakistan 92 D4 31 10N 71 30 E
Thala, Tunisia 108 A1 35 35N 8 40 E
Thalabarivat, Cambodia 86 F5 13 33N 105 57 E
Thalkirch, Switz. 33 C8 46 39N 9 17 E
Thallon, Australia 127 D4 28 39 S 148 49 E
Thalmann, U.S.A. 152 D8 31 18N 81 41W
Thalwil, Switz. 33 B7 47 17N 8 35 E
Thamarīt, Oman 99 C6 17 39N 54 2 E
Thames, N.Z. 130 D4 37 7 S 175 34 E
Thames →, Canada 140 D3 42 20N 82 25W
Thames →, U.K. 21 F8 51 29N 0 34 E
Thames →, U.S.A. 151 E12 41 18N 72 5W
Thames, Firth of, N.Z. 130 D4 37 0 S 175 25 E
Thames Estuary, U.K. 21 F8 51 29N 0 52 E
Thamesford, Canada 150 C4 43 4N 81 0W
Thamesville, Canada 150 D3 42 33N 81 59W
Thāmit, W. →, Libya 108 B3 31 11N 16 8 E
Thamūd, Yemen 99 C5 17 18N 49 55 E
Than Uyen, Vietnam 86 B4 22 0N 103 54 E
Thanbyuzayat, Burma 90 G6 15 58N 97 44 E
Thane, India 94 E1 19 12N 72 59 E
Thanesar, India 92 D7 30 1N 76 52 E
Thanet, I. of, U.K. 21 F9 51 21N 1 20 E

292 Torrevieja

Torrevieja, *Spain* 41 H4 37 59N 0 42W
Torrey, *U.S.A.* 159 G8 38 18N 111 25W
Torridge ➝, *U.K.* 21 G3 51 0N 4 13W
Torridon, L., *U.K.* 22 D3 57 35N 5 50W
Torrijos, *Phil.* 80 E4 13 19N 122 5 E
Torrijos, *Spain* 42 F6 39 59N 4 18W
Tørring, *Denmark* 17 J3 55 52N 9 29 E
Torrington, *Conn., U.S.A.* 151 E11 41 48N 73 7W
Torrington, *Wyo., U.S.A.* 154 D2 42 4N 104 11W
Torroella de Montgrì, *Spain* 40 C8 42 2N 3 8 E
Torrox, *Spain* 43 J7 36 46N 3 57W
Torsås, *Sweden* 17 H9 56 24N 16 0 E
Torsby, *Sweden* 16 D6 60 7N 13 0 E
Torshälla, *Sweden* 16 E10 59 25N 16 28 E
Tórshavn, *Færoe Is.* 14 E9 62 5N 6 56W
Torslanda, *Sweden* 17 G5 57 44N 11 45 E
Torsö, *Sweden* 17 F7 58 48N 13 45 E
Tortola, *Virgin Is.* 165 C7 18 19N 64 45W
Tórtoles de Esgueva, *Spain* 42 D6 41 49N 4 2W
Tortolì, *Italy* 46 C2 39 55N 9 39 E
Tortona, *Italy* 44 D5 44 54N 8 52 E
Tortorici, *Italy* 47 D7 38 2N 14 49 E
Tortosa, *Spain* 40 E5 40 49N 0 31 E
Tortosa, C., *Spain* 40 E5 40 41N 0 52 E
Tortosendo, *Portugal* 42 E3 40 15N 7 31W
Tortue, I. de la, *Haiti* 165 B5 20 5N 72 57W
Tortum, *Turkey* 101 B9 40 19N 41 35 E
Ţorūd, *Iran* 97 C7 35 25N 55 5 E
Torul, *Turkey* 101 B8 40 34N 39 18 E
Toruń, *Poland* 55 E5 53 2N 18 39 E
Toruń □, *Poland* 55 E5 53 20N 19 0 E
Tory I., *Ireland* 23 A3 55 16N 8 14W
Torysa ➝, *Slovak Rep.* 35 C14 48 39N 21 21 E
Torzhok, *Russia* 58 D8 57 5N 34 55 E
Torzym, *Poland* 55 F2 52 19N 15 5 E
Tosa, *Japan* 70 D5 33 24N 133 23 E
Tosa-Shimizu, *Japan* 70 E4 32 52N 132 58 E
Tosa-Wan, *Japan* 70 D5 33 15N 133 30 E
Tosa-yamada, *Japan* 70 D5 33 36N 133 38 E
Toscana □, *Italy* 44 E8 43 25N 11 0 E
Toscano, Arcipelago, *Italy* 44 F7 42 30N 10 30 E
Toshkent, *Uzbekistan* 63 C4 41 20N 69 10 E
Tosno, *Russia* 58 C6 59 38N 30 46 E
Tossa de Mar, *Spain* 40 D7 41 43N 2 56 E
Tösse, *Sweden* 17 F6 58 58N 12 39 E
Tostado, *Argentina* 174 B3 29 15 S 61 50W
Tostedt, *Germany* 30 B5 53 17N 9 42 E
Tostón, Pta. de, *Canary Is.* 39 F5 28 42N 14 2W
Tosu, *Japan* 70 D2 33 22N 130 31 E
Tosya, *Turkey* 100 B6 41 1N 34 2 E
Toszek, *Poland* 55 H5 50 27N 18 32 E
Totak, *Norway* 18 E5 59 40N 8 0 E
Totana, *Spain* 41 H3 37 45N 1 30W
Totebo, *Sweden* 17 G10 57 38N 16 12 E
Toteng, *Botswana* 116 C3 20 22 S 22 58 E
Tôtes, *France* 26 C8 49 41N 1 2 E
Tótkomlós, *Hungary* 52 D5 46 24N 20 45 E
Totland, *Norway* 18 C2 61 56N 5 23 E
Totma, *Russia* 56 C7 60 0N 42 40 E
Totnes, *U.K.* 21 G4 50 26N 3 42W
Totness, *Surinam* 169 B6 5 53N 56 19W
Totonicapán, *Guatemala* 164 D1 14 58N 91 12W
Totora, *Bolivia* 173 D4 17 42 S 65 9W
Totoya, I., *Fiji* 133 B3 18 57 S 179 50W
Totskoye, *Russia* 62 E4 52 32N 52 45 E
Totten Glacier, *Antarctica* 7 C8 66 45 S 116 10 E
Tottenham, *Australia* 129 B7 32 14 S 147 21 E
Tottenham, *Canada* 150 B5 44 1N 79 49W
Tottori, *Japan* 70 B6 35 30N 134 15 E
Tottori □, *Japan* 70 B6 35 30N 134 12 E
Touat, *Algeria* 111 C5 27 27N 0 30 E
Touba, *Ivory C.* 112 D3 8 22N 7 40W
Toubkal, Djebel, *Morocco* 110 B3 31 0N 8 0W
Toucy, *France* 27 E10 47 44N 3 15 E
Tougan, *Burkina Faso* 112 C4 13 11N 2 58W
Touggourt, *Algeria* 111 B6 33 6N 6 4 E
Tougué, *Guinea* 112 C2 11 25N 11 50W
Touho, *N. Cal.* 133 T19 20 47 S 165 14 E
Toukmatine, *Algeria* 111 D6 24 49N 7 11 E
Toul, *France* 27 D12 48 40N 5 53 E
Toulepleu, *Ivory C.* 112 D3 6 32N 8 24W
Toulon, *France* 29 E9 43 10N 5 55 E
Toulon, *U.S.A.* 156 C7 41 6N 89 52W
Toulouse, *France* 28 E5 43 37N 1 27 E
Toummo, *Niger* 108 D2 22 45N 14 8 E
Toummo Dhoba, *Niger* 108 D2 22 30N 14 31 E
Toumodi, *Ivory C.* 112 D3 6 32N 5 4W
Tounan, *Taiwan* 77 F13 23 41N 120 28 E
Tounassine, Hamada, *Algeria* 110 C4 28 48N 5 0W
Toungoo, *Burma* 90 F6 19 0N 96 30 E
Touques ➝, *France* 26 C7 49 22N 0 8 E
Touraine, *France* 26 E7 47 20N 0 30 E
Tourane = Da Nang, *Vietnam* 86 D7 16 4N 108 13 E
Tourcoing, *France* 27 B10 50 42N 3 10 E
Touriñán, C., *Spain* 42 B1 43 3N 9 17W
Tourine, *Mauritania* 110 D2 22 23N 11 50W
Tournai, *Belgium* 24 D3 50 35N 3 25 E
Tournan-en-Brie, *France* 27 D9 48 44N 2 46 E
Tournay, *France* 28 E4 43 13N 0 13 E
Tournon-St-Martin, *France* 26 F7 46 45N 0 58 E
Tournon-sur-Rhône, *France* 29 C8 45 4N 4 50 E
Tournus, *France* 27 F11 46 35N 4 54 E
Touros, *Brazil* 170 C4 5 12 S 35 28W
Tours, *France* 26 E7 47 22N 0 40 E
Touside, Pic, *Chad* 109 D3 21 1N 16 29 E
Touwsrivier, *S. Africa* 116 E3 33 20 S 20 2 E
Tovar, *Venezuela* 168 B3 8 20N 71 46W
Tovarkovskiy, *Russia* 58 F10 53 40N 38 14 E
Tovdal, *Norway* 18 F5 58 47N 8 10 E
Tovdalselva ➝, *Norway* 18 F5 58 11N 8 7 E
Tovuz, *Azerbaijan* 61 K7 41 0N 45 40 E
Towada, *Japan* 68 D10 40 37N 141 13 E
Towada-Ko, *Japan* 68 D10 40 28N 140 55 E
Towamba, *Australia* 129 D8 37 6 S 149 43 E
Towanda, *Ill., U.S.A.* 157 D8 40 36N 88 53W
Towanda, *Pa., U.S.A.* 151 E8 41 46N 76 27W
Tower, *U.S.A.* 154 B8 47 48N 92 17W
Towerhill Cr. ➝, *Australia* 126 C3 22 28 S 144 35 E
Towner, *U.S.A.* 154 A4 48 21N 100 25W
Towns, *U.S.A.* 152 D7 32 0N 82 45W
Townsend, *Ga., U.S.A.* 152 D8 31 33N 81 31W
Townsend, *Mont., U.S.A.* 158 C8 46 19N 111 31W
Townshend I., *Australia* 126 C5 22 10 S 150 31 E
Townsville, *Australia* 126 B4 19 15 S 146 45 E
Towson, *U.S.A.* 148 F7 39 24N 76 36W
Towuti, Danau, *Indonesia* 82 B2 2 45 S 121 32 E
Toya-Ko, *Japan* 68 C10 42 35N 140 51 E
Toyah, *U.S.A.* 155 K3 31 19N 103 48W
Toyahvale, *U.S.A.* 155 K3 30 57N 103 47W
Toyama, *Japan* 71 A9 36 40N 137 15 E
Toyama □, *Japan* 71 A9 36 45N 137 30 E
Toyama-Wan, *Japan* 69 F8 37 0N 137 30 E
Tōyō, *Japan* 70 D6 33 26N 134 16 E
Toyohashi, *Japan* 71 C9 34 45N 137 25 E
Toyokawa, *Japan* 71 C9 34 48N 137 27 E
Toyonaka, *Japan* 71 C7 34 50N 135 28 E
Toyooka, *Japan* 70 B6 35 35N 134 48 E
Toyota, *Japan* 71 B9 35 3N 137 7 E
Toyoura, *Japan* 70 C2 34 6N 130 57 E
Toytepa, *Uzbekistan* 63 C4 41 3N 69 20 E
Tozeur, *Tunisia* 108 B1 33 56N 8 8 E
Tqibuli, *Georgia* 61 J6 42 26N 43 0 E
Tqvarcheli, *Georgia* 61 J5 42 47N 41 42 E
Trá Li = Tralee, *Ireland* 23 D2 52 16N 9 42W
Tra On, *Vietnam* 87 H5 9 58N 105 55 E
Trabancos ➝, *Spain* 42 D5 41 36N 5 15W
Traben-Trarbach, *Germany* 31 F3 49 57N 7 7 E
Trabzon, *Turkey* 101 B8 41 0N 39 45 E
Tracadie, *Canada* 141 C7 47 30N 64 55W
Tracy, *Calif., U.S.A.* 160 H5 37 44N 121 26W
Tracy, *Minn., U.S.A.* 154 C7 44 14N 95 37W
Tradate, *Italy* 44 C5 45 43N 8 54 E
Traer, *U.S.A.* 156 B4 42 12N 92 28W
Trafalgar, *Australia* 129 E7 38 14 S 146 12 E
Trafalgar, C., *Spain* 43 J4 36 10N 6 2W
Trāghān, *Libya* 108 C2 26 0N 14 30 E
Tragowel, *Australia* 128 C6 35 50 S 144 0 E
Traian, *Brăila, Romania* 53 E12 45 11N 27 44 E
Traian, *Tulcea, Romania* 53 E13 45 2N 28 15 E
Traiguén, *Chile* 176 A2 38 15 S 72 41W
Trail, *Canada* 142 D5 49 5N 117 40W
Traill Ø, *Greenland* 10 C8 72 30N 23 0W
Trainor L., *Canada* 142 A4 60 24N 120 17W
Traíra ➝, *Brazil* 168 D4 1 4 S 69 26W
Trákhonas, *Cyprus* 38 D12 35 12N 33 21 E
Tralee, *Ireland* 23 D2 52 16N 9 42W
Tralee B., *Ireland* 23 D2 52 17N 9 55W
Tramelan, *Switz.* 32 B4 47 13N 7 7 E
Tramore, *Ireland* 23 D4 52 10N 7 10W
Tramore B., *Ireland* 23 D4 52 9N 7 10W
Tran Ninh, Cao Nguyen, *Laos* 86 C4 19 30N 103 10 E
Tranås, *Sweden* 17 F8 58 3N 14 59 E
Tranbjerg, *Denmark* 17 H4 56 6N 10 9 E
Tranby, *Norway* 18 E7 59 49N 10 15 E
Trancas, *Argentina* 174 B2 26 11 S 65 20W
Trancoso, *Portugal* 42 E3 40 49N 7 21W
Tranebjerg, *Denmark* 17 J4 55 51N 10 36 E
Tranemo, *Sweden* 17 G7 57 30N 13 20 E
Trang, *Thailand* 87 J2 7 33N 99 38 E
Trangahy, *Madag.* 117 B7 19 7 S 44 31 E
Trangan, *Indonesia* 83 C4 6 40 S 134 20 E
Trangie, *Australia* 129 B8 32 4 S 148 0 E
Trångsviken, *Sweden* 16 A7 63 19N 13 59 E
Trani, *Italy* 47 A9 41 17N 16 25 E
Tranoroa, *Madag.* 117 C8 24 42 S 45 4 E
Tranquebar, *India* 95 J4 11 1N 79 54 E
Tranqueras, *Uruguay* 175 C4 31 13 S 55 45W
Trans Nzoia □, *Kenya* 118 B4 1 0N 35 0 E
Transantarctic Mts., *Antarctica* 7 E12 85 0 S 170 0W
Transcaucasia = Zakavkazye, *Asia* 57 F7 42 0N 44 0 E
Transcona, *Canada* 143 D9 49 55N 97 0W
Transilvania, *Romania* 53 D9 46 30N 24 0 E
Transilvanian Alps = Carpații Meridionali, *Romania* 53 E9 45 30N 25 0 E
Transtrand, *Sweden* 16 C7 61 6N 13 20 E
Transtrandsfjällen, *Sweden* 16 C6 61 8N 13 0 E
Transylvania = Transilvania, *Romania* 53 D9 46 30N 24 0 E
Trápani, *Italy* 46 D5 38 1N 12 29 E
Trapper Pk., *U.S.A.* 158 D6 45 54N 114 18W
Traralgon, *Australia* 129 E7 38 12 S 146 34 E
Trarza, *Mauritania* 112 B2 17 30N 15 0W
Trás-os-Montes, *Angola* 115 E3 10 17 S 19 5 E
Trasacco, *Italy* 45 G10 41 57N 13 32 E
Trăscău, Munții, *Romania* 53 D8 46 14N 23 14 E
Trasimeno, L., *Italy* 45 E9 43 8N 12 6 E
Träslövsläge, *Sweden* 17 G6 57 4N 12 16 E
Trasvase Tajo-Segura, Canal de, *Spain* 40 E2 40 15N 2 55W
Trat, *Thailand* 87 F4 12 14N 102 33 E
Traun, *Austria* 34 C7 48 14N 14 15 E
Traunreut, *Germany* 31 H8 47 57N 12 36 E
Traunsee, *Austria* 34 D6 47 55N 13 50 E
Traunstein, *Germany* 31 H8 47 52N 12 37 E
Traveller's L., *Australia* 128 B5 33 20 S 142 0 E
Travemünde, *Germany* 30 B6 53 57N 10 52 E
Travers, Mt., *N.Z.* 131 C7 42 1 S 172 45 E
Traverse City, *U.S.A.* 148 C3 44 46N 85 38W
Travnik, *Bos.-H.* 52 F2 44 17N 17 39 E
Trayning, *Australia* 125 F2 31 7 S 117 40 E
Trbovlje, *Slovenia* 45 B12 46 12N 15 5 E
Treasure Island, *U.S.A.* 153 H7 27 46N 82 46W
Treasury Is., *Solomon Is.* 133 L8 7 22 S 155 37 E
Trébbia ➝, *Italy* 44 C6 45 4N 9 41 E
Trebel ➝, *Germany* 30 B9 53 54N 13 2 E
Trébeurden, *France* 26 D3 48 46N 3 35W
Třebíč, *Czech Rep.* 34 B8 49 14N 15 55 E
Trebinje, *Bos.-H.* 50 D2 42 44N 18 22 E
Trebisacce, *Italy* 47 C9 39 52N 16 32 E
Trebišnjica ➝, *Bos.-H.* 50 D2 42 47N 18 8 E
Trebišov, *Slovak Rep.* 35 C14 48 38N 21 41 E
Trebižat ➝, *Bos.-H.* 45 E14 43 15N 17 30 E
Trebnje, *Slovenia* 45 C12 45 54N 15 1 E
Třeboň, *Czech Rep.* 34 B7 49 1N 14 48 E
Trebujena, *Spain* 43 J4 36 52N 6 11W
Trecate, *Italy* 44 C5 45 26N 8 44 E
Trece Martires, *Phil.* 80 D3 14 20N 120 50 E
Tregaron, *U.K.* 21 E4 52 14N 3 56W
Tregnago, *Italy* 45 C8 45 31N 11 10 E
Tregrosse Is., *Australia* 126 B5 17 41 S 150 43 E
Tréguier, *France* 26 D3 48 47N 3 16W
Trégunc, *France* 26 E3 47 51N 3 51W
Treherne, *Canada* 143 D9 49 38N 98 42W
Tréia, *Italy* 45 E10 43 19N 13 19 E
Treignac, *France* 28 C5 45 32N 1 48 E
Treinta y Tres, *Uruguay* 175 C5 33 16 S 54 17W
Treis-karden, *Germany* 31 E3 50 10N 7 18 E
Treklyano, *Bulgaria* 50 D6 42 33N 22 36 E
Trélazé, *France* 26 E6 47 26N 0 30W
Trelew, *Argentina* 176 B3 43 10 S 65 20W
Trélissac, *France* 28 C4 45 11N 0 47 E
Trelleborg, *Sweden* 17 J7 55 20N 13 10 E
Tremadog Bay, *U.K.* 20 E3 52 51N 4 18W
Trémiti, *Italy* 45 F12 42 8N 15 30 E
Tremonton, *U.S.A.* 158 F7 41 43N 112 10W
Tremp, *Spain* 40 C5 42 10N 0 52 E
Trenche ➝, *Canada* 140 C5 47 46N 72 53W
Trenčiansky □, *Slovak Rep.* 35 C11 48 45N 18 20 E
Trenčín, *Slovak Rep.* 35 C11 48 52N 18 4 E
Trengereid, *Norway* 18 D2 60 26N 5 37 E
Trenggalek, *Indonesia* 85 D4 8 3 S 111 43 E
Trenque Lauquen, *Argentina* 174 D3 36 5 S 62 45W
Trent ➝, *U.K.* 20 D7 53 41N 0 42W
Trentino-Alto Adige □, *Italy* 45 B8 46 30N 11 20 E
Trento, *Italy* 44 B8 46 4N 11 8 E
Trenton, *Canada* 140 D4 44 10N 77 34W
Trenton, *Fla., U.S.A.* 153 F7 29 37N 82 49W
Trenton, *Mich., U.S.A.* 157 B13 42 8N 83 11W
Trenton, *Mo., U.S.A.* 156 D3 40 5N 93 37W
Trenton, *N.J., U.S.A.* 151 F10 40 14N 74 46W
Trenton, *Nebr., U.S.A.* 154 E4 40 11N 101 1W
Trenton, *S.C., U.S.A.* 152 B9 33 45N 81 51W
Trenton, *Tenn., U.S.A.* 155 H10 35 59N 88 56W
Trepassey, *Canada* 141 C9 46 43N 53 25W
Trepuzzi, *Italy* 47 B11 40 24N 18 4 E
Tres Arroyos, *Argentina* 174 D3 38 26 S 60 20W
Três Corações, *Brazil* 171 F2 21 44 S 45 15W
Três Lagoas, *Brazil* 171 F1 20 50 S 51 43W
Tres Lagos ➝, *Argentina* 176 C2 49 35 S 71 25W
Tres Marías, *Mexico* 162 C3 21 25N 106 28W
Três Marias, Represa, *Brazil* 171 E2 18 12 S 45 15W
Tres Montes, C., *Chile* 176 C1 46 50 S 75 30W
Tres Pinos, *U.S.A.* 160 J5 36 48N 121 19W
Três Pontas, *Brazil* 171 F2 21 23 S 45 29W
Tres Puentes, *Chile* 174 B1 27 50 S 70 15W
Tres Puntas, C., *Argentina* 176 C3 47 0 S 66 0W
Três Rios, *Brazil* 171 F3 22 6 S 43 15W
Tres Valles, *Mexico* 163 D5 18 15N 96 8W
Tresco, *U.K.* 21 H1 49 57N 6 20W
Tresfjord, *Norway* 18 B4 62 31N 7 7 E
Treska ➝, *Macedonia* 50 E5 42 0N 21 20 E
Treskavica, *Bos.-H.* 52 G3 43 40N 18 20 E
Trespaderne, *Spain* 42 C7 42 47N 3 24W
Trets, *France* 29 E9 43 27N 5 41 E
Treuchtlingen, *Germany* 31 G6 48 58N 10 54 E
Treuenbrietzen, *Germany* 30 C8 52 5N 12 52 E
Treungen, *Norway* 18 E5 59 1N 8 31 E
Trevi, *Italy* 45 F9 42 52N 12 45 E
Treviglio, *Italy* 44 C6 45 31N 9 35 E
Trevínca, Peña, *Spain* 42 C4 42 15N 6 46W
Treviso, *Italy* 45 C9 45 40N 12 15 E
Trévoux, *France* 29 C8 45 57N 4 47 E
Trgovište, *Serbia, Yug.* 50 D6 42 20N 22 10 E
Triabunna, *Australia* 126 G4 42 30 S 147 55 E
Triánda, *Greece* 38 C10 36 25N 28 10 E
Triaucourt-en-Argonne, *France* 27 D12 48 59N 5 2 E
Tribly, *U.S.A.* 153 G7 28 28N 82 12W
Tribsees, *Germany* 30 A8 54 5N 12 44 E
Tribulation, C., *Australia* 126 B4 16 5 S 145 29 E
Tribune, *U.S.A.* 154 F4 38 28N 101 45W
Tricárico, *Italy* 47 B9 40 37N 16 9 E
Tricase, *Italy* 47 C11 39 56N 18 22 E
Trichinopoly = Tiruchchirappalli, *India* 95 J4 10 45N 78 45 E
Trichur, *India* 95 J3 10 30N 76 18 E
Trida, *Australia* 129 B6 33 1 S 145 1 E
Trier, *Germany* 31 F2 49 45N 6 38 E
Trieste, *Italy* 45 C10 45 40N 13 46 E
Trieste, G. di, *Italy* 45 C10 45 40N 13 35 E
Trieux ➝, *France* 26 D3 48 43N 3 9W
Triggiano, *Italy* 47 A9 41 4N 16 55 E
Triglav, *Slovenia* 45 B10 46 21N 13 50 E
Trigno ➝, *Italy* 45 F11 42 4N 14 48 E
Trigueros, *Spain* 43 H4 37 24N 6 50W
Tríkeri, *Greece* 48 B5 39 6N 23 5 E
Trikhonis, Límni, *Greece* 48 C3 38 34N 21 30 E
Tríkkala, *Greece* 48 B3 39 34N 21 47 E
Tríkkala □, *Greece* 48 B3 39 41N 21 30 E
Trikomo, *Cyprus* 38 D12 35 17N 33 52 E
Trikora, Puncak, *Indonesia* 83 B5 4 15 S 138 45 E
Trilj, *Croatia* 45 E13 43 38N 16 42 E
Trillo, *Spain* 40 E2 40 42N 2 35W
Trim, *Ireland* 23 C5 53 33N 6 48W
Trincomalee, *Sri Lanka* 95 K5 8 38N 81 15 E
Trindade, *Brazil* 171 E2 16 40 S 49 30W
Trindade, I., *Atl. Oc.* 9 L6 20 20 S 29 50W
Třinec, *Czech Rep.* 35 B11 49 41N 18 39 E
Trinidad, *Bolivia* 173 C5 14 46 S 64 50W
Trinidad, *Colombia* 168 B3 5 25N 71 40W
Trinidad, *Cuba* 164 B4 21 48N 80 0W
Trinidad, *Uruguay* 174 C4 33 30 S 56 50W
Trinidad, *U.S.A.* 155 G2 37 10N 104 31W
Trinidad, *W. Indies* 165 D7 10 30N 61 15W
Trinidad ➝, *Mexico* 163 D5 17 49N 95 9W
Trinidad, G., *Chile* 176 C1 49 55 S 75 25W
Trinidad, I., *Argentina* 176 A4 39 10 S 62 0W
Trinidad & Tobago ■, *W. Indies* 165 D7 10 30N 61 20W
Trinitápoli, *Italy* 47 A9 41 21N 16 5 E
Trinity, *Canada* 141 C9 48 59N 53 55W
Trinity, *U.S.A.* 155 K7 30 57N 95 22W
Trinity ➝, *Calif., U.S.A.* 158 F2 41 11N 123 42W
Trinity ➝, *Tex., U.S.A.* 155 L7 29 45N 94 43W
Trinity B., *Canada* 141 C9 48 20N 53 10W
Trinity Is., *U.S.A.* 144 H9 56 33N 154 25W
Trinity Range, *U.S.A.* 158 F4 40 15N 118 45W
Trinkitat, *Sudan* 106 D4 18 45N 37 51 E
Trino, *Italy* 44 C5 45 12N 8 18 E
Trion, *U.S.A.* 149 H3 34 33N 85 19W
Trionto, C., *Italy* 47 C9 39 37N 16 43 E
Triora, *Italy* 44 D4 44 1N 7 46 E
Tripoli = Tarābulus, *Lebanon* 103 A4 34 31N 35 50 E
Tripoli = Tarābulus, *Libya* 108 B2 32 49N 13 7 E
Tripoli, *U.S.A.* 156 B4 42 49N 92 16W
Trípolis, *Greece* 48 D4 37 31N 22 25 E
Tripolitania, *Libya* 108 B2 31 0N 12 0 E
Tripp, *U.S.A.* 154 D6 43 13N 97 58W
Tripura □, *India* 90 D4 24 0N 92 0 E
Tripylos, *Cyprus* 38 E11 34 59N 32 41 E
Trischen, *Germany* 30 A4 54 4N 8 40 E
Tristan da Cunha, *Atl. Oc.* 9 M7 37 6 S 12 20W
Trivandrum, *India* 95 K3 8 41N 77 0 E
Trivento, *Italy* 45 G11 41 47N 14 33 E
Trnava, *Slovak Rep.* 35 C10 48 23N 17 35 E
Trnavský □, *Slovak Rep.* 35 C10 48 30N 17 45 E
Troarn, *France* 26 C6 49 11N 0 11W
Trobriand Is., *Papua N. G.* 132 E6 8 30 S 151 0 E
Trochu, *Canada* 142 C6 51 50N 113 13W
Trodely I., *Canada* 140 B4 52 15N 79 26W
Troezen, *Greece* 48 D5 37 25N 23 15 E
Trogir, *Croatia* 45 E13 43 32N 16 15 E
Troglav, *Croatia* 45 E13 43 56N 16 36 E
Tróia, *Italy* 47 A8 41 22N 15 18 E
Troilus, L., *Canada* 140 B5 50 50N 74 35W
Troina, *Italy* 47 E7 37 47N 14 36 E
Trois Fourches, Cap des, *Morocco* 111 A4 35 26N 2 58W
Trois-Pistoles, *Canada* 141 C6 48 5N 69 10W
Trois-Rivières, *Canada* 140 C5 46 25N 72 34W
Troisdorf, *Germany* 30 E3 50 48N 7 11 E
Troitsk, *Russia* 62 D8 54 10N 61 35 E
Troitskiy, *Russia* 62 C9 57 5N 63 43 E
Troitsko Pechorsk, *Russia* 56 B10 62 40N 56 10 E
Trölladyngja, *Iceland* 11 C9 64 54N 17 16W
Trollhättan, *Sweden* 17 F6 58 17N 12 20 E
Trollheimen, *Norway* 14 E13 62 46N 9 1 E
Trombetas ➝, *Brazil* 169 D6 1 55 S 55 35W
Tromelin I., *Ind. Oc.* 121 F4 15 52 S 54 25 E
Tromøy, *Norway* 18 F5 58 27N 8 51 E
Tromsø, *Norway* 14 B18 69 40N 18 56 E
Trona, *U.S.A.* 161 K9 35 46N 117 23W
Tronador, *Argentina* 176 B2 41 10 S 71 50W
Trøndelag, *Norway* 14 D14 64 17N 11 50 E
Trondheim, *Norway* 14 E14 63 36N 10 25 E
Trondheimsfjorden, *Norway* 14 E14 63 35N 10 30 E
Trondheimsleia, *Norway* 18 A5 63 25N 8 30 E
Trönninge, *Sweden* 17 H6 56 37N 12 51 E
Tronto ➝, *Italy* 45 F10 42 54N 13 55 E
Troodos, *Cyprus* 38 E11 34 55N 32 52 E
Troon, *U.K.* 22 F4 55 33N 4 39W
Tropea, *Italy* 47 D8 38 41N 15 54 E
Tropic, *U.S.A.* 159 H7 37 37N 112 5W
Tropoja, *Albania* 50 D4 42 23N 20 10 E
Trosa, *Sweden* 17 F11 58 54N 17 33 E
Trostan, *U.K.* 23 A5 55 3N 6 10W
Trostberg, *Germany* 31 G8 48 1N 12 33 E
Trostyanets, *Ukraine* 59 G8 50 33N 34 59 E
Troup, *U.S.A.* 155 J7 32 9N 95 7W
Trout ➝, *Canada* 142 A5 61 19N 119 51W
Trout L., *N.W.T., Canada* 142 A4 60 40N 121 14W
Trout L., *Ont., Canada* 143 C10 51 20N 93 15W
Trout Lake, *Mich., U.S.A.* 140 C2 46 12N 85 1W
Trout Lake, *Wash., U.S.A.* 160 E5 46 0N 121 32W
Trout River, *Canada* 141 C8 49 29N 58 8W
Trouville-sur-Mer, *France* 26 C7 49 21N 0 5 E
Trowbridge, *U.K.* 21 F5 51 18N 2 12W
Troy, *Turkey* 49 B8 39 57N 26 12 E
Troy, *Ala., U.S.A.* 152 D4 31 48N 85 58W
Troy, *Idaho, U.S.A.* 158 C5 46 44N 116 46W
Troy, *Ill., U.S.A.* 156 F7 38 44N 89 54W
Troy, *Ind., U.S.A.* 157 G10 37 59N 86 55W
Troy, *Kans., U.S.A.* 154 F7 39 47N 95 5W
Troy, *Mich., U.S.A.* 157 B13 42 37N 83 9W
Troy, *Mo., U.S.A.* 156 F6 38 59N 90 59W
Troy, *Mont., U.S.A.* 158 B6 48 28N 115 53W
Troy, *N.Y., U.S.A.* 151 D11 42 44N 73 41W
Troy, *Ohio, U.S.A.* 157 D12 40 2N 84 12W
Troy, *S.C., U.S.A.* 152 B7 33 59N 82 17W
Troyan, *Bulgaria* 51 D8 42 57N 24 43 E
Troyes, *France* 27 D11 48 19N 4 3 E
Trpanj, *Croatia* 45 E14 43 1N 17 15 E
Trstenik, *Serbia, Yug.* 50 C5 43 36N 21 0 E
Trubchevsk, *Russia* 59 F7 52 33N 33 47 E
Trucial States = United Arab Emirates ■, *Asia* 97 F7 23 50N 54 0 E
Truckee, *U.S.A.* 160 F6 39 20N 120 11W
Trudfront, *Russia* 61 H8 45 56N 47 40 E
Trudovoye, *Russia* 68 C6 43 17N 132 5 E
Trujillo, *Colombia* 168 C2 4 10N 76 19W
Trujillo, *Honduras* 164 C2 16 0N 86 0W
Trujillo, *Peru* 172 B2 8 6 S 79 0W
Trujillo, *Spain* 43 F5 39 28N 5 55W
Trujillo, *U.S.A.* 155 H2 35 32N 104 42W
Trujillo, *Venezuela* 168 B3 9 22N 70 38W
Trujillo □, *Venezuela* 168 B3 9 25N 70 30W
Truk, *Pac. Oc.* 134 G7 7 25N 151 46 E
Trumann, *U.S.A.* 155 H9 35 41N 90 31W
Trumbull, Mt., *U.S.A.* 159 H7 36 25N 113 8W
Trŭn, *Bulgaria* 50 D6 42 51N 22 38 E
Trun, *France* 26 D7 48 50N 0 2 E
Trun, *Switz.* 33 C7 46 45N 8 59 E
Trundle, *Australia* 129 B7 32 53 S 147 35 E
Trung-Phan = Annam, *Vietnam* 86 E7 16 0N 108 0 E
Truro, *Australia* 128 C3 34 24 S 139 9 E
Truro, *Canada* 141 C7 45 21N 63 14W
Truro, *U.K.* 21 G2 50 16N 5 4W
Truskavets, *Ukraine* 59 H2 49 17N 23 30 E
Truslove, *Australia* 125 F3 33 20 S 121 45 E
Trŭstenik, *Bulgaria* 51 C8 43 31N 24 28 E
Trustrup, *Denmark* 17 H4 56 20N 10 46 E
Truth or Consequences, *U.S.A.* 159 K10 33 8N 107 15W
Trutnov, *Czech Rep.* 34 A8 50 37N 15 54 E
Truyère ➝, *France* 28 D6 44 38N 2 34 E
Tryavna, *Bulgaria* 51 D9 42 54N 25 25 E
Tryon, *U.S.A.* 149 H4 35 13N 82 14W
Tryonville, *U.S.A.* 150 E5 41 42N 79 48W
Trysilelva ➝, *Norway* 18 C9 61 2N 12 35 E
Trzcianka, *Poland* 55 E3 53 3N 16 25 E
Trzciel, *Poland* 55 F2 52 23N 15 50 E
Trzcińsko Zdrój, *Poland* 55 F1 52 58N 14 35 E
Trzebiatów, *Poland* 54 D2 54 3N 15 18 E
Trzebiez, *Poland* 54 E1 53 38N 14 31 E
Trzebnica, *Poland* 55 G4 51 20N 17 1 E
Trzemeszno, *Poland* 55 F4 52 33N 17 48 E
Tržič, *Slovenia* 45 B11 46 22N 14 18 E
Tsagan Aman, *Russia* 61 G8 47 34N 46 43 E
Tsala Apopka L., *U.S.A.* 153 G7 28 53N 82 19W
Tsamandás, *Greece* 48 B2 39 46N 20 21 E
Tsaratanana, *Madag.* 117 B8 16 47 S 47 39 E
Tsaratanana, Mt. de, *Madag.* 117 A8 14 0 S 49 0 E
Tsarevo = Michurin, *Bulgaria* 51 D11 42 9N 27 51 E
Tsarevo, *Bulgaria* 51 D9 42 28N 25 52 E
Tsaritsáni, *Greece* 48 B4 39 53N 22 14 E
Tsau, *Botswana* 116 C3 20 8 S 22 22 E
Tsebrykove, *Ukraine* 59 J6 47 9N 30 10 E
Tselinograd = Aqmola, *Kazakstan* 64 D8 51 10N 71 30 E
Tsetserleg, *Mongolia* 72 B5 47 36N 101 32 E
Tshabong, *Botswana* 116 D3 26 2 S 22 29 E
Tshane, *Botswana* 116 C3 24 5 S 21 54 E
Tshela, *Dem. Rep. of the Congo* 115 C2 4 57 S 13 4 E
Tshesebe, *Botswana* 117 C4 21 51 S 27 32 E
Tshibeke, *Dem. Rep. of the Congo* 118 C2 2 40 S 28 35 E
Tshibinda, *Dem. Rep. of the Congo* 118 C2 2 23 S 28 43 E
Tshikapa, *Dem. Rep. of the Congo* 115 D4 6 28 S 20 48 E
Tshilenge, *Dem. Rep. of the Congo* 115 D4 6 17 S 23 48 E
Tshinsenda, *Dem. Rep. of the Congo* 119 E2 12 20 S 28 0 E
Tshofa, *Dem. Rep. of the Congo* 115 D5 5 13 S 25 16 E
Tshwane, *Botswana* 116 C3 22 24 S 22 1 E
Tsigara, *Botswana* 116 C4 20 22 S 25 54 E
Tsihombe, *Madag.* 117 D8 25 10 S 45 41 E

Uberaba 293

Tsimlyansk, *Russia* **61 G6** 47 40N 42 6 E
Tsimlyansk Res. =
Tsimlyanskoye Vdkhr.,
Russia **61 F6** 48 0N 43 0 E
Tsimlyanskoye Vdkhr.,
Russia **61 F6** 48 0N 43 0 E
Tsinan = Jinan, *China* **74 F9** 36 38N 117 1 E
Tsineng, *S. Africa* **116 D3** 27 5 S 23 5 E
Tsínga, *Greece* **51 E8** 41 23N 24 44 E
Tsinghai = Qinghai □, *China* **72 C4** 36 0N 98 0 E
Tsingtao = Qingdao, *China* **75 F11** 36 5N 120 20 E
Tsinjomitondraka, *Madag.* **117 B8** 15 40 S 47 8 E
Tsiroanomandidy, *Madag.* **117 B8** 18 46 S 46 2 E
Tsiteli-Tsqaro, *Georgia* **61 K8** 41 33N 46 0 E
Tsivilsk, *Russia* **60 C8** 55 50N 47 25 E
Tsivory, *Madag.* **117 C8** 24 4 S 46 5 E
Tskhinvali, *Georgia* **61 J7** 42 14N 44 1 E
Tsna →, *Russia* **60 C6** 54 55N 41 58 E
Tsnori, *Georgia* **61 K7** 41 40N 45 57 E
Tso Moriri, L., *India* **93 C8** 32 50N 78 20 E
Tsodilo Hill, *Botswana* **116 B3** 18 49 S 21 43 E
Tsogttsetsiy, *Mongolia* **74 C3** 43 43N 105 35 E
Tsolo, *S. Africa* **117 E4** 31 18 S 28 37 E
Tsomo, *S. Africa* **117 E4** 32 0 S 27 42 E
Tsu, *Japan* **71 C8** 34 45N 136 25 E
Tsu L., *Canada* **142 A6** 60 40N 111 52W
Tsuchiura, *Japan* **71 A12** 36 5N 140 15 E
Tsugaru-Kaikyō, *Japan* **68 D10** 41 35N 141 0 E
Tsukumi, *Japan* **70 D3** 33 4N 131 52 E
Tsukushi-Sanchi, *Japan* **70 D2** 33 25N 130 30 E
Tsumeb, *Namibia* **116 B2** 19 9 S 17 44 E
Tsumis, *Namibia* **116 C2** 23 39 S 17 29 E
Tsuna, *Japan* **70 C6** 34 28N 134 56 E
Tsuno-Shima, *Japan* **70 C2** 34 21N 130 52 E
Tsuru, *Japan* **71 B10** 35 31N 138 57 E
Tsuruga, *Japan* **71 B8** 35 45N 136 2 E
Tsuruga-Wan, *Japan* **71 B8** 35 50N 136 3 E
Tsurugi, *Japan* **71 A8** 36 29N 136 37 E
Tsurugi-San, *Japan* **70 D6** 33 51N 134 6 E
Tsurumi-Saki, *Japan* **70 E4** 32 56N 132 5 E
Tsuruoka, *Japan* **68 E9** 38 44N 139 50 E
Tsurusaki, *Japan* **70 D3** 33 14N 131 41 E
Tsushima, *Gifu, Japan* **71 B8** 35 10N 136 43 E
Tsushima, *Nagasaki, Japan* **70 C1** 34 20N 129 20 E
Tsvetkovo, *Ukraine* **59 H6** 49 8N 31 33 E
Tsyelyakhany, *Belarus* **59 F3** 52 30N 25 46 E
Tu →, *Burma* **90 E6** 21 50N 96 15 E
Tua →, *Portugal* **42 D3** 41 13N 7 26W
Tuai, *N.Z.* **130 E6** 38 47 S 177 10 E
Tuakau, *N.Z.* **130 D3** 37 16 S 174 59 E
Tual, *Indonesia* **83 C4** 5 38 S 132 44 E
Tuam, *Ireland* **23 C3** 53 31N 8 51W
Tuamarina, *N.Z.* **131 B8** 41 25 S 173 59 E
Tuamotu Arch. = Tuamotu
Is., *Pac. Oc.* **135 J13** 17 0 S 144 0W
Tuamotu Is., *Pac. Oc.* **135 J13** 17 0 S 144 0W
Tuamotu Ridge, *Pac. Oc.* **135 K14** 20 0 S 138 0W
Tuanfeng, *China* **77 B10** 30 38N 114 52 E
Tuanxi, *China* **76 D6** 27 28N 107 8 E
Tuao, *Phil.* **80 C3** 17 55N 121 22 E
Tuapse, *Russia* **61 H4** 44 5N 39 10 E
Tuas, *Singapore* **84 B2** 1 19N 103 39 E
Tuatapere, *N.Z.* **131 G2** 46 8 S 167 41 E
Tuba City, *U.S.A.* **159 H8** 36 8N 111 14W
Tuban, *Indonesia* **85 D4** 6 54 S 112 3 E
Tubarão, *Brazil* **175 B6** 28 30 S 49 0W
Tūbās, *West Bank* **103 C4** 32 20N 35 22 E
Tubau, *Malaysia* **85 B4** 3 10N 113 40 E
Tübingen, *Germany* **31 G5** 48 31N 9 4 E
Tubruq, *Libya* **108 B4** 32 7N 23 55 E
Tubuai Is., *Pac. Oc.* **135 K13** 25 0 S 150 0W
Tuburan, *Phil.* **81 H4** 10 44N 123 49 E
Tuc Trung, *Vietnam* **87 G6** 11 1N 107 12 E
Tucacas, *Venezuela* **168 A4** 10 48N 68 19W
Tucano, *Brazil* **170 D4** 10 58 S 38 48W
T'uch'ang, *Taiwan* **77 E13** 24 42N 121 25 E
Tuchodi →, *Canada* **142 B4** 58 17N 123 42W
Tuchola, *Poland* **54 E4** 53 33N 17 52 E
Tuchów, *Poland* **55 J8** 49 54N 21 1 E
Tucker, *U.S.A.* **152 B5** 33 51N 84 13W
Tucson, *U.S.A.* **159 K8** 32 13N 110 58W
Tucumán □, *Argentina* **174 B2** 26 48 S 66 2W
Tucumcari, *U.S.A.* **155 H3** 35 10N 103 44W
Tucunaré, *Brazil* **173 B6** 5 18 S 55 51W
Tucupido, *Venezuela* **168 B4** 9 17N 65 47W
Tucupita, *Venezuela* **169 B5** 9 2N 62 3W
Tucuruí, *Brazil* **170 B2** 3 42 S 49 44W
Tucuruí, Reprêsa de, *Brazil* **170 B2** 4 0 S 49 30W
Tuczno, *Poland* **55 E3** 53 13N 16 10 E
Tuddal, *Norway* **18 E5** 59 46N 8 49 E
Tudela, *Spain* **40 C3** 42 4N 1 39W
Tudmur, *Syria* **101 E8** 34 36N 38 15 E
Tudor, L., *Canada* **141 A6** 55 50N 65 25W
Tudora, *Romania* **53 C11** 47 31N 26 45 E
Tuella →, *Portugal* **42 D3** 41 30N 7 12W
Tuen, *Australia* **127 D4** 28 33 S 145 37 E
Tueré →, *Brazil* **170 B1** 2 48 S 50 59W
Tufi, *Papua N. G.* **132 E5** 9 8 S 149 19 E
Tufsingdalen, *Norway* **18 B8** 62 18N 11 44 E
Tugela →, *S. Africa* **117 D5** 29 14 S 31 30 E
Tuguegarao, *Phil.* **80 C3** 17 35N 121 42 E
Tugur, *Russia* **65 D14** 53 44N 136 45 E
Tui, *Spain* **42 C2** 42 3N 8 39W
Tuineje, *Canary Is.* **39 F5** 28 19N 14 3W
Tukangbesi, Kepulauan,
Indonesia **82 C2** 6 0 S 124 0 E
Tukarak I., *Canada* **140 A4** 56 15N 78 45W
Tukayyid, *Iraq* **96 D5** 29 47N 45 36 E
Tûkh, *Egypt* **106 H7** 30 21N 31 12 E
Tukituki →, *N.Z.* **130 F5** 39 36 S 176 56 E
Tukobo, *Ghana* **112 D4** 5 1N 2 47W
Tūkrah, *Libya* **108 B4** 32 30N 20 37 E
Tuktoyaktuk, *Canada* **138 B6** 69 27N 133 2W
Tukums, *Latvia* **15 H20** 56 58N 23 10 E
Tukums □, *Latvia* **54 B10** 56 55N 23 0 E
Tukuyu, *Tanzania* **119 D3** 9 17 S 33 35 E
Tula, *Hidalgo, Mexico* **163 C5** 20 5N 99 20W
Tula, *Tamaulipas, Mexico* **163 C5** 23 0N 99 40W
Tula, *Nigeria* **113 D7** 9 51N 11 27 E
Tula, *Russia* **58 E9** 54 13N 37 38 E
Tulak, *Afghan.* **91 B1** 33 55N 63 40 E
Tulancingo, *Mexico* **163 C5** 20 5N 99 22W
Tulangbawang →, *Indonesia* **84 C3** 4 24 S 105 52 E
Tulare, *Serbia, Yug.* **50 D5** 42 48N 21 28 E
Tulare, *U.S.A.* **160 J7** 36 13N 119 21W
Tulare Lake Bed, *U.S.A.* **160 K7** 36 0N 119 48W
Tularosa, *U.S.A.* **159 K10** 33 5N 106 1W
Tulbagh, *S. Africa* **116 E2** 33 16 S 19 6 E
Tulcán, *Ecuador* **168 C2** 0 48N 77 43W
Tulcea, *Romania* **53 E13** 45 13N 28 46 E
Tulcea □, *Romania* **53 E13** 45 0N 28 30 E
Tulchyn, *Ukraine* **59 H5** 48 41N 28 49 E
Tūleh, *Iran* **97 C7** 34 35N 52 33 E
Tulemalu L., *Canada* **143 A9** 62 58N 99 25W
Tulgheş, *Romania* **53 D10** 46 58N 25 45 E
Tuli, *Indonesia* **82 B2** 1 24 S 122 26 E
Tuli, *Zimbabwe* **119 G2** 21 58 S 29 13 E
Tulia, *U.S.A.* **155 H4** 34 32N 101 46W
Tuliszków, *Poland* **55 F5** 52 5N 18 18 E
Tulita, *Canada* **138 B7** 64 57N 125 30W
Ţūlkarm, *West Bank* **103 C4** 32 19N 35 2 E
Tulla, *Ireland* **23 D3** 52 53N 8 46W
Tullahoma, *U.S.A.* **149 H2** 35 22N 86 13W
Tullamore, *Australia* **129 B7** 32 39 S 147 36 E
Tullamore, *Ireland* **23 C4** 53 16N 7 31W
Tulle, *France* **28 C5** 45 16N 1 46 E
Tullibigeal, *Australia* **129 B7** 33 25 S 146 44 E
Tulln, *Austria* **34 C9** 48 20N 16 4 E
Tullow, *Ireland* **23 D5** 52 49N 6 45W
Tullus, *Sudan* **107 E1** 11 7N 24 31 E
Tully, *Australia* **126 B4** 17 56 S 145 55 E
Ţulmaythah, *Libya* **108 B4** 32 40N 20 55 E
Tulmur, *Australia* **126 C3** 22 40 S 142 20 E
Tulnici, *Romania* **53 E11** 45 51N 26 38 E
Tulovo, *Bulgaria* **51 D9** 42 33N 25 32 E
Tulsa, *U.S.A.* **155 G7** 36 10N 95 55W
Tulsequah, *Canada* **142 B2** 58 39N 133 35W
Tulu Milki, *Ethiopia* **107 F4** 9 55N 38 20 E
Tulu Welel, *Ethiopia* **107 F3** 8 56N 34 47 E
Tulua, *Colombia* **168 C2** 4 6N 76 11W
Tulucești, *Romania* **53 E13** 45 34N 28 2 E
Tuluksak, *U.S.A.* **144 F7** 61 6N 160 58W
Tulun, *Russia* **65 D11** 54 32N 100 35 E
Tulungagung, *Indonesia* **85 D4** 8 5 S 111 54 E
Tuma, *Russia* **58 E11** 55 10N 40 30 E
Tuma →, *Nic.* **164 D3** 13 6N 84 35W
Tumaco, *Colombia* **168 C2** 1 50N 78 45W
Tumaco, Ensenada,
Colombia **168 C2** 1 55N 78 45W
Tumatumari, *Guyana* **169 B6** 5 20N 58 55W
Tumauini, *Phil.* **80 C3** 17 17N 121 49 E
Tumba, *Sweden* **16 E11** 59 12N 17 48 E
Tumba, L.,
Dem. Rep. of the Congo **114 C3** 0 50 S 18 0 E
Tumbarumba, *Australia* **129 C8** 35 44 S 148 0 E
Tumbaya, *Argentina* **174 A2** 23 50 S 65 26W
Tumbes, *Peru* **172 A1** 3 37 S 80 27W
Tumbes □, *Peru* **172 A1** 3 50 S 80 30W
Tumbwe,
Dem. Rep. of the Congo **119 E2** 11 25 S 27 15 E
Tumby Bay, *Australia* **128 C2** 34 21 S 136 8 E
Tumd Youqi, *China* **74 D6** 40 30N 110 30 E
Tumen, *China* **75 C15** 43 0N 129 50 E
Tumen Jiang →, *China* **75 C16** 42 20N 130 35 E
Tumeremo, *Venezuela* **169 B5** 7 18N 61 30W
Tumiritinga, *Brazil* **171 E3** 18 58 S 41 38W
Tumkur, *India* **95 H3** 13 18N 77 6 E
Tump, *Pakistan* **91 D1** 26 7N 62 16 E
Tumpat, *Malaysia* **87 J4** 6 11N 102 10 E
Tumsar, *India* **94 D4** 21 26N 79 45 E
Tumu, *Ghana* **112 C4** 10 56N 1 56W
Tumucumaque, Serra, *Brazil* **169 C7** 2 0N 55 0W
Tumupasa, *Bolivia* **172 C4** 14 9 S 67 55W
Tumut, *Australia* **129 C8** 35 16 S 148 13 E
Tumwater, *U.S.A.* **158 C2** 47 1N 122 54W
Tunadal, *Sweden* **16 B11** 62 26N 17 22 E
Tunas de Zaza, *Cuba* **164 B4** 21 39N 79 34W
Tunbridge Wells =
Tunbridge Wells, *U.K.* **21 F8** 51 7N 0 16 E
Tunçbilek, *Turkey* **49 B11** 39 37N 29 29 E
Tunceli, *Turkey* **101 C8** 39 6N 39 31 E
Tuncurry, *Australia* **129 B10** 32 17 S 152 29 E
Tunduru, *Tanzania* **119 E4** 11 8 S 37 25 E
Tunduru □, *Tanzania* **119 E4** 11 5 S 37 22 E
Tundzha →, *Bulgaria* **51 E10** 41 40N 26 35 E
Tunga →, *India* **95 G2** 15 0N 75 50 E
Tunga Pass, *India* **90 A5** 29 0N 94 14 E
Tungabhadra →, *India* **95 G4** 15 57N 78 15 E
Tungabhadra Dam, *India* **95 G2** 15 0N 75 50 E
Tungaru, *Sudan* **107 E3** 10 9N 30 52 E
Tungi, *Bangla.* **90 D3** 23 53N 90 24 E
Tungla, *Nic.* **164 D3** 13 24N 84 21W
Tungnafellsjökull, *Iceland* **11 C9** 64 45N 17 55W
Tungshih, *Taiwan* **77 E13** 24 12N 120 43 E
Tungsten, *Canada* **142 A3** 61 57N 128 16W
Tungufell, *Iceland* **11 C6** 64 17N 20 10W
Tungurahua □, *Ecuador* **168 D2** 1 15 S 78 35W
Tunguska, Nizhnyaya →,
Russia **65 C9** 65 48N 88 4 E
Tunguska,
Podkamennaya →, *Russia* **65 C10** 61 50N 90 13 E
Tuni, *India* **94 F6** 17 22N 82 36 E
Tunia, *Colombia* **168 C2** 2 41N 76 31W
Tunica, *U.S.A.* **155 H9** 34 41N 90 23W
Tunis, *Tunisia* **108 A2** 36 50N 10 11 E
Tunis, Golfe de, *Tunisia* **108 A2** 37 0N 10 30 E
Tunisia ■, *Africa* **108 B1** 33 30N 9 10 E
Tunja, *Colombia* **168 B3** 5 33N 73 25W
Tunkhannock, *U.S.A.* **151 E9** 41 32N 75 57W
Tunliu, *China* **74 F7** 36 13N 112 52 E
Tunnsjøen, *Norway* **14 D15** 64 45N 13 25 E
Tunø, *Denmark* **17 J4** 55 57N 10 27 E
Tununak, *U.S.A.* **144 F6** 60 37N 165 15W
Tunungayualok I., *Canada* **141 A7** 56 0N 61 0W
Tunuyán, *Argentina* **174 C2** 33 35 S 69 0W
Tunuyán →, *Argentina* **174 C2** 33 33 S 67 30W
Tunxi, *China* **77 C12** 29 42N 118 25 E
Tuo Jiang →, *China* **76 C5** 28 50N 105 35 E
Tuolumne, *U.S.A.* **160 H6** 37 58N 120 15W
Tuolumne →, *U.S.A.* **160 H5** 37 36N 121 13W
Tūp Āghāj, *Iran* **101 D12** 36 3N 47 50 E
Tupã, *Brazil* **175 A5** 21 57 S 50 28W
Tupaciguara, *Brazil* **171 E2** 18 35 S 48 42W
Tupelo, *U.S.A.* **149 H1** 34 16N 88 43W
Tupik, *Russia* **58 E7** 55 42N 33 22 E
Tupinambaranas, *Brazil* **169 D6** 3 0 S 58 0W
Tupirama, *Brazil* **170 C2** 8 58 S 48 12W
Tupiratins, *Brazil* **170 C2** 8 23 S 48 8W
Tupiza, *Bolivia* **174 A2** 21 30 S 65 40W
Tupižnica, *Serbia, Yug.* **50 C6** 43 43N 22 10 E
Tupman, *U.S.A.* **161 K7** 35 18N 119 21W
Tupper, *Canada* **142 B4** 55 32N 120 1W
Tupper Lake, *U.S.A.* **151 B10** 44 14N 74 28W
Tupungato, Cerro, *S. Amer.* **174 C2** 33 15 S 69 50W
Tuquan, *China* **75 B11** 45 18N 121 38 E
Túquerres, *Colombia* **168 C2** 1 5N 77 37W
Tura, *India* **90 C3** 25 30N 90 16 E
Tura, *Russia* **65 C11** 64 20N 100 17 E
Turabah, *Si. Arabia* **96 D4** 28 20N 43 15 E
Turagua, Serranía, *Venezuela* **169 B5** 7 20N 64 35W
Turaiyur, *India* **95 J4** 11 9N 78 38 E
Turakina, *N.Z.* **130 G4** 40 3 S 175 16 E
Turakina →, *N.Z.* **130 G4** 40 5 S 175 8 E
Turakirae Hd., *N.Z.* **130 H3** 41 26 S 174 56 E
Tūrān, *Iran* **97 C8** 35 39N 56 42 E
Turan, *Russia* **65 D10** 51 55N 95 0 E
Turayf, *Si. Arabia* **96 D3** 31 41N 38 39 E
Turbacz, *Poland* **55 J7** 49 30N 20 8 E
Turbe, *Bos.-H.* **52 F2** 44 15N 17 35 E
Turbenthal, *Switz.* **33 B7** 47 27N 8 51 E
Turbeville, *U.S.A.* **152 B9** 33 54N 80 1W
Turčianske Teplice,
Slovak Rep. **35 C11** 48 52N 18 52 E
Turcoaia, *Romania* **53 E13** 45 7N 28 11 E
Turda, *Romania* **53 D8** 46 34N 23 47 E
Turek, *Poland* **55 F5** 52 3N 18 30 E
Turen, *Venezuela* **168 B4** 9 17N 69 6W
Turfan = Turpan, *China* **72 B3** 43 58N 89 10 E
Turfan Depression = Turpan
Hami, *China* **66 E12** 42 40N 89 25 E
Tŭrgovishte, *Bulgaria* **51 C10** 43 17N 26 38 E
Turgut, *Turkey* **49 D10** 37 22N 28 4 E
Turgutlu, *Turkey* **57 G4** 38 30N 27 43 E
Turhal, *Turkey* **100 B7** 40 24N 36 5 E
Turia →, *Spain* **41 F4** 39 27N 0 19W
Turiaçu, *Brazil* **170 B2** 1 40 S 45 19W
Turiaçu →, *Brazil* **170 B2** 1 36 S 45 19W
Turiec →, *Slovak Rep.* **35 B11** 49 7N 18 55 E
Turin = Torino, *Italy* **44 C4** 45 3N 7 40 E
Turin, *Canada* **142 D6** 49 58N 112 31W
Turinsk, *Russia* **62 B9** 58 3N 63 42 E
Turkana □, *Kenya* **118 B4** 3 0N 35 30 E
Turkana, L., *Africa* **118 B4** 3 30N 36 5 E
Türkeli, *Turkey* **51 F11** 40 30N 27 30 E
Turkestan = Türkistan,
Kazakstan **63 B4** 43 17N 68 16 E
Turkestanskiy, Khrebet,
Tajikistan **63 D4** 39 35N 69 0 E
Túrkeve, *Hungary* **52 C5** 47 6N 20 44 E
Turkey ■, *Eurasia* **100 C7** 39 0N 36 0 E
Turkey →, *U.S.A.* **156 B5** 42 43N 91 2W
Turkey Creek, *Australia* **124 C4** 17 2 S 128 12 E
Turki, *Russia* **60 D6** 52 0N 43 15 E
Türkistan, *Kazakstan* **63 B4** 43 17N 68 16 E
Türkmenbashi, *Turkmenistan* **57 G9** 40 5N 53 5 E
Turkmenistan ■, *Asia* **64 F6** 39 0N 59 0 E
Türkmenli, *Turkey* **49 B8** 39 45N 26 30 E
Türkoğlu, *Turkey* **100 D7** 37 23N 36 50 E
Turks & Caicos Is. ■,
W. Indies **165 B5** 21 20N 71 20W
Turks Island Passage,
W. Indies **165 B5** 21 30N 71 30W
Turku, *Finland* **15 F20** 60 30N 22 19 E
Turkwel →, *Kenya* **118 B4** 3 6N 36 6 E
Turlock, *U.S.A.* **160 H6** 37 30N 120 51W
Turnagain →, *Canada* **142 B3** 59 12N 127 35W
Turnagain, C., *N.Z.* **130 G5** 40 28 S 176 38 E
Turneffe Is., *Belize* **163 D7** 17 20N 87 50W
Turner, *Australia* **124 C4** 17 52 S 128 16 E
Turner, *U.S.A.* **158 B9** 48 51N 108 24W
Turner Pt., *Australia* **126 A1** 11 47 S 133 32 E
Turner Valley, *Canada* **142 C6** 50 40N 114 17W
Turners Falls, *U.S.A.* **151 D12** 42 36N 72 33W
Turnhout, *Belgium* **24 C4** 51 19N 4 57 E
Türnitz, *Austria* **34 D8** 47 55N 15 29 E
Turnor L., *Canada* **143 B7** 56 35N 108 35W
Turnov, *Czech Rep.* **34 A8** 50 34N 15 10 E
Tŭrnovo = Veliko Tŭrnovo,
Bulgaria **51 C9** 43 5N 25 41 E
Turnu Măgurele, *Romania* **53 G9** 43 46N 24 56 E
Turnu Roşu, P., *Romania* **53 E9** 45 33N 24 17 E
Turobin, *Poland* **55 H9** 50 50N 22 44 E
Turon, *U.S.A.* **155 G5** 37 48N 98 26W
Tuross Head, *Australia* **129 D9** 36 3 S 150 8 E
Turpan, *China* **72 B3** 43 58N 89 10 E
Turpan Hami, *China* **66 E12** 42 40N 89 25 E
Turrës, Kalaja e, *Albania* **50 E3** 41 10N 19 28 E
Turriff, *U.K.* **22 D6** 57 32N 2 27W
Tursāq, *Iraq* **101 F11** 33 27N 45 47 E
Tursi, *Italy* **47 B9** 40 15N 16 28 E
Turtle Head I., *Australia* **126 A3** 10 56 S 142 37 E
Turtle Is., *Phil.* **81 H2** 6 7N 118 14 E
Turtle L., *Canada* **143 C7** 53 36N 108 38W
Turtle Lake, *N. Dak.,*
U.S.A. **154 B4** 47 31N 100 53W
Turtle Lake, *Wis., U.S.A.* **154 C8** 45 24N 92 8W
Turtleford, *Canada* **143 C7** 53 23N 108 57W
Turua, *N.Z.* **130 D4** 37 14 S 175 35 E
Turugart, Pereval,
Kyrgyzstan **63 C7** 40 32N 75 24 E
Turukhansk, *Russia* **65 C9** 65 21N 88 5 E
Turzovka, *Slovak Rep.* **35 B11** 49 25N 18 35 E
Tuscaloosa, *U.S.A.* **149 J2** 33 12N 87 34W
Tuscánia, *Italy* **45 F8** 42 25N 11 52 E
Tuscany = Toscana □, *Italy* **44 E8** 43 25N 11 0 E
Tuscola, *Ill., U.S.A.* **157 E8** 39 48N 88 17W
Tuscola, *Tex., U.S.A.* **155 J5** 32 12N 99 48W
Tuscumbia, *Ala., U.S.A.* **149 H2** 34 44N 87 42W
Tuscumbia, *Mo., U.S.A.* **156 F4** 38 14N 92 28W
Tuskegee, *U.S.A.* **152 C4** 32 25N 85 42W
Tustin, *U.S.A.* **161 M9** 33 44N 117 49W
Tustna, *Norway* **18 A5** 63 10N 8 5 E
Tuszyn, *Poland* **55 G6** 51 36N 19 33 E
Tutak, *Turkey* **101 C10** 39 31N 42 46 E
Tutayev, *Russia* **58 D10** 57 53N 39 32 E
Tuticorin, *India* **95 K4** 8 50N 78 12 E
Tutin, *Serbia, Yug.* **50 D4** 42 58N 20 20 E
Tutóia, *Brazil* **170 B3** 2 45 S 42 20W
Tutong, *Brunei* **85 B4** 4 47N 114 40 E
Tutova →, *Romania* **53 D12** 46 7N 27 30 E
Tutrakan, *Bulgaria* **51 B10** 44 2N 26 40 E
Tutshi L., *Canada* **142 B2** 59 56N 134 30W
Tuttle, *U.S.A.* **154 B5** 47 9N 100 0W
Tuttlingen, *Germany* **31 H4** 47 58N 8 48 E
Tutuala, *Indonesia* **82 C3** 8 25 S 127 15 E
Tutuila, *Amer. Samoa* **133 X24** 14 19 S 170 50W
Tutuko Mt., *N.Z.* **131 E3** 44 35 S 168 1 E
Tututepec, *Mexico* **163 D5** 16 9N 97 38W
Tutye, *Australia* **128 C4** 35 12 S 141 29 E
Tuva □, *Russia* **65 D10** 51 30N 95 0 E
Tuvalu ■, *Pac. Oc.* **134 H9** 8 0 S 178 0 E
Tūwal, *Si. Arabia* **98 B2** 22 17N 39 6 E
Tuxer Alpen, *Austria* **34 D4** 47 10N 11 45 E
Tuxpan, *Mexico* **163 C5** 20 58N 97 23W
Tuxtla Gutiérrez, *Mexico* **163 D6** 16 50N 93 10W
Tuy = Tui, *Spain* **42 C2** 42 3N 8 39W
Tuy An, *Vietnam* **86 F7** 13 17N 109 16 E
Tuy Duc, *Vietnam* **87 F6** 12 15N 107 27 E
Tuy Hoa, *Vietnam* **86 F7** 13 5N 109 10 E
Tuy Phong, *Vietnam* **87 G7** 11 14N 108 43 E
Tuya L., *Canada* **142 B2** 59 7N 130 35W
Tuyen Hoa, *Vietnam* **86 D6** 17 50N 106 10 E
Tuyen Quang, *Vietnam* **86 B5** 21 50N 105 10 E
Tuymazy, *Russia* **62 D5** 54 36N 53 42 E
Tūysarkān, *Iran* **97 C6** 34 33N 48 27 E
Tuz Gölü, *Turkey* **100 C5** 38 42N 33 18 E
Ţūz Khurmātū, *Iraq* **101 E11** 34 56N 44 38 E
Tuzi, *Montenegro, Yug.* **50 D3** 42 22N 19 20 E
Tuzkan, Ozero, *Uzbekistan* **63 C3** 40 35N 67 28 E
Tuzla, *Bos.-H.* **52 F3** 44 34N 18 41 E
Tuzlov →, *Russia* **59 J10** 47 17N 39 57 E
Tuzluca, *Turkey* **101 B10** 40 3N 43 39 E
Tvååker, *Sweden* **17 G6** 57 4N 12 25 E
Tvardiţa, *Moldova* **53 D13** 46 9N 28 58 E
Tvedestrand, *Norway* **18 F5** 58 38N 8 58 E
Tver, *Russia* **58 D8** 56 55N 35 55 E
Tvrdošin, *Slovak Rep.* **35 B12** 49 21N 19 35 E
Tvrdošovce, *Slovak Rep.* **35 C11** 48 6N 18 4 E
Tvŭrditsa, *Bulgaria* **51 D9** 42 42N 25 53 E
Twain, *U.S.A.* **160 E5** 40 1N 121 3W
Twain Harte, *U.S.A.* **160 G6** 38 2N 120 14W
Twardogóra, *Poland* **55 G4** 51 23N 17 28 E
Tweed, *Canada* **150 B7** 44 29N 77 19W
Tweed →, *U.K.* **22 F6** 55 45N 2 0W
Tweed Heads, *Australia* **127 D5** 28 10 S 153 31 E
Tweedsmuir Prov. Park,
Canada **142 C3** 53 0N 126 20W
Twentynine Palms, *U.S.A.* **161 L10** 34 8N 116 3W
Twillingate, *Canada* **141 C9** 49 42N 54 45W
Twin Bridges, *U.S.A.* **158 D7** 45 33N 112 20W
Twin City, *U.S.A.* **152 C7** 32 35N 82 10W
Twin Falls, *U.S.A.* **158 E6** 42 34N 114 28W
Twin Hills, *U.S.A.* **144 G8** 59 23N 159 58W
Twin Lakes, *U.S.A.* **152 E6** 30 43N 83 13W
Twin Valley, *U.S.A.* **154 B6** 47 16N 96 16W
Twinnge, *Burma* **90 D6** 23 10N 96 2 E
Twisp, *U.S.A.* **158 B3** 48 22N 120 7W
Twistringen, *Germany* **30 C4** 52 48N 8 37 E
Two Harbors, *U.S.A.* **154 B9** 47 2N 91 40W
Two Hills, *Canada* **142 C6** 53 43N 111 52W
Two Rivers, *U.S.A.* **148 C2** 44 9N 87 34W
Two Thumbs Ra., *N.Z.* **131 D5** 43 45 S 170 44 E
Twofold B., *Australia* **129 D8** 37 8 S 149 59 E
Ty Ty, *U.S.A.* **152 D6** 31 28N 83 39W
Tyachiv, *Ukraine* **59 H2** 48 1N 23 35 E
Tychy, *Poland* **55 H5** 50 9N 18 59 E
Tyczyn, *Poland* **55 J9** 49 58N 22 2 E
Tydal, *Norway* **18 A8** 63 4N 11 34 E
Tyin, *Norway* **18 C5** 61 18N 8 15 E
Tykocin, *Poland* **55 E9** 53 13N 22 46 E
Tyler, *U.S.A.* **147 D7** 32 18N 95 17W
Tyler, *Minn., U.S.A.* **154 C6** 44 18N 96 8W
Tyler, *Tex., U.S.A.* **155 J7** 32 21N 95 18W
Tyligul →, *Ukraine* **59 J6** 47 4N 30 57 E
Tylldal, *Norway* **18 B7** 62 8N 10 48 E
Týn nad Vltavou, *Czech Rep.* **34 B7** 49 13N 14 26 E
Tynda, *Russia* **65 D13** 55 10N 124 43 E
Tyne →, *U.K.* **20 C6** 54 59N 1 32W
Tyne & Wear □, *U.K.* **20 B6** 55 6N 1 17W
Týnec nad Sázavou,
Czech Rep. **34 B7** 49 50N 14 36 E
Tynemouth, *U.K.* **20 B6** 55 1N 1 26W
Tynset, *Norway* **18 B7** 62 17N 10 47 E
Tyonek, *U.S.A.* **144 F10** 61 4N 151 8W
Tyonek Indian Reservation,
U.S.A. **144 F10** 61 5N 151 10W
Tyre = Sūr, *Lebanon* **103 B4** 33 19N 35 16 E
Tyrifjorden, *Norway* **15 F14** 60 2N 10 8 E
Tyringe, *Sweden* **17 H7** 56 9N 13 35 E
Tyristrand, *Norway* **18 D7** 60 4N 10 6 E
Tyrnyauz, *Russia* **61 J6** 43 21N 42 45 E
Tyrol = Tirol □, *Austria* **34 D3** 47 3N 10 43 E
Tyrone, *U.S.A.* **150 F6** 40 40N 78 14W
Tyrone □, *U.K.* **23 B4** 54 38N 7 11W
Tyrrell →, *Australia* **128 C5** 35 26 S 142 51 E
Tyrrell, L., *Australia* **128 C5** 35 20 S 142 50 E
Tyrrell Arm, *Canada* **143 A9** 62 27N 97 30W
Tyrrell L., *Canada* **143 A7** 63 7N 105 27W
Tyrrhenian Sea, *Medit. S.* **12 G8** 40 0N 12 30 E
Tysfjorden, *Norway* **14 B17** 68 7N 16 25 E
Tysnes, *Norway* **18 D2** 60 1N 5 30 E
Tysnesøya, *Norway* **18 D2** 60 0N 5 35 E
Tysse, *Norway* **18 D2** 60 23N 5 47 E
Tyssedal, *Norway* **18 D3** 60 7N 6 35 E
Tystberga, *Sweden* **17 F11** 58 51N 17 15 E
Tytuvėnai, *Lithuania* **54 C10** 55 36N 23 12 E
Tyub Karagan, Mys,
Kazakstan **61 H10** 44 40N 50 19 E
Tyuleni, Ostrova, *Kazakstan* **61 H10** 45 2N 50 16 E
Tyuleniy, *Russia* **61 H8** 44 28N 47 30 E
Tyuleniy, Mys, *Azerbaijan* **61 K10** 40 12N 50 22 E
Tyulgan, *Russia* **62 E6** 52 22N 56 12 E
Tyumen, *Russia* **64 D7** 57 11N 65 29 E
Tyumen-Aryk, *Kazakstan* **63 A3** 44 2N 67 1 E
Tyup, *Kyrgyzstan* **63 B9** 42 45N 78 20 E
Tywi →, *U.K.* **21 F3** 51 48N 4 21W
Tywyn, *U.K.* **21 E3** 52 35N 4 5W
Tzaneen, *S. Africa* **117 C5** 23 47 S 30 9 E
Tzermiádhes, *Greece* **38 D7** 35 12N 25 29 E
Tzoumérka, Óros, *Greece* **48 B3** 39 30N 21 26 E
Tzukong = Zigong, *China* **76 C5** 29 15N 104 48 E

U

U Taphao, *Thailand* **86 F3** 12 35N 101 0 E
U.S.A. = United States of
America ■, *N. Amer.* **146 C7** 37 0N 96 0W
Uacalla Iero, *Somali Rep.* **120 D2** 1 48N 42 38 E
Uachadi, Sierra, *Venezuela* **169 C4** 4 54N 65 18W
Uainambi, *Colombia* **168 C4** 1 43N 69 51W
Uanda, *Australia* **126 C3** 21 37 S 144 55 E
Uanle Uen, *Somali Rep.* **120 D2** 2 37N 44 54 E
Uarsciek, *Somali Rep.* **120 D3** 2 28N 45 55 E
Uascen, *Somali Rep.* **120 D2** 4 11N 43 13 E
Uasin □, *Kenya* **118 B4** 0 30N 35 20 E
Uato-Udo, *Indonesia* **82 C3** 9 7 S 125 36 E
Uatumã →, *Brazil* **169 D6** 2 26 S 57 37W
Uauá, *Brazil* **170 C4** 9 50 S 39 28W
Uaupés, *Brazil* **168 D4** 0 8 S 67 5W
Uaupés →, *Brazil* **168 C4** 0 2N 67 16W
Uaxactún, *Guatemala* **164 C2** 17 25N 89 29W
Ub, *Serbia, Yug.* **50 B4** 44 28N 20 6 E
Ubá, *Brazil* **171 F3** 21 8 S 43 0W
Ubaitaba, *Brazil* **171 D4** 14 18 S 39 20W
Ubangi = Oubangi →,
Dem. Rep. of the Congo **114 C3** 0 30 S 17 50 E
Ubaté, *Colombia* **168 B3** 5 19N 73 49W
Ubauro, *Pakistan* **92 E3** 28 15N 69 45 E
Ubay, *Phil.* **81 F5** 10 3N 124 28 E
'Ubaydīyah, *Yemen* **98 D3** 13 7N 43 20 E
Ubaye →, *France* **29 D10** 44 28N 6 18 E
Ubayyiḍ, W. al →, *Iraq* **101 F10** 32 34N 43 48 E
Ube, *Japan* **70 D3** 33 56N 131 15 E
Úbeda, *Spain* **43 G7** 38 3N 3 23W
Uberaba, *Brazil* **171 E2** 19 50 S 47 55W

Vaygach, Ostrov 295

Ústica, *Italy* **46 D6** 38 42N 13 11 E
Ustinov = Izhevsk, *Russia* **62 C4** 56 51N 53 14 E
Ustka, *Poland* **54 D3** 54 35N 16 55 E
Ustroń, *Poland* **55 J5** 49 43N 18 48 E
Ustrzyki Dolne, *Poland* **55 J9** 49 27N 22 40 E
Ustyurt Plateau, *Asia* **64 E6** 44 0N 55 0 E
Ustyuzhna, *Russia* **58 C9** 58 50N 36 32 E
Usu, *China* **72 B3** 44 27N 84 40 E
Usuki, *Japan* **70 D3** 33 8N 131 49 E
Usulután, *El Salv.* **164 D2** 13 25N 88 28W
Usumacinta →, *Mexico* **163 D6** 17 0N 91 0W
Usumbura = Bujumbura, *Burundi* **118 C2** 3 16 S 29 18 E
Usure, *Tanzania* **118 C3** 4 40 S 34 22 E
Usva, *Russia* **62 B6** 58 41N 57 37 E
Uta, *Indonesia* **83 B5** 4 33 S 136 0 E
'Uta Vava'u, *Tonga* **133 P14** 18 36 S 174 0W
Utah □, *U.S.A.* **158 G8** 39 20N 111 30W
Utah L., *U.S.A.* **158 F8** 40 10N 111 58W
Utansjö, *Sweden* **16 B11** 62 46N 17 55 E
Ute Creek →, *U.S.A.* **155 H3** 35 21N 103 50W
Utebo, *Spain* **40 D3** 41 43N 1 0W
Utena, *Lithuania* **15 J21** 55 27N 25 40 E
Utete, *Tanzania* **118 D4** 8 0 S 38 45 E
Uthai Thani, *Thailand* **86 E3** 15 22N 100 3 E
Uthal, *Pakistan* **92 G2** 25 44N 66 40 E
Utiariti, *Brazil* **173 C6** 13 0 S 58 10W
Utica, *N.Y., U.S.A.* **151 C9** 43 6N 75 14W
Utica, *Ohio, U.S.A.* **150 F2** 40 14N 82 27W
Utiel, *Spain* **41 F3** 39 37N 1 11W
Utik L., *Canada* **143 B9** 55 15N 96 0W
Utikuma L., *Canada* **142 B5** 55 50N 115 30W
Utinga, *Brazil* **171 D3** 12 6 S 41 5W
Utne, *Norway* **18 D3** 60 25N 6 37 E
Uto, *Japan* **70 E2** 32 41N 130 40 E
Utö, *Sweden* **16 F12** 58 56N 18 16 E
Utrecht, *Neths.* **24 B5** 52 5N 5 8 E
Utrecht, *S. Africa* **117 D5** 27 38 S 30 20 E
Utrecht □, *Neths.* **24 B5** 52 6N 5 7 E
Utrera, *Spain* **43 H5** 37 12N 5 48W
Utsira, *Norway* **18 E1** 59 19N 4 53 E
Utsjoki, *Finland* **14 B22** 69 51N 26 59 E
Utsunomiya, *Japan* **71 A11** 36 30N 139 50 E
Uttar Pradesh □, *India* **93 F9** 27 0N 80 0 E
Uttaradit, *Thailand* **86 D3** 17 36N 100 5 E
Uttoxeter, *U.K.* **20 E6** 52 54N 1 52W
Utva →, *Kazakstan* **62 F4** 51 28N 52 40 E
Uummannarsuaq = Farvel, Kap, *Greenland* **10 F6** 59 48N 43 55W
Uusikaarlepyy, *Finland* **14 E20** 63 32N 22 31 E
Uusikaupunki, *Finland* **15 F19** 60 47N 21 25 E
Uva, *Russia* **60 B11** 56 59N 52 13 E
Uvá →, *Colombia* **168 C3** 3 41N 70 3W
Uvac →, *Serbia, Yug.* **50 C3** 43 35N 19 30 E
Uvalda, *U.S.A.* **152 C7** 32 2N 82 31W
Uvalde, *U.S.A.* **155 L5** 29 13N 99 47W
Uvarovo, *Russia* **60 E6** 51 59N 42 14 E
Uvat, *Russia* **64 D7** 59 5N 68 50 E
Uvdal, *Norway* **18 D5** 60 17N 8 48 E
Uvéa, I., *Vanuatu* **123 E12** 20 30 S 166 35 E
Uvelskiy, *Russia* **62 D8** 54 26N 61 22 E
Uvinza, *Tanzania* **118 D3** 5 5 S 30 24 E
Uvira, *Dem. Rep. of the Congo* **118 C2** 3 22 S 29 3 E
Uvs Nuur, *Mongolia* **72 A4** 50 20N 92 30 E
Uwa, *Japan* **70 D4** 33 22N 132 31 E
Uwajima, *Japan* **70 D4** 33 10N 132 35 E
'Uwayfī, *Oman* **99 B7** 22 15N 56 59 E
Uweinat, Jebel, *Sudan* **106 C1** 21 54N 24 58 E
Uxbridge, *Canada* **150 B5** 44 6N 79 7W
Uxin Qi, *China* **74 E5** 38 50N 109 5 E
Uxmal, *Mexico* **163 C7** 20 22N 89 46W
Uyak, *U.S.A.* **144 H9** 57 38N 154 0W
Uyo, *Nigeria* **113 D6** 5 1N 7 53 E
Uyu →, *Burma* **90 C5** 24 51N 94 57 E
Uyuk, *Kazakstan* **63 B5** 43 36N 71 16 E
Üyüklü Tepe, *Turkey* **49 D9** 37 5N 27 21 E
Uyûn Mûsa, *Egypt* **103 F1** 29 53N 32 40 E
Uyuni, *Bolivia* **172 E4** 20 28 S 66 47W
Uzbekistan ■, *Asia* **63 C2** 41 30N 65 0 E
Uzen, *Kazakstan* **57 F9** 43 29N 52 54 E
Uzen, Bolshoi →, *Kazakstan* **61 F9** 49 6N 49 56 E
Uzen, Mal →, *Kazakstan* **61 F9** 49 4N 49 44 E
Uzerche, *France* **28 C5** 45 25N 1 34 E
Uzès, *France* **29 D8** 44 1N 4 26 E
Uzgen = Özgön, *Kyrgyzstan* **63 C6** 40 46N 73 18 E
Uzh →, *Ukraine* **59 G6** 51 15N 30 12 E
Uzhgorod = Uzhhorod, *Ukraine* **59 H2** 48 36N 22 18 E
Uzhhorod, *Ukraine* **59 H2** 48 36N 22 18 E
Užice, *Serbia, Yug.* **50 C3** 43 55N 19 50 E
Uzlovaya, *Russia* **58 F10** 54 0N 38 5 E
Üzümlü, *Turkey* **49 E11** 36 44N 29 14 E
Uzun-Agach, *Kazakstan* **63 B8** 43 35N 76 20 E
Uzunköprü, *Turkey* **51 E10** 41 16N 26 43 E
Uzunkuyu, *Turkey* **49 C8** 38 17N 26 33 E
Uzwil, *Switz.* **33 B8** 47 26N 9 9 E

V

Vaal →, *S. Africa* **116 D3** 29 4 S 23 38 E
Vaal Dam, *S. Africa* **117 D4** 27 0 S 28 14 E
Vaalwater, *S. Africa* **117 C4** 24 15 S 28 8 E
Vaasa, *Finland* **14 E19** 63 6N 21 38 E
Vabre, *France* **28 E6** 43 42N 2 24 E
Vác, *Hungary* **52 C4** 47 49N 19 10 E
Vacaria, *Brazil* **175 B5** 28 31 S 50 52W
Vacaville, *U.S.A.* **160 G5** 38 21N 121 59W
Vaccarès, Étang de, *France* **29 E8** 43 32N 4 34 E
Vach → = Vakh →, *Russia* **64 C8** 60 45N 76 45 E
Vache, Î. à, *Haiti* **165 C5** 18 2N 73 35W
Väckelsång, *Sweden* **17 H8** 56 37N 14 58 E
Väddö, *Sweden* **16 D12** 60 0N 18 50 E
Väderstad, *Sweden* **17 F8** 58 19N 14 55 E
Vadheim, *Norway* **18 C2** 61 13N 5 49 E
Vadnagar, *India* **92 H5** 23 47N 72 40 E
Vado Lígure, *Italy* **44 D5** 44 17N 8 26 E
Vadodara, *India* **92 H5** 22 20N 73 10 E
Vadsø, *Norway* **14 A23** 70 3N 29 50 E
Vadstena, *Sweden* **17 F8** 58 28N 14 54 E
Vaduz, *Liech.* **33 B9** 47 8N 9 31 E
Værlandet, *Norway* **18 C1** 61 18N 4 44 E
Værøy, *Norway* **14 C15** 67 40N 12 40 E
Vågåmo, *Norway* **18 C6** 61 52N 9 6 E
Vágar, *Færoe Is.* **14 E9** 62 5N 7 15W
Vaggeryd, *Sweden* **17 G8** 57 30N 14 10 E
Vagney, *France* **27 D13** 48 1N 6 43 E
Vagnhärad, *Sweden* **17 F11** 58 57N 17 33 E
Vagos, *Portugal* **42 E2** 40 33N 8 42W
Vågsfjorden, *Norway* **14 B17** 68 50N 16 50 E
Váh →, *Slovak Rep.* **35 D11** 47 43N 18 7 E
Vahsel B., *Antarctica* **7 D1** 75 0 S 35 0W
Vaï, *Greece* **38 D8** 35 15N 26 18 E
Vaigach, *Russia* **64 B6** 70 10N 59 0 E
Vaigai →, *India* **95 K4** 9 15N 79 10 E
Vaiges, *France* **26 D6** 48 2N 0 30W
Vaihingen, *Germany* **31 G4** 48 54N 8 57 E
Vaijapur, *India* **94 E2** 19 58N 74 45 E
Vaikam, *India* **95 K3** 9 45N 76 25 E
Vailly-sur-Aisne, *France* **27 C10** 49 24N 3 31 E
Vaippar →, *India* **95 K4** 9 0N 78 25 E
Vaison-la-Romaine, *France* **29 D9** 44 14N 5 4 E
Vajpur, *India* **94 D1** 21 24N 73 17 E
Vakarel, *Bulgaria* **50 D7** 42 35N 23 40 E
Vakfikebir, *Turkey* **101 B8** 41 2N 39 17 E
Vakh →, *Russia* **64 C8** 60 45N 76 45 E
Vakhsh →, *Tajikistan* **63 E4** 37 6N 68 18 E
Vakhtan, *Russia* **60 B8** 57 53N 46 47 E
Vaksdal, *Norway* **18 D2** 60 29N 5 45 E
Vál, *Hungary* **52 C3** 47 22N 18 40 E
Val-de-Marne □, *France* **27 D9** 48 45N 2 28 E
Val-d'Isère, *France* **29 C10** 45 27N 6 59 E
Val-d'Oise □, *France* **27 C9** 49 5N 2 10 E
Val-d'Or, *Canada* **140 C4** 48 7N 77 47W
Val Marie, *Canada* **143 D7** 49 15N 107 45W
Valaam, *Russia* **58 B6** 61 22N 30 57 E
Valadares, *Portugal* **42 D2** 41 5N 8 38W
Valahia, *Romania* **53 F9** 44 35N 25 0 E
Valais □, *Switz.* **32 D5** 46 12N 7 45 E
Valais, Alpes du, *Switz.* **32 D5** 46 5N 7 35 E
Valandovo, *Macedonia* **50 E6** 41 19N 22 34 E
Valašské Meziříčí, *Czech Rep.* **35 B10** 49 29N 17 59 E
Valáxa, *Greece* **48 C6** 38 50N 24 29 E
Vålberg, *Sweden* **16 E7** 59 23N 13 11 E
Valbo, *Sweden* **16 D10** 60 40N 17 0 E
Valbondione, *Italy* **44 B7** 46 2N 10 1 E
Vâlcani, *Romania* **52 D5** 46 0N 20 26 E
Vâlcea □, *Romania* **53 F9** 45 0N 24 10 E
Valcheta, *Argentina* **176 B3** 40 40 S 66 8W
Valdagno, *Italy* **45 C8** 45 39N 11 18 E
Valdahon, *France* **27 E13** 47 8N 6 21 E
Valday, *Russia* **58 D7** 57 58N 33 9 E
Valdayskaya Vozvyshennost, *Russia* **58 D7** 57 0N 33 30 E
Valdeazogues →, *Spain* **43 G6** 38 45N 4 55W
Valdecañas, Embalse de, *Spain* **42 F5** 39 45N 5 25W
Valdemarsvik, *Sweden* **17 F10** 58 14N 16 40 E
Valdemoro, *Spain* **42 E7** 40 12N 3 40W
Valdepeñas, *Spain* **43 G7** 38 43N 3 25W
Valderaduey →, *Spain* **42 D5** 41 31N 5 42W
Valdérice, *Italy* **46 D5** 38 4N 12 37 E
Valderrobres, *Spain* **40 E5** 40 53N 0 9 E
Valdés, Pen., *Argentina* **176 B4** 42 30 S 63 45W
Valdez, *Ecuador* **168 C2** 1 15N 79 0W
Valdez, *U.S.A.* **138 B5** 61 7N 146 16W
Valdivia, *Chile* **176 A2** 39 50 S 73 14W
Valdivia, *Colombia* **168 B2** 7 11N 75 27W
Valdivia □, *Chile* **176 B2** 40 0 S 73 0W
Valdobbiádene, *Italy* **45 C8** 45 54N 12 0 E
Valdosta, *U.S.A.* **152 E6** 30 50N 83 17W
Valdoviño, *Spain* **42 B2** 43 36N 8 8W
Valdres, *Norway* **15 F13** 61 5N 9 5 E
Vale, *Georgia* **61 K6** 41 30N 42 58 E
Vale, *U.S.A.* **158 E5** 43 59N 117 15W
Vale of Glamorgan □, *U.K.* **21 F4** 51 28N 3 25W
Valea lui Mihai, *Romania* **52 C7** 47 32N 22 11 E
Valea Mărului, *Romania* **53 E12** 45 49N 27 42 E
Valença, *Brazil* **171 D4** 13 20 S 39 5W
Valença, *Portugal* **42 C2** 42 1N 8 34W
Valença do Piauí, *Brazil* **170 C3** 6 20 S 41 45W
Valençay, *France* **27 E8** 47 9N 1 34 E
Valence = Valence d'Agen, *France* **28 D4** 44 6N 0 53 E
Valence, *France* **29 D8** 44 57N 4 54 E
Valence d'Agen, *France* **28 D4** 44 6N 0 53 E
Valencia, *Phil.* **81 H5** 7 57N 125 3 E
Valencia, *Spain* **41 F4** 39 27N 0 23W
Valencia, *Venezuela* **168 A4** 10 11N 68 0W
Valencia □, *Spain* **41 F4** 39 20N 0 40W
Valencia, G. de, *Spain* **41 F5** 39 30N 0 20 E
Valencia de Alcántara, *Spain* **43 F3** 39 25N 7 14W
Valencia de Don Juan, *Spain* **42 C5** 42 17N 5 31W
Valencia I., *Ireland* **23 E1** 51 54N 10 22W
Valenciennes, *France* **27 B10** 50 20N 3 34 E
Văleni, *Romania* **53 F9** 44 15N 24 45 E
Vălenii de Munte, *Romania* **53 E11** 45 11N 26 2 E
Valensole, *France* **29 E9** 43 50N 5 59 E
Valentigney, *France* **27 E13** 47 27N 6 51 E
Valentim, Sa. do, *Brazil* **170 C3** 6 0 S 43 30W
Valentin, *Russia* **68 C7** 43 8N 134 17 E
Valentine, *Nebr., U.S.A.* **154 D4** 42 52N 100 33W
Valentine, *Tex., U.S.A.* **155 K2** 30 35N 104 30W
Valenza, *Italy* **44 C5** 45 1N 8 38 E
Våler, *Hedmark, Norway* **18 D8** 60 41N 11 50 E
Våler, *Østfold, Norway* **18 E7** 59 29N 10 51 E
Valera, *Venezuela* **168 B3** 9 19N 70 37W
Valestrand, *Norway* **18 E2** 59 40N 5 26 E
Valga, *Estonia* **15 H22** 57 47N 26 2 E
Valguarnera Caropepe, *Italy* **47 E7** 37 30N 14 23 E
Valier, *U.S.A.* **158 B7** 48 18N 112 16W
Valinco, G. de, *France* **29 G12** 41 40N 8 52 E
Valjevo, *Serbia, Yug.* **50 B3** 44 18N 19 53 E
Valka, *Latvia* **15 H21** 57 42N 25 57 E
Valkeakoski, *Finland* **15 F20** 61 16N 24 2 E
Valkenswaard, *Neths.* **24 C5** 51 21N 5 29 E
Vall de Uxó = La Vall d'Uixó, *Spain* **40 F4** 39 49N 0 15W
Valla, *Sweden* **16 E10** 59 2N 16 20 E
Valladolid, *Mexico* **163 C7** 20 40N 88 11W
Valladolid, *Spain* **42 D6** 41 38N 4 43W
Valladolid □, *Spain* **42 D6** 41 38N 4 43W
Vallata, *Italy* **47 A8** 41 2N 15 15 E
Valldemossa, *Spain* **39 B9** 39 43N 2 37 E
Valle, *Norway* **18 E4** 59 13N 7 33 E
Valle d'Aosta □, *Italy* **44 C4** 45 45N 7 15 E
Valle de Arán, *Spain* **40 C5** 42 50N 0 55 E
Valle de la Pascua, *Venezuela* **168 B4** 9 13N 66 0W
Valle de las Palmas, *Mexico* **161 N10** 32 20N 116 43W
Valle de Santiago, *Mexico* **162 C4** 20 25N 101 15W
Valle de Suchil, *Mexico* **162 C4** 23 38N 103 55W
Valle de Zaragoza, *Mexico* **162 B3** 27 28N 105 49W
Valle del Cauca □, *Colombia* **168 C2** 3 45N 76 30W
Valle Fértil, Sierra del, *Argentina* **174 C2** 30 20 S 68 0W
Valle Hermoso, *Mexico* **163 B5** 25 35N 97 40W
Valledupar, *Colombia* **168 A3** 10 29N 73 15W
Vallehermoso, *Canary Is.* **39 F2** 28 10N 17 15W
Vallejo, *U.S.A.* **160 G4** 38 7N 122 14W
Vallenar, *Chile* **174 B1** 28 30 S 70 50W
Vallentuna, *Sweden* **16 E12** 59 32N 18 5 E
Valleraugue, *France* **28 D7** 44 6N 3 39 E
Vallet, *France* **26 E5** 47 10N 1 15W
Valletta, *Malta* **38 D2** 35 54N 14 31 E
Valley Center, *U.S.A.* **161 M9** 33 13N 117 2W
Valley City, *U.S.A.* **154 B6** 46 55N 98 0W
Valley Falls, *U.S.A.* **158 E3** 42 29N 120 17W
Valley Park, *U.S.A.* **156 F6** 38 33N 90 29W
Valley Springs, *U.S.A.* **160 G6** 38 12N 120 50W
Valley Station, *U.S.A.* **157 F11** 38 6N 85 52W
Valley Wells, *U.S.A.* **161 K11** 35 27N 115 46W
Valleyview, *Canada* **142 B5** 55 5N 117 17W
Valli di Comácchio, *Italy* **45 D9** 44 40N 12 15 E
Vallimanca, Arroyo, *Argentina* **174 D4** 35 40 S 59 10W
Vallo della Lucánia, *Italy* **47 B8** 40 14N 15 16 E
Vallon-Pont-d'Arc, *France* **29 D8** 44 24N 4 24 E
Vallorbe, *Switz.* **32 C2** 46 42N 6 20 E
Valls, *Spain* **40 D6** 41 18N 1 15 E
Valmaseda = Balmaseda, *Spain* **40 B1** 43 11N 3 12W
Valmeyer, *U.S.A.* **156 F6** 38 18N 90 19W
Valmiera, *Latvia* **15 H21** 57 37N 25 29 E
Valnera, *Spain* **42 B7** 43 9N 3 40W
Valognes, *France* **26 C5** 49 30N 1 28W
Valona = Vlóra, *Albania* **50 F3** 40 32N 19 28 E
Valongo, *Portugal* **42 D2** 41 8N 8 30W
Valozhyn, *Belarus* **58 E4** 54 3N 26 30 E
Valpaços, *Portugal* **42 D3** 41 36N 7 17W
Valparaíso, *Chile* **174 C1** 33 2 S 71 40W
Valparaíso, *Mexico* **162 C4** 22 50N 103 32W
Valparaiso, *Fla., U.S.A.* **153 E3** 30 29N 86 30W
Valparaiso, *Ind., U.S.A.* **157 C9** 41 28N 87 4W
Valparaíso □, *Chile* **174 C1** 33 2 S 71 40W
Valpovo, *Croatia* **52 E3** 45 39N 18 25 E
Valréas, *France* **29 D9** 44 24N 5 0 E
Vals, *Switz.* **33 C8** 46 39N 9 11 E
Vals →, *S. Africa* **116 D4** 27 23 S 26 30 E
Vals, Tanjung, *Indonesia* **83 C5** 8 26 S 137 25 E
Vals-les-Bains, *France* **29 D8** 44 42N 4 24 E
Valsad, *India* **94 D1** 20 40N 72 58 E
Valtellina, *Italy* **44 B6** 46 11N 9 55 E
Valþjófsstaður, *Iceland* **11 B12** 65 1N 14 59W
Valuyki, *Russia* **59 G10** 50 10N 38 5 E
Valverde, *Canary Is.* **39 G2** 27 48N 17 55W
Valverde del Camino, *Spain* **43 H4** 37 35N 6 47W
Valverde del Fresno, *Spain* **42 E4** 40 15N 6 51W
Vama, *Romania* **53 C10** 47 34N 25 42 E
Vamdrup, *Denmark* **17 J3** 55 25N 9 17 E
Våmhus, *Sweden* **16 C8** 61 7N 14 29 E
Vammala, *Finland* **15 F20** 61 20N 22 54 E
Vámos, *Greece* **38 D6** 35 24N 24 13 E
Vamsadhara →, *India* **94 E7** 18 21N 84 8 E
Van, *Turkey* **101 C10** 38 30N 43 20 E
Van, L. = Van Gölü, *Turkey* **101 C10** 38 30N 43 0 E
Van Alstyne, *U.S.A.* **155 J6** 33 25N 96 35W
Van Blommestein Meer, *Surinam* **169 C6** 4 45N 55 5W
Van Bruyssel, *Canada* **141 C5** 47 56N 72 9W
Van Buren, *Canada* **141 C6** 47 10N 67 55W
Van Buren, *Ark., U.S.A.* **155 H7** 35 26N 94 21W
Van Buren, *Maine, U.S.A.* **149 B11** 47 10N 67 58W
Van Buren, *Mo., U.S.A.* **155 G9** 37 0N 91 1W
Van Canh, *Vietnam* **86 F7** 13 37N 109 0 E
Van Diemen, C., *N. Terr., Australia* **124 B5** 11 9 S 130 24 E
Van Diemen, C., *Queens., Australia* **126 B2** 16 30 S 139 46 E
Van Diemen G., *Australia* **124 B5** 11 45 S 132 0 E
Van Gölü, *Turkey* **101 C10** 38 30N 43 0 E
Van Horn, *U.S.A.* **155 K2** 31 3N 104 50W
Van Horne, *U.S.A.* **156 B4** 42 1N 92 4W
Van Ninh, *Vietnam* **86 F7** 12 42N 109 14 E
Van Rees, Pegunungan, *Indonesia* **83 B5** 2 35 S 138 15 E
Van Tassell, *U.S.A.* **154 D2** 42 40N 104 5W
Van Tivu, *India* **95 K4** 8 51N 78 15 E
Van Wert, *U.S.A.* **157 D12** 40 52N 84 35W
Van Yen, *Vietnam* **86 B5** 21 4N 104 42 E
Vanadzor, *Armenia* **61 K7** 40 48N 44 30 E
Vanavara, *Russia* **65 C11** 60 22N 102 16 E
Vance, *U.S.A.* **152 B9** 33 26N 80 25W
Vanceburg, *U.S.A.* **157 F13** 38 36N 83 19W
Vancouver, *Canada* **142 D4** 49 15N 123 10W
Vancouver, *U.S.A.* **160 E4** 45 38N 122 40W
Vancouver, C., *Australia* **125 G2** 35 2 S 118 11 E
Vancouver, Mt., *U.S.A.* **144 F13** 60 20N 139 41W
Vancouver I., *Canada* **142 D3** 49 50N 126 0W
Vandalia, *Ill., U.S.A.* **156 F7** 38 58N 89 6W
Vandalia, *Mo., U.S.A.* **156 E5** 39 19N 91 29W
Vandalia, *Ohio, U.S.A.* **157 E12** 39 54N 84 12W
Vandavasi, *India* **95 H4** 12 30N 79 30 E
Vandeloos B., *Sri Lanka* **95 L5** 8 0N 81 45 E
Vandenburg, *U.S.A.* **161 L6** 34 35N 120 33W
Vanderbijlpark, *S. Africa* **117 D4** 26 42 S 27 54 E
Vandergrift, *U.S.A.* **150 F5** 40 36N 79 34W
Vanderhoof, *Canada* **142 C4** 54 0N 124 0W
Vanderkloof Dam, *S. Africa* **116 E3** 30 4 S 24 40 E
Vanderlin I., *Australia* **126 B2** 15 44 S 137 2 E
Vandyke, *Australia* **126 C4** 24 10 S 147 51 E
Vänern, *Sweden* **17 F7** 58 47N 13 30 E
Vänersborg, *Sweden* **17 F6** 58 26N 12 19 E
Vang, *Norway* **18 C5** 61 7N 8 34 E
Vang Vieng, *Laos* **86 C4** 18 58N 102 32 E
Vanga, *Kenya* **118 C4** 4 35 S 39 12 E
Vangaindrano, *Madag.* **117 C8** 23 21 S 47 36 E
Vangsnes, *Norway* **18 C3** 61 9N 6 39 E
Vanguard, *Canada* **143 D7** 49 55N 107 20W
Vangunu, *Solomon Is.* **133 M10** 8 40 S 158 5 E
Vanier, *Canada* **140 C4** 45 27N 75 40W
Vanimo, *Papua N. G.* **132 B1** 2 42 S 141 21 E
Vanino, *Russia* **65 E15** 48 50N 140 5 E
Vanivilasa Sagara, *India* **95 H3** 13 45N 76 30 E
Vaniyambadi, *India* **95 H4** 12 46N 78 44 E
Vânju Mare, *Romania* **52 F7** 44 25N 22 52 E
Vankleek Hill, *Canada* **140 C5** 45 32N 74 40W
Vanna, *Norway* **14 A18** 70 6N 19 50 E
Vännäs, *Sweden* **14 E18** 63 58N 19 48 E
Vannes, *France* **26 E4** 47 40N 2 47W
Vanoise, *France* **29 C10** 45 25N 6 40 E
Vanrhynsdorp, *S. Africa* **116 E2** 31 36 S 18 44 E
Vanrook, *Australia* **126 B3** 16 57 S 141 57 E
Vansbro, *Sweden* **16 D8** 60 32N 14 15 E
Vanse, *Norway* **18 F3** 58 5N 6 41 E
Vansittart B., *Australia* **124 B4** 14 3 S 126 17 E
Vantaa, *Finland* **15 F21** 60 18N 24 58 E
Vanthli, *India* **92 J4** 21 28N 70 25 E
Vanua Levu, *Fiji* **133 A2** 16 33 S 179 15 E
Vanuatu ■, *Pac. Oc.* **133 E6** 15 0 S 168 0 E
Vanwyksvlei, *S. Africa* **116 E3** 30 18 S 21 49 E
Vanzylsrus, *S. Africa* **116 D3** 26 52 S 22 4 E
Vapnyarka, *Ukraine* **59 H5** 48 32N 28 45 E
Var □, *France* **29 E10** 43 27N 6 18 E
Var →, *France* **29 E11** 43 39N 7 12 E
Vara, *Sweden* **17 F6** 58 16N 12 55 E
Varada →, *India* **95 G2** 15 0N 75 40 E
Varades, *France* **26 E5** 47 25N 1 1W
Varáita →, *Italy* **44 D4** 44 9N 7 53 E
Varaldsøy, *Norway* **18 D2** 60 6N 5 59 E
Varallo, *Italy* **44 C5** 45 49N 8 15 E
Varanasi, *India* **93 G10** 25 22N 83 0 E
Varanger-halvøya, *Norway* **14 A23** 70 25N 29 30 E
Varangerfjorden, *Norway* **14 A23** 70 3N 29 25 E
Varano, Lago di, *Italy* **45 G12** 41 53N 15 45 E
Varaždin, *Croatia* **45 B13** 46 20N 16 20 E
Varazze, *Italy* **44 D5** 44 22N 8 34 E
Varberg, *Sweden* **17 G6** 57 6N 12 20 E
Vardar = Axiós →, *Greece* **50 F6** 40 57N 22 35 E
Varde, *Denmark* **17 J2** 55 38N 8 29 E
Varde Å →, *Denmark* **17 J2** 55 35N 8 19 E
Vardø, *Norway* **14 A24** 70 23N 31 5 E
Varel, *Germany* **30 B4** 53 23N 8 8 E
Varella, Mui, *Vietnam* **86 F7** 12 54N 109 26 E
Varėna, *Lithuania* **15 J21** 54 12N 24 30 E
Varennes-sur-Allier, *France* **27 F10** 46 19N 3 24 E
Varennes-Vauzelles, *France* **27 E10** 47 3N 3 9 E
Vareš, *Bos.-H.* **52 F3** 44 12N 18 23 E
Varese, *Italy* **44 C5** 45 48N 8 50 E
Vârfurile, *Romania* **52 D7** 46 19N 22 31 E
Vårgårda, *Sweden* **17 F6** 58 2N 12 49 E
Vargem Bonita, *Brazil* **171 F2** 20 20 S 46 22W
Vargem Grande, *Brazil* **170 B3** 3 33 S 43 56W
Varginha, *Brazil* **175 A6** 21 33 S 45 25W
Vargön, *Sweden* **17 F6** 58 22N 12 20 E
Varhaug, *Norway* **18 F2** 58 37N 5 41 E
Variadero, *U.S.A.* **155 H2** 35 43N 104 17W
Varillas, *Chile* **174 A1** 24 0 S 70 10W
Varkaus, *Finland* **15 E22** 62 19N 27 50 E
Varmahlíð, *Iceland* **11 B7** 65 33N 19 28W
Värmdölandet, *Sweden* **16 E12** 59 20N 18 33 E
Värmeln, *Sweden* **16 E6** 59 35N 12 54 E
Värmlands Bro, *Sweden* **16 E7** 59 11N 13 0 E
Värmlands län □, *Sweden* **16 E6** 60 0N 13 20 E
Varna, *Bulgaria* **51 C11** 43 13N 27 56 E
Varna, *Russia* **62 E8** 53 24N 60 58 E
Varna, *U.S.A.* **156 C7** 41 2N 89 14W
Varna □, *Bulgaria* **51 C11** 43 20N 27 30 E
Varna →, *India* **94 F2** 16 48N 74 32 E
Värnamo, *Sweden* **17 G8** 57 10N 14 3 E
Varnsdorf, *Czech Rep.* **34 A7** 50 55N 14 35 E
Varnville, *U.S.A.* **152 C8** 32 51N 81 5W
Várpalota, *Hungary* **52 C3** 47 12N 18 8 E
Vars, *Canada* **151 A9** 45 21N 75 21W
Vars, *France* **29 D10** 44 37N 6 42 E
Vartdal, *Norway* **18 B3** 62 20N 6 8 E
Varto, *Turkey* **101 C9** 39 10N 41 27 E
Varvarin, *Serbia, Yug.* **50 C5** 43 43N 21 20 E
Varzaneh, *Iran* **97 C7** 32 25N 52 40 E
Várzea Alegre, *Brazil* **170 C4** 6 47 S 39 17W
Várzea da Palma, *Brazil* **171 E3** 17 36 S 44 44W
Várzea Grande, *Brazil* **173 D6** 15 39 S 56 8W
Varzi, *Italy* **44 D6** 44 49N 9 12 E
Varzo, *Italy* **44 B5** 46 12N 8 15 E
Varzy, *France* **27 E10** 47 22N 3 20 E
Vas □, *Hungary* **52 C1** 47 10N 16 55 E
Vasa Barris →, *Brazil* **170 D4** 11 10 S 37 10W
Vásárosnamény, *Hungary* **52 B7** 48 9N 22 19 E
Vascão →, *Portugal* **43 H3** 37 31N 7 31W
Vaşcău, *Romania* **52 D7** 46 28N 22 30 E
Vascongadas = País Vasco □, *Spain* **40 C2** 42 50N 2 45W
Vāshīr, *Afghan.* **91 B1** 32 16N 63 51 E
Vasht = Khāsh, *Iran* **97 D9** 28 15N 61 15 E
Vasilevichi, *Belarus* **59 F5** 52 15N 29 50 E
Vasilikón, *Greece* **48 C5** 38 25N 23 40 E
Vasilkov = Vasylkiv, *Ukraine* **59 G6** 50 7N 30 15 E
Vaslui, *Romania* **53 D12** 46 38N 27 42 E
Vaslui □, *Romania* **53 D12** 46 30N 27 45 E
Väsman, *Sweden* **16 D9** 60 9N 15 5 E
Vassar, *Canada* **143 D9** 49 10N 95 55W
Vassar, *U.S.A.* **148 D4** 43 22N 83 35W
Vassfaret, *Norway* **18 D6** 60 38N 9 28 E
Västerås, *Sweden* **16 E10** 59 37N 16 38 E
Västerbotten, *Sweden* **14 D18** 64 36N 20 4 E
Västerdalälven →, *Sweden* **16 D8** 60 30N 14 7 E
Västergötland, *Sweden* **17 F7** 58 0N 13 10 E
Västerhaninge, *Sweden* **16 E12** 59 7N 18 6 E
Västervik, *Sweden* **17 G10** 57 43N 16 33 E
Västmanland, *Sweden* **15 G16** 59 45N 16 20 E
Västmanlands län □, *Sweden* **16 E10** 59 45N 16 20 E
Vasto, *Italy* **45 F11** 42 8N 14 40 E
Vasvár, *Hungary* **52 C1** 47 3N 16 47 E
Vasylkiv, *Ukraine* **59 G6** 50 7N 30 15 E
Vatan, *France* **27 E8** 47 4N 1 50 E
Vaté = Efate, I., *Vanuatu* **133 G6** 17 40 S 168 25 E
Vatersay, *U.K.* **22 E1** 56 55N 7 32W
Váthia, *Greece* **48 E4** 36 29N 22 29 E
Vatican City ■, *Europe* **45 G9** 41 54N 12 27 E
Vaticano, C., *Italy* **47 D8** 38 37N 15 50 E
Vatili, *Cyprus* **38 D12** 35 6N 33 40 E
Vatin, *Serbia, Yug.* **52 E6** 45 12N 21 20 E
Vatnajökull, *Iceland* **11 C10** 64 30N 16 48W
Vatnås, *Norway* **18 E6** 59 58N 9 37 E
Vatneyri, *Iceland* **11 B2** 65 35N 24 0W
Vatólakkos, *Greece* **38 D5** 35 27N 23 53 E
Vatoloha, *Madag.* **117 B8** 17 52 S 47 48 E
Vatomandry, *Madag.* **117 B8** 19 20 S 48 59 E
Vatra-Dornei, *Romania* **53 C10** 47 22N 25 22 E
Vättern, *Sweden* **17 F8** 58 25N 14 30 E
Vättis, *Switz.* **33 C8** 46 55N 9 27 E
Vatulele, *Fiji* **133 B1** 18 33 S 177 37 E
Vaucluse, *U.S.A.* **152 B9** 33 37N 81 49W
Vaucluse □, *France* **29 E9** 43 50N 5 20 E
Vaucouleurs, *France* **27 D12** 48 37N 5 40 E
Vaud □, *Switz.* **32 C3** 46 35N 6 30 E
Vaughn, *Mont., U.S.A.* **158 C8** 47 33N 111 33W
Vaughn, *N. Mex., U.S.A.* **159 J11** 34 36N 105 13W
Vaulruz, *Switz.* **32 C3** 46 38N 6 58 E
Vaupés = Uaupés →, *Brazil* **168 C4** 0 2N 67 16W
Vaupes □, *Colombia* **168 C3** 1 0N 71 0W
Vauvert, *France* **29 E8** 43 42N 4 17 E
Vauxhall, *Canada* **142 C6** 50 5N 112 9W
Vava'u, *Tonga* **133 P14** 18 36 S 174 0W
Vavoua, *Ivory C.* **112 D3** 7 23N 6 29W
Vawkavysk, *Belarus* **59 F3** 53 9N 24 30 E
Vaxholm, *Sweden* **16 E12** 59 25N 18 20 E
Växjö, *Sweden* **17 H8** 56 52N 14 50 E
Våxtorp, *Sweden* **17 H7** 56 25N 13 8 E
Vaygach, Ostrov, *Russia* **64 C7** 70 0N 60 0 E

296 Váyia

Váyia, *Greece* 48 C5 38 19N 23 11 E
Váyia, Ákra, *Greece* 38 C10 36 15N 28 11 E
Veadeiros, *Brazil* 171 D2 14 7 S 47 31W
Vechelde, *Germany* 30 C6 52 16N 10 22 E
Vechta, *Germany* 30 C4 52 44N 8 17 E
Vechte →, *Neths.* 24 B6 52 34N 6 6 E
Vecsés, *Hungary* 52 C4 47 26N 19 19 E
Vedaranniyam, *India* 95 J4 10 25N 79 50 E
Vedavågen, *Norway* 18 E2 59 17N 5 12 E
Veddige, *Sweden* 17 G6 57 17N 12 20 E
Vedea →, *Romania* 53 G10 43 42N 25 41 E
Vedia, *Argentina* 174 C3 34 30 S 61 31W
Vedum, *Sweden* 17 F7 58 11N 13 0 E
Veendam, *Neths.* 24 A6 53 5N 6 52 E
Veenendaal, *Neths.* 24 B5 52 2N 5 34 E
Vefsna →, *Norway* 14 D15 65 48N 13 10 E
Vega, *Norway* 14 D14 65 40N 11 55 E
Vega, *U.S.A.* 155 H3 35 15N 102 26W
Vegadeo, *Spain* 42 B3 43 27N 7 4W
Vegårshei, *Norway* 18 F5 58 44N 8 51 E
Veggli, *Norway* 18 D6 60 3N 9 9 E
Vegorrítis, Límni, *Greece* 50 F5 40 45N 21 45 E
Vegreville, *Canada* 142 C6 53 30N 112 5W
Veidholmen, *Norway* 18 A4 63 31N 7 58 E
Veinge, *Sweden* 17 H7 56 33N 13 4 E
Veisiejai, *Lithuania* 54 D10 54 6N 23 42 E
Veitch, *Australia* 128 C4 34 39 S 140 31 E
Vejbystrand, *Sweden* 17 H6 56 19N 12 45 E
Vejen, *Denmark* 17 J3 55 30N 9 9 E
Vejer de la Frontera, *Spain* 43 J5 36 15N 5 59W
Vejle, *Denmark* 17 J3 55 43N 9 30 E
Vejle Amtskommune □, *Denmark* 17 J3 55 45N 9 20 E
Vejle Fjord, *Denmark* 17 J3 55 40N 9 50 E
Vela Luka, *Croatia* 45 F13 42 59N 16 44 E
Velanai I., *Sri Lanka* 95 K4 9 45N 79 45 E
Velas, C., *Costa Rica* 164 D2 10 21N 85 52W
Velasco, Sierra de, *Argentina* 174 B2 29 20 S 67 10W
Velay, Mts. du, *France* 28 D7 45 0N 3 40 E
Velbert, *Germany* 30 D3 51 20N 7 3 E
Velddrif, *S. Africa* 116 E2 32 42 S 18 11 E
Velebit Planina, *Croatia* 45 D12 44 50N 15 20 E
Velebitski Kanal, *Croatia* 45 D11 44 45N 14 55 E
Veleka →, *Bulgaria* 51 D11 42 4N 27 58 E
Velencei-tó, *Hungary* 52 C3 47 13N 18 36 E
Velenje, *Slovenia* 45 B12 46 23N 15 8 E
Veles, *Macedonia* 50 E5 41 46N 21 47 E
Velestínon, *Greece* 48 B4 39 23N 22 43 E
Vélez, *Colombia* 168 B3 6 1N 73 41W
Vélez-Málaga, *Spain* 43 J6 36 48N 4 5W
Vélez Rubio, *Spain* 41 H2 37 41N 2 5W
Velhas →, *Brazil* 171 E3 17 13 S 44 49W
Velika, *Croatia* 52 E2 45 27N 17 40 E
Velika Gorica, *Croatia* 45 C13 45 44N 16 5 E
Velika Kapela, *Croatia* 45 C12 45 10N 15 5 E
Velika Kladuša, *Bos.-H.* 45 C12 45 11N 15 48 E
Velika Morava →, *Serbia, Yug.* 50 B5 44 43N 21 3 E
Velika Plana, *Serbia, Yug.* 50 B5 44 20N 21 4 E
Velikaya →, *Russia* 58 D5 57 48N 28 10 E
Velikaya Kema, *Russia* 68 B8 45 30N 137 12 E
Velikaya Lepetikha, *Ukraine* 59 J7 47 2N 33 58 E
Veliké Kapušany, *Slovak Rep.* 35 C15 48 34N 22 5 E
Velike Lašče, *Slovenia* 45 C11 45 49N 14 45 E
Veliki Jastrebac, *Serbia, Yug.* 50 C5 43 25N 21 30 E
Veliki Kanal, *Serbia, Yug.* 52 E4 45 45N 19 15 E
Veliki Popović, *Serbia, Yug.* 50 B5 44 8N 21 18 E
Veliki Ustyug, *Russia* 56 B8 60 47N 46 20 E
Velikiye Luki, *Russia* 58 D6 56 25N 30 32 E
Veliko Gradište, *Serbia, Yug.* 50 B5 44 46N 21 29 E
Veliko Tŭrnovo, *Bulgaria* 51 C9 43 5N 25 41 E
Velikonda Range, *India* 95 G4 14 45N 79 10 E
Velingrad, *Bulgaria* 50 D7 42 4N 23 58 E
Velino, Mte., *Italy* 45 F10 42 9N 13 23 E
Velizh, *Russia* 58 E6 55 36N 31 11 E
Velké Karlovice, *Czech Rep.* 35 B11 49 20N 18 17 E
Velke Meziříčí, *Czech Rep.* 34 B9 49 21N 16 1 E
Vel'ký Javorník, *Slovak Rep.* 35 B11 49 19N 18 22 E
Vel'ký Krtíš, *Slovak Rep.* 35 C12 48 12N 19 21 E
Vel'ký Meder, *Slovak Rep.* 35 D10 47 52N 17 46 E
Vel'ký Tribeč, *Slovak Rep.* 35 C11 48 28N 18 15 E
Vella G., *Solomon Is.* 133 M9 8 0 S 156 50 E
Vella Lavella, *Solomon Is.* 133 L9 7 45 S 156 40 E
Vellar →, *India* 95 J4 11 30N 79 36 E
Velletri, *Italy* 46 A5 41 41N 12 47 E
Vellinge, *Sweden* 17 J6 55 29N 13 0 E
Vellmar, *Germany* 30 D5 51 22N 9 28 E
Vellore, *India* 95 H4 12 57N 79 10 E
Velsk, *Russia* 58 B11 61 10N 42 5 E
Velten, *Germany* 30 C9 52 42N 13 10 E
Velva, *U.S.A.* 154 A4 48 4N 100 56W
Velvendós, *Greece* 50 F6 40 15N 22 6 E
Vemb, *Denmark* 17 H2 56 21N 8 21 E
Vembanad L., *India* 95 K3 9 36N 76 15 E
Vemdalen, *Sweden* 16 B7 62 27N 13 51 E
Ven, *Sweden* 17 J6 55 55N 12 45 E
Venaco, *France* 29 F13 42 14N 9 11 E
Venado Tuerto, *Argentina* 174 C3 33 50 S 62 0W
Venafro, *Italy* 47 A7 41 29N 14 2 E
Venarey-les-Laumes, *France* 27 E11 47 32N 4 26 E
Venaría, *Italy* 44 C4 45 8N 7 38 E
Venčane, *Serbia, Yug.* 50 B4 44 24N 20 28 E
Vence, *France* 29 E11 43 43N 7 6 E
Vendas Novas, *Portugal* 43 G2 38 39N 8 27W
Vendée □, *France* 26 F5 46 50N 1 35W
Vendée →, *France* 26 F5 46 20N 1 10W
Vendéen, Bocage, *France* 28 B2 46 40N 1 20W
Vendeuvre-sur-Barse, *France* 27 D11 48 14N 4 28 E
Vendôme, *France* 26 E8 47 47N 1 3 E
Vendrell = El Vendrell, *Spain* 40 D6 41 10N 1 30 E
Vendsyssel, *Denmark* 17 G4 57 22N 10 0 E
Venelles, *France* 29 E9 43 35N 5 28 E
Véneta, L., *Italy* 45 C9 45 23N 12 25 E
Venetie Indian Reservation, *U.S.A.* 144 C11 67 20N 146 0W
Véneto □, *Italy* 45 C9 45 30N 12 0 E
Venev, *Russia* 58 E10 54 22N 38 17 E
Venézia, *Italy* 45 C9 45 27N 12 21 E
Venézia, G. di, *Italy* 45 C10 45 15N 13 0 E
Venezuela ■, *S. Amer.* 168 B4 8 0N 66 0W
Venezuela, G. de, *Venezuela* 168 A3 11 30N 71 0W
Vengurla, *India* 95 G1 15 53N 73 45 E
Vengurla Rocks, *India* 95 G1 15 55N 73 22 E
Venice = Venézia, *Italy* 45 C9 45 27N 12 21 E
Venice, *U.S.A.* 153 H7 27 6N 82 27W
Vénissieux, *France* 29 C8 45 42N 4 53 E
Venjansjön, *Sweden* 16 D8 60 54N 14 0 E
Venkatagiri, *India* 95 H4 14 0N 79 35 E
Venkatapuram, *India* 94 E5 18 20N 80 30 E
Venlo, *Neths.* 24 C6 51 22N 6 11 E
Vennesla, *Norway* 15 G12 58 15N 7 59 E
Venosa, *Italy* 47 B8 40 58N 15 49 E
Venray, *Neths.* 24 C6 51 31N 6 0 E
Venta, *Lithuania* 54 B9 56 12N 22 42 E
Venta →, *Latvia* 54 A8 57 24N 21 33 E
Venta de Baños, *Spain* 42 D6 41 55N 4 30W
Venta de Cardeña = Cardeña, *Spain* 43 G6 38 16N 4 20W
Ventana, Punta de la, *Mexico* 162 C3 24 4N 109 48W
Ventana, Sa. de la, *Argentina* 174 D3 38 0 S 62 30W
Ventersburg, *S. Africa* 116 D4 28 7 S 27 9 E
Venterstad, *S. Africa* 116 E4 30 47 S 25 48 E
Ventimíglia, *Italy* 44 E4 43 47N 7 36 E
Ventnor, *U.K.* 21 G6 50 36N 1 12W
Ventoténe, *Italy* 46 B6 40 47N 13 25 E
Ventoux, Mt., *France* 29 D9 44 10N 5 17 E
Ventspils, *Latvia* 15 H19 57 25N 21 32 E
Ventspils □, *Latvia* 54 A8 57 20N 21 50 E
Ventuarí →, *Venezuela* 168 C4 3 58N 67 2W
Ventucopa, *U.S.A.* 161 L7 34 50N 119 29W
Ventura, *U.S.A.* 161 L7 34 17N 119 18W
Venus, *U.S.A.* 153 H8 27 4N 81 22W
Venus B., *Australia* 129 E6 38 40 S 145 42 E
Vera, *Argentina* 174 B3 29 30 S 60 20W
Vera, *Spain* 41 H3 37 15N 1 51W
Veracruz, *Mexico* 163 D5 19 10N 96 10W
Veracruz □, *Mexico* 163 D5 19 0N 96 15W
Veraval, *India* 92 J4 20 53N 70 27 E
Verbánia, *Italy* 44 C5 45 56N 8 33 E
Verbicaro, *Italy* 47 C8 39 45N 15 55 E
Verbier, *Switz.* 32 D4 46 6N 7 13 E
Vercelli, *Italy* 44 C5 45 19N 8 25 E
Verchovchevo, *Ukraine* 59 H8 48 32N 34 10 E
Verdalsøra, *Norway* 14 E14 63 48N 11 30 E
Verde →, *Argentina* 176 B3 41 56 S 65 5W
Verde →, *Goiás, Brazil* 171 E1 19 11 S 50 44W
Verde →, *Goiás, Brazil* 171 E1 18 1 S 50 14W
Verde →, *Mato Grosso, Brazil* 173 C6 11 54 S 55 48W
Verde →, *Mato Grosso do Sul, Brazil* 173 E7 21 25 S 52 20W
Verde →, *Chihuahua, Mexico* 162 B3 26 29N 107 58W
Verde →, *Oaxaca, Mexico* 163 D5 15 59N 97 50W
Verde →, *Veracruz, Mexico* 162 C4 21 10N 102 50W
Verde →, *Paraguay* 174 A4 23 9 S 57 37W
Verde, Cay, *Bahamas* 164 B4 23 0N 75 5W
Verde Grande →, *Brazil* 171 E3 16 13 S 43 49W
Verde I., *Phil.* 80 E3 13 33N 121 5 E
Verde Island Pass, *Phil.* 80 E3 13 34N 120 51 E
Verde Pequeno →, *Brazil* 171 D3 14 48 S 43 31W
Verden, *Germany* 30 C5 52 55N 9 14 E
Verdery, *U.S.A.* 152 A7 34 7N 82 15W
Verdhikoúsa, *Greece* 48 B3 39 47N 21 59 E
Verdi, *U.S.A.* 160 F7 39 31N 119 59W
Verdigre, *U.S.A.* 154 D6 42 36N 98 2W
Verdon →, *France* 29 E9 43 43N 5 46 E
Verdun, *France* 27 C12 49 9N 5 24 E
Verdun-sur-le-Doubs, *France* 27 F12 46 54N 5 2 E
Vereeniging, *S. Africa* 117 D4 26 38 S 27 57 E
Vérendrye, Parc Prov. de la, *Canada* 140 C4 47 20N 76 40W
Vereshchagino, *Russia* 62 B5 58 5N 54 40 E
Verga, C., *Guinea* 112 C2 10 30N 14 10W
Vergato, *Italy* 44 D8 44 17N 11 7 E
Vergemont, *Australia* 126 C3 23 33 S 143 1 E
Vergemont Cr. →, *Australia* 126 C3 24 16 S 143 16 E
Vergennes, *U.S.A.* 151 B11 44 10N 73 15W
Vergt, *France* 28 C4 45 2N 0 43 E
Verín, *Spain* 42 D3 41 57N 7 27W
Verkhnedvinsk = Vyerkhnyadzvinsk, *Belarus* 58 E4 55 45N 27 58 E
Verkhneuralsk, *Russia* 62 E7 53 53N 59 13 E
Verkhnevilyuysk, *Russia* 65 C13 63 27N 120 18 E
Verkhniy Avzyan, *Russia* 62 E6 53 32N 57 33 E
Verkhniy Baskunchak, *Russia* 61 F8 48 14N 46 44 E
Verkhniy Tagil, *Russia* 62 C7 57 22N 59 56 E
Verkhniy Ufaley, *Russia* 62 C8 56 4N 60 14 E
Verkhniye Kigi, *Russia* 62 D7 55 25N 58 37 E
Verkhnyaya Salda, *Russia* 62 C8 58 2N 60 33 E
Verkhnyaya Tura, *Russia* 62 B7 58 22N 59 50 E
Verkhoturye, *Russia* 62 B8 58 52N 60 48 E
Verkhovye, *Russia* 59 F9 52 55N 37 15 E
Verkhoyansk, *Russia* 65 C14 67 35N 133 25 E
Verkhoyansk Ra. = Verkhoyanskiy Khrebet, *Russia* 65 C13 66 0N 129 0 E
Verkhoyanskiy Khrebet, *Russia* 65 C13 66 0N 129 0 E
Verlo, *Canada* 143 C7 50 19N 108 35W
Verma, *Norway* 18 B5 62 21N 8 3 E
Vermenton, *France* 27 E10 47 40N 3 42 E
Vermilion, *Canada* 143 C6 53 20N 110 50W
Vermilion →, *Alta., Canada* 143 C6 53 22N 110 51W
Vermilion →, *Qué., Canada* 140 C5 47 38N 72 56W
Vermilion →, *Ill., U.S.A.* 156 C7 41 19N 89 4W
Vermilion →, *Ind., U.S.A.* 157 E9 39 57N 87 27W
Vermilion, B., *U.S.A.* 155 L9 29 45N 91 55W
Vermilion Bay, *Canada* 143 D10 49 51N 93 34W
Vermilion Chutes, *Canada* 142 B6 58 22N 114 51W
Vermilion L., *U.S.A.* 154 B8 47 53N 92 26W
Vermillion, *U.S.A.* 154 D6 42 47N 96 56W
Vermont, *U.S.A.* 156 D6 40 18N 90 26W
Vermont □, *U.S.A.* 151 C12 44 0N 73 0W
Vermosh, *Albania* 50 D3 42 35N 19 42 E
Verna, *U.S.A.* 153 H7 27 23N 82 16W
Vernal, *U.S.A.* 158 F9 40 27N 109 32W
Vernalis, *U.S.A.* 160 H5 37 36N 121 17W
Vernayaz, *Switz.* 32 D4 46 8N 7 3 E
Vernazza, *Italy* 44 D6 44 10N 9 45 E
Verner, *Canada* 140 C3 46 25N 80 8W
Verneuil-sur-Avre, *France* 26 D7 48 45N 0 55 E
Verneukpan, *S. Africa* 116 E3 30 0 S 21 0 E
Vernier, *Switz.* 32 D2 46 13N 6 5 E
Vérnio, *Italy* 44 D8 44 3N 11 9 E
Vernon, *Canada* 142 C5 50 20N 119 15W
Vernon, *France* 26 C8 49 5N 1 30 E
Vernon, *Ill., U.S.A.* 156 F7 38 48N 89 5W
Vernon, *Ind., U.S.A.* 157 F11 38 59N 85 36W
Vernon, *Tex., U.S.A.* 155 H5 34 9N 99 17W
Vernonia, *U.S.A.* 160 E3 45 52N 123 11W
Vernouillet, *France* 26 D8 48 43N 1 22 E
Vero Beach, *U.S.A.* 149 M5 27 38N 80 24W
Véroia, *Greece* 50 F6 40 34N 22 12 E
Véroli, *Italy* 45 G10 41 41N 13 25 E
Verona, *Italy* 44 C7 45 27N 10 59 E
Verona, *U.S.A.* 156 B7 42 59N 89 32W
Verrès, *Italy* 44 C4 45 40N 7 42 E
Versailles, *France* 27 D9 48 48N 2 8 E
Versailles, *Ill., U.S.A.* 156 E6 39 53N 90 39W
Versailles, *Ind., U.S.A.* 157 E11 39 4N 85 15W
Versailles, *Ky., U.S.A.* 157 F12 38 3N 84 44W
Versailles, *Mo., U.S.A.* 156 F4 38 26N 92 51W
Versailles, *Ohio, U.S.A.* 157 D12 40 13N 84 29W
Versalles, *Bolivia* 173 C5 12 44 S 63 18W
Versmold, *Germany* 30 C4 52 2N 8 9 E
Versoix, *Switz.* 32 D2 46 17N 6 10 E
Vert, C., *Senegal* 112 C1 14 45N 17 30W
Vertou, *France* 26 E5 47 10N 1 28W
Vertus, *France* 27 D11 48 54N 4 2 E
Verulam, *S. Africa* 117 D5 29 38 S 31 2 E
Verviers, *Belgium* 24 D5 50 37N 5 52 E
Vervins, *France* 27 C10 49 50N 3 53 E
Veržej, *Slovenia* 45 B13 46 34N 16 13 E
Verzy, *France* 27 C11 49 9N 4 10 E
Vescovato, *France* 29 F13 42 30N 9 27 E
Veselí nad Lužnicí, *Czech Rep.* 34 B7 49 12N 14 43 E
Veselie, *Bulgaria* 51 D11 42 18N 27 38 E
Veselovskoye Vdkhr., *Russia* 61 G5 46 58N 41 25 E
Veshenskaya, *Russia* 60 F5 49 35N 41 44 E
Vesle →, *France* 27 C10 49 23N 3 28 E
Veslyana →, *Russia* 62 A4 60 20N 54 0 E
Vesoul, *France* 27 E13 47 40N 6 11 E
Vessigebro, *Sweden* 17 H6 56 58N 12 40 E
Vest-Agder □, *Norway* 18 F4 58 30N 7 15 E
Vesta, *U.S.A.* 152 B7 33 58N 82 56W
Vestbygd, *Norway* 18 F3 58 6N 6 34 E
Vestdalseyri, *Iceland* 11 B13 65 17N 13 59W
Vesterålen, *Norway* 14 B16 68 45N 15 0 E
Vestfjorden, *Norway* 14 C15 67 55N 14 0 E
Vestfold □, *Norway* 18 E7 59 15N 10 0 E
Vestfossen, *Norway* 18 E6 59 44N 9 52 E
Vestgrønland □, *Greenland* 10 C6 70 0N 47 0W
Vestmannaeyjar, *Iceland* 11 D6 63 27N 20 15W
Vestnes, *Norway* 18 B4 62 39N 7 5 E
Vestre Gausdal, *Norway* 18 C7 61 12N 10 8 E
Vestre Slidre, *Norway* 18 C5 61 5N 8 58 E
Vestsjællands Amtskommune □, *Denmark* 17 J5 55 30N 11 20 E
Vestspitsbergen, *Svalbard* 6 B8 78 40N 17 0 E
Vestur-skaftafellssýsla □, *Iceland* 11 D8 63 50N 18 30W
Vestvågøy, *Norway* 14 B15 68 18N 13 50 E
Vesuvio, *Italy* 47 B7 40 49N 14 26 E
Vesuvius, Mt. = Vesuvio, *Italy* 47 B7 40 49N 14 26 E
Vesyegonsk, *Russia* 58 C9 58 40N 37 16 E
Veszprém, *Hungary* 52 C2 47 8N 17 57 E
Veszprém □, *Hungary* 52 C2 47 5N 17 55 E
Vésztő, *Hungary* 52 D6 46 55N 21 16 E
Vetapalem, *India* 95 G5 15 47N 80 18 E
Vetlanda, *Sweden* 17 G9 57 24N 15 3 E
Vetluga, *Russia* 60 B7 57 53N 45 45 E
Vetlugu →, *Russia* 60 B8 56 36N 46 4 E
Vetluzhskiy, *Kostroma, Russia* 60 A7 58 23N 45 26 E
Vetluzhskiy, *Nizhniy Novgorod, Russia* 60 B7 57 17N 45 12 E
Vetovo, *Bulgaria* 51 C10 43 42N 26 16 E
Vetralla, *Italy* 45 F9 42 20N 12 2 E
Vetren, *Bulgaria* 51 D8 42 15N 24 3 E
Vettore, Mte., *Italy* 45 F10 42 49N 13 16 E
Veurne, *Belgium* 24 C2 51 5N 2 40 E
Vevay, *U.S.A.* 157 F11 38 45N 85 4W
Vevey, *Switz.* 32 D3 46 28N 6 51 E
Vévi, *Greece* 50 F5 40 47N 21 38 E
Veynes, *France* 29 D9 44 32N 5 49 E
Veys, *Iran* 97 D6 31 30N 49 0 E
Vézelay, *France* 27 E10 47 27N 3 45 E
Vézelise, *France* 27 D13 48 30N 6 5 E
Vézère →, *France* 28 D4 44 53N 0 53 E
Vezhen, *Bulgaria* 51 D8 42 50N 24 20 E
Vezirköprü, *Turkey* 100 B6 41 8N 35 27 E
Vezzani, *France* 29 F13 42 10N 9 15 E
Vi Thanh, *Vietnam* 87 H5 9 42N 105 26 E
Viacha, *Bolivia* 172 D4 16 39 S 68 18W
Viadana, *Italy* 44 D7 44 56N 10 31 E
Viamão, *Brazil* 175 C5 30 5 S 51 0W
Viana, *Brazil* 170 B3 3 13 S 44 55W
Viana, *Spain* 40 C2 42 31N 2 22W
Viana do Alentejo, *Portugal* 43 G3 38 17N 7 59W
Viana do Bolo, *Spain* 42 C3 42 11N 7 6W
Viana do Castelo, *Portugal* 42 D2 41 42N 8 50W
Viana do Castelo □, *Portugal* 42 D2 41 50N 8 30W
Vianden, *Lux.* 24 E6 49 56N 6 12 E
Vianópolis, *Brazil* 171 E2 16 40 S 48 35W
Viar →, *Spain* 43 H5 37 36N 5 50W
Viaréggio, *Italy* 44 E7 43 52N 10 14 E
Viaur →, *France* 28 D5 44 8N 1 58 E
Vibank, *Canada* 143 C8 50 20N 103 56W
Vibble, *Sweden* 17 G12 57 37N 18 16 E
Vibo Valéntia, *Italy* 47 D9 38 40N 16 6 E
Viborg, *Denmark* 17 H3 56 27N 9 23 E
Viborg Amtskommune □, *Denmark* 17 H3 56 30N 9 30 E
Vibraye, *France* 26 D7 48 3N 0 44 E
Vic, *Spain* 40 D7 41 58N 2 19 E
Vic, Étang de, *France* 28 E7 43 29N 3 50 E
Vic-en-Bigorre, *France* 28 E4 43 24N 0 3 E
Vic-Fézensac, *France* 28 E4 43 47N 0 19 E
Vic-le-Comte, *France* 27 G10 45 39N 3 14 E
Vic-sur-Cère, *France* 28 D6 44 59N 2 38 E
Vícar, *Spain* 41 J2 36 50N 2 38W
Vicenza, *Italy* 45 C8 45 33N 11 33 E
Vich = Vic, *Spain* 40 D7 41 58N 2 19 E
Vichada □, *Colombia* 168 C4 5 0N 69 30W
Vichada →, *Colombia* 168 C4 4 55N 67 50W
Vichuga, *Russia* 60 B5 57 12N 41 55 E
Vichy, *France* 27 F10 46 9N 3 26 E
Vicksburg, *Ariz., U.S.A.* 161 M13 33 45N 113 45W
Vicksburg, *Mich., U.S.A.* 157 B11 42 7N 85 32W
Vicksburg, *Miss., U.S.A.* 155 J9 32 21N 90 53W
Vico, *France* 29 F12 42 10N 8 49 E
Vico, L. di, *Italy* 45 F9 42 19N 12 10 E
Vico del Gargano, *Italy* 45 G12 41 54N 15 57 E
Viçosa, *Brazil* 170 C4 9 28 S 36 14W
Viçosa do Ceará, *Brazil* 170 B3 3 34 S 41 5W
Vicosoprano, *Switz.* 33 D9 46 22N 9 38 E
Vicovu de Sus, *Romania* 53 C10 47 56N 25 41 E
Victor, *India* 92 J4 21 0N 71 30 E
Victor, *Colo., U.S.A.* 154 F2 38 43N 105 9W
Victor, *N.Y., U.S.A.* 150 D7 42 58N 77 24W
Victor Emanuel Ra., *Papua N. G.* 132 C2 5 20 S 142 15 E
Victor Harbor, *Australia* 127 F2 35 30 S 138 37 E
Victoria = Labuan, *Malaysia* 85 A5 5 20N 115 14 E
Victoria, *Argentina* 174 C3 32 40 S 60 10W
Victoria, *Canada* 142 D4 48 30N 123 25W
Victoria, *Chile* 176 A2 38 13 S 72 20W
Victoria, *Guinea* 112 C2 10 50N 14 32W
Victoria, *Malta* 38 C1 36 2N 14 14 E
Victoria, *Mindoro, Phil.* 80 E3 13 12N 121 12 E
Victoria, *Tarlac, Phil.* 80 D3 15 35N 120 41 E
Victoria, *Romania* 53 E9 45 44N 24 41 E
Victoria, *Seychelles* 121 E4 5 0 S 55 40 E
Victoria, *Ill., U.S.A.* 156 C6 41 2N 90 6W
Victoria, *Kans., U.S.A.* 154 F5 38 52N 99 9W
Victoria, *Tex., U.S.A.* 155 L6 28 48N 97 0W
Victoria □, *Australia* 128 D6 37 0 S 144 0 E
Victoria →, *Australia* 124 C4 15 10 S 129 40 E
Victoria, Grand L., *Canada* 140 C4 47 31N 77 30W
Victoria, L., *Africa* 118 C3 1 0 S 33 0 E
Victoria, L., *Australia* 128 B4 33 57 S 141 15 E
Victoria, Mt., *Burma* 90 E4 21 15N 93 55 E
Victoria, Mt., *Papua N. G.* 132 E4 8 55 S 147 32 E
Victoria Beach, *Canada* 143 C9 50 40N 96 35W
Victoria de Durango = Durango, *Mexico* 162 C4 24 3N 104 39W
Victoria de las Tunas, *Cuba* 164 B4 20 58N 76 59W
Victoria Falls, *Zimbabwe* 119 F2 17 58 S 25 52 E
Victoria Harbour, *Canada* 140 D4 44 45N 79 45W
Victoria I., *Canada* 138 A8 71 0N 111 0W
Victoria Ld., *Antarctica* 7 D11 75 0 S 160 0 E
Victoria Nile →, *Uganda* 118 B3 2 14N 31 26 E
Victoria Peaks, *Phil.* 81 G2 9 22N 118 20 E
Victoria Ra., *N.Z.* 131 C7 42 12 S 172 7 E
Victoria Res., *Canada* 141 C8 48 20N 57 27W
Victoria River Downs, *Australia* 124 C5 16 25 S 131 0 E
Victoria West, *S. Africa* 116 E3 31 25 S 23 4 E
Victorias, *Phil.* 81 F4 10 54N 123 5 E
Victoriaville, *Canada* 141 C5 46 4N 71 56W
Victorica, *Argentina* 174 D2 36 20 S 65 30W
Victorville, *U.S.A.* 161 L9 34 32N 117 18W
Vicuña, *Chile* 174 C1 30 0 S 70 50W
Vicuña Mackenna, *Argentina* 174 C3 33 53 S 64 25W
Vidal, *U.S.A.* 161 L12 34 7N 114 31W
Vidal Junction, *U.S.A.* 161 L12 34 11N 114 34W
Vidalia, *U.S.A.* 152 C7 32 13N 82 25W
Vidauban, *France* 29 E10 43 25N 6 27 E
Videbæk, *Denmark* 17 H2 56 6N 8 38 E
Videle, *Romania* 53 F10 44 17N 25 31 E
Videseter, *Norway* 18 C4 61 57N 7 14 E
Vidette, *U.S.A.* 152 B7 33 2N 82 15W
Víðirhóll, *Iceland* 11 B10 65 44N 16 2W
Vídho, *Greece* 38 A3 39 38N 19 55 E
Vidigueira, *Portugal* 43 G3 38 12N 7 48W
Vidin, *Bulgaria* 50 C6 43 59N 22 50 E
Vidio, C., *Spain* 42 B4 43 35N 6 14W
Vidisha, *India* 92 H7 23 28N 77 53 E
Vidra, *Romania* 53 E11 45 56N 26 55 E
Viduša, *Bos.-H.* 50 D2 42 55N 18 21 E
Vidzy, *Belarus* 15 J22 55 23N 26 37 E
Viechtach, *Germany* 31 F8 49 4N 12 53 E
Viedma, *Argentina* 176 B4 40 50 S 63 0W
Viedma, L., *Argentina* 176 C2 49 30 S 72 30W
Vieira do Minho, *Portugal* 42 D2 41 38N 8 8W
Vielha, *Spain* 40 C5 42 43N 0 44 E
Viella = Vielha, *Spain* 40 C5 42 43N 0 44 E
Vielsalm, *Belgium* 24 D5 50 17N 5 54 E
Vienenburg, *Germany* 30 D6 51 57N 10 34 E
Vieng Pou Kha, *Laos* 86 B3 20 41N 101 4 E
Vienna = Wien, *Austria* 35 C9 48 12N 16 22 E
Vienna, *Ga., U.S.A.* 152 C6 32 6N 83 47W
Vienna, *Ill., U.S.A.* 155 G10 37 25N 88 54W
Vienna, *Mo., U.S.A.* 156 F5 38 11N 91 57W
Vienne, *France* 29 C8 45 31N 4 53 E
Vienne □, *France* 28 B4 46 30N 0 42 E
Vienne →, *France* 26 E7 47 13N 0 5 E
Vientiane, *Laos* 86 D4 17 58N 102 36 E
Vientos, Paso de los, *Caribbean* 165 C5 20 0N 74 0W
Viernheim, *Germany* 31 F4 49 31N 8 35 E
Viersen, *Germany* 30 D2 51 15N 6 23 E
Vierwaldstättersee, *Switz.* 33 C7 47 0N 8 30 E
Vierzon, *France* 27 E9 47 13N 2 5 E
Vieste, *Italy* 45 G13 41 53N 16 10 E
Vietnam ■, *Asia* 86 C6 19 0N 106 0 E
Vieux-Boucau-les-Bains, *France* 28 E2 43 48N 1 23W
Vif, *France* 29 C9 45 5N 5 41 E
Vigan, *Phil.* 80 C3 17 35N 120 28 E
Vigévano, *Italy* 44 C5 45 19N 8 51 E
Vigia, *Brazil* 170 B2 0 50 S 48 5W
Vigía Chico, *Mexico* 163 D7 19 46N 87 35W
Víglas, Ákra, *Greece* 38 D9 35 54N 27 51 E
Vignemale, *France* 28 F3 42 47N 0 10W
Vigneulles-lès-Hattonchâtel, *France* 27 D12 48 59N 5 43 E
Vignola, *Italy* 44 D8 44 29N 11 1 E
Vigo, *Spain* 42 C2 42 12N 8 41W
Vigo, Ría de, *Spain* 42 C2 42 15N 8 45W
Vigrestad, *Norway* 18 F2 58 34N 5 42 E
Vigsø Bugt, *Denmark* 17 G2 57 8N 8 47 E
Vihiers, *France* 26 E6 47 10N 0 30W
Vijayadurg, *India* 94 F1 16 30N 73 25 E
Vijayawada, *India* 95 F5 16 31N 80 39 E
Vík, *Iceland* 11 D7 63 25N 19 1W
Vik, *Norway* 18 C3 61 21N 6 6 E
Vika, *Sweden* 16 D8 60 57N 14 28 E
Vikarbyn, *Sweden* 16 D9 60 55N 15 1 E
Vike, *Norway* 18 D2 60 42N 5 35 E
Vikedal, *Norway* 18 E2 59 30N 5 55 E
Vikeke, *Indonesia* 82 C3 8 52 S 126 23 E
Vikeland, *Norway* 18 F4 58 5N 7 18 E
Viken, *Malmöhus, Sweden* 17 H6 56 9N 12 34 E
Viken, *Skaraborg, Sweden* 17 F8 58 39N 14 20 E
Vikersund, *Norway* 18 E6 59 58N 9 59 E
Vikeså, *Norway* 18 F3 58 38N 6 3 E
Vikevåg, *Norway* 18 E2 59 5N 5 41 E
Viking, *Canada* 142 C6 53 7N 111 50W
Vikmanshyttan, *Sweden* 16 D9 60 18N 15 50 E
Vikna, *Norway* 14 D14 64 55N 10 58 E
Vikramasingapuram, *India* 95 K3 8 40N 76 47 E
Viksøyri, *Norway* 18 C3 61 4N 6 34 E
Vila da Maganja, *Mozam.* 119 F4 17 18 S 37 30 E
Vila de João Belo = Xai-Xai, *Mozam.* 117 D5 25 6 S 33 31 E
Vila de Rei, *Portugal* 42 F2 39 41N 8 9W
Vila do Bispo, *Portugal* 43 H2 37 5N 8 53W
Vila do Chibuto, *Mozam.* 117 C5 24 40 S 33 33 E
Vila do Conde, *Portugal* 42 D2 41 21N 8 45W
Vila Franca de Xira, *Portugal* 43 G2 38 57N 8 59W
Vila Gamito, *Mozam.* 119 E3 14 12 S 33 0 E
Vila Gomes da Costa, *Mozam.* 117 C5 24 20 S 33 37 E
Vila Machado, *Mozam.* 119 F3 19 15 S 34 14 E
Vila Mouzinho, *Mozam.* 119 E3 14 48 S 34 25 E
Vila Nova de Famalicão, *Portugal* 42 D2 41 25N 8 32W

W

West Baines 299

Wangjiang, *China* **77 B11** 30 10N 116 42 E
Wangmo, *China* **76 E6** 25 11N 106 5 E
Wangqing, *China* **75 C15** 43 12N 129 42 E
Wankaner, *India* **92 H4** 22 35N 71 0 E
Wanless, *Canada* **143 C8** 54 11N 101 21W
Wannian, *China* **77 C11** 28 42N 117 4 E
Wanning, *Taiwan* **86 C8** 23 15N 121 17 E
Wanon Niwat, *Thailand* **86 D4** 17 38N 103 46 E
Wanquan, *China* **74 D8** 40 50N 114 40 E
Wanrong, *China* **74 G6** 35 25N 110 50 E
Wanshan, *China* **76 D7** 27 30N 109 12 E
Wanshengchang, *China* **76 C6** 28 57N 106 53 E
Wanstead, *N.Z.* **130 G5** 40 8 S 176 30 E
Wantage, *U.K.* **21 F6** 51 35N 1 25W
Wanxian, *China* **76 B7** 30 42N 108 20 E
Wanyin, *Burma* **90 E6** 20 23N 97 15 E
Wanyuan, *China* **76 A7** 32 4N 108 3 E
Wanzai, *China* **77 C10** 28 7N 114 30 E
Wapakoneta, *U.S.A.* **157 D12** 40 34N 84 12W
Wapato, *U.S.A.* **158 C3** 46 27N 120 25W
Wapawekka L., *Canada* **143 C8** 54 55N 104 40W
Wapello, *U.S.A.* **156 C5** 41 11N 91 11W
Wapikopa L., *Canada* **140 B2** 52 56N 87 53W
Wappingers Falls, *U.S.A.* **151 E11** 41 36N 73 55W
Wapsipinicon →, *U.S.A.* **156 C6** 41 44N 90 19W
Warabi, *Japan* **71 B11** 35 49N 139 41 E
Warangal, *India* **94 F4** 17 58N 79 35 E
Waratah, *Australia* **126 G4** 41 30 S 145 30 E
Waratah B., *Australia* **127 F4** 38 54 S 146 5 E
Warburg, *Germany* **30 D5** 51 28N 9 11 E
Warburton, *Vic., Australia* **129 D6** 37 47 S 145 42 E
Warburton, *W. Austral., Australia* **125 E4** 26 8 S 126 35 E
Warburton Ra., *Australia* **125 E4** 25 55 S 126 28 E
Ward, *N.Z.* **131 B9** 41 49 S 174 11 E
Ward →, *Australia* **127 D4** 26 28 S 146 6 E
Ward Cove, *U.S.A.* **142 B2** 55 25N 132 43W
Ward Hunt, C., *Papua N. G.* **132 E5** 8 2 S 148 10 E
Ward Hunt Str., *Papua N. G.* **132 E6** 9 30 S 150 0 E
Ward Mt., *U.S.A.* **160 H8** 37 12N 118 54W
Warden, *S. Africa* **117 D4** 27 50 S 29 0 E
Wardha, *India* **94 D4** 20 45N 78 39 E
Wardlow, *Canada* **142 C6** 50 56N 111 31W
Wards River, *Australia* **129 B9** 32 11 S 151 56 E
Ware, *Canada* **142 B3** 57 26N 125 41W
Ware, *U.S.A.* **151 D12** 42 16N 72 14W
Waregem, *Belgium* **24 D3** 50 53N 3 27 E
Wareham, *U.S.A.* **151 E14** 41 46N 70 43W
Waremme, *Belgium* **24 D5** 50 43N 5 15 E
Waren, *Germany* **30 B8** 53 31N 12 40 E
Warendorf, *Germany* **30 D4** 51 57N 8 1 E
Waresboro, *U.S.A.* **152 D7** 31 15N 82 29W
Warialda, *Australia* **127 D5** 29 29 S 150 33 E
Wariap, *Indonesia* **83 B4** 1 30 S 134 5 E
Warin Chamrap, *Thailand* **86 E5** 15 12N 104 53 E
Warka, *Poland* **55 G8** 51 47N 21 12 E
Warkopi, *Indonesia* **83 B4** 1 12 S 134 9 E
Warkworth, *N.Z.* **130 C3** 36 24 S 174 41 E
Warm Springs, *Ga., U.S.A.* **152 C5** 32 53N 84 41W
Warm Springs, *Nev., U.S.A.* **159 G5** 38 10N 116 20W
Warman, *Canada* **143 C7** 52 19N 106 30W
Warmbad, *Namibia* **116 D2** 28 25 S 18 42 E
Warmbad, *S. Africa* **117 C4** 24 51 S 28 19 E
Warminster, *U.K.* **21 F5** 51 12N 2 10W
Warnambool Downs, *Australia* **126 C3** 22 48 S 142 52 E
Warnemünde, *Germany* **30 A8** 54 10N 12 4 E
Warner, *Canada* **142 D6** 49 17N 112 12W
Warner Mts., *U.S.A.* **158 F3** 41 40N 120 15W
Warner Robins, *U.S.A.* **152 C6** 32 37N 83 36W
Warnes, *Bolivia* **173 D5** 17 30 S 63 10W
Warnow →, *Germany* **30 A8** 54 6N 12 9 E
Waroona, *Australia* **125 F2** 32 50 S 115 58 E
Warora, *India* **94 D4** 20 14N 79 1 E
Warracknabeal, *Australia* **128 D5** 36 9 S 142 26 E
Warragul, *Australia* **129 E6** 38 10 S 145 58 E
Warrawagine, *Australia* **124 D3** 20 51 S 120 42 E
Warrego →, *Australia* **127 E4** 30 24 S 145 21 E
Warrego Ra., *Australia* **126 C4** 24 58 S 146 0 E
Warren, *Australia* **129 A7** 31 42 S 147 51 E
Warren, *Ark., U.S.A.* **155 J8** 33 37N 92 4W
Warren, *Ill., U.S.A.* **156 B7** 42 29N 90 0W
Warren, *Mich., U.S.A.* **157 B13** 42 30N 83 0W
Warren, *Minn., U.S.A.* **154 A6** 48 12N 96 46W
Warren, *Ohio, U.S.A.* **150 E4** 41 14N 80 49W
Warren, *Pa., U.S.A.* **150 E5** 41 51N 79 9W
Warrenpoint, *U.K.* **23 B5** 54 6N 6 15W
Warrensburg, *Ill., U.S.A.* **156 E7** 39 56N 89 4W
Warrensburg, *Mo., U.S.A.* **154 F8** 38 46N 93 44W
Warrenton, *S. Africa* **116 D3** 28 9 S 24 47 E
Warrenton, *Ga., U.S.A.* **152 B7** 33 24N 82 40W
Warrenton, *Mo., U.S.A.* **156 F5** 38 49N 91 9W
Warrenton, *Oreg., U.S.A.* **160 D3** 46 10N 123 56W
Warrenville, *Australia* **127 D4** 25 48 S 147 22 E
Warrenville, *U.S.A.* **152 B9** 33 32N 81 48W
Warri, *Nigeria* **113 D6** 5 30N 5 41 E
Warrina, *Australia* **127 D2** 28 12 S 135 50 E
Warrington, *N.Z.* **131 F5** 45 43 S 170 35 E
Warrington, *U.K.* **20 D5** 53 24N 2 35W
Warrington, *U.S.A.* **149 K2** 30 23N 87 17W
Warrington □, *U.K.* **20 D5** 53 24N 2 35W
Warrnambool, *Australia* **128 E5** 38 25 S 142 30 E
Warroad, *U.S.A.* **154 A7** 48 54N 95 19W
Warsa, *Indonesia* **83 B5** 0 47 S 135 55 E
Warsaw = Warszawa, *Poland* **55 F8** 52 13N 21 0 E
Warsaw, *Ill., U.S.A.* **156 D5** 40 22N 91 26W
Warsaw, *Ind., U.S.A.* **157 C11** 41 14N 85 51W
Warsaw, *Ky., U.S.A.* **157 F12** 38 47N 84 54W
Warsaw, *Mo., U.S.A.* **156 F3** 38 15N 93 23W
Warsaw, *N.Y., U.S.A.* **150 D6** 42 45N 78 8W
Warsaw, *Ohio, U.S.A.* **150 F3** 40 20N 82 0W
Warstein, *Germany* **30 D4** 51 26N 8 22 E
Warszawa, *Poland* **55 F8** 52 13N 21 0 E
Warszawa □, *Poland* **55 F7** 52 30N 21 0 E
Warta, *Poland* **55 G5** 51 43N 18 38 E
Warta →, *Poland* **55 F1** 52 35N 14 39 E
Warthe = Warta →, *Poland* **55 F1** 52 35N 14 39 E
Warthen, *U.S.A.* **152 B7** 33 6N 82 48W
Waru, *Indonesia* **83 B4** 3 30 S 130 36 E
Warud, *India* **94 D4** 21 30N 78 16 E
Warwick, *Australia* **127 D5** 28 10 S 152 1 E
Warwick, *U.K.* **21 E6** 52 18N 1 35W
Warwick, *Ga., U.S.A.* **152 D6** 31 50N 83 57W
Warwick, *R.I., U.S.A.* **151 E13** 41 42N 71 28W
Warwickshire □, *U.K.* **21 E6** 52 14N 1 38W
Wasaga Beach, *Canada* **150 B4** 44 31N 80 1W
Wasatch Ra., *U.S.A.* **158 F8** 40 30N 111 15W
Wasbank, *S. Africa* **117 D5** 28 15 S 30 9 E
Wasco, *Calif., U.S.A.* **161 K7** 35 36N 119 20W
Wasco, *Oreg., U.S.A.* **158 D3** 45 36N 120 42W
Waseca, *U.S.A.* **154 C8** 44 5N 93 30W
Wasekamio L., *Canada* **143 B7** 56 45N 108 45W
Wash, The, *U.K.* **20 E8** 52 58N 0 20 E
Washago, *Canada* **150 B5** 44 45N 79 20W
Washburn, *Ill., U.S.A.* **156 D7** 40 55N 89 17W
Washburn, *N. Dak., U.S.A.* **154 B4** 47 17N 101 2W
Washburn, *Wis., U.S.A.* **154 B9** 46 40N 90 54W
Washim, *India* **94 D3** 20 3N 77 0 E
Washington, *U.K.* **20 C6** 54 55N 1 30W
Washington, *D.C., U.S.A.* **148 F7** 38 54N 77 2W
Washington, *Ga., U.S.A.* **152 B7** 33 44N 82 44W
Washington, *Ind., U.S.A.* **157 F9** 38 40N 87 10W
Washington, *Iowa, U.S.A.* **156 C5** 41 18N 91 42W
Washington, *Mo., U.S.A.* **156 F5** 38 33N 91 1W
Washington, *N.C., U.S.A.* **149 H7** 35 33N 77 3W
Washington, *N.J., U.S.A.* **151 F10** 40 46N 74 59W
Washington, *Pa., U.S.A.* **150 F4** 40 10N 80 15W
Washington, *Utah, U.S.A.* **159 H7** 37 8N 113 31W
Washington □, *U.S.A.* **158 C3** 47 30N 120 30W
Washington, Mt., *U.S.A.* **151 B13** 44 16N 71 18W
Washington Court House, *U.S.A.* **157 E13** 39 32N 83 26W
Washington I., *U.S.A.* **148 C2** 45 23N 86 54W
Washington Land, *Greenland* **10 A4** 80 30N 66 0W
Washougal, *U.S.A.* **160 E4** 45 35N 122 21W
Washuk, *Pakistan* **91 D2** 27 42N 64 45 E
Wasian, *Indonesia* **83 B4** 1 47 S 133 19 E
Wasilków, *Poland* **55 E10** 53 12N 23 13 E
Wasilla, *U.S.A.* **144 F10** 61 35N 149 26W
Wasior, *Indonesia* **83 B4** 2 43 S 134 30 E
Waskaganish, *Canada* **140 B4** 51 30N 78 40W
Waskaiowaka, L., *Canada* **143 B9** 56 33N 96 23W
Waskesiu Lake, *Canada* **143 C7** 53 55N 106 5W
Wassaw I., *U.S.A.* **152 D9** 31 53N 80 58W
Wassaw Sd., *U.S.A.* **152 D9** 31 55N 80 55W
Wassen, *Switz.* **33 C7** 46 42N 8 36 E
Wasserburg, *Germany* **31 G8** 48 3N 12 14 E
Wasserkuppe, *Germany* **30 E5** 50 29N 9 55 E
Wassy, *France* **27 D11** 48 30N 4 58 E
Waswanipi, *Canada* **140 C4** 49 40N 76 29W
Waswanipi, L., *Canada* **140 C4** 49 35N 76 40W
Watampone, *Indonesia* **82 B2** 4 29 S 120 25 E
Watansoppeng, *Indonesia* **82 B1** 4 10 S 119 56 E
Water Park Pt., *Australia* **126 C5** 22 56 S 150 47 E
Water Valley, *U.S.A.* **155 H10** 34 10N 89 38W
Waterberge, *S. Africa* **117 C4** 24 10 S 28 0 E
Waterbury, *Conn., U.S.A.* **151 E11** 41 33N 73 3W
Waterbury, *Vt., U.S.A.* **151 B12** 44 20N 72 46W
Waterbury L., *Canada* **143 B8** 58 10N 104 22W
Waterdown, *Canada* **150 C5** 43 20N 79 53W
Wateree →, *U.S.A.* **152 B9** 33 45N 80 37W
Waterford, *Canada* **150 D4** 42 56N 80 17W
Waterford, *Ireland* **23 D4** 52 15N 7 8W
Waterford, *Calif., U.S.A.* **160 H6** 37 38N 120 46W
Waterford, *Wis., U.S.A.* **157 B8** 42 46N 88 13W
Waterford □, *Ireland* **23 D4** 52 10N 7 40W
Waterford Harbour, *Ireland* **23 D5** 52 8N 6 58W
Waterhen L., *Man., Canada* **143 C9** 52 10N 99 40W
Waterhen L., *Sask., Canada* **143 C7** 54 28N 108 25W
Waterloo, *Belgium* **24 D4** 50 43N 4 25 E
Waterloo, *Ont., Canada* **140 D3** 43 30N 80 32W
Waterloo, *Qué., Canada* **151 A12** 45 22N 72 32W
Waterloo, *S. Leone* **112 D2** 8 26N 13 8W
Waterloo, *Ill., U.S.A.* **156 F6** 38 20N 90 9W
Waterloo, *Ind., U.S.A.* **157 C11** 41 26N 85 1W
Waterloo, *Iowa, U.S.A.* **156 B4** 42 30N 92 21W
Waterloo, *N.Y., U.S.A.* **150 D8** 42 54N 76 52W
Waterloo, *Wis., U.S.A.* **156 A8** 43 11N 88 59W
Waterman, *U.S.A.* **157 C8** 41 46N 88 47W
Watersmeet, *U.S.A.* **154 B10** 46 16N 89 11W
Waterton-Glacier International Peace Park, *U.S.A.* **158 B7** 48 45N 115 0W
Watertown, *Conn., U.S.A.* **151 E11** 41 36N 73 7W
Watertown, *Fla., U.S.A.* **152 E7** 30 11N 82 36W
Watertown, *N.Y., U.S.A.* **151 C9** 43 59N 75 55W
Watertown, *S. Dak., U.S.A.* **154 C6** 44 54N 97 7W
Watertown, *Wis., U.S.A.* **154 D10** 43 12N 88 43W
Waterval-Boven, *S. Africa* **117 D5** 25 40 S 30 18 E
Waterville, *Canada* **151 A13** 45 16N 71 54W
Waterville, *Maine, U.S.A.* **141 D6** 44 33N 69 38W
Waterville, *N.Y., U.S.A.* **151 D9** 42 56N 75 23W
Waterville, *Pa., U.S.A.* **150 E7** 41 19N 77 21W
Waterville, *Wash., U.S.A.* **158 C3** 47 39N 120 4W
Watervliet, *Mich., U.S.A.* **157 B10** 42 11N 86 18W
Watervliet, *N.Y., U.S.A.* **151 D11** 42 44N 73 42W
Wates, *Indonesia* **85 D4** 7 51 S 110 10 E
Watford, *Canada* **150 D3** 42 57N 81 53W
Watford, *U.K.* **21 F7** 51 40N 0 24W
Watford City, *U.S.A.* **154 B3** 47 48N 103 17W
Wathaman →, *Canada* **143 B8** 57 16N 102 59W
Watheroo, *Australia* **125 F2** 30 15 S 116 0 E
Wating, *China* **74 G4** 35 40N 106 38 E
Watkins Glen, *U.S.A.* **150 D8** 42 23N 76 52W
Watkinsville, *U.S.A.* **152 B6** 33 52N 83 25W
Watling I. = San Salvador, *Bahamas* **165 B5** 24 0N 74 40W
Watonga, *U.S.A.* **155 H5** 35 51N 98 25W
Watrous, *Canada* **143 C7** 51 40N 105 25W
Watrous, *U.S.A.* **155 H2** 35 48N 104 59W
Watsa, *Dem. Rep. of the Congo* **118 B2** 3 4N 29 30 E
Watseka, *U.S.A.* **157 D9** 40 47N 87 44W
Watson, *Australia* **125 F5** 30 29 S 131 31 E
Watson, *Canada* **143 C8** 52 10N 104 30W
Watson Lake, *Canada* **142 A3** 60 6N 128 49W
Watsonville, *U.S.A.* **160 J5** 36 55N 121 45W
Wattenwil, *Switz.* **32 C5** 46 46N 7 30 E
Wattiwarriganna Cr. →, *Australia* **127 D2** 28 57 S 136 10 E
Wattwil, *Switz.* **33 B8** 47 18N 9 6 E
Watuata = Batuata, *Indonesia* **82 C2** 6 12 S 122 42 E
Watubela, Kepulauan, *Indonesia* **83 B4** 4 28 S 131 35 E
Watubela Is. = Watubela, Kepulauan, *Indonesia* **83 B4** 4 28 S 131 35 E
Wau, *Papua N. G.* **132 D4** 7 21 S 146 47 E
Wau, *Sudan* **105 F6** 7 45N 28 1 E
Waubamik, *Canada* **150 A4** 45 27N 80 1W
Waubay, *U.S.A.* **154 C6** 45 20N 97 18W
Waubra, *Australia* **128 D5** 37 21 S 143 39 E
Wauchope, *Australia* **129 A10** 31 28 S 152 45 E
Wauchula, *U.S.A.* **149 M5** 27 33N 81 49W
Waugh, *Canada* **143 D9** 49 40N 95 11W
Waukarlycarly, L., *Australia* **124 D3** 21 18 S 121 56 E
Waukeenah, *U.S.A.* **152 E6** 30 25N 83 57W
Waukegan, *U.S.A.* **157 B9** 42 22N 87 50W
Waukesha, *U.S.A.* **157 B8** 43 1N 88 14W
Waukon, *U.S.A.* **154 D9** 43 16N 91 29W
Wauneta, *U.S.A.* **154 E4** 40 25N 101 23W
Waupaca, *U.S.A.* **154 C10** 44 21N 89 5W
Waupun, *U.S.A.* **154 D10** 43 38N 88 44W
Waurika, *U.S.A.* **155 H6** 34 10N 98 0W
Wausau, *Fla., U.S.A.* **152 E4** 30 38N 85 35W
Wausau, *Wis., U.S.A.* **154 C10** 44 58N 89 38W
Wauseon, *U.S.A.* **157 C12** 41 33N 84 8W
Wautoma, *U.S.A.* **154 C10** 44 4N 89 18W
Wauwatosa, *U.S.A.* **157 A9** 43 3N 88 0W
Wave Hill, *Australia* **124 C5** 17 32 S 131 0 E
Waveland, *U.S.A.* **157 E9** 39 53N 87 3W
Waveney →, *U.K.* **21 E9** 52 35N 1 39 E
Waverley, *N.Z.* **130 F3** 39 46 S 174 37 E
Waverley Hall, *U.S.A.* **152 C5** 32 41N 84 44W
Waverly, *Ala., U.S.A.* **152 C4** 32 44N 85 35W
Waverly, *Fla., U.S.A.* **153 H8** 27 59N 81 37W
Waverly, *Ga., U.S.A.* **152 D8** 31 6N 81 43W
Waverly, *Ill., U.S.A.* **156 E7** 39 36N 89 57W
Waverly, *Iowa, U.S.A.* **156 B4** 42 44N 92 29W
Waverly, *Mo., U.S.A.* **156 E3** 39 13N 93 31W
Waverly, *N.Y., U.S.A.* **151 E8** 42 1N 76 32W
Wavre, *Belgium* **24 D4** 50 43N 4 38 E
Wâw, *Sudan* **107 F2** 7 45N 28 1 E
Wāw al Kabīr, *Libya* **108 C3** 25 20N 16 43 E
Wāw an Nāmūs, *Libya* **108 D3** 24 55N 17 46 E
Wawa, *Canada* **140 C3** 47 59N 84 47W
Wawa, *Nigeria* **113 D5** 9 54N 4 27 E
Wawa, *Sudan* **106 C3** 20 30N 30 22 E
Wawanesa, *Canada* **143 D9** 49 36N 99 40W
Wawasee, L., *U.S.A.* **157 C11** 41 24N 85 42W
Wawoi →, *Papua N. G.* **132 D2** 7 48 S 143 16 E
Wawona, *U.S.A.* **160 H7** 37 32N 119 39W
Waxahachie, *U.S.A.* **155 J6** 32 24N 96 51W
Way, L., *Australia* **125 E3** 26 45 S 120 16 E
Wayabula Rau, *Indonesia* **82 A3** 2 29N 128 17 E
Wayatinah, *Australia* **126 G4** 42 19 S 146 27 E
Waycross, *U.S.A.* **152 D7** 31 13N 82 21W
Wayi, *Sudan* **107 F3** 5 8N 30 10 E
Wayland, *U.S.A.* **157 B11** 42 40N 85 39W
Wayne, *Nebr., U.S.A.* **154 D6** 42 14N 97 1W
Wayne, *W. Va., U.S.A.* **148 F4** 38 13N 82 27W
Wayne City, *U.S.A.* **157 F8** 38 21N 88 35W
Waynesboro, *Ga., U.S.A.* **152 B7** 33 6N 82 1W
Waynesboro, *Miss., U.S.A.* **149 K1** 31 40N 88 39W
Waynesboro, *Pa., U.S.A.* **148 F7** 39 45N 77 35W
Waynesboro, *Va., U.S.A.* **148 F6** 38 4N 78 53W
Waynesburg, *U.S.A.* **148 F5** 39 54N 80 11W
Waynesville, *Mo., U.S.A.* **156 G4** 37 50N 92 12W
Waynesville, *N.C., U.S.A.* **149 H4** 35 28N 82 58W
Waynesville, *Ohio, U.S.A.* **157 E12** 39 32N 84 5W
Waynoka, *U.S.A.* **155 G5** 36 35N 98 53W
Wayside, *U.S.A.* **152 B6** 33 4N 83 37W
Wazay, *Afghan.* **91 B3** 33 22N 69 26 E
Wāzin, *Libya* **108 B2** 31 58N 10 40 E
Wazirabad, *Pakistan* **92 C6** 32 30N 74 8 E
Wda →, *Poland* **54 E5** 53 25N 18 29 E
We, *Indonesia* **84 A1** 5 51N 95 18 E
Weald, The, *U.K.* **21 F8** 51 4N 0 20 E
Wear →, *U.K.* **20 C6** 54 55N 1 23W
Weatherford, *Okla., U.S.A.* **155 H5** 35 32N 98 43W
Weatherford, *Tex., U.S.A.* **155 J6** 32 46N 97 48W
Weaubleau, *U.S.A.* **156 G3** 37 54N 93 32W
Weaverville, *U.S.A.* **158 F2** 40 44N 122 56W
Webb City, *U.S.A.* **155 G7** 37 9N 94 28W
Weber, *N.Z.* **130 G5** 40 24 S 176 20 E
Webo = Nyaake, *Liberia* **112 E3** 4 52N 7 37W
Webster, *Fla., U.S.A.* **153 G7** 28 37N 82 3W
Webster, *Mass., U.S.A.* **151 D13** 42 3N 71 53W
Webster, *N.Y., U.S.A.* **150 C7** 43 13N 77 26W
Webster, *S. Dak., U.S.A.* **154 C6** 45 20N 97 31W
Webster, *Wis., U.S.A.* **154 C8** 45 53N 92 22W
Webster City, *U.S.A.* **156 B3** 42 28N 93 49W
Webster Green, *U.S.A.* **154 F9** 38 38N 90 20W
Webster Springs, *U.S.A.* **148 F5** 38 29N 80 25W
Weda, *Indonesia* **82 A3** 0 21N 127 50 E
Weda, Teluk, *Indonesia* **82 A3** 0 30N 127 50 E
Weddell I., *Falk. Is.* **176 D4** 51 50 S 61 0W
Weddell Sea, *Antarctica* **7 D1** 72 30 S 40 0W
Wedderburn, *Australia* **128 D5** 36 26 S 143 33 E
Wedel, *Germany* **30 B5** 53 34N 9 42 E
Wedemark, *Germany* **30 C5** 52 32N 9 43 E
Wedgeport, *Canada* **141 D6** 43 44N 65 59W
Wedowee, *U.S.A.* **152 B4** 33 19N 85 29W
Wedza, *Zimbabwe* **119 F3** 18 40 S 31 33 E
Wee Elwah, *Australia* **129 B6** 33 2 S 145 14 E
Wee Waa, *Australia* **127 E4** 30 11 S 149 26 E
Weed, *U.S.A.* **158 F2** 41 25N 122 23W
Weed Heights, *U.S.A.* **160 G7** 38 59N 119 13W
Weedsport, *U.S.A.* **151 C8** 43 3N 76 35W
Weedville, *U.S.A.* **150 E6** 41 17N 78 30W
Weemelah, *Australia* **127 D4** 29 2 S 149 15 E
Weenen, *S. Africa* **117 D5** 28 48 S 30 7 E
Weener, *Germany* **30 B3** 53 9N 7 20 E
Weert, *Neths.* **24 C5** 51 15N 5 43 E
Weggis, *Switz.* **33 B6** 47 2N 8 26 E
Węgierska-Górka, *Poland* **55 J6** 49 36N 19 7 E
Wegliniec, *Poland* **55 G2** 51 18N 15 10 E
Węgorzewo, *Poland* **54 D8** 54 13N 21 43 E
Węgorzyno, *Poland* **54 E2** 53 32N 15 33 E
Węgrów, *Poland* **55 F9** 52 24N 22 0 E
Wei He →, *Hebei, China* **74 F8** 36 10N 115 45 E
Wei He →, *Shaanxi, China* **74 G6** 34 38N 110 15 E
Weichang, *China* **75 D9** 41 58N 117 49 E
Weichuan, *China* **74 G7** 34 20N 113 59 E
Weida, *Germany* **30 E8** 50 46N 12 2 E
Weiden, *Germany* **31 F8** 49 41N 12 10 E
Weifang, *China* **75 F10** 36 44N 119 7 E
Weihai, *China* **75 F12** 37 30N 122 6 E
Weil, *Germany* **31 H3** 47 35N 7 37 E
Weilburg, *Germany* **30 E4** 50 28N 8 17 E
Weilheim, *Germany* **31 H7** 47 50N 11 9 E
Weimar, *Germany* **30 E7** 50 58N 11 19 E
Weinan, *China* **74 G5** 34 31N 109 29 E
Weinfelden, *Switz.* **33 A8** 47 34N 9 6 E
Weingarten, *Germany* **31 F4** 49 3N 8 31 E
Weinheim, *Germany* **31 F4** 49 32N 8 39 E
Weining, *China* **76 D5** 26 50N 104 17 E
Weipa, *Australia* **126 A3** 12 40 S 141 50 E
Weir →, *Australia* **127 D4** 28 20 S 149 50 E
Weir →, *Canada* **143 B10** 56 54N 93 21W
Weir, L., *U.S.A.* **153 F8** 29 0N 81 57W
Weir River, *Canada* **143 B10** 56 49N 94 6W
Weirsdale, *U.S.A.* **153 G8** 28 59N 81 55W
Weirton, *U.S.A.* **150 F4** 40 24N 80 35W
Weisen, *Switz.* **33 C9** 46 42N 9 43 E
Weiser, *U.S.A.* **158 D5** 44 10N 117 0W
Weishan, *Shandong, China* **75 G9** 34 47N 117 5 E
Weishan, *Yunnan, China* **76 E3** 25 12N 100 20 E
Weissenburg, *Germany* **31 F6** 49 2N 10 58 E
Weissenfels, *Germany* **30 D7** 51 11N 12 0 E
Weisshorn, *Switz.* **32 D5** 46 7N 7 43 E
Weissmies, *Switz.* **32 D6** 46 8N 8 1 E
Weisstannen, *Switz.* **33 C8** 46 59N 9 22 E
Weisswasser, *Germany* **30 D10** 51 30N 14 36 E
Weitra, *Austria* **34 C7** 48 41N 14 54 E
Weixi, *China* **76 D2** 27 10N 99 10 E
Weixin, *China* **76 D5** 27 48N 105 3 E
Weiyuan, *China* **74 G3** 35 7N 104 10 E
Weiz, *Austria* **34 D8** 47 13N 15 39 E
Weizhou Dao, *China* **76 G7** 21 0N 109 5 E
Wejherowo, *Poland* **54 D5** 54 35N 18 12 E
Wekusko L., *Canada* **143 C9** 54 40N 99 50W
Welbourn Hill, *Australia* **127 D1** 27 21 S 134 6 E
Welch, *U.S.A.* **148 G5** 37 26N 81 35W
Weldya, *Ethiopia* **107 E4** 11 50N 39 34 E
Welega □, *Ethiopia* **107 F3** 9 25N 34 20 E
Welkite, *Ethiopia* **107 F4** 8 15N 37 42 E
Welkom, *S. Africa* **116 D4** 28 0 S 26 46 E
Welland, *Canada* **140 D4** 43 0N 79 15W
Welland →, *U.K.* **21 E7** 52 51N 0 5W
Wellesley Is., *Australia* **126 B2** 16 42 S 139 30 E
Wellingborough, *U.K.* **21 E7** 52 19N 0 41W
Wellington, *Australia* **129 B8** 32 35 S 148 59 E
Wellington, *Canada* **140 D4** 43 57N 77 20W
Wellington, *N.Z.* **130 H3** 41 19 S 174 46 E
Wellington, *S. Africa* **116 E2** 33 38 S 19 1 E
Wellington, *Somst., U.K.* **21 G4** 50 58N 3 13W
Wellington, *Telford & Wrekin, U.K.* **21 E5** 52 42N 2 30W
Wellington, *Colo., U.S.A.* **154 E2** 40 42N 105 0W
Wellington, *Kans., U.S.A.* **155 G6** 37 16N 97 24W
Wellington, *Mo., U.S.A.* **156 E3** 39 8N 93 59W
Wellington, *Nev., U.S.A.* **160 G7** 38 45N 119 23W
Wellington, *Ohio, U.S.A.* **150 E2** 41 10N 82 13W
Wellington, *Tex., U.S.A.* **155 H4** 34 51N 100 13W
Wellington □, *N.Z.* **130 G4** 41 0 S 175 30 E
Wellington, I., *Chile* **176 C2** 49 30 S 75 0W
Wellington, L., *Australia* **129 E7** 38 6 S 147 20 E
Wells, *U.K.* **21 F5** 51 13N 2 39W
Wells, *Maine, U.S.A.* **151 C14** 43 20N 70 35W
Wells, *Minn., U.S.A.* **154 D8** 43 45N 93 44W
Wells, *Nev., U.S.A.* **158 F6** 41 7N 114 58W
Wells, L., *Australia* **125 E3** 26 44 S 123 15 E
Wells Gray Prov. Park, *Canada* **142 C4** 52 30N 120 15W
Wells-next-the-Sea, *U.K.* **20 E8** 52 57N 0 51 E
Wells River, *U.S.A.* **151 B12** 44 9N 72 4W
Wellsboro, *U.S.A.* **150 E7** 41 45N 77 18W
Wellsburg, *U.S.A.* **150 F4** 40 16N 80 37W
Wellsford, *N.Z.* **130 C3** 36 16 S 174 32 E
Wellsville, *Mo., U.S.A.* **156 E5** 39 4N 91 34W
Wellsville, *N.Y., U.S.A.* **150 D7** 42 7N 77 57W
Wellsville, *Ohio, U.S.A.* **150 F4** 40 36N 80 39W
Wellsville, *Utah, U.S.A.* **158 F8** 41 38N 111 56W
Wellton, *U.S.A.* **159 K6** 32 40N 114 8W
Welmel, Wabi →, *Ethiopia* **107 F5** 5 38N 40 47 E
Welo, *Somali Rep.* **120 C3** 9 25N 48 55 E
Welo □, *Ethiopia* **107 E4** 11 50N 39 48 E
Wels, *Austria* **34 C7** 48 9N 14 1 E
Welshpool, *U.K.* **21 E4** 52 39N 3 8W
Welwyn Garden City, *U.K.* **21 F7** 51 48N 0 12W
Wem, *U.K.* **20 E5** 52 52N 2 44W
Wembere →, *Tanzania* **118 C3** 4 10 S 34 15 E
Wemindji, *Canada* **140 B4** 53 0N 78 49W
Wen Xian, *Gansu, China* **74 H3** 32 43N 104 36 E
Wen Xian, *Henan, China* **74 G7** 34 55N 113 5 E
Wenatchee, *U.S.A.* **158 C3** 47 25N 120 19W
Wenchang, *China* **86 C8** 19 38N 110 42 E
Wencheng, *China* **77 D13** 27 46N 120 4 E
Wenchi, *Ghana* **112 D4** 7 46N 2 8W
Wenchow = Wenzhou, *China* **77 D13** 28 0N 120 38 E
Wenchuan, *China* **76 B4** 31 22N 103 35 E
Wendell, *U.S.A.* **158 E6** 42 47N 114 42W
Wenden, *U.S.A.* **161 M13** 33 49N 113 33W
Wendeng, *China* **75 F12** 37 15N 122 5 E
Wendesi, *Indonesia* **83 B4** 2 30 S 134 17 E
Wendo, *Ethiopia* **107 F4** 6 40N 38 27 E
Wendover, *U.S.A.* **158 F6** 40 44N 114 2W
Weng'an, *China* **76 D6** 27 5N 107 25 E
Wengcheng, *China* **77 E9** 24 22N 113 50 E
Wengen, *Switz.* **32 C5** 46 37N 7 55 E
Wengyuan, *China* **77 E10** 24 20N 114 9 E
Wenjiang, *China* **76 B4** 30 44N 103 55 E
Wenling, *China* **77 C13** 28 21N 121 20 E
Wenlock →, *Australia* **126 A3** 12 2 S 141 55 E
Wenona, *U.S.A.* **156 C7** 41 3N 89 3W
Wenshan, *China* **76 F5** 23 20N 104 18 E
Wenshang, *China* **74 G9** 35 45N 116 30 E
Wenshui, *Guizhou, China* **76 C6** 28 27N 106 28 E
Wenshui, *Shanxi, China* **74 F7** 37 26N 112 1 E
Wensleydale, *U.K.* **20 C6** 54 17N 2 0W
Wensu, *China* **72 B3** 41 15N 80 10 E
Wensum →, *U.K.* **20 E8** 52 40N 1 15 E
Wentworth, *Australia* **128 C4** 34 2 S 141 54 E
Wentzville, *U.S.A.* **156 F6** 38 49N 90 51W
Wenut, *Indonesia* **83 B4** 3 11 S 133 19 E
Wenxi, *China* **74 G6** 35 20N 111 10 E
Wenzhou, *China* **77 D13** 28 0N 120 38 E
Weott, *U.S.A.* **158 F2** 40 20N 123 55W
Wepener, *S. Africa* **116 D4** 29 42 S 27 3 E
Werda, *Botswana* **116 D3** 25 24 S 23 15 E
Werdau, *Germany* **30 E8** 50 44N 12 22 E
Werder, *Ethiopia* **120 C3** 6 58N 45 1 E
Werder, *Germany* **30 C8** 52 23N 12 55 E
Werdohl, *Germany* **30 D3** 51 15N 7 46 E
Wereilu, *Ethiopia* **107 E4** 10 40N 39 28 E
Weri, *Indonesia* **83 B4** 3 10 S 132 38 E
Werneck, *Germany* **31 F6** 49 58N 10 5 E
Wernigerode, *Germany* **30 D6** 51 50N 10 47 E
Werra →, *Germany* **30 D5** 51 24N 9 39 E
Werribee, *Australia* **128 D6** 37 54 S 144 40 E
Werrimull, *Australia* **128 C4** 34 25 S 141 38 E
Werris Creek, *Australia* **129 A9** 31 18 S 150 38 E
Wersar, *Indonesia* **83 B4** 1 30 S 131 55 E
Wertach →, *Germany* **31 G6** 48 22N 10 54 E
Wertheim, *Germany* **31 F5** 49 45N 9 32 E
Wertingen, *Germany* **31 G6** 48 33N 10 41 E
Wesel, *Germany* **30 D2** 51 39N 6 37 E
Weser →, *Germany* **30 B4** 53 36N 8 28 E
Weser-Ems □, *Germany* **30 C3** 53 0N 7 30 E
Weserbergland, *Germany* **30 C5** 52 12N 9 7 E
Wesiri, *Indonesia* **82 C3** 7 30 S 126 30 E
Wesley, *U.S.A.* **152 C7** 32 29N 82 20W
Wesley Vale, *U.S.A.* **159 J10** 35 3N 106 2W
Wesleyville, *Canada* **141 C9** 49 8N 53 36W
Wesleyville, *U.S.A.* **150 D4** 42 9N 80 0W
Wessel, C., *Australia* **126 A2** 10 59 S 136 46 E
Wessel Is., *Australia* **126 A2** 11 10 S 136 45 E
Wesselburen, *Germany* **30 A4** 54 13N 8 54 E
Wessington, *U.S.A.* **154 C5** 44 27N 98 42W
Wessington Springs, *U.S.A.* **154 C5** 44 5N 98 34W
West, *U.S.A.* **155 K6** 31 48N 97 6W
West Allis, *U.S.A.* **157 A10** 43 1N 88 0W
West B., *Fla., U.S.A.* **152 E4** 30 10N 85 45W
West B., *La., U.S.A.* **155 L10** 29 3N 89 22W
West Baines →, *Australia* **124 C4** 15 38 S 129 59 E

Y

Yerólakkos, *Cyprus* **38 D12** 35 11N 33 15 E
Yeropol, *Russia* **65 C17** 65 15N 168 40 E
Yeropótamos →, *Greece* **38 D6** 35 3N 24 50 E
Yeroskipos, *Cyprus* **38 E11** 34 46N 32 28 E
Yershov, *Russia* **60 E9** 51 23N 48 27 E
Yerunaja, Cerro, *Peru* **172 C2** 10 16 S 76 55W
Yerushalayim = Jerusalem, *Israel* **103 D4** 31 47N 35 10 E
Yerville, *France* **26 C7** 49 40N 0 53 E
Yes Tor, *U.K.* **21 G4** 50 41N 4 0W
Yesagyo, *Burma* **90 E5** 21 38N 95 14 E
Yesan, *S. Korea* **75 F14** 36 41N 126 51 E
Yeşilhisar, *Turkey* **100 C6** 38 20N 35 5 E
Yeşilırmak →, *Turkey* **100 B7** 41 22N 36 37 E
Yeşilkent, *Turkey* **100 D7** 36 57N 36 12 E
Yeşilköy, *Turkey* **51 F12** 40 57N 28 49 E
Yeşilova, *Turkey* **49 D11** 37 31N 29 46 E
Yeşilyurt, *Manisa, Turkey* **49 C10** 38 22N 28 40 E
Yeşilyurt, *Muğla, Turkey* **49 D10** 37 10N 28 20 E
Yesnogorsk, *Russia* **58 E9** 54 32N 37 38 E
Yeso, *U.S.A.* **155 H2** 34 26N 104 37W
Yessentuki, *Russia* **61 H6** 44 5N 42 53 E
Yessey, *Russia* **65 C11** 68 29N 102 10 E
Yeste, *Spain* **41 G2** 38 22N 2 19W
Yeu, Î. d', *France* **26 F4** 46 42N 2 20W
Yevlakh = Yevlax, *Azerbaijan* **61 K8** 40 39N 47 7 E
Yevlax, *Azerbaijan* **61 K8** 40 39N 47 7 E
Yevpatoriya, *Ukraine* **59 K7** 45 15N 33 20 E
Yeya →, *Russia* **59 J10** 46 40N 38 40 E
Yeysk, *Russia* **59 J10** 46 40N 38 12 E
Yezd = Yazd, *Iran* **97 D7** 31 55N 54 27 E
Yezerishche, *Belarus* **58 E5** 55 50N 29 59 E
Yhati, *Paraguay* **174 B4** 25 45 S 56 35W
Yhú, *Paraguay* **175 B4** 25 0 S 56 0W
Yi →, *Uruguay* **174 C4** 33 7 S 57 8W
Yi 'Allaq, G., *Egypt* **103 E2** 30 22N 33 32 E
Yi He →, *China* **75 G10** 34 10N 118 8 E
Yi Xian, *Anhui, China* **77 C11** 29 55N 117 57 E
Yi Xian, *Hebei, China* **74 E8** 39 20N 115 30 E
Yi Xian, *Liaoning, China* **75 D11** 41 30N 121 22 E
Yialí, *Greece* **49 E9** 36 41N 27 11 E
Yialiás →, *Cyprus* **38 D12** 35 9N 33 44 E
Yi'allaq, G., *Egypt* **106 H8** 30 21N 33 31 E
Yialousa, *Cyprus* **38 D13** 35 32N 34 10 E
Yiáltra, *Greece* **48 C4** 38 51N 22 59 E
Yianisádhes, *Greece* **38 D8** 35 20N 26 10 E
Yiannitsa, *Greece* **50 F6** 40 46N 22 24 E
Yibin, *China* **76 C5** 28 45N 104 32 E
Yichang, *China* **77 B8** 30 40N 111 20 E
Yicheng, *Henan, China* **77 B9** 31 41N 112 12 E
Yicheng, *Shanxi, China* **74 G6** 35 42N 111 40 E
Yichuan, *China* **74 F6** 36 2N 110 10 E
Yichun, *Heilongjiang, China* **73 B7** 47 44N 128 52 E
Yichun, *Jiangxi, China* **77 D10** 27 48N 114 22 E
Yidu, *Hubei, China* **77 B8** 30 25N 111 27 E
Yidu, *Shandong, China* **75 F10** 36 43N 118 28 E
Yidun, *China* **76 B2** 30 22N 99 21 E
Yihuang, *China* **77 D11** 27 30N 116 12 E
Yijun, *China* **74 G5** 35 28N 109 8 E
Yıldız Dağları, *Turkey* **51 E11** 41 48N 27 36 E
Yıldızeli, *Turkey* **100 C7** 39 51N 36 36 E
Yilehuli Shan, *China* **73 A7** 51 20N 124 20 E
Yiliang, *Yunnan, China* **76 D5** 27 38N 104 2 E
Yiliang, *Yunnan, China* **76 E4** 24 56N 103 11 E
Yilong, *China* **76 B6** 31 34N 106 23 E
Yimen, *China* **76 E4** 24 40N 102 10 E
Yimianpo, *China* **75 B15** 45 7N 128 2 E
Yinchuan, *China* **74 E4** 38 30N 106 15 E
Yindarlgooda, L., *Australia* **125 F3** 30 40 S 121 52 E
Ying He →, *China* **74 H9** 32 30N 116 30 E
Ying Xian, *China* **74 E7** 39 32N 113 10 E
Yingcheng, *China* **77 B9** 30 56N 113 35 E
Yingde, *China* **77 E9** 24 10N 113 25 E
Yingjiang, *China* **76 E1** 24 41N 97 55 E
Yingjing, *China* **76 C4** 29 41N 102 52 E
Yingkou, *China* **75 D12** 40 37N 122 18 E
Yingshan, *Henan, China* **77 B9** 31 35N 113 50 E
Yingshan, *Hubei, China* **77 B10** 30 41N 115 32 E
Yingshan, *Sichuan, China* **76 B6** 31 4N 106 35 E
Yingshang, *China* **77 A11** 32 38N 116 12 E
Yining, *China* **64 E9** 43 58N 81 10 E
Yinjiang, *China* **76 C7** 28 1N 108 21 E
Yinnietharra, *Australia* **124 D2** 24 39 S 116 12 E
Yiofiros →, *Greece* **38 D7** 35 20N 25 6 E
Yioúra, *Nótios Aiyaíon, Greece* **48 D6** 37 32N 24 40 E
Yioúra, *Thessalía, Greece* **48 B6** 39 23N 24 10 E
Yipinglang, *China* **76 E3** 25 10N 101 52 E
Yirga Alem, *Ethiopia* **107 F4** 6 48N 38 22 E
Yishan, *China* **76 E7** 24 28N 108 38 E
Yishui, *China* **75 G10** 35 47N 118 30 E
Yíthion, *Greece* **48 E4** 36 46N 22 34 E
Yitiaoshan, *China* **74 F3** 37 5N 104 2 E
Yitong, *China* **75 C13** 43 13N 125 20 E
Yiwu, *China* **77 C13** 29 20N 120 3 E
Yixing, *China* **77 B12** 31 21N 119 48 E
Yiyang, *Henan, China* **74 G7** 34 27N 112 10 E
Yiyang, *Hunan, China* **77 C9** 28 35N 112 18 E
Yiyang, *Jiangxi, China* **77 C11** 28 22N 117 20 E
Yizhang, *China* **77 E9** 25 27N 112 57 E
Yizheng, *China* **77 A12** 32 18N 119 10 E
Yli-Kitka, *Finland* **14 C23** 66 8N 28 30 E
Ylitornio, *Finland* **14 C20** 66 19N 23 39 E
Ylivieska, *Finland* **14 D21** 64 4N 24 28 E
Yngaren, *Sweden* **17 F10** 58 50N 16 35 E
Yoakum, *U.S.A.* **155 L6** 29 17N 97 9W
Yobe □, *Nigeria* **113 C7** 12 0N 11 30 E
Yobuko, *Japan* **70 D1** 33 32N 129 54 E
Yog Pt., *Phil.* **80 D5** 14 6N 124 12 E
Yogan, *Togo* **113 D5** 6 23N 1 30 E
Yoğuntaş, *Turkey* **51 E11** 41 50N 27 4 E
Yogyakarta, *Indonesia* **85 D4** 7 49 S 110 22 E
Yogyakarta □, *Indonesia* **85 D4** 7 48 S 110 22 E
Yoho Nat. Park, *Canada* **142 C5** 51 25N 116 30W
Yojoa, L. de, *Honduras* **164 D2** 14 53N 88 0W
Yŏju, *S. Korea* **75 F14** 37 20N 127 35 E
Yokadouma, *Cameroon* **114 B2** 3 26N 14 55 E
Yōkaichiba, *Japan* **71 B12** 35 42N 140 33 E
Yokkaichi, *Japan* **71 C8** 34 55N 136 38 E
Yoko, *Cameroon* **113 D7** 5 32N 12 20 E
Yokohama, *Japan* **71 B11** 35 27N 139 28 E
Yokosuka, *Japan* **71 B11** 35 20N 139 40 E
Yokote, *Japan* **68 E10** 39 20N 140 30 E
Yola, *Nigeria* **113 D7** 9 10N 12 29 E
Yolaina, Cordillera de, *Nic.* **164 D3** 11 30N 84 0W
Yolombo, *Dem. Rep. of the Congo* **114 C4** 1 36 S 23 12 E
Yombi, *Gabon* **114 C2** 1 26 S 10 37 E
Yonago, *Japan* **70 B5** 35 25N 133 19 E
Yonaguni-Jima, *Japan* **69 M1** 24 27N 123 0 E
Yŏnan, *N. Korea* **75 F14** 37 55N 126 11 E
Yonezawa, *Japan* **68 F10** 37 57N 140 4 E
Yong Peng, *Malaysia* **87 M4** 2 0N 103 3 E
Yong Sata, *Thailand* **87 J2** 7 8N 99 41 E
Yongampo, *N. Korea* **75 E13** 39 56N 124 23 E
Yong'an, *China* **77 E11** 25 59N 117 25 E
Yongcheng, *China* **74 H9** 33 55N 116 20 E
Yŏngchŏn, *S. Korea* **75 G15** 35 58N 128 56 E
Yongchuan, *China* **76 C5** 29 17N 105 55 E
Yongchun, *China* **77 E12** 25 16N 118 20 E
Yongdeng, *China* **74 F2** 36 38N 103 25 E
Yongding, *China* **77 E11** 24 43N 116 45 E
Yŏngdŏk, *S. Korea* **75 F15** 36 24N 129 22 E
Yŏngdŭngpo, *S. Korea* **75 F14** 37 31N 126 54 E
Yongfeng, *China* **77 D10** 27 20N 115 22 E
Yongfu, *China* **76 E7** 24 59N 109 59 E
Yonghe, *China* **74 F6** 36 46N 110 38 E
Yŏnghŭng, *N. Korea* **75 E14** 39 31N 127 18 E
Yongji, *China* **74 G6** 34 52N 110 28 E
Yŏngju, *S. Korea* **75 F15** 36 50N 128 40 E
Yongkang, *Yunnan, China* **76 E2** 24 9N 99 20 E
Yongkang, *Zhejiang, China* **77 C13** 28 55N 120 2 E
Yongnian, *China* **74 F8** 36 47N 114 29 E
Yongning, *Guangxi Zhuangzu, China* **76 F7** 22 44N 108 28 E
Yongning, *Ningxia Huizu, China* **74 E4** 38 15N 106 14 E
Yongping, *China* **76 E2** 25 27N 99 38 E
Yongqing, *China* **74 E9** 39 25N 116 28 E
Yongren, *China* **76 D3** 26 4N 101 40 E
Yongshan, *China* **76 C4** 28 11N 103 35 E
Yongsheng, *China* **76 D3** 26 38N 100 46 E
Yongshun, *China* **76 C7** 29 2N 109 51 E
Yongtai, *China* **77 E12** 25 49N 118 58 E
Yŏngwŏl, *S. Korea* **75 F15** 37 11N 128 28 E
Yongxin, *China* **77 D10** 26 58N 114 15 E
Yongxing, *China* **77 D9** 26 9N 113 8 E
Yongxiu, *China* **77 C10** 29 2N 115 42 E
Yonibana, *S. Leone* **112 D2** 8 30N 12 19W
Yonkers, *U.S.A.* **151 F11** 40 56N 73 54W
Yonne □, *France* **27 E10** 47 50N 3 40 E
Yonne →, *France* **27 D9** 48 23N 2 58 E
York, *Australia* **125 F2** 31 52 S 116 47 E
York, *U.K.* **20 D6** 53 58N 1 6W
York, *Ala., U.S.A.* **149 J1** 32 29N 88 18W
York, *Nebr., U.S.A.* **154 E6** 40 52N 97 36W
York, *Pa., U.S.A.* **148 F7** 39 58N 76 44W
York, C., *Australia* **126 A3** 10 42 S 142 31 E
York, City of □, *U.K.* **20 D6** 53 58N 1 6W
York, Kap, *Greenland* **10 B4** 75 55N 66 25W
York, Vale of, *U.K.* **20 C6** 54 15N 1 25W
York Sd., *Australia* **124 C4** 15 0 S 125 5 E
Yorke Pen., *Australia* **128 C2** 34 50 S 137 40 E
Yorkshire Wolds, *U.K.* **20 C7** 54 8N 0 31W
Yorkton, *Canada* **143 C8** 51 11N 102 28W
Yorktown, *U.S.A.* **155 L6** 28 59N 97 30W
Yorkville, *Calif., U.S.A.* **160 G3** 38 52N 123 13W
Yorkville, *Ga., U.S.A.* **152 B5** 33 55N 84 58W
Yorkville, *Ill., U.S.A.* **157 B5** 41 38N 88 27W
Yornup, *Australia* **125 F2** 34 2 S 116 10 E
Yoro, *Honduras* **164 C2** 15 9N 87 7W
Yoron-Jima, *Japan* **69 L4** 27 2N 128 26 E
Yos Sudarso, Pulau = Dolak, Pulau, *Indonesia* **83 C5** 8 0 S 138 30 E
Yosemite National Park, *U.S.A.* **160 H7** 37 45N 119 40W
Yosemite Village, *U.S.A.* **160 H7** 37 45N 119 35W
Yoshii, *Japan* **70 D1** 33 16N 129 46 E
Yoshimatsu, *Japan* **70 F2** 32 0N 130 47 E
Yoshkar Ola, *Russia* **60 B8** 56 38N 47 55 E
Yŏsu, *S. Korea* **75 G14** 34 47N 127 45 E
Yotala, *Bolivia* **173 D4** 19 10 S 65 17W
Yotvata, *Israel* **103 F4** 29 55N 35 2 E
You Xian, *China* **77 D9** 27 1N 113 17 E
Youbou, *Canada* **142 D4** 48 53N 124 13W
Youghal, *Ireland* **23 E4** 51 56N 7 52W
Youghal B., *Ireland* **23 E4** 51 55N 7 49W
Youkounkoun, *Guinea* **112 C2** 12 35N 13 11W
Young, *Australia* **129 C8** 34 19 S 148 18 E
Young, *Canada* **143 C7** 51 47N 105 45W
Young, *Uruguay* **174 C4** 32 44 S 57 36W
Young Ra., *N.Z.* **131 E4** 44 10 S 169 30 E
Younghusband, L., *Australia* **128 A2** 30 50 S 136 5 E
Younghusband Pen., *Australia* **128 D3** 36 0 S 139 25 E
Youngstown, *Canada* **143 C6** 51 35N 111 10W
Youngstown, *Fla., U.S.A.* **152 E4** 30 22N 85 26W
Youngstown, *N.Y., U.S.A.* **150 C5** 43 15N 79 3W
Youngstown, *Ohio, U.S.A.* **150 E4** 41 6N 80 39W
Youngsville, *U.S.A.* **150 E5** 41 51N 79 19W
Youssoufia, *Morocco* **110 B3** 32 16N 8 31W
Youxi, *China* **77 D12** 26 10N 118 13 E
Youyang, *China* **76 C7** 28 47N 108 42 E
Youyu, *China* **74 D7** 40 10N 112 20 E
Yoweragabbie, *Australia* **125 E2** 28 14 S 117 39 E
Yowrie, *Australia* **129 D8** 36 17 S 149 46 E
Yozgat, *Turkey* **100 C6** 39 51N 34 47 E
Ypané →, *Paraguay* **174 A4** 23 29 S 57 19W
Yport, *France* **26 C7** 49 45N 0 15 E
Ypres = Ieper, *Belgium* **24 D2** 50 51N 2 53 E
Ypsilanti, *U.S.A.* **157 B13** 42 14N 83 37W
Yreka, *U.S.A.* **158 F2** 41 44N 122 38W
Ysabel Chan., *Papua N. G.* **132 B6** 2 0 S 150 0 E
Ysleta, *U.S.A.* **159 L10** 31 45N 106 24W
Yssingeaux, *France* **29 C8** 45 9N 4 8 E
Ystabø, *Norway* **18 E2** 59 3N 5 23 E
Ystad, *Sweden* **17 J7** 55 26N 13 50 E
Ysyk-Köl, *Kyrgyzstan* **63 B8** 42 26N 76 12 E
Ysyk-Köl, Ozero, *Kyrgyzstan* **63 B8** 42 25N 77 15 E
Ythan →, *U.K.* **22 D7** 57 19N 1 59W
Ytre Arna, *Norway* **18 D2** 60 27N 5 25 E
Ytre Enebakk, *Norway* **18 E8** 59 44N 11 0 E
Ytre Rendal, *Norway* **18 C8** 61 46N 11 8 E
Ytterhogdal, *Sweden* **16 B8** 62 12N 14 56 E
Ytyk Kyuyel, *Russia* **65 C14** 62 30N 133 45 E
Yu Jiang →, *China* **73 D6** 23 22N 110 3 E
Yu Xian, *Hebei, China* **74 E8** 39 50N 114 35 E
Yu Xian, *Henan, China* **74 G7** 34 10N 113 28 E
Yu Xian, *Shanxi, China* **74 E7** 38 5N 113 20 E
Yuan Jiang →, *Hunan, China* **77 C8** 28 55N 111 50 E
Yuan Jiang →, *Yunnan, China* **76 F4** 22 20N 103 59 E
Yuan'an, *China* **77 B8** 31 3N 111 34 E
Yuanjiang, *Hunan, China* **77 C9** 28 47N 112 21 E
Yuanjiang, *Yunnan, China* **76 F4** 23 32N 102 0 E
Yüanli, *Taiwan* **77 E13** 24 27N 120 39 E
Yüanlin, *Taiwan* **77 F13** 23 57N 120 33 E
Yuanlin, *Taiwan* **77 F13** 23 58N 120 30 E
Yuanling, *China* **77 C8** 28 29N 110 22 E
Yuanmou, *China* **76 E3** 25 42N 101 53 E
Yuanqu, *China* **74 G6** 35 18N 111 40 E
Yuanyang, *Henan, China* **74 G7** 35 3N 113 58 E
Yuanyang, *Yunnan, China* **76 F4** 23 10N 102 43 E
Yuat →, *Papua N. G.* **132 C2** 4 10 S 143 52 E
Yuba →, *U.S.A.* **160 F5** 39 8N 121 36W
Yuba City, *U.S.A.* **160 F5** 39 8N 121 37W
Yūbari, *Japan* **68 C10** 43 4N 141 59 E
Yūbetsu, *Japan* **68 B11** 44 13N 143 50 E
Yucatán □, *Mexico* **163 C7** 21 30N 86 30W
Yucatán, Canal de, *Caribbean* **164 B2** 22 0N 86 30W
Yucatán, Península de, *Mexico* **136 H11** 19 30N 89 0W
Yucatán Basin, *Cent. Amer.* **136 H11** 19 0N 86 0W
Yucatan Str. = Yucatán, Canal de, *Caribbean* **164 B2** 22 0N 86 30W
Yucca, *U.S.A.* **161 L12** 34 52N 114 9W
Yucca Valley, *U.S.A.* **161 L10** 34 8N 116 27W
Yucheng, *China* **74 F9** 36 55N 116 32 E
Yuci, *China* **74 F7** 37 42N 112 46 E
Yudu, *China* **77 E10** 25 59N 115 30 E
Yuendumu, *Australia* **124 D5** 22 16 S 131 49 E
Yueqing, *China* **77 C13** 28 9N 120 59 E
Yueqing Wan, *China* **77 C13** 28 5N 121 20 E
Yuexi, *Anhui, China* **77 B11** 30 50N 116 20 E
Yuexi, *Sichuan, China* **76 C4** 28 37N 102 26 E
Yueyang, *China* **77 C9** 29 21N 113 5 E
Yufu-Dake, *Japan* **70 D3** 33 17N 131 33 E
Yugan, *China* **77 C11** 28 43N 116 37 E
Yugoslavia ■, *Europe* **50 C4** 43 20N 20 0 E
Yuhuan, *China* **77 C13** 28 9N 121 12 E
Yujiang, *China* **77 C11** 28 10N 116 43 E
Yukhnov, *Russia* **58 E8** 54 44N 35 15 E
Yūki, *Japan* **71 A11** 36 18N 139 53 E
Yukon →, *U.S.A.* **138 B3** 62 32N 163 54W
Yukon Territory □, *Canada* **138 B6** 63 0N 135 0W
Yüksekova, *Turkey* **101 D11** 37 34N 44 16 E
Yukta, *Russia* **65 C11** 63 26N 105 42 E
Yukuhashi, *Japan* **70 D2** 33 44N 130 59 E
Yule →, *Australia* **124 D2** 20 41 S 118 17 E
Yulee, *U.S.A.* **152 E8** 30 38N 81 36W
Yuli, *Nigeria* **113 D7** 9 44N 10 12 E
Yulin, *Guangxi Zhuangzu, China* **77 F8** 22 40N 110 8 E
Yulin, *Shaanxi, China* **74 E5** 38 20N 109 30 E
Yulin, *Shensi, China* **86 C7** 38 15N 109 30 E
Yuma, *Ariz., U.S.A.* **161 N12** 32 43N 114 37W
Yuma, *Colo., U.S.A.* **154 E3** 40 8N 102 43W
Yuma, B. de, *Dom. Rep.* **165 C6** 18 20N 68 35W
Yumali, *Australia* **128 C3** 35 32 S 139 45 E
Yumbe, *Uganda* **118 B3** 3 28N 31 15 E
Yumbi, *Dem. Rep. of the Congo* **118 C2** 1 12 S 26 15 E
Yumbo, *Colombia* **168 C2** 3 35N 76 28W
Yumen, *China* **72 C4** 39 50N 97 30 E
Yumurtalık, *Turkey* **100 D6** 36 45N 35 43 E
Yun Ho →, *China* **75 E9** 39 10N 117 10 E
Yun Xian, *Hubei, China* **77 A8** 32 50N 110 46 E
Yun Xian, *Yunnan, China* **76 E3** 24 27N 100 8 E
Yunak, *Turkey* **100 C4** 38 49N 31 43 E
Yunan, *China* **77 F8** 23 12N 111 30 E
Yuncheng, *Henan, China* **74 G8** 35 36N 115 57 E
Yuncheng, *Shanxi, China* **74 G6** 35 2N 111 0 E
Yundamindra, *Australia* **125 E3** 29 15 S 122 6 E
Yunfu, *China* **77 F9** 22 50N 112 5 E
Yungas, *Bolivia* **173 D4** 17 0 S 66 0W
Yungay, *Chile* **174 D1** 37 10 S 72 5W
Yungay, *Peru* **172 B2** 9 2 S 77 45W
Yunhe, *China* **77 C12** 28 8N 119 33 E
Yunlin, *Taiwan* **77 F13** 23 42N 120 30 E
Yunling, *China* **76 D2** 27 0N 99 20 E
Yunlong, *China* **76 E2** 25 57N 99 13 E
Yunmeng, *China* **77 B9** 31 2N 113 43 E
Yunnan □, *China* **76 E4** 25 0N 102 0 E
Yunomae, *Japan* **70 E2** 32 12N 130 59 E
Yunotso, *Japan* **70 B4** 35 5N 132 21 E
Yunquera de Henares, *Spain* **40 E1** 40 47N 3 11W
Yunt Dağı, *Turkey* **49 C9** 38 56N 27 13 E
Yunta, *Australia* **128 B3** 32 34 S 139 36 E
Yunxi, *China* **74 H6** 33 0N 110 22 E
Yunxiao, *China* **77 F11** 23 59N 117 18 E
Yunyang, *China* **76 B7** 30 58N 108 54 E
Yuping, *China* **76 D7** 27 13N 108 56 E
Yupukarri, *Guyana* **169 C6** 3 45N 59 20W
Yupyongdong, *N. Korea* **75 D15** 41 49N 128 53 E
Yuqing, *China* **76 D6** 27 13N 107 53 E
Yurga, *Russia* **64 D9** 55 42N 84 51 E
Yurimaguas, *Peru* **172 B2** 5 55 S 76 7W
Yurla, *Russia* **62 B5** 59 22N 54 21 E
Yurya, *Russia* **62 B2** 59 1N 49 13 E
Yuryev-Polskiy, *Russia* **58 D10** 56 30N 39 40 E
Yuryevets, *Russia* **60 B6** 57 25N 43 2 E
Yuryung Kaya, *Russia* **65 B12** 72 48N 113 23 E
Yuryuzan, *Russia* **62 D7** 54 57N 58 28 E
Yuscarán, *Honduras* **164 D2** 13 58N 86 45W
Yushanzhen, *China* **76 C7** 29 28N 108 22 E
Yushe, *China* **74 F7** 37 4N 112 58 E
Yushu, *Jilin, China* **75 B14** 44 43N 126 38 E
Yushu, *Qinghai, China* **72 C4** 33 5N 96 55 E
Yusufeli, *Turkey* **101 B9** 40 50N 41 33 E
Yutai, *China* **74 G9** 35 0N 116 45 E
Yutian, *China* **75 E9** 39 53N 117 45 E
Yuxarı Qarabağ = Nagorno-Karabakh, *Azerbaijan* **101 C12** 39 55N 46 45 E
Yuxi, *China* **76 E4** 24 30N 102 35 E
Yuyao, *China* **77 B13** 30 3N 121 10 E
Yuzawa, *Japan* **68 E10** 39 10N 140 30 E
Yuzha, *Russia* **60 B6** 56 34N 42 1 E
Yuzhno-Sakhalinsk, *Russia* **65 E15** 46 58N 142 45 E
Yuzhno-Surkhanskoye Vdkhr., *Uzbekistan* **63 E3** 37 53N 67 42 E
Yuzhnouralsk, *Russia* **62 D8** 54 26N 61 15 E
Yvelines □, *France* **27 D8** 48 40N 1 45 E
Yverdon, *Switz.* **32 C3** 46 47N 6 39 E
Yvetot, *France* **26 C7** 49 37N 0 44 E
Yvonand, *Switz.* **32 C3** 46 48N 6 44 E
Yzeure, *France* **27 F10** 46 33N 3 22 E

Z

Zaalayskiy Khrebet, *Asia* **63 D6** 39 20N 73 0 E
Zaanstad, *Neths.* **24 B4** 52 27N 4 50 E
Zab, Monts du, *Algeria* **111 B6** 34 55N 5 0 E
Zab al Kabīr →, *Iraq* **101 D10** 36 1N 43 24 E
Zāb aş Şagīr →, *Iraq* **101 E10** 35 17N 43 29 E
Žabalj, *Serbia, Yug.* **52 E5** 45 21N 20 5 E
Žabari, *Serbia, Yug.* **50 B5** 44 22N 21 15 E
Zabarjad, *Egypt* **106 C4** 23 40N 36 12 E
Zabaykalsk, *Russia* **65 E12** 49 40N 117 25 E
Zabid, *Yemen* **98 D3** 14 0N 43 10 E
Zabīd, W. →, *Yemen* **98 D3** 14 7N 43 6 E
Ząbki, *Poland* **55 F8** 52 17N 21 7 E
Ząbkowice Śląskie, *Poland* **55 H3** 50 35N 16 50 E
Žabljak, *Montenegro, Yug.* **50 C3** 43 18N 19 5 E
Zabłudów, *Poland* **55 F9** 53 0N 23 19 E
Žabno, *Poland* **55 H7** 50 9N 20 53 E
Zābol, *Iran* **97 D9** 31 0N 61 32 E
Zābol □, *Afghan.* **91 C2** 32 0N 67 0 E
Zābolī, *Iran* **97 E9** 27 10N 61 35 E
Zabré, *Burkina Faso* **113 C4** 11 12N 0 36W
Zábřeh, *Czech Rep.* **35 B9** 49 53N 16 52 E
Zabrze, *Poland* **55 H5** 50 18N 18 50 E
Zacapa, *Guatemala* **164 D2** 14 59N 89 31W
Zacapu, *Mexico* **162 D4** 19 50N 101 43W
Zacatecas, *Mexico* **162 C4** 22 49N 102 34W
Zacatecas □, *Mexico* **162 C4** 23 30N 103 0W
Zacatecoluca, *El Salv.* **164 D2** 13 29N 88 51W
Zacoalco, *Mexico* **162 C4** 20 14N 103 33W
Zacualtipán, *Mexico* **163 C5** 20 39N 98 36W
Zadar, *Croatia* **45 D12** 44 8N 15 14 E
Zadawa, *Nigeria* **113 C7** 11 33N 10 19 E
Zadetkyi Kyun, *Burma* **87 H2** 10 0N 98 25 E
Zadonsk, *Russia* **59 F10** 52 25N 38 56 E
Zafarqand, *Iran* **97 C7** 33 11N 52 29 E
Zafora, *Greece* **49 E8** 36 5N 26 24 E
Zafra, *Spain* **43 G4** 38 26N 6 30W
Żagań, *Poland* **55 G2** 51 39N 15 22 E
Žagarė, *Lithuania* **54 B10** 56 21N 23 15 E
Zagazig, *Egypt* **106 H7** 30 40N 31 30 E
Zāgheh, *Iran* **97 C6** 33 30N 48 42 E
Zaghouan, *Tunisia* **108 A2** 36 23N 10 10 E
Zagliverion, *Greece* **50 F7** 40 36N 23 15 E
Zaglou, *Algeria* **111 C4** 27 17N 0 3W
Zagnanado, *Benin* **113 D5** 7 18N 2 28 E
Zagorá, *Greece* **48 B5** 39 27N 23 6 E
Zagora, *Morocco* **110 B3** 30 22N 5 51W
Zagorje, *Slovenia* **45 B11** 46 8N 15 0 E
Zagórów, *Poland* **55 F4** 52 10N 17 54 E
Zagorsk = Sergiyev Posad, *Russia* **58 D10** 56 20N 38 10 E
Zagórz, *Poland* **55 J9** 49 30N 22 14 E
Zagreb, *Croatia* **45 C12** 45 50N 15 58 E
Zāgros, Kūhhā-ye, *Iran* **97 C6** 33 45N 48 5 E
Zagros Mts. = Zāgros, Kūhhā-ye, *Iran* **97 C6** 33 45N 48 5 E
Žagubica, *Serbia, Yug.* **50 B5** 44 15N 21 47 E
Zaguinaso, *Ivory C.* **112 C3** 10 1N 6 14W
Zagyva →, *Hungary* **52 C5** 47 5N 20 4 E
Zāhedān, *Fārs, Iran* **97 D7** 28 46N 53 52 E
Zāhedān, *Sīstān va Balūchestān, Iran* **97 D9** 29 30N 60 50 E
Zahirabad, *India* **94 F3** 17 43N 77 37 E
Zahlah, *Lebanon* **103 B4** 33 52N 35 50 E
Zahna, *Germany* **30 D8** 51 55N 12 49 E
Záhony, *Hungary* **52 B7** 48 25N 22 11 E
Zahrez Chergui, *Algeria* **111 B5** 35 11N 3 31 E
Zahrez Rharbi, *Algeria* **111 B5** 34 50N 2 55 E
Zailiyskiy Alatau, Khrebet, *Kazakstan* **63 B8** 43 5N 77 0 E
Zainsk, *Russia* **60 C11** 55 18N 52 4 E
Zaïre = Congo, Dem. Rep. of the ■, *Africa* **115 C4** 3 0S 23 0E
Zaire □, *Angola* **115 D2** 7 0S 14 0E
Zaïre → = Congo →, *Africa* **115 D2** 6 4 S 12 24 E
Zaječar, *Serbia, Yug.* **50 C6** 43 53N 22 18 E
Zakamensk, *Russia* **65 D11** 50 23N 103 17 E
Zakani, *Dem. Rep. of the Congo* **114 B4** 2 33N 23 16 E
Zakataly = Zaqatala, *Azerbaijan* **61 K8** 41 38N 46 35 E
Zakavkazye, *Asia* **57 F7** 42 0N 44 0 E
Zakháro, *Greece* **48 D3** 37 30N 21 39 E
Zakhodnaya Dzvina = Daugava →, *Latvia* **15 H21** 57 4N 24 3 E
Zākhū, *Iraq* **101 D10** 37 10N 42 50 E
Zákinthos, *Greece* **48 D2** 37 47N 20 57 E
Zákinthos □, *Greece* **48 D2** 37 47N 20 57 E
Zakopane, *Poland* **55 J6** 49 18N 19 57 E
Zakroczym, *Poland* **55 F7** 52 26N 20 38 E
Zákros, *Greece* **38 D8** 35 6N 26 10 E
Zala, *Angola* **115 D2** 7 52 S 13 42 E
Zala □, *Hungary* **52 D1** 46 42N 16 50 E
Zala →, *Hungary* **52 D2** 46 43N 17 16 E
Zalaegerszeg, *Hungary* **52 D1** 46 53N 16 47 E
Zalakomár, *Hungary* **52 D2** 46 33N 17 10 E
Zalalövő, *Hungary* **52 D1** 46 51N 16 35 E
Zalamea de la Serena, *Spain* **43 G5** 38 40N 5 38W
Zalamea la Real, *Spain* **43 H4** 37 41N 6 38W
Zalău, *Romania* **52 C8** 47 12N 23 3 E
Zalazna, *Russia* **62 B4** 58 39N 52 31 E
Žalec, *Slovenia* **45 B12** 46 16N 15 10 E
Zaleshchiki = Zalishchyky, *Ukraine* **59 H3** 48 45N 25 45 E
Zalew Wiślany, *Poland* **54 D6** 54 20N 19 50 E
Zalewo, *Poland* **54 E6** 53 50N 19 41 E
Zalim, *Si. Arabia* **98 B3** 22 43N 42 10 E
Zalingei, *Sudan* **109 F4** 12 51N 23 29 E
Zalishchyky, *Ukraine* **59 H3** 48 45N 25 45 E
Zaltan, Jabal, *Libya* **108 C3** 28 46N 19 45 E
Zambales □, *Phil.* **80 D3** 15 20N 120 10 E
Zambales Mts., *Phil.* **80 D3** 15 45N 120 5 E
Zambeke, *Dem. Rep. of the Congo* **118 B2** 2 8N 25 17 E
Zambeze →, *Africa* **119 F4** 18 35 S 36 20 E
Zambezi = Zambeze →, *Africa* **119 F4** 18 35 S 36 20 E
Zambezi, *Zambia* **115 E4** 13 30 S 23 15 E
Zambezia □, *Mozam.* **119 F4** 16 15 S 37 30 E
Zambia ■, *Africa* **119 F2** 15 0S 28 0 E
Zamboanga, *Phil.* **81 H4** 6 59N 122 3 E
Zamboanga del Norte □, *Phil.* **81 H4** 8 0N 123 0 E
Zamboanga del Sur □, *Phil.* **81 H4** 7 40N 123 0 E
Zamboanguita, *Phil.* **81 G4** 9 6N 123 12 E
Zambrano, *Colombia* **168 B3** 9 45N 74 49W
Zambrów, *Poland* **55 F9** 52 59N 22 14 E
Zametchino, *Russia* **60 D6** 53 30N 42 30 E
Zamora, *Ecuador* **168 D2** 4 4 S 78 58W
Zamora, *Mexico* **162 D4** 20 0N 102 21W
Zamora, *Spain* **42 D5** 41 30N 5 45W
Zamora □, *Spain* **42 D5** 41 30N 5 46W
Zamora-Chinchipe □, *Ecuador* **168 D2** 4 15 S 78 50W
Zamość, *Poland* **55 H10** 50 43N 23 15 E
Zamość □, *Poland* **55 H10** 50 40N 23 10 E
Zamuro, Sierra del, *Venezuela* **169 C5** 4 0N 62 30W
Zamzam, W. →, *Libya* **108 B2** 30 58N 14 48 E
Zan, *Ghana* **113 D4** 9 26N 0 17W
Zanaga, *Congo* **114 C2** 2 48 S 13 48 E
Záncara →, *Spain* **41 F1** 39 18N 3 18W
Zandvoort, *Neths.* **24 B4** 52 22N 4 32 E

304 Zanesville

Zanesville, *U.S.A.* **150 G2** 39 56N 82 1W
Zangābād, *Iran* **96 B5** 38 26N 46 44 E
Zangue →, *Mozam.* **119 F4** 17 50 S 35 21 E
Zanjān, *Iran* **97 B6** 36 40N 48 35 E
Zanjān □, *Iran* **97 B6** 37 20N 49 30 E
Zannone, *Italy* **46 B6** 40 58N 13 3 E
Zante = Zákinthos, *Greece* **48 D2** 37 47N 20 57 E
Zanthus, *Australia* **125 F3** 31 2 S 123 34 E
Zanzibar, *Tanzania* **118 D4** 6 12 S 39 12 E
Zanzūr, *Libya* **108 B2** 32 55N 13 1 E
Zaouiet El-Kala = Bordj Omar Driss, *Algeria* **111 C6** 28 10N 6 40 E
Zaouiet Reggane, *Algeria* **111 C5** 26 32N 0 3 E
Zaoyang, *China* **77 A9** 32 10N 112 45 E
Zaozhuang, *China* **75 G9** 34 50N 117 35 E
Zap Suyu = Zab al Kabīr →, *Iraq* **101 D10** 36 1N 43 24 E
Zapadna Morava →, *Serbia, Yug.* **50 C5** 43 38N 21 30 E
Zapadnaya Dvina, *Russia* **58 D7** 56 15N 32 3 E
Zapadnaya Dvina → = Daugava →, *Latvia* **15 H21** 57 4N 24 3 E
Západné Beskydy, *Europe* **35 B12** 49 30N 19 0 E
Zapadni Rodopi, *Bulgaria* **50 E7** 41 50N 24 0 E
Západočeský □, *Czech Rep.* **34 B6** 49 35N 13 0 E
Zapala, *Argentina* **176 A2** 39 0 S 70 5W
Zapaleri, Cerro, *Bolivia* **174 A2** 22 49 S 67 11W
Zapata, *U.S.A.* **155 M5** 26 55N 99 16W
Zapatón →, *Spain* **43 F4** 39 0N 6 49W
Zapiga, *Chile* **172 D4** 19 40 S 69 55W
Zapolyarnyy, *Russia* **56 A5** 69 26N 30 51 E
Zaporizhzhya, *Ukraine* **59 J8** 47 50N 35 10 E
Zaporozhye = Zaporizhzhya, *Ukraine* **59 J8** 47 50N 35 10 E
Zaqatala, *Azerbaijan* **61 K8** 41 38N 46 35 E
Zara, *Turkey* **100 C7** 39 58N 37 43 E
Zaragoza, *Colombia* **168 B3** 7 30N 74 52W
Zaragoza, *Coahuila, Mexico* **162 B4** 28 30N 101 0W
Zaragoza, *Nuevo León, Mexico* **163 C5** 24 0N 99 46W
Zaragoza, *Spain* **40 D4** 41 39N 0 53W
Zaragoza □, *Spain* **40 D4** 41 35N 1 0W
Zarand, *Kermān, Iran* **97 D8** 30 46N 56 34 E
Zarand, *Markazī, Iran* **97 C6** 35 18N 50 25 E
Zărandului, Munții, *Romania* **52 D7** 46 14N 22 7 E
Zaranj, *Afghan.* **91 C1** 30 55N 61 55 E
Zarasai, *Lithuania* **15 J22** 55 40N 26 20 E
Zarate, *Argentina* **174 C4** 34 7 S 59 0W
Zarautz, *Spain* **40 B2** 43 17N 2 10W
Zaraysk, *Russia* **58 E10** 54 48N 38 53 E
Zaraza, *Venezuela* **169 B4** 9 21N 65 19W
Zāreh, *Iran* **97 C6** 35 7N 49 9 E
Zarembo I., *U.S.A.* **142 B2** 56 20N 132 50W
Ẓaria, *Nigeria* **113 C6** 11 0N 7 40 E
Żarki, *Poland* **55 H6** 50 38N 19 23 E
Zárkon, *Greece* **48 B4** 39 38N 22 6 E
Zarneh, *Iran* **96 C5** 33 55N 46 10 E
Zărnești, *Romania* **53 E10** 45 33N 25 18 E
Ẓarós, *Greece* **38 D6** 35 8N 24 54 E
Żarów, *Poland* **55 H3** 50 56N 16 29 E
Zarqā', Nahr az →, *Jordan* **103 C4** 32 10N 35 37 E
Zarrīn, *Iran* **97 C7** 32 46N 54 37 E
Ẓaruma, *Ecuador* **168 D2** 3 40 S 79 38W
Żary, *Poland* **55 G2** 51 37N 15 10 E
Zarza de Granadilla, *Spain* **42 E4** 40 14N 6 3W
Zarzaîtine, *Algeria* **111 C6** 28 15N 9 34 E
Zarzal, *Colombia* **168 C2** 4 24N 76 4W
Zarzis, *Tunisia* **108 B2** 33 31N 11 2 E
Zas, *Spain* **42 B2** 43 4N 8 53W
Zaskar →, *India* **93 B7** 34 13N 77 20 E
Zaskar Mts., *India* **93 C7** 33 15N 77 30 E
Ẓastron, *S. Africa* **116 E4** 30 18 S 27 7 E
Żatec, *Czech Rep.* **34 A6** 50 20N 13 32 E
Zaterechnyy, *Russia* **61 H7** 44 48N 45 11 E
Zator, *Poland* **55 J6** 49 59N 19 28 E
Zavala, *Bos.-H.* **50 D1** 42 50N 17 59 E
Zavāreh, *Iran* **97 C7** 33 29N 52 28 E
Zavetnoye, *Russia* **61 G6** 47 13N 43 50 E
Zavidovići, *Bos.-H.* **52 F3** 44 27N 18 10 E
Zavitinsk, *Russia* **65 D13** 50 10N 129 20 E
Zavodovski, I., *Antarctica* **7 B1** 56 0 S 27 45W
Zavolzhsk, *Russia* **60 B6** 57 30N 42 0 E
Zavolzhye, *Russia* **60 B6** 56 37N 43 26 E
Zawadzkie, *Poland* **55 H5** 50 37N 18 28 E
Zawichost, *Poland* **55 H8** 50 48N 21 51 E
Zawidów, *Poland* **55 G2** 51 1N 15 1 E
Zawiercie, *Poland* **55 H6** 50 30N 19 24 E
Zāwiyat al Bayḑā = Al Bayḑā, *Libya* **108 B4** 32 50N 21 44 E
Zāwiyat Masūs, *Libya* **108 B4** 31 35N 21 1 E
Zawyet Shammâs, *Egypt* **106 A2** 31 30N 26 37 E
Zâwyet Um el Rakham, *Egypt* **106 A2** 31 18N 27 1 E
Zâwyet Ungeîla, *Egypt* **106 A2** 31 23N 26 42 E
Zāyā, *Iraq* **96 C5** 33 33N 44 13 E
Zaymah, *Si. Arabia* **98 B3** 21 37N 40 6 E
Zaysan, *Kazakstan* **64 E9** 47 28N 84 52 E
Zaysan, Oz., *Kazakstan* **64 E9** 48 0N 83 0 E
Zayü, *China* **76 C1** 28 48N 97 27 E
Zāzamt, W. →, *Libya* **108 B2** 30 29N 14 30 E
Zazir, O. →, *Algeria* **111 D6** 22 0N 5 40 E
Zázrivá, *Slovak Rep.* **35 B12** 49 16N 19 7 E
Zbarazh, *Ukraine* **59 H3** 49 43N 25 44 E
Zbąszyń, *Poland* **55 F2** 52 14N 15 56 E
Zbąszynek, *Poland* **55 F2** 52 16N 15 51 E
Zblewo, *Poland* **54 E5** 53 56N 18 19 E
Ždár nad Sázavov, *Czech Rep.* **34 B8** 49 34N 15 57 E
Zdolbuniv, *Ukraine* **59 G4** 50 30N 26 15 E
Ždrelo, *Serbia, Yug.* **50 B5** 44 16N 21 28 E
Zduńska Wola, *Poland* **55 G5** 51 37N 18 59 E
Zduny, *Poland* **55 G4** 51 39N 17 21 E
Zearing, *U.S.A.* **156 B3** 42 10N 93 18W
Zeballos, *Canada* **142 D3** 49 59N 126 50W
Zebediela, *S. Africa* **117 C4** 24 20 S 29 17 E
Zebulon, *U.S.A.* **152 B5** 33 6N 84 21W
Zeebrugge, *Belgium* **24 C3** 51 19N 3 12 E
Zeehan, *Australia* **126 G4** 41 52 S 145 25 E
Zeeland, *U.S.A.* **157 B10** 42 49N 86 1W
Zeeland □, *Neths.* **24 C3** 51 30N 3 50 E
Zeerust, *S. Africa* **116 D4** 25 31 S 26 4 E
Ẕefat, *Israel* **103 C4** 32 58N 35 29 E
Zegdou, *Algeria* **110 C4** 29 51N 4 45W
Zege, *Ethiopia* **107 E4** 11 43N 37 18 E
Zégoua, *Mali* **112 C3** 10 32N 5 35W
Zehdenick, *Germany* **30 C9** 52 58N 13 20 E
Zeigler, *U.S.A.* **156 G7** 37 54N 89 3W
Zeil, Mt., *Australia* **124 D5** 23 30 S 132 23 E
Zeila, *Somali Rep.* **120 B2** 11 21N 43 30 E
Zeist, *Neths.* **24 B5** 52 5N 5 15 E
Ẓeitz, *Germany* **30 D8** 51 2N 12 7 E
Żelechów, *Poland* **55 G8** 51 49N 21 54 E
Zelee, C., *Solomon Is.* **133 M11** 9 44 S 161 34 E
Zelengora, *Bos.-H.* **50 C2** 43 22N 18 30 E
Zelenodolsk, *Russia* **60 C9** 55 55N 48 30 E
Zelenogorsk, *Russia* **58 B5** 60 12N 29 43 E
Zelenograd, *Russia* **58 D9** 56 1N 37 12 E
Zelenogradsk, *Russia* **15 J19** 54 53N 20 29 E
Zelenokumsk, *Russia* **61 H6** 44 24N 44 0 E
Železná Ruda, *Czech Rep.* **34 B6** 49 8N 13 15 E
Železnik, *Serbia, Yug.* **50 B4** 44 43N 20 23 E
Želiezovce, *Slovak Rep.* **35 C11** 48 3N 18 40 E
Zelina, *Croatia* **45 C13** 45 57N 16 16 E
Zell, *Baden-W., Germany* **31 H3** 47 42N 7 52 E
Zell, *Rhld-Pfz., Germany* **31 E3** 50 1N 7 10 E
Zell am See, *Austria* **34 D5** 47 19N 12 47 E
Zella-Mehlis, *Germany* **30 E6** 50 39N 10 40 E
Zelów, *Poland* **55 G6** 51 28N 19 14 E
Zeltweg, *Austria* **34 D7** 47 11N 14 45 E
Zembra, I., *Tunisia* **108 A2** 37 5N 10 56 E
Zémio, *C.A.R.* **114 A5** 5 2N 25 5 E
Zemmora, *Algeria* **111 A5** 35 44N 0 51 E
Zemmur, *W. Sahara* **110 C2** 25 5N 12 0W
Zemoul, O. →, *Algeria* **110 C3** 29 15N 7 0W
Zempléni-hegység, *Hungary* **52 B6** 48 25N 21 25 E
Zemplínska šírava, *Slovak Rep.* **35 C15** 48 48N 22 0 E
Zemun, *Serbia, Yug.* **50 B4** 44 51N 20 25 E
Zendeh Jān, *Afghan.* **91 B1** 34 21N 61 45 E
Zengbe, *Cameroon* **113 D7** 5 46N 13 4 E
Zengcheng, *China* **77 F9** 23 13N 113 52 E
Zenica, *Bos.-H.* **52 F2** 44 10N 17 57 E
Zentsūji, *Japan* **70 C5** 34 14N 133 47 E
Zenza do Itombe, *Angola* **115 D2** 9 16 S 14 13 E
Žepče, *Bos.-H.* **52 F3** 44 28N 18 2 E
Zephyrhills, *U.S.A.* **153 G7** 28 14N 82 11W
Zeraf, Bahr ez →, *Sudan* **107 F3** 9 42N 30 52 E
Zeravshan, *Tajikistan* **63 D4** 39 10N 68 39 E
Zeravshanskiy, Khrebet, *Tajikistan* **63 D4** 39 20N 69 0 E
Ẓerbst, *Germany* **30 D8** 51 58N 12 5 E
Żerków, *Poland* **55 F4** 52 4N 17 32 E
Zermatt, *Switz.* **32 D5** 46 2N 7 46 E
Zernez, *Switz.* **33 C10** 46 42N 10 7 E
Zernograd, *Russia* **61 G5** 46 52N 40 19 E
Zerqani, *Albania* **50 E4** 41 27N 20 20 E
Zestaponi, *Georgia* **61 J6** 42 6N 43 0 E
Zetel, *Germany* **30 B3** 53 25N 7 58 E
Zeulenroda, *Germany* **30 E7** 50 39N 11 59 E
Zeven, *Germany* **30 B5** 53 17N 9 16 E
Zevenaar, *Neths.* **24 C6** 51 56N 6 5 E
Zévio, *Italy* **44 C8** 45 22N 11 8 E
Zeya, *Russia* **65 D13** 53 48N 127 14 E
Zeya →, *Russia* **65 D13** 51 42N 128 53 E
Zeytinbaği, *Turkey* **51 F12** 40 24N 28 47 E
Zeytindağ, *Turkey* **49 C9** 38 58N 27 4 E
Zghartā, *Lebanon* **103 A4** 34 21N 35 53 E
Zgierz, *Poland* **55 G6** 51 50N 19 27 E
Zgorzelec, *Poland* **55 G2** 51 10N 15 0 E
Zgurița, *Moldova* **53 B13** 48 8N 28 1 E
Zhabasak, *Kazakstan* **62 F8** 50 22N 61 41 E
Zhabinka, *Belarus* **59 F3** 52 13N 24 2 E
Zhailma, *Kazakstan* **62 F8** 51 37N 61 33 E
Zhalanash, *Kazakstan* **63 B9** 43 3N 78 38 E
Zhambyl, *Kazakstan* **63 B5** 42 54N 71 22 E
Zhambyl, Gora, *Kazakstan* **63 A6** 44 54N 73 0 E
Zhanadarya, *Kazakstan* **63 A2** 44 45N 64 40 E
Zhanatas, *Kazakstan* **63 B4** 43 35N 69 35 E
Zhangaly, *Kazakstan* **61 G10** 47 1N 50 37 E
Zhangaqazaly, *Kazakstan* **64 E7** 45 48N 62 6 E
Zhangbei, *China* **74 D8** 41 10N 114 45 E
Zhangguangcai Ling, *China* **75 B15** 45 0N 129 0 E
Zhangjiakou, *China* **74 D8** 40 48N 114 55 E
Zhangping, *China* **77 E11** 25 17N 117 23 E
Zhangpu, *China* **77 E11** 24 8N 117 35 E
Zhangwu, *China* **75 C12** 42 43N 123 52 E
Zhangye, *China* **72 C5** 38 50N 100 23 E
Zhangzhou, *China* **77 E11** 24 30N 117 35 E
Zhanhua, *China* **75 F10** 37 40N 118 8 E
Zhanjiang, *China* **77 G8** 21 15N 110 20 E
Zhannetty, Ostrov, *Russia* **65 B16** 76 43N 158 0 E
Zhanyi, *China* **76 E4** 25 38N 103 48 E
Zhanyu, *China* **75 B12** 44 30N 122 30 E
Zhao Xian, *China* **74 F8** 37 43N 114 45 E
Zhao'an, *China* **77 F11** 23 41N 117 10 E
Zhaocheng, *China* **74 F6** 36 22N 111 38 E
Zhaojue, *China* **76 C4** 28 1N 102 49 E
Zhaoping, *China* **77 E8** 24 11N 110 48 E
Zhaoqing, *China* **77 F9** 23 0N 112 20 E
Zhaotong, *China* **76 D4** 27 20N 103 44 E
Zhaoyuan, *Heilongjiang, China* **75 B13** 45 27N 125 0 E
Zhaoyuan, *Shandong, China* **75 F11** 37 20N 120 23 E
Zharkovskiy, *Russia* **58 E7** 55 56N 32 19 E
Zhashkiv, *Ukraine* **59 H6** 49 15N 30 5 E
Zhashui, *China* **74 H5** 33 40N 109 8 E
Zhayyq →, *Kazakstan* **57 E9** 47 0N 51 48 E
Zhdanov = Mariupol, *Ukraine* **59 J9** 47 5N 37 31 E
Zhecheng, *China* **74 G8** 34 7N 115 20 E
Zhegao, *China* **77 B11** 31 46N 117 45 E
Zhejiang □, *China* **77 C13** 29 0N 120 0 E
Zheleznodorozhnyy, *Russia* **56 B9** 62 35N 50 55 E
Zheleznogorsk, *Russia* **59 F8** 52 22N 35 23 E
Zheleznogorsk-Ilimskiy, *Russia* **65 D11** 56 34N 104 8 E
Zheltyye Vody = Zhovti Vody, *Ukraine* **59 H7** 48 21N 33 31 E
Zhen'an, *China* **74 H5** 33 27N 109 9 E
Zhenfeng, *China* **76 E5** 25 22N 105 40 E
Zheng'an, *China* **76 C6** 28 32N 107 27 E
Zhengding, *China* **74 E8** 38 8N 114 32 E
Zhenghe, *China* **77 D12** 27 20N 118 50 E
Zhengyang, *China* **77 A10** 32 37N 114 22 E
Zhengyangguan, *China* **77 A11** 32 30N 116 29 E
Zhengzhou, *China* **74 G7** 34 45N 113 34 E
Zhenhai, *China* **77 C13** 29 59N 121 42 E
Zhenjiang, *China* **77 A12** 32 11N 119 26 E
Zhenlai, *China* **75 B12** 45 50N 123 5 E
Zhenning, *China* **76 D5** 26 4N 105 45 E
Zhenping, *Henan, China* **74 H7** 33 10N 112 16 E
Zhenping, *Shaanxi, China* **76 B7** 31 59N 109 31 E
Zhenxiong, *China* **76 D5** 27 27N 104 50 E
Zhenyuan, *Gansu, China* **74 G4** 35 35N 107 30 E
Zhenyuan, *Guizhou, China* **76 D7** 27 4N 108 21 E
Zherdevka, *Russia* **60 E5** 51 56N 41 29 E
Zherong, *China* **77 D12** 27 15N 119 52 E
Zhetiqara, *Kazakstan* **62 E8** 52 11N 61 12 E
Zhezqazghan, *Kazakstan* **64 E7** 47 44N 67 40 E
Zhidan, *China* **74 F5** 36 48N 108 48 E
Zhigansk, *Russia* **65 C13** 66 48N 123 27 E
Zhigulevsk, *Russia* **60 D9** 53 28N 49 30 E
Zhijiang, *Hubei, China* **77 B8** 30 28N 111 45 E
Zhijiang, *Hunan, China* **76 D7** 27 27N 109 42 E
Zhijin, *China* **76 D5** 26 37N 105 45 E
Zhilinda, *Russia* **65 C12** 70 0N 114 20 E
Zhirnovsk, *Russia* **60 E7** 50 57N 44 49 E
Zhitomir = Zhytomyr, *Ukraine* **59 G5** 50 20N 28 40 E
Zhizdra, *Russia* **58 F8** 53 45N 34 40 E
Zhlobin, *Belarus* **59 F6** 52 55N 30 0 E
Zhmerinka = Zhmerynka, *Ukraine* **59 H5** 49 2N 28 2 E
Zhmerynka, *Ukraine* **59 H5** 49 2N 28 2 E
Zhob, *Pakistan* **91 C3** 31 20N 69 31 E
Zhodino = Zhodzina, *Belarus* **58 E5** 54 5N 28 17 E
Zhodzina, *Belarus* **58 E5** 54 5N 28 17 E
Zhokhova, Ostrov, *Russia* **65 B16** 76 4N 152 40 E
Zhong Xian, *China* **76 B7** 30 21N 108 1 E
Zhongdian, *China* **76 D2** 27 48N 99 42 E
Zhongdong, *China* **76 F6** 22 48N 107 47 E
Zhongdu, *China* **76 E7** 24 40N 109 40 E
Zhongning, *China* **74 F3** 37 29N 105 40 E
Zhongshan, *Guangdong, China* **77 F9** 22 26N 113 20 E
Zhongshan, *Guangxi Zhuangzu, China* **77 E8** 24 29N 111 18 E
Zhongtiao Shan, *China* **74 G6** 35 0N 111 10 E
Zhongwei, *China* **74 F3** 37 30N 105 12 E
Zhongxiang, *China* **77 B9** 31 12N 112 34 E
Zhongyang, *China* **74 F6** 37 20N 111 11 E
Zhoucun, *China* **75 F9** 36 47N 117 48 E
Zhouning, *China* **77 D12** 27 12N 119 20 E
Zhoushan Dao, *China* **77 C14** 28 5N 122 10 E
Zhouzhi, *China* **74 G5** 34 10N 108 12 E
Zhovti Vody, *Ukraine* **59 H7** 48 21N 33 31 E
Zhovtneve, *Ukraine* **59 J7** 46 54N 32 3 E
Zhovtnevoye = Zhovtneve, *Ukraine* **59 J7** 46 54N 32 3 E
Zhuanghe, *China* **75 E12** 39 40N 123 0 E
Zhuantobe, *Kazakstan* **63 B9** 43 43N 78 18 E
Zhucheng, *China* **75 G10** 36 0N 119 27 E
Zhugqu, *China* **74 H3** 33 40N 104 30 E
Zhuhai, *China* **77 F9** 22 15N 113 30 E
Zhuji, *China* **77 C13** 29 40N 120 10 E
Zhukovka, *Russia* **58 F7** 53 35N 33 50 E
Zhumadian, *China* **74 H8** 32 59N 114 2 E
Zhuo Xian, *China* **74 E8** 39 28N 115 58 E
Zhuolu, *China* **74 D8** 40 20N 115 12 E
Zhuozi, *China* **74 D7** 41 0N 112 25 E
Zhushan, *China* **77 A8** 32 15N 110 13 E
Zhuxi, *China* **76 A7** 32 25N 109 40 E
Zhuzhou, *China* **77 D9** 27 49N 113 12 E
Zhytomyr, *Ukraine* **59 G5** 50 20N 28 40 E
Zi Shui →, *China* **77 C9** 28 40N 112 40 E
Žiar nad Hronom, *Slovak Rep.* **35 C11** 48 35N 18 53 E
Zīārān, *Iran* **97 B6** 36 7N 50 32 E
Ziarat, *Pakistan* **92 D2** 30 25N 67 49 E
Zibo, *China* **75 F10** 36 47N 118 3 E
Zichang, *China* **74 F5** 37 18N 109 40 E
Zidarovo, *Bulgaria* **51 D11** 42 20N 27 24 E
Ziębice, *Poland* **55 H4** 50 37N 17 2 E
Zielona Góra, *Poland* **55 G2** 51 57N 15 31 E
Zielona Góra □, *Poland* **55 G2** 51 57N 15 30 E
Zierikzee, *Neths.* **24 C3** 51 40N 3 55 E
Ziesar, *Germany* **30 C8** 52 16N 12 17 E
Zifta, *Egypt* **106 H7** 30 43N 31 14 E
Zigey, *Chad* **109 F3** 14 43N 15 50 E
Zigong, *China* **76 C5** 29 15N 104 48 E
Zigui, *China* **77 B8** 31 0N 110 40 E
Ziguinchor, *Senegal* **112 C1** 12 35N 16 20W
Zihuatanejo, *Mexico* **162 D4** 17 38N 101 33W
Zijin, *China* **77 F10** 23 33N 115 8 E
Zile, *Turkey* **100 B6** 40 15N 35 52 E
Žilina, *Slovak Rep.* **35 B11** 49 12N 18 42 E
Žilinský □, *Slovak Rep.* **35 B12** 49 10N 19 0 E
Zillah, *Libya* **108 C3** 28 30N 17 33 E
Zillertaler Alpen, *Austria* **34 D4** 47 6N 11 45 E
Zima, *Russia* **65 D11** 54 0N 102 5 E
Zimane, Adrar in, *Algeria* **111 D5** 22 10N 4 30 E
Zimapán, *Mexico* **163 C5** 20 54N 99 20W
Zimba, *Zambia* **119 F2** 17 20 S 26 11 E
Zimbabwe, *Zimbabwe* **119 G3** 20 16 S 30 54 E
Zimbabwe ■, *Africa* **119 F3** 19 0 S 30 0 E
Zimnicea, *Romania* **53 G10** 43 40N 25 22 E
Zimovniki, *Russia* **61 G6** 47 10N 42 25 E
Zinal, *Switz.* **32 D5** 46 8N 7 38 E
Zinder, *Niger* **109 F1** 13 48N 9 0 E
Zinga, *Tanzania* **119 D4** 9 16 S 38 49 E
Zingst, *Germany* **30 A8** 54 26N 12 39 E
Ziniaré, *Burkina Faso* **113 C4** 12 35N 1 18W
Zinnowitz, *Germany* **30 A9** 54 4N 13 54 E
Zion, *U.S.A.* **157 B9** 42 27N 87 50W
Zion National Park, *U.S.A.* **159 H7** 37 15N 113 5W
Zionsville, *U.S.A.* **157 E10** 39 57N 86 16W
Zipaquirá, *Colombia* **168 C3** 5 0N 74 0W
Zirbitzkogel, *Austria* **34 D7** 47 4N 14 34 E
Zirc, *Hungary* **52 C2** 47 17N 17 52 E
Žiri, *Slovenia* **45 B11** 46 5N 14 5 E
Žirje, *Croatia* **45 E12** 43 39N 15 42 E
Zirl, *Austria* **34 D4** 47 17N 11 14 E
Zirndorf, *Germany* **31 F6** 49 27N 10 57 E
Ziros, *Greece* **38 D8** 35 5N 26 8 E
Zisterdorf, *Austria* **35 C9** 48 33N 16 45 E
Zitácuaro, *Mexico* **162 D4** 19 28N 100 21W
Žitava →, *Slovak Rep.* **35 C11** 48 14N 18 21 E
Žitište, *Serbia, Yug.* **52 E5** 45 30N 20 32 E
Zítsa, *Greece* **48 B2** 39 47N 20 40 E
Zittau, *Germany* **30 E10** 50 53N 14 48 E
Zitundo, *Mozam.* **117 D5** 26 48 S 32 47 E
Živinice, *Bos.-H.* **52 F3** 44 27N 18 36 E
Ziway, L., *Ethiopia* **107 F4** 8 0N 38 50 E
Zixi, *China* **77 D11** 27 45N 117 4 E
Zixing, *China* **77 E9** 25 59N 113 21 E
Ziyang, *Shaanxi, China* **74 H5** 32 32N 108 31 E
Ziyang, *Sichuan, China* **76 B5** 30 6N 104 40 E
Ziyun, *China* **76 E6** 25 45N 106 5 E
Ziz, Oued →, *Morocco* **110 B4** 31 40N 4 15W
Zizhixian, *China* **77 E8** 25 0N 111 47 E
Zizhong, *China* **76 C5** 29 48N 104 47 E
Zlarin, *Croatia* **45 E12** 43 42N 15 49 E
Zlatar, *Croatia* **45 B13** 46 5N 16 3 E
Zlatar, *Serbia, Yug.* **50 C3** 43 25N 19 47 E
Zlataritsa, *Bulgaria* **51 C9** 43 2N 25 55 E
Zlaté Moravce, *Slovak Rep.* **35 C11** 48 23N 18 24 E
Zlatibor, *Serbia, Yug.* **50 C3** 43 45N 19 43 E
Zlatitsa, *Bulgaria* **51 D8** 42 41N 24 7 E
Zlatna, *Romania* **53 D8** 46 8N 23 11 E
Zlatna Panega, *Bulgaria* **51 C8** 43 5N 24 9 E
Zlatni Pyasŭtsi, *Bulgaria* **51 C12** 43 17N 28 3 E
Zlatograd, *Bulgaria* **51 E9** 41 22N 25 7 E
Zlatoust, *Russia* **62 D7** 55 10N 59 40 E
Zletovo, *Macedonia* **50 E6** 41 59N 22 17 E
Zlín, *Czech Rep.* **35 B10** 49 14N 17 40 E
Ẕlītan, *Libya* **108 B2** 32 32N 14 35 E
Złocieniec, *Poland* **54 E3** 53 30N 16 1 E
Złoczew, *Poland* **55 G5** 51 24N 18 35 E
Zlot, *Serbia, Yug.* **50 B5** 44 1N 21 58 E
Złotoryja, *Poland* **55 G2** 51 8N 15 55 E
Złotów, *Poland* **54 E4** 53 22N 17 2 E
Ẓmeinogorsk, *Kazakstan* **64 D9** 51 10N 82 13 E
Żmigród, *Poland* **55 G3** 51 28N 16 53 E
Zmiyev, *Ukraine* **59 H9** 49 39N 36 27 E
Znamenka = Znamyanka, *Ukraine* **59 H7** 48 45N 32 30 E
Ẓnamyanka, *Ukraine* **59 H7** 48 45N 32 30 E
Żnin, *Poland* **55 F4** 52 51N 17 44 E
Znojmo, *Czech Rep.* **34 C9** 48 50N 16 2 E
Zobeyrī, *Iran* **96 C5** 34 10N 46 40 E
Zobia, *Dem. Rep. of the Congo* **118 B2** 3 0N 25 59 E
Zoetermeer, *Neths.* **24 B4** 52 3N 4 30 E
Zofingen, *Switz.* **32 B5** 47 17N 7 56 E
Zogang, *China* **76 C1** 29 55N 97 42 E
Zogno, *Italy* **44 C6** 45 48N 9 40 E
Zogqên, *China* **76 A2** 32 13N 98 47 E
Zolfo Springs, *U.S.A.* **153 H8** 27 30N 81 48W
Zollikofen, *Switz.* **32 C4** 47 0N 7 28 E
Zollikon, *Switz.* **33 B7** 47 21N 8 34 E
Zolochev = Zolochiv, *Ukraine* **59 H3** 49 45N 24 51 E
Zolochiv, *Ukraine* **59 H3** 49 45N 24 51 E
Zolotonosha, *Ukraine* **59 H7** 49 39N 32 5 E
Zomba, *Malawi* **119 F4** 15 22 S 35 19 E
Zongo, *Dem. Rep. of the Congo* **114 B3** 4 20N 18 35 E
Zonguldak, *Turkey* **100 B4** 41 28N 31 50 E
Zonqor Pt., *Malta* **38 D2** 35 51N 14 34 E
Zonza, *France* **29 G13** 41 45N 9 11 E
Zorgo, *Burkina Faso* **113 C4** 12 15N 0 35W
Zorita, *Spain* **43 F5** 39 17N 5 39W
Zorleni, *Romania* **53 D12** 46 14N 27 44 E
Zornitsa, *Bulgaria* **51 D10** 42 23N 26 58 E
Ẓorritos, *Peru* **172 A1** 3 43 S 80 40W
Żory, *Poland* **55 H5** 50 3N 18 44 E
Zorzor, *Liberia* **112 D3** 7 46N 9 28W
Zossen, *Germany* **30 C9** 52 13N 13 27 E
Zou Xiang, *China* **74 G9** 35 30N 116 58 E
Zouar, *Chad* **109 D3** 20 30N 16 32 E
Zouérate = Zouîrât, *Mauritania* **110 D2** 22 44N 12 21W
Zouîrât, *Mauritania* **110 D2** 22 44N 12 21W
Zousfana, O. →, *Algeria* **111 B4** 31 28N 2 17W
Zoushan Dao, *China* **77 B14** 30 5N 122 10 E
Zoutkamp, *Neths.* **24 A6** 53 20N 6 18 E
Zrenjanin, *Serbia, Yug.* **52 E5** 45 22N 20 23 E
Zuarungu, *Ghana* **113 C4** 10 49N 0 46W
Zuba, *Nigeria* **113 D6** 9 11N 7 12 E
Zubayr, *Yemen* **98 D3** 15 3N 42 10 E
Zubtsov, *Russia* **58 D8** 56 10N 34 34 E
Zudáñez, *Bolivia* **173 D5** 19 6 S 64 44W
Zuénoula, *Ivory C.* **112 D3** 7 34N 6 3W
Zuera, *Spain* **40 D4** 41 51N 0 49W
Zuetina = Az Zuwaytīnah, *Libya* **108 B4** 30 58N 20 7 E
Ẓufār, *Oman* **99 C6** 17 40N 54 0 E
Zug, *Switz.* **33 B7** 47 10N 8 31 E
Zug □, *Switz.* **33 B7** 47 9N 8 35 E
Zugdidi, *Georgia* **61 J5** 42 30N 41 55 E
Zugersee, *Switz.* **33 B7** 47 7N 8 35 E
Zugspitze, *Germany* **31 H6** 47 25N 10 59 E
Zuid-Holland □, *Neths.* **24 C4** 52 0N 4 35 E
Zuidbeveland, *Neths.* **24 C3** 51 30N 3 50 E
Zuidhorn, *Neths.* **24 A6** 53 15N 6 23 E
Zújar, *Spain* **43 H8** 37 34N 2 50W
Ẓújar →, *Spain* **43 F5** 39 1N 5 47W
Żukowo, *Poland* **54 D5** 54 21N 18 22 E
Zula, *Eritrea* **107 D4** 15 17N 39 40 E
Zulia □, *Venezuela* **168 B3** 10 0N 72 10W
Zülpich, *Germany* **30 E2** 50 41N 6 39 E
Zumaia, *Spain* **40 B2** 43 19N 2 15W
Zumárraga, *Spain* **40 B2** 43 5N 2 19W
Zumbo, *Mozam.* **119 F3** 15 35 S 30 26 E
Zummo, *Nigeria* **113 D7** 9 51N 12 59 E
Zumpango, *Mexico* **163 D5** 19 48N 99 6W
Zungeru, *Nigeria* **113 D6** 9 48N 6 8 E
Zunhua, *China* **75 D9** 40 18N 117 58 E
Zuni, *U.S.A.* **159 J9** 35 4N 108 51W
Zunyi, *China* **76 D6** 27 42N 106 53 E
Zuoquan, *China* **74 F7** 37 5N 113 22 E
Zuozhou, *China* **76 F6** 22 42N 107 27 E
Županja, *Croatia* **52 E3** 45 4N 18 43 E
Žur, *Serbia, Yug.* **50 D4** 42 13N 20 34 E
Zura, *Russia* **62 C4** 57 36N 53 24 E
Zurbātīyah, *Iraq* **101 F12** 33 9N 46 3 E
Zürich, *Switz.* **33 B7** 47 22N 8 32 E
Zürich □, *Switz.* **33 B7** 47 26N 8 40 E
Ẓürichsee, *Switz.* **33 B7** 47 18N 8 40 E
Żuromin, *Poland* **55 E6** 53 4N 19 51 E
Zuru, *Nigeria* **113 C6** 11 20N 5 11 E
Ẓurzach, *Switz.* **33 A6** 47 35N 8 18 E
Žut, *Croatia* **45 E12** 43 52N 15 17 E
Zutphen, *Neths.* **24 B6** 52 9N 6 12 E
Zuwārah, *Libya* **108 B2** 32 58N 12 1 E
Zuyevka, *Russia* **62 B3** 58 27N 51 10 E
Zūzan, *Iran* **97 C8** 34 22N 59 53 E
Žužemberk, *Slovenia* **45 C11** 45 52N 14 56 E
Zvenigorodka = Zvenyhorodka, *Ukraine* **59 H6** 49 4N 30 56 E
Zvenyhorodka, *Ukraine* **59 H6** 49 4N 30 56 E
Zverinogolovskoye, *Russia* **64 D7** 54 26N 64 50 E
Zvezdets, *Bulgaria* **51 D11** 42 6N 27 26 E
Zvishavane, *Zimbabwe* **119 G3** 20 17 S 30 2 E
Zvolen, *Slovak Rep.* **35 C12** 48 33N 19 10 E
Zvonce, *Serbia, Yug.* **50 D6** 42 57N 22 34 E
Zvornik, *Bos.-H.* **52 F4** 44 26N 19 5 E
Zwedru = Tchien, *Liberia* **112 D3** 5 59N 8 15W
Zweibrücken, *Germany* **31 F3** 49 15N 7 21 E
Zwenkau, *Germany* **30 D8** 51 13N 12 20 E
Zwettl, *Austria* **34 C8** 48 35N 15 9 E
Zwickau, *Germany* **30 E8** 50 44N 12 30 E
Zwierzyniec, *Poland* **55 H9** 50 36N 22 58 E
Zwiesel, *Germany* **31 F9** 49 1N 13 13 E
Zwoleń, *Poland* **55 G8** 51 21N 21 36 E
Zwolle, *Neths.* **24 B6** 52 31N 6 6 E
Ẓwolle, *U.S.A.* **155 K8** 31 38N 93 39W
Żychlin, *Poland* **55 F6** 52 15N 19 37 E
Zymoetz →, *Canada* **142 C3** 54 33N 128 31W
Żyrardów, *Poland* **55 F7** 52 3N 20 28 E
Zyryan, *Kazakstan* **64 E9** 49 43N 84 20 E
Zyryanka, *Russia* **65 C16** 65 45N 150 51 E
Zyryanovsk = Zyryan, *Kazakstan* **64 E9** 49 43N 84 20 E
Żywiec, *Poland* **55 J6** 49 42N 19 10 E
Zyyi, *Cyprus* **38 E12** 34 43N 33 20 E

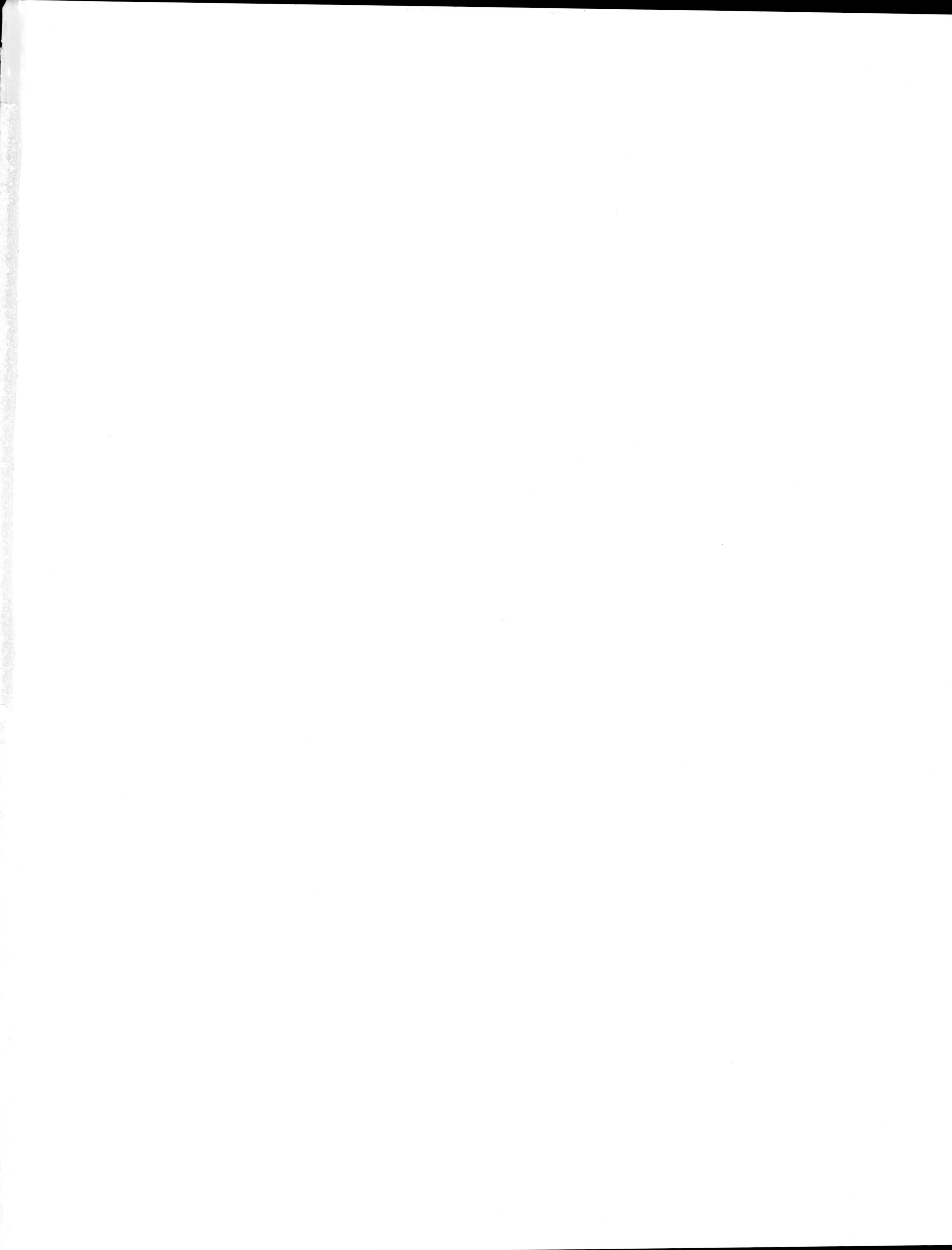

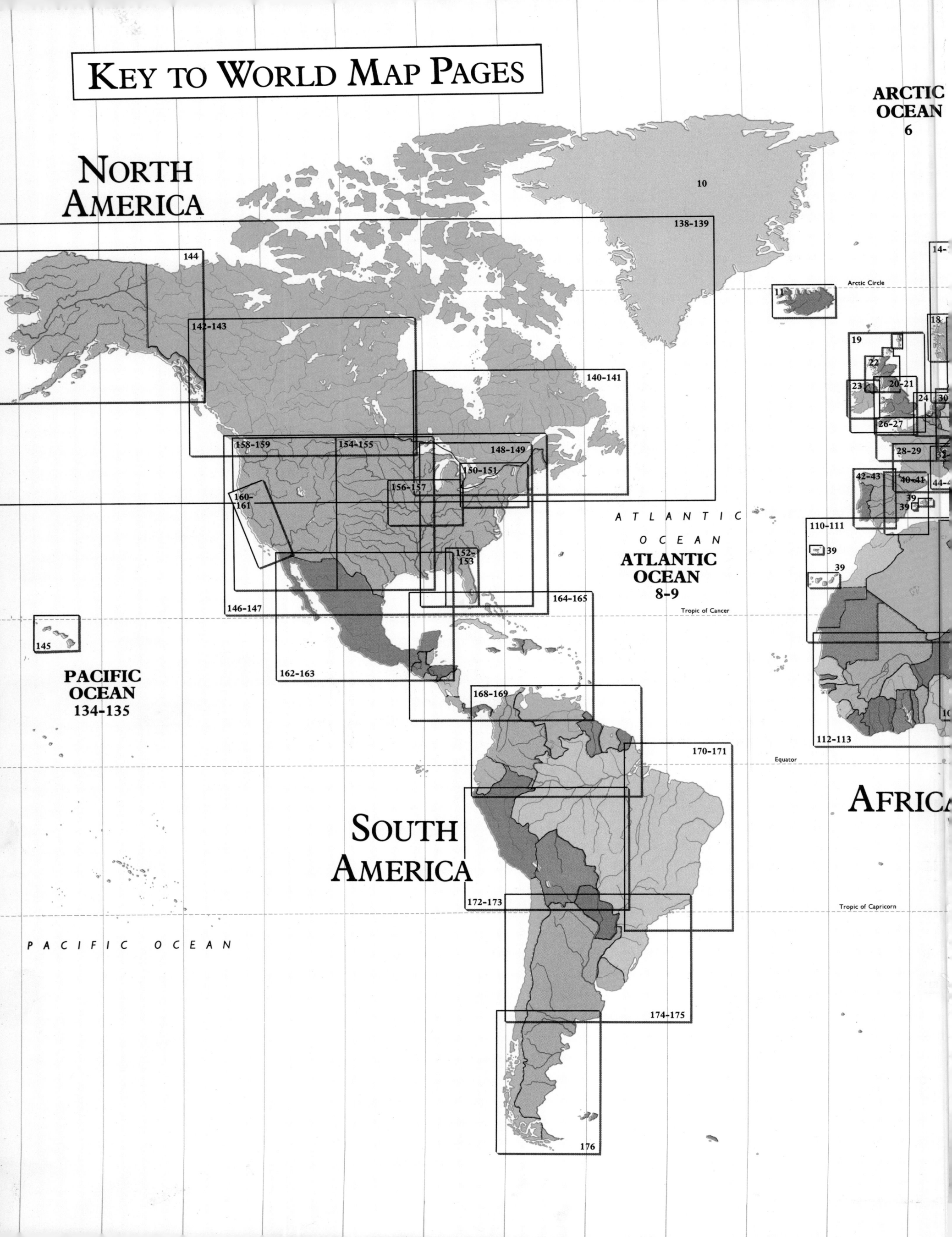

KEY TO WORLD MAP PAGES

NORTH AMERICA

SOUTH AMERICA

AFRICA

ARCTIC OCEAN
6

ATLANTIC OCEAN

ATLANTIC OCEAN
8-9

PACIFIC OCEAN
134-135

PACIFIC OCEAN

Arctic Circle

Tropic of Cancer

Equator

Tropic of Capricorn

10

138-139

11

14-

18

19

22

23

20-21

24

30

26-27

28-29

42-43

40-41

44-

39
39

110-111

39

39

112-113

10

144

142-143

140-141

158-159

154-155

148-149

150-151

156-157

160-161

152-153

164-165

146-147

162-163

168-169

170-171

172-173

174-175

176

145